Food Biochemistry and Food Processing

Second Edition

Food Biochemistry and Food Processing

Second Edition

Edited by
Benjamin K. Simpson

Associate Editors
Leo M.L. Nollet
Fidel Toldrá
Soottawat Benjakul
Gopinadhan Paliyath
Y.H. Hui

WILEY-BLACKWELL
A John Wiley & Sons, Ltd., Publication

Wiley-Blackwell is an imprint of John Wiley & Sons, formed by the merger of Wiley's global Scientific, Technical and Medical business with Blackwell Publishing.

Editorial offices: 2121 State Avenue, Ames, Iowa 50014-8300, USA
The Atrium, Southern Gate, Chichester, West Sussex, PO19 8SQ, UK
9600 Garsington Road, Oxford, OX4 2DQ, UK

For details of our global editorial offices, for customer services and for information about how to apply for permission to reuse the copyright material in this book please see our website at www.wiley.com/wiley-blackwell.

Library of Congress Cataloging-in-Publication Data
Food biochemistry and food processing. – 2nd ed. / edited by Benjamin Simpson . . . [et al.].
p. cm.
Includes bibliographical references and index.
ISBN 978-0-8138-0874-1 (hardcover : alk. paper) 1. Food industry and trade–Research.
2. Food–Analysis. 3. Food–Composition. 4. Food–Packaging. I. Simpson, Benjamin K.
TP370.8.F66 2012
664–dc23

2011052397

A catalogue record for this book is available from the British Library.

Wiley also publishes its books in a variety of electronic formats. Some content that appears in print may not be available in electronic books.

Cover image: © George Muresan/Shutterstock.com
Cover design by Meaden Creative

Set in 9.5/11.5 pt Times by Aptara® Inc., New Delhi, India
Printed and bound in Malaysia by Vivar Printing Sdn Bhd

1 2012

Contents

Contributor List

Joseph Alakali
Department of Food Science and Technology,
University of Agriculture
Makurdi, Benue State, Nigeria
E-mail: jsalakali@yahoo.com

Isaac N.A. Ashie, Ph.D.
Novozymes North America, Inc.
Franklinton, NC 27525
Phone: 919 637 3868
E-mail: Ikeashie@yahoo.com

Irene A. Axarli, Ph.D.
Laboratory of Enzyme Technology
Department of Agricultural Biotechnology
Agricultural University of Athens
Iera Odos 75, 11855-Athens, Greece
email: axarli_irene@yahoo.com

Sammy S.S. Bajwa
Department of Bioresource Engineering
McGill University
21,111 Lakeshore Road
Ste-Anne-de-Bellevue
Quebec, H9X 3V9, Canada
E-mail: Sammy.bajwa@mail.mcgill.ca

Soottawat Benjakul, Ph.D., (Associate Editor)
Department of Food Technology, Faculty of Agro-Industry
Prince of Songkla University
Hat Yai, Songkhla, 90112, Thailand
e-mail: soottawat.b@psu.ac.th

Joyce I. Boye, Ph.D.
Food Research and Development Centre
Agriculture and Agri-Food Canada
3600 Casavant Blvd. West
St-Hyacinthe, Quebec, J2S 8E3, Canada
Phone: 450-768-3232; Fax 450-773-8461
E-mail: joyce.boye@agr.gc.ca

Terri D. Boylston, Ph.D.
Department Food Science & Human Nutrition
2312 Food Sciences Building
Iowa State University
Ames, IA 50011-1061
Phone: 515-294-0077
Email: tboylsto@iastate.edu

Brian Bryksa, Ph.D.
Department of Food Science
University of Guelph
50 Stone Road East
Guelph, Ontario, N1G 2W1, Canada
Phone: 519-824-4120 x56585
E-mail: bbryksa@uoguelph.ca

Barry Byrne, Ph.D.
Biomedical Diagnostics Institute (BDI)
Dublin City University
Dublin 9, Ireland
E-mail: Barry.Byrne@dcu.ie

Claude P. Champagne, Ph.D.
Agriculture and Agri-Food Canada
3600 Casavant
St-Hyacinthe, Quebec, J2S 8E3, Canada
Phone: 450-768-3238
Fax: 450-773-8461
E-mail: Claude.Champagne@agr.gc.ca

Euggelia G. Chronopoulou, Ph.D.
Laboratory of Enzyme Technology
Department of Agricultural Biotechnology
Agricultural University of Athens
Iera Odos 75, 11855-Athens, Greece
E-mail: exronop@gmail.com

Nieves Corzo, Ph.D.
Institute of Food Science Research (CIAL) (CSIC-UAM)
c/Nicolás Cabrera, 9, Campus of Universidad Autónoma de Madrid,
28049-Madrid (Spain)
Phone: + 34 91 001 79 54
Fax: + 34 91 001 79 05
E-mail: nieves.corzo@csic.es

Justine R. Cottam, Ph.D.
Biomolecular Interaction Centre and School of Biological Sciences, University of Canterbury, Christchurch, New Zealand

and

Fonterra Research Centre, Palmerston North, New Zealand
E-mail: justine.cottam@canterbury.ac.nz

Angelina O. Danquah, Ph.D.
Department of Home Science
University of Ghana, Legon, Ghana
E-mail: adanquah@ug.edu.gh

M[a] Dolores del Castillo, Ph. D.
Institute of Food Science Research (CIAL) (CSIC-UAM)
c/Nicolás Cabrera, 9 Campus of Universidad Autónoma de Madrid
28049-Madrid (Spain)
Phone: + 34 91 001 79 53
Fax: + 34 91 001 79 05
E-mail: mdolores.delcastillo@csic.es

Juana Fernandez-Lopez, Ph.D.
IPOA Research Group. AgroFood Technology Department.
Orihuela Polytechnical High School
Miguel Hernandez University
Ctra. a Beniel. km. 3,2 Orihuela (Z.C. 03312) Alicante (Spain)
Phone: +34 966749784
Fax: +34 966749677
E-mail: j.fernandez@umh.es

Patrick Fox, Ph.D.
School of Food and Nutritional Sciences
University College Cork
Cork, Ireland
E-mail: PFF@ucc.ie

Juliet A. Gerrard, Ph.D.
IRL Industry and Outreach Fellow
Co-Director, Biomolecular Interaction Centre (BIC), and
School of Biological Sciences, University of Canterbury,
Christchurch, New Zealand
Phone: 64 3 3642987 extn 7302
Fax: 64 3 3642590
E-mail: juliet.gerrard@canterbury.ac.nz

Jissy K. Jacob
Nestle PTC
809 Collins Ave
Marysville, Ohio 43040, USA
Phone: 937-642-2132
E-mail: jissyjacob@gmail.com

Yueming Jiang, Ph.D.
South China Botanical Garden
The Chinese Academy of Science
Guangzhou, China
Phone. 86-20-37252525
E-mail: ymjiang@scib.ac.cn

Mary S. Kalamaki, Ph.D.
Technological Educational Institute of Thessaloniki
Department of Food Technology
P.O. Box 141
574 00 Thessaloniki, Greece
Phone: +30 231 041 2238
Fax: +30 231 041 2238
E-mail: kalamaki@vet.auth.gr

Christos Karamitros, Ph.D.
Laboratory of Enzyme Technology
Department of Agricultural Biotechnology
Agricultural University of Athens
Iera Odos 75, 11855-Athens, Greece
E-mail: karamitroschristos@yahoo.gr

Alan Kelly, Ph.D.
School of Food and Nutritional Sciences
University College Cork
Cork, Ireland
E-mail: a.kelly@ucc.ie

Phanat Kittiphattanabawon, Ph.D.
Department of Food Technology, Faculty of Agro-Industry
Prince of Songkla University
Hat Yai, Songkhla, 90112, Thailand
E-mail: 5111030006@email.psu.ac.th

Sappasith Klomklao, Ph.D.
Department of Food Science and Technology,
Faculty of Technology and Community Development,
Thaksin University, Phatthalung Campus,
Phatthalung, 93110, Thailand
E-mail: sappasith@tsu.ac.th

Georgia A. Kotzia, Ph.D.
Laboratory of Enzyme Technology
Department of Agricultural Biotechnology
Agricultural University of Athens
Iera Odos 75, 11855-Athens, Greece
E-mail: tzinakotz@yahoo.com

Hordur G. Kristinsson, Ph.D.
Acting CEO & Director
Matis - Icelandic Food & Biotech R&D
Biotechnology and Biomolecules Division
Biotechnology and Biomolecules Labs
Vinlandsleid 12, 113 Reykjavík

(Alternate Address: Matis Biotechnology Centre: Haeyri 1, 550 Saudarkrokur)
Phone: +354 422-5063 - Fax: +354 422-5002
E-mail: hordur.g.kristinsson@matis.is

Nikolaos Labrou, Ph.D.
Director of Division C (Biochemistry, Enzyme Technology, Microbiology and Molecular Biology),
Laboratory of Enzyme Technology,
Department of Agricultural Biotechnology,
Agricultural University of Athens,
Iera Odos 75, Gr-118 55, Athens, Greece
Phone: +30 210 5294308,
E-mail: lambrou@aua.gr

Catherine M. Logue, Ph.D.
Assistant Director, NDAES
Department of Veterinary and Microbiological Sciences
PO Box 6050, Dept 7690
North Dakota State University
Fargo, ND 58108
Phone: 701 231 7692
Fax: 701 231 9692
E-mail: Catherine.Logue@ndsu.edu

Yanjun Li,
Research and Development Center
Hangzhou Wahaha Group Co. Ltd.
Hangzhou, China
Phone : 86-571-86796066
E-mail: lyj@wahaha.com.cn

Miang H. Lim, Ph.D.
Honorary Associate Professor
University of Auckland
New Zealand
E-mail: mianglim@gmail.com.

Ying Ma, Ph.D.
College of Food Science and Engineering
Harbin Institute of Technology
Harbin, China
Phone: 86-451-86282903
E-mail: maying@hit.edu.cn

Samuel A. M. Martin, Ph.D.
School of Biological Sciences
University of Aberdeen
Aberdeen, AB24 3FX, UK.
E-mail: sam.martin@abdn.ac.uk

Marta Corzo Martínez, Ph.D.
Institute of Food Science Research (CIAL) (CSIC-UAM)
c/ Nicolás Cabrera, 9, Campus of Universidad Autónoma de Madrid,
28049-Madrid (Spain)
Phone: +34 647048116
E-mail: m.corzo@csic.es

Sitthipong Nalinanon, Ph.D.
Faculty of Agro-Industry
King Mongkut's Institute of Technology Ladkrabang
Chalongkrung Rd., Ladkrabang, Bangkok, 10520, Thailand
E-mail: knsitthi@kmitl.ac.th

Michael O. Ngadi, Ph.D.
Department of Bioresource Engineering
McGill University, Macdonald Campus
21,111 Lakeshore Road
Ste-Anne-de-Bellevue
Quebec, H9X 3V9, Canada
Phone: (514) 398-7779 Fax: (514) 398-8387
E-mail: michael.ngadi@mcgill.ca

Lisa K. Nolan, Ph.D.
College of Veterinary Medicine,
1600 S. 16th Street, 2506 Veterinary Administration
Iowa State University
Ames IA 50011-1250, USA
E-mail: lknolan@iastate.edu

Leo M. L. Nollet, Ph.D. (Associate Editor)
Hogeschool Gent
Department of Engineering Sciences
Schoonmeersstraat 52
B9000 Gent, Belgium
Phone: 00-329-242-4242 Fax: 00 329 243 8777
E-mail: leo.nollet@hogent.be

Moustapha Oke, Ph.D.
Ministry of Agriculture Food and Rural Affairs
Food Safety and Environment Division
Food Inspection Branch / Food Safety Science Unit
1 Stone Road West, 5th Floor NW
Guelph, Ontario N1G 4Y2, Canada
Phone: 519-826-3246; Fax: +519-826-3233
E-mail: Moustapha.oke@ontario.ca

Richard O'Kennedy, Ph.D.
Biomedical Diagnostics Institute (BDI)
and National Centre for Sensor Research (NCSR)
Dublin City University
Dublin 9, Ireland
E-mail: Richard.OKennedy@dcu.ie

Gopinadhan Paliyath, Ph.D., (Associate Editor)
Plant Agriculture
Edmond C. Bovey Bldg
University of Guelph
50 Stone Road East
Guelph, Ontario N1G 2W1, Canada
Phone: 519-824-4120 x 54856
E-mail: gpaliyat@uoguelph.ca

Jose Angel Perez-Alvarez, Ph.D.
IPOA Research Group. AgroFood Technology Department.
Orihuela Polytechnical High School
Miguel Hernandez University
Ctra. a Beniel. km. 3,2 Orihuela (Z.C. 03312), Alicante (Spain)
Phone: +34 966749739
Fax: +34 966749677
E-mail: ja.perez@umh.es

Dimitris Platis, Ph.D.
Laboratory of Enzyme Technology
Department of Agricultural Biotechnology
Agricultural University of Athens
Iera Odos 75, 11855-Athens, Greece,
E-mail: dimitris_platis@hotmail.com

Hosahalli S. Ramaswamy, Ph.D.
Department of Food Science & Agricultural Chemistry
McGill University, Macdonald Campus
21,111 Lakeshore Road
Ste-Anne-de-Bellevue
Quebec, H9X 3V9, Canada
Phone: 514-398-7919; Fax: 514-398-7977
E-mail: Hosahalli.Ramaswamy@Mcgill.Ca

Joe M. Regenstein, Ph.D.
Department of Food Science
Stocking Hall, Cornell University
Ithaca, NY, USA 14853-7201
E-mail: jmr9@cornell.edu

Milagro Reig, Ph.D.
Institute of Food Engineering for Development
Universidad Politécnica de Valencia, Ciudad Politécnica de la Innovación, ed.8E, Camino de Vera s/n, 46022, Valencia (Spain)
E-mail: mareirie@doctor.upv.es

Xin Rui
Department of Bioresource Engineering
McGill University, Macdonald Campus
21,111 Lakeshore Road
Ste-Anne-de-Bellevue
Quebec, H9X 3V9, Canada
Phone: (514) 398-7779 Fax: (514) 398-8387
E-mail: xin.rui@mail.mcgill.ca

Fereidoon Shahidi, Ph.D.
Department of Biochemistry
Memorial University of Newfoundland
St. John's
Newfoundland, A1B 3X9, Canada
E-mail: fshahidi@gmail.com

John Shi, Ph.D.
Guelph Food Research Center
Agriculture and Agri-Food Canada
Ontario, N1G 5C9, Canada
Phone: 519 780-8035
E-mail: John.Shi@AGR.GC.CA

Benjamin K. Simpson, Ph.D., Editor-in-Chief
Department of Food Science & Agricultural Chemistry
McGill University, Macdonald Campus
21,111 Lakeshore Road
Ste-Anne-de-Bellevue
Quebec, H9X 3V9, Canada
Phone: 514 398-7737
Fax: 514 398-7977
E-mail: benjamin.simpsom@mcgill.ca

Lee F. Siow, Ph.D.
Malaysia School of Science
Monash University
Sunway Campus, Malaysia
E-mail: siow.lee.fong@sci.monash.edu.my

Ana Cristina Soria, Ph.D.
Institute of General Organic Chemistry (CSIC)
c/ Juan de la Cierva, 3, 28006-Madrid (Spain)
Phone: +34 912587451
E-mail: acsoria@iqog.csic.es

Edwina Stack, Ph.D.
National Centre for Sensor Research (NCSR)
Dublin City University
Dublin 9, Ireland
E-mail: Edwina.Stack3@mail.dcu.ie

Nikolaos G. Stoforos, Ph.D.
Agricultural University of Athens
Department of Food Science and Technology
Iera Odos 75, 11855 Athens, Greece
Phone: +30-210 529 4706
Fax: +30 210 529 4682
E-mail: stoforos@aua.gr

Hólmfríður Sveinsdóttir, Ph.D.
Icelandic Food and Biotech R&D
Sauðárkrókur, Iceland
E-mail: holmfridur.sveinsdottir@matis.is

Petros Taoukis, Ph.D.
National Technical University of Athens
School of Chemical Engineering
Division IV- Product and Process Development
Laboratory of Food Chemistry and Technology
Iroon Polytechniou 5, 15780 Athens, Greece
Phone: +30-210-7723171
Fax: +30-210-7723163
E-mail: taoukis@chemeng.ntua.gr

Cin Lam Thang
Department of Animal Science
McGill University (Macdonald Campus)
21,111 Lakeshore Road
Ste. Anne de Bellevue
Quebec, H9X 3V9, Canada
E-mail: cin.thang@mail.mcgill.ca

Yetenayet Tola
Department of Food Science & Agricultural Chemistry
McGill University, Macdonald Campus
21,111 Lakeshore Road
Ste-Anne-de-Bellevue
Quebec, H9X 3V9, Canada
E-mail: yetenayet.tola@mail.mcgill.ca

Fidel Toldrá, Ph.D. (Associate Editor)
Department of Food Science
Instituto de Agroquímica y Tecnología de Alimentos (CSIC)
Avenue Agustín Escardino 7, 46980 Paterna, Valencia (Spain)
E-mail: ftoldra@iata.csic.es

Therese Uniacke-Lowe, Ph.D.
School of Food and Nutritional Sciences
University College Cork

Cork, Ireland
E-mail: t.uniacke@ucc.ie

Oddur T. Vilhelmsson, Ph.D.
Department of Natural Resource Sciences
University of Akureyri, IS-600 Akureyri, Iceland.
Phone: +354 460 8514 / +354 697 4252
E-mail: oddurv@unak.is

Mar Villamiel, Ph.D.
Institute of Food Science Research (CIAL) (CSIC-UAM)
c/Nicolás Cabrera, 9, Campus of Universidad Autónoma de Madrid,
28049-Madrid (Spain)
Phone: + 34 91 001 79 51
Fax: + 34 91 001 79 05
E-mail : m.villamiel@csic.es

Wonnop Visessanguan, Ph.D.
National Center for Genetic Engineering and Biotechnology (BIOTEC)
113 Thailand Science Park, Phahonyothin Road
Klong 1, Klong Luang
Pathumthani 12120, Thailand
Phone: +66 (0) 2564 6700
Fax: +66 (0) 2564 6701-5
E-mail : wonnop@biotec.or.th

Ronnie G. Willaert, Ph.D.
Department of Bioengineering Sciences
Vrije Universiteit Brussel
Pleinlaan 2
B-1050 Brussels, Belgium
E-mail: Ronnie.Willaert@vub.ac.be

Sophia Jun Xue, Ph.D.
Guelph Food Research Center
Agriculture and Agri-Food Canada
Ontario, N1G 5C9, Canada
Phone: 519 780-8096
E-mail: Jun.Xue@AGR.GC.CA

Rickey Yada, Ph.D.
Department of Food Science, Food Science, Rm. 224
University of Guelph
50 Stone Road East
Guelph, Ontario, N1G 2W1, Canada
Phone: 519.824.4120 x56585
Fax: 519.824.6631
E-mail: ryada@uoguelph.ca

Xingqian Ye, Ph.D.
Department of Food Science and Nutrition
School of Biosystems Engineering and Food Science
Zhejiang University
Hangzhou, China
Phone: 86-571- 88982155
E-mail: psu@zju.edu.cn

Fatemeh Zare
Department of Food Science & Agricultural Chemistry
McGill University, Macdonald Campus
21,111 Lakeshore Road
Ste-Anne-de-Bellevue
Quebec, H9X 3V9, Canada
E-mail: fatemeh.zare@mail.mcgill.ca

Xin Zhao, Ph.D.
Department of Animal Science
McGill University (Macdonald Campus)
21,111 Lakeshore Road
Ste. Anne de Bellevue
Quebec, H9X 3V9, Canada
Phone: 514 398-7975
E-mail: xin.zhao@mcgill.ca

Xianzhe Zheng, Ph.D.
College of Engineering
Northeast Agricultural University
Harbin, China
Phone : 86-451-55191606
E-mail: zhengxz2008@gmail.com

Preface

Food biochemistry principles and knowledge have become indispensable in practically all the major disciplines of food science, such as food technology, food engineering, food biotechnology, food processing, and food safety within the past few decades. Knowledge in these areas has grown exponentially and keeps growing, and is disseminated through various media in both printed and electronic forms, and entire books are available for almost all the distinct specialty areas mentioned above. The two areas of food biochemistry and food processing are becoming closely interrelated. Fundamental knowledge in food biochemistry is crucial to enable food technologists and food processing engineers to rationalize and develop more effective strategies to produce and preserve food in safe and stable forms. Nonetheless, books combining food biochemistry and food processing/engineering principles are rare, and the first edition of this book was designed to fill the gap by assembling information on following six broad topics in the two areas:

1. Principles of food biochemistry.
2. Advances in selected areas of food biochemistry.
3. Food biochemistry and the processing of muscle foods and milk.
4. Food biochemistry and the processing of fruits, vegetables and cereals.
5. Food biochemistry and the processing of fermented foods.
6. Food microbiology and food safety.

These topics were spread over 31 chapters in the first edition.

The second edition of the book provides an update of several chapters from the first edition and expands the contents to encompass eight broad topics as follows:

1. Principles and analyses
2. Biotechnology and enzymology
3. Muscle foods (meats, poultry, and fish)
4. Milk and dairy
5. Fruits, vegetables, and cereals
6. Health and functional foods
7. Food processing
8. Food safety and food allergens

These eight broad topics are spread over 45 chapters in the second edition, and represents close to 50% increase in content over the previous version. In addition, abstracts capturing the salient features of the different chapters are provided in this new edition of the book.

The book is the result of the combined efforts of more than 65 professionals with diverse expertise and backgrounds in food biochemistry, food processing, and food safety who are affiliated with industry, government research institutions, and academia from over 18 countries. These experts were led by an international editorial team of six members from four countries in assembling together the different topics in food biochemistry, food commodities, food processing, and food safety in this one book.

The end product is unique, both in depth and breadth, and is highly recommended both as an essential reference book on food biochemistry and food processing for professionals in government, industry, and academia; and as classroom text for undergraduate courses in food chemistry, food biochemistry, food commodities, food safety, and food processing principles.

We wish to thank all the contributing authors for sharing their knowledge and expertise for their invaluable contribution and their patience for staying the course and seeing this project through.

B.K. Simpson
L.M.L. Nollet
F. Toldrá
S. Benjakul
G. Paliyath
Y.H. Hui

Part 1
Principles/Food Analysis

1
An Introduction to Food Biochemistry

Rickey Y. Yada, Brian Bryksa, and Wai-kit Nip

Abstract: Compared to the siloed commodity departments of the past, the multi-disciplinary field of food science and technology has increasingly adopted a less segregated and more synergistic approach to research. At their most fundamental levels, all food-related processes from harvest to digestion are ways of bringing about, or preventing, biochemical changes. We contend that there is not a single scientific investigation of a food-related process that can avoid biochemical considerations. Even food scientists studying inorganic materials used in processing equipment and/or packaging must eventually consider potential reactions with biomolecules encountered in food systems. Moreover, since the food that we eat plays a central role in our overall well-being, it follows that tomorrow's food scientists and technologists must have a solid foundation in food biochemistry if they are to be innovators and visionaries. Introductions to biochemical topics are provided in this chapter, under the categories of carbohydrates, proteins, lipids, DNA, and toxicants. Within these broad divisions, general and specific food biochemical concepts are introduced, many of which are explored in detail in the chapters that follow.

INTRODUCTION

Many biochemical reactions and their products are the basis of much of food science and technology. Food scientists must be interdisciplinary in their approaches to studying and solving problems that require the integration of several disciplines, such as physics, chemistry, biology and various social sciences (e.g. sensory science, marketing, consumer attitude/acceptability). For example, in the development of food packaging materials, one must consider microbiological, environmental, biochemical (flavour/nutrient) and economic questions in addition to material/polymer science. In today's market, product development considerations may include several of the following: nutritional, environmental, microbiological (safety and probiotic), nutraceutical and religious/cultural questions in addition to cost/marketing and formulation methods. An ideal food product would promote healthy gut microflora, contain 20 g of vegetable

Food Biochemistry and Food Processing, Second Edition. Edited by Benjamin K. Simpson, Leo M.L. Nollet, Fidel Toldrá, Soottawat Benjakul, Gopinadhan Paliyath and Y.H. Hui.

protein with no limiting amino acids and have 25% of the daily fibre requirement. It would be lactose-free, nut-free, trans-fat-free, antibiotic- and pesticide-free, artificial colour-free, no sugar added and contain certified levels of phytosterols. The product would contain tasteless, odourless, mercury-free, cold-pressed, bioactive omega 3-rich fish oil harvested using animal-friendly methods. Furthermore, it would be blood sugar-stabilising and heart disease-preventing, boost energy levels, not interfere with sleep, be packaged in minimal, compostable packaging and manufactured using 'green' energy, transported by biodiesel-burning trucks and be available to the masses at a reasonable price.

At their most fundamental levels, growing crops and raising food animals, storing or ageing foods, processing via fermentation, developing food products, preparing and/or cooking, and finally ingesting food are all ways of bringing about, or preventing, biochemical changes. Furthermore, methods to combat both pathogenic and spoilage organisms are based upon biochemical effects, including acidifying their environments, heat denaturing their membrane proteins, oxygen depriving, water depriving and/or biotin synthesis inhibiting. Only recently have the basic mechanisms behind food losses and food poisoning begun to be unravelled.

Food scientists recognised long ago the importance of a biochemistry background, demonstrated by the recommendation of a general biochemistry course requirement at the undergraduate level by the Institute of Food Technologists (IFT) in the United States more than 40 years ago. Many universities in various countries now offer a graduate course in food biochemistry as an elective or have food biochemistry as a specialised area of expertise in their undergraduate and graduate programs. The complexity of this area is very challenging; a content-specific journal, the Journal of Food Biochemistry, has been available since 1977 for scholars to report their food biochemistry-related research results.

Our greater understanding of food biochemistry has followed developments in food processing technology and biotechnology, resulting in improved nutrition and food safety. For example, milk-intolerant consumers can ingest nutritious dairy products that are either lactose-free or by taking pills that contain an enzyme to reduce or eliminate lactose. People can decrease gas production resulting from eating healthy legumes by taking α-galactosidase (produced by *Aspergillus niger*) supplements with meals. Shark meat is made more palatable by controlling the action of urease on urea. Tomato juice production is improved by proper control of its pectic enzymes. Better colour in potato chips results from removal of sugars from the cut potato slices. More tender beef results from proper aging of carcasses or at the consumer level, the addition of instant marinades containing protease(s). Ripening inhibition of bananas during transport is achieved by controlling levels of the ripening hormone, ethylene, in packaging. Proper chilling of caught tuna minimises histamine production by inhibiting the activities of certain bacteria, thereby avoiding scombroid or histamine poisoning. Beyond modified atmosphere packaging, 'intelligent' packaging materials that respond to and delay certain deteriorative biochemical reactions are being developed and utilised. The above are just a few of the examples that will be discussed in more detail in this chapter and in the commodity chapters in this book.

The goal of this introductory chapter is to provide the reader with an overview of both basic and applied biochemistry as they relate to food science and technology, and to act as a segue into the following chapters. Readers are strongly encouraged to consult the references provided for further detailed information.

BIOCHEMISTRY OF FOOD CARBOHYDRATES

'Carbohydrate' literally means 'carbon hydrate,' which is reflected in the basic building block unit of simple carbohydrates, i.e. $(CH_2O)_n$. Carbohydrates make up the majority of organic mass on earth having the biologically important roles of energy storage (e.g. plant starch, animal glycogen), energy transmission (e.g. ATP, many metabolic intermediates), structural components (e.g. plant cellulose, arthropod chitin), and intra- and extracellular communication (e.g. egg-sperm binding, immune system recognition). Critical for the food industry, carbohydrates serve as the primary nutritive energy sources from foods like grains, fruits and vegetables, as well as being important ingredients for many formulated or processed foods. Carbohydrates are used to sweeten, gel, emulsify, encapsulate or bind flavours, can be altered to produce colour and flavour via various browning reactions, and are used to control humidity and water activity.

Structures

The basic unit of a carbohydrate is a monosaccharide; 2 monosaccharides bound together are called a disaccharide; 3 are called a trisaccharide, 2–10 monosaccharides in a chain are termed an oligosaccharide, and 10 or more are termed a polysaccharide. The simplest food-related carbohydrates, monosaccharides, are glucose, mannose, galactose and fructose.

Carbohydrate structures contain several hydroxyl groups (–OH) per molecule, a structural feature that imparts a high capacity for hydrogen bonding, making them very hydrophilic. This property allows them to serve as a means of moisture control in foods. The ability of a substance to bind water is termed humectancy, one of the most important properties of carbohydrates in foods. Maltose and lactose (discussed later) have relatively low humectancies; therefore, they allow for sweetness while resisting adsorption of environmental moisture. Hygroscopic (water absorbing) sugars like corn syrup and invert sugar (hydrolysed table sugar) help prevent water loss, e.g. in baked goods. In addition to the presence of hydroxyl groups, humectancy is also dependent on the overall structures of carbohydrates, e.g. fructose binds more water than glucose.

Sugar Derivatives – Glycosides

Most chemical reactions of carbohydrates occur via their hydroxyl and carbonyl groups. Under acidic conditions, the carbonyl carbon of a sugar can react with the hydroxyl of an alcohol, e.g. methanol (wood alcohol) to form *O-glycosidic bonds*. Other examples of glycosidic bonds are those between sugar carbonyl groups and amines (e.g. some amino acids as well as molecules such as DNA and RNA building blocks), as well as sugar carbonyl bonding with phosphate (e.g. phosphorylated metabolic intermediates). A glycosidic bond between a carbonyl carbon and the nitrogen of an amine group (R–NH) is termed an *N-glycosidic bond*. Similarly, reactions of carbonyl carbons with thiols (R–SH) produce *thio-glycosides*.

Food Disaccharides

The food carbohydrates sucrose, lactose and maltose (see structures below) are disaccharides (two monosaccharides joined by an *O*-glycosydic bond) and are three of the principal disaccharides used in the food industry. Used as a sweetening agent and fermentation carbon source, sucrose exists naturally in high amounts only in cane and beet, is composed of one glucose and one fructose and is a *non-reducing sugar* since it contains no free aldehyde. The enzymes responsible for catalysing the hydrolysis of sucrose to glucose and fructose are sucrases and invertases, which catalyse the hydrolysis of the sucrose glycosidic bond. Acid and heat also cause hydrolysis of sucrose, which is important in the commercial production of invert sugar, where sucrose is partially converted to glucose and fructose, thereby producing increased sweetness and water-binding ability.

The disaccharide lactose is made up of galactose and glucose, and is often referred to as milk sugar. Lactose is hydrolysed by lactase (EC 3.2.1.108), an enzyme of the β-galactosidase enzyme class, produced by various mammals, bacteria and fungi. By adulthood, some humans produce insufficient amounts of lactase, thereby restricting the consumption of dairy product in significant quantities. In deficient persons, a failure to hydrolyse lactose in the upper intestine results in this simple sugar passing into the large intestine, which in turn results in an influx of water as well as fermentation by lower gut bacteria, leading to bloating, cramping and diarrhoea. Lactose levels in dairy products can be reduced by treatment with lactase or by lactic acid bacteria fermentation. 'Lactose-free' milk products are widely sold and marketed with most of the lactose hydrolysed by treatment with lactase. Also, fermented dairy foods like yoghurt and cheese contain less lactose compared to the starting materials where the lactose is converted into lactic acid by bacteria, e.g. old cheddar cheese contains virtually no remaining lactose.

Maltose is composed of two glucose units and is derived from starch by treatment with β-amylase, thereby increasing the sweetness of the reaction mixture. The term *malt* (beer making) refers to the product where β-amylase, produced during the germination, has acted on the starch of barley or other grains when steeped in water.

Carbohydrate Browning Reactions

There are three general categories of browning reactions in foods: oxidative/enzymatic browning, *caramelisation* and non-oxidative/non-enzymatic/*Maillard browning*. Oxidative browning is discussed later in the section on proteins. The latter two types of browning involve carbohydrate reactions. Caramelisation involves a complex group of reactions that are the result of direct heating of carbohydrates, particularly sugars. Dehydration reactions result in the formation of double bonds along with the polymerisation of ring structures that absorb different light wavelengths, hence the flavour development, darkening and colour formation in such mixtures. Two important roles of caramelisation in the food industry are caramel flavour and colour production, a processes in which sucrose is heated in solution with acid or acid ammonium salts to produce a variety of products in food, candies and beverages (Ko et al. 2006).

The Maillard reaction is one of the most important reactions encountered in food systems, and it is also called non-enzymatic or non-oxidative browning. Reducing sugars and amino acids or other nitrogen-containing compounds react to produce *N*-glycosides displaying red-brown to very dark brown colours, caramel-like aromas, and colloidal and insoluble melanoidins. There are a complex array of possible reactions that can take place via Maillard chemistry, and the aromas, flavours and colours can be desirable or undesirable (BeMiller and Whistler 1996). Lysine is a nutritionally essential amino acid and its side chain can react during the Maillard reaction, thereby lowering the nutritional value of foods. Other amino acids that may be lost due to the Maillard reaction include the basic amino acids L-arginine and L-histidine.

Starch

Polysaccharides, or glycans, are made up of glycosyl units in a linear or branched structure. The three major food-related glycans are amylose, amylopectin and cellulose (cellulose is discussed below), which are all chains of D-glucose, but are structurally distinct based on the types of glycosidic linkages that join the glucose units and the amount of branching in their respective structures. Both amylose and amylopectin are components of starch, the energy storage molecules of plants, and cellulose is the structural carbohydrate that provides structural rigidity to plants. Starch is a critical nutritional component of many foods, especially flour-based foods, tubers, cereal grains, corn and rice. Starch can be both linear (amylose) and branched (amylopectin). Amylose glucose units are joined only by α-1,4-linkages and it usually contains 200–3000 units. Amylopectin also contains α-1,4-linkages, but additionally it has branch points at α-1,6 linkages that occur approximately every 20–30 α-linkages. The branched molecules of amylopectin produce bulkier structures than amylose. Most starches contain approximately 25% amylose, although amylose contents as high as 85% are possible. Starches containing only amylopectin are termed waxy starches.

Starch exists in *granules* that are deposited in organelles called amyloplasts. Granule size and shape vary with plant source, and they contain a cleft called the hilium, which serves as a nucleation point around which the granule develops as part of plant energy storage. Granules vary in size from 2 to 130 μ and they have a crystalline structure such that the starch molecules align radially within the crystals.

Metabolism of Carbohydrates

The characteristics of carbohydrates in both their natural states and as processed food ingredients determine the properties of many foods as well as their utilisation as nutrients. *Glycolysis* is a fundamental pathway of metabolism consisting of a series of reactions in the cytosol, where glucose is converted to pyruvate via nine enzymatically catalysed reactions (see Figure 1.1).

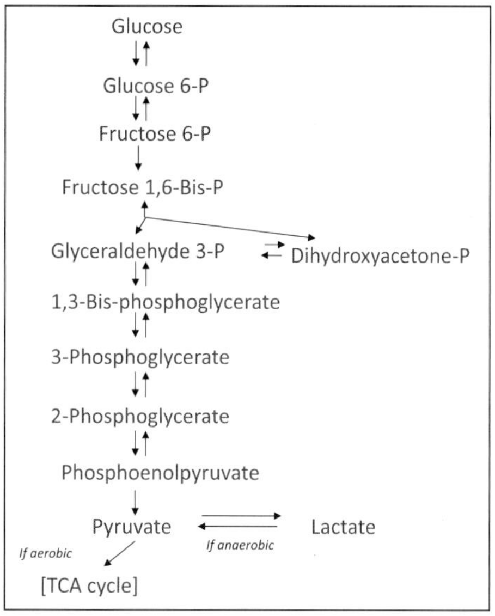

Figure 1.1. Outline of glycolysis; 3-C reactions beginning with glyceraldehyde 3-P occur twice per glucose.

Glycolytic processing of each glucose molecule results in a modest gain of only two ATPs (the universal biochemical energy currency). Although the gain of two ATPs is small, the creation of pyruvate feeds another metabolic pathway, the tricarboxylic acid (TCA) cycle, which yields two more ATPs. More importantly, glycolysis and the TCA cycle also generate the reduced forms of nicotinamide adenine dinucleotide (NADH) and flavin adenine dinucleotide ($FADH_2$), which drive subsequent oxidative phosphorylation of ADP. An overview of the TCA cycle is depicted in Figure 1.2. NADH and $FADH_2$ generated by glycolysis and the TCA cycle are subsequently part of oxidative phosphorylation, where they transfer their electrons to O_2 in a series of electron transfer reactions whose high energy potential is used to drive phosphorylation. Overall, a net yield of 30 ATPs is gained per glucose molecule as the result of glycolysis, the TCA cycle and oxidative phosphorylation.

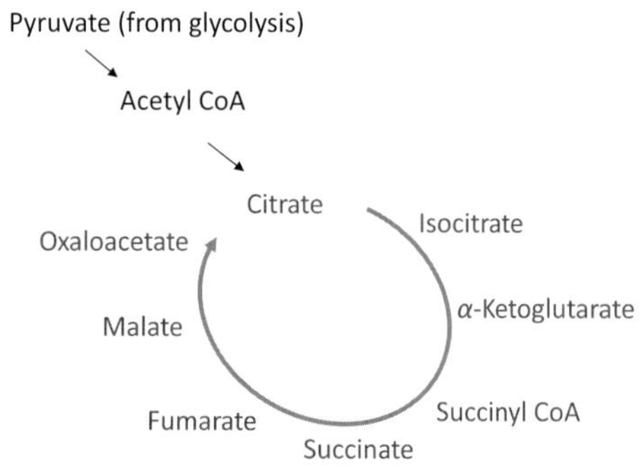

Figure 1.2. An overview of the TCA cycle.

Intermediates of carbohydrate metabolism play an important role in many food products. The conversion products of glycogen in fish and mammalian muscles are now known to utilise different pathways, but ultimately result in glucose-6-phosphate, leading into glycolysis. Lactic acid formation is an important phenomenon in rigor mortis, and souring and curdling of milk as well as in manufacturing sauerkraut and other fermented vegetables. Ethanol is an important end product in the production of alcoholic beverages, bread making and in some overripe fruits to a lesser extent. The TCA cycle is also important in alcoholic fermentation, cheese maturation and fruit ripening. In bread making, α-amylase (added or present in the flour itself) partially hydrolyses starch to release glucose units as an energy source for yeast growth and development, which is important for the dough to rise during fermentation.

During germination of cereal grains, glucose and glucose phosphates or fructose phosphates are produced from starch. Some of the relevant biochemical reactions are summarised in Table 1.1. The sugar phosphates are then converted to pyruvate via glycolysis, which is utilised in various biochemical reactions. The glucose and sugar phosphates can also be used in the building of various plant structures, e.g. cellulose is a glucose polymer and is the major structural component of plants.

In addition to starch, plants also possess complex carbohydrates, e.g. cellulose, β-glucans and pectins. Both cellulose and β-glucans are composed of glucose units bound by β-glycosidic linkages that cannot be metabolised in the human body. They are important carbohydrate reserves in plants that can be metabolised into smaller molecules for utilisation during seed germination. Pectic substances (pectins) act as the "glue" among cells in plant tissue and also are not metabolised in the human body. Together with cellulose and β-glucans, pectins are classified as dietary fibre.

Derivatives of cellulose can be made through chemical modification under strongly basic conditions where side chains such as methyl and propylene react and bind at sugar hydroxyl groups. The resultant derivatives are ethers (oxygen bridges) joining sugar residues and the side chain groups (Coffee et al. 1995). A major function of cellulose derivatives is to act as a bulking agent in food products. Two examples of important food-related derivatives of cellulose are carboxymethylcellulose and methylcellulose.

Pectin substances include polymers composed mainly of α-(1,4)-D-galacturonopyranosyl units and constitute the middle lamella of plant cells. Pectins exist in the propectin form in unripe (green) fruit, contributing to firm, hard structures. Upon ripening, propectins are metabolised into smaller molecules, giving ripe fruits a soft texture. Controlling enzymatic activity against propectin is commercially important in fruits such as tomatoes, apples and persimmons. Research and the development of genetically modified tomatoes allowed uniform ripening prior to processing and consumption. Fuji apples can be kept in the refrigerator for a much longer time than other varieties of apples before reaching the soft, grainy texture stage due to a lower pectic enzyme activity. Persimmons are hard in the unripe stage, but can be ripened to a very soft texture as a result of pectic- and starch-degrading enzymes. Table 1.2

Table 1.1. Starch Degradation During Cereal Grain Germination

Enzyme	Reaction
α-Amylase (EC 3.2.1.1)	Starch → Glucose + Maltose + Maltotriose + α-Limit dextrins + Linear maltosaccharides
Hexokinase (EC 2.7.1.1)	Glucose + ATP → Glucose-6-phosphate + ADP
α-Glucosidase (maltase, EC 3.2.1.20)	Hydrolysis of terminal, non-reducing 1,4-linked α-D-glucose residues releasing α-D-glucose
Oligo-1,6 glucosidase (limit dextrinase, isomaltase, EC 3.2.1.10)	α-Limit dextrin → Linear maltosaccharides
β-Amylase (EC 3.2.1.2)	Linear maltosaccharides → Maltose
Phosphorylase (EC 2.4.1.1)	Linear maltosaccharides + Phosphate → α-D-Glucose-1-phosphate
Phosphoglucomutase (EC 5.4.2.2)	α-D-Glucose-1-phosphaate → α-D-Glucose-6-phosphate
Glucosephosphate isomerase (EC 5.3.1.9)	D-Glucose-6-phosphate → D-Fructose-6-phosphate
UTP-glucose-1-phosphate uridyl transferase (UDP-glucose pyrophosphorylase, EC 2.7.7.9)	UTP + α-D-Glucose-1-phosphate → UDP-glucose + Pyrophosphate
Sucrose phosphate synthetase (EC 2.4.1.14)	UDP-glucose + D-Fructose-6-phosphate → Sucrose phosphate + UDP
Sugar phosphate phosphohydrolase (sugar phosphatase, EC 3.1.3.23)	Sugar phosphate (fructose-6-phosphate) → Sugar (fructose) + Inorganic phosphate
Sucrose phosphatase (EC 3.1.3.24)	Sucrose-6-F-phosphate → Sucrose + Inorganic phosphate
Sucrose synthetase (EC2.4.1.13)	NDP-glucose + D-Fructose → Sucrose + NDP
β-Fructose-furanosidase (invertase, succharase, EC 3.2.1.26)	Sucrose → Glucose + Fructose

Source: Duffus 1987, Kruger and Lineback 1987, Kruger et al. 1987, Eskin 1990, Hoseney 1994, IUBMB-NC website (www.iubmb.org).

Table 1.2. Degradation of Complex Carbohydrates

Enzyme	Reaction
Cellulose degradation during seed germination[a]	
Cellulase (EC 3.2.1.4)	Endohydrolysis of 1,4-β-glucosidic linkages in cellulose and cereal β-D-glucans
Glucan 1,4-β-glucosidase (exo-1,4, β-glucosidase, EC 3.2.1.74)	Hydrolysis of 1,4 linkages in 1,4-β-D-glucan so as to remove successive glucose units
Cellulose 1,4-β-cellubiosidase (EC 3.2.1.91)	Hydrolysis of 1,4-β-D-glucosidic linkages in cellulose and cellotetraose releasing cellubiose from the non-reducing ends of the chains
β-Galactosan degradation[a]	
β-Galactosidase (EC 3.21.1.23)	β-(1→4)-linked galactan → D-Galactose
β-Glucan degradation[b]	
Glucan endo-1,6, β-glucosidase (EC 3.2.1.75)	Random hydrolysis of 1,6 linkages in 1,6-β-D-glucans
Glucan endo-1,4, β-glucosidase (EC 3.2.1.74)	Hydrolysis of 1,4 linkages in 1,4-β-D-glucans so as to remove successive glucose units
Glucan endo-1,3-β-D-glucanase (EC 3.2.1.58)	Successive hydrolysis of β-D-glucose units from the non-reducing ends of 1,3-β-D-glucans, releasing α-glucose
Glucan 1,3-β-glucosidase (EC 3.2.1.39)	1,3-β-D-glucans → α-D-glucose
Pectin degradation[b]	
Polygalacturonase (EC 3.2.1.15)	Random hydrolysis of 1,4-α-D-galactosiduronic linkages in pectate and other galacturonans
Galacturan 1,4-α-galacturonidase (EC 3.2.1.67)	$(1,4\text{-}\alpha\text{-D-Galacturoniside})_n + H_2O \rightarrow (1,4\text{-}\alpha\text{-D-Galacturoniside})_{n-1} +$ D-Galacturonate
Pectate lyase (pectate transeliminase, EC 4.2.2.2)	Eliminative cleavage of pectate to give oligosaccharides with 4-deoxy-α-D-galact-4-enuronosyl groups at their non-reducing ends
Pectin lyase (EC 4.2.2.10)	Eliminative cleavage of pectin to give oligosaccharides with terminal 4-deoxy-6-methyl-α-D-galact-4-enduronosyl groups

Source: [a]Duffus 1987, Kruger and Lineback 1987, Kruger et al. 1987, Smith 1999, [b]Eskin 1990, IUBMB-NC website (www.iubmb.org).

Table 1.3. Changes in Carbohydrates in Cheese Manufacturing

Action, Enzyme or Enzyme System	Reaction
Formation of lactic acid	
Lactase (EC 3.2.1.108)	Lactose + H_2O→ D-Glucose + D-Galactose
Tagatose pathway	Galactose-6-P → Lactic acid
Embden–Meyerhoff pathway	Glucose → Pyruvate → Lactic acid
Formation of pyruvate from citric acid	
Citrate *(pro-3S)* lyase (EC 4.1.3.6)	Citrate → Oxaloaceate
Oxaloacetate decarboxylase (EC 4.1.1.3)	Oxaloacetate → Pyruvate + CO_2
Formation of propionic and acetic acids	
Propionate pathway	3 Lactate → 2 Propionate + 1 Acetate + CO_2 + H_2O
	3 Alanine → Propionic acid + 1 Acetate + CO_2 + 3 Ammonia
Formation of succinic acid	
Mixed acid pathway	Propionic acid + CO_2 → Succinic acid
Formation of butyric acid	
Butyric acid pathway	2 Lactate → 1 Butyrate + CO_2 + $2H_2$
Formation of ethanol	
Phosphoketolase pathway	Glucose → Acetylaldehyde → Ethanol
Pyruvate decarboxylase (EC 4.1.1.1)	Pyruvate → Acetylaldehyde + CO_2
Alcohol dehydrogenase (EC 1.1.1.1)	Acetylaldehyde + NAD + H^+ → Ethanol + NAD^+
Formation of formic acid	
Pyruvate-formate lyase (EC 2.3.1.54)	Pyruvate + CoA → Formic acid + Acetyl CoA
Formation of diacetyl, acetoin, 2,3-butylene glycol	
Citrate fermentation pathway	Citrate → Pyruvate → Acetyl CoA → Diacetyl → Acetoine → 2,3-Butylene glycol
Formation of acetic acid	
Pyruvate-formate lyase (EC 2.3.1.54)	Pyruvate + CoA → Formic acid + Acetyl CoA
Acetyl-CoA hydrolase (EC 3.1.2.1)	Acetyl CoA + H_2O→ Acetic acid + CoA

lists some of the enzymes and their reactions related to these complex carbohydrates.

Metabolism of Lactose in Cheese Production

Milk does not contain high-molecular weight carbohydrates; however, it does contain lactose. Lactose can be enzymatically degraded to glucose and galactose-6-phosphate by the enzyme lactase, which can be produced by lactic acid bacteria. Glucose and galactose-6-phosphate are then further metabolised to various smaller molecules through various biochemical reactions that are important in the flavour development of various cheeses, e.g. butyric acid via lactic acid decarboxylation. Table 1.3 lists some of these enzymatic reactions.

Removal of Glucose in Egg Powder

Glucose is present in very small quantities in egg albumen and egg yolk; however, it can undergo non-enzymatic reactions, e.g. Maillard reactions, which lower the quality of the final products. This problem can be overcome by using the glucose oxidase–catalase system. Glucose oxidase converts glucose to gluconic acid and hydrogen peroxide, which then decomposes into water and oxygen by the action of catalase. This process is used almost exclusively for whole egg and other yolk-containing products. However, for dehydrated egg albumen, bacterial and/or yeast fermentation is used to remove glucose.

Production of Starch Sugars and Syrups

The hydrolysis of starch by means of enzymes (α- and β-amylases) and/or acid to produce glucose (dextrose, D-glucose) and maltose syrups has resulted in the availability of various starch syrups, maltodextrins, maltose and glucose for the food, pharmaceutical and other industries. In the 1950s, researchers discovered that some xylose isomerase (D-xylose-keto-isomerase, EC 5.3.1.5) preparations possessed the ability to convert D-glucose to D-fructose. In the early 1970s, researchers developed immobilised enzyme technology for various applications. Since fructose is sweeter than glucose, xylose isomerase was successfully applied to this new technology with the production of high-fructose syrup (called high-fructose corn syrup in the United States; Carasik and Carroll 1983). High-fructose syrups have since replaced most of the glucose syrups in the soft drink industry.

FOOD PROTEIN BIOCHEMISTRY

PROPERTIES OF AMINO ACIDS

Proteins are polymers of *amino acids* joined by *peptide bonds*. Twenty amino acids commonly exist and their possible combinations result in the potential for an incredibly large number of sequence and 3D structural protein variants. Amino acids consist of a carbon atom (C_α) that is covalently bonded to an amino group and a carboxylic acid group. Thus, they have an N–C_α–C 'backbone'. In addition, the C_α is bound to a hydrogen atom and one of 20 'R groups' or 'side chains', hence the general formula:

$$+NH_3 - CHR - COO^-$$

The above general formula describes 19 of the 20 amino acids except proline, whose side chain is irregular, in that it is covalently bound to both the α-carbon and the backbone nitrogen. The covalent bonds among different amino acids in a protein are called peptide bonds. The 20 amino acids can be divided into three categories based on R-group differences: non-polar, polar and charged polar. The functional properties of food proteins are directly attributable to the amino acid R-group properties: structural (size, shape and flexibility), ionic (charge and acid–base character) and polarity (hydrophobicity/hydrophilicity).

At neutral pH, most free amino acids are zwitterionic, i.e. they are dipolar ions, carrying both a positive and negative charge, as shown in the general formula above. Since the primary amino and carboxyl groups of amino acids are involved in peptide bonds within a protein, it is only the R groups (and the ends of the peptide chains) that contribute to charge; the charge of a protein being determined by the charge states of the ionisable amino acid R groups that make up the polypeptide, namely aspartic acid (Asp), glutamic acid (Glu), histidine (His), lysine (Lys), arginine (Arg), cysteine (Cys) and tyrosine (Tyr). The acidic amino acids are Tyr, Cys, Asp and Glu (note: Tyr and Cys require pH above physiologic pH to act as acids, and therefore, are less important charge contributors in living systems). Lys, Arg and His are basic amino acids.

The amino acid sequence and properties determine overall protein structure. Some examples are as follows: Two residues of opposite charges can form a salt bridge. For example, Lys and Asp typically have opposite charges under the same conditions, and if the side chains are proximate, then the negatively charged carboxylate of Asp can salt-bridge to the positively charged ammonium of Lys. Another important inter-residue interaction is covalent bonding between Cys side chains. Under oxidising conditions, the sulfhydryl groups of Cys side chains (–S–H) can form a thiol covalent bond (–S–S–) also known as a *disulphide bond*. Lastly, hydrophobic (non-polar) amino acids are generally sequestered away from the solvent in aqueous solutions (which is not always the case for food products), since interaction with polar molecules is not energetically stable.

PROTEIN NUTRITIONAL CONSIDERATIONS

In terms of survival and good health, the contributions of protein in the diet are to provide adequate levels of what are referred to as 'essential amino acids' (Lys, methionine, phenylalanine, threonine, tryptophan, valine, leucine and isoleucine), amino acids that are either not produced in sufficient quantities or not at all by the body to support building/repairing and maintaining tissues as well as protein synthesis. Single source plant proteins are referred to as incomplete proteins since they do not have sufficient quantities of the essential amino acids in contrast to animal proteins, which are complete. For example, cereals are deficient in Lys, while oilseeds and nuts are deficient in Lys as well as methionine. In order for plant proteins to become 'complete', complementary sources of proteins must be consumed, i.e. the deficiency of one source is complemented by an excess from another source, thus making the combined protein 'complete'.

Although some amino acids are *gluconeogenic*, meaning that they can be converted to glucose, proteins are not a critical source of energy. Dietary protein breakdown begins with cooking (heat energy) and chewing (mechanical energy) followed by acid treatment in the stomach (chemical energy) as well as the mechanical actions of the upper GI. The 3D structures of proteins are partially lost due to such forces and are said to 'unfold' or denature.

In addition to protein denaturation, the stomach and upper intestine produce two types of *proteases* (enzymes that hydrolyse peptide bonds) that act on dietary proteins. *Endopeptidases* are proteases that cleave interior peptide bonds of polypeptide chains, while *exopeptidases* are proteases that cleave at the ends of proteins exclusively. Pepsin, an acid protease that functions optimally at extremely low pH of the stomach, releases peptides from muscle and collagen proteins. In the upper intestine, serine proteases trypsin and chymotrypsin further digest peptides, yielding free amino acids for absorption into the blood (Champe et al. 2005). An important consideration regarding the nutritional quality of proteins is the effect of processing. Heat and chemical treatments can serve to unfold proteins, thereby aiding to increase enzymatic hydrolysis, i.e. unfolded proteins have a larger surface area for enzymes to act. This may increase the amino acid bioavailability, but it can also lead to degradative/transformative reactions of amino acids, e.g. deamidation of asparagine and glutamine, reducing these amino acids as nutrient sources.

ANIMAL PROTEIN STRUCTURE AND PROTEOLYSIS IN FOOD SYSTEMS

Animal tissues have similar structures despite minor differences between land and aquatic (fish and shellfish) animal tissues. Post-mortem, meat structure breaks down slowly, resulting in desirable tenderisation and eventual undesirable degradation/spoilage. Understanding meat structure is critical to understanding these processes, and Table 1.4 lists the location and major functions of myofibrillar proteins associated with the contractile apparatus and cytoskeletal framework of animal tissues. Individual muscle fibres are composed of myofibrils, which are the basic units of muscular contraction. The skeletal muscle of fish differs from that of mammals, in that the fibres arranged between the sheets of connective tissue are much shorter. The connective tissue appears as short, transverse sheets

Table 1.4. Locations and Major Functions of Myofibrillar Proteins Associated with Contractile Apparatus and Cytoskeletal Framework

Location	Protein	Major Function
Contractile apparatus		
A-band	Myosin	Muscle contraction
	c-Protein	Binds myosin filaments
	F-, H-, I-Proteins	Binds myosin filaments
M-line	M-Protein	Binds myosin filaments
	Myomesin	Binds myosin filaments
	Creatine kinase	ATP synthesis
I-band	Actin	Muscle contraction
	Tropomyosin	Regulates muscle contraction
	Troponins T, I, C	Regulates muscle contraction
	β-, γ-Actinins	Regulates actin filaments
Cytoskeletal framework		
GAP filaments	Connectin (titin)	Links myosin filaments to Z-line
N_2-line	Nebulin	Unknown
By sarcolemma	Vinculin	Links myofibrils to sarcolemma
Z-line	α-Actinin	Links actin filaments to Z-line
	Eu-actinin, filamin	Links actin filaments to Z-line
	Desmin, vimmentin	Peripheral structure to Z-line
	Synemin, Z-protein, Z-nin	Lattice structure of Z-line

Source: Eskin 1990, Lowrie 1992, Huff-Lonergar and Lonergan 1999, Greaser 2001.

(myocommata) that divide the long fish muscles into segments (myotomes) corresponding in numbers to those of the vertebrae. A fine network of tubules, the sarcoplasmic reticulum, separates the individual myofibrils, and within each fibre is a liquid matrix referred to as the sarcoplasm-containing enzymes, mitochondria (cellular powerhouse), glycogen (carbohydrate storage form in animals), adenosine triphosphate (energy currency), creatine (part of energy transfer in muscle) and myoglobin (oxygen transport molecule). The basic unit of the myofibril is the sarcomere, which is made up of a thick set of filaments consisting mainly of myosin, a thin set containing primarily of F-actin and a filamentous 'cytoskeletal structure' composed of connectin and desmin. Meat tenderisation is a very complex, multi-factorial process involving glycolysis and the actions of both endogenous proteases (e.g. cathepsins and calpains) as well as intentionally added enzymes. Table 1.5 lists some of the more common enzymes used in meat tenderisation. Papain, ficin and bromelain are proteases of plant origin that efficiently break down animal proteins applied in meat tenderisation industrially or at the household/restaurant levels. Enzymes such as pepsins, trypsins and cathepsins cleave animal tissues at various sites of peptide chains, while enteropeptidase (enterokinase) is also known to activate trypsinogen by cleaving its Lys6-Ile7 peptide bond. Plasmin, pancreatic elastase and collagenase are responsible for the breakdown of animal connective tissues.

Chymosin (rennin) is the primary protease critical for the initial milk clotting step in cheese making and is traditionally obtained from calf stomach. Lactic acid bacteria (starter) gradually acidify milk to the pH 4.7, the optimal pH for coagulation by chymosin. Most lactic acid starters have limited proteolytic activities, i.e. product proteins are not degraded fully as in the case of GI tract breakdown of dietary proteins. The proteases and peptidases breakdown milk caseins to smaller protein molecules that, combined with milk fat, provide the cheese structure. Other enzymes such as decarboxylases, deaminases and transaminases are responsible for the degradation of amino acids into secondary amines, indole, α-keto acids and other compounds that give the typical flavour of cheeses (see Table 1.6 for enzymes and their reactions).

Germinating seeds also undergo proteolysis, although in a much lower amount relative to meat. Aminopeptidase and carboxypeptidase A are the main, known enzymes (Table 1.7) here that produce peptides and amino acids needed in the growth of the plant.

In beer production, a small amount of protein is dissolved from the wheat and malt into the wort. The protein fraction extracted from the wort may precipitate if present in the resulting beer due to its limited solubility at lower temperatures, resulting in hazing. Proteases of plant origin such as papain, ficin and bromelain break down such proteins to reduce this 'chill-haze' problem in the brewing industry.

Protein Modifications

A protein's amino acid sequence is critical to its physicochemical properties, and it follows that changes made to individual amino acids may alter its functionality. In addition to the many chemical alterations that may occur to amino acids during

Table 1.5. Proteases in Animal Tissues and Their Degradation

Enzyme	Reaction
Aspartic proteases	
Pepsin A (pepsin, EC 3.4.23.1)	Preferential cleavage, hydrophobic, preferably aromatic, residues in P1 and P'1 positions
Gastricsin (pepsin C, EC 3.4.23.3)	More restricted specificity than pepsin A; high preferential cleavage at Tyr bond
Cathepsin D (EC 3.4.23.5)	Specificity similar to, but narrower than that of pepsin A
Serine proteases	
Trypsin (a- and b-trypsin, EC 3.4.21.4)	Cleavage to the C-terminus of Arg and Lys
Chymotrypsin (Chymotrypsin A and B, EC 3.4.21.1)	Preferential cleavage: Tyr-, Trp-, Phe-, Leu-
Chymotrysin C (EC 3.4.21.2)	Preferential cleavage: Leu-, Tyr-, Phe-, Met-, Trp-, Gln-, Asn-
Pancreatic elastase (pancreato-peptidase E, pancreatic elastase I, EC 3.4.21.36)	Hydrolysis of proteins, including elastin. Preferential cleavage: Ala
Plasmin (fibrinase, fibrinolysin, EC 3.4.21.7)	Preferential cleavage: Lys >Arg; higher selectivity than trypsin
Enteropeptidase (enterokinase, EC 3.4.21.9)	Activation of trypsinogen by selective cleavage of Lys6-Ile7 bond
Collagenase	Hydrolysis of collagen into smaller molecules
Thiol/cysteine proteases	
Cathepsin B (cathepsin B1, EC 3.4.22.1)	Broad speicificity, Arg–Arg bond preference in small peptides
Papain (EC 3.4.22.2)	Broad specificity; preference for large, hydrophobic amino acid at P2; does not accept Val at P1′
Fiacin (ficin, EC 3.4.23.3)	Similar to that of papain
Bromelain (EC 3.4.22.4)	Broad specificity similar to that of pepsin A
γ-Glutamyl hydrolase (EC 3.4.22.12 changed to 3.4.1.99)	Hydrolyses γ-glutamyl bonds
Cathepsin H (EC 3.4.22.16)	Protein hydorlysis; acts also as an aminopeptidase and endopeptidase (notably cleaving Arg bond)
Calpain-1 (EC 3.4.22.17 changed to 3.4.22.50)	Limited cleavage of tropinin I, tropomyosin, myofibril C-protein, cytoskeletal proteins; activates phosphorylase, kinase, and cyclic-nucleotide-dependent protein kinase
Metalloproteases	
Procollagen *N*-proteinase (EC 3.4.24.14)	Cleaves *N*-propeptide of pro-collagen chain α1(I) at Pro+Gln and α1(II), and α2(I) at Ala+Gln

Source: Eskin 1990, Haard 1990, Lowrie 1992, Huff-Lonergan and Lonergan 1999, Gopakumar 2000, Jiang 2000, Simpson 2000, Greaser 2001, IUBMB-NC website (www.iubmb.org).

food processing, e.g. deamidation, natural, enzymatic protein modifications collectively known as *post-translational modifications* may also occur upon their expression in cells. Some examples are listed in Table 1.8.

Protein Structure

Protein folding largely occurs as a means to minimise the energy of the system where hydrophobic groups are maximally shielded from aqueous environments and while the exposure of hydrophilic groups to aqueous environments is maximised. Protein structures follow a hierarchy: primary, secondary, tertiary and quaternary structures. Primary structure refers to the amino acid sequence; secondary structures are the structures formed by amino acid sequences (e.g. α-helix, β-sheet, random coil); tertiary structures are the 3D structures made up of secondary structures (the way that helices, sheets and random coils pack together) and quaternary structure refers to the association of tertiary structures (e.g. two subunits of an enzyme) in oligomeric proteins. The overall shapes of proteins fall into two general types: globular and fibrous. Enzymes, transport proteins and receptor proteins are examples of globular proteins having a compact, spherical shape. Hair keratin and muscle myosin are examples of fibrous proteins having elongated structures that are simple compared to globular proteins.

Oxidative Browning

Oxidative browning, also called enzymatic browning, involves the actions of a group of enzymes generally referred to as polyphenol oxidase (PPO) or phenolase. PPO is normally compartmentalised in tissue such that oxygen is unavailable. Injury or cutting of plant material, especially apples, bananas, pears and lettuce, results in decompartmentalisation, making O_2 available

Table 1.6. Proteolytic Changes in Cheese Manufacturing

Enzyme	Reaction
Coagulation	
Chymosin (rennin, EC 3.4.23.4)	κ-Casein → Para-κ-casein + Glycopeptide, similar to pepsin A
Proteolysis	
Proteases	Proteins → High-molecular-weight peptides + Amino Acids
Amino peptidases, dipeptidases, tripeptidases	Low molecular weight peptides → Amino acids
Proteases, endopeptidases, aminopeptidases	High-molecular weight peptides → Low molecular weight peptides
Decomposition of amino acids	
Aspartate transaminase (EC 2.6.1.1)	L-Asparate + 2-Oxoglutarate → Oxaloacetate + L-Glutamate
Methionine γ-lyase (EC 4.4.1.11)	L-Methionine → Methanethiol + NH_3 + 2-Oxobutanolate
Tryptophanase (EC 4.1.99.1)	L-Tryptophan + H_2O → Indole + Pyruvate + NH_3
Decarboxylases	Lysine → Cadaverine
	Glutamate → Aminobutyric acid
	Tyrosine → Tyramine
	Tryptophan → Tryptamine
	Arginine → Putrescine
	Histidine → Histamine
Deaminases	Alanine → Pyruvate
	Tryptophan → Indole
	Glutamate → α-Ketoglutarate
	Serine → Pyruvate
	Threonine → α-Ketobutyrate

Source: Schormuller 1968, Kilara and Shahani 1978, Law 1984a, 1984b, Grappin et al. 1985, Gripon 1987, Kamaly and Marth 1989, Khalid and Marth 1990, Steele 1995, Walstra et al. 1999 (www.iubmb.org).

Table 1.7. Protein Degradation in Germinating Seeds

Enzyme	Reaction
Aminopeptidase (EC 3.4.11.xx)	Neutral or aromatic aminoacyl-peptide + H_2O → Neutral or aromatic amino acids + Peptide
Carboxypeptidase A (EC 3.4.17.1)	Release of a C-terminal amino acid, but little or no action with -Asp, -Glu, -Arg, -Lys or -Pro

Source: Stauffer 1987a, 1987b, Bewley and Black 1994, IUBMB-NC website (www.iubmb.org).

Table 1.8. Amino Acid Modifications

Amino Acid	Modification	Product
Arginine	Deamination	Citrulline
Glutamine	Deamination	Glutamic acid
Asparagine	Deamination	Aspartic acid
Various	C-terminal amidation	Amidated amino acid/protein
Serine	Phosphorylation	Phosphoserine
Histidine	Phosphorylation	Phosphohistidine
Tyrosine	Phosphorylation	Phosphotyrosine
Threonine	Phosphorylation	Phosphothreonine
Asparagine	Glycosylation	Various; *N*-linked glycoprotein
Serine	Glycosylation	Various; *O*-linked glycoprotein
Threonine	Glycosylation	Various; *O*-linked glycoprotein

to PPO for subsequent action on the phenolic ring of Tyr. Phenolics are hydroxylated, thus producing diphenols that are then subsequently oxidised to quinones:

$$2\,\text{Tyrosine}\ O_2{+}BH_2 \rightarrow 2\,\text{Dihydroxylphenylamine} + O_2$$
$$\rightarrow 2\,\text{o-Benzoquinone} + 2H_2O$$

The action of PPO can be desirable in various food products, such as raisins, prunes, dates, cider and tea; however, the extent of browning needs to be controlled. The use of reducing compounds is the most effective control method for PPO browning. The most widespread anti-browning treatment used by the food industry was the addition of sulfiting agents; however, due to safety concerns (e.g. allergenic-type reactions), other methods have been developed, including the use of other reducing agents (ascorbic acid and analogues, Cys, glutathione), chelating agents (phosphates, EDTA), acidulants (citric acid, phosphoric acid), enzyme inhibitors, enzyme treatment and complexing agents (e.g. copolymerised β-cyclodextrin or polyvinylpolypyrrolidone; Sapers et al. 2002). Application of these PPO activity inhibitors is strictly regulated in different countries (Eskin 1990, Gopakumar 2000, Kim et al. 2000). Oxidative browning is one of three types of browning reactions important in food colour, the other two being non-oxidative/Maillard browning and caramelisation (covered above and extensively in food chemistry texts; see Damodaran et al. 2008).

Enzymatic Texture Modifications

Transglutaminase (TGase, EC 2.3.2.13, protein-glutamine-y-glutamyltransferase) catalyses acyl transfer between R group carboxyamides of glutamine residues in proteins, peptides and various primary amines; the ε-amino group of Lys acts as acyl acceptor, resulting in polymerisation and inter- or intra-molecular cross-linking of proteins via formation of ε-(-y-glutamyl) Lys linkages via exchange of the Lys ε-amino group for ammonia at the carboxyamide group of a glutamine residue. Formation of covalent cross-links between proteins is the basis for TGase-based modification of food protein physical properties. The primary applications of TGase in seafood processing have been for cold restructuring, cold gelation of pastes and gel-strength enhancement through myosin cross-linking.

Quality Index

Trimethylamine and its *N*-oxide have long been used as indices for freshness in fishery products. Degradation of trimethylamine and its *N*-oxide leads to the formation of ammonia and formaldehyde with undesirable odours. The pathway on the production of formaldehyde and ammonia from trimethylamine and its *N*-oxide is shown in Figure 1.3.

Most live pelagic and scombroid fish (e.g. tunas, sardines and mackerel) contain an appreciable amount of His in the free state. In post-mortem scombroid fish, the free His is converted by the bacterial enzyme His decarboxylase into free histamine. Histamine is produced in fish caught 40–50 hours after death when

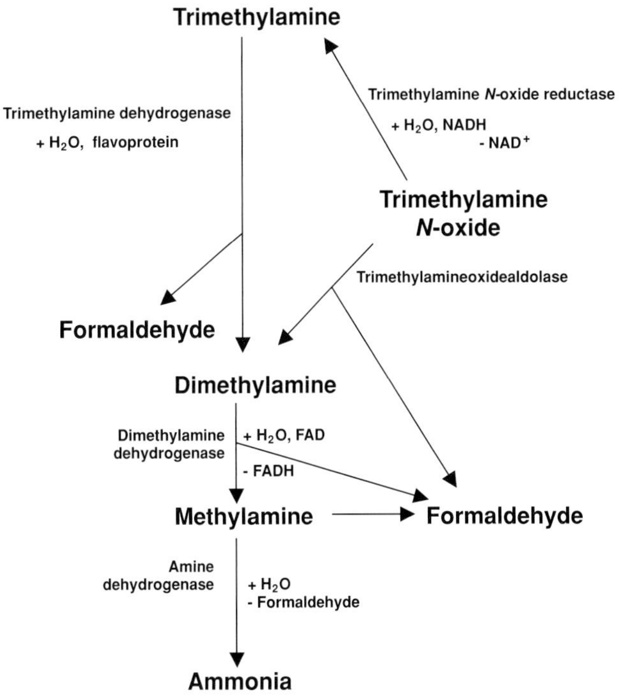

Figure 1.3. Degradation of trimethylamine and its *N*-oxide. Trimethylamine *N*-oxide reductase (EC 1.6.6.9), trimethylamine dehydrogenase (EC 1.5.8.2), dimethylamine dehydrogenase (EC 1.5.8.1), amine dehydrogenase (EC 1.4.99.3). (From Haard et al. 1982, Gopakumar 2000, Stoleo and Rehbein 2000, IUBMB-NC website (www.iubmb.org).)

fish are not properly chilled. Improper handling of tuna and mackerel after harvest can produce enough histamine to cause food poisoning (called scombroid or histamine poisoning), resulting in facial flushing, rashes, headache and gastrointestinal disorder. These disorders seem to be strongly influenced by other related biogenic amines, such as putrescine and cadaverine, produced by similar enzymatic decarboxylation (Table 1.9). The presence of putrescine and cadaverine is more significant in shellfish, such as shrimp. The detection and quantification of histamine is fairly simple and inexpensive; however, the detection and quantification of putrescine and cadaverine are more complicated and expensive. Despite the possibility that histamine may not be the main cause of poisoning (histamine is not stable under strong acidic conditions such as the stomach), it is used as an index of freshness of raw materials due to the simplicity of histamine analysis (Gopakumar 2000).

Urea is hydrolysed by the enzyme urease (EC 3.5.1.5), producing ammonia, which is one of the components measured by total volatile base (TVB). TVB nitrogen has been used as a quality index of seafood acceptability by various agencies (Johnson and Linsay 1986, Cadwallader 2000, Gopakumar 2000). Live shark contains relatively high amounts of urea, thus under improper handling urea is converted to ammonia, giving shark meat an ammonia odour, which is a quality defect.

Table 1.9. Secondary Amine Production in Seafood

Enzyme	Reaction
Histidine decarboxyalse (EC 4.1.1.22)	L-Histidine → Histamine + CO_2
Lysine decarboxylase (EC 4.1.1.18)	L-Lysine → Cadaverine + CO_2
Ornithine decarboxylase (EC 4.1.1.17)	L-Ornithine → Putrescine + CO_2

Source: Gopakumar 2000, IUBMB-NC website (www.iubmb.org).

Fruit Ripening

Ethylene, a compound produced as a result of fruit ripening, acts as an initiator and accelerator of fruit ripening. Its concentration is low in green fruits, but can accumulate inside the fruit and subsequently activate its own production (positive feedback). Table 1.10 lists enzymes in the production of ethylene starting from methionine. Ethylene is commonly used to intentionally ripen fruit. During shipping of green bananas, ethylene is removed through absorption by potassium permanganate to render a longer shelf life.

Analytical Protein Biochemistry

The ability to isolate food proteins from complex materials/matrices is critical to their biochemical characterisation. Differing biochemical characteristics such as charge, isoelectric point, mass, molecular shape and size, hydrophobicity, ligand affinity and enzymatic activity all offer opportunities for separation. Charge and size/mass are commonly exploited parameters used to separate different proteins using ion exchange and size exclusion chromatography, respectively. In ion exchange chromatography, proteins within a sample are bound to a stationary phase having an opposite charge (e.g. anionic proteins are bound to a cationic column matrix). A gradient of competing ions is then passed through the column matrix such that weakly bound proteins will elute at low gradient concentration and thus are separated from strongly bound proteins. In the case of size-exclusion chromatography, protein mixtures are passed through an inert stationary phase containing beads having pores of known size. Thus, larger molecules will take a more direct (and hence faster) path relative to smaller molecules, which can fit into the beads, resulting in a more 'meandering', longer path, and therefore, elute slower (in comparison to larger molecules).

Perhaps the most basic biochemical analysis is the determination of molecular mass. Polyacrylamide gel electrophoresis (PAGE) in the presence of a detergent and a reducing agent is typically used for this purpose (mass spectrometry is used for precise measurement where possible). Detergent (sodium dodecyl sulphate; SDS) is incorporated to negate charge effects and to ensure all proteins are completely unfolded, thereby leaving only mass as the sole determinant for rate of travel through the gel and hence the term SDS-PAGE. Proteins denatured with SDS have a negative charge and, therefore, will migrate through the gel in an electric field (see Figure 1.4). A standard curve for proteins with known molecular masses consisting of relative mobility versus log [molecular mass] can be generated to calculate masses of unknowns run on the same gel.

Food Allergenicity

The primary role of the immune system is to distinguish between self and foreign biomolecules in order to defend the host against invading organisms. Antibody proteins that are produced in response to the foreign compounds are specifically referred to as immunoglobulins (Ig), where five classes exist, which share common structural motifs: IgG, IgA, IgM, IgD and IgE. IgE, the least abundant class of antibodies, are the immunogenic proteins important to protection against parasites as well as the causative proteins in allergic reactions (Berg 2002).

Allergic diseases, particularly in industrialised countries, have significantly increased in the last two decades (Mine and Yang 2007). In the United States, food-induced allergies occur in an estimated 6% of young children and 3–4% of adults (Sicherer and Sampson 2006). The most common causes of food allergic reactions for the young are cow's milk and egg, whereas adults are more likely to develop sensitivity to shellfish

Table 1.10. Ethylene Biosynthesis

Enzyme	Reaction
Methionine adenosyltransferase (EC 2.5.1.6)	L-Methionine + ATP + H_2O → *S*-adenosyl-γ-methionine + Diphosphate + Pi
Aminocyclopropane carboxylate synthetase (EC 4.4.1.14)	*S*-adenosyl-γ-methionine → 1-Aminocyclopropane-1-carboxylate + 5′-Methylthio-adenosine
Aminocyclopropane carboxylate oxidase (EC 4.14.17.4)	1-Aminocyclopropane-1-carboxylate + Ascorbate + $\frac{1}{2}O_2$ → Ethylene +Dedroascorbate + CO_2 + HCN + H_2O

Source: Eskin 1990, Bryce and Hill 1999, Crozier et al. 2000, Dangl et al. 2000, IUBMB-NC website (www.iubmb.org).

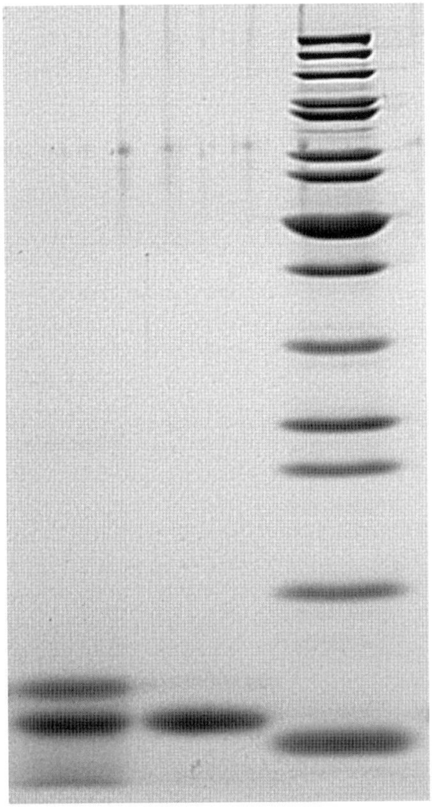

Figure 1.4. SDS-PAGE gel with protein standards of known masses (right lane), impure target protein (left lane) and pure target protein (middle lane; ***). The target protein was calculated to have a molecular mass of 11,700 Daltons (Da) or 11.7 kDa.

(Sampson 2004). Such reactions are thought to result from an abnormal response of the mucosal immune system towards normally harmless dietary proteins (antigens; Bischoff and Crowe 2005). Allergic reactions are distinct from food intolerances that do not involve the immune system (Sampson and Cooke 1990).

One of the biggest problems with food allergy management is that avoidance of antigen is the primary means of preventing allergic reactions; however, minute amounts of so-called 'hidden allergens' in the form of nut, milk and egg contaminants occur in many processed foods. To avoid allergens entirely may pose the risk of avoiding nutritionally important foods, resulting in malnutrition, especially in the young, a problem highlighting the need for control of allergens in foods including the making of hypoallergenic food products (Mine and Yang 2007). Allergic reactions to food components that are mediated by IgE are the best understood and most common type (Type I; Ebo and Stevens 2001).

In general, glycosyl biomolecules are often important elicitors of immunogenic responses (Berg 2002), e.g. bacterial lipopolysaccharides. Many proteins contain carbohydrate moieties and are termed 'glycoproteins'. IgE specific to glycans has been reported (van Ree et al. 1995), and it was originally reported that the carbohydrate portion of ovomucoid contributed to binding human IgE (Matsuda et al. 1985); however, subsequent investigation suggested that it did not participate in protein allergenicity (Besler et al. 1997).

As a means of reducing the allergenicity of egg proteins, enzymatic treatments have been studied. The major limitations or potential hurdles to such an approach are the need for the allergen epitope(s) to be directly impacted, i.e. cleavage upon enzyme treatment, and the retaining of the unique functional properties of egg proteins in foods, e.g. foaming and gelling (Mine et al. 2008). A combination of thermal treatments and enzymatic hydrolyses resulted in a hydrolysed liquid egg product with 100 times less IgE-binding activity than the starting material, determined by analysis of human subjects having egg allergies (Hildebrt et al. 2008). Flavour and texturising properties were not altered when incorporated into various food products, suggesting potential to manufacture customised products accessible to egg-allergy sufferers (Mine et al. 2008).

Enzyme Biotechnology in Foods

Various enzymes are used as processing aids in the food industry. Examples include acetolactate decarboxylase, α-amylase, amylo-1,6-glucosidase, chymosin, lactase and maltogenic α-amylase (Table 1.11), many of which are produced as recombinant proteins using genetic engineering techniques. Recombinant expression has the advantage of providing consistent enzyme preparations since expression cultures can be maintained indefinitely, and it is not dependent on natural sources (e.g. chymosin from calf stomachs). In addition to recombinant enzymes, microbial enzymes are also used, e.g. microbial rennets are used in cheese production from several organisms: *Rhizomucor pusillus, R. miehei, Endothia parasitica, Aspergillus oryzae and Irpex lactis*. Trade names of microbial milk-clotting enzymes include Rennilase, Fromase, Marzyme and Hanilase. Other enzymes include lactase, which is well accepted by the dairy industry for the production of lactose-free milk for lactose-intolerant consumers, and amylases, which are used for the production of high-fructose corn syrup and as an anti-stalling agent for bread.

FOOD LIPID BIOCHEMISTRY

Fatty Acids

Lipids are organic compounds characterised by little or no solubility in water and are the basic units of all organisms' membranes, the substituent of lipoproteins and the energy storage form of all animals. The basic units of lipids are fatty acids (FAs), simple hydrocarbon chains of varying length with a carboxylic acid group at one end.

$$CH_3 - [CH_2]_n - COOH$$

The carboxylic carbon is designated as carbon 1 (C1). FAs are characteristically named using hydrocarbon chain length. For example, a four carbon FA is butanoic acid, a five carbon FA is pentanoic acid, a six carbon FA is hexanoic acid etc.,

Table 1.11. Selected Commercial Biotechnology-Derived Food Enzymes

Enzyme	Application
Acetolactate decarboxylase (EC 4.1.1.5)	Beer aging and diacetyl reduction
α-Amylase (EC 3.2.1.1)	High-fructose corn syrup production
Amylo-1,6-glucosidase (EC 3.2.1.33)	High-fructose corn syrup production
Chymosin (EC 3.4.23.4)	Milk clotting in cheese manufacturing
Lactase (EC 3.2.1.108)	Lactose hydrolysis
Glucan-1,4-α-maltogenic α-amylase (EC 3.2.1.133)	Anti-stalling in bread

Source: Roller and Goodenough 1999, Anonymous 2004, IUBMB-NC website (www.iubmb.org).

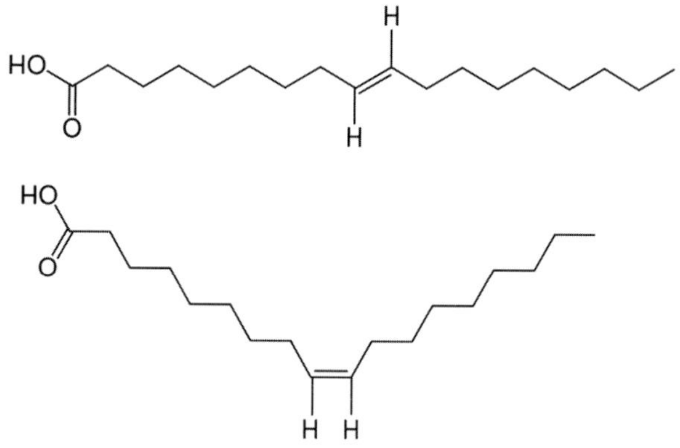

Figure 1.5. The structures of *cis*- and *trans*-oleic acid.

with designations 4:0, 5:0 and 6:0, respectively, with the first digit indicating the number of carbons and the second giving the number of double bonds. For FAs containing double bonds, hexadecanoic acid (16:0) becomes hexadecenoic acid (16:1), 16:2 is termed hexadecadienoic acid, indicating two double bonds, and 16:3 is hexadecatrienoic acid. Following this convention and in order to indicate the position of the double bond, if 16:1 has its double bond between C7 and C8, then 7-hexadecenoic acid is used.

Currently, the term 'omega' is often used. In this case, the methyl end of the FA is termed the omega carbon (ω); therefore, 9,12-octadecadienoic acid (18:2) becomes 18:2 ω-6 since the first double bond is six carbons from the ω carbon. Table 1.12 shows some common FAs, their lengths, and double bond characteristics. When discussing FAs, the term double bond simply indicates the lack of hydrogen across a hydrocarbon bond:

—CH=CH— (double bond; unsaturated)
vs.—CH2—CH2— (saturated)

Geometrically, double bonds can be either *cis* or *trans*, with the cis configuration being the naturally occurring form, bulky and susceptible to oxidation. The trans configuration is more linear, has properties similar to a saturated FA and is not found in nature (see Figure 1.5).

Triglycerides and Phospholipids

In foods, most lipids exist as triglycerides (TGs), making up 98% of food lipids. The name triglyceride refers to its biochemical structure consisting of a glycerol having three FAs bound at its hydroxyl groups. TGs are the primary energy storage form in animals, seeds and certain fruits (e.g. avocado and olive). In comparision to carbohydrates, TGs provide more than double the energy, on a dry basis (9 kcal/g vs. 4 kcal/g). Food TGs also provide mouthfeel and satiety as well as aid in the provision and absorption of fat-soluble vitamins (i.e. A, D, E and K). In terms of food structure, TGs play critical roles in the structures of emulsified food products (oil and water mixtures) like ice cream and chocolate. In ice cream, liquid fat globules partially

Table 1.12. Selected Common Fatty Acids

Fatty Acid	Systematic Name	Number of Carbons	Abbreviation
Butyric	Butanoic	4	4:0
Lauric	Dodecanoic	12	12:0
Myristic	Tetradecanoic	14	14:0
Palmitic	Hexadecanoic	16	16:0
Stearic	Octadecanoic	18	18:0
Oleic	9-octadecenoic	18	18:1 (*n*-9)
Linoleic	9,12-octadecadienoic	18	18:2 (*n*-6)
Linolenic	9,12,15-octadecatrienoic	18	18:3 (*n*-3)
Arachidonic	5,8,11,15-eicosatetraenoic	20	20:4 (*n*-6)
EPA	5,8,11,14,17-eicosapentaenoic	20	20:5 (*n*-3)
DHA	4,7,10,13,16,19-docosahexaenoic	22	22:6 (*n*-6)

Table 1.13. Lipid Degradation in Seed Germination

Enzyme	Reaction
Lipase (oil body)	Triacylglycerol → Diacylglycerol + Fatty acid Triacylglycerol → Monoacylglycerol + Fatty acids Diacylglycerol → Monoacylglycerol + Fatty acid Fatty acid + CoA → Acyl CoA
β-Oxidation (glyoxysome)	Acyl CoA → Acetyl CoA
Glyoxylate cycle (glyoxysome)	Acetyl CoA → Succinate
Mitochondrion	Succinate → Phosphoenol pyruvate
Reverse glycolysis (cytosol)	Phosphoenol pyruvate → Hexoses → Sucrose

Source: Bewley and Black 1994, Murphy 1999.

aggregate until a continuous network forms (partial coalescence) yielding a 'solid' product. TG globules cement to one another at globule interfaces due to interacting fat crystal networks (Goff 1997).

The physical and functional properties of TGs are highly variable and are dependent on FA chain length, number of FA double bonds and position/order of FAs on the glycerol backbone (e.g. oil seeds tend to contain more double bonds in the middle position, whereas animal TGs contain saturated FAs at this position). Furthermore, reactions such as oxidation or lipolysis affect TG behaviour.

Phospholipids

Another class of lipids are phospholipids (PLs), most of which consist of a glycerol backbone with two FAs and the third backbone position containing various substituents such as serine, choline and ethanolamine. These polar substituents render PLs amphipathic (one end hydrophilic, the other FA end hydrophobic). An example of a PL used in food is lecithin, a natural emulsifier and surfactant. In aqueous environments, PLs spontaneously form phospholipid bilayers in spherical-shaped structures called liposomes or lipid vesicles. Liposomes have been used for the study of membrane properties and substance permeabilities, drug delivery, and can be used for microencapsulation of food ingredients. Encapsulation of vitamin C significantly improves shelf life by about 2 months when degradative components like copper, Lys and ascorbate oxidase are present. Also, ingredients can be sequestered within liposomes that have well-defined melting temperatures, thus releasing the contents in a controlled fashion (Gouin 2004).

Food Lipid Degradation

Hydrolysis and oxidation reactions are the principal ways in which TGs are degraded in foods. Lipolysis refers to the hydrolysis of the ester linkage between the glycerol backbone and FAs, thereby releasing free FAs. Lipases are enzymes that release FAs from the outer TG positions by hydrolysis. Free FA's are more susceptible to oxidation and they are more volatile compared to FAs within TGs. Lipase activity is potentially problematic post-slaughter, and controlling temperature minimises their activities. By contrast, oil seeds 'naturally' undergo lipolysis pre-harvest and, therefore, contain a significant amount of free FAs, resulting in acidity that should be neutralised in the extracted oil. In terms of dairy products such as cheeses, yoghurts and bread, controlled lipolysis is used as a means of producing desired odours and flavours via microbial and endogenous lipases. However, lipolysis is also responsible for development of rancid flavour in milk, resulting from the release of short chain FAs, e.g. butyric acid. Deep frying also produces undesirable lipolysis due to the high heat and introduction of water from foods cooked in the oil medium.

During seed germination, lipids are degraded enzymatically to serve as an energy source for plant growth and development. Because of the presence of a considerable amount of seed lipids in oilseeds, they have attracted the most attention, and various pathways in the conversion of FAs have been reported (Table 1.13). The FAs hydrolysed from the oilseed glycerides are further metabolised by β-oxidation followed by the citric acid cycle to produce energy. Seed germination is important in the production of malted barley flour for bread making and brewing.

Autoxidation

The principal cause of lipid oxidation is *autoxidation*. This process takes place via the action of free radicals on the FA hydrocarbon chain in a chain reaction. Initiation of the chain reaction is the creation of free radicals by metal catalysis, light or peroxide decomposition. The initial free radicals then act on the FA by abstracting a hydrogen atom from the hydrocarbon chain, thereby making a new free radical, which then reacts with O_2, resulting in hydroperoxy free-radical formation of the FA. This FA-free radical then acts on other FAs abstracting a hydrogen atom, creating a new free radical and the formation of a stable hydroperoxide. The chain reaction terminates when two free radicals react together.

Initiation -- creation of $R^\bullet$

Propagation -- $R^\bullet + O^2 \rightarrow ROO^\bullet$

$ROO^\bullet + RH \rightarrow ROOH + R^\bullet$

(the new free radical can react with O_2 anew)

Termination -- $R^\bullet + R^\bullet$

$R^\bullet + ROO^\bullet$

$ROO^\bullet + ROO^\bullet$

Table 1.14. Enzymatic Lipid Oxidation in Food Systems

Enzyme	Reaction
Arachidonate-5-lipoxygenase (5-lipoxygenase, EC 1.13.11.34)	Arachidonate + O_2 → (6*E*,8*Z*,11*Z*,14*Z*)-(5*S*)-5-hydroperoxyicosa-6–8-11,14-tetraenoate
Arachidonate-8-lipoxygenase (8-lipoxygenase, EC 1.13.11.40)	Arachidonate + O_2 → (5*Z*,9*E*,11*Z*,14*Z*)-(8*R*)-8-hydroperoxyicosa-5,9,11,14-tetraenoate
Arachidonate 12-lipoxygenase (lipoxygenase, EC 1.13.11.31)	Arachidonate + O_2 → (5*Z*,8*Z*,10*E*,14*Z*)-(12*S*)-12-hydroperoxyicosa-(12–5,8,10,14-tetraenoate)
Arachidonate 15-lipoxygenase (15-lipoxygenase, EC 1.13.11.33)	Arachidonate + O_2 → (5*Z*,8*Z*,11*Z*,13*E*)-(15*S*)-15-hydroperoxyicosa-5,8,11,13-tetraenoate
Lipoxygenase (EC 1.13.11.12)	Linoleate + O_2 → (9*Z*,11*E*)-(13*S*)-13-hydroperoxyoctadeca-9,11-dienoate

Source: Lopez-Amaya and Marangoni 2000a, 2000b, Pan and Kuo 2000, Kolakowska 2003, IUBMB-NC website (www.iubmb.org).

Another mode of lipid oxidation is enzyme-catalysed lipid oxidation by a group of enzymes termed lipoxygenases (LOXs), whose activities are important in legumes and cereals. In LOX activity, a free radical of a FA initially is formed, followed by its reaction with O_2, yielding a hydroperoxide product (Klinman 2007). Linoleic acid and arachidonic acid are important FAs from the health perspective and are quite common in many food systems (Table 1.12). Because of the number of double bonds in arachidonic acid, enzymatic oxidation can occur at various sites, and the responsible LOXs are labelled according to these sites (Table 1.14). In addition to rancid off-flavours, LOX can also cause deleterious effects on vitamins and colour compounds. However, LOX activity also causes the pleasant odours associated with cut tomatoes and cucumbers.

The main enzymes involved in the generation of the aroma in fresh fish are also LOXs, specifically the 12- and 15-LOXs (Table 1.14) and hydroperoxide lyase. The 12-LOX acts on specific polyunsaturated FAs and produces *n*-9-hydroperoxides. Hydrolysis of the 9-hydroperoxide of eicosapentenoic acid by specific hydroperoxide lyases leads to the formation of aldehydes that undergo reduction to their corresponding alcohols, a significant step in the general decline of the aroma intensity due to alcohols having higher odour detection thresholds than the aldehydes (Johnson and Linsay 1986, German et al. 1992).

Milk contains a considerable amount of lipids and these milk lipids are subjected to enzymatic oxidation during cheese ripening. Under proper cheese maturation conditions, these enzymatic reactions starting from milk lipids create the desirable flavour compounds for these cheeses. These reactions are numerous and not completely understood, thus only general reactions are provided (Table 1.15). Readers should refer to Chapters 19, 20 and 26 in this book for a detailed discussion.

ELECTED PHYTOCHEMICAL FLAVOUR AND COLOUR COMPOUNDS

Many fruits and vegetables produce flavours that are significant in their acceptance and handling. There are a few well-known examples (Table 1.16). Garlic is well known for its pungent odour due to the enzymatic breakdown of its alliin to the thiosulfonate allicin, with the characteristic garlic odour. Strawberries have a very recognisable and pleasant odour when they ripen. Biochemical production of the key compound responsible for strawberry flavour, the furan 2,5-dimethyl-4-hydroxy-2H-furan-3-one (DMHF), is also known as furaneol, which results from hydrolysis of terminal, non-reducing ~-D-glucose residues from DMHF-glucoside with release of ~-D-glucose and DMHF. Lemon and orange seeds contain limonin, a furanolactone that is bitter and is hydrolysable to limonate, which is less bitter. Many cruciferous vegetables such as cabbage and broccoli have a sulfurous odour due to the production of a thiol compound (R–SH) after enzymatic hydrolysis of its glucoside. Brewed tea darkens

Table 1.15. Changes in Lipids in Cheese Manufacturing

Enzyme	Reaction
Lipolysis	
Lipases, esterases	Triglycerides → β-Keto acids, acetoacetate, fatty acids
Acetoacetate decarboxylase (EC 4.1.1.4)	Acetoacetate + H^+ → Acetone + CO_2
Acetoacetate-CoA ligase (EC 6.2.1.16)	Acetoacetate + ATP + CoA → Acetyl CoA + AMP + Diphosphate
Esterases	Fatty acids → Esters
Conversion of fatty acids	
β-Oxidation and decarboxylation	β-Keto acids → Methyl ketones

Source: Schormuller 1968, Kilara and Shahani 1978, IUBMB-NC website (www.iubmb.org).

Table 1.16. Selected Enzyme-induced Flavour Reactions

Enzyme	Reaction
Alliin lyase (EC 4.4.1.4; garlic, onion)	An *S*-alkyl-L-cysteine *S*-oxide → An alkyl sufenate + 2-Aminoacrylate
β-Glucosidase (EC 3.2.1.21; strawberry)	Hydrolysis of terminal non-reducing *β*-D-glucose residues with release of *β*-D-glucose (2,5-Dimethyl-4-hydroxy-2H-furan-3-one (DMHF)-glucoside → DMHF)
Catechol oxidase (EC 1.10.3.1; tea)	2 Catechol + O_2 → 2 1,2-Benzoquinone + 2 H_2O
Limonin-D-ring-lactonase (EC 3.1.1.36; lemon and orange seeds)	Limonoate-D-ring-lactone + H_2O → Limonate
Thioglucosidase (EC 3.2.1.147; cruciferous vegetables)	A thioglucoside + H_2O → A thiol + A sugar

Source: Wong 1989, Eskin 1990, Chin and Lindsay 1994, Orruno et al. 2001, IUBMB-NC website (www.iubmb.org).

after it is exposed to air due to enzymatic oxidation (discussed earlier). Flavours from cheese fermentation and fresh-fish odour have already been described earlier, and formation of fishy odour will be described later.

Colour is an intrinsic property of foods, and therefore, a change in colour is often caused by a change in quality. Vision is the most important sensory perception in selecting food and appreciating its quality (Diehl 2008). Chlorophylls are the most abundant natural pigments and are responsible for the green colour of plants (Marquez Ursula and Sinnecker 2008) and are the biomolecules responsible for capturing light energy in its transformation into chemical energy during photosynthesis. Light is able to be absorbed very efficiently due to the presence of many conjugated double bonds within the large, multi-ring chlorophyll structure. Disappearance of chlorophyll during fruit ripening and leaf senescence indicates slowing of photosynthesis. The loss of green colour is due to a loss of chlorophyll structure via two main stages: First, various reactions produce greenish chlorophyll derivatives and second, oxidative reactions result in opening of ring structures, thereby causing colourless products (Diehl 2008).

Carotenoids are the most widely distributed group of pigments, and although they are not produced by the human body, they are essential to human health. Vitamin A/*β*-carotene are carotenoids critical to a healthy diet. Additionally, carotenoids may help reduce the risk of cancer and heart disease (Bertram 1999), are important natural colourants in foods and are used as sources of red, yellow and orange food colouring (Otles and Cagindi 2008). Carotenoids are fat-soluble pigments that provide the colour for many common fruits such as yellow peaches, papayas and mangoes. During post-harvest maturation, these fruits show intense yellow to yellowish orange colours due to synthesis of carotenoids from its precursor isopentyl diphosphate, which is derived from (R)-mevalonate. Isopentyl diphosphate is a key building block for carotenoids (Croteau et al. 2000), and Table 1.17 lists the sequence of reactions in the formation of (R)-mevalonate from acetyl-CoA and from (R)-mevalonate to isopentyl diphosphate.

Flavonoids are not only a group of compounds responsible for various red, blue or violet colours of fruits and vegetables, they are also related to the group of bioactive, anti-plant pathogen compounds called stilbenes. Stilbenes have a common precursor of *trans-cinnamate* branching out into two routes, one that leads to flavonoids and the other leading to stilbenes (Table 1.18). Table 1.18 gives the series of reactions in the biosynthesis of naringenin chalcone, the building block for flavonoid

Table 1.17. Mevalonate and Isopentyl Diphosphate Biosyntheses

Enzyme	Reaction
Acetyl-CoA *C*-acetyltransferase (EC 2.3.1.9)	2 Acetyl-CoA → Acetoacetyl-Co-A + CoA
Hydroxymethylglutaryl-CoA-synthase (EC 2.3.3.10)	Acetoacetyl-CoA + Acetyl-CoA + H_2O → (*S*)-3-Hydroxy-3-methylglutaryl CoA + CoA
Hydroxymethylglutaryl-CoA reductase (EC 1.1.1.34)	(*S*)-3-Hydroxy-3-methylglutaryl-CoA + 2 $NADPH_2$ → (*R*)-Mevalonate + CoA + 2 NADP
Mevaldate reductase (EC 1.1.1.32)	(*R*)-Mevalonate + NAD → Mevaldate + $NADH_2$
Mevalonate kinase (EC 2.7.1.36)	(*R*)-Mevalonate + ATP → (*R*)-5-Phosphomevalonate + ADP
Phosphomevalonate kinase (EC 2.7.4.2)	(*R*)-5-Phosphomevalonate + ATP → (*R*)-5-Diphosphomevalonate + ADP
Diphosphomevalonate decarboxylase (EC 4.1.1.33)	(*R*)-5-Diphosphomevalonate + ATP → Isopentyl diphosphate + ADP + Pi + CO2

Source: Croteau et al. 2000, IUBMB-NC Enzyme website (www.iubmb.org).

Table 1.18. Naringenin Chalcone Biosynthesis

Enzyme	Reaction
Phenylalanine ammonia-lyase (EC 4.3.1.5)	L-phenylalanine → *Trans*-cinnamate + NH_3
Trans-cinnamate 4-monoxygenase (EC 1.14.13.11)	*Trans*-cinnamate + $NADPH_2$ + O_2 → 4-hydroxycinnamate + NADP + H_2O
4-Coumarate-CoA ligase (EC 6.2.1.12)	4-Hydroxycinnamate (4-coumarate) + ATP + CoA → 4-Coumaroyl-CoA + AMP + Diphosphate
Naringinin-chalcone synthase (EC 2.3.1.74)	Naringinin chalcone + 4 CoA + 3 CO_2

Source: Eskin 1990, Croteau et al. 2000, IUBMB-NC Enzyme (www.iubmb.org).

biosynthesis. Considerable interest has been given to the stilbene trans 3,5,4′ -trihydroxystilbene commonly called resveratrol in red grapes and red wine that may have human nutraceutical and/or pharmaceutical applications (Narayanan et al. 2009). Table 1.19 summarises various phytochemicals and their sources that are thought to confer health benefits (Gropper 2009).

Cholesterol

Cholesterol is an important lipid in animal biochemical processes that controls the fluidity of membranes and is the precursor for all steroid hormones. Cholesterol's structure is made up of adjacent carbon rings, making it hydrophobic (fat soluble), and thus it must be transported through the blood complex as a lipoprotein in structures called chylomicrons. The densities of lipoproteins vary greatly; an excessive level of low-density lipoprotein (LDL) is a critical risk factor for heart disease. In contrast, plants produce related molecules called phytosterols, which have various nutraceutical applications (Kritchevsky and Chen 2005).

Terpenoids

Terpenoids, a diverse and complex chemical group, are lipid-type molecules important to flavours and aromas of seasonings, herbs and fruits. Most terpenoids are multi-cyclic compounds made via successive polymerisation and cyclisation reactions derived from the 5-carbon building block isoprene, which contains two double bonds. Terpenoids are components of the flavour profiles of most soft fruit (Maarse 1991). In some fruit species, they are of great importance for the characteristic flavour and aroma

Table 1.19. Phytochemicals and Their Sources

	Phytochemical	Source
Anthocyanins	Cyanidin	Berries, cherries, plums, red wine
Carotenoids	β-carotene, α-carotene, lutein, lycopene	Tomato, pumpkin, squash, carrot, watermelon, papaya, guava
Flavanols	Catechins, epicatechins	Green tea, pear, wine, apple
Flavones	Apigenin, luteolin	Parsley, some cereals
Flavonols	Quercetin, kaempferol, myricetin	Onions, tea, olive, kale, leaf lettuce, cranberry, tomato, apple, turnip green, endive, *Gingko biloba*
Glucosinolates	Gucobrassicin, gluconapin, sinigrin, glucoiberin	Broccoli, cabbage, Brussels sprouts, mustard, watercress
Isoflavones	Genistein, daidzein, equol	Soy, nuts, milk, cheese, flour, miso, legumes
Isothiocyanates	Allylisothiocyanates, indoles, sulforaphane	Broccoli, cabbage, Brussels sprouts, mustard, watercress
Lignans	Secoisolariciresinol, mataresinol	Berries, flaxseed, nuts, rye bran
Organosulfides	Diallyl sulphide, allyl methyl sulphide, *S*-allylcysteine	Garlic, onions, leeks, broccoli, cabbage, brussel sprouts, mustard, watercress
Phenolic acids	Hydroxycinnamic acids (caffeic, ferukic, chlorogenic, curcumin) and hydroxybenzoic acids (ellagic, gallic)	Blueberry, cherry, pear, apple, orange, grapefruit, white potato, coffee bean, St. John's wort, *Echinacea*, raspberry, strawberry, grape juice
Phytosterols	β-Sitosterol, campesterol, stigmasterol	Oils: soy, rapeseed, corn, sunflower
Saponins	Panaxadiol, panaxatriol	Alfalfa sprouts, potato, tomato, ginseng
Terpenes	Limonene, carvone	Citrus fruits, cherries, *Gingko biloba*

Source: Adapted from Gropper et al. 2009.

(e.g. citrus fruits are high in terpenoids as is mango; Aharoni et al. 2004).

NUCLEIC ACIDS AND FOOD SCIENCE

DNA Structure

Although nucleic acids are not generally important components of foods, they are perhaps (and ironically) the most important biomolecule class of all for the simple reason that everything we eat was at some point alive, and every process and structure in those organisms were determined via enzymes encoded for by DNA. Indeed, the entire purpose of breeding programs is to control the propagation of DNA between successive generations. Understanding the basics of DNA (deoxyribonucleic acid) and its biochemical forms is important for food scientists to gain an appreciation of the basis for genetic manipulations of food-related proteins as well as DNA-based food authentication techniques.

DNA is composed of three main chemical components: a nitrogenous base, a sugar and a phosphate. There are four bases: Adenine (A) and guanine (G) are purines, while thymine (T) and cytosine (C) are pyrimidines. The bases bound to both the sugar and phosphate moieties make up nucleotides and the four building blocks of DNA are deoxyadenosine triphosphate, deoxyguanosine triphosphate, deoxythymidine triphosphate and deoxycytidine triphosphate (dATP, dGTP, dTTP and dCTP, respectively). These building blocks are typically referred to as 'bases' or simply A, T, G and C; however, common usage of these terms are meant to imply nucleotides as opposed to bases alone. The nucleotides covalently bond together forming a DNA strand that is synthesised by DNA polymerase. Additionally, each nucleotide's base moiety can bond via hydrogen bonds to other bases; A–T and G–C, termed base pairs and are said to be complementary. In fact, DNA exists in two-stranded form, consisting of two complementary strands. For example, a strand ATCG would be paired to its complement TAGC.

```
5′-ATCG-3′
3′-TAGC-5′
```

Two-stranded DNA spontaneously forms a helix, hence the term double helix, which can contain many thousands of base pairs with molecular weights in the millions, or billions, of Daltons. By comparison, the largest known protein is a mere 3 million Daltons. *Genes* are stretches of DNA that encode for the synthesis of proteins; DNA is transcribed, yielding messenger RNA (mRNA), which is then translated at ribosomes, yielding specific sequences of amino acids (proteins).

Genetic Modification

The advances in how to copy DNA, modify its sequence, and transfer genes between organisms efficiently and at low cost has produced a tremendous ability to study the roles of specific proteins in organisms as well as the roles of specific amino acids within proteins. This ability has produced an alternate to traditional breeding programs in the search for food plants and animals that have desired traits, such as increased yield, increased pesticide tolerance, lower pesticide requirements/higher pest resistance, longer shelf life post-harvest, etc. This alternate strategy is the basis of genetic modification (GM).

Critical to the study, transfer and manipulation of genes was the advent of the polymerase chain reaction (PCR), which allows for the easy and accurate copying and, equally important, the amplification of DNA sequences. Briefly, PCR works by inducing repeated copying of a given DNA sequence by the enzyme DNA polymerase via repeated temperature cycles such that exponential amplification results, i.e. the first round yields only a doubling, but the second round then makes new copies of each of the first round's copies and the originals; 2, 4, 8, 16, 32, 64, 128, etc. After 25–30 PCR rounds, millions of copies result. Thus, a gene encoding for a useful gene in organism A (e.g. an anti-freeze protein) can be copied, amplified and subsequently transferred to organism B (e.g. a fruit).

An example of GM is that of the Flavr savr^{TM} tomato, originally available for consumption in 1994 (Martineau 2001). A non-sense gene is a DNA sequence that encodes for complementary mRNA, which base-pair matches and binds to a natural gene transcript, thereby suppressing its translation. A 'non-sense', gene acting against the polygalacturonase gene, an enzyme responsible for the breakdown of a cell-wall component during ripening, was incorporated into a strain of tomato. The result was slowed softening of the texture of the engineered tomato compared to normal tomatoes, thus allowing producers to vine-ripen the Flavr savr^{TM}, reducing losses (e.g. bruising) during subsequent transport to market. Superior flavour and appearance relative to natural tomatoes picked green, as well as equivalent micro- and macronutrient content, pH, acidity and sugar content relative to non-transgenic tomatoes resulted.

In terms of food processing, lactic acid bacteria and yeast have been developed to solve problems in the dairy, baking and brewing industries (Tables 1.20 and 1.21). As with biotechnology-derived food enzymes, the use of genetically modified organisms is governed by laws of nations or regions (e.g. the European Union).

Safety assessments of GM foods before being released to market are done. These comparisons to non-engineered, conventional counterparts include proximate analysis as well as analyses of nutritional components, toxins, toxicants, anti-nutrients and other components relevant to given cases. As well, animal feeding trials are conducted to determine if any adverse health effects are observable (Institute of Medicine and National Research Council of the National Academies 2004). Ideally, the reference food for the above comparisons is the isogenic food (i.e. non-transformed) from which the GM version was derived.

Food Authentication and the Role of DNA Technologies

Another area utilising DNA technology is *food authentication*. Analysing processed food and ingredients for the presence of fraudulent or foreign components by DNA technologies can be

Table 1.20. Selected Commercial Biotechnology-derived Food Enzymes

Enzyme	Application
Acetolactate decarboxylase (EC 4.1.1.5)	Beer aging and diacetyl reduction
α-Amylase (EC 3.2.1.1)	High-fructose corn syrup production
Amylo-1,6-glucosidase (EC 3.2.1.33)	High-fructose corn syrup production
Chymosin (EC 3.4.23.4)	Milk clotting in cheese manufacturing
Lactase (EC 3.2.1.108)	Lactose hydrolysis
Glucan-1,4-α-maltogenic α-amylase (EC 3.2.1.133)	Anti-stalling in bread

Source: Roller and Goodenough 1999, Anonymous 2004, IUBMB-NC website (www.iubmb.org).

very complex due to degradation of genetic material as the result of processing conditions. Detecting misrepresentation of food components is a task well-suited for DNA-based detection in the case of meat species of origin, especially when the meat product is processed such that appearance and physical traits cannot easily cue experts as to product identity. Gene sequences within mitochondrial DNA are usually used as the PCR targets for such purposes (Wiseman 2009). For example, different tuna species can be differentiated (Michelini et al. 2007). Likewise, the presence of meat in cattle feed can be assayed (Rensen 2005). Such tests can be done for high-heat-processed samples at a low cost, and such tests can be highly automated for steps post sample collection. Lastly, another example of a food authentication application for PCR is the case of detecting the presence and/or the quantity of genetically modified ingredients in a food. For the purposes of quantification, real-time PCR must be used (Wiseman 2009), the most modern type of PCR technology, because it is capable of highly precise, quantitative determinations. This capability is due to real-time PCR's ability to very quickly and accurately modulate PCR reaction temperature via the use of low-volume reactions. Jurisdictions that have GM labelling laws (e.g. the European Union) can thus verify the correct labelling of incoming ingredients.

Table 1.21. Selected Genetically Modified Microorganisms Useful in Food Processing

Microorganism	Application
Lactobacillus lactis	Phage resistance, lactose metabolism, proteolytic activity, bacteriocin production
Saccharomyces cerevisiae (Baker's yeast, Brewer's yeast)	Gas (carbon dioxide) production in sweet, high-sugar dough
	Manufacture of low-calorie beer (starch degradation)

Source: Hill and Ross 1999, Roller and Goodenough 1999, Anonymous 2004.

In the authentication process, DNA within foods can also be characterised and differentiated by use of *DNA probes*, an earlier food authentication technology that dates back to at least 1990, where it was used to distinguish between heat-processed ruminant (goat, sheep and beef), chicken and pig meat (Chicuni et al. 1990). DNA probes are short pieces of DNA (Rensen et al. 2005) that contain a detectable feature (usually fluorescence) and that are complementary to a DNA sequence of interest. Thus, a signal is produced only if the probed sequence is present in the food, since the probe will bind only to its complement in a manner that yields a signal. In order to probe a food product, genetic material is isolated and immobilised on a piece of nitrocellulose (blotting), which is then 'probed' with a DNA complementary probe. The information from the above types of DNA-based authentication tests can aid in processing quality control to ensure both authenticity and safety at relatively cheap cost.

NATURAL TOXICANTS

The contamination of various foods by toxicants may occur as a result of microbial production, crop plant production or ingestion by animals for human consumption. Microbial sources of toxins are mycotoxin-producing fungi and toxin-producing bacteria. Notable examples of microbially derived toxins are botulinum toxin produced by *Clostridium botulinum* and the *Staphylococcus aureus* toxin. These toxins are produced in the food itself and result in food poisoning. Both toxins are heat-labile; however, the extreme toxicity of the botulinum toxin, potent at 10^{-9} g per kg body weight, makes it of particular concern for food processing of anaerobically stored foods.

Mycotoxins are extremely toxic compounds produced by certain filamentous fungi in many crop plants (Richard 2007). Ingestion of mycotoxins can be harmful to humans via contaminated foods or feed animals via their feed, particularly in maize, wheat, barley, rye and most oilseeds. Mycotoxins produce symptoms that include nervous system disorders, limb loss and death. Aflatoxins are a well-studied type of mycotoxin that are known carcinogens. An example of a mycotoxin structure, aflatoxin B1, is shown in Figure 1.6. Mycotoxin poisoning can take place either directly or indirectly, through consumption of a contaminated food or by a food ingredient that may be contaminated and subsequently eaten as part of a final product.

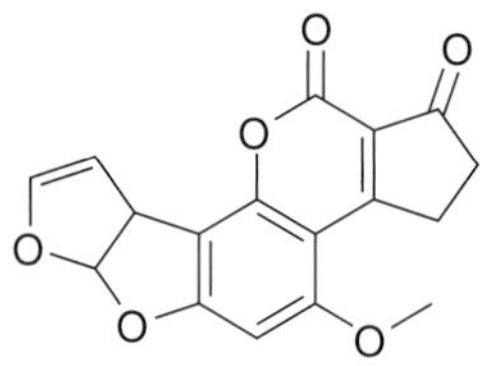

Figure 1.6. The structure of aflatoxin B1, produced by *Aspergillus flavus* and *A. parasiticus*.

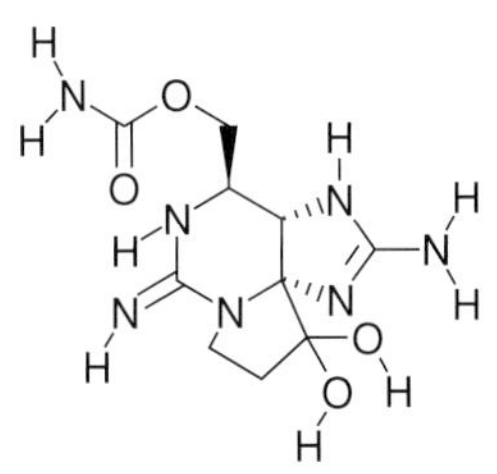

Figure 1.7. The structure of saxitoxin responsible for paralytic shellfish poisoning.

An important example of food-related mycotoxins is 'ergot', the common name for fungi of the genus *Claviceps*. Ergot species produce ergot alkaloids, which are derivatives of lysergic acid, isolysergic acid or dimethylergoline. When ingested, the various alkaloids produce devastating symptoms such as vasoconstriction, convulsion, gastrointestinal upset and central nervous system effects (Peraica et al. 1999)

Another type of natural toxins is that of shellfish toxins (phycotoxins), which are produced by certain species of marine algae and cyanobacteria (blue-green algae). The best known algal toxins are saxitoxins, responsible for paralytic shellfish poisoning (PSP; see Figure 1.7 for chemical structure). Saxitoxins block voltage-gated sodium channels of nerve cells. Thus, this group of 20 toxins induces extreme symptoms, including numbness, tingling and burning of the lips and skin, giddiness, ataxia and fever; severe poisoning may lead to muscular incoordination, respiratory distress or failure (Garthwaite 2000). The poisoning results from bioaccumulation of the algal toxins in algae-eating organisms and toxins can be step-wise passed up the food chain, e.g. from marine algae to shellfish to crabs to humans.

CONCLUSION

In 1939, the newly formed IFT was the world's first organisation to organise and coordinate those working in food processing, chemistry, engineering, microbiology and other sub-disciplines in order to better understand food systems. Food science began mainly within commodity departments, such as animal science, dairy science, horticulture, cereal science, poultry science and fisheries. Now, most of these programs have evolved into food science, or food science and human nutrition, departments. Many food science departments with a food biochemistry emphasis are now available all over the world. In addition, food science departments developed not only applied research programs, but also basic research programs that seek to understand foods at the atomic, molecular and/or cellular levels. The 'cook 'n' look, approach is not food science. Food science is a respected part of academic programmes worldwide, whose researchers often form collaborations with physicists, chemists, parasitologists, etc. Over the past several decades, food biochemistry research has led to industry applications, e.g. lactase supplements, lactose-free dairy products, Beano^{TM}, application of transglutaminase to control seafood protein restructuring, protease-based meat marinades for home, production of high-fructose syrups, rapid pathogen detection tests, massively diversified natural and artificial flavourings and many more. Thus, food biochemistry will continue to play critical roles in developments in food microbiology, packaging, product development, processing, crop science, nutrition and nutraceuticals.

REFERENCES

Aharoni A et al. 2004. Gain and loss of fruit flavor compounds produced by wild and cultivated strawberry species. *Plant Cell* 16: 3110–3131.

Institute of Medicine and National Research Council of the National Academies. 2004. Methods for Predicting and Assessing Unintended Effects on Human Health. In: *Safety of Genetically Engineered Foods: Approaches to Assessing Unintended Health Effects*. National Academies Press, Washington, DC, pp. 127–174.

BeMiller JN, Whistler RL. 1996. Carbohydrates. In: OR Fennema (ed.) *Food Chemistry,* 3rd edn. Marcel Dekker, New York, pp. 216–217.

Berg JM. 2002. *Biochemistry*. W.H. Freeman, New York.

Bertram JS. 1999. Carotenoids and gene regulation. *Nutr Rev* 57: 182–191.

Besler M et al. 1997. Allergenicity of hen's egg-white proteins: IgE-binding of native and deglycosylated ovomucoid. *Food Agric Immunol* 9: 227–328.

Bewley JD, Black M. 1994. *Physiology of Development and Germination*, 2nd edn. Plenum Press, New York, pp. 293–344.

Bischoff S, Crowe SE. 2005. Gastrointestinal food allergy: new insights into pathophysiology and clinical perspectives. *Gastroenterology* 128: 1089–1113.

Bryce JH, Hill SA. 1999. Energy production and plant cells In: PJ Lea, RC Leegood (eds.) *Plant Biochemistry and Molecular Biology*. John Wiley & Sons, Chichester, pp. 1–28.

Cadwallader KR. 2000. Enzymes and flavour biogenesis. In: NF Haard, BK Simpson (eds.) *Seafood Enzymes*, Marcel Dekker, New York, pp. 365–383.

Carasik W, Carroll JO. 1983. Development of immobilized enzymes for production of high-fructose corn syrup. *Food Technol* 37: 85–92.

Champe PC et al. 2005. *Biochemistry,* 3rd edn. Lippincott Williams and Wilkins, Baltimore, MD, pp. 243–251.

Chicuni K et al. 1990. Species identification of cooked meats by DNA hybridization assay. *Meat Sci.* 27: 119–128.

Chin HW, Lindsay RC. 1994. Modulation of volatile sulfur compounds in cruciferous vegetables. In: CJ Mussinan, ME Keelan (eds.) *Sulfur Compounds in Foods*. American Chemical Society, Washington, DC, pp. 90–104.

Coffee DG et al. 1995. Cellulose and cellulose derivatives. In: AM Steven (ed.) *Food Polysaccharides and Their Applications*. Marcel Dekker, New York, pp. 127–139.

Croteau R et al. 2000. Natural products (secondary metabolites). In: BB Buchenan, et al. (eds.) *Biochemistry and Molecular Biology of Plants*. American Society of Plant Physiologists, Rockwell, MD, pp. 1250–1318.

Crozier A et al. 2000. Biosynthesis of hormone and elicitor molecules). In: BB Buchenan, et al. (eds.) *Biochemistry and Molecular Biology of Plants*, American Society of Plant Physiologists, Rockwell, MD, pp. 850–929.

Damodaran S et al. 2008. *Fennema's Food Chemistry*. CRC Press, Boca Raton, FL.

Dangl JL et al. 2000. Senescence and programmed cell death). In: BB Buchenan, et al. (eds.) *Biochemistry and Molecular Biology of Plants*. American Society of Plant Physiologists, Rockwell, MD, pp. 1044–1100.

Diehl HA. 2008. Physics of color. In:C Socaciu (ed.) *Food Colorants: Chemical and Functional Properties*, CRC Press, Boca Raton, FL, pp. 3–21.

Duffus CM. 1987. Physiological aspects of enzymes during grain development and germination. In: Kruger JE, et al. (eds.) *Enzymes and Their Role in Cereal Technology*. American Association of Cereal Chemists, St. Paul, MN, pp. 83–116.

Ebo, DG, Stevens WJ. 2001. IgE-mediated food allergy – extensive review of the literature. *Acta Clin Belg* 56: 234–247.

Eskin NAM. 1990. *Biochemistry of Foods*, 2nd edn. Academic Press, San Diego, CA, pp. 268–272.

Garthwaite I. 2000. Keeping shellfish safe to eat: a brief review of shellfish toxins, and methods for their detection. *Trends Food Sci Technol* 11(7): 235–244.

German JB et al. 1992. Role of lipoxygenases in lipid oxidation in foods. In: AJ St-Angelo (ed.) *Lipid Oxidation in Food*, American Chemical Society, Washington, DC, pp. 74–92.

Goff HD. 1997. Partial coalescence and structure formation in dairy emulsions. In: S Damodran (ed.) *Food Proteins and Lipids*. Plenum Press, New York, pp. 137–147.

Gopakumar K. 2000. Enzymes and enzyme products as quality indices. In: NF Haard, BK Simpson (eds.) *Seafood Enzymes*. Marcel Dekker, New York, pp. 337–363.

Gouin S. 2004. Microencapsulation: industrial appraisal of existing technologies and trends. *Trends Food Sci Technol* 15: 330–347.

Grappin R et al. 1985. Primary proteolysis of cheese proteins during ripening. *J Dairy Sci* 68: 531–540.

Greaser M. 2001. Postmortem muscle chemistry. In: YH Hui, et al. (eds.) *Meat Science and Applications*. Marcel Dekker, New York, pp. 21–37.

Gripon JC. 1987. Mould-ripened cheeses. In: PF Fox (ed.) *Cheese: Chemistry, Physics and Microbiology*. Elsevier Applied Science, London, UK, pp. 121–149.

Gropper S. 2009. JL Groff (ed.) *Advanced Nutrition and Human Metabolism*. Wadsworth Cengage Learning, Belmont, CA.

Haard, NF, Lee, YZ. 1982. Hypobaric storage of Atlantic salmon in carbon dioxide atmospheres. *Can Inst Food Sci Technol J* 15: 68–71.

Haard NF. 1990. Biochemical reactions in fish muscle during frozen storage. In: EG Bligh (ed.) *Seafood Science and Technology*. Fishing News Books (Blackwell Scientific Publications, Ltd.), London, UK, pp. 176–209.

Hildebrt S, et al. 2008. In Vitro Determination of the Allergenic Potential of Technologically Altered Hen's Egg. *J Agric Food Chem* 56(5): 1727–1733.

Hill C, Ross RP. 1999. Starter cultures for the dairy industry. In: S Roller, S Harlander (eds.) *Genetic Modification in the Food Industry*, Blackie and Academic Professional, London, UK, pp. 174–192.

Hoseney RC. 1994. *Principles of Cereal Science and Technology*. American Association of Cereal Chemists, St. Paul, MN, pp. 177–195 (malting and brewing), 229–273 (yeast-leavened products).

Huff-Lonergan E, Lonergan SM. 1999. Postmortem mechanisms of meat tenderization. In: YL Xiong, CT Ho, F Shahidi (eds.) *Quality Attributes of Muscle Foods*. Kluwer Academic Plenum Publishers, New York, pp. 229–251.

Jiang ST. 2000. Enzymes and their effects on seafood texture. In: NF Haard, BK Simpson (eds.) *Seafood Enzymes*. Marcel Dekker, New York, pp. 411–450.

Johnson DB, Linsay RC. 1986. Enzymatic generation of volatile aroma compounds from fresh fish. In: TH Parliament, R Croteau (eds.) *Biogeneration of Aromas*, American Chemical Society, Washington, DC, pp. 201–219.

Kamaly KM, Marth EH. 1989. Enzyme activities of *Lactic streptococci* and their role in maturation of cheese: a review. *J Dairy Sci* 72: 1945–1966.

Khalid NM, Marth EH. 1990. Lactobacilli – their enzymes and role in ripening and spoilage of cheese: a review. *J Dairy Sci* 73: 2669–2684.

Kilara A, Shahani KM. 1978. Lactic fermentations of dairy foods and their biological significance. *J Dairy Sci* 61: 1793–1800.

Kim J et al. 2000. Polyphenoloxidase. In: NF Haard, BK Simpson (eds.) *Seafood Enzymes*. Marcel Dekker, New York, pp. 271–315.

Klinman, J. 2007. How do enzymes activate oxygen without inactivating themselves? *Acc Chem Res* 40: 325–333.

Ko HS et al. 2006. Aroma active compounds of bulgogi. *J Food Sci* 70: 517–522.

Kolakowska A. 2003. Lipid oxidation in food systems. In: ZE Sikorski, A Kolakowska (eds.) *Chemical and Functional Properties of Food Lipids*. CRC Press, Boca Raton, FL, pp. 133–166.

Kritchevsky D, Chen SC. 2005. Phytosterols – health benefits and potential concerns: a review. *Nutrition Res* 25: 413–428.

Kruger JE, Lineback DR. 1987. Carbohydrate-degrading enzymes in cereals. In: JE Kruger, et al. (eds.) *Enzymes and Their Role in Cereal Technology*. American Association of Cereal Chemists, St. Paul, MN, pp. 117–139.

Kruger JE et al. 1987. *Enzymes and Their Roles in Cereal Technology*. American Association of Cereal Chemists, St. Paul, MN.

Law BA. 1984a. Microorganisms and their enzymes in the maturation of cheeses. In: ME Bushell (ed.) *Progress in Industrial Microbiology*. Elsevier, London, UK, pp. 245–283.

Law BA. 1984b. Flavour development in cheeses. In: FL Davies, BA Law (ed.) *Advances in the Microbiology and Biochemistry of Cheese and Fermented Milks*. Elsevier Applied Science Publishers, London, UK, pp. 187–208.

Lopez-Amaya C, Marangoni AG. 2000a. Lipases. In: NF Haard, BK Simpson (eds.) *Seafood Enzymes*. Marcel Dekker, New York, pp. 121–146.

Lopez-Amaya C, Marangoni AG. 2000b. Phospholipases. In: NF Haard, BK Simpson (eds.) *Seafood Enzymes*. Marcel Dekker, New York, pp. 91–119.

Lowrie RA. 1992. Conversion of muscle into meat: biochemistry. In: DA Ledward, et al. (eds.) *The Chemistry of Muscle-Based Foods*. Royal Society of Chemistry, Cambridge, pp. 43–61.

Maarse H. 1991. *Volatile Compounds in Foods and Beverages*. Marcel Dekker, New York.

Marquez Ursula ML, Sinnecker P. 2008. Chlorophylls: properties, biosynthesis, degradation and functions. In: C Socaciu (eds.) *Food Colorants: Chemical and Functional Properties*. CRC Press, Boca Raton, FL.

Martineau B. 2001. *First Fruit: The Creation of the Flavr savr^{TM} Tomato and the Birth of Genetically Engineered Food*. McGraw-Hill, New York.

Matsuda T et al. 1985. Human IgE antibody to the carbohydrate-containing third domain of chicken ovomucoid. *Biochem Biophys Res Commun* 129: 505–510.

Michelini E et al. 2007. One-step triplex-polymerase chain reaction assay for the authentication of yellowfin (*Thunnus albacares*), bigeye (*Thunnus obesus*), and skipjack (*Katsuwonus pelamis*) tuna DNA from fresh, frozen, and canned tuna samples. *J Agric Food Chem* 55: 7638–7647.

Mine Y, Yang M. 2007. Concepts of hypoallegenicity. In: R Huopalahti (ed.) *Bioactive Egg Compounds*. Springer, New York, pp. 145–158.

Mine Y, Yang M. 2008. Recent advances in the understanding of egg allergens: basic, industrial, and clinical perspectives. *J Agric Food Chem* 56(13): 4874:4900.

Murphy OJ. 1999. Plant lipids – their metabolism, function, and utilization. In: PJ Lea, RC Leegood (eds.) *Plant Biochemistry and Molecular Biology*. John Wiley & Sons, Chichester, pp. 119–135.

Narayanan NK et al. 2009. Liposome encapsulation of curcumin and resveratrol in combination reduces prostate cancer incidence in PTEN knockout mice. *Int J Cancer* 125: 1–8.

Orruno E. 2001. The role of beta-glucosidase in the biosynthesis of 2,5-diemthyl-4-hydroxy-3(H)-furanone in strawberry *(Fragaria X ananassa* cv. *Elsanta). Flavour Frag J* 16: 81–84.

Otles S, Cagindi O. 2008. Carotenoids as natural colorants. In: C Socaciu (ed.) *Food Colorants: Chemical and Functional Properties*. CRC Press, Boca Raton, FL, pp. 51–70.

Pan BS, Kuo JM. 2000. Lipoxygenases. In: NF Haard, BK Simpson (eds.) *Seafood Enzymes*. Marcel Dekker, New York, pp. 317–336.

Peraica M, et al. 1999. Toxic effects of mycotoxins in humans. *Bulletin of the World Health Organization* 77(9): 754–766.

Richard JL. 2007. Some major mycotoxins and their mycotoxicoses – an overview. *Int J Food Microbiol* 119: 3–10.

Rensen G et al. 2005. Development and evaluation of a real-time fluorescent polymerase chain reaction assay for the detection of bovine contaminates in cattle feed. *Foodborne Pathog Dis* 2: 152–159.

Roller S, Goodenough PW. 1999. Food enzymes. In: S Roller, S Harlander (eds.) *Genetic Modification in the Food Industry*. Kluwer, London, UK, pp. 101–128.

Sampson HA. 2004. Update on food allergy. *J Allergy Clin Immunol* 113: 805–819.

Sampson HA, Cooke SK. 1990. Food allergy and the potential allerginicity–antigenicity of microparticulated egg and cow's milk proteins. *J Am Coll Nutr* 9: 410–417.

Sapers GM et al. 2002. Antibrowning agents. In: AL Branen (ed.) *Food Additives*. Marcel Dekker New York, pp. 543–561.

Schormuller J. 1968. The chemistry and biochemistry of cheese ripening. *Adv Food Res* 16: 231–334.

Sicherer SH, Sampson HA. 2006. Food allergy. *J Allergy Clin Immunol* 117: S470–S475.

Simpson B.K. 2000. Digestive proteases from marine animal. In N.F. Haard and B.K. Simpson (eds.) *Seafood enzymes*. Marcel Dekker, New York, pp. 191–213.

Smith CJ. 1999. Carbohydrate biochemistry. In: PJ Lea, RC Leegood (eds.) *Carbohydrate Biochemistry, Plant Biochemistry and Molecular Biology*. John Wiley & Sons, Chichester, pp. 81–118.

Stauffer CE. 1987a) Proteases, peptidases, and inhibitors. In: JE Kruger, et al. (eds.) *Enzymes and Their Role in Cereal Technology*. American Association of Cereal Chemists, St. Paul, MN, pp. 201–237.

Stauffer CE. 1987b) Ester hydrolases. In: JE Kruger, et al. (eds.) *Enzymes and Their Role in Cereal Technology*. American Association of Cereal Chemists, St. Paul, MN, pp. 265–280.

Steele JL. 1995. Contribution of lactic acid bacteria to cheese ripening. In: EL Malin, MH Tunick (eds.) *Chemistry of Structure–Function Relationships in Cheese*. Plenum Press, New York, pp. 209–220.

Stoleo CG, Rehbein H. 2000. TMAO-degrading enzymes. In: NF Haard, BK Simpson (eds.) *Seafood Enzymes*. Marcel Dekker, New York, pp. 167–190.

van Ree R, Aalberse RC. 1995. Demonstration of carbohydrate-specific immunoglobulin g4 antibodies in sera of patients receiving grass pollen immunotherapy. *Int Arch Allergy Immunol* 106(2): 148.

Walstra TJ et al. 1999. *Dairy Technology: Principles of Milk Properties and Processes*. Marcel Dekker, New York, pp. 94–97 (enzymes), 325–362 (lactic fermentation), 541–553 (cheese making).

Wong WSW. 1989. *Mechanisms and Theory in Food Chemistry*. Van Nostrand Reinhold, New York, pp. 242–263.

Wiseman G. 2009. Real-time PCR: Application to food authenticity and legislation. In: J Logan, K Edwards, N Saunders (eds.) *Real-Time PCR: Current Technology and Applications*. Caister Academic Press, Norfolk.

2
Analytical Techniques in Food Biochemistry

Massimo Marcone

Abstract: Food is a very complex heterogeneous "material" composed of thousands of different nutritive and nonnutritive compounds embedded in a variety of different plant and animal matrices. Nondesirable biochemical compounds such as environmental contaminants, microbial and plant toxins, and veterinary drugs are also present with their presence posing a danger to human health. In analytical food chemistry, the isolation, identification, and quantification of both desirable and undesirable compounds continue to pose immense challenges to food analysts. Methods and tests used to isolate, identify, and quantify must be precise, accurate, be increasing sensitive to satisfy the rigors of investigative and applicable science, have minimal interfering factors, use minimal hazardous chemicals, and produce minimal/no hazardous wastes.

This chapter is mainly concerned with the analytical methods used to determine the presence, identity, and quantity of all compounds of interest in a food. Although it is impossible to address the quantitative analysis of all food components, the major techniques used in food analysis will be addressed in detail. The food analyst has a variety of available tests but the test choice is primarily dependent on the goal of the analysis and the use of the final data. Many of the traditional analytical biochemical tests such as Kjeldahl digestion for protein determination are still regarded as the gold standard and is still used in many laboratories.

While useful, traditional analytical methods are continually being challenged by technological and instrumental developments. Technology is moving chemical analysis toward the use of more sophisticated instruments (either individually or in tandem) as both instrumental specificity and sensitivity are continually being "pushed" to new limits. Gas chromatography (GC), high performance liquid chromatography (HPLC), spectroscopy including near infrared (NIR), and mass spectroscopy (MS) are now considered to be basic laboratory equipment used in food analysis. Tandem hybrid analytical equipment such as GC-MS is also gradually becoming common although it is more sophisticated than the single analytical tests.

PROTEIN ANALYSIS

Proteins are a large diverse group of nitrogenous organic compounds that are indispensable constituents in the structure and function of all living cells. They contribute to a wide variety of functions within each cell, ranging from structural materials such as chitin in exoskeletons, hair, and nails to mechanical functions such as actin and myosin in muscular tissue. Chemically, they influence pH as well as catalyze thousands of critical reactions, producing a variety of essential substances that are involved in functions such as cell-to-cell signaling, immune responses, cell adhesion, cell reproduction, etc.

Proteins are essentially polymers of 20 L-∞-amino acids bonded together by covalent peptide bonds between adjacent carboxyl and amino functional groups. The sequence of these amino acids, and thus the function and structure of the protein, is determined by the base pairs in the gene that encodes it. These 20 different amino acids interact with each other within their own chain and/or with other molecules in their external environment, which also greatly influences their final structure and function. Proteins are by definition relatively "heavy" organic molecules ranging in weight from approximately 5000 to more than a million Daltons and, over the years, many of these food proteins have been purified, identified, and characterized. Chemically speaking, nitrogen is the most distinguishing element in proteins, varying in amounts from 13% to 19% due to the variations in the specific amino acid composition of proteins (Chang 1998).

Food Biochemistry and Food Processing, Second Edition. Edited by Benjamin K. Simpson, Leo M.L. Nollet, Fidel Toldrá, Soottawat Benjakul, Gopinadhan Paliyath and Y.H. Hui.

For the past several decades, protein analysis from food products has been performed by determining the nitrogen content after complete acid hydrolysis and digestion by the Kjeldahl method followed by an analytical step in which the resulting ammonium ion is quantified by titrimetry, colorimetry, or by the use of an ion-specific electrode. The result is then multiplied by a pre-established protein conversion factor that determines the final protein content of the sample (Chang 1998, Dierckx and Huyghebaert 2000). While this "wet" analytical technique is still the gold standard in protein analysis, it is time-consuming and involves the use of many dangerous chemicals both to the analyst and to the environment. Its main advantage is that the food sample used in this analytical procedure is considered large enough to be a genuine representative of the entire product. On the other hand, Dumas combustion is a more recent and faster "dry" analytical instrumental method of determining the protein content in foods and is based on the combustion of a very small sample at 900°C in the presence of oxygen. The resulting liberated nitrogen gas is analyzed in three minutes by the equipment through built-in programmed processes with the resultant value also multiplied by pre-determined conversion factors, requiring no further analysis or the use of dangerous chemicals. Both of these methods assume that all the nitrogenous compounds in the sample are proteins, but other organic molecules such as nucleotides, nucleic acids, some vitamins, and pigments (e.g., chlorophyll) also contains nitrogen, overestimating the actual protein content of the sample. Both techniques measure crude protein content, not actual protein content. While the Kjeldahl method is the internationally accepted method of protein determination for legal purposes, Dumas combustion is slowly becoming more acceptable as its accuracy and repeatability will soon be superior to that of the Kjeldahl method (Schmitter and Rihs 1989, Simonne et al. 1997).

As far back as the turn of the century, colorimetric methods for protein determination became available with procedures such as the Biuret, Lowry (original and modified), bicinchoninic acid (BCA), Bradford, and ultraviolet (UV) absorption at 280 nm (Bradford 1976). These colorimetric methods exploit the properties of specific proteins, the presence of specific amino acid functional groups, or the presence of peptide bonds. All require the extraction, isolation, and sometimes purification of the protein molecule of interest to attain an accurate absorbance reading. Considering that the nutrients in foods exist in complex matrices, these colorimetric methods are not practical for food analysis. Additionally, only a few dye-binding methods (official methods 967–12 and 975–17) have been approved for the direct determination of protein in milk (AOAC 1995).

The Biuret procedure measures the development of a purplish color produced when cupric salts in the reagent complex react with two or more peptide bonds in a protein molecule under alkaline conditions. The resultant color absorbance is measured spectroscopically at 540 nm with the color intensity (absorbance) being proportional to the protein content (Chang 1998) and with a sensitivity of 1–10 mg protein/mL. This method "measures" the peptide bonds that are common to cellular proteins, not just the presence of specific side-groups. While it is less sensitive compared to other UV methods, it is considered to be a good general protein assay for which yield is not an important issue. Likewise, the presence of interfering agents during absorbance measurement is not an issue as these substances usually absorb at lower wavelengths. While the color is stable, it should be measured within 10 minutes for best results. The main disadvantage of this method is that it consumes more material as well as it requires a 20-minute incubation period.

Over the years, further modifications to the colorimetric measurement of protein content have been made with the development of the Lowry method (Lowry et al. 1951, Peterson 1979), which combines both the Biuret reaction with the reduction of the Folin-Ciocalteu (F-C) phenol reagent (phosphomolybdic–phosphotungstic acid). The divalent cupric cations form a complex with the peptide bonds in the protein molecules, which cause them to be reduced to monovalent cations. The radical side groups of tyrosine, tryptophan, and cysteine then react with the Folin reagent, producing an unstable molybdenum/tungsten blue color when reduced under alkaline conditions. The resulting bluish color is read at both 500 nm and 750 nm wavelengths, which are highly sensitive to both high and low protein concentrations with a sensitivity of 20–100 ug, respectively. The modified Lowry method requires the absorbance measurement within 10 minutes, whereas the original Lowry method needs precise timing due to color instability. Additionally, the modified Lowry method is more sensitive to protein than the original method but less sensitive to interfering agents.

The BCA protein assay is used to determine the total protein content in a solution being similar to the Biuret, Lowry, and Bradford colorimetric protein assays. The peptide bonds in the protein molecules reduce the cupric cations in the BCA in the reagent solution to cuprous cations, a reaction that is dependent upon temperature. Afterwards, two molecules of BCA chelate the curprous ions, changing the solution color from green to purple, which strongly absorbs at 562 nm. The amount of cupric ions reduced is dependent upon the amount of protein present, which can be measured by comparing the results with protein solutions with known concentrations. Incubating the BCA assay at temperatures of 37–60°C and for longer time periods increases the assay's sensitivity as the cuprous cations complex with the cysteine, cystine, tyrosine, and tryptophan side-chains in the amino acid residues, while minimizing the variances caused by unequal amino acid composition (Olsen and Markwell 2007).

Other methods exploit the tendency of proteins to absorb strongly in the UV spectrum, that is, 280 nm primarily due to the presence of tryptophan and tyrosine amino acid residues. Since tryptophan and tyrosine content in proteins are generally constant, the absorbance at 280 nm has been used to estimate the concentration of proteins using Beer's law. As each protein has a unique aromatic amino acid composition, the extinction coefficient (E_{280}) must be determined for each individual protein for protein content estimation.

Although these methods are appropriate for quantifying the actual amounts of protein, they do have the ability to differentiate and quantify the actual types of proteins within a mixture. The most currently used methods to detect and/or quantify specific protein components belong to the field of spectrometry, chromatography, electrophoresis, or immunology or a combination of these methods (VanCamp and Huyghebaert 1996).

Electrophoresis is defined as the migration of ions (electrically charged molecules) in a solution through an electrical field (Smith 1998). Although several forms of this technique exist, zonal electrophoresis is perhaps the most common. Proteins are separated from a complex mixture into bands by migrating in aqueous buffers through a polyacrylamide gel with a pre-determined pore size (i.e., a solid polymer matrix). In nondenaturing/native electrophoresis, proteins are separated based on their charge, size, and hydrodynamic shape, while in denaturing polyacrylamide gel electrophoresis (PAGE), an anionic detergent sodium dodecyl sulfate (SDS) is used to separate protein subunits by size (Smith 1998). This method was used to determine the existence of differences in the protein subunits between control coffee beans and those digested by an Asian palm civet as well as between digested coffee beans from both the Asian and African civets (Marcone 2004). These protein subunits differences lead to differences in the final Maillard browning products during roasting and therefore flavor and aroma profiles. During the analysis of white and red bird's nests, Marcone determined that SDS–PAGE might be a useful analytical technique for differentiating between the more expensive red bird nest and the less expensive white bird (Marcone 2005). Additionally, this technique could possibly be used to determine if the red nest is adulterated with the less expensive white nest. Using SDS-PAGE analysis, Marcone was the first to report the presence of a 77 kDa ovotransferrin-like protein in both red and white nests, being similar in both weight and properties to ovotransferrin in chicken eggs. In isoelectric focusing, a modification of electrophoresis, proteins are separated by charge in an electrophoretic field on a gel matrix in which a pH gradient has been generated using ampholytes (molecules with both acidic and basic groups, that is, amphoteric, existing as zwitterions in specific pH ranges). The proteins migrate to the location in the pH gradient that equals the isoelectric point (pI) of the protein. Resolution is among the highest of any protein separation technique and can separate proteins with pI differences as small as 0.02 pH units (Chang 1998, Smith 1998). More recently, with the advent of capillary electrophoresis, proteins can be separated on the basis of charge or size in an electric field within a very short period of time. The primary difference between capillary electrophoresis and conventional electrophoresis (described above) is that a capillary tube is used in place of a polyacrylamide gel. The capillary tube can be used over and over again unlike a gel, which must be made and cast each time. Electrophoresis flow within the capillary also can influence separation of the proteins in capillary electrophoresis (Smith 1998).

HPLC is another extremely fast analytical technique having excellent precision and specificity as well as the proven ability to separate protein mixtures into individual components. Many different kinds of HPLC techniques exist depending on the nature of the column characteristics (chain length, porosity, etc.) and elution characteristics such as mobile phase, pH, organic modifiers, etc. In principle, proteins can be analyzed on the basis of their polarity, solubility, or size of their constituent components.

Reversed-phase chromatography was introduced in the 1950s (Howard and Martin 1950, Dierckx and Huyghebaert 2000) and has become a widely applied HPLC method for the analysis of both proteins and a wide variety of other biological compounds. Reversed-phase chromatography is generally achieved on an inert packed column, typically covalently bonded with a high density of hydrophobic functional groups such as linear hydrocarbons 4, 8, or 18 residues in length or with the relatively more polar phenyl group. Reversed-phase HPLC has proven itself useful and indispensable in the field of varietal identification. It has been shown that the processing quality of various grains depends on their physical and chemical characteristics, which are at least partially genetic in origin, and a wide range of qualities within varieties of each species exist (Osborne 1996). The selection of the appropriate cultivar is an important decision for a farmer, since it largely influences the return he receives on his investment (Dierckx and Huyghebaert 2000).

Size-exclusion chromatography separates protein molecules on the basis of their size or, more precisely, their hydrodynamic volume, and has in recent years become a very useful separation technique. Size-exclusion chromatography utilizes uniform rigid particles whose pre-determined pore size determines which protein molecules can enter and travel through the pores. Large molecules do not enter the pores of the column particles and are excluded, that is, they are eluted in the void volume of the column (i.e., elute first), whereas smaller molecules enter the column pores and therefore take longer to elute from the column. An application example of size-exclusion chromatography is the separation of soybean proteins (Oomah et al. 1994). In one study, nine peaks were eluted for soybean, corresponding to different protein size fraction, with one peak showing a high variability for the relative peak area and could serve as a possible differentiation among different cultivars. Differences, qualitatively and quantitatively, in peanut seed protein composition were detected by size-exclusion chromatography and contributed to genetic differences, processing conditions, and seed maturity. In 1990, Basha demonstrated that size-exclusion chromatography was an excellent indicator of seed maturity in peanuts as the area of one particular component (peak) was inversely proportional to increasing peanut seed maturity, which also remained unchanged toward later stages of seed maturity (Basha 1990). The peak was present in all studied cultivars, all showing a mature seed protein profile with respect to this particular protein, which was subsequently named Maturin.

LIPID ANALYSIS

Compared to most other food components, lipids are a group of relatively small, naturally occurring molecules containing carbon, hydrogen, and oxygen atoms, but with much less oxygen than carbohydrates. This large group of organic molecules includes fats, waxes, cholesterol, sterols, glycerides, phospholipids, etc. The most simplistic definition of a lipid is based on its solubility, that is, it is soluble in organic solvents (e.g., alcohol) but insoluble in water. Lipid molecules are hydrophobic, but this generality is sometimes not totally correct as some lipids are amphiphilic, that is, partially soluble in water and partially soluble in organic solvents. Lipids are widely distributed in nature and play many important biological roles, including signaling (e.g., cholesterol), acting as structural materials in cellular membranes, and energy storage. The amphiphilic nature of some lipids gives them the ability to form cellular structures such as vesicles and liposomes within cellular aquatic environments.

Analytically, lipid insolubility in water becomes an important distinguishing characteristic that can be maximally exploited in separating lipids from other nutritional components in the food matrix such as carbohydrates and proteins (Min and Steenson 1998). Classically, lipids are divided into two groups based upon the types of bonding between the carbon atoms in the backbone, which influences the lipid's physical characteristics. Fats having a greater number of fatty acids with single carbon to carbon bonds (saturated fatty acids) cause it to solidify at 23°C (room temperature). On the other hand, the fatty acids in oils have a greater proportion of one or more double bonds between the carbon atoms (unsaturated fatty acids) and hence are liquid at room temperature. Structurally, glycerides are composed primarily of one to three fatty acids (a mixture of saturated and unsaturated fatty acids of various carbon lengths) bonded to the backbone of a glycerol molecule, forming mono-, di-, and triglycerides (the most predominant), respectively. Animal fats in the milk (from mammals), meat, and from under the skin (blubber), including pig fat, butter, ghee, and fish oil (an oil), are generally more solid than liquid at room temperature. Plant lipids tend to be liquids (i.e., they are oils) and are extracted from seeds, legumes, and nuts (e.g., peanuts, canola, corn, soybean, olive, sunflower, safflower, sesame seeds, vegetable oils, coconut, walnut, grape seed, etc.). Margarine and vegetable shortening are made from the above plant oils that are solidified through a process called hydrogenation. This method of solidifying oils increases the melting point of the original substrate but produces a type of fat called *trans* fat, whose content can be as high as 45% of the total fat content of the product. The process, however, is greatly discouraged around the world as these *trans* fats have a detrimental effect on human health (Mozaffarian et al. 2006). *Trans* fats are so detrimental to human health that the amount of *trans* fat in food has been legislated even to the point of being banned in some countries. In 2003, Denmark became the first country in the world to strictly regulate the amount of *trans* fat in foods, and since then many countries have followed this lead.

The total lipid content of a food is commonly determined using extraction methods using organic solvents, either singularly or in combinations. Unfortunately, the wide relative hydrophobicity range of lipids makes the choice of a single universal solvent for lipid extraction and quantitation nearly impossible (Min and Steenson 1998). In addition to various solvents that can be used in the solvent extraction methods, non-solvent wet extraction methods and other instrumental methods also exist, which utilize the chemical and physical properties of lipids for content determination.

One of the most frequent and easiest methods to determine the crude fat content in a food sample is the Soxhlet method, a semicontinuous extraction method using relatively small amounts of various organic solvents. In this method, the solid food sample (usually, dried and pulverized) is placed in a thimble in the chamber and is completely submerged in the hot solvent for 10 minutes or more before both the extracted lipid and solvent are siphoned back into a boiling flask reservoir. The whole process is repeated numerous times until all the fat is removed, which usually requires at least two hours. The lipid content is determined by measuring either the weight loss of the sample in the thimble or by the weight gain of the flask reservoir. The main advantage of this method is that it is relatively quick and specific for fat as other food components such as proteins and carbohydrates are water soluble. Some preparation of the sample, such as drying, pulverizing, and weighing, may be needed to increase the extraction efficiency as extraction rates are influenced by the size of the food particles in the sample, food matrix, etc. If the fat component needs to be removed from the sample before further analysis, the method also allows for the sample in the thimble to remain intact without destruction.

Another excellent method for total fat determination is supercritical fluid extraction. In this method, a compressed gas (e.g., CO_2) is brought to a specific pressure–temperature point that allows it to attain supercritical solvent properties for the selective extraction of a lipid from a food matrix by diffusing through it (Mohamed and Mansoori 2002). This method permits the selective extraction of lipids, while other lipids remain in the food matrix (Min and Steenson 1998). The dissolved fat is then separated from the compressed/liquefied gas by reducing the pressure, and the precipitated lipid is then quantified as a percent lipid by weight (Min and Steenson 1998).

A third often used method for total lipid quantitation is infrared, which is based on the absorption of infrared energy by fat at a wavelength of 5.73 um (Min and Steenson 1998). In general, there is a direct proportional relationship between the amount of energy absorbed at this wavelength and the lipid content in the material. Near-infrared spectroscopy has been successfully used to measure the lipid content of various oilseeds, cereals, and meats. The added advantage is that it maintains the integrity of the sample, which is in contrast to the other previously reviewed methods.

Although these three cited methods are appropriate to quantify the actual amounts of lipids in a given sample, they are not able to determine the types of fatty acids within a lipid sample. In order to determine the composition of the lipid, GC offers the ability to characterize these lipids in terms of their fatty acid composition (Pike 1998). The first step is to isolate all mono-, di-, and triglycerides needed if a mixture exists usually by simple adsorption chromatography on silica or by using a one- or two-phase solvent–water extraction method. The isolated glycerides are then hydrolyzed to release the individual fatty acids and subsequently converted to their ester form, that is, the glycerides are saponified and the liberated fatty acids are esterified to form fatty acid methyl esters, that is, they are derivatized. The method of derivatization is dependent upon the food matrix, and the choice is an important consideration as some methods can produce undesired artifacts. The fatty acids are now volatile and can be separated chromatographically using various gases, various packed or capillary columns, and a variety of temperature–time gradients.

The separation of the actual mono-, di-, and triglycerides is usually much more problematic than determining their individual fatty acid constituents or building blocks. Although GC has also been used for this purpose, such methods result in insufficient information about the complete triglyceride composition in a complex mixture. Such analyses are important for the edible oil industry for process and product quality control purposes as well as for the understanding of triglyceride biosynthesis and deposition in plant and animal cells (Marini 2000).

Using HPLC analysis, Plattner et al. (1977) were able to establish that, under isocratic conditions, the logarithm of the elution volume of a triacylglycerol (TAG) was directly proportional to the total number of carbon atoms (CN) and inversely proportional to the total number of double bonds (X) in the three fatty acyl chains (Marini 2000). The elution behavior is controlled by the equivalent carbon number (ECN) or a TAG, which may be defined as ECN = CN − $X.n$ where n is the factor for double bond contribution, normally close to 2.

The IUPAC Commission on Oils, Fats, and Derivatives undertook the development of a method for the determination of triglycerides in vegetable oils by liquid chromatography. Materials studied included various oils extracted from plant materials such as soybeans, almonds, sunflowers, olives, canola, and blends of palm and sunflower oils, and almond and sunflower oils (Fireston 1994, Marini 2000). The method for the determination of triglycerides (by partition numbers) in vegetable oils by liquid chromatography was adopted by AOAC International as an official IUPAC-AOC-AOAC method. In this method, triglycerides in vegetable oils are separated according to their ECN by reversed-phase HPLC and detected by differential refractometry. Elution order is determined by calculating the ECNs, ECN = s and CN − 2_n, where CN is the carbon number and n is the number of double bonds (Marini 2000).

CARBOHYDRATE ANALYSIS

Carbohydrates comprise approximately 70% of the total caloric intake in many parts of the world, being the major source of energy for most of the world (BeMiller and Low 1998). In foods, these macromolecules have various important roles, including imparting important physical properties to foods such as sensory characteristics and viscosity.

The vast majority of carbohydrates are of plant origin mostly in the form of polysaccharides (BeMiller and Low 1998). Most of these polysaccharides are non-digestible by humans, the only digestible polysaccharide being the starch. The non-digestible polysaccharides are divided into two groups, soluble and insoluble, forming what is referred to as dietary fiber. Each of these two groups has specific functions not only in the plant tissues but also in human and animal digestive systems, subsequently influencing the health of the entire organism.

For many years, the total carbohydrate content was determined by exploiting their tendency to condense with phenolic-type compounds, including pheno, orcinol, resorcinol, napthoresorcinol, and α-naphthol (BeMiller and Low 1998). The most widely used condensation reaction was with phenol, which was used to determine virtually all types of carbohydrates including mono-, di-, oligo-, and polysaccharides. The analytic method is a rapid, simple, and specific determination for carbohydrates. After reaction with phenol in acidic conditions in the presence of heat, a stable color is produced, which is read spectrophotometrically. A standard curve is usually prepared with a carbohydrate similar to the one being measured.

Although the above method was, and is still, used to quantify the total amount of carbohydrate in a given sample, it does not offer the ability to determine the actual types of individual carbohydrates in a sample. Earlier methods such as paper chromatography, open column chromatography, and thin-layer chromatography have largely been replaced by either HPLC and/or GC (Peris-Tortajada 2000). GC has become established as an important method in carbohydrate determinations since the early 1960s (Sweeley et al. 1963, Peris-Tortajada 2000) and several unique applications have since then been reported (El Rassi 1995).

For GC analysis, carbohydrates must first be converted into volatile derivatives, the most common derivatization agent being trimethylsilyl (TMS). In this analytic technique, the aldonic acid forms of carbohydrates are converted into their TMS ethers, which are then injected directly into the chromatograph having a flame ionization detector. Temperature programming is utilized to maximally optimize the separation and identification of individual components. Unlike GC, HPLC analysis of carbohydrates requires no prior derivatization of carbohydrates and gives both qualitative (identification of peaks) and quantitative information of complex mixtures of carbohydrates. HPLC is an excellent method for the separation and analysis of a wide variety of carbohydrates ranging from the smaller and relatively structurally simpler monosaccharides to the larger and more structurally complex oligosaccharides. For the analysis of the larger polysaccharides and oligosaccharides, a hydrolysis step is needed prior to chromatographic analysis. A variety of different columns can be used with bonded amino phases used to separate carbohydrates with molecular weights up to about 2500 kDa depending upon carbohydrate composition and their specific solubility properties (Peris-Tortajada 2000). The elution order on amine-bonded stationary phases is usually monosaccharide and sugar alcohols followed by disaccharides and oligosaccharides. Such columns have been successfully used to analyze carbohydrates in fruits and vegetables as well as processed foods such as cakes, confectionaries, beverages, and breakfast cereals (BeMiller and Low 1998). With larger polysaccharides, gel filtration becomes the preferred chromatographic technique as found in the literature. Gel filtration media such as Sephadex and Bio-Gel have successfully been used to characterize polysaccharides according to molecular weights.

MINERAL ANALYSIS

A (dietary) mineral is any inorganic chemical element in its ionic form, excluding the four elements, namely carbon, nitrogen, oxygen, and hydrogen. These minerals are important as they are required for a wide range of physiological functions, including energy generation, enzyme production, providing structure (skeleton), circulation fluids (blood), movement (muscle and nerve function), digestion (stomach acid production), and hormone production (thyroxine), just to name a few. It has been estimated that 98% of the calcium and 80% of the phosphorous in the human body are bound-up within the skeleton (Hendriks 1998). Those that are directly involved in physiological function (such as in muscle contraction) include sodium, calcium, potassium, and magnesium.

Most minerals (calcium, chloride, cobalt, copper, iodine, iron, magnesium, manganese, molybdenum, nickel, phosphorous, potassium, selenium, sodium, sulfur, and zinc) are considered to be essential nutrients in the classical sense as they are needed for physiological functions and must be obtained from dietary sources. Additionally, there is a much smaller group of minerals (arsenic, boron, chromium, and silicon) that have a speculated role in human health, but their roles have not been conclusively established, yet. Some minerals are referred to as macro-minerals as they are required by humans in amounts greater than 100 mg per day, including sodium, potassium, magnesium, phosphorous, calcium, chlorine, and sulfur. A further ten minerals are classified as micro or trace minerals as they are required in milligram quantities per day, including silica, selenium, fluoride, molybdenum, manganese, chromium, copper, zinc, iodine, and iron (Hendriks 1998). In humans, both the macro- and micro- minerals can contribute to disease development in either excessive or deficient amounts.

Although many minerals are naturally found in most raw food sources, some are added to foods during processing for different purposes. An example of this is the addition of salt (sodium) during processing to decrease water activity and to act as a preservative, for example, pickles and cheddar cheese (Hendriks 1998). It should also be noted that food processing can also cause a decrease in the mineral content, for example, the milling of wheat removes the mineral-rich bran layer. During the actual washing and blanching of various foods, important minerals are often lost in the water. It can therefore be concluded that accurate and specific methods for mineral determination are in fact important for nutritional purposes as well as in properly processing food products for both human and animal consumption.

Many countries have laws specifying which minerals must be added to specific foods in legally specified amounts and forms. Iron is added to fortify white flour and various other minerals, including calcium, iron, and zinc, are added to various breakfast cereals. In North America, salt itself is fortified with iodine to inhibit the development of goiter. Additionally, minerals are affected by legal considerations. A food's nutrition label must declare the mineral content of the food in certain cases, such as baked goods. In some countries such as the United States, the link between a mineral in specified amounts and its health effects is permitted, whereas in countries such as Canada, legally predetermined statements may be put on the food product packaging.

To determine the total mineral content in a food, the ashing procedure is usually the method of choice. The word "ash" refers to the inorganic residue that remains after either ignition or, in some cases, complete oxidation of the organic material (Harbers 1998). Ashing can be divided into three main types—dry ashing (most commonly used), wet ashing (oxidation) for samples with a high fat content such as meat products or for preparation for elemental analysis, and plasma ashing (low temperature) when volatile elemental analysis is conducted.

In dry ashing, food samples are incinerated in a muffle furnace at temperatures of 500–600°C. During this process, most minerals are converted to either oxides, phosphates, sulfates, chlorides, or silicates. Unfortunately, some minerals such as mercury, iron, selenium, and lead may be partially volatized using this high-temperature procedure. Wet ashing involves the use of various acids that oxidize the organic materials while the minerals are subsequently solubilized without volatilization. Nitric or perchloric acids that are often used along with reagent blanks are carried throughout the entire procedure and then subtracted from sample results. In the low-temperature plasma ashing, the food sample is treated in a similar manner as in dry ashing but under a partial vacuum, with samples being oxidized by nascent oxygen formed by an electromagnetic field.

Although the above three ashing methods have been proven to be adequate for quantifying the total amount of minerals in a sample, they do not possess the ability to differentiate nor to quantify the actual mineral elements in a food sample. Atomic absorption spectrometers became widely used in the 1960s and 1970s and paved the way for measuring the presence and trace amounts of minerals in various biological samples (Miller 1998). Essentially, atomic absorption spectroscopy is an analytical technique based on the absorption of UV or visible radiation by free atoms in the gaseous state. However, the sample must first be ashed and then diluted in weak acid. The solution is then atomized into a flame. According to Beer's law, the absorption is directly related to the concentration of a particular element in the sample.

Atomic emission spectroscopy differs from atomic absorption spectroscopy in that the source of the radiation is in fact the excited atoms or ions in the sample rather than an external source has in part taken over. Atomic emission spectroscopy does have the advantages with regard to sensitivity, interference, and multielement analysis (Miller 1998).

Recently, the use of ion-selective electrodes has made online testing of the mineral composition of samples a reality. In fact, many different electrodes have been developed to directly measure various anions and cations, such as calcium, bromide, fluoride, chloride, potassium, sulfide, and sodium (Hendriks 1998). Typically, levels as low as 0.023 pm can be measured. When working with ion-selective electrodes, it is common procedure to establish a calibration curve.

VITAMIN ANALYSIS

Vitamins are low-molecular weight organic compounds obtained from external sources in the diet and are critical for normal physiological and metabolic functions (Russell 2000). Vitamins are divided into two groups based on liquid solubility, that is, those that are water soluble (B vitamins and vitamin C) and those that are fat soluble (vitamins A, D, E, and K). Since the vast majority of vitamins cannot be synthesized by humans, they must be obtained either from food sources or from manufactured dietary supplements.

Vitamin analysis of foods are completed for numerous reasons, including regulatory compliance, nutrient labeling, investigating the changes in vitamin content attributable to food processing, packaging, and storage (Ball 2000). As such, numerous analytical methods have been developed to determine their levels during processing and in their final product.

The scientific literature contains numerous analytical methods for the quantitation of water-soluble vitamins, including several bioassays, calorimetric, and fluorescent assays, all of which have been proven to be accurate, specific, and reproducible for both raw and processed food products (Eitenmiller et al. 1998). By far, HPLC has become the most popular technique for quantifying water-soluble vitamins, and the scientific literature contains an abundance of HPLC-based methodologies (Russell 2000). In general, it is the nonvolatile and hydrophilic natures of these vitamins that make them excellent candidates for reversed-phase HPLC analysis (Russell 1998). The ability to automate using autosamplers and robotics makes HPLC an increasingly popular technique. Since the vast majority of vitamins occur in trace amounts in foods, detection and sensitivity are critical considerations. Although UV absorbance is the most common detection method, both fluorescence and electrochemical detection are also used in specific cases. Refractive index detection is seldom used for vitamin detection due to its inherent lack of specificity and sensitivity.

During the 1960s, GC with packed columns was widely used for determining the presence and concentrations of various fat-soluble vitamins, especially vitamins D and E. Unfortunately, additional preliminary techniques such as thin-layer chromatography and open-column techniques were still necessary for the preliminary separation of the vitamins, followed by derivatization to increase their thermal stability and volatility (Ball 2000). More recently, the development of fused-silica open tubular capillary columns has revived the used of GC, leading to a number of recent applications for the determination of fat-soluble vitamins, especially vitamin E (Marks 1988, Ulberth 1991, Kmostak and Kurtz 1993, Mariani and Bellan 1996). Even so, HPLC is also the method of choice for the fat-soluble vitamins in foods (Ball 1998). The main advantage of this technique is that the vitamins need not be derivatized, and the method allows for both greater separation and detection selectivity. Various analytical HPLC methods have been introduced for the first time in the 1995 edition of the Official Methods of Analysis of AOAC International, including vitamin A in milk (AOAC 992.04 1995) and vitamin A (AOAC 992.26 1995), vitamin E (AOAC 992.03 1995), and vitamin K (AOAC 992.27 1995) in various milk-based infant formulas (Ball 1998).

At the present time, there are no universally recognized standardized methods that can be applied to all food types for determining the presence, amount, or identity of the fat-soluble vitamins (Ball 1998).

PIGMENT ANALYSIS

The color of food is a very important sensory characteristic as it is often used as an indicator of both food quality and safety (Schwartz 1998). A vast number of natural and synthetic pigments are obtained from both plant and animal sources as well as are developed during processes, for example, carmelization and Maillard browning reactions. The majority of the naturally occurring pigments in foods are divided into five major classes, four of which are found in plant tissues while the fifth is found in animal tissues (Schwartz 1998). Of those found in plants, two are lipid soluble, that is, the chlorophylls (green) and the carotenoids (yellow, orange, and red), with the other two are water soluble, that is, anthocyanins (color usually depending upon pH, red in acidic conditions and blue in basic) and betalains (yellow or red indole-derived). Carotenoids are found in both plant (e.g., carotenoids (carrots), lycopene (tomatoes)) and animal tissues (e.g., milk, egg yolk), but in animals they are derived from dietary sources, for example, the yellow color in milk/egg yolk is due to the consumption of green plants during pasture feeding with various grasses, clovers, etc. (Schwartz 1998), not from biosynthesis in the animal itself.

Several analytical methods exist for the analysis of chlorophylls in a wide variety of foods. Early spectrophotometric methods permitted the quantification of both chlorophyll a and b by measuring the absorbance maxima of both chlorophyll types. Unfortunately, only fresh plant material could be assayed as no pheophytin could be determined. This became the basis for the AOAC International spectrophotometric procedure (Method 942.04), which provides results for total chlorophyll content as well as for chlorophyll a and b quantitation.

Schwartz et al. (1981) described a simple reversed-phase HPLC method for the analysis of chlorophyll and their derivatives in fresh and processed plant tissues. This method simplified the determination of chemical alterations in chlorophyll during food processing and allowed for the determination of both pheophytins and pyropheophytins.

Numerous HPLC methods have been developed for carotenoid analysis, specifically for the separation of the different carotenoids found in fruits and vegetables (Bureau and Bushway 1986). Both normal and reversed-phase methods have been used with the reversed-phase methods predominating (Schwartz 1998). Reversed-phase chromatography on C-18 columns using isocratic elution procedures with mixtures of methanol and acetonitrile containing ethyl acetate, chloroform or tetrahydrofuran have been determined to be satisfactory (Schwartz 1998). Carotenoids are usually detected in wavelengths ranging from 430 nm to 480 nm. Since β-carotene in hexane has an absorption maximum at 453 nm, many methods have detected a wide variety of carotenoids in this region (Schwartz 1998).

Measurements of the water-soluble but pH color-dependent anthocyanins (red, purple, or blue) have been performed by determining absorbance of diluted samples acidified to approximately pH 1.0 at wavelengths between 510 nm and 540 nm. Unfortunately, absorbance measurements of anthocyanins provide only total quantification and any further information about the presence and amounts of various other individual anthocyanins must be performed by other methods. HPLC methods using reversed-phase methods employing C-18 columns have been the methods of choice as the anthocyanins are water soluble. Mixtures of water, acetic, formic, or phosphoric acids usually are used as part of the mobile phase.

Charged pigments such as the red-violet colored betalains can be separated by electrophoresis, but reversed-phase HPLC provides more rapid resolution and quantification (Cai et al. 2005). Betalains have been separated on reversed-phase columns using various ion-paring or ion-suppression techniques (Schwartz and Von Elbe 1980, Huang and Von Elbe 1985). This procedure allows for more interaction of the individual betalains molecules with the stationary phase and better separation between individual components.

ANTIOXIDANTS

An increasingly important group of nonnutritive food components are the antioxidants, a large group of biologically active chemicals in plant and animal tissues usually in relatively small concentrations (compared to their substrates; Halliwell and Gutteridge 1995). As the name implies, an antioxidant mitigates/prevents the oxidation of other molecules that are more easily oxidized, for example, vitamin E in oil prevents the auto-oxidation of lipids, and antioxidants are "strong" reducing agents. Antioxidants can accomplish this task via three possible mechanisms. The first is to prevent free-radical generating reactions (oxidation reactions) from occurring, the second involves inhibiting/scavenging for reactive species, and the third involves the chelation/sequestering of metals that can act as catalysts. Whichever pathway, antioxidants "protect" susceptible molecules such as membrane lipids from being oxidized, thereby maintaining their structural and functional integrity (Dilis and Trichopoulou 2010).

These naturally occurring antioxidants consist of vitamins, enzymes (proteins), minerals, pigments, organic acids, and hormones (Prior et al. 2005), which include vitamins C (ascorbic acid) and E and its isomers (α-tocopherol and tocotrienols); selenium and manganese (minerals); α-carotene, β-carotene, lutein, lycopene, etc. (carotenoid terpenoids); catalases, glutathione (a thiol), peroxidases, reductases, and superoxide dismutases (enzymes); flavones, flavenols, flavanones, flavanols and their polymers, isoflavone phytoestrogens, stilbenoids, and anthocyanins (all polyphenolic flavenoids); chlorogenic acid, cinnamic acid (and its derivatives), gallic acid, etc. (all phenolic acids and their esters); xanthones and eugenol (non-flavenoid phenolics); lignin, bilirubin, citric, oxalic, phytic acids, etc. (organic molecules); and melatonin, angiotensin, and estrogen (hormones) (Sies 1997, Prior et al. 2005, Dilis and Trichopoulou 2010).

Chemically speaking, an oxidation reaction involves the transfer of one or more electron(s) from one molecule to another (a redox reaction), thereby creating a charged molecule/molecular fragment called a free radical (essentially, an ion with positive or negative charge), which can further promote subsequent reactions, leading to the generation of more free radicals. By their very nature, free radicals are very reactive as they have an unpaired electron in their outer shell configuration and, in an attempt to reach stability, they donate/receive electrons to/from other molecules such as lipids, proteins, and genetic material (DNA). This interaction damages these more oxidizable molecules, which impairs their functions. Antioxidants can either donate hydrogen atoms or donate/receive electrons to/from free radicals. While the "intervention" of the antioxidants causes the antioxidants to become radicals themselves, these radicals are more stable, eventually returning to their original state.

The human body produces free radicals through normal metabolic processes such as aerobic respiration, while inflammation and disease processes, exercise, exposure to exogenous environmental chemicals, tobacco (including tobacco smoke), sun-exposure (radiation), and chemicals (cleaning products and cosmetics) are also sources of free radicals (Freeman and Capro 1982, Southorn and Powis 1988). The most abundant group of free radicals produced in the human body involve oxygen, and they are often referred to reactive oxygen species (ROS), a group of peroxyl radicals including superoxide anions ($O_2{}^{-}$), hydroxyl radicals ($OH\cdot$), singlet oxygen radicals ($^1O_2\cdot$), hydrogen peroxide (H_2O_2), and ozone (O_3). Other free radical producers/enablers include iron and copper (transition metals acting as catalysts), which promote the production of more aggressive free radicals (Valko et al. 2005, Apak et al. 2007) as well as nitric oxide. Biologically, some of the free radicals have critical biological functions, for example, some of the ROS species are used by the immune system to attack and kill pathogens and nitric oxide is involved in cell (redox) signaling (Somogyi et al. 2007).

A diet high in refined sugars and grains, foods cured with nitrites, pesticides, and hydrogenated vegetable oils coupled with decreased consumption of whole grains, fruits, vegetables, and non-/minimally processed foods, and non-healthy lifestyle generates more free radicals compared to a totally opposite diet (Bruce et al. 2000). It has been hypothesized that free-radical generating diets can lead to oxidative stress, an unbalanced state characterized by increased free-radical concentrations coupled with decreased antioxidant concentrations. This state has been implicated in the aging process as well as in many diseases/conditions such as Alzheimer's disease, Parkinson's disease, cardiovascular disease (including heart attack and stroke), various cancers, cellular DNA damage, rheumatoid arthritis, and cataracts (Etherton et al. 2002, Stanner et al. 2004, Shenkin 2006, Nunomura et al. 2006). Epidemiological studies have suggested the benefits of a high antioxidant diet, but clinical trials have provided mixed results (Pérez et al. 2009, Dilis and Trichopoulou 2010). This would seem to suggest that antioxidants may only be partially responsible or work in synergism with other biologically active molecules (Cherubini et al. 2005, Seeram et al. 2005). However, some clinical investigations have demonstrated that antioxidants can also act as pro-oxidants in some diseases (Bjelakovic et al. 2007). Because of their possible varied functionalities in disease prevention, mitigation, or even treatment, the food industry is very much interested in promoting the presence of either naturally occurring or added antioxidants in their products. Pharmaceutical companies too are interested in antioxidants through the creation and marketing of various dietary supplements or pharmaceutical drugs for the treatment of diseases.

Currently, antioxidant assays are divided into two groups, the division being based on the underlying reaction mechanisms, hydrogen atom transfer (HAT) and electron transfer (ET; Prior et al. 2005), which can work in tandem in some assays. Details of these two reaction mechanisms, the descriptions of the various assays, and their comparisons have been published extensively over the past 20 years (Halliwell 1994, Prior and Cao 1999, Huang et al. 2005, Prior et al. 2005, Apak et al. 2007, Phipps et al. 2007, Moon and Shibamoto 2009). HAT mechanisms measure the ability of antioxidants to quench free radicals by donating a hydrogen atom. This ability is critical as the transfer of a hydrogen atom is a critical step in preventing/stopping radical chain reactions (Huang et al. 2005). HAT-based assays uses a synthetic free-radical generator, an oxidizable probe (usually fluorescent), and the antioxidant (Huang et al. 2005), monitoring the competitive reaction kinetics with final quantification derived from kinetic curves. ET-based assays, on the other hand, are colorimetric assays that measure the change in color as the oxidant is reduced. The mechanism itself involves only one redox reaction, with the result being a quantifiable measurement of the antioxidant's reducing capacity (Huang et al. 2005).

Common HAT assays include oxygen radical absorbance capacity (ORAC), total radical trapping antioxidant parameter (TRAP; and some of its variants), and total oxidant scavenging capacity (TOSC). ORAC assays are the most widely used in research, clinical, and food laboratories for antioxidant quantification. This assay represents the classical antioxidant mechanism by hydrogen transfer and can measure both in vivo hydrophilic and lipophilic antioxidants, the latter by using selective combinations of solvents with relatively small polarities. It is widely used to measure the antioxidant capacity (AOC) of food samples ranging from pure compounds (e.g., melatonin) to complex matrices such as fruits and vegetables (Prior and Cao 1999). Peroxyl radicals generated with an azo-initiator compound are added to microplate wells along with a fluorescent probe and the extracted antioxidant. As the reaction progresses, fewer and fewer of the antioxidants are available to donate hydrogen atoms to the peroxyl radicals, which lead to radicals combining with the fluorescent probe and forming a nonfluorescent molecule. The reaction is quantified using a fluorometer over a period of 35 minutes, which is the time taken for the free radical action to complete. The length of the ORAC assay also minimizes the possibility of underestimating the total antioxidant concentration as some secondary antioxidant products have slower reaction kinetics. The effectiveness of the antioxidant is calculated from the area between two curves generated when the assay is performed with a sample and a blank, thereby combining both length of inhibition and percentage inhibition into a single quantity (Cao et al. 1995), with the results expressed as Trolox (a vitamin E analogue) equivalents, that is, μM Trolox equivalents (TE) per gram. One of the great advantages of the ORAC method is other radical sources can be used (Prior et al. 2005) and has also been developed for automation, which reduces the traditional time-consuming analysis as well as human reaction time and other sources of human error. In 2007, the United States Department of Agriculture published a report entitled "Oxygen Radical Absorbance Capacity (ORAC) of Selected Foods—2007" (USDA 2007), which lists the ORAC values (hydrophilic and lipophilic) as well as ferric-ion reducing antioxidant power (FRAP) and Trolox equivalence antioxidant capacity (TEAC) assays for 59 different foods in the American diet.

The second HAT mechanism is the TRAP assay, which measures the ability of an antioxidant to interfere with the reaction between a hydrophilic peroxyl radical generator such as 2,2′-azobis(2-amidinopropane) dihydrocholride and a target probe. The effectiveness of the antioxidant to interfere in this reaction is measured by the consumption of oxygen during the reaction with the oxidation of the probe followed optically or by fluorescence (Prior et al. 2005). In the initial lag phase, the oxidation is inhibited by the antioxidants and is compared to the internal Trolox standard. TRAP values are reported as a lag time or reaction time compared to the corresponding times for Trolox (Prior et al. 2005). The TRAP antioxidant assay is most useful for measuring serum or plasma AOC, that is, in vivo nonenzymatic antioxidants such as glutathione and nutritive antioxidant (pro-) vitamins such as vitamin A (β-carotene), C (ascorbic acid), and E (α-tocopherol). This assay is also sensitive to all antioxidants that are capable of breaking the reaction generating peroxyl radicals. The major disadvantage is that the assay assumes all antioxidants have a lag phase, but this is untrue. The total AOC is underestimated as the antioxidant value contributed after the lag phase (if one is present) is not taken into account and also assumes that the lag phase is proportional to AOC.

The third assay to measure the HAT mechanism is TOSC, which quantifies the absorbance capacity of three oxidants—hydroxyl radicals, peroxyl radicals, and peroxynitrite—thereby making it applicable to evaluate antioxidants from different biological sources (Prior et al. 2005). The substrate is α-keto-γ-methiolbutyric acid, which forms ethylene when oxidized and is measured over time by headspace analysis by GC. The AOC of the antioxidant is determined by its ability to inhibit ethylene formation. Like the ORAC, this assay is determined by the area between the curve generated by the sample and that of the control. The major advantage is that it can quantify the effects of three common radicals of interest, but it is time-consuming as the GC analysis can take 300 minutes, requiring multiple injections of ethylene that has been collected from a single sample into the GC. Additionally, the percentage antioxidant inhibition and its concentration are not linear, which requires a graph generated by various dilution factors with the DT_{50} calculated by the first derivation of this dose–response curve.

The only major assays measuring ET mechanisms (reducing power) is the FRAP, which was initially developed to measure the OAC of plasma but has been adapted for samples of botanical origins (Prior et al. 2005). The low pH reaction measures the reduction of ferric 2,4,6-tripyridyl-*s*-triazine by antioxidants in a redox-linked colorimetric method with a redox potential < 0.7 V (Phipps et al. 2007). The bright blue color caused by the presence of the ferrous ion is measured spectrophotometrically at 593 nm. The major advantage is that the assay is inexpensive, the results are producible, and the assay is short, leading to automated or semi-automated high throughput. The major drawback with the FRAP assay is that a chemical reductant and a biological antioxidant are treated identically, but the assay neither directly use a pro-oxidant nor use an oxidizable substrate (Prior and Cao 1999). The ferric ion (Fe^{3+}) is not necessarily a pro-oxidant such as the ferrous ion (Fe^{2+}) that reacts with hydrogen peroxide (a peroxyl free radical). Additionally, neither ion is able to cause direct oxidative damage to easily oxidized biological molecules such as lipids, proteins, and DNA. Additionally, the assay is unable to measure the effects of antioxidants that act strictly on HAT (mechanisms such as thiols and proteins), leading to underestimation of serum AOC but may be useful in combination with other assays to differentiate between antioxidant mechanisms.

Both the TEAC and other ABTS (2,2′-azinobis[3-ethylbenzothiazoline-6-sulfonic acid] di-ammonium salt) assay as well as DPPH assay (1,1-diphenyl-2-picryl-hydrazyl) use indicator radicals that may be neutralized by either HAT or ET mechanisms. In the TEAC (and its related assays), the long life anion ABTS molecule is oxidized to its cation radical form ($ABTS^{\cdot+}$) by peroxyl radicals, forming an intensely colored liquid. The AOC is determined by the ability of the antioxidant to decrease the color by reacting with the cation, which is measured with a spectrophotometer at 415, 645, 734, and 815 nm (Prior et al. 2005). As with ORAC, the results are expressed in TE. This assay offers advantages over some of the other assays described previously. Its major advantage is that antioxidant mechanisms as influenced by pH can be studied over a wide pH range. Additionally, the generation cation radical is soluble in both polar and nonpolar solvents and is not affected by ionic strength. While it measures both hydrophilic and lipophilic antioxidants, it does tend to favor the hydrophilic species, but this can be overcome by using buffered media or partitioning the antioxidants between hydrophilic and lipophilic solvents. In terms of effort, the assay is short (30 minutes) as the antioxidants react rapidly with the generated cation radical The major objection to this assay is that the ABTS radical is not found in mammalian tissues and thus represents a synthetic radical source, but any molecule, such as phenolic compounds, that have low redox potentials can react with $ABTS^{+}$ (Prior et al. 2005). As with some of the other assays, TEAC may have long initial lag phases, and the assay may not go to completion before stopping itself, resulting in OAC underestimation.

The theory of the DPPH assay is based on the ability of the antioxidants to reduce it to its radical form, DPPH·, a stable nitrogen radical producing a deep purple color with the decrease in color monitored by absorbance at 515 nm or by electron spin resonance. The results are reported as EC_{50}, which measures a 50% decrease in the DPPH· concentration, which is proportional to the antioxidant concentration (Prior et al. 2005). This assay is also quick and only requires the use of a spectrophotometer, but results are difficult to interpret as compounds such as the carotenoids also absorb maximally at 515 nm. The major disadvantage is that DPPH acts both as the radical probe and as the oxidant. As a radical, DPPH is not involved in lipid peroxidation. Many antioxidants react with peroxyl radicals, but DPPH may be inaccessible to the antioxidants or may interact with them very slowly.

The F-C assay, aka total phenolics assay, is a measure of the AOC of the total phenolic content of a (food) sample. Since its development in 1927 for tyrosine analysis, it has been developed and improved to measure the AOC of all the phenolic molecules of a sample (Singleton and Rossi 1965). These phenolic compounds are oxidized by a molybdotungsto-phosphoric heteropolyanion reagent with the reduced phenols, producing a colored product that is measured maximally at 765 nm. While the method is simple, it has a number of drawbacks. The method must be followed exactly and explicitly to minimize variability and erratic results. The assay suffers from a large number of interfering but naturally occurring substances such as fructose, ascorbic acid, aromatic amino acids, and other compounds that are listed by Prior et al. (2005). However, this limitation was overcome in a study on the phenolic content of urine by using solid-phase extraction to remove the water-soluble compounds from the sample (Roura et al. 2006). It should also be taken into consideration that the phenolic composition in different samples would react differently as is demonstrated by the F-C assay of different foods (cf. Figure 7, Prior et al. 2005).

For analytical chemists, it becomes increasingly important to understand the composition of the food matrix as the different antioxidant assays may not be identical in their underlying mechanisms. Additionally, the results must also take into account the effect of geography, soils, climatic conditions, variety, seasonality, transportation, processing, etc. of a particular food, as these can influence resulting antioxidant values (Dewanto et al. 2002, Tsao et al. 2006, Xu and Chang 2008).

In light of the different underlying reaction mechanisms, the different antioxidants, their target free radicals, and their environmental conditions, the food matrix, etc., the dilemma becomes apparent as to which assay to choose as much has to be taken into consideration. Extraction methodologies, detection, quantification, measuring the effectiveness of the vast array of antioxidants against their target free radicals, as well as interpreting the results (Prior et al. 2005) are just some considerations. As can also be deduced, the different assays have their advantages/disadvantages, and the result of one cannot be compared to the results of another. None of the assays can be considered as measurements of total antioxidant activity as antioxidants are indeed varied.

As a result of all these challenges and concerns, 144 scientists met at the First International Congress on Antioxidant Methods in 2004 to standardize these issues across all laboratories worldwide, with all resultant works are to be published in *The Journal of Agricultural Food Chemistry* (Prior et al. 2005). We look forward to the fruit of this enormous and invaluable effort.

GAS CHROMATOGRAPHY—MASS SPECTROSCOPY

GC–MS is a common and powerful analytical technique used in both agricultural and food industries, which combines the features of both GC and MS. Developed in the 1950s by Roland Gohike and Fred McLaffert, GC–MS first separates the prepared sample (containing a mixture of compounds) into separate and distinct compounds (by GC) and then positively identifies and quantifies these compounds based on their ion fragment pattern (by MS). This approach is advantageous as it identifies a molecule by both its unique retention time as well as its unique fragmentation pattern. The results are graphically illustrated in a mass spectrum graph that yields both qualitative and quantitative information about the substance(s) of interest.

GC separates (partitions) and analyzes molecules of interest such as lipids and carbohydrates, which can be volatilized (naturally or chemically) without decomposition or creation of chemical artifacts. The volatilized compounds are injected into the GC, vaporized, and carried through a column by an inert gas, where they differentially interact with the inner lining of the column. Their rate of release from the column lining is based on their attraction with the column lining, and after being released, they are carried to the detector that electronically records the amount of time each substance has taken to travel through the column. MS, on the other hand, determines the elemental composition of a sample, such as the gaseous effluent from a gas chromatograph. Molecules are ionized, creating both charged molecules as well as their various fragments. These charged substances are accelerated through an electromagnetic field that filters the ions based on a predetermined mass size. The ions are then counted by a detector that sends the information to a computer, which generates a "graph" showing both the number of the various ion fragments with their different masses and their relative amounts. Each chemical has a specific fragmentation pattern, just like a human fingerprint or iris pattern.

GC–MS is used for substances that are thermally stable (up to 300°C) and relatively volatile at temperatures less than 300°C, whereas liquid chromatography–MS (LC–MS or HPLC–MS) is used for chemicals that are not volatile and/or degrade at temperatures above room temperatures. Nonvolatile chemicals can be made volatile with derivatizing agents, but not all derivatizing agents and solvents are compatible with various components of GC or HPLC equipment. Both GC–MS and LC–MS provide the much-needed sensitivity than either equipment used alone. Conventional tests such as Kjeldahl or Dumas can determine if a sample contains any substances that can be categorized as proteins, carbohydrate, lipid, etc., but these single methods are not able to specifically identify the substance of interest. The main advantage of using these two analytical tools in tandem (GC–MS) is that together they positively identify (confirm) the presence of any chemical in a sample with no false positive results even to levels as low as 5 ppb (Tareke et al. 2002). Using the analogy of a puzzle, single tests such as Kjeldahl or Dumas provide one piece to the puzzle, whereas GC–MS provides the entire picture. Both single and tandem analytical approaches are valid but before starting any analysis, one has to determine what information is needed and which analytical test(s) would best provide that information to the meet the criteria of identification, accuracy, and sensitivity.

Currently, food applications of GC–MS is primarily limited to detect the presence of contaminants such as pesticides, insecticides, etc., in fruits and vegetables produced around the world, as their use in the agricultural industry is not globally standardized. This reality has made GC–MS a valuable tool in the area of food safety and quality assurance.

Foods and beverages contain many aromatic compounds such as esters, fatty acids, alcohols, aldehydes, terpenes, etc., which may be naturally present or formed during processing, and the presence or absence of specific compounds may be of interest to researchers, food product developers, food inspectors etc. As stated above, GC–MS is mainly used to detect the presence or absence of pesticides in foods, but it can easily be expanded to detect contaminants arising from the spoilage or adulteration of food. GC–MS detected the presence of pesticides in pet food from China in 2007 (FDA 2007) as well as the presence of melanine in Chinese baby food in 2008. Both Kjeldahl and Dumas would have shown the presence of much higher than expected nitrogen content, and hence protein content, in the contaminated pet food and melamine adulterated baby food. The higher than expected protein content would have raised the suspicions of analysts, but further expensive and time-consuming tests would need to have been completed. In this case, GC–MS clearly demonstrated the presence of pesticides in the pet food and was also able to differentiate between the nitrogenous milk proteins and the nitrogen from the melanine.

In the world of expensive foods, GC–MS is an effective tool in the detection of food adulteration. Some food delicacies such as caviar, bird's nests, saffron, and truffles as well as other expensive food commodities such as coffee, tea, olive oil, balsamic vinegar, honey, maple syrup, etc., are commonly adulterated to levels that may not be detectable by conventional tests, and visual inspection may not raise the suspicions of food inspectors or consumers. Ruiz-Matate et al. (2007) reported the development of a GC–MS method to detect the adulteration of honey with commercial syrups. Honey can be adulterated with inexpensive sugar syrups such as high-fructose corn syrup or inverted syrups. Using GC–MS, they discovered, and were also the first to report, the presence of di-fructose in the four syrup samples but not in any of the honey samples. Hence, the presence of di-fructose in any honey sample would indicate possible adulteration of the honey. Likewise, Plessi et al. (2006) used GC–MS to discover the presence of nine phenolic antioxidant compounds in balsamic vinegar, an internationally and nationally recognized commodity that has country of origin status in the European Union (European Council Regulation 813/2000). These phenolic acids certainly contribute to the unique sensory qualities of the balsamic vinegar from Modena and may very well be used as specific identification markers indicating country, and region, of origin.

As adulteration can be fatally dangerous as well as compromise the integrity of the genuine food product, constant vigilance by advanced analytical methods such as GC–MS is critically necessary.

REFERENCES

AOAC International. 1995. *Official Methods of Analysis*, 16th edn. AOAC International, Gaithersburg, MD.

AOAC Official Method 992.03. 1995. Vitamin E activity (all-*rac*-α"-tocopherol) in milk and milk-based infant formula. Liquid chromatographic method. In: MP Bueno (ed.) *Official Methods of Analysis of AOAC International*, 16th edn. AOAC International, Arlington, VA, pp. 50-4–50-5.

AOAC Official Method 992.04. 1995. Vitamin A (retinol isomers) in milk and milk-based infant formula. Liquid chromatographic method. In: MP Bueno (ed.) *Official Methods of Analysis of AOAC International*, 16th edn. AOAC International, Arlington, VA, pp. 50-1–50-2.

AOAC Official Method 992.06. 1995. Vitamin A (retinol isomers) in milk and milk-based infant formula. Liquid chromatographic method. In: MP Bueno(ed.) *Official Methods of Analysis of AOAC International*, 16th edn. AOAC International, Arlington, VA, pp. 50-2–50-3.

AOAC Official Method 992.26. 1995. Vitamin D_3 (cholecalciferol) in ready-to-feed milk-based infant formula. Liquid chromatographic method. In: MP Bueno (ed.) *Official Methods of Analysis of AOAC International*, 16th edn. AOAC International, Arlington, VA, pp. 50-5–50-6.

AOAC Official Method 992.27. 1995. *Trans*-vitamin K_1 (Phylloquinone) in ready-to-feed milk-based infant formula. Liquid chromatographic method. In: MP Bueno (ed.) *Official Methods of Analysis of AOAC International*, 16th edn. AOAC International, Arlington, VA, pp. 50-6–50-8.

Apak R et al. 2007. Comparative evaluation of various total antioxidant capacity. Assays applied to phenolic compounds with the CUPRAC assay. *Molecules* 12: 1496–1547.

Ball GF. 2000. The fat-soluble vitamins. In: LML Nollet (ed.) *Food Analysis by HPLC*. Marcel Dekker, New York, pp. 321–402.

Basha SM. 1990. Protein as an indicator of peanut seed maturity. *J Agric Food Chem* 38: 373–376.

BeMiller JN, Low NH. 1998. Carbohydrate analysis. In: SS Nielsen (ed.) *Food Analysis*, 2nd edn. Aspen Publication Inc., Gaithersburg, MA, pp. 167–187.

Bjelakovic G et al. 2007. Mortality in randomized trials of antioxidant supplements for primary and secondary prevention: systematic review and meta-analysis. *JAMA* 297(8): 842–857.

Bradford MM. 1976. A rapid and sensitive method for the quantitation of microgram quantitites of protein utilizing the principle of protein-dye binding. *Anal Biochem* 72: 248–254.

Bruce B et al. 2000. A diet high in whole and unrefined foods favorably alters lipids, antioxidant defenses, and colon function. *J Am Coll Nutr* 19(1): 61–67.

Bureau JL, Bushway RJ. 1986. HPLC determination of carotenoids in fruits and vegetables in the United States. *J Food Sci* 51: 128–130.

Cai Y et al. 2005. HPLC characterization of betalains from plants in the Amaranthaceae. *J Chromatogr Sci* 43(9): 454–460.

Cao G. et al. 1995. Automated assay of oxygen radical absorbance capacity with the COBAS FARA II. *Clin Chem* 41: 1738–1744.

Chang SKC. 1998. Protein analysis. In: SS Nielsen (ed.) *Food Analysis*, 2nd edn. Aspen Publication Inc., Gaithersburg, MA, pp. 237–249.

Cherubini A et al. 2005. Role of antioxidants in atherosclerosis: epidemiological and clinical update. *Curr Pharm Des* 11(16): 2017–2032.

Dewanto V et al. 2002. Thermal processing enhances the nutritional value of tomatoes by increasing total antioxidant activity. *J Agric Food Chem* 50: 3010–3014.

Dierckx S, Huyghebaert A. 2000. HPLC of food proteins. In: LML Nollet (ed.) *Food Analysis by HPLC*, Marcel Dekker, New York, pp. 127–167.

Dilis V, Trichopoulou A. 2010. Assessment of antioxidants in foods and biological samples; a short critique. *Inter J Food Sci Nutr*, Early online only, 1–8. Accessed March 23, 2010.

El Rassi Z. 1995. *Carbohydrate Analysis High Performance Liquid Chromotography and Capillary Electrophoresis*. Elsevier, Amsterdam.

Eitenmiller RR et al. 1998. Vitamin analysis. In: SS Nielsen (ed.) *Food Analysis*, 2nd edn. Aspen Publication Inc., Gaithersburg, MA, pp. 281–291.

Etherton TD. 2002. Bioactive compounds in foods: their role in the prevention of cardiovascular disease and cancer. *Am J Med* 113(9B), 71S–88S.

Food and Drug Administration (FDA). 2007. GC-MS method for screening and confirmation of melamine and related analogs, version 2. Available at www.fda.gov/cvm/GCMSscreen.

Fireston D. 1994. Liquid chromatographic method for determination of triglycerides in vegetable oils in terms of their partition numbers: summary of collaborative study. *J AOAC Inter* 77, 954–961.

Freeman BA, Capro JD. 1982. Biology of disease: free radicals and tissue injury. *Lab Invest* 47(5), 412–426.

Halliwell B. 1994. Free radicals and antioxidants: a personal view. *Nutr Rev* 52(8), 253–265.

Halliwell B, Gutteridge JM. 1995. The definition and measurement of antioxidants in biological systems. *Free Rad Biol Med* 18, 125–126.

Harbers LH. 1998. Ash analysis. In: SS Nielsen (ed.) *Food Analysis*, 2nd edn. Aspen Publication Inc., Gaithersburg, MA, pp. 141–149.

Hendriks DG. 1998. Mineral analysis. In: *Food Analysis*, 2nd edn. Aspen Publication Inc., Gaithersburg, MA, pp. 151–165.

Howard GA, Martin AJP. 1950. The separation of the C12-C18 fatty acids by reversed-phase partition chromatography. *Biochem J* 46, 532.

Huang D et al. 2005. The chemistry behind antioxidant capacity assays. *J Agric Food Chem* 53, 1841–1856.

Huang HS, Von Elbe JH. 1985. Kinetics of degradation and regeneration of betanine. *J Food Sci* 50, 1115–1120, 1129.

Kmostak S, Kurtz DA. 1993. Rapid determination of supplemental vitamin E acetate in feed premixes by capillary gas chromatography. *J Assoc Off Anal Chem Inter* 76, 735–741.

Lowry O H et al. 1951. Protein measurement with the Folin phenol reagent. *J Biol Chem* 193: 265–275.

Marcone M. 2004. Composition and properties of Indonesian palm civet coffee (*Kopi luwak*) and Ethiopian civet coffee. *Food Res Inter* 37(9): 901–912.

Marcone M. 2005. Characterization of the Edible Bird's Nest, the "Caviar of the East." *Food Res Inter* 38(10): 1125–1134.

Mariani C, Bellan G. 1996. Content of tocopherols, deidrotocopherols, tocodienols, tocotrienols in vegetable oils. *Riv Ital Sostanze Grasse* 73: 533–543 (In Italian).

Marini D. 2000. HPLC of lipids. In: LML Nollet (ed.) *Food Analysis by HPLC*. Marcel Dekker, New York, pp. 169–249.

Marks C. 1988. Determination of free tocopherols in deodorizer by capillary gas chromatography. *JOAC* 65: 1936–1939.

Miller DD. 1998. Atomic absorption and emission spectroscopy. In: SS Nielsen (ed.) *Food Analysis*, 2nd edn. Aspen Publication, Gaithersburg, MA, pp. 425–442.

Min DB, Steenson DF. 1998. Crude fat analysis. In: SS Nielsen (ed.) *Food Analysis*, 2nd edn. Aspen Publication Inc., Gaithersburg, MA, pp. 201–215.

Mohamed RS, Mansoori GA. 2002 (June). The use of supercritical fluid extraction technology in food processing. *Food Tech Mag*.

Moon J-K and Shibamoto T. 2009. Antioxidant assays for plant and food components. *J Agric Food Chem* 57: 1665–1666.

Mozaffarian D, et al. 2006. Trans fatty acids and cardiovascular disease. *New Eng J Med* 354(15): 1601–1613.

Navindra P et al. 2005. In vitro antiproliferative, apoptotic and antioxidant activities of punicalagin, ellagic acid and a total pomegranate tannin extract are enhanced in combination with other polyphenols as found in pomegranate juice. *J Nutr Biochem* 16: 360–367.

Nunomura A et al. 2006. Involvement of oxidative stress in Alzheimer disease. *J Neuropathol Exp Neurol* 65(7): 631–641.

Olsen BJ, Markwell J. 2007. Assays for the determination of protein concentration. *Cur Protoc Prot Sci* 3: 14–17.

Oomah BD et al. 1994. Characterization of soybean proteins by HPLC. *Plant Foods Hum Nutr* 45: 251–263.

Osborne BG. 1996. Authenticity of cereals. In: PR Ashurst, MJ Dennis (ed.) *Food Authentication*. Chapman and Hall, London, England, pp. 171–197.

Pérez VI et al. 2009. Is the oxidative stress theory of aging dead? *Biochimica et Biophysica Acta (BBA)—General Subjects* 1790(10): 1005–1014.

Peris-Tortajada M. 2000. HPLC determination of carbohydrates in foods. In: LML Nollet (ed.) *Food Analysis by HPLC*. Marcel Dekker, New York, pp. 287–302.

Pike OA. 1998. Fat characterization. In: SS Nielsen (ed.) *Food Analysis*, 2nd edn. Publication Inc., Gaithersburg, MA, pp. 217–235.

Phipps SM et al. 2007. Assessing antioxidant activity in botanicals and other dietary supplements. *Pharmacopeial Forum* 33(4): 810–814.

Plattner RD et al. 1977. Triglyceride separation by reverse phase high performance liquid chromatography. *J Am Oil Chem Soc* 54: 511–519.

Plessi RL et al. 2006. Extraction and identification by GC-MS of phenolic acids in traditional balsamic vinegar from Modena. *J Food Comp Anal* 19: 49–54.

Prior RL, Cao G. 1999. In vivo total antioxidant capacity comparison of different analytical methods. *Free Rad Biol Med* 27(11–12): 1173–1181.

Ronald L et al. 2005. Standardized methods for the determination of antioxidant capacity and phenolics in foods and dietary supplements. *J Agric Food Chem* 53: 4290–4302.

Roura E et al. 2006. Total polyphenol intake estimated by a modified Folin–Ciocalteu assay of urine. *Clin Chem* 52(4): 749–752.

Ruiz-Matate A et al. 2007. A new method based on GC-MS to detect honey adulteration with commercial syrups. *J Agric Food Chem* 55(18): 7264–7269.

Russell LF. 2000. Quantitative determination of watersoluble vitamins. In: LML Nollet (ed.) *Food Analysis by HPLC*. Marcel Dekker, New York, pp. 403–476.

Schmitter BM, Rihs T. 1989. Evaluation of a macro-combustion method for total nitrogen determination in feedstuff. *J Agric Food Chem* 37: 992–994.

Schwartz SJ. 1998. Pigment analysis. In: SS Nielsen (ed.) *Food Analysis*, 2nd edn. Aspen Publication Inc., Gaithersburg, MA, pp. 293–304.

Schwartz SJ, Von Elbe JH. 1980. Quantitative determination of individual betacyanin pigments by high-performance liquid chromatography. *J Agric Food Chem* 28: 540–543.

Schwartz SJ et al. 1981. High performance liquid chromatography of chlorophylls and their derivatives in fresh and processed spinach. *J Agric Food Chem* 29: 533–535.

Shenkin A. 2006. The key role of micronutrients. *Clin Nutr* 25(1): 1–13.

Sies H. 1997. Oxidative stress: oxidants and antioxidants. *Exp Physiol* 82(2): 291–295.

Simonne AH et al. 1997. Could the Dumas method replace the Kjeldahl digestion for nitrogen and crude protein determinations in foods? *J Sci Food Agric* 73, 39–45.

Singleton VL, Rossi JA. 1965. Colorimetry of total phenolics with phosphomolybdic–phosphotungstic acid reagents. *Am J Enol Vitic* 16: 144–158.

Smith DM. 1998. Protein separation and characterization procedures. In: SS Nielsen (ed.) *Food Analysis*, 2nd edn. Aspen Publication Inc., Gaithersburg, MA, pp. 251–263.

Somogyi A et al. 2007. Antioxidant measurements. *Physiol Meas* 28: R41–R55.

Southorn PA, Powis G. 1988. Free radicals in medicine. I. Chemical nature and biologic reactions. *Mayo Clin Proc* 63(4): 381–389.

Stanner SA et al. 2004. Review of the epidemiological evidence for the “antioxidant hypothesis.” *Public Health Nutr* 7(3): 407–422.

Sweeley CC et al. 1963. Gas–liquid chromatography of trimethylsilyl derivatives of sugars and related substances. *J Am Chem Soc* 85: 2497–2502.

Tareke E et al. 2002. Analysis of acrylamide, a carcinogen formed in heated foodstuffs. *J Agric Food Chem* 50(17): 4998–5006.

Tsao R et al. 2006. Designer fruits and vegetables with enriched phytochemicals for human health. *Can J Plant Sci* 86: 773–786.

Ulberth F. 1991. Simultaneous determination of vitamin E isomers and cholesterol by GLC. *J High Res Chrom* 14: 343–344.

United States Department of Agriculture (USDA). *Oxygen Radical Absorbance Capacity (ORAC) of Selected Foods—2007*. Nutrient Data Laboratory, Beltsville Human Nutrition Research Center (BHNRC), Agricultural Research Service (ARS), in collaboration with Arkansas Children’s Nutrition Center. Little Rock, AR. 2007.

VanCamp J, Huyghebaert A. 1996. Proteins. In: Nollet LML (ed.) *Handbook of Food Analysis. Vol. 1, Physical Characterization and Nutrient Analysis*. Marcel Dekker, New York, pp. 277–310.

Valko H et al. 2005. Metals, toxicity and oxidative stress. *Curr Med Chem* 12(10): 1161–1208.

Xu B, Chang SK. 2008. Effect of soaking, boiling, and steaming on total phenolic content and antioxidant activities of cool season food legumes. *Food Chem* 110: 1–13.

3
Enzymes in Food Analysis

Isaac N. A. Ashie

Abstract: Foods are primarily very complex matrices and therefore present an enormous technical challenge to the analyst whose responsibility is to determine and establish the levels of various components therein. A number of analytical tools, including enzymes, have therefore been developed through the years to meet this challenge. The basis for enzyme use is its specificity that enables selective, sensitive, and accurate determination of both macro- and micronutrients in foods. The application of enzymes as an analytical tool has, in recent years, been given a further boost by the development of biosensors with improved robustness and versatility. In this chapter, the application of enzymes (e.g., proteases, amylases, lipases, oxidases, etc.) and the underlying principles for analysis of food macro- and micronutrients including proteins, carbohydrates, lipids, antioxidants, cholesterol, toxins, food contaminants (pharmaceuticals, and food quality) are discussed.

INTRODUCTION

For centuries and prior to enzyme discovery and scientific understanding of its catalytic activity, humans had been inadvertently tapping into the benefits of enzymes for the production of a variety of foods. It is hard to think of food products like cheese, bread, wine, and other alcoholic beverages today without considering the role of enzymes in their production. Since their discovery, the unique characteristics of enzymes, such as their natural origins and specificity, have been exploited in the processing of a plethora of foods. These processes include (1) starch liquefaction and saccharification using alpha amylases and amyloglucosidases, (2) conversion of dextrins to fermentable sugars in brewing, (3) chill-haze prevention in the brewing industry using proteases, (4) juice clarification using pectinases, and (5) production of high-fructose corn syrups by conversion of glucose to its much sweeter keto-form, fructose, using glucose isomerase, just to mention a few. Recent developments in the field of molecular biology and protein engineering have enabled cost-effective production of "customized" enzymes with tailored characteristics that have further broadened the range of food applications involving enzymes. For example, phospholipases for increasing cheese yield and oil degumming, lipases for interesterification of lipids as an alternative to chemical processes, carbohydrate oxidases for lactose conversion to lactobionic acid used in the transplant industry for organ preservation, and asparaginases for acrylamide mitigation in foods are some of the emerging applications of enzymes in the food manufacturing process.

The broad range of enzyme classes used in the manufacture of food products is a very good indicator of the challenges presented to the food analyst, whose responsibility is to analyze the ingredients or components of different foods. Foods are basically a complex matrix of edible ingredients, some of which are present in high amounts (e.g., macronutrients such as

Food Biochemistry and Food Processing, Second Edition. Edited by Benjamin K. Simpson, Leo M.L. Nollet, Fidel Toldrá, Soottawat Benjakul, Gopinadhan Paliyath and Y.H. Hui.

proteins, carbohydrates, and lipids) and others in extremely low amounts (e.g., flavor components, micronutrients such as vitamins, antioxidants, etc.). Analysis of these different components requires techniques that are not only simple and affordable but also, more importantly, very specific and sensitive to ensure that the "needle in the haystack" can be picked up (identified) and its size (amount) determined. These challenges have been largely met by technological advancements in chromatography, electrophoresis, mass spectrometry, etc. The specificity and relative sensitivity of enzymes make them ideally suited and versatile candidates for addressing these challenges as well. The protein nature of enzymes, however, means their efficacy for such applications is influenced by factors such as pH, temperature, and ionic strength of the analytes and the catalytic milieu. These factors tend to limit the scope of enzyme applications for this purpose as they may destabilize or denature the enzyme, compromise specificity, or reduce affinity of the enzyme for the substrate or analyte of interest. Some of these limitations have been addressed using immobilization techniques. Furthermore, the significant developments in biosensor technology provide an even greater opportunity for use of enzymes for this purpose by providing the necessary robustness, versatility, and affordability. Basically, biosensors are analytical tools consisting of two main components: (i) a bioreceptor for selective analyte or food ingredient detection and (ii) a transducer that transmits the resulting effect of the bioreceptor–analyte interaction to a measurable signal (Viswanathan et al. 2009). These signals may be amperometric, piezoelectric, potentiometric, fluorometric, or colorimetric. Furthermore, developments in biotechnology have made possible the commercial production, and thus availability, of a broad range of enzymes with the desired characteristics for use in such applications. This chapter is aimed at discussing the application of enzymes in analysis of foods, with emphasis on the principles underlying their role in these analyses. Obviously, any attempt at reviewing the analytical role of enzymes for all food components will be a massive endeavor that cannot be captured in one chapter, and there have been a number of reviews on the subject in the past, several of which have been appropriately cited. Therefore, the emphasis of this chapter is on the enzymatic methods for analysis of a few selected topics related to food components, including the major food components (i.e., proteins, carbohydrates, and lipids), toxins, contaminants, and food quality.

ANALYSIS OF THE MAJOR FOOD COMPONENTS

Proteins and Amino Acids

Proteins are major components of foods, and while the primary interest in proteins is due to their nutritional and functional impact on foods, the allergic reactions induced in sensitized individuals following consumption of certain proteinaceous foods have been a key driver for the development of highly selective and sensitive tools for protein detection and analysis to ensure that food products conform to regulatory standards. For example, lysozyme (muramidase) is an enzyme derived from hen egg white and is approved for use as an antimicrobial agent for preservation of the quality of ripened cheese. However, it is a potential allergen and the joint Food and Agricultural Organization/World Health Organization Codex Alimentarius Commission require its presence in foods to be declared on the label as an egg product (Codex Standard I-1985). Enzyme-linked immunosorbent assay (ELISA) has been a major technique for analysis of these and other proteins with limits of detection ranging from 0.05 to 10 mg/kg (Schubert-Ulrich et al. 2009). The widespread application of ELISA for these analyses has been attributed to its specificity, sensitivity, simple sample handling, and high potential for standardization and automation. Schneider et al. (2010) recently reported an ELISA for analysis of lysozyme in different kinds of cheese. Their method involved use of a peroxidase-labeled antibody that undergoes a redox reaction with tetramethylbenzidine dihydrochloride to form a product that can be measured by the absorbance change at 450 nm. A number of ELISA methods have been used for analysis of lysozyme in hen egg white, wines, cheese, and other foods (Rauch et al. 1990, Yoshida et al. 1991, Rauch et al. 1995, Vidal et al. 2005, Kerkaert and De Meulenaer 2007, Weber et al. 2007). Other proteins that have been analyzed by ELISA include those found in hazelnuts (Holzhauser and Vieths 1999), peanuts (Pomes et al. 2003), walnuts (Doi et al. 2008), soy (You et al. 2008), sesame (Husain et al. 2010), etc. In soybeans, as many as 16 allergenic proteins have been detected, of which β-conglycinin is considered to be the major culprit (Maruyama et al. 2001, Xiang et al. 2002). In the method of You et al. (2008) for β-conglycinin analysis, they had used an epitope corresponding to amino acid residues 78–84 of the protein to synthesize an immunogen by linking it (the epitope) to chicken ovalbumin and bovine serum albumin as carrier proteins. Horseradish peroxidase was used as the enzyme for detection by measuring absorbance at 492 nm. Table 3.1 shows a list of ELISA-based methods commercially available for analysis of a number of food proteins.

ELISA is a major enzyme-based analytical tool for a broad range of food components and other ingredients, and therefore, an explanation of the underlying principles is appropriate here. Figures 3.1 and 3.2 illustrate the general principles underlying the use of sandwich and competitive ELISA (both of which may be direct or indirect) for protein analysis. In the direct sandwich ELISA, a capture antibody specific for the protein being analyzed is first immobilized onto a solid phase. When analytes containing the protein of interest are introduced or present in the system or food material, they bind specifically to the immobilized antibody. Detection of such binding is achieved by introducing a second enzyme-labeled analyte-specific antibody into the system, thereby forming a sandwich. In the indirect sandwich ELISA, a second analyte-specific antibody binds to a secondary site on the protein of interest (analyte) to form the sandwich before the enzyme-labeled antibody binding. The requirement for a secondary binding site or more than one epitope for specific binding makes this type of ELISA primarily applicable to large molecules like proteins. In both types, the binding reaction may be linked to a chromophore, with the color intensity being proportional to the protein concentration.

Table 3.1. Some Commercial ELISA Kits for Protein or Allergen Analysis

Protein Source	Target Protein	ELISA Test Kit	Supplier
Milk	β-Lactoglobulin, caseins	RIDASCREEN β-Lactoglobulin	R-Biopharm
		RIDASCREEN Casein	R-Biopharm
		Milk Protein ELISA Kit	Crystal Chem Inc
		Allergens ELISA Kits	Astori Labs
		FASTKIT SLIM Milk Allergens	Cosmo Bio USA
		ELISA Systems β-Lactoglobulin	ELISA Systems
Nuts	Peanut (Ara h1 and Ara h2)	RIDASCREEN FAST Peanut	R-Biopharm
		Peanut Protein ELISA Kit	Morinaga Inst.
		Allergens ELISA Kits	Crystal Chem Inc
		FASTKIT SLIM Peanut Allergens	Astori Labs
		Peanut Residue ELISA	Cosmo Bio USA
			ELISA Systems
Eggs	Ovomucoid, ovalbumin, ovotransferrin, lysozyme	RIDASCREEN FAST Egg Protein	R-Biopharm
		Egg Protein ELISA Kit	Crystal Chem Inc
		Allergens ELISA Kits	Diagnostic Automation
		FASTKIT SLIM Egg Allergens	Astori Labs
		Enhanced Egg Residue Assay	Cosmo Bio USA
			ELISA Systems
Cereals	Gliadins, secalin, hordein, prolamin, gluten	RIDASCREEN FAST Gliadin	R-Biopharm
		Gliadin ELISA Kit	Morinaga Inst.
		Wheat Protein ELISA Kit	Crystal Chem Inc
		Gluten ELISA Kits	ELISA Technologies
		FASTKIT SLIM Wheat Allergens	Astori Labs
		ELISA Systems Gliadin	Cosmo Bio USA
			ELISA Systems
Mustard	Mustard protein	RIDASCREEN FAST Senf/Mustard	R-Biopharm
		Allergens ELISA Kits	Astori Labs
		Veratox for Mustard Allergen	Neogen
		Mustard Seed Protein Assay	ELISA Systems
Wine, juice	Lupine (e.g., γ-conglutin)	RIDASCREEN FAST Lupine	R-Biopharm
		Allergens ELISA Kits	Astori Labs
		ELISA Systems Lupine Residue	ELISA Systems
Crustaceans	Tropomyosin	RIDASCREEN FAST Crustacean	R-Biopharm
		Allergens ELISA Kits	Astori Labs
		Crustacean Residue ELISA	ELISA Systems
Soybean	Soy trypsin inhibitor	Soy Protein ELISA Kit	US Biologicals
		Allergens ELISA Kits	Diagnostic Automation
		FASTKIT SLIM Soybean Allergens	Astori Labs
			Cosmo Bio USA
			ELISA Systems

Potentiometric, amperometric, electrochemical, and piezoelectric detection methods have since been developed and widely used for these analyses.

Competitive ELISA is more flexible and applicable to analysis of small as well as large molecules. In the direct form of the assay, the analyte of interest is first incubated together with the enzyme-labeled antibody and the mixture added to the analyte-specific capture antibody with which they react competitively and in a concentration-dependent manner for detection (Figure 3.2). The indirect form of competitive ELISA involves co-incubation of the bound analyte of interest with an analyte-specific antibody. Analyte detection is then achieved by introducing a second enzyme-labeled antibody into the mix, which binds to the analyte-specific antibody. In another form of indirect competitive ELISA, the analyte-specific antibody is first bound to the surface, and the analyte or protein of interest is made to compete with an enzyme-labeled analyte or a tracer for binding to the antibody. In all cases, the binding is concentration dependent and can be measured by the changes in absorbance or some other signal.

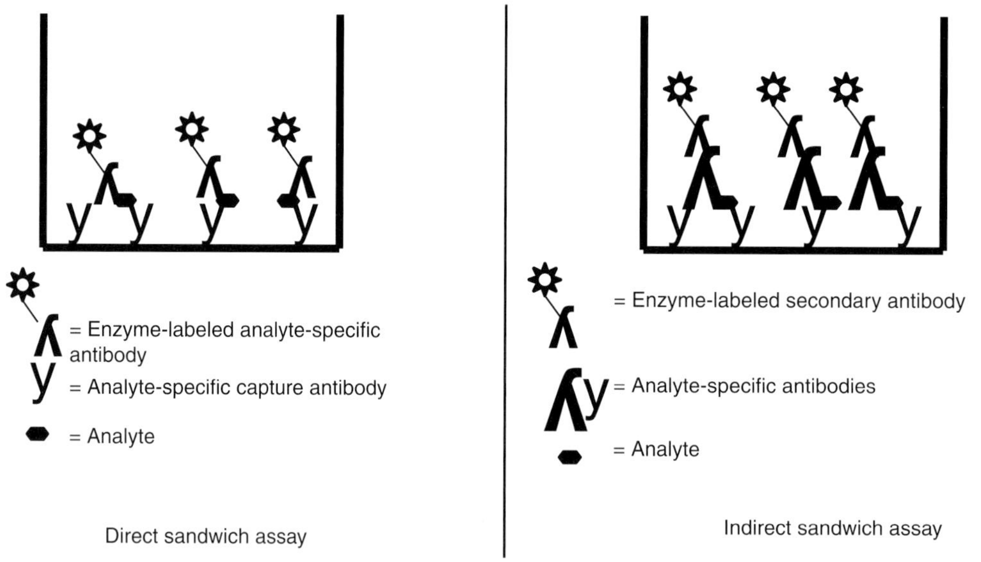

Figure 3.1. Sandwich ELISA with direct and indirect enzyme-labeled antibody detection.

Other techniques have been developed in recent years for protein or allergen analysis, which utilize enzyme-labeled antibodies for detection. These include lateral flow assays, which are immunochromatographic tests involving movement of immunoreactants along a test strip, and dipstick tests, which involve immobilized capture antibodies on a test strip with analyte detection by enzyme-labeled antibodies. Both techniques are generally inexpensive, rapid, and portable, thus making them ideal tools for online monitoring. For a detailed review of these techniques and other methods for protein and allergen analysis, the reader is referred to a recent review by Schubert-Ulrich et al. (2009). The growing field of nanotechnology has opened up other opportunities for development of enzyme-based biosensors for protein analysis. A recent report captured the essence of

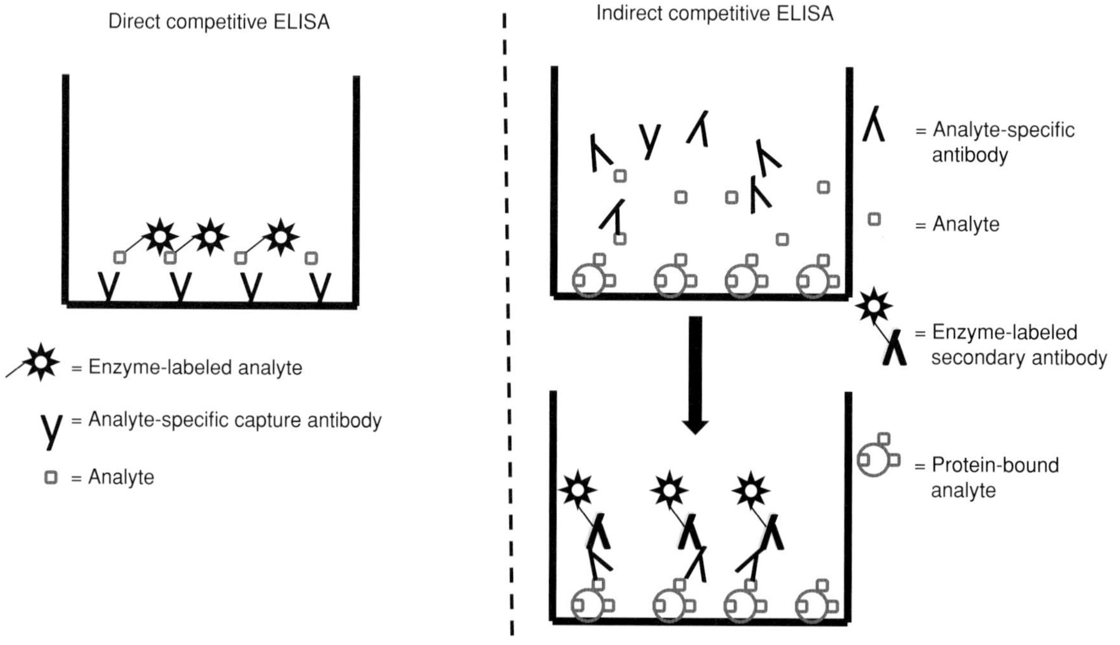

Figure 3.2. Competitive ELISA with direct and indirect enzyme-labeled antibody detection.

this using gold-nanoparticle–carbon-nanotube hybrid linked to horseradish peroxidase as a biosensor for measuring the amount of protein in serum (Cui et al. 2008). In this method, the hybrid-bound peroxidase catalyzes oxidation of o-phenylenediamine in the presence of hydrogen peroxide to produce diaminobenzene, which, being electroactive, generates an amperometric response. Comparison of this method with ELISA showed no significant differences between the two methods. Table 3.1 shows a list of some commercially available ELISA kits for analysis of food proteins.

The free amino acid content as well as amino acid composition of food proteins constitutes very important food components due to their impact on general protein metabolism and a number of food qualities. For example, the presence of glutamate is generally associated with umami or savory flavor, while high content of hydrophobic amino acids tend to cause bitterness in protein hydrolysates. Similarly, high content of free amino acids provide free amino groups that promote Maillard reactions particularly when these foods are subjected to high-temperature treatments. The products or intermediates of the Maillard reaction may undergo other reactions that eventually influence flavor and color of the food product. In the age of protein-fortified sports drinks, free amino acid and peptide content of foods also provides the athlete the "boost" needed for recovery from intense muscle activity during fitness training. A number of analytical methods for determination of amino acid content in foods involve application of enzymes. For instance, one of the early methods for amino acid analysis involved the use of amino acid decarboxylases to decarboxylate amino acids. The stoichiometric release of carbon dioxide was then analyzed using glass electrodes containing sodium bicarbonate (Wiseman 1981). In the analysis of tryptophan, tryptophanase was used to catalyze hydrolysis of the amino acid to produce pyruvate. Coupling of the pyruvate production to lactate dehydrogenase converted NADH to NAD^+ and the corresponding change in absorbance at 340 nm was recorded as the index of tryptophan content (Wiseman 1981). The principle underlying the method is shown in Figure 3.3. A similar method involving coupling to NADH has been exploited in the aspartate aminotransferase analysis of aspartic acid by conversion of aspartic acid to oxaloacetate in the presence of malate dehydrogenase (Wiseman 1981). A number of other enzyme-based methods have been developed for analysis of amino acids, with a significant number of these also using coupling to NADH–NAD^+ conversion (see Table 3.2).

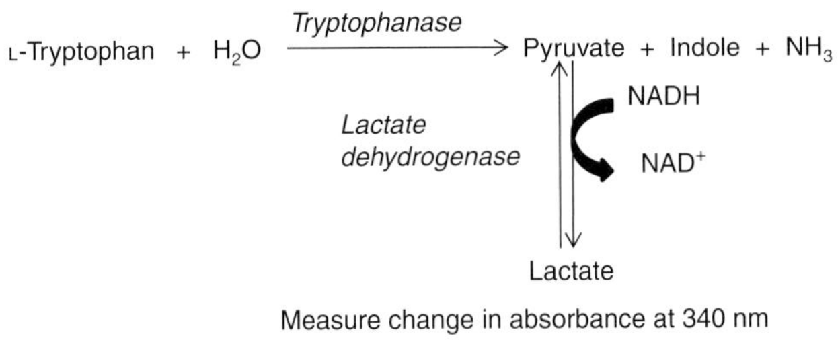

Figure 3.3. Principle underlying tryptophan analysis in foods using the coupled tryptophanase and lactate dehydrogenase (LDH) reaction.

Sugars

Glucose oxidase is by far the most predominant and widely used of enzymes for detection of glucose in foods. The underlying principle is based on hydrolysis of glucose to form gluconic acid and hydrogen peroxide. The latter is then coupled to peroxidase catalysis, which oxidizes a colored precursor, resulting in absorbance change that is proportional to the glucose content (Wiseman 1981). Other methods using enzyme complexes or combinations for analysis of glucose are available. One such method was based on the glucose-6-phosphate transferase catalyzed transfer of an acyl phosphate from a donor molecule to form glucose 6-phosphate. Coupling this reaction to NADP conversion to NADPH resulted in absorbance change that could be measured at 340 nm. Glucose oxidase has also been applied to the determination of sucrose (common sugar). The analysis involves invertase-catalyzed conversion of sucrose to D-fructose and α-D-glucose. This is followed by mutarotase-catalyzed isomerization of the α-D-glucose to β-D-glucose, which in turn is oxidized by glucose oxidase to D-glucono-δ-lactone and hydrogen peroxide (Morkyavichene et al. 1984, Matsumoto et al. 1988). As discussed earlier, the amount of peroxide released by the coupled reaction with peroxidase is proportional to the sucrose content. Fructose is another monosaccharide with widespread use in several foods and thus of significant interest in the food industry. The level of fructose in foods provides an indication of the stage of ripening, adulteration of products such as honey, and the sweetness of various foods. There have been several analytical methods based on the highly specific pyrroloquinoline quinone (PQQ) enzyme, D-fructose dehydrogenase (Paredes et al. 1997). The enzyme catalyzes oxidation of D-fructose to 5-keto-D-fructose with corresponding reduction of the covalently bound cofactor PQQ to $PQQH_2$. Coupling reoxidation of the reduced cofactor to electrodes produces electric current in direct proportion to the amount of fructose. Trivedi et al. (2009) recently reported another method involving use of D-fructose dehydrogenase with ferricyanide as electron acceptor. Figure 3.4 is a schematic representation of the principle underlying this method. The enzyme oxidizes fructose to keto-fructose with the reduction of ferricyanide. The reduced ferricyanide is subsequently re-oxidized, producing electrical current proportional to fructose concentration.

In a recent review, Zeravik et al. (2009) discussed the use of biosensors referred to as bioelectronic tongues for detection and analysis of sugars and other food components or contaminants. Glucose and ascorbic acid content of fruit juices have been analyzed using one of these "tongues" involving glucose oxidase and metal catalysts (Pt, Pd, and Au–Pd). The metal catalysts improved the sensitivity of the tongue by reducing the oxidation potential of the hydrogen peroxide released as a result of the glucose oxidase activity (Gutes et al. 2006). Similar simultaneous detection of glucose, fructose, and sucrose has been achieved by amperometric flow injection analysis using an immobilized

Table 3.2. Enzymatic Methods for Food Component Analysis

Food Component	Enzymes Involved	Commercial Kit	Reference
Amino Acids			
Amino acids	Amino acid decarboxylase, amino acid oxidase		Wiseman 1981, Mello and Kubota 2002, Prodromidis and Karayannis 2002, Cock et al. 2009
Tryptophan	Tryptophanase		
Aspartic acid	Aspartate aminotransferase		
Glutamic acid	Glutamate dehydrogenase, glutamate oxidase, diaphorase	Megazyme International	
Lysine	Lysine oxidase		
Carbohydrates			
Sucrose	Invertase, hexokinase, glucose-6-phosphate dehydrogenase	Megazyme International Sigma Aldrich, Inc.	
Fructose	Hexokinase, phosphoglucose isomerase, glucose-6-phosphate dehydrogenase, fructose dehydrogenase	Sigma Aldrich, Inc. Megazyme International	
Glucose	Glucose oxidase, peroxidase, hexokinase, glucose-6-phosphate dehydrogenase	Cayman Chemical Co. Sigma Aldrich, Inc. Megazyme International	Wiseman 1981, Mello and Kubota 2002, Prodromidis and Karayannis 2002, Cock et al. 2009
Lactose	β-Galactosidase	Megazyme International	
Lactulose	Fructose dehydrogenase, β-galactosidase		
Raffinose	α-Galactosidase, galactose mutarotase, β-galactose dehydrogenase	Megazyme International	
Starch	α-Amylase, amyloglucosidase, glucose oxidase, peroxidase, amyloglucosidase, hexokinase, glucose-6-phosphate dehydrogenase	Megazyme International Sigma Aldrich Company	
Dietary fiber, dextran, and hemicellulose	α-Amylase, protease, amyloglucosidase, dextranase, hemicellulase	Sigma Aldrich, Inc.	
Pectin	Pectate lyase	Megazyme International	
Lipids			
Triglycerides	Lipase, glycerol kinase, glycerol phosphate oxidase, peroxidase, lipoprotein lipase, glycerol kinase, glycerol phosphate oxidase	Cayman Chemical Co. Sigma Aldrich, Inc.	
Cholesterol	Cholesterol esterase, cholesterol oxidase, peroxidase	Cayman Chemical Co.	
Glycerol	Glycerol kinase, glycerol phosphate oxidase, peroxidase, glycerol dehydrogenase, glycerol-1-phosphate dehydrogenase, pyruvate kinase, lactate dehydrogenase	Megazyme International Cayman Chemical Co. Sigma Chemical Co.	Wiseman 1981, Mello and Kubota 2002, Prodromidis and Karayannis 2002, Cock et al. 2009
Phosphatidyl choline (lecithin)	Phospholipase D, choline oxidase, peroxidase	Cayman Chemical Co.	
Fatty acids	Lipases		

Table 3.2. *(Continued)*

Food Component	Enzymes Involved	Commercial Kit	Reference
Other Components			
Citric acid	Citrate lyase		Wiseman 1981, Mello and Kubota 2002, Prodromidis and Karayannis 2002, Cock et al. 2009
Catechol/polyphenols	Polyphenol oxidase, tyrosinase		
Oxalate	Oxalate oxidase		
Lactate	Transaminase, lactate dehydrogenase		
Aspartame	Carboxyl esterase, alcohol oxidase, aspartase, carboxypeptidase, chymotrypsin, glutamate oxidase		
Alcohol	Alcohol dehydrogenase	Megazyme International	Wiseman 1981, Mello and Kubota 2002, Cock et al. 2009
Ascorbic acid	Ascorbic acid oxidase	Megazyme International	
Xylitol	Alkaline phosphatase (ELISA)		Sreenath and Venkatesh 2010

enzyme reactor (Matsumoto et al. 1988). In the analysis of wine components, Zeravik et al. (2009) have discussed the use of (a) PQQ-bound dehydrogenases for the determination of glucose, alcohol, and glycerol; (b) tyrosinase, laccase, and peroxidase for phenolic compounds; and (c) sulfite oxidase for sulfite content.

For analysis of starch, a complex polysaccharide, the underlying principles of starch liquefaction and saccharification are utilized. The α-amylases hydrolyze starch, releasing maltose units and limit dextrin. This is characterized by viscosity reduction or what is more commonly referred to by the starch industry as liquefaction. Further treatment with pullulanase and amyloglucosidases (glucoamylases) results in hydrolysis of the maltose units and limit dextrin, ultimately producing glucose units that are subsequently analyzed by some of the methods discussed earlier. Figure 3.5 illustrates the salient features of the amylase, pullulanase, and amyloglucosidase reaction. Analyses of other polysaccharides have been conducted in similar fashion by initial breakdown to simpler sugars using different carbohydrases depending on the nature of the substrate. Dextranases have been used for dextrans, cellulases for cellulose, hemicellulases for hemicelluloses in dietary fiber, and pectinases and polygalacturonases for polysaccharides in soy sauce and other plant-based foods (Wiseman 1981). Table 3.2 provides a list of some of the enzymatic methods for food component analysis along with some of the commercially available test kits.

Other Food Components

Alcohol

Alcohol (ethanol) content in beer, wine, and other alcoholic beverages is a very important quality attribute and is also required for regulatory classification. The NAD-dependent alcohol dehydrogenase is widely used for analysis of alcohol content in foods and alcoholic beverages. The enzyme catalyzes oxidation of ethanol to acetaldehyde in an equilibrium reaction that also produces NADH. The reaction is made to proceed in the forward direction by acetaldehyde reaction with semicarbazine or hydrazine. The absorbance change resulting from NAD^+ to NADH conversion is proportional to alcohol content. Similar NAD-dependent assays have been used for measuring glycerol content in foods. Figure 3.6 shows the principle underlying the method for glycerol analysis.

Cholesterol

Consumption of cholesterol-rich foods and their association with atherosclerosis, hypertension, and high incidence of other

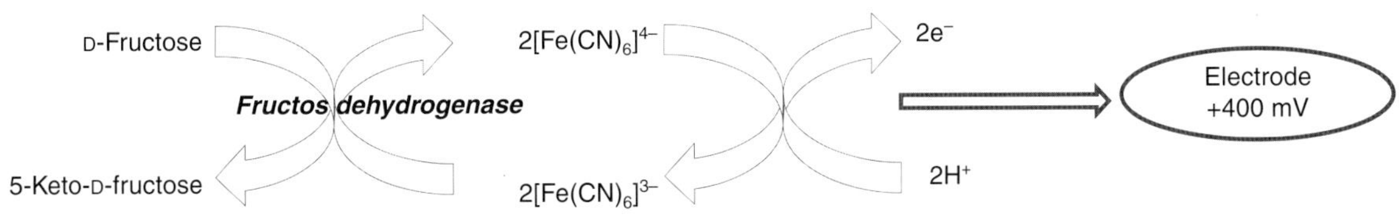

Figure 3.4. Schematic representation of the principle of fructose sensor based on fructose dehydrogenase.

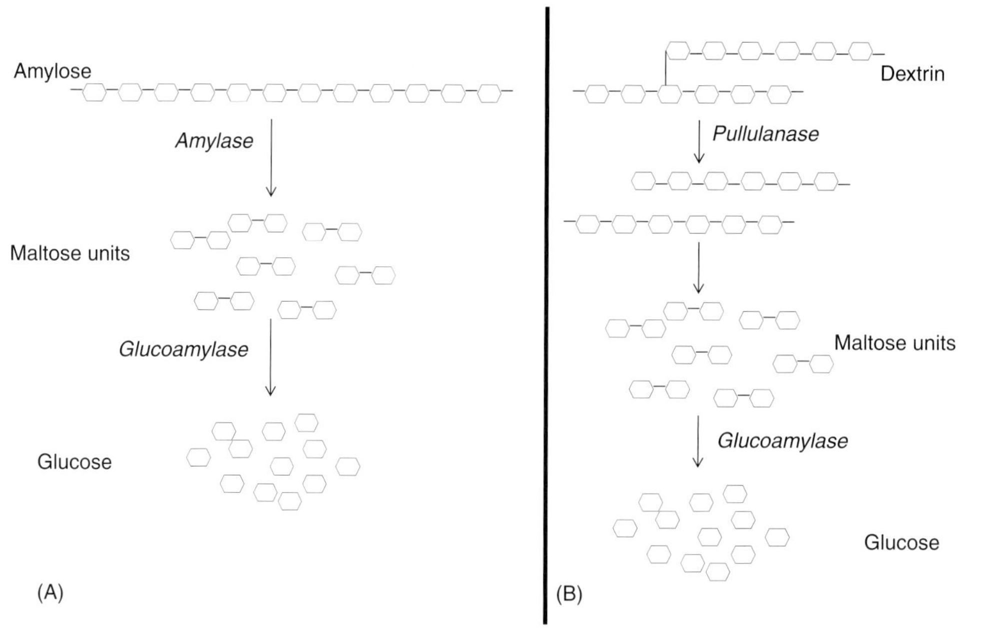

Figure 3.5. Schematic of enzymatic breakdown of starch (amylose and amylopectin) to glucose. (**A**) With amylose as substrate, amylase hydrolyzes the starch to maltose units, which are further simplified to glucose units. (**B**) With amylopectin substrate, pullulanase hydrolyzes the limit dextrin resulting from amylase activity. The products are further simplified by the amylase and glucoamylase to release glucose units.

cardiovascular disease has generated strong interest in the level of cholesterol in foods. Some foods characterized by high levels of cholesterol are cheese, egg yolk, beef, poultry, shrimps, and pork. The earliest enzymatic methods for cholesterol analysis were based on colorimetric assays involving initial de-esterification of cholesterol esters by cholesterol esterase to release free cholesterol, which is further oxidized by cholesterol oxidase to generate hydrogen peroxide. The peroxide produced is subsequently coupled to a colored or fluorescent reagent in a peroxidase-catalyzed reaction (Satoh et al. 1977, Hamada et al. 1980, Omodeo et al. 1984, Van Veldhoven et al. 2002). A commercial kit (Boehringer's Monotest cholesterol or CHOD-PAP method) for the colorimetric assay is available, in which 4-aminoantipyrine and phenol are converted to a red cyanogen-imine dye in the presence of the peroxide released. New and more sensitive electrochemical biosensors have since been developed involving immobilization of the enzymes (cholesterol esterase and cholesterol oxidase) on electrode surfaces and carbon nanotubes (Wisitsoraat et al. 2009). Similarly, a number of other enzyme-immobilized biosensors originally developed for analysis of serum cholesterol may also be applicable to food analysis. These include immobilization on polyvinyl chloride with glutaraldehyde as a coupling agent (Hooda et al. 2009), agarose gel and activation with cyanogen bromide (Karube et al. 1982), and pyrrole membrane coupled with flow injection for hydrogen peroxide analysis (Wolfgang et al. 1993).

Antioxidants

Oxidation of lipids in foods has been an age-old problem of the food industry due to the free-radical-induced off-flavors and the implication of these free radicals in cell damage and a number of pathogenetic conditions. Controlling such oxidation has been largely achieved by a number of process modifications as well

$$\text{Glycerol} \xrightarrow[\text{ATP}]{\textit{Glycerokinase}} \text{Glycerol-1-phosphate} \xrightarrow[\text{NAD+}]{\textit{Glycerol-1-phosphate dehydrogenase}} \text{NADH} + \text{Dihydroxyacetone phosphate} \;\ldots\ldots\ldots\ldots\;(1)$$

$$\text{Glycerol} \xrightarrow[\text{NAD}^+]{\textit{Glycerol dehydrogenase}} \text{Dihydroxyacetone phosphate} + \text{NADH} \;\ldots\ldots\ldots(2)$$

Figure 3.6. Glycerolkinase and glycerol dehydrogenase for glycerol analysis in foods.

as incorporation of antioxidants in various food product formulations. Though generally added for their ability to maintain food quality during processing and storage, antioxidants are also considered to protect consumers against some of the potential health risks associated with free radicals. An enzymatic assay recently developed for antioxidant analysis involves oxidation of the yellowish compound syringaldazine by laccase to produce the purple-colored tetramethoxy azobismethylene quinone (TMAMQ) with maximum absorbance at 530 nm. In the presence of antioxidants, the free-radical intermediates are unavailable or converted, thereby proportionately blocking formation of the purplish TMAMQ (Prasetyo et al. 2010). This method has been used to measure the level of antioxidant activity in a wide range of fruits and vegetables (e.g., apples, carrots, garlic, kiwi, lettuce, spinach, tomatoes), beverages (e.g., green tea, coffee), and oils. While in most cases, enzymes have been used for specific identification and analysis of these ingredients, they have also been used in extraction of some compounds for analysis. Analysis of flavanoids, which are often present in foods as the glycosylated or sulfated derivatives, generally involves enzymatic hydrolysis to the corresponding aglycones and desulfated forms during their extraction. Typically used in the extraction process are the β-glucuronidase or β-glycosidases and aryl sulfatase, as demonstrated in a recent report for determination of hesperitin and naringenin in foods (Shinkaruk et al. 2010). The extracted products were then subjected to peroxidase-linked ELISA.

Organic Acids

The presence and content of organic acids has long been known to exert significant influence on quality attributes of a wide variety of food products, including wines, milk, vinegar, soy sauce, and fermented products. Therefore, a number of enzymatic methods have been developed to detect and quantify these acids such as lactic, acetic, citric, malic, ascorbic, and succinic acids in a number of food products (Wiseman 1981). Also see Table 3.2.

The most widely studied of these acids is lactic acid due to its impact on a broad range of food products. In yogurt production and cheese maturation, lactic acid is a major intermediate of *Lactobacillus* fermentation and determines the flavor of the final product. In wine production, organic acids impact flavor, protect against bacterial diseases, and may slow down ripening. Excessive lactic acid in wines has a negative effect on wine taste due to formation of acetate, diacetyl, and other intermediates. These irreversible transformations designated by the French term "piqure lactique" have been attributed to the activities of some heterolactic bacteria during malo-lactic fermentation, which compromises wine race (Avramescu et al. 2001, Shkotova et al. 2008). Lima et al. (1998) developed a method for the simultaneous measurement of both lactate and malate levels in wines. This involved injection of enzymes (malate dehydrogenase and lactate dehydrogenase) into a buffer carrier stream flowing to a dialysis unit, where they react with the wine donor containing the two acids. Detection is based on absorbance change at 340 nm due to reduction of NAD^+ to NADH. Similar NAD^+-dependent lactate dehydrogenase application for lactate analysis has been developed using amperometric biosensors (Avramescu et al. 2001, 2002). Another enzymatic method for lactate determination in wines is the immobilized FAD-bound lactate oxidase. The enzyme catalyzes conversion of lactate to pyruvate and peroxide. Decomposition of the hydrogen peroxide results in generation of electrons that are recorded by an amperometric transducer. Table 3.2 lists a number of enzymatic methods and commercial test kits used for analysis of various acids in foods.

ANALYSIS OF FOOD CONTAMINANTS

Pharmaceuticals

The growing use of antibiotics and other drugs for treatment of animals meant for food (e.g., cattle, chicken, pigs) is of great concern to regulatory authorities due to the potential carry-over effect on consumers as these drugs enter the food supply. This has led to strict regulations on the prophylactic and therapeutic use of these drugs being introduced in many countries. Therefore, a number of methods have been developed for analysis of these drugs to ensure food safety. For instance, the high incidence of aplastic anemia in some countries has been attributed to the use of chloramphenicol in the treatment of food-producing animals, particularly aquaculture (Bogusz et al. 2004). Clenbuterol, a β-adrenergic agonist widely used in the treatment of some pulmonary diseases, also promotes muscle growth, and therefore, it is sometimes abused in animal feed to boost the lean meat-to-fat ratio and as a doping agent by athletes (Clarkson and Thompson 1997). Prolonged use of clenbuterol is known to cause adverse physiological effects, and incidents of poisoning have been reported in many countries following consumption of foods containing high concentrations of the drug (Pulce et al. 1991). ELISA methods based on monoclonal and polyclonal antibodies are the most widely used of the enzymatic methods for analysis (Petruzzelli et al. 1996, Matsumoto et al. 2000). He et al. (2009) recently reported development of a polyclonal indirect ELISA with high sensitivity for clenbuterol analysis in milk, animal feed, and liver samples. Another set of drugs subject to European Union Maximum Residual Limits are the nonsteroidal anti-inflammatory drugs (NSAIDs) used for treatment of some food-producing animals (porcine and bovine). Consumption of meat products containing these drugs has been shown to cause gastrointestinal problems such as diarrhea, nausea, and vomiting by irritation of the gastric mucosa. A simple enzymatic method involving cyclooxygenase has been developed for analysis of NSAIDs in milk and cheese (Campanella et al. 2009). In absence of NSAIDs, cyclooxygenase catalyzes oxidative conversion of arachidonic acid to prostaglandins. However, the presence of NSAIDs results in competitive inhibition of this activity, thereby reducing prostaglandin synthesis. The NSAID concentration is detected by coupling the catalytic reaction with an amperometric electrode for oxygen. Chloramphenicol, a potent broad-spectrum antibiotic banned in Europe and the United States due to potential toxic effects (aplastic anemia), is generally analyzed using various conventional methods such as gas chromatography, immunoassay, and mass spectrometry. The drug tends to be conjugated in biological systems, and Bogusz et al. (2004) recently developed an assay to distinguish between the free and conjugated forms. They used β-glucuronidase in the extraction process to hydrolyze the conjugated to the free form, enabling selective analysis of the drug in honey, shrimps, and chicken.

Fluoroquinolones, antimicrobial agents used for treatment and disease prevention in food-producing animals and sometimes used as feed additives to build up animal body mass, is another group of drugs that has come under scrutiny. This is due to their misuse, potentially leaving residues in meats, which could lead to drug resistance and allergic hypersensitivity (Martinez et al. 2006). Therefore, maximum residue limits have been instituted in the European Union, and a number of ELISA methods have been developed for their detection in milk, chicken, pork, eggs, and other foods (Coillie et al. 2004, Huet et al. 2006, Lu et al. 2006, Scortichini et al. 2009, Sheng et al. 2009). Euro Diagnostica B.V. (Arnhem, Netherlands) has a commercial kit currently available for analysis of quinolones.

Melamine is a triazine compound that contains 67% nitrogen and is used in the production of plastics, construction materials, and furniture. This compound was brought to public attention in the fall of 2008 following growing incidents of infants coming down with kidney stones in China as a result of adulteration of infant formula with the compound. Similar melamine adulteration in an effort to raise the apparent protein content of pet foods resulted in several pet deaths. These incidents were presumably possible due to protein analysis of foods being routinely conducted with the Kjeldahl method, which is nonspecific and does not distinguish between protein and nonprotein nitrogen. Several alternative methods for detection of melamine and related compounds (cyromazine, ammelide, cyanuric acid, and ammeline), including ELISA, have been developed and implemented (Tittlemier 2010) to ensure food safety. Romer Labs and Beacon Analytical Systems have commercially available ELISA test kits for melamine analysis in a broad range of foods.

Biotechnology, and particularly genetic modification, is undoubtedly a key factor for addressing the challenge of boosting food supplies to meet the demands of an exploding global population. This is clearly evident by the increase in the global area cultivated for biotech crops from 1.7 million hectares in 1996 to 125 million hectares in 2008 with the major food crops being soybean, canola, and maize (Lee et al. 2009). However, there is a deep sense of cynicism over the potential health and environmental impact of GMO-based foods. There have been calls for controls on GMO, and strict regulations on traceability and labeling of GMO and foods or feed made from GMO technology have been established in many countries (Council Regulation 2000, Lee et al. 2009). Hence, a number of detection and quantification methods have been developed based on analysis of two main components characteristic to GMO: (i) a specific transgenic DNA and (ii) a specific novel protein expressed by the GMO. The basic principle underlying these methods is the use of the polymerase chain reaction (PCR) to establish the presence of the most common recombinant elements in the vast majority of genetically modified crops (Vollenhofer et al. 1999, Michelini et al. 2008, Lee et al. 2009). These are the 35S promoter and nopalin synthetase (NOS) terminator, both of which are transcription sequences (Barbau-Piednoir et al. 2010). Variations of this method have been used to detect and quantify GMO in soy, maize, and other food crops (Matsuoka et al. 2001, James et al. 2003, Meric et al. 2004, Yang et al. 2006, 2007, Wu et al. 2007, Lee et al. 2009, Costa et al. 2010). In one of these methods, a disposable genosensor for detecting the NOS terminator (Meric et al. 2004) was used. It involved immobilization of a synthetic single stranded 25-base oligonucleotide that is complimentary to the NOS terminator to a screen-printed carbon electrode. In the presence of exogenous genetic materials, DNA polymerase catalyzes the PCR to synthesize the complimentary base sequence. Detection of hybridization is achieved by monitoring the reduction signal from the high-affinity binding of methylene blue to guanine.

In the alternative method based on detection of expressed proteins by the genetically modified organisms, PCR-ELISA has been the preferred approach and has been used for detecting and quantifying various GMO materials. This is demonstrated by the use of a sandwich ELISA for the detection of the Cry1Ab endotoxin characteristic of some transgenic maize. A Cry1Ab protein-specific antibody was first immobilized on a microtiter plate to capture the Cry1Ab toxin in the sample. This was followed by a second anti-Cry1Ab antibody and a horseradish peroxidase-labeled antibody for detection of the toxin using a chemiluminescent probe (Roda et al. 2006). A similar sandwich ELISA has been developed for the detection of phosphinotricin-*N*-acetyltransferase, which is encoded by a gene in genetically modified pepper (Shim et al. 2007). Using two commercially available ELISA kits, Whitaker et al. (2001) detected and quantified the distribution and variability of the Cry9C insecticidal protein produced by the genetically modified corn (StarLink) in a bulk of corn flour and meal. There are a number of other reports of PCR-ELISA application for GMO analysis in soy and maize products (Landgraf et al. 1991, Brunnert et al. 2001, Petit et al. 2003).

Food Toxins

Food toxicity is of high importance due to their potential impact on safety and risk to the environment. These toxins may be naturally present (endogenously), process induced, or from the environment. Food toxins such as phytotoxins (terpenoids, glycoalkaloids, phytosterols, flavonoids, lignans) and mycotoxins (aflatoxins, fumonisins, ochratoxins, citrinin, mycophenolic acid, etc.) are all naturally occurring and their toxic effects have been well noted. This section discusses the enzymatic methods for analyses of these and other toxins.

Endogenous Toxins

Mycotoxins are toxic secondary metabolites produced by various fungi (*Aspergillus*, *Fusarium*, *Penicillium*, etc.) and excreted in their substrates. Aflatoxins first came to the attention of researchers following the outbreak of the so-called Turkey X disease among turkeys in the 1960s and is by far the most widely studied of the mycotoxins. It has since been shown to occur mainly in the tropics in such food ingredients as peanuts, pistachios, Brazil nuts, walnuts, cottonseed, maize, and other cereal grains such as rice and wheat. Aflatoxins are considered to induce hepatotoxicity, mutagenicity, teratogenicity, immunosuppressant, and carcinogenic effects, and their presence in foods is highly regulated at a level of 0.05–2 μg/kg in the European Union (EC Report 2006). With such low limits, the analytical methods need to be of high sensitivity and specificity, and ELISA has developed to be one of the main methods for aflatoxin analysis. Some of the enzymes that have been used include horseradish peroxidase, which oxidizes tetramethylbenzidene to a product that can be measured by its electrical properties, and alkaline phosphatase for dephosphorylation of naphthyl phosphate to naphthol that is measured by pulse voltametry (Ammida et al. 2004, Parker and Tothill 2009). Detection limits for these methods were in the range of 20–30 pg/mL. Other ELISA methods for aflatoxin analysis have been reported (Lee et al. 2004, Liu et al. 2006, Jin et al. 2009, Piermarini et al. 2009, Tan et al. 2009). In their method, Liu et al. (2006) developed a biosensor by immobilizing horseradish peroxidase and aflatoxin antibodies onto microelectrodes. The presence of aflatoxin in the food resulted in immunocomplex formation with the antibody, which blocked electron transfer between the enzyme and the electrode, thereby modifying conductivity in proportion to aflatoxin concentration. Similar ELISA methods involving alkaline phosphatase and horseradish peroxidase has been used for analysis of ochratoxin A in wine (Prieto-Simon and Campas 2009). A fast and sensitive spectrophotometric method has also been reported based on acetylcholinesterase inhibition by aflatoxin (Arduini et al. 2007).

Unlike aflatoxins, which are mainly associated with foods from warm tropical regions, trichothecene mycotoxins are commonly found in cereals like wheat, barley, maize, oats, and rye, particularly in cold climates. Like the aflatoxins, these have also been found to cause immunosuppressive and cytotoxic effects as well as other disorders due to their inhibition of protein, DNA, and RNA synthesis. AOAC-approved methods (www.aoac.org/testkits/testedmethods.html) based on ELISA kits are commercially available for analysis of trichothecenes. These include the RIDASCREEN FAST and AgraQuant manufactured by Biopharm GmbH and Romer Labs, respectively, using horseradish peroxidase (Lattanzio et al. 2009).

Linamarin and lotaustralin are cyanogenic glycosides considered toxic and found in food crops like lima beans and cassava. The endogenous linamarase released on tissue damage hydrolyzes the glucosides to produce hydrogen cyanide, a highly toxic compound (Cooke 1978). In the analysis of these toxins, the endogenous linamarase is first inactivated, followed by introduction of the exogenous enzyme. The cyanide produced as a result of the enzyme activity is then measured by a spectrophotometer as index of toxin content. The linamarase-catalyzed reaction normally exists in equilibrium but is made to proceed in the forward direction by alkali treatment (Cooke 1978).

Biogenic amines are organic bases found in a broad range of foods such as meat, fish, wine, beer, chocolate, nuts, fruits, dairy products, sauerkraut, and some fermented foods, but are potentially toxic when consumed. These amines, most of which are produced by microbial decarboxylation of the corresponding amino acids, include putrescine, cadaverine, tyramine, spermine, spermidine, and agmatine. That is, putrescine, histamine, tryptamine, tyramine, agmatine, and cadaverine are formed from ornithine, histidine, tryptophan, tyrosine, arginine, and lysine, respectively (Teti et al. 2002). Histamine is the most regulated of these amines. In Germany, the maximum limit in fish products is set at 200 mg/kg, whereas in Canada, Finland, and Switzerland, it is set at 100 mg/kg. Various methods have been developed for analysis of these compounds (Onal 2007). For histamine determination, one of the methods has been based on the sequential activities of diamine oxidase and horseradish peroxidase (Landete et al. 2004). The diamine oxidase catalyzes breakdown of histamine, releasing imidazole acetaldehyde, ammonia, and hydrogen peroxide. The hydrogen peroxide produced is oxidized by the peroxidase, resulting in color change of a chromogen, which is measured by a colorimeter. Muresan et al. (2008) recently developed an amperometric biosensor involving use of amine oxidase and horseradish peroxidase for amine detection. In this method, the amines were first separated on a weak acid cation exchange column followed by enzymatic reactions that produce a potential difference, which is measured by a bioelectrochemical detector. An enzyme sensor array for simultaneous detection of putrescine, cadaverine, and histamine has also been reported by Lange and Wittman (2002). The principle underlying the method is illustrated in Figure 3.7. Changes in the levels of the various biogenic amines during ageing of salted anchovies were determined using immobilized diamine oxidase coupled to electrochemical detection of the hydrogen peroxide produced by the enzyme activity (Draisci et al. 1998).

Process-Induced Toxins

Acrylamide is a process-induced toxin that has gained significant global attention in the past decade following a publication by the Swedish National Food Administration (2002), which indicated its widespread presence in numerous food products and the risk posed to humans from consumption of such foods. Foods considered to have high levels of acrylamide include baked and fried products like potato chips (or crisps), biscuits, crackers, breakfast cereals, and French fries. The mechanism of acrylamide formation in these foods is attributed to the high levels of asparagine in the raw materials (especially wheat and potatoes) used in their production. Initiation of the Maillard reaction in presence of reducing sugars with asparagine as amino group donor channels the complex series of reactions through a pathway that results in the production of acrylamide (Mottram et al. 2002). While methods involving gas chromatography, liquid chromatography, and mass spectrometry still dominate analysis of this compound, a recent report has demonstrated the potential application of ELISA for analyzing acrylamide content in foods (Preston et al. 2008). To overcome the problem with the small size of acrylamide, which has limited the use of immunoassay for its analysis, these authors derivatized acrylamide with 3-mercaptobenzoic acid (3-MBA) and conjugated that to a carrier protein (bovine thyroglobulin) to form an

Enzymatic reaction:

$$R\text{-}CH_2\text{-}NH_2 + O_2 + H_2O \xrightarrow{\text{Amine oxidase}} R\text{-}CHO + NH_3 + H_2O_2$$

Color reaction:

$$H_2O_2 + \text{Reduced dye (colorless)} \xrightarrow{\text{Peroxidase}} H_2O + \text{Oxidized dye (colored)}$$

Electrochemical reaction:

$$H_2O_2 \longrightarrow O_2 + 2H^+ \quad 2e^-$$

Figure 3.7. Principle underlying amine detection by coupling amine oxidase and peroxidase reactions.

immunogen. Antisera raised against the immunogen and shown to have high affinity for the 3-MBA-derivatized acrylamide was used in the development of a peroxidase-based ELISA detection system with about 66 μg/kg limit of detection. A commercial ELISA kit has been recently introduced by Abraxis (www.abraxis.com) for acrylamide analysis.

Environmental Toxins

Certain toxin-producing phytoplanktons present in the aquatic environment have generally been recognized as the source of marine toxins in the human food chain following consumption of contaminated fish and shellfish. Safety concerns have led to regulatory limits being set for the presence of these toxins in seafoods.

Okadaic acid and its derivatives, which are considered responsible for diarrheic shellfish poisoning, are very potent inhibitors of protein phosphatase, and a number of methods have been developed around inhibition of the enzyme for its analysis (Tubaro et al. 1996, Della Loggia et al. 1999, Campas and Marty 2007, Volpe et al. 2009). Analysis of this toxin has also been accomplished by ELISA methods (Kreuzer et al. 1999, Campas et al. 2008). The toxic effects of yessotoxins remain to be elucidated even though in vitro studies indicate cardiotoxicity. Analytical methods based on ELISA (Garthwaite et al. 2001, Briggs et al. 2004) and yessotoxin interaction with pectinesterase enzymes have been developed (Alfonso et al. 2004, Pazos et al. 2005, Fonfria et al. 2008).

Another group of marine toxins of relevance to food are the brevetoxins responsible for neurotoxic shellfish poisoning. The inhibition by brevetoxins of the desulfo-yessotoxin interaction with phosphodiesterase provided the basis for an analytical method (Mouri et al. 2009), and also reported are various ELISA methods (Baden et al. 1995, Garthwaite et al. 2001, Naar et al. 2002). Saxitoxins, responsible for paralytic shellfish poisoning, have been analyzed in foods by ELISA (Chu and Fan 1985, Garthwaite et al. 2001). The paralytic effect is attributed to their high-affinity binding to specific sites on sodium channels, thereby blocking sodium flux across excitable cells. Like the saxitoxin group of toxins, amnesic shellfish poisoning is associated with consumption of foods contaminated with the domoic acid group of water-soluble neurotoxins produced by the genera *Pseudonitzschia*, *Nitzschia*, and *Chondria armata* (Vilarino et al. 2009). Ciguatoxin poisoning is characterized by symptoms of nausea, diarrhea, abdominal cramps, memory loss, dizziness, headaches, and death in extreme cases. An ELISA method approved by AOAC has been adopted in many countries (Kleivdahl et al. 2007). Tsai et al. (2009) have reported the identification of moray eel species responsible for ciguatera poisoning based on analysis of the cyt b gene in mitochondrial DNA using the PCR and restriction enzymes. In a recent study, Garet et al. (2010) compared relative sensitivities of ELISA and conventional methods (HPLC-UV and mouse bioassay) for detection of amnesic shellfish poisoning and paralytic shellfish poisoning toxins in naturally contaminated fresh, frozen, boiled, and canned fish and shellfish. Their findings showed lower limits of detection (50 μg saxitoxin and 60 μg per kg shellfish meat for saxitoxin and domoic acid, respectively) compared to 350 μg saxitoxin and 1.6 mg domoic acid per kg shellfish meat, respectively, when analyzed by conventional methods.

Detection of marine palytoxins and tetrodotoxins using ELISA techniques have also been reported (Bignami et al. 1992, Lau et al. 1995, Rivera et al. 1995, Kawatsu et al. 1997, Kreuzer et al. 2002, Neagu et al. 2006). Palytoxins and their analogues elicit their toxic effects by binding to the Na^+/K^+ ATPase pumps, thereby altering the selectivity of ion flux and membrane potential. Much like the saxitoxins, the tetrodotoxins elicit their neurotoxicity by inhibiting voltage-gated sodium channels. The method of Kreuzer et al. (2002) for analysis of tetrodotoxins involves an alkaline phosphatase-labeled antibody, the activity of which produces p-aminophenol, which can be detected by the amount of current generated. Alkaline phosphatase activity is competitively inhibited by tetrodotoxin, thereby reducing the current produced.

The presence of pesticides in foods and their impact on health is an issue of critical concern to the food industry, and therefore, regulatory measures are in place establishing maximum residual limits. Inhibition of acetylcholinesterase by pesticides has formed the basis for a broad range of techniques for pesticide analysis (see reviews by Mulchandani et al. 2001, Amine et al. 2006). These have used:

1. amperometric transducers to measure thiocholine and p-aminophenol produced by acetylcholinesterase hydrolysis of butyrylthiocholine, and p-aminophenyl acetate or hydrogen peroxide produced by oxidation of choline following acetylcholine hydrolysis in presence of choline oxidase;
2. potentiometric transducers for measuring pH changes as a result of acetic acid production from enzyme activity;
3. fiber optics for monitoring pH changes using a fluorescein-labeled enzyme.

These acetylcholinesterase inhibitory methods, however, are nondiscriminating between the organophoshates, carbamates, and other pesticides. In their review, Mulchandani et al. (2001) have discussed the use of organophosphorus hydrolase integrated with electrochemical and optical transducers to specifically measure organophosphate pesticides. The enzyme hydrolyzes organophosphates to produce acids or alcohols, which generally tend to be either electroactive or chromophores that can be monitored.

Bioelectronic tongues have also been proposed for the detection of pesticides such as organophosphates, carbamates, as well as phenols using an amperometric bioelectronic tongue involving cholinesterase, tyrosinase, peroxidases, and cellobiose dehydrogenase. The corresponding substrates for the enzymes were acetylcholine, phenols, hydrogen peroxide, and cellobiose, respectively, with detection limits in the nanomolar and micromolar range (Solna et al. 2005). A similar bioelectronic tongue with acetylcholinesterase as biosensor has been used to resolve pesticide mixtures of dichlorvos and methylparaoxon (Valdez-Ramirez et al. 2009).

Botrytis, a fungal disease responsible for significant losses of agricultural produce particularly in vineyards (e.g., wine grapes, strawberries), is controlled by preharvest or postharvest treatment with the fungicide fenhexamid, which inhibits growth of the germ tube and mycelia of the fungus. Detection and analysis of fenhexamid levels in must and wines have been accomplished by direct competitive ELISA containing horseradish peroxidase (Mercader and Abad-Fuentes 2009). Atrazine, a herbicide with widespread contamination of waterways and drinking water supplies and implicated in certain human health defects (e.g., birth defects, menstrual problems, low birth weight), is tightly regulated in the European Union with a limit of 0.1 g/L in potable water and fruit juices. Conductimetric and amperometric biosensors based on tyrosinase inhibition have been developed for the analysis of atrazine and its metabolites (Hipolito-Moreno et al. 1998, Vedrine et al. 2003, Anh et al. 2004). The method is based on deactivation of tyrosinase under aqueous conditions. The enzyme normally catalyzes oxidation of monophenols to o-diphenols, which are eventually dehydrogenated to the corresponding o-quinones. In aqueous environments, these form polymers that inhibit the enzyme (Hipolito-Moreno et al. 1998). Therefore, atrazine inhibition of the enzyme is conducted in nonaqueous media, where the quinone effect is eliminated, thus ensuring the inhibition is primarily due to the herbicide. Some recent reviews have also captured the application of enzymatic biosensors for determination of fertilizers (nitrates, nitrites, and phosphates), as well as several pesticides and heavy metals that are potentially toxic to humans when they enter the food chain (Amine et al. 2006, Cock et al. 2009). Some of the enzymes applied to these analyses include (i) urease, glucose oxidase, acetylcholinesterase, alkaline phosphatase, ascorbate oxidase, alcohol oxidase, glycerol-3-phosphate oxidase, invertase, and peroxidase for heavy metals and (ii) nitrate reductase, polyphenol oxidase, phosphorylase, glucose-6-phosphaste dehydrogenase, and phosphoglucomutase for nitrates, nitrites, and phosphates.

ENZYMATIC ANALYSIS OF FOOD QUALITY

While enzymes generally exert positive influence on food quality, the activities of some enzymes need to be controlled postharvest or postprocessing to avoid compromising the quality of food products. For instance, protease activity is desirable for meat tenderization during ageing, but left uncontrolled, meats become mushy and overtenderized. Similarly, excessive lipase and oxidase activities in foods during storage, especially under temperature-abuse conditions, may cause rancidity and impact quality attributes such as color, flavor, and texture. Therefore, a number of preservation techniques developed through the years have been aimed at controlling such undesirable postharvest enzymatic as well as microbial activities to ensure extended shelf life of food products.

The activities of a number of enzymes as well as the content of certain food components and secondary metabolites resulting from postharvest or postprocessing biochemical and microbial activities have been recognized as quality indices for various foods and consequently monitored. An example is the milk endogenous alkaline phosphatase, which is inactivated following heat treatment at 60°C for 5 seconds. With milk generally subjected to HTST/flash pasteurization (i.e., 71.5–74°C/160–165°F for 15–30 seconds) or ultra-high-temperature treatment (i.e., 135°C/275°F for 1–2 seconds), residual alkaline phosphatase activity following such treatment provides a good indication of efficacy of the treatment. The phosphatase test is based on hydrolysis of disodium phenyl phosphate to liberate phenol, which reacts with dichloroquinonechloroimide to form a blue indophenol that is measured colorimetrically at 650 nm (Murphy et al. 1992). An alternative fluorimetric method for alkaline phosphatase analysis has also been developed and commercialized (Rocco 1990). The adequacy of blanching, an essential step in vegetable processing, is also established by measuring residual lipoxygenase or peroxidase activities (Powers 1998). Amylase activity in malt, also referred to as its diastatic power, is a very essential quality parameter that influences dextrinizing time and, therefore, closely monitored (Powers 1998). In most seafoods, freshness is rapidly compromised postharvest, particularly when handled under temperature-abuse conditions that promote endogenous enzyme activity. The sequence of reactions (Figure 3.8) involved in ATP breakdown in postmortem fish

ATP ⇒ ADP ⇒ AMP ⇒ IMP ⇒ HxR ⇒ Hx ⇒ X ⇒ U

Figure 3.8. Sequence of reactions involved in postmortem ATP breakdown. ADP, adenosine diphosphate; AMP, adenosine-5′-phosphate; IMP, inosine-5-phosphate; HxR, inosine; Hx, hypoxanthine; X, xanthine; U, uric acid.

muscle has been very well investigated, and the relative amounts of the different intermediates are generally accepted as fish freshness indicator.

These key freshness indicators such as ornithine, amines, and hypoxanthine, which rapidly build up in fresh seafoods, have been analyzed by various enzymes. These include ornithine carbamoyl transferase, nucleoside phosphorylase, xanthine oxidase, and diamine oxidase for ornithine and amines; and xanthine oxidase for hypoxanthine (Cock et al. 2009). In meats, lactic acid levels tend to increase postmortem, and thus their levels have been monitored as freshness indicators using xanthine oxidase, diamine oxidase, and polyamide oxidase.

In addition to the indirect indicators of food quality or safety described, there are other enzymatic methods for direct detection of contaminating microflora. The latter has gained even greater significance for controlling not only food-borne diseases but also the potential for bioterrorism. The USDA Pathogen Reduction Performance Standards has set limits for the presence of *Salmonella* for all slaughter facilities and raw ground meat products (Alocilja and Radke 2003). Other well-known pathogens recognized as being responsible for food-borne diseases are *Escherichia coli* 0157:H7, *Campylobacter jejuni*, *Listeria monocytogenes*, *Bacillus cereus*, *Staphylococcus aureus*, *Streptococci*, *etc*. While there are a number of conventional methods for detection of such contaminating microflora or pathogens, they tend to be labor intensive and results are usually not available until after a couple of days. A number of ELISA methods with high sensitivity have since been developed that address this drawback (Croci et al. 2001, Delibato et al. 2006, Salam and Tothill 2009). Some of these ELISA kits (e.g., LOCATE SALMONELLA produced by R-Biopharm) are also commercially available. In their sandwich ELISA method for *Salmonella* detection in precooked chicken, Salam and Tothill (2009) immobilized the capture antibody onto a gold electrode surface, and a second antibody conjugated to horseradish peroxidase was used as the detection system for recognition of captured microbial cells. Detection or binding of the enzyme label is then conducted by an electrochemical system using tetramethylbenzidine dihydrochloride as electron transfer mediator and hydrogen peroxide as substrate.

CONCLUDING REMARKS

It is evident from the foregoing that enzymes play a tremendous role in food analysis and their application for this purpose will likely increase in importance, taking into consideration the developments in the food industry. With consumers becoming increasingly aware of the health implications of foods and the growing demand for functional foods and nutraceuticals, new ingredients are being pervasively introduced into the food supply at a rapid pace. Analysis of these novel ingredients will invariably require novel analytical tools with the sensitivities and specificities that are inherent in enzymes. The convergence of these unique characteristics of enzymes with developments in biotechnology and biosensors certainly make the application of enzymes for food analysis an easier proposition.

REFERENCES

Alfonso A et al. 2004. A rapid microplate fluorescence method to detect yessotoxins based on their capacity to activate phosphodiesterases. *Anal Biochem* 326: 93–99.

Alocilja EC, Radke SM. 2003. Market analysis of biosensors for food safety. *Biosens Bioelectron* 18: 841–846.

Ammida NHS et al. 2004. Electrochemical immunosensor for determination of aflatoxin B1 in barley. *Anal Chim Acta* 520: 159–164.

Amine A et al. 2006. Enzyme inhibition-based biosensors for food safety and environmental monitoring. *Biosens Bioelectron* 21: 1405–1423.

Anh TM et al. 2004. Conductometric tyrosinase biosensor for the detection of diuron, atrazine and its metabolites. *Talanta* 63: 365–370.

Arduini F et al. 2007. Enzymatic spectrophotometric method for aflatoxin B detection based on acetylcholinesterase inhibition. *Anal Chem* 79: 3409–3415.

Avramescu A et al. 2001. Chronoamperometric determination of D-lactate using screen-printed enzyme electrodes. *Anal Chim Acta* 433: 81–88.

Avramescu A et al. 2002. Screen-printed biosensors for the control of wine quality based on lactate and acetaldehyde determination. *Anal Chim Acta* 458: 203–213.

Baden DG et al. 1995. Modified immunoassays for polyether toxins: implications of biological matrices, metabolic states, and epitope recognition. *J AOAC Int* 78: 499–508.

Barbau-Piednoir E et al. 2010. SYBR Green qPCR screening methods for the presence of "35S promoter" and "NOS terminator" elements in food and feed products. *Eur Food Res Technol* 230: 383–393.

Bignami GS et al. 1992. Monoclonal antibody-based enzyme-linked immunoassays for the measurement of palytoxin in biological samples. *Toxicon* 30: 687–700.

Bogusz MJ et al. 2004. Rapid determination of chloramphenicol and its glucuronide in food products by liquid chromatography—electrospray negative ionization tandem mass spectrometry. *J Chromatogr B* 807: 343–356.

Briggs LR et al. 2004. Enzyme-linked immunosorbent assay for the detection of yessotoxin and its analogues. *J Agric Food Chem* 52: 5836–5842.

Brunnert HJ et al. 2001. PCR-ELISA for the CaMV-35S promoter as a screening method for genetically modified Roundup Ready soybeans. *Eur Food Res Technol* 213: 366–371.

Campanella L et al. 2009. Determination of nonsteroidal drugs in milk and fresh cheese bread on the inhibition of cyclooxygenase. *Food Technol Biotechnol* 47(2): 172–177.

Campas M, Marty JL. 2007. Enzyme sensor for the electrochemical detection of the marine toxin okadaic acid. *Anal Chim Acta* 605: 87–93.

Campas M et al. 2008. Enzymatic recycling-based amperometric immunosensor for the ultrasensitive detection of okadaic acid in shellfish. *Biosens Bioelectron* 24: 716–722.

Chu FS, Fan TS. 1985. Indirect enzyme-linked immunosorbent assay for saxitoxin in shellfish. *J Assoc Anal Chem* 68: 13–16.

Clarkson PM, Thompson HS. 1997. Drugs and sport. Research findings and limitations. *Sports Med* 24: 366–384.

Cock LS et al. 2009. Use of enzymatic biosensors as quality indices: a synopsis of present and future trends in the food industry. *Chilean J Agric Res* 69(2): 270–280.

Coillie EV et al. 2004. Development of an indirect ELISA for flumequine residues in raw milk using chicken egg yolk antibodies. *J Agric Food Chem* 52: 4975–4978.

Cooke RD. 1978. An enzymatic assay for the total cyanide content of cassava (*Manihot esculenta*). *J Sci Food Agric* 29: 345–352.

Costa J et al. 2010. Monitoring genetically modified soybean along the industrial soybean oil extraction and refining processes by polymerase chain reaction techniques. *Food Res Int* 43: 301–306.

Council Regulation (EC) No. 49/2000 of the European Parliament and of the council of 10 January 2000. 2000. *Official J* 11 January, L006, 15.

Croci L et al. 2001. A rapid electrochemical ELISA for the determination of *Salmonella* in meat samples. *Anal Lett* 34: 2597–2607.

Cui R et al. 2008. Horseradish peroxidase-functionalized gold nanoparticle label for amplified immunoanalysis based on gold nanoparticles/carbon nanotubes hybrids modified biosensor. *Biosens Bioelectron* 23: 1666–1673.

Delibato E et al. 2006. *Anal Lett* 39: 1611–1625.

Della Loggia R et al. 1999. Methodological improvement of the protein phosphatase inhibition assay for the detection of okadaic acid in mussels. *Nat Toxins* 7: 387–391.

Doi H et al. 2008. Reliable enzyme-linked immunosorbent assay for the determination of walnut proteins in processed foods. *J Agric Food Chem* 56: 7625–7630.

Draisci R et al. 1998. Determination of biogenic amines with an electrochemical biosensor and its application to salted anchovies. *Food Chem* 62: 225–232.

European Commission Regulation (EC) No 1881/2006 of 19 December 2006. 2006. *Off J Eur Union* L364/5.

Fonfria ES et al. 2008. Feasibility of using a surface plasmon resonance-based biosensor to detect and quantify yessotoxin. *Anal Chim Acta* 617: 167–170.

Garet E et al. 2010. Comparative evaluation of enzyme-linked immunoassay and reference methods for the detection of shellfish hydrophilic toxins in several presentations of seafood. *J Agric Food Chem* 58: 1410–1415.

Garthwaite I et al. 2001. Integrated ELISA screening system for amnesic, neurotoxic, diarrhetic and paralytic shellfish poisoning toxins found in New Zealand. *J Assoc Anal Chem Int* 84: 1643–1648.

Gutes A et al. 2006. Automated SIA e-tongue employing a voltammetric biosensor array for the simultaneous determination of glucose and ascorbic acid. *Electroanalysis* 18: 82–88.

Hamada C et al. 1980. A sensitive enzymatic method for the fluorimetric assay of cholesterol in serum and high-density lipoprotein. *Chem Pharm Bull* 28: 3131–3134.

He L et al. 2009. Development of a polyclonal indirect ELISA with sub-ng g-1 sensitivity for the analysis of clenbuterol in milk, animal feed, and liver samples and a small survey of residues in retail animal products. *Food Add Contamin* 26(8): 1153–1161.

Hipolito-Moreno A et al. 1998. Non-aqueous flow-injection determination of atrazine by inhibition of immobilized tyrosinase. *Anal Chim Acta* 362: 187–192.

Holzhauser T, Vieths S. 1999. Quantitative sandwich ELISA for determining traces of hazelnut (*Corylus avellana*) protein in complex food matrices. *J Agric Food Chem* 47: 4209–4218.

Hooda V et al. 2009. Biosensor based on enzyme coupled PVC reaction cell for electrochemical measurement of serum cholesterol. *Sensor Actuat B* 136: 235–241.

Huet AC et al. 2006. Simultaneous determination of fluoroquinolone antibiotics in kidney, marine products, eggs and muscle by enzyme-linked immunosorbent assay. *J Agric Food Chem* 54: 2822–2827.

Husain FT et al. 2010. Development and validation of an indirect competitive enzyme-linked immunosorbent assay for the determination of potentially allergenic sesame (*Sesamum indicum*) in food. *J Agric Food Chem* 58: 1434–1441.

James D et al. 2003. Reliable detection and identification of genetically modified maize, soybean, and canola by multiplex PCR analysis. *J Agric Food Chem* 51: 5829–5834.

Jin, X et al. 2009. Piezoelectric immunosensor with gold nanoparticles enhanced competitive immunoreaction technique for quantification of aflatoxin B1. *Biosens Bioelectron* 24: 2580–2585.

Karube I et al. 1982. Amperometric determination of total cholesterol in serum with use of immobilized cholesterol esterase and cholesterol oxidase. *Anal Chim Acta* 139: 127–132.

Kawatsu K et al. 1997. Rapid and highly sensitive enzyme immunoassay for quantitative determination of tetrodotoxin. *Jpn J Med Sci Biol* 50: 133–150.

Kerkaert B, De Meulenaer B. 2007. Detection of hen's egg white lysozyme in food: comparison between a sensitive HPLC and a commercial ELISA method. *Commun Agric Appl Biol Sci* 72: 215–218.

Kleivdahl H et al. 2007. Determination of domoic acid toxins in shellfish by biosense ASP ELISA—a direct competitive enzyme-linked immunosorbent assay: collaborative study. *J AOAC Int* 90: 1011–1027.

Kreuzer MP et al. 1999. Development of an ultrasensitive immunoassay for rapid measurement of okadaic acid and its isomers. *Anal Chem* 71: 4198–4202.

Kreuzer MP et al. 2002. Novel electrochemical immunosensors for seafood toxin analysis. *Toxicon* 40: 1267–1274.

Landete JM et al. 2004. Improved enzymatic method for the rapid determination of histamine in wine. *Food Addit Contamin* 21: 1149–1154.

Landgraf A et al. 1991. Determination of GMO in soybean. *Anal Biochem* 198: 86–91.

Lange J, Wittman C. 2002. Enzyme sensor array for the determination of biogenic amines in food samples. *Anal Bioanal Chem* 372: 276–283.

Lattanzio VMT et al. 2009. Current analytical methods for trichothecene mycotoxins in cereals. *Trends Anal Chem* 28(6): 758–768.

Lau CO et al. 1995. Lophozozymus pictor toxin: a fluorescent structural isomer for palytoxin. *Toxicon* 33: 1373–1377.

Lee NA et al. 2004. A rapid aflatoxin B1 ELISA: development and validation with reduced matrix effects for peanuts, corn, pistachio, and soybeans. *J Agric Food Chem* 52: 2746–2755.

Lee S-H et al. 2009. Event-specific analytical methods for biotech maize MIR 604 and DAS-59122-7. *J Sci Food Agric* 89: 2616–2624.

Lima JLFC et al. 1998. Enzymatic determination of lactic and malic acids in wines by flow injection spectrophotometry. *Anal Chim Acta* 366: 187–191.

Liu Y et al. 2006. Immune biosensor for aflatoxin B1-based bioelectrocatalytic reaction on micro-comb electrode. *Biochem Eng J* 32: 211–217.

Lu S et al. 2006. Preparation of antiperfloxacin antibody and development of an indirect competitive enzyme-linked immunosorbent assay for detection of perfloxacin residue in chicken liver. *J Agric Food Chem* 54: 6995–7000.

Martinez M et al. 2006. Pharmacology of the fluoroquinolones: a perspective for the use in domestic animals. *Vet J* 172: 10–28.

Maruyama N et al. 2001. Crystal structures of recombinant and native soybean β-conglycinin homotrimers. *Eur J Biochem* 268: 3595–3604.

Matsumoto K et al. 1988. Simultaneous determination of glucose, fructose, and sucrose in mixtures by amperometric flow injection analysis with immobilized enzyme reactors. *Anal Chem* 60: 147–151.

Matsumoto M et al. 2000. Residue analysis of clenbuterol in bovine and horse tissues and cow's milk by ELISA. *J Food Hyg Soc Jpn* 41: 48–53.

Matsuoka T et al. 2001. A multiplex PCR method of detecting recombinant DNAs from five lines of genetically modified maize. *J Food Hyg Soc Jpn* 42: 24–32.

Mello LD, Kubota LT. 2002. Review of the use of biosensors as analytical tools in the food and drink industries. *Food Chem* 77: 237–256.

Mercader JV, Abad-Fuentes A. 2009. Monoclonal antibody generation and direct competitive enzyme-linked immunosorbent assay evaluation for the analysis of the fungicide fenhexamid in must and wine. *J Agric Food Chem* 57: 5129–5135.

Meric B et al. 2004. Disposable genosensor, a new tool for detection of NOS terminator, a genetic element present in GMOs. *Food Control* 15: 621–626.

Michelini E et al. 2008. New trends in bioanalytical tools for the detection of genetically modified organisms: an update. *Anal Bioanal Chem* 392: 355–367.

Morkyavichene MV et al. 1984. Choice of initial ratio of enzymes for immobilization in multienzyme systems. *Appl Biochem Microbiol* 20: 60–63.

Mottram D et al. 2002. Acrylamide is formed in the Maillard reaction. *Nature* 419: 448–449.

Mouri R et al. 2009. Surface plasmon resonance-based detection of ladder-shaped polyethers by inhibition detection method. *Bioorg Med Chem Lett* 19: 2824–2828.

Mulchandani A et al. 2001. Biosensors for direct determination of organophosphate pesticides. *Biosens Bioelectron* 16: 225–230.

Muresan L et al. 2008. Amine oxidase amperometric biosensor coupled to liquid chromatography for biogenic amines determination. *Microchim Acta* 163: 219–225

Murphy GK et al. 1992. Phosphatase methods. In: GH Richardson (ed.) *Standard Methods for the Examination of Dairy Products*. American Public Health Association, Washington, DC, p. 413.

Naar J et al. 2002. A competitive ELISA to detect brevetoxins from *Karenia brevis* (formerly *Gymnodinium breve*) in seawater, shellfish, and mammalian body fluid. *Environ Health Perspect* 110: 179–185.

Neagu D et al. 2006. Study of a toxin-alkaline phosphatase conjugate for the development of an immunosensor for tetrodotoxin determination. *Anal Bioanal Chem* 385: 1068–1074.

Omodeo SF et al. 1984. A sensitive enzymatic assay for determination of cholesterol in lipid extracts. *Anal Biochem* 142: 347–350.

Onal A. 2007. A review: current analytical methods for the determination of biogenic amines in foods. *Food Chem* 103: 1475–1486.

Paredes PA et al. 1997. Amperometric mediated carbon paste sensor based on D-fructose dehydrogenase for the determination of fructose in food analysis. *Biosens Bioelectron* 12: 1233–1243.

Parker CO, Tothill IE. 2009. Development of an electrochemical immunosensor for aflatoxin M1 in milk with focus on matrix interference. *Biosens Bioelectron* 24: 2452–2457.

Pazos MJ et al. 2005. Kinetic analysis of the interaction between yessotoxin and analogues and immobilized phosphodiesterases using a resonant mirror optical biosensor. *Chem Res Toxicol* 18: 1155–1160.

Petit L et al. 2003. Screening of genetically modified organisms and specific detection of Bt176 maize in flours and starches by PCR-enzyme-linked immunosorbent assay. *Eur Food Res Technol* 217: 83–89.

Petruzzelli E et al. 1996. Preparation and characterization of a monoclonal antibody specific for the beta-agonist clenbuterol. *Food Agric Immunol* 8: 3–10.

Piermarini S et al. 2009. An ELIME-array for detection of aflatoxin B1 in corn samples. *Food Control* 20: 371–375.

Pomes A et al. 2003. Monitoring peanut allergen in food products by measuring Ara h1. *J Allergy Clin Immunol* 111: 640–645.

Powers JR. 1998. Application of enzymes in food analysis. In: SS Nielsen (ed.) *Food Analysis*. Aspen Publication Inc., Gaithersburg, MD, pp. 349–365.

Prasetyo EN et al. 2010. Laccase-generated tetramethoxy azobismethylene quinone as a tool for antioxidant activity measurement. *Food Chem* 118: 437–444.

Preston A et al. 2008. Development of a high throughput enzyme-linked immunosorbent assay for the routine detection of the carcinogen in food, via rapid derivatization pre-analysis. *Anal Chim Acta* 608: 178–185.

Prieto-Simon B, Campas M. 2009. Immunochemical tools for mycotoxin detection in food. *Monatsh Chem* 140: 915–920.

Prodromidis MI, Karayannis MI. 2002. Enzyme based amperometric biosensors for food analysis. *Electroanalysis* 14: 241–261.

Pulce C et al. 1991. Collective human food poisonings by clenbuterol residues in veal liver. *Vet Hum Toxicol* 52: 311–318.

Rauch P et al. 1990. Sandwich enzyme immunoassay of hen egg lysozyme in foods. *J Food Sci* 55: 103–105.

Rauch P et al. 1995. Enhanced chemiluminiscent sandwich enzyme assay for hen egg lysozyme. *J Biochim Chemilumin* 10: 35–40.

Roda A et al. 2006. Development and validation of a sensitive and fast chemiluminescent enzyme immunoassay for the detection of genetically modified maize. *Anal Bioanal Chem* 384: 1269–1275

Rivera VR et al. 1995. Prophylaxis and treatment with a monoclonal antibody of tetrodotoxin poisoning in mice. *Toxicon* 33: 1231–1237.

Rocco R. 1990. Fluorometric determination of alkaline phosphatase in fluid dairy products: collaborative study. *J Assoc Off Anal Chem* 73: 842–849

Salam F, Tothill IE. 2009. Detection of *Salmonella typhimurium* using an electrochemical immunosensor. *Biosens Bioelectron* 24: 2630–2636.

Satoh I et al. 1977. Enzyme electrode for free cholesterol. *Biotech Bioeng* 19: 1095–1099.

Schneider N et al. 2010. Development and validation of an ELISA for quantification of lysozyme in cheese. *J Agric Food Chem* 58: 76–81.

Schubert-Ulrich P et al. 2009. Commercialized rapid immunoanalytical tests for determination of allergenic food proteins: an overview. *Anal Bioanal Chem* 395: 69–81.

Scortichini G et al. 2009. Validation of an enzyme-linked immunosorbent assay screening for quinolones in egg, poultry muscle and feed samples. *Anal Chim Acta* 637: 273–278.

Sheng W et al. 2009. Determination of marbofloxacin residues in beef and pork with an enzyme-linked immunosorbent assay. *J Agric Food Chem* 57: 5971–5975.

Shim YY et al. 2007. Quantitative analysis of phosphinothricin-*N*-acetyltransferase in genetically modified herbicide tolerant pepper by an enzyme-linked immunosorbent assay. *J Microbiol Biotechnol* 17: 681–684.

Shinkaruk S et al. 2010. Development and validation of two new sensitive ELISAs for hesperetin and naringenin in biological fluids. *Food Chem* 118: 472–481.

Shkotova LV et al. 2008. Amperometric biosensor for lactate analysis in wine and must during fermentation. *Mat Sci Eng* C28: 943–948.

Solna R et al. 2005. Amperometric screen-printed biosensor arrays with co-immobilized oxidoreductases and cholinesterases. *Anal Chim Acta* 528: 9–19.

Sreenath K and Venkatesh YP. 2010. Quantification of xylitol in foods by an indirect competitive immunoassay. *J Agric Food Chem* 58(2): 1240–1246.

Swedish National Food Administration (Livsmedelverket). 2002. Analytical methodology and survey results for acrylamide in foods. Available at http://www.slv.se/templates/SLV_Page.aspx?id=15789&epslanguage=EN-GB.

Tan Y et al. 2009. A signal-amplified electrochemical immunosensor for aflatoxin B1 determination in rice. *Anal Biochem* 387: 82–86.

Teti D et al. 2002. Analysis of polyamines as markers of pathophysiological conditions. *J Chromatogr B* 781: 107–149.

Tittlemier SA. 2010. Methods for the analysis of melamine and related compounds in foods: a review. *Food Add Contamin* 27(2): 129–145.

Trivedi UB et al. 2009. Amperometric fructose biosensor based on fructose dehydrogenase enzyme. *Sens Actuat B* 136: 45–51.

Tsai W-L et al. 2009. A potential methodology for differentiation of ciguatera-carrying species of moray eel. *Food Control* 20: 575–579.

Tubaro A et al. 1996. A protein phosphatase 2A inhibition assay for a fast and sensitive assessment of okadaic acid contamination in mussels. *Toxicon* 34: 743–752.

Valdez-Ramirez G et al. 2009. Automated resolution of dichlorvos and methylparaoxon pesticide mixtures employing a flow injection system with an inhibition electronic tongue. *Biosens Bioelectron* 24: 1103–1108.

Van Veldhoven PP et al. 2002. Enzymatic quantitation of cholesterol esters in lipid extracts. *Anal Biochem* 258: 152–155.

Vedrine C et al. 2003. Amperometric tyrosinase based biosensor using an electrogenerated polythiophene film as an entrapment support. *Talanta* 59: 535–544.

Vidal ML et al. 2005. Development of an ELISA for quantifying lysozyme in hen egg white. *J Agric Food Chem* 53: 2379–2385.

Vilarino N et al. 2009. Use of biosensors as alternatives to current regulatory methods for marine biotoxins. *Sensors* 9: 9414–9443.

Viswanathan S et al. 2009. Electrochemical biosensors for food analysis. *Monatsch Chem* 140: 891–899.

Vollenhofer S et al. 1999. Genetically modified organisms in food—screening and specific detection by polymerase chain reaction. *J Agric Food Chem* 47: 5038–5043.

Volpe G et al. 2009. A bienzyme electrochemical probe for flow injection analysis of okadaic acid based on protein phosphatase-2A inhibition: an optimization study. *Anal Biochem* 385: 50–56.

Weber P et al. 2007. Investigation of the allergenic potential of wines fined with various proteinogenic fining agents by ELISA. *J Agric Food Chem* 55: 3127–3133.

Whitaker TB et al. 2001. Sampling grain shipments to detect genetically modified seed. *J Assoc Anal Chem Int* 84: 1941–1946.

Wisitsoraat A et al. 2009. High sensitivity electrochemical cholesterol sensor utilizing a vertically alligned carbon nanotube electrode with electropolymerized enzyme immobilization. *Sensors* 9: 8658–8668.

Wiseman A. 1981. Enzymes in food analysis. In: GG Birch et al. (eds.) *Enzymes and Food Processing*. Applied Science Publishers, London, pp. 275–288.

Wolfgang T et al. 1993. Cholesterol biosensors prepared by electropolymerization of pyrrole. *Electroanalysis* 5: 753–763.

Wu Y et al. 2007. Event-specific qualitative and quantitative PCR detection methods for transgenic rapeseed hybrids. *J Agric Food Chem* 55: 8380–8389.

Xiang P et al. 2002. Identification and analysis of a conserved immunoglobulin E-binding epitope in soybean G1a and G2a and peanut Ara h3 glycinins. *Arch Biochem Biophys* 408: 51–57.

Yang L et al. 2006. Event-specific qualitative and quantitative polymerase chain reaction analysis for genetically modified canola T45. *J Agric Food Chem* 54: 9735–9740.

Yang L et al. 2007. Event-specific quantitative detection of nine genetically modified maize using one novel standard reference molecule. *J Agric Food Chem* 55: 15–24.

Yoshida A et al. 1991. Enzyme immunoassay for hen egg white lysozyme used as a food additive. *J Assoc Off Anal Chem* 74: 502–505.

You J et al. 2008. Development of a monoclonal antibody-based competitive ELISA for detection of β-conglycinin, an allergen from soybean. *Food Chem* 106: 352–360.

Zeravik J et al. 2009. State of the art in the field of electronic and bioelectronic tongues towards the analysis of wines. *Electroanalysis* 21(23): 2509–2520.

4
Browning Reactions

Marta Corzo-Martínez, Nieves Corzo, Mar Villamiel, and M Dolores del Castillo

Abstract: Browning is one of the most important reactions taking place during food processing and storage. Both, enzymatic and nonenzymatic browning can affect the quality of food in either positive or negative ways, depending on the type of food. Special attention has been paid in this chapter to chemistry of the reaction. During enzymatic browning, phenolic compounds are oxidized to quinones by polyphenol oxidase, this reaction being of particular importance in fruits, vegetables, and seafoods. Nonenzymatic browning is referred to as the Maillard reaction (MR), when it takes place between free amino groups from amino acids, peptides, or proteins and the carbonyl group of reducing sugars. Ascorbic acid, its oxidation products and oxidized lipids can also react *via* MR. It has been also described that aminophospholipids and reducing carbohydrates can interact by MR. Food browning can be also due to degradation of carbohydrates also called caramelization reaction and it is hereby discussed.

INTRODUCTION

Browning reactions are some of the most important phenomena occurring in food during processing and storage. They represent an interesting research for the implications in food stability and technology as well as in nutrition and health. The major groups of reactions leading to browning are enzymatic phenol oxidation and so-called nonenzymatic browning (Manzocco et al. 2001).

ENZYMATIC BROWNING

Enzymatic browning is one of the most important color reactions that affect fruits, vegetables, and seafood. It is catalyzed by the enzyme polyphenol oxidase (PPO; 1,2 benzenediol; oxygen oxidoreductase, EC 1.10.3.1), which is also referred to as phenoloxidase, phenolase, monophenol oxidase, diphenol oxidase (DPO), and tyrosinase. Phenoloxidase enzymes (PPOs) catalyze the oxidation of phenolic constituents to quinones, which finally polymerize to colored melanins (Marshall et al. 2000). Significant advances have been made on biochemistry, molecular biology, and genetics of PPO (Yoruk and Marshall 2003, Queiroz et al. 2008).

Enzymatic browning reactions may affect fruits, vegetables, and seafood in either positive or negative ways. These reactions, for instance, may contribute to the overall acceptability of foods such as tea, coffee, cocoa, and dried fruits (raisins, prunes, dates, and figs). Products of enzymatic browning play key physiological roles. Melanins, produced as a consequence of PPO activity, may exhibit antibacterial, antifungal, anticancer, and antioxidant properties. PPOs impart remarkable physiological functions for the development of aquatic organisms, such as wound healing and hardening of the shell (sclerotization), after molting in insects and in crustaceans such as shrimp and lobster. The mechanism of wound healing in aquatic organisms is similar to

Food Biochemistry and Food Processing, Second Edition. Edited by Benjamin K. Simpson, Leo M.L. Nollet, Fidel Toldrá, Soottawat Benjakul, Gopinadhan Paliyath and Y.H. Hui.

that which occurs in plants in which the compounds produced as a result of the polymerization of quinones, melanins, or melanoidins exhibit both antibacterial and antifungal activities. In addition, enzymatic reactions are considered desirable during fermentation (Fennema 1976). Despite these positive effects, enzymatic browning is considered one of the most devastating reactions for many exotic fruits and vegetables, in particular tropical and subtropical varieties. Enzymatic browning is especially undesirable during processing of fruit slices and juices. Lettuce, other green leafy vegetables; potatoes and other starchy staples, such as sweet potato, breadfruit, and yam; mushrooms; apples; avocados; bananas; grapes; olive (Sciancalepore 1985); peaches; pears (Vamosvigyazo and Nadudvarimarkus 1982); and a variety of other tropical and subtropical fruits and vegetables are susceptible to browning. Crustaceans are also extremely vulnerable to enzymatic browning. Since enzymatic browning can affect color, flavor, and nutritional value of these foods, it can cause tremendous economic losses (Marshall et al. 2000, Queiroz et al. 2008).

A better understanding of the mechanism of enzymatic browning in fruits, vegetables, and seafood; the properties of the enzymes involved; and their substrates and inhibitors may be helpful for controlling browning development, avoiding economic losses, and providing high-quality foods. On the basis of this knowledge, new approaches for the control of enzymatic browning have been proposed. These subjects will be reviewed in this chapter.

Properties of PPO

PPO (EC 1.10.3.1; *o*-diphenol oxidoreductase) is an oxidoreductase able to oxidize phenol compounds, employing oxygen as a hydrogen acceptor. The abundance of phenolics in plants may be the reason for naming this enzyme PPO (Marshall et al. 2000). The molecular weight for PPO in plants ranges between 57 and 62 kDa (Hunt et al. 1993, Newman et al. 1993). PPO catalyzes two basic reactions: hydroxylation to the *o*-position adjacent to an existing hydroxyl group of the phenolic substrate (monophenol oxidase activity) and oxidation of diphenol to *o*-benzoquinones (DPO activity; Figure 4.1).

In plants, the ratio of monophenol to DPO activity is usually in the range of 1:40–1:10 (Nicolas et al. 1994). Monophenol oxidase activity is considered more relevant for insect and crustacean systems due to its physiological significance in conjunction with diphenolase activity (Marshall et al. 2000).

PPO is also referred to as tyrosinase to describe both monophenol and DPO activities in either animals or plants. The monophenol oxidase acting in plants is also called cresolase because of its ability to employ cresol as substrate (Marshall et al. 2000). This enzyme is able to metabolize aromatic amines and *o*-aminophenols (Toussaint and Lerch 1987).

The oxidation of diphenolic substrates to quinones in the presence of oxygen is catalyzed by DPO activity (Figure 4.1). The *p*-DPO catalyzed reaction follows a Bi Bi mechanism (Whitaker 1972). Diphenolase activity may be due to two different enzymes: catecholase (catechol oxidase) and laccase (Figure 4.1). Laccase (*p*-DPO,EC 1.10.3.2) is a type of copper-containing PPO. It has the unique ability to oxidize *p*-diphenols. Phenolic substrates, including polyphenols, methoxy-substituted phenols, diamines, and a considerable range of other compounds, serve as substrates for laccase (Marshall et al. 2000). Laccases occur in many phytopathogenic fungi, higher plants (Mayer and Harel 1991), peaches (Harel et al. 1970), and apricots (Dijkstra and Walker 1991).

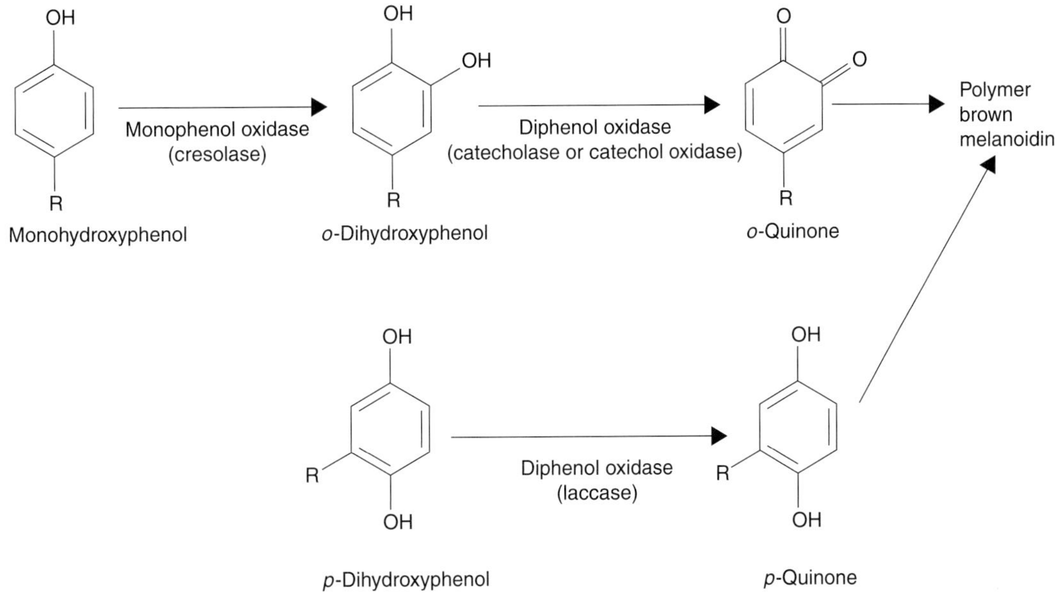

Figure 4.1. Polyphenol oxidase pathway.

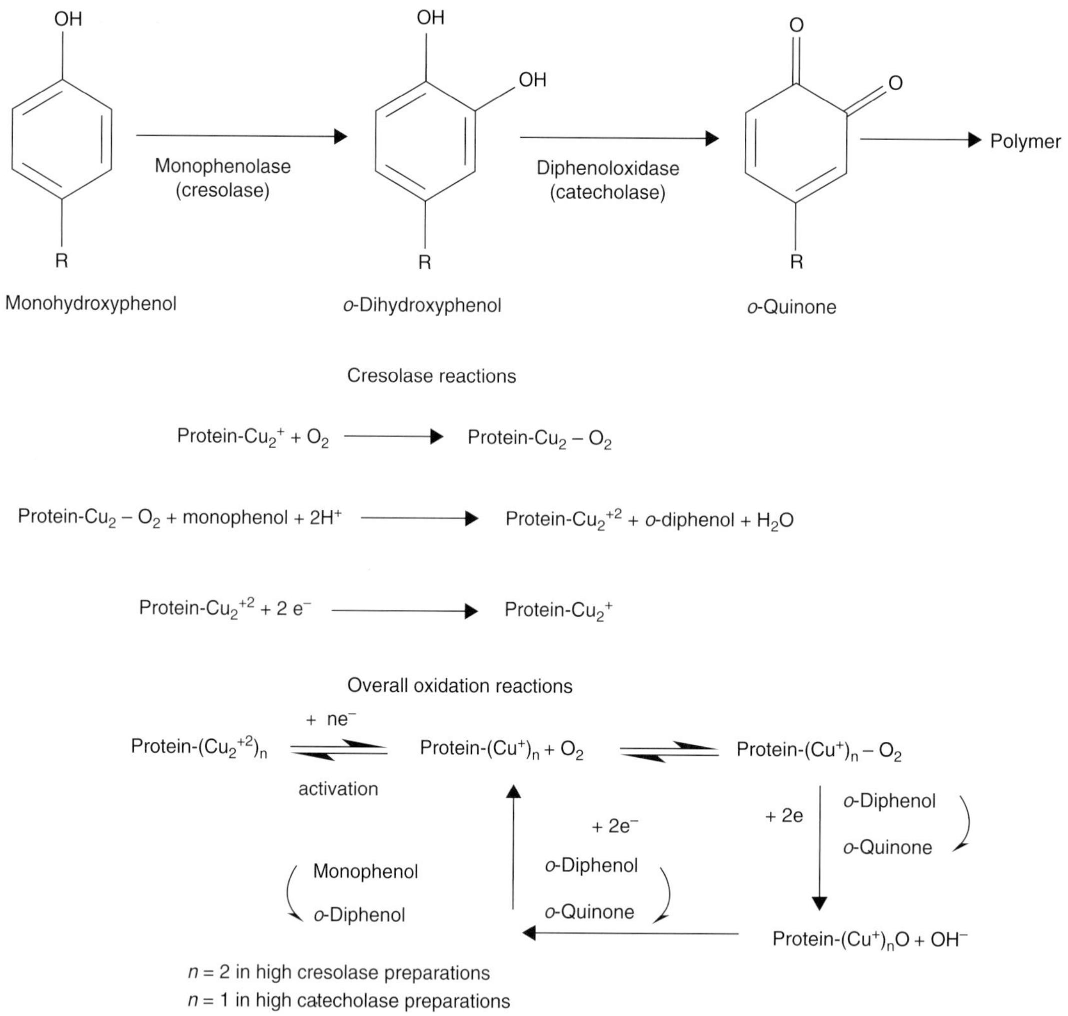

Figure 4.2. Simplified mechanism for the hydroxylation and oxidation of diphenol by phenoloxidase.

Catechol oxidase and laccase are distinguishable both on the basis of their phenolic substrates (Figure 4.1), and their inhibitor specificities (Marshall et al. 2000).

Catecholase activity is more important than cresolase action in food because most of the phenolic substrates in food are dihydroxyphenols (Mathew and Parpia 1971).

This bifunctional enzyme, PPO, containing copper in its structure, has been described as an oxygen and four electron-transferring phenol oxidase (Jolley et al. 1974). Figure 4.2 shows a simplified mechanism for the hydroxylation and oxidation of phenols by PPO. Both mechanisms involve the two copper moieties on the PPO.

PPO, active between pH 5 and 7, does not have a very sharp pH optimum. At lower pH values of approximately 3, the enzyme is irreversibly inactivated. Reagents that complex or remove copper from the prosthetic group of the enzyme inactivate the enzyme (Fennema 1976).

Substrates

Although tyrosine is the major substrate for certain phenolases, other phenolic compounds such as caffeic acid and chlorogenic acid also serve as substrate (Fennema 1976). Structurally, they contain an aromatic ring bearing one or more hydroxyl groups, together with a number of other substituents (Marshall et al. 2000). Structures of common phenolic compounds present in foods are shown in Figure 4.3. Phenolic subtrates of PPO in fruits, vegetables, and seafood are listed in Table 4.1. The substrate specificity of PPO varies in accordance with the source of the enzyme. Phenolic compounds and PPO are, in general, directly responsible for enzymatic browning reactions in damaged fruits during postharvest handling and processing. The relationship of the rate of browning to phenolic content and PPO activity has been reported for various fruits. In addition to serving as PPO substrates, phenolic compounds act as inhibitors of PPOs (Marshall et al. 2000).

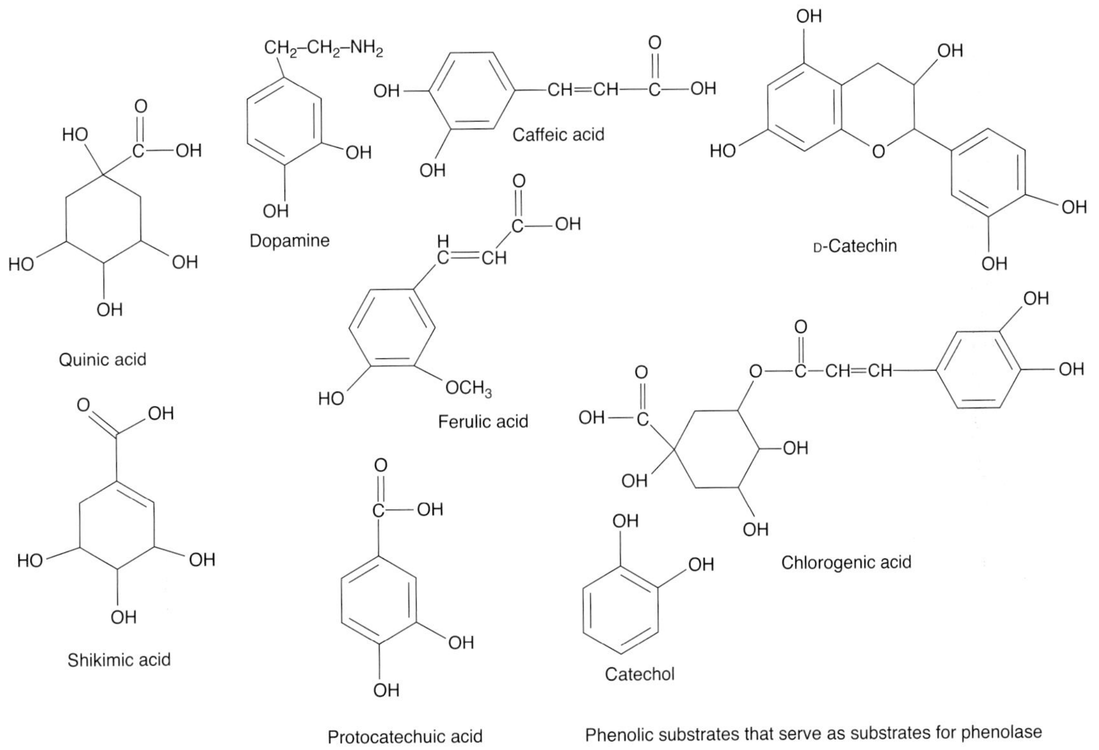

Figure 4.3. Structure of common phenols present in foods.

Table 4.1. Phenolic Substrates of PPO in Foods

Source	Phenolic Substrates
Apple	Chlorogenic acid (flesh), catechol, catechin (peel), caffeic acid, 3,4-dihydroxyphenylalanine (DOPA), 3,4-dihydroxy benzoic acid, p-cresol, 4-methyl catechol, leucocyanidin, *p*-coumaric acid, flavonol glycosides
Apricot	Isochlorogenic acid, caffeic acid, 4-methyl catechol, chlorogenic acid, catechin, epicatechin, pyrogallol, catechol, flavonols, *p*-coumaric acid derivatives
Avocado	4-Methyl catechol, dopamine, pyrogallol, catechol, chlorogenic acid, caffeic acid, DOPA
Banana	3,4-Dihydroxyphenylethylamine (Dopamine), leucodelphinidin, leucocyanidin
Cacao	Catechins, leucoanthocyanidins, anthocyanins, complex tannins
Coffee beans	Chlorogenic acid, caffeic acid
Eggplant	Chlorogenic acid, caffeic acid, coumaric acid, cinnamic acid derivatives
Grape	Catechin, chlorogenic acid, catechol, caffeic acid, DOPA, tannins, flavonols, protocatechuic acid, resorcinol, hydroquinone, phenol
Lettuce	Tyrosine, caffeic acid, chlorogenic acid derivatives
Lobster	Tyrosine
Mango	Dopamine-HCl, 4-methyl catechol, caffeic acid, catechol, catechin, chlorogenic acid, tyrosine, DOPA, *p*-cresol
Mushroom	Tyrosine, catechol, DOPA, dopamine, adrenaline, noradrenaline
Peach	Chlorogenic acid, pyrogallol, 4-methyl catechol, catechol, caffeic acid, gallic acid, catechin, dopamine
Pear	Chlorogenic acid, catechol, catechin, caffeic acid, DOPA, 3,4-dihydroxy benzoic acid, *p*-cresol
Plum	Chlorogenic acid, catechin, caffeic acid, catechol, DOPA
Potato	Chlorogenic acid, caffeic acid, catechol, DOPA, *p*-cresol, *p*-hydroxyphenyl propionic acid, *p*-hydroxyphenyl pyruvic acid, *m*-cresol
Shrimp	Tyrosine
Sweet potato	Chlorogenic acid, caffeic acid, caffeylamide
Tea	Flavanols, catechins, tannins, cinnamic acid derivatives

Source: Reproduced from Marshall et al. 2000.

Control of Browning

Enzymatic browning may cause a decrease in the market value of food products originating from plants and crustaceans (Kubo et al. 2000, Perez-Gilabert and García-Carmona 2000, Subaric et al. 2001, Yoruk and Marshall 2003, Queiroz et al. 2008). Processing such as cutting, peeling, and bruising is enough to cause enzymatic browning. The rate of enzymatic browning is governed by the active PPO content of the tissues, the phenolic content of the tissue, and the pH, temperature, and oxygen availability within the tissue. Table 4.2 gives a list of procedures and inhibitors that may be employed for controlling enzymatic browning in foods. A review on conventional and alternative methods to inactivate PPO has been recently published by Queiroz et al. (2008). The inhibition of enzymatic browning generally proceeds via direct inhibition of the PPO, nonenzymatic reduction of *o*-quinones, and chemical modification or removal of phenolic substrates of PPO. Among all these methods, inhibition of PPO is preferable. According to Marshall et al. (2000), there are six categories of PPO inhibitors applicable to control enzymatic browning: reducing agents, acidulants, chelating agents, complexing agents, enzyme inhibitors, and enzyme treatments. Tyrosinase inhibitors from natural and synthetic sources have been reported by Chang (2009).

Sulfites are the most efficient multifunctional agents in the control of enzymatic browning of foods. The use of sulfites has become increasingly restricted because they can produce adverse reaction in some consumers. L-ascorbic and its stereoisomer erythorbic acid have been considered as the best alternative to sulfite in controlling browning. 4-Hexylresorcinol is also a good substitute to sulfite. Sulphydryl amino acids such as cysteine and reduced glutathione, and inorganic salts like sodium chloride, kojic acid, and oxalic acid (Pilizota and Subaric 1998, Son et al. 2000, Burdock et al. 2001, Yoruk and Marshall 2009) cause an effective decrease in undesirable enzymatic browning in foods. An equivalent of hypotaurine can be an extract of a foodstuff that has a high hypotaurine level such clam, oyster, mussels, squid, octopus, or any combination thereof (Marshall and Schulbach 2009). The application of this naturally occurring compound provides a longer-lasting protection from enzymatic browning in comparison with ascorbic acid or citric acid. On the other hand, taurine addition may provide nutritional benefits to the supplemented foods because of its health-promoting properties.

A decrease in enzymatic browning is achieved by the use of various chelating agents, which either directly form complexes with PPO or react with its substrates. For example, *p*-cyclodextrin is an inhibitor that reacts with the copper-containing prosthetic group of PPO (Pilizota and Subaric 1998). Crown compounds have potential to reduce the enzymatic browning caused by catechol oxidase because of their ability to complex with copper present in its prosthetic group. The inhibition effect of crown compounds, macrocyclic esters, benzo-18-crown-6 with sorbic acid, and benzo-18-crown-6 with potassium sorbate have been proved for fresh-cut apples (Subaric et al. 2001).

Agro-chemical processes may also be employed for achieving an effective control of enzymatic browning. In vitro studies based on the evaluation of the effect of a range of commonly used pesticides on the activity of purified quince (*Cydonia oblonga* Miller) PPO have indicated that PPO enzyme is competitively inhibited by pesticides such as benomyl, carbaryl, deltamethrine, and parathion methyl (Fattouch et al. 2010). Salinity also affects PPO activity of the fresh-cut vegetables. Increasing salinity conditions may allow fresh-cut vegetables possessing low PPO activity, high phenol content and high antioxidant capacity (Chisari et al. 2010).

Thiol compounds like 2-mercaptoethanol may act as an inhibitor of the polymerization of *o*-quinone and as a reductant involved in the conversion of *o*-quinone to *o*-dihydroxyphenol (Negishi and Ozawa 2000).

Because of safety regulations and the consumer's demand for natural food additives, much research has been devoted to the search for natural and safe anti-browning agents. Honey (Chen et al. 2000, Gacche et al. 2009); papaya latex extract (De Rigal et al. 2001); banana leaf extract either alone or in combination with ascorbic acid and 4-hexylresorcinol (Kaur and Kapoor 2000); onion juice (Hosoda and Iwahashi 2002); onion oil (Hosoda et al. 2003); onion extracts (Kim et al. 2005); onion by-products (residues and surpluses; Roldan et al. 2008); rice bran extract (Boonsiripiphat and Theerakulkait 2009); solutions containing citric acid, calcium chloride, and garlic extract (Ihl et al. 2003); Maillard reaction products (MRPs) obtained by heating of hexoses in presence of cystein or glutathione (Billaud et al. 2003, 2004); *N*-acetylcysteine and glutathione (Rojas-Grau et al. 2008); resveratrol, a natural ingredient of red wine possessing several biological activities, and other hydroxystilbene compounds, including its analogous oxyresveratrol (Kim et al. 2002); and hexanal (Corbo et al. 2000), Brassicacaea processing water (Zocca et al. 2010); and chitosan (Martin-Diana et al. 2009) are some examples of natural inhibitors of PPO. In most cases, the inhibiting activity of plant extract is due to more than one component. Moreover, a good control of enzymatic browning may involve endogenous antioxidants (Mdluli and Owusu-Apenten 2003).

Regulation of the biosynthesis of polyphenols (Hisaminato et al. 2001) and the use of commercial glucose oxidase–catalase enzyme system for oxygen removal (Parpinello et al. 2002) have been described as essential and effective ways of controlling enzymatic browning.

Commonly, an effective control of enzymatic browning can be achieved by a combination of anti-browning agents (Zocca et al. 2010). A typical combination might consist of a chemical reducing agent such as ascorbic acid, an acidulant such as citric acid, and a chelating agent like EDTA (Marshall et al. 2000)

A great emphasis is put on research to develop methods for preventing enzymatic browning especially in fresh-cut (minimally processed) fruits and vegetables. The most efficient way to control this problem is the combination of physical and chemical methods, by avoiding the use of more severe individual treatments, which could harm the appearance and texture of vegetables. Technological processing, including microwave blanching either alone or combined with chemical anti-browning agents (Severini et al. 2001, Premakumar and Khurduya 2002, Yadav et al. 2008, Guan and Fan 2010); CO_2 treatments (Rocha and

Table 4.2. Inhibitors and Processes Employed in the Prevention of Enzymic Browning

Inhibition Targeted Toward the Enzyme		Inhibition Targeted Toward the Substrate		Inhibition Targeted Toward the Products
Processing	Enzymes Inhibitors	Removal of Oxygen	Removal of Phenols	
Heating 1. Steam and water blanching (70–105°C) 2. Pasteurization (60–85°C)	Chelating agents 1. Sodium azide 2. Cyanide 3. Carbon monoxide 4. Halide salts ($CaCl_2$, NaCl) 5. Tropolone 6. Ascorbic acid 7. Sorbic acid 8. Polycarboxylic acids (citric, malic, tartaric, oxalic, and succinic acids) 9. Polyphosphates (ATP and pyrophosphate) 10. Macromolecules (porphyrins, proteins, polysaccharides) 11. EDTA 12. Kojic acid	Processing 1. Vaccum treatment 2. Immersion in water, syrup, brine	Complexing agents 1. Cyclodextrins 2. Sulphate polysaccharides 3. Chitosan	Reducing agents 1. Sulphites(SO_2, SO_3^{2-}, HSO_3^-, $S_2O_5^{2-}$) 2. Ascorbic acid and analogs 3. Cysteine and other thiol compounds
Cooling 1. Refrigeration 2. Freezing (−18°C) Dehydration *Physical methods* 1. Freeze drying 2. Spray drying 3. Radiative drying 4. Solar drying 5. Microwave drying *Chemical methods* 1. Sodium chloride and other salts 2. Sucrose and other sugars 3. Glycerol 4. Propylene glycol 5. Modified corn syrup	Aromatic carboxylic acids 1. Benzoic acids 2. Cinnamic acids Aliphatic alcohol Peptides and amino acids	Reducing agents 1. Ascorbic acid 2. Erythorbic acid 3. Butylated hydroxyanisole (BHA) 4. Butylated hydroxytoluene (BTH) 5. Tertiarybutyl hydroxyquinone 6. Propyl gallate	Enzymatic modification 1. *O*-methyltransferase 2. Protocatechuate 3, 4-dioxigenase	Amino acids, peptides, and proteins Chitosan Maltol
Irradiation 1. Gamma rays up to 1 kGy (Cobalt 60 or Cesium 137) 2. X-rays 3. Electron beams 4. Combined treatments using irradiation and heat	Substituted resorcinols			

(*Continued*)

Table 4.2. *(Continued)*

Inhibition Targeted Toward the Enzyme		Inhibition Targeted Toward the Substrate		Inhibition Targeted Toward the Products
Processing	Enzymes Inhibitors	Removal of Oxygen	Removal of Phenols	
High pressure (600–900 Mpa)	Honey (peptide ~600 Da and antioxidants)			
Supercritical carbon dioxide (58 atm, 43°C)	Proteases			
Ultrafiltration	Acidulants			
	Citric acid (0.5–2% w/v)			
	Malic acid			
	Phosphoric acid			
Ultrasonication	Chitosan			
Employment of edible coating				

Source: Adapted from Marshall et al. 2000.

Morais 2001, Kaaber et al. 2002, Valverde et al. 2010); pretreatments employing sodium or calcium chloride and lactic acid followed by conventional blanching (Severini et al. 2003); combination of sodium chlorite and calcium propionate (Guan and Fan 2010); high-pressure treatments combined with thermal treatments and chemical anti-browning agents such as ascorbic acid (Prestamo et al. 2000, Ballestra et al. 2002) or natural anti-browning agents like pineapple juice (Perera et al. 2010); and UV-C light treatment of fresh-cut vegetables under nonthermal conditions (Manzocco et al. 2009) have been employed to prevent enzymatic browning in foods.

The use of edible coating from whey protein isolate-beeswax (Perez-Gago et al. 2003); edible coatings enriched with natural plant extracts (Ponce et al. 2008); oxygen-controlled atmospheres (Jacxsens et al. 2001, Soliva-Fortuny et al. 2001, Duan et al. 2009), and the use of active films (Endo et al. 2008) seem to improve the shelf life of foods by inhibition of PPO.

Biotechnological approaches may be also employed for the control of PPO (Rodov 2007). Transgenic fruits carrying an antisense PPO gene show a reduction in the amount and activity of PPO, and the browning potential of transgenic lines are reduced compared with the non-transgenic ones (Murata et al. 2000, Murata et al. 2001). These procedures may be used to prevent enzymatic browning in a wide variety of food crops without the application of various food additives (Coetzer et al. 2001).

NONENZYMATIC BROWNING

The Maillard Reaction

Nonenzymatic browning is the most complex reaction in food chemistry because a large number of food components are able to participate in the reaction through different pathways, giving rise to a complex mixture of products (Olano and Martínez-Castro 2004). It is referred to as the Maillard reaction when it takes place between free amino groups from amino acids, peptides, or proteins and the carbonyl group of a reducing sugar. The historical perspective showed by Finot (2005) shows the great importance of Maillard reaction on food science and nutrition.

The Maillard reaction is one of the main reactions causing deterioration of proteins during processing and storage of foods. This reaction can promote nutritional changes such as loss of nutritional quality (attributed to the destruction of essential amino acids) or reduction of protein digestibility and amino acid availability (Malec et al. 2002).

The Maillard reaction covers a whole range of complex transformations (Figure 4.4) that produces a large number of the so-called Maillard reaction products (MRPs) such as aroma compounds, ultraviolet absorbing intermediates, and dark-brown polymeric compounds named melanoidins (Kim and Lee 2008a). It can be divided into three major phases: the early, intermediate, and advanced stages. The early stage (Figure 4.5) consists of the condensation of primary amino groups of amino acids, peptides, or proteins with the carbonyl group of reducing sugars (aldose), with loss of a molecule of water, leading, via formation of a Schiff's base and Amadori rearrangement, to the so-called Amadori product (1-amino-1-deoxi-2-ketose), a relatively stable intermediate (Feather et al. 1995). The Heyns compound is the analogous compound when a ketose is the starting sugar. In many foods, the ε-amino group of the lysine residues of proteins is the most important source of reactive amino groups, but because of blockage, these lysine residues are not available for digestion, and consequently the nutritive value decreases (Brands and van Boekel 2001, Machiels and Istasse 2002). Amadori compounds are precursors of numerous compounds that are important in the formation of characteristic flavors, aromas, and brown polymers. They are formed before the occurrence of sensory changes; therefore, their determination provides a very sensitive indicator for early detection of quality changes caused by the Maillard reaction (Olano and Martínez-Castro 2004).

The intermediate stage leads to breakdown of Amadori compounds (or other products related to the Schiff's base) and the formation of degradation products, reactive intermediates (3-deoxyglucosone), and volatile compounds (formation of flavor).

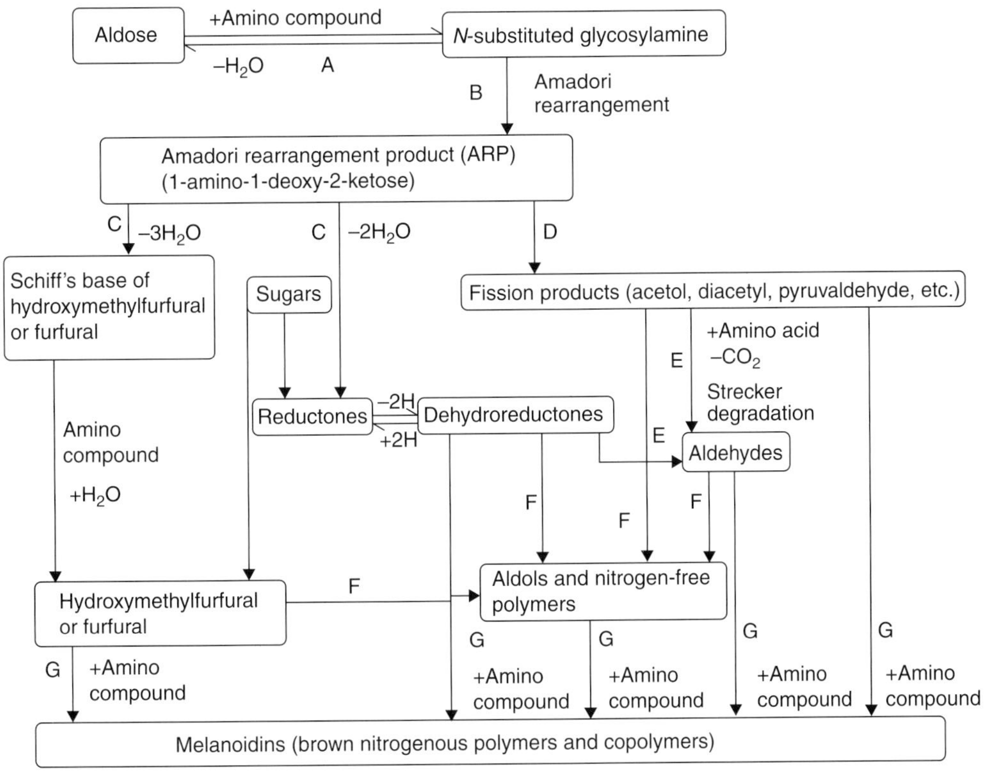

Figure 4.4. Scheme of different stages of Maillard reaction (Hodge 1953, Ames 1990).

The 3-deoxyglucosone participates in cross-linking of proteins at much faster rates than glucose itself, and further degradation leads to two known advanced products: 5-hydroxymethyl-2-furaldehyde and pyraline (Feather et al. 1995).

The final stage is characterized by the production of nitrogen-containing brown polymers and copolymers known as melanoidins (Badoud et al. 1995). The structure of melanoidins is largely unknown, but in the last few years, more data became available. In particular, it was shown that they can have a different structure according to the starting material. Melanoidins have been described as low-molecular weight (LMW) colored substances that are able to cross-link proteins via ε-amino groups of lysine or arginine to produce high-molecular weight (HMW) colored melanoidins. Also, it has been postulated that they are polymers consisting of repeating units of furans and/or pyrroles, formed during the advanced stages of the Maillard reaction and linked by polycondensation reactions (Martins and van Boekel 2003). Chemical structure of melanoidins can be mainly formed by a carbohydrate skeleton with few unsaturated rings and small nitrogen components; in other cases, they can have a protein structure linked to small chromophores (Borrelli and Fogliano 2005).

In foods, predominantly glucose, fructose, maltose, lactose, and to some extent reducing pentoses are involved with amino acids and proteins in forming fructoselysine, lactuloselysine or maltuloselysine. In general, primary amines are more important than secondary ones, because the concentration of primary amino acids in foods is usually higher than that of secondary amino acids (an exception is the high amount of proline in malt and corn products) (Ledl 1990).

Factors Affecting Maillard Reaction

The rate of the Maillard reaction and the nature of the products formed depend on the chemical environment of food, including water activity (a_w), pH, and chemical composition of the food system, temperature being the most important factor (Carabasa-Giribert and Ibarz-Ribas 2000). In order to predict the extent of chemical reactions in processed foods, knowledge of kinetic reactions is necessary to optimize the processing conditions. Since foods are complex matrices, these kinetic studies are often carried out using model systems in which sugars and amino acids react under simplified conditions. Model system studies may provide guidance regarding the directions in which to modify the food process and to find out which reactants may produce specific effects of the Maillard reaction (Lingnert 1990).

The reaction rate is significantly affected by the pH of the system; it generally increases with pH (Namiki et al. 1993, Ajandouz and Puigserver 1999). Bell (1997) studied the effect of buffer type and concentration on initial degradation of amino acids and formation of brown pigments in model systems of glycine and glucose that are stored for long periods at 25°C. The

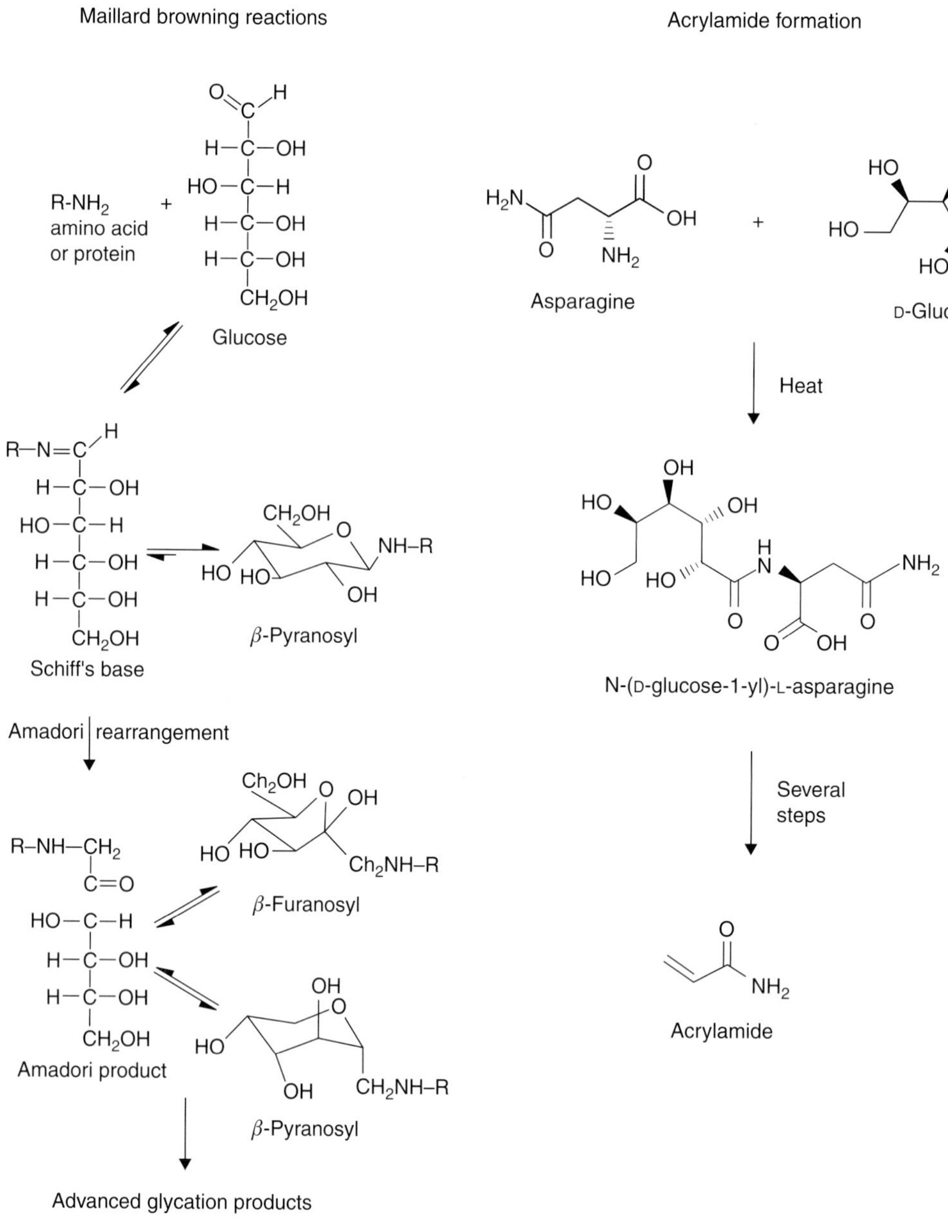

Figure 4.5. Initial stages of the Maillard reaction and acrylamide formation (Friedman 2003).

loss of glycine was faster at high phosphate buffer concentration (Figure 4.6), showing its catalytic effect on the Maillard reaction. The effects of salt concentration on the rates of browning reaction of amino acid, peptides, and proteins have also been studied by Yamaguchi et al. (2009). High concentration of sodium chloride retarded the reaction rate of glucose with amino acids but did not change the browning rate of glucose with peptides.

The type of reducing sugar has a great influence on Maillard reaction development. Pentoses (e.g., ribose) react more readily than hexoses (e.g., glucose), which, in turn, are more reactive than disaccharides (e.g., lactose; Ames 1990). A study on brown development (absorbance at 420 nm) in a heated model of fructose and lysine showed that browning was higher than in model systems with glucose (Ajandouz et al. 2001). A study using heated galactose/glycine model systems found that the rate of color development in Maillard reaction followed first-order kinetics on galactose concentration (Liu et al. 2008). The effects of tagatose on the Maillard reaction has been also investigated in aqueous model systems containing various sugars (glucose, galactose, fructose, and tagatose) and amino acids through volatile Maillard products determination (Cho et al. 2010).

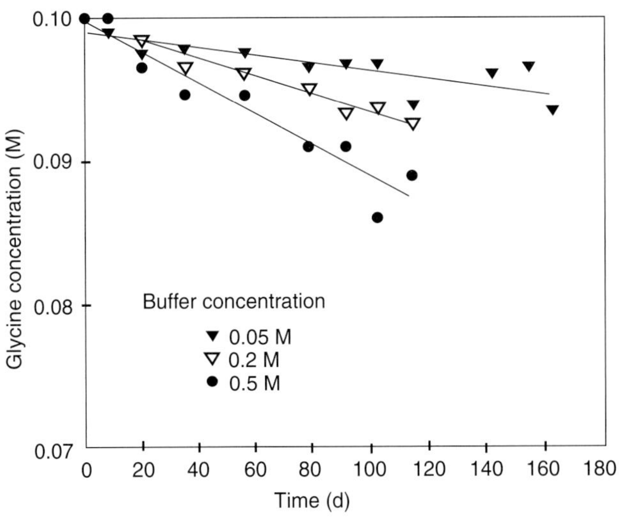

Figure 4.6. Effect of phosphate buffer concentration on the loss of glycine in 0.1 M glucose/glycine solutions at pH 7 and 25°C. (Bell 1997).

Participation of amino acids in the Maillard reaction is variable; thus lysine was the most reactive amino acid (Figure 4.7) in the heated model system of glucose and lysine, threonine, and methionine in phosphate buffer at different pH values (4–12) (Ajandouz and Puigserver 1999). The influence of type of amino acid and sugar in the Maillard reaction development has also been studied (Carabasa-Giribert and Ibarz-Ribas 2000, Mundt and Wedzicha 2003). The reaction between free amino acids and carbonyl compounds has been studied extensively; however, only a minor part of the Maillard reaction studies focused on peptides and proteins. Van Lancker et al. (2010) have studied the formation of flavor compounds from model systems of lysine-containing peptides and glucose, methylglyoxal, and glyoxal. The main compounds found were pyrazines, which contribute significantly to the roasted aroma of many heated food products.

Studies on the effect of time and temperature of treatment on the Maillard reaction development have also been conducted in different model systems, and it has been shown that an increase in temperature increases the rate of Maillard browning (Martins and van Boekel 2003, Ryu et al. 2003).

Concentration and ratio of reducing sugar to amino acid have a significant impact on the reaction. Browning reaction increased with increasing glycine to glucose ratios in the range 0.1:1–5:1 in a model orange juice system at 65°C (Wolfrom et al. 1974). In a model system of intermediate moisture ($a_w = 0.52$), Warmbier et al. (1976) observed an increase of browning reaction rate when the molar ratio of glucose:lysine increased from 0.5:1 to 3.0:1.

a_w is another important factor influencing the Maillard reaction development; thus, this reaction occurs less readily in foods with high a_w value due to the reactants are diluted. However, at low a_w values, the mobility of reactants is limited despite their presence at increased concentrations (Ames 1990). Numerous studies have demonstrated a browning rate maximum at a_w value from 0.5 to 0.8 in dried and intermediate-moisture foods (Warmbier et al. 1976, Tsai et al. 1991, Buera and Karel

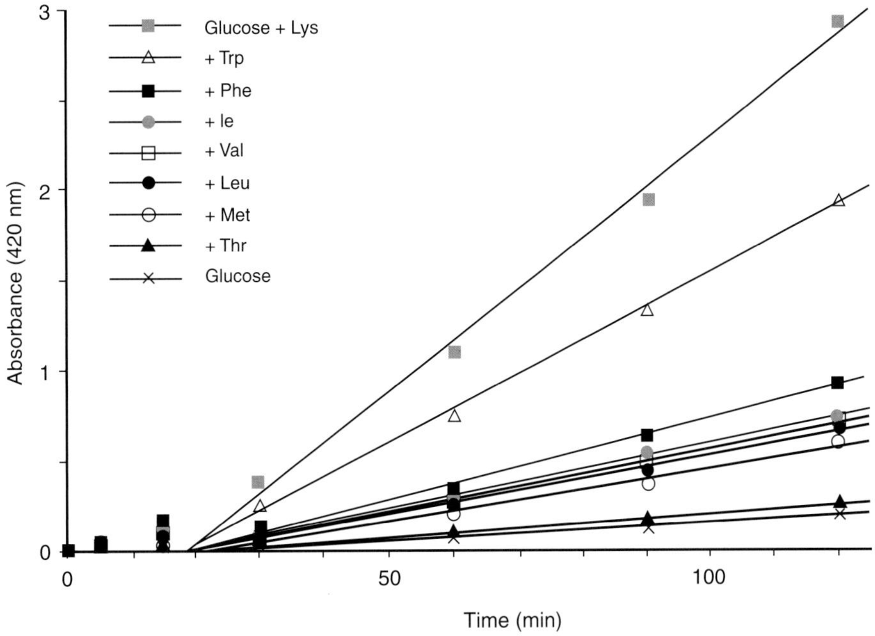

Figure 4.7. Brown color development in aqueous solutions containing glucose alone or in the presence of an essential amino acid when heated to 100°C at pH 7.5 as a function of time (Ajandouz and Puigserver 1999).

1995). The effect of a_w (0.33, 0.43, 0.52, 0.69, 0.85, and 0.98) on the rate of loss of lysine by mild heat treatment or during storage of milk powder was only significant at high a_w values (Pereyra-Gonzales et al. 2010).

Because of the complex composition of foods, it is unlikely that Maillard reaction involves only single compound (mono- or disaccharides and amino acids). For this reason, several studies on factors (pH, T, a_w) that influence the Maillard reaction development have been carried out using more complex model systems: heated starch–glucose–lysine systems (Bates et al. 1998), milk resembling model systems (lactose or glucose–caseinate systems) (Morales and van Boekel 1998), and lactose–casein model system (Malec et al. 2002, Jiménez-Castaño et al. 2005). Brands and van Boekel (2001) studied the Maillard reaction using heated monosaccharide (glucose, galactose, fructose, and tagatose)–casein model systems in order to quantify and identify the main reaction products and to establish the reaction pathways.

Studies on mechanisms of degradation, via the Maillard reaction, of oligosaccharides in a model system with glycine were performed by Hollnagel and Kroh (2000, 2002). The reactivity of di- and trisaccharides under quasi water-free conditions decreased in comparison to that of glucose due to the increasing degree of polymerization.

Study of Maillard Reaction in Foods

During food processing, the Maillard reaction produces desirable and undesirable effects. Processing such as baking, frying, and roasting are based on the Maillard reaction for flavor, aroma, and color formation (Lingnert 1990). Maillard browning may be desirable during manufacture of meat, coffee, tea, chocolate, nuts, potato chips, crackers, and beer and in toasting and baking bread (Weenen 1998, Burdulu and Karadeniz 2003). In other processes such as pasteurization, sterilization, drying, and storage, the Maillard reaction often causes detrimental nutritional (lysine damage) and organoleptic changes (Lingnert 1990). Available lysine determination methods has been used to assess the Maillard reaction extension in different types of foods: breads, breakfast cereals, pasta, infant formula (Erbersdobler and Hupe 1991), dried milks (El and Kavas 1997), heated milks (Ferrer et al. 2003), and infant cereals (Ramírez-Jimenez et al. 2004).

Sensory changes in foods due to the Maillard reaction have been studied in a wide range of foods, including honey (Gonzales et al. 1999), apple juice concentrate (Burdulu and Karadeniz 2003), and white chocolate (Vercet 2003).

Other types of undesirable effects produced in processed foods by the Maillard reaction may include the formation of mutagenic and cancerogenic compounds (Lingnert 1990, Chevalier et al. 2001). Frying or grilling of meat and fish may generate low (ppb) levels of mutagenic/carcinogenic heterocyclic amines via the Maillard reaction. The formation of these compounds depends on cooking temperature and time, cooking technique and equipment, heat, mass transport, and/or chemical parameters. Tareke et al. (2002) reported their findings on the carcinogen acrylamide in a range of cooked foods. Moderate levels of acrylamide (5–50 μg/kg) were measured in heated protein-rich foods, and higher levels (150–4000 μg/kg) were measured in carbohydrate-rich food, such as potato, beet root, certain heated commercial potato products, and crisp bread. Ahn et al. (2002) tested different types of commercial foods and some foods heated under home cooking conditions, and they observed that acrylamide was absent in raw or boiled foods, but it was present at significant levels in fried, grilled, baked, and toasted foods. Although the mechanism of acrylamide formation in heated foods is not yet clear, several authors have put forth the hypothesis that the reaction of asparagine (Figure 4.5), a major amino acid of potatoes and cereals, with reducing sugars (glucose, fructose) via the Maillard reaction, at temperatures above 120°C, could be the pathway (Mottram et al. 2002, Weiβhaar and Gutsche 2002, Friedman 2003, Yaylayan and Stadler 2005). Other amino acids have also been found to produce low amounts of acrylamide, including alanine, arginine, aspartic acid, cysteine, glutamine, threonine, valine, and methionine (Stadler et al. 2002, Tateo et al. 2007). Over the years, much work have been done to study the factor influencing the acrylamide formation during processing (Granda and Moreira 2005, Gökmen and Senyuva 2006, Tateo et al. 2007, Burch et al. 2008, Jom et al. 2008, Carrieri et al. 2010), the mechanism of acrylamide formation (Elmore et al. 2005, Hamlet et al. 2008, Zamora et al. 2010), the development of robust and sensitive analytical methods that provide reliable data in the different food categories (Zhang et al. 2005, Kaplan et al. 2009, Preston et al. 2009), and the content in different processed foods (Bermudo et al. 2006, Tateo et al. 2007, Burch et al. 2008, EFSA 2010). On the basis of the large number of existing studies, the International Agency for Research on Cancer (IARC 1994) has classified acrylamide as "probably carcinogenic" to humans. Currently, researchers are looking around for other strategies to obtain finished foods without acrylamide. In this sense, Anese et al. (2010) have proposed the possibility to remove acrylamide from foods by exploiting its chemical and physical properties. Thus, processed foods were subjected to vacuum treatments at different combinations of pressure, temperature, and time. Removal of acrylamide was achieved only in samples previously hydrated at a_w values higher than 0.83, and maximum removed amount was between 5 and 15 minutes of vacuum treatment at 6.67 Pa and 60°C.

Beneficial properties of Maillard products have also been described. Resultant products of the reaction of different amino acid and sugar model systems presented different properties: antimutagenic (Yen and Tsai 1993); antimicrobial (Chevalier et al. 2001, Rufián-Henares and Morales 2007) and antioxidative (Manzocco et al. 2001, Wagner et al. 2002, Rufián-Henares and Morales 2007). In foods, antioxidant properties of MRPs have been found in honey (Antony et al. 2000) and in tomato purees (Anese et al. 2002). Rufián-Henares and de la Cueva (2009) found antimicrobial activity of coffee melanoidins against different pathogenic bacteria.

On the other hand, the Maillard reaction is one of the safest and most efficient methods to generate new functional proteins with great potential as novel ingredients. Miralles et al. (2007) investigated the occurrence of the Maillard reaction between β-lactoglobulin and a LMW chitosan. Under the studied conditions, MRPs originated improved antibacterial activity against

Escherichia coli and emulsifying properties. Recently, Corzo-Martínez et al. (2010) observed an improvement in gelling and viscosity properties of glycosylated sodium caseinate via Maillard reaction with lactose and galactose.

Control of the Maillard Reaction in Foods

For a food technologist, one of the most important objectives must be to limit nutritional damage of food during processing. In this sense, many studies have been performed and found useful heat-induced markers derived from the Maillard reaction, and most of them have been proposed to control and check the heat treatments and/or storage of foods. There are many indicators of different stages of the Maillard reaction, but this review cites one of the most recent indicators proposed to control early stages of this reaction during food processing: the 2-furoylmethyl amino acids as an indirect measure of the Amadori compound formation.

Determination of the level of Amadori compounds provides a very sensitive indicator for early detection (before detrimental changes occur) of quality changes caused by the Maillard reaction as well as for retrospective assessment of the heat treatment or storage conditions to which a product has been subjected (del Castillo et al. 1999, Olano and Martínez-Castro 2004).

Evaluating for Amadori compounds can be carried out through furosine [ε-*N*-(2-furoylmethyl)-L-lysine] measurement. This amino acid is formed by acid hydrolysis of the Amadori compound ε-*N*-(1-deoxy-D-fructosyl)-L-lysine. It is considered a useful indicator of the damage in processed foods or foods stored for long periods: milks (Resmini et al. 1990, Villamiel et al. 1999); eggs (Hidalgo et al. 1995); cheese (Villamiel et al. 2000); honey (Villamiel et al. 2001, Morales et al. 2009); infant formula (Guerra-Hernandez et al. 2002, Cattaneo et al. 2009); milk-cereal-based baby foods (Bosch et al. 2008); jams and fruit-based infant foods (Rada-Mendoza et al. 2002); fresh filled pasta (Zardetto et al. 2003); prebaked breads (Ruiz et al. 2004), cookies, crackers, and breakfast cereals (Rada-Mendoza et al. 2004, Gökmen et al. 2008); fiber-enriched breakfast cereals (Delgado-Andrade et al. 2007); flour used for formulations of cereal-based-products (Rufián-Henares et al. 2009); and sauces and sauces-treated foods (Chao et al. 2009). Erbersdobler and Somoza (2007) have written an interesting review on 40 years of use of furosine as a reliable indicator of thermal damage in foods.

In the case of foods containing free amino acids, free Amadori compounds can be present, and acid hydrolysis gives rise to the formation of the corresponding 2-furoylmethyl derivatives. For the first time, 2-furoylmethyl derivatives of different amino acids (arginine, asparagine, proline, alanine, glutamic acid, and ϒ-amino butyric acid) have been detected and have been used as indicators of the early stages of the Maillard reaction in stored dehydrated orange juices (del Castillo et al. 1999). These compounds were proposed as indicators to evaluate quality changes either during processing or during subsequent storage. Later, most of these compounds were also detected in different foods: commercial orange juices (del Castillo et al. 2000), processed tomato products (Sanz et al. 2000), dehydrated fruits (Sanz et al. 2001) and carrots (Soria et al. 2010), commercial honey samples (Sanz et al. 2003), and infant formula (Penndorf et al. 2007).

Caramelization

During nonenzymatic browning of foods, various degradation products are formed via caramelization of carbohydrates, without amine participation (Ajandouz and Puigserver 1999, Ajandouz et al. 2001). Caramelization occurs when surfaces are heated strongly (e.g., during baking and roasting), during the processing of foods with high sugar content (e.g., jams and certain fruit juices), or in wine production (Kroh 1994). Caramelization is used to obtain caramel-like flavor and/or development of brown color in certain types of foods. Caramel flavoring and coloring, produced from sugar with different catalysts, is one of the most widely used additives in the food industry. However, caramelization is not always a desirable reaction because of the possible formation of mutagenic compounds (Tomasik et al. 1989) and the excessive changes in the sensory attributes that could affect the quality of certain foods.

More recently, Kitts et al. (2006) also observed the clastogenic activity of caramelized sucrose, this property being mainly due to the volatile and nonvolatile compounds of LMW derived from sucrose caramelization.

Caramelization is catalyzed under acidic or alkaline conditions (Namiki 1988) and many of the products formed are similar to those resulting from the Maillard reaction.

Caramelization of carbohydrates starts with the opening of the hemiacetal ring followed by enolization, which proceeds via acid- and base-catalyzed mechanisms, leading to the formation of isomeric carbohydrates (Figure 4.8). The interconversion of sugars through their enediols increases with increasing pH and is called the Lobry de Bruyn-Alberda van Ekenstein transformation (Kroh 1994). In acid media, low amounts of isomeric carbohydrates are formed; however, dehydration is favored, leading to furaldehyde compounds: 5-(hydroxymethyl)-2-furaldehyde (HMF) from hexoses (Figure 4.9) and 2-furaldehyde from pentoses. With unbuffered acids as catalysts, higher yields of HMF are produced from fructose than from glucose. Also, only the fructose moiety of sucrose is largely converted to HMF under unbuffered conditions, which produce the highest yields. The enolization of glucose can be greatly increased in buffered acidic solutions. Thus, higher yields of HMF are produced from

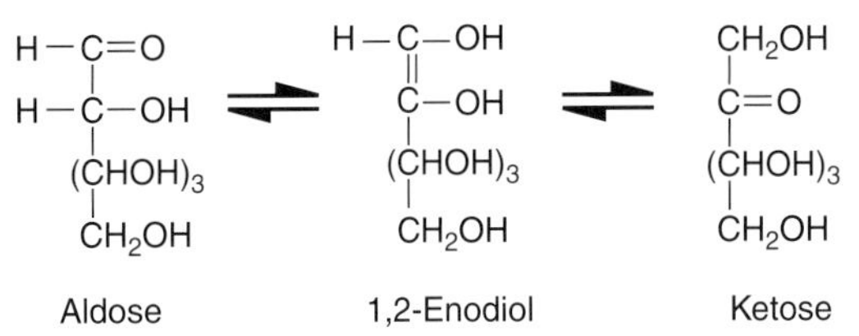

Figure 4.8. The Lobry de Bruyn-Alberda van Ekenstein transformation.

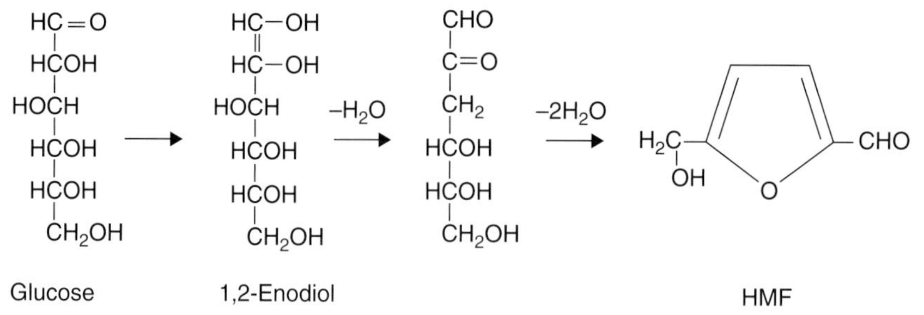

Figure 4.9. 1,2 Enolization and formation of hydroxymethyl furfural (HMF).

glucose and sucrose when a combination of phosphoric acid and pyridine is used as catalysts than when phosphoric acid is used alone (Fennema 1976).

In alkaline media, dehydration reactions are slower than in neutral or acid media, but fragmentation products such as acetol, acetoin, and diacetyl are detected. In the presence of oxygen, oxidative fission takes place, and formic, acetic, and other organic acids are also formed. All of these compounds react to produce brown polymers and flavor compounds (Olano and Martínez-Castro 2004).

In general, caramelization products (CPs) consist of volatile and nonvolatile (90–95%) fractions of LMWs and HMWs that vary depending on temperature, pH, duration of heating, and starting material (Defaye et al. 2000). Although it is known that caramelization is favored at temperatures higher than 120°C and at a pH greater than 9 and less than 3, depending on the composition of the system (pH and type of sugar), caramelization reactions may also play an important role in color formation in systems heated at lower temperatures. Thus, some studies have been conducted at the temperatures of accelerated storage conditions (45–65°C) and pH values from 4 to 6 (Buera et al. 1987a, 1987b). These authors studied the changes of color due to caramelization of fructose, xylose, glucose, maltose, lactose, and sucrose in model systems of 0.9 a_w and found that fructose and xylose browned much more rapidly than other sugars and lowering of pH inhibited caramelization browning of sugar solutions.

In a study on the kinetics of caramelization of several monosaccharides and disaccharides, Diaz and Clotet (1995) found that at temperatures of 75–95°C, browning increased rapidly with time and to a higher final value, with increasing temperature, this effect being more marked in the monosaccharides than in the disaccharides. In all sugars studied, increase of browning was greater at $a_w = 1$ than at $a_w = 0.75$. Likewise, results of a more recent study on the kinetics of browning development showed that at temperatures of 75–95°C and pH 3.0, color development increased linearly with a first-order kinetics on fructose concentration (Chen et al. 2009).

The effect of sugars, temperature, and pH on caramelization was evaluated by Park et al. (1998). Reaction rate was highest with fructose, followed by sucrose. As reaction temperature increased from 80°C to 110°C, reaction rate was greatly increased. With respect to pH, the optimum value for caramelization was 10. In agreement with this, more recently, Kim and Lee (2008b) reported that degradation/enolization of sugars rapidly increased at a high alkaline pH (10.0–12.0) and with increasing heating times, fructose being degraded/enolized more rapidly than glucose.

Although most studies on caramelization have been conducted in model systems of mono- and disaccharides, a number of real food systems contain oligosaccharides or even polymeric saccharides; therefore, it is also of great interest to know the contribution of these carbohydrates to the flavor and color of foods. Kroh et al. (1996) reported the breakdown of oligo- and polysaccharides to nonvolatile reaction products. Homoki-Farkas et al. (1997) studied, through an intermediate compound (methylglyoxal), the caramelization of glucose, dextrin 15, and starch in aqueous solutions at 170°C under different periods of time. The highest formation of methylglyoxal was in glucose and the lowest in starch systems. The authors attributed the differences to the number of reducing end groups. In the case of glucose, when all molecules are degraded, the concentration of methylglyoxal reached a maximum and began to transform, yielding LMW and HMW color compounds. Hollnagel and Kroh (2000, 2002) investigated the degradation of malto-oligosaccharides at 100°C through α-dicarbonyl compounds such as 1,4-dideoxyhexosulose, and they found that this compound is a reactive intermediate and precursor of various heterocyclic volatile compounds that contribute to caramel flavor and color. More recently, it has been observed that galactomannans of roasted coffee infusions are HMW supports of LMW brown compounds derived from caramelization reactions (Nunes et al. 2006).

Perhaps, as mentioned above, the most striking feature of caramelization is its contribution to the color and flavor of certain food products under controlled conditions. In addition, it is necessary to consider other positive characteristics of this reaction, such as the antioxidant activity of the CPs. This property has been found to vary depending on temperature, pH, duration of heating, and starting material, main variables affecting the caramelization kinetics. Thus, several studies (Benjakul et al. 2005, Phongkanpai et al. 2006, Kim and Lee 2008b) have reported that antioxidative activity of CPs from different sugars (fructose, glucose, ribose, and xylose) increased with increasing pH levels (7.0–12.0), heating temperature (80–180°C), and heating time (0–180 minutes). The CPs from fructose exhibited the highest antioxidant activity as evidenced by the greatest scavenging effect and reducing power. In addition, this increase in antioxidative activity was coincidental with the browning development and intermediate formation. This suggested that both colorless intermediates, such as reductones and dehydroreductones, produced in the earlier stages of the caramelization, (Rhee and Kim 1975) and HMW and colored pigments, produced in the advanced stages, might play an important role in the antioxidant activity of CPs (Kirigaya et al. 1968)

In addition, the effect of caramelized sugars on enzymatic browning has been studied by several authors. Thus, Pitotti et al. (1995) reported that the anti-browning effect of some CPs is in part related to their reducing power. Lee and Lee (1997) obtained CPs by heating a sucrose solution at 200°C under various conditions to study the inhibitory activity of these products on enzymatic browning. The reducing power of CPs and their inhibitory effect on enzymatic browning increased with prolonged heating and with increased amounts of CPs. Caramelization was investigated in solutions of fructose, glucose, and sucrose heated at temperatures up to 200°C for 15–180 minutes. Browning intensity increased with heating time and temperature. The effect of the caramelized products on PPO was evaluated, and the greatest PPO inhibitory effect was demonstrated by sucrose solution heated to 200°C for 60 minutes (Lee and Han 2001). More recently, Billaud et al. (2003) found CPS from hexoses with mild inhibitory effects on PPO, particularly after prolonged heating at 90°C.

Antioxidative activity of CPs from higher molecular weight carbohydrates has also been reported. Thus, Mesa et al. (2008), in a study on antioxidant properties of hydrolyzates of soy protein-fructooligosaccharide glycation systems, found that the most neoantioxidants products are able to prevent LDL oxidation and to scavenge peroxyl-alkyl radicals derived from the thermal degradation (95°C for 1 hour) of fructooligosaccharides rather than from the Maillard reaction.

Ascorbic Acid Browning

Ascorbic acid (vitamin C) plays an important role in human nutrition as well as in food processing (Chauhan et al. 1998). Its key effect as an inhibitor of enzymatic browning has been previously discussed in this chapter.

Browning of ascorbic acid can be briefly defined as the thermal decomposition of ascorbic acid under both aerobic and anaerobic conditions, by oxidative or non-oxidative mechanisms, either in the presence or absence of amino compounds (Wedzicha 1984).

Nonenzymatic browning is one of the main reasons for the loss of commercial value in citrus products (Manso et al. 2001). These damages, degradation of ascorbic acid followed by browning, also concern noncitrus foods such as asparagus, broccoli, cauliflower, peas, potatoes, spinach, apples, green beans, apricots, melons, strawberries, corn, and dehydrated fruits (Belitz and Grosch 1997).

Pathway of Ascorbic Acid Browning

The exact route of ascorbic acid degradation is highly variable and dependent upon the particular system. Factors that can influence the nature of the degradation mechanism include temperature, salt and sugar concentration, pH, oxygen, enzymes, metal catalysts, amino acids, oxidants or reductants, initial concentration of ascorbic acid, and the ratio of ascorbic acid to dehydroascorbic acid (DHAA; Fennema 1976).

Figure 4.10 shows a simplified scheme of ascorbic acid degradation. When oxygen is present in the system, ascorbic acid is degraded primarily to DHAA. DHAA is not stable and spontaneously converts to 2,3-diketo-L-gulonic acid (Lee and Nagy 1996). Under anaerobic conditions, ascorbic acid undergoes the generation of diketogulonic acid via its keto tautomer, followed by β elimination at C-4 from this compound and decarboxylation to give rise to 3-deoxypentosone, which is further degraded to furfural. Under aerobic conditions, xylosone is produced by simple decarboxylation of diketogulonic acid and that is later converted to reductones. In the presence of amino acids, ascorbic acid, DHAA, and their oxidation products furfural, reductones, and 3-deoxypentosone may contribute to the browning of foods by means of a Maillard-type reaction (Fennema 1976, Belitz and Grosch 1997). Formation of Maillard-type products has been detected both in model systems and foods containing ascorbic acid (Kacem et al. 1987, Ziderman et al. 1989, Loschner et al. 1990, 1991, Mölnar-Perl and Friedman 1990, Yin and Brunk 1991, Davies and Wedzicha 1992, 1994, Pischetsrieder et al. 1995, 1997, Rogacheva et al. 1995, Koseki et al. 2001).

The presence of metals, especially Cu^{2+} and Fe^{3+}, causes great losses of vitamin C. Catalyzed oxidation goes faster than the spontaneous oxidation. Anaerobic degradation, which occurs slower than uncatalyzed oxidation, is maximum at pH 4 and minimum at pH 2 (Belitz and Grosch 1997).

Ascorbic acid oxidation is nonenzymatic in nature, but oxidation of ascorbic acid is sometimes catalyzed by enzymes. Ascorbic acid oxidase is a copper-containing enzyme that catalyzes oxidation of vitamin C. The reaction is catalyzed by copper ions. The enzymatic oxidation of ascorbic acid is important in the citrus industry. Reaction takes place mainly during extraction of juices. Therefore, it becomes important to inhibit the ascorbic oxidase by holding juices for only short times and at low temperatures during the blending stage, by de-aerating the juice to remove oxygen, and finally by pasteurizing the juice to inactivate the oxidizing enzymes.

Enzymatic oxidation has also been proposed as a mechanism for the destruction of ascorbic acid in orange peels during the preparation of marmalade. Boiling the grated peel in water substantially reduces the loss of ascorbic acid (Fennema 1976).

Tyrosinase (PPO) may also possess ascorbic oxidase activity. A possible role of the ascorbic acid-PPO system in the browning of pears has been proposed (Espin et al. 2000).

In citrus juices, nonenzymatic browning is from reactions of sugars, amino acids, and ascorbic acid (Manso et al. 2001). In freshly produced commercial juice, filled into TetraBrik cartons, it has been demonstrated that nonenzymatic browning was mainly due to carbonyl compounds formed from L-ascorbic acid

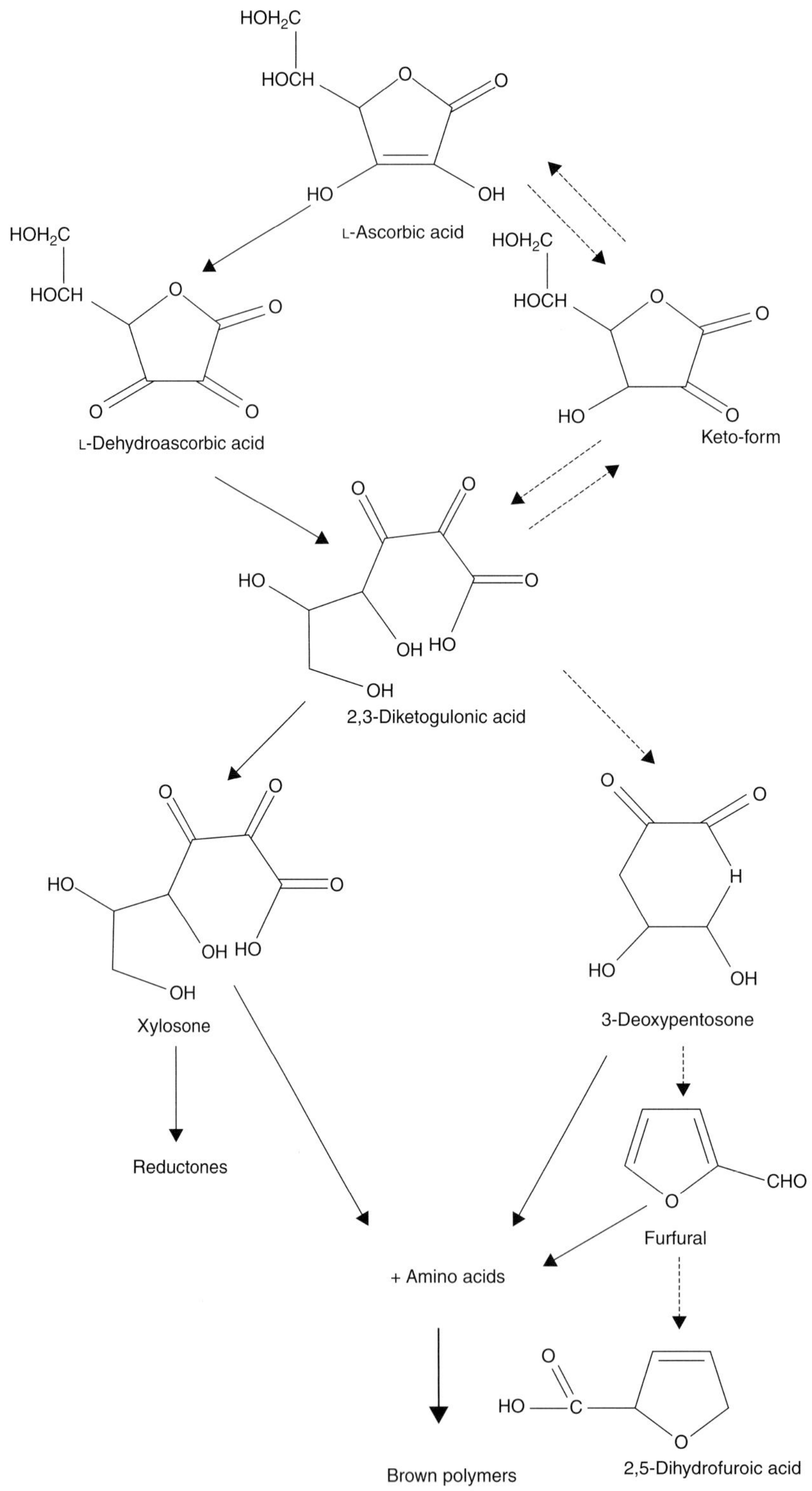

Figure 4.10. Pathways of ascorbic acid degradation (solid line, anaerobic route; dashed line, aerobic route).

degradation. Contribution from sugar-amine reactions is negligible, as evident from the constant total sugar content of degraded samples. The presence of amino acids and possibly other amino compounds enhance browning (Roig et al. 1999).

Both oxidative and nonoxidative degradation pathways are operative during storage of citrus juices. Since large quantities of DHAA are present in citrus juices, it can be speculated that the oxidative pathway must be dominant (Lee and Nagy 1996, Rojas and Gerschenson 1997a). A significant relationship between DHAA and browning of citrus juice has been found (Kurata et al. 1973, Sawamura et al. 1991, Sawamura et al. 1994). The rate of non-oxidative loss of ascorbic acid is often one-tenth or up to one-thousandth the rate of loss under aerobic conditions (Lee and Nagy 1996). In aseptically packed orange juice, the aerobic reaction dominates first and it is fairly rapid, while the anaerobic reaction dominates later and it is quite slow (Nagy and Smoot 1977, Tannenbaum 1976). A good prediction of ascorbic acid degradation and the evolution of the browning index of orange juice stored under anaerobic conditions at 20–45°C may be performed by employing the Weibull model (Manso et al. 2001).

Furfural, which is formed during anaerobic degradation of ascorbic acid, has a significant relationship to browning (Lee and Nagy 1988); its formation has been suggested as an adequate index for predicting storage temperature abuse in orange juice concentrates and as a quality deterioration indicator in single-strength juice (Lee and Nagy 1996). However, furfural is a very reactive aldehyde that forms and decomposes simultaneously; therefore, it would be more difficult to use as an index for predicting quality changes in citrus products (Fennema 1976). In general, ascorbic acid would be a better early indicator of quality.

Control of Ascorbic Acid Browning

Sulfites (Wedzicha and Mcweeny 1974, Wedzicha and Imeson 1977), thiols compounds (Naim et al. 1997), maltilol (Koseki et al. 2001), sugars, and sorbitol (Rojas and Gerschenson 1997b) may be effective in suppressing ascorbic acid browning. Doses to apply these compounds highly depend on factors such as concentration of inhibitors and temperatures. L-Cysteine and sodium sulfite may suppress or accelerate ascorbic acid browning as a function of their concentration (Sawamura et al. 2000). Glucose, sucrose, and sorbitol protect L-ascorbic acid from destruction at low temperatures (23°C, 33°C, and 45°C), while at higher temperatures (70°C, 80°C, and 90°C), compounds with active carbonyls promoted ascorbic acid destruction. Sodium bisulfite was only significant in producing inhibition at lower temperature ranges (23°C, 33°C, and 45°C; Rojas and Gerschenson 1997b).

Although the stability of ascorbic acid generally increases as the temperature of the food is lowered, certain investigations have indicated that there may be an accelerated loss on freezing or frozen storage. However, in general, the largest losses for noncitrus foods will occur during heating (Fennema 1976).

The rapid removal of oxygen from the packages is an important factor in sustaining a higher concentration of ascorbic acid and lower browning of citrus juices over long-term storage. The extent of browning may be reduced by packing in oxygen scavenging film (Zerdin et al. 2003).

Modified-atmosphere packages (Howard and Hernandez-Brenes 1998), microwave heating (Villamiel et al. 1998, Howard et al. 1999), ultrasound-assisted thermal processing (Zenker et al. 2003), pulsed electric field processing (Min et al. 2003), and carbon dioxide-assisted high-pressure processing (Boff et al. 2003) are some examples of technological processes that allow ascorbic acid retention and consequently prevent undesirable browning.

Lipid Browning

Protein-Oxidized Fatty Acid Reactions

The organoleptic and nutritional characteristics of several foods are affected by lipids, which can participate in chemical modifications during processing and storage. Lipid oxidation occurs in oils and lard, and also in foods with low amounts of lipids, such as products of vegetable origin. This reaction occurs in both unprocessed and processed foods, and although in some cases it is desirable, such as in the production of typical cheeses or fried-food aromas (Nawar 1985), in general, it can lead to undesirable odors and flavors (Nawar 1996). Quality properties such as appearance, texture, consistency, taste, flavor, and aroma can be adversely affected (Erikson 1987). Moreover, toxic compound formation and loss of nutritional quality can also be observed (Frankel 1980, Gardner 1989, Kubow 1990, 1992).

Although the lipids can be oxidized by both enzymatic and nonenzymatic reactions, the latter is the main involved reaction. This reaction arises from free radical or reactive oxygen species (ROS) generated during food processing and storage (Stadtman and Levine 2003), hydroperoxides being the initial products. As these compounds are quite unstable, a network of dendritic reactions, with different reaction pathways and a diversity of products, can take place (Gardner 1989). The enzymatic oxidation of lipids occurs sequentially. Lipolytic enzymes can act on lipids to produce polyunsaturated fatty acids that are then oxidized by either lipoxygenase or cyclooxygenase to form hydroperoxides or endoperoxides, respectively. Later, these compounds suffer a series of reactions to produce, among other compounds, long-chain fatty acids responsible for important characteristics and functions (Gardner 1995).

Via polymerization, brown-colored oxypolymers can be produced subsequently from the lipid oxidation derivatives (Khayat and Schwall 1983). However, interaction with nucleophiles such as the free amino group of amino acids, peptides, or proteins can also take place, because of the electrophilic character of free radicals produced during lipid oxidation, including hydroperoxides, peroxyl and alkoxyl radicals, carbonyl compounds and epoxides. As a result of this, end products different from those formed during oxidation of pure lipids can be also produced (Gillat and Rossell 1992, Schaich 2008). Both lipid oxidation and oxidized lipid/amino acid reactions occur simultaneously (Hidalgo and Zamora 2002).

Radical transfer occurs early in lipid oxidation, and this process underlies the antioxidant effect for lipids. In addition, protein radicals can also transfer radicals to other proteins, lipids, carbohydrates, vitamins, and other molecules, especially in the presence of metal ions such as iron and copper (Schaich 2008). Reactions between proteins and free radicals and ROS suggest that proteins could protect lipids from oxidation if they are oxidized preferentially to unsaturated fatty acids (Elias et al. 2008). A study of continuous phase β-lactoglobulin in oil-in-water emulsion showed that tryptophan and cysteine side chains, but not methionine, oxidized before lipids (Elias et al. 2005).

The interaction between oxidized fatty acids and amino groups has been related to the browning detected during the progressive accumulation of lipofuscin (age-related yellow-brown pigments) in lysosomes of men and animals (Yin 1996). In foods, evidence of this reaction has been found during storage and processing of some fatty foods (Hidalgo et al. 1992, Nawar 1996), fermented alcoholic and nonalcoholic beverages (Herraiz 1996), in cocoa, chocolate (Herraiz 2000), salted sun-dried fish (Smith and Hole 1991), boiled and dried anchovy (Takiguchi 1992), cuttlefish (*Sepia pharaonis*) (Thanonkaew et al. 2007), meat and meat products (Mottram 1998, Herraiz and Papavergou 2004), smoked foodstuffs such as sausages, cheeses, and fish (Zotos et al. 2001, Herraiz et al. 2003, Papavergou and Herraiz 2003), and in rancid oils and fats with amino acids or proteins (Yamamoto and Kogure 1969, Okumura and Kawai 1970, Gillat and Rossell 1992, Guillen et al. 2005). For instance, interaction between different carbonyl compounds, mainly aldehydes, derived from lipid oxidation and lysine, tryptophan, methionine, and cysteine side chains of whey proteins has been shown to occur in dairy products such as raw and different heat-treated milks (pasteurized, UHT, and sterilized), as well as in infant formula (Nielsen et al. 1985, Meltretter et al. 2007, 2008, Meltretter and Pischetsrieder 2008).

Several studies have been carried out in model systems with the aim to investigate the role of lipids in nonenzymatic browning. The role of lipids in these reactions seems to be similar to that of the role of carbohydrates during the Maillard reaction (Hidalgo and Zamora 2000). Similar to the Maillard reaction, oxidized lipid/protein interactions comprise a huge number of several related reactions. The isolation and characterization of the involved products is very difficult, mainly in the case of intermediate products, which are unstable and are present in very low concentrations.

According to the mechanism proposed for the protein browning caused by acetaldehyde, the carbonyl compounds derived from unsaturated lipids readily react with protein-free amino groups, following the scheme of Figure 4.11, to produce, by repeated aldol condensations, the formation of brown pigments (Montgomery and Day 1965, Gardner 1979, Belitz and Grosch 1997).

More recently, another mechanism based on the polymerization of the intermediate products 2-(1-hydroxyalkyl) pyrroles has been proposed (Zamora and Hidalgo 1994, 1995). These authors, studying different model systems, tried to explain, at least partially, the nonenzymatic browning and fluorescence produced when proteins are present during the oxidation of lipids (Figure 4.12). 2-(1-Hydroxyalkyl) pyrroles (I) have been found to be originated from the reaction of 4,5-epoxy-2-alkenals (formed during lipid peroxidation) with the amino group of amino acids and/or proteins, and their formation is always accompanied by the production of N-substituted pyrroles (II). Compounds derived from reaction of 4,5-epoxy-2-alkenals and phenylalanine have been found to be flavor compounds analogous to those of the Maillard reaction. Therefore, flavors traditionally connected to the Maillard reaction may also be produced as a result of lipid oxidation (Hidalgo and Zamora 2004, Zamora et al. 2006). N-substituted pyrroles are relatively stable and have been found in 22 fresh food products (cod, cuttlefish, salmon, sardine, trout, beef, chicken, pork, broad bean, broccoli, chickpea, garlic, green pea, lentil, mushroom, soybean, spinach, sunflower, almond, hazelnut, peanut, and walnut; Zamora et al. 1999). However, the N-substituted 2-(1-hydroxyalkyl) pyrroles are unstable and polymerize rapidly and spontaneously to produce brown macromolecules with fluorescent melanoidin-like characteristics (Hidalgo and Zamora 1993). Zamora et al. (2000) observed that the formation of pyrroles is a step immediately prior to the formation of color and fluorescence. Pyrrole formation and perhaps some polymerization finished before maximum color and fluorescence was achieved.

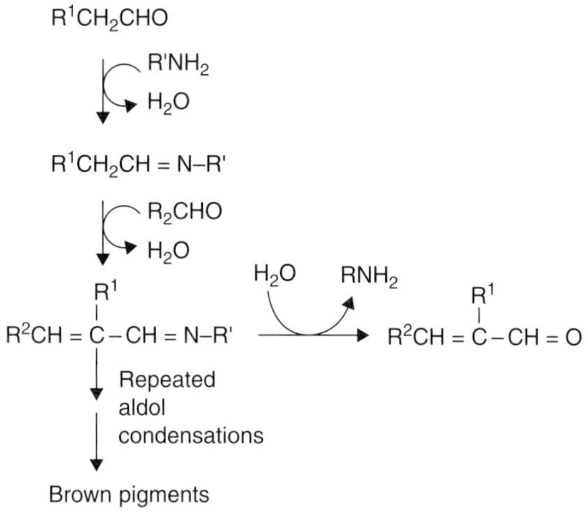

Figure 4.11. Formation of brown pigments by aldolic condensation (Hidalgo and Zamora 2000).

Although melanoidins starting from either carbohydrates or oxidized lipids would have analogous chemical structures, carbohydrate–protein and oxidized lipid–protein reactions are produced under different conditions. Hidalgo et al. (1999) studied the effect of pH and temperature in two model systems: (i) ribose and bovine serum albumin and (ii) methyl linoleate oxidation products and bovine serum albumin; they observed that from 25°C to 50°C, the latter exhibited higher browning than the former. Conversely, the browning produced in carbohydrate

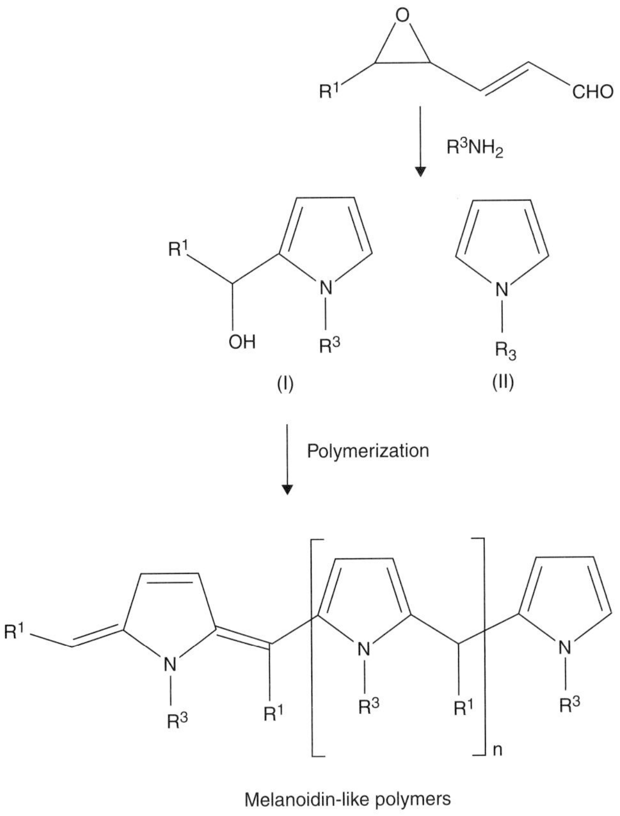

Figure 4.12. Mechanism for nonenzymatic browning produced as a consequence of 2-(1-hydroxyalkyl)pyrrole polymerization (Hidalgo et al. 2003).

model system was increased at temperatures in the range of 80–120°C. The effect of pH on browning was similar in both oxidized lipid–protein and carbohydrate–protein model systems.

Some of the oxidized lipid–amino acid reaction products have been shown to have antioxidant properties when they are added to vegetable oils (Zamora and Hidalgo 1993, Alaiz et al. 1995, Alaiz et al. 1996). All of the pyrrole derivatives, with different substituents in the pyrrole ring, play an important role in the antioxidant activity of foods, being the sum of the antioxidant activities of the different compounds present in the sample (Hidalgo et al. 2003). Moreover, others studies have suggested that tetrahydro-β-carbolines derived from oxidation of tryptophan might act as antioxidants when they are absorbed and accumulated in the body, contributing to the antioxidant effect of fruit products naturally containing these compounds (Herraiz and Galisteo 2003). Likewise, Hidalgo et al. (2006a), in a study carried out with amino phospholipids (phosphatidylethanolamine (PE) and phosphatidylcholine (PC)), lysine (Lys), and mixtures of them in edible oils, demonstrated the *in situ* formation of oxidized lipid–amino compound reaction products (PE–Lys and PC–Lys) with antioxidative activities. More recently, studies have shown that antioxidative activity of such carbonyl-amine products may be greatly increased with the addition of tocopherols and phytosterols such as β-sitosterol (Hidalgo et al. 2007, Hidalgo et al. 2009). Alaiz et al. (1997), in a study on the comparative antioxidant activity of both Maillard reaction compounds and oxidized lipid–amino acid reaction products, observed that both reactions seem to contribute analogously to increase the stability of foods during processing and storage.

Zamora and Hidalgo (2003a) studied the role of the type of fatty acid (methyl linoleate and methyl linolenate) and the protein (bovine serum albumin)–lipid ratio on the relative progression of the process involved when lipid oxidation occurs in the presence of proteins. These authors found that methyl linoleate was only slightly more reactive than the methyl linolenate for bovine serum albumin, producing an increase of protein pyrroles in the protein and an increase in the development of browning and fluorescence. In relation to the influence of the protein–lipid ratio on the advance of the reaction, the results observed in this study pointed out that a lower protein–lipid ratio increases sample oxidation and protein damage as a consequence of the antioxidant activity of the proteins. These authors also concluded that the changes produced in the color of protein–lipid samples were mainly due to oxidized lipid–protein reactions and not as consequence of polymerization of lipid oxidation products.

Analogous to the Maillard reaction, oxidized lipid and protein interaction can cause a loss of nutritional quality due to the destruction of essential amino acids, such as tryptophan, lysine, and methionine, and essential fatty acids. Moreover, a decrease in digestibility and inhibition of proteolytic and glycolytic enzymes can also be observed. In a model system of 4,5(*E*)-epoxy-2(*E*)-heptenal and bovine serum albumin, Zamora and Hidalgo (2001) observed denaturation and polymerization of the protein, and the proteolysis of this protein was impaired as compared with the intact protein. These authors suggested that the inhibition of proteolysis observed in oxidized lipid-damaged proteins may be related to the formation and accumulation of pyrrolized amino acid residues.

To date, although most of the studies have been conducted using model systems, the results obtained confirm that there is an interaction between lipid oxidation and the Maillard reaction. In fact, both reactions are so interrelated that they should be considered simultaneously to understand their consequences on foods when lipids, carbohydrates, and amino acids or proteins are present and should be included in one general pathway that can be initiated by both lipids and carbohydrates (Zamora et al. 2005a, 2005b). The complexity of the reaction is attributable to several fatty acids that can give rise to a number of lipid oxidation products that are able to interact with free amino groups. As summary, Figure 4.13 shows an example of a general pathway of pyrrole formation during polyunsaturated fatty acid oxidation in the presence of amino compounds.

Nonenzymatic Browning of Aminophospholipids

In addition to the above-mentioned studies on the participation of lipids in the browning reactions, several reports have been addressed on the amine-containing phospholipid interactions with carbohydrates. Because of the role of these membranous

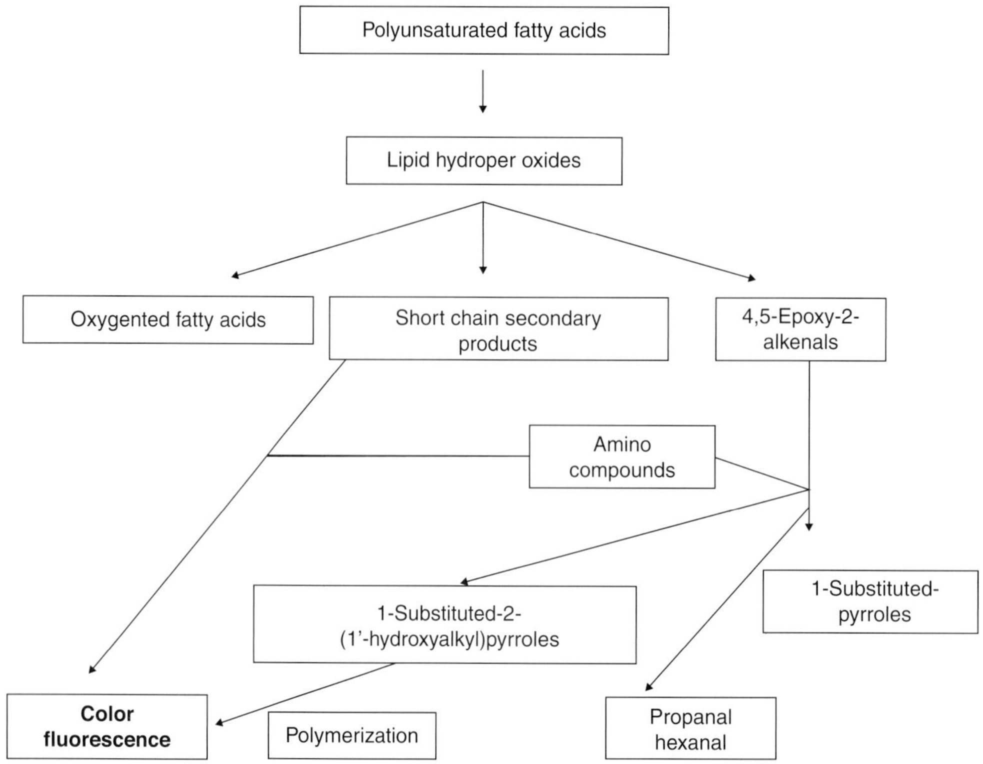

Figure 4.13. General pathways of pyrrole formation during polyunsaturated fatty acids oxidation (Zamora and Hidalgo 1995).

functional lipids in the maintenance of cellular integrity, most of the studies have been conducted in biological samples (Bucala et al. 1993, Requena et al. 1997, Lertsiri et al. 1998, Oak et al. 2000, Oak et al. 2002). The glycation of membrane lipids can cause inactivation of receptors and enzymes, cross-linking of membrane lipids and proteins, membrane lipid peroxidation, and consequently, cell death. Amadori compounds derived from the interaction between aminophospholipids and reducing carbohydrates are believed to be key compounds for generating oxidative stress, causing several diseases.

In foods, this reaction can be responsible for deterioration during processing. Although nonenzymatic browning of aminophospholipids was detected for the first time in dried egg by Lea (1957), defined structures from such reaction were reported later by Utzmann and Lederer (2000). These authors demonstrated the interaction of PE with glucose in model systems; moreover, they found the corresponding Amadori compound in spray-dried egg yolk powders and lecithin products derived therefrom and proposed the Amadori compound content as a possible quality criterion. Oak et al. (2002) detected the Amadori compounds derived from glucose and lactose in several processed foods with high amounts of carbohydrates and lipids, such as infant formula, chocolate, mayonnaise, milk, and soybean milk; of these, infant formula contained the highest levels of Amadori-PEs. However, these compounds were not detected in other foods due to differences in composition and relatively low temperatures used during the processing.

As an example, Figure 4.14 shows a scheme for the formation of the Amadori compounds derived from glucose and lactose with PE. Carbohydrates react with the amino group of PE to form an unstable Schiff's base, which undergoes an Amadori rearrangement to yield the stable PE-linked Amadori product (Amadori-PE).

On the other hand, similar to proteins, phospholipids, such as PE and ethanolamine, can react with secondary products of lipid peroxidation, such as 4,5-epoxyalkenals. Zamora and Hidalgo (2003b) studied this reaction in model systems, and they characterized different polymers responsible for brown color and fluorescence development, confirming that lipid oxidation products are able to react with aminophospholipids in a manner analogous to their reactions with protein amino groups; therefore, both amino phospholipids and proteins might compete for lipid oxidation products. Subsequent studies, besides to corroborate these findings (Zamora et al. 2004, Hidalgo et al. 2006b), showed the capacity of amino phospholipids to remove the cito- and genotoxic aldehydes produced in foods during lipid oxidation when its fatty acid chains were oxidized in the presence of an oxidative stress inducer. These results suggested that, in the presence of amino phospholipids, the oxidation of polyunsaturated fatty acid chains is not likely to produce toxic aldehydes in

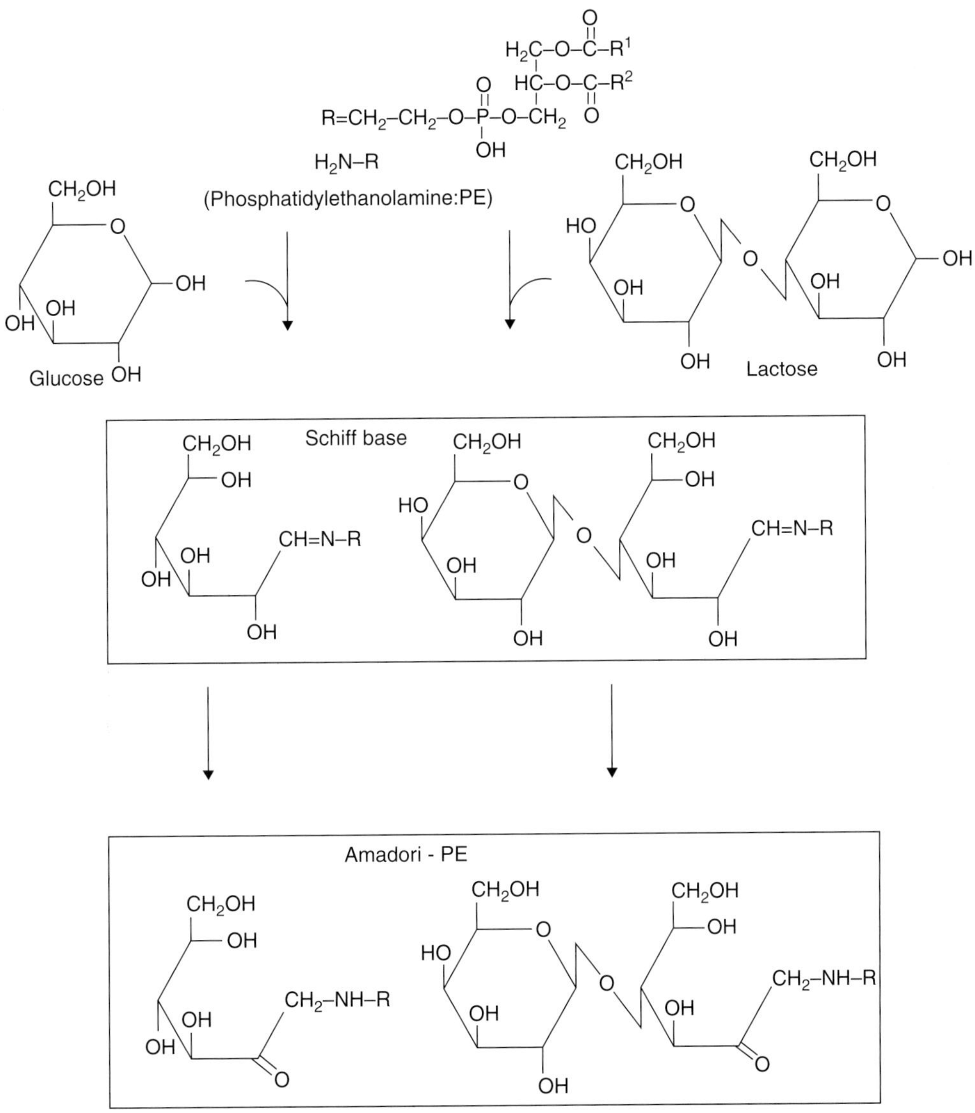

Figure 4.14. Scheme for the glycation with glucose and lactose of phosphatidylethanolamine (PE) (Oak et al. 2002).

enough concentration to pose a significant risk for human health (Hidalgo et al. 2008).

REFERENCES

Ahn JS et al. 2002. Verification of the findings of acrylamide in heated foods. *Food Addit Contam* 19(12): 1116–1124.

Ajandouz EH, Puigserver A. 1999. Non-enzymatic browning reaction of essential amino acids: effect of pH on caramelization and Maillard reaction kinetics. *J Agric Food Chem* 47(5): 1786–1793.

Ajandouz EH et al. 2001. Effects of pH on caramelization and Maillard reaction kinetics in fructose-lysine model systems. *J Agric Food Chem* 66(7): 926–931.

Alaiz M et al. 1995. Antioxidative activity of (E)-2-octenal/amino acids reaction products. *J Agric Food Chem* 43(3): 795–800.

Alaiz M et al. 1996. Antioxidative activity of pyrrole, imidazole, dihydropyridine, and pyridinium salt derivatives produced in oxidized lipid/amino acid browning reactions. *J Agric Food Chem* 44(3): 686–691.

Alaiz M et al. 1997. Comparative antioxidant activity of Maillard- and oxidized lipid-damaged bovine serum albumin. *J Agric Food Chem* 45(8): 3250–3254.

Ames JM. 1990. Control of the Maillard reaction in food systems. *Trends Food Sci Technol* 1(July): 150–154.

Anese, M et al. 2002. Effect of equivalent thermal treatments on the colour and the antioxidant activity of tomato purees. *J Food Sci* 67(9): 3442–3446.

Anese M et al. 2010. Acrylamide removal from heated foods. *Food Chem* 119(2): 791–794.

Antony SM et al. 2000. Antioxidative effect of Maillard products formed from honey at different reaction times. *J Agric Food Chem* 48(9): 3985–3989.

Badoud R et al. 1995. Periodate oxidative degradation of Amadori compounds. Formation of N^{ε}-carboxymethyllysine and *N*-carboxymethylamino acids as markers of the early Maillard reaction. In: T Ch Lee, HJ Kim (eds.) *Chemical Markers for Processed and Stored Foods*, ACS Symposium Series 631. Chicago, IL, pp. 208–220.

Ballestra P et al. 2002. Effect of high-pressure treatment on polyphenoloxidase activity of *Agaricus bisporus* mushroom. *High Press Res* 22(3–4, Sp. Iss): 677–680.

Bates L et al. 1998. Laboratory reaction cell to model Maillard colour developments in a starch-glucose-lysine system. *J Food Sci* 63(6): 991–996.

Belitz H-D, Grosch W. 1997. *Food Chemistry*, 2nd edn. Springer, Berlin, pp. 232–236.

Bell LN. 1997. Maillard reaction as influenced by buffer type and concentration. *Food Chem* 59(1): 143–147.

Benjakul S et al. 2005. Antioxidative activity of caramelisation products and their preventive effect on lipid oxidation in fish mince. *Food Chem* 90(1–2): 231–239.

Bermudo E et al. 2006. Determination of acrylamide in foodstuffs by liquid chromatography ion-trap tandem mass-spectrometry using an improved clean-up procedure. *Anal Chim Acta* 559(2): 207–214.

Billaud C et al. 2003. Inhibitory effect of unheated and heated D-glucose, D-fructose and L-cysteine solutions and Maillard reaction product model systems on polyphenoloxidase from apple. I. Enzymic browning and enzyme activity inhibition using spectrophotometric and polarographic methods. *Food Chem* 81(1): 35–50.

Billaud C et al. 2004. Effect of glutathione and Maillard reaction products prepared from glucose or fructose with glutathione on polyphenoloxidase from apple. I: Enzymic browning and enzyme activity inhibition. *Food Chem* 84(29): 223–233.

Boff JM et al. 2003. Effect of thermal processing and carbon dioxide-assisted high-pressure processing on pectin methylesterase and chemical changes in orange juice. *J Food Sci* 68(4): 1179–1184.

Boonsiripiphat K, Theerakulkait C. 2009. Extraction of rice bran extract and some factors affecting its inhibition of polyphenol oxidase activity and browning in potato. *Prep Biochem Biotechnol* 39(2): 147–158.

Borrelli RC, Fogliano V. 2005. Bread crust melanoidins as potential prebiotic ingredients. *Mol Nutr Food Res* 49(7): 673–678.

Bosch L et al. 2008. Effect of storage conditions on furosine formation in milk-ceral based baby foods. *Food Chem* 107(4): 1681–1686.

Brands CMJ, van Boekel MAJS. 2001. Reaction of monosaccharides during heating of sugar-casein systems: building of a reaction network model. *J Agric Food Chem* 49(10): 4667–4675.

Bucala R et al. 1993. Lipid advanced glycosylation: pathway for lipid oxidation in vivo. *Proc Natl Acad Sci USA* 90(July): 6434–6438.

Buera MP, Karel M. 1995. Effect of physical changes on the rates of non-enzymic browning and related reactions. *Food Chem* 52(2): 167–173.

Buera MP et al. 1987a. Non-enzymic browning in liquid model systems of high water activity: kinetics of colour changes due to caramelization of various single sugars. *J Food Sci* 52(4): 1059–1062, 1073.

Buera MP et al. 1987b. Non-enzymic browning in liquid model systems of high water activity: kinetics of colour changes due to Maillard's reaction between different single sugars and glycine and comparison with caramelization browning. *J Food Sci* 52(4): 1063–1067.

Burch RS et al. 2008. The effects of low-temperature potato storage and washing and soaking pre-treatments on the acrylamide content of French fries. *J Sci Food Agric* 88(6): 989–995.

Burdock FA et al. 2001. Evaluation of health aspects of kojic acid in food. *Regul Toxicol Pharmacol* 33(1): 80–101.

Burdulu HS, Karadeniz F. 2003. Effect of storage on nonenzymatic browning of apple juice concentrates. *Food Chem* 80(1): 91–97.

Carabasa-Giribet M, Ibarz-Ribas A. 2000. Kinetics of color development in aqueous glucose systems at high temperatures. *J Food Eng* 44(3): 181–189.

Carrieri G et al. 2010. Evaluation of acrylamide formation in potatoes during deep-frying: the effect of operation and configuration. *J Food Eng* 98(2): 141–149.

Cattaneo S et al. 2009. Liquid infant formulas: technological tools for limiting heat damage. *J Agric Food Chem* 57(22): 10689–10694.

Chang TS. 2009. An updated review of tyrosine inhibitors. *Int J Mol Sci* 10(6): 2440–2475.

Chao PC et al. 2009. Analysis of glycative products in sauces and sauce-treated foods. *Food Chem* 113(1): 262–266.

Chauhan AS et al. 1998. Properties of ascorbic acid and its application in food processing: a critical appraisal. J Food Sci *Technol Mysore* 35(5): 381–392.

Chen L et al. 2000. Honeys from different floral sources as inhibitors of enzymic browning in fruit and vegetables homogenates. *J Agric Food Chem* 48(10): 4997–5000.

Chen SL et al. 2009. Effect of hot acidic fructose solution on caramelisation intermediates including colour, hydroxymethylfurfural and antioxidative activity changes. *Food Chem* 114(2): 582–588.

Chevalier F et al. 2001. Scavenging of free radicals, antimicrobial, and cytotoxic activities of the Maillard reaction products of beta-lactoglobulin glycated with several sugars. *J Agric Food Chem* 49(10): 5031–5038.

Chisari M et al. 2010. Salinity effects on enzymatic browning and antioxidant capacity of fresh-cut baby Romaine lettuce (*Lactuca sativa* L. cv. *Duende*). *Food Chem* 119(4): 1502–1506.

Cho IH, Lee S, Jun HR, et al. 2010. Comparison of volatile Maillard reaction products from tagatose and other reducing sugars with amino acids. *Food Sci Biotechnol* 19(2): 431–438.

Coetzer C et al. 2001. Control of enzymic browning in potato (*Solanum tuberosum* L.) by sense and antisense RNA from tomato polyphenol oxidase. *J Agric Food Chem* 49(2): 652–657.

Corbo MR et al. 2000. Effects of hexanal, trans-2-hexanal, and storage temperature on shelf life of fresh sliced apples. *J Agric Food Chem* 48(6): 2401–2408.

Corzo-Martínez M et al. 2010. Characterization and improvement of rheological properties of sodium caseinate glycated with galactose, lactose and dextran. *Food Hydrocolloid* 24(1): 88–97.

Davies CGA, Wedzicha BL. 1992. Kinetics of the inhibition of ascorbic acid browning by sulfite. *Food Addit Contam* 9(5): 471–477.

Davies CGA, Wedzicha BL. 1994. Ascorbic acid browning. The incorporation of C(1) from ascorbic acid into melanoidins. *Food Chem* 49(2): 165–167.

De Rigal D et al. 2001. Inhibition of endive (*Cichorium endivia* L.) polyphenoloxidase by a *Carica papaya* latex preparation. *Int J Food Sci Technol* 36(69): 677–684.

Defaye J et al. 2000. The molecules of caramelization: structure and methodologies of detection and evaluation. *Actual Chim* 11(Nov): 24–27.

del Castillo MD et al. 1999. Early stages of Maillard reaction in dehydrated orange juice. *J Agric Food Chem* 47(10): 4388–4390.

del Castillo MD et al. 2000. Use of 2-furoylmethyl derivatives of GABA and arginine as indicators of the initial steps of Maillard reaction in orange juice. *J Agric Food Chem* 48(9): 4217–4220.

Delgado-Andrade C et al. 2007. Lysine availability is diminished in commercial fibre-enriched breakfast cereals. *Food Chem* 100(2): 725–731.

Diaz N, Clotet R. 1995. Kinetics of the caramelization of simple sugar solutions. *Alimentaria* (259): 35–38.

Dijkstra L, Walker JRL. 1991. Enzymic browning in apricots (*Prunus armeniaca*). *J Sci Food Agric* 54: 229–234.

Duan XW et al. 2009. Effect of low and high oxygen-controlled atmospheres on enzymatic browning of litchi fruit. *J Food Chem* 33(4): 572–586.

El SN, Kavas A. 1997. Available lysine in dried milk after processing. *Int J Food Sci Nutr* 48(2): 109–111.

Elias RJ et al. 2005. Antioxidant activity of cysteine, tryptophan, and methionine residues in continuous phase beta-lactoglobulin in oil-in-water emulsions. *J Agric Food Chem* 53: 10248–10253.

Elias RJ et al. 2008. Antioxidant activity of proteins and peptides. *Crit Rev Food Sci Nutr* 48: 430–441.

Elmore JS et al. 2005. Measurement of acrylamide and its precursors in potato, wheat, and rye model systems. *J Agric Food Chem* 53(4): 1286–1293.

Endo E et al. 2008. Use of active films in the minimally processed potato conservation. *Semina-Ciencias Agrarias* 29(2): 349–359.

Erbersdobler HF, Hupe A. 1991. Determination of lysine damage and calculation of lysine bioavailability in several processed foods. *Zeitschrift für Ernährungswissenschfat* 30(1): 46–49.

Erbersdobler HF, Somoza V. 2007. Forty years of furosine—forty years of using Maillard reaction products as indicators of the nutritional quality of foods. *Mol Nutr Food Res* 51(4): 423–430.

Erikson CE. 1987. Oxidation of lipids in food systems. In: HWS Chan (ed.), *Autoxidation of Unsaturated Lipids*. Academic Press, London, pp. 207–231.

Espin JC et al. 2000. The oxidation of L-ascorbic acid catalysed by pear tyrosinase. *Physiol Plant* 109(1): 1–6.

European Food Safety Authority (EFSA). 2010. Results on acrylamide levels in food from monitoring year 2008. *EFSA J* 8(5): 1599.

Fattouch S et al. 2010. Concentration-dependent effects of commonly used pesticides on activation versus inhibition of the quince (*Cydonia oblonga*) polyphenol oxidase. *Food Chem Toxicol* 48(3): 957–963.

Feather MS et al. 1995. The use of aminoguanidine to trap and measure decarbonyl intermediates produced during the Maillard reaction. In: T Ch Lee, HJ Kim (eds.) *Chemical Markers for Processed and Stored Foods*. ACS Symposium Series 631. Chicago, Illinois, pp. 24–31.

Fennema OR (ed.). 1976. *Principles of Food Science*. Marcel Dekker, New York.

Ferrer E et al. 2003. Fluorometric determination of chemically available lysine: adaptation, validation and application to different milk products. *Nahrung-Food* 47(6): 403–407.

Finot PA. 2005. Historical perspective of the Maillard reaction in food science. *Ann NY Acad Sci* 1043: 1–8.

Frankel EN. 1980. Lipid oxidation. *Prog Lipid Res* 19(1/2): 1–22.

Friedman M. 2003. Chemistry, biochemistry, and safety of acrylamide. A review. *J Agric Food Chem* 51(16): 4504–4526.

Gacche RN et al. 2009. Evaluation of the floral honey for inhibition of polyphenol oxidase-meadiated browning, antioxidant and antibacterial activities. *J Food Biochem* 33(5): 693–709.

Gardner HW. 1979. Lipid hydro peroxide reactivity with proteins and amino acids: a review. *J Agric Food Chem* 27(2): 220–229.

Gardner HW. 1989. Oxygen radical chemistry of polyunsaturated fatty acids. *Free Radic Biol Med* 7(1): 65–86.

Gardner HW. 1995. Biological roles and biochemistry of the lipoxygenase pathway. *HortScience* 30(2): 197–205.

Gillat PN, Rossell JB. 1992. The interaction of oxidized lipids with proteins. In: FB Padley (ed.) *Advances in Applied Lipid Research*. JAI Press, Greenwich, CT, pp. 65–118.

Gökmen V, Senyuva H. 2006. Study of colour and acrylamide formation in coffee, wheat flour and potato chips during heating. *Food Chem* 99(2): 238–243.

Gökmen V et al. 2008. Significance of furosine as heat-induced marker in cookies. J Cereal Sci 48(3): 843–847.

Gonzales AP et al. 1999. Color changes during storage of honeys in relation to their composition and initial color. *Food Res Int* 32(3): 185–191.

Granda C, Moreira RG. 2005. Kinetics of acrylamide formation during traditional and vacuum frying of potato chips. *J Food Process Eng* 28(5): 478–493.

Guan WQ, Fan XT. 2010. Combination of sodium chlorite and calcium propionate reduces enzymatic browning and microbial population of fresh-cut "Granny Smith" apples. *J Food Sci* 75(2): M72–M77.

Guerra-Hernandez E et al. 2002. Effect of storage on non-enzymic browning of liquid infant milk formulae. *J Sci Food Agric* 82(5): 587–592.

Guillen MD et al. 2005. Study of both sunflower oil and its headspace throughout the oxidation process. Occurrence in the headspace of toxic oxygenated aldehydes. *J Agric Food Chem* 53: 1093–1101.

Hamlet CG et al. 2008. Correlations between the amounts of free asparagine and saccharides present in commercial cereal flours in the United Kingdom and the generation of acrylamide during cooking. *J Agric Food Chem* 56(15): 6145–6153.

Harel E et al. 1970. Changes in the levels of catechol oxidase and laccase activity in developing peaches. *J Sci Food Agric* 21: 542–544.

Herraiz T. 1996. Occurrence of tetrahydro-beta-carboline-3-carboxylic acids in commercial foodstuffs. *J Agric Food Chem* 44: 3057–3065.

Herraiz T. 2000. Tetrahydro-beta-carbolines, potential neuroactive alkaloids, in chocolate and cocoa. *J Agric Food Chem* 48: 4900–4904.

Herraiz T, Galisteo J. 2003. Tetrahydro-beta-carboline alkaloids occur in fruits and fruit juices. Activity as antioxidants and radical scavengers. *J Agric Food Chem* 51: 7156–7161.

Herraiz T, Papavergou E. 2004. Identification and occurrence of tryptamine- and tryptophan-derived tetrahydro-beta-carbolines in commercial sausages. *J Agric Food Chem* 52: 2652–2658.

Herraiz T et al. 2003. L-tryptophan reacts with naturally occurring and food-occurring phenolic aldehydes to give phenolic tetrahydro-beta-carboline alkaloids: activity as antioxidants and free radical scavengers. *J Agric Food Chem* 51: 2168–2173.

Hidalgo FJ, Zamora R. 1993. Fluorescent pyrrole products from carbonyl-amine reactions. *J Biol Chem* 268(22): 16190–16197.

Hidalgo FJ, Zamora R. 1995. Characterization of the products formed during microwave irradiation of the non-enzymic browning lysine (E)-4,5-epoxy-(E)-2-heptanal model system. *J Agric Food Chem* 43(4): 1023–1028.

Hidalgo FJ, Zamora R. 2000. The role of lipids in non-enzymic browning. *Grasas y Aceites* 51(1–2): 35–49.

Hidalgo FJ, Zamora R. 2002. Methyl linoleate oxidation in the presence of bovine serum albumin. *J Agric Food Chem* 50(19): 5463–5467.

Hidalgo FJ, Zamora R. 2004. Strecker-type degradation produced by the lipid oxidation products 4,5-epoxy-2-alkenals. *J Agric Food Chem* 52: 7126–7131.

Hidalgo FJ et al. 1992. Modificaciones producidas en las proteínas alimentarias por su interacción con lípidos peroxidados. II. Mecanismos conocidos de la interacción lípido (oxidado)-proteína. *Grasas y Aceites* 43(1): 31–38.

Hidalgo A et al. 1995. Furosine as a freshness parameter of shell egg. *J Agric Food Chem* 43(6): 1673–1677.

Hidalgo FJ et al. 1999. Effect of pH and temperature on comparative non-enzymic browning of proteins produced by oxidized lipids and carbohydrates. *J Agric Food Chem* 47(2): 742–747.

Hidalgo FJ et al. 2003. Effect of the pyrrole polymerization mechanism on the antioxidative activity of no enzymatic browning reactions. *J Agric Food Chem* 51(19): 5703–5708.

Hidalgo FJ et al. 2006a. Antioxidative activity of amino phospholipids and phospholipid/amino acid mixtures in edible oils as determined by the rancimat method. *J Agric Food Chem* 54(15): 5461–5467.

Hidalgo FJ et al. 2006b. Nonenzymatic browning, fluorescence development, and formation of pyrrole derivatives in phosphatidylethanolamine/ribose/lysine model systems. *J Food Scis* 70(6): 387–391.

Hidalgo FJ et al. 2007. Effect of tocopherols in the antioxidative activity of oxidized lipid−amine reaction products. *J Agric Food Chem* 55(11): 4436–4442.

Hidalgo FJ et al. 2008. The role of amino phospholipids in the removal of the cito- and geno-toxic aldehydes produced during lipid oxidation. *Food Chem Toxicol* 46(1): 43–48.

Hidalgo FJ et al. 2009. Effect of β-sitosterol in the antioxidative activity of oxidized lipid–amine reaction products. *Food Res Int* 42(8): 1215–1222.

Hisaminato H et al. 2001. Relationship between the enzymatic browning and phenylalanine ammonia-lyase activity of cut lettuce, and the prevention of browning by inhibitors of polyphenol biosynthesis. *Biosci Biotechnol Biochem* 65(5): 1016–1021.

Hodge JE. 1953. Dehydrated foods. Chemistry of browning reactions in model systems. *J Agric Food Chem* 1(15): 928–943.

Hollnagel A, Kroh LW. 2002. 3-Deoxypentosulose: an alpha-dicarbonyl compound predominating in non-enzymic browning of oligosaccharides in aqueous solutions. *J Agric Food Chem* 50(6): 1659–1664.

Hollnagel A, Kroh LW. 2000. Degradation of oligosaccharides in non-enzymic browning by formation of α-dicarbonyl compounds via a "peeling off" mechanism. *J Agric Food Chem* 48(12): 6219–6226.

Homoki-Farkas P et al. 1997. Methylglyoxal determination from different carbohydrates during heat processing. *Food Chem* 59(1): 157–163.

Hosoda H, Iwahashi I. 2002. Inhibition of browning if apple slice and juice by onion juice. *J Jpn Soc Hortic Sci* 71(3): 452–454.

Hosoda H et al. 2003. Inhibitory effect of onion oil on browning of shredded lettuce and its active components. *J Jpn Soc Hortic Sci* 75(5): 451–456.

Howard LA et al. 1999. Beta-carotene and ascorbic acid retention in fresh and processed vegetables. *J Food Sci* 64(5): 929–936.

Howard LR, Hernandez-Brenes C. 1998. Antioxidant content and market quality of jalapeno pepper rings as affected by minimal processing and modified atmosphere packaging. *J Food Qual* 21(4): 317–327.

Hunt MD et al. 1993. cDNA cloning and expression of potato polyphenol oxidase. *Plant Mol Biol* 21(1): 59–68.

Ihl M et al. 2003. Effect of immersion solutions on shelf-life of minimally processed lettuce. *Food Sci Technol* 36(6): 591–599.

International Agency for Research on Cancer (IARC). 1994. Some Industrial Chemicals. In: *IARC Monographs on the Evaluation for Carcinogenic Risk of Chemical to Humans*. IARC, Lyon, vol. 60, pp. 389–433, 435–453.

Jacxsens L et al. 2001. Effect of high oxygen modified atmosphere packaging on microbiological growth and sensorial qualities of fresh cut produce. *Int J Food Microbiol* 71(2–3): 197–210.

Jiménez-Castaño L et al. 2005. Effect of the dry-heating conditions on the glycosylation of β-lactoglobulin with dextran through the Maillard reaction. *Food Hydrocolloid* 19(5): 831–837.

Jolley RL et al. 1974. Oxytyrosinase. *J Biol Chem* 249(2): 335–345.

Jom KN et al. 2008. Investigation of acrylamide in curries made from coconut milk. *Food Chem Toxicol* 46(1): 119–124.

Kaaber L et al. 2002. Browning inhibition and textural changes of pre-peeled potatoes caused by anaerobic conditions. *Food Sci Technol* 35(6): 526–531.

Kacem B et al. 1987. Non-enzymatic browning in aseptically packaged orange drinks: effect of ascorbic acid, amino acids and oxygen. *J Food Sci* 52(6): 1668–1672.

Kaplan O et al. 2009. Acrylamide concentrations in grilled foodstuffs of Turkish kitchen by high performance liquid chromatography-mass spectrometry. *Microchem J* 93(2): 173–179.

Kaur C, Kapoor HC. 2000. Inhibition of enzymic browning in apples, potatoes and mushrooms. *J Sci Ind Res* 59(5): 389–394.

Khayat A, Schwall D. 1983. Lipid oxidation in seafood. *Food Technol* 37(7): 130–140.

Kim JS, Lee YS. 2008a. Effect of pH on the enolization of sugars and antioxidant activity of caramelisation browning. *Food Sci Biotechnol* 17(5): 931–939.

Kim JS, Lee YS. 2008b. Effect of reaction pH on enolization and racemization reactions of glucose and fructose on heating with amino acid enantiomers and formation of melanoidins as result of the Maillard reaction. *Food Chem* 108(2): 582–592.

Kim YM et al. 2002. Oxyresveratrol and hydroxystilbene compounds. Inhibitory effect on tyrosinase and mechanism of action. *J Biol Chem* 277(18): 16340–16344.

Kim YM et al. 2005. Prevention of enzymatic browning pear by onion extract. *Food Chem* 89: 181–184.

Kirigaya N et al. 1968. Studies on the antioxidant of nonenzymatic browning reaction products. Part 1: Relation of colour intensity and reductones with antioxidant activity of browning reaction products. *Agric Biol Chem* 32(3): 287.

Kitts DD et al. 2006. Chemistry and genotoxicity of caramelized sucrose. *Mol Nutr Food Res* 50(12): 1180–1190.

Koseki H et al. 2001. Effect of sugars on decomposition and browning of vitamin C during heating storage. *J Jpn Soc Food Sci Technol* 48(4): 268–276.

Kroh LW. 1994. Caramelization in foods and beverages. *Food Chem* 51(4): 373–379.

Kroh LW et al. 1996. Non-volatile reaction products by heat-induced degradation of alpha-glucans. Part I: Analysis of oligomeric maltodextrins and anhydrosugars. *Starch/Stärke* 48(11/12): 426–433.

Kubo I et al. 2000. Flavonols from Heterotheca inuloides: tyrosinase inhibitory activity and structural criteria. *Biorganic Med Chem* 8(7): 1749–1755.

Kubow S. 1990. Toxicity of dietary lipid peroxidation products. *Trends Food Sci Technol* 1(3): 67–70.

Kubow S. 1992. Routes of formation and toxic consequences of lipid oxidation products in foods. *Free Radic Biol Med* 12(1): 63–81.

Kurata T et al. 1973. Red pigment produced by reaction of dehydro-L-ascorbic acid with alpha-amino-acids. *Agric Biol Chem* 37(6): 1471–1477.

Lea CH. 1957. Deteriorative reactions involving phospholipids and lipoproteins. *J Sci Food Agric* 8: 1–13.

Ledl F. 1990. Chemical pathways of the Maillard reaction. In: PA Finot, HU Aeschbacher, RF Hurrell, R Liardon (eds.) *The Maillard Reaction in Food Processing, Human Nutrition and Physiology*. Birkhäuser Verlag, Basel, pp. 19–42.

Lee GC, Han SC. 2001. Inhibition effects of caramelization products from sugar solutions subjected to different temperature on polyphenol oxidase. *J Korean Soc Food Sci Nutr* 30(6): 1041–1046.

Lee GC, Lee CY. 1997. Inhibitory effect of caramelization products on enzymic browning. *Food Chem* 60(2): 231–235.

Lee HS, Nagy S. 1988. Quality changes and non-enzymic browning intermediates in grapefruit juice during storage. *J Food Sci* 53(1): 168–172, 180.

Lee HS, Nagy S. 1996. Chemical degradative indicators to monitor the quality of processed and stored citrus products. In: T-C Lee, H-J Kim (eds.) *Chemical Markers for Processed Foods*. ACS Symposium Series 631, American Chemical Society, Washington, DC, pp. 86–106.

Lertsiri S et al. 1998. Identification of deoxy-D-fructosyl phosphatidylethanolamine as a non-enzymatic glycation product of phosphatidylethanolamine and its occurrence in human blood plasma and red blood cells. *Biosci Biotechnol Biochem* 62(5): 893–901.

Lingnert H. 1990. Development of the Maillard reaction during food processing. In: PA Finot et al. (eds.) *The Maillard Reaction in Food Processing, Human Nutrition and Physiology*. Birkhäuser Verlag, Basel, pp. 171–185.

Liu S Ch et al. 2008. Kinetics of color development, pH decreasing and anti-oxidative activity reduction of Maillard reaction in galactose/glycine model systems. *Food Chem* 108(2): 533–541.

Loschner J et al. 1990. L-Ascorbic acid—a carbonyl component on non-enzymic browning reactions. Browning of L-ascorbic acid in aqueous model system. *Zeitschrift fur Lebensmittel Untersuchung und Forschung* 191(4–5): 302–305.

Loschner J et al. 1991. L-Ascorbic-acid—A carbonyl component of non-enzymic browning reactions. 2. Amino-carbonyl reactions of L-ascorbic acid. *Zeitschrift fur Lebensmittel Untersuchung und Forschung* 192(4): 323–327.

Machiels D, Istasse L. 2002. Maillard reaction: importance and applications in food chemistry. *Annales de Medecine veterinaire* 146(6): 347–352.

Malec LS et al. 2002. Influence of water activity and storage temperature on lysine availability of a milk like system. *Food Res Int* 35(9): 849–853.

Manso MC et al. 2001. Modelling ascorbic acid thermal degradation and browning in orange juice under aerobic conditions. *Int J Food Sci Technol* 36(3): 303–312.

Manzocco L et al. 2001. Review of non-enzymic browning and antioxidant capacity in processed foods. *Trends Food Sci Technol* 11(9–10): 340–346.

Manzocco L et al. 2009. Polyphenoloxidase inactivation by light exposure in model systems and apple derivatives. *Innovative Food Sci Emerg Technol* 10(4): 506–511.

Marshall MR, Schulbach KF. 2009. Method of inhibition of enzymatic browning in food using hypotaurine and equivalents. Patent No. WO2009131996-A2; WO2009131996-A3.

Marshall MR et al. 2000. *Enzymatic Browning in Fruits, Vegetables and Seafoods*. FAO, Rome. Available at http://www.fao.org/waicent/faoinfo/agricult/ags/Agsi/ENZYMEFINAL/COPYRIGH.HTM.

Martin-Diana AB et al. 2009. Orange juices enriched with chitosan: optimisation for extending the shelf-life. *Innovat Food Sci Emerg Technol* 10(4): 590–600.

Martins SIFS, van Boekel MAJS. 2003. Melanoidins extinction coefficient in the glucose/glycine Maillard reaction. *Food Chem* 83(1): 135–142.

Mathew AG, Parpia HA. 1971. Food browning as a polyphenol reaction. *Adv Food Res* 19: 75–145.

Mayer AM, Harel E. 1991. Phenoloxidase and their significance in fruit and vegetables. In: PF Fx (ed.) *Food Enzymology*. Elsevier, London, p. 373.

Mdluli KM, Owusu-Apenten R. 2003. Enzymic browning in marula fruit. 1: Effect of endogenous antioxidants on marula fruit polyphenol oxidase. *J Food Biochem* 27(1): 67–82.

Meltretter J, Pischetsrieder M. 2008. Application of mass spectrometry for the detection of glycation and oxidation products in milk proteins. *Maillard Reaction: Recent Adv Food Biomed Sci* 1126: 134–140.

Meltretter J et al. 2007. Site-specific formation of Maillard, oxidation, and condensation products from whey proteins during reaction with lactose. *J Agric Food Chem* 55: 6096–6103.

Meltretter J et al. 2008. Identification and site-specific relative quantification of beta-lactoglobulin modifications in heated milk and dairy products. *J Agric Food Chem* 56: 5165–5171.

Mesa MD et al. 2008. Antioxidant properties of soy protein-fructooligosaccharide glycation systems and its hydrolyzates. *Food Res Int* 41(6): 606–615.

Min S et al. 2003. Commercial-scale pulsed electric field processing of orange juice. *J Food Sci* 68(4): 1265–1271.

Miralles B et al. 2007. The occurrence of a Maillard-type protein-polysaccharide reaction between β-lactobglobulin and chitosan. *Food Chem* 100(3): 1071–1075.

Mölnar-Perl I, Friedman M. 1990. Inhibition of browning by sulfur amino acids. 3. Apples and potatoes. *J Agric Food Chem* 38: 1652–1656.

Montgomery MW, Day EA. 1965. Aldehyde-amine condensation reaction: a possible fate of carbonyl in foods. *J Food Sci* 30(5): 828–832.

Morales FJ, van Boekel MAJS. 1998. A study on advanced Maillard reaction in heated casein/sugar solutions: color formation. *Int Dairy J* 8(10–11): 907–915.

Morales V et al. 2009. Combined use of HMF and furosine to assess fresh honey quality. *J Sci Food Agric* 89(8): 1332–1338.

Mottram DS. 1998. Flavour formation in meat and meat products: a review. *Food Chem* 62(4): 415–424.

Mottram DS et al. 2002. Acrylamide is formed in the Maillard reaction. *Nature* 419(3 October): 448.

Mundt S, Wedzicha BL. 2003. A kinetic model for the glucose-fructose-glycine browning reaction. *J Agric Food Chem* 51(12): 3651–3655.

Murata M et al. 2000. Transgenic apple (*Malus* x *domestica*) shoot showing low browning potential. *J Agric Food Chem* 48(11): 5243–5248.

Murata M et al. 2001. A transgenic apple callus showing reduced polyphenol oxidase activity and lower browning potential. *Biosci Biotechnol Biochem* 65(2): 383–388.

Nagy S, Smoot J. 1977. Temperature and storage effects on percent retention and percent United States' recommended dietary allowance of vitamin C in canned single-strength orange juice. *J Agric Food Chem* 25(1): 135–138.

Naim M et al.. 1997. Effects of orange juice fortification with thiols on *p*-vinylguaiacol formation, ascorbic acid degradation, browning, and acceptance during pasteurisation and storage under moderate conditions. *J Agric Food Chem* 45(5): 1861–1867.

Namiki M. 1988. Chemistry of Maillard reactions: recent studies on the browning reaction. Mechanism and the development of antioxidant and mutagens. *Adv Food Res* 32: 116–170.

Namiki M et al. 1993. Weak chemiluminiscence at an early-stage of the Maillard reaction. *J Agric Food Chem* 41(10): 1704–1709.

Nawar WW. 1985. Lipids. In: OR Fennema (ed.) *Food Chemistry*. Dekker, New York, pp. 139–244.

Nawar WW. 1996. Lipids. In: OR Fennema (ed.) *Food Chemistry*, 3rd edn. Marcel Dekker, New York, pp. 225–319.

Negishi O, Ozawa T. 2000. Inhibition of enzymic browning and protection of sulfhydryl enzymes by thiol compounds. *Phytochemistry* 54(5): 481–487.

Newman SM et al. 1993. Organization of the tomato polyphenol oxidase gene family. *Plant Mol Biol* 21(6): 1035–1051.

Nicolas JJ et al. 1994. Enzymic browning reactions in apple and apple products. *Crit Rev Food Sci Nutr* 32(2): 109–157.

Nielsen HK et al. 1985. Reaction of proteins with oxidized lipids. 1. Analytical measurement of lipid oxidation and amino acid losses in a whey protein-methyl lineolate model system. *Br J Nutr* 53(1): 61–73.

Nunes FM et al. 2006. Characterization of galactomannan derivatives in roasted coffee beverages. *J Agric Food Chem* 54(9): 3428–3439.

Oak J-H et al. 2000. Synthetically prepared Amadori-glycated phosphatidylethanolamine can trigger lipid peroxidation via free radical reactions. *FEBS Lett* 481(1): 26–30.

Oak J-H et al. 2002. UV analysis of Amadori-glycated phosphatidylethanolamine in foods and biological samples. *J Lipids Res* 43(3): 523–529.

Okumura S, Kawai H. 1970. Amino acids for seasoning food. Japanese Patent 7039622.

Olano A, Martínez-Castro I. 2004. Nonenzymatic browning. In: ML Leo (ed.) *Handbook of Food Analysis*, vol. 3. Nollet Marcel Dekker, New York , pp. 1855–1890.

Park CW et al. 1998. Effects of reaction conditions for improvement of caramelization rate. *Korean J Food Sci Technol* 30(4): 983–987.

Papavergou E, Herraiz T. 2003. Identification and occurrence of 1,2,3,4-tetrahydrobeta-carboline-3-carboxylic acid: the main beta-carboline alkaloid in smoked foods. *Food Res Int* 36: 843–848.

Parpinello GP et al. 2002. Preliminary study on glucose oxidase-catalase enzyme system to control the browning of apple and pear purees. *Food Sci Technol* 35(3): 239–243.

Penndorf I et al. 2007. Studies on N-terminal glycation of peptides in hypoallergenic infant formulas: quantification of alpha-*N*-(2-furoylmethyl)amino acids. *J Agric Food Chem* 55(3): 723–727.

Perera N et al. 2010. Colour and texture of apples high pressure processed in pineapple juice. *Innovative Food Sci Emerg Technol* 11(1): 39–46.

Pereyra-Gonzales AS et al. 2010. Maillard reaction kinetics in milk powder: effect of water activity. *Int Diary J* 20(1): 40–45.

Perez-Gago MB et al. 2003. Effect of solid content and lipid content of whey protein isolate-beeswax edible coatings on color change of fresh cut apples. *J Food Sci* 68(7): 2186–2191.

Perez-Gilabert M, García-Carmona F. 2000. Characterization of catecholase and cresolase activities of eggplant polyphenol oxidase. *J Agric Food Chem* 48(3): 695–700.

Phongkanpai V et al. 2006. Effect of pH on antioxidative activity and other characteristics of caramelization products. *J Food Biochem* 30(2): 174–186.

Pilizota V, Subaric D. 1998. Control of enzymic browning of foods. *Food Technol Biotechnol* 36(3): 219–227.

Pischetsrieder M et al. 1995. Reaction of ascorbic with aliphatic amines. *J Agric Food Chem* 43(12): 3004–3006.

Pischetsrieder M et al. 1997. Immunochemical detection of oxalic acid monoamides that are formed during the oxidative reaction of L-ascorbic and proteins. *J Agric Food Chem* 45(6): 2070–2075.

Pitotti A et al. 1995. Effect of caramelization and Maillard reaction products on peroxide activity. *J Food Biochem* 18(6): 445–457.

Ponce AG et al. 2008. Antimicrobial and antioxidant activities of edible coatings enriched with natural plant extracts: in vitro and in vivo studies. *Postharvest Biol Technol* 49(2): 294–300.

Premakumar K, Khurduya DS. 2002. Effect of microwave blanching on nutritional qualities of banana puree. *J Food Sci Technol-Mysore* 39(3): 258–260.

Prestamo G, et al. 2000. Fruit preservation under high hydrostatic pressure. *High Press Res* 19(1–6): 535–542.

Preston A et al. 2009. Monoclonal antibody development for acrylamide-adducted human haemoglobin. A biomarker of dietary acrylamide exposure. *J Immunol Methods* 341(1–2): 19–29.

Queiroz C et al. 2008. Polyphenol oxidase: characteristic and mechanisms of browning control. *Food Rev Int* 24(4): 361–375.

Rada-Mendoza M et al. 2002. Furosine as indicator of Maillard reaction in jams and fruit-based infant foods. *J Agric Food Chem* 50(14): 4141–4145.

Rada-Mendoza M et al. 2004. Study on nonenzymatic browning in cookies, crackers and breakfast cereals by maltulose and furosine determination. *J Cereal Sci* 39(2): 167–173.

Ramírez-Jimenez A et al. 2004. Effect to storage conditions and inclusión of milk on available lysine in infant cereals. *Food Chem* 85(2): 239–244.

Requena JR et al. 1997. Carboxymethylethanolamine, a biomarker of phospholipid modification during the Maillard reaction in vivo. *J Biol Chem* 272(28): 14473–17479.

Resmini P et al. 1990. Accurate quantification of furosine in milk and dairy products by direct HPLC method. *Italian J Food Sci* 2(3): 173–183.

Rhee C, Kim DH. 1975. Antioxidant activity of acetone extracts obtained from a caramelization-type browning reaction. *J Food Sci* 40(3): 460–462.

Rocha AMCN, Morais AMMB. 2001. Influence of controlled atmosphere storage on polyphenoloxidase activity in relation to color changes of minimally processed "Jonagored" apple. *Int J Food Sci Technol* 36(4): 425–432.

Rodov V. 2007. Biotechnological approaches to improving quality and safety of fresh-cut fruit and vegetable products. *Proc Int Conf Qual Manag Fresh Cut Produce* 746: 181–193.

Rogacheva SM et al. 1995. L-ascorbic acid in no enzymatic reactions. 1. Reaction with glycine. *Zeitschrift fur Lebensmittel Untersuchung und Forschung* 200(1): 52–58.

Roig MG et al. 1999. Studies on the occurrence of non-enzymic browning during storage of citrus juice. *Food Res Int* 32(9): 609–619.

Rojas AM, Gerschenson LN. 1997a. Ascorbic acid destruction in sweet aqueous model systems. *Food Sci Technol* 30(6): 567–572.

Rojas AM, Gerschenson LN. 1997b. Influence of system composition on ascorbic acid destruction at processing temperatures. *J Sci Food Agric* 74(3): 369–378.

Rojas-Grau MA et al. 2008. Effect of natural antibrowning agents on color and related enzymes in fresh-cut Fuji apples as an alternative to use of ascorbic acid. *J Food Sci* 73(6): S267–S272.

Roldan E et al. 2008. Characterisation of onion (*Allium cepa* L.) by products as food ingredients with antioxidant and antibrowning properties. *Food Chem* 108(3): 907–916.

Rufián-Henares JA, Morales FJ. 2007. Functional properties of melanoidins: in vitro antioxidant, antimicrobial and antihypertensive activities. *Food Res Int* 40(8): 995–1002.

Rufián-Henares JA, de la Cueva SP. 2009. Antimicrobial activity of coffee melanoidins. A study of their metal-chelating properties. *J Agric Food Chem* 57(2): 432–438.

Rufián-Henares JA et al. 2009. Assessing the Maillard reaction development during the toasting process of common flours employed by the cereal products industry. *Food Chem* 114(1): 93–99.

Ruiz JC et al. 2004. Furosine is a useful indicator in pre-baked breads. *J Sci Food Agric* 84(4): 336–370.

Ryu SY et al. 2003. Effects of temperature and pH on the nonenzymic browning reaction of tagatose-glycine model system. *Food Sci Biotechnol* 12(6): 675–679.

Sanz ML et al. 2000. Presence of 2-furoylmethyl derivatives in hydrolyzates of processed tomato products. *J Agric Food Chem* 48(2): 468–471.

Sanz ML et al. 2001. Formation of Amadori compounds in dehydrated fruits. *J Agric Food Chem* 49(11): 5228–5231.

Sanz ML et al. 2003. 2-Furoylmethyl amino acids and hydroxymethylfurfural as indicators of honey quality. *J Agric Food Chem* 51(15): 4278–4283.

Sawamura M et al. 1991. C-14 studies on browning of dehydroascorbic acid in and aqueous solution. *J Agric Food Chem* 39(10): 1735–1737.

Sawamura M et al. 1994. Identification of 2 degradation products from aqueous dehydroascorbic acid. *J Agric Food Chem* 42(5): 1200–1203.

Sawamura M et al. 2000. The effect of antioxidants on browning and on degradation products caused by dehydroascorbic acid. *J Food Sci* 65(1): 20–23.

Schaich KM. 2008. Co-oxidation of proteins by oxidizing lipids. In: A Kamal-Eldin, DB Min (eds.) *Lipid Oxidation Pathways*, vol. 2. AOCS Press, Urbana, IL, pp. 181–272.

Sciancalepore V. 1985. Enzymic browning in 5 olive varieties. *J Food Sci* 54(4): 1194–1195.

Severini C et al. 2001. Preventing enzymic browning of potato by microwave blanching. *Sci des Aliments* 21(2): 149–160.

Severini C et al. 2003. Prevention of enzymic browning in sliced potatoes by blanching in boiling saline solutions. *Food Sci Technol* 36(7): 657–665.

Smith G, Hole M. 1991. Browning of salted sun-dried fish. *J Sci Food Agric* 55(2): 291–301.

Soliva-Fortuny RC et al. 2001. Browning evaluation of ready-to-eat-apples as affected by modifies atmosphere packaging. *J Agric Food Chem* 49(8): 3685–3690.

Son SM et al. 2000. Kinetic study of oxalic inhibition on enzymic browning. *J Agric Food Chem* 48(6): 2071–2074.

Soria AC et al. 2010. 2-Furoylmethyl amino acids, hydroxymethylfurfural, carbohydrates and beta-carotene as quality markers of dehydrated carrots. *J Sci Food Agric* 89(2): 267–273.

Stadler RH et al. 2002. Acrylamide from Maillard reaction products. *Nature* 419(3 October): 449.

Stadtman ER, Levine RL. 2003. Free radical-mediated oxidation of free amino acids and amino acid residues in proteins. *Amino Acids* 25: 207–218.

Subaric D et al. 2001. Effectiveness of some crown compounds on inhibition of polyphenoloxidase in model systems an in apple. *Acta Alimentaria* 30(1): 81–87.

Takiguchi A. 1992. Lipid oxidation and brown discoloration in niboshi during storage at ambient temperature and low temperatures. *Bull Jpn Soc Sci Fish* 58(3): 489–494.

Tannenbaum SR. 1976. In: O Fennema (ed.) *Principles of Food Chemistry*, vol. 1. Marcel Dekker, New York and Basel, pp. 357–360.

Tareke A et al.. 2002. Analysis of acrylamide, a carcinogen formed in heated foodstuffs. *J Agric Food Chem* 50(17): 4998–5006.

Tateo F et al. 2007. Acrylamide levels in cooked rice, tomato sauces and some fast food on the Italian market. *J Food Composition Anal* 20(3–4): 232–235.

Thanonkaew A et al. 2007. Yellow discoloration of the liposome system of cuttlefish (*Sepia pharaonis*) as influenced by lipid oxidation. *Food Chem* 102(1): 219–224.

Tomasik P et al. 1989. The thermal decomposition of carbohydrates. Part I: the decomposition of mono-, di- and oligosaccharides. *Adv Carbohydr Chem Biochem* 47: 203–270.

Toussaint O, Lerch K. 1987. Catalytic oxidation of 2-aminophenols and ortho hydroxylation of aromatic amines by tyrosinase. *Biochemistry* 26(26): 8567–8571.

Tsai ChH et al. 1991. Water activity and temperature effects on nonenzymic browning of amino acids in dried squid and simulated model system. *J Food Sci* 56(3): 665–670, 677.

Utzmann CM, Lederer MO. 2000. Identification and quantification of aminophospholipid-linked Maillard compounds in model systems and egg yolk products. *J Agric Food Chem* 48(4): 1000–1008.

Valverde MT et al. 2010. Inactivation of *Saccharomyces cerevisiae* in conference pear with high pressure carbon dioxide and effect on pear quality. *J Food Eng* 98(4): 421–428.

Vamosvigyazo L, Nadudvarimarkus V. 1982. Enzymic browning, polyphenol content, polyphenol oxidase and preoxidase-activities in pear cultivars. *Acta Aliment* 11(2): 157–168.

Van Lancker F et al. 2010. Formation of pyrazines in Maillard model systems of lysine-containing peptides. *J Agric Food Chem* 58(4): 2470–2478.

Vercet A. 2003. Browning of white chocolate during storage. *Food Chem* 81(3): 371–377.

Villamiel M et al. 1998. Assessment of the thermal treatment of orange juice during continuous microwave and conventional heating. *J Sci Food Agric* 78(2): 196–200.

Villamiel M et al. 1999. Use of different thermal indices to assess the quality of pasteurized milks. *Zeitschrift fur Lebensmittel Untersuchung und Forschung* 208(3): 169–171.

Villamiel M et al. 2000. Survey of the furosine content in cheeses marketed in Spain. *J Food Prot* 63(7): 974–975.

Villamiel M et al. 2001. Presence of furosine in honeys. *J Sci Food Agric* 81(8): 790–793.

Wagner KH et al. 2002. Antioxidative potential of melanoidins isolated from a roasted glucose-glycine model. *Food Chem* 78(3): 375–382.

Warmbier HC et al. 1976. Nonenzymatic browning kinetics in an intermediate moisture model system. Effect of glucose to lysine ratio. *J Food Sci* 41(5): 981–983.

Wedzicha BL. 1984. *Chemistry of Sulfur Dioxide in Foods*. Elsevier Applied Science Publishers, London.

Wedzicha BL, Imeson AP. 1977. Yield of 3-deoxy-4-sulfopentulose in sulfite inhibited non-enzymic browning of ascorbic acid. *J Sci Food Agric* 28(8): 669–672.

Wedzicha BL, Mcweeny DJ. 1974. Non-enzymic browning reactions of ascorbic acid and their inhibition. Identification of 3-deoxy-4-sulphopentosulose in dehydrated, sulfited cabbage after storage. *J Sci Food Agric* 25(5): 589–593.

Weenen H. 1998. Reactive intermediates and carbohydrate fragmentation in Maillard chemistry. *Food Chem* 62(4): 339–401.

Wei*β*haar R, Gutsche B. 2002. Formation of acrylamide in heated potato products. Model experiments pointing to asparagines as precursor. *Deutsche Lebbensmittel* 98(11): 397–400.

Whitaker JR. 1972. Polyphenol oxidase. In: JR Whitaker (ed.) *Principles of Enzymology for the Food Sciences*. Marcel Dekker, New York, pp. 571–582.

Wolfrom ML et al. 1974. Factors affecting the Maillard browning reaction between sugars and amino acids studies on the nonenzymatic browning of dehydrated orange juice. *J Agric Food Chem* 22(5): 796–799.

Yadav DN et al. 2008. Effect of microwave heating of wheat grains on the browning of dough and quality chapattis. *Int J Food Sci* 43(7): 1217–1225.

Yaylayan VA, Stadler RH. 2005. Acrylamide formation in food: a mechanistic perspective. *J AOAC Int* 88(1): 262–267.

Yamaguchi K et al. 2009. Effects of salt concentration on the reaction rate of glucose with amino acids, peptides and proteins. *Biosci Biotechnol Biochem* 71(11): 2379–2383

Yamamoto A, Kogure M. 1969. Studies on new applications of amino acids in food industries. 1. Elimination of rancid odor in rice by L-lysine. *J Food Sci Technol, Nihon Shokuhin Kogyo Gakkai-shi* 16(9): 414–419.

Yen GC, Tsai LC. 1993. Antimutagenicity of a partially fractionated Maillard reaction-product. *Food Chem* 47(1): 11–15.

Yin D. 1996. Biochemical basis of lipofuscin, ceroid and age pigment-like fluorophores. *Free Radic Biol Med* 21(6): 871–888.

Yin DZ, Brunk UT. 1991. Oxidized ascorbic acid and reaction-products between ascorbic acid and amino acids might constitute part of age pigments. *Mech Ageing Dev* 61(1): 99–112.

Yoruk R, Marshall MR. 2003. Physicochemical properties and function of plant polyphenol oxidase: a review. *J Food Biochem* 27(5): 361–422.

Yoruk R, Marshall MR. 2009. Importance of pH on antibrowning activity of oxalic acid. *J Food Biochem* 33(4): 522–534.

Zamora R, Hidalgo FJ. 1993. Antioxidant activity of the compounds produced in the nonenzymatic browning reaction of E-4,5-epoxy-E-2-heptenal and lysine. In: C Benedito de Barber et al. (eds.) *Progress in Food Fermentation*. IATA, CSIC, Valencia, pp. 540–545.

Zamora R, Hidalgo FJ. 1994. Modification of lysine amino groups by the lipid peroxidation product 4,5(E)-epoxy-2(E)-heptenal. *Lipids* 29(4): 243–249.

Zamora R, Hidalgo FJ. 1995. Linoleic acid oxidation in the presence of amino compounds produces pyrroles by carbonyl amine reactions. *Biochim Biophys Acta* 1258(3): 319–327.

Zamora R, Hidalgo FJ. 2001. Inhibition of proteolysis in oxidized lipid-damaged proteins. *J Agric Food Chem* 49(12): 6006–6011.

Zamora R, Hidalgo FJ. 2003a. Comparative methyl linoleate and methyl linolenate oxidation in the presence of bovine serum albumin at several lipid/protein ratios. *J Agric Food Chem* 51(16): 4661–4667.

Zamora R, Hidalgo FJ. 2003b. Phosphatidylethanolamine modification by oxidative stress product 4,5(E)-epoxy-2(E)-heptenal. *Chem Res Toxicol* 16(12): 1632–1641.

Zamora R et al. 1999. Determination of ε-*N*-pyrrolylnorleucine in fresh food products. *J Agric Food Chem* 47(5): 1942–1947.

Zamora R et al. 2000. Contribution of pyrrole formation and polymerization to the nonenzymatic browning produced by amino-carbonyl reactions. *J Agric Food Chem* 48(8): 3152–3158.

Zamora R et al. 2004. Contribution of phospholipid pyrrolization to the color reversion produced during deodorization of poorly degummed vegetable oils. *J Agric Food Chem* 52(13): 4166–4171.

Zamora R et al. 2005a. Strecker-type degradation of phenylalanine by methyl 9,10-epoxy-13-oxo-11-octadecenoate and methyl 12,13-epoxy-9-oxo-11-octadecenoate. *J Agric Food Chem* 53(11): 4583–4588.

Zamora R et al. 2005b. Phospholipid oxidation and nonenzymatic browning development in phosphatidylethanolamine/ribose/lysine model systems. *Eur Food Res Technol* 220: 459–465.

Zamora R et al. 2006. Chemical conversion of alpha-amino acids into alpha-keto acids by 4,5-epoxy-2-decenal. *J Agric Food Chem* 54: 6101–6105.

Zamora R et al. 2010. Model reactions of acrylamide with selected amino compounds. *J Agric Food Chem* 58(3): 1708–1713.

Zardetto S et al. 2003. Effects of different heat treatment on the furosine content in fresh filata pasta. *Food Res Int* 36(9–10): 877–883.

Zenker M et al. 2003. Application of ultrasound-assisted thermal processing for preservation and quality retention of liquid foods. *J Food Prot* 66(9): 1642–1649.

Zerdin K et al. 2003. The vitamin C content of orange juice packed in an oxygen scavenger material. *Food Chem* 82(3): 387–395.

Zhang Y et al. 2005. Occurrence and analytical methods of acrylamide in heat-treated foods. Review and recent developments. *J Chromatogr A* 1075(1–2): 1–21.

Ziderman II et al. 1989. Thermal interaction of ascorbic acid and sodium ascorbate with proteins in relation to nonenzymatic browning and Maillard reaction of foods. *J Agric Food Chem* 37(6): 1480–1486.

Zocca F et al. 2010. Antibrowning potential of Brassicacaea. *Bioresour Technol* 101(10): 3791–3795.

Zotos A et al. 2001. Production and quality assessment of a smoked tuna (*Euthynnus affinis*) product. *J Food Sci* 66(8): 1184–1190.

5
Water Chemistry and Biochemistry

C. Chieh

INTRODUCTION

Water, the compound H_2O, is the most common food ingredient. Its rarely used chemical names are hydrogen oxide or dihydrogen monoxide. So much of this compound exists on the planet earth that it is often taken for granted. Water is present in solid, liquid, and gas forms in the small range of temperatures and pressures near the surface of the earth. Moreover, natural waters always have substances dissolved in them, and only elaborate processes produce pure water.

The chemistry and physics of water are organized studies of water: its chemical composition, formation, molecular structure, rotation, vibration, electronic energies, density, heat capacity, temperature dependency of vapor pressure, and its collective behavior in condensed phases (liquid and solid). In a broader sense, the study of water also includes interactions of water with atoms, ions, molecules, and biological matter. The knowledge of water forms the foundation for biochemistry and food chemistry. Nearly every aspect of biochemistry and food chemistry has something to do with water, because water is intimately linked to life, including the origin of life.

Science has developed many scientific concepts as powerful tools for the study of water. Although the study of water reveals a wealth of scientific concepts, only a selection of topics about water will be covered in the limited space here.

Biochemistry studies the chemistry of life at the atomic and molecular levels. Living organisms consist of many molecules. Even simple bacteria consist of many kinds of molecules. The interactions of the assembled molecules manifest life phenomena such as the capacity to extract energy or food, respond to stimuli, grow, and reproduce. The interactions follow

Food Biochemistry and Food Processing, Second Edition. Edited by Benjamin K. Simpson, Leo M.L. Nollet, Fidel Toldrá, Soottawat Benjakul, Gopinadhan Paliyath and Y.H. Hui.

chemical principles, and water chemistry is a key for the beginning of primitive life forms billions of years ago. The properties of water molecules give us clues regarding their interactions with other atoms, ions, and molecules. Furthermore, water vapor in the atmosphere increases the average temperature of the atmosphere by 30 K (Wayne 2000), making the earth habitable. Water remains important for human existence, for food production, preservation, processing, and digestion.

Water is usually treated before it is used by food industries. After usage, wastewaters must be treated before it is discharged into the ecological system. After we ingest foods, water helps us to digest, dissolve, carry, absorb, and transport nutrients to their proper sites. It further helps hydrolyze, oxidize, and utilize the nutrients to provide energy for various cells, and eventually, it carries the biological waste and heat out of our bodies. Oxidations of various foods also produce water. How and why water performs these functions depend very much on its molecular properties.

THE COMPOUND WATER

A *compound* is a substance that is made up of two or more basic components called *chemical elements* (e.g., hydrogen, carbon, nitrogen, oxygen, iron) commonly found in food. *Water* is one of the tens of millions of compounds in and on earth.

The chemical equation and thermal dynamic data for the formation of water from hydrogen and oxygen gas is

$$2H_2(g) + O_2(g) + 2H_2O(l), \quad \Delta H^0 = -571.78 \text{ kJ}$$

The equation indicates that 2 mol of gaseous hydrogen, $H_2(g)$, react with 1 mol of gaseous oxygen, $O_2(g)$, to form 2 mol of liquid water. If all reactants and products are at their *standard states* of 298.15 K and 101.325 kPa (1.0 atm), formation of 2 mol of water releases 571.78 kJ of energy, as indicated by the negative sign for ΔH^0. Put in another way, the **heat of formation** of water, ΔH_f^0, is −285.89 (= −571.78/2) kJ/mol. Due to the large amount of energy released, the water vapor formed in the reaction is usually at a very high temperature compared with its standard state, liquid at 298.15 K. The heat of formation includes the heat that has to be removed when the vapor is condensed to liquid and then cooled to 298.15 K.

The reverse reaction, that is, the decomposition of water, is endothermic, and energy, a minimum of 285.89 kJ per mole of water, must be supplied. More energy is required by electrolysis to decompose water because some energy will be wasted as heat.

A hydrogen-containing compound, when fully oxidized, also produces water and energy. For example, the oxidation of solid (s) sucrose, $C_{12}H_{22}O_{11}$, can be written as

$$C_{12}H_{22}O_{11}(s) + 12O_2(g) = 12CO_2(g) + 11H_2O(l),$$
$$\Delta H^0 = -5640 \text{ kJ}$$

The amount of energy released, −5640 kJ, is called the *standard enthalpy of combustion* of sucrose. The chemical energy derived this way can also be used to produce high-energy biomolecules. When oxidation is carried out in human or animal bodies, the oxidation takes place at almost constant and low temperatures. Of course, the oxidation of sucrose takes place in many steps to convert each carbon to CO_2.

THE POLAR WATER MOLECULES

During the twentieth century, the study of live organisms evolved from physiology and anatomy to biochemistry and then down to the molecular level of intermolecular relations and functions. Atoms and molecules are the natural building blocks of matter, including that of living organisms. Molecular shapes, structures, and properties are valuable in genetics, biochemistry, food science, and molecular biology, all involving water. Thus, we have a strong desire to know the shape, size, construction, dimension, symmetry, and properties of water molecules, because they are the basis for the science of food and life.

The structure of water molecules has been indirectly studied using X-ray diffraction, spectroscopy, theoretical calculations, and other methods. Specific molecular dimensions from these methods differ slightly, because they measure different properties of water under different circumstances. However, the O—H bond length of 95.72 pm (1 pm = 10^{-12} m) and the H—O—H bond angle of 104.52° have been given after careful review of many recent studies (Petrenko and Whitworth 1999). The atomic radii of H and O are 120 pm and 150 pm, respectively. The bond length is considerably shorter than the sum of the atomic radii. Sketches of the water molecule are shown in Figure 5.1 (spherical atoms assumed).

Bonding among the elements H, C, N, and O is the key for biochemistry and life. Each carbon atom has the ability to form four chemical bonds. Carbon atoms can bond to other carbon atoms as well as to N and O atoms, forming chains, branched chains, rings, and complicated molecules. These carbon-containing compounds are called *organic compounds*, and they include foodstuff. The elements N and O have one and two more electrons, respectively, than carbon, and they have the capacity to form only three and two chemical bonds with other atoms. *Quantum mechanics* is a theory that explains the structure, energy, and properties of small systems such as atoms and molecules.

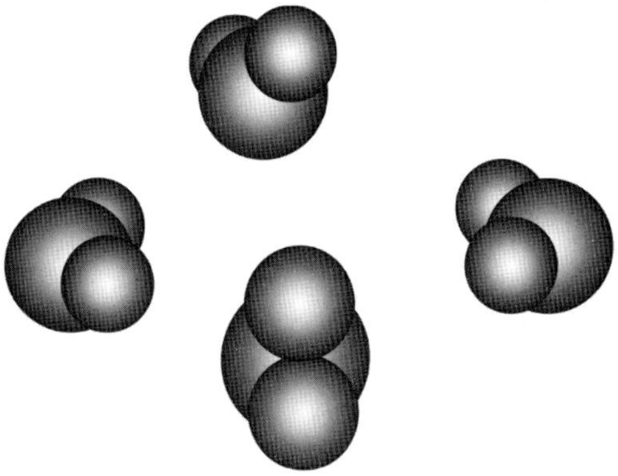

Figure 5.1. Some imaginative models of the water molecule, H_2O.

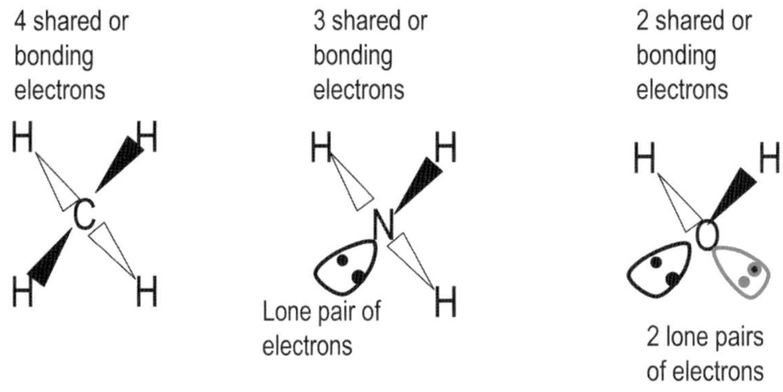

Figure 5.2. Molecules of CH_4, NH_3, and H_2O.

It provides excellent explanations for the electrons in the atom as well as the bonding of molecules, making the observed hard facts appear trivial (Bockhoff 1969).

The well-known inert elements helium (He) and neon (Ne) form no chemical bonds. The elements C, N, and O have four, five, and six electrons more than He, and these are called *valence electrons* (VE). Quantum mechanical designation for the VE of C, N, and O are $2s^2 2p^2$, $2s^2 2p^3$, and $2s^2 2p^4$, respectively. The C, N, and O atoms share electrons with four, three, and two hydrogen atoms, respectively. The formation of methane (CH_4), ammonia (NH_3), and water (H_2O) gave the C, N, and O atoms in these molecules eight VE. These molecules are related to each other in terms of bonding (Fig. 5.2).

The compounds CH_4, NH_3, and H_2O have zero, one, and two *lone pairs* (electrons not shared with hydrogen), respectively. Shared electron pairs form single bonds. The shared and lone pairs dispose themselves in space around the central atom symmetrically or slightly distorted when they have both bonding and lone pairs.

The lone pairs are also the negative sites of the molecule, whereas the bonded H atoms are the positive sites. The discovery of protons and electrons led to the idea of charge distribution in molecules. Physicists and chemists call NH_3 and H_2O *polar molecules* because their centers of positive and negative charge do not coincide. The polarizations in nitrogen and oxygen compounds contribute to their important roles in biochemistry and food chemistry. Elements C, N, and O play important and complementary roles in the formation of life.

Electronegativity is the ability of an atom to attract bonding electrons toward itself, and electronegativity increases in the order of H, C, N, and O (Pauling 1960). Chemical bonds between two atoms with different electronegativity are polar because electrons are drawn toward the more electronegative atoms. Thus, the polarity of the bonds increases in the order of H—C, H—N, and H—O, with the H atoms as the positive ends. The directions of the bonds must also be taken into account when the polarity of a whole molecule is considered. For example, the four slightly polar H—C bonds point toward the corner of a regular tetrahedron, and the polarities of the bonds cancel one another. The symmetric CH_4 molecules do not have a net dipole moment, and CH_4 is nonpolar. However, the asymmetric NH_3 and H_2O molecules are polar. Furthermore, the lone electron pairs make the NH_3 and H_2O molecules even more polar. The lone pairs also make the H—N—H, and H—O—H angles smaller than the 109.5° of methane. The chemical bonds become progressively shorter from CH_4 to H_2O as well. These distortions cause the dipole moments of NH_3 and H_2O to be 4.903×10^{-30} Cm (= 1.470 D) and 6.187×10^{-30} Cm (= 1.855 D), respectively. The tendency for water molecules to attract the positive sites of other molecules is higher than that of the NH_3 molecule, because water is the most polar of the two.

Bond lengths and angles are based on their equilibrium positions, and their values change as water molecules undergo vibration and rotation or when they interact with each other or with molecules of other compounds. Thus, the bond lengths, bond angles, and dipole moments change slightly from the values given above. Temperature, pressure, and the presence of electric and magnetic fields also affect these values.

Using atomic orbitals, valence bond theory, and molecular orbital theory, quantum mechanics has given beautiful explanations regarding the shapes, distortions, and properties of these molecules. Philosophers and theoreticians have devoted their lives to providing a comprehensive and artistic view of the water molecules.

WATER VAPOR CHEMISTRY AND SPECTROSCOPY

Spectroscopy is the study of the absorption, emission, or interaction of electromagnetic radiation by molecules in solid, liquid, and gaseous phases. The spectroscopic studies of vapor, in which the H_2O molecules are far apart from each other, reveal a wealth of information about individual H_2O molecules.

Electromagnetic radiation (light) is the transmission of energy through space via no medium by the oscillation of mutually perpendicular electric and magnetic fields. The oscillating *electromagnetic waves* move in a direction perpendicular to both fields at the speed of light ($c = 2.997925 \times 10^8$ m/s). Max Planck

(1858–1947) thought the waves also have particle-like properties except that they have no mass. He further called the light particles *photons,* meaning bundles of light energy. He assumed the *photon's energy, E,* to be proportional to its frequency. The proportional constant h ($= 6.62618 \times 10^{-34}$ J/s), now called the *Planck constant* in his honor, is universal. The validity of this assumption was shown by Albert Einstein's photoelectric-effect experiment.

Max Planck theorized that a bundle of energy converts into a light wave. His theory implies that small systems can be only at certain energy states called *energy levels.* Due to quantization, they can gain or lose only specific amounts of energy. Spectroscopy is based on these theories. Water molecules have quantized energy levels for their rotation, vibration, and electronic transitions. Transitions between energy levels result in the emission or absorption of photons.

The electromagnetic spectrum has been divided into several regions. From low energy to high energy, these regions are long radio wave, short radio wave, microwave, infrared (IR), visible, ultraviolet (UV), X rays, and gamma rays. Visible light of various colors is actually a very narrow region within the spectrum. On the other hand, IR and UV regions are very large, and both are often further divided into near and far, or A and B, regions.

Microwaves in the electromagnetic spectrum range from 300 MHz (3×10^{8} cycles/s) to 300 GHz (3×10^{11} cycles/s). The water molecules have many rotation modes. Their pure rotation energy levels are very close together, and the transitions between pure rotation levels correspond to microwave photons. *Microwave spectroscopy* studies led to, among other valuable information, precise bond lengths and angles.

Water molecules vibrate, and there are some fundamental vibration modes. The three fundamental vibration modes of water are symmetric stretching (for $^{1}H_2{}^{16}O$), ν_1, 3657 cm^{-1}, bending ν_2, 1595 cm^{-1}, and asymmetric stretching ν_3, 3756 cm^{-1}. These modes are illustrated in Figure 5.3. Vibration energy levels are represented by three integers, ν_1, ν_2, and ν_3, to represent the combination of the basic modes. The frequencies of fundamental vibration states differ in molecules of other isotopic species (Lemus 2004).

Water molecules absorb photons in the IR region, exciting them to the fundamental and combined overtones. As pointed out earlier, water molecules also rotate. The rotation modes combine with any and all vibration modes. Thus, transitions corresponding to the vibration-rotation energy levels are very complicated, and they occur in the *infrared* (frequency range 3×10^{11}–4×10^{14} Hz) region of the electromagnetic spectrum. High-resolution *IR spectrometry* is powerful for the study of water in the atmosphere and for water analyses (Bernath 2002a).

Visible light spans a narrow range, with wavelengths between 700 nm (red) and 400 nm (violet) (frequency 4.3–7.5 $\times 10^{14}$ Hz, wave number 14,000–25,000 cm^{-1}, photon energy 2–4 eV). It is interesting to note that the sun surface has a temperature of about 6000 K, and the visible region has the highest intensity of all wavelengths. The solar emission spectrum peaks at 630 nm (16,000 cm^{-1}, 4.8×10^{14} Hz), which is orange (Bernath 2002b).

Water molecules that have energy levels corresponding to very high overtone vibrations absorb photons of visible light, but the absorptions are very weak. Thus, visible light passes through water vapor with little absorption, resulting in water being transparent. On the other hand, the absorption gets progressively weaker from red to blue (Carleer et al. 1999). Thus, large bodies of water appear slightly blue.

Because visible light is only very weakly absorbed by water vapor, more than 90% of light passes through the atmosphere and reaches the earth's surface. However, the water droplets in clouds (water aerosols) scatter, refract, and reflect visible light, giving rainbows and colorful sunrises and sunsets.

Like the IR region, the *ultraviolet* (UV, 7×10^{14}–1×10^{18} Hz) region spans a very large range in the electromagnetic spectrum. The photon energies are rather high, more than 4 eV, and they are able to excite the electronic energy states of water molecules in the gas phase.

There is no room to cover the molecular orbitals (Gray 1964) of water here, but by analogy to electrons in atomic orbitals one can easily imagine that molecules have molecular orbitals or energy states. Thus, electrons can also be promoted to higher empty molecular orbitals after absorption of light energy. Ultraviolet photons have sufficiently high energies to excite electrons into higher molecular orbitals. Combined with vibrations and rotations, these transitions give rise to very broad bands in the UV spectrum. As a result, gaseous, liquid, and solid forms of water strongly absorb UV light (Berkowitz 1979). The

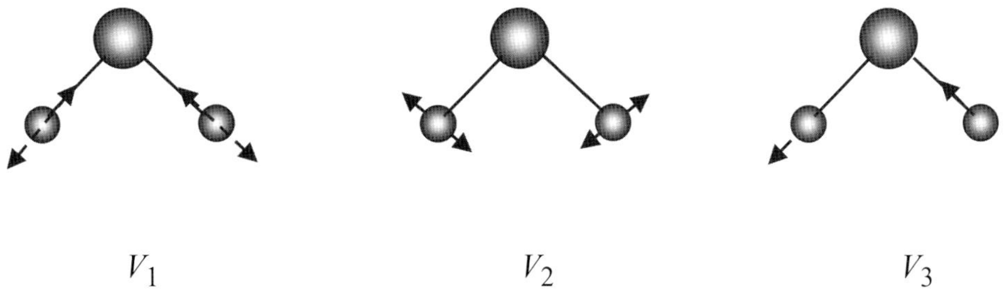

Figure 5.3. The three principal vibration modes of the water molecule, H_2O: ν_1, symmetric stretching; ν_2, bending; and ν_3, asymmetric stretching.

absorption intensities and regions of water vapor are different from those of ozone, but both are responsible for UV absorption in the atmosphere. Incidentally, both triatomic water and ozone molecules are bent.

Hydrogen Bonding and Polymeric Water in Vapor

Attraction between the lone pairs and hydrogen among water molecules is much stronger than any dipole–dipole interactions. This type of attraction is known as the hydrogen bond (O—H—O), a very prominent feature of water. Hydrogen bonds are directional and are more like covalent bonds than strong dipole–dipole interactions. Each water molecule has the capacity to form four hydrogen bonds, two by donating its own H atoms and two by accepting H atoms from other molecules. In the structure of ice, to be described later, all water molecules, except those on the surface, have four hydrogen bonds.

Attractions and strong hydrogen bonds among molecules form *water dimers* and polymeric *water clusters* in water vapor. Microwave spectroscopy has revealed their existence in the atmosphere (Goldman et al. 2001, Huisken et al. 1996).

As water dimers collide with other water molecules, trimer and higher polymers form. The directional nature of the hydrogen bond led to the belief that water clusters are linear, ring, or cage-like rather than aggregates of molecules in clusters (see Fig. 5.4). Water dimers, chains, and rings have one and two hydrogen-bonded neighbors. There are three neighbors per molecule in cage-like polymers. Because molecules are free to move in the gas and liquid state, the number of nearest neighbors is between four and six. Thus, water dimers and clusters are entities between water vapor and condensed water (Bjorneholm et al. 1999).

By analogy, when a few water molecules are intimately associated with biomolecules and food molecules, their properties would be similar to those of clusters.

CONDENSED WATER PHASES

Below the critical temperature of 647 K (374°C) and under the proper pressure, water molecules condense to form a liquid or solid—*condensed water*. Properties of water, ice, and vapor must be considered in freezing, pressure-cooking, and microwave heating. In food processing, these phases transform among one another. The transitions and the properties of condensed phases are manifestations of microscopic properties of water molecules. However, condensation modifies microscopic properties such as bond lengths, bond angles, vibration, rotation, and electronic energy levels. The same is true when water molecules interact with biomolecules and food molecules. All phases of water play important parts in biochemistry and food science.

Water has many anomalous properties, which are related to polarity and hydrogen bonding. The melting point (mp), boiling point (bp), and critical temperature are abnormally high for water. As a rule, the melting and boiling points of a substance are related to its molecular mass; the higher the molar mass, the higher the mp and bp. Mp and bp of water (molar mass 18, mp 273 K, bp 373 K) are higher than those of hydrogen compounds of adjacent elements of the same period, NH_3 (molar mass 17,

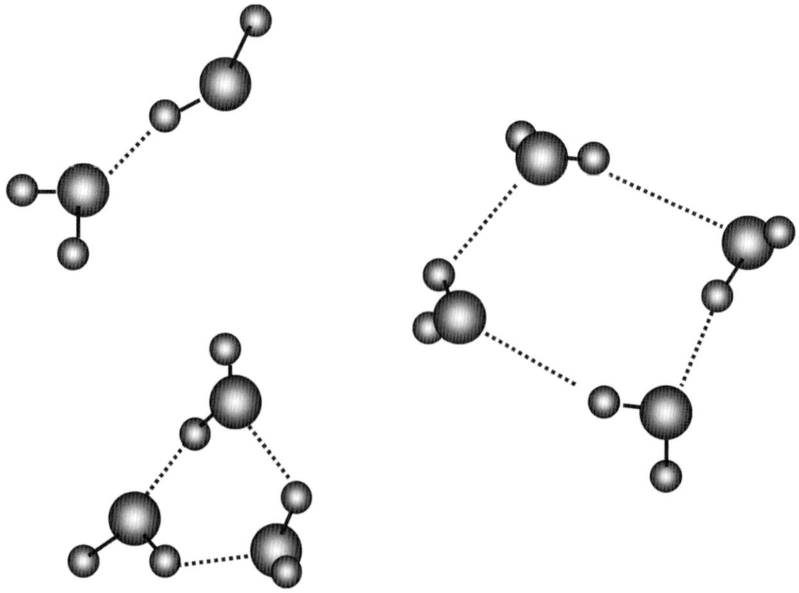

Figure 5.4. Hydrogen bonding in water dimers and cyclic forms of trimer and tetramer. Linear and transitional forms are also possible for trimers, tetramers, and polymers.

Table 5.1. Properties of Liquid Water at 298 K

Heat of formation ΔH_f	285.89 kJ mol^{-1}
Density at 3.98°C	1.000 g cm^3
Density at 25°C	0.9970480 g cm^3
Heat capacity	4.17856 J g^{-1}K^{-1}
$\Delta H_{vaporization}$	55.71 kJ mol^{-1}
Dielectric constant	80
Dipole moment	6.24×10^{-30} C m
Viscosity	0.8949 mPa s
Velocity of sound	1496.3 m s
Volumetric thermal expansion coefficient	0.0035 cm^3 g^{-1}K^{-1}

mp 195 K, bp 240 K) and HF (molar mass 20, mp 190 K, bp 293 K). If we compare the hydrogen compounds of elements from the same group (O, S, Se, and Te), the normal bp of H_2O (373 K) is by far the highest among H_2S, H_2Se, and H_2Te. Much energy (21 kJ mol^{-1}) is required to break the hydrogen bonds. The strong hydrogen bonds among water molecules in condensed phase result in anomalous properties, including the high enthalpies (energies) of fusion, sublimation, and evaporation given in Table 5.1. Internal energies and entropies are also high.

Densities of water and ice are also anomalous. Ice at 273 K is 9% less dense than water, but solids of most substances are denser than their liquids. Thus, ice floats on water, extending 9% of its volume above water. Water is the densest at 277 K (4°C). Being less dense at the freezing point, still water freezes from the top down, leaving a livable environment for aquatic organisms. The hydrogen bonding and polarity also lead to aberrant high surface tension, dielectric constant, and viscosity.

We illustrate the phase transitions between ice, liquid (water), and vapor in a phase diagram, which actually shows the equilibria among the common phases. Experiments under high pressure observed at least 13 different ices, a few types of amorphous solid water, and even the suggestion of two forms of liquid water (Klug 2002, Petrenko and Whitworth 1999). If these phases were included, the phase diagram for water would be very complicated.

Solid H_2O

At 273.16 K, ice, liquid H_2O, and H_2O vapor at 611.15 Pa coexist and are at equilibrium; the temperature and pressure define the *triple-point* of water. At the normal pressure of 101.3 kPa (1 atm), ice melts at 273.15 K. The temperature for the equilibrium water vapor pressure of 101.3 kPa is the bp, 373.15 K.

Under ambient pressure, ice often does not begin to form until it is colder than 273.15 K, and this is known as *supercooling*, especially for ultrapure water. The degree of supercooling depends on volume, purity, disturbances, the presence of dust, the smoothness of the container surface, and similar factors. Crystallization starts by *nucleation*, that is, formation of ice-structure clusters sufficiently large that they begin to grow and become crystals. Once ice begins to form, the temperature will return to the freezing point. At 234 K (−39°C), tiny drops of ultrapure water would suddenly freeze, and this is known as *homogeneous nucleation* (Franks et al. 1987). Dust particles and roughness of the surface promote nucleation and help reduce supercooling for ice and frost formation.

At ambient conditions, *hexagonal ice* (**Ih**) is formed. Snowflakes exhibit the hexagonal symmetry. Their crystal structure is well known (Kamb 1972). Every oxygen atom has four hydrogen bonds around it, two formed by donating its two H atoms, and two by accepting the H atoms of neighboring molecules. The hydrogen bonds connecting O atoms are shown in Figure 5.5. In normal ice, Ih, the positions of the H atoms are random or disordered. The hydrogen bonds O—H—O may be slightly bent, leaving the H—O—H angle closer to 105° than to 109.5°, the ideal angle for a perfect tetrahedral arrangement. Bending the hydrogen bond requires less energy than opening the H—O—H angle. Bending of the hydrogen bond and the exchange of H atoms among molecules, forming H_3O^+ and OH^- in the solids,

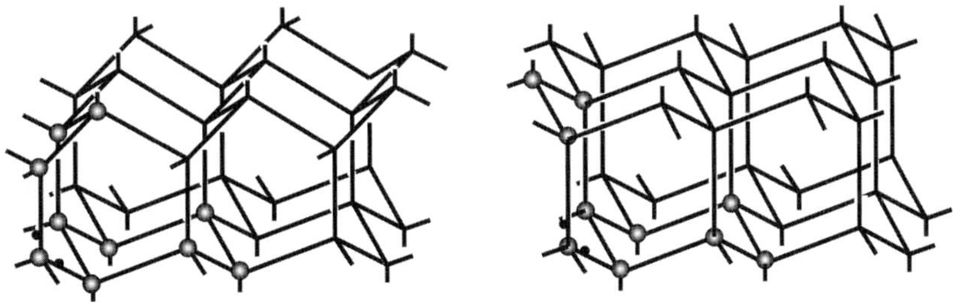

(Ih) (Ic)

Figure 5.5. The crystal structures of ice hexagonalice (**Ih**) and cubic ice (**Ic**). Oxygen atoms are placed in two rings in each to point out their subtle difference. Each line represents a hydrogen bond O—H—O, and the H atoms are randomly distributed such that on average, every oxygen atom has two O—H bonds of 100 pm. The O—H—O distance is 275 pm. The idealized tetrahedral bond angles around oxygen are 1095°.

give rise to the disorder of the H atoms. These rapid exchanges are in a dynamic equilibrium.

In the structure of Ih, six O atoms form a ring; some of them have a chair form, and some have a boat form. Two configurations of the rings are marked by spheres representing the O atoms in Figure 5.5. Formation of the hydrogen bond in ice lengthens the O—H bond distance slightly from that in a single isolated water molecule. All O atoms in Ih are completely hydrogen bonded, except for the molecules at the surface. Maximizing the number of hydrogen bonds is fundamental to the formation of solid water phases. Pauling (1960) pointed out that formation of hydrogen bonds is partly an electrostatic attraction. Thus, the bending of O—H—O is expected. Neutron diffraction studies indicated bent hydrogen bonds.

Since only four hydrogen bonds are around each O atom, the structure of Ih has rather large channels at the atomic scale. Under pressure, many other types of structures are formed. In liquid water, the many tetrahedral hydrogen bonds are formed with immediate neighbors. Since water molecules constantly exchange hydrogen-bonding partners, the average number of nearest neighbors is usually more than four. Therefore, water is denser than Ih.

Other Phases of Ice

Under high pressures, water forms many fascinating H_2O solids. They are designated by Roman numerals (e.g., ice XII; Klug 2002, Petrenko and Whitworth 1999). Some of these solids were known as early as 1900. Phase transitions were studied at certain temperatures and pressures, but metastable phases were also observed.

At 72 K, the disordered H atoms in Ih transform into an ordered solid called ice XI. The oxygen atoms of Ih and ice XI arrange in the same way, and both ices have a similar density, 0.917 Mg m^{-3}.

Under high pressure, various denser ices are formed. Ice II was prepared at a pressure about 1 GPa (1 GPa = 10^9 Pa) in 1900, and others with densities ranging from 1.17 to 2.79 Mg m^{-3} have been prepared during the twentieth century. These denser ices consist of hydrogen bond frameworks different from Ih and XI, but each O atom is hydrogen-bonded to four other O atoms.

Cubic ice, Ic, has been produced by cooling vapor or droplets below 200 K (Mayer and Hallbrucker 1987, Kohl et al. 2000). More studies showed the formation of Ic between 130 K and 150 K. Amorphous (glassy) water is formed below 130 K, but above 150 K Ih is formed. The hydrogen bonding and intermolecular relationships in Ih and Ic are the same, but the packing of layers and symmetry differ (see Fig. 5.5). The arrangement of O atoms in Ic is the same as that of the C atoms in the diamond structure. Properties of Ih and Ic are very similar. Crystals of Ic have cubic or octahedral shapes, resembling those of salt or diamond. The conditions for their formation suggest their existence in the upper atmosphere and in the Antarctic.

As in all phase transitions, energy drives the transformation between Ih and Ic. Several forms of amorphous ice having various densities have been observed under different temperatures and pressures. Unlike crystals, in which molecules are packed in an orderly manner, following the symmetry and periodic rules of the crystal system, the molecules in *amorphous ice* are immobilized from their positions in liquid. Thus, amorphous ice is often called frozen water or glassy water.

When small amounts of water freeze suddenly, it forms amorphous ice or glass. Under various temperatures and pressures, it can transform into high-density (1.17 mg/m^3) amorphous water, and very high-density amorphous water. Amorphous water also transforms into various forms of ice (Johari and Anderson 2004). The transformations are accompanied by energies of transition. A complicated phase diagram for ice transitions can be found in *Physics of Ice* (Petrenko and Whitworth 1999).

High pressures and low temperatures are required for the existence of other forms of ice, and currently these conditions are seldom involved in food processing or biochemistry. However, their existence is significant for the nature of water. For example, their structures illustrate the deformation of the ideal tetrahedral arrangement of hydrogen bonding presented in Ih and Ic. This feature implies flexibility when water molecules interact with foodstuffs and with biomolecules.

Vapor Pressure of Ice Ih

The equilibrium vapor pressure is a measure of the ability or potential of the water molecules to escape from the condensed phases to form a gas. This potential increases as the temperature increases. Thus, vapor pressures of ice, water, and solutions are important quantities. The ratio of equilibrium vapor pressures of foods divided by those of pure water is called the *water activity*, which is an important parameter for food drying, preservation, and storage.

Ice sublimes at any temperature until the system reaches equilibrium. When the vapor pressure is high, molecules deposit on the ice to reach equilibrium. Solid ice and water vapor form an equilibrium in a closed system. The amount of ice in this equilibrium and the free volume enclosing the ice are irrelevant, but the water vapor pressure or partial pressure matters. The equilibrium pressure is a function of temperature, and detailed data can be found in handbooks, for example, the *CRC Handbook of Chemistry and Physics* (Lide 2003). This handbook has a new edition every year. More recent values between 193 K and 273 K can also be found in *Physics of Ice* (Petrenko and Whitworth 1999).

The equilibrium vapor pressure of ice Ih plotted against temperature *(T)* is shown in Figure 5.6. The line indicates equilibrium conditions, and it separates the pressure–temperature *(P–T)* graph into two domains: vapor tends to deposit on ice in one, and ice sublimes in the other. This is the ice Ih—vapor portion of the phase diagram of water.

Plot of ln *P* versus 1/*T* shows a straight line, and this agrees well with the Clausius-Clapeyron equation. The negative slope gives the enthalpy of the phase transition, ΔH, divided by the gas constant R (= 8.3145 J/K/mol).

$$\frac{(d(\ln P))}{(d(1/T))} = \frac{-\Delta H}{R}$$

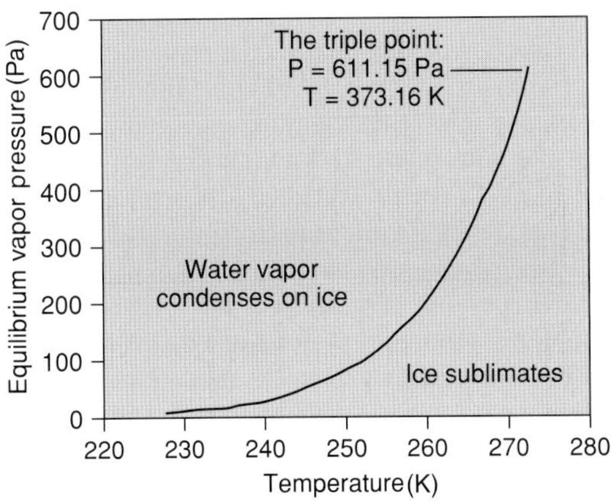

Figure 5.6. Equilibrium vapor pressure (Pa) of ice as a function of temperature (K).

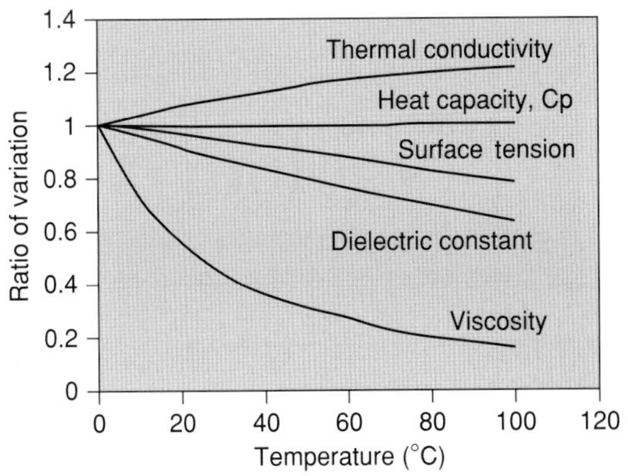

Figure 5.7. Variation of viscosity (1.793 mPa s), dielectric constant (87.90), surface tension (75.64 mN/m), heat capacity Cp (4.2176 J $g-^1K-^1$), and thermal conductivity (561.0 W $K-^1m-^1$) of water from their values at 273.16 K to 373.16 K (0 and 100°C). Values at 273.15°K are given.

This simplified equation gives an estimate of the vapor pressure (in Pa) of ice at temperatures in a narrow range around the triple point.

$$P = 611.15 \exp\left(-6148\left(\frac{1}{T} - \frac{1}{273.16}\right)\right)$$

The value 6148 is the enthalpy of sublimation ($\Delta_{sub}H = 51.1$ J/mol, varies slightly with temperature) divided by the gas constant R. Incidentally, the enthalpy of sublimation is approximately the sum of the enthalpy of fusion (6.0 J/mol) for ice and the heat of vaporization (45.1 J/mol, varies with temperature) of water at 273 K.

Liquid H_2O—Water

We started the chapter by calling the compound H_2O water, but most of us consider *water* the liquid H_2O. In terms of food processing, the liquid is the most important state. Water is contained in food, and it is used for washing, cooking, transporting, dispersing, dissolving, combining, and separating components of foods. Food drying involves water removal, and fermentation uses water as a medium to convert raw materials into commodities. Various forms of water ingested help digest, absorb, and transport nutrients to various part of the body. Water further facilitates biochemical reactions to sustain life. The properties of water are the basis for its many applications.

Among the physical properties of water, the heat capacity (4.2176 J $g^{-1}K^{-1}$ at 273.15 K) varies little between 273.15 and 373.15 K. However, this value decreases, reaches a minimum at about 308 K, and then rises to 4.2159–4.2176 J $g^{-1}K^{-1}$ at 373.15 K (see Fig. 5.7).

The viscosity, surface tension, and dielectric constant of liquid H_2O decrease as temperature increases (see Fig. 5.7). These three properties are related to the extent of hydrogen bonding and the ordering of the dipoles. As thermal disorder increases with rising temperature, these properties decrease. To show the variation, the properties at other temperatures are divided by the same property at 273 K. The ratios are then plotted as a function of temperature. At 273.15 K (0°C), all the ratios are unity (1). The thermal conductivity, on the other hand, increases with temperature. Thus, the thermal conductivity at 373 K (679.1 W $K^{-1}m^{-1}$) is 1.21 times that at 273 K (561.0 W $K^{-1}m^{-1}$). Warm water better conducts heat. Faster moving molecules transport energy faster. The variations of these properties play important roles in food processing or preparation. For example, as we shall see later, the dielectric constant is a major factor for the microwave heating of food, and heat conductivity plays a role cooking food.

Densities of other substances are often determined relative to that of water. Therefore, density of water is a primary reference. Variation of density with temperature is well known, and accurate values are carefully measured and evaluated especially between 273 K and 313 K (0–40°C). Two factors affect water density. Thermal expansion reduces its density, but the reduced number of hydrogen bonds increases its density. The combined effects resulted in the highest density at approximately 277 K (4°C). Tanaka et al. (2001) has developed a formula to calculate the density within this temperature range, and the *CRC Handbook of Chemistry and Physics* (Lide 2003) has a table listing these values. The variation of water density between the freezing point and the bp is shown in Figure 5.8. The densities are 0.9998426, 0.9999750, and 0.9998509 Mg m^{-3} at 0, 4, and 8°C, respectively. The decrease in density is not linear, and at 100°C, the density is 0.95840 Mg m^{-3}, a decrease of 4% from its maximum.

Vapor Pressure of Liquid H_2O

Equilibrium vapor pressure of water (Fig. 5.9) increases with temperature, similar to that of ice. At the triple point, the vapor

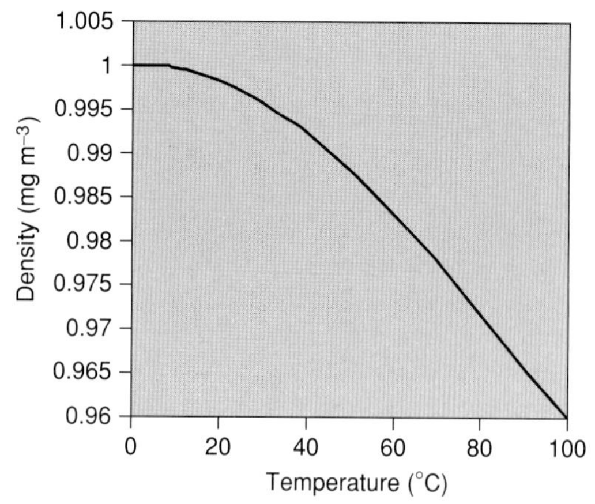

Figure 5.8. Density of water mg m−3 as a function of temperature (°C).

pressures of ice Ih and water are the same, 0.611 kPa, and the boiling point (373.15 K, 100°C) is the temperature at which the vapor pressure is 101.325 kPa (1 atm). At slightly below 394 K (121°C), the vapor pressure is 202.65 kPa (2.00 atm). At 473 and 574 K, the vapor pressures are 1553.6 and 8583.8 kPa, respectively. The vapor pressure rises rapidly as temperature increases. The lowest pressure to liquefy vapor just below the critical temperature, 373.98°C, is 22,055 kPa (217.67 atm), and this is known as the critical pressure. Above 373.98°C, water cannot be liquefied, and the fluid is called supercritical water.

The partial pressure of H_2O in the air at any temperature is the **absolute humidity**. When the partial pressure of water vapor in the air is the equilibrium vapor pressure of water at the same temperature, the **relative humidity** is 100%, and the air is saturated with water vapor. The partial vapor pressure in the air divided by the equilibrium vapor pressure of water at the temperature of the air is the **relative humidity**, expressed as a percentage. The temperature at which the vapor pressure in the air becomes saturated is the **dew point**, at which dew begins to form. Of course, when the dew point is below 273 K or 0°C, ice crystals (frost) begin to form. Thus, the relative humidity can be measured by finding the dew point and then dividing the equilibrium vapor pressure at the dew point by the equilibrium vapor pressure of water at the temperature of the air. The transformations between solid, liquid, and gaseous water play important roles in hydrology and in transforming the earth's surface. Solar energy causes phase transitions of water that make the weather.

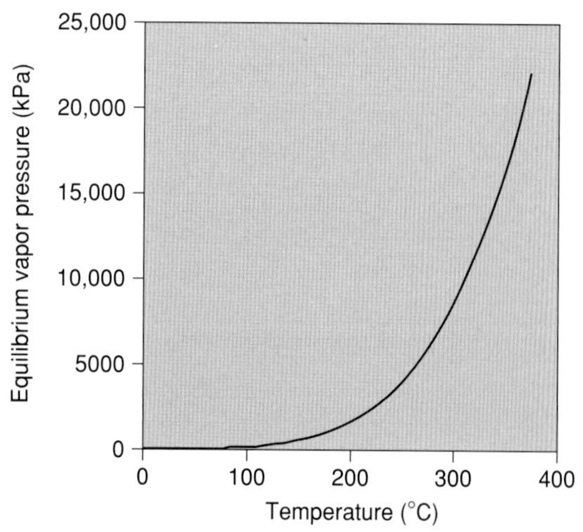

Figure 5.9. Equilibrium vapor pressure of water as a function of temperature.

Transformation of Solid, Liquid, and Vapor

Food processing and biochemistry involve transformations among solid, liquid, and vapor of water. Therefore, it is important to understand ice–water, ice–vapor, and water–vapor transformations and their equilibria. These transformations affect our daily lives as well. A map or diagram is helpful in order to comprehend these natural phenomena. Such a map, representing or explaining these transformations, is called a *phase diagram* (see Fig. 5.10). A sketch must be used because the range of pressure involved is too large for the drawing to be on a linear scale.

The curves representing the equilibrium vapor pressures of ice and water as functions of temperature meet at the *triple point* (see Fig. 5.10). The other end of the vapor pressure curve is the critical point. The mps of ice Ih are 271.44, 273.15, and 273.16 K at 22,055 kPa (the critical pressure), 101.325 kPa, and 0.611 kPa (the triple point), respectively. At a pressure of 200,000 kPa, Ih melts at 253 K. Thus, the line linking all these points represents the mp of Ih at different pressures. This line divides the conditions (pressure and temperature) for the formation of solid and liquid. Thus, the phase diagram is roughly divided into regions of solid, liquid, and vapor.

Ice Ih transforms into the ordered ice XI at low temperature. In this region and under some circumstance, Ic is also formed. The transformation conditions are not represented in Figure 5.10, and neither are the transformation lines for other ices. These occur at much higher pressures in the order of gigapascals. A box at the top of the diagram indicates the existence of these phases, but the conditions for their transformation are not given. Ices II–X, formed under gigapascals pressure, were mentioned earlier. Ice VII forms at greater than 10 GPa and at a temperature higher than the boiling point of water.

Formation and existence of these phases illustrate the various hydrogen bonding patterns. They also show the many possibilities of H_2O-biomolecule interactions.

Subcritical and Supercritical Waters

Water at temperatures between the boiling and critical points (100–373.98°C) is called subcritical water, whereas the phase

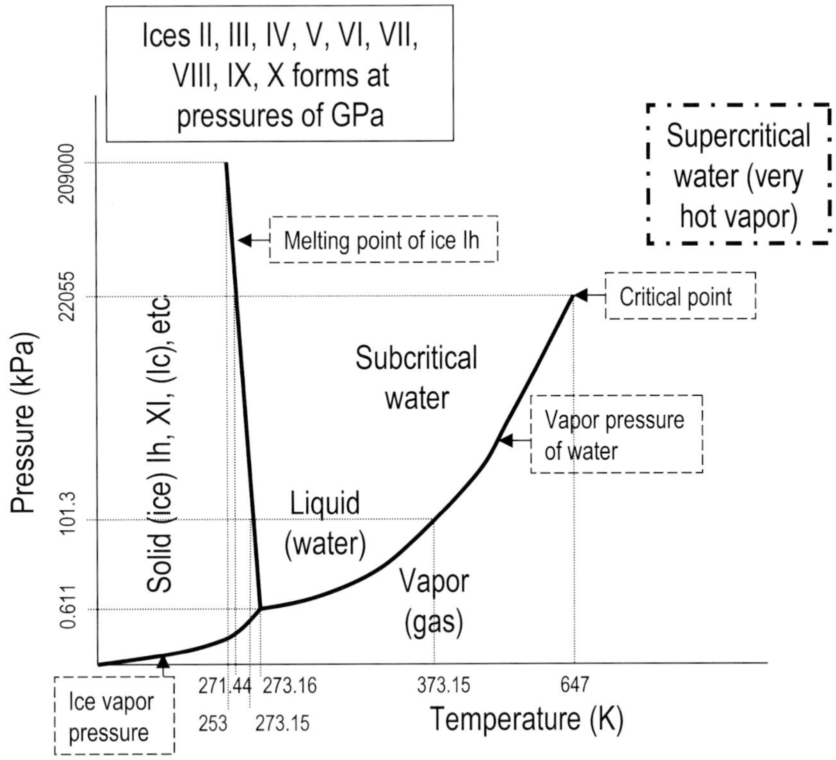

Figure 5.10. A sketch outlining the phase diagram of ice, water, and vapor.

above the critical point is *supercritical water*. In the seventeenth century, Denis Papin (a physicist) generated high-pressure steam using a closed boiler, and thereafter pressure canners have been used to preserve food. Pressure cookers were popular during the twentieth century. Pressure cooking and canning use subcritical water. Plastic bags are gradually replacing cans, and food processing faces new challenges.

Properties of subcritical and supercritical water such as dielectric constant, polarity, surface tension, density, viscosity, and others, differ from those of normal water. These properties can be tuned by adjusting water temperature. At high temperature, water is an excellent solvent for nonpolar substances such as those for flavor and fragrance. Supercritical water has been used for wastewater treatment to remove organic matter, and this application will be interesting to the food industry if companies are required to treat their wastes before discharging them into the environment. The conditions for supercritical water cause polymers to depolymerize and nutrients to degrade. More research will tell whether supercritical water will convert polysaccharides and proteins into useful products.

Supercritical water is an oxidant, which is desirable for the destruction of substances. It destroys toxic material without the need of a smokestack. Water is a "green" solvent and reagent, because it causes minimum damage to the environment. Therefore, the potential for supercritical water is great.

AQUEOUS SOLUTIONS

Water dissolves a wide range of natural substances. Life began in natural waters, *aqueous solutions*, and continuation of life depends on them.

Water, a polar solvent, dissolves polar substances. Its polarity and its ability to hydrogen bond make it a nearly universal solvent. Water-soluble polar substances are *hydrophilic*, whereas nonpolar substances insoluble in water are *hydrophobic* or *lipophilic*. Substances whose molecules have both polar and nonpolar parts are *amphiphilic*. These substances include detergents, proteins, aliphatic acids, alkaloids, and some amino acids.

The high dielectric constant of water makes it an ideal solvent for ionic substances, because it reduces the attraction between positive and negative ions in *electrolytes*: acids, bases, and salts. Electrolyte solutions are intimately related to food and biological sciences.

Colligative Properties of Aqueous Solutions

Vapor pressure of an aqueous solution depends only on the concentration of the solute, not on the type and charge of the solute. Vapor pressure affects the mp, the bp, and the osmotic pressure, and these are *colligative properties*.

In a solution, the number of moles of a component divided by the total number of moles of all components is the *mole fraction* of that component. This is a useful parameter, and the mole fraction of any pure substance is unity. When 1 mol of sugar dissolves in 1.0 kg (55.56 mol) of water, the mole fraction of water is reduced to 0.9823, and the mole fraction of sugar is 0.0177. Raoult's law applies to aqueous solutions. According to it, the vapor pressure of a solution at temperature T is the product of mole fraction of the solvent and its vapor pressure at T. Thus, the vapor pressure of this solution at 373 K is 99.53 kPa (0.9823 atm). In general, the vapor pressure of a solution at the normal bp is lower than that of pure water, and a higher temperature is required for the vapor pressure to reach 101.3 kPa (1 atm). The net increase in the boiling temperature of a solution is known as the *boiling point elevation*.

Ice formed from a dilute solution does not contain the solute. Thus, the vapor pressure of ice at various temperatures does not change, but the vapor pressure of a solution is lower than that of ice at the freezing point. Further cooling is required for ice to form, and the net lowering of vapor pressure is the *freezing point depression*.

The freezing points and bps are temperatures at ice–water and water–vapor (at 101.3 kPa) equilibria, not necessarily the temperatures at which ice begins to form or boiling begins. Often, the temperature at which ice crystals start to form is lower than the mp, and this is known as *supercooling*. The degree of supercooling depends on many other parameters revealed by a systematic study of heterogeneous nucleation (Wilson et al. 2003). Similarly, the inability to form bubbles results in *superheating*. Overheated water boils explosively due to the sudden formation of many large bubbles. Supercooling and superheating are nonequilibrium phenomena, and they are different from freezing point depression and bp elevation.

The temperature differences between the freezing points and bps of solutions and those of pure water are proportional to the concentration of the solutions. The proportionality constants are *molar freezing point depression* ($K_f = -1.86$ K mol^{-1} kg) and the *molar boiling point elevation* ($K_b = 0.52$ K mol^{-1} kg), respectively. In other words, a solution containing one mole of nonvolatile molecules per kilogram of water freezes at 1.86° below 273.15 K, and boils 0.52° above 373.15 K. For a solution with concentration C, measured in moles of particles (molecules and positive and negative ions counted separately) per kilogram of water, the bp elevation or freezing point depression ΔT can be evaluated:

$$\Delta T = KC,$$

where K represents K_f or K_b for freezing point depression or bp elevation, respectively. This formula applies to both cases. Depending on the solute, some aqueous solutions may deviate from Raoult's law, and the above formulas give only estimates.

In freezing or boiling a solution of a nonvolatile solute, the solid and vapor contain only H_2O, leaving the solute in the solution. A solution containing volatile solutes will have a total vapor pressure due to all volatile components in the solution. Its bp is no longer that of water alone. Freezing point depression and bp elevation are both related to the vapor pressure. In general, a solution of nonvolatile substance has a lower vapor pressure than that of pure water. The variations of these properties have many applications, some in food chemistry and biochemistry.

It is interesting to note that some fish and insects have antifreeze proteins that depress the freezing point of water to protect them from freezing in the arctic sea (Marshall et al. 2004).

Osmotic pressure is usually defined as the pressure that must be applied to the side of the solution to prevent the flow of the pure solvent passing a semipermeable membrane into the solution. Experimental results show that osmotic pressure is equal to the product of total concentration of molecules and ions (C), the gas constant (R = 8.3145 J/mol/K), and the temperature (T in K):

$$\text{Osmotic pressure} = CRT.$$

The expression for osmotic pressure is the same as that for ideal gas. Chemists calculate the pressure using a concentration based on mass of the solvent. Since solutions are never ideal, the formula gives only estimates. The unit of pressure works out if the units for C are in mol/m^3, since 1 J = 1 Pa m^3. Concentration based on mass of solvent differs only slightly from that based on volume of the solution.

An *isotonic solution* (*isosmotic*) is one that has the same osmotic pressure as another. Solutions with higher and lower osmotic pressures are called *hypertonic* (*hyperosmotic*), and *hypotonic* (*hypoosmotic*), respectively. Raw food animal and plant cells submerged in isotonic solutions with their cell fluids will not take up or lose water even if the cell membranes are semipermeable. However, in plant, soil, and food sciences, *water potential gradient* is the driving force or energy directing water movement. Water moves from high-potential sites to low-potential sites. Water moves from low osmotic pressure solutions to high osmotic pressure solutions. For consistency and to avoid confusion, negative osmotic pressure is defined as *osmotic potential*. This way, osmotic potential is a direct component of water potential for gradient consideration. For example, when red blood cells are placed in dilute solutions, their osmotic potential is negative. Water diffuses into the cells, resulting in the swelling or even bursting of the cells. On the other hand, when cells are placed in concentrated (hypertonic) saline solutions, the cells will shrink due to water loss.

Biological membranes are much more permeable than most man-made phospholipid membranes because they have specific membrane-bound proteins acting as water channels. Absorption of water and its transport throughout the body is more complicated than osmosis.

Solution of Electrolytes

Solutions of acids, bases, and salts contain ions. Charged ions move when driven by an electric potential, and electrolyte solutions conduct electricity. These ion-containing substances are called *electrolytes*. As mentioned earlier, the high dielectric constant of water reduces the attraction of ions within ionic solids and dissolves them. Furthermore, the polar water molecules surround ions, forming *hydrated ions*. The concentration of all ions and molecular substances in a solution contributes to the osmotic potential.

Water can also be an acid or a base, because H_2O molecules can receive or provide a proton (H^+). Such an exchange by water molecules in pure water, forming *hydrated protons* (H_3O^+ or $(H_2O)_4H^+$), is called *self-ionization*. However, the extent of ionization is small, and pure water is a very poor conductor.

Self-Ionization of Water

The self-ionization of water is a dynamic equilibrium,

$$H_2O(l) \leftrightarrow H^+(aq) + OH^-(aq),$$

$$K_w = [H^+][OH^-] = 10^{-14} \text{ at 298 K and 1 atm}$$

where $[H^+]$ and $[OH^-]$ represent the molar concentrations of H^+ (or H_3O^+) and OH^- ions, respectively, and K_w is called the *ion product of water*. Values of K_w under various conditions have been evaluated theoretically (Marshall and Franck 1981, Tawa and Pratt 1995). Solutions in which $[H^+] = [OH^-]$ are said to be *neutral*. Both pH and pOH, defined by the following equations, have a value of 7 at 298 K for a neutral solution

$$pH = -\log_{10}[H^+] = pOH = -\log_{10}[OH^-] = 7.\text{(at 298 K)}$$

The H^+ represents a hydrated proton (H_3O^+), which dynamically exchanges a proton with other water molecules. The self-ionization and equilibrium are present in water and all aqueous solutions.

Solutions of Acids and Bases

Strong acids perchloric acid ($HClO_4$), Chloric acid ($HClO_3$), hydrochoric acid (HCl), Aitric acid (HNO_3), and Sulphuric acid H_2SO_4 completely ionize in their solutions to give H^+ (H_3O^+) ions and anions ClO_4^-, ClO_3^-, Cl^-, NO_3^-, and HSO_4^-, respectively. Strong bases NaOH, KOH, and $Ca(OH)_2$ also completely ionize to give OH^- ions and Na^+, K^+, and Ca^{2+} ions, respectively. In an acidic solution, $[H^+]$ is greater than $[OH^-]$. For example, in a 1.00 mol/L HCl solution at 298 K, $[H^+] = 1.00$ mol/L, pH = 0.00, $[OH^-] = 10^{-14}$ mol/L.

Weak acids such as formic acid (HCOOH), acetic acid (HCH_3COO), ascorbic acid ($H_2C_6H_6O_6$), oxalic acid ($H_2C_2O_4$), carbonic acid (H_2CO_3), benzoic acid (HC_6H_5COO), malic acid ($H_2C_4H_4O_5$), lactic acid ($HCH_3CH(OH)COO$), and phosphoric acid (H_3PO_4) also ionize in their aqueous solutions, but not completely. The ionization of acetic acid is represented by the equilibrium

$$HCH_3COO(aq) \leftrightarrow H^+(aq) + CH_3COO^-(aq),$$

$$K_a = \frac{[H^+][CH_3COO^-]}{[HCH_3COO]} = 1.75 \times 10^{-5} \text{ at 298 K}$$

where K_a, as defined above, is the *acid dissociation constant*.

The solubility of CO_2 in water increases with partial pressure of CO_2, according to Henry's law. The chemical equilibrium for the dissolution is:

$$H_2O(l) + CO_2(g) \leftrightarrow H_2CO_3(aq)$$

Of course, H_2CO_3 dynamically exchanges H^+ and H_2O with other water molecules, and this weak *diprotic acid* ionizes in two stages with acid dissociation constants K_{a1} and K_{a2}:

$$H_2O + CO_2(aq) \leftrightarrow H^+(aq) + HCO_3^-(aq),$$
$$K_{a1} = 4.30 \times 10^{-7} \text{ at 298 K}$$
$$HCO_3^-(aq) \leftrightarrow H^+(aq) + CO_3^{2-}(aq),\ K_{a2} = 5.61 \times 10^{-11}.$$

Constants K_{a1} and K_{a2} increase as temperature rises, but the solubility of CO_2 decreases. At 298 K, the pH of a solution containing 0.1 mol/L H_2CO_3 is 3.7. At this pH, acidophilic organisms survive and grow, but most pathogenic organisms are neutrophiles, and they cease growing. Soft drinks contain other acids—citric, malic, phosphoric, ascorbic, and others. They lower the pH further.

All three hydrogen ions in phosphoric acid (H_3PO_4) are ionizable, and it is a *triprotic acid*. Acids having more than one dissociable H^+ are called *polyprotic acids*.

NH_3 and many nitrogen-containing compounds are weak bases. The ionization equilibrium of NH_3 in water and the *base dissociation constant* K_b are

$$NH_3 + H_2O \leftrightarrow NH_4O \leftrightarrow NH_4^+(aq) + OH^-(aq),$$

$$K_b = \frac{[H^+][OH^-]}{[NH_4OH]} = 1.70 \times 10^{-5} \text{ at 298 K.}$$

Other weak bases react with H_2O and ionize in a similar way.

The ionization or dissociation constants of inorganic and organic acids and bases are extensive, and they have been tabulated in various books (for example Perrin 1965, 1982, Kortüm et al. 1961).

Titration

Titration is a procedure for quantitative analysis of a solute in a solution by measuring the quantity of a reagent used to completely react with it. This method is particularly useful for the determination of acid or base concentrations. A solution with a known concentration of one reagent is added from a burette to a definite amount of the other. The *end point* is reached when the latter substance is completely consumed by the reagent from the burette, and this is detected by the color change of an indicator or by pH measurements. This method has many applications in food analysis.

The titration of strong acids or bases utilizes the rapid reaction between H^+ and OH^-. The unknown quantity of an acid or base may be calculated from the amount used to reach the end point of the titration.

The variation of pH during the titration of a weak acid using a strong base or a weak base using a strong acid is usually monitored to determine the end point. The plot of pH against the amount of reagent added is a *titration curve*. There are a number of interesting features on a titration curve. Titration of a weak acid HA using a strong base NaOH is based on two rapid equilibria:

$$H^+ + OH^- = H_2O,\ K = \frac{[H_2O]}{K_w} = 5.56 \times 10^{15} \text{ at 298 K.}$$

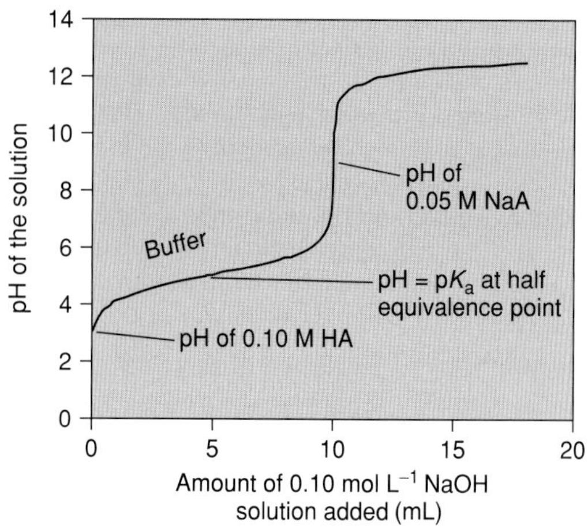

Figure 5.11. Titration curve of a 0.10 mol/L (or M) weak acid HA ($K_a = 1 \times 10{-}^{5}$) using a 0.10 mol/L strong base NaOH solution.

As the H^+ ions react with OH^-, more H^+ ions are produced due to the equilibrium

$$HA = H^+ + A^-, \ K_a \text{ (ionization constant of the acid).}$$

Before any NaOH solution is added, HA is the dominant species; when half of HA is consumed, [HA] = [A^-], which is called the *half equivalence point*. At this point, the pH varies the least when a little H^+ or OH^- is added, and the solution at this point is the *most effective buffer solution*, as we shall see later. The pH at this point is the same as the pK_a of the weak acid. When an equivalent amount of OH^- has been added, the A^- species dominates, and the solution is equivalent to a salt solution of NaA. Of course, the salt is completely ionized.

Polyprotic acids such as ascorbic acid H_2-($H_6C_6O_6$) (Vitamin C; $K_{a1} = 7.9 \times 10^{-5}$, $K_{a2} = 1.6 \times 10^{-12}$) and phosphoric acid H_3PO_4 ($K_{a1} = 6.94 \times 10^{-3}$, $K_{a2} = 6.2 \times 10^{-8}$, K_{a3} 2.1×10^{-12}) have more than one mole of H^+ per mole of acid. A titration curve of these acids will have two and three end points for ascorbic and phosphoric acids, respectively, partly due to the large differences in their dissociation constants (K_{a1}, K_{a2}, etc.). In practice, the third end point is difficult to observe in the titration of H_3PO_4. Vitamin C and phosphoric acids are often used as food additives.

Many food components (e.g., amino acids, proteins, alkaloids, organic and inorganic stuff, vitamins, fatty acids, oxidized carbohydrates, and compounds giving smell and flavor) are weak acids and bases. The pH affects their forms, stability, and reactions. When pH decreases by 1, the concentration of H^+, [H^+], increases 10-fold, accompanied by a 10-fold decrease in [OH^-]. The H^+ and OH^- are very active reagents for the esterification and hydrolysis reactions of proteins, carbohydrates, and lipids, as we shall see later. Thus, the acidity, or pH, not only affects the taste of food, it is an important parameter in food processing.

Solutions of Amino Acids

Amino acids have an amino group (NH_3^+), a carboxyl group (COO^-), a H, and a side chain (R) attached to the asymmetric alpha carbon. They are the building blocks of proteins, polymers of amino acids. At a pH called the *isoelectric point*, which depends on the amino acid in question, the dominant species is a *zwitterion*, $RHC(NH_3^+)(COO^-)$, which has a positive and a negative site, but no net charge. For example, the isoelectric point for glycine is pH = 6.00, and its dominant species is $H_2C(NH_3^+)COO^-$. An amino acid exists in at least three forms due to the following ionization or equilibria:

$$RHC(NH_3^+)(COOH) = RHC(NH_3^+)(COO^-) + H^+, \ K_{a1}$$
$$RHC(NH_3^+)(COO^-) = RHC(NH_2)(COO^-) + H^+, \ K_{a2}.$$

Most amino acids behave like a diprotic acid with two dissociation constants, K_{a1} and K_{a2}. A few amino acids have a third ionizable group in their side chains.

Among the 20 common amino acids, the side chains of eight are nonpolar, and those of seven are polar, containing —OH, >C=O, or —SH groups. Aspartic and glutamic acid contain acidic —COOH groups in their side chains, whereas arginine, histidine, and lysine contain basic —NH or —NH_2 groups. These have four forms due to adding or losing protons at different pH values of the solution, and they behave as triprotic acids. For example, aspartic acid [Asp = $(COOH)CH_2C(NH_3^+)(COO^-)$] has these forms:

$$AspH^+ = Asp + H^+$$
$$Asp = Asp^- + H^+$$
$$Asp^- = Asp^{2-} + H^+$$

Proteins, amino acid polymers, can accept or provide several protons as the pH changes. At its *isoelectric point* (a specific pH), the protein has no net charge and is least soluble because electrostatic repulsion between its molecules is lowest, and the molecules coalesce or precipitate, forming a solid or gel.

Solutions of Salts

Salts consist of positive and negative ions, and these ions are hydrated in their solutions. Positive, hydrated ions such as $Na(H_2O)_6{}^+$, $Ca(H_2O)_8{}^{2+}$, and $Al(H_2O)_6{}^{3+}$ have six to eight water molecules around them. Figure 5.12 is a sketch of the interactions of water molecules with ions. The water molecules point the negative ends of their dipoles toward positive ions, and their positive ends toward negative ions. Molecules in the hydration sphere constantly and dynamically exchange with those around them. The number and lifetimes of hydrated water molecules have been studied by various methods. These studies reveal that the hydration sphere is one layer deep, and the lifetimes of these hydrated water molecules are in the order of picoseconds (10^{-12} seconds). The larger negative ions also interact with the polar water molecules, but not as strongly as do cations. The presence of ions in the solution changes the ordering of water molecules even if they are not in the first hydration sphere.

The hydration of ions releases energy, but breaking up ions from a solid requires energy. The amount of energy needed

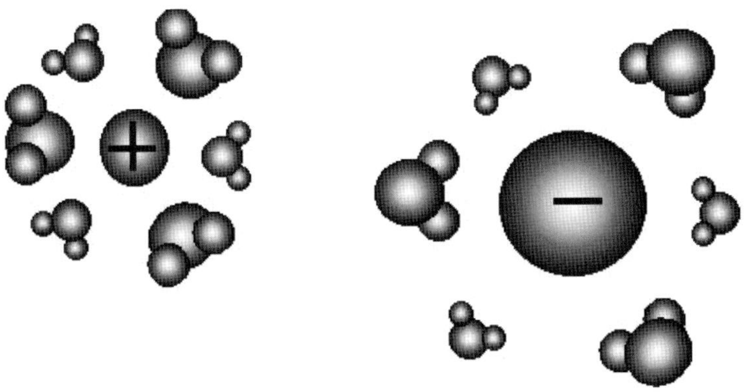

Figure 5.12. The first hydration sphere of most cations $M(H_2O)_6{}^+$, and anions $X(H_2O)_6-^1$. Small water molecules are below the plane containing the ions, and large water molecules are above the plane.

depends on the substance, and for this reason, some substances are more soluble than others. Natural waters in oceans, streams, rivers, and lakes are in contact with minerals and salts. The concentrations of various ions depend on the solubility of salts (Moeller and O'Connor 1972) and the contact time.

All salts dissolved in water are completely ionized, even those formed in the reaction between weak acids and weak bases. For example, the common food preservative sodium benzoate (NaC_6H_5COO) is a salt formed between a strong base NaOH and weak benzoic acid ($K_a = 6.5 \times 10^{-5}$). The benzoate ions, $C_6H_5COO^-$, in the solution react with water to produce OH^- ions giving a slightly basic solution:

$$C_6H_5COO^- + H_2O \leftrightarrow C_6H_5COOH + OH^-,$$
$$K = 1.6 \times 10^{-10}, \text{ at 298 K.}$$

Ammonium bicarbonate, NH_4HCO_3, was a leavening agent before modern baking powder was popular. It is still called for in some recipes. This can be considered a salt formed between the weak base NH_4OH and the weak acid H_2CO_3. When NH_4HCO_3 dissolves in water, the ammonium and bicarbonate ions react with water:

$$NH_4^+ + H_2O \leftrightarrow NH_3(aq) + H_3O^+, \; K = 5.7 \times 10^{-10}, \text{ at 298 K}$$
$$HCO_3^- + H_2O \leftrightarrow H_2CO_3 + OH^-, \; K = 2.3 \times 10^{-8}, \text{ at 298 K}$$
$$H_2CO_3 = H_2O + CO_2(g).$$

Upon heating, NH_3 and carbon dioxide (CO_2) become gases for the leavening action. Thus, during the baking or frying process, NH_3 is very pungent and unpleasant. When sodium bicarbonate is used, only CO_2 causes the dough to rise. Phosphoric acid, instead of $NH_4{}^+$, provides the acid in baking powder.

Buffer Solutions

A solution containing a weak acid and its salt or a weak base and its salt is a buffer solution, since its pH changes little when a small amount acid or base is added. For example, nicotinic acid ($HC_6H_4NO_2$, niacin, a food component) is a weak acid with $K_a = 1.7 \times 10^{-5}$ ($pK_a = -\log_{10}K_a = 4.76$):

$$HC_6H_4NO_2 = H^+ + C_6H_4NO_2^-,$$
$$K_a = \frac{[H^+][C_6H_4NO_2^-]}{[HC_6H_4NO_2]}$$
$$pH = pK_a + \log_{10}\left\{\frac{[C_6H_4NO_2^-]}{[HC_6H_4NO_2]}\right\}$$

(Henderson-Hesselbalch equation).

In a solution containing niacin and its salt, $[C_6H_4NO_2{}^-]$ is the concentration of the salt, and $[HC_6H_4NO_2]$ is the concentration of niacin. The pair, $HC_6H_4NO_2$ and $C_6H_4NO_2{}^-$, are called **conjugate acid and base**, according to the Bronsted-Lowry definition for acids and bases. So, for a general acid and its conjugate base, the pH can be evaluated using the Henderson-Hesselbalch equation:

$$pH = pK_a = \log_{10}\left\{\frac{[base]}{[acid]}\right\}$$

(Henderson-Hesselbalch Equation).

Adding H^+ converts the base into its conjugate acid, and adding OH^- converts the acid into its conjugate base. Adding acid and base changes the ratio [base]/[acid], causing a small change in the pH if the initial ratio is close to 1.0. Following this equation, the most effective buffer solution for a desirable pH is to use an acid with a pK_a value similar to the desired pH value and to adjust the concentration of the salt and acid to obtain the ratio that gives the desired pH. For example, the pK_a for $H_2PO_4{}^-$ is 7.21, and mixing KH_2PO_4 and K_2HPO_4 in the appropriate ratio will give a buffer solution with pH 7. However, more is involved in the art and science of making and standardizing buffer solutions. For example, the ionic strength must be taken into account.

The pH of blood from healthy persons is 7.4. The phosphoric acid and bicarbonate ions in blood and many other soluble biomaterials in the intercellular fluid play a buffering role in keeping the pH constant. The body fluids are very complicated buffer solutions, because each conjugate pair in the solution has an equilibrium of its own. These equilibria plus the equilibrium due to the self-ionization of water stabilize the pH of the solution. Buffer solutions abound in nature: milk, juice, soft drinks, soup, fluid contained in food, and water in the ocean, for example.

Hydrophilic and Hydrophobic Effects

The *hydrophilic effect* refers to the hydrogen bonding, polar-ionic and polar–polar interactions with water molecules, which lower the energy of the system and make ionic and polar substances soluble. The lack of strong interactions between water molecules and lipophilic molecules or the nonpolar portions of amphiphilic molecules is called the *hydrophobic effect*, a term coined by Charles Tanford (1980).

When mixed with water, ionic and polar molecules dissolve and disperse in the solution, whereas the nonpolar or hydrophobic molecules huddle together, forming groups. At the proper temperature, groups of small and nonpolar molecules surrounded by water cages form stable phases called *hydrates* or *clathrates*. For example, the clathrate of methane forms stable crystals at temperatures below 300 K (Sloan 1998). The hydrophobic effect causes the formation of micelles and the folding of proteins in enzymes so that the hydrophobic parts of the long chain huddle together on the inside, exposing the hydrophilic parts to the outside to interact with water.

Hydrophilic and hydrophobic effects together stabilize three-dimensional structures of large molecules such as enzymes, proteins, and lipids. Hydrophobic portions of these molecules stay together, forming pockets in globular proteins. These biopolymers minimize their hydrophobic surface to reduce their interactions with water molecules. Biological membranes often have proteins bonded to them, and the hydrophilic portions extend to the intra- and intercellular aqueous solutions. These membrane-bound proteins often transport specific nutrients in and out of cells. For example, water, amino acid, and potassium-sodium ion transporting channels are membrane-bound proteins (Garrett and Grisham 2002).

Hydrophilic and hydrophobic effects, together with the ionic interaction, cause long-chain proteins called *enzymes* to fold in specific conformations (three-dimensional structures) that catalyze specific reactions. The pH of the medium affects the charges of the proteins. Therefore, the pH may alter enzyme conformations and affect their functions. At a specific pH, some enzymes consist of several subunits that aggregate into one complex structure in order to minimize the hydrophobic surface in contact with water. Thus, the chemistry of water is intimately mingled with the chemistry of life.

During food processing, proteins are denatured by heat, acid, base, and salt. These treatments alter the conformation of the proteins and enzymes. Denatured proteins lose their life-maintaining functionality. Molecules containing hydrophilic and hydrophobic parts are emulsifiers that are widely used in the food industry.

Hydrophilic and hydrophobic effects cause nonpolar portions of phospholipids, proteins, and cholesterol to assemble into micelles and bilayers, or biological membranes (Sloan 1998). The membrane conformations are stable due to their low energy, and they enclose compartments with components to perform biological functions. Proteins and enzymes attached to the membranes communicate and transport nutrients and wastes for cells, keeping them alive and growing.

Hard Waters and Their Treatments

Waters containing dissolved CO_2 (same as H_2CO_3) are acidic due to the equilibria

$$H^+(aq) + HCO_3^-(aq) \leftrightarrow H_2CO_3(aq)\Delta H_2O + CO_2(g)$$

$$HCO_3^-(aq) \leftrightarrow H^+(aq) + CO_3^{2-}(aq).$$

Acidic waters dissolve $CaCO_3$ and $MgCO_3$, and waters containing Ca^{2+}, Mg^{2+}, HCO_3^-, and CO_3^{2-} are *temporary hard waters*, as the hardness is removable by boiling, which reduces the solubility of CO_2. When CO_2 is driven off, the solution becomes less acidic due to the above equilibria. Furthermore, reducing the acidity increases the concentration of CO_3^{2-}, and solids $CaCO_3$ and $MgCO_3$ precipitate:

$$Ca^{2+}(aq) + CO_3^{2-}(aq) \leftrightarrow CaCO_3(s)$$

$$Mg^{2+}(aq) + CO_3^{2-}(aq) \leftrightarrow MgCO_3(s).$$

Water containing less than 50 mg/L of these substances is considered soft; 50–150 mg/L moderately hard; 150–300 mg/L hard; and more than 300 mg/L very hard.

For water softening by the *lime treatment*, the amount of dissolved Ca^{2+} and Mg^{2+} is determined first; then an equal number of moles of lime, $Ca(OH)_2$, is added to remove them, by these reactions:

$$Mg^{2+} + Ca(OH)_2(s) \leftrightarrow Mg(OH)_2(s) + Ca^{2+}$$

$$Ca^{2+} + 2HCO_3^- + Ca(OH)_2(s) \leftrightarrow 2CaCO_3(s) + 2H_2O.$$

Permanent hard waters contain sulfate (SO_4^{2-}), Ca^{2+}, and Mg^{2+} ions. Calcium ions in the sulfate solution can be removed by adding sodium carbonate due to the reaction:

$$Ca^{2+} + Na_2CO3 \leftrightarrow CaCO_3(s) + 2Na^+.$$

Hard waters cause scales or deposits to build up in boilers, pipes, and faucets—problems for food and other industries. Ion exchange using resins or zeolites is commonly used to soften hard waters. The calcium and magnesium ions in the waters are taken up by the resin or zeolite that releases sodium or hydrogen ions back to the water. Alternatively, when pressure is applied to a solution, water molecules, but not ions, diffuse through the semipermeable membranes. This method, called *reverse osmosis*, has been used to soften hard waters and desalinate seawater.

However, water softening replaces desirable calcium and other ions with sodium ions. Thus, soft waters are not suitable for drinking. Incidentally, calcium ions strengthen the gluten proteins in dough mixing. Some calcium salts are added to the dough by bakeries to enhance bread quality.

Ionic Strength and Solubility of Foodstuff

Ions are attracted to charged or polar sites of large biomolecules. Cations strongly interact with large molecules such as proteins. At low concentrations, they may neutralize charges on large organic molecules, stabilizing them. At high concentrations, ions compete with large molecules for water and destabilize them, resulting in decreased solubility. The concentration of electrolytes affects the solubility of foodstuffs.

One of the criteria for concentration of electrolytes is *ionic strength*, *I*, which is half of the sum (Σ) of all products of the concentration (C_i) of the *i*th ion and the square of its charge (Z_i^2):

$$I = \frac{1}{2}\Sigma C_i Z_i^2.$$

However, solubility is not only a function of ionic strength; it also depends very much on the anions involved.

The *salting-in phenomenon* refers to increases of protein solubility with increased concentrations of salt at low ionic strength. The enhancement of broth flavor by adding salt may be due to an increase of soluble proteins or amino acids in it. At high ionic strength, however, the solubilities of some proteins decrease; this is the *salting-out phenomenon*. Biochemists often use potassium sulfate, K_2SO_4, and ammonium sulfate, $(NH_4)_2SO_4$, for the separation of amino acids or proteins because the sulfate ion is an effective salting-out anion. The sulfate ion is a *stabilizer*, because the precipitated proteins are stable. Table salt is not an effective salting-out agent. Damodaran (1996) and Voet and Voet (1995) discuss these phenomena in much more detail.

WATER AS REAGENT AND PRODUCT

Water is the product from the oxidation of hydrogen, and the standard cell potential (ΔE^o) for the reaction is 1.229 V:

$$2H_2(g) + O_2(g) = 2H_2O(l),\ \Delta E^o = 1.229\ V.$$

Actually, all hydrogen in any substance produces water during combustion and oxidation. On the other hand, water provides protons (H^+), hydroxide ions (OH^-), hydrogen atoms (H), oxygen atoms (O), and radicals (H·, ·OH) as reagents. The first two of these (H^+ and OH^-) also exhibit acid-base properties, as described earlier. Acids and bases promote hydrolysis and condensation reactions.

In *esterification* and *peptide synthesis*, two molecules are joined together, or condensed, releasing a water molecule. On the other hand, water breaks ester, peptide, and glycosidic bonds in a process called *hydrolysis*.

Esterification, Hydrolysis, and Lipids

Organic acids and various alcohols present in food react to yield esters in aqueous solutions. Esters, also present in food, hydrolyze to produce acids and alcohols. Water is a reagent and a product in these reversible equilibria. Figure 5.13 shows the Fisher esterification and hydrolysis reactions and the role of water in the series of intermediates in these equilibria. In general, esterification is favored in acidic solutions, and hydrolysis is favored in neutral and basic solutions.

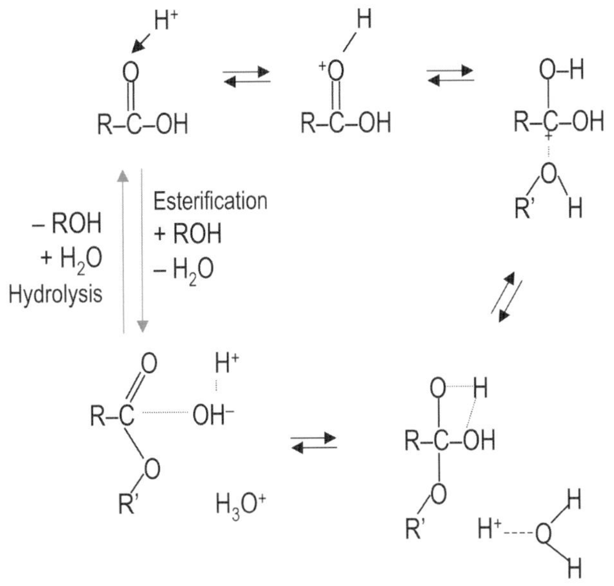

Figure 5.13. Esterification and hydrolysis in aqueous solutions.

The protonation of the slightly negative carbonyl oxygen (>C=O) of the carboxyl group (C(=O)OH) polarizes the C=O bond, making the carbon atom positive, to attract the alcohol group R′OH. Water molecules remove protons and rearrange the bonds in several intermediates for the simple overall reaction

$$RC(=O)OH + HOR' \leftrightarrow RC(=O)OR' + H_2O$$

In basic solutions, the OH^- ions are attracted to the slightly positive carbon of the carbonyl group. The hydrolysis is the reverse of esterification.

Hydrolysis of glycerol esters (glycerides: fat and oil) in basic solutions during soap making is a typical example of hydrolysis. Triglycerides are hydrophobic, but they can undergo partial hydrolysis to become amphiphilic diglycerides or monoglycerides. Esterification and hydrolysis are processes in metabolism.

Lipids, various water-insoluble esters of fatty acids, include glycerides, phospholipids, glycolipids, and cholesterol. Oils and fats are mostly triglycerides, which is a glycerol molecule (CH_2OH–CHOH–CH_2OH) esterified with three fatty acids [$CH_3(-CH_2)_nCOOH$, $n = 8$–16]. Some of the triglycerides are partially hydrolyzed in the gastrointestinal tract before absorption, but most are absorbed with the aid of bile salts, which emulsify the oil and facilitate its absorption. Many animals biosynthesize lipids when food is plentiful, as lipids provide the highest amount of energy per unit mass. Lipids, stored in fat cells, can be hydrolyzed, and upon further oxidation, they produce lots of energy and water. Some animals utilize fat for both energy and water to overcome the limitation of food and water supplies during certain periods of their lives.

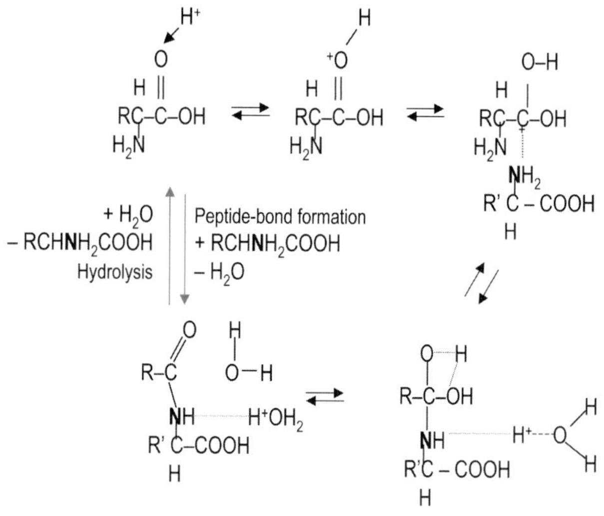

Figure 5.14. Hydrolysis and peptide-bond formation (polymerization).

Catalyzed by enzymes, esterification, and hydrolysis in biological systems proceed at much faster rates than when catalyzed by acids and bases.

Water in Digestion and Syntheses of Proteins

The digestion and the formation of many biopolymers (such as proteins and carbohydrates) as well as the formation and breakdown of lipids and esters involve reactions very similar to those of esterification and hydrolysis.

The digestion of proteins, polymers of amino acids, starts with chewing, followed by hydrolysis with the aid of protein-cleaving enzymes (proteases) throughout the gastrointestinal tract. Then, the partially hydrolyzed small peptides and hydrolyzed individual amino acids are absorbed in the intestine. Water is a reagent in these hydrolyses (Fig. 5.14).

The *deoxyribonucleic acids* (DNAs) store the genetic information, and they direct the synthesis of *messenger ribonucleic acids* (mRNAs), which in turn direct the protein synthesis machinery to make various proteins for specific body functions and structures. This is an oversimplified description of the biological processes that carry out the polymerization of amino acids.

Proteins and amino acids also provide energy when fully oxidized, but carbohydrates are the major energy source in normal diets.

Water in Digestion and Synthesis of Carbohydrates

On earth, water is the most abundant inorganic compound, whereas carbohydrates are the most abundant class of organic compounds. Carbohydrates require water for their synthesis and provide most of the energy for all life on earth. They are also part of the glycoproteins and the genetic molecules of DNA. *Carbohydrates* are compounds with a deceivingly simple general formula $(CH_2O)_n$, $n \geq 3$, that appears to be made up of carbon and water, but although their chemistry fills volumes of thick books and more, there is still much for carbohydrate chemists to discover.

Energy from the sun captured by plants and organisms converts CO_2 and water to high-energy carbohydrates,

$$6CO_2 + 12H_2O^* \rightarrow (CH_2O)_6 + 6H_2O + 6O_2^*$$

The stars (*) indicate the oxygen atoms from water released as oxygen gas (O_2^*). Still, this is an oversimplified equation for photosynthesis, but we do not have room to dig any deeper. The product $(CH_2O)_6$ is a *hexose*, a six-carbon simple sugar, or *monosaccharide,* that can be fructose, glucose, or another simple sugar. Glucose is the most familiar simple sugar, and its most common structure is a cyclic structure of the chair form. In glucose, all the OH groups around the ring are at the equatorial positions, whereas the small H atoms are at the axial locations (Fig. 5.15). With so many OH groups per molecule, glucose molecules are able to form several hydrogen bonds with water molecules, and thus most monosaccharides are soluble in water.

The *disaccharides* sucrose, maltose, and lactose have two simple sugars linked together, whereas starch and fiber are polymers of many glucose units. A disaccharide is formed when two OH groups of separate monosaccharides react to form an –O– link, called a *glycosidic bond*, after losing a water molecule:

$$C_6H_{12}O_6 + C_6H_{12}O_6 = C_6H_{11}O_5\text{–O–}C_6H_{11}O_5 + H_2O.$$

Disaccharides are soluble in water due to their ability to form many hydrogen bonds.

Plants and animals store glucose as long-chain *polysaccharides* in *starch* and *glycogen*, respectively, for energy. Starch is divided into amylose and amylopectin. *Amylose* consists of linear chains, whereas *amylopectin* has branched chains. Due to the many interchain hydrogen bonds in starch, hydrogen bonding to water molecules develops slowly. Small starch molecules are soluble in water. Suspensions of large starch molecules thicken

Figure 5.15. Chair form cyclic structure of glucose $C_6H_{12}O_6$. For glucose, all the OH groups are in the equatorial position, and these are possible H-donors for hydrogen bonding with water molecules. They are also possible sites to link to other hexoses.

soup and gravy, and starch is added to food for desirable texture and appearance. Water increases the molecular mobility of starch, and starch slows the movement of water molecules. Food processors are interested in a quantitative relationship between water and starch and the viscosity of the suspension.

Glycogen, animal starch, is easily hydrolyzed to yield glucose, which provides energy when required. In the hydrolysis of polysaccharides, water molecules react at the glycosidic links. Certain enzymes catalyze this reaction, releasing glucose units one by one from the end of a chain or branch.

Polysaccharide chains in cellulose are very long, 7000–15,000 monosaccharides, and interchain hydrogen bonds bind them into fibers, which further stack up through interfiber hydrogen bonds. Many interchain hydrogen bonds make the penetration of water molecules between chains a time-consuming process. Heating speeds up the process.

WATER, MINERALS, AND VITAMINS

Most minerals are salts or electrolytes. These are usually ingested as aqueous solutions of electrolytes, discussed earlier. Ions (Ca^{2+}, Mg^{2+}, Na^+, K^+, Fe^{2+}, Zn^{2+}, Cu^{2+}, Mn^{2+}, Cl^-, I^-, S^{2-}, Se^{2-}, $H_2PO_4{}^-$, etc.) present in natural water are leached from the ground. Some of them are also present in food, because they are essential nutrients for plants and animals that we use as food. A balance of electrolytes in body fluid must be maintained. Otherwise, shock or fainting may develop. For example, the drinking water used by sweating athletes contains the proper amount of minerals. In food, mineral absorption by the body may be affected by the presence of other molecules. For example, vitamin D helps the absorption of calcium ions.

Small amounts of a group of organic compounds not synthesized by humans, but essential to life, are called vitamins; their biochemistry is very complicated and interesting; many interact with enzymes, and others perform vital functions by themselves. Regardless of their biological function and chemical composition, vitamins are divided into water-soluble and fat-soluble groups. This division, based on polarity, serves as a guide for food processing. For example, food will lose water-soluble vitamins when washed or boiled in water, particularly after cutting (Hawthorne and Kubatova 2002).

The water-soluble vitamins consist of a complex group of vitamin Bs, vitamin C, biotin, lipoic acid, and folic acid. These molecules are either polar or have the ability to form hydrogen bonds. Vitamins A (retinal), D2, D3, E, and K are fat soluble, because major portions of their molecules are nonpolar organic groups.

Vitamin C, L-ascorbic acid or 3-oxo-L-gulofuranolactone, has the simplest chemical formula ($C_6H_8O_6$) among vitamins. This diprotic acid is widely distributed in plants and animals, and only a few vertebrates, including humans, lack the ability to synthesize it.

Vitamin B complex is a group of compounds isolated together in an aqueous solution. It includes thiamine (B1), riboflavin (B2), niacin (or nicotinic acid, B3), pantothenic acid (B5), cyanocobalamin (B12), and vitamin B6 (any form of pyridoxal, pyridoxine, or pyridoxamine). Biotin, lipoic acid, and folic acid are also part of the water-soluble vitamins. These vitamins are part of enzymes or coenzymes that perform vital functions.

FOOD CHEMISTRY OF WATER

Water ingestion depends on the individual, composition of the diet, climate, humidity, and physical activity. A nonexercising adult loses the equivalent of 4% of his or her body weight in water per day (Brody 1999). Aside from ingested water, water is produced during the utilization of food. It is probably fair to suggest that food chemistry is the chemistry of water, since we need a constant supply of water as long as we live.

Technical terms have special meanings among fellow food scientists. Furthermore, food scientists deal with dynamic and nonequilibrium systems, unlike most natural scientists who deal with static and equilibrium systems. There are special concepts and parameters useful only to food scientists. Yet, the fundamental properties of water discussed above lay a foundation for the food chemistry of water. With respect to food, water is a component, solvent, acid, base, and dispersing agent. It is also a medium for biochemical reactions, for heat and mass transfer, and for heat storage.

Food chemists are very concerned with water content and its effects on food. They need reliable parameters for references, criteria, and working objectives. They require various indicators to correlate water with special properties such as perishability, shelf life, mobility, smell, appearance, color, texture, and taste.

WATER AS A COMMON COMPONENT OF FOOD

Water is a food as well as the most common component of food. Even dry foods contain some water, and the degree of water content affects almost every aspect of food: stability, taste, texture, and spoilage.

Most food molecules contain OH, C=O, NH, and polar groups. These sites strongly interact with water molecules by hydrogen bonding and dipole-dipole interactions. Furthermore, dipole-ion, hydrophilic, and hydrophobic interactions also occur between water and food molecules. The properties of hydrogen-bonded water molecules differ from those in bulk water, and they affect the water molecules next to them. There is no clear boundary for affected and unaffected water molecules. Yet it is convenient to divide them into *bound water* and *free water*. This is a vague division, and a consensus definition is hard to reach. Fennema and Tannenbaum (1996) give a summary of various criteria for them, indicating a diverse opinion. However, the concept is useful, because it helps us understand the changes that occur in food when it is heated, dried, cooled, or refrigerated. Moreover, when water is the major ingredient, interactions with other ingredients modify the properties of the water molecules. These aspects were discussed earlier in connection with aqueous solutions.

WATER ACTIVITY

Interactions of water and food molecules mutually change their properties. Water in food is not pure water. Water molecules in vapor, liquid, and solid phases or in solutions and food react and interchange in any equilibrium system. The tendency to react and interchange with each other is called the *chemical potential*, μ. At equilibrium, the potential of water in all phases and forms must be equal at a given temperature, T. The potential, μ, of the gas phase may be expressed as:

$$\mu = \mu_w + RT \ln(p/p_w),$$

where R is the gas constant (8.3145 J $mol^{-1}K^{-1}$), p is the partial water vapor pressure, and p_w is the vapor pressure of pure water at T. The ratio p/p_w is called the *water activity* a_w ($= p/p_w$), although this term is also called *relative vapor pressure* (Fennema and Tannenbaum 1996). The difference is small, and for simplicity, a_w as defined is widely used for correlating the stability of foods. For ideal solutions and for most moist foods, a_w is less than unity ($a_w < 1.0$; Troller 1978).

The Clausius-Clapeyron equation, mentioned earlier, correlates vapor pressure, P, heat of phase transition, ΔH, and temperature, T. This same relationship can be applied to water activity. Thus, the plot of $\ln(a_w)$ versus $1/T$ gives a straight line, at least within a reasonable temperature range. Depending on the initial value for a_w or moisture content, the slope differs slightly, indicating the difference in the heat of phase transition due to different water content.

Both water activity and relative humidity are fractions of the pure-water vapor pressure. Water activity can be measured in the same way as humidity. Water contents have a sigmoidal relationship (Fig. 5.16). As water content increases, a_w increases: $a_w = 1.0$ for infinitely dilute solutions, $a_w > 0.7$ for dilute solutions and moist foods, and $a_w < 0.6$ for dry foods. Of course, the precise relationship depends on the food. In general, if the water vapor in the atmosphere surrounding the food is greater than the water activity of the food, water is adsorbed; otherwise, desorption takes place. Water activity reflects the combined effects of water-solute, water-surface, capillary, hydrophilic, and hydrophobic interactions.

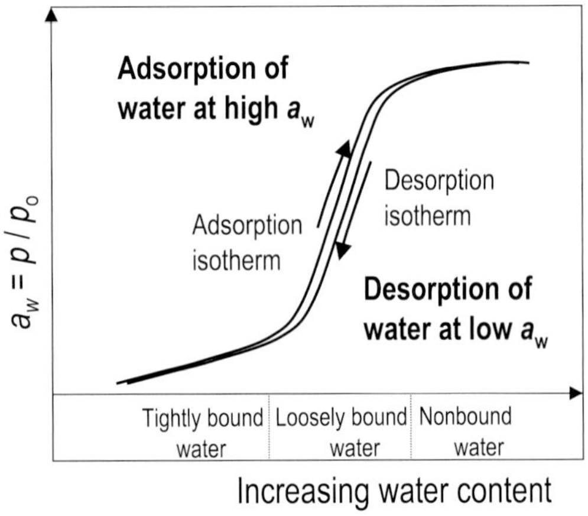

Figure 5.16. Nonequilibrium or hysteresis in desorption and adsorption of water by foodstuff. Arbitrary scales are used to illustrate the concept for a generic pattern.

Water activity is a vital parameter for food monitoring. A plot of a_w versus water content is called an *isotherm*. However, desorption and adsorption isotherms are different (Fig. 5.16) because this is a nonequilibrium system. Note that isotherms in most other literature plot water content against a_w, the reverse of the axes of Figure 5.16, which is intended to show that a_w is a function of water content.

Water in food may be divided into tightly bound, loosely bound, and nonbound waters. Dry foods contain tightly bound (monolayer) water, and a_w rises slowly as water content increases; but as loosely bound water increases, a_w increases rapidly and approaches 1.0 when nonbound water is present. Crisp crackers get soggy after adsorbing water from moist air, and soggy ones can be dried by heating or exposure to dry air.

Water activity affects the growth and multiplication of microorganisms. When $a_w < 0.9$, growth of most molds is inhibited. Growth of yeasts and bacteria also depends on a_w. Microorganisms cease growing if $a_w < 0.6$. *In a system, all components must be in equilibrium with one another, including all the microorganisms. Every type of organism is a component and a phase of the system, due to its cells or membranes. If the water activity of an organism is lower than that of the bulk food, water will be absorbed, and thus the species will multiply and grow. However, if the water activity of the organism is higher, the organism will dehydrate and become dormant or die.* Thus, it is not surprising that water activity affects the growth of various molds and bacteria. By this token, humidity will have the same effect on microorganisms in residences and buildings. Little packages of drying agent are placed in sealed dry food to reduce vapor pressure and prevent growth of bacteria that cause spoilage.

Aquatic Organisms and Drinking Water

Life originated in the water or oceans eons ago, and vast populations of the earliest unicellular living organisms still live in water today. Photosynthesis by algae in oceans consumes more CO_2 than the photosynthesis by all plants on land. Diversity of phyla (divisions) in the kingdoms of Fungi, Plantae, and Animalia live in water, ranging from single-cell algae to mammals.

All life requires food or energy. Some living organisms receive their energy from the sun, whereas others get their energy from chemical reactions. For example, the bacteria *Thiobacillus ferrooxidans* derive energy by catalyzing the oxidation of iron sulfide, FeS_2, using water as the oxidant (Barret et al. 1939). Chemical reactions provide energy for bacteria to sustain their lives and to reproduce. Many organisms feed on other organisms, forming a food chain. Factors affecting life in water include minerals, solubility of the mineral, acidity (pH), sunlight, dissolved oxygen level, presence of ions, chemical equilibria, availability of food, and electrochemical potentials of the material, among others.

Water used directly in food processing or as food is *drinking water*, and aquatic organisms invisible to the naked eye can be beneficial or harmful. The *Handbook of Drinking Water Quality* (De Zuane 1997) sets guidelines for water used in food services and technologies. Wastewater from the food industry needs treatment, and the technology is usually dealt with in industrial chemistry (Lacy 1992).

When food is plentiful, beneficial and pathogenic organisms thrive. Pathogenic organisms present in drinking water cause intestinal infections, dysentery, hepatitis, typhoid fever, cholera, and other diseases. Pathogens are usually present in waters that contain human and animal wastes that enter the water system via discharge, runoffs, flood, and accidents at sewage treatment facilities. Insects, rodents, and animals can also bring bacteria to the water system (Coler 1989, Percival et al. 2000). Testing for all pathogenic organisms is impossible, but some organisms have common living conditions. These are called *indicator bacteria*, because their absence signifies safety.

Water and State of Food

When a substance and water are mixed, they mutually dissolve, forming a homogeneous solution, or they partially dissolve in each other, forming solutions and other phases. At ambient pressure, various phases are in equilibrium with each other in isolated and closed systems. The equilibria depend on temperature. A plot of temperature versus composition showing the equilibria among various phases is a *phase diagram* for a two-component system. Phase diagrams for three-component systems are very complicated, and foods consist of many substances, including water. Thus, a strict phase diagram for food is almost impossible. Furthermore, food and biological systems are open, with a steady input and output of energy and substances. Due to time limits and slow kinetics, phases are not in equilibrium with each other. However, the changes follow a definite rate, and these are *steady states*. For these cases, plots of temperature against the composition, showing the existences of states (phases), are called *state diagrams*. They indicate the existence of various phases in multicomponent systems.

Sucrose (sugar, $C_{12}H_{22}O_{11}$) is a food additive and a sweetener. Solutions in equilibrium with excess solid sucrose are saturated, and their concentrations vary with temperature. The saturated solutions contain 64.4 and 65.4% at 0 and 10°C, respectively. The plot of saturated concentrations against temperature is *the equilibrium solubility curve,* ES, in Figure 5.17. The freezing curve, FE, shows the variation of freezing point as a function of temperature. Aqueous solution is in equilibrium with ice Ih along FE. At the *eutectic point,* E, the intersection of the solubility and freezing curves, solids Ih and sucrose coexist with a saturated solution. The eutectic point is the lowest mp of water–sugar solutions. However, viscous aqueous sugar solutions or syrups may exist beyond the eutectic point. These conditions may be present in freezing and tempering (thawing) of food.

Dry sugar is stable, but it spoils easily if it contains more than 5% water. The changes that occur as a sugar solution is chilled exemplify the changes in some food components when foods freeze. Ice Ih forms when a 10% sugar solution is cooled below the freezing point. As water forms Ih, leaving sugar in the solution, the solution becomes more concentrated, decreasing the freezing point further along the FE toward the E. However, when cooled, this solution may not reach equilibrium and yield

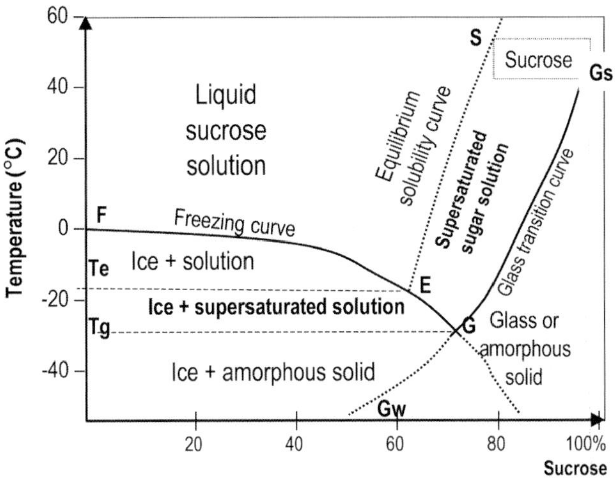

Figure 5.17. A sketch showing the phase and state diagram of water-sucrose binary system.

sugar crystals at the eutectic point. Part of the reason for not having sucrose crystals is the high viscosity of the solution, which prevents molecules from moving and orienting properly to crystallize. The viscous solution reaches a glassy or amorphous state at the *glass transition temperature (Tg)*, point G. The glass state is a frozen liquid with extremely high viscosity. In this state, the molecules are immobile. The temperature, Tg, for glass transition depends on the rate of cooling (Angell 2002). The freezing of sugar solution may follow different paths, depending on the experimental conditions.

In lengthy experiments, Young and Jones (1949) warmed glassy states of water-sucrose and observed the warming curve over hours and days for every sample. They observed the eutectic mixture of 54% sucrose (Te $= -13.95$°C). They also observed the formation of phases $C_{12}H_{22}O_{11}\cdot 2.5H_2O$ and $C_{12}H_{22}O_{11}\cdot 3.5H_2O$ hydrated crystals formed at temperatures higher than the Te, which is for anhydrous sucrose. The water-sucrose binary system illustrates that the states of food components during freezing and thawing can be very complicated. Freshly made ice creams have wonderful texture, and the physics and chemistry of the process are even more interesting.

INTERACTION OF WATER AND MICROWAVE

Wavelengths of microwave range from meters down to a millimeter, their frequencies ranging from 0.3 to 300 GHz. A typical domestic oven generates 2.45 GHz microwaves, wavelength 0.123 m, and energy of photon 1.62×10^{-24} J (10 μeV). For industrial applications, the frequency may be optimized for the specific processes.

Percy L. Spencer (1894–1970), the story goes, noticed that his candy bar melted while he was inspecting magnetron testing at the Raytheon Corporation in 1945. As a further test, he microwaved popping corns, which popped. A team at Raytheon developed microwave ovens, but it took more than 25 years and much more effort to improve them and make them practical and popular. Years ago, boiling water in a paper cup in a microwave oven without harming the cup amazed those who were used to see water being heated in a fire-resistant container over a stove or fire. Microwaves simultaneously heat all the water in the bulk food.

After the invention of the microwave oven, many offered explanations on how microwaves heat food. Water's high dipole moment and high dielectric constant caused it to absorb microwave energy, leading to an increase in its temperature. Driven by the oscillating electric field of microwaves, water molecules rotate, oscillate, and move about faster, increasing water temperature to sometimes even above its bp. In regions where water has difficulty forming bubbles, the water is overheated. When bubbles suddenly do form in superheated water, an explosion takes place. Substances without dipole moment cannot be heated by microwaves. Therefore, plastics, paper, and ceramics won't get warm. Metallic conductors rapidly polarize, causing sparks due to arcing. The oscillating current and resistance of some metals cause a rapid heating. In contrast, water is a poor conductor, and the heating mechanism is very complicated. Nelson and Datta (2001) reviewed microwave heating in the *Handbook of Microwave Technology for Food Applications*.

Molecules absorb photons of certain frequencies. However, microwave heating is due not only to absorption of photons by the water molecules, but also to a combination of polarization and dielectric induction. As the electric field oscillates, the water molecules try to align their dipoles with the electric field. Crowded molecules restrict one another's movements. The resistance causes the orientation of water molecules to lag behind that of the electric field. Since the environment of the water molecules is related to their resistance, the heating rate of the water differs from food to food and region to region within the same container. Water molecules in ice, for example, are much less affected by the oscillating electric field in domestic microwave ovens, which are not ideal for thawing frozen food. The outer thawed layer heats up quickly, and it is cooked before the frozen part is thawed. Domestic microwave ovens turn on the microwave intermittently or at very low power to allow thermal conduction for thawing. However, microwaves of certain frequencies may heat ice more effectively for tempering frozen food. Some companies have developed systems for specific purposes, including blanching, tempering, drying, and freeze-drying.

The electromagnetic wave form in an oven or in an industrial chamber depends on the geometry of the oven. If the wave forms a standing wave in the oven, the electric field varies according to the wave pattern. Zones where the electric field varies with the largest amplitude cause water to heat up most rapidly, and the nodal zones where there are no oscillations of electric field will not heat up at all. Thus, uniform heating has been a problem with microwave heating, and various methods have been developed to partly overcome this problem. Also, foodstuffs attenuate microwaves, limiting their penetration depth into foodstuff. Uneven heating remains a challenge for food processors and microwave chefs, mostly due to the short duration of microwaving. On the other hand, food is also seldom evenly heated when conventionally cooked.

Challenges are opportunities for food industries and individuals. For example, new technologies in food preparation, packaging, and sensors for monitoring food temperature during microwaving are required. There is a demand for expertise in microwaving food. Industries microwave-blanche vegetables for drying or freezing to take advantage of its energy efficiency, time saving, decreased waste, and retention of water-soluble nutrients. The ability to quickly temper frozen food in retail stores reduces spoilage and permits selling fresh meat to customers.

Since water is the heating medium, the temperature of the food will not be much higher than the boiling point of the aqueous solutions in the food. Microwave heating does not burn food; thus, the food lacks the usual color, aroma, flavor, and texture found in conventional cooking. The outer layer of food is dry due to water evaporation. Retaining or controlling water content in microwaved food is a challenge.

When microwaved, water vapor is continually removed. Under reduced pressure, food dries or freeze-dries at low temperature due to its tendency to restore the water activity. Therefore, microwaving is an excellent means for drying food because of its savings in energy and time. Microwaves are useful for industrial applications such as drying, curing, and baking or parts thereof.

Microwave ovens have come a long way, and their popularity and improvement continue. Food industry and consumer attitudes about microwavable food have gone up and down, often due to misconceptions. Microwave cooking is still a challenge. The properties of water affect cooking in every way. Water converts microwave energy directly into heat, attenuates microwave radiation, transfers heat to various parts of the foodstuff, affects food texture, and interacts with various nutrients. All properties of water must be considered in order to take advantage of microwave cooking.

WATER RESOURCES AND THE HYDROLOGICAL CYCLE

Fresh waters are required to sustain life and maintain living standards. Therefore, fresh waters are called *water resources*. Environmentalists, scientists, and politicians have sounded alarms about limited water resources. Such alarms appear unwarranted because the earth has so much water that it can be called a water planet. Various estimates of global water distribution show that about 94% of earth's water lies in the oceans and seas. These salt waters sustain marine life and are ecosystems in their own right, but they are not fresh waters that satisfy human needs. Of the remaining 6%, most water is in solid form (at the poles and in high mountains before the greenhouse effect melts them) or underground. Less than 1% of earth's water is in lakes, rivers, and streams, and waters from these sources flow into the seas or oceans. A fraction of 1% remains in the atmosphere, mixed with air (Franks 2000).

A human may drink only a few liters of water in various forms each day, but ten times more water is required for domestic usages such as washing and food preparation. A further equal amount is needed for various industries and social activities that support individuals. Furthermore, much more is required for food production, maintaining a healthy environment, and supporting lives in the ecosystems. Thus, one human may require more than 1000 L of water per day. In view of these requirements, a society has to develop policies for managing water resources both near and far as well as in the short and long terms. This chapter has no room to address the social and political issues, but facts are presented for readers to formulate solutions to these problems, or at least to ask questions regarding them. Based on these facts, is scarcity of world water resources a reality or not?

A major threat to water resources is climate change, because climate and weather are responsible for the *hydrologic cycle* of salt and fresh waters. Of course, human activities influence the climate in both short and long terms.

Based on the science of water, particularly its transformations among solid, liquid, and vapor phases under the influence of energy, we easily understand that heat from the sun vaporizes water from the ocean and land alike. Air movement carries the moisture (vapor) to different places than those from which it evaporated. As the vapor ascends, cooling temperature condenses the vapor into liquid drops. Cloud and rain eventually develop, and rain erodes, transports, shapes the landscape, creates streams and rivers, irrigates, and replenishes water resources. However, too much rain falling too quickly causes disaster in human life. On the other hand, natural water management for energy and irrigation has brought prosperity to society, easing the effects on humans of droughts and floods, when water does not arrive at the right time and place.

Water is a resource. Competition for this resource leads to "water war." Trade in food and food aid is equivalent to flow of water, because water is required for food production. Food and water management, including wastewater treatment, enable large populations to concentrate in small areas. Urban dwellers take these commodities for granted, but water enriches life both physically and mentally.

ACKNOWLEDGMENTS

The opportunity to put a wealth of knowledge in a proper perspective enticed me to a writing project, for which I asked more questions and found their answers from libraries and the Internet. I am grateful to all who have contributed to the understanding of water. I thank Professors L. J. Brubacher and Tai Ping Sun for their reading of the manuscript and for their suggestions. I am also grateful to other scholars and friends who willingly shared their expertise. The choice of topics and contents indicate my limitations, but fortunately, readers' curiosity and desire to know are limitless.

REFERENCES

Angell CA. 2002. Liquid fragility and the glass transition in water and aqueous solutions. *Chem Rev* 102: 2627–2650.

Barret J. et al. 1939. *Metal Extraction by Bacterial Oxidation of Minerals*. Ellis Horwood, New York.

Berkowitz J. 1979. *Photoabsorption, Photoionization, and Photoelectron Spectroscopy*. San Diego: Academic Press.

Bernath PF. 2002a. The spectroscopy of water vapour: Experiment, theory, and applications. *Phys Chem Chem Phys*. 4: 1501–1509.

———. 2002b. Water vapor gets excited. *Science* 297(Issue 5583): 943–945.

Bjorneholm O et al. 1999. Between vapor and ice: Free water clusters studied by core level spectroscopy, *J Chem Phys* 111(2): 546–550.

Bockhoff FJ. 1969. *Elements of Quantum Theory*. Reading, Masachusetts: Addison-Wesley, Inc.

Brody T. 1999. *Nutritional Biochemistry*, 2nd ed. San Diego: Academic Press.

Carleer M et al. 1999. The near infrared, visible, and near ultraviolet overtone spectrum of water. *J Chem Phys* 111: 2444–2450.

Coler RA. 1989. *Water pollution biology: A laboratory/field handbook*. Lancaster, Pennsylvania: Technomic Publishing Co.

Damodaran S. 1996. Chapter 6, Amino acids, peptides and proteins. In: OR Fennema (ed.) *Food Chemistry*, 3rd ed. New York: Marcel Dekker, Inc.

De Zuane J. 1997. *Handbook of Drinking Water Quality*, 2nd ed.. Van Nostrand, Reinheld.

Fennema OR, Tannenbaum SR. 1996. Chapter 2, Water and ice. In: OR Fennema (ed.) *Food Chemistry*, 3rd ed. New York: Marcel Dekker, Inc.

Franks F. 2000. *Water—a Matrix of Life*, 2nd ed.. Cambridge: Royal Society of Chemistry.

Franks F et al. 1987. Antifreeze activity of antarctic fish glycoprotein and a synthetic polymer. *Nature* 325: 146–147.

Garrett RH, Grisham CM. 2002. *Principles of Biochemistry, with a Human Focus*. Orlando, Florida: Harcourt College Publishers.

Goldman N et al. 2001. Water dimers in the atmosphere: Equilibrium constant for water dimerization from the VRT(ASP-W) potential surface. *J Phys Chem* 105: 515–519.

Gray HB. 1964. Chapter 7, *Angular triatomic molecules. In: Electrons and Chemical Bonding*. New York: Benjamin, pp. 142–154.

Hawthorne SB, Kubatova A. 2002. Hot (subcritical) water extraction. In: J Pawliszyn (ed.) *A Comprehensive Analytical Chemistry XXXVII, Sampling and Sample Preparation for Field and Laboratory*. New York: Elsevier, pp. 587–608.

Huisken F, Kaloudis M, Kulcke A. 1996. Infrared spectroscopy of small size-selected water clusters. *J Chem Phys* 104: 17–25.

Johari GP, Anderson O. 2004. Water's polyamorphic transitions and amorphization of ice under pressure. *J Chem Phys* 120: 6207–6213.

Kamb B. 1972. Structure of the ice. In: *Water and Aqueous Solutions—Structure, Thermodynamics and Transport Processes*, New York: Wiley-Interscience, pp. 9–25.

Klug DD. 2002. Condensed-matter physics: Dense ice in detail. *Nature* 420: 749–751.

Kohl I et al. 2000. The glassy water–cubic ice system: A comparative study by X-ray diffraction and differential scanning calorimetry. *Phys Chem Chem Phys* 2: 1579–1586.

Kortüm G et al. 1961. *Dissociation Constants of Organic Acids in Aqueous Solution*. London: Butterworths.

Lacy WJ. 1992. Industrial wastewater and hazardous material treatment (ed.) technology. In: JA Kent, *Riegel's Handbook of*

Industrial Chemistry, 9th edition. New York: Van Nostrand Reinhold, pp. 31–82.

Lemus R. 2004. Vibrational excitations in H_2O in the framework of a local model. *J Mol Spectrosc* 225: 73–92.

Lide DR (ed.) 2003. *CRC Handbook of Chemistry and Physics*, 83rd ed. Cleveland, Ohio: CRC Press (There is an Internet version).

Marshall CB et al. 2004. Hyperactive antifreeze protein in a fish. *Nature*, 429: 153–154.

Marshall WL, Franck EU. 1981. Ion product of water substance, 0–1000°C, 1–10,000 bars, new international formulation and its background, *J Phys Chem Ref Data* 10: 295–306.

Mayer E, Hallbrucker A. 1987. Cubic ice from liquid water. *Nature* 325: 601–602.

Moeller T, O'Connor R. 1972. *Ions in aqueous systems; an Introduction to Chemical Equilibrium and Solution Chemistry*. New York: McGraw-Hill.

Nelson SO, Datta AK. 2001. Dielectric properties of food materials and electric field interactions. *I:* Handbook of Microwave Technology for Food Applications, ed. AK Datta, RC Anantheswaran. New York: Marcel Dekker.

Pauling L. 1960. *The Nature of the Chemical Bond*. Ithaca, New York: Cornell University Press.

Percival SL et al. 2000. *Microbiological Aspects of Biofilms and Drinking Water*. Boca Raton: CRC Press.

Perrin DD. 1965. *Dissociation Constants of Organic bases in Aqueous Solution*. London: Butterworths.

———. 1982. *Ionisation Constants of Inorganic Acids and Bases in Aqueous Solution*. Toronto: Pergamon Press.

Petrenko VF, Whitworth RW. 1999. *Physics of Ice*. New York: Oxford University Press.

Sloan DE. 1998. *Clathrate Hydrates of Natural Gases*. New York: Marcel Dekker.

Tanaka M et al. 2001. Recommended table for the density of water between 0°C and 40°C based on recent experimental reports. *Metrologia* 38(4): 301–309.

Tanford C. 1980. *The Hydrophobic Effect: Formation of Micelles and Biological Membranes*, 2nd ed. New York: John Wiley & Sons.

Tawa GJ, Pratt LR. 1995. Theoretical calculation of the water ion product K_w, *J Am Chem Soc* 117: 1625–1628.

Troller JA. 1978. *Water Activity and Food*. New York: Academic Press.

Wayne RP. 2000. *Chemistry of Atmospheres*, 3rd ed. New York: Oxford University Press.

Wilson PW et al. 2003. Ice nucleation in nature: Supercooling point (SCP) measurements and the role of heterogeneous nucleation. *Cryobiology* 46: 88–98.

Young FE, Jones FT. 1949. Sucrose Hydrates. The sucrose-water phase diagram. *J of Physical and Colloid Chemistry* 53: 1334–1350.

Voet D, Voet JG. 1995. Chapter 5. *Biochemistry*, 2nd ed. New York: John Wiley & Sons, Inc.

Part 2
Biotechnology and Ezymology

6
Enzyme Classification and Nomenclature

H. Ako and W. K. Nip

INTRODUCTION

Before 1961, researchers reported on enzymes or enzymatic activities with names of their own preference. This situation caused confusion to others as various names could be given to the same enzyme. In 1956, the International Union of Biochemistry (IUB, later changed to International Union of Biochemistry and Molecular Biology, IUBMB) created the International Commission on Enzymes in consultation with the International Union of Pure and Applied Chemistry (IUPAC) to look into this situation. This Commission (now called the Nomenclature Committee of the IUBMB, NC-IUBMB) subsequently recommended classifying enzymes into six divisions (classes) with subclasses and sub-subclasses. General rules and guidelines were also established for classifying and naming enzymes. Each enzyme accepted to the Enzyme List was given a recommended name (trivial or working name; now called the common name), a systematic name, and an Enzyme Commission, or Enzyme Code (EC) number. The enzymatic reaction is also provided. A common name (formerly called recommended name) is assigned to each enzyme. This is normally the name most widely used for that enzyme, unless that name is ambiguous or misleading. A newly discovered enzyme can be given a common name and a systematic name, but not the EC number, by the researcher. EC numbers are assigned only by the authority of the NC-IUBMB.

The first book on enzyme classification and nomenclature was published in 1961. Some critical updates were announced as newsletters in 1984 (IUPAC-IUB and NC-IUB Newsletters 1984). The last (sixth) revision was published in 1992. Another update in electronic form was published in 2000 (Boyce and Tipton 2000). With the development of the Internet, most updated information on enzyme classification and nomenclature is now available through the website of the IUBMB (http://www.chem.qmul.ac.uk/iubmb/enzyme.html). This chapter should be considered as an abbreviated version of enzyme classification and nomenclature, with examples of common enzymes related to food processing. Readers should visit the IUBMB enzyme nomenclature website for the most up-to-date details on enzyme classification and nomenclature.

CLASSIFICATION AND NOMENCLATURE OF ENZYMES

General Principles

- *First principle*: Names purporting to be names of enzymes, especially those ending in *-ase* should be used only for single enzymes, that is, single catalytic entities. They should not be applied to systems containing more than one enzyme.
- *Second principle*: Enzymes are classified and named according to the reaction they catalyze.
- *Third principle*: Enzymes are divided into groups on the basis of the type of reactions catalyzed, and this, together

Food Biochemistry and Food Processing, Second Edition. Edited by Benjamin K. Simpson, Leo M.L. Nollet, Fidel Toldrá, Soottawat Benjakul, Gopinadhan Paliyath and Y.H. Hui.

with the name(s) of the substrate(s), provides a basis for determining the systematic name and EC number for naming individual enzymes.

Common and Systematic Names

- The common name (recommended, trivial, or working name) follows immediately after the EC number.
- While the common name is normally that used in the literature, the systematic name, which is formed in accordance with definite rules, is more precise chemically. It should be possible to determine the reaction catalyzed from the systematic name alone.

Scheme of Classification and Numbering of Enzymes

The first EC, in its report in 1961, devised a system for the classification of enzymes that also serves as a basis for assigning EC numbers to them. These code numbers (prefixed by EC), which are now widely in use, contain four elements separated by periods (e.g., 1.1.1.1), with the following meaning:

1. The first number shows to which of the six divisions (classes) the enzyme belongs.
2. The second figure indicates the subclass.
3. The third figure gives the sub-subclass.
4. The fourth figure is the serial number of the enzyme in its sub-subclass.

The main classes are as follows:

- *Class 1*: Oxidoreductases (dehydrogenases, reductases, or oxidases).
- *Class 2*: Transferases.
- *Class 3*: Hydrolases.
- *Class 4*: Lyases.
- *Class 5*: Isomerases (racemases, epimerases, cis-trans-isomerases, isomerases, tautomerases, mutases, cycloisomerases).
- *Class 6*: Ligases (synthases).

Class 1: Oxidoreductases

Enzymes catalyzing oxidoreductions belong to this class. The reactions are of the form $AH_2 + B = A + BH_2$ or $AH_2 + B^+ = A + BH + H^+$. The substrate oxidized is regarded as the hydrogen or electron donor. All reactions within a particular sub-subclass are written in the same direction. The classification is based on the order "donor:acceptor oxidoreductase." The common name often takes the form "substrate dehydrogenase," wherever this is possible. If the reaction is known to occur in the opposite direction, this may be indicated by a common name of the form "acceptor reductase" (e.g., the common name of EC 1.1.1.9 is D-xylose reductase). "Oxidase" is used only in cases where O_2 is an acceptor. Classification is difficult in some cases because of the lack of specificity toward the acceptor.

Class 2: Transferases

Transferases are enzymes transferring a group (e.g., the methyl group or a glycosyl group), from one compound (generally regarded as donor) to another compound (generally regarded as acceptor). The classification is based on the scheme "donor:acceptor group transferase." The common names are normally formed as "acceptor grouptransferase." In many cases, the donor is a cofactor (coenzyme) carrying the group to be transferred. The aminotransferases constitute a special case (subclass 2.6): the reaction also involves an oxidoreduction.

Class 3: Hydrolases

These enzymes catalyze the hydrolysis of various bonds. Some of these enzymes pose problems because they have a very wide specificity, and it is not easy to decide if two preparations described by different authors are the same, or if they should be listed under different entries.

While the systematic name always includes "hydrolase," the common name is, in most cases, formed by the name of the substrate with the suffix *-ase*. It is understood that the name of this substrate with the suffix means a hydrolytic enzyme. The peptidases, subclass 3.4, are classified in a different manner from other enzymes in this class.

Class 4: Lyases

Lyases are enzymes cleaving C–C, C–O, C–N, and other bonds by means other than hydrolysis or oxidation. They differ from other enzymes in that two substrates are involved in one reaction direction, but only one in the other direction. When acting on the single substrate, a molecule is eliminated, leaving an unsaturated residue. The systematic name is formed according to "substrate group-lyase." In common names, expressions like decarboxylase, aldolase, and so on are used. "Dehydratase" is used for those enzymes eliminating water. In cases where the reverse reaction is the more important, or the only one to be demonstrated, "synthase" may be used in the name.

Class 5: Isomerase

These enzymes catalyze changes within one molecule.

Class 6: Ligases

Ligases are enzymes catalyzing the joining of two molecules with concomitant hydrolysis of the diphosphate bond in ATP or a similar triphosphate. The bonds formed are often high-energy bonds. "Ligase" is commonly used for the common name, but in a few cases, "synthase" or "carboxylase" is used. Use of the term "synthetase" is discouraged.

General Rules and Guidelines for Classification and Nomenclature of Enzymes

Table 6.1 shows the classification of enzymes by class, subclass, and sub-subclass, as suggested by the Nomenclature Committee

Table 6.1. Classification of Enzymes by Class, Subclass, and Sub-subclass[a]

1. Oxidoreductases		
1.1. Acting on the CH–OH group of donors		
	1.1.1	With NAD^+ or $NADP^+$ as acceptor
	1.1.2	With a cytochrome as acceptor
	1.1.3	With oxygen as acceptor
	1.1.4	With a disulfide as acceptor
	1.1.5	With a quinone or similar compound as acceptor
	1.1.99	With other acceptors
1.2. Acting on the aldehyde or oxo group of donors		
	1.2.1	With NAD^+ or $NADP^+$ as acceptor
	1.2.2	With a cytochrome as acceptor
	1.2.3	With oxygen as acceptor
	1.2.4	With a disulfide compound as acceptor
	1.2.7	With an iron–sulfur protein as acceptor
	1.2.99	With other acceptors
1.3. Acting on the CH–CH group of donors		
	1.3.1	With NAD^+ or $NADP^+$ as acceptor
	1.3.2	With a cytochrome as acceptor
	1.3.3	With oxygen as acceptor
	1.3.5	With a quinone or related compound as acceptor
	1.3.6	With an iron–sulfur protein as acceptor
	1.3.99	With other acceptor
1.4. Acting on the CH–NH_2 of donors		
	1.4.1	With NAD^+ or $NADP^+$ as acceptor
	1.4.2	With a cytochrome as acceptor
	1.4.3	With oxygen as acceptor
	1.4.4	With a disulfide as acceptor
	1.4.7	With an iron–sulfur protein as acceptor
	1.4.99	With other acceptors
1.5. Acting on the CH–NH group of donors		
	1.5.1	With NAD^+ or $NADP^+$ as acceptor
	1.5.3	With oxygen as acceptor
	1.5.4	With disulfide as acceptor
	1.5.5	With a quinone or similar compound as acceptor
	1.5.8	With a flavin as acceptor
	1.5.99	With other acceptors
1.6. Acting on NADH or NADPH		
	1.6.1	With NAD^+ or $NADP^+$ as acceptor
	1.6.2	With a heme protein as acceptor
	1.6.3	With oxygen as acceptor
	1.6.4	With a disulfide compound as acceptor
	1.6.5	With a quinone or similar compound as acceptor
	1.6.6	With a nitrogenous group as acceptor
	1.6.8	With a flavin as acceptor
	1.6.99	With other acceptors
1.7. Acting on other nitrogenous compounds as donors		
	1.7.1	With NAD^+ or $NADP^+$ as acceptor
	1.7.2	With a cytochrome as acceptor
	1.7.3	With oxygen as acceptor
	1.7.7	With an iron–sulfur protein as acceptor
	1.7.99	With other acceptors
1.8. Acting on a sulfur group of donors		
	1.8.1	With NAD^+ or $NADP^+$ as acceptor
	1.8.2	With a cytochrome as acceptor
	1.8.3	With oxygen as acceptor
	1.8.4	With a disulfide as acceptor

(Continued)

Table 6.1. *(Continued)*

	1.8.5	With a quinone or related compound as acceptor
	1.8.7	With an iron–sulfur protein as acceptor
	1.8.98	With other, known, acceptors
	1.8.99	With other acceptors
	1.9.3	With oxygen as acceptor
	1.9.5	With a nitrogenous group as acceptor
	1.9.99	With other acceptors
1.10.	*Acting on diphenols and related substances as donors*	
	1.10.1	With NAD^+ or $NADP^+$ as acceptor
	1.10.2	With a cytochrome as acceptor
	1.10.3	With oxygen as acceptor
	1.10.99	With other acceptors
1.11.	*Acting on hydrogen peroxide as acceptor*	
	1.11.1	The peroxidases
1.12.	*Acting on hydrogen as donor*	
	1.12.1	With NAD^+ or $NADP^+$ as acceptor
	1.12.2	With a cytochrome as acceptor
	1.12.5	With a quinone or similar compound as acceptor
	1.12.7	With an iron–sulfur protein as acceptor
	1.12.98	With other known acceptors
	1.12.99	With other acceptors
1.13.	*Acting on single donors with incorporation of molecular oxygen (oxygenases)*	
	1.13.11	With incorporation of two atoms of oxygen
	1.13.12	With incorporation of one atom of oxygen (internal monoxygenase or internal mixed-function oxidases)
	1.13.99	Miscellaneous
1.14.	*Acting on paired donors with incorporation of molecular oxygen*	
	1.14.11	With 2-oxoglutarate as one donor, and incorporation of one atom of oxygen into both donors
	1.14.12	With NADH or NADPH as one donor, and incorporation of two atoms of oxygen into one donor
	1.14.13	With NADH or NADPH as one donor, and incorporation of one atom of oxygen
	1.14.14	With reduced flavin or flavoprotein as one donor, and incorporation of one atom of oxygen
	1.14.15	With reduced iron–sulfur protein as one donor, and incorporation of one atom of oxygen
	1.14.16	With reduced pteridine as one donor, and incorporation of one atom of oxygen
	1.14.17	With reduced ascorbate as one donor, and incorporation of one atom of oxygen
	1.14.18	With another compound as one donor, and incorporation of one atom of oxygen
	1.14.19	With oxidation of a pair of donors resulting in the reduction of molecular oxygen to two molecules of water
	1.14.20	With 2-oxoglutarate as one donor, and the other dehydrogenated
	1.14.21	With NADH or NADPH as one donor, and the other dehydrogenated
	1.14.99	Miscellaneous (requires further characterization)
1.15.	*Acting on superoxide as acceptor*	
1.16.	*Oxidizing metal ions*	
	1.16.1	With NAD^+ or $NADP^+$ as acceptor
	1.16.2	With oxygen as donor
1.17.	*Acting on -CH_2- groups*	
	1.17.1	With NAD^+ or $NADP^+$ as acceptor
	1.17.2	With oxygen as acceptor
	1.17.4	With a disulfide compound as acceptor
	1.17.99	With other acceptors
1.18.	*Acting on reduced ferredoxin as donor*	
	1.18.1	With NAD^+ or $NADP^+$ as acceptor
	1.18.3	With H^+ as acceptor (now EC 1.18.99)
	1.18.6	With dinitrogen as acceptor
	1.18.96	With other, known, acceptors
	1.18.99	With H^+ as acceptor
1.19.	*Acting on reduced flavodoxin as donor*	
	1.19.6	With dinitrogen as acceptor

Table 6.1. *(Continued)*

1.20.	*Acting on phosphorus or arsenic in donors*	
	1.20.1	With $NAD(P)^+$ as acceptor
	1.20.4	With disulfide as acceptor
	1.20.98	With other, known, acceptors
	1.20.99	With other acceptors
1.21.	*Acting on X-H and Y-H to form an X-Y bond*	
	1.21.3	With oxygen as acceptor
	1.21.4	With disulfide as acceptor
	1.21.99	With other acceptors
1.97.	*Other oxidoreductases*	
2. Transferases		
2.1.	*Transferring one-carbon groups*	
	2.1.1	Methyltransferases
	2.1.2	Hydroxymethyl-, formyl-, and related transferases
	2.1.3	Carboxyl- and carbamoyl transferases
	2.1.4	Amidinotransferases
2.2.	*Transferring aldehyde or ketonic groups*	
	2.2.1	Transketolases and transaldolase
2.3.	*Acyltransferases*	
	2.3.1	Transferring groups other than amino-acyl groups
	2.3.2	Aminoacyltransferases
	2.3.3	Acyl groups converted into alkyl on transfer
2.4.	*Glycosyltransferases*	
	2.4.1	Hexosyltransferases
	2.4.2	Pentosyltransferases
	2.4.99	Transferring other glycosyl groups
2.5.	*Transferring alkyl or aryl groups, other than methyl groups*	
	(There is no subdivision in this section.)	
2.6.	*Transferring nitrogenous groups*	
	2.6.1	Transaminases
	2.6.2	Amidinotransferases
	2.6.3	Oximinotransferases
	2.6.99	Transferring other nitrogenous groups
2.7.	*Transferring phosphorus-containing groups*	
	2.7.1	Phosphotransferases with an alcohol group as acceptor
	2.7.2	Phosphotransferases with a carboxy group as acceptor
	2.7.3	Phosphotransferases with a nitrogenous group as acceptor
	2.7.4	Phosphotransferases with a phosphate group as acceptor
	2.7.6	Diphosphotransferases
	2.7.7	Nucleotidyltransferases
	2.7.8	Transferases for other substituted phosphate groups
	2.7.9	Phosphotransferases with paired acceptors
2.8.	*Transferring sulfur-containing groups*	
	2.8.1	Sulfurtransferases
	2.8.2	Sulfotransferases
	2.8.3	CoA-transferases
	2.8.4	Transferring alkylthio groups
2.9.	*Transferring selenium-containing groups*	
	2.9.1	Selenotransferases
3. Hydrolases		
3.1.	*Acting on ester bonds*	
	3.1.1	Carboxylic ester hydrolases
	3.1.2	Phosphoric monoester hydrolases
	3.1.3	Phosphoric diester hydrolases
	3.1.4	Triphosphoric monoester hydrolases
	3.1.5	Sulfuric ester hydrolases

(Continued)

Table 6.1. *(Continued)*

	3.1.6	Diphosphoric monoester hydrolases
	3.1.7	Phosphoric triester hydrolases
	3.1.11	Exodeoxyribonucleases producing 5′-phosphomonoesters
	3.1.13	Exoribonucleases producing 5′-phosphomonoesters
	3.1.14	Exoribonucleases producing 3′-phosphomonoesters
	3.1.15	Exonucleases active with either ribo- or deoxyribonucleic acids and producing 5′-phosphomonoesters
	3.1.16	Exonucleases active with either ribo- or deoxyribonucleic acids and producing 3′-phosphomonoesters
	3.1.21	Endodeoxyribonucleases producing 5′-phosphomonoesters
	3.1.22	Endodeoxyribonucleases producing 3′-phosphomonoesters
	3.1.25	Site-specific endodeoxyribonucleases: specific for altered bases
	3.1.26	Endoribonucleases producing 5′-phosphomonoesters
	3.1.27	Endoribonucleases producing other than 5′-phosphomonoesters
	3.1.30	Endonucleases active with either ribo- or deoxyribonucleic acids and producing 5′-phosphomonoesters
	3.1.31	Endonulceases active with either ribo- or deoxyribonucleic acids and producing 3′-phophomonoesters
3.2.	*Glycosylases*	
	3.2.1	Glycosidases, i.e., enzymes hydrolyzing *O*- and *S*-glycosyl compounds
	3.2.2	Hydrolyzing *N*-glycosyl compounds
	3.2.3	Hydrolyzing *S*-glycosyl compounds
3.3.	*Acting on ether bonds*	
	3.3.1	Trialkylsulfonium hydrolases
	3.3.2	Ether hydrolases
3.4.	*Acting on peptide bonds (peptidases)*	
	3.4.11	Aminopeptidases
	3.4.13	Dipeptidases
	3.4.14	Dipeptidyl-peptidases and tripeptidyl-peptidases
	3.4.15	Peptidyl-dipeptidases
	3.4.16	Serine-type carboxypeptidases
	3.4.17	Metallocarboxypeptidases
	3.4.18	Cysteine-type carboxypeptidases
	3.4.19	Omega peptidases
	3.4.21	Serine endopeptidases
	3.4.22	Cysteine endopeptidases
	3.4.23	Aspartic endopeptidases
	3.4.24	Metalloendopeptidases
	3.4.25	Threonine endopeptidases
	3.4.99	Endopeptidases of unknown catalytic mechanism
3.5.	*Acting on carbon-nitrogen bonds, other than peptide bonds*	
	3.5.1	In linear amides
	3.5.2	In cyclic amides
	3.5.3	In linear amidines
	3.5.4	In cyclic amidines
	3.5.5	In nitriles
	3.5.99	In other compounds
3.6.	*Acting on acid anhydrides*	
	3.6.1	In phosphorus-containing anhydrides
	3.6.2	In sulfonyl-containing anhydrides
	3.6.3	Acting on acid anhydrides; catalyzing transmembrane movement of substances
	3.6.4	Acting on acid anhydrides; involved in cellular and subcellular movements
	3.6.5	Acting on GTP; involved in cellular and subcellular movements
3.7.	*Acting on carbon-carbon bonds*	
	3.7.1	In ketonic substances
3.8.	*Acting on halide bonds*	
	3.8.3	In C-halide compounds

Table 6.1. *(Continued)*

- *3.9. Acting on phosphorus-nitrogen bonds*
- *3.10. Acting on sulfur-nitrogen bonds*
- *3.11. Acting on carbon-phosphorus bonds*
- *3.12. Acting on sulfur-sulfur bonds*
- *3.13. Acting on carbon-sulfur bonds*

4. Lyases
    - *4.1. Carbon-carbon lyases*
        - 4.1.1 Carboxy-lyases
        - 4.1.2 Aldehyde-lyases
        - 4.1.3 Oxo-acid-lyases
        - 4.1.99 Other carbon-carbon lyases
    - *4.2. Carbon-oxygen lyases*
        - 4.2.1 Hydro-lyases
        - 4.2.2 Acting on polysaccharides
        - 4.2.3 Acting on phosphates
        - 4.2.99 Other carbon-oxygen lyases
    - *4.3. Carbon-nitrogen lyases*
        - 4.3.1 Ammonia-lyases
        - 4.3.2 Amidine-lyases
        - 4.3.3 Amine-lyases
        - 4.3.99 Other carbon-nitrogen lyases
    - *4.4. Carbon-sulfur lyases*
    - *4.5. Carbon-halide lyases*
    - *4.6. Phosphorus-oxygen lyases*
    - *4.99. Other lyases*
5. Isomerases
    - *5.1. Racemases and epimerases*
        - 5.1.1 Acting on amino acids and derivatives
        - 5.1.2 Acting on hydroxy acids and derivatives
        - 5.1.3 Acting on carbohydrates and derivatives
        - 5.1.99 Acting on other compounds
    - *5.2. Cis-trans-isomerases*
    - *5.3. Intramolecular oxidoreductases*
        - 5.3.1 Interconverting aldoses and ketoses
        - 5.3.2 Interconverting keto and enol groups
        - 5.3.3 Transposing C=C bonds
        - 5.3.4 Transposing S-S bonds
        - 5.3.99 Other intramolecular oxidoreductases
    - *5.4. Intramolecular transferases*
        - 5.4.1 Transferring acyl groups
        - 5.4.2 Phosphotransferases (phosphomutases)
        - 5.4.3 Transferring amino groups
        - 5.4.4 Transferring hydroxy groups
        - 5.4.99 Transferring other groups
    - *5.5. Intramolecular lyases*
        - 5.5.1 Catalyze reactions in which a group can be regarded as eliminated from one part of a molecule, leaving a double bond, while remaining covalently attached to the molecule
    - *5.99. Other isomerases*
        - 5.99.1 Miscellaneous isomerases
6. Ligases
    - *6.1. Forming carbon-oxygen bonds*
        - 6.1.1 Ligases forming aminoacyl-tRNA and related compounds
    - *6.2. Forming carbon-sulfur bonds*
        - 6.2.1 Acid-thiol ligases

(Continued)

Table 6.1. *(Continued)*

6.3.	*Forming carbon-nitrogen bonds*	
	6.3.1	Acid-ammonia (or amine) ligases (amide synthases)
	6.3.2	Acid-amino-acid ligases (peptide synthases)
	6.3.3	Cyclo-ligases
	6.3.4	Other carbon-nitrogen ligases
	6.3.5	Carbon-nitrogen ligases with glutamine as amido-*N*-donor
6.4.	*Forming carbon-carbon bonds*	
6.5.	Forming phosphoric ester bonds	
6.6.	Forming nitrogen-metal bonds	

Sources: Ref. NC-IUBMB. 1992. Enzyme Nomenclature, 6th ed. San Diego (CA): Academic Press, Inc. With permission. NC-IUBMB Enzyme Nomenclature Website (www.iubmb.org).
[a]Readers should refer to the website: http://www.chem.qmul.ac.uk/iubmb/enzyme/rules.html for the most up-to-date changes.

of the ICUMB (NC-IUBMB). The information is reformatted in table form instead of text form for easier reading and comparison.

Table 6.2 shows the rules for systematic names and guidelines for common names as suggested by NC-IUBMB. Table 6.3 shows the rules and guidelines for particular classes of enzymes as suggested by NC-IUBMB. The concept on reformatting used in Table 6.1 is also applied.

EXAMPLES OF COMMON FOOD ENZYMES

The food industry likes to use terms more easily understood by its people and is slow to adopt changes. For example, some commonly used terms related to enzymes are very general terms and do not follow the recommended guidelines established by NC-IUBMB. Table 6.4 is a list of enzyme groups commonly used by the food industry and researchers (Nagodawithana and Reed 1993).

Table 6.2. General Rules for Generating Systematic Names and Guidelines for Common Names[a]

Rules and Guidelines No.	Descriptions
1.	*Common names*: Generally accepted trivial names of substances may be used in enzyme names. The prefix D-should be omitted for all D-sugars and L-for individual amino acids, unless ambiguity would be caused. In general, it is not necessary to indicate positions of substitutes in common names, unless it is necessary to prevent two different enzymes having the same name. The prefix *keto-* is no longer used for derivatives of sugars in which -CHOH- has been replaced by –CO–; they are named throughout as dehydrosugars. *Systematic names*: To produce usable systematic names, accepted names of substrates forming part of the enzyme names should be used. Where no accepted and convenient trivial names exist, the official IUPAC rules of nomenclature should be applied to the substrate name. The 1, 2, 3 system of locating substitutes should be used instead of the α β γ system, although group names such as β-aspartyl-, γ-glutamyl- and also β-alanine-lactone are permissible; α and β should normally be used for indicating configuration, as in α-D-glucose. For nucleotide groups, adenylyl (not adenyl), etc. should be the form used. The name oxo acids (not keto acids) may be used as a class name, and for individual compounds in which –CH2– has been replaced by –CO–, oxo should be used.
2.	Where the substrate is normally in the form of an anion, its name should end in *-ate* rather than *-ic*, e.g., *lactate dehydrogenase*, not "lactic acid dehydrogenase."
3.	Commonly used abbreviations for substrates, e.g., ATP, may be used in names of enzymes, but the use of new abbreviations (not listed in recommendations of the IUPAC-IUB Commission on Biochemical Nomenclature) should be discouraged. Chemical formulae should not normally be used instead of names of substrates. Abbreviations for names of enzymes, e.g., GDH, should not be used.
4.	Names of substrates composed of two nouns, such as glucose phosphate, which are normally written with a space, should be hyphenated when they form part of the enzyme names, and thus become adjectives, e.g., *glucose-6-phosphate dehydrogenase* (EC 1.1.1.49). This follows standard practice in phrases where two nouns qualify a third; see e.g., *Handbook of Chemical Society Authors*, 2nd ed., p. 14 (The Chemical Society, London, 1961).

Table 6.2. *(Continued)*

Rules and Guidelines No.	Descriptions
5.	The use as enzyme names of descriptions such as *condensing enzyme, acetate-activating enzyme,* and *pH 5 enzyme* should be discontinued as soon as the catalyzed reaction is known. The word *activating* should not be used in the sense of converting the substrate into a substance that reacts further; all enzymes act by activating their substrates, and the use of the word in this sense may lead to confusion.
6.	*Common names*: If it can be avoided, a common name should not be based on a substance that is not a true substrate, e.g., enzyme EC 4.2.1.17 *(Enoyl-CoA hydratase)* should not be called "crotonase," since it does not act on crotonate.
7.	*Common names*: Where a name in common use gives some indication of the reaction and is not incorrect or ambiguous, its continued use is recommended. In other cases, a common name is based on the same principles as the systematic name (see later), but with a minimum of detail, to produce a name short enough for convenient use. A few names of proteolytic enzymes ending in *-in* are retained; all other enzyme names should end in *-ase*.
	Systematic names: Systematic names consist of two parts. The first contains the name of the substrate or, in the case of a bimolecular reaction, of the two substrates separated by a colon. The second part, ending in *-ase*, indicates the nature of the reaction.
8.	A number of generic words indicating a type of reaction may be used in either common or systematic names: *oxidoreductase, oxygenase, transferase* (with a prefix indicating the nature of the group transferred), *hydrolase, lyase, racemase, epimerase, isomerase, mutase, ligase.*
9.	*Common names*: A number of additional generic names indicating reaction types are used in common names, but not in the systematic nomenclature, e.g., *dehydrogenase, reductase, oxidase, peroxidase, kinase, tautomerase, dehydratase,* etc.
10.	Where additional information is needed to make the reaction clear, a phrase indicating the reaction or a product should be added in parentheses after the second part of the name, e.g., *(ADP-forming), (dimerizing), (CoA-acylating).*
11.	*Common names*: The direct attachment of *-ase* to the name of the substrate will indicate that the enzyme brings about hydrolysis.
	Systematic names: The suffix *-ase* should never be attached to the name of the substrate.
12.	*Common names*: The name "dehydrase," which was at one time used for both dehydrogenating and dehydrating enzymes, should not be used. *Dehydrogenase* will be used for the former and *dehydratase* for the latter.
13.	*Common names*: Where possible, common names should normally be based on a reaction direction that has been demonstrated, e.g., *dehydrogenase* or *reductase, decarboxylase* or *carboxylase.*
	Systematic names: In the case of reversible reactions, the direction chosen for naming should be the same for all the enzymes in a given class, even if this direction has not been demonstrated for all. Thus, systematic names may be based on a written reaction, even though only the reverse of this has been actually demonstrated experimentally.
14.	*Systematic names*: When the overall reaction included two different changes, e.g., an oxidative demethylation, the classification and systematic name should be based, whenever possible, on the one (or the first one) catalyzed by the enzyme: the other function(s) should be indicated by adding a suitable participle in parentheses, as in the case of *sarcosine:oxygen oxidoreductase (demethylating)* (EC 1.5.3.1); D-*aspartate:oxygen oxidoreductase (deaminating)* (EC 1.4.3.1); L-*serine hydro-lyase (adding indole glycerol-phosphatase)* (EC 4.2.1.20).
15.	When an enzyme catalyzes more than one type of reaction, the name should normally refer to one reaction only. Each case must be considered on its merits, and the choice must be, to some extent, arbitrary. Other important activities of the enzyme may be indicated in the list under "reaction" or "comments."
	Similarly, when any enzyme acts on more than one substrate (or pair of substrates), the name should normally refer only to one substrate (or pair of substrates), although in certain cases it may be possible to use a term that covers a whole group of substrates, or an alternative substrate may be given in parentheses.
16.	A group of enzymes with closely similar specificities should normally be described by a single entry. However, when the specificity of two enzymes catalyzing the same reactions is sufficiently different (the degree of difference being a matter of arbitrary choice) two separate entries may be made, e.g., EC 1.2.1.4 (Aldehyde dehydrogenase ($NADP^+$)) and EC 1.2.1.7 (Benzylaldehyde ($NADP^+$)).

Source: NC-IUBMB. 1992. Enzyme Nomenclature. 6th ed. San Diego, California: Academic Press, Inc. With permission.

[a]Readers should refer to the website: http://www.chem.qmul.ac.uk/iubmb/enzyme/rules.html for the most recent changes.

Table 6.3. Rules and Guidelines for Particular Classes of Enzymes[a]

Rules and Guidelines No.	Description
Class 1: Oxidoreductases	
1.	*Common names*: The terms *dehydrogenase* or *reductase* will be used much as hitherto. The latter term is appropriate when hydrogen transfer from the substance mentioned as donor in the systematic name is not readily demonstrated. *Transhydrogenase* may be retained for a few well-established cases. *Oxidase* is used only for cases where O_2 acts as an acceptor, and *oxygenase* only for those cases where the O_2 molecule (or part of it) is directly incorporated into the substrate. *Peroxidase* is used for enzymes using H_2O_2 as acceptor. *Catalase* must be regarded as exceptional. Where no ambiguity is caused, the second reactant is not usually named; but where required to prevent ambiguity, it may be given in parentheses, e.g., EC 1.1.1.1, *alcohol dehydrogenase* and EC 1.1.1.2 *alcohol dehydrogenase (NADP$^+$)*.
	Systematic names: All enzymes catalyzing oxidoreductions should be *oxidoreductases* in the systematic nomenclature, and the names formed on the pattern *donor:acceptor oxidoreductase.*
2.	*Systematic names*: For oxidoreductases using NAD$^+$ or NADP$^+$, the coenzyme should always be named as the acceptor except for the special case of Section 1.6 (enzymes whose normal physiological function is regarded as reoxidation of the reduced coenzyme). Where the enzyme can use either coenzyme, this should be indicated by writing NAD(P)$^+$.
3.	Where the true acceptor is unknown and the oxidoreductase has only been shown to react with artificial acceptors, the word *acceptor* should be written in parentheses, as in the case of EC 1.3.99.1, *succinate:(acceptor)oxidoreductase.*
4.	*Common names*: Oxidoreductases that bring about the incorporation of molecular oxygen into one donor or into either or both of a pair of donors are named *oxygenase.* If only one atom of oxygen is incorporated, the term *monooxygenase* is used; if both atoms of O_2 are incorporated, the term *dioxygenase* is used.
	Systematic names: Oxidoreductases that bring about the incorporation of oxygen into one pair of donors should be named on the pattern *donor, donor:oxygen oxidoreductase (hydroxylating).*
Class 2: Transferases	
1.	*Common names*: Only one specific substrate or reaction product is generally indicated in the common names, together with the group donated or accepted. The forms *transaminase,* etc. may be used, if desired, instead of the corresponding forms *aminotransferase,* etc. A number of special words are used to indicate reaction types, e.g., *kinase* to indicate a phosphate transfer from ATP to the named substrate (not "phosphokinase"), *diphospho-kinase* for a similar transfer of diphosphate.
	Systematic names: Enzymes catalyzing group-transfer reactions should be named *transferase,* and the names formed on the pattern *donor:acceptor group-transferred-transferase,* e.g., *ATP:acetate phosphotransferase* (EC 2.7.2.1). A figure may be prefixed to show the position to which the group is transferred, e.g., *ATP:*D*-fructose 1-phospho-transferase* (EC 2.7.1.3). The spelling "transphorase" should not be used. In the case of the phosphotransferases, ATP should always be named as the donor. In the case of the transaminases involving 2-oxoglutarate, the latter should always be named as the acceptor.
2.	*Systematic names*: The prefix denoting the group transferred should, as far as possible, be noncommittal with respect to the mechanism of the transfer, e.g., *phospho-* rather than *phosphate-*.
Class 3. Hydrolases	
1.	*Common names*: The direct addition of *-ase* to the name of the substrate generally denotes a hydrolase. Where this is difficult, e.g., EC 3.1.2.1 (acetyl-CoA hydrolase), the word *hydrolase* may be used. Enzymes should not normally be given separate names merely on the basis of optimal conditions for activity. The *acid and alkaline phosphatases* (EC 3.1.3.1–2) should be regarded as special cases and not as examples to be followed. The common name *lysozyme* is also exceptional.
	Systematic names: Hydrolyzing enzymes should be systematically named on the pattern *substrate hydrolase*. Where the enzyme is specific for the removal of a particular group, the group may be named as a prefix, e.g., *adenosine aminohydrolase* (EC 3.5.4.4). In a number of cases, this group can also be transferred by the enzyme to other molecules, and the hydrolysis itself might be regarded as a transfer of the group to water.
Class 4. Lyases	
1.	*Common names*: The names *decarboxylase, aldolase,* etc. are retained; and *dehydratase* (not "dehydrase") is used for the hydro-lyases. "Synthetase" should not be used for any enzymes in this class. The term *synthase* may be used instead for any enzyme in this class (or any other class) when it is desired to emphasize the synthetic aspect of the reaction.

Table 6.3. *(Continued)*

Rules and Guidelines No.	Description
	Systematic names: Enzymes removing groups from substrates nonhydrolytically, leaving double bonds (or adding groups to double bonds) should be called *lyases* in the systematic nomenclature. Prefixes such as *hydro-*, *ammonia-* should be used to denote the type of reaction, e.g., (*S*)-*malate hydro-lyase* (EC 4.2.1.2). Decarboxylases should be regarded as *carboxy-lyases*. A hyphen should always be written before *lyase* to avoid confusion with hydrolases, carboxylases, etc.
2.	*Common names*: Where the equilibrium warrants it, or where the enzyme has long been named after a particular substrate, the reverse reaction may be taken as the basis of the name, using *hydratase, carboxylase*, etc., e.g., *fumarate hydratase* for EC 4.2.1.2 (in preference to "fumarase,"which suggests an enzyme hydrolyzing fumarate).
	Systematic names: The complete molecule, not either of the parts into which it is separated, should be named as the substrate. The part indicated as a prefix to *-lyase* is the more characteristic and usually, but not always, the smaller of the two reaction products. This may either be the removed (saturated) fragment of the substrate molecule, as in *ammonia-, hydro-, thiol-lyase,* or the remaining unsaturated fragment, e.g., in the case of *carboxy-, aldehyde- or oxo-acid-lyases*.
3.	Various subclasses of the lyases include a number of strictly specific or group-specific pyridoxal-5-phosphate enzymes that catalyze *elimination* reactions of β- or γ-substituted α-amino acids. Some closely related pyridoxal-5-phosphate-containing enzymes, e.g., *tryptophan synthase* (EC 4.2.1.20) and *cystathionine-synthase* (4.2.1.22) catalyze *replacement* reactions in which a β-, or γ-substituent is replaced by a second reactant without creating a double bond. Formally, these enzymes appeared to be transferases rather than lyases. However, there is evidence that in these cases the elimination of the β- or γ-substituent and the formation of an unsaturated intermediate is the first step in the reaction. Thus, applying rule 14 of the general rules for systematic names and guidelines for common names (Table 6.2), these enzymes are correctly classified lyases.
Class 5. Isomerases	
In this class, the common names are, in general, similar to the systematic names that indicate the basis of classification.	
1.	*Isomerase* will be used as a general name for enzymes in this class. The types of isomerization will be indicated in systematic names by prefixes, e.g., *maleate cis-trans-isomerase* (EC 5.2.1.1), *phenylpyruvate keto-enol-isomerase* (EC 5.3.2.1), *3-oxosteroid* Δ^5-Δ^4-isomerase (EC 5.3.3.1). Enzymes catalyzing an aldose-ketose interconversion will be known as *ketol-isomerases*, e.g., L-*arabinose ketol-isomerase* (EC 5.3.1.4). When the isomerization consists of an intramolecular transfer of a group, the enzyme is named a *mutase*, e.g., EC 5.4.1.1 (lysolecithin acylmutase) and the *phosphomutases* in sub-subclass 5.4.2 (Phosphotransferases); when it consists of an intramolecular lyase-type reaction, e.g., EC 5.5.1.1 (muconate cycloisomerase), it is systematically named a *lyase (decyclizing)*.
2.	Isomerases catalyzing inversions at asymmetric centers should be termed *racemases* or *epimerases*, according to whether the substrate contains one, or more than one, center of asymmetry: compare, e.g., EC 5.1.1.5 (lysine racemase) with EC 5.1.1.7 (diaminopimelate epimerase). A numerical prefix to the word *epimerase* should be used to show the position of the inversion.
Class 6. Ligases	
1.	*Common names*: Common names for enzymes of this class were previously of the type *XP synthetase*. However, as this use has not always been understood, and synthetase has been confused with synthase (see Class 4, item 1), it is now recommended that as far as possible the common names should be similar in form to the systematic name.
	Systematic names: The class of enzymes catalyzing the linking together of two molecules, coupled with the breaking of a diphosphate link in ATP, etc. should be known as *ligases*. These enzymes were often previously known as "synthetase"; however, this terminology differs from all other systematic enzyme names in that it is based on the product and not on the substrate. For these reasons, a new systematic class name was necessary.
2.	*Common name*: The common names should be formed on the pattern *X-Y ligase*, where X-Y is the substance formed by linking X and Y. In certain cases, where a trivial name is commonly used for XY, a name of the type *XY synthase* may be recommended (e.g., EC 6.3.2.11, *carnosine synthase*).
	Systematic names: The systematic names should be formed on the pattern *X:Y ligase (ADP-forming)*, where X and Y are the two molecules to be joined together. The phrase shown in parentheses indicates both that ATP is the triphosphate involved and that the terminal diphosphate link is broken. Thus, the reaction is $X + Y + ATP = X–Y + ADP + P_i$.
3.	*Common name*: In the special case where glutamine acts an ammonia donor, this is indicated by adding in parentheses (*glutamine-hydrolyzing*) to a ligase name.
	Systematic names: In this case, the name *amido-ligase* should be used in the systematic nomenclature.

Source: NC-IUBMB. 1992. Enzyme Nomenclature. San Diego, California: Academic Press, Inc. With permission.

[a]Readers should refer to the web version of the rules, which are currently under revision to reflect the recent changes. Available at http://www.chem.qmul.ac.uk/iubmb/enzyme/rules.html

Table 6.4. Common Enzyme Group Terms Used in the Food Industry

Group Term	Selected Enzymes in This Group
Carbohydrases	Amylases (EC 3.2.1.1, 3.2.1.2, 3.2.1.3)
	Pectic enzymes (see later)
	Lactases (EC 3.2.1.108, 3.2.1.23)
	Invertase (EC 3.2.1.26)
	α-galactosidases (EC 3.2.1.22, 3.2.1.23)
	Cellulase (EC 3.2.1.4)
	Hemicellulases (EC not known)
	Dextranase (EC 3.2.1.11)
Proteases	Serine proteases (EC 3.4.21)
(Endopeptidases)	Cysteine proteases (EC 3.4.22)
	Aspartic proteases (EC 3.4.23)
	Metalloproteases (EC 3.4.24)
Proteases	Aminopeptidases (EC 3.4.11)
(Exopeptidase)	
Proteases	Carboxypeptidases (EC 3.4.16)
(Carboxypeptidases)	Metallocarboxypeptidases (EC 3.4.17)
	Cysteine carboxypeptidases (EC 3.4.18)
	Dipeptide hydrolases (EC 3.4.13)
Oxidoreductases	Polyphenol oxidase (EC 1.10.3.1)
	Peroxidase (EC 1.11.1.7)
	Lactoperoxidase (EC 1.11.1.7)
	Catalase (EC 1.11.1.6)
	Sulfhydryl oxidase (EC 1.8.3.2)
	Glucose oxidase (EC 1.1.3.4)
	Pyranose oxidase (EC 1.1.3.10)
	Xanthine oxidase (EC 1.1.3.22)
	Lipoxygenase (EC 1.13.11.12)
	Dehydrogenases (see below)
	Alcohol oxidase (EC 1.1.3.13)
Pectolytic enzymes	Pectin (methyl) esterase (PE, EC 3.1.11.1)
(Pectic enzymes)	Polygalacturonase (PG, EC 3.2.2.15 or 3.2.1.67)
	Pectate lyases or pectic acid lyases (PAL)
	(Endo-type, EC 3.2.1.15 or 4.2.2.2)
	(Exo-type, EC 3.2.1.67 or 4.2.2.9)
	Pectic lyase (PL, EC 4.2.2.10)
Heme enzymes	Catalase (EC 1.11.1.6)
	Peroxidase (EC 1.11.1.7)
	Phenol oxidase (Monophenol monooxygenase, EC 1.14.18.1 ?)
	Lipoperoxidase (EC 1.13.11.12)
Hemicellulase	Glucanases (EC 3.2.1.6, 3.2.1.39, 3.2.1.58, 3.2.1.75, 3.2.1.59, 3.2.1.71, 3.2.1.74, 3.2.1.84)
	Xylanase (EC 3.2.1.32)
	Galactanases (EC 3.2.1.89, 3.2.1.23, 3.2.1.145)
	Mannanase (EC 3.2.1.25)
	Galactomannanases (EC not known)
	Pentosanases (EC not known)
Nucleolytic enzymes	Nucleases
	(Endonucleases, EC 3.1.30 or 3.1.27)
	(Exonucleases, EC 3.1.15 or 3.1.14)
	Phosphatases
	(Nonspecific phosphatases, EC 3.1.3)
	(Nucleotidases, EC 3.1.3)
	Nucleosidases (EC 3)
	(Inosinate nucleosidase, EC 3.2.2.12)
	Nucleodeaminases
	(Adenine deaminase, EC 3.5.4.2)
Dehydrogenase	Alcohol dehydrogenase (EC 1.1.1.1)
	Shikimate dehydrogenase (EC 1.1.1.25)
	Maltose dehydrogenase (EC 3.1 to 3.3)
	Isocitrate dehydrogenase (EC 1.1.1.41, 1.1.1.42)
	Lactate dehydrogenase (EC 1.1.1.3, 1.1.1.4, 1.1.1.27)

Source: Nagodawithana and Reed. 1993, NC–IUBMB 1992, NC–IUBMB. Available at www.iubmb.org.

Table 6.5. Recommended Names, Systematic Names, and Enzyme Codes (EC) for Some Common Food Enzymes

Recommended Name	Systematic Name	Enzyme Code (EC)
5′-nulceotidase	5′-ribonucleotide phosphohydrolase	3.1.3.5
α-amylase (Glycogenase)	1,4-α-D-glucan glucanohydrolase	3.2.1.1
α-galactosidase (Melibiase)	α-D-galactoside galactohydrolase	3.2.1.22
α-glucosidase (Maltase, glucoinvertase, glucosidosucrase, maltase-glucoamylase)	α-D-glucoside glucohydrolase	3.2.1.20
D-amino acid oxidase	D-amino-acid:oxygen oxidoreductase (deaminating)	1.4.3.3
D-lactate dehydrogenase (Lactic acid dehydrogenase)	(R)-lactate:NAD(+) oxidoreductase	1.1.1.28
L-amino acid oxidase	L-amino-acid:oxygen oxidoreductase (deaminating)	1.4.3.2
L-lactate dehydrogenase (Lactic acid dehydrogenase)	(S)-lactate:NAD(+) oxidoreductase	1.1.1.27
β-amylase (Saccharogen amylase, glycogenase)	1,4-α-D-glucan maltohydrolase	3.2.1.2
β-galactosidase (Lactase)	β-D-galactoside galactohydrolase	3.2.1.23
β-glucosidase (Gentiobiase, cellobiase, amyygdalase)	β-D-glucoside glucohydrolase	3.2.1.21
β-glucanase, see Hemicellulase		
Acid phosphatase (Acid phosphomonoesterase, phosphomonoesterase, glycerol phosphatase)	Orthophosphoric-monoester phosphohydrolase (acid optimum)	3.1.3.2
Adenine deaminase (Adenase, adenine aminase)	Adenine aminohydrolase	3.5.4.2
Adenosine phosphate deaminase	Adenosine-phosphate aminohydrolase	3.5.4.17
AMP deaminase (Adenylic acid deaminase, AMP aminase)	AMP aminohydrolase	3.5.4.6
Alcalase, see Subtilisin Carlsberg		
Alcohol dehydrogenase (Aldehyde reductase)	Alcohol:NAD(+) oxidoreductase	1.1.1.1
Alcohol oxidase	Alcohol:oxygen oxidoreductase	1.1.3.13
Alkaline phosphatase (Alkaline phosphomono-esterase, phosphomonoesterase, glycerolphosphatase)	Orthophosphoric-monoester phosphohydrolase(alkaline optimum)	3.1.3.1
Alliin lyase (Alliinase)	Alliin alkyl-sulfenate-lyase	4.4.1.4
Aminoacylase (Dehydropeptidase II, histozyme, acylase I, hippuricase, benzamidase)	*N-acyl-L-amino-acid amidohydrolase*	3.5.1.14
Aspartate amino-lyase (fumaric aminase)	L-amino-acid:oxygen oxidoreductase (deaminating)	4.3.1.1
Aspergillus nuclease S (1) [Endonuclease S (1) (aspergillus), single-stranded-nucleate endo-nuclease, deoxyribonuclease S (1)]	None, see comments in original reference	3.1.30.1
Bacillus thermoproteolyticus neutral proteinase (Thermolysin)	None, see comments in original reference	3.4.24.4
Bromelain	None, see comments in original reference	3.4.22.4

(Continued)

Table 6.5. *(Continued)*

Recommended Name	Systematic Name	Enzyme Code (EC)
Carboxy peptidase α (Carboxypolypeptidase)	Peptidyl-L-amino-acid hydrolase	3.4.17.1
Catalase	Hydrogen-peroxide:hydrogen-peroxide oxidoreductase	1.11.1.6
Catechol oxidase (Diphenol oxidase, o-diphenolase, phenolase, polyphenol oxidase, tyrosinase)	1,2-benzenediol:oxygen oxidoreductase	1.10.3.1
Cellulase (Endo-1,4-β-glucanase)	1,4-(1,3:1,4)-β-D-glucan 4-glucanohydrolase	3.2.1.4
Chymosin (Rennin)	None, see comments in original reference	3.4.23.4
Chymotrypsin (Chymotrypsin α and β)	None, see comments in original reference	3.4.21.1
Debranching enzyme, see α-dextrin endo-1,6-α-glucosidase		
Dextranase	1,6-α-D-glucan 6-glucanohydrolase	3.2.1.11
α-dextrin endo-1,6-α-glucosidase (Limit dextrinase, debranching enzyme, amylopectin 6-gluco-hydrolase, pullulanase)	α-dextrin 6-glucanohydrolase	3.2.1.41
Diastase (obsolete term)		3.2.1.1 (obsolete)
Dipeptidase	Dipeptide hydrolase	3.4.13.11
Dipeptidyl peptidase I (Cathepsin C, dipeptidyl-amino-peptidase I, dipeptidyl transferase)	Dipeptidyl-peptide hydrolase	
Endopectate lyase, see Pectate lyase		
Exoribonuclease II (Ribonuclease II)	None, see comments in original reference	3.1.13.1
Exoribonuclease H	None, see comments in original reference	3.1.13.2
Ficin	None, see comments in original reference	3.4.22.3
Galactose oxidase	D-galactose-1,4-lactone:oxygen 2-oxidoreductase	1.1.3.9
Glucan-1,4-α-glucosidase (Glucoamylase, amyloglucosidase, γ-amylase, acid maltase, lysosomal α-glucosidase, exo-1,4-α-glucosidase)	1,4-α-D-glucan glucohydrolase	3.2.1.3
Glucose-6-phosphate isomerase [Phosphohexose isomerase, phosphohexomutase, oxoisomerase, hexosephosphate isomerase phosphosaccharomutase, phosphoglucoisomerase, phosphohexoisomerase, glucose isomerase (deleted, obsolete)]	D-glucose-6-phosphate keto-isomerase	5.3.1.9
Glucokinase	ATP:D-gluconate 6-phosphotransferase	2.7.1.12
Glucose oxidase (Glucose oxyhydrase)	β-D-glucose:oxygen 1-oxidoreductase	1.1.3.4
Glycerol kinase	ATP:glycerol 3-phosphotransferase	2.7.1.30
Glycogen (starch) synthase (UDPglucose-glycogen glucosyl-transferase)	UDPglucoase:glycogen glucosyltransferase	2.4.1.11
Inosinate nucleosidase (Inosinase)	5′-inosinate phosphoribohydrolase	3.2.2.2
β-fructofuranosidase (invertase, saccharase)	β-D-fructofuranoside fructohydrolase	3.2.1.26
Lactase	Lactose galactohydrolase	3.2.1.108
D-lactate dehydrogenase (Lactic acid dehydrogenase)	(R)-lactate:NAD(+) oxidoreductase	1.1.1.28

Table 6.5. *(Continued)*

Recommended Name	Systematic Name	Enzyme Code (EC)
L-lactate dehydrogenase (Lactic acid dehydrogenase)	(*S*)-lactate:NAD(+) oxidoreductase	1.1.1.27
Lipoxygenase (Lipoxidase, carotene oxidase)	Linoleate:oxygen oxidoreductase	1.13.11.12
Lysozyme (Muramidase)	Peptidoglycan N-acetylmuramoylhydrolase	3.2.1.17
Malate dehydrogenase (Malic dehydrogenase)	(S)-malate:NAD(+) oxidoreductase	1.1.1.37
Microbial aspartic proteinase (Trypsinogen kinase)	None, see comments in original reference	3.4.23.6
Monophenol monooxygenase (Tyrosinase, phenolase, monophenol oxidase, cresolase)	Monophenol, L-dopa:oxygen oxidoreductase	1.14.18.1
Palmitoyl-CoA hydrolase (Long-chain fatty-acyl-CoA hydrolase)	Palmitoyl-CoA hydrolase	3.1.2.2
Papain (Papaya peptidase I)	None, see comments in original reference	3.4.22.2
Pectate lyase (Pectate transliminase)	Poly (1,4-α-D-galacturonide) lyase	4.2.2.2
Pectic esterase (Pectin demethoxylase, pectin methoxylase, pectin methylesterase)	Pectin pectylhydrolase	3.1.1.11
Pepsin A (Pepsin)	None, see comments in original reference	3.4.23.1
Peroxidase	Donor:hydrogen-peroxide oxidoreductase	1.11.1.7
Phosphodiesterase I (5′-exonuclease)	Oligonucleate 5’-nucleotidohydrolase	3.1.4.1
Phosphoglucomutase (Glucose phosphomutase)	α-D-glucose 1,6-phosphomutase	5.4.2.2
Phosphorylase (Muscle phosphorylase α and β, amylophosphorylase, polyphosphorylase)	1,4-α-D-glucan:orthophosphate α-D-glucosyltransferase	2.4.1.1
Plasmin (Fibrinase, fibrinolysin)	None, see comments in original reference	3.4.21.7
Polygalacturonase (Pectin depolymerase, pectinase)	Poly (1,4-α-D-galacturonide) glycanohydrolase	3.2.1.15
Pyranose oxidase (Glucoase-2-oxidase)	Pyranose:oxygen 2-oxidoreductase	1.1.3.10
Rennet, see Chymosin		
Pancreatic ribonuclease (RNase, RNase I, RNase A, pancreatic RNase, ribonuclease I)	None, see comments in original reference	3.1.2.75
Serine carboxypeptidase	Peptidyl-L-amino-acid hydrolase	3.4.16.1
Shikimate dehydrogenase	Shikimate;NADP(+) 3-oxidoreductase	1.1.1.25
Staphylococcal cysteine protease (Staphylococcal proteinase II)	None, see comments in original reference	3.4.22.13
Starch (bacterial glycogen) synthase (ADPglucose-starch glucosyltransferase)	ADPglucose:1,4-α-D-glucan 4-α-D-glucosyltransferase	2.4.1.21
Streptococcal cysteine protease (Streptococcal proteinase, streptococcus peptidase a)	None, see comments in original reference	3.4.22.10
Subtilisin	None, see comments in original reference	3.4.21.14

(Continued)

Table 6.5. *(Continued)*

Recommended Name	Systematic Name	Enzyme Code (EC)
Sucrose-phosphate synthase (UDPglucaose-fructose phosphate glucosyltransferase, sucrose-phosphate-UDP glucosyl-transferase)	UDPglucose:D-fructose-6-phosphate 2-α-D-glucosyl-transferase	2.4.1.14
Sucrose synthase (UDPglucoase-fructose glucosyltransferase, sucrose-UDP glucosyltransferase)	UDPglucose:D-fructose 2-α-D-glucosyltransferase	2.4.1.13
Thiol oxidase	Thiol:oxygen oxidoreductase	1.8.3.2
Triacylglycerol lipase (Lipase, tributyrase, triglyceride lipase)		3.1.1.3
Trimethylamine-*N*-oxide aldolase	Trimethylamine-N-oxide formaldehydelase	4.1.2.32
Trypsin (α- and β-trypsin)	None, see comments in original reference	3.4.21.4
Tyrosinase, see Catechol oxidase		1.10.3.1
Xanthine oxidase (Hypoxanthin oxidase)	Xanthine:oxygen oxidoreductase	1.1.3.22

Sources: Nagadodawithana and Reed 1993, NC-IUBMB Enzyme Nomenclature. Available at www.iubmb.org.

Table 6.5 lists some common names, systematic names, and EC numbers for some common food enzymes.

Enzyme classification and nomenclature are now standardized procedures. Some journals already require that enzyme codes be used in citing or naming enzymes. Other journals are following the trend. It is expected that all enzymes will have enzyme codes in new research articles as well as (one hopes) in new reference books. However, for older literature, it is still difficult to identify the enzyme codes. It is hoped that Tables 6.4 and 6.5 will be useful as references.

ACKNOWLEDGMENTS

The authors thank the International Union of Biochemistry and Molecular Biology for permission to use of some of the copyrighted information on Enzyme Classification and Nomenclature. The authors also want to thank Prof. Keith Tipton, Department of Biochemistry, Trinity College, Dublin 2, Ireland, for his critical review of this manuscript.

REFERENCES

Nagodawithana T, Reed G. 1993. *Enzymes in Food Processing*. San Diego, California: Academic Press.

IUPAC-IUB Joint Commission on Biochemical Nomenclature (JCBN), and Nomenclature Commission of IUB (NC-IUB) Newsletters. 1984. *Arch Biochem Biophys* 229:237–245; Biochem J 217:I–IV; *Biosci Rep* 4:177–180; Chem. Internat (3): 24–25; *Eur J Biochem* 138:5–7; *Hoppe-Seyler's Z Physiol Chem* 365:I–IV.

NC-IUBMB. 1992. Enzyme Nomenclature 1992: *Recommendations of the Nomenclature Committee of the International Union of Biochemistry and Molecular Biology*, 6th ed. San Diego, California: Academic Press.

NC-IUBMB Enzyme Nomenclature. Available at http://www.chem.qmul.ac.uk/iubmb/enzyme/ or at www.iubmb.org.

Boyce S, Tipton KF. 2000. Enzyme classification and nomenclature. In: *Nature Encyclopedia of Life Sciences*. London: Nature Publishing Group. (http://www.els.net/[doi:10.1038/npg.els.0000710]).

7
Biocatalysis, Enzyme Engineering and Biotechnology

G. A. Kotzia, D. Platis, I. A. Axarli, E. G. Chronopoulou, C. Karamitros, and N. E. Labrou

Abstract: Enzymes are biocatalysts evolved in nature to achieve the speed and coordination of nearly all the chemical reactions that define cellular metabolism necessary to develop and maintain life. The application of biocatalysis is growing rapidly, since enzymes offer potential for many exciting applications in industry. The advent of whole genome sequencing projects enabled new approaches for biocatalyst development, based on specialised methods for enzyme heterologous expression and engineering. The engineering of enzymes with altered activity, specificity and stability, using site-directed mutagenesis and directed evolution techniques are now well established. Over the last decade, enzyme immobilisation has become important in industry. New methods and techniques for enzyme immobilisation allow for the reuse of the catalysts and the development of efficient biotechnological processes. This chapter reviews advances in enzyme technology as well as in the techniques and strategies used for enzyme production, engineering and immobilisation and discuss their advantages and disadvantages.

PREFACE

Enzymes are proteins with powerful catalytic functions. They increase reaction rates sometimes by as much as one million fold, but more typically by about one thousand fold. Catalytic activity can also be shown, to a limited extent, by biological molecules other than the 'classical' enzymes. For example, antibodies raised to stable analogues of the transition states of a number of enzyme-catalysed reactions can act as effective

Food Biochemistry and Food Processing, Second Edition. Edited by Benjamin K. Simpson, Leo M.L. Nollet, Fidel Toldrá, Soottawat Benjakul, Gopinadhan Paliyath and Y.H. Hui.

catalysts for those reactions (Hsieh-Wilson et al. 1996). In addition, RNA molecules can also act as a catalyst for a number of different types of reactions (Lewin 1982). These antibodies and RNA catalysts are known as abzymes and ribozymes, respectively.

Enzymes have a number of distinct advantages over conventional chemical catalysts. Among these are their high productivity, catalytic efficiency, specificity and their ability to discriminate between similar parts of molecules (regiospecificity) or optical isomers (stereospecificity). Enzymes, in general, work under mild conditions of temperature, pressure and pH. This advantage decreases the energy requirements and therefore reduces the capital costs. However, there are some disadvantages in the use of enzymes, such as high cost and low stability. These shortcomings are currently being addressed mainly by employing protein engineering approaches using recombinant DNA technology (Stemmer 1994, Ke and Madison 1997). These approaches aim at improving various properties such as thermostability, specificity and catalytic efficiency. The advent of designer biocatalysts enables production of not only process-compatible enzymes, but also novel enzymes able to catalyse new or unexploited reactions (Schmidt-Dannert et al. 2000, Umeno and Arnold 2004). This is just the start of the enzyme technology era.

ENZYME STRUCTURE AND MECHANISM

Nomenclature and Classification of Enzymes

Enzymes are classified according to the nature of the reaction they catalyse (e.g. oxidation/reduction, hydrolysis, synthesis, etc.) and sub-classified according to the exact identity of their substrates and products. This nomenclature system was established by the *Enzyme Commission* (a committee of the International Union of Biochemistry). According to this system, all enzymes are classified into six major classes:

1. *Oxidoreductases*, which catalyse oxidation–reduction reactions.
2. *Transferases*, which catalyse group transfer from one molecule to another.
3. *Hydrolases*, which catalyse hydrolytic cleavage of C–C, C–N, C–O, C–S or O–P bonds. These are group transfer reactions but the acceptor is always water.
4. *Lyases*, which catalyse elimination reactions, resulting in the cleavage of C–C, C–O, C–N, C–S bonds or the formation of a double bond, or conversely adding groups to double bonds.
5. *Isomerases*, which catalyse isomerisation reactions, e.g., racemisation, epimerisation, *cis-trans*-isomerisation, tautomerisation.
6. *Ligases*, which catalyse bond formation, coupled with the hydrolysis of a high-energy phosphate bond in ATP or a similar triphosphate.

The Enzyme Commission system consists of a numerical classification hierarchy of the form 'E.C. a.b.c.d' in which 'a' represents the class of reaction catalysed and can take values from 1 to 6 according to the classification of reaction types given above. 'b' denotes the sub-class, which usually specifies more precisely the type of the substrate or the bond cleaved, e.g. by naming the electron donor of an oxidation–reduction reaction or by naming the functional group cleaved by a hydrolytic enzyme. 'c' denotes the sub-subclass, which allows an even more precise definition of the reaction catalysed. For example, sub-subclasses of oxidoreductases are denoted by naming the acceptor of the electron from its respective donor. 'd' is the serial number of the enzyme within its sub-subclass. An example will be analysed. The enzyme that oxidises D-glucose using molecular oxygen catalyses the following reaction:

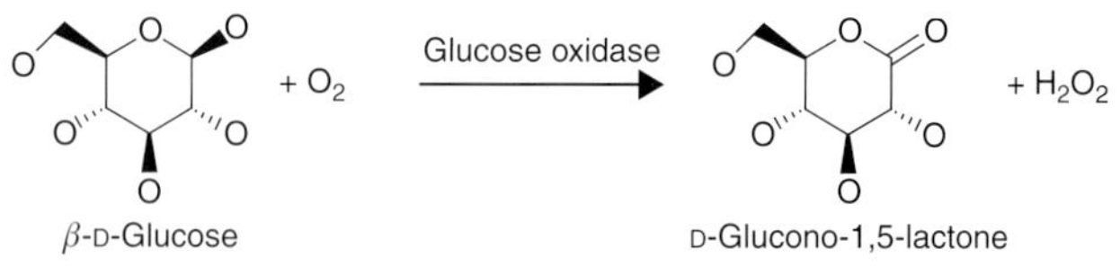

Scheme 7.1.

Hence, its systematic name is D-glucose: oxygen oxidoreductase, and its systematic number is EC 1.1.3.4.

The systematic names are often quite long, and therefore, short trivial names along with systematic numbers are often more convenient for enzyme designation. These shorter names are known as *recommended names*. The recommended names consist of the suffix '–ase' added to the substrate acted on. For example for the enzyme mentioned above, the recommended name is glucose oxidase.

It should be noted that the system of nomenclature and classification of enzymes is based only on the reaction catalysed and takes no account of the origin of the enzyme, that is from the species or tissue it derives.

Basic Elements of Enzyme Structure

The Primary Structure of Enzyme

Enzymes are composed of L-α-amino acids joined together by a *peptide bond* between the carboxylic acid group of one amino acid and the amino group of the next.

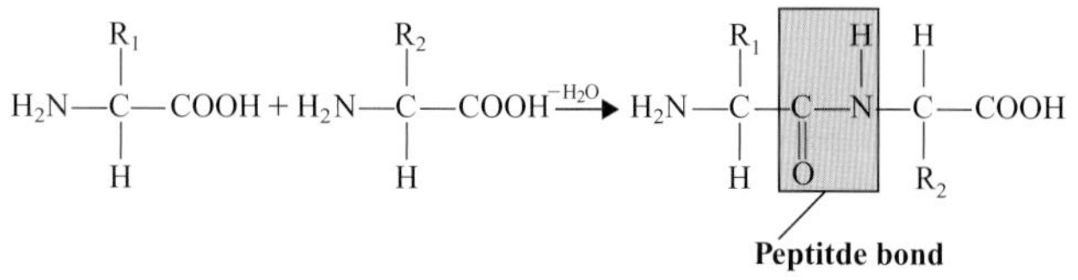

Scheme 7.2.

There are 20 common amino acids in proteins, which are specified by the genetic code; in rare cases, others occur as the products of enzymatic modifications after translation. A common feature of all 20 amino acids is a central carbon atom (C_α) to which a hydrogen atom, an amino group ($-NH_2$) and a carboxy group (–COOH) are attached. Free amino acids are usually zwitterionic at neutral pH, with the carboxyl group deprotonated and the amino group protonated. The structure of the most common amino acids found in proteins is shown in Table 7.1. Amino acids can be divided into four different classes depending on the structure of their side chains, which are called R groups: nonpolar, polar uncharged, negatively charged (at neutral pH) and positively charged (at neutral pH; Richardson 1981). The properties of the amino acid side chains determine the properties of the proteins they constitute. The formation of a succession of peptide bonds generates the 'main chain' or 'backbone'.

The primary structure of a protein places several constrains on how it can fold to produce its three-dimensional structure (Cantor 1980, Fersht 1999). The backbone of a protein consists of a carbon atom C_α to which the side chain is attached, a NH group bound to C_α and a carbonyl group $C = O$, where the carbon atom C is attached to C_α (Fig. 7.1) The peptide bond is planar since it has partial (~40%) double bond character with π electrons shared between the C–O and C–N bonds (Fig. 7.1). The peptide bond has a *trans*-conformation, that is the oxygen of the carbonyl group and the hydrogen of the NH group are in the *trans* position; the *cis* conformation occurs only in exceptional cases (Richardson 1981).

Enzymes have several 'levels' of structure. The protein's sequence, that is the order of amino acids, is termed as its 'primary structure'. This is determined by the sequence of nucleotide bases in the gene that codes for the protein. Translation of the mRNA transcript produces a linear chain of amino acids linked together by a peptide bond between the carboxyl carbon of the first amino acid and the free amino group of the second amino acid. The first amino acid in any polypeptide sequence has a free amino group and the terminal amino acid has a free carboxyl group. The primary structure is responsible for the higher levels of enzyme's structure and therefore for the enzymatic activity (Richardson 1981, Price and Stevens 1999).

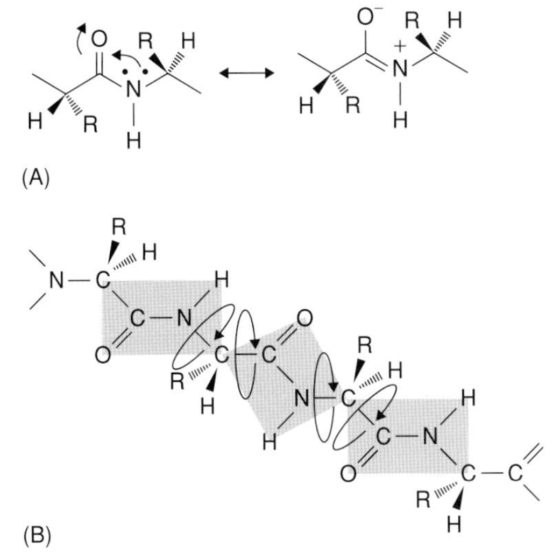

Figure 7.1. (**A**) The amide bond showing delocalisation of electrons. (**B**) A tripeptide unit of a polypeptide chain showing the planar amide units and the relevant angles of rotation about the bonds to the central α-carbon atom.

The Three-Dimensional Structure of Enzymes

Enzymes are generally very closely packed globular structures with only a small number of internal cavities, which are normally filled by water molecules. The polypeptide chains of enzymes are organised into ordered, hydrogen bonded regions, known as *secondary structure* (Andersen and Rost 2003). In these ordered structures, any buried carbonyl oxygen forms a hydrogen bond with an amino NH group. This is done by forming α-helices and β-pleated sheets, as shown in Figure 7.2. The α-helix can be thought of as having a structure similar to a coil or spring (Surewicz and Mantsch 1988). The β-sheet can be visualised as a series of parallel strings laid on top of an accordion-folded piece of paper. These structures are determined by the protein's primary structure. Relatively small, uncharged, polar amino acids readily organise themselves into α-helices, while relatively small, non-polar amino acids form β-sheets. Proline is a special amino acid because of its unique structure (Table 7.1). Introduction of proline into the sequence creates a permanent bend at that position (Garnier et al. 1990). Therefore, the presence of proline in an α-helix or β-sheet disrupts the secondary structure at that point. The presence of a glycine residue confers greater than normal flexibility on a polypeptide chain. This is due to the absence of a bulky side chain, which reduces steric hindrance.

Another frequently observed structural unit is the β-turn (Fang and Shortle 2003). This occurs when the main chain sharply changes direction using a bend composed of four successive residues, often including proline and glycine. In these units, the $C = O$ group of residue i is hydrogen bonded to the NH of residue i+3 instead of i+4 as in α-helix. Many different types of β-turn have been identified, which differ in terms of the number of amino acids and in conformation (e.g. Type I, Type II, Type III; Sibanda et al. 1989).

The three-dimensional structure of a protein composed of a single polypeptide chain is known as its *tertiary structure*. Tertiary structure is determined largely by the interaction of R groups on the surface of the protein with water and with other R groups on the protein's surface. The intermolecular non-covalent attractive forces that are involved in stabilising the enzyme's structure are usually classified into three types: ionic bonds, hydrogen bonds and van der Waals attractions (Matthews 1993). Hydrogen bonding results from the formation of hydrogen bridges between appropriate atoms; electrostatic bonds are due to the attraction of oppositely charged groups located on two amino acid side chains. Van der Waals bonds are generated by the interaction between electron clouds. Another important weak force is created by the three-dimensional structure of water, which tends to force hydrophobic groups together in order to minimise their disruptive effect on hydrogen-bonded

Table 7.1. Names, Symbols (One Letter and Three Letters Code) and Chemical Structures of the 20 Amino Acids Found in Proteins

Non-polar amino acids		
Alanine	Ala (A)	$CH_3-CH(NH_2)-COOH$
Valine	Val (V)	$(H_3C)(H_3C)>CH-CH(NH_2)-COOH$
Leucine	Leu (L)	$(G_2B)(G_2B)>CH-CH_2-CH(NH_2)-COOH$
Isoleucine	Ile (I)	$(H_3C-CH_2)(H_3C)>CH-CH(NH_2)-COOH$
Methionine	Met (M)	$CH_3-S-CH_2-CH(NH_2)-COOH$
Tryptophan	Trp (W)	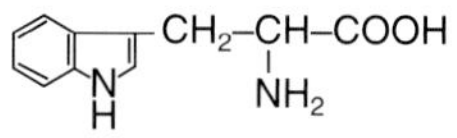
Phenylalanine	Phe (P)	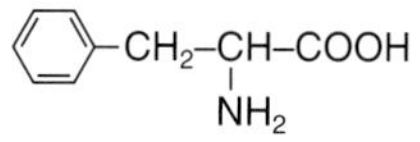
Proline	Pro (P)	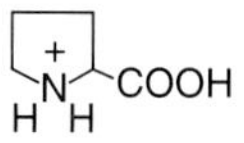
Polar uncharged amino acids		
Glycine	Gly (G)	$H-CH(NH_2)-COOH$
Asparagine	Asp (D)	$H_2N-C(=O)-CH_2-CH(NH_2)-COOH$
Glutamine	Gln (Q)	$H_2N-C(=O)-CH_2-CH_2-CH(NH_2)-COOH$
Serine	Ser (S)	$HO-CH_2-CH(NH_2)-COOH$
Threonine	Thr (T)	$HO-CH_2(CH_3)-CH(NH_2)-COOH$
Cysteine	Cys (C)	$HS-CH_2-CH(NH_2)-COOH$
Tyrosine	Tyr (Y)	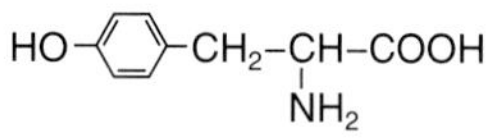
Negatively charged amino acids		
Glutamate	Glu (E)	$HOOC-CH_2-CH_2-CH(NH_2)-COOH$
Aspartate	Asn (N)	$HOOC-CH_2-CH(NH_2)-COOH$
Positively charged amino acids		
Arganine	Arg (R)	$HN(C(=NH)NH_2)-CH_2-CH_2-CH_2-CH(NH_2)-COOH$
Lysine	Lys (K)	$H_2N-(CH_2)_4-CH(NH_2)-COOH$
Histidine	His (H)	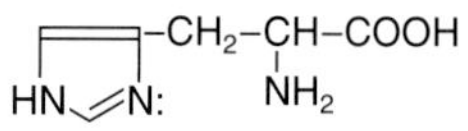

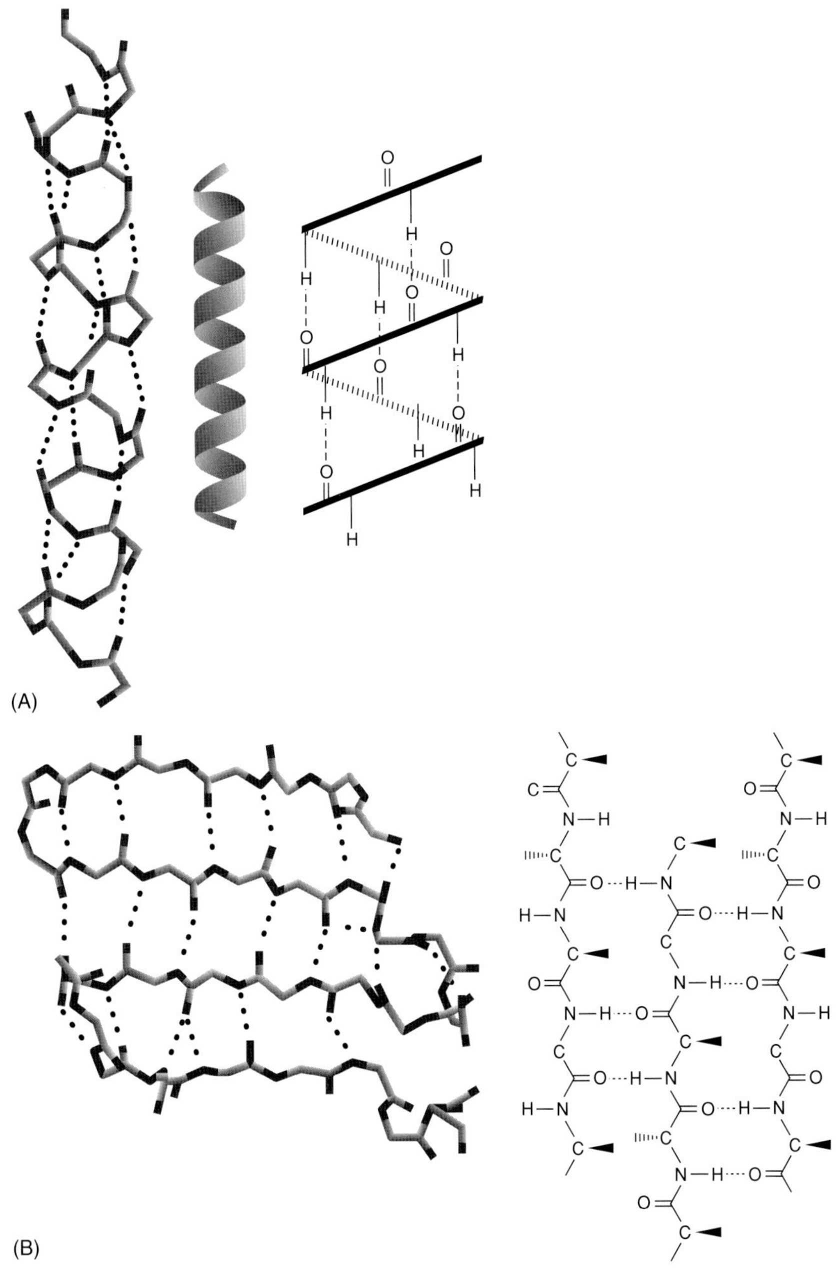

Figure 7.2. Representation of α-helix and β-sheet. Hydrogen bonds are depicted by doted lines.

network of water molecules. Apart from the peptide bond, the only other type of covalent bond involved in linking amino-acids is the disulphide (–S–S–) bond, which can be formed between two cysteine side-chains under oxidising conditions. The disulphide bond contributes significantly to the structural stability of an enzyme and more precisely in tertiary structural stabilisation (see below for details; Matthews 1993, Estape et al. 1998).

Detailed studies have shown that certain combinations of α-helices or β-sheet together with turns occur in many proteins. These often-occurring structures have been termed *motifs*, or super-secondary structure (Kabsch and Sander 1983, Adamson et al. 1993). Examples of motifs found in several enzymes are shown in Figure 7.3. These protein folds represent highly stable structural units, and it is believed that they may form nucleating centres in protein folding (Richardson 1981).

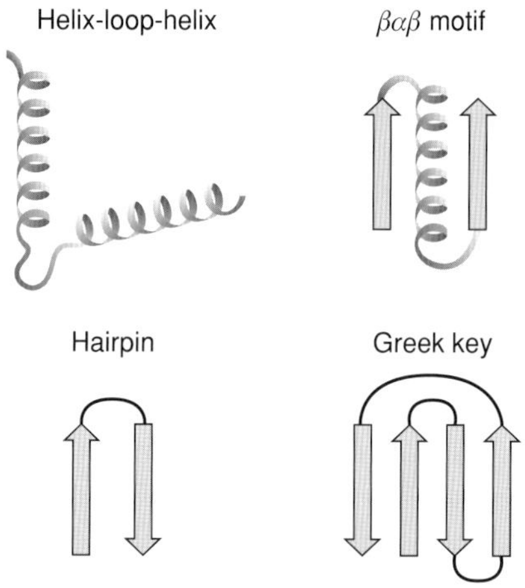

Figure 7.3. Common examples of motifs found in proteins.

Certain combinations of α-helices and β-sheets pack together to form compactly folded globular units, each of which is called protein domain with molecular mass of 15,000 to 20,000 Da (Orengo et al. 1997). Domains may be composed of secondary structure and identifiable motifs and therefore represent a higher level of structure than motifs. The most widely used classification scheme of domains has been the four-class system (Fig. 7.4; Murzin et al. 1995). The four classes of protein structure are as follows:

1. All-α-proteins, which have only α-helix structure.
2. All-β-proteins, which have β-sheet structure.
3. α/β-proteins, which have mixed or alternating segments of α-helix and β-sheet structure.
4. $\alpha+\beta$ proteins, which have α-helix and β-sheet structural segments that do not mix but are separated along the polypeptide chain.

While small proteins may contain only a single domain, larger enzymes contain a number of domains. In fact, most enzymes are dimers, tetramers or polymers of several polypeptide chains. Each polypeptide chain is termed 'subunit' and it may be identical or different to the others. The side chains on each polypeptide chain may interact with each other, as well as with water molecules to give the final enzyme structure. The overall organisation of the subunits is known as the *quaternary structure* and therefore the quaternary structure is a characteristic of multi-subunit enzymes. The four levels of enzyme structure are illustrated in Figure 7.5.

Theory of Enzyme Catalysis and Mechanism

In order for a reaction to occur, the reactant molecules must possess sufficient energy to cross a potential energy barrier, which is known as the *activation energy* (Fig. 7.6; Hackney 1990). All reactant molecules have different amounts of energy, but only a small proportion of them have sufficient energy to cross the activation energy of the reaction. The lower the activation energy, the more substrate molecules are able to cross the activation energy. The result is that the reaction rate is increased.

Enzyme catalysis requires the formation of a specific reversible complex between the substrate and the enzyme. This complex is known as the *enzyme–substrate complex* (ES) and provides all the conditions that favour the catalytic event (Hackney 1990, Marti et al. 2004). Enzymes accelerate reactions by lowering the energy required for the formation of a complex of reactants that is competent to produce reaction products. This complex is known as the *transition state complex* of the reaction and is characterised by lower free energy than it would be found in the uncatalysed reaction.

$$E + S \rightleftharpoons ES \rightleftharpoons ES^{*} \rightleftharpoons EP \rightleftharpoons E + P$$

Scheme 7.3.

The ES must pass to the transition state (ES*). The transition state complex must advance to an *enzyme–product complex* (EP), which dissociates to free enzyme and product (P). This reaction's pathway goes through the transition states TS_1, TS_2 and TS_3. The amount of energy required to achieve the transition state is lowered; hence, a greater proportion of the molecules in the population can achieve the transition state and cross the activation energy (Benkovic and Hammes-Schiffer 2003, Wolfenden 2003). Enzymes speed up the forward and reverse reactions proportionately, so that they have no effect on the equilibrium constant of the reactions they catalyse (Hackney 1990).

Substrate is bound to the enzyme by relatively weak non-covalent forces. The free energy of interaction of the ES complex ranges between −12 to −36 kJ/mole. The intermolecular attractive forces between enzyme–substrate, in general, are of three types: ionic bonds, hydrogen bonds and van der Waals attractions.

Specific part of the protein structure that interacts with the substrate is known as the *substrate binding site* (Fig. 7.7). The substrate binding site is a three-dimensional entity suitably designed as a pocket or a cleft to accept the structure of the substrate in three-dimensional terms. The binding residues are defined as any residue with any atom within 4 Å of a bound substrate. These binding residues that participate in the catalytic event are known as the *catalytic-residues* and form the *active-site*. According to Bartlett et al. (Bartlett et al. 2002), a residue is defined as catalytic if any of the following take place:

1. Direct involvement in the catalytic mechanism, for example as a nucleophile.
2. Exerting an effect, that aids catalysis, on another residue or water molecule, which is directly involved in the catalytic mechanism.
3. Stabilisation of a proposed transition-state intermediate.
4. Exerting an effect on a substrate or cofactor that aids catalysis, for example by polarising a bond that is to be broken.

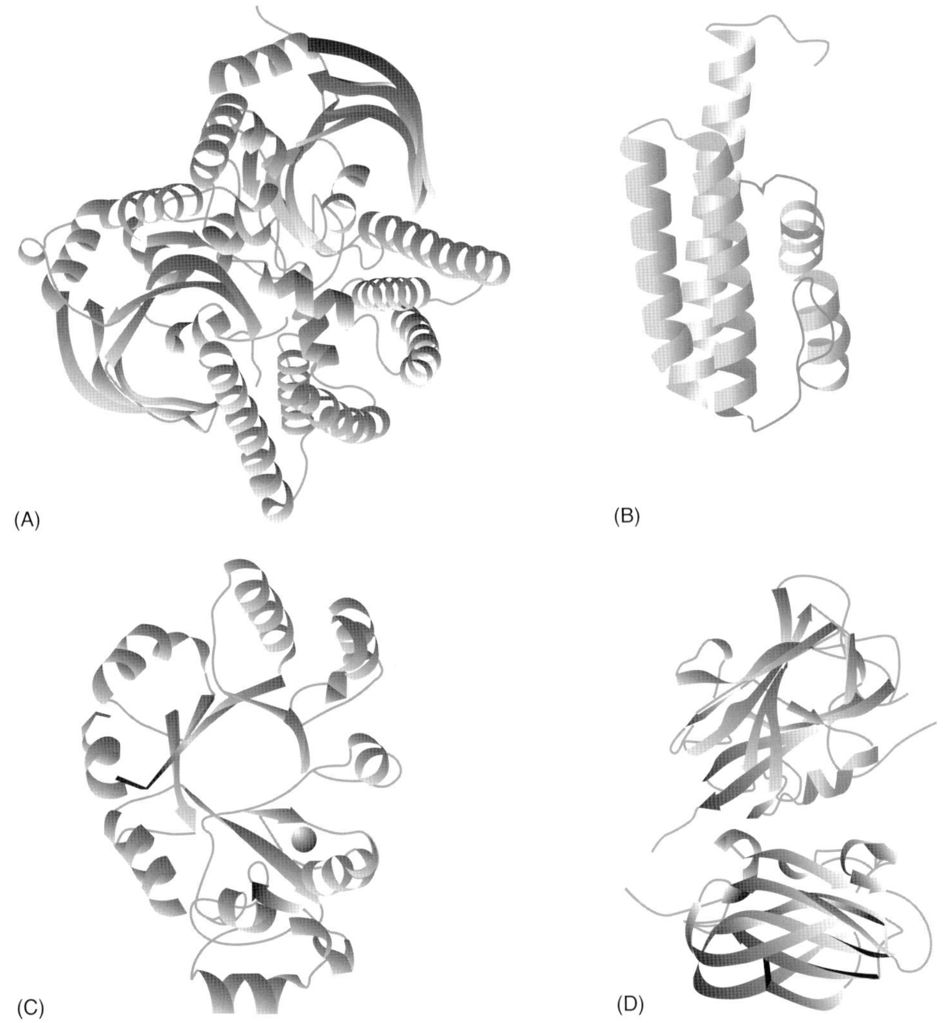

Figure 7.4. The four-class classification system of domains. (**A**) The $\alpha + \beta$ class (structure of glycyl-tRNA synthetase α-chain). (**B**) the all α class (structure of the hypothetical protein (Tm0613) from *Thermotoga maritima*). (**C**) The α/β class (structure of glycerophosphodiester phosphodiesterase). (**D**) The all β class (structure of allantoicase from *Saccharomyces cerevisiae*).

Despite the impression that the enzyme's structure is static and locked into a single conformation, several motions and conformational changes of the various regions always occur (Hammes 2002). The extent of these motions depends on many factors, including temperature, the properties of the solvating medium, the presence or absence of substrate and product (Hammes 2002). The conformational changes undergone by the enzyme play an important role in controlling the catalytic cycle. In some enzymes, there are significant movements of the binding residues, usually on surface loops, and in other cases, there are larger conformational changes. Catalysis takes place in the closed form and the enzyme opens again to release the product. This favoured model that explains enzyme catalysis and substrate interaction is the so-called *induced fit hypothesis* (Anderson et al. 1979, Joseph et al. 1990). In this hypothesis, the initial interaction between enzyme and substrate rapidly induces conformational changes in the shape of the active site, which results in a new shape of the active site that brings catalytic residues close to substrate bonds to be altered (Fig. 7.8). When binding of the substrate to the enzyme takes place, the shape adjustment triggers catalysis by generating transition-state complexes. This hypothesis helps to explain why enzymes only catalyse specific reactions (Anderson et al. 1979, Joseph et al. 1990). This basic cycle has been seen in many different enzymes, including triosephosphate isomerase, which uses a small hinged loop to close the active site (Joseph et al. 1990) and kinases, which use two large lobes moving towards each other when the substrate binds (Anderson et al. 1979).

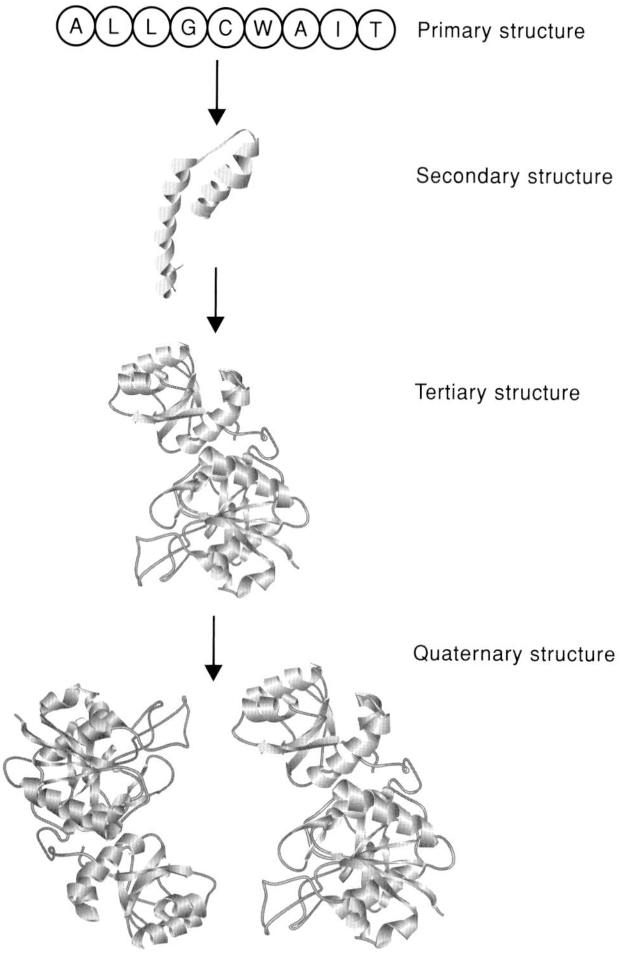

Figure 7.5. Schematic representation of the four levels of protein structure.

Coenzymes, Prosthetic Groups and Metal Ion Cofactors

Non-protein groups can also be used by enzymes to affect catalysis. These groups, called *cofactors*, can be organic or inorganic and are divided into three classes: coenzymes, prosthetic groups and metal ion cofactors (McCormick 1975). Prosthetic groups are tightly bound to an enzyme through covalent bond. Coenzymes bind to enzyme reversibly and associate and dissociate from the enzyme during each catalytic cycle and therefore may be considered as co-substrates. An enzyme containing a cofactor or prosthetic group is termed as *holoenzyme*. Coenzymes can be broadly classified into three main groups: coenzymes that transfer groups onto substrate, coenzymes that accept and donate electrons and compounds that activate substrates (Table 7.2). Metal ions such as Ca^{+2}, Mg^{+2}, Zn^{+2}, Mn^{+2}, Fe^{+2} and Cu^{+2} may in some cases act as cofactors. These may be bound to the enzyme by simple coordination with electron-donating atoms of side chains (imidazole of His, –SH group of Cys, carboxylate O^- of Asp and Glu). In some cases, metals, such as Mg^{+2}, are associated with the substrate rather than the enzyme. For example, Mg-ATP is the true substrate for kinases (Anderson et al. 1979). In other cases, metals may form part of a prosthetic group in which they are bound by coordinate bonds (e.g. heme; Table 7.2) in addition to side-chain groups. Usually, in this case, metal ions participate in electron transfer reactions.

Kinetics of Enzyme-Catalysed Reactions

The term enzyme kinetics implies a study of the velocity of an enzyme-catalysed reaction and of the various factors that may affect this (Moss 1988). An extensive discussion of enzyme kinetics would stay too far from the central theme of this chapter, but some general aspects will be briefly considered.

The concepts underlying the analysis of enzyme kinetics continue to provide significant information for understanding in vivo function and metabolism and for the development and clinical use of drugs aimed at selectively altering rate constants and interfering with the progress of disease states (Bauer et al. 2001). Central scope of any study of enzyme kinetics is knowledge of the way in which reaction velocity is altered by changes in the concentration of the enzyme's substrate and of the simple mathematics underlying this (Wharton 1983, Moss 1988, Watson and Dive 1994). As we have already discussed, the enzymatic reactions proceed through an intermediate ES in which each molecule of enzyme is combined, at any given instant during the reaction, with one substrate molecule. The reaction between enzyme and substrate to form the ES is reversible. Therefore, the overall enzymatic reaction can be shown as

$$E + S \underset{k_{-1}}{\overset{k_{+1}}{\rightleftharpoons}} ES \xrightarrow{k_{+2}} E + P$$

Scheme 7.4.

where k_{+1}, k_{-1} and k_{+2} are the respective rate constants. The reverse reaction concerning the conversion of product to substrate is not included in this scheme. This is allowed at the beginning of the reaction when there is no, or little, product present. In 1913, biochemists Michaelis and Menten suggested that if the reverse reaction between E and S is sufficiently rapid, in comparison with the breakdown of ES complex to form product, the latter reaction will have a negligible effect on the concentration of the ES complex. Consequently, E, S and ES will be in equilibrium, and the rates of formation and breakdown of ES will be equal. On the basis of these assumptions Michaelis and Menten produced the following equation:

$$v = \frac{V_{max} \cdot [S]}{K_m + [S]}$$

This equation is a quantitative description of the relationship between the rate of an enzyme-catalysed reaction (v) and the concentration of substrate $[S]$. The parameters V_{max} and K_m are

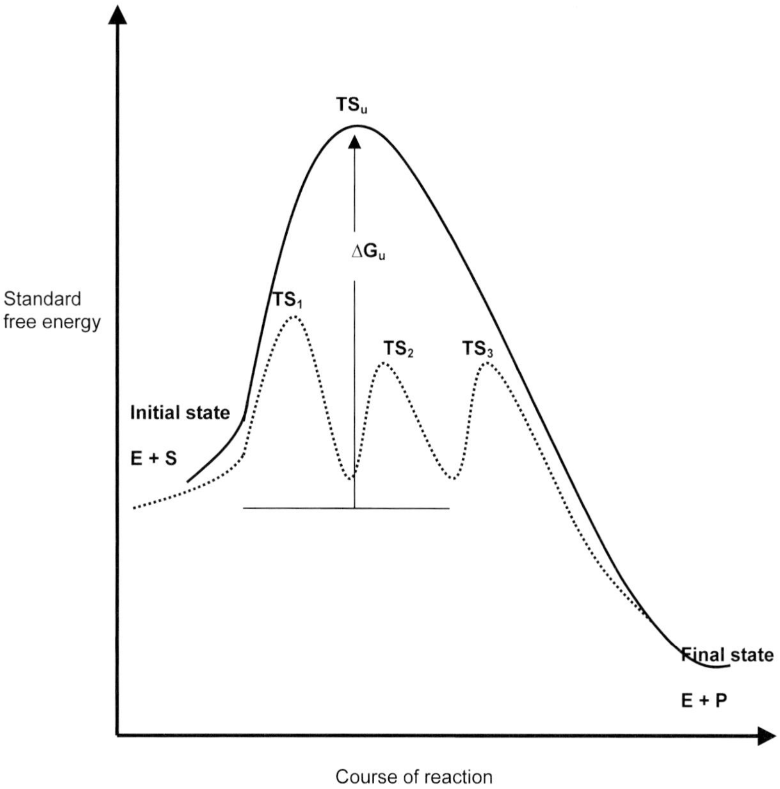

Figure 7.6. A schematic diagram showing the free energy profile of the course of an enzyme catalysed reaction involving the formation of enzyme–substrate (ES) and enzyme–product (EP) complexes. The catalysed reaction pathway goes through the transition states TS_1, TS_2, and TS_3, with standard free energy of activation ΔG_c, whereas the uncatalysed reaction goes through the transition state TS_u with standard free energy of activation ΔG_u.

constants at a given temperature and a given enzyme concentration. The K_m or Michaelis constant is the substrate concentration at which $v = V_{max}/2$ and its usual unit is M. The K_m provides us with information about the substrate *binding affinity* of the enzyme. A high K_m indicates a low affinity and vice versa (Moss 1988, Price and Stevens 1999).

The V_{max} is the maximum rate of the enzyme-catalysed reaction and it is observed at very high substrate concentrations where all the enzyme molecules are saturated with substrate, in the form of ES complex. Therefore

$$V_{max} = k_{cat}[E_t]$$

where $[E_t]$ is the total enzyme concentration and k_{cat} is the rate of breakdown of the ES complex (k_{+2} in the equation), which is known as the *turnover number*. k_{cat} represents the maximum number of substrate molecules that the enzyme can convert to product in a set time. The K_m depends on the particular enzyme and substrate being used and on the temperature, pH, ionic strength, etc. However, note that K_m is independent of the enzyme concentration, whereas V_{max} is proportional to enzyme concentration. A plot of the initial rate (v) against initial substrate concentration ($[S]$) for a reaction obeying the Michaelis–Menten kinetics has the form of a rectangular hyperbola through the origin with asymptotes $v = V_{max}$ and $[S] = -K_m$ (Fig. 7.9A). The term *hyperbolic kinetics* is also sometimes used to characterise such kinetics.

There are several available methods for determining the parameters from the Michaelis–Menten equation. A better method for determining the values of V_{max} and K_m was formulated by Hans Lineweaver and Dean Burk and is termed the Lineweaver-Burk (LB) or *double reciprocal plot* (Fig. 7.9B). Specifically, it is a plot of $1/v$ versus $1/[S]$, according to the equation:

$$\frac{1}{v} = \frac{K_m}{V_{max}} \cdot \frac{1}{[S]} + \frac{1}{V_{max}}$$

Such a plot yields a straight line with a slope of K_m/V_{max}. The intercept on the $1/v$ axis is $1/V_{max}$ and the intercept on the $1/[S]$ axis is $-1/K_m$.

The rate of an enzymatic reaction is also affected by changes in pH and temperature (Fig. 7.10). When pH is varied, the velocity of reaction in the presence of a constant amount of enzyme is typically greatest over a relatively narrow range of pH. Since enzymes are proteins, they possess a large number of ionic groups, which are capable of existing in different ionic forms (Labrou et al. 2004a). The existence of a fairly narrow pH-optimum for

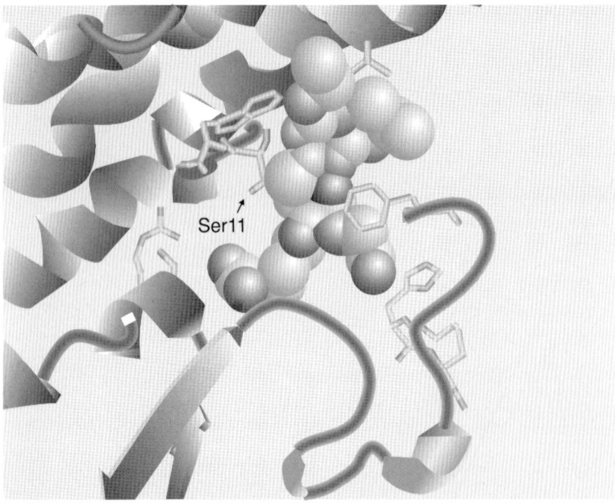

Figure 7.7. The substrate binding site of maize glutathione *S*-transferase. The binding residues are depicted as sticks, whereas the substrate is depicted in a space fill model. Only Ser 11 is involved directly in catalysis and is considered as catalytic residue.

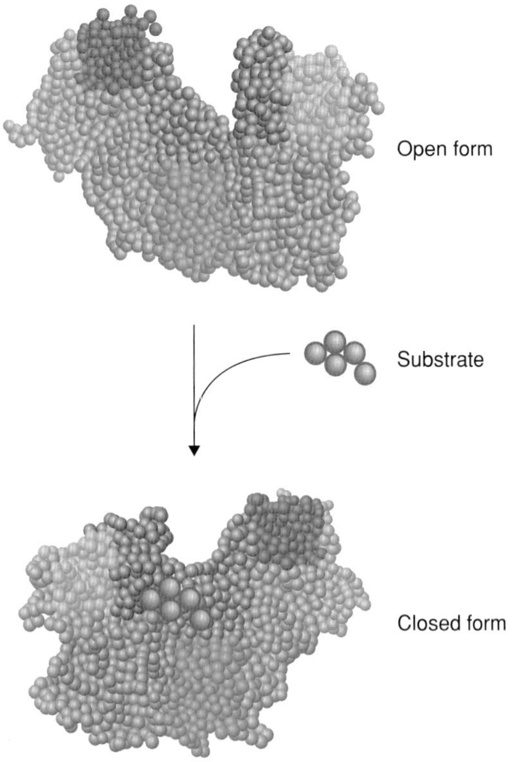

Figure 7.8. A schematic representation of the induced fit hypothesis.

most enzymes suggests that one particular ionic form of the enzyme molecule, out of the many that it can potentially exist, is the catalytically active one. The effect of pH changes on v is reversible, except after exposure to extremes of pH at which denaturation of the enzyme may occur.

The rate of an enzymatic reaction increases with increasing temperature. Although there are significant variations from one enzyme to another, on average, for each 10°C rise in temperature, the enzymatic activity is increased by an order of two. After exposure of the enzyme to high temperatures (normally greater that 65°C), denaturation of the enzyme may occur and the enzyme activity decreased. The Arrhenius equation

$$Log V_{\max} = \frac{-E_a}{2.303RT} + A$$

provides a quantitative description of the relationship between the rate of an enzyme-catalysed reaction ($V_{\max}$) and the temperature (T). Where E_a is the activation energy of the reaction, R is the gas constant, and A is a constant relevant to the nature of the reactant molecules.

The rates of enzymatic reactions are affected by changes in the concentrations of compounds other than the substrate. These modifiers may be activators, that is they increase the rate of reaction or their presence may inhibit the enzyme's activity. Activators and inhibitors are usually small molecules or even ions. Enzyme inhibitors fall into two broad classes: those causing irreversible inactivation of enzymes and those whose inhibitory effects can be reversed. Inhibitors of the first class bind covalently to the enzyme so that physical methods of separating the two are ineffective. Reversible inhibition is characterised by the existence of equilibrium between enzyme and inhibitor (I):

$$\mathrm{E + I \rightleftharpoons EI}$$

Scheme 7.5.

The equilibrium constant of the reaction, K_i, is given by the equation:

$$K_i = \frac{[ES]}{[E][I]}$$

K_i is a measure of the affinity of the inhibitor for the enzyme. Reversible inhibitors can be divided into three main categories: competitive inhibitors, non-competitive inhibitors and uncompetitive inhibitors. The characteristic of each type of inhibition and their effect on the kinetic parameters K_m and $V_{\max}$ are shown in Table 7.3.

Thermodynamic Analysis

Thermodynamics is the science that deals with energy. Chemical bonds store energy, as the subatomic particles are being attracted and chemical reactions are occurred with energy changes. Originally, thermodynamic laws were used to analyse and characterise mechanical systems, but the kinetic molecular theory

Table 7.2. The Structure of Some Common Coenzymes: Adenosine Triphosphate (ATP), Coenzyme A (CoA), Flavin Adenine Dinucleotide (FAD), Nicotinamide Adenine Dinucleotide (NAD^+) and Heam c

Cofactor	Type of Reactions Catalysed	Structure
Adenosine triphosphate (ATP)	Phosphate transfer reactions (e.g. kinases)	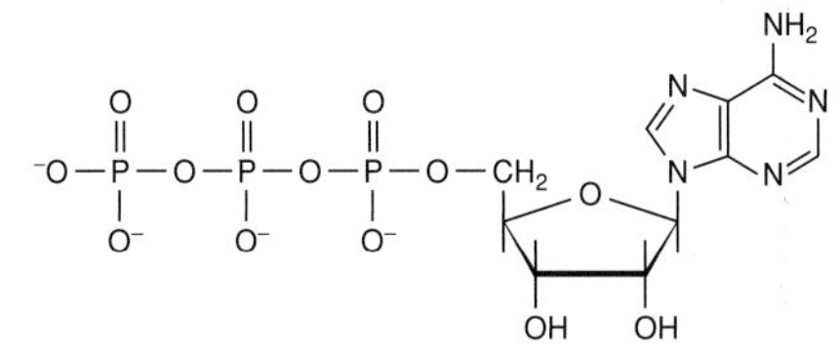
Coenzyme A (CoA)	Acyl transfer reactions (transferases)	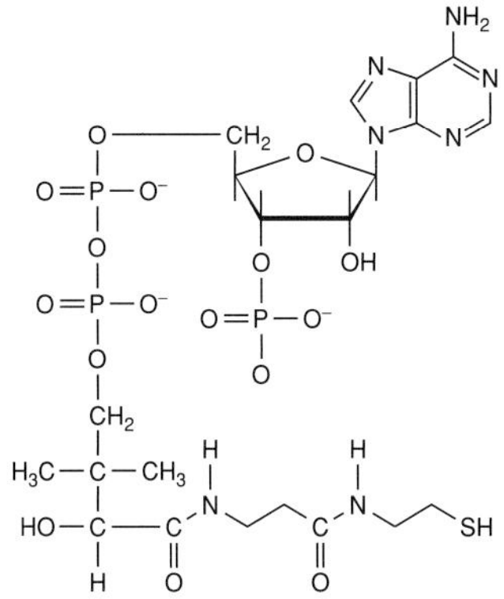
Flavin adenine dinucleotide (FAD)	Redox reactions (reductases)	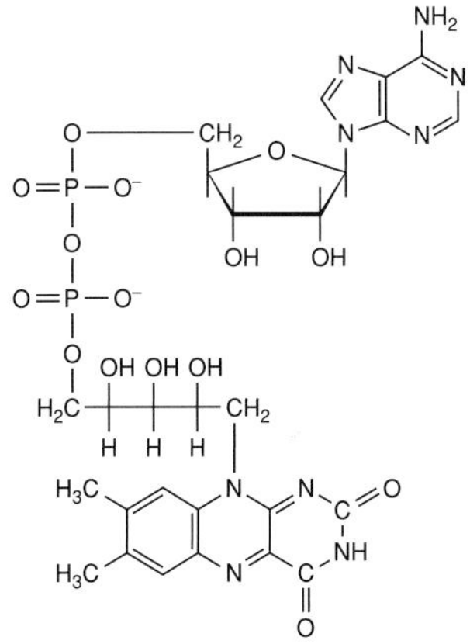
Nicotinamide adenine dinucleotide (NAD^+)	Redox reactions (e.g. dehydrogenases)	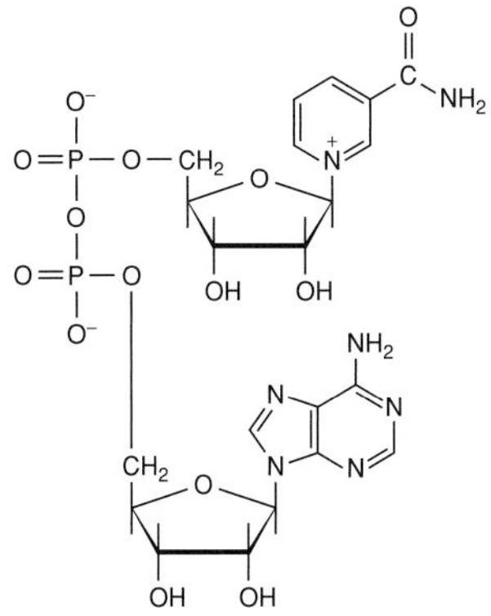
Heam c	Activate substrates	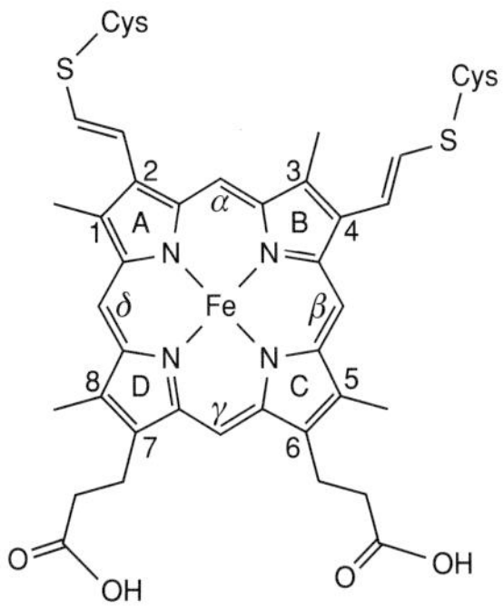

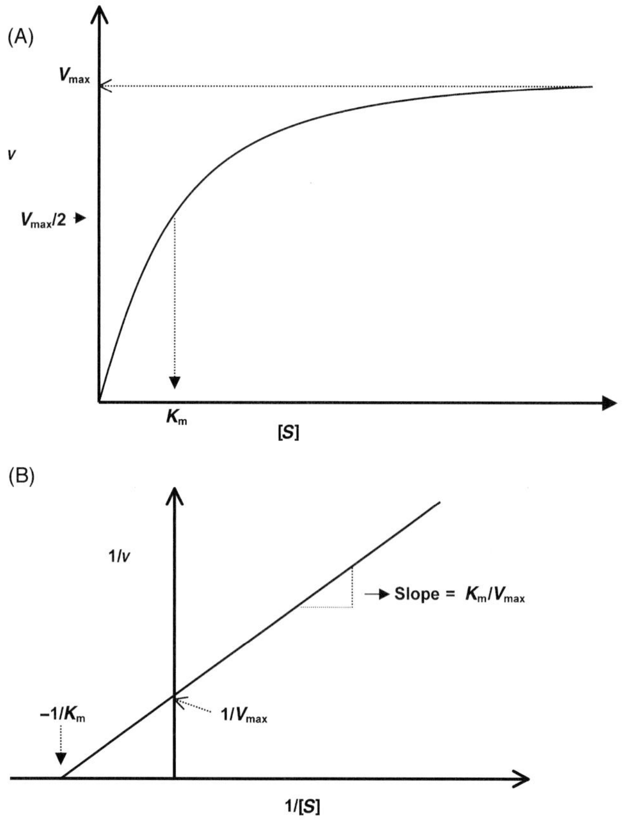

Figure 7.9. (**A**) A plot of the initial rate (v) against initial substrate concentration ([S]) for a reaction obeying the Michaelis-Menten kinetics. The substrate concentration, which gives a rate of half the maximum reaction velocity, is equal to the K_m. (**B**) The Lineweaver-Burk plot. The intercept on the $1/v$ axis is $1/V_{max}$, the intercept on the 1/[S] axis is $-1/K_m$ and the slope is K_m/V_{max}.

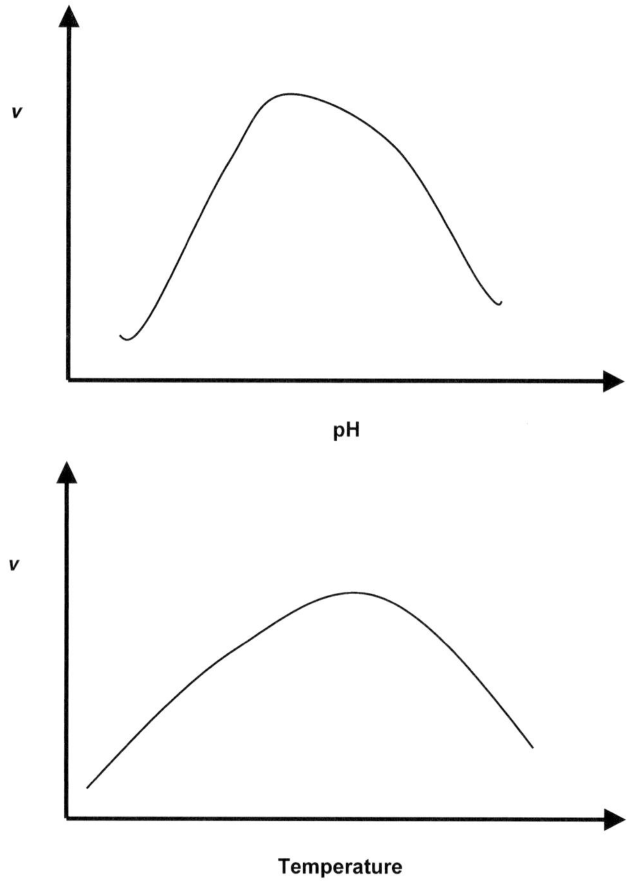

Figure 7.10. Relationship of the activity–pH and activity–temperature for a putative enzyme.

(Eblin 1964) suggests that molecules are in constant motion, which obey the same laws of mechanics as macroscopic objects. Since enzymes are biomolecules, which catalyse chemical reactions, they contribute to the energetic background of a chemical system. As a result, enzymes are directly related to thermodynamic parameters.

In this section, we will focus on the most representative thermodynamic parameters that are useful, in order to describe experimental data. The fundamental thermodynamic parameters are the Helmholtz free energy (F), the entropy (S) and the enthalpy (H). Free energy (F) correlates the energy (E) of a system with its entropy (S): $F = E - TS$, where T is the temperature. The entropy (S) of a system is the range of its disorder and the thermodynamic definition is (Rakintzis 1994) (the symbol q is used to represent the amount of heat absorbed by a system):

$$S = \int \frac{q}{T} \mathrm{d}T \Leftrightarrow S = q \ln(T)$$

Furthermore, enthalpy is a constitutive function of pressure (P) and temperature (T), and energy (E) is a constitutive function of volume (V) and temperature (T), respectively (Rakintzis 1994):

$$H = f(P, T)$$

$$\mathrm{d}H = \left[\frac{\partial H}{\partial P}\right]_T \mathrm{d}P + \left[\frac{\partial H}{\partial T}\right]_P \mathrm{d}T$$

$$E = f(V, T)$$

$$\mathrm{d}E = \left[\frac{\partial E}{\partial V}\right]_T \mathrm{d}V + \left[\frac{\partial E}{\partial T}\right]_V \mathrm{d}T$$

At this point, we have to emphasise that in biological systems, a normal experiment deals with a constant pressure (atmospheric pressure) and not with a constant volume. This means, that we determine not a change of energy (E) of a studied body (e.g. enzyme), but a change in its enthalpy: $H = E + PV$. However, the quantity (PV) is considered insignificant in case that we study a protein for instance, because the volume per molecule is too small. Hence, there is no difference between H and E, and we can refer both of them simply as "energy' (Finkelstein and Ptitsyn 2002). Likewise, there is no difference between Helmholtz free energy ($F = E - TS$) and Gibb's free energy ($G = H - TS$), since we consider that H is equivalent to E. As a result, we

Table 7.3. The Characteristic of Each Type of Inhibition and Their Effect on the Kinetic Parameters K_m and V_{max}

Inhibitor Type	Binding Site on Enzyme	Kinetic Effect
Competitive Inhibitor	The inhibitor specifically binds at the enzyme's catalytic site, where it competes with substrate for binding: $E \underset{-S}{\overset{+S}{\rightleftharpoons}} ES \rightarrow E + P$ $+I \Downarrow\Uparrow -I$ EI	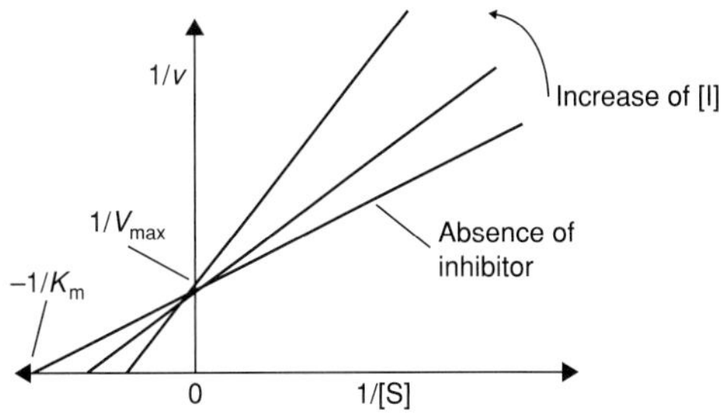 K_m is increased; V_{max} is unchanged. LB equation: $\frac{1}{v} = \frac{K_m}{V_{max}}\left(1 + \frac{[I]}{Ki}\right)\frac{1}{[S]} + \frac{1}{V_{max}}$
Non-competitive Inhibitor	The inhibitor binds to E or to the ES complex (may form an ESI complex) at a site other than the catalytic. Substrate binding is unchanged, whereas ESI complex cannot form products. $E \underset{-S}{\overset{+S}{\rightleftharpoons}} ES \rightarrow E + P$ $+I \Downarrow\Uparrow -I \quad +I \Downarrow\Uparrow -I$ $EI \underset{-S}{\overset{+S}{\rightleftharpoons}} ES \rightleftharpoons E + P$	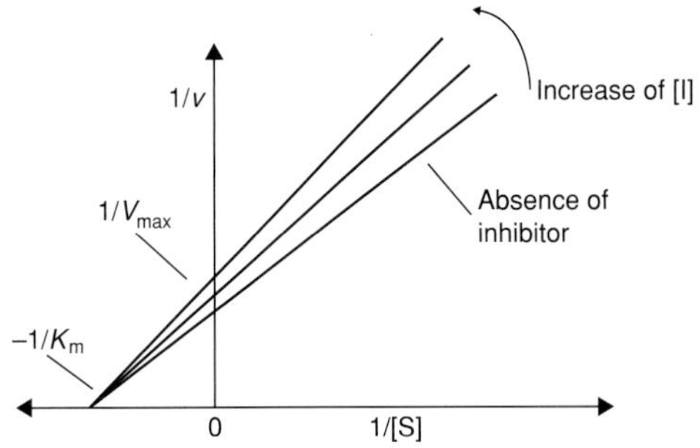 K_m is unchanged; V_{max} is decreased proportionately to inhibitor concentration. LB equation: $\frac{1}{v} = \frac{K_m}{V_{max}} \cdot \left(1 + \frac{[I]}{K_i}\right) \cdot \frac{1}{[S]} + \frac{1}{V_{max}}\left(1 + \frac{[I]}{K_i}\right)$
Uncompetitive Inhibitor	Binds only to ES complexes at a site other than the catalytic site. Substrate binding alters enzyme structure, making inhibitor-binding site available: $E \underset{-S}{\overset{+S}{\rightleftharpoons}} ES \rightarrow E + P$ $+I \Downarrow\Uparrow -I$ EIS	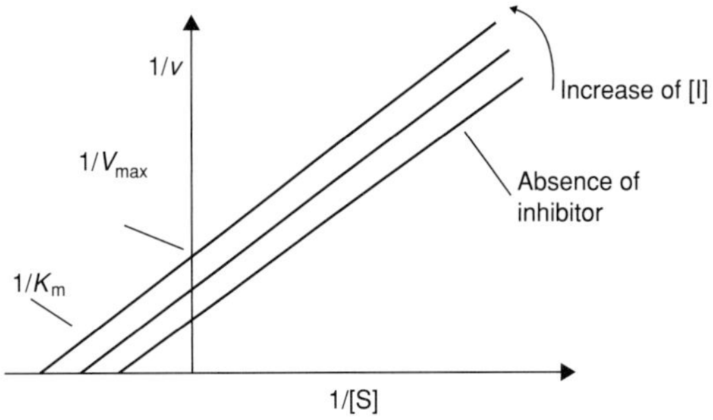 K_m and V_{max} are decreased. LB equation: $\frac{1}{v} = \frac{K_m}{V_{max}} \cdot \frac{1}{[S]} + \frac{1}{V_{max}}\left(1 + \frac{[I]}{K_i}\right)$

can refer both F and G as 'free energy' (Finkelstein and Ptitsyn 2002). Eventually, when we study the thermodynamic behaviour of an enzyme, regarding its structural stability or its functionality, we have to do with Gibb's free energy (G), entropy (S) and enthalpy (H).

As far as experimental procedures are concerned, someone can use two equations in order to measure the thermodynamic parameters, applying temperature differentiation experimental protocols (Kotzia and Labrou 2005). These equations are the Arrhenius (Equation 4) and the Eyring equations, which basically describe the temperature dependence of reaction state. So, considering an enzymic reaction as a chemical reaction, which is characterised by a transition state, we can efficiently use the Eyring's equation. This equation is based on transition state model, whose principles are (i) there is a thermodynamic equilibrium between the transition state and the state of reactants at the top of the energy barrier and (ii) the rate of chemical reaction is proportional to the concentration of the particles in the high-energy transition state.

Combining basic kinetic expressions, which describe a simple chemical reaction:

$$\mathrm{A} + \mathrm{B} \longrightarrow \mathrm{C}$$

Scheme 7.6.

and thermodynamic equations, such as (Finkelstein and Ptitsyn 2002)

$$\Delta G = -RT \ln k$$

$$\Delta G = \Delta H - T\Delta S$$

Eyring's equation is formed:

$$k = \frac{k_\beta T}{h} e^{-(\frac{\Delta G}{RT})} = \frac{k_\beta T}{h} e^{-(\frac{\Delta H}{RT} - \frac{\Delta S}{R})}$$

where k is the reaction velocity, R is the universal gas constant ($8.3145\ \mathrm{J \cdot mol^{-1}\ K}$), k_B the Boltzmann's constant ($1.381 \cdot 10^{-23}\ \mathrm{J \cdot K^{-1}}$), h is the Plank constant ($6.626 \cdot 10^{-34}\ \mathrm{J \cdot s}$) and T is the absolute temperature in degrees Kelvin (K). If we compare Eyring's equation and Arrhenius equation (Laidler 1984), we can realise that E and ΔH are parallel quantities, as we concluded above. Eyring's equation is an appropriate and useful tool, which allows us to simplify the complicated meaning of thermodynamics and, simultaneously, to interpret experimental data to physical meanings.

Enzyme Dynamics During Catalysis

Multiple conformational changes and intramolecular motions appear to be a general feature of enzymes (Agarwal et al. 2002). The structures of proteins and other biomolecules are largely maintained by non-covalent forces and are therefore subject to thermal fluctuations ranging from local atomic displacements to complete unfolding. These changes are intimately connected to enzymatic catalysis and are believed to fulfil a number of roles in catalysis: enhanced binding of substrate, correct orientation of catalytic groups, removal of water from the active site and trapping of intermediates. Enzyme conformational changes may be classified into four types (Gutteridge and Thornton 2004): (i) domain motion, where two rigid domains, joined by a flexible hinge, move relative to each other; (ii) loop motion, where flexible surface loops (2–10 residues) adopt different conformations; (iii) side chain rotation, which alters the position of the functional atoms of the side chain and (iv) secondary structure changes.

Intramolecular motions in biomolecules are usually very fast (picosecond–nanosecond) local fluctuations. The flexibility associated with such motions provides entropic stabilisation of conformational states (Agarwal et al. 2002). In addition, there are also slower- (microsecond–millisecond) and larger-scale, thermally activated, transitions. Large-scale conformational changes are usually key events in enzyme regulation.

ENZYME PRODUCTION

In the past, enzymes were isolated primarily from natural sources, and thus a relatively limited number of enzymes were available to the industry (Aehle 2007). For example, of the hundred or so enzymes being used industrially, over one half are from fungi and yeast and over a third are from bacteria, with the remainder divided between animal (8%) and plant (4%) sources (Panke and Wubbolts 2002, van Beilen and Li 2002, Bornscheuer 2005). Today, with the recent advances of molecular biology and genetic engineering, several expression systems were developed and exploited and used for the commercial production of several therapeutic (Mcgrath 2005), analytical or industrial enzymes (van Beilen and Li 2002, Kirk et al. 2002, Otero and Nielsen 2010). These systems have not only improved the efficiency, availability and cost with which enzymes can be produced, but they have also improved their quality (Labrou and Rigden 2001, Andreadeli et al. 2008, Kapoli et al. 2008, Kotzia and Labrou 2009, Labrou 2010).

Enzyme Heterologous Expression

There are two basic steps involved in the assembly of every heterologous expression system:

1. The introduction of the DNA, encoding the gene of interest, into the host cells, which requires: (i) the identification and isolation of the gene of the protein we wish to express, (ii) insertion of the gene into a suitable expression vector and (iii) introduction of the expression vector into the selected cell system that will accommodate the heterologous protein.
2. The optimisation of protein expression by taking into account the effect of various factors such as growing medium, temperature, and induction period.

A variety of vectors able to carry the DNA into the host cells are available, ranging from plasmids, cosmids, phagemids, viruses as well as artificial chromosomes of bacterial, yeast or human origin (BAC, YAC or HAC, respectively; Grimes and Monaco 2005, Monaco and Moralli 2006, Takahashi and Ueda 2010). The vectors are either integrated into the host chromosomal DNA or remain in an episomal form. In general, expression vectors have the following characteristics (Fig. 7.11):

1. Polylinker: it contains multiple restriction sites that facilitate the insertion of the desired gene into the vector.
2. Selection marker: encodes for a selectable marker, allowing the vector to be maintained within the host cell under conditions of selective pressure (i.e. antibiotic).
3. Ori: a sequence that allows for the autonomous replication of the vector within the cells.
4. Promoter: inducible or constitutive, that regulates RNA transcription of the gene of interest.

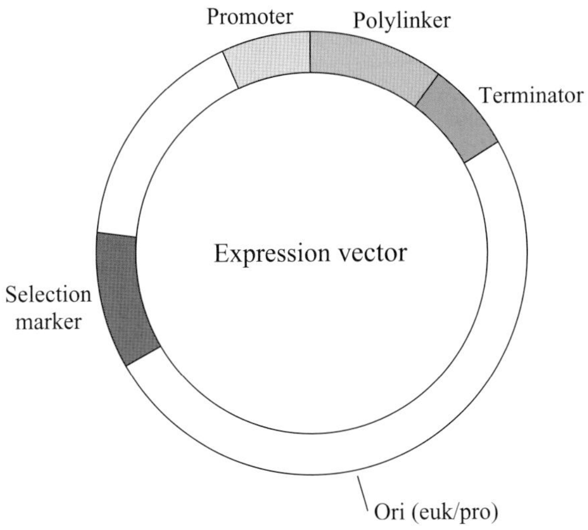

Figure 7.11. Generalised heterologous gene expression vector. Polylinker sequence (multiple cloning site (MCS)), promoter ($P^{euk/pro}$), terminator ($T^{euk/pro}$), selectable marker (prokaryotic and/or eukaryotic – $M^{euk/pro}$), and origin of replication (eukaryotic and/or prokaryotic – $Ori^{euk/pro}$).

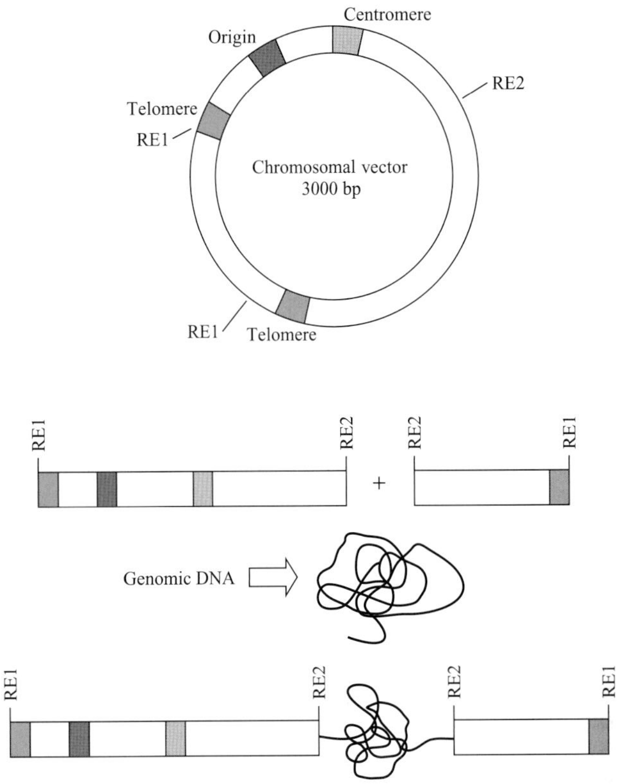

Figure 7.12. Artificial chromosome cloning system. Initially, the circular vector is digested with restriction endonucleases (RE) for linearisation and then ligated with size-fractionated DNA (≈100 kb). The vector contains centromeric and telomeric sequences, which assure chromosome-like propagation within the cell, as well as selection marker sequences for stable maintenance. Origin of replication (eukaryotic or prokaryotic – $Ori^{euk/pro}$), selectable marker (eukaryotic or prokaryotic – $M^{euk/pro}$), centromeric sequence (C), and telomeric sequence (T).

5. Terminator: a strong terminator sequence that ensures that the RNA polymerase disengages and does not continue to transcribe other genes located downstream.

Vectors are usually designed with mixed characteristics for expression in both prokaryotic and eukaryotic host cells (Brondyk 2009, Demain and Vaishnav 2009). Artificial chromosomes are designed for cloning of very large segments of DNA (100 kb), usually for mapping purposes, and contain host-specific telomeric and centromeric sequences. These sequences permit the proper distribution of the vectors to the daughter cells during cell division and increase chromosome stability (Fig. 7.12).

The Choice of Expression System

There are two main categories of expression systems: eukaryotic and prokaryotic (Demain and Vaishnav 2009). The choice of a suitable expression system involves the consideration of several important factors, such as protein yield, proper folding, post-translational modifications (e.g. phosphorylation, glycosylation), industrial applications of the expressed protein, as well as economic factors. For these reasons, there is no universally applied expression system. A comparison of the most commonly used expression systems is shown in Table 7.4.

Bacterial Cells

Expression of heterologous proteins in bacteria remains the most extensively used approach for the production of heterologous proteins (Zerbs et al. 2009), such as cytokines (Tang et al. 2006, Rabhi-Essafi et al. 2007), membrane proteins (Butzin et al. 2009, Zoonens and Miroux 2010), enzymes (Melissis et al. 2006, Kotzia and Labrou 2007, Melissis et al. 2007, Andreadeli et al. 2008, Wu et al. 2009, Okino et al. 2010), antibodies (Kwong and Rader 2009, Xiong et al. 2009) and antigens (viral or non-viral; Donayre-Torres et al. 2009, Liu et al. 2009a, Vahedi et al. 2009, Ebrahimi et al. 2010) at both laboratory and industrial scale (Koehn and Hunt 2009, Sahdev et al. 2008, Peti and Page 2007, Jana and Deb 2005). Bacteria can be grown inexpensively and genetically manipulated very easily. They can reach very high densities rapidly and express high levels of recombinant proteins, reaching up to 50% of the total protein. However, in many cases, high-level expression correlates with poor quality. Often, the expressed protein is accumulated in the form of insoluble inclusion bodies (misfolded protein aggregate) and additional, sometimes labour-intensive, genetic manipulation or re-solubilisation/refolding steps are required (Burgess 2009). Bacterial cells do not possess the eukaryotes' extensive post-translational modification system (such as *N*- or

Table 7.4. Comparison of the Main Expression Systems

	Bacteria	Yeast	Fungi	Insect Cells	Mammalian Cells	Transgenic Plants	Transgenic Animals
Developing time	Short	Short	Short	Intermediate	Intermediate	Intermediate	Long
Costs for downstream processing	+++	++	++	++	++	++	++
Levels of expression	High	Intermediate	Intermediate	Intermediate	Low	Low	Low
Recombinant protein stability	+/−	+/−	+/−	+/−	+/−	+++	+/−
Production volume	Limited	Limited	Limited	Limited	Limited	Unlimited	Unlimited
Postranslational modifications (disulphide bond formation, glycosylation, etc.)	No	Yes	Yes	Yes	Yes	Yes	Yes
'Human-type' glycosylation	No	No	No	No	Yes	No	Yes
Folding capabilities	−	++	+++	+++	+++	+++	+++
Contamination level (pathogens, EPL, etc.)	++	−	−	−	++	−	++

O-glycosylation), a serious disadvantage, when post-translational modifications are essential to the protein's function (Zhang et al. 2004). However, they are capable of a surprisingly broad range of covalent modifications such as acetylation, amidation and deamidation, methylation, myristylation, biotinylation and phosphorylation.

Mammalian Cells

Mammalian cells are the ideal candidate for expression hosts (Engelhardt et al. 2009, Hacker et al. 2009) when post-translational modifications (*N*- and *O*-glycosylation, disulfide bond formation) are a critical factor for the efficacy of the expressed protein (Baldi et al. 2007, Werner et al. 2007, Durocher and Butler 2009, Geisse and Fux 2009, Hacker et al. 2009). Despite substantial limitations, such as high cost, low yield, instability of expression and time-consuming, a significant number of proteins (e.g. cytokines; Fox et al. 2004, Sunley et al. 2008, Suen et al. 2010), antibodies (Kim et al. 2008, Chusainow et al. 2009), enzymes (Zhuge et al. 2004), viral antigens (Holzer et al. 2003), blood factors and related proteins (Halabian et al. 2009, Su et al. 2010) are produced in this system because it offers very high product fidelity. However, oligosaccharide processing is species- and cell type-dependent among mammalian cells and differences in the glycosylation pattern have been reported in rodent cell lines and human tissues. The expressed proteins are usually recovered in a bioactive, properly folded form and secreted into the cell culture fluids.

Yeast

Yeast is a widely used expression system with many commercial, medical and basic research applications. The fact that the yeast is the most intensively studied eukaryote at the genetic/molecular level makes it an extremely advantageous expression system (Idiris et al. 2010). Being unicellular organism, it retains the advantages of bacteria (low cultivation cost, high doubling rate, ease of genetic manipulation, ability to produce heterologous proteins in large-scale quantities) combined with the advantages of higher eukaryotic systems (post-translational modifications). The vast majority of yeast expression work has focused on the well-characterised baker's yeast *Saccharomyces cerevisiae* (Holz et al. 2003, Terpitz et al. 2008, Joubert et al. 2010), but a growing number of non-*Saccharomyces* yeasts are becoming available as hosts for recombinant polypeptide production, such as *Hansenula polymorpha*, *Candida boidinii*, *Kluyveromyces lactis*, *Pichia pastoris* (Cregg et al. 2000, Jahic et al. 2006, van Ooyen et al. 2006, Yurimoto and Sakai 2009), *Schizosaccharomyces pombe* (Alberti et al. 2007, Ahn et al. 2009, Takegawa et al. 2009), *Schwanniomyces occidentalis* and *Yarrowia lipolytica* (Madzak et al. 2004, 2005, Bankar et al. 2009). As in bacteria, expression in yeast relies on episomal or integrated multicopy plasmids with tightly regulated gene expression. Despite these advantages, expressed proteins are not always recovered in soluble form and may have to be purified from inclusion bodies. Post-translational modifications in yeast differ greatly from mammalian cells (Jacobs and Callewaert 2009, Hamilton and Gerngross 2007). This has sometimes proven to be a hindrance when high fidelity of complex carbohydrate modifications found in eukaryotic proteins appears to be important in many medical applications. Yeast cells do not add complex oligosaccharides and are limited to the high-mannose-type carbohydrates. These higher order oligosaccharides are possibly immunogenic and could potentially interfere with the biological activity of the protein.

Filamentous Fungi

Filamentous fungi have been extensively used for studies of eukaryotic gene organisation, regulation and cellular differentiation. Additionally, fungi belonging to the genus *Aspergillus* and *Penicillium* are of significant industrial importance because of their applications in food fermentation and their ability to secrete a broad range of biopolymer degrading enzymes and to produce primary (organic acids) and secondary metabolites (antibiotics, vitamins). The extensive genetic knowledge as well as the already well-developed fermentation technology has allowed for the development of heterologous protein expression systems (Nevalainen et al. 2005, Wang et al. 2005) expressing fungal (e.g. glucoamylase (Verdoes et al. 1993), propyl aminopeptidase (Matsushita-Morita et al. 2009)) or mammalian (e.g. human interleukin-6 (Contreras et al. 1991), antitrypsin (Chill et al. 2009), lactoferrin (Ward et al. 1995), bovine chymosin (Ward et al. 1990, Cardoza et al. 2003)) proteins of industrial and clinical interest using filamentous fungi as hosts. However, the expression levels of mammalian proteins expressed in *Aspergillus* and *Trichoderma* species are low compared to homologous proteins. Significant advances in heterologous protein expression have dramatically improved the expression efficiency by fusing the heterologous gene to the 3′-end of a highly expressed homologous gene (mainly glucoamylase). Even so, limitations in protein folding, post-translational modifications, translocation and secretion as well as secretion of extracellular proteases could pose a significant hindrance for the production of bioactive proteins (Gouka et al. 1997, Nevalainen et al. 2005, Lubertozzi and Keasling 2009).

Insect Cells

Recombinant baculoviruses are widely used as a vector for the expression of recombinant proteins in insect cells (Hitchman et al. 2009, Jarvis 2009, Trometer and Falson 2010), such as enzymes (Zhao et al. 2010), immunoglobulins (Iizuka et al. 2009), viral antigens (Takahashi et al. 2010) and transcription factors (Fabian et al. 1998). The recombinant genes are usually expressed under the control of the polyhedrin or *p10* promoter of the *Autographa californica* nuclear polyhedrosis virus (AcNPV) in cultured insect cells of *Spodoptera frugiperda* (Sf9 cells) or in insect larvae of *Lepidopteran* species infected with the recombinant baculovirus containing the gene of interest. The polyhedron and p10 genes possess very strong promoters and are highly transcribed during the late stages of the viral cycle. Usually, the recombinant proteins are recovered from the infected insect cells in soluble form and targeted in the proper cellular environment (membrane, nucleus and cytoplasm). Insect cells have many post-translational modification, transport and processing mechanisms found in higher eukaryotic cells (Durocher and Butler 2009), although their glycosylation efficiency is limited and they are not able to process complex-type oligosaccharides containing fucose, galactose and sialic acid.

Dictyostelium discoideum

Recently, the cellular slime mould *Dictyostelium discoideum*, a well-studied single-celled organism, has emerged as a promising eukaryotic alternative system (Arya et al. 2008b) for the expression of recombinant proteins (e.g. human antithrombin III; Dingermann et al. 1991, Tiltscher and Storr 1993, Dittrich et al. 1994) and enzymes (e.g. phosphodiesterase; Arya et al. 2008a). Its advantage over other expression systems lies in its extensive post-translational modification system (glycosylation, phosphorylation, acylation), which resembles that of higher eukaryotes (Jung and Williams 1997, Jung et al. 1997, Slade et al. 1997). It is a simple organism with a haploid genome of 5×10^7 bp and a life cycle that alternates between single-celled and multicellular stages. Recombinant proteins are expressed from extrachromosomal plasmids (*Dictyostelium discoideum* is one of a few eukaryotes that have circular nuclear plasmids) rather than being integrated in the genome (Ahern et al. 1988). The nuclear plasmids can be easily genetically manipulated and isolated in a one-step procedure, as in bacteria. This system is ideally suited for the expression of complex glycoproteins and although it retains many of the advantages of the bacterial (low cultivation cost) and mammalian systems (establishment of stable cell lines, glycosylation), the development of this system at an industrial scale is hampered by the relatively low productivity, compared to bacterial systems.

Trypanosomatid Protozoa

A newly developed eukaryotic expression system is based on the protozoan lizard parasites of the *Leishmania* and *Trypanosoma* species (Basile and Peticca 2009). Its gene and protein regulation and editing mechanisms are remarkably similar to those of higher eukaryotes and include the capability of 'mammalian-like' glycosylation. It has a very rapid doubling time and can be grown to high densities in relatively inexpensive medium. The recombinant gene is integrated into the small ribosomal subunit rRNA gene and can be expressed to high levels. Increased expression levels and additional promoter control can be achieved in T7 polymerase-expressing strains. Being a lizard parasite, it is not pathogenic to humans, which makes this system invaluable and highly versatile. Proteins and enzymes of significant interest, such as human tissue plasminogen activator (Soleimani et al. 2007), EPO (Breitling et al. 2002), IFNγ (Tobin et al. 1993) and IL-2 (La Flamme 1995), have been successfully expressed in this system.

Transgenic Plants

The current protein therapeutics market is clearly an area of enormous interest from a medical and economic point of view. Recent advances in human genomics and biotechnology have made it possible to identify a plethora of potentially important drugs or drug targets. Transgenic technology has provided an alternative, more cost-effective bioproduction system than the previously used (*E. coli*, yeast, mammalian cells; Larrick and Thomas 2001, Demain and Vaishnav 2009). The accumulated knowledge on plant genetic manipulation has been recently applied to the development of plant bioproduction systems (Twyman et al. 2005, Boehm 2007, Lienard et al. 2007, Faye and Champey 2008, Sourrouille et al. 2009). Expression in plants could be either constitutive or transient and directed to a specific tissue of the plant (depending on the type of promoter used). Expression of heterologous proteins in plants offers significant advantages, such as low production cost, high biomass production, unlimited supply and ease of expandability. Plants also have high-fidelity expression, folding and post-translational modification mechanisms, which could produce human proteins of substantial structural and functional equivalency compared to proteins from mammalian expression systems (Gomord and Faye 2004, Joshi and Lopez 2005). Additionally, plant-made human proteins of clinical interest (Schillberg et al. 2005, Twyman et al. 2005, Vitale and Pedrazzini 2005, Tiwari et al. 2009), such as antibodies (Hassan et al. 2008, Ko et al. 2009, De Muynck et al. 2010, Lai et al. 2010), vaccines (Hooper 2009, Alvarez and Cardineau 2010), cytokines (Elias-Lopez et al. 2008) and enzymes (Kermode et al. 2007, Stein et al. 2009), are free of potentially hazardous human diseases, viruses or bacterial toxins. However, there is considerable concern regarding the potential hazards of contamination of the natural gene pool by the transgenes and possible additional safety precautions will raise the production cost.

Transgenic Animals

Besides plants, transgenic technology has also been applied to many different species of animals (mice, cows, rabbits, sheep, goats and pigs; Niemann and Kues 2007, Houdebine 2009). The DNA containing the gene of interest is microinjected into the pronucleus of a single-cell fertilised zygote and integrated into the genome of the recipient; therefore, it can be faithfully passed on from generation to generation. The gene of interest is coupled with a signal targeting protein expression towards specific tissues, mainly the mammary gland, and the protein can therefore be harvested and purified from milk. The proteins produced by transgenic animals are almost identical to human proteins, greatly expanding the applications of transgenic animals in medicine and biotechnology. Several human protein of pharmaceutical value have been produced in transgenic animals, such as haemoglobin (Swanson 1992, Logan and Martin 1994), lactoferrin (Han et al. 2008, Yang et al. 2008), antithrombin III (Yang et al. 2009), protein C (Velander 1992) and fibrinogen (Prunkard 1996), and there is enormous interest for the generation of transgenic tissues suitable for transplantation in humans (only recently overshadowed by primary blastocyte technology (Klimanskaya et al. 2008)). Despite the initial technological expertise required to produce a transgenic animal, the subsequent operational costs are low and subsequent inbreeding ensures that the ability to produce the transgenic protein will be passed on to its offspring. However, certain safety issues have arisen concerning the potential contamination of transgenically produced proteins by animal viruses or prions, which could possibly be passed on to the human population. Extensive testing required by the FDA substantially raises downstream costs.

Enzyme Purification

Once a suitable enzyme source has been identified, it becomes necessary to design an appropriate purification procedure to isolate the desired protein. The extent of purification required for an enzyme depends on several factors, the most important of which being the degree of enzyme purity required as well as the starting material, for example the quantity of the desired enzyme present in the initial preparation (Lesley 2001, Labrou and Clonis 2002). For example, industrial enzymes are usually produced as relatively crude preparations. Enzymes used for therapeutic or diagnostic purposes are generally subjected to the most stringent purification procedures, as the presence of compounds other than the intended product may have an adverse clinical impact (Berthold and Walter 1994).

Table 7.5. Protein Properties Used During Purification

Protein Property	Technique
Solubility	Precipitation
Size	Gel filtration
Charge	Ion exchange
Hydrophobicity	Hydrophobic interaction chromatography
Biorecognition	Affinity chromatography

Purification of an enzyme usually occurs by a series of independent steps in which the various physicochemical properties of the enzyme of interest are utilised to separate it progressively from other unwanted constituents (Labrou and Clonis 2002, Labrou et al. 2004b). The characteristics of proteins that are utilised in purification include solubility, ionic charge, molecular size, adsorption properties and binding affinity to other biological molecules. Several methods that exploit differences in these properties are listed in Table 7.5.

Precipitation methods (usually employing $(NH_4)_2SO_4$, polyethyleneglycol or organic solvents) are not very efficient method of purification (Labrou and Clonis 2002). They typically give only a few fold purification. However, with these methods, the protein may be removed from the growth medium or from cell debris where harmful proteases and other detrimental compounds may affect protein stability. On the other hand, *chromatography* is a highly selective separation technique (Regnier 1987, Fausnaugh 1990). A wide range of chromatographic techniques has been used for enzyme purification: size-exclusion chromatography, ion-exchange, hydroxyapatite, hydrophobic interaction chromatography, reverse-phase chromatography and affinity chromatography (Labrou 2003). Of these, ion-exchange and affinity chromatography are the most common and probably the most important (Labrou and Clonis 1994).

Ion-Exchange Chromatography

Ion-exchange resins selectively bind proteins of opposite charge; that is, a negatively charged resin will bind proteins with a net positive charge and vice versa (Fig. 7.13). Charged groups are classified according to type (cationic and anionic) and strength (strong or weak); the charge characteristics of strong ion exchange media do not change with pH, whereas with weak ion-exchange media they do. The most commonly used charged groups include diethylaminoethyl, a weakly anionic exchanger; carboxymethyl, a weakly cationic exchanger; quaternary ammonium, a strongly anionic exchanger; and methyl sulfonate,

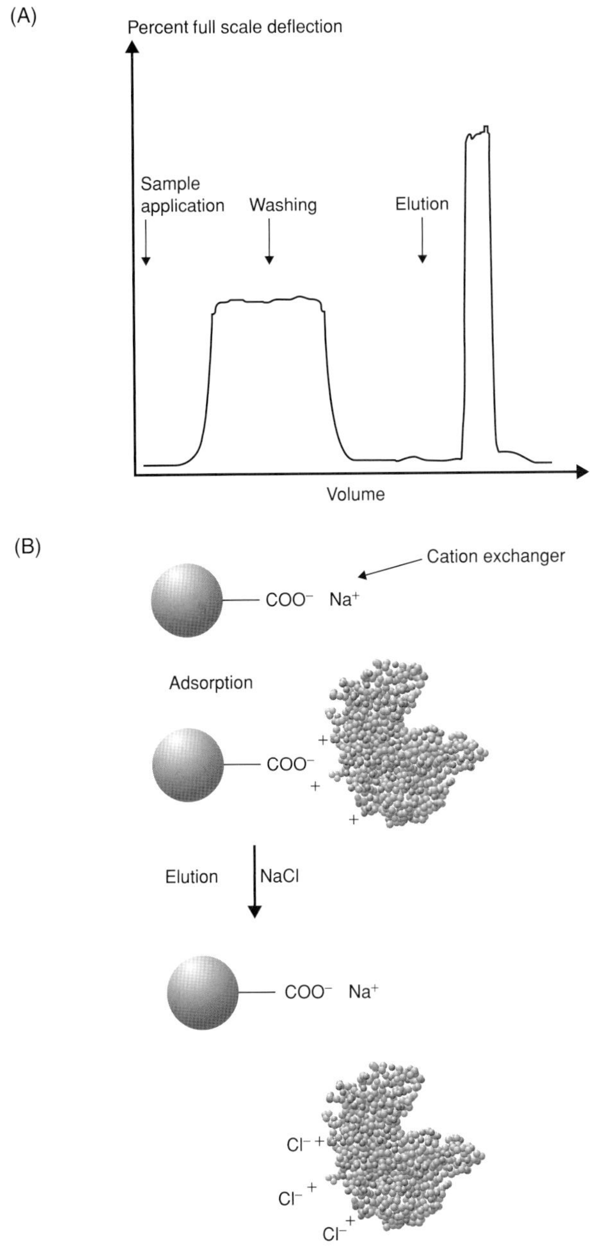

Figure 7.13. (**A**) Schematic diagram of a chromatogram showing the steps for a putative purification. (**B**) Schematic diagram depicting the principle of ion-exchange chromatography.

a strongly cationic exchanger (Table 7.6; Levison 2003). The matrix material for the column is usually formed from beads of carbohydrate polymers, such as agarose, cellulose or dextrans (Levison 2003).

The technique takes place in five steps (Labrou 2000; Fig. 7.13): *equilibration* of the column to pH and ionic strength conditions suitable for target protein binding; protein *sample application* to the column and reversible adsorption through counter-ion displacement; *washing* of the unbound contaminating proteins, enzymes, nucleic acids and other compounds; introduction of *elution* conditions in order to displace bound proteins; and *regeneration* and *re-equilibration* of the adsorbent for subsequent purifications. Elution may be achieved either by increasing the salt concentration or by changing the pH of the irrigating buffer. Both methods are used in industry, but raising the salt concentration is by far the most common because it is easier to control (Levison 2003). Most protein purifications are done on anion exchange columns because most proteins are negatively charged at physiological pH values (pH 6–8).

Affinity Chromatography

Affinity chromatography is potentially the most powerful and selective method for protein purification (Fig. 7.14; Labrou and Clonis 1994, Labrou 2003). According to the International Union of Pure and Applied Chemistry, affinity chromatography is defined as a liquid chromatographic technique that makes use of a 'biological interaction' for the separation and analysis of specific analytes within a sample. Examples of these interactions include the binding of an enzyme with a substrate/inhibitor or of an antibody with an antigen or in general the interaction of a protein with a binding agent, known as the 'affinity ligand' (Fig. 7.14; Labrou 2002, 2003, Labrou et al. 2004b). The development of an affinity chromatography-based purification step involves the consideration of the following factors: (i) selection of an appropriate ligand and (ii) immobilisation of the ligand onto a suitable support matrix to make an *affinity adsorbent*. The selection of the immobilised ligand for affinity chromatography is the most challenging aspect of preparing an affinity adsorbent. Certain factors need to be considered when selecting a ligand (Labrou and Clonis 1995, 1996): (i) the specificity of the ligand for the protein of interest, (ii) the reversibility of the interaction with the protein, (iii) its stability against the biological and chemical operation conditions and (iv) the affinity of the ligand for the protein of interest. The binding site of a protein is often located deep within the molecule and adsorbents prepared by coupling the ligands directly to the support exhibit low binding capacities. This is due to steric interference between the support matrix and the protein's binding site. In these circumstances, a 'spacer arm' is inserted between the matrix and ligand to facilitate effective binding (Fig. 7.14). A hexyl spacer is usually inserted between ligand and support by substitution of 1,6-diaminohexane (Lowe 2001).

The ideal matrix should be hydrophilic, chemically and biologically stable and have sufficient modifiable groups to permit an appropriate degree of substitution with the enzyme. Sepharose is the most commonly used matrix for affinity chromatography on the research scale. Sepharose is a commercially available beaded polymer, which is highly hydrophilic and generally inert to microbiological attack (Labrou and Clonis 2002). Chemically, it is an agarose (poly-{β-1,3-D-galactose-α-1,4-(3,6-anhydro)-L-galactose}) derivative.

The selection of conditions for an optimum affinity chromatographic purification involves the study of the following factors: (1) choice of adsorption conditions (e.g. buffer composition,

Table 7.6. Functional Groups Used on Ion Exchangers

	Functional Group
Anion exchangers	
Diethylaminoethyl (DEAE)	$-O-CH_2-CH_2-N^+H(CH_2CH_3)_2$
Quaternary aminoethyl (QAE)	$-O-CH_2-CH_2-N^+(C_2H_5)_2-CH_2-CHOH-CH_3$
Quaternary ammonium (Q)	$-O-CH_2-CHOH-CH_2O-CH_2-CHOH-CH_2N^+(CH_3)_2$
Cation exchangers	
Carboxymethyl (CM)	$-O-CH_2-COO^-$
Sulphopropyl (SP)	$-O-CH_2-CHOH-CH_2-O-CH_2-CH_2-CH_2SO_3^-$

pH, ionic strength) to maximise the conditions required for the formation of strong complex between the ligand and the protein to be purified, (2) choice of washing conditions to desorb non-specifically bound proteins and (3) choice of elution conditions to maximise purification (Labrou and Clonis 1995). The elution conditions of the bound macromolecule should be both tolerated by the affinity adsorbent and effective in desorbing the biomolecule in good yield and in the native state. Elution of bound proteins is performed in a non-specific or biospecific manner. Non-specific elution usually involves (1) changing the ionic strength (usually by increasing the buffer's molarity or including salt, e.g. KCl or NaCl) and the pH (adsorption generally weakens with increasing pH), (2) altering the polarity of the irrigating buffer by employing, for example ethylene glycol or other organic solvents, if the hydrophobic contribution in the protein-ligand complex is large. Biospecific elution is achieved by inclusion in the equilibration buffer of a suitable ligand, which usually competes with the immobilised ligand for the same binding site on the enzyme/protein (Labrou 2000). Any competing ligand may be used. For example, substrates, products, cofactors, inhibitors or allosteric effectors are all potential candidates as long as they have higher affinity for the macromolecule than the immobilised ligand.

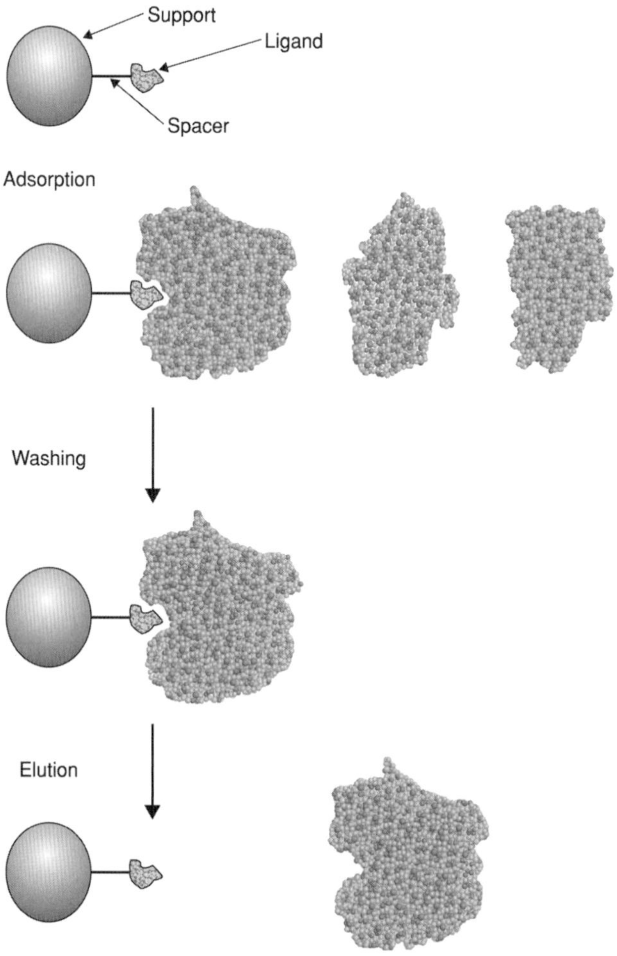

Figure 7.14. Schematic diagram depicting the principle of affinity chromatography.

Dye-ligand affinity chromatography represents a powerful affinity-based technique for enzyme and protein purification (Clonis et al. 2000, Labrou 2002, Labrou et al. 2004b). The technique has gained broad popularity due to its simplicity and wide applicability to purify a variety of proteins. The employed dyes as affinity ligands are commercial textile chlorotriazine polysulfonated aromatic molecules, which are usually termed as triazine dyes (Fig. 7.15). Such dye-ligands have found wide applications over the past 20 years as general affinity ligands in the research market to purify enzymes, such as oxidoreductases, decarboxylases, glycolytic enzymes, nucleases, hydrolases, lyases, synthetases and transferases (Scopes 1987). Anthraquinone triazine dyes are probably the most widely used dye-ligands in enzyme and protein purification. Especially the triazine dye Cibacron Blue F3GA (Fig. 7.18) has been widely exploited as an affinity chromatographic tool to separate and purify a variety of proteins (Scopes 1987). With the aim of increasing the specificity of dye-ligands, the biomimetic dye-ligand concept was introduced. According to this concept, new dyes that mimic natural ligands of the targeted proteins are designed by substituting the terminal 2-aminobenzene sulfonate moiety of the dye Cibacron Blue 3GA (CB3GA) for (ή with) a substrate-mimetic moiety (Clonis et al. 2000, Labrou 2002, 2003, Labrou et al. 2004b). These biomimetic dyes exhibit increased purification ability and specificity and provide useful tools for designing simple and effective purification protocols.

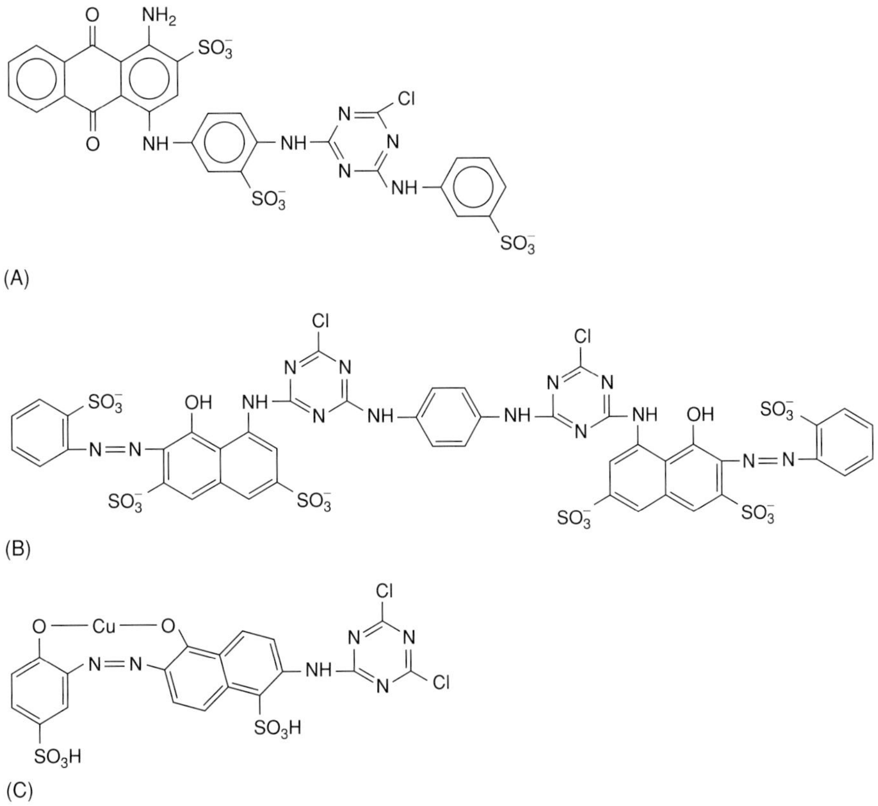

Figure 7.15. Structure of several representative triazine dyes: (**A**) Cibacron Blue 3GA, (**B**) Procion Red HE-3B and (**C**) Procion Rubine MX-B.

The rapid development of recombinant DNA technology since the early 1980s has changed the emphasis of a classical enzyme purification work. For example, epitope tagging is a recombinant DNA method for inserting a specific protein sequence (affinity tag) into the protein of interest (Terpe 2003). This allows the expressed tagged protein to be purified by affinity interactions with a ligand that selectively binds to the affinity tag. Examples of affinity tags and their respective ligands used for protein and enzyme purification are shown in Table 7.7.

ENZYME ENGINEERING

Another extremely promising area of enzyme technology is enzyme engineering. New enzyme structures may be designed and produced in order to improve existing ones or create new activities. Over the past two decades, with the advent of protein engineering, molecular biotechnology has permitted not only the improvement of the properties of these isolated proteins, but also the construction of 'altered versions' of these 'naturally occurring' proteins with novel or 'tailor-made' properties (Ryu

Table 7.7. Adsorbents and Elution Conditions of Affinity Tags

Affinity Tag	Matrix	Elution Condition
Poly-His	Ni^{2+}-NTA	Imidazole 20–250 mM or low pH
FLAG	Anti-FLAG monoclonal antibody	pH 3.0 or 2–5 mM EDTA
Strep-tag II	Strep-Tactin (modified streptavidin)	2.5 mM desthiobiotin
c-myc	Monoclonal antibody	Low pH
S	S-fragment of RNaseA	3 M guanidine thiocyanate, 0.2 M citrate pH 2, 3 M magnesium chloride
Calmodulin-binding peptide	Calmodulin	EGTA or EGTA with 1 M NaCl
Cellulose-binding domain	Cellulose	Guanidine HCl or urea > 4 M
Glutathione *S*-transferase	Glutathione	5–10 mM reduced glutathione

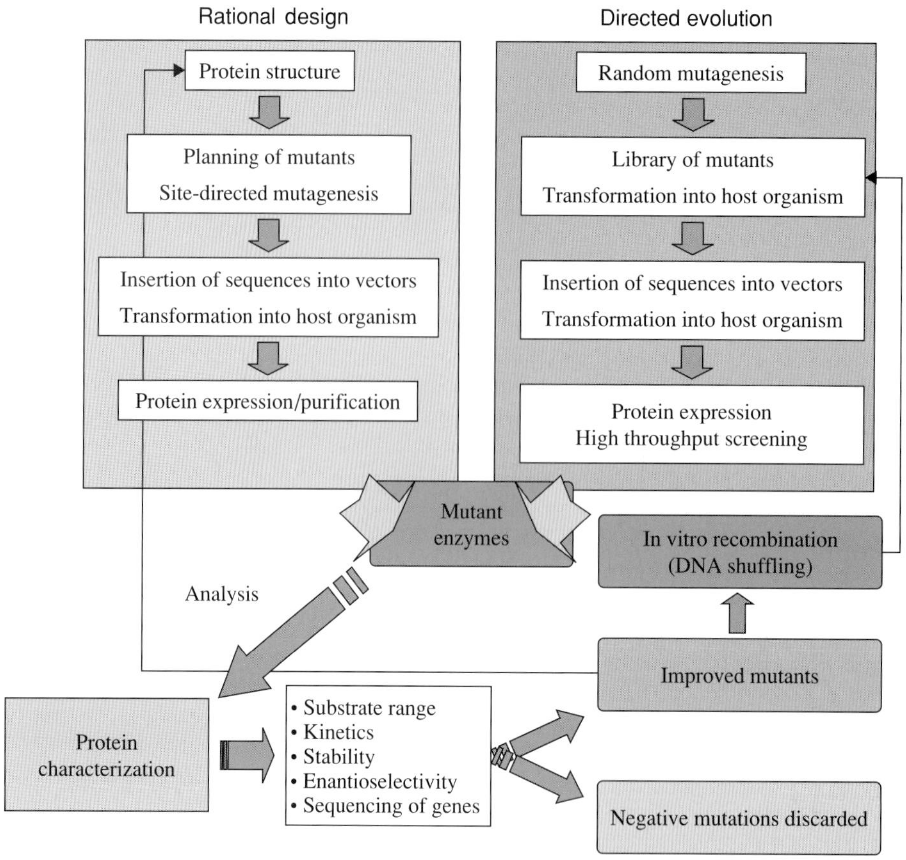

Figure 7.16. Comparison of rational design and directed evolution.

and Nam 2000, Gerlt and Babbitt 2009, Tracewell and Arnold 2009).

Tailor-Made Enzymes by Protein Engineering

There are two main intervention approaches for the construction of tailor-made enzymes: rational design and directed evolution (Chen 2001, Schmidt et al. 2009; Fig. 7.16).

Rational design takes advantage of knowledge of the three-dimensional structure of the enzyme, as well as structure/function and sequence information to predict, in a 'rational/logical' way, sites on the enzyme that when altered would endow the enzyme with the desired properties (Craik et al. 1985, Wells et al. 1987, Carter et al. 1989, Scrutton et al. 1990, Cedrone et al. 2000). Once the crucial amino acids are identified, site-directed mutagenesis is applied and the expressed mutants are screened for the desired properties. It is clear that protein engineering by rational design requires prior knowledge of the 'hot spots' on the enzyme. Directed evolution (or molecular evolution) does not require such prior sequence or three-dimensional structure knowledge, as it usually employs random-mutagenesis protocols to engineer enzymes that are subsequently screened for the desired properties (Tao and Cornish 2002, Dalby 2003, Jaeger and Eggert 2004, Jestin and Kaminski 2004, Williams et al. 2004). However, both approaches require efficient expression as well as sensitive detection systems for the protein of interest (Kotzia et al. 2006). During the selection process, the mutations that have a positive effect are selected and identified. Usually, repeated rounds of mutagenesis are applied until enzymes with the desired properties are constructed. For example, it took four rounds of random mutagenesis and DNA shuffling of *Drosophila melanogaster* 2′-deoxynucleoside kinase, followed by FACS analysis, in order to yield an orthogonal ddT kinase with a 6-fold higher activity for the nucleoside analogue and a 20-fold k_{cat}/K_m preference for ddT over thymidine, an overall 10,000-fold change in substrate specificity (Liu et al. 2009b).

Usually, a combination of both methods is employed by the construction of combinatorial libraries of variants, using random mutagenesis on selected (by rational design) areas of the parental 'wild-type' protein (typically, binding surfaces or specific amino acids; Altamirano et al. 2000, Arnold 2001, Saven 2002, Johannes and Zhao 2006). For example, Park et al. rationally manipulated several active site loops in the ab/ba metallohydrolase scaffold of glyoxalase II through amino acid insertion, deletion, and substitution, and then used directed evolution to introduce random point-mutations to fine-tune the enzyme activity (Park et al. 2006). The resulting enzyme completely lost its original activity and instead showed β-lactamase activity.

The industrial applications of enzymes as biocatalysts are numerous. Recent advances in genetic engineering have made possible the construction of enzymes with enhanced or altered properties (change of enzyme/cofactor specificity and enantioselectivity, altered thermostability, increased activity) to satisfy the ever-increasing needs of the industry for more efficient catalysts (Bornscheuer and Pohl 2001, Zaks 2001, Jaeger and Eggert 2004, Chaput et al. 2008, Zeng et al. 2009).

Rational Enzyme Design

The rational protein design approach is mainly used for the identification and evaluation of functionally important residues or sites in proteins. Although the protein sequence contains all the information required for protein folding and functions, today's state of technology does not allow for efficient protein design by simple knowledge of the amino acid sequence alone. For example, there are 10^{325} ways of rearranging amino acids in a 250-amino-acid-long protein, and prediction of the number of changes required to achieve a desired effect is an obstacle that initially appears impossible. For this reason, a successful rational design cycle requires substantial planning and could be repeated several times before the desired result is achieved. A rational protein design cycle requires the following:

1. Knowledge of the amino acid sequence of the enzyme of interest and availability of an expression system that allows for the production of active enzyme. Isolation and characterisation (annotation) of cDNAs encoding proteins with novel or pre-observed properties has been significantly facilitated by advances in genomics (Schena et al. 1995, Zweiger and Scott 1997, Schena et al. 1998, Carbone 2009) and proteomics (Anderson and Anderson 1998, Anderson et al. 2000, Steiner and Anderson 2000, Xie et al. 2009) and is increasing rapidly. These cDNA sequences are stored in gene (NCBI) and protein databanks (UniProtKB/Swiss-Prot; Release 57.12 of 15 Dec 09 of UniProtKB/Swiss-Prot contains 513,77 protein sequence entries; Apweiler et al. 2004, The UniProt Consortium 2008). However, before the protein design cycle begins, a protein expression system has to be established. Introduction of the cDNA encoding the protein of interest into a suitable expression vector/host cell system is nowadays a standard procedure (see above).
2. Structure/function analysis of the initial protein sequence and determination of the required amino acids changes. As mentioned before, the enzyme engineering process could be repeated several times until the desired result is obtained. Therefore, each cycle ends where the next begins. Although, we cannot accurately predict the conformation of a given protein by knowledge of its amino acid sequence, the amino acid sequence can provide significant information. Initial screening should therefore involve sequence comparison analysis of the original protein sequence to other sequence homologous proteins with potentially similar functions by utilising current bioinformatics tools (Andrade and Sander 1997, Fenyo and Beavis 2002, Nam et al. 2009, Yen et al. 2009, Zhang et al. 2009a, 2009b). Areas of conserved or non-conserved amino acids residues can be located within the protein and could possibly provide valuable information, concerning the identification of binding and catalytic residues. Additionally, such methods could also reveal information pertinent to the three-dimensional structure of the protein.
3. Availability of functional assays for identification of changes in the properties of the protein. This is probably the most basic requirement for efficient rational protein design. The expressed protein has to be produced in a bioactive form and characterised for size, function and stability in order to build a baseline comparison platform for the ensuing protein mutants. The functional assays should have the required sensitivity and accuracy to detect the desired changes in the protein's properties.
4. Availability of the three-dimensional structure of the protein or capability of producing a reasonably accurate three-dimensional model by computer modelling techniques. The structures of thousands of proteins have been solved by various crystallographic techniques (X-ray diffraction, NMR spectroscopy) and are available in protein structure databanks. Current bioinformatics tools and elaborate molecular modeling software (Wilkins et al. 1999, Gasteiger et al. 2003, Guex et al. 2009) permit the accurate depiction of these structures and allow the manipulation of the aminoacid sequence. For example, they are able to predict, with significant accuracy, the consequences of a single aminoacid substitution on the conformation, electrostatic or hydrophobic potential of the protein (Guex and Peitsch 1997, Gasteiger et al. 2003, Schwede et al. 2003). Additionally, protein–ligand interactions can, in some cases, be successfully simulated, which is especially important in the identification of functionally important residues in enzyme–cofactor/substrate interactions (Saxena et al. 2009). Finally, in allosteric regulation, the induced conformational changes are very difficult to predict. In last few years, studies on the computational modelling of allostery have also began (Kidd et al. 2009).

Where the three-dimensional structure of the protein of interest is not available, computer modelling methods (homology modelling, fold recognition using threading and ab initio prediction) allow for the construction of putative models based on known structures of homologous proteins (Schwede et al. 2003, Kopp and Schwede 2004, Jaroszewski 2009, Qu et al. 2009). Additionally, comparison with proteins having homologous three-dimensional structure or structural motifs could provide clues as to the function of the protein and the location of functionally important sites. Even if the protein of interest shows no homology to any other known protein, current amino acid sequence analysis software could provide putative tertiary structural models. A generalised approach to predict protein structure is shown in Figure 7.17.

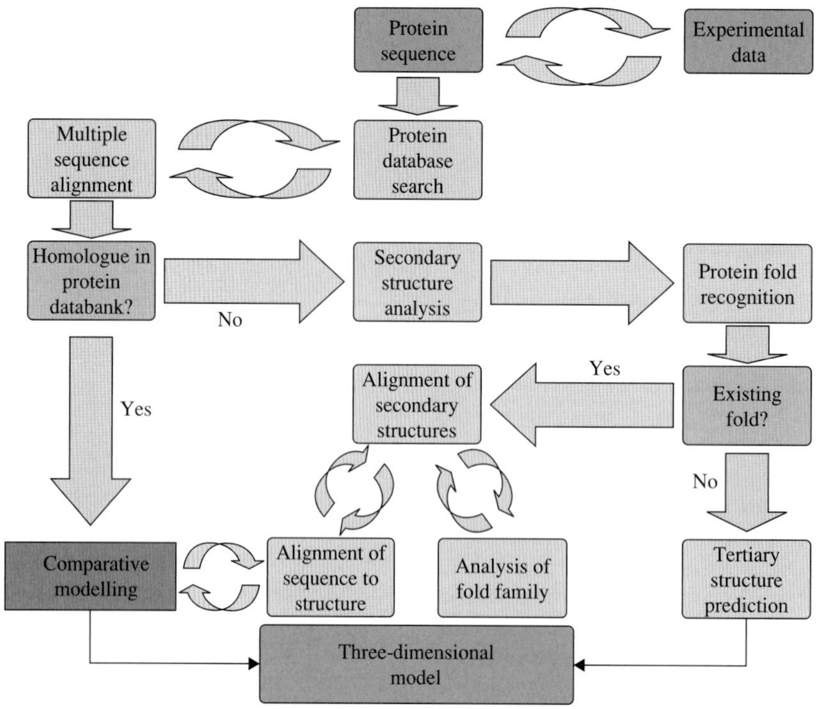

Figure 7.17. A generalised schematic for the prediction of protein three-dimensional structure.

5. Genetic manipulation of the wild-type nucleotide sequence. A combination of previously published experimental literature and sequence/structure analysis information is usually necessary for the identification of functionally important sites in the protein. Once an adequate three-dimensional structural model of the protein of interest has been constructed, manipulation of the gene of interest is necessary for the construction of mutants. Polymerase chain reaction (PCR) mutagenesis is the basic tool for the genetic manipulation of the nucleotide sequences. The genetically redesigned proteins are engineered by the following:
   a. Site-directed mutagenesis: alteration of specific amino acid residues. There are a number of experimental approaches designed for this purpose. The basic principle involves the use of synthetic oligonucleotides (oligonucleotide-directed mutagenesis) that are complementary to the cloned gene of interest but contain a single (or sometimes multiple) mismatched base(s) (Balland et al. 1985, Garvey and Matthews 1990, Wagner and Benkovic 1990). The cloned gene is either carried by a single-stranded vector (M13 oligonucleotide-directed mutagenesis) or a plasmid that is later denatured by alkali (plasmid DNA oligonucleotide-directed mutagenesis) or heat (PCR-amplified oligonucleotide-directed mutagenesis) in order for the mismatched oligonucleotide to anneal. The latter then serves as a primer for DNA synthesis catalysed by externally added DNA polymerase for the creation of a copy of the entire vector, carrying, however, a mutated base. PCR mutagenesis is the most frequently used mutagenesis method (Fig. 7.18). For example, substitution of specific amino acid positions by site-directed mutagenesis (S67D/H68D) successfully converted the coenzyme specificity of the short-chain carbonyl reductase from NADP(H) to NAD(H) as well as the product enantioselectivity without disturbing enzyme stability (Zhang et al. 2009). In another example, engineering of the maize GSTF1–1 by mutating selected G-site residues resulted in substantial changes in the pH-dependence of kinetic parameters of the enzyme (Labrou et al. 2004a). Mutation of a key residue in the H-site of the same enzyme (Ile118Phe) led to a fourfold improved specificity of the enzyme towards the herbicide alachlor (Labrou et al. 2005).

   So far, substitution of a specific amino acid by another has been limited by the availability of only 20 naturally occurring amino acids. However, it is chemically possible to construct hundreds of designer-made amino acids. Incorporation of these novel protein building blocks could help shed new light into the cellular and protein functions (Wang and Schultz 2002, Chin et al. 2003, Deiters et al. 2003, Arnold 2009).

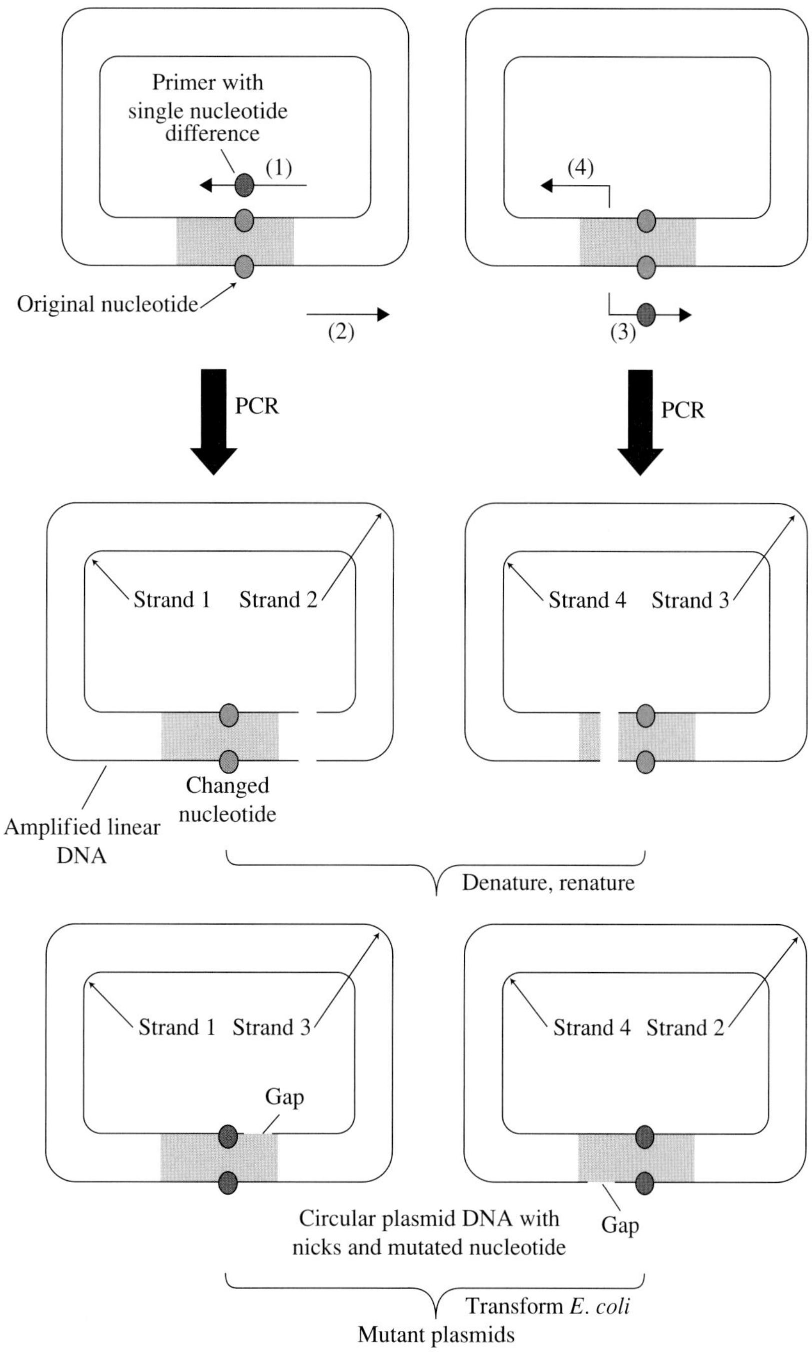

Figure 7.18. PCR oligonucleotide-directed mutagenesis. Two sets of primers are used for the amplification of the double-stranded plasmid DNA. The primers are positioned as shown and only one contains the desired base change. After the initial PCR step, the amplified PCR products are mixed together, denatured and renatured to form, along with the original amplified linear DNA, nicked circular plasmids containing the mutations. Upon transformation into *E. coli*, the nicked are repaired by host cell enzymes and the circular plasmids can be maintained.

b. Construction of deletion mutants: deletion of specified areas within or at the 5′/3′ ends (truncation mutants) of the gene.
c. Construction of insertion/fusion mutants: insertion of a functionally/structurally important epitope or fusion to another protein fragment.
   There are numerous examples of fusion proteins designed to facilitate protein expression and purification, display of proteins on surfaces of cells or phages, cellular localisation, metabolic engineering as well as protein–protein interaction studies (Nixon et al. 1998).
d. Domain swapping: exchanging of protein domains between homologous or heterologous proteins.
   For example, exchange of a homologous region between *Agrobacterium tumefaciens* β-glucosidase (optimum at pH 7.2–7.4 and 60°C) and *Cellvibrio gilvus* β-glucosidase (optimum at pH 6.2–6.4 and 35°C) resulted in a hybrid enzyme with optimal activity at pH 6.6–7.0 and 45–50°C (Singh et al. 1995). Also, domain swapping was used to clarify the control of electron transfer in nitric-oxide synthases (Nishida and Ortiz de Montellano 2001). In another example, domain swapping was observed in the structurally unrelated capsid of a rice yellow mottle virus, a member of the plant icosahedral virus group, where it was demonstrated to increase stability of the viral particle (Qu et al. 2000).

Although site-directed mutagenesis is widely used, it is not always feasible due to the limited knowledge of protein structure–function relationship and the approximate nature of computer-graphic modelling. In addition, rational design approaches can fail due to unexpected influences exerted by the substitution of one or more amino acid residues (Cherry and Fidantsef 2003, Johannes and Zhao 2006). Irrational approaches can therefore be preferable alternatives to direct the evolution of enzymes with highly specialised traits (Hibbert and Dalby 2005, Chatterjee and Yuan 2006, Johannes and Zhao 2006).

Directed Enzyme Evolution

Directed evolution by DNA recombination can be described as a mature technology for accelerating protein evolution. Evolution is a powerful algorithm with proven ability to alter enzyme function and especially to 'tune' enzyme properties (Cherry and Fidantsef 2003, Williams et al. 2004, Hibbert and Dalby 2005, Roodveldt et al. 2005, Chatterjee and Yuan 2006). The methods of directed evolution use the process of natural selection but in a directed way (Altreuter and Clark 1999, Kaur and Sharma 2006, Wong et al. 2006, Glasner et al. 2007, Gerlt and Babbitt 2009, Turner 2009). The major step in a typical directed enzyme evolution experiment is first to make a set of mutants and then to find the best variants through a high-throughput selection or screening procedure (Kotzia et al. 2006). The process can be iterative, so that a 'generation' of molecules can be created in a few weeks or even in a few days, with large numbers of progeny subjected to selective pressures not encountered in nature (Arnold 2001, Williams et al. 2004).

There are many methods to create combinatorial libraries, using directed evolution (Labrou 2010). Some of these are random mutagenesis using mainly error-prone PCR (Ke and Madison 1997, Cirino et al. 2003), DNA shuffling (Stemmer 1994, Crameri et al. 1998, Baik et al. 2003, Bessler et al. 2003, Dixon et al. 2003, Wada et al. 2003), StEP (staggered extension process; Zhao et al. 1998, Aguinaldo and Arnold 2003), RPR (random-priming in vitro recombination; Shao et al. 1998, Aguinaldo and Arnold 2003), incremental truncation for the creation of hybrid enzymes (ITCHY; Lutz et al. 2001), RACHITT (random chimeragenesis on transient templates; Coco et al. 2001, Coco 2003), ISM (iterative saturation mutagenesis; Reetz 2007), GSSM (gene site saturation mutagenesis; DeSantis et al. 2003, Dumon et al. 2008), PDLGO (protein domain library generation by overlap extension; Gratz and Jose 2008) and DuARCheM (dual approach to random chemical mutagenesis; Mohan and Banerjee 2008). The most frequently used methods for DNA shuffling are shown in Figure 7.19.

Currently, directed evolution has gained considerable attention as a commercially important strategy for rapid design of molecules with properties tailored for the biotechnological and pharmaceutical market. Over the past four years, DNA family shuffling has been successfully used to improve enzymes of industrial and therapeutic interest (Kurtzman et al. 2001, Chiang 2004, Dai and Copley 2004, Yuan et al. 2005). For example, by applying the DNA family shuffling approach, the catalytic properties of cytochrome P450 enzymes were further extended in the chimeric progeny to include a new range of blue colour formations. Therefore, it may be possible to direct the new enzymes towards the production of new dyes (Rosic 2009).

IMMOBILISED ENZYMES

The term 'immobilised enzymes' describes enzymes physically confined, localised in a certain region of space or attached on a support matrix (Abdul 1993). The main advantages of enzyme immobilisation are listed in Table 7.8.

There are at least four main areas in which immobilised enzymes may find applications, that is industrial, environmental, analytical and chemotherapeutic (Powell 1984, Liang et al. 2000). Environmental applications include waste water treatment and the degradation of chemical pollutants of industrial and agricultural origin (Dravis et al. 2001). Analytical applications include biosensors. Biosensors are analytical devices, which have a biological recognition mechanism (most commonly enzyme) that transduce it into a signal, usually electrical, and can be detected by using a suitable detector (Phadke 1992). Immobilised enzymes, usually encapsulated, are also being used for their possible chemotherapeutic applications in replacing enzymes that are absent from individuals with certain genetic disorders (DeYoung 1989).

Methods for Immobilisation

There are a number of ways in which an enzyme may be immobilised: adsorption, covalent coupling, cross-linking, matrix

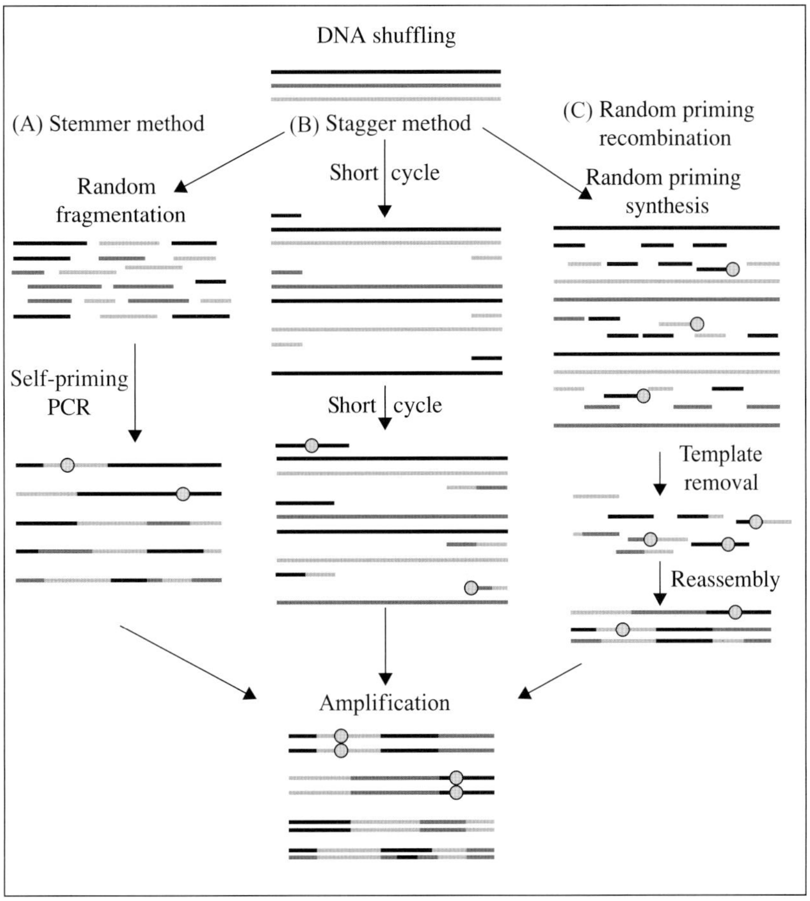

Figure 7.19. A schematic representation of the most frequently used methods for DNA shuffling.

entrapment or encapsulation (Podgornik and Tennikova 2002; Fig. 7.20). These methods will be discussed in the following sections.

Adsorption

Adsorption is the simplest method and involves reversible interactions between the enzyme and the support material (Fig. 7.20A). The driving force causing adsorption is usually the formation of several non-covalent bonds such as salt links, van der Waals, hydrophobic and hydrogen bonding (Calleri et al. 2004). The methodology is easy to carry out and can be applied to a wide range of support matrices such as alumina, bentonite, cellulose, anion and cation exchange resins, glass, hydroxyapatite, kaolinite, etc. The procedure consists of mixing together the enzyme and a support under suitable conditions of pH, ionic strength, temperature, etc. The most significant advantages of this method are (i) absence of chemicals resulting to a little damage to enzyme and (ii) reversibility, which allows regeneration with fresh enzyme. The main disadvantage of the method is the leakage of the enzyme from the support under many conditions of changes in the pH, temperature and ionic strength. Another disadvantage is the non-specific, adsorption of other proteins or other substances to the support. This may modify the properties of the support or of the immobilised enzyme.

Table 7.8. Advantages of Immobilised Enzymes

1. Repetitive use of a single batch of enzymes.
2. Immobilisation can improve enzyme's stability by restricting the unfolding of the protein.
3. Product is not contaminated with the enzyme. This is very important in the food and pharmaceutical industries.
4. The reaction is controlled rapidly by removing the enzyme from the reaction solution (or vice versa).

Covalent Coupling

The covalent coupling method is achieved by the formation of a covalent bond between the enzyme and the support

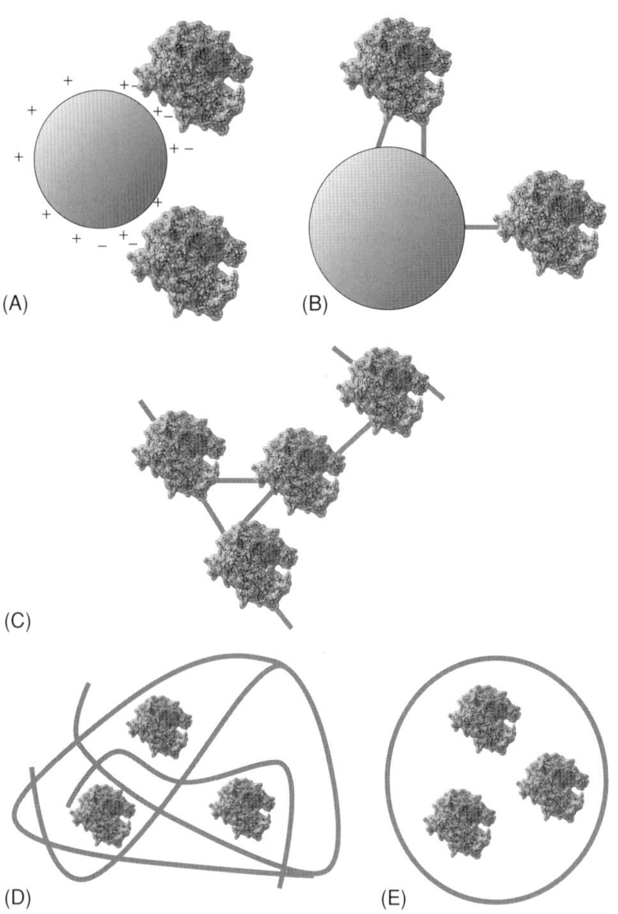

Figure 7.20. Representation of the methods by which an enzyme may be immobilised: adsorption, covalent coupling, cross-linking, matrix entrapment and encapsulation.

(Fig. 7.20B). The binding is very strong and therefore little leakage of enzyme from the support occurs (Calleri et al. 2004). The bond is formed between reactive electrophile groups present on the support and nucleophile side chains on the surface of the enzyme. These side-chains are usually the amino group ($-NH_2$) of lysine, the imidazole group of histidine, the hydroxyl group ($-OH$) of serine and threonine, and the sulfydryl group ($-SH$) of cysteine. Lysine residues are found to be the most generally useful groups for covalent bonding of enzymes to insoluble supports due to their widespread surface exposure and high reactivity, especially in slightly alkaline solutions.

It is important that the amino acids essential to the catalytic activity of the enzyme are not involved in the covalent linkage to the support (Dravis et al. 2001). This may be difficult to achieve, and enzymes immobilised in this fashion generally lose activity upon immobilisation. This problem may be prevented if the enzyme is immobilised in the presence of saturating concentrations of substrate, product or a competitive inhibitor to protect active site residues. This ensures that the active site remains 'unreacted' during the covalent coupling and reduces the occurrence of binding in unproductive conformations.

Various types of beaded supports have been used successfully as for example, natural polymers (e.g. agarose, dextran and cellulose), synthetic polymers (e.g. polyacrylamide, polyacryloyl trihydroxymethylacrylamide, polymethacrylate), inorganic (e.g. silica, metal oxides and controlled pore glass) and microporous flat membrane (Calleri et al. 2004).

The immobilisation procedure consists of three steps (Calleri et al. 2004): (i) activation of the support, (ii) coupling of ligand and (iii) blocking of residual functional groups in the matrix. The choice of coupling chemistry depends on the enzyme to be immobilised and its stability. A number of methods are available in the literature for efficient immobilisation of enzyme through a chosen particular functional side chain's group by employing glutaraldehyde, oxirane, cyanogen bromide, 1,1-carbonyldiimidazole, cyanuric chloride, trialkoxysilane to derivatise glass, etc. Some of them are illustrated in Figure 7.21.

Cross-linking

This type of immobilisation is achieved by cross-linking the enzymes to each other to form complex structures as shown in Figure 7.20C. It is therefore a support-free method and less costly than covalent linkage. Methods of cross-linking involve covalent bond formation between the enzymes using bi- or multifunctional reagent. Cross-linking is frequently carried out using glutaraldehyde, which is of low cost and available in industrial quantities. To minimise close proximity problems associated with the cross-linking of a single enzyme, albumin and gelatin are usually used to provide additional protein molecules as spacers (Podgornik and Tennikova 2002).

Entrapment and Encapsulation

In the immobilisation by entrapment, the enzyme molecules are free in solution, but restricted in movement by the lattice structure of the gel (Fig. 7.20D; Balabushevich et al. 2004). The entrapment method of immobilisation is based on the localisation of an enzyme within the lattice of a polymer matrix or membrane (Podgornik and Tennikova 2002). It is done in such a way as to retain protein while allowing penetration of substrate. Entrapment can be achieved by mixing an enzyme with chemical monomers that are then polymerised to form a cross-linked polymeric network, trapping the enzyme in the interstitial spaces of lattice. Many materials have been used, such as alginate, agarose, gelatin, polystyrene and polyacrylamide. As an example of this latter method, the enzymes' surface lysine residues may be derivatised by reaction with acryloyl chloride ($CH_2 = CH–CO–Cl$) to give the acryloyl amides. This product may then be copolymerised and cross-linked with acrylamide ($CH_2 = CH–CO–NH_2$) and bisacrylamide ($H_2N–CO–CH = CH–CH = CH–CO–NH_2$) to form a gel.

Encapsulation of enzymes can be achieved by enveloping the biological components within various forms of semipermeable

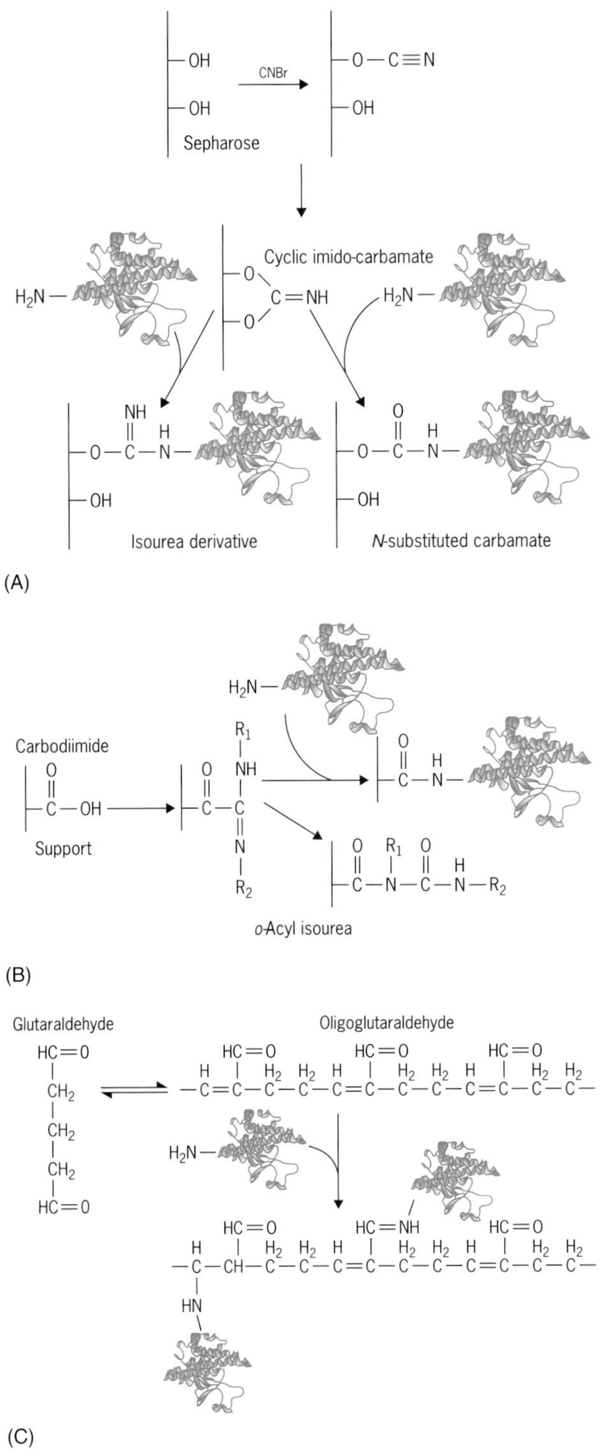

Figure 7.21. Commonly used methods for the covalent immobilisation of enzymes. (**A**) Activation of hydroxyl support by cyanogen bromide. (**B**) Carbodiimides may be used to attach amino groups on the enzyme to carboxylate groups on the support or carboxylate groups on the enzyme to amino groups on the support. (**C**) Glutaraldehyde is used to cross-link enzymes or link them to supports. The product of the condensation of enzyme and glutaraldehyde may be stabilised against dissociation by reduction with sodium borohydride.

membranes as shown in Figure 7.20E. Encapsulation is most frequently carried out using nylon and cellulose nitrate to construct microcapsules varying from 10 to 100 μM. In general, entrapment methods have found more application on the immobilisation of cells.

New Approaches for Oriented Enzyme Immobilisation: The Development of Enzyme Arrays

With the completion of several genome projects, attention has turned to the elucidation of functional activities of the encoded proteins. Because of the enormous number of newly discovered open reading frames, progress in the analysis of the corresponding proteins depends on the ability to perform characterisation in a parallel and high throughput format (Cahill and Nordhoff 2003). This typically involves construction of protein arrays based on recombinant proteins. Such arrays are then analysed for their enzymatic activities and the ability to interact with other proteins or small molecules, etc. The development of enzyme array technology is hindered by the complexity of protein molecules. The tremendous variability in the nature of enzymes and consequently in the requirement for their detection and identification makes the development of protein chips a particularly challenging task. Additionally, enzyme molecules must be immobilised on a matrix in a way that they preserve their native structures and are accessible to their targets (Cutler 2003). The immobilisation chemistry must be compatible with preserving enzyme molecules in native states. This requires good control of local molecular environments of the immobilised enzyme molecule (Yeo et al. 2004). There is one major barrier in enzyme microarray development: the immobilisation chemistry has to be such that it preserves the enzyme in native state and with optimal orientation for substrate interaction. This problem may be solved by the recently developed in vitro protein ligation methodology. Central to this method is the ability of certain protein domains (inteins) to excise themselves from a precursor protein (Lue et al. 2004). In a simplified intein expression system, a thiol reagent induces cleavage of the intein–extein bond, leaving a reactive thioester group on the C-terminus of the protein of interest. This group can then be used to couple essentially any polypeptide with an N-terminal cysteine to the thioester tagged protein by restoring the peptide bond. In another methodology, optimal orientation is based on the unique ability of protein prenyl-transferases to recognise short but highly specific C-terminal protein sequences (Cys–A–A–X–), as shown in Figure 7.22. The enzyme accepts a spectrum of phosphoisoprenoid analogues while displaying a very strict specificity for the protein substrate. This feature is explored for protein derivatisation. Several types of pyrophosphates (biotin analogues, photoreactive aside and benzophenone analogues; Fig. 7.22) can be covalently attached to the protein tagged with the Cys-A-A-X motif. After modification, the protein can be immobilised directly either reversibly through biotin–avidin interaction on avidin modified support or covalently through the photoreactive group on several supports.

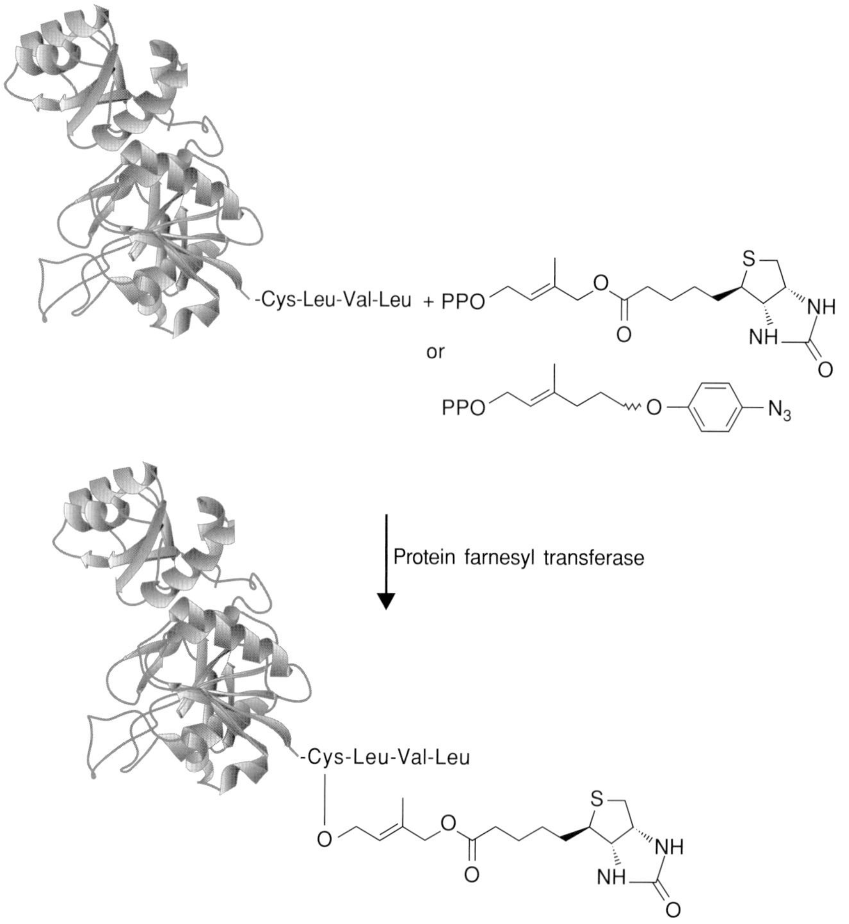

Figure 7.22. Principal scheme of using CAAX-tagged proteins for covalent modification with prenyl transferases.

ENZYME UTILISATION IN INDUSTRY

Enzymes offer potential for many exciting applications in industry. Some important industrial enzymes and their sources are listed in Table 7.9. In addition to the industrial enzymes listed above, a number of enzyme products have been approved for therapeutic use. Examples include tissue plasminogen activator and streptokinase for cardiovascular disease, adenosine deaminase for the rare severe combined immunodeficiency disease, β-glucocerebrosidase for Type 1 Gaucher disease, L-asparaginase for the treatment of acute lymphoblastic leukemia, DNAse for the treatment of cystic fibrosis and neuraminidase which is being targeted for the treatment of influenza (Cutler 2003).

There are also thousands of enzyme products used in small amounts for research and development in routine laboratory practice and others that are used in clinical laboratory assays. This group also includes a number of DNA- and RNA-modifying enzymes (DNA and RNA polymerase, DNA ligase, restriction endonucleases, reverse transcriptase, etc.), which led to the development of molecular biology methods and were a foundation for the biotechnology industry (Yeo et al. 2004). The clever application of one thermostable DNA polymerase led to the PCR and this has since blossomed into numerous clinical, forensic and academic embodiments. Along with the commercial success of these enzyme products, other enzyme products are currently in commercial development.

Another important field of application of enzymes is in metabolic engineering. Metabolic engineering is a new approach involving the targeted and purposeful manipulation of the metabolic pathways of an organism, aiming at improving the quality and yields of commercially important compounds. It typically involves alteration of cellular activities by manipulation of the enzymatic functions of the cell using recombinant DNA and other genetic techniques. For example, the combination of rational pathway engineering and directed evolution has been successfully applied to optimise the pathways for the production of isoprenoids such as carotenoids (Schmidt-Dannert et al. 2000, Umeno and Arnold 2004).

Table 7.9. Some Important Industrial Enzymes and Their Sources

Enzyme	EC Number	Source	Industrial Use
Rennet	3.4.23.4	Abomasum	Cheese
α-Amylase	3.2.1.1	Malted barley, *Bacillus, Aspergillus*	Brewing, baking
α-Amylase	3.2.1.2	Malted barley, *Bacillus*	Brewing
Bromelain	3.4.22.4	Pineapple latex	Brewing
Catalase	1.11.1.6	Liver, *Aspergillus*	Food
Penicillin amidase	3.5.1.11	*Bacillus*	Pharmaceutical
Lipoxygenase	1.13.11.12	Soybeans	Food
Ficin	3.4.22.3	Fig latex	Food
Pectinase	3.2.1.15	*Aspergillus*	Drinks
Invertase	3.2.1.26	*Saccharomyces*	Confectionery
Pectin lyase	4.2.2.10	*Aspergillus*	Drinks
Cellulase	3.2.1.4	*Trichoderma*	Waste
Chymotrypsin	3.4.21.1	Pancreas	Leather
Lipase	3.1.1.3	Pancreas, *Rhizopus, Candida*	Food
Trypsin	3.4.21.4	Pancreas	Leather
α-Glucanase	3.2.1.6	Malted barley	Brewing
Papain	3.4.22.2	Pawpaw latex	Meat
Asparaginase	3.5.1.1	*Erwinia chrisanthemy, Erwinia carotovora, Escherichia coli*	Human health
Glucose isomerase	5.3.1.5	*Bacillus*	Fructose syrup
Protease	3.4.21.14	*Bacillus*	Detergent
Aminoacylase	3.5.1.14	*Aspergillus*	Pharmaceutical
Raffinase	3.2.1.22	*Saccharomyces*	Food
Glucose oxidase	1.1.3.4	*Aspergillus*	Food
Dextranase	3.2.1.11	*Penicillium*	Food
Lactase	3.2.1.23	*Aspergillus*	Dairy
Glucoamylase	3.2.1.3	*Aspergillus*	Starch
Pullulanase	3.2.1.41	*Klebsiella*	Starch
Raffinase	3.2.1.22	*Mortierella*	Food
Lactase	3.2.1.23	*Kluyveromyces*	Dairy

ENZYMES INVOLVED IN XENOBIOTIC METABOLISM AND BIOCHEMICAL INDIVIDUALITY

The term xenobiotic metabolism refers to the set of metabolic pathways that chemically modify xenobiotics, which are compounds foreign to an organism's normal biochemistry, such as drugs and poisons. The term biochemical individuality of xenobiotic metabolism refers to variability in xenobiotic metabolism and drug responsiveness among different people. Biochemical individuality is a significant factor that can improve public health, drug therapy, nutrition and health impacts such as cancer, diabetes 2 and cardiovascular disease.

Most xenobiotics are lipophilic and able to bind to lipid membranes and be transported in the blood (Hodgson 2004). The enzymes that are involved in xenobiotic metabolism (Table 7.10) comprise one of the first defense mechanism against environmental carcinogens and xenobiotic compounds (Zhang et al. 2009a). Xenobiotic metabolism follows mainly three phases (I, II, III). In Phase I, the original compound obtain increased hydrophilicity and constitute an adequate substrate for phase II enzymes, by the introduction of a polar reactive group (–OH, $-NH_2$, –SH or –COOH). In Phase II, the products of Phase I can be conjugated to substrates such as GSH, which result in a significant increase of water solubility of xenobiotic, promoting its excretion (Hodgson 2004). The ATP-dependent transporters that facilitate the movement of the polar conjugates (by phase I and II) across biological membranes and their excretion from the cell constitute Phase III proteins (Josephy and Mannervik 2006). In general, the enzymes that are involved in xenobiotic metabolism are genetically polymorphic, affecting the individual delicacy to environmental pollutants (Zhang et al. 2009a).

Phase I

Human cytochrome P450, is one of the most important enzymes that takes part in xenobiotic metabolism; therefore, its genetic polymorphisms have been studied in depth. For example, P450 2A6 (CYP2A6) catalyses nicotine oxidation, and it has been found to have inter-individual and inter-ethnic variability. Genetic polymorphisms of this gene impact smoking behaviour (Xu et al. 2002). Another example of genetic polymorphism's impact of this enzyme came from Siraj et al., who suggested that CYP1A1 phenotype AA showed association with increased

Table 7.10. The Main Enzymes Involved in Xenobiotic Metabolism

	Reactions
Phase I enzymes	
Cytochrome P450 (CYPs)	Epoxidation/hydroxylation *N*-, *O*-, *S*-dealkylation *N*-, *S*-, *P*-oxidation Desulfuration Dehalogenation Azo reduction Nitro reduction
Flavin-containing monooxygenases (FMOs)	*N*-, *S*-, *P*-oxidation Desulfuration
Alcohol dehydrogenase	Oxidation Reductions
Aldehyde dehydrogenase	Oxidation
Prostaglandin synthetase co-oxidation	Dehydrogenation *N*-dealkylation Epoxidation/hydroxylation Oxidation
Molybdenum hydroxylases	Oxidation Reductions
Esterases and amidases	Hydrolysis
Epoxide hydrolase	Hydrolysis
Phase II enzymes	
UDP (uridine diphospho) glucuronosyl transferase (UGT)	Glucuronide conjugation
Sulfotransferases	Sulfation reaction
Sulfatases	Hydrolysis of sulfate esters
Methyltransferases	*N*-, *O*-, *S*-methylation,
Glutathione *S*-transferase	Alkyltransferase, aryltransferase, aralkyltransferase, alkenetransferase, epoxidetransferase
γ-Glutamyltranspeptidase	Hydrolysis Transpeptidation
N-acetyltransferase	Acetylation
Aminopeptidases	Hydrolysis of peptides
N,*O*-Acyltransferase	Acylation
Cysteine conjugate β-lyase	Methylation
Phase III enzymes	
MRP (multi-drug resistance – associated protein) MDR (multi-drug resistance family/*P*-glycoprotein) MXR (mitoxantrone-resistance protein) efflux	Transport and excretion of soluble products from phase I and II metabolic pathways

Source: Rommel and Richard 2002, Hodgson 2004, Josephy and Mannervik 2006.

risk of developing papillary thyroid cancer in Middle Eastern population (Siraj et al. 2008).

Another studied genetic polymorphic enzyme is superoxide dismutase-2 (SOD). The heterozygosity 9Val-allele of MnSOD is associated with a higher risk and severity of orofaciolingual dyskinesias (TDof) in Russian psychiatric inpatients from Siberia (Al Hadithy et al. 2010). In addition, there are several studies that have reported and showed that the 9Val/9Val genotype confer great susceptibility to tardive dyskinesia (Zhang et al. 2003, Akyol et al. 2005, Galecki et al. 2006, Hitzeroth et al. 2007). In addition, human MnSOD gene encoding alanine (A) or valine at codon 16 (Shimoda-Matsubayashi et al. 1996) can be a risk factor for several malignancies (Iguchi et al. 2009).

Phase II

Polymorphisms of human UGT correlate with diseases and side effects of drugs, for example the isoform UGT1A1 is associated with diseases of bilirubin metabolism (Hodgson 2004). *N*-Acetyltransferases (NAT) are important enzymes that participate in metabolic activation of carcinogenic aromatic and heterocyclic amines that are present in cigarette smoke (Wang et al. 1999). An association was found between smokers with MnSOD AA genotype and prostate cancer risk, especially in case of rapid NAT1 subjects (Iguchi et al. 2009). Tamini et al. found that the AA genotype of MnSOD in women smokers have an elevated risk for breast cancer (Tamini et al. 2004).

Tobacco smoke (Lioy and Greenberg 1990) as well as smoked foods, cereals, leafy green vegetables and fossil fuels combustion by-products (Waldman et al. 1991) are the sources of exposure to polycyclic aromatic hydrocarbons (PAHs). PAHs have been considered as potential carcinogens for human (Shimada 2006). According to McCarty et al., the lack of dose–response relationship of PAHs and breast cancer may be due to genetic differences in metabolic activation and detoxification of PAHs (McCarty et al. 2009).

Phase III

The proteins of phase III are membrane transporters. These proteins seem to be significantly implicated in the absorption, distribution and discard of drugs (Rommel and Richard 2002). There are genetic polymorphisms of drug transporters, which appear to have clinical impact and have been detected in multiple clinical and in vitro studies (Maeda and Sugiyama 2008).

ACKNOWLEDGEMENTS

This work was partially supported by the following grants: HRAKLEITOS II and THALIS ((fall under the Operational Programme "Education and Lifelong Learning") co-funded by the European Union – European Social Fund & National Resources), AquaPhage (funded by the European Union), BioExplore (co-funded by the European Union – European Social Fund & National Resources).

REFERENCES

Abdul MM. 1993. Biocatalysis and immobilized enzyme/cell bioreactors. Promising techniques in bioreactor technology. *Biotechnology (NY)* 11: 690–695.

Adamson JG et al. 1993. Structure, function and application of the coiled-coil protein folding motif. *Curr Opin Biotechnol* 4: 428–437.

Aehle W. 2007. *Enzymes in Industry: Production and Applications*, 3rd edn. Wiley-VCH, Weinheim.

Agarwal PK et al. 2002. Network of coupled promoting motions in enzyme catalysis. *Proc Natl Acad Sci USA* 99: 2794–2799.

Aguinaldo AM, Arnold FH. 2003. Staggered extension process (StEP) in vitro recombination. *Methods Mol Biol* 231: 105–110.

Ahern KG et al. 1988. Identification of regions essential for extrachromosomal replication and maintenance of an endogenous plasmid in *Dictyostelium*. *Nucleic Acids Res* 16(14B): 6825–6837.

Ahn J et al. 2009. Generation of expression vectors for high-throughput functional analysis of target genes in *Schizosaccharomyces pombe*. *J Microbiol* 47(6): 789–795.

Akyol O et al. 2005. Association between Ala-9Val polymorphism of Mn-SOD gene and schizophrenia. *Prog Neuropsychopharmacol Biol Psychiatry* 29: 123–131.

Al Hadithy AFY et al. 2010. Missense polymorphisms in three oxidative-stress enzymes (GSTP1, SOD2, and GPX1) and dyskinesias in Russian psychiatric inpatients from Siberia. *Hum Psychopharmacol Clin Exp* 25: 84–91.

Alberti S et al. 2007. A suite of Gateway cloning vectors for high-throughput genetic analysis in *Saccharomyces cerevisiae*. *Yeast* 24(10): 913–919.

Altamirano MM et al. 2000. Directed evolution of new catalytic activity using the α/β-barrel scaffold. *Nature* 403: 617–622.

Altmann F et al. 1999. Insect cells as hosts for the expression of recombinant glycoproteins. *Glycoconj J* 16: 109–123.

Altreuter DH, Clark DS. 1999. Combinatorial biocatalysis: taking the lead from nature. *Curr Opin Biotechnol* 10: 130–136.

Alvarez ML, Cardineau GA. 2010. Prevention of bubonic and pneumonic plague using plant-derived vaccines. *Biotechnol Adv* 28(1): 184–196.

Andersen CA, Rost B. 2003. Secondary structure assignment. *Methods Biochem Anal* 44: 341–363.

Anderson CM et al. 1979. Space-filling models of kinase clefts and conformation changes. *Science* 204: 375–380.

Anderson NL et al. 2000. Proteomics: applications in basic and applied biology. *Curr Opin Biotechnol* 11: 408–412.

Anderson NL, Anderson NG. 1998. Proteome and proteomics: new technologies, new concepts, and new words. *Electrophoresis* 19: 1853–1861.

Andrade MA, Sander C. 1997. Bioinformatics: from genome data to biological knowledge. *Curr Opin Biotechnol* 8: 675–683.

Andreadeli A et al. 2008. Structure-guided alteration of coenzyme specificity of formate dehydrogenase by saturation mutagenesis to enable efficient utilization of $NADP^+$. *FEBS J* 275(15): 3859–3869.

Apweiler R. 2001. Functional information in SWISS-PROT: the basis for large-scale characterisation of protein sequences. *Brief Bioinform* 2: 9–18.

Apweiler R et al. 2004. Protein sequence databases. *Curr Opin Chem Biol* 8: 76–80.

Archer DB. 1994. Enzyme production by recombinant *Aspergillus*. *Bioprocess Technol* 19: 373–393.

Archer DB, Peberdy JF. 1997. The molecular biology of secreted enzyme production by fungi. *Crit Rev Biotechnol* 17: 273–306.

Arnold FH. 2001. Combinatorial and computational challenges for biocatalyst design. *Nature* 409: 253–257.

Arnold U. 2009. Incorporation of non-natural modules into proteins: structural features beyond the genetic code. *Biotechnol Lett* 31(8): 1129–1139.

Arya R et al. 2008a. Production and characterization of pharmacologically active recombinant human phosphodiesterase 4B in *Dictyostelium discoideum*. *Biotechnol J* 3(7): 938–947.

Arya R et al. 2008b. *Dictyostelium discoideum* – a promising expression system for the production of eukaryotic proteins. *FASEB J* 22(12): 4055–4066.

Baik SH et al. 2003. Significantly enhanced stability of glucose dehydrogenase by directed evolution. *Appl Microbiol Biotechnol* 61: 329–335.

Balabushevich NG et al. 2004. Encapsulation of catalase in polyelectrolyte microspheres composed of melamine formaldehyde, dextran sulfate, and protamine. *Biochemistry (Mosc.)* 69: 763–769.

Balbas P. 2001. Understanding the art of producing protein and nonprotein molecules in *Escherichia coli*. *Mol Biotechnol* 19: 251–267.

Baldi L et al. 2007. Recombinant protein production by large-scale transient gene expression in mammalian cells: state-of-the-art and future perspectives. *Biotechnol Lett* 29(5): 677–684.

Balland A et al. 1985. Use of synthetic oligonucleotides in gene isolation and manipulation. *Biochimie* 67: 725–736.

Bankar AV et al. 2009. Environmental and industrial applications of *Yarrowia lipolytica*. *Appl Microbiol Biotechnol* 84(5): 847–865.

Bartlett GJ et al. 2002. Analysis of catalytic residues in enzyme active sites. *J Mol Biol* 324: 105–121.

Basile G, Peticca M. 2009. Recombinant protein expression in *Leishmania tarentolae*. *Mol Biotechnol* 43(3): 273–278.

Bauer C et al. 2001. A unified theory of enzyme kinetics based upon the systematic analysis of the variations of k(cat), $K(M)$, and k(cat)/$K(M)$ and the relevant Delta$G$(0 not equal) values-possible implications in chemotherapy and biotechnology. *Biochem Pharmacol* 61: 1049–1055.

Baumert TF et al. 1998. Hepatitis C virus structural proteins assemble into virus-like particles in insect cells. *J Virol* 72: 3827–3836.

Bendig MM. 1988. The production of foreign proteins in mammalian cells. *Genet Eng* 91–127.

Benkovic SJ, Hammes-Schiffer S. 2003. A perspective on enzyme catalysis. *Science* 301: 1196–1202.

Berka RM, Barnett CC. 1989. The development of gene expression systems for filamentous fungi. *Biotechnol Adv* 7: 127–154.

Berthold W, Walter J. 1994. Protein purification: aspects of processes for pharmaceutical products. *Biologicals* 22: 135–150.

Bessler C et al. 2003. Directed evolution of a bacterial alpha-amylase: toward enhanced pH-performance and higher specific activity. *Protein Sci* 12: 2141–2149.

Boehm R. 2007. Bioproduction of therapeutic proteins in the 21st century and the role of plants and plant cells as production platforms. *Ann NY Acad Sci* 1102: 121–134.

Bornscheuer UT. 2005. Trends and challenges in enzyme technology. *Adv Biochem Eng Biotechnol* 100: 181–203.

Bornscheuer UT, Pohl M. 2001. Improved biocatalysts by directed evolution and rational protein design. *Curr Opin Chem Biol* 5: 137–143.

Breitling R et al. 2002. Non-pathogenic trypanosomatid protozoa as a platform for protein research and production. *Protein Expr Purif* 25(2): 209–218.

Brondyk WH. 2009. Selecting an appropriate method for expressing a recombinant protein. *Methods Enzymol* 463: 131–147.

Burgess RR. 2009. Refolding solubilized inclusion body proteins. *Methods Enzymol* 463: 259–282.

Butzin NC et al. 2009. A new system for heterologous expression of membrane proteins: *Rhodospirillum rubrum*. *Protein Expr Purif* 70: 88–94.

Cahill DJ, Nordhoff E. 2003. Protein arrays and their role in proteomics. *Adv Biochem Eng Biotechnol* 83: 177–187.

Calleri E et al. 2004. Penicillin G acylase-based stationary phases: analytical applications. *J Pharm Biomed Anal* 35: 243–258.

Cantor CR. 1980. *The Conformation of Biological Macromolecules*. W.H. Freeman, San Francisco, CA.

Carbone R. 2009. An advanced application of protein microarrays: cell-based assays for functional genomics. *Methods Mol Biol* 570: 339–352.

Cardoza RE et al. 2003. Expression of a synthetic copy of the bovine chymosin gene in *Aspergillus awamori* from constitutive and pH-regulated promoters and secretion using two different pre-pro sequences. *Biotechnol Bioeng* 83(3): 249–259.

Carter P et al. 1989. Engineering subtilisin BPN' for site-specific proteolysis. *Proteins* 6: 240–248.

Cedrone F et al. 2000. Tailoring new enzyme functions by rational redesign. *Curr Opin Struct Biol* 10: 405–410.

Cereghino GP et al. 2002. Production of recombinant proteins in fermenter cultures of the yeast *Pichia pastoris*. *Curr Opin Biotechnol* 13: 329–332.

Chaput JC et al. 2008. Creating protein biocatalysts as tools for future industrial applications. *Expert Opin Biol Ther* 8(8): 1087–1098.

Chatterjee R, Yuan L. 2006. Directed evolution of metabolic pathways. *Trends Biotechnol* 24: 28–38.

Chen R. 2001. Enzyme engineering: rational redesign versus directed evolution. *Trends Biotechnol* 19: 13–14.

Chen W et al. 1999. Engineering of improved microbes and enzymes for bioremediation. *Curr Opin Biotechnol* 10: 137–141.

Cherry JR, Fidantsef AL. 2003. Directed evolution of industrial enzymes: an update. *Curr Opin Biotechnol* 14: 438–443.

Chiang SJ. 2004. Strain improvement for fermentation and biocatalysis processes by genetic engineering technology. *J Ind Microbiol Biotechnol* 31: 99–108.

Chica RA et al. 2005. Semi-rational approaches to engineering enzyme activity: combining the benefits of directed evolution and rational design. *Curr Opin Biotechnol* 16: 378–384.

Chill L et al. 2009. Production, purification, and characterization of human alpha1 proteinase inhibitor from *Aspergillus niger*. *Biotechnol Bioeng* 102(3): 828–844.

Chin JW et al. 2003. An expanded eukaryotic genetic code. *Science* 301: 964–967.

Choi JH, Lee SY. 2004. Secretory and extracellular production of recombinant proteins using *Escherichia coli*. *Appl Microbiol Biotechnol* 64: 625–635.

Chusainow J et al. 2009. A study of monoclonal antibody-producing CHO cell lines: what makes a stable high producer? *Biotechnol Bioeng* 102(4): 1182–1196.

Cirino PC et al. 2003. Generating mutant libraries using error-prone PCR. *Methods Mol Biol* 231: 3–9.

Clonis YD et al. 2000. Biomimetic dyes as affinity chromatography tools in enzyme purification. *J Chromatogr A* 891: 33–44.

Coco WM. 2003. RACHITT: gene family shuffling by random chimeragenesis on transient templates. *Methods Mol Biol* 231: 111–127.

Coco WM et al. 2001. DNA shuffling method for generating highly recombined genes and evolved enzymes. *Nat Biotechnol* 19: 354–359.

Contreras R et al. 1991. Efficient KEX2-like processing of a glucoamylase-interleukin-6 fusion protein by *Aspergillus nidulans* and secretion of mature interleukin-6. *Biotechnology (NY)* 9(4): 378–381.

Craik CS et al. 1985. Redesigning trypsin: alteration of substrate specificity. *Science* 228: 291–297.

Cramer CL et al. 1996. Bioproduction of human enzymes in transgenic tobacco. *Ann NY Acad Sci* 792: 62–71.

Crameri A et al. 1998. DNA shuffling of a family of genes from diverse species accelerates directed evolution. *Nature* 391: 288–291.

Cregg JM et al. 2000. Recombinant protein expression in *Pichia pastoris*. *Mol Biotechnol* 16(1): 23–52.

Cutler P. 2003. Protein arrays: the current state-of-the-art. *Proteomics* 3: 3–18.

Dai M, Copley SD. 2004. Genome shuffling improves degradation of the anthropogenic pesticide pentachlorophenol by *Sphingobium chlorophenolicum* ATCC 39723. *Appl Environ Microbiol* 70: 2391–2397.

Dalby PA. 2003. Optimising enzyme function by directed evolution. *Curr Opin Struct Biol* 13: 500–505.

Daniell H et al. 2001. Medical molecular farming: production of antibodies, biopharmaceuticals and edible vaccines in plants. *Trends Plant Sci* 6: 219–226.

De Muynck B et al. 2010. Production of antibodies in plants: status after twenty years. *Plant Biotechnol J* (in press).

Deiters A et al. 2003. Adding amino acids with novel reactivity to the genetic code of *Saccharomyces cerevisiae*. *J Am Chem Soc* 125: 11782–11783.

Demain AL, Vaishnav P. 2009. Production of recombinant proteins by microbes and higher organisms. *Biotechnol Adv* 27(3): 297–306.

DeSantis G et al. 2003. Creation of a productive, highly enantioselective nitrilase through gene site saturation mutagenesis (GSSM). *J Am Chem Soc* 125(38): 11476–11477.

DeYoung JL. 1989. Development of pancreatic enzyme microsphere technology and US findings with pancrease in the treatment of chronic pancreatitis. *Int J Pancreatol* 5 Suppl: 31–36.

Dingermann T et al. 1991. Expression of human antithrombin III in the cellular slime mould *Dictyostelium discoideum*. *Appl Microbiol Biotechnol* 35(4): 496–503.

Dittrich W et al. 1994. Production and secretion of recombinant proteins in *Dictyostelium discoideum*. *Biotechnology (NY)* 12: 614–618.

Dixon DP et al. 2003. Forced evolution of a herbicide detoxifying glutathione trasnsferase. *J Biol Chem* 278: 23930–23935.

Donayre-Torres AJ et al. 2009. Production and purification of immunologically active core protein p24 from HIV-1 fused to ricin toxin B subunit in *E. coli*. *Virol J* 6: 17.

Dracheva S et al. 1995. Expression of soluble human interleukin-2 receptor alpha-chain in *Escherichia coli*. *Protein Expr Purif* 6: 737–747.

Dravis BC et al. 2001. Haloalkane hydrolysis with an immobilized haloalkane dehalogenase. *Biotechnol Bioeng* 75: 416–423.

Dumon C et al. 2008. Engineering hyperthermostability into a GH11 xylanase is mediated by subtle changes to protein structure. *J Biol Chem* 283(33): 22557–22564.

Durocher Y, Butler M. 2009. Expression systems for therapeutic glycoprotein production. *Curr Opin Biotechnol* 20(6): 700–707.

Eblin LP. 1964. An introduction to molecular kinetic theory (Hildebrand, Joel H.). *J Chem Educ* 41(3): 171.

Ebrahimi F et al. 2010. Production and characterization of a recombinant chimeric antigen consisting botulinum neurotoxin serotypes A, B and E binding subdomains. *J Toxicol Sci* 35(1): 9–19.

Edmunds T et al. 1998. Transgenically produced human antithrombin: structural and functional comparison to human plasma-derived antithrombin. *Blood* 91: 4561–4571.

Eisenmesser EZ et al. 2002. Enzyme dynamics during catalysis. *Science* 295: 1520–1523.

Elias-Lopez AL et al. 2008. Transgenic tomato expressing interleukin-12 has a therapeutic effect in a murine model of progressive pulmonary tuberculosis. *Clin Exp Immunol* 154(1): 123–133.

Engelhardt EM et al. 2009. Suspension-adapted Chinese hamster ovary-derived cells expressing green fluorescent protein as a screening tool for biomaterials. *Biotechnol Lett* 31(8): 1143–1149.

Estape D et al. 1998. Susceptibility towards intramolecular disulphide-bond formation affects conformational stability and folding of human basic fibroblast growth factor. *Biochem J* 335(Pt 2): 343–349.

Fabian JR et al. 1998. Reconstitution and purification of eukaryotic initiation factor 2B (eIF2B) expressed in Sf21 insect cells. *Protein Expr Purif* 13(1): 16–22.

Fang Q, Shortle D. 2003. Prediction of protein structure by emphasizing local side-chain/backbone interactions in ensembles of turn fragments. *Proteins* 53(Suppl 6): 486–490.

Fausnaugh JL. 1990. Protein purification and analysis by liquid chromatography and electrophoresis. *Bioprocess Technol* 7: 57–84.

Faye L, Champey Y. 2008. Plants, medicine and genetics, which applications for tomorrow? *Med Sci (Paris)* 24(11): 939–945.

Fenyo D, Beavis RC. 2002. Informatics and data management in proteomics. *Trends Biotechnol* 20: S35–S38.

Fersht A. 1999 *Structure and Mechanism in Protein Science: a Guide to Enzyme Catalysis and Protein Folding*. W.H. Freeman, New York.

Finkelstein AV, Ptitsyn OB. 2002. Elements of thermodynamics. In: *Protein Physics*. Academic Press, New York, pp. 43–55.

Fischer R, Emans N. 2000. Molecular farming of pharmaceutical proteins. *Transgenic Res* 9: 279–299.

Fischer R et al. 1999d. Towards molecular farming in the future: transient protein expression in plants. *Biotechnol Appl Biochem* 30(Pt 2): 113–116.

Fischer R et al. 1999b. Towards molecular farming in the future: *Pichia pastoris*-based production of single-chain antibody fragments. *Biotechnol Appl Biochem* 30(Pt 2): 117–120.

Fischer R et al. 1999a. Towards molecular farming in the future: moving from diagnostic protein and antibody production in microbes to plants. *Biotechnol Appl Biochem* 30(Pt 2): 101–108.

type="Periodical">Fischer R et al. 1999c. Molecular farming of recombinant antibodies in plants. *Biol Chem* 380: 825–839.

Fischer R et al. 2000. Antibody production by molecular farming in plants. *J Biol Regul Homeost Agents* 14: 83–92.

Fox SR et al. 2004. Maximizing interferon-gamma production by Chinese hamster ovary cells through temperature shift optimization: experimental and modeling. *Biotechnol Bioeng* 85(2): 177–184.

Galecki P et al. 2006. Manganese superoxide dismutase gene (MnSOD) polimorphism in schizophrenics with tardive dyskinesia from central Poland. *Psychiatr Pol* 40: 937–948.

Garnier J et al. 1990. Secondary structure prediction and protein design. *Biochem Soc Symp* 57: 11–24.

Garvey EP, Matthews CR. 1990. Site-directed mutagenesis and its application to protein folding. *Biotechnology* 14: 37–63.

Gasteiger E et al. 2001. SWISS-PROT: connecting biomolecular knowledge via a protein database. *Curr Issues Mol Biol* 3: 47–55.

Gasteiger E et al. 2003. ExPASy: the proteomics server for in-depth protein knowledge and analysis. *Nucleic Acids Res* 31: 3784–3788.

Geisse S, Fux C. 2009. Recombinant protein production by transient gene transfer into mammalian cells. *Methods Enzymol* 463: 223–238.

Gerlt JA, Babbitt PC. 2009. Enzyme (re)design: lessons from natural evolution and computation. *Curr Opin Chem Biol* 13(1): 10–18.

Giga-Hama Y, Kumagai H. 1999. Expression system for foreign genes using the fission yeast *Schizosaccharomyces pombe*. *Biotechnol Appl Biochem* 30(Pt 3): 235–244.

Glasner ME et al. 2007. Mechanisms of protein evolution and their application to protein engineering. *Adv Enzymol Relat Area Mol Biol* 75: 193–239.

Gomord V, Faye L. 2004. Posttranslational modification of therapeutic proteins in plants. *Curr Opin Plant Biol* 7(2): 171–181.

Gouka RJ et al. 1997. Efficient production of secreted proteins by *Aspergillus*: progress, limitations and prospects. *Appl Microbiol Biotechnol* 47: 1–11.

Gratz A, Jose J. 2008. Protein domain library generation by overlap extension (PDLGO): a tool for enzyme engineering. *Anal Biochem* 378(2): 171–176.

Grimes BR, Monaco ZL. 2005. Artificial and engineered chromosomes: developments and prospects for gene therapy. *Chromosoma* 114(4): 230–241.

Guex N, Peitsch MC. 1997. SWISS-MODEL and the Swiss-PdbViewer: an environment for comparative protein modeling. *Electrophoresis* 18: 2714–2723.

Guex N et al. 2009. Automated comparative protein structure modeling with SWISS-MODEL and Swiss-PdbViewer: a historical perspective. *Electrophoresis* (Suppl 1): S162–S173.

Gutteridge A, Thornton J. 2004. Conformational change in substrate binding, catalysis and product release: an open and shut case? *FEBS Lett* 567: 67–73.

Hacker DL et al. 2009. 25 years of recombinant proteins from reactor-grown cells – where do we go from here? *Biotechnol Adv* 27(6): 1023–1027.

Hackney D. 1990. Binding energy and catalysis. In: *The Enzymes*, vol XIX. Academic Press, p. 136.

Halabian R et al. 2009. Expression and purification of recombinant human coagulation factor VII fused to a histidine tag using Gateway technology. *Blood Transfus* 7(4): 305–312.

Hamilton SR, Gerngross TU. 2007. Glycosylation engineering in yeast: the advent of fully humanized yeast. *Curr Opin Biotechnol* 18(5): 387–392.

Hamilton SR et al. 2003. Production of complex human glycoproteins in yeast. *Science* 301: 1244–1246.

Hammes GG. 2002. Multiple conformational changes in enzyme catalysis. *Biochemistry* 41: 8221–8228.

Han ZS et al. 2008. Adenoviral vector mediates high expression levels of human lactoferrin in the milk of rabbits. *J Microbiol Biotechnol* 18(1): 153–159.

Hasemann CA, Capra JD. 1990. High-level production of a functional immunoglobulin heterodimer in a baculovirus expression system. *Proc Natl Acad Sci USA* 87: 3942–3946.

Hassan S et al. 2008. Considerations for extraction of monoclonal antibodies targeted to different subcellular compartments in transgenic tobacco plants. *Plant Biotechnol J* 6(7): 733–748.

Hibbert EG, Dalby PA. 2005. Directed evolution strategies for improved enzymatic performance. *Microb Cell Fact* 4: 29.

Hitchman RB et al. 2009. Baculovirus expression systems for recombinant protein production in insect cells. *Recent Pat Biotechnol* 3(1): 46–54.

Hitzeroth A et al. 2007. Association between the MnSOD Ala-9Val polymorphism and development of schizophrenia and abnormal involuntary movements in the Xhosa population. *Prog Neuropsychopharmacol Biol Psychiatry* 31: 664–672.

Hodgson E. 2004. *A Textbook of Modern Toxicology*, 3rd edn. Wiley-Interscience, Hoboken, NJ, Chapter 7, pp. 111–148.

Holz C et al. 2003. Establishing the yeast *Saccharomyces cerevisiae* as a system for expression of human proteins on a proteome-scale. *J Struct Funct Genomics* 4: 97–108.

Holzer GW et al. 2003. Overexpression of hepatitis B virus surface antigens including the preS1 region in a serum-free Chinese hamster ovary cell line. *Protein Expr Purif* 29: 58–69.

Hooper DC. 2009. Plant vaccines: an immunological perspective. *Curr Top Microbiol Immunol* 332: 1–11.

Houdebine LM. 2009a. Genetically modified animals. Introduction. *Comp Immunol Microbiol Infect Dis* 32(2): 45–46.

Hsieh-Wilson LC et al. 1996. Insights into antibody catalysis: structure of an oxygenation catalyst at 1.9-angstrom resolution. *Proc Natl Acad Sci USA* 93: 5363–5367.

Humphreys DP. 2003. Production of antibodies and antibody fragments in *Escherichia coli* and a comparison of their functions, uses and modification. *Curr Opin Drug Discov Devel* 6: 188–196.

Idiris A et al. 2010. Engineering of protein secretion in yeast: strategies and impact on protein production. *Appl Microbiol Biotechnol* (in press).

Iguchi T et al. 2009. MnSOD genotype and prostate cancer risk as a function of NAT genotype and smoking status. *In Vivo* 23: 7–12.

Iizuka M et al. 2009. Production of a recombinant mouse monoclonal antibody in transgenic silkworm cocoons. *FEBS J* 276(20): 5806–5820.

Ikeno M et al. 1998. Construction of YAC-based mammalian artificial chromosomes. *Nat Biotechnol* 16: 431–439.

Jacobs PP, Callewaert N. 2009. *N*-glycosylation engineering of biopharmaceutical expression systems. *Curr Mol Med* 9(7): 774–800.

Jaeger KE, Eggert T. 2004. Enantioselective biocatalysis optimized by directed evolution. *Curr Opin Biotechnol* 15: 305–313.

Jahic M et al. 2006. Process technology for production and recovery of heterologous proteins with *Pichia pastoris*. *Biotechnol Prog* 22(6): 1465–1473.

Jana S, Deb JK. 2005. Strategies for efficient production of heterologous proteins in *Escherichia coli*. *Appl Microbiol Biotechnol* 67(3): 289–298.

Janne J et al. 1998. Transgenic bioreactors. *Biotechnol Annu Rev* 4: 55–74.

Jaroszewski L. 2009. Protein structure prediction based on sequence similarity. *Methods Mol Biol* 569: 129–156.

Jarvis DL. 2009. Baculovirus-insect cell expression systems. *Methods Enzymol* 463: 191–222.

Jestin JL, Kaminski PA. 2004. Directed enzyme evolution and selections for catalysis based on product formation. *J Biotechnol* 113: 85–103.

Johannes TW, Zhao H. 2006. Directed evolution of enzymes and biosynthetic pathways. *Current Opinion in Microbiology* 9: 261–267.

Joseph D et al. 1990. Anatomy of a conformational change: hinged "lid" motion of the triosephosphate isomerase loop. *Science* 249: 1425–1428.

Josephy PD, Mannervik B. 2006. *Molecular Toxicology*, 2nd edn. Oxford University Press, New York, Chapter 9, pp. 303–332.

Joshi L, Lopez LC. 2005. Bioprospecting in plants for engineered proteins. *Curr Opin Plant Biol* 8(2): 223–226.

Joubert O et al. 2010. Heterologous expression of human membrane receptors in the yeast *Saccharomyces cerevisiae*. *Methods Mol Biol* 601: 87–103.

Jung E, Williams KL. 1997. The production of recombinant glycoproteins with special reference to simple eukaryotes including *Dictyostelium discoideum*. *Biotechnol Appl Biochem* 25(Pt 1): 3–8.

Jung E et al. 1997. An in vivo approach for the identification of acceptor sites for *O*-glycosyltransferases: motifs for the addition of *O*-GlcNAc in *Dictyostelium discoideum*. *Biochemistry* 36(13): 4034–4040.

Kabsch W, Sander C. 1983. Dictionary of protein secondary structure: pattern recognition of hydrogen-bonded and geometrical features. *Biopolymers* 22: 2577–2637.

Kakkis ED et al. 1994. Overexpression of the human lysosomal enzyme alpha-L-iduronidase in Chinese hamster ovary cells. *Protein Expr Purif* 5: 225–232.

Kapoli P et al. 2008. Engineering sensitive glutathione transferase for the detection of xenobiotics. *Biosens Bioelectron* 24(3): 498–503.

Kaszubska W et al. 2000. Expression, purification, and characterization of human recombinant thrombopoietin in Chinese hamster ovary cells. *Protein Expr Purif* 18: 213–220.

Kaur J, Sharma R. 2006. Directed evolution: an approach to engineer enzymes. *Crit Rev Biotechnol* 26(3): 165–199.

Ke SH, Madison EL. 1997. Rapid and efficient site-directed mutagenesis by single-tube 'megaprimer' PCR method. *Nucleic Acids Res* 25: 3371–3372.

Kermode AR et al. 2007. Ectopic expression of a conifer abscisic acid insensitive 3 transcription factor induces high-level synthesis of recombinant human alpha-L-iduronidase in transgenic tobacco leaves. *Plant Mol Biol* 63(6): 763–776.

Kidd BA et al. 2009. Computation of conformational coupling in allosteric proteins. *PLoS Comput Biol* 5(8): e1000484.

Kim SH et al. 2008. Neutralization of hepatitis B virus (HBV) by human monoclonal antibody against HBV surface antigen (HBsAg) in chimpanzees. *Antiviral Res* 79(3): 188–191.

Kirk O et al. 2002. Industrial enzyme applications. *Curr Opin.Biotechnol* 13(4): 345–351.

Klimanskaya I et al. 2008. Derive and conquer: sourcing and differentiating stem cells for therapeutic applications. *Nat Rev Drug Discov* 7(2): 131–142.

Ko K et al. 2009. Production of antibodies in plants: approaches and perspectives. *Curr Top Microbiol Immunol* 332: 55–78.

Koehn J, Hunt I. 2009. High-throughput protein production (HTPP): a review of enabling technologies to expedite protein production. *Methods Mol Biol* 498: 1–18.

Kopp J, Schwede T. 2004. Automated protein structure homology modeling: a progress report. *Pharmacogenomics* 5: 405–416.

Kost TA, Condreay JP. 1999. Recombinant baculoviruses as expression vectors for insect and mammalian cells. *Curr Opin Biotechnol* 10: 428–433.

Kost TA, Condreay JP. 2002. Recombinant baculoviruses as mammalian cell gene-delivery vectors. *Trends Biotechnol* 20: 173–180.

Kotzia GA, Labrou NE. 2005. Cloning, expression and characterization of *Erwinia caratovora* L-asparaginase. *Journal of Biotechnology* 119(2005): 309–323.

Kotzia GA, Labrou NE. 2007. L-Asparaginase from *Erwinia chrysanthemi* 3937: cloning, expression and characterization. *J Biotechnol* 127(4): 657–669.

Kotzia GA, Labrou NE. 2009. Engineering thermal stability of L-asparaginase by *in vitro* directed evolution. *FEBS J* 276(6): 1750–1761.

Kotzia GA et al. 2006. Evolutionary methods in enzyme technology: high-throughput screening and selection of new enzyme variants. In: SG Pandalai (ed.) *Recent Research Developments in Biotechnology & Bioengineering*, vol. 7. Research Signpost, Kerala, pp. 85–104.

Kukuruzinska MA et al. 1987. Protein glycosylation in yeast. *Annu Rev Biochem* 56: 915–944.

Kurtzman AL et al. 2001. Advances in directed protein evolution by recursive genetic recombination: applications to therapeutic proteins. *Curr Opin Biotechnol* 12: 361–370.

Kwong KY, Rader C. 2009. *E. coli* expression and purification of Fab antibody fragments. *CurrProtocProtein Sci* Chapter 6: Unit 6.10.

La Flamme AC et al. 1995. Expression of mammalian cytokines by *Trypanosoma cruzi* indicates unique signal sequence requirements and processing. *Mol Biochem Parasitol* 75(1): 25–31.

Labrou NE. 2000. Dye-ligand affinity chromatography for protein separation and purification. *Methods Mol Biol* 147: 129–139.

Labrou NE. 2002. Affinity chromatography. In: MN Gupta (ed.) *Methods for Affinity-Based Separations of Enzymes and Proteins*. Birkhiuser Verlag AG, Switzerland, pp. 16–18.

Labrou NE. 2003. Design and selection of ligands for affinity chromatography. *J Chromatogr B Analyt Technol Biomed Life Sci* 790: 67–78.

Labrou NE. 2010. Random mutagenesis methods for *in vitro* directed enzyme evolution. *Curr Protein Pept* 11: 91–100.

Labrou NE, Rigden DJ. 2001. Active-site characterization of *Candida boidinii* formate dehydrogenase. *Biochem J* 354(Pt 2): 455–463.

Labrou N, Clonis YD. 1994. The affinity technology in downstream processing. *J Biotechnol* 36: 95–119.

Labrou NE, Clonis YD. 1995. Biomimetic dye affinity chromatography for the purification of bovine heart lactate dehydrogenase. *J Chromatogr A* 718: 35–44.

Labrou NE, Clonis YD. 1996. Biomimetic-dye affinity chromatography for the purification of mitochondrial L-malate dehydrogenase from bovine heart. *J Biotechnol* 45: 185–194.

Labrou NE, Clonis YD. 2002. Immobilised synthetic dyes in affinity chromatography. In: M.A. Vijayalakshmi (ed.) *Biochromatography – Theory and Practice*. Taylor and Francis Publishers, London, Chapter 8, pp. 235–251.

Labrou NE, Rigden DJ. 2001. Active-site characterization of *Candida boidinii* formate dehydrogenase. *Biochem J* 354: 455–463.

Labrou NE et al. 2004a. Engineering the pH-dependence of kinetic parameters of maize glutathione *S*-transferase I by site-directed mutagenesis. *Biomol Eng* 21: 61–66.

Labrou NE et al. 2001. The conserved Asn49 of maize glutathione *S*-transferase I modulates substrate binding, catalysis and intersubunit communication. *Eur J Biochem* 268: 3950–3957.

Labrou NE et al. 2005. Kinetic analysis of maize glutathione *S*-transferase I catalyzing the detoxification from chloroacetanilide herbicides. *Planta* 222: 91–97.

Labrou NE et al. 2004b. Dye-ligand and biomimetic affinity chromatography. In: DS Hage (ed.) *Handbook of Affinity Chromatography*. Marcel Dekker, New York.

Lai H et al. 2010. Monoclonal antibody produced in plants efficiently treats West Nile virus infection in mice. *Proc Natl Acad Sci USA* 107(6): 2419–2424.

Laidler KJ. 1984. The development of the Arrhenius equation. *J Chem Educ* 61(6): 494.

Larrick JW, Thomas DW. 2001. Producing proteins in transgenic plants and animals. *Curr Opin Biotechnol* 12(4): 411–418.

Lesley SA. 2001. High-throughput proteomics: protein expression and purification in the postgenomic world. *Protein Expr Purif* 22: 159–164.

Lesley SA et al. 2002. Gene expression response to misfolded protein as a screen for soluble recombinant protein. *Protein Eng* 15: 153–160.

Levison PR. 2003. Large-scale ion-exchange column chromatography of proteins. Comparison of different formats. *J Chromatogr B Analyt Technol Biomed Life Sci* 790: 17–33.

Lewin R. 1982. RNA can be a catalyst. *Science* 218: 872–874.

Liang JF et al. 2000. Biomedical application of immobilized enzymes. *J Pharm Sci* 89: 979–990.

Lienard D et al. 2007. Pharming and transgenic plants. *Biotechnol Annu Rev* 13: 115–147.

Lioy PJ, Greenberg A. 1990. Factors associated with human exposures to polycyclic aromatic hydrocarbons. *Toxicol Ind Health* 6: 209–223.

Liu D et al. 2009a. High-level expression and large-scale preparation of soluble HBx antigen from *Escherichia coli*. *Biotechnol Appl Biochem* 54(3): 141–147.

Liu L et al. 2009b. Directed evolution of an orthogonal nucleoside analog kinase via fluorescence-activated cell sorting. *Nucleic Acids Res* 37(13): 4472–4481.

Logan JS, Martin MJ. 1994. Transgenic swine as a recombinant production system for human hemoglobin. *Methods Enzymol* 231: 435–445.

Lowe CR. 2001. Combinatorial approaches to affinity chromatography. *Curr Opin Chem Biol* 5: 248–256.

Lubertozzi D, Keasling JD. 2009. Developing *Aspergillus* as a host for heterologous expression. *Biotechnol Adv* 27(1): 53–75.

Lue RY et al. 2004. Versatile protein biotinylation strategies for potential high-throughput proteomics. *J Am Chem Soc* 126: 1055–1062.

Lutz S et al. 2001. Rapid generation of incremental truncation libraries for protein engineering using alpha-phosphothioate nucleotides. *Nucleic Acids Res* 29: E16.

Madzak C et al. 2004. Heterologous protein expression and secretion in the non-conventional yeast *Yarrowia lipolytica*: a review. *J Biotechnol* 109(1–2): 63–81.

Madzak C et al. 2005. Heterologous production of a laccase from the basidiomycete *Pycnoporus cinnabarinus* in the dimorphic yeast *Yarrowia lipolytica*. *FEMS Yeast Res* 5(6–7): 635–646.

Maeda K, Sugiyama Y. 2008. Impact of genetic polymorphisms of transporters on the pharmacokinetic, pharmacodynamic and toxicological properties of anionic drugs. *Drug Metab Pharmacokinet* 23: 223–235.

Marti S et al. 2004. Theoretical insights in enzyme catalysis. *Chem Soc Rev* 33: 98–107.

Mason HS, Arntzen CJ. 1995. Transgenic plants as vaccine production systems. *Trends Biotechnol* 13: 388–392.

Matsushita-Morita M et al. 2009. Characterization of recombinant prolyl aminopeptidase from *Aspergillus oryzae*. *J Appl Microbiol* 109(10): 156–165.

Matsuura Y et al. 1987. Baculovirus expression vectors: the requirements for high level expression of proteins, including glycoproteins. *J Gen Virol* 68(Pt 5): 1233–1250.

Matthews BW. 1993. Structural and genetic analysis of protein stability. *Annu Rev Biochem* 62: 139–160.

McCarty KM et al. 2009. PAH–DNA adducts, cigarette smoking, GST polymorphisms, and breast cancer risk. *Environ Health Perspect* 117: 552–558.

McCormick DB. 1997. *Methods in Enzymology. Vitamins and coenzymes, Part L*. Academic Press, San Diego, p. 282.

Mcgrath BM. 2005. *Directory of Therapeutic Enzymes*, 1st edn. CRC Press.

Melissis S et al. 2006. Nucleotide-mimetic synthetic ligands for DNA-recognizing enzymes one-step purification of Pfu DNA polymerase. *J Chromatogr A* 1122(1–2): 63–75.

Melissis S et al. 2007. One-step purification of Taq DNA polymerase using nucleotide-mimetic affinity chromatography. *Biotechnol J* 2(1): 121–132.

Mohan U, Banerjee UC. 2008. Molecular evolution of a defined DNA sequence with accumulation of mutations in a single round by a dual approach to random chemical mutagenesis (DuARCheM). *Chembiochem* 9(14): 2238–2243.

Monaco ZL, Moralli D. 2006. Progress in artificial chromosome technology. *Biochem Soc Trans* 34(Pt 2): 324–327.

Moss DW. 1988. Theoretical enzymology: enzyme kinetics and enzyme inhibition. In: DL Williams, V Marks (eds.) *Principles of Clinical Biochemistry*, 2nd edn. Bath Press, Bath, UK, pp. 423–440.

Murzin AG et al. 1995. SCOP: a structural classification of proteins database for the investigation of sequences and structures. *J Mol Biol* 247: 536–540.

Nam HJ et al. 2009. Bioinformatic approaches for the structure and function of membrane proteins. *BMB Rep* 42(11): 697–704.

Nevalainen KM et al. 2005. Heterologous protein expression in filamentous fungi. *Trends Biotechnol* 23(9): 468–474.

Niemann H, Kues WA. 2007. Transgenic farm animals: an update. *Reprod Fertil Dev* 19(6): 762–770.

Nishida CR, Ortiz de Montellano PR. 2001. Control of electron transfer in nitric-oxide synthases. Swapping of autoinhibitory elements among nitric-oxide synthase isoforms. *J Biolog Chem* 276(23): 20116–20124.

Nixon AE et al. 1998. Hybrid enzymes: manipulating enzyme design. *Trends Biotechnol* 16: 258–264.

Okino N et al. 2010. Expression, purification, and characterization of a recombinant neutral ceramidase from *Mycobacterium tuberculosis*. *Biosci Biotechnol Biochem* 74(2): 316–321.

Orengo CA et al. 1997. CATH – a hierarchic classification of protein domain structures. *Structure* 5: 1093–1108.

Otero JM, Nielsen J. 2010. Industrial systems biology. *Biotechnol Bioeng* 105(3): 439–460.

Ozturk DH, Erickson-Viitanen S. 1998. Expression and purification of HIV-I p15NC protein in *Escherichia coli*. *Protein Expr Purif* 14: 54–64.

Panda AK. 2003. Bioprocessing of therapeutic proteins from the inclusion bodies of *Escherichia coli*. *Adv Biochem Eng Biotechnol* 85: 43–93.

Panke S, Wubbolts MG. 2002. Enzyme technology and bioprocess engineering. *Curr Opin Biotechnol* 13(2): 111–116.

Park HS et al. 2006. Design and evolution of new catalytic activity with an existing protein scaffold. *Science* 311: 535–538.

Peti W, Page R. 2007. Strategies to maximize heterologous protein expression in *Escherichia coli* with minimal cost. *Protein Expr Purif* 51(1): 1–10.

Phadke RS. 1992. Biosensors and enzyme immobilized electrodes. *Biosystems* 27: 203–206.

Piefer AJ, Jonsson CB. 2002. A comparative study of the human T-cell leukemia virus type 2 integrase expressed in and purified from *Escherichia coli* and *Pichia pastoris*. *Protein Expr Purif* 25: 291–299.

Platis D, Foster GR. 2003. High yield expression, refolding, and characterization of recombinant interferon alpha2/alpha8 hybrids in *Escherichia coli*. *Protein Expr Purif* 31: 222–230.

Podgornik A, Tennikova TB. 2002. Chromatographic reactors based on biological activity. *Adv Biochem Eng Biotechnol* 76: 165–210.

Powell LW. 1984. Developments in immobilized-enzyme technology. *Biotechnol Genet Eng Rev* 2: 409–438.

Price N, Stevens L. 1999. *Fundamentals of Enzymology*, 3rd edn. Oxford University Press, New York, pp. 79–92.

Prunkard D et al. 1996. High-level expression of recombinant human fibrinogen in the milk of transgenic mice. *Nat Biotechnol* 14(7): 867–871.

Qu C et al. 2000. 3D domain swapping modulates the stability of members of an icosahedral virus group. *Struct Fold Des* 8: 1095–1103.

Qu X et al. 2009. A guide to template based structure prediction. *Curr Protein Pept Sci* 10(3): 270–285.

Rabhi-Essafi I et al. 2007. A strategy for high-level expression of soluble and functional human interferon alpha as a GST-fusion protein in *E. coli*. *Protein Eng Des Sel* 20(5): 201–209.

Rakintzis TN. 1994. *Physical Chemistry, First Thermodynamic Law*. Chapter C, pp. 71–103.

Reetz MT. 2007. Controlling the selectivity and stability of proteins by new strategies in directed evolution: the case of organocatalytic enzymes. *Ernst Schering Found Symp Proc* 2: 321–340.

Regnier FE. 1987. Chromatography of complex protein mixtures. *J Chromatogr* 418: 115–143.

Richardson JS. 1981. The anatomy and taxonomy of protein structure. *Adv Protein Chem* 34: 167–339.

Rommel GT, Richard BK. 2002. Pharmacogenomics of drug transporters. In: J Licinio, M Wong (eds.) *Pharnacogenomics: The Search for Individualized Therapies*. Wiley VCH, Weinheim, Chapter 9, pp. 179–213.

Roodveldt C et al. 2005. Directed evolution of proteins for heterologous expression and stability. *Curr Opin Struct Biol* 15: 50–56.

Rosic NN. 2009. Versatile capacity of shuffled cytochrome P450s for dye production. *Appl Microbiol Biotechnol* 82(2): 203–210.

Roy P et al. 1994. Long-lasting protection of sheep against bluetongue challenge after vaccination with virus-like particles: evidence for homologous and partial heterologous protection. *Vaccine* 12: 805–811.

Rudolph NS. 1999. Biopharmaceutical production in transgenic livestock. *Trends Biotechnol* 17: 367–374.

Russell DA. 1999. Feasibility of antibody production in plants for human therapeutic use. *Curr Top Microbiol Immunol* 240: 119–138.

Ryu DD, Nam DH. 2000. Recent progress in biomolecular engineering. *Biotechnol Prog* 16: 2–16.

Sahdev S et al. 2008. Production of active eukaryotic proteins through bacterial expression systems: a review of the existing biotechnology strategies. *Mol Cell Biochem* 307(1–2): 249–264.

Sala F et al. 2003. Vaccine antigen production in transgenic plants: strategies, gene constructs and perspectives. *Vaccine* 21: 803–808.

Saven JG. 2002. Combinatorial protein design. *Curr Opin Struct Biol* 12(4): 453–458.

Saxena A et al. 2009. The basic concepts of molecular modeling. *Methods Enzymol* 467: 307–334.

Schatz SM et al. 2003. Higher expression of Fab antibody fragments in a CHO cell line at reduced temperature. *Biotechnol Bioeng* 84: 433–438.

Schena M et al. 1995. Quantitative monitoring of gene expression patterns with a complementary DNA microarray. *Science* 270: 467–470.

Schena M et al. 1998. Microarrays: biotechnology's discovery platform for functional genomics. *Trends Biotechnol* 16: 301–306.

Schillberg S et al. 2003a. 'Molecular farming' of antibodies in plants. *Naturwissenschaften* 90: 145–155.

Schillberg S et al. 2003b. Molecular farming of recombinant antibodies in plants. *Cell Mol Life Sci* 60: 433–445.

Schillberg S et al. 2005. Opportunities for recombinant antigen and antibody expression in transgenic plants–technology assessment. *Vaccine* 23(15): 1764–1769.

Schmidt M et al. 2009. Protein engineering of carboxyl esterases by rational design and directed evolution. *Protein Pept Lett* 16(10): 1162–1171.

Schmidt-Dannert C et al. 2000. Molecular breeding of carotenoid biosynthetic pathways. *Nat Biotechnol* 18: 750–753.

Schwede T et al. 2003. SWISS-MODEL: an automated protein homology–modeling server. *Nucleic Acids Res* 31: 3381–3385.

Scopes RK. 1987. Dye-ligands and multifunctional adsorbents: an empirical approach to affinity chromatography. *Anal Biochem* 165: 235–246.

Scrutton NS et al. 1990. Redesign of the coenzyme specificity of a dehydrogenase by protein engineering. *Nature* 343: 38–43.

Sgaramella V, Eridani S. 2004. From natural to artificial chromosomes: an overview. *Methods Mol Biol* 240: 1–12.

Shao Z et al. 1998. Random-priming in vitro recombination: an effective tool for directed evolution. *Nucleic Acids Res* 26: 681–683.

Shimada T. 2006. Xenobiotic-metabolizing enzymes involved in the activation and detoxification of carcinogenic polycyclic aromatic hydrocarbons. *Drug Metab Pharmacokinet* 24: 257–276.

Shimoda-Matsubayashi S et al. 1996. Structural dimorphism in the mitochondrial targeting sequence in the human manganese superoxide dismutase gene. A predictive evidence for conformational change to influence mitochondrial transport and a study of allelic association in Parkinson's disease. *Biochem Biophys Res Commun* 226: 561–565.

Sibanda BL et al. 1989. Conformation of beta-hairpins in protein structures. A systematic classification with applications to modelling by homology, electron density fitting and protein engineering. *J Mol Biol* 206: 759–777.

Singh A et al. 1995. Construction and characterization of a chimeric beta-glucosidase. *Biochem J* 305(Pt 3): 715–719.

Siraj AK et al. 2008. Polymorphisms of selected xenobiotic genes contribute to the development of papillary thyroid cancer susceptibility in Middle Eastern population. *BMC Medical Genetics* 9: 61–69.

Slade MB et al. 1997. Expression of recombinant glycoproteins in the simple eukaryote *Dictyostelium discoideum*. *Biotechnol Genet Eng Rev* 14: 1–35.

Soleimani M et al. 2007. Expression of human tissue plasminogen activator in the trypanosomatid protozoan *Leishmania tarentolae*. *Biotechnol Appl Biochem* 48(Pt 1): 55–61.

Sourrouille C et al. 2009. From Neanderthal to nanobiotech: from plant potions to pharming with plant factories. *Methods Mol Biol* 483: 1–23.

Stein H et al. 2009. Production of bioactive, post-translationally modified, heterotrimeric, human recombinant type-I collagen in transgenic tobacco. *Biomacromolecules* 10(9): 2640–2645.

Steiner S, Anderson NL. 2000. Pharmaceutical proteomics. *Ann NY Acad Sci* 919: 48–51.

Stemmer WP. 1994. Rapid evolution of a protein *in vitro* by DNA shuffling. *Nature* 370: 389–391.

Stoger E et al. 2002. Plantibodies: applications, advantages and bottlenecks. *Curr Opin Biotechnol* 13: 161–166.

Su D et al. 2010. Glycosylation-modified erythropoietin with improved half-life and biological activity. *Int J Hematol* 91(2): 238–244.

Suen KF et al. 2010. Transient expression of an IL-23R extracellular domain Fc fusion protein in CHO vs. HEK cells results in improved plasma exposure. *Protein Expr Purif* 71(1): 96–102.

Sunley K et al. 2008. CHO cells adapted to hypothermic growth produce high yields of recombinant beta-interferon. *Biotechnol Prog* 24(4): 898–906.

Surewicz WK, Mantsch HH. 1988. New insight into protein secondary structure from resolution-enhanced infrared spectra. *Biochim Biophys Acta* 952: 115–130.

Swanson ME et al. 1992. Production of functional human hemoglobin in transgenic swine. *Biotechnology (NY)* 10: 557–559.

Swartz JR. 2001. Advances in *Escherichia coli* production of therapeutic proteins. *Curr Opin Biotechnol* 12: 195–201.

Takahashi R, Ueda M. 2010. Generation of transgenic rats using YAC and BAC DNA constructs. *Methods Mol Biol* 597: 93–108.

Takahashi T et al. 2010. Binding of sulfatide to recombinant hemagglutinin of influenza A virus produced by a baculovirus protein expression system. *J Biochem* 174(4): 459–462.

Takegawa K et al. 2009. Production of heterologous proteins using the fission-yeast (*Schizosaccharomyces pombe*) expression system. *Biotechnol Appl Biochem* 53(Pt 4): 227–235.

Tamini RM et al. 2004. Manganese superoxide dismutase polymorphism, plasma antioxidants, cigarette smoking, and risk of breast cancer. *Cancer Epidemiol Biomarkers Prev* 13: 989–996.

Tang W et al. 2006. Expression, purification and identification of recombinant mouse interleukin 21 protein in *E. coli*. *Cell Mol Immunol* 3(4): 311–315.

Tao H, Cornish VW. 2002. Milestones in directed enzyme evolution. *Curr Opin Chem Biol* 6: 858–864.

Terpe K. 2003. Overview of tag protein fusions: from molecular and biochemical fundamentals to commercial systems. *Appl Microbiol Biotechnol* 60: 523–533.

Terpitz U et al. 2008. Electrofused giant protoplasts of *Saccharomyces cerevisiae* as a novel system for electrophysiological studies on membrane proteins. *Biochim Biophys Acta* 1778(6): 1493–1500.

The UniProt Consortium. 2008. The universal protein resource (UniProt). *Nucleic Acids Res* 36: D190–D195.

Tiltscher H, Storr M. 1993. Immobilization of the slime mould *Dictyostelium discoideum* for the continuous production of recombinant human antithrombin III. *Appl Microbiol Biotechnol* 40(2–3): 246–250.

Tiwari S et al. 2009. Plants as bioreactors for the production of vaccine antigens. *Biotechnol Adv* 27(4): 449–467.

Tobin JF et al. 1993. Transfected *Leishmania* expressing biologically active IFN-gamma. *J Immunol* 150(11): 5059–5069.

Tracewell CA, Arnold FH. 2009. Directed enzyme evolution: climbing fitness peaks one amino acid at a time. *Curr Opin Chem Biol* 13(1): 3–9.

Trometer C, Falson P. 2010. Mammalian membrane protein expression in baculovirus-infected insect cells. *Methods Mol Biol* 601: 105–117.

Trueman LJ. 1995. Heterologous expression in yeast. *Methods Mol Biol* 49: 341–354.

Turner NJ. 2009. Directed evolution drives the next generation of biocatalysts. *Nat Chem Biol* 5(8): 567–573.

Twyman RM et al. 2003. Molecular farming in plants: host systems and expression technology. *Trends Biotechnol* 21: 570–578.

Twyman RM et al. 2005. Transgenic plants in the biopharmaceutical market. *Expert Opin Emerg Drugs* 10(1): 185–218.

Umeno D, Arnold FH. 2004. Evolution of a pathway to novel long-chain carotenoids. *J Bacteriol* 186: 1531–1536.

Vahedi F et al. 2009. Characterization of *Bacillus anthracis* spores isolates from soil by biochemical and multiplex PCR analysis. *East Mediterr Health J* 15(1): 149–156.

van Beilen JB, Li Z. 2002. Enzyme technology: an overview. *Curr Opin Biotechnol* 13(4): 338–344.

van Berkel PH et al. 2002. Large scale production of recombinant human lactoferrin in the milk of transgenic cows. *Nat Biotechnol* 20: 484–487.

van den Hombergh JP et al. 1997. Aspergillus as a host for heterologous protein production: the problem of proteases. *Trends Biotechnol* 15: 256–263.

van Ooyen AJ et al. 2006. Heterologous protein production in the yeast *Kluyveromyces lactis*. *FEMS Yeast Res* 6(3): 381–392.

Velander WH et al. 1992. Production of biologically active human protein C in the milk of transgenic mice. *Ann NY Acad Sci* 665: 391–403.

Verdoes JC et al. 1993. Glucoamylase overexpression in *Aspergillus niger*: molecular genetic analysis of strains containing multiple copies of the glaA gene. *Transgenic Res* 2: 84–92.

Vitale A, Pedrazzini E. 2005. Recombinant pharmaceuticals from plants: the plant endomembrane system as bioreactor. *Mol Interv* 5(4): 216–225.

Wada M et al. 2003. Directed evolution of *N*-acetylneuraminic acid aldolase to catalyze enantiomeric aldol reactions. *Bioorg Med Chem* 11: 2091–2098.

Wagner CR, Benkovic SJ. 1990. Site directed mutagenesis: a tool for enzyme mechanism dissection. *Trends Biotechnol* 8: 263–270.

Waldman JM et al. 1991. Analysis of human exposure to benzo(a)pyrene via inhalation and food ingestion in the Total Human Environmental Exposure Study (THEES). *J Expo Anal Environ Epidemiol* 1: 193–225.

Walsh G. 2003. Pharmaceutical biotechnology products approved within the European Union. *Eur J Pharm Biopharm* 55: 3–10.

Wang CY et al. 1999. *N*-Acetyltransferase expression and DNA binding of *N*-hydroxyheterocyclic amines in human prostate epithelium. *Carcinogenesis* 20: 1591–1595.

Wang L, Schultz PG. 2002. Expanding the genetic code. *Chem Commun (Camb.)* 1: 1–11.

Wang L et al. 2005. Bioprocessing strategies to improve heterologous protein production in filamentous fungal fermentations. *Biotechnol Adv* 23(2): 115–129.

Ward M et al. 1990. Improved production of chymosin in *Aspergillus* by expression as a glucoamylase-chymosin fusion. *Biotechnology (NY)* 8: 435–440.

Ward PP et al. 1995. A system for production of commercial quantities of human lactoferrin: a broad spectrum natural antibiotic. *Biotechnology (NY)* 13: 498–503.

Wardell AD et al. 1999. Characterization and mutational analysis of the helicase and NTPase activities of hepatitis C virus full-length NS3 protein. *J Gen Virol* 80(Pt 3): 701–709.

Watson JV, Dive C. 1994. Enzyme kinetics. *Methods Cell Biol* 41: 469–507.

Wells JA et al. 1987. Designing substrate specificity by protein engineering of electrostatic interactions. *Proc Natl Acad Sci USA* 84: 1219–1223.

Werner RG et al. 2007. Glycosylation of therapeutic proteins in different production systems. *Acta Paediatr* 96(455): 17–22.

Wharton CW. 1983. Some recent advances in enzyme kinetics. *Biochem Soc Trans* 11: 817–825.

Wilkins MR et al. 1999. Protein identification and analysis tools in the ExPASy server. *Methods Mol Biol* 112: 531–552.

Williams GJ et al. 2004. Directed evolution of enzymes for biocatalysis and the life sciences. *Cell Mol Life Sci* 61: 3034–3046.

Wolfenden R. 2003. Thermodynamic and extrathermodynamic requirements of enzyme catalysis. *Biophys Chem* 105: 559–572.

Wong TS et al. 2006. The diversity challenge in directed protein evolution. *Comb Chem High Throughput Screen* 9(4): 271–288.

Wu ZL et al. 2009. Enhanced bacterial expression of several mammalian cytochrome P450s by codon optimization and chaperone coexpression. *Biotechnol Lett* 31(10): 1589–1593.

Xie S et al. 2009. Emerging affinity-based techniques in proteomics. *Expert Rev Proteomics* 6(5): 573–583.

Xiong H et al. 2009. *E. coli* expression of a soluble, active single-chain antibody variable fragment containing a nuclear localization signal. *Protein Expr Purif* 66(2): 172–180.

Xu C et al. 2002. CYP2A6 genetic variation and potential consequences. *Adv Drug Deliv Rev* 54: 1245–1256.

Yang H et al. 2009. Recombinant human antithrombin expressed in the milk of non-transgenic goats exhibits high efficiency on rat DIC model. *J Thromb Thrombolysis* 28(4): 449–457.

Yang P et al. 2008. Cattle mammary bioreactor generated by a novel procedure of transgenic cloning for large-scale production of functional human lactoferrin. *PLoS One* 3(10): e3453.

Yen MR et al. 2009. Bioinformatic analyses of transmembrane transport: novel software for deducing protein phylogeny, topology, and evolution. *J Mol Microbiol Biotechnol* 17(4): 163–176.

Yeo DS et al. 2004. Strategies for immobilization of biomolecules in a microarray. *Comb Chem High Throughput Screen* 7: 213–221.

Yeung PK. 2000. Technology evaluation: transgenic antithrombin III (rhAT-III), Genzyme Transgenics. *Curr Opin Mol Ther* 2: 336–339.

Yuan L et al. 2005. Laboratory-directed protein evolution. *Microbiol Mol Biol Rev* 69: 373–392.

Yurimoto H, Sakai Y. 2009. Methanol-inducible gene expression and heterologous protein production in the methylotrophic yeast *Candida boidinii*. *Biotechnol Appl Biochem* 53(Pt 2): 85–92.

Zaks A. 2001. Industrial biocatalysis. *Curr Opin Chem Biol* 5: 130–136.

Zeng QK et al. 2009. Reversal of coenzyme specificity and improvement of catalytic efficiency of *Pichia stipitis* xylose reductase by rational site-directed mutagenesis. *Biotechnol Lett* 31(7): 1025–1029.

Zerbs S et al. 2009. Bacterial systems for production of heterologous proteins. *Methods Enzymol* 463: 149–168.

Zhang R et al. 2009. Ser67Asp and His68Asp substitutions in *Candida parapsilosis* carbonyl reductase alter the coenzyme specificity and enantioselectivity of ketone reduction. *Appl Envir Micr* 25(7): 2176–2183.

Zhang Y et al. 2009a. Genetic variations in xenobiotic metabolic pathway genes, personal hair dye use, and risk of non-Hodgkin lymphoma. *Am J Epidemiol* 170: 1222–1230.

Zhang Y et al. 2009b. Bioinformatics analysis of microarray data. *Methods Mol Biol* 573: 259–284.

Zhang Z et al. 2004. A new strategy for the synthesis of glycoproteins. *Science* 303(5656): 371–373.

Zhang ZJ et al. 2003. Interaction between polymorphisms of the dopamine D3 receptor and manganese superoxide dismutase genes in susceptibility to tardive dyskinesia. *Psychiatr Genet* 13: 187–192.

Zhao P et al. 2010. Heterologous expression, purification, and biochemical characterization of a greenbug (*Schizaphis graminum*) acetylcholinesterase encoded by a paralogous gene (ace-1). *J Biochem Mol Toxicol* 24(1): 51–59.

Zhao H et al. 1998. Molecular evolution by staggered extension process (StEP) in vitro recombination. *Nat Biotechnol* 16: 258–261.

Zhuge J et al. 2004. Stable expression of human cytochrome P450 2D6*10 in HepG2 cells. *World J Gastroenterol* 10(2): 234–237.

Zoonens M, Miroux B. 2010. Expression of membrane proteins at the *Escherichia coli* membrane for structural studies. *Methods Mol Biol* 601: 49–66.

Zweiger G, Scott RW. 1997. From expressed sequence tags to 'epigenomics': an understanding of disease processes. *Curr Opin Biotechnol* 8: 684–687.

8
Enzyme Activities

D. J. H. Shyu, J. T. C. Tzen, and C. L. Jeang

This manuscript was reviewed by Dr. Hsien-Yi Sung, Department of Biochemical Science and Technology, National Taiwan University, Taipei, Taiwan, Republic of China, and Dr. Wai-Kit Nip, Department of Molecular Biosciences and Bioengineering, University of Hawaii at Manoa, Honolulu, Hawaii, USA.

INTRODUCTION

Long before human history, our ancestors, chimpanzees, might have already experienced the mild drunk feeling of drinking wine when they ate the fermented fruits that contained small amounts of alcohol. Archaeologists have also found some sculptured signs on 8000-year-old plates describing the beer-making processes. Chinese historians wrote about the lavish life of Tsou, a tyrant of Shang dynasty (around 1100 BC), who lived in a castle with wine-storing pools. All these indicated that people grasped the wine fermentation technique thousands of years ago. In the Chin dynasty (around 220 BC) China, a spicy paste or sauce made from fermentation of soybean and/or wheat was mentioned. It is now called soybean sauce.

Although people applied fermentation techniques and observed the changes from raw materials to special products, they did not realize the mechanisms that caused the changes. This mystery was uncovered by the development of biological sciences. Louis Pasteur claimed the existence and function of organisms that were responsible for the changes. In the same era, Justus Liebig observed the digestion of meat with pepsin, a substance found in stomach fluid, and proposed that a whole organism is not necessary for the process of fermentation.

In 1878, Kuhne was the first person to solve the conflict by introducing the word "enzyme," which means "in yeast" in Greek.

Food Biochemistry and Food Processing, Second Edition. Edited by Benjamin K. Simpson, Leo M.L. Nollet, Fidel Toldrá, Soottawat Benjakul, Gopinadhan Paliyath and Y.H. Hui.

This word emphasizes the materials inside or secreted by the organisms to enhance the fermentation (changes of raw material). Buchner (1897) performed fermentation by using a cellfree extract from yeast of significantly high light, and these were molecules rather than lifeforms that did the work. In 1905, Harden and Young found that fermentation was accelerated by the addition of some small dialyzable molecules to the cellfree extract. The results indicated that both macromolecular and micromolecular compounds are needed. However, the nature of enzymes was not known at that time. A famous biochemist, Willstátter, was studying peroxidase for its high catalytic efficiency; since he never got enough samples, even though the reaction obviously happened, he hesitated (denied) to conclude that the enzyme was a protein. Finally, in 1926, Summer prepared crystalline urease from jack beans, analyzed the properties of the pure compound, and drew the conclusion that enzymes are proteins. In the following years, crystalline forms of some proteases were also obtained by Northrop et al. The results agreed with Summer's conclusion.

The studies on enzyme behavior were progressing in parallel. In 1894, Emil Fischer proposed a "lock and key" theory to describe the specificity and stereo relationship between an enzyme and its substrate. In 1902, Herri and Brown independently reported a saturation-type curve for enzyme reactions. They revealed an important concept in which the enzyme substrate complex was an obligate intermediate formed during the enzyme-catalyzed reaction. In 1913, Michaelis and Menten derived an equation describing quantitatively the saturation-like behavior. At the same time, Monod and others studied the kinetics of regulatory enzymes and suggested a concerted model for the enzyme reaction. In 1959, Fischer's hypothesis was slightly modified by Koshland. He proposed an induced-fit theory to describe the moment the enzyme and substrate are attached, and suggested a sequential model for the action of allosteric enzymes.

For the studies on enzyme structure, Sanger et al. were the first to announce the unveiling of the amino acid sequence of a protein, insulin. After that, the primary sequences of some hydrolases with comparatively small molecular weights, such as ribonuclease, chymotrypsin, lysozyme, and others, were defined. Twenty years later, Sanger won his second Nobel Prize for the establishment of the chain-termination reaction method for nucleotide sequencing of DNA. On the basis of this method, the deduction of the primary sequences of enzymes blossomed; to date, the primary structures of 55,410 enzymes have been deduced. Combining genetic engineering technology and modern computerized X-ray crystallography and/or Nucleai magnetic resonance (NMR), about 15,000 proteins, including 9268 proteins and 2324 enzymes, have been analyzed for their three-dimensional structures (Protein Data Bank (PDB): http://www.rcsb.org/pdb). Computer software was created, and protein engineering on enzymes with demanded properties was carried on successfully (Fang and Ford 1998, Igarashi et al. 1999, Pechkova et al. 2003, Shiau et al. 2003, Swiss-PDBViewer(spdbv): http://www.expasy.ch/spdbv/mainpage.htm, SWISSMODELserver: http://www.expasy.org/swissmod/SWISS-MODEL.htm).

FEATURES OF ENZYMES

Most of the Enzymes are Proteins

Proteins are susceptible to heat, strong acids and bases, heavy metals, and detergents. They are hydrolyzed to amino acids by heating in acidic solution and by the proteolytic action of enzymes on peptide bonds. Enzymes give positive results on typical protein tests, such as the Biuret, Millions, Hopkins-Cole, and Sakaguchi reactions. X-ray crystallographic studies revealed that there are peptide bonds between adjacent amino acid residues in proteins. The majority of the enzymes fulfill the above criterion; therefore, they are proteins in nature. However, the catalytic element of some well-known ribozymes is just RNAs in nature (Steitz and Moore 2003, Raj and Liu 2003).

Chemical Composition of Enzymes

For many enzymes, protein is not the only component required for its full activity. On the basis of the chemical composition of enzymes, they are categorized into several groups, as follows:

1. Polypeptide, the only component, for example, lysozyme, trypsin, chymotrypsin, or papain.
2. Polypeptide plus one to two kinds of metal ions, for example, α-amylase containing Ca^{2+}, kinase containing Mg^{2+}, and superoxide dismutase having Cu^{2+} and/or Zn^{2+}.
3. Polypeptide plus a prosthetic group, for example, peroxidase containing a heme group.
4. Polypeptide plus a prosthetic group and a metal ion, for example, cytochrome oxidase ($a + a_3$) containing a heme group and Cu^{2+}.
5. Polypeptide plus a coenzyme, for example, many dehydrogenases containing NAD^+ or $NADP^+$.
6. Combination of polypeptide, coenzyme, and a metal ion, for example, succinate dehydrogenase containing both the FAD and nonheme iron.

Enzymes are Specific

In the life cycle of a unicellular organism, thousands of reactions are carried out. For the multicellular higher organisms with tissues and organs, even more kinds of reactions are progressing in every moment. Less than 1% of errors that occur in these reactions will cause accumulation of waste materials (Drake 1999), and sooner or later the organism will not be able to tolerate these accumulated waste materials. These phenomena can be explained by examining genetic diseases; for example, phenylketonuria (PKU) in humans, where the malfunction of phenylalanine hydroxylase leads to an accumulation of metabolites such as phenylalanine and phenylacetate, and others, finally causing death (Scriver 1995, Waters 2003). Therefore, enzymes catalyzing the reactions bear the responsibility for producing desired metabolites and keeping the metabolism going smoothly.

Enzymes have different types of specificities. They can be grouped into the following common types:

1. *Absolute specificity:* For example, urease (Sirko and Brodzik 2000) and carbonate anhydrase (Khalifah 2003) catalyze only the hydrolysis and cleavage of urea and carbonic acid, respectively. Those enzymes having small molecules as substrates or that work on biosynthesis pathways belong to this category.
2. *Group specificity:* For example, hexokinase prefers D-glucose, but it also acts on other hexoses (Cardenas et al. 1998).
3. *Stereo specificity:* For example, D-amino acid oxidase reacts only with D-amino acid, but not with L-amino acid; and succinate dehydrogenase reacts only with the fumarate (*trans* form), but not with maleate (*cis* form).
4. *Bond specificity:* For example, many digestive enzymes. They catalyze the hydrolysis of large molecules of food components, such as proteases, amylases, and lipases. They seem to have broad specificity in their substrates; for example, trypsin acts on all kinds of denatured proteins in the intestine, but prefers the basic amino acid residues at the *C*-terminal of the peptide bond. This broad specificity brings up an economic effect in organisms; they are not required to produce many digestive enzymes for all kinds of food components.

The following two less common types of enzyme specificity are impressive: (1) An unchiral compound, citric acid is formed by citrate synthase with the condensation of oxaloacetate and acetyl-CoA. An aconitase acts only on the moiety that comes from oxaloacetate. This phenomenon shows that, for aconitase, the citric acid acts as a chiral compound (Barry 1997). (2) If people notice the rare mutation that happens naturally, they will be impressed by the fidelity of certain enzymes, for example, amino acyltRNA synthethase (Burbaum and Schimmel 1991, Cusack 1997), RNA polymerase (Kornberg 1996), and DNA polymerase (Goodman 1997); they even correct the accidents that occur during catalyzing reactions.

Enzymes are Regulated

The catalytic activity of enzymes is changeable under different conditions. These enzymes catalyze the set steps of a metabolic pathway, and their activities are responsible for the states of cells. Intermediate metabolites serve as modulators on enzyme activity; for example, at high concentrations of adenosine diphosphate, the catalytic activity of the phosphofructokinase, pyruvate kinase, and pyruvate dehydrogenase complex are enhanced; as adenosine triphosphate becomes high, they will be inhibited. These modulators bind at allosteric sites on enzymes (Hammes 2002). The structure and mechanism of allosteric regulatory enzymes such as aspartate transcarbamylase (Cherfils et al. 1990) and ribonucleoside diphosphate reductase (Scott et al. 2001) have been well studied. However, allosteric regulation is not the only way that the enzymatic activity is influenced. Covalent modification by protein kinases (Langfort et al. 1998) and phosphatases (Luan 2003) on enzymes will cause large fluctuations in total enzymatic activity in the metabolism pool. In addition, sophisticated tuning phenomena on glycogen phosphorylase and glycogen synthase through phosphorylation and dephosphorylation have been observed after hormone signaling (Nuttall et al. 1988, Preiss and Romeo 1994).

Enzymes are Powerful Catalysts

The compound glucose will remain in a bottle for years without any detectable changes. However, when glucose is applied in a minimal medium as the only carbon source for the growth of *Salmonella typhimurium* (or *Escherichia coli*), phosphorylation of this molecule to glucose-6-phosphate is the first chemical reaction that occurs as it enters the cells. From then on, the activated glucose not only serves as a fuel compound to be oxidized to produce chemical energy, but also goes through numerous reactions to become the carbon skeleton of various micro- and macrobiomolecules. To obtain each end product, multiple steps have to be carried out. It may take less than 30 minutes for a generation to go through all the reactions. Only the existence of enzymes guarantees this quick utilization and disappearance of glucose.

ENZYMES AND ACTIVATION ENERGY

Enzymes Lower the Activation Energy

Enzymes are mostly protein catalysts; except for the presence of a group of ribonucleic acid-mediated reactions, they are responsible for the chemical reactions of metabolism in cells. For the catalysis of a reaction, the reactants involved in this reaction all require sufficient energy to cross the potential energy barrier, the activation energy (E_A), for the breakage of the chemical bonds and the start of the reaction. Few have enough energy to cross the reaction energy barrier until the reaction catalyst (the enzyme) forms a transition state with the reactants to lower the E_A (Fig. 8.1). Thus, the enzyme lowers the barrier that usually prevents the chemical reaction from occurring and facilitates the reaction proceeding at a faster rate to approach equilibrium. The substrates (the reactants), which are specific for the enzyme involved in the reaction, combine with enzyme to lower the E_A of the reaction. One enzyme type will combine with one specific type of substrate, that is, the active site of one particular enzyme will only fit one specific type of substrate, and this leads to the formation of and enzyme–substrate (ES) complex. Once the energy barrier is overcome, the substrate is then changed to another chemical, the product. It should be noted that the enzyme itself is not consumed or altered by the reaction, and that the overall free-energy change (ΔG) or the related equilibrium also remain constant. Only the activation energy of the uncatalyzed reaction (E_{Au}) decreases to that of the enzyme-catalyzed reaction (E_{Ac}).

One of the most important mechanisms of the enzyme function that decreases the E_A involves the initial binding of the enzyme to the substrate, reacting in the correct direction, and closing of the catalytic groups of the ES complex. The binding energy, part of the E_A, is required for the enzyme to bind to the substrate and is determined primarily by the structure complementarities

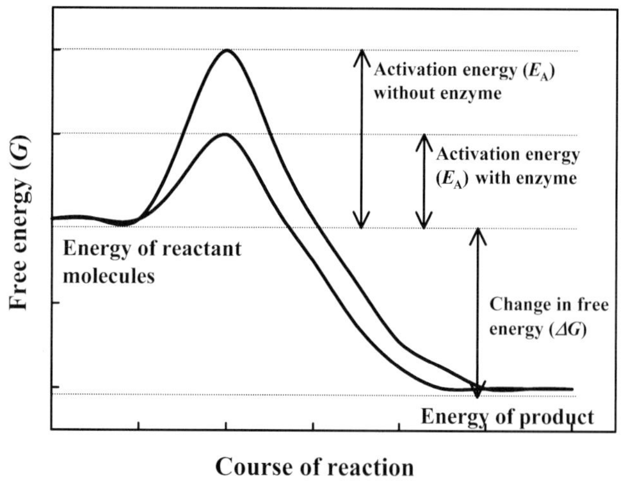

Figure 8.1. A transition state diagram, also called reaction coordinate diagram, shows the activation energy profile of the course of an enzyme-catalyzed reaction. One curve depicts the course of the reaction in the absence of the enzyme, while the other depicts the course of reaction in the presence of the enzyme that facilitates the reaction. The difference in the level of free energy between the beginning state (ground state) and the peak of the curve (transition state) is the activation energy of the reaction. The presence of enzyme lowers the activation energy (E_A) but does not change the overall free energy (ΔG).

between the enzyme and the substrate. The binding energy is used to reduce the free energy of the transition-state ES complex but not to form a stable, not easily separated, complex. For a reaction to occur, the enzyme acts as a catalyst by efficiently binding to its substrate, lowering the energy barrier, and allowing the formation of product. The enzyme stabilizes the transition state of the catalyzed reaction, and the transition state is the rate-limiting state in a single-step reaction. In a two-step reaction, the step with the highest transition-state free energy is said to be the rate-limiting state. Though the reaction rate is the speed at which the reaction proceeds toward equilibrium, the speed of the reaction does not affect the equilibrium point.

How Does an Enzyme work?

An enzyme carries out the reaction by temporarily combining with its specific kind of substrate, resulting in a slight change of their structures to produce a very precise fit between the two molecules. The introduction of strains into the enzyme and substrate shapes allows more binding energy to be available for the transition state. Two models of this minor structure modification are the induced fit, in which the binding energy is used to distort the shape of the enzyme, and the induced strain, in which binding energy is used to distort the shape of the substrate. These two models, in addition to a third model, the nonproductive binding model, have been proposed to describe conformational flexibility during the transition state. The chemical bonds of the substrate break and form new ones, resulting in the formation of new product. The newly formed product is then released from the enzyme, and the enzyme combines with another substrate for the next reaction.

ENZYME KINETICS AND MECHANISM

Regulatory Enzymes

For the efficient and precise control of metabolic pathways in cells, the enzyme, which is responsible for a specific step in a serial reaction, must be regulated.

Feedback Inhibition

In living cells, a series of chemical reactions in a metabolic pathway occurs, and the resulting product of one reaction becomes the substrate of the next reaction. Different enzymes are usually responsible for each step of catalysis. The final product of each metabolic pathway will inhibit the earlier steps in the serial reactions. This is called feedback inhibition.

Noncompetitive Inhibition

An allosteric site is a specific location on the enzyme where an allosteric regulator, a regulatory molecule that does not directly block the active site of the enzyme, can bind. When the allosteric regulator attaches to the specific site, it causes a change in the conformation of enzyme active site, and the substrate therefore will not fit into the active site of enzyme; this results in the inhibition phenomenon. The enzyme cannot catalyze the reaction, and not only the amount of final product but also the concentration of the allosteric regulator decrease. Since the allosteric regulator (the inhibitor) does not compete with the substrate for binding to the active site of enzyme, the inhibition mechanism is called noncompetitive inhibition.

Competitive Inhibition

A regulatory molecule is not the substrate of the specific enzyme but shows affinity for attaching to the active site. When it occupies the active site, the substrate cannot bind to the enzyme for the catalytic reaction, and the metabolic pathway is inhibited. Since the regulatory molecule will compete with the substrate for binding to the active site on enzyme, this inhibitory mechanism of the regulatory molecule is called competitive inhibition.

Enzyme Kinetics

During the course of an enzyme-catalyzed reaction, the plot of product formation over time (product formation profile) reveals an initial rapid increase, approximately linear, of product, and then the rate of increase decreases to zero as time passes (Fig. 8.2A). The slope of initial rate (v_i or v), also called the steady state rate, appears to be initially linear in a plot where the product concentration versus reaction time follows the

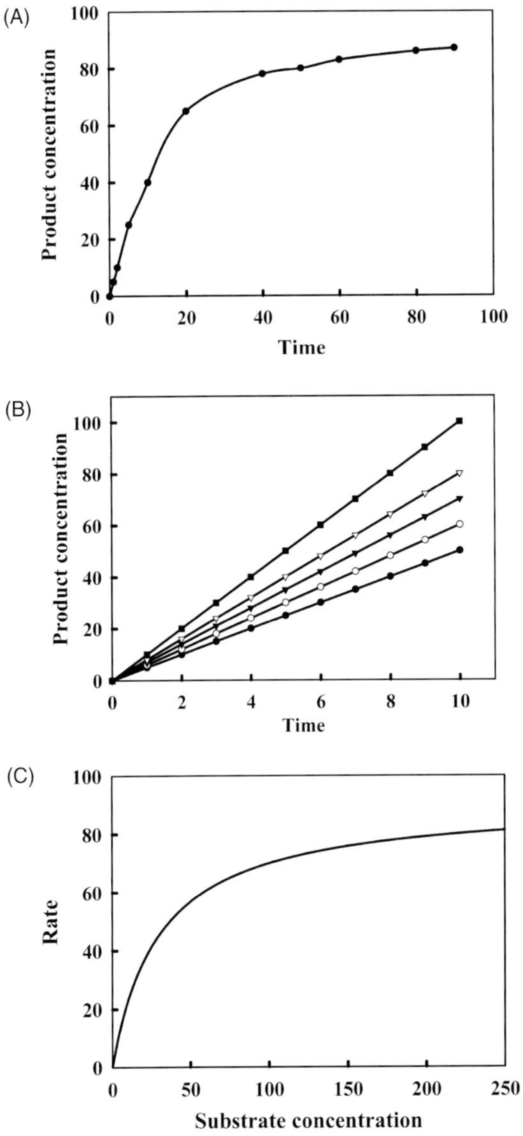

Figure 8.2. (A) Plot of progress curve during an enzyme-catalyzed reaction for product formation. **(B)** Plot of progress curves during an enzyme-catalyzed reaction for product formation with a different starting concentration of substrate. **(C)** Plot of reaction rate as a function of substrate concentration measured from slopes of lines from **(B)**.

establishment of the steady state of a reaction. When the reaction is approaching equilibrium or the substrate is depleted, the rate of product formation decreases, and the slope strays away from linearity. As is known by varying the conditions of the reaction, the kinetics of a reaction will appear linear during a time period in the early phase of the reaction, and the reaction rate is measured during this phase. When the substrate concentrations are varied, the product formation profiles will display linear substrate dependency at this phase (Fig. 8.2B). The rates for each substrate concentration are measured as the slope of a plot of product formation over time. A plot of initial rate as a function of substrate concentration is demonstrated as shown in Figure 8.2C. Three distinct portions of the plot will be noticed: an initial linear relationship showing first-order kinetic at low substrate concentrations; an intermediate portion showing curved linearity dependent on substrate concentration, and a final portion revealing no substrate concentration dependency, i.e., zero-order kinetic, at high substrate concentrations. The interpretation of the phenomenon can be described by the following scheme:

> The initial rate will be proportional to the ES complex concentration if the total enzyme concentration is constant and the substrate concentration varies. The concentration of the ES complex will be proportional to the substrate concentration at low substrate concentrations, and the initial rate will show a linear relationship and substrate concentration dependency. All of the enzyme will be in the ES complex form when substrate concentration is high, and the rate will depend on the rate of ES transformation into enzyme-product and the subsequent release of product. Adding more substrate will not influence the rate, so the relationship of rate versus substrate concentration will approach zero (Brown 1902).

However, a lag phase in the progress of the reaction will be noticed in coupled assays due to slow or delayed response of the detection machinery (see below). Though v can be determined at any constant concentration of substrate, it is recommended that the value of substrate concentration approaching saturation (high substrate concentration) be used so that it can approach its limiting value V_{max} with greater sensitivity and prevent errors occurring at lower substrate concentrations.

Measurement of the rate of enzyme reaction as a function of substrate concentration can provide information about the kinetic parameters that characterize the reaction. Two parameters are important for most enzymes—K_m, an approximate measurement of the tightness of binding of the substrate to the enzyme; and V_{max}, the theoretical maximum velocity of the enzyme reaction. To calculate these parameters, it is necessary to measure the enzyme reaction rates at different concentrations of substrate, using a variety of methods, and analyze the resulting data. For the study of single-substrate kinetics, one saturating concentration of substrate in combination with varying substrate concentrations can be used to investigate the initial rate (v), the catalytic constant (k_{cat}), and the specific constant (k_{cat}/K_m). For investigation of multisubstrate kinetics, however, the dependence of v on the concentration of each substrate has to be determined, one after another; that is, by measuring the initial rate at varying concentrations of one substrate with fixed concentrations of other substrates.

A large number of enzyme-catalyzed reactions can be explained by the Michaelis-Menten equation:

$$v = \frac{V_{max}[S]}{(K_m + [S])} = \frac{k_{cat}[E][S]}{(K_m + [S])},$$

where [E] and [S] are the concentrations of enzyme and substrate, respectively (Michaelis and Menten 1913). The equation describes the rapid equilibrium that is established between the reactant molecules (E and S) and the ES complex, followed by slow formation of product and release of free enzyme. It assumes that $k_2 \ll k_{-1}$ in the following equation, and K_m is the value of [S] when $v = V_{max}/2$:

$$E + S \underset{k_{-1}}{\overset{k_1}{\rightleftarrows}} ES \xrightarrow{k_2} E + P.$$

Briggs and Haldane (1924) proposed another enzyme kinetic model, called the steady state model; steady state refers to a period of time when the rate of formation of ES complex is the same as rate of formation of product and release of enzyme. The equation is commonly the same as that of Michaelis and Menten, but it does not require $k_2 \ll k_{-1}$, and $K_m = (k_{-1} + k_2)/k_1$, because [ES] now is dependent on the rate of formation of the ES complex (k_1) and the rate of dissociation of the ES complex (k_{-1} and k_2). Only under the condition that $k_2 \ll k_{-1}$ will $K_s = k_{-1}/k_1$ and be equivalent to K_m, the dissociation constant of the ES complex $\{K_s = k_{-1}/k_1 = ([E] - [ES])[S]/[ES]\}$, otherwise $K_m > K_s$.

The K_m is the substrate concentration at half of the maximum rate of enzymatic reaction, and it is equal to the substrate concentration at which half of the enzyme active sites are saturated by the substrate in the steady state. So, K_m is not a useful measure of ES binding affinity in certain conditions when K_m is not equivalent to K_s. Instead, the specific constant (k_{cat}/K_m) can be substituted as a measure of substrate binding; it represents the catalytic efficiency of the enzyme. The $k_{cat,}$ the catalytic constant, represents conversion of the maximum number of reactant molecules to product per enzyme active site per unit time, or the number of turnover events occurring per unit time; it also represents the turnover rate whose unit is the reciprocal of time. In the Michaelis-Menten approach, the dissociation of the EP complex is slow, and the k_{cat} contribution to this rate constant will be equal to the dissociation constant. However, in the Briggs-Haldane approach, the dissociation rate of ES complex is fast, and the k_{cat} is equal to k_2. In addition, the values of k_{cat}/K_m are used not only to compare the efficiencies of different enzymes but also to compare the efficiencies of different substrates for a specified enzyme.

Deviations of expected hyperbolic reaction are occasionally found due to such factors as experimental artifacts, substrate inhibition, existence of two or more enzymes competing for the same substrate, and the cooperativity. They show nonhyperbolic behavior that cannot fit well into the Michaelis-Menten approach. For instance, a second molecule binds to the ES complex and forms an inactive ternary complex that usually occurs at high substrate concentrations and is noticed when rate values are lower than expected. It leads to breakdown of the Michaelis-Menten equation. It is not only the Michaelis-Menten plot that is changed to show a nonhyperbolic behavior; it is also the Lineweaver–Burk plot that is altered to reveal the nonlinearity of the curve (see later).

A second example of occasionally occurring nonhyperbolic reactions is the existence of more than one enzyme in a reaction that competes for the same substrate; the conditions usually are realized when crude or partially purified samples are used. Moreover, the conformational change in the enzymes may be induced by ligand binding for the regulation of their activities, as discussed earlier in this chapter; many of these enzymes are composed of multimeric subunits and multiple ligand-binding sites in cases that do not obey the Michaelis-Menten equation. The sigmoid, instead of the hyperbolic, curve is observed; this condition, in which the binding of a ligand molecule at one site of an enzyme may influence the affinities of other sites, is known as cooperativity. The increase and decrease of affinities of enzyme binding sites are the effects of positive and negative cooperativity, respectively. The number of potential substrate binding sites on the enzymes can be quantified by the Hill coefficient, h, and the degree of cooperativity can be measured by the Hill equation (see page 164). More detailed illustrations and explanations of deviations from Michaelis-Menten behavior can be found in the literatures (Bell and Bell 1988, Cornish-Bowden 1995).

Data Presentation

Untransformed Graphics

The values of K_m and V_{max} can be determined graphically using nonlinear plots by measuring the initial rate at various substrate concentrations and then transforming them into different kinetic plots. First, the substrate stock concentration can be chosen at a reasonably high level; then, a two-fold (stringently) or five- or ten-fold (roughly) serial dilution can be made from this stock. After the data obtained has been evaluated, a Michaelis-Menten plot of rate v_i as a function of substrate concentration [S] is subsequently drawn, and the values of both K_m and V_{max} can be estimated. Through the plot, one can check if a hyperbolic curve is observed, and if the values estimated are meaningful. Both values will appear to be infinite if the range of substrate concentrations is low; on the other hand, although the value of V_{max} can be approximately estimated, that of K_m cannot be determined if the range of substrate concentrations is too high. Generally, substrate concentrations covering 20–80% of V_{max}, which corresponds to a substrate concentration of 0.25–5.0 K_m, is appreciated.

Lineweaver–Burk Plots

Though nonlinear plots are useful in determining the values of K_m and V_{max}, transformed, linearized plots are valuable in determining the kinetics of multisubstrate enzymes and the interaction between enzymes and inhibitors. One of the best known plots is the double-reciprocal or Lineweaver–Burk plot (Lineweaver and Burk 1934). Inverting both sides of the Michaelis-Menten equation gives the Lineweaver–Burk plot: $1/v = (K_m/V_{max})(1/[S]) + 1/V_{max}$ (Fig. 8.3). The equation is for a linear curve when one plots $1/v$ against 1/[S] with a slope of K_m/V_{max} and a y intercept of $1/V_{max}$. Thus, the kinetic values can also be determined from the slope and intercept values of the plot. However, caution should be taken due to small experimental errors that may be amplified by taking the reciprocal (the distorting effect), especially in the measurement of v values at low substrate concentrations. The problem can be solved by preparing a series of substrate concentrations evenly spaced along the 1/[S] axis, that

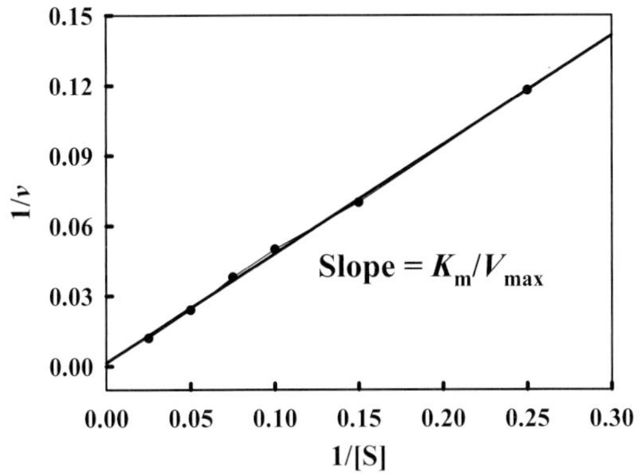

Figure 8.3. The Lineweaver–Burk double-reciprocal plot.

is, diluting the stock solution of substrate by two-, three-, four-, five-, . . . fold (Copeland 2000).

Eadie–Hofstee Plots

The Michaelis-Menten equation can be rearranged to give $v = -K_m\ (v/[S]) + V_{max}$ by first multiplying both sides by $K_m + [S]$, and then dividing both sides by [S]. The plot will give a linear curve with a slope of $-K_m$ and a y intercept of v when v is plotted against $v/[S]$ (Fig. 8.4). The Eadie–Hofstee plot has the advantage of decompressing data at high substrate concentrations and being more accurate than the Lineweaver–Burk plot, but the values of v against [S] are more difficult to determine (Eadie 1942, Hofstee 1959). It also has the advantage of observing the range of v from zero to V_{max} and is the most effective plot for revealing deviations from the hyperbolic reaction when two or more components, each of which follows the Michaelis-Menten equation, although errors in v affect both coordinates (Cornish-Bowden 1996).

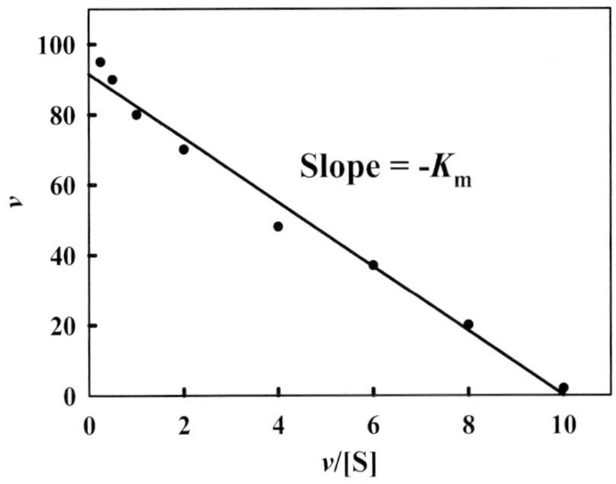

Figure 8.4. The Eadie–Hofstee plot.

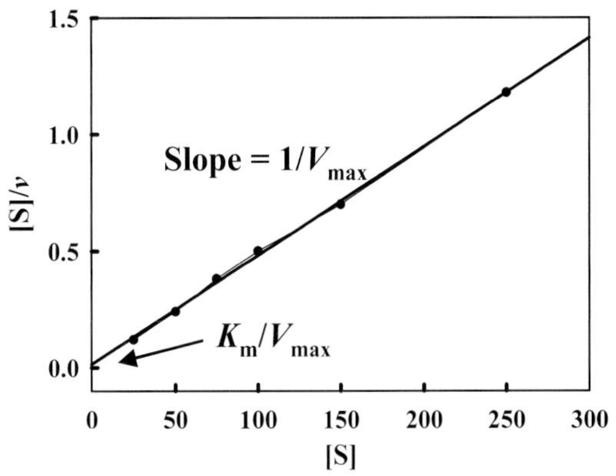

Figure 8.5. The Hanes–Wolff plot.

Hanes–Wolff Plots

The Lineweaver–Burk equation can be rearranged by multiplying both sides by [S]. This gives a linear plot $[S]/v = (1/V_{max})[S] + K_m/V_{max}$, with a slope of $1/V_{max}$, the y intercept of K_m/V_{max}, and the x intercept of $-K_m$ when $[S]/v$ is plotted as a function of [S] (Fig. 8.5). The plot is useful in obtaining kinetic values without distorting the data (Hanes 1932).

Eisenthal–Cornish-Bowden Plots (the Direct Linear Plots)

As in the Michaelis-Menten plot, pairs of rate values of v and the negative substrate concentrations [S] are applied along the y and x axis, respectively. Each pair of values is a linear curve that connects two points of values and extrapolates to pass the intersection. The location (x, y) of the point of intersection on the two-dimensional plot represents K_m and V_{max}, respectively (Fig. 8.6). The direct linear plots (Eisenthal and Cornish-Bowden

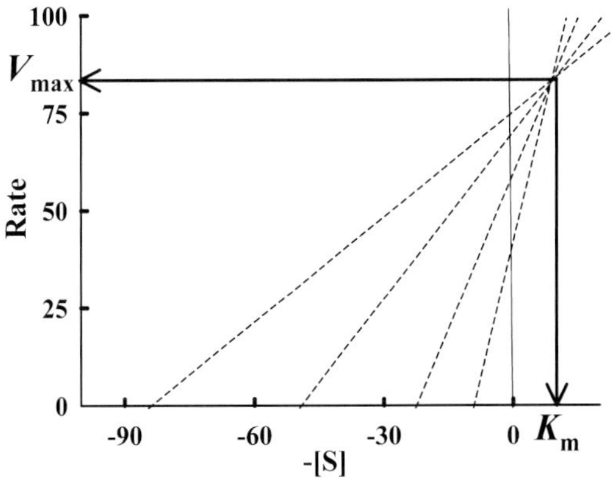

Figure 8.6. The Eisenthal–Cornish-Bowden direct plot.

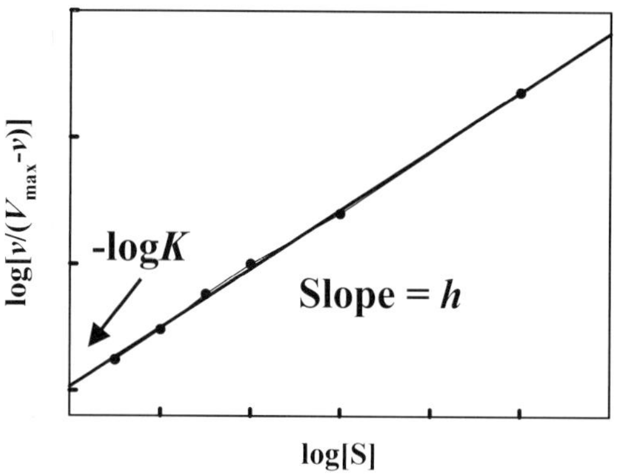

Figure 8.7. The Hill plot.

1974) are useful in estimating the Michaelis-Menten kinetic parameters and are the better plots for revealing deviations from the reaction behavior.

Hill Plot

The Hill equation is a quantitative analysis measuring non–Michaelis-Menten behavior of cooperativity (Hill 1910). In a case of completely cooperative binding, an enzyme contains h binding sites that are all occupied simultaneously with a dissociation constant $K = [\mathrm{E}][\mathrm{S}]^h/[\mathrm{ES}_h]$, where h is the Hill coefficient, which measures the degree of cooperativity. If there is no cooperativity, $h = 1$; there is positive cooperativity when $h > 1$, negative cooperativity when $h < 1$. The rate value can be expressed as $v = V_{\mathrm{max}}[\mathrm{S}]^h/(K + [\mathrm{S}]^h)$, and the Hill equation gives a linear $\log(v_i/V_{\mathrm{max}} - v) = h \log[\mathrm{S}] - \log K$, when $\log(v/V_{\mathrm{max}} - v)$ is plotted as a function of log[S] with a slope of h and a y intercept of $-\log K$ (Fig. 8.7). However, the equation will show deviations from linearity when outside a limited range of substrate concentrations, that is, in the range of $[\mathrm{S}] = K$.

FACTORS AFFECTING ENZYME ACTIVITY

As discussed earlier in this chapter, the rate of an enzymatic reaction is very sensitive to reaction conditions such as temperature, pH, ionic strength, buffer constitution, and substrate concentration. To investigate the catalytic mechanism and the efficiency of an enzyme with changes in the parameters, and to evaluate a suitable circumstance for assaying the enzyme activity, the conditions should be kept constant to allow reproducibility of data and to prevent a misleading interpretation.

Enzyme, Substrate, and Cofactor Concentrations

In general, the substrate concentration should be kept much higher than that of enzyme in order to prevent substrate concentration dependency of the reaction rate at low substrate concentrations. The rate of an enzyme-catalyzed reaction will also show linearity to the enzyme concentration when substrate concentration is constant. The reaction rate will not reveal the linearity until the substrate is depleted, and the measurement in initial rate will be invalid. Preliminary experiments must be performed to determine the appropriate range of substrate concentrations over a number of enzyme concentrations to prevent the phenomenon of substrate depletion. In addition, the presence of inhibitor, activator, and cofactor in the reaction mixture also influence enzyme activity. The enzymatic activity will be low or undetectable in the presence of inhibitors or in the absence of activators or cofactors. In the case of cofactors, the activated enzymes will be proportional to the cofactor concentration added in the reaction mixture if enzymes are in excess. The rate of reaction will not represent the total amount of enzyme when the cofactor supplement is not sufficient, and the situation can be avoided by adding excess cofactors (Tipton 1992). Besides, loss of enzyme activity may be seen before substrate depletion, as when the enzyme concentration is too low and leads to the dissociation of the dimeric or multimeric enzyme. To test this possibility, addition of the same amount of enzyme can be applied to the reaction mixture to measure if there is the appearance of another reaction progression curve.

Effects of pH

Enzymes will exhibit maximal activity within a narrow range of pH and vary over a relatively broad range. For an assay of enzymatic catalysis, the pH of the solution of the reaction mixture must be maintained in an optimal condition to avoid pH-induced protein conformational changes, which lead to diminishment or loss of enzyme activity. A buffered solution, in which the pH is adjusted with a component with $\mathrm{p}K_{\mathrm{a}}$ at or near the desired pH of the reaction mixture, is a stable environment that provides the enzyme with maximal catalytic efficiency. Thus, the appropriate pH range of an enzyme must be determined in advance when optimizing assay conditions.

To determine if the enzymatic reaction is pH dependent and to study the effects of group ionizations on enzyme kinetics, the rate as a function of substrate concentration at different pH conditions can be measured to simultaneously obtain the effects of pH on the kinetic parameters. It is known that the ionizations of groups are of importance, either for the active sites of the enzyme involved in the catalysis or for maintaining the active conformation of the enzyme. Plots of k_{cat}, K_{m}, and $k_{\mathrm{cat}}/K_{\mathrm{m}}$ values at varying pH ranges will reflect important information about the roles of groups of the enzyme. The effect of pH dependence on the value of k_{cat} reveals the steps of ionization of groups involved in the ES complex and provides a $\mathrm{p}K_{\mathrm{a}}$ value for the ES complex state; the value of K_{m} shows the ionizing groups that are essential to the substrate-binding step before the reaction. Moreover, the $k_{\mathrm{cat}}/K_{\mathrm{m}}$ value reveals the ionizing groups that are essential to both the substrate-binding and the ES complex–forming steps, and provides the $\mathrm{p}K_{\mathrm{a}}$ value of the free reactant molecules state (Brocklehurst and Dixon 1977, Copeland 2000). From a plot of the $k_{\mathrm{cat}}/K_{\mathrm{m}}$ value on the logarithmic scale as a function of pH,

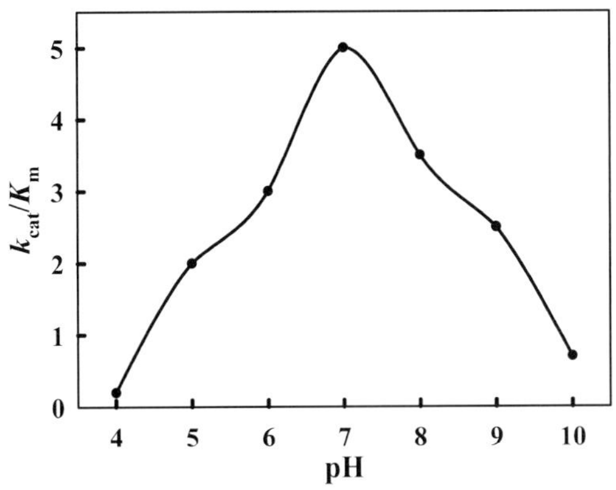

Figure 8.8. The effect of pH on the k_{cat}/K_m value of an enzymatic reaction that is not associated with the Henderson-Hasselbalch equation.

the pK_a value can be determined from the point of intersection of lines on the plot, especially for an enzymatic reaction that is not associated with the Henderson-Hasselbalch equation $pH = pK_a + \log([A^-]/[HA])$ (Fig. 8.8). Thus, the number of ionizing groups involved in the reaction can be evaluated (Dixon and Webb 1979, Tipton and Webb 1979).

Effects of Temperature

Temperature is one of the important factors affecting enzyme activity. For a reaction to occur at room temperature without the presence of an enzyme, small proportions of reactant molecules must have sufficient energy levels to participate in the reaction (Fig. 8.9A). When the temperature is raised above room temperature, more reactant molecules gain enough energy to be involved in the reaction (Fig. 8.9B). The E_A is not changed, but the distribution of energy-sufficient reactants is shifted to a higher average energy level. When an enzyme is participating in the reaction, the E_A is lowered significantly, and the proportion of reactant molecules at an energy level above the activation energy is also greatly increased (Fig. 8.9C). That means the reaction will proceed at a much higher rate.

Most enzyme-catalyzed reactions are characterized by an increase in the rate of reaction and increased thermal instability of the enzyme with increasing temperature. Above the critical temperature, the activity of the enzyme will be reduced significantly; while within this critical temperature range, the enzyme activity will remain at a relatively high level, and inactivation of the enzyme will not occur. Since the rate of reaction increases due to the increased temperature by lowering the activation energy E_A, the relationship can be expressed by the Arrhenius equation: $k = A\exp(-E_A/RT)$, where A is a constant related to collision probability of reactant molecules, R is the ideal gas constant (1.987 cal/mol—deg), T is the temperature in degrees Kelvin (K = °C + 273.15), and k represents the specific rate constant for any rate, that is, k_{cat} or V_{max}. The equation can be

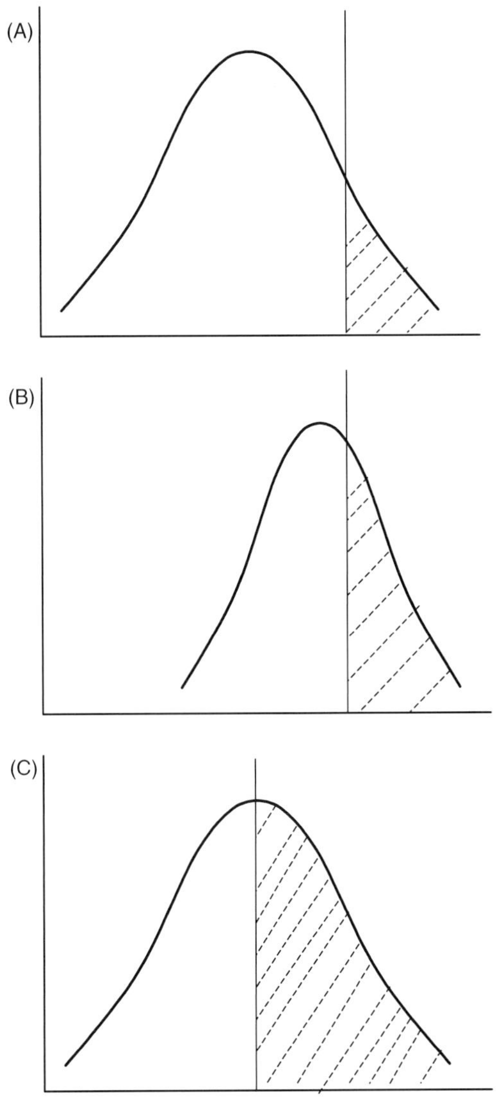

Figure 8.9. Plots of temperature effect on the energy levels of reactant molecules involved in a reaction. **(A)** The first plot depicts the distribution of energy levels of the reactant molecules at room temperature without the presence of an enzyme. **(B)** The second plot depicts that at the temperature higher than room temperature but in the absence of an enzyme. **(C)** The third plot depicts that at room temperature in the presence of an enzyme. The vertical line in each plot indicates the required activation energy level for a reaction to occur. The shaded portion of distribution in each plot indicates the proportion of reactant molecules that have enough energy levels to be involved in the reaction. The *x*-axis represents the energy level of reactant molecules, while the *y*-axis represents the frequency of reactant molecules at an energy level.

transformed into: $\ln K = \ln A - E_A/RT$, where the plot of $\ln K$ against $1/T$ usually shows a linear relationship with a slope of $-E_A/R$, where the unit of E_A is cal/mol. The calculated E_A of a reaction at a particular temperature is useful in predicting the E_A of the reaction at another temperature. And the plot is useful in judging if there is a change in the rate-limiting step of a sequential reaction when the line of the plot reveals bending of different slopes at certain temperatures (Stauffer 1989).

Taken together, the reaction should be performed under a constantly stable circumstance, with both temperature and pH precisely controlled from the start to the finish of the assay, for the enzyme to exhibit highly specific activity at appropriate acidbase conditions and buffer constitutions. Whenever possible, the reactant molecules should be equilibrated at the required assay condition following addition of the required components and efficient mixing to provide a homogenous reaction mixture. Because the enzyme to be added is usually stored at low temperature, the reaction temperature will not be significantly influenced when the enzyme volume added is at as low as 1–5% of the total volume of the reaction mixture. A lag phase will be noticed in the rate measurement as a function of time when the temperature of the added enzyme stock eventually influences the reaction. Significant temperature changes in the components stored in different environments and atmospheres should be avoided when the reaction mixture is mixed and the assay has started.

METHODS USED IN ENZYME ASSAYS

General Considerations

A suitable assay method is not only a prerequisite for detecting the presence of enzyme activity in the extract or the purified material: it is also an essential vehicle for kinetic study and inhibitor evaluation. Selection of an assay method that is appropriate to the type of investigation and the purity of the assaying material is of particular importance. It is known that the concentration of the substrate will decrease and that of product will increase when the enzyme is incubated with its substrate. Therefore, an enzyme assay is intended to measure either the decrease in substrate concentration or the increase in product concentration as a function of time. Usually, the latter is preferred because a significant increase of signal is much easier to monitor. It is recommended that a preliminary test be performed to determine the optimal conditions for a reaction including substrate concentration, reaction temperature, cofactor requirements, and buffer constitutions, such as pH and ionic strength, to ensure the consistency of the reaction. Also, an enzyme blank should be performed to determine if the nonenzymatic reaction is negligible or could be corrected. When a crude extract, rather than purified enzyme, is used for determining enzyme activity, one control experiment should be performed without adding substrate to determine if there are any endogenous substrates present and to prevent overcounting of the enzyme activity in the extract.

Usually, the initial reaction rate is measured, and it is related to the substrate concentration. Highly sufficient substrate concentration is required to prevent a decrease in concentration during the assay period of enzyme activity, thus allowing the concentration of the substrate to be regarded as constant.

Types of Assay Methods

On the basis of measuring the decrease of the substrate or the increase of the product, one common characteristic property that is useful for distinguishing the substrate and the product is their absorbance spectra, which can be determined using one of the *spectrophotometric methods*. Observation of the change in absorbance in the visible or near UV spectrum is the method most commonly used in assaying enzyme activity. $NAD(P)^+$ is quite often a cofactor of dehydrogenases, and NAD(P)H is the resulting product. The absorbance spectrum is 340 nm for the product, NAD(P)H, but not for the cofactor, NAD $(P)^+$. A continuous measurement at absorbance 340 nm is required to continuously monitor the disappearance of the substrate or the appearance of the product. This is called a *continuous assay*, and it is also a *direct assay* because the catalyzed reaction itself produces a measurable signal. The advantage of continuous assays is that the progress curve is immediately available, as is confirmation that the initial rate is linear for the measuring period. They are generally preferred because they give more information during the measurement. However, methods based on fluorescence, the *spectrofluorometric methods*, are more sensitive than those based on the changes in the absorbance spectrum. The reaction is accomplished by the release or uptake of protons and is the basis of performing the assay. Detailed discussions on the assaying techniques are available below.

Sometimes, when a serial enzymatic reaction of a metabolic pathway is being assayed, no significant products, the substrates of the next reaction, can be measured in the absorbance spectra due to rapid changes in the reaction. Therefore, there may be no suitable detection machinery available for this single enzymatic study. Nevertheless, there is still a measurable signal when one or several enzymatic reactions are coupled to the desired assay reaction. These are *coupled assays*, one of the particular *indirect assays* whose coupled reaction is not enzymatic. This means that the coupled enzymes and the substrates have to be present in excess to make sure that the rate-limiting step is always the reaction of the particular enzyme being assayed. Though a lag in the appearance of the preferred product will be noticed at the beginning of the assay, before reaching a steady state, the phenomenon will only appear for a short period of time if the concentration of the coupling enzymes and substrates are kept in excess. So, the V_{max} of the coupling reactions will be greater than that of the preferred reaction. Besides, not only the substrate concentration but also other parameters may affect the preferred reaction (see previously). To prevent this, control experiments can be performed to check if the V_{max} of the coupling reactions is actually much higher than that of the preferred reaction. For other indirect assays in which the desired catalyzed reaction itself does not produce a measurable signal, a suitable reagent that has no effect on enzyme activity can be added to the reaction mixture to react with one of the products to form a measurable signal.

Nevertheless, sometimes no easily readable differences between the spectrum of absorbance of the substrate and that of the product can be measured. In such cases, it may be possible to measure the appearance of the colored product, by the *chromogenic method* (one of the spectrofluorometric methods). The reaction is incubated for a fixed period of time. Then, because the development of color requires the inhibition of enzyme activity, the reaction is stopped and the concentration of colored product of the substrate is measured. This is the *discontinuous* or *end point assay*. Assays involving product separation that cannot be designed to continuously measure the signal change, such as electrophoresis, high performance liquid chromatography (HPLC), and sample quenching at time intervals, are also discontinuous assays. The advantage of discontinuous assays is that they are less time consuming for monitoring a large number of samples for enzyme activity. However, additional control experiments are needed to ensure that the initiation rate is linear for the measuring period of time. Otherwise, the radiolabeled substrate of an enzyme assay is a highly sensitive method that allows the detection of radioactive product. Separation of the substrate and the product by a variety of extraction methods may be required to accurately assay the enzyme activity.

Detection Methods

Spectrophotometric Methods

Both the spectrophotometric method and the spectrofluorometric method discussed below use measurements at specific wavelengths of light energy (in the wavelength regions of 200–400 nm (UV, ultraviolet) and 400–800 nm (visible)) to determine how much light has been absorbed by a target molecule (resulting in changes in electronic configuration). The measured value from the spectrophotometric method at a specific wavelength can be related to the molecule concentration in the solution using a cell with a fixed path length (l cm) that obeys the Beer-Lambert law:

$$A = -\log T = \varepsilon c l,$$

where A is the absorbance at certain wavelength, T is the transmittance, representing the intensity of transmitted light, ε is the extinction coefficient, and c is the molar concentration of the sample. Using this law, the measured change in absorbance, ΔA, can be converted to the change in molecule concentration, Δc, and the change in rate, v_i, can be calculated as a function of time, Δt, that is, $v_i = \Delta A / \varepsilon l \Delta t$.

Choices of appropriate materials for spectroscopic cells (cuvettes) depend on the wavelength used; quartz cuvettes must be used at wavelengths less than 350 nm because glass and disposable plastic cuvettes absorb too much light in the UV light range. However, the latter glass and disposible plastic cuvettes can be used in the wavelength range of 350–800 nm. Selection of a wavelength for the measurement depends on finding the wavelength that produces the greatest difference in absorbance between the reactant and product molecules in the reaction. Though the wavelength of measurement usually refers to the maximal wavelength of the reactant or product molecule, the most meaningful analytical wavelength may not be the same as the maximum wavelength because significant overlap of spectra may be found between the reactant and product molecules. Thus, a different spectrum between two molecules can be calculated to determine the most sensitive analytical wavelengths for monitoring the increase in product and the decrease in substrate.

Spectrofluorometric Methods

When a molecule absorbs light at an appropriate wavelength, an electronic transition occurs from the ground state to the excited state; this short-lived transition decays through various high-energy, vibrational substrates at the excited electronic state by heat dissipation, and then relaxes to the ground state with a photon emission, the fluorescence. The emitted fluorescence is less energetic (longer wavelength) than the initial energy that is required to excite the molecules; this is referred to as the Stokes shift. Taking advantage of this, the fluorescence instrument is designed to excite the sample and detect the emitted light at different wavelengths. The ratio of quanta fluoresced over quanta absorbed offers a value of quantum yield (Q), which measures the efficiency of the reaction leading to light emission. The light emission signals vary with concentration of fluorescent material by the Beer-Lambert law, where the extinction coefficient is replaced by the quantum yield, and gives: $I_f = 2.31\ I_0 \varepsilon c l Q$, where I_f and I_0 are the intensities of light emission and of incident light that excites the sample, respectively. However, conversion of light emission values into concentration units requires the preparation of a standard curve of light emission signals as a function of the fluorescent material concentration, which has to be determined independently, due to the not strictly linear relationship between I_f and c (Lakowicz 1983).

Spectroflurometric methods provide highly sensitive capacities for detection of low concentration changes in a reaction. However, quenching (diminishing) or resonance energy transfer (RET) of the measured intensity of the donor molecule will be observed if the acceptor molecule absorbs light and the donor molecule emits light at the same wavelength. The acceptor molecule will either decay to its ground state if quenching occurs or fluoresce at a characteristic wavelength if energy is transferred (Lakowicz 1983). Both situations can be overcome by using an appropriate peptide sequence, up to 10 or more amino acid residues, to separate the fluorescence donor molecule from the acceptor molecule if peptide substrates are used. Thus, both effects can be relieved by cleaving the intermolecular bonding when protease activity is being assayed.

Care should be taken in performing the spectrofluorometric method. For instance, the construction of a calibration curve as the experiment is assayed is essential, and the storage conditions for the fluorescent molecules are important because of photodecomposition. In addition, the quantum yield is related to the temperature, and the fluorescence signal will increase with decreasing temperature, so a temperature-constant condition is required (Bashford and Harris 1987, Gul et al. 1998).

Radiometric Methods

The principle of the radiometric methods in enzymatic catalysis studies is the quantification of radioisotopes incorporated into the substrate and retained in the product. Hence, successful radiometric methods rely on the efficient separation of the radiolabled product from the residual radiolabeled substrate and on the sensitivity and specificity of the radioactivity detection method. Most commonly used radioisotopes, for example, ^{14}C, ^{32}P, ^{35}S, and ^{3}H, decay through emission of β particles; however, ^{125}I decays through emission of γ particles, whose loss is related to loss of radioactivity and the rate of decay (the half-life) of the isotope. Radioactivity expressed in Curies (Ci) decays at a rate of 2.22×10^{12} disintegrations per minute (dpm); expressed in Becquerels (Bq), it decays at a rate of 1 disintegrations per second (dps). The experimental units of radioactivity are counts per minute (cpm) measured by the instrument; quantification of the specific activity of the sample is given in units of radioactivity per mass or per molarity of the sample (e.g., μCi/mg or dpm/μmol).

Methods of separation of radiolabeled product and residual radiolabeled substrate include chromatography, electrophoresis, centrifugation, and solvent extraction (Gul et al. 1998). For the detection of radioactivity, one commonly used instrument is a scintillation counter that measures light emitted when solutions of *p*-terphenyl or stilbene in xylene or toluene are mixed with radioactive material designed around a photomultiplier tube. Another method is the autoradiography that allows detecting radioactivity on surfaces in close contact either with X-ray film or with plates of computerized phosphor imaging devices. Whenever operating the isotopescontaining experiments, care should always be taken for assuring safety.

Chromatographic Methods

Chromatography is applied in the separation of the reactant molecules and products in enzymatic reactions and is usually used in conjunction with other detection methods. The most commonly used chromatographic methods include paper chromatography, column chromatography, thin-layer chromatography (TLC), and HPLC. Paper chromatography is a simple and economic method for the readily separation of large numbers of samples. By contrast, column chromatography is more expensive and has poor reproducibility. The TLC method has the advantage of faster separation of mixed samples, and like paper chromatography, it is disposable and can be quantified and scanned; it is not easily replaced by HPLC, especially for measuring small, radiolabeled molecules (Oldham 1992).

The HPLC method featured with low compressibility resins is a versatile method for the separation of either low molecular weight molecules or small peptides. Under a range of high pressures up to 5000 psi (which approximates to 3.45×10^{7} Pa), the resolution is greatly enhanced with a faster flow rate and a shorter run time. The solvent used for elution, referred to as the mobile phase, should be an HPLC grade that contains low contaminants; the insoluble media is usually referred to as the stationary phase. Two types of mobile phase are used during elution; one is an isocratic elution whose composition is not changed, and the other is a gradient elution whose concentration is gradually increased for better resolution. The three HPLC methods most commonly used in the separation steps of enzymatic assays are reverse phase, ion-exchange, and size-exclusion chromatography.

The basis of reverse phase HPLC is the use of a nonpolar stationary phase composed of silica covalently bonded to alkyl silane, and a polar mobile phase used to maximize hydrophobic interactions with the stationary phase. Molecules are eluted in a solvent of low polarity (e.g., methanol, acetonitrile, or acetone mixed with water at different ratios) that is able to efficiently compete with molecules for the hydrophobic stationary phase.

The ion-exchange HPLC contains a stationary phase covalently bonded to a charged functional group; it binds the molecules through electrostatic interactions, which can be disrupted by the increasing ionic strength of the mobile phase. By modifying the composition of the mobile phase, differential elution, separating multiple molecules, is achieved.

In the size-exclusion HPLC, also known as gel filtration, the stationary phase is composed of porous beads with a particular molecular weight range of fractionation. However, this method is not recommended where molecular weight differences between substrates and products are minor, because of overlapping of the elution profiles (Oliver 1989).

Selection of the HPLC detector depends on the types of signals measured, and most commonly the UV/visible light detectors are extensively used.

Electrophoretic Methods

Agarose gel electrophoresis and polyacrylamide gel electrophoresis (PAGE) are widely used methods for separation of macromolecules; they depend, respectively, on the percentage of agarose and acrylamide in the gel matrix. The most commonly used method is the sodium dodecyl sulfate (SDS)-PAGE method; under denaturing conditions, the anionic detergent SDS is coated on peptides or proteins giving them equivalently the same anionic charge densities. Resolving of the samples will thus be based on molecular weight under an electric field over a period of time. After electrophoresis, peptides or proteins bands can be visualized by staining the gel with Coomassie Brilliant Blue or other staining reagents, and radiolabeled materials can be detected by autoradiography. Applications of electrophoresis assays are not only for detection of molecular weight and radioactivity differences, but also for detection of charge differences. For instance, the enzyme-catalyzed phosphorylation reactions result in phosphoryl transfer from substrates to products, and net charge differences between two molecules form the basis for separation by electrophoresis. If radioisotope ^{32}P-labeled phosphate is incorporated into the molecules, the reactions can be detected by autoradiography, by monitoring the radiolabel transfer after gel electrophoresis, or by immunological blotting with antibodies that specifically recognize peptides or proteins containing phosphate-modified amino acid residues.

Native gel electrophoresis is also useful in the above applications, where not only the molecular weight but also the charge density and overall molecule shape affect the migration of molecules in gels. Though SDS-PAGE causes denaturing to peptides and proteins, renaturation in gels is possible and can be applied to several types of in situ enzymatic activity studies such as activity staining and zymography (Hames and Rickwood

1990). Both methods assay enzyme activity after electrophoresis, but zymography is especially intended for proteolytic enzyme activity staining in which gels are cast with high concentrations of proteolytic enzyme substrates, for example, casein, gelatin, bovine serum albumin, collagen, and others. Samples containing proteolytic enzymes can be subjected to gel electrophoresis, but the renaturation step has to be performed if a denaturing condition is used; then the reaction is performed under conditions suitable for assaying proteolytic enzymes. The gel is then subjected to staining and destaining, but the entire gel background will not be destained because the gel is polymerized with protein substrates, except in clear zones where the significant proteolysis has occurred; the amount of staining observed will be greatly diminished due to the loss of protein. This process is also known as reverse staining. Otherwise, reverse zymography is a method used to assay the proteolytic enzyme inhibitor activity in gel. Similar to zymography, samples containing proteolytic enzyme inhibitor can be subjected to gel electrophoresis. After the gel renaturation step is performed, the reaction is assayed under appropriate conditions in the presence of a specific type of proteolytic enzyme. Only a specific type of proteolytic enzyme inhibitor will be resistant to the proteolysis, and after staining, the active proteolytic enzyme inhibitor will appear as protein band (Oliver et al. 1999).

Other Methods

The most commonly used assay methods are the spectrophotometric, spectrofluorometric, radiometric, chromatographic, and electrophoretic methods described above, but a variety of other methods are utilized as well. Immunological methods make use of the antibodies raised against the proteins (Harlow and Lane 1988). Polarographic methods make use of the change in current related to the change in concentration of an electroactive compound that undergoes oxidation or reduction (Vassos and Ewing 1983). Oxygen-sensing methods make use of the change in oxygen concentration monitored by an oxygen-specific electrode (Clark 1992), and pH-stat methods use measurements of the quantity of base or acid required to be added to maintain a constant pH (Jacobsen et al. 1957).

Selection of an Appropriate Substrate

Generally, a low molecular mass, chromogenic substrate containing one susceptible bond is preferred for use in enzymatic reactions. A substrate containing many susceptible bonds or different functional groups adjacent to the susceptible bond may affect the cleavage efficiency of the enzyme, and this can result in the appearance of several intermediate-sized products. Thus, they may interfere with the result and make interpretation of kinetic data difficult. Otherwise, the chromophore-containing substrate will readily and easily support assaying methods with absorbance measurement. Moreover, substrate specificity can also be precisely determined when an enzyme has more than one recognition site on both the preferred bond and the functional groups adjacent to it. Different sized chromogenic substrates can then be used to determine an enzyme's specificity and to quantify its substrate preference.

Unit of Enzyme Activity

The unit of enzyme activity is usually expressed as either micromoles of substrate converted or product formed per unit time, or unit of activity per milliliter under a standardized set of conditions. Though any unit of enzyme activity can be used, the Commission on Enzymes of the International Union of Biochemistry and Molecular Biology (IUBMB) has recommended that a unit of enzyme, Enzyme Unit or International Unit (U), be used. An Enzyme Unit is defined as that amount that will catalyze the conversion of 1 micromole of substrate per minute, $1\ U = 1\ \mu mol/min$, under defined conditions. The conditions include substrate concentration, pH, temperature, buffer composition, and other parameters that may affect the sensitivity and specificity of the reaction, and usually a continuous spectropotometric method or a pH stat method is preferred. Another enzyme unit that now is not widely used is the International System of Units (SI unit) in which 1 katal (kat) = 1 mol/sec, so $1\ kat = 60\ mol/min = 6 \times 10^7\ U$.

In ascertaining successful purification of a specified enzyme from an extract, it is necessary to compare the specific activity of each step to that of the original extract; a ratio of the two gives the fold purification. The specific activity of an enzyme is usually expressed as units per milligram of protein when the unit of enzyme per milliliter is divided by milligrams of protein per milliliter, the protein concentration. The fold purification is an index reflecting only the increase in the specific activity with respect to the extract, not the purity of the specified enzyme.

REFERENCES

Barry MJ. 1997. Emzymes and symmetrical molecules. *Trends Biochem Sci* 22: 228–230.

Bashford CL, Harris DA. 1987. *Spectrophotometry and Spectrofluorimetry: A Practical Approach*. IRL Press, Washington, DC.

Bell JE, Bell ET. 1988. *Proteins and Enzymes*. New Jersey: Prentice-Hall.

Briggs GE, Haldane JBS. 1925. A note on the kinetics of enzyme action. *Biochem J* 19: 338–339.

Brocklehurst K, Dixon HBF. 1977. The pH dependence of second-order rate constants of enzyme modification may provide free-reactant pKa values. *Biochem J* 167: 859–862.

Brown AJ. 1902. Enzyme action. *J Chem Soc* 81: 373–386.

Burbaum JJ, Schimmel P. 1991. Structural relationships and the classification of aminoacyl-tRNA synthetases. *J Biol Chem* 266: 16965–16968.

Cardenas ML. et al. 1998. Evolution and regulatory role of the hexokinases. *Biochim Biophys Acta* 1401: 242–264.

Cherfils J. et al. 1990. Modelling allosteric processes in *E. coli* aspartate transcarbamylase. *Biochimie* 72(8): 617–624.

Clark JB. 1992. In: Eisenthal R, Danson MJ (eds.) *Enzyme Assays, A Practical Approach*. New York: Oxford University Press.

Copeland RA. 2000. *Enzymes*, 2nd ed. New York: Wiley-VCH Inc.

Cornish-Bowden A. 1995. *Fundamentals of Enzyme Kinetics*, 2nd ed. London: Portland Press.

Cornish-Bowden A. 1996. In: Engel PC, (ed.) *Enzymology Labfax*. San Diego: BIOS Scientific Publishers, Oxford and Academic Press, pp. 96–113.

Cusack S. 1997. Aminoacyl-tRNA synthetases. *Curr Opin Struct Biol* 7(6): 881–889.

Dixon M, Webb EC. 1979. *Enzymes*. 3rd ed. New York: Academic Press.

Drake JW. 1999. The distribution of rates of spontaneous mutation over viruses, prokaryotes, and eukaryotes. *Ann N Y Acad Sci*. 870: 100–107.

Eadie GS. 1942. The inhibition of cholinesterase by physostigmine and prostigmine. *J Biol Chem* 146: 85–93.

Eisenthal R, Cornish-Bowden A. 1974. The direct linear plot. A new graphical procedure for estimating enzyme parameters. *Biochem J* 139: 715–720.

Fang TY, Ford C. 1998. Protein engineering of *Aspergillus awamori* glucoamylase to increase its pH optimum. *Protein Eng* 11: 383–388.

Goodman MF. 1997. Hydrogen bonding revisited: Geometric selection as a principal determinant of DNA replication fidelity. *Proc Natl Acad Sci USA* 94: 10493–10495.

Gul S. et al. 1998. *Enzyme Assays*. Oxford: John Wiley & Sons Ltd.

Hames BD, Rickwood D. 1990. *Gel Electrophoresis of Proteins: A Practical Approach*, 2nd ed. Oxford: IRL Press.

Hammes GG. 2002. Multiple conformational changes in enzyme catalysis. *Biochemistry* 41(26): 8221–8228.

Hanes CS. 1932. Studies on plant amylases. I. The effect of starch concentration upon the velocity of hydrolysis by the amylase of germinated barley. *Biochem J* 26: 1406–1421.

Harlow E, Lane D. 1988. *Antibodies: A Laboratory Manual*. NewYork. Cold Spring Harbor: CSHL Press.

Hill AV. 1910. The possible effects of the aggregation of molecules of hemoglobin on its dissociation curves. *J Physiol* 40, 4–7.

Hofstee BHJ. 1959. Non-inverted versus inverted plots in enzyme kinetics. *Nature* 184: 1296–1298.

Igarashi K. et al. 1999. Thermostabilization by proline substitution in an alkaline, liquefying α-amylase from *Bacillus* sp. strain KSM-1378. *Biosci Biotechnol Biochem* 63: 1535-1540.

Jacobsen CF et al. 1957. In: Glick D, editor, *Methods of Biochemical Analysis*, vol IV. New York: Interscience Publishers, pp. 171–210.

Khalifah RG. 2003. Reflections on Edsall's carbonic anhydrase: Paradoxes of an ultra fast enzyme. *Biophys Chem*. 100: 159–170.

Kornberg RD. 1996. RNA polymerase II transcription control. *Trends Biochem Sci* 21: 325–327.

Lakowicz JR. 1983. *Principles of Fluorescence Spectroscopy*. New York: Plenum Publishing.

Langfort J. et al. 1998. Hormone-sensitive lipase expression and regulation in skeletal muscle. *Adv Exp Med Biol* 441: 219–228.

Luan S. 2003. Protein phosphatases in plants. *Annu Rev Plant Biol* 54: 63–92.

Lineweaver H, Burk D. 1934. The determination of enzyme dissociation constants. *J Am Chem Soc* 56: 658–666.

Michaelis L, Menten ML. 1913. Die Kinetik der Invertinwirkun. *Biochem Z* 49: 333–369.

Nuttall FQ. et al. 1988. Regulation of glycogen synthesis in the liver. *Am J Med* 85(Supp. 5A): 77–85.

Oldham KG. 1992. In: Eisenthal R, Danson MJ (eds.) *Enzyme Assays, A Practical Approach*. New York: Oxford University Press, pp. 93–122.

Oliver GW. et al. 1999. In: Sterchi EE, Stöcker W (ed.) *Proteolytic Enzymes: Tools and Targets*. Berlin Heidelberg: Springer-Verlag Berlin, pp. 63–76.

Oliver RWA. 1989. *HPLC of Macromolecules: A Practical Approach*. Oxford: IRL Press.

Pechkova E. et al. 2003. Threedimensional atomic structure of a catalytic subunit mutant of human protein kinase CK2. *Acta Crystallogr D Biol Crystallogr* 59(Pt 12): 2133–2139.

Preiss J, Romeo T. 1994. Molecular biology and regulation aspects of glycogen biosynthesis in bacteria. *Prog Nucl Acid Res Mol Biol* 47: 299–329.

Raj SML, Liu F. 2003. Engineering of RNase P ribozyme for gene-targeting application. *Gene* 313: 59–69.

Scott CP. et al. 2001. A quantitative model for allosteric control of purine reduction by murine ribonucleotide reductase. *Biochemistry* 40(6): 1651–1661.

Scriver CR. 1995. Whatever happened to PKU? *Clin Biochem* 28(2): 137–144.

Shiau RJ. et al. 2003. Improving the thermostability of raw-starch-digesting amylase from a *Cytophaga* sp. by site-directed mutagenesis. *Appl Environ Microbiol* 69(4): 2383–2385.

Sirko A, Brodzik R. 2000. Plant ureases: Roles and regulation. *Acta Biochim Pol* 47(4): 1189–1195.

Stauffer CE. 1989. *Enzyme Assays for Food Scientists*. New York: Van Nostrand Reinhold.

Steitz TA, Moore PB. 2003. RNA, the first macromolecular catalyst: The ribosome is a ribozyme. *Trends Biochem Sci* 28(8): 411–418.

Tipton KF. 1992. Principles of enzyme assay and kinetic studies. In: Eisenthal R, Danson MJ, editors, *Enzyme Assays, A Practical Approach*. New York: Oxford University Press, pp. 1–58.

Tipton KF, Webb HBF. 1979. Effects of pH on enzymes. *Methods Enzymol* 63: 183–233.

Vassos BH, Ewing GW. 1983. *Electroanalytical Chemistry*. New York: John Wiley & Sons.

Waters PJ. 2003. How PAH gene mutation cause hyperphenylalaninemia and why mechanism matters: Insights from in vitro expression. *Hum Mutat* 21(4): 357–369.

9
Enzymes in Food Processing

Benjamin K. Simpson, Xin Rui, and Sappasith Klomklao

Abstract: Enzymes have been used as food processing aids since time immemorial. Enzyme action in foods manifests as sensory, textural, and color changes that may or may not be desirable in the products, depending on the nature of the transformation. Their use as food processing aids is desirable, perhaps preferable to other food processing methods, because they are perceived as natural components of food; they are also more specific and can be used in low concentrations under mild reaction conditions. The products formulated with enzymes tend to be more uniform, and their action in

Food Biochemistry and Food Processing, Second Edition. Edited by Benjamin K. Simpson, Leo M.L. Nollet, Fidel Toldrá, Soottawat Benjakul, Gopinadhan Paliyath and Y.H. Hui.

foods can be stopped relatively easily after the desired transformation is attained. The chapter is organized in six sections and covers the major groups and sources of food enzymes, their properties and modes of action, the rationale for their use in food, their use in the manufacture of various foods, and the different strategies employed to control their undesirable effects in foods. The chapter also provides information on the future prospects of enzymes.

INTRODUCTION

Enzymes are biological molecules that enable biological reactions proceed at perceptible rates in living organisms (plants, animals, and microorganisms) and the products derived from these sources. Living organisms produce basically the same functional classes of enzymes to enable them carry out similar metabolic processes in their cells and tissues. Thus, it is to be expected that foodstuffs would also possess endogenous enzymes to catalyze biological reactions that occur in them both pre- and postharvest. The molecules that enzymes act upon are called substrates, while the resulting compounds from the enzymatic conversions are called products. Enzymes are able to speed up biological reactions by lowering the free energy of activation (ΔG^*) of the reaction (van Oort 2010). Unlike chemical catalysts, enzymes are much more specific and active under mild reaction conditions of pH, temperature, and ionic strength.

Humans have intentionally or unintentionally used enzymes to modify foodstuffs to alter their functional properties, extend their storage life, improve flavors, and provide variety and delight, and the use of enzymes in food processing has been increasing for reasons such as consumer preferences for their use in food modification instead of chemical treatments, their capacity to retain high nutritive value of foods, and the advances in biotechnology that are enabling the discoveries of new enzymes that are much more efficient in transforming foods. The recent advances in molecular biotechnology that are permitting the discoveries of new and better food enzymes augur well both for the need and capacity for food technologists and food manufacturers to produce food products in highly nutritious, safe, and stable forms as part of the overall strategy to achieve food security.

However, not all reactions catalyzed by enzymes in foods are useful; naturally present enzymes in fresh foods and their counterparts that survive food-processing operations can induce undesirable changes in foods, some of which may even be toxic. Thus, there is the need to also develop novel and more effective strategies to control the deleterious effects of enzymes in foods to reduce postharvest food losses and/or spoilage.

Some Terms and Definitions

Several useful terms are encountered in the study of enzymes. The active enzyme molecule may comprise a protein part exclusively, or the enzyme protein may require as essential nonprotein part for its functional activity. The essential nonprotein part that some enzymes require for functional activity is known as a cofactor or prosthetic group. For those enzymes requiring cofactors for activity, the enzyme–cofactor complex is known as the holoenzyme, and the protein part that has no functional activity without the cofactor is known as apoenzyme. Thus, cofactors or prosthetic groups may be regarded as "helper molecules" for the apoenzymes. Cofactors may be inorganic (e.g., metal ions like Zn, Mg, Mn, and Cu), or organic (e.g., riboflavin, thiamine, or folic acid) materials. Examples of food enzymes that do not require cofactors or prosthetic groups for activity include lysozyme, pepsin, trypsin, and chymotrypsin; examples of those enzymes that require cofactors or prosthetic groups for functional activity include carboxypeptidase, carbonic anhydrase and alcohol dehydrogenase (all have Zn^{2+} as cofactor), cytochrome oxidase (has Cu^{2+} as cofactor), glutathione peroxidase (which has Se cofactor), catalase (with either Fe^{2+} or Fe^{3+} as cofactor), and pyruvate carboxylase (that has thiamine in the form of thiamin pyrophosphate or TPP as cofactor).

Enzymes have a catalytic region known as the active site where substrate molecules (S) bind prior to their transformation into products (P). The active site is a very small region of the very large enzyme molecule, and the transformation of substrates to products may be measured by either the rate of disappearance of the substrate ($-\delta S/\delta t$) or the rate of appearance of the products ($\delta P/\delta t$) in the reaction mixture. The activity of the enzyme is usually expressed in units, where a unit of activity may be defined as the amount of enzyme required to convert 1 μmol of the substrate per unit time under specified conditions (e.g., temperature, pH, ionic strength, and/or substrate concentration). Enzyme activity may also be expressed in terms of specific activity, which may be defined as the number of enzyme units per unit amount (e.g., milligram) of enzyme; or it may be denoted by the molecular activity or turnover number, defined as the number of enzyme units per mol of the enzyme at optimal substrate concentration. The efficiency of an enzyme (catalytic efficiency) may be derived from knowledge of the binding capacity of the enzyme for the substrate (K'_m, which is the apparent Michaelis–Menten constant) and the subsequent transformation of the substrate into products (K_{cat} or V_{max}). The catalytic efficiency of an enzyme is defined as the ratio of the maximum velocity (V_{max}) of the transformation of the substrate(s) into product(s) to the binding capacity of the enzyme (or the apparent Michaelis–Menten constant, K'_m, i.e., V_{max}/K'_m or K_{cat}/K'_m). The catalytic efficiency is a useful parameter in selecting more efficient enzymes from a group of homologous enzymes for carrying out particular operations.

Rationale for Interest in Food Enzymes

Beneficial Effects

Enzymes have several advantages for food use compared to conventional chemical catalysts. They are relatively more selective and specific in their choice and action on substrates, thus obviating side reactions that could lead to the formation of undesirable coproducts in the finished products. They have higher efficiency and can conduct reactions several times faster than other catalysts. They are active in low concentrations and perform well under relatively mild reaction conditions (e.g., temperature and pH); thus, their use in food processing helps to preserve the integrity of heat-labile essential nutrients. They can also be immobilized onto stationary support materials to permit their reuse and thereby reduce processing costs. Most of them are quite heat labile; thus, they can be readily inactivated by mild heat treatments after they have been used to achieve the desired transformation in foods, and they are natural and relatively innocuous components of agricultural materials that are considered "safe" for food and other nonfood uses (e.g., drugs and cosmetics). Their action on food components other than their substrates are negligible, and more gentle, thus resulting in the formation of purer products with more consistent properties; and they are also more environmentally friendly and produce less residuals (or processing waste that must be disposed of at high costs) compared to traditional chemical catalysts (van Oort 2010). Because of these beneficial effects, enzymes are used in the food industry for a plethora of applications including: baking and milling, production of (both alcoholic and nonalcoholic beverages), cheese and other dairy products manufacture, as well as the manufacture of eggs and egg products, fish and fish products, meats and meat products, cereal and cereal products, and in confectionaries.

Undesirable Effects

Certain food enzymes cause undesirable autolytic changes in food products, such as excessive proteolysis to produce bitterness in cheeses and protein hydrolysates, or excessive texture softening in meats and fish products (e.g., canned tuna). Enzymes like proteases, lipases, and carbohydrases break down biological molecules (proteins, fats, and carbohydrates, respectively) to adversely impact flavor, texture, and keeping qualities of the products. Decarboxylases and deaminases degrade biomolecules (e.g., free amino acids, peptides, and proteins) to form undesirable and/or toxic components, e.g., biogenic amines, in foods. Some others, for example, polyphenol oxidases (PPO) and lipoxygenases (LOX), promote oxidations and undesirable discolorations and/or color loss in fresh vegetables, fruits, crustacea, and salmonids, and others like thiaminase and ascorbic acid oxidase cause destruction of essential components (vitamins) in foods. Thus, more effective treatments must be developed and implemented to safeguard against such undesirable effects of enzymes in foods.

SOURCES OF FOOD ENZYMES (PLANT, ANIMAL, MICROBIAL, AND RECOMBINANT)

Enzymes have been used inadvertently or deliberately in food processing since ancient times to make a variety of food products, such as breads, fermented alcoholic beverages, fish sauces, and cheeses, and for the production of several food ingredients. Enzymes have been traditionally produced by extraction and fermentation processes from plant and animal sources, as well as from a few cultivatable microorganisms.

Industrial enzymes have traditionally been derived from plant, animal, and microbial sources (Table 9.1). Examples of plant enzymes include α-amylase, β-amylase, bromelain, β-glucanase, ficin, papain, chymopapain, and LOX; examples of animal enzymes are trypsins, pepsins, chymotrypsins, catalase, pancreatic amylase, pancreatic lipase, and rennet; and examples of microbial enzymes are α-amylase, β-amylase, glucose isomerase, pullulanase, cellulase, catalase, lactase, pectinases, pectin lyase, invertase, raffinose, microbial lipases, and proteases.

Microorganisms constitute the foremost enzyme source because they are easier and faster to grow and take lesser space to cultivate, and their use as enzyme source is not affected by seasonal changes and inclement climatic conditions and are thus more consistent. Their use as sources of enzymes is also not affected by various political and agricultural policies or decisions that regulate the slaughter of animals or felling of trees or plants. Industrial microbial enzymes are obtained from bacteria (e.g., α- and β-amylases, glucose isomerase, pullulanase, and asparaginase), yeasts (e.g., invertase, lactase, and raffinose), and fungi (e.g., glucose oxidase (GOX), catalase, cellulase, dextranase, glucoamylase, pectinases, pectin lyase, and fungal rennet). Even though all classes of enzymes are expected to occur in all or most microorganisms, in practice, the great majority of industrial microbial enzymes are derived from only a very few GRAS (generally recognized as safe) species, the predominant ones being types like *Aspergillus* species, *Bacillus* species, and *Kluyveromyces* species. This is because microorganisms can coproduce harmful toxins, and therefore need to be stringently evaluated for safety at high cost before they can be put to use for food production. As such, there are only very few microorganisms currently used as safe sources of enzymes and most of these strains have either been used by the food industry for several years or been derived from such strains by genetic mutation and selection.

The traditionally produced enzymes are invariably not well suited for efficient use in food-processing applications for reasons such as sensitivity to processing temperature and pH, inhibitory reaction components naturally present in foods, as well as availability, consistency, and cost. Recent developments in industrial biotechnology (including recombinant DNA and fermentation technologies) have permitted molecular biologists and food manufacturers to design and manufacture novel enzymes tailor-made to suit particular food-processing applications (Olempska-Beer et al. 2006). These new microbial enzymes are adapted to have specific characteristics that make them better suited to function under extreme environmental conditions, e.g., pH and temperature. The production of recombinant enzymes by recombinant DNA technology entails the introduction of the genes that encode for those enzymes from traditional sources into special vectors to produce large amounts of the particular enzyme. Recombinant enzymes are useful because they can be produced to a high degree of purity, and they can be produced in high yields and made available on a continuing and consistent basis at much reduced cost. As well, their purity and large-scale production reduce extensive quality assurance practices; and the technique itself enables useful (food) enzymes from unsafe organisms to be produced in useful forms as recombinant enzymes in safer microorganisms.

Examples of recombinant enzymes available for food use include various amylases, lipases, chymosins, and GOX.

Table 9.1. Sources of Industrial Food Enzymes—Selected Traditional and Recombinant Forms

Source	Enzyme	Some Food Applications
Plant sources		
Papaya tree	Papain	Meats, baked goods, brewing
Pineapple stem	Bromelain	Meats, baked goods
Animal sources		
Stomachs of ruminants	Chymosin (rennet)	Dairy (cheese)
Mammals (guts)	Lipase	Dairy products, oleo products, baked goods
Microbial sources		
Bacillus spp.	Alcalase	Waste utilization, bioactive compounds
	α-Amylase	Baked goods, beverages, and high-glucose syrups
	β-Amylase	Baked goods, beverages, and high-glucose syrups
	Pectinase (alkaline)	Fruit juices, coffee, and tea
	α-D-Galactosidase	Soybean pretreatment
Aspergillus niger	Protease	Soy sauce
	Pectinase (acidic)	Fruit juices
	Lipase	Dairy products, oleo products, baked goods
	Glucose oxidase	Baked goods, beverages, dried food mixes
Aspergillus niger, Candida guilliermondii, Kluyveromyces marxianus	Fructosidase	Fructose production
Candida cylindracea, Penicillium spp., Rhizopus spp., Mucor spp.	Lipase	Dairy products, oleo products, baked goods
Aspergillus spp., Penicillium spp., Aureobasidium spp.	Fructosyl transferase	Fructo-oligosaccharides production
Streptoverticillium mobaraense	Transglutaminase (TGase)	Meats, seafood, dairy products
Pleurotus ostreatus	Laccase	Beverages (wine, fruit juice, beer), jams, baked goods
Recombinant enzymes[a]		
Aspergillus oryzae	Aspartic proteinase	Meats (potential as tenderizer)
	Chymosin	Dairy (cheese)
	Glucose oxidase	Baked goods, beverages, dried food mixes
	Laccase	Beverages (wine, fruit juice, beer), jams, baked goods
	Lipase	Dairy products, oleo products, baked goods
	Pectin esterase	Beverages (wine, fruit juice, beer)
	Phytase	Baked goods, animal feeds
Aspergillus oryzae, Pichia pastoris	Phospholipase A_1	Refining of vegetable oils
Bacillus lichenformis	α-Amylase	Baked goods, beverages, and high-glucose syrups
	Pullulanase	Baked goods, beverages, and high-glucose syrups
Bacillus subtilis	α-Acetolactate dehydrogenase	Dairy, flavor generation
Escherichia coli K-12	Chymosin	Dairy (cheese)
Fusarium venenatum	Xylanase	Cereals, baked goods, animal feed
Kluyveromyces maxianus var. *lactis*	Chymosin	Dairy (cheese)
Pseudomonas fluorenscens Biovar 1	α-Amylase	Baked goods, beverages, and high-glucose syrups
Trichoderma reesei	Pectin lyase	Beverages (wine, fruit juice, beer)

[a]Olempska-Beer et al. (2006).

Table 9.1 also lists other recombinant food enzymes, their source microorganisms, and their food applications, as well as those of their traditional counterparts.

MAJOR FOOD ENZYMES BY GROUPS

The major enzymes used in the food industry include various members from the oxidoreductase, transferase, hydrolase, and isomerase families of enzymes. The oxidoreductases catalyze oxidation–reduction reactions in their substrates. Examples of the oxidoreductases are catalases, GOXs, LOX, PPO, and peroxidases. The transferases catalyze the transfer of groups between molecules to results in new molecules or form inter-/intramolecular cross-linkages in their substrates. An example of transferase used in the food industry is transglutaminase (TGase). The hydrolases catalyze the hydrolytic splitting of large molecules (e.g., proteins, carbohydrates, lipids, and nucleic acids) into smaller ones (e.g., amino acids and peptides; mono-, di-, and oligosaccharides; free fatty acids, mono-, and diglycerides; nucleotides and purines). Examples of the hydrolases include the proteases, carbohydrases, lipases, and nucleases. The isomerases potentiate intramolecular rearrangements in their substrate molecules. An example of isomerase used in food processing is xylose isomerase (also known as glucose isomerase).

OXIDOREDUCTASES

Oxidoreductases are the group of enzymes that catalyze oxidation–reduction reactions in their substrates. They occur widely in plants, animals, and microorganisms. In the food industry, the oxidoreductases of economic importance include GOX, PPO, peroxidase and catalase, lipoxygenase, xanthine oxidase (XO), ascorbic acid oxidase, and sulfhydryl oxidase (SO). Other well-known oxidoreductases are alcohol dehydrogenase, aldehyde dehydrogenase, lactate dehydrogenase (LDH), glutathione dehydrogenase, laccases, and lactoperoxidase.

Glucose Oxidase

GOX is produced commercially from the fungi *Aspergillus niger*, *Penicillium chrysogenum* (formerly known as *P. notatum*), *P. Amagaskinese,* and *P. vitale*. GOX catalyzes the oxidation of β-D-glucose initially to D-glucono-δ-lactone and hydrogen peroxide (H_2O_2); the gluconolactone is subsequently hydrolyzed to D-gluconic acid. In the food industry, GOX is used to improve bread dough texture (in place of oxidants like ascorbate or bromate), and the H_2O_2 formed as coproduct in the oxidation of D-glucose facilitates formation of a network of disulfide bonds through the oxidation of thiol groups in proteins and cysteine to result in stronger doughs. GOX is used in the food industry to desugar egg white and prevent Maillard-type browning reactions, and is also used in combination with catalase to eliminate O_2 from the head spaces of beverages (e.g., wines) and dried food packages to curtail enzymatic browning reactions and off-flavor development.

Polyphenol Oxidase

The term PPO is used here to encompass all those enzymes that have variously been referred to as catecholase, cresolase, diphenolase, laccase, phenolase, polyphenolase, or tyrosinase. PPO is widespread in nature and occurs in plants, animals, and microorganisms, and is produced on a large scale from mushrooms and by fermentation from fungi such as *Alternaria tenius, Chaetomium thermophile,* and *Thermomyces lanuginosus*, and from bacteria, particularly from the genus *Bacillus* such as *B. licheniformis, B. Natto,* and *B. sphaericus*. They are Cu^{++}-containing enzymes and enhance enzymatic browning, whereby phenolic compounds are hydroxylated and/or oxidized via intermediates such as quinones that polymerize into dark-colored pigments known as melanins. These browning reactions are responsible for the undesirable darks discoloration in bruised or freshly cut fruits (e.g., apples, avocadoes, and bananas) and vegetables (e.g., lettuce and potatoes), as well as in melanosis or "blackspot" formation in raw crustacea (e.g., shrimps, lobsters, crabs). Nonetheless, PPO-induced browning is desirable in other products such as chocolate, cocoa, coffee, prunes, raisins, tea, and tobacco.

In the food industry, PPO is used to enhance the color and flavor in coffee, tea, and cocoa. PPO is also put to nonfood uses in the pharmaceutical and biomedical industries: as components of biosensors for distinguishing between codeine and morphine; for the treatment of Parkinson's disease, phenylketonuria, and leukemia; and in wastewater treatment for the removal of phenolic pollutants.

Lipoxygenase

LOX, also known as lipoxidases, are non-heme, iron-containing enzymes that are found in plants, animals, and microorganisms. LOX is obtained for commercial use from soy flour. The enzyme causes oxidization of essential unsaturated fatty acids, e.g., oleic, linoleic, linolenic, arachidonic, and other polyunsaturated fatty acids such as eicosapentaenoic acid and docosahexaenoic acid, as well as carotenoids, to form their breakdown products (e.g., hydrocarbons, alcohols, carbonyl compound, and epoxides). As a result of the activities of LOX, off flavors develop in foodstuffs (oxidative rancidity); as well, there is formation of free radical intermediates that cause destruction of biomolecules and essential food components (e.g., proteins, astaxanthin, β-carotene, and vitamin A).

In general, the action of LOX in foods is undesirable, particularly the generation of harmful free radicals, the destruction of essential biomolecules like carotenoids and other essential food components that manifest in effects such as gradual color loss in salmonids during frozen storage, and off flavor development in fatty foods. However, LOX is put to some beneficial uses in bread making for the bleaching of flour and for improving the viscoelastic properties of dough. In plants, LOX participate in flavor formation through the formation of the aldehyde 2-hexenal. Some of the aldehyde thus formed may be oxidized to the corresponding carboxylic acid by the enzyme aldehyde dehydrogenase or reduced by alcohol dehydrogenase to the corresponding alcohol. Esters form from the carboxylic acids and alcohols to impart flavors to the plant material.

Peroxidase and Catalase

These are heme iron-containing enzymes that occur naturally in plants, animals, and microorganisms. They use H_2O_2 as coreactant to catalyze the oxidation of other peroxides to form H_2O and oxidized products, such as alcohols. Examples of peroxidases are horseradish peroxidase, glutathione peroxidase, cytochrome c peroxidase, and catalase. Catalase is usually mentioned separately from the rest of the peroxidases as it displays absolute specificity for H_2O_2. In the food industry, thermal inactivation of peroxidases is used as a measure of the efficiency of blanching treatment. Catalase is used for the removal of H_2O_2 from food products (e.g., to remove residual H_2O_2 from cold pasteurization of milk); it is also used in conjunction with GOX for desugaring eggs and removing head space O_2. Peroxidases may also be used to remove phenolic pollutants from industrial waste waters. However, some other effects of peroxidases in foods are undesirable, such as browning in fruits and off flavors in vegetables.

Xanthine Oxidase

XO is a molybdo-flavoprotein enzyme that is found largely in the milk and liver of mammals, and it catalyzes the oxidation of hypoxanthine and xanthine (breakdown products of nucleic acids) to uric acid.

$$\text{Hypoxanthine} + H_2O + {}^1/_2O_2 \Leftrightarrow \text{Xanthine} + {}^1/_2H_2O_2 \quad (1)$$

$$\text{Xanthine} + H_2O + {}^1/_2O_2 \Leftrightarrow \text{Uric acid} + {}^1/_2H_2O_2 \quad (2)$$

It is variously suggested that XO is involved in the spontaneous development of oxidized flavors in raw milks. This defect caused by XO in raw milk may be prevented by treatment with the digestive enzyme trypsin (Lim and Shipe 1972).

Ascorbic Acid Oxidase

Ascorbic acid oxidase (or more precisely, L-ascorbic acid oxidase) catalyzes the oxidation of L-ascorbic acid (vitamin C) to dehydroascorbic acid.

$$2\,\text{L-ascorbate} + O_2 \Leftrightarrow 2\,\text{Dehydroascorbate} + 2\,H_2O$$

The above reaction is important as the presence of the enzyme can lead to the destruction of vitamin C in fruits and vegetables. The same reaction serves as the basis for its use for the enzymatic determination of vitamin C (in fruit juices) and for eliminating the interferences due to the vitamin in clinical analysis. Sources of ascorbic acid oxidase include squash (plant source) and the bacterium *Aerobacter aerogenes*.

Lactate Dehydrogenase

LDH occurs in plants, animals, and microorganisms (e.g., lactic acid bacteria such as *Lactococcus*, *Enterococcus*, *Streptococcus*, *Pediococcus*). LDH catalyzes the interconversions between lactate and pyruvate with the participation of nicotinamide adenine dinucleotide (NAD^+) as hydrogen acceptor (i.e., from lactate to pyruvate) or NADH as hydrogen donor (i.e., in the reverse reaction from pyruvate to lactate).

$$\text{Lactate} + NAD^+ \Leftrightarrow \text{Pyruvate} + NADH + H^+$$

Lactate formed by the enzymes is an essential component of dairy products such as sour milk products, yogurt, ititu, kefir, and cottage cheese. The formation of lactate lowers the pH and promotes curdling of casein in fermented milks; it also contributes to the sour flavor of sourdough breads and is used in beer brewing to lower the pH and add "body" to beer.

In general, the breakdown of body tissues is accompanied by an increase in the levels of LDH, thus LDH levels are used as part of the diagnosis for hemolysis and other health disorders, including cancer, meningitis, acute pancreatitis, and HIV.

Sulfhydryl Oxidase

SO enzymes catalyze the oxidation of thiol groups into disulfides as follows:

$$2RSH + O_2 \rightarrow RS - SR + H_2O_2$$

The enzymes have been described in animal milk (Burgess and Shaw 1983) and egg white (Hoober et al. 1996, 1999, Thorpe et al. 2002), and in microorganisms (de la Motte and Wagner 1987). SO enzymes from bovine milk, pancreas, and kidney have been reported to be Fe dependent (Schmelzer et al. 1982, Sliwkowski et al. 1984, Clare et al. 1988), while others from the small intestine and the skin have been shown to contain Cu^{++} ions (Yamada 1989).

The enzyme has been used in the food industry for flavor control in ultra-high temperature (UHT)-processed milk (Kaufman and Fennema 1987).

Transferases

The transferases catalyze the transfer of groups from one substrate molecule (donor molecule) to another (acceptor molecule). The reaction is typified by the equation:

$$A\text{–}X + B \rightarrow A + B\text{–}X$$

where A is the donor and B is the acceptor. Examples include TGases, transglycosidases, transacetylases, transmethylases, and transphosphorylases. Some of them like TGases, and various glucanotransferases (e.g., fructosyl transferase, cyclodextrin glycosyl transferase (CGTase), and amylomaltase), are used extensively in the food industry.

Fructosyl Transferase

Fructosyl transferases catalyze the synthesis of fructose oligomers, also known as fructose oligosaccharides (FOS). The enzymes have been found in plants, e.g., onions (Robert and Darbyshire 1980) and asparagus (Shiomi et al. 1979), and in microorganisms, such as *Streptococcus mutans* (Wenham et al. 1979), *Fusarium oxysporum* (Gupta and Bhatia 1980, 1982), *B. subtilis* (Cheetham et al. 1989, Homann and Seibel 2009), *Aspergillus* spp, *Penicillium* spp, and *Aureobasidium* spp (Prapulla et al. 2000).

Fructo-oligosaccharides account for about 30% of the sweetness of sucrose, but are non-digestible in the gastrointestinal tracts of humans (Oku et al. 1984, Yun et al. 1995), thus they serve as dietary fiber and render the gut milieu more conducive for the growth and proliferation of beneficial intestinal microflora. So, they are applied as low calorie sweeteners and as functional food ingredients with health benefits (Lee et al. 1992). FOS are also used as sugar substitutes in foods such as light jam products, ice cream, and confectionery. The use of FOS in these foods helps to reduce the caloric intake and contributes to a lowering of blood glucose levels and related health issues with diabetes. The enzyme from *B. subtilis* has been used to synthesize fructose disaccharides such as xylsucrose, galactosucrose, and 6-deoxysucrose for use as antigens, in glycosylated proteins, or as sweeteners or bulking agents in food products (Cheetham et al. 1989).

Cyclodextrin Glycosyl Transferase

CGTase is also known by names like cyclodextrin glucanotransferase and cyclodextrin glucosyltransferase. CGTases have been characterized from various genera of bacteria, including *Bacillus*, *Klebsiella*, *Pseudomonas*, *Brevibacterium*, *Micrococus*, and *Clostridium* (Bonilha et al. 2006, Menocci et al. 2008). CGTases are known to catalyze at least four different reactions; namely coupling, cyclization, disproportionation, and hydrolysis. The most important of these reactions is intramolecular cyclization, whereby linear polysaccharide chains in molecules such as amylose are cleaved, and the two ends of the resulting fragments joined together to produce cyclic dextrins or cyclodextrins or CDs for short. CDs with varying units of glucose that are linked together by α(1"4) glycosidic bonds have been described, namely the 6 glucose CD (or α-CD), the 7 glucose CD (or β-CD), and the 8 glucose CD (or γ-CD). By virtue of their ringed structures, CDs are able to trap or encapsulate other molecules to form inclusion composites with altered physicochemical properties that have great potential for applications in the (functional) foods, cosmetics, and fine chemicals (Shahidi and Xiao-Qing 1993).

Amylomaltase

The enzyme amylomaltase, also known as 4-α-glucanotransferase, catalyzes the transfer of glucans from one α-1,4-glucan to another α-1,4-glucan or to glucose. The amylomaltase enzyme has been found in microorganisms, e.g., *Clostridium butyricum, E. coli, S. pneumonia* (Goda et al. 1997, Pugsley and Dubreuil 1988, Tafazoli et al. 2010) and in plants, e.g., potato (Jones and Whelan 1969, Sträter et al. 2002) and barley (Yoshio et al. 1986). The enzyme acts on both malto-oligosaccharides and amylose to form medium to highly polymerized cycloamyloses that are highly water-soluble (Takaha et al. 1996, Terada et al. 1997). These products are also expected to be of use in the (functional) food, pharmaceuticals, and fine chemical industries.

Transglutaminase

TGases are transferase enzymes that catalyze the acyl transfer reactions between free amino groups furnished by $-NH_2$ groups (e.g., the ε-NH_2 group of lysine) of amino acids and proteins, and the γ-carboxyamide groups of glutamines (Motoki and Seguro 1998). The TGase-assisted acyl transfer results in the extensive formation of strong and resistant covalent bonds between biomolecules like proteins. The products formed invariably have altered properties in terms of mechanical strength and solubility in aqueous systems. This feature of the enzyme is exploited in commercial food processing to improve the texture of gluten-free breads for individuals suffering from celiac disease (Moore et al. 2004, 2006), for making imitation crabmeat and fish balls, and for binding small pieces of meat into larger chunks of meat. The enzyme is also used as a binding agent to improve the physical qualities (e.g., texture, firmness, and elasticity) of protein-rich foods such as surimi or ham, and emulsified meat products, such as sausages and hot dogs. They are also used to improve the texture of low-grade meats such as the so-called pale, soft, exudative meat or PSE meats; to improve the consistency of dairy products (milk and yogurt) to make them thicker and creamier; and to make noodles firmer (Kim and Kim 2009). TGases have been found to be present in animals (e.g., threadfin bream, big-eye snapper, white croaker (Benjakul et al. 2010), guinea pig (Huang et al. 2008), in birds, amphibians, and invertebrates (Lantto et al. 2005), and in plants and microorganisms (e.g., *Bacillus*, *Streptoverticullium*, and *Streptomyces* spp (Yokoyama et al. 2004).

PROTEASES

The term protease is used to represent the group of enzymes that catalyze the cleavage of peptide bonds in proteins and peptide molecules with the participation of water as co-reactant. Other names for this group of enzymes include proteinases or proteolytic enzymes. The majority of industrial enzymes currently in use are proteases, and it is estimated that they constitute about 40% of the total global enzyme production and use (Layman 1986, Godfrey and West 1996). These enzymes are used in the food processing, animal feed, detergent, leather and textiles, photographic, and other industries (Kalisz 1988). Proteolytic enzymes are also used as a digestive aid to facilitate digestion and use of proteins by some people who otherwise cannot readily digest these molecules.

On the basis of their mode of action on the protein/peptide molecules, proteases are classified as either endopeptidases (or endoproteases) or exopeptidases (or exoproteases). Endopeptidases are those proteases that cleave peptide bonds randomly within protein molecules to result in smaller peptides. Examples include trypsin, chymotrypsin, pepsin, chymosin, elastase, and thermolysin. Exopeptidases cleave peptide bonds by successively removing terminal amino acid residues to produce free amino acids and the residual polypeptide molecule. Thus, two types of exoproteases are distinguished, the carboxypeptidases, that is, those that cleave amino acids at the carboxyl or C-terminal of the protein or polypeptide molecule, and the aminopeptidases that cleave amino acids from the amino or N-terminal of the molecule. Examples of carboxypeptidases include depeptidase, cathepsin A, cathepsin C, metallo-carboxypeptidases (e.g., carboxypeptidases A, A2, B, and C), and angiotensin-converting enzyme; examples of aminopeptidase are glutamyl aminopeptidase (a.k.a. aminopeptidase A), arginyl aminopeptidase (also known as aminopeptidase B), alanine aminopeptidase, leucine aminopeptidase, and the methionine aminopeptidases 1 & 2 (also known as METAP 1 & 2).

Proteases are traditionally classified into four sub-groups based on the nature of their active sites as acid, serine, sulfhydryl, and metallo-proteases. Their distinctive features are summarized in the following subsections.

Acid Proteases

The acid proteases (also known as aspartate proteases) are so called because the enzymes in this group tend to have one or more side chain carboxyl groups from aspartic acid as essential component(s) of their active sites, and they are optimally active within the acid pH range. Examples include pepsins, chymosins, rennets, gastricins (from animal sources), cyprosin, and cardosin (from plant sources), and several other aspartic acid proteases from microorganisms such as *Rhizomucor miehei*, *R. pusillus*, *R. racemosus*, *Cryphonectria parasitica*, and *A. niger* (Mistry 2006). Most of the known acid proteases are inhibited by the hexapeptide pepstatin. The major food industry use of acid proteases is in dairy processing, such as for the milk coagulation step in cheese making. In cheese making, the acid proteases added (e.g., rennet of chymosin) act to cleave critical peptide bonds between phenylalanine and leucine residues in κ-casein of the milk to form micelles that aggregate and precipitate out as the curd. The characteristic specificity of these proteases in these dairy products is important to restrict hydrolysis to the peptide bonds between the critical phenylalanine and leucine residues in κ-casein; otherwise, excessive proteolysis could elicit undue texture softening, bitterness, and/or "off-odors" in the product(s).

Serine Proteases

The serine proteases are the most abundant group of the known proteases (Hedstrom 2002). The characteristic features of this group of enzymes include the presence of a seryl residue in their active sites, as well as a general susceptibility to inhibition by serpins, organophosphates (e.g., di-isopropylphosphofluoridate), aprotinin (also known as trasylol), and phenyl methyl sulfonyl fluoride. Examples of the animal serine proteases are chymotrypsin, trypsin, thrombin, and elastase; examples of plant serine proteases are cucumisin (from melon; Kaneda and Tominaga 1975), macluralisin, and taraxilisin (from the osage orange fruit and dandelion, respectively; Rudenskaya et al. 1995); and examples of microbial serine proteases are proteinase K and subtilisin A (Couto et al. 1993). Several of the serine proteases function optimally within neutral to alkaline pH range (pH 7–11), and they find extensive use in the food and animal feed industries. They are used to produce highly nutritive protein hydrolysates from protein substrates such as whey, casein, soy, keratinous materials, and scraps from meat processing, as well as from fish processing discards; for the development of flavor during ripening of dairy products; and for the production of animal fodder from keratinous waste materials generated from meat and fish processing (Dalev 1994, Wilkinson and Kilcawley 2005). Other industrial applications of serine proteases include their use for the regioselective esterification of sugars, resolution of racemic mixtures of amino acids, and production of synthetic peptides to serve as pharmaceutical drugs and vaccines (Chen et al. 1999, Guzmán et al. 2007, Barros et al. 2009). Serine proteases are also used commercially for the treatment and bating of leather to remove undesirable pigments and hair from leather (Valera et al. 1997, Adıgüzel et al. 2009). They are also used for the treatment of industrial effluents and household waste, to lower the biological oxygen demand, and viscosity, to increase the flow properties of solid proteinaceous waste (e.g., feather, hair, hooves, skin, and bones), as well as wastewater from animal slaughterhouses and fish processing plants (Dalev and Simeonova 1992). Serine proteases also find some use in the degumming and the detergent industries (Godfrey and West 1996), and in medicine for the treatment of burns, wounds, and abscesses (Lund et al. 1999).

Sulfhydryl Proteases

The sulfhydryl (also known as thiol or cysteine) proteases are so called because they contain intact sulfhydryl groups in their active sites and are inhibited by thiol reagents such as alkylating agents and heavy metal ions. Examples include plant types like actinidain, papain, bromelain, chymopapain, and ficin; animal types like cathepsins B and C, calpains, and caspases; and clostripain, protease 1 and others from various microorganisms such as *Streptococcus* sp (Whitaker 1994), thermophilic *Bacillus* sp (Almeida do Nascimento and Martins 2004), and *Pyrococcus* sp (Morikawa et al. 1994). The activities of this group of enzymes are enhanced by dithiothreitol and its epimer dithioerythritol, cysteine, and 2-mercaptoethanol, and they are inhibited by compounds such as iodoacetate, *p*-chloromercuribenzoic acid, cystatin, leupeptin, and *N*-ethylmaleimide. The enzymes in this group tend to be optimally active around neutral pH (i.e., pH 6–7.5) and are relatively heat stable, which accounts for their use in meat tenderizers. Thiol proteases act to degrade myofibril protein and collagen fibers to make meats tender. Meat tenderness is a very important factor in meat quality evaluation. Ficin tends to have the greatest effect on myofibril protein hydrolysis due to its relatively higher thermal stability and broader specificity, followed by bromelain and papain in that order, although bromelain tends to degrade collagen fibers more extensively (Miyada and Tappel 1956). In commercial application, papain and bromelain are more frequently used because the higher hydrolytic capacity of ficin often results in excessive softening and mushiness in the treated meat (Wells 1966). These proteases may be applied in various ways to achieve meat tenderization, as by blending, dipping, dusting, soaking, spraying, injection, and vascular pumping (Etherington and Bardsley 1991). A common drawback with most of these approaches is the unequal distribution of the enzyme that could result in excessive tenderization or inadequate tenderization of different parts of the meat. The problem of uneven distribution of the enzyme may be minimized by using an additional tumbling step after enzyme treatment to increase distribution or by pre-slaughter injections of the animal with the enzyme (Beuk et al. 1962, Etherington and Bardsley 1991).

Sulfhydryl proteases are also used in the brewing industry to control haziness and improve clarity of the beverage. Haziness may form in the product during the aging process via aggregation of larger peptide and polyphenol molecules to form high molecular weight complexes. This is particularly enhanced during chilling of the beer, where solubility is minimal, and this group of enzymes is also used in the baking industry to improve the stretchability and firmness of the dough, through the modification of gluten. The enzymes are able to do this by increasing the viscoelastic, gas-retaining, and thermosetting properties of the dough. The action of these enzymes results in benefits such as shorter time of mixing, better flow, improved shaping, and increased gas retention of the flour or dough (Mathewson 1998).

Metalloproteases

The term metalloproteases is used to encompass those proteases that have metal ions in their active sites. These enzymes typically require an essential metal ion (e.g., Ni^{2+}, Mg^{2+}, Mn^{2+}, Ca^{2+}, Zn^{2+}, and Co^{2+}) for functional activity. Urease is an example of the nickel (Ni^{2+})-containing metalloenzymes. It catalyzes the breakdown of urea into carbon dioxide (CO_2) and ammonia (NH_3). The urease-catalyzed reaction occurs as follows:

$$(NH_2)_2\,CO + H_2O \rightarrow CO_2 + 2NH_3$$

The NH_3 thus produced functions to neutralize gastric acid in the stomach. Urease is found in bacteria, yeast, and several higher plants. The major sources of urease include the Jack bean (*Canavalia ensiformis*) plant and the bacteria *B. pasteurii*. The enzyme is also found in large quantities in soybeans and other plant seeds, in certain animal tissues (e.g., liver, blood, and muscle), in intestinal bacteria (e.g., *Eubacterium aerofaciens*, *E. lentum*, and *Peptostreptococcus productus*), and in yeasts (e.g., *Candida curvata*, *C. humicola*, *C. bogoriensis*, *C. diffluens*, and *Rhodotorula* and *Cryptococcus* species), as well as the blue-green algae (*Anabaena doliolum* and *Anacystis nidulans*). Examples of the metalloproteases that are Mg^{2+} dependent include inorganic diphosphatase that catalyzes the cleavage of diphosphate into inorganic phosphates, that is,

$$\text{Diphosphate} + H_2O \Leftrightarrow 2\ \text{Phosphate}$$

and phosphatidylinositol triphosphate 3-phosphatase that catalyzes the hydrolysis of phosphatidylinositol 3,4,5-trisphosphate, that is,

$$\text{Phosphatidylinositol 3,4,5-trisphosphate} + H_2O$$
$$\Leftrightarrow \text{Phosphatidylinositol 4,5-bisphosphate} + \text{Phosphate}$$

An example of metalloproteases containing manganese (Mn^{2+}) metal ions is phosphoadenylylsulfatase, which participates in sulfur metabolism and catalyzes the hydrolytic cleavage of 3′-phosphoadenylyl sulfate into adenosine 3′,5′-bisphosphate and sulfate:

$$3'\text{-Phosphoadenylyl sulfate} + H_2O$$
$$\Leftrightarrow \text{Adenosine } 3',5'\text{-Bisphosphate} + \text{Sulfate}$$

A well-known calcium (Ca^{2+})-containing metalloprotease is lactonase, which hydrolyzes lactones into their corresponding hydroacids, that is,

$$\text{Lactone(s)} + H_2O \Leftrightarrow \text{4-Hydroxyacid(s)}$$

The enzyme lactonase also participates in the metabolism of galactose and ascorbic acid.

Other metalloproteases contain zinc (Zn^{2+}) metal ions, e.g., glutamate carboxypeptidase and glutamyl aminopeptidase. Glutamate carboxypeptidase catalyzes the hydrolysis of *N*-acetylaspartylglutamic acid into glutamic acid and *N*-acetylaspartate as shown in the following reaction:

$$N\text{-Acetylaspartylglutamate} + H_2O$$
$$\Leftrightarrow \text{Glutamic acid} + N\text{-Acetylaspartate}$$

while glutamyl aminopeptidase (also known as aminopeptidase A) catalyzes the hydrolysis of glutamic and aspartic acid residues from the N-termini of polypeptides, e.g.,

$$\text{Aspartic acid-polypeptide} + H_2O$$
$$\Leftrightarrow \text{Aspartic acid} + \text{Polypeptide}$$

Glutamyl aminopeptidase is important because of its capacity to degrade angiotensin II into angiotensin III, which promote vasoconstriction and high blood pressure. Both glutamate carboxypeptidase and glutamyl aminopeptidase are present in membranes. Another important Zn^{2+}-containing metalloprotease is the *insulin-degrading enzyme* that is found in humans and degrades short chain polypeptides such as the β-chain of insulin, glucagon, transforming growth factor, and endorphins. Collagenases are also Zn^{2+} dependent and they act to break down the peptide bonds in collagen. Cobalt (Co^{2+})-containing metalloproteases include methylmalonyl coenzyme A mutase that converts methyl malonyl-CoA to succinyl-CoA, which is involved in the extraction of energy from proteins and fats; other Co^{2+}-containing metalloenzymes are methionine aminopeptidase and nitrile hydratase (Kobayashi and Shimizu 1999).

Sources of metalloproteases include those derived from animals, like carboxypeptidases A and B, collagenases, gelatinases, and procollagen proteinase; microbial ones such as thermolysin, lysostaphin (from *Staphylococcus* sp), acidolysin (from *Clostridium acetobutylicum*), and neutral metalloproteases (from *Bacillus* sp); as well as plant types like the zinc-dependent endoproteases in the seeds of grass pea (*Lathyrus sativus* L.). Agents or substances that are capable of binding or interacting with the catalytic metal ions tend be potent inhibitors of this group of enzymes. Examples of these inhibitors include phosphoramidon, ethylenediaminetetraacetic acid (EDTA), methylamine, vitamin B-12, dimercaprol, porphine, phosphonates, and 1,10-phenanthroline.

Common commercially available metalloenzymes are Neutrase (Novo Nordisk 1993) and Thermoase (Amano Enzymes) and some of the industrial applications of metalloenzymes include the synthesis of peptides for use as low calorie sweetener and drugs for the food and pharmaceutical industries, respectively. Thermolysin and vimelysin (from *Vibrio* sp) are both used for the synthesis of aspartame (Murakami et al. 1998); in this regard, vimelysin is superior to thermolysin for synthesis of the dipeptide at low temperatures. The metalloprotease thermolysin is also used in combination with other proteases (protease cocktails) for the production of food protein hydrolysates and flavor-enhancing peptides and for accelerating the ripening of dry sausages (Fernandez et al. 2000). They are also used for the hydrolysis of slaughterhouse waste, and in the brewing industry to facilitate filtration of beer and production of low calorie beers.

Carbohydrases

Carbohydrases are a group of enzymes that catalyze the hydrolysis or synthesis of polysaccharides. Examples include the amylases that hydrolyze starch molecules, maltases that break down maltose into glucose units, and lactases that degrade lactose into galactose and glucose. Other examples include invertases that hydrolyze sucrose into glucose and fructose, glucanases that catalyze the hydrolysis of glucans, cellulases that break down celluloses into low molecular weight cellodextrins, cellobiose, and/or glucose. Carbohydrases are ubiquitous in nature and are widespread in animals, plants, and microorganisms. For example, the amylases, pectinases, chitinases, and β-galactosidases are all found in plants, animals, and microorganisms (bacteria, fungi, and yeast). Microbial sources of commercially available carbohydrases are bacteria (e.g., *B. licheniformis*), various fungi (e.g., *A. niger*, *A. oryzae*, *A. aculeatus*, *Trichoderma reesei*), and yeasts (e.g., *Kluyveromyces lactis*, *Saccharomyces cerevisiae*). They are also present in algae, e.g., marine algae.

Two types of carbohydrases may be distinguished based on their modes of action as the endo- versus exo-carbohydrases. The endo-hydrolases (e.g., α-amylase) catalyze the cleavage of the internal glycosidic bonds of carbohydrate molecules, while the exo-hydrolases (e.g., β-amylase) catalyze the hydrolysis of carbohydrates by removing the maltose units from the nonreducing end of complex carbohydrate molecules. Carbohydrases may also be classified into the starch-hydrolyzing types and the nonstarch-hydrolyzing types. The starch-hydrolyzing carbohydrases (amylases, glucanases) act on amylose and amylopectin, and their principal applications are in the production of glucose syrups, ethanol, as well as in baking and brewing. The nonstarch-hydrolyzing carbohydrases are those that break down polymers of hexoses other than amylose and amylopectin, e.g., cellulases, lactases, maltases, galactosidases, pectic enzymes, invertases, chitinases, and fructosidases.

Amylases, Glucanases, and Pullulanases

Some of the more common carbohydrases and their uses in the food industry are the amylases that are used in the starch, alcoholic beverages, and the sugar industries. For example, Novozymes produces a heat-stable α-amylase (known as termamyl) from a genetically modified strain of *B. licheniformis* for the continuous liquefaction of starch at high temperatures (105–110°C). This same heat-stable enzyme is used to liquefy or "thin" starch during brewing and mash distillation. It is also used in sugar manufacture to hydrolyze the starch in sugar cane juice. Nonfood applications of α-amylases include their use in the textiles industry for de-sizing of fabrics and in detergent formulations for the removal of starch-based stains from fabrics. Amyloglucosidase (AMG) is a type of amylase that cleaves the α-1,4- and α-1,6-glycosidic bonds and removes glucose units from liquefied starch sequentially. The commercially available AMG enzyme is derived from a strain of the fungus *A. niger*, and it is used to produce sugar syrups from starches and dextrins.

Glucanases are a group of hydrolytic enzymes that catalyze the cleavage of glucans into various glucose oligomers, trisaccharides (e.g., maltotriose and nigerotriose), disaccharides (e.g., maltose, trehalose, and cellobiose), and the monosaccharide glucose. The term glucans encompasses those carbohydrate compounds made up of repeating glucose units linked together by glycosidic bonds. Examples of glucans include amylose starch (with α-1,4-glycosidic bonds), cellulose (β-1,4-glycosidic bonds), zymosan (β-1,3-glycosidic bonds), glycogen (with both α-1,4- and α-1,6-glycosidic bonds), amylopectin starch (with both α-1,4- and α-1,6-glycosidic bonds), and pullulan (with both α-1,4- and α-1,6-glycosidic bonds). They occur in plants, animals, and microorganisms; in plants and fungi, in particular, they serve as structural, defense, and growth functions, and as storage forms of chemical energy (Thomas et al. 2000).

Several microorganisms (fungi, bacteria (Kikuchi et al. 2005), and yeasts (Abd-El-Al and Phaff 1968) produce glucanases that enable them to digest glucans for use as nutrients and energy source. Glucanases are also widespread in plants, e.g., in seed plants (Elortza et al. 2003), rice (Thomas et al. 2000), tobacco (Leubner-Metzger 2003), acacia (González-Teuber et al. 2010), and in animal tissues, e.g., the embryo of sea urchins (Bachman and McClay 1996), nematodes (Kikuchi et al. 2005), mollusks (Kozhemyako et al. 2004), crayfish (Lee et al. 1992), earthworms (Beschin et al. 1998), shrimp (Sritunyalucksana et al. 2002), and insects (Hrassnigg and Crailsheim 2005). Bacteria are used industrially to produce glucanases for use in applications such as brewing to break down glucans in barley, thereby facilitating filtration and preventing clogging of filters.

Pullulanase (also known as "debranching enzyme") is a specific glucanase that degrades pullulan, a carbohydrate made up of repeating units of the trisaccharide maltotriose linked together by α-1,6 glycosidic bonds. Depending on the nature of the glycosidic bond cleaved, two types of pullulanases are distinguished. The type I pullulanases cleave the α-1,6 bonds, while the type II pullulanases hydrolyze both the α-1,4 and α-1,6 linkages. Pullulanases are produced by bacteria (e.g., *B. cereus* (Nair et al 2006) and *Klebsiella* sp. (Pugsley et al. 1986)), and they are also found in plants (e.g., rice (Yamasaki et al. 2008), wheat (Liu et al. 2009), maize (Dinges et al. 2003)).

By virtue of their capacity to break down glucans into glucose and low molecular weight oligomers of glucose, glucanases and pullulanases are used extensively in industry for the manufacture of animal feeds, alcoholic (beer and wine) and nonalcoholic beverages (tea, fruit pulps), baked goods (breads), leather, and textiles, as well as for waste treatment, oil extraction, and in detergents.

Various amylase mixes are used as multienzyme cocktails, which are employed to hydrolyze carbohydrates. An example is the product Viscozyme, produced and marketed by Novozymes from the fungus *A. aculeatus*. Viscozyme is used widely in the food industry to degrade nonstarch polysaccharides, including branched pectin-like substances often associated with starch in plant materials, thereby reducing the viscosity of such materials to facilitate the extraction of useful products from plant materials. This way, the enzyme can also make starches more readily available for fermentation into various liquefied products.

Other carbohydrases available under various trade names for food uses include a fungal amylase powder derived from *A.*

oryzae known as asperzyme and a liquid saccharifying amylase for the production of fermentable sugars known as adjuzyme, which is derived from *A. niger*. An Enzeco glucoamylase powder derived from *A. niger* is also commercially available for use in baked goods to improve proofing in lean or frozen doughs, and brewers amylase produced in powder form from *Aspergillus sp* and *Rhizopus oryzae* is used for the manufacture of light beers. Others are Brewers Fermex derived from *Aspergillus sp.* and *Rhizopus oryzae* in powder form for use in increasing the fermentability of wort and cookerzyme derived from *B. subtilis* for accelerated hydrolysis of starch adjuncts.

Galactosidases and Lactases

Galactosidases are those carbohydrases that catalyze the hydrolysis of galactosides such as lactose, lactosylceramides, ganglioside, and various glycoproteins into their constituent monosaccharide units. Two types of galactosidases are distinguished based on their mode of action as α- and β-galactosidases. The α-galactosidases (also known as α-gal) catalyze the hydrolytic cleavage of the terminal α-galactosyl residues from α-galactosides (including certain glycoproteins and glycolipids) and they can catalyze the hydrolysis of the disaccharide melibiose into galactose and glucose. α-Gal is used as a dietary supplement (e.g., beano) to facilitate the breakdown of complex sugars into simpler forms to permit easier digestion and reduce intestinal gas and bloating. The enzyme may also be used to reduce the complex carbohydrate content of the beer to increase the simple sugar content and hence the alcohol content of beer. The β-gals facilitate the hydrolysis of β-galactosides such as lactose, various glycoproteins, and ceramides into simpler forms.

Lactases belong to the β-galactosidase family of enzymes; they catalyze the breakdown of the milk sugar lactose into its constituent monosaccharides, galactose and glucose units, that is,

$$\text{Lactose} + H_2O \Leftrightarrow \text{Galactose} + \text{Glucose}$$

Lactases are produced in commercial quantities from yeasts such as *K. fragilis* and *K. lactis*, and from fungi such as *A. niger* and *A. oryza* (Mustranta et al. 1981). Other lactase-producing microorganisms include the bacteria *Streptococcus cremoris*, *Lactobacillus bulgaricus*, and *Leuconostoc citrovorum* (Ramana Rao and Dutta 1978). Lactases are also found in the mucous lining of the small intestines of cows and rabbits, and in the walls of the small intestines of humans from as early as the fifth month of intrauterine life. However, lactase levels tend to be higher in younger animals and human infants, and the levels generally decrease with age (Heilskov 1951). The enzyme is also found in plants, e.g., from the calli and roots of gherkin, pea, poppy, and eastern blue star seedlings (Neubert et al. 2004).

Lactase is commercially produced by Novozymes from the dairy yeast *K. lactis* and marketed as Lactozym. Lactase as lactozym is widely used in the dairy industry to produce fermented milks, ice-cream, milk drinks, and lactose-reduced milk for people who are lactose-intolerant and for household pets (e.g., cats) that cannot utilize lactose. Milks treated with lactose tend to be sweeter than their untreated counterparts because the monosaccharides glucose and galactose produced in the treated milks are sweeter than the disaccharide lactose in untreated milks. Crystallization of lactose in condensed milk can also be prevented by treating the milk first with lactase. Lactase-treated milks used for the preparation of cultured milk products (e.g., cottage cheese or yogurt) ferment faster than untreated milks, because lactose in the latter ferment at a slower pace than the glucose and galactose in the treated milks. As well, the cultured milk products from the lactase-treated milks tend to be sweeter than the products prepared from the nontreated milks. Milk whey may also be treated with lactase and concentrated to form syrups for use as a sweetener in food products. Ice cream made from lactase-treated milk (or whey) also has little or no lactose crystals, thus the undesirable "sandy" feel due to lactose crystallization is averted.

Maltases

Maltase is the carbohydrase that breaks down the disaccharide maltose into glucose units as summarized in the reaction below:

$$\text{Maltose} + H_2O \Leftrightarrow 2\,\alpha\text{-Glucose}$$

Maltase is found in humans and other vertebrate animals, plants, and microorganisms (bacteria and yeast). In humans, digestion of starch commences in the mouth with ptyalin (saliva enzyme) and pancreatic amylases breaking down the starch into maltose. The maltose is subsequently hydrolyzed by intestinal maltase into glucose for utilization by the body to produce energy or for storage in the liver as glycogen. Maltase (from barley malt) is used in brewing to increase the glucose content for conversion into alcohol. Maltase from *A. oryzae* is used in the beet-sugar and molasses distilleries, and in the industrial production of glucose syrups (McWethy and Hartman 1979).

Cellulases

Cellulases are enzymes that potentiate the breakdown of the β-1,4-glycosidic bonds in cellulose. They are produced commonly by bacteria (e.g., *Clostridium thermocellum*, *Acidothermus cellulolyticus* (Tucker et al. 1988)), fungi (*Trichoderma reesei*, *T. hamatum*, *Fusarium roseum*, *Curvularia lunata* (Sidhu et al. 1986)), and protozoa (e.g., *Trichomitopsis termopsidis* (Odelson and Breznak 1985), *Eudiplodinium maggii*, *Epidinium ecaudatum caudatum*, and *Ostracodinium obtusum bilobum* (Coleman 1985)), although the enzymes are also found in plants (e.g., cress and pepper (Ferrarese et al. 29)) and animals (e.g., termites, crustacea, insects, and nematodes (Watanabe and Tokuda 2001)). A simplified equation for the hydrolysis of cellulose by cellulases may be summarized as

$$\text{Cellulose} + H_2O \Leftrightarrow n\beta\text{-D-Glucose}$$

Native cellulose is highly crystalline in nature and its complete hydrolysis entails stepwise reactions catalyzed by a complex of cellulase enzymes. It entails the preliminary unraveling of the crystalline cellulose molecules into simple (or linear) cellulose chains by endocellulases. The linear cellulose chains then serve as substrates for exocellulases to produce smaller oligosaccharides (such as cellobiose and cellotetrose), and the latter molecules are hydrolyzed further by cellobiases into β-D-glucose units.

Cellulase is used in coffee processing to hydrolyze cellulose during drying of the beans. They are also used extensively in laundry detergents, in textile, pulp, and paper industries, as digestive supplements in pharmaceutical applications, and to facilitate in the biotransformation of feedstock into biofuels, although this process is presently in its infancy. The enzyme is marketed commercially as *Celluclast* by Novozymes and is used to digest cellulose into fermentable sugars in the form of glucose oligomers, cellobiose, and glucose, or to reduce the viscosity of soluble cellulosic materials.

Invertases and Sucrases

Invertases, also known as β-fructofuranosidases, are enzymes that hydrolyze the disaccharide sucrose into its corresponding monosaccharide units, fructose and glucose, that is,

Sucrose + $H_2O \Leftrightarrow$ Fructose + Glucose

Sucrases also spilt sucrose into equimolar mixtures of glucose and fructose, the difference between the two being that invertases split the O–C (fructose) bond, while sucrases break the O–C (glucose) bond. The equimolar mixture of glucose and fructose thus produced is known as invert sugar or artificial honey. Invertase is produced for commercial use from the yeast *S. cerevisiae* or *S. carlsbergensis*, although bees also synthesize the enzyme to make honey from nectar. The enzymes are also present in bacteria (Bogs and Geider 2000), in the cell walls and vacuoles of higher plants like carrots (Quiroz-Castaneda et al. 2009), tomatoes (Elliot et al. 1993), as well as in the seeds, stems, roots, and tubers of several higher plants (Cooper and Greenshields 1961, 1964).

Invertase is produced and marketed commercially under the brand name Maxinvert by Novozymes and is used extensively in the confectionery industry to make products such as chocolate-covered berries and nuts, soft-centered candies, marshmallows, and fudges. Invertase derived from yeast is also commercially available from Enzeco for the inversion of sucrose into a mixture of fructose and glucose.

Pectinases

Pectinase or pectic enzymes are a group of carbohydrases that hydrolyze pectins, large and complex heteropolysaccharides comprising of 1,4-α-D-galactosyluronic acid residues. Pectins occur naturally in fruits and can contribute to the viscosity and haziness of alcoholic and nonalcoholic beverages. Pectic enzymes act to break down pectins into smaller molecules to enhance their solubility. By this action, pectic enzymes clear hazes in beverages that form from residual pectins in the juices and also increase the yields of fruit juices. Members of the pectinase family of enzymes include pectinesterase (also known as pectyl hydrolase, pectin methyl esterase, or PME) and polygalacturonase. They are widespread in the cell walls of higher plants (e.g., avocado, carrots, tomato), as well as in some bacteria (e.g., *B. polymyxa*, *Closytidium multifermentans*, and *Erwina aroideae*), fungi (e.g., *A. niger*, *Penicillium digitatum*, *Geotricum candidum*, and *Fusarium oxysporum*), and yeasts (e.g., *S. fragilis*). PME catalyzes the hydrolysis of pectins via the removal of methoxyl groups to form pectic acids, that is:

Pectin + $H_2O \Leftrightarrow$ Pectic acid + Methanol

Pectin lyase also known as pectolyase hydrolyzes pectins via eliminative cleavage of (1,4)-α-D-galacturonan methyl esters in pectins with 4-deoxy-6-*O*-methyl-α-D-galact-4-enuronosyl residues at the nonreducing termini. Food-grade pectolyase is produced on a commercial scale from fungi and is used to remove residual pectins in wines and ciders. The enzyme is also used in combination with cellulase in plant cell culture to degrade plant cell walls to generate protoplasts.

Polygalacturonases (PGases) break down the glycosidic bonds linking the galacturonic acid residues in the polymer, and they may be either endo or exo in action. The endo-PGases cleave the 1,4-α-D-galactosyluronic acid residues of pectins in a random fashion, while the exo-PGases degrade pectins by terminal action to produce galacturonic acid residues.

A variety of pectic enzymes are commercially available from Novozymes for applications in the food industry. Examples include Pectinex (a mixture of pectinases produced from *A. niger*) for fruit and vegetable juice processing and the maceration of plant tissues, Peelzym (produced from *A. aculeatus*) for hands-free peeling of citrus fruits, and NovoShape (produced from *A. aculeatus* and *A. oryzae*) for maintaining the shape and structure of whole chunks or pieces of fruits in a finished product.

Fructosidases

Fructosidases, also known as inulases, are a group of enzymes that catalyze the hydrolysis of inulin or fructans into simpler components. Inulin is a linear polysaccharide naturally present in plants. The linear chains consist of about 35 fructose molecules linked together by β-1,2-glycosidic bonds and terminating in a glucose unit. Interest in this group of enzymes has been growing in recent years because of their capacity to form industrial fructose from fructans. These enzymes are commonly found in plants and are also produced by microorganisms such as *Aspergillus* sp., *S. fragilis*, and *S. marxianus*. Fructosidases produce fructose from inulin in high yield for commercial use (Singh et al. 2007, Sirisansaneeyakul et al. 2007, Muñoz-Gutiérrez et al. 2009).

Lipases

The term lipases is used to refer to a large group of hydrolytic enzymes that catalyze the breakdown of ester bonds in biomolecules such as triglycerides (TGs), phospholipids, cholesterol esters, and vitamin esters. On the basis of the nature of their substrate molecules, lipases are classified into triacylglycerol lipase (or true lipases), phospholipases, sterol esterase, and retinyl-palmitate esterase. For TGs, arguably the most abundant food lipids, lipases hydrolyze the ester bonds in TGs to form free fatty acids and glycerol. Lipases belong to the α/β

hydrolase family of enzymes that are characterized by a particular α-helix and β-structure topology, that is, a fan-like pattern of the central β-sheets, and also include esterases, thioesterases, and several other intracellular hydrolases. The catalytic triad of the α/β hydrolase proteins consists of a nucleophilic serine residue, an aspartic or glutamic acid residue, and a histidine group and differs from the catalytic triad in serine proteases only in the sequence in which the amino acids in the triad occur (Wang and Hartsuck 1993, Wong and Schotz 2002). Homologous lipases exhibit subtle differences from one another, and four major types are distinguished based on structural, functional, or sequence homologies: (i) the consensus sequence GxSxG around the active site serine, where x can be any amino acid, (ii) those with the strand-helix motif also around the active site serine; (iii) those with a buried active site covered by a region known in different lipases as active site loop, lid domain, or flap, and (iv) those with the order of the catalytic triad residues Ser . . . Asp/Glu . . . His (Carriere et al. 2000, Svendsen 2000, Hui and Howles 2002).

Lipase Specificity, Mechanism of Action, and Some General Properties

There are several categories of lipase specificity. Fatty acid-specific lipases tend to prefer specific fatty acids or classes of fatty acids (e.g., short-chain, polyunsaturated, etc.). Some lipases exhibit positional specificity and are regioselective in discriminating between the external, primary (*sn*-1 and *sn*-3 positions), and internal, secondary (*sn*-2 position), ester bonds. For example, several lipases of microbial origin, as well as gastric and phospholipases, show a preference for the external ester bonds; some exhibit partial stereospecificity, that is, a preference for either *sn*-1 or *sn*-3 position of TG, and some others exhibit enantioselectivity and can differentiate between enantiomers of chiral molecules (Anthonsen et al. 1995, Hou 2002).

The most accepted kinetic model of lipase activity is that the lipase exists in both the active and inactive conformation in solution. In aqueous solutions, the inactive conformation is predominant, thus the enzyme activity is low. However, at the lipid–water interface, the active conformation is favored, and water-insoluble substrates are hydrolyzed by the enzyme. The active forms of the enzyme in solution that are adsorbed at the interface may have different conformations (Martinelle and Hult 1995). Interfacial activation of lipases is well documented. Excess substrate induces a drastic increase in lipase activity, while esterases display a hyperbolic, Michaelis–Menten relationship between the enzyme activity and substrate concentration, lipases show a sigmoidal response (Anthonsen et al. 1995). Binding of lipase to the emulsified substrate at the lipid–water interface elicits a conformational change that greatly enhances its catalytic activity: the lid domain flips backwards to expose the active site; as well, the $\beta5$ loop orients in a way to enhance access of the substrate to the enzyme's active site.

Lipases are water-soluble proteins; nonetheless, their substrates are water-insoluble. Pancreatic lipases have the following features: temperature optima around 37°C and thermal stabilities up to approximately 50°C (Gjellesvik et al. 1992), and pH optima in the range of 6.5–8.5 and pH stabilities between pH 6 and 10 (Wang and Hartsuck 1993). The molecular weights range between 45 and 100 kDa (Wong and Schotz 2002, Wang and Hartsuck 1993), and they are anionic with isoelectric points typically ranging from 4.5 to 7 under physiological conditions (Armand 2007). They are inhibited by organophosphorous compounds like di-isopropyl fluorophosphates via irreversible phosphorylation of active site serine. High concentrations of both cationic (e.g., quaternary ammonium salts) and anionic (e.g., SDS) surfactants also inhibit lipase activity. Other inhibitors of lipases include metal ions (e.g., $Fe^{2+/3+}$ and mercury derivatives). Calcium and zinc ions may either activate lipases or produce no effect (Patkar and Bjorkling 1994, Anthonsen et al. 1995).

Conventional Sources of Lipases—Mammalian, Microbial, and Plant

The pancreas and serous glands of ruminants, especially young calf and lamb, and the pancreas of pigs have served as important sources of lipases for flavor development in dairy products for quite some time now. In subsequent years, microbial lipases have received most of the attention in lipase research and have been considerably exploited in numerous applications as a consequence of the relative ease of their production and genetic manipulation, as well as the wider range in properties. Microbial lipases constitute the most significant commercially produced lipases.

Lipases have also been sourced from plants. However, there is a dearth of information on plant lipases compared to mammalian and microbial lipases. Plant lipases thus far characterized include the triacylglyceride lipases, various phospholipases, as well as glycolipase and sulpholipase. The appeal of plant lipases derives from their unique capacity to cleave all three fatty acids from TGs (e.g., as shown by oat lipase) and their distinct substrate specificities and abilities to synthesize "structured lipids" (as demonstrated with rape-seed lipase that esterifies fatty acids to primary alcohols exclusively). Several distinctive features of plant lipases have been imputed to phospholipases, such as the use of phospholipase D from cabbage to synthesize phosphatidylglycerols for use as an artificial lung surfactant. In food processing and preservation, an understanding of the properties of plant lipases is crucial for maintaining quality and freshness, particularly in oilseeds, cereals, and oily fruits, and for the production of high-quality edible oils (Mukherjee and Hills 1994, Mukherjee and Kiewit 2002).

Phospholipases are a broad class of enzymes catalyzing mainly water-insoluble substrates and are involved in digestive, regulatory, and signal transduction pathways, among others. Their nomenclature (phospholipase A_1, A_2, B, C, and D) is based on which phospholipid ester bond they attack. The majority of these enzymes require Ca^{2+} for (maximal) catalysis, have many disulfide bonds, and are relatively small with molecular weights ranging from 13 to 30 kDa, but there are few above 80 kDa (Lopez-Amaya and Marangoni 2000a). The amino acid sequences of most phospholipases are totally different from those of TG lipases (Svendsen 2000). Like lipase activity, phospholipase activity has also been associated with the deterioration of seafood postharvest (Lopez-Amaya and Marangoni 2000b).

Sphingomyelin (SM) phosphodiesterase (also known as sphingomyelinase or SMase) is a hydrolytic enzyme involved in sphingolipid metabolism and is responsible for the formation of ceramide (and phosphocholine) from sphingomyelin in response to cellular stresses. However, SMase is usually treated separately from "regular" lipases.

Some Applications of Lipases

Dairy Products Lipases occur naturally in milk and have been shown to be present in the milk of the cow, goat, sheep, sow, donkey, horse, camel, and humans. Perhaps, the natural role of lipases in mammalian milks is to aid digestion for the young suckling. However, in dairy products, lipases play an important role in flavor development: for example, "membrane" lipases can produce undesirable rancidity in freshly drawn milks. In processed milk and milk products, this may not be expected to pose a problem due to the high pH range for milk lipase activity (pH 6–9), the low thermal stability of the lipases, and their inability to survive pasteurization treatments. In some dairy products, post-process lipolysis is desirable to produce the flavors normally associated with such products, such as butterfat, cheddar cheese, blue cheese, and Roquefort cheese. In these products, supplementation with microbial lipases is done to facilitate the ripening process. During "ripening," the added lipases break down milk fat into free fatty acids. Different types of lipases contribute to distinctive flavors. In this regard, lipases that release short-chain fatty acids (i.e., C4–C6) produce "sharp" flavors in food products, while the lipases furnishing long-chain fatty acid (> C12) tend to elicit more subtle tastes (Schmidt and Verger 1998). For example, cheddar cheese has higher levels of butyric acid (C4:0) versus milk fat (Bills and Day 1964). In this regard, lipases from *Penicillium roqueforti*, *A. niger*, *Rhizopus arrhizus*, and *Candida cylindracea* have been found to be very useful (Ha and Lindsay 1993). Although free fatty acids produce distinctive flavors in cheeses and other dairy products, controlling of lipolysis after the desired transformation has been achieved in the product is necessary because excessive lipolysis could induce objectionable rancid flavors in the products (McSweeney and Sousa 2000).

Oleo Products Oleo products are spreads prepared mostly from vegetable oils and used as a substitute for butter. The term covers a broad spectrum of products ranging from margarines to specialty fats, shortenings, and lard substitutes. The making of oleo products takes advantage of the capacity of lipases to catalyze interesterification reactions that enable modifications in acylglycerols from fats and oils. An example is the making of cocoa butter equivalents. Cocoa butter has 1-palmitoyl-2-oleoyl-3-stearoyl-glycerol (POS) or 1,3-distearoyl-2-oleoyl-glycerol (SOS) as its key components and is widely used in making chocolates and candies. Palm oil mid fraction with its predominant TG as 1,3-dipalmitoyl-2-oleoyl-glycerol has been used as common substrates to mimic cocoa butter by mixing it with co-substrates such as tristearin. In the presence of lipase, interesterification reactions proceed, whereby fatty acids in *sn*-1 and *sn*-3 positions are interchanged to produce POS and/or SOS. Some commercially available fungal lipases, e.g., lipozyme produced by Novozymes from *Mucor miehei* and Enzymatix F3 produced from *Rhizopus sp.* by Gist-brocades have been shown to be able to carry out such product-simulating interesterification reactions (Bloomer et al. 1983). Immobilized *Rhizopus arrhizus* lipase has been utilized to produce cocoa butter equivalent from oils, and the substrate conversion rate was significantly increased by addition of defatted soy lecithin (Mojović et al. 1993).

Lipase-catalyzed interesterifications are also employed to produce shortenings devoid of *trans*-fatty acids. Shortenings are usually produced by the hydrogenating of vegetable oils and the process forms *trans*-fatty acids as common components. Lipozyme TLIM from *Thermomyces lanuginosus* produces *trans*-fatty acids-free shortening from high oleic acid sunflower oil and fully hydrogenated soybean oil (Li et al. 2010). *Trans*-fatty acids-free shortening was also formed from palm stearin and rice bran oil by using the same lipozyme.

Lipases in Human Health and Disease

In humans, lipase breaks down dietary fats in food so they can be absorbed in the intestines. The enzyme is primarily produced in the pancreas but is also produced in the mouth and stomach. Proper digestion of dietary fat is important prior to their absorption because the end products must be carried in an aqueous milieu (blood and lymph) in which fats are not soluble.

The importance of lipase in the gastrointestinal tract is due to the fact that it is the primary agent for splitting fats into free fatty acids and glycerol. Gastric lipases digest pre-emulsified fats in foods such as egg yolk and cream; however, emulsification in the small intestine followed by lipolysis is the basis for the proper digestion of dietary fats and oils. Emulsification is effected by the action of bile salts produced by the liver and results in the breakdown of the large fat molecules into tiny droplets, which provide lipases with an increased surface area to act on. If bile is insufficient or when the liver is not stimulated to produce bile, fats remain in such large particles that enzymes cannot readily combine with; hence, fat digestion is incomplete and fat absorption is markedly reduced. Most people produce enough pancreatic lipase for this purpose; however, individuals with cystic fibrosis, Crohn's disease, or celiac disease may not have sufficient lipase to assure proper absorption and assimilation of the lipid breakdown products.

Lipases control cell permeability so that nutrients can enter while waste exit. Lipase-deficient individuals have reduced cell permeability; thus, nutrients cannot enter into cells nor can waste metabolites leave. For example, diabetics are lipase-deficient and glucose entry into their cells is impaired, while waste or unwanted substances tend to accumulate in the cells. When lipase is deficient, individuals may suffer health defects including a tendency toward high cholesterol, high TGs, difficulty in losing weight, diabetes or glucosuria, and ultimately cardiovascular diseases.

In humans, inhibition of digestive lipases is important for the control of obesity. The inhibitors act by targeting gastric and pancreatic lipases. In this regard, an understanding of lipase structure and function relationship as well as lipase catalytic mechanism is crucial for the development of inhibitors for pharmacological and medical use. The well-known drug Orlistat (also known as xenical or tetrahydrolipstatin) used to treat obesity acts by blocking digestive lipase action, leaving consumed fat undigested (Halford 2006). The basic polypeptide, protamine, is a natural lipase inhibitor and acts to hinder fat absorption (Tsujita et al. 1996). Several soy proteins including lipoxygenase-1 are able to inhibit pancreatic and microbial lipases (Satouchi et al. 2002). The inhibitory capacity of soybean proteins has been attributed to interactions with lipids and the ability to transform the substrate emulsion. There are also reports of other promising natural inhibitors of lipases, e.g., medicinal plant extracts (Sharma et al. 2005), polyphenolics, e.g., resveratrol, catechins, and 3-*O*-*trans*-*p*-coumaroyl actinidic acid from red wine, green tea, and kiwi fruit (Armand 2007, Koo and Noh 2007, Jang et al. 2008).

While high dietary fat can cause severe adverse health effects (including cardiovascular diseases and cancer), some amount of fat is absolutely required to serve as components of biological membranes, hormones, vitamins, etc. Thus, an adequate supply of essential fats in the diet is important to sustain human health.

Isomerases

This group of enzymes catalyzes structural rearrangement of molecules into their corresponding isomers. Thus, for the general reaction A→B, catalyzed by an isomerase, A and B are isomers. While several isomerases have been characterized in the literature, the one with food use is glucose isomerase, or more specifically, xylose isomerase (Chaplin and Burke 1990). The enzyme is produced from several microorganisms, such as *Actinoplanes missouriensis*, *B. coagulans*, *Streptomyces albus*, *S. olivaceus*, and *S. olivochromogenes*, and it normally isomerizes D-xylose to D-xylulose, that is,

D-Xylose ⇔ D-Xylulose

However, this same enzyme can act on D-glucose and isomerize it to D-fructose, that is,

D-Glucose ⇔ D-Fructose

D-fructose is about 30% sweeter than both D-glucose and D-xylose, and plays an important role in the food industry as a sweetener or component of sweeteners; hence, the enzyme is commonly referred to as glucose isomerase instead of its more apt name xylose isomerase.

The free form of the enzyme is quite expensive; thus, it is produced in the immobilized form to permit reuse and reduce cost for the commercial production of fructose from D-glucose. Thus, high-fructose corn syrup (HFCS) is produced through isomerization by glucose isomerase from D-glucose derived from the hydrolysis of corn starch. To make HFCS, corn starch is first converted to liquefied starch by α-amylase and subsequently to glucose by glucoamylase before the isomerization step. The fructose yield in HFCS is temperature dependent, with much higher yields (about 55%) recovered at 95°C versus 42% at the conventional 60–70°C production temperature. The high thermal stability of glucose isomerase permits its use at high temperatures.

ENZYMES IN FOOD AND FEED MANUFACTURE

Enzymes from plants, animals, and microorganisms have been used since time immemorial to modify foodstuffs. Examples of the food enzymes of plant origin include papain, bromelain, and ficin; examples of animal enzymes include rennets, pepsins, trypsins, chymotrypsins, pancreatic lipases, lysozyme, and cathepsins; and examples of microbial enzymes are various amylases, cellulases, proteases, lipases, TGases, glucose isomerase, and GOX derived from selected bacteria, yeasts, and fungi. The uses of these enzymes in foods result in desirable changes with respect to improvements in appearance, consistency, flavor, solubility, texture, and yield, among others. In this regard, enzymes have been used as processing aids in the manufacture of various baked goods, dairy products, meats and fish products, beverages (alcoholic and nonalcoholic), etc. (Table 9.2).

Baked Goods

Baked goods are prepared from flours such as wheat flour, which has starch as its main constituent. Amylolytic enzymes break down flour starch into small dextrins that become better substrates for yeast to act upon in the bread-making process. The use of enzymes in the baking industry is expanding to replace the use of chemicals in making high-quality products, in terms of better dough handling, anti-staling properties, as well as texture, color, taste, and volume. For example, enzymes like GOX and lipase are used in baked goods to strengthen dough texture and enhance elasticity in place of chemicals such as potassium bromate and ascorbate. Other enzymes used in the baking industry include LOX for bleaching flour and strengthening doughs; proteases for degrading and weakening gluten to impart "plastic" properties to doughs and make them more suitable for the making of biscuits; pentosanases (also known as xylanases) for enhancing dough machinability to result in more flexible, easier-to-handle stable doughs that provide better crumb texture and increased loaf volume; and asparaginases employed to curtail the formation of acrylamide in baked goods derived from cereal- and potato-based flours. The enzyme asparaginase breaks down asparagine in the flours to reduce its availability for reaction with reducing sugars to form acrylamide at high temperature.

Dairy Products

The formulation of dairy products entails the use of several enzymes, such as proteases (e.g., microbial rennets or their homologs from calf and other ruminants), lipases, and lactases. The proteases are used to precipitate milk caseins as "clots" or "curds" during cheese-making; the proteases and lipases also promote flavor development during ripening of cheese and in products like coffee whiteners, snack foods, soups, and spreads (margarines); while lactases are used to produce lactose-free

Table 9.2. Selected Food Products and the Enzymes Involved in Their Transformations

Food Products	Group(s) of Enzyme Used in Transformations	Nature of Transformation(s) Induced
Baked goods	• Amylases	• To effect break down of starches to increase reducing sugars to improve fermentation, increase loaf volume, prevent staling, and increase product shelf life
	• Cellulases and xylanases	• To improve conditioning, mixing time, and machinability of doughs
	• Oxidase	• To improve texture/structure of dough, increase dough volume, and reduce the need for emulsifiers and chemical oxidants (e.g. bromates)
	• Proteases	• To break down protein molecules in the dough and improve dough handling; reduce dough mixing time and control dough retraction, enhance flavor development; and to break down gluten and protect individuals that are gluten intolerant
	• Lipase	• To modify fat to reduce dough stickiness, improve stability, and increase volume of doughs; and also enhance flavor
Dairy products	• Proteases	• To act on milk proteins to modify texture and solubility properties of milk and other dairy products; accelerate cheese ripening and improve flavor intensity
	• Lipases	• To transform the fat in milk and other dairy products to produce creamy texture, accelerate ripening, enhance flavor, and increase emulsifying properties
	• Lactase	• To hydrolyze lactose and prevent development of "sandy" taste in frozen ice cream and yoghurts, and protect individuals that are lactose intolerant
	• Transglutaminases (TGases)	• To improve texture and curtail syneresis in yoghurts, and to improve the mouthfeel properties of sugar-free low-calorie food products
	• Catalase	• To break down residual H_2O_2 used for cold pasteurization of milk in certain countries
Starch products	• Amylases, glucoamylases and pullulanases	• To liquefy starch to produce low-viscosity dextrose syrups comprised of maltodextrins and other oligosaccharides; and to saccharify the maltodextrins/oligosaccharides into glucose syrups that are high in reducing sugars (glucose, maltose, and isomaltose)
	• Glucose isomerase (immobilized)	• To isomerize some of the high-glucose syrups (HGS) into fructose—to result in high-fructose syrups (HFS) that are sweeter than glucose and HGS. HFS and HGS are both used as sweeteners in the food and beverage industries
	• Carbohydrases and protease mixtures	• For the break down and removal of starches and proteins from oil seeds during oils extraction, and to facilitate the recovery and improved oil yields
	• Proteases	• To modify grain proteins and improve their utility as animal feed
Alcoholic beverages	• Amylases including glucoamylases (heat-stable forms)	• To liquefy grain starches to increase fermentable sugar (glucose and maltose) content and wort yields, ease clarification and filtration of the beverage, enhance yeast growth and accelerate fermentation rates
	• Cellulases and pectinases	• To break down celluloses and pectic substances in plant materials to facilitate liquefaction, increase fermentable sugar contents, and reduce viscosity for easier handling by yeasts for faster fermentation
	• Proteases	• To break down proteins in plant materials to improve solubility and flavor, facilitate filtration, and prevent post-product haze formation for increased clarity and stability
	• Xylanases and glucanases	• To break down complex carbohydrates (xylans and glucans) to improve filtration and clarity, and reduce residual carbohydrate levels during the manufacture of light beers
	• Mix of glucanases and proteases	• To enhance beer clarity and storage stability, increase solubility and nutritional value, and reduce viscosity haziness in beverages

Table 9.2. *(Continued)*

Food Products	Group(s) of Enzyme Used in Transformations	Nature of Transformation(s) Induced
• Fruit juices (non-alcoholic beverages) • Coffee and tea products	• Pectinases	• To break down and remove pectins to increase yields, facilitate clarification, and improve stability; for peeling operations in some citrus products
	• Mix of cellulases and glucosidases	• To facilitate maceration of high carbohydrate-containing fruits (e.g., mangoes, bananas, and papaya) and vegetables (e.g., carrots and ginger) for better recovery of juices
	• Carbohydrase enzyme mix	• To improve maceration to facilitate and enhance juice extraction, increase juice yields, and enhance color and flavor of juices
	• Naringinase	• To break down naringin and reduce bitterness in citrus juices
	• Glucose oxidase	• For the removal of glucose and head space O_2 from bottled/carton packaged soft drinks and dry mixes
	• Pectinase	• To assist in maceration of tea leaves to improve strength and color
	• Tannases	• To remove tannins
	• Polyphenoloxidases	• To improve enzymatic browning and enhance color of the products
Meats and seafood products	• Proteases (heat stable forms, e.g., papain, ficin, and bromelain)	• To modify texture and induce tenderness in meats and squid, chewability, and digestibility; to reduce bitterness and improve flavor as well as nutritive value, produce hydrolysates from meat scraps, underutilized fish species and fish processing discards; enhanced flavors in fermented herring (matjes)
	• TGases	• To improve texture in meats and seafood products, form restructured meats from trimmings and surimi-type products, some soft-fleshed fish species, form "umami" flavors for use as additives to meat products
	• Mix of proteases and carbohydrases	• To de-skin fish and aid the break down of roe sacs to improve roe yields
Egg products	• Lipases	• To eliminate egg (whole, liquid, or yolk) lipids to preserve its formability
	• Phospholipases	• To modify whole egg or egg yolk to increase its emulsifying and gelling properties
	• Glucose oxidase (GOX)	• To de-sugar powdered eggs and curtail enzymatic browning
	• Catalase	• To break down residual H_2O_2 in egg products or egg ingredients (into H_2O and O_2) after partial sterilization of the eggs with H_2O_2. used to sterilize the product
Dietetic foods	• Pancreatin and glutaminase	• To facilitate digestion and utilization of wheat proteins (gluten) by individuals that are intolerant to gluten
	• Lactase	• To break down milk lactose to enhance safe utilization by lactose intolerants
Protein products	• Proteases	• To hydrolyze proteins derived from yeasts (e.g., single cell proteins), meats, and vegetable such as corn, soy, pulses (e.g., peas and lentils), and wheat to produce highly soluble and low viscosity protein hydrolysates with intense/improved flavors and high nutritive value for use as food ingredients and flavor enhancers; proteases (e.g., chymotrypsins) are also used in food to reduce bitterness due to bitter peptides
	• Carbohydrase mix	• To break down complex carbohydrates in high protein-containing vegetables, and facilitate their removal for the extraction of vegetable proteins in "pure" form and in high yields

(Continued)

Table 9.2. (*Continued*)

Food Products	Group(s) of Enzyme Used in Transformations	Nature of Transformation(s) Induced
Candies and confectioneries	• Amylases and invertases	• To produce high-maltose and high-glucose syrups for the manufacture of hard candies, soft drinks, and caramels; and to increase sweetness and aid soft cream candy manufacture; and for the recovery of sugars from candy scraps
	• Amylase, invertase and protease mix	• To improve sweetness, texture, and stability of candied fruits
	• Lipases	• To modify butterfat to increase buttery flavors and reduce sweetness in candies and caramels
Animal feed and pet care	• Cellulases and pectinases	• To facilitate silage production
	• Alkaline proteases	• To decolorize whole blood from abattoirs for use as animal feed
	• Carbohydrase, phytase, and protease mix	• To increase metabolizable energy and protein utilization, reduce viscosity, and liberate bound phosphorous
	• Proteases	• To reduce pain and inflammation and increase wound healing

milk and dairy products for the benefit of individuals that are lactose-intolerant. Catalase from *A. niger* is used to decompose residual H_2O_2 (that is used in some countries as a preservative for milk and whey; Burgess and Shaw 1983) into H_2O and O_2 that are nontoxic unlike H_2O_2.

Meat and Fish Products

Meat Tenderization

Tenderness in meats relates to the ease of these products to chomping or chewing. Meat tenderness arises from modifications in muscle proteins and entails breakdown of connective tissue proteins by proteases. These hydrolytic enzymes may be naturally present in the post-mortem animal (e.g., cathepsins) or they may be added exogenously. In some meats, the action of the endogenous enzymes may be inadequate to achieve tenderization within a time period that is useful. Exogenous enzymes that are used for household or commercial purposes include thermostable proteases from microbial and plant sources. Some peculiar features of these proteases used as meat tenderizers include their broader specificity and thermostability. Examples are papain, ficin, and bromelain (all from plant sources). The enzymes may be applied by sprinkling over the animal surface or dipping chunks of the meat in a protease solution (mostly in household practices). The disadvantage with these approaches is that the enzymes get unevenly distributed in the meat and results in unequal tenderness in different parts of the meat (e.g., the interior of the meat may invariably be tougher than the outside). Commercial practices include spreading the enzyme throughout the meat either by repeated pre-mortem injections of the protease solutions into the animal prior to slaughter or via post-mortem injections spread over several parts of the carcass.

Fish Processing

Enzymes are used in the fishing industry to achieve a variety of results such as their use in de-skinning fish species that are normally difficult to de-skin by manual or mechanical operations (e.g., squid, skate, and tuna). De-skinning is achieved using a mixture of proteolytic and glycolytic enzymes (Haard and Simpson 1994, Gildberg et al. 2000, Shahidi and Janak Kamil 2001). Heat-stable fungal and plant proteases are also used to accelerate fish sauce fermentation (Raksakulthai and Haard 1992a,1992b, Haard and Simpson 1994), and proteases are also used to enhance flavor development during the preparation of fermented herring (Matjes; Simpson and Haard 1984) and to tenderize squid meat (Raksakulthai and Haard 1992a,1992b).

Enzymes in Beverages

A number of enzymes are used in the manufacture of alcoholic and nonalcoholic beverages. Enzymes such as heat-stable bacterial amylases, glucanases, and pullanases are used to liquefy starches from cereals during mashing to increase fermentable sugar levels in the wort; heat-stable bacterial phytases are used to break down insoluble phytins into phytic acid, which is more soluble and lowers the pH for improved microbial and enzymatic activity in the fermentation medium; and various proteases are used to digest large-molecular weight and water-insoluble proteins (e.g., albumins and globulins) into smaller and soluble peptides in order to curtail haze formation in the products.

The nonalcoholic beverages include juices derived from various fruits (citrus, apples, apricots; and tropical fruits such as avocadoes, bananas, mangoes, and papayas) and vegetables (e.g., carrots), as well as tea, cocoa, and coffee. The enzymes that are used to facilitate the processing of fruits and vegetables into juices include a range of pectinases for peeling the fruits, maceration, removal of pectins, viscosity reduction, and clarification of the juice); amylases and AMGs are applied for breaking down starches in high starch-containing fruits (e.g., unripe apples, bananas) to prevent post-product haze development in the products; cellulases and hemicellulases are used for digesting cellulose and fruit cell wall hemicellulosic materials (i.e., polysaccharides made up of a number of different monosaccahride units such as xylose, mannose, galactose, rhamnose, and arabinose, and sugar acids such as mannuronic acid and galacturonic acid); and the enzyme naringinase is used for bitterness reduction in citrus fruit juice (e.g., grapefruit juice) processing.

Tea, cocoa, and coffee contain varying levels of caffeine and its related compound theobromine. Thus, intake of beverages derived from these plant materials elicits a stimulant effect in the consumer. Enzymes like polyphenoloxidases and peroxidases promote oxidation and browning reactions during "fermentation" of these products (tea, coffee, and cocoa) as part of the overall darker colors normally associated with these products; in the particular case of coffee, the dark color is enhanced by the roasting of the beans. Amylases are used in cocoa and chocolates to liquefy starches to permit free flow, and in the case of teas, the enzyme tannase breaks down tannins to enhance the solubilization of tea solids. Other enzymes such as invertase, LOX, peroxidases, and proteases also play an important role in flavor development in the beverages derived from these products.

The enzyme GOX is also used to remove head-space O_2 from bottled beverages.

Enzymes in Candies and Confectioneries

Various hydrolytic enzymes are used as processing aids for the manufacture of chocolate-covered soft cream candies, e.g., amylases and invertase, and for the recovery of sugars from candy scraps, e.g., amylases, invertase, and proteases (Cowan 1983). A mixture of cellulases, invertase, pectinases, and proteases are used in making candied fruits (Mochizuki et al. 1971). Amylases are also used to produce high-maltose and high-glucose syrups for use as sweetener in the manufacture of hard candies, soft drinks, and caramels. Lipases are used to modify butterfat to increase buttery flavors in candies and caramels and to reduce sweetness as well (Burgess and Shaw 1983).

Enzymes in Animal Feed and Pet Care

Agricultural (used here to encompass plant, livestock, and fish) harvesting and processing discards are high in useful nutrients such as carbohydrates (both starch and nonstarch types), proteins, and lipids, some of which cannot be fully digested and utilized directly by animals. In general, the discards are used as feed for animals. Different kinds of enzymes are added to these source materials to modify them into forms that are more amenable to digestion and utilization for improved feed efficiency. Examples of enzymes used for this purpose include microbial amylases, cellulases, glucanases, xylanases, phytases, and proteases, to degrade simple and complex carbohydrates, as well as proteins. Cellulases and pectinases also play an important role in silage production, while alkaline proteases (from bacteria and fungi) are employed for enzymatic decolorization of whole blood from abattoirs for use as animal feed protein. The enzymatic action results in an increase in available metabolizable energy and protein utilization, reduced viscosity, and liberation of bound phosphorous.

Enzymes are also used in pet care to provide relief from the effects of dry or scaly hair coats, skin problems, digestive and immune disorders, weight problems, allergies, bloating, and other disorders that afflict household pets. Enzymes are also used to enhance wound healing and rectify soreness, pain, and inflammations suffered by highly active as well as arthritic pets.

A summary of various foods produced commercially (and/or in households) via enzyme-assisted transformations is provided in Table 9.2.

CONTROLLING ENZYMATIC ACTIVITY IN FOODS

The action of enzymes in foods may not always be desirable. For example, continued enzymatic activity in foods after they have been used to attain the desired transformation could adversely affect food quality. The natural presence of certain enzymes in agricultural materials (e.g., LOX, PPO, peroxidases, and lipases) may induce undesirable changes (e.g., color loss, dark discolorations, and rancidity) in foods; the action of some naturally occurring enzymes (e.g., histidine decarboxylases) may produce toxic compounds (biogenic amines) in food products; and yet other enzymes (e.g., thiaminase and ascorbic acid oxidase) may act to destroy essential food components (e.g., vitamins B_1 and C). Thus, it is necessary to control enzymatic activity in food stuffs to obviate the potential deleterious effects they cause in these products. There are several factors that are known to affect enzyme activity and influence their behavior in foods. The factors include temperature, pH, water activity (A_w), and chemicals (inhibitors, chelating agents, and reducing agents), and advantage is taken of this fact by food technologists and food manufacturers to develop or rationalize procedures to control enzymatic activity in foods.

Temperature Effects

Temperature can affect the activity and stability of enzymes (as well as the substrate the enzymes act on). Thus, temperature effects on enzyme activity are mixed. Temperatures modulate the motion of biomolecules (both enzymes and their substrates) as well as interactions between molecules.

Heat Treatments

An increase in temperature (up to the temperature optimum of the enzyme) generally elevates the average kinetic energy of enzyme molecules and their substrates, and manifests as increased rates of enzyme catalyses. However, beyond the optimum temperature, further increases in temperature do not increase the average kinetic energy of molecules, rather they disrupt the forces that maintain the conformation and structural integrity of the molecules that are crucial for their stability and normal catalytic activity. Under those circumstances, enzyme molecules undergo denaturation, inactivation, and lose catalytic activity. The heat denaturation may be reversible or irreversible depending on the severity and/or duration of the heat treatment. Advantage is taken of this heat inactivation of enzymes in processes such as sterilization, blanching, UHT, high temperature short term treatment, pasteurization, and related thermal treatments to curtail undesirable enzymatic reactions in food and protect foods from enzyme-induced postharvest and post-processing spoilage that could arise from enzymes naturally occurring in food materials, or from enzymes deliberately added to foods to bring about particular transformations. Nevertheless, thermal processing of foods has the disadvantage of destroying heat-labile essential components in foods such as some vitamins and essential oils, thus the need for effective nonthermal techniques.

Low Temperature Treatments

A decrease in temperature slows down the average kinetic energy of biomolecules such as enzymes and their substrates. The molecules are sluggish at low temperatures and collide less frequently and effectively with one another, thus reaction rates are relatively slow. Hitherto, advantage is taken of lower thermal energy in approaches such as refrigeration, iced storage, refrigerated sea water storage, and frozen storage to slow down the undesirable effects of enzymes in foods after harvest. Low temperature treatments are generally used to slow down the deleterious effects of enzymes in fresh foods (vegetables, eggs, seafood, and meats); however, other processed food products such as high-fat food spreads (margarine and butter), vacuum-packaged meats and seafood products, pasteurized milk, cheeses, and yoghurt also benefit from the desirable effects derived from low temperature treatments.

It must be noted here that there are some enzymes that are quite stable and active in extreme temperatures (high- or low-temperature-adapted enzymes—known as extremophiles), which may survive the traditional temperature treatments and induce autolysis and spoilage in foods. For these enzymes, other techniques are needed to stop their undesirable effects.

Low-temperature treatments (refrigeration, chilling, and freezing) all slow down enzyme activity but do not completely inactivate the enzymes. Enzymatic activity still takes place in foods thus treated, albeit at much reduced rates. Once the food material is out of the source of the low temperature, once the frozen food material is thawed, enzyme activity may restart and cause undesirable autolytic changes in foods. Refrigerated and chilled storage in particular may not be considered as effective methods for long-term storage of fresh foods in high quality.

Effect of pH

Enzyme activity in foods is pH dependent (Table 9.3). In general, enzymes tend to be destabilized and irreversibly inactivated at extreme pH values, and advantage is taken of this property of pH to control enzymatic activity in foods. For example, the enzyme PPO is known to cause undesirable browning in fresh fruits and vegetables. By adding acids such as citrate, lactate, or ascorbate, acidic conditions are created (around pH 4) where the PPO enzyme is inactive. A similar effect is achieved when glucose/catalase/GOX cocktail is used on raw crustacea (e.g., shrimp); the GOX oxidizes the glucose to gluconic acid and this is accompanied by a drop in pH, which inactivates the PPO.

Table 9.3. pH Stabilities of Selected Enzymes

Enzyme	pH Stability Optimum
Pepsin	1.5
Malt amylase	4.5–5.0
Invertase	4.5
Gastric lipase	4.0–5.0
Pancreatic lipase	8.0
Maltase	6.5
Catalase	7.0
Trypsin	8.0

In household food preparations, vinegar or lemon juice is often sprinkled on fresh foods (e.g., vegetable salads) to prevent dark discolorations. Lowering pH of citrus juice with HCl has been used to achieve irreversible inactivation of pectin esterase (PE; Owusu-Yaw et al. 1988), and LOX activity in soy flour is drastically reduced at pH $\leq$ 5.0 (Thakur and Nelson 1997).

Effect of Inhibitors

Enzyme inhibitors are substances that slow down or prevent catalytic activities of enzymes. They do this by either binding directly to the enzyme or by removing co-substrates (e.g., O_2 in enzymatic browning) in the reaction catalyzed by particular enzymes. Enzyme inhibitors fall into several categories, such as reducing agents that remove co-substrates such as O_2 (e.g., sulfites and cysteine), metal chelators that bind or remove essential metal cofactors from enzymes (e.g., EDTA, polycarboxylic acids, and phosphates), acidulants that reduce pH and cause enzyme inactivation (e.g., phosphoric acids, citric acid, sorbates, benzoates), and enzyme inhibitors that bind directly to the enzymes and prevent their activity (e.g., polypeptides, organic acids, and resorcinols). Food-grade protein inhibitors from eggs, bovine/porcine plasma, and potato flour are all used to control undesirable proteolytic activity in foods. Egg white contains ovomucoid, a serine protease inhibitor; bovine and porcine plasmas have the broad spectrum protease inhibitor α_2-macroglobulin; and potato flour has a serine protease inhibitor (potato serine protease inhibitor or PSPI). Fractions containing these inhibitors have been used to prevent surimi texture softening due to proteolysis (Weerasinghe et al. 1996).

Effect of Water Activity

Enzymatic activity depends to the availability of water. Water availability, and hence water activity (A_w), in foods may be modified by using procedures that remove moisture, such as drying, freezing, addition of water-binding agents or humectants (such as salt, sugar, honey, glycerol, and other polyols), and by lyophilization (Tejada et al. 2008). Most enzyme-catalyzed reactions require moisture to progress effectively for several reasons: (i) a thin film of moisture (bound water) is required to maintain proper enzyme conformational integrity for functional activity, (ii) to solubilize substrates and also serve as the reaction medium, and (iii) to participate in the reaction as co-reactant. When moisture content is reduced by dehydration, there is conformational destabilization and loss of catalytic activity. For example, most enzymes such as proteases, carbohydrases, PPO, GOX, and peroxidases require $A_w \geq 0.85$ to have functional activity, the well-known exceptions being lipases that may actually gain in activity and remain active at A_w of 0.3, perhaps as low as 0.1 (Loncin et al. 1968). The unusual behavior of lipases is observed in lipase-catalyzed reactions with their water-insoluble substrates, lipids, and is due to the interfacial phenomenon that proceeds better in a reduced moisture milieu. Foodstuffs prepared based on A_w reduction to control the undesirable effects of enzymes (and microorganisms) include the use of sugar in food spreads (jams and jellies), salt in pickled vegetables, glycerol in cookies and liqueurs, and gelatin in candies and confectioneries.

Effect of Irradiation

Irradiation of foods (also known as cold pasteurization) is a process during which foods are subjected to ionizing radiations to preserve them. Food irradiation methods entail the use of gamma rays, X-rays, and accelerated electron beams, and can preserve food by curtailing enzymatic (and microbial) activities. However, the irradiation dosage needed to achieve complete and irreversible inactivation of enzymes may be too high and could elicit undesirable effects of their own (e.g., nutrient loss) in food materials. The technique is used to some extent in meats, seafood, fruits, and vegetables (especially, cereal grains) for long-term preservation; however, the procedure has low appeal and acceptability to consumers and, therefore, not extensively used in food processing.

Effects of Pressure

High-pressure treatment, also known as high-pressure processing (HPP) or ultra high-pressure processing, is a nonthermal procedure based on the use of elevated pressures (400–700 MPa) for processing foods. At the elevated pressures, there is inactivation of both enzymes and microorganisms, the two foremost causative agents of food spoilage, and the inactivation arises from conformational changes in the 3D structure of the enzyme protein molecules (Cheftel 1992). Enzymes in foodstuffs display different sensitivities to pressure; while some may be inactivated at relatively low pressures (few hundred MPa), others can tolerate pressures up to a thousand MPa. The pressure effects may also be reversible in some enzymes and irreversible in others, depending on the pressure intensity and the duration of the treatment. Those enzymes that can tolerate extreme pressures may be deactivated using appropriate pressure treatments in combination with other barriers to enzyme activity such as such as temperature, pH, and/or inhibitors (Ashie and Simpson 1995, Ashie et al. 1996, Sareevoravitkul et al. 1996, Katsaros et al. 2010).

HPP has minimal effects on the attributes of food, such as flavor, appearance, and nutritive value, compared with other procedures like thermal processing, dehydration, or irradiation. In contrast to those other processing methods, high-pressure treatments keep foods fresher and highly nutritious, improve food texture, make foods look better and taste better. The HPP approach also retains the native flavors associated with foods, protects heat-labile essential food components, and extends product shelf life with minimum need for chemical preservatives. The technology has been used mostly by companies in Japan and the United States to process foods such as ready-to-eat meats and meat products, food spreads (jams), fresh juices and beverages (sake), processed fruits and vegetables, fresh salads, and dips. HPP is also used to shuck and retrieve meats from shellfish (oysters, clams, and lobster).

CONCLUDING REMARKS: FUTURE PROSPECTS

Foodstuffs have naturally present enzymes as well as intentionally added ones that produce significant effects on food-processing operations. The actions of these enzymes may improve food quality or promote food quality deterioration. The use of enzymes as food-processing aids has increased steadily for several years now, and this trend is expected to continue for the foreseeable future due to the interest and need for more effective strategies and greener technologies to curtail the reliance on existing technologies and protect the environment. The factors that augur well for the expanded use of enzymes as food-processing aids include: consumer preferences for their use instead of chemicals, their use as food-processing aids is perceived to be more innocuous and more environmental friendly, their capacity to selectively and specifically remove toxic components in foods (e.g., glucosinolates with sulfatases, acrylamide with asparaginase, or phytates with phytases), and recent advances in enzyme engineering, which is permitting the discovery and design of new and superior enzymes tailored to suit specific applications.

Recent developments in enzyme engineering are enabling yields of particular enzymes to be improved by increasing the number of gene copies that code for enzyme proteins in safe host organisms, and this feature is particularly significant because they are permitting useful enzymes from plant and animal sources, and even their counterparts from uncertified microorganisms (including hazardous ones) to be produced in high yields much more consistently, rapidly, and safely for food use. It is also significant because the cultivation and growth of microorganisms for the purpose of producing recombinant enzymes is not dependent of weather conditions, proceeds much faster, and require much lesser space and regulations compared with plants or animals.

Enzyme engineering is also facilitating the design of new enzyme structures with superior performance characteristics with respect to catalytic activity, thermal stability, tolerance to pH and inhibitors, and other properties to make them more suited for applications in several fields of endeavor. This capacity is also expected to enhance the design and creation of synthetic or artificial enzymes (Chaplin and Burke 1990) with superior properties for both food and nonfood uses. Recombinant enzymes with improved stability would facilitate their use in producing more useful biosensors for food analysis and other analytical work (Chaplin and Burke 1990), stimulate the manufacture of more immobilized enzymes that are also endowed with superior properties for reuse and cost savings in food processing, produce new enzymes that can function well in nonaqueous milieu, and facilitate the synthesis of new molecules with predetermined structures and functions (as was achieved with the synthesis of aspartame by thermolysin) for use in foods. Recombinant enzyme technology is also expected to facilitate the discovery of newer applications for enzymes as food-processing aids and also produce new enzymes to assist the incorporation of specific essential molecules in food products to meet specific dietary needs, as use as dietary supplements to manipulate ingested carbohydrates, lipids, and cereal proteins (e.g., gluten) for improved human health and wellness.

The undesirable effects of enzymes in foods are controlled to some extent by traditional practices such as thermal treatments, cold storage, water activity (A_w) reduction, pH control, and treatment with chemicals. However, there are various limitations with these methods, such as destruction of heat-labile essential components in foods, continued enzymatic activity (albeit at a reduced rate) even under iced, refrigerated, or frozen storage; and the adverse effects of A_w reducing agents such as salt or sugar as well as chemicals on human health. Novel approaches based on nonthermal treatments for controlling enzymes (such as HPP and PEF) are promising, and studies to optimize their use either exclusively or in combination with each other or other barriers would be useful. Enzyme engineering provides a unique opportunity for exploiting the distinct capacity of enzymes from animals with no anatomical stomachs to inactivate native protein (and enzymes) molecules (Pfleiderer et al. 1967, Simpson and Haard 1987, Guizani et al. 1992, Brown 1995) in food processing. Thus, new enzymes designed to specifically recognize and bind undesirable native enzyme molecules (e.g., PME in fruit juices) could have a major impact on the use of nonthermal strategies to control the deleterious effects elicited by certain food enzymes.

REFERENCES

Abd-El-Al A, Pfaff HF. 1968. Exo-beta-glucanases in yeast. *Biochem J* 109: 347–360.

Adıgüzel AC et al. 2009. Sequential secretion of collagenolytic, elastolytic, and keratinolytic proteases in peptide-limited cultures of two *Bacillus cereus* strains isolated from wool. *J Appl Microbiol* 107(1): 226–234.

Almeida do Nascimento WC, Martins MLL. 2004. Production and properties of an extracellular protease from thermophilic *Bacillus* sp. *Brazilian J Microbiol* 35: 91–96.

Anthonsen HW et al. 1995. Lipases and esterases—a review of their sequences, structure and evolution. In: MR Gewely (ed.) *Biotechnology Annual Review*, vol. 1. Elsevier, Amsterdam, pp. 315–371.

Armand M. 2007. Lipases and lipolysis in the human digestive tract: where do we stand? *Curr Opin Clin Nutr Metab Car* 10(2): 156–164.

Ashie INA, Simpson BK. 1995. Effects of hydrostatic pressure on alpha 2-macroglobulin and selected proteases. *J Food Biochem* 18: 377–391.

Ashie INA et al. 1996. Control of endogenous enzyme activity by inhibitors and hydrostatic pressure using RSM. *J Food Sci* 61: 350–356.

Bachman ES, McClay DR. 1996. Molecular cloning of the first metazoan beta-1,3-glucanase from eggs of the sea urchin *Strongylocentrotus purpuratus*. *Proc Natl Acad Sci USA* 93: 6808–6813.

Barros TG et al. 2009. Pseudo-peptides derived from isomannide: inhibitors of serine proteases. *Amino Acids* 38(3): 701–709.

Benjakul S et al. 2010. Enzymes in fish processing. In: RJ Whitehurst, MV Oort (eds.) *Enzymes in Food Technology*, 2nd edn. Wiley-Blackwell, Ames, IA, Chapter 10, pp. 211–235.

Beschin A et al. 1998. Identification and cloning of a glucan- and lipopolysaccharide-binding protein from *Eisenia fetida* earthworm involved in the activation of prophenoloxidase cascade. *J Biol Chem* 273: 24948–24954.

Beuk JF et al. 1962. Improvement in or relating to meat products. UK Patent No. 913202.

Bills DD, Day EA. 1964. Determination of the major free fatty acids of cheddar cheese. *J Dairy Sci* 47: 733–738.

Bloomer S et al. 1983. Triglyceride interesterification by lipases. 1. Cocoa butter equivalents from a fraction of palm oil. *J Am Oil Chem Soc* 67(8): 519–524.

Bogs J, Geider K. 2000. Molecular analysis of sucrose metabolism of *Erwinia amylovora* and influence on bacterial virulence. *J Bacteriol* 182: 5351–5358.

Bonilha PRM et al. 2006. Cyclodextrin glycosyltransferase from *Bacillus licheniformis*: optimization of production and its properties. *Brazilian J Microbiol* 37: 317–323.

Brown PB. 1995. Physiological adaptations in the gastrointestinal tract of crayfish. *Am Zool* 35: 20–17.

Burgess K, Shaw M. 1983. Dairy. In: T Godfrey and J Reichelt (eds.) *Industrial Enzymology. The Application of Enzymes in Industry*. The Nature Press, New York, Chapters 4.6, pp. 260–283.

Carriere F et al. 2000. The specific activities of human digestive lipases measured from the in vivo and in vitro lipolysis of test meals. *Gastroenterology* 119: 949–960.

Chaplin MF, Bucke C. 1990. Enzyme preparation and use. In: MF Chaplin, C Bucke (eds.) *Enzyme Technology*, Cambridge University Press, Cambridge, Chapter 2, pp. 40–79.

Cheetham PSJ et al. 1989. Synthesis of novel disaccharides by a newly isolated fructosyltransferase from *Bacillus subtilis*. *Enzyme Microbiol Technol* 11: 212–219.

Cheftel JC. 1992. Effects of high hydrostatic pressure on food constitutes: an overview. In: C Balny et al.(eds.) *High Pressure and Biotechnology*, pp. 195–209.

Chen Y-X et al. 1999. Kinetically controlled synthesis catalyzed by proteases in reverse micelles and separation of precursor dipeptides of RGD. *Enzyme Microbiol Technol* 25(3): 310–315.

Clare DA, Blakistone BA, Swaisgood HE, et al. 1981. Sulfhydryl oxidase-catalyzed conversion of xanthine dehydrogenase to xanthine oxidase. *Arch Biochem Biophys* 11: 44–47.

Coleman GS. 1985. The cellulase content of 15 species of entodiniomorphid protozoa, mixed bacteria and plant debris isolated from the ovine rumen. *J Agric Sci* 104: 349–360.

Cooper RA, Greenshields RN. 1961. Sucrases in *Phaseolus vulgaris*. *Nature* 5(191): 601–602.

Cooper RA, Greenshields RN. 1964. The partial purification and some properties of two sucrases of *Phaseolus vulgaris*. *Biochem J* 92(2): 357–364.

Couto MA et al. 1993. Selective inhibition of microbial serine proteases by eNAP-2, an antimicrobial peptide from equine neutrophils. *Infect Immun* 61(7): 2991–2994.

Cowan D. 1983. "Proteins." In industrial enzymology. In: T Godfrey, J Reichelt (eds.) *The Application of Enzymes in Industry*. The Nature Press, New York, Chapter 4.14, pp. 352–374.

Dalev PG. 1994. Utilization of waste feathers from poultry slaughter for production of a protein concentrate. *Bioresour Technol* 48: 265–267.

Dalev PG, Simeonova LS. 1992. An enzyme biotechnology for the total utilization of leather wastes. *Biotechnol Lett* 14: 531–534.

de la Motte RS, Wagner FW. 1987. *Aspergillus niger* sulfhydryl oxidase. *Biochemistry* 26: 7363–7371.

Dinges JR et al. 2003. Mutational analysis of the pullulanase-type debranching enzyme of maize indicates multiple functions in starch metabolism. The plant cell. *Am Soc Plant Biol* 15: 666–680.

Elliot KJ et al. 1993. Isolation and characterization of fruit vacuolar invertase genes from two tomato species and temporal differences in mRNA levels during fruit ripening. *Plant Mol Biol* 21: 515–524.

Elortza F et al. 2003. Proteomic analysis of glycosylphosphatidylinositol-anchored membrane proteins. *Mol Cell Proteomics* 2: 1261–1270.

Etherington DJ, Bardsley RG. 1991. Enzymes in the meat industry. In: GA Tucker, LFJ Woods (eds.) *Enzymes in Food Processing*. Blackie Academic & Professional, London, pp. 144–191.

Fernandez M et al. 2000. Accelerated ripening of dry fermented sausages. *Trends Food Sci Technol* 11: 201–209.

Ferrarese L et al. 1995. Differential ethylene-inducible expression of cellulase in pepper plants. *Plant Mol Biol* 29: 735–747.

Gildberg A et al. 2000. Uses of enzymes from marine organisms. In: NF Haard, BK Simpson (eds.) *Seafood Enzymes. Utilization and Influence on Postharvest Seafood Quality*. Marcel Dekker, New York, Chapter 22, pp. 619–639.

Gjellesvik DR et al. 1992. Pancreatic bile salt dependent lipase from cod (*Gadus morhua*): purification and properties. *Biochimt Biophys Acta* 1124: 123–134.

Goda SK et al. 1997. Molecular analysis of a *Clostridium butyricum* NCIMB 7423 gene encoding 4-alpha-glucanotransferase and characterization of the recombinant enzyme produced in *Escherichia coli*. *Microbiology* 143: 3287–3294.

Godfrey T, West S. 1996 Introduction to industrial enzymology. In: T Godfrey, S West (eds.) *Industrial Enzymology*. Cambridge University Press, Cambridge, pp. 3–8.

González-Teuber M et al. 2010. Glucanases and chitinases as causal agents in the protection of acacia extrafloral nectar from infestation by phytopathogens. *Plant Physiol* 152: 1705–1715.

Guizani N et al. 1992. Purification and characterization of a trypsin-like enzyme from the hepatopancreas of crayfish (*Procambarus clarkii*). *Comp Biochem Physiol B* 103(4): 809–815.

Gupta AK, Bhatia IS. 1980. Glucofructosan biosynthesis in *Fusarium oxysporum*. Phytochemistry 19: 2557–2563.

Gupta AK, Bhatia IS. 1982. Glucofructosan biosynthesis in *Fusarium oxysporum*. *Phytochemistry* 21: 1249–1253.

Guzmán F et al. 2007. Peptide synthesis: chemical or enzymatic. *Electron J Biotechnol* 10(2): Issue of April 15.

Ha KJ, Lindsay RC. 1993. Release of volatile branched-chain and other fatty acids from ruminant milk fats by various lipases. *J Dairy Sci* 76: 677–690.

Haard NF, Simpson BK. 1994. Proteases from aquatic organisms and their uses in the seafood industry. In: AM Martin (ed.) *Fisheries Processing: Biotechnological Applications*. Chapman and Hall, London, Chapter 6, pp. 132–154.

Halford J. 2006. Obesity drugs in clinical development. Current opinion in investigational drugs. 7: 312–318.

Hedstrom L. 2002. Serine protease mechanism and specificity. *Chem Rev* 102: 4501–4524.

Heilskov NSC. 1951. Studies on animal lactase. *Acta Physiol Scand* 24: 84–89.

Homann A, Seibel J. 2009. Towards tailor-made oligosaccharides—chemo-enzymatic approaches by enzyme and substrate engineering. *Appl Microbiol Biotechnol* 83(2): 209–216.

Hoober KL et al. 1996. A sulfhydryl oxidase from chicken egg white. *J Biol Chem* 271: 30510–30516.

Hoober KL et al. 1999. Sulfhydryl oxidase from egg white: a facile catalyst for disulfide bond formation in proteins and peptides. *J Biol Chem* 274: 22147–22150.

Hou CT. 2002. Industrial uses of lipase. In: TM Kuo, HW Gardner (eds.) *Lipid Biotechnology*, Marcel Dekker, New York, pp. 387–397.

Hrassnigg N, Crailsheim K. 2005. Differences in drone and worker physiology in honeybees (*Apis mellifera*). *Apidologie* 36: 255–277.

Huang WN et al. 2008. Effects of transglutaminase on rheology, microstructure, and baking properties of frozen dough. *Cereal Chem* 85(3): 301–306.

Hui DY, Howles PN. 2002. Carboxyl ester lipase: structure–function relationship and physiological role in lipoprotein metabolism and atherosclerosis. *J Lipid Res* 43: 2017–2030.

Jang DS et al. 2008. A new pancreatic lipase inhibitor isolated from the roots of *Actinidia arguta*. *Arch Pharm Res* 31(5): 666–670.

Jones G, Whelan WJ. 1969. The action pattern of D-enzyme, a transmaltodextrinylase from potato. *Carbohydr Res* 9: 483–490.

Kalisz HM. 1988. Microbial proteinases. *Adv Biochem Eng Biotechnol* 36: 1–65.

Kaneda M, Tominaga N. 1975. Isolation and characterization of a protease from the sarcocarp of melon fruit. *J Biochem (Tokyo)* 78: 1287–1296.

Katsaros GI et al. 2010. Kinetic study of high pressure microbial and enzyme inactivation and selection of pasteurisation conditions for valencia orange juice. *Int J Food Sci Technol* 45(6): 1119–1129.

Kaufman SP, Fennema O. 1987. Evaluation of sulfhydryl oxidase as a strengthening agent for wheat flour dough. *Cereal Chem* 64: 172–176.

Kikuchi T et al. 2005. Molecular and biochemical characterization of an endo-beta-1,3-glucanase from the pinewood nematode *Bursaphelenchus xylophilus* acquired by horizontal gene transfer from bacteria. *Biochem J* 389: 117–125.

Kim SM, Kim EJ. 2009. Development of chicken breast noodles adding rubus coreanum miquel and opuntia ficus-indica var. saboten. *J Korean Soc Food Sci Nutr* 38(8): 1111–1117.

Kobayashi M, Shimizu S. 1999. Cobalt proteins. *Eur J Biochem* 261(1): 1–9.

Koo SI, Noh SK. 2007. Green tea as inhibitor of the intestinal absorption of lipids: potential mechanism for its lipid-lowering effect. *J Nutr Biochem* 18: 179–183.

Kozhemyako VB et al. 2004. Molecular cloning and characterization of an endo-1,3-b-D-glucanase from the mollusk *Spisula sachalinensis*. *Comp Biochem Physiol B* 137: 169–178.

Lantto R et al. 2005. Enzyme-aided modification of chicken-breast myofibril proteins: effect of laccase and transglutaminase on gelation and thermal stability. *J Agric Food Chem* 53: 9231–9237.

Layman PL. 1986. Industrial enzymes: battling to remain specialties. *Chem Eng News* 64: 11–24.

Lee KJ et al. 1992. Purification and properties of intracellular fructosyl transferase from *Aureobasidium pullulans*. *World J Microbiol Biotechnol* 8: 411–415.

Leubner-Metzger G. 2003. Functions and regulation of β-1,3-glucanases during seed germination, dormancy release and after-ripening. *Seed Sci Res* 13: 17–34.

Li D et al. 2010. Lipase-catalyzed interesterification of high oleic sunflower oil and fully hydrogenated soybean oil comparison of batch and continuous reactor for production of zero trans shortening fats. *LWT—Food Sci Technol* 43(3): 458–464.

Lim D, Shipe WF. 1972. Proposed mechanism for the antioxygenic action of trypsin in milk. *J Dairy Sci* 55: 753–758.

Liu B et al. 2009. Identification and antifungal assay of a wheat β-1,3-glucanase. *Biotechnol Lett* 31(7): 1005–1010.

Loncin N et al. 1968. Influence of water activity on the spoilage of foodstuffs. *J Food Technol* 3: 131–142.

Lopez-Amaya, C, Marangoni AG. (2000a). Lipases. In: NF Haard, BK Simpson (eds.) *Seafood Enzymes. Utilization and Influence on Postharvest Seafood Quality.* Marcel Dekker, New York, pp. 121–146.

Lopez-Amaya C, Marangoni AG. (2000b). Phospholipases. In: NF Haard, BK Simpson (eds.) *Seafood Enzymes. Utilization and Influence on Postharvest Seafood Quality.* Marcel Dekker, New York, pp. 91–120.

Lund LR et al. 1999. Functional overlap between two classes of matrix-degrading proteases in wound healing. *The EMBO J* 18: 4645–4656.

Martinelle M, Hult K. 1995. Kinetics of acyl transfer reactions in organic media catalyzed by *Candida antarctica* lipase B. *Biochim Biophys Acta* 1251(2): 191–197.

Mathewson PR. 1998. Enzymes—practical guides for the food industry. In: PR Mathewson (ed.) *Enzymes*, Eagan Press Handbook Series, American Chemical Society, St. Paul.

McSweeney PLH, Sousa MJ. 2000. Biochemical pathways for the production of flavour compounds in cheeses during ripening: a review. *Le Lait. Dairy Sci Technol* 80: 293–324.

McWethy SJ, Hartman P. 1979. Extracellular maltase of *Bacillus brevis*. *Appl Environ Microbiol* 37: 1096–1102.

Menocci V et al. 2008. Cyclodextrin glycosyltransferase production by new *Bacillus* sp. strains isolated from Brazilian soil. *Brazilian J Microbiol* 39(4): 682–688.

Mistry VV. 2006. Chymosin in cheese making. In: YH Hui et al. (eds.) *Food Biochemistry and Food Processing.* Blackwell Publishing, Ames, IA, pp. 241–253.

Miyada DS, Tappel AL. 1956. The hydrolysis of beef proteins by various proteolytic enzymes. *J Food Sci* 21: 217–225.

Mochizuki K et al. 1971. Method for producing candied fruit. US Patent No. 3615687.

Mojović L et al. 1993. *Rhizopus arrhizus* lipase-catalyzed interesterification of the midfraction of palm oil to a cocoa butter equivalent fat. *Enzyme Microbial Technol* 15: 438–443.

Moore MM et al. 2006. Network formation in gluten-free bread with application of transglutaminase. *Cereal Chem* 83(1): 28–36.

Moore MM et al. 2004. Textural comparison of gluten-free and wheat based doughs, batters and breads. *Cereal Chem* 81: 567–575.

Morikawa M et al. 1994. Purification and characterization of a thermostable thiol protease from a newly isolated hyperthermophilic *Pyrococcus* sp. *Appl Environ Microbiol* 60(12): 4559–4566.

Motoki M, Seguro K. 1998. Transglutaminase and its use for food processing. *Trends Food Sci Technol* 9(5): 204–210.

Mukherjee KD, Hills HJ. 1994. Lipases from plants. In: P Wooley, SB Petersen (eds.) *Lipases: Their Structure, Biochemistry and Application*, Cambridge University Press, Cambridge, pp. 49–75.

Mukherjee KD, Kiewit I. 1998. Structured triacylglycerols resembling human milk fat by transesterification catalyzed by papaya (*Carica papaya*) latex. *Biotechnol Lett* 20(6): 613–616.

Muñoz-Gutiérrez I et al. 2009. Kinetic behaviour and specificity of [beta]-fructosidases in the hydrolysis of plant and microbial fructans. *Process Biochem* 44: 891–898.

Murakami Y et al. 1998. Enzymatic synthesis of *N*-formyl-L-aspartyl-L-phenylalanine methyl ester (aspartame precursor) utilizing an extractive reaction in aqueous/organic biphasic medium. *Biotechnol Lett* 20(8): 767–769.

Mustranta A et al. 1981. Production of mold lactase. *Biotechnol Lett* 3: 333–338.

Nair SU et al. 2006. Enhanced production of thermostable pullulanase type 1 using *Bacillus cereus* FDTA 13 and its mutant. *Food Technol Biotechnol* 44: 275–282.

Neubert K et al. 2004. Study of extracellular lactase. *Eng Life Sci* 1: 281–283.

Novo Nordisk. 1993. *Annual Report Novo Nordisk A/S*, Bagsaverd, Denmark.

Odelson DA, Breznak JA 1985. Cellulase and other polymer-hydrolyzing activities of *Trichomitopsis termopsidis*, a symbiotic protozoan from termites. *Appl Environ Microbiol* 49(3): 622–626.

Oku T et al. 1984. Nondigestibity of a new sweetener "neo sugars" in the rat. *J Nutr* 114: 1574–1581.

Olempska-Beer ZS et al. 2006. Food-processing enzymes from recombinant microorganisms—a review. *Regul Toxicol Pharm* 45(2): 144–158.

Owusu-Yaw J et al. 1988. Low pH inactivation of pectinesterase in single-strength orange juice. *J Food Sci* 53(2): 504–507.

Patkar S, Björkling F. 1994. Lipase inhibitors. In: P. Woolley, S.B. Petersen (eds.) *Lipases: Their Structure, Biochemistry and Application.* Cambridge University Press, Cambridge, pp. 207–224.

Pfleiderer VG et al. 1967. Zur evolution der endopeptidasen III eine protease vom molekulargewicht 11 000 und eine trypsinähnliche fraktion aus *Astacus fluviatilis* (Fabr.). hoppe-seyler's Z. *Physiol Chem* 348: 1319–1331.

Prapulla SG et al. 2000. Microbial production of oligosaccharides: a review. In: SL Neidleman et al. (eds.). *Advances in Applied Microbiology*, Academic Press, New York, pp. 299–337.

Pugsley AP et al. 1986. Extracellular pullulanase of *Klebsiella pneumoniae* is a lipoprotein. *J Bacteriol* 166(3): 1083–1088.

Pugsley AP, Dubrevil C. 1988. Molecular characterization of malQ, the structural gene for the *Escherichia coli* enzyme amylomaltase. *Mol Microbiol* 2: 473–479.

Quiroz-Castaneda RE et al. 2009. Characterization of cellulolytic activities of *Bjerkandera adusta* and *Pycnoporus sanguineus* on solid wheat straw medium. *Electron J Biotechnol* [online]. 12(4): October 15. Available at http://www.ejbiotechnology.cl/content/vol12/issue4/full/3/index.html (accessed on December 24, 2010)

Raksakulthai N, Haard NF. 1992a. Correlation between the concentration of peptides and amino acids and the flavour of fish sauce. *ASEAN Food J* 7: 86–90.

Raksakulthai N, Haard NF. 1992b. Fish sauce from male capelin *(Mallotus villosus)*: contribution of cathepsin C to the fermentation. *ASEAN Food J* 7: 147–151.

Ramana Rao MV, Dutta SM. 1978. Lactase activity of microorganisms. *Folia Microbial* 23: 210–215.

Robert JH, Darbyshire B. 1980. Fructan synthesis in onion. *Phytochem* 19: 1017–1020.

Rudenskaya GN et al. (1995). Macluralisin—a serine protease from fruits of *Maclura pomifera* (Raf.) *Schneid*. *Planta* 196: 174–179.

Sareevoravitkul R et al. 1996. Comparative properties of bluefish (*Pomatomus saltatrix*) gels formulated by high hydrostatic pressure and heat. *J Aquat Food Prod Tech* 5: 65–79.

Satouchi K et al. 2002. A lipase-inhibiting protein from lipoxygenase-deficient soybean seeds. *Biosci Biotechnol Biochem* 66: 2154–2160.

Schmelzer C et al. 1982. Identification of splicing signals in introns of yeast mitochondrial split genes: mutational alterations in intron bit and secondary structures in related introns. *Nucl Acids Res* 10: 6797–6808.

Schmid RD, Verger R. 1998. Lipases: interfacial enzymes with attractive applications. *Angew Chem Int Ed* 37: 1608–1633.

Shahidi F, Han X-Q. 1993. Encapsulation of food ingredients. *Crit Rev Food Sci Nutr* 33(6): 501–547.

Shahidi F, Janak Kamil YVA. 2001. Enzymes from fish and aquatic invertebrates and their application in the food industry. *Trends Food Sci Technol* 12(12): 435–464.

Sharma N et al. 2005. Screening of some medicinal plants for antilipase activity. *J Ethnopharmacol* 97: 453–456.

Shiomi N et al. 1979. Synthesis of several fructo-oligosaccharides by asparagus fructosyl transferases. *Agric Biol Chem* 43: 2233–2244.

Sidhu MS et al. 1986. Purification and characterization of cellulolytic enzymes from *Trichoderma harzianum*. *Folia Microbiologica* 31(4): 293–302.

Simpson BK, Haard NF. 1984. Trypsin from Greenland cod as a food-processing aid. *J Appl Biochem* 6: 135–143.

Simpson BK, Haard NF. 1987 Cold-adapted enzymes from fish. In: D Knorr (ed.) *Food Biotechnology*, Marcel Dekker, New York, Chapter 19, pp. 495–527.

Singh R et al. 2007. Production of high fructose syrup from asparagus inulin using immobilized exoinulinase from *Kluyveromyces marxianus* YS-1. *J Ind Microbiol Biotechnol* 34: 649–655.

Sirisansaneeyakul S et al. 2007. Production of fructose from inulin using mixed inulinases from *Aspergillus niger* and *Candida guilliermondii*. *World J Microbiol Biotechnol* 23: 543–552.

Sliwkowski MX et al. 1984. Kinetic mechanism and specificity of bovine milk sulfhydryl oxidase. *Biochem J* 220: 51–55.

Sritunyalucksana K et al. 2002. A β-1,3-glucan binding protein from the black tiger shrimp, *Penaeus monodon*. *Dev Comp Immun* 26(3): 237–245.

Sträter N et al. 2002. Structural basis of the synthesis of large cycloamyloses by amylomaltase. *Biologia (Bratisl)* 11(57/Suppl): 93–99.

Svendsen A. (2000). Lipase protein engineering. *Biochim Biophys Acta* 1543: 223–238.

Tafazoli S et al. (2010). Safety evaluation of amylomaltase from *Thermus aquaticus*. *Regulat Toxicol Pharmacol* 57(1): 62–69.

Takaha T et al. 1996. Potato D-enzyme catalyzes the cyclization of amylose to produce cycloamylose, a novel cyclic glucan. *J Biol Chem* 271: 2902–2908.

Tejada L et al. 2008. Effect of lyophilisation, refrigerated storage and frozen storage on the coagulant activity and microbiological quality of *Cynara cardunculus* L. extracts. *J Sci Food Agric* 88(8): 1301–1306.

Terada Y et al. 1997. Cyclodextrins are not the major cyclic α-1,4-glucans produced by the initial action of cyclodextrin glucanotransferase on amylose. *J Biol Chem* 272: 15729–15733.

Thakur BR, Nelson PE. 1997. Inactivation of lipoxygenase in whole soy flour suspension by ultrasonic cavitation. *Food/Nahrung* 41(5): 299–301.

Thomas BR et al. 2000. Endo-1,3;1,4-β-glucanase from coleoptiles of rice and maize: role in the regulation of plant growth. *Int J Biol Macromol* 27: 145–149.

Thorpe C et al. 2002. Sulfhydryl oxidases: emerging catalysts of protein disulfide bond formation in eukaryotes. *Arch Biochem Biophys* 405: 1–12.

Tsujita T et al. 1996. Studies on the inhibition of pancreatic and carboxylester lipases by protamine. *J. Lipid Res* 37: 1481–1487.

Tucker ML et al. 1988. Bean abscission cellulase. *Plant Physiol* 88: 1257–1262.

Valera H et al. 1997. Skin unhairing proteases of *Bacillus subtilis*: production and partial characterization. *Biotechnol Lett* 19: 755–758.

Van Oort M. 2010. Enzymes in food technology—introduction. In: RJ Whitehurst, M Van Oort (eds.) *Enzymes in Food Technology*, Wiley-Blackwell, Ames, IA, Chapter 1, pp. 1–16.

Wang CS, Hartsuck JA. 1993. Bile salt-activated lipase. A multiple function lipolytic enzyme. *Biochim Biophys Acta* 1166(1): 1–19.

Watanabe H, Tokuda G. 2001. Animal cellulases. *Cell Mol Life Sci* 58: 1167–1178.

Weerasinghe VC et al. 1996. Characterization of active components in food-grade proteinase inhibitors for surimi manufacture *J Agric Food Chem* 44(9): 2584–2590.

Wells GH. 1966. Tenderness of freeze-dried chicken with emphasis on enzyme treatments. PhD Thesis, Michigan State University, University Microfilms, Ann Arbor, MI.

Wenham DG et al. 1979. Regulation of glucosyl and fructosyl transferase synthesis by continuous cultures of *Streptococcus mutans*. *J Gen Microbiol* 114: 117–124.

Whitaker JR. 1994. The sulfhydryl proteases. In: JR Whitaker (ed.) *Principles of Enzymology for the Food Sciences*, 2nd edn. Marcel Dekker, New York, pp. 367–385.

Wilkinson MG, Kilcawley KN. 2005. Mechanism of enzyme incorporation and release of enzymes in cheese during ripening. *Intl Dairy J* 15: 817–830.

Wong H, Schotz MC. 2002. The lipase gene family. *J Lipid Res* 43: 993–999.

Yamada H. 1989. Localization in skin, activation and reaction mechanisms of skin sulfhydryl oxidase. *Nippon Hifuka Gakkai Zasshi* 99: 861–869.

Yamasaki Y et al. 2008. Pullulanase from rice endosperm. *Acta Biochim Polonica* 55(3): 507–510.

Yokoyama K et al. 2004. Properties and applications of microbial transglutaminase. *Appl Microbiol Biotechnol* 64: 227–454.

Yoshio N et al. 1986. Purification and properties of D-enzyme from malted barley. *J Jpn Soc Starch Sci* 33: 244–252.

Yun JW et al. 1995. Continuous production of fructooligosaccharides from sucrose by immobilized fructosyltransferase. *Biotechnol Tech* 9(11): 805–808.

10

Protein Cross-linking in Food – Structure, Applications, Implications for Health and Food Safety

Juliet A. Gerrard and Justine R. Cottam

Abstract: The capacity of proteins to cross-link with each other has important implications for food quality. Protein cross-linking can influence food texture and appearance, or permit the incorporation of useful nutrients such as essential amino acids or essential oils in foods to accrue crucial benefits. Protein cross-linking in foods may be achieved by enzymatic means with transglutaminase or with cross-linking agents like glutaraldehyde. This chapter discusses the different cross-linking effects and how they may be exploited by food processor.

INTRODUCTION

Protein cross-links play an important role in determining the functional properties of food proteins. Manipulation of the number and nature of protein cross-links during food processing offers a means by which the food industry can manipulate the functional properties of food, often without damaging the nutritional quality. This chapter updates the chapter in the previous edition of this book and discusses advances in our understanding of protein cross-linking over the last two decades, as well as examining current and future applications of this chemistry in food processing and its implications for health and food safety. It draws on, and updates, two reviews in this area (Gerrard 2002, Miller and Gerrard, 2005) in addition to earlier reviews on this subject (Matheis and Whitaker 1987, Feeney and Whitaker 1988, Singh 1991).

The elusive relationship between the structure and the function of proteins presents a particular challenge for the food technologist. Food proteins are often denatured during processing, so there is a need to understand the protein both as a biological entity with a predetermined function and as a randomly coiled biopolymer. To understand and manipulate food proteins requires a knowledge of both protein biochemistry and polymer science. If the protein undergoes chemical reaction during processing, both the 'natural' function of the molecule and the properties of the denatured polymeric state may be influenced. One type of chemical reaction that has major consequences for protein function in either their native or denatured states is protein cross-linking. It is, therefore, no surprise that protein cross-linking can have profound effects on the functional properties of food proteins.

This chapter sets out to define the different types of protein cross-links that can occur in food, before and after processing, and the consequences of these cross-links for the functional and nutritional properties of the foodstuff. Methods that have been employed to introduce cross-links into food deliberately are then reviewed, and future prospects for the use of this chemistry for the manipulation of food during processing are surveyed, including a consideration of health and food safety aspects.

Food Biochemistry and Food Processing, Second Edition. Edited by Benjamin K. Simpson, Leo M.L. Nollet, Fidel Toldrá, Soottawat Benjakul, Gopinadhan Paliyath and Y.H. Hui.

PROTEIN CROSS-LINKS IN FOOD

Protein cross-linking refers to the formation of covalent bonds between polypeptide chains within a protein (intramolecular cross-links) or between proteins (intermolecular cross-links) (Feeney and Whitaker 1988). In biology, cross-links are vital for maintaining the correct conformation of certain proteins and may control the degree of flexibility of the polypeptide chains. As biological tissues age, further protein cross-links may form, which often have deleterious consequences throughout the body and play an important role in the many conditions of ageing (Zarina et al. 2000, Ahmed 2005, Nass et al. 2007, Gul et al. 2009). Chemistry similar to that which occurs during ageing may take place if biological tissues are removed from their natural environment – for example, when harvested as food for processing.

Food processing often involves high temperatures, extremes in pH, particularly alkaline, and exposure to oxidising conditions and uncontrolled enzyme chemistry. Such conditions can result in the introduction of protein cross-links, producing substantial changes in the structure of proteins, and therefore the functional (Singh 1991) and nutritional (Friedman 1999a, 1999b, 1999c) properties of the final product. A summary of protein cross-linking in foods is given in Figure 10.1, in which the information is organised according to the amino acids that react to form the cross-link. Not all amino acids participate in protein cross-linking, no matter how extreme the processing regime. Those that react do so with differing degrees of reactivity under various conditions.

DISULFIDE CROSS-LINKS

Disulfide bonds are the most common and well-characterised types of covalent cross-link in proteins in biology. They are formed by the oxidative coupling of two cysteine residues that are close in space within a protein. A suitable oxidant accepts the hydrogen atoms from the thiol groups of the cysteine residues, producing disulfide cross-links. The ability of proteins to form intermolecular disulfide bonds during heat treatment is considered to be vital for the gelling of some food proteins, including milk proteins, surimi, soybeans, eggs, meat and some vegetable proteins (Zayas 1997). Gels are formed through the cross-linking of protein molecules, generating a three-dimensional solid-like network, which provides food with desirable texture (Dickinson 1997).

Disulfide bonds are thought to confer an element of thermal stability to proteins and are invoked, for example, to explain the stability of hen egg white lysozyme, which has four intramolecular disulfide cross-links in its native conformation (Masaki et al. 2001). This heat stability influences many of the properties of egg white observed during cooking. Similarly, the heat treatment of milk promotes the controlled interaction of denatured β-lactoglobulin with κ-casein through the formation of a disulfide bond. This increases the heat stability of milk and milk products, preventing precipitation of β-lactoglobulin (Singh 1991). Disulfide bonds are also important in the formation of dough. Disulfide interchange reactions during the mixing of flour and water result in the production of a protein network with the viscoelastic properties required for bread making (Lindsay and Skerritt 1999). The textural changes that occur in meat during cooking have also been attributed to the formation of intermolecular disulfide bonds (Singh 1991).

CROSS-LINKS DERIVED FROM DEHYDROPROTEIN

Alkali treatment is used in food processing for a number of reasons, such as the removal of toxic constituents and the solubilisation of proteins for the preparation of texturised products. However, alkali treatment can also cause reactions that are undesirable in foods, and its safety has come into question (Savoie et al. 1991, Shih 1992, Friedman 1999a, 1999c). Exposure to alkaline conditions, particularly when coupled to thermal processing, induces racemisation of amino acid residues and the formation of covalent cross-links, such as dehydroalanine, lysinoalanine and lanthionine (Friedman 1999a, 1999b, 1999c). Dehydroalanine is formed from the base-catalysed elimination of persulfide from an existing disulfide cross-link. The formation of lysinoalanine and lanthionone cross-links occurs through β-elimination of cysteine and phosphoserine protein residues, thereby yielding dehydroprotein residues. Dehydroprotein is very reactive with various nucleophilic groups, including the ε-amino group of lysine residues and the sulfhydryl group of cysteine. In severely heat- or alkali-treated proteins, imidazole, indole and guanidino groups of other amino acid residues may also react (Singh 1991). The resulting intra- and intermolecular cross-links are stable, and food proteins that have been extensively treated with alkali are not readily digested, reducing their nutritional value. Mutagenic products may also be formed (Friedman 1999a, 1999c).

CROSS-LINKS DERIVED FROM TYROSINE

Various cross-links formed between two or three tyrosine residues have been found in native proteins and glycoproteins, for example in plant cell walls (Singh 1991). Dityrosine cross-links have recently been identified in wheat and are proposed to play a role in the formation of the cross-linked protein network in gluten (Tilley et al. 2001). They have also been formed indirectly by treating proteins with hydrogen peroxide or peroxidase (Singh 1991) and are implicated in the formation of caseinate films by gamma irradiation (Mezgheni et al. 1998). Polyphenol oxidase can also lead indirectly to protein cross-linking, due to reaction of cysteine, tyrosine, or lysine with reactive benzoquinone intermediates generated from the oxidation of phenolic substrates (Matheis and Whitaker 1987, Feeney and Whitaker 1988). Such plant phenolics have been used to prepare cross-linked gelatin gels to develop novel food ingredients (Strauss and Gibson 2004).

CROSS-LINKS DERIVED FROM THE MAILLARD REACTION

The Maillard reaction is a complex cascade of chemical reactions, initiated by the deceptively simple condensation of an amine with a carbonyl group, often within a reducing sugar or fat

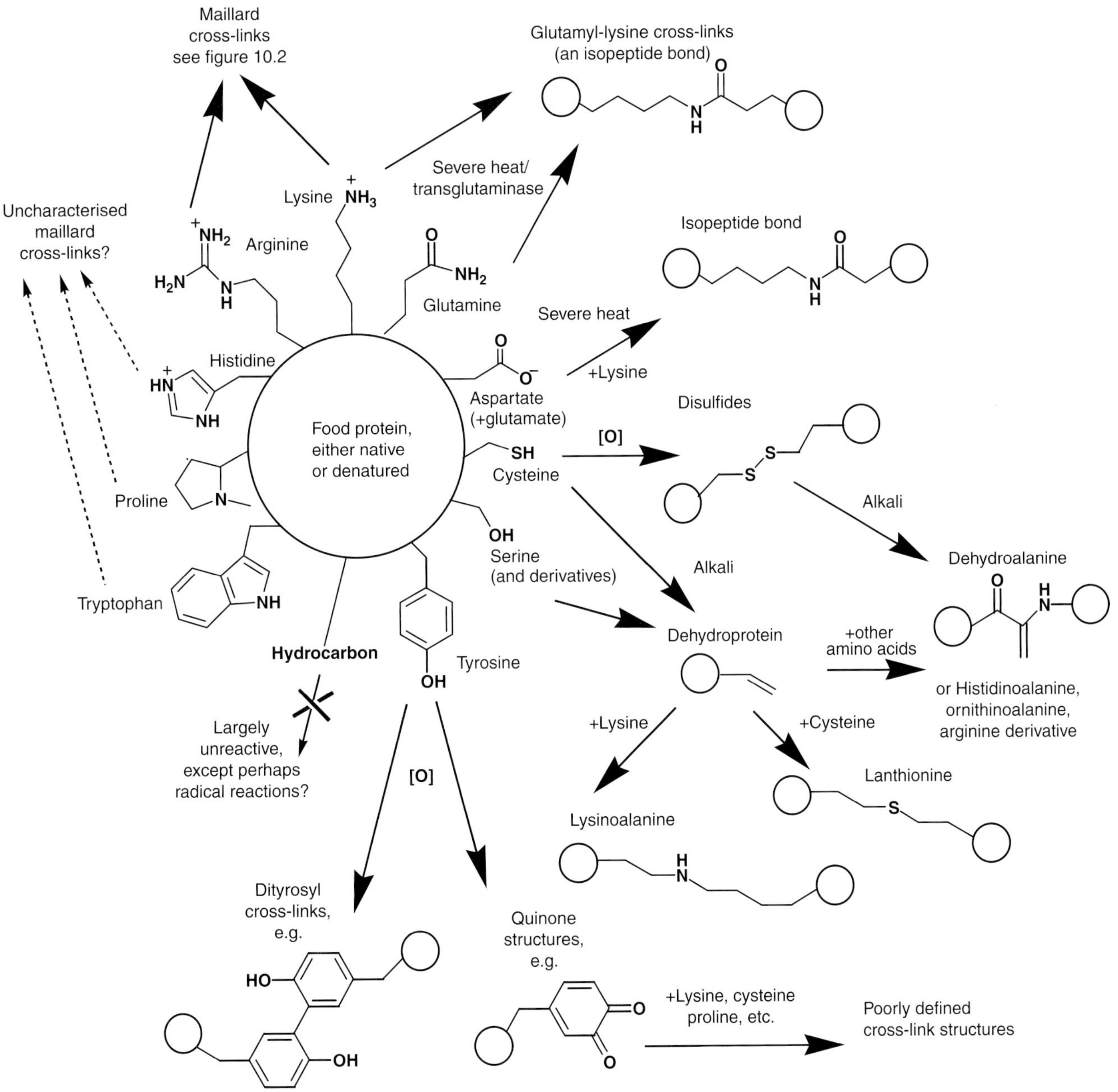

Figure 10.1. A summary of the cross-linking reactions that can occur during food processing, from Gerrard 2002. Further details are given in the text.

breakdown product (Fayle and Gerrard 2002, Friedman 1996a). During the course of the Maillard reaction, reactive intermediates, such as α-dicarbonyl compounds and deoxysones, are generated and lead to the production of a wide range of compounds, including polymerised brown pigments called melanoidins, furan derivatives, nitrogenous, and heterocyclic compounds (e.g. pyrazines; Fayle and Gerrard 2002). Protein cross-links form a subset of the many reaction products, and the cross-linking of food proteins by the Maillard reaction during food processing is well established (Gerrard 2002, Miller and Gerrard, in press, 2006). The precise chemical structures of these cross-links in food, however, are less well understood. Thus, surprisingly little is known about the extent of Maillard cross-linking in processed foods, the impact of this process on food quality, and how the reaction might be controlled to maximise food quality.

In biology and medicine, where the Maillard reaction is important during the ageing process, several cross-link structures have

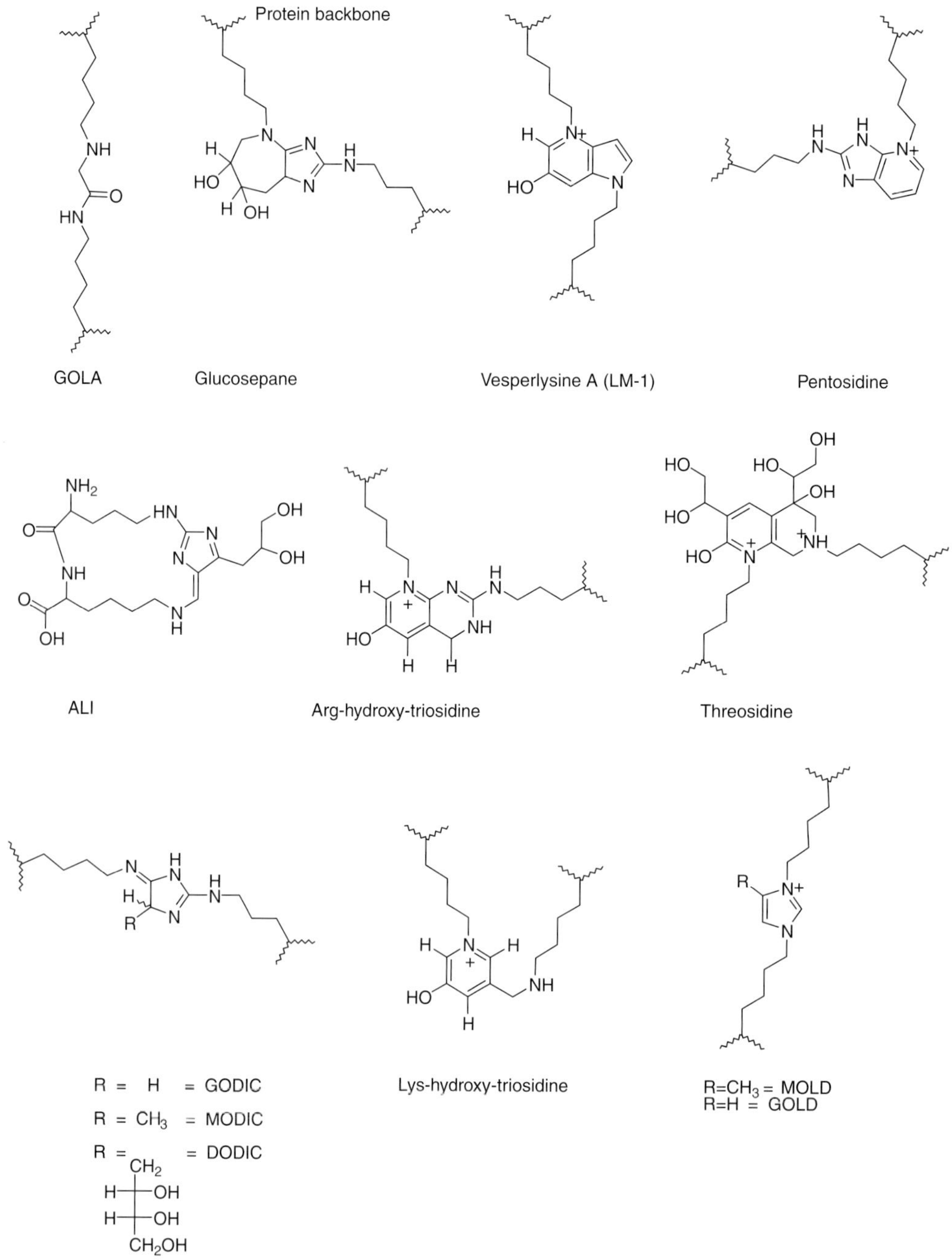

Figure 10.2. A selection of known protein cross-links derived from the Maillard reaction in the body, from Miller and Gerrard (2004). (Sell and Monnier 1989, Nakamura et al. 1997, Prabhakaram et al. 1997, Brinkmann-Frye et al. 1998, Lederer and Buhler 1999, Lederer and Klaiber 1999, Al-Abed and Bucala 2000, Glomb and Pfahler 2001, Biemel et al. 2002 Tessier et al. 2003). The extent to which these cross-links are present in food is largely undetermined. GOLA, N^6-{2-[(5-amino-5-carboxypentyl)amino]-2-oxoethyl}lysine; ALI, arginine–lysine imidazole; GODIC, N^6-(2-{[(4*S*)-4-ammonio-5-oxido-5-oxopentyl]amino}-3,5-dihydro-4*H*-imidazol-4-ylidene)-L-lysinate; MODIC, N^6-(2-{[(4*S*)-4-ammonio-5-oxido-5-oxopentyl]amino}-5-methyl-3,5-dihydro-4*H*-imidazol-4-ylidene)-L-lysinate; DOGDIC, N^6-{2-{[(4*S*)-4-ammonio-5-oxido-5-oxopentyl]amino}-5[(2*S*,3*R*)-2,3,4-trihydroxybutyl]-3,5-dihydro-4*H*-imidazol-4-ylidene)-L-lysinate; MOLD, methylglyoxal lysine dimer; GOLD, glyoxal lysine dimer.

been identified, including those shown in Figure 10.2 (Ames 1992, Hill et al. 1993, Easa et al. 1996a, 1996b, Gerrard et al. 1998a, 1999, 2002a, 2003a, 2003b, Hill and Easa 1998, Fayle et al. 2000, 2001, Mohammed et al. 2000). One of the first protein-derived Maillard reaction products (MRPs) isolated and characterised was the cross-link pentosidine (Sell and Monnier 1989, Dyer et al. 1991), a fluorescent moiety that is believed to form through the condensation of a lysine residue with an arginine residue and a reducing sugar. The exact mechanism of formation of pentosidine remains the subject of considerable debate (Chellan and Nagaraj 2001, Biemel et al. 2001).

Some of the earliest studies that assessed the effect of protein cross-linking on food quality examined the digestibility of a model protein following glycation. Kato et al. (1986a) observed that following incubation of lysozyme with a selection of dicarbonyl compounds for 10 days, the digestibility of lysozyme by a pepsin-pancreatin solution was reduced to up to 30% relative to the non-glycated sample. This trend of decreasing digestibility was concomitant with an observed increase in cross-linking of lysozyme. The increased resistance of cross-linked proteins to enzymes commonly involved in the digestion of proteins in the body is unfavourable from a nutritional standpoint. It has also been reported that the digestion process can be inhibited by the Maillard reaction (Friedman 1996b).

AGE PROTEIN CROSS-LINKS ISOLATED TO DATE IN FOOD

Information regarding the presence of specific Maillard protein cross-links in food is, to date, limited, with only a handful of studies in this area (Henle et al. 1997, Schwarzenbolz et al. 2000, Biemel et al. 2001). For example, compared with the extensive literature on the Maillard chemistry in vivo, relatively little has been reported on the existence of pentosidine in food. In a study by Henle et al. (1997), the pentosidine content of a range of foods was examined, with the highest values observed in roasted coffee. Overall concentrations, however, were considered low, amongst most of the commercial food products tested. It was, therefore, concluded that pentosidine does not have a major role in the polymerisation of food proteins. Schwarzenbolz et al. (2000) showed that pentosidine formation in a casein-ribose reaction is carried out under high hydrostatic pressure, which could be relevant to some areas of food processing. Iqbal et al. have investigated the role of pentosidine in meat tenderness in broiler hens (Iqbal et al. 1997, Iqbal et al. 1999a, 1999b, Iqbal et al. 2000).

Other Maillard cross-links have recently been detected in food, and attempts have been made to quantify the levels found. Biemel et al. (2001) examined the content of the lysine–arginine cross-links, GODIC, MODIC, DODIC and glucosepane in food. These cross-links were found to be present in proteins extracted from biscuits, pretzels, salt stick and egg white in the range of 7–151 mg/kg. The lysine–lysine imidazolium cross-links MOLD and GOLD were also isolated but were present at a lower concentration than MODIC and GODIC.

HEALTH AND FOOD SAFETY ASPECTS OF MRPS

The complexity of the Maillard reaction means that during any cross-linking reaction in the food matrix, a variety of other products are likely to form. Over the last decade, there has been an ongoing debate over the health and safety impacts of dietary MRPs, particularly advanced glycation end products (AGEs) generated in the later stages of the Maillard reaction in foods, with regards to their potential risks and/or benefits towards human health when consumed in the diet. To date, numerous research groups have argued that MRPs pose a risk to human health, particularly certain heterocyclic products, when consumed in the diet (Felton and Knize 1998, Faist and Erbersdobler 2002, Shin 2003, Shin et al. 2003, Bordas et al. 2004, Murkovic 2004, Taylor et al. 2004, Sebekova and Somoza 2007, Garcia et al. 2009). Other research has provided evidence to the contrary, highlighting the potential of MRPs to act as antioxidants or have other beneficial effects (Manzocco et al. 2000, Faist and Erbersdobler 2002, Lee and Shibamoto 2002, Morales and Babbel 2002, Dittrich et al. 2003, Morales and Jimenez-Perez 2004, Ames 2007), with a few groups considering both the positive and negative effects of MRPs when consumed in the diet (Faist and Erbersdobler 2002, Somaza 2005, Gerrard 2006, Henle 2007).

This ongoing debate was discussed by Henle (2007) in a review paper, in which he explored whether dietary MRPs are a risk to human health based on the arguments from two prominent papers on the topic by Ames (2007), who provided evidence against the MRPs being a risk to human health, and Sebekova and Somoza (2007), who argued the opposite. Both sides of the debate are outlined here.

Sebekova and Somoza (2007) argued that there was sufficient evidence concerning the negative impact of dietary AGEs on health. Their argument is based on research from both experimental studies in rodents and clinical studies in humans, including healthy people and those suffering from diseases that are associated with AGEs, such as diabetes and cardiovascular disease. Their analysis of the current literature revealed that a high intake of thermally processed foods in the diet may have significant nephrotoxic effects, aggravate low levels of inflammation and exert oxidative stress. There is also significant evidence to show that a high AGE-diet can influence the development of complications that are associated with diabetes and have detrimental effects on, for example, wound healing (Peppa et al. 2003, 2009) and atherosclerois (Lin et al. 2003, Walcher and Marx 2009). This is supported by further studies that show potential negative effects of an AGE-rich diet to health. The first of these studies was an animal study in which Wistar rats were fed either an AGE-rich diet or an AGE-poor diet over a 6-week period, where wheat starch was replaced by bread crusts (Sebekova et al. 2005, Somoza et al. 2005). The results obtained were interesting, with the rats on the AGE-rich diet gaining weight more rapidly than those on the AGE-poor diet, with a trend towards higher glucose and albumin levels. There was also noticeable weight gain in specific organs such as the heart, kidney, liver and lungs, but not the spleen, intestine or brain. These results suggest the idea that some organs are more susceptible to

MRP-rich diets. Similar results were found in a crossover study carried out by Wittmann et al. (2001) on 21 healthy volunteers, where those on the AGE-rich diet gained more weight. In this study, 21 healthy volunteers were fed on either an AGE-rich or AGE-poor diet of heated or unheated high protein (3 g/kg/day). However, to date, the majority of knowledge on the effects of AGE-rich diets is mainly based on experimental studies of rodents rather than human studies.

In contrast, Ames (2007) supports the idea that dietary AGEs are not a risk to human health and provides evidence to support this claim based on information obtained from both animal and human volunteer studies on the bioavailability and metabolic fate of dietary AGEs. The majority of the literature to date on the bioavailability and metabolic fate of AGEs is centred on the compounds CML and pyrraline, due to their abundance in food systems. Various studies have investigated the metabolic fate of dietary CML. One such study involved feeding rats on a diet of glycated casein (either 1.8 g of CML/kg or 6.2 g of CML/kg) against the control of native casein (Faist et al. 2000). The results found that approximately 50% of the initial CML consumed was excreted through urine and faeces, with a small portion being deposited on the kidney and liver. Although the majority of the CML consumed was readily excreted, further investigations would need to be carried out to determine the metabolic fate of the rest of the CML, with suggestions that it could have been degraded by the colonic microflora or metabolised post-absorption. Another study looked at distribution and elimination of AGEs, including CML, CEL (N^{ϵ}-carboxyethyllysine) and lysine (Bergmann et al. 2001). These AGEs were studied via fluorine-18-labelled analogues by [18F]-fluorobenzoylation of the α-amino group of the AGE and then intravenously injected into the tail vein of rats, which were subsequently sacrificed within 30 minutes of the injection. The results obtained showed all three compounds (CML, CEL and lysine) to be distributed swiftly through the body and be rapidly excreted via the kidneys with >87% of the radioactivity being excreted within 2 hours of injection (Bergmann et al. 2001).

Many studies have been carried out investigating the distribution and elimination of the other abundant AGE, pyrraline, which is excreted via the kidneys within 48 hours (Förster and Henle 2003). Other studies on excretion of pentosidine, another MRP, in the body, as well pyrraline and fructoselysine (Miyata et al. 1998, Förster et al. 2005) have found significant differences in the excretion rate of the individual MRPs, according to individual metabolic fates and whether the MRPs are free or protein bound. However, the weight of evidence appears to suggest that the majority of the ARPs and AGEs that are absorbed by the body are subsequently, rapidly excreted and are therefore considered not to be harmful to the body. Seiquer et al. (2008) carried out experiments on healthy adolescent males by feeding some of the participants a MRP-rich diet and the others a low MRP diet over a two-week period. Surprisingly, they found there was no obvious damage to the oxidative status or antioxidant defence in the MRP-rich diet participants. Lindenmeier et al. (2002) and Somoza et al. (2005) have suggested beneficial effects of AGEs. Definitive conclusions await detailed research on the bioactivity of dietary AGEs (Lopez-Garcia et al. 2004).

Ames (2009) suggests there needs to be a more holistic approach to the research of MRPs. The balance of evidence, based on many centuries of eating cooked food, is that MRPs/AGEs are not harmful, but further studies should be done to eliminate the possibility and provide reassurance to food manufacturers and consumers. An interdisciplinary approach incorporating food science, biology, chemistry and medicine is required for a better understanding into the biological consequences of thermally processed foods and more carefully defined human trials should resolve this. Food manufacturers wishing to harness the Maillard reaction for protein cross-linking or other uses are advised to keep a watching brief on this research.

MELANOIDINS

Very advanced glycation end-products that form as a result of food processing are dubbed melanoidins. This is a structurally diverse class of compounds, which until recently were very poorly characterised. However, a subset of food melanoidins undoubtedly includes those that cross-link proteins.

Melanoidins can have molecular weights of up to 100,000 Da (Hofmann 1998a). Because of their sheer chemical complexity, it has been difficult to isolate and characterise these molecules (Fayle and Gerrard 2002, Miller and Gerrard, 2005). However, some recent work in this area has lead to some new hypotheses. Hofmann proposed that proteins may play an important role in the formation of these complex, high-molecular weight melanoidins (Hofmann 1998b). Thus, it was proposed that a low-molecular weight carbohydrate-derived colourant reacts with protein-bound lysine and/or arginine, forming a protein cross-link. Formation of colour following polymerisation of protein has indeed been observed following incubation with carbohydrates (Cho et al. 1984, 1986a, 1986b, Okitani et al. 1984). Hofmann tested this hypothesis by the reaction of casein with a pentose-derived intermediate furan-2-carboxaldehyde and subsequent isolation of a melanoidin-type colourant compound from this reaction mixture (Figure 10.3). Although non-cross-linking in nature, the results encouraged further studies to isolate protein cross-links that are melanoidins, and possible deconvolution of the chemistry of melanoidin from these data. Further studies by Hofmann (1998a) proposed a cross-link structure BISARG, which was formed on reaction of N-protected arginine with glyoxal and furan-2-carboxaldehyde (Figure 10.3). This cross-link was proposed as plausible in food systems due to the large amount of protein, furan-2-carboxaldehyde and glyoxal that may be present in food (Hofmann 1998a).

A lysine–lysine radical cation cross-link, CROSSPY (Figure 10.3C), has been isolated from a model protein cross-link system involving bovine serum albumin and glycoladehyde, followed by thermal treatment (Hofmann et al. 1999). CROSSPY was also shown to form from glyoxal, but only in the presence of ascorbic acid, supporting the suggestion that reductones are able to initiate radical cation mechanisms resulting in cross-links (Hofmann et al. 1999). EPR spectroscopy of dark-coloured bread crust revealed results that suggested CROSSPY was most likely associated with the browning bread crust (Hofmann et al. 1999).

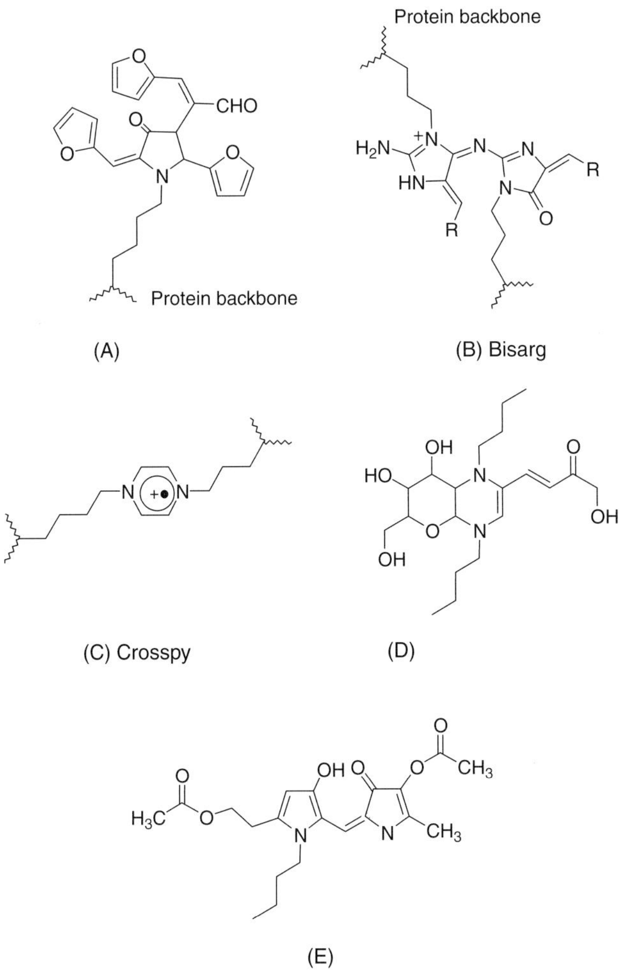

Figure 10.3. Formation of a melanoidin-type colourants on reaction of (**A**) protein-bound lysine with furan-2-aldehyde (Hofmann 1998b); (**B**) two protein-bound arginine residues with glyoxal in the presence of carbonyl compound (Hofmann 1998a); (**C**) two protein-bound lysine residues with glycolaldehyde or glyoxal (in the presence of ascorbate). R, remainder of carbonyl compound (Hofmann et al. 1999); (**D**) product of reaction of propylamine with glucose (Knerr et al. 2001); and (**E**) butylaminammonium acetate with glucose (Lerche et al. 2003).

Encouragingly, when browning was inhibited, radical formation was completely blocked.

A model system containing propylamine and glucose was found to yield a yellow cross-link product under food-processing conditions (Fig. 10.3D; Knerr et al. 2001); however, this product is still to be isolated from foodstuffs. In another model study undertaken by Lerche et al., reacting butylaminammonium acetate (a protein bound lysine mimic) with glucose resulted in the formation of a yellow product (Fig. 10.3E); this molecule is also still to be isolated from foodstuffs (Lerche et al. 2003).

The hypothesis that formation of melanoidins involves reaction of protein with carbohydrate is supported by Brands et al., who noted that following dialysis of a casein-sugar (glucose or fructose) system with 12,000 Da cutoff tubing, around 70% of the brown-coloured products were present in the retentate (Brands et al. 2002).

MAILLARD-RELATED CROSS-LINKS

Although not strictly classified under the heading of Maillard chemistry, animal tissues such as collagen and elastin contain complex heterocyclic cross-links, formed from the apparently spontaneous reaction of lysine and derivatives with allysine, an aldehyde formed from the oxidative deamination of lysine catalysed by the enzyme lysine oxidase (Feeney and Whitaker 1988). A selection of cross-links that form from this reaction is outlined in Figure 10.4. The extent to which these cross-links occur in food has not been well-studied, although their presence in gelatin has been discussed, along with the presence of pentosidine in these systems (Cole and Roberts 1996, 1997).

CROSS-LINKS FORMED VIA TRANSGLUTAMINASE CATALYSIS

An enzyme that has received extensive recent attention for its ability to cross-link proteins is transglutaminase. Transglutaminase catalyses the acyl-transfer reaction between the

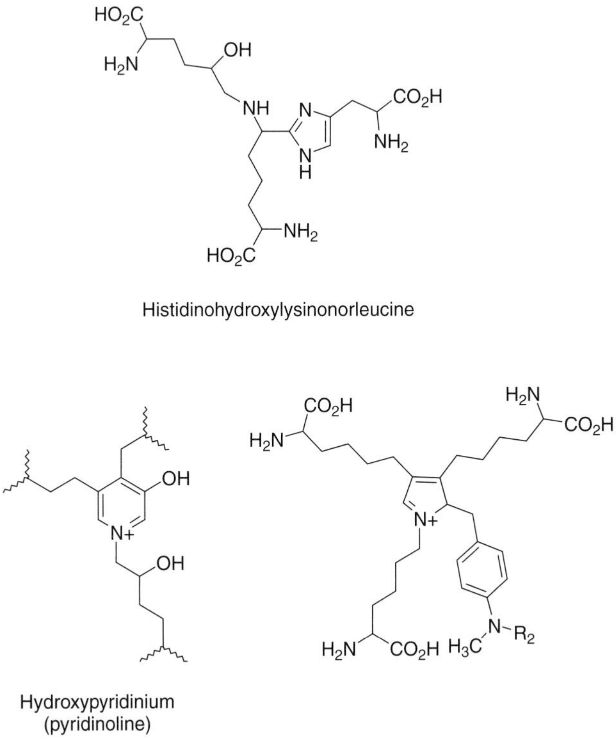

Figure 10.4. Trifunctional crosslinks reported to result from the spontaneous reaction of allysine with lysine (Eyre et al. 1984, Yamauchi et al. 1987, Brady and Robins 2001).

γ-carboxyamide group of peptide-bound glutamine residues and various primary amines. As represented in Figure 10.1, the ε-amino groups of lysine residues in proteins can act as the primary amine, yielding inter- and intramolecular ε-*N*-(γ-glutamyl)lysine cross-links (Zhu et al. 1995, Motoki and Seguro 1998, Yokoyama et al. 2004, Jaros et al. 2006, Özrenk 2006). The formation of this cross-link does not reduce the nutritional quality of the food as the lysine residue remains available for digestion (Seguro et al. 1996).

Transglutaminase is widely distributed in most animal tissues and body fluids and is involved in biological processes such as blood clotting and wound healing. ε-*N*-(γ-glutamyl)lysine cross-links can also be produced by severe heating (Motoki and Seguro 1998), but are most widely found where a food is processed from material that contains naturally high levels of the enzyme. The classic example here is the gelation of fish muscle in the formation of surumi products, a natural part of traditional food processing of fish by the Japanese suwari process, although the precise role of endogenous transglutaminase in this process is still under debate (An et al. 1996, Motoki and Seguro 1998) and an area of active research (Benjakul et al. 2004a, 2004b). ε-*N*-(γ-Glutamyl)lysine bonds have been found in various raw foods including meat, fish and shellfish. Transglutaminase-cross-linked proteins have thus long been ingested by man (Seguro et al. 1996). The increasing applications of artificially adding this enzyme to a wide range of processed foods, along with safety aspects, are discussed in detail below.

Other Isopeptide Bonds

In foods of low carbohydrate content, where Maillard chemistry is inaccessible, severe heat treatment can result in the formation of isopeptide cross-links during food processing, via condensation of the ε-amino group of lysine, with the amide group of an asparagine or glutamine residue (Singh 1991). This chemistry has not been widely studied in the context of food.

MANIPULATING PROTEIN CROSS-LINKING DURING FOOD PROCESSING

A major task of modern food technology is to generate new food structures with characteristics that please the consumer, using only a limited range of ingredients. Proteins are one of the main classes of molecule available to confer textural attributes, and the cross-linking and aggregation of protein molecules is an important mechanism for engineering food structures with desirable mechanical properties (Dickinson 1997, Oliver et al. 2006). The cross-linking of food proteins can influence many properties of food, including texture, viscosity, solubility, emulsification and gelling properties (Kuraishi et al. 2000, Motoki and Kumazawa 2000, Oliver et al. 2006). Many traditional food textures are derived from a protein gel, including those of yoghurt, cheese, sausage, tofu and surimi. Cross-linking provides an opportunity to create gel structures from protein solutions, dispersions, colloidal systems, protein-coated emulsion droplets or protein-coated gas bubbles and create new types of food or improve the properties of traditional ones (Dickinson 1997). In addition, judicious choice of starting proteins for cross-linking can produce food proteins of higher nutritional quality through cross-linking of different proteins containing complementary amino acids (Kuraishi et al. 2000).

Chemical Methods

An increasing understanding of the chemistry of protein cross-linking opens up opportunities to control these processes during food processing. Many commercial cross-linking agents are available, for example from Pierce (2001). These are usually double-headed reagents developed from molecules that derivatise the side chains of proteins (Matheis and Whitaker 1987, Feeney and Whitaker 1988) and generally exploit the lysine and/or cysteine residues of proteins in a specific manner. Doubt has been cast as to the accuracy with which reactivity of these reagents can be predicted (Green et al. 2001) but they remain widely used for biochemical and biotechnological applications.

Unfortunately, these reagents are expensive and not often approved for food use, so their use has not been widely explored (Singh 1991). They do, however, prove useful for 'proof of principle' studies to measure the possible effects of introducing specific new cross-links into food. If an improvement in functional properties is seen after treatment with a commercial cross-linking agent, then further research effort is merited to find a food approved, cost-effective means by which to introduce such cross-links on a commercial scale. Such 'proof of principle' studies include the use of glutaraldehyde to demonstrate the potential effects of controlled Maillard cross-linking on the texture of wheat-based foods (Gerrard et al. 2002, 2003a, 2003b). The cross-linking of hen egg white lysozyme with a double-headed reagent has also been used to show that the protein is rendered more stable to heat and enzyme digestion, with the foaming and emulsifying capacity reduced (Matheis and Whitaker 1987, Feeney and Whitaker 1988). Similarly, milk proteins cross-linked with formaldehyde showed greater heat stability (Singh 1991).

Food preparation for consumption often involves heating, which can result in a deterioration of the functional properties of these proteins (Bouhallab et al. 1999, Morgan et al. 1999a, Shepherd et al. 2000). In an effort to protect them from denaturation, particularly in the milk processing industry, some have harnessed the Maillard reaction to produce more stable proteins following incubation with monosaccharide (Aoki et al. 1999, Bouhallab et al. 1999, 2001, Handa and Kuroda 1999, Morgan et al. 1999a, Shepherd et al. 2000, Chevalier et al. 2001, Matsudomi et al. 2002). Increases in protein stability at high temperatures and improved emulsifying activity have been observed (Aoki et al. 1999, Bouhallab et al. 1999, Shepherd et al. 2000), and dimerisation and oligomerisation of these proteins has been noted (Aoki et al. 1999, Bouhallab et al. 1999, Morgan et al. 1999a, Pellegrino et al. 1999, Chevalier et al. 2001, French et al. 2002). In β-lactoglobulin, it has been suggested that oligomerisation is initiated by glycation of a β-lactoglobulin monomer, resulting in a conformational change in this protein that engenders a propensity to form stable covalent homodimers (Morgan et al. 1999b). From this point, it is thought that polymerisation occurs *via* hydrophobic interactions between unfolded homodimers and modified monomers (Morgan et al. 1999b). Experiments undertaken by Bouhallab et al. (1999) showed that this final polymerisation process was not due to intermolecular cross-linking via the Maillard reaction, but did confirm that increased solubility of the protein at high temperatures (65–90°C) was associated with polymerisation, presumably invoking disulfide linkages. Some recent data, however, suggests that Maillard cross-links may also be important in the polymerisation process (Chevalier et al. 2001). Interestingly, Pellegrino et al. have observed that an increase in pentosidine formation coincided with heat-induced covalent aggregation of β-casein, which lacks cysteine, but could not account for the extent of aggregation observed, suggesting the presence of other cross-links that remain undefined (Pellegrino et al. 1999).

The glycation approach has also been also employed as a tool to improve gelling properties of dried egg white (Handa and Kuroda 1999, Matsudomi et al. 2002), a protein which is used in the preparation of surimi and meat products (Weerasinghe et al. 1996, Chen 2000, Hunt et al. 2009). Some have suggested that the polymerisation may occur via protein cross-links formed as a result of the Maillard reaction, but disulfide bonds may still play some role (Handa and Kuroda 1999); the exact mechanism remains undefined (Matsudomi et al. 2002).

The effect of the Maillard reaction and particularly protein cross-linking on food texture has received some attention (Gerrard et al. 2002a). Introduction of protein cross-links into baked products has been shown to improve a number of properties that are valued by the consumer (Gerrard et al. 1998b, 2000). In situ studies revealed that following addition of glutaraldehyde to dough, albumin and globulin fraction of the extracted wheat proteins were cross-linked (Gerrard et al. 2003a). Inclusion of glutaraldehyde during bread preparation resulted in the formation of a dough with an increased dough relaxation time, relative to the commonly used flour improver ascorbic acid (Gerrard et al. 2003b). These results confirm that chemical cross-links are important in the process of dough development and suggest that they can be introduced via Maillard-type chemistry.

The Maillard reaction has also been used to modify properties in tofu. Kaye et al. (2001) reported that following incubation with glucose, a Maillard network formed within the internal structure of tofu, resulting in a loss in tofu solubility and a reduction in tofu weight loss. Furthermore, Kwan and Easa (2003) employed low levels of glucose for the preparation of retort tofu, which resulted in the production of firmer tofu product. Changes in tofu structure have also been observed in our laboratory when including glutaraldehyde, glyceraldehyde or formaldehyde in tofu preparation (Yasir et al. 2007a, 2007b). However, these results suggested that protein cross-linking agents may change the functional properties of tofu via non-cross-linking modifications of the side chains of the amino acid residues, perhaps by changing their isoelectric point and gelation properties.

Caillard et al. (2008) have also observed interesting results with soy protein that forms hydrogels in the presence of the cross-linking agents glutaraldehyde and glyceraldehyde. Their studies found that glutaraldehyde was the more efficient cross-linker, because it was able to form stronger hydrogels with modifiable properties than glyceraldehyde. More recently, Caillard et al. (2009) investigated the photophysical and microstructural properties of soy protein hydrogels cross-linked with glutaraldehyde in the absence or presence of salts.

The covalent polymerisation of milk during food processing has been reported (Singh and Latham 1993). This phenomenon was shown to be sugar dependent, as determined in model studies with β-casein. Further, pentosidine formation paralleled protein aggregation over time at 70°C (Pellegrino et al. 1999).

The Maillard reaction has been studied under dry conditions to gain an understanding of the details of the chemistry (Kato et al. 1988, French et al. 2002, Oliver et al. 2006). Kato et al. (1986b) reported the formation of protein polymers following incubation of galactose or talose with ovalbumin under desiccating conditions: an increase in polymerisation, relative to the protein only control, was observed. This study was extended using the milk sugar lactose, as milk can often be freeze dried for shipping and storage purposes. In this study, it was shown, in a dry reaction mixture, that ovalbumin was polymerised following incubation with lactose and glucose (Kato et al. 1988). In model studies with the milk protein β-lactoglobulin, the formation of protein dimers has been observed on incubation with lactose (French et al. 2002).

Proof of principle has thus been obtained in several systems, demonstrating that reactive cross-linking molecules are able to cross-link food proteins within the food matrix and lead to a noticeable change in the functional properties of the food. However, much work remains to be done in order to generate sufficient quantities of cross-linking intermediates during food processing to achieve a controlled change in functionality.

Enzymatic Methods

The use of enzymes to modify the functional properties of foods is an area that has attracted considerable interest, since consumers perceive enzymes to be more "natural" than chemicals. Enzymes are also favoured as they require milder conditions, have high specificity, are only required in catalytic quantities, and are less likely to produce toxic products (Singh 1991). Thus, enzymes are becoming commonplace in many industries for improving the functional properties of food proteins (Chobert et al. 1996, Poutanen 1997, Oliver et al. 2006, Hiller and Lorenzen 2009).

Because of the predominance of disulfide cross-linkages in food systems, enzymes that regulate disulfide interchange reactions are of interest to food researchers. One such enzyme is protein disulfide isomerase (PDI; Hatahet and Ruddock 2009). PDI catalyses thiol/disulfide exchange, rearranging 'incorrect' disulfide cross-links in a number of proteins of biological interest (Hillson et al. 1984, Singh 1991, Hatahet and Ruddock 2009). The reaction involves the rearrangement of low molecular sulfhydryl compounds (e.g. glutathione, cysteine and dithiothreitol) and protein sulfhydryls. It is thought to proceed by the transient breakage of the protein disulfide bonds by the enzyme and the reaction of the exposed active cysteine sulfhydryl groups with other appropriate residues to reform native linkages (Singh 1991).

PDI has been found in most vertebrate tissues and in peas, cabbage, yeast, wheat and meat (Singh 1991). It has been shown to catalyse the formation of disulfide bonds in gluten proteins synthesised in vitro. Early reports suggested that a high level of activity corresponded to a low bread-making quality (Grynberg et al. 1977). The use of oxidoreduction enzymes, such as PDI, to improve product quality is an area of interest to the baking industry (Watanabe et al. 1998, van Oort 2000, Joye et al. 2009a, 2009b) and the food industry in general (Hjort 2000). The potential of enzymes such as PDI to catalyse their interchange has been extensively reviewed by Shewry and Tatham (1997).

Sulfhydryl (or thiol) oxidase catalyses the oxidative formation of disulfide bonds from sulfhydryl groups and oxygen and occurs in milk (Matheis and Whitaker 1987, Singh 1991). Immobilised sulfhydryl oxidase has been used to eliminate the 'cooked' flavour of ultra-high temperature-treated milk (Swaisgood 1980). It is not a well-studied enzyme, although the enzyme from chicken egg white has received recent attention in biology (Hoober 1999). Protein disulfide reductase catalyses a further sulfhydryl–disulfide interchange reaction and has been found in liver, pea and yeast (Singh 1991).

Peroxidase, lipoxygenase and catechol oxidase occur in various plant foods and are implicated in the deterioration of foods during processing and storage. They have been shown to cross-link several food proteins, including bovine serum albumin, casein, β-lactoglobulin and soy, although the uncontrolled nature of these reactions casts doubt on their potential for food improvement (Singh 1991). Lipoxygenase, in soy flour, is used in the baking industry to improve dough properties and baking performance. It acts on unsaturated fatty acids, yielding peroxy free radicals and starting a chain reaction. The cross-linking action of lipoxygenase has been attributed to both the free radical oxidation of free thiol groups to form disulfide bonds and to the generation of reactive cross-linking molecules such as malondialdehyde (Matheis and Whitaker 1987).

Altering the functional properties of milk proteins by cross-linking with transglutaminase (see below), as well as the enzymes lactoglutaminase, lactoperoxidase, laccase and glucose oxidase has recently been studied by Hiller and Lorenzen (2009). In this study, a qualitative and quantitative overview of the enzymatic oligomerisation of milk proteins by various enzymes in relation to transglutaminase was presented, to fill a gap in understanding of enzymatic oligomerisation of proteins. A variety of milk proteins were chosen for this study, such as sodium caseinate, whey protein isolate and skim milk powder. Interestingly, the results from this study suggest it may be possible to specifically select the appropriate modified protein to use for specific industrial applications, depending upon the requirements of the food proteins and the desired functional protein properties.

All enzymes discussed so far have been over-shadowed in recent years by the explosion in research on the enzyme transglutaminase. Due largely to its ability to induce the gelation of protein solutions, transglutaminase has been investigated for uses in a diverse range of foods and food-related products. The use of this enzyme has been the subject of a series of recent reviews, covering both the scientific and patent literature (Nielson 1995, Zhu et al. 1995, Motoki and Seguro 1998, Kuraishi et al. 2001, Yokoyama et al. 2004, Jaros et al. 2006, Özrenk 2006). These are briefly highlighted below.

Transglutaminase

The potential of transglutaminase in food processing was hailed for many years before a practical source of the enzyme became widely available. The production of a microbially derived enzyme by Ajinomoto Inc. proved pivotal in paving the way for industrial applications (Motoki and Seguro 1998). In addition, the transglutaminases that were discovered in the early years were calcium-ion dependent, which imposed a barrier for their use in foods that did not contain a sufficient level of calcium. The commercial preparation is not calcium dependent and thus finds much wider applicability (Motoki and Seguro 1998). The production of microbial transglutaminase, derived from *Streptoverticillium mobaraense,* is described by Zhu et al. (1995) and methods with which to purify and assay the enzyme are reviewed by Wilhelm et al. (1996). The commercial enzyme operates effectively over the pH range 4–9, from 0°C to 50°C (Motoki and Seguro 1998).

There is a seemingly endless list of foods in which the use of transglutaminase has been successfully used (De Jong and Koppelman 2002): seafood, surimi, meat, dairy (Hiller and Lorenzen 2009), baked goods, sausages (as a potential replacement for phosphates and other salts), gelatin (Kuraishi et al. 2001), spaghetti (Aalami and Leelavathi 2008), noodles and pasta (Larre et al. 1998, 2000). It is finding increasing use in restructured products, such as those derived from scallops and pork (Kuraishi et al. 1997). In all cases, transglutaminase is reported to improve firmness, elasticity, water-holding capacity and heat stability (Kuraishi et al. 2001). It also has potential to alleviate the allergenicity of some proteins (Watanabe et al. 1994). Dickinson (1997) reviewed the application of transglutaminase to cross-link different kinds of colloidal structures in food and enhance their solid-like character in gelled and emulsified systems, controlling rheology and stability. These applications are particularly appealing in view of the fact that cross-linking by transglutaminase is thought to protect nutritionally valuable lysine residues in food from various deteriorative reactions (Seguro et al. 1996). Furthermore, the use of transglutaminase potentially allows production of food proteins of higher nutritional quality, through cross-linking of different proteins containing complementary amino acids (Zhu et al. 1995).

The use of transglutaminase in the dairy industry has been explored extensively. The enzyme has been trialled in many cheeses, from Gouda to Quark, and the use of transglutaminase in ice cream is reported to yield a product that is less icy and more easily scooped (Kuraishi et al. 2001). Milk proteins form emulsion gels that are stabilised by cross-linking, opening new opportunities for protein-based spreads, desserts and dressings (Dickinson and Yamamoto 1996). The use of transglutaminase has been explored in an effort to improve the functionality of whey proteins (Truong et al. 2004, Gauche et al. 2008, 2010), for example recent developments include the use of transglutaminase to incorporate whey protein into cheese (Cozzolino et al. 2003, Pereira et al. 2009) and yoghurt (Ozer et al. 2007, Gauche et al. 2009).

Soy products have also benefited from the introduction of transglutaminase, with the enzyme providing manufacturers with a greater degree of texture control. The enzyme is reported to enhance the quality of tofu made from old crops, giving a product with increased water-holding capacity, a good consistency, a silky and firmer texture and one that is more robust in the face of temperature change (Kuraishi et al. 2001, Soeda 2003). Transglutaminase has also been used to incorporate soy protein into new products, such as chicken sausages (Muguruma et al. 2003). More recent studies have found that transglutaminase can potentially reduce the extent of the Maillard reaction and cross-links (Gan et al. 2009). Soy protein isolate that had been heated in the presence of transglutaminase to initiate the ε-(γ-glutamyl)lysine bonds formed were then heated in the presence of ribose to initiate Maillard reaction. It was found that the ε-(γ-glutamyl)lysine bonds formed during incubation of soy protein isolate with transglutaminase reduced the number of free amino groups able to take part in the following Maillard reaction with ribose. Consequently, the use of transglutaminase is most likely to be most beneficial in products that contain reducing sugars, where it is advantageous for the Maillard reaction to be limited.

New foods are being created using transglutaminase, for example, imitation shark fin for the South East Asian market has been generated by cross-linking gelatin and collagen (Zhu et al. 1995). Cross-linked proteins have also been tested as fat substitutes in products such as salami and yoghurt (Nielson 1995) and the use of transglutaminase-cross-linked protein films as edible films has been patented (Nielson 1995).

Not surprisingly, the rate of cross-linking by transglutaminase depends on the particular structure of the protein acting as substrate. Most efficient cross-linking occurs in proteins that contain a glutamine residue in a flexible region of the protein or within a reverse turn (Dickinson 1997). Casein is a very good substrate, but globular proteins such as ovalbumin and β-lactoglobulin are poor substrates (Dickinson 1997). Denaturation of proteins increases their reactivity, as does chemical modification by disruption of disulfide bonds or by adsorption at an oil–water interface (Dickinson 1997). Many of the reported substrates of transglutaminase have actually been acetylated and/or denatured with reagents such as dithiothreitol under regimes that are not food approved (Nielson 1995). More work needs to be done to find ways to modify certain proteins in a food-allowed manner in order to render them amenable to cross-linking by transglutaminase in a commercial setting.

Whilst the applications of transglutaminase have been extensively reported in the scientific and patent literature, the precise mode of action of the enzyme in any one food processing situation remains relatively unexplored. The specificity of the enzyme suggests that in mixtures of food proteins, certain proteins will react more efficiently than others, and there is value in understanding precisely which protein modifications exert the most desirable effects. Additionally, transglutaminase has more than one activity: as well as cross-linking, the enzyme may catalyse the incorporation of free amines into proteins by attachment to a glutamine residue. Furthermore, in the absence of free amine, water becomes the acyl acceptor and the γ-carboxamide groups are deamidated to glutamic acid residues (Ando et al. 1989). The extent of these side reactions in foods, and the consequences of any deamidation to the functionality of food proteins, has yet to be fully explored. However, there have been suggestions that in the case of wheat proteins, these side reactions may lead to unwanted consequences, specifically associated with the celiac response (Gerrard and Sutton 2005, Leszczynska et al. 2006, Cabrera-Chavez et al. 2008), and the use in wheat products is not advised until further research has been done to investigate this possibility.

Perhaps the most advanced understanding of the specific molecular effects of transglutaminase in a food product is seen in yoghurt, where the treated product has been analysed by gel electrophoresis and specific functional effects correlated to the loss of β-casein, with the α-casein remaining. The specificity of the reaction was found to alter according to the exact transglutaminase source (Kuraishi et al. 2001). The specific effects of transglutaminase in baked goods (Gerrard et al. 1998b, 2000) have also been analysed at a molecular level. In particular, the enzyme produced a dramatic increase in the volume of croissants and puff pastries, with desirable flakiness and crumb texture (Gerrard et al. 2001). These effects were later correlated with cross-linking of the albumins and globulins and high-molecular weight glutenin fractions by transglutaminase (Gerrard et al. 2001). Subsequent research suggested that the dominant effect was attributable to cross-linking of the high-molecular weight glutenins (Gerrard et al. 2002). However, as noted above, the use of transglutaminase for wheat-based produce is not recommended (Gerrard and Sutton 2005).

FUTURE APPLICATIONS OF PROTEIN CROSS-LINKING

Although protein cross-linking is often considered to be detrimental to the quality of food, it is increasingly clear that it can also be used as a tool to improve food properties. The more we understand of the chemistry and biochemistry that take place during processing, the better placed we are to exploit it – minimising deleterious reactions and maximising beneficial ones.

Food chemists are faced with the task of understanding the vast array of reactions that occur during the preparation of food. Food often has complex structures and textures, and through careful manipulation of specific processes that occur during food preparation, these properties can be enhanced to generate a highly marketable product (Dickinson 1997). The formation of a structural network within food is critical for properties such as food texture. From a biopolymer point of view, this cross-linking process has been successfully applied in the formation of protein films, ultimately for use as packaging. Enhancing textural properties, emulsifying and foaming properties, by protein cross-linking has been a subject of interest of many working in this area, as exemplified by the number of studies detailed above.

Cross-linking using chemical reagents remains challenging, both in terms of controlling the chemistry and gaining consumer acceptance. However, as the extensive recent use of transglutaminase dramatically illustrates, protein cross-linking using enzymes has huge potential for the improvement of traditional products and the creation of new ones. Transglutaminase itself will no doubt find yet more application as its precise mode of action becomes better understood, especially if variants of the enzyme are found with a broader substrate specificity.

Whether other enzymes, which cross-link by different mechanisms, can find equal applicability remains open to debate. The thiol exchange enzymes, such as PDI, may offer advantages to food processors if their mechanisms can be unravelled, and then controlled within a foodstuff. Other enzymes, such as lysyl oxidase, have not yet been used, but have potential to improve foods (Dickinson 1997), especially in the light of recent work characterising this class of enzymes (Buffoni and Ignesti 2000), which may allow their currently unpredictable effects to be better understood.

ACKNOWLEDGEMENTS

We thank all the postgraduates and postdoctoral fellows who have worked with us on the Maillard reaction and protein cross-linking of food, in particular Dr Siân Fayle, Paula Brown, Dr Indira Rasiah, Dr Susie Meade, Dr Antonia Miller and Dr Suhaimi Yasir.

REFERENCES

Aalami M, Leelavathi K. 2008. Effect of microbial transglutaminase on spaghetti quality. *J Food Sci* 73: C306–C312.

Ahmed N. 2005. Advanced glycation endproducts – role in pathology of diabetic complications. *Diabetes Res Clin Pract* 67: 3–21.

Al-Abed Y, Bucala R. 2000. Structure of a synthetic glucose derived advanced glycation end product that is immunologically cross-reactive with its naturally occurring counterparts. *Bioconjug Chem* 11: 39–45.

Ames JM. 1992. The Maillard reaction in foods. In: BJF Hudson (ed.) *Biochemistry of Food Proteins*. Elsevier Applied Science, London, pp. 99–153.

Ames JM. 2007. Evidence against dietary advanced glycation endproducts being a risk to human health. *Mol Nutr Food Res* 51: 1085–1090.

Ames JM. 2009. Dietary Maillard reaction products: implication for human health and disease. *Czech J Food Sci* 27(SI): S66–S68.

An HJ et al. 1996. Roles of endogeneous enzymes in surimi gelation. *Trends Food Sci Technol* 7: 321–327.

Ando H et al. 1989. Purification and characteristics of a novel transglutaminase derived from microorganisms. *Agric Biol Chem* 53: 2613–2617.

Aoki T et al. 1999. Improvement of heat stability and emulsifying activity of ovalbumin by conjugation with glucuronic acid through the Maillard reaction. *Food Res Intern* 32: 129–133.

Aoki T et al. 2001. Modification of ovalbumin with oligogalacturonic acids through the Maillard reaction. *Food Res Intern* 34: 127–132.

Benjakul S et al. 2004a. Cross-linking activity of sarcoplasmic fraction from bigeye snapper (Priacanthus tayenus) muscle. *Food Sci Technol* 37: 79–85.

Benjakul S et al. 2004b. Effect of porcine plasma protein and setting on gel properties of surimi produced from fish caught in Thailand. *Food Sci Technol* 37: 177–185.

Bergmann R et al. 2001. Radio fluorination and positron emission tomography (PET) as a new approach to study the in vivo distribution and elimination of the advanced glycation endproducts *N*-epsilon-carboxymethyllysine (CML) and *N*-epsilon-carboxyethyllysine (CEL). *Nahrung* 45: 182–188.

Biemel KM et al. 2001. Identification and quantitative evaluation of the lysine-arginine crosslinks GODIC, MODIC, DODIC, and glucosepan in foods. *Nahrung* 45: 210–214.

Biemel KM et al. 2002. Identification and quantification of major Maillard-crosslinks in human serum albumin and lens protein: evidence for glucosepane as the dominant compound. *J Biol Chem* 277: 24907–24915.

Bordas M et al. 2004. Formation and stability of heterocycloc amines in a meat flavour model system – Effect of temperature, time and precursors. *J Chromatogr B* 802: 11–17.

Bouhallab S et al. 1999. Formation of stable covalent dimer explains the high solubility at pH 4.6 of lactose-β-lactoglobulin conjugates heated near neutral pH. *J Agric Food Chem* 47: 1489–1494.

Brady JD, Robins SP. 2001. Structural characterization of pyrrolic cross-links in collagen using a biotinylated Ehrlich's reagent. *J Biol Chem* 276: 18812–18818.

Brands CMJ et al. 2002. Quantification of melanoidin concentration in sugar-casein systems. *J Agric Food Chem* 50: 1178–1183.

Brinkmann-Frye E et al. 1998. Role of the Maillard reaction in aging of tissue proteins. *J Biol Chem* 273: 18714–18719.

Buffoni F, Ignesti G. 2000. The copper-containing amine oxidases: biochemical aspects and functional role. *Mol Genet Metab* 71: 559–564.

Cabrera-Chavez F et al. 2008. Transglutaminase treatment of wheat and maize prolamins of bread increases the serum IgA reactivity of celiac disease patients. *J Agric Food Chem* 56: 1387–1391.

Caillard R et al. 2008. Characterization of amino cross-linked soy protein hydrogels. *J Food Chem* 73: C283–C291.

Caillard R et al. 2009. Physiochemical properties and microstructure of soy protein hydrogels co-induced by Maillard-type cross-linking and salts. *Food Res Int* 42: 90–106.

Chellan P, Nagaraj RH. 2001. Early glycation products produce pentosidine cross-links on native proteins: novel mechanism of pentosidine formation and propagation of glycation. *J Biol Chem* 276: 3895–3903.

Chen HH. 2000. Effect of non muscle protein on the thermogelation of horse mackerel surimi and the resultant cooking tolerance of kamaboko. *Fish Sci* 66: 783–788.

Chevalier F et al. 2001. Improvement of functional properties of β-lactoglobulin glycated through the Maillard reaction is related to the nature of the sugar. *Int Dairy J* 11: 145–152.

Chevalier F et al. 2002. Maillard glycation of β-lactoglobulin induces conformation changes. *Nahrung* 46: 58–63.

Cho RK et al. 1984. Chemical properties and polymerizing ability of the lysozyme monomer isolated after storage with glucose. *Agric Biol Chem* 48: 3081–3089.

Cho RK et al. 1986a. Polymerization of acetylated lysozyme and impairment of their amino acid residues due to α-dicarbonyl and α-hydroxycarbonyl compounds. *Agric Biol Chem* 50: 1373–1380.

Cho RK et al. 1986b. Polymerization of proteins and impairment of their arginine residues due to intermediate compounds in the Maillard reaction. *Dev Food Sci* 13: 439–448.

Chober JM et al. 1996. Recent advances in enzymatic modifications of food proteins for improving their functional properties. *Nahrung* 40: 177–182.

Cole CGB, Roberts JJ. 1996. Gelatine fluorescence and its relationship to animal age and gelatine colour. *SA J Food Sci Nutr* 8: 139–143.

Cole CGB, Roberts JJ. 1997. The fluorescence of gelatin and its implications. *Imaging Sci J* 45: 145–149.

Cozzolino A et al. 2003. Incorporation of whey proteins into cheese curd by using transglutaminase *Biotechnol Appl Biochem* 38: 289–295.

De Jong GAH, Koppelman SJ. 2002. Transglutaminase catalyzed reactions: impact on food applications. *J Food Sci* 67: 2798–2806.

Dickinson E. 1997. Enzymic crosslinking as a tool for food colloid rheology control and interfacial stabilization. *Trends Food Sci Technol* 8: 334–339.

Dickinson E, Yamamoto Y. 1996. Rheology of milk protein gels and protein-stabilized emulsion gels cross-linked with transglutaminase. *J Agric Food Chem* 44: 1371–1377.

Dittrich R et al. 2003. Maillard reaction products inhibit oxidation of human low-density lipoproteins in vitro. *J Agric Food Chem* 51: 3900–3904.

Dyer DG et al. 1991. Formation of pentosidine during nonenzymatic browning of proteins by glucose. Identification of glucose and other carbohydrates as possible precursors of pentosidine in vivo. J Biol Chem 266: 11654–11660.

Easa AM et al. 1996a. Maillard induced complexes of bovine serum albumin – a dilute solution study. *Int J Biol Macromol* 18: 297–301.

Easa AM et al. 1996b. Bovine serum albumin gelation as a result of the Maillard reaction. *Food Hydrocolloids* 10: 199–202.

Eyre D et al. 1984. Cross-linking in collagen and elastin. *Ann Rev Biochem* 53: 717–748.

Faist V, Erbersdobler HF. 2002. Health impact of food-derived Maillard reaction products. *Kieler Milchwirtschaftliche Forschungsberichte* 54: 137–147.

Faist V et al. 2000. In vitro and in vivo studies on the metabolic transit of N^{ϵ}-carboxymethyllsine. *Czech J Food Sci* 18: 116–119.

Fayle SE, Gerrard JA. 2002. *The Maillard Reaction*. Royal Society of Chemistry, Cambridge.

Fayle SE et al. 2000. Crosslinkage of proteins by dehydroascorbic acid and its degradation products. *Food Chem* 70: 193–198.

Fayle SE et al. 2001. Novel approaches to the analysis of the Maillard reaction of proteins. *Electrophoresis* 22: 1518–1525.

Feeney RE, Whitaker JR. 1988. Importance of cross-linking reactions in proteins. Adv Cereal Sci Tech *IX*: 21–43.

Felton JS, Knize MG. 1998. Carcinogens in cooked foods: how do they get there and do they have an impact on human health? In: J O'Brian et al. (eds.) The Maillard Reaction in Foods and Medicine. *The Royal Society of Chemistry, Special Publication No. 223*, Cambridge, UK, pp. 11–18.

Förster A, Henle T. 2003. Glycation in food and metabolic transit of dietary AGEs (advanced glycation endproducts): studies on the urinary excretion of pyrraline. *Biochem Soc Trans* 31: 1383–1385.

Förster A et al. 2005. JB Baynes, VM Monnier, JM Ames, SR Thorpe (eds.) *The Maillard reaction: chemistry at the interface of nutrition, aging and disease. Ann NY Acad Sci* 1043: pp. 474–481.

Friedman M. 1996a. Food browning and its prevention: an overview. *J Agric Food Chem* 44: 631–653.

Friedman M. 1996b. The impact of the Maillard reaction on the nutritional value of food proteins. In: R Ikan (ed.) *The Maillard Reaction: Consequences for the Chemical and Life Sciences.* Wiley, Chichester, p. 119.

Friedman M. 1999a. Chemistry, biochemistry, nutrition, and microbiology of lysinoalanine, lanthionine, and histidinoalanine in food and other proteins. *J Agric Food Chem* 47: 1295–1319.

Friedman M. 1999b. Lysinoalanine in food and in antimicrobial proteins. *Adv Exp Med Biol* 459: 145–159.

Friedman M. 1999c Chemistry, nutrition, and microbiology of D-amino acids. *J Agric Food Chem* 47: 3457–3479.

French SJ et al. 2002. Maillard reaction induced lactose attachment to bovine β-lactoglobulin: electrospray ionization and matrix-assisted laser desorption/ionization examination. *J Agric Food Chem* 50: 820–823.

Gan CY et al. 2009. Assessment of cross-linking in combined cross-linked soy protein isolate gels by microbial tranglutaminase and Maillard reaction. *J Food Sci* 74: C141–C146.

Garcia MM et al. 2009. Intake of Maillard reaction products reduces iron bioavailability in male adolescents. *Mol Nutr Food Res* 53: 1551–1560.

Gauche C et al. 2008. Crosslinking of milk whey proteins by transglutaminase. *Process Biochem* 43: 788–794.

Gauche C et al. 2009. Physical properties of yoghurt manufactured with milk whey and transglutaminase. *Food Sci Technol* 42: 239–243.

Gauche C et al. 2010. Effect of thermal treatment on whey protein polymerization by transglutaminase: implications for functionality in processed dairy foods. *Food Sci Technol* 45: 213–219.

Gerrard JA. 2002. Protein–protein crosslinking in food: methods, consequences, applications. *Trends Food Sci Technol* 13: 391–399.

Gerrard JA. 2006. The Maillard reaction in food: progress made, challenges ahead – conference report from the eighth international symposium on the Maillard reaction. *Trends Food Sci Technol* 17: 324–330.

Gerrard JA, Sutton, KH. 2005. Addition of transglutaminase to cereal products may generate the epitope responsible for coelic disease. *Trends Food Sci Technol* 16: 510–512.

Gerrard JA et al. 1999. Covalent protein adduct formation and protein crosslinking resulting from the Maillard reaction between cyclotene and a model food protein. *J Agric Food Chem* 47: 1183–1188.

Gerrard JA et al. 1998a. Dehydroascorbic acid mediated crosslinkage of proteins using Maillard chemistry – relevance to food processing. In: JO'Brien, HE Nursten, MJC Crabbe, JM Ames (eds.) *The Maillard Reaction in Foods and Medicine*. Woodhead Publishing, Cambridge, pp. 127–132.

Gerrard JA et al. 1998b. Dough properties and crumb strength of white pan bread as affected by microbial transglutaminase. *J Food Sci* 63: 472–475.

Gerrard JA et al. 2000. Pastry lift and croissant volume as affected by microbial transglutaminase. *J Food Sci* 65: 312–314.

Gerrard JA et al. 2001. Effects of microbial transglutaminase on the wheat proteins of bread and croissant dough. *J Food Sci* 66: 782–786.

Gerrard JA et al. 2002. Maillard crosslinking of food proteins I: the reaction of glutaraldehyde, formaldehyde and glyceraldehyde with ribonuclease. *Food Chem* 79: 343–349.

Gerrard JA et al. 2003a. Maillard crosslinking of food proteins II: the reactions of glutaraldehyde, formaldehyde and glyceraldehyde with wheat proteins in vitro and in situ. *Food Chem* 80: 35–43.

Gerrard JA et al. 2003b. Maillard crosslinking of food proteins III: the effects of glutaraldehyde, formaldehyde and glyceraldehyde upon bread and croissants. *Food Chem* 80: 45–50.

Glomb MA, Pfahler C. 2001. Amides are novel protein modifications formed by physiological sugars. *J Biol Chem* 276: 41638–41647.

Green NS et al. 2001. Quantitative evaluation of the lengths of homobifunctional protein cross-linking reagents used as molecular rulers. *Protein Sci* 10: 1293–1304.

Grynberg A et al. 1977. De la presence d'une proteine disulfide isomerase dans l'embryon de ble. *CR Hebd Seances Acad Sci Paris* 284: 235–238.

GuI A et al. 2009. Advanced glycation end products in senile diabetic and nondiabetic patients with cataract. *J Diabetes Complications* 23: 343–348

Handa A, Kuroda N. 1999. Functional improvements in dried egg white through the Maillard reaction. *J Agric Food Chem* 47: 1845–1850.

Hatahet F, Ruddock LW. 2009. Protein disulfide isomerase: a critical evaluation of its function in disulfide bond formation. *Antioxid Redox Signal* 11: 2807–2850.

Henle T. 2007. Dietary advanced glycation end products – a risk to human health? A call for an interdisciplinary debate. *Mol Nutr Food Res* 51: 1075–1078.

Henle T et al. 1997. Detection and quantification of pentosidine in foods. *Z Lebensm – Unters Forsch A* 204: 95–98.

Hill S, Easa AM. 1998. Linking proteins using the Maillard reaction and the implications for food processors. In: J O'Brien et al. (eds.) *The Maillard Reaction in Foods and Medicine*. Woodhead Publishing, Cambridge, pp. 133–138.

Hill SE et al. 1993. The use and control of chemical reactions to enhance gelation of macromolecules in heat-processed foods. *Food Hydrocolloids* 7: 289–294.

Hiller B, Lorenzen PC. 2009. Functional properties of milk proteins as affected by enzymatic oligomerisation. *Food Res Int* 42: 899–908.

Hillson DA et al. 1984. Formation and isomerisation in proteins: protein disulfide isomerase. *Meth Enzymol* 107: 281–304.

Hjort CM. 2000. Protein disulfide isomerase variants that preferably reduce disulfide bonds and their manufacture. In *PCT Int Appl*, 82.

Hofmann T. 1998a. 4-alkylidene-2-imino-5-[4-alkylidene-5-oxo-1,3-imidazol-2- inyl]azamethylidene-1,3-imidazolidine – A novel colored substructure in melanoidins formed by Maillard reactions of bound arginine with glyoxal and furan-2-carboxaldehyde. *J Agric Food Chem* 46: 3896–3901.

Hofmann T. 1998b. Studies on melanoidin-type colorants generated from the Maillard reaction of protein-bound lysine and furan-2-carboxaldehyde. Chemical characterization of a red colored domain. *Z Lebens Unters Forsch A* 206: 251–258.

Hofmann T et al. 1999. Radical-assisted melanoidin formation during thermal processing of foods as well as under physiological conditions. *J Agric Food Chem* 47: 391–396.

Hoober KL et al. 1999. Sulfhydryl oxidase from egg white – a facile catalyst for disulfide bond formation in proteins and peptides. *J Biol Chem* 274: 22147–22150.

Hunt A et al. 2009. Effect of various types of egg white on characteristics and gelation of fish myofibrillar proteins. *J Food Sci* 74: C683–C692.

Iqbal M et al. 1997. Effect of dietary aminoguanidine on tissue pentosidine and reproductive performance in broiler breeder hens. *Poult Sci* 76: 1574–1579.

Iqbal M et al. 1999a. Age-related changes in meat tenderness and tissue pentosidine: effect of diet restriction and aminoguanidine in broiler breeder hens. *Poult Sci* 78: 1328–1333.

Iqbal M et al. 1999b. Protein glycosylation and advanced glycosylated endproducts (AGEs) accumulation: an avian solution? *J Gerontol A Biol Sci Med Sci* 54: B171–B176.

Iqbal M et al. 2000. Relationship between mechanical properties and pentosidine in tendon: effects of age, diet restriction, and aminoguanidine in broiler breeder hens. *Poult Sci* 79: 1338–1344.

Jaros D et al. 2006. Transglutaminase in dairy products: chemistry, physics, applications. *J Texture Stud* 37: 113–155.

Joye IJ et al. 2009a. Endogenous redox agents and enzymes that affect protein network formation during breadmaking – a review. *J Cer Sci* 50: 1–10.

Joye IJ et al. 2009b. Use of chemical redox agents and exogenous enzymes to modify the protein network during breadmaking – a review. *J Cer Sci* 50: 11–21.

Kato H et al. 1986a. Changes of amino acids composition and relative digestibility of lysozyme in the reaction with α-dicarbonyl compounds in aqueous system. *J Nutr Sci Vitaminol* 32: 55–65.

Kato Y et al. 1986b. Browning and insolubilization of ovalbumin by the Maillard reaction with some aldohexoses. *J Agric Food Chem* 34: 351–355.

Kato Y et al. 1988. Browning and protein polymerization induced by amino carbonyl reaction of ovalbumin with glucose and lactose. *J Agric Food Chem* 36: 806–809.

Kaye YK et al. 2001. Reducing weight loss of retorted soy protein tofu by using glucose- and microwave-pre-heating treatment. *Int J Food Sci Technol* 36: 387–392.

Knerr T et al. 2001. Formation of a novel colored product during the Maillard reaction of D-glucose. *J AgricFood Chem* 49: 1966–1970.

Kuraishi C et al. 1997. Production of restructured meat using microbial transglutaminase without salt or cooking. *J Food Sci* 62: 488–490.

Kuraishi C et al. 2000. Application of transglutaminase for food processing. *Hydrocolloids* 2: 281–285.

Kuraishi C et al. 2001. Transglutaminase: its utilization in the food industry. *Food Rev Int* 17: 221–246.

Kwan SW, Easa AM 2003. Comparing physical properties of retort-resistant glucono-δ-lactone tofu treated with commercial transglutaminase enzyme or low levels of glucose. *Lebensmittel-Wissenschaft und-Technologie* 3: 643–646.

Larre C et al. 1998. Hydrated gluten modified by a transglutaminase. *Nahrung Food* 42: 155–157.

Larre C et al. 2000. Biochemical analysis and rheological properties of gluten modified by transglutaminase. *Cereal Chem* 77: 32–38.

Lederer MO, Buhler HP. 1999. Cross-linking of proteins by Maillard processes – Characterization and detection of a lysine-arginine cross-link derived from D-glucose. *Bioorg Med Chem* 7: 1081–1088.

Lederer MO, Klaiber RG. 1999. Cross-linking of proteins by Maillard processes: characterization and detection of lysine-arginine cross-links derived from glyoxal and methylglyoxal. *Bioorg Med Chem* 7: 2499–2507.

Lee KG, Shibamoto T. 2002. Toxicology and antioxidant activities of non-enzymatic browning reaction products: review. *Food Rev Int* 18: 151–175.

Lerche H et al. 2003. Maillard reaction of D-glucose: identification of a colored product with hydroxypyrrole and hydroxypyrrolinone rings connected by a methine group. *J Agric Food Chem* 51: 4424–4426.

Leszczynska J et al. 2006. The use of transglutaminase in the reduction of immunoreactivity of wheat flour. *Food Agric Immunol* 17: 105–113.

Lin RY et al. 2003. Dietary glycotoxins promote diabetic atherosclerosis in apolipoprotein E-deficient mice. *Atherosclerosis* 168: 213–220.

Lindenmeier M et al. 2002. Structural and functional characterization of pronyl-lysine, a novel protein modification in bread crust melanoidins showing in vitro antioxidative and phase I/II enzyme modulating activity. *J Agric Food Chem* 50: 6997–7006.

Lindsay MP, Skerritt JH. 1999. The glutenin macropolymer of wheat flour doughs: structure–function perspectives. *Trends Food Sci Technol* 10: 247–253.

Lopez-Garcia E et al. 2004. Major dietary patterns are related to plasma concentrations of markers of inflammation and endothelial dysfunction. *Am J Clin Nutr* 80: 1029–1035.

Manzocco L et al. 2000. Review of non-enzymatic browning and antioxidant capacity in processed foods. *Trends Food Sci Tech* 11: 340–346.

Masaki K et al. 2001. Thermal stability and enzymatic activity of a smaller lysozyme from silk moth (*Bombyx mori*). *J Protein Chem* 20: 107–113.

Matheis G, Whitaker JR. 1987. Enzymic cross-linking of proteins applicable to food. *J Food Biochem* 11: 309–327.

Matsudomi N et al. 2002. Improvement of gel properties of dried egg white by modification with galactomannan through the Maillard reaction. *J Agric Food Chem* 50: 4113–4118.

Mezgheni E et al. 1998. Formation of sterilized edible films based on caseinates – effects of calcium and plasticizers. *J Agric Food Chem* 46: 318–324.

Miller AG, Gerrard JA. 2005. The Maillard reaction and food protein crosslinking. *Prog Food Biopol Res* 1: 69–86.

Miyata T et al. 1998. Renal catabolism of advanced glycation end-products. *Kidney Int* 53: 416–422.

Mohammed ZH et al. 2000. Covalent crosslinking in heated protein systems. *J Food Sci* 65: 221–226.

Morales FJ, Babbel MB. 2002. Antiradical efficiency of Maillard reaction mixtures in a hydrophilic media. *J Agric Food Chem* 50: 2788–2792

Morales FJ, Jimenez-Perez S. 2004. Peroxyl radical scavenging activity of melanoidins in aqueous systems. *Eur Food Res Technol* 218: 515–520.

Morgan F et al. 1999a. Modification of bovine β-lactoglobulin by glycation in a powdered state or in an aqueous solution: effect on association behaviour and protein conformation. *J Agric Food Chem* 47: 83–91.

Morgan F et al. 1999b. Glycation of bovine β-lactoglobulin: effect on the protein structure. *Int J Food Sci Technol* 34: 429–435.

Motoki M, Kumazawa Y. 2000. Recent trends in transglutaminase technology for food processing. *Food Sci Technol Res* 6: 151–160.

Motoki M, Seguro K. 1998. Transglutaminase and its use for food processing. *Trends Food Sci Technol* 9: 204–210.

Muguruma M et al. 2003. Soybean and milk proteins modified by transglutaminase improves chicken sausage texture even at reduced levels of phosphate. *Meat Sci* 63: 191–197.

Murkovoic M. 2004. Formation of heterocyclic aromatic amines in model systems. *J Chromatogr B* 802: 3–10.

Nakamura K et al. 1997. Acid-stable fluorescent advanced glycation end products: vesperlysines a, b and c are formed as crosslinked products in the Maillard reaction between lysine or protein with glucose. *Biochem Biophys Res Commun* 232: 227–230.

Nass N et al. 2007. Advanced glycation end products, diabetes and ageing. *Zeitschrift fur Gerontologie und Geriatrie* 40: 349–356.

Nielsen PM. 1995. Reactions and potential industrial applications of transglutaminase – review of literature and patents. *Food Biotechnol* 9: 119–156.

Okitani A et al. 1984. Polymerization of lysozyme and impairment of its amino acid residues caused by reaction with glucose. *Agric Biol Chem* 48: 1801–1808.

Oliver CM et al. 2006. Creating proteins with novel functionality via the Maillard reaction: a review. *Crit Rev Food Sci Nutr* 46: 337–350.

Ozer B et al. 2007. Incorportion of microbial transglutaminase into non-fat yoghurt production. *Int Dairy J* 17: 199–207.

Özrenk E. 2006. The use of transglutaminase in dairy products. *Int J Dairy Res* 59: 1–7.

Pellegrino L et al. 1999. Heat-induced aggregation and covalent linkages in β-casein model systems. *Int Dairy J* 9: 255–260.

Peppa M et al. 2003. Adverse effects of dietary glycotoxins on wound healing in genetically diabetic mice. *Diabetes* 52: 2805–2813.

Peppa M et al. 2009. Advanced glycoxidation products and impaired diabetic wound healing. *Wound Repair Regen* 17: 461–472.

Pereira CL et al. 2009. Microstructure of cheese: processing, technological and microbiological considerations. *Trends Food Sci Technol* 20: 213–219.

Pierce. 2001. Crosslinker selection guide. *Rockford Illinois Pierce Chemical Company.* Available at http://www.piercenet.com.

Poutanen K. 1997. Enzymes: an important tool in the improvement of the quality of cereal foods. *Trends Food Sci Technol* 8: 300–306.

Prabhakaram M et al. 1997. Structural elucidation of a novel lysine-lysine crosslink generated in a glycation reaction with L-threose. *Amino Acids* 12: 225–236.

Savoie L et al. 1991. Effects of alkali treatment on in vitro digestibility of proteins and release of amino acids. *J Sci Food Agric* 56: 363–372.

Schwarzenboltz U et al. 2000. Maillard-type reactions under high hydrostatic pressure: formation of pentosidine. *Eur Food Res Technol* 211: 208–210.

Sebekova K et al. 2005. Renel effects of oral Maillard reaction product load in form of bread crusts in healthy and subtotally nephrectomized rats. *Ann NY Acad Sci* 1043: 482–491.

Sebekova K, Somoza V. 2007. Dietary advanced glycation endproducts (AGEs) and their health effects – PRO. *Mol Nutr Food Res* 51: 1079–1084.

Seguro K et al. 1996. The epsilon-(gamma-glutamyl)lysine moiety in crosslinked casein is an available source of lysine for rats. *J Nutr* 126: 2557–2562.

Seiquer I et al. 2008. The antioxidant effect of a diet rich in Maillard reaction products is attenuated after consumption by healthy male adolescents. In vitro and in vivo comparative study. *J Sci Food Agri* 88: 1245–1252.

Sell DR, Monnier VM. 1989. Structure elucidation of a senescence crosslink from human extracellular matrix. *J Biol Chem* 264: 21597–21602.

Shepherd R et al. 2000. Dairy glycoconjugate emulsifiers: casein-maltodextrins. *Food Hydrocolloids* 14: 281–286.

Shewry PR, Tatham AS. 1997. Disulfide bonds in wheat gluten proteins. *J Cer Sci* 25: 207–222.

Shih FF. 1992. Modification of food proteins by non-enzymatic methods. In: BJF Hudson (ed.) *Biochemistry of Food Proteins*, Elsevier Applied Science, London, pp. 235–248.

Shin HS. 2003. Factors that affect the formation and mutagenicity of heterocyclic aromatic amines in food. Food Sci Biotech 12: 588–596.

Shin HS et al. 2003. Influence of different oligosaccharides and inulin on heterocyclic aromatic amine formation and overall mutagenicity in fored ground beef patties. *J Agri Food Chem* 51: 6726–6730.

Singh H. 1991. Modification of food proteins by covalent crosslinking. *Trends Food Sci Technol* 2: 196–200.

Singh H, Latham J. 1993. Heat stability of milk: aggregation and dissociation of protein at ultra-high temperatures. *Int Dairy J* 3: 225–235.

Soeda T. 2003. Effects of microbial transglutaminase for gelation of soy protein isolate during cold storage. *Food Sci Technol Res* 9: 165–169.

Somaza V. 2005. Five years of research on health benefits of Maillard reaction products. An update. *Mol Nutr Food Res* 49: 663–672.

Somoza V et al. 2005. Influence of feeding malt, bread crust, and a pronylated protein on the activity of chemopreventive enzymes and antioxidative defense parameters in vivo. *J Agric Food Chem* 53: 8176–8182.

Strauss G, Gibson SA. 2004. Plant phenolics as cross-linkers of gelatin gels and gelatin-based coacervates for use as food ingredients. *Food Hydrocolloids* 18: 81–89.

Swaisgood HE. 1980. Sulfhydryl oxidase: properties and applications. *Enzyme Micro Technol* 2: 265–272.

Taylor JLS et al. 2004. Genotoxicity of melanoidin fractions derived from a standard glucose/glycine model. *J Agric Food Chem* 52: 318–323.

Tessier FJ et al. 2003. Triosidines: novel Maillard reaction products and cross-links from the reaction of triose sugars with lysine and arginine residues. *Biochem J* 369: 705–719.

Tilley KA et al. 2001. Tyrosine cross-links: molecular basis of gluten structure and function. *J Agric Food Chem* 49: 2627–2632.

Truong VD et al. 2004. Cross-linking and rheological changes of whey proteins treated with microbial transglutaminase. *J Agric Food Chem* 52: 1170–1176.

van Oort M. 2000. Peroxidases in breadmaking. In: JD Schofield (ed.) *Wheat Structure, Biochemistry, Functionality*. Royal Society of Chemistry, Cambridge.

Walcher D, Marx N. 2009. Advanced glycation end products and C-peptide-modulators in diabetic vasculopathy and atherogenesis. *Semin Immunopathol* 31: 102–111.

Watanabe M et al. 1994. Controlled enzymatic treatment of wheat proteins for production of hypoallergenic flour. *Biosci Biotechnol Biochem* 58: 388–390.

Watanabe E et al. 1998. The effect of protein disulfide isomerase on dough rheology assessed by fundamental and empirical testing. *Food Chem* 61: 481–486.

Weerasinghe VC et al. 1996. Characterization of active components in food-grade proteinase inhibitors for surimi manufacture. *J Agric Food Chem* 44: 2584–2590.

Wilhelm B et al. 1996. Transglutaminases – purification and activity assays. *J Chromatogr B: Biomed Appl* 684: 163–177.

Wittmann I et al. 2001. Foods rich in advanced glycation end products (AGEs) induce microalbuminuria in healthy persons. *Nephrol Dial Transplant* 16: 106A (Abstract).

Yamauchi M et al. 1987. Structure and formation of a stable histidine-based trifunctional cross-link in skin collagen. *J Biol Chem* 262: 11428–11434.

Yasir S et al. 2007a. The impact of transglutaminase on soy proteins and tofu texture. *Food Chem* 104: 1491–1501.

Yasir S et al. 2007b. The impact of Maillard cross-linking on soy proteins and tofu texture. *Food Chem* 104: 1502–1508.

Yokoyama K et al. 2004. Properties and applications of microbial transglutaminase. *Appl Microbiol Biotechnol* 64: 447–454.

Zarina S et al. 2000. Advanced glycation end products in human senile and diabetic cataractous lenses. *Mol Cell Biochem* 210: 29–34.

Zayas JF. 1997. *Functionality of Proteins in Food*. Springer-Verlag, Berlin.

Zhu Y et al. 1995. Microbial transglutaminase. A review of its production and application in food processing. *Appl Microbiol Biotechnol* 44: 277–282.

11
Chymosin in Cheese Making

V. V. Mistry

INTRODUCTION

Cheeses are classified into those that are aged for the development of flavor and body over time and those that are ready for consumption shortly after manufacture. Many cheese varieties are also expected to develop certain functional characteristics either immediately after manufacture or by the end of the aging period. It is fascinating that what starts in the cheese vat, as a white liquid that is composed of many nutrients, emerges as a completely transformed compact mass, cheese, sometimes colored yellow, sometimes white, and sometimes with various types of mold on the surface or within the cheese mass (Kosikowski and Mistry 1997). Some cheeses may also exhibit a unique textural character that turns stringy when heated, and others may have shiny eyes. Each cheese also has unique flavor qualities. Thus, cheese is a complex biological material whose characteristics are specifically tailored by the cheese maker through judicious blends of enzymes, starter bacteria, acid, and temperature.

Milk is a homogeneous liquid in which components exist in soluble, emulsion, or colloidal form. The manufacture of cheese involves a phase change from liquid (milk) to solid (cheese). This phase change occurs under carefully selected conditions that alter the physicochemical environment of milk such that the milk system is destabilized and is no longer homogeneous. In many fresh cheeses, this conversion takes place with the help of acid through isoelectric precipitation of casein and subsequent temperature treatments and draining for the conversion of liquid to solid. In the case of most ripened cheeses, this conversion occurs enzymatically at higher pH and involves the transformation of the calcium paracaseinate complex through controlled lactic acid fermentation. Later on, the products of this reaction and starter bacteria interact for the development of flavor and texture. All this is the result of the action of the enzyme chymosin (Andren 2003).

CHYMOSIN

Chymosin, rennet, and rennin are often used interchangeably to refer to this enzyme. The latter, rennin, should not be confused with renin, which is an enzyme associated with kidneys and does not clot milk. Chymosin is the biochemical name of the enzyme that was formerly known as rennin. It belongs to the group of aspartic acid proteinases, Enzyme classification (EC) 3.4.23, that have a high content of dicarboxylic and hydroxyamino acids and

Reviewers of this chapter were Dr. Ashraf Hassan, Department of Dairy Science, South Dakota State University, Brookings, and Dr. Lloyd Metzger, Department of Food Science and Nutrition, University of Minnesota, St. Paul.

Food Biochemistry and Food Processing, Second Edition. Edited by Benjamin K. Simpson, Leo M.L. Nollet, Fidel Toldrá, Soottawat Benjakul, Gopinadhan Paliyath and Y.H. Hui.

a low content of basic amino acids. Its molecular mass is 40 kDa (Andren 2003, Foltmann 1993).

Rennet is the stomach extract that contains the enzyme chymosin in a stabilized form that is usable for cheese making (Green 1977). While the amount of chymosin that is required for cheese making is very small, this enzyme industry has undergone an interesting transformation over time. It is believed that in the early days of cheese making, milk coagulation occurred either by filling dried stomachs of calves or lambs with milk or by immersing pieces of such stomachs in milk (Kosikowski and Mistry 1997). The chymosin enzyme embedded within the stomach lining diffused into the milk and coagulated it. This crude process of extracting coagulating enzymes was eventually finessed into an industry that employed specific methods to extract and purify the enzyme and develop an extract from the fourth stomach of the calf or lamb. Extracts of known enzyme activity and predictable milk clotting properties then became available as liquids, concentrates, powders, and blends of various enzymes. Live calves were required for the manufacture of these products. Because of religious and economic reasons, another industry also had emerged for manufacturing alternative milk clotting enzymes from plants, fungi, and bacteria. These products remain popular and meet the needs of various Kosher and other religious needs. Applications of genetic technology in rennet manufacture were then realized, and in 1990 a new process utilizing this technology was approved in the United States for manufacturing rennet.

Rennet Manufacture

For the manufacture of traditional rennet, calves, lambs, or kids that are no more than about 2-weeks-old and fed only milk are used (Kosikowski and Mistry 1997). As calves become older and begin eating other feeds, the proportion of bovine pepsin in relation to chymosin increases. Extracts from milk-fed calves that are 3-weeks-old contain over 90% chymosin, and the balance is pepsin. As the calves age and are fed other feeds such as concentrates, the ratio of chymosin to pepsin drops to 30:70 by 6 months of age. In a full-grown cow, there are only traces of chymosin.

Milk-fed calves are slaughtered, and the unwashed stomachs (vells) are preserved for enzyme extraction by emptying their contents, blowing them into small balloons and drying them. The vells may also be slit opened and dry salted or frozen for preservation. Air-dried stomachs give lower yields of chymosin than frozen stomachs; 12–13 air-dried stomachs make 1 L of rennet standardized to 1:10,000 strength, but only 7–8 frozen stomachs would be required for the same yield.

Extraction of chymosin and production of rennet begin by extracting, for several days, chopped or macerated stomachs with a 10% sodium chloride solution containing about 5% boric acid or glycerol. Additional salt up to a total of 16–18% is introduced followed by filtration and clarification. Mucine and grass particles in suspension are removed by introducing 1% of potash alum, followed by an equal amount of potassium phosphate. The suspension is adjusted to pH 5.0 to activate prochymosin (zymogen) to chymosin, and the enzyme strength is standardized, so that one part coagulates 15,000 or 10,000 parts of milk. Sodium benzoate, propylene glycol, and salt are added as preservatives for the final rennet. Caramel color is also usually added. The finished rennet solutions must be kept cold and protected from light.

Powdered rennet is manufactured by saturating a rennet suspension with sodium chloride or acidifying it with a food-grade acid. Chymosin precipitates and secondary enzymes such as pepsin remain in the original suspension. The chymosin-containing precipitate is dried to rennet powder (Kosikowski and Mistry 1997).

A method has been developed for manufacturing rennet without slaughtering calves. A hole (fistula) is surgically bored in the side of live calves, and at milk feeding time, excreted rennet is removed. After the calf matures, the hole is plugged, and the animal is returned to the herd. This method has not been commercialized but may be of value where religious practices do not allow calf slaughter.

Rennet paste, a rich source of lipolytic enzymes, has been a major factor in the flavor development of ripened Italian cheeses such as Provolone and Romano. Conceivably, 2100 years ago, Romans applied rennet paste to cheese milks to develop a typical "picante" flavor in Romano and related cheeses, for even then calf rennet was used to set their cheese milks. In more modern times, many countries, including the United States, applied calf rennet paste to induce flavor, largely for Italian ripened cheeses.

Farntiam et al. (1964) successfully produced, from goats, dried preparations of a pregastric-oral nature, also rich in lipolytic enzymes. They applied them to milk for ripened Italian cheese with excellent results. Since then, lipase powders, of various character and strength, have largely replaced rennet paste.

Chymosin Production by Genetic Technology

Genetic technology has been used for the commercial production of a 100% pure chymosin product from microbes. This type of chymosin is often called fermentation-produced chymosin. Uchiyama et al. (1980) and Nishimori et al. (1981) published early on the subject, and in 1990 Pfizer, Inc., received US Food and Drug Administration approval to market a genetically transformed product, Chy-Max, with GRAS stature (Duxbury 1990). Other brands using other microorganisms have also been approved. Table 11.1 lists some examples of such products.

The microbes used for this type of rennet include nonpathogenic microorganisms *Escherichia coli* K-12, *Kluyveromyces marxianus* var. *lactis* and *Aspergillus niger* var. *awamori*. Prochymosin genes obtained from young calves are transferred through DNA plasmid intervention into microbial cells. Fermentation follows to produce prochymosin, cell destruction, activation of the prochymosin to chymosin, and harvesting/producing large yields of pure, 100% chymosin. This product, transferred from an animal, is considered a plant product, as microbes are of the plant kingdom. Thus, they are acceptable to various religious groups.

These products have been widely studied to evaluate their impact on the quality and yield of cheese. An initial concern was that because of the absence of pepsin from these types of

Table 11.1. Examples of Some New Chymosin Products

Product Name	Company and Type
AmericanPure	Sanofi Bio-Industries Calf rennet purified via ion exchange
Chy-max	Pfizer (Now Chr. Hansen) Fermentation—using *Escherichia coli* K-12
Chymogen	Chr. Hansen Fermentation—using *Aspergillus niger* var. *awamori*
ChymoStar	Rhône-Poulenc (Now Danisco) Fermentation—using *Aspergillus niger* var. *awamori*
Maxiren	Gist-Brocades Fermentation—using *Kluyveromyces marxianus* var. *lactis*
Novoren[a]	Novo Nordisk
(Marzyme GM)	Fermentation—using *Aspergillus oryzae*

[a]Novoren is not chymosin: it is the coagulating enzyme of *Mucor miehei* cloned into *A. oryzae*. The other products listed above are all 100% chymosin.

rennets, proper cheese flavor might not develop. Most studies have generally concluded that there are no significant differences in flavor, texture, composition, and yield compared with calf rennet controls (Banks 1992, Barbano and Rasmussen 1992, Biner et al. 1989, Green et al. 1985, Hicks et al. 1988, IDF 1992, Ortiz de Apodaca et al. 1994).

When these products were first introduced, it was expected that a majority of cheese makers would convert to these types of rennets because they are virtually identical to traditional calf rennet and much cheaper. Except for some areas, this has not generally happened because of some concerns toward genetically modified products. Further, some protected cheeses, such as those under French AOC regulations or Italian DOP regulations, require the use of only calf rennet. A method has been published by the International Dairy Federation to detect fermentation-produced chymosin (Collin et al. 2003). This method uses immunochemical techniques (ELISA) to detect such chymosin in rennet solutions. It cannot be used for cheese.

Thus, traditional calf rennet and rennet substitutes derived from *Mucor Miehei, Cryphonectria, Parasitica*, and others are still used, especially for high cooking temperature cheeses such as Swiss and Mozzarella. Since the introduction of fermentation chymosin, the cost of traditional calf rennet has dropped considerably, and they both now cost approximately the same. Fungal rennets are still available at approximately 65% of the cost of the recombinant chymosin.

A coagulating enzyme of *Mucor miehei* cloned into *Aspergillus oryzae* was developed by Novo Nordisk of Denmark. This product hydrolyzes kappa-casein only at the 105-106 bond, and reduces protein losses to whey.

Rennet Production by Separation of Bovine Pepsin

Traditional calf rennet contains about 5% pepsin and almost 95% chymosin. Industrial purification of standard rennet to 100% chymosin rennet by the removal of pepsin was developed by Sanofi-Bioingredients Co. The process involves separation of the pepsin by ion exchange. Chymosin has no charge at pH 4.5, but bovine pepsin is negatively charged. Traditional commercial calf rennet is passed through an ion-exchange column containing positively charged ions. Bovine pepsin, due to its negative charge, is retained, and chymosin passes through and is collected, resulting in a 100% chymosin product (Pszczola 1989).

RENNET SUBSTITUTES

Improved farming practices, including improved genetics, have led over the years to significant increases in the milk production capacity of dairy cattle. As a result, while total milk production in the world has increased, the total cow population has declined. Hence, there has been a reduction in calf populations and the availability of traditional calf rennet. Furthermore, the increased practice of raising calves to an older age for meat production further reduced the numbers available for rennet production. It is for these reasons that a shortage of calf rennet occurred and substitutes were sought. Various proteolytic enzymes from plant, microbial, and animal sources were identified and developed for commercial applications. These enzymes should possess certain key characteristics to be successful rennet substitutes: (1) the clotting-to-proteolytic ratio should be similar to that of chymosin (i.e., the enzyme should have the capacity to clot milk without being excessively proteolytic); (2) the proteolytic specificity for beta-casein should be low because otherwise bitterness will occur in cheese; (3) the substitute product should be free of contaminating enzymes such as lipases; and (4) cost should be comparable to or lower than that of traditional rennet.

Animal

Pepsin derived from swine shows proteolytic activity between pH 2 and 6.5, but by itself it has difficulty in satisfactorily coagulating milk at pH 6.6. For this reason, it is used in cheese making as a 1:1 blend with rennet. Pepsin, used alone as a milk coagulator, shows high sensitivity to heat and is inclined to create bitter cheese if the concentrations added are not calculated and measured exactly (Kosikowski and Mistry 1997).

Plant

Rennet substitutes from plant sources are the least widely used because of their tendency to be excessively proteolytic and to cause formation of bitter flavors. Most such enzymes are also heat stable and require higher setting temperatures in milk. Plant sources include ficin from the fig tree, papain from the papaya tree, and bromelin from pineapple. A notable exception to these problems is the proteolytic enzymes from the flower of thistle *(Cynara Cardunculus)*, as reported by Vieira de Sa and Barbosa

(1970) and *Cynara humilis* (Esteves and Lucey 2002). These enzymes have been used successfully for many years in Portugal to make native ripened Serra cheese with excellent flavor and without bitterness.

In India, enzymes derived from *Withania coagulans* have been used successfully for cheese making, but commercialization has been minimal, especially with the development of genetically derived chymosin.

Microbial

Substitutes from microbial sources have been very successful and continue to be used. Many act like trypsin and have an optimum pH activity between 7 and 8 (Green 1977, Kosikowski and Mistry 1997). Microorganisms, including *Bacillus subtilis*, *B. cereus*, *B. polymyxa*, *Cryphonectria parasitica* (formerly *Endothia parasitica*), and *Mucor pusillus* Lindt (also known as *Rhizomucor pusillus)* and *Rh. miehei* have been extracted for their protease enzymes. The bacilli enzyme preparations were not suited for cheese making because of excessive proteolytic activity while the fungal-derived enzymes gave good results, but not without off flavors such as bitter. Commercial enzyme preparations isolated from *Cr. parasitica* led to good quality Emmental cheese without bitter flavor, but some cheeses, such as Cheddar, that use lower cooking temperatures reportedly showed bitterness. Enzyme preparations of *Rh. Miehei*, at recommended levels, produced Cheddar and other hard cheeses of satisfactory quality without bitter flavor. Fungal enzyme preparations are in commercial use, particularly in North America. In cottage cheese utilizing very small amounts of coagulating enzymes, shattering of curd was minimum in starter-rennet set milk and maximum in starter–microbial milk coagulating enzyme set milk (Brown 1971).

Enzymes derived from *Rh. miehei* and *Rh. pusillus* are not inactivated by pasteurization. Any residual activity in whey led to hydrolysis of whey proteins during storage of whey powder. This problem has been overcome by treating these enzymes with peroxides to reduce heat stability. Commercial preparations of *Rh. miehei* and *Rh. pusillus* are inactivated by pasteurization. The heat stability of *Cr. parasitica*–derived enzyme is similar to that of chymosin.

These fungal enzymes are also milk-clotting aspartic enzymes, and all except *Cr. parasitica* clot milk at the same peptide bond as chymosin. *Cryphonectria* clots kappa-casein at the 104-105 bond.

CHYMOSIN ACTION ON MILK

Chymosin produces a smooth curd in milk, and it is relatively insensitive to small shifts in pH that may be found in milk due to natural variations, does not cause bitterness over a wide range of addition, and is not proteolytically active if the cheese supplements other foods. Chymosin coagulates milk optimally at pH 6.0–6.4 and at 20–30°C in a two-step reaction, although the optimum pH of the enzyme is approximately 4 (Kosikowski and Mistry 1997, Lucey 2003). Optimum temperature for coagulation is approximately 40°C, but milk for cheese making is coagulated with rennet at 31–32°C because at this temperature the curd is rheologically most suitable for cheese making. Above pH 7.0, activity is lost. Thus, mastitic milk is only weakly coagulable, or does not coagulate at all.

Chymosin is highly sensitive to shaking, heat, light, alkali, dilutions, and chemicals. Stability is highest when stored at 7°C and pH 5.4–6.0 under dark conditions. Liquid rennet activity is destroyed at 55°C, but rennet powders lose little or no activity when exposed to 140°C. Standard single-strength rennet activity deteriorates at about 1% monthly when held cold in dark or plastic containers.

Single-strength rennet is usually added at 100–200 mL per 1000 kg of milk. It serves to coagulate milk and to hydrolyze casein during cheese ripening for texture and flavor development.

It should be noted that bovine chymosin has greater specificity for cow's milk than chymosin derived from kid or lamb. Similarly, kid chymosin is better suited for goat milk.

Milk contains fat, protein, sugar, salts, and many minor components in true solution, suspension, or emulsion. When milk is converted into cheese, some of these components are selectively concentrated as much as ten fold, but some are lost to whey. It is the fat, casein, and insoluble salts that are concentrated. The other components are entrapped in the cheese serum or whey, but only at about the levels at which they existed in the milk. These soluble components are retained, depending on the degree that the serum or whey is retained in the cheese. For example, in a fresh Cheddar cheese, the serum portion is lower in volume than in milk. Thus, the soluble component percentage of the cheese is smaller.

In washed curd cheeses such as Edam or Brick, the above relationship does not hold, and lactose, soluble salts, and vitamins in the final cheese are reduced considerably. Approximately 90% of living bacteria in the cheese-milk, including starter bacteria, are trapped in the cheese curd. Natural milk enzymes and others are, in part, preferentially absorbed on fat and protein, and thus a higher concentration remains with the cheese. In rennet coagulation of milk and the subsequent removal of much of the whey or serum, a selective separation of the milk components occurs, and in the resulting concentrated curd mass, many biological agents become active, marking the beginning of the final product, cheese.

Milk for ripened cheese is coagulated at a pH above the isoelectric point of casein by special proteolytic enzymes, which are activated by small amounts of lactic acid produced by added starter bacteria. The curds are sweeter and more shrinkable and pliable than those of fresh, unripened cheeses, which are produced by isoelectric precipitation. These enzymes are typified by chymosin that is found in the fourth stomach, or abomasum, of a young calf.

The isoelectric condition of a protein is that at which the net electric charge on a protein surface is zero. In their natural state in milk, caseins are negatively charged, and this helps maintain the protein in suspension. Lactic acid neutralizes the charge on the casein. The casein then precipitates as a curd at pH 4.6, the isoelectric point, as in cottage or cream cheese and yogurt. For some types of cheeses, this type of curd is not desirable because it would be too acid, and the texture would be too firm and grainy. For ripened cheeses such as Cheddar and Swiss, a sweeter curd is desired, so casein is precipitated as a curd near neutrality, pH 6.2, by chymosin in 30–45 minutes at 30–32°C. This type of curd is significantly different than isoelectric curd and much more suitable for ripened cheeses. Milk proteins other than casein are not precipitated by chymosin. The uniform gel formed is made up of modified casein with fat entrapped within the gel.

$$\text{kappa-Casein} \xrightarrow{\text{chymosin}} \textit{para}\text{-kappa-casein} + \text{glycomacropeptide (soluble in whey)}$$

$$\textit{para}\text{-kappa-Casein} \xrightarrow[\text{pH 6.4-6.0}]{Ca^{++}} \text{dicalcium para-kappa-casein}$$

Figure 11.1. Destabilization of the casein micelle by introduction of chymosin.

Milk Coagulation and Protein Hydrolysis by Chymosin

The process of curd formation by chymosin is a complex process and involves various interactions of the enzyme with specific sites on casein, temperature and acid conditions, and calcium. Milk is coagulated by chymosin into a smooth gel capable of extruding whey at a uniformly rapid rate. Other common proteolytic enzymes such as pepsin, trypsin, and papain coagulate milk too, but may cause bitterness and loss of yield and are more sensitive to pH and temperature changes.

An understanding of the structure of casein in milk has contributed to the explanation of the mechanism of chymosin action in milk. Waugh and von Hippel (1956) began to unravel the heterogeneous nature of casein with their observations on kappa-casein. Casein comprises 45–50% alpha-s-casein, 25–35% beta-casein, 8–15% kappa-casein, and 3–7% gamma-casein. Various subfractions of alpha-casein have also been identified, as have genetic variants. Each casein fraction differs in sensitivity to calcium, solubility, amino acid makeup, and electrophoretic mobility.

An understanding of the dispersion of casein in milk has provoked much interest. Several models of casein micelles have been proposed including the coat-core, internal structure, and subunit models. With the availability of newer analytical techniques such as three-dimensional X-ray crystallography, work on casein micelle modeling is continuing, and additional understanding is being gained. The subunit model (Rollema 1992) is widely accepted. It suggests that the casein micelle is made up of smaller submicelles that are attached by calcium phosphate bonds. Hence, the micelle consists of not only pure casein (~93%), but also minerals (~7%), primarily calcium and phosphate. Most of the kappa-casein is located on the surface of the micelle. This casein has few phosphoserine residues and hence is not affected by ionic calcium. It is located on the surface of the micelle and provides for stability of the micelle and protects the calcium-sensitive and hydrophobic interior, including alpha- and beta- caseins from ionic calcium. These two caseins have phosphoserine residues and high calcium-binding affinity. The carbohydrate, *N*-acetyl neuramic acid, which is attached to kappa-casein, makes this casein hydrophilic. Thus, the casein micelle is suspended in milk as long as the properties of kappa-casein remain unchanged. More recently the dual-bonding model has been proposed (Horne 1998).

Introducing chymosin to the cheese-milk, normally at about 32°C, destabilizes the casein micelle in a two-step reaction, the first of which is enzymatic (or primary) and the second, nonenzymatic (or secondary) (Kosikowski and Mistry 1997, Lucey 2003). These two steps are separate but cannot be visually distinguished: only the appearance of a curd signifies the completion of both steps. The primary phase was probably first observed by Hammersten in the late 1800s (Kosikowski and Mistry 1997) and must occur before the secondary phase begins.

In the primary phase, chymosin cleaves the phenylalanine-methionine bond (105–106) of kappa-casein, thus eliminating its stabilizing action on calcium-sensitive alpha-s- and beta-caseins. In the secondary phase, the micelles without intact kappa-casein aggregate in the presence of ionic calcium in milk and form a gel (curd). This mechanism may be summarized as follows (Fig. 11.1):

The macropeptide, also known as caseinmacropeptide, contains approximately 30% amino sugar, hence the name glycomacropeptide (Lucey 2003). In the ultrafiltration processes of cheese making, this macropeptide is retained with whey proteins to increase cheese yields significantly. In the conventional cheese-making process, the macropeptide is found in whey.

The two-step coagulation does not fully explain the presence of the smooth curd or coagulum. Beau, many years ago, established that a strong milk gel arose because the fibrous filaments of paracasein cross-linked to make a lattice. Bonding at critical points between phosphorus, calcium, free amino groups, and free carboxyl groups strengthened the lattice, with its entrapped lactose and soluble salts. A parallel was drawn between the gel formed when only lactic acid was involved as in fresh, unripened cheese and when considerable chymosin was involved as in hard, ripened cheese. Both gels were considered originally as starting with a fibrous protein cross bonding, but the strictly lactic acid curd was considered weaker, because not having been hydrolyzed by a proteolytic enzyme, it possessed less bonding material. The essential conditions for a smooth gel developing in either case were sufficient casein, a quiescent environment, moderate temperature, and sufficient time for reaction.

FACTORS AFFECTING CHYMOSIN ACTION IN MILK

Rennet coagulation in milk is influenced by many factors that ultimately have an impact on cheese characteristics (Kosikowski and Mistry 1997). The cheese maker is usually able to properly control these factors, some of which have a direct affect on the primary phase and others on the secondary.

Chymosin Concentration

Chymosin concentration affects gel firmness (Lomholt and Qvist 1999), and enzyme kinetics have been used to study these effects. An increase in the chymosin concentration (larger amounts of rennet added to milk) reduces the total time required for rennet clotting, measured by the appearance of the first curd particle. As a result, the secondary phase of rennet action will also proceed much earlier, with the net result of an increase in the rate of increase in gel firmness. This property of chymosin is sometimes used to control the rennet coagulation in concentrated milks, which have a tendency to form firmer curd because of the higher protein content. Lowering the amount of rennet reduces the rate of curd firming. It should also be noted that with an alteration in the amount of rennet added to milk, there also will be a change in the amount of residual chymosin in the cheese.

Temperature

The optimal condition for curd formation in milk with chymosin is 40–45°C, but this temperature is not suitable for cheese making. Rennet coagulation for cheese making generally occurs at 30–35°C for proper firmness. At lower temperatures, rennet clotting rate is significantly reduced, and at refrigeration temperatures, virtually no curd is obtained.

pH

Chymosin action in milk is optimal at approximately pH 6. This is obtained with starter bacteria. If the pH is lowered further, rennet clotting occurs at a faster rate, and curd firmness is reduced. Lowering pH also changes the water-holding capacity, which will have an impact on curd firmness.

Calcium

Calcium has a significant impact on rennet curd, though it does not have a direct effect on the primary phase of rennet action. Addition of ionic calcium, as in the form of calcium chloride, for example, reduces the rate of rennet clot formation time and also increases rennet curd firmness. Similarly, if the calcium content of milk is lowered by approximately 30%, coagulation does not occur. Milks that have a tendency to form weak curd may be fortified with calcium chloride prior to the addition of rennet.

Milk Processing

Heating milk to temperatures higher than approximately 70°C leads to delayed curd formation and weak rennet curd (Vasbinder et al. 2003). In extreme cases, there may be no curd at all. These effects are a result of the formation of a complex involving disulphide linkages between kappa-casein and beta-lactoglobulin under high heat treatment. Under these conditions, the 105-106 bond in kappa-casein is inaccessible for chymosin action. This effect is generally not reversible. Under mild overheating conditions, the addition of 0.02% calcium chloride may help obtain firm rennet curd. Maubois et al. have developed a process for reversing this effect by the use of ultrafiltration. Increasing the protein content by ultrafiltration before or after UHT treatment restores curd-forming ability (Maubois et al. 1972). According to Ferron-Baumy et al. (1991), such a phenomenon results from lowering the zeta potential of casein micelles on ultrafiltration.

Homogenization has a distinct impact on milk rennet coagulation properties. Homogenized milk produces softer curd, but when only the cream portion of milk is homogenized, rennet curd becomes firmer (Nair et al. 2000). This is possibly because of the reduction of fat globule size due to homogenization and coating of the fat globule surface with casein. These particles then act as casein micelles.

Concentrating milk prior to cheese making is now a common practice. Techniques include evaporative concentration, ultrafiltration, and microfiltration. Each of these procedures increases the casein concentration in milk; thus, the rate of casein aggregation during the secondary phase increases. Ren-net coagulation usually occurs at a lower degree of kappa-casein hydrolysis, and rennet curd is generally firmer. For example, in unconcentrated milk, approximately 90% kappa-casein must be hydrolyzed by chymosin before curd formation, but in ultrafiltered milk only 50% must be hydrolyzed. (Dalgleish 1992). Under high concentration conditions, rennet curd is extremely firm, and difficulties are encountered in cutting curd using traditional equipment. In cheeses manufactured from highly concentrated milks, rennet should be properly mixed in the milk mixture to prevent localization of rennet action.

Genetic Variants

Protein polymorphism (genetic variants) of kappa-casein has been demonstrated to have an effect on rennet coagulation (Marzialli and Ng-Kwai-Hang 1986). Protein polymorphism refers to a small variation in the makeup of proteins due to minor differences in the amino acid sequence. Examples include kappa-casein AA, AB, or BB, beta-lactoglobulin AA, AB, and so on. Milk with kappa-casein BB variants forms a firmer rennet curd because of increased casein content associated with this variant of kappa-casein. Some breeds of cows, Jerseys in particular, have larger proportions of this variant. The BB variant of beta-lactoglobulin is also associated with higher casein content in milk.

EFFECT OF CHYMOSIN ON PROTEOLYSIS IN CHEESE

Once rennet curd has been formed and a fresh block of cheese has been obtained, the ripening process begins. During ripening, numerous biochemical reactions occur and lead to unique flavor and texture development. During this process, residual chymosin that is retained in the cheese makes important contributions to the ripening process (Kosikowski and Mistry 1997, Sousa et al. 2001). As the cheese takes form, chymosin hydrolyzes the paracaseins into peptides optimally at pH 5.6 and creates a peptide pool for developing flavor complexes (Kosikowski and Mistry 1997).

The amount of rennet required to coagulate milk within 30 minutes can be considerably below that required to properly break down the paracaseins during cheese ripening. Consequently, using too little rennet in cheese making retards ripening, as discerned by the appearance of the cheese and its end products. A lively, almost translucent looking cheese, after ripening several months, indicates adequate rennet; an opaque, dull-looking cheese indicates inadequate amounts. Apparently, conversion to peptides and later hydrolysis to smaller molecules by bacterial enzymes give the translucent quality. Residual rennet in cheese hydrolyzes alpha-s-1-casein to alpha-s-1-I-casein, which leads to a desirable soft texture in the aged cheese (Lawrence et al. 1987).

Electrophoretic techniques (Ledford et al. 1966) have demonstrated that in most ripening Cheddar cheese from milk coagulated by rennet, *para*-beta-casein largely remains intact while *para*-alpha-casein is highly degraded. According to Fox et al. (1993), if there is too much moisture in cheese or too little salt, the residual chymosin will produce bitter peptides due to excessive proteolysis. In cheeses with high levels of beta-lactoglobulin, as in cheeses made by ultrafiltration, proteolysis by residual chymosin is retarded because of partial inhibition of chymosin by the whey protein (Kosikowski and Mistry 1997).

Edwards (1969) found that milk-coagulating enzymes from the mucors hydrolyze the paracaseins somewhat similarly to rennet, while enzymes from *Cr. parasitica* and the papaya plant completely hydrolyze para-alpha- and para-beta-caseins during ripening.

Proteolytic activity of rennet substitutes, especially microbial and fungal substitutes, is greater than that of chymosin. Cheese yield losses, though small, do occur with these rennets. Barbano and Rasmussen (1992) reported that fat and protein losses to whey were higher with *Rh. miehei* and *Rh. pusillus* compared with those for fermentation-derived chymosin. As a result, the cheese yield efficiency was higher with the latter than with the former two microbial rennets.

EFFECT OF CHYMOSIN ON CHEESE TEXTURE

The formation of a rennet curd is the beginning of the formation of a cheese mass. Thus, a ripened cheese assumes its initial biological identity in the vat or press. This is where practically all of the critical components are assembled and the young cheese becomes ready for further development.

Undisturbed fresh rennet curd is not cheese because it still holds considerable amounts of water and soluble constituents that must be removed before any resemblance to a cheese occurs. This block of curd is cut into thousands of small cubes. Traditionally, this is done with wire knives (Fig. 11.2) but in large automated vats, built-in blades rotate in one direction at a selected speed to systematically cut the curd. Then, the whey, carrying with it lactose, whey proteins, and soluble salts, streams from the cut pores, accelerated by gentle agitation, slowly increasing temperature, and a rising rate of lactic acid production. These factors are manipulated to the degree desired by the cheese maker. Eventually, the free whey is separated from curd.

Figure 11.2. Cutting of rennet curd.

Optimum acid development is essential for forming rennet curd and creating the desired cheese mass. This requires viable, active microbial starters to be added to milk before the addition of rennet. These bacteria should survive the cheese making process.

Beta-D-galactosidase (lactase) and phosphorylating enzymes of the starter bacteria hydrolyze lactose and initiate the glucose-phosphate energy cycles for the ultimate production of lactic acid through various intermediary compounds. Lactose is hydrolyzed to glucose and galactose. Simultaneously, the galactose is converted in the milk to glucose, from which point lactic acid is produced by a rather involved glycolytic pathway. In most cheese milks, the conversion of glucose to lactic acid is conducted homofermentatively by lactic acid bacteria, and the conversions provide energy for the bacteria. For each major cheese type, lactic acid must develop at the correct time, usually not too rapidly, nor too slowly, and in a specific concentration.

The cheese mass, which evolves as the whey drains and the curds coalesce, contains the insoluble salts, $CaHPO_4$ and $MgHPO_4$, which serve as buffers in the pressed cheese mass, and should the acid increase too rapidly during cooking, they will dissolve in the whey. In the drained curds or pressed cheese mass under these circumstances, since no significant pool of insoluble divalent salt remains for conversion into buffering salts, the pH will lower to 4.7–4.8 from an optimum 5.2, leading to a sour acid-ripened cheese. Thus, conservation of a reserve pool of insoluble salts is necessary until the cheese mass is pressed. Thereafter, the insoluble salts are changed into captive soluble salts by the lactic acid and provide the necessary buffering power to maintain optimum pH at a level that keeps the cheese sweet. Such retention also is aided or controlled by the skills of the cheese maker. For Cheddar, excess wetness of the curd before pressing or too rapid an acid development in the whey causes sour cheese. For a Swiss or Emmental, the time period for acid development differs; most of it takes place in the press, but the principles remain the same (Kosikowski and Mistry 1997).

The development of proper acid in a curd mass controls the microbial flora. Sufficient lactic acid produced at optimum rates favors lactic acid bacteria and discriminates against spoilage or food poisoning bacteria such as coliforms, clostridia, and coagulase-positive staphylococci. But its presence does more than that, for it transforms the chemistry of the curd to provide the strong bonding that is necessary for a smooth, integrated cheese mass.

As discussed earlier, chymosin action on milk results in curd mass involving dicalcium paracasein. Dicalcium paracasein is not readily soluble, stretchable, or possessive of a distinguished appearance. However, if sufficient lactic acid is generated, this compound changes. The developing lactic acid solubilizes considerable calcium, creating a new compound, monocalcium paracasein. The change occurs relatively quickly, but there is still a time requirement. For example, a Cheddar cheese curd mass, salted prior to pressing, may show an 8:2 ratio of dicalcium paracasein to monocalcium paracasein; after 24 hours in the press, it is reversed, 2:8. Monocalcium paracasein has interesting properties. It is soluble in warm 5% salt solution, it can be stretched and pulled when warmed, and it has a live, glistening appearance. The buildup of monocalcium paracasein makes a ripened cheese pliable and elastic. Then, as more lactic acid continues to strip off calcium, some of the monocalcium paracasein changes to free paracasein as follows:

Dicalcium paracasein + lactic acid
→ monocalcium paracasein + calcium lactate

Monocalcium paracasein + lactic acid
→ free paracasein + calcium lactate

Free paracasein is readily attacked by many enzymes, contributing to a well-ripened cheese. The curd mass becomes fully integrated upon the uniform addition of sodium chloride, the amount of which varies widely for different cheese types. Salt directly influences flavor and arrests sharply the acid production by lactic acid starter bacteria. Also, salt helps remove excess water from the curd during pressing and lessens the chances for a weak-bodied cheese. Besides controlling the lactic acid fermentation, salt partially solubilizes monocalcium paracasein. Thus, to a natural ripened cheese or to one that is heat processed, salt helps give a smoothness and plasticity of body that is not fully attainable in its absence.

Texture of cheese is affected significantly during ripening, especially during the first 1–2 weeks, according to Lawrence et al. (1987). During this period, the alpha-s-1-casein is hydrolyzed to alpha-s-1-I by residual chymosin, making the body softer and smoother. Further proteolysis during ripening continues to influence texture.

REFERENCES

Andren A. 2003. Rennets and coagulants. In: H Roginski, JW Fuquay, PF Fox, editors. *Encyclopedia of Dairy Sciences*. London: Academic Press, London, pp. 281–286.

Banks JM. 1992. Yield and quality of Cheddar cheese produced using a fermentation-derived calf chymosin. *Milchwissenschaft* 47: 153–201.

Barbano DM, Rasmussen RR. 1992. Cheese yield performance of fermentation-produced chymosin and other milk coagulation. *J Dairy Sci* 75: 1–12.

Biner VEP, Young D, Law BA. 1989. Comparison of Cheddar cheese made with a recombinant calf chymosin and with standard calf rennet. *J Dairy Res* 56: 657–664.

Brown GD. 1971. Microbial enzymes in the production of Cottage cheese. MS Thesis. Ithaca NY: Cornell University.

Collin J-C. et al. 2003. Detection of fermentation produced chymosin. In: Bulletin 380. Brussels Belgium: Int Dairy Fed. pp. 21–24.

Dalgleish DG. 1992. Chapter 16. The enzymatic coagulation of milk. In: PF Fox (ed.) *Advanced Dairy Chemistry*. Vol 1, Proteins. Essex England: Elsevier Sci Publ Ltd.

Duxbury DD. 1990. Cheese enzyme is first rDNA technology derived food ingredients granted GRAS approved by FDA. Food Proc (June): 46.

Edwards JL. 1969. Bitterness and proteolysis in Cheddar cheese made with animal, microbial, or vegetable rennet enzymes. PhD Thesis. Ithaca NY: Cornell University.

Esteves CLC, Lucey JA. 2002. Rheological properties of milk gels made with coagulants of plant origin and chymosin. *Int Dairy J* 12: 427–434.

Ferron-Baumy C. et al. 1991. Coagulation présure du lait et des rétentats d'ultrafiltration. Effets de divers traitements thermiques. *Lait* 71: 423–434.

Farntiam MG, inventor. 1964, September 27. Cheese modifying enzyme product. U.S. patent 3,531,329.

Foltmann B. 1993. General and molecular aspects of rennets. In: PF Fox (ed.) *Cheese: Chemistry, Physics and Microbiology* Vol. 1, 2nd edition. New York: Chapman and Hall, pp. 37–68.

Fox PF. et al. 1993. Biochemistry of cheese ripening. In: PF Fox (ed.) *Cheese: Chemistry, Physics and Microbiology*, 2nd edition, vol. 1. New York: Chapman and Hall.

Green ML. 1977. Review of progress of dairy science: Milk coagulants. *J Dairy Res* 44: 159–188.

Green ML. et al. 1985. Cheddar cheesemaking with recombinant calf chymosin synthesized in Escherichia. *J Dairy Res* 52: 281.

Hicks CL, O'Leary J, Bucy J. 1988. Use of recombinant chymosin in the manufacture of Cheddar and Colby cheeses. *J Dairy Sci* 71: 1127–1131.

Horne DS. 1998. Casein interactions: Casting light on the black boxes, the structure in dairy products. *Int Dairy J* 8: 171–177.

IDF. 1992. Fermentation produced enzymes and accelerated ripening in cheesemaking. Bulletin No. 269. Brussels Belgium: Int Dairy Fed.

Kosikowski FV, Mistry VV. 1997. *Cheese and Fermented Milk Foods*. Vol 1, Origins and Principles. Great Falls, Virginia: FV Kosikowski.

Lawrence RC. et al. 1987. Texture development during cheese ripening. *J Dairy Sci* 70: 1748–1760.

Ledford RA. et al. 1966. Residual casein fractions in ripened cheese determined by polyacrylamide-gel electrophoresis. *J Dairy Sci* 49: 1098–1101.

Lomholt SB, Qvist KB. 1999. Gel firming rate of rennet curd as a function of rennet concentration. *Int Dairy J* 9: 417–418.

Lucey JA. 2003. Rennet coagulation in milk. In: H Roginski, JW Fuquay, PF Fox (ed.) *Encyclopedia of Dairy Sciences*, pp. 286–293. London: Academic Press.

Marzialli AS, Ng-Kwai-Hang KF. 1986. Relationship between milk protein polymorphisms and cheese yielding capacity. *J Dairy Sci* 69: 1193–1201.

Maubois JL. et al. 1972. Procédé traitement du lait et de sous-produits laitiers. *French Patent no.* 2 166 315.

Nair MG. et al. 2000. Yield and functionality of Cheddar cheese as influenced by homogenization of cream. *Int Dairy J* 10: 647–657.

Nishimori K. et al. 1981. Cloning in *Escherichia coli* of the structured gene of pro-rennin the precursor of calf milk—clotting rennin. *J Biochem* 90: 901–904.

Ortiz de Apodaca MJ. et al. 1994. Study of milk-clotting and proteolytic activity of calf rennet, fermentation-produced chymosin, vegetable and microbial coagulants. *Milchwissenschaft* 49: 13–16.

Pszczola DF. 1989. Rennet containing 100% chymosin increases cheese quality and yield. *Food Technol* 43(6): 84–89.

Rollema HS. 1992. Chapter 3. Casein association and micelle formation. In: PF Fox (ed.) *Advanced Dairy Chemistry*. Vol. 1, Proteins. Essex England: Elsevier Sci Publ Ltd.

Sousa M J. et al. 2001. Advances in the study of proteolysis during cheese ripening. *Int Dairy J* 11: 327–345.

Uchiyama H. et al. 1980. Purification of pro rennin MRNA and its translation in vitro. *Agric Biol Chem* 44: 1373–1381.

Vasbinder AJ. et al. 2003. Impaired rennetability of heated milk; study of enzymatic hydrolysis and gelation kinetics. *J Dairy Sci* 86: 1548–1555.

Vieira de sa F, Barbosa M. 1970. Cheesemaking experiments using a clotting enzyme from cardon *(Cynara cardunculus)* Proceed 18th Int Dairy Cong 1E:288.

Waugh DG, Von Hippel PH. 1956. Kappa-casein and the stabilization of casein micelles. *J Amer Chem Soc* 78: 4576–4582.

12
Pectic Enzymes in Tomatoes

Mary S. Kalamaki, Nikolaos G. Stoforos, and Petros S. Taoukis

Abstract: In this chapter, the enzymes that act on the pectic fraction of the tomato cell wall are described. In a brief introduction, the structure of the pectin network in cell walls of higher plants and a methodology for fractionation of pectic substances are presented. Following, the four most significant pectic enzymes, polygalacturonase, pectin methylesterase, β-galactosidase, and pectate lyase are described in terms of the genes that encode them in tomato, their expression profiles during fruit ripening, as well as the structure and activities of the corresponding proteins. Cases of genetic modification of these enzymes aiming in increasing cell wall integrity, postharvest shelf life, and the processing characteristics of tomatoes are also presented. Finally, the inactivation kinetics of tomato pectic enzymes by heat and high hydrostatic pressure are discussed.

INTRODUCTION

Pectin is a structurally and functionally diverse group of polysaccharides in the plant cell wall. Apart from its role in cell wall architecture, it functions in signaling, cell-to-cell adhesion, determining wall porosity, pH and ionic content, cell growth and differentiation, plant protection, and it is a major contributor to fruit texture. Pectin disassembly during ripening is associated with the development of soft fruit texture, which in turn determines the textural and rheological characteristics of the processed fruit products. In addition, pectin is used in the food industry as a gelling and stabilizing agent in order to improve food rheological behavior. Among the most studied pectic polysaccharides are those of tomato fruit. In this chapter, the enzymes that act on the pectin fraction of tomato cell walls are presented. In the introduction, background information on the structure of pectin in cell walls of higher plants as well as a method used for biochemical analysis of pectic substances are given. Next, pectic enzymes in tomato in terms of genes that encode pectin modifying enzymes, their expression profiles during fruit ripening, and the activities of the respective proteins are discussed. Furthermore, attempts to alter the expression of these enzymes using genetic engineering in order to retard fruit deterioration at the later stages of ripening, increase fruit firmness and shelf life, and improve processing characteristics of tomato products are described. Finally, the inactivation kinetics (by heat and high hydrostatic pressure) of two pectic enzymes, polygalactronase and pectin methylesterase, are discussed.

Structure of Pectin Network in Cell Walls of Higher Plants

The primary cell wall of higher plants constitutes a complex network, mainly constructed of polysaccharides, structural proteins, and some phenolics. Two co-extensive structural networks can

Food Biochemistry and Food Processing, Second Edition. Edited by Benjamin K. Simpson, Leo M.L. Nollet, Fidel Toldrá, Soottawat Benjakul, Gopinadhan Paliyath and Y.H. Hui.

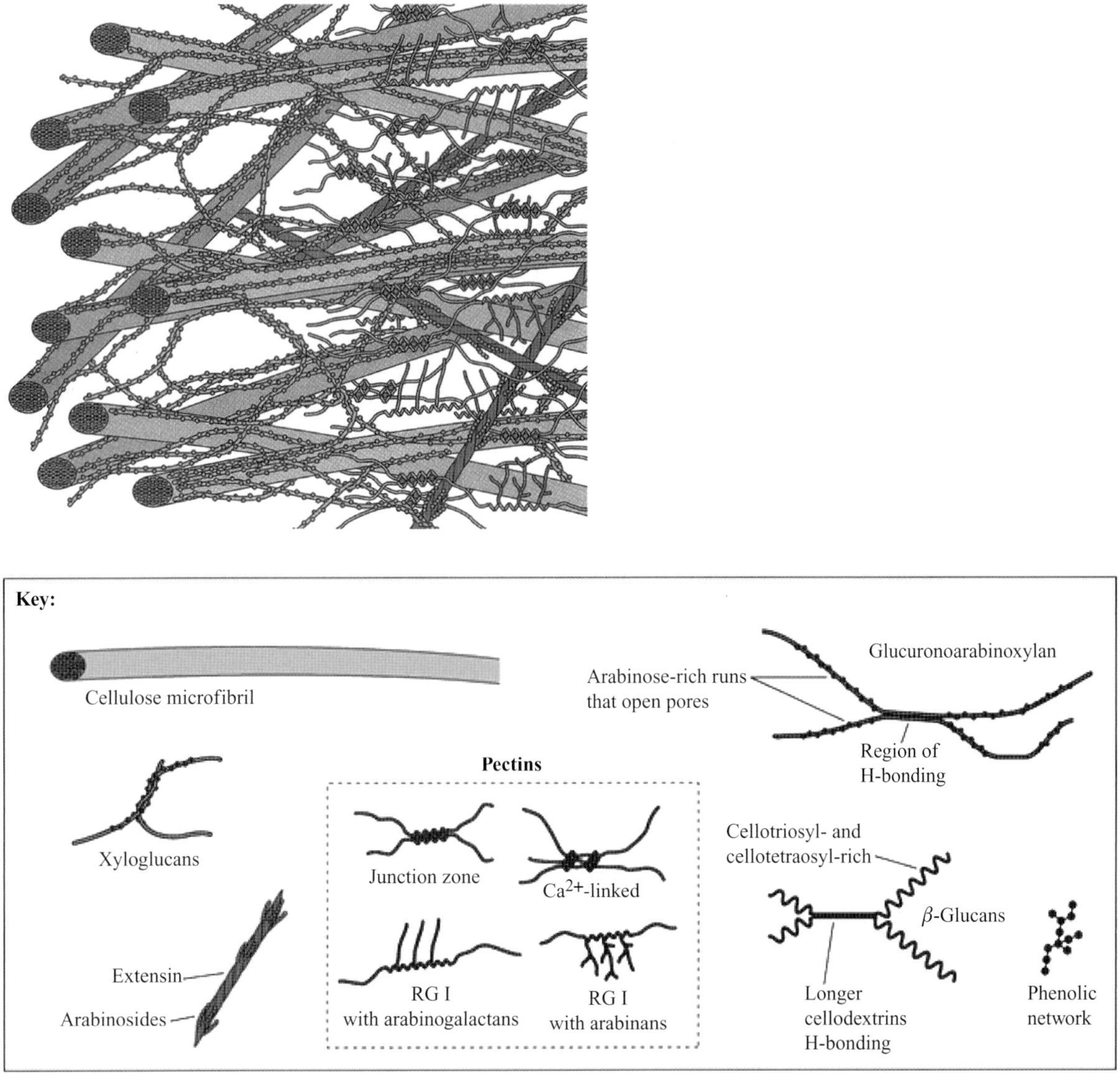

Figure 12.1. Current structural model of cell walls of flowering plants. (Reprinted from Buchanan et al. 2000).

be identified (Fig. 12.1), the cellulose–matrix glycan network, consisting of cellulose microfibrils held together by matrix glycans (formerly referred to as hemicelluloses), and the pectin network (Carpita and Gibeaut 1993). The way by which these polymers interact with each other is still unclear. Several cell wall models are proposed in an attempt to explain measured physical properties of the wall (Cosgrove 2000, Vincken et al. 2003, Thompson 2005). The cell wall provides shape and structural integrity to the cell, functions as a protective barrier to pathogen invasion and environmental stress, and provides signal molecules important in cell-to-cell signaling, plant–symbiot, and plant–pathogen interactions.

Pectin is an elaborate network of highly hydrated polysaccharides rich in galacturonic acid (GalA). Pectins can be classified in four major groups: homogalacturonan (HGA), rhamnogalacturonan I (RGI), rhamnogalacturonan II (RGII), and xylogalacturonan (XGA; Mohnen 2008). HGA is a homopolymer of 1,4-α-D-GalA (Fig. 12.2A) usually forming long chains of approximately 100 GalA residues. Galacturonosyl residues are esterified to various extents at the carboxyl group with methanol and may be acetylated at *O*-2 or *O*-3. Nonesterified regions of HGA chains can be associated with each other by ionic interactions via calcium ion bridges, forming egg-box-like structures (Fig. 12.2A). At least ten contiguous unesterified galacturonyl residues are required in order to form a stable cross link between HGA chains (Liners et al. 1992). RGI is composed of repeating disaccharide units of 1,2-α-D-rhamnose–1,4-α-D-GalA. Rhamnosyl residues on the backbone of this polymer can be further substituted with β-D-galactose- and α-L-arabinose-rich side chains (Fig. 12.2B). RGII is a minor

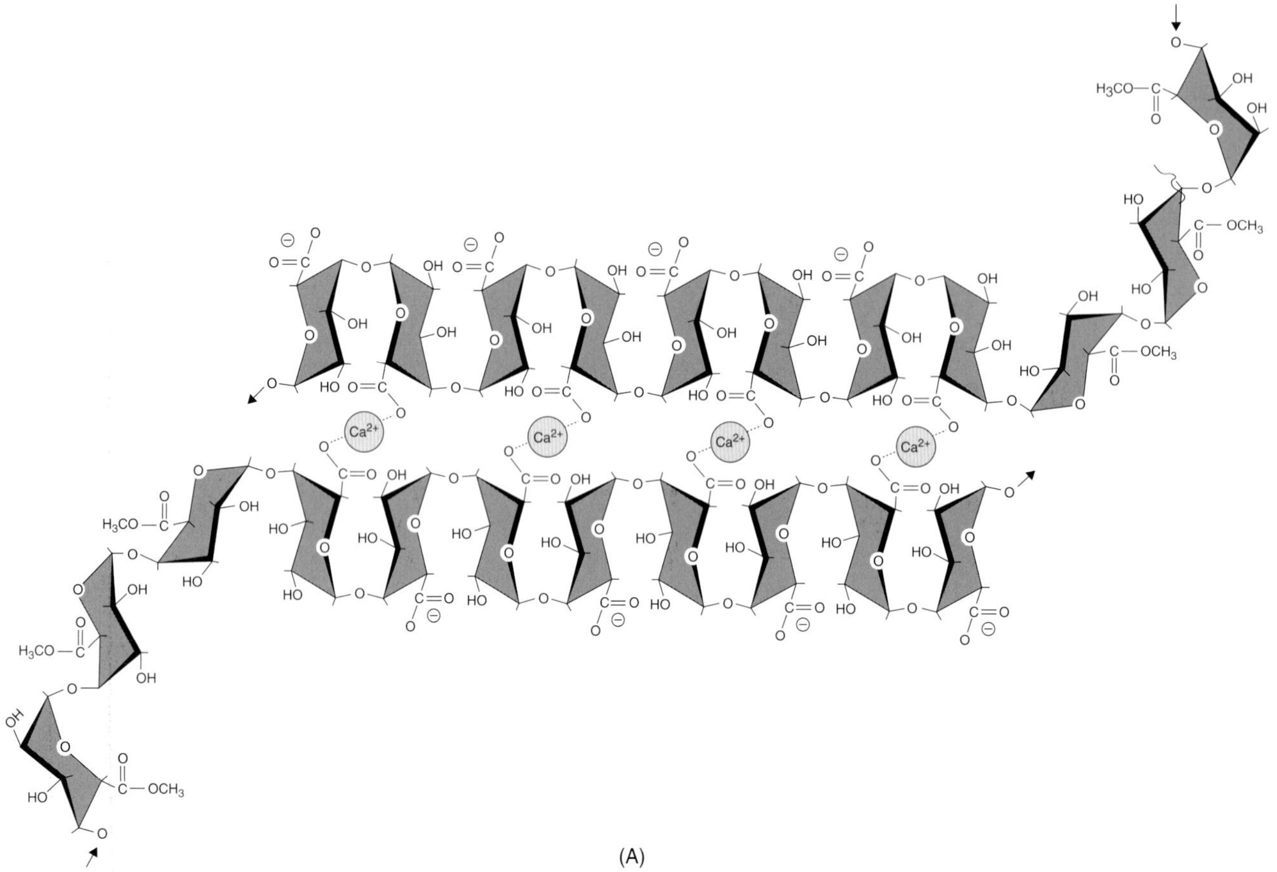

Figure 12.2. (A) Structure of homogalacturonan showing calcium ion associations. (Continued)

constituent of the cell wall (about 10% of pectin) consisting of an HGA backbone that exhibits a complex and highly conserved substitution pattern with many diverse sugars, linked together with more than 20 different linkages. RGII often is found in the wall as dimers cross-linked via borate-diol esters at the apiose residues (Ishii et al. 1999). Another, less-abundant pectic side chain is XGA, which is composed of a galacturonan backbone substituted at *O*-3 with xylosyl side units (Willats et al. 2004). Pectins are covalently interlinked, in part via diferulic acid, and further cross-linked to cellulose, matrix glycans, and other cell wall polymers (Thompson and Fry 2000, Zykwinska et al. 2005, Coenen et al. 2007). The relative abundance and distribution of these polymers in the primary cell wall is not constant. Variation in structure occurs within cell types and plant species, even within the cell wall of a single cell (Knox 2008, Jarvis 2011).

Fractionation of Pectic Substances

Methods for analyzing cell wall polymeric substances are based on deconstruction of the cell wall by sequentially extracting its components based on solubility. Tomato pericarp is boiled in ethanol or extracted with tris-buffered phenol to inactivate enzymes (Huber 1992), and cell wall polysaccharides are precipitated with ethanol. Alcohol-insoluble solids are sequentially extracted with 50 mM *trans*-1,2-diaminocyclohexane-*N,N,N′,N′*-tetraacetic acid (CDTA) and 50 mM sodium acetate buffer (pH = 6.5) to obtain the chelator-soluble (ionically bound) pectin. Next, the pellet is extracted with 50 mM sodium carbonate and 20 mM sodium borohydride, which results in breaking of ester bonds, to obtain the carbonate-soluble (covalently bound pectin via ester bonds) pectin. The remaining residue consists primarily of cellulose and matrix glycans with only about 5% remaining pectin. Extracts are dialyzed exhaustively against water at 4°C, and uronic acid (UA) content of dialyzed extracts (extractable polyuronide) is determined using the method of Blumenkrantz and Asboe-Hansen (1973), with GalA as a standard.

BIOCHEMISTRY OF PECTIC ENZYMES IN RIPENING TOMATO FRUIT

Ripening is a genetically coordinated process during which several biochemical and physiological modifications of the fruit

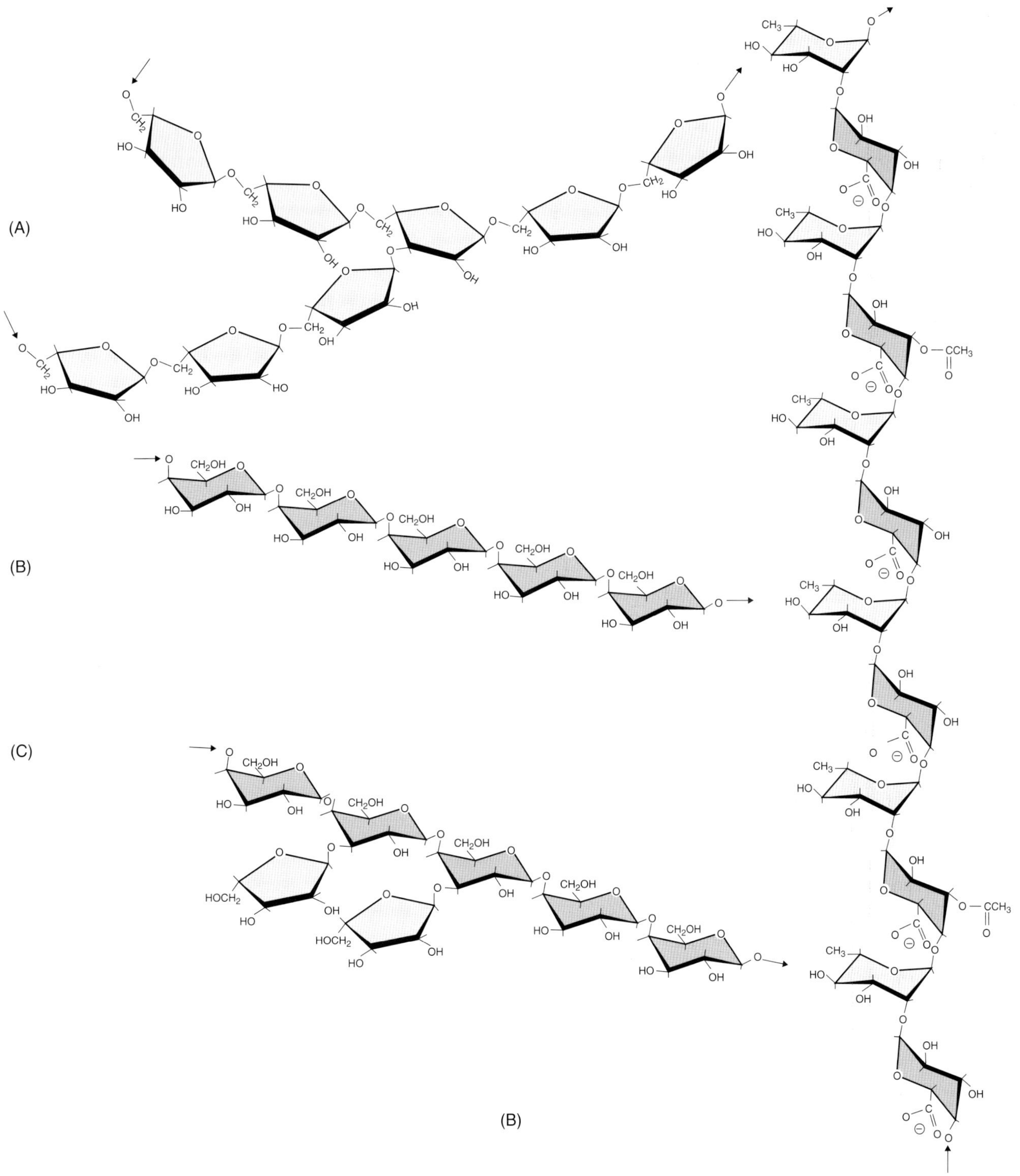

Figure 12.2. (*Continued*) (**B**) Structure of rhamnogalacturonan I. 5-Arabinan (**A**), 4-Galactan (**B**), and Type I Arabinogalactan (**C**) side chains (Reprinted from Buchanan et al. 2000).

occur. These include the conversion of starch to sugars, accumulation of pigments, organic acids, and aroma volatiles, as well as fruit softening. Pectin often consists more than 50% of the fruit cell wall. During ripening, pectin is extensively modified by de novo synthesized enzymes that bring about fruit softening (Brummell 2006, Cantu et al. 2008). The pectin backbone is de-esterified, allowing calcium associations and gelling to occur (Blumer et al. 2000). Pectin is solubilized from the wall complex, followed by a decrease in the average polymer size (Brummell and Labavitch 1997, Chun and Huber 1997). These changes are accompanied by loss of galactosyl and arabinosyl moieties from pectin side chains (Gross 1984). The dissolution of the pectin-rich middle lamella increases cell separation and contributes to softening. Cell wall disassembly in ripening tomato fruit has been extensively studied and a battery of polysaccharide-degrading enzymes involved in this process has been biochemically and genetically characterized (for review, see Brummell and Harpster 2001, Giovannoni 2001, Brummell 2006, Goulao and Oliveira 2008). The most thoroughly studied enzymes that act on the pectic polysaccharides of fruit are polygalacturonase (PG), pectin methylesterase (PME), β-galactosidase (β-GALase), and pectate lyase (PL).

Polygalacturonase

PGs are enzymes that act on the pectin fraction of cell walls and can be divided into endo-PGs and exo-PGs. Exo-PG (EC 3.2.1.67) removes a single GalA residue from the nonreducing end of HGA. Endo-PG (poly(1,4-α-D-galacturonide) glycanohydrolase, EC 3.2.1.15) is responsible for pectin depolymerization by hydrolyzing glycosidic bonds in demethylesterified regions of HGA (Fig. 12.3).

Endo-PGs are encoded by large multigene families with distinct temporal and spatial expression profiles (Hadfield and Bennett 1998). In ripening tomato fruit, the mRNA of one gene is accumulated at high levels, constituting 2% of the total polyadenylated RNA, in a fashion paralleling PG protein accumulation (DellaPenna et al. 1986). PG gene expression during ripening is ethylene responsive, with very low levels of ethylene (0.15 μL/L) being sufficient to induce PG mRNA accumulation and PG activity (Sitrit and Bennett 1998). PG mRNA levels continue to increase as ripening progresses and persist at the overripe stage. Immunodetectable PG protein accumulation follows the same pattern as mRNA accumulation during ripening and fruit senescence. This mRNA encodes a predicted polypeptide 457 amino acids long containing a signal sequence of 24 amino acids targeting it to the cell's endomembrane system for further processing and secretion. The mature protein is produced by cleaving a 47 amino acids amino-terminal prosequence and a sequence of 13 amino acids from the carboxyl terminus followed by glycosylation. Two isoforms, differing only with respect to the degree of glycosylation, are identified: PG2A, with a molecular mass of 43 kDa, and PG2B with a molecular mass of 46 kDa (Sheehy et al. 1987, DellaPenna and Bennett 1988, Pogson et al. 1991). Another isoform (PG1) accumulates during the early stages of tomato fruit ripening, when PG2 levels are low. PG1 has a molecular mass of 100 kDa and is composed of one or possibly two PG2A or PG2B subunits and the PG β-subunit. Recent work has identified yet another member of the PG1 multiprotein complex; a nonspecific lipid transfer protein of a molecular mass of 8 kDa and no detectable PG activity (Tomassen et al. 2007). When this 8 kDa protein was co-incubated with PG2, an increase in PG2 activity was observed. Moreover, heating of the 8 kDa–PG2 mixture resulted in increased heat stability of the PG2 protein. This report is the first to implicate non-specific lipid transfer proteins in PG-mediated HGA depolymerization in tomato.

The PG β-subunit is an acidic, heavily glycosylated protein. The precursor protein contains a 30 amino acid signal peptide, an N-terminus propeptide of 78 amino acids, the mature protein domain, and a large C-terminus propeptide. The mature protein presents a repeating motif of 14 amino acids and has a size of approximately 37–39 kDa. This difference in size results from different glycosylation patterns or different posttranslational processing at the carboxyl terminus of the protein (Zheng et al. 1992). The PG β-subunit protein is encoded by a single

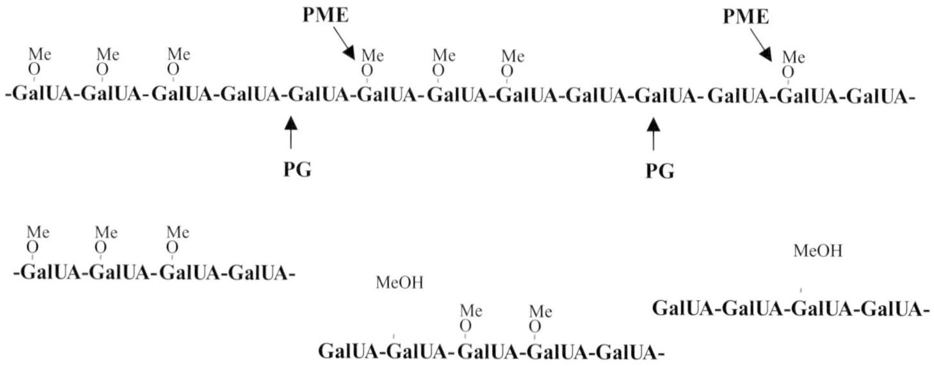

Figure 12.3. Action pattern of PG and PME on pectin backbone. Endo-PG hydrolyzes glycosidic bonds in demethylesterified regions of homogalacturonan. PME catalyzes the removal of methyl ester groups (OMe) from galacturonic acid (GalUA) residues of pectin.

gene. β-subunit mRNA levels increase during fruit development and reach a maximum at 30 days after pollination (i.e., just before the onset of ripening), then decrease to undetectable levels during ripening, whereas immunodetectable β-subunit protein persists throughout fruit development and ripening (Zheng et al. 1992, 1994). The presence of the PG β-subunit alters the physicochemical properties of PG2. Although the PG β-subunit does not possess any glycolytic activity, binding to PG2 leads to modification of PG2 activity with respect to pH optima, heat stability, and Ca^{++} requirements (Knegt et al. 1991, Zheng et al. 1992). Since the β-subunit is localized at the cell wall long before PG2 starts accumulating, it may serve as an anchor to localize PG2 to certain areas of the cell wall (Moore and Bennett 1994, Watson et al. 1994). Another proposed action for the β-subunit is that of limiting access of PG to its substrate or restricting PG activity by binding to the PG protein (Hadfield and Bennett 1998).

Generally, PG-mediated pectin disassembly contributes to fruit softening at the later stages of ripening and during fruit deterioration. Overall, PG activity is neither sufficient nor necessary for fruit softening, as it is evident from data using transgenic tomato lines with suppressed levels of PG mRNA accumulation (discussed later in this chapter) and studies on ripening-impaired tomato mutants. It is evident from data in other fruit, especially melon (Hadfield and Bennett 1998), persimmon (Cutillas-Iturralde et al. 1993), and apple (Wu et al. 1993), that very low levels of PG may be sufficient to catalyze pectin depolymerization in vivo.

Pectin Methylesterase

PME is a de-esterifying enzyme (EC 3.1.1.11) catalyzing the removal of methyl ester groups from GalA residues of pectin, thus leaving negatively charged carboxylic residues on the pectin backbone (Fig. 12.3). Demethylesterification of galacturonan residues leads to a change in the pH and charge density on the HGA backbone. Free carboxyl groups from adjacent polygalacturonan chains can then associate with calcium or other divalent ions to form gels (Fig. 12.2A). PMEs are encoded by large multigene families in many plant species (Pelloux et al. 2007). In tomato, PME protein is encoded by at least four genes (Turner et al. 1996, Gaffe et al. 1997). The PME polypeptide is 540–580 amino acids long and contains a signal sequence targeting it to the apoplast. The mature protein has a molecular mass of 34–37 kDa and is produced by cleaving an amino-terminal prosequence of approximately 22 kDa (Gaffe et al. 1997). PME is found in multiple isoforms in fruit and other plant tissues (Gaffe et al. 1994). PME isoforms have pIs in the range of 8–8.5 in fruit and around 9.0 in vegetative tissues (Gaffe et al. 1994). Tomato fruit PME is active throughout fruit development and influences accessibility of PG to its substrate. PME transcript accumulates early in tomato fruit development and peaks at the mature green fruit stage, followed by a decline in transcript levels. In contrast, PME protein levels increase in developing fruit at the early stages of ripening and then decline (Harriman et al. 1991, Tieman et al. 1992). Pectin is synthesized in highly methylated form, which is then demethylated by the action of PME. In ripening tomato fruit, the methylester content of pectin is reduced from an initial 90% at the mature green stage to about 35% in red ripe fruit (Koch and Nevins 1989). Evidence supports the hypothesis that demethylesterification of pectin, which allows Ca^{++} cross-linking to occur, may restrict cell expansion. Constitutive expression of a petunia PME gene in potato resulted in diminished PME activity in some plants. A decrease in PME activity in young stems of these transgenic potato plants correlated with an increased growth rate (Pilling et al. 2000). The action of PME is required for PG action on the pectin backbone (Wakabayashi et al. 2003), indicating a major role of PME in pectin remodeling during ripening.

β-Galactosidase

β-Galactosidase (EC 3.2.1.23) is an exo-acting enzyme that catalyzes the cleavage of terminal galactose residues from pectin β-(1,4)-D-galactan side chains (Fig. 12.2B). Loss of galactose from wall polysaccharides occurs throughout fruit development, and it accelerates with the onset of ripening (Gross 1984, Seymour et al. 1990). In tomato fruit, β-galactosidase is encoded by a small multigene family of at least seven members, *TBG1* through *7* (Smith et al. 1998, Smith and Gross 2000). On the basis of sequence analysis, it was found that β-galactosidase genes encode putative polypeptides with a predicted molecular mass between 89.8 and 97 kDa except for TBG4, which is predicted to be shorter by 100 amino acids at its carboxyl terminus (Smith and Gross 2000). A signal sequence targeting these proteins to the apoplast was only identified in *TBG4*, *TBG5*, and *TBG6*, whereas *TBG7* is predicted to be targeted to the chloroplast. These seven genes showed distinct expression patterns during fruit ripening, and transcripts of all clones except *TBG2* were detected in other tomato plant tissues (Smith and Gross 2000). β-galactosidase transcripts were also detected in the ripening impaired *ripening-inhibitor* (*rin*), *nonripening* (*nor*), and *never-ripe* (*nr*) mutant tomato lines, which do not soften during ripening; however, the expression profile differed from that in wild-type fruit. *TGB4* mRNA accumulation was impaired in comparison to wild-type levels, whereas accumulation of *TBG6* transcript persisted up to 50 days after pollination in fruit from the three mutant lines (Smith and Gross 2000). Presence of β-galactosidase transcripts in these lines suggests that β-galactosidase activity alone cannot lead to fruit softening. Expression of the *TBG4* clone, which encodes the β-galactosidase isoform II, in yeast resulted in production of active protein. This recombinant protein was able to hydrolyze synthetic substrates (*p*-nitro-phenyl-β-D-galactopyranoside) and lactose, as well as galactose-containing wall polymers (Smith and Gross 2000). Heterologous expression of *TBG1* in yeast also resulted in the production of a protein with exo-galactanase/β-galactosidase activity, also active in cell wall substrates (Carey et al. 2001).

Pectate Lyase

PLs (pectate transeliminase, EC 4.2.2.2) are a family of enzymes that catalyze the random cleavage of demethylesterified polygalacturonate by β-elimination generating oligomers with 4,5-unsaturated reducing ends (Yoder et al. 1993). PLs were thought to be of microbial origin and were commonly isolated from macerated plant tissue infected with fungal or bacterial pathogens (Yoder et al. 1993, Barras et al. 1994, Mayans et al. 1997, Yadav et al. 2009). In plants, presence of PL-like sequences was first identified in mature tomato flowers, anthers, and pollen (Wing et al. 1989). Presence of PL in pollen of other species (Albani et al. 1991, Taniguchi et al. 1995, Wu et al. 1996, Kulikauskas and McCormick 1997) as well as in other tissues (Domingo et al. 1998, Milioni et al. 2001, Pilatzke-Wunderlich and Nessler 2001) has been well documented. Recently, PL-like sequences were identified in ripening banana (Pua et al. 2001, Marín-Rodríguez et al. 2003) and strawberry fruit (Medina-Escobar et al. 1997, Benítez-Burraco et al. 2003). Transgenic strawberry plants that expressed a pectin lyase antisense construct were firmer than controls, with the level of PL suppression being correlated to internal fruit firmness (Jiménez-Bermúdez et al. 2002). Additional studies using transgenic strawberries suppressed in the expression of PL support a key role of this enzyme in strawberry fruit softening (Santiago-Doménech et al. 2008, Youssef et al. 2009) and strawberry juice viscosity (Sesmero et al. 2009). PL activity has been detected in ripening cherry tomato fruit grown under different environmental conditions (Rosales et al. 2009) and several putative PL sequences exist in the tomato expressed sequence tag databases (Marín-Rodríguez et al. 2002), suggesting that these enzymes may also contribute to cell wall disassembly during tomato fruit ripening.

GENETIC ENGINEERING OF PECTIC ENZYMES IN TOMATO

Cell wall disassembly in ripening fruit is an important contributor to the texture of fresh fruit. Modification of cell wall enzymatic activity during ripening, using genetic engineering, can impact cell wall polysaccharide metabolism, which in turn can influence texture. Texture of fresh fruit, in turn, influences processing characteristics and final viscosity of processed tomato products. The development of transgenic tomato lines in which the expression of single or multiple genes is altered has allowed the role of specific enzymes to be evaluated both in fresh fruit and processed products.

Single Transgenic Tomato Lines

Polygalacturonase

PG was the first target of genetic modification aiming to retard softening in tomatoes (Sheehy et al. 1988, Smith et al. 1988). Transgenic plants were constructed in which expression of PG was suppressed to 0.5–1% of wild-type levels by the expression of an antisense PG transgene under the control of the cauliflower mosaic virus 35S (CaMV35S) promoter. Fruit with suppressed levels of PG ripened normally as wild-type plants did. An increase in the storage life of the fruit was observed in the PG-suppressed fruit (Schuch et al. 1991, Kramer et al. 1992, Langley et al. 1994). The influence of PG suppression on individual classes of cell wall polymers was investigated (Carrington et al. 1993, Brummell and Labavitch 1997). Sequential extractions of cell wall constituents revealed a decrease in the extent of depolymerization of CDTA soluble polyuronides during fruit ripening. Naturally occurring mutant tomato lines exist, which carry mutations, affecting the normal process of ripening. One of these lines, the *ripening-inhibitor* (*rin*), possesses a single locus mutation that arrests the response to ethylene. *rin* fruit fail to synthesize ethylene and remain green and firm throughout ripening (DellaPenna et al. 1989). Transgenic *rin* plants that expressed a functional PG gene driven by the ethylene-inducible E8 promoter were obtained. After exposure to propylene, PG mRNA accumulation as well as active PG protein was attained, but the fruit did not soften. In these fruit, depolymerization and solubilization of cell wall pectin was at near wild-type levels (Giovannoni et al. 1989, DellaPenna et al. 1990). Altogether, results obtained from analyzing transgenic tomato fruit with altered levels of PG indicate that PG-mediated depolymerization of pectin does not influence softening at the early stages of ripening, and it is not enough to cause fruit softening. However, its outcome is manifested at the later stages of ripening and in fruit senescence.

Suppression of β-subunit expression by the introduction of a β-subunit cDNA in the antisense orientation resulted in reduction of immunodetectable β-subunit protein levels and PG1 protein levels to $<1\%$ of PG1 in control fruit. In transgenic lines, PG2 expression remained unaltered (Watson et al. 1994). Suppression of β-subunit expression resulted in increased CDTA-extractible polyuronide levels in transgenic fruit during ripening. The size distribution profiles of CDTA-extractable polyuronides of transgenic and control fruit were nearly identical at the mature green stage. However, at subsequent ripening stages, an increased number of smaller size polyuronides were observed in suppressed β-subunit fruit compared to control fruit (Watson et al. 1994).

Apart from the influence of transgenic modification of PG gene expression on the ripening characteristics and shelf life of fresh tomato fruit, the processing characteristics of these fruit were also modified by the introduction of the PG transgene. A shift in the partitioning of polyuronides in water and CDTA extracts of cell walls from pastes prepared from control and suppressed PG lines was observed. The majority of polyuronides were solubilized in the water extract in control pastes, whereas in pastes from PG-suppressed fruit, a greater proportion was solubilized in CDTA. Water and CDTA extracts from control paste contained a larger amount of smaller sized polyuronides compared with the corresponding extracts from PG-suppressed fruit (Fig. 12.4). However, the molecular size profiles from carbonate-soluble pectin were indistinguishable in the two lines. Tomato paste prepared from PG-suppressed fruit exhibited enhanced gross and serum viscosity (Brummell and Labavitch 1997, Kalamaki 2003). The enhancement of viscosity characteristics of pastes from transgenic PG fruit was attributed to the presence of pectin polymers of higher molecular size and wider size distribution in the serum. Although PG seems to play only a small role in fruit softening, the viscosity of juice and reconstituted paste prepared from antisense PG lines is substantially increased, largely due to reduced break down of soluble

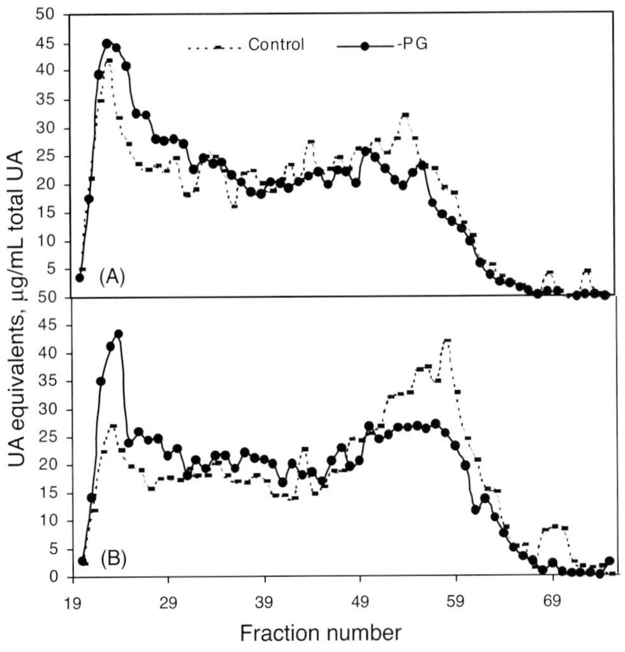

Figure 12.4. Size exclusion chromatography of CDTA-soluble pectin fractions from juice (**A**) and paste (**B**) prepared from control (-) and suppressed PG (.) fruit. Larger polysaccharide molecules elute close to the void volume of the column (fraction 20), whereas small molecules elute at the total column volume (fraction 75). (Based on data presented by Kalamaki 2003.)

pectin during processing (Schuch et al. 1991, Kramer et al. 1992, Brummell and Labavitch 1997, Kalamaki et al. 2003a).

Pectin Methylesterase

PME protein is accumulated during tomato fruit development, and protein levels increase with the onset of ripening. Production of transgenic tomato lines in which the expression of PME has been suppressed via the introduction of an antisense PME gene driven by the CaMV35S promoter has been reported (Tieman et al. 1992, Hall et al. 1993). PME activity was reduced to less than 10% of wild-type levels, and yet fruit ripened normally. The degree of methylesterification increased significantly in antisense PME fruit by 20–40%, compared to wild-type fruit. Additionally, the amount of chelator-soluble polyuronides decreased in transgenic lines, a result that was expected since demethylesterification is necessary for calcium-mediated cross-linking of pectin. Furthermore, chelator-soluble polyuronides from red ripe tomato pericarp exhibited a larger amount of intermediate size polymers in transgenic lines than controls, indicating a decrease in pectin depolymerization of the chelator-soluble fraction. As described above, PME action precedes that of PG. An increase of about 15% in soluble solids was also observed in transgenic fruit pericarp cell walls (Tieman et al. 1992). Expression of the antisense PME construct also influenced the ability of the fruit tissue to bind divalent cations. Calcium accumulation in the wall was not influenced; however, two-thirds of the total calcium in transgenic fruit was present as soluble calcium, compared to one-third in wild-type fruit (Tieman and Handa 1994). Pectin with a higher degree of methylesterification was found in PME-suppressed fruit, thus presenting fewer sites for calcium binding in the cell wall.

From a food-processing standpoint, the effect of these modifications in PME activity in tomato cell walls during ripening was further investigated. Juice prepared from transgenic antisense PME fruit had about 35–50% higher amount of total UAs than control fruit. Pectin extracted from juice processed either by a hot break, microwave break, or a cold-break method was methylesterified at a higher degree in PME-suppressed juices. In addition, pectin from transgenic fruit was of larger molecular mass compared to controls (Thakur et al. 1996a). Juice and paste prepared from antisense PME fruit had higher viscosity and lower serum separation (Thakur et al. 1996b).

β-Galactosidase

In tomato fruit, a considerable loss of galactose is observed during the course of ripening. This is attributed to the action of enzymes with exo-galactanase/β-galactosidase activity. In order to further investigate their role in ripening, β-galactosidase *TBG1* gene expression was suppressed in transgenic tomato by the expression of a sense construct of 376 bp of the *TBG1* gene driven by the CaMV35S promoter (Carey et al. 2001). Although plants with different expression levels of the transgene were obtained, there was no effect on β-galactosidase protein activity and fruit softening. The results suggest that the *TBG1* gene product may not be readily involved in fruit softening but may act on a specific cell wall substrate (Carey et al. 2001). The β-galactosidase isoform II, encoded by *TBG4*, has been found to accumulate in ripening tomato fruit (Smith et al. 1998). The activity of TBG4, has been suppressed by the expression of an 1.5 kb antisense construct of *TBG4* (Smith et al. 2002). Antisense lines exhibited various degrees of *TBG4* mRNA suppression that was correlated to reduced extractable exo-galactanase activity. In one antisense line, a 40% increase in fruit firmness was observed, which correlated with the highest—among transgenic lines—*TBG4* mRNA suppression, lowest exo-galactanase activity levels, and highest galactosyl content during the early stages of ripening, suggesting a possible role for *TBG4* in the early events of fruit softening. (Smith et al. 2002).

Double Transgenic Tomato Lines

As reported above, modifying the expression of individual cell wall hydrolases did not alter fruit softening substantially, suggesting that cell-wall-modifying enzymes may act as a consortium to bring about fruit softening during ripening. Generation of double transgenic lines further aids in understanding the influence of cell wall deconstruction during ripening to the texture of ripe fruit and on the physicochemical properties of processed tomato products.

PG and Expansin

The influence of the suppression of the expression of two cell-wall-related proteins in the same transgenic line has been reported (Kalamaki et al. 2003a, Powell et al. 2003). Tomato lines with simultaneous suppression of the expression of both PG and the ripening-associated expansin have been produced and their chemical and physical properties characterized (Kalamaki et al. 2003a, Powell et al. 2003). Expansins are proteins that lack wall hydrolytic activity and are proposed to act by disrupting the hydrogen bonding between cellulose microfibrils and the cross-linking glycan (xyloglucan) matrix (see Fig. 12.1). Expansins are involved in cell wall disassembly during ripening (Brummell et al. 1999a). Their role in ripening is proposed to be that of loosening the wall and increasing the accessibility of wall polymers to hydrolytic enzymes (Rose and Bennett 1999). In tomato, the product of one expansin gene (*Exp1*) is accumulated exclusively during tomato fruit ripening (Rose et al. 1997, Brummell et al. 1999b). Juices and pastes prepared from fruit with modified levels of *Exp1* exhibit enhanced viscosity attributes (Kalamaki et al. 2003b).

Double transgenic lines were generated by crossing single transgenic homozygous lines. Fruit was harvested at the mature green stage and allowed to ripen at 20°C, and texture was determined as the force required to compress the blossom end of the tomato by 2 mm. Fruit from the double suppressed line were firmer than controls and single transgenic lines at all ripening stages. The same results were observed in fruit ripened on vine. At the red ripe stage, fruit from the double suppressed line were 20% firmer than controls. The flow properties of juice prepared from control, single transgenic, and double transgenic lines at the mature green/breaker, pink, and red ripe stages were evaluated using a Bostwick consistometer. Average juice viscosity decreased as ripening progressed in all genotypes. Juice prepared from fruit of the suppressed PG and the double suppressed line at the red ripe stage exhibited higher viscosity than control (Powell et al. 2003).

In another experiment, an elite processing variety was used as the wild-type background to suppress the expression of PG, Exp1 and both PG and Exp1 in the same line (Kalamaki et al. 2003a). Juice and paste were prepared from control, single suppressed and double suppressed fruit, and their flow properties were characterized. In paste diluted to 5 Brix, an increase in Bostwick consistency of about 18% was observed for the PG-suppressed and the Exp1-suppressed genotypes. Diluted paste from the double suppressed genotype exhibited the highest viscosity; however, this increase in consistency of the double transgenic line was only by an additional 4% compared to the single transgenic lines. Analysis of particle size distributions at 5 Brix in juices and pastes of the different genotypes indicated that suppression of PG or Exp1 results in small increases in the number of particles below a diameter of 250 μm (Fig. 12.5). Suppression of Exp1 also results in the appearance of some larger particles, above 1400 μm in diameter. The presence of these larger fragments shows that suppression of Exp1 affects the way the cells and cell walls rupture during processing. Biochemical characterization of cell wall polymers did not show large differences in the amounts of extractable polyuronide or neutral sugar in sequential cell wall extracts of juices and diluted concentrates of the various genotypes. Only minor size differences appear in individual polymer sizes in sequential extracts. In suppressed PG genotypes, larger pectin sizes were observed in the CDTA extract. The most dramatic difference was in the carbonate extracts where the double suppressed-line contained larger pectin polymers in all paste concentration levels relative to other genotypes (Kalamaki 2003). Concurrent suppression of PG and Exp1 in the same transgenic line showed an overall increase in viscosity compared to control, but only a 4% additional increase compared with the single transgenic lines. Particle size distribution in juices and pastes from the double transgenic line are more poly-disperse (Fig. 12.5), and particles appear to be more rigid. In this line, an increase in the size of

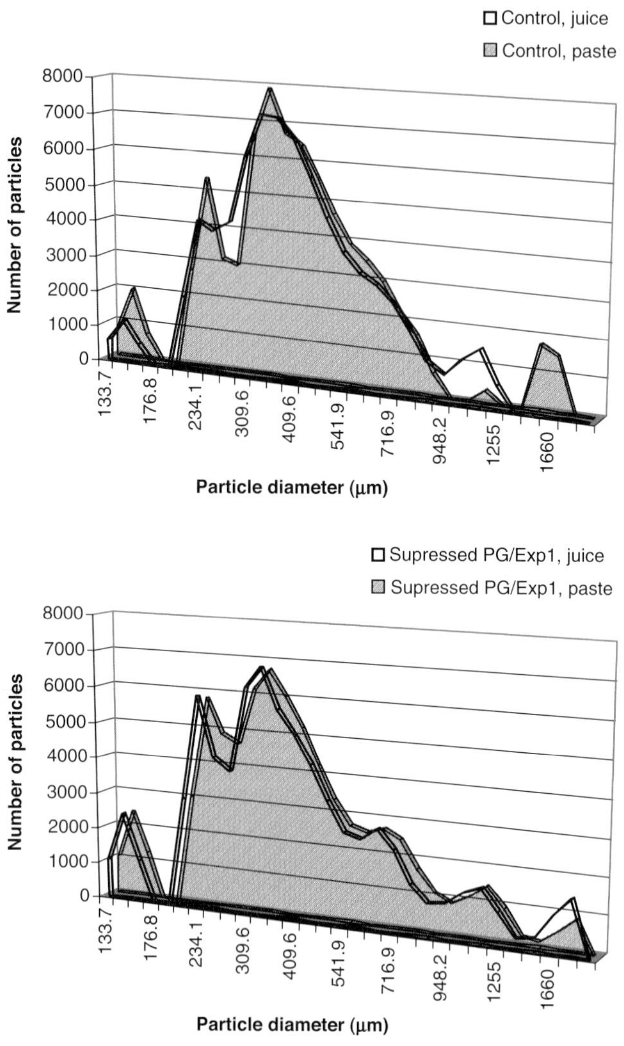

Figure 12.5. Particle size determination in juice and paste from control and suppressed PG/Exp1 fruit. (Based on data presented by Kalamaki et al. 2003a.)

carbonate-soluble pectin polymers was observed. This size increase could suggest differences in the degree of polymer cross-linking in the particles, resulting in altered particle properties. By modifying cell wall metabolism during ripening, the processing qualities of the resulting juices and pastes were influenced. Hence, reducing the activity of these cell wall enzymes may be a route to the selection of improved processing tomato varieties.

PG and PME

The activities of both PG and PME were suppressed in the same line, and juices were evaluated and compared to single transgenics and controls (Errington et al. 1998). Small differences in Bostwick consistency were observed in hot-break juices with the suppressed PG juices having the highest viscosity. In cold-break juices from PG-suppressed fruit, a time-dependent increase in viscosity was observed. Since a similar increase was not observed in the double transgenic line, it was concluded that in order for the increase in viscosity of cold-break juice to occur, both absence of PG and continued action of PME is required. Absence of PG activity will lead to larger size of pectin, whereas continuing demethylesterification of pectin chains by PME will increase calcium associations between pectin chains, leading to gel formation and thus improved viscosity. In conclusion, simultaneous transgenic suppression of PG and PME expression did not result in an additive effect.

TOMATO PROCESSING

The majority of tomatoes are consumed in a processed form, such as juice, paste, pizza and pasta sauce, and various diced or sliced products. Most of the products are concentrated to different degrees and stored in a concentrated form until ready to use. Industrial concentrates are then diluted to reach the desired final product consistency.

Textural properties of tomato fruit are important contributors to the overall quality in both fresh market and processing tomatoes (Barrett et al. 1998). In some processed products, the most important quality attribute is viscosity (Alviar and Reid 1990). It was recognized early that the structure most closely associated with viscosity is the cell wall (Whittenberger and Nutting 1957). Both the concentration and type of cell wall polymers in the serum fraction and the pulp (particle fraction) are important contributors to viscosity. In serum, the amount and size of the soluble cell wall polymers influence serum viscosity (Beresovsky et al. 1995), whereas in pulp, the size distribution, the shape, and the degree of deformability of cell wall fragments influence viscosity (Den Ouden and Van Vliet 1997). Viscosity is influenced partly by factors that dictate the chemical composition and physical structure of the juice such us fruit variety, cultivation conditions, and the ripening stage of the fruit at harvest. However, it is also influenced by processing factors such as break temperature (Xu et al. 1986), finisher screen size (Den Ouden and Van Vliet 1997), mechanical shearing during manufacture, and degree of concentration (Marsh et al. 1978). These processes result in changes in the microstructure of fruit cell wall that are manifested as changes in viscosity (Xu et al. 1986). Since the integrity of cell wall polymers in the juice is imperative, plant pectic enzymes usually have to be inactivated during processing in order to diminish their activity and prevent pectin degradation. Pectin degradation could lead to increased softening in, for example, pickled vegetable production or peach canning, and loss of viscosity in processed tomato products (Crelier et al. 2001). Although several enzymes are reported to act on cell wall polymers, the main depolymerizing enzyme is PG. Therefore, inactivation of PG activity during tomato processing is essential for viscosity retention. However, there are cases where residual activity of other pectic enzymes is desirable, as for example, with PME. Retention of PME activity with concurrent and complete inactivation of PG could lead to products with higher viscosity (Errington et al. 1998, Crelier et al. 2001). Selective inactivation of PG is not possible with conventional heat treatment, since PME is rather easily inactivated by heat at ambient pressure, while PG requires much more severe heat treatment for complete inactivation. PME is inactivated by heating at 82.2°C for 15 seconds at ambient pressure, while for PG inactivation, a temperature of 104.4°C for 15 seconds is required in canned tomato pulp (Luh and Daoud 1971). In order to achieve this selective inactivation, a combination of heating and high hydrostatic pressure can be used during processing or alternatively, the expression of a particular enzyme can be suppressed using genetic engineering in fruit.

Thermal Inactivation

Thermal processing (that is, exposure of the food to elevated temperatures for relatively short times) has been used for almost 200 years to produce shelf-stable products by inactivating microbial cells, spores, and enzymes in a precisely defined and controlled procedure. Kinetic description of the destructive effects of heat on both desirable and undesirable attributes is essential for proper thermal process design.

At constant temperature and pressure conditions, PME thermal inactivation follows first-order kinetics (Crelier et al. 2001, Fachin et al. 2002, Stoforos et al. 2002).

$$-\frac{dA}{dt} = kA, \quad k = f(T, P, \ldots) \tag{1}$$

where A is the enzyme activity at time t, k the reaction rate constant, and P and T the pressure and temperature process conditions.

Ignoring the pressure dependence, Equation 1 leads to

$$A = A_o e^{-k_T t} \tag{2}$$

where A_o is the initial enzyme activity and the reaction rate constant k_T, function of temperature, is adequately described by Arrhenius kinetics through Equation 3.

$$k_T = k_{T_{ref}} \exp\left[-\frac{E_a}{R}\left(\frac{1}{T} - \frac{1}{T_{ref}}\right)\right] \tag{3}$$

where $k_{T_{ref}}$ is the reaction rate constant at a constant reference temperature T_{ref}, E_a is the activation energy, and R is the universal gas constant (8.314 J/(mol · K)).

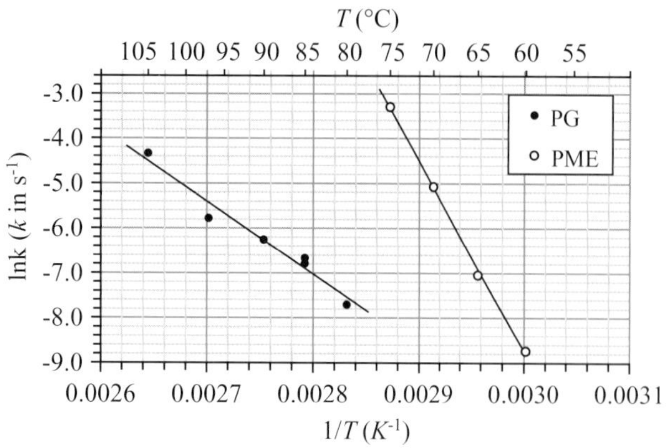

Figure 12.6. Effect of processing temperature on PG and PME thermal inactivation rates at ambient pressure. (Based on data presented by Crelier et al. 2001.)

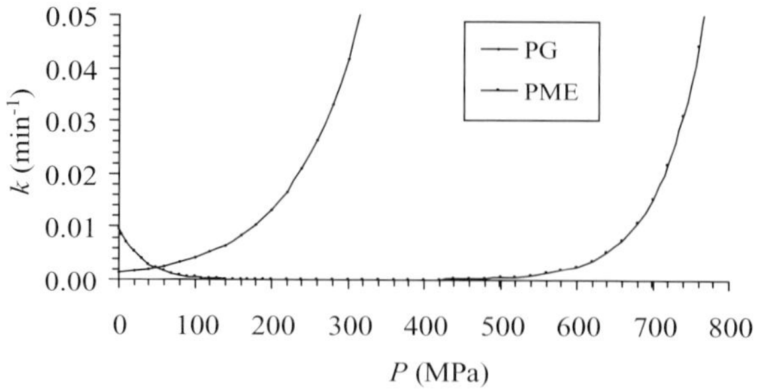

Figure 12.7. Schematic representation of the effect of processing pressure on PG and PME inactivation rates during high-pressure treatment (at 60°C).

PG thermal inactivation follows first-order kinetics (Crelier et al. 2001, Fachin et al. 2003) as suggested by Equation 2 above, or a fractional conversion model (Equation 4), which suggests a residual enzyme activity at the end of the treatment.

$$A = A_\infty + (A_o - A_\infty)e^{k_T} \qquad (4)$$

where A_∞ is the residual enzyme activity after prolonged heating.

From the data presented by Crelier et al. (2001), the effect of processing temperature on crude tomato juice PG and PME thermal inactivation rates at ambient pressure is illustrated in Figure 12.6. The higher resistance of PG, compared with PME, to thermal inactivation is evident (Fig. 12.6). Furthermore, PG inactivation is less sensitive to temperature changes, compared with PME inactivation (as can be seen by comparing the slopes of the corresponding curves in Fig. 12.6). Values for the activation energies (E_a, see Equation 3) equal to 134.5±15.7 kJ/mol for the case of PG and 350.1±6.0 kJ/mol for the case of PME thermal inactivation have been reported (Crelier et al. 2001). It must be noted that the literature values presented here are restricted to the system used in the particular study and are mainly reported here for illustrative purposes. Thus, for example, the origin and the environment (e.g., pH) of the enzyme can influence the heat resistance (*k* or *D*—the decimal reduction time—values) of the enzyme as well as the temperature sensitivity (E_a or *z*—the temperature difference required for 90% change in *D*—values) of the enzyme thermal inactivation rates.

High Pressure Inactivation

High hydrostatic pressure processing of foods (i.e., processing at elevated pressures (up to 1000 MPa) and low to moderate temperatures (usually less than 100°C)) has been introduced as an alternative nonthermal technology that causes inactivation of microorganisms and denaturation of several enzymes with minimal destructive effects on the quality and the organoleptic characteristics of the product. The improved product quality attained during high-pressure processing of foods, and the potential for production of a variety of novel foods, in particular, desirable characteristics, have made the high pressure technology attractive (Farr 1990, Knorr 1993).

As far as high pressure enzyme inactivation goes, PG is easily inactivated at moderate pressure and temperatures (Crelier et al. 2001, Shook et al. 2001, Fachin et al. 2003), while PME inactivation at elevated pressures reveals an antagonistic (protective) effect between pressure and temperature (Crelier et al. 2001, Shook et al. 2001, Fachin et al. 2002, Stoforos et al. 2002). Depending on the processing temperature, PME inactivation rate at ambient pressure (0.1 MPa) is high, rapidly decreases as pressure increases, practically vanishes at pressures of 100–500 MPa, and thereafter starts increasing again, as illustrated on Figure 12.7.

High pressure inactivation kinetics for both PG and PME follow first-order kinetics (Crelier et al. 2001, Fachin et al. 2002, Stoforos et al. 2002, 2003). Values for the reaction rate constants, *k*, for high pressure inactivation of PME and PG as a function of processing temperature, at selected conditions, are given in Table 12.1 (Crelier et al. 2001).

Through the activation volume concept (Johnson and Eyring 1970), the pressure effects on the reaction rate constants can be expressed as:

$$k_p = k_{P_{ref}} \exp\left[-\frac{V_a}{R}\frac{(p - P_{ref})}{T}\right] \qquad (5)$$

where $k_{P_{ref}}$ is the reaction rate constant at a constant reference pressure, P_{ref}, and V_a is the activation volume.

Models to describe the combined effect of pressure and temperature on tomato PME or PG inactivation have been presented in the literature (Crelier et al. 2001, Stoforos et al. 2002, Fachin et al. 2003). On the basis of the literature data (Crelier et al. 2001), a schematic representation of high pressure inactivation of tomato PG and PME is presented in Figure 12.7. From data like these, one can see the possibilities of selective inactivation

Table 12.1. First-Order Reaction Rate Constants, k (sec^{-1}), for High Pressure PME and PG Inactivation as a Function of Processing Temperature (Based on Data Presented by Crelier et al. 2001)

	First-Order Reaction Rate Constants, k (sec^{-1})							
	PME				PG			
P (MPa)	60°C	65°C	70°C	75°C	30°C	40°C	50°C	60°C
0.1	$161 \cdot 10^{-6}$	$872 \cdot 10^{-6}$	$6250 \cdot 10^{-6}$	$3680 \cdot 10^{-6}$				
100	$7.58 \cdot 10^{-6}$	$163 \cdot 10^{-6}$	$883 \cdot 10^{-6}$	$574 \cdot 10^{-6}$				$52.6 \cdot 10^{-6}$
300	$9.70 \cdot 10^{-6}$	$112 \cdot 10^{-6}$	$51.4 \cdot 10^{-6}$	$18.3 \cdot 10^{-6}$				$309 \cdot 10^{-6}$
400				$38.6 \cdot 10^{-6}$	$193 \cdot 10^{-6}$	$351 \cdot 10^{-6}$	$701 \cdot 10^{-6}$	$2510 \cdot 10^{-6}$
500				$103 \cdot 10^{-6}$	$3890 \cdot 10^{-6}$	$4090 \cdot 10^{-6}$	$3620 \cdot 10^{-6}$	$7020 \cdot 10^{-6}$
600	$43.8 \cdot 10^{-6}$	$44.3 \cdot 10^{-6}$	$70.2 \cdot 10^{-6}$	$52.5 \cdot 10^{-6}$		$1720 \cdot 10^{-6}$	$4810 \cdot 10^{-6}$	$7530 \cdot 10^{-6}$
800	$1500 \cdot 10^{-6}$	$2330 \cdot 10^{-6}$	$2920 \cdot 10^{-6}$	$2710 \cdot 10^{-6}$				

of the one or the other enzyme by appropriately optimizing the processing conditions (pressure, temperature, and time).

FUTURE PERSPECTIVES

Texture in ripe tomato fruit is largely dictated by cell wall disassembly during the ripening process. Cell wall polysaccharides are depolymerized, and their composition is changed as ripening progresses. The coordinated and synergistic activities of many proteins are responsible for fruit softening, and although expansins and PGs are among the more abundantly expressed proteins in ripening tomato fruit, the modification of their expression has not been sufficient to account for all of the cell wall changes associated with softening or for the overall extent of fruit softening. The texture of fresh fruit directly influences the rheological characteristics of processed tomato products. Viscosity is influenced by modifications of the cell wall architecture but also by changes that may affect polysaccharide mobility and interactions of different components in juices and pastes. Since cell wall metabolism in ripening fruit is a complex process and many enzymes have been identified, which contribute to this process, further insight to the effect of these enzymes on fruit texture and processing attributes can be gained by the simultaneous suppression or over-expression of combinations of enzymes. Of course, the quest is on for identification of enzymes acting on RGI, RGII, and XGA of plant pectins similar to those found in fungi and bacteria (Wong 2008). Moreover, as our knowledge on cell wall biosynthesis and deconstruction increases, the ability to genetically engineer pectin structure in target plants would further improve fruit texture and processing characteristics (Ramakrishna et al. 2003, Vicente et al. 2007, Matas et al. 2009). Manipulation of the expression of multiple ripening-associated genes and/or transcription factors that regulate ripening (Vrebalov et al. 2009) in transgenic tomato lines will further shed light to the ripening physiology, postharvest shelf life and processing qualities of tomato fruit. Finally, the interest exists and efforts are made to design optimal processes for selective enzyme inactivation and thus for production of products with desirable characteristics, by introducing pressure as an additional (to time and temperature) processing variable.

ACKNOWLEDGMENTS

The authors would like to thank Asst. Professor P. Christakopoulos, National Technical University of Athens, School of Chemical Engineering, Biosystems Technology Laboratory, and Dr. C. Mallidis, Institute of Technology of Agricultural Products, National Agricultural Research Foundation of Greece, for reviewing the manuscript and their constructive comments.

REFERENCES

Albani D et al. 1991. A gene showing sequence similarity to pectin esterase is specifically expressed in developing pollen of *Brassica napus* sequences in its 5′ flanking region are conserved in other pollen-specific promoters. *Plant Mol Biol* 16: 501–513.

Alviar MSB, Reid DS. 1990. Determination of rheological behavior of tomato concentrates using back extrusion. *J Food Sci* 55: 554–555.

Barras F et al. 1994. Extracellular enzymes and soft-rot Erwinia. *Ann Rev Phytopathol* 32: 201–234.

Barrett DM et al. 1998. Textural modification of processing tomatoes. *Crit Rev Food Sci Nutr* 38: 173–258.

Benítez-Burraco A et al. 2003. Cloning and characterization of two ripening-related strawberry (*Fragaria x ananassa* cv *Chandler*) pectate lyase genes. *J Exp Bot* 54: 633–645.

Beresovsky N et al. 1995. The role of pulp interparticle interaction in determining tomato juice viscosity. *J Food Process Pres* 19: 133–146.

Blumenkrantz N, Asboe-Hansen G. 1973. New method for quantitative determination of uronic acids. *Ann Biochem* 54: 484–489.

Blumer JM et al. 2000. Characterization of changes in pectin methylesterase expression and pectin esterification during tomato fruit ripening. *Can J Bot* 78: 607–618.

Brummell DA. 2006. Cell wall disassembly in ripening fruit. *Funct Plant Biol* 33: 103–119.

Brummell DA, Labavitch JM. 1997. Effect of antisense suppression of endopolygalacturonase activity on polyuronide molecular weight in ripening tomato fruit and in fruit homogenates. *Plant Physiol* 115: 717–725.

Brummell DA, Harpster MH. 2001. Cell wall metabolism in fruit softening and quality and its manipulation in transgenic plants. *Plant Mol Biol* 47: 311–340.

Brummell DA et al. 1999a. Modification of expansin protein abundance in tomato fruit alters softening and cell wall polymer metabolism during ripening. *The Plant Cell* 11: 2203–2216.

Brummell DA et al. 1999b. Differential expression of expansin gene family members during growth and ripening of tomato fruit. *Plant Mol Biol* 39: 161–169.

Buchanan BB et al. 2000. *Biochemistry and Molecular Biology of Plants*. American Society of Plant Physiologists, Rockville, MD, p. 1367.

Cantu D et al. 2008. The intersection between cell wall disassembly, ripening, and fruit susceptibility to *Botrytis cinerea*. *Proc Natl Acad Sci USA* 105: 859–864.

Carey AT et al. 2001. Down-regulation of a ripening-related β-galactosidase gene (TBG1) in trasgenic tomato fruits. *Journal of Experimental Botany* 52: 663–668.

Carpita NC, Gibeaut DM. 1993. Structural models of primary cell walls in flowering plants consistency of molecular structure with the physical properties of the walls during growth. *Plant J* 3: 1–30.

Carrington CMS et al. 1993. Cell wall metabolism in ripening fruit. 6. Effect of the antisense polygalacturonase gene on cell wall changes accompanying ripening in transgenic tomatoes. *Plant Physiol* 103: 429–434.

Chun JP, Huber DJ. 1997. Polygalacturonase isoenzyme 2 binding and catalysis in cell walls from tomato fruit: pH and β-subunit effects. *Physiol Plant* 101: 283–290.

Coenen GJ et al. 2007. Identification of the connecting linkage between homo- or xylogalacturonan and rhamnogalacturonan type I. *Carbohydr Polym* 70: 224–235.

Cosgrove DJ. 2000. Expansive growth of plant cells. *Plant Physiol Biochem* 38: 109–124.

Crelier S et al. 2001. Tomato (*Lycopersicon esculentum*) pectin methylesterase and polygalacturonase behaviors regarding heat- and pressure-induced inactivation. *J Agric Food Chem* 49: 5566–5575.

Cutillas-Iturralde A et al. 1993. Metabolism of cell wall polysaccharides from persimmon fruit: pectin solubilization during fruit ripening occurs in apparent absence of polygalacturonase activity. *Physiol Plant* 89: 369–375.

DellaPenna D et al. 1986. Molecular cloning of tomato fruit polygalacturonase: analysis of polygalacturonase mRNA levels during ripening. *Proc Natl Acad Sci USA* 83: 6420–6424.

DellaPenna D, Bennett AB. 1988 *In vitro* synthesis and processing of tomato fruit polygalacturonase. *Plant Physiol* 86: 1057–1063.

DellaPenna D et al. 1989. Transcriptional analysis of polygalacturonase and other ripening associated genes in Rutgers, *rin*, *nor*, and *nr* tomato fruit. *Plant Physiol* 90: 1372–1377.

DellaPenna D et al. 1990. Polygalacturonase isoenzymes and pectin depolymerization in transgenic *rin* tomato fruit. *Plant Physiol* 94: 1882–1886.

Den Ouden FWC, Van Vliet T. 1997. Particle size distribution in tomato concentrate and effects on rheological properties. *J Food Sci* 62: 565–567.

Domingo C et al. 1998. A pectate lyase from *Zinnia elegans* is auxin inducible. *Plant J* 13: 17–28.

Errington N et al. 1998. Effect of genetic down-regulation of polygalacturonase and pectin esterase activity on rheology and composition of tomato juice. *J Sci Food Agric* 76: 515–519.

Fachin D et al. 2002. Comparative study of the inactivation kinetics of pectinmethylesterase in tomato juice and purified form. *Biotechnol Prog* 18: 739–744.

Fachin D et al. 2003. Inactivation kinetics of polygalacturonase in tomato juice. *Innovat Food Sci Emerg Tech* 4: 135–142.

Farr D. 1990. High pressure technology in the food industry. *Trends Food Sci Technol* 1: 14–16.

Gaffe J et al. 1994. Pectin methylesterase isoforms in tomato *Lycopersicon esculentum* tissue. Effects of expression of a pectin methylesterase antisense gene. *Plant Physiol* 105: 199–203.

Gaffe J et al. 1997. Characterization and functional expression of a ubiquitously expressed tomato pectin methylesterase. *Plant Physiol* 114: 1547–1556.

Giovannoni J. 2001. Molecular biology of fruit maturation and ripening. *Ann Rev Plant Physiol Plant Mol Biol* 52: 725–749.

Giovannoni JJ et al. 1989. Expression of a chimeric polygalacturonase gene in transgenic *rin* (ripening inhibitor) tomato fruit results in polyuronide degradation but not fruit softening. *Plant Cell* 1: 53–63.

Goulao LF, Oliveira CM. 2008. Cell wall modifications during fruit ripening: when the fruit is not the fruit. *Trends Food Sci Technol* 19: 4–25.

Gross KC. 1984. Fractionation and partial characterization of cell walls from normal and non-ripening mutant tomato fruit. *Physiol Plant* 62: 25–32.

Hadfield KA, Bennett AB. 1998. Polygalacturonases—many genes in search of a function. *Plant Physiol* 117: 337–343.

Hall LN et al. 1993. Antisense inhibition of pectin esterase gene expression in transgenic tomatoes. *Plant J* 3: 121–129.

Harriman RW et al. 1991. Molecular cloning of tomato pectin methylesterase gene and its expression in Rutgers, ripening inhibitor, nonripening, and never ripe tomato fruits. *Plant Physiol* 97: 80–87.

Huber DJ. 1992. Inactivation of pectin depolymerase associated with isolated tomato fruit cell wall: implications for the analysis of pectin solubility and molecular weight. *Physiol Plant* 86: 25–32.

Ishii T et al. 1999. The plant cell wall polysaccharide rhamnogalacturonan II self-assembles into a covalently cross-linked dimmer. *J Biol Chem* 274: 13098–13104.

Jarvis MC. 2011. Plant cell walls: supramolecular assemblies. *Food Hydrocolloid* 25: 257–262.

Jiménez-Bermúdez S et al. 2002. Manipulation of strawberry fruit softening by antisense expression of a pectate lyase gene. *Plant Physiol* 128: 751–759.

Johnson FH, Eyring H. 1970. The kinetic basis of pressure effects in biology and chemistry. In: AM Zimmerman (ed.) *High Pressure Effects on Cellular Processes*. Academic Press, New York, pp. 1–44.

Kalamaki MS. 2003. The influence of transgenic modification of gene expression during ripening on physicochemical characteristics of processed tomato products. D.Phil. Dissertation. University of California, Davis, CA. UMI Digital Dissertations, Ann Arbor, MI. p. 155.

Kalamaki MS et al. 2003a. Simultaneous transgenic suppression of LePG and LeExp1 influences rheological properties of juice and concentrates from a processing tomato variety. *Journal of Agricultural and Food Chemistry* 51: 7456–7464.

Kalamaki MS et al. 2003b. Transgenic over-expression of expansin influences particle size distribution and improves viscosity of tomato juice and paste. *J Agric Food Chem* 51: 7465–7471.

Koch JL, Nevins DJ. 1989. Tomato fruit cell wall. I. Use of purified tomato polygalacturonase and pectinmethylesterase to identify developmental changes in pectins. *Plant Physiol* 91: 816–822.

Knegt E et al. 1991. Function of the polygalacturonase convertor in ripening tomato fruit. *Physiol Plant* 82: 237–242.

Knorr D. 1993. Effects of high-hydrostatic-pressure processes on food safety and quality. *Food Technology* 47(6): 156–161.

Knox JP. 2008. Revealing the structural and functional diversity of plant cell walls. *Curr Opin Plant Biol* 11: 308–313.

Kramer M et al. 1992. Postharvest evaluation of transgenic tomatoes with reduced levels of polygalacturonase: Processing, firmness, and disease resistance. *Postharvest Biol Technol* 1: 241–255.

Kulikauskas R, McCormick S. 1997. Identification of the tobacco and Arabidopsis homologues of the pollen-expressed LAT59 gene of tomato. *Plant Mol Biol* 34: 809–814.

Langley KR et al. 1994. Mechanical and optical assessment of the ripening of tomato fruit with reduced polygalacturonase activity. *J Sci Food Agric* 66: 547–554.

Liners F et al. 1992. Influence of the degree of polymerization of oligogalacturonates and of esterification pattern of pectin on their recognition by monoclonal antibodies. *Plant Plysiology* 99: 1099–1104.

Luh BS, Daoud HN. 1971. Effect of break temperature and holding time on pectin and pectic enzymes in tomato pulp. *J Food Sci* 36: 1039–1043.

Marín-Rodríguez MC et al. 2002. Pectate lyases, cell wall degradation and fruit softening. *J Exp Bot* 53: 2115–2119.

Marín-Rodríguez MC et al. 2003. Pectate lyase gene expression and enzyme activity in ripening banana fruit. *Plant Mol Biol* 51: 851–857.

Marsh GL et al. 1978. Effect of degree of concentration and of heat treatment on consistency of tomato pastes after dilution. *J Food Process Pres* 1: 340–346.

Matas AJ et al. 2009. Biology and genetic engineering of fruit maturation for enhanced quality and shelf-life. *Curr Opin Biotechnol* 20: 197–203.

Mayans O et al. 1997. Two crystal structures of pectin lyase A from *Aspergillus niger* reveal a pH driven conformational change and striking divergence in the substrate-binding clefts of pectin and pectate lyases. *Structure* 5: 677–689.

Medina-Escobar N et al. 1997. Cloning molecular characterisation and expression pattern of a strawberry ripening-specific cDNA with sequence homology to pectate lyase from higher plants. *Plant Mol Biol* 34: 867–877.

Milioni D et al. 2001. Differential expression of cell-wall-related genes during the formation of tracheary elements in the *Zinnia* mesophyll cell system. *Plant Mol Biol* 47: 221–238.

Mohnen D. 2008. Pectin structure and biosynthesis. *Curr Opin Plant Biol* 11: 266–277.

Moore T, Bennett AB. 1994. Tomato fruit polygalacturonase isozyme 1. *Plant Physiology* 106: 1461–1469

Pelloux J et al. 2007. New insights into pectin methylesterase structure and function. *Trends Plant Sci* 12: 267–277.

Pilatzke-Wunderlich I, Nessler CL. 2001. Expression and activity of cell-wall-degrading enzymes in the latex of opium poppy, *Papaver somniferum* L. *Plant Mol Biol* 45: 567–576.

Pilling J et al. 2000. Expression of a *Petunia inflata* pectin methyl esterase in *Solanum tuberosum* L. enhances stem elongation and modifies cation distribution. *Planta* 210: 391–399.

Pogson BJ et al. 1991. On the occurrence and structure of subunits of endopolygalacturonase isoforms in mature-green and ripening tomato fruits. *Aus J Plant Physiol* 18: 65–79.

Powell ALT et al. 2003. Simultaneous transgenic suppression of LePG and LeExp1 influences fruit texture and juice viscosity in a fresh-market tomato variety. *J Agric Food Chem* 51: 7450–7455.

Pua E-C et al. 2001. Isolation and expression of two pectate lyase genes during fruit ripening of banana (*Musa acuminata*). *Physiol Plant* 113: 92–99.

Ramakrishna W et al. 2003. A novel small heat shock protein gene, *vis1*, contributes to pectin depolymerization and juice viscosity in tomato fruit. *Plant Physiol* 131: 725–735.

Rosales MA et al. 2009. Environmental conditions affect pectin solubilization in cherry tomato fruits grown in two experimental Mediterranean greenhouses. *Environ Exp Bot* 67: 320–327.

Rose JKC, Bennett AB. 1999. Cooperative disassembly of the cellulose–xyloglucan network of plant cell walls: parallels between cell expansion and fruit ripening. *Trends Plant Sci* 4: 176–183.

Rose JKC et al. 1997. Expression of a divergent expansin gene is fruit-specific and ripening-regulated. *Proc Natl Acad Sci USA* 94: 5955–5960.

Santiago-Doménech N et al. 2008. Antisense inhibition of a pectate lyase gene supports a role for pectin depolymerization in strawberry fruit softening. *J Exp Bot* 59: 2769–2779.

Sesmero R et al. 2009. Rheological characterisation of juices obtained from transgenic pectate lyase-silenced strawberry fruits. *Food Chem* 116: 226–232.

Seymour GB et al. 1990. Composition and structural features of cell wall polysaccharides from tomato fruits. *Phytochemistry* 29: 725–731.

Schuch W et al. 1991. Fruit quality characteristics of transgenic tomato fruit with altered polygalacturonase activity. *Hortscience* 26: 1517–1520.

Sheehy RE et al. 1987. Molecular characterization of tomato fruit polygalacturonase. *Mol Gen Genet* 208: 30–36.

Sheehy RE et al. 1988. Reduction of polygalacturonase activity in tomato fruit by antisense RNA. *Proc Natl Acad Sci USA* 85: 8805–8809.

Shook CM et al. 2001. Polygalacturonase, pectinmethylesterase, and lipoxygenase activities in high-pressure-processed diced tomatoes. *J Agric Food Chem* 49: 664–668.

Sitrit Y, Bennett AB. 1998. Regulation of tomato fruit polygalacturonase mRNA accumulation by ethylene: a re-examination. *Plant Physiol* 116: 1145–1150.

Smith CJS et al. 1988. Antisense RNA inhibition of polygalacturonase gene expression in transgenic tomatoes. *Nature* 334: 724–726.

Smith DL et al. 1998. A gene coding for tomato fruit β-galactosidase II is expressed during fruit ripening. *Plant Physiol* 117: 417–423.

Smith DL, Gross KC. 2000. A family of at least seven β-galactosidase genes is expressed during tomato fruit development. *Plant Physiol* 123: 1173–1183.

Smith DL et al. 2002. Down-regulation of tomato beta-galactosidase 4 results in decreased fruit softening. *Plant Physiol* 129: 1755–1762.

Stoforos NG et al. 2002. Kinetics of tomato pectin methylesterase inactivation by temperature and high pressure. *J Food Sci* 67: 1026–1031.

Taniguchi Y et al. 1995. *Cry j* I, a major allergen of Japanese cedar pollen, has a pectate lyase enzyme activity. *Allergy* 50: 90–93.

Thakur BR et al. 1996a. Effect of an antisense pectin methylesterase gene on the chemistry of pectin in tomato (*Lycopersicon esculentum*) juice. *J Agric Food Chem* 44: 628–630.

Thakur BR et al. 1996b. Tomato product quality from transgenic fruits with reduced pectin methylesterase. *J Food Sci* 61: 85–108.

Thompson DS. 2005. How do cell walls regulate plant growth? *J Exp Bot* 56: 2275–2285.

Thompson JE, Fry SC. 2000. Evidence for covalent linkage between xyloglucan and acidic pectins in suspension-cultured rose cells. *Planta* 211: 275–286.

Tieman DM, Handa AK. 1994. Reduction in pectin methylesterase activity modifies tissue integrity and cation levels in ripening tomato (*Lycopersicon esculentum* Mill.) fruits. *Plant Physiol* 106: 429–436.

Tieman DM et al. 1992. An antisence pectin methylesterase gene alters pectin chemistry and soluble solids in tomato fruit. *Plant Cell* 4: 667–679.

Tomassen MMM et al. 2007. Isolation and characterization of a tomato non-specific lipid transfer protein involved in polygalacturonase-mediated pectin degradation. *J Exp Bot* 58: 1151–1160.

Turner LA et al. 1996. Isolation and nucleotide sequence of three tandemly arranged pectin methylesterase genes (Accession Nos. U70675, U70676 and U70677) from tomato. *Plant Physiol* 112: 1398.

Vicente AR et al. 2007. The linkage between cell wall metabolism and fruit softening: looking to the future. *J Sci Food Agric* 87: 1435–1448

Vincken J-P et al. 2003. If homogalacturonan were a side chain of rhamnogalacturonan I. Implications for cell wall architecture. *Plant Physiol* 132: 1781–1789.

Vrebalov J et al. 2009. Fleshy fruit expansion and ripening are regulated by the tomato SHATTERPROOF gene *TAGL1*. *Plant Cell* 21: 3041–3062.

Wakabayashi K et al. 2003. Methyl de-esterification as a major factor regulating the extent of pectin depolymerization during fruit ripening: a comparison of the action of avocado (*Persea americana*) and tomato (*Lycopersicon esculentum*) polygalacturonases. *J Plant Physiol* 160: 667–673.

Watson CF et al. 1994. Reduction of tomato polygalacturonase b-subunit expression affects pectin solubilization and degradation during fruit ripening. *Plant Cell* 6: 1623–1634.

Whittenberger RT, Nutting GC. 1957. Effect of tomato cell structures on consistency of tomato juice. *Food Technol* 11: 19–22.

Willats WGT et al. 2004. A xylogalacturonan epitope is specifically associated with plant cell detachment. *Planta* 218: 673–681.

Wing RA et al. 1989. Molecular and genetic characterization of two pollen-expressed genes that have sequence similarity to pectate lyases of the plant pathogen *Erwinia*. *Plant Mol Biol* 14: 17–28.

Wong D. 2008. Enzymatic deconstruction of backbone structures of the ramified regions in pectins. *Protein J* 27: 30–42.

Wu Q et al. 1993. En-dopolygalacturonase in apples (*Malus domestica*) and its expression during fruit ripening. *Plant Physiol* 102: 219–225.

Wu Y et al. 1996. PO149 a new member of pollen pectate lyase-like gene family from alfalfa. *Plant Mol Biol* 32: 1037–1042.

Xu S-Y et al. 1986. Effect of break temperature on rheological properties and microstructure of tomato juices and pastes. *J Food Sci* 51: 399–407.

Yadav S et al. 2009. Pectin lyase: a review. *Process Biochem* 44: 1–10.

Yoder MD et al. 1993. New domain motif: the structure of pectate lyase C, a secreted plant virulence factor. *Science* 260: 1503–1507.

Youssef SM et al. 2009. Fruit yield and quality of strawberry plants transformed with a fruit specific strawberry pectate lyase gene. *Sci Hor* 119: 120–125.

Zheng L et al. 1992. The β-subunit of tomato fruit polygalacturonase 1: isolation characterization, and identification of unique structural features. *Plant Cell* 4: 1147–1156.

Zheng L et al. 1994. Differential expression of the two subunits of tomato polygalacturonase isoenzyme 1 in wild-type and *rin* tomato fruit. *Plant Physiol* 105: 1189–1195.

Zykwinska AW et al. 2005. Evidence for in vitro binding of pectin side chains to cellulose. *Plant Physiol* 139: 397–407.

13
Seafood Enzymes

M. K. Nielsen and H. H. Nielsen

INTRODUCTION

"Freshness" serves as a basic parameter of quality to a greater extent in seafood than in muscle foods of other types. The attention directed at the freshness of seafood and seafood products is appropriate since certain changes in seafood occur soon after capture and slaughter, often leading to an initial loss in prime quality. Such changes are not necessarily related to microbial spoilage, but may be due to the activity of endogenous enzymes. However, not all postmortem endogenous enzyme activities in seafood lead to a loss in quality; many autolytic changes are important for obtaining the desired texture and taste.

Despite the biological differences between aquatic and terrestrial animals, the biochemical reactions of most animals are identical and are catalyzed by homologous enzymes; only few reactions are unique for certain species. Few reactions are catalyzed by analog enzymes that differ in structure and evolutionary origin. The aquatic habitat can be extreme in various respects including temperature, hydrostatic pressure, and osmotic pressure, compared with the environment of terrestrial animals. These all conditions have thermodynamic implications and can affect the kinetics and stability of marine enzymes. Also, because of the large biological diversity of marine animals and such seasonally determined factors as spawning and the availability of food, seafood species can differ markedly in enzyme activity.

Seafood is a gastronomic term that in principle includes all edible aquatic organisms. In this chapter, it is mainly the "classical" seafood groups—fish, shellfish, and molluscs—that are dealt with. Muscle enzymes are mostly considered, since muscle is the organ of which most seafood products primarily consist. Various other types of seafood enzymes, such as those used in

Food Biochemistry and Food Processing, Second Edition. Edited by Benjamin K. Simpson, Leo M.L. Nollet, Fidel Toldrá, Soottawat Benjakul, Gopinadhan Paliyath and Y.H. Hui.

digestion, are only dealt with to a lesser extent. Further information on seafood enzymes can be found in the book *Seafood Enzymes*, edited by Haard and Simpson (2000), and the references therein.

Cold Adaptation of Enzymes

Most seafood species are poikilotherm and are thus forced to adapt to the temperature of their habitat, which may range from the freezing point of oceanic seawater, of approximately −1.9°C, to temperatures similar to those of the warm-blooded species. Since most of the aquatic environment has a temperature of below 5°C, the enzyme systems of seafood species are often adapted to low temperatures. Although all enzyme reaction rates are positively correlated with temperature, the enzyme systems of most cold-adapted species exhibit, at low temperatures, catalytic activities similar to those that homologous mesophilic enzymes of warm-blooded animals exhibit at higher in vivo temperatures. This is primarily due to the higher turnover number (k_{cat}) of the cold-adapted enzymes; substrate affinity, estimated by the Michaelis-Menten constant (K_M), and other regulatory mechanisms are generally conserved (Somero 2003).

Several studies reviewed by Simpson and Haard (1987) have found the level of activity of cold-adapted marine enzymes to be higher than that of homologous bovine enzymes; they have also shown marine enzymes to be much lower in thermal stability. Although the attempt has been made to explain the temperature adaptation of the enzymes of marine species in terms of molecular flexibility in a general sense, studies of the molecular structure of such enzymes have not revealed any common structural features that can provide a universal explanation of temperature adaptation. Rather, there appear to be various minor structural adjustments of varying character that provide the necessary flexibility of the protein structure for achieving cold adaption (Smalås et al. 2000, Fields 2001, de Backer et al. 2002).

Since the habitat temperature of most marine animals used as seafood tends to be near to the typical cold storage temperature of approximately 0–5°C, the enzymatic reaction rates are not thermally depressed at storage temperatures as they usually would be in the meat of warm-blooded animals. With a habitat temperature of seafood of about 5°C and Q_{10} values in the normal range of 1.5–2.0, the enzyme reactions during ice storage can thus be expected to proceed at rates four to nine times as high as in warm-blooded animals under similar conditions.

Effects of Pressure on Enzyme Reactions

Although the oceans, with their extreme depth, represent the largest habitat on earth, it is often forgotten that hydrostatic pressure is an important biological variable, affecting the equilibrium constants of all reactions involving changes in volume. Regarding enzyme reactions, pressure affects the molecular stability of enzymes and the intermolecular interactions between enzymes and substrates, cofactors, and macromolecules. Therefore, pressure alterations may change both the concentration of catalytically active enzymes and their kinetic parameters, such as the Michaelis-Menten constant, K_M.

Siebenaller and Somero (1978) showed that the apparent K_M values of a lactate dehydrogenase (LDH) enzyme from a deep-sea fish species are less sensitive to pressure changes than its homologous enzyme from a closely related fish species living in shallow water. Similar relations between the pressure sensitivity of LDH and habitat pressure have been found for the LDH of other marine species (Siebenaller and Somero 1979, Dahlhoff and Somero 1991). In this case, only the LDH substrate interaction was found to be affected by pressure, whereas the turnover number appeared to be unaffected by pressure.

Very high pressures, of 10^8 Pa (ca. 1000 bar), corresponding to the pressure in the deepest trenches on earth can lead to denaturization of proteins. Some deep-sea fish have been found to accumulate particularly high levels of trimethylamine-*N*-oxide (TMAO). TMAO stabilizes most protein structures and may counteract the effect of pressure on proteins. Adaptation to high pressure thus seems to be achieved both by evolution of pressure-resistant proteins and by adjustments of the intracellular environment.

The metabolic rates of deep-sea fish tend to be lower than those of related species living in shallow waters. Several lines of evidence suggest that the decrease in metabolic rate with increasing depth is largely due to the food consumption levels being lower and to a reduced need for locomotor capability in the darkness (Childress 1995, Gibbs 1997). The latter makes intuitive sense if one considers the often far from streamlined, and sometimes bizarre, anatomy of deep-sea fish species.

Osmotic Adaptation of Enzymes

Marine teleosts (bonefish) are hypoosmotic, having intracellular solute concentrations that are similar to those of terrestrial vertebrates. By contrast, elasmobranches (sharks and rays) are isoosmotic, employing the organic compounds urea and TMAO as osmolytes. Although the concentration of osmolytes can be high, the stability and functionality of enzymes remain mostly unaffected due to the compatibility and balanced counteracting effects of the osmolytes (Yancey et al. 1982).

ENZYMATIC REACTIONS IN THE ENERGY METABOLISM OF SEAFOOD

Most key metabolic pathways such as the glycolysis are found in all animal species, and the majority of animal enzymes are thus homologous. Nevertheless, the energy metabolism of seafood species shows a number of differences from that of terrestrial animals, and these differences contribute to the characteristics of seafood products.

The red and white fish muscle cells are anatomically separated into minor portions of red muscle and major portions of white muscle tissues (see Fig. 13.2 in the Section Postmortem Proteolysis in Fresh Fish). The energy metabolism and the histology of the two muscle tissues differ, the anaerobic white muscle having only few mitochondria and blood vessels. When blood circulation stops after death, the primary effect on the white muscle metabolism is therefore not the seizure of oxygen supply but the accumulation of waste products that are no longer removed.

Adenosine-5′-triphosphate (ATP) is the predominant nucleotide in the muscle cells of rested fish and represents a readily available source of metabolic energy. It regulates numerous biochemical processes and continues to do so during postmortem processes. Both the formation and degradation of ATP continue postmortem, the net ATP level being the result of the difference between the rates of the two processes. In the muscle of rested fish, the ATP content averages 7–10 μmol/g tissue (Cappeln et al. 1999, Gill 2000). This concentration is low in comparison with the turnover of ATP, and the ATP pool would thus be quickly depleted if not continuously replenished.

Postmortem Generation of ATP

In anaerobic tissues such as postmortem muscle tissue, ATP can be generated from glycogen by glycolysis. Smaller amounts of ATP may also be generated by other substances, including creatine or arginine phosphate, glucose, and adenosine diphosphate (ADP).

Creatine phosphate provides an energy store that can be converted into ATP by a reversible transfer of a high-energy phosphoryl group to ADP catalyzed by sarcoplasmic creatine kinase Enzyme Commision number (EC 2.7.3.2). The creatine phosphate in fish is usually depleted within hours after death (Chiba et al. 1991). In many invertebrates, including shellfish, creatine phosphate is often substituted by arginine phosphate, and ATP is thus generated by arginine kinase (EC 2.7.3.3) (Ellington 2001, Takeuchi 2004).

Glucose can enter glycolysis after being phosphorylated by hexokinase (EC 2.7.1.1). Although glucose represents an important energy resource in living animals, it is in scarce supply in postmortem muscle tissue due to the lack of blood circulation, and it contributes only little to the postmortem formation of ATP.

ATP is also formed by the enzyme adenylate kinase (EC 2.7.4.3), which catalyzes the transfer of a phosphoryl group between two molecules of ADP, producing ATP and adenosine monophosphate (AMP). As AMP is formed, adenylate kinase parti-cipates also in the degradation of adenine nucleotides, as will be described later.

The glycogen deposits in fish muscle are generally smaller than the glycogen stores in the muscle of rested mammals. Nevertheless, muscle glycogen is normally responsible for most of the postmortem ATP formation in fish. The degradation of glycogen and the glycolytic formation of ATP and lactate proceed by pathways that are similar to those in mammals. Glycogen is degraded by glycogen phosphorylase (EC 2.4.1.1), producing glucose-1-phosphate. The highly regulated glycogen phosphorylase is inhibited by glucose-1-phosphate and glucose and is activated by AMP. Glucose-1-phosphate is subsequently transformed into glucose-6-phosphate by phosphoglucomutase (EC 5.4.2.2). Glucose-6-phosphate is further degraded by the glycolysis, leading to the formation of ATP, pyruvate, and Nicotineamide adenine dinucleotide (NADH). NADH originates from the reduction of NAD^+ during glycolysis. In the absence of oxygen, NAD^+ is regenerated by LDH (EC 1.1.1.27), as pyruvate is reduced to lactate. The formation of lactate forms a metabolic blind end and causes pH to fall, the final pH level being reached when the formation of lactate has stopped. The postmortem concentration of lactate reflects, by and large, the total degradation of ATP. Since in living fish, however, removal of lactate from the large white muscle often proceeds only slowly, most lactate formed just before death may remain in the muscle postmortem. Thus, the final pH level in the fish is affected only a little by the stress to which the fish was exposed and the struggle it went through during capture (Foegeding et al. 1996).

The accumulation of lactate in fish is limited by the amount of glycogen in the muscle. Since glycogen is present in fish muscle at only low concentrations, the final pH level of the fish meat is high in comparison to that of other types of meat, typically 6.5–6.7. In species that have more muscular glycogen, this substance rarely limits the formation of lactate, which may continue until the catalysis stops for other reasons, at a pH of approximately 5.5–5.6 in beef (Hultin 1984). The final pH depends not only on the lactate concentration, but also on the pH buffer capacity of the tissue. In the neutral and slightly acidic pH range, the buffer capacity of the muscle tissue in most animals, including fish, originates primarily from histidyl amino acid residues and phosphate groups (Somero 1981, Okuma and Abe 1992). Some fish species, in order to decrease muscular acidosis, have evolved particularly high pH buffer capacities. For example, pelagic fish belonging to the tuna family (genus *Thunnus*) possess a very high glycolytic capacity (Storey 1992, and references therein) and also accumulate high concentrations of histidine and anserine (*N*-(β-alanyl)-*N*-methylhistidine) (van Waarde 1988, Dickson 1995). Since tuna and a few other fish species can effectively retain metabolic heat, their core temperature at the time of capture may significantly exceed the water temperature. The high muscle temperature of such species expectedly accelerates autolytic processes, including proteolysis, and may, in combination with the lowering of pH, promote protein denaturation (Haard 2002).

Some species of carp, including goldfish, are able to survive for several months without oxygen. The metabolic adaptation this requires includes substituting ethanol for lactate as the major end product of glycolysis. Ethanol has the advantage over lactate that it can be easily excreted by the gills and does not lead to acidosis (Shoubridge and Hochachka 1980). Under anoxic conditions, NAD^+ is thus regenerated by muscular alcohol dehydrogenase (EC 1.1.1.1).

Postmortem Degradation of Nucleotides

ATP can be metabolized by various ATPases. In muscle tissues, most of the ATPase activity is represented by Ca^{2+}-dependent myosin ATPase (EC 3.6.1.32) that is directly involved in contraction of the myofibrils. Thus, release of Ca^{2+} from its containment in the sarcoplasmic reticulum dramatically increases the rate of ATP breakdown.

The glycogen consumption during and after slaughter is highly dependent on the fishing technique and the slaughtering process employed. In relaxed muscle cells, the sarcoplasmic Ca^{2+} levels are low, but upon activation by the neuromuscular junctions, Ca^{2+} is released from the sarcoplasmic reticulum, activating myosin ATPase. Since the muscular work performed during capture is often intensive, the muscular glycogen stores may be almost depleted even before slaughtering takes place. The activation of myosin ATPase by spinally mediated reflexes continues after slaughter, but the extent to which it occurs can be reduced by employing rested harvest techniques such as anesthetization (Jerrett and Holland 1998, Robb et al. 2000); by destroying the nervous system by use of appropriate methods such as *iki jime,* where the brain is destroyed by a spike (Lowe 1993), or by use of still other techniques (Chiba et al. 1991, Berg et al. 1997).

ATPases hydrolyze the terminal phosphate ester bond, forming ADP. ADP is subsequently degraded by adenylate kinase, forming AMP. The degradation of ADP is favored by the removal of AMP by AMP deaminase (EC 3.5.4.6), producing inosine monophosphate (IMP) and ammonia (NH_3). AMP deaminase is a key regulator of the intracellular adenine nucleotide pool and as such is a highly regulated enzyme inhibited by inorganic phosphate, IMP and NH_3 and activated by ATP.

It is widely accepted that IMP contributes to the desirable taste of fresh seafood (Fluke 1994, Gill 2000, Haard 2002). IMP is present in small amounts in freshly caught relaxed fish but builds up during the depletion of ATP (Berg et al. 1997). IMP can be hydrolyzed to form inosine, yet the rate of hydrolysis in fish is generally low, following zero-order kinetics, indicating that the IMP-degrading enzymes are fully substrate saturated (Tomioka et al. 1987, Gill 2000, Itoh and Kimura 2002). However, a large species variation in the degradation of IMP has been shown, and thus IMP may persist during storage for several days or even weeks (Dingle and Hines 1971, Gill 2000, Haard 2002). The formation of inosine is often rate limiting for the overall breakdown of nucleotides and is considered to constitute the last purely autolytic step in the nucleotide catabolism of chilled fish. The formation of inosine usually indicates a decline in the prime quality of seafood (Surette et al. 1988, Gill 2000, Haard 2002). IMP can be hydrolyzed by various enzymes, 5'-nucleotidase (EC.3.1.3.5) being regarded as the most important of these in the case of chilled fish (Gill 2000, Haard 2002). Several forms of 5'-nucleotidase exist, of which a soluble sarcoplasmic form has been documented in fish (Itoh and Kimura 2002).

The degradation of inosine continues via hypoxanthine and its oxidized products xanthine and uric acid. These reactions, all of them related to spoilage, are catalyzed by both endogenous and microbial enzymes, the activity of the latter depending upon the spoilage flora that are present and thus differing much between species and products. Inosine can be degraded by either nucleoside phosphorylase (EC 2.4.2.1) or inosine nucleosidase (EC 3.2.2.2), leading to the formation of different by-products (see Fig. 13.1). The final two-step oxidation of hypoxanthine is catalyzed by one and the same enzymes, xanthine oxidase (EC 1.1.3.22). The oxidation requires a reintroduction of oxygen in the otherwise anaerobic pathway, which often limits the extent of the xanthine oxidase reaction and causes inosine and hypoxanthine to accumulate. Hypoxanthine is an indicator of spoilage and may contribute to bitter taste (Hughes and Jones 1966). The hydrogen peroxide produced by the oxidation of hypoxanthine can be expected to cause oxidation, but the technological

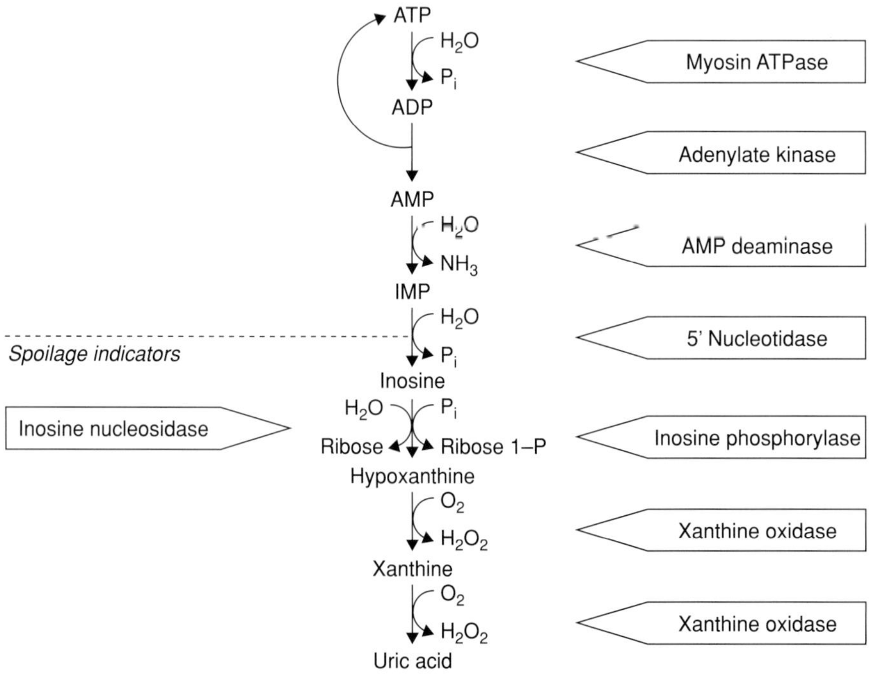

Figure 13.1. The successive reactions of the main pathway for nucleotide degradation in fish. Substances listed below the dotted line are generally considered to be indicators of spoilage. P_i is inorganic phosphate.

implications of this are minor, since it is only after the quality of the fish has declined beyond the level of acceptability that these reactions take place on a large scale.

The relation between the accumulation of breakdown products of ATP and the postmortem storage period of seafood products has resulted in various freshness indicators being defined. Saito et al. (1959) defined a freshness indicator K as a simplified way of expressing the state of nucleotide degradation by a single number:

$$K = 100 \times \frac{[\text{Inosine}] + [\text{Hypoxanthine}]}{\left(\begin{array}{c}[\text{ATP}] + [\text{ADP}] + [\text{AMP}] + [\text{IMP}] \\ + [\text{Inosine}] + [\text{Hypoxanthine}]\end{array}\right)}$$

Higher K values correspond to a lower quality of fish. The K value often increases linearly during the first period of storage on ice, but the K value at sensory rejection differs between species. An alternative and simpler K_i value correlates well with the K value of wild stock fish landed by traditional methods (Karube et al. 1984):

$$K_i = 100 \times \frac{[\text{Inosine}] + [\text{Hypoxanthine}]}{[\text{IMP}] + [\text{Inosine}] + [\text{Hypoxanthine}]}$$

Both K values reflect fish quality well, provided the numerical values involved are not compared directly with the values from other species. Still, K values necessarily remain less descriptive than the concentrations of the degradation products themselves.

ATP is an important regulator of biochemical processes in all animals and continues to be so during the postmortem processes. One of the most striking postmortem changes is that of rigor mortis. When the concentration of ATP is low, the contractile filaments lock into each other and cause the otherwise soft and elastic muscle tissue to stiffen. Rigor mortis is thus a direct consequence of ATP depletion. The biochemical basis and physical aspects of rigor mortis are reviewed by Hultin (1984) and Foegeding et al. (1996).

ATP depletion and the onset of rigor mortis in fish are highly correlated with the residual glycogen content at the time of death and with the rate of ATP depletion. The onset of rigor mortis may occur only a few minutes after death or may be postponed for up to several days when rested harvest techniques, gentle handling, and optimal storage conditions are employed (Azam et al. 1990, Lowe et al. 1993, Skjervold et al. 2001). The strength of rigor mortis is greater when it starts soon after death (Berg et al. 1997). Although the recommended temperature for chilled storage is generally close to 0°C, in some fish from tropical and subtropical waters, the degradation of nucleotides has been found to be slower, and the onset of rigor mortis to be further delayed when the storage temperature is elevated to 5°C, 10°C, and even 20°C (Saito et al. 1959, Iwamoto et al. 1987, 1991).

Rigor mortis impedes filleting and processing of fish. In the traditional fishing industry, therefore, filleting is often postponed until rigor mortis has been resolved. In aquaculture, however, it has been shown that rested harvesting techniques can postpone rigor mortis sufficiently to allow prerigor filleting to be carried out (Skjervold et al. 2001). Fillets from prerigor filleted fish shorten, in accordance with the rigor contraction of the muscle fibers, by as much as 8%. Prerigor filleting has few problems with gaping, since the rigor contraction of the fillet is not hindered by the backbone (Skjervold et al. 2001). Gaping, which represents the formation of fractures between segments of the fillets, is described further in the Section Postmortem Proteolysis in Fresh Fish.

All enzyme reactions are virtually stopped during freezing at low temperatures. The ATP content of frozen prerigor fish can thus be stabilized. During the subsequent thawing of prerigor fish muscle, the leakage of Ca^{2+} from the organelles results in a very high level of myosin Ca^{2+}–ATPase activity and a rapid consumption of ATP. This leads to a strong form of rigor mortis called thaw rigor. Thaw rigor results in an increase in drip loss, in flavor changes, and in a dry and tough texture (Jones 1965) and gaping (Jones 1969). Thaw rigor can be avoided by controlled thawing, holding the frozen products at intermediate freezing temperatures above −20°C for a period of time (Mcdonald and Jones 1976, Cappeln et al. 1999). During this holding period, ATP is degraded at moderate rates, allowing a slow onset of rigor mortis in the partially frozen state.

Rigor mortis is a temporary condition, even though in the absence of ATP, the actin-myosin complex is locked. The resolution of rigor mortis is due to structural decay elsewhere in the muscular structure, as will be discussed later.

ENZYMATIC DEGRADATION OF TRIMETHYLAMINE-*N*-OXIDE

The substance TMAO is found in all marine seafood species but occurs in some freshwater fish as well (Anthoni et al. 1990, Parab and Rao 1984, Niizeki et al. 2002). It contributes to cellular osmotic pressure, as previously described, but several other physiological functions of TMAO have also been suggested. TMAO itself is a harmless and nontoxic constituent, yet it forms a precursor of undesirable breakdown products. In seafood products, TMAO can be degraded enzymatically by two alternative pathways as described later.

THE TRIMETHYLAMINE-*N*-OXIDE REDUCTASE REACTION

Many common spoilage bacteria reduce TMAO to trimethylamine (TMA) by means of the enzyme TMAO reductase (EC 1.6.6.9):

$$\begin{array}{c}CH_3 \\ | \\ O=N-CH_3 \\ | \\ CH_3\end{array} + NADH \rightarrow \begin{array}{c}CH_3 \\ | \\ N-CH_3 \\ | \\ CH_3\end{array} + NAD^+ + H_2O$$

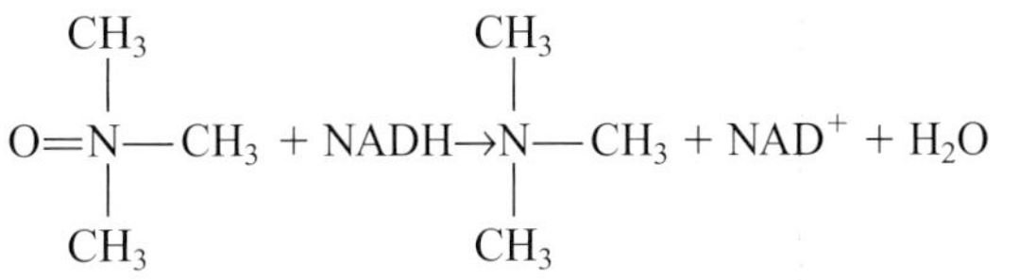

TMA has a strong fishy odor, and TMAO reductase activity is responsible for the typical off-odor of spoiled fish. Since TMAO reductase is of microbial origin, the formation of TMA occurs primarily under conditions such as cold storage that allow microbial growth to take place. TMA is thus an important spoilage indicator of fresh seafood products. The formation of

TMA is further discussed in the food microbiology literature (e.g., Barrett and Kwan 1985, Dalgaard 2000).

The Trimethylamine-*N*-Oxide Aldolase Reaction

TMAO is also the precursor of the formation of dimethylamine (DMA) and formaldehyde:

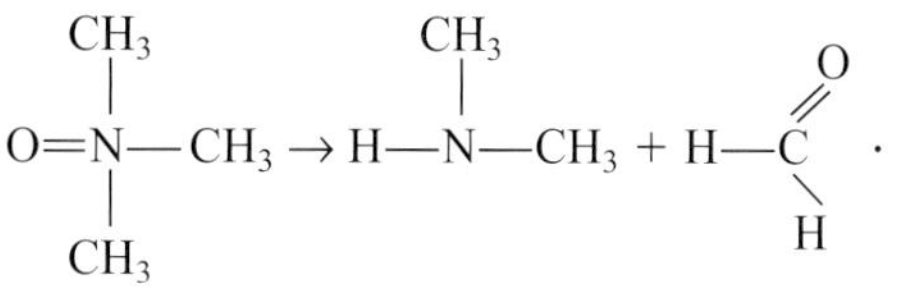

This reaction is catalyzed by trimethylamine-*N*-oxide aldolase (TMAOase or TMAO demethylase, EC 4.1.2.32) but may also to some extent be nonenzymatic, catalyzed by iron and various reductants (Vaisey 1956, Spinelli and Koury 1981, Nitisewojo and Hultin 1986, Kimura et al. 2002). In most cases in which significant amounts of formaldehyde and DMA are accumulated, however, species possessing TMAOase enzyme activity are involved. The reaction leads to cleavage of a C-N bond and the elimination of an aldehyde, resulting in the classification of TMAOase as a lyase (EC 4.1.2.32) and the International Union of Biochemistry and Molecular Biology name aldolase.

DMA is a reactive secondary amine with a milder odor than TMA. Formaldehyde is highly reactive and strongly affects the texture of fish meat by making it tougher, harder, more fibrous, and less juicy, as well as increasing the drip loss in the thawed products. The quality changes are associated with a loss in protein solubility and in particular the solubility of the myofibrillar proteins. For reviews, see Sikorski and Kostuch (1982), Hultin (1992), Mackie (1993), Sikorski and Kolakowska (1994), and Sotelo et al. (1995).

Formaldehyde can react with a number of chemical groups, including some protein amino acid residues and terminal amino groups, resulting in denaturation and possibly in the cross-linking of proteins, both of which are believed to be the cause of the observed effects on seafood products. The formaldehyde concentration in severely damaged seafood products may reach 240 μg/g (Nielsen and Jørgensen 2004). These concentrations are generally considered nontoxic, but may still exceed various national trade barrier limits.

Whereas TMAO is widespread, TMAOase is only found in a limited number of animals, many of which belong to the order of gadiform fish (pollock, cod, etc.). In gadiform species, the highest levels of enzyme activity are found in the inner organs (kidney, spleen, and intestine), while the enzyme activity in the large white muscle is low. The formation of DMA and formaldehyde in various tissues of certain nongadiform fish, as well as of crustaceans and mollusks, has also been reported, as reviewed by Sotelo and Rehbein (2000), although the taxonomic distribution of the enzyme has not been investigated systematically. The TMAOase content of gadiform fish exhibits large individual variation, probably due to the influence of biological factors that have not yet been adequately studied (Nielsen and Jørgensen 2004).

The enzyme is stable and tolerates both high salt concentrations and freezing. The accumulation of DMA and formaldehyde progresses only slowly, and most of the accumulation is produced during prolonged frozen storage or cold storage of the salted fish (e.g., salted cod and bacalao). During freezing, the enzymatic reaction proceeds as long as liquid water and substrate are available. In practice, the formation of DMA and formaldehyde is found at temperatures down to approximately −30°C (Sotelo and Rehbein 2000). At higher temperatures, the rate of formation of formaldehyde is also higher; thus, freezing of gadiform fish at insufficiently low temperatures may result in dramatic changes in the solubility of myofibrillar proteins within a few weeks (Nielsen and Jørgensen 2004).

Despite the TMAOase concentration of the white muscle being low, it is the enzyme activity of the white muscle that is responsible for the accumulation of formaldehyde in whole fish and in fillets (Nielsen and Jørgensen 2004). In the case of minced products, even minor contamination by TMAOase-rich tissues leads to a marked rise in the rate of accumulation (Dingle et al. 1977, Lundstrøm et al. 1982, Rehbein 1988, Rehbein et al. 1997). Despite its dependency upon other factors, such as cofactors, the rate of accumulation of formaldehyde can to a large extent be predicted from the TMAOase enzyme activity of fish meat alone (Nielsen and Jørgensen 2004).

The physiological function of TMAOase remains unknown. The formation of formaldehyde could have a digestive function, yet this would not explain the high TMAOase content of the kidney and spleen. Accordingly, it has been speculated that TMAO may not even be its primary natural substrate (Sotelo and Rehbein 2000).

The in vitro reaction rate is low without the presence of a number of redox-active cofactors. Although the overall TMAOase reaction is not a redox reaction, its dependency on redox agents shows that the reaction mechanism must include redox steps. Little is known about the in vivo regulation of TMAOase, but studies suggest the importance of nucleotide coenzymes and iron (Hultin 1992). TMAOase appears to be membrane bound and is only partly soluble in aqueous solutions. This has complications for its purification and is one of the main reasons why the complete characterization of the enzyme remains yet to be done.

POSTMORTEM PROTEOLYSIS IN FRESH FISH

Postmortem proteolysis is an important factor in many changes in seafood quality. During cold storage, the postmortem proteolysis of myofibrillar and connective tissue proteins contributes to deterioration in texture. Despite the natural tenderness of seafood, texture is an important quality parameter, in fish and shellfish alike. Deterioration in seafood quality through proteolysis involves a softening of the muscle tissue. This reduces the cohesiveness of the muscle segments in fillets, which promotes gaping, a formation of gaps and slits between muscle segments. The negative character of this effect contrasts with the effect of postmortem proteolysis on the meat of cattle and

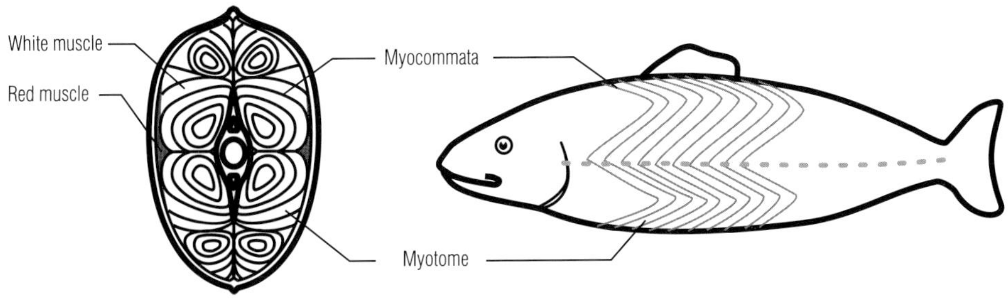

Figure 13.2. Gross anatomy of fish muscle. (Drawing courtesy of A. S. Matforsk, Norwegian Food Research Institute.)

pigs, in which the degradation of myofibrillar proteins produces a highly desired tenderization in the conversion of muscle to meat.

The proteolytic enzymes that cause a softening of seafood are of basically the same classes as those found in terrestrial animals. The special effects of proteolysis on seafood result from the combined action of the homologous proteolytic enzymes on the characteristic muscle structures of seafood. The muscle of fish differs on a macrostructural level from that of mammals, fish muscle being segmented into muscle blocks called myotomes. A myotome consists of a single layer of muscle fibers arranged side by side and separated from the muscle fibers of the adjacent myotomes by collagenous sheets termed the myocommata (Bremner and Hallett 1985). See Figure 13.2.

A myotome resembles a mammalian skeletal muscle in terms of its intracellular structure (see Fig. 13.3) and the composition of the extracellular matrix. Each muscle fiber is surrounded by fine collagen fibers, the endomysium, joined with a larger network of collagen fibers, the perimysium, which is contiguous with the myocommata (Bremner and Hallet 1985). Although the endomysium, perimysium, and myocommata are considered to be discrete areas of the extracellular matrix, they join to form a single weave.

Such phenomena in seafood as softening, gaping, and the resolution of rigor are believed to be caused by hydrolysis of the myofibrillar and the extracellular matrix proteins. Initially, the disintegration of the attachment between the myocommata and muscle fibers leads to the resolution of rigor (Taylor et al. 2002). Endogenous proteolytic enzymes that cleave muscle proteins under physiological conditions and at neutral pH can be a factor in the resolution of rigor mortis. The continuation of this process is regarded as one of the main causes of gaping (Taylor et al. 2002, Fletcher et al. 1997).

The softening of fish muscle appears to be the result of multiple changes in muscle structure. Histological studies have shown that the attachment between muscle fibers in fish muscle is broken during ice storage and has been associated with a loss of "hardness" as measured by instrumental texture analysis (Taylor et al. 2002). During cold storage, the junction between the myofibrils and connective tissue of the myocommata is hydrolyzed (Bremner 1999, Fletcher et al. 1997, Taylor et al. 2002), and the collagen fibers of the perimysium surrounding the bundles of muscle fibers are degraded (Ando et al. 1995, Sato et al. 2002).

It appears that cleavage of the costameric proteins that link the myofibrils (the sarcomere) with the sarcolemma (the cell membrane), and of the basement membrane, which is attached

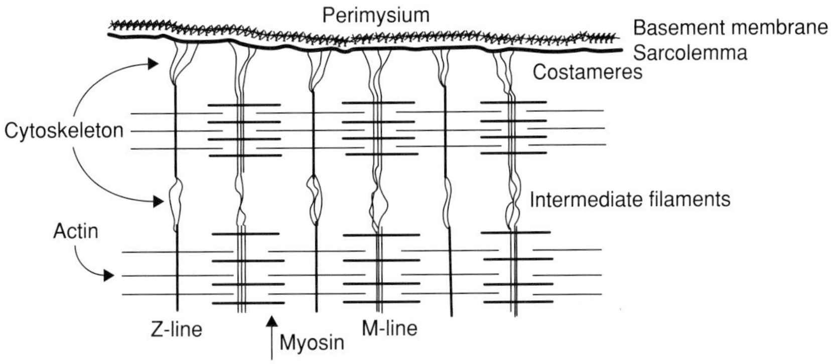

Figure 13.3. A longitudinal view of the intracellular structure of a skeletal muscle fiber showing the contractile elements (actin and myosin), the cytoskeleton, and the attachment to the extracellular matrix (endomysium and perimysium). (Adapted from Lødemel 2004.)

to the fine collagen fibers of the endomysium (see Fig. 13.3), leads to muscle fibers being detached, which results in softening. Studies have shown that the costameric proteins found in fish muscle are already degraded 24 hours postmortem (Papa et al. 1997). This emphasizes the importance of early postmortem changes.

It has been shown that changes in the collagen fraction of the extracellular matrix and the softening of fish are related. Collagen V, a minor constituent of the pericellular connective tissue of fish muscle, becomes soluble when rapid softening takes place, whereas no solubilization is observed in fish that do not soften (Sato et al. 2002). Recent studies also indicate a degradation of the most abundant collagen in fish muscle, collagen I, during 24 hours of cold storage (Shigemura et al. 2004). This indicates that the degradation of collagen and the resulting weakening of the intramuscular pericellular connective tissue also play a role in the textural changes in fish muscle that occur early postmortem.

A number of studies have revealed structural changes in the myofibrillar proteins in fish muscle during cold storage. Analysis of myofibrillar proteins from the muscle of salmon stored at 0°C for as long as 23 days has shown that several new protein fragments form (Lund and Nielsen 2001). Other studies suggest that predominantly proteins of the cytoskeletal network, such as the high molecular weight proteins titin and nebulin, are degraded (Busconi et al. 1989, Astier et al. 1991). The extent of degradation of the muscle myofibrillar proteins varies among species. It has been found, for example, that the intermediate filament protein desmin is clearly degraded during the cold storage of turbot and sardines but that no degradation occurs during the cold storage of sea bass and brown trout (Verrez-Bagnis et al. 1999).

As research shows, both myofibrillar and extracellular matrix proteins in the muscle of many fish are degraded during storage, and textural changes can be expected to result from degradation of proteins from both structures.

Proteases in Fish Muscle

The structural and biochemical changes just described can be considered to largely represent the concerted action of different endogenous proteolytic enzymes. Proteolytic enzymes of all major classes have been documented in the muscle of various fish species. An overview of the different proteases believed to play a role in the postmortem proteolysis of seafood is presented later. Further information on the proteases found in fish and marine invertebrates can be found in reviews by Kolodziejska and Sikorski (1995, 1996).

Matrix Metalloproteinases

Matrix metalloproteinases are extracellular enzymes involved in the in vivo catabolism (degradation) of the extracellular matrix (degradation of the helical regions of the collagens). They have been isolated from rainbow trout (Saito et al. 2000), Japanese flounder (Kinoshita et al. 2002) and Pacific rockfish (Bracho and Haard 1995). Collagenolytic and gelatinolytic activities have also been detected in the muscle of winter flounder (Teruel and Simpson 1995), yellowtail (Kubota et al. 1998), ayu (Kubota et al. 2000), salmon, and cod (Lødemel and Olsen 2003). Type I collagen is solubilized and degraded by matrix metalloproteinases in rainbow trout (Saito et al. 2000), Japanese flounder (Kubota et al. 2003), and Pacific rockfish (Bracho and Haard 1995), suggesting that these proteases can participate in postmortem textural changes.

Cathepsins

Lysosomal proteinases such as cathepsins B, D, and L have been isolated from a number of fish species, including herring (Nielsen and Nielsen 2001), mackerel (Aoki and Ueno 1997), and tilapia (Jiang et al. 1991). In vitro studies show that certain cathepsins are capable of cleaving myofibrillar proteins (Ogata et al. 1998, Nielsen and Nielsen 2001) and may also participate in degradation of the extracellular matrix since they can cleave both nonhelical regions of the collagen (Yamashita and Konagaya 1991) and collagen that has already been partly degraded by matrix metalloproteinases. Since cathepsins in the muscle of living fish are located in the lysosomes, they are not originally in direct contact with either the myofibrils or the extracellular matrix, although it has been shown that the enzymes leak from the lysosomes in fish muscle postmortem, and from lysosomes in bovine muscles postmortem as well (Geromel and Montgomery 1980, Ertbjerg et al. 1999).

Calpains

Calpains have been reported in the muscle of various fish species (Wang and Jiang 1991, Watson et al. 1992). They are only active in the neutral pH range, although research shows that they also remain active at the slightly acidic postmortem pH (Wang and Jiang 1991) and are capable of degrading myofibrillar proteins in vitro (Verrez-Bagnis et al. 2002, Geesink et al. 2000). This suggests a participation in the postmortem hydrolysis of fish muscle. Watson et al. (1992) found evidence for calpains being involved in the degradation of myofibrillar proteins in tuna, leading to the muscle becoming pale and grainy, which is referred to as "burnt tuna."

20S Proteasome

The 20S proteasome enzyme is a 700 kDa multicatalytic proteinase with three major catalytic sites. The term 20S refers to its sedimentation coefficient. In the eukaryotic cells, the enzyme exists in the cytoplasm, either in a free state or associated with large regulatory complexes. In vivo, it is involved in nonlysosomal proteolysis and apoptosis. Research strongly indicates that this proteasome is involved in meat tenderization in cattle (Sentandreu et al. 2002). Although proteasomes have been detected in the muscle of such fish species as carp (Kinoshita et al. 1990a), white croaker (Busconi et al. 1992), and salmon (Stoknes and Rustad 1995), their postmortem activity in fish muscle has not been clarified.

Heat-Stable Alkaline Proteinases

Kinoshita et al. (1990b) reported the existence of up to four distinct heat-stable alkaline proteinase (HAP) in fish muscle: two sarcoplasmic proteinases activated at 50°C and 60°C, and two myofibril-associated proteinases activated at 50°C and 60°C, respectively. The distribution of the four proteinases was found to be quite diverse among the 12 fish species that were studied. The mechanisms activating these proteinases in vivo and their precise physiological functions are not clear.

The participation of one or the other protease in the many different degradation scenarios that occur has still been only partially elucidated. However, various studies have found a close relationship between protein degradation in fish muscle and the activity of specific proteases. A high level of activity of cathepsin L has been found in chum salmon during spawning, a period during which the fish exhibit an extensive softening in texture (Yamashita and Konagaya 1990). Similarly, Kubota et al. (2000) found an increase in gelatinolytic activity in the muscle of ayu during spawning, a period involving a concurrent marked decrease in muscle firmness. Also, in the muscle of hake, a considerably higher level of proteolytic activity during the prespawning period than in the postspawning period was found (Perez-Borla et al. 2002).

The possible existence of a direct link between protease activity and texture has been explored in situ by perfusing protease inhibitors into fish muscle and later measuring changes in texture during cold storage. The activity of metalloproteinase and of trypsin-like serine protease was found in this way to play a role in the softening of flounder (Kubota et al. 2001) and of tilapia (Ishida et al. 2003).

POSTMORTEM HYDROLYSIS OF LIPIDS IN SEAFOOD DURING FROZEN AND COLD STORAGE

Changes in the lipid fraction of fish muscle during storage can lead to changes in quality. Both the content and the composition of the lipids in fish muscle can vary considerably between species and from one time of year to another, and also differ greatly depending upon whether white or red muscle fibers are involved. As already mentioned, these two types of muscle fibers are separated from each other, the white fibers generally constituting most of the muscle as a whole, although fish species vary considerably in the amounts of dark meat, which has a higher myoglobin and lipid content. In species like tuna and in small and fatty pelagic fish, the dark muscle can constitute up to 48% of the muscle as a whole, whereas in lean fish such as cod and flounder, the dark muscle constitutes only a small percentage of the muscle (Love 1970). Since triglycerides are deposited primarily in the dark muscle, providing fatty acids as substrate to aerobic metabolism, whereas the phospholipids represent most of the lipid fraction of the white muscle, phospholipids constitute a major part of the lipid fraction in lean fish (Lopez-Amaya and Marangoni 2000b).

Not much research on lipid hydrolysis in fresh fish during ice storage has been carried out, research having concentrated more on changes in the lipid fraction during frozen storage. This could be due to freezing being the most common way of storing and processing seafood and to lipid hydrolysis playing no appreciable role in ice-stored fish before microbial spoilage becomes extreme. Knudsen (1989), however, detected an increase in free fatty acids in cod muscle during 11 days of ice storage, indicative of the occurrence of lipolytic enzyme activity, an increase that was most pronounced during days 5 to 11. Ohshima et al. (1984) reported a similar delay in the increase in free fatty acids in cod muscle stored in ice for 30 days. Both results are basically consistent with the observation of Geromel and Montgomery (1980) of no lysosomal lipase activity being evident in trout muscle after seven days on ice. In contrast, the authors reported that slow freezing and fluctuations in temperature during frozen storage were found to result in the release of acid lipase from the lysosomes of the dark muscle of rainbow trout.

Several researchers have reported an increase in free fatty acids during frozen storage of muscle of different fish species such as trout (Ingemansson et al. 1995), salmon (Refsgaard et al. 1998, 2000), rayfish (Fernandez-Reiriz et al. 1995), tuna, cod, and prawn (Kaneniwa et al. 2004). The release of free fatty acids during frozen storage can induce changes in texture by stimulation of protein denaturation and through off-flavors being produced by lipid oxidation (Lopez-Amaya and Marangoni 2000b). Refsgaard et al. (1998, 2000) observed a marked increase in free fatty acids in salmon stored at −10°C and −20°C. This increase in free fatty acid content was connected with changes in sensory attributes, suggesting that lipolysis plays a significant role in deterioration of the quality of salmon during frozen storage.

Kaneniwa et al. (2004) detected a large variation in the formation of free fatty acids among nine species of fish and shellfish stored at −10°C for 30 days. Once again, this demonstrates the large variation in enzyme activity in seafood species. Findings of Ben-gigirey et al. (1999) indicate that the temperature at which fish are stored has a clear influence on the lipase activity occurring in the muscle. They noted, for example, that the formation of free fatty acids in the muscle of albacore tuna during storage for the period of a year was considerably higher at −18°C than at −25°C.

Two classes of lipases, the lysosomal lipases and the phospholipases, are apparently involved in the hydrolysis of lipids in fish muscle during storage. Nayak et al. (2003) found significant differences among four fish species (rohu, oil sardine, mullet, and Indian mackerel) in the degree of red muscle lipase activity. This is quite in line with differences in lipid hydrolysis among species that Kaneniwa et al. (2004) reported.

Although both the lipase activity and the formation of free fatty acids in fish muscle are well documented, only a few studies have actually isolated and characterized the muscle lipases and phospholipases involved. Aaen et al. (1995) have isolated and characterized an acidic phospholipase from cod muscle, and Hirano et al. (2000) a phospholipase from the white muscle of bonito. Similarly, triacylglycerol lipase from salmon and from rainbow trout has been isolated and characterized by Sheridan and Allen (1984) and by Michelsen et al. (1994), respectively. Knowledge of the properties of the lipolytic enzymes in the muscle in seafood and of the responses the enzymes show to various processing parameters, however, is sparse. Extensive reviews of work done on the lipases and phospholipases in seafood has been presented by Lopez-Amaya and Marangoni (2000a,b).

ENDOGENOUS ENZYMATIC REACTIONS DURING THE PROCESSING OF SEAFOOD

During seafood processing such as salting, heating, fermentation, and freezing, endogenous enzymes can be active and contribute to the sensory characteristics of the final product. Such endogenous enzyme activities are sometimes necessary in order to obtain the desired taste and texture.

SALTING OF FISH

Salting has been used in many countries for centuries as a means of preserving fish. Today, the primary purpose of salting fish is no longer only to preserve them. Instead, salting enables fish products with sensory attributes that are sought after, such as salted herring and salted cod, as produced on the northern European continent and in Scandinavia. Enzymatic degradation of muscle proteins during salting is a factor that contributes to the development of the right texture and taste of the products.

Ripening of Salt-Cured Fish

Spiced sugar–salted herring in its traditional form is made by mixing approximately 100 kg of headed, ungutted herring with 15 kg of salt and 7 kg of sugar in barrels, usually adding spices as well. After a day or two, after a blood brine has been formed, saturated brine is added, after which the barrels are stored at 0–5°C for up to a year. During this period, a ripening of the herring takes place, and it achieves its characteristic taste and texture (Stefánsson et al. 1995).

During the ripening period, both intestinal and muscle proteases participate in the degradation of muscle protein, contributing to the characteristic softening of the fillet and liberating free amino acids and small peptides that help create the characteristic flavor of the product (Nielsen 1995, Olsen and Skåra 1997). Studies have shown that intestinal trypsin- and chymotrypsin-like enzymes migrate into the fillet, where they play an active role in the degradation of muscle proteins during storage (Engvang and Nielsen 2000, Stefánsson et al. 2000). Furthermore, Nielsen (1995) shows that muscle amino peptidases are also active in the salted herring during storage.

In southern Europe, a similar product based on the use of whole sardines or anchovies is produced. In contrast to the salted herring from Scandinavia, these salted sardines and anchovies are stored at ambient temperature, which can vary between 18°C and 30°C (Nunes et al. 1997). Nunes et al. (1997) found that proteases from both the intestines and the muscles participated in the ripening of sardines *(Sardina pilchardus)*. Hernadez-Herrero et al. (1999) reported an increase in proteinase activity as well as in protein hydrolysis during the storage of salted anchovies *(Engraulis encrasicolus)* and found a close relationship between proteolysis and the development of the sensory characteristics of the product.

PRODUCTION OF FISH SAUCE AND FISH PASTE

Fish sauce and fish paste are fermented fish products produced mainly in Southeast Asia, where they are highly appreciated food flavorings. Fish sauce is the liquefied protein fraction, and fish paste is the "solid" protein fraction obtained from the prolonged hydrolysis of heavily salted small pelagic fish. Production takes place in closed tanks at ambient tropical temperatures during a period of several months (Gildberg 2001, Saishiti 1994). The hydrolysis represents the combined action of the fishes' own digestive proteases and of enzymes from halotolerant lactic acid bacteria (Saishiti 1994). Orejani and Liston (1981) concluded, on the basis of inhibitor studies, that a trypsin-like protease is one of the enzymes responsible for the hydrolysis. Vo et al. (1984) detected a high and stable level of activity of intestinal amino peptidase during the production of fish sauce. Del Rosario and Maldo (1984) measured the activity of four different proteases in fish sauce produced from horse mackerel. During a four-month period of measurement, they found the activity of cathepsins A, C, and B to be stable and that of cathepsin D to decrease. Raksakulthai and Haard (1992) found that cathepsin C obtained from capelin was active in the presence of 20–25% salt, suggesting that this enzyme is involved in the hydrolysis of capelin fish sauce. These studies indicate that several different proteases need to be active and that their concerted action is necessary to achieve the pronounced hydrolysis required for production of such fish sauces.

It is notable that the similarly high storage temperatures present in southern Europe do not lead to a solubilization of salted sardines and anchovies. Ishida et al. (1994) reported that at 35°C salted Japanese anchovies *(Eriobotrya japonica)* degraded to a marked degree, whereas at this temperature salted anchovies from southern Europe *(E. encrasicolus)* were structurally stable. Also, they detected a thermostable trypsin-like proteinase in the muscle of both salted Japanese anchovies and European anchovies, but its activity was much higher in the Japanese anchovies. An explanation of the difference between the European and the Asian products might be a large difference between the hydrolytic enzyme activity of the respective raw materials.

PRODUCTION OF SURIMI

Surimi is basically a myofibrillar protein concentrate that forms a gel due to cross-linking of its actomyosin molecules (An et al. 1996). It is made from minced fish flesh obtained mainly from pelagic white fish of low fat content, such as Alaskan pollack and Pacific whiting. The mince is washed several times with water to remove undesired elements such as connective tissue and lipids. The particulate is then stabilized by cryoprotectants before being frozen (Park and Morrisey 2000).

Surimi is used as a raw material for the manufacture of various products, such as imitation crabmeat and shellfish substitutes.

The manufacture of surimi products involves the use of a slow temperature-setting process, which can be in the temperature range of 4–40°C, followed by a heating process involving temperatures of 50–70°C, which results in the gel strength being enhanced (Park 2000). Surimi-based products are very important fish products in the Asian and Southeast Asian countries, where Japan has the largest production and marketing of surimi-based products. The quality and the price are closely dependent on the gel strength (Park 2000).

The enzyme transglutaminase plays an important role in the gelation process through catalyzing the cross-linking of the actomyosin (An et al. 1996). Some of the differences in gelling capability among different species are due to the properties and levels of activity of muscle transglutaminase (An et al. 1996, Lanier 2000). Studies have shown that the addition of microbial transglutaminase can increase the gel strength obtained in fish species having low transglutaminase activity (Perez-Mateos et al. 2002, Sakamoto et al. 1995). The effect of transglutaminase on the processing of surimi has been reviewed extensively by An et al. (1996) and by Ashie and Lanier (2000).

A softening of the gel during the temperature-setting and heating processes can occur due to auto-lysis of the myosin and actomyosin through the action of endogenous heat-stable proteinases. Whether or not this occurs is partly a function of the species used for the surimi production. Two groups of proteinases have been identified as being responsible for the softening: cysteine cathepsin and HAP.

The presence of HAP in the muscle of different fish species used for surimi production and the effects it has on degradation of the fish gel during the heating process have been taken up in a number of studies (Makinodan et al. 1985, Toyahara et al. 1990, Cao et al. 1999). The finding of Kinoshita et al. (1990b) that the fish species differ in the amount of HAP in muscle, could partly explain why softening of the gel is more pronounced in some fish species than in others.

A study by An et al. (1994) indicated that the cysteine proteinase cathepsin L contributes to degradation of the myofibrils in surimi at a temperature of about 55°C during the heating process, a result substantiated by Ho et al. (2000), who also measured the softening of mackerel surimi upon the addition of mackerel cathepsin L.

TECHNOLOGICAL APPLICATIONS OF ENZYMES FROM SEAFOOD

Utilization of enzymes from seafood as technical aids in both seafood processing and other areas of food and feed processing has been an area of active research for many years. There are two factors that have provided the primary motivation for such research: (1) the cold-adaptation properties of seafood enzymes and (2) the increasing production of marine by-products used as potential sources of enzymes. Although the results have been promising, many of the potential technologies are still in their initial stages of development and are not yet fully established industrially.

Improved Processing of Roe

Roe is considered by many to be a seafood delicacy. Russian caviar, produced from the roe of sturgeon, is the form best known, although roe produced from a variety of other species, such as salmon, trout, herring, lumpfish, and cod, has also gained wide acceptance. The roe is originally covered by a two-layer membrane (chorion) termed the roe sack. In some species, mechanical or manual separation of the roe from the sack results in damage to the eggs and in yields as low as 50% (Gildberg 1993). Pepsin-like proteases isolated from the intestines of seafood species, as well as collagenases from the hepatopancreas of crabs, have been shown to cleave the linkages between the sac and the eggs without damaging the eggs. Such enzyme treatment has been reported to increase the yield from 70% to 90% (Gildberg et al. 2000, and references therein).

Production of Fish Silage

Fish silage is a liquid nitrogenous product made from small pelagic fish or fish by-products mixed with acid. It is used as a source of protein in animal feed (Aranson 1994, Gildberg 1993). In its manufacture, the fish material is mixed with 1–3.5% formic acid solution, reducing pH to 3–4. This is optimal for the intestinal proteases and aspartic muscle proteases contained in the fish material, allowing the solubilization of the fish material to proceed as an autolytic process driven by both types of protease (Gildberg et al. 2000). Gildberg and Almas (1986) have reported the existence of two very active pepsins (I and II) in silage manufactured from cod viscera. They were able to show that fish by-products having low protease activity could be hydrolyzed and used for silage by adding protease-rich cod viscera.

Deskinning and Descaling of Fish

Deskinning fish enzymatically can increase the edible yield as compared with that achieved by mechanical deskinning (Gildberg et al. 2000). It also provides the possibility of utilizing alternative species such as skate, the skin of which is very difficult to remove mechanically without ruining the flesh (Stefánsson and Steingrimsdottir 1990). It has been shown that herring can be deskinned enzymatically by use of acid proteases obtained from cod viscera (Joakimsson 1984). Enzymatic removal of the skin of other species has been reported as well. Kim et al. (1993) have described removal of the skin of filefish by use of collagenase extracted from the intestinal organs of the fish. Crude protease extract obtained from minced arrowtooth flounder has been found to be effective in solubilizing the skin of pollock (Tschersich and Choudhury 1998).

Removing squid skin can be a difficult task. Skinning machines only remove the outer skin of the squid tubes, leaving the tough rubbery inner membrane. Strom and Raa (1991) reported a gentler and more efficient enzymatic method of deskinning the squid, using digestive enzymes from the squid itself. Also, a method for deskinning the squid by making use of squid liver extract has been developed by Leuba et al. (1987).

For certain markets, such as Japanese sashimi restaurants and fresh fish markets, skin-on fillets without the scales are demanded. Also, fish skin is used in the leather industry, where descaling is likewise necessary. Obtaining a prime quality product requires gentle descaling. Mechanical descaling can be difficult to accomplish without the fish flesh being damaged, especially in the case of certain soft-fleshed species, where enzymatic descaling results in a gentle descaling (Svenning et al. 1993). Digestive enzymes of fish have proved to be useful for removing scales gently (Gildberg et al. 2000).

Improved Production of Fish Sauce

As already indicated, the original process for manufacturing fish sauce is carried out at high ambient temperature, involving the use of an autolytic process catalyzed by endogenous proteases. The rate of the hydrolysis depends on the content of digestive enzymes in the fish. There has been an obvious interest, however, in shortening the time required for producing the fish sauce, and use has also been made of other fish species, such as Arctic capelin and Pacific whiting. Arctic capelin is usually caught during the winter, when the fish has a low feed intake and its digestive enzyme content thus is low. Research has shown, however, that supplementing Arctic capelin with cod intestines or squid pancreas, both of which are rich in digestive enzymes, allows an acceptable fish sauce to be produced during the winter (Gildberg 2001, Raksakulthai et al. 1986). Tungkawachara et al. (2003) showed that fish sauce produced from a mixture of Pacific whiting and surimi by-products (head, bone, guts, and skin from Pacific whiting) has the same sensory quality as a commercial anchovy fish sauce.

Seafood Enzymes Used in Biotechnology

The poor temperature stability of seafood enzymes is a useful property that has led to the production of enzymes useful in gene technology, where only very small amounts of enzymes are needed. Alkaline phosphatase from cold-water shrimp *(Pandalus borealis)* is more heat labile than alkaline phosphatases from mammals and can be denaturated at 65°C for 15 minutes (Olsen et al. 1991). The heat-labile enzyme is therefore more suitable as a DNA-modifying enzyme in gene-cloning technology, where higher temperatures can denaturate the DNA. The enzyme is recovered for commercial use from shrimp-processing wastewater in Norway. Other seafood enzymes with heat-labile properties, such as Uracil-DNA *N*-glycosylase from cod (Lanes et al. 2000), and shrimp nuclease, are likewise produced commercially as recombinant enzymes for gene-cloning technology.

Potential Applications of Seafood Enzymes in the Dairy Industry

Research has shown that digestive proteases from fish, due to their specificity, can be useful as rennet substitutes for calf chymosin in cheese making (Brewer et al. 1984, Tavares et al. 1997, Shamsuzzaman and Haard 1985). Due to their heat lability, they can be useful for preventing oxidized flavor from developing in milk (Simpson and Haard 1984). Simpson and Haard (1984) found that cod and bovine trypsin are equally effective in preventing copper-induced off-flavors from developing in milk. But the cod enzyme has the advantage of being completely inactivated after pasteurization at 70°C for 45 minutes, whereas 47% of the bovine trypsin is still active. These studies show that cold-adaption properties of marine enzymes can be an advantage in the processing of different foods.

REFERENCES

Aaen B. et al. 1995. Partial purification and characterization of a cellular acidic phospholipase A_2 from cod *(Gadus morhua)* muscle. *Comp Biochem Physiol* 110B: 547–554.

An H. et al. 1996. Roles of endogenous enzymes in surimi gelation. *Trends in Food Sci Technol* 7: 321–327.

An H. et al. 1994. Cathepsin degradation of Pacific whiting surimi proteins. *J Food Sci* 59(5): 1013–1033.

Ando M. et al. 1995. Post-mortem change of the three-dimensional structure of collagen fibrillar network in fish muscle pericellular connective tissues corresponding to post-mortem tenderization. *Fish Sci* 61: 327–330.

Anthoni U. et al. 1990. Is trimethylamine oxide a reliable indicator for the marine origin of fish. *Comp Biochem Phys* 97B: 569–571.

Aoki T, Ueno R. 1997. Involvement of cathepsins B and L in the *post-mortem* autolysis of mackerel muscle. *Food Res Intern* 30(8): 585–591.

Arason S. 1994. Production of fish silage. In: AM Martin (ed.) *Fisheries processing–Biotechnological Applications*. London, Chapman and Hall, pp. 244–269.

Ashie INA, Lanier TC. 2000. Transglutaminases in seafood processing. In: NF Haard, BK Simpson (eds.) *Seafood Enzymes*. New York: Marcel Dekker, pp. 147–166.

Astier C. et al. 1991. Sarcomeric disorganization in post-mortem fish muscles. *Comp Biochem Physiol* 100B: 459–465.

Azam K. et al. 1990. Effect of slaughter method on the progress of rigor of rainbow trout Salmo gairdneri as measured by an image processing system. *Int J Food Sci Technol* 25: 477–482.

Barrett EL, Kwan HS. 1985. Bacterial reduction of trimethylamine oxide. *Ann Rev Microbiol* 39: 131–149.

Ben-gigirey B. et al. 1999. Chemical changes and visual appearance of albacore tuna as related to frozen storage. *J Food Sci* 64(1): 20–24.

Berg T. et al. 1997. Rigor mortis assessment of Atlantic salmon *(Salmo salar)* and effects of stress. *J Food Sci* 62(3): 439–446.

Bracho GE, Haard NF. 1995. Identification of two matrix metalloproteinase in the skeletal muscle of Pacific rockfish (*Sebastes* sp.). *J Food Biochem* 19: 299–319.

Bremner HA. 1999. Gaping in fish flesh. In: K Sato, M Sakaguchi, HA Bremner (eds.) *Extra cellular matrix of Fish and Shellfish*. Trivandrum India: India Research Signpost, pp. 81–94.

Bremner HA, Hallett IC. 1985. Muscle fiber–connective tissue junctions in the fish blue grenadier *(Macruronus novaezelandia)*. A scanning electron microscope study. *J Food Sci* 50: 975–980.

Brewer P. et al. 1984. Atlantic cod pepsin—Characterization and use as a rennet substitute. *Can Inst Food Sci Technol J* 17(1): 38–43.

Busconi L. et al. 1989. Postmortem changes in cytoskeletal elements of fish muscle. *J of Food Biochem* 13: 443–451.

Busconi L. et al. 1992. Purification and characterization of a latent form of multicatalytical proteinase from fish muscle. *Comp Biochem Physiol* 102B: 303–309.

Cappeln G. et al. 1999. Synthesis and degradation of adenosine triphosphate in cod *(Gadus morhua)* at subzero temperatures. *J Sci Food Agric* 79(8): 1099–1104.

Cao MJ. et al. 1999. Myofibril-bound serine proteinase (MBP) and its degradation of myofibrillar proteins. *J Food Sci* 64(4): 644–647.

Chiba A. et al. 1991. Quality evaluation of fish meat by ^{31}phosphorus-nuclear magnetic-resonance. *J Food Sci* 56(3): 660–664.

Childress JJ. 1995. Are there physiological and biochemical adaptations of metabolism in deep-sea animals? *Trends Ecol Evol* 10(1): 30–36.

Dahlhoff, Somero GN. 1991. Pressure and temperature adaptation of cytosolic malate dehydrogenase of shallow and deep-living marine invertebrates. Evidence for high body temperatures in hydrothermal vent animals. *J Exp Biol* 159: 473–487.

Dalgaard P. 2000. Fresh and lightly preserved seafood. In: CMD Man, AA Jones, AN Aspen, editors. *Shelf-life Evaluation of Foods*, 2nd edn. Gaithersburg, Maryland: Aspen Publishers, Inc, pp. 110–139.

de Backer M. et al. 2002. The 1.9 Å crystal structure of heat-labile shrimp alkaline phosphatase. *J Mol Biol* 318: 1265–1274.

Del Rosario RR, Maldo SM. 1984. Biochemistry of patis formation I. Activity of cathepsins in patis hydrolysates. *Phil Agric* 67: 167–175.

Dickson KA. 1995. Unique adaptations of the metabolic biochemistry of tunas and billfishes for the life in the pelagic environment. *Env Biol Fish* 42: 65–97.

Dingle JR, Hines JA. 1971. Degradation of inosine 5′ monophosphate in the skeletal muscle of several North Atalantic fishes. *J Fish Res Bd Can* 28(8): 1125–1131.

Dingle JR. et al. 1977. Protein instability in frozen storage induced in minced muscle of flatfishes by mixture with muscle of red hake. *Can Inst Food Sci Technol J* 10: 143–146.

Ellington WR. 2001. Evolution and physiological roles of phosphagen systems. *Ann Rev Phys* 63: 289–325.

Engvang K, Nielsen HH. 2000. *In situ* activity of chymotrypsin in sugar-salted herring during cold storage. *J Sci Food Agric* 80: 1277–1283.

Ertbjerg P. et al. 1999. Effect of prerigor lactic acid treatment on lysosomal enzyme release in bovine muscle. *J Sci Food Agric* 79: 98–100.

Fernandez-Reiriz MJ. et al. 1995. Changes in lipids of whole and minced rayfish *(Raja clavata)* muscle during frozen storage. *Z Lebensm Unters Forsch.* 200(6): 420–424.

Fields PA. 2001. Protein function at thermal extremes: balancing stability and flexibility. *Comp Biochem Physiol* 129A: 417–431.

Fletcher GC. et al. 1997. Changes in the fine structure of the myocommata-muscle fibre junction related to gaping in rested and exercised muscle from king salmon *(Oncorrhynchus tshawytscha) Lebensm-Wiss Technol* 30(3): 246–252.

Fluke S. 1994. Taste active compounds of seafoods with a special reference to umani substances. In: F Shahidi, JR Botta (eds.) *Seafoods: Chemistry, Processing Technology and Quality*. Glasgow, Great Britain: Blackie Academic and Professional, pp. 115–139.

Foegeding EA. et al. 1996. Characteristics of edible muscle tissues. In: OR Fenema (ed.) *Food Chemistry*. 3rd edn. New York: Marcel Dekker, pp. 879–942.

Geromel EJ, Montgomery MW. 1980. Lipase release from lysosomes of rainbow trout (*Salmo gairdneri*) muscle subjected to low temperatures. *J Food Sci* 45: 412–419.

Geesink GH. et al. 2000. Partial purification and characterization of Chinook salmon *(Oncorhynchus tshawytscha)* calpains and an evaluation of their role in postmortem proteolysis. *J Food Sci* 77: 2685–2692.

Gibbs, AG. 1997. Biochemistry at depth. In: DJ Randall, AP Farrell (eds.) *Deep-Sea Fishes. Fish Physiology*, vol. 16. San Diego, California: Academic Press. pp. 239–271.

Gildberg A. 1993. Enzymic processing of marine raw materials. *Process Biochem* 28: 1–15.

——. 2001. Utilisation of male Arctic capelin and Atlantic cod intestines for fish sauce production—Evaluation of fermentation conditions. *Biores. Technol.* 76: 119–123.

Gildberg A, Almas KA. 1986. Utilization of fish viscera. In: ML Maguer, P Jelen (eds.) *Food Engineering and Process Applications 2, Unit Operations*. London: Elsevier Applied Science, pp. 425–435.

Gildberg A. et al. 2000 Uses of enzymes from marine organisms. In: NF Haard, BK Simpson (eds.) *Seafood Enzymes*. New York: Marcel Dekker, Inc, pp. 619–639.

Gill T. 2000. Nucleotide-degrading enzymes. In: NF Haard, BK Simpson (eds.) *Seafood Enzymes*. New York: Marcel Dekker, pp. 37–68.

Haard N. 2002. The role of enzymes in determining seafood color, flavor and texture. In: HA Bremner (ed.) *Safety and Quality Issues in Fish Processing*. Cambridge: Woodhead Publishing Limited and CRC Press LLC, pp. 220–253.

Haard NF, Simpson BK (eds.). 2000. *Seafood Enzymes*. New York: Marcel Dekker, p. 681.

Hernandez-Herrero MM. et al. 1999. Protein hydrolysis and proteinase activity during the ripening of salted anchovy (*Engraulis encrasicholus* L.): A microassay method for determing the protein hydrolysis. *J Agric Food Chem* 47: 3319–3324.

Hirano K. et al. 2000. Purification and regiospecificity of multiple enzyme activities of phospholipase A_1 from bonito muscle. *Biochim Biophys Acta* 1483: 325–333.

Ho ML. et al. 2000. Effects of mackerel cathepsins L and L-like, and calpain on the degradation of mackerel surimi. *Fish Sci* 66: 558–568.

Hughes RB, Jones NR. 1966. Measurement of hypoxanthine concentration in canned herring as an index of freshness of raw material with a comment on flavour relations. *J Sci Food Agric* 17(9): 434–436.

Hultin HU. 1984. Postmortem biochemistry of meat and fish. *J Chem Ed* 61(4): 289–298.

Hultin HO. 1992. Trimethylamine-*N*-oxide (TMAO) demethylation and protein denaturation in fish muscle. In: GJ Flick, Jr, RE Martin (eds.) *Advances in Seafood Biochemistry. Composition and Quality*. Lancaster, Pennsylvania: Technomic Publishing Company, Inc, pp. 25–42.

Ingemansson T. et al. 1995. Multivariate evaluation of lipid hydrolysis and oxidation data from light and dark muscle of frozen stored rainbow trout *(Oncorhynchus mykiss)*. *J Agric Food Chem* 43: 2046–2052.

Ishida M. et al. 1994. Thermostable proteinase in salted anchovy muscle. *J Food Sci* 59(4): 781–785.

Ishida N. et al. 2003. Inhibition of post-mortem muscle softening following in situ perfusion of protease inhibitors in tilapia. *Fish Sci* 69: 632–638.

Itoh R, Kimura K. 2002. Occurrence of IMP-GMP 5'-nucleotidase in three fish species: A comparative study on *Trachurus japonicus, Oncorhynchus masou* and *Triakis scyllium*. *Comp Biochem Physiol* 132B(2): 401–408.

Iwamoto M. et al. 1987. Effect of storage temperature on rigor-mortis and ATP degradation in plaice *Paralichthys olivaceus* muscle. *J Food Sci* 52(6): 1514–1517.

——. 1991. Changes in ATP and related breakdown compounds in the adductor muscle of itayagai scallop *Pecten albicans* during storage at various temperatures. *Nipp Suis Gakkai* 57(1): 153–156.

Jiang ST. et al. 1991. Purification and characterization of a proteinase identified as cathepsin D from tilapia muscle (*Tilapia nilotica* × *Tilapia aurea*). *J Agric Food Chem* 39: 1597–1601.

Jerrett AR. et al. 1998. Rigor contractions in "rested" and "partially exercised" chinook salmon white muscle as affected by temperature. *J Food Sci* 63(1): 53–56.

Joakimsson K. 1984. Enzymatic deskinning of herring. [DPhil thesis]. Tromsø, Norway: Institute of Fisheries, University of Tromsø.

Jones NR. 1965. Freezing Fillets at Sea. In: *Fish Quality at Sea*. London: Grampian Press Ltd, pp. 81–95.

——. 1969. Fish as a raw material for freezing. Factors influencing the quality of products frozen at sea. In: R Kreuzer (ed.) *Freezing and Irradiation of Fish*. London: Fishing News (Books) Limited, pp. 31–39.

Kaneniwa M. et al. 2004. Enzymatic hydrolysis of lipids in muscle of fish and shellfish during cold storage. In: F Shahidi, AM Spanier, CT Ho, T Braggins (eds.) *Quality of Fresh and Processed Foods*. New York: Kluwer Academic Publishers, pp. 113–119.

Karube I. et al. 1984. Determination of fish freshness with an enzyme sensor system. *J Agric Food Chem* 32(2): 314–319.

Kim SK. et al. 1993. Removal of skin from filefish using enzymes. *Bull Korean Fish Soc* 26: 159–172.

Kimura M. et al. 2002. Enzymic and nonenzymic cleavage of trimethylamine-*N*-oxide *in vitro* at subzero temperatures. *Nipp Suis Gakkai* 68: 85–91.

Kinoshita M. et al. 1990a. Induction of carp muscle multicatalytic proteinase activities by sodium dodecyl sulfate and heating. *Comp Biochem Physiol* 96B: 565–569.

——. 1990b. Diverse distribution of four distinct types of modori (gel degradation)-inducing proteinases among fish species. *Bull Japan Soc Sci Fish* 56(9): 1485–1492.

Kinoshita M. et al. 2002. cDNA cloning and characterization of two gelatinases from Japanese flounder. *Fish Sci* 68: 618–626.

Knudsen LB, 1989. Protein-lipid interaction in fish. [DPhil thesis]. Lyngby: Danish Institute for Fisheries Research, Technical University of Denmark. p. 188.

Kolodziejska I, Sikorski ZE. 1995. Muscle cathepsins of marine fish and invertebrates. *Pol J Food Nutr Sci* 4/45(3): 3–10.

——. 1996. Neutral and alkaline muscle proteases of marine fish and invertebrates A review. *J Food Biochem* 20: 349–363.

Kubota S. et al. 1998. Occurrence of gelatinolytic activities in yellowtail tissues. *Fish Sci* 64: 439–442.

Kubota S. et al. 2000. Induction of gelatino activities in ayu muscle at the spawning stage. *Fish Sci* 66: 574–578.

Kubota M. et al. 2001. Possible implication of metalloproteinases in post-mortem tenderization of fish muscle. *Fish Sci* 67: 965–968.

Kubota M. et al. 2003. Solubilization of type I collagen from fish muscle connective tissue by matrix metalloproteinase-9 at chilled temperature. *Fish Sci* 69: 1053–1059.

Lanes O. et al. 2000. Purification and characterization of a cold-adapted uracil-DNA N glycosylase from Atlantic cod (Gadus morhua). *Comp Biochem Physiol* 127B: 399–410.

Lanier TC. 2000. Surimi gelation chemistry. In: JW Park (ed.) *Surimi and Surimi Seafood*. New York: Marcel Dekker, pp. 237–266.

Leuba JL. et al. 1987 (Oct 20). Method for dissolving squid membranes. U.S. patent 4,701,339.

Lødemel JB. 2004. Gelatinolytic activities, matrix metalloproteinases and tissue inhibitors of metalloproteinases in fish [Dphil thesis]. Tromsø, Norway: University of Tromsø, Dept. of Marine Biotechnology. p. 46.

Lødemel JB, Olsen RL. 2003. Gelatinolytic activities in muscle of Atlantic cod (Gadus morhua), spotted wolffish (Anarhichas minor) and Atlantic salmon (Salmo salar). *J Sci Food Agric* 83: 1031–1036.

Lopez-Amaya C, Marangoni AG. 2000a. *Phospholipases*. In seafood processing. In: NF Haard, BK Simpson (eds.) *Seafood Enzymes*. New York: Marcel Dekker, pp. 91–119.

——. 2000b. Phospholipases in seafood processing. In: NF Haard, BK Simpson (eds.) *Seafood Enzymes*. New York: Marcel Dekker, pp. 121–146.

Love RM. 1970. *The Chemical Biology of Fishes*. New York: Academic Press, p. 547.

Lowe TE. et al. 1993. Flesh quality in snapper, Pagrus auratus, affected by capture stress. *J Food Sci* 58: 770–773.

Lund KE, Nielsen HH. 2001. Proteolysis in salmon (Salmo salar) during cold storage: Effects of storage time and smoking process. *J Food Biochem* 25: 379–395.

Lundstrøm RC. et al. 1982. Enzymatic dimethylamine and formaldehyde production in minced American plaice and blackback flounder mixed with a red hake TMAOase active fraction. *J Food Sci* 47: 1305–1310.

Mackie IM. 1993. The effects of freezing on flesh proteins. *Food Rev Int* 9: 575–610.

Mcdonald I, Jones NR. 1976. Control of thaw rigor by manipulation of temperature in cold store. *J Food Technol* 11: 69–76.

Makinodan Y. et al. 1985. Implication of muscle alkaline proteinase in the textural degradation of fish meat gel. *J Food Sci* 50: 1351–1355.

Michelsen KG. et al. 1994. Adipose tissue lipolysis in rainbow trout, Onco-rhynchus mykiss, is modulated by phosphorylation of triacylglycerol lipase. *Comp Biochem Physiol* 107B: 509–513.

Nayak J. et al. 2003. Lipase activity in different tissues of four species of fish: Rohu (Labeo rohita Hamilton), oil sardine (Sardinella longceps linnaeus), mullet (Liza subviridis Valenciennes) and India mackerel (Rastrelliger kanagurta Cuvier). *J Sci Food Agric* 83: 1139–1142.

Nielsen HH. 1995. Proteolytic enzyme activites in salted herring during cold storage. [DPhil thesis]. Lyngby: Dept. of Biotechnology, Technical University of Denmark. p. 131.

Nielsen LB, Nielsen HH. 2001. Purification and characterization of cathepsin D from herring muscle (Clupea harengus). *Comp Biochem Physiol* 128B: 351–363.

Nielsen MK, Jorgensen BM. 2004. Quantitative relationship between trimethylamine oxide aldolase activity and formaldehyde accumulation in white muscle from gadiform fish during frozen storage. *J Agric Food Chem* 52(12): 3814–3822.

Niizeki N. et al. 2002. Mechanism of biosynthesis of trimethylamine oxide from choline in the teleost tilapia, Oreochromis niloticus, under freshwater conditions. *Comp Biochem Physiol* 131B: 371–386.

Nitisewojo P, Hultin HO. 1986. Characteristics of TMAO degrading systems in Atlantic short finned squid (Illex illecebrosus). *J Food Biochem* 10: 93–106.

Nunes ML. et al. 1997. Sardine ripening: Evolution of enzymatic, sensorial and biochemical aspects. In: JB Luten et al. (eds.) *Seafood from Producer to Consumer, Integrated Approach to Quality*, Amsterdam: Elsevier Science B.V, pp. 319–330.

Ogata H. et al. 1998. Proteolytic degradation of myofibrillar components by carp cathepsin L. *J Sci Food Agric* 76: 499–504.

Ohshima T. et al. 1984. Enzymatic hydrolysis of phospholipids in cod flesh during storage in ice. *Bull Japan Soc Sci Fish* 50(1): 107–114.

Okuma E, Abe H. 1992. Major buffering constituents in animal muscle. *Comp Biochem Physiol* 102A(1): 37–41.

Olsen RL. et al. 1991. Alkaline phosphatase from the hepatopancreas of shrimp (Pandalus borealis): A dimeric enzyme with catalytically active subunits. *Comp Biochem Physiol* 99B: 755–761.

Olsen SO, Skåra T. 1997. Chemical changes during ripening of North Sea herring. In: JB Luten et al. (eds.) *Seafood from Producer to Consumer, Integrated Approach to Quality*. Amsterdam: Elsevier Science B.V, pp. 305–318.

Orejani FL, Liston J. 1981. Agents of proteolysis and its inhibition in patis (fish sauce) fermentation. *J Food Sci* 47: 198–203.

Papa I. et al. 1997. Dystrophin cleavage and sarcolemna detachment are early postmortem changes of bass (Dicentrachus labrax) white muscle. *J Food Sci* 62: 917–921.

Parab SN, Rao SB. 1984. Distribution of trimethylamine oxide in some marine and freshwater fish. *Curr Sci* 53: 307–309.

Park JW. 2000. Surimi Seafood—Products, Market and Manufacturing. In: JW Park (ed.) *Surimi and Surimi Seafood*. New York: Marcel Dekker, pp. 201–235.

Park JW, Morrissey MT. 2000. Manufacturing of surimi from light muscle fish. In: JW Park (ed.) *Surimi and Surimi Seafood*. New York: Marcel Dekker, pp. 23–58.

Perez-Mateos M. et al. 2002. Addition of microbial transglutaminase and protease inhibitors to improve gel properties of frozen squid muscle. *Eur Food Res Technol* 214: 377–381.

Perez-Borla O. et al. 2002. Proteolytic activity of muscle in pre- and post-spawning hake (Merluccius hubbsi marini) after frozen storage. *Lebensm-Wiss Technol* 35: 325–330.

Raksakulthai N, Haard NF. 1992. Fish sauce from capelin (Mallotus villosus): Contribution of cathepsin C to the fermentation. *ASEAN Food J* 7: 147–151.

Raksakulthai N. et al. 1986. Effect of enzyme supplements on the production on the production of fish sauce prepared from male capelin (Mallotus villosus). *Can Inst Food Sci Technol J* 19: 28–33.

Refsgaard HHF. et al. 1998. Sensory and chemical changes in farmed Atlantic salmon (Salmo salar) during frozen storage. *J Agric Food Chem* 46: 3473–3479.

——. 2000. Free polyunsaturated fatty acids cause taste deterioration of salmon during frozen storage. *J Agric Food Chem* 48: 3280–3285.

Rehbein H. 1988. Relevance of trimethylamine oxide demethylase activity and haemoglobin content to formaldehyde production and texture deterioration in frozen stored minced fish muscle. *J Sci Food Agric* 43: 261–276.

Rehbein H. et al. 1997. Relation between TMAOase activity and content of formaldehyde in fillet minces and belly flap minces from gadoid fishes. *Inf Fischwirtsch* 44: 114–118.

Robb DHF. et al. 2000. Muscle activity at slaughter: I. Changes in flesh colour and gaping in rainbow trout. *Aquaculture* 182(3–4): 261–269.

Saishiti P. 1994. Traditional fermented fish: Fish sauce production. In: AM Martin (ed.) *Fisheries Processing—Biotechnological Applications*. London: Chapman and Hall, pp. 111–129.

Saito T. et al. 1959. A new method for estimating the freshness of fish. *Bull Jpn Soc Sci Fish* 24(9): 749–750.

Saito M. et al. 2000. Characterization of a rainbow trout matrix metalloproteinase capable of degrading type I collagen. *Eur J Biochem* 267: 6943–6950.

Sakamoto H. et al. 1995. Gel strength enhancement by addition of microbial transglutaminase during onshore surimi manfacture. *J Food Sci* 60(2): 300–304.

Sato K. et al. 2002. Effect of slaughter method on degradation of intramuscular type V collagen during short-term chilled storage of chub mackerel Scomber japonicus. *J Food Biochem* 26: 415–429.

Sentandreu MA. et al. 2002. Role of muscle endopeptidases and their inhibitors in meat tenderness. *Trends Food Sci Technol* 13: 398–419.

Shamsuzzaman K, Haard NF. 1985. Milk clotting and cheese making properties of a chymosin-like enzyme from harp seal mucosa. *J Food Biochem* 9: 173–192.

Sheridan MA, Allen WV. 1984. Partial purification of a triacylglycerol lipase isolated from steelhead trout (Salmo gairdneri) adipose tissue. *Lipids* 19(5): 347–352.

Shigemura Y. et al. 2004. Possible degradation of type I collagen in relation to yellowtail muscle softening during chilled storage. *Fish Sci* 70: 703–709.

Shoubridge EA, Hochachka PW. 1980. Ethanol—Novel end product of vertebrate anaerobic metabolism. *Science* 209(4453): 308–309.

Siebenaller JF, Somero GN. 1978. Pressure-adaptive differences in lactate dehydrogenases of congeneric marine fishes living at different depths. *Science* 201: 255–257.

——. 1979. Pressure-adaptive differences in the binding and catalytic properties of muscle-type (M_4) lactate dehydrogenase of shallow- and deep-living marine fishes. *J Comp Physiol* 129: 295–300.

Sikorski Z, Kostuch S. 1982. Trimethylamine N-oxide demethylase—its occurrence, properties, and role in technological changes in frozen fish. *Food Chem* 9: 213–222.

Sikorski ZE, Kolakowska A. 1994. Changes in proteins in frozen stored fish. In: ZE Sikorski et al. (eds.) *Seafood Proteins*. New York: Chapman and Hall, pp. 99–231.

Simpson BK, Haard NF. 1984. Trypsin from Greenland cod as a food-processing aid. *J Appl Biochem* 6: 135–143.

——. 1987. Cold-adapted enzymes from fish. In: D Knorr (ed.) *Food Biotechnology*. New York: Marcel Dekker, pp. 495–527.

Skjervold PO. et al. 2001. Live-chilling and crowding stress before slaughter of Atlantic salmon (Salmo salar). *Aquaculture* 192(2-4): 265–280.

Smalås AO. et al. 2000. Cold adapted enzymes. *Biotechnol Ann Rev* 6: 1–57.

Somero GN. 1981. pH-temperature interactions on proteins—Principles of optimal pH and buffer system-design. *Mar Biol Lett* 2(3): 16–78.

——. 2003. Protein adaptations to temperature and pressure: complementary roles of adaptive changes in amino acid sequence and internal milieu. *Comp Biochem Physiol* 136B: 577–591.

Sotelo CG. et al. 1995. Denaturation of fish proteins during frozen storage: role of formaldehyde. *Z Lebensm Unters Forsch* 200: 14–23.

Sotelo CG, Rehbein H. 2000. TMAO-degrading enzymes. Utilization and influence on post harvest seafood quality. In: NF Haard, BK Simpson (eds.) *Seafood Enzymes*. New York: Marcel Dekker, pp. 167–190.

Spinelli J, Koury BJ. 1981. Some new observations on the pathways of formation of dimethylamine in fish muscle and liver. *J Agric Food Chem* 29: 327–331.

Stefánsson G. et al. 1995. Ripening of spice-salted herring. TemaNord 613. Copenhagen, Denmark: Nordic Council of Ministers.

Stefánsson G. et al. 2000. Frozen herring as raw material for spice-salting. *J Sci Food Agric* 80: 1319–1324.

Stefánsson G, Steingrimsdottir U. 1990. Application of enzymes for fish processing in Iceland—Present and future aspects. In: MN Voight, JR Botta (eds.) *Advances in Fisheries Technology and Biotechnology for Increased Profitability*. Lancaster: Technomic Publishing Co., pp. 237–250.

Stoknes I, Rustad T. 1995. Purification and characterization of a multicatalytic proteinase from Atlantic salmon (Salmo salar) muscle. *Comp Biochem Physiol* 111B(4): 587–596.

Storey KB. 1992. The basis of enzymatic adaptation. In: EE Bittar (ed.) *Chemistry of the Living Cell. Fundamentals of Medical Cell Biology*, vol. 3A. Amsterdam: Elsevier, pp. 137–156.

Strom T, Raa J. 1991. From basic research to new industries within marine biotechnology: Successes and failures in Norway. In: HK Kuanf et al. (eds.) *Proceeding of a Seminar on Advances in Fishery Post-Harvest Technology in Southeast Asia*. Singapore: Changi Point, pp. 63–71.

Svenning R. et al. 1993. Biotechnological descaling of fish. *INFOFISH Int* 6/93: 30–31.

Surette M. et al. 1988. Biochemical basis of postmortem nucleotide catabolism in cod (Gadus morhua) and its relationship to spoilage. *J Agric Food Chem* 36: 19–22.

Takeuchi M. et al. 2004. Unique evolution of Bivalvia arginine kinases. *Cellular and Molecular Life Sciences* 61(1): 110–117.

Tavares JFP. et al. 1997. Milk-coagulating enzymes of tuna fish waste as a rennet substitute. *Int J Food Sci Nutr* 48(3): 169–176.

Taylor RG. et al. 2002. Salmon fillet texture is determined by myofiber-myofiber and myofiber myocommata attachment. *J Food Sci* 67: 2067–2071.

Teruel SRL, Simpson BK. 1995. Characterization of the collagenolytic enzyme fraction from winter flounder (Pseudopleuronectes americanus). *Comp Biochem Physiol* 112B: 131–136.

Tomioka K. et al. 1987. Effect of storage temperature on the dephosphorylation of mucleotides in fish muscle. *Nipp Suis Gakkai* 53(3): 503–507.

Toyohara H. et al. 1990. Degradation of oval-filefish meat gel caused by myofibrillar proteinase(s). *J Food Sci* 52: 364–368.

Tschersich P, Choudhury GS. 1998. Arrowtooth flounder (Atheresthes stomias) protease as a processing aid. *J Aqua Food Prod Technol* 7(1): 77–89.

Tungkawachara S. et al. 2003. Biochemical properties and consumer acceptance of Pacific whiting fish sauce. *J. Food Sci* 68(3): 855–860.

Vaisey EB. 1956. The non-enzymatic reduction of trimethylamine oxide to trimethylamine, dimethylamine, and formaldehyde. *Can J Biochem Physiol* 34: 1085–1090.

van Waarde A. 1988. Biochemistry of non-protein nitrogenous compounds in fish including the use of amino acids for anaerobic energy production. *Comp Biochem Physiol* 91B: 207–228.

Verrez-Bagnis V. et al. 2002. In vitro proteolysis of myofibrillar and sarcoplasmic proteins of European sea bass (Dicentrarchus labrax L.) by an endogenous m-calpain. *J Sci Food Agric* 82: 1256–1262.

Verrez-Bagnis V. et al. 1999. Desmin degration in postmortem fish muscle. *J Food Sci* 64: 240–242.

Vo VT. et al. 1984. The aminopeptidase activity in fish sauce. *Agric Biol Chem* 48(2): 525–527.

Wang JH, Jiang ST. 1991. Properties of calpain II from tilapia muscle (Tilapia nilotica × Tilapia aurea) *Agric Biol Chem* 55: 339–345.

Watson CL. et al. 1992. Proteolysis of skeletal muscle in yellowfin tuna (Thunnus albacares): Evidence of calpain activation. *Comp Biochem Physiol* 103B: 881–887.

Yamashita M, Konagaya S. 1990. Participation of cathepsin L into extensive softening of the muscle of chum salmon caught during spawning migration. *Bull Japan Soc Sci Fish* 56: 1271–1277.

——. 1991. Hydrolytic action of salmon cathepsins B and L to muscle structural proteins in respect of muscle softening. *Bull Japan Soc Sci Fish* 57: 1917–1922.

Yancey PH. et al. 1982. Living with water stress: Evolution of osmolyte systems. *Science* 217: 1214–1222.

14
Seafood Enzymes: Biochemical Properties and Their Impact on Quality

Sappasith Klomklao, Soottawat Benjakul, and Benjamin K. Simpson

Abstract: Recovery and characterization of enzymes from seafood has taken place and this has led to the emergence of some interesting new applications of these enzymes in food processing. The major enzymes isolated from marine organism and used for industrial application are protease, transglutaminase, and lipase. However, some enzymes found in the fish and aquatic invertebrates have exhibited an adverse effect on seafood quality such as polyphenoloxidase and TMAOase. Therefore, appropriate approach to tackle the problem associated with undesirable enzyme has become the technical concern for processors to maintain the prime quality of seafood and seafood products. On the other hand, the maximization of desirable enzyme activities is the means to come across the fulfillment of seafood processing. In addition, the cost of enzyme from other sources could be lowered, leading to the sustainability of seafood industry, in which all resources could be fully utilized.

INTRODUCTION

The marine environment contains a wide variety of genetic material, offering enormous potential as a source of enzymes. Because of the biological diversity of marine species, a wide array of enzymes with unique properties can be recovered and effectively utilized. Most enzymes from fish and aquatic invertebrates are present in terrestrial organisms. However, the enzymes from different marine species, as well as their organs or environment habitat display differences in molecular weight (MW), amino acid composition, pH and temperature optimum, stability, and inhibition characteristics and kinetic properties. Recovery and characterization of enzymes from seafood has been studied extensively, and this has led to the emergence of some interesting new applications of these enzymes in food processing. The major enzymes isolated from marine organisms and used for industrial application are proteases, transglutaminases, and lipases. However, some enzymes found in fish and aquatic invertebrates exhibit adverse effects on seafood quality such as polyphenoloxidases (PPO) and trimethylamine-*N*-oxide demethylase (TMAOases). Therefore, appropriate approaches to tackle the problem associated with undesirable enzymes in seafood products have become a technical concern for processors to maintain the prime quality of seafood and seafood products. Conversely, the activities of desirable enzyme should be maximized to fully exploit the marine resources. Furthermore, the need of commercial enzymes can be reduced, thereby lowering the operation cost associated with those enzymes.

Food Biochemistry and Food Processing, Second Edition. Edited by Benjamin K. Simpson, Leo M.L. Nollet, Fidel Toldrá, Soottawat Benjakul, Gopinadhan Paliyath and Y.H. Hui.

PROTEASES

Proteases play an important role in the growth and survival of all living organisms. Protease catalyzes the hydrolysis of peptide bonds in polypeptides and protein molecules (Garcia-Carreno and Hernandez-Cortes 2000). On the basis of their in vitro properties, proteases have been classified in a number of ways such as the pH range over which they are active (acid, neutral, or alkaline proteases), their ability to hydrolyze specific proteins, and their mechanism of catalysis. Proteases may be classified based on their similarities to well-characterized proteases such as trypsin-like, chymotrypsin-like, chymosin-like or cathepsin-like, and so on (Haard 1990). In addition, proteases are classified according to their catalytic action (endopeptidase or exopeptidase) and the nature of the catalytic site. In Enzyme Commission (EC) system for enzyme nomenclature, all proteases (peptide hydrolases) belong to subclass 3.4, which is further divided into 3.4.11–19, the exopeptidases and 3.4.21–24, the endopeptidases or proteinases (Nissen 1993). Endopeptidases cleave the polypeptide chain at particularly susceptible peptide bonds distributed along the chain, whereas exopeptidases hydrolyze one amino acid unit sequentially from the *N*-terminus (aminopeptidases) or from *C*-terminus (carboxypeptidases) (Fig. 14.1). Exopeptidases, especially aminopeptidases, are ubiquitous, but less readily available as commercial products, since many of them are intracellular or membrane bound (Simpson 2000). For endopeptidases or proteinases, the four major classes of endopeptidases can be distinguished according to the chemical group of their active site, including serine proteinases (EC 3.4.21), thiol or cysteine proteinases (EC 3.4.22), acid or aspartic proteinases (EC 3.4.23), and metalloproteinases (EC 3.4.24) (Simpson 2000). The enzymes in the different classes are differentiated by various criteria, such as the nature of the groups in their catalytic sites, their substrate specificity, and their response to inhibitors or by their activity/stability under acid or alkaline conditions (Nissen 1993).

For marine animals, proteases are mainly produced by the digestive glands. Like the proteases from plants, animals, and microorganisms, digestive proteases from marine animals are polyfunctional enzymes catalyzing the hydrolytic degradation of proteins (Garcia-Carreno and Hernandez-Cortes 2000). For some fish species, proteases are present at high levels in the muscle (Kolodziejska and Sikorski 1996), and are associated with the induced changes of proteins during postmortem storage or processing. Marine animals have adapted to different environmental conditions, and these adaptations, together with inter- and intraspecies genetic variations, are associated with certain unique properties of their proteinases, compared with their counterpart from land animals, plants, and microorganisms (Simpson 2000). Some of these distinctive properties include higher catalytic efficiency at low temperature and lower thermal stability (Klomklao et al. 2009a).

Digestive Proteases from Marine Animals

Digestive proteases have been studied in several species of fish and decapods (De-Vecchi and Coppes 1996). Proteases found in the digestive organs of fish include pepsin, gastricsin, trypsin, chymotrypsin, collagenase, elastase, carboxypeptidase, and carboxyl esterase (Simpson 2000). Pepsin, chymotrypsin, and trypsin are three main groups of proteases found in fish viscera. Pepsin is localized in fish stomach (Klomklao et al. 2007a), while chymotrypsin and trypsin are concentrated in tissues such as the pancreas, pyloric ceca, and intestine (Klomklao et al. 2004). The distribution and properties of protease vary depending on species and organs (Table 14.1). According to the International Union of Applied Biochemists classification, digestive proteases from fish and aquatic invertebrates may be classified into four major groups such as acidic/aspartic protease, serine protease, thiol or cysteine protease, and metalloprotease (Simpson 2000).

Acid/Aspartic Proteases

The acid or aspartic proteases are a group of endopeptidases characterized by high activity and stability at acidic pH. They are referred to as "aspartic" proteases (or carboxyl proteases) because their catalytic sites are composed of the carboxyl group of two aspartic acid residues (Whitaker 1994). On the basis of the EC system, all the acid/aspartic proteases from marine animals have the first three digits in common as EC 3.4.23. Three common aspartic proteases that have been isolated and

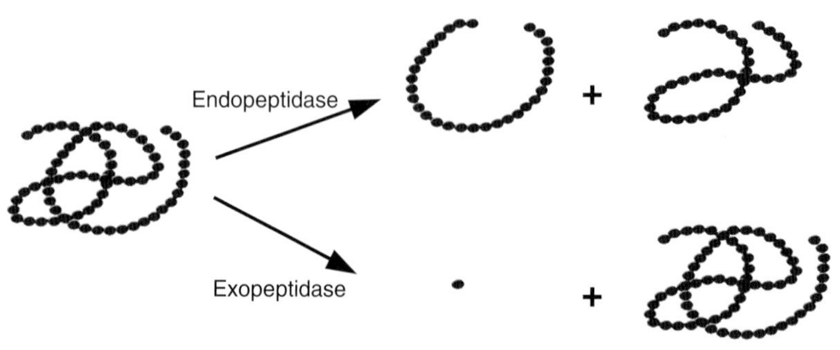

Figure 14.1. Cleavage of proteins by endopeptidases and exopeptidases.

Table 14.1. Proteases from Digestive Organs of Fish and Aquatic Invertebrates

Family	Enzyme	Identified Species	Optimum		Reference
			pH	Temperature (°C)	
Aspartic protease	Pepsin	Pectoral rattail	3.0–3.5	45	Klomklao et al. (2007a)
		Smooth hound	2.0	40	Bougatef et al. (2008)
		Sea bream	3.0–3.5	45–50	Zhou et al. (2007)
		European eel	2.5–3.5	35–40	Wu et al. (2009)
		African coelacanth	2.0–2.5	–	Tanji et al. (2007)
	Chymosin	Harp seal	2.2–3.5	–	Shamsuzzaman and Haard (1984)
	Gastricsin	Hake	3.0	–	Sanchez-Chiang and Ponce (1981)
Serine protease	Trypsin	Skipjack tuna	9.0	55–60	Klomklao et al. (2009a)
		Tongol tuna	8.5	65	Klomklao et al. (2006a)
		Yellowfin tuna	8.5	55–65	Klomklao et al. (2006b)
		Pectoral rattail	8.5	45	Klomklao et al. (2009b)
		Atlantic bonito	9.0	65	Klomklao et al. (2007b)
		Cuttlefish	8.0	70	Balti et al. (2009)
		Bluefish	9.5	55	Klomklao et al. (2007c)
		Yellow tail	8.0	60	Kishimura et al. (2006a)
		Brown hekeling	8.0	50	Kishimura et al. (2006a)
		Spotted mackerel	8.0	60	Kishimura et al. (2006b)
		Jacopever	8.0	60	Kishimura et al. (2007)
		Elkhorn sculpin	8.0	50	Kishimura et al. (2007)
		Sardine	8.0	60	Bougatef et al. (2007)
		Walleye pollock	8.0	50	Kishimura et al. (2008)
	Chymotrypsin	Scallop	8.0–8.5	50–55	Chevalier et al. (1995)
		Anchovy	8.0	45	Heu et al. (1995)
		Monterey sardine	8.0	50	Castillo-Yanez et al. (2006)
		Crucian carp	7.5–8.0	45–50	Yang et al. (2009)
Cysteine protease	Cathepsin B	Carp	6.0	45	Aranishi et al. (1997a)
	Cathepsin L	Jumbo squid	4.5	55	Cardenas-Lopez and Haard (2009)
		Carp	5.5–60	50	Aranishi et al. (1997b)
	Cathepsin S	Carp	7.0	37	Pangkey et al. (2000)
Melloprotease	Collagenolytic melloprotease	Carb	7–7.5	25	Sivakumar et al. (1999)

characterized from the stomach of marine animals are pepsin, chymosin, and gastricsin (Simpson 2000).

Pepsin, one of the major proteases found in fish viscera, has an extracellular function as the major gastric protease. Pepsin, secreted as a zymogen (pepsinogen), is activated by the acid in stomach to an active form (Klomklao et al. 2007a). The activation of pepsin is initiated by the cleavage of proenzyme either by single step or multiple steps. Nalinanon et al. (2010) reported that pepsinogen from the stomach of albacore tuna was activated within 30–60 minutes under acidic condition, in which protein with MW of 36.8 kDa was found as the intermediate during activation process. Bovine pepsin is composed of a single polypeptide chain of 321 amino acids and has a MW of about 35 kDa (Simpson 2000). However, pepsins from marine animals have been reported to have MWs ranging from 27 to 42 kDa. The MWs of two pepsins (I and II) from orange roughy stomach were estimated to be approximately 33.5 kDa and 34.5 kDa, respectively (Xu et al. 1996). Sanchez-Chiang et al. (1987) reported that the MWs of two pepsins from stomach of salmon to be approximately 32 kDa and 27 kDa by gel filtration. MWs of two pepsins from polar cod stomachs were estimated by sodium dodecyl sulfate-polyacrylamide gel electrophoresis (SDS-PAGE) to be approximately 42 kDa and 40 kDa (Arunchalam and Haard 1985). The MWs of four pepsins from sea bream stomach were determined as approximately 30 kDa by SDS-PAGE (Zhou et al. 2007). Bougatef et al. (2008) reported that pepsin from smooth hound stomach were estimated to be 35 kDa using SDS-PAGE and gel filtration. The MWs of three pepsins from the stomach of European eel were determined as 30 kDa (Wu et al. 2009). Klomklao et al. (2007a) reported that pepsin A and B from stomach of pectoral rattail (*Coryphaenoides pectoralis*) had apparent MWs of 35 kDa and 31 kDa, respectively,

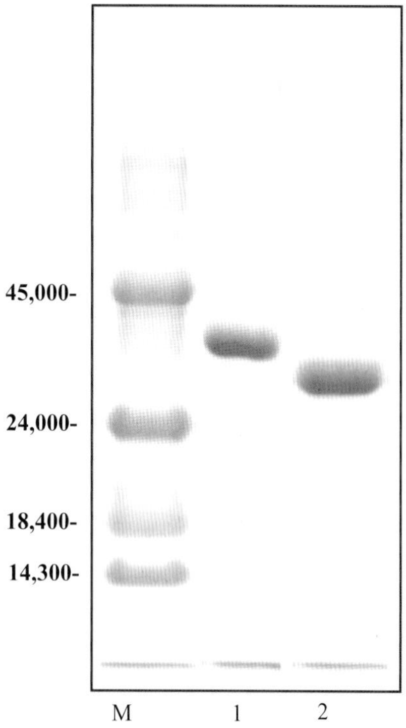

Figure 14.2. Protein pattern of purified pepsins A and B from pectoral rattail stomach determined by sodium dodecyl sulfate-polyacrylamide gel electrophoresis. M, MW standard; lane 1, pepsin A; lane 2, pepsin B (Klomklao et al. 2007a).

as estimated by SDS-PAGE (Fig. 14.2) and gel filtration on Sephacryl S-200.

Pepsins and pepsin-like enzymes can be extracted from the stomach of marine animals such as capelin (*Mallotus villosus*) (Gildberg and Raa 1983), polar cod (*Boreogadus saida*) (Arunchalam and Haard 1985) sardine (*Sardinops melanostica*) (Noda and Murakami 1981), Monterey sardine (*Sardinops sagax caerulea*) (Castillo-Yanez et al. 2004), sea bream (*Sparus latus* Houttuyn) (Zhou et al. 2007), pectoral rattail (*Coryphaenoides pectoralis*) (Klomklao et al. 2007a), smooth hound (*Mustelus mustelus*) (Bougatef et al. 2008), and European eel (*Anguilla anguilla*) (Wu et al. 2009).

Several methods have been described in the literature for the isolation and purification of pepsins from marine animals. Gildberg et al. (1990) purified pepsin from stomach of Atlantic cod (*Gadus morha*) by ammonium sulfate fractionation (20–70% saturation), followed by ion-exchange chromatography using S-Sepharose column. Klomklao et al. (2007a) purified two pepsins, A and B, from the stomach of pectoral rattail by acidification, ammonium sulfate precipitation (30–70% saturation), followed by a series of column chromatographies including Sephacryl S-200, diethylaminoethyl (DEAE)–cellulose and Sephadex G-50. Pepsin from the stomach of smooth hound was purified by 20–70% ammonium sulfate precipitation, Sephadex G-100 gel filtration, and DEAE–cellulose anion exchange chromatography (Bougatef et al. 2008). Recently, Wu et al. (2009) isolated pepsins from the stomach of freshwater fish European eel by ammonium sulfate fractionation, column chromatography on anion exchange (DEAE–Sephacel) and gel filtration (Sephacryl S-200 and Superdex G-75).

Pepsin activity is very dependent on pH values, temperatures, and the type of substrate. Hemoglobin is the substrate most commonly used for determination of pepsin activity (De-Vecchi and Coppes 1996, Klomklao et al. 2004). Pepsin from polar cod stomach exhibited a maximal activity against hemoglobin at pH 2.0 and 37°C (Arunchalam and Haard 1985). Gildberg et al. (1990) reported that the optimal pH of Atlantic cod pepsin for hemoglobin hydrolysis was 3.0. Fish pepsins were shown to hydrolyze hemoglobin much faster than casein (Gildberg and Raa 1983). Most fish species contain two or three major pepsins with an optimum hemoglobin digestion at pH between 2 and 4 (Gildberg and Raa 1983). Gildberg et al. (1990) found that the affinity of cod pepsin, especially pepsin I toward hemoglobin, was lower at pH 2 than at pH 3.5. Klomklao et al. (2007a) found that pepsins A and B from pectoral rattail stomach had maximal activity at pH 3.0 and 3.5, respectively, and had the same optimal temperature at 45°C using hemoglobin as the substrate. Pepsins I, II, and III purified from the stomach of European eel showed maximal activity at pH 3.5, 2.5, and 2.5, respectively, with hemoglobin as substrate (Wu et al. 2009). Bougatef et al. (2008) found that the optimal pH and temperature for the pepsin activity of smooth hound stomach were pH 2.0 and 40°C, respectively, using hemoglobin as a substrate.

Pepsins from marine species are quite stable from pH 2 to about 6, but rapidly lose activity at pH 6 and above due to the denaturation (Simpson 2000). Pepsin from sardine stomach was stable between pH 2 and 6 and showed drastic loss of activity at pH 7.0 (Noda and Murakami 1981). Castillo-Yanez et al. (2004) found that Monterey sardine acidic pepsin-like enzymes were stable at pH ranging from 3.0 to 6.0. Smooth hound pepsin was stable within the pH range of 1.0–4.0 (Bougatef et al. 2008). Klomklao et al. (2007a) also reported that both pepsins A and B from the stomach of pectoral rattail were stable in the pH range of 2.0–6.0. Pepsin from the stomach of albacore tuna was stable in the pH range of 2–5 (Nalinanon et al. 2010).

Chymosins (EC 3.4.23.4), commonly known as rennins, have also been described as acid proteases with some characteristics distinct from other acid proteases (Shahidi and Kamil 2001). For example, these enzymes are most active and stable around pH 7.0 unlike other acid proteases. They also have relatively narrower substrate specificity, compared with other acid proteases such as pepsin (Simpson 2000). Digestive proteases with chymosin-like activity were purified as zymogens from the gastric mucosa of young and adult seals (*Pagophilus groenlandicus*) (Shamsuzzaman and Haard 1984) by a series of chromatographies including DEAE–Sephadex A-50, Sephadex G-100, and Z-D-Phe-T-Sepharose gel. The enzyme had optimal pH of 2.2–3.5 for hemoglobin hydrolysis. The chymosins from marine animals did not hydrolyze the specific synthetic substrate for pepsin (i.e., *N*-acetyl-L-phenylalanine diiodotyrosine) and were also more susceptible to inactivation by urea (Simpson 2000).

Gastricsins, like pepsins, are aspartic proteases that have many properties in common with other enzymes in the family of gastric proteases. However, they differ from pepsins in structure and certain catalytic properties (Simpson 2000). Sanchez-Chiang and Ponce (1981) isolated and characterized two gastricsin isozymes from the gastric juices of hake (*Merluccius gayi*). The optimum pH for the hydrolysis of hemoglobin by hake gastricsins was 3.0, which is similar to that of mammalian gastricsins. Hake gastricsins were stable up to pH 10 but were rapidly inactivated at higher pH values (Sanchez-Chiang and Ponce 1981). The latter property would appear to distinguish the gastricsins from pepsins (Simpson 2000).

Serine Proteases

Serine proteases have been described as a group of endopeptidases with a serine residue, together with an imidazole group and aspartyl carboxyl group in their catalytic sites (Simpson 2000). The proteases in serine subclass all have the same first three digits: EC 3.1.21. A survey of proteolytic enzymes in various species of fish digestive tracts has revealed that serine proteases distributed in fish intestine possess high activity under alkaline rather than neutral pH (Shahidi and Kamil 2001). Among the serine proteases, trypsin, and chymotrypsin are the major serine proteases found in the digestive glands of marine animals.

Trypsins (EC 3.4.21.4) specifically hydrolyze proteins and peptides at the carboxyl side of arginine and lysine residues (Klomklao et al. 2006a). Trypsins play major roles in biological processes including digestion, activation of zymogens of chymotrypsin and other enzymes (Cao et al. 2000). Trypsins from marine animals resemble mammalian trypsins with respect to their molecular size (22–30 kDa), amino acid composition, and sensitivity to inhibitors. Their pH optima for the hydrolysis of various substrates were from 7.5 to 10.0, while their temperature optima for hydrolysis of those substrates ranged from 35°C to 65°C (De-Vecchi and Coppes 1996).

A number of studies on trypsins from fish viscera have been carried out. Two anionic trypsins (trypsin A and trypsin B) from the hepatopancreas of carp were purified using a series of chromatographies including DEAE–Sephacel, Ultrogel AcA54, and Q-Sepharose (Cao et al. 2000). Trypsin A was purified to homogeneity with a MW of 28 kDa, while trypsin B showed two close bands of 28.5 kDa and 28 kDa on SDS-PAGE. Trypsins A and B had optimal activity at 40°C and 45°C, respectively, and had the optimum pH of 9.0 using Boc-Phe-Ser-Arg-MCA as a substrate. Both enzymes were effectively inhibited by trypsin inhibitors. Trypsin from the pyloric ceca of Monterey sardine (*Sardinops sagax caerulea*) with MW of 25 kDa was purified and characterized by Castillo-Yanez et al. (2005). The optimum pH for activity was 8.0 and the maximal activity was found at 50°C. The purified enzyme was partially inhibited by 1.4 mg/mL phenylmethylsulfonyl fluoride and fully inhibited by 0.5 mg/mL soybean trypsin inhibitor or 2.0 mg/mL benzamidine, but was not inhibited by the metalloprotease inhibitor, 0.25 mg/mL ethylenediaminetetraacetic acid (EDTA.) In addition, trypsin was reported to be the major form of protease in the spleen of tongol tuna (*Thunnus tongol*) based on the MW, the inhibition by *N-p*-tosyl-L-lysine chloromethyl ketone (TLCK) and the activity toward specific substrates (Klomklao et al. 2006a). Two anionic trypsins (A and B) were purified from yellowfin tuna (*Thunnus albacores*) spleen. Trypsins A and B exhibited the maximal activity at 55°C and 65°C, respectively, and had the same optimal pH at 8.5 using *N-p*-tosyl-L-arginine methyl ester hydrochloride (TAME) as a substrate (Klomklao et al. 2006b). Klomklao et al. (2007d) purified trypsins from skipjack tuna (*Katsuwonus pelamis*) spleen by a series of chromatographies including Sephacryl S-200, Sephadex G-50, and DEAE-cellulose. Skipjack tuna spleen contained three trypsin isoforms, trypsins A, B, and C. The MW of all trypsin isoforms was estimated to be 24 kDa by size exclusion chromatography on Sephacryl S-200 and SDS-PAGE. The optimum pH and temperature for TAME hydrolysis of all trypsin isoforms were 8.5°C and 60°C, respectively.

Recently, Klomklao et al. (2009a) purified two isoforms of trypsin (A and B) from the intestine of skipjack tuna (*Katsuwonus pelamis*) by Sephacryl S-200, Sephadex G-50, and DEAE–cellulose. The MWs of both trypsins were 24 kDa as estimated by size exclusion chromatography and SDS-PAGE. Trypsin A and B exhibited maximal activity at 55°C and 60°C, respectively, and had the same optimal pH at 9.0. Trypsin from the pyloric ceca of pectoral rattail (*Coryphaenoides pectoralis*) was purified and characterized (Klomklao et al. 2009b). Purification was carried out by ammonium sulfate precipitation, followed by column chromatographies on Sephacryl S-200, DEAE–cellulose, and Sephadex G-50. Purified trypsin had an apparent MW of 24 kDa when analyzed using SDS-PAGE and size exclusion chromatography. Optimal profiles of pH and temperature of the enzyme were 8.5°C and 45°C, respectively, using TAME as substrate (Fig. 14.3).

Trypsins from marine animals tend to be more stable at alkaline pH, but are unstable at acidic pH, and in this respect differ from trypsins from mammals that are most stable under acidic condition (Simpson 2000, Klomklao et al. 2006a). Trypsin from tongol tuna spleen showed high pH stability within the range of pH 6–11, but the inactivation was more pronounced at pH values below 6 (Klomklao et al. 2006a). Klomklao et al. (2007d) reported that skipjack tuna spleen trypsins were stable in the pH ranging from 6.0 to 11.0 but were unstable at pH below 5.0. Also, trypsin purified from the pyloric ceca of pectoral rattail was stable in a wide pH range of 6–11 but unstable at pH below 5.0 (Klomklao et al. 2009b). The stability of trypsins at a particular pH might be related to the net charge of the enzyme at that pH (Castillo-Yanez et al. 2005). Trypsin might undergo the denaturation under acidic conditions, where the conformational change took place and enzyme could not bind to the substrate properly (Klomklao et al. 2006a, 2006b). For thermal stability, the thermal stability of fish trypsin varies with species as well as with incubation conditions (De-Vecchi and Coppes 1996).

N-terminal amino acid sequences are useful as tools to identify the type of enzymes and may be useful for designing primers for cDNA cloning of enzyme (Cao et al. 2000, Klomklao et al. 2007b). Table 14.2 shows the *N*-terminal amino acid sequences of fish trypsins compared with those of mammals. Generally, fish trypsins had a charged Glutamic acid residue at position 6, where threonine (Thr) is most common in mammalian pancreatic

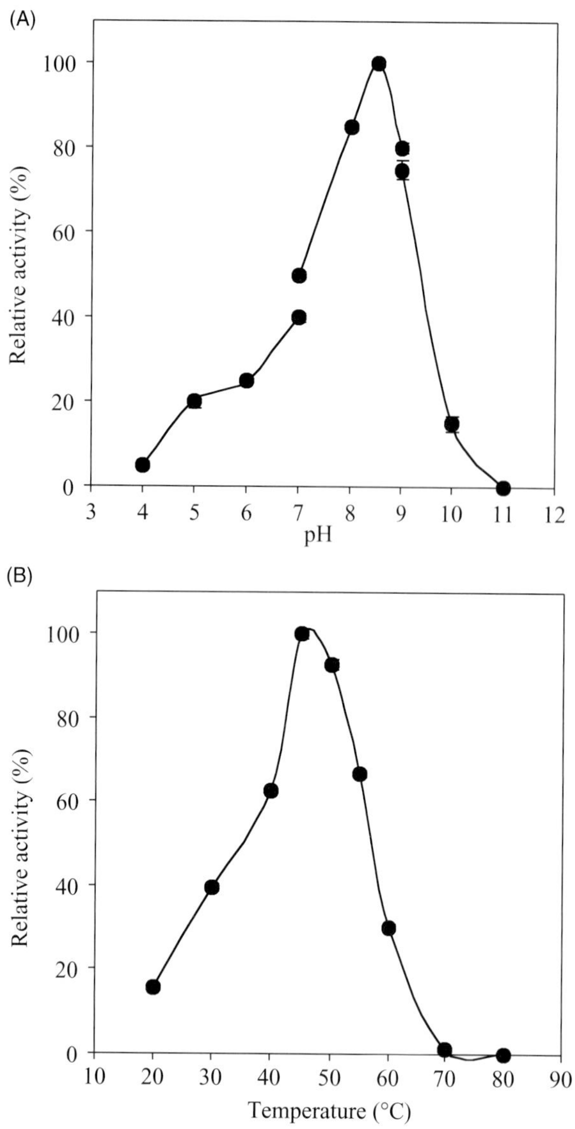

Figure 14.3. pH **(A)** and temperature **(B)** profiles of purified trypsin from the pyloric ceca of pectoral rattail (Klomklao et al. 2009b).

trypsins (Table 14.2). Moreover, the sequence of all trypsins starts with IVGG after limited proteolysis of inactive trypsinogen into the active trypsin (Cao et al. 2000, Klomklao et al. 2006b).

Chymotrypsin is another member of a large family of serine proteases functioning as a digestive enzyme. Chymotrypsins have been isolated and characterized from marine species such as anchovy (Heu et al. 1995), Atlantic cod (Asgeirsson and Bjarnason 1991), scallop (Chevalier et al. 1995), Monterey sardine (Castillo-Yanez et al. 2006), and crucian carp (Yang et al. 2009). In general, these enzymes are single-polypeptide molecules with MWs between 25 kDa and 28 kDa. They are most active within the pH range of 7.5–8.5 and are most stable at around pH 9.0 (Simpson 2000). Chymotrypsins have a much broader specificity than trypsins. They cleave peptide bonds involving amino acids with bulky side chains and nonpolar amino acids such as tyrosine (Tyr), phenylalanine, tryptophan, and leucine (Simpson 2000).

Thiol/Cysteine Proteases

Thiol or cysteine proteases are a group of endopeptidases that have cysteine and histidine residues as the essential groups in their catalytic sites. These enzymes require the thiol (-SH) group furnished by the active site cysteine residue to be intact, hence, this group is named "thiol" or "cysteine" proteases (Simpson 2000). The thiol proteases are inhibited by heavy metal ions and their derivatives, as well as by alkylating agents and oxidizing agents (Simpson 2000). The first three digits common to thiol proteases are EC 3.4.22.

Digestive cysteine proteases have been found in the viscera of fish and aquatic invertebrates. Digestive cysteine proteases from marine animals are most active at acidic pH and inactive at alkaline pH. Cathepsin B, cathepsin L, and cathepsin S are common examples of digestive thiol protease from marine animals (Simpson 2000). Various researchers have described different procedures for isolating marine cysteine or thiol proteases from the digestive glands of marine animals.

Cathepsin B was isolated from a few aquatic animals including the horse clam (Reid and Rauchert 1976), mussel (Zeef and Dennison 1988), and carp (Aranishi et al. 1997a). Generally, cathepsin B from marine animals is a single polypeptide chain with molecular sizes ranging from 13.6 to 25 kDa. Cathepsins from different species show maximum activity over a broad pH range of 3.5–8.0. Cathepsin B is activated by Cl^- ions, and requires sulfhydryl-reducing agents or metal-chelating agents for activity (Zeef and Dennison 1988).

Cathepsin L from carp hepatopancreas was purified by using ammonium sulfate precipitation and a series of chromatographies, in which the enzyme had an affinity toward Concanavalin A and Cibacron Blue F3GA (Aranishi et al. 1997b). Its homogeneity was established by a native-PAGE. Two protein bands corresponding to MWs of 30 kDa and 24 kDa were found on SDS-PAGE. The enzyme exhibited a maximal activity against Z-Phe-Arg-MCA at pH 5.5–6.0 and 50°C. All tested cysteine protease inhibitors, TLCK and chymostatin, markedly inhibited its activity, whereas the other serine protease inhibitors and metalloprotease inhibitors showed no inhibitory effects on the enzyme (Aranishi et al. 1997b). Recently, Cardenas-Lopez and Haard (2009) isolated cathepsin L from jumbo squid hepatopancreas by a two-step procedure involving ammonium sulfate precipitation and gel filtration chromatography. The MW of the enzyme was 24 kDa determined by SDS-PAGE and 23.7 kDa by mass spectrometry. The activity had an optimum pH of 4.5 and optimum temperature of 55°C using benzyloxycarbonyl-Phe-Arg-7-amino-4-trifluoro methyl coumarin as substrate.

Cathepsin S from hepatopancreas of carp (*Cyprinus carpio*) was purified by ammonium sulfate fractionation, followed by SP-Sepharose, Sephacryl S-200, and Q-Sepharose, respectively, (Pangkey et al. 2000). The MW of purified protease was 37 kDa estimated by SDS-PAGE. It hydrolyzed Z-Phe-Arg-MCA but not Z-Arg-MCA. The optimal pH and

Table 14.2. *N*-Terminal Amino Acid Sequence of Fish and Mammalian Trypsins

Source of Trypsin	*N*-Terminal Sequence	Reference
Vertebrate		
Pectoral rattail	IVGGYECQEHSQ	Klomklao et al. (2009a)
Skipjack tuna	IVGGYECQAHSQPPQVSLNA	Klomklao et al. (2009b)
Sardine	IVGGYECQKYSQ	Bougatef et al. (2007)
Yellowfin tuna	IVGGYECQAHSQPHQVSLNA	Klomklao et al. (2006b)
Tongol tuna	IVGGYECQAHSQPHQVSLNA	Klomklao et al. (2006a)
True sardine	IVGGYECKAYSQPWQVSLNS	Kishimura et al. (2006c)
Japanese anchovy	IVGGYECQAHSQPHTVSLNS	Kishimura et al. (2005)
Arabaesque greenling	IVGGYECTPHTQAHQVSLDS	Kishimura et al. (2006c)
Walleye pollok	IVGGYECTKHSQAHQVSLNS	Kishimura et al. (2008)
Spotted mackerel	IVGGYECTAHSQPHQVSLNS	Kishimura et al. (2006b)
Bluefish	IVGGYECKPKSAPVQVSLNL	Klomklao et al. (2007c)
Atlantic bonito	IVGGYECQAHSQPWQPVLNS	Klomklao et al. (2007b)
Cod	IVGGYECTKHSQAHQVSLNS	Gudmundsdottir et al. (1993)
Salmon	IVGGYECKAYSQTHQVSLNS	Male et al. (1995)
Invertebrate		
Cuttlefish	IVGGJESSPYNQ	Balti et al. (2009)
Starfish	IVGGKESSPHSR	Kishimura et al. (2002)
Mammal		
Porcine	IVGGYTCAANSVPYQVSLNS	Hermodson et al. (1973)
Bovine	IVGGYTCGANTVPYQVSLNS	Walsh (1970)

temperature for the hydrolysis of Z-Phe-Arg-MCA were 7.0 and 37°C, respectively. This protease activity was inhibited by E-64, leupeptin, 5–5'-dithiobis (2-nitro-benzoic acid), and *p*-tosyl-lys-chloromethylketone.

Metalloproteases

The metalloproteases are hydrolytic enzymes whose activity depends on the presence of bound divalent cations (Simpson 2000). There may be at least one tyrosyl residue and one imidazole residue associated with the catalytic sites of metalloproteases (Whitaker 1994). The metalloproteases are inhibited by chelating agents such as 1, 10-phenanthroline, EDTA, and sometimes by the simple process of dialysis. The metalloproteases have been characterized from marine animals (e.g., rockfish, carp, and squid mantle) but have not been found in the digestive glands except in the muscle tissue (Simpson 2000). Metalloproteases do not seem to be common in marine animals (Simpson 2000). However, Sivakumar et al. (1999) purified collagenolytic metalloprotease with gelatinase activity from carp hepatopancreas by ammonium sulfate fractionation and gel filtration chromatography. The enzyme had a MW of 55 kDa and was active against native type I collagen. Optimum temperature and pH were 25°C and 7–7.5, respectively. The enzyme was strongly inactivated by 10 mM EDTA.

Muscle Proteases from Marine Animals

Muscles of marine animals contain a variety of proteases. These enzymes perform different metabolic functions in the living organisms. Their activity depends on the species, life cycle, and the feeding status of the animal (Kolodziejska and Sikorski 1996). Endogenous proteases in fish muscle are found in the sarcoplasmic (or soluble) component of muscle tissues, in association with cellular organelles, connective tissues, and myofibrils and in the interfiber space (Kolodziejska and Sikorski 1996). A number of endogenous proteases have been investigated as enzymes contributing to postmortem softening of fish flesh, which also participate to a different extent in the degradation of myofibrillar protein (Visessanguan et al. 2001). The action of these enzymes exerts detrimental effects on the sensory quality and functional properties of muscle food (Visessanguan et al. 2001). The muscles of marine fish and invertebrates contain lysosomal cathepsins, as well as different proteases active in neutral and alkaline range of pH (Table 14.3).

Cathepsins

Lysosomal proteases such as cathepsin A (carboxypeptidase A), cathepsin B (carboxypeptidase B), cathepsin D, cathepsin H, and L have been recovered and identified from various fish and aquatic invertebrates (Shahidi and Kamil 2001). Except for cathepsin D, all other lysosomal proteases belong to serine or cysteine (thiol). However, cathepsin D, which is an aspartic protease, has its pH optimum within the acidic range (Simpson 2000).

Cathepsin A (EC 3.4.16.1) is now called carboxypeptidase A. The enzyme has a pH optimum of 5–6 and is readily inactivated by heat and alkaline, and has a molecular mass of about 35 kDa. Cathepsin A has been purified about 1700-fold to homogeneity from carp muscle (Toyohara et al. 1982). The molecular mass was estimated to be 36 kDa and the pH optimum was 5.0.

Table 14.3. Muscle Proteases from Marine Animals

Major Group	Family	Enzyme	Purified Organism	Reference
Lysosomal cathepsin	Cysteine	Cathepsin A	Carp	Toyohara et al. (1982)
			Milkfish	Jiang et al. (1990)
			Atlantic cod	McLay (1980)
		Cathepsin B	Carp	Makinodan and Ikeda (1971)
			Grey mullet	Bonete et al. (1984)
			Tilapia	Sherekar et al. (1988)
			Mackerel	Aoki et al. (2002)
		Cathepsin C	Atlantic squid	Hameed and Haard (1985)
			Pacific whiting	Erickson et al. (1983)
		Cathepsin L	Chum salmon	Yamashita and Konagaya (1990)
			Spotted mackerel	Lee et al. (1993)
			Anchovy	Heu et al. (1997)
			Pacific whiting	Visessanguan et al. (2001)
	Aspartic	Cathepsin D	True sardine	Klomklao et al. (2008)
			Herring	Nielsen and Nielsen (2001)
Alkaline proteases	Serine	Trypsin-like	Atlantic menhaden	Choi et al. (1999)
			Bigeye snapper	Benjakul et al. (2003a)
Neutral proteases	Cysteine	Calpains	Tilapia	Wang et al. (1993)
			Octopus	Hatzizisis et al. (1996)
			Sea bass	Ladrat et al. (2000)
			Rainbow trout	Saito et al. (2007)
	Metalloprotease	Neutral metalloprotease	Crucian carp	Kinoshita et al. (1990)

Two cathepsin A-like proteases (15.6 kDa and 35.6 kDa) have been purified from milkfish muscle and had optimal activity with carbobenzoyl-L-Gly-L-Phe at pH 7.0 (Jiang et al. 1990). Cathepsin A, partially purified from Atlantic cod muscle, hydrolyzed carbobenzoyl-α-Glutamyl-L-Tyr with an optimum of pH 5.0 (McLay 1980). Cathepsin A-like activity has also been detected in muscle of tilapia, Bombay duck, pomfret, and shrimp by Sherekar et al. (1988) but was not detected in cod muscle (Shahidi and Kamil 2001).

Cathepsin B (EC 3.4.22.1) is the best known and most thoroughly investigated lysosomal thiol protease. It is now recognized that at least two enzymes have this activity including cathepsin B (B_1) and B_2. Cathepsin B_1 is a thiol endopeptidase, having a MW of 24–28 kDa with a pH of 5.0–5.2, and maximal activity at pH 6.0. In contrast, cathepsin B_2 is an exopeptidase also called carboxypeptidase B, with a MW ranging between 47 kDa and 52 kDa (Kang and Lanier 2000). Cathepsin B_3, now called cathepsin H, is a 25-kDa glycoprotein that is very heat stable. Cathepsin B purified from carp (Makinodan and Ikeda 1971), grey mullet (Bonete et al. 1984), tilapia (Sherekar et al. 1988), and mackerel (Aoki et al. 2002) had the MWs of 23–29 kDa and optimal pH range of 5.5–6.0. Aoki et al. (2002) reported that cathepsin B purified from the white muscle of common mackerel had a MW of 23 kDa. The optimal pH for activity was 5.5.

Cathepsin C (EC 3.4.4.9), an exopeptidase, is also called dipeptidyl transferase and dipeptidyl-aminopeptidase. It was first recognized as an enzyme that deaminates Gly-Phe-NH_2. It is unlikely to act on intact proteins directly, but rather has the highest specific activity among all the lysosomal peptidase, suggesting that it further digests the resulting peptide fragments by cathepsin D. It is an oligomeric thiol protease with a MW of 200 kDa requiring Cl^- ions for activity (Ashie and Simpson 1997). The enzyme is optimally active within a pH range of 5–6 (Ashie and Simpson 1997). Hameed and Haard (1985) reported that cathepsin C extracted from Atlantic squid muscle had MW of 25 kDa and was Cl^- and sulfhydryl dependent, being inhibited by sulfhydryl enzyme inhibitors, such as iodoacetate, PCMB, and $HgCl_2$. For cathepsin C from sarcoplasmic fluid of Pacific whiting, it showed the maximal activity at pH 7.0, which was greater than the maximal activity obtained at pH 6.0 in cod (Erickson et al. 1983).

Cathepsin D (EC 3.4.4.23) is a major lysosomal aspartic protease, which is involved in the cellular degradation of proteins (Nielsen and Neilsen 2001). Cathepsin D is a 42–60 kDa endopeptidase that is active within a pH range of 3–5. It exists in several isomeric forms with pH ranging between 5.7 and 6.8, and exhibits a broad range of activity on muscle tissue between pH 3.0 and 6.0 (Ashie and Simpson 1997). The enzyme preferentially cleaves peptide bonds between hydrophobic amino acid residues (Ashie and Simpson 1997). Cathepsin D is strongly inhibited by pepstatin A, a specific inhibitor of carboxyl proteases, whereas thiol protease inhibitors have negligible effects on enzyme activity (Bonete et al. 1984). Cathepsin D was purified from herring muscle. The purified enzyme is a monomer with a MW of 38–39 kDa. It is inhibited by pepstatin and has an optimal activity at pH 2.5 when hemoglobin is used as the substrate. The isoelectric point is 6.8 (Nielsen and Nielsen 2001). Klomklao et al. (2008) reported that the predominant enzyme responsible for autolysis in true sardine was a cathepsin D, which exhibited the maximal activity at 55°C, pH 3.5 and was effectively inhibited by pepstatin A.

Cathepsin L (EC 3.4.22.15), is a typical cysteine protease found in lysosomes and is considered to have major biological roles in proteolysis. It is a 24-kDa molecule, existing in multiple forms with pH between 5.9 and 6.1 and it exhibits the activity in a wide range of pH (3.0–6.5). It is strongly inhibited by iodoacetate, leupeptin, and antipain (Kang and Lanier 2000). Visessanguan et al. (2003) purified cathepsin L from arrowtooth flounder muscle by heat treatment, followed by a series of chromatographic separations. The apparent molecular mass of the purified enzyme was 27 kDa by size exclusion chromatography and SDS-PAGE. The enzyme had high affinity and activity toward Z-Phe-Arg-NMec. Activity was inhibited by sulfhydryl reagents and activated by reducing agents. The purified enzyme displayed optimal activity at pH 5.0–5.5 and 60°C, respectively. Cathepsin L from the muscle of anchovy was purified and characterized (Heu et al. 1997). The enzyme was activated by thiol reagents and inhibited by thiol-blocking reagents. The MW was estimated to be 25.8 kDa by SDS-PAGE. The enzyme exhibited its maximal activity at pH 6.0 and 50°C. Cathepsin L is a major protease, which degrades myofibrillar proteins in antemortem or postmortem muscle of chum salmon (Yamashita and Konagaya 1990), spotted mackerel (Lee et al. 1993), and Pacific whiting (Visessanguan et al. 2001). Cathepsin L is capable of hydrolyzing a broad range of proteins including myosin, actin, nebulin, cytosolic proteins, collagen, and elastin (Visessanguan et al. 2001).

Alkaline Proteases

Alkaline proteases are located in the muscle sarcoplasm, microsomal fraction or are bound to myofibrils (Makinodan and Ikeda 1971). Alkaline proteases are active at neutral to slightly alkaline pH and are unstable under acidic conditions (Simpson 2000). They are sometimes characterized as trypsin-like serine enzyme, which show great capacity of degrading intact myofibrils in vitro (Busconi et al. 1987). Generally, alkaline proteases are heat stable and also become active at the neutral pH of fish meat paste. Properties of alkaline proteases vary with sources. Alkaline proteases are oligomeric protease with high MWs ranging from 560 to 920 kDa (Kolodziejska and Sikorski 1996). However, an alkaline protease isolated from Atlantic croaker was much smaller with MW of 80–84 kDa (Lin and Lanier 1980). These enzymes normally exhibit little or no catalytic activity unless assayed at a nonphysiologically high temperature (60–65°C) or activated by protein denaturing agents such as urea, fatty acids, or detergents (Visessanguan et al. 2001). Two alkaline proteases (A and B) were purified from Atlantic menhaden muscle (Choi et al. 1999). The MWs of purified enzymes were 707 kDa and 450 kDa, respectively. Both are probably serine proteases, and optimum caseinolytic activity was observed at pH 8 and 55°C. Purification of a novel myofibril-bound serine protease from the ordinary muscle of the carp was carried out by solubilizing it from the myofibril fraction with acid treatment, followed by Ultrogel AcA 54 and Arginine-Sepharose 4B chromatography (Osatomi et al. 1997). The MW of purified enzyme was 30 kDa by SDS-PAGE and gel filtration. The optimum pH and temperature of the enzyme were 8.0 and 55°C, respectively. Benjakul et al. (2003a) reported that heat stable alkaline proteases purified from bigeye snapper had the optimum pH and temperature for casein hydrolysis at 8.5 and 60°C, respectively. The MW of purified enzyme was estimated to be 72 kDa by gel filtration.

Neutral Ca^{2+}-Activated Proteases

Neutral proteases in skeletal muscle have been classified with the Ca^{2+}-activated, neutral endopeptidases known as CANP, and recently as calpains (EC 3.4.33.17) (Kolodziejska and Sikorski 1996). These enzymes are further subclassified into μ-calpain and *m*-calpain, which differ in sensitivity to calcium ions. Both are heterodimers: the large subunit and the small subunit for μ- and *m*-calpain, which have MWs near 80 kDa and 28 kDa, respectively (Cheret et al. 2007). The calcium requirement of μ- and *m*-calpain was 50–70 μM and 1–5 mM, respectively. Calpastatin is known to be the endogenous specific inhibitor of calpain (Kolodziejska and Sikorski 1996). Calpains have been isolated and characterized from the muscles of seafood species like carp, American lobster, and scallop (Toyohara et al. 1982, Kolodziejska and Sikorski 1996). Wang et al. (1993) purified *m*-calpains from tilapia muscle by Phenyl-Sepharose CL-4B, Sephacryl S-200, Q-Sepharose, and Superose 12 HR column chromatography. The enzyme had a MW of 110,000 containing two subunits of 80 kDa and 28 kDa. It has optimal temperature and pH of 30°C and 7.5, respectively. The protease was activated by dithiothreitol, glutathione, and β-mercaptoethanol and was inhibited by calpain inhibitor I, calpain inhibitor II, leupeptin, antipain, iodoacetic acid, but not affected by pepstatin A and *N*-ethylmaleimide. Calpain was purified to homogeneity from the arm muscle of *Octopus vulgaris* (Hatzizisis et al. 1996). The enzyme consists of a 65-kDa protein when separated by SDS-PAGE. The Ca^{2+} requirements for its half maximal and maximal activities are 1.5 mM and 7 mM, respectively. It is strongly inhibited by thiol protease inhibitors such as leupeptin, E-64, and antipain and by alkylating agents such as iodoacetic acid and iodoacetamide. The enzyme displayed maximal activity at 30°C and showed broad pH optimum between 6.5 and 7.5.

Fish Protease Applications

Proteases are by far the most studied enzymes for industrial bioprocessing. Almost half of all industrial enzymes are proteases, mostly used in the detergent, leather, and food industries (Klomklao et al. 2005). The food industry uses proteases as processing aids for many products including baked goods, beer, wine, cereals, milk, fish products, and legumes. Production of protein hydrolysates and flavor extracts has been known to be successful by using selected proteases (Table 14.4) (Klomklao 2008).

Commercial supplies of food processing enzymes are presently derived from various plants, animal, and microbial sources. Compared with these enzymes used in food processing,

Table 14.4. Some Uses of Proteolytic Enzymes in Food Industry

Commodity	Application
Cereals, baked goods	Increase drying rate of proteins; improve product handling. Characteristics: decrease dough mixing time; improve texture and loaf volume of bread; and decrease dough mixing time
Egg and egg products	Improve quality of dried products
Meats	Tenderization; recover protein from bones; hydrolysis of blood proteins
Fish	Fish protein hydrolysates, viscosity reduction, skin removal, roe processing, acceleration of fish sauce production, recover carotenoprotein from shrimp shell
Pulses	Tofu; soy sauce; protein hydrolysis; off-flavor removal soy milk
Dairy	Cheese curd formation; accelerate cheese aging; rennet puddings
Brewing	Fermentation and filtration aid; chill proofing
Wine	Clarification; decrease foaming, promote malolactic fermentation
Coco	Facilitate fermentation for chocolate production

the use of proteases from marine fish is still rare. Fish digestive enzymes can be recovered as by-products from fish processing by-products (Haard 1992). However, most proteases from marine animals are extracellular digestive enzymes with characteristics differing from homologous proteases from warm-blooded animals (De-Vecchi and Coppes 1996). They are more active catalysts at relatively low temperature, compared with similar enzymes from mammals, thermophilic organisms, and plants (Klomklao 2008). Low temperature processing could provide various benefits, such as low thermal costs, protection of substrates or products from thermal degradation and/or denaturation, and minimization of unwanted side reactions (Haard 1992). Certain fish enzymes are excellent catalysts at low temperature, which is advantageous in some food processing operations (Simpson 2000). For example, cold-adopted pepsins are very effective for cold renneting milk clotting because of the high activity at low reaction temperatures (Simpson 2000). Moreover, fish digestive enzymes possessing other unique properties might make them better suited as food processing aids.

Recently, the use of alkaline proteases from marine digestive organs, especially trypsin, has increased remarkably since they are both stable and active under harsh conditions, such as at temperatures of 50–60°C, high pHs, and in the presence of surfactants or oxidizing agents (Klomklao et al. 2005). Trypsin can be used for extraction of carotenoprotein from shrimp processing wastes. Cano-Lopez et al. (1987) reported that using trypsin from Atlantic cod pyloric ceca in conjunction with a chelating agent (EDTA) in the extraction medium increased the efficacy in recovering both protein and pigment from crustacean wastes. This method has facilitated the recovery of as much as 80% of astaxanthin and protein from shrimp wastes as carotenoprotein complex. Recently, Klomklao et al. (2009c) also recovered carotenoproteins from black tiger shrimp waste by using trypsin from the pyloric ceca of bluefish. The product obtained contained higher protein and pigment content than those of black tiger shrimp waste and had low contents of chitin and ash.

The enzymes recovered from fish have also been successfully used as seafood processing aids including the acceleration of fish sauce fermentation. Chaveesuk et al. (1993) reported that the supplementation with trypsin and chymotrypsin significantly increased protein hydrolysis of fish sauce. Fish sauce prepared from herring with enzyme supplementation contained significantly more total nitrogen, soluble protein, free amino acid content, and total amino acid content, compared to fish sauce with no added enzyme (Chaveesuk et al. 1993). By supplementing minced capelin with 5–10% enzyme-rich cod pyloric ceca, a good recovery of fish sauce protein (60%) was obtained after 6 months of storage (Gildberg 2001). In addition, Klomklao et al. (2006c) reported that fish sauce prepared from sardine with spleen supplementation contained greater total nitrogen, amino nitrogen, and ammonia nitrogen contents than those without spleen supplementation throughout the fermentation. Therefore, the addition of spleen can accelerate the liquefaction of sardine for fish sauce production.

Proteolytic enzymes also show profound effects on collagen extraction. Generally, use of pepsin in combination with acid extraction increased in the yield of collagen. Nagai et al. (2002) reported that the yield of pepsin-solubilized collagen was higher (44.7%) than acid-solubilized collagen (10.7%). Nagai and Suzuki (2002) found that the collagen extracted from the outer skin of the paper nautilus was hardly solubilized in 0.5 M acetic acid. The insoluble matter was easily digested by 10% pepsin (w/v), and a large amount of collagen was obtained with 50% yield (pepsin-solubilized collagen). Collagen from the outer skin of cuttlefish (*Sepia lycidas*) was also extracted by Nagai et al. (2001). The initial extraction of the cuttlefish outer skin in acetic acid yielded only 2% of collagen (dry weight basis). With a subsequent digestion of the residue with 10% pepsin (w/v), a solubilized collagen was obtained with a yield of 35% (dry weight basis). Pepsin solubilized collagen was extracted from the skin of grass carp (*Ctenopharyngodon idella*) with a yield of 35% (dry weight basis) (Zhang et al. 2007). Recently, Nalinanon et al. (2007) studied the use of fish pepsin as processing aid to increase the yield of collagen extracted from fish skin. Addition of bigeye snapper pepsin resulted in the increased content of collagen extracted from bigeye snapper skin. The yields of collagen from bigeye snapper skin extracted for 48 hours with acid and with bigeye snapper pepsin were 5.31% and 18.74%

(dry basis), respectively. With pre-swelling in acid for 24 hours, collagen extracted with bigeye snapper pepsin for 48 hours had the yield of 19.79%, which was greater than that of collagen extracted using porcine pepsin at the same level (13.03%) (Nalinanon et al. 2007).

TRANSGLUTAMINASE

Transglutaminase (TGase; EC 2.3.2.13) is a transferase whose systematic name is protein-glutamine γ-glutamyltransferase. The enzyme catalyzes the acyl transfer reaction in which the γ-carboxyamine groups of glutamine residues in proteins, peptides, and various primary amines, act as acyl donors and primary amino groups including ε-amino groups of lysine residues, either as peptide-proteins bound or free lysine, act as the acyl acceptor (Folk 1980, Worratao and Youngsawatdigul 2005). When acceptors are ε-amino groups of lysine residues, the formation of ε-(γ-glutamyl)lysine linkages occur both intra-and intermolecularly (Folk 1980). This occurs through the exchange of the ε-amino groups of the lysine residues for the ammonia at the γ-carboxyamide group of a glutamine residue in a protein molecule (Fig. 14.4). Formation of covalent cross-links between proteins is the basis of the ability of TGase to modify the physical characteristics of protein foods (Ashie and Lanier 2000). These bonds are stable and resistant to proteolysis (Worratao and Youngsawatdigul 2005). Amino acid sequence and the charge of amino acids surrounding the susceptible glutamine residue and local secondary structures affected the cross-linking capability (Folk 1980).

TGase has been widely used to improve functional properties of various food proteins (Benjakul et al. 2004a). TGase is an endogenous fish enzyme responsible for cross-linking of proteins and change textural properties of fish sol during processing, especially surimi production (Ashie and Lanier 2000). This is known to reinforce the thermal myosin gelation that occurs during surimi production by the formation of intra- and intermolecular cross-links between myosin heavy chains known as setting. This setting phenomenon, called suwari, is one of the important processes for manufacturing surimi-based products (Shahidi and Kamil 2001).

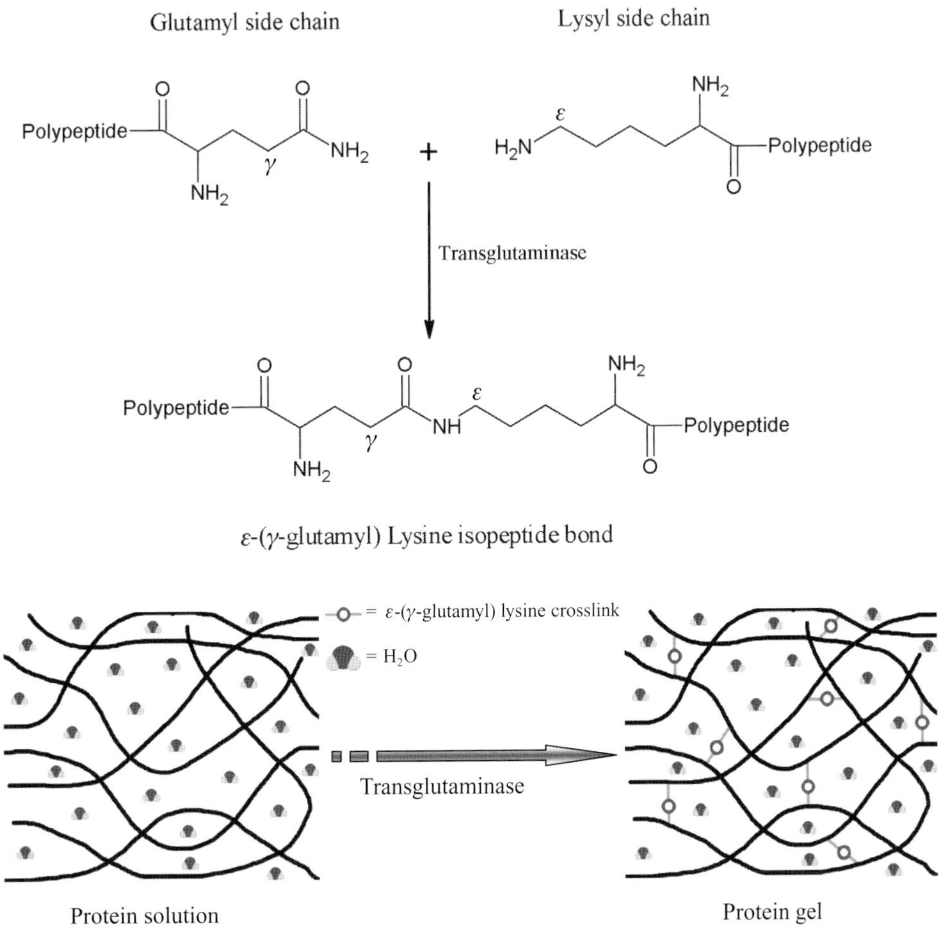

Figure 14.4. Cross-linking of proteins by transglutaminase.

Endogenous TGases from the muscle tissues of different fish species vary in properties such as different MWs, pH, and temperature optima. TGase activity has been found in muscle of red sea bream (*Pagrus major*), rainbow trout (*Oncorhynchus mykiss*), atka mackerel (*Pleurogrammus azonus*) (Muruyama et al. 1995), walleye pollock liver (*Theragra chalcogramma*) (Kumazawa et al. 1996), muscle of scallop (*Patinopecten yessoensis*), botan shrimp (*Pandalus nipponensis*), and squid (*Todarodes pacificus*) (Nozawa et al. 1997). Worratao and Yongsawatdigul (2005) demonstrated that tropical tilapia (*Oreochromis niloticus*) muscle contained high TGase activity. Generally, the optimum temperature of TGase ranged from 25°C to 55°C, depending on fish species. Nozawa et al. (1997) described TGase in fish muscle as calcium dependent. Worratao and Youngsawatdigul (2005) also reported that purified tilapia TGase had an absolute requirement for calcium ions. TGase activity increased with Ca^{2+} concentration and reached the maximum at 1.25 mM. Optimal Ca^{2+} concentrations for TGase from red sea bream liver, Japanese oyster, scallop, and pollock liver were at 0.5 mM, 25 mM, 10 mM, and 3 mM, respectively (Yasueda et al. 1994, Kumazawa et al. 1996,, Nozawa et al. 2001). It was postulated that the calcium ions induced conformational changes of the enzyme that consequently exposed the cysteine located at the active site to a substrate (Ashie and Lanier 2000). Noguchi et al. (2001) reported that the calcium ions attach to a binding site of red sea bream TGase molecule, resulting in conformational changes. Subsequently, Tyr covering the catalysis cystine (Cys) was removed. Then, the acyl donor binds with the Cys at the active site, forming an acyl-enzyme intermediate.

POLYPHENOLOXIDASE

Polyphenoloxidase (PPO, EC 1.14.18.1) is a copper enzyme (Goulart et al. 2003), whose active site contains two copper atoms, each of which is liganded to three histidine residues (Fig. 14.5) (Whitaker 1995, Kim et al. 2000). This copper pair is the site of interaction of PPO with both molecular oxygen and its phenolic substrates (Kim et al. 2000). PPO is also known as phenoloxidasc (PO), polyphenolase, phenolase, tyrosinase, and cresolase (Whitaker 1995). This enzyme catalyzes the hydroxylation of monophenols to *o*-diphenols and the oxidation of *o*-diphenols to *o*-quinones. Quinones are highly reactive products and can polymerize spontaneously to form brown pigments (melanin), or react with amino acids and proteins that enhance the brown color produced (Fig. 14.6) (Goulart et al. 2003, Martinez-Alvarez et al. 2008). Enzyme isolated from various sources exhibited different activities depending on the substrates used. Tyr has been reported to be a natural substrate for PO activity in crustacea. Hydroxylation of Tyr leads to the formation of dihydroxyphenylalanine (DOPA) (Kim et al. 2000), which is then converted to dopaquinone and water and can be readily converted to the orange-red pigment, dopachrome (Simpson et al. 1987).

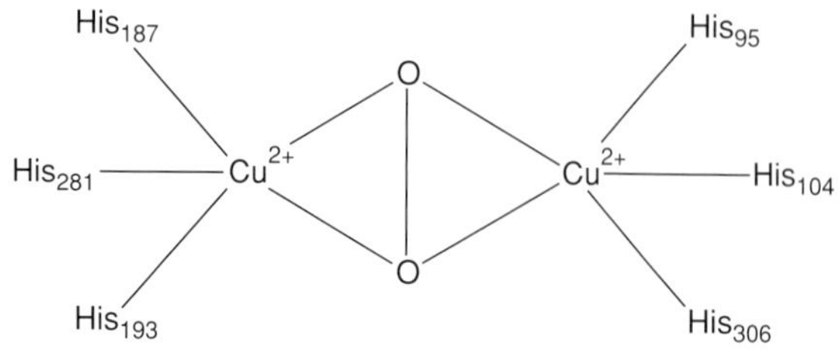

Figure 14.5. Coordination of copper to six histidine residues in active site of polyphenoloxidase.

PO has been shown to play a central role in the innate immune system of crustacea (Williams et al. 2003). This enzyme is expressed from their hemocytes as its precursor, prophenoloxidase (proPO) and functions in various phases of the live organism such as sclerotization, pigmentation, wound healing on cuticles, and defense reaction (Adachi et al. 2001). PO has been detected in the hemocyte, hemolymph, cuticle, accessory gland, egg case, salivary gland, and midgut in insects (Adachi et al. 1999). In crustaceans, PO is localized in different parts. It is found on the exoskeleton, chiefly on the region of the pleopods' connection (Montero et al. 2001).

The activation of proPO in hemocyte, or hemolymph, is initiated by minute amounts of bacterial or fungal cell wall components (β-1,3-glucans, lipopolysaccharides, and peptidoglycans) and mediated by several factors including protease cascade reaction (Adachi et al. 1999, Williams et al. 2003). The serine protease that exhibits trypsin-like activity cleaves the proPO between an arginine residue and a Thr residue to yield active PO (Williams et al. 2003). The proPO of kuruma prawn homocyte was activated by SDS and methanol but not by trypsin (Adachi et al. 1999).

PPO has been found in crustaceans such as shrimp (Chen et al. 1997), lobster (Chen et al. 1991), pink shrimp (Simpson et al. 1988), white shrimp (Simpson et al. 1987), black tiger shrimp (Rolle et al. 1991), Florida prawn (Ali et al. 1994), kuruma prawn (Benjakul et al. 2005), imperial tiger prawn (Montero et al. 2001), and deep water pink shrimp (Zamorano et al. 2009). PPO is distributed in many parts of shrimps with different levels of activity (Montero et al. 2001). PO is localized in the carapace of the cephalothorax, in the caudal zone and in the cuticle of the abdomen, mainly where the cuticle segments are joined and where the cuticle is connected to the pleopods (Ogawa et al. 1984). Zamorano et al. (2009) studied the tissue distribution of PPO in deepwater pink shrimp (*Parapenaeus longirostris*) postmortem. PPO activity was highest in the carapace, followed by that in the abdomen exoskeleton, cephalotorax, pleopods, and telson. No PPO activity was found in the abdominal muscle and in the pereopods and maxillipeds. Storage of whole shrimps and of the different organs showed that melanosis required the presence of the cephalotorax to be initiated, indicating that its development depends on other factions in addition to the PPO levels.

Factors Affecting PPO Activity

PPO differs in isoforms, latency, catalytic behavior, MW, specificity, and hydrophobicity. In addition, optimum pH, thermal

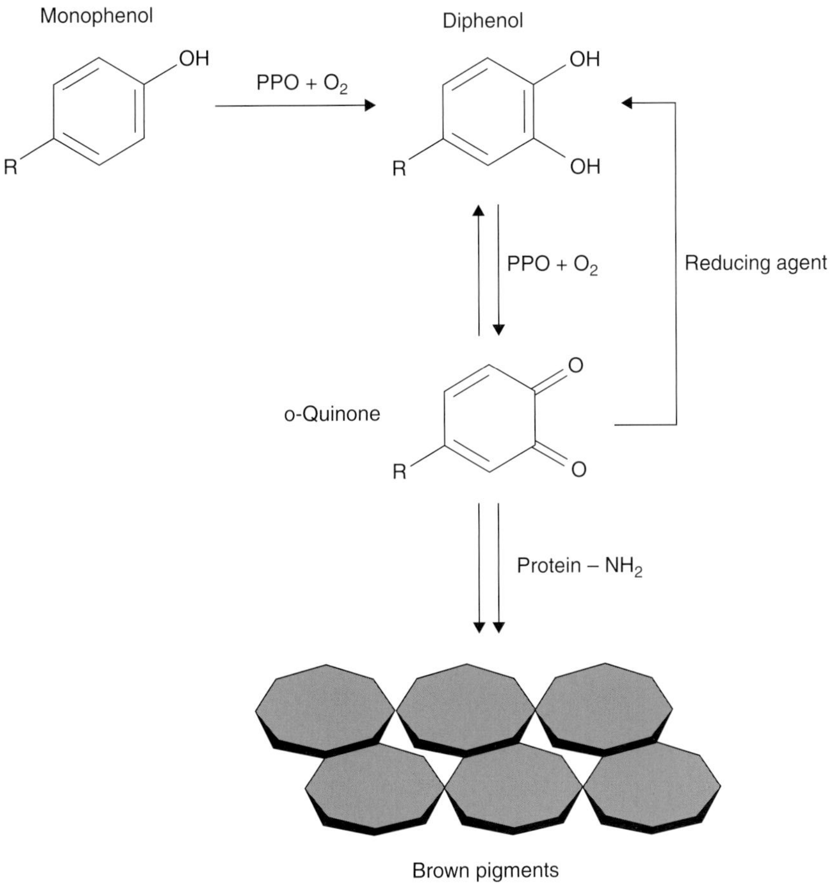

Figure 14.6. Enzymatic oxidation induced by PPO.

stability, and kinetic parameters can vary with species, body part, and so forth. (Montero et al. 2001).

Temperature

PPO shows different optimal temperatures among various animals or species. Montero et al. (2001) reported that the optimum temperature of PPO from imperial tiger prawn was between 40°C and 60°C. Thermal stability was considerably reduced when the enzyme extract was subjected to temperatures up to 35°C. Rolle et al. (1991) observed that the PPO from Taiwanese tiger shrimp (*Penaeus monodon*) exhibited a temperature optimum of 45°C and showed stability over the temperature range of 30–35°C. The optimum temperature for PPO from the imperial tiger prawn carapace was between 40°C and 60°C; however, thermal stability was greatest at temperatures below 35°C (Montero et al. 2001). Zamorano et al. (2009) found that the maximal activity of PPO of deepwater pink shrimp was observed in the 15–60°C range. Benjakul et al. (2005) reported that the PO from the kuruma prawn cephalothorax showed the maximum activity at 35°C and the enzyme was unstable at temperatures above 50°C.

The determination of the temperature/time profiles required to activate the proenzyme and to prevent melanosis and improve postprocessing quality and acceptability is a means to regulate or control the quality of crustaceans deteriorated by melanosis (Williams et al. 2003).

pH

The optimum pH depends largely on the physiological pH in which the enzyme activity occurs in nature. For example, the pH of the carapace of the cephalothorax was 7.16, while in the abdominal cuticle, it was 8.76. Therefore, the enzymes would probably present different optimum pH characteristics, depending on the locus of extraction. Montero et al. (2001) reported that high PPO activity occurred at pH 5 and 8 in *Penaeus japonicus*. PPO from deepwater pink shrimp had the highest activity at pH 4.5 and was most stable at pH 4.5 and 9.0. Benjakul et al. (2005) found that PO from kuruma prawn cephalotorax showed the maximal activity at pH 6.5 and it was stable in a wide pH range of 3–10. The stability of the PO was very different depending on a number of factors such as temperatures, pH, substrate used, ionic strength, buffer system, and time of incubation (Kim et al.

Table 14.5. Characteristic of polyphenoloxidase from Crustaceans

Source	MW (kDa)	pH Optimum	pH Stability	Temperature Optimum (°C)	Reference
Kuruma prawn	160	6.5	3–10	35	Benjakul et al. (2005)
Imperial tiger prawn	–	5–8	Basic pH	40–60	Montero et al. (2001)
Deepwater pink shrimp	200	4.5	4.5–9.0	15–60	Zamorano et al. (2009)
Clam	76.9	7.0	-	40	Cong et al. (2005)
White shrimp	30	6.5–7.5	8.0	45	Simpson et al. (1987)
Lobster	–	6.5	7.5	30	Opoku-Gyamfua et al. (1992)
Crab	64.5	6.0	–	40	Liu et al. (2006)

2000). The characteristics of PPO from different crustaceans are summarized in Table 14.5.

Activators/Inhibitors

One characteristic of PPO is its ability to exist in an inactive or latent state. PPO can be released from latency by several agents or treatments such as incubation with ammonium sulfate, anionic detergents, carboxymethylcellulose, urea, guanidine salts, short-time acid or base treatment or a limited proteolysis (Whitaker 1995), and alcohol (methanol) (Adachi et al. 1999).

Several factors have been reported to influence PPO activity. They include: (a) a proenzyme, (b) a PPO-bound inhibitor, and (c) a conformation change (Gauillard and Richard-Forget 1997). Chazarra et al. (1997) reported that the presence of SDS in the assay medium provokes a conformation change in the latent iceberg lettuce PPO, leading to an increase in PPO activity with increasing concentration. Adachi et al. (1999) and Mazzafera and Robinson (2000) also found that proPPO from kuruma prawn was activated by SDS. Some ion or salts affect the PPO activity. $MnSO_4$ and $CaCl_2$ enhanced PPO activity (Adachi et al. 1999). Kim et al. (2000) observed that the activity was increased by addition of 1 mM Mg^{2+}, K^+, or Cu^{2+}. However, the activity of PPO was dependent on the concentrations of divalent cations (Adachi et al. 1999). Adachi et al. (1999) reported that the highest activity of PPO from kuruma prawn was obtained at a concentration of 50 mM for both Mg^{2+} and Ca^{2+}. Ascorbic acid, cysteine, glutathione, thiourea, and hexylresorcinol were demonstrated to control enzymatic browning induced by PPO (Guerrero-Beltran et al. 2005). Benjakul et al. (2005) reported that phenylthiourea, a copper-chelating agent, and cysteine exhibited inhibitory effects against PPO from kuruma prawn. Total inhibition of the extract from imperial tiger prawn was achieved with ascorbic acid and citric acid at pH 3.0 (Montero et al. 2001). Nirmal and Benjakul (2009a) found that ferulic acid and oxygenated ferulic acid inhibited PPO from Pacific white shrimp, and ferulic acid was the more effective inhibitor than oxygenated ferulic acid. Recently, Nirmal and Benjakul (2009b) reported that catechin showed inhibitory activity toward PPO of Pacific white shrimp in a dose-dependent manner.

Melanosis

Melanosis is a process that is triggered by the oxidation of phenols to quinones induced by PPO or PO (Montero et al. 2001). PO is responsible for a discoloration in crustacean species such as lobster, shrimp, and crab. The postmortem dark discoloration in crustaceans, called melanosis or blackspot, connotes spoilage (Kim et al. 2000). Melanosis reduces the market value of these foods and causes large financial losses to the food industry (Williams et al. 2003).

Melanosis or blackspot appears very frequently during postmortem storage of crustacea prior to the onset of spoilage. Melanosis starts in refrigerated crustaceans within a few hours of capture, and the susceptibility is often greater in autumn and winter, coinciding with molting and lower food availability (Montero et al. 2001). Marine crustaceans form the black spots on their tail, cephalothoraxes, and podites during storage (Adachi et al. 1999). In chilled prawns and shrimps, the melanosis reaction begins from the head region and spreads to the tail (Simpson et al. 1987), and the rate of spread of melanosis differs among various species. This could be related to differences in levels of substrate or levels of enzyme concentration or enzymatic activity in each species (Simpson et al. 1987). Pink shrimp PPO oxidized DOPA 1.8-fold more rapidly than white shrimp PPO. Pink shrimp carapace has relatively higher levels of free Tyr and phenylalanine, the natural substrates of melanosis, than white shrimp carapace (Rolle et al. 1991). The enzymatic oxidation to form melanin is associated with the stages of molting cycle, sex, and occurrence of injury or trauma (Ogawa et al. 1984). To prevent or retard melanosis, several PPO inhibitors including reducing agents or antioxidants inhibitors such as sulfiting agent, cysteine, acidulants, and metal ion chelators have been used. Because of the susceptibility of certain individuals to the residual level of sulfiting agents commonly used for melanosis prevention in crustaceans, natural additives such as plant phenols have been used to reduce melanosis formation in

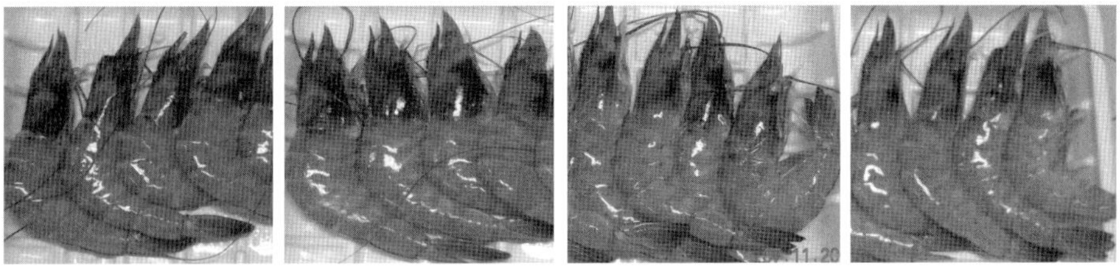

Figure 14.7. Pacific white shrimp without and with different treatments at day 10 of iced storage (Nirmal and Benjakul 2009b). SMS, sodium metabisulfite.

crustaceans during postmortem storage or handling. Whole Pacific white shrimp treated with catechin solution (0.05% or 0.1%) kept in ice for 10 days retarded melanosis in comparison with the control and those treated with 1.25% sodium metabisulfite (Nirmal and Benjakul 2009b) (Fig. 14.7).

TRIMETHYLAMINE-*N*-OXIDE DEMETHYLASE

TMAOase (EC 4.1.2.32) catalyzes the conversion of trimethylamine oxide (TMAO) to dimethylamine (DMA) and formaldehyde (FA) (Gill and Paulson 1982). TMAOase is one of the important enzymes in fish exhibiting an adverse effect on fish quality. TMAOase has been found at different degrees in various fish species. The inhibition or lowering of TMAOase activity in the postharvest animal could be an effective approach to maintain fish quality.

Distribution of TMAOase

Although TMAO is distributed among all kinds of marine fish and invertebrates, and in some fresh water fish, TMAOase activity has been identified in only 30 species of marine fish belonging to 10 families and 8 species of invertebrates, mainly mollusk (Sikorski and Kostuch 1982). Accumulation of DMA and FA has been described in species belonging to Gadidae, Merluccidae, and Myctophidae (Sikorski and Kostuch 1982). Among the fish, Gadiform (families Gadidae and Merluccidae) is a group of fish with high TMAOase activity (Sotelo and Rehein 2000).

The level of TMAOase in a particular species and tissue exhibits variations (Rehbein and Schreiber 1984) and seems to be influenced by factors such as gender, maturation stage, temperature of the habitat, and feeding status, or size (Sotelo and Rehein 2000).

Depending on the source of the enzyme, the activity may be located in the soluble fraction or in the particulate matter of homogenates of organs or tissues (Sotelo and Rehein 2000). The organs and tissues including viscera and skin contain TMAOase. Harada (1975) isolated TMAOase from the liver of lizardfish (*Saurida tumbil*). Distribution of TMAOase in various internal organs of lizardfish (*Saurida micropectoralis*) were studied (Benjakul et al. 2004b). Rey-Mansilla et al. (2002) studied the activities of TMAOase from several internal organs of hake (kidney, spleen, liver, heart, bile, and gallbladder). It was found that kidney and spleen showed the highest activities while liver, heart, bile, and gallbladder activities were much lower. Moreover, TMAOase has been found in muscle of some fish species, such as red hake (Phillipy and Hultin 1993) and walleye pollack (Kimura et al. 2000).

Purification and Characterization of TMAOase

Harada (1975) partially purified TMAOase from liver of lizardfish (*Saurida tumbil*) and found that the enzyme had an optimum pH of 5.0 and required methylene blue for activation. Tomioka et al. (1974) found that TMAOase activity from pyloric ceca of Alaska pollack was stimulated by Fe^{2+}, L-ascorbic acid, flavin mononucleotide (FMN) and inhibited by Fe^{3+}, Cu^{2+}, EDTA, trimethylamine, and choline. TMAOase associated with Alaska pollack myofibrils has been characterized (Kimura et al. 2000). The enzyme requires cofactors for full activity. The system of nicotinamide adenine dinucleotide and FMN requires anaerobic conditions while the system of iron and cysteine and/or ascorbate functions in the presence or absence of oxygen (Parkin and Hultin 1986). Benjakul et al. (2003b) purified TMAOase from lizardfish (*Saurida micropectoralis*) kidney by acidification followed by DEAE–cellulose column chromatography, in which a purification fold of 82 with a yield of 65.4% were obtained. The molecular mass of the partially purified enzyme was estimated to be 128 kDa based on activity staining. Optimum pH and temperature were 7.0 and 50°C, respectively. The activation energy was calculated to be 30.5 kJ $mol^{-1}K^{-1}$. Combined cofactors ($FeCl_2$, ascorbate, and cysteine) were required for full activation. Fe^{2+} exhibited a higher stimulating effect on TMAOase activity than Fe^{3+}. At concentrations less than 2 mM, ascorbate was more stimulatory of TMAOase activity than cysteine. TMAOase was tolerant to salt at concentrations up to 0.5 M. The enzyme had a K_m of 16.24 mM and V_{max} of 0.35 μmol/min, and was able to convert TMAO to DMA and FA. Leelapongwattana et al. (2008a) partially purified TMAOase from lizardfish kidney by heat treatment, ammonium sulfate precipitation, and a series of chromatographies including Sephacryl S-300 and DEAE–cellulose. The addition of partially purified TMAOase from lizardfish kidney to haddock natural actomyosin in the presence of cofactors ($FeCl_2$, ascorbate, and cysteine) accelerated formaldehyde formation throughout the storage either at 4°C or −10°C.

Prevention of TMAOase Activity

TMAOase inhibition has been focused on preventing loss of quality in frozen fish, especially in some species such as cod. The inhibition of the TMAO demethylation can be achieved using several inhibitors. Most of them are general inhibitors of redox reactions, being strong oxidants, such as H_2O_2, which inhibits the reaction (Phillipy and Hultin 1993). The presence of oxygen or oxidants can promote inhibition of enzymatic TMAO breakdown in frozen red hake (Racicot et al. 1984). Iodoacetamide, cyanide, and azide have been shown to inhibit TMAOase activity (Sotelo and Rehein 2000); while sodium citrate and H_2O_2 have been found to slow down the rate of DMA and FA formation in frozen gadoid mince (Racicot et al. 1984). Treating Alaska pollack fillets with sodium citrate and sodium pyruvate before freezing resulted in a less texture toughening (Krueger and Fennema 1989). Furthermore, the addition of 0.5% alginate could lower the DMA and FA formation in minced fillet of mackerel stored at −20°C for 2 months (Ayyad and Aboel-Niel 1991).

Apart from reducing protein denaturation, cryoprotectants have also been reported to prevent FA formation. Sotelo and Mackie (1993) found a protective role of different cryoprotectants (amino acids and sugars) in preventing FA-promoted aggregation of bovine serum albumin during frozen storage. The binding of FA during frozen storage was dependent on protein rearrangements in the way that reactive groups become available. The constraints of cryostabilizers on molecular diffusion reduced the exposure of these groups (Herrera et al. 1999). Herrera et al. (2000) also reported that the addition of maltodextrins to minced blue whiting muscle inhibited FA production during storage at −10°C and −20°C. Sucrose, however, was effective only at −20°C. Herrera et al. (2002) studied the effects of various cryostabilizers on inhibiting FA production in frozen-stored minced blue whiting muscle. Several maltodextrins (dextrose equivalent 9, 12, 18, and 28) or sucrose minimized the decrease in protein solubility during storage at −10°C and −20°C. These effects were greater at −20°C, and DE 18 maltodextrin seemed to be the most effective treatment at both temperatures. Conversely, sucrose was as effective as maltodextrins at −20°C, but showed hardly any effect at −10°C. A high correlation was found between the effectiveness of cryostabilizers in preventing protein alterations and their effectiveness in inhibiting FA production, particularly at −10°C. In addition, a sigmoidal relationship between protein solubility and FA content was found at this temperature, which supports the hypothesis of the cooperative nature of the effect of FA on protein alterations (Herrera et al. 2002).

Addition of TMAOase inhibitors could be a means to retard the loss in textural properties of some fish species associated with FA formation. Leelapongwattana et al. (2008b) reported that sodium citrate and pyrophosphate could inhibit TMAOase activity from lizardfish muscle in a concentration dependent manner, most likely due to their chelating property. Sodium alginate is a hydrocolloid possessing inhibitory activity toward TMAOase. During the storage of lizardfish mince at −20°C for 24 weeks, the addition of 0.5% sodium alginate and 0.3% pyrophosphate in combination with 4% sucrose and 4% sorbitol as the cryoprotectants slowed down TMAOase activities and lowered the formation of DMA and FA. The addition of TMAOase inhibitor (0.3% pyrophosphate and 0.5% alginate) in haddock mince containing TMAOase retarded FA formation throughout storage at −10°C for 6 weeks. On the other hand, antioxidants (250 mg ascorbic acid/kg and 250 mg tocopherol/kg) induced FA formation (Leelapongwattana et al. 2008a, 2008b).

Moreover, TMAO has also been found to exert a protective role over proteins. It has been shown that TMAO stabilizes fish muscle proteins and some enzymes during the frozen storage (LeBlanc and LeBlanc 1989).

LIPASE

Lipases (triacylglycerol acylhydrolases; EC 3.1.1.3) constitute a group of enzymes defined as carboxylesterases that catalyze the hydrolysis (and synthesis) of long-chain acyl-glycerols at the lipid–water interface (Cherif and Gargouri 2009). The two main categories in which lipase-catalyzed reactions may be classified are given in Figure 14.8.

From Figure 14.8, the last three reactions are often grouped together into a single term, namely, transesterification.

Lipases are important because of the role they play in the postmortem quality deterioration of seafood (and other foodstuffs) during handling, chilled, and frozen storage. Lipases are finding increasing uses as food and others industrial processing aids (Aryee et al. 2007). Compared with other hydrolytic enzymes (e.g., proteases and carbohydrases), lipases are relative less well studied and in this regard, lipases from aquatic animals are even less well-known versus their counterparts from mammalian, plant, and microbial sources (Aryee et al. 2007). The presence of a lipolytic activity and partial characterization of a lipase in marine organisms have been reported in few studies. The pH and temperature optima of most lipases lie between 7 and 9 and 35°C and 60°C, respectively. Lipases are active over a very wide temperature range of −20°C to 65°C, but the more usual range is 30–45°C (Shahidi and Kamil 2001).

The presence of a lipase activity has been described for some fish and aquatic organisms. Mukundan et al. (1985) described the purification of a lipase from the hepatopancreas of oil sardine using anion-exchange chromatography and gel filtration. The glycosylated enzyme (6.1% carbohydrate) showed a MW of 54–57 kDa (by gel filtration), was marginally activated by approximately 1 mM Ca^{2+}, was completely inhibited by approximately 1 mM F^-, and hydrolyzed tributyrin most efficiently among C2–C18 monoacid triglycerides (TGs). Sardine lipase exhibited optimum activity against tributyrin at pH 8–8.5 and

(A) Hydrolysis:

$$RCOOR' + H_2O \leftrightarrow RCOOH + R'OH$$

(B) Synthesis:

Reactions under this category can be further separated:

(a) Esterification

$$RCOOH + R'OH \leftrightarrow RCOOR' + H_2O$$

(b) Interesterification

$$RCOOR' + R''COOR^{\bullet} \leftrightarrow RCOOR^{\bullet} + R''COOR'$$

(c) Alcoholysis

$$RCOOR' + R''OH \leftrightarrow RCOOR'' + R'OH$$

(d) Acidolysis

$$RCOOR' + R''COOH \leftrightarrow R''COOR' + RCOOH$$

Figure 14.8. Lipase reactions.

37°C. The enzyme was stable up to 45°C for 15 minutes and in the pH range 5–9.5 after a 1-hour incubation period. A lipase from the hepatopancreas of red sea bream was purified and characterized by Iijima et al. (1998). The enzyme was extracted by a combination of anion exchange, hydrophobic interaction, and gel filtration. The MW of the lipase was 64 kDa and the enzyme exhibited the maximal activity at pH 7–9. The lipolytic activity of the red sea bream enzyme is Ca^{2+} independent. Lipase was purified from tilapia intestine using ion exchange, chromatofocusing, and gel filtration chromatography techniques (Taniguchi et al. 2001). The temperature optimum was 35°C and it was stable below 40°C. The pH stability fell within the relatively narrow range of 6.5–8.5 and the highest activity was at pH 7.5, typical of corresponding mammalian lipases and those from other fish species. The inclusion of bile salts caused an approximate 9-fold increase in the enzyme activity. The tilapia lipase's MW of 46 kDa, pH of 4.9 and efficient hydrolysis of soybean and coconut oils are characteristics similar to mammalian lipase. A lipase was partially purified from the viscera of grey mullet by ammonium sulfate fractionation, ultrafiltration, and cholate-Sepharose affinity chromatography (Aryee et al. 2007). Optimum activity against *p*-nitrophenyl palmitate was displayed at pH 8. The temperature optimum was 50°C. Grey mullet lipase is completely stable in several water-immiscible organic solvents, suggesting that it may have potential applications for synthesis reactions in organic media.

Recently, Cherif and Gargouri (2009) purified crab digestive lipase from crab hepatopancreas by heat treatment, DEAE–cellulose, ammonium sulfate precipitation, S-200, Mono-Q, and S-200. The enzyme had a molecular mass of 65 kDa. The maximum activity of crab lipase appeared at pH 8. Lipase activity was compatible with the presence of organic solvents, except for butanol. Furthermore, the hydrolysis was found to be specifically dependent on the presence of Ca^{2+} to trigger the hydrolysis of tributyrin emulsion.

Fish Lipase Applications

Lipases have been widely used for biotechnological applications in detergents, and dairy and textile industries, production of surfactants and oil processing. Recently, lipases have received considerable attention with regard to the preparation of enantiomerically pure pharmaceuticals, since they have a number of unique characteristics: substrate specificity, regiospecificity, and chiral selectivity (Cherif and Gargouri 2009). Because of their high affinity for long chain fatty acids, specificity for particular fatty acids and regiospecificity (Kurtovic et al. 2009), fish digestive lipases may find applications in the synthesis of structured lipids. Using lipases, TGs could be enriched with certain beneficial fatty acids, like eicosapentaenoic acid or docosahexaenoic acid at a specific position along the glycerol backbone. Enzymatic synthesis of modified lipids using lipases could be economically more attractive than chemical synthesis because of the lower energy requirements and specificity. In addition, long-chain PUFAs are highly labile and lipase-catalyzed lipid modifications could prevent detrimental oxidation and *cis-trans* isomerization processes. Lipases have the important physiological role of preparing the fatty acids of water-insoluble triacylglycerols (TAG) for absorption into and transport through membranes by converting the TAG to the more polar diacylglycerols, monoacylglycerols, free-fatty acids, and glycerol (Shahidi and Kamil 2001). Marine lipases are likely to have advantages over their microbial or mammalian counterparts because they may operate more efficiently at lower temperatures. Termination of enzymatic reactions could be achieved by smaller changes in temperature as marine lipases often have lower temperature optima and stabilities than lipases from other sources. The lipases do not require potentially hazardous high or low pH media in contrast to the requirements of many chemically driven reactions. Fish digestive lipases could also carry out lipid interesterification reactions (e.g., transesterification) with minimal or no by-products, a problem often encountered in chemical transformations. The screening for fish digestive lipases with alkaline pH optima would allow the identification of suitable enzymes for use in cleaners and household detergents.

Some fish digestive lipases are stereospecific as well as enantiospecific and these features could allow the lipases to be used in the synthesis of specialty pharmaceuticals and agrochemicals where enantiomerically pure products are desired. Specialty esters for personal care products and environmentally safe surfactants for applications in detergents and as food emulsions may also await the application of fish digestive lipases (Kurtovic et al. 2009).

REFERENCES

Adachi K et al. 1999. Purification and characterization of prophenoloxidase from kuruma prawn Penaeus japonicus. *Fish Sci* 65: 919–925.

Adachi K et al. 2001. Hemocyanin a must likely inducer of black spots in kuruma prawn *Penaeus japonicus* during storage. *J Food Sci* 66: 130–136.

Ali MT et al. 1994. Activation mechanism of Pro-polyphenoloxidase on melanosis development in Florida spiny lobster (*Panulirus argns*) cuticle. *J Food Sci* 59: 1024–1031.

Aoki T et al. 2002. A cathepsin B-like enzyme from mackerel white muscle is a precursor of cathepsin B. *Comp Biochem Physiol* 133B: 307–316.

Aranishi F et al. 1997a. Purification and characterization of cathepsin B from hepatopancreas of carp *Cyprinus carpio*. *Comp Biochem Physiol* 117B: 579–587.

Aranishi F et al. 1997b. Purification and characterization of cathepsin L from hepatopancreas of carp (*Cyprinus carpio*). *Comp Biochem Physiol* 118B: 531–537.

Arunchalam K, Haard NF. 1985. Isolation and characterization of pepsin isoenzymes from polar cod (*Boreogadus saida*). *Comp Biochem Physiol* 80B: 467–473.

Aryee ANA et al. 2007. Lipase fraction from the viscera of grey mullet (*Mugil cephalus*): Isolation, partial purification and some biochemical characteristics. *Enzyme Microb Tech* 40: 394–402.

Asgeirsson B, Bjarnason JB. 1991. Structural and kinetic properties of chymotrypsin from Atlantic cod (*Gadus morhua*). Comparison with bovine trypsin. *Comp Biochem Physiol* 9B: 327–335.

Ashie INA, Lanier TC. 2000. Transglutaminase in seafood processing. In: NF Haard, BK Simpson (eds.) *Seafood Enzymes Utilization and Influence on Postharvest Seafood Quality*. Marcel Dekker, New York, pp. 147–166.

Ashie INA, Simpson BK. 1997. Proteolysis in food myosystems: a review. *J Food Biochem* 21: 91–123.

Ayyad K, Aboel-Niel E. 1991. Effect of addition of some hydrocolloids on the stability of frozen minced fillet of mackerel. *Carbohydr Polym* 15: 143–149.

Balti R et al. 2009. A heat-stable trypsin from the hepatopancreas of the cuttlefish (*Sepia officinalis*): Purification and characterization. *Food Chem* 113: 146–154.

Benjakul S et al. 2003a. Heat-activated proteolysis in lizardfish (*Saurida tumbil*) muscle. *Food Res Internat* 36: 1021–1028.

Benjakul S et al. 2003b. Partial purification and characterization of trimethylamine-*N*-oxide demethylase from lizardfish kidney. *Comp Biochem Physiol* 135B: 359–371.

Benjakul S et al. 2004a. Suwari gel properties as affected by transglutaminase activator and inhibitors. *Food Chem* 85: 91–99.

Benjakul S et al. 2004b. Induced formation of dimethylamine and formaldehyde by lizardfish kidney trimethylamine-*N*-oxide demethylase. *Food Chem* 94: 297–305.

Benjakul S et al. 2005. Properties of phenoloxidase isolated from the cephalothorax of kuruma prawn (Penaeus japonicus). *J Food Biochem* 29: 470–485.

Bonete MJ et al. 1984. Acid proteinase activity in fish II. Purification and characterization of cathepsin B and D from *Mujil Auratus* muscle. *Comp Biochem Physiol* 78B: 207–213.

Bougatef A et al. 2007. Purification and characterization of trypsin from the viscera of sardine (*Sardina pilchardus*). *Food Chem* 102: 343–350.

Bougatef A et al. 2008. Pepsinogen and pepsin from the stomach of smooth hound (*Mustelus mustelus*): Purification, characterization and amino acid terminal sequences. *Food Chem* 107: 777–784.

Busconi L et al. 1987. Action of serine proteinases from fish skeletal muscle on myofibrils. *Arch Biochem Biophys* 252: 329–333.

Cano-Lopez A et al. 1987. Extraction of carotenoprotein from shrimp processing wastes with the aid of trypsin from Atlantic cod. *J Food Sci* 52: 503–506.

Cao MJ et al. 2000. Purification and characterization of two anionic trypsin from the hepatopancreas of carp. *Fish Sci* 66: 1172–1179.

Cardenas-Lopez JL, Haard N. 2009. Identification of a cysteine proteinase from jumbo squid (*Dosidicus gigas*) hepatopancreas as cathepsin L. *Food Chem* 112: 442–447.

Castillo-Yanez FJ et al. 2004. Characterization of acidic proteolytic enzymes from Monterey sardine (*Sardinops sagax caerulea*) viscera. *Food Chem* 85: 343–350.

Castillo-Yanez FJ et al. 2005. Isolation and characterization of trypsin from the pyloric ceca of Monterey sardine *Sardinops sagax caerulea*. *Comp Biochem Physiol* 140B: 91–98.

Castillo-Yanez FJ et al. 2006. Purification and biochemical characterization of chymotrypsin from the viscera of Monterey sardine (*Sardinops sagax caeruleus*). *Food Chem* 99: 252–259.

Chaveesuk R et al. 1993. Production of fish sauce and acceleration of sauce fermentation using proteolytic enzymes. *J Aquatic Food Prod Tech* 2(3): 59–77.

Chazarra S et al. 1997. Kinetic study of the suicide inactivation of latent polyphenoloxidase from iceberg lettuce (*Lactuca sativa*) induced by 4-*tert*-butylcatechol in the presence of SDS. *Biochim Biophys Acta* 1339: 297–303.

Chen JS et al. 1991. Comparison of phenoloxidase activity from Florida spiny lobster and western Australian lobster. *J Food Sci* 51: 154–160.

Chen JS et al. 1997. Comparison of two treatment methods on the purification of shrimp polyphenol oxidase. *J Sci Food Agri* 75: 12–18.

Cheret R et al. 2007. Calpain and cathepsin activities in post mortem fish and meat muscles. *Food Chem* 101: 1474–1479.

Cherif S, Gargouri Y. 2009. Thermoactivity and effects of organic solvents on digestive lipase from hepatopancreas of the green crab. *Food Chem* 116: 82–86.

Chevalier PL et al. 1995. Purification and partial characterization of chymotrypsin-like proteases from the digestive gland of the scallop *Pecten maximus*. *Comp Biochem Physiol* 110B: 777–784.

Choi YJ et al. 1999. Purification and characterization of alkaline proteinase from Atlantic menhaden muscle. *J Food Sci* 64: 768–771.

Cong R et al. 2005. Purification and characterization of phenoloxidase from clam *Ruditapes philippinarum*. *Fish Shellfish Immunol* 18: 61–70.

De-Vecchi SD, Coppes Z. 1996. Marine fish digestive proteases-relevance to food industry and south-west Atlantic region-a review. *J Food Biochem* 20: 193–214.

Erickson MC et al. 1983. Proteolytic activities in the sarcoplasmic fluids of parasitized Pacific whiting (*Merluccius productus*) and unparasitized true cod (*Gadus macrocephalus*). *J Food Sci* 48: 1315–1319.

Folk JE. 1980. Transglutaminase. *Annu Rev Biochem* 17: 517–531.

Garcia-Carreno FC, Hernandez-Cortes P. 2000. Use of protease inhibitors in seafood products. In: NF Haard, BK Simpson (eds.) *Seafood Enzymes: Utilization and Influence on Postharvest Seafood Quality*. Marcel Dekker, New York, pp. 531–540.

Gauillard F, Richard-Forget F. 1997. Polyphenoloxidase from William pear (*Pyrus communis* L, CV Williams): Activation, purification and some properties. *J Sci Food Agric* 74: 49–56.

Gildberg A. 2001. Utilization of male Arctic capelin and Atlantic cod intestines for fish sauce production-evaluation of fermentation conditions. *Bioresour Technol* 76: 119–123.

Gildberg A, Raa J. 1983. Purification and characterization of pepsins from the Arctic fish capelin (*Mallotus villosus*). *Comp Biochem Physiol* 75A: 337–342.

Gildberg A et al. 1990. Catalytic properties and chemical composition of pepsin from Atlantic cod (*Gadus morhua*). *Comp Biochem Physiol* 69B: 323–330.

Gill TA, Paulson AT. 1982. Localization, characterization and partial purification of TAMOase. *Comp Biochem Physiol* 71B: 49–56.

Goulart PFP et al. 2003. Purification of polyphenoloxidase from coffee fruits. *Food Chem* 83: 7–11.

Guerrero-Beltran JA et al. 2005. Inhibition of polyphenoloxidase in mango puree with 4-hexylresorcinal, cysteine and ascorbic acid. *LWT-Food Sci Technol* 38: 625–630.

Gudmundsdottir A et al. 1993. Isolation and characterization of cDNAs from Atlantic cod encoding two different forms of trypsinogen. *Eur J Biochem* 217: 1091–1097.

Haard NF. 1990. Enzymes from myosystems. *J Muscle Foods* 1: 293–338.

Haard NF. 1992. A review of protolytic enzymes from marine organisms and their application in the food industry. *J Aq Food Prod Technol* 1(1): 17–35.

Hameed KS, Haard NF. 1985. Isolation and characterization of cathepsin C from Atlantic short finned squid, *Illexex illecebrosus*. *Comp Biochem Physiol* 110B: 777–784.

Harada K. 1975. Studies on the enzyme catalyzing the formation of formaldehyde and dimethylamine in tissues of fishes and shells. *J Shimonoseki Univ Fish* 23: 163–241.

Hatzizisis D et al. 1996. Purification and properties of a calpain II-like proteinase from Octopus vulgaris arm muscle. *Comp Biochem Physiol* 113B: 295–303.

Hermodson MA et al. 1973. Determination of the amino sequence of procine trypsin by sequenator analysis. *Biochem* 12: 3146–3153.

Herrera JJ et al. 1999. Effect of various cryostabilizers on the production and reactivity of formaldehyde in frozen-stored minced blue whiting muscle. *J Agric Food Chem* 47: 2386–2397.

Herrera JJ et al. 2000. Inhibition of formaldehyde production in frozen-stored minced blue whiting (*Micromesistius poutassou*) muscle by cryostabilizers: an approach from the glassy state theory. *J Agric Food Chem* 48: 5256–5262.

Herrera JJ et al. 2002. Effects of various cryostabilisers on protein functionality in frozen-stored minced blue whiting muscle: the importance of inhibiting formaldehyde production. *Eur J Biochem* 214: 382–387.

Heu MS et al. 1997. Purification and characterization of cathepsin L-like enzyme from the muscle of anchovy, *Engraulis japonica*. *Comp Biochem Physiol* 118B: 523–529.

Heu MS et al. 1995. Comparison of trypsin and chymotrypsin from the viscera of anchovy, *Engraulis japonica*. *Comp Biochem Physiol* 112B: 557–567.

Iijima N et al. 1998. Purification and characterization of bile salt-activated lipase from the hepatopancreas of red sea bream, *Pagrus major*. *Fish Physiology Biochemistry* 18: 59–69.

Jiang ST et al. 1990. Purification and properties of proteases from milkfish muscle (*Chanos chanos*). *J Agric Food Chem* 38: 1458–1463.

Kang IS, Lanier TC. 2000. Heat-induced softening of surimi gels by proteinases. In: JW Park (ed.) *Surimi and Surimi Seafood*. Marcel Dekker, New York, pp. 445–474.

Kim J et al. 2000. Polyphenoloxidase. In: NF Haard, BK Simpson (eds.) *Seafood Enzymes: Utilization and Influence on Postharvest Seafood Quality*. Marcel Dekker, New York, pp. 271–316.

Kimura M et al. 2000. Purification and characterization of trimethylamine-*N*-oxide demethylase from walleye pollack muscle. *Fish Sci* 66: 967–973.

Kinoshita M et al. 1990. Characterization of two distinct latent proteinases associated with myofibrils of crucian carp (*Carassius auratus cuvieri*). *Comp Biochem Physiol* 97B: 315–319.

Kishimura H, Hayashi K. 2002. Isolation and characteristics of trypsin from the pyloric ceca of the starfish *Asterina pectinifera*. *Comp Biochem Physiol* 132B: 485–490.

Kishimura H et al. 2005. Characteristics of two trypsin isozymes from the viscera of Japanese anchovy (*Engraulis japonica*). *J Food Biochem* 29: 459–469.

Kishimura H et al. 2006a. Comparative study on enzymatic characteristics of trypsins from the pyloric ceca of yellow tail (*Seriola quinqueradiata*) and brown hakeling (*Physiculus japonicus*). *J Food Biochem* 30: 521–534.

Kishimura H et al. 2006b. Enzymatic characteristics of trypsin from the pyloric ceca of spotted mackerel (*Scomber australasicus*). *J Food Biochem* 30: 466–477.

Kishimura H et al. 2006c. Characteristics of trypsins from the viscera of true sardine (*Sardinops melanostictus*) and the pyloric ceca of arabesque greenling (*Pleuroprammus azonus*). *Food Chem* 97: 65–70.

Kishimura H et al. 2007. Trypsins from the pyloric ceca of jacopever (*Sebastes schlegeli*) and elkhorn sculpin (*Alcichthys alcicornis*): Isolation and characterization. *Food Chem* 100: 1490–1495.

Kishimura H et al. 2008. Characteristics of trypsin from the pyloric ceca of walleye pollok (*Theragra chalcogramma*). *Food Chem* 106: 194–199.

Klomklao S. 2008. Digestive proteinases from marine organisms and their applications. *Songklanakarin J Sci Technol* 30: 37–46.

Klomklao S et al. 2004. Comparative studies on proteolytic activity of spleen extracts from three tuna species commonly used in Thailand. *J Food Biochem* 28: 355–372.

Klomklao S et al. 2005. Partitioning and recovery of proteinases from tuna spleen by aqueous two-phase systems. *Process Biochem* 40: 3061–3067.

Klomklao S et al. 2006a. Purification and characterization of trypsin from the spleen of tongol tuna (*Thunnus tonggol*). *J Agric Food Chem* 54: 5617–5622.

Klomklao S et al. 2006b. Trypsins from yellowfin tuna (*Thunnus albacores*) spleen: Purification and characterization. *Comp Biochem Physiol* 144B: 47–56.

Klomklao S et al. 2006c. Effects of the addition of spleen of skipjack tuna (*Katsuwonus pelamis*) on the *liquefaction and characteristics of fish sauce made from sardine (Sardinella gibbosa). Food Chem* 98: 440–452.

Klomklao S et al. 2007a. Purification and characterization of two pepsins from the stomach of pectoral rattail (*Coryphaenoides pectoralis*). *Comp Biochem Physiol* 147B: 682–689.

Klomklao S et al. 2007b. 29 kDa trypsin from the pyloric ceca of Atlantic bonito (*Sarda sarda*): Recovery and characterization. *J Agric Food Chem* 55: 4548–4553.

Klomklao S et al. 2007c. Trypsin from the pyloric ceca of bluefish (*Pomatomus saltatrix*). *Comp Biochem Physiol* 148B: 382–389.

Klomklao S et al. 2007d. Purification and characterization of trypsins from the spleen of skipjack tuna (*Katsuwonus pelamis*). *Food Chem* 100: 1580–1589.

Klomklao S et al. 2008. Endogenous proteinases in true sardine (*Sardinops melanostictus*). *Food Chem* 107: 213–220.

Klomklao S et al. 2009a. Biochemical properties of two isoforms of trypsin purified from the intestine of skipjack tuna (*Katsuwonus pelamis*). *Food Chem* 115: 155–162.

Klomklao S et al. 2009b. Trypsin from the pyloric ceca of pectoral rattail (*Coryphaenoides pectoralis*): Purification and characterization. *Journal of Agricultural and Food Chem* 57: 7097–7103.

Klomklao S et al. 2009c. Extraction of carotenoprotein from black tiger shrimp shell with the aid of bluefish trypsin. *J Food Biochem* 33: 201–217.

Kolodziejska I, Sikorski ZE. 1996. Neutral and alkaline muscle proteases of marine fish and aquatic invertebrates. A review. *J Food Biochem* 20: 349–363.

Krueger DJ, Fennema OR. 1989. Effect of chemical additives on toughening of fillets of frozen storage Alaska pollack (*Theragra chacogramma*). *J Food Sci* 54: 1101–1106.

Kumazawa Y et al. 1996. Purification and characterization of transglutaminase from walleye Pollock liver. *Fish Sci* 62: 959–964.

Kurtovic I et al. 2009. Lipases from mammals and fishes. *Fish Sci* 17: 18–40.

Ladrat C et al. 2000. Neutral calcium-activated proteases from European sea bass (*Dicentrarchus labrax* L.) muscle: polymorphism and biochemical studies. *Comp Biochem Physiol* 125B: 83–95.

LeBlanc EL, LeBlanc RJ. 1989. Separation of cod (*Gadus morhua*) fillet proteins by electrophoresis and HPLC after various frozen storage treatments. *J Food Sci* 53: 328–340.

Lee JJ et al. 1993. Purification and characterization of proteinases identified as cathepsin L and L-like (58 kDa) proteinase from mackerel (*Scomber australasicus*). *Biosci Biotech Biochem* 57: 1470–1476.

Leelapongwattana K et al. 2008a. Effect of trimehtylamine-*N*-oxide demethylase from lizardfish kidney on biochemical changes of haddock natural actomyosin stored at 4 and -10°C. *Eur J Biochem* 226: 833–841.

Leelapongwattana K et al. 2008b. Effect of some additives on the inhibition of lizardfish trimehtylamine-*N*-oxide demethylase and frozen storage stability of minced fish. *Int J Food Sci Technol* 43: 448–455.

Lin TS, Lanier TC. 1980. Properties of alkaline protease from the skeletal muscle of Atlantic croaker. *J Food Biochem* 4: 17–28.

Liu G et al. 2006. Purification and characterization of phenoloxidase from crab *Charybdis japonica*. *Fish Shellfish Immunol* 20: 47–57.

Makinodan Y, Ikeda S. 1971. Studies on fish muscle protease. Part V. The existence of protease active in neutral pH range. *Bulletin of the Japanese Society of Scientific Fisheries* 49: 1153–1161.

Male R et al. 1995. Molecular cloning and characterization of anionic and cationic variants of trypsin from Atlantic salmon. *Eur J Biochem* 232: 677–685.

Martinez-Alvarez O et al. 2008. Evidence of an active laccase-like enzyme in deepwater pink shrimp (*Parapenaeus longirostris*). *Food Chem* 108: 624–632.

Mazzafera P, Robinson SP. 2000. Characterization of polyphenol oxidase in coffee. *Phytochem* 55: 285–296.

McLay R. 1980. Activities of cathepsin A and D in cod muscle. *J Sci Food Agric* 31: 1050–1054.

Montero P et al. 2001. Characterization of polyphenoloxidase of prawns (*Penaeus japonicus*). Alternatives to inhibition: additives and high-pressure treatment. *Food Chem* 75: 317–324.

Mukundan MK et al. 1985. Purification of a lipase from the hepatopancreas of oil sardine (*Sardinella longiceps* Linnaeus) and its characteristics and properties. *J Sci Food Agric* 36: 191–203.

Muruyama N et al. 1995. Transglutaminase induced polymerization of a mixture of different fish myosins. *Fish Sci* 61: 495–500.

Nagai T, Suzuki N. 2002. Preparation and partial characterization of collagen from paper nautilus (*Argonauta argo*, Linnaeus) outer skin. *Food Chem* 76: 149–153.

Nagai T et al. 2002. Collagen of the skin of ocellate puffer fish (*Takifugu rubripes*). *Food Chem* 78: 173–177.

Nagai T et al. 2001. Isolation and characterization of collagen from the outer skin waste material of cuttlefish (*Sepia lycidas*). *Food Chem* 72: 425–429.

Nalinanon S et al. 2007. Use of pepsin for collagen extraction from the skin of bigeye snapper (*Priccanthus tayenus*). *Food Chem* 104: 593–601.

Nalinanon S et al. 2010. Biochemical properties of pepsinogen and pepsin from the stomach of albacore tuna (*Thunnus alalunga*). *Food Chem* 121: 49–55.

Nielsen LB, Nielsen HH. 2001. Purification and characterization of cathepsin D from herring muscle (*Clupea harengus*). *Comp Biochem Physiol* 128B: 351–363.

Nissen JA. 1993. Proteases. In: T Nagodawithana, G Reed (eds.) *Enzymes in Food Processing*. Academic Press, New York, pp. 159–203.

Nirmal NP, Benjakul S. 2009a. Effect of ferulic acid on inhibition of polyphenoloxidase and quality changes of Pacific white shrimp (*Litopenaeus vannamei*) during iced storage. *Food Chem* 116: 323–331.

Nirmal NP, Benjakul S. 2009b. Melanosis and quality changes of Pacific white shrimp (*Litopenaeus vannamei*) treated with catechin during iced storage. *Journal of Agricultural and Food Chem* 57: 3578–3586.

Noda M, Murakami K. 1981. Studies on proteinases from the digestive organs of sardine. Purification and characterization of two acid proteinases from the stomach. *Biochimica Biophysica Acta* 658: 27–32.

Noguchi K et al. 2001. Cystal structure of red sea bream transglutaminase. *J Biologic Chem* 276: 12055–12059.

Nozawa H, Seki N. 2001. Purification of transglutaminase from scallop striated adductor muscle and NaCl-induced inactivation. *Fish Sci* 67: 493–499.

Nozawa H et al. 1997. Partial purification and characterization of six transglutaminases from ordinary muscles of various fishes and marine invertebrates. *Comp Biochem Physiol* 118B: 313–317.

Ogawa M et al. 1984. Incidence of melanosis in the integumentary tissue. *Bulletin of the Japanese Society of Scientific Fisheries* 50: 471–475.

Opoku-Gyamfua A et al. 1992. Comparative studies on the polyphenol oxidase fraction from lobster and tyrosinase. *J Agric Food Chem* 40: 772–775.

Osatomi K et al. 1997. Purification and characterization of myofibril-bound serine proteinase from carp *Cyprinus carpio* ordinary muscle. *Comp Biochem Physiol* 116B: 183–190.

Pangkey H et al. 2000. Purification and characterization of cathepsin S from hepatopancreas of carp *Cyprinus carpio*. *Fish Sci* 66: 1130–1137.

Parkin KL, Hultin HO. 1986. Characterization of tirmethylamine-N-oxide (TMAO) demethylase activity from fish muscle microsomes. *J Biochem* 100: 77–86.

Phillipy BQ, Hultin HO. 1993. Distribution and some characteristics of trimethylamine-*N*-oxdie (TMAO) demethylase activity of red hake muscle. *J Food Biochem* 17: 235–250.

Racicot LD et al. 1984. Effect of oxidizing and reducing agents on trimethylamine oxide demethylase activity in red hake muscle. *J AgricFood Chem* 32: 459–464.

Rehbein H, Schreiber W. 1984. TMAOase activity in tissues of fish species from the Northeast Atlantic. *Comp Biochem Physiol* 79B: 447–452.

Reid RGB, Rauchert K. 1976. Catheptic endopeptidases and protein digestion in the horse clam *Tresus capax* (Gould). *Comp Biochem Physiol* 54B: 467–472.

Rey-Mansilla MM et al. 2002. TMAOase activity of European hake (*Merluccius merluccius*) organs: influence of biological condition and season. *J Food Sci* 67: 3242–3251.

Rolle RS et al. 1991. Purification and characterization of phenoloxidase from Taiwanese black tiger shrimp (*Penaeus monodon*). *J Food Biochem* 15: 17–32.

Saito M et al. 2007. Purification and characterization of calpain and calpastatin from rainbow trout, *Oncorhynchus mykiss*. *Comp Biochem Physiol* 146B: 445–455.

Sanchez-Chiang L et al. 1987. Partial purification of pepsins from adult and juvenile salmon fish (*Oncorrhynchus keta*). Effect on NaCl on proteolytic activities. *Comp Biochem Physiol* 87B: 793–797.

Sanchez-Chiang L, Ponce O. 1981. Gastric sinogens and gastrisins from *Merluccaius gayi*-purification and properties. *Comp Biochem Physiol* 68B: 251–257.

Shahidi F, Kamil YVAJ. 2001. Enzymes from fish and aquatic invertebrates and their application in the food industry. *Trends Food Sci Technol* 12: 435–464.

Shamsuzzaman K, Haard NF. 1984. Purification and characterization of a chymosin-like protease from gastric mucosa of harp seal (*Paophilus groenlandicus*). *Canadian J BiochemCell Biol* 62: 699–708.

Sherekar SV et al. 1988. Purification and characterization of cathepsin B from the skeletal muscle of fresh water fish, *Tilapia mossambia*. *J Food Sci* 53: 1018–1023.

Sikorski ZE, Kostuch S. 1982. Trimethylamine-N-oxide demethylase: its occurrence, properties, and role in technological changes in frozen fish. *Food Chem* 9: 213–222.

Simpson BK. 2000. Digestive proteinases from marine animals. In: NF Haard, BK Simpson (eds.) *Seafood Enzymes: Utilization and Influence on Postharvest Seafood Quality*. Marcel Dekker, New York, pp. 191–214.

Simpson BK et al. 1987. Phenoloxidase from shrimp (*Penaeus setiferus*): Purification and some properties. *J Agric Food Chem* 35: 918–921.

Simpson BK et al. 1988. Phenoloxidase from pink and white shrimp: Kinetic and other properties. *J Food Biochem* 12: 205–217.

Sivakumar P et al. 1999. Collagenolytic metalloprotease (gelatinase) from the hepatopancreas of the marine carb, *Scylla serrata*. *Comp Biochem Physiol* 123B: 273–279.

Sotelo CG, Mackie IM. 1993. The effect of formaldehyde on the aggregation behavior of bovine serum albumin during storage in the frozen and unfrozen states in the presence and absence of cryoprotectants and other low MW hydrophilic compounds. *Food Chem* 47: 263–270.

Sotelo CG, Rehein H. 2000. TMAO-degrading enzymes. In: NF Haard, BK Simpson (eds.) *Seafood Enzymes: Utilization and Influence on Postharvest Seafood Quality*. Marcel Dekker, New York, pp. 167–190.

Taniguchi A et al. 2001. Purification and properties of lipase from Tilapia intestine-VI. *Nippon Suisan Gakkaishi* 67: 78–84.

Tanji M et al. 2007. Purification and characterization of pepsinogens from the gastric mucosa of African coelacanth, *Latimeria chalumnae*, and properties of the major pepsins. *Comp Biochem Physiol* 146B: 412–420.

Tomioka K et al. 1974. Studies on dimethylamine in foods II. Enzymatic formation of dimethylamine from trimethylamine oxide. *Bulletin of the Japanese Society of Scientific Fisheries* 40: 1021–1026.

Toyohara H et al. 1982. Purification and properties of carp muscle cathepsin A. *Bulletin of the Japanese Society of Scientific Fisheries* 48: 1145–1150.

Visessanguan W et al. 2003. Purification and characterization of cathepsin L in arrowtooth flounder (*Atheresthes stomias*) muscle. *Comp Biochem Physiol* 134B: 477–487.

Visessanguan W et al. 2001. Cathepsin L: A predominant heat-activated proteinase in arrowtooth flounder muscle. *J Agric Food Chem* 49: 2633–2640.

Walsh KA. 1970. Trypsinogens and trypsins of various species. *Methods Enzymol* 19: 41–63.

Wang JH et al. 1993. Comparison of properties of m-calpain from tilapia and shrimp muscle. *J Agric Food Chem* 41: 1379–1384.

Whitaker JR. 1994. Classification and nomenclature of enzymes. In: *Principles of Enzymology for the Food Sciences*. Marcel Dekker, New York, pp. 367–385.

Whitaker JR. 1995. Polyphenoloxidase. In: DWS Wong (ed.) *Food Enzymes, Structure and Mechanism*. Chapman and Hall. New York, pp. 271–303.

Williams HG et al. 2003. Heat-induced activation of polyphenoloxidase in Western rock lobster (*Panulirus Cygnus*) hemolymph: implications for heat processing. *J Food Sci* 68: 1928–1932.

Worratao A, Yongsawatdigul J. 2005. Purification and characterization of transglutaminase from tropical tilapia (*Oreochromis niloticus*). *Food Chem* 93: 651–658.

Wu T et al. 2009. Identification of pepsinogens and pepsins from the stomach of European eel (*Anguilla anguilla*). *Food Chem* 115: 137–142.

Xu RA et al. 1996. Purification and characterization of acidic proteases from the stomach of the deep water finfish orange roughy (*Hoplostethus atlanticus*). *J Food Biochem* 20: 31–48.

Yamashita M, Konagaya S. 1990. Participitation of cathepsin L into extensive softening of the muscle of chum salmon caught during spawning migration. *Nippon Suisan Gakkaishi* 56: 1271–1277.

Yang FY et al. 2009. Purification and characterization of chymotrypsins from the hepatopancreas of crucian carp (*Carassius auratus*). *Food Chem* 116: 860–866.

Yasueda H et al. 1994. Purification and characterization of a tissue-type transglutaminase from red sea bream (*Pagrus major*). *Biosci Biotech Biochem* 58: 2041–2045.

Zamorano JP et al. 2009. Characterization and tissue distribution of polyphenol oxidase of deepwater pink shrimp (*Parapenaeus longirostris*). *Food Chem* 112: 104–111.

Zeef AH, Dennison C. 1988. A novel cathepsin from mussel (*Perna perna Linne*). *Comp Biochem Physiol* 90B: 204–210.

Zhang Y et al. 2007. Isolation and partial characterization of pepsin-soluble collagen from the skin of grass carp (*Ctenopharyngodon idella*). *Food Chem* 103: 906–912.

Zhou Q et al. 2007. Purification and characterization of sea bream (*Sparus latus Houttuyn*) pepsinogens and pepsins. *Food Chem* 103: 795–801.

Part 3
Meat, Poultry and Seafoods

15
Biochemistry of Raw Meat and Poultry

Fidel Toldrá and Milagro Reig

Abstract: The consumers' concern for safety and health is driving the consumption patterns in accordance to meat composition, especially for some key substances. In addition, a significant percentage of meat is used as raw material for further processing into different products such as cooked, fermented, and dry-cured meats and one of the most relevant problems of the meat industry is that genetics, age and sex, intensive or extensive production systems, type of feeding and slaughter procedures including preslaughter handling, stunning methods, and postmortem treatment have an influence on important biochemical traits with a direct effect on the meat quality and its aptitude as raw material. All these factors are described in this chapter.

BACKGROUND INFORMATION

In recent years, there has been a decline in consumption of beef, accompanied by a slight increase in pork and an increased demand for chicken. There are many reasons behind these changes including the consumer concern for safety and health, changes in demographic characteristics, changes in consumer lifestyles, availability and convenience, price, and so on (Resurreccion 2003). The quality perception of beef changes depending on the country. So, Americans are concerned with cholesterol, calorie content, artificial ingredients, convenience characteristics, and price (Resurreccion 2003). In the United States, this changing demand has influenced the other meat markets. Beef has been gradually losing market share to pork and, especially, chicken (Grunert 1997).

But a significant percentage of meat is used as raw material for further processing into different products such as cooked, fermented, and dry-cured meats. Some of the most well-known meat products are bacon, cooked ham, fermented and dry-fermented sausages, and dry-cured ham (Flores and Toldrá 1993, Toldrá 2002).

The processing meat industry faces various problems, but one of the most important is the variability in the quality of meat as a raw material. Genetics, age and sex, intensive or extensive production systems, type of feeding and slaughter procedures including pre-slaughter handling, stunning methods, and postmortem treatment have an influence on important biochemical traits with a direct effect on the meat quality and its aptitude as raw material.

Food Biochemistry and Food Processing, Second Edition. Edited by Benjamin K. Simpson, Leo M.L. Nollet, Fidel Toldrá, Soottawat Benjakul, Gopinadhan Paliyath and Y.H. Hui.

STRUCTURE OF MUSCLE

A good knowledge of the structure of muscle is essential for the use of muscle as meat. The structure is important for the properties of the muscle, and its changes during postmortem events influence the quality properties of the meat.

Muscle has different colors within the range of white to red, depending on the proportion of fibers. There are different classifications for fibers. On the basis of color, they can be classed as red, white, or intermediate (Moody and Cassens 1968): (1) red fibers are characterized by a higher content of myoglobin and higher numbers of capillaries and mitochondria, and they exhibit oxidative metabolism; (2) white fibers contain low amounts of myoglobin and exhibit glycolytic metabolism; and (3) intermediate fibers exhibit intermediate properties. Red muscles contain a high proportion of red fibers and are mostly related to locomotion, while white muscles contain a higher proportion of white fibers and are engaged in support tasks (Urich 1994). Other classifications are based on the speed of contraction (Pearson and Young 1989): Type I for slow-twitch oxidative fibers, type IIA for fast-twitch oxidative fibers, and IIB for fast-twitch glycolytic fibers.

The skeletal muscle contains a great number of fibers. Each fiber, which is surrounded by connective tissue, contains around 1000 myofibrils, all of them arranged in a parallel way and responsible for contraction and relaxation. They are embedded in a liquid known as sarcoplasm, which contains the sarcoplasmic (water-soluble) proteins. Each myofibril contains clear dark lines, known as Z-lines, regularly located along the myofibril (see Fig. 15.1). The distance between two consecutive Z-lines is known as a sarcomere. In addition, myofibrils contain thick and thin filaments, partly overlapped and giving rise to alternating dark (A band) and light (I band) areas. The thin filaments extend into the Z-line that serves as linkage between consecutive sarcomeres. All these filaments are composed of proteins known as myofibrillar proteins.

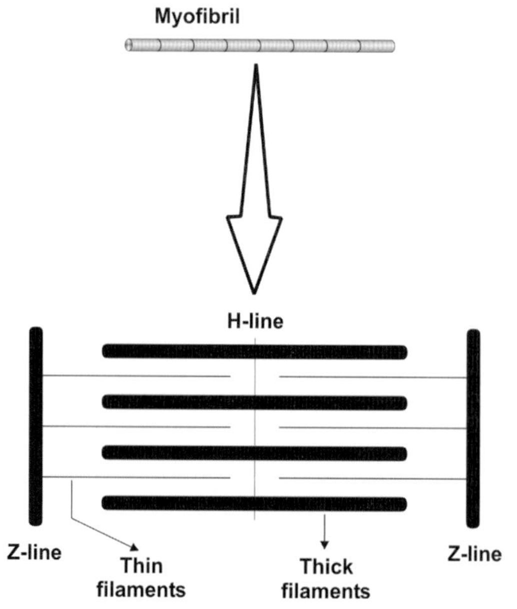

Figure 15.1. Structure of a myofibril with details of main filaments in the sarcomere.

Muscle, fat, bones, and skin constitute the main components of carcasses; muscle is the major compound and is, furthermore, associated to the term of meat. The average percentage of muscle in relation to live weight varies depending on the species, degree of fatness, and dressing method: 35% for beef, 32% for veal, 36% for pork, 25% for lamb, 50% for turkey, and 39% for broiler chicken. The muscle to bone ratio is also an important parameter representative of muscling: 3.5 for beef, 2.1 for veal, 4.0 for pork, 2.5 for lamb, 2.9 for turkey, and 1.8 for poultry (Kauffman 2001).

MUSCLE COMPOSITION

Essentially, meat is basically composed of water, protein, lipids, minerals, and carbohydrates. Lean muscle tissue contains approximately 72–74% moisture, 20–22% protein, 3–5% fat, 1% ash, and 0.5% of carbohydrate. An example of typical pork meat composition is shown in Table 15.1. These proportions are largely variable, especially in the lipid content, which depends on species, amount of fattening, inclusion of the adipose tissue, and so on. There is an inverse relationship between the percentages of protein and moisture and the percentage of fat so that

Table 15.1. Example of the Approximate Composition of Pork Muscle *Longissimus Dorsi*. Expressed as Means and Standard Deviation (SD)

	Units	Mean	SD
Gross composition			
Moisture	g/100g	74.5	0.5
Protein	g/100g	21.4	2.1
Lipid	g/100g	2.7	0.3
Carbohydrate	g/100g	0.5	0.1
Ash	g/100g	0.9	0.1
Proteins			
Myofibrillar	g/100g	9.5	0.8
Sarcoplasmic	g/100g	9.1	0.6
Connective	g/100g	3.0	0.4
Lipids			
Phospholipids	g/100g	0.586	0.040
Triglycerides	g/100g	2.12	0.22
Free fatty acids	g/100g	0.025	0.005
Some minor compounds			
Cholesterol	mg/100g	46.1	6.1
Haem content	mg/100g	400	30
Dipeptides	mg/100g	347.6	35.6
Free amino acids	mg/100g	90.2	5.8

Source:Aristoy and Toldrá 1998, Hernández et al. 1998, Toldrá 1999, Unpublished.

Table 15.2. Type, Localization and Main Role of Major Muscle Proteins

Cellular Localization	Main Proteins	Main Role in Muscle
Myofibrillar	Myosin and actin	Cytoskeletal proteins providing support to the myofibril and responsible for contraction–relaxation of the muscle.
	Tropomyosin	Regulatory protein associated to troponins than cause its movement toward the F-actin helix to permit contraction.
	Troponins T, C, I	Regulatory proteins. Tn-T binds to tropomyosin, Tn-C binds Ca^{2+} and initiates the contractile process and Tn-I inhibits the actin–myosin interaction in conjunction with tropomyosin and Tn-T and Tn-C.
	Titin and nebulin	Large proteins located between Z-lines and thin filaments. They provide myofibrils with resistance and elasticity.
	α-, β-, γ-, and eu-actinin	Proteins regulating the physical state of actin. α-actinin acts as a cementing substance in the Z-line. γ-actinin inhibits the polymerization of actin at the nucleation step. β-actinin is located at the free end of actin filaments, preventing them to bind each other. Eu-actinin interacts with both actin and α-actinin.
	Filamin, synemin, vinculin, zeugmatin, Z-nin, C, H, X, F, I proteins	Proteins located in the Z-line that contribute to its high density.
	Desmin	Protein that links adjacent myofibrils through Z-lines.
	Myomesin, creatine kinase and M-protein	Proteins located in the center of the sarcomere forming part of the M-line.
Sarcoplasm	Mitochondrial enzymes	Enzymes involved in the respiratory chain.
	Lysosomal enzymes	Digestive hydrolases very active at acid pH (cathepsins, lipase, phospholipase, peptidases, glucohydrolases, etc.).
	Other cytosolic enzymes	Neutral proteases, lipases, glucohydrolases, ATPases, etc.
	Myoglobin	Natural pigment of meat.
	Hemoglobin	Protein present from remaining blood within the muscle.
Connective tissue	Collagen	Protein giving support, strength, and shape to the fibers,
	Elastin	Protein that gives elasticity to tissues like capillaries, nerves, tendons, etc.

Source: Bandman 1987, Pearson and Young 1989.

meats with high content of fat have lower content of moisture and proteins. A brief description of proteins and lipids as major components of meat is given later.

Muscle Proteins

Proteins constitute the major compounds in the muscle, approximately 15–22%, and have important roles for the structure, normal function, and integrity of the muscle. Proteins experience important changes during the conversion of muscle to meat that mainly affect tenderness; and additional changes occur during further processing, through the generation of peptides and free amino acids as a result of the proteolytic enzymatic chain. There are three main groups of proteins in the muscle (see Table 15.2): myofibrillar proteins, sarcoplasmic proteins, and connective tissue proteins.

Myofibrillar Proteins

These proteins are responsible for the basic myofibrillar structure, and thus they contribute to the continuity and strength of the muscle fibers. They are soluble in high ionic strength buffers, and the most important of them are listed in Table 15.2. Myosin and actin are by far the most abundant and form part of the structural backbone of the myofibril. Tropomyosin and troponins C, T, and I are considered as regulatory proteins because they play an important role in muscle contraction and relaxation (Pearson 1987). There are many proteins in the Z-line region (although in a low percentage) that serve as bridges between the thin filaments of adjacent sarcomeres. Titin and nebulin are two very large proteins, present in a significant proportion, that are located in the void space between the filaments and the Z-line and contribute to the integrity of the muscle cells (Robson et al. 1997). Desmin is located on the external area of the Z-line and connects adjacent myofibrils at the level of the Z-line.

Sarcoplasmic Proteins

These are water-soluble proteins, comprising about 30–35% of the total protein in muscle. Sarcoplasmic proteins contain a high diversity of proteins (summarized in Table 15.2), mainly metabolic enzymes (mitochondrial, lysosomal, microsomal, nuclear, or free in the cytosol) and myoglobin. Some of these enzymes play a very important role in postmortem meat and during further processing, as described in Chapter 16. Minor amounts of hemoglobin may be found in the muscle if blood has not been drained properly. Myoglobin is the main sarcoplasmic protein, responsible for the red color of meat as well as the typical pink color of drippings. The amount of myoglobin depends on many factors. Red fibers contain higher amounts of myoglobin than white fibers. The species is very important, and thus beef and lamb contain more myoglobin than pork and poultry. For a given species, the myoglobin content in the muscle increases with the age of the animal.

Connective Tissue Proteins

Collagen, reticulin, and elastin constitute the main stromal proteins in connective tissue. There are several types (I–V) of collagen containing different polypeptide chains (up to ten α-chains). Type I collagen is the major component of the epimysium and perimysium that surround the muscles. Types III, IV, and V collagen are found in the endomysium, which provides support to the muscle fiber (Eskin 1990). There are a high number of cross-linkages in the collagen fibers that increase with age, and this is why meat tougher in older animals. Elastin is found in lower amounts, usually in capillaries, tendons, nerves, and ligaments.

Enzymes

Muscle contains a large number of enzymes, mainly peptidases like calpains, cathepsins, proteasome, tri- and dipeptidylpeptidases, aminopeptidases, carboxypeptidases, and dipeptidases, but also lipases, glycohydrolase, nucleotidases, and so on (Toldrá 2007). All these enzymes have a special relevance for meat processing and they are covered in-depth in Chapter 16.

Non-protein Compounds

Free Amino Acids

The action of muscle aminopeptidases contributes to the generation of free amino acids in living muscle. An example of the typical content of free amino acids, at less than 45 minutes postmortem, in glycolytic and oxidative porcine muscles is shown in Table 15.3. It can be observed that most of the amino acids are present in significantly higher amounts in the oxidative muscle (Aristoy and Toldrá 1998). The free amino acid content is relatively low just postmortem, but it is substantially increased during postmortem storage due to the action of the proteolytic chain, which is very active and stable during meat aging.

Table 15.3. Example of the Composition in Free Amino Acids (Expressed as mg/100g of Muscle) of the Glycolytic Muscle *Longissimus Dorsi* and Oxidative Muscle *Trapezius*.

Amino Acids	Muscle Longissimus dorsi	Muscle Trapezius
Essential		
Histidine	2.90	4.12
Threonine	2.86	4.30
Valine	2.78	2.09
Methionine	0.90	1.01
Isoleucine	1.52	1.11
Leucine	2.43	1.82
Phenylalanine	1.51	1.25
Lysine	2.57	0.22
Nonessential		
Aspartic acid	0.39	0.74
Glutamic acid	2.03	5.97
Serine	2.02	4.43
Asparagine	0.91	1.63
Glycine	6.01	12.48
Glutamine	38.88	161.81
Alanine	11.29	26.17
Arginine	5.19	5.51
Proline	2.83	4.45
Tyrosine	2.11	1.63
Ornithine	0.83	0.83

Source: Aristoy and Toldrá 1998.

Dipeptides

Muscle contains three natural dipeptides: carnosine (β-alanyl-L-histidine), anserine (β-alanyl-L-1-methylhistidine), and balenine (β-alanyl-L-3-methylhistidine). These dipeptides perform some physiological functions in muscle, for example, as buffers, antioxidants, neurotransmitters, and modulators of enzyme action (Chan and Decker 1994, Gianelli et al. 2000). Dipeptide content is especially higher in muscles with glycolytic metabolism (see Table 15.4), but it varies with the animal species, age, and diet (Aristoy and Toldrá 1998, Aristoy et al. 2004). Beef and pork have a higher content of carnosine and are lower in anserine, lamb has similar amounts of carnosine and anserine, and poultry is very rich in anserine (see Table 15.3). Balenine is present in minor amounts in pork muscle but at very low concentrations in other animal muscle, except in marine mammalians such as dolphins and whales (Aristoy and Toldrá 2004).

Muscle and Adipose Tissue Lipids

Skeletal muscle contains a variable amount of lipids, between 1% and 13%. Lipid content mainly depends on the degree of fattening and the amount of adipose tissue. Lipids can be found within the muscle (intramuscular), between muscles (intermuscular), and in adipose tissue. Intramuscular lipids are mainly composed of triacylglycerols, which are stored in fat cells, and phospholipids, which are located in cell membranes. The amount of cholesterol in lean meat is around 50–70 mg/100 g. Intermuscular and adipose tissue lipids are mainly composed of triacylglycerols and small amounts of cholesterol, around 40–60 mg/100g (Toldrá and Flores 2004).

Table 15.4. Example of the Composition in Dipeptides (Expressed as mg/100g of Muscle) of the Porcine Glycolytic Muscle *Longissimus Dorsi* and Oxidative Muscle *Trapezius*.

	Carnosine	Anserine
Effect of muscle metabolism		
Glycolytic (muscle *Longissimus dorsi)*	313	14.6
Oxidative *(muscle. Trapezius)*	181.0	10.7
Animal species		
Pork (loin)	313.0	14.5
Beef (top loin)	372.5	59.7
Lamb (neck)	94.2	119.5
Chicken (pectoral)	180.0	772.2

Source: Aristoy and Toldrá 1998, Aristoy and Toldrá 2004.

Triacylglycerols

Triacylglycerols are the major constituents of fat, as shown in Table 15.1. The fatty acid content mainly depends on age, production system, type of feed, and environment (Toldrá et al. 1996b). Monogastric animals such as swine and poultry tend to reflect the fatty acid composition of the feed in their fat. In the case of ruminants, the nutrients and fatty acid composition are somehow standardized due to biohydrogenation by the microbial population of the rumen (Jakobsen 1999). The properties of the fat will depend on its fatty acid composition. A great percentage of the triacylglycerols are esterified to saturated and monounsaturated fatty acids (see neutral muscle fraction and adipose tissue data in Table 15.5). When triacylglycerols are rich in polyunsaturated fatty acids (PUFA) such as linoleic and linolenic acids, fats tend to be softer and prone to oxidation. These fats may even have an oily appearance when kept at room temperature.

Phospholipids

These compounds are present in cell membranes, and although present in minor amounts (see Table 15.1), they have a strong relevance to flavor development due to their relatively high proportion of PUFA (see polar fraction in Table 15.5). Major constituents are phosphatidylcholine (lecithin) and phosphatidylethanolamine. The phospholipid content may vary depending on the genetic type of the animal and the anatomical location of the muscle (Hernández et al. 1998, Armero et al. 2002). For instance, red oxidative muscles have a higher amount of phospholipids than white glycolytic muscles.

CONVERSION OF MUSCLE TO MEAT

A great number of chemical and biochemical reactions take place in living muscle. Some of these reactions continue, while others are altered due to changes in pH, the presence of inhibitory compounds, the release of ions into the sarcoplasm, and so on during the early postmortem time. In a few hours, these reactions are responsible for the conversion of muscle to meat; this process is basically schematized in Figure 15.2 and consists of the following steps: Once the animal is slaughtered, the blood circulation is stopped, and the importation of nutrients and the removal of metabolites to the muscle cease. This fact has very important

Table 15.5. Example of Fatty Acid Composition (Expressed as Percentage of Total Fatty Acids) of Muscle *Longissimus Dorsi* and Adipose Tissue in Pigs Feeded with a Highly Unsaturated Feed. Neutral and Polar Fractions of Muscle Lipids are also Included

Fatty Acid	Total	Muscle Neutral	Polar	Adipose Tissue
Myristic acid (C 14:0)	1.55	1.97	0.32	1.40
Palmitic acid (C 16:0)	25.10	26.19	22.10	23.78
Stearic acid (C 18:0)	12.62	11.91	14.49	11.67
Palmitoleic acid (C 16:1)	2.79	3.49	0.69	1.71
Oleic acid (C 18:1)	36.47	42.35	11.45	31.64
C 20:1	0.47	0.52	0.15	0.45
Linoleic acid (C 18:2)	16.49	11.38	37.37	25.39
C 20:2	0.49	0.43	0.66	0.78
Linolenic acid (C 18:3)	1.14	1.17	0.97	2.64
C 20:3	0.30	0.10	1.04	0.10
Arachidonic acid (C 20:4)	2.18	0.25	9.83	0.19
C 22:4	0.25	0.08	0.84	0.07
Total SFA	39.42	40.23	37.03	37.02
Total MUFA	39.74	46.36	12.26	33.81
Total PUFA	20.84	13.41	50.70	29.17
Ratio MUFA/SFA	1.01	1.15	0.33	0.91
Ratio PUFA/SFA	0.53	0.33	1.37	0.79

SFA, saturated fatty acids; MUFA, monosaturated fatty acids; PUFA, polysaturated fatty acid.

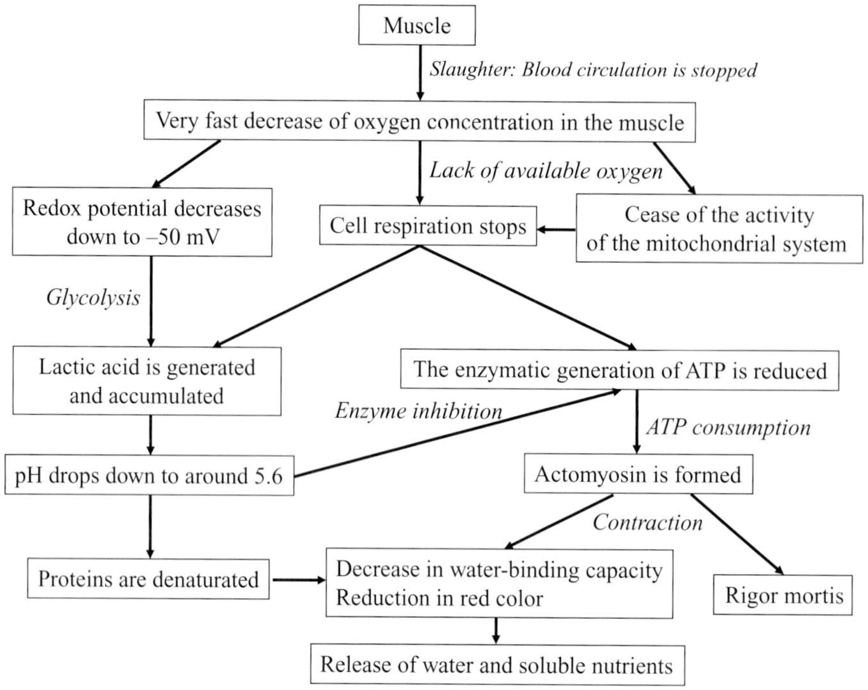

Figure 15.2. Summary of main changes during conversion of muscle to meat.

and drastic consequences. The first consequence is the reduction of the oxygen concentration within the muscle cell because the oxygen supply has stopped. An immediate consequence is a reduction in mitochondrial activity and cell respiration (Pearson 1987). Under normal aerobic values (see an example of resting muscle in Fig. 15.3), the muscle is able to produce 12 moles of adenosine triphosphate (ATP) per mole of glucose, and thus the ATP content is kept around 5–8 μmol/g of muscle (Greser 1986). ATP constitutes the main source of energy for the contraction and relaxation of the muscle structures as well as other biochemical reactions in postmortem muscle. As the redox potential is reduced toward anaerobic values, ATP generation is more costly. So, only 2 moles of ATP are produced per mole of glucose under anaerobic conditions (an example of a stressed muscle is shown in Fig. 15.3). The extent of anaerobic glycolysis depends on the reserves of glycogen in the muscle (Greaser 1986). Glycogen is converted to dextrins, maltose, and finally, glucose through a phosphorolytic pathway; glucose is then converted into lactic acid with the synthesis of 2 moles of ATP (Eskin 1990). In addition, the enzyme creatine kinase may generate some additional ATP from adenosine diphosphate (ADP) and creatine phosphate at very early postmortem times, but only while creatine phosphate remains. The contents of creatine have been reported to vary depending on the type of muscle (Mora et al. 2008). The main steps in glycolysis are schematized in Figure 15.4.

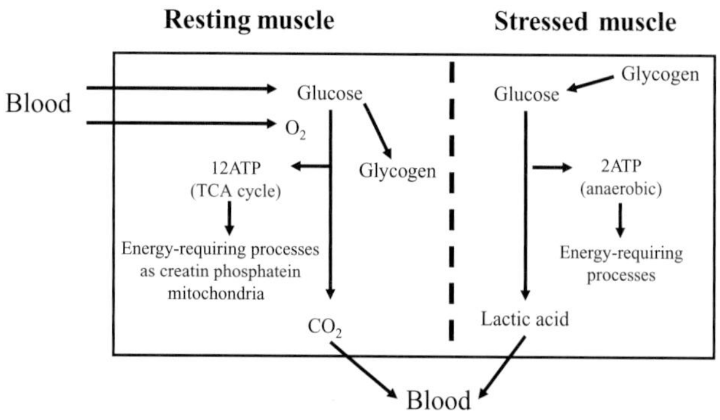

Figure 15.3. Comparison of energy generation between resting and stressed muscles.

The generation of ATP is strictly necessary in the muscle to supply the required energy for muscle contraction and relaxation and to drive the sodium-potassium pump of the membranes and the calcium pump in the sarcoplasmic reticulum. The initial situation in postmortem muscle is rather similar to that in the stressed muscle, but with an important change: the absence of blood circulation. Thus, there is a lack of nutrient supply and waste removal (see Fig. 15.5). Initially, the ATP content in postmortem muscle does not drop substantially because some ATP may be formed from ceratin phosphate through the action of the enzyme creatine kinase and through anaerobic glycolysis. As mentioned earlier, once creatine phosphate and glycogen are exhausted, ATP drops within a few hours to negligible

Enzymes	Reactions	Other	Comments
Hexokinase	Glucose → glucose-6-P	ATP → ADP	Requires Mg^{2+}
Phosphoglucoisomerase	Glucose-6-P → fructose-6-P		Requires Mg^{2+}
Phosphofructokinase	Fructose-6-P → fructose-1,6-biP	ATP → ADP	Inhibited by excess of ATP Requires Mg^{2+}
Aldolase	Fructose-1,6-biP → dihydroxyacetone-3-P ↓↑ Fructose-1,6-biP → glyceraldehyde-3-P		
Triose phosphate dehydrogenase	Glyceraldehyde-3-P → 1,3diphosphoglycerol	$2NAD^+$ → 2NADH	
Phosphoglycerokinase	1,3-Diphosphoglycerol → 3-phosphoglycerol	2ADP → 2ATP	Requires Mg^{2+}
Phosphoglyceromutase	3-Phosphoglycerol → 2-phosphoglycerol		Requires Mg^{2+}
Enolase	2-Phosphoglycerol → phosphoenolpyruvate	+ H_2O	
Pyruvate kinase	Phosphoenolpyruvate → pyruvic acid	2ADP → 2ATP	If O_2 available, produces CO_2 via TCA cycle Requires K^+, Mg^{2+}
Lactate dehydrogenase	Pyruvic acid → lactic acid	NADH → NAD^+	Only under anaerobic conditions

Figure 15.4. Main steps in glycolysis during early postmortem. (Adapted from Greaser 1986.)

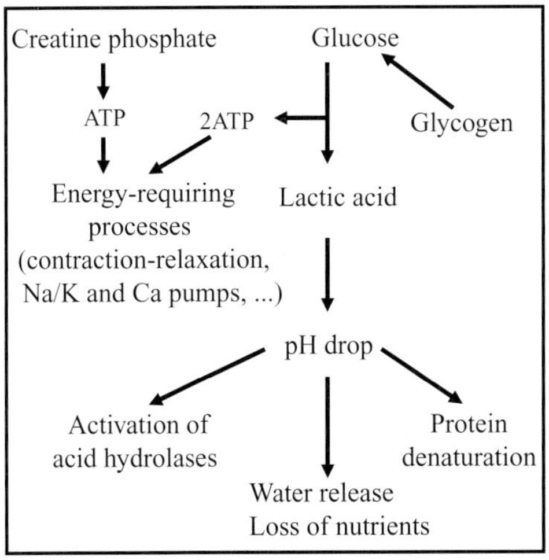

Figure 15.5. Scheme of energy generation in postmortem muscle.

values by conversion into ADP, adenosine monophosphate, and other derived compounds such as 5′-inosine monophosphate, 5′-guanosine monophosphate, and inosine (see Fig. 15.6). An example of the typical content of ATP breakdown products in pork at 2 hours and 24 hours postmortem is shown in Table 15.6. The reaction rates depend on the metabolic status of the animal prior to slaughter. For instance, reactions proceed very quickly in pale, soft, exudative (PSE) muscle, where ATP can be almost fully depleted within few minutes. The rate is also affected by the pH and temperature of the meat (Batlle et al. 2000, 2001). For instance, the ATP content in beef *Sternomandibularis* kept at 10–15°C is around 5 μmol/g at 1.5 hours postmortem and decreases to 3.5 μmol/g at 8–9 hours postmortem. However, when that muscle is kept at 38°C, ATP content is below 0.5 μmol/g at 6–7 hours postmortem.

Once the ATP concentration is exhausted, the muscle remains contracted, as no more energy is available for relaxation. The muscle develops a rigid condition known as rigor mortis, in which the crossbridge of myosin and actin remains locked, forming actomyosin (Greaser 1986). The postmortem time necessary for the development of rigor mortis is variable, depending on the animal species, size of carcass, amount of fat cover, and environmental conditions such as the temperature of the chilling

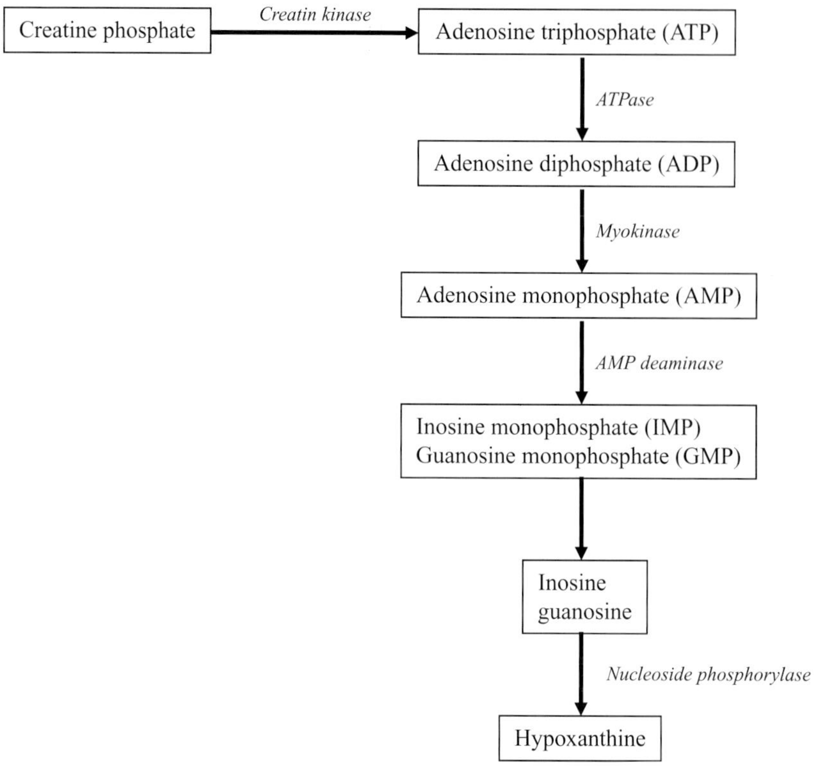

Figure 15.6. Main adenosine triphosphate (ATP) breakdown reactions in early postmortem muscle.

tunnel and the air velocity (see pork pieces after cutting in a slaughterhouse in Fig. 15.7). The rates of enzymatic reactions are strongly affected by temperature. In this sense, the carcass-cooling rate will affect glycolysis rate, pH drop rate, and the time course of rigor onset (Faustman 1994). The animal species and size of carcass have a great influence on the cooling rate of the carcass. Furthermore, the location in the carcass is also important because surface muscles cool more rapidly than deep muscles (Greaser 2001). So, when carcasses are kept at 15°C, the time required for rigor mortis development may be about 2–4 hours in poultry, 4–18 hours in pork, and 10–24 hours in beef (Toldrá 2006).

Table 15.6. Example of Nucleotides and Nucleosides Content (Expressed as μmol/g Muscle) in Pork Postmortem Muscle at 2 hours and 24 hours

Compound	2 h Postmortem Time	24 h Postmortem Time
ATP	4.39	–
ADP	1.08	0.25
AMP	0.14	0.20
ITP + GTP	0.18	–
IMP	0.62	6.80
Inosine	0.15	1.30
Hypoxanthine	0.05	0.32

Source: Batlle et al. 2001.
ATP, adenosine triphosphate; ADP, adenosine diphosphate; AMP, adenosine monophosphate; ITP, inosine triphosphate; GMP, guanosine monophosphate; IMP, inosine monophosphate.

Muscle glycolytic enzymes hydrolyze the glucose to lactic acid, which is accumulated in the muscle because muscle waste substances cannot be eliminated due to the absence of blood circulation. This lactic acid accumulation produces a relatively rapid (in a few hours) pH drop to values of about 5.6–5.8. The pH drop rate depends on the glucose concentration, the temperature of the muscle, and the metabolic status of the animal previous to slaughter. Water binding decreases with pH drop because of the change in the protein's charge. Then, some water is released out of the muscle as a drip loss (DL). The amount of released water depends on the extent and rate of pH drop. Soluble compounds such as sarcoplasmic proteins, peptides, free amino acids, nucleotides, nucleosides, B vitamins, and minerals may be partly lost in the drippings, affecting nutritional quality (Toldrá and Flores 2004).

The pH drop during early postmortem has a great influence on the quality of pork and poultry meats. The pH decrease is very fast, below 5.8 after 2 hours postmortem, in muscles from animals with accelerated metabolism. This is the case of the PSE pork meats and red, soft, exudative pork meats. ATP breakdown

Figure 15.7. Pork hams after cutting in a slaughterhouse, ready for submission to a processing plant. (Courtesy of Industrias Cárnicas Vaquero SA, Madrid, Spain.)

also proceeds very quickly in these types of meats, with almost full ATP disappearance in less than 2 hours (Batlle et al. 2001). Red, firm, normal meat experiences a progressive pH drop down to values around 5.8–6.0 at 2 hours postmortem. In this meat, full ATP breakdown may take up to 8 hours. Finally, the dark, firm, dry pork meat (DFD) and dark cutting beef meat are produced when the carbohydrates in the animal are exhausted from before slaughter, and thus almost no lactic acid can be generated during early postmortem due to the lack of a substrate. Very low or almost negligible glycolysis is produced, and the pH remains high in these meats, which constitutes a risk from the microbiological point of view. These meats constitute a risk because they are prone to contamination by foodborne pathogens and must be carefully processed, with extreme attention to good hygienic practices.

Protein oxidation is another relevant change during postmortem aging. Some amino acid residues may be converted into carbonyl derivatives and cause the formation of inter- and intraprotein disulfide links that can reduce the functionality of proteins (Huff-Lonergan 2010).

FACTORS AFFECTING BIOCHEMICAL CHARACTERISTICS

Effect of Genetics

Genetic Type

The genetic type has an important relevance for quality, not only due to differences among breeds, but also to differences among animals within the same breed. Breeding strategies have been focused toward increased growth rate and lean meat content and decreased backfat thickness. Although grading traits are really improved, poorer meat quality is sometimes obtained. Usually, large ranges are found for genetic correlations between production and meat quality traits, probably due to the reduced number of samples when analyzing the full quality of meat, or to a large number of samples but with few determinations of quality parameters. This variability makes it necessary to combine the results from different research groups to obtain a full scope (Hovenier et al. 1992).

Current pig breeding schemes are usually based on a backcross or on a three- or four-way cross. For instance, a common cross in the European Union is a three-way cross, where the sow is a Landrace × Large White (LR × LW) crossbreed. The terminal sire is chosen depending on the desired profitability per animal, and there is a wide range of possibilities. For instance, the Duroc terminal sire grows faster and shows a better food conversion ratio but accumulates an excess of fat; Belgian Landrace and Pietrain are heavily muscled but have high susceptibility to stress and thus usually present a high percentage of exudative meats; or a combination of Belgian Landrace × Landrace gives good conformation and meat quality (Toldrá 2002).

Differences in tenderness between cattle breeds have also been observed. Brahman cattle is used extensively in the southwest of the United States since it is tolerant to adverse environmental conditions but may give some tenderness issues (Brewer 2010). Even toughness was associated to an increased amount of calpastatin, the endogenous muscle calpain inhibitor (Ibrahim 2008). Studies have been performed for cattle breeding. For instance, after 10 days of aging, the steaks from an Angus breed were more tender than those from a 1/2Angus–1/2Brahman breed, and more tender than steaks from a 1/4Angus–3/4Brahman breed (Johnson et al. 1990). In other studies, it was found that meat from Hereford cattle was more tender than that from Brahman cattle (Wheeler et al. 1990). Differences in the activity of proteolytic enzymes, especially calpains, that are deeply involved in the degradation of Z-line proteins appear to have a major role in the tenderness differences between breeds.

The enzyme fingerprints, which include the assay of many different enzymes such as endo- and exoproteases as well as lipases and esterases, are useful for predicting the expected proteolysis and lipolysis during further meat processing (Armero et al. 1999a, 1999b). These enzymatic reactions, extensively described in Chapter 14, are very important for the development of sensory characteristics such as tenderness and flavor in meat and meat products.

Genes

Some genes have been found to have a strong correlation to certain positive and negative characteristics of meat. The dominant RN^- allele, also known as the Napole gene, is common in the Hampshire breed of pigs and causes high glycogen content and an extended pH decline. The carcasses are leaner, and the eating quality is better in terms of tenderness and juiciness, but the more rapid pH fall increases DL by about 1%, while the technological yield is reduced by 5–6% (Rosenvold and Andersen 2003, Josell et al. 2003). The processing industry is not interested in pigs with this gene because most pork meat is used for further processing, and the meat from carriers of the RN^- allele gives such a low technological yield (Monin and Sellier 1985).

Pigs containing the halothane gene are stress susceptible, a condition also known as porcine stress syndrome (PSS). These pigs are very excitable in response to transportation and environmental situations, have a very high incidence of PSE, and are susceptible to death due to malignant hyperthermia. These stress-susceptible pigs may be detected through the application of the halothane test, observing their reaction to inhalation anesthesia with halothane (Cassens 2000). These pigs give a higher carcass yield and leaner carcasses, which constitutes a direct benefit for farmers. However, the higher percentage of PSE, with high DL, poor color, and deficient technological properties, makes it unacceptable to the meat processing industry. These negative effects convinced major breeding companies to remove the halothane gene from their lines (Rosenvold and Andersen 2003).

Incidence of Exudative Meats

The detection of exudative meats at early postmortem time is of primary importance for meat processors to avoid further losses during processing. It is evident that PSE pork meat is not appealing to the eye of the consumer because it has a pale color, abundant dripping in the package, and a loose texture (Cassens 2000). Exudative pork meat also generates a loss of the nutrients that are solubilized in the sarcoplasm and lost in the drip and an economic loss due to the lower weight as a consequence of its poor binding properties if the meat is further processed.

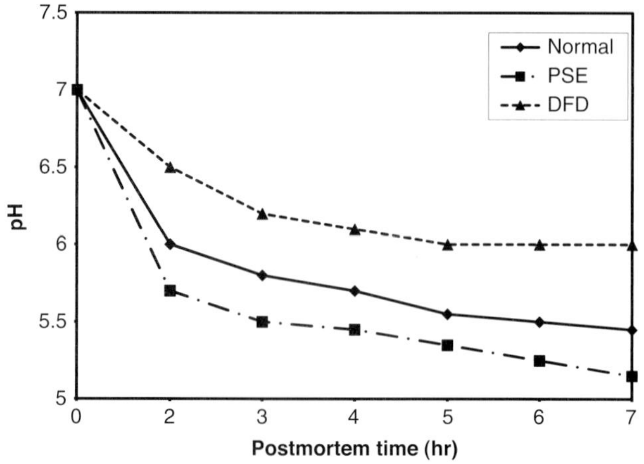

Figure 15.8. Typical postmortem pH drop of normal, pale, soft, exudative (PSE) and dark, firm, dry (DFD) pork meats (Toldrá, unpublished).

PSE meat is the result of protein denaturation at acid pH and relatively high postmortem temperatures. There are several classification methodologies, for example, the measurement of pH or conductivity at 45 minutes postmortem. Other methodologies involve the use of more data and thus give a more accurate profile (Warner et al. 1993, 1997, Toldrá and Flores 2000). So, exudative meats are considered when pH measured at 2 hours postmortem (pH_{2h}) is lower than 5.8 and DL is higher than 6%. DL, which is usually expressed as a percent, gives an indication of water loss (difference in weight between 0 and 72 hours): a weighed muscle portion is hung within a sealed plastic bag for 72 hours under refrigeration, then reweighed (Honikel 1997). The color parameter, L, is higher than 50 (pale color) for PSE meats and between 44 and 50 for red exudative (RSE) meats. Meats are considered normal when pH_{2h} is higher than 5.8, L is between 44 and 50, and DL is below 6%. Meats are classified as DFD when L is lower than 44 (dark red color), DL is below 3%, and pH measured at 24 hours postmortem (pH_{24h}) remains high. Typical pH drops are shown in Figure 15.8.

There are some measures such as appropriate transport and handling, adequate stunning, and chilling rate of carcasses that can be applied to prevent, or at least reduce, the incidence of negative effects in exudative meats. Even though the problem is well known and there are some available corrective measures, exudative meats still constitute a problem. A survey carried out in the United States in 1992 revealed that 16% of pork carcasses were PSE, 10% DFD, and about 58% of questionable quality, mainly RSE, indicating little progress in the reduction of the problem (Cassens 2000). A similar finding was obtained in a survey carried out in Spain in 1999, where 37% of carcasses were PSE, 12% RSE, and 10% DFD (Toldrá and Flores 2000).

Effect of the Age and Sex

The content in intramuscular fat content increases with the age of the animal. In addition, the meat tends to be more flavorful and colorful, due to an increased concentration of volatiles and myoglobin, respectively (Armero et al. 1999a). Some of the muscle proteolytic and lipolytic enzymes are affected by age. Muscles from heavy pigs (11 months old) are characterized by a greater peptidase to proteinase ratio and a higher lipase, dipeptidylpeptidase IV, and pyroglutamyl aminopeptidase activity. On the other hand, the enzyme activity in light pigs (7–8 months old) shows two groups. The larger one is higher in moisture content and cathepsins B and B+L and low in peptidase activity, while the minor one is intermediate in cathepsin B activity and high in peptidase activity (Toldrá et al. 1996a). In general, there is a correlation between the moisture content and the activity of cathepsins B and B+L (Parolari et al. 1994). So, muscles with higher moisture content show higher levels of cathepsins B and B+L activity. This higher cathepsin activity may produce an excess of proteolysis in processed meat products with long processing times (Toldrá 2002).

A minor effect of sex is observed. Meats from barrows contain more fat than those from gilts. They present higher marbling, and the subcutaneous fat layer is thicker (Armero et al. 1999a). In the case of muscle enzymes, only very minor differences have been found. Sometimes, meats from entire males may give some sexual odor problems due to high contents of androstenone or escatol.

Effect of the Type of Feed

A great research effort has been exerted since the 1980s for the manipulation of the fatty acid composition of meat, to achieve nutritional recommendations, especially an increase in the ratio between PUFA and saturated fatty acids (SFA) (PUFA:SFA ratio). More recently, nutritionists recommend that PUFA composition should be manipulated toward a lower *n*-6:*n*-3 ratio.

Fats with a higher content of PUFA have lower melting points that affect the fat firmness. Softer fats may raise important problems during processing if the integrity of the muscle is disrupted by any mechanical treatment (chopping, mincing, stuffing, etc.). The major troubles are related to oxidation and generation of off-flavors (rancid aromas) and color deterioration (trend toward yellowness in the fat) (Toldrá and Flores 2004).

Pigs and poultry are monogastric animals that incorporate part of the dietary fatty acids practically unchanged into the adipose tissue and cellular membranes, where desaturation and chain elongation processes may occur (Toldrá et al. 1996b, Jakobsen 1999). The extent of incorporation may vary depending on the specific fatty acid and the type of feed. Different types of cereals as well as dietary oils and their effects on the proportions in fatty acid composition have been studied. The use of canola or linseed oils produces a substantial increase in the content of linolenic acid (C 18:3), which is an *n*-3 fatty acid (Jiménez-Colmenero et al. 2006). In this way, the *n*-6:*n*-3 ratio can be reduced from 9 to 5 (Enser et al. 2000). Other dietary oils such as soy, peanut, corn, and sunflower increase the content of linoleic acid (C 18:2), an *n*-6 fatty acid. Although it increases the total PUFA content, this fatty acid does not contribute to decrease the *n*-6:*n*-3 ratio, just the reverse. A similar trend is observed in the case of poultry, where the feeds with a high content of linoleic acid such as grain, corn, plant seeds, or oils also increase the *n*-6:*n*-3 ratio (Jakobsen 1999). As in the case of pork, the use of feeds containing fish oils or algae, enriched in *n*-3 fatty acids such as eicosapentaenoic (C 22:5 *n*-3) and docosahexanoic (C 22:6 *n*-3) acids, can enrich the poultry meat in *n*-3 fatty acids and reduce the *n*-6:*n*-3 ratio from around 8.4 to 1.7 (Jakobsen 1999).

The main problem arises from oxidation during heating, because some volatile compounds such as hexanal are typically generated, producing rancid aromas (Larick et al. 1992). The rate and extent of oxidation of muscle foods mainly depends on the level of PUFA, but they are also influenced by early postmortem events such as pH drop, carcass temperature, aging, and other factors. It must be pointed out that the increased linoleic acid content is replacing the oleic acid to a large extent (Monahan et al. 1992). Feeds rich in saturated fats such as tallow yield the highest levels of palmitic, palmitoleic, stearic, and oleic acids in pork loin (Morgan et al. 1992). Linoleic and linolenic acid content may vary as much as 40% between the leanest and the fattest animals (Enser et al. 1988). An example of the effect of feed type on the fatty acid composition of subcutaneous adipose tissue of pigs is shown in Table 15.7. The PUFA content is especially high in phospholipids, located in subcellular membranes such as mitochondria, microsomes, and so on, making them vulnerable to peroxidation because of the proximity of a range of pro-oxidants such as myoglobin, cytochromes, nonheme iron, and trace elements (Buckley et al. 1995). Muscle contains several antioxidant systems, for example, those of superoxide dismutase and glutathione peroxidase, and ceruloplasmin and transferrin, although they are weakened during postmortem storage.

An alternative for effective protection against oxidation consists of the addition of natural antioxidants like vitamin E (alpha-tocopheryl acetate); this has constituted a common practice in the last decade. This compound is added in the feed as an antioxidant and is accumulated by animals in tissues and subcellular structures, including membranes, substantially increasing its effect. The concentration and time of supplementation are important. Usual levels are around 100–200 mg/kg in the feed for several weeks prior to slaughter. The distribution of vitamin E in the organism is variable, being higher in the muscles of the thoracic limb, neck, and thorax and lower in the muscles of the pelvic limb and back (O'Sullivan et al. 1997). Dietary supplementation with this lipid-soluble antioxidant improves the oxidative stability of the meat. Color stability in beef, pork, and poultry is improved by protection of myoglobin against oxidation (Houben et al. 1998, Mercier et al. 1998). The water-holding capacity in pork is improved by protecting the membrane phospholipids against oxidation (Cheah et al. 1995, Dirinck et al. 1996). The reduction in DL by vitamin E is observed even in frozen pork meat, upon thawing. Oxidation of membrane phospholipids causes a loss in membrane integrity and affects its function as a semipermeable barrier. As a consequence, there is an increased passage of sarcoplasmic fluid through the membrane, known as DL.

Table 15.7. Effect of Type of Feed on Total Fatty Acid Composition (Expressed as Percentage of Total Fatty Acids) of Pork muscle Lipids

Fatty Acid\Feed Enriched	Barley + Soybean Meal[a]	Safflower Oil[b]	Tallow Diet[c]	High Oleic Sunflower Oil[d]	Canola Oil[e]
Myristic acid (C 14:0)	–	–	1.37	0.05	1.6
Palmitic acid (C 16:0)	23.86	27.82	24.15	6.35	20.6
Stearic acid (C 18:0)	10.16	12.53	11.73	4.53	9.8
Palmitoleic acid (C 16:1)	3	3.56	3.63	0.45	3.6
Oleic acid (C 18:1)	39.06	37.81	46.22	71.70	45.9
C 20:1	–	0.01	0.29	0.26	
Linoleic acid (C 18:2)	17.15	14.60	8.95	15.96	12.3
C 20:2	–	0.01	0.44	–	0.4
Linolenic acid (C 18:3)	0.91	0.01	0.26	0.71	3.0
C 20:3	0.21	0.01	0.25	–	0.1
Arachidonic acid (C 20:4)	4.26	2.14	2.13	–	0.74
Eicosapentaenoic acid (C 22:5)	0.64	0.01	–	–	–
Hexadecanoic acid (C22:6)	0.75	0.01	–	–	–
Total SFA	34.02	40.35	37.83	10.93	33.6
Total MUFA	42.06	42.38	50.26	72.41	49.5
Total PUFA	23.92	16.79	11.91	16.67	16.6
Ratio MUFA/SFA	1.24	1.05	1.33	6.62	1.47
Ratio PUFA/SFA	0.70	0.42	0.32	1.53	0.49

Sources: [a]Morgan et al. (1992), [b]Larick et al. (1992), [c]Leszczynski et al. (1992), [d]Rhee et al. (1988), [e]Miller et al. (1990).
SFA, saturated fatty acids; MUFA, monosaturated fatty acids; PUFA, polysaturated fatty acid.

The fatty acid profile in ruminants is more saturated than in pigs, and thus the fat is firmer (Wood et al. 2003). The manipulation of fatty acids in beef is more difficult due to the rumen biohydrogenation. More than 90% of the PUFA are hydrogenated, leaving a low margin for action to increase the PUFA:SFA ratio above 0.1. However, meats from ruminants are rich in conjugated linoleic acid (CLA), mainly 9-*cis*,11-*trans*-octadecadienoic acid, which exerts important health-promoting biological activity (Belury 2002).

In general, a good level of nutrition increases the amount of intramuscular fat. On the other hand, food deprivation may result in an induced lipolysis that can be rapidly detected (in just 72 hours) through a higher content of free fatty acids and monoacylglycerols, especially in glycolytic muscles (Fernandez et al. 1995). Fasting within 12–15 hours preslaughter is usual to reduce the risk of microbial cross-contamination during slaughter.

CARCASS CLASSIFICATION

Current Grading Systems

Meat grading constitutes a valuable tool for the classification of a large number of carcasses into classes and grades with similar characteristics such as quality and yield. The final purpose is to evaluate specific characteristics to determine carcass retail value. In addition, the weight and category of the carcass are useful for establishing the final price to be paid to the farmer. Carcasses are usually evaluated for conformation, carcass length, and back-fat thickness. The carcass yields vary depending on the degree of fatness and the degree of muscling. The grade is determined based on both degrees. The grading system is, thus, giving information on quality traits of the carcass that help producers, processors, retailers, and consumers.

Official grading systems are based on conformation, quality, and yield. Yield grades indicate the quantity of edible meat in a carcass. In the United States, beef carcasses receive a grade for quality (prime, choice, good, standard, commercial, utility, and cutter) and a grade for predicted yield of edible meat (1–5). There are four grades for pork carcasses (US No. 1 to US No. 4) based on back-fat thickness and expected lean yield. The lean yield is predicted by a combination of back-fat thickness measured at the last rib and the subjective estimation of the muscling degree. In the case of poultry, there are three grades (A C) based on the bilateral symmetry of the sternum, the lateral convexity and distal extension of the pectoral muscles, and fat cover on the pectoral muscles (Swatland 1994).

In Europe, beef, pork, and lamb carcasses are classified according to the EUROP scheme (Council regulation 1208/81, Commission directives 2930/81, 2137/92, and 461/93). These European Union Directives are compulsory for all the member states. Carcass classification is based on the conformation according to profiles, muscle development, and fat level. Each carcass is classified by visual inspection, based on photos corresponding to each grade (see Fig. 15.9). The six conformation classification ratings are S (superior), E (excellent), U (very good), R (good), O (fair), and P (poor). S represents the highest quality level and must not present any defect in the main pieces. The five classification ratings for fat level are 1 (low), 2 (slight), 3 (average), 4 (high), and 5 (very high). The grading system for each carcass consists in a letter for conformation, which is given

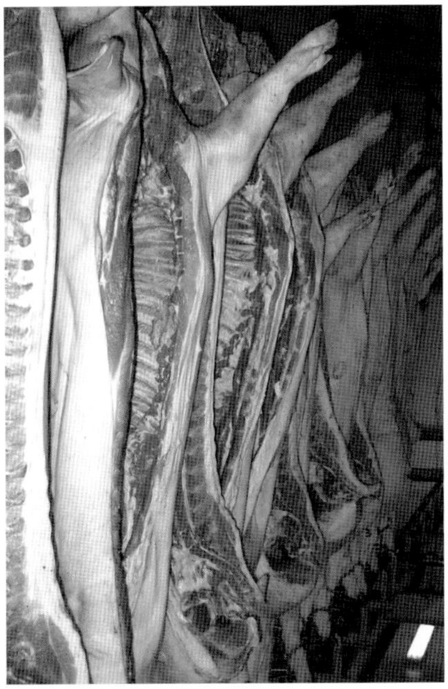

Figure 15.9. Classification of pork carcasses in the slaughterhouse. (Courtesy of Industrias Cárnicas Vaquero SA, Madrid, Spain.)

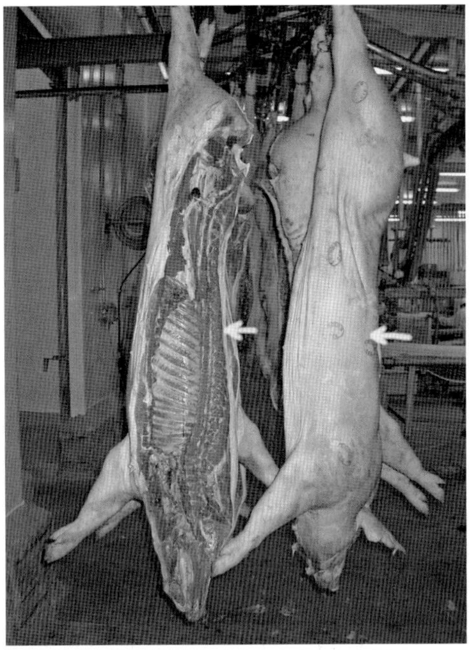

Figure 15.10. Example of application of physical-based methods for the online evaluation of meat yield and quality. At the slaughterhouse, the hand-held devices are applied to the carcass between the last three and four ribs (place marked with an *arrow*). The signal is then received in the unit and computer-processed. The carcass quality is estimated and, depending on the technique, is also classified by yield. (Courtesy of Industrias Cárnicas Vaquero SA, Madrid, Spain.)

first, and a number for the fat level. In some countries, additional grading systems may be added. For instance, in France, the color is measured and rated 1 to 4 from white to red color.

New Grading Systems

New grading systems are being developed that take advantage of the rapid developments in video image analysis and other new physical techniques as well as those in biochemical assay tests. New methodologies based on physical methods for the online evaluation of meat yield and meat quality, applied as schematized in Figure 15.10, include near infrared reflectance (NIR), video image analysis, ultrasound, texture analysis, nuclear magnetic resonance, and magnetic resonance spectroscopy techniques.

Physical Techniques

NIR is applied for the rapid and nondestructive analysis of meat composition in fat, protein, and water (Byrne et al. 1998, Rodbotten et al. 2000). This technique has also been applied to aged meat, giving a good correlation with texture, and thus constituting a good predictor for meat tenderness. In addition, NIR spectroscopy allows for a minimal or nonnecessary sample preparation (Niemöller and Behmer 2008). Video image analysis is very useful for the measurement of carcass shape, marbling, and meat color. Conductivity has been in use for several years to predict meat composition and quality.

Ultrasounds are based on the measurement of different parameters such as velocity, attenuation, and backscattering intensity and may constitute a valuable tool for the measurement of meat composition (Got et al. 1999, Abouelkaram et al. 2000). Texture analysis, as the image processing of the organization of gray pixels of digitized images, can be used for the classification of photographic images of meat slices (Basset et al. 2000). This technique appears to give good correlation with fat and collagen, which are especially visible under ultraviolet light, and would allow classification according to three factors, muscle type, age, and breed (Basset et al. 1999).

Nuclear magnetic resonance has good potential as a noninvasive technique for better characterization and understanding of meat features. Thus, magnetic resonance imaging can give a spatial resolution that characterizes body composition. This technique is well correlated to important meat properties such as pH, cooking yield, and water-holding capacity (Laurent et al. 2000). Magnetic resonance spectroscopy may be useful to determine the fatty acid composition of animal fat. This technique may have further applications; for example, in the possible use of ^{23}Na imaging to follow brine diffusion in cured meat products (Renou et al. 1994).

The use of bioimpedance spectroscopy has shown good performance for the assessment of meat quality, especially to distinguish among PSE, DFD, and normal pork meats (Castro-Giráldez et al. 2009, 2010a). The amount of ATP is characteristic for each type of pork meat quality and the ion activity of ATP may be correlated with the ionic dispersion in the dielectric spectra (Castro-Giráldez et al. 2010b).

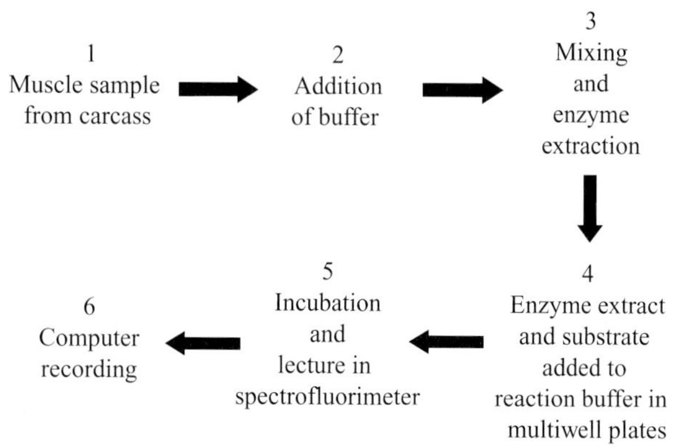

Figure 15.11. Example of application of biochemical-based methods for evaluation of meat quality. A sample of the carcass (1) is mixed with buffer (2) for enzyme extraction and homogenization (3). Enzyme extracts are placed in the wells of a multiwell plate and synthetic substrates, previously dissolved in reaction buffer, are added (4). The released fluorescence, which is proportional to the enzyme activity, is read by a multiwell plate spectrofluorometer (5) and computer-recorded (6).

Biochemical Assay Techniques

Biochemical assay methodologies are based on biochemical compounds that can be used as markers of meat quality. The mode of operation is essentially schematized in Figure 15.11. Some of the most promising techniques include assay of proteolytic muscle enzymes and use of peptides as biochemical markers.

The assay of certain proteolytic muscle enzymes such as calpain I, alanyl aminopeptidase, or dipeptidylpeptidase IV at just 2 hours postmortem, has shown good ability to predict the water-holding capacity of the meat (Toldrá and Flores 2000). Modified assay procedures, based on the use of synthetic fluorescent substrates, have been developed to allow relatively fast and simple measurements of enzyme activity with enough sensitivity.

Some peptides have been proposed as biochemical markers for meat tenderness, which is particularly important in beef. These are peptides with molecular masses ranging from 1282 to 5712 kDa, generated from sarcoplasmic and myofibrillar proteins (Stoeva et al. 2000). The isolation and identification of these peptides are tedious and time consuming, but once the full sequence is known, Enzyme-linked immunosorbent assay (ELISA) test kits can be developed, and this would allow rapid assay and online detection at the slaughterhouse.

BIOACTIVE COMPOUNDS

It is well known that meat contains a large amount of proteins with high biological value. However, in recent years, the bioactivities of certain compounds that are present in meat, even though in minor amounts, have received increased attention because they may have an important nutritional role.

CLA is a group of isomers of octadecadienoic acid that is abundant in the fat of ruminants like cattle and sheep. The content may change depending on the breed, feed, and age. CLA has been reported to reduce the risk for certain types of cancer like the colorectal cancer as well as having other activities like antiartherioesclerotic, antioxidative, and also playing certain role in controlling obesity (Arihara and Motoko 2010).

Some of these substances are small peptides, which are present in meat. Dipeptides carnosine and anserine are present in meat although their content varies depending on the animal species and the type of muscle metabolism (Aristoy et al. 2004, Aristoy and Toldrá 2004, Mora et al. 2008). Both peptides have shown good resistance to proteases and a good antioxidant activity. Other bioactive peptides are derived from meat proteins after digestion. These peptides exert an antihypertensive effect through the inhibitory action on the angiotensin I converting enzyme. So, several bioactive peptides have been identified and sequenced after simulated in vitro gastrointestinal digestion of pork meat proteins (Escudero et al. 2010a, 2010b).

In summary, there are many biochemical and chemical reactions of interest in raw meats that contribute to important changes in meat and affect its quality. Most of these changes, as described in this chapter, have an important role in defining the aptitude of the meat as raw material for further processing.

REFERENCES

Abouelkaram S et al. 2000. Effects of muscle texture on ultrasonic measurements. *Food Chem* 69: 447–455.

Arihara K, Motoko O. 2010. Functional meat products. In: F Toldrá (ed.) *Handbook of Meat Processing*. Wiley-Blackwell, Ames, IA, pp. 423–439.

Aristoy MC, Toldrá F. 1998. Concentration of free amino acids and dipeptides in porcine skeletal muscles with different oxidative patterns. *Meat Sci* 50: 327–332.

Aristoy MC, Toldrá F. 2004. Histidine dipeptides HPLC-based test for the detection of mammalian origin proteins in feeds for ruminants. *Meat Sci* 67: 211–217.

Aristoy MC et al. 2004. A simple, fast and reliable methodology for the analysis of histidine dipeptides as markers of the presence of animal origin proteins in feeds for ruminants. *Food Chem* 94: 485–491

Armero E et al. 1999a. Effect of the terminal sire and sex on pork muscle cathepsin (B, B+L and H), cysteine proteinase inhibitors and lipolytic enzyme activities. *Meat Sci* 51: 185–189.

Armero E et al. 1999b. Effects of pig sire types and sex on carcass traits, meat quality and sensory quality of dry-cured ham. *J Sci Food Agric* 79: 1147–1154.

Armero E et al. 2002. Lipid composition of pork muscle as affected by sire genetic type. *J Food Biochem* 26: 91–102.

Bandman E. 1987. Chemistry of animal tissues. Part 1-Proteins. In: JF Price, BS Schweigert (eds.) *The Science of Meat and Meat Products*. Food and Nutrition Press, Inc, Trumbull, CT, pp. 61–102.

Basset O et al. 1999. Texture image analysis: application to the classification of bovine muscles from meat slice images. *Optical Eng* 38: 1956–1959.

Basset O et al. 2000. Application of texture image analysis for the classification of bovine meat. *Food Chem* 69: 437–445.

Batlle N et al. 2000. Early postmortem detection of exudative pork meat based on nucleotide content. *J Food Sci* 65: 413–416.

Batlle N et al. 2001. ATP metabolites during aging of exudative and non exudative pork meats. *J Food Sci* 66: 68–71.

Belury, MA. 2002. Dietary conjugated linoleic acid in health: physiological effects and mechanisms of action. *Ann Rev Nutr* 22: 505–531.

Brewer S. 2010. Technological quality of meat for processing. In: F Toldrá (ed.) *Handbook of Meat Processing*. Wiley-Blackwell, Ames, IA, pp. 25–42.

Buckely DJ et al. 1995. Influence of dietary vitamin E on the oxidative stability and quality of pig meat. *J Anim Sci* 73: 3122–3130.

Byrne CE et al. 1998. Non-destructive prediction of selected quality attributes of beef by near infrared reflectance spectroscopy between 750 and 1098 nm. *Meat Sci* 49: 399–409.

Cassens RG. 2000. Historical perspectives and current aspects of pork meat quality in the USA. *Food Chem* 69: 357–363.

Castro-Giráldez M et al. 2009. Physical sensors and techniques. In: LML Nollet, F Toldrá (eds.) *Handbook of Processed Meats and Poultry Analysis*. CRC Press, Boca Raton, FL, pp. 7–34.

Castro-Giráldez M et al. 2010a. Physical sensors for quality control during processing. In: F Toldrá (ed.) *Handbook of Meat Processing*. Wiley-Blackwell, Ames, IA, pp. 443–456.

Castro-Giráldez M et al. 2010b. Low-frequency dielectric spectrum to determine pork meat quality. *Innov. Food Sci. Emerging Technol* 11: 376–386.

Chan KM, Decker EA. 1994. Endogenous skeletal muscle antioxidants. *Crit Rev Food Sci Nutr* 34: 403–426.

Cheah KS et al. 1995. Effect of dietary supplementation of vitamin E on pig meat quality. *Meat Sci* 39: 255–264.

Dirinck P et al. 1996. Studies on vitamin E and meat quality: 1. Effect of feeding high vitamin E levels on time-related pork quality. *J Agric Food Chem* 44: 65–68.

Enser M et al. 1998. Fatty acid content and composition of UK beef and lamb muscle in relation to production system and implications for human nutrition. *Meat Sci* 49: 329–341.

Enser M et al. 2000. Feeding linseed to increase the n-3 PUFA of pork: fatty acid composition of muscle, adipose tissue, liver and sausages. *Meat Sci* 55: 201–212.

Escudero E et al. 2010a. Angiotensin I converting enzyme inhibitory peptides generated from in vitro gastrointestinal digestion of pork meat. *J Agric Food Chem* 58: 2895–2901.

Escudero E et al. 2010b. Characterization of Peptides Released by *in vitro* Digestion of Pork Meat. *J Agric Food Chem* in press.

Eskin NAM. 1990. Biochemical changes in raw foods: meat and fish. In: *Biochemistry of Foods*, 2nd edn. Academic Press, San Diego, pp. 3–68

Faustman LC. 1994. Postmortem changes in muscle foods. In: DM Kinsman, AW Kotula, BC Breidenstein (eds.) *Muscle Foods*. Chapman and Hall, New York, pp. 63–78.

Fernández X et al. 1995. Effect of muscle type and food deprivation for 24 hours on the composition of the lipid fraction in muscles of large white pigs. *Meat Sci* 41: 335–343.

Flores J, Toldra F. 1993. Curing: processes and applications. In: R MacCrae et al. (eds.) *Encyclopedia of Food Science, Food Technology and Nutrition*. Academic Press, London, pp. 1277–1282.

Gianelli MP et al. 2000. Effect of carnosine, anserine and other endogenous skeletal peptides on the activity of porcine muscle alanyl and arginyl aminopeptidases. *J Food Biochem* 24: 69–78.

Got F et al. 1999. Effects of high intensity high frequency ultrasound on aging rate, ultrastructure and some physico-chemical properties of beef. *Meat Sci* 51: 35–42.

Greaser ML. 1986. Conversion of muscle to meat. In: PJ Bechtel (ed.) *Muscle as Food*. Academic Press, Orlando, pp. 37–102.

Greaser ML. 2001. Postmortem muscle chemistry. In: YH Hui, et al. (eds.) *Meat Science and Applications*. Marcel Dekker, New York, pp. 21–37.

Grunert KG. 1997. What's in a steak? A cross-cultural study on the quality perception of beef. *Food Qual Pref* 8: 157–174.

Hernández P et al. 1998. Lipid composition and lipolytic enzyme activities in porcine skeletal muscles with different oxidative pattern. *Meat Sci* 49: 1–10.

Honikel KO. 1997. Reference methods supported by OECD and their use in Mediterranean meat products. *Food Chem* 59: 573–582.

Houben JH et al. 1998. Effect of the dietary supplementation with vitamin E on colour stability and lipid oxidation in packaged, minced pork. *Meat Sci* 48: 265–273.

Hovenier R et al. 1992. Genetic parameters of pig meat quality traits in a halothane negative population. *Meat Sci* 32: 309–321.

Huff-Lonergan E. 2010. Chemistry and Biochemistry of meat. In: F Toldrá, (ed.), *Handbook of Meat Processing*. Wiley-Blackwell, Ames, IA, pp 5–24.

Ibrahim RM et al. 2008. Effect of two dietary concentrate levels on tenderness, calpain and calpastatin activities and carcass merit in Waguli and Brahman sterrs. *J Anim Sci* 86: 1426–1433.

Jakobsen K. 1999. Dietary modifications of animal fats: status and future perspectives. *Fett/Lipid* 101: 475–483.

Jiménez Colmenero F et al. 2006. New approaches for the development of functional meat products. In: LML Nollet, F Toldrá (eds.) *Advanced technologies for meat processing*. CRC Press, Boca Raton, FL, pp. 275–308.

Johnson MH et al. 1990. Differences in cathepsin B+L and calcium-dependent protease activities among breed type and their relationship to beef tenderness. *J Anim Sci* 68: 2371–2379.

Josell A et al. 2003. Sensory and meat quality traits of pork in relation to post-slaughter treatment and RN genotype. *Meat Sci* 66: 113–124.

Kauffman RG. 2001. Meat composition. In: YH Hui, WK Nip, RW Rogers, OA Young, (eds.) *Meat Science and Applications*. Marcel Dekker, New York, pp. 1–19.

Larick DK et al. 1992. Volatile compound contents and fatty acid composition of pork as influenced by linoleic acid content of the diet. *J Anim Sci* 70: 1397–1403.

Laurent W et al. 2000. Muscle characterization by NMR imaging and spectroscopic techniques. *Food Chem* 69: 419–426.

Leszczynski et al. 1992. Characterization of lipid in loin and bacon from finishing pigs fed full-fat soybeans or tallow. *J Anim Sci* 70: 2175–2181.

Mercier Y et al. 1998. Effect of dietary fat and vitamin E on colour stability and on lipid and protein oxidation in turkey meat during storage. *Meat Sci* 48: 301–318.

Miller MF et al. 1990. Determination of the alteration in fatty acid profiles, sensory characteristics and carcass traits of swine fed elevated levels of monounsaturated fats in the diet. *J Anim Sci* 68: 1624–1631.

Monahan FJ et al. 1992. Influence of dietary vitamin E (alpha tocopherol) on the color stability of pork chops. Proc 38th Int Congress of Meat Science and Technology, pp. 543–544, Clermont-Ferrand, France, August 1992.

Monin G, Sellier P. 1985. Pork of low technological quality with a normal rate of muscle pH fall in the immediate post-mortem period: the case of the Hampshire breed. *Meat Sci* 13: 49–63.

Moody WG, Cassens RG. 1968. Histochemical differentiation of red and white muscle fibers. *J Anim Sci* 27: 961–966.

Mora L et al. 2008. Contents of creatine, creatinine and carnosine in pork muscles of different metabolic type. *Meat Sci* 79: 709–715.

Morgan CA et al. 1992. Manipulation of the fatty acid composition of pig meat lipids by dietary means. *J Sci Food Agric* 58: 357–368.

Niemöller A, Behmer D. 2008. Use of near infrared spectroscopy in the food industry. In J Irudayaraj, C Reh (eds.) *Nondestructive Testing of Food Quality*. Blackwell Publishing, Ames, IA, pp. 67–118.

O'Sullivan MG. 1997. The distribution of dietary vitamin E in the muscles of the porcine carcass. *Meat Sci* 45: 297–305.

Parolari G et al. 1994. Relationship between cathepsin B activity and compositional parameters in dry-cured ham of normal and defective texture. *Meat Sci* 38: 117–122.

Pearson AM. 1987. Muscle function and postmortem changes. In: JF Price, BS Schweigert (eds.) *The Science of Meat and Meat Products*. Food and Nutrition Press, Westport, CT, pp. 155–191.

Pearson AM, Young RB. 1989. *Muscle and Meat Biochemistry*. Academic Press, San Diego, pp. 1–261.

Renou JP et al. 1994. 23Na magnetic resonance imaging: distribution of brine in muscle. *Magn Reson Imaging* 12: 131–137.

Resurreccion AVA. 2003. Sensory aspects of consumer choices for meat and meat products. *Meat Sci* 66: 11–20.

Rhee KS et al. 1988. Effect of dietary high-oleic sunflower oil on pork carcass traits and fatty acid profiles of raw tissues. *Meat Sci* 24: 249–260.

Robson RM et al. 1997. Postmortem changes in the myofibrillar and other cytoskeletal proteins in muscle. Proceedings of the 50th Annual Reciprocal Conference, Ames, IA. 50: 43–52.

Rodbotten R et al. 2000. Prediction of beef quality attributes from early post mortem near infrared reflectance spectra. *Food Chem* 69: 427–436.

Rosenvold A, Andersen HJ. 2003. Factors of significance for pork quality: A review. *Meat Sci* 64: 219–237.

Stoeva S et al. 2000. Isolation and identification of proteolytic fragments from TCA soluble extracts of bovine M. *Longissimus dorsi*. *Food Chem* 69: 365–370.

Swatland HJ. 1994. *Structure and Development of Meat Animals and Poultry*. Technomic Publishing, Lancaster, pp. 143–199.

Toldrá F. 1992. The enzymology of dry-curing of meat products. In: FJM Smulders, et al. (eds.) *New Technologies for Meat and Meat Products*. Audet, Nijmegen, pp. 209–231.

Toldrá F. 2002. *Dry-Cured Meat Products*. Food and Nutrition Press, Trumbull, CT, pp. 1–238.

Toldrá F. 2006. Meat: chemistry and biochemistry. In: YH Hui et al. (eds.) *Handbook of Food Science, Technology and Engineering*, Vol 1. CRC Press, Boca Raton, FL, pp. 28–1 a 28–18.

Toldrá F. 2007. Biochemistry of muscle and fat. In: F Toldrá et al. (eds.) *Handbook of Fermented Meat and Poultry*. Blackwell Publishing, Ames, IA, pp. 51–58.

Toldrá F, Flores, M. 2000. The use of muscle enzymes as predictors of pork meat quality. *Food Chem* 69: 387–395.

Toldrá F, Flores M. 2004. Analysis of meat quality. In: LML Nollet (ed.) *Handbook of Food Analysis*. Marcel Dekker, New York, pp. 1961–1977.

Toldrá F et al. 1996a. Pattern of muscle proteolytic and lipolytic enzymes from light and heavy pigs. *J Sci Food Agric* 71: 124–128.

Toldrá F et al. 1996b. Lipids from pork meat as related to a healthy diet. *Recent Res Devel Nutr* 1: 79–86.

Urich K. 1994. *Comparative Animal Biochemistry*. Springer, Berlin, pp. 526–623.

Warner RD et al. 1993. Quality attributes of major porcine muscles: a comparison with the *Longissimus lumborum*. *Meat Sci* 33: 359–372.

Warner RD et al. 1997. Muscle protein changes post mortem in relation to pork quality traits. *Meat Sci* 45: 339–372.

Wheeler TL et al. 1990. Mechanisms associated with the variation in tenderness of meat from Brahman and Hereford cattle. *J Anim Sci* 68: 4206–4220.

Wood JD et al. 2003. Effects of fatty acids on meat quality. *Meat Sci* 66: 21–23.

16
Biochemistry of Processing Meat and Poultry

Fidel Toldrá

Abstract: The biochemical changes happening during meat conditioning (aging) were abundantly reported during the 1970s and 1980s. It has been in recent decades that more information has been available for the biochemical changes in other products such as cooked, dry-fermented, and dry-cured meats. The processing and quality of these meat products have been improved based on a better knowledge of the biochemical mechanisms involved in the generation of color, flavor, and texture. The endogenous enzyme systems that play important roles in these processes mainly through proteolysis and lipolysis reactions are described in this chapter. Other biochemical reactions like oxidation, glycolysis, and nucleotides breakdown are also described.

BACKGROUND INFORMATION

There are a wide variety of meat products that are attractive to consumers because of their characteristic color, flavor, and texture. This perception varies depending on local traditions and heritage. Most of these products have been produced for many years or even centuries based on traditional practices. For instance, cured meat products reached America with settlers. Pork was cured in New England for consumption in the summer. Curers expanded these products by trying different recipes based on the use of additives such salt, sugar, pepper, spices, and so forth, and smoking (Toldrá 2002).

Although scientific literature on biochemical changes during meat conditioning (aging) and in some meat products were abundantly reported during the 1970s and 1980s, little information was available on the origin of the biochemical changes in other products such as cooked, dry-fermented, and dry-cured meats. The need to improve the processing and quality of these meat products prompted research in the last decades on endogenous enzyme systems that play important roles in these processes, which has been later demonstrated (Toldrá 2007). It is important to remember that the potential role of a certain enzyme in a specific observed or reported biochemical change can only be established if all the following requirements are met (Toldrá 1992): (1) the enzyme is present in the skeletal muscle or adipose tissue, (2) the enzyme is able to degrade in vitro the natural substance (i.e., a protein in the case of a protease, a triacylglycerol in the case of a lipase, etc.), (3) the enzyme and substrate

Food Biochemistry and Food Processing, Second Edition. Edited by Benjamin K. Simpson, Leo M.L. Nollet, Fidel Toldrá, Soottawat Benjakul, Gopinadhan Paliyath and Y.H. Hui.

are located close enough in the real meat product for an effective interaction, and (4) the enzyme exhibits enough stability during processing for the changes to be developed.

DESCRIPTION OF THE MUSCLE ENZYMES

There are a wide variety of enzymes in the muscle. Most of them have an important role in the in vivo muscle functions, but they also serve an important role in biochemical changes such as the proteolysis and lipolysis that occur in postmortem meat, and during further processing of meat. Some enzymes are located in the lysosomes, while others are free in the cytosol or linked to membranes. The muscle enzymes with most important roles during meat processing are grouped by families and are described in the succeeding sections.

MUSCLE PROTEASES

Proteases are characterized by their ability to degrade proteins, and they receive different names depending on respective mode of action (see Fig. 16.1). They are endoproteases or proteinases, when they are able to hydrolyze internal peptide bonds, but they are exopeptidases, when they hydrolyze external peptide bonds, either at the amino termini or the carboxy termini.

Neutral Proteinases: Calpains

Calpains are cysteine endopeptidases consisting of heterodimers of 110 kDa, composed of an 80 kDa catalytic subunit and a 30 kDa subunit of unknown function. They are located in the cytosol, around the Z-line area. Calpains have received different names in the scientific literature, such as calcium-activated neutral proteinase, calcium-dependent protease, and calcium-activated factor. Calpain I is also called μ-calpain because it needs micromolar amounts (50–70 μM) of Ca^{2+} for activation. Similarly, calpain II is called *m*-calpain because it requires millimolar amounts (1–5 mM) of Ca^{2+}. Both calpains show maximal activity around pH 7.5. Calpain activity decreases very quickly when pH decreases to 6.0, or even reaches ineffective activity at pH 5.5 (Etherington 1984). Calpains have shown good ability to degrade important myofibrillar proteins, such as titin, nebulin, troponins T and I, tropomyosin, C-protein, filamin, desmin, and vinculin, which are responsible for the fiber structure. On the other hand, they are not active against myosin, actin, α-actinin and troponin C (Goll et al. 1983, Koohmaraie 1994).

The stability of calpain I in postmortem muscle is very poor because it is readily autolyzed, especially at high temperatures, in the presence of the released Ca^{2+} (Koohmaraie 1994). Calpain II appears as more stable, just 2–3 weeks before losing its activity (Koohmaraie et al. 1987). In view of this rather poor stability, the importance of calpains should be restricted to short-term processes. A minor contribution, just at the beginning, has been observed in long processes such as dry curing of hams (Rosell and Toldrá 1996) or in fermented meats where the acid pH values makes calpain activity rather unlikely (Toldrá et al. 2001).

Calpastatin is a polypeptide (between 50 and 172 kDa) acting as an endogenous reversible and competitive inhibitor of calpain in the living muscle. In postmortem muscle, calpastatin regulates the activity of calpains, through a calcium-dependent interaction, although only for a few days, because it is destroyed by autolysis (Koohmaraie et al. 1987). The levels of calpastatin vary with animal species, and pork muscle has the lowest level (Valin and Ouali 1992).

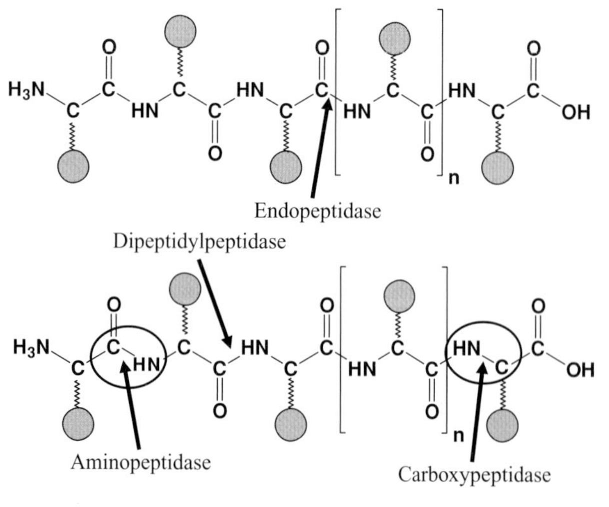

Figure 16.1. Mode of action of the different types of muscle proteases.

Lysosomal Proteinases: Cathepsins

There are several acid proteinases in the lysosomes that degrade proteins in a nonselective way. The most important are cathepsins B, H, and L, which are cysteine proteinases, and cathepsin D, which is an aspartate proteinase. The optimal pH for activity is slightly acid (pH around 6.0) for cathepsins B and L, acid (pH around 4.5) for cathepsin D, and neutral (pH 6.8) for cathepsin H (Toldrá et al. 1992). Cathepsins require a reducing environment such as that found in postmortem muscle to express their optimal activity (Etherington 1987). All of them are of small size, within the range 20–40 kDa, and are thus able to penetrate into the myofibrillar structure. Cathepsins have shown a good ability to degrade different myofibrillar proteins. Cathepsins D and L are very active against myosin heavy chain, titin, M and C proteins, tropomyosin, and troponins T and I (Matsukura et al. 1981, Zeece and Katoh 1989). Cathepsin L is extremely active in degrading both titin and nebulin. Cathepsin B is able to degrade myosin heavy chain and actin (Schwartz and Bird 1977). Cathepsin H exhibits both endo- and aminopeptidase activity, and this is the reason for its classification as an aminoendopeptidase (Okitani et al. 1981). In the muscle, there are endogenous inhibitors against cysteine peptidases. These inhibitory compounds, known as cystatins, are able to inhibit cathepsins B, H, and L. Cystatin C and chicken cystatin are the most well-known cystatins.

Proteasome Complex

The proteasome is a multicatalytic complex with different functions in living muscle, even though its role in postmortem muscle is still not well understood. The 20S proteasome has a large molecular mass, 700 kDa, and a cylinder structure with several subunits. Its activity is optimal at pH above 7.0, but it rapidly decreases when pH decreases, especially below 5.5. It exhibits three major activities: (1) chymotrypsin-like activity, (2) trypsin-like activity, and (3) peptidyl-glutamyl hydrolyzing activity (Coux et al. 1996). This multiple activity behavior is the reason why there was originally some confusion among laboratories over its identification. The 20S proteasome concentration is higher in oxidative muscles than in glycolytic ones (Dahlmann et al. 2001). This enzyme has shown degradation of some myofibrillar proteins such as troponin C and myosin light chain and could be involved in postmortem changes in slow twitch oxidative muscles or in high-pH meat, where an enlargement of the Z-line with more or less density loss is observed (Sentandreu et al. 2002).

A new family of peptidases, named as caspases or apoptosis-generating peptidases, are cysteine peptidases that have been recently proposed to be involved in cell death and thus immediate postmortem changes in proteins having some impact on the phases of rigor and meat aging. Three main pathways of cellular death development have been proposed (Herrera-Méndez et al. 2006). These peptidases are active at neutral pH, and one of the limitations to its activity in postmortem meat is the acid pH in the postmortem muscle.

Exoproteases: Peptidases

There are several peptidases in the muscle with the ability to release small peptides of importance for taste. Tripeptidylpeptidases (TPPs) are enzymes capable of hydrolyzing different tripeptides from the amino termini of peptides, while dipeptidylpeptidases (DPPs) are able to hydrolyze different dipeptide sequences. There are two TPPs and four DPPs, and their molecular masses are relatively high, between 100 and 200 kDa, or even as high as 1000 kDa for TPP II, and have different substrate specificities. TPP I is located in the lysosomes, has an optimal acid pH (4.0), and is able to hydrolyze tripeptides Gly-Pro-X, where X is an amino acid, preferentially of hydrophobic nature. TPP II has optimal neutral pH (6.5–7.5) and wide substrate specificity, except when Pro is present on one of both sides of the hydrolyzed bond. DPPs I and II are located in the lysosomes and have optimal acid pH (5.5). DPP I has a special preference for hydrolyzing the dipeptides Ala-Arg and Gly-Arg, while DPP II prefers a terminal Gly-Pro sequence. DPP III is located in the cytosol and has special preference for terminal Arg-Arg and Ala-Arg sequences. DPP IV is linked to the plasma membrane and prefers a terminal Gly-Pro sequence. Both DPP III and IV have an optimal pH around 7.5–8.0. All these peptidases have been purified and fully characterized in porcine skeletal muscle (Toldrá 2002).

Exoproteases: Aminopeptidases and Carboxypeptidases

There are five aminopeptidases, known as leucyl, arginyl, alanyl, pyroglutamyl, and methionyl aminopeptidases, based on their respective preference or requirement for a specific *N*-terminal amino acid. They are able, however, to hydrolyze other amino acids, although at a lower rate (Toldrá 1998). Aminopeptidases are metalloproteases with a very high molecular mass and complex structures. All of them are active at neutral or basic pH. Alanyl aminopeptidase, also known as the major aminopeptidase because it exhibits very high activity, is characterized by its preferential hydrolysis of alanine, but it is also able to act against a wide spectrum of amino acids such as aromatic, aliphatic, and basic aminoacyl bonds. Methionyl aminopeptidase has preference for methionine, alanine, lysine, and leucine, but also has a wide spectrum of activity. This enzyme is activated by calcium ions. Arginyl aminopeptidase, also known as aminopeptidase B, hydrolyzes basic amino acids such as arginine and lysine (Toldrá and Flores 1998).

Carboxypeptidases are located in the lysosomes and have optimal acid pH. They are able to release free amino acids from the carboxy termini of peptides and proteins. Carboxypeptidase A has preference for hydrophobic amino acids, whereas carboxypeptidase B has a wide spectrum of activity (McDonald and Barrett 1986).

Lipolytic Enzymes

Lipolytic enzymes are characterized by their ability to degrade lipids, and they receive different names depending on their mode of action (see Fig. 16.2). They are known as lipases when they are able to release long-chain fatty acids from triacylglycerols, while they are know as esterases when they act on short-chain fatty acids. Lipases and esterases are located either in the skeletal muscle or in the adipose tissue. Phospholipases, mainly found in the skeletal muscle, hydrolyze fatty acids at positions 1 or 2 in phospholipids.

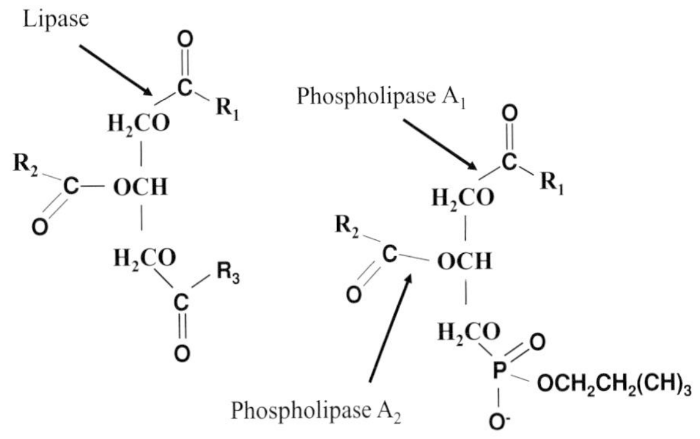

Figure 16.2. Mode of action of muscle lipases and phospholipases.

Muscle Lipases

Lysosomal acid lipase and acid phospholipase are located in the lysosomes. Both have an optimal acid pH (4.5–5.5) and are responsible for the generation of long-chain free fatty acids. Lysosomal acid lipase has preference for the hydrolysis of triacylglycerols at positions 1 or 3 (Fowler and Brown 1984). This enzyme also hydrolyzes di- and monoacylglycerols but at a lower rate (Imanaka et al. 1985, Negre et al. 1985). Acid phospholipase hydrolyzes phospholipids at position 1 at the water-lipid interface.

Phospholipase A and lysophospholipase have optimal pH in the basic region and regulate the hydrolysis of phospholipids, at positions 1 and 2, respectively. The activities of these enzymes are higher in oxidative muscles than in glycolytic muscles, and this fact would explain the high content of free fatty acids in oxidative muscles. The increase in activity is about 10- to 25-fold for phospholipase A and 4- to 5-fold for lysophospholipase (Alasnier and Gandemer 2000).

Acid and neutral esterases are located in the lysosomes and cytosol, respectively, and are quite stable (Motilva et al. 1992). Esterases are able to hydrolyze short-chain fatty acids from tri-, di-, and monoacylglycerols, but they exert poor action due to the lack of adequate substrate.

Adipose Tissue Lipases

Hormone-sensitive lipase is the most important enzyme present in adipose tissue. This enzyme is responsible for the hydrolysis of stored adipocyte lipids. It has a high specificity and preference for the hydrolysis of long-chain tri- and diacylglycerols (Belfrage et al. 1984). This enzyme has positional specificity since it hydrolyzes fatty acids at positions 1 or 3 in triacylglycerols four times faster than it hydrolyzes fatty acids in position 2 (Belfrage et al. 1984). The hormone-sensitive lipase has a molecular mass of 84 kDa and neutral optimal pH, around 7.0. The monoacylglycerol lipase is mainly present in the adipocytes, and very little is present in stromal and vascular cells. It has a molecular mass of 32.9 kDa and hydrolyzes medium- and long-chain monoacylglycerols resulting from previous hydrolysis by the hormone-sensitive lipase (Tornquist et al. 1978). Lipoprotein lipase is located in the capillary endothelium and is able to hydrolyze the acylglycerol components at the luminal surface of the endothelium (Smith and Pownall 1984), with preference for fatty acids at position 1 over those at position 3 (Fielding and Fielding 1980). Lipoprotein lipase is an acylglycerol lipase responsible for the degradation of lipoprotein triacylglycerol. Its molecular mass is around 60 kDa and it has an optimal basic pH. Unsaturated monoacylglycerols are more quickly hydrolyzed than saturated compounds (Miller et al. 1981).

The lipolysis phenomenon in adipose tissue is not so complex as in muscle. The hormone-sensitive lipase hydrolyzes tri- and diacylglycerols, as a rate-limiting step (Xiao et al. 2010). The resulting monoacylglycerols from this reaction or from lipoprotein lipase (Belfrage et al. 1984) are then further hydrolyzed by the monoacylglycerol lipase. The end products are glycerol and free fatty acids.

Acid and neutral esterases are also present in adipose tissue (Motilva et al. 1992). During mobilization of depot lipids, esterases can participate by mobilizing stored cholesteryl esters. Esterases can also degrade lipoprotein cholesteryl esters taken up from the plasma (Belfrage et al. 1984).

Muscle Oxidative and Antioxidative Enzymes

Oxidative Enzymes

Lipoxygenase contains iron and catalyzes the incorporation of molecular oxygen into polyunsaturated fatty acids, especially arachidonic acid, and esters containing a Z,Z-1,4-pentadien (Marczy et al. 1995). They receive different names, 5-, 12-, or 15-lipoxygenase, depending on the position where oxygen is introduced. The final product is a conjugated hydroperoxide. They usually require millimolar concentrations of Ca^{+2}, and their activity is stimulated by ATP (Yamamoto 1992). Lipoxygenase has been found to be stable during frozen storage and is responsible for rancidity development in chicken, especially in the muscle *Gastrocnemius* (Grossman et al. 1988).

Antioxidative Enzymes

Antioxidative enzymes and their regulation in the muscle constitute a defense system against oxidative susceptibility (i.e., an increased concentration of polyunsaturated fatty acids) and physical stress (Young et al. 2003). Glutathione peroxidase contains a covalently bound selenium atom that is essential for its activity. This enzyme catalyzes the dismutation of alkyl hydroperoxides by reducing agents like phenols. Its activity has been reported to be lower in oxidative muscles than in glycolytic muscles (Daun et al. 2001). Superoxide dismutase is a copper metalloenzyme, and catalase an iron metalloenyzme. Both enzymes catalyze the dismutation of hydrogen peroxide to less harmful hydroxides. These enzymes influence the shelf life of the meat and protect against the pro-oxidative effects of chloride during further processing.

PROTEOLYSIS

Proteolysis constitutes an important group of reactions during the processing of meat and meat products. In fact, proteolysis has a high impact on texture, and thus meat tenderness, because it contributes to the breakdown of the myofibrillar proteins responsible for muscle network, but proteolysis also generates peptides and free amino acids that have a direct influence on taste and also act as substrates for further reactions contributing to aroma (Toldrá 1998, 2002). In general, proteolysis has several consecutive stages (see Fig. 16.3) as follows: (1) action of calpains and cathepsins on major myofibrillar proteins, generating protein fragments and intermediate size polypeptides; (2) these generated fragments and polypeptides are further hydrolzyed to small peptides by DPPs and TPPs; and (3) dipeptidases, aminopeptidases, and carboxypeptidases are the last proteolytic enzymes that act on previous polypeptides and peptides to generate free amino acids. The progress of proteolysis varies depending on

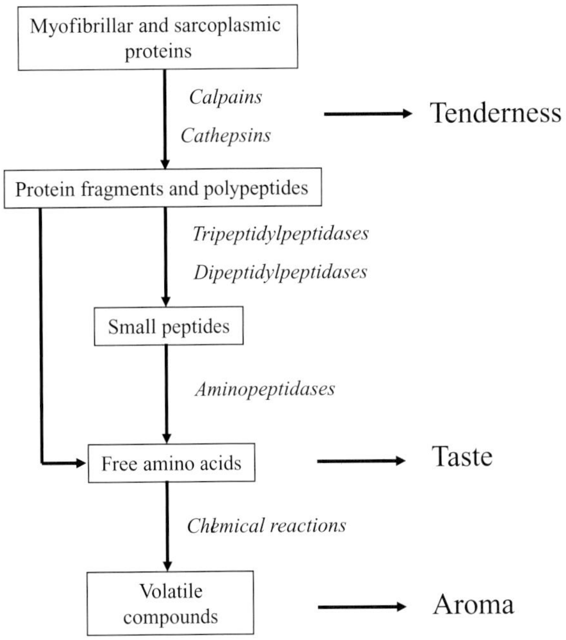

Figure 16.3. General scheme of proteolysis during the processing of meat and meat products.

the processing conditions, the type of muscle, and the amount of endogenous proteolytic enzymes, as described later.

Proteolysis in Aged Meat and Cooked Meat Products

During meat aging, there is proteolysis of important myofibrillar proteins such as troponin T by calpains, with the associated release of a characteristic 30 kDa fragment, which is associated with meat tenderness, nebulin, desmin, titin, troponin I, myosin heavy chain, and proteins at the Z-line level (Yates et al. 1983). A 95 kDa fragment is also characteristically generated (Koohmaraie 1994). Examples of cathepsin and aminopeptidase activity during beef aging are shown in Figures 16.4 and 16.5, respectively.

The fastest aging rate is observed in chicken, followed by pork, lamb, and beef. For instance, it has been reported that chicken myofibrils are easily damaged by cathepsin L, whereas beef myofibrils are much more resistant (Mikami et al. 1987). The reasons for this difference are the differences between species in enzymatic activity, inhibitor content, and susceptibility to proteolysis of the myofibrillar structure. These factors are also strongly linked to the type of muscle metabolism, which has a strong influence on the aging rate (Toldrá 2006a). In fact, proteolysis is faster in fast-twitch white fibers (which contract rapidly) than in slow-twitch red fibers (which contract slowly), that is, aging rate increases with increasing speed of contraction but decreases with the level of heme iron (Ouali 1991). Even though the effect of muscle type is estimated to be tenfold lower

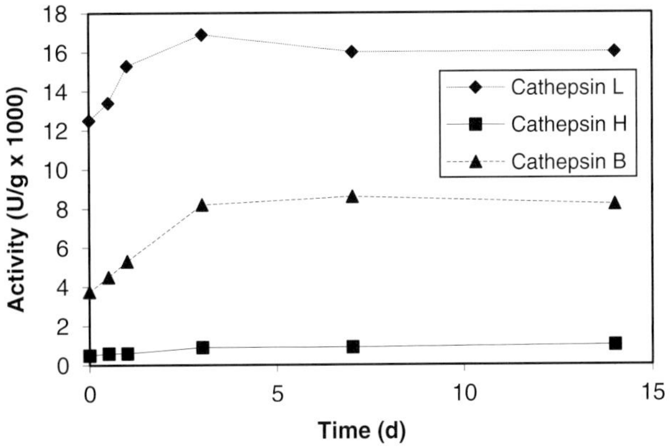

Figure 16.4. Evolution of muscle cathepsins during the aging of beef (Toldrá, unpublished data).

than the effect of temperature, it is threefold higher than animal effects (Dransfield 1980–1981). There are also some physical and chemical conditions in postmortem muscle, listed in Tables 16.1 and 16.2, respectively, that can affect enzyme activity. Of these conditions, the most significant are pH, which decreases once the animal is slaughtered, and osmolality, which increases from 300 to around 550 mOsm within 2 days, due to the release of ions to the cytosol. The osmolality has been observed to be higher in fast-twitch muscle, which also experiences a faster aging rate (Valin and Ouali 1992). Age of the animal also decreases the aging rate, because collagen content as well as the cross-links that make the muscle more heat stable and mechanically resistant, increase with age.

Proteolysis in Fermented Meats

The progressive pH decrease by lactic acid (generated during the fermentation by lactic acid bacteria), the added salt (2–3%),

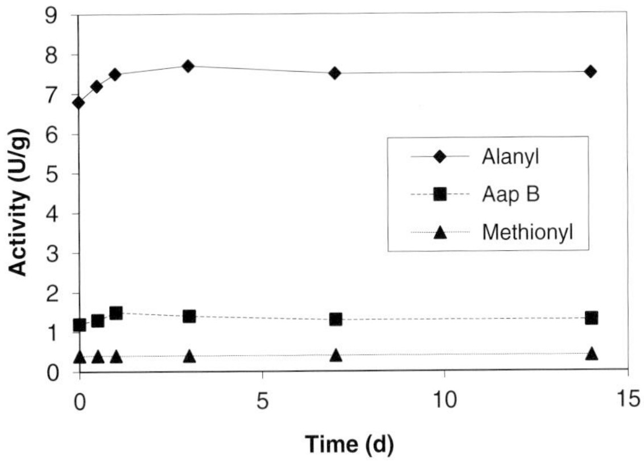

Figure 16.5. Evolution of muscle aminopeptidases during the aging of beef (Adapted from Goransson et al. 2002).

Table 16.1. Physical Factors Affecting Proteolytic Activity During Meat and Meat Products Processing

Factor	Typical Trend	Effect on Proteases
pH	Near neutral pH in dry-cured ham and cooked meat products.	Favors the activity of calpain, cathepsins B and L, DPP III and IV, TPP II and aminopeptidases.
	Slightly acid in aged meat	Favors the activity of cathepsins B, H and L
	Acid pH in dry-fermented sausages	Favors the activity of cathepsin D, DPP I, DPP II, and TPP I
Time	Short in aged meat, cooked meat products, and fermented products	Short action for enough enzyme action except for calpains that contribute to tenderness
	Medium in dry-fermented sausages	Time allows significant biochemical changes
	Long in dry-cured ham	Very long time for important biochemical changes
Temperature	High increase in cooking and smoking	Enzymes are strongly activated by cooking temperatures although stability decreases rapidly and time of cooking is short
	Mild temperatures in fermentation and dry-curing	Enzymes have time enough for its activity even though the use of mild temperatures
Water activity	Slightly reduced in cooked meats	Enzymes have good conditions for activity
	Substantially reduced in dry-meats.	Restricted enzyme activity as aw drops
Redox potential	Anaerobic values in postmortem meat.	Most of the muscle enzymes need reducing conditions for activity
Osmolality	High in fast-twitch muscle. It may increase from 300 to 550 mOsm within 2 d	Enhances proteases activity and these muscles are aged at faster rate

and the heating/drying conditions during the fermentation and ripening/drying affect protein solubility (Toldrá and Reig 2007). The reduction may reach 50–60% in the case of myofibrillar proteins and 20–47% for sarcoplasmic proteins (Klement et al. 1974). There is a variable degree of contribution to proteolysis by both endogenous and proteases and those of microbial origin. This contribution mainly depends on the raw materials, the type of product, and the processing conditions. The contribution of microbial proteases to proteolysis also depends on the type of strains. For instance, strains *Staphylococcus carnosus* and *Staphylococcus simulans* were reported to hydrolyze sarcoplasmic but not myofibrillar proteins; however, no protease activity was detected in another staphylococci but instead, some low aminopeptidase and high esterase activity was found (Casaburi et al. 2006).

The pH drop during the fermentation stage is very important. So, when pH drops below 5.0, the proteolytic activity of endogenous cathepsin D becomes very intense (Toldrá et al. 2001). Several myofibrillar proteins, such as myosin and actin, are degraded, and some fragments of 135, 38, 29 kDa, and 13 kDa are formed. The major role of cathepsin D has been confirmed in model systems using antibiotics and specific protease inhibitors in order to inhibit bacterial proteinases or other endogenous muscle proteinases (Molly et al. 1997). A minor role is played by other muscle cathepsins (B, H, and L) and bacterial proteinases. Peptides and small protein fragments are produced during fermentation, heating (smoking), and ripening. The generation of free amino acids, as final products of proteolysis, depends on the pH reached in the product (as aminopeptidases are affected by low pH values), concentration of salt (these enzymes are

Table 16.2. Chemical Factors Affecting Proteolytic Activity During Meat and Meat Products Processing

Factor	Typical Trend	Effect on Proteases
NaCl	Low in aged meat	No effect on enzyme activity
	Medium concentration in cooked meat products	Partial inhibition of most proteases except calpain and aminopeptidase B that are chloride-activated at low NaCl concentration.
	High concentration in dry-cured hams	Strong inhibition of almost all proteases
Nitrate and nitrite	Concentration around 125 ppm in cured meat products	Slight inhibitory effect on the enzyme activity, except cathepsin B that is activated
Ascorbic acid	Concentration around 500 ppm in cured meat products	Slight inhibition of *m*-calpain, cathepsin H, leucyl aminopeptidase, and aminopeptidase B
Glucose	Poor concentration in aged meat and dry-cured ham	No effect
	High concentration (up to 2 gL^{-1}) in fermented meats	Slight activation of leucyl aminopeptidase and cathepsins B, H, and D

inhibited or activated by salt), and processing conditions (time, temperature, and water activity (a_w) that affect the enzyme activity) (Toldrá 1998). All these factors affect the contribution of the different aminopeptidases, as described in Tables 16.1 and 16.2; thus, the pH reached at the fermentation stage (pH $<$ 5.0) is decisive because it reduces substantially both muscle and microbial aminopeptidase activity (Sanz et al. 2002). So, the processing environmental conditions at the beginning of drying are very important for the accumulation of free amino acids in the final products (Roseiro et al. 2010).

Finally, it must be taken into account that some microorganisms, grown during fermentation, might have decarboxylase (an enzyme able to generate amines from amino acids) activity.

Proteolysis in Dry-Cured Ham

The analysis of muscle sarcoplasmic proteins and myofibrillar proteins by sodium dodecyl sulfate–polyacrylamide gel electrophoresis reveals an intense proteolysis during the process. This proteolysis appears to be more intense in myofibrillar than in sarcoplasmic proteins (Toldrá and Aristoy 2010). The patterns for myosin heavy chain, myosin light chains 1 and 2, and troponins C and I show a progressive disappearance during the processing (Toldrá et al. 1993). Several fragments of 150, 95, and 16 kDa and in the ranges of 50–100 kDa and 20–45 kDa are formed (Toldrá 2002). The analysis of ultrastructural changes by both scanning and transmission electron microscopy shows weakening of the Z-line as well as important damage to the fibers, especially at the end of salting (Monin et al. 1997). An excess of proteolysis may create unpleasant textures because of intense structural damage. The result is a poor firmness that is poorly rated by sensory panelists and consumers. This excess of proteolysis is frequently due to the breed type and/or age, which have a marked influence on some enzymes, or just a higher level of cathepsin B activity (Toldrá 2004a). A high residual cathepsin B activity and/or low salt content, a strong inhibitor of cathepsin activity, are correlated with the increased softness. The action of calpains is restricted to the initial days of processing due to their poor stability. Cathepsin D would contribute during the initial 6 months, and cathepsins B, L, and H, which are very stable and have an optimal pH closer to that in ham, would act during the full process (Toldrá 1992). An example of the evolution of these enzymes is shown in Figure 16.6.

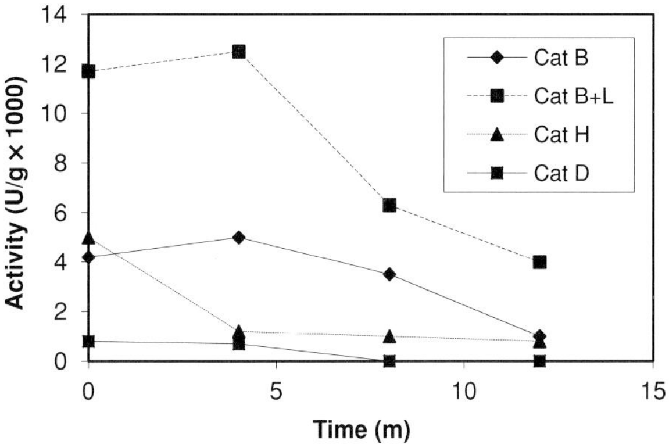

Figure 16.6. Evolution of cathepsins during the processing of dry-cured ham (Toldrá, unpublished data).

Numerous peptides are generated during processing: mainly in the range 2700–4500 Da during postsalting and early ripening, and below 2700 Da during ripening and drying (Aristoy and Toldrá 1995). Some of these peptides have been generated from specific myofibrillar and sarcoplasmic proteins degradation. Proteomic tools have been used for the identification of long-chain peptides resulting from proteolysis of actin (Sentandreu et al. 2007), titin, and light-chain myosin I (Mora et al. 2009a) and creatine kinase (Mora et al. 2009b). These peptides are object of further research for its potential bioactivity and contribution to the nutritional properties of dry-cured ham (Jiménez-Colmenero et al. 2010). Some of the smaller tri- and dipeptides recently have been sequenced. DPP I and TPP I appear to be the major enzymes involved in the release of di- and tripeptides, respectively, due to their good activity, stability, and an optimal pH near to that in ham. The other peptidases would play a minor role (Sentandreu and Toldrá 2002). The generation of free amino acids during the processing of dry-cured ham is very high (Toldrá 2004b). Alanine, leucine, valine, arginine, lysine, and glutamic and aspartic acids are some of the amino acids generated in higher amounts. An example of generation is shown in Figure 16.7. The final concentrations depend on the length of the process and the type of ham (Toldrá et al. 2000). On the basis of the specific enzyme characteristics and the process conditions, alanyl and methionyl aminopeptidases appear to be the most important enzymes involved in the generation of free amino acids, while arginyl aminopeptidase would mostly generate arginine and lysine (Toldrá 2002).

NUCLEOTIDE BREAKDOWN

The disappearance of ATP is very fast; in fact, it only takes a few hours to reach negligible levels. Many enzymes are involved in the degradation of nucleotides and nucleosides, as described in Chapter 15. The main changes in the nucleotide breakdown products occur during a few days postmortem, as shown in Figure 16.8. So, adenosine triphosphate (ATP), adenosine diphosphate, and adenosine monophosphate, which are intermediate degradation compounds, also disappear within 24 hours postmortem. Inosine monophosphate reaches a maximum by 1 day postmortem, but some substantial amount is still recovered after 7 days postmortem. On the other hand, inosine and hypoxanthine, as final products of these reactions, increase up to 7 days postmortem (Batlle et al. 2001).

GLYCOLYSIS

Glycolysis consists in the hydrolysis of carbohydrates, mainly glucose, either that remaining in the muscle or that formed from glycogen, to give lactic acid as the end product. As lactic acid

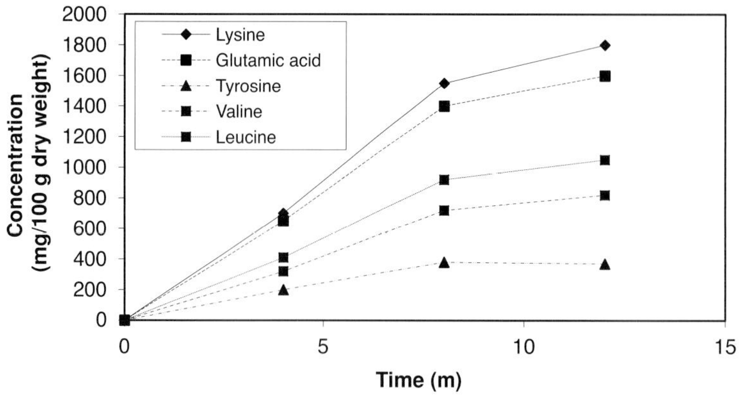

Figure 16.7. Example of the generation of some free amino acids during the processing of dry-cured ham (Adapted from Toldrá et al. 2000).

accumulates in the muscle, pH falls from neutral values to acid values around 5.3–5.8. The glycolytic rate, or speed of pH fall, depends on the animal species and the metabolic status. On the other hand, the glycolytic potential, which depends on the amount of stored carbohydrates, gives an indication of ultimate pH. The pH drop due to the lactic acid accumulation is perhaps the main consequence of glycolysis, and it has very important effects on meat processing because pH affects numerous chemical and biochemical (all the enzymes) reactions (Toldrá 2006a). There are many enzymes involved in the glycolytic chain; some of the most important are phosphorylase, phosphofructokinase, pyruvic kinase, and lactate dehydrogenase (Demeyer and Toldrá 2004). Lactate dehydrogenase is involved in the last step, which consists in the conversion of pyruvic acid into lactic acid. The contribution of glycolysis is restricted to a few hours postmortem, although it is also important in fermented meats where sugar is added for microorganism growth (see Chapter 18). The evolution of some glycolytic enzymes is shown in Figure 16.9.

LIPOLYSIS

Lipolysis makes an important contribution to the quality of meat products by the generation of free fatty acids, some of which have a direct influence on flavor, and others that, with polyunsaturations, may be oxidized to volatile aromatic compounds, acting as flavor precursors. The general scheme for lipolysis in muscle and adipose tissue is shown in Figure 16.10. In addition, the breakdown of triacylglycerols affects the texture of the adipose tissue, and an excess of lipolysis/oxidation may contribute to the development of rancid aromas or yellowish colors in fat (Toldrá 1998).

Lipolysis in Aged Meat and Cooked Meat Products

The relative amounts of released free fatty acids depends not only on the enzyme preference, but also on many other factors such as raw materials (especially affected by feed composition), type

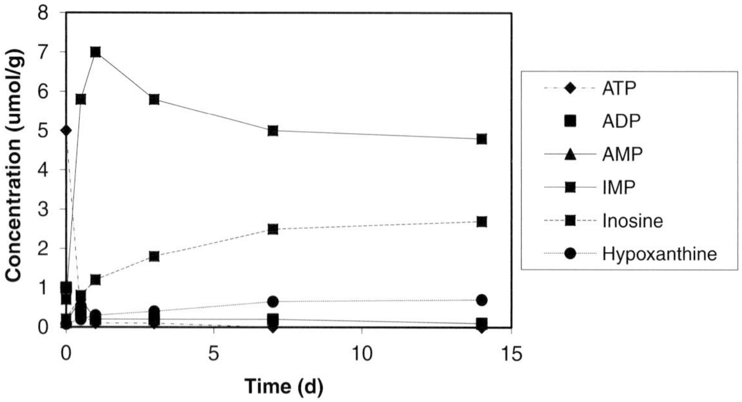

Figure 16.8. Evolution of nucleotides and nucleosides during the aging of pork meat. (Adapted from Batlle et al. 2001.)

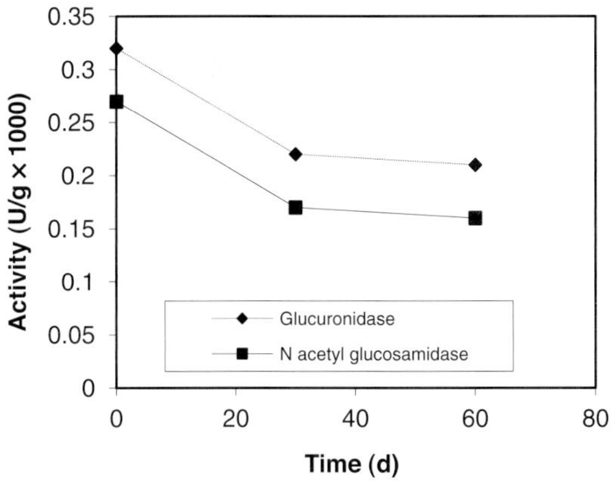

Figure 16.9. Evolution of β-glucuronidase and N-acetyl-β-glucosaminidase during the processing of a dry-fermented sausage (Toldrá, unpublished data).

of process, and extent of cooking (Toldrá 2002). The generation rate observed during in vitro incubations of muscle lipases with pure phospholipids is the following (in order of importance): linoleic > oleic > linolenic > palmitic > stearic > arachidonic acid. Lipolysis may also vary depending on the type of muscle, as oxidative and glucolytic muscles exhibit different lipase activity (Flores et al. 1996). The generation of free short-chain fatty acids is very low due to the lack of adequate substrate (Motilva et al. 1993b). Studies comparing pale, soft, exudative (PSE) versus normal pork reported that phospholipase A2 activity was higher and well correlated with the occurrence of the PSE syndrome (Chen et al. 2010). Reverse correlation was also reported for the antioxidant activity of glutathione peroxidase (Chen et al. 2010). The physical and chemical conditions in the muscle and adipose tissue, especially during cooking, may affect the enzyme activity (see Tables 16.3 and 16.4). The evolution of muscle lipases during aging of pork meat is shown in Figure 16.11.

Lipolysis in Fermented Meats

Depending on the raw materials, type of product, and processing conditions, the degree of contribution of endogenous and microbial origin lipases will vary (Toldrá 2007). The relative importance of both enzyme systems has been checked with mixtures of antibiotics and antimycotics used in sterile meat model systems (Molly et al. 1997). The results showed a minimal effect of antibiotics, and it was concluded that lipolysis is mainly brought about by muscle and adipose tissue lipases (60–80% of total free fatty acids generated), even though some variability might be found, depending on the batch and the presence of specific strains (Molly et al. 1997). Other authors have also observed that fatty acids are released in higher amounts when starters are added, but that there is significant lipolysis in the absence of microbial starters (Montel et al. 1993, Hierro et al. 1997). When pH drops during fermentation, the action of muscle lysosomal acid lipase and acid phospholipase becomes very important. Some strains are selected as starters based on their contribution to lipolysis. So, *Micrococcaceae* present a highly variable amount of extra- and intracellular lipolytic enzymes, dependent on the strain and type of substrate (Casaburi et al. 2008). The action of the extracellular enzymes on the hydrolysis of triacylglycerols becomes more important after 15–20 days of ripening (Ordoñez et al. 1999). Other microorganisms used as starters are *Staphylococcus warneri*, which gives the highest lipolytic activity, and

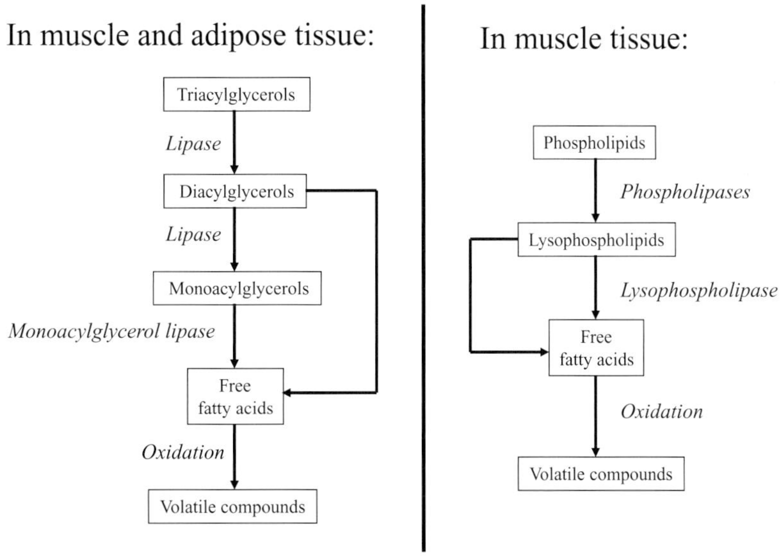

Figure 16.10. General scheme of lipolysis during the processing of meat and meat products.

Table 16.3. Physical Factors Affecting Lipolytic Activity During Meat and Meat Products Processing

Factor	Typical Trend	Effect on Lipolytic Enzymes
pH	Near neutral pH in dry-cured ham and cooked meat products.	Favors the activity of neutral lipase, neutral esterase and hormone-sensitive lipase.
	Slightly acid in aged meat	Favors the activity of lysosomal acid lipase, acid esterase and acid phospholipase
	Acid pH in dry-fermented sausages	Favors the activity of lysosomal acid lipase, acid esterase and acid phospholipase
Time	Short in aged meat, cooked meat products and fermented products	Short action for enough enzyme action but significant oxidation of the released fatty acids
	Medium in dry-fermented sausages and long in dry-cured ham	Time allows significant biochemical changes in dry sausages and very important in dry-cured ham
Temperature	High increase in cooking and smoking	Enzymes are strongly activated by cooking temperatures although for short time
	Mild temperatures in fermentation and dry-curing	Enzymes have time enough for its activity even though the use of mild temperatures. Lipases are active even during freezing storage of the raw meats
Water activity	Slightly reduced in cooked meats	Enzymes have very good conditions for activity
	Substantially reduced in dry-meats.	Muscle neutral lipase and esterases are affected. Rest of lipolytic enzymes remain unaffected by a_w drop
Redox potential	Anaerobic values in postmortem meat.	Slight effect on lysosomal acid lipase and phospholipase. More intense for rest of lipases

Staphylococcus saprophyticus; *S. carnosus* and *Staphylococcus xylosus* present poor and variable lipolytic activity (Montel et al. 1993). Lactic acid bacteria have poor lipolytic activity, mostly intracellular. Many molds and yeasts such as *Candida*, *Debaryomyces*, *Cryptococcus*, and *Trichosporum* have been isolated from fermented sausages, and all of them exhibit lipolytic activity (Ordoñez et al. 1999, Ludemann et al. 2004, Sunesen and Stahnke 2004).

The increases in the levels of free fatty acids show a great variability depending on the raw materials, type of sausage, and processing conditions (Toldrá et al. 2002). The increase may reach 2.5–5% of the total fatty acids, and the rate of release decreases in the following order: oleic > palmitic > stearic > linoleic (Demeyer et al. 2000). The release of polyunsaturated fatty acids from phospholipids is more pronounced during ripening (Navarro et al. 1997). Some short-chain fatty acids such as acetic acid may increase, especially during early stages of ripening. The generation of volatile compounds with impact on the aroma of fermented sausages depends on the type of processing and the starter culture added (Talon et al. 2002, Tjener et al. 2004). Some of the most important are aliphatic saturated and unsaturated aldehydes, ketones, methyl-branched aldehydes and acids, free short-chain fatty acids, sulfur compounds, some alcohols, terpenes (from spices), and some nitrogen-derived volatile compounds (Stahnke 2002).

Lipolysis in Dry-Cured Ham

The generation of free fatty acids in the muscle is correlated with the period of maximal phospholipid degradation (Motilva

Table 16.4. Chemical Factors Affecting Lipolytic Activity During Meat and Meat Products Processing

Factor	Typical Trend	Effect on Lipolytic Enzymes
NaCl	Low in aged meat	No effect on enzyme activity
	Medium concentration in cooked meat products	Partial inhibition of muscle neutral lipase and esterases. Adipose tissue lipases not affected. Activation of lysosomal acid lipase and acid phospholipase
	High concentration in dry-cured hams	Slight activation of lysosomal acid lipase and acid phospholipase
Nitrate and nitrite	Concentration around 125 ppm in cured meat products	No significant effect
Ascorbic acid	Concentration around 500 ppm in cured meat products	Slight inhibition of all lipolytic enzymes
Glucose	Poor concentration in aged meat and dry-cured ham	No effect
	High concentration (up to 2 gL^{-1}) in fermented meats	No effect

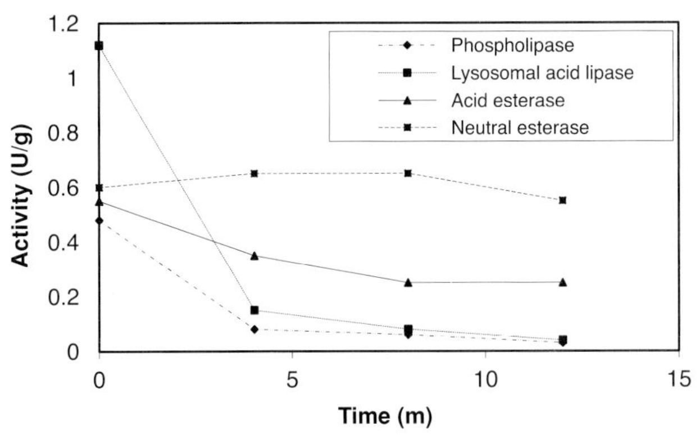

Figure 16.11. Evolution of muscle lipases along the processing of dry-cured ham (Toldrá, unpublished data).

et al. 1993b, Buscailhon et al. 1994). Furthermore, a decrease in linoleic, arachidonic, oleic, palmitic, and stearic acids from phospholipids is observed at early stages of processing (Martin et al. 1999). This fact corroborates muscle phospholipases as the most important enzymes involved in muscle lipolysis. The amount of generated free fatty acids increases with aging time, up to 6 months of processing (see Fig. 16.12) and is higher in the external muscle *(Semimembranosus),* which contains more salt and is more dehydrated, than in the internal muscle *(Biceps femoris).* As lipase activity is mainly influenced by pH, salt concentration, and a_w (Motilva and Toldrá 1993), it appears that the observed lipid hydrolysis is favored by the same variables (salt increase and a_w reduction), as shown in Tables 16.3 and 16.4, that enhance enzyme activity in vitro (Vestergaard et al. 2000, Toldrá et al. 2004).

In the case of the adipose tissue, triacylglycerols form the major part of this tissue (around 90%) and are mostly hydrolyzed by neutral lipases to di- and monoacylglycerols and free fatty acids (see Fig. 16.12), especially up to 6 months of processing (Motilva et al. 1993a). Hormone-sensitive lipase was reported to be more stable than the adipose tissue neutral triglyceride lipase (Xiao et al. 2010). The amount of triacylglycerols decreases from about 90% to 76% (Coutron-Gambotti and Gandemer 1999). There is a preferential hydrolysis of polyunsaturated fatty acids, although some of them may not accumulate due to further oxidation during processing. Triacylglycerols that are rich in oleic and linoleic acids and are liquid at 14–18°C more hydrolyzed than triacylglycerols that are rich in saturated fatty acids such as palmitic acid and are solid at those temperatures (Coutron-Gambotti and Gandemer 1999). This means that the physical state of the triacylglycerols would increase the lipolysis rate by favoring the action of lipases at the water–oil interface. The rate of release of individual fatty acids is as follows: linoleic > oleic > palmitic > stearic > arachidonic (Toldrá 1992). The generation rate remains high up to 10 months of processing, when the accumulation of free fatty acids remains asymptotic or even decreases as a consequence of further oxidative reactions. Oleic, linoleic, stearic, and palmitic acids are those accumulated in higher amounts because they are present in great amounts in the triacylglycerols and have a better stability against oxidation. Similar results for generation of oleic, linoleic and palmitic acids were reported for Chinese Xuanwei ham (Xiao et al. 2010).

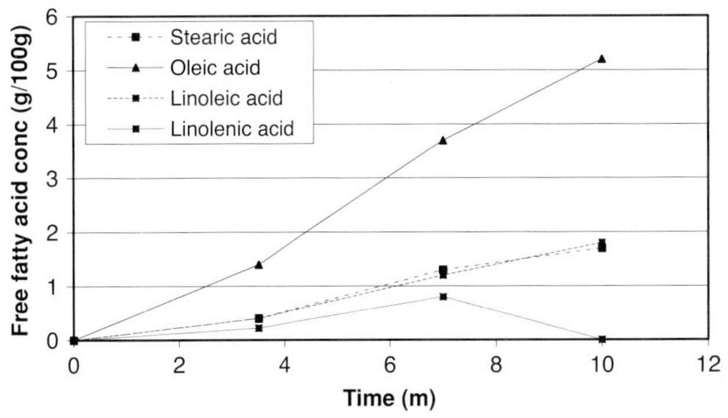

Figure 16.12. Example of the generation of some free fatty acids in the adipose tissue during the processing of dry-cured ham (Adapted from Motilva et al. 1993b).

OXIDATIVE REACTIONS

The lipolysis and generation of free polyunsaturated fatty acids, susceptible to oxidation, constitute a key stage in flavor generation. The susceptibility of fatty acids to oxidation and the rate of oxidation depend on their unsaturation (Shahidi 1998a). So, linolenic acid (C 18:3) is more susceptible than linoleic acid (C 18:2), which is more susceptible than oleic acid (C 18:1). The animal species have different susceptibility to autoxidation in the following order: poultry > pork > beef > lamb (Tichivagana and Morrisey 1985). Oxidation has three consecutive stages (Shahidi 1998b). The first stage, initiation, consists in the formation of a free radical. This reaction can be enzymatically catalyzed by muscle lipoxygenase or chemically catalyzed by light, moisture, heat, and/or metallic cations. The second stage, propagation, consists in the formation of peroxide radicals by reaction of the free radicals with oxygen. When peroxide radicals react with double bonds, they form primary oxidation products, or hydroxyperoxides, that are very unstable. Their breakdown produces many types of secondary oxidation products by a free radical mechanism. Some of them *are* potent flavor-active compounds that can impart off-flavor to meat products during cooking or storage. The oxidative reactions finish by inactivation of free radicals when they react with each other (last stage). Thus, the result of these oxidative reactions consists in the generation of volatile compounds responsible of final product aroma. It is important to have a good control of these reactions because sometimes, oxidation may give undesirable volatile compounds with unpleasant off-flavors.

Muscle proteins may also experience some oxidative reactions during the processing of dry-cured ham. Dry-cured ham presented an intermediate oxidation when compared to other dry-cured and cooked meat products when based on total protein carbonyls and fluorescence corresponding to protein carbonyls. But a higher oxidation was reported in dry-cured ham when using α-aminoadipic semialdehyde and γ-glutamic semialdehyde as indicators of protein oxidation (Armenteros et al. 2009a).

Oxidation to Volatile Compounds

As mentioned previously, some oxidation is needed to generate volatile compounds with desirable flavor properties. For instance, a characteristic aroma of dry-cured meat products is correlated with the initiation of lipid oxidation (Buscailhon et al. 1994, Flores et al. 1998). However, an excess of oxidation may lead to off-flavors, rancidity, and yellow colors in fat.

The primary oxidation products, or hydroperoxides, are flavorless, but the secondary oxidation products have a clear contribution to flavor. There are a wide variety of volatile compounds formed by oxidation of the unsaturated fatty acids. The most important are (1) aliphatic hydrocarbons that result from autoxidation of the lipids; (2) alcohols, mainly originated by oxidative decomposition of certain lipids; (3) aldehydes, which can react with other components to produce flavor compounds; and (4) ketones produced through either β-keto acid decarboxylation or fatty acid β-oxidation. Other compounds, like esters, may contribute to characteristic aromas (Shahidi et al. 1987).

Oxidation rates may vary depending on the type of product or the processing conditions. For instance, TBA (thiobarbituric acid), a chemical index used as an indication of oxidation, increases more markedly in products such as Spanish chorizo than in French saucisson or Italian salami (Chasco et al. 1993). On the other hand, processing conditions such as curing or smoking also give a characteristic flavor to the product (Toldrá 2006b).

Antioxidants

The use of spices such as paprika and garlic, which are rich in natural antioxidants, protects the product from certain oxidations. The same applies to antioxidants such as vitamin E that are added in the feed to prevent undesirable oxidative reactions in polyunsaturated fatty acids. Nitrite constitutes a typical curing agent that generally retards the formation of off-flavor volatiles that can mask the flavor of the product, and allows extended storage of the product (Shahidi 1998). Nitrite acts against lipid oxidation through different mechanisms: (1) binding of heme and prevention of the release of the catalytic iron, (2) binding of heme and nonheme iron and inhibition of catalysis, and (3) stabilization of lipids against oxidation. Smoking also contains some antioxidant compounds such as phenols that protect the external part of the product against undesirable oxidations. The muscle antioxidative enzymes also exert some contribution to the lipid stability against oxidation. In the case of fermented meats, the microbial enzyme catalase degrades the peroxides formed during the processing of fermented sausages (Toldrá et al. 2001). Thus, this enzyme contributes to stabilizing the color and flavor of the final sausage. Catalase increases its activity with cell growth to a maximum at the onset or during the stationary phase, but it is mainly formed during the ripening stage. Large amounts of salt exert an inhibitory effect on catalase activity, especially at low pH values. The catalase activity is different depending on the strain. For instance, *S. carnosus* has a high catalase activity in anaerobic conditions, while *S. warneri* has a low catalase activity (Talon et al. 1999).

REFERENCES

Alasnier C, Gandemer G. 2000. Activities of phospholipase A and lysophospholipases in glycolytic and oxidative skeletal muscles in the rabbit. *J Sci Food Agric* 80: 698–704.

Aristoy MC, Toldrá F. 1995. Isolation of flavor peptides from raw pork meat and dry-cured ham. In: G Charalambous (ed.) *Food Flavors: Generation, Analysis and Process Influence*. Elsevier Science Pub., Amsterdam, pp. 1323–1344

Armenteros M et al. 2009a. Analysis of protein carbonyls in meat products by using the DNPH method, fluorescence spectroscopy and liquid chromatography-electrospray lonisation-mass spectrometry (LC-ESI-MS). *Meat Sci* 83: 104–112.

Batlle N et al. 2001. ATP metabolites during aging of exudative and nonexudative pork meats. *J Food Sci* 66: 68–71.

Belfrage P et al. 1984. Adipose tissue lipases. In: B Borgström, HL Brockman (eds.) *Lipases*. Elsevier Science Pub., London, pp. 365–416.

Buscailhon S et al. 1994. Time-related changes in intramuscular lipids of French dry-cured ham. *Meat Sci* 37: 245–255.

Casaburi A et al. 2006. Protease and esterase activity of staphylococci. *Int J Food Microbiol* 112: 223–229.

Casaburi A et al. 2008. Proteolytic and lipolytic starter cultures and their effect on traditional fermented sausages ripening and sensory traits. *Food Microbiol* 25: 335–347.

Chasco J et al. 1993. A study of changes in the fat content of some varieties of dry sausage during the curing process. *Meat Sci* 34: 191–204.

Chen T, et al. 2010. Phospholipase A2 and antioxidant enzyme activities in normal and PSE pork. *Meat Sci* 84: 143–146.

Coutron-Gambotti C, Gandemer G. 1999. Lipolysis and oxidation in subcutaneous adipose tissue during dry-cured ham processing. *Food Chem* 64: 95–101.

Coux O et al. 1996. Structure and function of the 20S and 26S proteasomes. *Ann Rev Biochem* 65: 801–847.

Dahlmann et al. 2001. Subtypes of 20S proteasomes from skeletal muscle. *Biochimie* 83: 295–299.

Daun C et al. 2001. Glutathione peroxidase activity, tissue and soluble selenium content in beef and pork in relation to meat aging and pig RN phenotype. *Food Chem* 73: 313–319.

Demeyer DI, Toldrá F. 2004. Fermentation. In: W Jensen, C Devine, M Dikemann (eds.) *Encyclopedia of Meat Sciences*. Elsevier Science, Ltd., London, pp. 467–474

Demeyer DI et al. 2000. Control of bioflavor and safety in fermented sausages: First results of a European project. *Food Research Int* 33: 171–180.

Dransfield E et al. 1980–1981. Quantifying changes in tenderness during storage of beef meat. *Meat Sci* 5: 131–137.

Etherington DJ. 1984. The contribution of proteolytic enzymes to postmortem changes in muscle. *J Anim Sci* 59: 1644–1650.

Etherington DJ. 1987. Conditioning of meat factors influencing protease activity. In: A Romita, C Valin, AA Taylor (eds.) *Ac-*

celerated Processing of Meat. Elsevier Applied Sci., London, pp. 21–28.

Fielding CJ, Fielding PE. 1980. Characteristics of triacylglycerol and partial acylglycerol hydrolysis by human plasma lipoprotein lipase. *Biochim Biophys Acta* 620: 440–446.

Flores M et al. 1996. Activity of aminopeptidase and lipolytic enzymes in five skeletal muscles with various oxidative patterns. *J Sci Food Agric* 70: 127–130.

Flores M et al. 1998. Flavour analysis of dry-cured ham. In: F Shahidi (ed.) *Flavor of Meat, Meat Products and Seafoods*. Blackie Academic and Professional, London, pp. 320–341.

Fowler SD, Brown WJ. 1984. Lysosomal acid lipase. In: B Borgström, HL Brockman (eds.) *Lipases*. Elsevier Science Pub., London, pp. 329–364.

Goll DE et al. 1983. Role of muscle proteinases in maintenance of muscle integrity and mass. *J Food Biochem* 7: 137–177.

Goransson A et al. 2002. Effect of electrical stimulation on the activity of muscle exoproteases during beef aging. *Food Sci Tech Int* 8: 285–289.

Grossman S et al. 1988. Lipoxygenase in chicken muscle. *J Agric Food Chem* 36: 1268–1270.

Herrera-Méndez CH et al. 2006. Meat aging: Reconsideration of the current concept. *Trends Food Sci Technol* 17: 394–405.

Hierro E et al. 1997. Contribution of microbial and meat endogenous enzymes to the lipolysis of dry fermented sausages. *J Agric Food Chem* 45: 2989–2995.

Imanaka T et al. 1985. Positional specificity of lysosomal acid lipase purified from rabbit liver. *J Biochem* 98: 927–931

Jiménez-Colmenero J et al. 2010. Nutritional composition of dry-cured ham and its role in a healthy diet. *Meat Sci* 84: 585–593.

Klement JT et al. 1974. The effect of bacterial fermentation on protein solubility in a sausage model system. *J Food Sci* 39: 833–835.

Koohmaraie M. 1994. Muscle proteinases and meat aging. *Meat Sci* 36: 93–104.

Koohmaraie M et al. 1987. Effect of postmortem storage on Ca^{2+}-dependent proteases, their inhibitor and myofibril fragmentation. *Meat Sci* 19: 187–196.

Ludemann V et al. 2004. Determination of growth characteristics and lipolytic and proteolytic activities of *Penicillium* strains isolated from Argentinian salami. *Int J Food Microbiol* 96: 13–18.

Marczy JS et al. 1995. Comparative study on the lipoxygenase activities of some soybean cultivars. *J Agric Food Chem* 43: 313–315.

Martin L et al. 1999. Changes in intramuscular lipids during ripening of Iberian dry-cured ham. *Meat Sci* 51: 129–134.

Matsukura U et al. 1981. Mode of degradation of myofibrillar proteins by an endogenous protease, cathepsin L. *Biochim Biophys Acta* 662: 41–47.

McDonald JK, Barrett AJ (eds.). 1986. *Mammalian proteases: A glossary and bibliography*, vol 2, Exopeptidases. Academic Press, London.

Mikami M et al. 1987. Degradation of myofibrils from rabbit, chicken and beef by cathepsin L and lysosomal lysates. *Meat Sci* 21: 81–87.

Miller CH et al. 1981. Specificity of lipoprotein lipase and hepatic lipase towards monoacylglycerols varying in the acyl composition. *Biochim Biophys Acta* 665: 385–392.

Molly K et al. 1997. The importance of meat enzymes in ripening and flavor generation in dry fermented sausages. First results of a European project. *Food Chem* 54: 539–545.

Monin G et al. 1997. Chemical and structural changes in dry-cured hams (Bayonne hams) during processing and effects of the dehairing technique. *Meat Sci* 47: 29–47.

Montel MC et al. 1993. Effects of starter cultures on the biochemical characteristics of French dry sausages. *Meat Sci* 35: 229–240.

Mora L et al. 2009a. Naturally generated small peptides derived from myofibrillar proteins in serrano dry-Cured Ham. *J Agric Food Chem* 57: 3228–3234

Mora L et al. 2009b. The degradation of creatin kinase in dry-Cured Ham. *J Agric Food Chem* 57: 8982–8988.

Motilva MJ, Toldrá F. 1993. Effect of curing agents and water activity on pork muscle and adipose subcutaneous tissue lipolytic activity. *Z Lebensm Unters Forsch* 196: 228–231.

Motilva MJ et al. 1992. Assay of lipase and esterase activities in fresh pork meat and dry-cured ham. *Z Lebensm Unters Forsch* 195: 446–450.

Motilva MJ et al. 1993a. Subcutaneous adipose tissue lipolysis in the processing of dry-cured ham. *J Food Biochem* 16: 323–335.

Motilva MJ et al. 1993b. Muscle lipolysis phenomena in the processing of dry-cured ham. *Food Chem* 48: 121–125.

Navarro JL et al. 1997. Lipolysis in dry-cured sausages as affected by processing conditions. *Meat Sci* 45: 161–168.

Negre AE et al. 1985. New fluorimetric assay of lysosomal acid lipase and its application to the diagnosis of Wolman and cholesteryl ester storage diseases. *Clin Chim Acta* 149: 81–88.

Okitani A et al. 1981. Characterization of hydrolase H, a new muscle protease possessing aminoendopeptidase activity. *Eur J Biochem* 115: 269–274.

Ordoñez JA et al. 1999. Changes in the components of dry-fermented sausages during ripening. *Crit Rev Food Sci Nutr* 39: 329–367.

Ouali A. 1991. Sensory quality of meat as affected by muscle biochemistry and modern technologies. In: LO Fiems, BG Cottyn, (eds.) *Animal Biotechnology and the Quality of Meat Production*. Elsevier Science Pub., B.V., Amsterdam, pp. 85–105.

Roseiro LC et al. 2010 Effect of processing on proteolysis and biogenic amines formation in a Portuguese traditional dry-fermented ripened sausage Chouriço Grosso Estremoz e Borba PGI. *Meat Sci* 84: 172–179.

Rosell CM, Toldrá F. 1996. Effect of curing agents on m-calpain activity throughout the curing process. *Z Lebensm Unters Forchs* 203: 320–325.

Sanz Y et al. 2002. Role of muscle and bacterial exopeptidases in meat fermentation. In: F Toldrá (ed.) *Research Advances in the Quality of Meat and Meat Products*. Research Signpost, Trivandrum, pp. 143–155

Schwartz WN, Bird JWC. 1977. Degradation of myofibrillar proteins by cathepsins B and D. *Biochem J* 167: 811–820.

Sentandreu MA, Toldrá F. 2002. Dipeptidylpeptidase activities along the processing of Serrano dry-cured ham. *Eur Food Res Technol* 213: 83–87.

Sentandreu MA et al. 2002. Role of muscle endopeptidases and their inhibitors in meat tenderness. *Trends Food Sci Technol* 13: 398–419.

Sentandreu MA et al. 2007. Proteomic identification of actin-derived oligopeptides in dry-cured ham. *J Agric Food Chem* 55: 3613–3619.

Shahidi F. 1987. Assessment of lipid oxidation and off-flcour development in meat, meat products and seafoods. In: F Shahidi (ed) *Flavour of meat, meat products and seafoods*, 2nd edn. London: Blackie Academic & Professional, pp. 373–419.

Shahidi F. 1998a. Assessment of lipid oxidation and off-flavour development in meat, meat products and seafoods. In: F Shahidi (ed.) *Flavor of Meat, Meat Products and Seafoods*. Blackie Academic and Professional, London, pp. 373–394.

Shahidi F. 1998b. Flavour of muscle foods—An overview. In: F Shahidi (ed.) *Flavor of Meat, Meat Products and Seafoods*. Blackie Academic and Professional, London, pp. 1–4.

Smith LC, Pownall HJ. 1984. Lipoprotein lipase. In: B Borgström, HL Brockman (eds.) *Lipases*. Elsevier Science Pub., London, pp. 263–305.

Stahnke LH. 2002. Flavour formation in fermented sausage. In: F Toldrá (ed.) *Research Advances in the Quality of Meat and Meat Products*. Research Signpost, Trivandrum, pp. 193–223.

Sunesen LO, Stahnke LH. 2004. Mould starter cultures for dry sausages: Selection, application and effects. *Meat Sci* 65: 935–948.

Talon R et al. 1999. Effect of nitrate and incubation conditions on the production of catalase and nitrate reductase by staphylococci. *Int J Food Microbiol* 52: 47–56.

Talon R et al. 2002. Bacterial starters involved in the quality of fermented meat products. In: F Toldrá (ed.) *Research Advances in the Quality of Meat and Meat Products*. Research Signpost, Trivandrum, pp. 175–191.

Tichivagana JZ, Morrisey PA. 1985. Metmyoglobin and inorganic metals as prooxidants in raw and cooked muscle systems. *Meat Sci* 15: 107–116.

Tjener K et al. 2004. Growth and production of volatiles by *Staphylococcus carnosus* in dry sausages: Influence of inoculation level and ripening time. *Meat Sci* 67: 447–452.

Toldrá F. 1992. The enzymology of dry-curing of meat products. In: FJM Smulders et al. (eds.) *New Technologies for Meat and Meat Products*. Audet, Nijmegen, pp. 209–231.

Toldrá F. 1998. Proteolysis and lipolysis in flavour development of dry-cured meat products. *Meat Sci* 49: s101–s110.

Toldrá F. 2002. *Dry-Cured Meat Products*. Food and Nutrition Press, Trumbull, CT, pp. 1–238.

Toldrá F. 2004a. Dry-cured ham. In: YH Hui et al. (eds.) *Handbook of Food and Beverage Fermentation Technology*. Marcel Dekker, New York, pp. 369–384.

Toldrá F. 2004b. Curing: Dry. In: W Jensen, (eds.) *Encyclopedia of Meat Sciences*. Elsevier Science, Ltd., London, pp. 360–365.

Toldrá F. 2006a. Meat: Chemistry and biochemistry. In: YH Hui et al. (eds.) *Handbook of Food Science, Technology and Engineering*. CRC Press, Boca Raton, FL, vol 4, pp. 28–1 a 28–18.

Toldrá F. 2006b. Dry-cured ham. In: YH Hui et al. (eds.) *Handbook of Food Science, Technology, and Engineering*. CRC Press, Boca Raton, FL, vol. 4, pp. 164–1 a 164–11.

Toldrá F. 2007. Biochemistry of muscle and fat. In: F Toldrá et al. (eds.) *Handbook of fermented meat and poultry*. Blackwell Publishing, Ames, IA, pp. 51–58.

Toldrá F, Aristoy MC. 2010. Dry-cured ham. In: F Toldrá (ed.) *Handbook of Meat Processing*. Wiley-Blackwell, Ames, IA, pp. 351–362.

Toldrá F, Flores M. 1998. The role of muscle proteases and lipases in flavor development during the processing of dry-cured ham. *CRC Crit Rev Food Sci Nutr* 38: 331–352.

Toldrá F, Reig M. 2007. Sausages. In: YH Hui et al. (eds.) *Handbook of Food Product Manufacturing*, vol 2. John Wiley Interscience, New York, pp. 249–262.

Toldrá F et al. 1992. Activities of pork muscle proteases in cured meats. *Biochimie* 74: 291–296.

Toldrá F et al. 1993. Cathepsin B, D, H and L activity in the processing of dry-cured-ham. *J Sci Food Agric* 62: 157–161.

Toldrá F, et al. 2000. Contribution of muscle aminopeptidases to flavor development in dry-cured ham. *Food Res Internat* 33: 181–185.

Toldrá F et al. 2001. Meat fermentation technology. In: YH Hui, WK Nip, RW Rogers, OA Young (eds.) *Meat Science and Applications*. Marcel Dekker, New York, pp. 537–561.

Toldrá F et al. 2004. Packaging and quality control. In: YH Hui et al. (eds.) *Handbook of Food and Beverage Fermentation Technology*. Marcel Dekker, New York, pp. 445–458.

Tornquist H et al. 1978. Enzymes catalyzing the hydrolysis of long-chain monoacylglycerols in rat adipose tissue. *Biochim Biophys Acta* 530: 474–486.

Valin C, Ouali A. 1992. Proteolytic muscle enzymes and post-mortem meat tenderisation. In: FJM Smulders et al. (eds.) *New Technologies for Meat and Meat Products*. Audet, Nijmegen, pp. 163–179.

Vestergaard CS et al. 2000. Lipolysis in dry-cured ham maturation. *Meat Sci* 55: 1–5.

Xiao S et al. 2010. Changes of hormone-sensitive lipase (HSL), adipose tissue triglyceride lipase (ATGL) and free fatty acids in subcutaneous adipose tissue throughout the ripening process of dry-cured ham. *Food Chem* 121: 191–195.

Yamamoto S. 1992. Mammalian lipoxygenases: Molecular structures and functions. *Biochim Biophys Acta* 1128: 117–131.

Yates LD et al. 1983. Effect of temperature and pH on the post-mortem degradation of myofibrillar proteins. *Meat Sci* 9: 157–179.

Young JF et al. 2003. Significance of preslaughter stress and different tissue PUFA levels on the oxidative status and stability of porcine muscle and meat. *J Agric Food Chem* 51: 6877–6881.

Zeece MG, Katoh K. 1989. Cathepsin D and its effects on myofibrillar proteins: A review. *J Food Biochem* 13: 157–161.

17
Chemical and Biochemical Aspects of Color in Muscle-Based Foods

José Angel Pérez-Alvarez and Juana Fernández-López

Abstract: The term muscle-based foods encompass food products derived from cattle, hogs, poultry, fish, and shellfish. Various factors influence the quality of muscle food products during various handling, processing, and storage. In this regard, color plays a crucial role in consumers' perception and acceptance of muscle-based food products. The purplish-red to bright red color in muscle foods is due to the myoglobin pigment, and the postmortem changes in its color are useful indicators of the state of freshness of the source material. This chapter discusses various postmortem biochemical and related effects of the pigment in foods derived from these animals during processing and storage.

GENERAL ASPECTS OF MUSCLE-BASED FOOD COLOR

The first impression that a consumer receives concerning a food product is established visually, and among the properties observed are color, form, and surface characteristics.

All meats and muscle-based foods have a number of visually perceived attributes that contribute to their color and appearance as well as to their overall quality. According to Lozano (2006), the appearance can be divided in three different categories: color, cesia, and spatial properties or spatiality. Color is related to optical power spectral properties of the stimulus detected by observers (Hutchings 2002). Cesia include transparency, translucence, gloss, luster, haze, lightness, opacity, and matt, and is related to the properties of reflecting, transmitting, or diffusing light by foods evaluated by human observation. Spatial properties are divided in two main groups: (i) modes of appearance in which color is modified depending on the angle of observation related to the light incidence angle, such as metallic, pearlescent, or iridescent materials, and (ii) modes of appearance related to optical properties of surfaces or objects in which effects of ordered patterns (textures) or finishing characteristics of food (as roughness, polish, etc.) (Lozano 2006).

Color is the main aspect that defines a food's quality and a product may be rejected simply because of its color, even before other properties, such as aroma, texture, and taste, can be evaluated. This is why the appearance (optical properties, physical form, and presentation) of meat products at the sales point is of such importance for the meat industry (Lanari et al. 2002). As regards the specific characteristics that contribute to the physical appearance of meat and muscle-based foods, color is the quality that most influences consumer choice (Krammer 1994).

The relation between meat color and quality has been the subject of study since the 1950s since, indeed, Urbain (1952) described how consumers had learnt through experience that the color of fresh meat is bright red; and any deviation from this color (nonuniform or anomalous coloring) is unacceptable (Diestre 1992). The color of fresh meat and associated adipose tissue is, then, of great importance for its commercial acceptability, especially in the case of beef and lamb (Cornforth 1994) and in

Food Biochemistry and Food Processing, Second Edition. Edited by Benjamin K. Simpson, Leo M.L. Nollet, Fidel Toldrá, Soottawat Benjakul, Gopinadhan Paliyath and Y.H. Hui.

certain countries like the United States and Canada, and there have been many studies to identify the factors controlling its stability. Adams and Huffman (1972) affirmed that consumers relate a meat's color with its freshness. In poultry, the consumers of many countries also associate the meat color with the way in which the animal was raised (intensive or extensive) and fed (cereals, animal feed, etc.).

Color as quality factor on meat can be appreciated in different ways in different countries, for example, in Denmark, pork meat color has the fifth place on consumers' purchase decision (Bryhni et al. 2002). Sensorial quality, especially color and appearance (Brewer and Mckeith 1999), of meat can be affected by internal and external factors.

Food technologists, especially those concerned with the meat industry, have a special interest in the color of food for several reasons. First, because of the need to maintain a uniform color throughout processing; second, to prevent any external or internal agent from acting on the product during processing, storage, and display; third, to improve or optimize a product's color and appearance; and lastly, to attempt bring the product's color into line with what the consumer expects.

Put simply, the color of meat is determined by the pigments present in the same. These can be classified into four types: biological (carotenes and hemopigments), which are accumulated or synthesized in the organism antemortem (Lanari et al. 2002); pigments produced as a result of damage during manipulation or inadequate processing conditions; pigments produced postmortem (through enzymatic or nonenzymatic reactions) (Montero et al. 2001); and finally, those resulting from the addition of natural or artificial colorants (Fernández-López et al. 2002).

As a quality parameter, color has been widely studied in fresh meat (MacDougall 1982, Cassens et al. 1995, Faustman et al. 1996) and cooked products (Anderson et al. 1990, Fernández-Ginés et al. 2003, Fernández-López et al. 2003a), while dry-cured meat products have received less attention (Pérez-Alvarez 1996, Pagán-Moreno et al. 1998, Aleson et al. 2003) because in this type of product color formation takes place during the different processing stages (Pérez Alvarez et al. 1997, Fernández-López et al. 2000, Pérez-Alvarez and Fernández-López 2009b); recently, new heme pigment has been identified in this type of products (Parolari et al. 2003, Wakamatsu et al. 2004a, 2004b).

From a practical point of view, color plays a fundamental role in the animal production sector, especially in meat production (beef and poultry, basically) (Zhou et al. 1993, Esteve 1994, Verdoes et al. 1999, Irie 2001), since in many countries of the European Union (Spain and Holland, for example) paleness receives a wholesale premium.

CHEMICAL AND BIOCHEMICAL ASPECTS OF COLOR IN MUSCLE-BASED FOODS

Of the major components of meat, proteins are the most important because they are only provided by essential amino acids (aa), which are very important for the organism's correct functioning, although they also make a technological contribution during processing, while some are also responsible for such important attributes as color. These are the so-called chromoproteins and they are mainly composed of a porphyrinic group conjugated with a transition metal, principally iron (metalloporphyrin), which forms conjugation complexes (heme group) (Whitaker 1972), this last being responsible for the color. However, other organic compounds exist alongside this, with isoprenoid-type conjugated systems (carotenes and carotenoproteins), also play an important part in meat color.

There are also some enzymatic systems whose coenzymes or prosthetic groups possess chromophoric properties (peroxidases, cytochromes, and flavins) (Faustman et al. 1996). However, their contribution to meat color is slight. In Section "Cytochromes", we describe the principal characteristics of the major compounds that impart color to meat.

CYTOCHROMES

Cytochromes are metalloproteins with a prosthetic heme group, whose role in meat coloration is undergoing revision (Boyle et al. 1994, Faustman et al. 1996), since initially they were not thought to play a very important role (Ledwar 1984). These compounds are found in low concentrations in the skeletal muscle; in poultry, they do not represent more than 4.23% of the total hemeoproteines present (Pikul et al. 1986). The role of cytochrome (especially its concentration) in poultry meat color is fundamental, when the animal is previously exposed to stress (Ngoka and Froning 1982, Pikul et al. 1986). Cytochromes are most concentrated in cardiac muscle, so that when this organ is included in meat products, heart contribution to color must be taken into consideration, not to mention the reactions that take place during elaboration processes (Pérez-Álvarez et al. 2000b).

CAROTENES

Carotenes are responsible for the color of beef, poultry meat and skin, fish and shellfish, in the latter case being of great economic importance. The color of the fat is also important in the carcasses grading. Also, carotenoids can be used as food coloring agents (Verdoes et al. 1999).

An important factor to be taken into account with these compounds is that they not synthesized by the animal's organism but by assimilation and during storage (Pérez-Álvarez et al. 2000b). On fat color, fatty acid composition can affect its color. When the ratio of cis-monounsaturated to saturated fatty acids is high, the fat exhibit a greater yellow color (Zhou et al. 1993). In the case of the carotenes present in fish tissues, these come from the ingestion of zooplankton and algae, and the levels are sometimes very high. The shell of many crustaceans also contains these compounds, for example, lobster (*Panilurus argus*).

The pigments responsible for color fish, particularly salmonids (trout, salmon among others), are astaxanthin and canthaxanthin, although they are also present in tunids and are one of the most important natural pigments of a marine origin.

In the case of shellfish, their color depends on the so-called carotenoproteins, which are proteins with a prosthetic group that may contain various types of carotene (Minguez-Mosquera 1997), which are themselves water soluble (Shahidi and Matusalach-Brown 1998).

In fish-derived products, the carotene content has previously been used as a quality parameter on its own; however, it has been demonstrated that this is not so and that other characteristics generally influence color (Little et al. 1979).

The carotene content and its influence on color is perhaps one of the characteristics that have received most attention (Swatland 1995). In the case of meat, especially beef, an excess of carotenes may actually lower the quality (Irie 2001), as occurs sometimes when classifying carcasses. The Japanese system for beef carcass classification classifies as acceptable fats with a white, slightly off-white, or slightly reddish-white color, while pink-yellowish and dark yellow are unacceptable (Irie 2001). It is precisely the carotenes that are responsible for these last two colorations.

However, in other animal species, such as chicken, the opposite effect is observed, since a high carotene (xanthophylls) concentration is much appreciated by consumers (Esteve 1994), yellow being associated with traditional or "home-reared" feeding (Pérez-Álvarez et al. 2000b).

The use of carotenoid canthaxanthin as a coloring agent in poultry feeds designed to result in the desired coloration of poultry meat skins. Although the carotenoids used mostly are citranaxanthin, capsanthin, and capsorubin, canthaxanthin shows superior pigmenting properties and stability during processing and storage (Blanch 1999). Zhu et al. (2009) reported that dietary supplementation with *Lactobacillus salivarius* increased the xanthophyll concentration in tissue and the Roche color fan scores of the shank skin of chickens.

Farmed fish, specially colored fish (salmon, rainbow trout, for example), is now of the main industries, for example, Norway exports a great part of its salmon. To improve its color and brilliance, 0.004–0.04 weight% proanthocyanidin is added to fish feed containing carotenoids (Sakiura 2001). For rainbow trout, carotenoid concentrations could be 10.7 or 73 mg/kg canthaxanthin, or 47 or 53 mg/kg astaxanthin.

Hemoproteins

Of the hemoproteins present in the muscle postmortem, myoglobin (Mb) is the one mainly responsible for color, since hemoglobin (Hb) arises from the red cells that are not eliminated during the bleeding process and are retained in the vascular system, basically in the capillaries (incomplete exsanguination, the average amount of blood remaining in meat joints being 0.3%) (Warris and Rodes 1977). However, their contribution to color does not usually exceed 5% (Swatland 1995). There was wide variation in amounts of Hb from muscle tissue of bled and unbled fish. Mb content was minimal as compared to Hb content in fish light muscle and white fish whole muscle. Hb made up 65% and 56% of the total hem protein by weight in dark muscle from unbled and bled fish (Richards and Hultin 2002).

Mb, on average, represents 1.5% in weight of the proteins of the skeletal muscle, while Hb represents about 0.5%, the same as the cytochromes and flavoproteins together. Mb is an intracellular (sarcoplasmic) pigment apparently distributed uniformly within muscles (Ledwar 1992, Kanner 1994). It is red in color and water soluble and is found in the red fibers of both vertebrates and invertebrates (Knipe 1993, Park and Morrisey 1994), where it fulfils the physiological role of intervening in the oxidative phosphorylation chain in the muscle (Moss 1992).

Structure of Myoglobin

Structurally, Mb can be described as a monomeric globular protein with a very compact, well-ordered structure that is specifically, almost triangularly, folded and bound to a hemo group (Whitaker 1972). It is structurally composed of two groups: a proteinaceous group and a heme group.

The protein group has only one polypeptide chain composed of 140–160 amino acid residues, measuring 3.6 nm and weighing 16,900 Da in vertebrates (Lehninger 1981). It is composed of eight relatively straight segments (where 70% of the aa is found), separated by curvatures caused by the incorporation in the chain of proline and other aa that do not form α-helices (such as serine and isoleucine). Each segment is composed of a portion of α-helix, the largest of 23 aa and the shortest of 7 aa, all dextrogyrating.

The high helicoidal content (forming an ellipsoid of $44 \times 44 \times 25$ Å) and the lack of disulphide bonds (there is no cysteine) means that Mb is an atypical globular protein. The absence of these groups makes the molecule highly stable (Whitaker 1972). Although the three-dimensional structure seems irregular and asymmetric, it is not totally anarchic, and all the molecules of Mb have the same conformation.

One very important aspect of the protein part of Mb is its lack of color. However, the variations presented by its primary structure and the aa composition of different animal and fish species destined for human consumption are the cause of the different colorations of meat and their stability when they are not displayed in the same conditions (Lorient 1982, Lee et al. 2003).

The heme group of Mb (as in Hb and other proteins) is, as mentioned in the preceding text, a metalloporphyrin. These molecules are characterized by their high degree of coloration as a result of their conjugated cyclic tetrapyrrole structure (Kalyanasundaram 1992). The heme group is composed of a complex, organic annular structure, protoporphyrin, to which an iron atom in ferrous state is united (Fe II). This atom has six coordination bonds, four with the flat protoporphyrin molecule (forming a flat square complex) and two perpendiculars to it. The sixth bond is open and acts as a binding site for the oxygen molecule.

Protoporphyrin is a system with a voluminous flat ring composed of four pyrrole units connected by methyl bridges (= C-). The Fe atom with a coordination number of six lies in the center of the tetrapyrrole ring and is complexed to four pyrrolic nitrogen's. The heme group is complexed to the polypeptide chain (globin) through a specific histidine residue (imadazole ring) occupying the fifth position of the Fe atom (Dadvison and Henry 1978).

The heme group is bound to the molecule by hydrogen bridges that are formed between the propionic acid side chains and other side chains. Other aromatic rings exist near, and almost parallel to, the heme group, which may also form pi (π) bonds (Stauton-West et al. 1969).

The Hb contains a porphyrinic heme group identical to that of Mb and equally capable of undergoing reversible oxygenation and deoxygenation. Indeed, it is functionally and structurally paired with Mb and its molecular weight is four times greater, since it contains four peptidic chains and four heme groups. The Hb, like Mb, has its fifth ligand occupied by the imadazole group of a histidine residue, while the sixth ligand may or may not be occupied. It should be mentioned that the fifth and sixth positions of other hemoproteins (cytochromes) are occupied by R groups of specific aa residues of the proteins and, therefore, cannot bind to oxygen (O_2), carbon monoxide (CO), or cyanide (CN^-), except a_3, which, in its biological role, usually binds to oxygen.

One of the main differences between fish and mammalian is that fish Mb had two distinct endothermic peaks indicating multiple states of structural unfolding, whereas mammalian Mb followed a two-state unfolding process. Changes in α-helix content and tryptophan fluorescence intensity with temperature were greater for fish Mb than for mammalian Mb. Fish Mb show labile structural folding, suggesting greater susceptibility to heat denaturation than that of mammalian Mb (Saksit et al. 1996).

Helical contents of frozen-thawed Mb were practically the same as those of unfrozen Mb, regardless of pH. Frozen-thawed Mb showed a higher autoxidation rate than unfrozen Mb. During freezing and thawing, Mb suffered some conformational changes in the no helical region, resulting in a higher susceptibility to both unfolding and autoxidation (Chow et al. 1989). In tuna fish Mb, its stability was in the order bluefin tuna (*Thunnus thynnus*) > yellowfin tuna (*Thunnus albacares*) > bigeye tuna (*Thunnus obesus*), with autoxidation rates in the reverse order. The pH dependency of Mb from skipjack tuna (*Katsuwonus pelamis*) and mackerel (*Scomber scombrus*) were similar. Lower Mb stability was associated with higher autoxidation rate (Chow 1991).

Chemical Properties of Myoglobin

The chemical properties of Mb center on its capacity to form ionic and covalent groups with other molecules. Its interaction with several gases and water depends on the oxidation state of the Fe of the heme group (Fox 1966), since this may be either in its ferrous (Fe II) state or in ferric (Fe III) state. Upon oxidation, the Fe of the heme group takes on a positive charge (Kanner 1994) and, typically, binds with negatively charged ligands, such as nitrites, the agents responsible for the nitrosation reactions in cured meat products.

When the sixth coordination ligand is free, the Mb is usually denominated deoxymyoglobin (DMb), which is purple in color. However, when this site is occupied by oxygen, a noncovalent complex is formed between this gas and the Mb, and it is denominated oxymyoglobin (OMb), which is cherry or bright red (Lanari and Cassens 1991). When the oxidation state of the iron atom is modified to the ferric state, the sixth position is occupied by a molecule of water; it is denominated metmyoglobin (MMb), which is brown.

Mb absorbs light in the ultraviolet region and through practically the complete visible region of light (Fox 1966). MMb, OMb, nitrosomyoglobin (NOMb), and DMb have maximum absorbance's (>400 nm) at about 410, 418, 419, and 434 nm, respectively (Millard et al. 1996). The absorbance band is typically much weaker at higher wavelengths (500–600 nm). Above 500 nm, OMb and NOMb have absorption maxima at around 545 and 585 nm both. The NOMb complex maintains Mb in ferrous state, but is somewhat unstable and can be displaced and oxidized if stored with excess oxygen and light (Kanner 1994).

There are several possible causes for MMb to be generated, and these may be due to the ways in which tunids, meat, and meat products are obtained, transformed, or stored (MacDougall 1982, Lee et al. 2003, Mancini et al. 2003). Among the most important factors are: low pH, the presence of ions and high temperatures during processing (Osborn et al. 2003), the growth and/or formation of metabolites from the microbiota (Renerre 1990), the activity of endogenous reducing enzymes (Arihara et al. 1995, Osborn et al. 2003), the levels of endogenous (Lanari et al. 2002) or exogenous antioxidants, such as ascorbic acid or its salts, tocopherols (Irie et al. 1999) or plant extracts (Fernández-López et al. 2003b, Sánchez-Escalante et al. 2003).

This change in the oxidation state of the heme group will result in the group being unable to bind with the oxygen molecule (Arihara et al. 1995).

DMb is able to react with other molecules to form colored complexes, many of which are of great economic relevance for the meat industry. The most characteristic example is the reaction of DMb with nitrite, since its incorporation generates a series of compounds with distinctive colors: red in dry-cured meat products or pink in heat-treated products. The products resulting from the incorporation of nitrite are denominated cured, and such products are of enormous economic importance worldwide (Pérez-Alvarez 1996).

The reaction mechanism is based on the property of nitric oxide (NO, generated in the reaction of nitrite in acid medium that readily gives up electrons) to form strong coordinated covalent bonds, which form an iron complex with the heme group without the oxidation state of its structure having any influence. The compound formed after the nitrification reaction is denominated NOMb.

As mentioned in the preceding text, the presence of reducing agents such as hydrosulfhydric acid (H_2S) and ascorbates, lead to the formation of undesirable pigments both in meat and meat products. These green pigments are called sulphomyoglobin (SMb) and colemyoglobin (ColeMb), respectively, and are formed as a result of bacterial activity and by an excess of reducing agents in the medium. The formation of SMb is reversible, but that of ColeMb is an irreversible mechanism, since it is rapidly oxidized between pH 5 and 7, releasing the different parts of the Mb (globin, iron, and the tetrapyrrole ring).

From a chemical point of view, it should be borne in mind that the color of Mb, and therefore of the meat or meat products, not only depends on the molecule that occupies the sixth coordination site, but also the oxidation state of the iron atom (ferrous or ferric), the type of bond formed between the ligand and the heme group (coordinated covalent, ionic, or none) and the state of the protein (native or denaturalized form), not to mention the state of the porphyrin of the heme group (intact, substituted or degraded) (Pérez-Alvarez 1996).

During heat treatment of fish flesh, aggregation of denatured fish proteins is generally accompanied by changes in light scattering intensity. Results demonstrate the use of changes in relative light scattering intensity for studying structural unfolding and aggregation of proteins under thermal denaturation (Saksit et al. 1998). When fatty fish meat like *Trachurus japonicus* was heat treated, MMb content increased linearly, and the percentages of denatured Mb and apomyoglobin increased rapidly when mince was exposed to heat, but when temperature reach 60°C, the linearity is broken. Results indicated that stability of the color was higher than that of Mb and that the thermal stability of heme was higher than that of apomyoglobin (Hui et al. 1998).

Both Mb and ferrous iron accelerated lipid oxidation of cooked water-extracted fish meat. EDTA inhibited the lipid oxidation accelerated by ferrous iron, but not that accelerated by Mb. Also, with cooked nonextracted mackerel meat, EDTA noticeably inhibited lipid oxidation. Nonheme iron-catalysis seemed to be related in part to lipid oxidation in cooked mackerel meat. Addition of nitrite in combination with ascorbate resulted in a marked inhibition of lipid oxidation in the cooked mackerel meat. From these results, it was postulated that nitric oxide ferrohemochromogen, formed from added nitrite and Mb, present in the mackerel meat in the presence of a reducing agent, possesses an antioxidant activity, which is attributable in part to the function as a metal chelator (Ohshima et al. 1988).

Tuna fish meat can be improved in its color when the flesh is treated with CO. The specific spectrum of carboxymyoglobin (COMb) within the visible range can be obtained. Penetration of CO into tuna muscle was very slow. After approximately 1–4 hours CO had penetrated 2–4 mm under the surface, and after 8 hours, CO had penetrated 4–6 mm. Mb extracts from tuna muscle treated with CO exhibited higher absorbance at 570 than at 580 nm (Chau et al. 1997). Jayasingh et al. (2001) reported that CO (0.5%) can penetrate to a depth of 15 mm (1 week) in ground beef. These authors also mentioned that when COMb is exposed to atmospheres free of CO, COMb slowly dissociate from Mb.

In dry-cured meat products, as Parma ham (produced without nitrite or nitrate), the characteristic bright red color (Wakamatsu et al. 2004a) is caused by Zn-protoporphyrin IX (ZPP) complex, a heme derivative. This type of pigment can be formed by endogenous enzymes as well as microorganisms (Wakamatsu et al. 2004b). Spectroscopic studies of Parma ham during processing all process, revealed a gradual transformation of muscle Mb, initiated by salting and continuing during aging. Pigments became increasingly lipophilic during processing, suggesting that a combination of drying and maturing yields a stable red color (Parolari et al. 2003). Electron spin resonance spectra showed that the pigment in dry-cured Parma ham is at no stage a nitrosyl complex of ferrous Mb as found in brine-cured ham and Spanish Serrano hams (Moller et al. 2003). These authors also establish that heme moiety is present in the acetone/water extract and that Parma ham pigment is gradually transformed from an Mb derivative into a nonprotein heme complex, thermally stable in acetone/water solution. Adamsen et al. (2003) also demonstrated that the heme moieties of Parma ham pigments have also antioxidative properties.

COLOR CHARACTERISTICS OF BLOOD PIGMENTS

Hb has function carrying oxygen to the tissues. Thus, the oxygenated hemoglobin $Hb(Fe(II))O_2$ is a very stable molecule but does slowly auto-oxidize at a rate of about 3% per day. This rate is accelerated at lower oxygen tensions if the Hb is partially oxygenated. The "blood pigments" chemistry is actually quite complex (Umbreit 2007).

Nagababu and Rifkind (2000) reported that the auto-oxygenation generates Hb(Fe(III)), called methemoglobin (MHb), and superoxide. At least in vitro, the superoxide undergoes dismutation to hydrogen peroxide and oxygen. The hydrogen peroxide is rapidly decomposed by catalase. The sixth coordinate of the MHb is occupied with water. If not immediately destroyed, the hydrogen peroxide would react with $Hb(Fe(II))O_2$ to produce ferrylhemoglobin, $Hb(Fe(IV)) = O$, with a rhombic heme that reacts with further hydrogen peroxide to produce free Fe(III) and porphyrin degradation products.

Jaffe (1981) reported that the MHb can be reduced by the NADH-cytochrome b5-MHb reductase or by direct reduction by ascorbate and glutathione.

The Hb can react with nitrites; the reaction with oxyhemoglobin (OHb) limits the half-life of the NO. The reaction with free Hb is so fast that any NO is consumed immediately (Vaughin et al. 2000, Huanf et al. 2001). According to Umbreit (2007), there are two major reactions of the nitrogenous compounds with Hb Fe(II), one binding NO to the heme and the other reducing nitrite to NO.

Muscles can retain several amounts of blood. Thus, Richard et al. (2005) reported that dark muscle from Atlantic mackerel contained roughly equal amounts of Hb and Mb, although in other fish species such as bluefin tuna (*T. thynnus*) Hb can be the major heme compound in dark muscle "sangacho" (Sánchez-Zapata et al. 2009a). Niewiarowicz et al. (1986) reported that in different poultry dark meat species the ratio between Hb and Mb varies from 20% to 40%. In mammals, this ratio has been reported to range from 7% to 35% (Han et al. 1994).

Animal blood is little used in the food industry, because of dark color it imparts to the products to which it is added. Attempts to solve food color related problems have employed several different processes and means, but they are not always completely satisfactory. The addition of 12% blood plasma to meat sausages lead to pale-colored products. Another means of solving color problems is the addition of discolored whole blood or globin by eliminating Hb's heme group. From blood, natural red pigments can be obtained without using coloring agents such as nitrous acid salts; these pigments have zinc protoporphyrin as the metalloporphyrin moiety, and can be used in producing beef products, whale meat products, and fish products (including fish pastes), and a favorable color (Numata and Wakamatsu 2003).

There was wide variation in amounts of Hb extracted from muscle tissue of bled and unbled fish, and the residual level in the muscle of bled fish was substantial. Mb content was minimal as compared to Hb content in mackerel light muscle and trout whole muscle. Hb made up 65% and 56% of the total heme protein by weight in dark muscle from unbled and bled mackerel. That blood-mediated lipid oxidation in fish muscle depends on various factors, including Hb concentration, Hb type, plasma volume, and erythrocyte integrity (Richards and Hultin 2002).

Alvarado et al. (2007) reported that the Hb content can vary a lot among animal species and muscle types; it is clear that the Hb levels present in muscle-based foods have the potential to substantially contribute to lipid oxidation. Also, the presence of blood, Mb, Fe^{2+}, Fe^{3+}, or Cu^{2+} can stimulate lipid oxidation in the fillets of ice-fish (Rehbein and Orlick 1990).

The formation of ferryl and/or perferryl species upon reaction with hydrogen peroxide or lipid hydroperoxides has been reported to be either truly initiators or important catalysts of lipid oxidation in raw muscle meat products (Baron and Andersen 2002). Various factors such as the ability of the Hb to auto-oxidize and release of hematin, the heme–iron moiety nonbound to the protein, have also been reported to be crucial in promoting lipid oxidation (Grunwald and Richards 2006).

As was mentioned in the preceding text, blood utilization by the food industry is minimal, but many blood sausages are elaborated around the world (Diez et al. 2009). Problem associated with blood uses is discoloration through the disruption of the heme group of the Hb moiety. Several attempts are made to find a possible solution to the color problems caused by blood addition, one of them is the Hb moiety conversion into the more stable and sensorial accepted carboxyhaemoglobin (COHb) by saturation with carbon monoxide (CO). This pigment is very stable and can be visualized by the reflectance spectra (400–700 nm) (Fontes et al. 2010).

Fontes et al. (2010) reported differences between the spectra of blood CO-treated and its untreated counterpart. When CO was bubbled into the blood samples, the spectra reflected the COHb color properties in the red region, whose maximum reflectance occurs at 700 nm. Greater color stability was also observed in CO-treated sample as indicated by small differences between reflectance curves.

In adipose tissues, residual Hb is associated with the presence of capillaries or with hemorrhage. Irie (2001) reported that internal fat is whiter, with less Hb and with harder fat, than subcutaneous fat. The Hb derivatives are MHb, OHb, and deoxyhemoglobin (DHb). This Hb derivatives show different reflectance and absorbance spectra in which each showed different absorbance bands (AB); thus, MHb showed AB at 406, 500, and 630 nm; OHb showed 418, 540–542, 516–578, 950 nm; and DHb showed 430, 555, 760, and 910 nm.

From technological point of view, COMb can be used in several types of meat processing; thus, CO can be used in (i) modified atmosphere packaging of several type of meats (Raines and Hunt 2010), (ii) pasteurized meat products (Fontes et al. 2004), (iii) dry blood (Fontes et al. 2010), and (iv) combination with injection-enhancement ingredient (lactate) for beef color stabilization (Suman et al. 2010). But several legal problems are associated with its use; thus, the industry must avoid its use to rejuvenate the color of spoiled meat.

Stiebing (1990) reported that during cooking, the oxygen partial pressure in blood sausages decreases to a point where oxidation occurs, instead of oxygenation. But the CO-treated blood can increase shelf life (refrigerated storage, 4 days).

FAT COLOR

From a technological point of view, fat fulfils several functions, although as regards color, its principal role is in the brightness of meat products. Processes such as "afinado" during the elaboration of dry-cured ham involve temperatures at which fat melts, so that it infiltrates the muscle mass and increases the brilliance (Sayas 1997). When the fat is finely chopped, it "dilutes" the red components of the color, thus decreasing the color intensity of the finished product (Pérez-Alvarez et al. 2000). However, fats do not play such an important role in fine pastes because, after emulsification, the fat is masked by the matrix effect of the emulsion, so that it contributes very little to the final color.

The color of fat basically depends on the feed that the live animal received (Esteve 1994, Irie 2001). In the case of chicken and ostrich, the fat has a "white" appearance (common in Europe) when the animal has been fed with "white" cereals or other ingredients not containing xanthophylls, since these are accumulated in subcutaneous fat and other fatty deposits. However, when the same species are fed on maize (rich in xanthophylls), the fatty deposits take on a yellow color.

Beef or veal fat that is dark, hard (or soft), excessively bright or shiny lowers the carcass and cut price. Fat with a yellowish color in healthy animals reflects a diet containing beta-carotene (Swatland 1988). While fat color evaluation has traditionally been a subjective process, modern methods include such techniques as optical fiber spectrophotometry (Irie 2001).

Another factor influencing fat color is the concentration of the Hb retained in the capillaries of the adipose tissues (Swatland 1995). As in meat, the different states of Hb may influence the color of the meat cut. OMb is responsible for the yellowish appearance of fat, since it affects different color components (yellow-blue and red-green).

The different states of Hb present in adipose tissue may react in a similar way as in meat, so that fat color should be measured as soon as possible to avoid possible color alterations.

When the Hb in the adipose tissue reacts with the nitrite incorporated in the form of salt, nitrosohemoglobin (NOHb) is generated, a pigment that imparts a pink color to the fat. This phenomenon occurs principally in dry-cured meat products with a degree of anatomical integrity, such as dry-cured ham or shoulder (Sayas 1997).

When fat color is measured, its composition should be kept in mind, since its relation with fatty acids modifies its characteristics, making it more brilliant or duller in appearance. The fat content of the conjunctive tissue must also be borne in mind, since collagen may present a glassy appearance because, at acidic pH, it is "swollen", imparting a transparent aspect to the product.

REDUCED NITRITE MEAT PRODUCTS

Health concerns relating to the use of nitrates and nitrites in cured meats (cooked and dry cured) have led to a tendency toward decreased usage to alleviate the potential risk of the formation of carcinogenic, teratogenic, mutagenic *N*-nitroso compounds (Karolyi 2003), cytotoxic effects of nitrosamines and by the toxic effects of nitrite per se (metmyoglobinaemia and fall of blood pressure). But it is well known that nitrate/nitrite is widely used as a curing agent in the meat industry, although in recent years its omission from meat processing has been proposed (Viuda-Martos et al. 2009a).

Meat scientists have to take into account that any technological strategy (Pérez-Alvarez 2008) or the use of other ingredients (meat, nonmeat, and functional ingredients) can modify color characteristics (Pérez-Alvarez and Fernández-López 2009a, Sánchez-Zapata et al. 2009b).

As soon as nitrite is added in the meat formulation, it starts to disappear (nitrite is reduced to nitric oxide (NO)) that reacts with Mb to form NO-OMb and often can no longer be detected analytically at a later time. The rate of depletion is dependent on various factors, such as pH, initial nitrite concentration, processing and storage temperatures, meat-to-water ratio, and the presence of reductants (Pérez-Alvarez et al. 1993).

According to Pérez-Alvarez (2006), two-thirds of the Mb present in the meat is transformed into NOMb, although it is possible that only 50% of the Mb reacts. The rest corresponds to the residual nitrite. Residual nitrite levels (10–20% of the originally added sodium nitrite) correspond to nitrite that has not reacted with Mb and it is available for other reactions in the organism (Fernández-López et al. 2007).

Recently, it has been found that plants and their extracts can be used as indirect sources of nitrate (Shahid-Umar and Iqbal 2007, Shahid-Umar et al. 2007, Parks et al. 2008) in the production of meat products (Sebranek and Bacus 2007a, 2007b).

Six "technological" pathways exist for reducing the nitrite present in the meat products:

1. Exposure to γ radiation, as described by Wei et al. (2009).
2. Reaction with compounds of a polyphenolic nature (Garrote et al. 2004) derived basically from spices added in the formulation of the meat products. Also, can react with polyphenols, mainly hydroxycinnamic acids, like caffeic or ferulic acid, or with glycosylated flavanones, like hesperidin or narirutin, that are natural compounds of citrus coproducts (Viuda-Martos et al. 2010a).
3. Reaction with reducing agents present in the formulation or with endogenous substances. For example, when the pH of the meat is less than six, the nitrite added or that arising from the microbial reduction of nitrates is transformed into nitrous acid (relatively unstable). This, in turn, reacts with endogenous (cysteine, reduced nicotinamide adenine dinucleotide, cytochromes, and quinines) or exogenous (ascorbic acid and its salts) reducing substances of the meat and is transformed into NO, completing the dismutation reaction, as described by Pérez-Alvarez (2006).
4. Action of bacteria with nitrite-reductase activity such as *Staphylococcus carnosus*, *Staphylococcus simulans*, or *Staphylococcus saprophyticus* (Gøtterup et al. 2007). The action mechanism through which strains of *Staphylococcus* use nitrate/nitrite is thought to be related with its capacity to act as alternative electron acceptor in the cell respiratory chain (Gøtterup et al. 2008). The synthesis of nitrite reductases only occurs in anaerobic conditions and is induced by the presence of nitrate or nitrite in the medium (Neubauer and Götz 1996).
5. Combined method, reducing the quantity of residual nitrite by interaction with natural ingredients (tomato paste, annatto). Thus, Deba et al. (2007) reported that the addition of tomato paste to frankfurters reduces the added nitrite level from 150 to 100 mg/kg without any negative effect on the processing and quality characteristics of the product during storage.
6. Reaction with compounds present in citric coproducts such as orange and lemon albedo (Aleson-Carbonell et al. 2003, Fernández-Ginés et al. 2004), orange dietary fiber (Fernández-López et al. 2008b, 2009), citrus fiber washing water (Viuda-Martos et al. 2009b), orange fiber plus essential oils (Viuda-Martos et al. 2010a, 2010b).

ALTERATIONS IN MUSCLE-BASED FOOD COLOR

The color of meat and meat products may be altered by several factors, including exposure to light (source and intensity), microbial growth, rancidity, and exposure to oxygen. Despite the different alterations in color that may take place, few have been studied, including the pink color of boiled uncured products, premature browning (PMB), and melanosis in crustaceans.

Pink Color of Uncured Meat Products

The normal color of a meat product that has been heat treated but not cured is "brown," although it has recently been observed that these products show an anomalous coloration (red or pink) (Hunt and Kropf 1987). This problem is of great economic importance in "grilled" products, since this type of color is not considered desirable.

This defect may occur both in meats with a high hemoprotein content such as beef and lamb (red) and in those with a low concentration, including chicken and turkey (pink) (Conforth et al. 1986).

One of the principal causes of this defect is the use of water rich in nitrates, which are reduced to nitrites by nitrate-reducing bacteria, which react with the Mb in meat to form NOMb (Nash et al. 1985). The same defect may occur in meat products containing paprika, which, according to Fernández-López (1998), contains nitrates that, once incorporated in the product, may be similarly reduced by the microorganisms. Conforth et al. (1991) mentions that several nitrogen oxides may be generated in gas and electric ovens used for cooked ham and that these will react with the Mb to generate nitrosohemopigments. Also produced in ovens is CO, which reacts with Mb during thermal treatment to form a pink-colored pigment, carboxyhemochrome. It has also been described how the use of adhesives formed from starchy substances produces the same undesirable color in cooked products (Scriven et al. 1987).

The same anomalous color may be generated when the pH of the meat is high (because of the addition of egg albumin to the ingredients) (Froning et al. 1968) and when the cooking temperature during processing is too low. These conditions favor the development of a reducing environment that maintains the iron of the Mb in its ferrous form, imparting a reddish/pink color (as a function of the concentration of hemopigments) instead of the typical grayish-brown color of heat-treated uncured meat products.

Cooking uncured meat products, such as roast beef, at low temperatures (less than 60°C) may produce a reddish color inside the product, which some consumers may like. This internal coloring is not related with the formation of nitrosopigments, but results from the formation of OMb, a phenomenon that occurs because MMb-reducing enzymatic systems exist in the muscle that are activated at temperatures below 60°C (Osborn et al. 2003).

Microbial growth may also cause the formation of a pink color in cooked meats, since these reduce the oxido-reduction potential of the product during their growth. This is important when the microorganisms that develop in the medium are anaerobes, since they may generate reducing substances that reduce the heme iron. When extracts of *Pseudomonas* cultures are applied, the MMb may be reduced to Mb (Faustman et al. 1990). Recently, Gallego-Restrepo et al. (2010) reported that microorganism can play an important role in the color of uncured meat products during its elaboration process. These authors recommended the use of microbiological starter culture with a higher metabolic action at refrigeration conditions to obtain excellent color characteristics on this type of products.

Melanosis

Melanosis or blackspot, involving the appearance of a dark, even black, color, may develop post mortem in certain shellfish during chilled and frozen storage (Slattery et al. 1995). Melanosis is of huge economic importance, since the coloration may suggest a priori in the eyes of the consumer that the product is in bad condition, despite the fact that the formation of the pigments responsible involves no health risk.

Melanosis is an objectionable surface discoloration of high valuable shellfish as lobsters caused by enzymic formation of precursors of phenolic pigments. Blackspot is a process regulated by a complex biochemical mechanism, whereby the phenols present in a food are oxidized to quinones in a series of enzymatic reactions caused by polyphenoloxidase (PO) (Ogawa et al. 1984). This is followed by a polymerization reaction, which produces pigments of a high molecular weight and dark in color. Melanosis is produced in the exoskeleton of crustaceans, first in the head and gradually spreading toward the tail. Melanosis of shell and hyperdermal tissue in some shellfish as lobsters was related to stage of molt. The molting fluid is considered as the source of the natural activator(s) of pro-PO. Polyphenol oxidase (catechol oxidase) can be isolated from shellfish cuticle (Ali et al. 1994) and still active during iced or refrigerated storage. The process can be controlled by using sulfites (Ferrer et al. 1989), although their use is prohibited in many countries. Also ficin (Taoukis et al. 1990), 4-hexylresorcinol, functioned as a blackspot inhibitor, alone and in combination with L-lactic acid (Benner et al. 1994). Recently, Encarnacion et al. (2010) reported that dietary supplementation of the *Flammulina velutipes* mushroom extract in shrimp could be a promising approach to control postmortem development of melanosis and lipid oxidation in shrimp muscles.

Premature Browning

Hard-to-cook patties show persistent internal red color and are associated with high pH (>6) raw meat. Pigment concentration affects red color intensity after cooking (residual undenatured Mb), so this phenomenon is often linked to high pH dark-cutting meat from older animals. PMB is a condition in which ground beef (mince) looks well done at lower than expected temperature (Warren et al. 1996).

PMB of beef mince is a condition in which Mb denaturation appears to occur on cooking at temperature lower than expected and, therefore, may indicate falsely that an appropriate internal core temperature of 71°C has been achieved (Suman et al. 2004). The relationship between cooked color and internal temperature of beef muscle is inconsistent and depends on pH and animal maturity. Increasing the pH may be of benefit in preventing PMB but may increase incidence of red color in well-cooked meat (cooked over internal temperature of 71.1°C) (Berry 1997). When PSE meat is used in patties processing and those containing OMb easily exhibited PMB. One of the reasons of this behavior is that percentage of Mb denaturation increased as cooking temperature increased (Lien et al. 2002). Recently, Mancini et al. (2010) reported that the use of several additives such as lactate improved raw color stability, but did not minimize PMB.

COLOR AND SHELF LIFE OF MUSCLE-BASED FOODS

Meat and meat products are susceptible to degradation during storage and throughout the retail process. In this respect, color is one of the most important quality attributes for indicating the state of preservation in meat.

Any energy received by food can initiate its degradation, the rate of any reactions depending on the exact composition of the product (Jensen et al. 1998), environmental factors (light, temperature, presence of oxygen), or the presence of additives.

Transition metals, such as copper or iron, are very important in the oxidative/antioxidative balance of meat. When the free ions of these two metals interact, they reduce the action of certain agents, such as cysteine, ascorbate and α-tocopherol, oxidizing them and significantly reducing the antioxidant capacity in muscle (Zanardi et al. 1998).

The iron in meat is 90% heme iron (HI), which is several times more absorbable than nonheme iron (NHI) present in other foods (Fernández-López et al. 2008a). Although the mechanism for heme iron release in meat has not been determined, oxidation of the porphyrin ring and denaturation of Mb (Kristensen and Purslow 2001) are probably involved.

Harel et al. (1988) reported that the interaction of MMb with H_2O_2 or lipid hydroperoxide results in the release of free ionic iron. Free ionic iron can serve as a catalyst in the production of àOH from H_2O_2 as well as in the degradation of lipid hydroperoxides to produce peroxyl and alkoxyl radicals, which can initiate lipid oxidation and/or be self-degraded to the secondary products of lipid oxidation (Min and Ahn 2005). However, reducing compounds are essential to convert ferric to ferrous ion, a catalyst for Fenton reaction.

Lee et al. (1998) reported that during refrigerated storage of meat, there is a release of iron from the heme group, with a consequent increase in NHI, which speeds up lipid oxidation. Also, the heme molecule can break down during cooking or storage (Gómez-Basauri and Regenstein 1992a, 1992b, Miller et al. 1994a, 1994b) and can generate NHI. The increase of NHI in meats and fish is considered to be a reflection of the decrease of HI and is linked to the oxidative deterioration (Schricker and Miller 1983).

Nonheme iron is considered the most important oxidation promoter in meat systems, and therefore, knowledge of the proportions of the chemical forms of iron is of great importance (Kanner et al. 1991). An increase in the amount of NHI as a result of thermal processes on meat systems has been demonstrated by several authors (Schricker et al. 1982, Lombardi-Boccia et al. 2002). Gomez-Basauri and Regenstein (1992a, 1992b), suggested cooking is not as important as the subsequent refrigerated storage of cooked meats for the release of NHI from Mb.

The relationship between oxidative processes and the release of iron from Mb and the effect of these chemical changes on color characteristics of cooked products is still not well understood.

The increase of NHI could have some important consequences, affecting both the nutritional and technological properties of liver pate. The degradation of heme iron would reduce the nutritional value of the pates in terms of bioavailability of iron, since HI is more available than NHI (Hunt and Roughead 2000). Furthermore, iron achieves enhanced ability for promoting oxidation processes when it is released from the heme molecule (Kanner et al. 1991), and therefore, pates with increasing amounts of NHI might also have increased oxidative susceptibility.

From a nutritional point of view, Ahn and Kim (1998) reported that the status of ionic iron is more important than the amount of iron.

Traditionally, researchers have determined the discoloration of meat using as criterion the brown color of the product, calculated as percent of MMb (Mancini et al. 2003). These authors demonstrated that in the estimation of the shelf life of beef or veal (considered as discoloration of the product), the diminution in the percent of OMb is a better tool than the increase in percent of MMb.

Occasionally, when the meat cut contains bone (especially in pork and beef), the haemopigments (mainly Hb) present in the medulla lose color because the erythrocytes are broken during cutting and accumulate on the surface of the bone Hb. When exposed to light and air, the color of the Hb changes from bright red (OHb) characteristic of blood to brown (MHb) and even black (Gill 1996). This discoloration basically takes place during long periods of storage and especially during shelf life display (Mancini et al. 2004). This characteristic is aggravated if the product is kept in a modified atmosphere rich in oxygen (Lanari et al. 1995). These authors also point out that the effect of bone marrow discoloration is minimized by the effect of bacterial growth in modified atmosphere packaging.

As in the case of fresh meat, the shelf life of meat products is limited by discoloration (Mancini et al. 2004). This phenomenon is important in this type of product because they are normally displayed in illuminated cabinets. Consequently, the possibility of nitrosated pigments photo-oxidation of (NOMb) needs to be taken into account. During this process, the molecule is activated because it absorbs light; this may subsequently deactivate the NOMb and give the free electrons to the oxygen to generate MMb and free nitrite.

In model systems of NOMb photo-oxidation, this effect can be diminished by adding solutions of dextrose, which is a very important component of the salts used for curing cooked products and in meat emulsions.

When a meat product is exposed to light or is stored in darkness, the use of ascorbic acid or its salts may help stabilize the product's color. Such behavior has been described both in model systems of NOMb (Walsh and Rose 1956) and in dry-cured meat products (e.g., "longanizas" and Spanish dry-fermented sausage). However, when sodium isoascorbate or erythorbate is used in "longaniza" production, color stability is much reduced during the retail process (Ruíz-Peluffo et al. 1994).

The discoloration of white meats like turkey is characterized by a color changes, which go from pink/yellow to yellow/brown, while in veal/beef the changes go from purple to grayish/brown. In turkey, it has been demonstrated that the presence or absence of lipid oxidation depends on, among other things, the concentration of vitamin E in the tissues. The color and lipid oxidation are interrelated, since it has been seen that lipid oxidation in red and white muscle depends on the predominant form of catalyzing iron, Mb, or free iron (Mercier et al. 1998).

Compared with red meat, tuna flesh tends to undergo more rapid discoloration during refrigerated storage. Discoloration due to oxidation of Mb in red fish presented a problem, even at low temperature. This low color stability might be related to the lower activity or poorer stability of MMb reductase in tuna flesh (Ching et al. 2000). One of the reasons of this behavior is that aldehydes known to be produced during lipid oxidation can accelerate tuna OMb oxidation in vitro (Lee et al. 2003). Also, tuna flesh could be immersed in MMb reductase solution that could extend the color stability of tuna fish. Also, the use of this enzyme can reduce metMb formation during refrigerated storage of tuna (Tze et al. 2001).

Yellowtail (*Seriola quinqueradiata*) fillets stored in gas barrier film packs filled with N_2 and placed in cold storage at 0–5°C, stayed fresh for 4–7 days. N_2 or CO_2 packaging did not prevent discoloration in frozen tuna fillets; better results were achieved by thawing the frozen tuna meat in an O_2 atmosphere (Oka 1989). Packaging in atmospheres containing 4% or 9% O_2 was inferior to packaging in air, as they promoted MMb formation. Packaging in 70% O_2 maintained the fresh red color of tuna dorsal muscle for storage periods less than 3 days (Tanaka et al. 1996). In order to change the dark brown color into a bright red color, processors sometimes treat with 100% carbon monoxide (CO) during modified atmosphere packaging. This may result in high CO residues in the flesh and may cause health problems, since Mb can react with CO rapidly even at low CO concentration (Chi et al. 2001).

MICROORGANISMS AND COLOR

Although the real limiting factor in the shelf life of fresh meat is the microbial load, the consumer chooses it according to its color. The bacterial load is usually the most important cause of discoloration in fresh meat and meat products (sausages and other cooked products), so slaughter, cutting, and packaging must be strictly controlled.

Bacterial contamination decisively affects the biochemical mechanisms responsible for the deterioration of meat (Renerre 1990). An important aspect to take into account is that, just as with the bacterial load, the effect is more pronounced in meats that are more strongly pigmented (beef) than in less pigmented meats like pork and chicken (Gobantes and Oliver 2000).

Another variable affecting color stability in meat is the quantity of microorganisms present (Houben et al. 1998), a concentration in excess of 10^6 per gram having a strong effect. Although antioxidants, such as ascorbic acid, slow down lipid oxidation and consequently improves color stability, they (antioxidants) do not alter color changes caused by bacterial growth (Zerby et al. 1999).

REFERENCES

Adams DC, Huffman RT. 1972. Effect of controlled gas atmospheres and temperature on quality of packaged pork. *J Food Sci* 37: 869–872.

Adamsen CE et al. 2003. Studies on the antioxidative activity of red pigments in Italian-type dry-cured ham. *Eur Food Res Technol* 217: 201–206.

Ahn DU, Kim SM. 1998. Prooxidant effects of ferrous iron, hemoglobin, and ferritin in oil emulsion and cooked meat homogenates are different from those in raw-meat homogenates. *Poult Sci* 77: 348–355.

Aleson L et al. 2003. Utilization of lemon albedo in dry-cured sausages. *J Food Sci* 68: 1826–1830.

Ali MT et al. 1994. Activation mechanisms of pro-phenoloxidase on melanosis development in Florida spiny lobster (*Panulirus argus*) cuticle. *J Food Sci* 59(5): 1024–1030.

Alvarado CZ et al. 2007. The effect of blood removal on oxidation and shelf life of broiler breast meat. *Poultry Sci* 86: 156–161.

Anderson HJ et al. 1990. Color and color stability of hot processed frozen minced beef. Result from chemical model experiments tested under storage conditions. *Meat Sci* 28(2): 87–97.

Arihara K et al. 1995. Significance of metmyoglobin reducing enzyme system in myocytes. *Proc. 41th International Congress of Meat Sci and Technology. San Antonio (Texas)*. C70: 378–379.

Baron CP, Andersen HJ. 2002. Myoglobin-induced lipid oxidation: a review. *J Agric Food Chem* 50: 3887–3897.

Benner RA et al. 1994. Lactic acid/melanosis inhibitors to improve shelf life of brown shrimp (*Penaeus aztecus*). *J Food Sci* 59(2): 242–245.

Berry BW. 1997. Colour of cooked beef patties as influenced by formulation and final internal temperature. *Food Res Int* 30(7): 473–478.

Blanch A. 1999. Getting the colour of yolk and skin right. *World Poult* 15(9): 32–33.

Boyle RC et al. 1994. Quantitation of heme proteins from spectra of mixtures. *J Agric Food Chem* 42: 100–104.

Brewer MS, Mckeith FK. 1999. Consumer-rated quality characteristics as related to purchase intent of fresh pork. *J Food Sci* 64: 171–174.

Bryhni EA et al. 2002. Consumer perceptions of pork in Denmark, Norway and Sweden. *Food Qual Pref* 13: 257–266.

Cassens RG et al. 1995. Recommendation of reference method for assessment of meat color. *Proc 41th Int Congr Meat Sci Technol*, San Antonio, TX, C86: 410–411.

Chau JC et al. 1997. Characteristics of reaction between carbon monoxide gas and myoglobin in tuna flesh. *J Food Drug Anal* 5(3): 199–206.

Chi CY et al. 2001. Effect of carbon monoxide treatment applied to tilapia fillet. *Taiwanese J Agric Chem Food Sci* 39(2): 117–121.

Ching YP et al. 2000. Effect of polyethylene package on the metmyoglobin reductase activity and color of tuna muscle during low temperature storage. *Fish Sci* 66(2): 384–389.

Chow CJ. 1991. Relationship between the stability and autoxidation of myoglobin. *J Agric Food Chem* 39(1): 22–26.

Chow CJ et al. 1989. Reduced stability and accelerated autoxidation of tuna myoglobin in association with freezing and thawing. *J Agric Food Chem* 37(5): 1391–1395.

Conforth D. 1994. Colour–its basis and importance. In: AM Pearson, TR Dutson (eds.) *Advances in Meat Research*, vol 9. Chapman and Hall, London, pp. 34–78.

Conforth DP et al. 1986. Role of reduced haemochromes in pink colour defect of cooked turkey. *J Food Sci* 51: 1132–1135.

Conforth DP et al. 1991. Methods for identification and prevention of pink color in cooked meats. *ProcReci Meat Conf* 44: 53–58.

Davidsson I, Henry JB. 1978. *Diagnóstico clínico por el laboratorio*. Salvat, Barcelona, pp. 310–316.

Diestre A. 1992. Principales problemas de la calidad de la carne en el porcino. *Alim Eq Tecnol* 98: 73–78.

Deba MS et al. 2007. Effect of tomato paste and nitrite level on processing and quality characteristics of frankfurters. *Meat Sci* 76(3): 501–508.

Diez AM et al. 2009. Effectiveness of combined preservation methods to extend the shelf life of Morcilla de Burgos. *Meat Sci* 81: 171–177.

Encarnacion AB et al. 2010. Effects of ergothioneine from mushrooms (*Flammulina velutipes*) on melanosis and lipid oxidation of kuruma shrimp (*Marsupenaeus Japonicus*). *J Agric Food Chem* 58: 2577–2585.

Esteve E. 1994. Alimentación animal y calidad de la carne. *Eurocarne* 31: 71–77.

Faustman C et al. 1990. Color reversion in beef: influence of psychotrophic bacteria. *Fleischwirt* 70: 676.

Faustman C et al. 1996. Strategies for increasing oxidative stability of (fresh) meat color. *Proc. 49th Annu Reci Meat Conf*, pp. 73–78, Provo, Utah.

Fernández-Ginés JM et al. 2003. Effect of storage conditions on quality characteristics of bologna sausages made with citrus fiber. *J Food Sci* 68: 710–715.

Fernández-Ginés JM et al. 2004. Lemon albedo as a new source of dietary fiber: application to bologna sausages. *Meat Sci* 67: 7–13.

Férnández-López J. 1998. Estudio del color por métodos objetivos en sistemas modelo de pastas de embutidos crudo-curados. [PhD Thesis]. Murcia, Spain: Universidad de Murcia, p. 310.

Fernández-López J et al. 2000. Effect on mincing degree on color properties in pork meat. *Color Res Appl* 25: 376–380.

Fernández-López J et al. 2002. Effect of paprika (Capsicum annum) on color of spanish-type sausages during the resting stage. *J Food Sci* 67: 2410–2414.

Fernández-López J et al. 2003a. Physical, chemical and sensory properties of bologna sausage made with ostrich meat. *J Food Sci* 68: 1511–1515.

Fernández-López J et al. 2003b. Evaluation of antioxidant potential of hyssop (*Hyssopus officinalis L.*) and rosemay (*Rosmarinus officinalis L*) extract in cooked pork meat. *J Food Sci* 68: 660–664.

Fernández-López J et al. 2007. Orange fibre as potential functional ingredient for dry-cured sausage. *Eur Food Res Technol* 226: 1–6.

Fernández-López J et al. 2008a. Effect of packaging conditions on shelf-life of ostrich steaks. *Meat Sci* 78: 143–152.

Fernández-López J et al. 2008b. Physico-chemical and microbiological profiles of "Salchichón" (Spanish dry fermented sausage) enriched with orange fiber. *Meat Sci* 80: 410–417.

Fernández-López J et al. 2009. Storage stability of a high dietary fibre powder from orange by-products. *Int J Food Sci Technol* 44(4): 748–756.

Ferrer OJ et al. 1989. Effect of bisulfite on lobster shell phenoloxidase. *J Food Sci* 54(2): 478–480.

Fontes PR et al. 2004. Color evaluation of carbon monoxide treated porcine blood. *Meat Sci* 68: 507–513.

Fontes PR et al. 2010. Composition and color stability of carbon monoxide treated dried porcine blood. *Meat Sci* Forthcoming.

Fox JB. 1966. The chemistry of meat pigments. *J Agric Food Chem* 14: 207–210.

Froning GW et al. 1968. Color and texture of ground turkey meat products as affected by dry egg white solids. *Poult Sci* 47: 1187.

Gallego-Restrepo JA et al. 2010. Use of starter cultures and alternative sources of sodium nitrite for processing of Medellin-type cooked ham: influence on ph and colour. IPOA research group internal report. Miguel Hernández University, Orihuela, Spain.

Garrote G et al. 2004. Antioxidant activity of by-products from the hydrolytic processing of selected lignocellulosic materials. *Trends Food Sci Technol* 15: 191–200.

Gill CO. 1996. Extending the storage life of raw chilled meats. *Meat Sci* 43(Suppl): S99–S109.

Gobantes IY, Oliver MA. 2000. Problemática de la estabilidad del color en carne de vacuno: la adición de vitamina E y el envasado bajo atmósfera protectora. *Eurocarne* 85: 57–62.

Gomez-Basauri JV, Regenstein JM. 1992a. Processing and frozen storage effects on the iron content of cod and mackerel. *J Food Sci* 57: 1332–1336.

Gomez-Basauri JV, Regenstein JM. 1992b. Vacuum packaging, ascorbic acid and frozen storage effects on heme and non-heme iron content of mackerel. *J Food Sci* 57: 1337–1339.

Gøtterup J et al. 2007. Relationship between nitrate/nitrite reductase activities in meat associated staphylococci and nitrosylmyoglobin formation in a cured meat model system. *Int J Food Microbiol* 120(3): 303–310.

Gøtterup J et al. 2008. Colour formation in fermented sausages by meat-associated staphylococci with different nitrite and nitrate-reductase activities. *Meat Sci* 78: 492–501.

Grunwald EW, Richards MP. 2006. Mechanisms of heme protein-mediated lipid oxidation using hemoglobin and myoglobin variants in raw and heated washed muscle. *J Agric Food Chem* 54: 8271–8280.

Han D et al. 1994. Hemoglobin, myoglobin, and total pigments in beef and chicken muscles chromatographic determination. *J Food Sci* 59: 1279–1282.

Harel S et al. 1988. Iron release from metmyoglobin, methaemoglobin and cytochrome c by a system generating hydrogen peroxide. *Free Rad Res Comm* 5: 11–19.

Houben JH et al. 1998. Effect of the dietary supplementation with vitamin E on colour stability and lipid oxidation in packaged, minced pork. *Meat Sci* 48(3/4): 265–273.

Huanf Z et al. 2001. Nitric oxide binding to oxygenated hemoglobin under physiological conditions. *Biochim Biophys Acta* 1568: 252–260.

Hui HC et al. 1998. Thermal stability of myoglobin and color in horse mackerel mince. *Food Science Taiwan* 25(4): 419–427.

Hunt JR, Roughead ZK. 2000. Adaption of iron absorption in men consuming diets with high or low iron bioavailability. *Am J Clin Nut* 71: 94–102.

Hunt MC, Kropf DH. 1987. Colour and appearance. In: AM Pearson, TR Dutson (eds.) *Restructured Meat and Poultry Products, Advances in Meat Research Series*, vol 3. Van Nostran Reinhold, New York, pp. 125–159.

Hutchings JB. 2002. The perception and sensory assessment of colour. In: DB MacDougall (ed.) *Colour in Food, Improving Quality*. Woodhead Publishing, Cambridge, pp. 9–32.

Irie M. 2001. Optical evaluation of factors affecting appearance of bovine fat. *Meat Sci* 57: 19–22.

Irie M et al. 1999. Changes in meta color and alpha-tocopherol concentrations in plasma and tissues from Japanese beef cattle fed by two methods of vitamin E supplementation. *Asian Aust J Animal Sci* 11: 810–814.

Jaffe ER. 1981. Methaemoglobinaemia. *Clin Hematol* 10: 99–122.

Jayasingh P et al. 2001. Evaluation of carbon monoxide treatment in modified atmosphere packaging or vacuum packaging to increase color stability of fresh beef. *Meat Sci* 59: 317–324.

Jensen C et al. 1998. Dietary vitamin E: quality and storage stability of pork and poultry. *Trends Food Sci Technol* 9: 62–72.

Kalyanasundaram K. 1992. *Photochemistry of Polypyridine and Porphyrin Complexes*. Academic Press, New York, pp. 25–56.

Kanner J. 1994. Oxidative processes in meat products: quality implications. *Meat Sci* 36: 169–189.
Kanner J et al. 1991. Catalytic ‘free’ iron ions in muscle foods. *J Agric Food Chem* 36: 412–415.
Karolyi D. 2003. Cured meat and consumer health. *Meso* 5: 16–18.
Knipe L. 1993. *Basic Science of Meat Processing. Cured Meat Short Course*. Meat Laboratory, Iowa State University, Ames, IA.
Krammer A. 1994. Use of color measurements in quality control of food. *Food Technol* 48(10): 62–71.
Kristensen L, Purslow PP. 2001. The effect of processing temperature and addition of mono- and di-valent salts on the heme nonheme-iron ratio in meat. *Food Chem* 73: 433–439.
Lanari MC, Cassens RG. 1991. Mitochondrial activity and beef muscle color stability. *J Food Sci* 56: 1476–1479.
Lanari MC et al. 1995. Dietary vitamin E supplementation and discoloration of pork bone and muscle following modified atmosphere packaging. *Meat Sci* 41(3): 237–250.
Lanari MC et al. 2002. Pasture and grain finishing affect the color stability of beef. *J Food Sci* 67: 2467–2473.
Ledward DA. 1984. Haemoproteins in meat and meat products. In: DA Ledward (ed.) *Developments in Food Proteins*. Applied Science, London, pp. 33–68.
Ledward DA. 1992. Colour of raw and cooked meat. In: DA Ledward el al. (eds.) *The Chemistry of Muscle Based Foods*. The Royal Society of Chemistry, Cambridge, pp. 128–144.
Lee BJ et al. 1998. Antioxidant effect of carnosine and phytic acid in a model beef system. *J Food Sci* 63: 394–398.
Lee S et al. 2003. Oxymyoglobin and lipid oxidation in yellowfin tuna (*Thunnus albacares*)loins. *J Food Sci* 68(5): 1164–1168.
Lehninger AL. 1981. *Bioquímica. Las bases Moleculares de la Estructura y Dunción Celular*. Omega, Barcelona, pp. 230–233.
Lien R et al. 2002. Effects of endpoint temperature on the internal color of pork patties of different myoglobin form, initial cooking state, and quality. *J Food Sci* 67(3): 1011–1015.
Little AC et al. 1979. Color assessment of experimentally pigmented rainbow trout. *Col Res Appl* 4(2): 92–95.
Lombardi-Boccia G et al. 2002. Total, heme and non-heme iron in raw and cooked meats. *J Food Sci* 67: 1738–1741.
Lorient D. 1982. Propiedades funcionales de las proteínas de origen animal. In: CM Bourgeois, P Le Roux (eds.) *Proteínas Animales*. El Manual Moderno, México, pp. 215–225.
Lozano D. 2006. A new approach to appearance characterization. *Color Res Appl* 31(3): 164–167.
MacDougall DB. 1982. Changes in the colour and opacity of meat. *Food Chem* 9(1/2): 75–88.
Mancini RA et al. 2003. Reflectance at 610 nm estimates oxymyoglobin content on the surface of ground beef. *Meat Sci* 64: 157–162.
Mancini RA et al. 2004 Ascorbic acid minimizes lumbar vertebrae discoloration. *Meat Sci* 68(3): 339–345.
Mancini RA et al. 2010. Effects of lactate and modified atmospheric packaging on premature browning in cooked ground beef patties. *Meat Sci* 85: 339–346.
Mercier Y et al. 1998. Effect of dietary fat and vitamin E on colour stability and on lipid and protein oxidation in turkey meat during storage. *Meat Sci* 48(3/4): 301–318.
Millar SJ et al. 1996. Some observations on the absorption spectra of various myoglobin derivatives found in meat. *Meat Sci* 42(3): 277–288.
Miller DK et al. 1994a. Dietary iron in swine rations affects nonheme iron and TBARS in pork skeletal muscles. *J Food Sci* 59: 747–749.
Miller DK et al. 1994b. Lipid oxidation and warmed-over aroma in cooked ground pork from swine fed increasing levels of iron. *J Food Sci* 59: 751–756.
Min B, Ahn DU. 2005. Mechanism of lipid peroxidation in meat and meat products—a review. *Food Sci Biotech* 14: 152–163.
Minguéz-Mosquera MI. 1997. Clorofilas y carotenoides en tecnología de alimentos. Sevilla: Secretariado de publicaciones de la Universidad de Sevilla, pp. 15–17.
Moller JKS et al. 2003. Spectral characterisation of red pigment in Italian-type dry-cured ham. Increasing lipophilicity during processing and maturation. *Eur Food Res Technol* 216(4): 290–296.
Montero P et al. 2001. Characterization of polyphenoloxidase of prawns (*Penaeus japonicus*). Alternatives to inhibition: additives and high pressure treatment. *Food Chem* 75: 317–324.
Moss BW. 1992. Lean meat, animal welfare and meat quality. In: DA Ledward el al. (eds.) *The Chemistry of Muscle Based Foods*. The Royal Society of Chemistry, Cambridge, pp. 62–76.
Nagababu E, Rifkind JM. 2000. Reaction of hydrogen peroxide with ferrylhemoglobin: superoxide production and heme degradation. *Biochem* 9: 12503–12511.
Nash DM et al. 1985. Pink discoloration in broiler chicken. *Poult Sci* 64: 917–919.
Neubauer H, Götz F. 1996. Physiology and interaction of nitrate and nitrite reduction in *Staphylococcus carnosus*. *J Bacteriol* 178: 2005–2009.
Ngoka DA, Froning GW. 1982. Effect of free struggle and preslaughter excitement on color of turkey breast muscles. *Poult Sci* 61: 2291–2293.
Niewiarowicz A et al. 1986. Myoglobin and haemoglobin content in the meat of different types of poultry. *Fleischw* 66: 1281–1282.
Numata M, Wakamatsu J. 2003. Natural red pigments and foods and food materials containing the pigments. Patent. WO 03/063615 A1.
Ogawa M et al. 1984. On physiological aspects of black spot appearance in shrimp. *Bull Jap Soc Sci Fish* 50(10): 1763–1769.
Ohshima T et al. 1988. Influences of haeme pigment, non-haeme iron, and nitrite on lipid oxidation in cooked mackerel meat. *Bull Jap Soc Sci Fish* 54(12): 2165–2171.
Oka H. 1989. Packaging for freshness and the prevention of discoloration of fish fillets. *Pack Technol Sci* 2(4): 201–213.
Osborn HM et al. 2003. High temperature reduction of metmyoglobin in aqueous muscle extracts. *Meat Sci* 65: 631–637.
Pagán-Moreno MJ et al. 1998. The evolution of colour parameters during "chorizo" processing. *Fleischwt* 78(9): 987–989.
Park JW, Morrisey MT. 1994. The need for developing the surimi standar. In: G Sylvia et al. (eds.) *Quality Control and Quality Assurance Seafood*. Oregon Sea Grant, Corvallis, OR, p. 265.
Parks S et al. 2008. Nitrate and nitrite in Australian leafy vegetables. *Austral J Agric Res* 59: 632–638.
Parolari G et al. 2003. Extraction properties and absorption spectra of dry cured hams made with and without nitrate. *Meat Sci* 64(4): 483–490.
Pérez-Alvarez JA. 1996. Contribución al estudio objetivo del color en productos cárnicos crudo-curados. PhD Thesis. Universidad Politécnica de Valencia, Valencia, Spain, p. 210.

Pérez-Alvarez JA. 2006. Aspectos tecnológicos de los productos crudo-curados. In: YH Hui et al. (eds.) *Ciencia y Tecnología de Carnes*. Limusa, Mexico D.F., pp. 463–492.

Pérez-Alvarez JA. 2008. Overview of meat products as functional foods. In: J Fernández-López, JA Pérez-Alvarez (eds.) *Technological Strategies for Functional Meat Products Development*. Transworld Research, Kerala, pp. 1–17.

Pérez-Alvarez JA, Fernández-López J. 2009a. Color characteristics of meat and poultry processing. In: *Handbook of processed meats and poultry analysis*. CRC Press, Boca Raton, FL, pp. 355–373.

Pérez-Alvarez JA, Fernández-López J. 2009b. Colour of meat products from a food engineering perspective. *Alim Eq Tecnol* 247: 34–37.

Pérez-Alvarez JA et al. 1993. Estudio del color durante la elaboración de paté de hígado de cerdo. *Cárnicas 2000* IV: 429–444.

Pérez-Alvarez JA et al. 1997. Chemical and color characteristics of "Lomo embuchado" during salting seasoning. *J Muscle Food* 8(4): 395–411.

Pérez-Alvarez JA et al. 1998. Utilización de vísceras como materias primas en la elaboración de productos cárnicos: güeña. *Alimentaria* 291: 63–70.

Pérez-Alvarez JA et al. 2000. Fundamentos físicos, químicos, ultraestructurales y tecnológicos en el color de la carne. In: M Rosmini et al. (eds.) *Nuevas Tendencias en la Tecnología e Higiene de la Industria Cárnica*. Miguel Hernández de Elche, Elche, pp. 51–71.

Pikul J et al. 1986. The cytochrome c content of various poultry meats. *J Sci Food Agric* 37: 1236–1240.

Raines CR, Hunt MC. 2010. Headspace volume and percentage of carbon monoxide affects carboxymyoglobin layer development of modified atmosphere packaged beef steaks. *J Food Sci* 75: C62–C65.

Rehbein H, Orlick B. 1990. Comparison of the contribution of formaldehyde and lipid oxidation products to protein denaturation and texture deterioration during frozen storage of minced icefish fillet (*Champsocephalus gunnari* and *Pseudochaenichthys georgianus*). *Int J Ref* 13(5): 336–341.

Renerre M. 1990. Review: factors involved in discoloration of beef meat. *Int J Food Sci Technol* 25: 613–630.

Richards MP, Hultin HO. 2002. Contributions of blood and blood components to lipid oxidation in fish muscle. *J Agric Food Chem* 50(3): 555–564.

Richards MP et al. 2005. Pro-oxidative characteristics of trout hemoglobin and myoglobin: a role for released heme in oxidation of lipids. *J Agric Food Chem* 53: 10231–10238.

Ruíz-Peluffo MC et al. 1994. Longaniza de pascua: influencia del ascorbato e isoascorbato de sodio sobre las propiedades fisicoquímicas y de color. In: P Fito et al. (eds.) *Anales de Investigación del Master en Ciencia e Ingeniería de Alimentos*, Vol. VI. Reproval, Valencia, pp. 675–688.

Sakiura T. 2001. Cultured fish carotenoid and polyphenol added feed for improving fish body color tone and fish meat brilliance. U.S. Patent 2001/0043982 A1.

Saksit C et al. 1998. Changes in light scattering intensity of fish holo-, and apo- and reconstituted myoglobins under thermal denaturation. *Fish Sci* 64(5): 846–847.

Saksit C et al. 1996. Studies on thermal denaturation of fish myoglobins using differential scanning calorimetry, circular dichroism, and tryptophan fluorescence. *Fish Sci* 62(6): 927–932.

Sánchez-Escalante A et al. 2003. Antioxidant action of borage, rosemary, oregano and ascorbic acid in beef patties packaged in modified atmospheres. *J Food Sci* 68: 339–344.

Sánchez-Zapata E et al. 2009a. Effect of orange fibre on the colour and heme iron content of paté elaborated with dark muscle ("sangacho") of bluefin tuna (*Thunnus thynnus*). *EULAFF/CYTED International Functional Foods Conference, Porto, Portugal*, p. 83.

Sánchez-Zapata EJ et al. 2009b. Non-meat ingredients. In: I Guerrero-Legarreta, YH Hui (eds.) *Handbook of Poultry Science and Technology*, vol. 2, Chapter 10. John Wiley & Sons, London, pp. 101–123.

Sayas ME. 1997. Contribuciones al proceso tecnológico de elaboración del jamón curado: aspectos físicos, fisicoquímicos y ultraestructurales en los procesos de curado tradicional y rápido. PhD Thesis. Universidad Politécnica de Valencia, Valencia, Spain, p. 175.

Sebranek J, Bacus J. 2007a. Cured meat products without direct addition of nitrate or nitrite: what are the issues? *Meat Sci* 77: 136–145.

Sebranek J, Bacus J. 2007b. Natural and organic cured meat products: regulatory, manufacturing, marketing, quality and safety issues. AMSA. White Paper Series. Number 1.

Schricker BR, Miller DD. 1983. Effects of cooking and chemical treatment on heme and nonheme iron in meat. *J Food Sci* 48: 1340–1343.

Schricker BR et al. 1982. Measurement and content of nonheme and total iron in muscle. *J Food Sci* 47: 470–473.

Scriven F et al. 1987. Investigations of nitrite and nitrate levels in paper materials used to package fresh meats. *J Agric Food Chem* 35: 188–192.

Shahidi F, Matusalach-Brown JA. 1998. Carotenoid pigments in seafoods and aquaculture. *Crit Rev Food Sci* 38(1): 1–67.

Shahid-Umar A, Iqbal M. 2007. Nitrate accumulation in plants, factors affecting the process, and human health implications: a review. *Agron Sustain Develop* 27: 45–57.

Shahid-Umar A et al. 2007. Are nitrate concentrations in leafy vegetables within safe limits? *Current Sci* 92(3): 355–360.

Slattery SL et al. 1995. A sulphite-free treatment inhibits blackspot formation in prawns. *Food-Australia* 47(11): 509–514.

Stauton-West E et al. 1969. *Bioquímica Médica*. Interamericana, México, pp. 616–623.

Stiebing A. 1990. Blood sausage technology. *Fleischw Int* 3: 34–40.

Suman SP et al. 2004. Effect of muscle source on premature browning in ground beef. *Meat Sci* 68(3): 457–461.

Suman SP et al. 2010. Color-stabilizing effect of lactate on ground beef is packaging-dependent. *Meat Sci* 84: 329–333.

Swatland HJ. 1988. Carotene reflectance and the yellowness of bovine adipose tissue measured with a portable fiber-optic spectrophotometer. *J Sci Food Agric* 46: 195–200.

Swatland HJ. 1995. Reversible pH effect on pork paleness in a model system. *J Food Sci* 60(5): 988–991.

Tanaka M et al. 1996. Gas-exchanged packaging of tuna fillets. *Nip Sui Gak* 62(5): 800–805.

Taoukis PS et al. 1990. Inhibition of shrimp melanosis (black spot) by ficin. *Lebens Wis Technol* 23(1): 52–54.

Tze KC et al. 2001. Effect of met-myoglobin reductase on the color stability of blue fin tuna during refrigerated storage. *Fish Sci* 67(4): 694–702.

Umbreit J. 2007. Methemoglobin—it's not just blue: a concise review. *Am J Hematol* 82: 134–144.

Urbain MW. 1952. Oxygen is key to the color of meat. *Nat Prov* 127: 140–141.

Vaughin MW et al. 2000. Erythrocytes posses an intrinsic barrier to nitric oxide consumption. *J Biol Chem* 275: 2342–2348.

Verdoes JC et al. 1999. Isolation and functional characterisation of a novel type of carotenoid biosynthetic gene from Xanthophyllomyces dendrorhous. *Mol Gen Genet* 262(3): 453–461.

Viuda-Martos M et al. 2009a. Citrus co-products as technological strategy to reduce residual nitrite content in meat products. *J Food Sci* 74: R93–R100.

Viuda-Martos M et al. 2009b. Physico-chemical characterisation of the orange juice wastewater of a citrus co-product. *J Food Process Preserv* 8: R93–R100.

Viuda-Martos M et al. 2010a. Effect of adding citrus fibre washing water and rosemary essential oil on the quality characteristics of a bologna sausage. LWT. *Food Sci Technol* 43: 958–963.

Viuda-Martos M et al. 2010b. Effect of added citrus fibre and spice essential oils on quality characteristics and shelf-life of Mortadella. *Meat Sci* 85(3): 568–576.

Wakamatsu J et al. 2004a. A Zn-porphyrin complex contributes to bright red colour in Parma ham. *Meat Sci* 67(1): 95–100.

Wakamatsu J et al. 2004b. Establishment of a model experiment system to elucidate the mechanism by which Zn-protoporphyrin IX is formed in nitrite-free dry-cured ham. *Meat Sci* 68(2): 313–317.

Walsh KA, Rose D. 1956. Factors affecting the oxidation of nitric oxide myoglobin. *J Agric Food Chem* 4(4): 352–355.

Warren KE et al. 1996. Chemical properties of ground beef patties exhibiting normal and premature brown internal cooked color. *J Muscle Foods* 7(3): 303–314.

Warris PD, Rhodes DN. 1977. Haemoglobin concentrations in beef. *J Sci Food Agric* 28: 931–934.

Wei F et al. 2009. Irradiated Chinese Rugao ham: changes in volatile N-nitrosamine, biogenic amine and residual nitrite during ripening and post-ripening. *Meat Sci* 81: 451–455.

Whitaker JR. 1972. *Principles of Enzymology for the Food Sciences*. Marcel Decker, New York, pp. 225–250.

Zanardi E et al. 1998. Oxidative stability and dietary treatment with vitamin E, oleic acid and copper of fresh and cooked pork chops. *Meat Sci* 49(3): 309–320.

Zerby HN et al. 1999. Display life of fresh beef containing different levels of vitamin E and initial microbial contamination. *J Muscle Foods* 10: 345–355.

Zhou GH et al. 1993. A relationship between bovine fat colour and fatty acid composition. *Meat Sci* 35(2): 205–212.

Zhu NH et al. 2009. Effects of Lactobacillus cultures on growth performance, xanthophyll deposition, and color of the meat and skin of broilers. *J App Poultry Res* 18: 570–578.

18
Biochemistry of Fermented Meat

Fidel Toldrá

Abstract: A wide variety of fermented sausages are produced, the type and composition depending on the raw materials, microbial population, and processing conditions. Undry and semidry sausages are fermented to reach low pH values, and are usually smoked and cooked before consumption. Dry-fermented sausages experience a long drying after fermentation reaching low water activity values. This chapter describes the raw ingredients including microbial starters used for the manufacture of fermented sausages as well as its main processing stages. The physical and chemical changes that take place during processing are also described.

BACKGROUND INFORMATION

The origin of fermented meats is quite far away in time. Ancient Romans and Greeks already manufactured fermented sausages, and, in fact, the origin of words like sausage and salami might have come from the Latin expressions *salsiccia* and *salumen*, respectively (Toldrá 2002). The production and consumption of fermented meats expanded throughout Europe in the Middle Ages, being adapted to climatic conditions (i.e., smoked in Northern Europe and dried in Mediterranean countries). The experience in manufacturing these meats went to America with settlers (e.g., states like Wisconsin still have a good number of typical Northern European sausages like Norwegian and German sausages).

Today, a wide variety of fermented sausages are produced, depending on the raw materials, microbial population, and processing conditions (Toldrá et al. 2007). For instance, Northern European-type sausages contain beef and pork as raw meats, are ripened for short periods (up to 3 weeks), and are usually subjected to smoking. In these sausages, shelf-life is mainly due to acid pH and smoking rather than drying. On the other hand, Mediterranean sausages mostly use only pork and are ripened for longer periods (several weeks or even months), and smoke is not so typically applied (Flores and Toldrá 1993). Examples for different types of fermented sausages, according to the intensity of drying, are shown in Table 18.1. Undry and semidry sausages are fermented to reach low pH values, and are usually smoked and cooked before consumption. Shelf-life and safety is mostly determined by pH drop and reduced water activity (a_w) as a consequence of fermentation and drying, respectively. The product may be considered stable at room temperature when pH < 5.0 and the moisture:protein ratio is below 3.1:1 (Sebranek 2004). Moisture:protein ratios are defined for the different dry and semidry fermented sausages in the Unites States, while a_w values are preferred in Europe.

Food Biochemistry and Food Processing, Second Edition. Edited by Benjamin K. Simpson, Leo M.L. Nollet, Fidel Toldrá, Soottawat Benjakul, Gopinadhan Paliyath and Y.H. Hui.

Table 18.1. Examples of Fermented Meats with Different Dryness Degrees

Product	Type	Examples	Weight Loss (%)	Drying/Ripening
Undry fermented sausages	Spreadable	German teewurst	<10	No drying
		Frische mettwurst	<10	No drying
Semidry fermented sausages	Sliceable	Summer sausage	<20	Short
		Lebanon bologna	<20	Short
		Saucisson dáAlsace	<20	Short
		Chinese Laap cháeung	<20	Short
		Chinese Xunchang	<20	Short
Dry fermented sausages	Sliceable	Hungarian and Italian salami	>30	Long
		Pepperoni	>30	Long
		Spanish chorizo	>30	Long
		Spanish salchichón	>30	Long
		French Saucisson	>30	Long

Source: Lücke 1985, Roca and Incze 1990, Campbell-Platt 1995, Toldrá 2002.

RAW MATERIAL PREPARATION

There are several considerations (listed in Table 18.2) that need to be taken into account when producing fermented meats. The selection of the different options, which are discussed in the later sections, facilitates the choice of the most adequate conditions for the correct processing, safety, and optimal final quality.

INGREDIENTS

Lean meats from pork and beef, in equal amounts, or only pork are generally used. Quality characteristics such as color, pH (preferably < 5.8), and water-holding capacity are very important. When the pH of pork meat is >6.0, the meat is known as DFD (dark, firm, and dry). This type of meat binds water tightly and spoils easily. Pork meat with another defect, known as PSE (pale, soft, and exudative), is not recommended because the color is pale, and the sausage would release water too fast, which could cause casings to wrinkle. Meat from older animals is preferred because of its more intense color, which is due to the accumulation of myoglobin, a sarcoplasmic protein that is the natural pigment responsible for color in meat (Toldrá and Reig 2007).

Pork back and belly fats constitute the main source for fats. Special care must be taken for the polyunsaturated fatty acid

Table 18.2. Some Decisions to Adopt and Options to Choose in the Processing of Fermented Meats

Aspects	Options
Type of meat	Pork, beef, etc.
Quality of meat	Choose good quality. Reject defective meats (pork PSE and DFD), abnormal colors, exudation, etc.
Origin of fat	Choose either chilled, or frozen (how long?) fats. Reject oxidized fats.
Type of fat	Control of fatty acids profile (excess of PUFA?).
Ratio	Choose desired meat:fat ratio.
Particle size	Choose adequate plate (grinder) or speeds (cutter).
Additives: Salt	Decide concentration.
Additives: Curing agent	Nitrite or nitrate depending on type and length of process.
Additives: Carbohydrates	Type and concentration depending on type of process and required pH drop.
Spices	Choose according to required specific flavor.
Microflora	Natural or added as starter?
Starters	Choose microorganisms depending on type of process and product.
Casing	Material and diameter depending on type of product.
Fermentation	Conditions depending on type of starter used and product.
Ripening/drying	Conditions depending on type of product.
Smoking	Optional application. Conditions depending on type of product and specific flavor.
Color	Depends on raw meat, nitrite and processing conditions.
Texture	Depends on meat:fat ratio, stuffing pressure and extent of drying.
Flavor	Choose adequate starter and process conditions.
Water activity	Depends on drying conditions and length of process.

PSE, pale, soft, and exudative; DFD, dark, firm, and dry; PUFA, polyunsaturated fatty acids.

profile, which should be lower than 12%; and the level of oxidation, measured as peroxide value, should be as low as possible (Demeyer 1992). Some rancidity may develop after long-term frozen storage since lipases present in adipose tissue are active even at temperatures as low as −18°C and are responsible for the continuous release of free fatty acids that are susceptible to oxidation (Hernández et al. 1999). So, extreme caution must be taken with fats stored for several months as they may develop a rancid flavor.

Other Ingredients and Additives

Salt is the oldest additive used in cured meat products since ancient times. Salt, at about 2–4%, serves several functions, including (1) an initial reduction in a_w, (2) providing a characteristic salty taste, and (3) contributing to increased solubility of myofibrillar proteins. Nitrite is a typical curing agent used as a preservative against pathogens, especially *Clostridium botulinum*. Nitrite is also responsible for the development of the typical cured meat color, prevention of oxidation, and contribution to the cured meat flavor (Gray and Pearson 1984). The reduction of nitrite to nitric oxide is favored by the presence of ascorbic and erythorbic acids or their sodium salts. They also exert antioxidative action and inhibit the formation of nitrosamines.

Carbohydrates like glucose and lactose are used quite often as substrates for microbial growth and development. Disaccharides, and especially polysaccharides, may delay the growth and pH drop rate because they have to be hydrolyzed to monosaccharides by microorganisms.

Sometimes, additional substances may be used for specific purposes (Demeyer and Toldrá 2004). This is the case of glucono-delta-lactone, added at 0.5%, which may simulate bacterial acidulation. In the presence of water, glucono-delta-lactone is hydrolyzed to gluconic acid and produces a rapid decrease in pH. The quality is rather poor because the rapid pH drop drastically reduces the activity of flavor-related enzymes such as exopeptidases and lipases. Phosphates may be added to improve stability against oxidation; vegetable proteins, such as soy isolates, to replace meat proteins; and manganese sulfate as a cofactor for lactic acid bacteria.

Spices, either in natural form or as extracts, are added to give a characteristic aroma or color to the fermented sausage. There are a wide variety of spices (pepper, paprika, oregano, rosemary, garlic, onion, etc.), each one giving a particular aroma to the product. Some spices also contain powerful antioxidants. The most important aromatic volatile compounds may vary depending on the geographical and/or plant origin. For instance, garlic, which gives a pungent and penetrating smell, is typically used in chorizo, and pepper is used in salchichón and salami. Paprika gives a characteristic flavor and color due to its high content of carotenoids (Ordoñez et al. 1999). The presence of manganese in some spices, like red pepper and mustard, is necessary for the activity of several enzymes involved in glycolysis and thus enhances the generation of lactic acid (Lücke 1985).

Starters

Typical fermented products were initially based on the development and growth of desirable indigenous flora, sometimes reinforced with backslopping, which consists in the addition of a previous ripened, fermented sausage with adequate sensory properties. However, this practice usually gave a high heterogeneity in the product quality. The microbiota in spontaneously fermented sausages have been identified in many different types of sausages worldwide. The most frequent microorganisms found were *L. curvatus, Lb. plantarum, Lb. sakei, St. carnosus, St. xylosus*, and *St. saprophyticus* (Paramithitis et al. 2010).

The use of microbial starters, as a way to standardize processing as well as quality and safety, is relatively new. In fact, the first commercial use was in the United States in the 1950s, followed by its use in Europe in the 1960s; since then, starters have seen extensive use. Today, most of the fermented sausages are produced with a combination of lactic acid bacteria, to get an adequate acidulation, and two or more cultures to develop flavor and facilitate other reactions, such as nitrate reduction.

In general, microorganisms used as starter cultures must satisfy several requirements according to the purposes of their use: nontoxicity for humans, good stability under the processing conditions (resistance to acid pH, low a_w, tolerance to salt, resistance to phage infections), intense growth at the fermentation temperature (i.e., 18–25°C in Europe or 35–40°C in the United States), generation of products with technological interest (i.e., lactic acid for pH drop, volatile compounds for aroma, nitrate reduction, secretion of bacteriocins, etc.), and lack of undesirable enzymes (e.g., decarboxylases responsible for amine generation). Thus, the most adequate strains must be carefully selected and controlled as they will have a very important role in the process and will be decisive for the final quality. The most important microorganisms used as starters belong to one of the following groups: lactic acid bacteria, coagulase-negative staphylococci, Micrococcaceae, yeasts, or molds (Leistner 1992). These microorganisms provide a good number of enzymes involved in relevant biochemical changes like the enzymatic breakdown of carbohydrates, proteins, and lipids that affect color, flavor, and texture (Flores and Toldrá 2010). It must be taken into account that these enzymes will be affected by the temperature of fermentation and ripening, a_w and pH decrease during the process, the length of ripening time, the content of salt, nitrate and nitrite, the type and content of carbohydrates, and the presence of spices (Stahnke and Tjener 2007). The main roles and functions for each group are shown in Table 18.3.

Lactic Acid Bacteria

The most important function of lactic acid bacteria consists in the generation of lactic acid from glucose or other carbohydrates through either homo- or heterofermentative pathway. The accumulation of lactic acid produces a pH drop in the sausage. However, some undesirable secondary products, like acetic acid, hydrogen peroxide, acetoin, etc., may be generated in case of certain species having heterofermentative pathways. *L. sakei* and *L. curvatus* grow at mild temperatures and are usual in

Table 18.3. Main Roles and Effects of Microorganisms in Fermented Meats

Group	Microorganisms	Primary Role	Secondary Role	Chemical Contribution	Effects
Lactic acid bacteria	*L. sakei, L. curvatus, L. plantarum, L. pentosus, P. acidilactici, P. pentosaceus*	Glycolysis	Proteolysis (endo and exo)	Lactic acid generation Generation of free amino acids	pH drop Safety Firmness Taste
Micrococcaceae	*K. varians, S. Xylosus, S. carnosus*	Nitrate reductase, lipolysis, Catalase	Proteolysis (exo)	Reduction of nitrate to nitrite Generation of free fatty acids, ready to oxidation Generation of free amino acids Degradation of hydrogen peroxide	Safety Aroma Taste
Yeasts	*Debaryomyces hansenii*	Lipolysis	Deamination/ deamidation Transamination	Generation of free fatty acids Transformation of amino acids Lactic acid consumption and generation of ammonia	Aroma Taste pH rise
Molds	*Penicillium nalgiovense, P. chrysogenum*	Lipolysis, proteolysis (exo)	Deamination/ deamidation Transamination	Generation of free fatty acids, ready to oxidation Generation of free amino acids and its transformation Generation of ammonia	Aroma Taste pH rise external aspect

the processing of European sausages while *L. plantarum* and *P. acidilactici* grow well at higher temperatures (30–35°C) closer to the fermentation conditions in the sausages produced in the United States. Lactic acid bacteria also have a proteolytic system, consisting in endo- and exopeptidases, that contributes to the generation of free amino acids during processing and most of them are also able to generate different types of bacteriocins with antimicrobial properties. Recently, an absence of effect of certain lactic acid bacteria like *L. curvatus, L. sakei, Str. Griseu*, and *P. pentosaceus* was reported in American dry sausages ripened for a short time and finally heat processed (Candogan et al. 2009).

Micrococcaceae

This group consists of *Staphylococcus* and *Kocuria* (formerly *Micrococcus*), which are major contributors to flavor due to their proteolytic and lipolytic activities (Casaburi et al. 2006). However, *L. curvatus* and *S. xylosus* were reported to have no significant effect on proteolysis and lipolysis (Casaburi et al. 2007), although the sensory characteristics were observed to improve in long-ripened sausages (Casaburi et al. 2008). Another important function consists of the nitrate reductase activity, necessary to reduce nitrate to nitrite and contribute to color formation and safety. However, these microorganisms must be added in high amounts because they grow little or even die just at the onset of fermentation, when low pH conditions prevail. Preferably, low pH-tolerant strains should be carefully selected. The species from this family also have an important catalase activity that contributes to color stability and, somehow, prevention of lipid oxidation.

Yeasts

Debaryomyces hansenii is the predominant yeast in fermented meats, mainly growing in the outer area of the sausage due to its aerobic metabolism. *D. hansenii* has a good lipolytic activity and is able to degrade lactic acid. In addition, it also exhibits an important deaminase/deamidase activity, using free amino acids as substrates and producing ammonia as a subproduct that raises the pH in the sausage (Durá et al. 2002). *D. hansenii* has also some proteases of interest like proteases A, B, and D, as well as prolyl and arginyl aminopeptidases (Bolumar et al. 2005, 2006a, 2008).

Molds

Some typical Mediterranean dry fermented sausages have molds on the surface. The most usual are *Penicillium nalgiovense* and *P. chrysogenum*. They contribute to flavor through their proteolytic and lipolytic activity, and to appearance in the form of a white coating on the surface (Sunesen and Stahnke 2003). They also generate ammonia through their deaminase and deamidase activities and contribute to pH rise. Inoculation of sausages with natural molds present in the fermentation room is dangerous because toxigenic molds might grow. So, fungal starter cultures are mainly used as a preventive measure against the growth of other mycotoxin-producing molds and give a typical white color on the surface as demanded in certain Mediterranean areas.

Casings

Casings may be natural, semisynthetic, or synthetic, but a common required characteristic is its permeability to water and air.

Natural casings are natural portions of the gastrointestinal tract of pigs, sheep, and cattle, and although irregular in shape, they have good elasticity, tensile strength, and permeability. Natural casings are typically used for traditional sausages because they give a homemade aspect to the product. Semisynthetic casings are based on collagen that shrinks with the product and is permeable, but cannot be overstuffed (Toldrá et al. 2002). Synthetic cellulose-based casings are nonedible but are preferred for industrial processes due to important advantages such as controlled and regular pore size, uniformity for standard products, and hygiene. These casings are easily peeled off.

A wide range of sizes, between 2 and 15 cm, may be used, depending on the type of product. Of course, the diameter strongly affects fermentation and drying conditions. So, pH drop is more important in large-diameter sausages, where drying is more difficult to achieve.

PROCESSING STAGE 1: COMMINUTION

An example of a flow diagram for the processing of fermented sausages is shown in Figure 18.1. Chilled meats, pork alone or mixtures of pork and beef, and porcine fats are submitted to comminution in a grinder (Fig. 18.2). There are several plates with different hole sizes depending on the desired particle size. Previous trimming for removal of connective tissue is recommended, especially when processing undry or semidry fermented sausages where no further hydrolysis of collagen will occur. Salt, nitrate and/or nitrite, carbohydrates, microbial starters, spices, sodium ascorbate, and optionally, other nonmeat proteins are added to the ground mass, and the whole mix is homogenized under vacuum to avoid bubbles and undesirable oxidations that affect color and flavor (Fig. 18.3). Grinding and mixing take several minutes, depending on the amount. Industrial processes may use a cutter, as an alternative to grinding and mixing, when the required particles sizes are small. The cutter consists in a slowly moving bowl, containing the meats, fat, and additives, which rotates against a set of knives operating with rapid rotation. The fat and meat must be prefrozen (−6°C to −7°C) to avoid smearing of fat particles during chopping. This phenomenon consists in a fine film of fat formed over the lean parts that may reduce the release of water during drying (Roca and Incze 1990). The cutter operates under vacuum to avoid any damage by oxygen,

Figure 18.2. Grinding of meats and fats. There are many sizes of grinder plates in accordance to the required particle size.

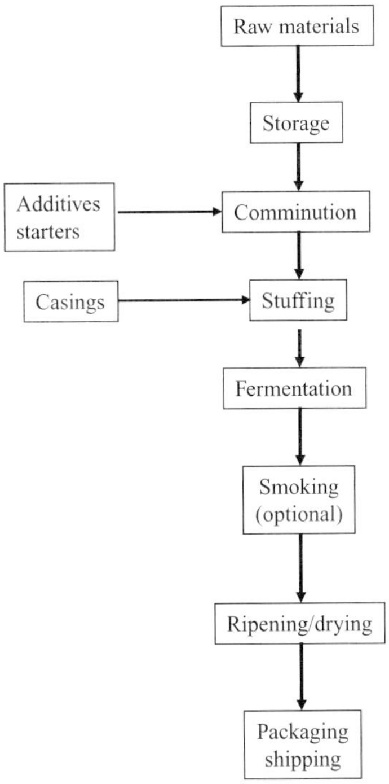

Figure 18.1. Flow diagram showing the most important stages in the processing of fermented sausages.

Figure 18.3. Detail of the batter after mixing in a vacuum mixer massager.

Figure 18.4. Sausage stuffed into a collagen casing, 80-mm diameter, and clipped on both extremes.

Figure 18.5. Example of a fermentation/drying chamber with computer control of temperature, relative humidity, and air rate. (By courtesy of Embutidos y Conservas Tabanera, Segovia, Spain.)

although the operation takes only a short time, just a few minutes, and the ratio of bowl speed to knife speed determines the desired particle size.

PROCESSING STAGE 2: STUFFING

The mixture is stuffed under vacuum into casings—natural, collagen-based, or synthetic—with both extremes clipped. The vacuum avoids the presence of bubbles within the sausage and disruptions in the casing. The stuffing must be adequate in order to avoid smearing of the batter, and temperature must be kept below 2°C to avoid this problem. Once stuffed (Fig. 18.4), the sausages are hung in racks and placed in natural or air-conditioned drying chambers.

PROCESSING STAGE 3: FERMENTATION

Fermentation Technology

Once sausages are stuffed, they are placed in computer-controlled air-conditioned chambers and left to ferment for microbial growth and development. A typical chamber is shown in Figure 18.5. Temperature, relative humidity, and air speed must be carefully controlled in order to have correct microbial growth and enzyme action. The whole process can be considered as a lactic acid solid-state fermentation in which several simultaneous processes take place: (1) microbial growth and development, (2) biochemical changes, mainly enzymatic breakdown of carbohydrates, proteins, and lipids, and (3) physical changes, mainly acid gelation of meat proteins and drying.

Meat fermentation technology differs between the United States and Europe. High fermentation temperatures (35–40°C) are typical in US sausages, followed by a mild heating process, as a kind of pasteurization, instead of drying, to kill any *Trichinella*. Thus, starters such as *Lactobacillus plantarum* or *Pediococcus acidilactici*, which grow well at those temperatures, are typically used. In Europe, different technologies may be found, depending on the location and climate. There is a historical trend toward short-processed, smoked sausages in cold and humid countries, as in Northern Europe, and long-processed, dried sausages in warmer and drier countries, as in the Mediterranean area. In the case of Northern European countries, sausages are fermented for about 3 days at intermediate temperatures (25–30°C), followed by short ripening periods (up to 3 weeks). These sausages are subjected to a rapid pH drop and are usually smoked for a specific flavor (Demeyer and Stahnke 2002). On the other hand, Mediterranean sausages require longer processing times. Fermentation takes place at milder temperatures (18–24°C) for about 4 days, followed by mild drying conditions for a longer time, usually several weeks or months. *L. sakei* or *L. curvatus* are the lactic acid bacteria most often used as starter cultures (Toldrá et al. 2001). The time required for the fermentation stage is a function of the temperature and the type of microorganisms used as starters.

The technology is quite different in China and other Asian countries. Sausages are first dried over charcoal at 48°C and 65% relative humidity for 36 hours and then at 20°C and 75% relative humidity for 3 days. Water activity rapidly drops below 0.80, although pH remains at about 5.9, which is a relatively high value. Fermentation is relatively poor, and the sour taste, which is considered undesirable, is reduced. Chinese raw sausage is consumed after heating (Leistner 1992).

Microbial Metabolism of Carbohydrates

The added carbohydrate is converted, during the fermentation, into lactic acid of either the D(−) or L(+) configuration, or a mixture of both, depending on the species of lactic acid bacteria used as starter. The ratio between the L- and D-enantiomers depends on the action of L- and D-lactate dehydrogenase, respectively, and the presence of lactate racemase. The rate of generation and the final amount of lactic acid depend on the type of lactic acid bacteria species used as starter, the type and content of carbohydrates, the fermentation temperature, and other processing parameters. The accumulation of lactic acid produces a pH drop, more or less, intense depending on its generation rate. Some secondary products such as acetic acid, acetoin, and others may be formed through heterofermentative pathways (Demeyer and Stahnke 2002). Acid pH favors protein coagulation, as it approaches its isoelectric point, and thus also favors water release. Acid pH also contributes to safety by contributing to the inhibition of undesirable pathogenic or spoilage bacteria. The pH drop favors initial proteolysis and lipolysis by stimulating the activity of muscle cathepsin D and lysosomal acid lipase, both active at acid pH, but an excessive pH drop does not favor later enzymatic reactions involved in the generation of flavor compounds (Toldrá and Verplaetse 1995).

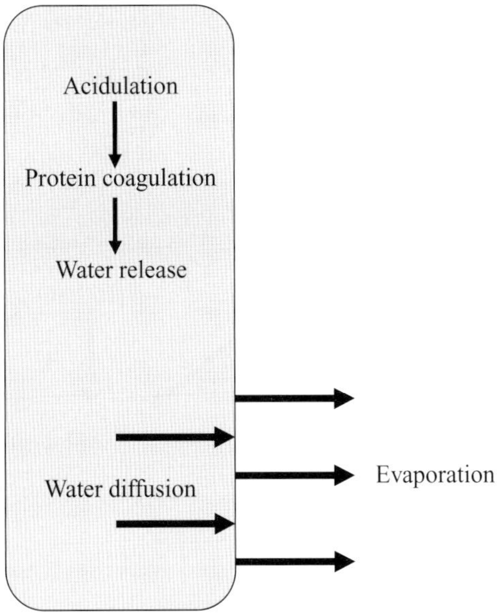

Figure 18.6. Scheme showing important physical changes during the processing of fermented meats.

PROCESSING STAGE 4: RIPENING AND DRYING

Temperature, relative humidity, and air flow have to be carefully controlled during fermentation and ripening to allow correct microbial growth and enzyme action while maintaining adequate drying progress. The air velocity is kept at around 0.1 m/s, which is enough for a good homogenization of the environment. Ripening and drying are important for enzymatic reactions related to flavor development and obtaining the required water loss and thus reduction in a_W. The length of the ripening/drying period takes from 7 to 90 days, depending on many factors, including the kind of product, its diameter, dryness degree, fat content, desired flavor intensity, and so on. The reduction in a_W is slower in beef-containing sausages. The casing must remain attached to the sausage when it shrinks during drying. In general, long-ripened products tend to be drier and more flavorful.

PHYSICAL CHANGES

The most important physical changes during fermentation and ripening/drying are summarized in Figure 18.6. The acidulation produced during the fermentation stage induces protein coagulation and thus some water release. The acidulation also reduces the solubility of sarcoplasmic and myofibrillar proteins, and the sausage starts to develop consistency. The drying process is a delicate operation that must achieve an equilibrium between two different mass transfer processes—diffusion and evaporation (Baldini et al. 2000). Water inside the sausage must diffuse to the outer surface and then evaporate to the environment. Both rates must be in equilibrium because a very fast reduction in the relative humidity of the chamber would cause excessive evaporation from the sausage surface that would reduce the water content on the outer parts of the sausage, causing hardening. This is typical of sausages of large diameter because of the slow water diffusion rate. The cross section of these sausages shows a darker, dry, hard outer ring. On the other hand, when the water diffusion rate is much higher than the evaporation rate, water accumulates on the surface of the sausage, causing a wrinkled casing. This situation may happen in small-diameter sausages being ripened in a chamber with high relative humidity. The progress in drying reduces the water content, up to 20% weight loss in semidry sausages and 30% in dry sausages (Table 18.1). The a_W decreases according to the drying rate, reaching values below 0.90 for long-ripened sausages.

CHEMICAL CHANGES

There are different enzymes, of both muscle and microbial origin, involved in reactions related to color, texture, and flavor generation. These reactions, which are summarized in Figure 18.7, are very important for the final sensory quality of the product. One of the most important groups of reactions, mainly affecting myofibrillar proteins and yielding small peptides and free amino acids as final products, is known as proteolysis (Toldrá 1998). An intense proteolysis during fermentation and ripening is mainly carried out by endogenous cathepsin D, an acid muscle proteinase that is very active at acid pH. This enzyme hydrolyses myosin and actin, producing an accumulation of polypeptides that are further hydrolyzed to small peptides by muscle and microbial peptidyl peptidases and to free amino acids by muscle and microbial aminopeptidases (Sanz et al. 2002). The generation of small peptides and free amino acids increases with the length of processing, although the generation rate is reduced at acid pH values because the conditions are far from optimal for enzyme activity. Free amino acids may be further transformed into other products, for example, volatile compounds through Strecker degradations and Maillard reactions; ammonia through

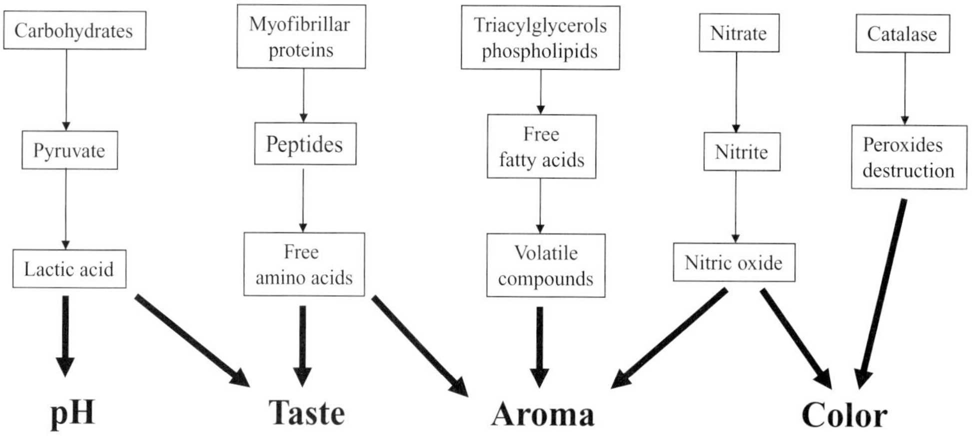

Figure 18.7. Scheme showing the most important reactions by muscle and microbial enzymes involved in chemical and biochemical changes affecting sensory quality of fermented meats.

deamination and/or deamidation reactions by deaminases and deamidases, respectively, present in yeasts and molds; or amines by microbial decarboxylases.

Another important group of enzymatic reactions, affecting muscle and adipose tissue lipids, is known as lipolysis (Toldrá 1998). Thus, a large amount of free fatty acids (between 0.5% and 7%) is generated through the enzymatic hydrolysis of triacylglycerols and phospholipids. Most of the observed lipolysis is attributed, after extensive studies on model sterile systems and sausages with added antibiotics, to endogenous lipases present in muscle and adipose tissue (e.g., lysosomal acid lipase, present in the lysosomes and very active at acid pH; Toldrá 1992, Hierro et al. 1997, Molly et al. 1997).

Catalases are mainly present in microorganisms such as *Kocuria* and *Staphylococcus;* they are responsible for peroxide reduction and thus contribute to color and flavor stabilization. Nitrate reductase, also present in these microorganisms, is also important for reducing nitrate to nitrite in slow-ripened sausages with an initial addition of nitrate. Recently, two strains of *Lactobacillus fermentum* have proved to be able to generate nitric oxide and give an acceptable color in sausages without nitrate/nitrite. This could be used to produce cured meats free of nitrate and nitrite (Moller et al. 2003).

PROCESSING STAGE 5: SMOKING

Smoking is mostly applied in Northern countries with cold and/or humid climates. Initially, it was used for preservation purposes, but today its contribution to flavor and color is more important (Ellis 2001). In some cases, smoking can be applied just after fermentation or even at the start of the fermentation. Smoking can be accompanied by heating at 60°C and has a strong impact on the final sensory properties. It has a strong antioxidative effect and gives a characteristic color and flavor to the product, which is the primary role of smoking. The antimicrobial effect of some smoking compounds, especially phenols, carboxylic acids, and formaldehyde inhibit the growth of certain bacteria even though some yeasts and molds may be resistant (Sikorski and Kolakowski 2010).

SAFETY

The stability of the sausage against pathogen and/or spoilage microorganisms is the result of successive hurdles (Leistner 1992). Initially, the added nitrite curing salt is very important for the microbial stability of the mix. During mixing under vacuum, oxygen is gradually removed, and redox potential is reduced. This effect is enhanced when ascorbic acid or ascorbate is added. Low redox potential values inhibit aerobic bacteria and make nitrite more effective as bactericide. During the fermentation, lactic acid bacteria can inhibit other bacteria, not only by the generation of lactic acid (and the subsequent pH drop), but also by generation of other metabolic products such as acetic acid and hydrogen peroxide and, especially, bacteriocins (low-molecular mass peptides synthetized in bacteriocin-positive strains; Lücke 1992). The drying of the sausage continues the reduction in a_w to low values (a_w below 0.92) that inhibit growth of spoilage and/or pathogenic microorganisms. Thus, the correct interaction of all these factors assures the stability of the product.

Some foodborne pathogens that might be found in fermented meats are briefly described. *Salmonella* is more usual in fresh, spreadable sausages (Lücke 1985), but can be inhibited by acidification to pH 5.0 and/or drying to $a_w < 0.95$ (Talon et al. 2002). Lactic acid bacteria exert an antagonistic effect against *Salmonella* (Roca and Incze 1990). *Staphylococcus aureus* may grow under aerobic or anaerobic conditions and requires $a_w <$ 0.91 for inhibition, but is sensitive to acid pH. So, it is important to control the elapsed time before reaching the pH drop in order to avoid toxin production. Furthermore, this toxin is produced only in aerobic conditions (Roca and Incze 1990). *Clostridium botulinum* and its toxin-producing capability are affected by a rapid pH drop and low a_w even more than by the addition of

Table 18.4. Safety Aspects: Generation of Undesirable Compounds in Dry Fermented Meats.

Compounds	Route of Formation	Origin	Concentrations (mg/100g)
Tyramine	Microbial decarboxylation	Tyrosine	<16.0
Tryptamine	Microbial decarboxylation	Trytophan	<6.0
Phenylethylamine	Microbial decarboxylation	Phenylalanine	<3.5
Cadaverine	Microbial decarboxylation	Lysine	<0.6
Histamine	Microbial decarboxylation	histidine	<3.6
Putrescine	Microbial decarboxylation	Ornithine	<10.0
Spermine	Microbial decarboxylation	Methionine	<3.0
Spermidine	Microbial decarboxylation	Methionine	<0.5
Cholesterol oxides	Oxidation	Cholesterol	<0.15

Source: Adapted from Maijala et al. (1995), Shalaby (1996), Hernández-Jover et al. (1997) and Demeyer et al. (2000).

lactic acid bacteria and nitrite (Lücke 1985). *Listeria monocytogenes* is limited in growth at $a_w < 0.90$ combined with low pH values and specific starter cultures (Hugas et al. 2002). *Escherichia coli* is rather resistant to low pH and a_w but is reduced when exposed to $a_w < 0.91$ (Nissen and Holck 1998). Adequate prevention measures consist in correct cooling and a hazard analysis critical control point (HACCP) plan with application of good manufacturing practices (GMP), sanitation, and strict hygiene control of personnel and raw materials.

In recent years, most attention has been paid to biopreservation as a way to enhance preservation against spoilage bacteria and foodborne pathogens. The bioprotective culture consists in a competitive bacterial strain that grows very fast or produces antagonistic substances like bacteriocins. Another precise way consists in the direct addition of purified bacteriocins. Those bacteriocins belonging to group IIa (also called pediocin-like) that display inhibition against *Listeria* have been reported to be the most interesting for the meat industry (Hugas et al. 2002).

Parasites like *Trichinella spiralis* are almost eliminated through modern breeding systems. Pork meat free of trichinae must be used as raw material for fermented sausages; otherwise, heat treatments of the sausage to reach internal temperatures above 62.2°C are required to inactivate them (Sebranek 2004).

The generation of undesirable compounds, listed in Table 18.4, depends on several factors. The most important factor is the hygienic quality of the raw materials. For instance, the presence of cadaverine and/or putrescine may be indicative of the presence of contaminating meat flora. Another factor is processing conditions, which may favor the generation of biogenic amines; however, the type of natural flora or microbial starters used for the process is the most important issue, because the presence of microorganisms with decarboxylase activity can induce the generation of biogenic amines. In general, tyramine is the amine generated in higher amounts, and it is formed by certain lactic acid bacteria that exhibit enzymatic activity for decarboxylation of tyrosine (Eerola et al. 1996). Tyramine releases noradrenaline from the sympathetic nervous system, and the peripheral vasoconstriction and increase in cardiac output result in higher blood pressure and risk for hypertensive crisis (Shalaby 1996). However, the estimated tolerance level for tyramine (100–800 mg/kg) is higher than for other amines (Nout 1994). The amines derived from foods are generally degraded in humans by the enzyme monoamine oxidase (MAO) through oxidative deamination reactions. Those consumers using MAO inhibitors are less protected against amines and are thus susceptible for risk situations such as hypertensive crisis when ingesting significant amounts of amines. Other amines may also cause problems; for example, phenylethylamine, which may cause migraine and an increase in blood pressure; or histamine, which excites the smooth muscles of the uterus, the intestine, and the respiratory tract. Biogenic amines may be typically found in traditional fermented meat products (Roseiro et al. 2010). One way to reduce health risks from amines consists in the use of starter cultures that are unable to produce amines but are competitive against amine-producing microorganisms. Additionally, the use of microorganisms that exhibit amine oxidase activity and are able to degrade amines, the selection of raw materials of high quality, and the use of GMP assures products of high quality and reduced risks (Talon et al. 2002). Finally, the generation of nitrosamines during the process is almost negligible due to the restricted amount of nitrate and/or nitrite that can be initially added and the low amount of residual nitrite remaining by the end of the process (Cassens 1997, Honikel 2010).

The processing conditions may favor the oxidation of cholesterol. Some generated oxides can be involved in cardiovascular-related diseases (e.g., 7-ketocholesterol and α-5,6-epoxycholesterol), but in general, the reported levels of all cholesterol oxides is very low, less than 0.15 mg/100 g, for exerting any toxic effect (Demeyer et al. 2000).

FINISHED PRODUCT

Once the product is finished, it is packaged and distributed. Fermented sausages can be sold as either entire or as thin slices (Fig. 18.8). The developed color, texture, and flavor depend on the processing and type of product. Main sensory properties are described in the forthcoming sections.

Color

The color of the sausage depends on the moisture and fat content as well as its content of hemoprotein, particularly myoglobin.

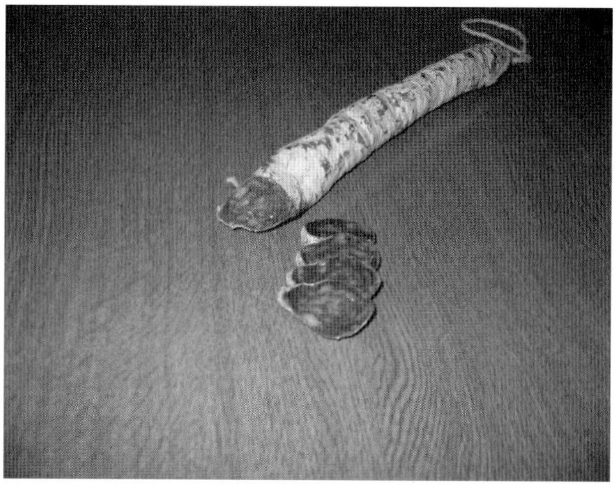

Figure 18.8. Picture of a typical small-diameter salchichón, showing its cross section.

Color is also influenced by pH drop rate and the ultimate pH, but it may be also affected by the presence of spices like red pepper. An excess of acid generation by lactobacilli may also affect color.

The characteristic color is due to the action of nitrite with myoglobin. Nitrite is reduced to nitric oxide, favored by the presence of ascorbate/erythorbate. Myoglobin and nitric oxide may then interact to form nitric oxide myoglobin, which gives the characteristics cured pinkish-red color (Pegg and Shahidi 1996). This reaction is favored at low pH. Long-processing sausages using nitrate need some time for the growth of Micrococcaceae before pH drops. Nitrate reductase, which is present in Micrococcaceae, reduces nitrate to nitrite, which is afterwards further reduced to nitric oxide, which can react with myoglobin. Oxidative discoloration consists in the conversion of nitrosylmyoglobin to nitrate and metmyoglobin, which affects the oxidative stability because of the pro-oxidant effect of ferric heme.

Texture

The consistency of fermented meats is initiated with the salt addition and pH reduction. The water-binding capacity of myofibrillar proteins decreases as pH approaches its isoelectric point and releases water. The solubility of myofibrillar proteins is also reduced, with a trend toward aggregation and coagulation, forming a gel. The consistency of this gel increases with water loss during drying. So, there is a continuous development of textural characteristics such as firmness, hardness, and cohesiveness of meat particles during drying (Toldrá 2002). The meat:fat ratio

Table 18.5. Quality Aspects: Generation or Presence of Desirable Nonvolatile Compounds Contributing to Taste in Fermented Meats

Group of Compounds	Main Representative Compounds	Routes of Generation	Presence in Final Product	Main Contribution	Expected Intensity
Peptides	Tri and dipeptides	Proteolysis	Increases with length of process	Taste	High
Free amino acids	Glutamic acid, aspartic acid, alanine, lysine, threonine	Proteolysis	Increases with length of process	Taste	High
Nucleotides and nucleosides	Inosine monophosphate, guanosine monophosphate, inosine, hypoxanthine	ATP degradation	Around 100 mg/100 g	Taste enhancement	Low
Long chain free fatty acids	Oleic acid, linoleic acid, linolenic acid, arachidonic acid, palmitic acid	Lipolysis	Increases with length of process	Taste	Low
Short chain fatty acids	Acetic acid, propionic acid	Microbial metabolism	Depends on microflora	Taste	Medium
Acids	Lactic acid	Glycolysis	Depends on initial amount of sugar and fermentation	Sour taste	High
Carbohydrates	Glucose, lactose	Remaining (nonconsumed through glycolysis)	Depends on initial amount of sugar and microflora	Sweet taste	Low
Inorganic compounds	Salt	Addition	Depends on initial amount	Salty taste	High

may affect some of these textural characteristics, but, in general, the final texture of the sausage will mainly depend on the extent of drying (Toldrá et al. 2004).

Flavor

Little or no flavor is usually detected before meat fermentation, although a large number of flavor precursors are present. As fermentation and further ripening/drying progress, the combined effect of endogenous muscle enzyme and microbial activity produces a high number of nonvolatile and volatile compounds with sensory impact. The longer the process, the more the accumulation of these compounds is increased and their sensory impact enhanced. Although not so important as in meat cooking, some compounds with sensory impact may be produced through further chemical reactions. The addition of spices also has an intense contribution to specific flavors.

Taste

The main nonvolatile compounds contributing to taste of fermented meats are summarized in Table 18.5. Sour taste, mainly resulting from lactic acid generation through microbial glycolysis, is the most relevant taste in fermented meats. Sourness is also correlated with other microbial metabolites such as acetic acid. Ammonia may be generated through deaminase and deamidase activities, usually present in yeasts and molds, reducing the intensity of the acid taste. Salty taste is usually perceived as a direct taste from salt addition. ATP (adenosine triphosphate)-derived compounds such as inosine monophosphate and guanosine monophosphate exert some taste enhancement, while hypoxanthine contributes to bitterness. Other taste contributors are those compounds resulting from protein hydrolysis. The generation and accumulation of small peptides and free amino acids contribute to taste perception, which increases with the length of process. Some of these small peptides (e.g., leucine, isoleucine, and valine) also act as aroma precursors, as described in Section "Aroma".

Aroma

The origin of aroma mainly depends on the ingredients and processing conditions. Different pathways are responsible for the formation of volatile compounds with aroma impact (Table 18.6). As mentioned previously, proteolysis gives rise to a large amount of small peptides and free amino acids. Microorganisms can convert the amino acids leucine, isoleucine, valine, phenylalanine, and methionine to important sensory compounds with low threshold values. Some of the most important are branched aldehydes such as 2- and 3-methylbutanal and 2-methylpropanal,

Table 18.6. Quality Aspects: Generation of Desirable Volatile Compounds Contributing to Aroma in Fermented Meats.

Group of Compounds	Main Representative Compounds	Routes of Generation	Main Aroma	Expected Contribution
Aliphatic aldehydes	Hexanal, pentanal, octanal, etc.	Oxidation of unsaturated fatty acids	Green	High
Strecker aldehydes	2- and 3-methylbutanal, etc.	Strecker degradation of free amino acids	Roasted cocoa, cheesy-green	High
Branched-chain acids	2- and 3-methyl butanoic acid	Secondary products of previous Strecker degradation	Sweaty	Medium
Alcohols	Ethanol, butanol, etc.	Oxidative decomposition of lipids	Sweet, alcohol, etc.	Low
Ketones	2-pentanone, 2-heptanone, 2-octanone, etc.	Lipid oxidation	Ethereal, soapy	Medium
Sulfides	Dimethyldisulfide	Strecker degradation of sulfur-containing amino acids (methionine)	Dirty socks	Low
Esters	Ethyl acetate, ethyl 2-methyl-butanoate	Interaction of carboxylic acids and alcohols	Pineapple, fruity	High
Hydrocarbons	Pentane, heptane, etc.	Lipids autoxidation	Alkane	Very low
Dicarbonyl products	Diacetyl, acetoin, acetaldehyde	Pyruvate microbial metabolism	Butter	Low
Nitrogen compounds	Ammonia	Deamination, deamidation	Ammonia	Variable, depends on growth of yeasts and molds

Source: Adapted from Viallon et al. (1996), Flores et al. (1997), Stahnke (2002), Toldrá (2002) and Talon et al. (2002).

branched alcohols such as 2- and 3-methylbutanol, acids such as 2- and 3-methylbutanoic and 2-methylpropanoic acids, and esters such as ethyl 2- and 3-methylbutanoate (Stahnke 2002). Some of these branched-chain aldehydes may also be formed through the Strecker degradation, consisting in the reaction of amino acids with diketones. However, conditions found in sausages are far from those optimal for this kind of reaction, which needs high temperature and low a_w (Talon et al. 2002).

Methyl ketones may be formed either by β-oxidation of free fatty acids or decarboxylation of free β-keto acids. Other non-branched aliphatic compounds generated by lipid oxidation are alkanes, alkenes, aldehydes, alcohols, and several furanic cycles.

A large number of volatile compounds are generated by chemical oxidation of the unsaturated fatty acids. These volatile compounds are mainly generated during ripening and further storage. Other low-molecular weight volatile compounds are generated by microorganisms from carbohydrate catabolism. The most usual compounds are diacetyl, acetoin, butanediol, acetaldehyde, ethanol, and acetic propionic and butyric acids. However, some of these compounds may be derived from pyruvate originated through other metabolic pathways than carbohydrate glycolysis (Demeyer and Stahnke 2002, Demeyer and Toldrá 2004). A comparison of the generation of certain volatile compounds between fat tissue and intramuscular fat in lean meat showed that major development took place in the intramuscular fat (Olivares et al. 2009). The flavor profile may have important variations depending on the type of microorganisms used as starters (Berdagué et al. 1993).

REFERENCES

Baldini P et al. 2000. Dry sausages ripening: influence of thermohygrometric conditions on microbiological, chemical and physicochemical characteristics. *Food Research Int* 33: 161–170.

Berdagué JL et al. 1993. Effects of starter cultures on the formation of flavour compounds in dry sausages. *Meat Sci* 35: 275–287.

Bolumar T et al. 2005. Protease B from *Debaryomyces hansenii*. Purification and biochemical properties. *Int J Food Microbiol* 98: 167–177.

Bolumar T et al. 2006a. Protease (PrA and PrB) and prolyl and arginyl aminopeptidase activities from *Debaryomyces hansenii* as a function of growth phase and nutrient sources. *Int J Food Microbiol* 107: 20–26.

Bolumar T et al. 2008. Purification and characterisation of proteases A and D from *Debaryomices hansenii*. *Int J Food Microbiol* 124: 135–141.

Campbell-Platt G. 1995. Fermented meats- a world perspective. In: G Campbell-Platt, PE Cook (eds.) *Fermented Meats*. Blackie Academic & Professional, London, pp. 39–51.

Candogan K et al. 2009. Effect of starter culture on proteolytic changes during processing of fermented beef sausages. *Food Chem* 116: 731–737.

Casaburi A et al. 2006. Protease and esterase activity of Staphylococci. *Int J Food Microbiol* 112: 223–229.

Casaburi A et al. 2007. Biochemical and sensory characteristics of traditional fermented sausages of Vallo di Diano (Southern Italy) as affected by use of starter cultures. *Meat Sci* 76: 295–307.

Casaburi A et al. 2008. Proteolytic and lipolytic starter cultures and their effect on traditional fermented sausages ripening and sensory traits. *Food Microbiol* 25: 335–347.

Cassens RG. 1997. Composition and safety of cured meats in the USA. *Food Chem* 59: 561–566.

Demeyer DI et al. 2000. Control of bioflavor and safety in fermented sausages: first results of a European project. *Food Research Int* 33: 171–180.

Demeyer D, Stahnke L. 2002. Quality control of fermented meat products. In: J Kerry et al. (eds.) *Meat Processing: Improving Quality*. Woodhead Pub. Co., Cambridge, pp. 359–393.

Demeyer D, Toldrá F. 2004. Fermentation. In: W Jensen et al. (eds.) *Encyclopedia of Meat Sciences*. Elsevier Science, London, pp. 467–474.

Demeyer D. 1992. Meat fermentation as an integrated process. In: FJM Smulders et al. (eds.) *New Technologies for Meat and Meat Products*. Audet, Nijmegen, pp. 21–36.

Durá A et al. 2002. Purification and characterization of a glutaminase from *Debaryomices* spp. *Int J Food Microbiol* 76: 117–126.

Eerola S et al. 1996. Biogenic amines in dry sausages as affected by starter culture and contaminant amine-positive *Lactobacillus*. *J Food Sci* 61: 1243–1246.

Ellis DF. 2001. Meat smoking technology. In: YH Hui et al. (eds.) *Meat Science and Applications*. Marcel Dekker, New York, pp. 509–519.

Flores J, Toldrá F. 1993. Curing: Processes and applications. In: R MacCrae et al. (eds.) *Encyclopedia of Food Science, Food Technology and Nutrition*. Academic Press, London, pp. 1277–1282.

Flores M et al. 1997. Correlations of sensory and volatile compounds of Spanish Serrano dry-cured ham as a function of two processing times. *J Agric Food Chem* 45: 2178–2186.

Flores M, Toldrá F. 2010. Microbial enzymes for improved fermented meats. *Trends Food Sci Technol*, submitted.

Gray JY, Pearson AM. 1984. Cured meat flavor. In: CO Chichester et al. (eds.) *Advances in Food Research*. Academic Press, Orlando, FL, pp. 2–70.

Hernández P et al. 1999. Effect of frozen storage on lipids and lipolytic activities in the *Longissimus dorsi* muscle of the pig. *Z Lebensm Unters Forsch A* 208: 110–115.

Hernández-Jover T et al. 1997. Effect of starter cultures on biogenic amine formation during fermented sausage production. *J Food Protec* 60: 825–830.

Hierro E et al. 1997. Contribution of microbial and meat endogenous enzymes to the lipolysis of dry fermented sausages. *J Agric Food Chem* 45: 2989–2995.

Honikel KO. 2010. Curing. In: F Toldrá (ed.) *Handbook of Meat Processing*. Wiley-Blackwell, Ames, IA, pp. 125–142.

Hugas M et al. 2002. Bacterial cultures and metabolites for the enhancement of safety and quality of meat products. In: F Toldrá (ed.) *Research Advances in the Quality of Meat and Meat Products*. Research Signpost, Trivandrum, pp. 225–247.

Leistner L. 1992. The essentials of producing stable and safe raw fermented sausages. In: FJM Smulders et al. (eds.) *New Technologies for Meat and Meat Products*. Audet, Nijmegen, pp. 1–19.

Lücke FK. 1985. Fermented sausages. In: BJB Wood (ed.) *Microbiology of Fermented Foods*. Elsevier Applied Science, London, pp. 41–83.

Lücke FK. 1992. Prospects for the use of bacteriocins against meat-borne pathogens. In: FJM Smulders et al. (eds.) *New Technologies for Meat and Meat Products*. Audet, Nijmegen, pp. 37–52.

Maijala R, et al. 1995. Formation of biogenic amines during ripening of dry sausages as affected by starter culture and thawing time of raw materials. *J Food Sci* 60: 1187–1190.

Moller JKS, et al. 2003. Microbial formation of nitrite-cured pigment, nitrosylmyoglobin, from metmyoglobin in model systems and smoked fermented sausages by Lactobacillus fermentum strains and a commercial starter culture. *Eur Food Res Technol* 216: 463–469.

Molly K et al. 1997. The importance of meat enzymes in ripening and flavor generation in dry fermented sausages. First results of a European project. *Food Chem* 54: 539–545.

Nissen H, Holck AL. 1998. Survival of Escherichia coli O157:H7, Listeria monocytogenes and Salmonella kentucky in Norwegian fermeted dry sausage. *Food Microbiol* 15: 273–279.

Nout MJR. 1994. Fermented foods and food safety. *Food Res Int* 27: 291–296.

Olivares A et al. 2009. Distribution of volatile compounds in lean and subcutaneous fat tissues during processing of dry fermented sausages. *Food Res Int* 42: 1303–1308.

Ordoñez JA et al. 1999. Changes in the components of dry-fermented sausages during ripening. *Crit Rev Food Sci Nutr* 39: 329–367.

Paramithitis S et al. 2010. Fermentation: microbiology and biochemistry. In: F Toldrá (ed.) *Handbook of Meat Processing*. Wiley-Blackwell, Ames, IA, pp. 185–198.

Pegg BR, Shahidi F. 1996. A novel titration methodology for elucidation of the structure of preformed cooked cured-meat pigment by visible spectroscopy. *Food Chem* 56: 105–110.

Roca M, Incze K. 1990. Fermented sausages. *Food Reviews Int* 6: 91–118.

Roseiro LC et al. 2010. Effect of processing on proteolysis and biogenic amines formation in a Portuguese traditional dry-fermented sausage Chouriço Grosso Estremoz a Borba PGI. *Meat Sci* 84: 172–179.

Sanz Y et al. 2002. Role of muscle and bacterial exopeptidases in meat fermentation.. In: F Toldrá (ed.) *Research Advances in the Quality of Meat and Meat Products*. Research Signpost, Trivandrum, pp. 143–155.

Sebranek JG 2004. Semi-dry fermented sausages. In: YH Hui, et al. (eds.) *Handbook of Food and Beverage Fermentation Technology*. Marcel Dekker, New York, pp. 385–396.

Shalaby, AR. 1996. Significance of biogenic amines to food safety and human health. *Food Res Int* 29: 675–690.

Sikorski ZE, Kolakowski E. 2010. Smoking. In: F. Toldrá (ed.) *Handbook of Meat Processing*.: Wiley-Blackwell, Ames, IA, pp. 231–245.

Stahnke L. 2002. Flavour formation in fermented sausage. In: F Toldrá (ed.) *Research Advances in the Quality of Meat and Meat Products*. Research Signpost, Trivandrum, pp. 193–223.

Stahnke LH, Tjener K. 2007. Influence of processing parameters on cultures performance. In: F Toldrá, et al. (eds.) *Handbook of Fermented Meat and Poultry*. Wiley-Blackwell, Ames, IA, pp. 187–194.

Sunesen LO, Stahnke LH. 2003. Mould starter cultures for dry sausages-selection, application and effects. *Meat Sci* 65: 935–948.

Talon R et al. 2002. Bacterial starters involved in the quality of fermented meat products. In: F Toldrá (ed.) *Research Advances in the Quality of Meat and Meat Products*. Research Signpost, Trivandrum, pp. 175–191.

Toldrá F. 1992. The enzymology of dry-curing of meat products. In: FJM Smulders et al.(eds.) *New Technologies for Meat and Meat Products*. Audet, Nijmegen, pp. 209–231.

Toldrá F, Verplaetse A. 1995. Endogenous enzyme activity and quality for raw product processing. In: K Lündstrom, et al. (eds.) *Compiosition of Meat in Relation to Processing, Nutritional and Sensory Quality*. Ecceamst, Uppsala, pp. 41–55.

Toldrá F. 1998. Proteolysis and lipolysis in flavour development of dry-cured meat products. *Meat Sci* 49: s101–s110.

Toldrá F et al. 2001. Meat fermentation technology. In: YH Hui et al. (eds.) *Meat Science and Applications*. Marcel Dekker, New York, pp. 537–561.

Toldrá F. 2002. *Dry-Cured Meat Products*. Food & Nutrition Press, Trumbull, CT, pp. 1–238.

Toldrá F et al. 2004. Packaging and quality control. In: YH Hui, et al. (eds.) *Handbook of Food and Beverage Fermentation Technology*. Marcel Dekker, New York, pp. 445–458.

Toldrá F, Reig M. 2007. Sausages. In: YH Hui et al. (eds.) *Handbook of Food Product Manufacturing*, vol 2. John Wiley Interscience, New York, pp. 249–262.

Toldrá F et al. 2007. Dry fermented sausages: An overview. In: F Toldrá et al. (eds.) *Handbook of Fermented Meat and Poultry*. Blackwell Publishing, Ames, IA, pp. 321–325.

Viallon C et al. 1996. The effect of stage of ripening and packaging on volatile content and flavour of dry sausage. *Food Res Int* 29: 667–674.

19
Biochemistry of Seafood Processing

Y. H. Hui, N. Cross, H. G. Kristinsson, M. H. Lim, W. K. Nip, L. F. Siow, and P. S. Stanfield

Yerlikaya P, Gokoglu N. 2010. Inhibition effects of green tea and grape seed extracts on lipid oxidation in bonito fillets during frozen storage. *Int J Food Sci Technol* 45(2): 252–257.

Abstract: Seafood is valued for its nutritive components and desirable sensory attributes. However, quality of seafood is vulnerable to rapid degradation if no appropriate postharvest handling or processing methods are used. It is important to understand the basic biochemical reactions of the relevant components in seafood in order that their quality can be best preserved. Biochemical changes in glycogen, protein, lipids, and pigments are discussed in detail and how such changes could affect the quality of seafood. Metabolic changes from the activity of enzymes can be used as indices of freshness and be monitored by biochemical or chemical methods. Three different manufacturing process techniques, (1) freezing, (2) dehydration, and (3) thermal are discussed with respect to the biochemical changes.

INTRODUCTION

Most of us like to eat seafood, especially when it is fresh, although processed products are also favorites of many consumers. In addition to sensory attributes, the preference for seafood now

Food Biochemistry and Food Processing, Second Edition. Edited by Benjamin K. Simpson, Leo M.L. Nollet, Fidel Toldrá, Soottawat Benjakul, Gopinadhan Paliyath and Y.H. Hui.

includes awareness of its health benefits. Thus, when processing seafood, it is important to understand the scientific and technical reasons that are responsible for the sensory and health attributes of seafood, in addition to how the manufacturing process can affect the basic quality of seafood. The same knowledge may lead to a reduction in the perishability of seafood, a problem that has existed since the beginning of time, though it has been partially solved by many modern techniques including canning, freezing, and dehydration.

The nutritive value of seafood is well known for its content of protein and some minerals and, recently, its fat quality. For the past 20 years, considerable attention has been paid to the content of polyunsaturated omega-3 fatty acids in fish, especially their prevalence in fatty fish. Of course, one must not forget the relatively high levels of cholesterol in some fish and shellfish, especially crustaceans and squid. The biochemical changes in macronutrients of seafood are of interest to nutritionists, food processors, and consumers. Their importance in public health cannot be overestimated if one considers how seafood contributes to the nutrition and health of the human body. A short discussion of the nutritive values of fish serves as an introduction.

Perishability of seafood is important economically as well as for its safety. The price of seafood can drop drastically if it is not handled properly and is handled without refrigeration or other forms of preservation. The biochemical aspects of seafood will be discussed in this chapter using the following approaches:

- *Glycogen in seafood*: Biochemical changes in its glycogen, which exists in small quantities, initiates the biochemical process, in the same way that it occurs in other animal products.
- *Nitrogenous compounds in seafood*: The degradation of protein will be discussed, with a special reference to sarcoplasmic, myofibrillar, and stromal protein. This discussion is accompanied by a brief mention of nonprotein nitrogenous substances. Such biochemical changes in seafood's protein and nonprotein components usually reduce its economic value and may also create safety problems.
- *Lipids in seafood*: Fat can undergo many biochemical changes, both qualitatively and quantitatively. Some changes can cause rancidity, especially during the storage of fatty fish.
- *Pigments in seafood*: Biochemical changes will be discussed with reference to epithelial discoloration, hemoglobin, hemocyanin, myoglobin, carotenoids, and melanosis.
- *Quality indices in seafood*: The monitoring of the quality of seafood has been the subject of intense scientific research. Currently, the techniques available are lactic acid formation with lowering of pH, nucleotide catabolism, degradation of myofibrillar proteins, collagen degradation, dimethylamine (DMA) formation, free fatty acid (FFA) accumulation, and tyrosine accumulation.
- *Processing methods for seafood*: In the last 30 years, great advances have taken place in the processing of seafood. The basic observation is simple. No matter what method is used or invented to process the products, their biochemistry is affected. Three types of processing techniques will be discussed in this chapter: freezing, drying, and heating.

When studying this chapter, the reader should refer to Chapter 1 of this book for more details on food biochemistry.

NUTRITIVE COMPOSITION OF THE MAJOR GROUPS OF SEAFOOD AND THEIR HEALTH ATTRIBUTES

As mentioned earlier, seafood is nutritious. It contains nutrients, some of which are essential to human health. The nutrients are divided into two groups: macronutrients and micronutrients. This first section covers the macronutrients (water, protein, lipids) and their changes during processing and preservation. This is followed by a discussion of the micronutrients (vitamins and minerals). The implications of undesirable metals in seafood are then discussed. Issues related to health attributes in seafood are presented last.

COMPOSITION

Macronutrients

The macronutrients found in seafood include protein, fats and oils, and water. All other nutrients found in seafood are considered micronutrients and are of minor significance. Fish contains 63–84% water, 14–24% protein, and 0.5–17% lipid by weight. Fish, like other muscle sources of protein, have only small amounts of carbohydrates (as muscle glycogen). The amount of protein is similar in pelagic and demersal fish. However, pelagic fish, commonly called fatty fish, are higher in lipids (9–17 g/100 g) than demersal fish that live at the bottom of the ocean (0.3–1.6 g/100 g).

The primary source of fish protein is muscle, and the protein quality is comparable to other animal protein from milk, eggs, and beef. Muscle from fish and shellfish has very little connective tissue and is readily hydrolyzed upon heating, resulting in a product that is tender and easy to chew.

The forms of lipid in fish are triglycerides or triacylglycerols. Triglycerides in pelagic fish contain the long-chain polyunsaturated fatty acids (PUFAs) 20:5v3 eicosapentoic acidand 22:6v3 docosahexanoic acid (DHA), which have many health benefits including normal development of the brain and retina in infants and prevention of heart diseases in adults.

Pelagic fish store lipids in the head and muscle, whereas lipids in demersal fish are stored primarily in the liver and peritoneal lining and under the skin, except in halibut, which has both liver and muscle stores. The lipid content of pelagic fish varies with the season, feeding ground, water salinity, and spawning. Fish do not feed during spawning but depend on lipid stores for energy. For example, the lipid content of salmon may be 13% when the salmon begins traveling upstream to spawning grounds in the spring and only 5% at the end of the spawning season in the fall. Herring show more seasonal variation in lipid content, with a peak level of 20% in the spring, which drops to 10–15% during

Table 19.1. Distribution of Water, Protein, and Lipids in 100 g of Raw Edible Fish including Bones and Cartilage

Fish	Water (g)	Protein (g)	Lipid (g)
Cod	79–81	17–19	0.6–0.8
Herring	78–80	18–20	0.5–0.7
Ling	79–80	17–19	0.6–0.8
Mackerel	63–65	17–19	15–17
Monkfish	82–84	14–16	0.3–0.5
Pilchard/sardine	66–68	19–21	8–10
Plaice	78–80	15–17	1.3–1.5
Saithe/coley	79–81	17–19	0.9–1.1
Salmon	66–68	19–21	10–12
Skate	79–81	14–16	0.3–0.5
Sole	80–82	16–18	1.4–1.6
Sprat	68–70	16–18	9–10
Trout (rainbow)	75–77	18–20	5–6
Tuna	69–71	22–24	4–5
Whiting	79–81	17–19	0.5–0.8

Table 19.2. Distribution of Water, Protein, and Lipid in 100 g of Cooked, Edible Fish

Fish	Water (g)	Protein (g)	Lipid (g)
Haddock (smoked)	71–72	23—24	0.8–1.0
Haddock (steamed)	78–79	20—21	0.5–0.7
Herring (grilled)	63–64	20—21	11–12
Mackerel (smoked)	47–48	18—20	30–31
Mackerel (grilled)	58–59	20—21	17–18
Salmon (smoked)	64–65	25—26	4–5
Salmon (steamed)	64–65	21—22	11–12
Salmon (breaded nuggets)	60–61	12—13	11–12
Tuna (canned in water)	75–76	25—26	0.9–0.9
Tuna (canned in oil)	64–65	26—27	0.8–0.9
Tuna (raw, fresh, or frozen)	70–71	23—24	4–5

the fall spawning season, and to 3–4% in the winter, because of colder water temperatures and a scarcity of food. Water replaces the lipid depleted from muscle tissue and results in fish of poor eating quality.

Livers of species of the Gadidae (cod) family contain 50% lipids and are a concentrated source of vitamins A and D and omega-3 fatty acids. Oil is commercially extracted from the liver and purified for medicinal use and animal feed. Livers from other species of fish (haddock, coley, ling, shark, huss, halibut, and tuna) are also commercially processed to extract the fish oil.

In general, processing does not change the nutrient content of fish. Drying and smoking fish reduces the water content. The process can also reduce the lipid content of the fish if extremely high temperatures are used. Drying also increases the concentration of all nutrients on a weight basis. Sodium nitrite (usually with sodium nitrate) that is added during the process of drying or smoking increases the content of sodium considerably. The salmon flesh used for smoking is lower in lipids than whole salmon.

Breading fish adds carbohydrate and yields a final product with a relatively lower concentrations of protein, vitamins, and minerals than unbreaded fish.

Freezing and canning do not change the nutrient composition of fish except for canned tuna. Most canned tuna is low in lipids, since the lighter colored tuna is lower in fat and is preferred by consumers over the deep brown-red-colored tuna. Tuna is also cooked prior to canning, which removes some of the lipid.

Tables 19.1 and 19.2 compare the water, protein, and lipid contents of different forms of fish.

The macronutrients in crustaceans and mollusks differ from those in bony fish, since all mollusks contain carbohydrate and lobster and scallops contain small amounts of carbohydrate. Carbohydrate in shellfish is in the form of glycogen. The protein content is more variable than in bony fish, and the lipid content is comparable to that in pelagic fish (see Table 19.3; refer to Tables 19.4 and 19.5 for comparison of micronutrients). Ranges are used in Tables 19.1–19.5, because data from many sources are so variable.

Micronutrients

Micronutrients cover vitamins and minerals. Seafood is an important dietary source of vitamins A and D and the B vitamins B_6 and B_{12}. Although seafood is not high in thiamin or riboflavin, the amounts are comparable to those found in beef. Fish and shellfish have little or no vitamin C or folate.

Cod liver oil has been used to supplement diets with vitamins A and D; one teaspoon (5 g) of cod liver oil provides 12.5 μg of vitamin D and 1500 μg of vitamin A.

Mineral contributions of fish and shellfish include iodine and selenium; a 100 g portion contains more than 50% of the dietary reference intake (DRI) for iodine (DRI 5-150 μg/day) and selenium (DRI 5-55 μg/day), respectively. Generally, meat, eggs, and dairy products are poor sources of iodine and selenium. Iron and zinc are found in moderate amounts in most seafood. Clams and oysters are concentrated sources of zinc, with 24 and 91 mg/100 g serving, respectively, compared with a DRI of 15 mg/day. Seafood, similar to other dietary protein sources, is also an excellent source of potassium. Fresh seafood contains varying amounts of sodium, from a low of 31 mg/100 g of trout to a high of 856 mg/100 g of crab. Sodium is also added during canning and smoking fish and other seafood. In general, seafood is a poor source of calcium, except for whole canned salmon and sardines: when their bones are included, they will provide 200–400 mg calcium/100 g portion.

Other Components

Seafood, like other animal products, contains cholesterol. The cholesterol content in fish and mollusks ranges from 40 to 100 mg/100 g portion and is lower than in other meat such as beef, pork, and chicken. However, the shellfish, shrimp, and prawns are quite high in cholesterol—195 mg /100 g portion

Table 19.3. Distribution of Water, Protein, and Lipid in 100 g of Raw or Cooked Nonfish Seafood

Edible Nonfish Seafood	Water (g)	Protein (g)	Lipid (g)	Carbohydrate (g)
Crustacean varieties				
Crab (steamed)	77–78	19–20	1.5–1.6	0
Crayfish (steamed)	79–80	16–17	1–2	0
Lobster (steamed)	66–67	26–27	1–2	3–4
Prawn (raw)	79–80	17–18	0.5–0.7	0
Scallops (steamed)	73–74	23–24	1–2	3–4
Shrimp (boiled)	62–63	23–24	2–3	Trace
Mollusk varieties				
Clam (steamed)	63–64	25–26	1–2	5–6
Cuttlefish (steamed)	61–62	32–33	1–2	1–2
Mussel (steamed)	61–62	23–24	4–5	7–8
Octopus (steamed)	60–61	29–30	2–3	4–5
Oyster (steamed)	64–65	18–19	4–5	9–10
Whelk (steamed)	32–33	47–48	0.8–0.9	15–16

of shrimp. The cholesterol content of shrimp is of interest to individuals with cardiovascular disease who are advised to limit dietary cholesterol to 300–400 mg/day. However, diets high in total fat and saturated fat also increase the risk for cardiovascular disease, and both shrimp and prawns contain negligible amounts of fat.

Health Attributes

In the history of mankind, fish has been considered to have special health attributes, especially in various ethnic populations. Recently, scientific research has confirmed certain claims and raised controversy in relation to others. Seafood is a rich source of omega-3 fatty acids and may be good for the heart. Fish is being recommended for people of all ages to prevent a variety of disorders. Fish is also being promoted for women during pregnancy, for children at risk for asthma, and in the elderly to reduce the risk of Alzheimer's disease. However, pregnant women and children have been advised to limit fish to prevent the risk of mercury toxicity. Natural habitats are becoming polluted with mercury that is taken up and stored in some varieties of fish. Fish farming is a rapidly growing enterprise and has the benefit of controlling the food and water supply to avoid possible contamination of fish with toxic materials.

Fish and shellfish are an important part of a healthy diet. Fish and shellfish contain high-quality protein and other essential

Table 19.4. Vitamin Content of Fish and Shellfish per 100 g of Edible Fish or Shellfish (Raw Unless Otherwise Stated)

	Vitamins					
Seafood	A (mg)	D (mg)	E (mg)	B_1 (mg)	B_2 (mg)	B_6 (mg)
Fish						
Cod	1–3	0.5–1	0.2–0.5	0.02–0.05	0.03–0.06	0.15–0.20
Haddock	0.02–0.04	0.02–0.04	0.3–0.5	0.03–0.06	0.05–0.08	0.3–0.5
Herring	39–50	17–21	0.5–0.9	0.0–0.02	0.18–0.30	0.30–0.50
Mackerel	40–50	3–6	0.3–0.6	0.08–0.16	0.22–0.30	0.3–0.5
Salmon (Atlantic)	11–15	6–10	1.7–2.0	0.21–0.30	0.09–0.15	0.6–0.9
Trout (rainbow)	45–55	8–11	0.5–0.9	0.1–0.3	0.05–0.15	0.3–0.5
Tuna	20–30	6–9	Trace	0.05–0.15	0.1–0.2	0.2–0.5
Crustacean						
Crab	Trace	Trace	Trace	0.05–0.10	0.7–0.9	0.05–0.20
Lobster (boiled)	Trace	Trace	1.0–1.6	0.05–0.10	0.5–1.0	0.01–0.10
Prawns	Trace	Trace	2–3	0.0–0.7	0.05–0.20	0.02–0.07
Shrimp (boiled)	Trace	Trace	Trace	Trace	Trace	0.05–1.0
Mollusk						
Mussel	Trace	Trace	0.6–0.9	Trace	0.2–0.4	0.00–1.0
Oyster	60–80	0.5–1.5	0.5–1.0	0.1–0.3	0.1–0.3	0.1–0.2

Table 19.5. Mineral Content of Fish and Shellfish per 100 g of Edible Fish or Shellfish (Raw Unless Otherwise Stated)

	Minerals						
Seafood	Sodium (mg)	Potassium (mg)	Calcium (mg)	Iron (mg)	Zinc (mg)	Iodine (mg)	Selenium (mg)
Fish							
Cod	50–70	300–400	5–10	0.0–0.2	0.2–0.5	100–110	20–32
Haddock	50–70	350–400	10–18	0.0–0.2	0.2–0.5	200–250	25–30
Herring	119–130	300–350	50–70	0.9–1.5	0.5–1.3	25–35	30–40
Mackerel	60–70	270–320	9–13	0.6–1.0	0.4–0.8	120–160	25–35
Salmon (Atlantic)	40–50	350–370	19–23	0.5–1.0	0.5–1.5	70–82	20–30
Trout (rainbow)	40–50	400–450	15–20	0.1–0.5	0.3–0.8	10–16	15–20
Tuna	45–50	350–450	12–20	1.0–1.6	0.5–1.0	25–35	55–60
Crustacean							
Crab	400–450	230–280	Trace	1.3–1.8	5,0–6,0	Trace	15–20
Lobster (boiled)	300–350	250–280	60–70	0.5–1.0	2.0–3.0	95–105	120–140
Shrimp (boiled)	3500–3900	350–450	300–350	1.5–2.0	2.0–2.5	90–110	40–50
Prawn	150–220	300–400	75–80	1–2	1–2	1–3	10–20
Mollusk							
Mussel	250–310	300–350	30–42	5–7	2–3	120–160	45–55
Oyster	500–550	250–300	120–160	5–7	55–65	55–65	20–30

nutrients, are low in saturated fat, and contain two omega-3 fatty acids, eicosapentoic acid and DHA. The American Heart Association recommends that healthy adults eat at least two servings of fish a week (12 oz), especially varieties that are high in omega-3 fatty acids, such as mackerel, lake trout, herring, sardines, albacore tuna, and salmon.

Fish oil can help reduce deaths from heart disease, according to evidence in reports from the Agency for Healthcare Research and Quality (AHRQ, www.ahrq.gov/clinic/epcindex.htm#dietsup). A systematic review of the available literature found evidence that long-chain omega-3 fatty acids, the beneficial component ingested by eating fish or taking a fish oil supplement, reduce not only the risk for a heart attack and other problems related to heart and blood vessel disease in persons who already have these conditions, but also their overall risk of death. Although omega-3 fatty acids do not alter total cholesterol, HDL cholesterol, or LDL cholesterol, evidence suggests that they can reduce levels of triglycerides—a fat in the blood that may contribute to heart disease.

The AHRQ review reported evidence that fish oil can help lower high blood pressure slightly, may reduce the risk of coronary artery reblockage after angioplasty, may increase exercise capability among patients with clogged arteries, and may possibly reduce the risk of irregular heart beats—particularly in individuals with a recent heart attack. α-Linolenic acid (ALA), a type of omega-3 fatty acid from plants such as flaxseed, soybeans, and walnuts, may help reduce deaths from heart disease, but to a much lesser extent than fish oil.

On the basis of the evidence to date, it is not possible to conclude whether omega-3 fatty acids help improve respiratory outcomes in children and adults who have asthma. Omega-3 fatty acids appear to have mixed effects on people with inflammatory bowel disease, kidney disease, and osteoporosis, and no discernible effect on rheumatoid arthritis.

Depending on the stage of life, consumers need to be aware of both the benefits and risks of eating fish. Fish may contain mercury, which can harm an unborn baby or young child's developing nervous system. For most people, the risk from mercury by eating fish and shellfish is not a health concern. However, the Food and Drug Administration (FDA) and the Environmental Protection Agency (EPA) are advising women who may become pregnant, pregnant women, nursing mothers, and young children to avoid some types of fish known to be high in mercury content.

The risks from mercury in fish and shellfish depend on the amount of fish and shellfish eaten and the levels of mercury in the fish and shellfish. Nearly all fish and shellfish contain traces of mercury, but two varieties of fish high in omega-3 fatty acids (mackerel and albacore tuna) may contain high amounts of mercury. Larger fish that have lived longer have the highest levels of mercury because they have had more time to accumulate it. These large fish (swordfish, shark, king mackerel, and tilefish) pose the greatest risk. Other types of fish and shellfish may be eaten in the amounts recommended by the FDA and EPA (www.FDA.gov).

By following these recommendations for selecting and eating fish or shellfish, women and young children will receive the benefits of eating fish and shellfish and be confident that they have reduced their exposure to the harmful effects of mercury:

- Avoid eating fish that contain high levels of mercury: shark, swordfish, king mackerel, or tilefish.
- Eat up to 12 oz (two average meals) a week of a variety of fish lower in mercury: shrimp, canned light tuna, salmon,

pollock, and catfish. However, limit intakes of albacore ("white") tuna to 6 oz (one average meal) per week because albacore has more mercury than canned light tuna.
- Check local advisories about the safety of fish caught by family and friends in local lakes, rivers, and coastal areas. Some kinds of fish and shellfish caught in local waters may have higher or much lower than average levels of mercury. This depends on the levels of mercury in the water in which the fish are caught. Those fish with much lower levels may be eaten more frequently and in larger amounts. If no advice is available, eat up to 6 oz (one average meal) per week of fish you catch from local waters, but do not consume any other fish during that week.

For young children, these same recommendations can be followed, except that a serving size for children is smaller than for adults, 2–3 oz for children instead of 6 oz.

BIOCHEMISTRY OF GLYCOGEN DEGRADATION

When the fish or crustacean animals are being caught, they struggle vigorously in the fishing gear and on board, causing antemortem exhaustion of energy reserves, mainly glycogen, and high-energy phosphates. Asphyxia phenomena set in with gradual formation of anoxial conditions in the muscle. Tissue enzymes continue to metabolize energy reserves. Degradation of high-energy phosphates eventually produces hypoxanthine, followed by formation of formaldehyde, ammonia, inorganic phosphate, and ribose phosphates. The degradation of glycogen in fish follows the Embden-Meyerhof-Parnas pathway via the amylolytic route, catalyzed by endogenous enzymes. This results in an accumulation of lactic acid and a reduction of pH (7.2 to 5.5), contracting the tissues and inducing rigor mortis (Hobbs 1982, Hultin 1992a, Sikorski et al. 1990a). The rate of glycolysis is temperature dependent and is slowed by a lower storage temperature.

BIOCHEMISTRY OF PROTEIN DEGRADATION

Approximately 11–27% of seafood (fish, crustaceans, and mollusks) consists of crude proteins. The types of seafood proteins, similar to other muscle foods, may be classified as sarcoplasmic, myofibrillar, and stromal. The sarcoplasmic proteins, mainly albumins, account for approximately 30% of the total muscle proteins. A large proportion of sarcoplasmic proteins are composed of hemoproteins. The myofibrillar proteins are myosin, actin, actomysin, and troponin; these account for 40–60% of the total crude protein in fish. The rest of the muscle proteins, classified as stromal, consist mainly of collagenous material (Shahidi 1994).

Sarcoplasmic Proteins

Sarcoplasmic proteins are soluble proteins in the muscle sarcoplasm. They include a large number of proteins such as myoglobin, enzymes, and other albumins. The enzymatic degradation of myoglobin is discussed in the section on pigment degradation. Sarcoplasmic enzymes are responsible for quality deterioration of fish after death and before bacterial spoilage. The significant enzyme groups are hydrolases, oxidoreductases, and transferases. Other sarcoplasmic proteins are the heme pigments, parvalbumins, and antifreeze proteins (Haard et al. 1994). Glycolytic deterioration in seafood was discussed in Chapter 1 of this book and will not be repeated here. Changes in the heme pigments will be covered in the Section on "Biochemical Changes in Pigments during Handling, Storage, and Processing". Hydrolytic deterioration of seafood myofibrillar and collagenous proteins are discussed in the Section on "Myofibrillar Protein Deterioration."

Myofibrillar Protein Deterioration

The most common myofibrillar proteins in the muscles of aquatic animals are myosin, actin, tropomyosin, and troponins C, I, and T (Suzuki 1981). Myofibrillar proteins undergo changes during rigor mortis, resolution of rigor mortis, and long-term frozen storage. The integrity of the myofibrillar protein molecules and the texture of fish products are affected by these changes. These changes have been demonstrated in various research reports and reviews (Kye et al. 1988, Sikorski et al. 1990a, Martinez 1992, Haard 1992a, 1992b, 1994, Sikorski and Pan 1994, Sikorski 1994a, 1994b, Jiang 2000). The degradation of myofibrillar proteins in seafood causes these proteins to lose their integrity and gelation power in ice-stored seafood. The cooked seafood will no longer possess the characteristic firm texture of very fresh seafood; it will show a mushy or soft (sometimes mislabeled as tender) mouthfeel. For frozen seafoods, these degradations are accompanied by a loss in the functional characteristics of muscle proteins, mainly solubility, water retention, gelling ability, and lipid emulsifying properties. This situation gets even worse when the proteins are cross-linked due to the presence of formaldehyde formed from trimethylamine degradation. The cooked products become tough, chewy, and stringy or fibrous. Repeated freezing and thawing make the situation even worse. Readers should refer to the review by Sikorski and Kolakowska (1994) for a detailed discussion of the topic.

Stromal Protein Deterioration

The residue remaining after extraction of sarcoplasmic and myofibrillar proteins is known as stromal protein. It is composed of collagen and elastin from connective tissues (Sikorski and Borderias 1994). Degradation causes textural changes in these seafoods (honeycombing in skipjack tuna and mackerel and mushiness in freshwater prawn) (Frank et al. 1984, Nip et al. 1985, Pan et al. 1986). Bremmer (1992) reviewed the role of collagen in fish flesh structure, postmortem aspects, and the implications for fish processing, using electron microscopic illustrations. Jiang (2000) reviewed the proteinases involved in the textural changes of postmortem muscle and surimi-based products. Microstructural changes in ice-stored freshwater prawns have been revealed (Nip and Moy 1988). These textural changes are due to the degradation of collagenous matter and definitely influence the quality of seafood. For example, in freshwater prawn, the development of mushy texture downgrades its quality. In tuna, the development of honeycombing is an undesirable defect. In the postcooking examination of tuna before filling into the cans, the appearance of honeycombing is a sign of mishandling of the raw material. In more extensive cases, it may even cause the rejection of the precooked fish for further processing into canned tuna.

BIOCHEMICAL CHANGES IN NONPROTEIN NITROGENOUS COMPOUNDS

Reviews on the postmortem degradation of nonprotein nitrogenous (NPN) compounds are available (Sikorski et al. 1990a, Haard et al. 1994, Sikorski and Pan 1994). Such compounds in the meat of marine animals vary among species, with the habitat, and with life cycle; more importantly, they play a role in the postmortem handling processes (Sikorski et al. 1990a, Sikorski 1994a). Bykowski and Kolodziejski (1983) reported that white meat generally contains less NPN compounds than dark meat. For example, in the meat of white fish, the NPN generally made up 9–15% of the total nitrogen, in clupeids 16–18%, in muscles of mollusks and crustaceans 20–50%, and in some sharks up to 55%. Ikeda (1979) showed that about 95% of the total amount of NPN in the muscle of marine fish and shellfish is composed of free amino acids, imidazole dipeptides, trimethylamine oxide (TMAO) and its degradation products, urea, guanidine compounds, nucleotides and the products of their postmortem changes, and betaines.

The content of free amino acids in the body of oysters is higher in the winter than it is in the summer (Sakaguchi and Murata 1989).

The endogenous enzymatic breakdown of TMAO to DMA and then to formaldehyde and the bacterial reduction of TMAO to TMA have been most extensively studied. It should be noted that the production of DMA and formaldehyde takes place mainly in anaerobic conditions (Lundstrom et al. 1982).

Shark muscles contain fairly high amounts of urea, and ammonia may accumulate due to the activity of endogenous urease after death. This problem renders shark meat with ammonia odor unacceptable. Fortunately, this problem can be overcome by complete bleeding of the shark at the tail, gutting, filleting, and thorough washing right after catch to reduce the amount of substrate (urea). This procedure has been successfully practiced to provide acceptable shark meat for consumers.

After death, fish muscle also produces large amounts of ammonia due to degradation of adenosine triphosphate (ATP) to adenosine monophosphate (AMP), followed by deamination of AMP.

It should be noted that both ammonia production and degradation of TMAO can be either endogenous or contributed by bacteria. Ammonia, TMA, small amounts of DMA, and methylamine constitute the "total volatile base (TVB)," an indicator of freshness commonly used for seafood.

Postmortem enzymatic breakdown of nucleotides in fish may have a positive or negative impact on the flavor of seafood. The production of inosine monophosphate (IMP) at a certain concentrations in dried fish product can enhance the flavor (Murata and Sakaguchi 1989). Hypoxanthine contributes to bitterness and may add undesirable notes to the product (Lindsay 1991).

LIPIDS IN SEAFOODS

Lipid Composition

Lipids play an important nutritional role in seafoods, but at the same time, they contribute greatly to quality changes in many species. Just as aquatic species vary widely biologically, the lipid content and fatty acid types are greatly variable between species and sometimes within the same species. This can create a challenge for the seafood harvester and processor since varying amounts of lipids and the presence (or absence) of certain fatty acids can lead to significantly different effects on quality and shelf life. Seafood can be roughly classified into four categories based on fat content (O'Keefe 2000). Lean species (e.g., cod, halibut, pollock, snapper, shrimp, and scallops) have fat contents below 2%; low-fat species (e.g., yellowfin tuna, Atlantic sturgeon, and smelt) have fat contents between 2% and 4%; medium-fat species (e.g., catfish, mullet, trout, and salmon) are classified as having 4–8% total fat; and fatty species (e.g., herring, mackerel, eel, and sablefish) are classified as having more than 8% total fat. However, this classification is not valid for many species, since they can fluctuate widely in composition between seasons. For example, Atlantic mackerel can have less than 4% fat during late spring shortly after spawning, while it can have about 24% fat during late November due to active feeding (Leu et al. 1981; Fig. 19.1).

Different parts of the aquatic animal will also have different fat contents. Ohshima et al. (1993a) reported the following descending order for the fat content of mackerel and sardines:

- Skin (including subcutaneous lipids),
- Viscera,
- Dark muscle, and
- White muscle.

Some species, most notably salmonids, have larger fat deposits in the belly flaps than in other parts of the fish. Distribution of fat in many fish seems to be in descending fat levels from the head to the tail (Icekson et al. 1998, Kolstad et al. 2004). Diet will have a significant impact on the fat content and fatty acid profiles of aquatic animals. Aquacultured fish on a controlled diet, for example, do not show the same seasonal variation in fat content as wild species. It is also possible to modify the fat content and fat type via diet. For example, adding fish oil to fish feed will increase the fat content of fish (Lovell and Mohammed 1988), usually proportionately to the level of fat/oil in the diet (Solberg 2004). It has also been shown with many species (salmon being the most investigated) that the fatty acid composition is a reflection of the fatty acid composition in the fish diet (Jobling 2004). Therefore, it is possible to selectively increase the nutritional value of aquacultured fish by increasing the level and improving the ratio of nutritionally beneficial fatty acids (Lovell and Mohammed 1988). Conversely, one can

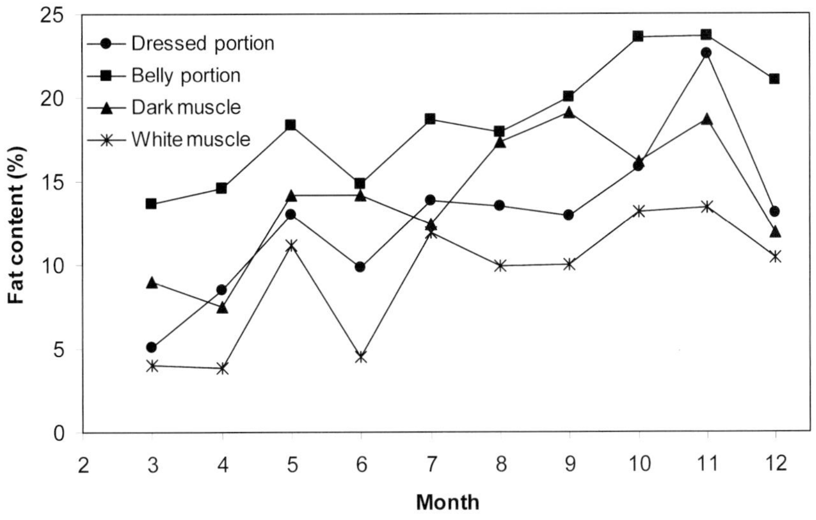

Figure 19.1. Seasonal variation in the fat content in different parts of Atlantic mackerel. (Adapted from Leu et al. 1981.)

decrease the level of unstable fatty acids to increase the shelf life of the product. A good example is catfish (Table 19.6), which in the wild has more unstable (but more nutritionally beneficial) fatty acids than grain-fed aquacultured catfish (Angelo 1996).

Table 19.6. Fatty Acid Composition of Extracted Lipids from Wild and Pond-Reared Channel Catfish

Fatty Acids	Wild (%)	Pond-Reared (%)
16:0	20.63	19.62
16:1v7	7.48	4.15
18:0	8.04	6.67
18:1v	23.72	42.45
18:2v6	2.9	12.35
18:3v3	2.78	1.66
18:4v3	0.6	0.96
20:1v9	0.83	0.97
20:3v6	0.36	0.98
20:4v6	6.82	1.88
20:4v3	0.62	0.16
20:5v3	7.02	1.49
24:1v9	0.34	0.16
22:4v6	0.68	0.14
22:5v6	1.34	0.51
22:5v3	3.05	0.98
22:6v6	13.56	4.97
Saturated	28.67	26.28
Monounsaturated	31.64	47.64
Polyunsaturated	39.77	26.07
v-6 PUFA	12.13	15.85
v-3 PUFA	27.64	10.22
v3/v6	2.54	0.62

Source: Adapted from Chanmugam et al. (1986).

Although varying levels of total fat content in fish can influence quality, fatty acid type rather than quantity appears to be more important. The major classes of lipids in fish are triglycerides and phospholipids. Triglycerides, the majority of lipids in fish (except very lean fish), are neutral lipids. They exist as either large droplets within the adipose tissue or smaller droplets between or within muscle cells (Undeland 1997). These lipids are deposited and stored as an energy source for fish (Huss 1994). The phospholipids, which are polar lipids, are one of the main building blocks of muscle cell membranes and are usually present only at low levels, 0.5–1.1% (O'Keefe 2000). Some muscles of very lean species such as cod, pollock, and whiting can contain less than 1% lipids, most of which are in the form of membrane phospholipids.

Compared with other animals, aquatic animals are especially rich in long-chain PUFAs, which are found in higher numbers in the membrane phospholipids than in storage triglycerides (Hultin and Kelleher 2000). Aquatic animals adapted to cold water also have higher levels of PUFAs than those adapted to warmer waters, since more unsaturation is essential to maintain membrane fluidity and function at low temperatures. It has been reported that about half of the membrane phospholipids in cold-water fish are the omega-3 fatty acids eicosapentoic acid (20:5v3) and DHA (22:6v3) (Shewfelt 1981). These fatty acids have been found to be highly active physiologically, for example, leading to decreased plasma triacylglycerols, cholesterol, and blood pressure (Wijendran and Hayes 2004) and modulating immunological activities (e.g., inflammation) (Klurfeld 2002), to name a few important functions.

Lipids and Quality Problems

Even though the phospholipids are at very low levels compared to neutral fats, they are believed to lead to more quality problems such as lipid oxidation, which is a major cause of quality deterioration in many aquatic foods. This is in part because of their higher level of unsaturation, but also is due to their significantly larger surface area compared with neutral lipids (Hultin and Kelleher 2000). Undeland et al. (2002) tested this hypothesis in a model system with cod membrane lipids and varying levels of neutral menhaden lipids and hemoglobin as a prooxidant. The oxidation was found to be independent of the amount of neutral lipids in the system. Their larger surface area also makes the unstable phospholipids more exposed to a variety of prooxidants (and antioxidants) such as heme proteins, radicals, iron, and copper. These prooxidants are especially rich in the metabolically more active dark muscle, which is in turn more susceptible to lipid oxidation than white muscle. These prooxidants, especially the heme proteins hemoglobin and myoglobin, can lead to high levels of lipid oxidation products, especially as pH of the muscle is reduced (which occurs postmortem) or when muscle is minced and these components are mixed in with lipids and oxygen (Hultin 1994, Undeland et al. 2004). The secondary products formed as a result of lipid oxidation have highly unpleasant off-odors and flavors, and they can also adversely affect the texture and color of the muscle. Several studies have suggested a link between lipid oxidation and protein oxidation (e.g., Srinivasan et al. 1996, King and Li 1999, Tironi et al. 2002). Oxidation of proteins can lead to protein cross-linking and thus textural defects, including muscle toughening and loss of water-holding capacity. King and Li (1999) proposed that lipid oxidation products may also induce protein denaturation, which in turn could cause textural problems.

Although oxidation is often delayed during frozen storage of fish muscle, thawed fish can oxidize more rapidly than fresh fish since muscle cells are disrupted, leading to an increase in free iron and copper, among other changes (Hultin 1994, Benjakul and Bauer 2001). Much frozen seafood has also been found to accumulate high levels of FFAs over time as a result of lipase and phospholipase activity (Reddy et al. 1992, Undeland 1997). These FFAs may or may not be more susceptible to lipid oxidation (Shewfelt 1981), but they can lead to protein denaturation and cross-linking and thus adverse effects on texture and water-holding capacity (Reddy et al. 1992).

Conflicting results have been reported on the effect of thermal processing on lipids. Canning has been reported to increase oxidation (Aubourg 2001). Short-term heating to 80°C, in contrast to long-term heating, has been reported to reduce oxidation due to inactivation of lipoxygenases, while long-term heating may accelerate nonenzymatic oxidation reactions (Wang et al. 1991). Shahidi and Spurvey (1996) reported that cooked dark muscle of mackerel oxidized more than raw dark muscle, while interestingly, the opposite was found for white muscle. In a study where herring was cooked by various methods (microwave, boiling, grilling, and frying), there was essentially no effect on lipid oxidation (Regulska-Ilow and Ilow 2002). That same study also demonstrated that fish lipids, including the nutritionally important omega-3 fatty acids, were highly stable under the different cooking methods (Fig. 19.2). However, very high temperatures (e.g., 550°C) can lead to decomposition of fish fatty acids and thus to decreased nutritional value (Sathivel et al. 2003). Changes in seafood fatty acid profiles can occur with frying. It has been reported that there may be an exchange of fatty acids from the frying medium and the muscle (Sebedio et al. 1993). This can either lower or increase the nutritional quality and stability of the seafood, depending on the type of frying oil and the type of seafood.

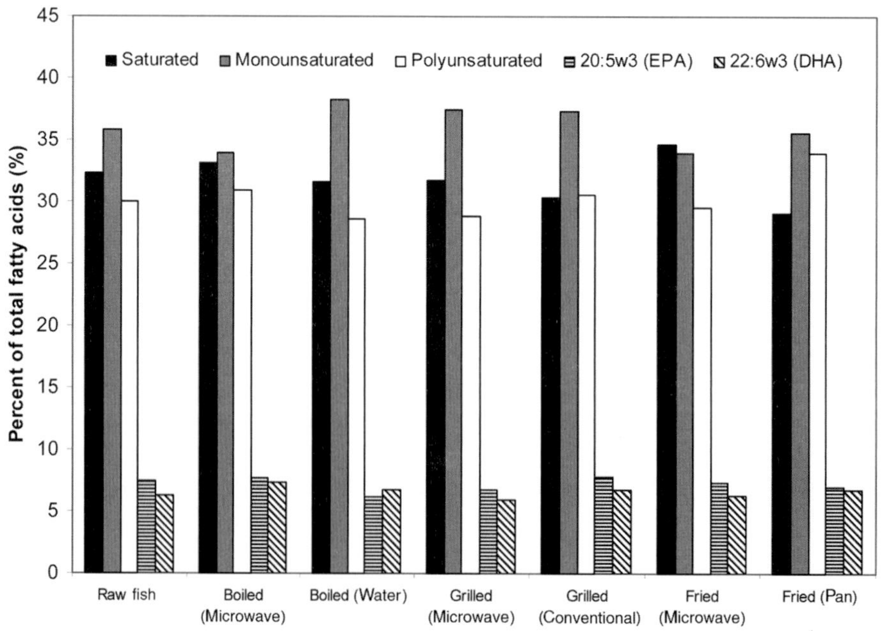

Figure 19.2. Effect of different thermal treatments on fatty acids in herring. (Adapted from Regulska-Ilow and Ilow 2002.)

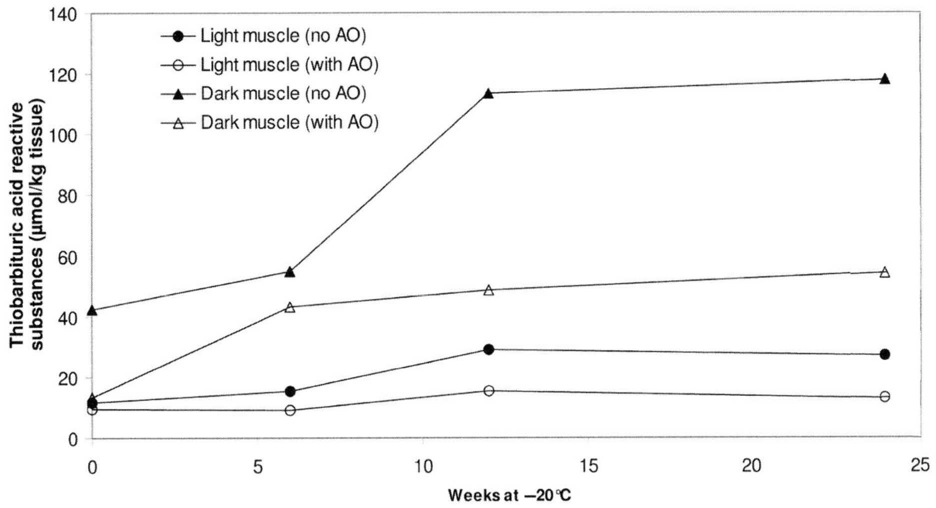

Figure 19.3. Effect of absence or presence of antioxidants (AO) on the oxidative stability of washed dark and white mackerel muscle during frozen storage (220°C). The antioxidant TBHQ was added during grinding, while ascorbate and EDTA were added during grinding and washing. (Adapted from Kelleher et al. 1992.)

Novel processing methods such as irradiation and high-pressure processing have been reported to lead to increased levels of lipid oxidation. High-pressure treatment has been reported to have relatively small effects on purified lipids from fish, while high-pressure treatment of fish muscle leads to high levels of lipid oxidation products (Ohshima et al. 1993a). This increase in oxidation is hypothesized, in part, to be via pressure-induced denaturation of heme proteins, which are potent prooxidants (Yagiz et al. unpublished data). Increased pressure and time also appears to lead to increased oxidation. Lipid hydrolysis has been found to be increased by medium levels of pressure, while it is reduced at high pressures, likely due to lipase inactivation (Ohshima et al. 1993b). Several studies have been conducted on the effect irradiation has on seafood quality, although this process is not allowed in many countries. It has been observed that low doses of gamma irradiation (5 kGy) may have an accelerating effect on the oxidative instability of some fish species and not others during postirradiation refrigerated storage. Al-Kahtani et al. (1996) reported that irradiation (1.5–10 kGy) led to more oxidation in tilapia and Spanish mackerel during refrigerated storage than in untreated fish. Fatty acid changes were also observed in that study. The likely cause for increased lipid oxidation on irradiation is the formation of radicals, more at higher levels of irradiation.

Minimization of Lipid-Derived Quality Problems

To maintain or extend the sensory and nutritional qualities of seafood, it is important to minimize or delay the undesirable reactions of lipids, such as oxidation and hydrolysis. Time is of the essence with these reactions, and thus interventions should be done as early as possible to be able to extend quality as much as possible. The simplest means of controlling oxidation is maintaining low temperatures, since enzymatic and nonenzymatic oxidation reactions are greatly influenced by temperature (Hultin 1994). As mentioned before, very low temperatures such as freezing will accelerate lipid hydrolysis. However, hydrolysis as well as oxidation will be reduced if products are kept at extremely low frozen storage temperatures (e.g., below −40°C) compared with conventional frozen storage (about −20°C or higher).

At the harvest level, bleeding fish can lead to significantly lower levels of oxidation, since heme proteins are reduced (Richards et al. 1998). Special care should be taken to prevent tissue disruption in storage, since both oxidation and hydrolysis will be increased. Washing fish fillets and fish mince will also remove significant amounts of heme proteins (Kelleher et al. 1992, Richards et al. 1998). Another effective way to reduce oxidation of unstable species is the removal of dark muscle, or deep skinning. This removes not only a large fraction of the heme proteins but also a large amount of oxidatively unstable lipids present in the dark muscle itself and below the skin. Protection from oxygen is another effective means for reducing oxidation. This can be achieved through modified atmosphere packaging, where oxygen is either reduced (for lean species) or completely removed (for fatty species). Vacuum packaging of seafood is also highly effective for reducing oxidation (Flick et al. 1992). Filleting fish under water (where O_2 is low) has also been reported to lead to less oxidation on storage than filleting in air (Richards et al. 1998). Special gases such as carbon monoxide have been found to significantly reduce oxidation of several species, even after fillets are removed from the gas, most likely since the gas reduces the prooxidative activities of heme proteins (Kristinsson et al. 2003).

Various antioxidative components can be highly effective in delaying lipid oxidation. All aquatic animals have a number of different indigenous antioxidants, and it is important to avoid conditions that can negatively affect or destroy these (Hultin 1994). It has been well researched and is well known that antioxidants can be added to seafood during processing to increase their oxidative stability (Fig. 19.3). Addition of antioxidants early during processing is also important. Since many of the prooxidants in seafood are in the aqueous phase and the lipids constitute a nonpolar phase, a combination of polar and nonpolar antioxidants has been found to be very effective. Tocopherol (vitamin E), a nonpolar antioxidant, and ascorbate (vitamin C), a polar antioxidant, are among the most commonly added antioxidants. Sometimes, metal chelators such as EDTA are added. However, it is worth mentioning that under certain conditions antioxidants can act as prooxidants. For example, this is true for ascorbic acid and EDTA, both of which can stimulate iron-mediated lipid oxidation. For example, when the ratio of EDTA: iron is 1, it is prooxidative, but when the ratio is .1, it is antioxidative (Halliwell and Gutteridge 1989). Since ascorbate is a reducing agent, it can also reduce iron under certain conditions and thus make it more prooxidative (Halliwell and Gutteridge 1988).

BIOCHEMICAL CHANGES IN PIGMENTS DURING HANDLING, STORAGE, AND PROCESSING

The main pigments in seafood can be classified as heme proteins (hemoglobin and myoglobin) in red-meat (warm-blooded) fish, hemocyanin in cold-blooded shellfish such as crustaceans and mollusks, and carotenoids in some important fish and shellfish products. The changes during handling, storage, and processing greatly affect the quality of these seafoods. Epithelial pigments and carotenoids also undergo changes during postmortem ice-chilled storage.

Epithelial Discoloration

The market value of squid is related to the contraction state of its epidermal chromatophores, called ommochromes. In recently harvested, prerigor squid, the pigment is dispersed throughout the chromatophores; hence, the dark red-brown appearance of the epithelial tissue. Following rigor mortis, the pigment cells contract, giving the skin a pale, light coloration dotted with dark flecks. The continued storage of squid results in a structural deterioration of the ommochrome membrane, leading to bleeding of pigment and downgrading of quality. The dark brown coloration characteristic of very fresh squid can be retained during processing, for example, by freezing or dehydration. However, improper storage of frozen or dried squid will result in chromatophore disruption and red discoloration of the meat (Hink and Stanley 1985).

The frozen storage of some fish may result in subcutaneous yellowing of flesh below the pigmented skin (Thompson and Thompson 1972). Apparently, freezing or other processes that disrupt chromatophores can lead to the release of carotenoids and their migration to the subcutaneous fat layer. Subcutaneous yellowing that occurs during the prolonged storage of frozen fish can originate from other causes, that is, yellowing associated with lipid oxidation and carbonyl-amine reactions.

Hemoglobin

Normally, hemoglobin contributes less to the appearance of seafood than myoglobin because it is lost easily during handling and storage, while myoglobin is retained in the intracellular structure. The amount of hemoglobin present also affects to a greater extent the color appearance of light red and dark red muscles of red-meat fish flesh (Wang and Amiro 1979). For example, in yellowfin tuna *Neothunnus macropterus*, hemoglobin concentrations ranged from 12 to 50 mg% in light red muscle and 50 to 380 mg% in dark red muscle (Livingston and Brown 1981). The amount of residual hemoglobin in fish muscle is obviously influenced by the bleeding efficacy at the time of catch. Method of catch also affects the residual hemoglobin in the fish muscle. For example, the percent of total heme as hemoglobin in the meat of yellowfin caught by bait boat and purse seine was 24% and 32%, respectively (Barrett et al. 1965).

The green meat of raw or frozen broadbill swordfish (*Xiphias gladus*) is believed to be due to the combination of hemoglobin with hydrogen sulfide generated from the fairly extensive decomposition of the meat (Amano and Tomiya 1953).

Hemocyanin

Hemocyanin, not hemoglobin, is present in the blood of shellfish, that is, crustaceans and mollusks. Hemocyanins are copper-containing proteins, as compared with iron-containing proteins in hemoglobins, and they combine reversely with oxygen. The contribution of hemocyanins to seafood quality is not very well understood. It is suspected that the blue discoloration of canned crabmeat is associated with a high content of hemocyanin. The average copper content of blue meat (e.g., 2.8 mg%) is higher than meat of normal color (e.g., 0.5 mg%) (Ghiretti 1956).

Myoglobin

Myoglobin in fish muscle is retained in the intracellular structure. In fish muscle, the red, white, and intermediate fibers tend to be more distinctively segregated than they are in muscle from land animals. The myoglobin content in muscle of yellowfin tuna was found to range from 37 to 128 mg% in the light-colored muscle and from 530 to 22,400 mg% in the dark-colored muscle (Wolfe et al. 1978). In cod, the deep-seated, dark-colored muscle is richer in myoglobin than superficial dark-colored muscle (Brown 1962, Love et al. 1977).

Myoglobin in fish is easily oxidized to a brown-colored metmyoglobin. The discoloration of tuna during frozen storage is associated with the formation of metmyoglobin, depending on temperature and location (Tichivangana and Morrissey 1985).

Greening is a discoloration problem associated with cooking various tunas. The problem arises from the formation of a sulfhydryl adduct of myoglobin in the presence of an oxidizing agent (e.g., TMAO). This greening problem can be effectively prevented by the use of a reducing agent (e.g., sodium hydrosulfite or another legally permitted additive) (Tomlinson 1966, Koizumi and Matsura 1967, Grosjean et al. 1969).

Carotenoids

The carotenoids contribute to the attractive yellow, orange, and red color of several important fish and shellfish products. The more expensive seafoods such as lobster, shrimp, salmon, and red snapper have orange-red integument and/or flesh from carotenoid pigments. For example, the red color of salmon is directly related to the price of the product. Carotenoids in seafoods may be easily oxidized by a lipogenase-like enzyme (Schwimmer 1981).

Melanosis (Melanin Formation)

Melanosis (melanin formation) or "blackspot" in shrimp during postmortem storage is caused by phenol oxidase and has economic implications (Simpson et al. 1987). Consumers consider shrimp with blackspot to be defective. This problem can be overcome by the use of appropriate reducing or inhibiting agent(s).

A comprehensive review of the biochemistry of color and color change in seafood was presented by Haard (1992a). Readers should refer to this reference for detailed information on this subject.

BIOCHEMICAL INDICES

Postharvest biochemical events in fish can be classified into two phases: metabolic (enzymatic) and microbial (Eskin et al. 1971). A discussion of microbial events is not the objective of this chapter. If interested, please consult references in this chapter or other sources. Abundant literature is available. Physical and instrumental methods for assessing seafood quality were reviewed by Sorenson (1992).

The metabolic (enzymatic) changes result from the activity of enzymes remaining in the fish flesh after death. Metabolites from these enzymatic changes can be used as indices of freshness and can be monitored by biochemical or chemical methods. The major metabolites coming from the actions of inherent enzymes in the fish themselves are lactic acid, nucleotide catabolites, collagen and myofibrillar protein degradation products, DMA formation, FFA accumulation, and tyrosine. Methodology for the analysis of these metabolites can be found in government publications such as Official Methods of Analysis of the American Association of Official Chemists in the United States.

Lactic Acid Formation with Lowering of pH

It is well known that, in animals, lactic acid accumulates during postmortem changes because of glycolytic conversion of storage glycogen in the muscle after cessation of respiration, and finfish, crustaceans, and shellfish are no exception. The metabolic pathway of glycogen degradation in animals was presented in Chapter 1 of this book. The accumulation of lactic acid can cause a drop in pH. Both lactic acid and pH can be measured without difficulty using modern instrumentation (Jacober and Rand 1982).

Nucleotide Catabolism

Postmortem dephosphorylation of nucleotides by autolysis in fish has been studied for many years as an index of quality. Nucleotide degradation commences with death and proceeds at a temperature-dependent rate (Spinelli et al. 1964, Eskin et al. 1971). Adenosine nucleotides are rapidly deaminated to IMP and further degraded from inosine to hypoxanthine during storage (Jones et al. 1964). There is no hypoxanthine in freshly caught fish and marine invertebrates. Accumulation of the metabolite hypoxanthine has attracted considerable attention for many years as an index of fish freshness. Hypoxanthine content can be determined by applying the colorimetric xanthine oxidase (EC 1.2.3.2) test. The rate of postmortem degradation of adenosine nucleotides in marine fish and invertebrates differs with the species and muscle type. For this reason, hypoxanthine analysis is of limited use for the evaluation of quality in certain species. It is most useful for the analysis of pelagics, redfish, salmon, and squid but is of little value for the estimation of lean fish quality (Gill 1992).

Because hypoxanthine is a metabolite fairly close to the end of the nucleotide degradation, Saito et al. (1959) proposed the use of K-value, defined as the ratio of inosine (HxR) plus hypoxanthine (Hx) to the total amount of ATP and related compounds (ADP, AMP, IMP, HxR, and Hx) in a fish muscle extract. Many researchers have demonstrated that K-value is related to fish freshness. However, its limitation is that it is difficult to analyze all these nucleotides and their metabolites by simple procedures. The use of high performance liquid chromatography offers accurate and reliable results but is not suitable for routine analysis. Since adenosine nucleotides rapidly break down and disappear within 24 hours postmortem, the K-value can be simplified as $(HxR + Hx)/(IMP + HxR + Hx)$. Karube et al. (1984) called it the K_1-value and reported that this value is strongly correlated to the K-value proposed by Saito et al. (1959). Like the K-value, the K_1-value is species dependent (Huynh et al. 1990).

Degradation of Myofibrillar Proteins

During chilled storage of fish and marine invertebrates, it is common to notice textural changes before bacterial spoilage. It is believed that these textural changes are caused by the autolytic degradation of myofibrillar proteins. This was demonstrated in ice-stored freshwater prawn (Kye et al. 1988) and in other finfish (Shewfelt 1980). Analysis of myofibrillar proteins is tedious and time-consuming using electrophoresis, but it will give a definite picture of protein degradation.

Collagen Degradation

In the tuna canning industry, it was long recognized that the appearance of honeycombing in precooked tuna was an index of deterioration or mishandling. It was demonstrated that the appearance of honeycombing was related to collagen degradation (Frank et al. 1984). Mackerel was also shown to develop honeycombing when poorly handled (Pan et al. 1986). Freshwater prawn developed mushiness rapidly due to poor handling after catch. It was also demonstrated that this mushiness problem was related to collagen degradation (Nip et al. 1985). It should be noted that all these problems are observable only after the fish or prawn have been cooked and are only detected visibly. It is believed that these problems developed before bacterial spoilage of the fish. Analysis of collagen degradation is also tedious, involving hydrolysis of the extracted collagen and analysis of hydroxyproline.

Dimethylamine Formation

TMAO is commonly found in large quantities in marine species of fish, especially the elasmobranch (Jiang and Lee 2004) and gadoid (Bonnell 1994) species. After death, TMAO is readily degraded to DMA through a series of reactions during iced and frozen storage. DMA is typically observed in frozen gadoid species such as cod, hake, haddock, whiting, red hake, and pollock (Castell et al. 1973). TMAO degradation with DMA formation was enhanced by the presence of an endogenous enzyme (TMAOase) in the fish tissues, as observed in cod muscle by Amano and Yamada (1965). It should be noted that TMAO degradation can also be bacterial. Readers should consult the reviews by Regenstein et al. (1982), Hebard et al. (1982), Hultin (1992b), or other literature elsewhere.

Free Fatty Acid Accumulation

Postmortem lipid degradation in seafood, especially fatty fish, proceeds mainly due to enzymatic hydrolysis, with the accumulation of FFAs. About 20% of lipids are hydrolyzed during the shelf life of iced fish. The amount of FFAs is more or less doubled during that period, mostly from phospholipids, followed by triglycerides, cholesterol esters, and wax esters (Haard 1990, Sikorski et al. 1990b).

Tyrosine Accumulation

The accumulation of lactic acid accompanied with a drop in pH causes the liberation and activation of inherent acid cell proteases, cathepsins (Eskin et al. 1971). Tyrosine has been reported to accumulate in stored fish due to autolysis. Its use as an index of freshness has been proposed by Shenouda et al. (1979) because of its simplicity in analysis. However, it was shown that the pattern of tyrosine accumulation was similar to that of the TVB, an indicator of bacterial spoilage. Therefore, its use as a biochemical index (before bacterial spoilage) is not sufficiently sensitive and specific enough to assess total fish quality (Simpson and Haard 1984).

BIOCHEMICAL AND PHYSICOCHEMICAL CHANGES IN SEAFOOD DURING FREEZING AND FROZEN STORAGE

Freezing is widely used to preserve and maintain the quality of food products for an extended period of time because microbial growth and enzymatic and biochemical reactions are reduced at low temperatures. However, ice crystals formed as a result of freezing may damage cells and disrupt the texture of food products, and the concentrated unfrozen matrix may result in changes in pH, osmotic pressure, and ionic strength. These changes can affect biochemical and physicochemical reactions such as protein denaturation, lipid oxidation, and enzymatic degradation of TMAO in frozen seafood. Therefore, it is essential to understand these reactions in order to extend the shelf life and improve the quality attributes of seafood.

Protein Denaturation

During freezing or frozen storage, changes in the physical state of water and the presence of lipids create an environment that induces protein denaturation. This denaturation can be caused by one or more of the following factors: (1) ice crystal formation, (2) dehydration effect, (3) increase in solute concentration, (4) interaction of protein with intact lipids and FFAs, and (5) interaction of protein with oxidized lipids (Shenouda 1980). Denaturation of protein changes the texture and functional properties of protein. The texture of fish may become more fibrous and tough as a result of the loss of protein solubility (Benjakul and Sutthipan 2009, Reza et al. 2009) and water-holding capacity. For example, the gel strength of rainbow fish (*Trichiurus savala*) protein significantly decreases after 100 days of storage at −20°C (Mishra and Dora 2010). These textural changes as a result of protein changes give rise to undesirable sensory attributes, which are often described as sponginess, dryness, rubbery texture, and loss of juiciness (Haard 1992a). Tseng et al. (2003) suggested that to retain good eating qualities, that is, to maintain tenderness and cooking yield, and reduce lipid oxidation, red claw crayfish (*Cherax quadricarinatus*) should not be subjected to more than three freeze-thaw cycles.

Ice Crystal Effect

Ice may form inter- and intracellularly during freezing, which ruptures membranes and changes the structure of the muscle cells (Mazur 1970, 1984, Friedler et al. 1988). At a slow freezing rate, fluids in the extracellular spaces freeze first, thus increasing the concentration of extracellular solutes and drawing water osmotically from the unfrozen cell through the semipermeable cellular membrane (Mazur 1970, 1984). The diffusion of water from the internal cellular spaces to the extracellular spaces results in drip, collected from frozen muscle tissues when thawed (Jiang and Lee 2004). Drip contains proteins, peptides, amino acids, lactic acid, purines, vitamin B complex, and various salts (Sulzbacher and Gaddis 1968), and their concentration in drip increases with storage time (Einen et al. 2002). On the other hand, when the tissue is frozen rapidly, the cellular fluids do not have enough time to migrate out to the extracellular spaces and will freeze as small crystals uniformly distributed throughout the tissue (Mazur 1984). Thus, rapid freezing has a less detrimental effect on the cell or tissue (Coggins and Chamul 2004). However, if the storage temperature fluctuates, the intracellular ice recrystallizes, forming large ice crystals and causing cell disruption, which leads to drip loss as well. Mechanical damage from ice crystals on the texture of food during freezing and frozen storage has been studied in kiwi fruit (Fuster et al. 1994), lamb meat (Payne and Young 1995), squid mantle muscle (Ueng and Chow 1998), pork (Ngapo et al. 1999), and beef (Farouk et al. 2003).

During storage, if the product is not wrapped or is improperly packaged, ice may sublime into the headspace and produce a product defect called freezer burn (Blond and Le Meste 2004, Coggins and Chamul 2004). The sublimation of ice crystals leaves behind small cavities and causes the surface of fish to appear grayish. It is more pronounced when the storage temperature is high (Blond and Le Meste 2004). Freezer burn increases the rate of rancidity and discoloration because of the greater exposed surface, resulting in a "woody" (Blond and Le Meste 2004), tough, and dry texture in fish, making and leaving the fish less acceptable (Coggins and Chamul 2004).

Dehydration Effect

Native protein is stabilized by the folding of hydrophobic chains into the protein molecules and is held together by many other forces, including hydrogen bonding, dipole–dipole interactions, electrostatic interactions, and disulfite linkages (Benjakul and Bauer 2000). The formation of ice during freezing removes water from the protein molecules, thus disrupting the hydrogen bonding network between the native proteins and water molecules and exposing the hydrophobic or hydrophilic sites of the protein molecules (Shenouda 1980). The exposed hydrophobic or hydrophilic sites of the protein molecules interact with each other to form hydrophobic–hydrophobic and hydrophilic–hydrophilic bonds, either within the protein molecules, resulting in deconformation of the native three-dimensional structure of protein, or between adjacent protein molecules, causing protein–protein interactions and resulting in aggregation (Shenouda 1980). Xiong (1997) proposed that protein aggregates to maintain its lowest free energy as water forms ice, thus resulting in protein denaturation. Matsumoto (1979) suggested that redistribution of water during freezing allows protein molecules to move closer together and aggregate through intermolecular interactions. Lim and Haard (1984) found that the loss of protein solubility as a result of protein denaturation in Greenland halibut during frozen storage was mostly due to the noncovalent, hydrophobic interactions in protein molecules. Buttkus (1970) proposed that the formation of intermolecular S–S bonds is the major cause of protein denaturation.

Solute Concentration Effect

As ice forms, the concentration of mineral salts and soluble organic substances in the unfrozen matrix increases. As a result, salts and other compounds that are only slightly soluble (such as phosphate) may precipitate out, which will change the pH (Einen et al. 2002) and ionic strength of the unfrozen matrix and cause conformational changes in proteins. Ions in the concentrated matrix will compete with the existing electrostatic bonds and cause the breakdown of some of the electrostatic bonds (Dyer and Dingle 1961, Shenouda 1980). Takahashi et al. (1993) found, in their freeze denaturation study of carp myofibrils with KCl or NaCl, that freeze denaturation above $-13^{\circ}C$ is caused by the concentrated salt solution. Effect of freeze-concentration in a cellular structure was demonstrated in a phospholipid liposomal model system (Siow et al. 2007).

Reaction of Protein with Intact Lipids

There are different views in the literature on the effect of intact lipids (i.e., lipids that have not been subjected to partial or total hydrolysis or oxidation) on fish proteins. On one hand, they seem to protect proteins; on the other, they form lipoprotein complexes, which affect protein properties (Shenouda 1980, Mackie 1993). Dyer and Dingle (1961) found that lean fish (fat content less than 1%) showed a rapid decrease in protein (actomyosin) extractability when compared with fatty fish species (3–10% lipids). Therefore, they hypothesized that moderate levels of lipids may reduce protein denaturation during frozen storage. In contrast, Shenouda and Piggot (1974) observed a detrimental effect of intact lipids on protein denaturation in their study of a model system, which involved incubating lipid and protein extracted from the same fish at $4^{\circ}C$ overnight. They showed that when fish actin (G-form) was incubated with fish polar or neutral lipids, high molecular weight protein aggregates formed. They suggested that during freezing, lipid and protein components form lipoprotein complexes, which change the textural quality of muscle tissue.

Reaction of Proteins with Oxidized Lipids

During frozen storage, lipid oxidation products cause proteins to become insoluble and harder (Takama 1974). When proteins are exposed to peroxidized lipids, peroxidized lipid–protein complexes will form through hydrophobic interactions or hydrogen bonds (Narayan et al. 1964), thus causing conformational changes in the protein. The unstable free radical intermediates of lipid peroxidation remove hydrogen from protein, forming a protein radical, which could initiate various reactions such as cross-linking with other proteins or lipids and formation of protein–protein and protein–lipid aggregates (Karel et al. 1975, Schaich and Karel 1975, Gardner 1979). Roubal and Tappel (1966) found that peroxidized protein cross-links into a range of oligomers, which are associated with protein insolubility. Careche and Tejada (1994) found that oleic and myristic acid had a detrimental effect on the ATPase activity, protein solubility, and viscosity of hake muscle during frozen storage.

Secondary products from lipid oxidation such as aldehydes react chemically with the amino groups of proteins through the formation of Shiff base adducts, which fluoresce (Leake and Karel 1985, Kikugawa et al. 1989). Ang and Hultin (1989) suggested that formaldehyde might interact with protein side chains and form aggregates without causing cross-linking.

Lipid Oxidation and Hydrolysis

Lipids degrade by two mechanisms: hydrolysis (lipolysis) and oxidation (Shenouda 1980, Shewfelt 1981). In frozen foods, enzymes generally catalyze lipid hydrolysis. Phospholipase A and lipases from muscle tissues are responsible for the hydrolysis of phospholipids and lipids in frozen fish (Shewfelt 1981, De Koning et al. 1987). As the storage time increases and the frozen storage temperature elevates, FFAs from the hydrolysis of lipids start to accumulate (Dyer and Dingle 1961). FFAs have a detrimental effect on both the textural properties and flavor of fish.

Lipid oxidation is considered one of the major factors that limit the shelf life of frozen seafood. Fish lipids are known for their susceptibility to oxidation, particularly during frozen storage. Oxidation of PUFAs yields various oxidative products including a mixture of aldehydes, epoxides, and ketones, which give fish a rancid flavor (Gardner 1979). The low flavor threshold of most aldehydes formed during lipid oxidation means they are easily perceived by the consumer and, therefore, reduce the acceptability of the products. Rancid flavor in salmon is caused by the formation of volatile products such as (E,Z)-2,6-nonadienal (cucumber odor), (Z)-3-hexanal (green odor), and (Z,Z)-3,6-nonadienal (fatty odor) (Milo and Grosch 1996). The characteristic "seaweed" odor of fresh fish tissue results from the volatile compounds formed during rapid degradation of site-specific hydroperoxides (Josephson and Lindsey 1986). The content of lipid hydroperoxides and FFAs in salmon increases during storage, and these changes are fastest when stored at -10°C (Refsgaard et al. 1998). Cod samples stored for 18 months at -15°C had hepta-trans-2-enal and hepta-trans-2,cis-dienal. These compounds were described as cold storage flavor (cardboard, musty) with a very low flavor threshold (Coggins and Chamul 2004). Brake and Fennema (1999) found that the rate of decrease for thiobarbituric acid reactive substances (TBARS) was abrupt below glass transition temperature (Tg′), whereas the rates of decrease for lipid hydrolysis and peroxide values were moderate to small, respectively, in frozen minced mackerel. They suggested that the TBARS reduction rate is more diffusion limited than those for lipid hydrolysis and peroxide.

Color changes are an indicator of food quality deterioration. Nonenzymic browning occurs as a consequence of chemical reactions between peroxidizing lipids in the presence of protein. Fluorescence Schiff base adducts formed as a result of chemical reactions between tetrameric dialdehyde and amino groups of protein (Haard 1992a, 1992b). Aubourg (1998) observed a higher fluorescence ratio in a formaldehyde and fatty fish model system compared with the model for formaldehyde and lean fish. In addition to flavor and color changes, lipid oxidation products, including FFAs and aldehydes, decrease protein solubility and cause undesirable changes in the functional properties of proteins (Sikorski et al. 1976).

Aquatic species are especially rich in long-chain PUFAs and are more susceptible to lipid oxidation. As a result, various antioxidants have been experimented with by adding antioxidants to seafood during processing, thus minimizing or delaying the oxidation process and prolonging shelf life of seafood. Natural antioxidants seem to be the preferred types for recent research studies. Rosemary extract added to sea salmon mince reduced lipid oxidation during storage up to 6 months (Tironi et al. 2010) and garlic and grape seed extract have also shown to have a lipid oxidation protective effect (Yerlikaya et al. 2010). However, grape seed extract did not score well in sensory aspects for appearance, odor, and taste. Coating of fish with chitosan film has been demonstrated to be successful in effectively inhibiting lipid oxidation during frozen storage. Examples are the coating of silver carp (Fan et al. 2009) and lingcod fillet (Duan et al. 2010) with chitosan prior to frozen storage.

Lipid oxidation stability may also be achieved through feeding commercially farmed fish with a diet enriched with natural antioxidants such as tocopherol isomers (Ortiz et al. 2009). α-Tocopheryl acetate was also used as a dietary vitamin E supplement in beluga sturgeon (*Huso huso*) because it is fairly stable in the feed and only becomes active as an antioxidant after the hydrolysis of the acetate group in the fish body. It was shown to slow down lipid oxidation of the fish muscle during frozen storage (Hosseini et al. 2010).

Degradation of Trimethylamine Oxide

TMAO, a source of formaldehyde, is present naturally in many marine animals as an osmoregulator and as a means of excreting nitrogen (Hebard et al. 1982). After death, TMAO is readily degraded to DMA and formaldehyde in the presence of the endogenous enzyme (TMAOase) in the fish tissues (Bremmer 1977, Hebard et al. 1982, Nielsen and Jorgensen 2004). It has been postulated that formaldehyde binds covalently to various functional groups in proteins and hence results in a deconformation of the protein, followed by cross-linking between the protein peptide chains via methylene bridges (Sikorski et al. 1976). The interaction of formaldehyde with muscle protein accelerates muscle protein denaturation (Crawford et al. 1979, Ciarlo et al. 1985, Sotelo et al. 1995). Research using both mechanical and sensory tests has shown that formaldehyde formed from the degradation of TMAO increases the firmness and decreases juiciness of mince prepared from white muscle or fillets of gadoid and nongadoid fish species (Rehbein 1988). The presence of formaldehyde also causes a noticeable decrease in the extractability of total proteins, particularly the myofibrillar group (Lim and Haard 1984, Benjakul and Bauer 2000). Ishikawa et al. (1978) suggested that depletion of TMAO accelerates the autoxidation reaction of lipids. Therefore, it is clear that the degradation of TMAO increases the toughness of muscle protein and accelerates the oxidation and hydrolysis of lipids.

Summary

During freezing, the formation of ice changes the order of water molecules in the food environment, causes dehydration and solute concentration, and thus disturbs the conformation of protein, leading to protein denaturation. Lipid degradation and enzymatic degradation of TMAO during freezing affect the textural and sensory properties of frozen seafood. Therefore, it is important to understand the mechanisms of various biochemical and physicochemical reactions occurring during freezing and frozen storage so that the quality of frozen food can be maintained.

BIOCHEMISTRY OF DRIED, FERMENTED, PICKLED, AND SMOKED SEAFOOD

Salting, fermenting, marinating (pickling), drying, and smoking of fish and marine invertebrates increase the shelf life and develop in the products' desirable sensory properties. Extension of storage life is achieved mainly through the combined effects of

- a reduction in water activity from the addition of salt;
- a decrease in microbial load by the application of heat;
- the presence of inherent preservatives such as acetic acid;
- the use of chemical preservatives such as ascorbic acid, BHA, and BHT; and
- the antibacterial and antioxidant activities of various smoke components.

The enzymatic and spoilage processes are controlled not only by these chemical components but also by the temperature, pH, and availability of oxygen. Desirable changes in sensory properties are thus developed as a result of these carefully controlled chemical and enzymatic processes (Shewan 1944, Doe and Olley 1990, Miler and Sikorski 1990, Shenderyuk and Bykowski 1990, Perez-Villarreal 1992, Fuke 1994, Haard 1994, Sikorski and Ruiter 1994).

There are some changes in proteins. In the salting of fish, salt will penetrate slowly into the tissues, affecting the stability of the native proteins and reducing their extractability. Heavily salted fish, when compared with less salted fish, has the following disadvantages:

- More water loss due to osmosis
- Tougher texture
- Less developed flavor

It should be remembered that enzymes are also proteins, and their activities are affected by salt concentration.

Another effect of salting is the changes in texture of the final product. It is believed that the calcium and magnesium ions present as impurities in salt may penetrate the fish, giving rise to a soft, "mushy" texture in the fillet. This may be undesirable for most salted fish, but this effect is considered highly desirable in some Chinese salted products (tender salted threadfin, tender salted mackerel, and others), and Scandinavian products (such as kryddersild, tidbit, and gaffelbiter) (Shewan 1944). Salting of whole fish should be controlled precisely, as overripening will result in excessively soft products with sensory properties that are undesirable to most consumers. However, changes in salted fish fillets depend mostly on endogenous muscle proteases.

Fermented fish paste and sauce are popular products prepared and consumed in southern China and Southeast Asian countries as a source of nutrients and as condiments. Generally, whole fish or shrimp are used as the raw materials for the preparation of these products. It is believed that extensive proteolysis occurs under carefully controlled conditions. It is difficult to differentiate the endogenous and bacterial actions of proteolysis because of the use of whole fish or shrimp and the way these products are produced. Sikorski and Ruiter (1994) summarized some of the work on proteolysis in fermented fish products. Cathepsins A and C as well as trypsin-like enzyme are endogenous proteases that appear to contribute to fish sauce production both in yield and quality (Orejana and Liston 1982, Rosario and Maldo 1984, Raksakulkthai et al. 1986).

Chiou et al. (1989) reported that cathepsin D-like and aminopeptidase activities release a large amount of free amino acids during roe processing and contribute to the flavor. In the drying of squid, endogenous cathepsin C appears to contribute to desirable qualities (texture and flavor) of traditional products (Haard 1983). Simpson and Haard (1984) reported that added trypsin appears to be a key enzyme contributing to the texture and flavor of matjes herring.

Marinating fish (mainly herring) by means of salt and acetic acid is one of the oldest ways of preserving food in European countries. The acid condition of the marinades, with pH 4–4.5, makes the tissue cathepsins much more active. This results in the degradation of muscle proteins into peptides and amino acids. This permits the marinade to create the proper flavor and texture in the product (Meyers 1965).

Smoke curing means the smoking of presalted fish. The action of the smoke constituents produces a unique smoky odor, taste, and color. The process also contributes to the tenderizing action on the fish tissue. In cold-smoked fish, this tenderization is caused predominately by the action of endogenous proteolytic enzymes because of their activity at such ambient temperatures.

BIOCHEMISTRY OF THERMAL-PROCESSED PRODUCTS

Exposure of the fish to elevated temperatures is detrimental to tissue structure, and results in very undesirable effects (Haard 1994). Extreme texture softening in fish has been attributed to cysteine proteases acting on the fish muscle at elevated temperatures (Konegaya 1984). Heat-stable alkaline proteases and neutral proteases (modori or gel-degradation) are active when the temperature reaches 60–70°C (Lin and Lanier 1980, Kinoshita et al. 1990). Degradation of connective tissue of mackerel appears to be associated with muscle proteases and pyloric ceca collagenase (Pan et al. 1986).

The rate of oxidation of desirable myoglobin and oxymyoglobin of the red muscles of tuna to brown metmyoglobin depends on the species of the fish and on the storage temperature (Mattews 1983). Color deterioration in iced and frozen stored bonito, yellowfin, and skipjack tuna caught in Seychelles waters was demonstrated in this study.

Food biochemistry plays a role in the production of thermally processed seafood. During the canning of seafood, it is a common practice to precook the raw materials, for example, fish and crustaceans. The accompanying thermal treatment coagulates the proteins in the muscles for easy handling and removal of nonedibles as well as for quality inspection in the later steps. It also inactivates the enzymes that can cause biochemical deterioration of the raw materials, when exposed to elevated temperatures and extended storage times. For example, in manufacturing canned tuna, the fish is precooked and permitted to cool completely, sometimes overnight, before the following steps: (1) removal of skin, bones, dark meats, and viscera and (2) inspection for the presence of defects, for example, honeycombed and/or burnt tissues.

It should be noted that honeycombed tissues are only detectable in the cooked fish. In the production of canned crabmeat, the precooked crab is cooled before the extraction of crabmeat. Without precooking to coagulate the crabmeat, it is almost impossible to separate the crabmeat from its shell efficiently. The precooking also inactivates those enzymes that can degrade the quality of the product. This is especially important in products like crab and shrimp, as the deteriorating enzymes can act very quickly on these tissues at elevated temperatures. However, there are exceptions where the raw fish is not precooked in order to preserve the premium quality. This includes the production of canned salmon, where sections of raw salmon with skin and bones are stuffed into the can before sealing and processing.

Thermal processing such as mild heating also is applied in the production of some dried seafood products such as fish, shrimp, and squid. The heating process inactivates the deteriorating enzymes in the raw materials. This stops the enzymatic reactions from occurring during drying or dehydrating processes that expose the intermediate products to ambient or elevated temperatures for extended periods. For example, in the production of dried, shaved bonito, the raw bonito is first precooked in brine before the processes of recovery of loin tissues, drying to coagulate the tissues, and shaving of the dried product. Without this precooking process, the fish tissue will deteriorate during the long drying process. For dried shrimp, the majority of the product is produced with the precooking process to inactivate the deteriorating enzymes before drying or dehydrating. However, a small amount of dried shrimp is produced without precooking to produce specialty products. For dried squid, it is usually produced without precooking to develop the unique flavor from enzymatic reactions during drying or dehydration. However, a small amount is produced by precooking the product prior to the drying process.

Thermal processing of seafood such as canning and mild heat treatment attempts to produce a final product with long shelf life and favorable consumer acceptance. For more details, refer to the review by Aubourg (2001).

REFERENCES

Al-Kahtani HA et al. 1996. Chemical changes after irradiation and post-irradiation storage in tilapia and Spanish mackerel. *J Food Sci* 61(4): 729–733.

Amano K, Tomiya F. 1953. Studies on the green discoloration of frozen swordfish (*Xiphia gladis* L.) II. Some experiments on the cause of discoloration. *Bull Jpn Soc Sci Fish* 19(5): 671–678.

Amano K, Yamada K. 1965. The biological formation of formaldehyde in cod fish. In: R Kreuzer (ed.) *The Technology of Fish Utilization.* Fishing News Books, London, pp. 73–87.

Ang JF, Hultin HO. 1989. Denaturation of cod myosin during freezing after modification with formaldehyde. *J Food Sci* 54: 814–818.

Angelo AJ. 1996. Lipid oxidation in foods. *Crit Rev Food Sci Hum Nutr* 36(3): 175–224.

Aubourg SP. 1998. Influence of formaldehyde in the formation of fluorescence related to fish deterioration. *Zeitschrift fur Lebensmittel-Untersuchung und-Forschung A—Food Research and Technology* 206: 29–32.

Aubourg SP. 2001. Review: Loss of quality during the manufacture of canned fish products. *Food Sci Technol Int* 7(3): 199–215.

Barrett I et al. 1965. Changes in tuna quality and associated biochemical changes during handling and storage aboard vessels. *Food Technol* 19(2): 108–117.

Benjakul S, Bauer F. 2000. Physicochemical and enzymatic changes of cod muscle proteins subjected to different freeze-thaw cycles. *J Sci Food Agric* 80: 1143–1150.

Benjakul S, Bauer F. 2001. Biochemical and physiochemical changes in catfish (*Silurus glanis* Linne) muscle as influenced by different freeze-thaw cycles. *Food Chem* 72(2): 207–217.

Benjakul S, Sutthipan N. 2009. Muscle changes in hard and soft shell crabs during frozen storage. *LWT—Food Sci Technol* 42: 723–729.

Blond G, Le Meste M. 2004. Principles of Frozen Storage. In: YH Hui et al. (eds.) *Handbook of Frozen Foods.* Marcel Dekker, New York, pp. 25–53.

Bonnell AD. 1994. *Quality Assurance in Seafood Processing: A Practical Guide.* Chapman and Hall, London, pp. 74–75.

Brake NC, Fennema OR. 1999. Lipolysis and lipid oxidation in frozen minced mackerel as related to Tg′, molecular diffusion, and presence of gelatin. *J Food Sci* 64: 25–32.

Bremmer HA. 1977. Storage trials on the mechanically separated flesh of three Australian mid-water fish species. 1. Analytical tests. *Food Technol Aust* 29: 89–93.

Bremmer HA. 1992. Fish flesh structure and the role of collagen–Its post-mortem aspects and implications for fish processing. In: HH Huss et al. (eds.) *Quality Assurance in the Fish Industry.* Elsevier, Amsterdam, pp. 39–62.

Brown WD. 1962. The concentration of myoglobin hemoglobin in tuna flesh. *J Food Sci* 27(1): 26–28.

Buttkus H. 1970. Accelerated denaturation of myosin in frozen solution. *J Food Sci* 35: 558–562.

Bykowski P, Kolodziejski W. 1983. Wlasciwosci Miesa z Kryla odskorupionego Metoda Rolkowa. *Bull Sea Fish Inst, Gdynia* 14(5–6): 53–57.

Careche M, Tejada M. 1994. Hake natural actomyosin interaction with free fatty acids during frozen storage. *J Sci Food and Agric* 64: 501–507.

Castell C et al. 1973. Comparison of changes in TMA, DMA, and extarctable protein in iced and frozen stored gadoid fillets. *J Fish Res Board Canada* 30: 1246–1250.

Chanmugam O et al. 1986. Differences in the v3 fatty acid contents in pond-reared and wild fish and shellfish. *J Food Sci* 51(6): 1556–1557.

Chiou TK et al. 1989. Proteolytic activities of Mullet and Alaska pollack roes and their changes during processing. *Nippon Suisan Gakkaishi* 55: 805–809.

Ciarlo AS et al. 1985. Storage life of frozen blocks of Patagonian Hake (Merluccius hubbsi) filleted and minced. *J Food Sci* 50: 723–726.

Coggins PC, Chamul RS. 2004. Food sensory attributes. In: YH Hui et al. (eds.) *Handbook of Frozen Foods*. Marcel Dekker, New York, pp. 93–147.

Crawford D et al. 1979. Comparative stability and desirability of frozen Pacific hake fillet and minced flesh blocks. *J Food Sci* 44: 363–367.

De Koning AJ et al. 1987. The origin of free fatty acids formed in frozen cape hake mince (*Merluccius capensis*, castelnau) during cold storage at 218°C. *J Sci Food Agric* 39: 79–84.

Doe P, Olley J. 1990. Drying and dried fish products. In: ZE Sikorski (ed.) *Seafood: Resources, Nutritional Composition, and Preservation*. CRC Press, Boca Raton, FL, pp. 125–145.

Duan J et al. 2010. Quality enhancement in fresh and frozen lingcod (Ophiodon elongates) fillets by employment of fish oil incorporated chitosan coatings. *Food Chem* 119: 524–532.

Dyer WJ, Dingle JR. 1961. Fish proteins with special reference to freezing. In: G Borgstrom (ed.) *Fish as Food*. Academic Press, New York, pp. 275–327.

Einen O et al. 2002. Freezing of pre-rigor fillets of Atlantic salmon. *Aquaculture* 212: 129–140.

Eskin NA et al. 1971. *Biochemistry of Foods*. Academic Press, New York.

Fan W et al. 2009. Effects of chitosan coating on quality and shelf life of silver carp during frozen storage. *Food Chem* 115: 66–70.

Farouk MM et al. 2003. Ultra-fast freezing and low storage temperatures are not necessary to maintain the functional properties of manufacturing beef. *Meat Sci* 66: 171–179.

Flick GJ et al. 1992. Lipid oxidation of seafood during storage. In: AJ Angelo (ed.) *Lipid Oxidation in Food*. American Chemical Society, Washington, DC, pp. 183–207.

Frank HA et al. 1984. Relationship between honeycombing and collagen breakdown in skipjack tuna. *Mar Fish Rev* 46(2): 40–42.

Friedler S et al. 1988. Cryopreservation of embryos and ova. *Fertil Steril* 49: 743–763.

Fuke S. 1994. Taste-active compounds of seafoods with special references to umami substances. In: F Shahidi, JR Botta (eds.) *Seafoods: Chemistry, Processing Technology and Quality*. Blackie Academic and Professional, London, pp. 115–139.

Fuster C et al. 1994. Drip loss, peroxidase and sensory changes in kiwi fruit slices during frozen storage. *J Sci Food Agric* 64: 23–29.

Gardner HW. 1979. Lipid hydroperoxide reactivity with proteins and amino acids: A review. *J Agric Food Chem* 27: 220–229.

Ghiretti F. 1956. The decomposition of hydrogen peroxide by hemocyanin and by its dissociation products. *Arch Biochem Physiol* 63: 165–176.

Gill TA. 1992. Biochemical and chemical indices of seafood quality. In: HH Huss et al. (eds.) *Quality Assurance in the Fish Industry*. Elsevier, Amsterdam, pp. 377–388.

Grosjean O et al. 1969. Formation of a green pigment from tuna myoglobin. *J Food Sci* 34: 404–407.

Haard NF. 1983. Dehydration of Atlantic short finned squid. In: BS Pan (ed.) *Properties and Processing of Marine Foods*. National Taiwan Ocean University, Keelung, Taiwan, Republic of China, pp. 36–60.

Haard NF. 1990. Biochemical reactions in fish muscle during frozen storage. In: EG Bligh (ed.) *Seafood Science and Technology*. Fishing News Books (Blackwell Scientific Publications), Oxford, pp. 46–57.

Haard NF. 1992a. Biochemistry and chemistry of color and color change in seafoods. In: GJ Flick, RE Martin (eds.) *Advances in Seafood Biochemistry*. Technomic Publishing Co., Lancester, PA, pp. 305–360.

Haard NF. 1992b. Biochemical reactions in fish muscle during frozen storage. In: EG Bligh (ed.) *Seafood Science and Technology*. Fishing New Books, Oxford, pp. 176–209.

Haard NF. 1994. Protein hydrolysis in seafoods. In: F Shahidi, JR Botta (eds.) *Seafoods: Chemistry, Processing Technology and Quality*. Blackie Academic and Professional, London, pp. 10–33.

Haard NF et al. 1994. Sarcoplasmic proteins and other nitrogenous compounds. In: ZE Skorski et al. (eds.) *Seafood Proteins*. Chapman and Hall, New York, pp. 13–39.

Halliwell B, Gutteridge JMC. 1989. *Free Radicals in Biology and Medicine*. Claredon Press, Oxford.

Hebard CE et al. 1982. Occurrence and significance of trimethylamine oxide and its derivatives in fish and shellfish. In: RE Martin et al. (eds.) *Chemistry and Biochemistry of Marine Food Products*. AVI Publishing Company, Westport, CT, pp. 149–304.

Hink MJ, Stanley DW. 1985. Colour measurement of the squid (Illex illecebrosus) and its relationship to quality and chromatophore structure. *Can Inst Food Sci Technol J* 18(3): 233–241.

Hobbs G. 1982. Changes in fish after catching. In: S Aitkin et al. (eds.) *Fish Handling and Processing*. Her Majesty's Stationery Office, Edinburgh, pp. 20–27.

Hosseini SV et al. 2010. Influence of the in vivo addition of alphatocopheryl acetate with three lipid sources on the lipid oxidation and fatty acid composition of Beluga sturgeon, *Huso huso*, during frozen storage. *Food Chem* 118: 341–348.

Hultin HO. 1992a. Biochemical deterioration of fish muscle. In: HH Huss et al. (eds.) *Quality Assurance in the Fish Industry*. Elsevier, Amsterdam, pp. 125–138.

Hultin HO. 1992b. Trimethylamine-N-oxide (TMAO) demethylation and protein denaturation in fish muscle. In: GJ Flick, RE Martin (eds.) *Advances in Seafood Biochemistry*. Technomic Publishing Co., Lancaster, CT, pp. 25–42.

Hultin HO. 1994. Oxidation of lipids in seafoods. In: F Shahidi, JR Botta (eds.) *Seafoods: Chemistry, Processing Technology and Quality*. Blackie Academic and Professional, Glasgow, pp. 49–74.

Hultin HO, Kelleher SD. 2000. Surimi processing from dark muscle fish. In: JW Park (ed.) *Surimi and Surimi Seafood*. Marcel Dekker, New York, pp. 59–77.

Huss HH. 1994. *Quality and Quality Changes in Fresh Fish*. Food and Agricultural Organization of the United Nations, Rome.

Huynh MD et al. 1990. Freshness assessment of Pacific fish species using K-value. In: EG Bligh (ed.) *Seafood Science and Technology*. Fishing News Books (Blackwell Scientific Publications), Oxford, pp. 258–268.

Icekson I et al. 1998. Lipid oxidation levels in different parts of the mackerel, *Scomber scombrus*. *J Aquat Food Prod Technol* 7(2): 17–29.

Ikeda S. 1979. Other organic components and inorganic components. In: JJ Connell, the staff of Torry Research Station (eds.) *Advances of Fish Science and Technology*. Fish News Books, Farnham, Surrey, pp. 111–112.

Ishikawa Y et al. 1978. Synergistic effect of trimethylamine oxide on the inhibition of the autoxidation of methyl linoleate by g-tocopherol. *Agric Biol Chem* 42: 703–709.

Jacober LF, Rand AG. 1982. Biochemical evaluation of seafood. In: RE Martin et al. (eds.) *Chemistry and Biochemistry of Marine Food Products*. AVI Publishing Company, Westport, CT, pp. 347–365.

Jiang ST. 2000. Enzymes and their effects on seafood texture. In: NF Haard, BK Simpson (eds.) *Seafood Enzymes*. Marcel Dekker, New York, pp. 411–450.

Jiang ST, Lee TC. 2004. Frozen seafood and seafood products: Principles and applications. In: YH Hui et al. (eds.) *Handbook of Frozen Foods*. Marcel Dekker, New York, pp. 245–294.

Jobling M. 2004. Are modifications in tissue fatty acid profiles following a change in diet the result of dilution? Test of a simple dilution model. *Aquaculture* 232(1–4): 551–562.

Jones NR et al. 1964. Rapid estimation of hypoxanthine concentrations as indices of freshness of chilled-stored fish. *J Sci Food Agric* 15(11): 763–773.

Josephson DB, Lindsey RC. 1986. Enzymic generation of fresh fish volatile aroma compounds. In: TH Parliment, R Croteau (eds.) *Biogenesis of Aromas*. ACS Symposium No. 317, p. 201.

Karel M et al. 1975. Interaction of peroxidizing methyl linoleate with some proteins and amino acids. *J Agric Food Chem* 23: 159–163.

Karube I et al. 1984. Determination of fish freshness with an enzyme sensor system. *J Agric Food Chem* 32: 314–319.

Kelleher SD et al. 1992. Inhibition of lipid oxidation during processing of washed, minced Atlantic mackerel. *J Food Sci* 57(5): 1103–1108, 1119.

Kikugawa K et al. 1989. A tetrameric dialdehyde formed in the reaction of butyraldehyde and benzylamine: a possible intermediary component for protein cross-linking induced by lipid oxidation. *Lipids* 24: 962–969.

King AJ, Li SJ. 1999. Association of malonaldehyde with rabbit myosin subfragment 1. In: YL Xiong et al. (eds.) *Quality Attributes of Muscle Foods*. Kluwer Academic/Plenum Publishers, New York, pp. 277–286.

Kinoshita M et al. 1990. Diverse distribution of four distinct types of modori (gel degradation)-inducing proteinases among fish species. *Nippon Suisan Gakkaishi* 56: 1485–1492.

Klurfeld DM. 2002. Dietary fats, ecosanoids, and the immune system. In: CC Akoh, DB Min (eds.) *Food Lipids: Chemistry, Nutrition and Biochemistry*, 2nd edn. Marcel Dekker, New York, pp. 589–601.

Koizumi C, Matsura F. 1967. Studies on 'green' tuna. IV. Effect of cysteine on greening of myoglobin in the presence of trimenthylamine oxide. *Bull Jpn Soc Sci Fish* 33(9): 839–842.

Kolstad K et al. 2004. Quantification of fat deposits and fat distribution in Atlantic halibut (*Hippoglossus hippoglossus* L.) using computerized X-ray tomography (CT). *Aquaculture* 229(1–4): 255–264.

Konegaya S. 1984. Studies on the jellied meat of fish, with special reference to that of yellowfin tuna. *Bull Tokai Reg Fish Lab* 116: 39–47.

Kristinsson HG et al. 2003. *The Effect of Carbon Monoxide and Filtered Smoke on the Properties of Aquatic Muscle and Selected Muscle Components*. Icelandic Fisheries Laboratory, Reykjavik, Iceland, pp. 27–29.

Kye HW et al. 1988. Changes in myofibrillar proteins and texture in freshwater prawn during ice-storage. *Marine Fish Rev* 50(1): 53–56.

Leake L, Karel M. 1985. Nature of fluorescent compounds generated by exposure of protein to oxidized lipids. *J Food Biochem* 9: 117–136.

Leu S-S et al. 1981. Atlantic mackerel (*Scomber scombrus* L.): seasonal variation in proximate composition and distribution of chemical nutritients. *J Food Sci* 46: 1635–1638.

Lim HK, Haard NF. 1984. Protein insolubilization in frozen Greenland halibut (*Reinhardtius Hippoglossoides*). *J Food Biochem* 8: 163–187.

Lin TS, Lanier TC. 1980. Properties of an alkaline protease from the skeletal muscle of Atlantic croaker. *J Food Biochem* 4(1): 17–28.

Lindsay RC. 1991. Chemical basis of the quality of seafood flavors and aromas. *Mar Technol Soc Jpn* 25: 16–22.

Livingston DJ, Brown WD. 1981. The chemistry of myoglobin and its reactions. *Food Technol* 35: 244–252.

Love RM et al. 1977. Adaptation of the dark muscle of cod to swimming activity. *J Fish Biol* 11: 431–436.

Lovell RT, Mohammed T. 1988. Content of omega-3 fatty acids can be increased in farm-raised catfish. *Highlights Agric Res* 35(3): 16.

Lundstrom RC et al. 1982. Dimethylamine production of fresh red hake (*Urophycis chuss*): the effect of packaging material, oxygen permeability and cellular damage. *J Food Biochem* 6: 229–241.

Mackie IM. 1993. The effects of freezing on flesh proteins. *Food Rev Int* 9: 575–610.

Martinez I. 1992. Fish myosin degradation upon storage. In: HH Huss et al. (eds.) *Quality Assurance in the Fish Industry*. Elsevier, Amsterdam, pp. 389–397.

Matsumoto JJ. 1979. Denaturation of fish muscle proteins during frozen storage. In: O Fennema (ed.) *Proteins at Low Temperatures*. American Chemical Society, Washington, DC, pp. 205–224.

Mattews AD. 1983. Muscle color deterioration in iced and frozen stored bonito, yellowfin and skipjack tuna caught in Seychelles waters. *J Food Technol* 18: 387–392.

Mazur P. 1970. The freezing of biological systems. *Science* 168: 939–949.

Mazur P. 1984. Freezing of living cells: Mechanisms and implications. *Am J Physiol* 247: C125–C142.

Meyers V. 1965. Chapter 5. Marinades. In: G Borgstrom (ed.), *Fish as Food*, Vol. III. Academic Press, New York, pp. 165–193.

Miler KBM, Sikorski ZE. 1990. Smoking. In: ZE Sikorski (ed.) *Seafood: Resources, Nutritional Composition, and Preservation*. CRC Press, Boca Raton, FL, pp. 163–180.

Milo C, Grosch W. 1996. Changes in the odorants of boiled salmon and cod as affected by the storage of the raw material. *J Agric Food Chem* 44: 2366–2371.

Mishra R, Dora KC. 2010. Effect of frozen storage on the functional property of Ribbon fish (*Trichiurus savala*) cuvier. *J Food Process Preserv* 34: 364–372.

Murata M, Sakaguchi M. 1989. The effects of phosphatase treatment of yellowtail muscle extracts and subsequent addition of

IMP on flavor intensity. *Nippon Suisan Gakkaishi* 55: 1599–1603.
Narayan KA et al. 1964. Complex formation between oxidized lipids 1 egg albumin. *JAOCS* 41: 254–259.
Ngapo TM et al. 1999. Freezing and thawing rate effects on drip loss from samples of pork. *Meat Sci* 53: 149–158.
Nielsen MK, Jorgensen BM. 2004. Quantitative relationship between trimethylamine oxide aldolase activity and formaldehyde accumulation in white muscle from gadiform fish during frozen storage. *J Agric Food Chem* 52: 3814–3822.
Nip WK et al. 1985. Partial characterization of a collagenolytic enzyme fraction from the hepatopancreas of the freshwater prawn, *Macrobrachium rosenbergii*. *J Food Sci* 50: 1187–1188.
Nip WK, Moy JH. 1988. Microstructural changes of iced-chilled and cooked freshwater prawn, *Macrobrachium rosenbergii*. *J Food Sci* 53: 319–322.
Ohshima T et al. 1993a. Oxidative stability of sardine and mackerel lipids with reference to synergism between phospholipid and a-tocopherol. *JAOCS* 7(3): 269–276.
Ohshima T et al. 1993b. High pressure processing of fish and fish products. *Trends Food Sci Technol* 4: 370–375.
O'Keefe SF. 2000. Fish and shellfish. In: GL Christen, JS Smith (eds.) *Food Chemistry: Principles and Applications*. Science Technology Systems, West Sacramento, CA, pp. 399–420.
Orejana FM, Liston J. 1982. Agents of proteolysis and its inhibition of patis (fish sauce) fermentation. *J Food Sci* 47(1): 198–203, 209.
Ortiz O et al. 2009. Rancidity development during the frozen storage of farmed coho salmon (*Oncorhynchus kisutch*): effect of antioxidant composition supplied in the diet. *Food Chem* 115: 143–148.
Pan BS et al. 1986. Effect of endogenous proteinases on histamine and honeycombing in mackerel. *J Food Biochem* 10: 305–319.
Payne SR, Young OA. 1995. Effects of pre-slaughter administration of antifreeze proteins on frozen meat quality. *Meat Sci* 41: 147–155.
Perez-Villarreal PR. 1992. Ripening of the salted anchovy (Engraulis encrasicholus): Study of the sensory, biochemical and microbiological aspects. In: HH Huss et al. (eds.) *Quality Assurance in the Fish Industry*. Elsevier, Amsterdam, pp. 157–167.
Raksakulkthai N et al. 1986. Influence of mincing and fermentation aids on fish sauce prepared from male, inshore capelin, *Mallotus villosus*. *Can Inst Food Sci Technol J* 19: 28–33.
Reddy GVS et al. 1992. Deteriorative changes in pink perch mince during frozen storage. *Int J Food Sci Technol* 27(3): 271–276.
Refsgaard HH et al. 1998. Sensory and chemical changes in farmed Atlantic salmon (*Salmo salar*) during frozen storage. *J Agric Food Chem* 46: 3473–3479.
Regenstein JM et al. 1982. Chemical changes of trimethylamine oxide during fresh and frozen storage of fish. In: RE Martin et al. (eds.) *Chemistry and Biochemistry of Marine Food Products*. AVI Publishing Company, Westport, CT, pp. 137–148.
Regulska-Ilow B, Ilow R. 2002. Comparison of the effects of microwave cooking and conventional cooking methods on the composition of fatty acids and fat quality indicators in herring. *Nahrung/Food* 46(6): 383–388.
Rehbein H. 1988. Relevance of trimethylamine oxide demethylase activity and haemoglobin content of formaldehyde production and texture deterioration in frozen stored minced fish muscle. *J Sci Food Agric* 43: 261–276.
Reza MS et al. 2009. Shelf life of several marine fish species of Bangladesh during ice storage. *Int J Food Sci Technol* 44: 1485–1494.
Richards MP et al. 1998. Effect of washing with or without antioxidants on quality retention of mackerel fillets during refrigerated and frozen storage. *J Agric Food Chem* 46(10): 4363–4371.
Rosario RR, Maldo SM. 1984. Biochemistry of patis formation. I. Activity of cathepsins in patis hydrolysis. *Philippine Agriculturist* 67(2): 167–175.
Roubal WT, Tappel AL. 1966. Damage to proteins enzymes and amino acids by peroxidizing lipids. *Arch Biochem Biophys* 113: 5–8.
Saito T et al. 1959. A new method for estimating the freshness of fish. *Bull Jpn Soc Sci Fish* 24: 749–750.
Sakaguchi M, Murata M. 1989. Seasonal variations of free amino acids in oyster whole body and adductor muscle. *Nippon Suisan Gakkaishi* 55: 2037–2041.
Sathivel S et al. 2003. Thermal degradation of FA and catfish and menhaden oils at different refining steps. JAOCS 80(11): 1131–1134.
Schaich KM, Karel M. 1975. Free radicals in lysozyme reacted with peroxidizing methyl linoleate. *J Food Sci* 40: 456–459.
Schwimmer S. 1981. *Source Book of Food Enzymology*. AVI Publishing Company, Westport, CT, p. 967.
Sebedio JL et al. 1993. Stability of polyunsaturated omega-3 fatty acids during deep fat frying of Atlantic mackerel (*Scomber scombrus* L.). *Food Res Int* 26: 163–172.
Shahidi F. 1994. Seafood proteins and preparation of protein concentrates. In: F Shahidi, JR Botta (eds.) *Seafoods: Chemistry, Processing Technology and Quality*. Blackie Academic and Professional, London, pp. 3–9.
Shahidi F, Spurvey SA. 1996. Oxidative stability of fresh and heat-processed dark and light muscles of mackerel (*Scomber scombrus*). *J Food Lipids* 3(1): 13–25.
Shenderyuk VI, Bykowski PJ. 1990. Sating and mari-nating of fish. In: ZE Sikorski (ed.) *Seafood: Resources, Nutritional Composition, and Preservation*. CRC Press, Boca Raton, FL, pp. 147–162.
Shenouda S et al. 1979. Technical studies on ocean pout, an unexploited fish species, for direct human consumption. *J Food Sci* 44: 164–168.
Shenouda SYK. 1980. Theories of protein denaturation during frozen storage of fish flesh. *Adv Food Res* 26: 275–311.
Shenouda SYK, Piggot GM. 1974. Lipid-protein interaction during aqueous extraction of fish protein: Myosin-lipid interaction. *J Food Sci* 39: 726–734.
Shewan JM. 1944. The effect of smoke curing and salt curing on the composition, keeping quality and culinary properties of fish. *Proc Nutr Soc* (1–2): 105–112.
Shewfelt R. 1980. Fish muscle hydrolysis—a review. *J Food Biochem* 5: 79–94.
Shewfelt RL. 1981. Fish muscle lipolysis—a review. *J Food Biochem* 5: 79–100.
Sikorski Z et al. 1976. Protein changes in frozen fish. *CRC Crit Rev Food Sci Nutr* 8: 97–129.
Sikorski ZE. 1994a. The contents of proteins and other nitrogenous compounds in marine animals. In: ZE Skorski et al. (eds.) *Seafood Proteins*. Chapman and Hall, New York, pp. 6–12.

Sikorski ZE. 1994b. The myofibrillar proteins in seafoods. In: ZE Skorski et al. (eds.) *Seafood Proteins*. Chapman and Hall, New York, pp. 40–57.

Sikorski ZE, Borderias JA. 1994. Collagen in the muscles and skin of marine animals. In: ZE Skorski et al. (eds.) *Seafood Proteins*. Chapman and Hall, New York, pp. 58–70.

Sikorski ZE, Kolakowska A. 1994. Changes in proteins in frozen stored fish. In: ZE Skorski et al. (eds.) *Seafood Proteins*. Chapman and Hall, New York, pp. 99–112.

Sikorski ZE, Pan BS. 1994. The involvement of proteins and nonprotein nitrogen in postmortem changes in seafoods. In: ZE Skorski et al. (eds.) *Seafood Proteins*. Chapman and Hall, New York, pp. 71–83.

Sikorski ZE, Ruiter A. 1994. Changes in proteins and nonprotein nitrogen compounds in cured, fermented, and dried seafoods. In: ZE Sikorski et al. (eds.) *Seafood Proteins*. Chapman and Hall, New York, pp. 113–126.

Sikorski ZE et al. 1990a. Postharvest biochemical and microbial changes. In: ZE Sikorski (ed.) *Seafood Resources: Nutritional Composition, and Preservation*. CRC Press, Boca Raton, FL, pp. 55–75.

Sikorski ZE et al. 1990b. The nutritive composition of the major groups of marine food organisms. In: ZE Sikorski (ed.) *Seafood Resources: Nutritional, Compositional and Preservation*. CRC Press, Boca Raton, FL, pp. 29–54.

Siow LF et al. 2007. Characterizing the freezing behavior of liposomes as a tool to understand the cryopreservation procedures. *Cryobiology* 55: 210–221.

Simpson BK, Haard NF. 1984. Trypsin from Greenland cod (*Gadus ogac*) as a food processing aid. *J Appl Biochem* 6: 135–143.

Simpson BK et al. 1987. Phenoloxidase from shrimp (*Paneus setiferus*): purification and some properties. *J Agric Food Chem* 35: 918–921.

Solberg C. 2004. The influence of dietary oil content on the growth and chemical composition of Atlantic salmon (Salmo salar). *Aquaculture Nutr* 10(1): 31–37.

Sorenson NK. 1992. Physical and instrumental methods for assessing seafood quality. In: M Jackobsen, J Listin (eds.) *Quality Assurance in the Fish Industry*. Elsevier, Amsterdam, pp. 321–332.

Sotelo CG et al. 1995. Denaturation of fish proteins during frozen storage-role of formaldehyde. *Zeitschrift fur Lebensmittel-Untersu chung Und-Forschung A—Food Research and Technology* 200: 14–23.

Spinelli J et al. 1964. Measurement of hypoxanthin in fish as a method of assessing freshness. *J Food Sci* 79: 710–714.

Srinivasan S et al. 1996. Inhibition of protein and lipid oxidation in beef heart surimi-like materials by antioxidants and combinations of pH, NaCl, and buffer type in the washing media. *J Agric Food Chem* 44(1): 119–125.

Sulzbacher WL, Gaddis AM. 1968. Meats: Preservation of quality by freezer storage. In: DK Tressler et al. (eds.) *The Freezing Preservation on Foods*, Vol. II. AVI Publishing Company, Westport, CT, pp. 159–178.

Suzuki T. 1981. *Fish and Krill Processing Technology*. Applied Science Publishers, London, pp. 10–13.

Takahashi K et al. 1993. Effect of storage temperature on freeze denaturation of Carp myofibrils with KCl or NaCl. *Nippon Suisan Gakkaishi* 59: 519–527.

Takama K. 1974. Insolubilization of rainbow trout actomyosin during storage at 220°C. 1. Properties of insolubilized proteins formed by reaction of propanal or caproic acid with actomyosin. *Bull Jpn Soc Sci Fish* 40: 585–588.

Thompson HC, Thompson MH. 1972. Inhibition of flesh browning and skin color fading in frozen fillets of yelloweye snapper (*Lutjanus vivanus*) NOAA Technical Report NMFS SSRF-544, p. 6.

Tichivangana JZ, Morrissey PA. 1985. Metmyoglobin and inorganic metals as pro-oxidants in raw and cooked muscle systems. *Meat Sci* 15(2): 107–116.

Tironi VA et al. 2002. Structural and functional changes in myofibrillar proteins of sea salmon (*Pseudopercis semifasciata*) by interaction with malonaldehyde. *J Food Sci* 67: 930–935.

Tironi VA et al. 2010. Quality loss during the frozen storage of sea salmon (*Pseudopercis semifasciata*). Effect of rosemary (*Rosmarinus officinalis* L.) extract. *LWT—Food Sci Technol* 43: 263–272.

Tomlinson N. 1966. *Bulletin 150: Greening in Tuna and Related Species*. Fisheries Research Board of Canada, Ottawa, p. 21.

Tseng YC et al. 2003. Quality changes in Australian red claw crayfish (*Cherax quadricarinatus*) subjected to multiple freezing-thawing cycles. *J Food Qual* 26: 285–298.

Ueng YE, Chow CJ. 1998. Textural and histological changes of different squid mantle muscle during frozen storage. *J Agric Food Chem* 46: 4728–4733.

Undeland I. 1997. Lipid oxidation in fish—Causes, changes and measurements. In: G Olafsdóttir, et al. (eds.) *Methods to Determine the Freshness of Fish in Research and Industry*. International Institute of Refrigeration, Paris, pp. 241–256.

Undeland I et al. 2002. Added triacylglycerols do not hasten hemoglobin-mediated lipid oxidation in washed minced cod muscle. *J Agric Food Chem* 50(23): 6847–6853.

Undeland I et al. 2004. Hemoglobin-mediated oxidation of washed minced cod muscle phospholipids: Effect of pH and hemoglobin source. *J Agric Food Chem* 52(14): 4444–4451.

Wang JCC, Amiro ER. 1979. A fluorometric method for the micro-quantitative determination of hemoglobin and myoglobin in fish muscle. *J Sci Food Agric* 30(11): 1089–1096.

Wang Y-J et al. 1991. Effect of heat inactivation of lipoxygenase on lipid oxidation in lake herring (*Coregonus artedii*). *JAOCS* 68(10): 752–757.

Wijendran V, Hayes KC. 2004. Dietary n-6 and n-3 fatty acid balance and cardiovascular health. *Annu Rev Nutr* 24: 597–615.

Wolfe SK et al. 1978. Analysis of myoglobin derivatives in meat or fish samples using absorption spectrophotometry. *J Agric Feed Chem* 26(1): 217–219.

Xiong YL. 1997. Protein denaturation and functionality losses. In: MC Erickson, YC Hung (eds.) *Quality in Frozen Food*. Chapman and Hall, New York, pp. 111–140.

20
Fish Collagen

Soottawat Benjakul, Sitthipong Nalinanon, and Fereidoon Shahidi

Abstract: Fish collagen has gained increasing interest as the alternative for mammalian counterpart. It can be generally produced from by-products generated during processing of fish and invertebrates. The potential raw materials include skin, bone, scale, and so on. Types and molecular properties of collagen vary with the source, habitat of fish, extraction process, and other factors. In general, collagen can be extracted from collagenous materials at low temperature with the aid of various acids to avoid thermal denaturation. To increase the extraction yield, pepsins from mammalian and fish origins, which specifically cleave at telopeptide region, have been used successfully without the changes in molecular properties. Fish collagen can be of food, biomedical, and pharmaceutical applications.

INTRODUCTION

Collagen is the fibrous protein of animal connective tissue, contributing to the unique physiological functions of tissues in skins, tendons, bones, cartilages, and so on and is associated with toughness in mammalian muscle (Foegeding et al. 1996, Ogawa et al. 2003, Muyonga et al. 2004, Yan et al. 2008, Kittiphattanabawon et al. 2010a). The collagen fibers are essentially inextensible and, therefore, provide mechanical strength and also allow flexibility between various organs of the body (Bailey et al. 1998).

Collagen is widely used in food, biomedical, pharmaceutical, and cosmetics, and its consumption has been increasing along with the development of new industrial application (Nalinanon et al. 2007, Regenstein and Zhou 2007, Woo et al. 2008). Collagen exhibits biodegradability, weak antigenecity, and superior biocompatibility compared with other natural polymers, such as albumin and gelatin (Lee et al. 2001). Generally, commercial collagens are produced from bovine and porcine hides and bones. Currently, the increasing attention of alternative sources for replacement of mammalian collagen has been paid, especially from seafood processing by-products. Fish collagen can be used for Halal and Kosher products for Muslim and Jewish communities, respectively. Because of the outbreak of bovine spongiform encephalopathy and bird flu, the increasing demand of fish collagen has been gained (Jongjareonrak et al. 2005, Nalinanon et al. 2007, 2008, Regenstein and Zhou 2007, Duan et al. 2009, Kittiphattanabawon et al. 2010a). Collagen can be extracted from fish skin, scale, and bone. Collagen is also found in the body walls and cuticles of invertebrates (Meena et al. 1999). With the appropriate extraction technology, collagen from fish or aquatic animals can be used as the potential alternative for mammalian counterpart.

Food Biochemistry and Food Processing, Second Edition. Edited by Benjamin K. Simpson, Leo M.L. Nollet, Fidel Toldrá, Soottawat Benjakul, Gopinadhan Paliyath and Y.H. Hui.

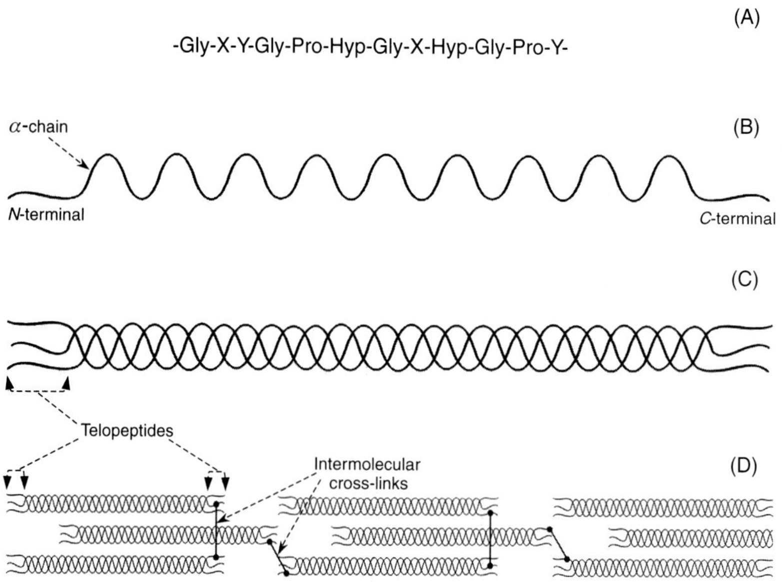

Figure 20.1. Arrangement of collagen fibril in collagen fiber. (**A**) Amino acid sequence of a collagen polypeptide. (**B**) Collagen polypeptide. (**C**) Tropocollagen. (**D**) Collagen fibril.

COLLAGEN COMPOSITION AND STRUCTURE

Collagen is typically used as a generic term to cover a large family of distinct proteins, each with specific structures, functions, and tissue distributions in the extracellular matrix (Ramshaw et al. 2009). The collagen monomer is a long cylindrical protein about 2800 Å long and 14–15 Å in diameter (Foegeding et al. 1996). Collagen is constructed from tropocollagen, a rod-shaped protein consisting of three polypeptides unit (called α-chains) intertwined to form a triple-helical structure. Each α-chain coils in a left-handed helix with three residues per turn, and the three chains are twisted right-handed to form the triple helix (Wong 1989) (Fig. 20.1). The three chains are held together primarily by hydrogen bonding between adjacent –CO and –NH group (Wong 1989, Lee et al. 2001). At least 19 types of collagen designated types I–XIX have been reported (Bailey et al. 1998, Lee et al. 2001, Woo et al. 2008). Different α-chains, designated α1, α2, or α3, within the same type of collagen differ in their amino acid composition. The distribution of α1-, α2-, and α3-chains in collagen molecules varies depending on the specific genetic variants (Xiong 1997). The most common collagen is type I collagen. It contains two α1(I)-chains and one α2(I)-chain, which has different amino acid sequence and composition. Each chain has a molecular mass about 100,000Da, yielding a total molecular mass of about 300,000 Da of collagen (Foegeding et al. 1996).

Type I collagen is the major of epimysium. Both types I and III are present in the perimysium in large amounts, whereas types II, IV, and V are primary species in epimysium (Table 20.1) (Xiong 1997). Types I, II, and III collagen as well as types V

Table 20.1. Collagens and Their Distribution

Type	Peptide Chains[a]	Molecular Composition	Occurrence
I	α1, α2	$[\alpha 1(\mathrm{I})]_2\ \alpha 2(\mathrm{I})$	Skin, tendon, bone, muscle (epimysium)
II	α1	$[\alpha 1(\mathrm{II})]_3$	Intervertebral disc, cartilage
III	α1	$[\alpha 1(\mathrm{III})]_3$	Fetal skin, cardiovascular vessel, uterus, synovial membranes, inner organ, muscle (perimysium)
IV	α1, α2	$[\alpha 1(\mathrm{IV})]_3$(?) [b]	Basement membrane, kidney glomeruli, lens capsule, glomeruli
V	AA, αB, αC (?)	$[\alpha B]_2\alpha A$ or $(\alpha B)_3 + (\alpha A)_3$ or $(\alpha C)_3$(?)	Placental membrane, cardiovascular system, lung, muscle (endomysium)

Sources: Wong (1989) and Belitz et al. (2004).
[a]Since the α-chains of various types of collagen differ, the are called α1(I), α1(II), αA, etc.
[b](?) Not completely elucidated.

and XI are built up of three chains and all are composed of the continuous triple-helical structure (Lee et al. 2001). Types I, II, III, and V are also called as fibril-forming collagens and have large sections of homologous sequences independent of species (Timpl 1984). Foegeding et al. (1996) reported that types IV and III collagens contain oxidizable cysteine residues and is rich in hydroxyproline and hydroxylysine. In type IV collagen (basement membrane), the triple-helical conformation is interrupted with large nonhelical domains as well as with the short nonhelical peptide interruption (Lee et al. 2001). Like type IV collagen, type V collagen is rich in hydroxyproline and hydroxylysine but contains no cysteine. Type VI is microfibrillar collagen, and type VII is anchoring fibril collagens (Samuel et al. 1998).

Based on its macromolecular structures, collagen can be divided into three major groups: (a) striated fibrous collagen, which includes types I, II, and III collagen, (b) nonfibrous collagen, which contains type IV or basement membrane collagen, and (c) microfibrillar collagen, which encompasses types VI and VII (the matrix microfibrils), types V, IX, and X (the pericellular collagen), and types VIII and XI, which are yet unclassified (Xiong 1997). Fibril-associated collagens (types IX, XI, XII, and XIV) have small chains, which contain some nonhelical domains (Lee et al. 2001). Each is characterized by having the triple-helical motif as part of its structure (Ramshaw et al. 2009).

In the extracellular space, collagen molecules align themselves into microfibrils in quarter stagger array. Cross-linking is initiated and larger diameter fibrils are formed either by the addition of microfibrils or by association with other fibrils (McCormick 1994). During maturation or aging, collagen fibers can be strengthened and are stabilized primarily by covalent cross-linkages (Belitz et al. 2004). Thus, the cross-links confer mechanical strength to collagen fiber. Belitz et al. (2004) proposed that cross-link formation involves (i) enzymatic oxidation of lysine and hydroxylysine to the corresponding ω-aldehydes, (ii) conversion of these aldehydes to aldols and aldimines, and (iii) stabilization of these primary products by additional reduction or oxidation reactions. The cross-linking stereochemistry derives from the reaction of specific peptidyl aldehydes, in the NH_2- and COOH-terminal nonhelical peptide, with vicinal ε-amino groups of specific peptidyl residues of Lys and Hyl, located in the triple-helical regions of molecules (Mechanic et al. 1987). Intermolecular cross-links are confined to the end-overlap region involving a lysine aldehyde in the telopeptide of one chain and a hydroxylysine of an adjacent chain (Foegeding et al. 1996). Furthermore, Avery and Bailey (2005) proposed that the slow turnover of mature collagen subsequently allows accumulation of the products of the adventitious nonenzymic reaction of glucose with the lysines in the triple helix to form glucosyl lysine and its Amadori product. These products are subsequently oxidized to a complex series of advanced glycation end-products; some of which are intermolecular cross-links between the triple helices, rendering the fiber too stiff for optimal functioning of the collagen fibers, and, consequently, of the particular tissue involved.

The number of cross-links in collagen increase with increasing age of the animal (Zayas 1997, Belitz et al. 2004). Most connective tissue in fish is renewed annually, and highly cross-linked protein is not generally found in fish (Foegeding et al. 1996). Collagenous tissue from older animals with more cross-linkages is more resistant to swelling and has a lower water-holding capacity (Zayas 1997). The steady increase in mature collagen cross-linking is due to progressive and ongoing cross-linking reactions that occur within fibrillar collagen and with the slowing of collagen synthesis rates as animals reach maturity (McCormick 1994). Collagen from young animals is more easily solubilized and produces structures with low tensile strength. In contrast, collagen from old animals is difficult to solubilize and produces a structure with high tensile strength (Miller et al. 1983). In starving fish, the sarcoplasmic and myofibrillar proteins undergo gradual degradation, while the connective tissues are not utilized, or extra collagen is even deposited in the myocommata and in the skin (Sikorski et al. 1990). Thus, the toughening process in fish seems to be much more reversible than that of higher animals, where the amount of cross-linking increases with age (Regenstein and Regenstein 1991). Foegeding et al. (1996) reported that starving fish produce more collagen, especially collagen with a greater degree of cross-linking, than the fish that are well fed. Furthermore, collagens in myocommata are thickened with more intermolecular cross-links of collagen during starvation (Love et al. 1976).

Glycine represents nearly one-third of the total residues, and it is distributed uniformly at every third position throughout most of the first collagen molecule. The repetitive occurrence of glycine is absent in the first 14 amino acid residues from N-terminus and the first ten residues from the C-terminus, in which these end portions are termed "telopeptides" (Foegeding et al. 1996). The presence of glycine at every three residues is a critical requirement for the collagen superhelix structure (Regenstein and Zhou 2007). Glycine contains no side chain, which allows it to come into the center of the superhelix without any steric hindrance to form a close packing structure (Te Nijenhuis 1997). The triple helix of collagen assembled from specific polypeptide chain (α-chain) with the Gly-X-Y repeat and contain frequent occurrence of proline (Pro) and hydroxyproline (Hyp) in the X- and Y-positions, respectively (Johnston-Banks 1990, Xu et al. 2002). Hydroxyproline is located at Y-position, as is hydroxylysine, while proline can be found in either the X- or Y-positions (Johnston-Banks 1990). The distribution of polar and nonpolar residues in the X-position determines the ordered aggregation of molecules into fibrils (Xiong 1997). The segments of the polypeptide chain consisting of repeating triplets with imino acid residues are the nonpolar regions, and the segments containing Gly-X-Y triplets without imino acids are mostly polar (Wong 1989). Collagen is a hydrophilic protein because of the greater content of acidic, basic, and hydroxylated amino acid residues than lipophilic residues. Therefore, it swells in polar liquids (Johnston-Banks 1990).

The amino acid composition of collagen is unique. The collagens are rich in glycine, proline and alanine, and contained low or no cysteine and tryptophan (Table 20.2). Relatively low contents of tyrosine and histidine are reported in collagen. Two amino acids that are not commonly present in other proteins include hydroxyproline and hydroxylysine (Wong 1989). The amino acid composition of collagen is nutritionally unbalanced,

Table 20.2. Amino Acid Composition of Collagen from Fish and Mammalian (Residues per 1000 residues)

	Cod [(a)]	Bigeye Snapper [(b)]		Skate [(c)]	Grass Carp [(d)]	*Pagus Major* [(e)]	Porcine [(e)]	Calf [(f)]
Amino Acids	Skin	Skin	Bone	Muscle	Skin	Scale	Dermis	Skin
Alanine	107	136	129	115	135	133	115	119
Arginine	54	60	46	51	57	49	48	50
Aspartic acid/ asparagine	53	51	47	36	42	43	44	45
Cysteine	0	0	0	0	4	0	0	0
Glutamic acid/ glutamine	80	78	74	78	61	71	72	75
Glycine	342	286	361	356	334	346	341	330
Histidine	8	10	6	8	5	5	7	5
Isoleucine	12	5	5	17	10	7	10	11
Leucine	22	24	25	22	22	18	22	23
Lysine	29	31	25	25	23	26	27	26
Hydroxylysine	7	10	20	6	8	7	7	7
Methionine	15	12	8	9	10	15	6	6
Phenylalanine	12	15	12	12	17	13	12	3
Hydroxyproline	51	77	68	74	65	73	97	94
Proline	103	116	95	83	121	107	123	121
Serine	59	36	34	46	39	41	33	33
Threonine	23	29	25	36	24	24	16	18
Tyrosine	4	4	2	2	2	3	1	3
Valine	19	22	17	25	31	19	22	21
Imino acid[a]	154	193	163	157	186	180	220	215

Source: (a) Duan et al. (2009) (b) Kittiphattanabawon et al. (2005), (c) Mizuta et al. (2002), (d) Zhang et al. (2007), (e) Ikoma et al. (2003), and (f) Giraud-Guille et al. (2000).
[a]Imino acids include proline and hydroxyproline.

which is almost devoid of tryptophan. Fish muscle, skin, bone, and scale collagen differ from bovine and porcine hide collagens in having significantly higher contents of seven essential amino acids (Sikorski et al. 1990). Collagen is the only protein that is rich in hydroxyproline (up to 10% in mammalian collagen) (Sikorski et al. 1990). On the other hands, hydroxyproline is present in negligible amounts in other proteins, thus it is often used as a measure of collagen content in foods (Foegeding et al. 1996). Proline and hydroxyproline, imino acids, contents vary with species and their living habitat (Foegeding et al. 1996, Jongjareonrak et al. 2005, Kittiphattanabawon et al. 2005). Fish collagens contain less imino acids than do mammalian counterpart (Table 20.2). Lower content of imino acids is generally found in cold-water fish, compared with warm-water fish. Imino acids contribute to the stability of helix structure of collagen (Ikoma et al. 2003, Jongjareonrak et al. 2005). Pyrrolidine rings of proline and hydroxyproline impose restrictions on the conformation of the polypeptide chain and help to strengthen the triple helix (Wong 1989, Bae et al. 2008). The hydroxyl group of the hydroxyproline plays a role in stabilizing the helix by interchain hydrogen bonding via a bridging water molecule as well as direct hydrogen bonding to a carbonyl group (Wong 1989). Interchains between α-chains of collagen, in which hydroxyproline is involved in bondings are illustrated in Figure 20.2. Hydroxylysine of 6–20 residues/1000 residues is found in collagens, suggesting the partial cross-linking of collagen via covalent bond (Mechanic et al. 1987). Transition temperature of collagen from tropical fish is generally higher than that from cold- or temperate-water fish (Muyonga et al. 2004, Nalinanon et al. 2010). Differential scanning calorimetric study revealed the differences in maximal transition temperature (T_m) among collagen from different sources (Table 20.5). Therefore, collagen from different sources might have the different molecular properties due to the difference in imino acid content and cross-linking.

Collagen contains carbohydrates such as glucose and galactose. Other sugars are also found in some fish and invertebrate collagens. These are attached to hydroxylysine residues of the peptide chain by *O*-glycosidic bonds. The presence of 2-*O*-α-D-glucosyl-*O*-β-D-galactosyl-hydroxylysine and of *O*-β-D-galactosyl-hydroxylysine has been confirmed (Belitz et al. 2004). Some fish and invertebrate collagens also contain small amounts of arabinose, xylose, and ribose residues (Kimura 1972). These hydroxylysine-linked carbohydrates may have an impact on the structure of the fibrils in the invertebrate collagen (Sikorski et al. 1990).

ISOLATION OF COLLAGEN

Isolation of collagen is generally separated into three main steps, including sample preparation, extraction, and recovery.

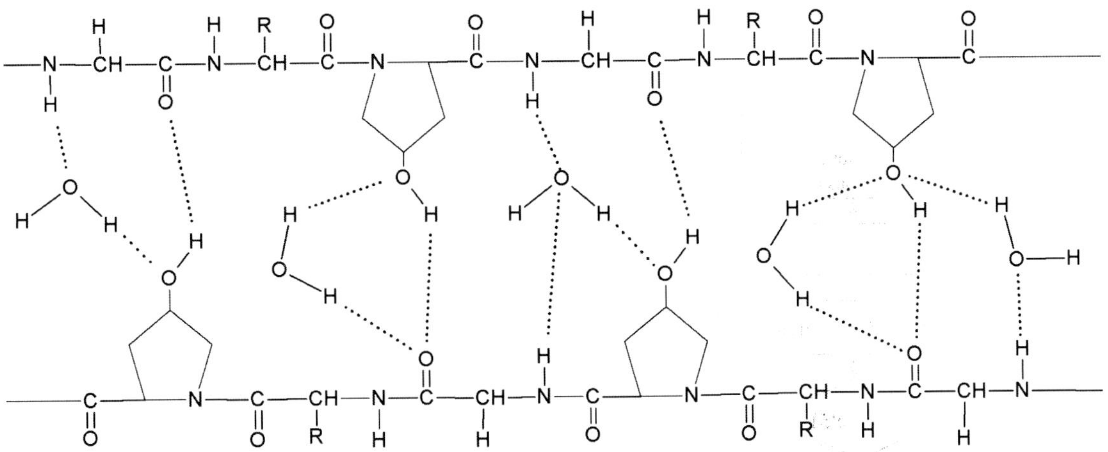

Figure 20.2. Contribution of hydroxyproline in hydrogen bonding between collagen polypeptides.

Generally, collagen can be extracted from fish skin, bone, scale, and so on. Those raw materials have been prepared by cleaning, size reduction, followed by appropriate pretreatment, prior to extraction. The extraction is carried out by the aid of acid, in which the resulting collagen is termed "acid-soluble collagen, ASC." Because of the lower yield of collagen obtained from this process, pepsin-aided process has been developed to maximize the extraction yield and the resultant collagen is named "pepsin-soluble collagen, PSC." Flow chart for the preparation of ASC and PSC is shown in Figure 20.3. All procedures are performed at low temperature (4°C) to avoid the thermal denaturation. All processes used for collagen extraction directly affect the extraction yield and properties of collagen obtained. The conditions for the extraction of collagen from different sources can be summarized in Table 20.4.

Preparation of Raw Materials

In general, the raw material for collagen preparation contains a number of contaminants including noncollagenous proteins, lipids, and pigments, and so on. Calcium or other inorganic matters are found in scale and bone. One or several pretreatments are used to remove the contaminants and increase the purity of extracted collagen. The removal of residual meats and cleaning are performed before further chemical pretreatments. Size reduction of raw material is also useful to facilitate the contaminant removal and collagen extraction. To remove noncollagenous proteins and pigments, alkaline pretreatment is used (Nagai and Suzuki 2002, Kittiphattanabawon et al. 2005, Hwang et al. 2007, Duan et al. 2009, Zhang et al. 2009, Benjakul et al. 2010). NaOH at a concentration of 0.1 M is widely used. However, NaCl and H_2O_2 have also been used to remove noncollagenous proteins and pigments, respectively (Wang et al. 2007, Zhuang et al. 2009). Some raw materials such as bones from carp (Duan et al. 2009) and bigeye snapper (Kittiphattanabawon et al. 2005) and scale from carp (Duan et al. 2009) contain high amount of calcium. Those raw materials are effectively decalcified by using ethylenediaminetetraacetic acid (EDTA) because of its chelating function. The decalcification can be also achieved by using inorganic acid, especially hydrochloric acid. As a consequence, the porous decalcified raw material with increased surface area can be readily subjected to collagen extraction.

The yield and properties of collagen would be influenced by the source of the raw material, the nature and concentration of acid or alkali used during pretreatment and the temperature and time of pretreatment and extraction (Regenstein and Zhou 2007). Prior to collagen extraction, skin, scale, and bone are generally subjected to noncollagenous protein removal. Thereafter, pretreated raw material is defatted or the pigments in the skin can be bleached to lower the color. In shark skin, different process was proposed by Yoshimura et al. (2000). For the alkali–acid extraction of collagen from the skin of great blue shark (*Prionace glauca*), a pretreatment with 0.5 or 1 M NaOH containing 15% Na_2SO_4 for 5 days at 20°C was required prior to acid extraction. For the alkaline direct extraction, a treatment with 0.5 M NaOH containing 10% NaCl at 4°C for 20–30 days was implemented.

Extraction

Acid Solubilization Process

The acid solubilization process has been widely used for collagen preparation. The collagen can be extracted using acid solution and the collagen obtained is referred to as "acid-soluble collagen, ASC." Extraction is conducted using acidic condition, in which the positive charge of collagen polypeptides becomes dominant. As a consequence, the enhanced repulsion among tropocollagen can be achieved, leading to the increased solubilization. Generally, tropocollagen is still in the form of triple helix with negligible changes. Fourier transform infrared (FTIR) spectra of collagens from the skins of carp and brownbanded bamboo shark and arabesque greenling extracted with acetic acid apparently revealed the triple-helical structure of collagen

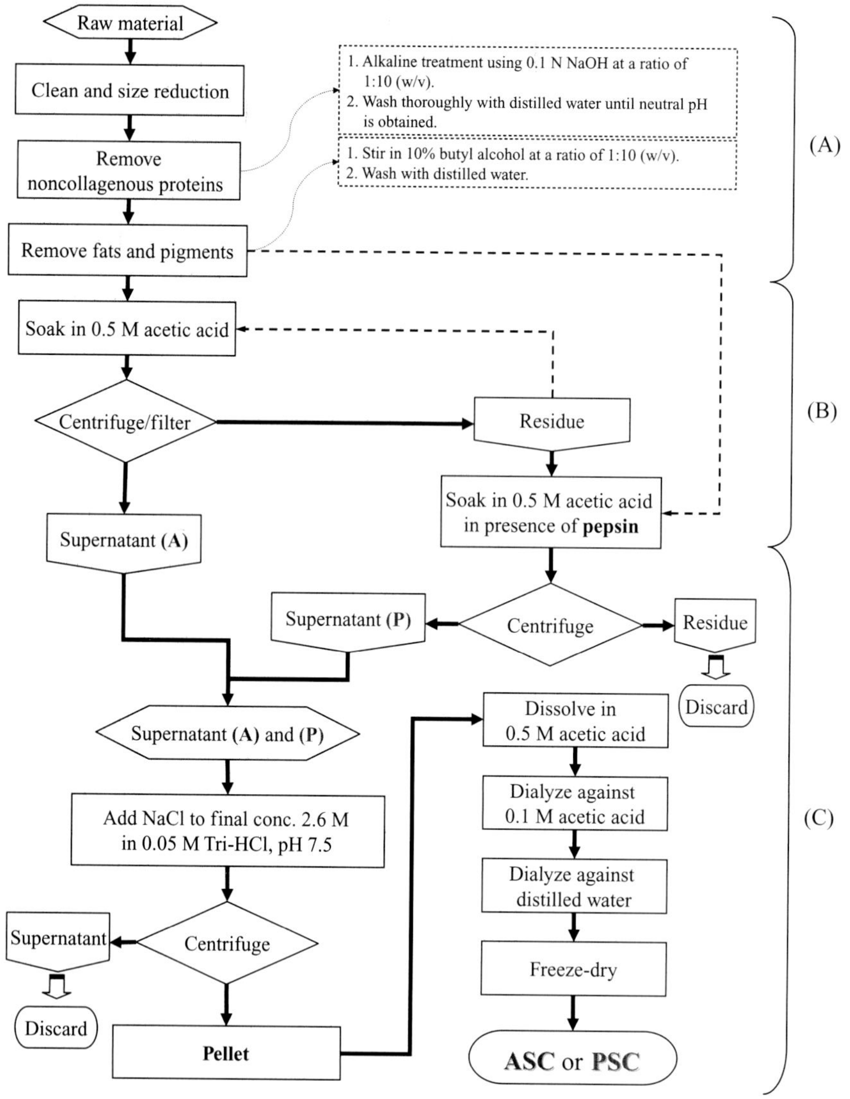

Figure 20.3. Flow chart for the preparation of acid soluble collagen (ASC) and pepsin soluble collagen (PSC). All procedures are performed at 4°C. (**A**) Preparation of raw material. (**B**) Extraction. (**C**) Recovery.

(Duan et al. 2009, Nalinanon et al. 2010, Kittiphattanabawon et al. 2010a). Collagen from total tissue can be isolated by direct extraction with organic acids (acetic, chloracetic, citric, lactic) or inorganic acid (hydrochloric). The yield of extracted collagen depends on the animal species used and the age and parameters of extraction (Skierka and Sadowska 2007). Collagen extraction from different species has been performed using acetic acid. Collagen extraction is generally achieved by 0.5 M acetic acid treatment at 4°C for 24–48 hours. Compared to simply prolonging the extraction time, the use of several consecutive extraction may get better ASC yield (Sadowska et al. 2003, Regenstein and Zhou 2007, Skierka and Sadowska 2007). Nalinanon et al. (2008) found that ASC yield from the skin of threadfin bream was increased from 12.32% to 34.90% (based on hydroxyproline content), when extended the extraction time from 6 to 48 hours. The similar result was observed in collagen extraction from bigeye snapper skin, in which the yield increased from 7.3% to 9.3% (based on hydroxyproline content) with increasing extraction time from 24 to 48 hours (Nalinanon et al. 2007). Increasing the concentration of acetic acid used from 0.1 to 0.5 M resulted in the increased yield of collagen extracted from minced cod skin, from 52% to 59% (dry weight). Re-extraction of residual matter with 0.5 M acetic acid can be performed to increase the yield of ASC, particularly in carp bone (Duan et al. 2009), bigeye snapper skin and bone (Kittiphattanabawon et al. 2005).

Skierka and Sadowska (2007) studied the influence of different acids on the extractability of collagen from the skin of Baltic cod (*Gadus morhua*). The maximal yield of collagen extracted with 0.5 M citric acid and acetic or lactic acids was 60% and 90%

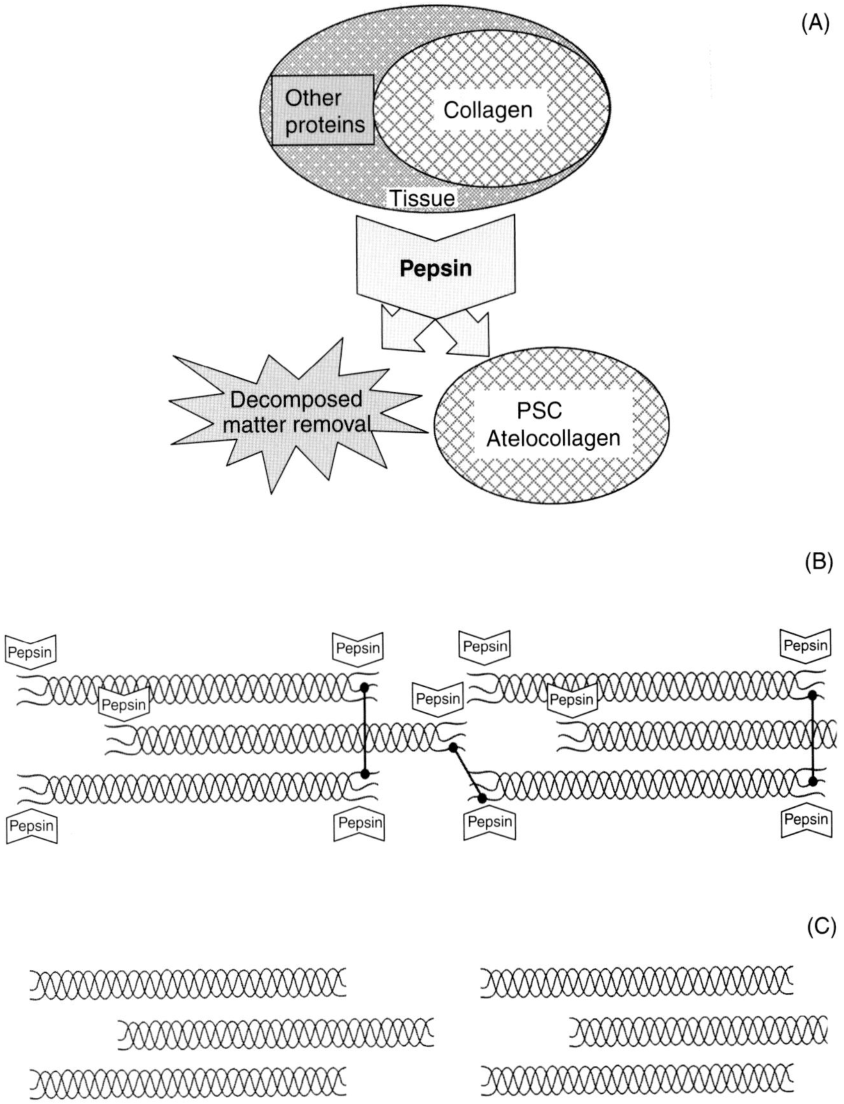

Figure 20.4. Collagen extraction with the aid of pepsin in 0.5 M acetic acid at 4°C. (**A**) Collagenous tissue degradation. (**B**) The cleavage of telopeptide region of collagen by pepsin. (**C**) Pepsin soluble collagen (PSC) or atelocollagen.

(based on hydroxyproline content), respectively, whereas collagen extracted with 0.15 M HCl had the yield of only 18%. The positively charged amine groups of protein can bind with anions (Cl^-), reducing electrostatic repulsive forces between charged groups. As a consequence, the structure of collagenous fibers is tightening, the ability of bonding water decreases, and the solubility of collagen is, therefore, reduced (Skierka and Sadowska 2007).

Pepsin Solubilization Process

Generally, typical acid solubilization process renders a low yield of collagen. To tackle the problem, pepsin has been applied because it is able to cleave peptides specifically in telopeptide region of collagen, leading to increased extraction efficiency (Fig. 20.4; Nagai et al. 2002a, Nalinanon et al. 2007). Use of pepsin as the aid for collagen extraction is a potential method for several reasons: (1) some of noncollagenous proteins are hydrolyzed and are easily removed by salt precipitation and dialysis, improving collagen purity; (2) hydrolyze those components and telopeptides of collagen to make the sample ready to solubilize in acid solution, resulting in improvement of extraction efficiency; (3) reduce antigenicity cause by telopeptide in the collagen which serve as the major problem in food and pharmaceutical applications (Werkmeister and Ramshaw 2000, Lee et al. 2001, Lin and Liu 2006, Cao and Xu 2008). Pepsin, particularly pepsin from porcine stomach, has been used to maximize the extraction efficiency of collagen from several species such as largefin

Table 20.3. Band Intensity Ratios of Cross-Linked Chain to Total Monomer Chains in Collagen from the Skin of Several Sources

Collagens	Sources	$\alpha 1/\alpha 2$	$\beta/(\alpha 1+\alpha 2)$	$\gamma/(\alpha 1+\alpha 2)$
ASC	Calf [(a)]	ND	1.26	0.75
	Black drum [(a)]	ND	1.59	1.32
	Sheepshead sea bream [(a)]	ND	1.30	0.62
	Deep-sea redfish [(b)]	2.47	1.52	1.10
PSC	Black drum [(a)]	ND	0.50	0.23
	Sheepshead sea bream [(a)]	ND	0.84	0.51
	Deep-sea redfish [(b)]	2.15	1.03	0.19

Source: (a) Ogawa et al. (2003) and (b) Wang et al. (2007).
ND, not determined.

langbarbel catfish (Zhang et al. 2009), seaweed pipefish (Khan et al. 2009), deep-sea redfish (Wang et al. 2007), brownstripe red snapper (Jongjareonrak et al. 2005), ocellate puffer fish (Nagai et al. 2002a), skate (Hwang et al. 2007), brownbanded bamboo shark (Kittiphattanabawon et al. 2010a), blacktip shark (Kittiphattanabawon et al. 2010c), cuttlefish (Nagai et al. 2001), octopus (Nagai et al. 2002b), scallop (Xuan Ri et al. 2007), paper nautilus (Nagai and Suzuki 2002), and common minke whale (Nagai et al. 2008). Skierka and Sadowska (2007) found that pepsin treatment of Baltic cod skin in acetic acid shortened the extraction time to 24 hours and the solubility of collagen, after pepsin digestion increased from 55% to 90%. After ASC extraction, the residues that represent the cross-linked molecules are further extracted in the presence of pepsin. The collagen obtained with pepsin treatment is referred to as "pepsin-soluble collagen, PSC" (Nagai et al. 2001, Ogawa et al. 2003, Nalinanon et al. 2007) or "atelocollagen" (Lee et al. 2001, Ikoma et al. 2003). The use of pepsin as the aid of collagen extraction resulted in the increased yield (see Table 20.4). The yield of extracted collagen also increased with increasing preswelling time and pepsin treatment time. From the skin of bigeye snapper (*Priacanthus tayenus*) subjected to preswelling with 0.5 M acetic acid and subsequent pepsin hydrolysis, those treated with these two steps with longer time led to the increase in yield of collagen extracted (Nalinanon et al. 2007). At the same preswelling time and hydrolysis time, bigeye snapper pepsin (BSP) exhibited greater extracting ability than did porcine pepsin. However, the highest yield (~65%) for collagen extraction from the skin of bigeye snapper was found when BSP was added with a reaction time of 48 hours, regardless of preswelling time (Nalinanon et al. 2007). Swollen skin, which was pretreated with acetic acid, possibly had a porous and loose structure caused by charge repulsion. As a result, the penetration of pepsin into the skin matrix could be enhanced. Thus, hydrolytic reaction of pepsin toward collagen was augmented. Miller (1972) proposed that the mechanism whereby the proteolytic activity of pepsin alters the solubility properties of cartilage collagen involves the degradation of the nonhelical region, thus effectively eliminating a site of intermolecular cross-linking. Drake et al. (1966) reported that most of intra- and intermolecular cross-links found in collagen occur through the telopeptide region. Some of the telopeptides of calfskin tropocollagen are vulnerable to pepsin action, since intramolecular cross-links are broken on pepsin digestion and fragments comprising a small fraction of the molecule (approximately 1%) become dialyzable (Drake et al. 1966). Pepsin treatment for collagen extraction has the impact on the proportion of compositions. In general, β- and γ-chains, dimmer, and trimer in PSC are decreased as compared with those found in ASC. Cleavage of telopeptide region more likely results in the conversion of cross-links, such as β- and γ-chains, to the α-chain, which can be of ease for extraction. The band intensity ratios of $\alpha 1$-chain to $\alpha 2$-chain and cross-linked chain (trimer form, γ, or dimmer form, β) to total monomer chain ($\alpha 1 + \alpha 2$) in collagens are shown in Table 20.3. Nalinanon et al. (2007) reported that higher molecular-weight (MW) components, including γ-chain, were found to a greater extent in ASC extracted from the skin of bigeye snapper than in PSC. Ogawa et al. (2003) found that the intra- and/or intermolecular cross-linking of collagens, β and γ components, were richer in ASC than in PSC for collagens from the skin of black drum and sheepshead sea bream. Similar results were also reported in collagens extracted from deep-sea redfish (Wang et al. 2007). Those high-MW components might be degraded to smaller components, such as β, $\alpha 1$, or $\alpha 2$, by pepsin action. Pepsin removes the cross-link-containing telopeptide, and the concomitantly one β-chain is converted to two α-chains (Nagai et al. 2001, Wang et al. 2007). Similar changes were also found in acid-solubilized calfskin tropocollagen when treated with pepsin. Native calfskin tropocollagen consisted of α:β:γ at a ratio of 32:65:3, whereas the components of pepsin-treated tropocollagen were changed to 72:23:3 (Drake et al. 1966). Therefore, the yield of collagen extraction was significantly increased because of the improvement of collagen extractability by pepsin (Ogawa et al. 2003, Nalinanon et al. 2007, Wang et al. 2007, Nalinanon et al. 2008, Benjakul et al. 2010).

Generally, commercial pepsin used for collagen extraction is isolated from porcine stomach. Owing to the limitation of porcine pepsin mostly associated with the religious constraint, pepsins from fish origin including bigeye snapper (Nalinanon et al. 2007), albacore tuna (Nalinanon et al. 2008), and tongol tuna (Nalinanon et al. 2008, Benjakul et al. 2010) have been used as the potential aid for collagen extraction from fish skin.

Nalinanon et al. (2007) reported that the addition of BSP at a level of 20 kUnits/g of defatted skin resulted in an increased content of collagen extracted from bigeye snapper skin. The yields of collagen from bigeye snapper skin extracted for 48 hours with acid and with BSP were 5.31% and 18.74% (dry basis), respectively. With preswelling in acid for 24 hours, collagen extracted with BSP at a level of 20 kUnits/g of defatted skin for 48 hours had a yield of 19.79%, which was greater than that of collagen extracted using porcine pepsin at the same level (13.03%). When tuna pepsin (10 units/g defatted skin) was used for collagen extraction from the skin of threadfin bream for 12 hours, the yield of collagen increased by 1.84- to 2.32-fold and albacore pepsin showed the comparable extraction efficacy to porcine pepsin (Nalinanon et al. 2008). Recently, PSCs from the skin of two species of bigeye snapper, *P. tayenus* and *Priacanthus macracanthus*, were extracted with the aid of tongol tuna (*Thunnus tonggol*) pepsin and porcine pepsin (Benjakul et al. 2010). Yields of PSCs extracted from the skin of *P. tayenus* with the aid of porcine pepsin (T-PP) and tongol tuna pepsin (T-TP) were 77.4 and 87.3 g kg^{-1} (based on wet weight of skin), respectively, while those of PSCs extracted from the skin of *P. macracanthus* with the aid of porcine pepsin (M-PP) and tongol tuna pepsin (M-TP) were 70.6 and 72.9 g kg^{-1}, respectively. For the same pepsin used, *P. tayenus* skin gave a higher yield than *P. macracanthus* skin. When the same skin was used for collagen extraction, tongol tuna pepsin treatment resulted in the higher yield (Benjakul et al. 2010). Therefore, tuna pepsin showed higher efficiency in collagen extraction. Pepsin from different sources might have a different cleavage site on the collagen. As a result, the peptides in the telopeptide region, which is susceptible to hydrolysis, were hydrolyzed to different degrees (Nalinanon et al. 2008, Benjakul et al. 2010). In addition, collagen from different sources was probably cleaved by the same pepsin in different ways (Benjakul et al. 2010). Furthermore, collagen from the two skins may have some differences in configuration and amino acid sequence. Fish pepsins used for collagen extraction from several sources exhibited the comparable extraction efficacy to porcine counterpart. Therefore, fish pepsin could be used as a replacer of mammalian pepsin.

The extraction conditions for ASC and PSC from different sources and their yield, type, and molecular compositions are summarized in Table 20.4.

Recovery of Collagen

After extraction, the collagen is generally recovered by salt precipitation prior to dialysis and freeze-drying. The collagen solution is generally precipitated by adding NaCl to a final concentration of 2.6 M in the presence of 0.05 M tris(hydroxylmethyl)aminomethane, pH 7.5 (Nagai and Suzuki 2000, Jongjareonrak et al. 2005, Kittiphattanabawon et al. 2005, Nalinanon et al. 2007, Benjakul et al. 2010). However, the concentrations of NaCl used for collagen precipitation in different methods are varied from 0.9 to 2.6 M. This can be adjusted to maximize the collagen recovery and removal of impurities. The resultant precipitate is then collected by centrifugation. The pellet is dissolved in 0.5 M acetic acid with minimal volume prior to dialysis against 0.1 M acetic acid and distilled water. The dialysate is finally freeze-dried and the powder obtain is used as collagen. Collagen can be isolated from fish skin, bone, and scale (Nomura et al. 1996, Nagai et al. 2002a, Ikoma et al. 2003, Mizuta et al. 2003, Muyonga et al. 2004, Ogawa et al. 2004, Nalinanon et al. 2007, Nalinanon et al. 2008). Nagai and Suzuki (2000) extracted type I collagen from underutilized resources including fish skin, bone, and fin with varying yields as follows: (1) skin collagen: 51.4% (Japanese sea bass), 49.8% (chub mackerel), and 50.1% (bullhead shark); (2) bone collagen: 42.3% (skipjack tuna), 40.7% (Japanese sea bass), 53.6% (ayu), 40.1% (yellow sea bream), and 43.5% (horse mackerel); (3) fin collagen: 5.2% (Japanese sea bass acid-soluble collagen) and 36.4% (Japanese sea bass acid-insoluble collagen) on the dry basis. Kittiphattanabawon et al. (2005) extracted ASC from the skin and bone of bigeye snapper (*P. tayenus*) with the yields of 10.94% and 1.59% on the basis of wet weight, respectively. The yield of ASC from the skins of young and adult Nile perch (*Lates niloticus*) extracted using 0.5 M acetic acid and precipitation with 0.9 M NaCl were 63.1% and 58.7% (on a dry weight basis), respectively (Muyonga et al. 2004). Sadowska et al. (2003) extracted the collagen from the skins of Baltic cod (*G. morhua*). The use of a one-stage, 24-hour extraction of whole skins with acetic at ratio of material to solvent of 1:6, yielded 20% collagen. Using three consecutive 24-hour extractions of whole skins with citric acid, 85% of collagen protein could be separated. Recently, ASC and PSC from the skin of brownbanded bamboo shark (*Chiloscyllium punctatum*) were isolated and characterized by Kittiphattanabawon et al. (2010a). The yield of ASC and PSC were 9.38% and 8.86% (wet weight basis), respectively. Collagen from the outer skin of cuttlefish (*Sepia lycidas*) was extracted by Nagai et al. (2001). The initial extraction of the cuttlefish outer skin in acetic acid yielded only 2% of collagen (dry weight basis). On subsequent digestion of the residue with 10% pepsin (w/v), a solubilized collagen (PSC) was obtained with a yield of 35% (dry weight basis). Nagai et al. (2000) found that 35.2% collagen was extracted from rhizostomous jellyfish (*Rhopilema asamushi*) by limited pepsin digestion.

CHARACTERISTICS AND PROPERTIES OF COLLAGENS

Collagen from different sources generally has the varying properties, especially the susceptibility to thermal denaturation or cleavage by proteases (Nalinanon et al. 2007, Nalinanon et al. 2010). Fish collagen, comprising lower imino acid content, is less stable than mammalian collagen. This property directly affected the application.

Mammalian Collagen

Bovine skin corium obtained from freshly slaughtered cattle of increasing biological age showed greater resistance to the action of acids, as measured by extractable collagen (Miller et al. 1983). Citrate-soluble content of corium from fetal skin was 30.9%, while that of 3–6-week-old calf was 6.4% and of

Table 20.4. The Conditions for the Extraction of Collagen from Different Sources and Their Yield, Type, and Molecular Compositions

Raw Materials	Pretreatment[a]	Extraction[a]	%Yield (Based on Wet Basis)	Collagen Type	Molecular Compositions	References
Bigeye snapper skin (*Priacanthus tayenus*, T and *Priacanthus macracanthus*, M)	0.1 M NaOH for 6 h, followed by 10% butanol for 18 h	0.5 M acetic acid containing pepsin from tongol tuna stomach extract (TP) or porcine (PP) for 48 h	7.74 (T-PP) 8.73 (T-TP) 7.06 (M-PP) 7.29 (M-TP)	Type I	$(\alpha 1)_2\alpha 2$	Benjakul et al. (2010)
Largefin longbarbel catfish skin (*Mystus macropterus*)	0.1 M NaOH containing 0.5% nonionic detergent for 24 h, followed by 15% (v/v) butanol for 24 h and 3% H_2O_2 for 24 h	0.5 M acetic acid for 24 h and re-extract with 0.5 M acetic acid for 12 h (ASC) 0.5 M acetic acid containing porcine pepsin for 30 h (PSC)	16.8[b] (ASC) 28.0[b] (PSC)	Type I	$\alpha 1\alpha 2\alpha 3$	Zhang et al. (2009)
Seaweed pipefish (*Syngnathus schlegeli*)	0.1 M NaOH for 36 h	0.5 M acetic acid for 3 days and re-extract with the same solution for 3 days (ASC) 0.5 M acetic acid containing porcine pepsin for 48 h (PSC)	5.5[b] (ASC) 33.2[b] (PSC)	Type I	$(\alpha 1)_2\alpha 2$- and $\alpha 1\alpha 2\alpha 3$ for ASC and PSC, respectively	Khan et al. (2009)
Ocellate puffer fish skin (*Takifugu rubripes*)	0.1 N NaOH for 3 days, followed by 10% butanol for 2 days	0.5 M acetic acid for 3 days (ASC) 0.5 M acetic acid containing porcine pepsin for 48 h (PSC)	10.7[b] (ASC) 44.7[b] (PSC)	NM[d]	$(\alpha 1)_2\alpha 2$	Nagai et al. (2002a)
Bigeye snapper skin (*P. tayenus*)	0.1 M NaOH for 6 h, followed by 10% (v/v) butanol for 18 h	0.5 M acetic acid for 24 h and re-extract with the same solution for 24 h (ASC)	10.9 (ASC)	Type I	At least two different α-chains	Kittiphattanabawon et al. (2005)
Deep-sea redfish (*Sebastes mentella*)	1.0 M NaCl for 24 h	0.5 M acetic acid for 24 h and re-extract with 0.5 M acetic acid for 24 h (ASC) 0.5 M acetic acid containing porcine pepsin for 48 h (PSC)	47.5 (ASC) 92.2 (PSC)	Type I	At least two different α-chains	Wang et al. (2007)
Brownstripe red snapper (*Lutjanus vitta*)	0.1 M NaOH for 24 h, followed by 10% (v/v) butanol for 24 h	0.5 M acetic acid for 24 h and re-extract with the same solution for 16 h (ASC) 0.5 M acetic acid containing porcine pepsin for 48 h (PSC)	9.0 (ASC) 4.7 (PSC)	Type I	At least two different α-chains	Jongjareonrak et al. (2005)
Skate skin (*Raja kenojei*)	0.1 M NaOH for 24 h	0.5 M acetic acid (ASC) 0.5 M acetic acid containing porcine pepsin for 24 h (PSC)	8.9 (NM[c])	Type I (major) and V (minor)	$(\alpha 1(I))_2\alpha 2(I)$ with cross-link of $\alpha 2$-chain and $(\alpha 1(V))_2\alpha 2(V)$	Hwang et al. (2007)

Nile tilapia skin (*Oreochromis niloticus*)	0.1 M NaOH for 48 h, followed by 10%(v/v) butanol for 24 h	0.5 M acetic acid for 3 days and re-extract with same solution for 2 days (ASC)	39.4[b] (ASC)	Type I	$\alpha 1\alpha 2\alpha 3$	Zhang et al. (2009)
Carp skin (*Cyprinus carpio*)	0.1 M NaOH for 6 h, followed by 1.0% detergent for 12 h	0.5 M acetic acid for 3 days (ASC)	41.3[b] (ASC)	Type I	$(\alpha 1)_2\alpha 2$	Duan et al. (2009)
Brownbanded bamboo shark skin (*Chiloscyllium punctatum*)	0.1 M NaOH for 6 h	0.5 M acetic acid for 48 h (ASC) 0.5 M acetic acid containing porcine pepsin for 48 h (PSC)	9.38 (ASC) 8.86 (PSC)	Type I	$\alpha 1$ (major) with the cross-link of $\alpha 2$-chain	Kittiphattanabawon et al. (2010a)
Blacktip shark skin (*Carcharhinus limbatus*)	0.1 M NaOH for 6 h	0.5 M acetic acid for 48 h (ASC) 0.5 M acetic acid containing porcine pepsin for 48 h (PSC)	20.01 (ASC) 0.86 (PSC)	Type I	$\alpha 1$ (major) with the cross-link of $\alpha 2$-chain	Kittiphattanabawon et al. (2010c)
Arabesque greenling skin (*Pleurogrammus azonus*)	0.1 M NaOH for 6 h, followed by 10% (v/v) butanol for 18 h	0.5 M acetic acid for 24 h and re-extract with the same solution for 24 h (ASC) 0.5 M acetic acid containing albacore tuna pepsin for 24 h (PSC)	30.3[b] (ASC) 14.0[b] (PSC	Type I	At least two different α-chains	Nalinanon et al. (2010)
Threadfin bream skin (*Nemipterus* spp.)	0.1 M NaOH for 6 h, followed by 10% (v/v) butanol for 18 h	0.5 M acetic acid for 12 h (ASC) 0.5 M acetic acid containing pepsin from albacore tuna (A-PSC) or skipjack tuna (S-PSC) or tongol tuna (T-PSC) or porcine (P-PSC) for 12 h	22.45[c] (ASC) 74.48[c] (A-PSC) 63.81[c] (S-PSC) 71.95[c] (T-PSC) 75.92[c] (P-PSC)	Type I	$(\alpha 1)_2\alpha 2$	Nalinanon et al. (2008)
Brownbanded bamboo shark cartilage (*C. punctatum*)	0.1 M NaOH for 6 h, followed by 0.5 M EDTA for 40 h	0.5 M acetic acid for 48 h (ASC) 0.5 M acetic acid containing porcine pepsin for 48 h (PSC)	1.27[b] (ASC) 9.59[b] (PSC)	Types I and II	$(\alpha 1(I))_2\alpha 2(I)$ and $\alpha 1(II)_3$	Kittiphattanabawon et al. (2010b)
Blacktip shark cartilage (*C. limbatus*)	0.1 M NaOH for 6 h, followed by 0.5 M EDTA for 40 h	0.5 M acetic acid for 48 h (ASC) 0.5 M acetic acid containing porcine pepsin for 48 h (PSC)	1.04[b] (ASC) 10.3[b] (PSC)	Types I and II	$\alpha 1(I))_2\alpha 2(I)$ and $\alpha 1(II)_3$	Kittiphattanabawon et al. (2010b)
Carp bone (*C. carpio*)	0.1 M NaOH for 24 h, followed by 0.5 M EDTA-2Na for 5 days and 1.0% detergent for 12 h	0.5 M acetic acid for 3 days and re-extract with 0.5 M acetic acid for 2 days (ASC)	1.06[b]	Type I	$(\alpha 1)_2\alpha 2$	Duan et al. (2009)

(*Continued*)

Table 20.4. (*Continued*)

Raw Materials	Pretreatment[a]	Extraction[a]	%Yield (Based on Wet Basis)	Collagen Type	Molecular Compositions	References
Bigeye snapper bone (*P. tayenus*)	0.1 M NaOH for 6 h, followed by 10% (v/v) butanol for 18 h and 0.5 M EDTA-4 Na for 40 h	0.5 M acetic acid for 24 h and re-extract with the same solution for 24 h (ASC)	1.6 (ASC)	Type I	At least two different α-chains	Kittiphattanabawon et al. (2005)
Pagrus major scale	10% (w/v) for 48 h, followed by 0.5 M EDTA for 24–48 h	0.01 M HCl for 48 h (ASC) 0.01 M HCl containing porcine pepsin for 72 h (PSC)	~2 for ASC and PSC	Type I	NM[d]	Ikoma et al. (2003)
Oreochromis niloticas scale	10% (w/v) for 48 h, followed by 0.5 M EDTA for 24–48 h	0.01 M HCl for 48 h (ASC) 0.01 M HCl containing porcine pepsin for 72 h (PSC)	~2 for ASC and PSC	Type I	NM[d]	Ikoma et al. (2003)
Carp scale (*C. carpio*)	0.1 M NaOH for 6 h, followed by 0.5 M EDTA-2Na for 24 h	0.5 M acetic acid for 4 days (ASC)	1.35[b]	Type I	$(\alpha 1)_2\alpha 2$	Duan et al. (2009)
Scallop mantle (*Patinopecten yessoensis*)	0.1 M NaOH for 12 h	0.5 M acetic acid containing porcine pepsin for 2 days	0.48 (PSC)	Type I (major) and V (minor)	Two different α-chains	Xuan Ri et al. (2007)
Cuttlefish outer skin (*Sepia lycidas*)	0.1 N NaOH for 3 days	0.5 M acetic acid for 3 days (ASC) days and re-extract with the same solution for 2 days (ASC) 0.5 M acetic acid containing porcine pepsin for 48 h (PSC)	2[b] (ASC) 35[b] (PSC)	NM[d]	$(\alpha 1)_2\alpha 2$	Nagai et al. (2001)
Rhizostomous jellyfish mesogloea (*Rhopilema asamushi*)	0.1M NaOH	0.5M acetic acid for 3 days (ASC) 0.5 M acetic acid containing porcine pepsin for 48 h (PSC)	32.5[b] (PSC)	NM[d]	$\alpha 1\alpha 2\alpha 3\alpha 4$ for PSC	Nagai et al. (2000)
Common minke whale (*Balaenoptera acutorostrata*) unesu	99.5% ethanol, followed by 0.1 M NaOH for 2 days	0.5 M acetic acid for 3 days (ASC) 0.5 M acetic acid containing porcine pepsin (PSC)	0.9 (ASC) 28.4 (PSC)	Type I	$(\alpha 1)_2\alpha 2$	Nagai et al. (2008)

[a]All procedures were performed at 4°C.
[b]Based on dry basis.
[c]Based on hydrohyproline content.
[d]NM, not mentioned.

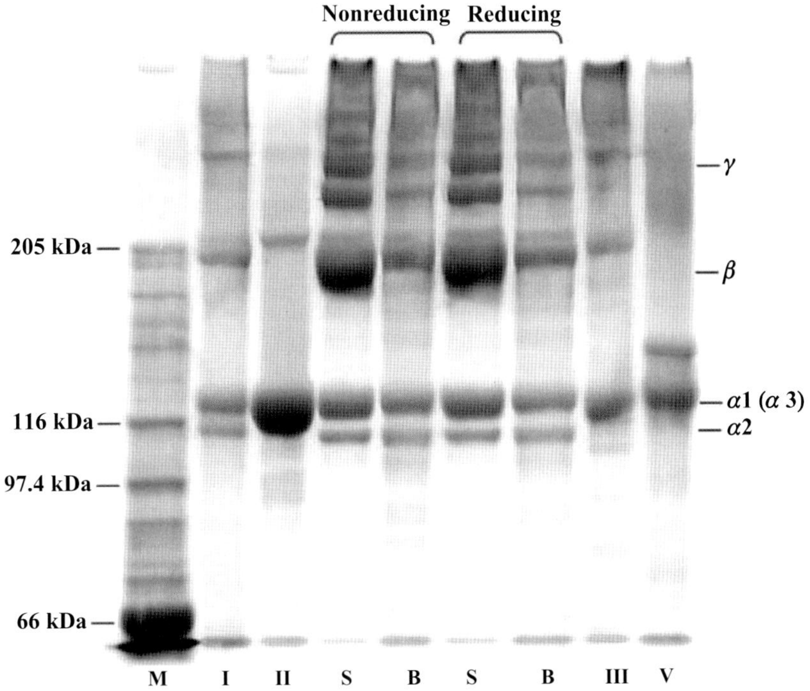

Figure 20.5. SDS–PAGE patterns of collagen from the skin (S) and bone (B) of bigeye snapper under reducing and nonreducing conditions. M, I, II, III, and V denote high molecular weight protein markers, and collagens type I, II, III, and V, respectively (Kittiphattanabawon et al. 2005).

18-month steer was 3.2%. Similar trends were also noted in neutral salt-soluble (NSC) and pepsin-treated, acid-soluble (PSC) fractions. The greater resistance to degradation in biologically older bovine collagen is thought to be directly related to increased cross-link formation. Gel permeation chromatography, using a newly developed 03BC-Bondagel column, was found to provide a rapid means for separation and determination of molecular size distribution (Miller et al. 1983). Chomarat et al. (1994) compared the effectiveness of two proteolytic enzymes, pepsin and proctase (isolated from *Aspergillus niger*), for the solubilization of collagen from bovine skin. Pepsin- and proctase-solubilized collagens had the similar yields (75% and 76% of total collagen as calculated from hydroxyproline). However, proctase-extracted collagen exhibited a decrease in high-MW components compared with pepsin-extracted collagen (Chomarat et al. 1994).

Collagen was prepared from common minke whale unesu and characterized (Nagai et al. 2008). The yield of collagen was high, about 28.4% (wet weight basis). By SDS–PAGE and TOYOPEARL® CM-650M column chromatography, the collagen was classified as type I collagen. The denaturation temperature of the collagen was 31.5°C, about 6–7°C lower than that of porcine collagen. Attenuated total reflectance-FTIR analysis indicated that ASC from common minke whale unesu held its triple-helical structure well, but the structures of porcine skin collagen and pepsin-solubilized collagen from common minke whale were changed slightly because of the loss of *N*- and *C*-terminus domains (Nagai et al. 2008).

Fish Collagen

Fish collagen from different sources can have different components. This may affect the properties of those collagens. Generally, fish collagens, from both finfish and shellfish, are classified to be types I and V as a major and minor component, respectively.

Protein Components

Both collagens, from the skin and bone of bigeye snapper (*P. tayenus*), comprised at least two different α-chains, $\alpha 1$ and $\alpha 2$ (see Fig. 20.5). The $\alpha 3$-chain, if present, could not be separated under the electrophoretic conditions employed because $\alpha 3$(I) migrates electrophoretically to the same position as $\alpha 1$(I) (Kimura and Ohno 1987, Matsui et al. 1991). Collagen of skin and bone from other fish species showed the similarity (Kimura and Ohno 1987, Matsui et al. 1991, Nagai and Suzuki 2000). The electrophoretic patterns of collagens from the skin and bone under nonreducing and reducing conditions were similar, indicating the absence of the disulfide bonds in those collagens. However, peptide maps of collagen from the skin and bone of bigeye snapper, digested by V8 protease and lysyl endopeptidase, revealed differences between collagens from skin and bone, and both were completely different from those of calfskin collagen (Kittiphattanabawon et al. 2005).

Nalinanon et al. (2008) reported that threadfin bream skin collagen (PSC) obtained with the aid of tuna pepsin showed similar protein patterns compared with those found in ASC. Nevertheless, pepsin from skipjack tuna caused the degradation of α- and β-components. All collagens were classified as type I with large portion of β-chain, consists of two identical α1-chains and one α2-chain. However, proteins with MW greater than 200 kDa were abundant in ASC. Recently, Benjakul et al. (2010) extracted PSCs from the skin of two species of bigeye snapper, *P. tayenus* and *P. macracanthus*, with the aid of tongol tuna (*T. tonggol*) pepsin and porcine pepsin. PSCs from the skin of both species extracted using porcine pepsin had a higher content of β-chain but a lower content of α-chains compared with those extracted using tuna pepsin.

The elution profiles of ASC and PSC from the skin of arabesque greenling on the TOYOPEARL CM-650M column after being denatured by heat treatment in the presence of 2 M urea showed that the primary structure was modified to some degree by albacore tuna pepsin digestion (Nalinanon et al. 2010). β12/β22 dimer of ASC and PSC was the major component, and high-MW component and γ-chain were also detected in the same fractions. It was suggested that some of α2-component might either dimerize covalently into β-component and form β12- or β22-dimer or polymerize into higher MW components. As a result, much lower band intensity of α2-chain was detected on SDS–PAGE and band intensity ratios of α1/α2 more than twofold were observed (Nalinanon et al. 2010). Collagens from alkali–acid extraction and alkali direct extraction of great blue shark skin were composed of α-chains (α1 and α2) (Yoshimura et al. 2000). Based on protein patterns and TOYOPEARL CM-650M column chromatography, ASC and PSC from the skin of brownbanded bamboo shark contained α- and β-chains as their major components and were characterized as type I collagen with the cross-link of α2-chain, which was different from type I collagen from teleost fish species (Kittiphattanabawon et al. 2010a). The similar result was observed in collagens from the skin of blacktip shark (*Carcharhinus limbatus*) (Kittiphattanabawon et al. 2010c). The chromatographic fractions indicated by the numbers were subjected to SDS–PAGE. The fractions of ASC (Fig. 20.6A) and PSC (Fig. 20.6B) were eluted as two major peaks. The α1-chain was found in the first peak (fraction nos. 28–40 and 23–36 for ASC and PSC, respectively), while the α2- and β-chains were found in the second peak (fraction nos. 41–59 and 37–55 for ASC and PSC, respectively). The results revealed that both collagens might be type I collagen, although the band intensity of α1-chain was not twofold higher than that of α2-chain. α2-component might dimerize into the β-component and form β_{12}-dimer (Hwang et al. 2007). As a result, much lower band intensity of α2-chain was detected on SDS–PAGE. Similar results have been reported for collagen type I from other elasmobranches (Hwang et al. 2007, Bae et al. 2008). Hwang et al. (2007) purified PSC from skate skin by differential salt precipitation, followed by phosphocellulose column chromatography. The collagen was purified as major collagen and minor collagen. The major collagen was defined as type I collagen, contained α1 and α2, with dimerization of α2-chain into β-chain and form β12-dimer. This result was in accordance with other elasmobranch type I collagen obtained from other reports. For the minor collagen, it contained α1 and α2 and was classified as type V collagen.

Fourier Transform Infrared Spectroscopy

Generally, the collagens from fish skin such as Nile perch (Muyonga et al. 2004), deep-sea redfish (Wang et al. 2007), channel catfish (Liu et al. 2007), yellowfin tuna (Woo et al. 2008), carp (Duan et al. 2009), arabesque greenling (Nalinanon et al. 2010), brownbanded bamboo shark (Kittiphattanabawon et al. 2010a), and blacktip shark (Kittiphattanabawon et al. 2010c) exhibited similar FTIR spectra in which the absorption bands were situated in the amide band region including amides A, B, I, II, and III. A slight difference in FTIR spectra between ASC and PSC from deep-sea redfish (Wang et al. 2007) and arabesque greenling (Nalinanon et al. 2010) has been reported, indicating slight differences in the secondary structure and functional groups of collagens. Amide A band (3400–3440 cm^{-1}) is generally associated with the N–H stretching vibration and shows the existence of hydrogen bonds (Woo et al. 2008, Kittiphattanabawon et al. 2010a). When the NH group of a peptide is involved in a hydrogen bond, the position is shifted to lower frequencies. Amide B band of collagens is observed at 2924–2926 cm^{-1}, which is related to CH_2 asymmetrical stretching vibration (Muyonga et al. 2004). The amide I and II bands of several fish collagens are observed at the wavenumber of 1640–1654 cm^{-1} and 1531–1555 cm^{-1}, respectively (Muyonga et al. 2004, Duan et al. 2009, Nalinanon et al. 2010, Kittiphattanabawon et al. 2010a, 2010c). The amide I band, with characteristic frequencies in the range of 1600–1700 cm^{-1}, is mainly associated with the stretching vibrations of the carbonyl groups (C = O bond) along the polypeptide backbone and is a sensitive marker of polypeptide secondary structure (Payne and Veis 1988, Wang et al. 2007, Kittiphattanabawon et al. 2010c). The shoulder of amide I band appearing at 1637 cm^{-1} could be attributed to the triple helix absorption of collagens (Petibois et al. 2006). Amide II (~1550 cm^{-1}) is associated with N-H bending coupled with C-N stretching (Muyonga et al. 2004, Woo et al. 2008). Amide III band of collagens is observed at 1234–1238 cm^{-1}, which is related to C-N stretching and N–H in plane bending from amide linkages as well as absorptions arising from wagging vibrations from CH_2 groups from the glycine backbone and proline side chains (Mohd Nasir et al. 2006) involved in triple-helical structure of collagen (Muyonga et al. 2004, Wang et al. 2007, Woo et al. 2008). The triple-helical structure of collagens is also confirmed from the absorption ratio between Amide III and 1450 cm^{-1} bands, which was approximately equal to 1.0 (Plepis et al. 1996, Wang et al. 2007, Kittiphattanabawon et al. 2010a). A slight difference in this ratio between ASC and PSC from deep-sea redfish (Wang et al. 2007) and arabesque greenling (Nalinanon et al. 2010) has been reported. The result suggested that pepsin might affect the triple-helical structure of collagen to some degree. The differences in those ratios and wavenumbers of amide bands between

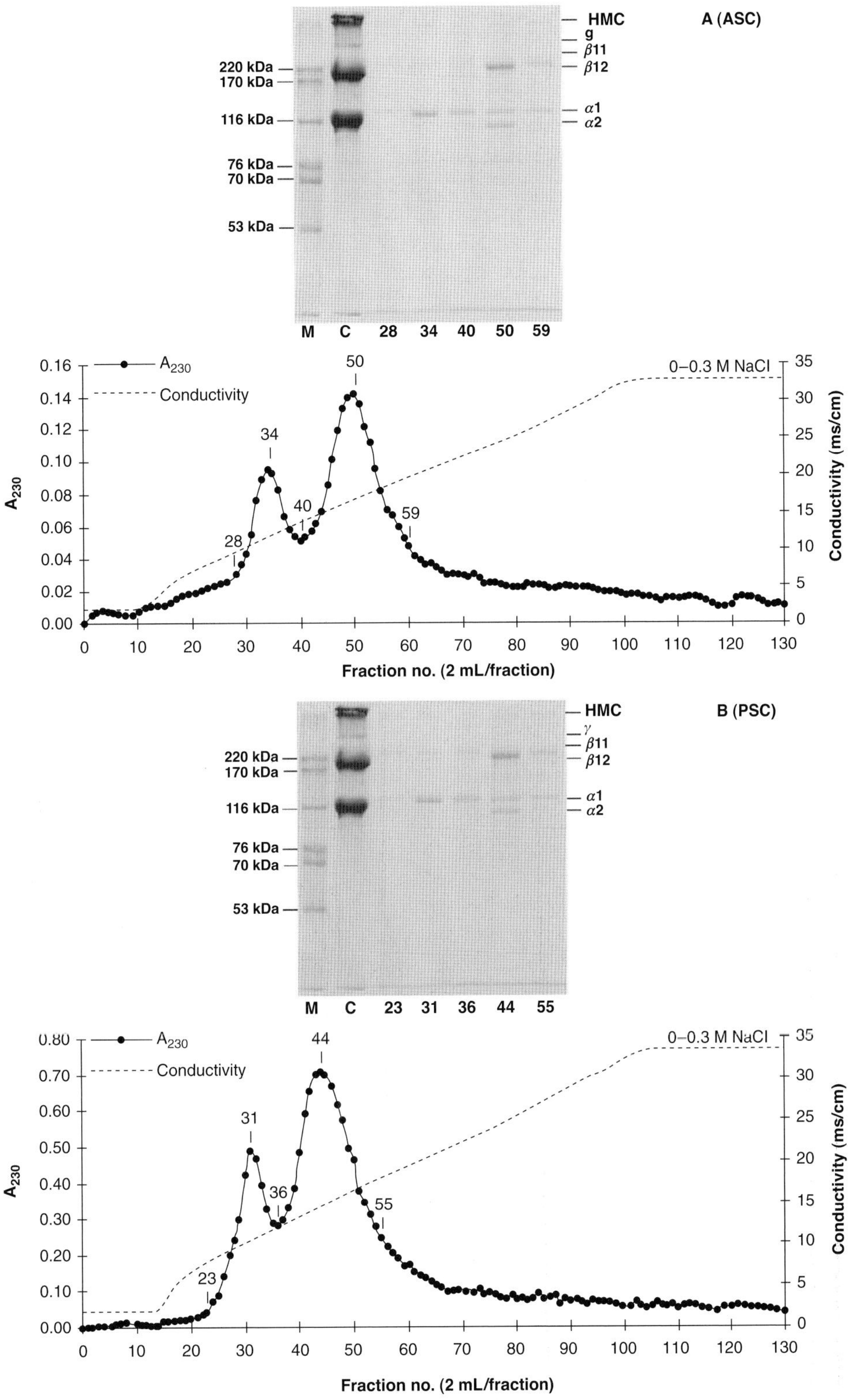

Figure 20.6. Elution profiles of ASC (A) and PSC (B) from the skin of brownbanded bamboo shark on the TOYOPEARL CM-650M ion-exchange column. The fractions indicated by numbers were examined by SDS–PAGE using 7.5% separating gel and 4% stacking gel. M, C, and HMC denote high-molecular-weight protein markers, collagen, and high-molecular-weight cross-linked components, respectively (Kittiphattanabawon et al. 2010a).

Table 20.5. The Maximum Transition Temperature (T_m) or Denaturation Temperature (T_d) of Collagens from the Skin of Several Fish Species

	Imino Acid[a] Content (%)		T_m or T_d (°C)		
Sources of Skin Collagen	ASC	PSC	ASC	PSC	References
Calf[b]	21.5	–	36.3	–	Ogawa et al. (2003)
Young Nile perch[b]	19.3	–	36.0	–	Muyonga et al. (2004)
Adult Nile perch[b]	20.0	–	36.0	–	Muyonga et al. (2004)
Black drum[b]	20.0	19.7	34.2	35.8	Ogawa et al. (2003)
Sheepshead sea bream[b]	20.5	19.8	34.0	34.3	Ogawa et al. (2003)
Brownstripe red snapper[c]	21.1	22.1	31.5	31.0	Jongjareonrak et al. (2005)
Bigeye snapper[c]	19.3	–	32.5	31.5	Nalinanon et al. (2007)
Carp[b]	19.0	–	28.0	–	Duan et al. (2009)
Channel catfish[b]	17.1	–	32.5	–	Liu et al. (2007)
Pacific whiting[b]	16.4	–	21.7	–	Kim and Park (2004)
Deep-sea redfish[b]	16.5	16.0	16.1	15.7	Wang et al. (2007)
Cod[b]	15.4	–	15.0	–	Duan et al. (2009)
Arabesque greenling[c]	15.9	15.7	15.7	15.4	Nalinanon et al. (2010)
Brownbanded bamboo shark[c]	20.4	20.7	34.5	34.5	Kittiphattanabawon et al. (2010a)
Blacktip shark[c]	19.7	20.3	34.2	34.4	Kittiphattanabawon et al. (2010c)

[a]Imino acids include proline and hydroxyproline.
[b]Thermal stability was expressed as T_d.
[c]Thermal stability was expressed as T_m.

collagens indicated some differences in the molecular structure. However, the triple-helical structure of PSC from other species, such as brownbanded bamboo shark (Kittiphattanabawon et al. 2010a) and blacktip shark (Kittiphattanabawon et al. 2010c), was not affected by pepsin digestion during collagen extraction. This confirms that pepsin specifically cleaves only at telopeptide regions of tropocollagen.

Thermal Stability

The contents of imino acids (proline and hydroxyproline) are related to thermal stability of collagen. Generally, collagens containing small concentrations of both imino acids denature at lower temperature than do those with the larger concentrations. The maximal transition temperatures (T_m) or denaturation temperatures (T_d) of collagen from different fish are shown in Table 20.5. A high content of imino acid is needed for stabilization of collagen (Xu et al. 2002). The thermal stability of the collagen triple helix is attributed to the hydrogen-bonded networks, mediated by water molecules, which connect the hydroxyl group of hydroxyproline in one strand to the main chain amide carboxyl of another chain (Babu and Ganesh 2001). Therefore, differences in hydroxyproline content might determine the denaturation temperatures of collagens from different species. Collagens derived from fish species living in cold environments have lower contents of hydroxyproline and they exhibit lower thermal stability than those from fish living in warm environment (Muyonga et al. 2004). Therefore, the imino acid content of fish collagens is associated with their thermal stability and correlates with the water temperature of their normal habitat (Foegeding et al. 1996). Norland (1990) reported that collagen from cold-water fish consisted of imino acid about 16–18%. Muyonga et al. (2004) found that ASC extracted from the skins of young and adult Nile perch contain higher content of imino acids (19.3% and 20.0%, respectively) than that from other fish species. The denaturation temperature for the collagens from the skins of young and adult Nile perch was determined to be 36°C, which was higher than that for most other fish species. The circular dichroism measurement of type I collagens from fish scales of *Pagrus major* and *Oreochromis niloticas* indicated that the denaturation temperatures were dependent on the amount of hydroxyproline rather than proline residues. Raman spectra also indicated that the relative intensities of Raman lines at 879 and 855 cm^{-1} assigned to Hyp and Pro rings were changed due to the contents of the imino acids (Ikoma et al. 2003). Nalinanon et al. (2010) reported that T_m of collagens from the skin of arabesque greenling was 15.4–15.7°C, which was in accordance with that of cold-water fish species such as cod (15°C) (Duan et al. 2009) and deep-sea redfish (16.1°C) (Wang et al. 2007). Denaturation temperatures of skin collagen from several cold- and temperate-water fish including carp (28°C) (Duan et al. 2009), grass carp (28.4°C) (Zhang et al. 2007), ocellate puffer fish (28°C) (Nagai et al. 2002a), tiger puffer fish (28.4°C) (Bae et al. 2008), Japanese seabass (26.5°C), chub mackerel (25.6°C), and bullhead shark (25°C) (Nagai and Suzuki 2000) have been reported. These collagens had much lower denaturation temperature than those of subtropical and tropical fish such as black drum (34.2°C) (Ogawa et al. 2003), sheephead sea bream (34°C) (Ogawa et al. 2003), bigeye snapper (32.5°C) (Nalinanon et al. 2007), brownstripe red snapper (30.5°C) (Jongjareonrak et al. 2005), brownbanded bamboo shark (34.5°C) (Kittiphattanabawon et al. 2010a), and blacktip shark (34.5°C) (Kittiphattanabawon et al. 2010c). The

Table 20.6. Isoelectric Point (pI) of Collagens from Different Sources

Collagens	Sources of Collagen	Sources of Pepsin Used	PI	References
ASC	Arabesque greenling skin	–	6.31	Nalinanon et al. (2010)
	Brownbanded bamboo shark skin	–	6.21	Kittiphattanabawon et al. (2010a)
	Blacktip shark skin	–	6.78	Kittiphattanabawon et al. (2010c)
	Brownbanded bamboo shark cartilage	–	6.53	Kittiphattanabawon et al. (2010b)
	Blacktip shark cartilage	–	6.96	Kittiphattanabawon et al. (2010b)
PSC	Arabesque greenling skin	Albacore tuna	6.38	Nalinanon et al. (2010)
	Brownbanded bamboo shark skin	Porcine	6.56	Kittiphattanabawon et al. (2010a)
	Blacktip shark skin	Porcine	7.02	Kittiphattanabawon et al. (2010c)
	P. tayenus skin	Porcine	7.15	Benjakul et al. (2010)
	Priacanthus tayenus skin	Tongol tuna	7.46	Benjakul et al. (2010)
	Priacanthus macracanthus skin	Porcine	5.97	Benjakul et al. (2010)
	P. macracanthus skin	Tongol tuna	6.44	Benjakul et al. (2010)
	Brownbanded bamboo shark cartilage	Porcine	7.03	Kittiphattanabawon et al. (2010b)
	Blacktip shark cartilage	Porcine	7.26	Kittiphattanabawon et al. (2010b)

differences in denaturation temperature of collagen from different sources might be governed by different contents of imino acids (proline and hydroxyproline) (Jongjareonrak et al. 2005, Zhang et al. 2007, Kittiphattanabawon et al. 2010a). The thermal stability of collagen is associated with the restriction of the secondary structure of the polypeptide chain governed by the pyrrolidine rings of proline and hydroxyproline and partially by the hydrogen bonding through the hydroxyl group of hydroxyproline (Benjakul et al. 2010).

Zeta (ζ) Potential

Charge of proteins plays a role in properties, especially solubility as well as interaction with other molecules via ionic interaction. Collagen from different fish possesses the zero net charge at pH range of 6.21 to 7.26 (Table 20.6). Generally, the pI of collagen is determined by amino acid composition. The pH where the ζ-potential is zero corresponds to the pI of the protein (Ma et al. 2009), in which a net electrical charge of zero at the surface is obtained. The collagens had a net positive or negative charge when pH values are below and above their pI, respectively (Nalinanon et al. 2010, Kittiphattanabawon et al. 2010a, b, c). At pI of proteins, hydrophobic–hydrophobic interaction increased, thereby promoting the precipitation and aggregation of protein molecules (Jongjareonrak et al. 2005). Nalinanon et al. (2010) determined the ζ-potential of ASC and PSC from arabesque greenling (*Pleurogrammus azonus*) at different pHs. Generally, ζ-potential profiles of both collagens were similar within pH range tested (Fig. 20.7). The similarity in ζ-potential profiles of ASC and PSC indicated that partial removal of telopeptides

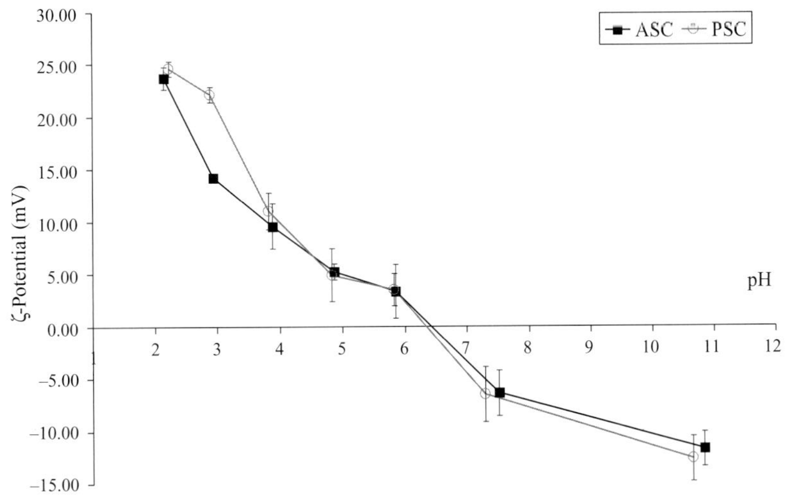

Figure 20.7. Zeta (ζ) potential of ASC and PSC from the skin of arabesque greenling at different pHs. Bars represent the standard deviation from triplicate determinations (Nalinanon et al. 2010).

in PSC by the albacore tuna pepsin did not significantly alter surface charge of collagen (Nalinanon et al. 2010). However, slight differences in ζ-potential profiles of ASC and PSC from shark skins were observed when porcine pepsin was used as extraction aid. The difference in pI and the rate of changes in ζ-potential between ASC and PSC from shark skins might be caused by the different amino acids mediated by partial removal of telopeptides by pepsin used (Kittiphattanabawon et al. 2010a, 2010c). It was noticed that PSCs from different sources showed different pIs. Benjakul et al. (2010) reported that PSC of the skin of *P. tayenus* showed a higher pI than those from the skin of *P. macracanthus*. For the same skin, PSCs prepared using porcine pepsin possessed a lower pI than those prepared using tongol tuna pepsin. Owing to the similar amino acid composition of all PSCs, the differences in net surface charge at different pHs were most likely governed by the different unfolding or exposure of charged amino acids, in which protonation or deprotonation could take place to different degrees (Benjakul et al. 2010). Thus, the different pepsins may cleave the telopeptide region at different sites, leading to differences in the ease of conformational changes at different pHs and to different charges at the surface (Benjakul et al. 2010, Kittiphattanabawon et al. 2010b). The pI of ASC and PSC of arabesque greenling was estimated to be 6.31 and 6.38, respectively. Both collagens had pI in a slight acidic pH. This might be due to higher content of acidic amino acids, aspartic acid, and glutamic acid compared with basic amino acids, including histidine, lysine, and arginine (Nalinanon et al. 2010). The pI of collagens from the skin of brownbanded bamboo shark and blacktip shark was reported in the range of 6.21–7.02 (Kittiphattanabawon et al. 2010a, 2010c). The pI of ASC and PSC from the cartilage of brownbanded bamboo shark and blacktip shark were estimated to be 6.53, 7.03, 6.96, and 7.26, respectively. The difference in pI between both shark cartilages might be caused by the variation in their amino acid compositions (Kittiphattanabawon et al. 2010b). Proteins arise from different genes may have a variable number of amino acids with charged side chains, resulting in different surface charge and a variety of pI values (Bonner 2007).

Invertebrate Collagen

Collagens with different types and contents have been found in invertebrates. The presence of homotrimetic collagen molecules $(\alpha 1)_3$ has been demonstrated in the muscle of abalone *Haliotis discus* and top shell *Turbo cornutus* (Yoshinaka and Mizuta 1999). Kimura and Matsuura (1974) reported that collagen extracted by limited digestion with pepsin from foot muscle of abalone consisted of $\alpha 3$ -chain, while collagens from the skins of octopus and squid, and the subcuticular membranes of swimming crab and spiny lobster had $(\alpha 1)_2\alpha 2$, commonly found in various vertebrate collagens. The squid and lobster collagens contained the greater amount of glycosylated hydroxylysines (mainly glucosylgalactosylhydroxylysine) than most vertebrate collagens (Kimura and Matsuura 1974). On the other hand, bivalves have been shown to possess $(\alpha 1)_2\alpha 2$ heterotrimers, which have similar compositional features to vertebrate type I collagen. The byssus threads of sea mussel (*Mytilus edulis*) and the adductor muscles of pearl oyster (*Pinctada fucata*) contained heterotrimer collagen $(\alpha 1)_2\alpha 2$ (Yoshinaka and Mizuta 1999). Xuan Ri et al. (2007) reported that PSC from scallop (*Patinopecten yessoensis*) mantle consisted of two different collagen molecules, with different profiles in molecules, amino acids, and peptide maps, and had different uronic acid contents and gel-forming abilities.

Collagen in the mantle muscle of common squid *Todarodes pacificus* consisted of two genetically distinct types of PSC (Yoshinaka and Mizuta 1999). The major collagen, named type SQ-I collagen, is insoluble in 0.5 M acetic acid containing 0.45 M NaCl, and has a similar amino acid composition to those of vertebrate type I collagen. Type SQ-I collagen is the major collagen in the cranial cartilage and skin of common squid *T. pacificus* and in the skin and arm muscle of octopus *Octopus vulgaris*. In addition, similar collagens have been isolated from the sucker of cuttlefish and octopus. The minor collagen, called type SQ-II, is very soluble in 0.5 M acetic acid containing 0.45 M NaCl. Type SQ-II collagen has some compositional characteristics similar to vertebrate type V collagen, showing a low level of alanine and high level of hydroxylysine. Both type SQ-I and SQ-II collagens are heterotrimers, of which the subunit compositions are represented as $[\alpha 1(\text{SQ-I}]_2\alpha 2(\text{SQ-I})$ and $[\alpha 1(\text{SQ-II}]_2\alpha 2(\text{SQ-II})$, respectively (Yoshinaka and Mizuta 1999). Octopus *Callistoctopus arakawai* arm consisted of ASC and PSC at levels of 10.4% and 62.9%, respectively. PSC had a chain composition of $\alpha 1\alpha 2\alpha 3$ heterotrimer, while ASC showed only a single α-chain, $\alpha 1$. The denaturation temperature of those collagens was lower than that of porcine collagen (Nagai et al. 2002b). Nagai and Suzuki (2002) found that the collagen extracted from the outer skin of the paper nautilus was hardly solubilized in 0.5 M acetic acid. The insoluble matter was easily digested by 10% pepsin (w/v), and a large amount of collagen (PSC) was obtained with about a 50% yield. PSC had a chain composition of $\alpha 1\ \alpha 2\ \alpha 3$ heterotrimer similar to *C. arakawai* arm collagen. PSC from skin of cuttlefish (*S. lycidas*) had a chain composition of $(\alpha 1)_2\alpha 2$ heterotrimer, which was similar to Japanese common squid PSC. The denaturation temperature of this collagen was 27°C, that is about 10°C lower than that of porcine collagen.

The primary structure of PSC from rhizostomous jellyfish was very similar to that of PSC from edible jellyfish mesogloea, but it was different from those of the collagen from edible jellyfish exumbrella and the ASC from its mesogloea. The rhizostomous jellyfish mesogloea collagen had a denaturation temperature (T_d) of 28.8°C. This collagen contained a large amount of a fourth subunit, designated as $\alpha 4$. This collagen may have the chain composition of an $\alpha 1\alpha 2\alpha 3\alpha 4$ heterotetramer. Mizuta et al. (1998) reported that collagen from muscle of Antarctic krill, *Euphausa superba*, referred to as $\alpha 1(\text{Kr})$ component, constituted more than 80% of the total pepsin-soluble collagen and showed typical compositional feature of crustacean major collagens, with low alanine and high hydroxylysine contents. The $\alpha 1(\text{Kr})$ component was mainly distributed in relatively thick connective tissues, epimysium and perimysium. $\alpha 1(\text{Kr})$ component may functionally correspond to the major α-component of collagen in decapods, $\alpha 1(\text{AR-I})$ (AR-I : Arthropod-type I), and comprise a major homotrimeric collagen molecule, $[\alpha 1(\text{AR-I})_3]$.

FACTOR AFFECTING COLLAGEN PROPERTIES

Some factors have been known to influence the collagen properties. pH and salt concentration were reported to affect water-binding capacity, viscosity, and emulsifying properties of collagen from fish muscle and skin connective tissues (Montero et al. 1991). Jongjareonrak et al. (2005) reported that maximum solubility in 0.5 M acetic acid of collagen from brownstripe red snapper (*Lutjanus vitta*) was observed at pH 3 and 4 for ASC and PSC, respectively. A sharp decrease in solubility was observed in the presence of NaCl above 2% and 3% (w/v) for ASC and PSC, respectively. Collagens from the skin and bone of bigeye snapper (*P. tayenus*) had the highest solubility at pH 2 and 5, respectively. No changes in solubility were observed in the presence of NaCl up to 3% (w/v). However, a sharp decrease in solubility was found with NaCl above 3% (w/v) (Kittiphattanabawon et al. 2005). Where pH was between 2 and 4, the solubility and water binding capacity of trout (*Salmo irideus* Gibb) collagen were highest, but in the addition of NaCl, functionality was reduced (Montero et al. 1991). Montero et al. (1999) also reported that solubility, apparent viscosity, and water-binding capacity of collagenous material from hake and trout showed the maximum values at pH levels between 2 and 4 and at concentrations less than 0.25 M NaCl. Furthermore, emulsifying capacity decreased as the NaCl concentration increased and was highest at pH levels between 1 and 3 (Montero et al. 1991).

Collagen properties are also affected by processing parameters such as freezing, heating, and so on. Montero et al. (1995) compared four stabilizing methods: (1) freezing, (2) freeze-drying, (3) partial solubilization with 0.05 M acetic acid then freezing, and (4) partial solubilization with 0.05 M acetic acid then freeze-drying on the functional properties of collagen from plaice skin. Only freeze-drying caused reduction in solubility and emulsifying capacity. Viscosity was greatest when samples were presolubilized. Emulsifying capacity was not changed when samples were frozen and decreased when they were either freeze-dried or presolubilized. Optimum water-holding capacity was observed in samples, which were previously solubilized. During the storage of fish on ice, a progressive change in solubility of muscle collagen was found. For insoluble collagen, significantly lower values were detected at day 15 compared with day 0. A little increase in ASC was found, while no changes were seen in PSC during storage (Eckhoff et al. 1998). Therefore, some cleavage of intermolecular cross-links seems to occur during storage on ice (Eckhoff et al. 1998).

APPLICATIONS OF COLLAGEN

Food Applications

The collagen protein can be refined so that it can be used in processed meat products to improve protein functionality through the immobilization of free water, increasing the stability of the finished product (Prabhu et al. 2004). Kenney et al. (1992) used pork collagen as an inexpensive adjunct to increase cooking yields and tensile strength in restructured beef. In addition, the use of 10% raw and preheated pork collagen reduced ($P < 0.05$) product initial cohesiveness and juiciness and tended ($P < 0.09$) to reduce beefiness. Utilization of pork collagen in boneless cured pork that incorporates pale, soft, and exudative (PSE) meat increases water-holding capacity and has the potential to improve protein functionality characteristics of the product (Schilling et al. 2003). Effects of pork collagen in emulsified and whole muscle products have also been evaluated. Prabhu et al. (2004) found that incorporation of pork collagen at 1% and above significantly ($P < 0.05$) increased cooked and chilled yields in frankfurters but did not have any effect in hams. Purge was significantly ($P < 0.05$) reduced in both frankfurters and hams after 4 weeks of storage. Sensory difference testing showed no significant difference up to 2% usage level of pork collagen in both frankfurters and hams ($P > 0.05$). The use of pork collagen as a binder and extender in emulsion sausages and hams provides for the absorption and binding of moisture, so that during thermal processing, cook losses are minimized. (Prabhu et al. 2004). Effects of raw material and the inclusion of chicken collagen on the protein functionality of chunked and formed chicken breast manufactured from 100% PSE like and 100% normal broiler breast with either 0% or 1.5% collagen were determined (Schilling et al. 2005). Inclusion of collagen was effective in decreasing cooking loss and increasing protein–protein bind. PSE like with no collagen had lower ($P < 0.05$) protein–protein bind than normal meat with collagen. Schilling et al. (2004) also found that inclusion of turkey collagen improved the texture of turkey breast deli products through increasing the protein's ability to bind with each other in pale meat and was effective in decreasing cooking and purge loss in both pale and normal samples. These results revealed that addition of collagens is effective in increasing water-holding capacity and has the potential to improve quality characteristics of products made from either pale or normal raw material.

Biomedical Applications

The primary reason for the usefulness of collagen in biomedical application is that collagen can form fibers with extra strength and stability through its self-aggregation and cross-linking (Lee et al. 2001). Out of the ten collagen types that have been characterized, types I, III, and V are the most desirable for biomedical applications because of their high biocompatibility and low immunogenicity (Antiszko et al. 1996). Uses of collagen as surgical suture, hemostatic agents, and tissue engineering including basic matrices for cell systems and replacement/substitutes for artificial blood vessels and valves were reported (Lee et al. 2001). As a commercial medical product, collagen can be part of natural, stabilized tissue that is used in the device, for example, as in a bioprosthetic heart valve, or it can be fabricated as a reconstituted, purified product from animal sources, for example, as in wound dressings (Ramshaw et al. 2009). Polymers of natural origin that function as regeneration templates have so far been synthesized as graft copolymers of collagen and chondroitin 6-sulfate, a glycosaminoglycan. Certain collagen-graft glycosaminoglycans can induce synthesis of dermis within a wound bed in guinea pigs in which the epidermis and the complete dermis have been removed by surgical excision (Meena et al. 1999). Wisser and Steffes (2003) described the successful wound management of a 19-year-old patient with 63% of the body surface burnt covered with Integra®, consists of a modified bovine collagen matrix and a silastic membrane facing toward the body surface. After 1 year of surgical development, a common feature of wound areas treated with Integra showed the good sliding characteristics of the skin on its support and the lack of scar lines. Taguchi et al. (2004) developed a novel tissue adhesive, consisting of type I collagen from pig skin and cross-linking by citric acid derivative. This adhesive is biodegraded under the skin of mice within 7 days after subcutaneous injection. Synthetic materials in the form of dressings are used primarily for treatment of superficial wounds to prevent moisture and heat loss from the skin as well as to prevent bacterial infiltration from the environment (Meena et al. 1999). Antiszko et al. (1996) developed collagen membranes as carriers of antiseptics. Microdressings prepared with collagen membranes containing povidone-iodine or chlorhexidine glycerin solution and packed into envelopes made of laminated aluminum foil were stable for 8 weeks concerning their antibacterial activity. After implantation beneath mouse skin, the dressings did not cause inflammation. Such dressings may be useful in a treatment of ulcers, superficial burns, and chronic granulated wounds (Antiszko et al. 1996).

Advantages of collagen for biomedical application include (Meena et al. 1999) (i) abundant sources of highly purified (medical grade) collagen, (ii) the ability to be reconstituted into high strength forms useful in surgery, (iii) the wealth of research literature on the characterization of collagen, (iv) improved processing techniques, (v) introduction of several commercial collagen products, and (vi) recent advances in the use of collagen as a delivery system.

Pharmaceutical Applications

Bovine collagen was approved by the Food and Drug Administration in 1982, and since that time, it has remained as a primary filler of choice for soft-tissue augmentation of dermal defects (Matarasso 2006). The advent of human-derived collagen from fibroblast cultures further increased the popularity of injectable products. Cultured autologous collagen and several other formulations of collagen have been undergoing clinical trials or are available outside the United States (Matarasso 2006). These advances in collagen technology may deliver products that are more persistent in duration and comfortable to administer than materials that are presently in use. It is widely known that the digestion of collagen proteins or hydrolysates in vivo will generate peptides that may be useful for organic biosynthetic processes (Cúneo et al. 2010). Thus, the consumption of collagen-rich foods might be beneficial for bone health. Several functional properties of enzymatic collagen hydrolysates have been investigated in osteoarthritis and suggest a beneficial stimulating effect with an increased synthesis of extracellular matrix macromolecules produced by chondrocytes (Bello and Oesser 2006, Cúneo et al. 2010). Wu et al. (2004) postulated that oral administered collagen peptide may provide beneficial effects on bone metabolism in rat, especially in the calcium-deficient condition, without obvious undesirable effects. In a randomized, double-blind, placebo-controlled trial with osteoporotic women, Adam et al. (1996) investigated the use of a diet rich in collagen hydrolysate (10 g/day) and observed that calcitonin associated with this type of diet had a greater effect in inhibiting bone collagen breakdown than calcitonin alone, as determined by a decrease in urinary levels of pyridinoline cross-links. Bello and Oesser (2006) postulated that collagen hydrolysate ingestion stimulates a statistically significant increase in synthesis of extracellular matrix macromolecules by chondrocytes compared with untreated controls. These findings suggest mechanisms that might help patients affected by joint disorders such as osteoarthritis. A growing body of evidence provides a rationale for the use of collagen hydrolysate for patients with osteoarthritis (Bello and Oesser 2006). The cell biological properties of collagen, gelatin, and collagen hydrolysate (<15,000 Da) were studied using murine keratinocytes (Li et al. 2005). Keratinocyte culture experiments demonstrated that only collagen had significant effects on cell attachment and proliferation, but the results of cells cultured on gelatin and collagen hydrolysate showed that the rates of adhesion and proliferation were similar to those of cells cultured on plastic as a control. It is concluded that collagen has better physiological effects than those of gelatin and collagen hydrolysate as skin-care cosmetic materials (Li et al. 2005).

In addition, collagen has served as an excellent carrier for drug delivery systems. Collagen can be extracted into an aqueous solution and molded into various forms of delivery systems. The main applications of collagen as drug delivery systems are collagen shield in ophthalmology, sponges for burns/wounds, minipellets and tablets for protein delivery, gel formulation in combination with liposomes for sustained drug delivery, controlling material for transdermal delivery, and nanoparticles for gene delivery (Lee et al. 2001). Atelocollagen, which is produced by elimination of the telopeptide moieties using pepsin, has demonstrated its potential as a drug carrier, especially for gene delivery (Kohmura et al. 1999, Ochiya et al. 1999, Lee et al. 2001). Ho et al. (1997) demonstrated that gel formulation prepared by gelling the vehicle mixture of citric acid solution, ethanol, and propylene glycol with 1% (w/w) of telopeptide-poor collagen (atelocollagen) sample was suitable for transdermal delivery. Takaoka et al. (1991) reported that bovine skin collagens prepared by use of conventional methods (acid-solubilized collagen, or collagen-digested once with pepsin) were also assessed as carriers for bone morphogenetic protein (BMP) in mice, but were found to be inferior in terms of consistency of bone formation and amount of induced bone mass. The results suggest that telopeptide-depleted collagen (atelocollagen) permitted a gradual release of purified BMP for induction of bone, with minimal immunogenic interference. Consequently, this collagen carrier represents an important development for future clinical application of BMP (Takaoka et al. 1991). Ochiya et al. (1999) described a new gene transfer method that allows prolonged release and expression of plasmid DNA in vivo in normal adult animals through the use of atelocollagen, a biocompatible polymer, as the carrier. Plasmid DNA was carried in the atelocollagen, resulting in a minipellet of cylindrical shape (0.6 mm in diameter and 10 mm in length) that contained 50 μg of plasmid DNA and human HST-1/FGF-4 cDNA. The minipellet was able to carry the bioactive macromolecules even when as much as 50% (w/w) of the atelocollagen was replaced by pharmaceutical ingredients and/or bioactive macromolecules. This new technique of controlled gene transfer using atelocollagen and allowing prolonged systemic circulation of target products will facilitate a more effective and long-term use of naked plasmid vectors for somatic gene therapy.

REFERENCES

Adam M et al. 1996. Postmenopausal osteoporosis: calcitonin treatment on a diet rich in collagen proteins. *C'asopis Lekaru Ceskych* 135(3): 74–78.

Antiszko M et al. 1996. Collagen membranes of increased absorption as carriers of antiseptics. *Polimery W Medycynie* 26(3–4): 21–28.

Avery NC, Bailey AJ. 2005. Enzymic and non-enzymic cross-linking mechanisms in relation to turnover of collagen: relevance to aging and exercise. *Scand J Med Sci Sports* 15(4): 231–240.

Babu IR, Ganesh KN. 2001. Enhanced triple helix stability of collagen peptides with 4R-aminoprolyl (Amp) residues: relative roles of electrostatic and hydrogen bonding effects [11]. *J Am Chem Soc* 123(9): 2079–2080.

Bae I et al. 2008. Biochemical properties of acid-soluble collagens extracted from the skins of underutilised fishes. *Food Chem* 108(1): 49–54.

Bailey AJ et al. 1998. Mechanisms of maturation and ageing of collagen. *Mech Ageing Dev* 106(1–2): 1–56.

Belitz H-D et al. 2004. Meat. In: H-D Belitz et al. (eds.) *Food Chemistry*, 3rd edn. Springer-Verlag, Berlin, pp. 566–616.

Bello AE, Oesser S. 2006. Collagen hydrolysate for the treatment of osteoarthritis and other joint disorders: a review of the literature. *Curr Med Res Opin* 22(11): 2221–2232.

Benjakul S et al. 2010. Extraction and characterisation of pepsin-solubilised collagens from the skin of bigeye snapper (*Priacanthus tayenus* and *Priacanthus macracanthus*). *J Sci Food Agric* 90(1): 132–138.

Bonner PLR. 2007. *Protein Purification*. Taylor & Francis Group, Cornwall.

Cao H, Xu S-Y. 2008. Purification and characterization of type II collagen from chick sternal cartilage. *Food Chem* 108(2): 439–445.

Chomarat N et al. 1994. Comparative efficiency of pepsin and proctase for the preparation of bovine skin gelatin. *Enzyme Microb Technol* 16(9): 756–760.

Cúneo F et al. 2010. Effect of dietary supplementation with collagen hydrolysates on bone metabolism of postmenopausal women with low mineral density. *Maturitas* 65(3): 253–257.

Drake MP et al. 1966. Action of proteolytic enzymes on tropocollagen and insoluble collagen. *Biochemistry* 5(1): 301–312.

Duan R et al. 2009. Properties of collagen from skin, scale and bone of carp (*Cyprinus carpio*). *Food Chem* 112(3): 702–706.

Eckhoff KM et al. 1998. Collagen content in farmed Atlantic salmon (*Salmo salar*, L.) and subsequent changes in solubility during storage on ice. *Food Chem* 62(2): 197–200.

Foegeding EA et al. 1996. Characteristics of edible muscle tissues. In: OR Fennema (ed.) *Food Chemistry*. Marcel Dekker, New York, pp. 902–906.

Giraud-Guille MM et al. 2000. Structural aspects of fish skin collagen which forms ordered arrays via liquid crystalline states. *Biomaterials* 21(9): 889–906.

Ho H-O et al. 1997. Characterization of collagen isolation and application of collagen gel as a drug carrier. *J Control Release* 44(2–3): 103–112.

Hwang J-H et al. 2007. Purification and characterization of molecular species of collagen in the skin of skate (*Raja kenojei*). *Food Chem* 100(3): 921–925.

Ikoma T et al. 2003. Physical properties of type I collagen extracted from fish scales of *Pagrus major* and *Oreochromis niloticas*. *Int J Biol Macromol* 32(3–5): 199–204.

Johnston-Banks FA. 1990. Gelatin. In: P Harris (ed.) *Food Gels*. Elsevier Applied Science, London, pp. 233–289.

Jongjareonrak A et al. 2005. Isolation and characterisation of acid and pepsin-solubilised collagens from the skin of brownstripe red snapper (*Lutjanus vitta*). *Food Chem* 93(3): 475–484.

Kenney PB et al. 1992. Raw and preheated epimysium and gelatin affect properties of low-salt, low-fat, restructured Beef. *J Food Sci* 57(3): 551–554.

Khan SB et al. 2009. Isolation and biochemical characterization of collagens from seaweed pipefish, *Syngnathus schlegeli*. *Biotechnol Bioprocess Eng* 14(4): 436–442.

Kim JS, Park JW. 2004. Characterization of acid-soluble collagen from Pacific whiting surimi processing byproducts. *J Food Sci* 69(8): 637–642.

Kimura S. 1972. Studies on marine invertibrate collagen type V. The neutral sugar composition and glucosylated hydroxylysine contents of several collagens. *Bull Jpn Soc Sci Fish* 38: 1153–1156.

Kimura S, Matsuura F. 1974. The chain compositions of several invertebrate collagens. *J Biochem* 75(6): 1231–1240.

Kimura S, Ohno Y. 1987. Fish type I collagen: tissue-specific existence of two molecular forms, $(\alpha 1)_2\alpha 2$ and $\alpha 1\alpha 2\alpha 3$, in Alaska pollack. *Comp Biochem Physiol B Biochem Mol Biol* 88(2): 409–413.

Kittiphattanabawon P et al. 2005. Characterisation of acid-soluble collagen from skin and bone of bigeye snapper (*Priacanthus tayenus*). *Food Chem* 89(3): 363–372.

Kittiphattanabawon P et al. 2010a. Isolation and characterisation of collagen from the skin of brownbanded bamboo shark (*Chiloscyllium punctatum*). *Food Chem* 119(4): 1519–1526.

Kittiphattanabawon P et al. 2010b. Isolation and characterization of collagen from the cartilages of brownbanded bamboo shark (*Chiloscyllium punctatum*) and blacktip shark (*Carcharhinus limbatus*). *LWT—Food Sci Technol* 43(5): 792–800.

Kittiphattanabawon P et al. 2010c. Isolation and properties of acid- and pepsin-soluble collagen from the skin of blacktip shark (*Carcharhinus limbatus*). *Eur Food Res Technol* 230(3): 475–483.

Kohmura E et al. 1999. BDNF atelocollagen mini-pellet accelerates facial nerve regeneration. *Brain Res* 849(1–2): 235–238.

Lee CH et al. 2001. Biomedical applications of collagen. *Int J Pharm* 221(1–2): 1–22.

Li GY et al. 2005. Comparative study of the physiological properties of collagen, gelatin and collagen hydrolysate as cosmetic materials. *Int J Cosmet Sci* 27(2): 101–106.

Lin YK, Liu DC. 2006. Effects of pepsin digestion at different temperatures and times on properties of telopeptide-poor collagen from bird feet. *Food Chem* 94(4): 621–625.

Liu H et al. 2007. Studies on collagen from the skin of channel catfish (*Ictalurus punctaus*). *Food Chem* 101(2): 621–625.

Love RM et al. 1976. The connective tissues and collagens of cod during starvation. *Comp Biochem Physiol B Biochem Mol Biol* 55(4): 487–492.

Ma H et al. 2009. Sodium caseinates with an altered isoelectric point as emulsifiers in oil/water systems. *J Agric Food Chem* 57(9): 3800–3807.

Matarasso SL. 2006. The use of injectable collagens for aesthetic rejuvenation. *Semin Cutan Med Surg* 25(3): 151–157.

Matsui R et al. 1991. Characterization of an $\alpha 3$ chain from the skin type I collagen of chum salmon (*Oncoorhynchus keta*). *Comp Biochem Physiol B Biochem Mol Biol* 99(1): 171–174.

McCormick RJ. 1994. Structure and properties of tissues. In: DM Kinsman et al. (eds.) *Muscle Foods; Meat Poultry and Seafood Technology*. Chapman and Hall, New York, pp. 54–58.

Mechanic GL et al. 1987. Locus of a histidine-based, stable trifunctional, helix to helix collagen cross-link: stereospecific collagen structure of type I skin fibrils. *Biochemistry* 26(12): 3500–3509.

Meena C et al. 1999. Biomedical and industrial applications of collagen. *Proc Indian Acad Sci, Chem Sci* 111(2): 319–329.

Miller EJ. 1972. Structural studies on cartilage collagen employing limited cleavage and solubilization with pepsin. *Biochem* 11(26): 4903–4909.

Miller AT et al. 1983. Age-related changes in the collagen of bovine corium: studies on extractability, solubility and molecular size distribution. *J Food Sci* 48(3): 681–685.

Mizuta S et al. 1998. Characterization of a major alpha component of collagen from muscle of Antarctic krill *Euphausa superba*. *Comp Biochem Physiol B Biochem Mol Biol* 120(3): 597–604.

Mizuta S et al. 2002. Molecular species of collagen from wing muscle of skate (*Raja kenojei*). *Food Chem* 76(1): 53–58.

Mizuta S et al. 2003. Molecular species of collagen in pectoral fin cartilage of skate (*Raja kenojei*). *Food Chem* 80(1): 1–7.

Mohd Nasir NF et al. 2006. The study of morphological structure, phase structure and molecular structure of collagen-PEO 600K blends for tissue engineering application. *Am J Biochem Biotechnol* 2(4): 175–179.

Montero P et al. 1991. Effect of pH and the presence of NaCl on some hydration properties of collagenous material from trout (*Salmo irideus* Gibb) muscle and skin. *J Sci Food Agric* 54(1): 137–146.

Montero P et al. 1995. Plaice skin collagen extraction and functional properties. *J Food Sci* 60(1): 1–3.

Montero P et al. 1999. Functional characterisation of muscle and skin collagenous material from hake (*Merluccius merluccius* L.). *Food Chem* 65(1): 55–59.

Muyonga JH et al. 2004. Characterisation of acid soluble collagen from skins of young and adult Nile perch (*Lates niloticus*). *Food Chem* 85(1): 81–89.

Nagai T, Suzuki N. 2000. Isolation of collagen from fish waste material - skin, bone and fins. *Food Chem* 68(3): 277–281.

Nagai T, Suzuki N. 2002. Preparation and partial characterization of collagen from paper nautilus (*Argonauta argo*, Linnaeus) outer skin. *Food Chem* 76(2): 149–153.

Nagai T et al. 2000. Isolation and characterization of collagen from rhizostomous jellyfish (*Rhopilema asamushi*). *Food Chem* 70(2): 205–208.

Nagai T et al. 2001. Isolation and characterisation of collagen from the outer skin waste material of cuttlefish (*Sepia lycidas*). *Food Chem* 72(4): 425–429.

Nagai T et al. 2002a. Collagen of the skin of ocellate puffer fish (*Takifugu rubripes*). *Food Chem* 78(2): 173–177.

Nagai T et al. 2002b. Collagen of octopus *Callistoctopus arakawai* arm. *Int J Food Sci Technol* 37(3): 285–289.

Nagai T et al. 2008. Collagen from common minke whale (*Balaenoptera acutorostrata*) unesu. *Food Chem* 111(2): 296–301.

Nalinanon S et al. 2007. Use of pepsin for collagen extraction from the skin of bigeye snapper (*Priacanthus tayenus*). *Food Chem* 104(2): 593–601.

Nalinanon S et al. 2008. Tuna pepsin: characteristics and its use for collagen extraction from the skin of threadfin bream (*Nemipterus* spp.). *J Food Sci* 73(5): C413–C419.

Nalinanon S et al. 2010. Collagens from the skin of arabesque greenling (*Pleurogrammus azonus*) solubilized with the aid of acetic acid and pepsin from albacore tuna (*Thunnus alalunga*) stomach. *J Sci Food Agric* 90(9): 1492–1500.

Nomura Y et al. 1996. Preparation and some properties of type I collagen from fish scales. *Biosci Biotechnol Biochem* 60(12): 2092–2094.

Norland RE. 1990. Fish gelatin. In: MN Voight, JK Botta (eds.) *Advance in Fisheries Technology and Biotechnology for Increased Profitability*. Techcomic Publishing, Lancaster, pp. 325–333.

Ochiya T et al. 1999. New delivery system for plasmid DNA in vivo using atelocollagen as a carrier material: the Minipellet. *Nat Med* 5(6): 707–710.

Ogawa M et al. 2003. Biochemical properties of black drum and sheepshead seabream skin collagen. *J Agric Food Chem* 51(27): 8088–8092.

Ogawa M et al. 2004. Biochemical properties of bone and scale collagens isolated from the subtropical fish black drum (*Pogonia cromis*) and sheepshead seabream (*Archosargus probatocephalus*). *Food Chem* 88(4): 495–501.

Payne KJ, Veis A. 1988. Fourier transform IR spectroscopy of collagen and gelatin solutions: deconvolution of the amide I band for conformational studies. *Biopolymers* 27(11): 1749–1760.

Petibois C et al. 2006. Analysis of type I and IV collagens by FT-IR spectroscopy and imaging for a molecular investigation of skeletal muscle connective tissue. *Anal Bioanal Chem* 386(7): 1961–1966.

Plepis AMDG et al. 1996. Dielectric and pyroelectric characterization of anionic and native collagen. *Polym Eng Sci* 36(24): 2932–2938.

Prabhu GA et al. 2004. Utilization of pork collagen protein in emulsified and whole muscle meat products. *J Food Sci* 69(5): C388–C392.

Ramshaw J et al. 2009. Collagens as biomaterials. *J Mater Sci Mater Med* 20(0): 3–8.

Regenstein JM, Regenstein CE. 1991. Current issues in Kosher foods. *Trends Food Sci Technol* 2(C): 50–54.

Regenstein JM, Zhou P. 2007. Collagen and Gelatin from Marine By-Products. In: F Shahidi (ed.) *Mazimising the Value of Marine By-Products*, 1st edn. Woodhead Publishing Limited, Cambridge, pp. 279–303.

Sadowska M et al. 2003. Isolation of collagen from the skins of Baltic cod (*Gadus morhua*). *Food Chem* 81(2): 257–262.

Samuel CS et al. 1998. Effects of relaxin, pregnancy and parturition on collagen metabolism in the rat pubic symphysis. *J Endocrinol* 159(1): 117–125.

Schilling MW et al. 2003. Utilization of pork collagen for functionality improvement of boneless cured ham manufactured from pale, soft, and exudative pork. *Meat Sci* 65(1): 547–553.

Schilling MW et al. 2004. Pale and normal turkey breast enhancement through the utilization of Turkey collagen in a chunked and formed Turky breast roll. *J Appl Poul Res* 13(3): 406–411.

Schilling MW et al. 2005. Effects of collagen addition on the functionality of PSE-like and normal broiler breast meat in a chunked and formed deli roll. *J Muscle Foods* 16(1): 46–53.

Sikorski ZE et al. 1990. The nutritive composition of the major groups of marine food organisms. In: ZE Sikorski (ed.) *Seafood: Resources, Nutritional Composition and Preservation*. CRC Press, Boca Raton, FL, pp. 29–54.

Skierka E, Sadowska M. 2007. The influence of different acids and pepsin on the extractability of collagen from the skin of Baltic cod (*Gadus morhua*). *Food Chem* 105(3): 1302–1306.

Taguchi T et al. 2004. Bonding of soft tissues using a novel tissue adhesive consisting of a citric acid derivative and collagen. *Mater Sci Eng C* 24(6–8): 775–780.

Takaoka K et al. 1991. Telopeptide-depleted bovine skin collagen as a carrier for bone morphogenetic protein. *J Orthop Res* 9(6): 902–907.

Te Nijenhuis K. 1997. *Thermoreversible Networks: Viscoelastic Properties and Structure of Gels*, vol 130. Springer, New York.

Timpl R. 1984. Immunology of the collagens. In: KA Piez, AH Reddi (eds.) *Extracellular Matrix Biochemistry*. Elsevier, New York, pp. 159–190.

Wang L et al. 2007. Isolation and characterization of collagen from the skin of deep-sea redfish (*Sebastes mentella*). *J Food Sci* 72(8): E450–E455.

Werkmeister JA, Ramshaw JAM. 2000. Immunology of collagen-based biomaterials. In: DL Wise (ed.) *Biomaterials and bioengineering handbook*. Marcel Dekker, New York, pp. 739–759.

Wisser D, Steffes J. 2003. Skin replacement with a collagen based dermal substitute, autologous keratinocytes and fibroblasts in burn trauma. *Burns* 29(4): 375–380.

Wong DWS. 1989. *Mechanism and Theory in Food Chemistry*. Van Nostrand Reinhold Company Inc., New York.

Woo J-W et al. 2008. Extraction optimization and properties of collagen from yellowfin tuna (*Thunnus albacares*) dorsal skin. *Food Hydrocolloids* 22(5): 879–887.

Wu J et al. 2004. Assessment of effectiveness of oral administration of collagen peptide on bone metabolism in growing and mature rats. *J Bone Miner Metab* 22(6): 547–553.

Xiong YL. 1997. Structure-function relationships of muscle proteins. In: S Damodaran, A Paraf (eds.) *Food Proteins and Their Applications*. Marcel Dekker, New York, pp. 341–392.

Xu Y et al. 2002. Characterization of the nucleation step and folding of a collagen triple-helix peptide. *Biochemistry* 41(25): 8143–8151.

Xuan Ri S et al. 2007. Characterization of molecular species of collagen in scallop mantle. *Food Chem* 102(4): 1187–1191.

Yan M et al. 2008. Characterization of acid-soluble collagen from the skin of walleye pollock (*Theragra chalcogramma*). *Food Chem* 107(4): 1581–1586.

Yoshimura K et al. 2000. Preparation and dynamic viscoelasticity characterization of alkali-solubilized collagen from shark skin. *J Agric Food Chem* 48(3): 685–690.

Yoshinaka R, Mizuta S. 1999. Collagen types in aquatic molluses and crustaceans. In: R Yoshinaka, S Mizuta (eds.) *Extracellular Matrix of Fish and Shellfish*. The Haworth Press, New York, pp. 31–43.

Zayas JF. 1997. *Functionality of Proteins in Food*. Springer, Berlin.

Zhang M et al. 2009. Isolation and characterisation of collagens from the skin of largefin longbarbel catfish (*Mystus macropterus*). *Food Chem* 115(3): 826–831.

Zhang Y et al. 2007. Isolation and partial characterization of pepsin-soluble collagen from the skin of grass carp (*Ctenopharyngodon idella*). *Food Chem* 103(3): 906–912.

Zhuang Y et al. 2009. Antioxidant and melanogenesis-inhibitory activities of collagen peptide from jellyfish (*Rhopilema esculentum*). *J Sci Food Agric* 89(10): 1722–1727.

21
Fish Gelatin

Soottawat Benjakul, Phanat Kittiphattanabawon, and Joe M. Regenstein

Abstract: Fish gelatin, partially hydrolyzed or denatured collagen, has received attention as the alternative of mammalian and avian gelatin. It can be extracted from the by-products rich in collagen such as fish skin, bone, scale, or the skin of some invertebrate, and so on. Composition and properties of fish gelatin can be governed by the sources of raw materials. Processing parameters such as pretreatment, extraction temperature, bleaching, drying, and so on influence the chemical and functional properties of gelatin. Generally, fish gelatin exhibited lower functional properties than those from land animals, thereby limiting the applications of fish gelatin. Several approaches have been therefore developed to improve the properties of fish gelatin via modification using chemical or enzymatic processes. Therefore, fish gelatin can be applied more widely in several industries.

INTRODUCTION

Gelatin is biopolymer obtained from partial denaturation of collagen. It has a wide range of applications in food and nonfood (photographic, cosmetic, and pharmaceutical) industries (Regenstein and Zhou 2007a). Generally, gelatin is obtained from mammals, especially pig and cow skins and bones. Recently, outbreaks of mad cow disease (bovine spongiform encephalopathy, BSE) have caused anxiety in customers. Additionally, the gelatin obtained from pigskin and pig bone cannot be used in Kosher and Halal foods due to religious constraints. Furthermore, an increasing attention to health issues by consumers has also resulted in a high demand of fish gelatin (Kittiphattanabawon et al. 2010). As a consequence, the alternative sources for gelatin production have been intensively increased, especially skin, bone, and scale from seafood processing by-products due to their abundance and low cost. However, fish gelatins have limited application, mainly owing to the lower gel strength and lower gel stability, compared with those from their mammalian counterparts (Leuenberger 1991). This chapter covers the literature reviews of gelatin, particularly fish gelatin, regarding their production, functional properties, and improvement of their functional properties as well as applications.

Gelatin is one of the commonly used food ingredients, obtained by the thermal denaturation of collagen (Bailey and Light 1989). The process involves the disruption of noncovalent

Food Biochemistry and Food Processing, Second Edition. Edited by Benjamin K. Simpson, Leo M.L. Nollet, Fidel Toldrá, Soottawat Benjakul, Gopinadhan Paliyath and Y.H. Hui.

bonds and is partially reversible in agreement with the gelling properties of gelatin (Bigi et al. 1998). Gelatin has been widely applied in the food industry as ingredients to improve the elasticity, consistency, and stability of foods. It can be used for medical and pharmaceutical purposes such as encapsulation, production of hard and soft capsules, wound dressing, adsorbent pads, and edible film formation. Furthermore, it can be applied for biomaterial-based packaging and photographic industries (Jongjareonrak et al. 2006c).

PRODUCTION OF GELATIN

Production of gelatin can be divided into three main steps, (1) pretreatment, (2) extraction, and (3) drying (Fig. 21.1). All processes used for gelatin extraction have direct impact on the yield and properties of gelatin obtained. The process has been optimized for different sources of raw materials (Cho et al. 2004, Zhou and Regenstein 2004, Cho et al. 2005, Yang et al. 2007).

Raw materials
↓
Noncollagenous protein removal process
↓
Fat removal process (optional)
↓
Demineralization process (optional)
↓
Swelling process
↓
Extraction with distilled water
↓
Filtration
↓
Clarification (optional)
↓
Drying
↓
Gelatin powder

Figure 21.1. Scheme for gelatin extraction.

PRETREATMENT OF RAW MATERIAL

Removal of Noncollagenous Protein

Prior to gelatin extraction from raw material, the pretreatment is practically implemented to increase purity of gelatin extracted. Alkaline solution has been used to remove considerable amounts of noncollagenous materials (Johns and Courts 1977, Zhou and Regenstein 2005) and break some interchain cross-links. Also, the process is able to inactivate proteases involved in degradation of collagen (Regenstein and Zhou 2007b). During alkaline pretreatment, the type of alkali does not make a significant difference, but the concentration of alkali is critical (Zhou and Regenstein 2005). Yoshimura et al. (2000) reported that alkali attacks predominantly the telopeptide region of the collagen molecule during pretreatment. Thus, some collagen can be solubilized by an alkali solution. Long time and high concentration of alkaline pretreatment decreased the yield of gelatin from skin of channel catfish (Yang et al. 2007). The concentration of alkali, time, and temperature used for pretreatment varied with raw materials (Table 21.1).

Table 21.1. Pretreatment Conditions for the Extraction of Gelatin from Different Fishes

Raw Materials	Alkaline Type/Concentration	Conditions: Temperature (°C)	Time (min)	Alkaline Changing (times)	References
Brownbanded bamboo shark and blacktip shark	0.1 M NaOH	15–20	40	3	Kittiphattanabawon et al. (2010)
Atlantic salmon skin	0.04 N NaOH	8	30	2	Arnesen and Gildberg (2007)
Cod skin	0.2%(w/v) NaOH	NM[a]	40	3	Gudmundsson and Hafsteinsson (1997)
Sin croaker skin and shortfin scad skin	0.2%(w/v) NaOH	NM	40	3	Cheow et al. (2007)
Yellowfin tuna skin	0.1 N NaOH	20	40	3	Rahman et al. (2008)
Megrim skin	0.2 N NaOH	5	30	3	Montero and Gomez-Guillen (2000)
Black tilapia and red tilapia skin	0.2%(w/v) NaOH	15–27	40	3	Jamilah and Harvinder (2002)
Shark cartilage	1.6 N NaOH	8	3.14 days	1	Cho et al. (2004)

[a]NM: not mentioned.

Removal of Fat

Raw material with high fat content can be associated with the negative effect on gelatin. Soaps may be formed during treatment with alkali prior to gelatin extraction. This soap can contaminate the resulting gelatin. Nevertheless, fat removal process was not included in the pretreatment process for gelatin extraction from some fish skin with low fat content (Gudmundsson and Hafsteinsson 1997, Gómez-Guillén et al. 2003, Zhou and Regenstein 2004Arnesen and Gildberg 2007, Cheow et al. 2007, Yang et al. 2007, Aewsiri et al. 2008, Nalinanon et al. 2008, Rahman et al. 2008). A simple defatting can be achieved by hot water (Waldner 1977). Muyonga et al. (2004) removed fat in bone of Nile perch (*Lates niloticus*) by tumbling in warm water (35°C).

Removal of Minerals

Demineralization of the raw material, especially bone or scale, aims to remove the calcium and other inorganic substances to facilitate the extraction of collagenous component (Waldner 1977). The inorganic substances in raw material can be removed by treatment with dilute hydrochloric acid solution, whereby the calcium phosphate is dissolved as acid phosphates (Waldner 1977). The fresh bone is commonly treated with hydrochloric acid solution, in which almost all minerals are completely removed. Depending on the nature of the material, temperature, and acid concentration, the demineralization time can be varied. The acid concentration used is in the range 2–6% HCl (Waldner 1977). Acid hydrolysis of the protein should be minimized during demineralization. High temperature should also be avoided since it can enhance the hydrolysis of protein. Bone from Nile perch (*Lates niloticus*) was demineralized with 3% HCl at room temperature prior to extraction using warm water (60°C) (Muyonga et al. 2004). Demineralization of tuna fin (*Katsuwonus pelamis*) was carried out at room temperature using 0.6 N HCl (Aewsiri et al. 2008).

Swelling of Pretreated Raw Materials

Prior to gelatin extraction, pretreatment is generally required to enhance the extraction efficiency. Swelling is important because it can favor protein unfolding by disruption of noncovalent bonding and predispose the collagen to subsequent extraction and solubilization (Stainby 1987). Pretreatments can be classified into two processes and are selectively used on the basis of the raw materials.

Acid Process Acid hydrolysis is a milder treatment that effectively solubilizes collagens of animals slaughtered at a young age such as pigs (Foegeding et al. 1996). The pretreatment is aimed to convert the collagen into a form suitable for extraction. The covalent cross-links in the collagen must be disrupted to enable the release of free α-chains during the extraction (Johnston-Banks 1990). The process is able to remove other organic substances. Sulfuric and hydrochloric acids are used, often with the addition of phosphoric acid to retard color development (Johnston-Banks 1990). Additionally, the acid pretreatment can partially inactivate endogenous proteases involved in degradation. As a result, the enzymatic breakage of intrachain peptide bonds of collagen during extraction can be lowered (Zhou and Regenstein 2005). Nalinanon et al. (2008) reported that gelatin extraction from bigeye snapper skin using a pepsin-aided process in combination with a protease inhibitor showed the higher bloom strength than the gelatin extracted by the conventional process, which had a substantial degradation of gelatin components.

Moreover, type of acid and concentration affected the yield and properties of gelatin. The concentration of H^+ used in processing of gelatin from cod skins affected yield and quality of resulting gelatin (Gudmundsson and Hafsteinsson 1997). Megrim skin was treated with 0.05 M acetic and 0.05 M propionic acid prior to gelatin extraction using distilled water at 45°C for 30 minutes. The gelatin obtained had the highest elastic modulus, viscous modulus, melting temperature, and gel strength. On the other hand, gelatin obtained from skin swollen with citric acid exhibited the lowest turbidity of gelatin solution, whereas propionic acid led to the most turbid gelatin solution (Gómez-Guillén and Montero 2001). Gimenez et al. (2005) reported that use of lactic acid (25 mM) could be an excellent substitute for acetic acid for the skin swelling process. The gelatin so obtained showed similar properties to that prepared by using 50 mM acetic acid without the negative organoleptic properties. Gelatin obtained from the acid process is known as type A gelatin.

Alkaline Process Type B gelatins are generally produced by alkali hydrolysis of bovine materials. This process results in deamidation as well as degradation, leading to the lowered chain length (Foegeding et al. 1996). Alkaline pretreatments are normally applied to bovine hide and ossein. Lime is most commonly used for this purpose; it is relatively mild and does not cause significant damage to the raw material by excessive hydrolysis. However, 8 weeks or more are required for complete treatment. Lime at concentrations of up to 3% is used in conjunction with small amounts of calcium chloride or caustic soda. Alkaline process using caustic soda takes 10–14 days for pretreatment (Johnston-Bank 1983). Gelatin obtained from the alkaline process is known as type B gelatin.

Cho et al. (2004) optimized the extraction condition for production of gelatin from shark cartilage using response surface methodology. Gelatin production has two important steps, (1) alkali treatment and (2) hot-water extraction. The alkali treatment removes noncollagenous protein. Hot-water extraction causes thermal hydrolysis, leading to the solubilization of gelatin. The predicted maximum yields of 79.9% for gelatin production were obtained when alkali treatment using 1.6 N NaOH for 3.16 days and hot-water extraction at 65°C for 3.4 hours were implemented.

Acid Process in Conjunction with Pepsin Traditionally, alkaline process has been used for gelatin production from raw material with high degree of cross-linking. These harsh conditions result in random hydrolysis of peptide bonds and decomposition of some amino acids, leading to a product of variable quality with a broad molecular weight distribution (Slade and

Levine 1987, Galea et al. 2000). Generally, gelatin quality is determined by the length of collagen chains: longer chains produce higher quality gelatin (Tomka 1982, Galea et al. 2000). To obtain the high-quality gelatin, acid process should be used for gelatin production since this process disrupts acid-labile collagen cross-links with negligible peptide bond hydrolysis and less amino acids are destroyed (Slade and Levine 1987, Galea et al. 2000). However, raw materials with high amount of molecular cross-links result in the lower extraction yield (Galea et al. 2000). To increase the extraction yield, some proteases capable of solubilizing the collagen or cleaving the cross-links have been applied during acid process. In general, triple-helical domain of collagen is relatively resistant to proteolytic attack, while the terminal telopeptide regions, which contain inter- and intramolecular cross-links, are readily cleaved (Galea et al. 2000). Chomarat et al. (1994) prepared bovine skin gelatin using pepsin or proctase (isolated from *Aspergillus niger*) to solublize the collagen prior to extraction at 90°C. Both enzymes solubilized collagen with a yield of 75–76% as calculated from hydroxyproline content. After heating, collagen was converted to gelatin with yields of 32% and 39% when pepsin and proctase were used as the extraction aid, respectively. Nalinanon et al. (2008) reported that the extraction yield of gelatin from bigeye snapper skin treated with pepsin during swelling process was approximately two fold higher than that of gelatin from skin without pepsin treatment.

Extraction of Gelatin

When the heat is applied, collagen fibrils shrink to less than one-third of their original length at a critical temperature known as the shrinkage temperature (Foegeding et al. 1996). This shrinkage involves a disassembly of fibers and a collapse of the triple-helical arrangement of polypeptide subunits in the collagen molecule. During the collagen to gelatin transition, many non-covalent bonds are broken along with some covalent inter- and intramolecular bonds (Schiff base and aldo-condensation bonds) and a few peptide bonds are cleaved. Heat applied at temperature higher than transition temperature (T_{max}) is able to disrupt the bonds, mainly H-bond, which stabilizes collagen structure (Fig. 21.2). This results in conversion of the helical collagen structure to a more amorphous form known as gelatin. Nevertheless, when the collagen molecule is completely destructured, glue is produced instead of gelatin (Foegeding et al. 1996). The conversion of collagen to gelatin yields molecules of varying mass, depending on the temperatures used, indigenous proteinases, and so on. Therefore, gelatin is a mixture of fractions varying in molecular weight from 15 to 400 kDa (Gelatin Manufacturers Institute of America 1993).

The maximum yield together with the desirable physical properties is the main objective for the processor. The pH of extraction can be selected either for the maximum extraction rate

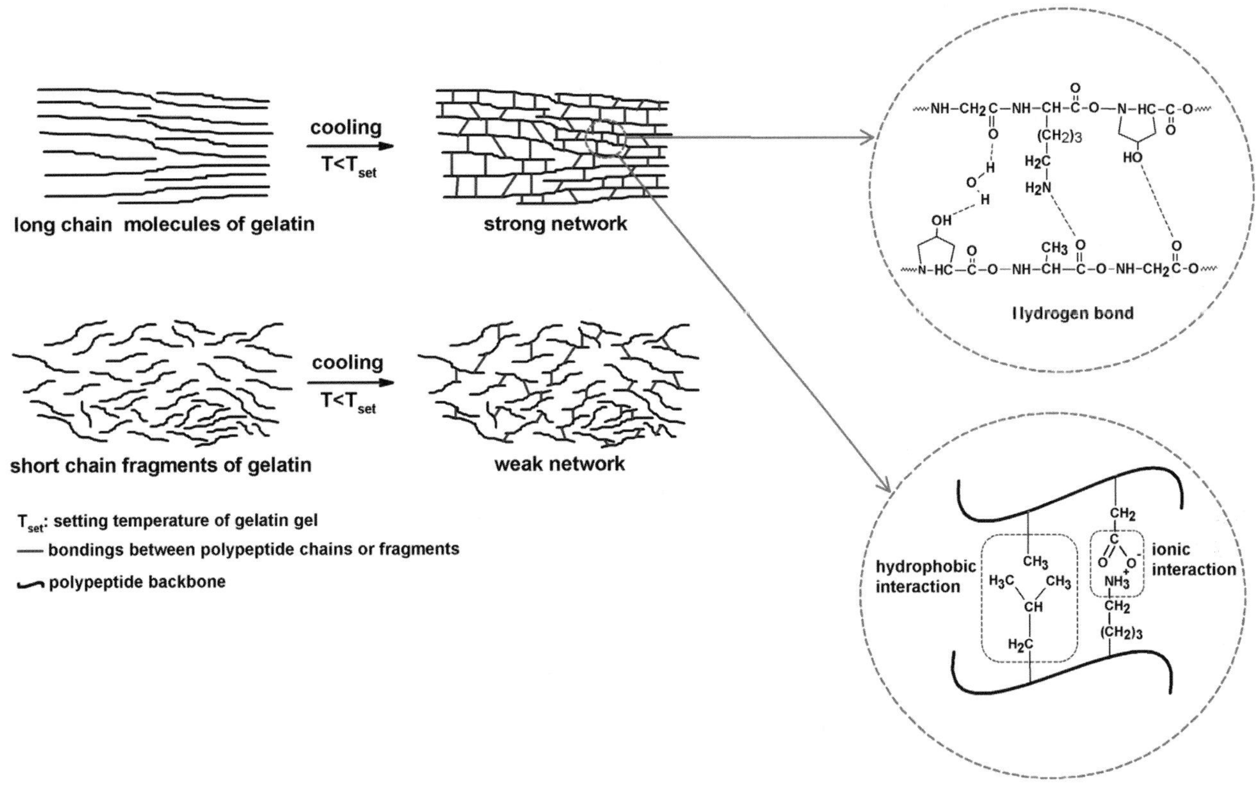

Figure 21.2. Gelatin network associated with hydrogen bond, hydrophobic interaction, and ionic interaction.

(low pH) or for the maximum physical properties (neutral pH) (Johnston-Banks 1990). Basically, more efficient pretreatment conditions allow the manufacturer to use lower extraction temperatures, and still obtain gelatins with high bloom strength (Johnston-Banks 1990). Shorter time is generally associated with higher extraction temperatures if neutral pH levels are chosen. However, lower gel strength is obtained (Johnston-Banks 1990). Muyonga et al. (2004) reported that Nile perch skin extracted at higher temperature yielded lower gel strength, melting point, setting temperature, and longer setting time. Yang et al. (2007) also reported that gelatin from channel catfish skin showed lower gel strength as extraction temperature was increased from 60°C to 75°C. Additionally, gelatin extracted from the skin of brownbanded bamboo shark and blacktip shark using higher temperature had decreasing band intensity of major components (α-, β- and γ-chains) as well as increasing low-molecular weight components. Such changes correlated well with lowered bloom strength. Shorter chain fragments of gelatin could not form the junction zone, in which the strong network could be developed (Kittiphattanabawon et al. 2010) (Fig. 21.2).

Drying

After extraction, the gelatins are filtered to remove suspended or insoluble matters including fat, unextracted collagen fibers, and other residues. Diatomaceous earth or activated carbon can be used to make gelatin solution clear. The final stage is evaporation, sterilization, and drying. These are performed as quickly as possible to minimize loss of properties (Johnston-Banks 1990). Kwak et al. (2009) extracted gelatin from shark cartilage using three drying methods, (1) freeze drying, (2) hot-air drying, and (3) spray drying. Freeze-dried gelatin showed the highest gel strength and foam formation ability, but its foam stability was the lowest. Nevertheless, spray-dried gelatin exhibited the best emulsion capacities.

FISH GELATIN

Fish gelatin from both cold- and warm-water fish has been extracted under different conditions (Table 21.2) (Montero and Gómez-Guillén 2000, Jamilah and Harvinder 2002, Cho et al. 2004, Muyonga et al. 2004, Zhou and Regenstein 2004, Cho et al. 2005, Jongjareonrak et al. 2006a, Arnesen and Gildberg 2007, Cheow et al. 2007, Yang et al. 2007, Nalinanon et al. 2008, Aewsiri et al. 2008, Giménez et al. 2009, Aewsiri et al. 2009a, Jongjareonrak et al. 2010, Kittiphattanabawon et al. 2010). However, gelatins from these sources have limited application as they have lower gel strength and their gels are less stable, compared with those from their mammalian counterparts. The differences between fish and mammalian gelatin mainly depend on their molecular weight distribution and amino acid content, especially the content of imino acid (proline and hydroxyproline) (Johnston-Banks 1990). However, gelatin from skin of brownbanded bamboo shark extracted under the appropriate conditions had properties similar to those of mammalian gelatin. The gel could be set at room temperature and had the bloom strength comparable to commercial bovine gelatin (Kittiphattanabawon et al. 2010). This shark skin gelatin might contain similar amino acids, especially imino acids, as well as could form gel comparable to that from mammalian source.

FUNCTIONAL PROPERTIES OF GELATIN

Gelatin is a gelling protein, which has widely been applied in the food and pharmaceutical industries (Cho et al. 2005). Gelatin gel has "melt-in-the-mouth" characteristic and shows an excellent release of flavor (Choi and Regenstein 2000). Moreover, its functional properties, including gelation, emulsifying properties, foam-forming properties, and film formation, are important for the food industry as it enhances the elasticity, consistency, and stability of food products, and it is also used as an outer film to protect foods against light and oxygen (Montero and Gómez-Guillén 2000).

Gelation

An aqueous solution of gelatin becomes slightly viscous at temperature above its melting temperature. On cooling, the gelatin solution starts to form transparent elastic thermoreversible gels when the temperature is below the setting temperature (Normand et al. 2000, Babin and Dickinson 2001). The interaction initiates a disorder-to-order transition, as the random coil gelatin molecules seek to return to the ordered triple helix conformation. Gelatin gel is a reversibly cross-linked biopolymer network stabilized mainly by hydrogen-bonded junction zones (Fig. 21.2). Furthermore, hydrophobic and ionic interactions are also involved in the gelation of gelatin (Fig. 21.2). The gelation of gelatin is dependent on many factors such as source of raw material for gelatin extraction, presence of endogenous protease in raw material, and the conditions for gelatin extraction, particularly temperature, play a role in gelation of resulting gelatin. High temperature generally renders high yield, the gelatin so obtained has shorter chains (Kittiphattanabawon et al. 2010). These fragments cannot form the junction zone effectively. As a result, a poor gel is formed or gel of these fragments cannot set, especially at room temperature (25°C) (Fig. 21.2).

Source of Raw Material

The sources of raw materials influence the composition of gelatin, especially amino acid composition of gelatin, which affect gelling properties. Generally, the imino acid content of gelatin obtained from mammals is higher than that of gelatin from both warm-water and cold-water fish (Table 21.3). Imino acid content correlates with gel strength of gelatin. Imino acid, especially hydroxyproline, is involved in gel formation by acting as a H-donor, in which hydrogen bond can be formed with an adjacent chain possessing H-acceptor (Fig. 21.2). The hydroxyl group of the hydroxyproline plays a part in the stability of the helix by interchain hydrogen bonding via a bridging water molecule as well as direct hydrogen bonding to a carbonyl group (Wong 1989).

Fish gelatin has been recognized to exhibit poorer gelling properties than its mammalian counterpart. Gelatin from

Table 21.2. Extraction Conditions, Yield, and Bloom/Gel Strength of Gelatin Extracted from Different Fish

	Pretreatments					
Raw Materials	Non-Collageneous Matter Removal	Swelling	Extraction	Yield (On Wet Basis)	Bloom/Gel Strength	References
Giant catfish skin (*Pangasianodon gigas*)	0.2 M NaOH at 4°C for 90 min	0.05 M acetic acid at 24–26°C for 3 h	Water at 45°C for 12 h	20.1%	153 g	Jongjareonrak et al. (2010)
Brownbanded bamboo shark (BBS, *Chiloscyllium punctatum*) and blacktip shark (BTS, *Carcharhinus limbatus*)	0.1 M NaOH at 15–20°C for 2 h, followed by 1 M HCl at 15–20°C for 1 h	0.2 M acetic acid at 25–26°C for 15 min	Water at 45°C, 60°C, and 75°C for 6 and 12 h	19.06–22.81% (BBS) and 21.17–24.76% (BTS)	56.53–217.26g (BBS) 10.43–207.83g (BTS)	Kittiphattanabawon et al. (2010)
Cuttlefish skin (*Sepia pharaonis*)	0.05 N NaOH at 26–28°C for 6 h, followed by 5% (v/v) h_2O_2 at 4°C for 16 h	None	Water at 60°C for 12 h	36.82%[a] (dorsal skin) and 59.69%[a] (ventral skin)	126 g (dorsal skin) and 137 g (ventral skin)	Aewsiri et al. (2009a)
Bigeye snapper skin (*Priacanthus tayenus*)	0.025 M NaOH at 26–28°C for 2 h	None	0.2 M acetic acid in the presence of pepsin (15 unit/g) at 4°C for 48 h, followed by incubation at 45°C for 12 h	40.3%[b]	138.6 g	Nalinanon et al. (2008)
Tuna fin (*Katsuwonus pelamis*)	0.025 N NaOH at 26–28°C for 1 h, followed by 0.6 N HCl at 26–28°C for 5 days	0.2 M acetic acid at 26–28°C for 3 h	Water at 50°C for 12 h	1.25%	126 g (at pH 6)	Aewsiri et al. (2008)
Channel catfish (*Ictalurus punctatus*)	0.2 M NaOH at 4°C for 84 min	0.115 M acetic acid at 4°C for 60 min	Water at 55°C for 3 h	19.2%[c]	252 g	Yang et al. (2007)
Atlantic salmon skin (*Salmo salar*)	0.04 N NaOH at 8°C for 1 h	0.12 M h_2SO_4 for 30 min, followed by 0.005 M citric acid for 30 min at 8°C	Two step-extractions with water at 56°C and 65°C, respectively, for 2 h each step	11.3% (56°C) and 4.0% (65°C)	108 g (56°C) 80 g (65°C)	Arnesen and gildberg (2007)
Bigeye snapper skin (*P. tayenus*)	0.2 M NaOH at 4°C for 90 min	0.05 M acetic acid at 25°C for 3 h	Water at 45°C for 12 h	6.5%	105.7 g	Jongjareonrak et al. (2006c)
Brownstripe red snapper skin (*Lutjanus vitta*)	0.2 M NaOH at 4°C for 90 min	0.05 M acetic acid at 25°C for 3 h	Water at 45°C for 12 h	9.4%	218.6 g	Jongjareonrak et al. (2006)
Giant squid inner and outer tunics (*Dosidicus gigas*)	Not mentioned	0.5 M acetic acid containing porcine pepsin at 2°C for 72 h	Water at 60°C for 18 h	12%	147 g	Giménez et al. (2009)

(*continued*)

Table 21.2. *(Continued)*

Raw Materials	Pretreatments: Non-Collageneous Matter Removal	Pretreatments: Swelling	Extraction	Yield (On Wet Basis)	Bloom/Gel Strength	References
Shortfin scad skin (*Decapterus macrosoma*)	0.2% (w/v) NaOH for 2 h	0.2% (w/v) h_2SO_4 for 2 h, followed by 1.0% (w/v) citric acid for 2 h	Water at 40–50°C for 12 h	7.25%	177 g	Cheow et al. (2007)
Sin croaker skin (*Johnius dussumieri*)	0.2% (w/v) NaOH for 2 h	0.2% (w/v) h_2SO_4 for 2 h, followed by 1.0% (w/v) citric acid for 2 h	Water at 40–50°C for 12 h	14.3%	125 g	Cheow et al. (2007)
Yellowfin tuna skin (*Thunnus albacares*)	1.89% (w/v) NaOH at 10°C for 2.87 days	None	Water at 58.15°C for 4.72 h	89.7%[b]	426 kPa	Cho et al. (2005)
Shark cartilage (*Isurus oxyrinchus*)	1.6 N NaOH at 8°C for 3.14 days	None	Water at 65°C for 3.4 h	17.34%	111.9 kPa	Cho et al. (2004)
Pollock skin	0.25 M $Ca(OH)_2$ at 2°C for 2 h	0.09 M acetic acid at at 2°C for 45 min	Water at 50°C for 180 min	18%[c]	460 g	Zhou and Regenstein (2004)
Nile perch (adult) bone (*Lates niloticus*)	3% HCl at 20–25°C for 9–12 h	None	Four-step extractions with water at 50°C, 60°C, 70°C and 100°C, respectively, for 5 h each step	2.4%	134 g (50°C) 151 g (60°C) 160 g (70°C)	Muyonga et al. (2004)
Nile perch (adult) skin (*L. niloticus*)	0.01 M h_2SO_4 at 20–25°C for 16 h	None	Four step-extractions with water at 50°C, 60°C, 70°C and 100°C, respectively, for 5 h each step	16%	229 g (50°C) 175 g (60°C) 134 g (70°C)	Muyonga et al. (2004)
Red tilapia skin (*Oreochromis nilotica*)	0.2% (w/v) NaOH for 40 min	0.2% (w/v) h_2SO_4, followed by 1.0% (w/v) citric acid	Water at 45°C for 12 h	7.81%	128 g	Jamilah and Harvinder (2002)
Black tilapia skin (*Oreochromis mossambicus*)	0.2% (w/v) NaOH for 40 min	0.2% (w/v) h_2SO_4, followed by 1.0% (w/v) citric acid	Water at 45°C for 12 h	5.39%	181 g	Jamilah and Harvinder (2002)
Megrim skin (*Lepidorhombus boscii*)	0.2 N NaOH at 5°C for 90 min	0.05 N acetic acid at room temperature for 3 h	Water at 45°C for 12 h	10.0%	360 g	Montero and gómez-Guillén (2000)

[a]Based on dry weight.
[b]Based on hydroxyproline content of the gelatin in comparison with that in the skin.
[c]Based on protein content of the gelatin in comparison with wet weight of raw material.

Table 21.3. Amino Acid Compositions of Cold-Water Fish, Warm-Water Fish, and Mammalian gelatin (Residues Per 1000 Residues)

	Sources of gelatin										
	Mammals		Tropical fish					Cold-water/temperate fish			
Amino Acid	Calfskin[a]	Porcine Skin[b]	Brcwnbanded Bamboo Shark Skin[c]	Blacktip Shark Skin[c]	Giant Catfish Skin[a]	Tilapia Skin[d]	Cutttlefish Skin[e]	Alaska Pollock Skin[f]	Baltic Cod Skin[g]	Hake Skin[g]	Salmon Skin[h]
Alanine	104	115	106	120	106	123	83	108	96	119	104
Arginine	57	38	51	54	63	47	60	51	56	54	53
Aspartic acid/ asparagine	30	51	40	40	15	48	72	51	52	49	54
Cysteine	–	–	1	1	–	0		0	0	–	–
Glutamine/ glu-tamic acid	63	83	76	76	62	69	92	74	78	74	74
Glycine	356	329	322	321	359	347	314	358	344	331	366
Histidine	3	4	7	7	4	6	5	8	8	10	13
Isoleucine	13	12	17	18	13	8	20	11	11	9	9
Leucine	25	24	22	23	23	23	26	20	22	23	19
Lysine	32	29	28	27	32	25	12	26	29	28	24
Hydroxylysine	7	–	6	5	5	8	12	6	6	5	–
Methionine	5	9	12	15	10	9	7	16	17	15	18
Phenylalanine	13	10	13	13	13	13	12	12	16	15	13
Hydroxyproline	93	108	95	91	87	79	91	55	50	59	60
Proline	126	114	113	110	124	119	103	95	106	114	106
Serine	34	35	41	29	36	35	40	63	64	49	46
Threonine	16	17	22	20	24	24	24	25	25	22	23
Tyrosine	3	1	2	2	3	2	7	3	3	4	3
Valine	21	21	24	25	22	15	19	18	18	19	15
Imino acid	**219**	**222**	**208**	**201**	**211**	**198**	**194**	**150**	**156**	**173**	**166**

[a]Jongjareonrak et al. (2010).
[b]Cho et al. (2004).
[c]Kittiphattanabawon et al. (2010).
[d]Sarabia et al. (2000).
[e]Hoque et al. (2010).
[f]Zhou et al. (2006).
[g]Gómez-Guillén et al. (2002).
[h]Arnesen and gildberg (2007).

brownbanded bamboo shark and blacktip shark extracted at 45°C for 6 hours exhibited higher bloom strength as compared to that of bovine gelatin. Additionally, gelatin from brownbanded bamboo shark can be set at room temperature within 3 hours (Kittiphattanabawon et al. 2010). This superior gelling ability is mainly attributed to the high imino acid content of gelatin from the skin of both sharks.

Presence of Endogenous Protease in Raw Material

Proteinases are enzymes inducing the cleavage of peptides. Collagenolytic enzyme is another proteinase, which catalyzes the hydrolysis of collagen and gelatin (Sikorski et al. 1990, Kristjánsson et al. 1995, Sovik and Rustad 2006). The major collagenolytic enzyme is collagenase, a group of matrix metalloproteinase in a family of zinc-dependent enzyme. Collagenase can cleave triple-helical collagen at a single site, resulting in the formation of fragments corresponding to one-fourth and three-fourth of its initial length (Sano et al. 2004). Metallocollagenase obtained from fibroblastic cells of rainbow trout was classified as MMP-13 (collagenase-3, EC 3.4.24) (Saito et al. 2000).

Recently, the presence of an endogenous protease—serine protease—in bigeye snapper skin has been reported (Jongjareonrak et al. 2006a, Intarasirisawat et al. 2007, Nalinanon et al. 2008). Intarasirisawat et al. (2007) reported that endogenous serine protease contributing to the autolysis of the skin of bigeye snapper had the maximal activity at 60°C, which is within the temperature range used for gelatin extraction. Nalinanon et al. (2008) reported that the gelatin extracted by the typical process in the absence of protease inhibitor—soybean trypsin inhibitor (SBTI)—showed lower gel strength as compared to that extracted in the presence of SBTI. When gelatin was extracted from the skin of bigeye snapper skin in the presence of SBTI, β- and α-chains along with the polymerized components were retained (Fig. 21.3). Gelatin molecules with the shorter chains generated by serine protease are not able to form the strong interjunction zone, especially via hydrogen bond or other weak bonds such as hydrophobic interaction or ionic interaction. As a consequence, a weaker network develops, in contrast to the gelatin molecules with longer chain length, which are capable of alignment or self-aggregation more effectively (Fig. 21.4).

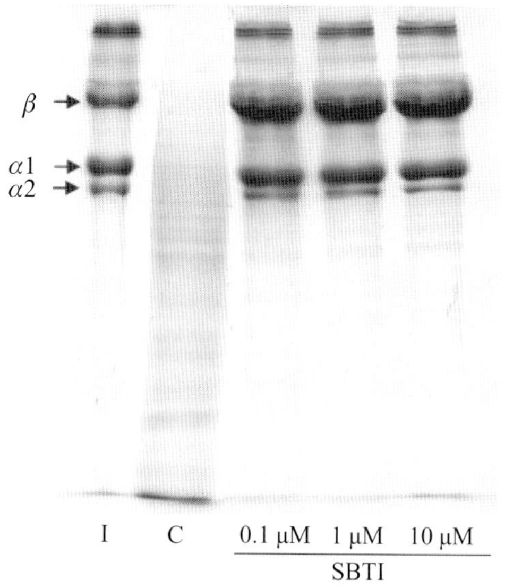

Figure 21.3. Protein patterns of gelatin extracted from bigeye snapper skin in the absence and the presence of soybean trypsin inhibitor (SBTI) at different concentrations. I and C denote calfskin collagen type I and gelatin extracted without protease inhibitor. (Adapted from Nalinanon et al. (2008).)

Extraction Conditions

Most of the processes involved in gelatin extraction directly or partially affect the functional properties of gelatin. Pretreatment of fish skin has been reported to have an impact on gel-forming ability of gelatin. Gómez-Guillén and Montero (2001) extracted the gelatin from megrim skins pretreated with different organic acids including formic acid, acetic acid, propionic acid, lactic acid, malic acid, tartaric acid, and citric acid at a concentration of 0.05 M. Gelatin extracted from the skin pretreated with acetic acid and propionic acid exhibited the highest gel strength. Furthermore, the skin from dover sole swollen with 50-mM acetic acid produced gelatin with the highest gel strength as compared to that swollen in 25- and 50-mM lactic acid (Giménez et al. 2005). Ahmad and Benjakul (2011) reported that gelatin from the skin of unicorn leatherjacket swollen with 0.2 M phosphoric acid (GPA) resulted in the decreasing major band intensity (α-, β-, and γ-chains) and the increasing low-molecular weight components occurred in comparison with that swollen with 0.2 M acetic acid (GAA). However, GPA exhibited higher bloom strength than GAA since some phosphate groups might attach with some amino acids during swelling process, leading to phosphorylation of GPA. Phosphorylation may introduce ionic interaction between phosphate groups and $-NH^+$ proton of amino acids, increasing the cross-links of proteins (Guo et al. 2005).

Extraction temperature and time are two other important factors affecting the gelation of gelatin. Generally, bloom strength of gelatin gel decreases as the extraction temperature and time increase. Normand et al. (2000) reported that higher extraction temperature caused protein degradation, producing protein fragments and lowering gelling ability. A weak gelatin gel was associated with the formation of small fragments (Gómez-Guillén et al. 2002). Degraded α-chains of gelatin are not able to anneal correctly during overnight stabilization by hindering the growth of the existing nucleation sites (Ledward 1986). Gómez-Guillén et al. (2002) found that squid gelatin extracted at 80°C showed very weak gel. For gelatin from Nile perch skin and bone, the gelatin extracted at 50°C exhibited higher gel strength than the corresponding bone gelatin. Gelatin extracted from skin of Nile perch at higher temperature exhibited lower gel strength, but temperature had no effect on gel strength of bone gelatin (Muyonga et al. 2004). Bloom strength of gelatin from brownbanded bamboo shark and blacktip shark decreased with increasing extraction temperature and time. The marked decrease in bloom

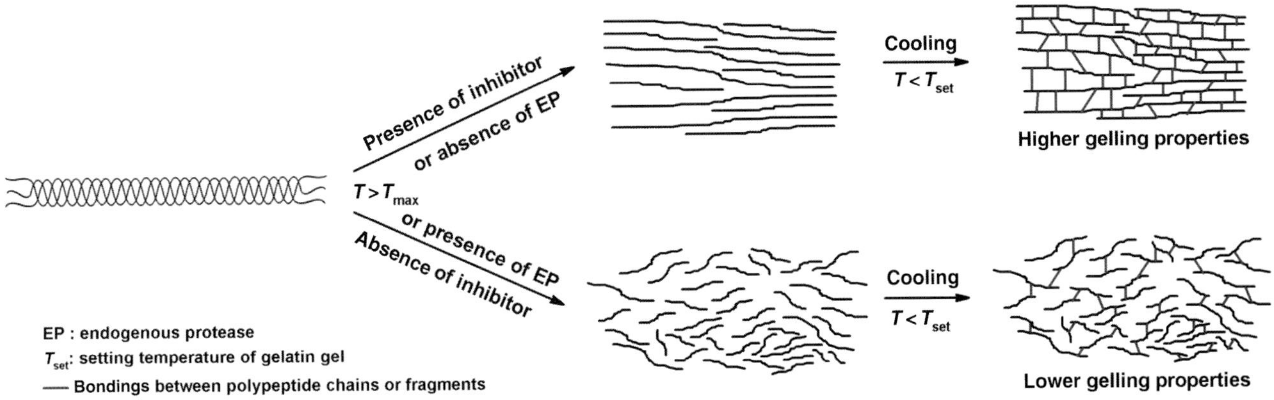

Figure 21.4. Impact of endogenous protease associated with fish skin on degradation of gelatin molecules.

strength was noticeable when extraction temperature of 75°C was used for extraction of gelatin from the skin of both sharks (Kittiphattanabawon et al. 2010).

Emulsifying and Foaming Properties

Emulsions and foams are heterogeneous systems consisting of one phase dispersed in another. An emulsion is a dispersion or suspension of two immiscible liquids, while a foam is a gas phase dispersed in liquid (Hill 1998). Proteins extracted from the different natural sources, for example, soy, milk, fish, meat, and plant, can be used as the emulsifier or foaming agents because of their ability to facilitate the formation and improve the stability of emulsion or foam (Surh et al. 2006). Gelatin is surface-active and is capable of acting as an emulsifier in oil-in-water emulsions and foaming agent (Lobo 2002, Cho et al. 2004, Surh et al. 2006, Aewsiri et al. 2008, Binsi et al. 2009, Kwak et al. 2009, Aewsiri et al. 2009a, Jongjareonrak et al. 2010). The hydrophobic areas on the peptide chain are responsible for giving gelatin its emulsifying and foaming properties (Galazka et al. 1999, Cole 2000). Gelatin concentration has an impact on interfacial properties, but these properties are also governed by the source of the raw material.

Aewsiri et al. (2008) reported that the higher emulsifying and foaming properties were observed in gelatin from precooked tuna fin when the concentration of gelatin was increased. In contrast, the emulsifying capacity of gelatin from bigeye snapper (*Priacanthus hamrur*) decreased with increasing gelatin concentration (Binsi et al. 2009). Additionally, Aewsiri et al. (2009a) studied the emulsifying properties, emulsion activity index (EAI) and emulsion stability index (ESI), and foaming properties, foam expansion (FE) and foam stability (FS) of gelatin from cuttlefish skin with and without bleaching using hydrogen peroxide (H_2O_2). Emulsions containing gelatin from bleached dorsal and ventral skin were more stable than those of gelatin without bleaching. A longer bleaching time and higher H_2O_2 concentration led to a lower ESI of gelatin for all samples, except for gelatin from dorsal skin in which the highest ESI was obtained when the skin was bleached with 5% H_2O_2 for 48 hours ($p < 0.05$). Surh et al. (2006) found that the oil-in-water emulsion prepared with high-molecular weight fish gelatin (~120 kDa) was more stable than that prepared with low-molecular weight fish gelatin (~50 kDa). For foam-forming ability, gelatin from unbleached skin, both dorsal and ventral, had a slightly lower FE than gelatin extracted from bleached skin, while bleaching had no effect on the FS of gelatin from ventral skin, but gelatin from dorsal skin bleached with 5% H_2O_2 for 48 hours showed the highest FS (Aewsiri et al. 2009a). Jongjareonrak et al. (2010) reported that foam capacity and foam stability of gelatin from farmed giant catfish were higher than that from calfskin. Gelatin from shark cartilage and precooked tuna fin showed lower foam capacity and foam stability than gelatin from porcine skin (Cho et al. 2004, Aewsiri et al. 2008).

Film formation

Film and coating from biopolymers such as proteins and polysaccharides have been receiving increasing attention since synthetic packaging films have led to serious ecological problems because of their nonbiodegradability. Proteins are important biopolymers possessing good film-forming ability (Benjakul et al. 2008). Moreover, their mechanical and barrier properties are generally superior to polysaccharide-based films (Cuq et al. 1998). Among all proteins, gelatin has attracted the attention for the development of edible films due to its abundance, biodegradability, and excellent film-forming properties (Gennadios et al. 1994, Bigi et al. 2002). Thus, it is one of the first materials applied to edible coatings and films. The main parameters affecting film-forming properties of gelatin are the source of raw material, extraction method, molecular weight, film preparation method, and degree of hydration or type and level of plasticizer used (Jongjareonrak et al. 2006a, 2006b).

Giménez et al. (2009) suggested that giant squid gelatin obtained from pepsin aid process during the swelling step showed good film-forming properties. Its puncture force, puncture deformation, and water permeability were 4.94 N, 46%, and

1.89×10^{-8} g mm h^{-1} cm^{-2} Pa^{-1}, respectively. Zhang et al. (2007) characterized edible film from channel catfish gelatin extracted using different pretreatment methods. The pretreatment method with 0.25 M NaOH and 0.09 M acetic acid, followed by extraction at 50°C for 3 hours was selected as the optimum extraction method. The resultant film had tensile strength (TS), percentage elongation (%E), and water vapor permeability (WVP) comparable to the film obtained from commercial mammalian gelatin. Additionally, Nile perch skin gelatin films were also found to exhibit film strength (stress at break) and percentage of strain (elongation at break) similar to that of bovine bone gelatin (Muyonga et al. 2004).

TS and %E of catfish gelatin films significantly increased from 27.1 MPa to 61.7 MPa and from 105.1% to 115.1%, respectively, when the gelatin contents were increased from 0.5% to 1%. WVP also increased with the increase in gelatin contents from 0.5% to 2.5% (Zhang et al. 2007). TS and %E of skin-gelatin films of bigeye snapper increased with increasing protein concentration from 2% to 3%, but the increase in protein concentration did not affect WVP (Jongjareonrak et al. 2006a). Additionally, the decrease in TS and increase in %E and WVP were observed in the gelatin-based film from catfish, bigeye snapper, and brownstripe red snapper, when the concentration of glycerol was increased (Jongjareonrak et al. 2006a, Zhang et al. 2007). Generally, less stiff and rigid, more extensible films may be obtained by increasing the plasticizer concentration in the film, due to the reduction in interactions between the biopolymer chains (Arvanitoyannis 2002). Hoque et al. (2010) studied the effect of heat treatment, at different temperatures (40–90°C), of film-forming solution (FFS) containing 3% gelatin from cuttlefish ventral skin and 25% glycerol (based on protein) on film properties. The gelatin film prepared from FFS heated at 60°C and 70°C showed the highest TS ($p<0.05$), while that from FFS heated at 90°C had the highest elongation at break (EAB). Additionally, increasing heating temperatures resulted in the decrease in WVP of films, while L^*-value and transparency value were not different. However, an increase in b^*-value was observed with increasing heating temperature, mainly mediated by Maillard reaction at higher temperature (Manzoccu et al. 2000, Chinabhark et al. 2007).

IMPROVEMENT OF FUNCTIONAL PROPERTIES OF FISH GELATIN

Because of the inferior gel-forming ability of fish gelatin to its mammalian counterpart, fish gelatin can be subjected to modifications for gel improvement or be used for other functional properties. To increase or maximize its functional properties, the modification of gelatin by several methods or the addition of selected compounds has been implemented.

USE OF PROTEIN CROSS-LINKERS

Aldehydes

The reaction of glutaraldehyde with an amine group involves the formation of an imine followed (or preceded) by aldol reactions (Monsan et al. 1975). Subsequent dehydration leads to stable unsaturated cross-links. The sugar-based counterpart of this process is known as the Maillard or browning reaction (Ellis 1959, Ledl and Schleicher 1990), the reaction between aldose sugars and amino groups. The first step in the Maillard reaction is the reversible formation of a glycosylamine that can undergo a so-called Amadori rearrangement (Hodge 1955) to 1-amino-1-deoxy-2-ketose, which will not hydrolyze under ambient conditions. A sugar aldehyde, therefore, can act as a protein cross-linker through glycation and Amadori rearrangement reactions (Schoevaart and Kieboom 2002). The aldehydes commonly used include glutaraldehyde, formaldehyde, and glyoxal (Marquie et al. 1995). However, there are indications of possible toxicity, which make it questionable to use as the modifying agents in foods (Lam et al. 1986, Galietta et al. 1998).

Phenolic Compounds

Phenolic compounds as food components represent more than 6000 identified substances and have been known as the largest group of secondary metabolites in plant foods (Rawel et al. 2007). They generally possess an aromatic ring bearing one or more hydroxy substituents (Robards et al. 1999). They are usually found in plants and are bound to sugars as glycosides (Hollman and Arts 2000). Phenolic compounds can interact with proteins in two different ways: via noncovalent (reversible) interactions and via covalent interactions, which in most cases are irreversible (Prigent 2005). Hydrogen bonds may involve the interactions between hydroxyl groups of phenolic compounds and the nitrogen or oxygen of lysine, arginine, histidine, asparagine, glutamine, serine, threonine, aspartic acid, glutamic acid, tyrosine, cysteine, and tryptophan. Hydrophobic interactions may occur between phenolic compounds and amino acids such as alanine, valine, isoleucine, leucine, methionine, phenylalanine, tyrosine, tryptophan, cysteine, and glycine residues (Prigent 2005).

Covalent interactions between phenolic compounds and proteins can occur via oxidation of phenolic compounds to radicals or quinones (Balange and Benjakul 2009). Monophenol and polyphenol are readily oxidized to *ortho*-quinone, either enzymatically as in plant tissues, or by molecular oxygen. The quinone forms a dimer through a secondary reaction, or reacts with amino or sulfhydryl side chains of polypeptides to form covalent C–N or C–S bonds with the phenolic ring (Fig. 21.5) (Strauss and Gibson 2004). However, gelatin contains a negligible content of cysteine. Thus, the cross-links formed are mainly via amino groups, in which covalent bonds can be formed. Lysine in gelatin molecule could provide ε-NH_3 as the binding site in this reaction. Fish gelatin had lysine ranging from 12 to 32 residues/1000 residues (Table 21.3). Polyphenols can be reoxidized and bind a second polypeptide. As a consequence, protein cross-links can be formed (Fig. 21.5).

Strauss and Gibson (2004) prepared oxidized phenolic compounds–gelatin mixtures by adjusting pH to 8 followed by mixing with gelatin in the presence of oxygen at 40°C. Gels so obtained had greater mechanical strength, reduced swelling,

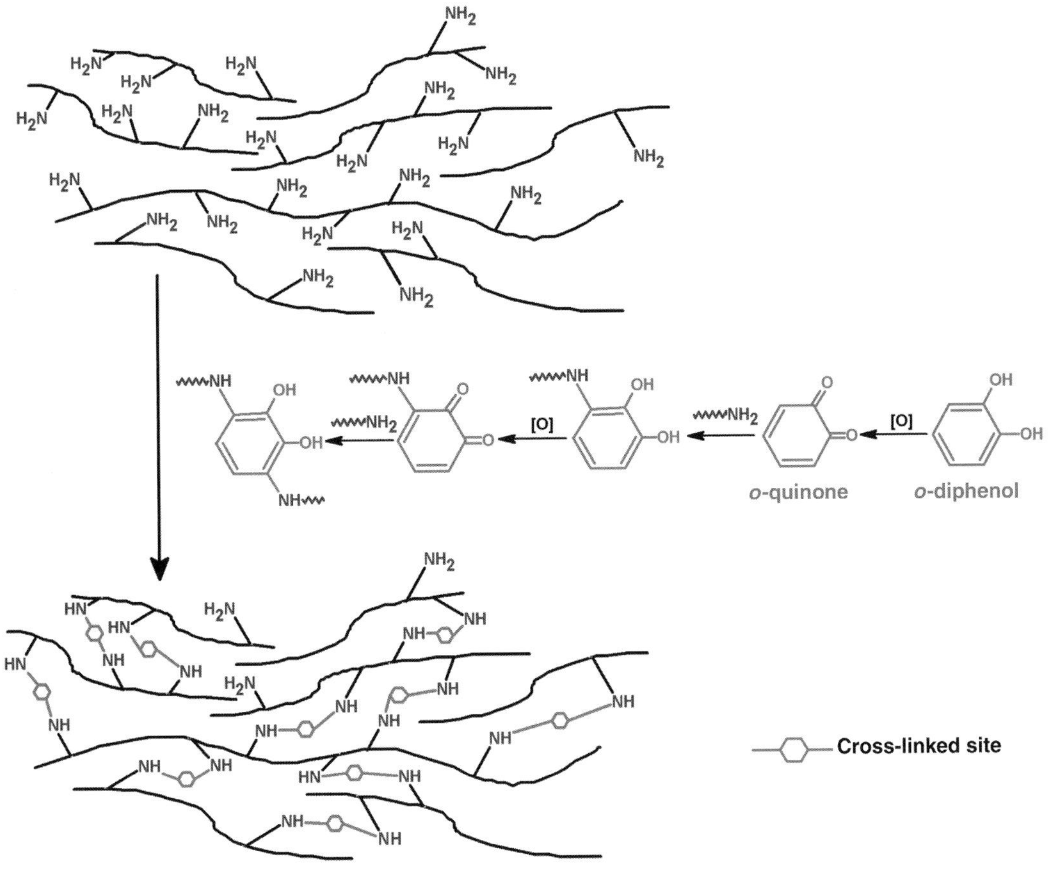

Figure 21.5. Reaction between phenolic compounds and side chain of gelatin.

and fewer free amino groups. Dynamic light scattering analyses showed that such cross-linking results in denser polymeric networks. Aewsiri et al. (2009b) reported that cuttlefish skin gelatin modified with oxidized phenolic compounds, including ferulic acid and tannic acid, at the level above 5% (w/w of protein) exhibited a decrease in the emulsion activity index (EAI). This was possibly associated with the decrease in surface hydrophobicity of the modified gelatin.

Genipin

Genipin is a hydrolytic product of geniposide and its related iridoid glucosides, which are extracted from the fruits of *Gardenia jasminoides Ellis* (Yao et al. 2004) via enzymatic hydrolysis with β-glucosidase (Butler et al. 2003). Genipin has been known as a naturally occurring cross-linking agent (Nickerson et al. 2006b).

Genipin participates in both short- and long-range covalent cross-linking of ε-amino groups in amine-containing polymers (Nickerson et al. 2006a). It can react with amino acid or proteins in the presence of oxygen to form dark blue pigments used in the fabrication of food dyes (Touyama et al. 1994). Additionally, it can form stable cross-linked products with resistance against enzymatic degradation, comparable to that of glutaraldehyde-fixed tissue (Huang et al. 1998, Sung et al. 1998). Furthermore, gelatin-derived bioadhesives display higher biocompatibility and less cytotoxicity when cross-linked with genipin than with other agents, such as formaldehyde, glutaraldehyde, and epoxy compounds (Sung et al. 1999). The mechanism of the reaction of amino acids or proteins with genipin is proposed by Touyama's group (Touyama et al. 1994a, 1994b). The formation of the genipin–methylamine monomer is through a nucleophilic attack by methylamine on the olefinic carbon at C-3 in genipin, followed by the opening of the dihydropyran ring and an attack by the secondary amino group on the resulting aldehyde group. The blue-pigment polymers are presumably formed through the oxygen radical-induced polymerization and dehydrogenation of several intermediary pigments.

Yao et al. (2004) reported that genipin is a good cross-linker for the gelatin. The rate of degradation and the degree of cross-linking of the genipin-fixed gelatin depend upon genipin concentration. Moreover, increasing genipin concentration makes the gelatin network gradually shift from being dominated by hydrogen bonds to covalent cross-linking (Nickerson et al. 2006b). However, the concentration of genipin should exceed 0.5% (w/v) of the overall weight of the gelatin-based material (25% (w/v)) if a complete cross-linking reaction between gelatin and genipin molecules is required (Yao et al. 2004).

Transglutaminase

Transglutaminase (TGase; protein-glutamine-glutamyltransferase, EC 2.3.2.13) catalyzes an acyl-transfer reaction between the γ-carboxamide group of peptide-bound glutamine residues (acyl donors) and a variety of primary amines (acyl acceptors), including the ε-amino group of lysine residues in certain proteins (Fig. 21.6). It results in polymerization and inter- or intramolecular cross-linking of protein via formation of ε-(γ-glutamyl)lysine linkages (Motoki and Seguro 1998, Ashie and Lanier 2000). In the absence of amine substrates, water may act as the acyl acceptor, resulting in deamidation of γ-carboxamide group of glutamine to form glutamic acid. Formation of covalent cross-links between proteins is the basis of the ability of TGase to modify the physical properties of food protein (Ashie and Lanier 2000). TGases are widely distributed in most animal tissues and body fluids, and are involved in several biological phenomena, such as blood clotting, wound healing, epidermal keratinization, and stiffening of the erythrocyte membrane (Aeschlimann and Paulsson 1994). They are responsible for the regulation of cellular growth, differentiation, and proliferation. TGases have also been discovered in plants (Icekson and Apelbaum 1987), fish (Araki and Seki 1993), and microorganisms (Ando et al. 1989, Klein et al. 1992).

Microbial transglutaminase (MTGase) are mass-produced at low cost by fermentation. MTGase catalyzes the cross-linking of most food proteins through the formation of an ε-(γ-glutamyl)lysine bond, in the same way as well-known mammalian enzymes. MTGase is calcium independent and its molecular weight is smaller than that of other known enzymes (Motoki and Seguro 1998). The results of many studies suggest that MTGase, as well as other transglutaminases, has many potential applications in food processing and other areas.

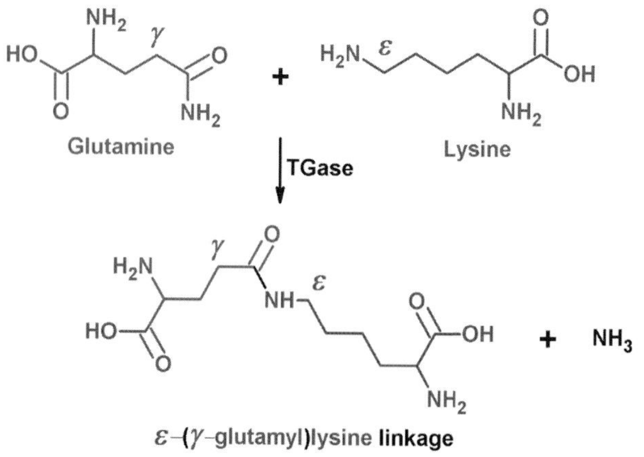

Figure 21.6. Cross-linking reaction between glutamine and lysine residues induced by transglutaminase.

Babin and Dickinson (2001) studied the influence of transglutaminase cross-linking on the rheology of gelatin gels. The gel strength might be either reduced or enhanced depending on whether covalent cross-linking occurred predominantly before or after the development of triple helix junction zones. The extensive covalent cross-linking occurred during cold-set process, and afterward in the molten state, thermoreversible character of gelatin gel was almost completely lost.

Jongjareonrak et al. (2006c) studied the effect of microbial transglutaminase (MTGase) on the gel properties of gelatin from bigeye snapper skin and brownstrip red snapper skin. The addition of MTGase at concentrations up to 0.005% and 0.01% (w/v) increased the bloom strength of gelatin gel from bigeye snapper and brownstripe red snapper, respectively. However, the bloom strength of gelatin gel from both fish species decreased with further increase in MTGase concentration. Gómez-Guillén et al. (2001) reported that the addition of microbial transglutaminase to a fish skin gelatin can considerably raise melting point, gel strength, and viscosity at 60°C, depending on the concentration of the enzyme and the incubation time. Increasing concentrations of TGase increase the elasticity and cohesiveness of the gels, but result in lower gel strength and hardness because they produce excessively rapid gel network formation (Gómez-Guillén et al. 2001). Recently, Norziah et al. (2009) reported that the addition of MTGase with the concentration of 0.5 and 1.0 mg/g gelatin into fish gelatin extracted from the wastes of fish herring species (*Tenualosa ilisha*) had higher gel strength than that without the addition of enzyme, while enzyme concentration above 1.0 mg/g gelatin resulted in a decreasing gel strength. Adding a high concentration of TGase or cross-linking before gelation would be detrimental to gel strength (Babin and Dickinson 2001). Regenstein and Zhou (2007a) also suggested that TGase should not be added over an appropriate enzyme concentration. The reaction should occur during or after gel development to obtain the desired cross-linking. Additionally, Fernandez-Diaz et al. (2001) reported that the gelatin from hake skin had the highest bloom strength with the addition of 10 mg transglutaminase/g.

Use of hydrocolloids

Food products are composed of a wide range of ingredients such as proteins and carbohydrate-based polysaccharides (Ye 2008).

Complex formation between proteins and polysaccharides occurs at pH values below the isoelectric point (IEP) of the proteins and at low ionic strength, usually <0.3 (Ye 2008). Protein molecules have a net positive charge and behave as polycations at pH values below the IEP (De Kruif et al. 2004). At mildly acidic and neutral pH values, which are typical of most foods, carboxyl-containing polysaccharides behave as polyanions (Ye 2008). Electrostatic complex formation between proteins and anionic polysaccharides generally occurs in the pH range between the pK_a value of the anionic groups (carboxyl groups) on the polysaccharide and the IEP of the protein (Tolstoguzov 1997).

The mixtures of fish gelatin and κ-carrageenan had the gels with varying degree of turbidity, depending on the concentration of polymers, pH, ionic strength, and the nature of the added salt (Haug et al. 2004). The highest gel strength was found when 20 mM KCl and 20 mM NaCl were added into the mixtures of 1% (w/v) κ-carrageenan with 2% and 5% (w/v) fish gelatin, respectively. However, these mixtures exhibited more turbid gel than κ-carrageenan or fish gelatin alone since the system undergoes associative phase separation promoted by the release of counterions (Piculell et al. 1995). Complexes of fish gelatin and κ-carrageenan at 60°C were probably stabilized by electrostatic interactions and the solutions were highly turbid, and at 4°C the strongest gel was obtained (Haug et al. 2004).

Liu et al. (2007a) determined the hardness of gelatin/pectin mixed gels with different ratios. The addition of pectin resulted in the increased gel hardness. The strong interactions between two polymers in the mixed solution resulted in a synergistic effect. At the same gelatin content, the lower pectin content added to the system resulted in a harder gel. The excess pectin content formed the repulsive forces in the junction zones, which might reduce the formation of linkages between the aggregated helices, leading to weakened gel structures (Liu et al. 2007a).

Oxidation process

Free radical-mediated protein modification affects protein functionalities. Protein oxidation can lead to the formation of several different kinds of protein cross-links (Stadtman 1998). For example, abstraction of H atoms from protein by radical could result in the reaction with one another to form –C–C– protein cross-linked products. The oxidation of protein sulfhydryl groups can lead to disulfide –S–S– cross-linked proteins. Additionally, the carbonyl groups obtained by the direct oxidation of amino acid side chains of one protein may react with the lysine amino groups of another protein to form Schiff-based cross-linked products.

Fenton reaction, involving a mixture of ferrous ion (Fe^{2+}) and H_2O_2, generates hydroxyl radicals ($HO^\bullet$) at room temperature (Walling 1975, Fenton 1984) as shown in equation below:

$$Fe^{2+} + H_2O_2 \rightarrow Fe^{3+} + HO^\bullet + OH^-$$

H_2O_2 is largely used in oxidation process. Its oxidizing properties are not only due to the presence of an active oxygen atom in the molecule, but also to its ability to participate in radical reactions, with the homolysis of the O–O bond, leading to the formation of the $HO^\bullet$ radical (Chedeville et al. 2005). $HO^\bullet$ radical can oxidize organic compounds in solution (Namkung et al. 2008).

Aewsiri et al. (2009a) reported that uses of H_2O_2 for bleaching of cuttlefish skin prior to gelatin extraction could improve the bloom strength of gelatin gel. H_2O_2 might induce the oxidation of protein with the concomitant formation of carbonyl groups. Those carbonyl groups might undergo Schiff base formation with the amino groups, in which the protein cross-links were most likely formed (Stadtman 1997). Moreover, $OH^\bullet$ can abstract H atoms from amino acid residues to form carbon-centered radical derivatives, which can react with one another to form C–C protein cross-linked products (Stadtman 1997). The larger protein aggregates were mostly associated with the improved bloom strength.

Additionally, free radical in protein could be generated by irradiation, such as UV-irradiation and gamma-irradiation. Irradiation makes protein oxidized, followed by the generation of free radical, resulting in either cross-linking (protein aggregation) or degradation of protein. Bhat and Karim (2009) reported that gelatin treated with UV-irradiation for 60 minutes prior to preparing the gel showed higher gel strength than the gelatin treated with UV-irradiation for 30 minutes as well as that without any UV-irradiation, respectively. Bessho et al. (2007) reported that insolubility due to cross-linking of the gelatin hydrogels was induced at doses above 8 kGy gamma-irradiation. Hydrocarbon groups (alkyl or a phenyl group of the side chains) are the cross-linking sites of gelatin hydrogels.

APPLICATIONS OF GELATIN

Gelatin has been used widely in many sections of the food industry. The major uses cover jellies, confectionery, meat products, and chilled dairy products. It can be used in the pharmaceutical industry, especially for hard and soft capsule manufacture and in the photographic industry, which uses the unique combination of gelling agent and surface activity to suspend particles of silver chloride or light-sensitive dyes (Kragh 1977, Fiszman et al. 1999, Soper 1999, Choi and Regenstein 2000, Park et al. 2007, Zhou and Regenstein 2007, Cheng et al. 2008, Binsi et al. 2009).

Due to the drawbacks of fish gelatin—that it does not gel at room temperature and requires temperature below 8–10°C for setting, it can also be used in other applications that do not require a high bloom value such as for prevention of syneresis and modification of food texture. Fiszman et al. (1999) studied the effect of the addition of gelatin on the microstructure of acid milk gels and yogurt and on their rheological properties. The addition of 1.5% gelatin developed fairly firm and deformable gels with almost total absence of syneresis. Dynamic rheology showed that the yogurts with added gelatin exhibited more solid-like behavior than the ones prepared without it.

Gelatins from tuna or tilapia skin (warm-water fish) have a melting point of 25–27°C (Choi and Regenstein 2000) and have a bloom value of 200–250 g. These gelatins more closely resemble bovine or pig gelatin, which melts at 32–35°C. Fish gelatin with lower gel melting temperatures had a better release of aroma and offered a stronger flavor (Choi and Regenstein 2000). By increasing the concentration of gelatin or by using gelatin mixtures, desserts made from fish gelatins would be more similar to desserts made from high bloom pork skin gelatin (Zhou and Regenstein 2007).

Cheng et al. (2008) observed that combinations of fish gelatin with pectin have been used to make a low-fat spread. A decrease in the fish gelatin to pectin ratio (3:0, 2:1, 1:1, and 1:2) resulted in an increase in bulk density, firmness, compressibility, adhesiveness, elasticity, and meltability. On the other hand, use of gelatin/sodium alginate blends to form casings could lower water losses and lipid oxidation during chilled storage as compared to pectin casing (Liu et al. 2007b). Recently, Binsi et al.

(2009) used gelatin from the skin of bigeye snapper (*Priacanthus hamrur*) to modify the texture of threadfin bream mince gel. The addition of fish gelatin (0.1–1%) to fish mince resulted in higher storage modulus (G′) values from the beginning of heating regime. A maximum G′ value of 443.7 kPa at 68.3°C was obtained in the presence of 0.5% gelatin, which was 42% higher than that of fish mince without the added gelatin.

Fish gelatins with low melting points could also be used in dry products. One of the major applications of fish gelatin is in the microencapsulation of vitamins and other pharmaceutical additives. Soper (1999) described a method for microencapsulation of food flavors such as vegetable oil, lemon oil, garlic flavor, apple flavor, or black pepper with warm-water fish gelatin (150–300 bloom). Park et al. (2007) patented a process describing the preparation of a film-forming composition for hard capsules composed of fish gelatin. Using transglutaminase for cross-linking circumvented the problems caused by the low gelling temperature property of fish gelatin.

REFERENCES

Aeschlimann D, Paulsson M. 1994. Transglutaminases: protein cross-linking enzymes in tissues and body fluids. *Thromb Haemost* 71(4): 402–415.

Aewsiri T et al. 2009a. Functional properties of gelatin from cuttlefish (*Sepia pharaonis*) skin as affected by bleaching using hydrogen peroxide. *Food Chem* 115(1): 243–249.

Aewsiri T et al. 2009b. Antioxidative activity and emulsifying properties of cuttlefish skin gelatin modified by oxidised phenolic compounds. *Food Chem* 117(1): 160–168.

Aewsiri T et al. 2008. Chemical compositions and functional properties of gelatin from pre-cooked tuna fin. *Int J Food Sci Technol* 43(4): 685–693.

Ahmad M, Benjakul S. 2011. Characteristics of gelatin from the skin of unicorn leatherjacket (*Aluterus monoceros*) as influenced by acid pretreatment and extraction time. *Food Hydrocolloids* 25(3): 381–388.

Ando H et al. 1989. Purification and characteristics of a novel transglutaminase derived from microorganisms. *Agric Biol Chem* 53(10): 2613–2617.

Araki H, Seki N. 1993. Comparison of reactivity of transglutaminase to various fish actomyosins. *Nippon Suisan Gakkaishi* 59(4): 711–716.

Arnesen JA, Gildberg A. 2007. Extraction and characterisation of gelatine from Atlantic salmon (*Salmo salar*) skin. *Bioresour Technol* 98(1): 53–57.

Arvanitoyannis IS. 2002. Formation and properties of collagen and gelatin films and coatings. In: A Gennadios (ed.) *Protein-Based Films and Coatings*. CRC Press, Boca Raton, FL, pp. 275–304.

Ashie INA, Lanier TC. 2000. Transglutaminases in seafood processing. In: NF Haard, BK Simpson (eds.) *Seafood Enzymes: Utilization and Influence on Postharvest Seafood Quality*. Marcel Dekker, New York, pp. 147–166.

Babin H, Dickinson E. 2001. Influence of transglutaminase treatment on the thermoreversible gelation of gelatin. *Food Hydrocolloids* 15(3): 271–276.

Bailey AJ, Light ND. 1989. *Connective Tissue in Meat and Meat Products*. Elsevier Applied Science, New York.

Balange A, Benjakul S. 2009. Enhancement of gel strength of bigeye snapper (*Priacanthus tayenus*) surimi using oxidised phenolic compounds. *Food Chem* 113(1): 61–70.

Benjakul S et al. 2008. Properties of protein-based film from round scad (*Decapterus maruadsi*) muscle as influenced by fish quality. *LWT - Food Sci Technol* 41(5): 753–763.

Bessho M et al. 2007. Radiation-induced cross-linking of gelatin by using γ-rays: insoluble gelatin hydrogel formation. *Bull Chem Soc Jpn* 80(5): 979–985.

Bhat R, Karim AA. 2009. Ultraviolet irradiation improves gel strength of fish gelatin. *Food Chem* 113(4): 1160–1164.

Bigi A et al. 1998. Drawn gelatin films with improved mechanical properties. *Biomaterials* 19(24): 2335–2340.

Bigi A et al. 2002. Stabilization of gelatin films by crosslinking with genipin. *Biomaterials* 23(24): 4827–4832.

Binsi PK et al. 2009. Rheological and functional properties of gelatin from the skin of bigeye snapper (*Priacanthus hamrur*) fish: influence of gelatin on the gel-forming ability of fish mince. *Food Hydrocolloids* 23(1): 132–145.

Butler MF et al. 2003. Mechanism and kinetics of the crosslinking reaction between biopolymers containing primary amine groups and genipin. *J Polym Sci, Part A: Polym Chem* 41(24): 3941–3953.

Chedeville O et al. 2005. Modeling of Fenton reaction for the oxidation of phenol in water. *J Autom Methods Manage Chem* 2005(2): 31–36.

Cheng LH et al. 2008. Using fish gelatin and pectin to make a low-fat spread. *Food Hydrocolloids* 22(8): 1637–1640.

Cheow CS et al. 2007. Preparation and characterisation of gelatins from the skins of sin croaker (*Johnius dussumieri*) and shortfin scad (*Decapterus macrosoma*). *Food Chem* 101(1): 386–391.

Chinabhark K et al. 2007. Effect of pH on the properties of protein-based film from bigeye snapper (*Priacanthus tayenus*) surimi. *Bioresour Technol* 98(1): 221–225.

Cho SM et al. 2005. Extracting optimization and physical properties of yellowfin tuna (*Thunnus albacares*) skin gelatin compared to mammalian gelatins. *Food Hydrocolloids* 19(2): 221–229.

Cho SM et al. 2004. Processing optimization and functional properties of gelatin from shark (*Isurus oxyrinchus*) cartilage. *Food Hydrocolloids* 18(4): 573–579.

Choi SS, Regenstein JM. 2000. Physicochemical and sensory characteristics of fish gelatin. *J Food Sci* 65(2): 194–199.

Chomarat N et al. 1994. Comparative efficiency of pepsin and proctase for the preparation of bovine skin gelatin. *Enzyme Microb Technol* 16(9): 756–760.

Cole B. 2000. Gelatin. In: FJ Francis (ed.) *Encyclopedia of Food Science and Technology 2*. Wiley, New York, pp. 1183–1188.

Cuq B et al. 1998. Proteins as agricultural polymers for packaging production. *Cereal Chem* 75(1): 1–9.

De Kruif CG et al. 2004. Complex coacervation of proteins and anionic polysaccharides. *Curr Opin Colloid Interface Sci* 9(5): 340–349.

Ellis GP. 1959. The Maillard reaction. In: ML Wolform (ed.) *Advances in Carbohydrate Chemistry*. Vol. 14 Academic Press, New York, pp. 63–134.

Fenton HJH. 1984. Oxidation of tartaric acid in the presence of iron. *J Chem Soc Trans* 65: 899–910.

Fernández-Díaz MD et al. 2001. Gel properties of collagens from skins of cod (*Gadus morhua*) and hake (*Merluccius*

merluccius) and their modification by the coenhancers magnesium sulphate, glycerol and transglutaminase. *Food Chem* 74(2): 161–167.

Fiszman SM et al. 1999. Effect of addition of gelatin on microstructure of acidic milk gels and yoghurt and on their rheological properties. *Int Dairy J* 9(12): 895–901.

Foegeding EA et al. 1996. Characteristics of edible muscle tissues. In: OR Fennema (ed.) *Food Chemistry*. Marcel Dekker, New York, pp. 879–942.

Galazka VB et al. 1999. Emulsifying behaviour of 11S globulin *Vicia faba* in mixtures with sulphated polysaccharides: comparison of thermal and high-pressure treatments. *Food Hydrocolloids* 13(5): 425–435.

Galea CA et al. 2000. Modification of the substrate specificity of porcine pepsin for the enzymatic production of bovine hide gelatin. *Protein Sci* 9(10): 1947–1959.

Galietta G et al. 1998. Mechanical and thermomechanical properties of films based on whey proteins as affected by plasticizer and crosslinking agents. *J Dairy Sci* 81(12): 3123–3130.

Gelatin Manufacturers Institute of America. 1993. *Gelatin*. GMIA, New York.

Gennadios A et al. 1994. Edible coating and films based on proteins. In: JM Krochta et al. (eds.) *Edible Coatings and Films to Improve Food Quality*. Technomic Publishing Co. Inc., Lancaster, NC, pp. 201–278.

Giménez B et al. 2009. Physico-chemical and film forming properties of giant squid (*Dosidicus gigas*) gelatin. *Food Hydrocolloids* 23(3): 585–592.

Giménez, B et al. 2005. Use of lactic acid for extraction of fish skin gelatin. *Food Hydrocolloids* 19(6): 941–950.

Gómez-Guillén MC et al. 2003. Method for the production of gelatin of marine origin and product thus obtained. US 2003/0022832 A1.

Gómez-Guillén MC, Montero P. 2001. Extraction of gelatin from megrim (*Lepidorhombus boscii*) skins with several organic acids. *J Food Sci* 66(2): 213–216.

Gómez-Guillén MC et al. 2001. Effect of microbial transglutaminase on the functional properties of megrim (*Lepidorhombus boscii*) skin gelatin. *J Sci Food Agric* 81(7): 665–673.

Gómez-Guillén MC et al. 2002. Structural and physical properties of gelatin extracted from different marine species: a comparative study. *Food Hydrocolloids* 16(1): 25–34.

Gudmundsson M, Hafsteinsson H. 1997. Gelatin from cod skins as affected by chemical treatments. *J Food Sci* 62(1): 37–39, 47.

Guo YT et al. 2005. Effect of the phosphate group with different negative charges on the conformation of phosphorylated Ser/Thr-Pro motif. *Int J Pept Res Ther* 11(2): 159–165.

Haug IJ et al. 2004. Physical behaviour of fish gelatin-κ-carrageenan mixtures. *Carbohydr Polym* 56(1): 11–19.

Hill SE. 1998. Emulsions and foams. In: SE Hill et al. (eds.) *Functional Properties of Food Macromolecules*. 2nd edn. Aspen Publishers, Gaithersburg, Maryland, pp. 302–334.

Hodge JE. 1955. The Amadori rearrangement. In: ML Wolforn (ed.) *Advances in Carbohydrate Chemistry*, vol 10 Academic Press, New York, pp. 169–205.

Hollman PCH, Arts ICW. 2000. Flavonols, flavones and flavanols - Nature, occurrence and dietary burden. *J Sci Food Agric* 80(7): 1081–1093.

Hoque MS et al. 2010. Effect of heat treatment of film-forming solution on the properties of film from cuttlefish (*Sepia pharaonis*) skin gelatin. *J Food Eng* 96(1): 66–73.

Huang LLH et al. 1998. Biocompatibility study of a biological tissue fixed with a naturally occurring crosslinking reagent. *J Biomed Mater Res* 42(4): 568–576.

Icekson I, Apelbaum A. 1987. Evidence for transglutaminase activity in plant tissue. *Plant Physiol* 84(4): 972–974.

Intarasirisawat R et al. 2007. Autolysis study of bigeye snapper (*Priacanthus macracanthus*) skin and its effect on gelatin. *Food Hydrocolloids* 21(4): 537–544.

Jamilah B, Harvinder KG. 2002. Properties of gelatins from skins of fish - Black tilapia (*Oreochromis mossambicus*) and red tilapia (*Oreochromis nilotica*). *Food Chem* 77(1): 81–84.

Johns R, Courts A. 1977. Relationship between collagen and gelatin. In:AG Ward, A Courts (eds.) *The Science and Technology of Gelatin*. Academic Press, London, pp. 137–177.

Johnston-Bank FA. 1983. From tannery to table: an account of gelatin production. *J Soc Leather Technol Chem* 68: 141–145.

Johnston-Banks FA. 1990. *Gelatin*. In: P Harris (ed.) *Food Gels*. Elsevier Applied Science, London, pp. 233–289.

Jongjareonrak A et al. 2006a. Characterization of edible films from skin gelatin of brownstripe red snapper and bigeye snapper. *Food Hydrocolloids* 20(4): 492–501.

Jongjareonrak A et al. 2006b. Effects of plasticizers on the properties of edible films from skin gelatin of bigeye snapper and brownstripe red snapper. *Eur Food Res Technol* 222(3–4): 229–235.

Jongjareonrak A et al. 2006c. Skin gelatin from bigeye snapper and brownstripe red snapper: chemical compositions and effect of microbial transglutaminase on gel properties. *Food Hydrocolloids* 20(8): 1216–1222.

Jongjareonrak A et al. 2010. Chemical compositions and characterisation of skin gelatin from farmed giant catfish (*Pangasianodon gigas*). *LWT - Food Sci Technol* 43(1): 161–165.

Kittiphattanabawon P et al. 2010. Comparative study on characteristics of gelatin from the skins of brownbanded bamboo shark and blacktip shark as affected by extraction conditions. *Food Hydrocolloids* 24(2–3): 164–171.

Klein JD et al. 1992. Purification and partial characterization of transglutaminase from *Physarum polycephalum*. *J Bacteriol* 174(8): 2599–2605.

Kragh AM. 1977. Swelling, adsorption and the photographic uses of gelatin. In: AG Ward, A Courts (eds.) *The Science and Technology of Gelatin*. Academic Press, London, pp. 439–474.

Kristjánsson MM et al. 1995. Characterization of a collagenolytic serine proteinase from the Atlantic cod (*Gadus morhua*). *Comp Biochem Physiol - B Biochem Mol Biol* 110(4): 707–717.

Kwak KS et al. 2009. Changes in functional properties of shark (*Isurus oxyrinchus*) cartilage gelatin produced by different drying methods. *Int J Food Sci Technol* 44(8): 1480–1484.

Lam CW et al. 1986. Decreased extractability of DNA from proteins in the rat nasal mucosa after acetaldehyde exposure. *Fundam Appl Toxicol* 6(3): 541–550.

Ledl F, Schleicher E. 1990. New aspects of the Maillard reaction in foods and in the human body. *Angew Chem Int Ed Eng* 29(6): 565–594.

Ledward DA. 1986. Gelation of gelatin. In: JR Mitchell, DA Ledward (eds.) *Functional Properties of Food Macromolecules*. Elsevier Applied Science, London, pp. 171–201.

Leuenberger BH. 1991. Investigation of viscosity and gelation properties of different mammalian and fish gelatins. *Food Hydrocolloids* 5(4): 353–361.

Liu L et al. 2007b. Application and assessment of extruded edible casings manufactured from pectin and gelatin/sodium alginate blends for use with breakfast pork sausage. *Meat Sci* 75(2): 196–202.

Liu H et al. 2007a. Rheological, textural and microstructure analysis of the high-methoxy pectin/gelatin mixed systems. *J Texture Stud* 38(5): 577–600.

Lobo L. 2002. Coalescence during emulsification: 3. Effect of gelatin on rupture and coalescence. *J Colloid Interface Sci* 254(1): 165–174.

Manzoccu L et al. 2000. Review of non-enzymatic browning and antioxidant capacity in processed foods. *Trends Food Sci Technol* 11(9–10): 340–346.

Marquie C et al. 1995. Biodegradable packaging made from cottonseed flour: formation and improvement by chemical treatments with gossypol, formaldehyde, and glutaraldehyde. *J Agric Food Chem* 43(10): 2762–2767.

Monsan P et al. 1975. On the mechanism of the formation of glutaraldehyde protein bonds. *Biochimie* 57(11–12): 1281–1292.

Montero P, Gómez-Guillén MC. 2000. Extracting conditions for megrim (*Lepidorhombus boscii*) skin collagen affect functional properties of the resulting gelatin. *J Food Sci* 65(3): 434–438.

Motoki M, Seguro K. 1998. Transglutaminase and its use for food processing. *Trends Food Sci Technol* 9(5): 204–210.

Muyonga JH et al. 2004. Extraction and physico-chemical characterisation of Nile perch (*Lates niloticus*) skin and bone gelatin. *Food Hydrocolloids* 18(4): 581–592.

Nalinanon S et al. 2008. Improvement of gelatin extraction from bigeye snapper skin using pepsin-aided process in combination with protease inhibitor. *Food Hydrocolloids* 22(4): 615–622.

Namkung KC et al. 2008. Advanced Fenton processing of aqueous phenol solutions: a continuous system study including sonication effects. *Ultrason Sonochem* 15(3): 171–176.

Nickerson MT et al. 2006a. Some physical and microstructural properties of genipin-crosslinked gelatin-maltodextrin hydrogels. *Int J Biol Macromol* 38(1): 40–44.

Nickerson MT et al. 2006b. Kinetic and mechanistic considerations in the gelation of genipin-crosslinked gelatin. *Int J Biol Macromol* 39(4–5): 298–302.

Normand V et al. 2000. Gelation kinetics of gelatin: a master curve and network modeling. *Macromolecules* 33(3): 1063–1071.

Norziah MH et al. 2009. Characterization of fish gelatin from surimi processing wastes: thermal analysis and effect of transglutaminase on gel properties. *Food Hydrocolloids* 23(6): 1610–1616.

Park HJ et al. 2007. Film-forming and composition for hard capsules comprising fish gelatin and its preparation method. Wipo Patent WO/2007/123350.

Piculell L et al. 1995. Factors determining phase behaviour of multi component polymer systems. In: SE Harding et al. (eds.) *Biopolymer Mixtures*. Nottingham University Press, Nottingham, pp 13–35.

Prigent S. 2005. *Interactions of Phenolic Compounds with Globular Proteins and their Effect on Food-Related Functional Properties*. The Degree of Doctor. Wageningen University.

Rahman MS et al. 2008. Thermal characterisation of gelatin extracted from yellowfin tuna skin and commercial mammalian gelatin. *Food Chem* 108(2): 472–481.

Rawel HM et al. 2007. Effect of non-protein components on the degradability of proteins. *Biotechnol Adv* 25(6): 611–613.

Regenstein JM, Zhou P. 2007a. Collagen and gelatin from marine by-products. In: F Shahidi (ed.) *Maximising the Value of Marine By-Products*, 1st edn. Woodhead Publishing Limited, Cambridge, pp. 279–303.

Regenstein JM, Zhou P. 2007b. Collagen and gelatin from marine by-products. In: F Shahidi (ed.) *Mazimising the Value of Marine By-Products*, 1st edn. Woodhead Publishing Limited, Cambridge, pp. 279–303.

Robards K et al. 1999. Phenolic compounds and their role in oxidative processes in fruits. *Food Chem* 66(4): 401–436.

Saito M et al. 2000. Characterization of a rainbow trout matrix metalloproteinase capable of degrading type I collagen. *Eur J Biochem* 267(23): 6943–6950.

Sano M et al. 2004. Assay of collagenase activity for native triple-helical collagen using capillary gel electrophoresis with laser-induced fluorescence detection. *J Chromatogr B Anal Technol Biomed Life Sci* 809(2): 251–256.

Sarabia AI et al. 2000. The effect of added salts on the viscoelastic properties of fish skin gelatin. *Food Chem* 70(1): 71–76.

Schoevaart R, Kieboom T. 2002. Galactose dialdehyde as potential protein cross-linker: proof of principle. *Carbohydr Res* 337(10): 899–904.

Sikorski ZE et al. 1990. The nutritive composition of the major groups of marine food organisms. In: ZE Sikorski (ed.) *Seafood: Resources, Nutritional Composition, and Preservation*. CRC Press, Boca Raton, FL:, pp. 29–54.

Slade L, Levine H. 1987. Polymer-chemical properties of gelatin in foods. In: AM Pearson, TR Dustson, AJ Bailey (eds.) *Advances in Meat Research*. Vol. 4. Van Nostrand Reinhold Company, New York, pp. 251–266.

Soper JC. 1999. Method of encapsulating food or flavor particles using warm water fish gelatin, and capsules produced therefrom. EP 0 797 394 B1.

Sovik SL, Rustad T. 2006. Effect of season and fishing ground on the activity of cathepsin B and collagenase in by-products from cod species. *LWT - Food Sci Technol* 39(1): 43–53.

Stadtman ER. 1997. Free radical mediated oxidation of proteins. In: T Özben (ed.) *Free Radicals, Oxidative Stress, and Antioxidants: Pathological and Physiological Significance*. Plenum Press, New York, pp. 51–65.

Stadtman ER. 1998. Free radical mediated oxidation of proteins. In: T Ozben (ed.) *Free Radicals, Oxidative Stress, and Antioxidants. Pathological and Physiological Significance. NATO ASI Series, Series A: Life Sciences*. Vol. 296. Plenum Press, New York pp. 51–64.

Stainby G. 1987. Gelatin gels. In: AM Pearson et al. (eds.) *Advances in Meat Research*. Vol. 4. Van Nostrand Reinhold Company, New York, pp. 209–222.

Strauss G, Gibson SM. 2004. Plant phenolics as cross-linkers of gelatin gels and gelatin-based coacervates for use as food ingredients. *Food Hydrocolloids* 18(1): 81–89.

Sung HW et al. 1999. Evaluation of gelatin hydrogel crosslinked with various crosslinking agents as bioadhesives: *in vitro* study. *J Biomed Mater Res* 46(4): 520–530.

Sung HW et al. 1998. Feasibility study of a natural crosslinking reagent for biological tissue fixation. *J Biomed Mater Res* 42(4): 560–567.

Surh J et al. 2006. Properties and stability of oil-in-water emulsions stabilized by fish gelatin. *Food Hydrocolloids* 20(5): 596–606.

Tolstoguzov VB. 1997. Protein-polysaccharide interactions. In: S Damodaran, A Paraf (eds.) *Food Proteins and Their Applications*. Marcel Dekker, New York, pp. 171–198.

Tomka I. 1982. Photographic material. 4360590.

Touyama R et al. 1994a. Studies on the blue pigments produced from genipin and methylamine. II. On the formation mechanisms of brownish-red intermediates leading to the blue pigment formation. *Chem Pharm Bull* 42(8): 1571–1578.

Touyama R et al. 1994b. Studies on the blue pigments produced from genipin and methylamine. I. Structures of the brownish-red pigments, intermediates leading to the blue pigments. *Chem Pharm Bull* 42(3): 668–673.

Touyama R et al. 1994. Studies on the blue pigments produced from genipin and methylamine. I. Structures of the brownish-red pigments, intermediates leading to the blue pigments. *Chem Pharm Bull* 42(3): 668–673.

Waldner RH. 1977. Raw material. In: AG Ward, A Courts (eds.) *The Science and Technology of Gelatin*. Academic Press, London, pp. 295–314.

Walling C. 1975. Fenton's reagent revisited. *Acc Chem Res* 8(4): 125–131.

Wong DWS. 1989. *Mechanism and Theory in Food Chemistry*. Van Nostrand Reinhold Company, New York.

Yang H et al. 2007. 2-Step optimization of the extraction and subsequent physical properties of channel catfish (*Ictalurus punctatus*) skin gelatin. *J Food Sci* 72(4): C188–C195.

Yao CH et al. 2004. Preparation of networks of gelatin and genipin as degradable biomaterials. *Mater Chem Phys* 83(2–3): 204–208.

Ye A. 2008. Complexation between milk proteins and polysaccharides via electrostatic interaction: principles and applications - a review. *Int J Food Sci Technol* 43(3): 406–415.

Yoshimura K et al. 2000. Preparation and dynamic viscoelasticity characterization of alkali-solubilized collagen from shark skin. *J Agric Food Chem* 48(3): 685–690.

Zhang S et al. 2007. Characterization of edible film fabricated with channel catfish (*Ictalurus punctatus*) gelatin extract using selected pretreatment methods. *J Food Sci* 72(9): C498–C503.

Zhou P et al. 2006. Properties of Alaska pollock skin gelatin: A comparison with tilapia and pork skin gelatins. *J Food Sci* 71(6): C313–C321.

Zhou P, Regenstein JM. 2004. Optimization of extraction conditions for pollock skin gelatin. *J Food Sci* 69(5): C393–C398.

Zhou P, Regenstein JM. 2005. Effects of alkaline and acid pretreatments on Alaska pollock skin gelatin extraction. *JournalFood Sci* 70(6): C393–C396.

Zhou P, Regenstein JM. 2007. Comparison of water gel desserts from fish skin and pork gelatins using instrumental measurements. *Journal Food Sci* 72(4): C196–C201.

22
Application of Proteomics to Fish Processing and Quality

Hólmfríður Sveinsdóttir, Samuel A. M. Martin, and Oddur T. Vilhelmsson

Abstract: Proteomics involves the study of proteins, with regards to proteins, their expression by genomes, their structures and functions. The entire set of proteins or proteome expressed by a genome display variations in tissues and organisms, and can be used as the basis for evaluating the status and changes in the proteins in living organisms including fish and shellfish. This feature can be useful for developing standards for fish and other food materials and assessing their quality and/or safety. This chapter discusses current uses of proteomics for establishing the attributes of fish and fish products.

Proteomics is most succinctly defined as "the study of the entire proteome or a subset thereof," the proteome being the expressed protein complement of the genome. Unlike the genome, the proteome varies among tissues, as well as with time in reflection of the organism's environment and its adaptation thereto. Proteomics can, therefore, give a snapshot of the organism's state of being and, in principle at least, map the entirety of its adaptive potential and mechanisms. As with all living matter, foodstuffs are in large part made up of proteins. This is especially true of fish and meat, where the bulk of the food matrix is constructed from proteins. Furthermore, the construction of the food matrix, both on the cellular and tissue-wide levels, is regulated and brought about by proteins. It stands to reason, then, that proteomics is a tool that can be of great value to the food scientist, giving valuable insight into the composition of the raw materials, quality involution within the product before, during, and after processing or storage, the interactions of proteins with one another or with other food components, or with the human immune system after consumption. In this chapter, a brief overview of "classical" proteomics methodology is presented, and their present and future application in relation to fish and seafood processing and quality is discussed.

PROTEOMICS METHODOLOGY

Unlike nucleic acids, proteins are an extremely variegated group of compounds in terms of their chemical and physical properties. It is not surprising, then, that a field that concerns itself with "the systematic identification and characterization of proteins for their structure, function, activity and molecular interactions" (Peng et al. 2003) should possess a toolkit containing a wide spectrum of methods that continue to be developed at a brisk pace. While high-throughput, gel-free methods, for example, based on liquid chromatography tandem mass spectrometry (LC-MS/MS) (Peng et al. 2003), surface-enhanced laser desorption/ionization (Hogstrand et al. 2002), or protein arrays (Lee and Nagamune 2004), hold great promise and are deserving of discussion in their own right, the "classic" process of

Food Biochemistry and Food Processing, Second Edition. Edited by Benjamin K. Simpson, Leo M.L. Nollet, Fidel Toldrá, Soottawat Benjakul, Gopinadhan Paliyath and Y.H. Hui.

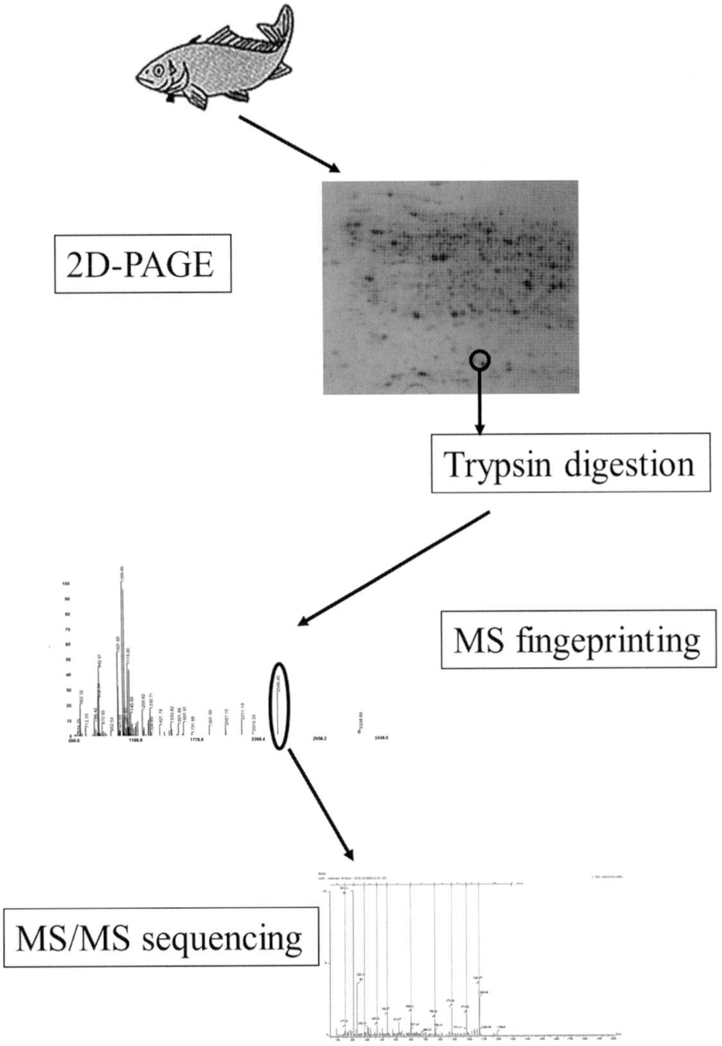

Figure 22.1. An overview over the "classic approach" in proteomics. First, a protein extract (crude or fractionated) from the tissue of choice is subjected to two-dimensional polyacrylamide gel electrophoresis (2D-PAGE). Once a protein of interest has been identified, it is excised from the gel, subjected to degradation by trypsin (or other suitable protease) and the resulting peptides analyzed by mass spectrometry (MS), yielding a peptide mass fingerprint. In many cases, this is sufficient for identification purposes, but if needed, peptides can be dissociated into smaller fragments and small partial sequences obtained by tandem mass spectrometry (MS/MS). See text for further details.

two-dimensional electrophoresis (2DE) followed by protein identification via peptide mass fingerprinting of trypsin digests (Fig. 22.1) remains the workhorse of most proteomics work, largely because of its high resolution, simplicity, and mass accuracy. This "classic approach" will, therefore, be the main focus of this chapter. A number of reviews on the advances and prospects of proteomics within various fields of study are available. Some recent ones include: Andersen and Mann (2006), Balestrieri et al. (2008), Beretta (2009), Bogyo and Cravatt (2007), Drabik et al. (2007), Ikonomou et al. (2009), Issaq and Veenstra (2008), Jorrin-Novo et al. (2009), Latterich et al. (2008), López (2007), Mamone et al. (2009), Malmstrom et al. (2007), Premsler et al. (2009), Smith et al. (2009), Wang et al. (2006), Wilm (2009), Yates et al. (2009).

Two-Dimensional Electrophoresis

2DE, the cornerstone of most proteomics research, is the simultaneous separation of hundreds, or even thousands, of proteins on a two-dimensional polyacrylamide slab gel. The potential of a two-dimensional protein separation technique was realized early on, with considerable development efforts taking place in the 1960s (Margolis and Kenrick 1969, Kaltschmidt and Wittmann 1970). The method most commonly used today was developed by Patrick O'Farrell. It is described in his seminal and thorough 1975 paper (O'Farrell 1975) and is outlined briefly later. It is worth emphasizing that great care must be taken that the proteome under investigation is reproducibly represented on the 2DE gels, and that individual variation in specific protein

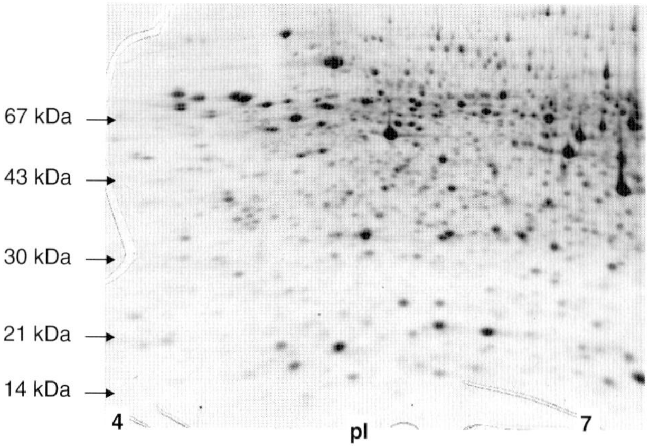

Figure 22.2. A two-dimensional electrophoresis protein map of rainbow trout (*Oncorhynchus mykiss*) liver proteins with pI between 4 and 7 and molecular mass about 10–100 (S. Martin, unpublished). The proteins are separated according to their pI in the horizontal dimension and according to their mass in the vertical dimension. Isoelectro focussing was by pH 4–7 immobilized pH gradient (IPG) strip and the second dimension was in a 10–15% gradient polyacrylamide slab gel.

abundance is taken into consideration by running gels from a sufficient number of samples and performing the appropriate statistics. Pooling samples may also be an option, depending on the type of experiment.

Basic 2DE Methods Overview

O'Farrell's original 2DE method first applies a process called isoelectric focusing (IEF), where an electric field is applied to a tube gel on which the protein sample and carrier ampholytes have been deposited. This separates the proteins according to their molecular charge. The tube gel is then transferred onto a polyacrylamide slab gel and the isoelectrically focused proteins are further separated according to their molecular mass by conventional sodium dodecyl sulfate–polyacrylamide gel electrophoresis (SDS-PAGE), yielding a two-dimensional map (Fig. 22.2) rather than the familiar banding pattern observed in one-dimensional SDS-PAGE. The map can be visualized and individual proteins quantified by radiolabeling or by using any of a host of protein dyes and stains, such as Coomassie blue, silver stains, or fluorescent dyes. By comparing the abundance of individual proteins on a number of gels (Fig. 22.3), upregulation or downregulation of these proteins can be inferred. Although a number of refinements have been made to 2DE since O'Farrell's paper, most notably, the introduction of immobilized pH gradients (IPGs) for IEF (Görg et al. 1988), the procedure

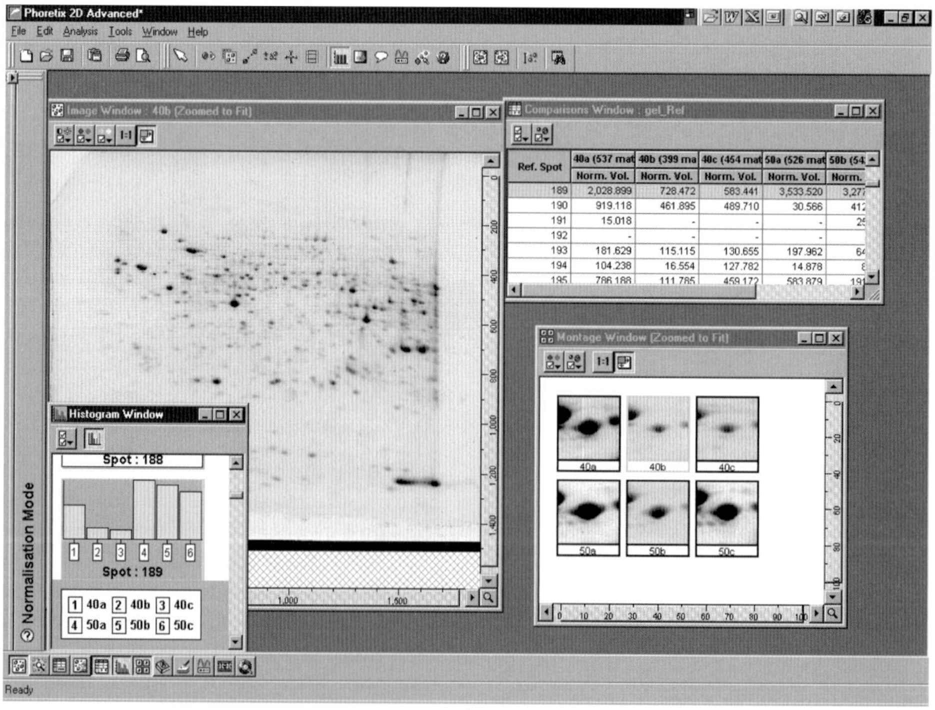

Figure 22.3. A screenshot from the two-dimensional electrophoresis analysis program Phoretix 2-D (NonLinear Dynamics, Gateshead, Tyne & Wear, UK) showing some steps in the analysis of a two-dimensional protein map. Variations in abundance of individual proteins, as compared with a reference gel, can be observed and quantified.

remains essentially as outlined earlier. In the following sections, a general protocol is outlined briefly with some notes of special relevance to the seafood scientist. For more detailed, up-to-date protocols, the reader is referred to any of a number of excellent reviews and laboratory manuals such as Berkelman and Stenstedt (1998), Görg et al. (2000, 2004), Kraj and Silberring (2008), Link (1999), Simpson (2003), Walker (2005) and Westermeier and Naven (2002).

Sample Extraction and Cleanup

For most applications, sample treatment prior to electrophoresis should be minimal in order to minimize in-sample proteolysis and other sources of experimental artifacts. We have found direct extraction into the gel reswelling buffer (7-M urea, 2-M thiourea, 4% (w/v) CHAPS [3-(3-chloramidopropyl)dimethylamino-1-propanesulfonate], 0.3% (w/v) DTT [dithiothreitol], 0,5% Pharmalyte ampholytes for the appropriate pH range), supplemented with a protease inhibitor cocktail, to give good results for proteome extraction from whole Atlantic cod larvae (Guðmundsdóttir and Sveinsdóttir 2006, Sveinsdóttir et al. 2008) and Arctic charr (*Salvelinus alpinus*) liver (Coe and Vilhelmsson 2008). Thorough homogenization is essential to ensure complete and reproducible extraction of the proteome. Cleanup of samples using commercial two-dimensional sample cleanup kits may be beneficial for some sample types.

First-Dimension Electrophoresis

The extracted proteins are first separated by IEF, which is most conveniently performed using commercial dry IPG gel strips. These strips consist of a dried IPG-containing polyacrylamide gel on a plastic backing. Ready-made IPG strips are currently available in a variety of linear and sigmoidal pH ranges. This method is thus suitable for most 2DE applications and has all but completely replaced the older and less reproducible method of IEF by carrier ampholytes in tube gels. Broad-range linear strips (e.g., pH 3–10) are commonly used for whole-proteome analysis of tissue samples, but for many applications narrow-range and/or sigmoidal IPG strips may be more appropriate as these will give better resolution of proteins in the fairly crowded pI 4–7 range. Narrow-range strips also allow for higher sample loads (since part of the sample will run off the gel) and thus may yield improved detection of low-abundance proteins.

Before electrophoresis, the dried gel needs to be reswelled to its original volume. A recipe for a typical reswelling buffer is presented earlier. Reswelling is normally performed overnight at 4°C. Application of a low-voltage current may speed up the reswelling process. Optimal conditions for reswelling are normally provided by the IPG strip manufacturer. If the protein sample is to be applied during the reswelling process, extraction directly into the reswelling buffer is recommended.

IEF is normally performed for several hours at high voltage and low current. Typically, the starting voltage is about 150 V, which is then increased step-wise to about 3500 V, usually totaling about 10,000 to 30,000 Vh, although this will depend on the IPG gradient and the length of the strip. The appropriate IEF protocol will depend not only on the sample and IPG strip, but also on the equipment used. The manufacturer's instructions should be followed. Görg et al. (2000) reviewed IEF for 2DE applications.

Equilibration

Before the isoelectrofocused gel strip can be applied to the second-dimension slab gel, it needs to be equilibrated for 30–45 minutes in a buffer containing SDS and a reducing agent such as DTT. During the equilibration step, the SDS–polypeptide complex that affords protein-size-based separation will form and the reducing agent will preserve the reduced state of the proteins. A tracking dye for the second electrophoresis step is also normally added at this point. A typical equilibration-buffer recipe is as follows: 50 mM Tris-HCl at pH 8.8, 6-M urea, 30% glycerol, 2% SDS, 1% DTT, and trace amount of bromophenol blue. A second equilibration step in the presence of 2.5% iodoacetamide and without DTT (otherwise identical buffer) may be required for some applications. This will alkylate thiol groups and prevent their reoxidation during electrophoresis, thus reducing vertical streaking (Görg et al. 1987).

Second-Dimension Electrophoresis

Once the gel strip has been equilibrated, it is applied to the top edge of an SDS-PAGE slab gel and cemented in place using a molten agarose solution. Optimal pore size depends on the size of the target proteins, but for most applications gradient gels or gels of about 10% or 12% polyacrylamide are appropriate. Ready-made gels suitable for analytical 2DE are available commercially. While some reviewers recommend alternative buffer systems (Walsh and Herbert 1999), the Laemmli method (Laemmli 1970), using glycine as the trailing ion and the same buffer (25-mM Tris, 192-mM glycine, 0.1% SDS) at both electrodes, remains the most popular one. The gel is run at a constant current of 25 mA until the bromophenol blue dye front has reached the bottom of the gel.

Staining

Visualization of proteins spots is commonly achieved through staining with colloidal Coomassie Blue G-250 due to its low cost and ease of use. A typical staining procedure includes fixing the gel for several hours in 50% ethanol/2% *ortho*-phosphoric acid, followed by several 30-minute washing steps in water, followed by incubation for 1 hour in 17% ammonium sulfate/34% methanol/2% *ortho*-phosphoric acid, followed by staining for several days in 0.1% Coomassie Blue G-250/17% ammonium sulfate/34% methanol/2% *ortho*-phosphoric acid, followed by destaining for several hours in water. There are, however, commercially available colloidal Coomassie staining kits that do not require fixation or destaining.

A great many alternative visualization methods are available, many of which are more sensitive than colloidal Coomassie and thus may be more suitable for applications where visualization of low-abundance proteins is important. These include radiolabeling, such as with [^{35}S]methionine, and staining with fluorescent dyes such as the SYPRO or Cy series of dyes. Multiple staining with dyes fluorescing at different wavelengths offers the possibility of differential display, allowing more than one proteome to be compared on the same gel, such as in difference gel electrophoresis (DIGE). Patton published a detailed review of visualization techniques for proteomics (Patton 2002).

Analysis

Although commercial 2DE image analysis software, such as ImageMaster (Amersham), PDQuest (BioRad), or Progenesis (Nonlinear Dynamics), has improved by leaps and bounds in recent years, analysis of the 2DE gel image, including protein spot definition, matching, and individual protein quantification, remains the bottleneck of 2DE-based proteome analysis and still requires a substantial amount of subjective input by the investigator (Barrett et al. 2005). In particular, spot matching between gels tends to be time consuming and has proved difficult to automate (Wheelock and Goto 2006). These difficulties arise from several sources of variation among individual gels, such as protein load variability due to varying IPG strip reswelling or protein transfer from strip to slab gel. Also, gene expression in several tissues varies considerably among individuals of the same species, and therefore individual variation is a major concern and needs to be accounted for in any statistical treatment of the data. Pooling samples may also be an option, depending on the type of experiment. These multiple sources of variation has led some investigators (Barrett et al. 2005, Karp et al. 2005, Wheelock and Goto 2006) to cast doubt on the suitability of univariate tests such as the Student's t-test, commonly used to assess the significance of observed protein expression differences. Multivariate analysis has been successfully used by several investigators in recent years (Gustafson et al. 2004, Karp et al. 2005, Kjaersgard et al. 2006b).

Some Problems and Their Solutions

The high resolution and good sensitivity of 2DE are what make it the method of choice for most proteomics work, but the method nevertheless has several drawbacks. The most significant of these have to do with the diversity of proteins and their expression levels. For example, hydrophobic proteins do not readily dissolve in the buffers used for isoelectrofocussing. This problem can be overcome, though, using nonionic or zwitterionic detergents, allowing for 2DE of membrane- and membrane-associated proteins (Chevallet et al. 1998, Herbert 1999, Henningsen et al. 2002, Babu et al. 2004). Vilhelmsson and Miller (2002), for example, were able to use "membrane protein proteomics" to demonstrate the involvement of membrane-associated metabolic enzymes in the osmoadaptive response of the foodborne pathogen *Staphylococcus aureus*. A 2DE gel image of *S. aureus* membrane-associated gels is shown in Figure 22.4.

Similarly, resolving alkaline proteins, particularly those with pI above 10, on two-dimensional gels has been problematic in the past. Although the development of highly alkaline, narrow-range IPGs (Bossi et al. 1994) allowed reproducible two-dimensional resolution of alkaline proteins (Görg et al. 1997), their representation on wide-range 2DE of complex mixtures such as cell extracts remained poor. Improvements in resolution and representation of alkaline proteins on wide-range gels have been made (Görg et al. 1999), but nevertheless an approach that involves several gels, each of a different pH range, from the same sample is advocated for representative inclusion of alkaline proteins when studying entire proteomes (Cordwell et al. 2000). Indeed, Cordwell and coworkers were able to significantly improve the representation of alkaline proteins in their study on the relatively highly alkaline *Helicobacter pylori* proteome using both pH 6–11 and pH 9–12 IPGs (Bae et al. 2003).

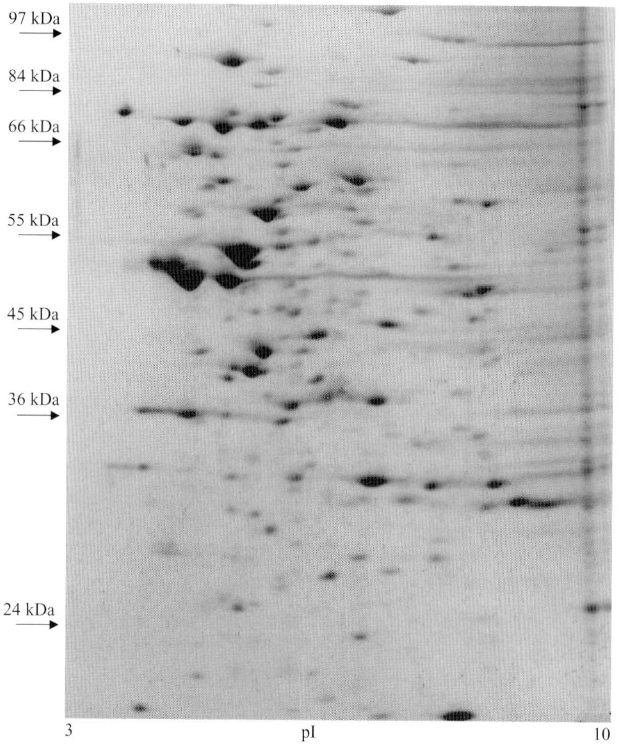

Figure 22.4. A two-dimensional electrophoresis membrane proteome map from *Staphylococcus aureus*, showing proteins with pI between 3 and 10 and molecular mass about 15–100 (O. Vilhelmsson and K. Miller, unpublished). Isoelectrofocussing was in the presence of a mixture of pH 5–7 and pH 3–10 carrier ampholytes and the second dimension was in a 10% polyacrylamide slab gel with a 4% polyacrylamide stacker.

A second drawback of 2DE has to do with the extreme difference in expression levels of the cell's various proteins, which can be as much as 10,000-fold. This leads to swamping of low-abundance proteins by high-abundance ones on the two-dimensional map, rendering analysis of low-abundance proteins difficult or impossible. For applications such as species identification or study of the major biochemical pathways, where the proteins of interest are present in relatively high abundance, this does not present a problem. However, when investigating, for example, regulatory cascades, the proteins of interest are likely to be present in very low abundance and may at times be undetectable because of the dominance of high-abundance ones. Simply increasing the amount of sample is usually not an option, as it will give rise to overloading artifacts in the gels (O'Farrell 1975). In transcriptomic studies, where a similar disparity can be seen in the abundance of RNA transcripts present, this problem can be overcome by amplifying the low-abundance transcripts using the polymerase chain-reaction, but no such technique is available for proteins. The remaining option, then, is fractionation of the protein sample in order to weed out the high-abundance proteins, allowing a larger sample of the remaining proteins to be analyzed. A large number of fractionation protocols, both specific and general, are available. Thus, Østergaard and coworkers used acetone precipitation to reduce the abundance of hordeins present in barley (*Hordeum vulgare*) extracts (Østergaard et al. 2002) whereas Locke and coworkers used preparative isoelectrofocussing to fractionate Chinese snow pea (*Pisum sativum macrocarpon*) lysates into fractions covering three pH regions (Locke et al. 2002). The fractionation method of choice will depend on the specific requirements of the study and on the tissue being studied. Discussion of some fractionation methods can be found in Ahmed (2009), Bodzon-Kulakowska et al. (2007), Butt et al. (2001), Canas et al. (2007), Corthals et al. (1997), Dreger (2003), Fortis et al. (2008), Issaq et al. (2002), Lee and Pi (2009), Lopez et al. (2000), Millea and Krull (2003), Pieper et al. (2003), Righetti et al. (2005a, b) Rothemund et al. (2003), von Horsten (2006).

Identification by Peptide Mass Fingerprinting

Identification of proteins on 2DE gels is most commonly achieved via mass spectrometry (MS) of trypsin digests. Briefly, the spot of interest is excised from the gel, digested with trypsin (or another protease), and the resulting peptide mixture is analyzed by MS. The most popular MS method is matrix-assisted laser desorption/ionization time-of-flight (MALDI-TOF) MS (Courchesne and Patterson 1999), where peptides are suspended in a matrix of small, organic, ultraviolet-absorbing molecules (such as 2,5-dihydroxybenzoic acid) followed by ionization by a laser at the excitation wavelength of the matrix molecules and acceleration of the ionized peptides in an electrostatic field into a flight tube where the time of flight of each peptide is measured, giving its expected mass.

The resulting spectrum of peptide masses (Fig. 22.5) is then used for protein identification by searching against expected peptide masses calculated from data in protein sequence databases, such as SwissProt or the National Centre for Biotechnology Information (NCBI) nonredundant protein sequences data base, using the appropriate software. Several programs are available, many with a web-based open-access interface. The ExPASy

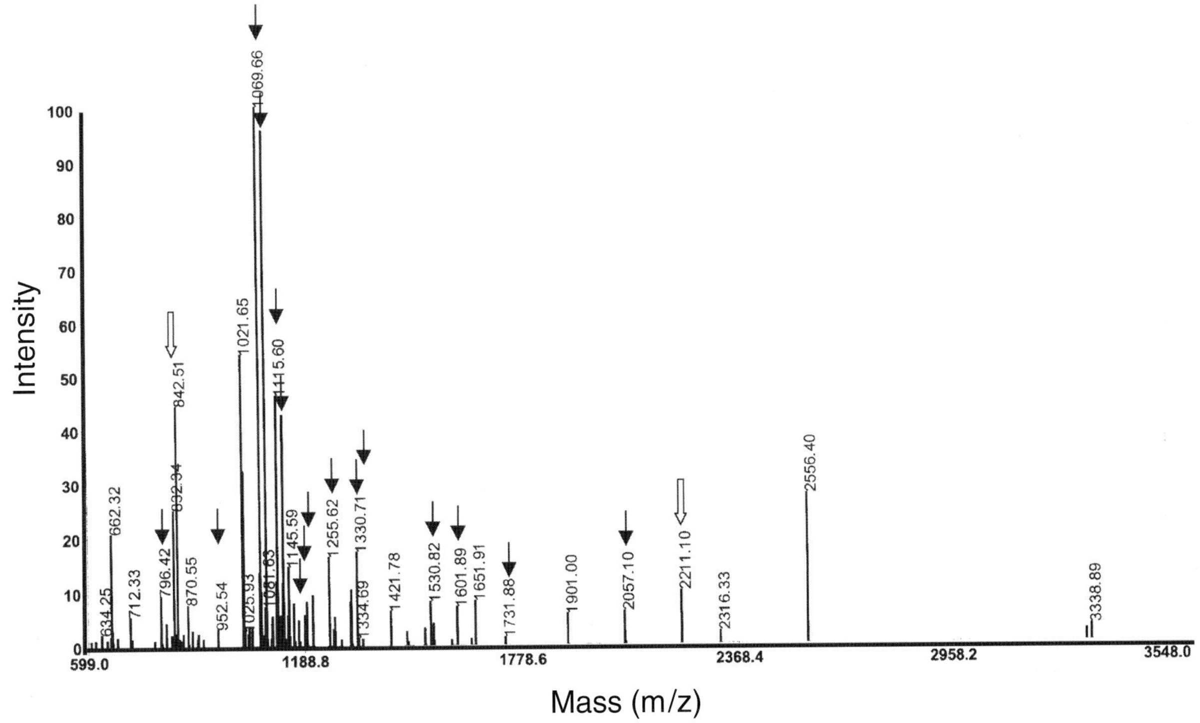

Figure 22.5. A trypsin digest mass spectrometry fingerprint of a rainbow trout liver protein spot, identified as apolipoprotein A I-1 (S. Martin, unpublished). The open arrows indicate mass peaks corresponding to trypsin self-digestion products and were, therefore, excluded from the analysis. The solid arrows indicate the peaks that were found to correspond to expected apolipoprotein A I-1 peptides.

Tools website (http://www.expasy.org/tools) contains links to most of the available software for protein identification and several other tools. Attaining a high identification rate is problematic in fish and seafood proteomics due to the relative paucity of available protein sequence data for these animals. As can be seen in Table 22.1, this problem is surprisingly acute for species of commercial importance. To circumvent this problem, it is possible to take advantage of the available nucleotide sequences, which in many cases is more extensive than the protein sequences available, to obtain a tentative identity. How useful this method is will depend on the length and quality of the available nucleotide sequences. It is important to realize, however, that an identity obtained in this manner is less reliable than that obtained through protein sequences and should be regarded only as tentative in the absence of corroborating evidence (such as two-dimensional immunoblots, correlated activity measurements, or transcript abundance). In their work on the rainbow trout (*Oncorhynchus mykiss*) liver proteome, Martin et al. (2003b) and Vilhelmsson et al. (2004) were able to attain an identification rate of about 80% using a combination of search algorithms that included the open-access Mascot program (Perkins et al. 1999) and a licensed version of Protein Prospector MS-Fit (Clauser et al. 1999), searching against both protein databases and a database containing all salmonid nucleotide sequences. In those cases where both the protein and nucleotide databases yielded results, a 100% agreement was observed between the two methods.

A more direct, if rather more time-consuming, way of obtaining protein identities is by direct sequence comparison. Until recently, this was accomplished by *N*-terminal or internal (after proteolysis) sequencing by Edman degradation of eluted or electroblotted protein spots (Kamo and Tsugita 1999, Erdjument-Bromage et al. 1999). Today, the method of choice is tandem mass spectrometry (MS/MS). In the peptide mass fingerprinting discussed earlier, each peptide mass can potentially represent any of a large number of possible amino acid sequence combinations. The larger the mass (and longer the sequence), the higher the number of possible combinations. In MS/MS, one or several peptides are separated from the mixture and dissociated into fragments that then are subjected to a second round of MS, yielding a second layer of information. Correlating this spectrum with the candidate peptides identified in the first round narrows down the number of candidates. Furthermore, several short stretches of amino acid sequence will be obtained for each peptide, which, when combined with the peptide and fragment masses obtained, enhances the specificity of the method even further (Wilm et al. 1996, Chelius et al. 2003, Yu et al. 2003b,). MS methods in proteomics are reviewed in Yates (1998, 2004) and Yates et al. (2009).

SEAFOOD PROTEOMICS AND THEIR RELEVANCE TO PROCESSING AND QUALITY

2DE-based proteomics have found a number of applications in food science. Among early examples are such applications as characterization of bovine caseins (Zeece et al. 1989), wheat flour baking quality factors (Dougherty et al. 1990), and soybean protein bodies (Lei and Reeck 1987). In recent years, proteomic investigations on fish and seafood products, as well as in fish physiology, have gained considerable momentum, as can be seen in recent reviews (Parrington and Coward 2002, Piñeiro et al. 2003). Herein, we consider recent and future developments in fish and seafood proteomics as related to issues of concern in fish processing or other quality considerations.

Early Development and Proteomics of Fish

Fishes go through different developmental stages (embryo, larva, and adult) during their lifespan that coincide with changes in the morphology, physiology, and behavior of the fish (O'Connell 1981, Govoni et al. 1986, Skiftesvik 1992, Osse and van den Boogaart 1995). The morphological and physiological changes that occur during these developmental stages are characterized by differential cellular and organelle functions (Einarsdóttir et al. 2006). This is reflected in variations of global protein expression and posttranslational modifications of the proteins that may cause alterations of protein function (Campinho et al. 2006).

Proteome analysis provides valuable information on the variations that occur within the proteome of organisms. These variations may, for example, reflect a response to biological perturbations or external stimuli (Anderson and Anderson 1998, Martin et al. 2001, Martin et al. 2003b, Vilhelmsson et al. 2004) resulting in different expression of proteins, posttranslational modifications, or redistribution of specific proteins within cells (Tyers and Mann 2003). To date, few studies on fish development exist in which proteome analysis techniques have been applied. Recent studies on global protein expression during early developmental stages of zebrafish (Tay et al. 2006) and Atlantic cod (Sveinsdóttir et al. 2008) revealed that distinctive protein profiles characterize the developmental stages of these fishes even though abundant proteins are largely conserved during the experimental period. In both these studies, the identified proteins consisted mainly of proteins located in the cytosol, cytoskeleton, and nucleus. Proteome analyses in developing organisms have shown that many of the identified proteins have multiple isoforms (Paz et al. 2006), reflecting either different gene products (Guðmundsdóttir et al. 1993) or posttranslationally modified forms of these proteins (Jensen 2004). Different isoforms generated by posttranslational modifications are largely overlooked by studies based on RNA expression. This fact further indicates the importance of the proteome approach to understand cellular mechanisms underlying fish development. Studies on various proteins have shown that during fish development sequential synthesis of different isoforms appear successively (Huriaux et al. 1996, Galloway et al. 1998, Focant et al. 1999, Galloway et al. 1999, Huriaux et al. 1999, Focant et al. 2000, Huriaux et al. 2003, Focant et al. 2003, Hall et al. 2003, Galloway et al. 2006, Campinho et al. 2006, Campinho et al. 2007). In this context, developmental stage-specific muscle protein isoforms have gained a special attention (Huriaux et al. 1996, Galloway et al. 1998, Focant et al. 1999, Galloway et al. 1999, Huriaux et al. 1999, Focant et al. 2000, Focant et al. 2003, Hall et al. 2003, Huriaux et al. 2003, Galloway et al. 2006, Campinho et al. 2007).

Table 22.1. Some Commercially or Scientifically Important Fish and Seafood Species and the Availability of Protein and Nucleotide Sequence Data as of February 18, 2010, According to the NCBI TaxBrowser[(a)]

	Protein Sequences	Nucleotide Sequences		Protein Sequences	Nucleotide Sequences
Actinopterygii (ray-finned fishes)	257,843	1,586,862	*Tetraodontiformes* (puffers and filefishes)	34,032	114,496
Elopomorpha	3,218	4,459	Pufferfish (*Takifugu rubripes*)	2,593	16,406
Anguilliformes (eels and morays)	2,898	4,144	Green pufferfish (*Tetraodon nigroviridis*)	28,483	126,200
European eel (*Anguilla anguilla*)	227	404	*Zeiformes* (dories)	138	126
Clupeomorpha	2,333	4,697	John Dory (*Zeus faber*)	102	100
Clupeiformes (herrings)	2,333	4,697	*Scorpaeniformes* (scorpionfishes/flatheads)	4,858	7,896
Atlantic herring (*Clupea harengus*)	105	166	Redfish (*Sebastes marines*)	13	17
European pilchard (*Sardina pilchardus*)	79	401	Lumpsucker (*Cyclopterus lumpus*)	20	30
Ostariophysii	104,088	211,491			
Cypriniformes (carps)	94,392	196,415	*Chondrichthyes* (cartilagenous fishes)	6,407	692,476
Zebrafish (*Danio rerio*)	70,151	163,995	*Carcharhiniformes* (ground sharks)	1,812	2,234
Siluriformes (catfishes)	7,149	10,472	Lesser spotted dogfish (*Scyliorhinus canicula*)	261	197
Channel catfish (*Ictalurus punctatus*)	1,890	1,996	Blue shark (*Prionace glauca*)	48	52
Protacanthopterygii	32,187	42,475	*Lamniformes* (mackrel sharks)	554	676
Salmoniformes (salmons)	31,244	42,278	Basking shark (*Cetorhinus maximus*)	66	72
Atlantic salmon (*Salmo salar*)	16,230	19,107	*Rajiformes* (skates)	848	1,050
Rainbow trout (*Oncorhynchus mykiss*)	6,216	8,361	Thorny skate (*Raja radiata*)	41	60
Arctic charr (*Salvelinus alpinus*)	257	465	Blue skate (*Raja batis*)	7	1
Paracanthopterygii	5,396	6,397	Little skate (*Raja erinacea*)	201	185
Gadiformes (cods)	4,517	5,573			
Atlantic cod (*Gadus morhua*)	2,106	1,964	*Mollusca* (mollusks)	52,622	154,100
Alaska pollock (*Theragra chalcogramma*)	382	303	*Bivalvia*	14,572	25,149
Saithe (*Pollachius virens*)	51	53	Blue mussel (*Mytilus edulis*)	1,307	1,266
Haddock (*Melanogrammus aeglefinus*)	189	151	Bay scallop (*Argopecten irradians*)	194	380
Lophiiformes (anglerfishes)	412	471	*Gastropoda*	32,489	121,832
Monkfish (*Lophius piscatorius*)	40	104	Common whelk (*Buccinum undatum*)	6	11
Acanthopterygii	104,101	1,369,306	Abalone (*Haliotis tuberculata*)	133	279
Perciformes (perch like)	44,027	619,647	*Cephalopoda*	3,964	5,045
Gilthead sea bream (*Sparus aurata*)	516	956	Northern European squid (*Loligo forbesi*)	34	49
European sea bass (*Dicentrachus labrax*)	466	36,916	Common cuttlefish (*Sepia officinalis*)	351	335
Atlantic mackrel (*Scomber scombrus*)	111	129	Common octopus (*Octopus vulgaris*)	169	204
Albacore (*Thunnus alalunga*)	116	227			
Bluefin tuna (*Thunnus thynnus*)	179	935	*Crustacea* (crustaceans)	44,586	70,954
Spotted wolffish (*Anarhichas minor*)	38	23	*Caridea*	5,103	7,630
Beryciformes (*sawbellies*)	794	616	Northern shrimp (*Pandalus borealis*)	19	26

Table 22.1. (*Continued*)

	Protein Sequences	Nucleotide Sequences		Protein Sequences	Nucleotide Sequences
Orange roughy (*Hoplostethus atlanticus*)	11	40	*Astacidea* (lobsters and crayfishes)	1,742	4,466
Pleuronectiformes (flatfishes)	4,109	11,851	American lobster (*Homerus americanus*)	195	173
Atlantic halibut (*Hippoglossus hippoglossus*)	218	2,958	European crayfish (*Astacus astacus*)	28	39
Witch (*Glyptocephalus cynoglossus*)	22	54	Langoustine (*Nephrops norvegicus*)	19	35
Plaice (*Pleuronectes platessa*)	88	285	Brachyura (short-tailed crabs)	5,143	8,238
Winter flounder (*Pseudopleuronectes americanus*)	139	196	Edible crab (*Cancer pagurus*)	37	38
Turbot (*Scophthalmus maximus*)	385	1,014	Blue crab (*Callinectes sapidus*)	123	133

Source: [a]http://www.ncbi.nlm.nih.gov/Taxonomy/taxonomyhome.html/.

The developmental changes in the composition of muscle protein isoforms have been tracked by proteome analysis in African catfish (*Heterobranchus longifilis*) (Focant et al. 1999), common sole (*Solea solea*) (Focant et al. 2003), and dorada (*Brycon moorei*) (Huriaux et al. 2003). These studies demonstrated that the muscle shows the usual sequential synthesis of protein isoforms in the course of development. For example, in the common sole, 2DE revealed two isoforms (larval and adult) of myosin light chain 2, and likewise in dorada larval and adult isoforms of troponin I were sequentially expressed during development.

Proteomic techniques have, thus, been shown to be applicable for investigating cellular and molecular mechanisms involved in the morphological and physiological changes that occur during fish development.

The major obstacle on the use of proteomics in embryonic fish has been the high proportion of yolk proteins. These interfere with any proteomic application that intends to target the cells of the embryo proper. In a recent study on the proteome of embryonic zebrafish, the embryos were deyolked to enrich the pool of embryonic proteins and to minimize ions and lipids found in the yolk prior to two-dimensional gel analysis (Tay et al. 2006). Despite this undertaking, a large number of yolk proteins remained prominently present in the embryonic protein profiles. Link et al. (2006) published a method to efficiently remove the yolk from large batches of embryos without losing cellular proteins. The success in the removal of yolk proteins by Link et al. (2006) is probably due to dechorionation prior to the deyolking of the embryos. By dechorionation, the embryos fall out of their chorions facilitating the removal of the yolk (Huriaux et al. 1996, Westerfield 2000).

Changes in the Proteome of Early Cod Larvae in Response to Environmental Factors

The production of good quality larvae is still a challenge in marine fish hatcheries. Several environmental factors can interfere with the protein expression of larvae affecting larval quality like growth and survival rate. Proteome analysis allows us to examine the effects of environmental factors on larval global protein expression, posttranslational modifications, and redistribution of specific proteins within cells (Tyers and Mann 2003), all important information for controlling factors influencing the aptitude to continue a normal development until adult stages.

A variety of environmental factors have shown to improve the health and survival of fish larvae, including probiotic bacteria (Tinh et al. 2008) and protein hydrolysates (Cahu et al. 1999). However, the beneficial effects of these treatments on fish larvae are poorly understood at the molecular level. Only a few proteome analysis studies on fish larvae have been published (Focant et al. 2003, Tay et al. 2006, Guðmundsdóttir and Sveinsdóttir 2006, Sveinsdóttir et al. 2008, Sveinsdóttir and Guðmundsdóttir 2008, Sveinsdóttir et al. 2009), of which two have focused on the changes in the whole larval proteome after treatment with probiotic bacteria (Sveinsdóttir et al. 2009) and protein hydrolysate (Sveinsdóttir and Gudmundsdóttir 2008). These studies provide protocols for the production of high-resolution two-dimensional gels of whole larval proteome, where peptide mass mapping (MALDI-TOF MS) and peptide fragment mapping (LC-MS/MS) allowed identification of ca. 85% of the of the selected cod protein spots (Guðmundsdóttir and Sveinsdóttir 2006, Sveinsdóttir and Gudmundsdóttir 2008, Sveinsdóttir et al. 2008, Sveinsdóttir et al. 2009).

The advantages of working with whole larvae versus distinct tissues is the ease of keeping the sample handling to a minimum in order to avoid loss or modification of the proteins. Nevertheless, there are several drawbacks when working with the whole larval proteome, like the overwhelming presence of muscle and skin proteins. These proteins may mask subtle changes in proteins expressed in other tissues or systems, such as the gastrointestinal tract or the central nervous system caused by various environmental factors. The axial musculature is the largest tissue in larval fishes as it constitutes approximately 40% of their body mass (Osse and van den Boogaart 1995). This is reflected in our studies on whole cod larval proteome, where the majority of the highly abundant proteins were identified as muscle proteins (Guðmundsdóttir and Sveinsdóttir 2006, Sveinsdóttir and Gudmundsdóttir 2008, Sveinsdóttir et al. 2008, 2009). Also, cytoskeletal proteins were prominent among the identified proteins. Removal of those proteins may increase detection of other proteins present at low concentrations. However, it may also result in a loss of other proteins, preventing identification of holistic alterations in the analyzed proteomes. Various strategies have been presented for the removal of highly abundant proteins (Ahmed et al. 2003) or enrichment of low-abundance proteins (Oda et al. 2001, Ahmed and Rice 2005).

Tracking Quality Changes Using Proteomics

A persistent problem in the seafood industry is postmortem degradation of fish muscle during chilled storage, which has deleterious effects on the fish flesh texture, yielding a tenderized muscle. This phenomenon is thought to be primarily due to autolysis of muscle proteins, but the details of this protein degradation are still somewhat in the dark. However, degradation of myofibrillar proteins by calpains and cathepsins (Ogata et al. 1998, Ladrat et al. 2000) and degradation of the extracellular matrix by the matrix metalloproteases and matrix serine proteases, capable of degrading collagens, proteoglycans, and other matrix components (Woessner 1991, Lødemel and Olsen 2003), are thought to be among the main culprits. Whatever the mechanism, it is clear that these quality changes are species dependent (Papa et al. 1996, Verrez-Bagnis et al. 1999) and, furthermore, appear to display seasonal variations (Ingólfsdóttir et al. 1998, Ladrat et al. 2000). For example, whereas desmin is degraded postmortem in sardine and turbot, no desmin degradation was observed in sea bass and brown trout (Verrez-Bagnis et al. 1999). Of further concern is the fact that several commercially important fish muscle processing techniques, such as curing, fermentation, and production of surimi and conserves occur under conditions conducive to endogenous proteolysis (Pérez-Borla et al. 2002). As with postmortem protein degradation during storage, autolysis during processing seems to be somewhat specific. Indeed, the myosin heavy chain of the Atlantic cod was shown to be significantly degraded during processing of "salt fish" (*bacalhau*) whereas actin was less affected (Thorarinsdottir et al. 2002). Problems of this kind, where differences are expected to occur in the number, molecular mass, and pI of the protein present in a tissue, are well suited to investigation using 2DE-based proteomics. It is also worth noting that protein isoforms other than proteolytic ones, whether they be encoded in structural genes or brought about by posttranslational modification, usually have different molecular weight or pI and can, therefore, be distinguished on 2DE gels. Thus, specific isoforms of myofibrillar proteins, many of which are correlated with specific textural properties in seafood products, can be observed using 2DE or other proteomic methods (Martinez et al. 1990, Piñeiro et al. 2003).

Several 2DE studies have been performed on postmortem changes in seafood flesh (Verrez-Bagnis et al. 1999, Morzel et al. 2000, Martinez et al. 2001a, Kjaersgard and Jessen 2003, 2004, Martinez and Jakobsen Friis 2004, Kjaersgard et al. 2006a, b) and have demonstrated the importance and complexity of proteolysis in seafood during storage and processing. For example, Martinez and Jakobsen Friis used a 2DE approach to demonstrate different protein composition of surimi made from prerigor versus postrigor cod (Martinez and Jakobsen Friis 2004). They found that 2DE could be used as a diagnostic tool to indicate the freshness of the raw material used for surimi production, a finding of considerable economic and public health interest. Kjaersgard and Jessen, who used 2DE to study changes in abundance of several muscle proteins during storage of the Atlantic cod (*Gadus morhua*), proposed a general model for postmortem protein degradation in fish flesh where initially calpains were activated due to the increase in calcium levels in the muscle tissue. Later, as pH decreases and ATP is depleted with the consequent onset of rigor mortis, cathepsins and the proteasome are activated sequentially (Kjaersgard and Jessen 2003).

Antemortem Effects on Quality and Processability

Malcolm Love started his 1980 review paper on biological factors affecting fish processing (Love 1980) with a lament for the easy life of poultry processors who, he said, had the good fortune to work on a product reared from hatching under strictly controlled environmental and dietary conditions "so that plastic bundles of almost identical foodstuff for man can be lined up on the shelf of a shop." Since the time of Love's review, the advent of aquaculture has made attainable, in theory at least, just such a utopic vision. As every food processor knows, the quality of the raw material is among the most crucial variables that affect the quality of the final product. In fish processing, therefore, the animal's own individual physiological status will to a large extent dictate where quality characteristics will fall within the constraints set by the species' physical and biochemical makeup. It is well known that an organism's phenotype, including quality characteristics, is determined by environmental as well as genetic factors. Indeed, Huss noted in his review (Huss 1995) that product quality differences within the same fish species can depend on feeding and rearing conditions, differences wherein can affect postmortem biochemical processes in the product, which, in turn, affect the involution of quality characteristics in the fish product. The practice of rearing fish in aquaculture, as opposed to wild-fish catching, therefore raises the tantalizing prospect of managing quality characteristics of the fish flesh antemortem, where individual physiological characteristics, such as those governing gaping tendency, flesh softening during storage, etc., are optimized. To achieve that goal, the interplay between these physiological parameters and environmental and dietary variables needs to be understood in detail. With the ever-increasing resolving power of molecular techniques, such as proteomics, this is fast becoming feasible.

In mammals, antemortem protease activities have been shown to affect meat quality and texture (Vaneenaeme et al. 1994, Kristensen et al. 2002). For example, an antemortem upregulation of calpain activity in swine (*Sus scrofa*) will affect postmortem proteolysis and, hence, meat tenderization (Kristensen et al. 2002).

In beef (*Bos taurus*), a correlation was found between antemortem and postmortem activities of some proteases, but not others (Vaneenaeme et al. 1994). As discussed in the earlier section, postmortem proteolysis is a matter of considerable importance in the fish and seafood industry and any antemortem effects thereupon are surely worth investigating.

In a recent study on the feasibility of substituting fish meal in rainbow trout (*O. mykiss*) diets with protein from plant sources, 2DE-based proteomics were among the techniques used (Martin et al. 2003a, b, Vilhelmsson et al. 2004). Concomitantly, various quality characteristics of fillet and body were also measured (De Francesco et al. 2004, Parisi et al. 2004). Among the findings was that, according to a triangular sensory test using a trained panel, cooked 5-day-old trout that had been fed the plant protein diet had higher hardness, lower juiciness, and lower odor intensity than those fed the fish meal-containing diet, indicating an effect of antemortem metabolism on product texture. The study identified a number of metabolic pathways sensitive to plant protein substitution in rainbow trout feed, such as pathways involved in cellular protein degradation, fatty acid breakdown, and NADPH (nicotinamide adenine dinucleotide phosphate) metabolism (Table 22.2). In the context of this chapter,

Table 22.2. Protein Spots Affected by Dietary Plant Protein Substitution in Rainbow Trout as Judged by Two-dimensional electrophoresis and Their Identities as Determined by Trypsin Digest Mass Fingerprinting

Spot Reference Number	pI	MW (kDa)	Normalized Volume Diet FM[a]	Normalized Volume Diet PP100[a]		Fold Difference	P
Downregulated:							
128	6.3	66	303 ± 57	60 ± 19	Vacuolar ATPase β-subunit	5	0.026
291	6.4	42	521 ± 37	273 ± 30	β-ureidopropionase	2	0.004
356	6.3	38	161 ± 37	44 ± 19	Transaldolase	4	0.031
747	5.6	43	101 ± 19	19 ± 11	β-actin	2	0.040
760	6.3	39	41 ± 6	21± 5	ND	2	0.040
766	4.8	27	12 ± 1	6 ± 1	ND	2	0.004
Upregulated:							
80	4.4	82	9 ± 4	47 ± 8	"Unknown protein"	5	0.007
87	5.7	75	58 ± 14	262 ± 21	Transferrin	5	<0.001
138	5.5	67	99 ± 16	267 ± 39	Hemopexin-like	3	0.009
144	5.4	63	26 ± 6	265 ± 66	L-Plastin	10	0.018
190	5.9	54	6 ± 2	50 ± 9	Malic enzyme	9	0.018
199	5.9	53	60 ± 16	156 ± 13	Thyroid hormone receptor	3	0.020
275	6.1	45	1 ± 0.6	11 ± 0.6	NSH	9	<0.001
387	5.6	35	97 ± 3	251 ± 49	Electron transferring flavoprotein	3	0.035
389	5.8	35	192 ± 45	414 ± 54	Electron transferring flavoprotein	2	0.027
399	6.8	33	59 ± 12	130 ± 10	Aldolase B	2	0.028
457	4.7	29	26 ± 7	57 ± 5	14–3-3 B2 Protein	2	0.021
461	4.7	27	75 ± 9	190 ± 12	Proteasome alpha 2	3	0.004
517	4.4	22	15 ± 6	135 ± 29	Cytochrome *c* oxidase	9	0.013
539	4.9	19	7 ± 3	18 ± 3	ND	3	0.033
551	4.1	17	40 ± 11	143 ± 28	ND	4	0.018
563	5.2	15	814 ± 198	3762 ± 984	Fatty acid-binding protein	5	0.039
639	6.4	84	10 ± 6	28 ± 5	NSH	3	0.047
648	6.1	55	17 ± 5	154 ± 46	Hydroxymethylglutaryl-CoA synthase	9	0.040
678	5.3	48	26 ± 7	69 ± 15	Proteasome 26S ATPase subunit 4	3	0.044
746	4.4	46	45 ± 13	107 ± 15	"Similar to catenin"	2	0.012
754	4.1	15	6 ± 2	36 ± 4	ND	7	<0.001
761	6.1	36	44 ± 21	204 ± 34	Transaldolase	5	0.006
764	6.2	65	0	102 ± 17	NSH	>10	NA
770	5.0	21	4 ± 1	18 ± 4	ND	4	0.026

[a]Values are mean normalized protein abundance (± SE). Data were analyzed by the Student's *t* test ($n = 5$). In the diet designated FM, protein was provided in the form of fish meal; in diet designated PP100, protein was provided by a cocktail of plant product with an equivalent amino acid composition to fish meal.

NSH, no significant homology detected; ND, identity not determined; NA, not available.

the effects on the proteasome are particularly noteworthy. The proteasome is a multisubunit enzyme complex that catalyzes proteolysis via the ATP-dependent ubiquitin–proteasome pathway, which, in mammals, is thought to be responsible for a large fraction of cellular proteolysis (Rock et al. 1994, Craiu et al. 1997). In rainbow trout, the ubiquitin–proteasome pathway has been shown to be downregulated in response to starvation (Martin et al. 2002) and to have a role in regulating protein deposition efficiency (Dobly et al. 2004). It, therefore, seems likely that the observed difference in texture is affected by antemortem proteasome activity, although further studies are needed to verify that statement.

In addition to having a hand in controlling autolysis determinants, protein turnover is a major regulatory engine of cellular structure, function, and biochemistry. Cellular protein turnover involves at least two major systems: (1) the lysosomal system and (2) the ubiquitin–proteasome system (Hershko and Ciechanover 1986, Mortimore et al. 1989). The 20S proteasome has been found to have a role in regulating the efficiency with which rainbow trout deposit protein (Dobly et al. 2004). It seems likely that the manner in which protein deposition is regulated, particularly in muscle tissue, has profound implications for quality and processability of the fish flesh.

Protein turnover systems, such as the ubiquitin–proteasome or the lysosome systems, are suitable for rigorous investigation using proteomic methods. For example, lysosomes can be isolated and the lysosome subproteome queried to answer the question whether and to what extent lysosome composition varies among fish expected to yield flesh of different quality characteristics. Proteomic analysis on lysosomes has been successfully performed in mammalian (human) systems (Journet et al. 2000, Journet et al. 2002).

An exploitable property of proteasome-mediated protein degradation is the phenomenon of polyubiquitination, whereby proteins are targeted for destruction by the proteasome by covalent binding to multiple copies of ubiquitin (Hershko and Ciechanover 1986, Ciechanover 1994). By targeting these ubiquitin-labeled proteins, it is possible to observe the ubiquitin–proteasome "degradome," that is, which proteins are being degraded by the proteasome at a given time or under given conditions. Gygi and coworkers have developed methods to study the ubiquitin–proteasome degradome in the yeast *Saccharomyces cerevisiae* using multidimensional LC-MS/MS (Peng et al. 2003).

Some proteolysis systems, such as that of the matrix metalloproteases, may be less directly amenable to proteomic study. Activity of matrix metalloproteases is regulated via a complex network of specific proteases (Brown et al. 1993, Okumura et al. 1997, Wang and Lakatta 2002). Monitoring of the expression levels of these regulatory enzymes, and how they vary with environmental or dietary variables, may be more conveniently carried out using transcriptomic methods.

A well-known quality issue when farmed fish are compared with wild catch is that of gaping, a phenomenon caused by cleavage of myocommatal collagen cross-links that results in weakening and rupturing of connective tissue (Børresen 1992, Foegeding et al. 1996). Gaping can be a serious quality issue in the fish processing industry as, apart from the obvious visual defect, it causes difficulties in mechanical skinning and slicing of the fish (Love 1992). Weakening of collagen and, hence, gaping is facilitated by low pH. Well-fed fish, such as those reared in aquaculture, tend to yield flesh of comparatively low pH, which, thus, tends to gape (Foegeding et al. 1996, Einen et al. 1999). Gaping is, therefore, a cause for concern with aquaculture-reared fish, particularly of species with high natural gaping tendency, such as the Atlantic cod. Gaping tendency varies considerably among wild fish caught in different areas (Love et al. 1974) and, thus, it is conceivable that gaping tendency can be controlled with dietary or other environmental manipulations. Proteomics and transcriptomics, with their capacity to monitor multiple biochemical processes simultaneously, are methodologies eminently suitable to finding biochemical or metabolic markers that can be used for predicting features such as gaping tendency of different stocks reared under different dietary or environmental conditions.

Species Authentication

Food authentication is an area of increasing importance, both economically and from a public health standpoint. Taking into account the large difference in market value of different fish species and the increased prevalence of processed product on the market, it is perhaps not surprising that species authentication is fast becoming an issue of supreme commercial importance. Along with other molecular techniques, such as DNA-based species identification (Sotelo et al. 1993, Mackie et al. 1999, Martinez et al. 2001b, Pascoal et al. 2008) and nuclear magnetic resonance-based techniques for determining geographical origin (Campana and Thorrold 2001), proteomics are proving to be a powerful tool in this area, particularly for addressing questions on health status of the organism, stresses or contamination levels at place of breeding, and postmortem treatment (Martinez et al. 2003, Martinez and Jakobsen Friis 2004). Since, unlike the genome, the proteome is not a static entity, but changes between tissues and with environmental conditions, proteomics can potentially yield more information than genomic ones, possibly indicating freshness and tissue information in addition to species. Therefore, although it is likely that DNA-based methods will remain the methods of choice for species authentication in the near term, proteomic methods are likely to develop rapidly and find commercial uses within this field. In many cases, the proteomes of even closely related fish species can be easily distinguishable by eye from one another on two-dimensional gels (Fig. 22.6), indicating that diagnostic protein spots may be used to distinguish closely related species.

From early on, proteomic methods have been recognized as a potential way of fish species identification. During the 1960s, one-dimensional electrophoretic techniques were developed to identify the raw flesh of various species (Tsuyuki et al. 1966, Cowie 1968, Mackie 1969), which was soon followed by methods to identify species in processed or cooked products (Mackie 1972, Mackie and Taylor 1972). These early efforts were reviewed in 1980 (Mackie 1980, Hume and Mackie 1980).

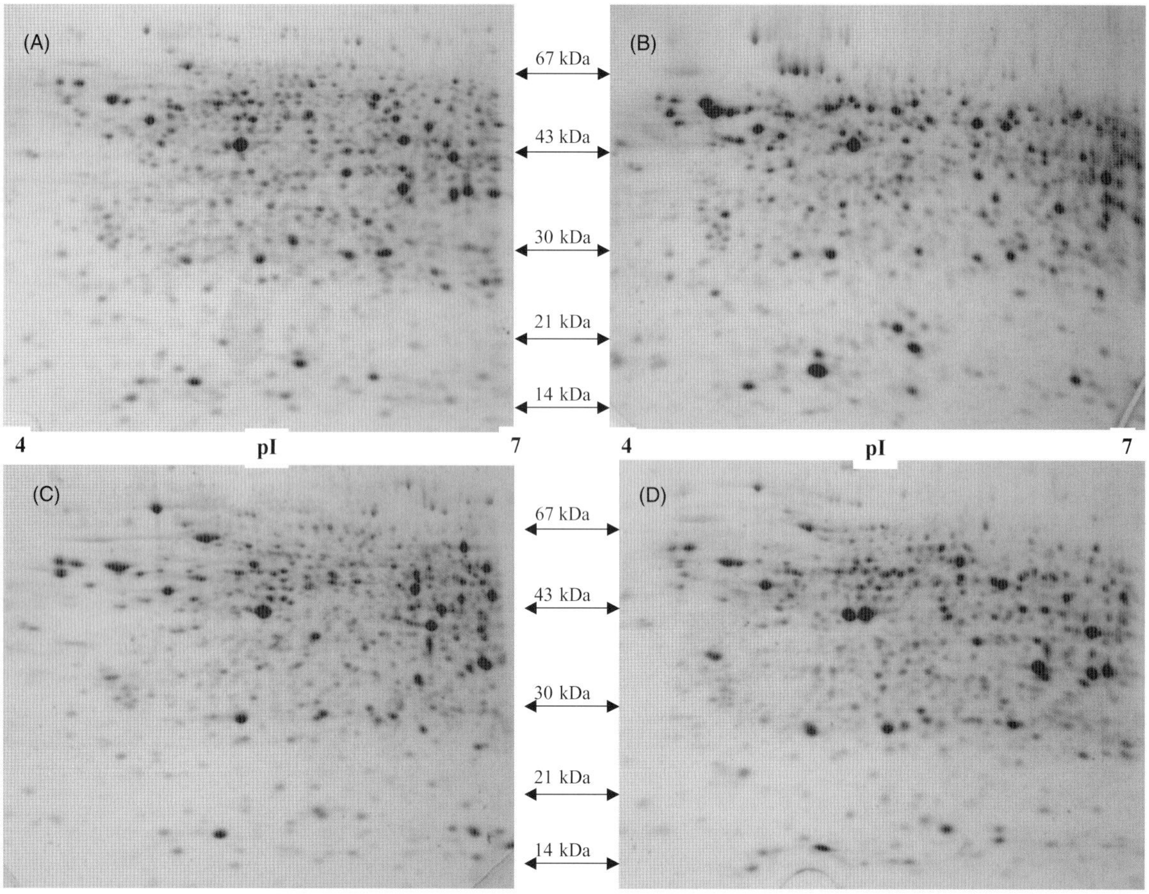

Figure 22.6. Two-dimensional electrophoresis liver proteome maps of four salmonid fish (S. Martin and O. Vilhelmsson, unpublished). Running conditions are as in Figure 22.2. (**A**) Brown trout (*Salmo trutta*), (**B**) Arctic charr (*Salvelinus alpinus*), (**C**) rainbow trout (*Oncorhynchus mykiss*), (**D**) Atlantic salmon (*Salmo salar*).

More recently, 2DE-based methods have been developed to distinguish various closely related species, such as the gadoids or several flat fishes (Piñeiro et al. 1998, 1999, 2001). Piñeiro and coworkers have found that Cape hake (*Merluccius capensis*) and European hake (*Merluccius merluccius*) can be distinguished on two-dimensional gels from other closely related species by the presence of a particular protein spot that they identified, using nanoelectrospray ionization MS, as nucleoside diphosphate kinase (Piñeiro et al. 2001). Lopez and coworkers, studying three species of European mussels: (1) *Mytilus edulis*, (2) *Mytilus galloprovincialis*, and (3) *Mytilus trossulus*, found that *M. trossulus* could be distinguished from the other two species on foot extract two-dimensional gels by a difference in a tropomyosin spot. They found the difference to be due to a single T to D amino acid substitution (Lopez et al. 2002). Martinez and Jakobsen Friis (2004) attempted to identify not only the species present, but also their relative ratios in mixtures of several fish species and muscle types. They concluded that such a strategy would become viable once a suitable number of markers have been identified, although detection of species present in very different ratios is problematic. Ortea et al. (2009a, b) used peptide mass fingerprinting of 2DE-separated proteins to discriminate between closely related shrimp species.

Recently, gel-free methods have received some attention. Mazzeo et al. (2008) found that specific protein biomarkers detectable with MALDI-TOF MS could discriminate among several commercially important fish species, such as those belonging to Gadidae and Pleuronectiformes.

Identification and Characterization of Allergens

Food safety is a matter of increasing concern to food producers and should be included in any consideration of product quality. Among issues within this field that are of particular concern to the seafood producer is that of allergenic potential. Allergic reactions to seafood affect a significant part of the population: about 0.5% of young adults are allergic to shrimp (Woods et al. 2002). Seafood allergies are caused by an immunoglobulin E-mediated response to particular proteins, including structural proteins such as tropomyosin (Lehrer et al. 2003). Proteomics provide a highly versatile toolkit to identify and characterize allergens. As yet, these have seen little use in the study of seafood allergies, although an interesting and elegant approach was reported by Yu and coworkers (Yu et al. 2003a) at National Taiwan University. These authors, studying the cause of shrimp allergy in humans, performed a 2DE on crude protein extracts from the tiger prawn, *Penaeus monodon*, blotted the two-dimensional gel onto a PVDF membrane and probed the membranes with serum from confirmed shrimp allergic patients. The allergens were then identified by MALDI-TOF MS of tryptic digests. The allergen was identified as a protein with close similarity to arginine kinase. The identity was further corroborated by cloning and sequencing the relevant cDNA. A final proof was obtained by purifying the protein, demonstrating that it had arginine kinase activity and reacted to serum IgE from shrimp allergic patients and, furthermore, induced skin reactions in sensitized shrimp allergic patients.

IMPACTS OF HIGH THROUGHPUT GENOMIC AND PROTEOMIC TECHNOLOGIES

In recent years, proteomics has moved on from technical issues related to protein separation and protein identity to highly reproducible gel-based or gel-free systems as discussed earlier. This increased ability to separate proteins and to perform peptide sequence analysis by MS has meant that the volume of data that is produced and the rate of identification of proteins is orders of magnitude greater than only a few years ago (reviewed by Seidler et al. 2010). Directly related to the increased volume of data generated means that extracting the relevant information is no longer a simple matter of discussing a list of protein identities, and interpretation of the proteins and their function is central to any medium- to large-scale proteome study (Malik et al. 2010).

The number of expressed sequence tags (ESTs) related to salmonid fish is currently in the order of 800,000 sequences, representing mRNAs encoding about 30–40,000 different proteins, other commercial species including cod, sea bass, sea bream, and catfish amongst others are quickly catching up (Martin et al. 2008). These sequences can be used to help identify amino acid sequences generated during proteomics studies, which means that now the majority of proteins can be identified rather than the minority.

There are now a considerable number of fish species whose whole genome is completely sequenced; zebrafish, two species of puffer fish, and stickleback have their genomes sequenced. And recently the sea bass, a major aquacultured species has been almost fully sequenced (Kuhl et al. 2010). The technology and tools used for these projects have led the way to expand the sequence data and interpretation of sequences from commercially important species (Forne et al. 2010)—the data that can be directly used for proteomic studies. A good example of this is zebrafish where 5716 proteins were identified with an estimated false discovery rate of 1.34% (De Souza et al. 2009).

As more and more genes and protein sequences are deposited in databases, they are automatically annotated; the quality of these annotations is probably one of the greatest hurdles in fully interpreting the output of either a transcriptomic study or proteomic studies. Currently, annotations include the nucleotide sequence, protein sequence, tissue distribution abundance of mRNA, gene ontology (GO) (Gaudet et al. 2009), and Kyoto Encyclopedia of Genes and Genomes (KEGG) (Okuda et al. 2008) pathways. A high proportion of the genes in human, mice, yeast, among others, have many of their protein annotated to this extent; in the near future, this will be the case for fish as well. GO and KEGG are two databases that are used to assign function to a particular protein. For GO, these are three different assignments: (1) biological process, (2) molecular function, and (3) cellular component, the same terms being used across all species. This gives a vocabulary of definitions for a particular protein; in fact, there are now thousands of GO terms. When proteins are identified they can be matched to these different GO terms; for example, glycolysis or protein metabolism as very basic examples. If a data set produced contains tens or hundreds of proteins identities, it is possible to use GO to show the global changes occurring in that tissue; if many GO terms are related to the same biological process or function, this can help interpret the data as opposed to a simple list of proteins. If there are two samples to compare, statistical analysis of the GO terms can indicate if there are major differences in cellular activity between the two samples, or by using GO cellular component, it can be indicated what are the particular organelles that are being affected. The KEGG pathways (http://www.genome.jp/kegg/) are similar to biochemical maps with interactive links to other nucleotide, protein, and literature databases, as with GO these help interpret the output data when a large number of proteins have been identified. New pathway analysis programs are continually being developed including Reactome (Vastrik et al. 2007) that allows networks of proteins to be explored and integration of data with literature through a platform called iHOP (Hoffmann and Valencia 2004). Although there is a requirement for significant manual analysis of the data when working with fish species (including the model species), clearly as the users become more familiar with the available tools and the databases become more integrated, the knowledge gap between those working on species with well-annotated genomes and fish will decrease.

In future studies, data from other "omic" platforms should be combined, that is, incorporating transcriptomic data from microarray and deep sequencing and from metabolomic data. The complementary techniques, although performed often in different laboratories, do ask the same questions and one of the future steps will be to perform meta-analysis across these high throughput technologies.

REFERENCES

Ahmed FE. 2009. Sample preparation and fractionation for proteome analysis and cancer biomarker discovery by mass spectrometry. *Journal of Separation Science* 32: 771–798.

Ahmed N et al. 2003. An approach to remove albumin for the proteomic analysis of low abundance biomarkers in human serum. *Proteomics* 3: 1980–1987.

Ahmed N, Rice GE. 2005. Strategies for revealing lower abundance proteins in two-dimensional protein maps. *Journal of Chromatography B* 815: 39–50.

Andersen JS, Mann M. 2006. Organellar proteomics: turning inventories into insights. *EMBO Reports* 7: 874–879.

Anderson NL, Anderson NG. 1998. Proteome and proteomics: new technologies, new concepts, and new words. *Electrophoresis* 19: 1853–1861.

Babu GJ et al. 2004. Solubilization of membrane proteins for two-dimensional gel electrophoresis: identification of sarcoplasmic reticulum membrane proteins. *Analytical Biochemistry* 325: 121–125.

Bae SH et al. 2003. Strategies for the enrichment and identification of basic proteins in proteome projects. *Proteomics* 3: 569–579.

Balestrieri ML et al. 2008. Proteomics and cardiovascular disease: an update. *Current Medicinal Chemistry* 15: 555–572.

Barrett J et al. 2005. Analysing proteomic data. *International Journal for Parasitology* 35: 543–553.

Beretta L. 2009. Comparative analysis of the liver and plasma proteomes as a novel and powerful strategy for hepatocellular carcinoma biomarker discovery. *Cancer Letters* 286: 134–139.

Berkelman T, Stenstedt T. 1998. *2-D Electrophoresis Using Immobilized pH Gradients. Principles and Methods.* Amersham Biosciences, Uppsala.

Bodzon-Kulakowska A et al. 2007. Methods for samples preparation in proteomic research. *Journal of Chromatography B* 849: 1–31.

Bogyo M, Cravatt BF. 2007. Genomics and proteomics: from genes to function: advances in applications of chemical and systems biology. *Current Opinion in Chemical Biology* 11: 1–3.

Bossi A et al. 1994. Focusing of alkaline proteases (subtilisins) in pH 10–12 immobilized pH gradients. *Electrophoresis* 15: 1535–1540.

Brown PD et al. 1993. Cellular activation of the 72 kDa type IV procollagenase/TIMP-2 complex. *Kidney International* 43: 163–170.

Butt A et al. 2001. Chromatographic separations as a prelude to two-dimensional electrophoresis in proteomics analysis. *Proteomics* 1: 42–53.

Børresen T. 1992. Quality aspects of wild and reared fish. In: HH Huss et al. (eds.) *Quality Assurance in the Fish Industry*. Elsevier Science Publishers, Amsterdam, pp. 1–17.

Cahu CL et al. 1999. Protein hydrolysate vs. fish meal in compound diets for 10-day old sea bass *Dicentrarchus labrax* larvae. *Aquaculture* 171: 109–119.

Campana SE, Thorrold SR. 2001. Otoliths, increments, and elements: keys to a comprehensive understanding of fish populations? *Canadian Journal of Fisheries and Aquatic Sciences* 58: 30–38.

Campinho MA et al. 2007. Molecular, cellular and histological changes in skin from a larval to an adult phenotype during bony fish metamorphosis. *Cell and Tissue Research* 327: 267–284.

Campinho MA et al. 2006. Regulation of troponin T expression during muscle development in sea bream *Sparus auratus* Linnaeus: the potential role of thyroid hormones. *Journal of Experimental Biology* 209: 4751–4767.

Canas B et al. 2007. Trends in sample preparation for classical and second generation proteomics. *Journal of Chromatography A* 1153: 235–258.

Chelius D et al. 2003. Global protein identification and quantification technology using two-dimensional liquid chromatography nanospray mass spectrometry. *Analytical Chemistry* 75: 6658–6665.

Chevallet M et al. 1998. New zwitterionic detergents improve the analysis of membrane proteins by two-dimensional electrophoresis. *Electrophoresis* 19: 1901–1909.

Ciechanover A. 1994. The ubiquitin-proteasome proteolytic pathway. *Cell* 79: 13–21.

Clauser KR et al. 1999. Role of accurate measurement (+/− 10 ppm) in protein identification strategies employing MS or MS/MS and database searching. *Analytical Chemistry* 71: 2871–2882.

Coe JE, Vilhelmsson O. 2008. Two-dimensional polyacrylamide gel electrophoresis of Arctic charr liver proteins: setup of equipment and standardization of protocols. Technical Report No. TS08:08. The University of Akureyri. Akureyri, Iceland.

Cordwell SJ et al. 2000. Subproteomics based upon protein cellular location and relative solubilities in conjunction with composite two-dimensional electrophoresis gels. *Electrophoresis* 21: 1094–1103.

Corthals GL et al. 1997. Prefractionation of protein samples prior to two-dimensional electrophoresis. *Electrophoresis* 18: 317–323.

Courchesne PL, Patterson SD. 1999. Identification of proteins by matrix-assisted laser desorption/ionization mass spectrometry using peptide and fragment ion masses. In: AJ Link (ed.) *2-D Proteome Analysis Protocols*, Chapter 51. Humana Press, Totowa, NJ, pp. 487–511.

Cowie WP. 1968. Identification of fish species by thin slab polyacrylamide gel electrophoresis. *Journal of the Science of Food and Agriculture* 19: 226–229.

Craiu A et al. 1997. Two distinct proteolytic processes in the generation of a major histocompatibility complex class I-presented peptide. *Proceedings of the National Academy of Sciences of the United States of America* 94: 10850–10855.

De Francesco M et al. 2004. Effect of long-term feeding with a plant protein mixture based diet on growth and body/fillet quality traits of large rainbow trout (*Oncorhynchus mykiss*). *Aquaculture* 236: 413–429.

De Souza AG et al. 2009. Large-scale proteome profile of the zebrafish (*Danio rerio*) gill for physiological and biomarker discovery studies. *Zebrafish* 6: 229–238.

Dobly A et al. 2004. Efficiency of conversion of ingested proteins into growth; protein degradation assessed by 20S proteasome activity in rainbow trout, *Oncorhynchus mykiss*. *Comparative Biochemistry and Physiology A* 137: 75–85.

Dougherty DA et al. 1990. Evaluation of selected baking quality factors of hard red winter wheat flours by two-dimensional electrophoresis. *Cereal Chemistry* 67: 564–569.

Drabik A et al. 2007. Proteomics in neurosciences. *Mass Spectrometry Reviews* 26: 432–450.

Dreger M. 2003. Subcellular proteomics. *Mass Spectrometry Reviews* 22: 27–56.

Einarsdóttir IE et al. 2006. Thyroid and pituitary gland development from hatching through metamorphosis of a teleost flatfish, the Atlantic halibut. *Anatatomy and Embryology* 211: 47–60.

Einen O et al. 1999. Feed ration prior to slaughter - a potential tool for managing product quality of Atlantic salmon (*Salmo salar*). *Aquaculture* 178: 149–169.

Erdjument-Bromage H et al. 1999. Characterizing proteins from 2-DE gels by internal sequence analysis of peptide fragments. In: AJ Link (ed.) *2-D Proteome Analysis Protocols*, Chapter 49. Humana Press, Totowa, NJ, pp. 467–472.

Focant B et al. 2000. Expression of myofibrillar proteins and parvalbumin isoforms in white muscle of the developing turbot *Scophthalmus maximus* (Pisces, Pleuronectiformes). *Basic and Applied Myology* 10: 269–278.

Focant B et al. 1999. Muscle parvalbumin isoforms of *Clarias gariepinus, Heterobranchus longifilis* and *Chrysichthys auratus*: isolation, characterization, and expression during development. *Journal of Fish Biology* 54: 832–851.

Focant B et al. 2003. Expression of myofibrillar proteins and parvalbumin isoforms during the development of a flatfish, the common sole *Solea solea*: comparison with the turbot *Scophthalmus maximus*. *Comparative Biochemistry and Physiology B* 135: 493–502.

Foegeding EA et al. 1996. Characteristics of edible muscle tissues. In: OR Fennema (ed.) *Food Chemistry*. Marcel Dekker, New York, pp. 879–942.

Forne I et al. 2010. Fish proteome analysis: model organisms and non-sequenced species. *Proteomics* 10: 858–872.

Fortis F et al. 2008. A pI-based protein fractionation method using solid-state buffers. *Journal of Proteomics* 71: 379–389.

Galloway TF et al. 2006. Somite formation and expression of MyoD, myogenin and myosin in Atlantic halibut (*Hippoglossus hippoglossus* L.) embryos incubated at different temperatures: transient asymmetric expression of MyoD. *Journal of Experimental Biology* 209: 2432–2441.

Galloway TF et al. 1998. Effect of temperature on viability and axial muscle development in embryos and yolk sac larvae of the Northeast Atlantic cod (*Gadus morhua*). *Marine Biology* 132: 547–557.

Galloway TF et al. 1999. Muscle growth in yolk sac larvae of the Atlantic halibut as influenced by temperature in the egg and yolk sac stage. *Journal of Fish Biology* 55: 26–43.

Garcia-Canas V et al. 2010. Advances in Nutrigenomics research: novel and future analytical approaches to investigate the biological activity of natural compounds and food functions. *Journal of Pharmaceutical and Biomedical Analysis* 51: 290–304.

Gaudet P et al. 2009. The gene ontology's reference genome project: a unified framework for functional annotation across species. *PLOS Computational Biology* 5: e1000431.

Görg A et al. 1987. Elimination of point streaking on silver stained two-dimensional gels by addition of iodoacetamide to the equilibration buffer. *Electrophoresis* 8: 122–124.

Görg A et al. 1988. The current state of two-dimensional electrophoresis with immobilized pH gradients. *Electrophoresis* 9: 531–546.

Görg A et al. 1997. Very alkaline immobilized pH gradients for two-dimensional electrophoresis of ribosomal and nuclear proteins. *Electrophoresis* 18: 328–337.

Görg A et al. 1999. Recent developments in two-dimensional gel electrophoresis with immobilized pH gradients: wide pH gradients up to pH 12, longer separation distances and simplified procedures. *Electrophoresis* 20: 712–717.

Görg A et al. 2000. The current state of two-dimensional elelctrophoresis with immobilized pH gradients. *Electrophoresis* 21: 1037–1053.

Görg A et al. 2004. Current two-dimensional electrophoresis technology for proteomics. *Proteomics* 19: 3665–3685.

Govoni JJ et al. 1986. The physiology of digestion in fish larvae. *Environmental Biology of Fishes* 16: 59–77.

Guðmundsdóttir Á et al. 1993. Isolation and characterization of cDNAs from Atlantic cod encoding two different forms of trypsinogen. *European Journal of Biochemistry* 217: 1091–1097.

Guðmundsdóttir Á, Sveinsdóttir H. 2006. Rýnt í próteinmengi lirfa Atlantshafsþorsks (*Gadus morhua*). In: GG Haraldsson (ed.) *Vísindin heilla: afmælisrit til heiðurs Sigmundi Guðbjarnasyni 75 ára*. Háskólaútgáfan, Reykjavík, pp. 425–440.

Gustafson JS et al. 2004. Statistical exploration of variation in quantitative two-dimensional gel electrophoresis data. *Proteomics* 4: 3791–3799.

Hall TH et al. 2003. Temperature and the expression of seven muscle-specific protein genes during embryogenesis in the Atlantic cod *Gadus morhua* L. *Journal of Experimental Biology* 206: 3187–3200.

Henningsen R et al. 2002. Application of zwitterionic detergents to the solubilization of integral membrane proteins for two-dimensional gel electrophoresis and mass spectrometry. *Proteomics* 2: 1479–1488.

Herbert B. 1999. Advances in protein solubilization for two-dimensional electrophoresis. *Electrophoresis* 20: 660–663.

Hershko A, Ciechanover A. 1986. The ubiquitin pathway for the degradation of intracellular proteins. *Progress in Nucleic Acid Research and Molecular Biology* 33: 19–56.

Hoffmann R, Valencia A. 2004. A gene network for navigating the literature. *Nature Genetics* 36: 664.

Hogstrand C et al. 2002. Application of genomics and proteomics for study of the integrated response to zinc exposure in a non-model fish species, the rainbow trout. *Comparative Biochemistry and Physiology B* 133: 523–535.

Hume A, Mackie I. 1980. The use of electrophoresis of the water-soluble muscle proteins in the quantitative analysis of the species components of a fish mince mixture. In: JJ Connell (ed.) *Advances in Fish Science and Technology*. Fishing News Books Ltd., Aberdeen, pp. 451–456.

Huriaux F et al. 2003. Expression of myofibrillar proteins and parvalbumin isoforms in white muscle of dorada during development. *Journal of Fish Biology* 62: 774–792.

Huriaux F et al. 1996. Parvalbumin isotypes in white muscle from three teleost fish: characterization and their expression during development. *Comparative Biochemistry and Physiology B* 113: 475–484.

Huriaux F et al. 1999. Myofibrillar proteins in white muscle of the developing catfish *Heterobranchus longifilis* (Siluriforms, Claridae). *Fish Physiology and Biochemistry* 21: 287–301.

Huss HH. 1995. Quality and quality changes in fresh fish. FAO Fisheries Technical Paper No. 348. FAO. Rome.

Ikonomou G et al. 2009. Proteomic methodologies and their application in colorectal cancer research. *Critical Reviews in Clinical Laboratory Sciences* 46: 319–342.

Ingólfsdóttir S et al. 1998. Seasonal variations in physicochemical and textural properties of North Atlantic cod (*Gadus morhua*) mince. *Journal of Aquatic Food Product Technology* 7: 39–61.

Issaq HJ et al. 2002. Methods for fractionation, separation and profiling of proteins and peptides. *Electrophoresis* 23: 3048–3061.

Issaq HJ, Veenstra TD. 2008. Two-dimensional polyacrylamide gel electrophoresis (2D-PAGE): advances and perspectives. *Biotechniques* 44: 697–698.

Jensen O. 2004. Modification-specific proteomics: characterization of post-translational modifications by mass spectrometry. *Current Opinion in Chemical Biology* 8: 33–41.

Jorrin-Novo JV et al. 2009. Plant proteomics update (2007–2008): Second-generation proteomic techniques, an appropriate experimental design, and data analysis to fulfill MIAPE standards, increase plant proteome coverage and expand biological knowledge. *Journal of Proteomics* 72: 285–314.

Journet A et al. 2000. Towards a human repertoire of monocytic lysosomal proteins. *Electrophoresis* 21: 3411–3419.

Journet A et al. 2002. Proteomic analysis of human lysosomes: application to monocytic and breast cancer cells. *Proteomics* 2: 1026–1040.

Kaltschmidt E, Wittmann H-G. 1970. Ribosomal proteins. VII. Two-dimensional polyacrylamide gel electrophoresis for fingerprinting of ribosomal proteins. *Analytical Biochemistry* 36: 401–412.

Kamo M, Tsugita A. 1999. N-terminal amino acid sequencing of 2-DE spots. In: AJ Link (ed.) *2-D Proteome Analysis Protocols*, Chapter 48. Humana Press, Totowa, NJ, pp. 461–466.

Karp NA et al. 2005. Application of partial least squared discriminant analysis to two-dimensional difference gel studies in expression proteomics. *Proteomics* 5: 81–90.

Kjaersgard IVH, Jessen F. 2003. Proteome analysis elucidating postmortern changes in cod (*Gadus morhua*) muscle proteins. *Journal of Agricultural and Food Chemistry* 51: 3985–3991.

Kjaersgard IVH, Jessen F. 2004. Two-dimensional gel electrophoresis detection of protein oxidation in fresh and tainted rainbow trout muscle. *Journal of Agricultural and Food Chemistry* 52: 7101–7107.

Kjaersgard IVH et al. 2006a. Identification of carbonylated protein in frozen rainbow trout (*Oncorhynchus mykiss*) fillets and development of protein oxidation during frozen storage. *Journal of Agricultural and Food Chemistry* 54: 9437–9446.

Kjaersgard IVH et al. 2006b. Changes in cod muscle proteins during frozen storage revealed by proteome analysis and multivariate data analysis. *Proteomics* 6: 1606–1618.

Kraj A, Silberring J. 2008. *Introduction to Proteomics* John Wiley & Sons, Hoboken, NJ.

Kristensen L et al. 2002. Dietary-induced changes of muscle growth rate in pigs: effects on in vivo and postmortem muscle proteolysis and meat quality. *Journal of Animal Science* 80: 2862–2871.

Kuhl H et al. 2010. The European sea bass *Dicentrarchus labrax* genome puzzle: comparative BAC-mapping and low coverage shotgun sequencing. *BMC Genomics* 11: 68.

Ladrat C et al. 2000. Neutral calcium-activated proteases from European sea bass (*Dicentrachus labrax* L.) muscle: polymorphism and biochemical studies. *Comparative Biochemistry and Physiology B* 125: 83–95.

Laemmli UK. 1970. Cleavage of structural proteins during the assembly of the head of bacteriophage T4. *Nature* 227: 680–685.

Latterich M et al. 2008. Proteomics: new technologies and clinical applications. *European Journal of Cancer* 44: 2737–2741.

Lee BH, Nagamune T. 2004. Protein microarrays and their applications. *Biotechnology and Bioprocess Engineering* 9: 69–75.

Lee K, Pi K. 2009. Proteomic profiling combining solution-phase isoelectric fractionation with two-dimensional gel electrophoresis using narrow-pH-range immobilized pH gradient gels with slightly overlapping pH ranges. *Analytical and Bioanalytical Chemistry* 396: 535–539.

Lehrer SB et al. 2003. Seafood allergy and allergens: a review. *Marine Biotechnology* 5: 339–348.

Lei MG, Reeck GR. 1987. Two dimensional electrophoretic analysis of isolated soybean protein bodies and of the glycosylation of soybean proteins. *Journal of Agricultural and Food Chemistry* 35: 296–300.

Link AJ. 1999. *2-D Proteome Analysis Protocols*. Humana Press, Totowa, NJ.

Link V et al. 2006. Proteomics of early zebrafish embryos. *BMC Developmental Biology* 6: 1–9.

Locke VL et al. 2002. Gradiflow as a prefractionation tool for two-dimensional electrophoresis. *Proteomics* 2: 1254–1260.

Lopez JL et al. 2002. Application of proteomics for fast identification of species-specific peptides from marine species. *Proteomics* 2: 1658–1665.

Lopez MF et al. 2000. High-throughput profiling of the mitochondrial proteome using affinity fractionation and automation. *Electrophoresis* 21: 3427–3440.

Love RM. 1980. Biological factors affecting processing and utilization. In: JJ Connell (ed.) *Advances in Fish Science and Technology*. Fishing News Books, Aberdeen, pp. 130–138.

Love RM. 1992. Biochemical dynamics and the quality of fresh and frozen fish. In: GM Hall (ed.) *Fish Processing Technology*. Blackie Academic and Professional, London, pp. 1–26.

Love RM et al. 1974. The texture of cod muscle. *Journal of Texture Studies* 5: 201–212.

López JL. 2007. Two-dimensional electrophoresis in proteome expression analysis. *Journal of Chromatography B* 849: 190–202.

Lødemel JB, Olsen RL. 2003. Gelatinolytic activities in muscle of Atlantic cod (*Gadus morhua*), spotted wolffish (*Anarhichas minor*) and Atlantic salmon (*Salmo salar*). *Journal of the Science of Food and Agriculture* 83: 1031–1036.

Mackie I. 1980. A review of some recent applications of electrophoresis and isoelectric focusing in the identification of species of fish in fish and fish products. In: JJ Connell (ed.) *Advances in Fish Science and Technology*. Fishing News Books Ltd., Aberdeen, pp. 444–450.

Mackie IM. 1969. Identification of fish species by a modified polyacrylamide disc electrophoresis technique. *Journal of the Association of Public Analysts* 5: 83–87.

Mackie IM. 1972. Some improvements in the polyacrylamide disc electrophoretic method of identifying species of cooked fish. *Journal of the Association of Public Analysts* 8: 18–20.

Mackie IM et al. 1999. Challenges in the identification of species of canned fish. *Trends in Food Science and Technology* 10: 9–14.

Mackie IM, Taylor T. 1972. Identification of species of heat-sterilized canned fish by polyacrylamide disc electrophoresis. *Analyst* 97: 609–611.

Malik R et al. 2010. From proteome lists to biological impact- tools and strategies for the analysis of large MS data sets. *Proteomics* 10: 1270–1283.

Malmstrom J et al. 2007. Advances in proteomic workflows for systems biology. *Current Opinion in Biotechnology* 18: 378–384.

Mamone G et al. 2009. Analysis of food proteins and peptides by mass spectrometry-based techniques. *Journal of Chromatography A* 1216: 7130–7142.

Margolis J, Kenrick KG. 1969. Two-dimensional resolution of plasma proteins by combination of poly-acrylamide disc and gradient gel electrophoresis. *Nature* 221: 1056–1057.

Martin SA et al. 2002. Ubiquitin-proteasome-dependent proteolysis in rainbow trout (*Oncorhynchus mykiss*): effect of food deprivation. *Pflugers Archive* 445: 257–266.

Martin SAM et al. 2001. Proteome analysis of rainbow trout (*Oncorhynchus mykiss*) liver proteins during short-term starvation. *Fish Physiology and Biochemistry* 24: 259–270.

Martin SAM et al. 2003a. Rainbow trout liver proteome - dietary manipulation and protein metabolism. In: WB Souffrant, CC Metges (eds.) *Progress in Research on Energy and Protein Metabolism*. Wageningen Academic Publishers, Wageningen, pp. 57–60.

Martin SAM et al. 2003b. Proteomic sensitivity to dietary manipulations in rainbow trout. *Biochimica et Biophysica Acta* 1651: 17–29.

Martin SAM et al. 2008. Genomic tools for examining immune gene function in salmonid fish. *Reviews in Fisheries Science* 16: 112–118.

Martinez I et al. 2003. Destructive and non-destructive analytical techniques for authentication and composition analyses of foodstuffs. *Trends in Food Science and Technology* 14: 489–498.

Martinez I, Jakobsen Friis T. 2004. Application of proteome analysis to seafood authentication. *Proteomics* 4: 347–354.

Martinez I et al. 2001a. *Post mortem* muscle protein degradation during ice-storage of Arctic (*Pandalus borealis*) and tropical (*Penaeus japonicus* and *Penaeus monodon*) shrimps: a comparative electrophoretic and immunological study. *Journal of the Science of Food and Agriculture* 81: 1199–1208.

Martinez I et al. 2001b. Requirements for the application of protein sodium dedecyl sulfate-polyacrylamide gel electrophoresis and randomly amplified polymorphic DNA analyses to product speciation. *Electrophoresis* 22: 1526–1533.

Martinez I et al. 1990. Electrophoretic study of myosin isoforms in white muscles of some teleost fishes. *Comparative Biochemistry and Physiology B* 96: 221–227.

Mazzeo MF et al. 2008. Fish authentication by MALDI-TOF mass spectrometry. *Journal of Agricultural and Food Chemistry* 56: 11071–11076.

Millea KM, Krull IS. 2003. Subproteomics in analytical chemistry: chromatographic fractionation techniques in the characterization of proteins and peptides. *Journal of Liquid Chromatography & Related Technologies* 26: 2195–2224.

Mortimore GE et al. 1989. Mechanism and regulation of protein degradation in liver. *Diabetes/Metababolism Reviews* 5: 49–70.

Morzel M et al. 2000. Use of two-dimensional electrophoresis to evaluate proteolysis in salmon (*Salmo salar*) muscle as affected by a lactic fermentation. *Journal of Agricultural and Food Chemistry* 48: 239–244.

O'Connell CP. 1981. Development of organ systems in the northern anchovy *Engraulis mordax* and other teleosts. *American Zoologist* 21: 429–446.

O'Farrell PH. 1975. High resolution two-dimensional electrophoresis of proteins. *Journal of Biological Chemistry* 250: 4007–4021.

Oda Y et al. 2001. Enrichment analysis of phosphorylated proteins as a tool for probing the phosphoproteome. *Nature Biotechnology* 19: 379–382.

Ogata H et al. 1998. Proteolytic degradation of myofibrillar components by carp cathepsin L. *Journal of the Science of Food and Agriculture* 76: 499–504.

Okuda S et al. 2008. KEGG Atlas mapping for global analysis of metabolic pathways. *Nucleic Acids Research* 36: W423–W426.

Okumura Y et al. 1997. Proteolytic activation of the precursor of membrane type 1 matrix metalloproteinase by human plasmin. A possible cell surface activator. *FEBS Letters* 402: 181–184.

Ortea I et al. 2009a. Closely related shrimp species identification by MALDI-ToF mass spectrometry. *Journal of Aquatic Food Product Technology* 18: 146–155.

Ortea I et al. 2009b. Arginine kinase peptide mass fingerprinting as a proteomic approach for species identification and taxonomic analysis of commercially relevant shrimp species. *Journal of Agricultural and Food Chemistry* 57: 5665–5672.

Osse JWM, van den Boogaart JGM. 1995. Fish larvae, development, allometric growth, and the aquatic environment. *ICES Marine Science Symposia* 201: 21–34.

Østergaard O et al. 2002. Initial proteome analysis of mature barley seeds and malt. *Proteomics* 2: 733–739.

Papa I et al. 1996. *Post mortem* release of fish white muscle a-actinin as a marker of disorganisation. *Journal of the Science of Food and Agriculture* 72: 63–70.

Parisi G et al. 2004. Effect of total replacement of dietary fish meal by plant protein sources on early *post mortem* changes in the biochemical and physical parameters of rainbow trout. *Veterinary Research Communications* 28: 237–240.

Parrington J, Coward K. 2002. Use of emerging genomic and proteomic technologies in fish physiology. *Aquatic Living Resources* 15: 193–196.

Pascoal A et al. 2008. Identification of shrimp species in raw and processed food products by means of a polymerase chain reaction-restriction fragment length polymorphism method targeted to cytochrome b mitochondrial sequences. *Electrophoresis* 29: 3220–3228.

Patton WE. 2002. Detection technologies in proteome analysis. *Journal of Chromatography B* 771: 3–31.

Paz M et al. 2006. Proteome profile changes during mouse testis development. *Comparative Biochemistry and Physiology D* 1: 404–415.

Peng J et al. 2003. A proteomics approach to understanding protein ubiquitination. *Nature Biotechnology* 21: 921–926.

Perkins DN et al. 1999. Probability-based protein identification by searching sequence databases using mass spectrometry data. *Electrophoresis* 20: 3551–3567.

Pérez-Borla O et al. 2002. Proteolytic activity of muscle in pre- and post-spawning hake (*Merluccius hubbsi* Marini) after frozen storage. *Lebensmittel-Wissenschaft und –Technologie* 35: 325–330.

Pieper R et al. 2003. Multi-component immunoaffinity subtraction chromatography: an innovative step towards a comprehensive survey of the human plasma proteome. *Proteomics* 3: 422–432.

Piñeiro C et al. 1998. Two-dimensional electrophoretic study of the water-soluble protein fraction in white muscle of gadoid fish species. *Journal of Agricultural and Food Chemistry* 46: 3991–3997.

Piñeiro C et al. 1999. The use of two-dimensional electrophoresis for the identification of commercial flat fish species. *Zeitschrift für Lebensmitteluntersuchung und –Forschung* 208: 342–348.

Piñeiro C et al. 2003. Proteomics as a tool for the investigation of seafood and other marine products. *Journal of Proteome Research* 2: 127–135.

Piñeiro C et al. 2001. Characterization and partial sequencing of species-specific sarcoplasmic polypeptides from commercial hake species by mass spectrometry following two-dimensional electrophoresis. *Electrophoresis* 22: 1545–1552.

Premsler T et al. 2009. Recent advances in yeast organelle and membrane proteomics. *Proteomics* 9: 4731–4743.

Righetti PG et al. 2005a. Prefractionation techniques in proteome analysis: the mining tools of the third millennium. *Electrophoresis* 26: 297–319.

Righetti PG et al. 2005b. How to bring the "unseen" proteome to the limelight via electrophoretic pre-fractionation techniques. *Bioscience Reports* 25: 3–17.

Rock KL et al. 1994. Inhibitors of the proteasome block the degradation of most cell proteins and the generation of peptides presented on MHC class I molecules. *Cell* 78: 761–771.

Rothemund DL et al. 2003. Depletion of the highly abundant protein albumin from human plasma using the Gradiflow. *Proteomics* 3: 279–287.

Seidler J et al. 2010. *De novo* sequencing of peptides by MS/MS. *Proteomics* 10: 634–649.

Simpson R. 2003. *Proteins and Proteomics: A Laboratory Manual*. Cold Spring Harbor Laboratory Press, Cold Spring Harbor, New York.

Skiftesvik AB. 1992. Changes in behaviour at onset of exogenous feeding in marine fish larvae. *Canadian Journal of Fisheries and Aquatic Sciences* 49: 1570–1572.

Smith MPW et al. 2009. Application of proteomic analysis to the study of renal diseases. *Nature Reviews Nephrology* 5: 701–712.

Sotelo CG et al. 1993. Fish species identification in seafood products. *Trends in Food Science and Technology* 4: 395–401.

Sveinsdóttir H, Gudmundsdóttir A. 2008. Proteome profile comparison of two differently fed groups of Atlantic cod (*Gadus morhua*) larvae. *Aquaculture Nutrition* 16: 662–670.

Sveinsdóttir H et al. 2008. Proteome analysis of abundant proteins in two age groups of early Atlantic cod (*Gadus morhua*) larvae. *Comparative Biochemistry and Physiology D* 3: 243–250.

Sveinsdóttir H et al. 2009. Differential protein expression in early Atlantic cod larvae (*Gadus morhua*) in response to treatment with probiotic bacteria. *Comparative Biochemistry and Physiology D* 4: 249–254.

Tay TL et al. 2006. Proteomic analysis of protein profiles during early development of the zebrafish, *Danio rerio*. *Proteomics* 6: 3176–3188.

Thorarinsdottir KA et al. 2002. Changes in myofibrillar proteins during processing of salted cod (*Gadus morhua*) as determined by electrophoresis and differential scanning calorimetry. *Food Chemistry* 77: 377–385.

Tinh NTN et al. 2008. A review of the functionality of probiotics in the larviculture food chain. *Marine Biotechnology* 10: 1–12.

Tsuyuki H et al. 1966. Comparative electropherograms of *Coregonis clupeoformis, Salvelinus namaycush, S. alpinus, S. malma* and *S. fontinalis* from the family Salmonidae. *Journal of the Fisheries Research Board of Canada* 23: 1599–1606.

Tyers M, Mann M. 2003. From genomics to proteomics. *Nature* 422: 193–197.

Vaneenaeme C et al. 1994. Postmortem proteases activity in relation to muscle protein-turnover in Belgian blue bulls with different growth-rates. *Sciences des Aliments* 14: 475–483.

Vastrik I et al. 2007. Reactome: a knowledge base of biologic pathways and processes. *Genome Biology* 8: R39.

Verrez-Bagnis V et al. 1999. Desmin degradation in postmortem fish muscle. *Journal of Food Science* 64: 240–242.

Vilhelmsson O, Miller KJ. 2002. Synthesis of pyruvate dehydrogenase in *Staphylococcus aureus* is stimulated by osmotic stress. *Applied and Environmental Microbiology* 68: 2353–2358.

Vilhelmsson OT et al. 2004. Dietary plant protein substitution affects hepatic metabolism in rainbow trout (*Oncorhynchus mykiss*). *British Journal of Nutrition* 92: 71–80.

von Horsten HH. 2006. Current strategies and techniques for the analysis of low-abundance proteins and peptides from complex samples. *Current Analytical Chemistry* 2: 129–138.

Walker JM. 2005. *The Proteomics Protocols Handbook*. Humana Press, Totowa, NJ.

Walsh BJ, Herbert BR. 1999. Casting and running vertical slab-gel electrophoresis for 2D-PAGE. In: AJ Link (ed.) *2-D Proteome Analysis Protocols*. Humana Press, Totowa, NJ, pp. 245–253.

Wang J et al. 2006. Proteomics and its role in nutrition research. *Journal of Nutrition* 136: 1759–1762.

Wang M, Lakatta EG. 2002. Altered regulation of matrix metalloproteinase-2 in aortic remodeling during aging. *Hypertension* 39: 865–873.

Westerfield M. 2000. *The Zebrafish Book. A Guide for the Laboratory Use of Zebrafish (Danio rerio)*, 4th edn. University of Oregon Press, Eugene, OR.

Westermeier R, Naven T. 2002. *Proteomics in Practice*. Wiley-VCH, Weinheim.

Wheelock AM, Goto S. 2006. Effects of post-electrophoretic analysis on variance in gel-based proteomics. *Expert Review of Proteomics* 3: 129–142.

Wilm M. 2009. Quantitative proteomics in biological research. *Proteomics* 9: 4590–4605.

Wilm M et al. 1996. Femtomole sequencing of proteins from polyacrylamide gels by nano-electrospray mass spectrometry. *Nature* 379: 466–469.

Woessner JF. 1991. Matrix metalloproteases and their inhibitors in connective-tissue remodelling. *FASEB Journal* 5: 2145–2154.

Woods RK et al. 2002. Prevalence of food allergies in young adults and their relationship to asthma, nasal allergies, and eczema. *Annals of Allergy, Asthma & Immunology* 88: 183–189.

Yates JR. 1998. Mass spectrometry and the age of the proteome. *Journal of Mass Spectrometry* 33: 1–19.

Yates JR. 2004. Mass spectral analysis in proteomics. *Annual Review of Biophysics and Biomolecular Structure* 33: 297–316.

Yates JR et al. 2009. Proteomics by mass spectrometry: approaches, advances, and applications. *Annual Review of Biomedical Engineering* 11: 49–79.

Yu C-J et al. 2003a. Proteomics and immunological analysis of a novel shrimp allergen, pen m 2. *Journal of Immunology* 170: 445–453.

Yu YL et al. 2003b. Proteomic studies of macrophage-derived foam cell from human U937 cell line using two-dimensional gel electrophoresis and tandem mass spectrometry. *Journal of Cardiovascular Pharmacology* 42: 782–789.

Zeece MG et al. 1989. High-resolution two-dimensional electrophoresis of bovine caseins. *Journal Of Agricultural and Food Chemistry* 37: 378–383.

Part 4
Milk

23
Dairy Products

Terri D. Boylston

Abstract: This chapter provides an overview of the biochemistry and processing of cultured dairy products. The role of lactic acid bacteria in developing cultured dairy products and coagulation of the milk proteins through either enzyme addition or acid production is emphasized. The biochemistry involved in the processing of natural and processed cheeses, butter, buttermilk, sour cream, yogurt, and acidophilus milk is discussed.

INTRODUCTION

Cultured dairy products were originally developed as a means to preserve milk through the production of lactic acid. However, these products are also recognized because of their desirable sensory characteristics and nutritional value. The term, cultured dairy products, is currently used to indicate that these products are prepared using lactic acid bacteria starter cultures and controlled fermentation. The lactic acid bacteria utilize the nutrients in the milk to support their growth. The production of lactic acid reduces the pH of these products to inhibit the growth of many pathogenic and spoilage microorganisms. In addition, the fermentation process develops a wide range of dairy products with a diversity of flavor and textural attributes, including cheese, yogurt, buttermilk, butter, acidophilus milk, and sour cream.

Chapters 24 and 25 provide detailed information on the biochemistry and processing of milk and milk products. This chapter will focus on the biochemistry and processing involved in the production of cultured dairy products. This chapter will include a general discussion of the biochemistry and processing of cultured dairy products, followed by a discussion of the processing of the individual dairy products.

BIOCHEMISTRY OF CULTURED DAIRY PRODUCTS

COMPOSITION OF MILK

The quality of the cultured dairy products is influenced by the composition and quality of the raw milk. Species, breed, nutritional status, health, and stage of lactation of the cow can have an impact on the fat, protein, and calcium content and overall composition and quality of the milk. On the average, cows' milk consists of 3.7% fat, 3.4% protein, 4.8% lactose, 0.7% ash, and has a pH of 6.6 (Fox et al. 2000). Good microbiological

Food Biochemistry and Food Processing, Second Edition. Edited by Benjamin K. Simpson, Leo M.L. Nollet, Fidel Toldrá, Soottawat Benjakul, Gopinadhan Paliyath and Y.H. Hui.

quality of the raw milk is also essential to the production of safe, high-quality cultured dairy products.

The predominant sugar in milk is lactose, a disaccharide of glucose and galactose. The fermentation of lactose by lactic acid bacteria in cultured dairy products provides the flavor and textural attributes that are desirable in cultured dairy products.

The fat is present in the milk in the form of fat globules, which are surrounded by a polar milk fat globule membrane (MFGM). Triacylglycerols are the predominant lipid fraction in milk, accounting for 98% of the total lipids. Diacylglycerols, monoacylglycerols, fatty acids, phospholipids, and sterols account for the remaining lipid fraction. The phospholipids are integral components of the MFGM. Approximately 65% of the fatty acids in milk fat are saturated, including 26% palmitic acid and 15% stearic acid. A significant amount of short- and middle-chain fatty acids, including 3.3% butyric acid are present. These fatty acids and the breakdown products of these fatty acids are important contributors to the flavor of many cultured dairy products.

Two major classes of milk proteins are caseins and whey proteins. The caseins, which make up 80% of the total protein in cow milk, are insoluble at a pH of 4.6, but are stable to heating. The whey proteins remain soluble at pH 4.6 and are heat sensitive.

The casein micelles exist in milk as a colloidal dispersion, with a diameter ranging from 40 to 300 nm and containing approximately 10,000 casein molecules. The principal casein proteins, α_{s1}, α_{s2}, β, and κ, present in the ratio 40:10:35:12, vary in the number of phosphate residues, calcium-sensitivity, and hydrophobicity. Within the casein micelle, the more hydrophobic proteins, such as β-casein, are located on the interior of the micelle, while the more hydrophilic proteins, such as κ-casein, are located on the surface of the micelle. The carboxyl end of κ-casein is dominated by glutamic acid residues and glycoside groups. These hydrophilic carboxyl-ends are represented as "hairy" regions in the model of the casein micelle and promote the stability of the casein micelle in solution (Walstra and Jenness 1984). Calcium phosphate further facilitates the association of individual calcium-sensitive casein proteins (α_{s1}-, α_{s2}-, and β-casein), within the casein micelle. Hydrogen bonds and hydrophobic interactions also play a critical role in stabilizing the casein micelle. Processing treatments applied during the formation of cultured dairy products, such as the addition of acid or enzymes destabilize the casein micelle causing the casein proteins to precipitate (Lucey 2002).

The whey proteins consist of four major proteins, β-lactoglobulin (50%), α-lactalbumin (20%), blood serum albumin (10%), and immunoglobulins (10%). These proteins have a significant number of cysteine and cysteine residues and are able to form disulfide linkages with other proteins following heat treatment.

Fresh cow's milk is characterized as having a distinctive subtle flavor. Classes of volatile flavor compounds that have been shown to have the greatest impact on milk flavor include nitrogen heterocyclics, linolenic acid oxidation products, γ-lactones, phenolics and phytol derivatives; many of these compounds are found in foods of plant origin. Differences in the milk flavor from cows fed different diets have been attributed to concentration differences of these flavor compounds rather than the presence of different compounds (Friedrich and Acree 1998, Bendall 2001). Although these flavor compounds are not significant contributors to the characteristic flavors of cultured dairy products, they do contribute to the background flavors of these products (Urbach 1995).

Lactic Acid Bacteria

The lactic acid bacteria used in the development of cultured dairy products include *Streptococcus*, *Lactococcus*, *Leuconostoc*, and *Lactobacillus* genera. These bacteria are gram-positive bacteria and belong to either the *Streptococcaceae* or *Lactobacillaceae* families, depending on the morphology of the bacteria as cocci or rods, respectively. These bacteria also differ in their optimal temperature for growth, with 20–30°C the optimal temperature for mesophilic bacteria and 35–45°C the optimal temperature for thermophilic bacteria. Although the lactic acid bacteria are quite diverse in growth requirements, morphology, and physiology, they all have the ability to metabolize lactose to lactic acid and reduce the pH of the milk to produce specific cultured dairy products. The heat treatment the cultured dairy products receive following inoculation is one of the factors that influences the selection of lactic acid bacteria for specific cultured dairy products. Table 23.1 summarizes the growth characteristics of common lactic acid bacteria.

KEY PROCESSING STEPS IN CULTURED DAIRY PRODUCTS

Many of the processing steps important in the production of cultured dairy products are not unique to a specific product. Therefore, the following discussion will provide an overview of the key processing steps that are used in the production of several cultured dairy products. Specific processing treatments and concerns will be highlighted within the discussion of the processing of the specific cultured dairy products.

Lactic Acid Fermentation

Lactic acid bacteria use the lactose in the milk to produce lactic acid and other important flavor compounds in the cultured dairy products. Many of these bacteria have lactase activity and hydrolyze lactose to its monosaccharide units, glucose and galactose, prior to further metabolism. The hydrolysis of lactose in most cultured dairy products is significant for individuals who are lactose intolerant, allowing them to consume dairy products without the undesirable effects of the inability to hydrolyze lactose.

The homofermentative lactic acid bacteria produce lactic acid from lactose according to the reaction

$$\text{Lactose} + 4\text{ADP} + 4\text{H}_3\text{PO}_4 \rightarrow 4\text{Lactic Acid} + 4\text{ATP} + 3\text{H}_2\text{O}$$

The glucose and galactose molecules are metabolized through the glycolytic and tagatose pathways, respectively. Key steps in

Table 23.1. Characteristics of Lactic Acid Bacteria Used in Fermented Dairy Products

	Morphology	Lactose Fermentation	Temperature for Growth (°C) Minimum	Optimal	Maximum	Lactic Acid Production[1]	Primary Metabolic Products
			Lactococcus lactis ssp.				
Lc. lactis ssp. *lactis*	Cocci	Homofermentative	8–10	28–32	40	~0.9	Lactate
Lc. lactis ssp. *cremoris*	Cocci	Homofermentative	8–10	22	37–39	~0.9	Lactate
			Leuconostoc				
Ln. mesenteroides ssp. *cremoris*	Cocci	Heterofermentative	4–10	20–25	~37	Trace	Lactate, diacetyl, carbon dioxide
Ln. lactis	Cocci	Heterofermentative	4–10	20–25	~37	~0.8	Lactate, diacetyl, carbon dioxide
			Streptococcus				
S. thermophilus	Cocci	Homofermentative	20	40	50	~0.9	Lactate, acetaldehyde
			Lactobacillus				
Lb. delbrueckii ssp. *bulgaricus*	Rods	Homofermentative	22	40–45	52	~2.5	Lactate, acetaldehyde
Lb. helveticus	Rods	Homofermentative	20–22	42	54	~2.5	Lactate
Lb. delbrueckii ssp. *lactis*	Rods	Homofermentative	18	40	50	~1.2	Lactate
Lb. acidophilus	Rods	Homofermentative	20–22	37	45—48	~1.0	

Sources: Walstra et al. 1999, Stanley 1998.
[1]Percentage in milk in 24 h at optimum temperature.

the formation of lactic acid from the monosaccharides include the hydrolysis of the hexose diphosphates to glyceraldehyde-3-phosphate by aldolases, the formation of pyruvate from phosphoenolpyruvate by pyruvate kinase, and the reduction of pyruvate to lactate by lactate dehydrogenase. Heterofermentative lactic acid bacteria metabolize lactose to produce carbon dioxide, acetic acid, and ethanol, in addition to lactic acid according to the reaction

$$\text{Lactose} + 2H_3PO_4 + 2\text{ADP} \rightarrow 2\text{Lactic Acid} + 2\text{Ethanol} + 2CO_2 + 2\text{ATP} + H_2O$$

The heterofermentative lactic acid bacteria lack aldolases, therefore, the sugars are metabolized through 6-phosphogluconate pathway rather than the glycolytic pathway. In this pathway, glucose-6-phosphate is oxidized to 6-phosphogluconate, which is decarboxylated to a pentose-5-phosphate and carbon dioxide by phosphoketolase. The pentose-5-phosphate is converted to glyceraldehyde-3-phosphate, which enters the glycolytic pathway to form lactic acid and acetyl-phosphate, which is metabolized to acetaldehyde. The acetaldehyde is reduced to ethanol by alcohol dehydrogenase. However, several bacteria, including *Lactococcus lactis* ssp. *lactis* biovar. *diacetylactis*, *Streptococcus thermophilus*, and *Lactobacillus delbrueckii* ssp. *bulgaricus* lack the dehydrogenase and accumulate acetaldehyde.

The production of acid by the lactic acid bacteria has a significant impact on the safety and quality of the cultured dairy products. The reduction in pH increases the shelf life and safety of the cultured dairy products through the inhibition of spoilage and pathogenic microorganisms. Acid production by these lactic acid bacteria is critical for the precipitation of the casein proteins in the formation of several cultured dairy products, such as yogurt, sour cream, and unripened cheeses. These bacteria may also contribute to the degradation of proteins and lipids through proteolytic and lipolytic reactions to further develop the unique texture and flavor characteristics of the cultured dairy products. These reactions are especially important in ripened cheeses. The selection of the appropriate type and level of starter cultures is imperative for the development of the appropriate flavor characteristics.

Metabolism of citrate by *Lactococcus*, *Leuconostoc*, and other lactic acid bacteria results in the formation of diacetyl

(2,3-butanedione), acetoin, and 2,3-butanediol, which are key volatile flavor compounds in cultured dairy products. Diacetyl is formed through the chemical oxidation of α-acetolactate. Diacetyl reductase (also referred to as acetoin reductase) catalyzes the irreversible reduction of diacetyl to acetoin and the reversible reduction of acetoin to butanediol in which NAD(P)H acts as the coenzyme (Hugenholtz 1993, Rattray et al. 2000, Østlie et al. 2003). Diacetyl contributes to the characteristic buttery flavor of many cultured dairy products. There is significant interest in adapting processing methods that will enhance the formation of diacetyl or decrease diacetyl reductase activity to maintain these desirable flavor characteristics (Rattray et al. 2000).

Many commercial dairy processors now use the direct vat inoculation (DVI) process for frozen or freeze-dried cultures (up to 10^{12} bacteria per gram of starter) in the processing of cultured dairy products. The use of DVI cultures allows the dairy processor to directly add the cultures to the milk and by-pass on-site culture preparation. This recent progress in the development of starter cultures has also increased phage resistance, minimized the formation of mutants which may alter the characteristics of the starter cultures, enhanced the ability to characterize the composition of the cultures, and improved the consistent quality of cultured dairy products. However, the DVI process is limited by the additional cost of these cultures, the dependence of the cheese plants on the starter suppliers for the selection and production of the starters, and the increased lag phase of these cultures in comparison to on-site culture preparation (Stanley 1998, Tamime and Robinson 1999a, Canteri 2000).

Coagulation of Milk Proteins

The production of lactic acid by lactic acid bacteria decreases the pH of the milk to cause coagulation of the casein. As the pH decreases to less than 5.3, colloidal calcium phosphate is solubilized from the casein micelle, causing the micelles to dissociate and the casein proteins to aggregate and precipitate at the isoelectric point of casein (pH 4.6). The resulting gel, which is somewhat fragile in nature, provides the structure for sour cream, yogurt, and acid-precipitated cheeses, such as cream cheese and cottage cheese (Lucey 2002).

The casein micelles are also susceptible to coagulation through enzymatic activity. Rennet, a mixture of chymosin and pepsin, obtained from calf stomach, is most commonly recognized as the enzyme for coagulation of casein. However, proteases from microorganisms and produced through recombinant DNA technologies have been successfully adapted as alternatives to calf rennet (Banks 1998). Chymosin, the major enzyme present in rennet, cleaves the peptide bond between Phe-105 and Met-106 of κ-casein, releasing the hydrophilic, charged casein macropeptide, while the *para*-κ-casein remains associated with the casein micelle. The loss of the charged macropeptide reduces the surface charge of the casein micelle and results in the aggregation of the casein micelles to form a gel network stabilized by hydrophobic interactions. Temperature influences both the rate of the enzymatic reaction and the aggregation of the casein proteins, with 40–42°C, the optimal temperature for casein coagulation. The use of rennet to hydrolyze the peptide bond and cause aggregation of the casein micelles is used in the manufacture of most ripened cheeses (Lucey 2002).

Homogenization

Milk fat globules have a tendency to coalesce and separate upon standing. Homogenization reduces the diameter of the fat globules from 1–10 μm to less than 2 μm and increases the total fat globule surface area. The physical change in the fat globule occurs through forcing the milk through a small orifice under high pressure. The decrease in the size of the milk fat globules reduces the tendency of the fat globules to aggregate during the gelation period. In addition, denaturation of the whey proteins and interactions of the whey proteins with casein or the fat globules can alter the physical and chemical properties of the milk proteins to result in a firmer gel with reduced syneresis (Tamime and Robinson 1999b, Fox et al. 2000). Milk to be used to process yogurt, cultured buttermilk, and unripened cheeses is commonly homogenized to improve the quality of the final product.

Pasteurization

The original cultured dairy products relied on the native microorganisms in the milk for the fermentation process. Current commercial methods for all cultured dairy products include a pasteurization treatment to kill the native microorganisms, followed by inoculation with starter cultures to produce the desired product. The heat process, which must be sufficient to inactivate alkaline phosphatase, also destroys many pathogenic and spoilage microorganisms, and enzymes that may have a negative impact on the quality of the finished products. The time-temperature treatments for the fluid milk pasteurization have been adapted for the milk to be used in the processing of cultured dairy products (62.8°C for 30 minutes or 71.1°C for 15 seconds). More severe heat treatments than characteristic of pasteurization causes denaturation of whey proteins and interactions between β-lactoglobulin and κ-casein. In cheeses, this interaction decreases the ability of chymosin to hydrolyze the casein molecule and initiate curd precipitation and formation.

Pasteurization has a significant effect of the flavor profile of the milk. Cultured dairy products produced from pasteurized milks tend to have less intense flavor characteristics due to the heat inactivation of the naturally occurring microorganisms and enzymes in the milk that contribute to flavor formation (Buchin et al. 1998). Lactones and heterocycles are also formed during the heat treatment of raw milk to contribute cooked flavors (Friedrich and Acree 1998).

Cooling

The processing of cultured dairy products relies on the metabolic activity of the starter cultures to contribute to acid formation and flavor and texture development. Once the desired pH or titratable acidity is reached for these products, the products are cooled to 5–10°C to slow the growth of the bacteria and limit further acid production and other biological reactions.

PROCESSING OF CULTURED DAIRY PRODUCTS

The following discussion highlights the unique processing steps that are involved in the production of cultured dairy products. These processing steps contribute to the unique flavor, texture, and overall sensory characteristics of these products.

CHEESE

Over 400 different varieties of cheese have been recognized throughout the world. The wide diversity in the flavor, texture, and appearance of these cheeses is attributed to differences in the milk source, starter cultures, ripening conditions, and chemical composition. Cheeses are frequently classified based on moisture content, method of precipitation of the cheese proteins, and the ripening process. Table 23.2 compares processing methods and composition of selected cheeses.

The coagulation of the casein proteins, separation of the curds from the whey, and ripening of the curd are the primary steps involved in the processing of cheese. The resulting product is a highly nutritious product in which the casein and fat from the milk are concentrated. The fat plays a critical role in the texture of the cheese by preventing the casein molecules from associating to form a tough structure. In general, most cheeses can be classified as natural or processed cheeses. The natural cheeses include ripened or unripened cheeses. The stages involved in processing these different types of cheeses and their unique characteristics will be discussed further.

Natural Cheeses

A simplified overview of the steps involved in processing fresh and ripened natural cheeses is presented in Figure 23.1. Fresh cheeses are consumed immediately after processing and are characterized as having a high moisture content and mild flavor. Most fresh cheeses are acid-coagulated cheeses, in which lactic acid produced by the starter cultures causes the precipitation of the caseins. The final pH of the acid-coagulated cheeses is 4.6. Rennet is primarily used for the coagulation of the casein proteins and curd formation in ripened cheeses. Starter cultures are added to produce acid and contribute enzymes for flavor and texture development during ripening. Ripened cheeses undergo a ripening period, ranging from 3 weeks to more than 2 years, following processing that contributes to the development of the flavor and texture of the cheese. The moisture content of these cheeses ranges from 30% to 55% and the pH ranges from 5.0 to 5.3.

Standardization of the Milk Casein and fat contents of the milk are standardized to minimize variations in the quality of the cheese due to seasonal effects and variation in the milk supply. The casein-to-fat ratio can be adjusted by the addition of skim milk, cream, milk powder, or evaporated milk or the removal of fat. Calcium chloride (0.1%) may also be added to improve coagulation of the milk by rennet and further processing of the cheese. The actual casein and fat content of the milk will vary for each cheese type and influence the curd formation, cheese yield, fat content, and texture of the cheese (Banks 1998).

Coagulation of the Milk Proteins Aggregation of the casein micelles to form a three-dimensional gel protein network is initiated through the addition of rennet or other proteolytic enzymes or the addition of acid. Fat and water molecules are also entrapped within this protein network. Enzymes and starter bacteria also tend to associate with the curds, and thus contribute to a number of biochemical changes that occur during the ripening process. The whey, which includes water, salts, lactose, and the soluble whey proteins, is expelled from the gel. The aggregation of the casein micelles by either enzyme or acid treatment results in gels with different characteristics.

In most natural, aged cheeses, coagulation of the casein proteins by the addition of rennet is most common. This process is temperature dependent, with no coagulation occurring below 10°C, and an increase in coagulation rate accompanying an increase in temperature until the optimal temperature for coagulation at 40–45°C. Above 65°C, the enzyme is inactivated (Fox 1969, Brulé et al. 2000). The aggregation of the casein micelles is influenced by temperature ($Q_{10} \sim 12$) to a greater degree than the enzymatic hydrolysis of κ-casein ($Q_{10} \sim 2$) (Cheryan et al. 1975). Aggregation of the micelles begins when approximately 70–85% of the κ-casein molecules are hydrolyzed, which reduces the steric hindrance between the micelles. Reducing the pH or increasing the temperature reduces the degree of casein hydrolysis necessary for coagulation (Fox and McSweeney 1997). The presence of Ca^{2+} ions further facilitates the aggregation of the casein micelles through the neutralization of the negative charge on the micelle and the formation of ionic bonds. The resulting gel has an irregular network, is highly elastic and porous, and exhibits a high degree of syneresis.

The production of acid by lactic acid bacteria or the direct addition of hydrochloric or lactic acid can also result in the aggregation of the casein micelles and formation of clots. As the pH of the milk is reduced, the casein micelles become insoluble and begin to aggregate. Because calcium phosphate is solubilized as the pH of the milk drops, the gel formed during acid coagulation is not stabilized by Ca^{2+} ions. The acid-coagulated gels are less cohesive and exhibit less syneresis than enzyme-coagulated cheeses. These cheeses generally have a high moisture content and low mineral content. Acid coagulation is most frequently used in the manufacture of cottage cheese and other unripened cheeses.

A few unique types of cheese are prepared through acid coagulation of whey or a blend of whey and skim milk in conjunction with heat treatment. Ricotta cheese is the most common cheese prepared in this manner.

Lactic acid bacteria cultures are added to the milk in conjunction with the rennet in ripened cheeses. Although the lactic acid bacteria cultures do not have a significant role in the coagulation of casein, they contribute to the changes that occur during the ripening process. The different strains of starter cultures differ in characteristics including growth rate, metabolic rate, phage interactions, proteolytic activity, and flavor promotion (Stanley 1998). Frequently, mixed-strain cultures, which

Table 23.2. Comparison of Processing Characteristics of Different Types of Ripened, Natural Cheeses

Cheese	Starter Cultures	Ripening	Scald Temperature	Other Treatments	Composition (%)		
					Moisture	Fat	Salt
Very hard							
Romano	Thermophilic cultures	18–24 mo at 16–18°C	45–48°C		23.0	24.0	5.5
Hard: Ripened by bacteria, without eyes							
Cheddar	Mesophilic D cultures	0.5–2 yr at 6–8°C	37–39°C	Curds are cheddared prior to salting, dry salting	37.0	32.0	1.5
Hard: Ripened by bacteria, with eyes							
Emmental	Thermophilic cultures and propionibacteria	10–14 d at 10–15°C; 3 wk–2 mo at 20–24°C; 1–2 mo at 7°C	55°C	Brine-salted	35.5	30.5	1.2
Swiss	Thermophilic cultures, *Pr. freudenreichii* ssp. *shermanii*	7–14 d at 21–25°C; 6–9 mo at 2–5°C	50–53°C	Brine-salted	37.1	27.8	0.5
Semisoft: Ripened principally by bacteria, without eyes							
Colby	Mesophilic D cultures	2–3 mo at 3–4°C	37–39°C	Curds and whey stirred after cooking, as whey is drained	38.2	32.1	1.5
Provalone	L. bulgaricus	1–12 mo at 15–16°C	44–50°C	Curd—knead and stretch in 85–90°C water	42.5	27.0	3.0
Mozzarella	Thermophilic cultures	<1 mo	41°C	Curd—kneaded and stretched in 70°C water	54.0	18.0	0.7
Feta	Thermophilic or mesophilic starter culture	7 d in salt brine at 14–16°C; 2 mo at 3–4°C	Not cooked	From sheep or sheep/goat milk	59.7	20.3	2.2

Semisoft: Ripened principally by bacteria, with eyes							
Gouda	Mesophilic DL culture	2–3 mo at 15°C	36–38°C	Curd washed with hot water, stirred for 30 min at 36–38°C, pressed under whey, brine salted	41.0	28.5	2.0
Edam	Mesophilic DL culture	6 wk–6 mo	36°C	Similar to Gouda, made from semiskimmed (2.5% fat) milk	43.0	24.0	2.0
Semisoft: Ripened by bacteria and surface microorganisms							
Limburger	Mesophilic DL culture	2–3 wk at 10–15°C; 3–8 wk at 4°C (wrapped in foil)	37°C	Whey drained and replaced with salt brine, surface smeared with microflora during initial ripening	45.0	28.0	2.0
Semisoft: Ripened principally by blue mold in the interior							
Roquefort	Indigenous microflora	3–5 mo in limestone caves at 5–10°C and 95% relative humidity	Not cooked	Ewe's milk. Curds mixed with *Penicillium roqueforti* spores. Cheeses are pierced during ripening	40.0	31.0	3.5
Soft: Ripened principally by surface molds							
Camembert	Mesophilic DL culture	10–12 d at 12°C, 7–10 d at 7°C	Not cooked	Surface mold growth by *Penicillium camemberti*	52.5	23.0	2.5

Sources: Kosikowski 1978, Fox et al. 2000.

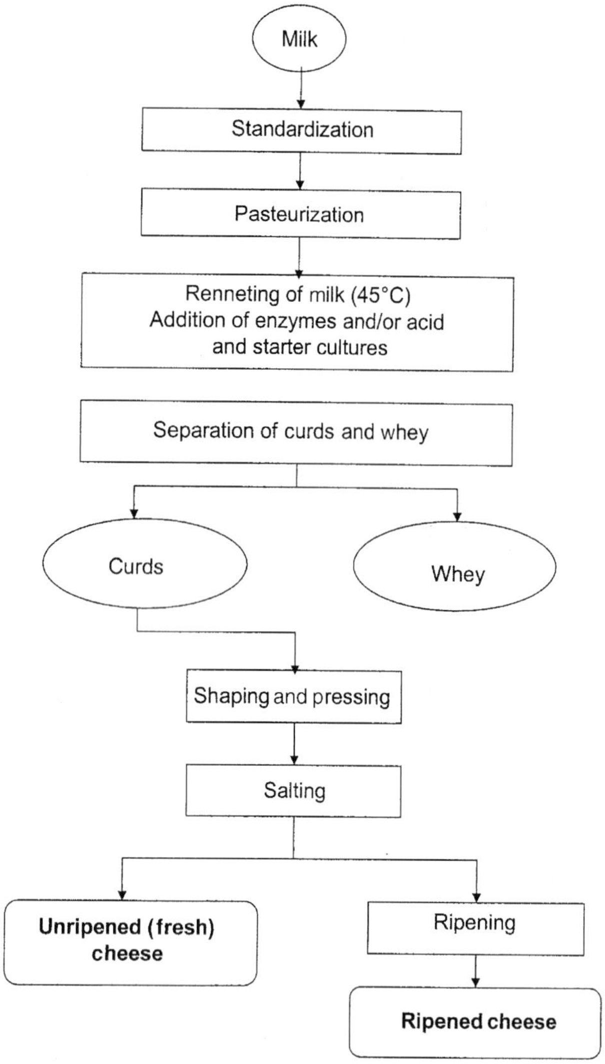

Figure 23.1. Processing scheme for natural cheeses.

contain unknown numbers of strains of the same species, are used. Mesophilic L cultures contain *Leuconostoc* species, including *Ln. mesenteroides* ssp. *cremoris* and *Ln. lactis*, while the main species in mesophilic D cultures *Lactococcus lactis* ssp. *cremoris* with lesser amounts of *Lc. lactis* ssp. *lactis*. Mesophilic DL cultures would consist of both lactococci and leuconostoc. The thermophilic cultures consist of *S. thermophilus*, and either *Lactobacillus helveticus*, *Lb. delbrueckii* ssp. *lactis* or *Lb. delbrueckii* ssp. *bulgaricus*.

Cutting the Coagulum Cutting the coagulum increases the drainage of the whey from the curds and contributes to a sharp decrease in the moisture content of the curds. Acid-coagulated and rennet-coagulated cheeses differ in the pH at which curd formation occurs, and subsequently the pH at which the coagulum is cut. For acid-coagulated cheeses, curd formation occurs at pH 4.6, the isoelectric point of casein. Curd formation of rennet-coagulated cheeses occurs at higher pHs, ranging from pH 6.3 to 6.6. Following the cutting of the coagulum, the curd and whey mixture is heated and agitated in a process called "scalding." The agitation is necessary to keep the curds suspended in the whey and to promote drainage of whey from the curds. The temperature during the scalding process is dependent on the type of the cheese and ranges from 20°C to 55°C. The temperature affects gel formation, gel viscoelasticity, and regulates the growth of the lactic acid bacteria. A high temperature results in greater drainage from the cheese and a firmer cheese.

The conversion of lactose to lactic acid by starter cultures decreases the pH of the curd, which contributes to the loss of the whey from the curd and a decrease in moisture content. While the curds are in the whey, diffusion of lactic acid into the whey and lactose into the curds occurs. The rate of acid production is affected by the amount and type of the starter cultures, composition of the milk, and temperature during acid production.

When the required acidity of the cheese curds is reached, the whey is drained to recover the curds. Following the separation of the curds from the whey, acid production continues at an increased rate, because of the lack of diffusion of the lactic acid from the curd. To minimize excess acid development following the drainage of the whey, some types of cheese include a washing step to reduce the lactose content. The moisture content of the curd is affected by the extent the coagulum is cut, the temperature of the curd after cutting, and agitation of the curd in the whey.

The handling of the curd following the cutting of the coagulum greatly affects the characteristics of the cheese. The cheddaring process, used in the manufacture of Cheddar, Colby, Monterey, and Mozzarella cheeses, involves piling blocks of curd on top of each other with regular turning to allow the curds to fuse together. In Colby and Monterey cheeses, vigorous stirring during the drainage of the whey inhibits the development of a curd structure and results in the softer cheese with a higher moisture content. Frequently, the whey may be partially drained off and replaced with water or salt brine to remove lactose and reduce the development of acidity, as is done in the processing of Gouda and Limburger cheeses.

Shaping and Pressing The resulting curds are shaped to form a coherent mass that is easy to handle. The curds are placed in a mold and are often pressed with an external force to cause the curds to deform and fuse. The pressure and time of pressing range from a few g/cm^2 for a few minutes for moist cheeses, to 200 to 500 g/cm^2 for up to 16–48 hours for cooked and hard cheeses. The temperature (20–27°C) and humidity (95% relative humidity) of the pressing room is controlled to optimize the growth of the lactic acid bacteria and facilitate the deformation and shaping of the curd. Acid production by lactic acid bacteria and additional drainage of whey from the curd continues during the shaping and pressing stage. The deformability is affected by the composition of the cheese, and increases until the curds reach a pH of 5.2–5.3. Deformability also increases with an increase in moisture content and temperature.

Salting The salting step reduces the moisture content of the curd, inhibits the growth of starter bacteria, and affects the flavor, preservation, texture, and rate of ripening of the cheese. Final salt contents of cheeses range from 0.7% to 4% (2–10% salt in moisture contents). The amount of salt, the method for application of the salt, and the timing of the salting is dependent on the specific type of cheese. The salt may be incorporated through (1) mixing with dry milled curd pieces, (2) rubbing onto the surface of the molded cheese, or (3) immersing the cheese in a salt brine. Following the salting step, the salt diffuses into the interior of the cheese, with the subsequent displacement of whey. Depending on the size of the cheese block and the composition of the cheeses, it may take from 7 days to over 4 months for the salt to equilibrate within the cheese.

Ripening Fresh, green cheese has a bland flavor and a smooth, rubbery texture. During the ripening process, the characteristic texture and flavor of the cheese develop through a complex series of biochemical reactions. Ripening starter cultures are selected to develop the texture and flavor characteristics of the specific cheese type. Enzymes released following lysis of the microorganisms catalyze the degradation of proteins, lipids, and lactose in the cheese to produce volatile flavor compounds, which contribute to the characteristic flavor of the cheeses. As the ripening time increases, the moisture content of the cheese decreases and the intensity of the flavor increases. The resulting quality attributes of the finished cheese depend on the initial composition of the milk and the starter cultures used, the water activity of the cheeses, and the temperature, time, and humidity during the ripening period. Depending on the type of cheese, the ripening period can range from 3 weeks to more than 2 years.

The ripening temperature influences the rate of the microbial growth and enzyme activity during the process and the equilibrium between the biochemical reactions that occur during ripening. Ripening temperatures generally range from 5°C to 20°C, which is well below the optimum temperatures for microbial growth and enzyme activity. Soft cheeses are often ripened at 4°C to slow the biochemical processes. An increase in ripening temperature for hard cheeses reduces the ripening time necessary for flavor development, with a 5°C increase in ripening temperature reducing the ripening time 2–3 months. However, caution must be exercised in altering ripening temperatures since not all microorganisms and enzymes respond to temperature changes in the same manner, resulting in an imbalance in flavor characteristics (Choisy et al. 2000).

The growth of most of the starter bacteria added to the milk in the initial stages of cheese-making is slowed as the pH of the cheese approaches 5.7 and following the addition of salt, however fermentation and a decrease in pH continue. The fermentation of lactose to lactic acid by the starter cultures provides an environment that prevents the growth of undesirable microorganisms through the reduced pH and the formation of an anaerobic environment. The reduced pH also has an influence on activity of the proteases and lipases, which contribute to the formation of critical volatile flavor compounds. The optimal activity for proteases is between pH 5.5 and 6.5 and for lipases is between pH 6.5 and 7.5. Lactose and citrate are precursors for a number of volatile flavor compounds, including diacetyl, acetoin, 2,3-butanediol, and acetaldehyde, which are also formed during ripening (Urbach 1995).

Protease and lipase activity during the ripening is probably most important to the development of the flavor and texture of the cheese. Enzymes of the starter bacteria, nonstarter lactic acid bacteria, and secondary cultures added during cheese-making are most important in the development of the flavor and texture of the cheese during ripening. These enzymes are released by the lysis of the cell wall of the bacteria. Rennet enzymes and endogenous milk enzymes, such as plasmin, also contribute to these hydrolytic reactions during ripening. The extent of these enzymatic reactions depends on the activity and specificity of the enzymes, the concentration of the substrates, pH, water activity, salt concentration, and ripening temperature and duration. The degradation of the amino acids and fatty acids, through enzymatic and nonenzymatic reactions, results in the formation of several important volatile flavor compounds, including sulfur-containing compounds, amines, aldehydes, alcohols, esters, and lactones.

Rennet and plasmin are associated with the primary phase of proteolysis and hydrolyze the caseins to large polypeptides. This proteolysis alters three-dimensional protein network of the cheese to form a less firm and less elastic cheese. Although these polypeptides do not have a direct impact on flavor, they do function as a substrate for the proteases associated with the starter and nonstarter bacteria. However, if the primary proteolysis is extensive, bitter peptides, with a high percentage of hydrophobic amino acids predominate. Free amino acids and short-chain peptides contribute sweet, bitter, and brothy-like taste characteristics to the cheese. Further degradation and chemical reactions of these peptides and amino acids through the action of decarboxylases, transaminases, or deaminases, contribute to the formation of amines, acids, ammonia, and thiols, which contribute to cheese flavor.

Lipases break down triacylglycerols into free fatty acids, and mono- and diacylglycerols. The short-chain free fatty acids contribute to the sharp, pungent flavor characteristics of the cheese. The degree of lipolysis that is acceptable without producing soapy and rancid flavors depends on the type of cheese. Several *Penicillium* strains form methyl ketones, lactones, and unsaturated alcohols through their enzymatic systems associated with β-oxidation and decarboxylation, β-oxidation and lactonization, and lipoxygenase activity. Aliphatic and aromatic esters are synthesized by esters present in a range of microorganisms, including mesophilic and thermophilic lactic acid bacteria (Choisy et al. 2000).

The texture of cheese is attributed to the three-dimensional protein network, which entraps fat and whey. This structure is altered through proteolysis during ripening to form a less firm and less elastic cheese.

Carbon dioxide produced by the metabolism of the bacteria and entrapped within the curd results in the formation of eyes in several types of cheeses. The small eyes characteristics of Edam, Gouda, and related cheese varieties are formed by carbon dioxide produced from citrate by *Leuconostoc* ssp. In Swiss-type cheeses, *Propionibacterium freudenreichii* ssp. *shermanii* metabolize lactate to form the carbon dioxide responsible for the eye formation. These cheeses are ripened in hot rooms (20–22°C) to optimize the growth of the propionibacteria and allow eye formation.

Ripening of cheese is a time-intensive process; thus, there is much interest in accelerating the ripening process without an adverse effect on cheese quality. Addition of enzymes, including the use of encapsulated enzymes; addition of attenuated or modified starter cultures and adjunct cultures; elevated ripening temperatures; and the addition of cheese slurry have been explored as possible means to shorten the ripening processing. An important consideration is that increased proteolysis or an imbalance of enzyme activity can result in higher contents of hydrophobic amino acids and peptides that contribute to bitterness (Choisey 2000, Law 2001, Azarnia et al. 2006).

Processed Cheese

Processed cheeses are produced by blending and heating several natural cheeses with emulsifiers, water, butterfat, whey powder and/or caseinates to form a homogeneous mixture. The proportion of cheese used in the formulation ranges from 51% for cheese spreads and cheese foods to 98% in processed cheeses. Different types of natural cheeses will produce processed cheeses with different flavor and textural characteristics. The formulation of these ingredients affects the consistency, texture, flavor, and melting characteristics of the processed cheese. The heating process inactivates the microorganisms and denatures enzymes to produce a stable product. The flavor of the processed cheese is generally milder than the natural cheeses, due to the effects of the heat. However, the melting properties of the processed cheese are much improved due to the addition of the emulsifiers. Processed cheeses range in consistency from block cheese to sliced cheese to cheese for spreading. (Banks 1998, Fox et al. 2000).

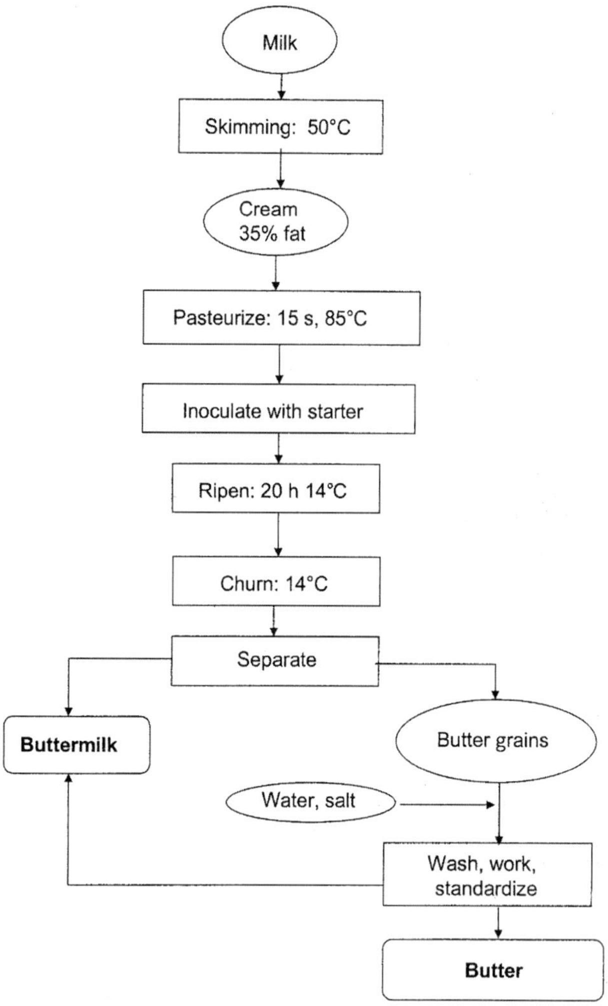

Figure 23.2. Processing scheme for butter processed from ripened cream.

Butter

Flavor and consistency are the most important quality attributes to consider during the processing of butter. The characteristic flavor of butter is a result of the formation of diacetyl by heterofermentative lactic acid bacteria. Off-flavors due to lipolysis, oxidation, or other contaminants should be avoided. The butter should be firm enough to hold its shape, yet soft enough to spread easily. Modification of the cow's diet to increase contents of polyunsaturated fatty acids and conjugated linoleic acids in the milk (Avramis et al. 2003, Bobe et al. 2007) or selecting cows that produce milk with a higher content of unsaturated fatty acids (Bobe et al. 2003) results in butter with a softer texture and the potential for improved cold spreadability due to decrease in the content of saturated fatty acids. Figure 23.2 outlines the process involved for making butter and the by-product buttermilk.

The first step in the processing of butter is skimming the milk to obtain cream with a fat content of at least 35%, in which the fat is dispersed as droplets within the aqueous phase. This step increases the efficiency of the process through increasing the yield of butter, reducing the yield of buttermilk, and reducing the size of machinery needed. The pasteurization step functions to kill microorganisms, to inactivate enzymes, and to make the butter less susceptible to oxidative degradation. Following pasteurization, the cream is inoculated with a starter culture mixture (1–2%) of *Lactococcus* ssp. and *Leuconostoc* ssp., which contributes to the development of the characteristics butter flavor. It is desirable that the starter cultures grow rapidly at low temperatures. During ripening, the cream begins to ferment and the fat begins to crystallize. The formation of lactic acid and diacetyl by the starter cultures contributes to flavor development in the butter and buttermilk. Crystallization of the fat is important to maximize the yield of the butter and minimize loss of fat into the buttermilk during the churning process. During the churning process, air is incorporated into the cream using dashboards or rotary agitators contributing to the aggregation of fat globules to form butter granules. The resulting butter grains may be washed with water to remove nonfat solids followed by working or kneading the butter grains. During the working step, a water-in-oil emulsion is formed as the small water droplets are dispersed into the fat matrix.

The churning process is probably most critical to the textural quality of the butter. As air is incorporated into the cream, the fat globules surround the air bubbles and coalesce with other fat-coated air bubbles to form clumps. These clumps continue to coalesce during the churning process and the volume of air bubbles decreases. The proportion of solid fat, which is influenced by temperature, is critical to the aggregation of fat clumps to form butter. If the rate of churning is too fast, the fat globules are less stable and less likely to coalesce, resulting in a greater loss of fat into the buttermilk.

Buttermilk

Natural buttermilk is a by-product of butter manufacture (Figure 23.2) and is categorized as sweet (nonfermented) or acidic (fermented). Sweet and acid buttermilk are obtained from the manufacture of butter from nonfermented, sweet cream and sour or cultured cream, respectively. In general, the composition of buttermilk is similar to that of skim milk, consisting of milk proteins, lactose, minerals, and some lipids. The fat content of natural buttermilk is approximately 0.4% and consists primarily of the membrane components of the fat globules. Churning disrupts the fat globules and the MFGM associates with the buttermilk fraction. The MFGM is characterized as having a higher content of phospholipids and unsaturated fatty acids than found in whole milk. The high phospholipid content provides emulsification ability and expands the functionality of buttermilk as a dairy ingredient. However, because of the high content of unsaturated fatty acids associated with the membrane lipids, natural buttermilk has a characteristic flavor and is also more susceptible to the development of oxidized off-flavors (Sodini et al. 2006).

Cultured buttermilk and cultured skim milk are more frequently produced as alternatives to the traditional buttermilk (Figure 23.3). The solids-not-fat and fat contents of cultured buttermilk range from 7.4% to 11.4% and 0.25% to 1.9%, respectively. The homogenized and pasteurized milk with the desired fat content is inoculated with a 1% starter culture of *Lc. lactis* ssp. *lactis* and *Lc. lactis* biovar. *diacetylactis* and *Ln. mesenteroides* ssp. *cremoris*. These bacteria metabolize citrate to diacetyl to contribute to the development of "buttery" flavor. Acid is produced by these bacteria at a relatively slow rate.

Only a minor percentage of the buttermilk produced is consumed directly. The commercial use of buttermilk includes applications in the baking and dairy industries (Sodini et al. 2006). For many of these applications, the buttermilk is concentrated by evaporation and spray dried to form a powder. Spray drying has been shown to result in a significant decrease in the phospholipid content, which could have important implications with respect to the functionality of buttermilk powders as an ingredient (Morin et al. 2007).

Sour Cream

Sour cream is produced from pasteurized, homogenized cream (20–30% fat) and has a pleasant acidic taste and buttery aroma

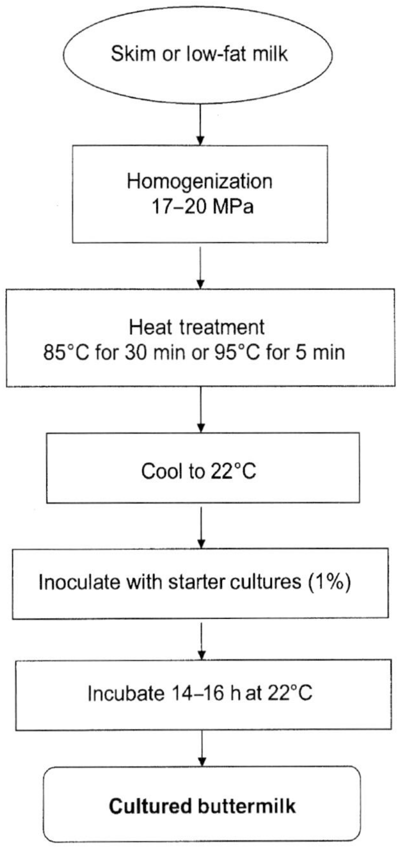

Figure 23.3. Processing scheme for cultured buttermilk.

(Fig. 23.4). Low-fat sour cream has a fat content ranging from 10% to 12%. Following standardization of the milk and cream to the desired fat content, the mixture is warmed and homogenized to improve the consistency of the final product. The starter cultures include *Lc. lactis* ssp. *lactis* and *Lc. lactis* biovar. *diacetylactis* and *Ln. mesenteroides* ssp. *cremoris*. As in buttermilk, these cultures contribute to the acid and buttery flavor of the sour cream. The sour cream may be packaged prior to or following fermentation at 20°C until the pH is reduced to 4.5. Packaging prior to fermentation results in a thicker product because the gel is not disturbed. As the pH decreases, the fat clusters aggregate to form a viscous cream. Rennet or thickening agents, such as starch, guar gum, carrageenan, and locust bean gum, are sometimes added to increase the firmness of the sour cream and stabilize the gel to minimize syneresis (Hunt and Maynes 1997).

Yogurt

The tremendous increase in the popularity of yogurt in recent decades has been attributed to its health food image and the wide diversity of flavors, compositions, and viscosities available to consumers. The manufacturing methods, raw materials, and

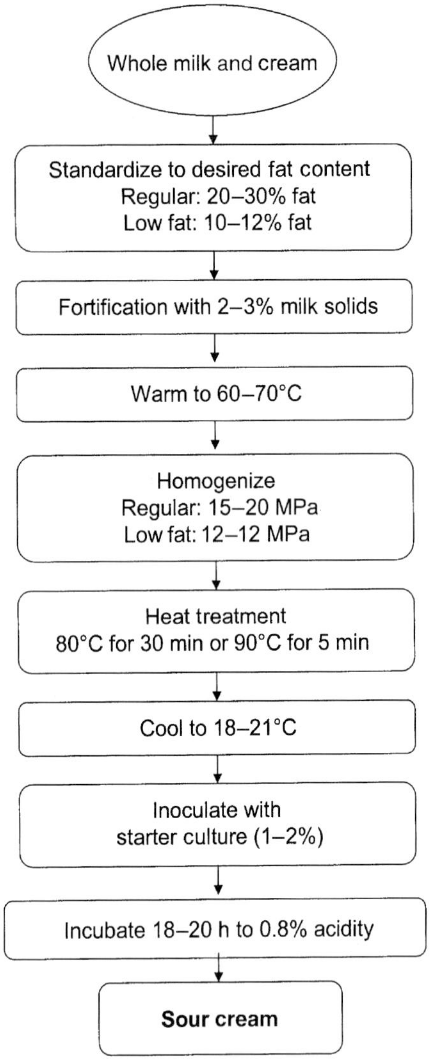

Figure 23.4. Processing scheme for sour cream.

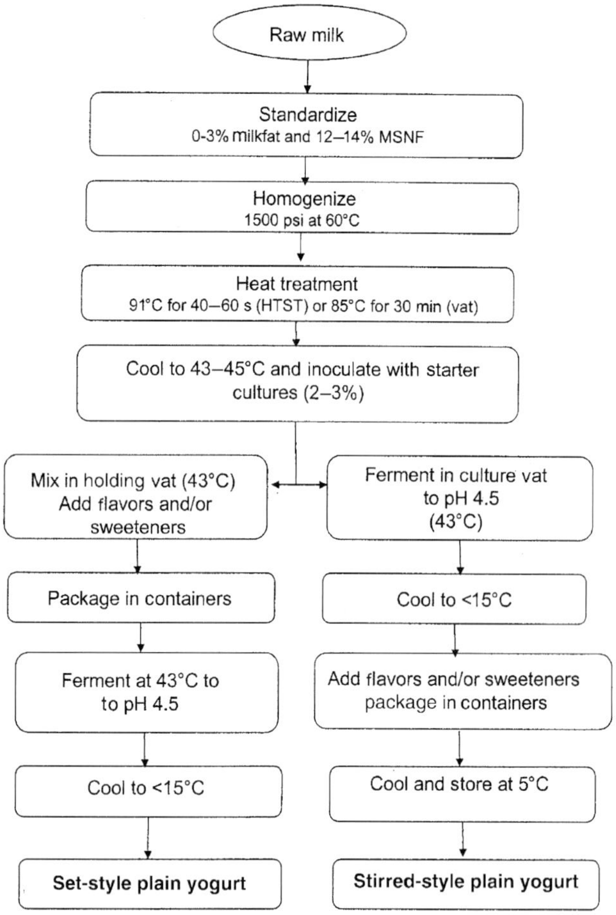

Figure 23.5. Processing scheme for set-style and stirred-style yogurt.

formulation vary widely from country to country resulting in products with a diversity of flavor and texture characteristics. While in many Western societies, yogurt is produced from cows' milk, other mammalian milk can be used to produce yogurt. Figure 23.5 outlines the steps involved in the processing of stirred-style and set-style yogurt.

The milk is initially standardized to the desired fat (0.5–3.5% fat) and milk solids-not-fat (12.5%) contents. The increase in the protein content is most commonly achieved through the addition of nonfat milk powder and improves the body and decreases syneresis of the final product. In addition to decreasing the size of the fat globule, homogenization of the milk alters the milk proteins to reduce syneresis and increase firmness (Tamime and Robinson 1999b). The subsequent heat treatment (85°C for 30 minutes) eliminates pathogenic microorganisms and reduces the oxygen in the milk to provide a good growth media for the starter cultures. Enzymes and the major whey proteins, including β-lactoglobulin and α-lactalbumin, but not the casein proteins are also denatured by the heat treatment. The denaturation of the whey proteins and subsequent interactions between the whey proteins and casein and/or fat globules improves the stability of the gel and increases water-binding capacity of the caseins to decrease syneresis (Tamime and Robinson 1999b).

Following heat treatment, the milk is cooled to 43–45°C for inoculation of the starter cultures. The fast acid-producing thermophilic lactic acid bacteria, *Streptococcus salivarius* ssp. *thermophilus* and *Lactobacillus delbrueckii* ssp. *bulgaricus* are the primary microorganisms used in the production of yogurt. These bacteria have a synergistic effect on each others' growth and should be present in approximately equal numbers for optimal flavor development. *Lb. delbrueckii* ssp. *bulgaricus* has important protease activity and hydrolyzes the milk proteins to small peptides and amino acids. These peptides and amino acids enhance the growth of *S. thermophilus*, which has limited proteolytic activity. *S. thermophilus* metabolizes pyruvic acid to formic acid and carbon dioxide, which in turn, stimulates the growth of *Lb. delbrueckii*. Initially, *S. thermophilus* grows faster than *Lb. delbrueckii*, however, at the later stages of the fermentation process, the growth of the *S. thermophilus* is inhibited by the reduced pH of the yogurt. The mutual stimulation of the yogurt cultures through their metabolic activity significantly increases the formation of lactic acid at a rate greater than would be possible by the individual cultures.

The acid produced by the yogurt cultures decreases the pH of the milk to cause the aggregation of the casein micelles. As the pH decreases to less than 6.0, the colloidal calcium phosphate that stabilizes the casein micelle structure is solubilized and micelle undergoes partial rearrangement. Solubilization of the colloidal calcium phosphate is essentially completed when the milk reaches a pH of 5.0. With further acid production, the pH of the milk reaches the isoelectric point of casein (pH 4.6) and electrostatic repulsion of the charged groups on the casein proteins decreases as these groups become protonated. The increased casein–casein association through hydrophobic interactions results in the aggregation of casein micelles and gel formation (Lucey 2002). Complete solubilization of colloidal calcium phosphate prior to the initiation of gel formation results in the formation of yogurt with stronger gels and reduced tendency for syneresis. Higher inoculation rates for the starter cultures (2%) and a lower incubation temperature (40°C) contribute to stronger interactions between casein molecules and a more stable gel structure and reduced syneresis (Lee and Lucey 2004).

Two types of yogurt, set-style and stirred-style are produced. Set-style yogurt is fermented following packaging, with color and flavors added to the container prior to the addition of the inoculated milk, resulting in a gel that forms a firm, unbroken coagulum. Stirred-style yogurt is fermented in a vat prior to packaging, with the gel structure broken following fermentation during the addition of flavors and colors, cooling and packaging stage. For both types of yogurt, either a short incubation period at 40–45°C for $2^1/_2$–3 hours or a long incubation period at 30°C for 16–18 hours may be used to attain a pH of 4.5 or titratable acidity of 0.9% lactic acid prior to cooling.

The metabolism of the yogurt cultures contributes to the characteristic flavor of yogurt. Both bacteria are heterofermentative and produce lactic acid from glucose, yet are unable to metabolize galactose. Acetaldehyde, a key flavor component of yogurt described as having a fruity aroma, is produced by the degradation of threonine to acetaldehyde and glycine by the enzyme threonine aldolase. Although *S. thermophilus* forms a majority of the acetaldehyde produced, the proteolytic activity of *Lb. delbrueckii* ssp. *bulgaricus* generates the precursors for the formation of acetaldehyde (Wilkins et al. 1986). Other volatile flavor compounds generated by the metabolism of lactic acid bacteria include diacetyl and 2,3-pentanedione, both of which have buttery aroma characteristics. The formation of lactic acid during the fermentation is also a critical contributor to the typical yogurt flavor (Ott et al. 1997, 2000).

Syneresis, the expelling of interstitial liquid due to association of the protein molecules and shrinkage of a gel network, is undesirable in yogurt. Syneresis increases with an increase in incubation temperature. In stirred yogurt, extensive amounts of syneresis results in a thin product. Therefore, incubation of the yogurt at a lower temperature, such as 32°C, is recommended to maintain the desirable viscosity. Stabilizers, such as gelatin, xanthan gum, locust bean gum, and κ-carrageenan, increase the firmness and viscosity of yogurt and minimize syneresis through their ability to bind with water and the milk proteins to stabilize the protein network (Soukoulis et al. 2007).

Acidophilus Milk

Acidophilus milk was developed as a fluid milk product that provides therapeutic benefits through the alleviation of intestinal disorders. Following homogenization and heat treatment of the milk to minimize contamination by other microorganisms, the milk is cooled to 37°C and inoculated with a 1% culture of *Lactobacillus acidophilus* (Fig. 23.6). The milk is incubated at this temperature for 18–24 hours until the titratable acidity reaches 0.63–0.72%. *Lb. acidophilus* is a thermophilic starter that grows slowly in milk and has a relatively high acid-tolerance. To provide the desired therapeutic effects, the milk

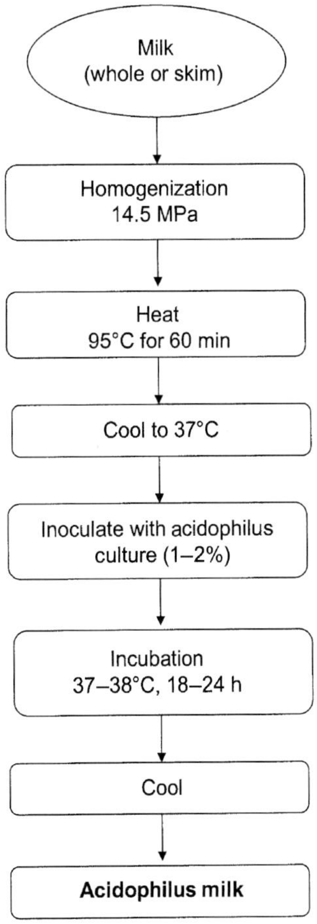

Figure 23.6. Processing scheme for acidophilus milk.

should contain 5 × 10^8 cfu/mL at the time of consumption (Tamime and Marshall 1997).

THE FUTURE—CULTURED DAIRY PRODUCTS WITH THERAPEUTIC BENEFITS

The availability of dairy products with positive nutritional benefits, in addition to yogurt and acidophilus milk, is expected to increase in the future. Probiotic bacteria, galactooligosaccharides and other lactose derivatives, prebiotics, bioactive peptides, whey proteins, and conjugated linoleic acids are among the components identified in cultured dairy products with enhanced nutritional benefits, ranging from improved balance of intestinal function, reduction of serum cholesterol, enhanced immune function, and anticarcinogenic and antimicrobial activity. In some cases, modifications of the traditional processing protocols will be necessary to increase the content and stability of these components in cultured dairy products without adversely affecting the flavor, texture, and other sensory quality attributes. The development of cultured dairy products with improved health benefits will provide numerous challenges and rewards to food scientists and consumers (McIntosh et al. 1998, Hilliam 2003, Korhonen and Pihlanto 2003, Parodi 2003, Playne et al. 2003, Stanton et al. 2003, Stevenson and Knowles 2003).

ACKNOWLEDGMENTS

This chapter of the Iowa Agriculture and Home Economics Experiment Station, Ames, Iowa, Project No. 3574, was supported by Hatch Act and State of Iowa funds.

FURTHER READING

Early R. 1998. *The Technology of Dairy Products*, 2nd edn. Blackie Academic and Professional, London, p. 446.

Kurmann JA et al. 1992. *Encyclopedia of Fermented Fresh Milk Products: An International Inventory of Fermented Milk, Cream, Buttermilk, Whey and Related Products*. Van Nostrand Reihnold, New York, p. 368.

Law BA. 1997. *Microbiology and Biochemistry of Cheese and Fermented Milk*, 2nd edn. Blackie Academic and Professional, London, p. 365.

Tamime AY, Robinson RK. 1999. *Yoghurt – Science and Technology*, 2nd edn. CRC Press, Boca Raton, FL, p. 619.

Walstra P. 1990. On the stability of casein micelles. *J Dairy Sci* 73: 1965–1979.

REFERENCES

Avramis CA et al. 2003. Physical and processing properties of milk, butter, and Cheddar cheese from cows fed supplemental fish meal. *J Dairy Sci* 86: 2568–2576.

Azarnia S et al. 2006. Biotechnological methods to accelerate Cheddar cheese ripening. *Crit Rev Biotechnol* 26: 121–143.

Banks JM. 1998. Chapter 3, Cheese. In: R Early (ed.) *The Technology of Dairy Products*, 2nd edn. Blackie Academic and Professional, London, pp. 81–122.

Bendall JG. 2001. Aroma compounds of fresh milk from New Zealand cows fed different diets. *J Agric Food Chem* 49: 4825–4832.

Bobe G et al. 2003. Texture of butter from cows with different milk fatty acid compositions. *J Dairy Sci* 86: 3122–3127.

Bobe G et al. 2007. Butter composition and texture from cows fed with different milk fatty acid compositions fed fish oil or roasted soybeans. *J Dairy Sci* 90: 2596–2603.

Brulé G et al. 2000. Chapter 1, The casein micelle and milk coagulation. In: A Eck, J-C Gillis (eds.) *Cheesemaking – From Science to Quality Assurance*, 2nd edn. Lavoisier Publishing, Paris, pp. 7–40.

Buchin S et al. 1998. Influence of pasteurization and fat composition of milk on the volatile compounds and flavor characteristics of semi-hard cheese. *J Dairy Sci* 81: 3097–3108.

Canterri G. 2000. Chapter 5.2, Lactic starters. In: A Eck, J-C Gillis (eds.) *Cheesemaking – From Science to Quality Assurance*, 2nd edn. Lavoisier Publishing, Paris, pp. 164–182.

Cheryan M et al. 1975. Secondary phase and mechanism of enzymic milk coagulation. *J Dairy Sci* 58: 477–481.

Choisy C et al. 2000. Chapter 4, The biochemistry of ripening. In: A Eck, J-C Gillis (eds.) *Cheesemaking – From Science to Quality Assurance*, 2nd edn. Lavoisier Publishing, Paris, pp. 82–151.

Fox PF. 1969. Milk-clotting and proteolytic activities of rennet, bovine pepsin, and porcine pepsin. *J Dairy Res* 36: 427–433.

Fox PF et al. 2000. *Fundamentals of Cheese Science*. Aspen Publishers, Gaithersburg, MD, p. 587.

Fox PF, McSweeney PLH. 1997. Chapter 1, Rennets: their role in milk coagulation and cheese ripening. In: BA Law (ed.) *Microbiology and Biochemistry of Cheese and Fermented Milk*, 2nd edn. Blackie Academic and Professional, London, pp. 1–49.

Friedrich JE, Acree TE. 1998. Gas chromatography olfactometry (GC/O) of dairy products. *Int Dairy J* 8: 235–241.

Hilliam M. 2003. Future for dairy products and ingredients in the functional foods market. *Aus J Dairy Technol* 58: 98–103.

Hugenholtz J. 1993. Citrate metabolism. *FEMS Microbiol Rev* 12: 165–178.

Hunt CC, Maynes JR. 1997. Current issues in the stabilization of cultured dairy products. *J Dairy Sci* 80: 2639–2643.

Korhonen H, Pihlanto A. 2003. Bioactive peptides: new challenges and opportunities for the dairy industry. *Aus J Dairy Technol* 58: 129–133.

Kosikowski F. 1978. *Cheese and Fermented Milk Foods*, 2nd edn. F.V. Kosikowski and Associates, Brooktondale, NY, p. 711.

Law BA. 2001. Controlled and accelerated cheese ripening: the research base for new technology. *Int Dairy J* 11: 383–398.

Lee WJ, Lucey JA. 2004. Structure and physical properties of yogurt gels: effect of inoculation rate and incubation temperature. *J Dairy Sci* 87: 3153–3164.

Lucey JA. 2002. Formation and physical properties of milk protein gels. *J Dairy Sci* 85: 281–294.

McIntosh GH, et al. 1998. Whey proteins as functional food ingredients. *Int Dairy J* 8: 425–434.

Morin P et al. 2007. Effect of processing on the composition and microstructure of buttermilk and its milk fat globule membranes. *Int Dairy J* 17: 1179–1187.

Østlie HM et al. 2003. Growth and metabolism of selected strains of probiotic bacteria in milk. *Int J Food Microbiol* 87: 17–27.

Ott A et al. 1997. Determination and origin of the aroma impact compounds of yogurt flavour. *J Agric Food Chem* 45: 850–858.

Ott A et al. 2000. Sensory investigation of yogurt flavour perception: mutual influence of volatiles and acidity. *J Agric Food Chem* 48: 441–450.

Parodi PW. 2003. Anticancer agents in milkfat. *Aus J Dairy Technol* 58: 114–118.

Playne MJ, et al. 2003. Functional dairy foods and ingredients. *Aus J Dairy Technol* 58: 242–264.

Rattray FP et al. 2000. Purification and characterization of a diacetyl reductase from *Leuconostoc pseudomesenteroides*. *Int Dairy J* 10: 781–789.

Sodini I et al. 2006. Compositional and functional properties of buttermilk: a comparison between sweet, sour, and whey buttermilk. *J Dairy Sci* 89: 525–536.

Soukoulis C et al. 2007. Industrial yogurt manufacture: monitoring of fermentation process and improvement of final product quality. *J Dairy Sci* 90: 2641–2654.

Stanley G. 1998. Chapter 2, Microbiology of fermented milk products. In: R Early (ed.) *The Technology of Dairy Products*, 2nd edn. Blackie Academic and Professional, London, pp. 50–80.

Stanton C, et al. 2003. Probiotic health benefits – reality or myth? *Aus J Dairy Technol* 58: 107–113.

Stevenson L, Knowles G. 2003. The role of dairy ingredients in enhancing immunity. *Aus J Dairy Technol* 58: 135–139.

Tamime AY, Marshall VME. 1997. Chapter 3, Microbiology and technology of fermented milks. In: BA Law (ed.) *Microbiology and Biochemistry of Cheese and Fermented Milk*, 2nd edn. Blackie Academic and Professional, London, pp. 57–152.

Tamime AY, Robinson RK. 1999a. Chapter 8, Preservation and production of starter cultures. In: *Yoghurt – Science and Technology*, 2nd edn. CRC Press, Boca Raton, FL, pp. 486–534.

Tamime AY, Robinson RK. 1999b. Chapter 2, Background to manufacturing practice. In: *Yoghurt – Science and Technology*, 2nd edn. CRC Press, Boca Raton, FL, pp. 11–128.

Urbach G. 1995. Contribution of lactic acid bacteria to flavour compound formation in dairy products. *Int Dairy J* 5: 877–903.

Walstra P et al. 1999. *Dairy Technology: Principles of Milk Properties and Processes*. Marcel Dekker, New York, p. 727.

Walstra P, Jenness R. 1984. *Dairy Chemistry and Physics*. John Wiley & Sons, New York, p. 467.

Wilkins DW et al. 1986. Threonine aldolase activity in yogurt bacteria as determined by headspace gas chromatography. *J Agric Food Chem* 34: 150–152.

24
Chemistry and Biochemistry of Milk Constituents

P.F. Fox and A.L. Kelly

Abstract: Mammalian milk is a highly complex physicochemical system, containing colloidal proteins (the casein micelle) and emulsified lipids, as well as dissolved lactose, minerals, vitamins and minerals. The properties of milk, and the products made or isolated from milk, are very much determined by the properties of its constituents. These properties are particularly relevant when milk is processed, for example through denaturation of proteins, oxidation, hydrolysis of proteins and lipids, or Maillard reactions involving lactose. In this chapter, the principal families of milk constituents, and their most significant characteristics, are described.

INTRODUCTION

Milk is a fluid secreted by female mammals, of which there are approximately 4500 species, to meet the complete nutritional, and some of the physiological, requirements of the neonate of the species. Because nutritional requirements are species-specific and change as the neonate matures, it is not surprising that the composition of milk shows very large interspecies differences, for example the concentrations of fat, protein and lactose range from 1% to 50%, 1% to 20% and 0% to 10%, respectively. Interspecies differences in the concentrations of many of

Food Biochemistry and Food Processing, Second Edition. Edited by Benjamin K. Simpson, Leo M.L. Nollet, Fidel Toldrá, Soottawat Benjakul, Gopinadhan Paliyath and Y.H. Hui.

the minor constituents of milk are even greater than those of the macro constituents. The composition of milk also changes markedly during lactation, reflecting the changing nutritional requirements of the neonate and during mastitis and physiological stress. In this chapter, the typical characteristics of the principal, and some of the minor, constituents of bovine milk will be described. The milk of the principal dairying species, cow, buffalo, sheep and goat are generally similar but differ in detail. The milks of several non-bovine species are described by Park and Haenlein (2006) and Fuquay et al. (2011); the latter includes articles on monotremes and marsupials, marine mammals and primates.

Sheep and goats were domesticated around 8000 BC and their milk has been used by humans since. However, cattle, especially breeds of *Bos taurus*, are now the dominant dairy animals. Total recorded world milk production is approximately 600 × 10^6 tonnes/annum, of which approximately 85% is bovine, 11% is buffalo and 2% each is from sheep and goats. Camels, mares, reindeer and yaks are important dairy animals in limited geographical regions with specific cultural and/or climatic conditions.

Milk is a very flexible raw material; several thousand dairy products are produced around the world in a great diversity of flavours and forms, including about 1400 varieties/variants of cheese. The principal dairy products and the percentage of milk used in their production are: liquid (beverage) milk, approximately 40%; cheese, approximately 35%; butter, approximately 32%; whole milk powder, approximately 6%; skimmed milk powder, approximately 9%; concentrated milk products approximately 2%; fermented milk products, approximately 2%; casein, approximately 2% and infant formulae, approximately 0.3%. The flexibility of milk as a raw material is a result of the properties, many of them unique, of its principal constituents; many of these are very easily isolated, permitting the production of valuable food ingredients. Milk is free of off-flavours, pigments and toxins, which greatly facilitates its use as a food or as a raw material for food production.

The processability and functionality of milk and milk products are determined by the chemical and physicochemical properties of its principal constituents, that is lactose, lipids, proteins, and salts, which will be described in this chapter. The exploitation and significance of the chemical and physico-chemical properties of milk constituents in the production and properties of the principal groups of dairy foods, that is liquid milk products, cheese, butter, fermented milks, functional milk proteins and lactose will be described in Chapter 25. Many of the principal problems encountered during the processing of milk are caused by variability in the concentrations and properties of the principal constituents arising from several factors, including breed, individuality of the animal (i.e., genetic factors), stage of lactation, health of the animal, especially mastitis, and nutritional status. Synchronised calving, as practised in New Zealand, Australia and Ireland, to avail of cheap grass as the principal component of the cow's diet, has a very marked effect on the composition and properties of milk (see O'Brien et al. 1999a, b, c, Mehra et al. 1999). However, much of the variability can be offset by standardising the composition of milk using various methods (e.g., centrifugation, ultrafiltration or supplementation) or by modifying the process technology.

The chemical and physico-chemical properties of the principal constituents of milk are well characterised and described. The very extensive literature includes the following textbooks: Walstra and Jenness (1984), Wong et al. (1988), Fox (1992, 1995, 1997, 2003a), Fox and McSweeney (1998, 2003, 2006), Walstra et al. (1999, 2005) and McSweeney and Fox (2009).

SACCHARIDES

LACTOSE

Introduction

Lactose is a reducing disaccharide comprising glucose and galactose, linked by a β1–4-*O*-glycosidic bond (Fig. 24.1). It is unique to milk and is synthesised in the mammary gland from glucose transported from the blood; one molecule of glucose is epimerised to galactose, as UDP-galactose (Gal), *via* the Leloir pathway and is condensed with a second molecule of glucose by a two-component enzyme, lactose synthetase. Component A is a general UDP-galactosyl transferase (UDP-GT; EC 2.4.1.2.2), which transfers galactose from UDP-Gal to a range of sugars, peptides or lipids. Component B is the whey protein, α-lactalbumin (α-La), in the presence of which, the K_M of UDP-GT for glucose is reduced 1000-fold and lactose is the principal product synthesised. There is a good correlation between the concentrations of lactose and α-La in milk. Lactose is responsible for approximately 50% of the osmotic pressure of milk, which is equal to that of blood and varies little; therefore, the concentration of lactose in milk is tightly controlled and is independent of breed, individuality and nutritional factors, but decreases as lactation advances and especially during mastitis, in both cases due to the influx of NaCl from the blood. The physiological function of α-La is probably to control the synthesis of lactose, and thus maintain the osmotic pressure of milk relatively constant.

The concentration of lactose in milk ranges from approximately 0, for some species of seal, to approximately 10% in the milk of some monkeys. The concentration of lactose in the milk of the principal dairy species is quite similar (cow, 4.8%; buffalo, 4.3%; sheep, 4.6%; goat, 4.9%; camel, 5.1%), exceptions are the horse (6.1%), donkey (6.9%) and reindeer (2.5%); human milk contains about 7.0% lactose. The lactose content of bulk herd milk from randomly calved cows varies little throughout the year but differences can be quite large when calving of cows is synchronised, for example in Ireland, the level of lactose in creamery milk varies from approximately 4.8% in May to approximately 4.2% in October.

Chemical and Physico-Chemical Properties of Lactose

Among sugars, lactose has a number of distinctive characteristics, some of which cause problems in milk products during processing and storage; however, some of its characteristics are exploited to advantage.

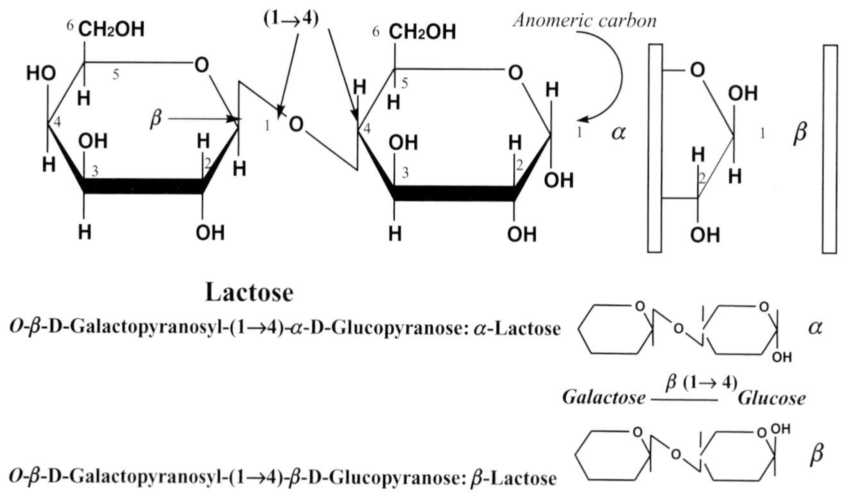

Figure 24.1. Structures of lactose.

The functional aldehyde group at the C-1 position of the glucose moiety exists mainly in the hemiacetal form, forming a cyclic structure and, consequently, C-1 is a chiral, asymmetric, carbon. Therefore, like all reducing sugars, lactose can exist as two anomers, α and β, which have markedly different properties. From a functional viewpoint, the most important of these properties are differences in solubility and crystallisation characteristics between the isomers; α-lactose crystallises as a monohydrate, while crystals of β-lactose are anhydrous. Since crystalline α-lactose contains 5% H_2O, the yield of this anomer is higher than that of β-lactose and this must be considered when expressing the concentration of lactose. The solubility of α- and β-lactose in water at 20°C is approximately 7 g/100 mL and 50 g/100 mL, respectively. However, the solubility of α-lactose is much more temperature-dependent than that of β-lactose and the solubility curves intersect at approximately 93.5°C (see Fox and McSweeney 1998).

At equilibrium in aqueous solution, lactose exists as a mixture of α and β anomers in the approximate ratio of 37:63. When an excess of α-lactose is added to water, approximately 7 g/100 mL dissolve immediately, some of which mutarotates to give an α:β ratio of 37:63, leaving the solution unsaturated with respect to both α- and β-lactose. Further α-lactose then dissolves, some of which mutarotates to β-lactose. Solubilisation and mutarotation continue until two conditions exist, that is approximately 7 g of dissolved α-lactose/100 mL of water and an α :β ratio of 37:63, giving a final solubility of approximately 18.2 g/100 mL. When β-lactose is added to water, approximately 50 g/100 mL dissolve initially but approximately 18.5 g of this mutarotates to α-lactose, which then exceeds its solubility and some lactose crystallises. This upsets the α:β ratio and more β-lactose mutarotates to α-lactose, which crystallises. Mutarotation of β-lactose and crystallisation of α-lactose continue until approximately 7 g and 11.2 g of α- and β-lactose, respectively, are in solution.

Although lactose has low solubility in comparison with other sugars, once dissolved, it crystallises with difficulty and forms supersaturated solutions. Highly supersaturated solutions (greater than twofold saturated) crystallise spontaneously but if the solution is only slightly supersaturated (one to twofold), lactose crystallises slowly and forms large, sharp, tomahawk-shaped crystals of α-lactose. If the dimensions of the crystals exceed approximately 15 μm, they are detectable on the tongue and palate as a sandy texture. Crystals of β-lactose are smaller and monoclinical in shape. In the metastable zone, crystallisation of lactose is induced by seeding with finely powdered lactose (Fig. 24.2). Since the solubility of α-lactose is lower than that of the β anomer below 93.5°C, α-lactose is the normal commercial form.

When concentrated milk is spray-dried, the lactose does not have sufficient time to crystallise during drying and an amorphous glass is formed. If the moisture content of the powder is kept low ($<$4%), the lactose glass is stable, but if the moisture content increases to about 6%, for example on exposure of the powder to a high-humidity atmosphere, the lactose will crystallise as α-lactose monohydrate. If extensive crystallisation occurs, an inter-locking mass of crystals is formed, resulting in 'caking' of the powder, which is a particularly serious problem in whey powders owing to the high content of lactose ~70%). The problem is avoided by pre-crystallising, as much as possible, of the lactose before drying, which is achieved by seeding the concentrated solution with finely powdered lactose.

Controlled agglomeration (caking) is used to improve the wettability of spray-dried milk powder This product has poor wettability because the small particles swell on contact with water, thereby blocking the channels between the particles (see Kelly et al. 2003 for review). The wettability (often incorrectly referred to as 'solubility') of spray-dried milk powder may be improved by controlling the drying process to produce milk

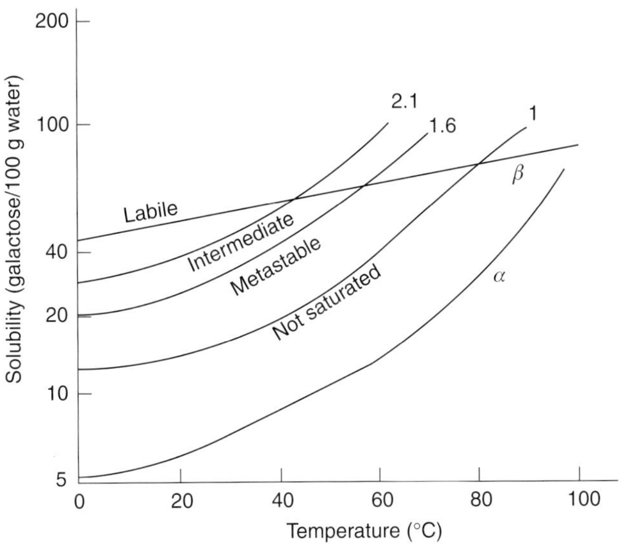

Figure 24.2. Initial solubility of α- and β-lactose, final solubility at equilibrium (line 1) and supersaturation by a factor of 1.6 and 2.1 (α-lactose excluding water of crystallisation) (Fox and McSweeney 1998).

powder with coarser, more easily wetted particles; such powders are said to be 'instantised' and are produced by agglomerating the fine powder particles, in effect by controlling the caking process. In the case of whole milk powder, instantisation processes must also overcome the intrinsic hydrophobic nature of milk fat; this is normally achieved by adding the amphiphilic agent, lecithin. Although lactose is hygroscopic when it crystallises, properly crystallised lactose has very low hygroscopicity and, consequently, is a very effective component of icing sugar.

The crystallisation of lactose in frozen milk products results in destabilisation of the casein, which aggregates when the product is thawed. In this case, the effect of lactose is indirect; when milk is frozen, pure water freezes and the concentration of solutes in the unfrozen water is increased. Since milk is supersaturated with calcium phosphate (~66% and ~57% of the Ca and PO_4, respectively, are insoluble and occur in the casein micelles, known as colloidal calcium phosphate (CCP); see Section 'Milk Salts'), when the amount of water becomes limiting, soluble $Ca(H_2PO_4)_2$ and $CaHPO_4$ crystallise as $Ca_3(PO_4)_2$, with the concomitant release of H^+ and a decrease in pH to approximately 5.8. During frozen storage, lactose crystallises as α-lactose monohydrate, thus, reducing the amount of solvent water and aggravating the problems of calcium phosphate solubility and pH decline. Thorough crystallisation of lactose before freezing alleviates, but does not eliminate, this problem. Preheating milk prior to freezing also alleviates the problem, but pre-hydrolysis of lactose to the more soluble sugars, glucose and galactose, using β-galactosidase, appears to be the best solution.

Lactose has low sweetness (16% as sweet as sucrose in a 1% solution). This limits its usefulness as a sweetener (the principal function of sugars in foods) but makes it is a very useful diluent, for example for food colours, flavours or enzymes, when a high level of sweetness is undesirable.

Being a reducing sugar, lactose can participate in the Maillard (non-enzymatic) browning reaction, with very undesirable consequences in all dairy products, for example brown colour, off-flavours, reduced solubility and reduced nutritional value. The Maillard reaction, and factors that affect it, has been studied extensively for around 100 years and the relevant literature has been reviewed frequently (see O'Brien 2009, Nursten 2011).

Food Applications of Lactose

Total milk production (~600 × 10^6 tonnes/annum) contains ~30 × 10^6 tonnes of lactose. Most of this lactose is consumed as a constituent of milk but whey, a by-product of the manufacture of cheese and, to a lesser extent, of casein, contains 8–9 × 10^6 tonnes of lactose. About 400,000 tonnes of lactose are isolated/prepared per annum. A number of high-lactose food products are also produced, for example approximately 2,000,000 tonnes of whey powder, electrodialysed whey powder and whey permeate powder; these serve as crude sources of lactose for several food products, including infant formulae. Thus, most available lactose is now utilised in some form and little is wasted. The production of lactose is basically similar to that for other sugars, involving concentration, crystallisation, recovery, washing and drying of the crystals (see Paterson 2009, 2011).

Although some of its properties, especially its low sweetness and low solubility, limit the usefulness of lactose as a sugar, other properties, that is very low hygroscopicity if properly crystallised, low sweetness and reducing properties, make it a valuable ingredient for the food and pharmaceutical industries. In the pharmaceutical industry, lactose is widely used as a diluent in pelleting operations. The principal application of lactose in the food industry is in the humanisation of infant formulae; human milk contains approximately 7% lactose, compared to approximately 4.8% in bovine milk. For this application, demineralised whey is widely used; it is cheaper and more suitable than purified lactose because it also supplies whey proteins, which help bring the casein:whey protein ratio of the formulae closer to the value 40:60, found in human milk, compared with 80:20 in bovine milk. It is necessary to demineralise the whey because bovine milk contains approximately 4 times as much inorganic salts as human milk. Demineralisation is accomplished by electrodialysis, ion exchange or nanofiltration or combinations of these.

Lactose is also used as an agglomerating/free-flowing agent in foods (e.g., butter powders), in the confectionery industry to improve the functionality of shortenings, as an anti-caking agent in icing mixtures at high humidity, or as a reducing sugar if Maillard browning is desired. The low sweetness of lactose limits its widespread use as a sugar but is advantageous in many applications. Lactose absorbs compounds and may be used as a diluent for food flavours and colours or to trap flavours.

Lactose Derivatives

A number of more useful and more valuable products may be produced from lactose (see Playne and Crittenden 2009). The most significant are the following:

- *Lactulose (galactosyl β-1–4 fructose)*: A sugar not found in nature, which is produced from lactose by heating, especially under slightly alkaline conditions. The concentration of lactulose in milk is a useful index of the severity of the heat treatment to which milk is subjected, for example in-container sterilisation > indirect ultra-high temperature (UHT) > direct UHT > high-temperature short-time (HTST) pasteurisation. Lactulose is not hydrolysed by intestinal β-galactosidase and enters the large intestine, where it promotes the growth of *Bifidobacterium* spp. It also has a laxative effect and is widely used for this purpose; more than 20,000 tonnes are produced annually.
- *Glucose-galactose syrups, produced by acid or enzymatic (β-galactosidase) hydrolysis* (see Chapter 25): The technology for the production of such hydrolysates has been developed but the product is not cost-competitive with other sugars (sucrose, glucose, glucose–fructose).
- *Tagatose*: Is the keto analogue of galactose. It occurs at a low level in the gum of the evergreen tree, *Sterculia setigera,* and in severely heated milk and stored milk powder. It can be produced by treating β-galactosidase–hydrolysed lactose with a weak alkali, for example $Ca(OH)_2$, which converts the galactose to tagatose, which can be purified by demineralisation and chromatography. Tagatose is nearly as sweet as sucrose, has a good quality taste and enhances the flavour of other sweeteners. It is absorbed poorly from the small intestine, serves as a probiotic and has little effect on blood glucose; it is fermented in the lower intestine to volatile short-chain acids that can be absorbed but provide only approximately 35% of the energy derived from sugars catabolised *via* the normal route. Tagatose has GRAS status and is produced commercially by SweetGredients, a company formed by Arla Foods and Nordzuker (Denmark).
- *Galacto-oligsaccharides*. β-Galactosidase has transferase as well as hydrolytic activity and under certain conditions, the former predominates, leading to the formation of galacto-oligosaccharides containing up to six monosaccharides linked by glycosidic bonds that are not hydrolysed by the enzymes secreted by the human small intestine. The undigested oligosaccharides enter the large intestine, where they have bifidogenic properties and are considered to have promising food applications. These oligosaccharides are quite distinct from the naturally occurring oligosaccharides referred to in Section 'Oligosaccharides'(see Ganzle 2011).
- *Ethanol*: Is produced commercially by the fermentation of lactose by *Kluyveromyces lactis*. Depending on local legislation, the ethanol may be used in alcoholic drinks, which are profitable. The current interest in renewable energy sources has created vast opportunities for lactose-derived ethanol but its commercial success will depend on local taxation policy.

Other derivatives that have limited but potentially important applications include lactitol, lactobionic acid, lactic acid, acetic acid, propionic acid, lactosyl urea and single-cell proteins. Most of these derivatives can be produced by fermentation of sucrose, which is cheaper than lactose, or by chemical synthesis. However, lactitol and lactobionic acids are derived specifically from lactose and may have economic potential. Lactitol is a synthetic sugar alcohol produced by reduction of lactose; it is not metabolised by higher animals but is relatively sweet, and hence has potential for use as a non-calorific sweetener. It has also been reported that lactitol reduces blood cholesterol level, reduces sucrose absorption and is anti-carcinogenic. Lactobionic acid has a sweet taste, which is unusual for an acid and therefore should have some interesting applications.

Nutritional Aspects of Lactose

Lactose is responsible for two enzyme deficiency syndromes: lactose intolerance and galactosemia. The former is due to a deficiency of intestinal β-galactosidase, which is rare in infants but common in adults, except North-Western Europeans and a few African tribes. Since humans are unable to absorb disaccharides, including lactose, from the small intestine, unhydrolysed lactose enters the large intestine where it is fermented by bacteria, leading to flatulence and cramp, and to the absorption of water from the intestinal mucosa, causing diarrhoea. These conditions cause discomfort and perhaps death. Lactose intolerance has been studied extensively since its discovery in 1959 and the literature has been reviewed regularly (see Ingram and Swallow 2009).

The problems caused by lactose intolerance can be avoided by:

- excluding lactose-containing products from the diet, which is the normal practice in regions of the world where lactose intolerance is widespread;
- removing lactose from milk, for example by ultrafiltration;
- hydrolysis of the lactose by adding β-galactosidase at the factory or in the home. The technology for the production of lactose-hydrolysed milk and dairy products is well developed but is of commercial interest mainly for lactose-intolerant individuals in Europe or North America. Because the consumption of milk is very limited in South-East Asia, the use of β-galactosidase is of little interest, although lactose intolerance is widespread.

Galactosemia is caused by the inability to catabolise galactose, owing to a deficiency of either of two enzymes, galactokinase or galactose-1P uridyltransferase (see Flynn 2003). A deficiency of galactokinase leads to the accumulation of galactose that is catabolised *via* alternative routes, one of which leads to the accumulation of galactitol in various tissues, including the eye, where it causes cataracts over a period of about 20 years. A deficiency of galactose-1P uridyltransferase leads to abnormalities in membranes of the brain and to mental retardation unless galactose is excluded from the diet within a few weeks of birth. Both forms of galactosemia occur at a frequency of 1 per approximately 50,000 births.

Lactose in Fermented Dairy Products

The fermentation of lactose to lactic acid, by lactic acid bacteria (LAB) is a critical step in the manufacture of all fermented dairy products (cheese, fermented milks and lactic butter). The fermentation pathways are well established (see Cogan and Hill 1993, Poolman 2002). Lactose is not a limiting factor in the manufacture of fermented dairy products; only approximately 20% of the lactose is fermented in the production of these products. Individuals suffering from lactose intolerance may be able to consume fermented milks without ill-effects, possibly because LAB produce β-galactosidase and emptying of the stomach is slower than that for fresh milk products, thus, releasing lactose more slowly into the intestine.

In the manufacture of cheese, most (96–98%) of the lactose is removed in the whey. The concentration of lactose in fresh curd depends on its concentration in the milk and on the moisture content of the curd and varies from approximately 1.7%, w/w, in fresh Cheddar curd to approximately 2.4%, w/w, in fresh Camembert. The metabolism of residual lactose in the curd to lactic acid has a major effect on the quality of mature cheese (Fox et al. 1990, 2000). The resultant lactic acid may remain essentially unchanged in the cheese during ripening (e.g., Cheddar cheese) or may be catabolised to other compounds, for example CO_2 and H_2O, by surface mould in Camembert, or to propionic acid, acetic acid, H_2O and CO_2 in Emmental-type cheeses. Excessive lactic acid in cheese curd may lead to a low pH and a number of defects, such as a strong, acid, harsh taste, an increase in brittleness and a decrease in firmness. The pH of full-fat Cheddar is inversely related to the lactose/lactic acid content of the curd. Excess residual lactose may also be fermented by heterofermentative lactobacilli, with the production of CO_2 leading to an open texture.

In the manufacture of some cheese varieties, for example Dutch cheese, the curds are washed to reduce the lactose content and thereby regulate the pH of the pressed curd at approximately 5.3. For Emmental, the curd-whey mixture is diluted with water by approximately 20%, again to reduce the lactose content of the curd, maintain the pH at approximately 5.3, and keep the calcium concentration high, which is important for the textural properties of this cheese. For Cheddar, the level of lactose, and hence lactic acid, in the curd is not controlled. Hence, changes in the concentration of lactose in milk, such as those occurring throughout lactation, can result in marked changes in the quality of such cheeses. To overcome seasonal variations in the lactose content of milk, the level of wash water used for Dutch-type cheeses is related to the concentrations of lactose and casein in the milk. Ideally, the lactose-to-protein ratio in any particular variety should be standardised, for example by washing the curd, to minimise variations in the concentration of lactic acid, pH and the quality of cheese.

Oligosaccharides

Lactose is the principal sugar in milk but the milk of most, if not all, species also contains oligosaccharides, up to hexasaccharides, derived from lactose (the reducing end of the oligosaccharides is lactose and many contain fucose and *N*-acetylneuraminic acid). About 130 oligosaccharides have been identified in human milk; the milk of elephant, bears and marsupials also contains high levels of oligosaccharides. The oligosaccharides are considered to be important sources of certain monosaccharides, especially fucose and *N*-acetylglucosamine, for neonatal development, especially of the brain (Urashima et al. 2001, 2009, 2011).

MILK LIPIDS

Definition and Variability

The lipid fraction of milk is defined as those compounds that are soluble in non-polar solvents (ethyl/petroleum ether or chloroform/methanol) and is comprised mainly of triglycerides (98%), with approximately 1% phospholipids and small amounts of diglycerides, monoglycerides, cholesterol, cholesterol esters and traces of fat-soluble vitamins and other lipids. The lipids occur as globules, 0.1–20 μm in diameter, each surrounded by a membrane, the milk fat globule membrane (MFGM), which serves as an emulsifier. The concentration of total and individual lipids varies with breed, individual animal, stage of lactation, mastitic infection, plane of nutrition, interval between milking and point during milking when the sample is taken. Among the principal dairy breeds, Friesian/Holsteins produce milk with the lowest fat content (~3.5%) and Jersey/Guernsey the highest (~6%). The fat content varies considerably throughout lactation; when synchronised calving is practised, the fat content of bulk Friesian milk varies from approximately 3% in early lactation to >4.5% in late lactation. Such large variations in lipid content obviously affect the economics of milk production and the composition of milk products, but can be modified readily by natural creaming, centrifugal separation or addition of cream, and hence need not affect product quality. Milk lipids also exhibit variability in fatty acid composition and in the size and stability of the globules. These variations, especially fatty acid profile, are essentially impossible to standardise and hence are responsible for considerable variations in the rheological properties, colour, chemical stability and nutritional properties of fat-containing dairy products.

Fatty Acid Profile

Ruminant milk fat contains a wider range of fatty acids than any other lipid system – up to 400 fatty acids have been reported in bovine milk fat; the principal fatty acids are the homologous series of saturated fatty acids with an even number of C-atoms, $C_{4:0}$–$C_{18:0}$, and $C_{18:1}$. The outstanding features of the fatty acid profile of bovine milk fat are a high concentration of short- and medium -chain acids (ruminant milk fats are the only natural lipids that contain butanoic acid, $C_{4:0}$) and a low concentration of polyunsaturated fatty acids.

In ruminants, the fatty acids for the synthesis of milk lipids are obtained from triglycerides in chylomicrons in the blood or synthesised de novo in the mammary gland from acetate or β-hydroxybutyrate produced in the rumen. The triglycerides in chylomicrons are derived from the animal's feed or synthesised in the liver. Butanoic acid ($C_{4:0}$) is produced by the reduction of β-hydroxybutyrate, which is synthesised from dietary roughage by bacteria in the rumen and therefore varies substantially with the animal's diet. All $C_{6:0}$–$C_{14:0}$ and 50% of $C_{16:0}$ are synthesised in the mammary gland *via* the malonylCoA pathway from acetylCoA produced from acetate synthesised in the rumen. Essentially 100% of $C_{18:0}$, $C_{18:1}$, $C_{18:2}$ and $C_{18:3}$ and 50% of $C_{16::0}$ are derived from blood lipids (chylomicrons) and represent approximately 50% of total fatty acids in ruminant milk fat. Unsaturated fatty acids in the animal's diet are saturated by bacteria in the rumen unless they are protected, for example by encapsulation.

Mainly due to saturation in the rumen, ruminant milk fats are quite saturated, approximately 65% of the fatty acids in bovine milk fat are saturated, and are considered nutritionally undesirable although not all to an equal extent. However, according to Parodi (2009), the case against saturated fatty acids as causative factors for coronary heart disease is not proven and further research is required.

When milk production is seasonal, for example Australia, New Zealand and Ireland, very significant changes occur in the fatty acid profile of milk fat throughout the production season (see Fox 1995, Fox and McSweeney 1998, 2006). These variations are reflected in the hardness of butter produced from such milk; the spreadability of butter produced in winter is much lower than that of summer butter. Owing to the lower degree of unsaturation, winter butter should be less susceptible to lipid oxidation than the more unsaturated summer product but the reverse appears to be the case, probably owing to higher levels of pro-oxidants, for example Cu and Fe, in winter milk.

Although a ruminant's diet, especially if grass-based, is rich in polyunsaturated fatty acids (PUFA), these are hydrogenated by bacteria in the rumen and, consequently, ruminant milk fat contains very low levels of PUFAs, for example bovine milk fat contains approximately 2.4% $C_{18:2}$ compared to approximately 13% in human or porcine milk fat. PUFAs are considered to be nutritionally desirable and consequently there has been interest in increasing the PUFA content of bovine milk fat. This can be done by feeding encapsulated PUFA-rich lipids or crushed PUFA-rich oil seed to the animal. Increasing the PUFA content also reduces the melting point (MP) of the fat and makes butter produced from it more spreadable. However, the lower MP fat may have undesirable effects on the rheological properties of cheese, and PUFA-rich dairy products are very susceptible to lipid oxidation. Although the technical feasibility of increasing the PUFA content of milk fat by feeding protected PUFA-rich lipids to the cow has been demonstrated, it is not economical to do so in most cases. Blending milk fat with PUFA-rich or $C_{18:1}$-rich vegetable oil appears to be much more viable and is now widely practised commercially.

Conjugated Linoleic Acid

Linoleic acid (*cis, cis* 9, 12-octadecadienoic acid) is the principal essential fatty acid and has been the focus of nutritional research for many years. However, conjugated isomers of linoleic acid (CLA) have attracted very considerable attention recently (for review, see Bauman and Lock 2006). CLA is a mixture of eight positional and geometric isomers of linoleic acid, which have a number of health-promoting properties, including anticarcinogenic and anti-atherogenic activities, reduction of the catabolic effects of immune stimulation and the ability to enhance growth and reduce body fat (see Parodi 1999, Yurawecz et al. 1999). Of the eight isomers of CLA, only the *cis*-9, *trans*-11 isomer is biologically active. This compound is effective at very low concentrations, 0.1 g/100 g diet.

Fat-containing foods of ruminant origin, especially milk and dairy products, are the principal sources of dietary CLA, which is produced as an intermediate during the biohydrogenation of linoleic acid by the rumen bacterium, *Butyrivibrio fibrisolvens*. Since CLA is formed from linoleic acid, it is not surprising that the CLA content of milk is affected by diet and season, being highest in summer when cows are on fresh pasture rich in PUFAs (Lock and Garnsworthy 2000, Lawless et al. 2000) and is higher in the fat of milk from cows on mountain pasture than on lowland pasture (Collomb et al. 2002). The concentration of CLA in milk fat can be increased five to seven folds by increasing the level of dietary linoleic acid, for example by duodenal infusion (Kraft et al. 2000) or by feeding a linoleic acid-rich oil, for example sunflower oil (Kelly et al. 1998)

A number of other lipids may have anticarcinogenic activity, for example sphingomyelin, butanoic acid and ether lipids, but little information is available on these to date (Parodi 1997, 1999)

Structure of Milk Triglycerides

Glycerol for milk lipid synthesis is obtained in part from hydrolysed blood lipids (free glycerol and monoglycerides), partly from glucose and a little from free blood glycerol. Synthesis of triglycerides within the cell is catalysed by enzymes located on the endoplasmic reticulum. Esterification of fatty acids is not random (Table 24.1). The concentrations of $C_{4:0}$ and $C_{18:1}$ appear to be rate-limiting because of the need to keep the lipid liquid at body temperature. Some notable features of the structure are as follows:

- Butanoic and hexanoic acids are esterified almost entirely, and octanoic and decanoic acids predominantly, at the *sn*-3 position.
- As the chain-length increases up to $C_{16:0}$, an increasing proportion is esterified at the *sn*-2 position; this is more marked for human than for bovine milk fat, especially in the case of palmitic acid ($C_{16:0}$).
- Stearic acid ($C_{18:0}$) is esterified mainly at *sn*-1.
- Unsaturated fatty acids are esterified mainly at the *sn*-1 and *sn*-3 positions, in roughly equal proportions.

The fatty acid distribution is significant from two viewpoints:

1. It affects the MP and hardness of the fat, which can be reduced by randomising the fatty acid distribution. Trans-esterification can be performed by treatment with $SnCl_2$ or enzymatically under certain conditions: increasing

Table 24.1. Composition of Fatty Acids (mol% of the Total) Esterified to Each Position of the Triacyl-*sn*-Glycerols in Bovine or Human Milk

Fatty Acid	Cow			Human		
	sn-1	*sn*-2	*sn*-3	*sn*-1	*sn*-2	*sn*-3
4:0	–	–	35.4	–	–	–
6:0	–	0.9	12.9	–	–	–
8:0	1.4	0.7	3.6	–	–	–
10:0	1.9	3.0	6.2	0.2	0.2	1.1
12:0	4.9	6.2	0.6	1.3	2.1	5.6
14:0	9.7	17.5	6.4	3.2	7.3	6.9
16:0	34.0	32.3	5.4	16.1	58.2	5.5
16:1	2.8	3.6	1.4	3.6	4.7	7.6
18:0	10.3	9.5	1.2	15.0	3.3	1.8
18:1	30.0	18.9	23.1	46.1	12.7	50.4
18:2	1.7	3.5	2.3	11.0	7.3	15.0
18:3	–	–	–	0.4	0.6	1.7

attention is being focused on the latter as an acceptable means of modifying the harness of butter.

2. Pancreatic and many other lipases are specific for the fatty acids at the *sn*-1 and *sn*-3 positions. Therefore, $C_{4:0}$–$C_{8:0}$ are released rapidly from milk fat; these are water-soluble and are readily absorbed from the intestine. Medium- and long-chain acids are absorbed more effectively as 2-monoglycerides than as fatty acids; this appears to be quite important for the digestion of lipids by human infants who have limited ability to digest lipids due to the absence of bile salts. Infants metabolise human milk fat more efficiently than bovine milk fat, apparently due to the very high proportion of $C_{16:0}$ esterified at *sn*-2 in the former. The effect of trans-esterification on the digestibility of milk fat by infants merits investigation.

Short-chain fatty acids ($C_{4:0}$–$C_{10:0}$) have a strong aroma and flavour and their release by indigenous lipoprotein lipase (LPL) and microbial lipases cause off-flavours in milk and many dairy products, referred to as hydrolytic rancidity.

Rheological Properties of Milk Fat

The melting characteristics of ruminant milk fat are such that, at low temperatures (e.g., *ex*-refrigerator), it contains a high proportion of solid fat and has poor spreadability. The rheological properties of milk lipids may be modified by fractional crystallisation, for example an effective treatment involves removing the middle MP fraction and blending high- and low-MP fractions. Fractional crystallisation is expensive and is practised in industry to only a limited extent; in particular, securing profitable outlets for the middle-MP fraction is a major economic problem.

Alternatively, the rheological properties of milk fat may be modified by increasing the level of PUFAs through feeding cows with protected PUFA-rich lipids, but this practise is also expensive. The melting characteristics of blends of milk fat and vegetable oils can be easily varied by changing the proportions of the different fats and oils in the blend. This procedure is economical and is widely practised commercially; blending also increases the level of nutritionally desirable PUFAs. The rheological properties of milk fat-based spreads can also be improved by increasing the moisture content of the product; obviously, this is economical and nutritionally desirable in the sense that the caloric value is reduced, but the resultant product is less microbiologically stable than butter.

The melting characteristics and rheological properties of milk fat can also be modified by inter- and trans-esterification. Chemically-catalysed inter- and trans-esterification are not permitted in the food industry but enzymatic catalysis may be acceptable. Lipases capable of such modifications on a commercial scale are available but their use is rather limited. Enzymatic trans-esterification allows modification of the nutritional as well as the rheological properties of lipids. The nutritional and rheological properties of lipids can also be modified by the use of a desaturase that converts $C_{18:0}$ to $C_{18:1}$ (these enzymes are a subject of ongoing research, see hppt://bioinfo.pbi.nrc.ca/covello/r-fattyacid.html; and Meesapyodsuk et al. 2000). However, this type of enzyme does not seem to be available commercially yet.

Milk Fat as an Emulsion

An emulsion consists of two immiscible, mutually insoluble liquids, usually referred to as oil and water, in which one of the liquids is dispersed as small droplets (globules; the dispersed phase) in the other (the continuous phase). If the oil is the dispersed phase, the emulsion is referred to as an oil-in-water (O/W) emulsion; if water is the dispersed phase, the emulsion is referred to as a water-in-oil (W/O) emulsion. The dispersed phase is usually, but not necessarily, the phase present in the smaller amount. An emulsion is prepared by dispersing one phase into the other. Since the liquids are immiscible, they will remain discrete and separate if they differ in density, as is the case with lipids and water, the density of which are 0.9 and 1.0, respectively; the lipid globules will float to the surface and coalesce. Coalescence is prevented by adding a compound which reduces the interfacial tension, γ, between the phases. Compounds capable of doing this have an amphipathic structure, that is hydrophobic and hydrophilic regions, for example phospholipids, monoglycerides, diglycerides, proteins, soaps and numerous synthetic compounds, and are known as emulsifiers or detergents. The emulsifier forms a layer on the surface of the globules with its hydrophobic region penetrating the oil phase and its hydrophilic region in the aqueous phase. An emulsion thus stabilised will cream if left undisturbed, but the globules remain discrete and can be redispersed readily by gentle agitation.

In milk, the lipids exist as an O/W emulsion in which the globules range in size from approximately 0.1 to 20 μm, with a mean of 3–4 μm. The mean size of the fat globules is higher in high-fat milk than in low-fat milk, for example Jersey compared to Friesian, and decreases with advancing lactation. Consequently, the separation of fat from milk is less efficient in winter than in summer, especially when milk production is seasonal, and it may not be possible to meet the upper limit for fat content in some products, for example casein, during certain periods.

Stability of Milk Fat Globules

In milk, the emulsifier is the MFGM. On the inner side of the MFGM is a layer of unstructured lipoproteins, acquired within the secretory cells as the triglycerides move from the site of synthesis in the rough endoplasmic reticulum (RER) in the basal region of the cell towards the apical membrane. The fat globules are excreted from the cells by exocytosis, that is they are pushed through and become surrounded by the apical cell membrane. Milk proteins and lactose are excreted from the cell by the reverse process: the proteins are synthesised in the RER and are transported to the Golgi region, where the synthesis of lactose occurs under the control of α-La. The milk proteins and lactose are encapsulated in Golgi membrane; the vesicles move towards, and fuse with, the apical cell membrane, open and discharge their contents into the alveolar lumen, leaving the vesicle (Golgi) membrane as part of the apical membrane, thereby replacing the membrane lost on the excretion of fat globules. Thus, the outer layer of the MFGM is composed of a trilaminar membrane, consisting of phospholipids and proteins, with a fluid mosaic structure.

Many of the proteins of the MFGM are strongly hydrophobic and difficult to isolate and characterise. Modern proteomic methods have shown that the MFGM contains about 100 proteins, of which the following are the principal and have been isolated and characterised: butyrophilin (BTN), xanthine dehydrogenase (XDH), acidophilin, PAS (periodic acid Schiff staining) 6/7, CD (cluster of differentiation) 36, fatty acid-binding protein, mucins 1 and 15 (MUC). BTN is a trans-membrane protein that complexes with XDH (located on the inner face of the membrane) that initiates the blebbing of the fat globule through the apical membrane of the cell (in this role, XDH does not act as an enzyme). MUC1, MUC 15, CD 36 and PAS6/7 are heavily glycosylated and are located mainly on the outer surface of the membrane, increasing its hydrophilicity. Very considerable progress has been made on the proteins during the past 10 years, which has been reviewed by Mather (2000, 2011) and Keenan and Mather (2006). In human and equine milk, the MUC form long (up to 50 μm) filaments that are lost easily; the filaments probably retard the passage of lipids through the small intestine, thereby improving digestibility, and prevent the adhesion of pathogens.

The MFGM contains many enzymes that originate mainly from the Golgi apparatus: in fact, most of the indigenous enzymes in milk are concentrated in the MFGM, notable exceptions being plasmin and LPL that are associated with the casein micelles. The trilaminar membrane is unstable and is shed during storage, and especially during agitation, into the aqueous phase, where it forms microsomes.

The stability of the MFGM is critical for many aspects of the milk fat system:

- The existence of milk as an emulsion depends on the effectiveness of the MFGM.
- Damage to the MFGM leads to the formation of non-globular (free) fat, which may be evident as 'oiling-off' on tea or coffee, cream plug or age thickening. An elevated level of free fat in whole milk powder reduces its wettability. Problems related to, or arising from, free fat are more serious in winter than in summer, probably due to the reduced stability of the MFGM. Homogenisation, which replaces the natural MFGM by a layer of proteins from the skim milk phase, principally caseins, eliminates problems caused by free fat.
- The MFGM protects the lipids in the core of the globule against lipolysis by LPL in the skim milk (adsorbed on the casein micelles). The MFGM may be damaged by agitation, foaming, freezing, for example on bulk tank walls, and especially by homogenization, allowing access for LPL to the core lipids and leading to lipolysis and hydrolytic rancidity. This is potentially a major problem in the dairy industry unless milking machines, especially pipeline milking installations, are properly installed and serviced.
- The MFGM appears to be less stable in winter/late lactation than in summer/mid lactation; therefore, hydrolytic rancidity is more likely to be a problem in winter than in summer. An aggravating factor is that less milk is usually produced in winter than in summer, especially in seasonal milk production systems, which leads to greater agitation and air incorporation during milking and, consequently, a greater risk of damage to the MFGM.

Creaming

Since the specific gravity of lipids and skim milk is 0.9 and 1.036, respectively, the fat globules in milk held under quiescent conditions will rise to the surface under the influence of gravity, a process referred to as creaming. The rate of creaming, V, of fat globules is given by Stoke's equation:

$$V = \frac{2r^2(\rho^1 - \rho^2)g}{9\eta}$$

where

$r =$ radius of the fat globules
$\rho^1 =$ specific gravity of skim milk
$\rho^2 =$ specific gravity of the fat globules
$g =$ acceleration due to gravity
$\eta =$ viscosity of milk

The typical values of r, ρ^1, ρ^2 and η suggest that a cream layer should form in milk after approximately 60 hours but milk creams in approximately 30 minutes. The rapid rate of creaming is due to the strong tendency of the fat globules to agglutinate (stick together) due to the action of indigenous immunoglobulin (Ig) M, which precipitates onto the fat globules when milk is cooled (hence, they are called cryoglobulins). Considering the effect of globule size (r) on the rate of creaming, large globules rise faster than smaller ones and collide with, and adhere to, smaller globules, an effect promoted by cryoglobulins. Owing to the larger value of r, the clusters of globules rise faster than individual globules, and therefore the creaming process accelerates as the globules rise and clump. Ovine, caprine and buffalo milk do not contain cryoglobulins and therefore cream much more slowly than bovine milk.

In the past, creaming was a very important physicochemical property of milk:

- The cream layer served as an index of fat content and hence of quality to the consumer.
- Creaming was the traditional method for preparing fat (cream) from milk for use in the manufacture of butter. Its significance in this respect declined after the development of the mechanical separator by Gustav de Laval in 1878 but natural creaming is still used to adjust the fat content of milk for some cheese varieties, for example Parmigiano-Reggiano. A high proportion (~90%) of the bacteria in milk become occluded in the clusters of fat globules.

Homogenisation of Milk

Today, creaming is of little general significance. In most cases, its effect is negative and for most dairy products, milk is homogenised, that is subjected to a high-shear pressure that reduces the size of the fat globules (average diameter <1 μm), increases the fat surface area (four to six fold), replaces the natural MFGM by a layer of caseins and denatures cryoglobulins, hence preventing the agglutination of globules.

Homogenization of milk is usually performed using a valve homogeniser (developed by August Gaulin in 1899), in which milk is forced at a pressure of approximately 20 MPa against a spring-loaded valve. The residence time in the valve is very short and many newly formed globules share emulsifier (probably casein micelles), leading to the formation of clusters that will cream rapidly and increase the viscosity of the system. The clusters are dispersed by a second homogenisation at approximately 3.5 MPa. Recently, a high-pressure homogeniser, operating at a pressure up to 300 MPa, has been developed. This type of homogeniser gives a very narrow distribution of fat globules, resulting in improved product quality.

Homogenisation has several very significant effects on the properties of milk:

- If properly executed, creaming is delayed indefinitely due to the reduced size of the fat globules, the denaturation of cryoglobulins and the increased density of the fat globules, due to the layer of adsorbed casein on the surface.
- Susceptibility to hydrolytic rancidity is markedly increased because indigenous LPL has ready access to the triglycerides; consequently, milk must be heated under conditions sufficiently severe to inactivate LPL before (usually) or immediately after homogenisation.
- Susceptibility to oxidative rancidity is reduced because prooxidants in the MFGM, for example metals and xanthine oxidase, are distributed throughout the milk.
- The whiteness of milk is increased, due to the greater number of light-scattering particles.
- The strength and syneretic properties of rennet-coagulated milk gels for cheese manufacture are reduced; hence, cheese with a higher moisture content is obtained. Consequently, milk for cheese manufacture is not normally homogenised; an exception is reduced-fat cheese, in which a higher moisture content improves texture.
- The heat stability of whole milk and cream is reduced, the magnitude of the effect increasing directly with fat content and homogenisation pressure; homogenisation has no effect on the heat stability of skimmed milk.
- The viscosity of whole milk and cream is increased by single-stage homogenisation due to the clustering of newly formed fat globules, caused by sharing of casein micelles; the clumps of globules are dispersed by a second homogenisation stage at a lower pressure, which may be omitted if an increased viscosity is desired.

Lipid Oxidation

The chemical oxidation of lipids is a major cause of instability in dairy products (and many other foods). Lipid oxidation is a free radical, autocatalytic process principally involving the methylene group between a pair of double bonds in PUFAs; oxygen is a primary reactant (see O'Brien and O'Connor 2011). The process is initiated and/or catalysed by polyvalent metals, especially Cu and Fe, ultraviolet (UV) light, ionising radiation or enzymes, such as lipoxygenase in the case of plant oils, and xanthine oxidase, which is a major component of the MFGM, in milk. The principal end-products are unsaturated carbonyls, which cause major flavour defects; the reaction intermediates, that is fatty acid free radicals, peroxy free radicals and hydroperoxides, have no flavour. Polymerisation of free radicals and other species leads to the formation of pigmented products and to an increase in viscosity but it is unlikely that polymerisation-related problems occur to a significant extent in dairy products.

Lipid oxidation can be prevented or controlled by the following:

- Avoiding metal contamination at all stages of processing through the use of stainless steel equipment.
- Avoiding exposure to UV light by using opaque packing (foil or paper).
- Packaging under an inert atmosphere, usually N_2.
- Use of O_2 or free-radical scavengers, for example glucose oxidase and superoxide dismutase (an indigenous enzyme in milk), respectively.
- Use of antioxidants which break the free radical chain reaction; synthetic antioxidants are not permitted in dairy products but the level of natural antioxidants, for example tocopherols (vitamin E), in milk may be increased by supplementing the animal's feed. Polyphenols are very effective antioxidants; their direct addition to dairy products is not permitted but it may also be possible to increase their concentration in milk by supplementing the animal's feed. Antioxidants are compounds which readily supply a $H^{\cdot}$ to fatty acid and peroxy radicals, leaving a stable oxidised radical. Many antioxidants are polyphenols, which give up a $H^{\cdot}$ and are converted to a quinone; examples are tocopherols (vitamin E) and catechins. At low concentrations, ascorbic acid is a good antioxidant but at high concentrations it

functions as a pro-oxidant, apparently as a complex with Cu. Sulphydryl groups of proteins are also effective antioxidants; cream for butter making is usually heated to a high temperature to denature proteins and expose and activate sulphydryl groups.

Fat-Soluble Vitamins

Since the fat-soluble vitamins (A, D, E and K) in milk are derived from the animal's diet, large seasonal variations can be expected in their concentration in milk. The breed of cow also has a significant effect on the concentration of fat-soluble vitamins in milk; high-fat milk (Jersey and Guernsey) has a higher content of these vitamins than Friesian or Holstein milk. Variations in the concentrations of fat-soluble vitamins in milk have a number of consequences:

- Nutritionally, milk contributes a substantial portion of the RDA (recommended daily allowance) for these vitamins to Western diets; it is common practice to fortify milk and butter with vitamins A and D.
- The yellow-orange colour of high-fat dairy products depends on the concentrations of carotenoids and vitamin A present, and hence on the diet of the animal. New Zealand butter is much more highly coloured than Irish butter, which in turn is much more yellow than American or German products. The differences are due in part to the greater dependence of milk production in New Zealand and Ireland on pasture and to the higher proportion of carotenoid-rich clover in New Zealand pasture and the higher proportion of Jersey cows in New Zealand herds.
- Goats, sheep and buffalo do not transfer carotenoids to their milk, which is, consequently, whiter than bovine milk. Products produced from these milks are whiter than corresponding products made from bovine milk. The darker colour of the latter may be unattractive to consumers accustomed to caprine or ovine milk products. If necessary, the carotenoids in bovine milk may be bleached (by benzoyl peroxide) or masked (by chlorophyll or T_iO_2).
- Vitamin E (tocopherols) is a potent antioxidant and contributes to the oxidative stability of dairy products. The tocopherol content of milk and meat can be readily increased by supplementing the animal's diet with tocopherols, which is sometimes practised.

MILK PROTEINS

Introduction

Technologically, the proteins of milk are its most important constituents. They play important, even essential, roles in all dairy products except butter and anhydrous milk fat. The roles played by milk proteins include the following:

- *Nutritional*: All milk proteins.
- *Physiological*: Igs, lactoferrin (Lf), lactoperoxidase, vitamin-binding proteins, protein-derived biologically-active peptides.
- *Physico-chemical*:
  - *Gelation*. Enzymatically, acid- or thermally induced gelation in all cheeses, fermented milks, whey protein concentrates and isolates.
  - *Heat stability*: All thermally processed dairy products,
  - *Surface activity*: Caseinates, whey protein concentrates and isolates.
  - *Rheological*: All protein-containing dairy products.
  - *Water sorption*: Most dairy products, comminuted meat products.

Milk proteins have been studied extensively and are very well characterised at molecular and functional levels (for reviews, see Fox and McSweeney 2003).

Heterogeneity of Milk Proteins

It has been known since 1830 that milk contains two types of protein which can be separated by acidification, to what we now know is pH 4.6. The proteins insoluble at pH 4.6 are called caseins and represent approximately 78% of the total nitrogen in bovine milk; the soluble proteins are called whey or serum proteins. As early as 1885, it was shown that there are two types of whey protein, globulins and albumins, which were thought to be transferred directly from the blood (the proteins of blood and whey have generally similar physico-chemical properties and are classified as albumins and globulins). Initially, the term casein was not restricted to the acid-insoluble proteins in milk but was used to describe all acid-insoluble proteins; however, it was recognised at an early stage that the caseins are unique milk-specific proteins.

The casein fraction of milk protein was considered initially to be homogeneous but from 1918 onwards, evidence began to accumulate that it is heterogeneous. Through the application of free boundary electrophoresis (FBE) in the 1930s and especially zone electrophoresis in starch or polyacrylamide gels (SGE, PAGE) containing urea and a reducing agent in the 1960s, it has been shown that casein is in fact very heterogeneous. Bovine casein consists of four families of caseins: α_{s1}-, α_{s2}- β- and κ-, which represent about 38%, 10%, 36% and 12%, respectively, of whole casein. Urea-PAGE showed that each of the casein families exhibits micro-heterogeneity due to:

- genetic polymorphism, usually involving substitution of one or two amino acids;
- variations in the degree of phosphorylation;
- variations in the degree of glycosylation of κ-casein;
- inter-molecular disulphide bond formation in α_{s2}- and κ-caseins;
- limited proteolysis, especially of β- and α_{s2}-caseins, by plasmin; the resulting peptides include the γ- and λ-caseins and proteose peptones.

In the 1930s, FBE showed that both the globulin and albumin fractions of whey protein are heterogeneous and, in the 1950s, the principal constituents were isolated and characterised. It is now known that the whey protein fraction of bovine milk comprises four main proteins: β-lactoglobulin (β-Lg), α-La, Igs and blood serum albumin (BSA), which represent about 40%, 20%, 10% and 10%, respectively, of the total whey protein in mature milk. The remaining 10% is mainly non-protein nitrogen and trace amounts of several proteins, including *ca.* 60 indigenous. Sixty indigenous enzymes. About 1% of total milk protein is part of the MFGM, including many enzymes. β-Lg and α-La are synthesised in the mammary gland and are milk-specific; they exhibit genetic polymorphism (in fact, the genetic polymorphism of milk proteins was first demonstrated for β-Lg in 1956). BSA, most of the Ig (i.e., IgG) and most of the minor proteins are transferred from the blood.

Methods for the isolation of the individual proteins were developed and gradually improved during the period 1950–1970, so that by around 1970 all the principal milk proteins had been purified to homogeneity.

The concentration of total protein in milk is affected by most of the factors that affect the concentration of fat, that is breed, individuality, nutritional status, health and stage of lactation but, with the exception of the last, the magnitude of most effects is less than in the case of milk fat. The concentration of protein in milk decreases very markedly during the first few days *post-partum*, mainly due to the decrease in Ig from approximately 10% in the first colostrum to 0.1% within about 1 week. The concentration of total protein continues to decline more slowly thereafter, to a minimum after about 4 weeks and then increases until the end of lactation. Data on variations in the groups of proteins throughout lactation have been published (see Mehra et al. 1999) but there are few data on variations in the concentrations of the individual principal proteins.

Molecular Properties of Milk Proteins

The six principal milk-specific proteins have been isolated and are very well characterised at the molecular level; their chemical composition is summarised in Table 24.2. The most notable features of the principal milk-specific proteins are discussed later.

They are quite small molecules, approximately 15–25 kDa. All the caseins are phosphorylated but to different and variable degrees, 1–13 mol P/mol protein; the phosphate groups are esterified as monoesters of serine residues. The primary structures of the caseins have a rather uneven distribution of polar and non-polar residues along their sequences. Clustering of the phosphoseryl residues is particularly marked, resulting in strong anionic patches in α_{s1}-, α_{s2}- and β-caseins. κ-Casein does not have a phosphoseryl cluster but the *N*-terminal two-thirds of the molecule is quite hydrophobic while its *C*-terminal is relatively hydrophilic – it contains no aromatic and no cationic residues. These features give the caseins an amphipathic structure, making them very surface-active, with good emulsifying and foaming properties. The amphipathic structure of κ-casein is particularly significant and is largely responsible for its micelle-stabilising properties. The distribution of amino acids in β-Lg and α-La is quite random.

The two principal caseins, α_{s1}- and β-caseins, are devoid of cysteine or cystine residues; the two minor caseins, α_{s2}- and κ-caseins, contain two inter-molecular disulphide bonds. β-Lg contains two intra-molecular disulphide bonds and one sulphydryl group that is buried and unreactive in the native protein but becomes exposed and reactive when the molecule is denatured; it reacts *via* sulphydryl-disulphide interactions with other proteins, especially κ-casein, with major consequences for many important properties of the milk protein system, especially heat stability and cheese making properties. α-La has four intra-molecular disulphide bonds.

All the caseins, especially β-casein, contain a high level of proline (in β-casein, 17 of the 209 residues are proline), which disrupts α- and β-structures; consequently, the caseins are rather unstructured molecules and are readily susceptible to proteolysis, which is as would be expected for proteins the putative function of which is as a source of amino acids for the neonate. However, theoretical calculations suggest that the caseins may have a considerable level of secondary and tertiary structure; to explain the differences between the experimental and theoretical indices of higher structures, it has been suggested that the caseins are very mobile, flexible, rheomorphic molecules (i.e., the casein molecules, in solution, are sufficiently flexible to adopt structures that are dictated by their environment; Holt and Sawyer 1993). In contrast, the whey proteins are highly compact and structured, with high levels of α-helices, β-sheets and β-turns. In β-Lg, the β-sheets are in an anti-parallel arrangement and form a β-barrel calyx.

Table 24.2. Characteristics of the Principal Proteins in Bovine Milk

Protein	Molecular Weight	Amino Acid Residues: Total	Proline	Cysteine	Phosphate Groups	Concentration (g/L)	Genetic Variants
α_{s1}-casein	23,164	199	17	0	8	10.0	A, B, C, D, E, F, G, H
α_{s2}-casein	25,388	207	10	2	10–13	2.6	A, B, C, D
β-casein	23,983	209	35	0	5	9.3	A^1, A^2. A^3, B, C, D, E, F, G
κ-casein	19,038	169	20	2	1	3.3	A, B, C, D, E, F^S, F^I, G^S, H, I, J
β-lactoglobulin	18,277	162	8	5	0	3.2	A, B, C, D, E, F, G, H, I, J
α-lactalbumin	14,175	123	2	8	0	1.2	A, B, C

Source: Adapted from Fox 2003.

The caseins are often regarded as very hydrophobic proteins but they are not particularly so; however, they do have a high surface hydrophobicity, owing to their open structures. Because of their hydrophobic patches they have a high propensity to yield bitter hydrolysates, even in cheese that undergoes relatively little proteolysis. In contrast, the highly structured whey proteins are very resistant to proteolysis in the native state and may traverse the intestinal tract of the neonate intact.

About 60% of κ-casein molecules are glycosylated. The sugar moieties present are galactose, galactosamine and *N*-acetylneuraminic acid (sialic acid), which occur as trisaccharides or tetrasaccharides. The oligosaccharides are attached to the polypeptide *via* threonine residues in the *C*-terminal region of the molecule and vary from 0 to 4 mol/mol of protein.

Probably because of their rather open structures, the caseins are extremely heat stable, for example sodium caseinate can be heated at 140°C for 1 hour without obvious physical effects. The more highly structured whey proteins are comparatively heat-labile, although compared to many other globular proteins, they are quite heat-stable; they are completely denatured on heating at 90°C for 10 minutes.

Under the ionic conditions in milk, α-La exists as monomers of molecular weight (MW) approximately 14.7 kDa. β-Lg exists as dimers (MW ~36 kDa) in the pH range 5.5–7.5; at pH values <3.5 or >7.5, it exists as monomers, while at pH 3.5–5.5, it exists as octamers. The caseins have a very strong tendency to associate. Even in sodium caseinate, the most soluble form of whole casein, the proteins exist mainly as decamers or larger aggregates at 20°C. At low concentrations, for example <0.6%, at 4°C, β-casein exists as monomers but associates to form micelles if the concentration or temperature is increased. In the presence of Ca, for example, calcium caseinate, casein forms large aggregates of several million Daltons. In milk, the caseins exist as very complex structures, known as casein micelles, which are described in Section 'Casein Micelles'.

The function of the caseins appears to be to supply amino acids to the neonate. They have no biological function *stricto sensu* but their Ca-binding properties enable a high concentration of calcium phosphate to be carried in milk in a 'soluble' form; the concentration of calcium phosphate far exceeds its solubility and, without the 'solubilising' influence of casein, would precipitate in the ducts of the mammary gland and cause atopic milk stones.

β-Lg has a hydrophobic pocket within which it can bind small hydrophobic molecules. It binds and protects retinol *in vitro* and perhaps functions as a retinol carrier *in vivo*. In the intestine, it exchanges retinol with a retinol-binding protein. It also binds fatty acids and thereby stimulates lipase; this is perhaps its principal biological function. β-Lg is a member of the lipocalin family, which now includes 14 proteins (for review see Akerstrom et al. 2000); all members of the lipocalin family have some form of binding function.

α-La is a metalloprotein; it binds one Ca atom per molecule in a peptide loop containing four aspartic acid residues. The apo-protein is quite heat-labile but the metalloprotein is rather heat-stable; when studied by differential scanning calorimetry, α-La is in fact observed to be quite heat-labile. The metallo-protein renatures on cooling, whereas the apo-protein does not. The difference in heat stability between the halo- and apo-protein may be exploited in the isolation of α-La on a large (i.e., potentially industrial) scale.

As discussed in Section 'Lactose', α-La is an enzyme modifier protein in the lactose synthesis pathway; it makes UDP-galactosyl transferase highly specific for glucose as an acceptor for galactose, resulting in the synthesis of lactose.

Interspecies Comparison of Milk Proteins

The milk of all species that have been studies contain caseins and whey proteins, but the protein systems differ in detail:

- The total concentration of protein ranges from approximately 1% in human milk and that of the other great apes to approximately 20% for lagomorphs; the protein content is directly related to the growth rate of the neonate.
- The ratio of casein to whey proteins ranges from approximately 80:20 in bovine, buffalo, ovine and caprine milk to approximately 30:70 for human milk.
- The types and proportions of the caseins and whey proteins show considerable inter-species differences. The milk of all species studied seems to contain α-, β- and κ- caseins but human milk and that of some other primates contains very little α-casein and equine milk contains little κ-casein. The milk of humans and that of many others, but not all, primates and rodents lacks β-Lg. The principal protein in human milk is α-La but the milk of some seals lacks this protein. The milk of some species contains a protein called acid whey protein, which is absent from the milk of many species, including cow, buffalo, sheep and goat. There are particularly large inter-species differences with respect to many minor proteins, including enzymes.
- All individual milk proteins show inter-species differences with respect to amino acid composition and perhaps other features. An interspecies comparison of the principal lactoproteins of several species has been made by Martin et al. (2003, 2011).

Casein Micelles

Three of the caseins, α_{s1}-, α_{s2}- and β-, which together represent approximately 85% of total casein, are precipitated by calcium at concentrations >6 mM at temperatures >20°C. Since milk contains approximately 30 mmol/L Ca, it would be expected that most of the caseins would precipitate in milk. However, κ-casein is soluble at high concentrations of Ca and it reacts with and stabilises the Ca-sensitive caseins through the formation of casein micelles.

Since the stability, or perhaps instability, of the casein micelles is responsible for many of the unique properties and processing characteristics of milk, their structure and properties have been studied intensively. It has been known for nearly 200 years that the casein in milk exists as large colloidal particles which are retained by porcelain-Chamberlain filters. During the early part of the twentieth century, there was interest in explaining the rennet coagulation of milk and hence in the structure and properties of the Ca-caseinate particles (for review, see Fox and Kelly 2003, Fox and McSweeney 2003). The term casein micelle appears to have been first used by Beau (1921). The idea that the rennet-induced coagulation of milk is due to the destruction of a protective colloid (Schütz colloid) dates from the 1920s; initially, it was suggested that the whey proteins were the 'protective colloid'. The true nature of the protective colloid, the structure of the casein micelle and the mechanism of rennet coagulation did not become apparent until the pioneering work on the identification, isolation and characterisation of κ-casein by Waugh and von Hippel (1956). Since then, the structure and properties of the casein micelle have been studied intensively. The evolution of views on the structure of the casein micelle was described by Fox and Kelly (2003), Horne (2003, 2011) and Fox and Brodkorb (2008). Current knowledge on the composition, structure and properties of the casein micelle and the key features thereof are summarised later.

The micelles are spherical colloidal particles, with a mean diameter of approximately 120 nm (range, 50–600 nm). They have a mean particle mass of approximately 10^8 Da, that is there are about 5000 casein molecules (20,000–25,000 Da each) in an average micelle. On a dry weight basis, the micelles contain approximately 94% protein and approximately 6% non-protein species, mainly calcium and phosphate, with smaller amounts of Mg and citrate and traces of other metals; these are collectively called colloidal (or micellar) calcium phosphate (CCP or MCP). Under the conditions that exist in milk, the micelles are hydrated to the extent of approximately 2 g H_2O/g protein. There are approximately 10^{15} micelles/mL milk, with a total surface area of approximately 5×10^4 cm^2; the micelles are about 240 nm apart. Owing to their very large surface area, the surface properties of the micelles are of major significance and because they are quite closely packed, even in unconcentrated milk, they collide frequently due to Brownian, thermal and mechanical motion.

The casein micelles scatter light; the white appearance of milk is due mainly to light scattering by the micelles, with a contribution from the fat globules. The micelles are generally stable to most processes and conditions to which milk is normally subjected and may be reconstituted from spray-dried or freeze-dried milk without major changes in their properties. Freezing milk destabilises the micelles, due to a decrease in pH and an increase in [Ca^{2+}]. On very severe heating, for example at 140°C, the micelles shatter initially, then aggregate and eventually, after approximately 20 minutes, the system coagulates (see Chapter 25)

The micelles can be sedimented by centrifugation; approximately 50% are sedimented by centrifugation at 20,000 × g for 30 minute and approximately 95% by 100,000 × g for 60 minute. The pelleted micelles can be redispersed by agitation, for example by ultrasonication, in natural or synthetic milk ultrafiltrate; the properties of the redispersed micelles are not significantly different from those of native micelles.

The casein micelles are not affected by regular (~20 MPa) or high-pressure (up to 250 MPa) homogenisation (Hayes and Kelly 2003). The average size is increased by high-pressure treatment at 200 MPa, but they are disrupted (i.e., average size reduced by ~50%) by treatment at a somewhat higher pressure, that is ≥400 MPa (Huppertz et al. 2002, 2004).

The micro-structure of the casein micelle has been the subject of considerable research during the past 40 years, that is since the discovery and isolation of the micelle-stabilising protein, κ-casein; however, there is still a lack of general consensus. Numerous models have been proposed, the most widely supported initially was the sub-micelle model, first proposed by Morr (1967) and refined several times since (Fox and Kelly 2003, Horne 2003, 2011, Fox and Brodkorb 2008). Essentially, this model proposes that the micelle is built up from sub-micelles (MW ~5×10^6 Da) held together by CCP and surrounded and stabilised by a surface layer, approximately 7 nm thick, rich in κ-casein but also containing some of the other caseins (Fig. 24.3A). It is proposed that the hydrophilic C-terminal region of κ-casein protrudes from the surface, creating a hairy layer around the micelle and stabilising it through a zeta potential of *ca* −20 mV and by steric stabilisation. The principal direct experimental evidence for this model is provided by electron microscopy, which indicates a non-uniform electron density; this has been interpreted as indicating sub-micelles.

However, several authors have expressed reservations about the sub-unit model and three alternative models have been proposed. Visser (1992) suggested that the micelles are spherical aggregates of casein molecules randomly aggregated and held together partly by salt bridges in the form of amorphous calcium phosphate and partly by other forces, for example hydrophobic interactions, with a surface layer of κ-casein. Holt (1992) proposed that the Ca-sensitive caseins are linked by micro-crystals of CCP and surrounded by a layer of κ-casein, with its C-terminal region protruding from the surface (Fig. 24.3B). In the dual-binding model of Horne (2003, 2011), it is proposed that individual casein molecules interact *via* hydrophobic regions in their primary structures, leaving the hydrophilic regions free and with the hydrophilic C-terminal region of κ-casein protruding into the aqueous phase (Fig. 24.3C). Thus, the key structural features of the sub-micelle model are retained in the three alternatives, that is the integrating role of CCP and a κ-casein-rich surface layer.

The micelles disintegrate:

- when the CCP is removed, for example by acidification to pH 4.6 and dialysis in the cold against bulk milk, or by addition of trisodium citrate, also followed by dialysis; removal of >60% of the CCP results in disintegration of the micelles;
- on raising the pH to approximately 9.0, which does not solubilise the CCP and presumably causes disintegration by increasing the net negative charge;
- on adding urea to >5 M, which suggests that hydrogen and/or hydrophobic bonds are important for micelle integrity.

The micelles are precipitated by ethanol or other low MW alcohols at levels of ≥ approximately 35% at 20°C. However, if the temperature is increased to ≥70°C, surprisingly, the precipitated casein dissolves and the solution becomes quite clear, indicating

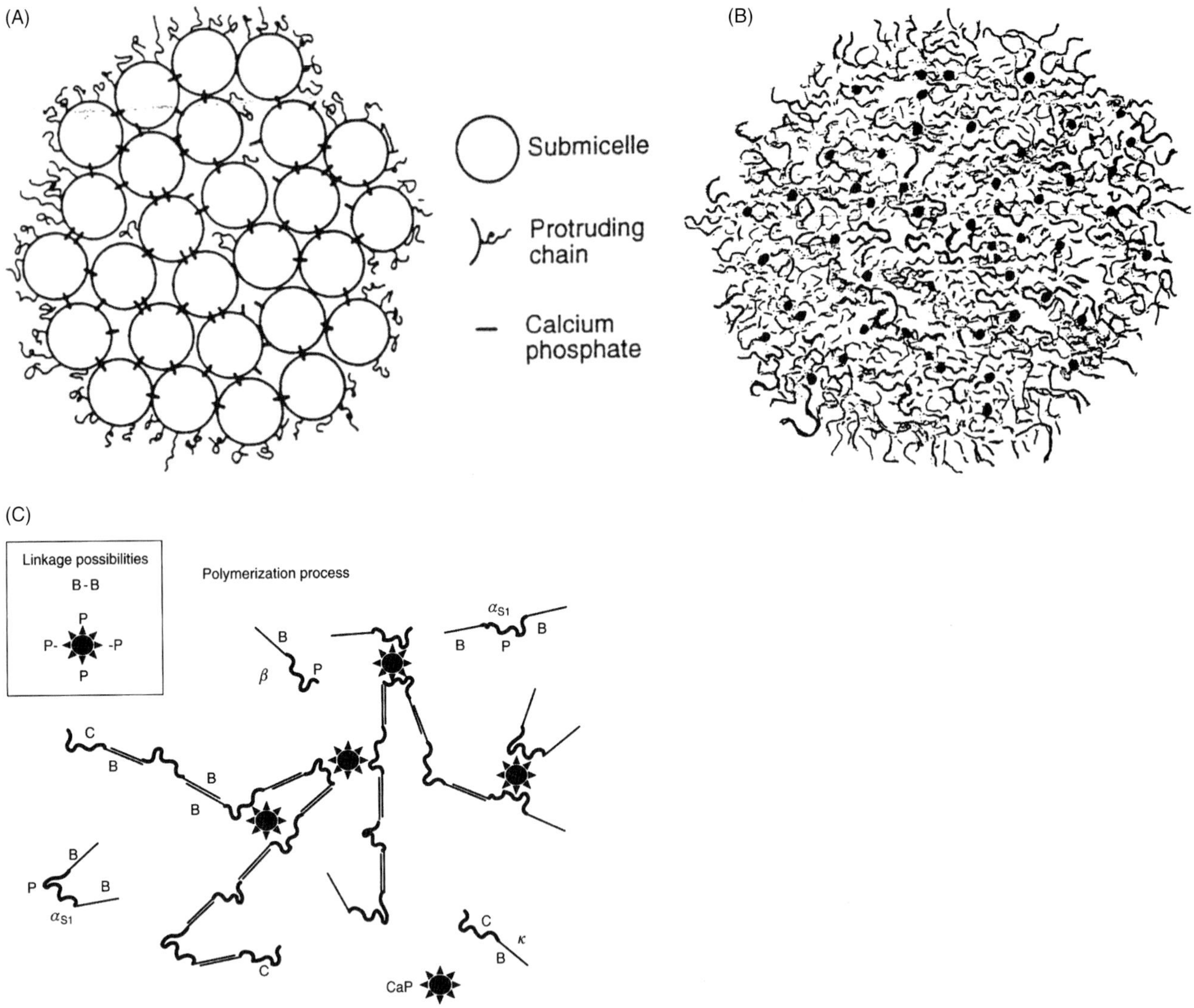

Figure 24.3. Models of casein micelle structure: (**A**) the submicelle model of Walstra (1999); (**B**) the Holt model (from Fox and McSweeney 1998); (**C**) the dual binding model, showing interactions between α_{s1}-, β- and κ-caseins (Horne 2003).

dissociation of the micelles (O'Connell et al. 2001a, b). Micelle-like particles reform on cooling and form a gel at ~4°C. It is not known if the sub-particles formed by any of these treatments correspond to casein sub-micelles.

The turbidity of skim milk increases on addition of SDS up to 21mM, at which concentration a gel is formed; however, at ≥28 mM (~0.8%, w/v) the micelles disperse (see Lefebvre-Cases et al. 2001a, b).

There have been few studies on variations in micelle size throughout lactation and these have failed to show consistent trends. No studies on variability in the micro-structure of the casein micelle have been published. We are not aware of studies on the effect of the nutritional or health status of the animal on the structure of the casein micelles, although their stability and behaviour are strongly dependent on pH, milk salts and whey proteins, all of which are affected by the health and nutritional status of the cow.

The milk of all species that have been studied is white, presumably due to light scattering by the casein micelles. However, the micelles of only a few species have been studied in detail; electron microscopy indicates that the micelles of all species are spherical and range in size from approximately 50 nm in human milk to ~600 nm in equine milk. Some of the large micelles in equine milk settle out under gravity or low centrifugal forces; the large micelles reflect the low level of κ-casein in equine milk. The milk of most species is supersaturated with calcium phosphate and the insoluble part (CCP) is present in the casein micelles, in which it acts as a cementing material; the micelles disintegrate when the CCP is removed. Human milk lacks CCP and its micelles have a rather porous structure.

Minor Proteins

In addition to the caseins and the two principal whey proteins, milk contains several proteins at low or trace levels. Many of these minor proteins are biologically active (see Schrezenmeir et al. 2000); some are regarded as highly significant and have attracted considerable attention as nutraceuticals. When ways of increasing the value of milk proteins are discussed, the focus is usually on these minor proteins but they are, in fact, of little economic value to the overall dairy industry. They are found mainly in the whey but some are also located in the fat globule membrane. Reviews on the minor proteins include Fox and Flynn (1992), Haggarty (2003) and Wynn (2011).

Immunoglobulins

Bovine colostrum contains approximately 10% (w/v) Igs, but this level declines rapidly to approximately 0.1% about 5 days *post-partum*. IgG1 is the principal Ig in bovine, caprine or ovine milk, with lesser amounts of IgG2, IgA and IgM; IgA is the principal Ig in human milk. The cow, sheep and goat do not transfer Ig to the foetus *in utero* and the neonate is born without Ig in its blood serum; consequently, it is very susceptible to bacterial infection, with a very high risk of mortality. The young of these species can absorb Ig from the intestine for several days after birth and thereby acquire passive immunity until they synthesise its own Ig, within a few weeks of birth. The human mother transfers Ig *in utero* and the offspring is born with a broad spectrum of antibodies. Although the human baby cannot absorb Ig from the intestine, the ingestion of colostrum is still very important because the Igs it contains prevent intestinal infection. Some species, for example the horse, transfer Ig both *in utero* and *via* colostrum.

The modern dairy cow produces colostrum far in excess of the requirements of her calf. Therefore, surplus colostrum is available for the recovery of Ig and other nutraceuticals (Paakanen and Aalto 1997, Marnila and Korhonen 2011). There is also considerable interest in hyper-immunising cows against certain human pathogens, for example rota virus, for the production of antibody-rich milk for human consumption, especially by infants; the Ig could be isolated from the milk and presented as a 'pharmaceutical' or consumed directly in the milk.

Blood Serum Albumin

About 1–2% of the protein in bovine milk is BSA, which enters from the blood by leakage through inter-cellular junctions. As befits its physiological importance, BSA is very well characterised (see Carter and Ho 1994). BSA has no known biological function in milk and, considering its very low concentration, it probably has no technological significance either.

Metal-Binding Proteins

Milk contains several metal-binding proteins: the caseins (which bind Ca, Mg, PO_4) are quantitatively the most important; others are α-La (Ca), xanthine oxidase (Fe, Mo), alkaline phosphatase (Zn, Mg), lactoperoxidase (Fe), catalase (Fe), ceruloplasmin (Cu), glutathione peroxidase (Se), Lf (Fe) and seroferrin (Fe).

Lf, a non-haem iron-binding glycoprotein (see Lonnerdal 2003, Korhonen and Marnila 2011), is a member of a family of iron-binding proteins, which includes seroferrin and ovotransferrin (conalbumin). It is present in several body fluids, including saliva, tears, sweat and semen. Lf has several potential biological functions; it improves the bioavailability of Fe, is bacteriostatic (by sequestering Fe and making it unavailable to intestinal bacteria), and has antioxidant, antiviral, anti-inflammatory, immunomodulatory and anti-carcinogenic activity. Human milk contains a much higher level of Lf (~20% of total N) than bovine milk and therefore there is interest in fortifying bovine milk-based infant formula with Lf. The pH of Lf is ~9.0, that is it is cationic at the pH of milk, whereas most milk proteins are anionic, and can be isolated on an industrial scale by adsorption on a cation-exchange resin. Hydrolysis of Lf by pepsin yields peptides called lactoferricins, which are more bacteriostatic than Lf and their activity is independent of iron status. Bovine milk also contains a low level of serum transferrin.

Milk contains a copper-binding glycoprotein, ceruloplasmin, also known as ferroxidase (EC 1.16.3.1) (see Wooten et al. 1996). Ceruloplasmin is an α_2-globulin with a MW of ~126,000 Da; it binds six atoms of copper per molecule and may play a role in delivering essential copper to the neonate.

β_2-Microglobulin

β_2-Microglobulin, initially called lactollin, was first isolated from acid-precipitated bovine casein by Groves et al. (1963). Lactollin, with a MW of 43, 000 Da, is a tetramer of β_2-microglobulin, which consists of 98 amino acids, with a calculated MW of 11, 636 Da. β_2-Microglobulin, a component of the immune system (see Groves and Greenberg 1982), is probably produced by proteolysis of a larger protein, mainly within the mammary gland; it has no known significance in milk.

Osteopontin

Osteopontin (OPN) is a highly phosphorylated acidic glycoprotein, consisting of 261 amino acid residues with a calculated MW of 29,283 (total MW of the glycoprotein, ~60, 000 Da). OPN has 50 potential calcium-binding sites, about half of which are saturated under normal physiological concentrations of calcium and magnesium. OPN occurs in bone (it is one of the major non-collagenous proteins in bone), in many other normal and malignant tissues, in milk and urine and can bind to many cell types. It is believed to have a diverse range of functions (Denhardt and Guo 1993, Bayless et al. 1997), but its role in milk is not clear.

Proteose Peptone 3

Bovine proteose peptone 3 (PP3) is a heat-stable phosphoglycoprotein that was first identified in the proteose-peptone (heat-stable, acid-soluble) fraction of milk. Unlike the other peptides in this fraction, which are proteolytic products of the caseins, PP3 is an indigenous milk protein, synthesised in the mammary gland. Bovine PP3 is a polypeptide of 135 amino acids, with five phosphorylation and three glycosylation sites. When isolated from milk, the PP3 fraction contains at least three components of MW ~28, 18 and 11 kDa; the largest of these is PP3, while the smaller components are fragments thereof generated by plasmin (see Girardet and Linden 1996). PP3 is present mainly in acid whey but some is present in the MFGM also. Girardet and Linden (1996) proposed that its name should be changed to *lactophorin* or *lactoglycoporin*. PP3 has also been referred to as *the hydrophobic fraction of proteose peptone*.

Owing to its strong surfactant properties (Campagna et al. 1998), PP3 can prevent contact between milk lipase and its substrates, thus preventing spontaneous lipolysis. Although its amino acid composition suggests that PP3 is not a hydrophobic protein, it behaves hydrophobically, possibly owing to the formation of an amphiphilic α-helix, one side of which contains hydrophilic residues while the other side is hydrophobic. The biological role of PP3 is unknown.

Vitamin-Binding Proteins

Milk contains binding proteins for at least the following vitamins: retinol (vitamin A, i.e., β-Lg, see Section 'Molecular Properties of Milk Proteins'), biotin, folic acid and cobalamine (vitamin B_{12}). The precise role of these proteins is not clear but they may improve the absorption of vitamins from the intestine or act as antibacterial agents by rendering vitamins unavailable to bacteria. The concentration of these proteins varies during lactation, but the influence of other factors such as individuality, breed and nutritional status is not known. The activity of these proteins is reduced or destroyed on heating at temperatures somewhat higher than HTST pasteurisation.

Angiogenins

Angiogenins induce the growth of new blood vessels, that is angiogenesis. They have high sequence homology with members of the RNase A superfamily of proteins and have RNase activity. Angiogenesis is a complex biological process of which the ribonuclease activity of angiogenins is one of a number of essential biochemical steps that lead to the formation of new blood vessels (Strydom 1998).

Two angiogenins (ANG-1 and ANG-2) have been identified in bovine milk and blood serum. Both strongly promote the growth of new blood vessels in a chicken membrane assay. Bovine ANG-1 has 64% sequence identity with human angiogenin and 34% identity with bovine RNase A. The amino acid sequence of bovine ANG-2 has 57% identity with that of bovine ANG-1; ANG-2 has lower RNase activity than ANG-1. The function(s) of the angiogenins in milk is unknown; they may be part of a repair system to protect either the mammary gland or the intestine of the neonate and/or part of the host-defence system.

Kininogen

Two forms of kininogen have been identified in bovine milk, a high (>68 kDa) and a low (16–17 kDa) MW form (Wilson et al. 1989). Bradykinin, a biologically-active peptide containing nine amino acids that is released from the high MW kininogen by the action of the enzyme, kallikrein, has been detected in the mammary gland, and is secreted into milk, from which it has been isolated. Plasma kininogen is an inhibitor of thiol proteases and has an important role in blood coagulation. Bradykinin affects smooth muscle contraction, induces hypertension and is involved in natriuresis and diuresis. The biological significance of bradykinin and kininogen in milk is unknown.

Glycoproteins

Many of the minor proteins discussed previously are glycoproteins; in addition, several other minor glycoproteins have been found in milk and colostrum but their identity and function have not been elucidated fully. Some of these glycoproteins belong to a family of closely related, highly acidic glycoproteins, called M-1 glycoproteins. Some glycoproteins stimulate the growth of bifidobacteria, presumably *via* their amino sugars. One of the M-1 glycoproteins in colostrum, but which has not been detected in milk, is orosomucoid (α_1-acid glycoprotein), a member of the lipocalin family, which is thought to modulate the immune system.

One of the high MW glycoproteins in bovine milk is prosaposin, a neurotrophic factor that plays an important role in the development, repair and maintenance of the nervous system (Patton et al. 1997). It is a precursor of saposins A, B, C and D, which are sphingolipid-activator proteins, but saposins have not been detected in milk. The physiological role of prosaposin in milk is not known, although the potent biological activity of saposin C, released by digestion, could be important for the growth and development of the young.

Proteins in the Milk Fat Globule Membrane

About 1% of the total protein in milk is in the MFGM. Most of the proteins are present at trace levels, including many of the indigenous enzymes in milk. The principal proteins in the MFGM include MUC1 and MUC15, adipophilin, BTN and XDH (Keenan and Mather 2003, 2003, 2006). BTN is a very hydrophobic protein and has similarities to the Igs (for review see Mather 2000, 2011).

Growth Factors

A great diversity of protein growth factors (hormones), including epidermal growth factor, insulin, insulin-like growth factors 1 and 2, three human milk growth factors (α_1, α_2 and β), two mammary-derived growth factors (I and II), colony-stimulating factor, nerve growth factor, platelet-derived growth factor and bombasin, are present in milk. It is not clear whether these factors play a role in the development of the neonate or in the development and functioning of the mammary gland, or both (Fox and Flynn 1992, Pihlanto-Leppala 2011, Wynn 2011).

Milk Protein-Derived Bioactive Peptides

The six lacto-proteins contain sequences, which when released on proteolysis display various biological activities (Pihlanto-Leppala 2011, Fosset and Tome 2011). Some of these peptides are formed in the GIT in vivo but it not known whether or not they are active in vivo.

Indigenous Milk Enzymes

Milk contains about 60 indigenous enzymes, which represent a minor but very important part of the milk protein system (for review, see Fox and Kelly 2006, Fox 2003b). The enzymes originate from the secretory cells or the blood. Many of the indigenous enzymes are concentrated in the MFGM and originate in the Golgi membranes of the cell or the cell cytoplasm, some of which occasionally becomes entrapped as crescents inside the encircling membrane during exocytosis. Plasmin and LPL are associated with the casein micelles and several enzymes are present in the milk serum, many of which are derived from the MFGM, which is shed during the storage of milk.

There has been interest in indigenous milk enzymes since lactoperoxidase was discovered in 1881. Although present at relatively low levels, the indigenous enzymes are significant for several reasons:

Nutritional and Protective Some indigenous enzymes have a protective effect on the neonate or perhaps on the mammary gland of the mother, for example lysozyme, lactoperoxidase, superoxide dismutase and xanthine oxidoreductase. Many of the enzymes have the potential to participate in digestion but probably the only enzyme significant in this regard is bile salts stimulated lipase (carboxyl ester hydrolase, CEH) which is believed to make a significant contribution to lipolysis; human milk is particularly rich in CEH but it is also present in the milk of several other species.

Technological

- Plasmin causes proteolysis in milk and some dairy products; it may be responsible for age gelation in UHT milk and contributes to proteolysis in cheese during ripening, especially in varieties that are cooked at a high temperature, in which the coagulant is extensively or completely denatured, for example Emmental, Parmesan and Mozzarella.
- LPL may cause hydrolytic rancidity in milk and butter but contributes positively to cheese ripening, especially that made from raw milk.
- Acid phosphatase can dephosphorylate casein and modify its functional properties; it may contribute to cheese ripening.
- Xanthine oxidoreductase is a very potent pro-oxidant and may cause oxidative rancidity in milk; it reduces nitrate, used to control the growth of clostridia in several cheese varieties, to nitrite.
- Lactoperoxidase is a very effective bacteriocidal agent in the presence of a low level of H_2O_2 and SCN^- and is exploited for the cold-sterilisation of milk.

Indices of Milk Quality and History

- The standard assay to assess the adequacy of HTST pasteurisation is the inactivation of alkaline phosphatase. Proposed assays for super-pasteurisation of milk are based on the inactivation of γ-glutamyltranspeptidase or lactoperoxidase.
- The concentration/activity of several enzymes in milk increases during mastitic infection and some have been used as indices of this condition, for example catalase, acid phosphatase and especially *N*-acetylglucosaminidase.

Antibacterial

- Milk contains several bactericidal agents, two of which are the enzymes, lysozyme and lactoperoxidase.

MILK SALTS

Milk contains six inorganic elements, Ca, Mg, Na, K, Cl and P, at substantial concentrations (macroelements) and about 20 more at trace levels (microelements). These elements are referred to as milk salts or, incorrectly, as minerals. In addition, milk contains citrate that interacts with and affects the state of inorganic salts. Although they are relatively minor constituents, the milk salts of milk have major effects on its technological and biological properties and there have been several reviews, including Holt (1997) and Lucey and Horne (2009).

When milk is heated in a muffle furnace at 500°C for approximately 5 hours, an ash derived mainly from the inorganic salts of milk and representing approximately 0.7%, w/w, of the milk, remains. However, the elements are changed from their original forms to oxides or carbonates and the ash contains P and S derived from caseins, lipids, sugar phosphates or high-energy phosphates. The organic salts, the most important of which is citrate, are oxidised and lost during ashing; some volatile metals, for example sodium, are partially lost. Thus, ash does not accurately represent the salts of milk. However, the principal inorganic and organic ions in milk can be determined directly by potentiometric, spectrophotometric or other methods. The typical concentrations of the principal elements, often referred to as macro-elements, are shown in Table 24.3. Considerable variability occurs, due, in part, to use of poor analytical methods and/or to samples from cows in very early or late lactation or suffering from mastitis. The micro-elements are very important from a nutritional viewpoint, some, for example Fe and Cu, are very potent lipid pro-oxidants and some, for example Fe, Mo, Zn, are enzyme co-factors.

Some of the salts in milk are fully soluble but others, especially calcium phosphate, exceed their solubility under the conditions in milk and occur partly in the colloidal state, associated with the

Table 24.3. Distribution of Organic and Inorganic Ions Between Soluble and Colloidal Phases of Bovine Milk

Species	Concentration (mg/L)	Soluble %	Soluble Form	Colloidal (%)
Sodium	500	92	Ionised	8
Potassium	1450	92	Ionised	8
Chloride	1200	100	Ionised	–
Sulphate	100	100	Ionised	–
Phosphate	750	43	10% bound to Ca^{+2} and Mg^{+2} 51% H_2PO4^- 39% $HPO_4{}^{2-}$	57
Citrate	1750	94	84% bound to Ca and Mg 14% $Citr^{3-}$ 1% $HCitr^{2-}$	6
Calcium	1200	34	35% Ca^{2+} 55% bound to citrate 10% bound to phosphate	66
Magnesium	130	67	Probably similar to calcium	33

Source: Modified from Fox and McSweeney 1998.

casein micelles; these salts are referred to as CCP, although some magnesium, citrate and traces of other elements are also present in the micelles. The typical distribution of the principal organic and inorganic ions between the soluble and colloidal phases is summarised in Table 24.3. The actual form of the principal species can be determined or calculated after making certain assumptions; typical values are shown in Table 24.3. Many of the micro-elements are bound by proteins, including enzymes, some of which are present in the MFGM.

The precise nature and structure of CCP are uncertain. It is associated with the caseins, probably *via* the casein phosphate residues; it probably exists as nanocrystals that include PO_4 residues of casein. The simplest stoichiometry is $Ca_3(PO_4)_2$ but spectroscopic data suggest that $CaHPO_4$ is the most likely form. CCP plays a major integrating role in the casein micelles, as discussed earlier.

The solubility and ionisation status of many of the principal ionic species are interrelated, especially H^+, Ca^{2+}, $PO_4{}^{3-}$ and citrate^{3-}. These relationships have major effects on the stability of the caseinate system and consequently on the processing properties of milk. The status of various species in milk can be modified by adding certain salts to milk, for example [Ca^{2+}] by $PO_4{}^{3-}$ or citrate^{3-}; addition of $CaCl_2$ to milk affects the distribution and ionisation status of calcium and phosphate and the pH of milk.

The distribution of species between the soluble and colloidal phases is strongly affected by pH and temperature. As the pH is reduced, CCP dissolves and is completely soluble < *ca.* pH 4.9; the reverse occurs when the pH is increased. These pH-dependent shifts mean that acid-precipitated products, for example acid casein and acid-coagulated cheeses, have a very low concentration of Ca.

The solubility of calcium phosphate decreases as the temperature is increased. Consequently, soluble calcium phosphate is transferred to the colloidal phase, with the release of H^+ and a decrease in pH:

$$CaHPO_4/Ca(H_2PO_4)_2 \leftrightarrow Ca_3(PO_4)_2 + 3H^+$$

These changes are quite substantial, but are at least partially reversible on cooling.

Since milk is supersaturated with calcium phosphate, concentration of milk by evaporation of water increases the degree of super-saturation and the transfer of soluble calcium phosphate to the colloidal state, with the concomitant release of H^+. Dilution has the opposite effect. The concentration of salts in milk may be modified by electrodialysis, ion exchangers or addition of chelators or selected salts; such operations are practised widely, for example for sterilised milks, processed cheese or nutritional products, such as infant formulae.

Milk salts equilibria are also shifted on freezing; as pure water freezes, the concentrations of solutes in unfrozen liquid are increased. Soluble calcium phosphate precipitates as $Ca_3(PO_4)_2$, releasing H^+ (the pH may decrease to 5.8). The crystallisation of lactose as a monohydrate aggravates the situation by reducing the amount of solvent water.

There are substantial changes in the concentrations of the macroelements in milk during lactation, especially at the beginning and end of lactation and during mastitic infection (see White and Davies 1958, Keogh et al. 1982, O'Keeffe 1984, O'Brien et al. 1999a). Changes in the concentration of some of the salts in milk, especially calcium phosphate and citrate, have major effects on the physico-chemical properties of the casein system and on the processability of milk, especially rennet coagulability and related properties and heat stability.

Table 24.4. Typical Concentrations of Vitamins in Milk and Proportion of Recommended Daily Allowance (RDA) Supplied by 1 L Milk

Vitamin	Average Level in 1 L Milk	RDA (%) in 1 L Milk
A	380 rational equivalents	38
B_1 (thiamine)	0.4 mg	33
B_2 (riboflavin)	1.8 mg	139
B_3 (niacin)	9.0 niacin equivalents	53
B_5 (pantothenic acid)	3.5 mg	70
B_6 (pyridoxine)	0.5 mg	39
B_7 (biotin)	30 g	100
B_{11} (folic acid)	50 g	13
B_{12} (cobalamin)	4 g	167
C	15 mg	25
D	0.5 g	10
E	1 mg	10
K	35 mg	44

Source: Adapted from Schaafma 2002.

VITAMINS

Milk contains all the vitamins in sufficient quantities to allow normal growth and maintenance of the neonate. Cow's milk is a very significant source of vitamins, especially biotin (B_7), riboflavin (B_2) and cobalamine (B_{12}), in the human diet. Typical concentrations of all the vitamins and the percentage of recommended daily allowance supplied by 1 L of milk are shown in Table 24.4. For specific aspects of vitamins in relation to milk and dairy products, including stability during processing and storage, the reader is referred to a set of articles in Roginski et al. (2002) or Fuquay et al. (2011).

In addition to their nutritional significance, four vitamins are of significance for other reasons, which have been discussed in Section 'Fat-Soluble Vitamins':

- Vitamin A (retinol), and especially carotenoids, are responsible for the yellow-orange colour of fat-containing products made from cows' milk.
- Vitamin E (tocopherols) is a potent antioxidant.
- Vitamin C (ascorbic acid) may act as an antioxidant or pro-oxidant, depending on its concentration.
- Vitamin B2 (riboflavin), which is greenish-yellow, is responsible for the colour of whey or UF permeate. It co-crystallises with lactose and is responsible for its yellowish colour, which may be removed by recrystallisation or bleached by oxidation. Riboflavin acts as a photocatalyst in the development of light-oxidised flavour in milk, which is due to the oxidation of methionine (not to the oxidation of lipids).

SUMMARY

Milk is a very complex fluid. It contains several hundred molecular species, mostly at trace levels. Most of the micro-constituents are derived from blood or mammary tissue but most of the macro-constituents are synthesised in the mammary gland and are milk-specific. The constituents of milk may be in true aqueous solution (e.g., lactose and most inorganic salts) or as a colloidal solution (proteins, which may be present as individual molecules or as large aggregates of several thousand molecules, called micelles) or as an emulsion (lipids). The macro-constituents can be fractionated readily and are used as food ingredients. The natural function of milk is to supply the neonate with its complete nutritional requirements for a period (sometimes, several months) after birth and with many physiologically important molecules, including carrier proteins, protective proteins and hormones.

The properties of milk lipids and proteins may be modified readily by biological, biochemical, chemical or physical means and thus converted into a wide range of dairy products. In this chapter, the chemical and physico-chemical properties of milk sugar (lactose), lipids, proteins and inorganic salts are discussed. The technology used to convert milk into a range of food products is described in Chapter 25.

REFERENCES

Akerstrom B et al. 2000. Lipocalins 2000. *Biochimica et Biophysica Acta* 1482: 1–356.

Bauman DE, Lock AL. 2006. Conjugated linoleic acid: Biosynthesis and significance. In: *Advanced Dairy Chemistry, Volume 3, Lactose, Water, Salts and Vitamins*, 2nd edn. Chapman & Hall, London, pp. 93–136.

Bayless KJ et al. 1997. Isolation and biological properties of osteopontin from bovine milk. *Protein Expr Purif* 9: 309–314.

Beau M. 1921. Les mátiéres albuminoides du lait. *Le Lait* 1: 16–26.

Campagna S et al. 1998. Conformational studies of a synthetic peptide from the putative lipid-binding domain of bovine milk component PP3. *J Dairy Sci* 81: 3139–3148.

Carter DC, Ho JX. 1994. Structure of serum albumin. *Adv Protein Chem* 45: 153–203.

Cogan TM, Hill C. 1993. Cheese starter cultures. In: PF Fox (ed.) *Cheese: Chemistry, Physics and Microbiology*, vol 1, 2nd edn. Chapman & Hall, London, pp. 193–255.

Collomb M et al. 2002. Composition of fatty acids in cow's milk fat produced in the lowlands, mountains and highlands of Switzerland using high-resolution gas chromatography. *Int Dairy J* 12: 649–659.

Denhardt DT, Guo X. 1993. Osteopontin: a protein with diverse functions. *FASEB Journal* 7: 1475–1482.

Flynn A. 2003. Nutritional significance of lactose II Metabolism and toxicity of galactose. In: PF Fox (ed.) *Advanced Dairy Chemistry, Volume 3, Lactose and Minor Constituents*. Chapman and Hall, London, pp. 133–141.

Fosset S, Tome D. 2011. Nutraceuticals from milk. In: J Fuquay et al. (eds.) *Encyclopedia of Dairy Sciences*, 2nd edn. Elsevier, Oxford, pp. 1062–1066.

Fox PF (ed.). 1992. *Advanced Dairy Chemistry, Volume 1, Proteins*, 2nd edn. Elsevier Applied Science, London.

Fox PF (ed.). 1995. *Advanced Dairy Chemistry, Volume 2, Lipids*, 2nd edn. Chapman and Hall, London.

Fox PF (ed.). 1997. *Advanced Dairy Chemistry, Volume 3, Lactose, Water, Salts and Vitamins*, 2nd edn. Chapman and Hall, London.

Fox PF. 2003a. The major constituents of milk. In: G Smit (ed.) *Dairy Processing: Improving Quality*. Woodhead Publishing Ltd., Cambridge, pp. 5–41.

Fox PF. 2003b. Indigenous enzymes in milk. In: PF Fox, PLH McSweeney (eds.) *Advanced Dairy Chemistry, Volume 1, Proteins*. Kluwer Academic-Plenum Publishers, New York, pp. 467–603.

Fox PF, Brodkorb A. 2008. The casein micelle: Historical Aspects, current concepts and significance. *Int Dairy J* 18: 677–684.

Fox PF, Flynn A. 1992. Biological properties of milk proteins. In: PF Fox (ed.) *Advanced Dairy Chemistry, Volume 1, Proteins*. Elsevier Applied Sciences, London, pp. 255–284.

Fox PF, Kelly AL. 2003. The caseins. In: R Yada (ed.) *Proteins in Food Processing*. Woodhead Publishing, New York, pp. 29–71.

Fox PF, Kelly AL. 2006. Indigenous enzymes in milk: Overview and historyical aspects, Parts I and II. Collomb M et al. 2002. Composition of fatty acids in cow's milk fat produced in the lowlands, mountains and highlands of Switzerland using high-resolution gas chromatography. *Int Dairy J* 12: 649–659 16: 500–516, 517–532.

Fox PF, McSweeney PLH. 1998. *Dairy Chem Biochem*. Chapman and Hall, London.

Fox PF, McSweeney PLH (eds.). 2003. *Advanced Dairy Chemistry, Volume 1, Proteins*. Kluwer Academic-Plenum Publishers, New York.

Fox PF, McSweeney PLH (eds.). 2006. *Advanced Dairy Chemistry, Volume 1, Lipids*. Springer, New York.

Fox PF et al. 2000. *Fundamentals of Cheese Science*. Aspen Publishers, Gaithersburg, MD.

Fox PF et al. 1990. Glycolysis and related reactions during cheese manufacture and ripening. *Crit Rev Food Sci Nutr* 29: 237–253

Fuquay J et al. (eds.). 2011. *Encyclopedia of Dairy Sciences*, 2nd edn. Elsevier, Oxford.

Ganzle M. 2011. Galacto-oligosaccharides. In: J Fuquay et al. (eds.) *Encyclopedia of Dairy Sciences*, 2nd edn. Elsevier, Oxford, pp. 29–71.

Girardet JM, Linden G. 1996. PP3 component of bovine milk: a phosphorylated whey glycoprotein. *J Dairy Res* 63: 333–350.

Groves ML, Greenberg R. 1982. β_2-Microglobulin and its relationship to the immune system. *J Dairy Sci* 65: 317–325.

Groves ML et al. 1963. Isolation, characterisation and amino acid composition of a new crystalline protein, lactollin, from milk. *Biochem* 2: 814–817.

Haggarty NW. 2003. Milk proteins; minor proteins, bovine serum albumin and vitamin-binding proteins. In: H Roginski et al. (eds.) *Encyclopedia of Dairy Science*. Academic Press, London, pp. 1939–1946.

Hayes MG, Kelly AL. 2003. High pressure homogenisation of raw whole bovine milk (a) Effects on fat globule size and other properties. *J Dairy Res* 70: 297–305.

Holt C. 1992. Structure and properties of bovine casein micelles. *Adv Protein Chem* 43: 63–151.

Holt C. 1997. The milk salts and their interaction with caseins. In: PF Fox (ed.) *Advanced Dairy Chemistry, Volume 2, Lactose, Salts, Water and Vitamins*. Chapman and Hall, London, pp. 233–256.

Holt C, Sawyer L. 1993. Caseins as rheomorphic proteins. Interpretation of primary and secondary structures of α_{s1}, β- and κ-caseins. *J Chemical Soc Faraday Trans* 89: 2683–2692.

Horne DS. 2003. Caseins, micellar structure. In: H Roginski et al. (eds.) *Encyclopedia of Dairy Science*. Academic Press, London, pp. 1902–1909.

Horne DS. 2011 Casein: Micellar structure. In: J Fuquay et al. (eds.) *Encyclopedia of Dairy Sciences*, 2nd edn. Elsevier, Oxford, pp. 772–779.

Huppertz T et al. 2002. Effects of high pressure on constituents and properties of milk: a review. *Int Dairy J* 561–572.

Huppertz T et al. 2004. High pressure treatment of bovine milk: effects on casein micelles and whey proteins. *J Dairy Res* 71: 87–106.

Ingram CJE, Swallow DM. 2009. Lactose malabsorption. In: PLH McSweeney, PF Fox (eds.) *Advanced Dairy Chemistry, Volume 3, Lactose, Salts, Water and Minor Constituents*. Springer, New York, pp. 203–229.

Keenan TW, Mather IH. 2003 Milk fat globule membrane. In: H Roginski et al. (eds.) *Encyclopedia of Dairy Science*. Academic Press, London, pp. 1568–1576.

Kelly ML et al. 1998. Dietary fatty acid content significantly affects conjugated linoleic acid concentrations in milk from lactating dairy cows. *J Nutr* 128: 881–885.

Kelly AL et al. 2003. Manufacture and properties of milk powders. In: PF Fox, PLH McSweeney (eds.) *Advanced Dairy Chemistry, Volume 1, Proteins*. Kluwer Academic-Plenum Publishers, New York, pp. 1027–1062.

Keenan TW, Mather IH. 2006. Intercellular origin of milk fat globules and the nature of the milk fat globule membrane. In: PF Fox, PLH McSweeney (eds.) *Advanced Dairy Chemistry, Volume 1, Lipids*. Springer, New York, pp. 137–171.

Keogh MK et al. 1982. Studies of milk composition and its relationship to some processing criteria. I. Seasonal variation in the mineral levels in milk. *Ir J Food Sci Tech* 6: 13–27.

Korhonen H, Marnila P. 2011. Lactoferrin. In: *Encyclopedia of Dairy Sciences*, 2nd edn. Elsevier, Oxford, pp. 801–806.

Kraft J et al. 2000. Duodenal infusion of conjugated linoleic acid mixture influences milk fat synthesis and milk conjugated linoleic acid content. In: *Milk Composition. British Society of Animal Science*. Occasional Publication No 25, pp. 143–147.

Lawless F et al. 2000. Dietary effect on bovine milk fat conjugated linoleic acid content. In: *Milk Composition. British Society of Animal Science*. Occasional Publication No 25, pp. 283–293.

Lefebvre-Cases E et al. 2001a. Effect of SDS on casein micelles: SDS-induced milk gel formation. *Journal of Food Science* 66: 38–42.

Lefebvre-Cases E et al. 2001b. Effect of SDS on acid milk coagulability. *J Food Sci* 66: 555–560.

Lock AL, Garnsworthy PC. 2000. Independent effects of dietary linoleic and linolenic fatty acids on the conjugated linoleic acid content of cow's milk. *J Anim Sci* 74: 163–176.

Lonnerdal B. 2003. Lactoferrin. In: PF Fox, PLH McSweeney (eds.) *Advanced Dairy Chemistry, Volume 1, Proteins*. Kluwer Academic-Plenum Publishers, New York, pp. 449–466.

Lucey JA, Horne DS. 2009. Milk salts: technological significance. In: PLH McSweeney, PF Fox (eds.) *Advanced Dairy Chemistry, Volume 3, Lactose, Salts, Water and Minor Constituents*. Springer, New York, pp. 351–389.

Marnila P, Korhonen H. 2011. Immunoglobulins. In: J Fuquay et al. (eds.) *Encyclopedia of Dairy Sciences*, 2nd edn. Elsevier, Oxford, pp. 807–815.

Martin P et al. 2003. Non-bovine caseins: Quantitive variability and molecular diversity. In: PF Fox, PLH McSweeney (eds.) *Advanced Dairy Chemistry, Volume 1, Proteins*. Kluwer Academic-Plenum Publishers, New York, pp. 277–317.

Martin P et al. 2011. Interspecies comparison of milk proteins: Quantitative variability and molecular diversity. In: J Fuquay et al. (eds.) *Encyclopedia of Dairy Sciences*, 2nd edn. Elsevier, Oxford, pp. 821–842.

Mather IH. 2000. A revised and proposed nomenclature for major proteins of the milk fat globule membrane. J Dairy Sci 83: 203–247.

Mather I. 2011. Milk fat globule membrane. In: J Fuquay et al. (eds.) *Encyclopedia of Dairy Sciences*, 2nd edn. Elsevier, Oxford, pp. 680–690.

McSweeney PLH, Fox PF (eds.). 2009. *Advanced Dairy Chemistry, Volume 3, Lactose, Water, Salts and Minor Constituents*, 3rd edn. Springer, New York.

Mehra R et al. 1999. Seasonal variation in the composition of Irish manufacturing and retail milk. 2. Nitrogen fractions. *Ir J Agr Food Res* 38: 65–74

Meesapyodsuk D et al. 2000. Substrate specificity, regioselectivity and cryptoregiochemistry of plant and animal omega-3-fatty acid desaturases. *Biochem Soc Trans* 28: 632–635.

Morr CV. 1967. Effect of oxalate and urea upon ultracentrifugation properties of raw and heated skim milk casein micelles. *J Dairy Sci* 50: 1744–1751.

Nursten H. 2011. Maillard reactions. In: J Fuquay et al. (eds.) *Encyclopedia of Dairy Sciences*, 2nd edn. Elsevier, Oxford, pp. 217–235.

O'Brien J. 2009. Non-enzymatic degradation pathways of lactose and their significance in dairy products. In: PLH McSweeney, PF Fox (eds.) *Advanced Dairy Chemistry, Volume 3, Lactose, Water, Salts and Minor Constituents*, 3rd edn. Springer, New York, pp. 231–294.

O'Brien NM, O'Connor TP. 2011. Lipid oxidation. In: J Fuquay et al. (eds.) *Encyclopedia of Dairy Sciences*, 2nd edn. Elsevier, Oxford, pp. 716–720.

O'Brien B et al. 1999a. Seasonal variation in the composition of Irish manufacturing and retail milk. 1. Chemical composition and renneting properties. *Ir J Agr Food Res* 38: 53–64

O'Brien B et al. 1999b. Seasonal variation in the composition of Irish manufacturing and retail milk. 3. Vitamins. *Ir J Agr Food Res* 38: 75–85

O'Brien B et al. 1999c. Seasonal variation in the composition of Irish manufacturing and retail milk. 4. Minerals and trace elements. *Ir J Agr Food Res* 38: 87–99.

O'Connell JE et al. 2001a. Ethanol-dependent heat-induced dissociation of casein micelles. *Ir J Agr Food Chem* 49: 4420–4423.

O'Connell JE et al. 2001b. Mechanism for the ethanol-dependent heat-induced dissociation of casein micelles. *J Agri Food Chem* 49: 4424–4428.

O'Keeffe AM. 1984. Seasonal and lactational influences on moisture content of Cheddar cheese. *Ir J Food Sci Tech* 8: 27–37.

Paakanen R, Aalto J. 1997. Growth factors and antimicrobial factors of bovine colostrums. *Int Dairy J* 7: 285–297.

Park YW, Haenlein FW (eds.). 2006. *Handbook of Non-bovine Mammals*. Blackwell Publishing, Oxford.

Parodi PW. 1997. Cows' milk fat components as potential anticarcinogenic agents. *J Nutr* 127: 1055–1060.

Parodi PW. 1999. Conjugated linoleic acid and other anticarcinogenic agents in bovine milk fat. J Dairy Sci 82: 1399–1349.

Parodi P. 2009. Has the association between saturated fatty acids, serum cholesterol and coronary heart disease been overemphasised? *Int Dairy J* 19: 345–361.

Paterson AHJ. 2009. Production and uses of lactose. In: PLH McSweeney, PF Fox (eds.) *Advanced Dairy Chemistry, Volume 3, Lactose, Water, Salts and Minor Constituents*, 3rd edn. Springer, New York, pp. 105–120.

Patterson AHJ. 2011. Lactose: Production and applications reactions. In: J Fuquay et al. (eds.) *Encyclopedia of Dairy Sciences*, 2nd edn. Elsevier, Oxford, pp. 196–201.

Patton S et al. 1997. Prosaposin, a neurotrophic factor: presence and properties in milk. *J Dairy Sci* 80: 264–272.

Pihlanto-Leppala A. 2011. Bioactive peptides. In: J Fuquay et al. (eds.) *Encyclopedia of Dairy Sciences*, 2nd edn. Elsevier, Oxford, pp. 879–886.

Playne MJ, Crittenden RG. 2009. Galacto-oligosaccharides and other products derived from lactose In: PLH McSweeney, PF Fox (eds.) *Advanced Dairy Chemistry, Volume 3, Lactose, Water, Salts and Minor Constituents*, 3rd edn. Springer, New York, pp. 121–201.

Poolman B. 2002. Transporters and their role in LAB cell physiology. *Antonie van Leewenhoek* 82: 147–164.

Roginski H et al. (eds.). 2002. *Encyclopedia of Dairy Science*. Academic Press, Oxford.

Schrezenmeir J et al. 2000. Beneficial natural bioactive substances in milk and colostrums. *Brit J Nutr* 84 (Suppl 1): S1–S166.

Strydom DJ. 1998. The angiogenins. *Cell Mol Life Sci* 54: 811–824.

Urashima T et al. 2001. Oligosaccharides of milk and colostrums in non-human mammals. *Glycoconj J* 18: 357–371.

Urashima T et al. 2009. Milk oligosaccharides. In: PLH McSweeney, PF Fox (eds.) *Advanced Dairy Chemistry, Volume 3, Lactose, Water, Salts and Minor Constituents*, 3rd edn. Springer, New York, pp. 295–349.

Urashima T et al. 2011. Indigenous oligosaccharides in milk. In: J Fuqua et al. (eds.) *Encyclopedia of Dairy Sciences*, 2nd edn. Elsevier, Oxford, pp. 241–273.

Visser H. 1992. A new casein micelle model and its consequences for pH and temperature effects on the properties of milk. In: H Visser (ed.) *Protein Interactions*. VCH Publishing, Weinheim, pp. 135–165.

Walstra P, Jenness R. 1984. *Dairy Chemistry and Physics*. John Wiley & Sons, New York.

Walstra P et al. 1999. *Dairy Technology: Principles of Milk Properties and Processing*. Marcel Dekker, New York.

Walstra P et al. 2005. *Dairy Science and Technology*, 2nd edn. CRC Press, Boca Raton, FL.

Waugh DF, von Hippel PH. 1956. κ-Casein and the stabilisation of casein micelles. *J Am Chem Soc* 78: 4576–4582.

White JCD, Davies DT. 1958. The relationship between the chemical composition of milk and the stability of the caseinate complex. I. General introduction, description of samples, methods and chemical composition of samples. *J Dairy Res* 25: 236–255.

Wilson WE et al. 1989. Bradykinin and kininogens in bovine milk. *J Biol Chem* 264: 17777–17783.

Wong NP et al. 1988. *Fundamentals of Dairy Chemistry*. Van Nostrand Reinhold Co., New York.

Wooten L et al. 1996. Ceruloplasmin is found in milk and amniotic fluid and may have a nutritional role. *J Nutr Biochem* 7: 632–639.

Wynn P. 2011. Minor proteins bovine serum albumin and vitamin-binding proteins. In: J Fuquay, PF Fox, PLH McSweeney (eds.) *Encyclopedia of Dairy Sciences*, 2nd edn. Elsevier, Oxford, pp. 795–800.

Yurawecz MP et al. 1999. *Advances in Conjugated Linoleic Acid Research*. American Oil Chemists Society, Minneapolis, MN.

25
Biochemistry of Milk Processing

A.L. Kelly and P.F. Fox

Abstract: Milk is subjected to a wide range of processes before consumption directly or in the form of the very wide range of modern dairy products. Primary objectives of processing involve rendering milk microbiologically safe or stable (e.g., pasteurisation, ultra-high temperature (UHT) milk), isolation of valuable components (e.g., whey proteins, caseins) and conversion to other forms (e.g., cheese, yoghurt, enzymatic hydrolysates). All such processes applied to milk have very significant effects on the properties and distribution of food constituents. The principal families of dairy products, which may be produced from milk, and their principles of production, are described in this chapter.

INTRODUCTION

As described in Chapter 24, milk is a very complex system; the continuous phase is an aqueous solution of a specific sugar, lactose, globular (whey/serum) proteins, inorganic salts and hundreds of minor constituents, for example vitamins, at trace levels. Dispersed in the aqueous phase as an oil-in-water (o/w) emulsion are lipids in the form of small globules and a second, unique group of proteins, the caseins, which exist as large colloidal

Food Biochemistry and Food Processing, Second Edition. Edited by Benjamin K. Simpson, Leo M.L. Nollet, Fidel Toldrá, Soottawat Benjakul, Gopinadhan Paliyath and Y.H. Hui.

particles known as casein micelles, with an average diameter of approximately 150 nm (range 50–600 nm) and containing, on average, approximately 5000 protein molecules.

Owing to its physical state, milk is an unstable system. The fat globules, which are less dense than the aqueous phase, rise to the surface where they form a cream layer. The fat-rich cream can be skimmed off and used for the manufacture of butter, butter oil, ghee or other fat-rich products. Alternatively, the fat-depleted lower layer (skimmed milk) may be run off, for example through a valve at the bottom of a separating vat. The skimmed milk may be consumed directly as a beverage or used for the manufacture of a wide range of products. Gravity creaming and butter manufacture have been used since prehistoric times, but gravity creaming has been largely replaced by centrifugal separation of the fat since the development of the cream separator by Gustav de Laval in 1878. Gravity separation is still used to prepare reduced-fat milk for the manufacture of Parmigiano-Reggiano cheese. The physics of fat separation by gravity or centrifugal separation and the process of demulsifying milk fat to butter or butter oil will be described briefly later in this chapter.

Creaming (fat separation) is undesirable in many dairy products (e.g., liquid/beverage milk, concentrated milks) and is prevented by a process known as homogenisation (the commonly used valve homogeniser was developed by Auguste Gaulin in 1899). Homogenisation prevents creaming by reducing the size of the fat globules and preventing their agglomeration (clustering), by denaturing a particular minor protein known as cryoglobulin (type M immunoglobulin).

The casein micelles are physicochemically stable but can be destabilised by a number of processes/treatments, which are exploited for the production of new dairy products. The most important of these are limited proteolysis and acidification, which are used in the production of cheese, fermented milks and functional milk proteins. The mechanism and consequences of proteolysis and acidification will be described briefly. Thus, the three main constituents/phases of milk, that is fat, casein and the aqueous solution (whey), can be separated easily.

Milk is remarkably heat stable; it can be sterilised by heat in standard or concentrated form, or dried to produce a wide range of dairy products. The principal processes and the resulting product families are summarised in Table 25.1. All these families of dairy products contain many products, for example approximately 1400 varieties of cheese are produced worldwide.

In addition to the above considerations, milk is a rich medium for the growth of a wide range of microorganisms. While this is exploited in the production of a range of fermented dairy products, it also means that milk can harbour microorganisms that may cause spoilage of products, or present health risks to the consumer. Largely for the latter reason, very little raw milk is now consumed, the vast majority being heat treated sufficiently severe to kill all pathogenic and food-poisoning bacteria.

Table 25.1. Diversity of Dairy Products Produced by Different Processes

Process	Primary (*italic*) and Secondary Products
Heat treatment (generally with homogenisation)	*Market milk*
Centrifugal separation	*Standardised milk*
	Cream
	Butter, butter oil, ghee
	Creams of various fat content (pasteurised or UHT treated); whipping, coffee, dessert creams
	Skim milk
	Skim milk powder, casein, cheese, protein concentrates
Concentration (e.g., evaporation or membrane separation)	In container or UHT-sterilised concentrated milks, sweetened condensed milk
Concentration and drying	Whole milk powders, infant formulae
Enzymatic coagulation	*Cheese*
	Numerous varieties, processed cheese, cheese sauces and dips
	Rennet casein; cheese analogues
	Whey
	Whey powders; demineralised whey powders; whey protein concentrates; fractionated whey proteins; whey protein hydrolysates; lactose and lactose derivatives
Acid coagulation	*Cheese (fresh)*
	Acid casein; functional proteins
	Whey (as rennet whey)
Fermentation	*Various fermented milk products*, e.g., yoghurt, buttermilk, acidophilus milk, bioyoghurt
Freezing, aeration and whipping	*Ice cream*

Source: Adapted From Fox and McSweeney 1998.
UHT, ultra-high temperature.

In this chapter, several of the main processes used in the modern dairy industry will be discussed and the relationships between the biochemical properties of milk and the processes applied, explored.

THERMAL PROCESSING OF MILK

Introduction

The most common process applied to most food products is probably heat treatment, principally used to kill microorganisms (e.g., bacteria, yeasts, moulds and viruses) responsible for foodborne disease, food poisoning or spoilage. Secondary objectives of heat-treating foods include inactivation of enzymes in the tissue or fluid, which would otherwise negatively influence product quality, and effecting changes in the structure of the food, for example through denaturation of proteins or gelatinisation of starch.

Today, a wide range of thermal processes are commonly applied to milk, as summarised in Table 25.2.

The most widely used thermal process for milk is pasteurisation. In 1864–1866, Louis Pasteur, the famous French scientist, discovered that the spoilage of wine and beer could be prevented by heating the product to around 60°C for several minutes; this process is now referred to as pasteurisation. The historical development of the thermal processing of milk was reviewed by Westhoff (1978).

Although pasteurisation was introduced to improve the stability and quality of food, it soon became apparent that it offered consumers protection against hazards associated with the consumption of raw milk (particularly the risk of transmission of tuberculosis from infected cows to humans); developments in technology and widespread implementation occurred early in the nineteenth century. The first commercial pasteurisers, in which milk was heated at 74°C–77°C for an unspecified time period, were made by Albert Fesca in Germany in 1882 (Westhoff 1978). The first commercially operated milk pasteuriser in the United States was installed in Bloomville, New York, in 1893; the first law requiring the pasteurisation of liquid (beverage) milk was passed by the authorities in Chicago in 1908.

Many early pasteurisation processes used conditions not very different from those proposed by Pasteur, generally heating milk to 62–65°C for at least 30 minutes, followed by rapid cooling to less than 10°C (now referred to as low-temperature long-time, LTLT, pasteurisation). However, high-temperature short-time (HTST) pasteurisation, in which milk is treated at 72–74°C for 15 seconds in a continuous-flow plate heat exchanger, was introduced by APV in 1923 and gradually became the standard industrial procedure for the heat treatment of liquid milk and cream. Compared to the LTLT process, the HTST process has the advantages of reduced heat damage and flavour changes, reduced cost and increased efficiency and throughput (Kelly and O'Shea 2002).

Later developments in the thermal processing of milk led to the development of ultra-high temperature (UHT) processes, in which milk is heated to a temperature in the range 135–140°C for 2–5 seconds (Lewis and Heppell 2000). UHT treatments can be applied using a range of heat exchanger technologies (e.g., indirect and direct processes), and essentially result in a sterile product that is typically shelf-stable for at least 6 months at room temperature; eventual deterioration of product quality generally results from physicochemical rather than microbiological or enzymatic processes.

Sterilised milk products may also be produced using in-container retort systems; in fact, such processes were developed before the work of Pasteur. In 1809, Nicholas Appert developed an in-container sterilisation process that he applied to the preservation of a range of food products, with varying degrees of success. Today, in-container sterilisation, as first successfully and consistently applied towards the end of the nineteenth century, is generally used for concentrated (condensed) milks; typical conditions involve heating at 115°C for 10–15 minutes. A related, although not sterile, class of product is preserved by adding a high level of sugar to concentrated milk; the sugar preserves the product through osmotic action (sweetened condensed milk was patented by Gail Borden in 1856).

The primary function of thermal processing of milk is to kill undesirable microorganisms; modern pasteurisation is a very

Table 25.2. Thermal Processes Commonly Applied to Liquid Milk Products

Process	Conditions	Heat Exchanger	Reason
Thermisation	63°C for 15 s	Plate heat exchanger	Killing psychrotrophic bacteria before cold storage of milk
Pasteurisation	72–74°C for 15–30 s	Plate heat exchanger	Killing of vegetative pathogenic bacteria and shelf-life extension
UHT (indirect)	135–140°C for 3–5 s	Plate heat exchanger Scraped-surface heat exchanger Tubular heat exchanger	Production of safe long shelf-life milk
UHT (direct)	135–140°C for 3–5 s	Steam infusion system Steam injection system	Production of safe long shelf-life milk
Sterilisation	118°C for 12 min	Batch or continuous Retort	Production of sterile Long-life milk

UHT, ultra-high temperature.

effective means for ensuring that liquid milk is free of the most heat-resistant pathogenic bacteria likely to be present in raw milk. The original target species was *Mycobacterium tuberculosis* but, a little later, *Coxiella burnetti* became the target. In recent years, evidence for the survival of *Listeria monocytogenes* and, in particular, *Mycobacterium avium* subsp. *paratuberculosis* in pasteurised milk is of concern and the subject of ongoing research (Grant et al. 2001, Ryser 2002). Most health risks linked to the consumption of pasteurised milk are probably due to post-pasteurisation contamination of the product.

Heat-Induced Changes in Milk Proteins

The caseins are very resistant to temperatures normally used for processing milk and, in general, unconcentrated milk is stable to the thermal processes to which it is exposed, for example HTST pasteurisation, UHT or in-container sterilisation. However, very severe heating (e.g., at 140°C for a prolonged period) causes simultaneous dissociation and aggregation of the micelles and eventually coagulation.

However, the whey proteins are typical globular proteins and are relatively susceptible to changes on heating milk. β-Lactoglobulin (β-lg) is susceptible to thermal denaturation, to an extent dependant on factors such as pH, protein concentration and ionic strength (Sawyer 2003); denaturation exposes a highly reactive sulphydryl (-SH) group. Denatured β-lg can react, *via* sulphydryl–disulphide interchange reactions, with other proteins including κ-casein, the micelle-stabilising protein that is concentrated at the surface of the casein micelles. Thus, heating milk at a temperature higher than that used for minimal HTST pasteurisation (which itself causes only low levels of denaturation of β-lg, although even slightly higher temperatures will cause progressively increased levels of denaturation) results in the formation of casein/whey protein complexes in milk, with major effects on many technologically important properties of milk.

Such complexes are found either at the surface of the micelles or in the serum phase, depending on the pH, which determines the likelihood of heat-induced dissociation of the κ-casein/β-lg complex from the casein micelles. Largely as a result of this, the heat stability of milk (expressed as the number of minutes at a particular temperature, for example 140°C, before visible coagulation of the milk occurs – the heat coagulation time, HCT) is highly dependent on pH, with large differences over a narrow range of pH values (O'Connell and Fox 2003).

Most milk samples show a type A heat stability–pH profile, with a pronounced minimum and maximum, typically around pH 6.7 and 7.0, respectively, with decreasing stability on the acidic side of the maximum and increasing stability on the alkaline side of the minimum. The rare type B HCT–pH profile has no minimum, and heat stability increases progressively throughout the pH range 6.4–7.4.

In addition to pH, a number of factors affect the heat stability of milk (O'Connell and Fox 2002), including:

- Reduction in the concentration of Ca^{2+} or Mg^{2+} increases stability in pH range 6.5–7.5
- Lactose hydrolysis increases heat stability throughout the pH range
- Addition of κ-casein eliminates the minimum in the type A profile
- Addition of β-lg to a type B milk converts it to a type A profile
- Addition of phosphates increases heat stability
- Reducing agents reduce heat stability and convert a type A to a type B profile
- Alcohols and sulphydryl-blocking agents reduce the heat stability of milk

Concentrated (e.g., evaporated) milk is much less thermally stable than unconcentrated milk and its HCT–pH profile, normally assayed at 120°C, is quite different (O'Connell and Fox 2003). It shows a maximum at approximately pH 6.4, with decreasing stability at higher and lower pH values (Fig. 25.1).

Severe heating has a number of effects on the casein molecules, including dephosphorylation, deamidation of glutamine and asparagine residues, cleavage and formation of covalent cross-links; these changes may result in protein–protein interactions that contribute to thermal instability.

The heat stability of milk of different species differ, primarily due to differences in protein and mineral content; for example equine, camel, buffalo and caprine milk are less heat stable than bovine milk (O'Connell and Fox 2003).

Stability of UHT Milk on Storage

UHT milk is stable to long-term storage at ambient temperatures if microbiological sterility has been achieved by the thermal process and maintained by aseptic packaging in hermetically sealed containers. The shelf life of UHT milk is often limited by

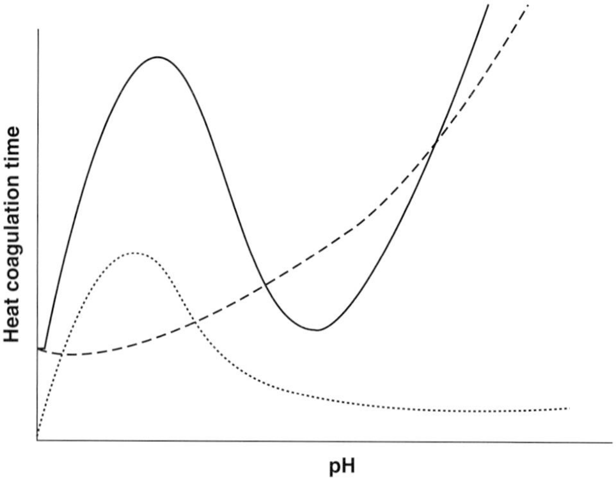

Figure 25.1. Heat coagulation time (HCT)–pH profiles of typical type A bovine milk (____), type B or serum protein-free milk (_ _ _ _), as determined at 140°C, or concentrated milk (......), as determined at 120°C.

age gelation, which refers to a progressive increase in viscosity during storage, followed by the formation of a gel that cannot be redispersed. The mechanism of age gelation is not completely understood, but is probably due to one or both of the following:

1. *Physicochemical factors:* For example dissociation of casein/whey protein complexes; changes in casein micelle structure and/or properties; cross-linking due to the Maillard reaction; removal or binding of Ca^{2+}.
2. *Biochemical factors:* For example action of proteolytic enzymes, such as plasmin or heat-stable bacterial proteinases (i.e., from *Pseudomonas* species), which may hydrolyse κ-casein, inducing micelle coagulation.

Recent reviews on this subject include Nieuwenhuijse (1995), Datta and Deeth (2001) and Nieuwenhuijse and van Boekel (2003).

Heat-Induced Changes in Lactose in Milk and the Maillard Reaction

Heating lactose causes several chemical modifications, the nature and extent of which depend on environmental conditions and the severity of heating; changes include degradation to acids (with a concomitant decrease in pH), isomerisation (e.g., to lactulose), production of compounds such as furfural and interactions with amino groups of proteins (Maillard reaction).

In the Maillard reaction, lactose or lactulose reacts with an amino group, such as the ε-amino group of lysine residues, in a complex (and not yet fully understood) series of reactions with a variety of end products (O'Brien 1995, van Boekel and Walstra 1998; O'Brien 2009). The early stages of the Maillard reaction result in the formation of a protein-bound Amadori product, lactulosyllysine, which then degrades to a range of advanced Maillard products, including hydroxymethylfurfural, furfurals and formic acid. The breakdown of lactose to organic acids reduces the pH of milk. The most obvious result of the Maillard reaction is a change in the colour of milk (browning), due to the formation of pigments called melanoidins, or advanced-stage Maillard products; extensive Maillard reactions also result in the polymerisation of proteins. The Maillard reaction also changes the flavour and nutritive quality of dairy products, in the latter case through reduced digestibility of the caseins and loss of available lysine.

Moderately intense heating also causes the isomerisation of lactose to lactulose. Lactulose, a disaccharide of galactose and fructose that does not occur naturally, is of interest as a bifidogenic factor and as a laxative. More severe treatments (e.g., sterilisation) will result preferentially in Maillard reactions.

There is particular interest in the use of products of heat-induced changes in lactose, such as lactulose and forosine, as indices of heat treatment of milk (Birlouez-Aragon et al. 2002).

Heat-Induced Changes in Milk Lipids

There are little effects of heating on milk triglycerides at the temperatures commonly encountered in dairy processing. Heating at modest temperatures can promote oxidation of lipids, but more severe heating inhibits oxidation (van Boekel and Walstra 1998). However, heat-induced denaturation of whey proteins can result in their interaction with the milk fat globule membrane (MFGM) (Huppertz et al. 2009). Because of denaturation of immunoglobulins, creaming due to cold agglutination of milk fat globules is prevented in heated milk.

Inactivation of Enzymes on Heating of Milk

Heat treatment of milk inactivates many enzymes, both indigenous and endogenous (i.e., of bovine or bacterial origin). The inactivation of enzymes is of interest, both for the stability of heated milk products and as indices of heat treatment. Thermal inactivation characteristics of a number of indigenous milk enzymes are summarised in Table 25.3.

Because of its importance, the thermal inactivation kinetics of several milk enzymes have been studied in detail (Andrews et al. 1987). A case of particular interest is that of alkaline phosphatase, which has been used since 1925 as an indicator of the adequacy of pasteurisation of milk, because its thermal inactivation kinetics in milk closely approximate those of

Table 25.3. Thermal Inactivation Characteristics of Some Indigenous Milk Enzymes

Enzyme	Characteristics
Alkaline phosphatase	Inactivated by pasteurisation; index of pasteurisation; may reactivate on storage
Lipoprotein lipase	Almost completely inactivated by pasteurisation
Xanthine oxidase	May be largely inactivated by pasteurisation, depending on whether milk is homogenised or not
Lactoperoxidase	Affected little by pasteurisation; inactivated rapidly around 80°C; used as an index of flash pasteurisation or pasteurisation of cream
Sulphydryl oxidase	About ~40% of activity survives pasteurisation, completely inactivated by UHT
Superoxide dismutase	Largely unaffected by pasteurisation
Catalase	Largely inactivated by pasteurisation but may reactivate during subsequent cold storage
Acid phosphatase	Very thermostable; survives pasteurisation
Cathepsin D	Largely inactivated by pasteurisation
Amylases (α and β)	Largely inactivated; β more stable than α
Lysozyme	Largely survives pasteurisation

Source: Adapted From Farkye and Imifadon 1995 UHT, ultra-high temperature.

M. tuberculosis. A negative result (residual activity below a set maximum value) in a phosphatase test (assay) is regarded as an indication that milk has been pasteurised correctly (Wilbey 1996). Detailed kinetic studies on the thermal inactivation of alkaline phosphatase under conditions similar to pasteurisation have been published (McKellar et al. 1994, Lu et al. 2001).

To test for over-pasteurisation (excessive heating) of milk, a more heat-stable enzyme, for example lactoperoxidase, has been used as the indicator enzyme (Storch test); lactoperoxidase may also be used as an index of the efficacy of pasteurisation of cream, which must be heated more severely than milk to ensure the killing of target bacteria, due to the protective effect of the fat therein.

Recently, other enzymes have been studied as indicators (sometimes called Time Temperature Integrators, TTIs) of heat treatments (particularly at temperatures above 72°C) that may be applied to milk; these include catalase, lipoprotein lipase, acid phosphatase, *N*-acetyl-β-glucosaminidase and γ-glutamyl-transferase (McKellar et al. 1996, Wilbey 1996). The latter has been a focus of particular attention due to its potential role as an indicator of the efficacy of heat treatments in the range 70–80°C for 16 seconds (McKellar et al. 1996).

The alkaline milk proteinase, plasmin, is resistant to pasteurisation; indeed, its activity may increase during storage of pasteurised milk, due to inactivation of inhibitors of plasminogen activators (Richardson 1983). Treatment under UHT conditions greatly reduces plasmin activity (Enright et al. 1999); there is some evidence that the low level of plasmin activity in UHT milk contributes to the destabilisation of the proteins, leading to defects such as age gelation, although this is not universally accepted (for review see Datta and Deeth 2001).

Since most indigenous milk enzymes are inactivated in UHT-sterilised milk products and in all in-container sterilised products, enzymes are not suitable indices of adequate processing and chemical indices are more usually used (e.g., the concentration of lactulose or extent of denaturation of β-lg). The use of a range of indicators for different heat treatments applied to milk was reviewed by Claeys et al. (2002).

The thermal inactivation of bacterial enzymes in milk is also of considerable significance. For example, psychrotrophic bacteria of the genus *Pseudomonas* produce heat-stable lipases and proteases during growth in refrigerated milk; while pasteurisation readily kills the bacteria, the enzymes survive such treatment and may contribute to the deterioration of dairy products made therefrom (Stepaniak and Sørhaug 1995). *Pseudomonas* enzymes partially survive in UHT milk and may be involved in its age gelation (Datta and Deeth 2001).

CHANGES ON EVAPORATION AND DRYING OF MILK

Dehydration, either partial (e.g., concentration to ~40–50% total solids, TS) or (almost) total (e.g., spray-drying to ~96–97% TS) is a common method for the preservation of milk. Reviews of the technology of drying of milk and its effects on milk quality include Caríc and Kalab (1987), Pisecký (1997), Early (1998) and Kelly et al. (2003).

Concentration of Milk

The most common technology for concentrating milk is thermal evaporation in a multi-effect (stage) falling-film evaporator, which is used either as a prelude to spray-drying, to increase the efficiency of the latter, or to produce a range of concentrated dairy products (e.g., sterilised concentrated milk, sweetened condensed milk).

In many processes, a key stage in milk evaporation is preheating the feed. The minimum requirements of this step are to pasteurise the milk (e.g., skim milk, whole milk, filled milk) to ensure food safety and to bring its temperature to the boiling point in the first effect of the evaporator, for thermal efficiency. However, the functionality of the final product can be determined and modified by this processing step; preheating conditions range from conventional pasteurisation to 90°C for 10 minutes or 120°C for 2 minutes. The most significant result of such heating is denaturation of β-lg, to a degree depending on the severity of heating. During subsequent evaporation and, if applied, spray-drying, relatively little denaturation of the whey proteins occurs, as the temperature of the milk generally does not exceed 70°C (Singh and Creamer 1991). However, further association of whey proteins with the casein micelles can occur during evaporation, probably because the decrease in pH reduces protein charge, facilitating association reactions (Oldfield 1998).

Evaporation also increases the concentrations of lactose and salts in milk and induces a partially reversible transfer of soluble calcium phosphate to the colloidal form, with a concomitant decrease in pH (Le Graet and Brule 1982, Nieuwenhuijse et al. 1988, Oldfield 1998). The extent of the transfer of phosphate to the colloidal phase, which is greater than that of calcium, depends on the temperature of preheating. The concentrations of soluble calcium and phosphate in reconstituted milk powder are generally lower than those in the original milk, due to irreversible shifts induced during drying (Le Graet and Brule 1982).

Holding milk concentrate at >60°C for an extended period before spray-drying can increase the viscosity of the concentrate, making it more difficult to atomise and thereby affecting the properties of the final powder.

Alternative technologies for concentrating milk are available, the most significant of which is probably membrane separation using reverse osmosis (RO). RO can achieve only a relatively low level of total solids (<20%) and has a relatively low throughput, but is far less thermally severe than evaporation (Pisecký 1997). Other membrane techniques may be used to concentrate or fractionate milk, such as nanofiltration (NF), which essentially removes only water, and ultrafiltration (UF), which allows standardisation of the composition (e.g., protein and lactose contents) of the final powder (Mistry and Pulgar 1996, Horton 1997). The most significant area of application for membrane processing in the dairy industry to date is probably in whey processing, as will be discussed below.

Spray-Drying of Milk

Today, milk powders are produced in large, highly efficient spray-dryers; the choice of dryer design (e.g., single versus multi-stage, with integrated or external fluidised bed drying steps, choice of atomiser, air inlet and outlet temperatures) depends on the final product characteristics required. Because of a wide range of applications of milk powders, from simple reconstitution to use in cheese manufacture or incorporation as an ingredient into complex food products, the precise functionality required for a powder can vary widely.

Skim milk powders (SMP) are usually classified on the basis of heat treatment (mostly meaning the intensity of preheating during manufacture), which influences their flavour and solubility (Caríc and Kalab 1987, Kyle 1993, Pellegrino et al. 1995). Instant powders (readily dispersible in cold or warm water) are desirable for applications requiring reconstitution in the home, and are produced by careful production of agglomerated powders containing an extensive network of air spaces that can fill rapidly with water on contact (Pisecký 1997). A number of techniques are used to produce agglomerated powders, including feeding fines (small powder particles) back to the atomisation zone during drying (in straight-through agglomeration processes) or wetting dry powder in chambers that promote aggregation of moistened particles that, when subsequently dried, produce porous agglomerated powder particles (rewet processes).

Milk powder contains occluded air (i.e., contained in vacuoles within individual powder particles) and interstitial air (i.e., entrapped between neighbouring powder particles). The amount of occluded air depends on the heat treatment applied to the feed (whey protein denaturation affects foaming properties), method of atomisation and outlet air temperature, while the interstitial air content of a powder depends on the size, shape and surface geometry of individual powder particles. The drying process can be manipulated (e.g., by using multiple stages or by returning fines to the atomisation zone) to increase the levels of interstitial and occluded air and instantise the powder (Kelly et al. 2003).

In the case of whole milk powders (WMP), a key characteristic is the state of the milk fat, whether fully emulsified or partially in free form (the latter is desirable for WMP used in chocolate manufacture). The production of instant WMP requires the use of an amphiphilic additive such as lecithin, as well as agglomeration, to overcome the intrinsic hydrophobicity of milk fat (Jensen 1975, Kim et al. 2009). Lecithin may be added either during multi-stage drying of milk (between the main drying chamber and the fluidised bed dryer) or separately in a rewet process.

During spray-drying of milk, there is relatively little change in milk proteins, apart from that caused by preheating and evaporation. The constituent most affected by the drying stage *per se* is probably lactose that, on the rapid removal of moisture from the matrix, assumes an amorphous glassy state, which is hygroscopic and can readily absorb moisture if the powder is exposed to a high relative humidity. This can result in the crystallisation of lactose, with a concomitant uptake of water, which can cause caking and plasticisation of the powder (Kelly et al. 2003).

Amorphous lactose is the principal constituent of both SMP and WMP and forms the continuous matrix in which proteins, fat globules and air vacuoles are dispersed. In products containing a very high level of lactose (e.g., whey powders, SMP), the lactose may be pre-crystallised to avoid the formation of an amorphous glass. Usually, small lactose crystals are added to milk or whey concentrate prior to drying to promote crystallisation under relatively mild conditions (i.e., in the liquid rather than the powder form); the added crystals act as nuclei for crystallisation.

Freezing of Milk

When milk is cooled to temperatures below approximately –0.54°C, pure ice crystals form, and the remainder of the milk, thus, becomes progressively more concentrated at lower temperatures. This results in changes in milk that are analogous to those occurring on concentration by other means, including decrease in pH and increase in Ca^{2+} activity. At temperatures below approximately –23°C, the non-frozen portion of milk is in a glassy state, and changes cease, due to diffusion coefficients being close to zero. On thawing of frozen milk, aggregation of casein micelles due to salting out may occur, as may clumping of fat globules, due to mechanical damage to globule membranes by ice crystals.

CHEESE AND FERMENTED MILKS

Introduction

About 35% of total world milk production is used to produce cheese ($\sim 16 \times 10^6$ tonnes per annum), mainly in Europe, North and South America, Australia and New Zealand. Cheese manufacture essentially involves coagulating the casein micelles to form a gel that entraps the fat globules, if present; when the gel is cut or broken the casein network contracts (synereses), expelling whey. The resulting curds may be consumed fresh as mild-flavoured products, or ripened for a period ranging from 2 weeks, for example for Mozzarella, to >2 years, for example for Parmigiano-Reggiano.

Cheese is undoubtedly the most biochemically complex dairy product. However, since the action of rennet on milk and dairy fermentations are described in Chapters 33 and 27, respectively, only a brief summary is given here. For more detailed discussion of the science and technology of cheesemaking, see also Robinson and Wilbey (1998), Law (1999), Eck and Gillis (2000) and Fox et al. (2004), McSweeney (2007) and a number of articles in the Encyclopedia of Dairy Science (2011).

The coagulation of milk for cheese is achieved by one of three methods:

1. Rennet-induced coagulation, which is used for most ripened cheeses, and accounts for approximately 75% of total cheese production,
2. Acidification to pH approximately 4.6 at 30–36°C by *in situ* production of acid by fermentation of lactose to lactic acid by lactic acid bacteria (LAB; *Lactococcus*, *Lactobacillus* or *Streptococcus*) or direct acidification with acid or acidogen (e.g., gluconic acid-δ-lactone). Most acid-coagulated cheeses are consumed fresh and represent approximately 25% of total cheese production. Major examples of acid-coagulated cheeses are Cottage, Quarg and Cream cheeses,

3. Acidification of milk, whey or mixtures thereof to pH approximately 5.2 and heating to approximately 90°C. These cheeses are usually consumed fresh; common examples include Ricotta and variants thereof, Manouri and some forms of Queso Blanco.

Rennet-Coagulated Cheeses

These cheeses, which represent approximately 75% of total production, are produced in a great diversity of shapes, flavours and textures (~1400 varieties worldwide). Their production can be divided into two phases: (1) conversion of milk to cheese curd and (2) ripening of the curd (Fig. 25.2B).

Cheese may be made from milk of many species, with bovine, ovine and caprine being most common globally. Cheese may be made from raw or pasteurised milk, and in large-scale industrial cheese production the latter is far more common, with some exceptions. The heat treatment applied is normally conventional HTST pasteurisation, as more severe treatments may impair the coagulation properties of milk due to heat-induced interaction of whey proteins and casein micelles. Other pre-treatments that may be applied to cheese milk include adjustment of the fat:protein or fat:casein ratio (i.e., standardisation) to achieve certain targets for cheese composition, and addition of $CaCl_2$, which may be used to adjust for seasonal differences in mineral balance of milk. For some varieties, milk pH may be reduced (e.g., by adding gluconic acid-δ-lactone) prior to coagulation.

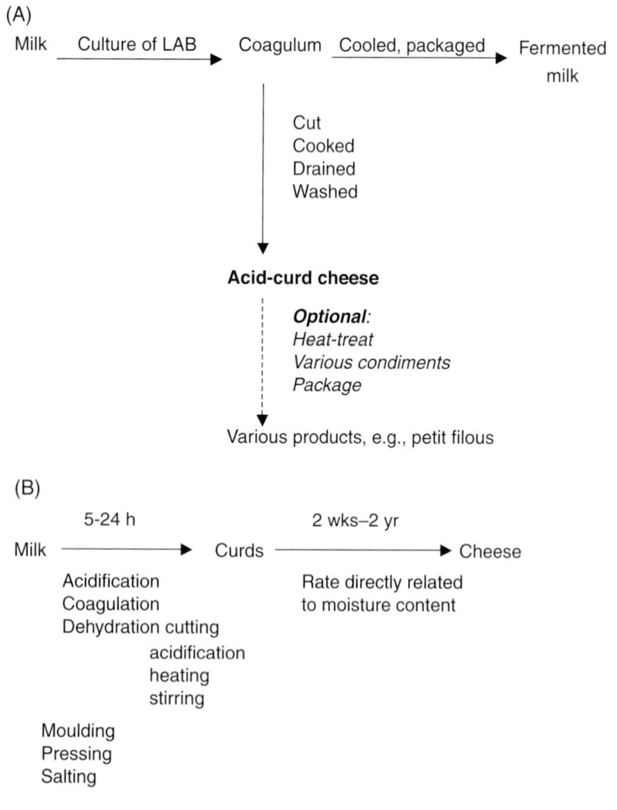

Figure 25.2. Principles of manufacture of (**A**) acid-coagulated and (**B**) rennet-coagulated cheese.

Coagulation

The coagulation of milk for the production of rennet-coagulated cheese exploits a unique characteristic of the casein system. As described in Chapter 24, the casein in milk exists as large (diameter 50–600 nm, mean 150 nm) colloidal particles, known as casein micelles. The micelles are stabilised by κ-casein, which is concentrated on the surface, with its hydrophobic *N*-terminal segment interacting with the α_{s1}–, α_{s2}– and β-caseins, and its hydrophilic *C*-terminal third protruding into the aqueous environment, forming a layer approximately 7 nm thick, which stabilises the micelles by a zeta potential of about –20 mV and by steric stabilisation. The stability of the micelles is lost when the surface κ-casein layer is destroyed by heat, alcohol or proteinases (known as rennets).

Several proteinases can coagulate milk but the traditional and the most effective rennets are NaCl extracts of the stomachs of young, milk-fed calves, kids or lambs. The active enzyme in these rennets is chymosin; as the animal ages, the secretion of chymosin decreases and is replaced by pepsin. The supply of chymosin has been inadequate for about 50 years due to the increased production of cheese and reduced availability of calf stomachs (due to the birth of fewer calves and the slaughter of calves at an older age). This shortage has led to a search for alternative coagulants (rennet substitutes).

Many, perhaps most, proteinases can coagulate milk under certain conditions, but almost all are unsuitable as rennets because they are too proteolytic, resulting in a reduced yield of cheese curd and off-flavoured cheese. Only five successful rennet substitutes have been identified: (1) bovine and (2) porcine pepsins and acid proteinases from (3) *Rhizomucor mehei,* (4) *Rhizomucor pusillus* and (5) *Cryphonectria parasitica*. Plant rennets (e.g., that from *Cynara cardunculus*) have also been studied as possible coagulants, but have high proteolytic activity relative to their milk-clotting activity, which may lead to bitterness in cheese. Plant rennets have been most successfully used for cheese varieties made in Spain and Portugal from ovine milk (e.g., Torta del Casar and Serra Da Estrela), in which bitterness does not seem to be a problem.

Rennet pastes, including those from kids or lambs, have been traditionally used for some hard Italian and Greek cheese varieties. These are less purified than conventional calf rennet, and include a lipase, pregastric esterase (PGE), which contributes to lipolysis during ripening of these varieties. The calf chymosin gene has been cloned in *Kluyveromyces lactis, Escherichia coli* and *Aspergillus niger* and fermentation-produced chymosin is now used widely.

Chymosin and most of the other commercially successful rennets hydrolyse κ-casein specifically at the Phe_{105}–Met_{106} bond; *C. parasitica* proteinase cleaves κ-casein at Ser_{104}–Phe_{105}. The liberated hydrophilic *C*-terminal segment, known as the (glyco) caseinomacropeptide (CMP), diffuses into the surrounding aqueous phase and the stability of the micelles is destroyed.

When approximately 85% of the κ-casein has been hydrolysed, the rennet-altered micelles aggregate to form a gel in the presence of a critical concentration of Ca^{2+} and at a temperature about $>18^{\circ}C$.

Acidification

The second characteristic step in cheesemaking is acidification of the milk and curd from pH approximately 6.7 to a value in the range 4.6–5.2, depending on the variety. Until relatively recently, and for some minor artisanal varieties still, acidification was due to the production of lactic acid from lactose by adventitious LAB. Today, cheese milk is inoculated with a culture (starter) of selected LAB for more controlled and reproducible acidification. If the cheese curds are cooked to <40°C, a culture of *Lactococcus lactis* and/or *cremoris* is used, but for high-cooked cheese (curds cooked to 50–55°C), a culture of *Lactobacillus delbrueckii* subsp. *bulgaricus,* possibly in combination with *Streptococcus thermophilus,* is used. Cheese starters have been refined progressively over the years, especially with respect to the rate of acidification, resistance to bacteriophage and cheese-ripening characteristics. Today, mixtures of highly selected defined strains of LAB are used widely.

Acidification at the correct rate and time is essential for successful cheesemaking; it affects at least the following aspects:

- Activity and stability of the coagulant during renneting
- Strength of the rennet-induced gel
- Rate and extent of syneresis of the gel when cut and hence the composition of the cheese
- Retention of rennet in the curd
- Dissolution of colloidal calcium phosphate; the concentration of calcium in the cheese has a major effect on the texture and functionality of cheese
- Inhibition of growth of undesirable microorganisms, especially pathogenic bacteria
- Activity of various enzymes in the cheese during ripening

Post-Coagulation Operations

If left undisturbed, a rennet-coagulated milk gel is quite stable, but if cut or broken, it synereses (contracts), producing curds and whey. By controlling the rate and extent of syneresis, the cheese maker controls the moisture content of cheese and thereby, the rate and pattern of ripening and the quality and stability of the cheese.

Traditionally, the point at which the gel was ready to be cut was judged by the cheese maker based on experience; however, in some modern cheesemaking systems, this decision may be made automatically based on sensors in the vat or rheological measurements (Fagan et al. 2008).

Syneresis is affected by:

- Concentrations of fat, protein and calcium
- Denaturation of whey protein
- pH
- Size of curd particles
- Temperature of cooking
- Stirring the curds–whey mixture and the curds after whey drainage
- Pressing of the curd
- Salting (2 kg H_2O lost per kg NaCl taken up; NaCl should not be used to control moisture content)

When the desired degree of syneresis has occurred, as judged subjectively by the cheese maker, the curds and whey are separated, usually on some form of perforated metal screen. The curds are subjected to various treatments that are more or less variety-specific. These include cheddaring, kneading–stretching, moulding, pressing and salting. Salt is added to essentially all cheese varieties, but may be added either in dry form (e.g., for Cheddar cheese) or, more commonly, through immersion of the cheese in brine for sufficient time for salt uptake to the desired final level, followed by enough time for diffusion throughout the cheese so as to achieve a uniform concentration. The functions of salt in cheese are in flavour and control of bacterial growth and metabolism.

The extent of physical force applied to cheese curd during the later stages of manufacture depends on the target final moisture content, and may range from significant pressures being applied through piling of curds in the vat followed by pressing in moulds (Cheddar), to very light drainage under gravity in perforated moulds (e.g., Camembert). In addition, variety-specific techniques such as heating–stretching may be used at the final stages to achieve specific desirable functional properties (e.g., for Mozzarella).

Ripening

Rennet-coagulated cheese curd may be consumed at the end of the manufacturing process, and a little is, for example Junket or Burgos cheese, but most is ripened (matured) for a period ranging from approximately 2 weeks (Mozzarella) to >2 years (Parmigiano-Reggiano, extra-mature Cheddar) during which the characteristic flavour, texture and functionality of the cheese develop. The principal changes in cheese during ripening are listed in Table 25.4.

Ripening is a very complex biochemical process, catalysed by:

- *Coagulant*: Depending on the coagulant used, the pH of the curd at whey drainage, the temperature to which the curds are cooked and the moisture content of the curds, 0–30% of the rennet added to the milk is retained in the curd.
- *Indigenous milk enzymes*: These are particularly important in raw-milk cheese but many indigenous enzymes are sufficiently heat stable to withstand HTST pasteurisation. Important indigenous enzymes include plasmin, xanthine oxidase, acid phosphatase and, in raw-milk cheese, lipoprotein lipase. Information on the significance of other indigenous enzymes is lacking.
- *Starter LAB and their enzymes*: These reach maximum numbers (approximately 10^9 CFU/g) at the end of curd manufacture (6–12 hours). They then die and lyse at strain-dependent rates, releasing their intracellular enzymes.

Table 25.4. Changes in Cheese During Ripening

Softening of the Texture, due to:
• Proteolysis of the protein matrix, the continuous solid phase in cheese
• Increase in pH due to the catabolism of lactic acid and/or production of NH_3 from amino acids
• Migration of Ca to the surface in some cheeses, e.g. Camembert
Decrease in Water Activity, due to:
• Uptake of NaCl
• Loss of water through evaporation
• Production of low molecular weight compounds
• Binding of water to newly formed charged groups, e.g., $-NH_4^+$ or COO^-
Changes in Appearance, e.g.:
• Mould growth
• Colour, due to growth of *Brevibacterium linens*
• Formation of eyes
Flavour Development, due to:
• Formation of numerous sapid and aromatic compounds
• Release of flavourful compounds from the cheese matrix due to proteolysis
Changes in Functionality, e.g., Meltability, Stretchability, Water-Binding Properties, Browning, due to:
• Proteolysis
• Catabolism of lactose

- *Non-starter LAB (NSLAB)*: These are adventitious LAB that contaminate the cheese milk at farm and/or factory. Traditionally, cheesemakers relied on NSLAB for acid production during curd manufacture; this is still the case with some minor artisanal varieties but starter LAB are now used in all factory and most artisanal cheesemaking operations. Most NSLAB are killed by HTST pasteurisation and curd made from good-quality pasteurised milk in modern enclosed equipment has only a few hundred NSLAB, mainly mesophilic lactobacilli, per gram at the start of ripening. Indeed, only a limited number of species of mesophilic lactobacilli grow in cheese curd. However, those NSLAB that are present, grow at a rate depending mainly on the temperature to 10^7–10^8 cfu/g within about 3 months. In cheese made from raw milk, the NSLAB microflora is more diverse and reaches a higher number than in cheese made from pasteurised milk; the difference is probably mainly responsible for the more intense flavour of the former compared to the latter. This situation applies to all cheese varieties that have been investigated and the NSLAB dominate the viable microflora of cheese ripened for >2 months.
- *Secondary cultures*: Most cheese varieties develop a secondary microflora, which was originally adventitious but now develops mainly from added cultures. Examples are *Propionibacteria freudenrichii* subsp. *shermanii* (Swiss cheeses), *Penicillium roqueforti* (Blue cheeses), *Penicillium camemberti* and *Geotrichium candidum* (Camembert and Brie), *Brevibacterium linens*, *Arthrobacter* spp., *Corynebacterium* and yeasts (surface smear-ripened cheese) and citrate-positive *Lc. lactis* and *Leuconostoc* spp. (Dutch-type cheeses). Most of the microorganisms used as secondary cultures are very active metabolically and many secrete very active proteolytic and lipolytic enzymes. Consequently, they dominate the ripening of varieties in which they are used. Traditionally, a secondary culture was not used for Cheddar-type cheese but it is becoming increasingly common to add an adjunct culture, usually selected mesophilic lactobacilli (equivalent to adding a NSLAB population), to accelerate ripening, intensify flavour and perhaps tailor-make flavour. Essentially, the objective is to reproduce the microflora, and consequently the flavour, of raw-milk cheese. It is also becoming more common to add *Str. thermophilus* as an adjunct culture for Cheddar cheese.
- *Exogenous enzymes*: For some varieties of Italian cheese, for example Provalone and Pecorino varieties, rennet paste is used as coagulant. The rennet paste contains a lipase, PGE, in addition to chymosin; PGE is responsible for extensive lipolysis and the characteristic piquant flavour of the cheese. Rennet extract used for most varieties lacks PGE.

During ripening, a very complex series of reactions, which fall into three groups, occur:

1. Glycolysis of lactose and modification and/or catabolism of the resulting lactate
2. Lipolysis and the modification and catabolism of the resulting fatty acids
3. Proteolysis and catabolism of the resulting amino acids

Glycloysis and Related Events Fresh cheese curd contains approximately 1% lactose, which is converted to lactic acid (mainly the L-isomer) by the starter LAB, usually within 24 hours. Depending on the variety, the L-lactic acid is racemised to DL-lactic acid by NSLAB, or catabolised to CO_2 and H_2O in mould-ripened and in smear-ripened cheese. In Swiss-type cheese, lactic acid is converted to propionic and acetic acids, CO_2 and H_2O by *P. freudenrichii* subsp. *shermanii*; the CO_2 is responsible for the characteristic eyes in such cheese.

Lipolysis Little lipolysis occurs in most cheese varieties, in which it is catalysed by the weakly-lipolytic LAB or NSLAB, and indigenous milk lipase if raw milk is used. Extensive lipolysis occurs in some hard Italian-type cheeses, for example Provalone and Pecorino varieties, for which PGE is responsible, and in Blue cheese, for which *P. roqueforti* is responsible. Fatty acids are major contributors to the flavour of Provalone and Pecorino cheese and some may be converted to lactones or esters, which have characteristic flavours. In blue-veined cheese, fatty acids are converted to methyl ketones by *P. roqueforti* and these are mainly responsible for the characteristic peppery taste of such cheese. Some of the methyl ketones may be reduced to secondary alcohols.

Proteolysis Proteolysis is the most complex and probably the most important of the three primary ripening reactions for the quality of cheese, especially internal bacterially ripened

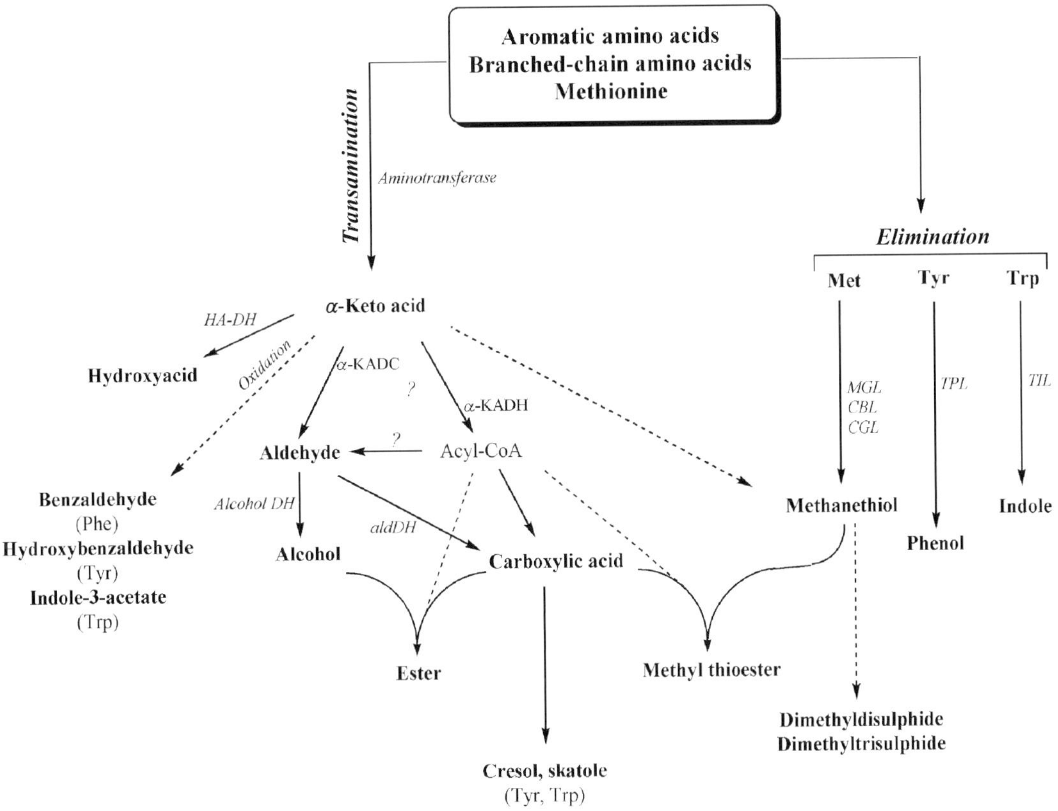

Figure 25.3. Summary of amino acid catabolism during cheese ripening. MGL, methionine-γ-lyase; CBL, cystathionine-β- lyase; CGL, cystathionine-γ-lyase; TPL, tyrosine-phenol lyase; aldDH, aldehyde dehydrogenase; KADH, α-keto acid dehydrogenase; HADH, hydroxyl acid dehydrogenase; α−KADC, α−keto acid decarboxylase. (From Curtin and McSweeney 2004).

varieties. Initially, the caseins are hydrolysed by chymosin and, to a lesser extent by plasmin, in the case of β-casein; chymosin is almost completely inactivated in high-cook varieties and in this case plasmin is the principal agent of primary proteolysis. The polypeptides produced by chymosin or plasmin are too large to affect flavour, but primary proteolysis has a major influence on the texture and functionality of cheese. The peptides produced by chymosin or plasmin are hydrolysed to smaller peptides and amino acids by proteinases and peptidases of starter LAB and NSLAB. Small peptides contribute positively to the background brothy flavour of cheese but some are bitter. Many amino acids also have a characteristic flavour but, more importantly, they serve as substrates for a great diversity of catabolic reactions catalysed by enzymes of LAB, NSLAB and the secondary culture. The flavour of internal bacterially ripened cheese is probably mainly due to compounds produced from amino acids (Fig. 25.3).

Acceleration of Cheese Ripening

As the cost of storing cheese under controlled conditions for ripening represents a significant proportion of the cost of manufacturing cheese, there have been many studies on factors that could accelerate the process of cheese ripening, that is achieve the desired texture and flavour in a shorter time. Key strategies that have been studied include increasing ripening temperature, manipulating starter or adding adjunct starter culture (e.g., increasing level of addition, or accelerating the rate at which starter cells lyse and release their enzymes into the cheese curd) and use of exogenous enzymes. Of these, the simplest is increasing ripening temperature (e.g., from 7°C for Cheddar to 13°C), which has been shown to increase rates of lipolysis and proteolysis. For the use of exogenous enzymes, the key is to get even distribution of enzyme in the cheese, which can most easily be achieved by adding enzymes to the cheese milk; however, much of the added enzyme is then lost in the whey. A general consideration applying to all methods for accelerating cheese ripening is that the overall balance of flavour should remain comparable to the reference product, which can be difficult to achieve.

Cheese-Related Products

In addition to the many types of natural cheese, as introduced previously, there is also a range of cheese-derived or substitute products, such as processed cheese, cheese analogues and enzyme-modified cheese (EMC). Processed cheese is produced by blending natural cheeses of differing stages of ripening, to balance structural and flavour properties, and adding water and emulsifying salts that convert the insoluble calcium para-caseinate to more soluble and amphiphilic sodium para-caseinate, which then produces a stable emulsion. Emulsification is carried out under high-shear conditions at 70–90°C, yielding a safe and stable product. Following this molten stage, the cheese can be set (on cooling) into a wide range of shapes and forms, including blocks and slices; in addition, by choosing the formulation and ingredients used, a range of textures from solid blocks to spreads, which can have a range of added food components and flavours, can be produced.

In the case of cheese analogues, natural cheese is not used, and the main source of casein is usually rennet casein; such products may be used as Mozzarella substitutes for pizza toppings. EMC is cheese curd that has been incubated with enzymes (e.g., proteinases, peptidases and lipases) to yield a concentrated product with an intense cheese flavour that can be used as an ingredient (often in dried form) in food formulations. EMC with flavour profiles matching different cheese varieties have been produced.

Acid-Coagulated Cheeses

Acid-coagulated cheeses are at one end of the spectrum of fermented dairy products, the production of which is summarised in Figure 25.2A. Depending on the desired fat content of the final product, the starting material may be low-fat cream, whole milk, semi-skimmed milk or skimmed milk. The milk for Cottage-type cheese is subjected to only mild heat treatment so that the syneretic properties of the coagulum are not impaired.

Fermented Milks

In addition to acid-coagulated cheese, a wide range of fermented milk products are produced worldwide today, including:

- products of fermentation by mesophilic or thermophilic LAB (e.g., yoghurt); and
- products obtained by alcohol-lactic fermentation by LAB and yeast (e.g., kefir, koumiss).

For yoghurt, as for acid-coagulated cheese, the milk base (with a fat content depending on the final product) is usually supplemented with milk solids to enhance the viscosity and rigidity of the final coagulum, in some cases to simulate the viscosity of yoghurt made from sheep's milk, which has a higher total solids content than cows' milk. The milk solids added may be WMP or SMP or whey powders, depending on the type of product required (González-Martínez et al. 2002, Remeuf et al. 2003). The most commonly used milk powder today is probably SMP, typically of medium heat classification. Buttermilk powder may be used also as a substitute for SMP, with the advantages of good emulsifying properties and potential biological activity.

Syneresis, which causes free whey in the product, is very undesirable in fermented milk products. Therefore, the milk or starting mix is subjected to a severe heat treatment (e.g., 90°C for 10 minutes) to denature the whey proteins, increase their water-binding capacity and produce a firmer gel network, consisting of casein–whey protein complexes. The heat treatment also kills vegetative pathogenic bacteria, rendering the product safe (Robinson 2002a). Homogenisation of the mix also increases the firmness of the gel, because the fat globules in homogenised milk are stabilised mainly by a surface layer of adsorbed casein micelles, through which they can become structural elements of the final acid-induced gel.

The relative proportion of caseins and whey proteins in yoghurt has a significant influence on the structure of the gel (Puvanenthiran et al. 2002). Hydrocolloid stabilisers (polysaccharides or proteins) may be added to the mix to modify the viscosity (by gelling or thickening) and/or prevent syneresis (Fizsman et al. 1999, Tamime and Robinson 1999). Pectin is probably the most commonly used hydrocolloid stabiliser for yoghurt, but carageenan, guar gum, alginate and various modified and natural starches are also widely used. Sweetening agents (e.g., sugars or synthetic sweeteners such as aspartame) may be added to the mix, and fruit purees or fruit essence may be added either at the start or end of fermentation.

The starter culture for yoghurt contains *Str. thermophilus* and *Lb. bulgaricus* subsp. *bulgaricus*. In recent years, many fermented milk products have been introduced with added probiotic bacteria such as *Bifidobacteria* spp. or *Lactobacillus acidophilus*, due to increasing consumer interest in the health benefits of such bacteria, and increasing clinical evidence of their efficacy (Surono and Hosono 2002, Robinson 2002b, Reid 2008, Sanchez et al. 2009). The texture and viscosity of yoghurt may be modified by using an exopolysaccharide (EPS)-producing starter culture (strains of both *Str. thermophilus* and *Lb. bulgaricus* subsp. *bulgaricus* have been identified that secrete capsular heteropolysaccharides; De Vuyst et al. 2001, Hassan et al. 2003).

During fermentation, the production of lactic acid destabilises the casein micelles resulting in their coagulation (acid-induced gelation) (Lucey and Singh 1998). The main point of difference between the production protocols for yoghurt and acid-coagulated cheese is that when the pH has reached 4.6 and the milk has set into a coagulum, this coagulum is cut or broken for the latter but not for the former. In addition, the heat treatment applied to milk for the manufacture of acid-coagulated cheese is less severe than for yoghurt, to facilitate syneresis in the former, while avoiding it in the latter.

Depending on the desired characteristics of the finished product, the milk may be fermented in the final package (to produce a firm set yoghurt) or in bulk tanks, followed by filling (to produce a more liquid stirred yoghurt) (Tamime and Robinson 1999, Robinson 2002a). The rate of acidification for set or stirred yoghurt differs due to differences in the level of inoculation and incubation temperature.

It is important to cool the yoghurt rapidly when pH 4.6 is reached, to preserve the gel structure and prevent further starter activity (Robinson 2002a). In some cases, the viscosity of the yoghurt may be reduced by post-fermentation homogenisation to produce drinking yoghurt.

Yoghurt has a relatively short shelf life; common changes during storage involve syneresis, with whey expulsion, and further slow acidification by starter bacteria. To prevent the latter, and thereby extend the shelf life of yoghurt, the final product may be pasteurised after acidification; to prevent damage to the gel structure at this point, the role of hydrocolloid or protein stabilisers is critical (Walstra et al. 1999).

Acid-Heat Coagulated Cheeses

These cheeses were produced initially in Southern European countries from whey as a means of recovering nutritionally valuable whey proteins; their production involves heating whey from rennet-coagulated cheese to approximately 90°C to denature and coagulate the whey proteins. Well-known examples are Ricotta and Manouri. Today, such cheeses are made from blends of milk and whey, and in this case, it is necessary to adjust the pH of the blend to approximately 5.2 using vinegar, citrus juice or fermented milk.

Acid-heat coagulated cheese may also be produced from whole milk by acidifying to pH 5.2 and heating to 90°C, for example US-style Queso Blanco, which does not melt on heating and hence has interesting functional properties for certain applications.

WHEY PROCESSING

Range of Whey Products

For many centuries, whey was often viewed as an unwanted and valueless by-product of cheese manufacture; however, since around 1970, whey has been repositioned as a valuable source of a range of food ingredients, rather than a disposal problem. While part of the impetus for this strategic re-evaluation of the potential of whey was driven by environmental and other concerns about the disposal of a product of such a high Biological Oxygen Demand, it has resulted in a previously unsuspected richness and diversity of products, and a significant new processing sector of the dairy industry (Sienkiewicz and Riedel 1990, Jelen 2002).

The first stage in the production of any whey product is preliminary purification of the crude whey, recovered after cheese or casein manufacture. Fresh whey is classified as either sweet, with a pH > 5.6 and a low calcium content, or acid whey, with a pH of around 4.6 and high calcium content; the difference in calcium content arises from the solubilisation of colloidal calcium phosphate as the pH of the milk decreases. Common pre-treatments include centrifugal clarification to remove curd particles (fines), and pasteurisation (to inactivate starter LAB). For many whey protein-based products, removal of all lipids is desirable; this can be achieved by centrifugal separation to produce whey cream, which may be churned into whey butter. To obtain a very low level of lipids, calcium chloride may be added to the whey, with adjustment of the pH to more alkaline values, followed by heating and cold storage to precipitate and remove lipoprotein complexes (thermocalcic aggregation; Karleskind et al. 1995).

The next level of processing technology applied to the pretreated whey depends on the end product to be manufactured. The simplest whey products are probably whey beverages, consisting of clarified whey, typically blended with natural or concentrated fruit juices. Whey beverages have a nutritionally beneficial amino acid profile; although not widely commercialised, such drinks have been successful in some European countries (e.g., Rivella in Switzerland).

Whey may be concentrated by evaporation and spray-dried to whey powders. The key consideration in spray-drying whey is the constraints imposed by the fact that the major constituent of whey is lactose (typically, >75% of whey solids). Thus, processes for drying whey must include controlled crystallisation of lactose to yield small crystals (30–50 μm). This is typically achieved by controlled cooling of concentrated (i.e., supersaturated with respect to lactose) whey under a programmed temperature regime, with careful stirring and addition of seed crystals, typically of α-lactose monohydrate, followed by holding under conditions sufficient to allow crystallisation to proceed. In general, only approximately 70% of the lactose present will crystallise at this stage, and processes often include a post-drying crystallisation stage, for example on a belt attached to the spray-dryer, where the remainder crystallises.

The simplest dried whey product is whey powder, which is produced in a single-stage spray-dryer. Additional care must be taken in drying acid whey products, which may be sticky; the corrosive nature of the acids (e.g., HCl) used in their manufacture may also present processing difficulties and such products may be neutralised prior to drying.

Whey powders may be unsuitable for use in certain food applications (e.g., infant formulae) due to their high mineral content; in such cases, whey is demineralised by ion exchange or electrodialysis before concentration and dialysis (for a discussion of the technologies involved, see Burling 2002). Demineralisation of whey is also desirable for use in ice cream production, to reduce the salty taste of normal whey powder.

Whey-Protein-Rich Products

Whey powder has a low (approximately 12–15%) protein content and a high (approximately 75%) lactose content; many whey products have been at least partially purified to increase the level of a particular constituent, usually protein (Matthews 1984). There is a family of protein-enriched whey-derived powders, which are differentiated based on the level of protein; whey protein concentrates (WPCs) with a protein content in the range 35–80%, and whey protein isolates (WPIs) with a protein content >88%. Typically, WPCs are produced by UF, with protein being progressively concentrated in the retentate, while lactose, salts and water are removed in the permeate (Ji and Haque 2003). Higher protein levels can be achieved by diafiltration, that is, dilution of the retentate followed by UF. The use of ion exchangers to adsorb the proteins from whey, followed by selective release into suitable buffer solutions, is required to achieve the high degree of purity required of WPIs.

Methods for the purification of individual whey proteins have been developed, usually exploiting differences in the stability of individual proteins under specific conditions of pH, temperature and ionic salts. In addition, methods involving organic extraction, selective enzymatic hydrolysis and ion-exchange resins have been described. On a commercial scale, the production of products containing purified but unfractionated whey proteins remains most common, although ion-exchange methods for the recovery of biologically significant (and hence high value) proteins from whey, such as lactoperoxidase and lactoferrin have been scaled up and are in commercial operation.

Lactose Processing

Whey or whey permeate, produced as a by-product of UF processing, contains a high level of lactose. This can be recovered by crystallisation from a concentrated (supersaturated, 60–62% TS) preparation of either whey or whey permeate (for review see Muir 2002, Peterson 2009). The crystalline lactose is usually recovered using a decanter centrifuge, dried in a fluidised bed dryer and ground to a fine powder, which is used in food applications, for example confectionery products and infant formulae. Lactose is also used widely in pharmaceutical applications (e.g., as a diluent in drug tablets); for such applications, the lactose is generally further refined by re-dissolving in hot water and mixing with activated carbon, followed by filtration, crystallisation and drying.

Many derivatives of lactose can be produced, some of which are more valuable and useful than lactose (Fig. 25.4, Playne and Crittenden 2009). For example, lactulose (β-D-galactosyl-D-fructose) and lactitol (β-D-galactosyl-sorbitol) are used in treatment of patients with chronic hepatic encephalopathy, and both have prebiotic effects. In addition, lactose in whey can be fermented to alcohol for industrial use in beverages.

CASEINS: ISOLATION, FRACTIONATION AND APPLICATIONS

In recent years, it has been increasingly recognised that milk proteins have functionalities that can be exploited in food systems other than conventional dairy products; these proteins have been increasingly recognised as desirable ingredients for a range of food products (Mulvihill and Ennis 2003). The very different properties of the two classes of milk proteins, the caseins and whey proteins, present different technological challenges for recovery, and are suitable for quite different applications. The whey proteins have been discussed previously, and the caseins will be discussed in the following section.

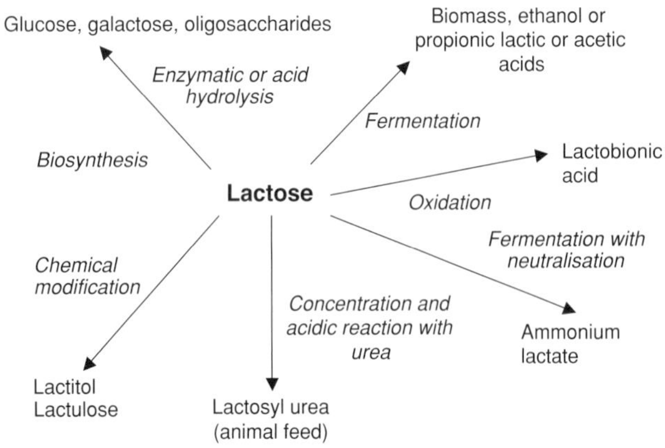

Figure 25.4. Food-grade derivatives of lactose.

Recovery and Application of Caseins

Technologies for the recovery of caseins from milk are based on the fact that relatively simple perturbations of the milk system can destabilise the caseins selectively, resulting in their precipitation and facilitating their recovery from milk (Mulvihill and Fox 1994, Mulvihill and Ennis 2003). As discussed for rennet-coagulated cheese, the stability of the caseins in a colloidal micellar form is possible due to the amphiphilic nature of κ-casein. To overcome this stability, and precipitate the caseins, merely requires that this stabilising effect be overcome.

The two key principles used to destabilise the micelles are (1) acidification to the isoelectric point of the caseins (pH 4.6) or (2) limited proteolysis, for example by chymosin, which hydrolyses κ-casein, removing the stabilising glycomacropeptide. In both cases, the starting material for casein production is skim milk. For acid casein, acidification can be achieved either by addition of a mineral acid, usually HCl, or fermentation of lactose to lactic acid by a culture of LAB. When the isoelectric point is reached, a precipitate (rapid acidification) or a gel (slower acidification) is formed; the latter is cut/broken to initiate syneresis and expel whey. The mixture of casein and whey is stirred and cooked to enhance syneresis, and the casein separated from the whey (either centrifugally or by sieving). To improve the purity of the casein, the curds are washed repeatedly with water to remove residual salts and lactose, and the final casein is dried, typically in specialised dryers, such as attrition or ring dryers. For rennet casein, the milk gel is formed by adding a suitable coagulant to skim milk, but subsequent stages are similar to those for acid casein.

Both acid and rennet casein are relatively insoluble in water, and are used in the production of cheese analogues (rennet casein) and convenience food products, as well as non-food applications, such as the manufacture of glues, plastics and paper glazing (acid casein).

To extend the range of applications of acid casein, and improve its functionality, it may be converted to a metal salt (e.g., Na, K, Ca caseinates), by mixing a suspension of acid casein with the appropriate hydroxide and heating, followed by drying (typically with a low, that is approximately 20%, total solids concentration in the feed, due to high product viscosity, in a spray-dryer). Caseinates, particularly sodium caseinate, have a range of useful functional properties, including emulsification and thickening; calcium caseinate has a micellar structure (Mulvihill and Ennis 2003).

Microfiltration (MF) technology may be used to produce powders enriched in micellar casein (Pouliot et al. 1996, Maubois 1997, Kelly et al. 2000, Garem et al. 2000). Casein recovered by MF has very different functionality and properties compared to acid or rennet casein, and may be more suitable for use in applications such as cheese manufacture, as it closely resembles the micellar casein in milk.

Whole casein may be resolved into β- and α_s/κ-casein-rich fractions by processes that exploit the dissociation of β-casein at a low temperature (reviewed by Mulvihill and Ennis 2003), or low-temperature dissociation after renneting of milk (Huppertz et al. 2006). Chromatographic or precipitation processes can be used to purify the other caseins, although such processes are generally not easy to scale up for industrial-level production (Coolbear et al. 1996, Farise and Cayot 1998).

EXOGENOUS ENZYMES IN DAIRY PROCESSING

The dairy industry represents one of the largest markets for commercial enzyme preparations. The use of some enzymes in the manufacture of dairy products, for example rennet for the coagulation of milk, is in fact probably the oldest commercial application of enzyme biotechnology.

PROTEINASES

The use of rennets to coagulate milk and the ripening of cheese, were discussed earlier. A further application of proteolytic enzymes in dairy products, that is the production of dairy protein hydrolysates, is discussed later in this chapter. The applications of other enzymes of significance for dairy products are discussed in this Section.

β-GALACTOSIDASE

Hydrolysis of lactose (4-*O*-β-D-galactopyranosyl-D-glucopyranose), either at a low pH or enzymatically (by β-galactosidases), yields two monosaccharides, D-glucose and D-galactose (Mahoney 1997, 2002). A large proportion of the world's population suffers from lactose intolerance, leading to varying degrees of gastrointestinal distress if lactose-containing dairy products are consumed. However, glucose and galactose are readily metabolised; they are also sweeter than lactose. Glucose/galactose syrups are one of the simplest and most common commercial products of lactose hydrolysis.

β-Galactosidases is available from several sources, including *E. coli, A. niger, Aspergillus oryzae and K. lactis* (Whitaker 1994; Mahoney 2002, Husain 2010). β-Galactosidase for use in dairy products must be isolated from a safe source and be acceptable to regulatory authorities; for this reason, genes for some microbial β-galactosidases have been cloned into safe hosts for expression and recovery. Of particular interest are enzymes that are active either at a low or a very high temperature, to avoid the microbiological problems associated with enzymatic treatment at temperatures that encourage the growth of mesophilic microorganisms (Vasiljevic and Jelen 2001). Microorganisms capable of producing thermophilic β-galactosidases include *Lb. bulgaricus* subsp. *bulgaricus* (Vasiljevic and Jelen 2001) and *Thermus thermophilus* (Maciunska et al. 1998), while cold-active β-galactosidase has been purified from psychrophilic bacteria (Nakagawa et al. 2003).

Enzymatic hydrolysis of lactose rarely leads to complete conversion to monosaccharides, due to feedback inhibition of the reaction by galactose, as well as concurrent side-reactions (due to transferase activity) that produce isomers of lactose and oligosaccharides (Chen et al. 2002). While initially considered to be undesirable by-products of lactose hydrolysis, galacto-oligosaccharides are now recognised to be bifidogenic factors, which enhance the growth of desirable probiotic bacteria in the intestine of consumers and suppress the growth of harmful anaerobic colonic bacteria (Shin et al. 2000).

Lactose has a number of other properties that cause difficulty in the processing of dairy products, such as its tendency to form large crystals on cooling of concentrated solutions of the sugar; lactose-hydrolysed concentrates are not susceptible to such problems. For example, hydrolysis of lactose in whey concentrates can preserve these products through increased osmotic pressure, while maintaining physical stability (Mahoney 1997, 2002).

For the hydrolysis of lactose in dairy products, β-galactosidase may be added in free solution, allowed sufficient time to react at a suitable temperature and inactivated by heating the product. To control the reaction more precisely and avoid uneconomical single use of the enzyme, immobilised enzyme technology (e.g., where the enzyme is immobilised on an inert support, such as glass beads) or systems where the enzyme is recovered by UF of the product after hydrolysis and reused have been studied widely (Obon et al. 2000). A further technique with potential for application in lactose hydrolysis is the use of permeabilised bacterial or yeast cells (e.g., *K. lactis*) with β-galactosidase activity. In such processes, the cells are treated with agents, for example ethanol, to damage their cell membrane and allow diffusion of substrate and reaction products across the damaged membrane; the cell itself becomes the immobilisation matrix and the enzyme is active in its natural cytoplasmic environment (Fontes et al. 2001, Becerra et al. 2001). This provides a crude but convenient and inexpensive enzyme-utilisation strategy. Overall, however, few immobilised systems for lactose hydrolysis are used commercially, due to technological limitations and high cost of such processes (Zadow 1993); one strategy of potential interest involves the recovery of soluble enzyme by UF, allowing separation from the product and substrate, and recovery and reuse of the enzyme.

One of the more common applications of lactose hydrolysis is the production of low-lactose liquid milk, suitable for consumption by lactose-intolerant consumers; this may be achieved in a number of ways, including adding a low level of β-galactosidase to packaged UHT milk, or the consumer may add β-galactosidase to milk during domestic refrigerated storage (Modler et al. 1993). In the case of ice cream, lactose hydrolysis reduces the incidence of sandiness (due to lactose crystallisation) during storage and, due to the enhanced sweetness, permits the reduction of sugar content. Lactose hydrolysis may also be applied in yoghurt manufacture, to make reduced-calorie products.

Overall, despite considerable interest, industrial use of lactose hydrolysis by β-galactosidases has not been widely adopted, although such processes have found certain niche applications (Zadow 1993).

Transglutaminase

Transglutaminases (TGase; protein-glutamine: amine γ-glutamyl-transferase) are enzymes that create inter- or intramolecular cross-links in proteins by catalysing an acyl-group transfer reaction between the γ-carboxyamide group of peptide-bound glutamine residues and the primary amino group of lysine residues in proteins. TGase treatment can modify the properties of many proteins through the formation of new cross-links, incorporation of amines or deamidation of glutamine residues.

TGases are widespread in nature; for example blood factor XIIIa, or fibrinoligase, is a TGase-type enzyme, and TGase-mediated cross-linking is involved in cellular and physiological phenomena such as cell growth and differentiation, as well as blood clotting and wound healing. TGases have been identified in animals, plants and microbes (e.g., *Streptoverticillium mobaraense*, which is the source of much of the TGase used in food studies). TGase may be either calcium-independent (most microbial enzymes) or calcium-dependent (typical for mammalian enzymes).

TGase-catalysed cross-linking can alter the solubility, hydration, gelation, rheological and emulsifying properties, rennetability and heat stability of a variety of food proteins (Motoki and Seguro 1998). The structure of individual proteins determines whether cross-linking by TGase is possible. Caseins are good substrates for TGase due to their open structure but the whey proteins, due to their globular structure, require modification, for example heat-induced denaturation, to allow cross-linking (Ikura et al. 1980, Sharma et al. 2001, O'Sullivan et al. 2002a, b). When the whey proteins in milk are in the native state, the principal cross-linking reactions involve the caseins; heat-induced denaturation renders the whey proteins susceptible to cross-linking, both to each other and to casein molecules.

TGase treatment has several significant effects on the properties of milk and dairy products. For example, TGase treatment of fresh raw milk increases its heat stability; if milk is preheated under conditions that denature the whey proteins prior to TGase treatment, the increase in heat stability is even more marked (O'Sullivan et al. 2002a). This is probably due to the formation of cross-links between the caseins and denatured whey proteins, brought into close proximity by the formation of disulphide bridges between β-lg and micellar κ-casein. There has also been considerable interest in the effects of TGase treatment on the cheesemaking properties of milk, in part due to the potential for increasing cheese yield. It has been suggested that TGase treatment of milk before renneting can achieve this effect. However, a number of studies (Lorenzen 2000, O'Sullivan et al. 2002a) have indicated that the rennet coagulation properties of milk and the syneretic properties of TGase-cross-linked renneted milk gels, as well as the proteolytic digestibility of casein, are impaired by cross-linking the proteins, which may alter cheese manufacture and ripening. Further studies are required to evaluate whether limited, targeted cross-linking may give desirable effects.

Other potential benefits of TGase treatment of dairy proteins include physical stablilisation and structural modification of products such as yoghurt, cream and liquid milk products, particularly in formulations with a reduced fat content, and the production of protein products, such as caseinates or whey protein products, with modified or tailor-made functional properties (Dickinson and Yamamoto 1996, Lorenzen and Schlimme 1998, Færgamand and Qvist 1998, Færgamand et al. 1998).

Overall, it is likely that TGase will have commercial applications in modifying the functional characteristics of milk and dairy products. However, some important issues remain to be clarified; for example the heat inactivation kinetics of the enzyme, to facilitate control of the reaction through inactivation at the desired extent of cross-linking. There is also a dearth of information on the effects of processing variables (e.g., temperature, pH) on the nature and rate of cross-linking reactions. Finally, a clear commercial advantage of using TGase over other methods for manipulating the structure and texture of dairy products (such as addition of proteins or hydrocolloids) must be established.

Lipases

Lipases, that is enzymes that produce free fatty acids (FFAs) and flavour precursors from triglycerides, diglycerides and monoglycerides, are produced by a wide range of plants, animals and microorganisms (Kilara 2003). For example, calves produce a salivary enzyme called PGE, which is present at high levels in the stomachs of calves, lambs or kids slaughtered immediately after suckling (Kilcawley et al. 1998). Microbial lipases have been isolated from *A. niger*, *A. oryzae*, *Pseudomonas fluorescens* and *P. roqueforti*. Exogenous lipases may be used for a range of applications (Kilara 2011), including:

- modification of milk fat to improve its physical properties and digestibility, reduce calorific value and enhance flavour (Balcão & Malcata 1998);
- production of EMC flavours (e.g., 'buttery', 'blue cheese' or 'yoghurt'); and
- production of lipolysed creams for bakery applications.

The correct choice of lipase for particular applications is very important, as the FFA profile differs with the type of enzyme used, and different lipases vary in their pH and temperature optima, heat stability and other characteristics. Calf PGE, for example generates a buttery and slightly peppery flavour, while kid PGE generates a sharp peppery flavour (Birsbach 1992).

The intense flavour of blue cheese results from lipolysis and catabolism of FFAs; most other cheese varieties undergo very little lipolysis during ripening. A method for the production of blue cheese flavour concentrates was described by Tomasini et al. (1995). It has been reported that the use of exogenous lipases (e.g., PGE) can enhance the flavour of Cheddar cheese and accelerate the development of perceived maturity (for review see Kilcawley et al. 1998). EMC preparations (e.g., slurries or powders) can also be produced by incubation of cheese emulsified in water with commercial enzymes (such as those produced by Novo Nordisk) that have 5–20 times the flavour intensity of mild Cheddar cheese.

Other Enzymes

Other enzymes that have been used in dairy processing include catalase, lactoperoxidase and lysozyme. All of these are added for preservative reasons, and their use is more common in some countries than others. Lysozyme may be added to milk for making certain cheese varieties (e.g., Edam, Emmental) to prevent growth of *Clostridium tyrobutyricum* and gas blowing, although its use is not widespread. Catalase and lactoperoxidase may be used in milk preservation strategies (particularly in tropical countries), with the latter (either the indigenous activity present or added purified enzyme) acting on hydrogen peroxide (which may be added to the milk) and the former being used to destroy excess hydrogen peroxide and terminate the enzymatic reaction (Brennan and McSweeney 2011).

MILK LIPIDS

Production of Fat-Based Dairy Products

Milk fat is a complex mixture of triglycerides that, in milk, is maintained as a stable o/w emulsion by the MFGM, which surrounds milk fat globules and protects the lipids therein from physical or enzymatic damage.

As mentioned earlier, separation of milk, either slowly under the influence of gravity or more rapidly under centrifugal force, readily produces a fraction (cream) enriched in the less dense milk fat globules, separated from the fat-depleted skim milk. Cream products of varying fat content (12–40%) are produced industrially for a range of applications, including dessert, coffee and as a food ingredient. The technology of cream products was reviewed by Smiddy et al. (2009).

One of the oldest dairy products is butter, a fat-continuous (water-in-oil, w/o) emulsion containing 80–81% fat, not more than 16% H_2O and, usually, 1.5% added salt. The production of butter from cream requires destabilisation of the emulsion and phase inversion, followed by consolidation of the fat and removal of a large part of the aqueous phase (Frede and Buchheim 1994, Frede 2002a). The manufacture of butter has been the subject of several reviews (e.g., Keogh 1995, Lane 1998; Ranjith and Rajah 2001, Frede 2002a, Wilbey 2009).

In traditional butter manufacture, cream is churned (mixed) in a large partially filled rotating cylindrical, conical or cubical churn, which damages the MFGM. Mechanically damaged fat globules become adsorbed on the surfaces of air bubbles (flotation churning) and gradually coalesce, being bound together by expressed free fat, to form butter grains. Eventually, the air entrapped within such grains is expelled and churning generally progresses until the grains have grown to approximately the size of a pea (Frede and Buchheim 1994, Vanapilli and Coupland 2001). In a batch process, growth of grains is monitored visually and audibly (impact of masses of grains on the churn walls); at this point, the buttermilk (essentially, skim milk but with a high content of sloughed off MFGM components such as phospholipids) is drained off and the butter worked by repeated falls and impaction within the rotating churn. After a set time, salt is added and worked throughout the butter.

With the exception of very small-scale plants, most modern dairy factories use continuous rather than batch buttermakers. In the most common system, the Fritz system, each stage of the process occurs in a separate horizontal cylindrical chamber, with product passing vertically between stages. In the churning cylinder, rotating impellers churn the cream very rapidly in the first stage, and in the second stage, butter grains are consolidated and the buttermilk drained off. In the final stage, the butter mass is worked; this chamber is sloped upwards and, while augers transport the butter and squeeze it through a series of perforated plates, the buttermilk drains off in the opposite direction. Midway through the working stage, salt (generally as a concentrated brine) is added, and becomes distributed in small moisture droplets with a high local salt concentration (~12.5% S/M), which acts to flavour and microbiologically stabilise the butter. The moisture present in the final droplets comes originally from milk serum, with a small contribution from water added in brine.

The by-product buttermilk is rich in MFGM materials, including phospholipids, and is commonly dried and used as an ingredient in many food products; however, due to its high content of polyunsaturated fatty acids, it deteriorates quite rapidly due to oxidation (O'Connell and Fox 2000). In recent years, there has been increasing interest in the fact that buttermilk contains high levels of biologically active (e.g., anticarcinogenic, anticholesterolemic) membrane-derived lipids, including glycosphingolipids and gangliosides (Jensen 2002, Dewettinck et al. 2008). Buttermilk isolates may also have potential application as emulsifying agents (Corrideg and Dalgleish 1997, Singh 2006). In addition, proteins in the MFGM such as butyrophilin may have anticancer and antimicrobial properties, and may play a role in diseases including multiple sclerosis and autism (Dewettinck et al. 2008).

Traditionally, butter was produced from cream that was naturally soured, and, today, in some countries, cream for butter manufacture is ripened with LAB (lactic or fermented butter). Bacterial acidification enhances the keeping quality of butter and changes the flavour, through production of diacetyl. However, the production of lactic butter leads to the production of acidic buttermilk as an unwanted by-product, and in recent years alternative technologies have been developed for the production of lactic butter. One of the most successful is the NIZO method, in which a sweet (non-lactic) cream is used, and a concentrated starter permeate is added to the butter grains midway through the process. This process leads to production of normal buttermilk, gives a well-flavoured product, which is very resistant to autoxidation (Walstra et al. 1999).

In the second half of the nineteenth century, the high cost of butter led to the development of alternatives to butter, such as margarine; consumer preference for reduced-fat (i.e., higher perceived 'healthiness') products has strengthened this trend. A further significant disadvantage of butter for domestic applications is the fact that at typical refrigeration temperatures, butter behaves essentially as a solid and has poor spreadability; moreover, at room temperature, it oils off and exudes water. The spreadability of butter may be improved by blending with vegetable oils, or by modifications such as interesterification (Marangoni and Rousseau 1998).

Today, there are two principal classes of non-butter milk-fat based spreads; (1) full-fat products with partial replacement of milk fat by another (e.g., vegetable) fat, with physicochemical, rheological, economic or dietary advantage, and (2) reduced-fat products containing varying levels of milk fat (Frede 2002b). Such products may be manufactured either in a modified continuous buttermaking system (butter technology) or in a votator scraped-surface cooler system (margarine technology). Generally, all products are made by preparation of aqueous (e.g., skim milk or cream) and lipid phases, each containing the appropriate ingredients, followed by mixing and phase inversion of the initial o/w emulsion to a final w/o emulsion. Reduction of the fat content of the mix requires increased attention to the structural characteristics of the aqueous phase, which must increasingly contribute to the texture and body of the product. Aqueous phase structuring agents, such as polysaccharides and/or proteins, may be added for this purpose (Keogh 1995, Frede 2002b).

Butter is a relatively stable product; the small size and high salt content of the water droplets make them an inhospitable environment for microbial growth. Lower fat spreads, while also being generally stable products, may have preservatives, such as sorbates, incorporated into their formulation to ensure shelf life and safety (Delamarre and Batt 1999).

Some new developments in the technology of spreads include the production of triple-emulsion (o/w/o) products (in which the droplets of aqueous phase in the product contain very small fat globules) (Frede 2002b). These systems have three distinct phases, and need two or more emulsifiers and/or the use of proteins or polysaccharides as stabilisers/thickeners, as they can be quite unstable due to the presence of at least two interfaces (Kanouni et al. 2002, O'Regan and Mulvhill 2010).

Production and Fractionation of Milk Fat

The most purified form of milk fat commercially produced is anhydrous milk fat (AMF), which consists of >99% triglycerides, and is produced in many countries. The production of AMF requires the removal of water and water-soluble components of milk, and may be achieved either by successive high-speed centrifugation, with phase-inversion at high fat levels, or by starting with butter, which is melted and centrifuged to recover the purified fat phase. Lipid oxidation is a significant concern during AMF manufacture, and the presence of oxygen during the process must be avoided.

AMF may be fractionated to yield fractions of milk fat with different desirable properties (Illingworth et al. 2009), usually on the basis of melting point. The most common industrial processes for milk fat fractionation involve cooling melted milk fat to temperatures where crystallisation of some of the fat occurs, facilitating the removal of a high melting point fraction (stearin crystals) from the low melting point fraction (olein). This removal may be by centrifugation or filtration through plate-and-frame membrane units. This basic fractionation process may be improved through the use of detergents or solvents, such as acetone.

There are also a number of methods for removal of cholesterol from dairy products, either through enzymatic, microbial, chemical (solid–liquid extraction, complexation) or physical means (distillation and crystallisation, supercritical fluid extraction). There have been several studies of the application of such methods for the removal of cholesterol from dairy products, including homogenised milk, butter and butter oil, and reduced-cholesterol dairy products have appeared on the marker since the 1990s; however, greater consumer preference for reduced-fat versus specifically reduced-cholesterol products may limit their widespread appeal (Sieber & Eyer 2011).

Lipid Oxidation

Lipids with double bonds (i.e., unsaturated fatty acids) are inherently susceptible to attack by active O_2, that is, lipid oxidation. Oxidation can give rise to a range of undesirable flavour compounds (such as aldehydes, ketones and alcohols that cause rancid off-flavours), and possibly result in toxic products. Oxidation is dependent on factors such as availability of oxygen, exposure to light, temperature, presence of pro- or anti-oxidants and the nature of the fat (O'Brien and O'Connor 1995, 2002, 2011).

Milk fat contains a high level of monounsaturated fatty acids (although less than many other fats), but a low level of polyunsaturated fatty acids. It is susceptible to oxidation but raw milk samples differ in susceptibility to oxidation as follows:

- *Spontaneous*: Will develop oxidised flavour within 48 hours without the addition of pro-oxidant metals, such as iron or copper
- *Susceptible*: Will not oxidise spontaneously, requires contamination by pro-oxidant metals
- *Non-susceptible*: Will not oxidise even in the presence of iron or copper

The reasons for the difference between milk samples are not clear, but are likely to be related to lactational and dietary factors, and the enzyme xanthine oxidase in milk may be critical.

Of particular current interest is the oxidation of the unsaturated alcohol, cholesterol, in milk; consumption of cholesterol oxidation products (COPs) in the diet is tentatively linked to the incidence of artherosclerosis (Kumar and Singhal 1991), and hence these products are regarded as potential health hazards. COPs are not found in all dairy product, but have been detected in WMP, baby foods, butter and certain cheese varieties, albeit at levels that probably do not pose a risk to consumers (Sieber et al. 1997, RoseSallin et al. 1997). The formation of COPs is influenced by factors such as temperature and exposure of the food to light (Angulo et al. 1997, RoseSallin et al. 1997, Hiesberger and Luf 2000).

Ice Cream

Ice cream, probably the most popular dairy dessert, is a frozen aerated emulsion. The continuous phase consists of a syrup containing dissolved sugars and minerals, while the dispersed phase consists of air cells, milk fat (or other kinds of fat) globules, ice crystals and insoluble proteins and hydrocolloids (Marshall 2002). The structure of ice cream is particularly complex, with several phases (e.g., ice crystals, fat globules and air bubbles, freeze-concentrated aqueous phase) coexisting in a single product.

The production technology for ice cream was reviewed by Goff (2002 2011), and will be summarised briefly here. The base for production of ice cream is milk blended with sources of milk solids non-fat (e.g., SMP) and fat (e.g., cream), added sugars or other sweeteners, emulsifying agents and hydrocolloid stabilisers. The exact formulation depends on the characteristics of the final product and, once blended, the mix is pasteurised, by either batch (e.g., 69°C for 30 minutes) or continuous (e.g., 80°C for 25 seconds) processes, and homogenised at 15.5–18.9 MPa, first stage, and 3.4 MPa, second stage. The mix is then cooled and stored at 2–4°C for at least 4 hours; this step is called ageing, and facilitates the hydration of milk proteins and stabilisers, and crystallisation of fat globules. During this period, emulsifiers generally displace milk proteins from the milk fat globule surface. Ageing improves the whipping quality of the mix and the melting and structural properties of the final ice cream.

After ageing, the ice cream is passed through a scraped-surface heat exchanger, cooled using a suitable refrigerant flowing in the jacket, under high-shear conditions with the introduction of air into the mix. These conditions result in rapid ice crystal nucleation and freezing, yielding small ice crystals, and the incorporation of air bubbles, resulting in a significant increase in the volume (over-run) of the product. The partially crystalline fat phase at refrigeration temperatures undergoes partial coalescence during the whipping and freezing stage, and a network of agglomerated fat develops, which partially surrounds the air bubbles and produces a solid-like structure (Hartel 1996, Goff 1997).

Flavourings and colourings may be added either to the mix before freezing, or to the soft semi-frozen mix exiting the heat exchanger. The mix typically exits the barrel of the freezer at –6°C, and is transferred immediately to a hardening chamber (–30°C or below) where the majority of the unfrozen water freezes.

Today, ice cream is available in a wide range of forms and shapes (e.g., stick, brick or tub, low- or full-fat varieties).

PROTEIN HYDROLYSATES

The bovine caseins contain several peptide sequences with specific biological activities when released by enzymatic hydrolysis (Table 25.5). Such enzymatic hydrolysis can occur either *in vivo* during the digestion of ingested food, or *in vitro* by treating the parent protein with appropriate enzymes under closely controlled conditions.

Casein-derived bioactive peptides have been the subject of considerable research for several years and the very extensive literature has been reviewed by Miesel (1998), Pihlanto-Lappälä (2002), Gobbetti et al. (2002) and FitzGerald & Meisel (2003). Several bioactive peptides are liberated during the digestion of bovine milk, as shown by studies of the intestinal contents of consumers, confirming that such peptides are liberated *in vivo*.

Table 25.5. Range and Properties of Casein-Derived Peptides with Potential Biological Activity

Peptides	Putative Biological Activities
Phosphopeptides	Metal binding
Caseinomacropeptide	Anticancerogenic action; inhibition of viral and bacterial adhesion; bifidogenic action; immunomodulatory activity; suppression of gastric secretions
Casomorphins	Opioid agonist and ACE inhibitors (antihypertensive action)
Immunomodulating peptides	Immunomodulatory activity
Blood platelet-modifying (antithrombic) peptides (e.g., casoplatelin)	Inhibition of aggregation of platelets
Angiotensin converting enzyme (ACE)	Anti-hypertension action; blood pressure inhibitors (casokinins) regulation; effects on immune and nervous systems
Bacteriocidal peptides (casocidins)	Antibiotic-like activity

Laboratory-scale processes for the production and purification (e.g., using chromatography, salt fractionation or UF) of many interesting peptides from the caseins have been developed; enzymes used for hydrolysis include chymotrypsin and pepsin (Pihlanto-Lappälä 2002).

Bioactive peptides may also be produced on enzymatic hydrolysis of whey proteins; α- and β-lactorphins, derived from α-lactalbumin and β-lg, respectively, are opioid agonists and possess angiotensin-converting enzyme (ACE) inhibitory activity. The whey proteins are also the source of lactokinins, which are probably ACE inhibitory.

Currently, few milk-derived biologically active peptides are produced commercially. Perhaps the peptides most likely to be commercially viable in the short term are the caseinophosphopeptides, which contain clusters of phosphoserine residues and are claimed to promote the absorption of metals (Ca, Fe, Zn) through chelation, and acting as passive transport carriers for the metals across the distal small intestine, although evidence for this is equivocal (Miquel and Farré 2008, Phelan et al. 2009). Caseinophosphopeptides are currently used in some dietary and pharmaceutical supplements, for example in the prevention of dental caries.

The CMP (κ-CN f106–169) is a product of the hydrolysis of the Phe_{105}–Met_{106} bond of κ-casein by rennet; during cheesemaking, it diffuses into the whey, while the *N*-terminal portion of κ-casein remains with the cheese curd. CMP has several interesting biological properties; for example it has no aromatic amino acids and is thus suitable for individuals suffering from phenylketonuria; however, it lacks several essential amino acids.

It also inhibits viral and bacterial adhesion, acts as a bifidogenic factor, suppresses gastric secretions, modulates immune system responses and inhibits the binding of bacterial toxins (e.g., toxins produced by cholera and *E. coli*). Of particular interest from the viewpoint of the commercial exploitation of CMP, relatively high levels of this peptide are present in whey (~4% of total casein, 15–20% of protein in cheese whey, an estimated 180 × 10^3 tonnes per annum are available globally in whey), and can be recovered therefrom quite easily.

Overall, detailed information is lacking regarding the physiological efficacy and mechanism of action of many milk protein-derived peptides, and possible adverse effects. Technological barriers also remain in terms of methods for industrial-scale production and purification of desired products.

CHOCOLATE

Milk powder is a key ingredient in many chocolate products, to which they contribute flavour and texture, for example through the role of milk fat in retarding the undesirable appearance of a white discolouration called chocolate bloom. The presence of milk powder (at levels up to 20% w/w) also influences processing characteristics of the molten chocolate during manufacture, for example flow properties. Certain characteristics of WMP that may be undesirable in other applications (e.g., high free fat content, low vacuole volume) are actually advantageous in chocolate. For this reason, roller-drying was often preferred to spray-drying for the production of milk powder for chocolate, although the latter process may be modified to yield powers with optimised properties for this application (Keogh et al. 2003). In addition, new approaches for tailoring powder functionality for chocolate, such as extrusion, have been described (Franke and Heinzelmann 2008).

INFANT FORMULAE

Today, a high proportion of infants in the developed world receive some or all of their nutritional requirements during the first year of life from prepared infant formulae, as opposed to breast milk. The raw material for such formulae is usually bovine milk or ingredients derived therefrom, but there are significant differences between the composition of bovine and human milk. This fact has led to the development of specialised processing strategies for transforming its composition to a product more nutritionally suitable for the human neonate. Today, most formulae are in fact prepared from isolated constituents of bovine milk (e.g., casein, whey proteins, lactose), blended with non-milk components. This, combined with the requirement for high hygienic standards and the absence of potentially harmful agents, makes the manufacture of infant formulae a highly specialised branch of the dairy processing industry, with almost pharmaceutical-grade quality control.

Most infant formulae are formulated by blending dairy proteins, vegetable (e.g., soya) proteins, lactose and other sugars, with vegetable oils and fats, minerals, vitamins, emulsifiers and micronutrients (O'Callaghan et al. 2011). The mixture of ingredients is then homogenised and heat treated to ensure microbiological safety. Subsequent processing steps differ in the case of dry or liquid formulae.

The dairy ingredients used are generally demineralised, as the mineral balance in bovine milk is very different from that of human milk (Burling 2002), and desired minerals are added back to the formula as required. Certain proteins (e.g., lactoferrin and α-lactalbumin) are present at higher levels in human than in bovine milk, and β-lg is absent from the former. There is interest in fortifying infant formulae with α-lactalbumin and/or lactoferrin, although technological challenges exist in the economical production of such proteins at acceptable purity.

The exact formulation of infant formulae differs based on the age and special requirements of the infant. Formulae for very young (<6-months old) babies generally have a high proportion of whey protein (e.g., 60% of total protein), whereas follow-on formulae for older infants contain a higher level of casein. In cases of infants with allergies to milk proteins or, in some cases, proteins in general, formulae in which milk proteins have been substituted by soya proteins or in which the proteins have been hydrolysed to small peptides and amino acids may be used. Soya-based formulae are also used in cases of intolerance to lactose.

The exact composition and level of added vitamins and minerals also differ based on nutritional requirements (O'Callaghan et al. 2011). Certain lipids are of interest as supplements for infant formulae, such as long-chain polyunsaturated fatty acids and conjugated linoleic acids. The final category of additives of interest for infant formulae is oligosaccharides, which are present at quite high levels (~15 g/L) in human milk (Urashima et al. 2009). Such oligosaccharides may be produced enzymatically, chemically synthesised or produced by fermentations; for example conversion of lactose to prebiotic oligosaccharides by β-galactosidase, and challenges involved in such processes, were reviewed by Gosling et al. (2010).

Two main categories of infant formulae are available in most countries: dry and liquid (UHT). For dry powder manufacture, the liquid mix is concentrated by evaporation and spray-dried to yield a highly agglomerated powder that will disperse readily in warm water (dissolving infant formulae is probably the most common dairy powder reconstitution operation practiced by consumers in the home). Some components may be dry-blended with the base powder, allowing flexibility in manufacture for different applications (e.g., infant age, dietary requirements, etc.). Powdered formulae are typically packaged in N_2/CO_2-flushed cans.

Liquid formulae (ready-to-feed) are generally subjected to far more severe thermal treatment than those intended for drying, for example UHT processing followed by aseptic packaging or retort sterilisation of product in screw-capped glass jars. These products, which are stable at room temperature, have the obvious advantage of convenience over their dry counterparts.

NOVEL TECHNOLOGIES FOR PROCESSING MILK AND DAIRY PRODUCTS

In recent years, a number of novel processing techniques have been developed for applications in food processing; major reasons for this trend include consumer demand for minimally processed food products, and ongoing challenges in satisfactorily processing certain food products without significant loss of quality using existing technologies (e.g., deterioration of nutrient content on heating fruit juices, problems with the safety of shellfish).

One new process that has received particular attention during the last decade is high pressure (HP) treatment. The principle of HP processing involves subjecting food products to a very high pressure (100–1000 MPa, or 1000–10,000 atm), typically at room temperature, for a fixed period (e.g., 1–30 minutes); under such conditions, microorganisms are killed, proteins denatured and enzymes either activated (at low pressures) or inactivated (at higher pressures). The principal advantage of HP processing, which does not rupture covalent bonds, is that low molecular weight substances, such as vitamins, are not affected and hence there are few detrimental effects on the nutritional or sensory characteristics of food.

Although it was first described for the inactivation of microorgansisms in milk in the 1890s, by Bert Hite at the Agricultural Research Station in Morganstown, West Virginia, lack of available processing equipment meant that HP remained unexploited in the food industry for most of the twentieth century. Only in the late 1990s were HP-processed foods launched, such as shellfish in the United States, fruit juice in France and meat in Spain.

To date, no HP-processed dairy products are available, probably due, at least in part, to the complexity of the effects of HP on dairy systems, which necessitate considerable fundamental research to underpin future commercial applications.

In short, HP affects the properties of milk in several, often unique, ways (for reviews, see Huppertz et al. 2002, Trujillo et al. 2002). Key effects include the following:

- Denaturation of the whey proteins, α-lactalbumin (α-la) and β-lg, at pressures $\geq$200 or $\geq$600 MPa, respectively, and interactions of denatured β-lg with the casein micelles.
- Increased casein micelle size (by ~25%) after treatment at 250 MPa for $\geq$15 minutes (Huppertz et al. 2004a) and reductions in micelle size by approximately 50% on treatment at 300–800 MPa.
- Increased levels of non-micellar α_{s1}–, α_{s2}–, β- and κ-caseins after HP treatment at $\geq$200 MPa.
- Inactivation of indigenous alkaline phosphatase and plasmin at pressures $\geq$400 MPa; lactoperoxidase is not inactivated by treatment at $\leq$700 MPa.
- Reduced rennet coagulation time of milk and time required for the gel to become firm enough for cutting, and enhanced strength of the rennet gel, following treatment at 200 MPa for up to 60 minutes or 400 MPa for $\leq$15 minutes.
- Increased yield of rennet-coagulated cheese curd, in particular after treatment $\geq$400 MPa.
- Increased rate and level of creaming in milk, by up to 70%, after HP treatment at 100–250 MPa, whereas treatment at 400 or 600 MPa reduces the rate and level of creaming by up to 60% (Huppertz et al. 2003).
- Reduced stability of milk to coagulation by ethanol (Johnston et al. 2002, Huppertz et al. 2004b),

HP treatment can also be applied to cheese (for review see O'Reilly et al. 2000a). A patent issued in Japan in 1992 suggested that the ripening of Cheddar cheese could be significantly accelerated by treatment at 50 MPa for 3 days; however, subsequent research (O'Reilly et al. 2001) has failed to substantiate this claim. There have been studies on the effects of HP on the ripening of many other cheese varieties; varying degrees of acceleration have been reported, for example Saldo et al. (2002) reported effects of HP treatment on ripening of cheese made from caprine milk. Of particular interest may be positive effects of HP treatment on the functionality of Mozzarella cheese (Johnston et al. 2002, O'Reilly et al. 2002a). HP may also be used to modify the microbial population (either starter or contaminants) in cheese (O'Reilly et al. 2000b, 2002b, Voigt et al. 2010).

In the last few years, the first commercial processes applying HP technology in the dairy sector have appeared, particularly for high-value functional foods (Kelly and Zeece 2009). Fonterra in New Zealand have proposed the commercial application of HP treatment for processing of bovine colostrum, with HP achieving a commercially relevant shelf life while retaining functional activity; to achieve a comparable shelf life using heat treatment would require heating of sufficient severity to result in loss of functionality of IgG through denaturation. In addition, Fonterra have patented a process whereby the shelf life of probiotic yoghurt may be extended by HP treatment to inactivate spoilage microorganisms while leaving active specially selected baro-resistant strains of probiotic bacteria, and a process for inactivating microorganisms in preparations containing lactoferrin.

Another technology that may be of potential interest for dairy processing in the future is high-pressure homogenisation (HPH), which works like conventional homogenisation, but at significantly higher pressures, up to 250 MPa; a related process, microfluidisation, is based on the principle of collisions between high-speed liquid jets. As well as reducing the size of oil droplets in an emulsion, HPH may inactivate enzymes and microorganisms and denature proteins in food. Many effects of HPH are due to the extremely high shear forces encountered by a fluid being processed; however, there is also a significant heating effect during the process. Recent studies have indicated that HPH significantly inactivates bacteria in raw bovine milk, and affects its rennet coagulation properties, fat globule size distribution and enzyme profile (Hayes and Kelly 2003a, b).

Another novel process that may be applied to dairy products is pulsed electric field treatment (which inactivates microorganisms but has relatively few other effects; Datta and Deeth 2002b). Some other processes, such as ultrasonication, irradiation, addition of antimicrobial peptides or enzymes and addition of carbon dioxide, have been studied for potential application to milk (Datta and Deeth 2002a). Arguably, HP processing is the most likely of the novel processes to be adopted by the dairy industry in the near future, due to the availability of equipment for commercial processing, but scale of available equipment combined with high cost will remain a significant hurdle (Patel et al. 2008).

CONCLUSION AND SUMMARY

Milk is a very complex raw material, with constituents and properties that are sensitive to applied stresses such as heat, changes in pH or concentration. Changes that can occur include inactivation of enzymes, mineral distribution and equilibria and denaturation of proteins. The extent of such changes and the consequences for the properties and stability of milk depend on the severity of the treatment.

In addition to the physicochemical changes that occur during processing, many dairy products involve complex enzymatic pathways, such as those involved in cheese ripening; exogenous enzymes may be added to milk to achieve a variety of end results. Milk may be fractionated by a range of complex technological processes to yield a broad portfolio of food ingredients.

In conclusion, the processing of milk represents perhaps one of the most complex fields in food science and technology and while many underpinning principles have been characterised, much research remains to be done in several areas.

REFERENCES

Andrews AT et al. 1987. A study of the heat stabilities of a number of indigenous milk enzymes. *J Dairy Res* 54: 237–246.

Angulo AJ et al. 1997. Determination of cholesterol oxides in dairy products. Effect of storage conditions. *J Agric Food Chem* 45: 4318–4323.

Balcão VM, Malcata FX. 1998. Lipase-catalysed modification of milkfat. *Biotechnol Adv* 16: 309–341.

Becerra M et al. 2001. Lactose bioconversion by calcium-alginate immobilisation of *Kluyveromyces lactis* cells. *Enzyme Microb Technol* 29: 506–512.

Birlouez-Aragon I et al. 2002. A new method of discriminating milk heat treatment. *Int Dairy J* 12: 59–67.

Birsbach P. 1992. Pregastric lipases. In: *Bulletin 269*. International Dairy Federation, Brussels, pp. 36–39.

Brennan NM, McSweeney PLH. 2011. Enzymes exogenous to milk in dairy technology: catalase, glucose oxidase, glucose isomerise and hexose oxidase. In: JW Fuquay et al. (eds.) *Encyclopedia of Dairy Sciences*, 2nd edn., vol 2. Elscivicr, London, pp. 301–303.

Burling H. 2002. Whey processing. Demineralisation. In: H Roginski et al. (eds.) *Encyclopedia of Dairy Science*. Academic Press, London, pp. 2745–2751.

Caríc M, Kalab M. 1987. Effects of drying techniques on milk powder quality and microstructure: a review. *Food Microstruct* 6: 171–180.

Chen CS et al. 2002. Optimization of the enzymic process for manufacturing low-lactose milk containing oligosaccharides. *Process Biochem* 38: 801–808.

Claeys WL et al. 2002. Intrinsic time temperature integrators for heat treatment of milk. *Trends Food Sci Technol* 13: 293–311.

Coolbear KP et al. 1996. Comparative study of methods for the isolation and purification of bovine κ-casein and its hydrolysis by chymosin. *J Dairy Res* 63: 61–71.

Corrideg M, Dalgleish DG. 1997. Isolates from industrial buttermilk: emulsifying properties of materials derived from the milk fat globule membrane. *J AgricFood Chem* 45: 4595–4600.

Curtin ÁC, McSweeney PLH. 2004. Catabolism of amino acids in cheese during ripening. In: PF Fox et al. (eds.) *Cheese: Chemistry, Physics and Microbiology, Volume 1, General Aspects*, 3rd edn. Elsevier Applied Science, Amsterdam, pp. 436–454.

Datta N, Deeth HC. 2001. Age gelation of UHT milk – a review. *Trans Am Inst Chem Eng* 71: 197–210.

Datta N, Deeth HC. 2002a. Other nonthermal technologies. In: H Roginski et al. (eds.) *Encyclopedia of Dairy Science*. Academic Press, London, pp. 1339–1346.

Datta N, Deeth HC. 2002b. Pulsed electric technologies. In: H Roginski et al. (eds.) *Encyclopedia of Dairy Science*. Academic Press, London, pp. 1333–1339.

Delamarre S, Batt CA. 1999. The microbiology and historical safety of margarine. *Food Microbiol* 16: 327–333.

De Vuyst L et al. 2001. Recent developments in the biosynthesis and applications of heteropolysaccharides from lactic acid bacteria. *Int Dairy J* 11: 687–707.

Dewettinck K et al. 2008. Nutritional and technological aspects of milk fat globule membrane material. *Int Dairy J* 18: 436–457.

Dickinson E, Yamamoto Y. 1996. Rheology of milk protein gels and protein-stabilised emulsion gels cross-linked with transglutaminase. *J Agric Food Chem* 44: 1371–1377.

Early R. 1998. Milk concentrates and milk powders. In: R Early (ed.) *The Technology of Dairy Products*, 2nd edn. Blackie Academic and Professional, London, pp. 228–300.

Eck A, Gilles JC. 2000. *Cheesemaking – from Science to Quality Assurance*, 2nd edn. Laviosier Technique and Documentation, Paris.

Enright E et al. 1999. Proteolysis and physicochemical changes in milk on storage as affected by UHT treatment, plasmin activity and KIO_3 addition. *Int Dairy J* 9: 581–591.

Færgamand M et al. 1998. Emulsifying properties of milk proteins cross-linked with microbial transglutaminase. *Int Dairy J* 8: 715–723.

Færgamand, M, Qvist KB. 1998. Transglutaminase: effect on rheological properties, microstructure and permeability of set-style acid skim milk gel. *Food Hydrocolloids* 11: 287–292.

Fagan CC et al. 2008. On-line prediction of cheese making using backscatter of near infrared indices light. *Int Dairy J* 18: 120–128.

Farise JF, Cayot P. 1998. New ultrarapid method for the separation of milk proteins by capillary electrophoresis. *J Agric Food Chem* 46: 2628–2633.

Farkye NY, Imafidon GI. 1995. Thermal denaturation of indigenous milk enzymes. In: PF Fox (ed.) *Heat-Induced Changes in Milk. Special Issue 9501*. International Dairy Federation, Brussels, pp. 331–348.

FitzGerald R, Meisel H. 2002. Milk protein hydrolyzates. In: PF Fox, PLH McSweeney (eds.) *Advanced Dairy Chemistry, Volume 1, Proteins*, 3rd edn. Kluwer Academic–Plenum Publishers, New York, pp. 625–698.

Fizsman SM et al. 1999. Effect of addition of gelatin on microstructure of acidic milk gels and yoghurt and on their rheological properties. *Int Dairy J* 9: 895–901.

Fontes EAF et al. 2001. A mechanistical mathematical model to predict lactose hydrolysis by β-galactosidase in a permeabilised cell mass of *Kluyveromyces lactis*: variability and sensitivity analysis. *Process Biochem* 37: 267–274.

Fox PF, McSweeney PLH. 1998. *Dairy Chemistry and Biochemistry*. Chapman and Hall, London.

Fox PF et al. 2004. *Cheese: Chemistry, Physics and Microbiology*. Elsevier, London.

Franke K, Heinzelmann K. 2008. Structure improvement of milk powder for chocolate processing. *Int Dairy J* 18: 928–931.

Frede E. 2002a. Butter: the product and its manufacture. In: H Roginski et al. (eds.) *Encyclopedia of Dairy Science*. Academic Press, London, pp. 220–226.

Frede E. 2002b. Milk fat products: milk fat-based spreads. In: H Roginski et al. (eds.) *Encyclopedia of Dairy Science*. Academic Press, London, pp. 1859–1868.

Frede E, Bucheim W. 1994. Buttermaking and the churning of blended fat emulsions. *Soc Dairy Technol* 47: 17–27.

Garem A et al. 2000. Cheesemaking properties of a new dairy-based powder made by a combination of microfiltration and ultrafiltration. *Lait* 80: 25–32.

Gobbetti M et al. 2002. Latent bioactive peptides in milk proteins: proteolytic activation and significance in dairy products. *CRC Crit Rev Food Sci Nutr* 42: 223–239.

Goff HD. 1997. Colloidal aspects of ice cream: a review. *Int Dairy J* 7: 363–373.

Goff HD. 2002. Ice cream and frozen desserts. Manufacture. In: H Roginski et al. (eds.) *Encyclopedia of Dairy Science*. Academic Press, London, pp. 1374–1380.

Goff HD. 2011. Ice cream and frozen desserts. In: JW Fuquay et al. (eds.) *Encyclopedia of Dairy Science*, 2nd edn., vol 2. Academic Press, London, pp. 899–904.

González-Martínez C et al. 2002. Influence of substituting milk powder for whey powder on yoghurt quality. *Trends Food Sci Technol* 13: 334–340.

Gosling A et al. 2010. Recent advances refining galactooligosaccharide production from lactose. *Food Chem* 121: 307–318.

Grant IR et al. 2001. *Mycobacterium avium* ssp. *paratuberculosis*: its incidence, heat resistance and detection in milk and dairy products. *Int J Dairy Technol* 54: 2–13.

Hartel RW. 1996. Ice crystallisation during the manufacture of ice cream. *Trends Food Sci Technol* 7: 315–321.

Hassan AN et al. 2003. Observation of bacterial exopolysaccharide in dairy products using cryo-scanning electron microscopy. *Int Dairy J* 13: 755–762.

Hayes MG, Kelly AL. 2003a. High pressure homogenisation of raw whole bovine milk (a) Effects on fat globule size and other properties. *J Dairy Res* 70: 297–305.

Hayes MG, Kelly AL. 2003b. High pressure homogenisation of milk (b) Effects on indigenous enzymatic activity. *J Dairy Res* 70: 307–313.

Hiesberger J, Luf W. 2000. Oxidation of cholesterol in butter during storage – effects of light and temperature. *Eur Food Res Technol* 211: 161–164.

Horton BS. 1997. What ever happened to the ultrafiltration of milk? *Aust J Dairy Technol* 52: 47–49.

Huppertz T et al. 2002. Effects of high pressure on constituents and properties of milk: a review. *Int Dairy J* 12: 561–572.

Huppertz T et al. 2003. High pressure-induced changes in the creaming properties of bovine milk. *Innovative Food Sci Emerg Technol* 4: 349–359.

Huppertz T et al. 2004a. High pressure treatment of bovine milk: effects on casein micelles and whey proteins. *J Dairy Res* 71: 98–106.

Huppertz T et al. 2004b. Heat and ethanol stability of high pressure-treated bovine milk. *Int Dairy J* 13: 125–133.

Huppertz T et al. 2006. A method for the large-scale isolation of β-casein. *Food Chem* 99: 45–40.

Huppertz T et al. 2009. Milk lipids – composition, origin and properties. In: AT Tamime (ed.) *Dairy Fats and Related Products*. Blackwell Publishing Limited, UK, pp. 1–27.

Husain Q. 2010. Beta-galactosidases and their potential applications: a review. *Crit Rev Biotechnol* 30: 41–62.

Ikura K et al. 1980. Crosslinking of casein components by transglutaminase. *Agric Biol Chem* 44: 1567–1573.

Illingworth D et al. 2009. Anhydrous milk fat manufacture and fractionation. In: AT Tamime (ed.) *Dairy Fats and Related Products*. Blackwell Publishing Limited, UK, pp. 108–166.

Jelen P. 2002. Whey processing: utilisation and products. In: Roginski H et al. (eds.) *Encyclopedia of Dairy Science*. Academic Press, London, pp. 2739–2745.

Jensen JD. 1975. Some recent advances in agglomeration, instantising and spray drying. *Food Technol* 29: 60–71.

Jensen RJ. 2002. The composition of bovine milk lipids: January 1995 to December 2000. *J Dairy Sci* 85: 295–350.

Ji T, Haque ZU. 2003. Cheddar whey processing and source: I. Effect on composition and functional properties of whey protein concentrates. *Int J Food Sci Technol* 38: 453–461.

Johnston DE et al. 2002. Effects of high-pressure treatment on the texture and cooking performance of half-fat Cheddar cheese. *Milchwissenschaft* 57: 198–201.

Johnston DE et al. 2002. Ethanol stability and chymosin-induced coagulation behaviour of high pressure treated milk. *Milchwissenschaft* 57: 363–366.

Kanouni M et al. 2002. Preparation of a stable double emulsion (W1/O/W2): role of the interfacial films on the stability of the system. *Adv Colloid Interface Sci* 99: 229–254.

Karleskind D et al. 1995. Chemical pre-treatment and microfiltration for making delipidised whey-protein concentrate. *J Food Sci* 60: 221–226.

Kelly AL et al. 2003. Manufacture and properties of milk powders. In: PF Fox, PLH McSweeney (eds.) *Advanced Dairy Chemistry, Volume 1, Proteins*, 3rd edn. Academic–Plenum Publishers, New York, pp. 1027–1062.

Kelly AL, O'Shea N. 2002. Pasteurisers: design and operation. In: H Roginski et al. (eds.) *Encyclopedia of Dairy Science*. Academic Press, London, pp. 2237–2244.

Kelly AL, Zeece M. 2009. Applications of novel technologies in processing of functional foods. *Aust J Dairy Technol* 64: 12–15.

Kelly PM et al. 2000. Implementation of integrated membrane processes for pilot-scale development of fractionated milk components. *Le Lait* 80: 139–153.

Keogh KK. 1995. Chemistry and technology of milk fat spreads. In: PF Fox (ed.) *Advanced Dairy Chemistry, Volume 2, Lipids*. Chapman and Hall, London, pp. 213–248.

Keogh MK et al. 2003. Effects of selected properties of ultrafiltered spray-dried milk powders on some properties of chocolate. *Int Dairy J* 12: 719–726.

Kilara A. 2003. Enzymes exogenous to milk. Lipases. In: H Roginski et al. (eds.) *Encyclopedia of Dairy Science*. Academic Press, London, pp. 914–918.

Kilara A. 2011. Lipases. In: JW Fuquay et al. (eds.) *Encyclopedia of Dairy Science*, 2nd edn., vol 2. Academic Press, London, pp. 284–288.

Kilcawley KN et al. 1998. Enzyme-modified cheese. *Int Dairy J* 8: 1–10.

Kim EHJ et al. 2009. Surface composition of industrial spray-dried milk powders. 1. Development of surface composition during manufacture. *J Food Eng* 94: 163–168.

Kumar N, Singhal OP. 1991. Cholesterol oxides and atherosclerosis – a review. *J Sci Food Agric* 55: 497–510.

Kyle WSA. 1993. Powdered milk. In: R McCrae et al. (eds.) *Encyclopedia of Food Science, Food Technology and Nutrition*. Academic Press, New York, pp. 3700–3713.

Lane R. 1998. Butter and mixed fat spreads. In: R Early (ed.) *The Technology of Dairy Products*. Blackie Academic and Professional, London, pp. 158–197.

Law BA. 1999. *Technology of Cheesemaking*. CRC Press, Boca Raton, FL.

Le Great Y, Brule G. 1982. Effect of concentration and drying on mineral equilibria in skim milk and retentates. *Lait* 62: 113–125.

Lewis M, Heppell N. 2000. Continuous thermal processing of foods. Pasteurisation and UHT sterilization. Aspen Publishers, Gaithersburg, MD.

Lorenzen PC. 2000. Renneting properties of transglutaminase-treated milk. *Milchwissenschaft* 55: 433–437.

Lorenzen PC, Schlimme E. 1998. Properties and potential fields of application of transglutaminase preparations in dairying. In: *Bulletin 332*. International Dairy Federation, Brussels, pp. 47–53.

Lu W et al. 2001. Modelling alkaline phosphatase inactivation in bovine milk during high-temperature short-time pasteurisation. *Food Sci Technol Int* 7: 479–485.

Lucey JA, Singh H. 1998. Formation and physical properties of acid milk gels: a review. *Food Res Int* 30: 529–542.

Maciunska J et al. 1998. Isolation and some properties of beta-galactosidase from the thermophilic bacterium *Thermus thermophilus*. *Food Chem* 63: 441–445.

Mahoney RR. 1997. Lactose: enzymatic modification. In: PF Fox (ed.) *Advanced Dairy Chemistry, Volume 3, Lactose, Salts, Water and Vitamins*. Chapman and Hall, London, pp. 77–126.

Mahoney RR. 2002. Enzymes exogenous to milk in dairy technology: β-D-Galactosidase. In: H Roginski et al. (eds.) *Encyclopedia of Dairy Science*. Academic Press, London, pp. 907–914.

Marangoni AG, Rousseau D. 1998. Chemical and enzymatic modification of butter fat and butterfat-canola oil blends. *Food Res Int* 31: 595–599.

Marshall RT. 2002. Ice cream and frozen desserts. Product types. In: H Roginski et al. (eds.) *Encyclopedia of Dairy Science*. Academic Press, London, pp. 1367–1373.

Matthews ME. 1984. Whey protein recovery processes and products. J Dairy Sci 67: 2680–2692.

Maubois JL. 1997. Current uses and future perspectives of MF technology in the dairy industry. In: *Bulletin 320*. International Dairy Federation, Brussels, pp. 154–159.

McKellar RC et al. 1994. Predictive modelling of alkaline phosphatase inactivation in a high-temperature short-time pasteurizer. *J Food Prot* 57: 424–430.

McKellar RC et al. 1996. Predictive modelling of lactoperoxidase and γ-glutamyl transpeptidase inactivation in a high-temperature short-time pasteurizer. *Int Dairy J* 6: 295–301.

McSweeney PLH. (ed.). 2007. *Cheese Problems Solved*. Woodhead Publishers, Cambridge.

Miesel H. 1998. Overview on milk protein-derived peptides. *Int Dairy J* 8: 363–373.

Miquel E, Farré R. 2008. Effects and future trends of casein phosphopeptides on zinc bioavailability. *Trends Food Sci Technol* 18: 139–143.

Mistry VV, Pulgar JB. 1996. Physical and storage properties of high milk protein powder. *Int Dairy J* 6: 195–203.

Modler HW et al. 1993. Production of fluid milk with a high degree of lactose hydrolysis. In: PF Fox (ed.) *Heat-Induced Changes in Milk*. International Dairy Federation, Brussels. pp. 57–61.

Motoki M, Seguro K. 1998. Transglutaminase and its use for food processing. *Trends Food Sci Technol* 9: 204–210.

Muir DD. 2002. Lactose; properties, production, applications. In: H Roginski et al. (eds.) *Encyclopedia of Dairy Science*. Academic Press, London, pp. 1525–1529.

Mulvihill DM, Ennis MP. 2003. Functional milk proteins: production and utilisation. In: PF Fox, PLH McSweeney (eds.) *Advanced Dairy Chemistry, Volume 1, Proteins*, 3rd edn. Kluwer Academic–Plenum Publishers, New York, pp. 1175–1128.

Mulvihill DM, Fox PF. 1994. Developments in the production of milk proteins. In: JBF Hansen (ed.) *New and Developing Sources of Food Proteins*. Plenum Press, Chapman and Hall, pp. 1–30.

Nakagawa T et al. 2003. Isolation and characterisation of psychrophiles producing cold-active beta-galactosidase. *Lett Appl Microbiol* 37: 154–157.

Nieuwenhuijse JA. 1995. Changes in heat-treated milk products during storage. In: PF Fox (ed.) *Heat-Induced Changes in Milk. Special Issue 9501*. International Dairy Federation, Brussels, pp. 231–255.

Nieuwenhuijse JA et al. 1988. Calcium and phosphate paritions during the manufacture of sterilised concentrated milk and their relations to heat stability. *Neth Milk Dairy J* 42: 387–421.

Nieuwenhuijse JA, van Boekel MAJS. 2003. Protein stability in sterilised milk and milk products. In: PF Fox, PLH McSweeney (eds.) *Advanced Dairy Chemistry, Volume 1, Proteins*, 3rd edn. Kluwer Academic–Plenum Publishers, New York, pp. 947–974.

Obon JM et al. 2000. β-Galactosidase immobilisation for milk lactose hydrolysis: a simple experimental and modelling study of batch and continuous reactors. *Biochem Educ* 28: 164–168.

O'Brien J. 1995. Heat-induced changes in lactose: isomerisation, degradation, Maillard browning. In: PF Fox (ed.) *Heat-Induced Changes in Milk. Special Issue 9501*. International Dairy Federation, Brussels, pp. 134–170.

O'Brien J. 2009. Non-enzymatic degradation pathways of lactose and their significance in dairy products. In: PLH McSweeney, PF Fox (eds.) *Advanced Dairy Chemistry, Volume 3, Lactose, Water, Salts and Minor Constituents*. Springer, New York, pp. 231–294.

O'Brien NM, O'Connor T. 1995. Lipid oxidation. In: PF Fox (ed.) *Advanced Dairy Chemistry, Volume 2, Lipids*. Chapman and Hall, London, pp. 309–348.

O'Brien NM, O'Connor T. 2002. Lipid oxidation. In: H Roginski et al. (eds.) *Encyclopedia of Dairy Science*. Academic Press, London, pp. 1600–1604.

O'Brien NM, O'Connor TP. 2011. Lipid oxidation. In: JW Fuquay et al. (eds.) *Encyclopedia of Dairy Science*, 2nd edn., vol 3. Academic Press, London, pp. 716–720.

O'Callaghan D et al. 2011. Infant formula. In: JW Fuquay et al. (eds.) Encyclopedia of Dairy Science, vol 2. Elsevier, Oxford, pp. 135–145.

O'Connell JE, Fox PF. 2000. Heat stability of buttermilk. *J Dairy Sci* 83: 1728–1732.

O'Connell JE, Fox PF. 2002. Heat stability of milk. In: H Roginski et al. (eds.) *Encyclopedia of Dairy Science*. Academic Press, London, pp. 1321–1326.

O'Connell JE, Fox PF. 2003. Heat-induced coagulation of milk. In: PF Fox, PLH McSweeney (eds.) *Advanced Dairy Chemistry, Volume 1, Proteins*, 3rd edn. Kluwer Academic–Plenum Publishers, New York, pp. 879–945.

Oldfield DJ. 1998. Heat induced whey protein reactions in milk: kinetics of denaturation and aggregation as related to milk powder manufacture. PhD Thesis, Massey University, Palmerstown North, New Zealand.

O'Regan J, Mulvihill DM. 2010. Sodium caseinate-maltodextrin conjugate stabilised double emulsions: encapsulation and stability. *Food Res Int* 43: 224–231.

O'Reilly CE et al. 2000a. Effect of high pressure on proteolysis during ripening of Cheddar cheese. *Innovative Food Sci Emerging Technol* 1: 109–117.

O'Reilly CE et al. 2000b. Inactivation of a number of microbial species in cheese by high hydrostatic pressure. *Appl Environ Microbiol* 66: 4890–4896.

O'Reilly CE et al. 2001. High pressure treatment: applications to cheese manufacture and ripening. *Trends Food Sci Technol* 12: 51–59.

O'Reilly CE et al. 2002a. The effect of high pressure treatment on the functional and rheological properties of Mozzarella cheese. *Innovative Food Sci Emerging Technol* 3: 3–9.

O'Reilly CE et al. 2002b. The effect of high pressure on starter viability and autolysis and proteolysis in Cheddar cheese. *Int Dairy J* 12: 915–922.

O'Sullivan MM et al. 2002a. Effect of transglutaminase on the heat stability of milk: a possible mechanism. *J Dairy Sci* 85: 1–7.

O'Sullivan MM et al. 2002b. Influence of transglutaminase treatment on some physicochemical properties of milk. *J Dairy Res* 69: 433–442.

Patel HA et al. 2008. Nonthermal preservation technologies for dairy applications. In: RC Chandan (ed.) *Dairy Processing and Quality Assurance*. Wiley-Blackwell, Ames, IA, pp. 465–482.

Pellegrino L et al. 1995. Assessment (indices) of heat treatment of milk. In: PF Fox (ed.) *Heat-Induced Changes in Milk. Special Issue 9501*. International Dairy Federation, Brussels, pp. 419–453.

Peterson AHK. 2009. Production and uses of lactose. In: PLH McSweeney, PF Fox (eds.) *Advanced Dairy Chemistry, Volume 3, Lactose, Water, Salts and Minor Constituents*. Springer, New York, pp. 105–120.

Phelan M et al. 2009. Casein-derived bioactive peptides: biological effects, industrial uses, safety aspects and regulatory status. *Int Dairy J* 19: 643–654.

Pihlanto-Lappäla A. 2002. Bioactive peptides. In: H Roginski et al. (eds.) *Encyclopedia of Dairy Science*. Academic Press, London, pp. 1960–1967.

Písecký J. 1997. *Handbook of Milk Powder Manufacture*. Niro A/S, Copenhagen, Denmark.

Playne MJ, Crittenden RG. 2009. Galacto-oligosaccharides and other products derived from lactose. In: PLH McSweeney, PF Fox (eds.) *Advanced Dairy Chemistry, Volume 3, Lactose, Water, Salts and Minor Constituents*. Springer, New York, pp. 121–202.

Pouliot M et al. 1996. On the conventional cross-flow microfiltration of skim milk for the production of native phosphocaseinate. *Int Dairy J* 6: 105–111.

Puvanenthiran A et al. 2002. Structure and visco-elastic properties of set yoghurt with altered casein to whey protein ratios. *Int Dairy J* 12: 383–391.

Ranjith HMP, Rajah KK. 2001. High fat content dairy products. In: AY Tamime, BA Law (eds.) *Mechanisation and Automation in Dairy Technology*. Sheffield Academic Press, Sheffield, pp. 119–151.

Reid G. 2008. Probiotics and prebiotics – progress and challenges. *Int Dairy J* 18: 969–975.

Remeuf F et al. 2003. Preliminary observations on the effects of milk fortification and heating on microstructure and physical properties of stirred yoghurt. *Int Dairy J* 13: 773–782.

Richardson BC. 1983. The proteinases of bovine milk and the effect of pasteurisation on their activity. *New Zealand J Dairy Sci Technol* 18: 233–245.

Robinson RK. 2002a. Fermented milks. Yoghurt types and manufacture. In: H Roginski et al. (eds.) *Encyclopedia of Dairy Science*. Academic Press, London, pp. 1055–1058.

Robinson RK. 2002b. Fermented milks. Yoghurt, role of starter culture. In: H Roginski et al. (eds.) *Encyclopedia of Dairy Science*. Academic Press, London, pp. 1059–1063.

Robinson RK, Wilby RA. 1998. *Cheesemaking Practice*, 3rd edn. Aspen Publishers: Gaithersburg, MD.

RoseSallin C et al. 1997. Effects of storage or heat treatment on oxysterols formation in dairy products. *Food Sci Technol(Lebensmittel-Wissenschaft Technologie)* 30: 170–177.

Ryser ET. 2002. Pasteurisation of liquid milk products. Principles, public health aspects. In: H Roginski et al. (eds.) *Encyclopedia of Dairy Science*. Academic Press, London, pp. 2232–2237.

Saldo J et al. 2002. Proteolysis in caprine milk cheese treated by high pressure to accelerate cheese ripening. *Int Dairy J* 12: 35–44.

Sanchez B et al. 2009. Probiotic fermented milks: present and future. *Int J Dairy Technol* 62: 472–483.

Sawyer L. 2003. β-Lactoglobulin. In: PF Fox, PLH McSweeney (eds.) *Advanced Dairy Chemistry, Volume 1, Proteins*, 3rd edn. Kluwer Academic–Plenum Publishers, New York, pp. 319–386.

Sharma R et al. 2001. Influence of transglutaminase treatment of skim milk on the formation of ε-(γ-glutamyl) lysine and the susceptibility of individual proteins towards crosslinking. *Int Dairy J* 11: 785–793.

Shin HS et al. 2000. Growth and viability of commercial *Bifidobacterium* spp. in skim milk containing oligosaccharides and inulin. *J Food Sci* 65: 884–887.

Sieber R, Eyer H. 2011. Removal of cholesterol from dairy products. In: JW Fuquay et al. (eds.) *Encyclopedia of Dairy Science*, vol 3. Elsevier, Oxford, pp. 734–740.

Sieber R et al. 1997. Cholesterol and its oxidation products in dairy products: analysis, effect of technological treatments, nutritional consequences. *Sci Aliments* 17: 243–252.

Sienkiewicz T, Riedel CL. 1990. *Whey and Whey Utilisation*. Verlag Th. Mann, Gelsenkirchen.

Singh H. 2006. The milk fat globule membrane – a biophysical system for food applications. *Curr Opin J Colloid Interface Sci* 11: 154–163.

Singh H, Creamer LK. 1991. Denaturation, aggregation and heat stability of milk protein during the manufacture of skim milk powder. *J Dairy Res* 58: 269–283.

Smiddy M et al. 2009. Cream and related products. In: AT Tamime (ed.) *Dairy Fats and Related Products*. Blackwell Publishing Limited, Oxford, UK, pp. 61–85.

Stepaniak L, Sørhaug T. 1995. Thermal denaturation of bacterial enzymes in milk In: PF Fox (ed.) *Heat-Induced Changes in Milk. Special Issue 9501*. International Dairy Federation, Brussels, pp. 349–363.

Surono I, Hosono A. 2002. Fermented milks. Starter cultures. In: H Roginski et al. (eds.) *Encyclopedia of Dairy Science*. Academic Press, London, pp. 1023–1028.

Tamime AY, Robinson RK. 1999. Yoghurt: Science and Technology. Woodhead Publishers, Cambridge.

Tomasini A et al. 1995. Production of blue cheese flavour concentrates from different substrates supplemented with lipolysed cream. *Int Dairy J* 5: 247–257.

Trujillo AJ et al. 2002. Applications of high-hydrostatic pressure on milk and dairy products: a review. *Innovative Food Sci Emerging Technol* 3: 295–307.

Urashima T et al. 2009. Milk oligosaccharides. In: PLH McSweeney, PF Fox (eds.) *Advanced Dairy Chemistry, Volume 3, Lactose, Water, Salts and Minor Constituents*. Springer, New York, pp. 295–349.

Van Boekel MAJS, Walstra P. 1998. Effect of heat treatment on chemical and physical changes to milkfat globules. In: PF Fox (ed.) *Heat Induced Changes in Milk*, 2nd edn. International Dairy Federation, Brussels, pp. 51–65.

Vanapalli SA, Coupland JN. 2001. Emulsions under shear – the formation and properties of partially coalesced lipid structures. *Food Hydrocolloids* 51: 507–512.

Vasiljevic T, Jelen P. 2001. Production of β-galactosidase for lactose hydrolysis in milk and dairy products using thermophilic lactic acid bacteria. *Innovative Food Sci Emerging Technol* 2: 75–85.

Voigt DD et al. 2010. Effect of high-pressure treatment on microbiology, proteolysis, lipolysis and levels of flavour compounds in mature blue-veined cheese. *Innovative Food Sci Emerging Technol* 11: 68–77.

Walstra P et al. 1999. Dairy technology. Principles of milk properties and processes. Marcel Dekker, New York.

Westhoff DC. 1978. Heating milk for microbial destruction: a historical outline and update. *J Food Prot* 41: 122–130.

Whitaker JR. 1994. β-Galactosidase. In: *Principles of Enzymology for the Food Sciences*. Marcel Dekker, New York, pp. 419–423.

Wilbey RA. 1996. Estimating the degree of heat treatment given to milk. *J Soc Dairy Technol* 49: 109–112.

Wilbey RA. 2009. Butter. In: AT Tamime (ed.) *Dairy Fats and Related Products*. Blackwell Publishing Limited, Oxford, pp. 86–107.

Zadow JG. 1993. Economic considerations related to the production of lactose and lactose by-products. In: *Lactose Hydrolysis, IDF Bulletin 289*. International Dairy Federation, Brussels, pp. 10–15.

26
Equid Milk: Chemistry, Biochemistry and Processing

T. Uniacke-Lowe and P.F. Fox

Abstract: The characteristics of equid milk of interest in human nutrition include a high concentration of polyunsaturated fatty acids, a low cholesterol content, high lactose and low protein levels, as well as high levels of vitamins A, B and C. Compared to bovine milk, the protein fraction of equid milk contains proportionally less

Food Biochemistry and Food Processing, Second Edition. Edited by Benjamin K. Simpson, Leo M.L. Nollet, Fidel Toldrá, Soottawat Benjakul, Gopinadhan Paliyath and Y.H. Hui.

casein and more whey proteins. The low fat and unique fatty acid profile of both equine and asinine milk result in low atherogenic and thrombogenic indices. The high lactose content of equid milk gives good palatability and improves the intestinal absorption of calcium that is important for bone mineralisation in children. The renal load of equid milk, based on levels of protein and inorganic substances, is equal to that of human milk, a further indication of its suitability as an infant food. The concentration of lysozyme, lactoferrin and *n*-3 fatty acids is exceptionally high in equid milk, suggesting a potential anti-inflammatory effect. Equine and asinine milk can be used for their prebiotic and probiotic activity and as alternatives for infants and children with cow milk protein allergy and other food intolerances.

While the gross composition of equine and asinine milk has been reasonably well established and the caseins have been fractionated and well characterised, the presence of κ-casein remains a contentious issue and there is little or no information on the physicochemical properties of equid milk. The only significant product from equine milk is the fermented product, koumiss.

The composition of equid milk suggests a product with interesting nutritional characteristics with potential use in dietetics and therapeutics, especially in diets for the elderly, convalescent and newborns, but the lack of scientific research must be addressed to develop the potential of equine and asinine milk in the health and nutritional markets.

INTRODUCTION

Approximately one-third of all mammalian genera are herbivores, more than half of which belong to two orders, (1) the Perissodactyla (odd-toed ungulates (hoofed animals)) and (2) the Artiodactyla (even-toed ungulates) (Savage and Long 1986). The horse and donkey belong to the order Perissodactyla that has three families: (1) Equidae (nine species of horses, donkey and zebras), (2) Tapiridae (four species of tapirs) and (3) Rhinocerotidae (five species of rhinoceros). Uniquely among species, all Equidae can interbreed, but the hybrid offspring are almost always infertile because horses have 64 chromosomes and donkeys have 62 chromosomes (Trujillo et al. 1962). Zebras have between 32 and 46 chromosomes, depending on breed (Burchelli's zebra, *Equus burchelli*, has 44 chromosomes, Carbone et al. (2006) and viable hybrids with donkeys have been produced where gene combinations have allowed for embryonic development to birth. Differences in chromosome numbers between horses and zebras are most likely due to horses having two longer chromosomes that contain a single gene content compared to four zebra chromosomes (Benirschke et al. 1964).

This chapter presents a brief historical overview of some aspects of the domestication of equid species, a review of the chemistry and biochemistry of the principal constituents of equine milk and, to a lesser extent, of asinine milk, with comparative data for bovine and human milk. The technological properties of equid milk, with reference to processing of the milk are examined and, finally, a synopsis of the nutritional and biological significance of equid milks in the human diet is presented.

EQUID EVOLUTION

Perissodactyla species evolved during the early Eocene ($\sim$55 million years ago) and were the dominant ungulate order until approximately 15 million years ago when numbers declined, probably due to climatic factors rather than competition with emerging Artiodactyla species (including the Ruminantia family) (MacDonald 2001). Nevertheless, wild equids ranged in vast herds across the grasslands of the Northern Hemisphere and in South America until the end of the ice age, approximately 10,000 years ago. Wild horses were the chief prey of increasing human populations and numbers decreased until they became extinct in North America approximately 7000 years ago and in Europe, herds of horses were pushed eastwards into Central Asia where the last few Przewalski horses (*Equus ferus przewalski*) survived until the twentieth century (Clutton-Brock 1992). It is generally assumed that equids were domesticated approximately 5000 years ago but archaeologists believe man did not ride horses until approximately 1000 BC, although horses replaced the ox as draft animals approximately 2000 BC (Clutton-Brock 1992). One of the early records of the domestic horse in Western Europe comes from horse remains found at the late Neolithic site at Newgrange, County Meath, Ireland (Clutton-Brock 1992). Outram (2009) demonstrated domestication of the horse in the Eneolithic Botai culture of Kazakhstan approximately 3500 years ago and analysis of organic residues based on characterisation of stable isotopes of carbon, $\delta^{13}C$ (which allows differentiation of non-ruminant and ruminant carcass and dairy fats) and deuterium δD, values of fatty acid analysis revealed processing of mares' milk and meat in ceramics at that time. The donkey (*Equus asinus*) is believed to have evolved from the Nubian and Somalian sub-species of African wild asses approximately 4000–5000 years ago and has since been an integral part of human life as a pack and riding animal. Today, perissodactyls are a poor second to artiodactyls in terms of numbers of species, geographical distribution, variety of form and ecological diversity (MacDonald 2001). For further details on equid evolution and genetic lineages of domestic horse species, see Vilà et al. (2001) and references therein.

EQUID DOMESTICATION

Only five major species of large, plant-eating mammals have been widely domesticated: sheep, goat, cattle, pig and horse. Nine additional minor species have been domesticated but are restricted to certain geographical areas: Arabian and Bactrian camels, llama, alpaca, donkey, reindeer, water buffalo, yak, Bali cattle of Southeast Asia and the gaur (mithan) of India and Burma (Bruns 1999). Despite taxonomic congruity and behavioural similarities between equid species, only two equids have been domesticated: the horse (*Equus ferus*) and the donkey (*Equus asinus africanus*) (Clutton-Brock 1992).

Because milk and products derived from it provide up to 30% of dietary protein in developed countries, to meet this demand, man has genetically improved some species for milk production (Mercier 1986). Domesticated animals have been modified from their wild ancestors through being kept and selectively bred for use by humans who control the animals' breeding and feeding. Domestication of dairy species has meant that different breeds of farm animals, for example dairy cows and goats, can be developed and maintained to optimise certain hereditary factors, for example, length of lactation, period of gestation and milk yield, as well as protein and fat levels in the milk produced. Genetic selection of breeds of horse and donkey for milk production has not occurred, as yet, and consequently there is high variability for milk yield and length of lactation, as well as high individual variability.

In some regions of the world, for example Mongolia and Southern Russia, Hungary, France and Belgium, the horse and, less frequently, the donkey, is an important source of meat and milk. Steppe Mongols, forest-steppe Kazakhs, the Hadza hunter–gatherers of Tanzania and the urban French all regard horses as a valuable food source and believe that horse flesh and milk have special nutritional and medicinal attributes (Levine 1998). The use of donkeys as a dairy species can be traced back to Roman times when the nutritional value of its milk and beneficial properties in skin care were first recognised (Salimei 2011).

The importance of the horse in leisure activities, especially racing, has led to the scientific breeding and nutrition of horses, including foals, and has created the need to characterise the composition and properties of equine milk, which are now relatively well known. The fact that the horse is spread throughout the world, has been domesticated, is relatively easily handled, and produces large amounts of milk, which can be obtained relatively easily, makes equine milk a relatively easy subject for research. Although the donkey is now less widely distributed than the horse, its milk can be obtained readily. The zebra has not been domesticated and although it is widespread in the wild in Africa and in captivity elsewhere, apparently it is very difficult to obtain zebra milk, even from captive animals. Studies of the social behaviour of equids have shown that there are no intrinsic reasons why the zebra has never been domesticated but it is believed that the peoples of Africa may have had cultural rather than biological reasons not to use zebras as pack animals (Clutton-Brock 1992).

Ruminants and Non-Ruminants

The horse and donkey are non-ruminant herbivores and digest portions of their feed first enzymatically in a foregut and then ferment it in a very large sacculated hind-gut. Limited digestion occurs in the equid stomach which liquefies incoming feed and secretes gastric acid and pepsin to initiate breakdown of feed components. The equid digestive system is designed to process small amounts of food frequently (Sneddon and Argenzio 1998). Equids rarely fast for more than 2–4 hours at a time and naturally forage for 16–18 hours/day. Horses in the wild roam widely, grazing both day and at night on immature, easily digested food and exhibit few digestive problems in comparison to domesticated horses (Sellnow 2006). Unlike ruminants, and in keeping with their status as prey animals, horses do not require periods of rest to stop and ruminate. Ruminants have very efficient digestive systems with microbial breakdown of fibrous food at the start of the gastrointestinal tract and nutrient absorption along the entire intestine. Ruminants can digest fiber and carbohydrate more completely than any other species.

Donkeys, like horses, generally survive on a diet high in fiber and low in soluble carbohydrate and protein but while donkeys are not as efficient as ruminants at digesting cell wall components, they are far more efficient than horses (Izraely et al. 1989). Donkeys are capable of consuming large amounts of forage and gain more digestible energy from it than even goats fed a similar diet (Smith and Sherman 2009). The donkey achieves this by substantially increasing its forage intake rate to compensate for a low-quality diet. A donkey uses its narrower muzzle and prehensile lips for greater selectivity of its food, thereby, maximising feed quality rather than quantity (Aganga et al. 2000). Donkeys have a much lower water requirement per unit weight than any other mammal, except the camel, and can rehydrate quickly with large volumes of water without complications. Donkeys can work while suffering from severe dehydration by reducing water and energy turnover rates, while maintaining feed intake and its plasma volume can be maintained by drawing on the substantial fluid reservoir in the hind-gut (Sneddon et al. 2006). The zebra, on the other hand, needs a constant source of water for survival (Aganga et al. 2000).

Why Equid Milk in Human Nutrition?

The benefits of equine milk for human health is an ancient idea and there is much literature from the former Soviet Union on this subject, although it is now accepted that the results of experimental work are dubious (Doreau and Martin-Rosset 2002). Because equine milk resembles human milk in many respects and is claimed to have special therapeutic properties, it is becoming increasingly important in Western Europe, especially in France, Italy, Hungary and the Netherlands. Equine milk (and koumiss, fermented equine milk) is often used for the treatment of a myriad of ailments including anaemia, nephritis, diarrhoea, gastritis disorders, cardiovascular disease and in post-operative care, as well as for stimulation of the immune system (Lozovich 1995). In Mongolia, where koumiss is the national drink, people have a saying that 'kumys cures 40 diseases' (Levine 1998). In Italy, equine milk has been recommended as a possible substitute for bovine milk for allergic children (Curadi et al. 2001). Equine milk is considered to be highly digestible, rich in essential nutrients and possesses an optimum whey protein:casein ratio, making it very suitable as a substitute for bovine milk in paediatric dietetics. Estimates suggest that more than 30 million people drink equine milk regularly, with this figure increasing significantly annually (Doreau and Martin-Rosset 2002).

The use of asinine milk by humans for alimentary and cosmetic purposes has been popular since Egyptian antiquity. Cleopatra is reputed to have bathed daily in asinine milk and kept a herd of 700 to fill her bath. Hippocrates (460–370 BC) was a strong advocate of the use of asinine milk as a medicine and used it to cure many ailments including, liver disease, oedema, nosebleed, poisoning and wounds. Today, asinine milk is consumed mainly in countries where donkeys were traditionally bred, Asia, Africa and Eastern Europe but more recently it has been used successfully as a substitute for human milk in Western Europe (Vincenzetti et al. 2008) and is the milk of choice in Italy for children with severe immunoglobulin E (IgE)-mediated cows' milk allergy. (Businco et al. 2000).

Production of Equid Milk

The production of equine milk and the factors that affect it have been the subject of several reviews, including Doreau and Boulot (1989), Doreau et al. (1990), Doreau (1994), Doreau and Martin-Rosset (2002) and Park et al. (2006) and, therefore, is considered only briefly here.

World milk production was approximately 695 million tonnes in 2009, of which 84% was bovine, 13% buffalo and 3% sheep, goat and other species (IDF 2009). Statistics for milk production from species other than cows and buffalo are not very reliable and are available only from countries where these milks are processed industrially although it is accepted that, while bovine milk production figures have changed little in recent years, the production of milk from buffalo, camels, horses and donkeys is increasing. In Europe, it is estimated that about 1–1.3 million litres of equine milk are currently produced per annum but in countries such as Mongolia the figure is considerably higher, probably approximately 9 million litres.

A decade ago, equine milk was produced only in isolated small holdings in parts of Eastern Europe and Mongolia, but now there are large-scale operations in France, Belgium, Germany, Austria and the Netherlands. Asinine milk is produced in large donkey farms in Italy, France, Spain, Belgium, Xinjiang and Shanxi provinces of China, Ethiopia and Pakistan (Salimei 2011).

For equine milk production, milking begins when the foal reaches approximately 8 weeks and is eating some hay and grass. The mare is separated from the foal by day and milked approximately five times at intervals of about 2.5 hours and produces 1–1.5 L of milk at each milking. At night, the foal feeds freely (van Laar, Orchid's Paardenmelkerij, the Netherlands, personal communication). Thus, milk production is very labour intensive and expensive, with the result that equine milk is priced as a delicacy, typically, €11/L. Milking schedules are similar for asinine milk, but the yield is lower than that of the horse, approximately 350–850 mL of milk per milking, depending on several factors including: foal and mare management, milking procedure, stage of lactation, body size and condition and feeding regime (Salimei 2011). Mastitis is rarely a problem with equids and occurs only if teat injury occurs during milking. Furthermore, equid species appear to be relatively resistant to brucellosis and tuberculosis, which is advantageous from a dairy farming point of view (Stoyanova et al. 1988).

COMPOSITION OF EQUID MILK

With the exception of the major domesticated dairy species and humans, information on milk composition is poor and of >4500 mammalian species in existence, milk compositional data are available for approximately 200 species, of which, data for only approximately 55 species appears to be reliable. The milk of all mammals contains the same principal components: water, salts, vitamins, fats, carbohydrate and proteins, but these constituents differ significantly both quantitatively and qualitatively between species (Table 26.1), although species from the same taxonomic order, for example equids, produce milk of similar composition (Table 26.1). Equid milk is similar in composition to human milk but considerably different from that of other dairy mammals, for example cow, buffalo, sheep, goat, camel, llama and yak (Table 26.1). Why equid milk is so similar in macro composition to that of human milk is unclear, especially as equids and humans are phylogenetically distantly related. Inter-species differences in milk composition reflect very divergent patterns of nutrient transfer to the young and presumably reflect adaptations in maternal rearing of offspring to physiological constraints and environmental conditions (Oftedal and Iverson 1995). In all situations, lactation must be effective in providing nourishment to the offspring without over-taxing maternal resources.

The protein content of milk varies considerably between species and reflects the growth rate of the young. Bernhart (1961) found a linear correlation between the percent of calories derived from protein and the logarithm of the days to double birth weight for 12 mammalian species. For humans, one of the slowest growing and slowest maturing species, it takes 120–180 days to double birth weigh and only 7% of calories come from protein. In contrast, carnivores can double their birth weigh in as little as 7 days and acquire >30% of their energy from protein. Equid species take between 30 and 60 days to double their birth weight and, like humans, have an exceptionally low level of protein in their milk (Table 26.1). The high metabolic needs of the foal are met through frequent feeding. Equid milks have a significantly lower energy value than human milk, owing to their low fat content (Table 26.1).

The fat content of milk across species shows large variation and ranges from approximately 0.6% for some breeds of donkey and less than 1% for rhinoceros to approximately 50% in the milk of some seals. High-fat milks are important for some species, for example desert mammals, when maternal water economy is important and energy needs to be transferred to the young in an efficient manner. Equid milk has one of the highest contents of carbohydrate, which is similar to that of human milk. In equid species, lactation lasts naturally for approximately 7 months.

Factors that Affect the Composition of Equid Milk

Stage of Lactation

Within species, the stage of lactation is the most important determinant of milk composition and the difference between colostrum and mid-lactation milk shows the most significant difference but absolute values and the direction of change vary among species (Casey 1989). Shorty after parturition, the mammary gland produces colostrum that is richer in dry matter, proteins, fat, vitamins and minerals (except calcium and phosphorus) but poorer in lactose than mature milk. One of the major biological benefits of colostrum is the presence of Igs, IgA, IgM and IgG, and high levels of some enzymes, including

Table 26.1. Gross Composition of the Milk of Equid Species and Some Dairy Species, with Human and Other Selected Species Included for Comparison

Species	Total Solids	Protein	Casein: Whey Ratio	Fat	Lactose	Ash	Gross Energy (kJ.kg^{-1} or kJ.L^{-1})	Days to Double Birth Rate
[a]Horse (*Equus caballus*)	102.0	21.4	1.1:1	12.1	63.7	4.2	1883	40–60
[a]Donkey (*Equus africanus asinus*)	88.4	17.2	1.28:1	14.0	68.8	3.9	1966	30–50
[a]Mountain zebra (*Equus zebra hartmannae*)	100.0	15.6	–	10.2	69.0	3.0	1800	–
[a]Plains zebra (*Equus burchelli*)	113.0	16.3	–	22.0	70.0	4.0	2273	–
[a]Przewalski horse (*Equus caballus przewalski*)	105.0	15.5	1.1:1	15.0	67.0	3.0	1946	–
[b]Cow (*Bos taurus*)	127.0	34.0	4.7:1	37.0	48.0	7.0	2763	30–47
[a]Buffalo (*Bubalus bubalis*)	172.0	46.5	4.6:1	81.4	48.5	8.0	4644	48–50
[b]Sheep (*Ovis aries*)	181.0	55.9	3.1:1	68.2	48.8	10.0	4309	10–15
[a]Goat (*Capra hircus*)[h]	122.0	35.0	3.5:1	38.0	41.0	8.0	2719	12–19
[b]Camel (*Camelus dromedarius*)	124.7	33.5	1.68:1	38.2	44.6	7.9	2745	250
[b] Llama (*Llama glama*)	131.0	34.0	3.1:1	27.0	65.0	5.0	2673	120
[a]Yak (*Bos grunniens*)	160.0	42.3	4.5:1	56.0	52.9	9.1	3702	60
[a]Man(*Homo sapiens*)	124.0	9.0	0.4:1	38.0	70.0	2.0	2763	120–180
[b]Pig (*Sus scrofa*)	188.0	36.5	1.4:1	65.8	49.6	10.0	3917	9
[b]Rabbit (*Oryctolagus cuniculus*)	328.0	139.0	2.0:1	183.0	21.0	18.0	9581	4–6
[b]Blue whale (*Balaenoptera musculus*)	550.0	119.0	2.0:1	409.0	13.0	14.0	17614	10
[b]Northern Fur Seal (*Callorhinus ursinus*)	633.0	103.0	1.1:1	507.0	1.0	5.0	20836	5
[a]Rat (*Rattus norvegicus*)	210.0	84.0	3.2:1	103.0	26.0	13.0	5732	2–5
[a]Elephant (*Loxodonta africana africana*)	176.9	47.3	0.61:1	60.7	38.8	7.0	3975	100–260
[a]Rhinoceros (*Ceratotherium simum*)	77.5	16.2	0.22:1	7.4	61.0	3.0	1589	25–35

Source: Modified from Uniacke et al. 2010.
Values are expressed as [a]g.kg^{-1} or [b]g.L^{-1} Milk.

catalase, lipase and proteinase. Figure 26.1 shows the affect of lactation on the main constituents of equine milk and indicates a very rapid transition from equine colostrum to mature equine milk, that is within the first 24 hours of lactation. The concentration of protein in equine milk is very high, >15 g/100 g milk, immediately post-partum, but decreases rapidly to <4 g/100 g milk after 24 hours of lactation and to less than 2 g/100 g milk after 4 weeks of lactation (Fig. 26.1A). The casein to whey protein ratio in equine colostrum is 0.2:1 immediately post-partum and this changes to approximately 1.1:1 within 1 week. The protein content of bovine milk decreases during the first 3 months of lactation, but increases subsequently (Walstra et al. 2006a). The concentration of lactose in equine more than doubles during the first 24 hours (Fig. 26.1C), and this is also observed for bovine milk (Walstra et al. 2006a). The concentration of lactose in equine milk subsequently increases steadily throughout further lactation (Fig. 26.1C), a trend that is different from that for bovine milk in which lactose content decreases progressively (Walstra et al. 2006a). Close agreement is observed between the data of various studies for the level of protein (Fig. 26.1A) and lactose (Fig. 26.1C) in equine milk but considerable differences are reported for the lipid content of equine milk between different studies (Csapó et al. 1995, Doreau et al. 1986). This may be due to an increase in the fat content of equine milk that occurs during a milking session and, in some cases, the use of the hormone, oxytocin, which promotes complete evacuation of the udder (Doreau et al. 1986). Hence, the volume of milk drawn and the degree of evacuation of the udder will significantly influence the lipid content of the milk and thus explain differences in lipid content observed between different studies. However, all studies indicated in Figure 26.1B show the same trend, that is a decrease in the lipid content of equine milk with

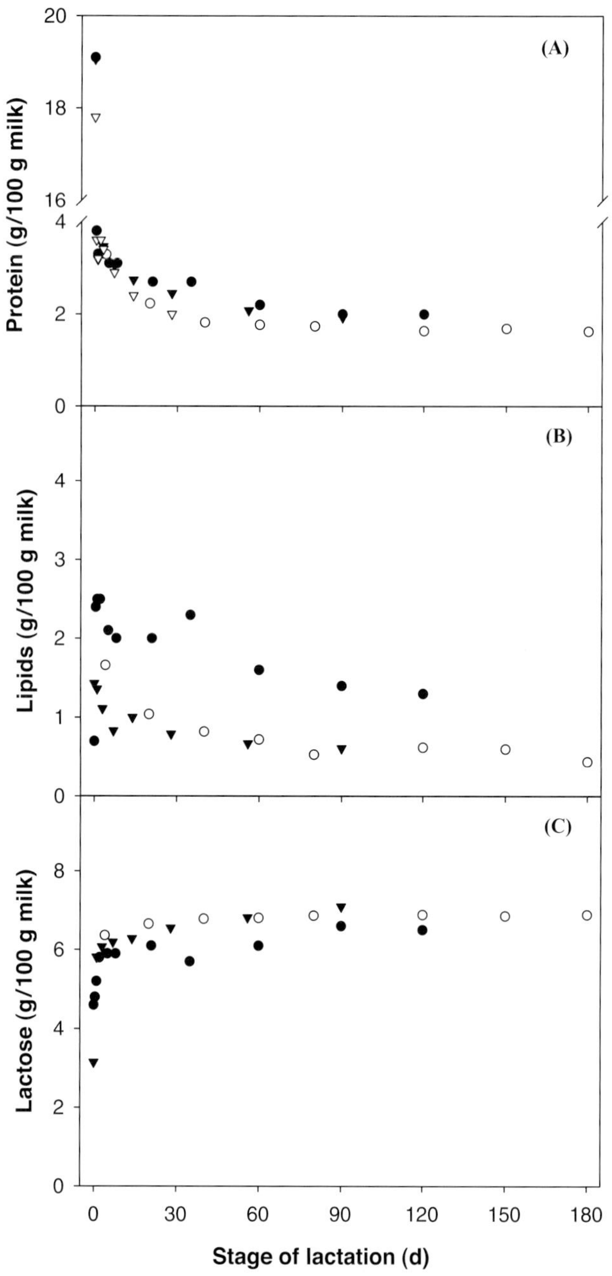

Figure 26.1. Influence of lactation stage on the concentration of (**A**) protein, (**B**) lipids, or (**C**) lactose in equine milk. (Data from Ullrey et al. (1966, •), Mariani et al. (2001, ○) Smolders et al. (1990, ▼) and Zicker and Lönnerdal (1994, ▽).)

advancing lactation (Fig. 26.1B), whereas that the lipid content in bovine milk shows a distinct minimum after approximately 3 months of lactation (Walstra et al. 2006a). The fat content of asinine milk increases from approximately 0.5% to 1.5% from days 15 to 105 but decreases sharply thereafter (Guo et al. 2007, Salimei et al. 2004). Piccione et al. (2008) reported a decrease in fat in the milk of Ragusana donkeys throughout lactation and observed a daily rhythmicity, similar to that found in bovine and human milk, for the levels of fat, lactose and protein in asinine milk, with fat and lactose peaking at night and protein reaching a maximum level during the daytime.

Effect of Equid Breed on Milk Composition

Data in the literature are not conclusive as to whether or not the breed of mare has an effect on the concentration of protein in milk. Kulisa (1977), Doreau et al. (1990), Csapó-Kiss et al. (1995) and Csapó et al. (1995) reported no effect of breed on the concentration of proteins or lipids in equine milk throughout lactation. On the other hand, Boulot (1987) and Formaggioni et al. (2003) have reported significant differences in protein content between breeds. Civardi et al. (2002), who compared Arabian, Haflinger, Trotter and Norico breeds, found that Norico milk had significantly lower α-lactalbumin (α-La), highest lysozyme (Lyz) and β-lactoglobulin (β-Lg) and highest thermal resistance of the breeds studied. Pelizzola et al. (2006), who compared the milk of Haflinger, Quarter horse, Sella/Salto and Rapid Heavy Draft, found that Quarter horse milk had significantly higher concentrations of the main constituents and higher concentrations of linoleic and α-linolenic fatty acids (ALA) than in the milk of the other species.

Asinine milk shows variability in fat content among breeds and is reported to be as low as 0.4% for Martina Franca mares, 0.6% in Ragusana mares and as high as 1.7% in Jiangyue donkeys (for these donkeys an increase from 0.5% to 1.7% was recorded in the fat content of the milk over 180 days of lactation) (Guo et al. 2007). Milk yield is significantly lower for Jiangyue donkeys than for Martina Franca and Ragusana breeds and the protein pattern of Jangyue milk is significantly different from the other breeds (Guo et al. 2007)

PROTEINS

While the protein content of mature equid milk is lower than that of bovine milk, there is a strong qualitative resemblance, the principal classes of proteins, that is caseins and whey proteins are similar in both milks. However, while the caseins are the predominant class of proteins in bovine milk (~80% of total milk protein), equid milk contains less casein and more whey proteins. The distribution of casein and whey proteins in equid milk is shown in Table 26.2, with comparative data for bovine and human milk.

Caseins

About 80% of the proteins in bovine milk are caseins that are primarily a source of amino acids, calcium, phosphate and bioactive peptides for neonates (Shekar et al. 2006). The low-casein concentration in mature equine milk (~55% of total protein) has many implications that will be discussed later. The traditional method for separating caseins from whey proteins is isoelectric precipitation of the caseins at pH approximately 4.6. The casein fraction of most milks consists of four gene products: α_{s1}-,

Table 26.2. Concentration of Caseins and Whey Proteins (g. kg^{-1}) in Equine, Asinine, Human and Bovine Milk

	Equine	Asinine	Human	Bovine
Total casein	13.56	7.8	2.4	26
α_{s1}-casein	2.4	Identified	0.77	10.7
α_{s2}-casein	0.20	Unknown	–	2.8
β-casein	10.66	Identified	3.87	8.6
κ-casein	0.24	Unknown	0.14	3.1
γ-casein	Identified	Unknown	–	0.8
Total whey protein	8.3	5.8	6.2	6.3
β-lactoglobulin	2.55	3.3	–	3.2
α-lactalbumin	2.37	1.9	2.5	1.2
Serum albumin	0.37	0.4	0.48	0.4
Proteose peptone	–	–	–	0.8
Immunoglobulins	1.63	1.30	0.96	0.80
$IgG_{1,2}$	0.38		0.03	0.65
IgA	0.47		0.96	0.14
IgM	0.03		0.02	0.05
Lactoferrin	0.58	0.37	1.65	0.10
Lysozyme	0.87	1.00	0.34	126×10^{-6}
NPN (mg.L^{-1})	375	455	454	266
Casein micelle size (nm)	255	~100–200	64	182

Source: Modified from Uniacke et al. 2010, with asinine data from Guo et al. 2007 and Salimei et al. 2004.
NPN, non-protein nitrogen.

α_{s2}-, β- and κ-caseins, of which the first three are calcium sensitive. All caseins lack secondary structure, which led Holt and Sawyer (1993) to consider them as rheomorphic proteins. The lack of secondary structure may be attributed, at least partially, to the relatively high level of proline residues in casein. As a result, caseins do not denature or associate on heating (Paulson and Dejmek 1990). The biological function of the caseins lies in their ability to form macromolecular structures, casein micelles, which transfer large amounts of calcium to the neonate with a minimal risk of pathological calcification of the mammary gland. The individual caseins will be discussed separately in the following sections with focus on their interactions to form casein micelles and the colloidal stability thereof.

Fractionation and characterisation of individual equine caseins has been poorly researched to date in comparison to those of bovine milk and it had been reported that equine, and presumably asinine, caseins exhibit greater heterogeneity and a higher level of post-translational modifications than those of bovine milk (Miranda et al. 2004). Table 26.3 shows the biochemical properties of individual casein proteins that are discussed later.

Table 26.3. Properties of Equine, Bovine and Human α_{s1}-, β- and κ-Caseins

Protein	Species	Primary Accession Number[a]	Amino Acid Residues	Molecular Weight (Da)	pH	GRAVY[b]	Cysteine Residues
α_{s1}-casein	Equine	Q8SPR1	205	24,614.4	5.47	−1.127	0
	Bovine	P02662	199	22,974.8	4.99	−0.704	0
	Human	P47710	170	20,089.4	5.17	−1.013	3
β-casein	Equine	Q9GKK3	226	25,511.4	5.78	−0.415	0
	Bovine	P02666	209	23,583.2	5.13	−0.355	0
	Human	P05814	211	23,857.8	5.33	−0.289	0
κ-casein	Equine	P82187	165	18,844.7	8.03	−0.313	2
	Bovine	P02668	169	18,974.4	5.93	−0.557	2
	Human	P07498	162	18,162.6	8.68	−0.528	1

Source: Modified from Uniacke et al. 2010.
[a]Primary accession number for the protein in SWISS-PROT database.
[b]Grand average hydropathy (GRAVY) score using the scale of Kyte and Doolittle (1982).

α_{S1}-Casein

The amino acid sequence of equine α_{s1}-casein has been deduced from its cDNA sequence (Lenasi et al. 2003). The protein contains 205 amino acids and has a molecular mass of 26,614.4 Da prior to post-translational modification, that is, it is considerably larger than its bovine or human counterpart (Table 26.3). Two smaller isoforms of α_{s1}-casein have been identified in equine milk, which probably result from the skipping of exons during transcription (Miranda et al. 2004). Equine α_{s1}-casein contains six potential phosphorylation sites (Lenasi et al. 2003), five of which are in very close proximity (Ser_{75}, Ser_{77}, Ser_{79}, Ser_{80}, Ser_{81}) and can thus form a phosphorylation centre, which is important in the structure of casein micelles. Mateos et al. (2009a) determined the different phosphorylation levels of the native isoforms of equine α_{s1}-casein and identified 36 different variants with several phosphate groups ranging from two to six or eight which, like equine β-casein, present a complex pattern on one dimensional and two dimensional electrophoresis. Bovine α_{s1}-casein contains eight or nine phosphorylation sites (Swaisgood 2003), which form two phosphorylation centres (De Kruif and Holt 2003). Bovine α_{s1}-casein contains three distinct hydrophobic regions, roughly including residues 1–44, 90–113 and 132–199 (Swaisgood 2003). These regions are characterised by positive values for hydropathy. Likewise, equine α_{s1}-casein has three domains with a high hydropathy value, that is, around residues 25–30, 95–105 and 150–205 and therefore it probably has association properties similar to those of bovine α_{s1}-casein. Furthermore, equine α_{s1}-casein contains two regions with very low hydropathy, that is, around residues 45–55 and 125–135, which are expected to behave hydrophilically. Human α_{s1}-casein does not appear to have distinct hydrophobic regions. Overall, equine and human α_{s1}-casein have comparable grand average hydropathy (GRAVY) scores, which are lower than that of bovine α_{s1}-casein (Table 26.3), indicating an overall higher hydrophobicity for the latter. GRAVY scores reflect the relative ratio of hydrophobic and hydrophilic amino acid residues in a protein, with a positive value reflecting an overall hydrophobic and a negative value an overall hydrophilic nature of the protein.

Prior to the mid-1990s, it was generally assumed that human milk contains mainly β- and κ-caseins with little or no α_s-casein (Kunz and Lönnerdal 1990). A minor casein component has since been identified and is considered to be the human equivalent of α_{s1}-casein, although this identification highlights several inconsistencies in comparison with the equivalent casein in other species. Uniquely, human α_{s1}-casein appears to contain at least two cysteine residues and exists as a multimer in complex with κ-casein (Cavaletto et al. 1994, Rasmussen et al. 1995). Johnsen et al. (1995) identified three cysteine residues in human α_{s1}-casein and provided a molecular explanation for α_{s1}-κ-casein complex formation. Martin et al. (1996) provided definitive evidence for the presence of a functional α_{s1}-casein locus in the human genome which is expressed in the mammary gland during lactation, while Sørensen et al. (2003) determined the phosphorylation pattern of human α_{s1}-casein. In bovine milk, α_{s1}- casein is a major structural component of the casein micelle and plays a functional role in curd formation (Walstra and Jenness 1984). The relatively low level of α_{s1}-casein in equine milk (Table 26.2), and similarly in human milk, may be significant and, coupled with the low protein content, could be responsible for the soft curd produced in the stomach of the infant or foal (Dr. Ursula Fogarty, National Equine Centre, Ireland – personal communication). Goat milk lacking α_{s1}-casein has poor coagulation properties compared to milk containing α_{s1}-casein (Clark and Sherbon 2000). Bevilaçqua et al. (2001), who assessed the capacity of goat's milk with a low or high α_{s1}-casein content to induce milk protein sensitisation in guinea pigs, found significantly less sensitisation in milk with low α_{s1}-casein. This may represent another important attribute of the low α_{s1}-casein content of equine milk for use in human allergology.

An α_{s1}-like protein of approximately 31–33 kDa has been identified in asinine milk although Criscione et al. (2009) reported its absence in one Ragusana donkey under investigation.

α_{S2}-Casein

The complete amino acid sequence of equine α_{s2}-casein is unknown, but Ochirkhuyag et al. (2000) published the sequence of the *N*-terminal 15 amino acid residues (Lys-His-Lys-Met-Glu-His-Phe-Ala-Pro-Xaa-Tyr-Xaa-Gln-Val-Leu, where Xaa is an unknown amino acid). Only five of these amino acids were confirmed by Miranda et al. (2004). Isoelectric focusing showed two major bands for equine α_{s2}-casein, with isoelectric points in the pH range 4.3–5.1 (Ochirkhuyag et al. 2000). Bovine α_{s2}-casein is the most highly phosphorylated casein, usually containing 11 phosphorylated serine residues, with lesser amounts containing 10, 12 or 13 phosphate groups (Swaisgood 2003). There are no reports on the presence of α_{s2}-casein in human milk. Using three different methods for protein identification, Criscione et al. (2009) could not detect α_{s2}- or κ-casein in asinine milk.

β-Casein

The amino acid sequence of equine β-casein, derived from the cDNA, has been reported by Lenasi et al. (2003), and revised by Girardet et al. (2006) with the insertion of eight amino acids (glutamic acid (Glu_{27}) to Lys_{34}). The theoretical molecular mass of this 226 amino acid polypeptide is 25,511.4 Da (Table 26.3). Bovine and human β-casein contain 209 and 211 amino acid residues, respectively (Table 26.3). Two smaller variants of equine β-casein, which probably result from casual exon-skipping during transcription, were reported by Miranda et al. (2004). The 28 *C*-terminal amino acids contain seven potential phosphorylation sites (Ser_9, Ser_{15}, Ser_{18}, Ser_{23}, Ser_{24}, Ser_{25}, Ser_{28}) and multiple-phosphorylated isoforms of equine β-casein containing three to seven phosphoserine residues have been reported, with the isoelectric point varying from pH 4.74–5.30 (Girardet et al. 2006, Mateos et al. 2009b). Bovine β-casein, which contains four or five phosphorylated serine residues, has an isoelectric point of 5.0–5.5 (Swaisgood 2003). Human β-casein has up to six levels of phosphorylation, that is, 0, 1, 2, 3, 4 or 5 phosphorylated serine residues (Sood and Slattery 2000).

Equine, bovine and human β-casein have a very hydrophilic N-terminus, followed by a relatively random hydropathy distribution in the rest of the protein, leading to an amphiphilic protein with a hydrophilic N-terminus and a hydrophobic C-terminus. In equine sodium caseinate, the Lys_{47}–Ile_{48}, bond of β-casein is hydrolysed readily by bovine plasmin, whereas no cleavage of the corresponding bond, Lys_{48}–Ile_{49} in bovine β-casein has been shown (Egito et al. 2003). In bovine β -casein, Lys_{28}-Lys_{29} is readily cleaved by plasmin but the equivalent, Lys_{28}–Leu_{29}, in equine β-casein is insensitive (Egito et al. 2002). Other plasmin cleavage sites in equine β-casein are Lys_{103}–Arg_{104}, Arg_{104}–Lys_{105} and Lys_{105}–Val_{196} (Egito et al. 2002). Equine β-casein is readily hydrolysed by chymosin at Leu_{190}–Tyr_{191} (Egito et al. 2001).

Equine β-casein and equine α-La undergo spontaneous deamidation under physiological conditions at Asn_{135}–Gly_{136} and Asn_{45}–Gly_{46}, respectively (Girardet et al. 2004), which has been reported also for canine milk Lyz (Nonaka et al. 2008) and human lactoferrin (Lf) (Belizy et al. 2001) but not, to our knowledge, for bovine or human β-casein or α-la. Recent research has shown that temperature may be an important factor controlling the spontaneous deamidation process and at 10°C, the phenomenon is strongly reduced (Matéos et al. 2009b). Spontaneous deamidation represents an important modification of equine milk proteins under certain conditions where bovine milk proteins, which do not contain a potential site for deamidation, remain unaffected. Equine Lf also contains the Asn–Gly sequence and may be susceptible to spontaneous deamidation (Girardet et al. 2006).

Unique to equine milk and apparently absent from the milk of other species, including ruminants, is a low-molecular weight (MW) multi-phosphorylated β-casein variant which accounts for 4% of the total casein (Miclo et al. 2007). This short protein (94 amino acid residues) is the result of a large deletion (residues 50–181) from full-length equine β-casein. No spontaneous deamidation of this low-MW form of β-casein has been found. Multi-phosphorylated isoforms of β-casein, approximately 34–35.4 kDa, have been identified in asinine milk but no further characterisation has been reported to date (Criscione et al. 2009).

κ-Casein

The presence of κ-casein in equine milk was an issue of debate for several years, with several authors (Visser et al. 1982, Ono et al. 1989, Ochirkhuyag et al. 2000) reporting its absence. However, other studies (Kotts and Jenness 1976, Malacarne et al. 2000, Iametti et al. 2001, Egito et al. 2001) showed its presence, albeit at a low concentration. The primary structure of equine κ-casein has been derived (Iametti et al. 2001, Lenasi et al. 2003, Miranda et al. 2004); it contains 165 amino acids residues, that is four less than bovine κ-casein but three more than human κ-casein (Table 26.3). The MW of equine κ-casein, prior to post-translational modification, is 18,844.7 Da. Equine and human κ-casein have a considerably higher isoelectric pH than bovine κ-casein (Table 26.3), and they have a net positive charge at physiological pH, whereas bovine κ-casein has a net negative charge. The GRAVY score of bovine κ-casein is considerably lower than that of equine κ-casein (Table 26.3), indicating that the latter is more hydrophilic. Bovine κ-casein is characterised by a hydrophilic C-terminus, which is very important for the manner in which bovine casein micelles are stabilised, but a comparison of the hydropathy distribution of bovine and equine κ-caseins indicates that the C-terminus of equine κ-casein is far less hydrophilic, particularly as a result of the absence of a strong hydrophilic region at residues 110–120. Human κ-casein appears to be more like equine than bovine κ-casein in terms of the distribution of hydropathy along the polypeptide chain. Studies on asinine milk have not found κ-casein (Vincenzetti et al. 2008, Chianese et al. 2010).

Glycosylation of κ-Casein κ-Casein, the only glycosylated member of the casein family, exhibits microheterogeneity due to the level of glycosylation (Saito and Itoh 1992). Tri- or tetrasaccharides consisting of N-acetylneuraminic acid (NANA), galactose and N-acetylgalactosamine are attached to κ-casein *via* O-glycosidic linkages to threonine residues in the C-terminal portion of the molecule (the glycomacropeptide region). About two-thirds of bovine κ-casein molecules are glycosylated at one of six threonyl residues, that is Thr_{121}, Thr_{131}, Thr_{133}, Thr_{135}, Thr_{136} (only in bovine κ-casein variant A) or Thr_{142} (Pisano et al. 1994); Ser $_{141}$ is also a potential glycosylation site (Kanamori et al. 1981). Human κ-casein has seven glycosylation sites, Thr_{113}, Thr_{123}, Thr_{128}, Thr_{131}, Thr_{137}, Thr_{147} and Thr_{149} (Fiat et al. 1980). Although no direct information is available, lectin-binding studies indicate that equine κ-casein is glycosylated (Iametti et al. 2001), possibly at residues Thr_{123}, Thr_{127}, Thr_{131}, Thr_{149} and Thr_{153} (Lenasi et al. 2003) (these glycosylation sites are not fully in agreement with those proposed by Egito et al. (2001)). To date, no non-glycosylated κ-casein has been identified in equine milk (Martuzzi and Doreau 2006).

κ-Casein is located mainly on the surface of the casein micelles and is responsible for their stability (Walstra 1990). The presence of a glycan moiety in the C-terminal region of κ-casein enhances its ability to stabilise the micelle, by electrostatic repulsion, and may increase the resistance by the protein to proteolytic enzymes and high temperatures (Minkiewicz et al. 1993, Dziuba and Minkiewicz 1996). Biologically, NANA residues have antibacterial properties and act as a bifidogenic factor (Dziuba and Minkiewicz 1996). κ-Casein is thought to play a major role in preventing the adhesion of *Helicobacter pylori* to human gastric mucosa (Strömqvist et al. 1995). It is likely that heavily glycosylated κ-casein provides some protection to breast-feeding infants due to its carbohydrate content which may be important especially as *H. pylori* infection is occurring at an increasingly younger age (Lönnerdal 2003).

Hydrolysis of κ-Casein The hydrolysis of bovine κ-casein by chymosin at Phe_{105}–Met_{106} leads to the production of the hydrophobic N-terminal para-κ-casein and the hydrophilic C-terminal caseinomacropeptide (CMP) (Walstra and Jenness 1984). Chymosin hydrolyses the Phe_{97}–Ile_{98} bond of equine κ-casein (Egito et al. 2001) and slowly hydrolyses the Phe_{105}–Ile_{106} bond of human κ-casein (Plowman et al. 1999). However, as

Table 26.4. Properties of Equine, Bovine and Human Para-κ-Casein and Caseinomacropeptide

Protein	Species	Residues	Amino Acid Residues	Molecular Weight (Da)	pI	GRAVY[a]
Para-κ-casein	Equine	1–97	97	11,693.3	8.96	−0.675
	Bovine	1–105	105	12,285.0	9.33	−0.617
	Human	1–105	97	11,456.9	9.63	−1.004
CMP	Equine	98–165	68	7,169.3	4.72	0.203
	Bovine	106–169	63	6,707.4	4.04	−0.370
	Human	106–162	65	6,723.7	4.24	0.182

[a]Grand average hydropathy (GRAVY) score using the scale of Kyte and Doolittle (1982).
Source: Modified from Uniacke et al. 2010.
CMP, *C*-terminal caseinomacropeptide.

summarised in Table 26.4, the CMPs released from equine and human κ-caseins are considerably less hydrophilic than bovine CMP. The sequence 97–116 of κ-casein is highly conserved across species, suggesting that the limited proteolysis of κ-casein and subsequent coagulation of milk are of major biological significance (Mercier et al. 1976, Martin et al. 2011)

A grouping system for mammals based on κ-casein structure and the site of cleavage by chymosin has been suggested (Mercier et al. 1976, Nakhasi et al. 1984). Group I species (cow, goat, sheep and buffalo) have a higher content of dicarboxylic amino acids and low hydrophobicity and carbohydrate content and κ-casein is cleaved at Phe_{105}–Met_{106}, while Group II species (horse, human, mouse, pig, rat) have a high proline content, less dicarboxylic amino acids and a much higher hydrophobicity and carbohydrate content and are cleaved at Phe_{97}–Ile_{98} or Phe_{105}–Leu_{106}. Marsupial κ-casein appears to form a separate group with a cleavage site different from that in eutherian mammals (Stasiuk et al. 2000). Cleavage of equine milk at Phe_{97}–Ile_{98}, as well as other characteristics of its κ-casein, place the horse in Group II. The divergence between species into Groups I and II could account for differences in the clotting mechanisms of ruminant and non-ruminant milks (Herskovits 1966). In addition to the differences in cleavage site, the grouping system also divides species based on the number of *O*-glycosylation sites in κ-caseins. As equine and human κ-casein are considerably more highly glycosylated than bovine κ-casein and non-glycosylated κ-casein has not been found in equine milk (Egito et al. 2001), equine and human κ-caseins belong to the same group. The level of glycosylation does not affect micelle structure but it does affect the susceptibility of κ-casein to hydrolysis by chymosin, with susceptibility decreasing as the level of glycosylation increases (Doi et al. 1979, Addeo et al. 1984, Van Hooydonk et al. 1984, Vreeman et al. 1986, Zbikowska et al. 1992). Therefore, equine milk probably has a different clotting mechanism by chymosin than bovine milk.

Equid Casein Micelles

In the milk of all species studied in sufficient detail, the caseins exist predominantly as micelles, which are hydrated spherical structures with dimensions in the sub-micron range. The dry matter of casein micelles consists predominantly (>90%) of proteins, with small amounts of inorganic matter, collectively referred to as micellar calcium phosphate (MCP). The structure and sub-structure of bovine casein micelles has been studied in detail and reviews include: Holt and Horne (1996), Horne (1998, 2006), De Kruif and Holt (2003), Phadungath (2005), Farrell et al. (2006), Qi (2007), Fox and Brodkorb (2008).

Equine casein micelles are larger than bovine or human micelles (Table 26.2) (Welsch et al. 1988, Buchheim et al. 1989) while those of asinine milk are similar in size to bovine micelles (Salimei 2011). Electron microscopy shows that bovine and equine micelles have a similar 'spongy' appearance, while human micelles seem to have a much 'looser', more open structure (Jasińska and Jaworska 1991). Such a loose open structure may affect the susceptibility to hydrolysis by pepsin. Jasińska and Jaworska (1991) reported that human micelles are much more susceptible to pepsin hydrolysis than either equine or bovine micelles. There are no specific reports on the sub-structure of equine casein micelles although equine milk does contain approximately 10.1 $mmol.L^{-1}$ micellar calcium and approximately 2.6 $mmol.L^{-1}$ micellar inorganic phosphate, suggesting a micellar calcium:casein ratio of >20:1 which, on a molar basis, far exceeds the calcium-binding capacity of equine casein molecules. Hence, it may be assumed that equine micelles, like bovine casein micelles, contain nanoclusters of calcium phosphate. Since equine milk contains little or no κ-casein, unphosphorylated β-casein may play a role in micellar stability (Ochirkhuyag et al. 2000, Doreau and Martin-Rosset 2002). A similar conclusion was reported by Dev et al. (1994) for the stabilisation of human casein micelles.

Both equine α_{s1}-casein (residues 75–81) and β-casein (residues 23–28) contain a phosphorylation centre, which is required for the formation of nanoclusters; furthermore, both proteins also contain distinct hydrophobic regions through which solvent-mediated protein–protein interactions may occur. Equine α_{s2}-casein may have similar properties to equine α_{s1}-casein, pending further characterisation. The ratio of micellar calcium:micellar inorganic phosphate is 2.0 in equine milk, but approximately 3:9 in bovine milk (Holt and Jenness 1984) and might indicate that either a smaller proportion of micellar calcium is incorporated into nanoclusters in equine milk, or that equine nanoclusters contain a higher proportion of casein-bound phosphate, which would imply smaller nanoclusters. However, unlike bovine κ-casein, equine κ-casein does not have a distinctly hydrophilic *C*-terminal domain; thus, it is unclear if this part of the protein is capable of protruding from the micellar surface to sterically stabilise the micelles. Furthermore, given that the size of casein micelles and the content of κ-casein are inversely related (Yoshikawa et al. 1982, Dalgleish 1998.), a low level of κ-casein would be expected in equine milk compared to bovine milk. Further research is required to elucidate the structure of equine and asinine casein micelles as destabilisation of the micelles is the basis for the successful conversion of milk into a range of dairy products, for example cheese or yoghurt.

Stability of Equid Casein Micelles

Coagulation of milk occurs when the colloidal stability of the casein micelles is destroyed and may be desirable or undesirable. Coagulation is desirable in the manufacture of yoghurt and cheese and is also important from a nutritional point of view, as clotting of the caseins in the stomach, and the type and structure of the resultant coagulum strongly affect digestibility. In contrast, heat-induced coagulation of casein micelles, which can occur at a temperature >120°C, is undesirable. In this section, common types of micellar instability are described.

Bovine casein micelles are sterically stabilised by a brush of predominantly κ-casein (De Kruif and Zhulina 1996), which protrudes from the micelle surface. Coagulation of casein micelles can occur only following collapse of the brush, which occurs on acidification of milk, that is in the manufacture of yoghurt or on removal of the brush that occurs on rennet-induced coagulation of milk. The combined process of enzyme- and acid-induced coagulation is likely to contribute to coagulation of casein micelles in the stomach.

Enzymatic Coagulation of Equid Milk

Enzymatic coagulation of milk is the first step in the manufacture of most cheese varieties and also plays an important role in the flocculation of casein micelles in the stomach. For cheese manufacture, the process involves the addition of a milk-clotting enzyme, for example chymosin, to the milk, followed by incubation at a temperature ≥30°C. During the incubation of bovine milk with rennet, chymosin hydrolyses the Phe_{105}–Met_{106} bond of κ-casein, leading to the formation of two fragments, the hydrophobic *N*-terminal fragment, f1–105, which remains attached to the casein micelles and is referred to as para-κ-casein, and the hydrophilic *C*-terminal fragment, f106–169, which is released into the milk serum and is referred to as the CMP. As a result, the micelles lose steric stabilisation and become susceptible to aggregation, particularly in the presence of Ca^{2+} (for reviews, see Walstra and Jenness 1984, Wong et al. 1988, Walstra 1990, Fox and McSweeney 1998, Walstra et al. 2006b). Equine κ-casein is hydrolysed slowly by chymosin at the Phe_{97}–Ile_{98} bond (Kotts and Jenness 1976, Egito et al. 2001), without gel formation, and it appears that either the chymosin-sensitive bond of equine κ-casein is located in the micelle in a manner that renders it inaccessible by chymosin, or that the equine casein micelle derives colloidal stability from constituents other than κ-casein. The high degree of glycosylation may also affect the ability of chymosin to hydrolyse equine κ-casein. Figure 26.2 illustrates the coagulation of equine, asinine and bovine milk by calf chymosin at 30°C. While it is clear that no gel is formed from equine milk, as judged by lack of an increase in storage modulus, G^1, asinine milk seems to form a gel, although it is very weak compared to the gel formed from bovine milk. Further investigation is warranted to determine if there are differences in the coagulation properties of asinine and equine milk.

Acid-induced Coagulation of Equid Milk

When bovine milk is acidified to a pH below 5.0, flocculation of casein micelles occurs, leading ultimately to gel formation. This process is the basis of the manufacture of yoghurt, in which acidification is induced by the production of lactic acid by lactic acid bacteria and also occurs at the low pH of the stomach (for

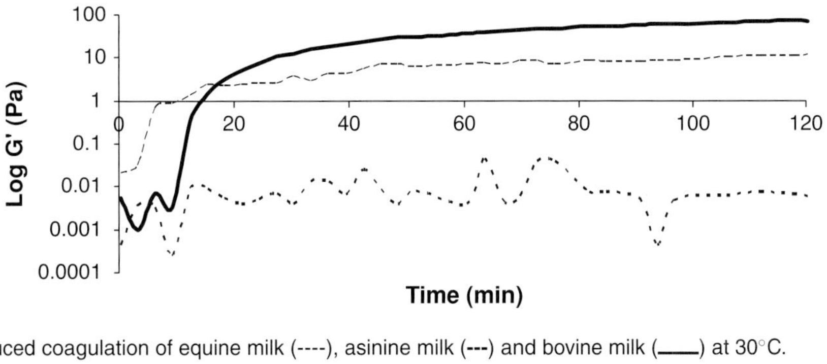

Figure 26.2. Rennet-induced coagulation of equine milk (----), asinine milk (---) and bovine milk (——) at 30°C.

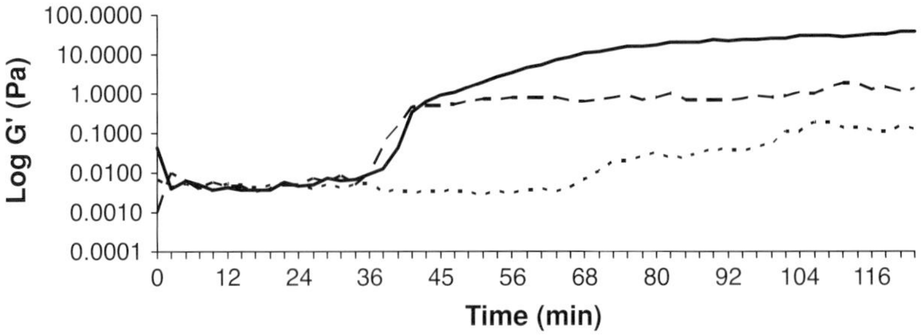

Figure 26.3. Coagulation of equine milk (----), asinine milk (---) and bovine milk (____) acidified with 3% glucono-δ-lactone at 30°C.

review, see Lucey and Singh 2003). Acid-induced flocculation of bovine casein micelles is believed to result from a reduction in the solvency of the κ-casein brush on the micellar surface due to protonation of the negatively charged carboxylic acid groups of Glu and aspartic acid (Asp). Equine casein micelles are considerably less susceptible to acid-induced flocculation. Di Cagno et al. (2004) reported that equine milk acidified at pH 4.2, the point of minimum solubility of equine caseins (Egito et al. 2001), had an apparent viscosity only approximately seven times higher than that of equine milk at its natural pH (Waelchli et al. 1990) and is probably indicative of micellar flocculation rather than gelation. By comparison, the viscosity of acidified bovine milk is approximately 100 times higher than that of bovine milk at natural pH. Differences in acid-induced flocculation between equine and bovine casein micelles may be related to differences in the mechanism by which they are sterically stabilised. Figure 26.3 shows the effect of acidification of bovine, equine and asinine milk at 30°C and 3% glucono-δ-lactone (GDL). Asinine milk appears to form a weak gel when treated with GDL, unlike equine milk that shows little or no gel formation. Elucidation of the mechanism of steric stabilisation of equine casein micelles is likely to shed further light on this subject.

Heat-induced Coagulation of Equine Milk

Although milk, compared to most other foods, is extremely heat-stable, coagulation does occur when heated for a sufficiently long time at >120°C. Unconcentrated bovine milk, usually assayed at 140°C, displays a typical profile, with a heat coagulation time (HCT) maximum (~20 minutes) at pH approximately 6.7 and a minimum at pH approximately 6.9 (O'Connell and Fox 2003). In contrast, the HCT of unconcentrated equine milk at 140°C increases with pH, that is, it has an almost sigmoidal pH-HCT profile (Fig. 26.4), from <2 minutes at pH 6.3–6.9 to >20 minutes at pH 6.9–7.1; a slight maximum is observed at pH 7.2. Pre-heating unconcentrated milk shifts the pH-HCT profile and reduces the HCT in the pH region around the maximum, similar to the effect reported for bovine milk (O'Connell and Fox 2003). The HCT of concentrated equine milk at 120°C increases up to pH 7.1 but decreases progressively at higher pH values. While the profile for concentrated bovine milk is somewhat similar, the maximum HCT occurs at a considerably lower pH, that is, approximately 6.6. Differences in heat stability between equine and bovine milk may be related to differences in steric stabilisation of the micelles and, while heat-induced complexation of β-Lg with κ-casein greatly affects the heat stability of bovine milk (O'Connell and Fox 2003), it is unlikely to do so in equine milk due to lack of a sulphydryl group in equine β-Lg. The lower protein, particularly casein, concentration in equine milk is also likely to contribute to its higher heat stability.

The colloidal stability of equine casein micelles differs considerably from that of bovine casein micelles, which may have significant implications for the conversion of equine milk into dairy products. On the basis of the evidence outlined, manufacture of cheese and yoghurt from equine milk is unlikely to be successful using conventional manufacturing protocols.

Stability of Equine Milk to Ethanol

The ethanol stability of bovine milk (for review, see Horne 2003), defined as the minimum concentration of added aqueous ethanol that causes it to coagulate at its natural pH (~6.7), is 70–75% (added 1:1 to milk), whereas the ethanol stability of equine milk (pH ~7.2) is 40–45% (Uniacke-Lowe, 2011). The high concentration of ionic calcium and low level of κ-casein in equine milk probably contribute to its low ethanol stability.

Whey Proteins

Similar to bovine milk, the major whey proteins in equine and asinine milk areβ-Lg, α-La, Igs, blood serum albumin (BSA), Lf and Lyz (Bell et al. 1981a, Salimei et al. 2004, Guo et al. 2007). Except for β-Lg, all these proteins are also present in human milk. However, the relative amounts of the whey proteins differ considerably between these milks (Table 26.2). Compared to bovine milk, equine milk contains less β-Lg but more α-La and Igs. The principal anti-microbial agent in equine milk is Lyz and to a lesser extent Lf (which predominates in human milk (Table 26.2). Both Lf and Lyz are present at low levels in bovine milk, in which Igs form the main defense against

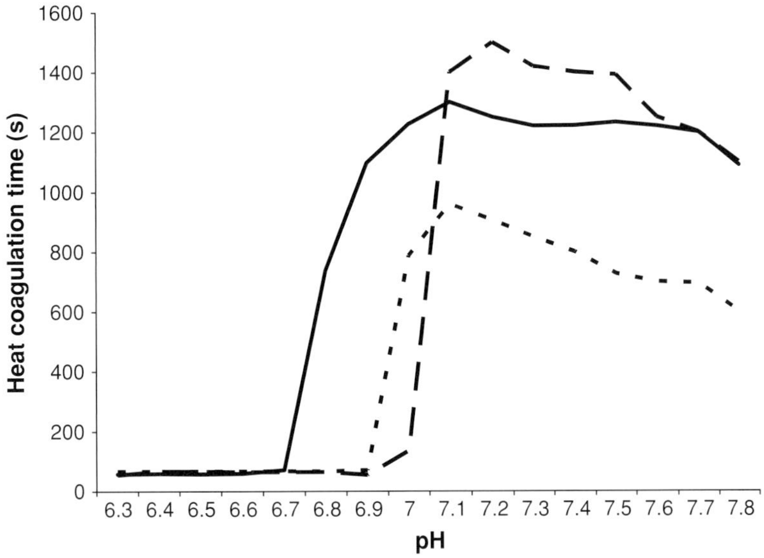

Figure 26.4. Heat coagulation time-pH profile of raw unconcentrated skimmed equine milk at 140°C (-----), preheated and unconcentrated milk (____) and concentrated milk at 120°C (---).

microbes (Malacarne et al. 2002). Together, Ig A, G, M, Lf and Lyz provide the neonate with immune and non-immune protection against infection (Baldi et al. 2005).

β-Lactoglobulin

β-Lg is the major whey protein in the milk of most ruminants and is also present in the milk of some monogastrics and marsupials, but is absent from the milk of humans, camels, lagomorphs and rodents. *β*-Lg is synthesised in the secretory epithelial cells of the mammary gland under the control of prolactin. Although several biological roles for *β*-Lg have been proposed, for example facilitator of vitamin A (retinol) uptake and an inhibitor, modifier or promoter of enzyme activity, conclusive evidence for a specific biological function of *β*-Lg is not available (Sawyer 2003, Creamer and Sawyer 2011). *β*-Lg of all species studied binds retinol; *β*-Lg of many species, but not equine or porcine, binds fatty acids also (Pérez et al. 1993). During digestion, milk lipids are hydrolysed by pre-gastric and pancreatic lipases, greatly increasing the amount of free fatty acids that could potentially bind to *β*-Lg, displacing any bound retinol, and implying that fatty acid metabolism, rather than retinol transport, is the more important function of *β*-Lg (Pérez and Calvo 1995). Bovine *β*-Lg is very resistant to peptic digestion and can cause allergenic reactions on consumption. Resistance to digestion is not consistent among species, with ovine *β*-Lg being far more digestible than bovine *β*-Lg (El-Zahar et al. 2005). The digestibility of equine *β*-Lg, which has, to our knowledge, not been studied, warrants research, particularly considering the potential applications of equine milk as a hypo-allergenic dairy product.

Two isoforms of equine *β*-Lg have been isolated, *β*-Lg I and II, which contain 162 and 163 amino acids, respectively. The extra amino acid in equine *β*-Lg II is a glycine residue inserted after position 116 of *β*-Lg I (Halliday et al. 1991). Asinine milk also has two forms of *β*-Lg, I and II (MW. 18.5 and 18.2 kDa, respectively); two variants of *β*-Lg I, that is, A and B, are known and four variants of *β*-Lg II, A, B, C and D (Cunsolo et al. 2007). Godovac-Zimmermann et al. (1985, 1988a,b) reported that *β*-Lg I from asinine milk has 162 amino acids, similar to equine *β*-Lg I (from which it differs by 5 amino acids). *β*-Lg II in both asinine and equine milks has 163 amino acids and shows substantial differences between both milks, with only six clusters of amino acid residues conserved (Godovac-Zimmermann et al. 1990). Criscione et al. (2009) reported the absence of *β*-Lg II from more than 23% of Ragusana donkeys in one study.

Bovine *β*-Lg occurs mainly as two genetic variants, A and B, both of which contain 162 amino acids and differ only at positions 63 (Asp in variant A, Gly in variant B) and 117 (valine (Val) in variant A, alanine (Ala) in variant B); a further 11, less common, genetic variants of bovine *β*-Lg have also been reported (Sawyer 2003). On the basis of its amino acid sequence, unmodified equine *β*-Lg I has a molecular mass of 18,500 Da and an isoelectric pH of 4.85, whereas equine *β*-Lg II, despite having one more amino acid, has a molecular mass of 18,262 Da (ExPASy ProtParam Tool 2009), and an isoelectric pH of 4.71 (Table 26.5). Bovine *β*-Lg A and B have a molecular mass of 18,367 and 18,281 Da, respectively, and an isoelectric pH of 4.76 and 4.83, respectively (Table 26.5). Using the hydropathy scale proposed by Kyte and Doolittle (1982), equine *β*-Lg I and II have a GRAVY score of

Table 26.5. Properties of Equine and Bovine β-Lactoglobulin (β-Lg) and Equine Bovine and Human α-Lactalbumin (α-La) and Lactoferrin.

Protein	Species	Variant	Primary Accession Number[a]	Amino Acid Residues	Molecular Mass (Da)	Isoelectric point	GRAVY Score[b]	Disulphide Bridges
β-Lg	Equine	I	P02758	162	18500.2	4.85	−0.386	2
		II	P07380	163	18261.6	4.71	−0.300	2
	Bovine	A	P02754	162	18367.3	4.76	−0.167	2
		B	P02754	162	18281.2	4.83	−0.162	2
α-La	Equine	A	P08334	123	14223.2	4.95	−0.416	4[c]
		B	P08896	123	14251.2	4.95	−0.503	4[c]
		C	P08896	123	14249.3	5.11	−0.438	4[c]
	Bovine		P00711	123	14186.0	4.80	−0.453	4
	Human		P00709	123	14078.1	4.70	−0.255	4
Lactoferrin	Equine		O77811	689	75420.4	8.32	−0.376	17
	Bovine		P24627	689	76143.9	8.67	−0.350	16[d]
	Human		P02788	691	76165.2	8.47	−0.415	16

Source: Modified from Uniacke et al. 2010.
Values were calculated from the amino acid sequences of the mature proteins provided on http://au.expasy.org.
[a]Primary accession number for the protein in SWISS-PROT database.
[b]Grand average hydropathy (GRAVY) score using the scale of Kyte and Doolittle (1982).
[c]Estimated from structural similarity with bovine and human α-La.
[d]Estimated from structural similarity with human lactoferrin.

−0.386 and −0.300, respectively (Table 26.5). Bovine β-Lg A and B have a GRAVY score of −0.167 and −0.162, respectively (Table 26.5), and are, therefore, considered to be less hydrophilic than equine β-Lg I and II. Both equine and bovine β-Lg contain two intramolecular disulphide bridges, linking Cys_{66} to Cys_{160} and Cys_{106} to Cys_{119} in equine β-Lg I, Cys_{66} to Cys_{161} and Cys_{106} to Cys_{120} in equine β-Lg II and Cys_{66} to Cys_{160} and Cys_{106} to Cys_{119} or Cys_{121} in bovine β-Lg A and B. Bovine β-Lg contains one sulphydryl group at Cys_{119} or Cys_{121}. Equine β-Lg contains only four cysteine residues and lacks a sulphydryl group that has major implications for denaturation and aggregation of the protein (see later).

At physiological conditions (neutral pH and β-Lg concentration >50 μM), bovine β-Lg occurs predominantly in dimeric form and at its isoelectric point (pH 3.7–5.2) the dimers associate into octamers but below pH 3.4 and above pH 8.0 the protein dissociates into its monomeric form (Gottschalk et al. 2003). Equine and asinine β-Lg I exist in the monomeric form only (Godovac-Zimmermann et al. 1990).

α-Lactalbumin

α-La, a unique milk protein, is homologous with the well-characterised C-type Lyz. It is a calcium metalloprotein, in which the Ca^{2+} plays a crucial role in folding and structure and has a regulatory function in the synthesis of lactose (Larson 1979, Brew 2003, Neville 2009).

Similar to the α-La of asinine, bovine, caprine, ovine, camelid and human milk, equine α-La contains 123 amino acids (Brew 2003). Equine α-La occurs as three genetic variants, A, B and C, which differ by only a few single amino acid replacements (Godovac-Zimmermann et al. 1987). Bovine α-La occurs as two, or possibly three, genetic variants (Bell et al. 1981b) and human α-La has two genetic variants, one of which has been identified only recently (Chowanadisai et al. 2005). The primary structure of equine, bovine and human α-La differ only by a few single amino acid replacements, and the proteins have similar properties (Table 26.5). Equine α-La A, B and C have an isoelectric point at pH 4.95, 4.95 and 5.11, respectively, whereas bovine and human α-La have isoelectric point at pH 4.80 and 4.70, respectively (Table 26.5). The GRAVY scores of equine and bovine α-La are comparable, whereas that of human α-La is distinctly higher (Table 26.5), indicating a lower hydrophobicity. The eight-cysteine residues of bovine and human α-La form four intramolecular disulphide bonds, linking Cys_6 to Cys_{120}, Cys_{28} to Cys_{111}, Cys_{61} to Cys_{77} and Cys_{73} to Cys_{93}. On the basis of the very high similarity between equine, bovine and human α-Las, as well as the α-La of other species, it is very likely that equine α-La also contains four intramolecular disulphide bridges, in the aforementioned positions. Equine, bovine or human α-La does not contain a sulphydryl group.

Three genetic variants of equine α-La have been reported but asinine milk has only one (123 amino acid residues, Mw approximately 14.2 kDa and four disulphide bonds), although some heterogeneity has been shown. Two isoforms, A and B, of asinine α-La (whose isoelectric points differ by 0.23 units) have been reported but subsequent analysis showed that the protein has only one form and misidentification in earlier work was probably due to differences in calcium binding by asinine α-La (Giuffrida et al. 1992). The primary structure of asinine α-La has been determined and differs from those of equine and bovine proteins with 39 and 40 amino acid substitutions, respectively (Godovac-Zimmermann et al. 1987).

Immunoglobulins

The concentration of whey proteins is significantly elevated in the colostrum of all ruminants and equids as maternal Igs are passed from mother to neonate after birth when the small intestine is capable of absorbing intact proteins. After a few days, the gut 'closes' and further significant passage of proteins is prevented and within 2–3 days, the serum level of IgG in the neonate is similar to adult levels (Widdowson 1984). In contrast, *in utero* transfer of Igs occurs in humans and in some carnivores Igs are passed to the newborn both before and after birth. The milk of species that provide prenatal passive immunisation tends to have relatively small differences in protein content between colostrum and mature milk compared to species that depend on post-natal passage of maternal Igs. In the latter cases, of which all ungulates are typical, colostrum is rich in Igs and there are large quantitative differences in protein content between colostrum and mature milk (Langer 2009).

Three classes of Igs, which form part of a mammal's natural defense against infection, are commonly found in milk, IgIgG, IgA and IgM; IgG is often sub-divided into two sub-classes, IgG_1 and IgG_2 (Hurley 2003, Madureira et al. 2007). All monomeric Igs consist of a similar basic structure of four polypeptides, two heavy chains and two light chains, linked by disulphide bridges, yielding a sub-unit with a molecular mass of approximately 160 kDa. IgG consists of one sub-unit, while IgA and IgM consist of two or five sub-units, with a molecular mass of approximately 400 or approximately 1000 kDa, respectively. The relative proportions of the Igs in milk differ considerably between species (Table 26.2). IgG is the principal Ig in equine colostrum, but IgA is the principal form in equine milk. In bovine milk and colostrum, IgG is the principal immunoglobulin, while IgA is the predominant Ig in human colostrum and milk.

Lactoferrin

Lf is an iron-binding glycoprotein, comprising of a single polypeptide chain of MW approximately 78 kDa (Conneely 2001). Lf is structurally very similar to transferrin (Tf), a plasma iron transport protein, but has a much higher (~300-fold) affinity for iron (Brock 1997). Lf is not unique to milk although it is especially abundant in colostrum, with small amounts in tears, saliva and mucus secretions and in the secondary granules of neutrophils. The expression of Lf in the bovine mammary gland is dependent on prolactin (Green and Pastewka 1978); its concentration is very high during early pregnancy and involution and is expressed predominantly in the ductal epithelium close to the teat (Molenaar et al. 1996). Human, equine, asinine and bovine milk contain 1.65 , 0.58 , 0.37 and 0.1 g Lf/kg, respectively (Table 26.2). The concentration of Lf in asinine milk, which comprises approximately 4% of total whey protein, is significantly lower than in equine milk (Table 26.2).

Shimazaki et al. (1994) purified Lf from equine milk and compared its iron-binding ability with that of human and bovine Lfs and with bovine Tf. The iron-binding capacity of equine Lf is similar to that of human Lf but higher than that of bovine Lf and Tf. Various biological functions have been attributed to Lf but the exact role of Lf in iron-binding in milk is unknown and there is no relationship between the concentrations of Lf and Tf and the concentration of iron in milk (human milk is very rich in Lf but low in iron) (Masson and Heremans 1971).

Lf is a bioactive protein with nutritional and health-promoting properties (Baldi et al. 2005). Bacterial growth is inhibited by its ability to sequester iron and also to permeabilise bacterial cell walls by binding to lipopolysaccharides through its *N*-terminus. Lf can inhibit viral infection by binding tightly to the envelope proteins of viruses and is also thought to stimulate the establishment of a beneficial microflora in the gastrointestinal tract (Baldi et al. 2005). Ellison and Giehl (1991) suggested that Lf and Lyz work synergistically to effectively eliminate Gram-negative bacteria; Lf binds oligosaccharides (OSs) in the outer bacterial membrane, thereby opening 'pores' for Lyz to hydrolyse glycosidic linkages in the interior of the peptidoglycan matrix. This synergistic process leads to inactivation of both Gram-negative, for example *E. coli* (Rainhard 1986) and Gram-positive bacteria, for example *Staphylococcus epidermidis* (Leitch and Willcox 1999) bacteria. Furthermore, a proteolytic digestion product of bovine and human Lf, lactoferricin, has bactericidal activity (Bellamy et al. 1992). Bovine and human Lf are reported to have antiviral activity and a role as a growth factor (Lönnerdal 2003). The specific biological function of equine Lf has not been studied, but is likely to be similar to that of bovine and human Lf.

Equine Lf contains 689 amino acid residues, which is similar to bovine Lf and two more than human Lf (Table 26.5). Compared to most other milk proteins, Lf has a high isoelectric point, that is, at pH 8.32, 8.67 or 8.47 for equine, bovine or human Lf (Table 26.5). As a result, the protein is positively charged at the pH of milk and may associate with negatively charged proteins *via* electrostatic interactions. GRAVY scores are comparable for equine, bovine and human Lf (Table 26.5). Equine and human Lf contain 17 and 16 intra-molecular disulphide bonds, respectively (Table 26.5). On the basis of structural similarities with human Lf, it has been assumed that bovine Lf contains 16 intra-molecular disulphide bonds (Table 26.5). The iron-binding capacity of equine, bovine and human Lfs are equivalent, although the pH-dependence of the iron-binding capacity of bovine Lf differs from that of equine and human Lf (Shimazaki et al. 1994).

All Lfs studied to date are glycosylated, but the location and number of potential glycosylation sites, as well as the number of sites actually glycosylated, vary. In bovine Lf, four out of five potential glycosylation sites, that is, Asn_{223}, Asn_{368}, Asn_{476} and Asn_{545}, are glycosylated (Moore et al. 1997), whereas in human Lf, two of three potential glycosylation sites, that is, Asn_{137} and Asn_{478}, are glycosylated (Haridas et al. 1995). Glycosylation of equine Lf has not been studied, but using the consensus sequence, Asn-Xaa-Ser/Thr (where Xaa is not Pro), for glycosylation, three potential glycosylation sites are likely in equine Lf, that is Asn_{137}, Asn_{281} and Asn_{476}.

Whey Protein Denaturation

Whey proteins are susceptible to heat-induced denaturation. The thermal stability of equine Lf and BSA is comparable to that of the bovine proteins but equine β-Lg and α-La are more heat stable than the bovine proteins (Bonomi et al. 1994). Equine β-Lg is more thermally stable than equine α-La, which is different from bovine milk, where α-La is the more thermally stable (Civardi et al. 2007). The high thermal stability of equine β-Lg may be related to its lack of a sulphydryl group. Thermal denaturation of bovine β-Lg is a two-stage process, unfolding of the polypeptide chain and exposure of the sulphydryl group, followed by self-association or interaction with other proteins *via* sulphydryl–disulphide interchange (Sawyer 2003). Owing to the lack of a sulphydryl group, equine β-Lg cannot undergo the second denaturation step and therefore its structure may refold on cooling. Denaturation of α-La is commonly a result of complex formation with β-Lg *via* sulphydryl–disulphide interchange and its higher thermal stability may therefore be a result of differences in its environment, rather than its molecular structure. Recent research suggests that equine α-La and β-Lg are also less susceptible to denaturation than their bovine counterparts under high pressure.

DIGESTIBILITY OF EQUID MILK

Milk is highly digestible and, because it is liquid, the gastrointestinal tract of mammals has mechanisms for delaying its passage; coagulation of milk in the stomach delays the degradation of proteins and improves their assimilation by the body. Caseins are precipitated by gastric acid and enzymes, forming a clot in the stomach that entraps fat. The hardness of this clot depends on the casein content of the milk; high casein-containing milks will produce firm clots. Generally, species that nurse their young at frequent intervals, for example equids and humans, tend to produce dilute milk in which <60% of total protein is casein and which form a soft clot, whereas those that nurse infrequently, for example cattle and sheep, produce milk that is high in fat and casein and has much longer gastric retention (Jenness 1986). Degradation of casein in the gastrointestinal tract is slow but extensive and while β-Lg is relatively resistant to gastric proteolysis, α-La is readily hydrolysed when the gastric pH is approximately 3.5 (Savalle et al. 1988).

The physico-chemical differences between human and bovine caseins result in the formation of different types of curd in the stomach (Hambræus 1982) and because the protein profile of equine milk is quite similar to that of human milk, equine milk may be more suitable in human nutrition than bovine milk. Turner (1945) compared the digestibility of equine, human and bovine milk based on the average percentage conversion of acid-insoluble protein to acid-soluble protein during digestion. Equine and human milk have a much lower buffering capacity than bovine milk and, while equine milk is very digestible, it is slightly less than human milk but significantly better than bovine milk. Turner (1945) concluded that both equine and human milk form soft curds in the stomach that pass through the digestive tract more quickly than bovine milk curd. Kalliala et al. (1951) also reported that the overall digestibility of equine and human milk (by *in vitro* experiments) appeared to be quite similar and both were easier to digest than bovine milk. Human milk forms fine, soft flocs in the stomach with an evacuation time of 2–2.5 hours, whereas bovine milk forms compact hard curds with a digestion time of 3–5 hours.

TOTAL AMINO ACIDS

Guo et al. (2007) investigated the total amino acid composition of asinine milk and expressed the results both in grams of individual amino acids per 100 grams of milk and per 100 grams of protein and compared values to those for equine, bovine and human milk (Table 26.6). Results expressed per 100 g of milk demonstrated differences related, most likely, to differences in the total protein content between the milks, but when expressed as g/100 g protein, the differences were not so apparent. It has been reported that glycine is exceptionally high in equine casein (Lauer and Baker 1977) and other studies have reported that the mean values of peptide-bound amino acids in equid milks are generally higher than those in camel and buffalo milks and may indicate that equid milks are more suitable for human consumption than other milks studied to date. Asinine milk has noticeably higher levels of serine, glutamate, arginine and valine and much less cystine and the percentage of seven of the eight essential amino acids (isoleucine, leucine, lysine, methionine, phenylalanine, threonine, tyrosine, valine) is also higher than those of equine and bovine milk (Guo et al. 2007). In equine, bovine and human milk cystine, glycine, serine, threonine and alanine decrease as lactation progresses while glutamate, methionine, isoleucine and lysine tend to increase (Davis et al. 1994).

NON-PROTEIN NITROGEN

The non-protein nitrogen (NPN) of milk consists primarily of urea, peptides, amino acids and ammonia. NPN constitutes 10–15% of the total nitrogen in mature equine milk which is intermediate between the values for human milk and ruminant milk, 25% and 5%, respectively (Hambræus 1984, Oftedal et al. 1983, Atkinson et al. 1989, Walstra et al. 2006b). In equine milk, NPN increases from <2% of total nitrogen at parturition to >10% after 2 weeks (Zicker and Lönnerdal 1994). The components of the NPN in human and bovine milk have been characterised (see Atkinson et al. 1989, Atkinson and Lönnerdal 1995, Rudloff and Kunz 1997, Carratù et al. 2003), but the NPN of equine milk has not been studied in detail.

Up to 50% of the NPN of human milk is urea and free amino acids, the exact function of which is, as yet, unknown. Asinine milk has a significantly higher level of NPN than equine milk and is close to that of human milk (Table 26.2). Equine and asinine milk have similar urea levels.

Free Amino Acids

The free amino acid content of equine, bovine and human milk are 1960, 578 and 3019 μmol.L^{-1}, respectively (Rassin et al. 1978, Agostini et al. 2000) (Table 26.7). Glutamine, glutamate,

Table 26.6. Amino Acid Composition of Asinine and Equine Milk Expressed as g amino acid per 100 g Protein, with Comparative Data for Bovine and Human Milk

Amino Acid	Asinine	Equine	Bovine	Human
Aspartic acid	8.9	10.4	7.8	8.3
Serine	6.2	6.2	4.8	5.1
Glutamic acid	22.8	20.1	23.2	17.8
Glycine	1.2	1.9	1.8	2.6
Histidine	2.3	2.4	3.0	2.3
Arginine	4.6	5.2	3.3	4.0
Threonine	3.6	4.3	4.5	4.6
Alanine	3.5	3.2	3.0	4.0
Proline	8.8	8.4	9.6	8.6
Cystine	0.4	0.6	0.6	1.7
Tyrosine	3.7	4.3	4.5	4.7
Valine	6.5	4.1	4.8	6.0
Methionine	1.8	1.5	1.8	1.8
Lysine	7.3	8.0	8.1	6.2
Isoleucine	5.5	3.8	4.2	5.8
Leucine	8.6	9.7	8.7	10.1
Phenylalanine	4.3	4.7	4.8	4.4
Tryptophan	–	1.2	1.5	1.8
Essential amino acids	38.2	36.7	37.5	40.7

Source: Modified from Guo et al. 2007.

Table 26.7. Free Amino Acids (μM.L^{-1}) of Equine, Bovine and Human Milk

Amino Acid	Equine[a]	Bovine[a]	Human[b]
Alanine	105	30.0	227.5
Arginine	14.0	10.0	35.4
Aspartic acid	40.0	15.0	183.2
Cystine	2.0	21.0	56.0
Glutamic acid	568.0	117.0	1184.1
Glutamine	485.0	12.0	284.8
Glycine	100.0	88.0	124.6
Histidine	46.0	9.0	7.7
Isoleucine	8.0	3.0	33.4
Leucine	16.0	3.0	55.6
Lysine	26.0	15.0	39.0
Methionine	~0	~0	8.8
Phenylalanine	5.0	3.0	23.6
Proline	1.61	–	64.3
Serine	175	23.0	273.7
Taurine	32.0	13.0	301.1
Threonine	137.0	16.0	97.6
Tyrosine	3.0	0.3	2.5
Valine	45.0	5.0	72.7
Total	~1960.0	578.0	3019.7

[a]Rassin et al. 1978.
[b]Agostini et al. 2000.

glycine, alanine and serine are the most abundant free amino acids in equine, bovine and human milk, and taurine also is exceptionally high in human milk (Rassin et al. 1978, Sarwar et al. 1998, Carratù et al. 2003). Taurine is an essential metabolite for the human infant and may be involved in the structure and function of retinal photoreceptors (Agostini et al. 2000). Compared to bovine milk, equine milk has an appreciable amount of taurine although it is ten times less than that of human milk (Table 26.7). In contrast to total amino acid composition, which is essentially similar in equine, bovine and human milks, free amino acids show a pattern characteristic of each species (Table 26.7), which may be important for early post-natal development in different animals. Free amino acids are more easily absorbed than protein-derived amino acids and glutamic acid and glutamine, which comprise >50% of the total free amino acids of human milk, are a source of α-ketoglutaric acid for the citric acid cycle and also act as neurotransmitters in the brain (Levy 1998, Agostini et al. 2000).

Bioactive Peptides

Both caseins and whey proteins are believed to contribute to human health through latent biological activity produced enzymatically during digestion, fermentation with specific starter cultures or enzymatic hydrolysis by microorganisms, resulting in the formation of bioactive peptides. These peptides are important for their physiological roles, their opioid-like features, as well as their immunostimulating and anti-hypertensive activities and their ability to enhance Ca^{2+} absorption and are released or activated during gastrointestinal digestion. Several peptides generated by the hydrolysis of milk proteins are known to regulate the overall immune function of the neonate (Baldi et al. 2005). A detailed discussion on bioactive peptides in milk is outside the scope of this chapter, for reviews, see Donnet-Hughes et al. (2000), Shah (2000), Malkoski et al. (2001), Fitzgerald and Meisel (2003), Silva and Malcata (2005), Fitzgerald and Murray (2006), Lopéz-Fandiño et al. (2006), Michaelidou and Steijns (2006), Thomä-Worringer et al. (2006) and Phelan et al. (2009).

Research on the bioactive peptides derived from equid milk is very limited. Peptides from the hydrolysis of equine β-casein may have a positive action on human health (Doreau and Martin-Rosset 2002). Chen et al. (2010) reported the presence of four novel angiotensin-converting enzyme-inhibitory peptides in koumiss which may enhance the beneficial effects of koumiss on cardiovascular health. Peptides with trophic or protective activity have been identified in asinine milk (Salimei 2011).

Hormones and Growth Factors

Leptin is a protein hormone of approximately 16kDa that has been discovered recently in human milk and plays a key role in the regulation of energy intake and energy expenditure, as well as functioning in mammary cell proliferation, differentiation and apoptosis. Human-like leptin has been isolated from asinine milk at a level of 3.2–5.4 ng.mL^{-1} which is similar to levels reported for other mammals and showed little variation throughout lactation (Salimei et al. 2002).

Levels of the bioactive peptides, ghrelin and insulin growth factor I, which play a direct role in metabolism, body composition and food intake, have also been reported for asinine milk at 4.5 pg.mL^{-1} and 11.5 ng.mL^{-1}, respectively, similar to levels in human milk (Salimei 2011).

Amyloid A

Amyloid A3 (AA3) is a protein produced in the mammary gland and is encoded by a separate gene from that for serum amyloid A (serum AA) (Duggan et al. 2008). AA3 is believed to prevent attachment of pathogenic bacteria to the intestinal cell wall (Mack et al. 2003) and may prevent necrotising enterocolitis in human infants (Larson et al. 2003). McDonald et al. (2001) demonstrated the presence of AA3 in the colostrum of cows, ewes, sows and horses. Bovine colostrum has a high concentration of AA3 but by approximately 3 days *post-partum* the levels decline. In bovine milk, the presence of serum AA in milk is an indicator of mastitic infection (Kaneko et al. 2004, Winter et al. 2006). In equine colostrum, the concentration of AA3 is considerably lower than in milk and consequently may play a crucial role in intestinal cell protection in the foal especially after gut closure (Duggan et al. 2008).

INDIGENOUS ENZYMES

Milk contains many indigenous enzymes that originate from the mammal's blood plasma, leucocytes (somatic cells), or cytoplasm of the secretory cells and the milk fat globule membrane (MFGM)(Fox and Kelly 2006). The indigenous enzymes in bovine and human milks have been studied extensively but the enzymes in the milk of other species have been studied only sporadically. Equine milk probably contains all the enzymes that have been identified in bovine milk but relatively few studies have been reported.

Lysozyme

Lyz (EC 3.1.2.17) occurs at high levels in equine, asinine and human milk (Table 26.2). Human, equine and asinine milk contain 3000, 6000 and >6000 times more Lyz, respectively, than bovine milk (Salimei et al. 2004, Guo et al. 2007) with levels as high as 0.4 g.100g^{-1} for Martina Franca donkeys (Coppola et al. 2002) although 0.1 g/100 g is reported and more commonly found in asinine milk (Vincenzetti et al. 2008). The concentration of Lyz in human milk increases strongly after the second month of lactation, suggesting that Lyz and Lf play major roles in fighting infection in breast-fed infants during late lactation, and protect the mammary gland (Montagne et al. 1998).

Equine milk Lyz is more stable to denaturation than human Lyz during pasteurisation at 62°C for 30 minutes but at 71°C for 2 minutes or 82°C for 15 seconds, the inactivation of both were similar (Jauregui-Adell 1975). It has been suggested, but research is scarce, that while the composition of breast milk varies widely between well-nourished and poorly nourished mothers, the amount of Lyz is conserved.

Lyz found in egg white, tears and saliva do not generally bind calcium but equine and canine milk Lyz do and this is believed to enhance the stability and activity of the enzyme (Nitta et al. 1987). The binding of a Ca^{2+} by Lyz is considered to be an evolutionary linkage between non-Ca^{2+}-binding Lyzs and α-La (Tada et al. 2002, Chowdhury et al. 2004). The conformation of the calcium-binding loop of equine Lyz is similar to that of α-La (Tsuge et al. 1992, Tada et al. 2002) and both equine Lyz and α-La form stable, partially folded, 'molten globules' under various denaturing conditions (Koshiba et al. 2001,) with that of equine Lyz being considerably more stable than α-La (Lyster 1992, Morozova-Roche 2007). The molten state of canine Lyz is significantly more stable than that of equine Lyz (Koshiba et al. 2000, Spencer et al. 1999). Equine milk Lyz is very resistant to acid (Jauregui-Adell 1975) and proteolysis (Kuroki et al. 1989), and may reach the gut relatively intact.

Asinine Lyz contains 129 amino acids, is a C-type Lyz, binds calcium strongly and has 51% homology to human Lyz (Godovac-Zimmermann et al. 1988b). Two genetic variants of Lyz, A and B, have been reported in asinine milk (Herrouin et al. 2000) but only one is found in equine milk. Asinine Lyz is remarkably heat stable and requires 121°C for 10 minutes for inactivation. The Lyz content of equid milks is one of the main attractions for use of these milks in cosmetology as it is reputed to have a smoothing effect on the skin and may reduce scalp inflammation when incorporated into shampoo. Equid milk has very good antibacterial activity, presumably due to its high level of Lyz.

Other Indigenous Enzymes in Equine Milk

Lactoperoxidase, catalase, amylase, proteinase (plasmin), lipase, lactate dehydrogenase and malate dehydrogenase have been reported in equine milk. Bovine milk is a rich source of xanthine oxidoreductase (XOR) but the milk of other species for which data are available have much lower XOR activity, because in non-bovine species, most (up to 98% in human milk) of the enzyme molecules lack Mo and are inactive. XOR has not been reported in equine milk, which is unusual considering the role of XOR in the excretion of fat globules from the secretory cells and also considering that equine milk contains quite a high level of molybdenum (Mo), which presumably is present exclusively in XOR. Chilliard and Doreau (1985) characterised the lipoprotein lipase activity of equine milk and reported that the milk has high lipolytic activity, comparable to that in bovine milk and higher that in caprine milk. There are no reports on hydrolytic rancidity in equine milk, which is potentially a serious problem in equine milk products and warrants investigation.

Plasmin, a serine proteinases, is one of a number of proteolytic enzymes in milk. Visser et al. (1982) and Egito et al. (2002) reported γ-caseins in equine milk and, it is therefore assumed, that equine milk contains plasmin. Humbert et al. (2005) reported that equine milk contains five times more plasmin activity than bovine milk and 90% of total potential plasmin activity was plasmin, with 10% as plasminogen; the plasmin:plasminogen ratio in bovine and human milk is 18:82 and 28:72, respectively.

Alkaline phosphatase (ALP) is regarded as the most important indigenous enzyme of bovine milk because ALP activity is used as the index of the efficiency of high-temperature short-time pasteurisation. About 40% of ALP activity in bovine milk is associated with the MFGM. Equine milk has 35–350 times less ALP activity than bovine milk and there are no reports on ALP in the equine MFGM. Because of the low level of ALP in equine milk it has been suggested that it is not suitable as an indicator of pasteurisation efficiency of equine milk (Marchand et al. 2009) although one would expect that once the exact initial concentration of ALP is known, the use of a larger sample size or a longer incubation period would overcome the low level of enzyme.

CARBOHYDRATES

Lactose and Glucose

The chemistry, properties and applications of lactose are described in Chapter 24 and have been reviewed extensively elsewhere for example Fox (1985, 1997) and McSweeney and Fox (2009) and will not be considered here. The concentration of lactose in asinine milk is high (51–72.5 $g.kg^{-1}$), probably marginally higher than equine milk (approximately 64 $g.kg^{-1}$) (Table 26.2), which is similar to the level in human milk and significantly higher than that in bovine milk. As an energy source, lactose is far less metabolically complicated than lipids but the latter provides significantly more energy per unit mass. As well as being a major energy source for the neonate, lactose affects bone mineralisation during the first few months *post-partum* as it stimulates the intestinal absorption of calcium (Schaafsma 2003). Equine milk contains a significant concentration of glucose, approximately 50 mg/L in colostrum which increases to approximately 150 mg/L 10 days *post-partum* and then decreases gradually to approximately 120 mg/L (Enbergs et al. 1999). Although the lactose content of equid milks is high, the physico-chemical properties of lactose that cause problems in the processing of bovine milk are of no consequence for equid milks, which are consumed either fresh or fermented. In Mongolia, where approximately 88% of the population is lactose intolerant (Yongfa et al. 1984), lactose intolerance is not a problem with fermented equine milk, koumiss, as approximately 30% of lactose is converted to lactic acid, ethanol and carbon dioxide during fermentation.

Oligosaccharides

The milk of all species examined contains OSs but the concentration varies markedly (see Urashima et al. 2009). The OSs in milk contain 3 to 10 monosaccharides and may be linear or branched; they contain lactose at the reducing end and also contain fucose, galactosamine and *N*-acetylneuraminic acid. The highest levels are in the milk of monotremes, marsupials, marine mammals, humans, elephants and bears. OSs are the third most abundant constituent of human milk that has an exceptionally high content (approximately 20 $g.L^{-1}$ in colostrum, which decreases to 5–10 $g.L^{-1}$ in milk) and structural diversity of OSs (>200 molecular species), which have a range of functions,

Table 26.8. Principal Oligosaccharides of Equine Colostrum

Oligosaccharide	(mg/L)
Acidic	
Neu5Ac(α2–3) Gal(β1—4)Glc	N/a
Gal(β1–4)GlcNAcα1-diphosphate (*N*-acetyllactosamine-α1-phospahte)	N/a
Neutral	
Gal(β1–3)Gal(β1–4)Glc (β3'-galactosyllactose)	7.8
Gal(β1–6)Gal(β1–4)Glc (β6'-galactosyllactose)	4.8
Gal(β1–4)GlcNAc(β1–3)Gal(β1–4)Glc (lacto-*N*-neotetraose)	N/a
Gal(β1–4)GlcNAc(β1–6)Gal(β1–4)Glc (iso-lacto-*N*-neotetraose)	0.5
Gal(β1–4)GlcNAc(β1–6)[Gal(β1–3)]Gal(β1–4)Glc (lacto-*N*-novopentanose 1)	1.1
Gal(β1–4)GlcNAc(β1–6)[Gal(β1–4)GlcNAc(β1–3)]Gal(β1–4)Glc (lacto-*N*-neohexaose)	1.1
Gal(β1–4)GlcNAc-1-phosphate (*N*-acetyllactosamine-1–0-phosphate)	N/a
Neu5Ac(α2–3)Gal(β1–4)Glc (3'-*N*-acetylneuraminyllactose)	N/a

Source: From Urashima et al. 1989, 2001, Nakamura et al. 2001.
Gal, D-galactose; Glc, D-glucose; GlcNAc, *N*-acetylglucosamine; Neu5A, *N*-acetylneuraminic acid; N/a, not available.

including as important components of our immune system and as prebiotics to promote a healthy gut microflora (Donovan 2009). Bovine, ovine, caprine and equine milk contain relatively low levels of OSs, which have been characterised (see Urashima et al. 2001). The OSs identified in equine colostrum are summarised in Table 26.8. The OSs in mature equine milk have not been reported but it can be assumed that the level is considerably lower than in colostrum which has approximately 18.6 g/L (Nakamura et al. 2001). The neutral OSs, lacto-*N*-neotetraose and lacto-*N*-neohexaose, in equine colostrum are also abundant in human milk, while iso-lacto-*N*-neotetraose and lacto-*N*-novopentanose 1 are not, but have been identified in bovine colostrum; the latter has been identified also in the milk of the Tammar wallaby and brown capuchin monkey (Urashima et al. 2009).

LIPIDS

Milk fat is important for the provision of energy to the newborn as well as being the vehicle for fat-soluble vitamins and essential fatty acids. From a practical point of view, milk lipids are important as they confer distinctive nutritional, textural and organoleptic properties on dairy products. Dietary composition is considered one of the major determinants of the fatty acid composition of equid milk and non-dietary factors such as stage of lactation, age and parity of the mare play minor roles.

Triglycerides (TGs) represent approximately 80–85% of the lipids in equine and asinine milk, while approximately 9.5% are free fatty acids (FFAs) and approximately 5–10% are phospholipids (Jahreis et al. 1999). In contrast, approximately 97–98% of the lipids in bovine and human milk are TGs, with low levels of phospholipids and free fatty acids, 1.3 and 1.5 $g.100g^{-1}$, respectively. The high level of free fatty acids in equid milk implies that rancidity is a problem with these milks and is dealt with in Section 'Stability of Equine Milk Fat'.

The relatively high content of phospholipids in equid milk is thought to contribute to its buffering properties. TGs, the primary transport and storage form of lipids, are synthesised in the mammary gland from fatty acids that originate from three sources: *de novo* synthesis (C8:0, C10:0 and C12:0), direct uptake from the blood (>14 carbons) and modification of fatty acids in the mammary gland by desaturation and/or elongation. Circulating fatty acids in the blood may originate from dietary fat or from lipids mobilised from body fat stores. The principal phospholipids of equid milk are phosphatidylcholine (19%), phosphatidylethanolamine (31%), phosphatidylserine (16%) and sphingomylin (34%); the corresponding values for bovine milk are 35%, 32%, 3% and 25% and for human milk are 28, 20, 8 and 39%, respectively. The high level of FFAs in equid milk implies considerable lipolysis but this has not been suggested; if lipolysis was responsible for the high level of FFAs, they should be accompanied by high levels of mono- and di-glycerides but these are reported to be quite low at approximately 1.8% of total lipids.

Table 26.9 shows the monounsaturated fatty acids (MUFA) and polyunsaturated fatty acids (PUFA) in the milk fat of some ruminant and non-ruminant species. The milk fat of non-ruminants contains substantially higher levels of PUFAs than ruminant milks due to the lack of biohydrogenation of fatty acids in the former, and for the two-equid species shown, the horse has considerably more PUFAs in its milk than the donkey. Saturated fatty acids are the dominant class in asinine milk and levels are significantly higher than those in equine or human milk (Table 26.9).

Fatty Acids in Equid Milks

The fatty acid profile of equid milk (Table 26.10) differs from that of bovine and human milk fat in a number of respects. Like human, and unlike bovine milk, equid milk is characterised by low proportions of saturated fatty acids with low or higher numbers of carbons, that is, $C_{4:0}$, $C_{6:0}$, $C_{16:0}$ and $C_{18:0}$ (Pikul and Wójtowski 2008). Butyric acid ($C_{4:0}$) is present at high levels in bovine and other ruminant milk fats, produced from

Table 26.9. Monounsaturated and Polyunsaturated Fatty Acids (Percent of Total Fatty Acids ± Standard Deviations) in the Milk Fat of Some Ruminants and Non-Ruminants

	MUFAs	PUFAs	CLA
Non-ruminants			
Equine	20.70	36.80	0.09
Asinine	15.30	16.00	–
Porcine	51.80	12.40	0.23
Human	33.20	12.50	0.39
Ruminants			
Caprine	26.90	2.58	0.65
Bovine	23.20	2.42	1.01
Ovine	23.00	3.85	1.08

Data from Jahreis et al. 1999, Salimei et al. 2004.
MUFA, monounsaturated fatty acids; PUFA, polyunsaturated fatty acids; CLA, conjugated linoleic acid.

3-hydroxybutanoic acid, which is synthesised by bacteria in the rumen (Pikul and Wójtowski 2008). Caprylic acid, $C_{8:0}$, is very high in equid milk compared to the level in human and bovine milk (Table 26.10). Levels of middle chain-length FAs, especially $C_{10:0}$ and $C_{12:0}$, are high in equid milk (20–35% of all FAs contain <16 C) and in all non-ruminant herbivores, suggesting that they arise from the use of glucose as the principal precursor for fatty acid synthesis (Palmquist 2006). Mammalian *de novo* fatty acid synthesis requires a carbon source (acetyl-CoA) and reducing equivalents in the form of NADPH + H^+. In ruminants, acetate and β-hydroxybutyrate are the primary sources of carbon while glucose and acetate are the primary sources of reducing equivalents; in non-ruminants, for example equid species, glucose is the primary source of both carbon and reducing equivalents and also supplies some of the glycerol for milk TGs (Dils 1986). When horses were infused with either glucose or acetate and palmitate, $C_{12:0}$ and $C_{14:0}$ were formed exclusively from acetate, as in ruminants, and $C_{16:0}$ was formed, partly from acetate and partly from palmitate; unlike ruminants, 44% of $C_{18:0}$ and 7% of $C_{18:1}$ are formed from acetate in the horse (Palmquist 2006). If this is so, acetate and 3-hydroxybutyrate are presumably produced by bacterial fermentation in the lower intestine of the horse and why 3-hydroxybutyrate is not converted to butanoic acid, as in ruminants, is unclear. Fatty acids from $C_{6:0}$ to $C_{16:0}$ are released from the fatty acid synthesis complex by acyl-specific thioesterases; presumably, the middle chain-length-specific thioesterases are particularly active in equids; investigation of this possibility is warranted.

Equine milk-fat contains a relatively high level of $C_{16:1}$ (2–10%, w/w) and $C_{18:1}$, reflecting high Δ-9 desaturase activity. Equid milk fats contain a very high level of *n*-3-octadecatrienoic acid (linolenic acid), which reflects the high level of PUFAs in the diet and the lack of biohydrogenation, as occurs with ruminants. In the rumen, extensive hydrogenation of double bonds occurs and most fatty acids taken up from the intestinal tract are saturated. The large intestine of equids shows significant differences in the relative rate of transport of volatile fatty acids compared to ruminants and *de novo* synthesis of $C_{18:0}$ fatty acids occurs with a further high proportion of $C_{6:0}$ to $C_{14:0}$ carbon fatty acids and some $C_{16:0}$ arising from products of their large bowel fermentation.

Equine and asinine milks have similar fatty acid profiles although the former has a higher content of monosaturated fatty acids (Tables 26.9 and 26.10). Both equine and asinine milk have characteristic low levels of stearic acid, and oleic acid is exceptionally low in asinine milk. Asinine and zebra milk fat contain a high level of PUFAs, although considerably lower than in equine milk. The well-balanced ratio of *n*-6: *n*-3 of 1.17:1 in asinine milk compared to 3.14:1 in equine milk makes it an interesting product for human nutrition. *n*-6 and *n*-3 fatty acids are essential in human metabolism as components of membrane phospholipids, precursors of eicosanoids, ligands for membrane receptors and transcription factors that regulate gene expression. The importance of *n*-6 $C_{18:2}$, linoleic acid (LA), has been known for many years but the significance of *n*-3 $C_{18:3}$, α-linolenic (ALA) was not recognised until the late 1980s and has since been identified as a key component in the diet for the prevention of atopic dermatitis (Horrobin 2000). LA and ALA are not inter-convertible but are the parent acids of the *n*-6 and *n*-3 series of long chain (LC) polyunsaturated fatty acids, respectively (e.g., *n*-6 $C_{20:4}$, arachidonic acid (AA); *n*-3 $C_{20:5}$, eicosapentaenoic acid (EPA) and *n*-3 $C_{22:6}$, docosahexaenoic acid (DHA)) which are components of cellular membranes and precursors of other essential metabolites such as prostaglandins and prostacyclins (Cuthbertson 1999, Innis 2007). DHA and AA are now recognised as being crucial for normal neurological development (Carlson 2001). Humans have evolved on a diet with a ratio of *n*-6 to *n*-3 fatty acids of approximately 1:1 but Western diets nowadays have a ratio of 15:1 to

Table 26.10. Typical Fatty Acid (Percent of Total Fatty Acids) Composition in the Milk of Equid Species; Bovine and Human Milk are Included for Comparison

Fatty Acid	Common Name	Equine	Asinine	Zebra	Bovine	Human
Saturates						
$C_{4:0}$	Butyric	0.09	0.60		3.90	0.19
$C_{6:0}$	Caproic	0.24	1.22		2.50	0.15
$C_{8:0}$	Caprylic	3.15	12.80	8.20	1.50	0.46
$C_{10:0}$	Capric	6.48	18.65	15.30	3.20	1.03
$C_{12:0}$	Lauric	6.65	10.67	9.20	3.60	4.40
$C_{13i:0}$			0.22			
$C_{13:0}$		0.17	3.92		0.19	0.06
$C_{14i:0}$			0.12			0.04
$C_{14:0}$	Myristic	7.04	5.77	6.50	11.10	6.27
$C_{15a:0}$						0.21
$C_{15\ i:0}$		0.16	0.07			
$C_{15:0}$	Pentadecanoic	0.39	0.32	0.50	1.20	0.43
$C_{16i:0}$			0.12			0.17
$C_{16:0}$	Palmitic	20.43	11.47	13.30	27.90	22.00
$C_{17i:0}$		0.31	0.20			0.23
$C_{17:0}$	Margaric	0.38	0.22		0.60	0.58
$C_{18i:0}$						0.11
$C_{18:0}$	Stearic	1.18	1.12	2.00	12.20	8.06
$C_{20:0}$			0.12		0.35	0.44
$C_{21:0}$					0.04	0.13
$C_{22:0}$			0.05		0.20	0.12
$C_{24:0}$					0.14	0.25
Total		*46.67*	*67.66*	*55.00*	*68.62*	*45.33*
Monounsaturates						
$C_{10:1}$		1.46	2.20		0.15	
$C_{12:1}$		0.20	0.25		0.06	
$C_{14:1c\ n\text{-}5}$	Myristoleic	0.52	0.22		0.80	0.41
$C_{14:1t\ n\text{-}5}$						0.07
$C_{15:1}$	Pentadecanoic	0.22			0.30	0.11
$C_{16:1c\ n\text{-}7}$	Palmitoleic	5.68	2.37	4.20	1.50	3.29
$C_{16:1c\ n\text{-}9}$		0.56				
$C_{16:1t\ n\text{-}7}$						0.36
$C_{17:1}$	Heptadecanoic	0.62	0.27		0.36	0.37
$C_{18:1c\ n\text{-}9}$	Oleic	20.26	9.65	20.40	17.20	31.30
$C_{18:1c\ n\text{-}11}$		1.31				
$C_{18:1t\ n\text{-}9}$						2.67
$C_{18:1t\ n\text{-}11}$	Vaccenic				3.90	
$C_{20:1\ n\text{-}9}$		0.40			0.32	0.67
$C_{20:1\ n\text{-}11}$			0.35			
$C_{22:1\ n\text{-}9}$					0.06	0.08
$C_{24:1\ n\text{-}9}$						0.12
Total		*31.23*	*15.31*	*24.65*	*24.65*	*39.45*
Polyunsaturates						
n-6 Series						
$C_{18:2cc}$	Linoleic	10.10	8.15	16.04	1.40	10.85
C18:2 conj.	Conjugated linoleic			0.07	1.10	

(*Continued*)

Table 26.10. *(Continued)*

Fatty Acid	Common Name	Equine	Asinine	Zebra	Bovine	Human
$C_{18:2tt}$						0.46
$C_{18:2ct}$						0.69
$C_{18:3}$	γ-linolenic		0.15	0.61	1.00	0.25
$C_{20:2}$			0.35	0.37	0.07	0.27
$C_{20:3}$	dihomo-γ-linolenic			0.10	0.10	0.32
$C_{20:4}$	Arachidonic			0.11	0.14	0.46
$C_{22:2}$					0.04	0.11
$C_{22:4}$					0.03	0.09
$C_{22:5}$					0.04	0.09
n-3 Series						
$C_{18:3}$	α-linolenic	8.00	6.32	5.31	1.80	1.03
$C_{18:4}$			0.22			
$C_{20:3}$			0.12			
$C_{20:4}$			0.07			0.09
$C_{20:5}$	Eicosapentaenoic		0.27	0.02	0.09	0.12
$C_{22:5}$			0.07	0.10		0.19
$C_{22:6}$	Docosahexaenoic		0.30	0.04	0.01	0.25
Total PUFA		*18.00*	*16.02*	*22.77*	*5.82*	*15.27*
Ratio n-6:n-3		*1.26*	*1.17*	*3.14*	*2.06*	*8.09*
Ratio $C_{18:2}$ to $C_{18:3}$		*1.26*	*1.28*	*2.72*	*1.55*	*9.37*

Source: Modified from Uniacke and Fox 2011.
c, cis; *t*,trans; *i*, iso; *a*, anteiso, PUFA, polyunsaturated fatty acids.

16.7:1. As a species, humans are generally deficient in *n*-3 fatty acids and have excessive levels of *n*-6 which is associated with the pathogenesis of cardiovascular, cancerous, inflammatory and autoimmune diseases (Simopoulos 2002).

Equine milk-fat contains a very low level of conjugated linoleic acid (CLA; rumenic acid, Tables 26.9 and 26.10) which is virtually absent from asinine milk but is high in ruminant milk-fats, being produced in the rumen by abortive biohydrogenation of *n*-6 octadecadienoic acid (LA) (see Whigham et al. 2000, Bauman and Lock 2006, Collomb et al. 2006). CLA has several desirable effects in the diet; some of the positive health effects attributed to it include: suppression of carcinogenesis, anti-obesity agent, modulator of the immune system and control of artherogenesis and diabetes. Small amounts of eicosapentaenoic (EPA) and docosahexaenoic acid (DHA) are present in asinine milk whereas equine milk has only trace amounts. EPA and DHA are especially important in infant nutrition but their absence from equine milk is not considered to be a problem as an infant's liver can desaturate linoleic and ALA to form EPA and DHA.

Structure of Triglycerides

The distribution of fatty acids in animal TGs is non-random, apparently so that the TGs will be liquid at body temperature. Inter-species comparison of the positional distribution of fatty acids has been determined, using fatty acid and stereospecific analysis, for 11 species: echidna, koala, Tammar wallaby, guinea pig, dog, cat, Weddell seal, horse, pig, cow and human (Parodi 1982). Generally, the positional distribution of fatty acids is similar, except for the echidna, with short chain fatty acids preferentially esterified at *sn*-3, saturated fatty acids at *sn*-1 and unsaturated fatty acids generally at *sn*-2. In equine milk fat, $C_{10:0}$ occurs at the *sn*-3 position, whereas in bovine milk fat, more $C_{10:0}$ is found at the *sn*-2 than at the *sn*-3 position. In human and equine milk, $C_{16:0}$ is located predominantly at the *sn*-2 position which is regarded as favourable for assimilation by infants and children, whereas in bovine milk, $C_{16:0}$ is equally distributed between the *sn*-1 and *sn*-2 positions. In the milk fat of the Weddell seal and horse, $C_{18:1}$ is esterified preferentially at *sn*-1 but for all other species studied it occurs mainly at *sn*-3 (Parodi 1982). No information is available on the stereospecific distribution of fatty acids in the TGs of asinine milk.

Equid Milk Fat Globules and MFGM

The fat in milk is emulsified as globules which are surrounded and stabilised by a very complex emulsifying layer, consisting of phospholipids and proteins, called the MFGM. Many of the indigenous enzymes in milk are concentrated in the MFGM. The glycoproteins in the MFGM of human, rhesus monkey, chimpanzee, dog, sheep, goat, cow, grey seal, camel, horse and alpaca have been studied; large intra- and inter-species differences have been found (see Keenan and Mather 2006). Very highly glycosylated proteins occur in the MFGM of primates, horse, donkey, camel and dog. Long (0.5–1 μm) filamentous structures, comprising of mucins (highly glycosylated proteins), extend from the surface of the fat globules in equine and human milk (Welsch et al. 1988). These filaments dissociate from the surface into the milk serum on cooling and are lost on heating. For unknown reasons, the filaments on bovine milk fat globules are lost much more easily than those in equine or human milk. The filaments facilitate the adherence of fat globules to the intestinal epithelium and probably improve the digestion of fat (Welsch et al. 1988). The mucins prevent bacterial adhesion and may protect mammary tissue against tumors (Patton 1999). The milk fat globules (MFGs) of asinine milk can be up to approximately 10 μm in diameter, whereas those of equine milk are generally smaller at 2–3 μm on average; the MFGs of bovine and human milk are 3–3.5 μm and approximately 4 μm, respectively. Little is known about the proteins of the MFGM of equids but they are known to play a major role in neonatal defense mechanisms in humans (see Mather 2000).

Butyrophilin, acidophilin and XOR have been identified in the equine MFGM and appear to be similar to the corresponding proteins of the human MFGM, as does lactadherin which shares 74% identity with that of the human lactadherin (Barello et al. 2008). Both XOR and acidophilin are involved in fat globule secretion with butyrophylin while lactadherin is thought to have a protective function against rotovirus in the intestinal tract (Barello et al. 2008). Like ovine and buffalo milk, equine milk does not cream due to the lack of cryoglobulin.

Rheology Equid Milk Fat

The temperature-dependent melting characteristics of bovine milk-fat have been studied thoroughly but since equid milk is not used for the production of butter, the spreadability, rheology and melting characteristics of these fats have not been studied in detail (see Chandan et al. 1971). Considering the rather unusual fatty acid profile of equid milk fats, they should have interesting melting and rheological properties.

Stability of Equine Milk Fat

Lipids generally are susceptible to two forms of chemical spoilage, lipid oxidation (oxidative rancidity) and lipolysis (hydrolytic rancidity) which are of great commercial significance to the dairy industry and have been studied in detail (see Fox and McSweeney 2006). No studies on the chemical spoilage of equid milk-fat have been reported. Considering the high content of PUFAs in these fats, they are probably quite susceptible to oxidation. Since equine milk contains a lipase, hydrolytic rancidity would be expected under certain conditions.

VITAMINS

The overall vitamin content of any milk depends on maternal vitamin status but water-soluble vitamins are more responsive to the maternal diet than fat-soluble vitamins. Vitamin levels in the milk of some species are shown in Table 26.11. The level of vitamin E is low in asinine milk (~0.05 mg.L^{-1}) and is reduced further if the milk is heated. Concentrations of fat-soluble vitamins are generally similar in equine and bovine milks (Table 26.12). The levels of vitamins A, D_3, K and C are significantly higher in equine colostrum than in equine milk, whereas the concentration

Table 26.11. Vitamin Levels (mg.L^{-1}) in the Milk of Some Species

Vitamin	Buffalo	Goat	Sheep	Donkey	Cow	Horse	Human
Water-soluble							
Thiamine, B_1	0.5	0.49	0.48	0.41	0.37	0.3	0.15
Riboflavin, B_2	1.0	1.5	2.3	0.64	1.8	0.3	0.38
Niacin, B_3	0.8	3.2	4.5	0.74	0.9	1.4	1.7
Pantothenic acid, B_5	3.7	3.1	3.5	–	3.5	3	2.7
Pyridoxine, B_6	0.25	0.27	0.27	–	0.64	0.3	0.14
Biotin, B_7	0.11	0.039	0.09	–	0.035	–	0.006
Folic acid, B_9					0.18		0.16
Cobalamin, B_{12}	3.0	0.7	0.007[a]	1.1	0.004	0.003	0.5
Ascorbic acid, C	–	9.0	4.25[b]	–	21	17.2[c]	43
Fat-soluble							
Vitamin A and β-carotene	–	0.5	0.5	–	0.32–0.50	0.12	2.0
Cholecalciferol, D_3	–	–	–	–	0.003	0.003	0.001
α-Tocopherol, E	–	–	–	0.05	0.98–1.28	1.128	6.6
Phylloquinone, K	–	–	–	–	0.011	0.020	0.002

Source: Modified from Walstra and Jenness 1984, Souci et al. 2000.
[a]Ramos et al. 1994.
[b]Recio et al. 2009.
[c]Csapó et al. 1995.

Table 26.12. Vitamins ($mg.L^{-1}$) in Equine Milk

Vitamin	Equine Colostrum	Equine Milk	Bovine Milk
A	0.88	0.34	0.352
D_3	0.0054	0.0032	0.0029
E	1.342	1.128	1.135
K_3	0.043	0.029	0.032
C	23.80	17.2	15.32

Source: Modified from Csapó et al. 1995.

of vitamin E remains unchanged throughout lactation (Table 26.12)(Csapó et al. 1995).

MINERALS

MACRO-ELEMENTS

The levels of ash in equine and asinine milk are similar (Table 26.2) and the levels of inorganic elements are close to those in human milk except for higher concentrations of Ca and P (Table 26.13) (Holt and Jenness 1984). The principal salts in equine milk are phosphates, chlorides, carbonates and citrates of potassium, sodium, calcium and magnesium. However, there are considerable quantitative inter-species differences in milk salts, (Table 26.13). The concentrations of all macro-elements, except potassium, are higher in equine and asinine milk than in human milk but all are considerably lower than in bovine, caprine, ovine or porcine milk. The low level of salts in equid milk reduces renal load, making it suitable in infant nutrition. Pieszka and Kulisa (2005) reported on the low tolerance of equine species to imbalances in mineral concentrations in milk post-partum; slight increases in some minerals can cause severe deformation of teeth and bones in horses and affect metabolism and protein synthesis.

The concentration of macro-elements in equine milk is comparable to that in zebra milk (*Equus zebra*), whereas in early lactation, the concentrations of calcium and phosphate are considerably higher in domestic horse milk (*Equus caballus*)(Table 26.14) than in zebra milk (Schryver et al. 1986). The concentrations of macro-elements in equine milk are strongly influenced

Table 26.13. Total Concentrations of Inorganic Elements ($mmol.L^{-1}$) and Citrate in the Milk of Eight Different Species

Species	Calcium	Magnesium	Sodium	Potassium	Phosphorus (Inorganic)	Citrate	Chloride
Horse	16.5	1.6	5.7	11.9	6.7	3.1	6.6
Cow	29.4	5.1	24.2	34.7	20.9	9.2	30.2
Man	7.8	1.1	5.0	16.5	2.5	2.8	6.2
Goat	23.1	5.0	20.5	46.6	15.6	5.4	34.2
Sheep	56.8	9.0	20.5	31.7	39.7	4.9	17.0
Pig	104.1	9.6	14.4	31.4	51.2	8.9	28.7
Rat	80.4	8.8	38.3	43.6	93.3	0.06	36.1
Rabbit	214.4	19.5	83.7	89.5	54.2	17.4	80.0

Source: Modified from Holt and Jenness 1984.

Table 26.14. Concentration of Inorganic Elements ($\mu g.g^{-1}$ Whole Milk) in Early and Late Lactation Milk of Various Equid Species

Species	Total Solids	Ash	Ca	P	Mg	Na	K	Cu	Zn	Fe
Early Lactation										
Przewalski horse	11.6	0.58	1380	790	104	220	590	0.42	4.1	1.3
Hartmann's zebra	11.3	0.50	1100	800	120	290	590	1.13	2.4	1.1
Domestic horse	11.6	0.58	1327	884	102	198	655	0.64	2.7	0.37
Domestic pony	9.6	0.48	1036	600	70	189	483	0.26	1.8	–
Late Lactation										
Przewalski horse	10.3	0.33	804	419	62	137	344	0.23	1.9	1.1
Hartmann's zebra	10.0	0.32	840	550	80	200	422	0.26	1.9	3.6
Grant's zebra	10.7	0.35	690	490	93	277	187	1.0	2.9	2.2
Domestic horse	10.2	0.36	811	566	53	140	410	0.25	1.9	0.27
Domestic pony	9.4	0.36	857	418	77	127	250	0.37	1.7	–

Source: Modified from Schryver et al. 1986.

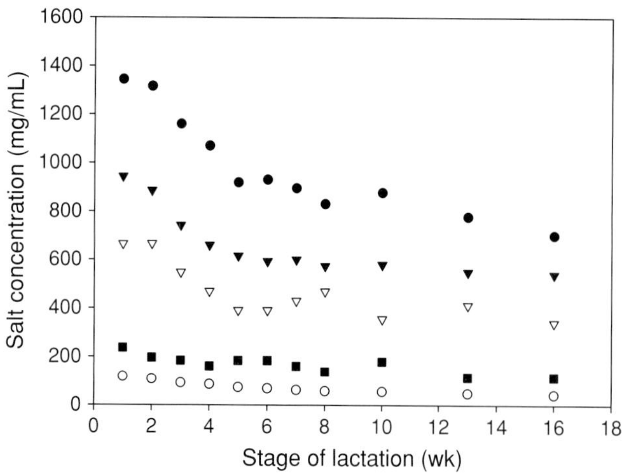

Figure 26.5. Influence of the stage of lactation on the concentration of calcium (•), magnesium (○), phosphate (▼), potassium (▽) or sodium (■) in equine milk. (Data from Schryver et al. 1986.)

Table 26.15. Concentrations of Trace Elements ($\mu g.L^{-1}$) in Equine, Bovine and Human Milk

	Species		
Element	Equine	Bovine	Human
Aluminum	123	98	125
Boron	97	333	273
Barium	76	188	149
Copper	155	52	314
Iron	224	194	260
Lithium	15	24	7
Molybdenum	16	22	17
Manganese	14	21	7
Silicon	161	434	472
Strontium	442	417	60
Titanium	145	111	25
Zinc	1835	3960	2150

Source: Modified from Anderson 1992.

by the stage of lactation (Fig. 26.5), which shows a progressive decrease in the concentrations of calcium, magnesium, phosphorus, sodium and potassium from the end of the first week of lactation. The concentration of calcium in equine milk increases during the first week of lactation, before decreasing steadily thereafter (Ullrey et al. 1966). As in the milk from all mammals which have been studied in sufficient detail, the concentrations of calcium and phosphate in equine milk far exceeds the solubility of calcium phosphate in milk. As a result, part of the calcium and phosphate exist in a non-ultrafiltrable, micellar, form, that is MCP. Holt and Jenness (1984) reported that approximately 60%, 20% or 40% of total calcium, magnesium or inorganic phosphorus, respectively, in equine milk are non-ultrafiltrable, compared to approximately 70%, 35% or 45% of total calcium, magnesium and inorganic phosphorus in bovine milk. On the basis of measurements of the distribution of salts between the ultrafiltrable and non-ultrafiltrable phase of milk, Holt and Jenness (1984) estimated the concentrations of ionic calcium and magnesium in equine milk at pH 7.0 are 2.5 and 0.6 mmol L^{-1}, respectively, compared to 2.0 or 0.8 mmol L^{-1}, respectively, in bovine milk at pH 6.7. For human infant nutrition, a Ca:P ratio of approximately 2:1 is considered optimal; for bovine milk, the ratio is approximately 1:1, but in equine milk it is about 2:1, and is very close to that in human milk.

Trace Elements

Data on concentrations of trace elements, that is, those elements present at concentrations less than 30 mg L^{-1}, in equine milk have been sporadic. The concentrations of trace elements in equine, bovine and human milk are compared in Table 26.15. Compared to bovine milk, equine milk contains markedly higher levels of aluminum, copper, iron and titanium but lower levels of boron, barium, lithium, molybdenum, manganese, silicon and zinc. Human milk contains more boron, barium, copper, iron, silicon and zinc, but less lithium, manganese, strontium and titanium than equine milk (Table 26.15). Concentrations of zinc, iron and copper in equine milk decrease progressively with advancing lactation, whereas the concentration of manganese increases during the first 5 days of lactation, after which it decreases progressively (Csapó-Kiss et al. 1995, Ullrey et al. 1974).

PHYSICAL PROPERTIES OF EQUID MILK

The physical properties of the milk of some species are compared in Table 26.16.

Density

The density (kg m^{-3}) of equine colostrum is higher than that of equine milk (Waelchli et al. 1990, Ullrey et al. 1966, Mariani et al. 2001), due primarily to its considerably higher protein content. Values for colostrum can reach up to approximately 1080 kg m^{-3} (Ullrey et al. 1966) and show a significant linear correlation with the IgG content of colostrum (Waelchli et al. 1990, LeBlanc et al. 1986). Density is highest immediately *post-partum* and decreases rapidly during the first 12 h (Ullrey et al. 1966); considerably smaller decreases in density are observed during the rest of lactation (Ullrey et al. 1966, Mariani et al. 2001). Density values reported for mature equine milk range from approximately 1028 to 1035 kg m^{-3}. The density of whole mature bovine milk normally ranges from 1027 to 1033 kg m^{-3} (Singh et al. 1997).

Refractive Index

The refractive index for equine colostrum ranges from 1.340 to 1.354, whereas that of mature equine milk is approximately

Table 26.16. Physical Properties of the Milk of Equid Species, with Comparative Data for Bovine and Human Milk

Property	Equine Milk	Equine Colostrum	Asinine Milk	Bovine Milk	Human milk
Freezing point (°C)	−0.525 – −0.554	–	−0.55 – −0.49	−0.512 – −0.55	–
pH (25°C)	7.1–7.3	–	7.01–7.35	6.5–6.7	6.8
Density ($kg.m^{-3}$), (20°C)	1032	1080	1029–1037	1027–1033	1031
Refractive index, n_D^{20}	1.3394	1.340–1.354	–	1.344–1.349	–
Viscosity (mPa s)	1.5031	–	–	1.6314	–
Zeta potential (mV)	−10.3	–	–	−20.0	–
Colour L*a*b*	86.52, −2.34, −0.15	–	80.88 −2.27 −3.53	79.12, −7.46 −2.31	–

Source: Modified from Uniacke and Fox 2011.

1.339 (Waelchli et al. 1990). The higher refractive index of colostrum than of mature milk is probably related to its higher total solids content, since the refractive index increases with increasing mass fraction of each solute. The refractive index of bovine milk is in the range 1.344–1.349 (Singh et al. 1997).

pH

There is considerable variation among reported values for the pH of equine milk. Mariani et al. (2001) reported that its pH 4 days *post-partum* is approximately 6.6 and increases to approximately 6.9 after 20 days and to approximately 7.1 at 180 days post-partum. A value of approximately 7.0 for the pH for mature equine milk was reported by Küçükcetin et al. (2003), but Pagliarini et al. (1993) reported an average value of approximately 7.2. The pH of bovine milk is generally between 6.5 and 6.7 (Singh et al. 1997) and increases during lactation (Tsioulpas et al. 2007). These differences are presumably related to differences in protein and salt composition of the milks.

Freezing Point

The freezing point of milk is directly related to the concentrations of water-soluble compounds therein. Fat globules and proteins have a negligible influence on freezing point, with the main effect arising from lactose and minerals. A freezing point of −0.554°C (Pagliarini et al. 1993) or −0.548°C (Neseni et al. 1958) has been reported for equine milk, whereas the vast majority of bovine milk samples have a freezing point in the range −0.512 to −0.550°C (Singh et al. 1997). The lower freezing point of equine milk is probably related to its higher lactose content.

Viscosity

Equid milks are less viscous than bovine milk due to a lower total solids content.

Colour

Equid milk may be expected to be less white than bovine milk due to its low protein content and large casein micelles but this is not the case and equid milk is considerably whiter due to the absence of β-carotene that confers a yellow colour to bovine milk.

PROCESSING OF EQUID MILK

Because of its unique physico-chemical properties outlined earlier, the processing of equine and asinine milk into traditional dairy products is not possible; cheese is not produced from these milks as a firm curd is not formed on renneting. Equid milk will form a weak coagulum under acidic conditions and this is exploited in the production of yoghurt-type products with reputed probiotic and therapeutic properties. Traditionally, and to date, the only significant product from equine milk is the fermented product, koumiss, the production and properties of which are described below in section 'Koumiss'. Interest in koumiss production has grown recently which may be attributed to the fact that, worldwide, the overall consumption of fermented milk products has grown faster than the consumption of fresh milk (IDF 2009).

Fermentation is one of the oldest methods for preserving milk and probably dates back approximately 10,000 years to the Middle East where the first organised agriculture occurred. Traditional fermented milk products have been developed independently worldwide and were, and continue to be, especially important in areas where transportation, pasteurisation and refrigeration facilities are inadequate. Worldwide, milk from eight species of domesticated mammals (cow, buffalo, sheep, goat, camel, horse, reindeer and yak) have been used to produce traditional fermented milk products. There are three categories of fermented milks, those resulting from lactic fermentations, yeast-lactic fermentations and mould (*Geotrichum candidum*)-lactic fermentations. Koumiss and kefir belong to the yeast-lactic fermentation group where alcoholic fermentation by yeasts is used in combination with a lactic acid fermentation (Tamine and Marshall 1984). The conversion of milk into a fermented product has several important advantages; as well as being a means of preservation, it also improves taste and digestibility and increases the variety of food. Current interest in the health benefits of fermented milks started with the theory of longevity proposed by the Russian microbiologist, Elie Metchnikoff (1845–1918); he claimed that people who consumed fermented milks lived longer as lactic acid bacteria from the fermented product colonised the intestine and inhibited 'putrefaction' caused by harmful bacteria (a probiotic effect), thereby slowing down the aging process. Nowadays, the principal effects of probiotics are thought to be: improved gastrointestinal transit time of digesta, bowel function and glycemic index; some reports claim that they have an anticarcinogenic effect (see McIntosh 1996).

Further details on the fermentation of milk can be found in many publications including: Tamine and Marshall (1984); Koroleva (1991); Kurmann et al. (1992); Nakazawa and Hosono (1992); Surono and Hosono (2002).

Koumiss

Koumiss (Kumys), fermented equine milk, is widely consumed in Russia, Mongolia and Kazakhstan, primarily for its therapeutic value. Russians, in particular, have long advocated the use of koumiss for a wide variety of illnesses but the variable microbiology of these products has made it difficult to confirm any theoretical basis for the claims (Tamine and Robinson 1999). In Mongolia, koumiss is the national drink (Airag) and a high-alcoholic drink made by distilling koumiss, called Arkhi, is also produced (Kanbe 1992). *Per caput* consumption of koumiss in Mongolia is estimated to be about 50 L/yr.

Koumiss is still manufactured in remote areas of Mongolia by traditional methods but with increased demand elsewhere it is now produced under more controlled and regulated conditions. Traditional koumiss (from fresh raw milk) was prepared by seeding milk with a mixture of bacteria and yeasts using part of the previous day's product as an inoculum. The milk was held in a leather sack called a 'turdusk' (also called a 'saba' or 'burduk'), made from smoked horsehide taken from the thigh area, that is, it has a broad bottom and a narrow, long, sleeve with a capacity of 25–30 L, fermentation takes from 3 to 8 hours. In the 1960s, the microbial population was analysed and found to consist mainly of *Lactobacillus delbrueckii* subsp. *bulgaricus*, *Lactobacillus casei*, *Lactococcus lactis* subsp. *lactis*, *Kluveromyces fragilis* and *Saccharomyces unisporus* (Tamine and Marshall 1984). The lactic acid bacteria are responsible for acid production and the yeasts for the production of ethanol and carbon dioxide. During the mixing and maturation stages of production, more equine milk is usually added to control the levels of acidity and ethanol. The whole process is poorly controlled and often results in a product with an unpleasant taste due to the presence of too much yeast or excessive acidification. Turdusks, often containing fermented caprine milk from the previous season, were stored in a cool place over winter and the culture was reactivated when required by gradually filling the turdusk with equine milk over about 5 days (Tamine and Marshall 1984).

Koumiss contains about 90% moisture, 2–2.5% protein (1.2% casein and 0.9% whey proteins), 4.5–5.5% lactose, 1–1.3% fat and 0.4–0.7% ash, as well as the end-products of microbial fermentation, that is, lactic acid (1.8%), ethanol (0.6–2.5%) and CO_2 (0.5–0.9%) and provides 37 to 40 kCal/100 mL). After production, koumiss contains between 0.6 and 3% ethanol and is effervescent. Koumiss is thought to be more effective than raw equine milk in disease treatment due to the additional peptides and bactericidal substances produced during microbial metabolism (Doreau and Martin-Rosset 2002).

In the last decade, technological advances have been made in the manufacture of koumiss, such as the development of blends of microorganisms in starter cultures that enhance flavour development and extend the shelf life up to 14 days. The production of koumiss and other fermented milk products is carried out using a more standardised protocol for manufacture and is of considerable interest for increasing the market and consumption of equine milk products in countries where it has not normally been consumed (Di Cagno et al. 2004). As well as pasteurising the raw equine milk, pure cultures of lactobacilli such as *Lb. delbrueckii* subsp. *bulgaricus* and yeasts are used for koumiss manufacture. The use of *Saccharomyces lactis* is considered best for ethanol production (2–5%) and *S. cartilaginosus* is sometimes used for its antibiotic activity against *Mycobacterium tuberculosis* (Park et al. 2006). Other microorganisms such as *Candida* spp., *Torula* spp., *Lb. acidophilus* and *Lb. lactis* may also be used in koumiss production (Surono and Hosono 2002). A protocol for the manufacture of commercial koumiss is presented in Figure 26.6. The characteristics of a good koumiss are optimal when lactic and alcoholic fermentations proceed simultaneously so that the products of fermentation occur in definite proportions. As well as lactic acid, ethanol and CO_2, volatile acids and other compounds are formed which are important for aroma and taste and approximately 10% of equine milk proteins are hydrolysed. Products with varying amounts of lactic acid and ethanol are produced and generally 3 categories of koumiss are recognised: mild (0.6–0.8% acidity, 0.7–1.0% alcohol; medium (0.8–1.0% acidity, 1.1–1.8% alcohol) and strong (1.0–1.2% acidity, 1.8–2.5% alcohol (Tamine and Marshall 1984).

The presence of a high level of thermostable Lyz in equine milk may interfere with the activity of some starter microorganisms in the production of fermented products and thus cause problems in the processing of equine milk. Di Cagno et al. (2004) who heated equine milk to 90°C for 3 minutes to inactivate Lyz, produced an acceptable fermented product. In sensory tests, fermented equine milk generally scores low and, in an attempt to enhance the rheological and sensory properties of fermented products made from equine milk, Di Cagno et al. (2004) fortified equine milk with bovine Na caseinate (1.5 $g.100^{-1}g$), pectin (0.25 $g.100^{-1}g$) and threonine (0.08 $g.100^{-1}g$). The resultant product had good microbiological, rheological and sensory characteristics after 45 days at 4°C. Fermented unmodified equine milk had an unacceptable viscosity and scored very low in comparison to fortified products for appearance, consistency and taste.

Research has turned also to producing koumiss-like products from bovine milk, which must be modified to make it suitable for

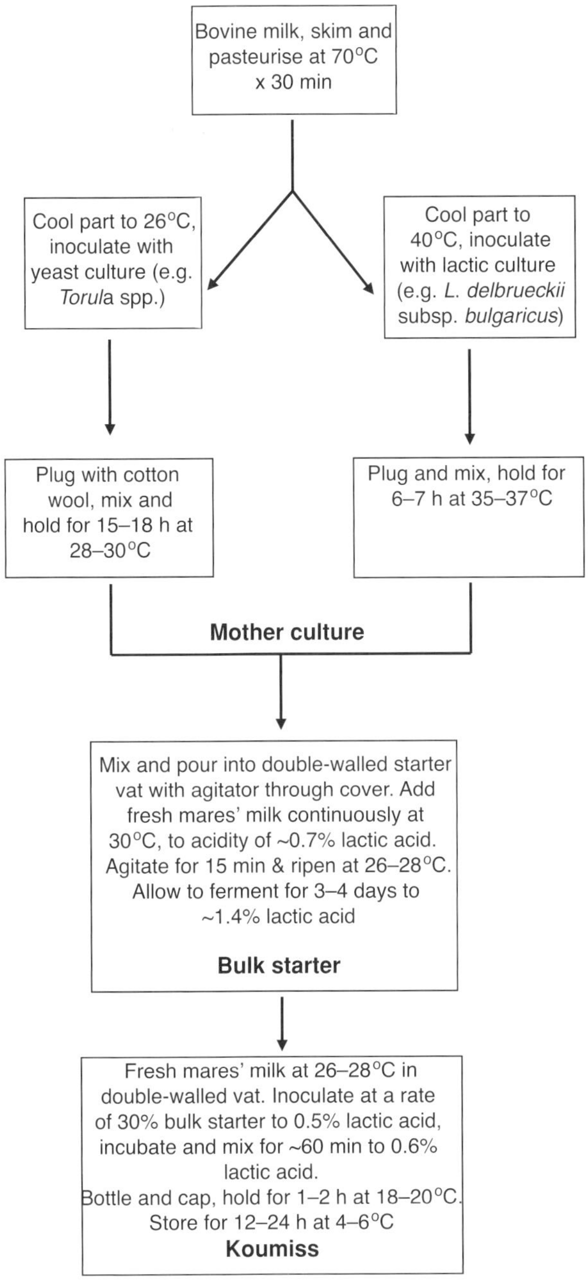

Figure 26.6. Schematic for the production of koumiss. (Based on Berlin 1962.)

koumiss production. Methods have been developed, with varying degrees of success, where a single constituent of bovine milk has been altered to resemble that of equine milk, for example the carbohydrate content has been increased or the protein content reduced but, until recently both had not been altered simultaneously. Koumiss of reasonable quality has been produced successfully from whole or skimmed bovine milk containing added sucrose using a mixture of *Lb. acidophilus*, *Lb. delbrueckii* ssp. *bulgaricus* and *Kluyveromyces marxianus* var. *marxianus* or var. *lactis* as starter culture (Kücükcetin et al. 2003). Koumiss has also been made from diluted bovine milk supplemented with lactose and, more successfully, from bovine milk mixed with concentrated whey using a starter culture of *Kluyveromyces lactis* (AT CC 56498), *Lb. delbrueckii* subsp. *bulgaricus* and *Lb. acidophilus*. Starter cultures for koumiss manufacture from bovine milk may also include *Saccharomyces lactis* (high antimicrobial activity against *Mycobacterium tuberculosis*) in order to retain the 'anti-tuberculosis image' of equine milk (Kücükcetin et al. 2003). More recently, bovine milk has been modified to approximate the composition of mares' milk using a series of membrane filtration steps and a starter culture (*Kluyveromyces lactis*, *Lb. delbrueckii* subsp. *bulgaricus* and *Lb. acidophilus*) that ensures consistent fermentation; the resulting product was found to be very similar to koumiss with respect to pH, titratable acidity, ethanol content, proteolytic activity, apparent viscosity and microbial composition, both when fresh or stored (15 days at 4°C) (Kücükcetin et al. 2003).

The physico-chemical and microbiological properties of asinine milk, similar to equine milk, such as low microbiological load and high Lyz make it a good substrate for the production of fermented products with probiotic *Lactobacillus* strains. Coppola et al. (2002) incubated asinine milk with the probiotic *Lb. rhamnosus* (AT 194, GTI/1, GT 1/3) and found that the strain is unaffected by the high Lyz activity in the milk and remained viable after 15 days at 4°C and pH 3.7–3.8. *Lb. rhamnosus* inhibits the growth of most harmful bacteria in the intestine and acts as a natural preservative in yoghurt-type products, considerably extending shelf life. Chiavari et al. (2005) produced fermented beverages from asinine milk using a mixed culture of *Lb. rhamnosus* (AT 194, CLT 2.2) and *Lb. casei* (LC 88) and in all cases found a high level of viable bacteria after 30 days storage. Some sensory differences were recorded for the fermented drinks and those made with the *Lb. casei* strain developed a more acceptable and balanced aroma than the boiled vegetable/acidic taste and aroma of the products made with *Lb. rhamnosus*.

Other Products from Equid Milk

As sales of equine milk have increased considerably in recent years, research is now focused on the development of new fermented products or new methods for extending the shelf life of existing products, while maintaining some of the unique components of equine milk. The ability of milk to withstand relatively high processing temperatures is very important from a technological point of view. The whey proteins in equine milk are much more thermostable than those of bovine milk. Heat treatment at 80°C × 80 s causes only a 10–15% decrease in non-casein nitrogen, with a marked decrease evident only when the temperature is increased above 100°C (Bonomi et al. 1994). Lf and equine BSA are the most heat-sensitive but are not completely denatured until the temperature reaches 130°C x 10 min. β-Lg and α-La are almost completely denatured at temperatures over 130°C and Lyz at temperatures greater than 110°C (68% residual Lyz activity after heating at 82°C for 15 minutes (Jauregui-Adell 1975).

NUTRITIONAL AND BIOMEDICAL PROPERTIES OF EQUID MILK

It is claimed that fresh and fermented equid milk relieve metabolic and intestinal problems while having a gut-cleansing effect coupled with 'repair' of the intestinal micro-flora. It is claimed to give relief from stomach ulcers, high blood pressure, high cholesterol and liver problems and is also recommended as an aid in the treatment of cancer patients. The recommended amount of equine milk is 250 mL/d. The most significant use of equid milk is as a substitute for bovine milk for patients with cows' milk protein allergy (CMPA); limited research has shown that both equine and asinine milks are generally tolerated well by CMPA patients. The use of equine milk in the production of cosmetics is relatively new and includes soaps, creams and moisturisers (Doreau and Martin-Rosset 2002). At present, equine milk is available in several forms: frozen milk, frozen yoghurt-type drink, lyopholised powder, shampoos and various cosmetic and medicinal creams. Lyopholised equine colostrum is available and used in the high-value horse industry to feed orphaned foals. Marconi and Panfili (1998) suggested that research is required to identify optimum drying and storage conditions for powdered equine milk for retention of some of the unique characteristics of raw milk, including high levels of whey proteins, PUFAs, lysine, tocopherols and vitamin C which are partially destroyed in the preparation of commercial powdered equine milk. Asinine milk is used to make ice cream and other desserts and also a fermented milk product. To improve the nutritive value of asinine milk and increase its overall energy content for human nutrition, it is frequently supplemented with approximately 4% medium-chain TGs (Salimei 2011).

Cow Milk Protein Allergy

Equine and asinine milk, with a composition close to that of human milk, may be good nutritional sources for the neonate when breast milk is not available. Bovine milk or bovine milk products are used traditionally as substitutes for human milk in infant nutrition but bovine milk is considerably different from human milk in terms of its macro- and micro-nutrients and the absorption rates of vitamins and minerals from the two milks are different, which can be problematic for infants. CMPA is an IgE-mediated type I allergy, which may be life-threatening, and is defined as a set of immunologically mediated adverse reactions which occur following the ingestion of milk, affecting from 2 to 6% of children in their first year of life. About 50% of affected children recover after the age of one and 80–90% of those affected recover by 5 years of age (Caffarelli et al. 2010). The high frequency of CMPA in infants and children is thought to be due to an incomplete gut mucosal barrier, increased gut permeability to large molecules and immature local and systemic responses which are aided by breast-milk which facilitates gut maturation and provides passive protection against bacteria and antigens (Hill 1994).

The difference between bovine milk protein allergy and lactose intolerance is of particular interest and it is an area which causes much confusion. CMPA is a food allergy, that is an adverse immune reaction to a food protein that is normally harmless to the non-allergic individual. Lactose intolerance is a non-allergic food hypersensitivity due to a deficiency of the enzyme β-galactosidase (lactase), required to hydrolyse lactose. Lactase deficiency manifests as abdominal symptoms and chronic diarrhoea after ingestion of milk (see Bindslev-Jensen 1998, Vesa et al. 2000). Lactose intolerance is not a disease or malady; 70% of the world's population is lactose-intolerant. Adverse effects of lactose intolerance occur at a much higher level of milk consumption than that which causes milk allergy.

CMPA is important because bovine milk is the first foreign antigen ingested in large quantities in early infancy. Reviews on CMPA include: Hill (1994); Hill and Hosking (1996); Taylor (1986); Høst (1988, 1991); Bindslev-Jensen (1998); Wal (2002, 2004); El-Agamy (2007); Apps and Beattie (2009). Because β-Lg is absent from human milk, it has commonly been considered to be the most important cows' milk allergen (Goldman et al. 1963, Ghosh et al. 1989) although other whey proteins (Jarvinen et al. 2001) and caseins (Savilahti and Kuitunen 1992, Restani et al. 1995) have also been implicated in allergic reactions. In children, β-Lg is the major allergen, whereas casein appears to be the most allergenic for adults. The resistance of β-Lg to proteolysis allows the protein to remain intact through the gastrointestinal tract with the possibility of being absorbed across the gut mucosa. Ingested β-Lg has been detected in human milk and could be responsible for colic in breast-fed infants and the sensitisation of infants, predisposing them to allergies (Jakobsson and Lindberg 1978, Kilshaw and Cant 1984, Stuart et al. 1984, Jakobsson et al. 1985, Axelsson et al. 1986).

The choice of substitute for cows' milk in cases of CMPA depends on two major factors, that is, nutritional adequacy and allergenicity; cost and taste must also be taken into account. Many soy or hydrolysate (casein-based, and more recently, whey protein-based) formulae are available for treatment of CMPA but they can themselves induce allergic reactions. Heat treatment of milk may destroy heat-labile proteins, especially BSA and Igs, and change the antigenic properties of other whey proteins, such as β-Lg and α-la, although caseins need severe heat treatment (121°C × 20 minutes) to reduce sensitising capacity (Hill 1994). Enzymatic treatment of milk proteins may result in products with an unacceptable taste due to bitterness arising from the production of peptides and amino acids and such peptides may, in fact, be allergenic (Schmidt et al. 1995, Sélo et al. 1999, El-Agamy 2007).

Many clinical studies have been carried out on the use of the milk of different species in infant nutrition, for example goat, sheep (Restani et al. 2002), camel (El-Agamy 2007), buffalo (El-Agamy 2007), horse and donkey (Iacono et al. 1992, Carroccio et al. 2000, El-Agamy et al. 1997, Businco et al. 2000, Monti et al. 2007). Results on the benefits of such milks are conflicting and infants with CMPA may suffer allergic reactions to buffalo, goat, sheep, donkey or mare milk proteins due to positive immunological cross-reaction with their counterparts in cows' milk (El-Agamy 2007). Lara-Villoslada et al. (2005) found that the balance between casein and whey proteins may be important in determining the allergenicity of bovine milk proteins in humans and that modification of this balance may reduce the allergenicity of bovine milk; a readjustment of the casein:whey proteins ratio to 40:60 was found to make the bovine milk significantly less allergenic. Presumably, equine and asinine milk, with a ratio of casein:whey proteins close to that in human milk, are potentially good substitutes for human milk. It is noteworthy that the study by Lara-Villoslada et al. (2005) was carried out using mice as subjects and the increased level of β-Lg (a major bovine milk allergen) in bovine milk adjusted to 40:60, casein:whey was not considered.

Cross-Reactivity of Milk Proteins

Cross-reaction occurs when two food proteins have similar amino acid sequences or when the three-dimensional conformation makes two molecules similar in their capacity to bind specific antibodies (Restani et al. 2002). Cross-reactivity of proteins from different species generally reflects the phylogenetic relations between animal species, for example homologous proteins from vertebrates often cross-react. A comprehensive study on the subject by Jenkins et al. (2007) highlights some interesting points, especially concerning the potential allergenicity of caseins from different species. The authors set out to determine how closely a foreign protein has to resemble a human homologue before it actually loses its allergenic affect. A high degree of similarity to human homologues would, presumably, imply that a foreign animal food protein would be much less likely than a protein with little or no similarity to its human homologue to be allergenic in human subjects. In addition, the study of potential animal allergens must take into account the ability of the human immune system to discriminate between its own proteins, that is an autoimmune response, and those from another species which have a high similarity, that is how closely a foreign protein has to resemble a human homologue before it loses its ability to act as an allergen? (Spitzauer 1999). Table 26.17 gives the percentage homology of α_{S1}-, α-$_{S2}$ and β-caseins from different species to bovine and human homologues. Known allergens are less than 53% identical to human sequences. Natale et al. (2004) found that 90% of a group of infants with CMPA had serum IgE against bovine α_{S2}-casein, 55% against bovine α_{S1}-casein and only 15% against bovine β-casein, which is closest in amino acid

Table 26.17. Homology (percentage) Between Milk Proteins from Different Species and Their Human Homologues

		Percent Identity to Closest	
Casein	Accession Code	Bovine Homolog	Human Homolog
α-$_{S1}$-Casein			
Cow	P02662	100	29
Goat	Q8M1H4	88	29
Sheep	P04653	88	28
Horse	Q8SPR1	39	44
Human	P47710	29	100
Rat	PO2661	22	27
Camel	097943	41	36
Rabbit	PO9115	37	37
α-$_{S2}$-Casein			
Cow	P02663	100	16
Goat	P33049	88	17
Sheep	P04654	89	17
Camel	097944	56	11
Rabbit	P50418	36	16
β-Casein			
Cow	P02666	100	53
Goat	Q712N8	91	54
Sheep	P11839	91	54
Horse	Q9GKK3	56	58
Human	P05814	53	100
Camel	Q9TVD0	66	58
Rabbit	PO9116	52	55

Source: Modified from Jenkins et al. 2007.

composition to human β-casein. Caprine and ovine milk proteins are more closely related to each other than either is to bovine milk proteins, thus explaining why an individual allergic to goats' milk cheese may exhibit high IgE cross-reactivity with sheep's milk proteins but could tolerate cow's milk and its products.

Allergy to equine milk appears to be rare and, to date, only two documented cases have been reported. Fanta and Ebner (1998) reported the case of an individual who experienced sensitisation to horse dander allergen and subsequently produced IgE antibodies on ingestion of equine milk which was prescribed to 'strengthen' her immune system. Gall et al. (1996) demonstrated the existence of an IgE-mediated equine milk allergy in one patient, caused by low MW heat-labile proteins, most likely α-La and β-Lg, without cross-reaction to the corresponding whey proteins from bovine milk. Presumably, the previous cases are not isolated incidents and as the consumption of equine milk and its products increases, it is likely that further cases will be reported.

Bevilaçqua et al. (2001) tested the capacity of goats' milk with low or high α_{s1}-casein content to induce milk protein sensitisation in guinea pigs and found significantly less sensitisation by milk with low α_{s1}-casein. This may represent an important attribute of the low α_{s1}-casein content of equine milk for use in human allergology. The absence of α_{s2}-casein (and lack of α_{s1} casein in one donkey) and β-lg II in donkey milk reported by Criscione et al. (2009) could be potentially interesting for future research on the allergenicity of asinine milk; α_{s1}-casein and β-lg are scarce or absent in human milk and are considered to be the most significant proteins causing allergic reactions in children and adults.

SUMMARY

The characteristics of equine and asinine milk of interest in human nutrition include an exceptionally high concentration of polyunsaturated fatty acids, low cholesterol content, high lactose and low protein levels (Solaroli et al. 1993, Salimei et al. 2004), as well as high levels of vitamins A, B and C. The low fat and unique fatty acid profile of both equine and asinine milk results in low atherogenic and thrombogenic indices. Research has shown that human health is considerably improved when dietary fat intake is reduced and, more importantly, when the ratio of saturated to unsaturated fatty acids is reduced. The high lactose content of equid milk gives good palatability and improves intestinal absorption of calcium, which is important for bone mineralisation in children. The renal load of equine milk, based on levels of protein and inorganic substances, is equal to that of human milk, a further indication of its suitability as an infant food. Equine and asinine milk can be used for their prebiotic and probiotic activity and as alternatives for infants and children with CMPA and multiple food intolerances (Iacono et al. 1992, Carroccio et al. 2000).

The invigorating effect of equine milk may be, at least partially, due to its immunostimulating ability. Lyz, Lf and n-3 fatty acids have long been associated with the regulation of phagocytosis of human neutrophils *in vitro* (Ellinger et al. 2002). The concentration of these compounds is exceptionally high in equine milk and the consumption of frozen equine milk significantly inhibits chemotaxis and respiratory burst, two important phases of the phagocytic process (Ellinger et al. 2002). This result suggests a potential anti-inflammatory effect by equine milk.

To be successful as a substitute for human milk in infant nutrition, equine milk must be capable of performing many biological functions associated with human milk. The presence of high concentrations of Lf, Lyz, n-3 and n-6 fatty acids in equine milk are good indicators of its potential role. However, the lack of research must be addressed to develop the potential of equine milk in the health and nutritional markets. Studies are required to bring the health claims for equine milk out of the realms of regional folklore. It seems reasonable to suggest that equine milk could be marketed as a dietary aid where the immune system is already depleted, that is, as a type of 'immuno-boost'. More than 30% of customers who purchase equine milk in the Netherlands are patients undergoing chemotherapy, who find equine milk helpful in counteracting the effects of the treatment. The composition of equine milk suggests a product with interesting nutritional characteristics with potential use in dietetics and therapeutics, especially in diets for the elderly, convalescent and newborns.

Future research should also include comprehensive characterisation of the proteins of different breeds of horse and donkey, with the possibility of selection of animals for specific proteins which could, in turn, optimise the nutritional and technological properties of the milks; the genetic control of α_s-, β- and κ-caseins as well as β-Lg in bovine and caprine milk has been researched, albeit in a somewhat limited manner, as has the effects of these proteins on the technological properties of both types of milk, for example colloidal stability, coagulation and curd strength. These characteristics can determine the physical and chemical behaviour of milk in the infant gastrointestinal tract with the result that digestion and availability of nutrients to the young may be affected (Cuthbertson 1999). Furthermore, genetic selection of certain breeds of horse and donkey may improve milk yield and lactation pattern and make the production of milk more cost effective.

REFERENCES

Addeo F et al. 1984. Susceptibility of buffalo and cow κ-caseins to chymosin action. *Milchwissenschaft* 39: 202–205.

Aganga AA et al. 2000. Feeding donkeys. *Livest Res Rural Dev* 12: 1–5.

Agostini C et al. 2000. Free amino acid content in standard infant formulas: comparison with human milk. *J Am Coll Nutr* 19: 434–438.

Anderson RR. 1992. Comparison of trace elements in milk of four species. *J Dairy Sci* 75: 3050–3055.

Apps JR, Beattie RM. 2009. Cows milk allergy in children. *BMJ* 339: 343–345.

Atkinson SA et al. 1989. Non-protein nitrogen components in human milk: biochemistry and potential roles. In: SA Atkinson,

B Lönnerdal (eds.) *Protein and Non-protein-nitrogen in Human Milk*, Chapter 12. CRC Press, Boca Raton, FL, pp. 117–136.

Atkinson SA, Lönnerdal B. 1995. Nitrogenous components of milk. Non protein nitrogen fractions of human milk. In: RG Jensen (ed.) *Handbook of Milk Composition*, Chapter 5B. Academic Press, San Diego, CA, pp. 369–387.

Axelsson I et al. 1986. Bovine β-lactoglobulin in human milk. *Acta Paediatrica Scandinavica* 75: 702–707.

Baldi A et al. 2005. Biological effects of milk proteins and their peptides with emphasis on those related to the gastrointestinal ecosystem. *J Dairy Res* 72: 66–72.

Barello C et al. 2008. Analysis of major proteins and fat fractions associated with mare's milk fat globules. *Mol Nutr Food Res* 58: 1448–1456.

Bauman DE, Lock AL. 2006. Conjugated linoleic acid: Biosynthesis and nutritional significance. In: PF Fox, PLH McSweeney (eds.) *Advanced Dairy Chemistry, vol 2: Lipids*, 3rd edn, Springer, New York, pp. 93–136.

Belizy S et al. 2001. Changes in antioxidative properties of lactoferrin from women's milk during deamidation. *Biochem (Moscow)* 66: 576–580.

Bell K et al. 1981a. Equine whey proteins. *Comp Biochem Physiol B Biochem Mol Biol* 68: 225–236.

Bell K et al. 1981b. Bovine alpha-lactalbumin C and alpha S1-, beta- and kappa-caseins of Bali (Banteng) cattle, *Bos* (Bibos) *javanicus*. *Aust J Biol Sci* 34: 149–159.

Bellamy W et al. 1992. Identification of the bactericidal domain of lactoferrin. *Biochimica et Biophysica Acta* 1121: 130–136.

Benirschke K et al. 1964. Chromosomes studies of a donkey-Grevy zebra hybrid. *Chromosoma* 15: 1–13.

Berlin PJ. 1962. Kumiss. In: *Annual Bulletin, IV*. International Dairy Federation, Brussels, pp. 4–16.

Bernhart FW. 1961. Correlations between growth-rate of the suckling of various species and the percentage of total calories from protein in the milk. *Nature* 191: 358–360.

Bevilaçqua C et al. 2001. Goats' milk of defective α_{s1}-casein genotype decreases intestinal and systemic sensitization to β-lactoglobulin in guinea pigs. *J Dairy Res* 68: 217–227.

Bindslev-Jensen C. 1998. Food allergy. *BMJ* 316: 1299–1302.

Bonomi F et al. 1994. Thermal sensitivity of mares' milk proteins. *J Dairy Res* 61: 419–422.

Boulot S. 1987. L'ingestion chez la jument. Etude de quelques facteurs de variation au tours du cycle gestation-lactation. Implications, nutritionelles et métaboliques. Thése Doct. Ing., ENSA Rennes, University of Rennes I, France.

Brew K. 2003. α-Lactalbumin. In: PF Fox, PLH McSweeney (eds.) *Advanced Dairy Chemistry, vol 1: Proteins*, 3rd edn, Kluwer Academic/Plenum Publishers, New York, pp. 387–419.

Brock JH. 1997. Lactoferrin structure – function relationships. In: TW Hutchens, B Lönnerdal (eds.) *Lactoferrin: Interactions and Biological Functions*, Chapter 1. Humana Press, NJ, pp. 3–38.

Bruns E. 1999. Commission on horse production. *Livest Prod Sci* 60: 197–198.

Buchheim W et al. 1989. Verrgleichende Untersuchungen zur Struktur und Grobe von Casein Micellen in der Milchverschiedener Species. *Kieler Milchwirtschaffliche Forschungsberichte* 41: 253–265.

Businco L et al. 2000. Allergenicity of mare's milk in children with cow's milk allergy. *J Allergy Clin Immunol* 105: 1031–1034.

Caffarelli C et al. 2010. Cow's milk protein allergy in children: a practical guide. *Ital J Pediatr* 36: 1–7.

Carbone L et al. 2006. Evolutionary movement of centromeres in horse, donkey, and zebra. *Genomics* 87: 777–782.

Carlson SE. 2001. Docosahexaenoic acid and arachidonic acid in infant development. *Semin Neonatol* 6: 437–449.

Carratù B et al. 2003. Nitrogeneous components of human milk: non-protein nitrogen, true protein and free amino acids. *Food Chem* 81: 357–362.

Carroccio A et al. 2000. Intolerence to hydrolysed cow's milk proteins in infants: clinical characteristics and dietary treatment. *Clin Exp Allergy* 30: 1597–1603.

Casey CE. 1989. The nutritive and metabolic advantages of homologous milk. *Proc Nutr Soc* 48: 271–281.

Cavaletto M et al. 1994. Human α_{s1}-casein like protein: purification and N-terminal sequence determination. *Biol Chem Hoppe Seyler* 375: 149–151.

Chandan RC et al. 1964. Lysozyme content of human milk. *Nature* 224: 76–77.

Chandan RC et al. 1971 Physicochemical analyses of bovine milk fat globule membrane. I. Differential thermal analysis. *J Dairy Sci* 54: 1744–1751.

Chen Y et al. 2010. Identification of angio 1-converting enzyme inhibitory peptides from koumiss, a traditional fermented mare's milk. *J Dairy Sci* 93: 884–892.

Chianese L et al. 2010. Proteomic characterization of donkey milk "caseome". *J Chromatogr A1* 217: 4834–4940.

Chiavari C et al. 2005. Use of donkey's milk for a fermented beverage with lactobacilli. *Lait* 85: 481–490.

Chilliard Y, Doreau M. 1985. Characterization of lipase in mare milk. *J Dairy Sci* 68: 37–39.

Chowanadisai W et al. 2005. Detection of a single nucleotide polymorphism in the human α-lactalbumin gene: implications for human milk proteins. *J Nutr Biochem* 16: 272–278.

Chowdhury FA et al. 2004. Protein dissection experiments reveal key differences in the equilibrium folding of alpha-lactalbumin and the calcium binding lysozymes. *Biochem* 43: 9961–9967.

Civardi G et al. 2002. Capillary electrophoresis (CE) applied to analysis of mare's milk. *Milchwissenschaft* 57: 515–517.

Civardi G et al. 2007. Mare's milk: monitoring the effect of thermal treatments on whey proteins stability by SDS capillary electrophoresis (CE-SDS). *Milchwissenschaft* 62: 32–35.

Clark S, Sherbon JW. 2000. Alpha$_{s1}$-casein, milk composition and coagulation properties of goat milk. *Small Ruminant Res* 38: 123–134.

Clutton-Brock J. 1992. *Horse Power: A History of the Horse and Donkey in Human Societies*. British Museum, London.

Collomb M et al. 2006. Conjugated linoleic acids in milk fat: Variation and physiological effects. *Intern Dairy J* 16: 1347–1361.

Conneely O. 2001. Antiinflammatory activities of lactoferrin. *J Am Coll Nutr* 20: 389S–395S.

Coppola R et al. 2002. Behaviour of *L. rhamnosus* strain in ass's milk. *Ann Microbiol* 52: 55–60.

Creamer L, Sawyer W. 2011. Beta-lactoglobulin. In: J Fuquay et al. (eds.) *Encyclopedia of Dairy Sciences*, 2nd edn. Elsevier, Oxford, in press.

Criscione A et al. 2009. Donkey's milk protein fraction investigated by electrophoretic methods and mass spectrometric analysis. *Intern Dairy J* 19: 190–197.

Csapó J et al. 1995. Composition of mare's colostrum and milk. Fat content, fatty acid composition and vitamin content. *Intern Dairy J* 5: 393–402.

Csapó-Kiss Zs et al. 1995. Composition of mare's colostrum and milk. Protein content, amino acid composition and contents of macro- and micro-elements. *Intern Dairy J* 5: 403–415.

Cunsolo V et al. 2007. Characterization of the protein profile of donkey's whey fraction. *Mass Spectr* 42: 1162–1174.

Curadi MC et al. 2001. Use of mare milk in pediatric allergology. Proceedings of the XIV Congress Associazione Scientifica Produzioni Animali. Firenze, Italy. Wageningen Academic, Wageningen, The Netherlands.

Cuthbertson WFJ. 1999. Evolution of infant nutrition. *Brit J Nutr* 81: 359–371.

Dalgleish DG. 1998. Casein micelles as colloids. Surface structures and stabilities. *J Dairy Sci* 81: 3013–3018.

Davis TA et al. 1994. Amino acid composition of human milk is not unique. *J Nutr* 124: 1126–1132.

De Kruif CG, Holt C. 2003. Casein micelle structure, functions and interactions. In: PF Fox, PLH McSweeney (eds.) *Advanced Dairy Chemistry, vol 1: Proteins*, 3rd edn, Chapter 5. Kluwer Academic/Plenum Publishers, New York, pp. 233–276.

De Kruif CG, Zhulina EB. 1996. κ-Casein as a polyelectrolyte brush on the surface of casein micelles. *Colloids Surf A Physicochem Eng Asp* 117: 151–159.

Dev BC et al. 1994. κ-Casein and β-casein in human milk micelles: structural studies. *Arch Biochem* 314: 329–336.

Di Cagno R et al. 2004. Uses of mares' milk in the manufacture of fermented milk. *Intern Dairy J* 14: 767–775.

Dils RR. 1986. Comparative aspects of milk fat synthesis. *J Dairy Sci* 69: 904–910.

Doi H et al. 1979. Heterogeneity of reduced bovine κ-casein. *J Dairy Sci* 62: 195–203.

Donnet-Hughes A et al. 2000. Bioactive molecules in milk and their role in health and disease: The role of transforming growth factor-beta. *Immunol Cell Biol* 78: 74–79.

Donovan S. 2009. Human milk oligosaccharides – the plot thickens. *Brit J Nutr* 101: 1267–1269.

Doreau M et al. 1986. Relationship between nutrient intake, growth and body composition of the nursing foal. *Reprod Nutr Dev* 26: 683–690.

Doreau M, Boulot S. 1989. Recent knowledge on mare milk production: A review. *Livest Prod Sci* 22: 213–235.

Doreau M et al. 1990. Yield and composition of milk from lactating mares: effect of lactation stage and individual differences. *J Dairy Res* 57: 449–454.

Doreau M. 1994. Le lait de jument et sa production: particularités et facteurs de variation. *Lait* 74: 401–418.

Doreau M, Martin-Rosset W. 2002. Dairy animals: "Horse". In: H Roginski et al.(eds.) *Encyclopedia of Dairy Sciences*. Academic Press, London, pp. 630–637.

Duggan VE et al. 2008. Amyloid A in equine colostrum and early milk. *Vet Immunol Immunopathol* 121: 150–155.

Dziuba J, Minkiewicz P. 1996. Influence of glycosylation on micelle-stabilizing ability and biological properties of C-terminal fragments of cow's κ-casein. *Intern Dairy J* 6: 1017–1044.

Egito AS et al. 2001. Susceptibility of equine κ- and β-caseins to hydrolysis by chymosin. *Intern Dairy J* 11: 885–893.

Egito AS et al. 2002. Separation and characterization of mares' milk α_{s1}-, β-, κ-caseins, γ-casein-like, and proteose peptone component 5-like peptides. *J Dairy Sci* 85: 697–706.

Egito AS et al. 2003. Action of plasmin on equine β-casein. *Intern Dairy J* 13: 813–820.

El-Agamy EI et al. 1997. A comparative study of milk proteins from different species. II. Electrophoretic patterns, molecular characterization, amino acid composition and immunological relationships. In: *Third Alexandria Conference on Food Science and Technology. Alexandria, Egypt*, pp. 51–62.

El-Agamy EI. 2007. The challenge of cow milk protein allergy. *Small Ruminant Res* 68: 64–72.

Ellinger S et al. 2002. The effect of mare's milk consumption on functional elements of phagocytosis of human neutrophils granulocytes from healthy volunteers. *Food Agric Immunol* 14: 191–200.

Ellison RT, Giehl TJ. 1991. Killing of Gram-negative bacteria by lactoferrin and lysozyme. *JClin Invest* 88: 1080–1091.

El-Zahar K et al. 2005. Peptic hydrolysis of ovine β-lactoglobulin and α-lactalbumin. Exceptional susceptibility of native ovine β-lactoglobulin to pepsinolysis. *Intern Dairy J* 15: 17–27.

Enbergs H et al. 1999. Eiweiss-, Glucose- und Lipidgehalt in der Milch von Zuchtstuten im Verlauf der Laktation. *Zuchtungskunde* 71: 245–266.

ExPASy ProtParam Tool, Swiss Institute of Bioinfomatics. 2009. Available at http://www.expasy.org/tools/pi_tool.html. Accessed on October 16, 2009.

Fanta C, Ebner C. 1998. Allergy to mare's milk. *Allergy* 53: 539–540.

Farrell HM et al. 2006. Casein micelle structure: what can be learned from milk synthesis and structural biology? *Curr Opin Colloid Interface Sci* 11: 135–147.

Fiat AM et al. 1980. Localisation and importance of the sugar part of human casein. *Eur J Biochem* 111: 333–339.

Fitzgerald RJ, Meisel H. 2003. Milk protein hydrolysates and bioactive peptides. In: PF Fox, PLH McSweeney (eds.). *Advanced Dairy Chemistry, vol 1: Proteins*, 3rd edn. Kluwer Academic/Plenum Publishers, New York, pp. 675–698.

Fitzgerald RJ. Murray BA. 2006. Bioactive peptides and lactic fermentations. *Int J Dairy Technol* 59: 118–125.

Formaggioni P et al. 2003. Casein number variability of mare milk from Haflinger and Italian Saddle horse breeds. *Annali della Facoltà di Medicina Veterinaria, Università di Parma* 13: 175–179.

Fox PF (ed.). 1985. *Developments in Dairy chemistry, vol 3: Lactose and Minor Constituents*. Elsevier Applied Science, London.

Fox PF (ed.). 1997. *Advanced Dairy Chemistry, vol 3: Lactose, Water, Salts and Vitamins*, 2nd edn. Chapman and Hall, London.

Fox PF, McSweeney PLH. 1998. *Dairy Chemistry and Biochemistry*. Blackie Academic and Professional, New York.

Fox PF, Kelly A. 2006. Indigenous enzymes in milk: Overview and historical aspects-Part 1. *Intern Dairy J* 16: 500–516.

Fox PF, McSweeney PLH (eds.). (2006). *Advanced Dairy Chemistry, vol 3: Lipids*, 3rd edn. Springer, New York.

Fox PF, Brodkorb A. 2008. The casein micelle: historical aspects, current concepts and significance. *Intern Dairy J* 18: 677–684.

Gall H et al. 1996. Allergy to the heat-labile proteins α-lactalbumin and β-lactoglobulin in mare's milk. *J Allergy Clin Immunol* 97: 1304–1307.

Ghosh J et al. 1989. Hypersensitivity of human subjects to bovine milk proteins: a review. *Indian J Dairy Sci* 42: 744–749.

Girardet J-M et al. 2004. Multiple forms of equine α-lactalbumin: evidence for *N*-glycosylated and deamidated forms. *Inter Dairy J* 14: 207–217.

Girardet J-M et al. 2006. Determination of the phosphorylation level and deamidation susceptibility of equine β-casein. *Proteomics* 6: 3707–3717.

Giuffrida MG et al. 1992. The amino acid sequence of two isoforms of alpha-lactalbumin from donkey (*Equus asinus*) milk is identical. *Biol Chem Hoppe Seyler* 373: 931–935.

Godovac-Zimmermann J et al. 1985. The amino acid sequence of β-lactoglobulin II from horse colostrum (*Equus caballus*, Perissodactyla): β-lactoglobulins are retinol-binding proteins. *Biol Chem Hoppe Seyler* 366: 601–608.

Godovac-Zimmermann J et al. 1987. Identification and the primary structure of equine α-lactalbumin B and C (*Equus caballus*, Perissodactyla). *Biol Chem Hoppe Seyler* 368: 427–433.

Godovac-Zimmermann J et al. 1988a. Microanalysis of the amino acid sequence of monomeric beta-lactoglobulin I from donkey (*Equus asinus*) milk. The primary structure and its homology with a superfamily of hydrophobic molecule transporters. *Biol Chem Hoppe Seyler* 369: 171–179.

Godovac-Zimmermann J et al. 1988b. The primary structure of donkey (*Equus asinus*) lysozyme contains the Ca(II) binding site of alpha-lactalbumin. *Biol Chem Hoppe Seyler* 369: 1109–1115.

Godovac-Zimmermann J et al. 1990. Covalent structure of the minor monomeric beta-lactoglobulin II component from donkey milk. *Biol Chem Hoppe Seyler* 371: 871–879.

Goldman AS et al. 1963. Milk allergy. 1. Oral challenge with milk and isolated milk proteins in allergic children. *Pediatrics* 32: 425–443.

Gottschalk M et al. 2003. Protein self-association in solution: the bovine β-lactoglobulin dimmer and octamer. *Protein Sci*12: 2404–2411.

Green MR, Pastewka JV. 1978. Lactoferrin is a marker for prolactin response in mouse mammary explants. *Endocrinology* 103: 1510–1513.

Guo HY et al. 2007. Composition, physicochemical properties, nitrogen fraction distribution, and amino acid profile of donkey milk. *J Dairy Sci* 90: 1635–1643.

Halliday JA et al. 1991. The complete amino acid sequence of feline β-lactoglobulin II and a partial revision of the equine β-lactoglobulin II sequence. *Biochimi et Biophys Acta* 1077: 25–30.

Hambræus L. 1982. Nutritional aspects of milk proteins. In: PF Fox (ed.) *Developments in Dairy Chemistry, vol 1: Proteins*. Applied Science Publishers, London, pp. 289–313.

Hambræus L. 1984. Human milk composition. *Nutrit Abstr Rev. Series A* 54: 219–236.

Haridas M et al. 1995. Structure of human diferric lactoferrin refined at 2.2 Å resolution. *Acta Crystallographica* D 51: 629–646.

Herrouin M et al. 2000. New genetic variants identified in donkey's milk whey proteins. *JProtein Chem* 19: 105–115.

Herskovits TT. 1966. On the conformation of caseins. Optical rotatory properties. *Biochem* 5: 1018–1026.

Hill DJ. 1994. Cow milk allergy within the spectrum of atopic disorders. *Clin Exp Allergy* 24: 1137–1143.

Hill DJ, Hosking CS. 1996. Cow milk allergy in infancy and early childhood. *J Allergy Clin Immunol* 97: 9–15.

Holt C, Horne DS. 1996. The hairy casein micelle: evolution of the concept and its implications for dairy technology. *Nether Milk Dairy J* 50: 85–111.

Holt C, Jenness R. 1984. Interrelationships of constituents and partition of salts in milk samples from eight species. *Comp Biochem Physiol* 77A: 175–282.

Holt C, Sawyer L. 1993. Caseins as rheomorphic proteins: interpretation of primary and secondary structures of the α_{s1}-, β- and κ-caseins. *J Chem Soc, Faraday Trans* 89: 2683–2692.

Horne DS. 1998. Casein interactions: casting light on the *black boxes*, the structure in dairy products. *Intern Dairy J* 8: 171–177.

Horne DS. 2003. Ethanol stability. In: PF Fox, PLH McSweeney (eds.) *Advanced Dairy Chemistry, vol. 1: Proteins*, 3rd edn. Kluwer Academic/Plenum Publishers, New York, pp. 975–1000.

Horne DS. 2006. Casein micelle structure: Models and muddles. *Curr Opin Colloid Interface Sci*11: 148–153.

Horrobin DF. 2000. Essential fatty acid metabolism and its modification in atopic eczema. *Am J Clin Nutr*71: 367S–372S.

Høst A. 1988. A prospective study of cow's milk allergy in exclusively breast-fed infants. Incidence, pathogenetic role of early inadvertent exposure to cow's milk formula, and characterization of bovine milk protein in human milk. *Acta Paediatrica Scandinavica* 77: 663–670.

Høst A. 1991. Importance of the first meal on the development of cow's milk allergy and intolerance. *Allergy Proc* 10: 227–232.

Humbert G et al. 2005. Plasmin activity and plasminogen activation in equine milk. *Milchwissenschaft* 60: 134–137.

Hurley WL. 2003. Immunoglobulins in mammary secretions. In: PF Fox, PLH McSweeney (eds.) *Advanced Dairy Chemistry, vol 1: Proteins*, 3rd edn. Kluwer Academic/Plenum Publishers, New York, pp. 422–447.

Iacono G et al. 1992. Use of ass' milk in multiple food allergy. *J Pediatr Gastroenterol Nutr* 14: 177–181.

Iametti S et al. 2001. Primary structure of κ-casein isolated from mares' milk. *J Dairy Res* 68: 53–61.

International Dairy Foundation (IDF). 2009. The World Dairy Situation 2009. *Bulletin of the International Dairy Federation N° 438*. International Dairy Federation, Brussels, Belgium.

Innis S. 2007. Fatty acids and early human development. *Early Hum Dev* 83: 761–766.

Izraely H et al. 1989. Factors determining the digestive efficiency of the domesticated donkey (*Equus asinus asinus*). *Q J Exp Physiol* 74: 1–6.

Jahreis G et al. 1999. The potential anticarcinogenic conjugated linoleic acid, cis-9, trans-11 C18:2, in milk of different species: cow, goat, ewe, cow, mare, women. *Nutr Res* 19: 1541–1549.

Jakobsson I, Lindberg T. 1978. Cow's milk as a cause of infantile colic in breast-fed infants. *Lancet* 11: 437–446.

Jakobsson I et al. 1985. Dietary bovine beta lactoglobulin is transferred to human milk. *Acta Paediatrica Scandinavia* 74: 342–345.

Jarvinen KM et al. 2001. IgE and IgG binding epitopes on α-lactalbumin and β-lactoglobulin cow's milk allergy. *Int Arch Allergy Immunol* 126: 111–118.

Jasińska B, Jaworska G. 1991. Comparison of structures of micellar caseins of milk of cows, goats and mares with human milk casein. *Anim Sci Pap Rep* 7: 45–55.

Jauregui-Adell J. 1975. Heat stability and reactivation of mare milk lysozyme. *J Dairy Sci* 58: 835–838.

Jenkins JA et al. 2007. Evolutionary distance from human homologs reflects allergenicity of animal food proteins. *J Allergy Clin Immunol* 120: 1399–1405.

Jenness R. 1986. Lactational performance of various mammalian species. *J Dairy Sci* 69: 869–885.

Johnsen LB et al. 1995. Characterization of three types of human α_{s1}-casein mRNA transcripts. *Biochem J* 309: 237–242.

Kalliala H et al. 1951. On the use of mare's milk in infant feeding. *Acta Paediatrica* 40: 97–117.

Kanamori M et al. 1981. Presence of O-glycosidic linkage through serine residue in kappa-casein component from bovine mature milk. *J Nutr Sci Vitaminol (Tokyo)* 27: 231–241.

Kanbe M. 1992. Traditional fermented milk of the world. In: Y Nakazawa, A Hosono (eds.) *Functions of Fermented Milk: Challenges for the Health Sciences*. Elsevier Applied Science, London.

Kaneko K et al. 2004. Bovine subclinical mastitis diagnosed on the basis of amyloid A protein in milk. *J Jpn Vet Med Assoc* 57: 515–518.

Keenan TW, Mather IH. 2006. Intracellular origin of milk fat globules and the nature of the milk fat globule membrane. In: PF Fox, PLH McSweeney (eds.) *Advanced Dairy Chemistry, vol 2: Lipids*, 3rd edn. Springer, New York, pp. 137–171.

Kilshaw PJ, Cant AJ. 1984. The passage of maternal dietary proteins into human breast milk. *Int Arch Allergy Appl Immunol* 75: 8–15.

Koshiba T et al. 2000. Structure and thermodynamics of the extraordinary stable molten globule state of canine milk lysozyme. *Biochem* 39: 3248–3257.

Koshiba T et al. 2001. Energetics of three-state unfolding of a protein: canine milk lysozyme. *Protein Eng* 14: 967–974.

Kotts C, Jenness R. 1976. Isolation of κ-casein-like proteins from milk of various species. *J Dairy Sci* 59: 816–822.

Koroleva NS. 1991. Products prepared with lactic acid bacteria and yeasts. In: RK Robinson (ed.) *Therapeutic Properties of Fermented Milk*. Elsevier Applied Science, Barking, pp. 159–179.

Kücükcetin A et al. 2003. Adaptation of bovine milk towards mares' milk composition by means of membrane technology for koumiss manufacture. *Intern Dairy J* 13: 945–951.

Kulisa M. 1977. The composition of mare's milk in three horse breeds with reference to *N*-acetylneuraminic acid. *Acta Agraria et Silvestria Series Zootechnica* 27: 25–37.

Kunz C, Lönnerdal B. 1990. Casein and casein subunits in preterm milk, colostrum, and mature human milk. *J Pediatr Gastroenterol Nutr.* 10: 454–461.

Kuroki R et al. 1989. Design and creation of a Ca^{2+} binding site in human lysozyme to enhance structural stability. *Proc Natl Acad Sci U S A* 86: 6903–6907.

Kurmann JA et al. 1992. *Encyclopedia of Fermented Fresh Milk Products: An International Inventory of Fermented Milk, Cream, Buttermilk, Whey, and Related Products*. Van Nostrand Reinhold, New York.

Kyte J, Doolittle RF. 1982. A simple method for displaying the hydropathic character of a protein. *J Mol Biol* 157: 105–132.

Langer P. 2009. Differences in the composition of colostrum and milk in eutherians reflect differences in immunoglobulin transfer. *J Mammal* 90: 332–339.

Lara-Villoslada F et al. 2005. The balance between caseins and whey proteins in cow's milk determines its allergenicity. *J Dairy Sci* 88: 1654–1660.

Larson BL. 1979. Biosynthesis and secretion of milk proteins: a review. *J Dairy Res* 46: 161–174.

Larson MA et al. 2003. Human serum amyloid A3 peptide enhances intestinal MUC3 expression and inhibits EPEC adherence. *Biochemi et Biophys Res Comm* 300: 531–540.

Lauer BH, Baker BE. 1977. Amino acid composition of casein isolated from the milks of different species. *Can J Zool* 55: 231–236.

LeBlanc MM et al. 1986. Relationships among serum immunoglobulin concentration in foals, colostral specific gravity, and colostral immunoglobulin concentration. *J Am Vet Med Assoc* 189: 57–60

Leitch EC, Willcox MD. 1999. Elucidation of the antistaphylococcal action of lactoferrin and lysozyme. *J Med Microbiol* 48: 867–871.

Lenasi T et al. 2003. Characterization of equine cDNA sequences for α_{s1}-, β- and κ-caseins. *J Dairy Res* 70: 29–36.

Levine MA. 1998. Eating horses: the evolutionary significance of hippophagy. *Antiquity* 72: 90–100.

Levy J. 1998. Immunonutrition: the pediatric experience. *Nutrition* 14: 641–647.

Lönnerdal B. 2003. Lactoferrin. In: PF Fox, PLH McSweeney (eds.) *Advanced Dairy Chemistry, vol 1: Proteins*, 3rd edn. Kluwer Academic/Plenum Publishers, New York, pp. 449–460.

Lopéz-Fandiño R et al. 2006. Physiological, chemical and technological aspects of milk-protein-derived peptides with antihypertensive and ACE-inhibitory activity. *Intern Dairy J* 16: 1277–1293.

Lozovich S. 1995. Medical uses of whole and fermented mare milk in Russia. *Cultured Dairy Products* 30: 18–21.

Lucey JA, Singh H. 2003. Acid coagulation of milk. In: PF Fox, PLH McSweeney (eds.) *Advanced Dairy Chemistry, vol 1: Proteins*, 3rd edn. Kluwer Academic/Plenum Publishers, New York, pp. 1001–1026.

Lyster RLJ. 1992. Effect of calcium on the stability of mares' milk lysozyme. *J Dairy Res* 59: 331–338.

MacDonald D. 2001. *The Encyclopedia of Mammals*. Barnes and Noble/Andromeda Oxford Ltd., Abingdon.

Mack DR et al. 2003. The conserved TFLK motif of mammary-associated serum amyloid A3 is responsible for up-regulation of intestinal MUC3 mucin expression *in vitro*. *Pediatr Res* 53: 137–142.

Madureira AR et al. 2007. Bovine whey proteins – Overview on their main biological properties. *Food Res Intel* 40: 1197–1211.

Malacarne M et al. 2000. Observations on percentage distribution of the main mare milk caseins separated by reversed-phase

HPLC. *Annali della Facoltà di Medicina Veterinaria, Università di Parma* (Italy) 20: 143–152.

Malacarne M et al. 2002. Protein and fat composition of mare's milk: some nutritional remarks with reference to human and cow's milk. *Intern Dairy J* 12: 869–897.

Malkoski M et al. 2001. Kappacin, a novel antibacterial peptide from bovine milk. *Antimicrob Agents Chemother* 45: 2309–2315.

Marchand S et al. 2009. Thermal inactivation kinetics of alkaline phosphatase in equine milk. *Intern Dairy J* 19: 763–767.

Marconi E, Panfili G. 1998. Chemical composition and nutritional properties of commercial products of mare milk powder. *J Food Compost Anal* 11: 178–187.

Mariani P et al. 2001. Physicochemical properties, gross composition, energy value and nitrogen fractions of Halflinger nursing mare milk throughout 6 lactation months. *Anim Res* 50: 415–425.

Martin P et al. 1996. The gene encoding α_{s1}-casein is expressed in human mammary epithelial cells during lactation. *Lait* 76: 523–535.

Martin P et al. 2011. Inter-species comparison of milk proteins: quantitative variability and molecular diversity. In: J Fuquay et al. (eds.) *Encyclopedia of Dairy Sciences*, 2nd edn. Elsevier, Oxford, 3: 821–842.

Martuzzi F, Doreau M. 2006. Mare milk composition: recent findings about protein fractions and mineral content. In: N Margilia, W Martin-Rosset (eds.) *Nutrition and Feeding of the Broodmare*. EAAP publication No.20, Wageningen Academic Publishers, The Netherlands, pp. 65–76.

Masson PL, Heremans JF. 1971. Lactoferrin in milk from different species. *Comp Biochem Physiol* 39B: 119–129.

Matéos A et al. 2009a. Equine α_{s1}-casein: Characterization of alternative splicing isoforms and determination of phosphorylation levels. *J Dairy Sci* 92: 3604–3615.

Matéos A et al. 2009b. Two-dimensional cartography of equine β-casein variants achieved by isolation of phosphorylation isoforms and control of the deamidation phenomenon. *J Dairy Sci* 92: 2389–2399.

Mather IH. 2000. A review and proposed nomenclature for major proteins of the milk fat globule membrane. *J Dairy Sci* 83: 203–247.

McDonald TL et al. 2001. Elevated extra hepatic expression and secretion of mammary-associated serum amyloid A 3 (M-SAA3) into colostrum. *Vet Immunol Immunopathol* 3: 203–211.

McIntosh GH. 1996. Probiotics and colon cancer prevention. *Asia Pac J Clin Nutr* 5: 48–52.

McSweeney PLH, Fox PF. 2009. *Advanced Dairy Chemistry, vol 3: Lactose, Water, Salts and Minor Constituents*, 3rd edn. Springer Science, New York.

Mercier J-C et al. 1976. Comparative study of the amino acid sequences of the caseinomacropeptides from seven species. *FEBS Letters* 72: 208–214.

Mercier JC. 1986. Genetic engineering applied to milk producing animals: some expectations. In: C Smith (ed.) *Exploiting New Technologies in Animal Breeding: Genetic Developments* Proceedings of a Seminar in the CEC Animal Husbandry Research Programme, Oxford University Press, Edinburgh, pp. 122–131.

Michaelidou A, Steijns J. 2006. Nutritional and technological aspects of minor bioactive components in milk and whey: Growth factors, vitamins and nucleotides. *Intern Dairy J* 16: 1421–1426.

Miclo L et al. 2007. The primary structure of a low-Mr multiphosphorylated variant of β-casein in equine milk. *Proteomics* 7: 1327–1335.

Minkiewicz P et al. 1993. The contribution of *N*-acetylneuraminic acid on the stabilization of micellar casein. *Pol J Food Nutr Sci* 2: 39–48.

Miranda G et al. 2004. Proteomic tools to characterize the protein fraction of Equidae milk. *Proteomics* 4: 2496–2509.

Molenaar AJ et al. 1996. Elevation of lactoferrin gene expression in developing, ductal, resting, and regressing parenchymal epithelium of the ruminant mammary gland. *J Dairy Sci* 79: 1198–1208.

Montagne P et al. 1998. Microparticle-enhanced nephelometric immunoassay of lysozyme in milk and other human body fluids. *Clin Chem* 44: 1610–1615.

Monti G et al. 2007. Efficacy of donkey's milk in treating highly problematic cow's milk allergic children : an *in vivo* and *in vitro* study. *Pediatr Allergy Immunol* 18: 258–264.

Moore SA et al. 1997. Three dimensional structure of diferric bovine lactoferrin at 2.8 Å resolution. *J Mol Biol* 274: 222–236.

Morozova-Roche L. 2007. Equine lysozyme: the molecular basis of folding, self-assembly and innate amyloid toxicity. *FEBS Letters* 581: 2587–2592.

Nakamura T et al. 2001. Occurrence of an unusual phosphorylated *N*-acetyllactosamine in horse colostrum. *Biochimi et Biophy Acta* 1525: 13–18.

Nakazawa T, Hosono A. 1992. *Functions of Fermented Milk: Challenges for the Health Sciences.* Elsevier Applied Science, London.

Nakhasi HL et al. 1984. Expression of κ-casein in normal and neoplastic rat mammary gland is under the control of prolactin. *J Biol Chem* 259: 14894–14898.

Natale M et al. 2004. Cow's milk allergens identification by two-dimensional immunoblotting and mass spectrometry. *Mol Nutr Food Res* 48: 363–369.

Neseni R et al. 1958. Milchleistung und Milchzusammenstzung von Studen im Verlauf der Laktation. *Archiv fur Tierzucht* 1: 191–129.

Neville MC. 2009. Introduction: alpha-lactalbumin, a multifunctional protein that specifies lactose synthesis in the Golgi. *J Mammary Gland Biol Neoplasia* 14: 211–212.

Nitta K et al. 1987. The calcium-binding properties of equine lysozyme. *FEBS Letters* 223: 405–408.

Nonaka Y et al. 2008. Spontaneous asparaginyl deamidation of canine milk lysozyme under mild conditions. *Proteins* 72: 313–322.

O'Connell JE, Fox PF. 2003. Heat-induced coagulation of milk. In: PF Fox, PLH McSweeney (eds.) *Advanced Dairy Chemistry, vol. 1: Proteins*, 3rd edn. Kluwer Academic/Plenum Publishers, New York, pp. 879–945.

Ochirkhuyag B et al. 2000. Characterization of mare caseins. Identification of α_{s1}- and α_{s2}-caseins. *Lait* 80: 223–235.

Oftedal OT et al. 1983. Lactation in the horse: milk composition and intake by foals. *J Nutr* 113: 2096–2106.

Oftedal OT, Iverson SJ. 1995. Comparative analysis of non-human milks. In: RG Jensen (ed.) *Handbook of milk composition.* Academic Press, California, pp. 749–827.

Ono T et al. 1989. Subunit components of casein micelles from bovine, ovine, caprine and equine milk. *J Dairy Res* 56: 61–68.

Outram AK. 2009. The earliest horse harnessing and milking. *Science* 323: 1332–1335.

Pagliarini E et al. 1993. Chemical and physical characteristics of mare's milk. *Ital J Food Sci* 4: 323–332.

Palmquist DL. 2006. Milk fat: origin of fatty acids and influence of nutritional factors thereon. In: PF Fox, PLH McSweeney (eds.) *Advanced Dairy Chemistry, vol 2: Lipids*, 3rd edn. Springer, New York, pp. 43–92.

Park YW et al. 2006. Mare milk. In: YW Park, FW Haenlein (eds.) *Handbook of Non-bovine Mammals*. Blackwell Publishing, Oxford, pp. 275–296.

Parodi PW. 1982. Positional distribution of fatty acids in triglycerides from milk of several species of mammals. *Lipids* 17: 437–442.

Patton S. 1999. Some practical implications of the milk mucins. *J Dairy Sci* 82: 1115–1117.

Paulson M, Dejmek P. 1990. Thermal denaturation of whey proteins in mixtures with caseins studied by differential scanning calorimetry. *J Dairy Sci* 73: 590–600.

Pelizzola V et al. 2006. Chemical-physical characteristics and fatty acid composition of mare's milk. *Milchwissenschaft* 61: 33–36.

Pérez MD, Calvo M. 1995. Interaction of β-lactoglobulin with retinol and fatty acids and its role as a possible biological function for this protein: A review. *J Dairy Sci* 78: 978–988.

Pérez MD et al. 1993. Comparison of the ability to bind ligands of β-lactoglobulin and serum albumin from ruminant and non-ruminant species. *J Dairy Res* 60: 55–63.

Phadungath C. 2005. Casein micelle structure: a concise review. *Songklanakarin J Sci Technol* 27: 201–212.

Phelan M et al. 2009. Casein-derived bioactive peptides: Biological effects, industrial uses, safety aspects and regulatory status. *Intern Dairy J* 19: 643–654.

Piccione G et al. 2008. Daily rhythmicity in nutrient content of asinine milk. *Livestock Sci* 116: 323–327.

Pieszka M, Kulisa M. 2005. Magnesium content in mares' milk and growth parameters of their foals. *J Elementology* 10: 985–990.

Pikul J, Wójtowski J. 2008. Fat and cholesterol content and fatty acid composition of mares' colostrums and milk during five lactation months. *Livestock Sci* 113: 285–290.

Pisano A et al. 1994. Characterization of *O*-linked glycosylation motifs in the glycopeptide domain of bovine κ-casein. *Glycobiology* 4: 837–844.

Plowman JE et al. 1999. Structural features of a peptide corresponding to human κ-casein residues 84–101 by 1H-nuclear magnetic resonance spectroscopy. *J Dairy Res* 66: 53–63.

Qi PX. 2007. Studies of casein micelle structure: the past and the present. *Lait* 87: 363–383.

Rainhard P. 1986. Bacteriostatic activity of bovine milk lactoferrin against mastitic bacteria. *Veter Microbiol* 11: 387–392.

Ramos JJ et al. 1994. Vitamin B12 levels in ewe colostrum and milk and in lamb serum. *Vet Res* 25: 405–409.

Rasmussen LK et al. 1995. Human α_{s1}-casein: purification and characterization. *Comp Biochem PhysiolB* 111: 75–81.

Rassin DK et al. 1978. Taurine and other free amino acids in milk of man and other mammals. *Early Hum Dev* 2/1: 1–13.

Restani P et al. 1995. Evaluation by SDS-PAGE and immunoblotting of residual antigenicity in hydrolysed protein formulas. *Clin Exp Allergy* 25: 651–658.

Recio I et al. 2009. Bioactive components in sheep milk. In: Y.W. Park (ed.), *Bioactive Components in Milk and Dairy Products*. Wiley-Blackwell Publishers. Ames, IA, pp. 83–104.

Restani P et al. 2002. Cross-reactivity between mammalian proteins. *Ann Allergy Asthma Immunol* 89: 11–15.

Rudloff S, Kunz C. 1997. Protein and nonprotein nitrogen components in human milk, bovine milk, and infant formula: Quantitative and qualitative aspects in infant nutrition. *J Pediatr Gastroenterol Nutr* 26: 328–344.

Saito T, Itoh T. 1992. Variations and distributions of O-glycosidically linked sugar chains in bovine κ-casein. *J Dairy Sci* 75: 1768–1774.

Salimei E. 2011. Animals that produce dairy foods: Donkey. In: W. Fuquay et al. (eds.) *Encyclopedia of Dairy Sciences, vol 1*, 2nd edn. Elsevier. Oxford, pp. 365–373.

Salimei E et al. 2002. Major constituents leptin, and non-protein nitrogen compounds in mares' colostum and milk. *Reprod Nutr Dev* 42: 65–72.

Salimei E et al. 2004. Composition and characteristics of ass' milk. *Anim Res* 53: 67–78.

Sarwar G et al. 1998. Free amino acids in milk of human subjects, other primates and non-primates. *Brit J Nutr* 79: 129–131.

Savage RJG, Long MR. 1986. *Mammalian Evolution: an Illustrated Guide*. Natural History Museum, London.

Savalle B et al. 1988. Modelization of gastric digestion of milk proteins. In: CA Barth, E Schlimme (eds.) *Milk Proteins. Nutritional, Clinical, Functional and Technological Aspects*. Steinkopff Publishers, Darmstad, pp. 143–149.

Savilahti E, Kuitunen M. 1992. Allergenicity of cow milk proteins. *J Pediatr* 121: S12–S20.

Sawyer L. 2003. β-Lactoglobulin. In: PF Fox, PLH McSweeney (eds.) *Advanced Dairy Chemistry, vol. 1: Proteins*, 3rd edn. Kluwer Academic/Plenum Publishers, New York, pp. 319–386.

Schaafsma G. 2003. Nutritional significance of lactose and lactose derivatives. In: H Roginski, JW Fuquay, PF Fox (eds.) *Encyclopedia of Dairy Science*. Academic Press, London, pp. 1529–1533.

Schmidt DG et al. 1995. Raising the pH of the pepsin-catalyzed hydrolysis of bovine whey proteins increases the antigenicity of the hydrolysates. *Clin Exp Allergy* 25: 1007–1017.

Schryver HF et al. 1986. Lactation in the horse: the mineral composition of mare milk. *J Nutr* 116: 2142–2147.

Sellnow L. 2006. Food Factory. Article 10, Equine Anatomy and Physiology. Available at http://www.thehorse.com/pdf/anatomy/anatomy10.pdf. Accessed on October 3, 2010.

Sélo I et al. 1999. Allergy to bovine β-lactoglobulin: specificity of human IgE to tryptic peptides. *Clin Exp Allergy* 29: 1055–1063.

Shah NP. 2000. Effects of milk-derived bioactives: an overview. *Brit J Nutr* 84: S3–S10.

Shekar CP et al. 2006. κ-Casein-deficient mice fail to lactate. *Proc Natl Acad Sci U S A* 103: 8000–8005.

Shimazaki KI et al. 1994. Comparative study of the iron-binding strengths of equine, bovine and human lactoferrins. *J Dairy Res* 61: 563–566.

Silva SV, Malcata FX. 2005. Caseins as source of bioactive peptides. *Intern Dairy J* 15: 1–15.

Simopoulos AP. 2002. The importance of the ratio of omega-6/omega-3 essential fatty acids. *Biomedical Pharmacotherapy* 56: 365–379.

Singh H et al. 1997. Physico-chemical properties of milk. In: PF Fox (ed.) *Advanced Dairy Chemistry vol. 3: Lactose, Water, Salts and Vitamins*, 2nd edn. Chapman and Hall, London, pp. 469–518.

Smith MC, Sherman DM. 2009. *Goat Medicine*, 2nd edn. Wiley-Blackwell, New York.

Smolders EAA et al. 1990. Composition of horse milk during the suckling period. *Livest Prod Sci* 25: 163–171.

Sneddon JC, Argenzio RA. 1998. Feeding strategy and water homeostasis in equids: the role of the hind-gut. *J Arid Envir* 38: 493–509.

Sneddon JC et al. 2006. Mucosal surface area and fermentation activity in the hind gut of hydrated and chronically dehydrated working donkeys. *J Anim Sci* 84: 119–124.

Solaroli G et al. 1993. Composition and nutritional quality of mare's milk. *Ital J Food Sci* 5: 3–10.

Sood SM, Slattery CW. 2000. Association of the quadruply phosphorylated β-casein from human milk with the nonphosphorylated form. *J Dairy Sci* 83: 2766–2770.

Sørensen ES et al. 2003. The phosphorylation pattern of human α_{s1}-casein is markedly different from the ruminant species. *Eur J Biochem* 270: 3651–3655.

Souci SW et al. 2000. *Food Composition and Nutrition Tables*, 6th edn. Medpharm Scientific Publishers, Stuttgart.

Spencer A et al. 1999. Expression, purification, and characterization of the recombinant calcium-binding equine lysozyme secreted by the filamentous fungus Aspergillus niger: comparisons with the production of hen and human lysozymes. *Protein Expression and Purification* 16: 171–180.

Spitzauer S. 1999. Allergy to mammalian proteins: at the borderline between foreign and self? *Int Arch Allergy Immunol* 120: 259–269.

Stasiuk SX et al. 2000. Cloning of a marsupial kappa-casein cDNA from the brushtail possum (*Trichosurus vulpecula*). *Reprod Fertil Dev* 12: 215–222.

Stoyanova LG et al. 1988. Freeze-dried mares' milk and its potential use in infant and dietetic food products. *Voprosy Pitaniia* 2: 64–67.

Strömqvist M et al. 1995. Human milk β-casein and inhibition of Helicobacter pylori adhesion to human gastric mucosa. *J Pediatr Gastroenterol Nutr.* 21: 288–296.

Stuart CA et al. 1984. Passage of cows' milk protein in breast milk. *Clin Allergy* 14: 533–535.

Surono IS, Hosono A. 2002. Fermented milks: Types and standards of identity. In: H Roginski, JA Fuquay, PF Fox (eds.) *Encyclopedia of Dairy Sciences*. Academic Press, London, UK, pp. 1018–1069.

Swaisgood HF. 2003. Chemistry of caseins. In: PF Fox, PLH McSweeney (eds.) *Advanced Dairy Chemistry, vol 1: Proteins*, 3rd edn. Kluwer Academic/Plenum Publishers, New York, pp. 139–202.

Tada M et al. (2002). Stabilization of protein by replacement of a fluctuating loop: Structural analysis of a chimera of bovine α-lactalbumin and equine lysozyme. *Biochem* 41: 13807–13813.

Tamine AY, Marshall VME. 1984. Microbiology and technology of fermented milk. In: FL Davies, BA Law (eds.) *Advances in the Microbiology and Biochemistry of Cheese and Fermented Milk*. Elsevier Applied Science Publishers, London, pp. 118–122.

Tamine AY, Robinson RK. 1999. *Yoghurt: Science and Technology*. Woodhead Publishing Ltd., Cambridge.

Taylor SL. 1986. Immunologic and allergic properties of cows' milk proteins in humans. *J Food Prot* 49: 239–250.

Thomä-Worringer C et al. 2006. Health effects and technological features of caseinomacropeptide. *Intern Dairy J* 16: 1324–1333.

Trujillo JM et al. 1962. Chromosomes of the horse, the donkey, and the mule. *Chromosoma* 13: 243–248.

Tsioulpas A et al. 2007. Changes in physical properties of bovine milk from the colostrum period to early lactation. *J Dairy Sci* 90: 5012–5017.

Tsuge H et al. 1992. Crystallographic studies on a calcium-binding lysozyme from equine milk at 2.5 Å resolution. *J Biochem* 111: 141–143.

Turner AW. 1945. Digestibility of milk as affected by various types of treatment. *Food Res Inter* 10: 52–59.

Ullrey DE et al. 1966. Composition of mare's milk. *J Anim Sci* 25: 217–222.

Ullrey DE et al. 1974. Iron, zinc and copper in mare's milk. *J Anim Sci* 38: 1276–1277.

Uniacke-Lowe T. 2011. Studies on equine milk and comparative studies on equine and bovine milk systems. PhD Thesis. The National University of Ireland, Cork.

Uniacke T, Fox PF. 2011. Equid milk. In: J Fuquay et al. (eds.). *Encyclopedia of Dairy Sciences*, vol 3, 2nd edn. Elsevier, Oxford, pp. 518–529.

Uniacke T et al. 2010. Equine milk proteins: Chemistry, structure and nutritional significance. *Intern Dairy J* 20: 609–629.

Urashima T et al. 1989. Structural determination of three neutral oligosaccharides obtained from horse colostrum. *Carbohydr Res* 194: 280–287.

Urashima T et al. 2001. Oligosaccharides of milk and colostrum in non-human mammals. *Glycoconj J* 18: 357–371.

Urashima T et al. 2009. Milk oligosaccharides. In: PLH McSweeney, PF Fox (eds.) *Advanced Dairy Chemistry, vol. 3: Lactose, Water, Salts and Minor Constituents*, 3rd edn. Springer, London, pp. 295–349.

Van Hooydonk ACM et al. 1984. Kinetics of the chymosin-catalysed proteolysis of κ-casein in milk. *Neth Milk Dairy J* 38: 207–222.

Vesa TH et al. 2000. Lactose intolerance. *J Am Coll Nutr* 19: 165S–175S.

Vilà C et al. 2001. Widespread origins of domestic horse lineages. *Science* 291: 474–477.

Vincenzetti S et al. 2008. Donkey's milk protein fractions characterization. *Food Chem* 106: 640–649.

Visser S et al. 1982. Isolation and characterization of β- and γ-caseins in horse milk. *Biochem J* 203: 131–139.

Vreeman HJ et al. 1986. Characterization of bovine κ-casein fractions and the kinetics of chymosin-induced macropeptide release from carbohydrate-free and carbohydrate-containing fractions determined by high-performance gel-permeation chromatography. *Biochem J* 240: 87–97.

Waelchli RO et al. 1990. Relationships of total protein, specific gravity, viscosity, refractive index and latex agglutination to immunoglobulin G concentration in mare colostrum. *Equine Vet J* 22: 39–42.

Wal J-M. 2002. Cow's milk proteins/allergens. *Ann Allergy Asthma Immunol* 89: 3–10.

Wal J-M. 2004. Bovine milk allergenicity. *Ann Allergy Asthma Immunol* 93: S2–S11.

Walstra P. 1990. On the stability of casein micelles. *J Dairy Sci* 73: 1965–1979.

Walstra P, Jenness R. 1984. *Dairy Chemistry and Physics*. John Wiley & Sons, New York.

Walstra P et al. 2006a. *Dairy Science and Technology*. CRC Press, Boca Raton, FL.

Walstra P et al. 2006b. *Dairy Technology:Principles of Milk Properties and Processes*. CRC/Taylor Francis, New York.

Welsch U et al. 1988. Structural, histochemical and biochemical observations on horse milk-fat-globule membranes and casein micelles. *Histochemistry* 88: 357–365.

Whigham LD et al. 2000. Conjugated linoleic acid: implications for human health. *Pharmacol Res* 42: 503–510.

Widdowson EM. 1984. Lactation and feeding patterns in different species. In: DLM Freed (ed.) *Health Hazards of Milk*. Baillière Tindall, London, pp. 85–90.

Winter P et al. 2006. The potential of measuring serum amyloid A in individual ewe milk and in farm bulk milk for monitoring udder health on sheep dairy farms. *Res Vet Sci* 81: 321–326.

Wong NP et al. (eds.) 1988. *Fundamentals of Dairy Chemistry*, 3rd edn. Van Nostrand Reinhold, New York.

Yongfa W et al. 1984. Prevalence of primary adult lactose malabsorption in three populations of northern China. *Human Genetics* 67: 103–106.

Yoshikawa M et al. 1982. Chemical characteristics of bovine casein micelles fractionated by size on CPG-10/3000 chromatography. *Agri Biol Chem* 46: 1043–1048.

Zbikowska A et al. 1992. The influence of casein micelle size on selected functional properties of bulk milk proteins. *Polish J Food Nutr Sci* 1: 23–32.

Zicker SC, Lönnerdal B. 1994. Protein and nitrogen composition of equine (Equus caballus) milk during early lactation. *Comp Biochem Physiol* 108A: 411–421.

Part 5
Fruits, Vegetables, and Cereals

27
Biochemistry of Fruits

Gopinadhan Paliyath, Krishnaraj Tiwari, Carole Sitbon, and Bruce D. Whitaker

Abstract: Fruits are major ingredients of human diet and provide several nutritional ingredients including carbohydrates, vitamins and functional food ingredients such as soluble and insoluble fibers, polyphenols and carotenoids. Biochemical changes during fruit ripening make the fruit edible by making them soft, changing the texture through the breakdown of cell wall, converting acids or stored starch into sugars and causing the biosynthesis of pigments and flavour components. Fruits are processed into several products to preserve these qualities.

INTRODUCTION

Because of various health benefits associated with the consumption of fruits and various products derived from fruits, these are at the centre stage of human dietary choices in recent days. The selection of trees that produce fruits with ideal edible quality has been a common process throughout human history. Fruits are developmental manifestations of the seed-bearing structures in plants, the ovary. After fertilisation, the hormonal changes induced in the ovary result in the development of the characteristic fruit that may vary in ontogeny, form, structure and quality. Pome fruits such as apple and pear are developed from the development of the thalamus in the flower. Drupe fruits, such as cherry, peach, plum, apricot and so on, are developed from the ovary wall (mesocarp) enclosing a single seed. Berry fruits such as tomato possess the seeds embedded in a jelly-like pectinaceous matrix, with the ovary wall developing into the flesh of the fruit. Cucumbers and melons develop from an inferior ovary. Citrus fruits belong to the class hesperidium, where the ovary wall develops as a protective structure surrounding the juice-filled locules that become the edible part of the fruit. In strawberry, the seeds are located outside the fruit, and it is the receptacle of the ovary (central portion) that develops into the edible part. The biological purpose of the fruit is to attract vectors that help in the dispersal of the seeds. For this, the fruits have developed various organoleptic (stimulatory to organs) characteristics that include attractive colour, flavour and taste. The biochemical characteristics and pathways in the fruits are developmentally structured to achieve these goals. The nutritional and food qualities of fruits arise as a result of the accumulation of components derived from these intricate biochemical pathways. In terms of production and

Food Biochemistry and Food Processing, Second Edition. Edited by Benjamin K. Simpson, Leo M.L. Nollet, Fidel Toldrá, Soottawat Benjakul, Gopinadhan Paliyath and Y.H. Hui.

volume, tomato, orange, banana and grape are the major fruit crops used for consumption and processing around the world (Kays 1997).

BIOCHEMICAL COMPOSITION OF FRUITS

Fruits contain a large percentage of water, which can often exceed 95% by fresh weight. During ripening, activation of several metabolic pathways often leads to drastic changes in the biochemical composition of fruits. Fruits such as banana store starch during development, and hydrolyse the starch to sugars during ripening, that also results in fruit softening. Most fruits are capable of photosynthesis, store starch and convert them to sugars during ripening. Fruits such as apple, tomato, grape and so on have a high percentage of organic acids, which decreases during ripening. Fruits also contain large amounts of fibrous materials such as cellulose and pectin. The degradation of these polymers into smaller water-soluble units during ripening leads to fruit softening as exemplified by the breakdown of pectin in tomato and cellulose in avocado. Secondary plant products are major compositional ingredients in fruits. Anthocyanins are the major colour components in grape, blueberry, apple and plum; carotenoids, specifically lycopene and carotene, are the major components that impart colour in tomato and watermelon. Aroma is derived from several types of compounds that include monoterpenes (as in lime, orange), ester volatiles (ethyl, methyl butyrate in apple, isoamyl acetate in banana), simple organic acids such as citric and malic acids (citrus fruits, apple) and small chain aldehydes such as hexenal and hexanal (cucumber). Fruits are also rich in vitamin C. Lipid content is quite low in fruits, the exceptions being avocado and olives, in which triacylglycerols (oils) form the major storage components. The amounts of proteins are usually low in most fruits.

Carbohydrates, Storage and Structural Components

As the name implies, carbohydrates are organic compounds containing carbon, hydrogen and oxygen. Basically, all carbohydrates are derived by the photosynthetic reduction of CO_2 to the pentoses (ribose, ribulose) and hexoses (glucose, fructose), which are also intermediates in the metabolic pathways. Polymerisation of several sugar derivatives leads to various storage (starch, inulin) and structural components (cellulose, pectin).

During photosynthesis, the glucose formed is converted to starch and stored as starch granules. Glucose and its isomer fructose, along with phosphorylated forms (glucose-6-phosphate, glucose-1,6- diphosphate, fructose-6-phosphate and fructose-1,6-diphosphate), can be considered to be the major metabolic hexose pool components that provide carbon skeleton for the synthesis of carbohydrate polymers. Starch is the major storage carbohydrate in fruits. There are two molecular forms of starch, amylose and amylopectin and both components are present in the starch grain. Starch is synthesised from glucose phosphate by the activities of a number of enzymes designated as ADP-glucose pyrophosphorylase, starch synthase and a starch-branching enzyme. ADP-glucose pyrophosphorylase catalyses the reaction between glucose-1-phosphate and ATP that generates ADP-glucose and pyrophosphate. ADP-glucose is used by starch synthase to add glucose molecules to amylose or amylopectin chain, thus increasing their degree of polymerisation. By contrast to cellulose that is made up of glucose units in β-1,4-glycosidic linkages, the starch molecule contains glucose linked by α-1,4-glycosidic linkages. The starch branching enzyme introduces glucose molecules through α-1,6- linkages which further gets extended into linear amylose units with α-1,4- glycosidic linkages. Thus, the added glucose branch points (α-1,6-linkages) serve as sites for further elongation by starch synthase, thus resulting in a branched starch molecule, also known as amylopectin.

Cell wall is a complex structure composed of cellulose and pectin, derived from hexoses such as glucose, galactose, rhamnose and mannose, and pentoses such as xylose and arabinose, as well as some of their derivatives such as glucuronic and galacturonic acids (Negi and Handa 2008). A model proposed by Keegstra et al. (1973) describes the cell wall as a polymeric structure constituted by cellulose microfibrils and hemicellulose embedded in the apoplastic matrix in association with pectic components and proteins. In combination, these components provide the structural rigidity that is characteristic to the plant cell. Most of the pectin is localised in the middle lamella. Cellulose is biosynthesised by the action of β-1,4-glucan synthase enzyme complexes that are localised on the plasma membrane. The enzyme uses uridine diphosphate glucose (UDPG) as a substrate, and by adding UDPG units to small cellulose units, extends the length and polymerisation of the cellulose chain. In addition to cellulose, there are polymers made of different hexoses and pentoses known as hemicelluloses, and based on their composition, they are categorised as xyloglucans, glucomannans and galactoglucomannans. The cellulose chains assemble into microfibrils through hydrogen bonds to form crystalline structures. In a similar manner, pectin is biosynthesised from UDP-galacturonic acid (galacturonic acid is derived from galactose, a six carbon sugar), as well as other sugars and derivatives and includes galacturonans and rhamnogalacturonans that form the acidic fraction of pectin. As the name implies, rhamnogalacturonans are synthesised primarily from galacturonic acid and rhamnose. The acidic carboxylic groups complex with calcium that provide the rigidity to the cell wall and the fruit. The neutral fraction of the pectin comprises polymers such as arabinans (polymers of arabinose), galactans (polymers of galactose) or arabinogalactans (containing both arabinose and galactose). All these polymeric components form a complex three-dimensional network stabilised by hydrogen bonds, ionic interactions involving calcium, phenolic components such as diferulic acid and hydroxyproline-rich glycoproteins (Fry 1986, Negi and Handa 2008). It is also important to visualise that these structures are not static and the components of cell wall are constantly being turned over in response to growth conditions.

Lipids and Biomembranes

By structure, lipids can form both structural and storage components. The major forms of lipids include fatty acids, diacyl- and triacylglycerols, phospholipids, sterols, and waxes that provide an external barrier to the fruits. Fruits in general are not rich in lipids with the exception of avocado and olives that store large amounts of triacylglycerols or oil. As generally observed in plants, the major fatty acids in fruits include palmitic (16:0), stearic (18:0), oleic (18:1), linoleic (18:2) and linolenic (18:3) acids. Among these, oleic, linoleic and linolenic acids possess an increasing degree of unsaturation. Olive oil is rich in triacylglycerols containing the monounsaturated oleic acid and is considered as a healthy ingredient for human consumption.

Compartmentalisation of cellular ingredients and ions is an essential characteristic of all life forms. The compartmentalisation is achieved by biomembranes, formed by the assembly of phospholipids and several neutral lipids that include diacylglycerols and sterols, the major constituents of the biomembranes. Virtually, all cellular structures include or are enclosed by biomembranes. The cytoplasm is surrounded by the plasma membrane, the biosynthetic and transport compartments such as the endoplasmic reticulum and Golgi bodies form an integral network of membranes within the cell. Photosynthesis, which converts light energy into chemical energy, occurs on the thylakoid membrane matrix in the chloroplast, and respiration, which further converts chemical energy into more usable forms, occurs on the mitochondrial cristae. All these membranes have their characteristic composition and enzyme complexes to perform their designated function.

The major phospholipids that constitute the biomembranes include phosphatidylcholine, phosphatidylethanolamine, phosphatidylglycerol and phosphatidylinositol. Their relative proportion may vary from tissue to tissue. In addition, metabolic intermediates of phospholipids such as phosphatidic acid, diacylglycerols, free fatty acids and so on are also present in the membrane in lower amounts. Phospholipids are integral functional components of hormonal and environmental signal transduction processes in the cell. Phosphorylated forms of phosphatidylinositol such as phosphatidylinositol-4- phosphate and phosphatidylinositol-4,5-bisphosphate are formed during signal transduction events, though their amounts can be very low. The membrane also contains sterols such as sitosterol, campesterol and stigmasterol, as well as their glucosides, and they are extremely important for the regulation of membrane fluidity and function (Whitaker 1988, 1991, 1993, 1994).

Biomembranes are bilamellar layers of phospholipids. The amphipathic nature of phospholipids having hydrophilic head groups (choline, ethanolamine, etc.) and hydrophobic fatty acyl chains, thermodynamically favour their assembly into bilamellar or micellar structures when exposed to an aqueous environment. In a biomembrane, the hydrophilic headgroups are exposed to the external aqueous environment. The phospholipid composition between various fruits may differ, and within the same fruit, the inner and outer lamella of the membrane may have a different phospholipids composition. Such differences may cause changes in polarity between the outer and inner lamellae of the membrane, and lead to the generation of a voltage across the membrane. These differences usually become operational during signal transduction events.

An essential characteristic of the membrane is its fluidity. The fluid-mosaic model of the membrane (Singer and Nicholson 1972) depicts the membrane as a planar matrix comprising phospholipids and proteins. The proteins are embedded in the membrane bilayer (integral proteins) or are bound to the periphery (peripheral proteins). The nature of this interaction results from the structure of the proteins. If the proteins have a much larger proportion of hydrophobic amino acids, they would tend to become embedded in the membrane bilayer. If the protein contains more hydrophilic amino acids it may tend to prefer a more aqueous environment, and thus remain as a peripheral protein. In addition, proteins may be covalently attached to phospholipids such as phosphatidylinositol. Proteins that remain in the cytosol may also become attached to the membrane in response to an increase in cytosolic calcium levels. The membrane is a highly dynamic entity. The semi-fluid nature of the membrane allows for the movement of phospholipids in the plane of the membrane, and between the bilayers of the membrane. The proteins are also mobile within the plane of the membrane. However, this process is not always random and is regulated by the functional assembly of proteins into metabolons (functional assembly of enzymes and proteins, e.g., photosynthetic units in thylakoid membrane, respiratory complexes in the mitochondria, cellulose synthase on plasma membrane, etc.), their interactions with the underlying cytoskeletal system (network of proteins such as actin and tubulin), and the fluidity of the membrane.

The maintenance of homeostasis (life processes) requires the maintenance of the integrity and function of discrete membrane compartments. This is essential for the compartmentalisation of ions and metabolites, which may otherwise destroy the cell. For instance, calcium ions are highly compartmentalised within the cell. The concentration of calcium is maintained at the millimolar levels within the cell wall compartment (apoplast), endoplasmic reticulum and the tonoplast (vacuole). This is achieved by energy-dependent transport of calcium from the cytoplasm into these compartments by ATPases. As a result, the cytosolic calcium levels are maintained at low micromolar ($<1\ \mu M$) levels. Maintenance of this concentration gradient across the membrane is a key requirement for the signal transduction events, as regulated entry of calcium into the cytosol can be achieved simply by opening calcium channels. Calcium can then activate several cellular biochemical reactions that mediate the response to the signal. Calcium is pumped back into the storage compartments when the signal diminishes in intensity. In a similar manner, cytosolic pH is highly regulated by the activity of proton ATPases. The pH of the apoplast and the vacuole is maintained near four, whereas the pH of the cytosol is maintained in the range of 6–6.5. The pH gradient across the membrane is a key feature that regulates the absorption, or extrusion of other ions and metabolites such as sugars. The cell could undergo senescence if this compartmentalisation is lost.

There are several factors that affect the fluidity of the membrane. The major factor that affects the fluidity is the type and proportion of acyl chain fatty acids of the phospholipids. At a given temperature, a higher proportion of unsaturated fatty acyl chains (oleic, linoleic, linolenic) in the phospholipids can increase the fluidity of the membrane. An increase in saturated fatty acids such as palmitic and stearic acids can decrease the fluidity. Other membrane components such as sterols, and degradation products of fatty acids such as fatty aldehydes, alkanes and so on can also decrease the fluidity. Based on the physiological status of the tissue, the membrane can exist in either a liquid crystalline state (where the phospholipids and their acyl chains are mobile) or a gel state where they are packed as rigid ordered structures and their movements are much restricted. The membrane usually has co-existing domains of liquid crystalline and gel phase lipids depending on growth conditions, temperature, ion concentration near the membrane surface and so on. The tissue has the ability to adjust the fluidity of the membrane by altering the acyl lipid composition of the phospholipids. For instance, an increase in the gel phase lipid domains resulting from exposure to cold temperature could be counteracted by increasing the proportion of fatty acyl chains having a higher degree of unsaturation, and, therefore, a lower melting point. Thus, the membrane will tend to remain fluid even at a lower temperature (Whitaker 1991, 1992, 1993, 1994). An increase in gel phase lipid domains can result in the loss of compartmentalisation. The differences in the mobility properties of phospholipid acyl chains can cause packing imperfections at the interface between gel and liquid crystalline phases, and these regions can become leaky to calcium ions and protons that are highly compartmentalised. The membrane proteins are also excluded from the gel phase into the liquid crystalline phase. Thus, during examinations of membrane structure by freeze fracture electron microscopy, the gel phase domains can appear as regions devoid of proteins (Paliyath and Thompson 1990).

Proteins

Fruits, in general, are not very rich sources of proteins. During the early growth phase of fruits, the chloroplasts and mitochondria are the major organelles that contain structural proteins. The structural proteins include the light-harvesting complexes in chloroplast or the respiratory enzyme/protein complexes in mitochondria. Ribulose-bis-phosphate carboxylase/oxygenase (Rubisco) is the most abundant enzyme in photosynthetic tissues. Fruits do not store proteins as an energy source. The green fruits such as bell peppers and tomato have a higher level of chloroplast proteins.

Organic Acids

Organic acids are major components of some fruits. The acidity of fruits arises from the organic acids that are stored in the vacuole, and their composition can vary depending on the type of fruit. In general, young fruits contain more organic acids, which may decline during maturation and ripening due to their conversion to sugars (gluconeogenesis; eg. conversion of malic acid into glucose during ripening of apple). Some fruit families are characterised by the presence of certain organic acids. For example, fruits of oxalidaceae members (e.g., starfruit, *Averrhoa carambola*) contain oxalic acid, and fruits of the citrus family, rutaceae, are rich in citric acid. Apples contain malic acid and grapes are characterised by the presence of tartaric acid. In general, citric and malic acids are the major organic acids of fruits. Grapes contain tartaric acid as the major organic acid. During ripening, these acids can enter the citric acid cycle and undergo further metabolic conversions.

L-(+)tartaric acid is the optically active form of tartaric acid in grape berries. A peak in acid content is observed before the initiation of ripening, and the acid content declines on a fresh weight basis during ripening. Tartaric acid can be biosynthesised from carbohydrates and other organic acids. Radiolabelled glucose, glycolate and ascorbate were all converted to tartarate in grape berries. Malate can be derived from the citric acid cycle or through carbon dioxide fixation of pyruvate by the malic enzyme (NADPH-dependent malate dehydrogenase). Malic acid, as the name implies, is also the major organic acid in apples.

FRUIT RIPENING AND SOFTENING

Fruit ripening is a physiological event that results from a very complex and interrelated biochemical changes that occur in the fruits. Ripening is the ultimate stage of the development of the fruit, which entails the development of ideal organoleptic characters such as taste, colour and aroma that are important features of attraction for the vectors (animals, birds, etc.) responsible for the dispersal of the fruit, and thus the seeds, in the ecosystem. Human beings have developed an agronomic system of cultivation, harvest and storage of fruits with ideal food qualities. In most cases, the ripening process is very fast, and the fruits undergo senescence resulting in the loss of desirable qualities. An understanding of the biochemistry and molecular biology of the fruit ripening process has resulted in developing biotechnological strategies for the preservation of postharvest shelf life and quality of fruits (Negi and Handa 2008, Paliyath et al. 2008a).

A key initiator of the ripening process is the gaseous plant hormone ethylene. In general, all plant tissues produce a low, basal, level of ethylene. Based on the pattern of ethylene production and responsiveness to externally added ethylene, fruits are generally categorised into climacteric and non-climacteric fruits. During ripening, the climacteric fruits show a burst in ethylene production and respiration (CO_2 production). Non-climacteric fruits show a considerably low level of ethylene production. In climacteric fruits (apple, pear, banana, tomato, avocado, etc.), ethylene production can reach levels of 30–500 ppm (parts per million, microlitre/L), whereas in non-climacteric fruits (orange, lemon, strawberry, pineapple, etc.) ethylene levels usually are in the range of 0.1–0.5 ppm. Ethylene can stimulate its own biosynthesis in climacteric fruits, known as autocatalytic ethylene production. As well, the respiratory carbon dioxide evolution increases in response to ethylene treatment, termed as the respiratory climacteric. Climacteric fruits respond to external ethylene treatment by accelerating the respiratory climacteric and time required for ripening, in a concentration-dependent manner. Non-climacteric fruits show increased respiration in response to increasing ethylene concentration without accelerating the time required for ripening.

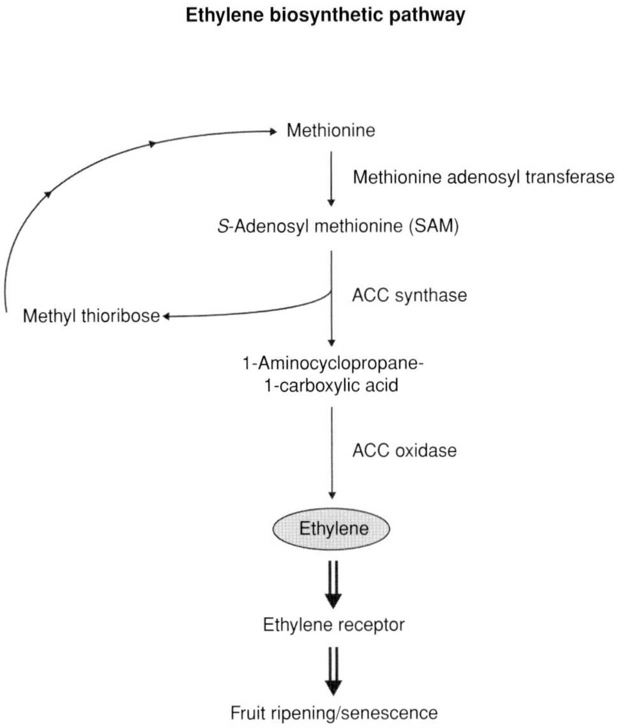

Figure 27.1. Summary of ethylene biosynthesis and action during fruit ripening. ACC, 1-Aminocyclopropane 1- Carboxylic Acid.

Ethylene is biosynthesised through a common pathway that uses the amino acid methionine as the precursor (Yang 1981, Fluhr and Mattoo 1996) (Fig. 27.1). The first reaction of the pathway involves the conversion of methionine to *S*-adenosyl methionine (SAM) mediated by the enzyme methionine adenosyl transferase. SAM is further converted into 1-aminocyclopropane-1-carboxylic acid (ACC) by the enzyme ACC synthase. The sulphur moiety of methylthioribose generated during this reaction is recycled back to methionine by the action of a number of enzymes. ACC is the immediate precursor of ethylene and is acted upon by ACC oxidase to generate ethylene. ACC synthase and ACC oxidase are the key control points in the biosynthesis of ethylene. ACC synthase is a soluble enzyme located in the cytoplasm, with a relative molecular mass of 50 kDa (kiloDalton). ACC oxidase is found to be associated with the vacuolar or mitochondrial membrane. Using molecular biology tools, a cDNA (complementary DNA representing the coding sequences of a gene) for ACC oxidase was isolated from tomato (Hamilton et al. 1991) and is found to encode a protein with a relative molecular mass of 35 kDa. There are several isoforms of ACC-synthase. These are differentially expressed in response to wounding, other stress factors and at the initiation of ripening. ACC oxidase reaction requires Fe^{2+}, ascorbate and oxygen.

Regulation of the activities of ACC synthase and ACC oxidase is extremely important for the preservation of shelf life and quality in fruits. Inhibition of the ACC synthase and ACC oxidase gene expression by the introduction of their respective antisense cDNAs resulted in delayed ripening and better preservation of the quality of tomato (Hamilton et al. 1990, Oeller et al. 1991) and apple (Hrazdina et al. 2000) fruits. ACC synthase, which is the rate-limiting enzyme of the pathway, requires pyridoxal-5-phosphate as a cofactor, and is inhibited by pyridoxal phosphate inhibitors such as aminoethoxyvinylglycine (AVG) and aminooxy acetic acid (AOA). Field application of AVG as a growth regulator (Retain™, Valent Biosciences, Chicago) has been used to delay ripening in fruits such as apples, peaches and pears. Also, commercial storage operations employ controlled atmosphere with very low oxygen levels (1–3%) for long-term storage of fruits such as apples to reduce the production of ethylene, as oxygen is required for the conversion of ACC to ethylene.

In response to the initiation of ripening, several biochemical changes are induced in the fruit, which ultimately results in the development of ideal texture, taste, colour and flavour. Several biochemical pathways are involved in these processes as described in the subsequent text.

Carbohydrate Metabolism

Cell Wall Degradation

Cell wall degradation is the major factor that causes softening of several fruits. This involves the degradation of cellulose components, pectin components or both. Cellulose is degraded by the enzyme cellulase or β-1,4-glucanase. Pectin degradation involves the enzymes pectin methylesterase, polygalacuronase (pectinase) and β-galactosidase (Negi and Handa 2008). The degradation of cell wall can be reduced by the application of calcium as a spray or drench in apple fruits. Calcium binds and cross-links the free carboxylic groups of polygalacturonic acid components in pectin. Calcium treatment, therefore, also enhances the firmness of the fruits.

The activities of both cellulase and pectinase have been observed to increase during ripening of avocado fruits and result in their softening. Cellulase is an enzyme with a relative molecular mass of 54.2 kDa and formed by extensive post-translational processing of a native 54 kDa protein involving proteolytic cleavage of the signal peptide and glycosylation (Bennet and Christopherson 1986). Further studies have shown three isoforms of cellulose ranging in molecular masses between 50 and 55 kDa. These forms are associated with the endoplasmic reticulum, the plasma membrane and the cell wall (Dallman et al. 1989). The cellulase isoforms are initially synthesised at the style end of the fruit at the initiation of ripening, and the biosynthesis progressively increases towards the stalk end of the fruit with the advancement of ripening. Degradation of hemicelluloses (xyloglucans, glucomannans and galactoglucomannans) is also considered as an important feature that leads to fruit softening. Degradation of these polymers could be achieved by cellulases and galactosidases.

Loss of pectic polymers through the activity of polygalacturonases (PG) is a major factor involved in the softening of fruits such as tomato. There are three major isoforms of PG responsible for pectin degradation in tomato, designated as PG1, PG2a and PG2b (Fischer and Bennet 1991). PG1 has a relative molecular mass of 100 kDa, and is the predominant form at the initiation of ripening. With the advancement of ripening, PG2a and PG2b isoforms increase, becoming the predominant isoforms in ripe fruit. The different molecular masses of the isozymes result from the post-translational processing and glycosylation of the polypeptides. PG2a (43 kDa) and PG2b (45 kDa) appear to be the same polypeptide with different degrees of glycosylation. PG1 is a complex of three polypeptides, PG2a, PG2b and a 38kDa subunit known as the β-subunit. The 38 kDa subunit is believed to exist in the cell wall space where it combines with PG2a and PG2b forming the PG1 isoform of PG. The increase in activity of PG1 is related to the rate of pectin solubilisation and tomato fruit softening during the ripening process.

Research into the understanding of the regulation of biosynthesis and activity of PG using molecular biology tools has resulted in the development of strategies for enhancing the shelf life and quality of tomatoes. PG mRNA was one of the first ripening-related mRNAs isolated from tomato fruits. All the different isoforms of PGs are encoded by a single gene. The PG cDNA which has an open reading frame of 1371 bases encodes a polypeptide having 457 amino acids, which includes a 24 amino acid signal sequence (for targeting to the cell wall space) and a 47 amino acid pro-sequence at the N-terminal end, which are proteolytically removed during the formation of the active PG isoforms. A 13 amino acid long C-terminal peptide is also removed resulting in a 373 amino acid long polypeptide, which undergoes different degrees of glycosylation resulting in the PG2a and PG2b isozymes. Complex formation among PG2a, PG2b and the 38-kDa subunit in the apoplast results in the PG1 isozyme (Grierson et al. 1986, Bird et al. 1988). In response to ethylene treatment of mature green tomato fruits which stimulates ripening, the levels of PG mRNA and PG are found to increase. These changes can be inhibited by treating tomatoes with silver ions, which interfere with the binding of ethylene to its receptor and initiation of ethylene action (Davies et al. 1988). Thus, there is a link between ethylene, PG synthesis and fruit softening.

Genetic engineering of tomato with the objective of regulating PG activity has yielded complex results. In the rin mutant of tomato which lacks PG and does not soften, introduction of a PG gene resulted in the synthesis of an active enzyme; however, this did not cause fruit softening (Giovannoni et al. 1989). As a corollary to this, introduction of the PG gene in the antisense orientation resulted in near total inhibition of PG activity (Smith et al. 1988). In both these cases, there was very little effect on fruit softening, suggesting that factors other than pectin de-polymerisation may play an integral role in fruit softening. Further studies using a tomato cultivar such as UC82B (Kramer et al. 1992) showed that antisense inhibition of ethylene biosynthesis or PG did indeed result in lowered PG activity, improved integrity of cell wall and increased fruit firmness during fruit ripening. As well, increased activity of pectin methylesterase, which removes the methyl groups from esterified galacturonic acid moieties, may contribute to the fruit softening process.

The activities of pectin degrading enzymes have been related to the incidence of physiological disorders such as "mealiness" or "wooliness" in mature unripened peaches that are stored at a low temperature. The fruits with such a disorder show a lack of juice and a dry texture. De-esterification of pectin by the activity of pectin methyl esterase is thought to be responsible for the development of this disorder. Pectin methyl esterase isozymes with relative molecular masses in the range of 32 kDa have been observed in peaches, and their activity increases after 2 weeks of low temperature storage. Polygalacuronase activity increases as the fruit ripens. The ripening fruits which possess both polygalacturonase and pectin methyl esterase do not develop mealy symptoms when stored at low temperature implicating the potential role of pectin degradation in the development of mealiness in peaches.

There are two forms of PG in peaches, the exo- and endo-PG. The endo-PG are the predominant forms in the freestone type of peaches, whereas the exo-PG are observed in the mesocarp of both freestone and clingstone varieties of peaches. As the name implies, exo-PG remove galacturonic acid moieties of pectin from the terminal reducing end of the chain, whereas the endo-PG can cleave the pectin chain at random within the chain. The activities of these enzymes increase during the ripening and softening of the fruit. Two exo-PG isozymes have been identified in peach, having a relative molecular mass of near 66 kDa. The exo-acting enzymes are activated by calcium. Peach endo-PG is observed to be similar to the tomato endo-PG. The peach endo-PG is inhibited by calcium. The freestone peaches possess enhanced activities of both exo-PG and endo-PG leading to a high degree of fruit softening. However, the clingstone varieties with low levels of endo-PG activity do not soften as the freestone varieties. In general, fruits such as peaches, tomatoes, strawberries, pears and so on, which soften extensively, possess high levels of endo-PG activity. Apple fruits which remain firm lack endo-PG activity.

Starch Degradation

Starch is the major storage form of carbohydrates. During ripening, starch is catabolised into glucose and fructose, which enters the metabolic pool where they are used as respiratory substrates or further converted to other metabolites (Fig. 27.2). In fruits such as banana, the breakdown of starch into simple sugars is associated with fruit softening. There are several enzymes involved in the catabolism of starch. α-amylase hydrolyses amylose molecules by cleaving the α-1,4-linkages between sugars providing smaller chains of amylose termed as dextrins. β-amylase is another enzyme that acts upon the glucan chain releasing maltose, which is a diglucoside. The dextrins as well as maltose can be further catabolised to simple glucose units by the action of glucosidases. Starch phosphorylase is another enzyme, which mediates the phosphorylytic cleavage of terminal glucose units at the non-reducing end of the starch molecule using inorganic phosphate, thus releasing glucose-1-phosphate. The amylopectin molecule is not only degraded in a similar manner to amylose but also involves the action of de-branching

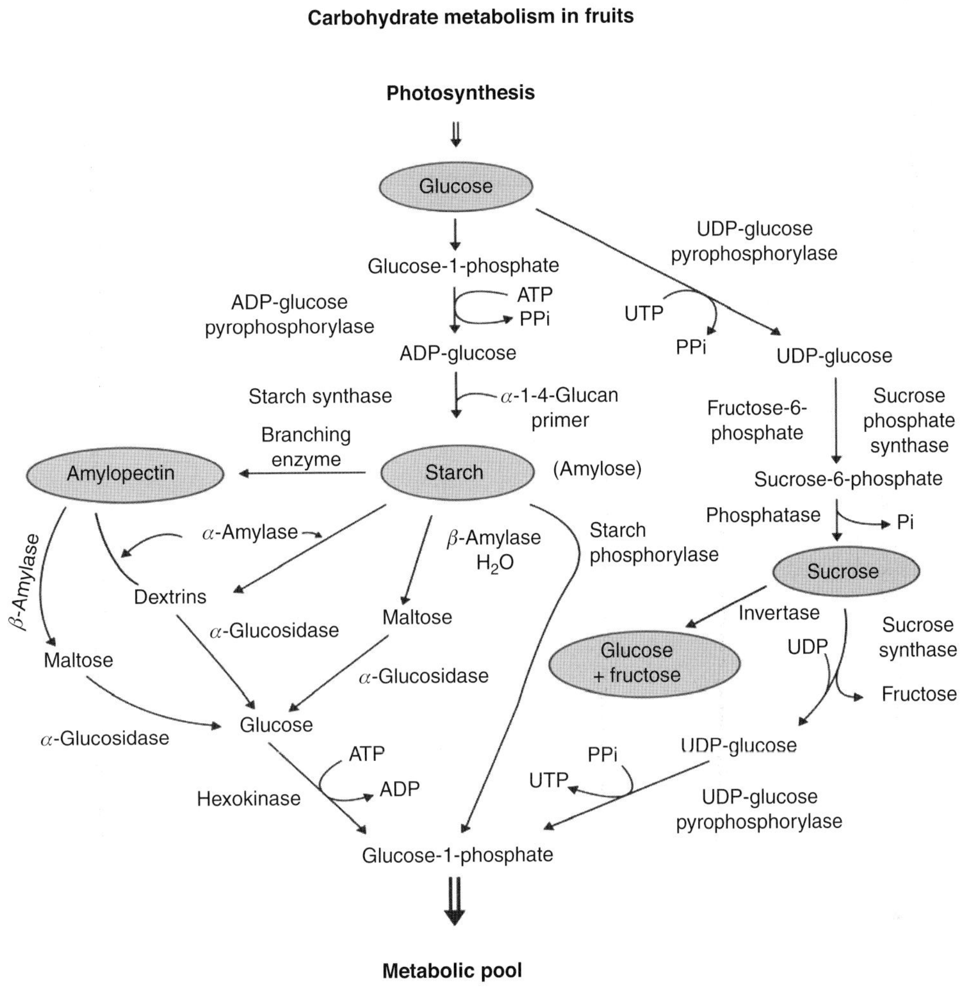

Figure 27.2. Carbohydrate metabolism in fruits. UDP, Uridine diphosphate; UTP, Uridine triphosphate.

enzymes which cleaves the α-1,6-linkages in amylopectin and releases linear units of the glucan chain.

In general, starch is confined to the plastid compartments of fruit cells, where it exists as granules made up of both amylose and amylopectin molecules. The enzymes that catabolise starch are also found in this compartment and their activities increase during ripening. The glucose-1-phosphate generated by starch degradation (Fig. 27.2) is mobilised into the cytoplasm where it can enter into various metabolic pools such as that of glycolysis (respiration), pentose phosphate pathway (PPP) or for turnover reactions that replenish lost or damaged cellular structures (cell wall components). It is important to visualise that the cell always tries to extend its life under regular developmental conditions (the exceptions being programmed cell death which occurs during hypersensitive response to kill invading pathogens, thus killing both the pathogen and the cell/tissue; formation of xylem vessels, secondary xylem tissues, etc.), and the turnover reactions are a part of maintaining the homeostasis. The cell ultimately succumbs to the catabolic reactions during senescence. The compartmentalisation and storage of chemical energy in the form of metabolisable macromolecules are all the inherent properties of life, which is defined as a struggle against increasing entropy.

The biosynthesis and catabolism of sucrose is an important part of carbohydrate metabolism. Sucrose is the major form of transport sugar and is translocated through the phloem tissues to other parts of the plant. It is conceivable that carbon dioxide fixed during photosynthesis in leaf tissues may be transported to the fruits as sucrose during fruit development. Sucrose is biosynthesised from glucose-1-phosphate by three major steps (Fig. 27.2). The first reaction involves the conversion of glucose-1-phosphate to UDP-glucose by UDP-glucose pyrophosphorylase in the presence of UTP (Uridine triphosphate). UDP-glucose is also an important substrate for the biosynthesis of cell wall components such as cellulose. UDP-glucose is converted to sucrose-6-phosphate by the enzyme sucrose phosphate synthase (SPS), which utilises fructose-6-phosphate during this reaction. Finally, sucrose is formed from sucrose-6-phosphate by the action of phosphatase with the liberation of the inorganic phosphate.

Even though sucrose biosynthesis is an integral part of starch metabolism, sucrose often is not the predominant sugar that accumulates in fruits. Sucrose is further converted into glucose and fructose by the action of invertase, which are characteristic to many ripe fruits. By the actions of sucrose synthase and UDP-glucose pyrophosphorylase, glucose-1-phosphate can be regenerated from sucrose. As well, sugar alcohols such as sorbitol and mannitol formed during sugar metabolism are major transport and storage components in apple and olive, respectively.

Biosynthesis and catabolism of starch has been extensively studied in banana, where prior to ripening, it can account for 20–25% by fresh weight of the pulp tissue. All the starch degrading enzymes, α-amylase, β-amylase, α-glucosidase and starch phosphorylase, have been isolated from banana pulp. The activities of these enzymes increase during ripening. Concomitant with the catabolism of starch, there is an accumulation of the sugars, primarily, sucrose, glucose and fructose. At the initiation of ripening, sucrose appears to be the major sugar component, which declines during the advancement of ripening with a simultaneous increase in glucose and fructose through the action of invertase (Beaudry et al. 1989). Mango is another fruit which stores large amounts of starch. The starch is degraded by the activities of amylases during the ripening process. In mango, glucose, fructose and sucrose are the major forms of simple sugars (Selvaraj et al. 1989). The sugar content is generally very high in ripe mangoes and can reach levels in excess of 90% of the total soluble solids content. By contrast to the bananas, the sucrose levels increase with the advancement of ripening in mangoes, potentially due to gluconeogenesis from organic acids (Kumar and Selvaraj 1990). As well, the levels of pentose sugars increase during ripening, and could be related to an increase in the activity of the PPP.

Glycolysis

The conversion of starch to sugars and their subsequent metabolism occur in different compartments. During the development of fruits, photosynthetically fixed carbon is utilised for both respiration and biosynthesis. During this phase, the biosynthetic processes dominate. As the fruit matures and begin to ripen, the pattern of sugar utilisation changes. Ripening is a highly energy-intensive process. And this is reflected in the burst in respiratory carbon dioxide evolution during ripening. As mentioned earlier, the respiratory burst is characteristic of some fruits which are designated as climacteric fruits. The post-harvest shelf life of fruits can depend on their intensity of respiration. Fruits such as mango and banana possess high level of respiratory activity and are highly perishable. The application of controlled atmosphere conditions having low oxygen levels and low temperature have thus become a routine technology for the long-term preservation of fruits.

The sugars and sugar phosphates generated during the catabolism of starch are metabolised through the glycolysis and citric acid cycle (Fig. 27.3). Sugar phosphates can also be channelled through the PPP, which is a major metabolic cycle that provides reducing power for biosynthetic reactions in the form of NADPH, as well as supplying carbon skeletons for the biosynthesis of several secondary plant products. The organic acids stored in the vacuole are metabolised through the functional reversal of respiratory pathway and is termed as gluconeogenesis. Altogether, sugar metabolism is a key biochemical characteristic of the fruits.

In the glycolytic steps of reactions (Fig. 27.3), glucose-6-phosphate is isomerised to fructose-6-phosphate by the enzyme hexose-phosphate isomerase. Glucose 6-phosphate is derived from glucose-1-phosphate by the action of glucose phosphate mutase. Fructose-6-phosphate is phosphorylated at the C1 position, yielding fructose-1,6- bisphosphate. This reaction is catalysed by the enzyme phosphofructokinase (PFK) in the presence of ATP. Fructose-1,6-bisphosphate is further cleaved into two three carbon intermediates, dihydroxyacetone phosphate and glyceraldehyde-3-phosphate, catalysed by the enzyme aldolase. These two compounds are interconvertible through an isomerisation reaction mediated by triose phosphate isomerase. Glyceraldehyde-3-phosphate is subsequently phosphorylated at the C1 position using orthophosphate, as well as oxidised using NAD, to generate 1,3-diphosphoglycerate and NADH. In the next reaction, 1,3-diphosphoglycerate is dephosphorylated by glycerate-3-phosphate kinase in the presence of ADP, along with the formation of ATP. Glycerate-3-phosphate formed during this reaction is further isomerised to 2-phosphoglycerate in the presence of phosphoglycerate mutase. In the presence of the enzyme enolase, 2-phosphoglycerate is converted to phosphoenol pyruvate (PEP). Dephosphorylation of phosphoenolpyruvate in the presence of ADP by pyruvate kinase yields pyruvate and ATP. Metabolic fate of pyruvate is highly regulated. Under normal conditions, it is converted to acetyl CoA, which then enters the citric acid cycle. Under anaerobic conditions, pyruvate can be metabolised to ethanol, which is a by-product in several ripening fruits.

There are two key regulatory steps in glycolysis, one mediated by PFK and the other by pyruvate kinase. In addition, there are other types of modulation involving cofactors and enzyme structural changes reported to be involved in glycolytic control. ATP levels increase during ripening. However, in fruits, this does not cause a feed back inhibition of PFK as observed in animal systems. There are two isozymes of PFK in plants, one localised in plastids and the other localised in the cytoplasm. These isozymes regulate the flow of carbon from the hexose phosphate pool to the pentose phosphate pool. PFK isozymes are strongly inhibited by PEP. Thus, any conditions that may cause the accumulation of PEP will tend to reduce the carbon flow through glycolysis. By contrast, inorganic phosphate is a strong activator of PFK. Thus, the ratio of PEP to inorganic phosphate would appear to be the major factor that regulates the activity of PFK and carbon flux through glycolysis. Structural

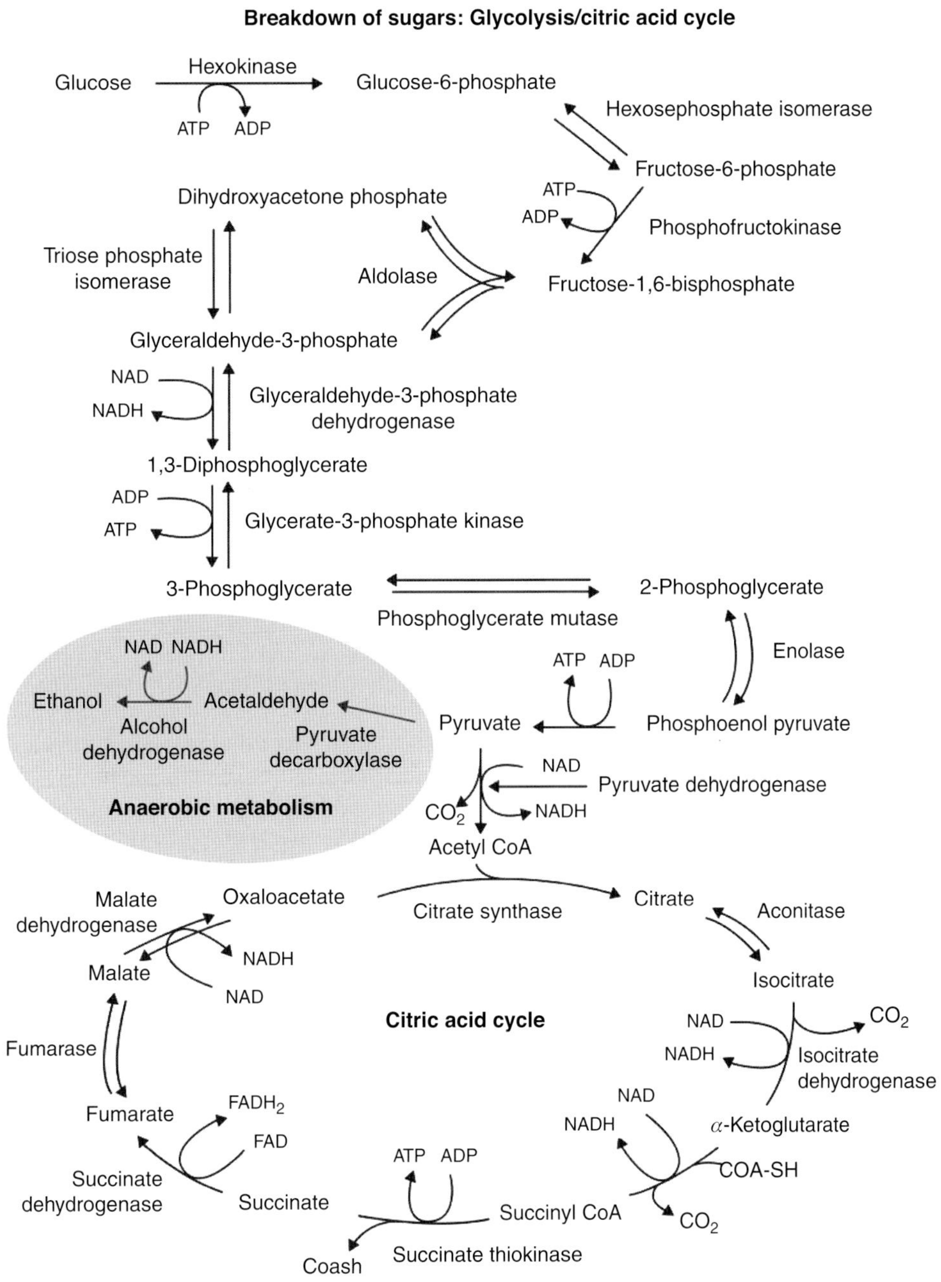

Figure 27.3. Catabolism of sugars through glycolytic pathway and citric acid cycle.

alteration of PFK, which increases the efficiency of utilisation of fructose-6-phosphate, is another means of regulation that can activate the carbon flow through the glycolytic pathway.

Other enzymes of the glycolytic pathway are involved in the regulation of starch/sucrose biosynthesis (Figs. 27.2 and 27.3). Fructose-1,6-bisphosphate is converted back to fructose-6-phosphate by the enzyme fructose-1,6-bisphosphatase, also releasing inorganic phosphate. This enzyme is localised in the cytosol and chloroplast. Fructose-6-phosphate is converted to fructose-2,6-bisphosphate by fructose-6-phosphate 2-kinase which can be dephosphorylated at the 2 position by fructose-2,6-bisphosphatase. Fructose-6-phosphate is an intermediary in sucrose biosynthesis (Fig. 27.2). SPS is regulated by reversible phosphorylation (a form of post-translational modification that involves addition of a phosphate moiety from ATP to an OH-amino acid residue in the protein, such as serine or threonine, mediated by a kinase, and dephosphorylation mediated by a phosphatase) by SPS kinase and SPS phosphatase. Phosphorylation of the enzyme makes it less active. Glucose-6-phosphate is an allosteric activator (a molecule that can bind to an enzyme and increase its activity through enzyme subunit association) of the active form of SPS (dephosphorylated). Glucose-6-phosphate is an inhibitor of SPS kinase and inorganic phosphate is an inhibitor of SPS phosphatase. Thus, under conditions when glucose-6-phosphate/inorganic phosphate ratio is high, the active form of SPS will dominate favouring sucrose phosphate biosynthesis. These regulations are highly complex and may be regulated by the flux of other sugars in several pathways.

The conversion of PEP to pyruvate mediated by pyruvate kinase is another key metabolic step in the glycolytic pathway and is irreversible. Pyruvate is used in several metabolic reactions. During respiration, pyruvate is further converted to acetyl coenzyme A (acetyl CoA), which enters the citric acid cycle through which it is completely oxidised to carbon dioxide (Fig. 27.3). The conversion of pyruvate to acetyl CoA is mediated by the enzyme complex pyruvate dehydrogenase, and is an oxidative step that involves the formation of NADH from NAD. Acetyl CoA is a key metabolite and starting point for several biosynthetic reactions (fatty acids, isoprenoids, phenylpropanoids, etc.).

Citric Acid Cycle

The citric acid cycle involves the biosynthesis of several organic acids, many of which serve as precursors for the biosynthesis of several groups of amino acids. In the first reaction, oxaloacetate combines with acetyl CoA to form citrate, and is mediated by citrate synthase (Fig. 27.3). In the next step, citrate is converted to isocitrate by the action of aconitase. The next two steps in the cycle involve oxidative decarboxylation. The conversion of isocitrate to α-ketoglutarate involves the removal of a carbon dioxide molecule and reduction of NAD to NADH. This step is catalysed by isocitrate dehydrogenase. α-ketoglutarate is converted to succinyl-CoA by α-ketoglutarate dehydrogenase, along with the removal of another molecule of carbon dioxide and the conversion of NAD to NADH. Succinate, the next product, is formed from succinyl CoA by the action of succinyl CoA synthetase that involves the removal of the CoA moiety and the conversion of ADP to ATP. Through these steps, the complete oxidation of the acetyl CoA moiety has been achieved with the removal of two molecules of carbon dioxide. Thus, succinate is a four-carbon organic acid. Succinate is further converted to fumarate and malate in the presence of succinate dehydrogenase and fumarase, respectively. Malate is oxidised to oxaloacetate by the enzyme malate dehydrogenase along with the conversion of NAD to NADH. Oxaloacetate then can combine with another molecule of acetyl CoA to repeat the cycle. The reducing power generated in the form of NADH and FADH (succinate dehydrogenation step) is used for the biosynthesis of ATP through the electron transport chain in the mitochondria.

Gluconeogenesis

Several fruits store large amounts of organic acids in their vacuole and these acids are converted back to sugars during ripening, a process termed as gluconeogenesis. Several irreversible steps in the glycolysis and citric acid cycle are bypassed during gluconeogenesis. Malate and citrate are the major organic acids present in fruits. In fruits such as grapes, where there is a transition from a sour to a sweet stage during ripening, organic acids content declines. Grape contains predominantly tartaric acid along with malate, citrate, succinate, fumarate and several organic acid intermediates of metabolism. The content of organic acids in berries can affect their suitability for processing. High acid content coupled with low sugar content can result in poor-quality wines. External warm growth conditions enhance the metabolism of malic acid in grapes during ripening and could result in a high tartarate/malate ratio, which is considered ideal for vinification.

The metabolism of malate during ripening is mediated by the malic enzyme, NADP-dependent malate dehydrogenase. Along with a decline in malate content, there is a concomitant increase in the sugars, suggesting a possible metabolic precursor product relationship between these two events. Indeed, when grape berries were fed with radiolabelled malate, the radiolabel could be recovered in glucose. The metabolism of malate involves its conversion to oxaloacetate mediated by malate dehydrogenase, the decarboxylation of oxaloacetate to PEP catalysed by PEP-carboxykinase, and a reversal of glycolytic pathway leading to sugar formation (Ruffner et al. 1983). The gluconeogenic pathway from malate may contribute only a small percentage (5%) of the sugars, and a decrease in malate content could primarily result from reduced synthesis and increased catabolism through the citric acid cycle. The inhibition of malate synthesis by the inhibition of the glycolytic pathway could result in increased sugar accumulation. Metabolism of malate in apple fruits is catalysed by NADP-malic enzyme, which converts malate to pyruvate. In apples, malate appears to be primarily oxidised through the citric acid cycle. Organic acids are important components of citrus fruits. Citric acid is the major form of the acid followed by malic acid and several less abundant acids such as acetate, pyruvate, oxalate, glutarate, fumarate and so on. In oranges, the acidity increases during maturation of the fruit and declines during the ripening phase. Lemon fruits, by contrast, increase their acid content through the accumulation of citrate. The citrate levels in various citrus fruits range from 75% to 88%, and malate levels range from 2% to 20%. Ascorbate is another major component of citrus fruits. Ascorbate levels can range from 20 to 60 mg/100 g juice in various citrus fruits. The orange skin may possess 150–340 mg/100 g fresh weight of ascorbate, which may not be extracted into the juice.

Anaerobic Respiration

Anaerobic respiration is a common event in the respiration of ripe fruits and especially becomes significant when fruits are exposed to low temperature. Often, this may result from oxygen-depriving conditions induced inside the fruit. Under anoxia, ATP production through the citric acid cycle and mitochondrial electron transport chain is inhibited. Anaerobic respiration is a means of regenerating NAD, which can drive the glycolyic pathway and produce minimal amounts of ATP (Fig. 27.3). Under anoxia, pyruvate formed through glycolysis is converted to lactate-by-lactate dehydrogenase using NADH as the reducing factor, and generating NAD. Accumulation of lactate in the cytosol could cause acidification, and under these low pH conditions, lactate dehydrogenase is inhibited. The formation of acetaldehyde by the decarboxylation of pyruvate is stimulated by the activation of pyruvate decarboxylase under low pH conditions in the cytosol. It is also likely that the increase in concentration of pyruvate in the cytoplasm may stimulate pyruvate decarboxylase directly. Acetaldehyde is reduced to ethanol by alcohol dehydrogenase using NADH as the reducing power. Thus, acetaldehyde and ethanol are common volatile components observed in the headspace of fruits indicative of the occurrence of anaerobic respiration. Cytosolic acidification is a condition that stimulates deteriorative reactions. By removing lactate through efflux and converting pyruvate to ethanol, cytosolic acidification can be avoided.

Anaerobic respiration plays a significant role in the respiration of citrus fruits. During early stages of growth, respiratory activity predominantly occurs in the skin tissue. Oxygen uptake by the skin tissue was much higher than the juice vesicles (Purvis 1985). With advancing maturity, a decline in aerobic respiration and an increase in anaerobic respiration was observed in Hamlin orange skin (Bruemmer 1989). In parallel with this, the levels of ethanol and acetaldehyde increased. As well, a decrease in the organic acid substrates pyruvate and oxaloacetate was detectable in Hamlin orange juice. An increase in the activity levels of pyruvate decarboxylase, alcohol dehydrogenase and malic enzyme was noticed in parallel with the decline in pyruvate and accumulation of ethanol. In apple fruits, malic acid is converted to pyruvate by the action of NADP-malic enzyme, and pyruvate subsequently converted to ethanol by the action of pyruvate decarboxylase and alcohol dehydrogenase. The alcohol dehydrogenase in apple can use NADPH as a cofactor, and NADP is regenerated during ethanol production, thus driving malate utilisation. Ethanol is either released as a volatile or can be used for the biosynthesis of ethyl esters of volatiles.

Pentose Phosphate Pathway

Oxidative PPP is a key metabolic pathway that provides reducing power (NADPH) for biosynthetic reactions as well as carbon precursors for the biosynthesis of amino acids, nucleic acids, secondary plant products and so on. The PPP shares many of the sugar phosphate intermediates with the glycolytic pathway (Fig. 27.4). The PPP is characterised by the interconversion of sugar phosphates with three (glyceraldehyde-3-phosphate), four (erythrose-4-phosphate), five (ribulose-, ribose-, xylulose-phosphates), six (glucose-6-phosphate, fructose-6-phosphate) and seven (sedoheptulose-7-phosphate) carbon long chains.

The PPP involves the oxidation of glucose-6-phosphate, and the sugar phosphate intermediates formed are recycled. The first two reactions of PPP are oxidative reactions mediated by the enzymes glucose-6-phosphate dehydrogenase and 6-phosphogluconate dehydrogenase (Fig. 27.4). In the first step, glucose-6-phosphate is converted to 6-phosphogluconate by the removal of two hydrogen atoms by NADP to form NADPH. In the next step, 6-phosphogluconate, a six-carbon sugar acid phosphate, is converted to ribulose-5-phosphate, a five-carbon sugar phosphate. This reaction involves the removal of a carbon dioxide molecule along with the formation of NADPH. Ribulose-5-phosphate undergoes several metabolic conversions to yield fructose-6-phosphate. Fructose-6-phosphate can then be converted back to glucose-6-phosphate by the enzyme glucose-6-phosphate isomerase and the cycle repeated. Thus, six complete turns of the cycle can result in the complete oxidation of a glucose molecule.

Despite the differences in the reaction sequences, the glycolytic pathway and the PPP intermediates can interact with one another and share common intermediates. Intermediates of both the pathways are localised in plastids, as well as the cytoplasm, and intermediates can be transferred across the plastid membrane into the cytoplasm and back into the chloroplast. Glucose-6-phosphate dehydrogenase is localised both in the chloroplast and cytoplasm. Cytosolic glucose-6-phosphate dehydrogenase activity is strongly inhibited by NADPH. Thus, the ratio of NADP to NADPH could be the regulatory control point for the enzyme function. The chloroplast-localised enzyme is regulated differently through oxidation and reduction, and related to the photosynthetic process. 6-Phosphogluconate dehydrogenase exists as distinct cytosol- and plastid-localised isozymes.

The PPP is a key metabolic pathway related to biosynthetic reactions, antioxidant enzyme function and general stress tolerance of the fruits. Ribose-5-phosphate is used in the biosynthesis of nucleic acids and erythrose-4- phosphate is channelled into phenyl propanoid pathway leading to the biosynthesis of the amino acids phenylalanine and tryptophan. Phenylalanine is the metabolic starting point for the biosynthesis of flavonoids and anthocyanins in fruits. Glyceraldehyde-3-phosphate and pyruvate serve as the precursors for the isoprenoid pathway localised in the chloroplast. Accumulation of sugars in fruits during ripening has been related to the function of PPP. In mangoes, an increase in the levels of pentose sugars observed during ripening has been related to increased activity of PPP. Increases in glucose-6-phosphate dehydrogenase and 6-phosphogluconate dehydrogenase activities were observed during ripening of mango.

NADPH is a key component required for the proper functioning of the antioxidant enzyme system (Fig. 27.4). During growth, stress conditions, fruit ripening and senescence, free radicals are generated within the cell. Activated forms of oxygen, such as superoxide, hydroxyl and peroxy radicals, can attack enzymes, proteins, nucleic acids and lipids, causing structural and functional alterations of these molecules. Under most conditions, these are deleterious changes, which are nullified by the action of antioxidants and antioxidant enzymes. Simple antioxidants such as ascorbate and vitamin E can scavenge the free radicals and protect the tissue. Anthocyanins and other polyphenols may also serve as simple antioxidants. In addition, the antioxidant enzyme system involves the integrated function of several enzymes. The key antioxidant enzymes are superoxide dismutase (SOD), catalase, ascorbate peroxidase and peroxidase. SOD converts superoxide into hydrogen peroxide. Hydrogen peroxide is immediately acted upon by catalase, generating water. Hydrogen peroxide can also be removed by the action of peroxidases.

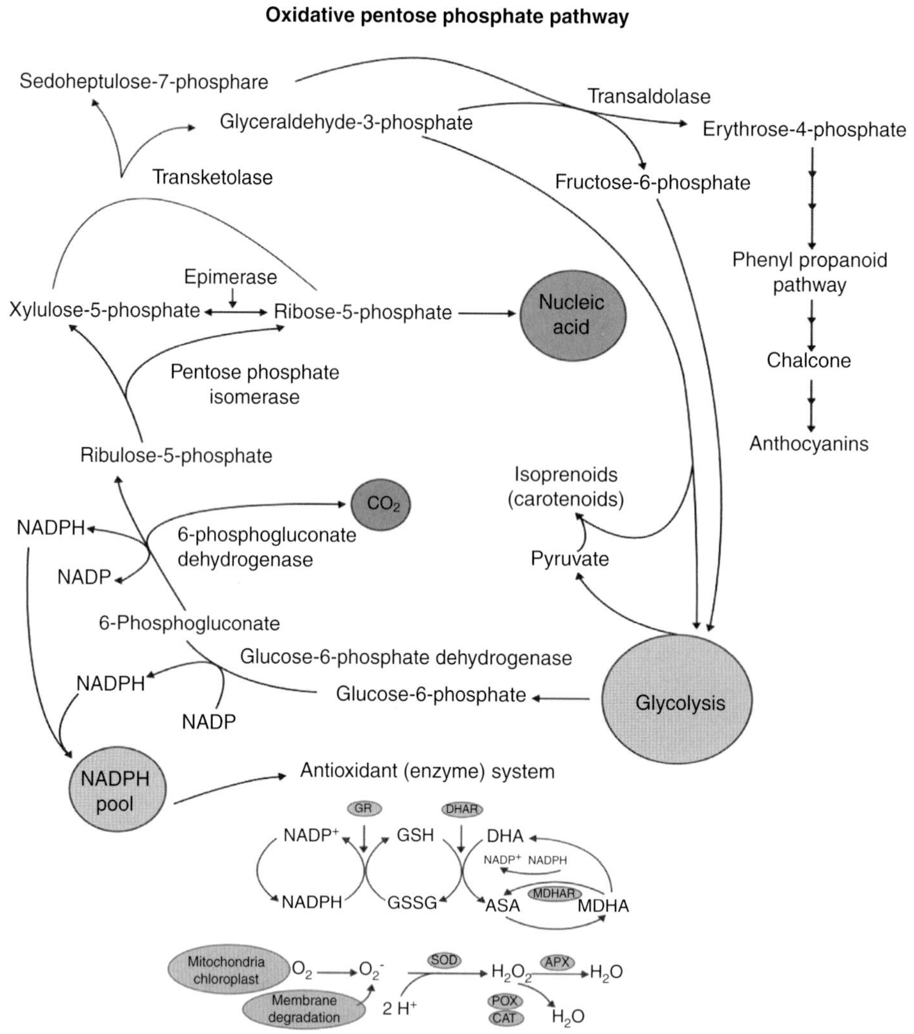

Figure 27.4. Oxidative pentose phosphate pathway in plants. NADPH generated from the pentose phosphate pathway is channeled into the antioxidant enzyme system, where the regeneration of oxidised intermediates requires NADPH. GSH, reduced glutathione; GSSG, oxidised glutathione; ASA, reduced ascorbate; MDHA, monodehydroascorbate; DHA, dehydroascorbate; GR, glutathione reductase; DHAR, dehydroascorbate reductase; MDHAR, monodehydroascorbate reductase; SOD, superoxide dismutase; CAT, catalase; POX, peroxidase; APX, ascorbate peroxidase.

A peroxidase uses the oxidation of a substrate molecule (usually having a phenol structure, C–OH, which becomes a quinone, C = O, after the reaction) to react with hydrogen peroxide, converting it to water. Hydrogen peroxide can also be acted upon by ascorbate peroxidase, which uses ascorbate as the hydrogen donor for the reaction, resulting in water formation. The oxidised ascorbate is regenerated by the action of a series of enzymes (Fig. 27.4). These include monodehydroascorbate reductase (MDHAR) and dehydroascorbate reductase (DHAR). Dehydroascorbate is reduced to ascorbate using reduced glutathione (GSH) as a substrate, which itself gets oxidised (GSSG) during this reaction. The oxidised GSH is reduced back to GSH by the activity of GSH reductase using NADPH. Antioxidant enzymes exist as several functional isozymes with differing activities and kinetic properties in the same tissue. These enzymes are also compartmentalised in chloroplast, mitochondria and cytoplasm. The functioning of the antioxidant enzyme system is crucial to the maintenance of fruit quality through preserving cellular structure and function (Meir and Bramlage 1988, Ahn et al. 2002).

Lipid Metabolism

Among fruits, avocado and olive are the only fruits that significantly store reserves in the form of lipid triglycerides. In avocado, triglycerides form the major part of the neutral lipid fraction, which can account for nearly 95% of the total lipids. Palmitic (16:0), palmitoleic (16:1), oleic (18:1) and linoleic (18:2) acids are the major fatty acids of triglycerides. The oil content progressively increases during maturation of the fruit, and the oils are compartmentalised in oil bodies or oleosomes. The biosynthesis of fatty acids occurs in the plastids, and the fatty acids are exported into the endoplasmic reticulum where they are esterified with glycerol-3-phosphate by the action of a number of enzymes to form the triglyceride. The triglyceride-enriched regions then are believed to bud off from the endoplasmic reticulum as the oil body. The oil body membranes are different from other cellular membranes, since they are made up of only a single layer of phospholipids. The triglycerides are catabolised by the action of triacylglycerol lipases with the release of fatty acids. The fatty acids are then broken down into acety CoA units through β-oxidation.

Even though phospholipids constitute a small fraction of the lipids in fruits, the degradation of phospholipids is a key factor that controls the progression of senescence. As in several senescing systems, there is a decline in phospholipids as the fruit undergoes senescence. With the decline in phospholipid content, there is a progressive increase in the levels of neutral lipids, primarily diacylglycerols, free fatty acids and fatty aldehydes. In addition, the levels of sterols may also increase. Thus, there is an increase in the ratio of sterol:phospholipids. Such changes in the composition of membrane can cause the formation of gel phase or non-bilayer lipid structures (micelles). These changes can make the membranes leaky, thus resulting in the loss of compartmentalisation, and ultimately, senescence (Paliyath and Droillard 1992).

Membrane lipid degradation occurs by the tandem action of several enzymes, one enzyme acting on the product released by the previous enzyme in the sequence. Phospholipase D (PLD) is the first enzyme of the pathway which initiates phospholipids catabolism and is a key enzyme of the pathway (Fig. 27.6). PLD acts on phospholipids liberating phosphatidic acid and the respective headgroup (choline, ethanolamine, glycerol, inositol). Phosphatidic acid, in turn, is acted upon by phosphatidate phosphatase which removes the phosphate group from phosphatidic acid with the liberation of diacylglycerols (diglycerides). The acyl chains of diacylglycerols are then de-esterified by the enzyme lipolytic acyl hydrolase liberating free fatty acids. Unsaturated fatty acids with a *cis*-1,4- pentadiene structure (linoleic acid, linolenic acid) are acted upon by lipoxygenase (LOX) causing the peroxidation of fatty acids. This step may also cause the production of activated oxygen species such as singlet oxygen, superoxide and peroxy radicals and so on. The peroxidation products of linolenic acid can be 9-hydroperoxy linoleic acid or 13-hydroperoxy linoleic acid. The hydroperoxylinoleic acids undergo cleavage by hydroperoxide lyase resulting in several products including hexanal, hexenal and ω-keto fatty acids (keto group towards the methyl end of the molecule). For example, hydroperoxide lyase action on 13-hydroperoxylinolenic acid results in the formation of *cis*-3-hexenal and 12-keto-cis-9- dodecenoic acid. Hexanal and hexenal are important fruit volatiles. The short-chain fatty acids may feed into catabolic pathway (β-oxidation) that results in the formation of short-chain acyl CoAs, ranging from acetyl CoA to dodecanoyl CoA. The short-chain acyl CoAs and alcohols (ethanol, propanol, butanol, pentanol, hexanol, etc.) are esterified to form a variety of esters that constitute components of flavour volatiles that are characteristic to fruits. The free fatty acids and their catabolites (fatty aldehydes, fatty alcohols, alkanes, etc.) can accumulate in the membrane causing membrane destabilisation (formation of gel phase, non-bilayer structures, etc.). An interesting regulatory feature of this pathway is the very low substrate specifity of enzymes that act downstream from PLD for the phospholipids. Thus, phosphatidate phosphatase, lipolytic acyl hydrolase and LOX do not directly act on phospholipids, though there are exceptions to this rule. Therefore, the degree of membrane lipid catabolism will be determined by the extent of activation of PLD (Fig. 27.5).

The membrane lipid catabolic pathway is considered as an autocatalytic pathway (Fig. 27.5). The destabilisation of the membrane can cause the leakage of calcium and hydrogen ions from the cell wall space, as well as the inhibition of calcium- and proton ATPases, the enzymes responsible for maintaining a physiological calcium and proton concentration within the cytoplasm (calcium concentration below micromolar range, pH in the 6–6.5 range). Under conditions of normal growth and development, these enzymes pump the extra calcium- and hydrogen ions that enter the cytoplasm from storage areas such as apoplast and the ER lumen in response to hormonal and environmental stimulation using ATP as the energy source. The activities of calcium- and proton ATPases localised on the plasma membrane, the endoplasmic reticulum and the tonoplast are responsible for pumping the ions back into the storage compartments. In fruits (and other senescing systems), with the advancement in ripening and senescence, there is a progressive increase in leakage of calcium and hydrogen ions. PLD is stimulated by low pH and calcium concentration over 10 μM. Thus, if the cytosolic concentrations of these ions progressively increase during ripening or senescence, the membranes are damaged as a consequence. However, this is an inherent feature of the ripening process in fruits, and results in the development of ideal organoleptic qualities that makes them edible. The uncontrolled membrane deterioration can result in the loss of shelf life and quality in fruits (Paliyath et al. 2008).

The properties and regulation of the membrane degradation pathway are increasingly becoming clear. Enzymes such as PLD and LOX are very well studied. There are several isoforms of PLD designated as PLD alpha, PLD beta, PLD gamma and so on. The expression and activity levels of PLD alpha are much higher than that of the other PLD isoforms. Thus, PLD alpha is considered as a housekeeping enzyme; however, it is also developmentally regulated (Pinhero et al. 2003). The regulation of PLD activity is an interesting feature. PLD is normally a soluble enzyme. The secondary structure of PLD shows the presence of a segment of around 130 amino acids at the *N*-terminal end, designated as the C2 domain. This domain is characteristic of several enzymes and proteins that are integral components of the hormone signal transduction system. In response to hormonal and environmental stimulation and the resulting increase in cytosolic calcium concentration, C2 domain binds calcium and transports

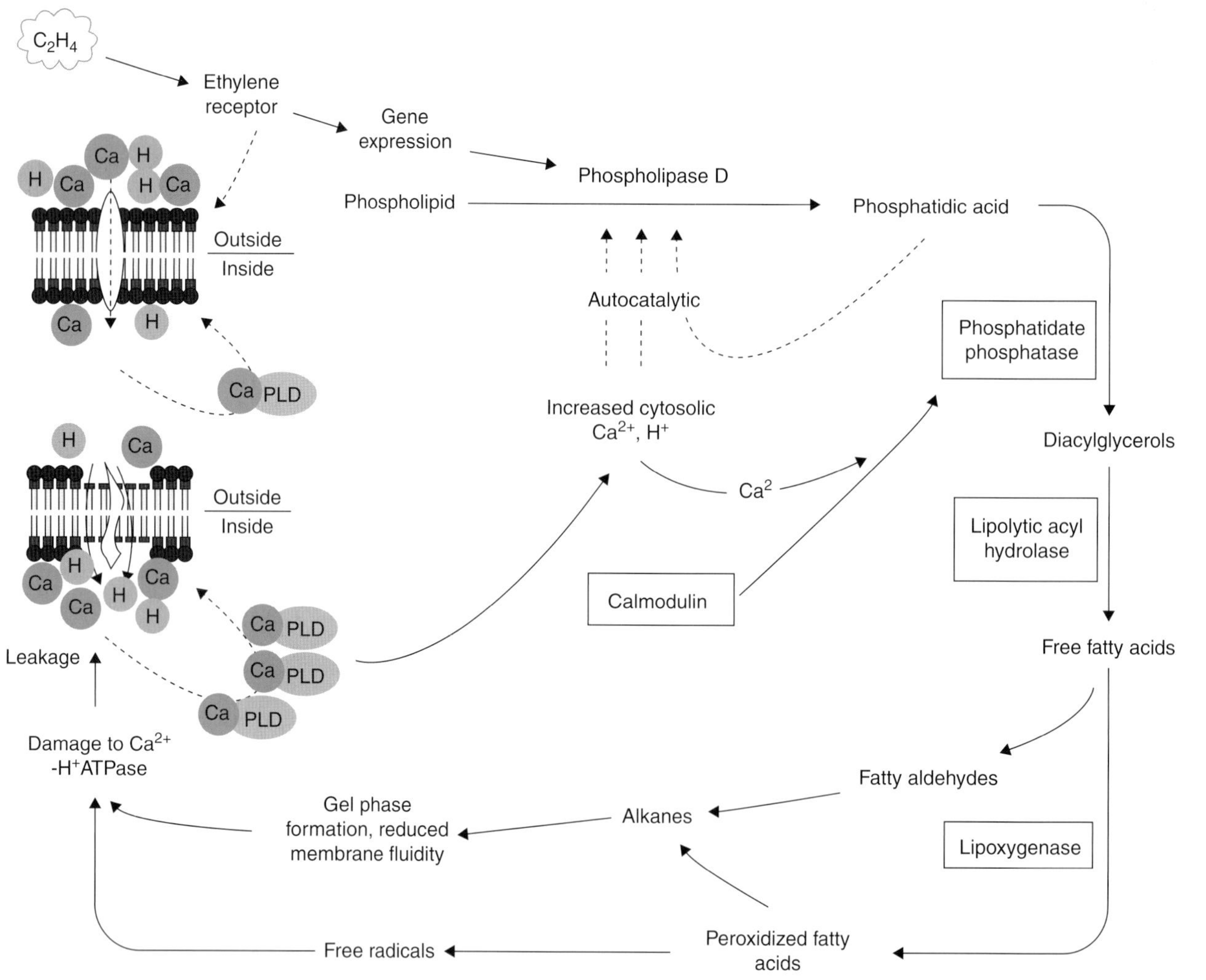

Figure 27.5. Diagrammatic representation of the autocatalytic pathway of phospholipid degradation that occur during fruit ripening/harvest stress in horticultural produce.

PLD to the membrane where it can initiate membrane lipid degradation. The precise relation between the stimulation of the ethylene receptor and PLD activation is not fully understood, but could involve the release of calcium and migration of PLD to the membrane, formation of a metabolising enzyme complex (metabolon) with other lipid degrading enzymes of the pathway as well as calmodulin. PLD alpha appear to be the key enzyme responsible for the initiation of membrane lipid degradation in tomato fruits (Pinhero et al. 2003). Antisense inhibition of PLD alpha in tomato fruits resulted in the reduction of PLD activity and consequently, an improvement in the shelf life, firmness, soluble solids and lycopene content of the ripe fruits (Whitaker et al. 2001, Pinhero et al. 2003, Oke et al. 2003, Paliyath et al. 2008a). There are other phospholipid degrading enzymes such as phospholipase C and phospholipase A_2. Several roles of these enzymes in signal transduction processes have been extensively reviewed (Wang 2001, Meijer and Munnik 2003).

LOX exists as both soluble and membranous forms in tomato fruits (Todd et al. 1990). Very little information is available on phosphatidate phosphatase and lipolytic acyl hydrolase in fruits.

Proteolysis and Structure Breakdown in Chloroplasts

The major proteinaceous compartment in fruits is the chloroplast which is distributed in the epidermal and hypodermal layers of fruits. The chloroplasts are not very abundant in fruits. During senescence, the chloroplast structure is gradually disassembled with a decline in chlorophyll levels due to the degradation and disorganisation of the grana lamellar stacks of the chloroplast. With the disorganisation of the thylakoid, globular structures termed as plastoglobuli accumulate within the chloroplast stroma, which are rich in degraded lipids. The degradation of chloroplasts and chlorophyll result in the unmasking of other coloured pigments and is a prelude to the state of ripening and development of organoleptic qualities. Mitochondria, which are also rich in protein, are relatively stable and undergo disassembly during the latter part of ripening and senescence.

Chlorophyll degradation is initiated by the enzyme chlorophyllase which splits chlorophyll into chlorophyllide and the phytol chain. Phytol chain is made up of isoprenoid units (methyl-1,3-butadiene), and its degradation products accumulate in the plastoglobuli. Flavour components such as 6-methyl-5-heptene-2-one, a characteristic component of tomato flavour, are also produced by the catabolism of phytol chain. The removal of magnesium from chlorophyllide results in the formation of pheophorbide. Pheophorbide, which possesses a tetrapyrole structure, is converted to a straight chain colourless tetrapyrrole by the action of pheophorbide oxidase. Action of several other enzymes is necessary for the full catabolism of chlorophyll. The protein complexes that organise the chlorophyll, the light-harvesting complexes, are degraded by the action of several proteases. The enzyme ribulose-bis-phosphate carboxylase/oxygenase (Rubisco), the key enzyme in photosynthetic carbon fixation, is the most abundant protein in chloroplast. Rubisco levels also decline during ripening/senescence due to proteolysis. The amino acids resulting from the catabolism of proteins may be translocated to regions where they are needed for biosynthesis. In fruits, they may just enrich the soluble fraction with amino acids.

SECONDARY PLANT PRODUCTS AND FLAVOUR COMPONENTS

Secondary plant products are regarded as metabolites that are derived from primary metabolic intermediates through well-defined biosynthetic pathways. The importance of the secondary plant products to the plant or organ in question may not readily be obvious, but these compounds appear to have a role in the interaction of the plant with the environment. The secondary plant products may include non-protein amino acids, alkaloids, isoprenoid components (terpenes, carotenoids, etc.), flavonoids and anthocyanins, ester volatiles and several other organic compounds with diverse structure. The number and types of secondary plant products are enormous, but, with the perspective of fruit quality, the important secondary plant products include isoprenoids, anthocyanins and ester volatiles.

Isoprenoid Biosynthesis

In general, isoprenoids possess a basic five-carbon skeleton in the form of 2-methyl-1,3-butadiene (isoprene), which undergoes condensation to form larger molecules. There are two distinct pathways for the formation of isoprenoids: the acetate/mevalonate pathway (Bach et al. 1999) localised in the cytosol and the DOXP pathway (Rohmer pathway, Rohmer et al. 1993) localised in the chloroplast (Fig. 27.6). The metabolic precursor for the acetate/mevalonate pathway is acetyl Coenzyme A. Through the condensation of three acetyl CoA molecules, a key component of the pathway, 3-hydroxy-3-methyl-glutaryl CoA (HMG CoA) is generated. HMG-CoA undergoes reduction in the presence of NADPH mediated by the key regulatory enzyme of the pathway HMG CoA reductase (HMGR), to form mevalonate. Mevalonate undergoes a two-step phosphorylation in the presence of ATP, mediated by kinases, to form isopentenyl pyrophosphate (IPP), the basic five carbon condensational unit of several terpenes. IPP is isomerised to dimethylallylpyrophosphate (DMAPP) mediated by the enzyme IPP isomerase. Condensation of these two components results in the synthesis of C10 (geranyl), C15 (farnesyl) and C20 (geranylgeranyl) pyrophosphates. The C10 pyrophosphates give rise to monoterpenes, C15 pyrophosphates give rise to sesquiterpenes and C20 pyrophosphates give rise to diterpenes. Monoterpenes are major volatile components of fruits. In citrus fruits, these include components such as limonene, myrcene, pinene and so on occurring in various proportions. Derivatives of monoterpenes such as geranial, neral (aldehydes), geraniol, linalool, terpineol (alcohols), geranyl acetate, neryl acetate (esters) and so on are also ingredients of the volatiles of citrus fruits. Citrus fruits are especially rich in monoterpenes and derivatives. Alpha-farnesene is a major sesquiterpene (C15) component evolved by apples. The catabolism of alpha-farnesene in the presence of oxygen into oxidised forms has been implicated as a causative feature in the development of the physiological disorder superficial scald (a type of superficial browning) in certain varieties of apples such as red Delicious, McIntosh, Cortland and so on (Rupasinghe et al. 2000, 2003).

HMGR is a highly conserved enzyme in plants and is encoded by a multigene family (Lichtenthaler et al. 1997). The HMGR genes (*hmg1*, *hmg2*, *hmg3*, etc.) are nuclear encoded and can be differentiated from each other by the sequence differences at the 3′-untranslated regions of the cDNAs. There are three distinct genes for HMGR in tomato and two in apples. The different HMGR end products may be localised in different cellular compartments and are synthesised differentially in response to hormones, environmental signals, pathogen infection and so on. In tomato fruits, the level of *hmg1* expression is high during early stage of fruit development when cell division and expansion processes are rapid, when it requires high levels of sterols for incorporation into the expanding membrane compartments. The expression of *hmg2* which is not detectable in young fruits increases during the latter part of fruit maturation and ripening.

HMGR activity can be detected in both membranous and cytosolic fractions of apple fruit skin tissue extract. HMGR is a membrane-localised enzyme, and the activity is detectable in the endoplasmic reticulum, plastid and mitochondrial membranes. It is likely that HMGR may have undergone proteolytic cleavage releasing a fragment into the cytosol, which also possesses enzyme activity. There is a considerable degree of interaction between the different enzymes responsible for the biosynthesis of isoprenoids, which may exist as multienzyme complexes. The enzyme Farnesyl pyrophosphate synthase, responsible for the synthesis of farnesyl pyrophosphate is a cytosolic enzyme. Similarly, farnesene synthase, the enzyme which converts farnesyl pyrophosphate to alpha-farnesene in apples, is a cytosolic enzyme. Thus, several enzymes may act in concert at the cytoplasm/endoplasmic reticulum boundary to synthesise isoprenoids.

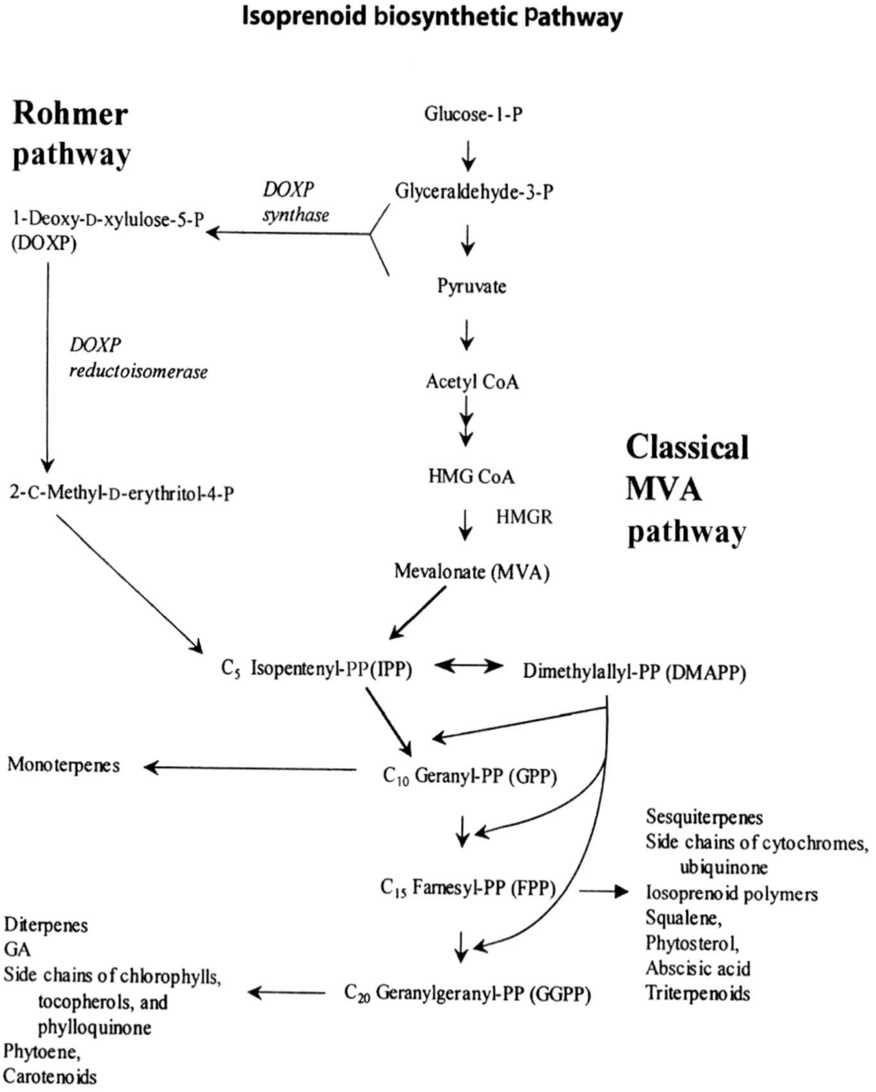

Figure 27.6. Isoprenoid biosynthetic pathway in plants.

HMG CoA reductase expression and activities in apple fruits are hormonally regulated (Rupasinghe et al. 2001, 2003). There are two genes for HMGR in apples designated as *hmg1* and *hmg2,* which are differentially expressed during storage. The expression of *hmg1* was constitutive, and the transcripts (mRNA) were present throughout the storage period. In contrast, the expression of *hmg2* increased during storage in parallel with the accumulation of alpha-farnesene. Ethylene production also increased during storage. Ethylene stimulates the biosynthesis of alpha-farnesene as evident from the inhibition of alpha-farnesene biosynthesis and the expression of *hmg2* by the ethylene action inhibitor 1-methylcyclopropene (MCP). Thus, biosynthesis of isoprenoids is a highly controlled process.

Carotenoids, which are major isoprenoid components of chloroplasts, are biosynthesised through the Rohmer pathway. The precursors of this pathway are pyruvate and glyceraldehyde-3-phosphate, and through a number of enzymatic steps, 1-deoxy-D-xylulose-5-phosphate (DOXP), a key metabolite of the pathway, is formed. NADPH-mediated reduction of DOXP leads ultimately to the formation of IPP. Subsequent condensation of IPP and DMAPP are similar as in the classical mevalonate pathway. Carotenoids have a stabilising role in the photosynthetic reactions. By virtue of their structure, they can accept and stabilise excess energy absorbed by the light-harvesting complex. During the early stages of fruit development, the carotenoids have primarily photosynthetic function. During fruit ripening, the composition of carotenoids changes to reveal the coloured xanthophylls pigments. In tomato, lycopene is the major carotenoid pigment that accumulates during ripening. Lycopene is an intermediate of the carotene biosynthetic pathway. In young fruits, lycopene formed by the condensation of two geranylgeranyl pyrophosphate (C20) moieties mediated by the enzyme phytoene synthase is converted to beta-carotene by the action of the enzyme sesquiterpene cyclase. However, as ripening proceeds, the levels and activity of sesquiterpene cyclase are reduced leading to the accumulation of lycopene in the stroma. This leads to the development of red colour in ripe tomato fruits. In yellow tomatoes, the carotene biosynthesis is not inhibited, and as the fruit ripens, the chlorophyll pigments are degraded exposing the yellow carotenoids. Carotenoids are also major components that contribute to the colour of melons. Beta-carotene is the major pigment in melons with an orange flesh. In addition, the contribution to colour is also provided by alpha-carotene, delta-carotene, phytofluene, phytoene, lutein and violaxanthin. In red-fleshed melons, lycopene is the major ingredient, whereas in yellow-fleshed melons, xanthophylls and beta-carotene predominate. Carotenoids provide not only a variety of colour to the fruits but also important nutritional ingredients in human diet. Beta-carotene is converted to vitamin A in the human body and thus serves as a precursor to vitamin A. Carotenoids are strong antioxidants. Lycopene is observed to provide protection from cardiovascular diseases and cancer (Giovanucci 1999). Lutein, a xanthophyll, has been proposed to play a protective role in the retina maintaining the vision and prevention of age-related macular degeneration.

Anthocyanin Biosynthesis

The development of colour is a characteristic feature of the ripening process, and in several fruits, the colour components are anthocyanins biosynthesised from metabolic precursors. The anthocyanins accumulate in the vacuole of the cell, and are often abundant in the cells closer to the surface of the fruit. Anthocyanin biosynthesis starts by the condensation of three molecules of malonyl CoA with *p*-coumaroyl CoA to form tetrahydroxychalcone, mediated by the enzyme chalcone synthase (Fig. 27.7). Tetrahydroxychalcone has the basic flavonoid structure C6-C3-C6, with two phenyl groups separated by a three-carbon link. Chalcone isomerase enables the ring closure of chalcone leading to the formation of the flavanone, naringenin that possesses a flavonoid structure having two phenyl groups linked together by a heterocyclic ring. The phenyl groups are designated as A and B and the heterocyclic ring is designated as ring C. Subsequent conversions of naringenin by flavonol hydroxylases result in the formation of dihydrokaempferol, dihydromyricetin and dihydroquercetin, which differ in their number of hydroxyl moieties. Dihydroflavonol reductase converts the dihydroflavonols into the colourless anthocyanidin compounds leucocyanidin, leucopelargonidin and leucodelphinidin. Removal of hydrogens and the induction of unsaturation of the C-ring at C2 and C3, mediated by anthocyanin synthase results in the formation of cyanidin, pelargonidin and delphinidin, the coloured compounds (Figs. 27.7 and 27.8). Glycosylation, methylation, coumaroylation and a variety of other additions of the anthocyanidins result in colour stabilisation of the diverse types of anthocyanins seen in fruits. Pelargonidins give orange, pink and red colour, cyanidins provide magenta and crimson colouration, and delphinidins provide the purple, mauve and blue colour characteristic to several fruits. The colour characteristics of fruits may result from a combination of several forms of anthocyanins existing together, as well as the conditions of pH and ions present in the vacuole.

Anthocyanin pigments cause the diverse colouration of grape cultivars resulting in skin colours varying from translucent, red and black. All the forms of anthocyanins along with those with modifications of the hydroxyl groups are routinely present in the red and dark varieties of grapes. A glucose moiety is attached at the 3 and 5 positions or at both in most grape anthocyanins. The glycosylation pattern can vary between the European (*Vitis vinifera*) and North American (*Vitis labrusca*) grape varieties. Anthocyanin accumulation occurs towards the end of ripening, and is highly influenced by sugar levels, light, temperature, ethylene and increased metabolite translocation from leaves to fruits. All these factors positively influence the anthocyanin levels. Most of the anthocyanin accumulation may be limited to epidermal cell layers and a few of the sub-epidermal cells. In certain high anthocyanin containing varieties, even the interior cells of the fruit may possess high levels of anthocyanins. In the red wine varieties such as merlot, pinot noir and cabernet sauvignon, anthocyanin content may vary between 1500 and 3000 mg/kg fresh weight. In some high-anthocyanin-containing varieties such as Vincent, Lomanto and Colobel, the anthocyanin levels can exceed 9000 mg/kg fresh weight. Anthocyanins are very strong antioxidants and are known to provide protection from the development of cardiovascular diseases and cancer.

Many fruits have a tart taste during early stage of development, which is termed as astringency, and is characteristic to fruits such as banana, kiwi, grape and so on. The astringency is due to the presence of tannins and several other phenolic components in fruits. Tannins are polymers of flavonoids such as catechin and epicatechin, phenolic acids (caffeoyl tartaric acid, coumaroyl tartaric acid, etc.). The contents of tannins decrease during ripening, making the fruit palatable.

Ester Volatile Biosynthesis

The sweet aroma characteristic to several ripe fruits are due to the evolution of several types of volatile components that include monoterpenes, esters, organic acids, aldehydes, ketones, alkanes and so on. Some of these ingredients specifically provide the aroma characteristic to fruits and are referred to as character impact compounds. For instance, the banana flavour is predominantly from isoamyl acetate, apple flavour from ethyl-2 methyl butyrate, and the flavour of lime is primarily due to the monoterpene limonene. As the name implies, ester volatiles are formed from an alcohol and an organic acid through the formation of an ester linkage. The alcohols and acids are, in general, products of lipid catabolism. Several volatiles are esterified with ethanol giving rise to ethyl derivatives of aliphatic acids (ethyl acetate, ethyl butyrate, etc.).

The ester volatiles are formed by the activity of the enzyme Acyl CoA: alcohol acyltransferase or generally called as alcoholacyltransferase (AAT). In apple fruits, the major aroma components are ester volatiles (Paliyath et al. 1997). The alcohol can vary from ethanol, propanol, butanol, pentanol, hexanol and

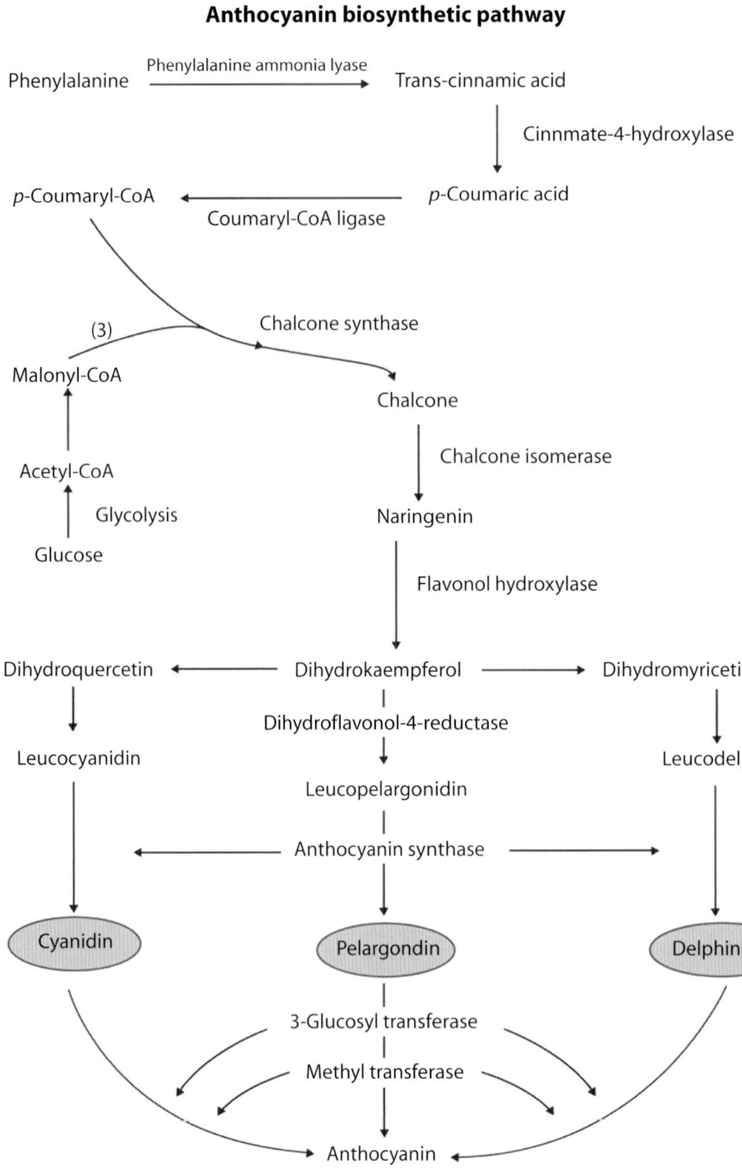

Figure 27.7. Anthocyanin biosynthetic pathway in plants.

so on. The organic acid moiety containing the CoA group can vary in chain length from C2 (acetyl) to C12 (dodecanoyl). AAT activity has been identified in several fruits that include banana, strawberry, melon, apple and so on. In banana, esters are the predominant volatiles enriched with esters such as acetates and butyrates. The flavour may result from the combined perception of amyl esters and butyl esters. Volatile production increases during ripening. The components for volatile biosynthesis may arise from amino acids and fatty acids. In melons, the volatile components comprise esters, aldehydes, alcohols, terpenes and lactones. Hexyl acetate, isoamyl acetate and octyl acetate are the major aliphatic esters. Benzyl acetate, phenyl propyl acetate and phenyl ethyl acetate are also observed. The aldehydes, alcohols, terpenes and lactones are minor components in melons. In mango fruits, the characteristic aroma of each variety is based on the composition of volatiles. The variety "Baladi" is characterised by the presence of high levels of limonene, other monoterpenes and sesquiterpenes, and ethyl esters of even numbered fatty acids. By contrast, the variety "Alphonso" is characterised by high levels of C6 aldehydes and alcohols (hexanal, hexanol) that may indicate a high level of fatty acid peroxidation in ripe fruits. C6 aldehydes are major flavour components of tomato fruits as well. In genetically transformed tomatoes (antisense PLD), the evolution of pentanal and hexenal/hexanal was much higher

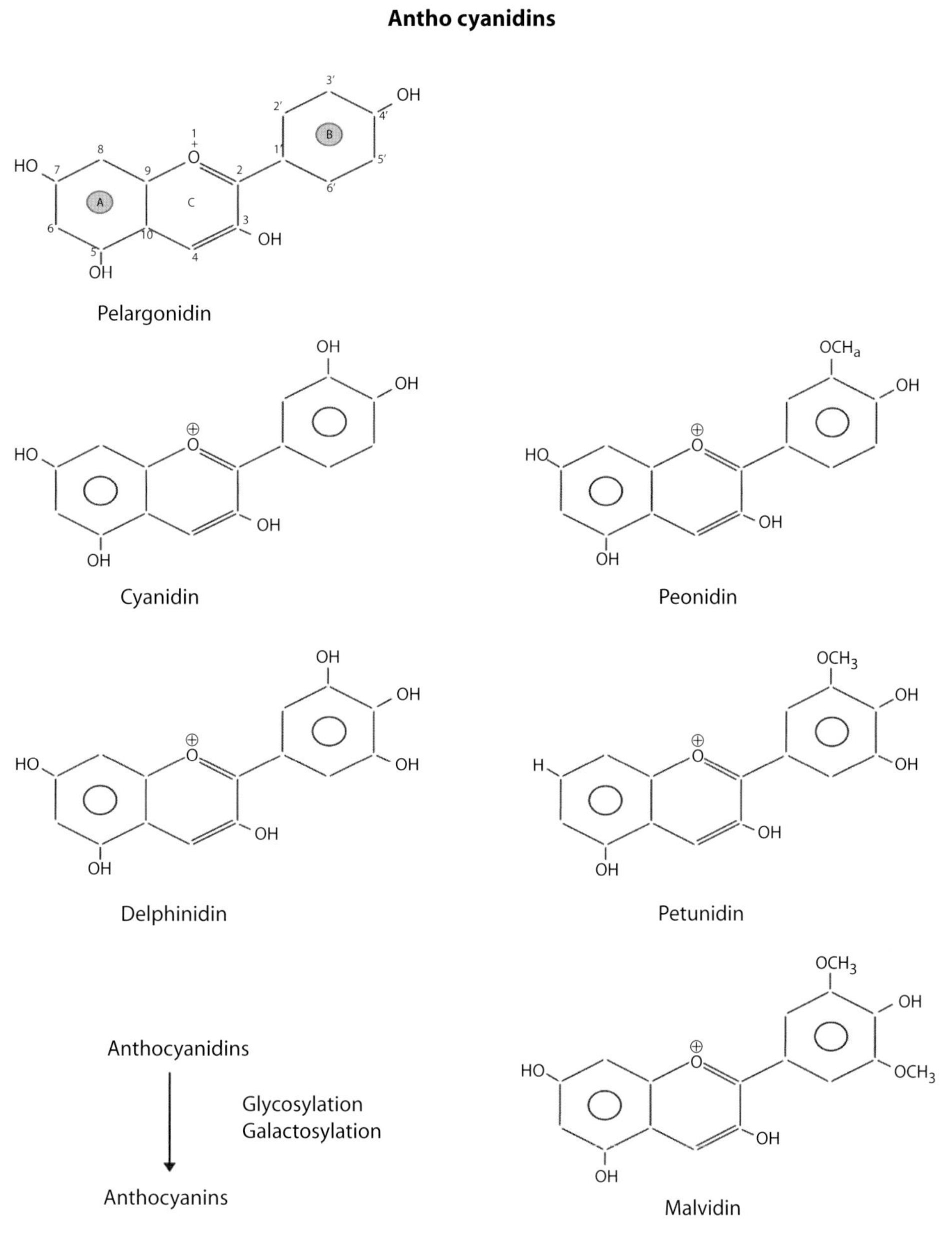

Figure 27.8. Some common anthocyanidins found in fruits and flowers.

after blending, suggesting the preservation of fatty acids in ripe fruits. Preserving the integrity of the membrane during ripening could help preserve the fatty acids that contribute to the flavour profile of the fruits and this feature may provide a better flavour profile for fruits.

GENERAL READING

Buchanan BB et al. (eds.) 2000. *Biochemistry and Molecular Biology of Plants*. American Society of Plant Physiologists, Bethesda, MD.

Kays SJ. 1997. *Postharvest Physiology of Perishable Plant Products*. Exon Press, Athens.

Paliyath G et al. (eds.) 2008a. *Postharvest Biology and Technology of Fruits, Vegetables and Flowers*. Wiley-Blackwell, Iowa.

Seymour GB et al. (eds.) 1991. *Biochemistry of Fruit Ripening*. Chapman and Hall, London.

REFERENCES

Ahn T et al. 2002. Changes in antioxidant enzyme activities during tomato fruit development. *Physiol Mol Biol Plants* 8: 241–249.

Bach TJ et al. 1999. Mevalonate biosynthesis in plants. *Crit Rev Biochem Mol Biol* 34: 107–122.

Beaudry RM et al. 1989. Banana ripening: implication of changes in glycolytic intermediate concentrations, glycolytic and gluconeogenic carbon flux, and fructose 2,6-bisphosphate concentration. *Plant Physiol* 91: 1436–1444.

Bennet AB, Christofferson RE. 1986. Synthesis and processing of cellulose from ripening avocado fruit. *Plant Physiol* 81: 830–835.

Bird CR et al. 1988. The tomato polygalacturonase gene and ripening specific expression in transgenic plants. *Plant Mol Biol* 11: 651–662.

Bruemmer JH. 1989. Terminal oxidase activity during ripening of Hamlin orange. *Phytochemistry* 28: 2901–2902.

Dallman TF et al. 1989. Expression and transport of cellulase in avocado mesocarp during ripening. *Protoplasma* 151: 33–46.

Davies KM et al. 1988. Silver ions inhibit the ethylene stimulated production of ripening related mRNAs in tomato fruit. *Plant Cell Env* 11: 729–738.

Fischer RL, Bennet AB. 1991. Role of cell wall hydrolases in fruit ripening. *Annu Rev Plant Physiol Plant Mol Biol* 42: 675–703.

Fluhr R, Mattoo AK. 1996. Ethylene- biosynthesis and perception. *Crit Rev Plant Sci* 15: 479–523.

Fry SC. 1986. Cross linking of matrix polymers in the growing cell walls of angiosperms. *Annu Rev Plant Physiol* 37: 165–186.

Giovannoni JJ et al. 1989. Expression of a chimeric PG gene in transgenic rin tomato fruit results in polyuronide degradation but not fruit softening. *The Plant Cell* 1: 53–63.

Giovanucci EL. 1999. Tomatoes, tomato based products, lycopene and cancer: a review of the epidemiologic literature. *J Natl Cancer Inst* 91: 317–329.

Grierson D et al. 1986. Sequencing and identification of a cDNA clone for tomato polygalacturonase. *Nucleic Acids Res* 14: 8595–8603.

Hamilton AJ et al. 1991. Identification of a tomato gene for the ethylene forming enzyme by expression in yeast. *Proc Natl Acad Sci (USA)* 88: 7434–7437.

Hamilton AJ et al. 1990. Antisense gene that inhibits synthesis of the hormone ethylene in transgenic plants. *Nature* 346: 284–287.

Hrazdina G et al. 2000. Down regulation of ethylene production in apples. In: G Hrazdina (ed.) *Use of Agriculturally Important Genes in Biotechnology*. IOS Press, Amsterdam, pp. 26–32.

Keegstra K et al. 1973. The structure of plant cell walls. III. A model of the walls of suspension cultured sycamore cells based on the interaction of the macromolecular components. *Plant Physiol* 51: 188–196.

Kramer M et al. 1992. Postharvest evaluation of transgenic tomatoes with reduced levels of polygalacturonase: processing, firmness and disease resistance. *Postharv Biol Technol* 1: 241–255.

Kumar R, Selvaraj Y. 1990. Fructose-1,6-bisphosphatase in ripening mango fruit. *Indian J Exp Biol* 28: 284–286.

Lichtenthaler HK et al. 1997. Two independent biochemical pathways for isopentenyl diphosphate and isoprenoid biosynthesis in higher plants. *Physiologia Plant* 101: 643–652.

Meijer HJG, Munnik T. 2003. Phospholipid-based signalling in plants. *Annu Rev Plant Biol* 54: 265–306.

Meir S, Bramlage WJ. 1988. Antioxidant activity in "Cortland" apple peel and susceptibility to superficial scald after storage. *J Amer Soc Hort Sci* 113: 412–418.

Negi PS, Handa AK. 2008. Structural deterioration of the produce: the breakdown of cell wall components. In: G Paliyath et al. (eds.) *Postharvest Biology and Technology of Fruits, Vegetables and Flowers*. Wiley-Blackwell, Iowa, pp. 162–194.

Oeller PW et al. 1991. Reversible inhibition of tomato fruit senescence by antisense RNA. *Science* 254: 437–439.

Oke M et al. 2003. The effects of genetic transformation of tomato with antisense phospholipase D cDNA on the quality characteristics of fruits and their processed products. *Food Biotechnol* 17: 163–182.

Paliyath G, Droillard MJ. 1992. The mechanism of membrane deterioration and disassembly during senescence. *Plant physiol Biochem* 30: 789–812.

Paliyath G, Thompson JE. 1990. Evidence for early changes in membrane structure during post harvest development of cut carnation flowers. *New Phytologist* 114: 555–562.

Paliyath G et al. 2008. Structural deterioration in produce: phospholipase D, membrane deterioration and senescence. In: G Paliyath et al. (eds.) *Postharvest Biology and Technology of Fruits, Vegetables and Flowers*. Wiley-Blackwell, Ames, IA, pp. 195–239.

Paliyath G et al. 1997. Volatile production and fruit quality during development of superficial scald in Red Delicious apples. *Food Res Int* 30: 95–103.

Pinhero RG et al. 2003. Developmental regulation of phospholipase D in tomato fruits. *Plant Physiol Biochem* 41: 223–240.

Purvis AC. 1985. Low temperature induced azide-insensitive oxygen uptake in grapefruit flavedo tissue. *J Amer Soc Hort Sci* 110: 782–785.

Rohmer M et al. 1993. Isoprenoid biosynthesis in bacteria: a novel pathway for early steps leading to isopentenyl diphosphate. *Biochem J* 295: 517–524.

Ruffner HP et al. 1983. The physiology of acid metabolism in grape berry ripening. *Acta Horticulturae* 139: 123–128.

Rupasinghe HPV et al. 2001. Cloning of *hmg1* and *hmg2* cDNAs encoding 3-hydroxy-3-methylglutaryl coenzyme A reductase and their expression and activity in relation to alpha-farnesene synthesis in apple. *Plant Physiol Biochem* 39: 933–947.

Rupasinghe HPV et al. 2000. Sesquiterpene alpha-farnesene synthase; partial purification, characterization and activity in relation to superficial scald development in "Delicious" apples. *J Amer Soc Hort Sci* 125: 111–119.

Rupasinghe HPV et al. 2003. Biosynthesis of isoprenoids in higher plants. *Physiol Mol Biol Plants* 9: 19–28.

Selvaraj Y et al. 1989. Changes in sugars, organic acids, amino acids, lipid constituents and aroma characteristics of ripening mango fruit. *J Food Sci Technol* 26: 308–313.

Singer SJ, Nicholson GL. 1972. The fluid mosaic model of the structure of cell membranes. *Science* 175: 720–731.

Smith CJS et al. 1988. Antisense RNA inhibition of polygalacturonase gene expression in transgenic tomatoes. *Nature* 334: 724–726.

Todd JF et al. 1990. Characteristics of a membrane-associated lipoxygenase in tomato fruit. *Plant Physiol* 94: 1225–1232.

Wang X. 2001. Plant phospholipases. *Annu Rev Plant Physiol Plant Mol Biol* 52: 211–231.

Whitaker BD. 1988. Changes in the steryl lipid content and composition of tomato fruit during ripening. *Phytochemistry* 27: 3411–3416.

Whitaker BD. 1991. Changes in lipids of tomato fruit stored at chilling and non-chilling temperatures. *Phytochemistry* 30: 757–761.
Whitaker BD. 1992. Changes in galactolipid and phospholipid levels of tomato fruits stored at chilling and non-chilling temperatures. *Phytochemistry* 31: 2627–2630.
Whitaker BD. 1993. Lipid changes in microsomes and crude plastid fractions during storage of tomato fruits at chilling and non-chilling temperatures. *Phytochemistry* 32: 265–271.
Whitaker BD. 1994. Lipid changes in mature green tomato fruit during ripening, during chilling, and after re-warming subsequent to chilling. *J Amer Soc Hort Sci* 119: 994–999.
Whitaker BD et al. 2001. Cloning, characterization and functional expression of a phospholipase D alpha from tomato fruit. *Physiol Plant* 112: 87–94.
Yang SF. 1981. Ethylene, the gaseous plant hormone and regulation of biosynthesis. *Trends Biochem Sci* 6: 161–164.

28
Biochemistry of Fruit Processing

Moustapha Oke, Jissy K. Jacob, and Gopinadhan Paliyath

Abstract: Processing of fruits is an important segment of food industry. As fruits are highly perishable, processing is an efficient way of ensuring their year-round availability. Fruits are processed into juice, sauce, infused and dried products. Several standardized techniques are available to preserve the quality of processed products, still, there are nutrient losses associated with processing. Common fruits processed include grape, orange, apples, pears, peach, and so on. Processed fruits can be enriched with the addition of nutraceutical components.

INTRODUCTION

Overall, the food and beverage processing industry is an important manufacturing sector all across the world. The United States is among the top producers and consumers of fruits and tree nuts in the world. Each year, fruit and tree nut production generates about 13% of US farm cash receipts for all agricultural crops. Annual US per capita use of fruit and tree nuts totals nearly 300 pounds fresh-weight equivalent. Oranges, apples, grapes, and bananas are the most popular fruits. The consumption of fruits and processed products has enjoyed an unprecedented growth during the past decade. Many factors motivate this increase, including consumers' awareness of the health benefits of fruit constituents such as the importance of dietary fiber, antioxidants, vitamins, minerals, and phytochemicals present in fruits. In food stores, one can buy fresh and processed exotic food items, either canned, frozen, dehydrated, fermented, pickled, or made into jams, jellies, and marmalades year-round. Several varieties of fruits are sold throughout the year in developed countries, and with the increase in international trade of fruits, even tropical fruits are available at a reasonable cost. The food processing industry uses fruits as ingredients in juice blends, snacks, baby foods, and many other processed food items. As a

Food Biochemistry and Food Processing, Second Edition. Edited by Benjamin K. Simpson, Leo M.L. Nollet, Fidel Toldrá, Soottawat Benjakul, Gopinadhan Paliyath and Y.H. Hui.

Table 28.1. World Production of Fruit in 2002 in Metric Tons

Production (Metric Tons)	Africa	Asia	Europe	America	Australia and New Zealand	World
Primary fruits	61,934,001	204,640,809	73,435,847	129,642,621	4,438,090	475,503,880
Apples	1,696,109	30,870,497	15,820,454	7,875,880	831,999	57,094,939
Bananas	7,140,629	35,766,134	446,600	15,372,622	250,000	69,832,378
Grapes	3,113,940	14,761,974	28,739,848	12,529,900	1,872,588	61,018,250
Mangoes	2,535,781	19,841,469	–	3,343,697	28,000	25,748,947
Oranges	4,920,544	14,177,590	6,180,846	38,402,953	444,500	64,128,523
Pears	530,547	10,996,503	3,626,020	1,759,538	202,597	17,115,205
Plums	173,946	5,385,634	2,661,194	1,069,953	24,000	9,314,727
Strawberries	171,132	325,670	1,276,192	1,157,540	21,300	3,237,533

Source: Calculated from FAOSTAT database. http://faostat.fao.org/default.aspx. Accessed on December 07, 2011.

result, the world production of primary fruits has increased from 384 million metric tons in 1992 to 475 million metric tons in 2002 (FAO). The world production of fruits in 2002 by regions is shown in Table 28.1.

Advances in fruit processing technologies mostly occur in response to consumer demands or improvement in the efficiency of technology. Traditional methods of canning, freezing, and dehydration are progressively being replaced by fresh-cut, ready-to-eat fruit preparations. The use of modified atmosphere and irradiation also helps in extending the shelf life of produce.

Fruits are essential source of minerals, vitamins, and dietary fiber. In addition to these components, they provide carbohydrates, and to a lesser extent, proteins. Fruits play an important role in the digestion of meat, cheese, and other high-energy foods by neutralizing the acids produced by the hydrolysis of lipids.

FRUIT CLASSIFICATION

Fruits can be broadly grouped into three categories according to their growth latitude, whether it is in temperate, subtropical, or tropical regions.

1. *Temperate zone fruits*: Four subgroups can be distinguished among temperate zone fruits that include small fruits and berries (e.g., grape, strawberry, blueberry, blackberry, cranberry), stone fruits (e.g., peach, plum, cherry, apricot, nectarine) and pome fruits, (e.g., apple, pear, Asian pear or Nashi, and European pear or quince).
2. *Subtropical fruits*: Two subgroups can be differentiated among subtropical fruits and include citrus fruits (e.g., orange, grapefruit, lemon, tangerine, mandarin, lime, and pummelo), and noncitrus fruits (e.g., avocado, fig, kiwifruit, olive, pomegranate, and cherimoya).
3. *Tropical fruits*: Major tropical fruits include banana, mango, papaya, and pineapple. Minor tropical fruit include passion fruit, cashew apple, guava, longan, lychee, mangosteen, carambola, rambutan, sapota, and so on.

CHEMICAL COMPOSITION OF FRUITS

CARBOHYDRATES

Fruits typically contain between 10% and 25% carbohydrates, less than 1% of protein and less than 0.5% of fat (Somogyi et al. 1996b). Carbohydrates, sugars, and starches are broken down into CO_2, water, and energy during catabolism. The major sources of carbohydrates are banana, plantain, date, raisin, breadfruit, and jackfruit. Proteins and amino acids are contained in dried apricot and fig, whereas fats are the major components of avocado and olives. Sugars are the major carbohydrate components of fruits and their composition vary from fruit to fruit. In general, sucrose, glucose, and fructose are the major sugar components present in fruits (Table 28.2). Several fruits also contain sugar alcohols such as sorbitol.

VITAMINS

Fruits and vegetables are major contributors to our daily vitamin requirements. The nutrient contribution from a specific fruit or vegetable is dependent on the amount of vitamins present in the fruit or vegetable, as well as the amount consumed. The approximate percentage contribution to daily vitamin intake from fruits and vegetables is: vitamin A, 50%; thiamine, 60%; riboflavin, 30%; niacin, 50%; and vitamin C, 100% (Somogyi et al. 1996b). Vitamins are sensitive to different processing conditions such as exposure to heat, cold, reduced and high levels of oxygen, light, free water, and mineral ions. Trimming, washing, blanching, and canning can also cause the loss in vitamin content of fruits and vegetables.

MINERALS

Minerals found in fruits in general may not be fully nutritionally available (e.g., calcium, found as calcium oxalate in certain produce is not nutritionally available). Major minerals in fruits are base-forming elements (K, Ca, Mg, Na) and acid-forming elements (P, Cl, S). Minerals often present in microquantities are Mn, Zn, Fe, Cu, Co, Mo, and I. Potassium is the most abundant mineral in fruits, followed by calcium and magnesium. High potassium is often associated with increased acidity and

Table 28.2. Sugar Composition of Selected Fruits

	Sugar, g/100 mL of Juice			
Fruit	Sucrose	Glucose	Fructose	Sorbitol
Apple	0.82±0.13	2.14±0.43	5.31±0.94	0.20±0.04
Cherry	0.08±0.02	7.50±0.81	6.83±0.74	2.95±0.33
Grape	0.29±0.08	9.59±1.03	10.53±1.04	ND
Nectarine	8.38±0.73	0.85±0.04	0.59±0.02	0.27±0.04
Peach	5.68±0.52	0.67±0.06	0.49±0.01	0.09±0.02
Pear	0.55±0.12	1.68±0.36	8.12±1.56	4.08±0.79
Plum	0.51±0.36	4.28±1.18	4.86±1.30	6.29±1.97
Kiwifruit	1.81±0.72	6.94±2.85	8.24±3.43	ND
Strawberry	0.17±0.06	1.80±0.16	2.18±0.19	ND

Source: Reprinted with permission from Van Gorsel et al. 1992. Copyright, American Chemical Society.
ND, not detected (<0.05 g/100 mL).

improved color, whereas high calcium content reduces the incidence of physiological disorders and improves quality of fruits. Phosphorus is a constituent of several metabolites and nucleic acids and plays an important role in carbohydrate metabolism and energy transfer.

Dietary Fiber

Dietary fiber consists of cellulose, hemicellulose, lignin, and pectic substances, which are derived from fruit cell walls and skins. The dietary fiber content of fruits ranges from 0.5% to 1.5% by fresh weight. Because of its properties that include a high water-holding capacity, dietary fiber plays a major role in the movement of digested food and reducing the incidence of colon cancer and cardiovascular disease.

Proteins

Fruits contain less than 1% protein (as opposed to 9–20% protein in nuts). In general, plant protein sources provide a significant portion of dietary protein requirement in countries where animal proteins are in short supply. Plant proteins, unlike animal proteins, are often deficient or limiting in one or more essential amino acids. The green fruits are important sources of proteins as they contain enzymes and proteins associated with the photosynthetic apparatus. Enzymes, which catalyze metabolic processes in fruits, are important in the reactions involved in fruit ripening and senescence. Some of the enzymes important to fruit quality are:

- *Ascorbic acid oxidase*: Catalyzes oxidation of ascorbic acid and results in loss of nutritional quality.
- *Chlorophyllase*: Catalyzes removal of phytol ring from chlorophyll; results in loss of green color.
- *Polyphenol oxidase*: Catalyzes oxidation of phenolics, resulting in the formation of brown colored polymers.
- *Lipoxygenase*: Catalyzes oxidation of unsaturated fatty acids; results in off-odor and off-flavor production.
- *Polygalacturonase*: Catalyzes hydrolysis of glycosidic bonds between adjacent polygalacturonic acid residues in pectin; results in tissue softening.
- *Pectinesterase*: Catalyzes deesterification of methyl groups in pectin; Act in conjunction with polygalacturonases, leading to tissue softening.
- *Cellulase*: Catalyzes hydrolysis of cellulose polymers in the cell wall and therefore is involved in fruit softening.
- *Phospholipase D*: Initiates the degradation of cell membrane.

Lipids

In general, the lipid content of fruits is very small amounting to 0.1–0.2%. Avocado, olive, and nuts are exceptions. Despite the relatively small amount of storage lipids in fresh fruits, they still possess cell membrane lipids. Lipids make up the surface wax, which contributes to the shiny appearance of fruits, and cuticle, which protects fruits against water loss and pathogens. The degree of fatty acid saturation establishes membrane fluidity with greater saturation, resulting in less fluidity. Lipids play a significant role in the characteristic aroma and flavor of fruits. For example, the characteristic aroma and flavor of cut fruits result from the action of the enzyme lipoxygenase on fatty acids (linoleic and linolenic acids), eventually leading to the production of volatile compounds. The action of lipoxygenase is increased after the fruit is cut because there is a greater chance for the enzyme and substrate to mix together. Lipoxygenase can also be responsible for the off-flavor and off-aroma in certain plant products (soybean oil).

Volatiles

The specific aroma of fruits is due to the amount and diversity of volatiles they contain. Volatiles are present in extremely small quantities (<100 ug/g fresh wt). Ethylene is the most abundant volatile in fruit, but has no typical fruit aroma. Characteristic flavors and aromas are a combination of various character impact compounds, mainly esters, terpenes, short chain aldehydes,

Table 28.3. Organic Acids of Selected Fruits

	Organic Acid, mg/100 mL of Juice				
Fruit	Citric	Ascorbic	Malic	Quinic	Tartaric
Apple	ND	Tr	518±32	ND	ND
Cherry	ND	Tr	727±20	ND	ND
Grape	Tr	Tr	285±58	ND	162±24
Nectarine	730±92	114±6	501±42	774±57	Tr
Peach	140±39	Tr	383±67	136±28	ND
Pear	109±16	Tr	358±72	121±11	Tr
Plum	ND	Tr	371±16	220±2	ND
Kiwifruit	ND	Tr	294±24	214±68	ND
Strawberry	207±35	56±4	199±26	ND	ND

Source: Reprinted with permission from Van Gorsel et al. 1992. Copyright, American Chemical Society.
ND, not detected; Tr, <10 mg/100 mL.

ketones, organic acids, and so on. Their relative importance depends upon the threshold concentration (sometimes, <1 ppb) and interaction with other compounds. For some fruits such as apple and bananas, the specific aroma is determined by the presence of a single compound, ethyl-2-methylbutyrate in apples, and isoamylacetate in bananas.

Water

Water is the most abundant single component of fruits and vegetables (up to 90% of total weight). The maximum water content of fruits varies due to structural differences. Agricultural conditions also influence the water content of plants. As a major component of fruits, water has an impact on both the quality and the deterioration. Turgidity is a major quality determinant factor in fruits. Loss of turgor is associated with loss of quality, and is a major problem during postharvest storage and transportation. Harvest should be done during the cool part of the day in order to keep the turgidity at its optimum.

Organic Acids

The role of organic acids in fruits is twofold.

1. Organic acids are an integral part of many metabolic pathways, especially Krebs (TCA) cycle. Tricarboxylic acid cycle (respiration) is the main channel for the oxidation of organic acids, serving as an important energy source for the cells. Organic acids are metabolized into many constituents including amino acids. Major organic acids present in fruits include citric acid, malic acid, and quinic acid.
2. Organic acids are important contributors to the taste and flavor of many fruits and vegetables. Total titratable acidity, the quantity and specificity of organic acids present in fruits and so one influence the buffering system and the pH. Acid content decreases during ripening, because part of it is used for respiration and another part is transformed into sugars (gluconeogenesis). The composition of organic acids in some fruits is given in (Table 28.3).

Pigments

Pigments are mainly responsible for the skin and flesh colors in fruits and vegetables. They undergo changes during maturation and ripening of the fruits including, loss of chlorophyll (green color), synthesis and/or revelation of carotenoids (yellow and orange), and biosynthesis of anthocyanins (red, blue, and purple) (Table 28.4.). Anthocyanins occur as glycosides and are water soluble, unstable, and easily hydrolyzed by enzymes to free

Table 28.4. Anthocyanins of Selected Fruits

	Anthocyanin	
Fruit	Identification	Peak Area
Apple	Cyanidin 3-arabinoside	22±6
	Cyanidin 7-arabinoside or Cyanidin 3-glucoside	27±15
	Cyanidin 3-galactoside	260±69
Cherry	Cyanidin 3-glucoside	189±40
	Cyanidin 3-rutinoside	1320±109
	Peonidin 3 rutinoside	47±36
Grape	Cyanidin 3-glucoside	121±33
	Delphinidin 3-glucoside	586±110
	Malvidin 3-glucoside	2157±375
	Peonidin 3-glucoside	478±92
Nectarine	Cyanidin 3-glucoside	322±51
Peach	Cyanidin 3-glucoside	180±43
Plum	Cyanidin 3-glucoside	42±5
Strawberry	Cyanidin 3-glucoside	70±18
	Pelargonidin 3-glucoside	1302±29
	Pelargonidin 3-glycoside	78±9

Source: Reprinted with permission from Van Gorsel et al. 1992. Copyright, American Chemical Society.

anthocyanins. The latter may be oxidized to brown products by phenoloxidases after wounding.

Phenolics

Fruit phenolics include chlorogenic acid, catechin, epicatechin, leucoanthocyanidins and anthocyanins, flavonoids, cinnamic acid derivatives, and simple phenols. The total phenolic content of fruits varies from 1 to 2 g/100 g fresh weight. Chlorogenic acid is the main substrate responsible for the enzymatic browning of damaged fruit tissue. Phenolic components are oxidized by polyphenol oxidase enzyme in the presence of O_2 to brown products according to the following reactions.

$$\text{Monophenol} + O_2 \xrightarrow{\text{Polyphenol oxidase}} \text{quinone} + H_2O$$

$$\text{1,4 diphenol} + O_2 \xrightarrow{\text{Polyphenol oxidase}} \text{1,4 quinone} + H_2O$$

Phenolic components are responsible for the astringency in fruits and decrease with maturity because of their conversion from soluble to insoluble forms, and metabolic conversions.

Cell Structure

The fruit is composed of three kinds of cells that include parenchyma, collenchyma, and sclerenchyma.

1. *Parenchyma cells*: Parenchyma cells are the most abundant cells found in fruits. These cells are mostly responsible for the texture resulting from turgor pressure.
2. *Collenchyma cells*: Collenchyma cells contribute to the plasticity of fruit material. The collenchyma cells are located usually closer to the periphery of fruits. Collenchyma cells have thickened lignified cell walls and are elongated in their form.
3. *Sclerenchyma cells*: Also called stone cells, for example, in pears, sclerenchyma cells are responsible for the gritty/sandy texture of pears. They have very thick lignified cell walls and are also responsible for the stringy texture in such commodities as asparagus. For some commodities such as asparagus, the number of sclerenchyma cells increases after harvest and during handling, and storage.

FRUIT PROCESSING

Harvesting and Processing of Fruits

Harvesting is one of the most important of fruit-growing activities. It represents about half of the cost involved in fruit production because most fruit crops are still harvested by hand. Improvement needs to be done for the development of mechanical harvesters. Production of fruits of nearly equal size, and resistant to mechanical damage during harvest, is another area addressed by breeding technology. Certain fruits that are not to be processed are harvested with stems in order to avoid microbial invasion. Examples of this include cherries and apples.

Harvested fruit is washed to remove soil, microorganisms and pesticide residues, and then sorted according to their size and quality. Fruit sorting can be done by hand or mechanically. There are two ways of mechanical sorting of fruits, sorting with water, which takes advantage of changes in density with ripening; and automatic high-speed sorting, in which compressed air jets separate fruit in response to differences in color and ripeness as measured by light reflectance or transmittance. When not marketed as fresh, fruits are processed in many ways including: canning, freezing, concentration, and drying.

Freezing and Canning of Fruits

For freezing and canning of fruits, the general sequence of operations includes (1) washing; (2) peeling; (3) cutting, trimming, coring; (4) blanching; (5) packaging and sealing; and (6) heat processing or freezing.

Freezing is generally superior to canning for preserving the firmness of fruits. In order to minimize postharvest changes, freezing process is done at the harvest site for certain fruits. Fruit to be frozen must be stabilized against enzymatic changes during frozen storage and on thawing. The principal enzymatic changes are oxidations, causing darkening of color and the production of off-flavor. An important color change is enzymatic browning in apples, peaches, and bananas. This is due to the oxidation of pigment precursors such as *o*-diphenols and tannins by enzymes of the group known as phenol oxidases and polyphenol oxidases. Frozen fruits are produced for home use, restaurants, baking manufacturers, and other food industries. Depending on the intended end use, various techniques are employed to prevent oxidation.

The objective of blanching is to inhibit various enzyme reactions that reduce the quality of fruits. Fruits generally are not heat blanched because the heat causes loss of turgor, resulting in sogginess and juice drainage after thawing. An exception to this is when the frozen fruit is to receive heating during the baking operation, in which case calcium salts are added to the blanching water or after blanching. Calcium salts increase the firmness of the fruit by forming calcium pectates. For the same purpose, pectin, carboxymethyl cellulose, alginates, and other colloidal thickeners can also be added to fruit prior to freezing. Instead of heat treatments, certain chemicals can be used in order to inactivate oxidative enzymes or to act as antioxidants. The major antioxidant treatments include ascorbic acid dip and treatment with sulfites. Other methods for the prevention of contact with oxygen using physical barriers (sugar syrup) can be used.

Commonly dissolved in sugar syrup, vitamin C acts as an antioxidant and protects the fruit from darkening by subjecting itself to oxidation. A concentration of 0.05–0.2% ascorbic acid in syrup is effective for apple and peach. Peaches subjected to this treatment may not darken during frozen storage at −18°C for up to 2 years. Because low pH also helps in delaying oxidative processes, vitamin C and citric acid may be used in combination. Furthermore, citric acid removes cofactors of oxidative enzymes such as copper ions in polyphenol oxidase.

Sulfite is a multiple function antioxidant. Bisulfites of sodium or calcium are used to stabilize the color of fresh fruits. Like vitamin C, sulfite is an oxygen acceptor and inhibits the activity of oxidizing enzymes. Furthermore, sulfite reduces the nonenzymatic Maillard-type browning reactions by reacting with aldehyde groups of sugars so that they are no longer free to combine with amino acids. Inhibition of browning is especially important in dried fruits such as apples, apricots, and pears. Sulfite also exhibits antimicrobial properties. The disadvantage of sulfite is that some people are allergic to this chemical, which makes its use restricted, for example, FDA prohibited the use of sulfite in fresh produce and limited the residual in processed products to 10 ppm, with appropriate label.

Sugar syrup has long been used to minimize oxidation before the mechanisms of browning reactions were understood, and still remains as a common practice today. Application of sugar syrup provides a coating to the fruit and thus prevents the contact of the cut surface and oxidizable components from contact with atmospheric oxygen. Sugar syrup also increases the sensory attributes of fruits, by reducing the loss of volatiles and improving the taste.

Vacuum treatment is used in combination with the chemical dips in order to improve their effects. While fruits are submerged in the dip, vacuum is applied to draw air from the fruit tissue, allowing better penetration.

In order to reduce the cost of handling and shipping, some high-moisture fruits are pureed and concentrated to two or three times their natural solids content. Many others are dried for various purposes to different moisture levels. The majority of fruit including apricot, apple, figs, pears, prunes, and raisins are sun dried. Sulfite is commonly used to preserve the color when fruits are dried under high temperature that does not inactivate the oxidative enzymes.

Nonenzymatic Browning

Also called Maillard reaction, the nonenzymatic browning is of great importance in fruit processing. During Maillard reaction, the amino groups of amino acids, peptides, or proteins, react with aldehyde groups of sugars resulting in the formation of brown nitrogenous polymers called melanoidins (Ellis 1959, deMan 1999). The velocity and pattern of the reaction depend on the nature of the reacting compounds, the pH of the medium, and the temperature. Each kind of food shows a different browning pattern. Lysine is the most reactive amino acid because it contains a free amino group. The destruction of lysine reduces the nutritional value of a food since it is a limiting essential amino acid. The major steps involved in the Maillard reaction are:

1. An aldose or ketose reacts with a primary amino group of amino acid, peptide or protein, to produce an *N*-substituted glycosylamine;
2. The Amadori reaction, which rearranges the glycosylamine to yield ketosamine or aldoseamine;
3. Rearrangement of the ketosamine with a second molecule of aldose to yield diketosamine, whereas the rearrangement of the aldoseamine with a second molecule of amino acid leads to a diamino sugar;
4. Amino sugars are degraded by losing water to give rise to amino or nonamino compounds;
5. The condensation of the obtained products with amino acid or with each other.

In Maillard reaction, the basic amino group is the reactive component and therefore, the browning is dependent on the initial pH, or the presence of a buffer system. Low pH results in the protonation of the basic amino group, and therefore it inhibits the reaction. The effect of pH is also heavily dependent on the moisture content of the product. When the moisture is high, the browning is caused by caramelization, whereas at low water content and pH of about 6, the Maillard reaction prevails. The flavors produced by the Maillard reaction in some cases are reminiscent of caramelization.

The nonenzymic browning can be prevented by closely monitoring the factors leading to its occurrence, including temperature, pH, and moisture content. The use of inhibitors such as sulfite is an effective way of controlling nonenzymic browning. It is believed that sulfite reacts with the degradation products of the amino sugars to prevent the condensation of these products into melanoidins. However, sulfite can react with thiamine, which prohibits its use in foods containing this vitamin (Figs. 28.1 and 28.2).

Fruit Juice Processing

The quality of the juice depends on the quality of the raw material, regardless of the process. Often, the quality of the fruit is dependent on the stage of maturity or the stage of ripening. The major physicochemical parameters used in assessing fruit ripening are sugar content, acidity, starch content, and firmness (Somogyi 1996a, 1996b).

The main steps involved in the processing of most type of juice include the extraction of the juice, clarification, juice deaeration, pasteurization, concentration, essence add-back, canning or bottling, and freezing (less frequent). Juice extractors for oranges and grapefruits, whose peel contain bitter oils, are designed to cause the peel oil to run down the outside of the fruit and not enter the juice stream. Since apples do not contain bitter oil, the whole apples are pressed. Juice extraction should be done as quickly as possible, in order to minimize oxidation of phenolic compounds in fruit juice by naturally present enzymes.

Apple Juice Processing

Apple juice is processed and sold in many forms including American apple cider, European apple cider, and shelf-stable apple juice. American apple cider is the product of sound, ripe fruit that has been pressed and bottled or packaged with no form of preservatives added and stored under refrigeration. This is a sweet type of cider, which has not been fermented. Because of an increased number of food-borne outbreaks of *E. coli* 0157:H7 and *Cryptosporidium parvum* in apple cider, many jurisdictions are working toward implementation of a kill step in the processing of apple cider. Several pathogen outbreaks make unpasteurized apple cider not suitable for general consumption, and especially for children, elderly, and immunocompromised people.

European type of apple cider is a naturally fermented apple juice, usually fermented to a specific gravity of one or less (Anonymous 1980). Shelf-stable apple juice includes clarified

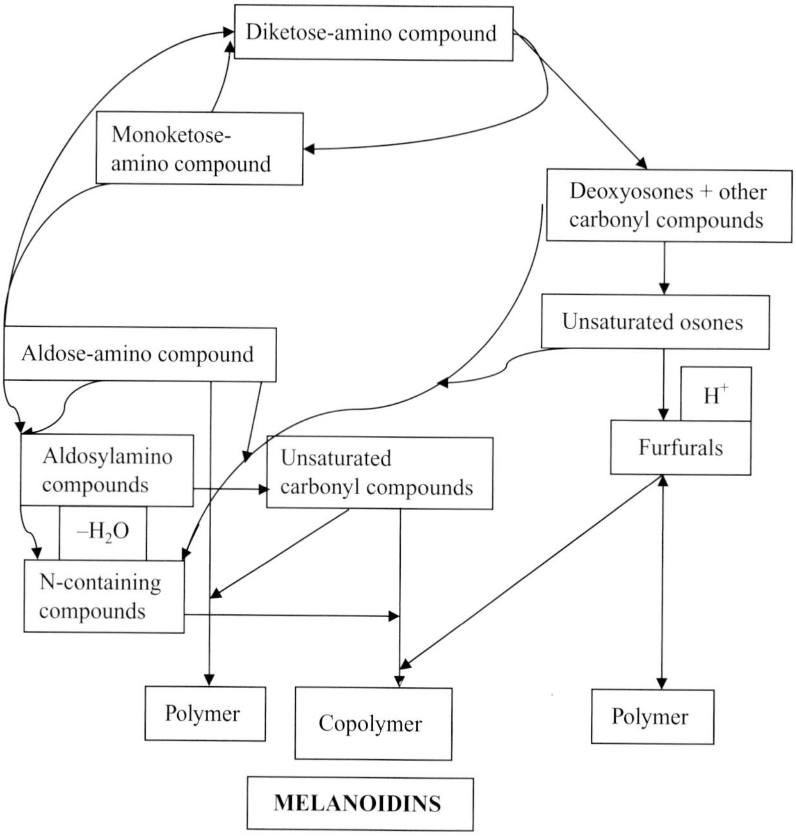

Figure 28.1. Mechanism of browning reaction.

and crushed apple juice and concentrates. The most popular type of apple juice is the one that has been clarified and filtered before pasteurization and bottling. Natural juice is a juice that comes from the press. Up to 2% of ascorbic acid can be added before pasteurization and bottling in order to preserve the color. Crushed apple juice is produced by passing coarsely ground apples through a pulper and desecrator before pasteurization.

Juice concentrates can be two types: frozen, when the Brix is about 42°, or regular (commercial) concentrate with a Brix of 70° or more. Regardless of the type of apple product that is to be produced, the quality of the apple juice is directly related to the quality of the raw material used in its production. Only sound ripe fruits should be used in the processing of apple juice. "Windfalls" or apples that are fallen and picked from the ground, should not be used for the juice due to the risk of contamination by pathogens such as *E. coli* O157:H7, *C. parvum*, *Penicillium expansium* (producing the mycotoxin patulin) and also the risk of pronounced musty or earthy off-flavor of the juice. Overmature apples are very difficult to process, whereas immature apples give a starchy and astringent juice with poor flavor.

The processing of apple juice starts with washing and sorting of the fruit in order to remove soil and other foreign material as well as decayed fruits. Any damaged or decayed fruit should be removed or trimmed in order to keep down the level of patulin in the finished juice. Patulin is an indicator, which tells if the juice was produced from windfalls or spoiled apples. The acceptable level of patulin in most countries is less than 50 ppb. Patulin is carcinogenic and teratogenic. Various methods are currently used to reduce the levels of patulin in apple juice, namely, charcoal treatment, chemical preservation (sulfur dioxide), gamma irradiation, fermentation, and trimming of fungus-infected apples. A recent study shows that pressing followed by centrifugation resulted in an average toxin reduction of 89%. Total toxin reduction using filtration, enzyme treatment, and fining were 70, 73, and 77%, respectively, in the finished juice. (Bissessur et al. 2001). Patulin reduction was due to the binding of the toxin to solid substrates such as the filter cake, pellet, and sediment.

Prior to pressing, apples are ground using disintegrators, hammers, or grating mills. The effectiveness of the pressing operation depends on the maturity level of the fruits, as more mature fruits are often difficult to press. A wide range of presses is used in juice extraction including hydraulic, pneumatic, and screw/basket type. The vertical hydraulic is a batch-type press and requires no press aid. The main disadvantage is that the press is labor intensive and produces juice with low solids content. Hydraulic presses are the oldest type of press and are still used worldwide. The newer versions are automated and require

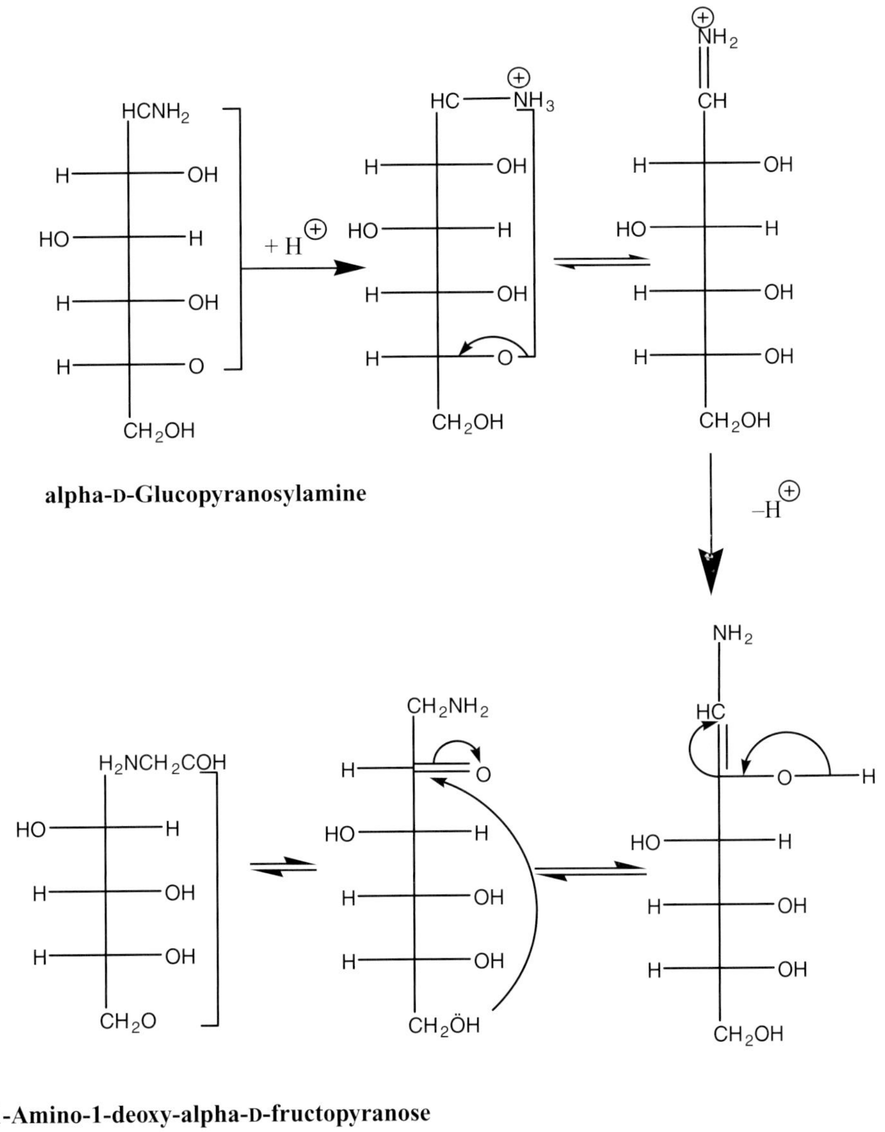

Figure 28.2. Amadori rearrangement in amino sugars.

press aid such as up to 1–2% paper pulp or rice hulls or both in order to reduce spillage and increase juice channel in the mash. The apple mash has many natural enzymes. Pectinolytic enzyme products contain the primary types of pectinases, pectinmethylesterase (PME), polygalacturonase (PG), pectin lyase, and pectin transeliminase (PTE). PME deesterifies methylated carboxylic acid moieties of pectin, liberating methanol from the side chain, after which PG can hydrolyze the long pectin chains (Fig. 28.3). Enzymatic mash treatment has been developed to improve the pressabiliy of the mash and, therefore the throughput and yield. An amount of 80–120 mL of enzyme per ton is added to the apple mash in order to break down the cell structure. High-molecular weight constituents of cell walls, like protopectin, are insoluble and inhibit the extraction of the juice from the fruit and keep solid particles suspended in the juice. Pectinase used in the apple juice processing is extracted from the fungus *Aspergillus niger*. Pectinase developed for apple mash pretreatment acts mainly on the cell wall, breaking the structure and freeing the juice. Also, the viscosity of the juice is lowered, and it can emerge more easily from the mash. The high content of pectin esterase (PE) causes the formation of deesterified pectin fragments, which has a low water-binding capacity and reduces the slipperiness. These pectins consist of chains of galacturonic acid joined by alpha-glycoside linkage. In addition, polymers of xylose, galactose, and arabinose (hemicelluloses) form a link with the cellulose. The entire system forms a gel that retains the juice in the mash. The extraction is easier, even if the pectins are partially broken by pectinesterase. The pomace acts like a

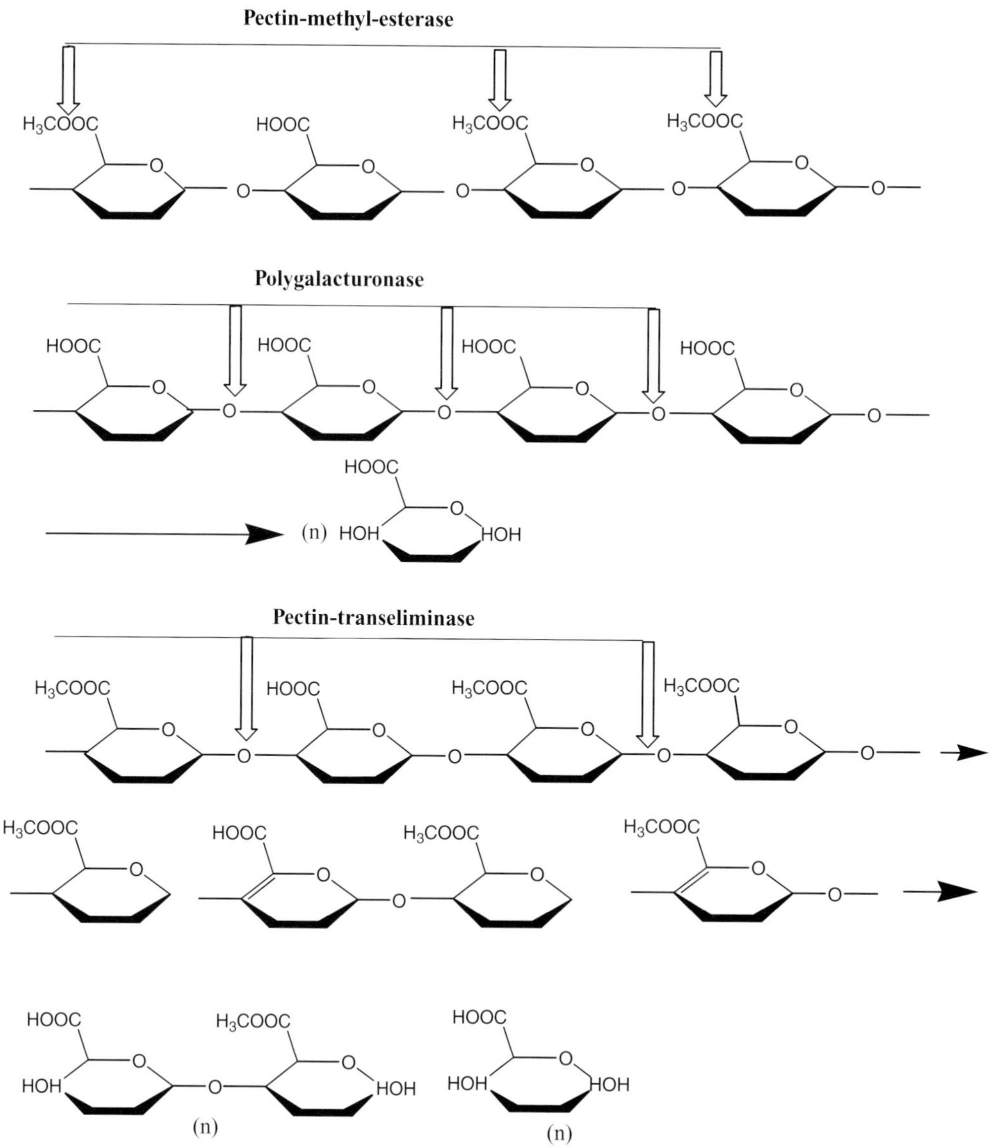

Figure 28.3. Diagrammatic representation of the site of action of various pectinolytic enzymes.

pressing aid, when used with mash predraining. The application of enzyme treatment can increase the press throughput by 30–40% and the juice yield by over 20%. Mash pretreatment also increases the flux rate of ultrafiltered apple juice by up to 50%. An important by-product of apple juice industry is pectin. Therefore, overtreatment of mash with pectinolytic enzymes could render the pomace unsuitable for the production of pectin. Inactivation of the enzyme after reaching the appropriate level of pectin degradation is an important step in the production of apple juice. Residual pectic enzymes in apple juice concentrate could cause setup problems, when used for making apple jelly. There is a recent development in apple juice extraction with the process of liquefaction. Liquefaction is a process of completely breaking down the mash by using an enzyme preparation, temperature, and time combination. The liquefied juice is extracted from the residual solid by the use of decanter centrifuges and rotary vacuum filters. It is a common practice, with the addition of cellulose enzyme to the mash to further degrade the cellulose to soluble solids, increasing the juice Brix nearly by 5°. The commercially available enzyme preparations contain more than 120 substrate specific enzyme components. Another popular extraction method is the countercurrent extraction method, developed in South Africa and refined in Europe in the 1970s. The principle of the system is as follows: the mash is heated, predrained, and counterwashed with water and recycled hot juice. A 90–95% recovery is possible when the throughput is about 3 tons per hour. The main disadvantage of the system is the lower soluble solids content of the obtained juice (6–8 vs. 11–12 from

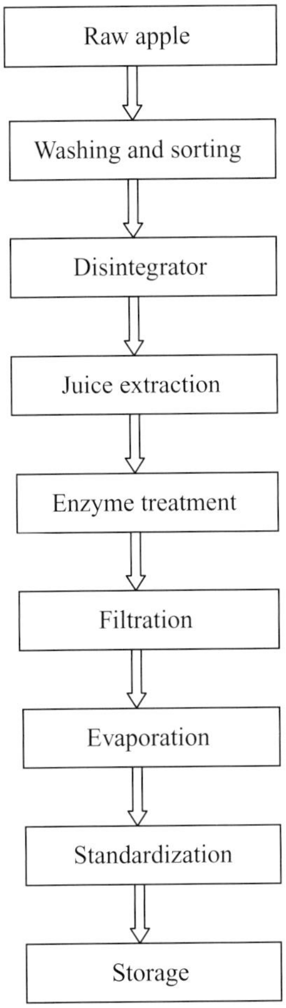

Figure 28.4. Apple juice and concentrate flow chart.

other methods). Juice yield from different types of extraction varies from 75% to 95%, and depends on many factors including the cultivar and maturity of the fruit, the type of extraction, the equipment and press aids, the time, temperature, and the addition and concentration of the enzyme to the apple mash (Fig. 28.4).

Enzyme Applications

Enzyme applications in the apple juice processing follow the extraction, especially when producing a clarified type of juice. The main objective is to remove the suspended particles from the juice (Smock and Neubert 1950). The soluble pectin in the juice has colloidal properties and inhibits the separation of the undissolved cloud particles from the clear juice. Pectinase hydrolyzes the pectin molecule so it can no longer hold juice. The treatment dosage of pectinase depends on the enzyme strength and varies from one manufacturer to another. A typical '3X' enzyme dosage is about 100 mL per 4000 liters of raw juice. Depectinization is important for the viscosity reduction and the formation of galacturonic acid groups that helps flocculate the suspended matter. This material, if not removed, binds to the filters, reduces the production, and can result in a haze formation in the final product. There are two methods of enzyme treatment commonly used in the juice industry including (1) hot treatment where the enzymes are added to 54°C juice, mixed, and held for 1–2 hours; and (2) cold treatment where the enzymes are added to the juice at reduced temperature (20°C) and held for 6–8 hours. The complete breakdown of the pectin is monitored by means of an acidified alcohol test. Five milliliters of juice are added to 15 mL of HCl-acidified, ethyl alcohol. Pectin is present if a gel develops in 3–5 minutes after mixing the juice with the ethanol solution. The absence of gel formation means the juice depectination is complete. In the cloud of the postprocessed juice, other polymers such as starch, and arabinans can be present and therefore, the clear juice is treated with alpha-amylase and hemicellulase enzymes, in order to partially or completely degrade them. Gelatin can be used to remove fragmented pectin chains and tannins from the juice. Best results are obtained when hydrating 1% gelatin in warm 60°C water. Gelatin can be added in combination with the enzyme treatment, bentonite, or by adding midway through the enzyme treatment period. The positively charged gelatin facilitates the removal of the negatively charged suspended colloidal material from the juice. Bentonite can be added in the range of 1.25–2.5 kg of rehydrated bentonite per 4000 L of juice. Bentonite is also added to increase the efficiency of settling, protein removal, and prevention of cloudiness caused by metal ions. For cloudy and natural apple juice, enzymes are usually not used. The enzyme-treated, refined, and settled apple juice is then pumped from the settled material (lees) and further clarified by filtration.

The filtration of apple juice is done with or without a filter aid. The major types of filters are pressure leaf, rotary vacuum, frame, belt, and Millipore filters. To obtain the desired product color and clarity, most juice manufacturers use a filter medium or filter aid in the filtration process. The filter media include diatomaceous earth, paper pulp pads, cloth pads or socks, and ceramic membranes to name a few. The filter aid helps to prevent the blinding of the filters and increase throughput. As the fruit matures, more filter aid will be required. Several types of filter aids are available, in which the most commonly used is diatomaceous earth or cellulose-type materials. Additional juice can be recovered from the tank bottoms or "lees" by centrifugation or filtration. This recovered juice can be added to the raw juice before filtration.

Diatomaceous earth, or kieselguhr, is a form of hydrated silica. It has also been called fossil silica or infusorial earth. Diatomaceous earth is made up of the skeletal remains of prehistoric diatoms that were single-cell plant life and is related to the algae that grow in the lakes and oceans. Diatomaceous earth filtration is a three-step operation: (1) first, a firm thin protective precoat layer of filter aid, usually of cellulose, is built up on the filter septum (which is usually a fine wire screen, synthetic cloth, or felt); (2) the use of the correct amount of a diatomite body feed or admix (about 4.54 kg per 3000 cm^2 of filter screen); and (3) the separation of the spent filter cake from the septum prior to the next filter cycle. Prior to the filtration, centrifugation may be used to remove high-molecular weight suspended solids.

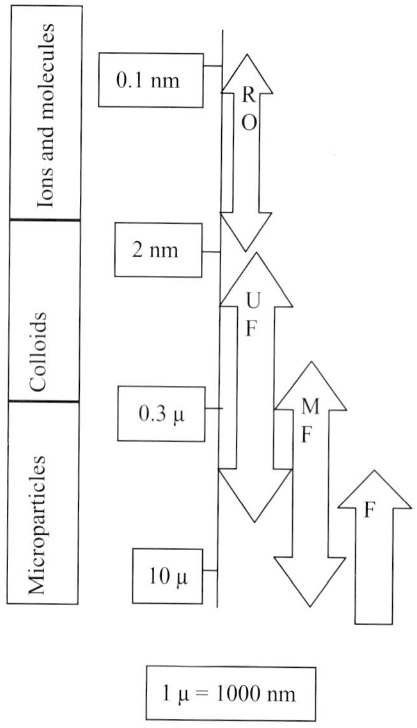

Figure 28.5. Membrane filtration. RO, reverse osmosis; UF, ultrafiltration; MF, microfiltration; F, filtration.

In some juice plants, high-speed centrifugation is used prior to the filtration. This centrifugation step reduces the solids by about 50%, thus minimizing the amount of filter aid required. The filtration process is critical, not only from the point of view of production, but also for the quality of the end product. Both pressure and vacuum filters have been used with success in juice production (Nelson and Tresler 1980). A recent development in the juice industry is the membrane ultrafiltration. Ultrafiltration, based on membrane separation, has been used with good results to separate, clarify, and concentrate various food products. Ultrafiltration of apple juice will not only clarify the products, but depending on the size of the membrane, will also remove the yeast and mold microorganisms common in apple juice (Fig. 28.5).

Apple Juice Preservation

Preservation of apple juice can be achieved by refrigeration, pasteurization, concentration, chemical treatment, membrane filtration, or irradiation. The most common method is pasteurization based on temperature and time of exposure. The juice is heated to over 83°C, held for 3 minutes, filled hot into the container (cans or bottles), and hermetically sealed. The apple juice is held for 1 minute at 83°C and then cooled to less than 37°C. When containers are closed hot and then cooled, vacuum develops, reducing the available oxygen that also aids in the prevention of microbial growth. After the heat treatment, the juice may also be stored in bulk containers. Aseptic packaging is another common process where, after pasteurization, the juice is cooled and packed in a closed, commercially sterile system under aseptic conditions. This process provides shelf-stable juice in laminated, soft-sided consumer cartons, bag-in-box cartons, or aseptic bags in 200–250 liter drums.

Apple juice concentration is another common method of preservation. The single-strength apple juice is concentrated by evaporation, preferably to 70° or 71° Brix. By an alternate method, the single-strength juice is preconcentrated by reverse osmosis to about 40° Brix, then further concentrated by evaporation methods. The reduced water content and natural acidity make the final concentrated apple juice shelf stable at room temperature. There are several evaporators used in apple juice production including rising film evaporators, falling film evaporators, and multiple effect tubular and plate evaporators. Due to the heat sensitivity of the apple juice, the multiple-effect evaporator with aroma recovery is the most commonly used. The general method in a multiple-effect evaporator is to heat the juice to about 90°C and capture the volatile (aroma) components by cooling and condensation. This is followed by reheating the 20–25° Brix juice concentrate in the first stage to about 100°C and then concentrate it to about 40–45° Brix. Another stage of heating and evaporation at about 50°C to 60° Brix, and final heating and concentration in the fourth stage to 71° Brix provides fully concentrated juice. The warm concentrate is then chilled to 4–5°C prior to adjusting the Brix to 70° before barreling or bulk storage.

Chemicals such as benzoic acid, sorbic acid, and sulfite are sometimes used to reduce spoilage of unpasteurized apple juice, either in bulk or as an aid in helping to preserve refrigerated products. Application of irradiation and ultrasonic sound are new emerging methods of preservation with high potential even though not fully accepted by consumers at this time.

Apple essence is recovered during the concentration of apple juice. The identification of volatile apple constituents, commonly known as essence or aroma, has been the subject of considerable research. In 1967, researchers at the USDA identified 56 separate compounds from apple essence. These compounds were further refined by organoleptic identification, using a trained panel of sensory specialists. These laboratory evaluations revealed 18 threshold compounds identified as "Delicious" apple compounds consisting of alcohols, aldehydes, and esters. Three of the eighteen compounds had "apple-like aromas" according to the taste panel. These were 1-hexanal, trans-2-hexenal, and ethyl, 2-methyl butyrate (Flath et al. 1967, Somogyi et al. 1996b).

Processed Apple Products

The popularity of blended traditional/tropical juice products has pushed the consumption of traditional fruit juices such as apple, orange, and grape juices to 25 liters in 2002, an increase of more than 24% from 1992 (Statistics Canada 2003). Apples are normally processed into a variety of products, although apple juice is the most popular processed apple product. With a production of 465,418 metric tons of apple in 2001, Canada exported

18,538 metric tons of concentrated apple juice. For the same year, United States exported 25,170 metric tons of concentrated apple juice (FAO 2003). Apples for processing should be of high quality, proper maturity, of medium size, and uniform in shape. Apples are processed into frozen, canned, dehydrated apple slices and dices, and different kinds of applesauce. Apples that are unsuitable for peeling are diverted to juice processing.

Apple Sauce

Diced or chopped apples with added sugar, preferably a sugar concentrate, are cooked at 93–98°C for 4–5 minutes in order to soften the fruit and inactivate polyphenol oxidase. Sauce with a good texture, color, and consistency is produced with high quality raw apples and a good combination of time and temperature treatment. Cooked applesauce is passed through a pulper of 1.65–3.2 mm finishing screen to remove unwanted debris and improve the texture. Applesauce is then heated to 90°C and immediately filled in glass jars or metal cans. The filled applesauce are seamed or capped at 88°C and cooled to 35–40°C after 1–2 minutes (Fig. 28.6). There are various types of applesauce that include natural, no sugar added, "chunky", cinnamon applesauce, and mixture of applesauce and other fruits such as apricot, peach, or cherry.

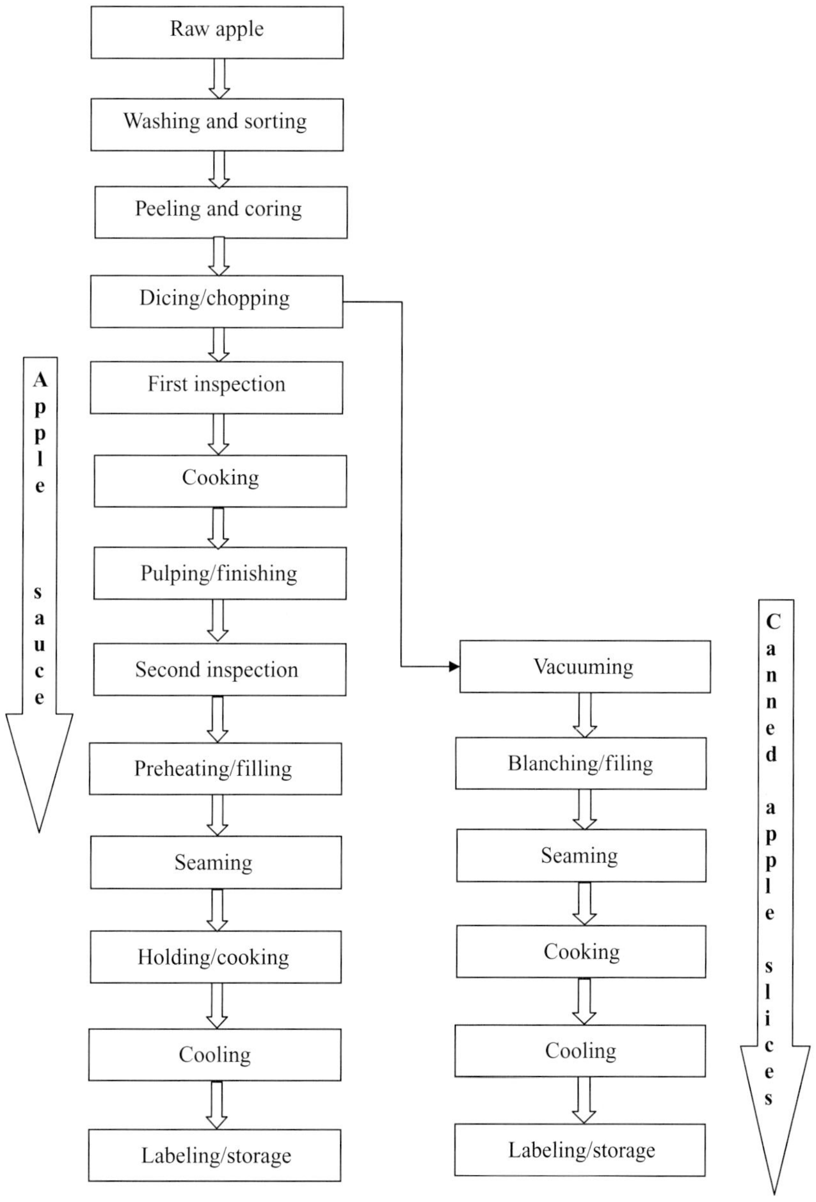

Figure 28.6. A flow chart depicting steps in apple processing.

Sliced Apples

Sliced apples have multiple uses and are preserved by many different methods such as canned, refrigerated, frozen, or dehydrated states. About 85% of sliced apples are processed, whereas only 15% are refrigerated, frozen or dehydrated and frozen. Apple slice texture is very important, as well as the consistency of the slice size, therefore, apples with firm flesh and high-quality, falling within a specified size range are desirable. Apples are sliced into 12–16 slices and blanched after inspection for eventual defects. The blanched apple slices are hot filled into cans and closed under steam vacuum after addition of hot water or sugar syrup. The canned apples are heated at 82.2°C and immediately cooled to 37–40°C.

When bulk frozen, apple slices are vacuum treated and blanched and filled into 13–15 kg tins or poly-lined boxes. The tins or boxes are then sealed, frozen, and stored at −17°C.

Individual quick frozen apple slices (IQF) are usually treated with sodium bisulfite after inspection. Nitrogen (N_2) and carbon dioxide (CO_2) are the most popular freezing media. From the vacuum tank, the apple slices pass through an IQF unit where the slices are individually frozen. From the freezing unit, the slices are filled into tins or poly-lined boxes and stored frozen at −17°C or below.

Dehydrofrozen apple slices are dehydrated to less than 50% of their original weight and then frozen. The dehydrofrozen slices are packed in cardboard containers or large metal cans with polyethylene liners and rapidly frozen before storing. Frozen slices are thawed and then soaked in a combined solution of sugar, $CaCl_2$, and ascorbic acid, or bisulfite. For fresh and refrigerated apple slices, the use of 0.1–0.2% calcium chloride, and ascorbate protects apple slices from browning and microbial spoilage. Blanched slices resist browning for approximately 2 days, but lose flavor, sugar, and acid. This type of product has a very short shelf life. Recently, the use of natural protein polymer coatings (NatureSeal) has shown promise in enhancing the shelf life and quality of fresh-cut apple products.

Dried Apple Products

Most processing cultivars are used for drying, but the best quality dried apple slices are obtained from "Red Delicious" and "Golden Delicious" apples. A desirable quality attribute of apples used for drying is a high sugar/water ratio. Color preservation and reduction in undesirable enzyme activities are achieved by the use of bisulfite. There are two types of dried apple products including evaporated and dehydrated apples. Evaporated apples are cut into rings, pie pieces, or dices and then dried to less than 24% moisture by weight. The dehydrated apples, on the other hand, are cut into dices, pie pieces, granules, and flakes prior to drying to 3–0.5% moisture content. Up to 300 ppm of bisulfite is used to prevent color deterioration. The maximum allowed SO_2 in dried apples in Europe is 500 ppm, whereas in United States, the limit is 1000 ppm.

Quality Control

Apples contain several organic acids and as such, only a limited number of microorganisms can grow in them. The most common microbes are molds, yeasts, aciduric bacteria, and certain pathogens such as *E. coli* O157:H7, capable of growing at low pH (Swanson 1989). Several approaches to quality management are available today including total quality management (TQM), statistical quality control (SQC), and hazard analysis critical control points (HACCP).

Biochemical Composition and Nutritional Value of Processed Apples

The nutritive value of most processed apple products is similar to the fresh raw product. Apple products are sources of potassium, phosphorus, calcium, vitamin A, and ascorbic acid. Glucose, sucrose, and fructose are the most abundant sugars. Dried or dehydrated apples have a higher energy value per gram tissue due to the concentration of sugars (Tables 28.4 and 28.5).

The nutritional value of apples, and fruits in general, is enhanced by the presence of flavonoids. More than 4000 flavonoids have been identified to date. There are many classes of flavonoids of which flavanones, flavones, flavonols, isoflavonoids, anthocyanins, and flavans are of interest. Flavanones occur predominantly in citrus fruits, anthocyanins, catechins, and flavonols are widely distributed in fruits, and isoflavonoids are present in

Table 28.5. Nutrients in Fruits and Processed Products (454 g)

Apples	Food Energy (kcal)	Protein (g)	Fat (g)	Carbohydrate (g)	Calcium (mg)	Phosphorus (mg)	Iron (mg)
Raw fresh	242	0.8	2.5	60.5	29	42	1.3
Applesauce[a]	413	0.9	0.5	108.0	18	23	2.3
Unsweetened Juice	186	0.9	0.9	49.0	18	23	2.3
Apple juice	213	0.5	0.1	54.0	27	41	2.7
Frozen sliced[a]	422	0.9	0.5	110.2	23	27	2.3
Apple butter[a]	844	1.8	3.6	212.3	64	163	3.2
Dried, 24%	1247	4.5	7.3	325.7	141	236	7.3
Dehydrated, 2%	1601	6.4	9.1	417.8	181	299	9.1

Source: Composition of foods. Agriculture Handbook No 8.
[a]With sugar.

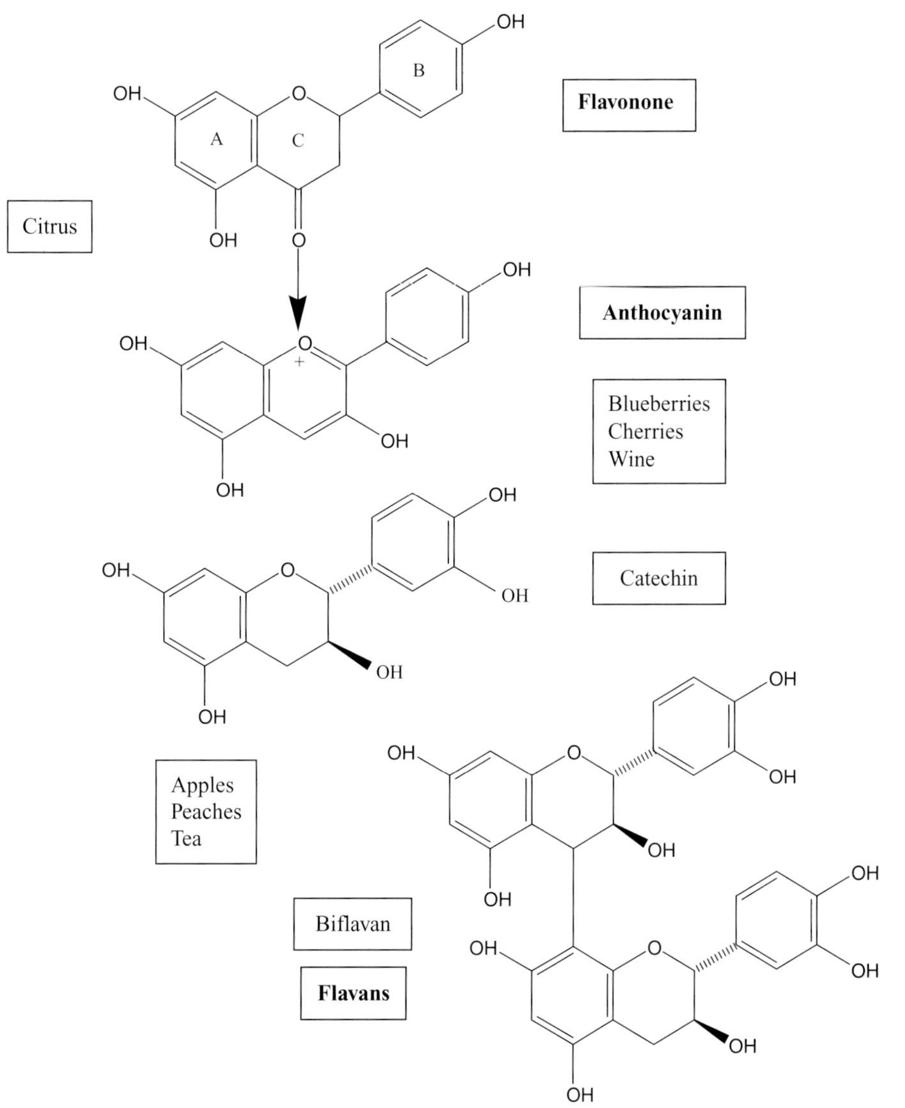

Figure 28.7. Chemical structures of fruit flavonoids. The propanoid structure consists of two fused rings. "A" is aromatic ring and "C" is a heterocyclic ring attached by carbon–carbon bond to "B" ring. The flavonoids differ in their structure at the "B" and "C" rings.

legumes. Structurally, flavonoids are characterized by a C_6-C_3-C_6 carbon skeleton (Fig. 28.7). They occur as aglycones (without sugar moieties) and as glycosides (with sugar moieties). Processing may decrease the content of flavonoids by up to 50%, due to leaching into water or removal of portions of the fruit such as the skin, that are rich in flavonoids and anthocyanins. It is estimated that the dietary intake of flavonoids may vary from 23 mg/d to 1000 mg/d in various populations (Peterson and Dwyer 1998). Flavonoids are increasingly recognized as playing potentially important roles in health including but not limited to their roles as antioxidants. Epidemiological studies suggest that flavonoids may reduce the risk of developing cardiovascular diseases and stroke.

Flavanones

The major source of flavanones is citrus fruits and juices. Flavanones contribute to the flavor of citrus, for example, naringin found in grapefruit provides the bitter taste, whereas hesperidin, found in oranges is tasteless.

Flavonols

The best-known flavonols are quercetin and kaempferol. Quercetin is ubiquitous in fruits and vegetables. Kaempferol is most common among fruits and leafy vegetables. In fruits, flavonols and their glycosides are found predominantly in the skin. Myricetin is found most often in berries.

Table 28.6. Nutrients in Fruit and Processed Products (454 g)

Apples	Sodium (mg)	Potassium (mg)	Vitamin A (IU)	Thiamin (mg)	Riboflavin (mg)	Niacin (mg)	Vitamin C (mg)
Raw fresh	4	459	380	0.12	0.08	0.3	16
Applesauce[a]	9	295	180	0.08	0.05	0.2	5
Unsweetened juice	9	354	180	0.08	0.05	0.2	5
Apple juice	5	458	–	0.03	0.07	0.4	4
Frozen sliced[a]	64	308	80	0.05	0.014	1.0	33
Apple butter[a]	9	1143	0	0.05	0.09	0.7	9
Dried, 24%	23	22,581	–	0.26	0.053	2.3	48
Dehydrated, 2%	32	3311	–	0.02	0.026	2.9	47

Source: Composition of foods. Agriculture Handbook No 8.
[a]With sugar.

Anthocyanins

Anthocyanins often occur as a complex mixture. Grape extracts can have glucosides, acetyl glucosides, and coumaryl glucosides of delphinidin, cyanidin, petunidin, peonidin, and malvidin. The color of anthocyanins is pH dependent. An anthocyanin is usually red at a pH of 3.5, becoming colorless and then shifting to blue as the pH increases. Fruit anthocyanin content increases with maturity. In stone fruits (peaches and plums) and pome fruits (apple and pears), anthocyanins are restricted to the skin, whereas in soft fruits (berries), they are present both in skin and flesh. Anthocyanins are used as coloring agents for beverages and other food products.

Flavans

Flavans are what was once called catechins, leucoanthocyanins, proanthocyanins, and tannins. They occur as monoflavans, biflavans, and triflavans. Monoflavans are found in ripe fruits and fresh leaves. Biflavans and triflavans are found in fruits such as apples, blackberries, blackcurrants, cranberries, grapes, peaches, and strawberries (Table 28.6).

FURTHER READING

Arthey D, Ashwurst PR. 2001. *Fruit Processing: Nutrition, Products and Quality Management*. Aspen Publishers, Gaithersburg, MD, p. 312.

Enachescu DM. 1995. *Fruit and Vegetable Processing*. FAO Agricultural Services Bulletin 119, FAO, p. 382.

Jongen WMF. 2002. *Fruit and Vegetable Processing: Improving Quality*. (Electronic Resource). CRC Press, Boca Raton, FL.

Salunkhe DK, Kadam SS. 1998. *Handbook of Vegetable Science and Technology: Production, Composition, storage and Processing*. Marcel Dekker, New York, p. 721.

REFERENCES

Anonymous. 1980. Cider. In: *National Association of Cider Makers*. NACM, Dorchester, Dorset.

Bissessur J et al. 2001. Reduction of patulin during apple juice clarification. *J Food Protection* 64: 1216–1222.

deMan JM. 1999. *Principles of food chemistry*, 3rd edn. Aspen Publishers, Gaithersburg, MD, pp. 120–137.

Ellis GP. 1959. The Maillard reaction. In: ML Wolfram, RS Tipson (eds.) *Advances in Carbohydrates Chemistry, Volume 14*. Academic Press, New York.

Flath RA et al. 1967. *J Agric Food Chem* 15: 29–38.

Canadian apple industry. http://www.ats.agr.gc.ca/can/4480-eng.htm. Accessed on December 06, 2011.

FAO. 2003. http://www.statcan.gc.ca/pub/82-003-x/2008004/article/6500821-eng.pdf. Accessed on December 06, 2011.

Statistics Canada. 2003. http://www.statcan.ca/english/ads/23F0001XCB/highlight.htm. Accessed on December 07, 2011.

Nelson PE, Tresler DK. 1980. *Fruit and Vegetable Juice Production Technology*, 3rd edn. AVI Co., Westport, CT.

Peterson J, Dwyer J. 1998. Flavonoids: dietary occurrence and biochemical activity. *Nutrition Research* 18: 1995–2018.

Somogyi LP et al. 1996a. *Processing Fruits: Science and Technology, Volume 2, Major Processed Products*. Technomic Publishing Co., Lancaster, PA, pp. 9–35.

Somogyi LP et al. 1996b. *Processing Fruits: Science and Technology, Volume 1, Biology, Principles and Applications*. Technomic Publishing Co., Lancaster, PA, pp. 1–24.

Smock RM, Neubert AM. 1950. *Apples and Apple Products*. Interscience Publishers, New York.

Swanson KMJ. 1989. Microbiology and preservation. In: DL Downing (ed.) *Processed Apple Products*. Van Nostrand Reinhold/AVI, New York, pp. 343–363.

Van Gorsel H et al. 1992. Compositional characterisation of prune juice. *J Agric Food Chem* 40: 784–789.

29
Biochemistry of Vegetable Processing

Moustapha Oke, Jissy K. Jacob, and Gopinadhan Paliyath

Abstract: Vegetables are very important components of diet. Vegetables are highly perishable, and transportation, storage, and distribution require low-temperature conditions. Vegetables are processed into juice, sauce, and canned under aseptic conditions. This enables long-term storage of processed products. Tomato is the major produce that is processed into juice and sauce. Preservation of nutritional components is compromised during processing. Stability of juice is influenced by the particle size distribution.

INTRODUCTION

Vegetables and fruits have many similarities with respect to their composition, harvesting, storage properties, and processing. In the true botanical sense, many vegetables are considered as fruits. Thus, tomatoes, cucumbers, eggplant, peppers, and so on could be considered as fruits, since they develop from ovaries or flower parts and are functionally designed to help the development and maturation of seeds. However, the important distinction between fruits and vegetables is based on their use. In general, most vegetables are immature or partially mature and are consumed with the main course of a meal, whereas fruits are generally eaten alone or as a dessert. The United States is one of the world's leading producers and consumers of

Food Biochemistry and Food Processing, Second Edition. Edited by Benjamin K. Simpson, Leo M.L. Nollet, Fidel Toldrá, Soottawat Benjakul, Gopinadhan Paliyath and Y.H. Hui.

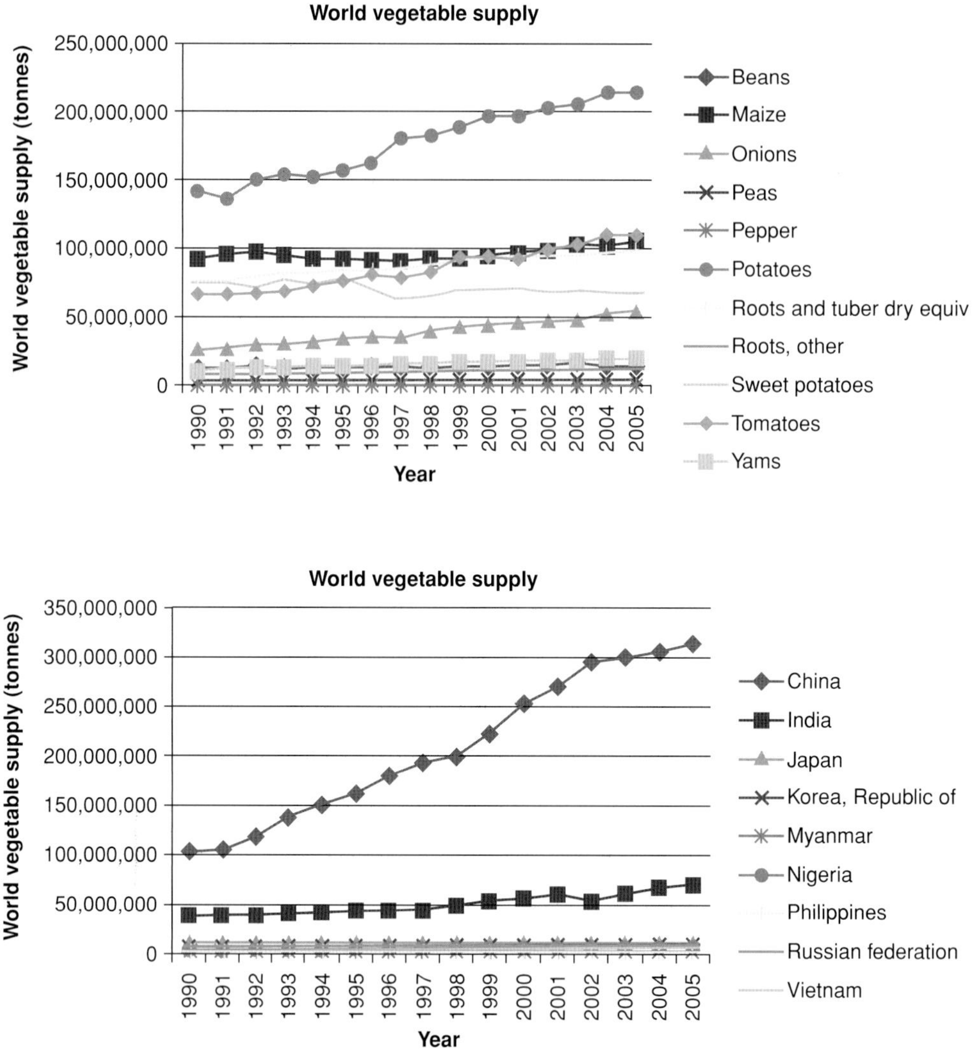

Figure 29.1. An estimate of world vegetable production by region, and type of vegetable. (*Source*: FAOSTAT 2010.)

vegetables and fruits. In 2002, the farm gate value of vegetables and melons (including mushrooms) sold in the United States reached $17.7 billion. Annual per capita use of fresh vegetables and melons rose 7% between 1990–1992 and 2000–2002, reaching 442 pounds, as fresh consumption increased and consumption of processed products decreased. According to Food and Agriculture Organization (FAO) data, the world production of vegetables has increased from 147 million metric tons (Mt) in 1992 to 350 million Mt in 2005 (Fig. 29.1). Potato, tomato, sweet potato, onions and so on are the major vegetables that are produced across the world (Fig. 29.1). Vegetables remain a popular choice for consumers worldwide. In general, the supply of vegetables per capita has increased in developing countries, but lags behind in certain parts of the world (Table 29.1). In 2002, each individual ate an average of 110

Table 29.1. Changes in Vegetable Supply in Different Parts of the World

Region	1979	2000
World	66.1	101.9
Developed countries	107.4	112.8
Developing countries	51.1	98.8
Africa	45.4	52.1
North and Central America	88.7	98.3
South America	43.2	47.8
Asia	56.6	116.2
Europe	110.9	112.5
Oceania	71.8	98.7

Source: Fresco LO, Baudoin WO 2002.

Table 29.2. World Production of Selected Vegetables in 2002 in Metric Tons (Mt)

Countries	Production in 2002	
	Fresh Vegetable	Tomatoes
World	233,223,758	108,499,056
Africa	12,387,390	12,428,174
Asia	199,192,449	53,290,273
Australia	80,000	400,000
Canada	124,000	690,000
European Union (15)	8,462,000	14,534,582
New Zealand	120,000	87,000
North and Central America	2,023,558	15,837,877
South America	3,503,611	6,481,410
United States of America	1,060,000	12,266,810

kilograms of vegetables (including potatoes), up from 106 kg a decade earlier. Potatoes represented 35% of all vegetables consumed. In 2008, greater than 50% of women and nearly 40% of men reported to have more than 5 servings of fruits and vegetables per day. (http://www.statcan.gc.ca/pub/82-229-x/2009001/deter/fvc-eng.htm.) The world production of certain vegetables is shown in Tables 29.2 and 29.3. From 2002 to 2005, the increase in vegetable production observed was primarily due to an increase in production of potatoes (Fig. 29.1).

In the United States, consumption of fresh vegetables (excluding potatoes) has increased from 60 kg in 1986 to 72 kg in 2003. In 2003, the most consumed fresh vegetable was head lettuce at 12 kg, followed by onions at 9 kg and fresh tomatoes at 8 kg. In the same year, the consumption of processed tomatoes per capita was 35 kg and processed sweet corn 9.5 kg. In Canada, the consumption of fresh vegetables (excluding potatoes) has been increasing steadily reaching 70.2 kg/capita in 1997 from 41 kg/capita in 1971. Also in 1997, the consumption of fresh vegetables has declined 2.4% from the amount of fresh vegetables consumed in 1996 (71.9 kg/capita). Lettuce (16%) is the most consumed fresh vegetable followed by onions (12.1%), carrots (12.0%), tomatoes (11.6%), and cabbage (8.1%). Brussels sprouts, parsnips, asparagus, beets, and peas each represent less than 1.0% of Canadian fresh vegetable diet. On a per capita basis, Canada has one of the highest consumption rates of fresh vegetables in the world. Frozen vegetable consumption has declined 3.4% to 5.6 kg/capita while canned vegetables and vegetable juices have increased 2.4% to 12.9 kg/capita for the same period. Vegetable purchases by consumers represent 6.6% of total food expenditures, virtually unchanged in the past 10 years.

Table 29.3. Production of Tomato in North America

Tomato Production (Mt)	Year 2002
World	108,499,056
Canada	690,000
United States of America	12,266,810
Mexico	2,083,558

Source: Food and Agricultural Organization (FAO).

Table 29.4. Classification of Vegetables

Types of Vegetables	Examples
Earth vegetables	
Roots	Sweet potatoes, carrots
Modified stems	
Corms	Taro
Tubers	Potatoes
Modified buds	
Bulbs	Onions, garlic
Herbage vegetables	
Leaves	Cabbage, spinach, lettuce
Petioles (leaf stalk)	Celery, rhubarb
Flower buds	Cauliflower, artichokes
Sprouts, shoots (young stems)	Asparagus, bamboo shoots
Fruit vegetables	
Legumes	Peas, green beans
Cereal	Sweet corn
Vine fruits	Squash, cucumber
Berry fruits	Tomato, eggplant
Tree fruits	Avocado, breadfruit

CLASSIFICATION OF VEGETABLES

Vegetables can be classified according to the part of the plant from which they are derived, such as leaves, roots, stems, and buds as shown in Table 29.4. They also can be classified into "wet" or "dry" crops. Wet crops such as celery or lettuce have water as their major component, whereas dry crops such as soybeans have carbohydrates, protein, and fat as their major constituents, and relatively low amount of water. Soybeans, for example, are composed of 36% protein, 35% carbohydrate, 19% fat, and 10% water. Wet crops tend to perish more rapidly compared to dry crops.

CHEMICAL COMPOSITION OF VEGETABLES

Fresh vegetables contain more than 70% water, and very frequently greater than 85%. Beans and other dry crops are exceptions. The protein content is often less than 3.5% and the fat content less than 0.5%. Vegetables are also important sources of digestible and indigestible carbohydrate, as well as of minerals and vitamins. They contain the precursor of vitamin A, beta-carotene, and other carotenoids. Carrots are one of the richest sources of beta-carotene (provitamin A).

Vitamins

Vegetables are major contributors to our daily vitamin requirements. The nutrient contribution from a specific vegetable is dependent on the amount of vitamins present in the vegetable, as well as the amount consumed. The approximate percentage that vegetables contribute to daily vitamin intake is: vitamin A-50%, thiamine-60%, riboflavin-30%, niacin-50%, and vitamin C-100%. Vitamins are sensitive to different processing conditions including; exposure to heat, oxygen, light, free water, and traces of certain minerals. Trimming, washing, blanching, and canning can cause loss in vitamin content of fruits and vegetables.

Minerals

The amount and types of minerals depend on the specific vegetable. Not all minerals in plant materials are readily available and are mostly in the form of complexes. An example is calcium found in vegetables as calcium oxalate. Green leafy vegetables are rich in magnesium and iron.

Dietary Fiber

The major polysaccharides found in vegetables include starch, and dietary fiber such as cellulose, hemicellulose, pectic substances, and lignin. Cell walls in young vegetables are composed of cellulose. As the produce ages, cell walls become higher in hemicellulose and lignin. These materials are tough and fibrous and their consistency is not affected by processing. Some vegetables such as potato also contain varying amounts of starch resistant to hydrolysis (resistant starch). Resistant starch is not digested as efficiently as regular starch, and reaches the colon where they undergo microbial fermentation. The short chain fatty acids liberated during the digestion of resistant starch is considered to be beneficial to the health. Roots of vegetables such as endive are rich in fructooligosaccharides such as inulin, which are considered to possess health regulatory function.

Proteins

Most vegetables contain less than 3.5% protein. Soybeans are an exception. In general, plant proteins are major sources of dietary protein in places where animal protein is in short supply. Plant proteins, however, are often deficient or limiting in one or more essential amino acids. Wheat protein is limiting in lysine, while soybean protein is limiting in methionine. Leafy green vegetables are also rich in proteins, especially photosynthetic proteins such as ribulose-bis-phosphate carboxylase oxygenase (RubisCO) and other chloroplast proteins. Multiple sources of plant proteins are recommended in the diet because of the absence of key amino acids.

Lipids

The lipid content of vegetables is less than 0.5% and primarily found in the cuticles, as constituents of the cell membrane, and in some cases, as a part of the internal cell structure (oleosomes). Even though lipids are a minor component of vegetables, they play an important role in the characteristic aroma and flavor of the vegetable. The characteristic aroma of cut tomato and cucumber results from components released from the lipoxygenase pathway, through the action of lipoxygenase upon linoleic and linolenic acids and the hydroperoxide lyase action on the peroxidized fatty acids to produce volatile compounds. This action is accentuated, when the tissue is damaged. The results of the action of lipoxygenase are sometimes deleterious to the quality—for example, the action of lipoxygenase on soybean oil leads to rancid flavors and aromas.

Volatiles

The specific aroma of vegetables is due to the amount and diversity of volatiles they contain. Volatiles are present in extremely small quantities (<100 ug/g fresh wt). Characteristic flavors and aromas are a combination of various compounds, mainly short chain aldehydes, ketones, and organic acids. Their relative importance depends upon threshold concentration (sometimes < 1 ppb), and interaction with other compounds. Of more than 400 volatiles identified in tomato, the following have been reported to play important roles in fresh tomato flavor: hexanal, trans-2-hexenal, cis-3-hexenal, cis-3-hexenol, trans-2-trans-4 decadienal, 2-isobutylthiazole, 6-methyl-5-hepten-2-one, 1-penten-3-one, and β-ionone (Petro-Turza 1986–1987).

Water

In general, water is the most abundant single component of most vegetables (up to 90% of total weight). The maximum water content of vegetables varies between individuals due to structural differences. Agricultural conditions also influence the water content of plants. As a major component of vegetables, water impacts both on the quality and the rate of deterioration. Harvest should be done during the cool part of the day in order to keep the turgidity to its optimum. Loss of turgor under post harvest storage is a major quality-reducing factor in vegetables (wilting of leaves such as spinach).

Organic Acids

Organic acids are important contributors to the taste and flavor of many vegetables (tomato). Total titratable acidity, the quantity and specificity of organic acids present in vegetables influence the buffering system and the pH. Acid content decreases during maturation, because of its use for respiration and transformation into sugars through gluconeogenesis. Certain fruits are used as vegetables when immature. For example, immature mango fruits, which are rich in organic acids, are used as vegetables.

Pigments

Pigments are mainly responsible for the skin and flesh colors in vegetables. The vegetables undergo changes during maturation and ripening of vegetables, including loss of chlorophyll (green color), synthesis and/or revelation of carotenoids (yellow and orange) and development of anthocyanins as in eggplant (red, blue and purple). Vegetables such as carrots are especially rich in beta-carotene, and red beets owe their red color to betacyanins. Tomatoes are a rich source of the red carotenoid lycopene. Anthocyanins belong to the group of flavonoids, occur as glycosides, and are water-soluble. They are unstable and easily hydrolyzed by enzymes to free anthocyanins. The latter are oxidized to brown products by phenoloxidases. The colors of anthocyanins are pH dependent. In basic medium, they are mostly violet or blue, whereas in acidic medium they tend to be red. Several exotic new vegetables with different colors have been introduced recently, that include purple carrots and tomatoes containing anthocyanins, yellow, and green cauliflower (broccoflower) and so on with added characters than their parental crops.

Phenolic Components

Phenolic compounds found in vegetables vary in structure from simple monomers to complex tannins. Under most circumstances they are colorless, but after reaction with metal ions, they assume red, brown, green, gray, or black coloration. The various shades of color depend on the particular tannins, the specific metal ion, pH, and the concentration of the phenolic complex. Phenolics, which are responsible for the astringency in vegetables, decrease with maturity, because of the conversion of astringent phenolics from soluble to insoluble form (egg plant, plantain).

Carbohydrates

The carbohydrate content of vegetables is between 3% and 27% (Table 29.5). Carbohydrates vary from low molecular weight sugars to polysaccharides such as starch, cellulose, hemicellulose, pectin, and lignin. The main sugars found in vegetables are glucose, fructose, and sucrose. In general, vegetables contain higher amount of polysaccharides (starch). Root vegetables are rich in starch (e.g., taro and colocasia, various yams, tapioca, sweet potato, potato). Plantains are also rich in starch. Fruits of the Artocarpaceae family (jack fruit, bread fruit) are also rich in starch prior to their ripening, when they are used as vegetables.

Turgor and Texture

The predominant structural feature of vegetables is the presence of parenchyma cells that are assembled into metabolically and functionally important regions for their function. In leaf tissues, the parenchymal cells are organized to obtain photosynthetic efficiency. In root tissues, the cells are loaded with starch granules. The texture of vegetables is largely related to the elasticity and permeability of the parenchyma cells. Cells with high content of water exhibit a crisp texture. The cell vacuoles contain most of the water of plant cells. The vacuolar solution contains dissolved sugars, acids, salts, amino acids, pigments and vitamins, and several other low-molecular-weight constituents. The osmotic pressure within the cell vacuole, and within the protoplast against the cell walls, causes them to stretch slightly in accordance with their elastic properties. These processes determine the specific appearance and crispness of vegetables. Damaged vegetables lose their turgor after processing and tend to become soft unless precautions are taken in packaging and storage.

VEGETABLE PROCESSING

Harvesting and Processing of Vegetables

The quality of the vegetables constantly varies depending on the growth conditions. Since the optimum quality is transient, harvesting and processing of several vegetables such as corn, peas, and tomatoes are strictly planned. Serious losses can

Table 29.5. Percentage Composition of Vegetables

	Component				
Food	Carbohydrate	Protein	Fat	Ash	Water
Cereals					
Maize (corn) whole grain	72.9	9.5	4.3	1.3	12
Earth vegetables					
Potatoes, white	18.9	2.0	0.1	1.0	78
Sweet potatoes	27.3	1.3	0.4	1.0	70
Other Vegetables					
Carrots	9.1	1.1	0.2	1.0	88.6
Radishes	4.2	1.1	0.1	0.9	93.7
Asparagus	4.1	2.1	0.2	0.7	92.9
Beans (snap−, green−)	7.6	2.4	0.2	0.7	89.1
Peas, fresh	17.0	6.7	0.4	0.9	75.0
Lettuce	2.8	1.3	0.2	0.9	94.8

Source: Food and Agriculture Organization (FAO).

occur when harvesting is done before or after the peak quality stage. The time of harvesting vegetables is very important to the quality of the raw produce and the manner of harvesting and handling is critical economically. A study on sweet corn showed that 26% of total sugars were lost in just 24 hours by storing the harvested corn at room temperature. Even when stored at low temperature, the sugar loss could reach 22% in 4 days. Peas and lima beans can lose up to 50% of their sugars in just 1 day. Losses are slower under refrigeration but the reduction in sweetness and freshness of the produce is an irreversible process. It is assumed that a part of the sugars is used for respiration and starch formation in commodities such as corn, whereas the sugars are converted to cellulose, hemicellulose, and lignin in the case of asparagus. Each type of vegetable has its optimum cold storage temperature, which may vary between 0°C and 10°C. Water loss is another problem, which reduces the quality of the produce. Continued water loss due to transpiration and drying of cut surfaces results in wilting of leafy vegetables. Hermetic packaging (anaerobic packaging) does not prevent water loss, but instead creates conditions that prevent deterioration due to an increase in the level of carbon dioxide and a decrease in the oxygen level. In order to keep consistent produce quality, many processors monitor the growing practices so that harvest and processing are programmed according to the capacity of the processing plant.

Preprocessing Operations

Vegetables can be processed in different ways including canning, freezing, freeze-drying, pickling, and dehydration. The operations involved in the processing depend on the type of vegetable and the method to be used. After harvest, the processing steps involved in canning are washing, sorting and grading, peeling, cutting and sizing, blanching, filling and brining (brining is very important for filling weight and heat transfer), exhausting (help maintaining high vacuum; exhaust temperature in the center of the can should be about 71°C), sealing, processing (heating cycle), cooling, labeling, and storage.

Harvesting

Decision to harvest should be based on experience and on objective testing method. It is recommended that the vegetables be harvested at the optimum maturity and processed promptly (Luh and Kean 1988, Woodroof 1988).

Sorting and Grading

This operation is done using roller grader, air blower, rod shaker, or any mechanical device, followed by sorting on conveyor belts. Electronic sorting is commonly used recently to remove vegetables affected by diseases and insects.

Washing

Vegetables are washed to remove not only field soil and surface microorganisms, but also fungicides, insecticides, and other pesticides. There are laws specifying the types and maximum levels of contaminants that are permitted. In order to remove dirt, insects, and small debris, vegetables are rinsed with water or with detergent in some cases. Mechanically harvested tomatoes, potatoes, red beets, and leafy vegetables are washed with fruit grade detergents. The choice of washing equipment depends on the size, shape, and fragility of the particular type of vegetable. Flotation cleaners can be used for peas and other small vegetables, whereas fragile vegetables such as asparagus may be washed by gentle spraying belt.

Peeling

Several methods are used to remove skins from vegetables including lye, steam, and direct flame. Lye peeling of mechanically harvested tomatoes and potatoes is a common practice. Vegetables with loosened skins are jet washed with water to remove skins and residual sodium hydroxide. Steam is used to peel vegetables with thick skin such as red beets and sweet potatoes, whereas for onions and pepper direct flame or hot gases in rotary tube flame peelers are used.

Cutting and Trimming

Cutting, stemming, pitting, or coring depends on the type of vegetable. Asparagus spears are cut to precise lengths. The most fibrous part is used for soup and other heated products where heat tenderizes them. Green beans are cut by machine into several different shapes along the length of the vegetable. Brussels sprouts are trimmed by hand by pressing the base against a rapidly rotating knife. Olives are pitted by aligning them in small cups, and mechanically pushing the plungers through the olives.

Blanching

The purpose of blanching is to inactivate the enzymes present in the vegetables. Since many vegetables do not receive a high-temperature heat treatment, heating to a minimal temperature before processing or storing inactivates the enzymes responsible for changes in texture, color, flavor and nutritional quality of the produce. Several enzymes are responsible for the loss of quality in vegetables. The deterioration of the cell membrane caused by the action of phospholipase D and lipoxygenases account for the flavor development in vegetables (Pinhero et al. 2003, Oke et al. 2003). Proteases and chlorophyllases contribute to the destruction of chloroplast and chlorophyll. Changes in texture occur due to the activity of pectic enzymes and cellulases. Color deterioration occurs due to the activity of polyphenol oxidase, chlorophyllase, and peroxidase (Robinson 1991). Changes in nutritional quality can occur by the activity of enzymes that destroy the vitamins. Ascorbic acid oxidase can cause a decline in the level of vitamin C.

The blanching process also reduces the microbial load of vegetables and renders packaging into containers easier. To evaluate the effectiveness of blanching, indicator enzymes such as catalase and peroxidase are traditionally used. The reason for using indicators is that blanching is not a process of indiscriminate

Table 29.6. Relative Content of Vitamins in Peas During Processing

	Vitamins, Percentage of Original			
Processing	C	B1	B2	Niacin
Fresh	100	100	100	100
Blanched	67	95	81	90
Blanched/frozen	55	94	78	76
Blanched, frozen, cooked	38	63	72	79

heating, meaning too little heating is ineffective, whereas heating too much on the other hand, negatively impacts the freshness of certain vegetables. The choice of an indicator depends on the vegetable being processed. For example, lipoxygenase may be an ideal indicator for peas and beans. The problem with using peroxidase as a universal indicator is that it sometimes overestimates heat requirements, which may vary from one product to another. Blanching prior to freezing has the advantages of stabilizing color, texture, flavor, and nutritional quality, as well as helping the destruction of microorganisms. Blanching, however, can cause deterioration of taste, color, texture, flavor, and nutritional quality, because of heating (Table 29.6.). There are three ways of blanching a produce including blanching using water, steam, or microwave. Blanchers need to be energy efficient, to give a uniform heat distribution and time, to have ability to keep quality of the produce while destroying enzymes and reducing microbial load. Blanching using water is done at 70–100°C for a specific time frame, giving a thermal energy transfer efficiency of about 60%, versus 5% for steam blanchers. Time-temperature combination is very important in order to inactivate enzymes and keeping the quality of vegetables. Effects of blanching on plant tissues include alteration of membranes, pectin demethylation, protein denaturation, and starch gelatinization. Microwave blanching gives similar result as water blanching, but the loss of vitamins is higher than in steam and water methods (Table 29.7.)

Canning Procedure

Depending on their pH, vegetables can be grouped into four categories (Banwart 1989), including high-acid vegetables with pH < 3.7, acid vegetables with pH 3.7–4.6, medium-acid vegetables with pH 4.6–5.3, and low-acid vegetables with pH > 5.3.

Table 29.7. Effect of Blanching Methods on Vitamin C Content

	Vitamin C (mg/100g Fresh Weight)			
	Asparagus	Beans	Peas	Corn
Water	35.7	22.5	15.6	15.8
Steam	35.3	23.3	11.0	13.6
Microwave	18.9	13.1	9.3	12.9

The purpose of canning is to ensure food safety and high quality to the product, as the growth of several pathogens may compromise these parameters. *Clostridium botulinum* (*C. botulinum*), an anaerobic and neuroparalytic toxin-forming bacterium, is a primary safety concern in hermetically sealed, canned vegetables. Other important spoilage organisms include *Clostridium sporogenes* group including putrefactive anaerobe 3679, *Clostridium thermosaccharolyticum*, *Bacillus stearothermophilus* and related species (Stumbo et al. 1975). Although *C. botulinum* spores do not produce toxin, the vegetative cells formed after germination produce a deadly neurotoxin in canned low-acid vegetables. Botulism, the poisoning caused by the *C. botulinum* is mostly associated with canned products including vegetables. Canned vegetables are commercially sterile, meaning that all pathogens and spoilage organisms of concern have been destroyed. The product may still contain a few microbial spores that could be viable under better conditions. During the processing of low-acid vegetables, it is necessary to provide a margin of safety in the methods schedule. This is achieved according to a "12D" process. D-value for a given temperature is taken from a thermal death time (TDT) curve. D-value is the time in minutes required to kill 90% of a bacterial population. The assumption is that by increasing this time by 12, any population of *C. botulinum* present in the canned product will decrease by 12 log cycles. This process time allows for adequate reduction of bacterial load to achieve a commercially sterile product.

$$t = D(\log a - \log b) \qquad (1)$$

where

t = heating time in minutes at a constant lethal temperature
D = time in minutes to kill 90% of a bacterial population
$\log a$ = log of initial number of viable cells
$\log b$ = log of number of viable cells after time t

For *C. botulinum*, with $D = 0.21$ at 121°C, the 12D is equal to 2.52 minutes (12×0.21 min), which means, if a can contains one spore of C. botulinum with this D value, then it can be seen from the above equation that $2.52 = 0.21$ ($\log 1 - \log b$), or $\log b = -2.52/0.21 = -12$; therefore $b = 10^{-12}$. The probability of survival of a single *C. botulinum* spore in the can is one in 10^{12} (Hersom and Hulland 1980). In canning low and medium-acid vegetables, where the destruction of spores of *C. botulinum* is the major concern, a 12D process (2.52 minutes or 3 minutes at 121°C) is the minimum safe standard for the "botulinum cook" (Banwart 1989).

Canned Tomatoes

Tomato, *Lycopersicon esculentum* Mill., belongs to the family of Solanaceae. Tomato is a major vegetable crop in North America. In North America alone, over 15 million Mt of processing tomatoes are produced. Canned tomatoes are prepared from red ripe tomatoes as whole, diced, sliced, or wedges. The fruit may or may not be peeled but stems and calices should be removed. Canned tomatoes may be packed with or without an added liquid. Calcium salts varying from 0.045% to 0.08% by weight of the finished products can also be added. Other ingredients such as organic acids, spices, oil, and flavorings can be added up to 10%. There are three categories of canned tomatoes. The label tomatoes are valid only for peeled and canned tomato. Unpeeled tomatoes are labeled accordingly. Stewed tomatoes are canned tomatoes containing onion, celery, and peppers (Anon 1993). The flowchart for the manufacture of canned tomatoes is as follows: Fresh tomatoes → Sorting → Washing → Resorting → Trimming → Peeling → Final Inspection → Cutting (except for whole tomatoes) → Filling → Exhausting → Steaming and Thermal processing. Most of the operations are similar to the ones described for canning in general.

Peeling

Tomatoes for canning are peeled with hot water, steam, or lye. Tomatoes are passed through boiling water bath or live steam chamber for 15–60 seconds, depending on the variety, size, and ripeness of the fruit. The temperature of 98°C or above is recommended. The hotter the water, the shorter is the scalding time. Boiled tomatoes are immediately cooled in cold water and then peeled by hand or with a machine. Hand peeling is labor intensive and is completed by removing the core with a knife. Machine peeling is faster and is done by scrubbing or cutting and squeezing. For lye treatment, a solution of hot 14–20% soda caustic (sodium hydroxide-NaOH) is normally used in two ways. Tomatoes could be immersed in the solution or sprayed. Then, a scrubber removes the disintegrated skins. Lye peeling is less labor intensive, but uses more water than hot water and steam peeling. Since the peeled tomatoes are thoroughly rinsed to remove the residual lye, treatment with up to 10% of citric acid is recommended. Lye treatment creates more pollutants, and loss of soluble solids compared to the other two methods.

Inspection

The final inspection after peeling is very important for the grading of the finished product. Canned tomatoes are graded in categories, A, B, C or 1st, 2nd, or 3rd class product. The purpose of the final inspection is to remove any visible defects in the products including residual peel fragments, extraneous vegetable materials, as well as checking the wholeness (when necessary).

Cutting

Peeled tomato may be cut into halves, slices, dices, wedges when necessary, using cutter, slicer, or dicer before filling in the can.

Filling

Filling of tomato in cans can be done by hand or by machine. The best quality whole tomatoes are filled by hand and filled up with tomato juice. Softening during heating can be avoided by addition of calcium chloride or calcium sulfate, in the form of tablets or mixed in tomato juice and dispensed in each can. In order to inhibit the action of *C. botulinum*, a required pH less than 4.6 is secured by the addition of citric acid in the form of tablets in each can. For stewed tomato, 3/4 filled cans with peeled tomatoes are spiced with dehydrated onion, garlic, chopped celery, green bell pepper dices, as well as tablets made of citric acid and a mixture of salt, sugar, and calcium chloride into the can. The can is then filled up with tomato juice until an acceptable level of headspace is reached.

Exhausting

The purpose of exhausting is to create enough vacuum in the can, in order to avoid fast deterioration of the canned product during the summer season. A minimum temperature of 71°C at the center of a can after the completion of exhaust is recommended (Luh and Kean 1988). Exhaust is normally done in steam chambers. The exhaust time in the chamber depends on the size of the can and varies from 3 minutes for 300 mm × 407 mm to 10 minutes for 603 mm × 700 mm cans. Exhaust is also done by mechanical vacuum closing machines (Lopez 1987, Gould 1992).

Processing Time of Canned Tomatoes

The processing time of canned tomato depends on many factors including the pH of the canned tomato, the major spoilage microorganisms of concern, the size of the can, and the type of retort (sterilizer). Organisms of concern in canned tomatoes are *Clostridium pasteurianum* and *Clostridium butyricum*, as well as *Bacillus coagulans* (*Bacillus thermoacidurans*). Spores of butyric acid anaerobes are destroyed at 93.3°C for 10 minutes when the pH is higher than 4.3, and after 5 minutes, when the pH is between 4 and 4.3. Canned tomatoes can be processed with a rotary sterilizer, a conventional stationary retort or in an open nonagitating cooker. In general, the processing time is longer if the cooling is done by air instead of with water. For a rotary sterilizer the processing time at 100°C for a can of 307 mm × 409 mm may be reduced by 9 minutes or shorter if air cooled, and by 13 minutes or shorter if water cooled. For the same size of can at 100°C an open nonagitating cooker or a conventional retort would require a processing time of 35–55 minutes. The can center temperature should reach 82°C, when air cooled, and 90°C when water cooled (Lopez 1987). The retort time depends on the size of the can, the fill weight, and the initial temperature of the canned product.

Tomato Juice Processing

Processing

Tomato juice is defined as the unconcentrated, pasteurized liquid containing a substantial portion of fine tomato pulp extracted from good quality, ripe, whole tomatoes from which all stems and unnecessary portions have been removed by any method that does not increase the water content, and may contain salt and a sweetening ingredient in dry form (Canadian Food and Drugs Act). Only high-quality tomatoes should be used for juice production. Tomatoes are important source of vitamins A and

C, and antioxidants such as lycopene. In tomato and tomato products, color serves as a measure of total quality. Consumers notice color first and their observation often supplements preconceived ideas about other quality attributes such as aroma and flavor. Color in tomato is due to carotenoids, a class of isoprenoid compounds varying from yellow to red color. Most carotenoids are tetraterpenes (C-40), derived from 2 C-20 isoprene units (Geranylgeranyl pyrophosphate). The most isolated and quantified carotenoids in tomato and tomato products include lycopene, lycope-5–6-diol, α-carotene, β-carotene, γ-carotene, δ-carotene, lutein, xanthophylls (carotenol), neurosporene, phytoene, and phytofluene. Lycopene is the major carotenoid of tomato and comprises about 83% of the total pigments present in the ripe fruit (Thakur et al. 1996). Therefore, the levels of lycopene are very important in determining the quality of processed tomato products. Not only does it determine the color of tomato products, but also provides antioxidant properties to it. Lycopene is considered as a preventive agent against coronary heart disease and cancers (Gerster 1991, Clinton 1998). The flowchart for making tomato juice is as follows: Fresh tomatoes → Washing → Sorting and Trimming → Comminution → Extraction → Deaeration → Homogenization → Salting and Acidification → Thermal Processing → Tomato juice.

Comminution

Comminution is a process of chopping or crushing tomatoes into small particles prior to extraction. The comminuted tomatoes are subjected to either cold-break or to hot-break processing. The cold-break processing produces tomato juice with a more natural color, fresh flavor, and higher vitamin C content than does hot-break process. The hot-break process on the other hand, produces tomato juice with higher consistency, lesser tendency to separate, but with a cooked flavor. During the cold-break process, the comminuted tomato is heated below 65°C to introduce rapid enzyme inactivation, and held at room temperature for a short time (a few seconds to many minutes), prior to extraction. In the hot-break process, the tomatoes are rapidly heated to above 82°C, immediately following comminution, in order to inactivate pectinesterase and enhance pectin extraction. Hot-break process can be done either in a rotary heat exchanger or in a rotary coil tank. The latter not only inactivates the enzymes fast enough to retain most of the serum viscosity, but also deaerates the juice. Low pH inhibits pectic enzymes and enhances the extraction of macromolecules such as pectin. Therefore, the addition of citric acid to the juice during comminution improves the consistency of the juice (Miers et al. 1970, Becker et al. 1972).

Extraction

Extraction is a process of separating the seeds and skins from the juice. There are two types of extractors: screw type and paddle type. The screw-type presses tomatoes between a screw and a screen with about 0.5–0.8 mm (Lopez 1987). Paddle-type extractors beat the tomato against screens. The extraction yields 3% of seeds and skins and 97% juice. Procedures that target a lower yield of 70–80% result in a better quality juice and a high quality residue for other purposes.

Deaeration

The deaeration process improves color, flavor, and ascorbic acid content of the juice. It is done immediately after the extraction by a vacuum deaerator. Removal of oxygen inhibits oxidative processes.

Homogenization

A stable juice is one in which the solid and liquid phases do not separate during storage for a long period of time. The ability of a juice to separate depends on many factors including the serum viscosity, the gross viscosity, the pH of the medium, the size and shape of the suspended solids and so on. Homogenization is a process of forcing the juice through narrow orifices at a pressure of 6.9–9.7 MPa and a temperature of about 65°C to break up the suspended particles to a fine consistency (Lopez 1987).

Salting

Sodium chloride is sometimes added to the juice at a rate of 0.5–1.25% by weight in order to improve the taste of the tomato juice. Citric acid is added to improve the color, flavor, and taste. In addition, citric acid inhibits polyphenoloxidase by removing the bound copper from the enzyme, reducing browning reactions and improving color.

Quality Attributes of Tomato and Processed Products

The major quality attribute of ripe tomato is its red color, which is due to the lycopene content of the fruit. Other important physicochemical parameters, which determine the quality of tomato are Brix, acidity, pH, vitamin C, ash, dry matter, firmness, fruit weight, and flavor volatiles. For processed tomato product, the required quality attributes are precipitate weight ratio, serum viscosity, total viscosity (Brookfield), and lycopene content of the product. Several quality attributes of tomato and tomato products can be improved by genetic modification of tomatoes (Oke et al. 2003). These comparisons were made between fruits obtained from untransformed and genetically transformed tomato plants carrying an antisense phospholipase D cDNA, and juice prepared from these fruits. The levels and activity of phospholipase D, the key enzyme involved in membrane lipid degradation, were considerably reduced by antisense transformation. These changes potentially resulted in increased membrane stability and function that also improved several quality parameters. The flavor profiles of blended antisense tomato fruits were different from the controls, being enriched in volatile aldehydes such as pentenal and hexenal. Increased membrane stability in transgenic fruits potentially resulted in lowered degradation of unsaturated fatty acids such as linoleic and linolenic acids and may have contributed to increased substrate availability for lipoxygenase pathway enzymes (lipoxygenase, hydroperoxide lyase, and so on) during blending with an increased evolution of volatile

Table 29.8. Physicochemical Parameters of Tomato Fruits and Processed Juice from Transgenic and Control Tomato

Properties	Control	Transgenic
Fruit weight (g)	95.14 ± 36.56^a	34.87 ± 21.57^b
Firmness (N)	4.97 ± 0.59^a	5.99 ± 1.03^b
Redness (a+)	29.6 ± 0.70^a	31.70 ± 1.67^a
Acidity (%)	0.36 ± 0.00^a	0.38 ± 0.05^a
Brix	4.55 ± 0.71^a	4.75 ± 0.71^b
Dry matter-NSS (%)	3.54 ± 0.05^a	5.15 ± 0.03^b
Ash (%)	0.63 ± 0.03^a	0.089 ± 0.13^b
Vitamin C (mg/100g)	3.9 ± 0.00^a	10.4 ± 0.016^b
PPT (%)	15.70 ± 0.33^a	16.17 ± 0.48^a
Serum viscosity (mpa·s)	1.0919 ± 0.04^a	1.2503 ± 0.010^b
Brookfield viscosity (mpa·s)	1075 ± 35^a	1400 ± 35^b
Lycopene (mg/100g)	11.73 ± 2.10^a	17.47 ± 0.58^b

[a,b]The values showing different superscripts are significantly different at $P < 0.05$.

aldehydes. Table 29.8 provides a comparison of various quality parameters between a genetically modified tomato and a control, which shows improvements in several quality parameters due to the transformation.

Physicochemical Stability of Juices

There are two categories of juices, clear and comminuted. Clear juice such as apple juice contains no visible vegetal particles, whereas a comminuted juice such as tomato juice contains mostly vegetal particles suspended in a liquid. To be stable, a clear juice needs to remain clear (without sediment) during its shelf life. On the other hand, a comminuted juice may not separate into distinct phases during the shelf life of the product. An important quality attribute of juices such as tomato juice is the stability of their disperse system. In order to be stable, a juice needs to be kinetically and physically stable.

Kinetic Stability

The ability of a poly-disperse system containing suspended particles to maintain its homogenous distribution without agglomeration is called kinetic or sedimentation stability. Kinetic stability depends on many factors, the most important of which include size of suspended vegetable particles, viscosity of the disperse medium and the intensity of the Brownian motion. In a liquid medium, heavier or larger particles sediment faster than the lighter ones in response to gravity. The sedimentation velocity of any particle is described by the following Stokes Equation:

$$V = 2/9r^2(\rho 1 - \rho 2)(1/\eta)g \qquad (2)$$

where V is the sedimentation velocity, m/s (meter/second); r is the radius of the suspended particles (m); $\rho 1$ and $\rho 2$ are the densities of the particles and the serum, respectively (kg/m^3; serum = liquid medium in which particles are suspended);

Table 29.9. Sedimentation of Spherical Mineral Particles in Water and in Juice With a Depth of 1 cm

Particle Size (μ)	Velocity (m/s)	Sedimentation Time in Water	Sedimentation Time in Juice
10	3.223×10^{-4}	31.03 s	2.29 min
0.1	3.223×10^{-8}	86.2 h	16 d
0.001	3.223×10^{-12}	100 yrs	436 yrs

η is the viscosity of the juice (Pa·s; Pascal·second); and g is the gravitational force (g = 9.81 m/s^2; meter/second^2).

The sedimentation time of a given particle is about 4 times longer in a juice than in water (Table 29.9.).

Particles larger than 10 μm will sediment in a few seconds. This is the reason why the sedimentation stability of juices, especially comminuted juices such as tomato juice is a serious processing issue.

The physical force that affects the sedimentation of particles in a juice is called normal force and can be calculated by:

$$f' = mg, \qquad (3)$$

where m is the mass (kg) of the suspended particle. In general, a particle with a spherical shape has a mass represented by:

$$m = 4/3\pi r^3 \rho 1 \qquad (4)$$

where $\pi = 3.14$, r is the radius of the particle and $\rho 1$ is the density of the particle. For particles with nonspherical shape (most of suspended particles), r is equal to the nearest equivalent value of a spherical particle with an identical mass and an identical density.

During particle sedimentation, another important force, friction, also comes into play. Friction between particles results in a reduction in their movement.

The frictional force $f'' = 6\pi\rho rv \qquad (5)$

where η is the viscosity of the medium; r is the radius of the particle in meters; and v is the velocity of the particle (m/s).

Friction between the particles increases depending on the density of the medium, higher the density, higher the friction. When $f' = f''$, sedimentation of the particles occur.

Kinetic stability of a heterogeneous dispersion system also depends on Brownian motion. The most dispersed particles have a very complex motion due to collisions from molecules in the dispersed medium. Because of this, the suspended particles are subjected to constant changes in their velocity and trajectory. Molecular kinetics shows that dispersed particles with colloidal size change their path 10^{20} times/second. These particles may also acquire a rotational Brownian motion. This is why colloidal particles have higher sedimentation stability than larger particles. With an increase in the mass of the suspended particles, their momentum also increases. Particles smaller than 5×10^{-4} cm in diameter that oscillate around a point do not sediment, whereas larger particles that do not experience as much Brownian motion as smaller particles easily sediment. Thus, when particles have reduced motion, they tend to aggregate and enhance sedimentation. Under ideal conditions, obtaining a colloidal particle size will enhance the stability of a juice preparation.

Physical Stability

Physical stability results from the property of a polydisperse system that inhibits the agglomeration of suspended particles. In a system with low physical stability, suspended particles agglomerate to form heavier particles ($> 5 \times 10^{-4}$ cm in diameter) that easily sediment. Physical stability depends on two opposing forces, attracting and repulsing forces.

Attracting forces between molecules are referred to as van der Vaals forces, which reduce the physical stability of heterogeneous colloidal system. The intensity of attractive forces increases as the distance between suspended particles decreases.

Repulsing forces between particles are caused by the charges surrounding the particles designated by their ζ- potential. When ζ-potential is zero, the net charge surrounding the particle is also zero, and the suspended particles are said to be at their isoelectric point. Agglomeration of particles begins at a given value of ζ-potential called as the critical potential. This is the point at which equilibrium is reached between van der Vaals forces (attraction) and repulsing forces. Different heterogeneous colloidal disperse systems have different values of critical potential. When the ζ-potential (repulsive forces) of a particle is higher than the critical potential, hydrophilic colloids are stable due to the repulsion between particles, whereas at a lower potential than the critical potential ζ, the particles tend to aggregate. The effective energy of interaction between particles of a heterogeneous colloidal system is expressed by:

$$\mathrm{E} = \mathrm{E_A} + \mathrm{E_R}, \tag{6}$$

where $\mathrm{E_A}$ is the attracting energy, and $\mathrm{E_R}$ is the repulsing energy.

Attracting energy is actually the integration of the sum of all attracting forces between molecules of two colloidal particles. For two particles with radius r, the potential energy is expressed by:

$$E_A = -\mathrm{A}r/12h \tag{7}$$

where, h is the distance between surfaces of the two particles, and A is the Hamaker constant (10^{-1}–10^{-21} J). Consequently, attracting forces decrease as the distance between particles increase.

On the other hand, particles can come closer to one another up to a certain distance after which they start repulsing each other because of their ζ-potential. When two particles are very close, their similar ionic charge (positive or negative) layers create a repulsive force, which keeps them apart. The repulsing energy between two particles with the same radius r and the same surface potential Ψ_0 is expressed by:

$$\mathrm{E_R} = -2\int \Delta P \, d\delta, \tag{8}$$

where ΔP is the increase in osmotic pressure; δ is thickness of the double ionic layer; and the number 2 relates to the energy changes between 2 particles.

$$\Delta P = \Delta n \mathrm{K} T \tag{9}$$

where Δn is the increase in the ionic charge between the 2 particles; K is the Boltzman's constant; and T is the temperature in degrees Kelvin (°K).

Assuming that the two particles are spherical, the formula (8) can be expressed by:

$$\mathrm{ER} \approx \left[\left(\varepsilon r \psi_0^2 \exp\{-\chi h\}\right)\right]/2, \tag{10}$$

where ε is the dielectric constant; r is the radius of particles; h is the distance between the two particles; ψ_0 is the surface potential, and χ is a constant which characterizes the double ionic layer.

The inferences from the Equation 10 are that:

1. repulsing energy exponentially decreases as the distance between the particles increases; and
2. repulsing energy quadratically increases as the surface potential increases and as the radius linearly increases.

For particles with radius r and a constant surface potential, $\mathrm{E_R}$ depends on χ. Figure 29.2 shows the interaction between two particles (1 and 2) under constantly increasing χ. As a result of decreasing the thickness of the double ionic layer and a decrease in ζ-potential, the distance between the two particles decreases

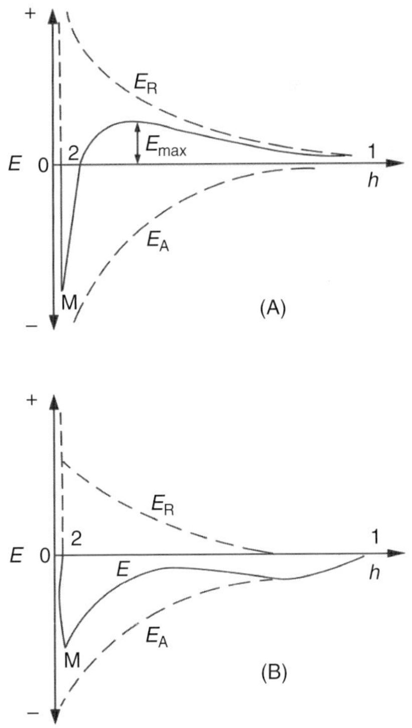

Figure 29.2. Interactive energy between two particles.

leading to an increase in the repulsing energy according to the Equation (10). For a given h, the residual energy reaches the maximum value (E_{max}), which must be overcome by particle 1 in order to aggregate with particle 2. Continual increase of χ and a decrease in the distance between particles h reduces the effect of repulsing energy E_R over the residual energy E ($E_R - E_A$ or E_{max}), which reaches its minimal value M. At this stage, the attracting energy E_A dominates and the particles aggregate. The speed of aggregation is an important physicochemical parameter for the stability of a microheterogeneous system. E_{max} is called coagulation energy and determines the speed of aggregation. When $E_{max} > 0$ (Fig. 29.2A), the aggregation process is slow, whereas when $E_{max} < 0$ (Fig. 29.2B) the process is quick, since the particles of the system do not have to overcome E_{max} (When $E_A > E_R$, $E_R - E_A$ is <0, Fig. 29.2A, B). The critical concentration of coagulation is reached at $E_{max} = 0$, when both the attractive and repulsive energies are equal.

In a real situation of polydisperse colloidal system such as juice, suspended particles have different sizes and different ψ_0 and ζ- potential. When two particles with different sizes and different surface ζ- potential approach each other, an opposite charge is induced upon the particle with lower ζ- potential, which activates the agglomeration. Another important factor in the stability of juices is the hydrated layer of suspended colloidal particles. The formation of hydrated layer is due to the orientation of water molecules towards the hydrophilic groups such as -COOH, -OH, and so on situated on the surface of the particles. A decrease in ζ- potential leads to a decrease in the hydrated layer of suspended particles, and therefore, increases the susceptibility to aggregation. The ability of the hydrated layer to increase stability is explained by Deryagin theory, which stipulates that when suspended particles are in close proximity, their hydrated layers are reduced, leading to an increase in repulsive forces between them. Only strong hydrated layers can affect the aggregative stability of a polydisperse system. The probability of the formation of such hydrated layers around vegetal particles in juices is very low, because of the weak energetic relationship between the particles and the disperse phase. In this case, hydrated layers play a secondary role to supplement the action of the double ionic layer.

Stability from particular aggregation in juice is also influenced by the viscosity of the disperse medium. The viscosity of the disperse phase or serum of juices is mainly composed of high molecular mass compounds such as pectin, protein, and starch. Practically, this is achieved by the addition of such compound to the juice. The mechanism of action is as follows. Upon addition of high molecular mass compounds (pectin), they adsorb on the surface of the hydrophobic vegetal particles to make a layer of hydrophilic molecules, circled by a thick hydrated layer, which gradually blends into the disperse medium. Pectin is the stabilizer of choice in the juice processing industry, compared to protein and starch. With an increase in the concentration of high molecular mass compounds, the sedimentation stability of the juice increases. It has been established that with the increase in the concentration of pectin in the juice, the stability increases to reach a plateau after which, a further increase does not provide an added beneficial effect (Idrissou 1992). The mechanism of the sedimentation stability of the juice is as follows:

- Low adsorption velocity: Lower adsorption velocity is observed under lower concentration of stabilizers.
- Irreversible adsorption: Stabilizers in most cases are bonded to the particular phase surface with strong adsorption bonds.
- Adsorption is described by the curve in Figure 29.3, following a typical saturation curve.
- Different configuration of adsorption by particles is shown in Figure 29.3. Molecules of linear polymers can be adsorbed in three different ways, depending on their affinity to the liquid phase and the surface of the suspended particles, adopting horizontal, vertical, and stitchlike configurations.

Enhancing Nutraceutical Quality of Juice Products

Health beneficial effects of enhanced fruit and vegetable consumption through enhanced intake of nutraceuticals (functional food ingredients) are well established. There are several types of juice blends from fruits and vegetables available in the market including those that are blends of fruits and vegetables. An additional way of enhancing the nutraceutical value of various juices is by the addition of specific ingredients. In a recent study, Oke et al. (2010) have described the benefits of adding soy lecithin to tomato juice and sauce preparations and its effect on stability characteristics of the juice and sauce. Soy lecithin is enriched in proteins as well as lipids such as phosphatidylcholine, a key component in the biomembrane. Choline is also a key component of nervous system function linked to the biosynthesis of acetyl choline, involved in nerve signal transmission, lipid secretion, enhancement in liver function, and so on. Juice/sauce preparations from fruits and vegetables are generally low in phospholipids. Thus, addition of soy lecithin to juice could be beneficial both in terms of nutrition and stability of juice. Addition of 0.5–1% (w/v) soy lecithin to tomato juice and sauce enhanced the stability characteristics of the processed products such as precipitate weight ratio and Brookfield viscosity. Commercially, soy lecithin is not added to juice or sauce preparations at present.

MINIMALLY PROCESSED VEGETABLES

Definitions

Rolle and Chism (1987) defined minimally processed refrigerated fruit and vegetable (MPR F & V) as produce which have undergone minimal processing such as washing, sorting, peeling, slicing, or shredding prior to packaging and sale to consumers. Odumeru et al. (1997, 1999) defined ready-to-use vegetables as fresh-cut, packaged vegetables requiring minimal or no processing prior to consumption. It is generally accepted that MPR F & V are products that contain live tissues or those that have been only slightly modified from the fresh condition and are freshlike in character and quality.

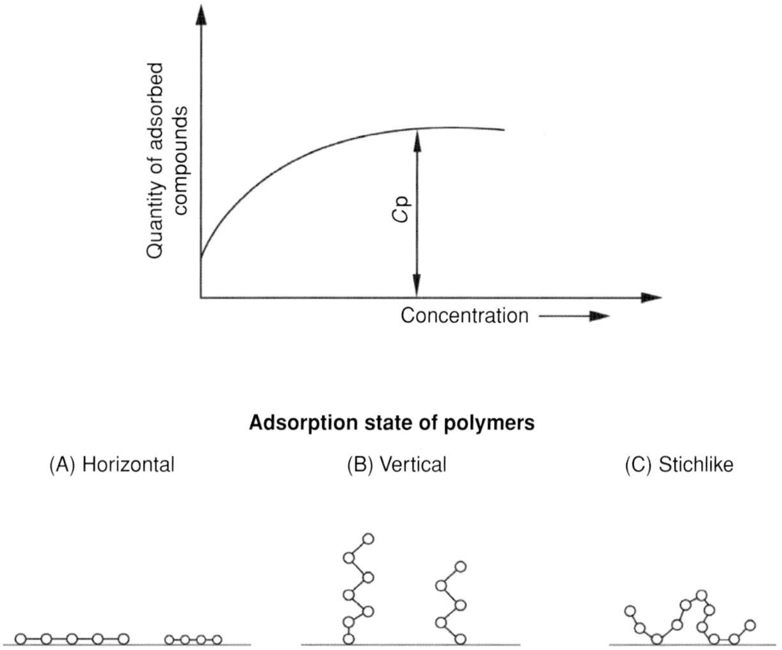

Figure 29.3. Isotherm of adsorption of polymers; polymer adsorption states.

The rapid growth of fast-food restaurants in the 1980s encouraged the development and use of fresh-cut products. As the demand for products increased, new technology and product innovation were developed to help make fresh-cut produce, one of the fastest growing segments of the food industry. It is estimated that in the US fresh-cut produce market, up to 8% of all produce is sold in retail grocery outlets and 20% sold in the food service industry. It is expected to continue growing in size and popularity as more products are introduced and consumers change their buying habits. In the first quarter of 2006, the sale of fresh cut produce in the United States accounted for \$1.3 billion, with major sales (~80%) coming from vegetables. In 2007, the fresh cut produce market was estimated at \$15.9 billion in the United States (www.agecon.ucdavis.edu). Freshly prepared ready to eat salads, now with cut tomatoes, will have a shelf life of up to 14 days utilizing unique Modified Atmosphere Packaging technology.

Processing

The major operations in processing of MPR produce include Sorting/Sizing/Grading, Cleaning/washing/disinfection, Centrifugation, and Packaging and Distribution. For certain vegetables such as eggplant, the peeling operation follows the sorting operation that results in a size reduction, and mixing, packaging and distribution. Washing of vegetables for minimal processing, uses a significant amount of water (5–10 L/kg) to achieve a reduction in the bacterial load. The main pathogen of concern in this type of products is *Listeria monocytogenes*. The water temperature of about 4°C is recommended. The safety of the product is best secured through disinfection with chlorine at a concentration of about 100 mg/L. Two forms of chlorine are commonly used: gaseous chlorine, which is 100% active chlorine, and calcium or sodium hypochlorites. The latter is most popular in MPR industry.

Centrifugation is one of the major operations in MPR processing. The purpose is to dry the vegetables rapidly. The efficacy of the centrifugation depends on the speed and time of rotation of the centrifuge. Certain antioxidants such as ascorbic acid and citric acid at a concentration of about 300 ppm in the wash solution would enhance the quality of the products. The processing techniques employed usually wound the tissue and may cause the liberation of "wound ethylene" that can activate deteriorative reactions within the tissue. Vegetables of high quality may also have a good complement of the antioxidants (vitamins C and E, reduced glutathione) and the antioxidant enzyme system that comprise the enzymes such as superoxide dismutase, catalase, peroxidase, ascorbate peroxidase, and so on. An active pentose phosphate pathway is required for the supply of reduced nicotinamide adenine dinucleotide phosphate-reduced form for the efficient functioning of the antioxidant enzyme system. Membrane phospholipid degradation mediated by phospholipase D, is activated in response to wounding the tissue. Damage to the tissue during processing can cause the leakage of ions such as Ca^{2+} and H^+ which can activate phospholipase D. Naturally occurring phospholipase D inhibitors such as hexanal can be used to reduce the activity of phospholipase D in processed tissues and enhance their shelf life and quality (Paliyath et al.

2003). The quality of MPR F & V can be increased by following Good Manufacturing Practices, the main points of which are: minimizing handling frequency, providing continued control of temperature, the relative humidity (% RH), modified atmosphere and controlled atmosphere (MA/CA storage etc.). The product is always transferred from truck to refrigerated storage immediately to minimize degradative and oxidative reactions. The products are replaced on a first-in/first-out basis, and the inventory on a weekly basis. Currently MPR include the following items: ready-to-eat fruits and vegetables, ready-to-cook fruits and vegetables, ready-to-cook mixed meals, and fresh ready-to-use herbs and sprouts.

Quality of MPR

Consumers expect fresh-cut products to be of optimum maturity, without defects, and in fresh condition. The most important physicochemical quality parameters targeted for preservation include good appearance, nutrients, and excellent sensory attributes such as texture, firmness, and taste. Minimally processed products are vulnerable to discoloration because of damaged cells and tissues that become dehydrated. Cutting and slicing of carrots with a very sharp blade reduces the amount of damaged cells and dehydration, when compared with those sliced with a regular culinary knife. To overcome the problem with dehydration, fresh-cut produce can be treated with calcium chloride and kept in a high humidity atmosphere. Enzymatic browning is a serious problem with minimally processed produce. During the processing of produce, several types of oxidative reactions may occur, leading to the formation of oxidized products. These reactions cause browning reactions resulting in the loss of nutritional value by the destruction of vitamins and essential fatty acids. In lipid-rich vegetables, oxidative processes lead to the development of rancid off-flavors and sometimes to toxic oxidative products (Dziezak 1986). There are a number of chemicals used to stabilize MPR produce including: (a) free radical scavengers such as tocopherols; (b) reducing agents and oxygen scavengers such as ascorbic acid and erythorbic acid; (c) chelating agents such as citric acid, and (d) other secondary antioxidants such as carotenoids. Ascorbic acid is commonly used either alone, or in combination with other organic acids. Temperature is the most important factor that influences the general quality of MPR produce. When temperature increases from 0°C to 10°C, respiration rate increases substantially, with Q_{10} (fold increase in respiration by a 10°C increase in temperature) ranging from 3.4 to 8.3 among various fresh-cut products. With the increase in respiration rate, deterioration rate also increases at a comparable rate; therefore, low temperature storage is essential for maintaining good quality (Watada et al. 1996). MA within MPR container or bags is useful in keeping the quality of the produce (Gorny 1997). Gas mixtures suitable for MA storage have been the same as those recommended for the whole commodity (Saltveit 1997). CA system is used to simulate the MA with similar gas composition. A mixture of 10% O_2 + 10% CO_2 has been shown to retard chlorophyll degradation in parsley and broccoli florets (Yamauchi and Watada 1993, 1998). An atmosphere of 3% O_2 + 10% CO_2 was beneficial for fresh-cut iceberg lettuce, slightly beneficial for romaine lettuce, and not beneficial for butterhead lettuce (Lopez-Galvez et al. 1996). The oxygen level can be allowed to drop to the level of respiratory quotient breakpoint. The O_2 level could be dropped to 0.25% for zucchini slices (Izumi et al. 1996), and 0.8% for spinach (Ko et al. 1996). Other technologies that involve inactivation of ethylene action using 1-Methylcyclopropene (Lurie and Paliyath 2008), or controlling gas movements using edible coating technologies (Fallik 2008) are also being employed increasingly.

FURTHER READING

Arthey D, Ashwurst PR. 2001. *Fruit Processing: Nutrition, Products and Quality Management*. Aspen Publishers, Gaithersburg, MD, p. 312.

Enachescu DM. 1995. *Fruit and Vegetable Processing*. FAO Agricultural Services Bulletin 119. FAO, Rome, p. 382.

Jongen WMF. 2002. *Fruit and Vegetable Processing: Improving Quality (Electronic Resource)*. CRC Press, Boca Raton, FL.

Paliyath G et al. 2008. *Postharvest Biology and Technology of Fruits, Vegetables and Flowers*. Wiley-Blackwell, Iowa, p. 482.

Salunkhe DK, Kadam SS. 1998. *Handbook of Vegetable Science and Technology: Production, Composition, Storage and Processing*. Marcel Dekker, New York, p. 721.

REFERENCES

Anonymous. 1993. Canned tomatoes. Tomato concentrates. Catsup. In: *Code of Federal Regulations*, Title 21, Sections 155.190–155.194.

Banwart GJ. 1989. Control of Microorganisms by destruction. In: *Basic Food Microbiology*, 2nd edn. AVI Press, New York, p. 773.

Becker R et al. 1972. Consistency of tomato products. Effects of acidification on cell walls and cell breakage. *J Food Sci* 37: 118–125.

Clinton SK. 1998. Lycopene: chemistry, biology and implications for human health and disease. *Nutr Rev* 56: 35–51.

Dziezak JD. 1986. Preservative systems in foods, antioxidants and antimicrobial agents. *Food Technol* 40: 94–136.

Fallik E. 2008. Postharvest treatments affecting sensory quality of fresh and fresh-cut products. In: G Paliyath et al. (eds.) *Postharvest Biology and Technology of Fruits, Vegetables and Flowers*. Wiley-Blackwell, Iowa, pp. 301–318.

FAOSTAT. 2010. *Food and Agricultural Commodities Production*. Available at http://faostat.fao.org/site/339/default.aspx. Accessed on January 25, 2010.

Fresco, LO and Baudoin, WO. 2002. Food and nutrition security towards human security. In: Proceedings of the International Conference on Vegetables, (ICV-2002), 11–14 November 2002, Bangalore, India. Dr Prem Nath Agricultural Science Foundation (in press).

Gerster H. 1991. Potential role of β-carotene in the prevention of cardiovascular disease. *Int J Vitamin Nutr Res* 61: 277–291.

Gorny JR. 1997. Summary of CA and MA requirements and recommendations for fresh-cut (minimally-processed) fruits and vegetables. In: JR Gorny (ed.) Proceedings of the Seventh International Controlled Atmosphere Conference, vol 5. Postharvest Outreach Program, University of California, Davis, CA, pp. 30–66.

Gould WA. 1992. Tomato production. In: *Processing and Technology*, 3rd edn. CTI Publications, Baltimore, MD, 437 p.

Hersom AC, Hulland ED. 1980. Principles of thermal processing. In: *Canned Foods*, 7th edn. Churchill Livingstone, New York, 394 p.

Idrissou M. 1992. Investigation and improvement of the technology of processing citrus bases and concentrates. Ph.D. thesis. Higher Institute for Food and Flavour Industries. Plovdiv, Bulgaria (In Bulgarian).

Izumi H et al. 1996. Low O_2 atmospheres affect storage quality of zucchini squash slices treated with calcium. *J Food Sci* 41: 317–321.

Ko NP et al. 1996. Storage of spinach under low oxygen atmosphere above the extinction point. *J Food Sci* 61: 398–400.

Lopez A. 1987. Canning of vegetables. In: *A Complete Course in Canning, Book III*, 12th edn. Canning Trade Inc., Maryland, pp. 11–143.

Lopez-Galvez G et al. 1996. The visual quality of minimally processed lettuce stored in air or controlled atmosphere with emphasis on romaine and iceberg types. *Postharvest Biol Technol* 8: 179–190.

Luh BS, Kean, CE. 1988. Canning of vegetables. In: BS Kuh, JG Woodroof (eds.) *Commercial Vegetable Processing*, 2nd edn. Van Nostrand Reinhold, New York, pp. 250–255.

Lurie S, Paliyath G. 2008. Enhancing postharvest shelf life and quality in horticultural commodities using 1-MCP technology. In: G Paliyath et al. (eds.) *Postharvest Biology and Technology of Fruits, Vegetables and Flowers*. Wiley-Blackwell, Ames, IA, pp. 139–161.

Miers JC et al. 1970. Consistency of tomato products. 6. Effects of holding temperature and pH. *Food Technol* 24: 1399–1403.

Odumeru JA. 1999. Microbiology of fresh-cut ready-to-use vegetables. *Recent Res Dev Microbiol* 3: 113–124.

Odumeru JA et al. 1997. Assessment of the microbiological quality of ready-to-use vegetables for health-care food services. *J Food Prot* 60: 954–960.

Oke M et al. 2003. The effects of genetic transformation of tomato with antisense phospholipase D cDNA on the quality characteristics of fruits and their processed products. *Food Biotechol* 17: 163–182.

Oke M et al. 2010. Effect of soy lecithin in enhancing fruit juice/sauce quality. *Food Res Int* 43: 232–240.

Paliyath G et al. 2003. Inhibition of phospholipase D. US patent # 6,514,914.

Petro-Turza M. 1986–1987. Flavour of tomato and tomato products. *Food Rev Int* 2: 309–351.

Pinhero RP et al. 2003. Developmental regulation of phospholipase D in tomato fruits. *Plant Physiol Biochem* 41: 223–240.

Robinson DS. 1991. Peroxidases and their significance in fruits and vegetables. In: PF Fox (ed.) *Food Enzymology*, vol 1, Chapter 10. Elsevier Applied Science, London, pp. 399–426.

Rolle RS, Chism GW III. 1987. Physiological consequences of minimally processed fruits and vegetables. *J Food Qual* 10: 157–177.

Saltveit ME. Jr. 1997. A summary of CA and MA requirements and recommendations for harvested vegetables. In: ME Saltveit (ed.) Proceedings of the seventh international controlled atmosphere conference, vol 4. Postharvest Outreach Program, University of California, Davis, CA, pp. 98–117.

Stumbo CR et al. 1975. Thermal process lethality guide for low-acid foods in metal containers. *J Food Sci* 40: 1316–1323.

Thakur BR et al. 1996. Quality attributes of processed tomato products: A review. *Food Rev Int* 12: 375–401.

Watada AE et al. 1996. Factors affecting quality of fresh-cut horticultural products. *Postharv Biol Technol* 9: 115–125.

Woodroof JG. 1988. Harvesting, handling, and storing vegetables. In: BS Luh, JG Woodroof (eds.) *Commercial Vegetable Processing*. Van Nostrand Reinhold, New York, pp. 135–174.

Yamauchi N, Watada AE. 1993. Pigment changes in parsley leaves during storage in controlled or ethylene containing atmosphere. *J Food Sci* 58: 616–618.

Yamauchi N, Watada AE. 1998. Chlorophyll and xanthophyll changes in broccoli florets stored under elevated CO_2 or ethylene containing atmosphere. *Hort Sci* 33: 114–117.

30
Non-Enzymatic Browning in Cookies, Crackers and Breakfast Cereals

A.C. Soria and M. Villamiel

Abstract: In this chapter, the authors have attempted to collect the main studies carried out up to 2010 on the most investigated quality indicators in cookies, crackers and breakfast cereals. These cereal-based foods, mainly consumed for breakfast, constitute a very important source of energy, particularly in the case of children. For this reason, their processing (baking or extrusion cooking) should be carefully carried out in order to keep their quality. One of the main reactions that takes place during the processing of these cereal-based foods is non-enzymatic browning including Maillard reaction and caramelisation, which might contribute to the development of colour and fluorescence. In this sense, the evaluation of chemical indicators such as available lysine, furosine, hydroxymethylfurfural, furfural, acrylamide and maltulose might afford important information on the processing and storage conditions to which cereal-based products have been submitted.

INTRODUCTION

Cereal-based products such as cookies, crackers and breakfast cereals represent a predominant source of energy in the human diet, especially for children consuming cereal derivatives for breakfast meals.

Cookies, crackers and breakfast cereals can be manufactured by means of traditional processes or by extrusion cooking. In general, during conventional treatment of flour products (usually, temperatures higher than 200°C for several minutes), more intense processing conditions are applied as compared to those used in the extrusion process (González-Galán et al. 1991, Manzaneque Ramos 1994, Huang 1998).

Extrusion cooking is a well-established industrial technology with a number of food applications since, in addition to the usual benefits of heat processing, extrusion has the possibility of changing the functional properties of food ingredients and/or of texturising them (Cheftel 1986). In the extruder, the mixture of ingredients is subjected to intense mechanical shear through the action of one or two rotating screws. The cooking can occur at high temperatures (up to 250°C), relatively short residence times (1–2 minutes), high pressures (up to 25 MPa), intense shear forces (100 rpm) and low moisture conditions (below 30%). In addition to the cooking step, cookies, crackers and breakfast cereals manufacture involves toasting and/or drying operations (Cheftel 1986, Camire and Belbez 1996, Manley 2000).

During these technological treatments, due to the elevated temperatures and low moisture conditions used, different chemical reactions such as the non-enzymatic browning can take place. Non-enzymatic browning includes Maillard reaction (MR) and caramelisation. The products resulting from both reactions depend on food composition, temperature, water activity and pH, and can occur simultaneously (Zanoni et al. 1995).

As it is known, MR occurs between reducing sugars such as glucose, fructose, lactose or maltose, and free amino groups of amino acids or proteins (usually the ε-amino group of lysine). The MR is favoured in foods with high protein and reducing carbohydrate content at intermediate moisture content, temperatures above 50°C and a pH in the range from 4 to 7. Caramelisation depends on direct degradation of carbohydrates due to heat and it needs more drastic conditions than those of the MR. Thus, at temperatures higher than 120°C, pH lower than 3 or higher than 9 and very low moisture content, caramelisation is favoured (Kroh 1994).

Food Biochemistry and Food Processing, Second Edition. Edited by Benjamin K. Simpson, Leo M.L. Nollet, Fidel Toldrá, Soottawat Benjakul, Gopinadhan Paliyath and Y.H. Hui.

Furthermore, other chemical reactions that might take place during processing of these products may also affect the extent of non-enzymatic browning. Thus, starch and non-reducing sugars such as sucrose can be hydrolysed into reducing sugars that can later be involved in other reactions, for instance, in the MR (Camire et al. 1990).

The chemical changes that take place during the technological process used in the elaboration of this type of cereal-based foods contribute, to a certain extent, to their typical organoleptic characteristics. However, important losses of lysine due to the formation of chemically stable and nutritionally unavailable derivatives of protein-bound lysine can be observed under MR conditions (Torbatinejad et al. 2005).

The amount of lysine and its biological availability are meaningful criteria for the nutritive quality of cereals, as foods processed from cereal grains are low in essential amino acids such as lysine and methionine (Horvatić and Eres 2002). As this deficiency can be further impaired by losses originated in browning reactions during processing, a compromise must be found where the objectives of heat treatment are reached with a minimal decrease in the nutritional quality of the food. Different indicators based on the evaluation of the extent of non-enzymatic browning have proved to be useful, not only for processing control, but for optimisation of operating conditions in the manufacture of cereal-based foods (Singh et al. 2000, Gökmen et al. 2008a, 2008b). Although some of the studies on cereal-based products reported in this review have not been directly carried out in cookies, crackers or breakfast cereals, they have been considered here since both the manufacture and composition could be similar and, consequently, the same chemical reactions might be involved.

NON-ENZYMATIC BROWNING INDICATORS

Available Lysine

The determination of available lysine has been used to evaluate the effect of heating on the protein quality of the following cereal-based products: pasta (Nepal-Sing and Chauhan 1989, Acquistucci and Quattrucci 1993), bread (Tsen et al. 1983, Ramírez-Jiménez et al. 2001), cereals (Fernández-Artigas et al. 1999a), cookies fortified with oilseed flours (Martinkus et al. 1977) and biscuits (Singh et al. 2000).

Since extrusion cooking is a well-established technology for the industrial elaboration of cereal-based foods, several authors have studied the effect of the initial composition and the different operating conditions on lysine loss in extruded materials. In samples of protein-enriched biscuits, Noguchi et al. (1982) found that the loss of reactive lysine is significant (up to 40% of the initial value) when the extrusion cooking is carried out at high temperature (190–210°C) with a relatively low water content (13%). When moisture is increased to 18%, the lysine loss is much less pronounced or even negligible.

Noguchi et al. (1982) also studied the effect of the decrease of pH on lysine loss in biscuits obtained by extrusion. These authors observed a higher lysine loss at low pH, since strong acidification markedly increases starch or sucrose hydrolysis and, consequently, the formation of reducing carbohydrates. Starch hydrolysis and corresponding formation of reducing sugars was also proposed as the main cause of lysine loss in extruded wheat flours (Bjorck et al. 1984).

An important decrease of available lysine content (40.9–69.2%) in wheat grain processed by flaking and toasting was observed by McAuley et al. (1987), and they attributed these results to the high temperature reached during toasting. In general, extrusion cooking of cereal-based foods appears to cause lysine losses that do not exceed those for other methods of food processing. In order to keep lysine losses within low levels (10–15%), it is necessary to avoid operating conditions above 180°C at water contents below 15% (even if a subsequent drying step is then necessary) (Cheftel 1986).

Phillips (1988) suggested that the MR is more likely to occur in expanded snack foods in which nutritional quality is not a major factor than in other extruded foods with higher moisture contents, if the processing conditions are controlled. Horvatić and Guterman (1997), in a study on the available lysine content during industrial cereal (wheat, rye, barley and oat) flake production, found that the effects of particular processing phases can result in a significant decrease of available lysine amounts in rye and oat flakes, whereas less influence can be observed in the case of wheat and barley flakes. Apart from the importance of processing conditions, the results obtained by these authors seem to indicate that the decrease of lysine availability is higher for cereals with greater available lysine contents in total proteins.

Other important consideration is to avoid the presence of reducing sugars during extrusion. Lysine loss and browning are more intense when reducing carbohydrates such as glucose, fructose or lactose (added as skim milk) are added to the food mix above 2–5% level (Cheftel et al. 1981). In agreement with these results, Singh et al. (2000) and Awasthi and Yadav (2000) found higher lysine loss and browning in traditionally elaborated biscuits enriched with whey or skim milk.

An investigation of the changes in available lysine content during industrial production of dietetic biscuits was performed by Horvatić and Eres (2002). Dough preparation did not significantly affect the available lysine content. However, after baking, a significant loss (27–47%) of available lysine was observed in the studied biscuits. The loss of available lysine was found to be significantly correlated with technological parameters, mainly baking temperature/time conditions.

The influence of storage on the lysine loss in protein-enriched biscuits was studied by Noguchi et al. (1982). Lysine loss was observed to increase when samples were stored at room temperature for a long period of time. Hozova et al. (1997) estimated, by measurement of lysine, the nutritional quality of amaranth biscuits and crackers stored during 4 months under laboratory conditions (20°C and 62% RH). Although a slight decrease in the level of lysine was detected, this was not significant. However, these authors suggested that lysine degradation can continue with prolonged storage and it is necessary to consider this fact in relation to consumers and the extension of storage time.

In a survey on 20 commercially available cereal-based breakfast foods, Torbatinejad et al. (2005) stated that reactive lysine is a more accurate measure of unmodified lysine since total lysine determination often includes lysine that has reverted from modified lysine derivatives (early Maillard products) during acid hydrolysis. In addition to the formation of unavailable lysine derivatives, it has also been described that the advance of MR as a result of processing is at least partly responsible for the reduction in the overall protein digestibility (Rutherfurd et al. 2006).

Furosine

The determination of furosine (ε-N-2-furoylmethyl-lysine), generated from the acid hydrolysis of Amadori compounds formed during the early stages of MR (Erbersdobler and Hupe 1991), has been used for the assessment of lysine loss in malt (Molnár-Perl et al. 1986), pasta (Resmini and Pellegrino 1991, García-Baños et al. 2004, Gallegos-Infante et al. 2010), baby cereals (Guerra-Hernández and Corzo 1996, Guerra-Hernández et al. 1999), baby biscuits (Carratù et al. 1993), cookies (Gökmen et al. 2008a), fibre-enriched breakfast cereals (Delgado-Andrade et al. 2007), flours employed for the formulation of cereal-based products (Rufián-Henares et al. 2009) and bread (Ramírez-Jiménez et al. 2001, Cárdenas-Ruiz et al. 2004). Other 2-furoylmethyl derivatives such as the ones corresponding to GABA (2-FM-GABA) have been, among others, suggested as indicators of the extent of the MR in different vegetable products (Del Castillo et al. 2000, Sanz et al. 2000, 2001, Soria et al. 2009).

Furosine determination was used by Rada-Mendoza et al. (2004) to evaluate the advance of MR in commercial cookies, crackers and breakfast cereals. Furosine was detected in all analysed samples within a wide range of variation: 25–982 (in cookies), 163–751 (in crackers) and 87–1203 (in cereals) mg/100 g protein. Figure 30.1 illustrates, as an example, the HPLC chromatogram of the 2- FM-amino acids of the acid hydrolysate of a cereal breakfast sample. Besides furosine, these authors found 2-FM-GABA in samples of crackers and breakfast cereals, which was absent in most cookie samples analysed. The presence of 2-FM-GABA in breakfast cereals and crackers may be attributed to the considerable amount of free GABA present in rice and corn used in their manufacture.

Recently, Delgado-Andrade et al. (2007) showed that regardless of the protein source (i.e., type of cereal), the higher the protein content, the higher the furosine level of breakfast cereal formulations. According to this, breakfast cereals enriched in dietary fibre showed the highest content of this quality marker. Regarding the physical form of the sample, higher furosine levels were found in puffed cereals as compared to flakes, probably due to the more pronounced heating during processing.

Contribution of milk components used during breakfast cereal manufacture to furosine content is still controversial (Rada-Mendoza et al. 2004, Delgado-Andrade et al. 2007). Whereas some authors partly attributed the wide variation of this heat-induced marker to the considerable amounts of furosine in dried milk (Corzo et al. 1994), others concluded that addition of these components was not directly related to the levels of furosine in the product. The effect of fibre, which was not taken into account in some of these studies, could be responsible for the discrepancies observed.

In cookies, furosine formation has been shown to be highly dependent on the dough formulation and baking conditions (Gökmen et al. 2008a). Therefore, furosine cannot be

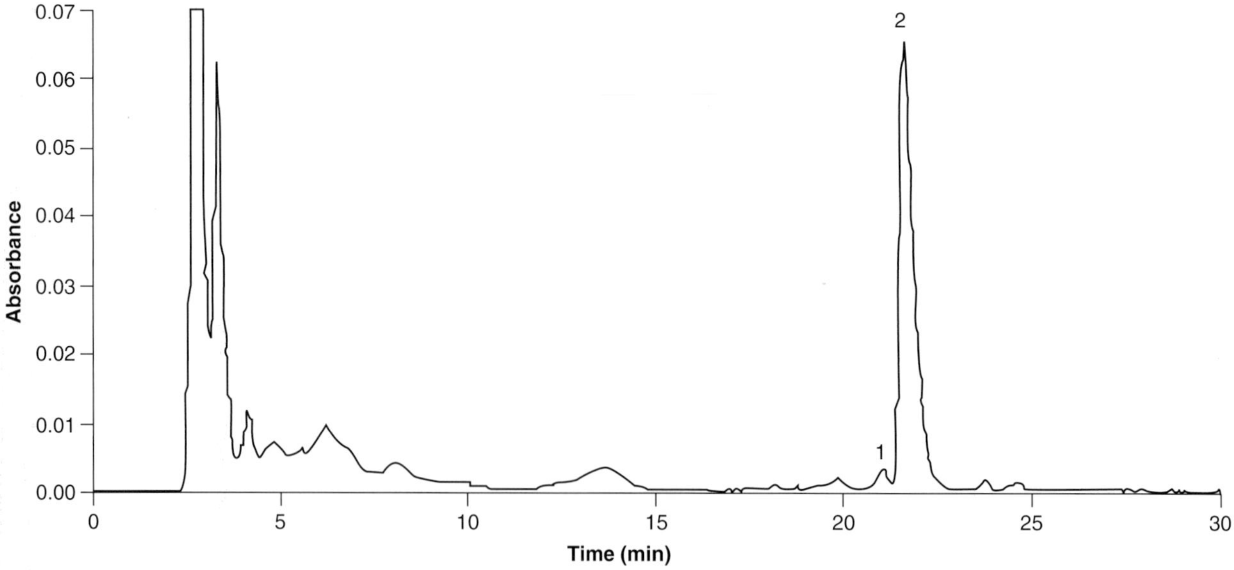

Figure 30.1. HPLC chromatogram of the acid hydrolysate of a breakfast cereal sample: (1) 2-furoylmethyl-GABA and (2) furosine (From Rada-Mendoza et al. 2004).

considered a reliable marker of the extent of the thermal treatment for comparison of different products or processing plants. In these cases, the measure of lysine availability is a better index of protein quality.

Similarly, the variable amounts of dried milk used in manufacture of cereal-based products and the different levels of free GABA in unprocessed cereals appear to be the major drawback for the use of furosine and 2-FM-GABA as suitable indicators to differentiate between commercial cereal-based products. However, in the cereal industry, where exact composition is known, measurement of 2-FM-GABA and furosine formed might be used as indicators to monitor processing conditions during the manufacture of cereal products.

Finally, the effect of the addition of common bean flour (15% and 30%) to semolina on furosine content, total phenolics and other parameters related to quality evaluation (moisture, optimal cooking time, cooking loss, water absorption capacity, colour and firmness) was determined by Gallegos-Infante et al. (2010) in spaghetti pasta obtained at different temperatures (60–80°C). Although an increase in furosine and phenolic contents was observed in pasta fortified with bean flour and dried at high temperature, further research is necessary to elucidate the actual phenolic content and the antioxidant capacity related to the MR products or polyphenols.

Hydroxymethylfurfural and Furfural

Hydroxymethylfurfural (HMF) and furfural are intermediate products in the MR, originated from the degradation in acid solution of hexoses and pentoses, respectively (Hodge 1953, Kroh 1994).

HMF is a classic indicator of browning in several foods such as milk (van Boekel and Zia-Ur-Rehman 1987, Morales et al. 1997), juices (Lee and Nagy 1988) and honey (Jeuring and Kuppers 1980, Sanz et al. 2003). HMF has also been detected in several cereal-based foods, including dried pasta (Acquistucci and Bassotti 1992, Resmini et al. 1993), baby cereals (Guerra-Hernández et al. 1992, Fernández-Artigas et al. 1999b), cookies (Ait-Ameur et al. 2006) and bread (Ramírez-Jiménez et al. 2000, Cárdenas-Ruiz et al. 2004).

In a study on the effect of various sugars on the quality of baked cookies, furfural (in cookies elaborated with pentoses) and HMF (in cookies elaborated with hexoses) were detected (Nishibori and Kawakishi 1992). HMF has also been found in model systems of cookies baked at 150°C for 10 minutes (Nishibori and Kawakishi 1995, Nishibori et al. 1998).

García-Villanova et al. (1993) proposed the determination of HMF to control the heating procedure in breakfast cereals as well as in other cereal derivatives. Birlouez-Aragon et al. (2001) detected HMF in commercial samples of breakfast cereals (cornflakes), in samples before and after processing using traditional cooking and roasting, and in a model system of wheat, oats and rice with two levels of sugars submitted to extrusion. Although only traces of HMF were detected in the raw material, HMF formation during extrusion increased proportionally as sugar concentration increased. Among all the analysed samples, the highest content of HMF was found in commercial samples.

The effect of recipe composition in terms of leavening agent (ammonium and sodium bicarbonates) and sugars (sucrose and glucose), and baking conditions (temperature and time) on HMF formation in cookies has been recently studied by Gökmen et al. (2008b). The type of leavening agent, affecting both pH of formulation and degradation of sugars, appeared as one of the most important parameters from the viewpoint of HMF formation. Water activity, which has also been reported to be highly influential on HMF formation, must reach levels lower than 0.4 for allowing a significant formation of this indicator in commercial cookies (Ait-Ameur et al. 2006).

Together with HMF and furfural, glucosylisomaltol (GIM, an intermediate product of the MR that is generated principally from the reaction between maltose and glutamine) has been determined by HPLC as a control index of the MR development in breakfast cereals (Rufián-Henares et al. 2006a), with the advantage over HMF that the MR, and not the addition of thermally damaged ingredients, is the sole pathway of formation of GIM.

Recently, Rufián-Henares and Delgado-Andrade (2009) studied the effect of digestive process on the bioaccessibility of furosine and HMF and on the antioxidant activity of corn-based breakfast cereals. Digestion allowed liberation of HMF and Amadori products (furosine) and increased the solubility of antioxidant compounds.

Colour

Colour is an important characteristic of cereal-based foods and, together with texture and aroma, contributes to consumer preference. Colour is another indication of the extent of MR and caramelisation. The kinetic parameters of these reactions are extremely complex for cereal products. As a consequence, the colouring reaction is always studied globally, without taking into account individual reaction mechanisms (Chevallier et al. 2002). Colour depends both on the physico-chemical characteristics of the raw dough (water content, pH, reducing sugars and amino acid content) and on the operating conditions during processing (Zanoni et al. 1995, Gökmen et al. 2008b).

Brown pigments (melanoidins) formation occurs at the advanced stages of browning reactions and, although is undesirable in milk, fruit juices and tomatoes, among other foods (De Man 1980), it is desirable during the manufacture of cookies, crackers and breakfast cereals. Colour development has been studied in cereal-based foods such as baby cereals (Fernández-Artigas et al. 1999b), bread (Ramírez-Jiménez et al. 2001), cookies (Kane et al. 2003, Gökmen et al. 2008b), breakfast cereals (Rufián-Henares et al. 2006b, Moreau et al. 2009) and gluten-free biscuits (Schober et al. 2003).

The formation of melanoidins and other products of the MR darkens a food, but a darker colour is not always attributed to the presence of these compounds, since the initial composition of the mixture can also afford colour. In a study on the protein nutritional value of extrusion-cooked wheat flours, Bjorck et al. (1984) determined the colour of the extruded material and found a correlation between the reflectance values and total lysine content of extruded wheat. As the reflectance of the raw flours was very different, comparison could not be made for the effect of extrusion.

Hunter 'L' values determined for wheat flours, commercial flaked-toasted, extruded-toasted, and extruded-puffed breakfast cereals were found to be positively correlated with available lysine (McAuley et al. 1987). Rufián-Henares et al. (2009) concluded that the colour difference (ΔE) obtained from the CIE Lab parameters L∗, a∗, b∗ between toasted and untoasted samples results in a reliable measurement of the visible colour production for the toasting step of different flours usually used for the formulation of cereal-based products. During industrial baking of cookies, the effect of time on colour development and other parameters (volume, structure, weight and crispness) was studied by Piazza and Masi (1997). The development of crispness increased with time and was found to be related to the other physical processes that occur during baking. Gallagher et al. (2003) observed different colour development in the production of a functional low-fat and low-sugar biscuit, depending upon the quantities of sugar and protein present. Regarding spaghetti pasta fortified with different levels of Mexican common bean flour, the difference in colour was associated with the use of common flour rather than the drying temperature assayed (Gallegos-Infante et al. 2010).

The manufacture of many breakfast cereals starts with the cooking of whole cereal grains in a rotary pressure cooker. During this operation, the grains absorb heat and moisture, and undergo chemical (browning reactions) and physico-chemical changes as a consequence. The cooking stage is thought to have a key influence on the properties of the final product, such as colour, flavour and texture. Horrobin et al. (2003) studied the interior and surface colour development during wheat grain steaming and the results obtained indicated the possible relationship between the colour development and moisture uptake during the cooking process. In a study on the effects of oven humidity on foods (bread, cakes and cookies) baked in gas convection ovens, Xue et al. (2004) observed that increased oven humidity results in products with lighter colour and reduced firmness.

As aforementioned, the formation of MR products and intense colour are responsible for the organoleptic properties of this type of products. Moreover, it is also important to consider that the brown pigments formed could present some biological activities. Thus, Bressa et al. (1996) observed a considerable antioxidant capacity in cookies during the first 20–30 minutes of cooking (when browning takes place). Whole grain breakfast cereals have also proved to be an important dietary source of antioxidants (Miller et al. 2000). Borrelli et al. (2003) studied in bread and biscuits the formation of coloured compounds and they examined the antioxidant activity and the potential cytotoxic effects of the formed products. Summa et al. (2006) also correlated acrylamide concentration and antioxidant activity with colour of cookies by means of a three-dimensional model in which the three coordinates of the CIE colour space were used as variables.

Bernussi et al. (1998) studied the effect of microwave baking on the moisture gradient and overall quality of cookies, and they observed that colour did not differ significantly from that of the control samples (cookies baked using the traditional process).

Broyart et al. (1998) carried out a study on the kinetics of colour formation during the baking of crackers in a static electrically heated oven. These authors observed that the darkening step starts when the product temperature reaches a critical value in the range of 105–115°C. A kinetic model was developed in order to predict the lightness variation of the cracker surface using the product temperature and moisture content variations during baking. The evolution of lightness appears to follow a first-order kinetic influenced by these two parameters.

Colour development has also been included, together with other parameters (temperature, water loss, etc) in a mathematical model that simulated the functioning of a continuous industrial-scale biscuit oven (Broyart and Trystram 2003).

The effect of raising agents and sugars on surface colour of cookies was studied by Gökmen et al. (2008b).

Changes in surface colour, water activity and pH of cookies were determined to understand the chemical mechanisms responsible for HMF formation during baking of cookies from different formulations.

Fluorescence

During the advanced stages of the non-enzymatic browning, compounds with fluorescence properties are also produced. Several analytical methods based on fluorescence measurements have been used to evaluate the extent of this reaction (Rizkallah et al. 2008, Trégoat et al. 2009, Calvarro et al. 2009). For instance, the FAST (fluorescence of advanced Maillard products and soluble tryptophan) method proposed by Birlouez-Aragon et al. (1998) is based on the fluorescence ratio between MR products and tryptophan determined when excited at 330–350 nm. This parameter is dependent on heat treatment of the product and is related to its protein nutritional loss. Thus, this method, firstly validated on milk samples, has been used in other foods modified by the MR such as breakfast cereals (Birlouez-Aragon et al. 2001) and rye bread (Michalska et al. 2008).

Birlouez-Aragon et al. (2001) studied the correlation between the FAST index, lysine loss and HMF formation during the manufacture of breakfast cereals by extrusion and in commercial samples. The FAST index was in good agreement with HMF formation. These authors also found that the relationship between FAST index and lysine loss indicates that, for severe treatments inducing lysine blockage higher than 30%, lysine loss is less rapid than the increase in the FAST index. Thus, the fluorimetric FAST method appears to be an interesting alternative to evaluate the nutritional damage on a great variety of cereal-based products submitted to heat treatment.

Delgado-Andrade et al. (2006) determined free and total (free + linked-to-protein backbone) fluorescence intermediate compounds (FIC) related to MR in commercial breakfast cereals. Levels of free and total FIC were also correlated with other well-established heat-induced markers of the extent of the MR or sugar caramelisation during cereal processing, such as HMF, furfural, GIM and furosine. Data support the usefulness of FIC measurements as an unspecific but adequate heat-induced marker of the extent of thermal processing in breakfast cereals, since statistically significant correlations were found between

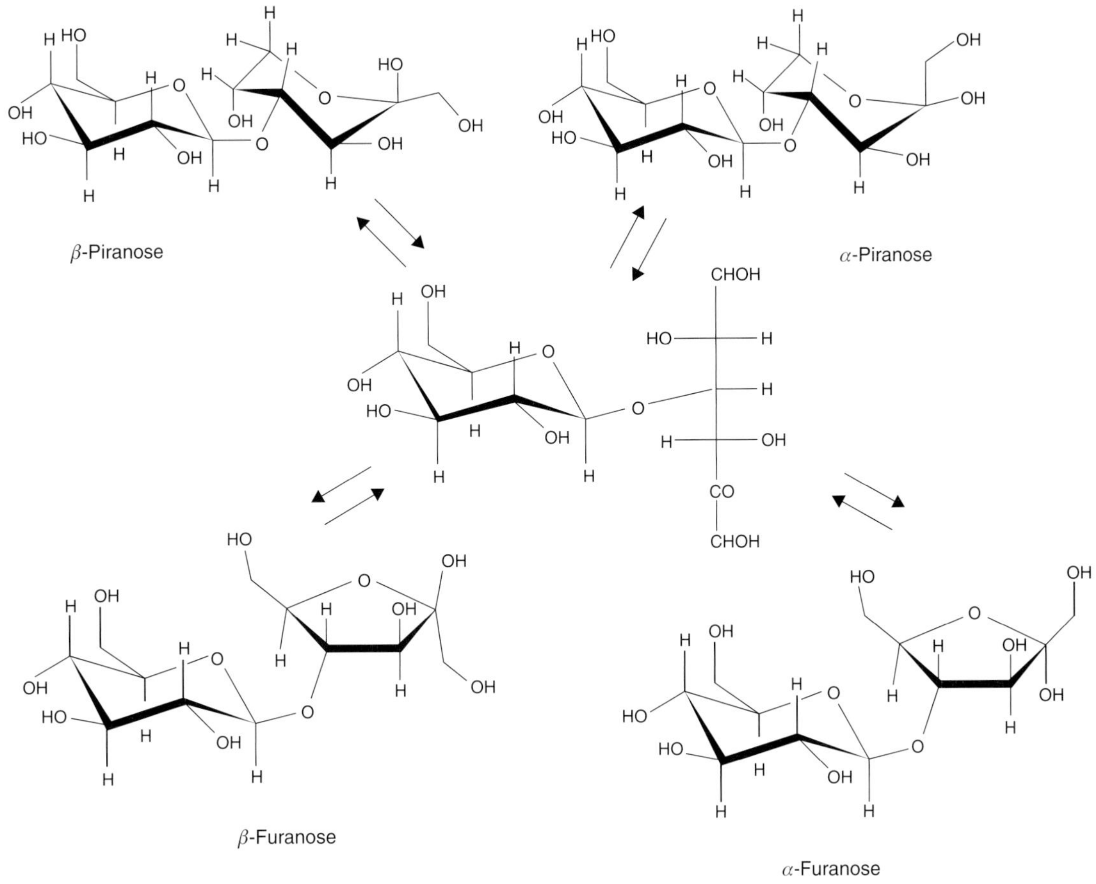

Figure 30.2. Anomeric forms of maltulose.

FIC values and other heat-induced markers. Recently, Calvarro et al. (2009) reported a new methodology to monitor Maillard-derived fluorescent compounds (FC) in cookies by flow-injection analysis. Ratio between total and free FC was proposed as reference for setting the limits to the appropriate range of processing and to avoid over-processing. Because of the significant correlation with acrylamide levels, this methodology is also proposed for estimation of acrylamide formed in cookies.

Acrylamide

For almost 8 years now, a significant amount of research activities has focused on the mitigation of the formation of the 'probable carcinogen to human', acrylamide (2-propenamide), in a number of fried and oven-cooked foods (IARC 1994, Jensen et al. 2008). One of the possible mechanisms involved in the acrylamide formation is the reaction between asparagine and reducing sugars such as glucose or fructose via MR (Mottram et al. 2002, Weisshaar and Gutsche 2002, Courel et al. 2009). Since asparagine is a major amino acid in cereals, the possible formation of acrylamide in cereal-based foods should be considered.

Variable amounts of acrylamide have been found in cereal-based foods such as bread, cookies, crackers, biscuits and breakfast cereals (Yoshida et al. 2002, Ono et al. 2003, Riediker and Stadler 2003, Delatour et al. 2004, Taeymans et al. 2004, Rufián-Henares et al. 2006b, Sadd et al. 2008).

The presence of acrylamide, together with that of other undesirable Maillard compounds, such as furosine, HMF, carboxymethyllysine, has been related to formulation and processing conditions of model systems of cookies (Courel et al. 2009) and crackers (Levine and Smith 2005). In general, baking conditions, the selected leavening agent and addition of asparagine to the formula were identified as the most significant sources of variability (including those contributing either to the formation or elimination) of acrylamide. In a similar approach by Rufián-Henares et al. (2006b), the relationship among levels of acrylamide and compositional parameters (type of cereal and protein and dietary fibre contents) of commercial breakfast cereals was shown. In addition, no significant relationship between acrylamide and classical browning indicators such as HMF, furosine or coloured compounds was found in the commercial samples analysed. However, it could be feasible that this relation become significant for samples processed at industrial/

pilot-scale under controlled formulation and processing conditions.

Sadd et al. (2008) studied different alternative mitigation strategies to reduce acrylamide in bakery products. Whereas yeast fermentation was an effective way of reducing asparagine and hence acrylamide, amino acid addition to dough gave modest reduction of this contaminant. Removing ammonium-based raising agents was also beneficial. Calcium supplementation for binding asparagine looked promising, but interactions with other ingredients need further investigation. Lowering the dough pH reduced acrylamide, but at the expense of higher levels of other process contaminants.

Alternatively, Anese et al. (2010) have recently evaluated the possibility to remove acrylamide from previously hydrated commercial biscuits by vacuum treatment. Selection of suitable temperature, time and pressure conditions has proved to be required to maximise acrylamide removal while minimising its formation and effect on the sensory properties.

On the other hand, the correlation between mitigation of acrylamide and its influence on beneficial properties of food, such as the antioxidant activity has been scarcely studied. Summa et al. (2006) reported a strong link between the composition of the pastries and baking parameters and the genesis of acrylamide as well as antioxidants. Therefore, suppressing the MR generally leads not only to acrylamide mitigation but also to losses of substances considered to be beneficial.

Maltulose

Besides MR, isomerisation of reducing carbohydrates may take place during processing of cookies, crackers and breakfast cereals. Maltulose (Fig. 30.2), an epimerisation product of maltose, has been found in the crust of bread (Westerlund et al. 1989). Maltulose formation has also been observed during the heating of a maltodextrin solution at high temperature (180°C) (Kroh et al. 1996). García-Baños et al. (2000, 2002) detected maltulose in commercial enteral products and proposed the ratio maltose:maltulose as a heat treatment and storage indicator of enteral formulas.

In cookies, crackers and breakfast cereals, Rada-Mendoza et al. (2004) detected maltulose (from traces to 842 mg/100 g) in all commercial samples analysed. Figure 30.3 shows the gas chromatographic profile of the trimethylsilyl oxime-derivatives of mono- and disaccharide fractions of a cookie sample. Similar patterns were obtained for crackers and breakfast cereals. The formation of maltulose depends mainly on initial maltose

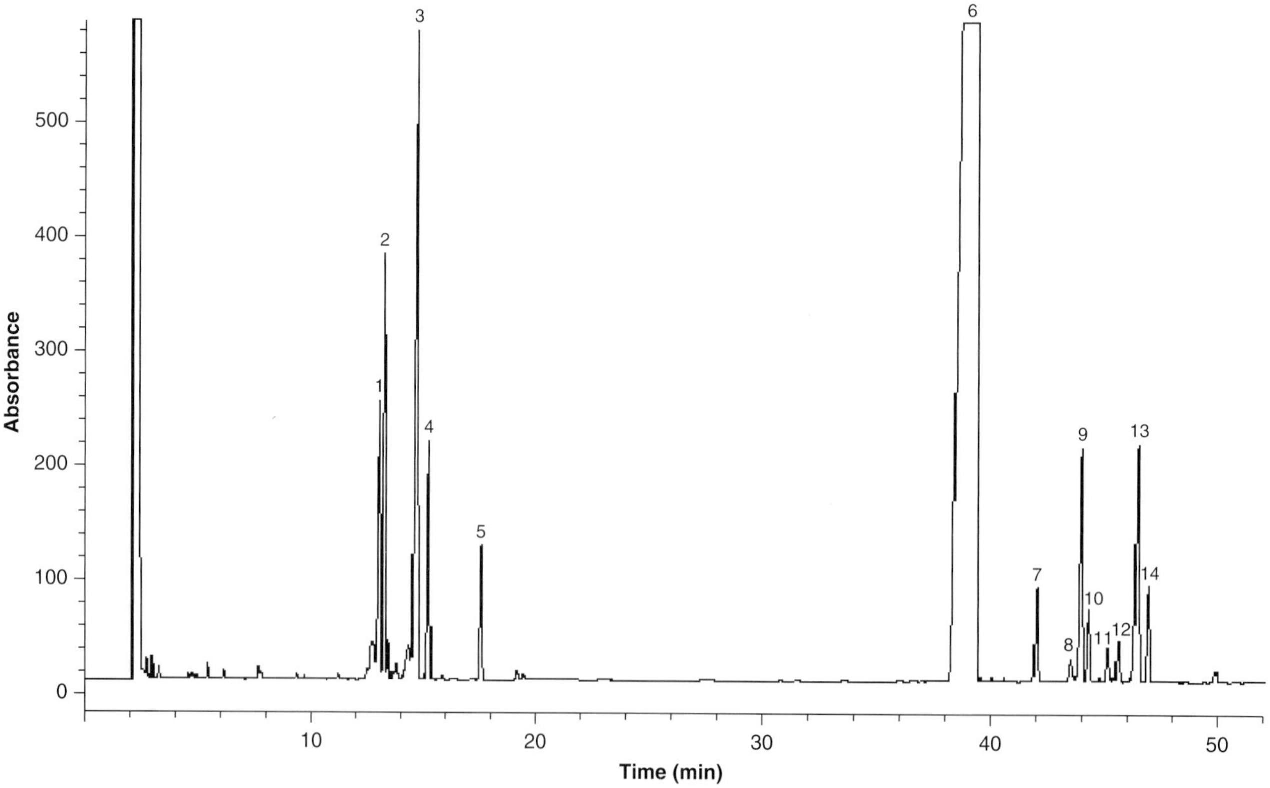

Figure 30.3. Gas chromatographic profile of the trimethylsilyl oxime-derivatives of fructose (1,2), glucose (3,4), sucrose (6), lactulose (8), lactose (9, 10), maltulose (11, 12) and maltose (13,14) of a cookie sample. Peaks 5 and 7 were the internal standards: *myo*-inositol and trehalose, respectively (from Rada-Mendoza et al. 2004).

content, pH and heat treatment intensity of the process. Since cookies, crackers and breakfast cereal samples may contain variable amounts of maltose, the usefulness of maltulose as indicator of heat treatment might be questionable. Previous studies on the formation of maltulose during heating of enteral formulas (García-Baños et al. 2002) pointed out that values of the ratio maltose:maltulose were similar in samples with different maltose content submitted to the same heat treatment. Therefore, the ratio maltose: maltulose is an adequate parameter for comparing samples with different initial maltose contents. Because maltose isomerisation increases with pH, only in samples with similar pH, differences in maltose: maltulose ratio can be due to different heat processing conditions. These results, shown by Rada-Mendoza et al. (2004), seem to indicate that the ratio maltose: maltulose can allow the differentiation among commercial cereal-based products and may serve as an indicator of the heat load during its manufacture.

REFERENCES

Acquistucci R, Bassotti G. 1992. Effects of Maillard reaction on protein in spaghetti samples. Chemical reaction in Foods II. Abstract Papers, Symposium Prague, September, p. 94.

Acquistucci R, Quattrucci E. 1993. In vivo protein digestibility and lysine availability in pasta samples dried under different conditions. Nutritional, chemical and food processing implications of nutrient availability. *Bioavailability* 1: 23–27.

Ait-Ameur L et al. 2006. Accumulation of 5-hydroxymethyl-2-furfural in cookies during the backing process: Validation of an extraction method. *Food Chem* 98: 790–796.

Anese M et al. 2010. Acrylamide removal from heated foods. *Food Chem* 119: 791–794.

Awasthi P, Yadav MC. 2000. Effect of incorporation of liquid dairy by-products on chemical characteristics of soy-fortified biscuits. *J Food Sci Technol-Mysore* 37(2): 158–161.

Bernussi ALM et al. 1998. Effects of production by microwave heating after conventional baking on moisture gradient and product quality of biscuits (cookies). *Cereal Chem* 75(5): 606–611.

Birlouez-Aragon I et al. 1998. A rapid fluorimetric method to estimate the heat treatment of liquid milk. *Int Dairy J* 8(9): 771–777.

Birlouez-Aragon I et al. 2001. The FAST method, a rapid approach of the nutritional quality of heat-treated foods. *Nahrung/Food* 45(3): 201–205.

Bjorck I et al. 1984. Protein nutritional value of extrusion-cooked wheat flours. *Food Chem* 15(8): 203–214.

Borrelli RC et al. 2003. Characterization of coloured compounds obtained by enzymatic extraction of bakery products. *Food Chem Toxicol* 41(10): 1367–1374.

Bressa F et al. 1996. Antioxidant effect of Maillard reaction products: Application to a butter cookie of a competition kinetics analysis. *J Agric Food Chem* 44(3): 692–695.

Broyart B et al. 1998. Predicting colour kinetics during cracker baking. *J Food Eng* 35(3): 351–368.

Broyart B, Trystram G. 2003. Modelling of heat and mass transfer phenomena and quality changes during continuous biscuit baking using both deductive and inductive (neural network) modelling principles. *Food Bioproducts Process* 81(C4): 316–326.

Calvarro J et al. 2009. A generic procedure to monitor Maillard-derived fluorescent compounds in cookies by flow-injection analysis. *Eur Food Res Technol* 229: 843–851.

Camire ME et al. 1990. Chemical and nutritional changes in foods during extrusion. *Crit Rev Food Sci Nutr* 29(1): 35–57.

Camire ME, Belbez EO. 1996. Flavor formation during extrusion cooking. *Cereal Foods World* 41(9): 734–736.

Cárdenas-Ruiz J et al. 2004. Furosine is a useful indicator in pre-baked bread. *J Sci Food Agric* 84(4): 366–370.

Carratù B et al. 1993. Influenza del trattamento tecnologico sugli alimenti: determinazione per cromatografía ionica di piridosina e furosina in prodotti dietetici ed alimentari. *La Rivista di Scienza dell'Alimentazione* 22(4): 455–457.

Cheftel JC et al. 1981. *In Progress Food Nutrition Science*, vol 5. Pergamon Press, London, p. 487.

Cheftel JC. 1986. Nutritional effects of extrusion-cooking. *Food Chem* 20: 263–283.

Chevallier S et al. 2002. Structural and chemical modifications of short dough during baking. *J Cereal Sci* 35(1): 1–10.

Corzo N et al. 1994. Ratio of lactulose to furosine as indicator of quality of commercial milks. *J Food Prot* 57(8): 737–739.

Courel M et al. 2009. Effects of formulation and baking conditions on neo-formed contaminants in model cookies. *Czech J Food Sci* 27: S93–S95.

Delatour Y et al. 2004. Improved sample preparation to determine acrylamide in difficult matrixes such as chocolate powder, cocoa, and coffee by liquid chromatography tandem mass spectroscopy. *J Agric Food Chem* 52(15): 4625–4631.

Del Castillo MD et al. 2000. Use of 2-furoylmethyl derivatives of GABA and arginine as indicators of the initial steps of Maillard reaction in orange juice. *J Agric Food Chem* 48(9): 4217–4220.

Delgado-Andrade C et al. 2006. Study on fluorescence of Maillard reaction compounds in breakfast cereals. *Mol Nutr Food Res* 50: 799–804.

Delgado-Andrade C et al. 2007. Lysine availability is diminished in commercial fibre-enriched breakfast cereals. *Food Chem* 100: 725–731.

De Man JM. 1980. Nonenzymatic browning. In: *Principles of Food Chemistry*. AVI Publishing, Westport, pp. 98–112.

Erbersdobler HF, Hupe A. 1991. Determination of the lysine damage and calculation of lysine bio-availability in several processed foods. *Zeitschrift für Ernaühhrungswissenchaft* 30(1): 46–49.

Fernández-Artigas P et al. 1999a. Blockage of available lysine at different stages of infant cereal production. *J Sci Food Agric* 79(6): 851–854.

Fernández-Artigas P et al. 1999b. Browning indicators in model systems and baby cereals. *J Agric Food Chem* 47(7): 2872–2878.

Gallagher E et al. 2003. Use of response surface methodology to produce functional short dough biscuits. *J Food Eng* 56(2–3): 269–271.

Gallegos-Infante JA et al. 2010. Quality of spaghetti pasta containing Mexican common vean flour (*Phaseolus vulgaris* L.). *Food Chem* 119: 1544–1549.

García-Baños JL et al. 2000. Determination of mono and disaccharide content of enteral formulations by gas chromatography. *Chromatographia* 52(3/4): 221–224.

García-Baños JL et al. 2002. Changes in carbohydrate fraction during manufacture and storage of enteral formulas. *J Food Sci* 67(9): 3232–3235.

García-Baños JL et al. 2004. Maltulose and furosine as indicators of quality of pasta products. *Food Chem* 88(1): 35–38.

García-Villanova B et al. 1993. Liquid Chromatography for the determination of 5-hydroxymethyl-2-furaldehyde in breakfast cereals. *J Agric Food Chem* 41(8): 1254–1255.

Gökmen V et al. 2008a. Significance of furosine as heat-induced marker in cookies. *J Cereal Sci* 48: 843–847.

Gökmen V et al. 2008b. Effect of leavening agents and sugars on the formation of hydroxymethylfurfural in cookies during baking. *Eur Food Res Technol* 226: 1031–1037.

González-Galán A et al. 1991. Sensory and nutritional properties of cookies based on wheat-rice-soybean flours baked in a microwave oven. *J Food Sci* 56(6): 1699–1701, 1706.

Guerra-Hernández E et al. 1992. Determination of hydroxymethylfurfural in baby cereals by High-Performance Liquid Chromatography. *J Liq Chromatogr* 15(14): 2551–2559.

Guerra-Hernández E, Corzo N. 1996. Furosine determination in baby cereals by ion-pair reversed phase liquid chromatography. *Cereal Chem* 73(6): 729–731.

Guerra-Hernández E et al. 1999. Maillard reaction evaluation by furosine determination during infant cereal processing. *J Cereal Sci* 29(2): 171–176.

Hodge JE. 1953. Dehydrated foods; chemistry of browning reactions in model systems. *J Agric Food Chem* 1(15): 928–943.

Horrobin DJ et al. 2003. Interior and surface color development during wheat grain steaming. *J Food Eng* 57(1): 33–43.

Horvatić M, Guterman M. 1997. Available lysine content during cereal flake production. *J Sci Food Agric* 74(3): 354–358.

Horvatić M, Eres M. 2002. Protein nutritive quality during production and storage of dietetic biscuits. *J Sci Food Agric* 82(14): 1617–1620.

Hozova B et al. 1997. Microbiological, nutritional and sensory aspects of stored amaranth biscuits and amaranth crackers. *Nahrung/Food* 41(3): 155–158.

Huang WN. 1998. Comparing cornflake manufacturing processes. *Cereal Foods World* 43(8): 641–643.

IARC. 1994. International Agency for Research on Cancer. Monographs on the Evaluation of Carcinogenic Risks to Humans: Some Industrial Chemicals. Volume 60. IARC, Lyon, France, p. 389.

Jensen BBB et al. 2008. Robust modelling of heat-induced reactions in an industrial food production process exemplified by acrylamide generation in breakfast cereals. *Food Bioproducts Process* 86(3): 154–162.

Jeuring HJ, Kuppers FJEM. 1980. High performance liquid chromatography of furfural and hydroxymethylfurfural in spirits and honey. *J Assoc Off Anal Chem* 63(6): 1215–1218.

Kane AM et al. 2003. Comparison of two sensory and two instrumental methods to evaluate cookie color. *J Food Sci* 68(5): 1831–1837.

Kroh LW. 1994. Caramelisation in Food and Beverages. *Food Chem* 51(4): 373–379.

Kroh LW et al. 1996. Non-volatile reaction products by heat-induced degradation of α-glucans. Part I: Analysis of oligomeric maltodextrins and anhydrosugars. *Starch* 48(11–12): 426–433.

Lee HS, Nagy S. 1988. Quality changes and nonenzymatic browning intermediates in grapefruit juice during storage. *J Food Sci* 53(1): 168–172.

Levine RA, Smith RE. 2005. Sources of variability of acrylamide levels in a cracker model. *J Agric Food Chem* 53: 4410–4416.

Manley D. 2000. In: *Technology of Biscuits, Crackers and Cookies*, 3rd edn. Woodhead Publishing and CRC Press, Cambridge, pp. 388–402.

Manzaneque Ramos A. 1994. Snacks y preparados especiales para desayuno. *Alimentación, Equipos y Tecnología* 2: 61–65.

Martinkus VB et al. 1977. Effect of heat-treatment on available lysine and other amino-acids in a cookie fortified with oilseed flours. *Fed Proc* 36(3): 1111.

McAuley JA et al. 1987. Relationships of available lysine to lignin, color and protein digestibility of selected wheat-based breakfast cereals. *J Food Sci* 52(6): 1580–1582.

Michalska A et al. 2008. Effect of bread making on formation of Maillard reaction products contributing to the overall antioxidant activity of rye bread. *J Cereal Sci* 48: 123–132.

Miller HE et al. 2000. Antioxidant content of whole grain breakfast cereals, fruits and vegetables. *J Am Coll Nutr* 19(3 Suppl): 312S–319S.

Molnár-Perl I et al. 1986. Optimum yield of pyridosine and furosine originating from Maillard reactions monitored by ion-exchange chromatography. *J Chromatogr* 361: 311–320.

Morales FJ et al. 1997. Chromatographic determination of bound hydroxymethylfurfural as an index of milk protein glycosylation. *J Agric Food Chem* 45(5): 1570–1573.

Moreau L et al. 2009. Influence of sodium chloride on colour, residual volatiles and acrylamide formation in model systems and breakfast cereals. *Int J Food Sci Technol* 44: 2407–2416.

Mottram DS et al. 2002. Acrylamide is formed in the Maillard reaction. *Nature* 419(6906): 448–449.

Nepal-Sing S, Chauhan GS. 1989. Some physico-chemical characteristics of defatted soy flour fortified noodles. *J Food Sci Technol* 26: 210–212.

Nishibori S, Kawakishi S. 1992. Effect of various sugars on the quality of baked cookies. *Cereal Chem* 69(2): 160–163.

Nishibori S, Kawakishi S. 1995. Effect of various amino-acids on formation of volatile compounds during baking in a low moisture food system. *J Jpn Soc Food Sci Technol-Nippon Shokuhin Kagaku Kogaku Kaishi* 42(1): 20–25.

Nishibori S et al. 1998. Volatile components formed from reaction of sugar and beta-alanine as a model system of cookie processing. *Process-Induced Chem Changes Food (Adv Exp Med Biol)* 434: 255–267.

Noguchi A et al. 1982. Maillard reactions during extrusion-cooking of protein-enriched biscuits. *Lebensmittel Wissenschaft und Technologie* 15(2): 105–110.

Ono H et al. 2003. Analysis of acrylamide by LC-MS/MS and GC-MS in processed Japanese foods. *Food Addit Contam* 20(3): 215–220.

Phillips RD. 1988. Effect of extrusion cooking on the nutritional quality of plant proteins. In: RD Phillips, JW Fionley (eds.) *Protein Quality and the Effects of Processing*. Marcel Dekker, New York.

Piazza L, Masi P. 1997. Development of crispness in cookies during baking in an industrial oven. *Cereal Chem* 74(2): 135–140.

Rada-Mendoza M et al. 2004. Study on nonenzymatic browning in cookies, crackers and breakfast cereals by maltulose and furosine determination. *J Cereal Sci* 39(2): 167–173.

Ramírez-Jiménez A et al. 2000. Browning indicators in bread. *J Agric Food Chem* 48(9): 4176–4181.

Ramírez-Jiménez A et al. 2001. Effect of toasting time on browning of sliced bread. *J Sci Food Agric* 81(5): 513–518.

Resmini P, Pellegrino L. 1991. Analysis of food heat damage by direct HPLC of furosine. *Int Chromatogr Lab* 6: 7–11.

Resmini P et al. 1993. Formation of 2-acetyl-3-D-glucopyranosylfuran (glucosylisomaltol) from nonenzymatic browning in pasta drying. *Italian J Food Sci* 5(4): 341–353.

Riediker S, Stadler RH. 2003. Analysis of acrylamide in food by isotope-dilution liquid chromatography coupled with electrospray ionisation tandem mass spectrometry. *J Chromatogr A* 1020(1): 121–130.

Rizkallah J et al. 2008. Front face fluorescence spectroscopy and multiway analysis for process control and NFC prediction in industrially processed cookies. *Chemom Intell Lab Syst* 93: 99–107.

Rufián-Henares JA et al. 2006a. Application of a fast high-performance liquid chromatography method for simultaneous determination of furanic compounds and glucosylisomaltol in breakfast cereals. *J AOAC Int* 89: 161–165.

Rufián-Henares JA et al. 2006b. Relationship between acrylamide and thermal-processing indexes in commercial breakfast cereals: a survey of Spanish breakfast cereals. *Mol Nutr Food Res* 50: 756–762.

Rufián-Henares JA, Delgado-Andrade C. 2009. Effect of digestive process on Maillard reaction indexes and antioxidant properties of breakfast cereals. *Food Res Int* 42: 394–400.

Rufián-Henares JA et al. 2009. Assessing the Maillard reaction development during the toasting process of common flours employed by the cereal products industry. *Food Chem* 114: 93–99.

Rutherfurd SM et al. 2006. Available (ileal digestible reactive) lysine in selected cereal-based food products. *J Agric Food Chem* 54: 9453–9457.

Sadd PA et al. 2008. Effectiveness of methods for reducing acrylamide in bakery products. *J Agric Food Chem* 56: 6154–6161.

Sanz ML et al. 2000. Presence of 2-furoylmethyl derivatives in hydrolysates of processed tomato products. *J Agric Food Chem* 48(2): 468–471.

Sanz ML et al. 2001. Formation of Amadori compounds in dehydrated fruits. *J Agric Food Chem* 49(11): 5228–5231.

Sanz ML et al. 2003. 2-Furoylmethyl amino acids and hydroxymethylfurfural as indicators of honey quality. *J Agric Food Chem* 51(15): 4278–4283.

Schober TJ et al. 2003. Influence of gluten-free flour mixes and fat powders on the quality of gluten-free biscuits. *Eur Food Res Technol* 216(5): 369–376.

Singh R et al. 2000. Nutritional evaluation of soy fortified biscuits. *J Food Sci Technol-Mysore* 37(2): 162–164.

Soria AC et al. 2009. 2-Furoylmethyl amino acids, hydroxymethylfurfural, carbohydrates and β-carotene as quality markers of dehydrated carrots. *J Sci Food Agric* 89: 267–273.

Summa C et al. 2006. Investigation of the correlation of the acrylamide content and the antioxidant activity of model cookies. *J Agric Food Chem* 54(3): 853–859.

Taeymans D et al. 2004. A review of acrylamide: an industry perspective on research, analysis, formation and control. *Crit Rev Food Sci Nutr* 44(5): 323–347.

Torbatinejad NM et al. 2005. Total and reactive lysine contents in selected cereal-based food products. *J Agric Food Chem* 53: 4454–4458.

Trégoat V et al. 2009. Immunofluorescence detection of advanced glycation end products (AGEs) in cookies and its correlation with acrylamide content and antioxidant activity. *Food Agric Immunol* 20(3): 253–268.

Tsen CC et al. 1983. Effects of the Maillard browning reaction on the nutritive value of breads and pizza crusts. In: GR Waller, MS Feather (eds.) *The Maillard Reaction in Food and Nutrition*. ACS, Washington, pp. 379–394.

van Boekel MAJS, Zia-Ur-Rehman. 1987. Determination of HMF in heated milk by HPLC. *Neth Milk Dairy J* 41(4): 297–306.

Weisshaar R, Gutsche B. 2002. Formation of acrylamide in heated potato products. Model experiments pointing to asparagines as precursor. *Deutsche Lebbensmittel* 98 Jahrgang(11): 397–400.

Westerlund E et al. 1989. Effects of baking on protein and ethanol-extractable carbohydrate in white bread fractions. A review of acrylamide: an industry perspective. *J Cereal Sci* 10(2): 139–147.

Xue J et al. 2004. Effects of oven humidity on foods baked in gas convection ovens. *J Food Process Preservation* 28(3): 179–200.

Yoshida M et al. 2002. Determination of acrylamide in processed foodstuffs in Japan. *J Jpn Soc Food Sci Technol-Nippon Shokuhin Kagaku Kogaku Kaishi* 49(12): 822–825.

Zanoni B et al. 1995. Modelling of browning kinetics of bread crust during baking. *Lebensmittel Wissenschaft und Technologie* 28(3): 604–609.

31
Bakery and Cereal Products

J. A. Narvhus and T. Sørhaug

INTRODUCTION

Cereals are the edible seeds of plants of the grass family. They can be grown in a large part of the world and provide the staple food for most of mankind. Maize, wheat, and rice contribute about equally to 85% of world cereal production, which is at present about 2000 million tons (FAO 1999).

Cereals in their dry state are not subjected to fermentation due to their low water content. Properly dried cereals contain less than 14% water, and this limits microbial growth and chemical changes during storage. However, on mixing grains or cereal flour with water or other water-based fluids, enzymatic changes occur that may be attributed to the enzymes inherent in the grain itself and/or to microorganisms. These microorganisms can either be those present as the natural contaminating flora of the cereal, or they can be added as a starter culture.

This chapter will be mainly devoted to fermented bakery products made from wheat. However, on a global basis, many fermented cereal products are derived wholly or in part from other grains such as rice, maize, sorghum, millet, barley, and rye. Different cereals differ not only in nutrient content, but also in the composition of the protein and carbohydrate polymers. The functional and sensory characteristics of products made from different cereals will therefore vary at the outset due to these factors. In addition to this, the opportunity to vary technological procedures and microbiological content and activity provides us with the vast range of fermented cereal products that are prepared and consumed in the world today.

CEREAL COMPOSITION

Carbohydrates are quantitatively the most important constituents of cereal grains, contributing 77–87% of the total dry matter. In wheat, the carbohydrate in the endosperm is mainly starch, whereas the pericarp, testa, and aleurone contain most of the

Food Biochemistry and Food Processing, Second Edition. Edited by Benjamin K. Simpson, Leo M.L. Nollet, Fidel Toldrá, Soottawat Benjakul, Gopinadhan Paliyath and Y.H. Hui.

crude and dietary fiber present in the grain. The pericarp, testa, and aleurone also contain over half of the total mineral matter. Whole meal flour is derived, by definition, from the whole grain and contains all its nutrients. When wheat is milled into flour, the yield of flour from the grain (extraction rate) reflects the extent to which the bran and the germ are removed and thereby determines not only the whiteness of the flour, but also its nutritive value and baking properties. Decreasing extraction rate results in a marked, and nutritionally important, decrease in fiber, fat, vitamins, and minerals (Kent 1983). The protein content of different cereal grains varies between 7% and 20%, governed not only by cereal genus, species, or variety (i.e., genetically regulated), but also by plant growth conditions such as temperature, availability of water during plant growth, and also of nitrogen and other minerals in the soil. There is an uneven distribution of different protein types in the different parts of the grain, so that although the protein concentration is not radically affected by milling, the proteins present in different milling fractions will vary.

Starch

The starch in cereals is contained in granules that vary in size, from 2–3 μm to about 30 μm according to grain species. Barley, rye, and wheat have starch granules with a bimodal size distribution, with large lenticular and small spherical grains. Almost 100% of the starch granule is composed of the polysaccharides amylose and amylopectin, and the relative proportions of these polymers vary not only according to species of cereal, but also according to variety within a species. However, both wheat and maize contain about 28% amylose, but in wheat the ratio of amylose to amylopectin does not vary (Fenema 1996, Hoseney 1998). The amylose molecule is essentially linear, with up to 5000 glucose molecules polymerized by α-1,4 linkages and only occasional α-1,6 linkages. Amylopectin is a much larger (up to 10^6 glucose units) and more highly branched molecule with approximately 4% of α-1,6 linkages that cause branching in the α-1,4 glucosidic chain.

During milling of grains, some of the starch granules become damaged, particularly in hard wheat. The starch exposed in these broken granules is more susceptible to attack by amylases and also absorbs water much more readily. Therefore, the degree of damage to the starch grains dictates the functionality of the flour in various baking processes. When water is added to starch grains, they absorb water, and soluble starch leaks out of damaged granules. Heating of this mixture results in an increase in viscosity and a pasting of the starch, which on further heating leads to gelatinization as the ordered crystalline structure is disrupted and water forms hydration layers around the separated molecules. The gelatinization temperature of starch from different cereals varies from 55°C to 78°C, partly due to the ratio of amylose to amylopectin. The gelatinization of wheat starts at about 60°C. Despite the fact that there is not sufficient water to totally hydrate the starch in most bakery foods, the heat causes irreversible changes to the starch. On cooling a heated cereal product, some starch molecules reassociate, causing firming of the product.

Nonstarch polysaccharides, the pentosans, which are principally arabinoxylans, comprise approximately 2–3% of the weight of flour. They are derived from the grain cell walls and are polymers that may contain both pentoses and hexoses. They are able to absorb many times their own weight in water and contribute in baking by increasing the viscosity of the aqueous phase, but they may also compete with the gluten proteins for available water. Cereals also contain small amounts (1–3%) of mono-, di- and oligosaccharides, and these are important as an energy source for yeast at the start of dough fermentation.

Protein

Cereal proteins contribute to the nutritional value of the diet, and therefore the composition and amount of protein present are inherently important. However, the protein content in cereals also has several important aspects in fermented bakery products. The amount and type of some of the proteins is important for the formation of an elastic dough and for its gas-retaining properties. Other proteins in cereals are enzymes with specific functions, not only for the developing germ, but also for various changes that take place from the processing of flour to bakery products.

Gluten Proteins

The unique storage proteins of wheat are also the functional proteins in baking. The gliadins and glutenins, collectively called gluten proteins, make up about 80% of the total protein in the grain and are mostly found in the endosperm. These proteins have very limited solubility in water or salt solutions, unlike albumins and globulins (Fig. 31.1). A good bread flour (known as "strong" flour) must contain adequate amounts of gluten proteins to give the desired dough characteristics, and extra gluten may be added to the bread formulation.

Enzyme Proteins

The albumin and globulin proteins are concentrated in the bran, germ, and aleurone.

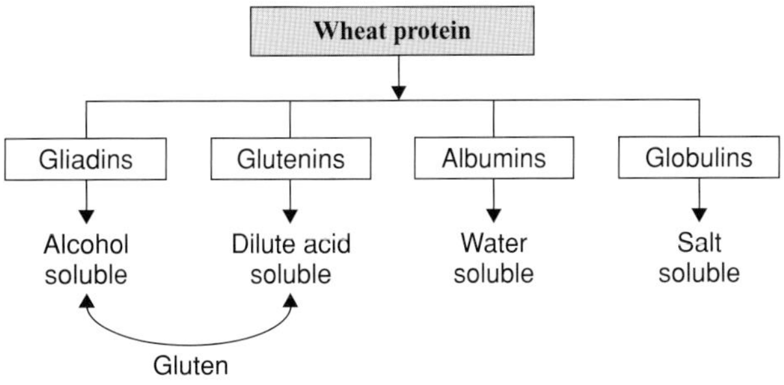

Figure 31.1. Wheat proteins.

Amylases The primary function of starch-hydrolyzing enzymes is to mobilize the storage polysaccharides to readily metabolized carbohydrates when the grain germinates (Hoseney 1998).

α-Amylase hydrolyzes α-1,4 glycosidic bonds at random in the starch molecule chain but is unable to attack the α-1,6 linkages at the branching points on the amylopectin molecule. The activity of α-amylase causes a rapid reduction in size of the large starch molecule, and the viscosity of a heated solution or slurry of starch is greatly decreased. It is most active on gelatinized starch, but granular starch is also slowly degraded.

β-Amylase splits off two glucose units (maltose) at a time from the nonreducing end of the starch chain, thus providing a large amount of fermentable carbohydrate. β-Amylase is also called a saccharifying enzyme since its action causes a marked increase in sweetness of the hydrated cereal. Neither the hydrolysis of amylopectin nor of amylose is completed by β-amylase, since the enzyme is not able to move past the branching points. The presence of both α- and β-amylases, however, leads to a much more comprehensive hydrolysis, since α-amylase produces several new reducing ends in each starch molecule.

The level of α-amylase is very low in intact grain but increases markedly on germination, whereas β-amylase levels in intact and germinated grain are similar.

Flour containing too much α-amylase absorbs less water and therefore results in heavy bread. In addition, the dough is sticky and hard to handle, and the texture of the loaf is usually faulty, having large open holes and a sticky crumb texture. However, some activity is required, and bakers may add amylase either as an enzyme preparation or as wheat or barley malt in order to slightly increase loaf volume and improve crumb texture. The thermal stability of amylases from different sources dictates their activity during the baking process. Microbial amylases with greater thermal stability have been used in bread to decrease firming (retrogradation) upon storage since these enzymes are not fully denatured during baking.

Proteases Proteinases and peptidases are found in cereals, and their primary function is to make small amino nitrogen compounds available for the developing seed embryo during germination, when the levels of these enzymes also increase. However, whether these enzymes have a role in bread baking is not certain. Peptidases may furnish the yeast with soluble nitrogen during fermentation, and a proteinase in wheat that is active at low pH may be important in acidic fermentations such as sourdough bread.

Lipases Lipases are present in all grains, but oats and pearl millet have a relatively high activity of lipase compared with wheat or barley (Linko et al. 1997). In flour of the former grain types, hydrolytic rancidity of the grain lipids and added baking fat may be a problem.

Lipids

Lipids are present in grains as a large number of different compounds, and they vary from species to species and also within each cereal grain. Most lipids are found within the germ. Wheat flour contains about 2.5% lipids, of which about 1% are polar lipids (tri- and diglycerides, free fatty acids, and sterol esters) and 1.5% are nonpolar lipids (phospholipids and galactosyl glycerides). During dough mixing, much of the lipid forms hydrophobic bonds to the gluten protein (Hoseney 1998).

BREAD

Many different types of bread are produced in the world. Bread formulations and technologies differ both within and between countries due to both traditional and technological factors including: (1) which cereals are traditionally grown in a country and their suitability for bread baking, (2) the status of bread in the traditional diet, (3) changes in lifestyle and living standards, (4) globalization of eating habits, and (5) economic possibilities for investing in new types of bread-making equipment.

The basic production of most bread involves the addition of water to wheat flour, yeast, and salt. Other cereal flours may be blended into the mixture, and other optional ingredients include sugar, fat, malt flour, milk and milk products, emulsifiers, and gluten (for further ingredients and their roles in bread, see Table 31.1). The mixture is worked into an elastic dough that is then leavened by the yeast to a soft and spongy dough that retains its shape and porosity when baked. An exception to this is the production of bread containing 20–100% rye flour, where the application of sourdough and low pH are required.

Bread Formulation

The formulation of bread is determined by several factors. In a simple bread, the baking properties of the flour are of vital importance in determining the characteristics of the loaf using a given technology. In addition, the bread obtained from using poor bread flour or suboptimum technology may be improved by using certain additives (Table 31.1).

The major methods used to prepare bread are summarized in Figure 31.2. In the straight dough method, all the ingredients are added together at the start of the process, which includes two fermentation steps and then two proofing steps. In the sponge and dough method, only part of the dry ingredients are added to the water, and this soft dough undergoes a fermentation of about 5 hours before the remainder of the ingredients are added and the dough is kneaded to develop the structure. Although these processes are time consuming, their advantages are that they develop a good flavor in the bread and that the timing and technology of the processes are less critical (Hoseney 1998). Mechanical dough development processes, such as the Chorleywood bread process developed in the United Kingdom in the 1960s, radically cut down the total bread-making time. The fermentation step is virtually eliminated, and dough formation is achieved by intense mechanical mixing and by various additives that hasten the process (Kent 1983). The resulting loaf has a high volume and a thin crust but lacks flavor and aroma. The trend is now away from this kind of process due to customer demand for more flavorful bread and reduced use of additives.

Table 31.1. Bread Additives

Dough Additive	Role
Cysteine. Sodium sulphite and metabisulphite[a]	Reducing agent. Aids optimal dough development during mixing by disrupting disulphide (–S–S–) bonds. A "dough relaxer."
Amylase	Releases soluble carbohydrate for yeast fermentation and Maillard browning reaction. Reduces starch retrogradation.
Ascorbic acid	Oxidizing agent. Strengthens gluten and increases bread volume by improving gas retention.
Potassium iodate, calcium iodate; calcium peroxide; azodicarbonamide[a]	Fast-acting oxidants; oxidizes flour lipids, carotene and converts sulphydryl (–SH) groups to disulphide (–S–S–) bonds.
Potassium and calcium bromates[a]	Delayed-acting oxidants: Develops dough consistency, reduces proofing stage.
Emulsifiers, strengtheners/conditioners and crumb softeners	Dispersion of fat in the dough. Increase dough extensibility. Interact with the gluten-starch complex and thereby retard staling.
Soy flour	Increases nutritional value, bleaches flour pigments, increases in loaf volume, increases crumb firmness and crust appearance, promotes a longer shelf life.
Vital wheat gluten and its derivatives	Increases gluten content, used especially when mixing time or fer- mentation time is reduced. Water adsorbant. Improves dough and loaf properties.
Hydrocolloids: Starch-based products from various plants	Regulates water distribution and water-holding capacity and thereby improves yield. Strengthens bread crumb structure and improves digestibility.
Cellulose and cellulose-based derivatives	Source of dietary fiber.
Salt	Enhances flavor (ca. 2% based on flour weight) and modifies mixing time for bread and rolls. Increases dough stability, firmness and gas retention properties. Raises starch gelatinization temperature

Source: Compiled from Stear 1990, Williams and Pullen 1998.
[a]Not allowed in all countries.

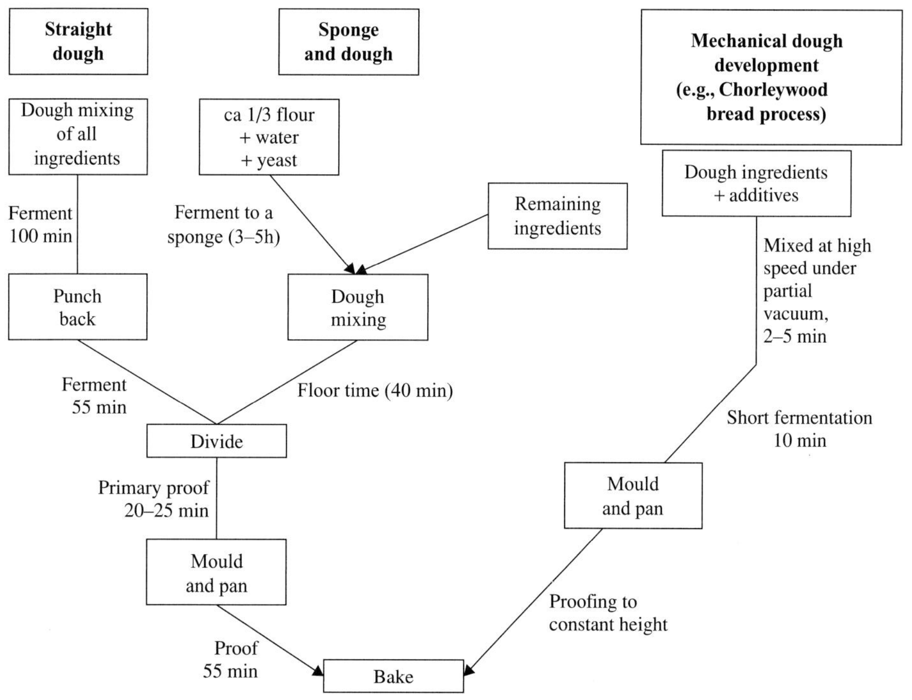

Figure 31.2. Bread-processing methods. (Adapted from Hoseney 1994.)

The Development of Dough Structure

When wheat flour and water are mixed together in an approximately 3:1 ratio and kneaded, a viscoelastic dough is formed that can entrap the gas formed during the subsequent fermentation. The amount of water absorbed by the flour is dependent upon, and therefore must be adjusted to, the integrity of the starch granules and the amount of protein present. A high proportion of damaged granules, as found in hard wheat flour, results in greater water absorption. The unique elastic property of the dough is due to the nature of the gluten proteins. Hydrated gliadin is sticky and extensible, whereas glutenin is cohesive and plastic. When hydrated during the mixing process, the gluten proteins unfold and bond with each other by forming a complex (gluten) as kneading proceeds, with an increasing number of cross-linkages between the protein molecules as they become aligned. Disulphide bonds (-S-S-) break and re-form within and between the protein molecules during mixing.

Gluten does not form spontaneously when flour and water are mixed; energy must be provided (i.e., in the actual mixing process) in order for the molecular bonds to break and re-form as the gluten structure. At this point, the dough stiffens and becomes smooth and shiny. The gluten is now composed of protein sheets in which the starch granules are embedded. In addition, free polar lipids and glycolipids are incorporated in the complex by hydrophobic and hydrogen bonds. The properties of the dough are determined by the amount of protein present and by the relative proportions of the gluten proteins.

Another important part of the dough formation is the incorporation of air, in particular nitrogen. This forms insoluble bubbles in the dough that become weak points where carbon dioxide collects during the subsequent fermentation step. In the Chorleywood bread process, the dough is mixed under partial vacuum so that the incorporated bubbles expand and are then split into many small ones as mixing continues, thus giving a fine-pored loaf crumb after baking.

Dough Fermentation

During the fermentation step, several processes happen simultaneously, and in order to produce a bread of the required quality characteristics, each of these processes must be optimized to that end.

Yeasts have been used to leaven bread for thousands of years, but only in comparatively recent times have pure cultures of the yeast *Saccharomyces cerevisiae* been added to the bread dough as a leavening agent. The commercial production of baker's yeast follows procedures similar to those used in the production of brewing, wine making, and distilling strains of this same species. Indeed, the baking industry was originally supplied with yeast waste from the brewing industry until about 1860 (Ponte and Tsen 1987). However, commercial production of yeast biomass specifically for the baking industry developed alongside an increasingly expanding manufacture of bread in commercial bakeries and the development of the technology that provided the great volumes required by the industry.

Commercial Production of Baker's Yeast

S. cerevisiae was originally produced commercially using grain mash as a growth substrate, but for economic reasons, it is now grown on sucrose-rich molasses, a by-product from the sugar cane or sugar beet refining industry. Nitrogen, phosphorous, and essential mineral ions such as magnesium are added to promote growth. The production of the yeast biomass for the baking industry is multistage and takes about 10–13 hours at 30°C. *S. cerevisiae* shows the Crabtree effect, as its metabolism favors fermentative metabolism at high levels of energy-giving substrate, thus resulting in a low production of biomass (Walker 1998a). To avoid this, molasses is added incrementally toward the end of the production of yeast biomass, and the mixture is vigorously aerated in order to promote respiration and avoid fermentative metabolism. At the end of the production, the yeast is allowed to "ripen" by aeration in the absence of nutrients. This step synchronizes the yeast cells into the stationary growth phase and also promotes an increase in the storage sugar trehalose in the cells, thus improving their viability and activity.

When the fermentation is complete, the amount of yeast is about 3.5–4% w/v. The biomass is separated and concentrated by centrifugation and filtering. The yeast cream is then processed into pressed yeast or is dried. The most usual types of commercial yeast preparations are (Stear 1990) the following:

- *Cream yeast* is a near liquid form of baker's yeast that must be kept at refrigerated temperatures. It may be added directly to the bakery product being made.
- *Compressed yeast* is formed by filtering cream yeast under pressure to give approximately 30% solids. It has a refrigerated shelf life of 3 weeks.
- *Active dry yeast* (ADY) is produced by extruding compressed yeast through a perforated steel plate. The resulting thin strands are dried and then broken into short lengths to give a free-flowing granular product after further drying. Depending on the subsequent treatment and packaging, ADY may have a shelf life of over a year. However, ADY requires rehydration before application in dough, and this can be a labor-intensive operation in a large bakery. The product rehydrates best using steam or in water with added sugar at 40°C. Rehydration in pure water promotes leaching of cell contents and a reduction in the activity of the yeast.
- *High activity dry yeast* (HADY) (instant ADY, IADY) is a similar product, where improved drying techniques are used to give a product with smaller particle size that does not need to be rehydrated before use and can therefore be incorporated directly into bread dough without prior treatment.

Desirable Properties of Baker's Yeast

Yeast plays a critically important role in leavened bread production, and over the decades of commercial production, strains have been selected that give improved performance. Desirable characteristics include the following:

- High CO_2 production during the dough fermentation due to high glycolytic rate.

- The ability to quickly commence maltose utilization when the glucose in the flour is depleted.
- The ability to store high concentrations of trehalose, which gives tolerance to freezing and to high sugar and salt concentrations.
- Tolerance to bread preservatives such as propionate.
- Viability and retained activity during various storage conditions.

In the future, strains will probably be developed with even more useful properties. In particular, the flavor-forming properties will receive special attention (Walker 1998b).

The Role of Yeast in Leavened Bread

When yeast is incorporated into the dough, conditions allow a resumption of metabolic activity, although there is little actual multiplication of the yeast during shorter bread-making processes such as the straight dough and Chorleywood processes (Fig. 31.2). The yeast has been produced under aerobic (respiratory) conditions and is therefore adapted to this metabolism, but conditions very quickly become anaerobic in bread dough since the oxygen incorporated in the dough is soon depleted. The sugars are metabolized to pyruvate by glycolysis; pyruvate is then decarboxylated to acetaldehyde, thus producing carbon dioxide; and then ethanol is formed by reduction of acetaldehyde by $NADH_2$ (Fig. 31.3). For each molecule of glucose (or half molecule of maltose) that is metabolized, two molecules each of ethanol and carbon dioxide are produced. This fermentative metabolism is the prevalent pathway in *S. cerevisiae* in dough due both to the absence of oxygen and to the nonlimiting supply of fermentable sugars (Maloney and Foy 2003).

The amount of maltose available is a complex interaction between the amount of damaged starch, the level of amylases in the flour and the stage, and length of the fermentation process. Maltose accumulates during the early stages of the fermentation because it is generated by amylase but is not metabolized by the yeast because the presence of glucose represses maltose utilization.

When readily fermentable sugars (glucose and fructose) are exhausted, the yeast shows a lag in fermentation and then turns its metabolism to the maltose produced from the action of β-amylase on starch. If sucrose has been added in the bread formulation (e.g., 4%), this is fermented in preference to maltose, and the lag in the fermentation may not be observed. High amounts of added sucrose (e.g., 20%) significantly retard fermentation due to the high osmotic stress on the yeast (Maloney and Foy 2003).

The products of yeast metabolism in dough fermentation vary considerably with pH. In bread, the pH is usually below 6.0, but above this, end products in addition to ethanol and CO_2 are formed, such as succinate, acetic acid, and glycerol, and less ethanol and CO_2 are formed. *S. cerevisiae* is also able to degrade proteins and lipids, and several flavor compounds are produced (Fig. 31.3).

It is generally not considered that the yeast fermentation is important for bread flavor and aroma development in traditional

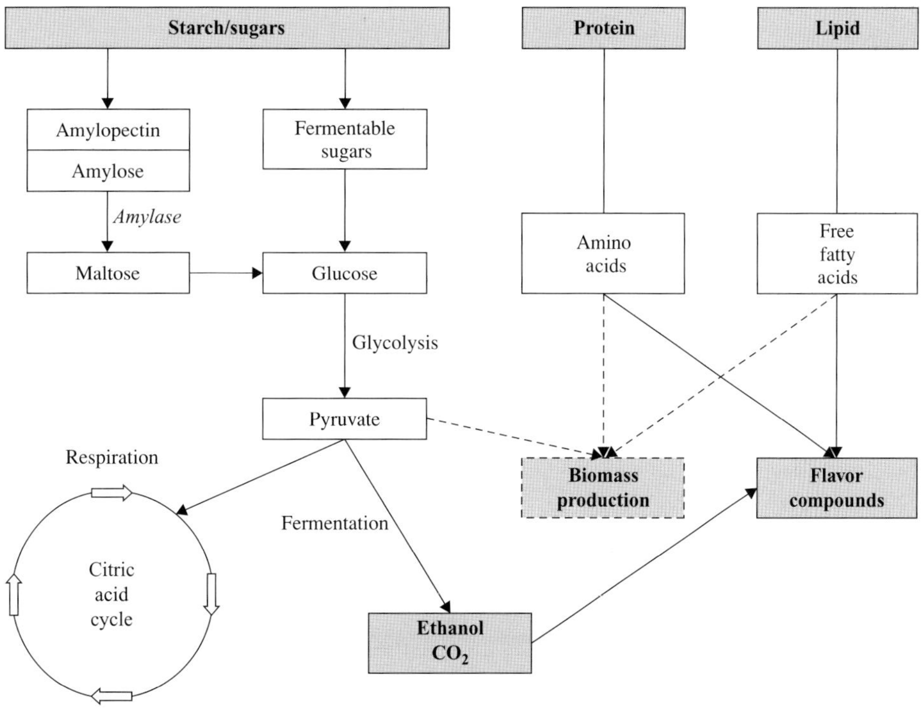

Figure 31.3. Biochemical changes during yeast fermentation of bread.

bread processes. However, the modern mechanical dough development processes, where the fermentation stage has been radically reduced, produce bread with a flavor that is inferior to that produced by the traditional straight dough process. This indicates that the yeast fermentation does make a positive contribution to bread flavor (Stear 1990). Zehentbauer and Grosch (1998) showed that yeast level and fermentation time and temperature affected aroma in the crust of baguettes, and they identified the flavor compounds 2-acetyl-1-pyrroline (roasty), methyl propanal and 2- and 3-methylbutanal (malty), and 1-octene-3-ol and (E)-2-nonenal (fatty). An increase in fermentation time allows for a development of flavor, but this trend is not really noticeable until much longer fermentation times are used, as in sourdough breads. There is not a clear borderline between regular bread and sourdough bread.

The production of ethanol and CO_2 is essential for the development of the desired bread crumb structure, and several factors affect both the development of the dough and its leavening (Fig. 31.4). During fermentation, some of the CO_2 is lost to the atmosphere, but most either collects in the small pockets of air incorporated during dough mixing or is dissolved in the dough's aqueous phase. The amount that can be dissolved in the aqueous phase is dependent on temperature, and is greater at lower temperatures. As the aqueous phase is already saturated with CO_2, it cannot escape from the bubbles by diffusion into the dough, so the bread begins to increase in volume. As the gas collects, the rheological properties of the dough allow it to expand in order to equalize the pressure that builds up. Ethanol reacts with the gluten to slightly soften it, allowing for easier expansion of the dough. It is important that CO_2 develop immediately after dough preparation and proceed at an adequate intensity. In addition, the dough must have the physical properties necessary to withstand dough manipulation and allow for gas retention, so that the optimal structure has been obtained for the final proof and baking (Stear 1990).

The Bread-Baking Process

When the bread has undergone the final proofing and is put in the oven, the outer surface rapidly starts to form the crust. A temperature gradient develops due to transfer of heat from the pan to the loaf, and if the loaf is to achieve optimal properties, then the heat of the oven and the state of the bread proof need to be synchronized (Stear 1990). Apart from the outer crust, no part of the bread ever becomes dry; therefore, despite oven temperatures of well over 200°C, the temperature in most of the loaf will not exceed 100°C. The primary rise in temperature increases the activity of the yeast, and its production of CO_2. At the same time, the solubility of CO_2 decreases, ethanol and water evaporate, and the gases increase in volume. This results in a marked increase in the volume of the dough, called "oven spring."

As the temperature in the loaf continues to rise, several other changes take place. The yeast is increasingly inhibited, and its enzymes are inactivated at about 65°C. The amylases in the dough are active until about 65–70°C is reached and a rapid increase in the amount of soluble carbohydrate takes place. Gelatinization of starch occurs at 55–65°C, and the water that this requires is taken from the gluten protein network, which then becomes more rigid, viscous, and elastic until a temperature is

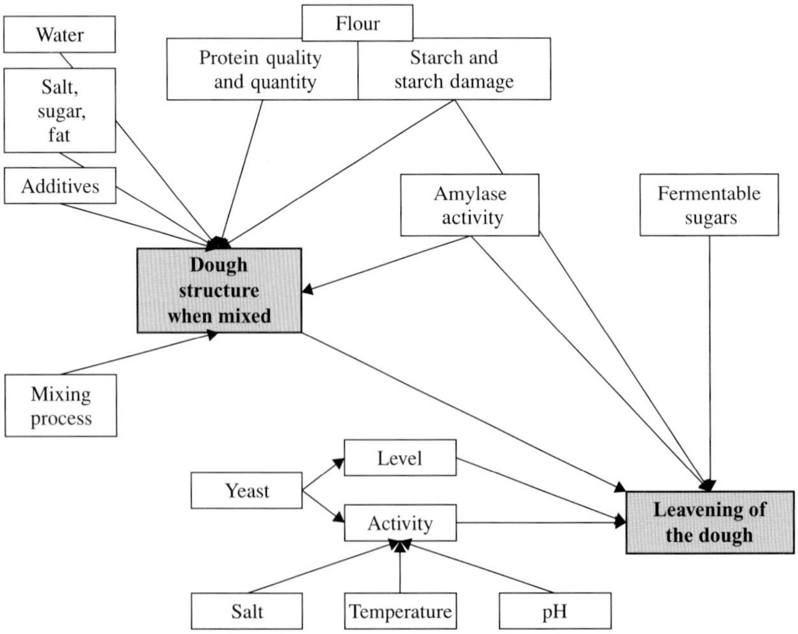

Figure 31.4. Important factors for bread leavening.

reached at which the protein begins to coagulate. At this stage, the structure of the dough has changed to a more rigid structure due to denatured protein and gelatinized starch. These changes first occur near the crust and gradually move into the crumb as the heat is transferred inward.

Toward the end of baking, the temperature at the crust is much higher than 100°C. The crust becomes brown, and aroma compounds, predominantly aldehydes and ketones, are formed, mainly from Maillard reactions. The formation of flavor compounds is two staged. First, compounds are formed from the fermentation itself, and then during baking some of these compounds may react with each other or with the bread components to form other flavor compounds. Some other flavor compounds formed during the fermentation may be lost due to the high temperature, but those that remain gradually diffuse into the crumb after cooling. On further storage, the levels of flavor compounds decrease due to volatilization. Over 200 different flavor and aroma compounds have been identified in bread (Stear 1990).

Staling

The two main components of staling (firming of the bread crumb) are loss of moisture, mainly due to migration of moisture from the crumb to the crust (which becomes soft and leathery), and the retrogradation of starch. The major chemical change that occurs during staling is starch retrogradation, but a redistribution of water between the starch and the gluten also has been proposed. The gelatinization of the starch that occurs during cooking or baking gradually reverses, and the starch molecules form intermolecular bonds and crystallize, expelling water molecules and resulting in the firming of crumb texture. Starch retrogradation is a time- and temperature-dependent process and proceeds fastest at low temperatures, just above freezing point. Since the rearrangement of the starch molecules is facilitated by a high water activity, staling is retarded by the addition of ingredients that lower water activity (e.g., salt and sugar) or bind water (e.g., hydrocolloids and proteins). The staling rate can also be slowed by the incorporation of surfactants, shortening, or heat-stable α-amylase. Freezing of baked goods also retards staling since the water activity is drastically lowered. Much of the firming of the loaf during cooling is due to retrogradation of the amylose whereas the slower reaction of staling is due, in addition, to retrogradation of amylopectin (Hoseney 1998, Stear 1990).

SOURDOUGH BREAD

When cereal flour does not contain gluten, it is not suitable for production of leavened bread in the manner described above. However, if rye flour, which is very low in gluten proteins, is mixed with water and incubated at 25–30°C for a day or two, there is a good possibility that first step of sourdough production will be started. This mixture will regularly develop fermentation with lactic acid bacteria (LAB) and yeasts. This forms the basis of sourdough production, and this low-pH dough is able to leaven.

The use of cereal flour and water as a basis for spontaneous or directed fermentation products is common in many countries. In Africa, fermented porridge and gruel as well as their diluted thirst quenchers, are the main products of these natural fermentations, whereas Europeans and Americans and their descendants enjoy a variety of sourdough breads. In all these areas beers are also produced.

This great variety of fermented products has an historic prototype in the earliest reported leavened breads in Egypt about 1500 BC. Considering the simplicity of the process and the ease with which it succeeds, it has been suggested that peoples in several places must have shared this experience independently. It may be surmised that the experience with gruels and porridge preceded the idea of making bread.

Common bread fermented only with yeast appeared later in our history, and it was a staple food in the Roman Empire. This also indicates that the Romans had wheat with sufficient gluten potential.

It is possible to make leavened bread without gluten using sourdough, and this bread has become a favorite among many peoples (Hammes and Gänzle 1998, Wood 2000).

Advantages of Making Sourdough Bread

- Sourdough bread does not have to contain high levels of gluten for successful leavening.
- Low pH inhibits amylase, and thereby, degradation of starch is avoided.
- Sourdough improves the water-binding capacity of starch and the swelling and solubility of pentosans.
- Sourdough bread has very good keeping quality and an excellent safety potential.
- Less costly cereal flours can be used.
- A different variety of flavor and taste attributes can be offered.
- Sourdough bread can nutritionally compete with regular bread.
- Phytic acid is degraded by phytase in flour and from lactic acid bacteria. This improves the availability of iron and other minerals.
- Bread volume is increased, crumb quality is improved, and staling is delayed.

Rye flour is very low in gluten proteins, and instead, starch and pentosans make an important contribution to bread structure. The swelling and solubility of pentosans increase when LAB fermentation lowers pH. Gelatinization of starch occurs at about 55–58°C. Considering that the flour amylase has a temperature optimum around 50–52°C, it is crucial that the amylase is actually inactivated in the pH range that is obtained during sourdough fermentation. When mixtures of wheat and rye flour are used for bread making, a sourdough process is necessary if the content of rye flour exceeds 20%.

Rye and wheat flour contain phytic acid that binds minerals, particularly iron, that then become nutritionally unavailable. However, these cereals also contain phytases with pH optima around 5.0–5.5; thus, phytate degradation is very good in fermented flour, where these phytate complexes are also more soluble. Lactic acid bacteria also appear to have some phytase activity.

As wheat flour is able to form gluten, some of the considerations about amylases and starch are not equally relevant when baking wheat bread. Nevertheless, wheat flour is often used in sourdough bread.

However, preferred qualities like improved keeping and safety potential as well as the increased variety of flavors appeal to many consumers. These desirable qualities are also praised because they represent an alternative natural preservation method (Gobbetti 1998, Hammes and Gänzle 1998, Wood 2000).

Microbiology of Sourdough

An established, "natural" sourdough is dominated by a few representatives of some bacteria and yeast species. This results from the selective ecological pressures exerted in the (rye) flour-water environment. Rye flour is an appropriate choice for this mixture because leavening of the dough is dependent on sourdough development. At the start of fermentation, a 50:50 (w/w) rye flour–water mixture at 25–30°C will harbor approximately the following:

Mesophilic microorganisms, aerobes	10^3–10^7 cfu/g
Lactic acid bacteria	<10 to 5×10^2 cfu/g
Yeast	10–10^3 cfu/g
Molds	10^2 to 5×10^4 cfu/g

Among the mesophiles at the start, members of the Enterobacteriaceae dominate. Microorganisms dominating in the sourdough after 1–2 days at 25–30°C are as follows:

Lactic acid bacteria	10^9 cfu/g
Yeast	10^6 to 5×10^7 cfu/g

Some important properties of the rye flour–water environment determine that certain LAB and yeasts will compete most favorably. Lactobacilli have a superior ability to ferment maltose, they thrive despite limited iron due to the presence of phytic acid, and they are able to grow at about pH 5.0 and lower.

Reports show that different LAB may be isolated from sourdoughs; however, *Lactobacillus sanfrancis censis* has been found most often. Table 31.2 presents some of the other *Lactobacillus* species that have been isolated from sourdough. The selection of yeasts may be even narrower, with *Candida milleri* often cited (Table 31.2). Several other species are isolated occasionally (Spicher 1983, Gobbetti and Corsetti 1997, Hammes and Gänzle 1998, Martinez-Anaya 2003, Stolz 2003).

Starters

It is traditional bakers' practice to maintain a good sourdough over time by regular transfer, for example, every 8 hours. This is called "rebuilding." Such established cultures are referred to as Type I sourdoughs (Type I process), and a three-stage fermentation procedure is considered necessary to obtain an optimal sourdough. Each step is defined by specific dough yield, temperature, and incubation time.

Dough yield is defined as:

$$\frac{\text{(Flour + Water) by weight}}{\text{Flour by weight}} = \times 100 \text{ Dough yield}$$

A high dough yield implies that a relatively large amount of water is used to make the dough; such a dough would conform with certain requirements in industrial production when there is a need to pump the dough.

The lactobacilli and yeasts in Table 31.2 are all common in sourdoughs; however, the composition of starter cultures for Type I processes have been continuously stably maintained for many years, and may be compared to certain mixed cultures in dairy technology. The LAB and yeasts in such cultures will be particularly well adjusted and adapted for the conditions in sourdoughs. In Germany, established natural sourdough starter cultures with a stable composition of *Lb. sanfranciscensis* and *Candida milleri* have been propagated for decades; they are marketed as "Reinzuchtsauerteig." When the sourdough process has been started through a one-stage or a three-stage procedure, a part of the optimized dough is withdrawn to start up the sourdough production for the next day. These sourdoughs are mentioned in different languages as Anstellgut (German), mother sponge (English), chef (French), masa madre (Spanish), madre (Italian).

Sourdoughs in Type II processes are used mainly for enhancing the flavor and taste of the regular bread. Addition of baker's yeast is required for efficient leavening. Type I sourdoughs are good alternatives as starters for the production of Type II sourdoughs. At the start of a production period, industrially large quantities of Type II sourdoughs may be stocked for portionwise use over time.

Sourdoughs for Type III processes are dried sourdough preparations.

Table 31.2. Common Representatives of Lactic Acid Bacteria and Yeasts Isolated from Mature Sourdoughs

	Heterofermentative	Homofermentative
Lactic acid bacteria	*Lactobacillus sanfranciscencis* (formerly *Lb.sanfrancisco, Lb. brevis* subsp. *lindneri*) *Lb. brevis, Lb. fructivorans, Lb. fermentum, Lb. pontis, Lb. sakei* (formerly *Lb. bavaricus, Lb. reuteri*)	*Lb. plantarum* *Lb. delbrueckii*
Yeasts	*Candida milleri, C. krusei, Saccharomyces cerevisiae, S. exiguus* *Torulopsis holmii, T. candida*	

Defined cultures have also been marketed; however, the suitability of Type I and Type II cultures appears to outcompete the alternatives offered. In addition to baker's yeast, the collection of defined cultures comprise at least pure cultures of *Lb. brevis* (heterofermentative) and *Lb. delbrueckii* and *Lb. plantarum* (homofermentative). The homofermentative cultures produce mainly lactic acid under anaerobic conditions, whereas the heterofermentative cultures will also produce acetic acid or ethanol and carbon dioxide. By controlling, if possible, the contributions from these different cultures, the relative amounts of acetic and lactic acids may be regulated. This important relationship:

$$\frac{\text{Lactic acid (Mole)}}{\text{Acetic acid (Mole)}}$$

is called the fermentation quotient (FQ).

Relatively mild acidity will have an FQ about 4–9, whereas a more strongly flavored rye bread, as produced in Germany, requires a much lower FQ, for example, 1.5–4.0 (Spicher 1983, Gobbetti and Corsetti 1997, Hammes and Gänzle 1998, Martinez-Anaya 2003, Stolz 2003).

Sourdough Processes

Several more or less traditional sourdough processes are practiced on a large scale in present-day bakery industry. (Fig. 31.5) One line comprises processes designed for baking with rye flour. They may be the Type I processes mentioned above. The Berliner short-sour process, the Detmolder one-stage process, and the Lönner one-stage process are typical for central and northern Europe. In every case, the process is initiated by a starter culture, 2–20% (often, 9–10%), in a rye flour–water (close to 50:50 w/w) mixture that is incubated for 3–24 hours at 20–35°C, depending on the process. When the resulting sourdough is ready, bread making starts with an "inoculation" of about 30% sourdough together with rye flour, wheat flour, baker's yeast, and salt. The final rye:wheat ratio is regularly 70:30 w/w. Dough yield is adjusted to satisfy handling (e.g., pumping) and microbial nutrition requirements.

Another line of sourdough processes comprises those for baking with wheat only. The San Francisco sourdough process is often mentioned in this connection; however, Italian wheat sourdough products, for example, Panettone, Colomba, and Pandoro, are also very important (Cauvain 1998, Gobbetti and Corsetti 1997, Spicher 1983, Wood 2000).

The starter for San Francisco sourdough is ideally rebuilt every 8 hours to maintain maximum activity. However, a mature sponge may be kept refrigerated for days with acceptable performance. For rebuilding the starter, sponge (40%) is mixed with high-gluten wheat flour (40%) and water (about 20%) to ferment at bakery temperature. Final bread making requires a dough consisting of the ripe sourdough (9.2%), regular wheat flour (45.7%), water (44.2%), and salt (0.9%). Proofing for 7 hours follows, during which the pH decreases from about 5.3 to about 3.9. A sourdough starter culture for San Francisco French bread production commonly contains *Lb. sanfranciscensis* and *Candida holmii*.

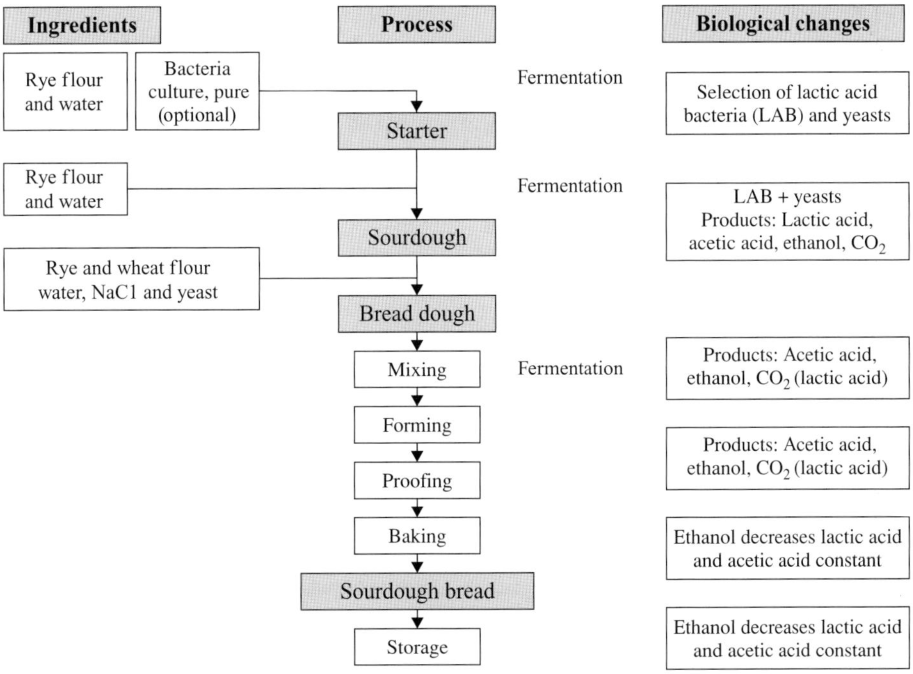

Figure 31.5. The sourdough process.

Selection and Biochemistry of Microorganisms in Sourdough

Sourdough breads based on rye or wheat flour or their mixtures enjoys a remarkable standing in many societies, either as established, traditional products or as "innovative" developments for more natural products and a wider choice of flavors. Well-functioning and popular sourdough starters that have been maintained by simple rebuilding for decades, and the reestablishment of the "same" stable starter over and over again from a constant quality flour, are both expressions of stable ecological conditions. The simplicity of the procedures may be somewhat deceiving with respect to the actual complexity of these biological systems. In the following sections, metabolic events and biochemical aspects of sourdough fermentation will be discussed. Attention will be drawn to some of the more clear points about the metabolic events and other biochemical facts. The near future should bring us closer to a comprehensive understanding.

Carbohydrate Metabolism

The development of a mixture (1:1) of rye flour or even wheat flour with water incubated some hours at 25–30°C will almost inevitably lead to a microbiological population consisting of lactic acid bacteria (LAB) and yeasts. It may need rebuilding several times in order to stabilize it, but from then on the composition of the microflora may be constant for years, provided the composition of the flour and the conditions for growth are not changed much. Representative LAB and yeasts have been presented in Table 31.2. These microorganisms have certain characteristics in common. First, the selected LAB are very efficient maltose fermenters, a prime reason why they competed so well in the first place. Several lactobacilli in sourdoughs, e.g. *Lb. sanfranciscensis, Lb. pontis, Lb. reuteri*, and *Lb. fermentum*, harbor a key enzyme, maltose phosphorylase, which cleaves maltose (the phosphorolytic reaction) to glucose-1-phosphate and glucose. Glucose-1-phosphate is metabolized heterofermentatively via the phosphogluconate pathway, while glucose is excreted into the growth medium. Glucose repression has not been observed with these lactobacilli. Most of the yeast species identified in sourdoughs are, per se, maltose negative, and will thus prefer to take up glucose when it is available. Other microorganisms may experience glucose repression of the maltose enzymes, to the benefit of the sourdough lactobacilli. Among the yeasts, *S. cerevisiae*, which is maltose positive and transports maltose and hexoses very efficiently, cannot take up maltose due to glucose repression and will, as a consequence, be defeated from the sourdough flora. *S. cerevisiae* as baker's yeast is, however, used at the bread-making stage, but as an addition in the recipe. Additional yeast cells may also be necessary for fast and efficient CO_2 production, because the yeasts are relatively sensitive to acids, particularly to acetic acid, which is excreted by the heterofermentative lactobacilli that often dominate the LAB flora of the sourdough. *Candida milleri* (syn. *S. exiguus, Torulopsis holmii*) is common in sourdoughs for San Francisco French bread. This yeast tolerates the acetic acid from heterolactic fermentation and thrives on glucose and sucrose in preference to maltose; it thus appears to be a near ideal partner for *Lb. sanfranciscensis* (Gobbetti and Corsetti 1997, Gobbetti 1998, Wood 2000, Hammes and Gänzle 1998).

Wheat and rye flour contain mainly maltose as a readily available carbohydrate, although rye flour has greater amylase activity and therefore has a greater potential for release of maltose. Early work in the United States on *Lb. sanfranciscensis* indicated that this organism would only ferment maltose (Kline and Sugihara 1971). However, strains isolated in Europe appeared more diversified, and some of them would ferment up to eight different sugars (Hammes and Gänzle 1998). Utilization of maltose by *Lb. sanfranciscensis, Lb. pontis, Lb. reuteri*, and *Lb. fermentum* through phosphorolytic cleavage with maltose phosphorylase is energetically very favorable (Stolz et al. 1993) and shows increased cell yield and excretion of glucose when maltose is available. In these conditions the cells have very low levels of hexokinase.

Co-metabolism

Lactobacilli in sourdough production are not only specialized for maltose fermentation they also exploit co-fermentations for optimized energy yield (Gobbetti and Corsetti 1997, Hammes and Gänzle 1998, Stolz et al. 1995, Romano et al. 1987). *Lactobacillus sanfranciscensis, Lb. pontis* and *Lb. fermentum* all have mannitol dehydrogenase. Thus, fructose may be used as an electron acceptor for the reoxidation of NADH in maltose or glucose metabolism, and then acetylphosphate may react on acetate kinase to yield ATP and acetate (Axelsson 1993). The lactobacilli gain energetically and more acetic acid may contribute to the desirable taste and flavor of bread. In practical terms addition of fructose is used to increase acetate in the products, that is, lower FQ (Spicher 1983). Comparable regulation of acetate production may be achieved by providing citrate, malate or oxygen as electron acceptors, resulting in products like succinate, glycerol and acetate (Gobbetti and Corsetti 1996, Condon 1987, Stolz et al. 1993).

Proteolysis and Amino Compounds

In a sourdough, the flour contributes considerable amounts of amino acids and peptides; however, in order to satisfy nutritional requirements of growing LAB and provide sufficient amino compounds, precursors, for flavor development, some proteolytic action is necessary. The LAB have been suspected as the main contributors of proteinase and peptidase activities for release of amino acids in sourdoughs (Spicher and Nierle 1984, Spicher and Nierle 1988, Gobbetti et al. 1996), although the flour enzymes may also have considerable input (Hammes and Gänzle 1998). In addition, lysis of microbial cells, particularly yeast cells, add to the pool of amino acids; a stimulant peptide containing aspartic acid, cysteine, glutamic acid, glycine, and lysine that appears in the autolytic process of *C. milleri* has also been identified (Berg et al. 1981). *Lactobacillus sanfranciscensis* has been found to have a regime of intracellular peptidases, endopeptidase, and proteinase, as well as a dipeptidase and proteinase in the cell envelope (Gobbetti et al. 1996). Limited autolysis of lactobacillus populations in sourdoughs may add to the repertoire of enzymes that will release amino acids from flour proteins, including those from proline-rich gluten in wheat. Some of the enzymes have been purified for further characterization (Gobbetti et al. 1996), and they express interesting activity levels at sourdough pH and temperatures.

The addition of exogenous microbial glucose oxidase, lipase, endoxylanase, α-amylase, or protease in the production of sourdough with 11 different LAB cultures showed positive effects on acidification rate and level for only three cultures, one *Leuconostoc citreum*, one *Lactococcus lactis* subsp. *lactis* and one *Lb. hilgardii*. *Lactobacillus hilgardii* with lipase, endoxylanase or α-amylase showed increased production of acetic acid. *Lactobacillus hilgardii* interacted with the different enzymes for higher stability and softening of doughs (Di Cagno et al. 2003).

Recent work with *Lb. sanfranciscensis, Lb. brevis*, and *Lb. alimentarius* in model sourdough fermentations showed, by using two-dimensional electrophoresis, that 37–42 polypeptides had been hydrolyzed. The polypeptides varied over wide ranges of pIs and molecular masses, and they originated from albumin, globulin, and gliadin, but not from glutenin. Free amino acid concentrations increased, in particular those of proline and glutamic and aspartic acid. Proteolysis by the lactobacilli had a positive effect on the softening of the dough. A toxic peptide for celiac patients, A-gliadin fragment 31–43, was degraded by enzymes from lactobacilli. The agglutination of human myelogenous leukemia–derived cells (K562) by toxic peptic-tryptic digest of gliadins was abolished by enzymes from lactobacilli (Di Cagno et al. 2002).

Volatile Compounds and Carbon Dioxide

Both yeasts and LAB contribute to CO_2 production in sourdough products, but the importance of the two varies. In bread production with only the (natural) sourdough microflora, the input from LAB may even be decisive for leavening because the counts and kinds of yeast may not be optimal for gas production. Relatively low temperature (e.g., 25°C) and low dough yield (e.g., 135) would select for LAB activities and less yeast metabolism. More complete volatile profiles were obtained at higher temperatures (e.g., 30°C) and with a more fluid dough. Of course, increasing the leavening time may give substantially richer volatile profiles (Gobbetti et al. 1995). If baker's yeast, *S. cerevisiae*, is added to optimize and speed up the production process, the contribution from yeasts will dominate (Gobbetti 1998, Hammes and Gänzle 1998).

Bread made with chemical acidification without fermentation starter failed in sensory analysis. This indicates that fermentation with yeasts and LAB is important for good flavor, although high quality raw materials and proofing and baking are also decisive factors. Flavor compounds distinguishing the different metabolic contributions in sourdough are as follows (Gobbetti 1998):

- *Yeast fermentation (alcoholic)*: 2-methyl-1-propanol, 2,3-methyl-1-butanol.
- *LAB homofermentative*: diacetyl, other carbonyls.
- *LAB heterofermentative*: ethyl acetate, other alcohols and carbonyls.

Antimicrobial Compounds from Sourdough LAB

The primary antimicrobial compounds produced by sourdough LAB are lactic and acetic acid, diacetyl, hydrogen peroxide, carbon dioxide, and ethanol, and among these, the two organic acids continue to be the most important contributions for beneficial effects in fermentations.

Researchers in the field, of course, also consider and test possibilities that LAB may produce bacteriocins and other antimicrobials. Thus antifungal compounds from *Lb. plantarum* 21B have been identified, for example, phenyl lactic acid and 4-hydroxyphenyl lactic acid (Lavermicocca et al. 2000). Caproic acid from *Lb. sanfranciscensis* also has some antifungal activity (Corsetti et al. 1998).

A real broad-spectrum antimicrobial from *Lb. reuteri* is reuterin (β-hydroxypropionic aldehyde), which comes as a monomer and a cyclic dimer (El-Ziney et al. 2000). Reutericyclin, which was isolated from *Lb. reuteri* LTH2584 after the screening of 65 lactobacilli, is a tetramic acid derivative. Reutericyclin inhibited Gram-positive bacteria (e.g., *Lactobacillus* spp., *Bacillus subtilis, B. cereus, Enterococcus faecalis, Staphylococcus aureus*, and *Listeria innocua*), and it was bactericidal towards *B. subtilis, S. aureus*, and *Lb. sanfranciscensis*. The ability to produce reutericyclin was stable in sourdough fermentations over a period of several years. Reutericyclin produced in sourdough was also active in the dough (Gänzle et al. 2000; Höltzel et al. 2000; Gänzle and Vogel 2003).

A few bacteriocins or bacteriocin-like compounds have also been identified, isolated, and characterized (Messens and De Vuyst 2002). Bavaricin A from *Lb. sakei* MI401 was selected by screening 335 LAB strains, including 58 positive strains (Larsen et al. 1993). Bavaricin A (and Bavaricin MN from *Lb. sakei* MN) have the *N*-terminal consensus motif of bacteriocin class IIA in common, comprise 41 and 42 amino acids, respectively, and have interesting sequence homologies and similar hydrophobic regions. Bavaricin A inhibits *Listeria* strains and some other Gram-positive bacteria but not *Bacillus* or *Staphylococcus* (Larsen et al. 1993, Kaiser and Montville 1996). Plantaricin ST31 is produced by *Lb. plantarum* ST; it contains 20 amino acids, and the activity spectrum includes several Gram-positive bacteria but not *Listeria* (Todorov et al. 1999). A bacteriocin-like compound, BLIS C57 from Lb. *sanfranciscensis* C57, was detected after screening 232 *Lactobacillus* isolates, including 52 strains expressing antimicrobial activity. BLIS C57 inhibits Gram-positive bacteria including bacilli and *Listeria* strains (Corsetti et al. 1996).

TRADITIONAL FERMENTED CEREAL PRODUCTS

Only two cereals, wheat and rye, contain gluten and are thereby suitable for the production of leavened bread, but many other food cereals are grown in the world. On a global basis, a great proportion of cereals are consumed as spontaneously fermented products, in particular in Africa, Asia, and Latin America. Most fermented cereals are dominated by lactic acid bacteria (LAB), and the microflora associated with the grains, flour, or any other ingredient, together with contamination from water, food-making equipment, and the producers themselves, represent the initial fermentation flora. Malted flour is also an important source of microorganisms. The changes that take place during the fermentation are due to both the metabolism of the microorganisms present and the activity of enzymes in the cereal, and these are in turn affected by the great variety of technologies that are used. The technology may be simple, involving little more than a mixing of flour with water and allowing it to ferment, or it may be extremely complex and involve many steps with obscure roles. Indigenous fermented foods are usually based upon raw materials that have a sustainable production in their country of origin and are therefore attracting increasing interest from researchers—both within pure and applied food science and also in anthropology. These ancient technologies often have deep roots in the culture of a country, and there is increasing awareness of the importance of preserving these traditional foods. Many products are not yet described in the literature, and knowledge of them is in danger of disappearing. It is therefore necessary to document the technologies used and to identify the fermenting organisms and the metabolic changes that are essential for the characteristics of the product. It is, however, often difficult to describe the sensory attributes of a product that is inherently variable.

In Africa, as much as 77% of the total caloric consumption is provided by cereals, of which rice, maize, sorghum, and millet are most important. Cereals are also significant sources of protein. Most of the cereal foods consumed in Africa are traditional fermented products and are very important both as weaning foods and as staple foods and beverages for adults. In Asia, many products are based on rice, and maize is most widely utilized in Latin America (FAO 1999).

Indigenous fermented cereals can be classified according to raw material, type of fermentation, technology used, product usage, or geographical location. They can range from quite solid products such as baked flat breads to sour, sometimes mildly alcoholic, refreshing beverages.

Many factors have an influence on the characteristics of an indigenous product (Fig. 31.6). The choice of raw material may be primarily influenced by price and availability rather than by preference.

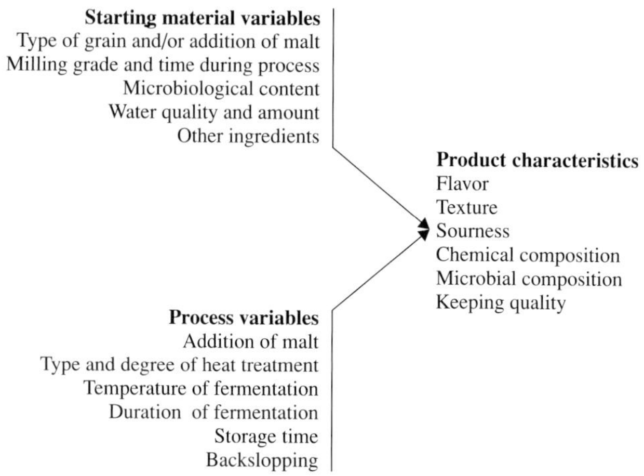

Figure 31.6. Important factors determining the characteristics of spontaneously fermented cereal products.

For instance Togwa, a Tanzanian fermented beverage, may be made from maize in the inland areas of Morogoro and Iringa, but from sorghum in the coastal areas of Dar es Salaam and Zanzibar (Mugula 2001). Similarly, the Ethiopian product borde may be made from several different grains according to availability—sorghum, maize, millet, barley and also the Ethiopian cereal tef (Abegaz et al. 2002). The use of different grains obviously affects the sensory characteristics of a product, and yet it may have the same name throughout the country. Some fermented cereal products also contain other ingredients. Idli is a leavened steamed cake made primarily from rice to which black gram dahl is added. This not only improves the nutritional quality, but in addition, the black gram imparts a viscosity, apparently specific for this legume, which may aid air entrapment during fermentation and thereby lighten the texture of the product (Soni and Sandu 1990). However, on a broader basis, the addition of legumes such as soybean flour to fermented cereals has been suggested as an economically feasible way to generally improve the nutritional quality of cereal foods.

Some fermented cereal products are made using unmalted grain, with no extra addition of amylase, but they tend to either be very thick or of low nutritional density. Malted flour is added to many indigenous fermented cereals, a traditional technology that has far-reaching effects on several product characteristics. The addition of malt provides amylases (in particular, α-amylase) that hydrolyze the starch, sweeten the product, and also cause a considerable decrease in viscosity of the product after heat treatment. The malting process, the germination of grain following steeping in water, is associated with colossal microbiological proliferation, and the organisms that develop during malting are a source of fermenting organisms. Many Asian products, for example koji, a Japanese fermented cereal or soybean product, are first inoculated with a fungus, as a source of amylase, in order to liberate fermentable sugars from the cereal starch (Lotong 1998).

Many fermented cereals are multipurpose. A single product may be prepared in varying thicknesses and used as a fermented gruel for both adults and children, or it may be watered down and used as a fermented thirst-quenching beverage. As Wood (1994) remarked, the latter type of product makes a meaningful contribution to nutrition; the potential of their replacement by cola-type beverages would result in a serious negative impact on the nutrition of people in developing countries.

The use of fermented cereals as weaning foods in developing countries raises several important issues. Unfermented gruels deteriorate very rapidly in unhygienic conditions, especially if refrigeration is not available. They then represent a significant source of foodborne infections that annually claim the lives of millions of young children (Adams 1998). Fermented malted cereal gruels have been shown on the whole to contain low numbers of pathogenic organisms since these are inhibited and killed by the low pH that rapidly develops in the product.

Fermented cereals are therefore usually regarded as safer than their unfermented counterparts (Nout and Motarjemi 1997). A weaning food made from unmalted cereals may be a cause of malnutrition because its thick viscosity limits the nutritional intake of a small child. Addition of malted flour decreases the viscosity so that more food can be ingested. If the fermentation flora includes yeasts in addition to LAB, a measurable reduction of carbohydrate will occur due to the production of CO_2 and other volatile compounds (Muyanja 2001). Analysis of fermented cereal products therefore shows that the protein:carbohydrate ratio is improved during the fermentation, and this obviously has nutritional benefits.

Milling of cereals into flour is usually done prior to fermentation, but in some products, for example, borde (Ethiopia), wet milling is used. This technique can be used when mechanical grain mills are not available and if the product is required to be smooth and without bits of suspended bran. The starch is also liberated from the grain more thoroughly when slurried with water and sieved than if it has been previously dry milled (Abegaz 2002).

A heat treatment step is found at some point in the production technology of most fermented cereal products and may involve boiling, steaming, or roasting. The type of heating employed is likely to have an effect on the flavor of the product, certainly if the temperature attained is sufficient to promote Maillard reactions. The heat also gelatinizes the starch, making it more susceptible to amylolytic enzymes, thus providing greater amounts of fermentable carbohydrates. However, at the same time, most of the natural contaminating (and potentially fermenting) flora and cereal enzymes are destroyed. Such products are also prone to contamination after the heat treatment step, and are thereby potentially unsafe should pathogenic organisms grow during the subsequent fermentation. The traditional solution to this is to use "backslopping," the addition of some of a previous batch of the product, and/or the addition of malted flour. Regular backslopping results in a selection of acid-tolerant organisms and functions as an empirical starter culture.

Fermentation usually takes place at ambient temperatures, and this may cause seasonal variations in products due to selection of different microorganisms at different temperatures. The duration of fermentation is largely a matter of personal choice, based on expected sensory attributes. Heat treatment after fermentation makes for a safer product, but it has the disadvantage of change of taste or loss of volatile flavor and aroma compounds.

The Microflora of Spontaneously Fermented Cereals

Spontaneously acid-fermented cereal products may contain a variety of microorganisms, but the flora in the final product is generally dominated by acid-tolerant LAB. Yeasts are also invariably present in large numbers when the fermentation is prolonged. A typical fermented cereal product contains approximately 10^9 and 10^7 cfu/g of product, of LAB and yeasts, respectively. However, since yeast cells are considerably larger than bacteria cells, their metabolic contribution to product characteristics is likely to be just as important as that of the LAB. The buffer capacity of cereal slurries is low, and the pH therefore drops quickly as acid is produced. Pathogenic organisms are inhibited by a fast acid production, so the addition of starter cultures, either as a pure culture or by "backslopping" promotes acid production and contributes to the safety of the fermented product (Nout et al. 1989).

The potential and the need for upgrading traditional fermentation technologies have initiated considerable research (Holzapfel 2002). In some recent studies of spontaneously fermented cereals, the LAB and yeasts have been isolated and identified as a first stage towards developing starter cultures for small-scale production of traditional fermented cereals. Muyanja et al. (2003) recorded that bushera, a traditional Ugandan fermented sorghum beverage that contains high numbers of LAB, was usually consumed by children after one day of fermentation as "sweet bushera." After 2–4 days, the product became sour and alcoholic and was consumed by adults. However, the sweet bushera showed very high counts of coliforms and had a reputation for causing diarrhea (Muyanja 2001). Clearly, the development of defined starter cultures would improve the safety of this and similar products.

Some recent examples of studies on the microbial flora of spontaneously fermented cereals are shown in Table 31.3. For each product, several different types of organisms have been isolated. In other words, a specific product is not produced from fermentation by a specific organism or organisms. *Lb. plantarum* seems to be the most commonly isolated *Lactobacillus* species in fermented cereals. In addition, heterofermentative LAB such as leuconostocs, *Lb. brevis*, and *Lb. fermentum* frequently occur. Yeasts are always present in spontaneously fermented products, but few studies have characterized the predominating species. However, Jespersen (2003) reported that *S. cerevisciae* is the predominant yeast in many African fermented foods and beverages.

All spontaneously fermented products contain, or have contained, many different types of microorganisms. These have grown in the product and will have metabolized some of the cereal components, thereby making a contribution (positive or negative) with their metabolites to the overall sensory characteristics of that product. However, studies on spontaneously fermented products have focused on LAB and yeasts since these organisms are often associated with other, better known, fermented products and have a history of safe use in food. Stanton (1998) proposed that the nature of the substrate (raw material) and the technology used to produce fermented foods are the predominating factors that determine the development of microorganisms and, thereby, the properties of a product.

Desirable Properties of the Fermenting Microflora

The most important property of a starter culture for a fermented cereal is the ability to quickly produce copious amounts of lactic acid in order to achieve a rapid decline in pH and retard the growth of pathogens and other undesirable organisms. Some workers (Sanni et al. 2002) have sought amylolytic LAB strains, as this could remove the need for using the highly contaminated

Table 31.3. Lactic Acid Bacteria and Yeasts Isolated from Some Traditional Fermented Cereal Products

Product (Country of Origin)	Cereal Basis	Most Prevalent		
		Species of LAB	Species of Yeast	Reference
Togwa (Tanzania)	Various	*Lb. brevis, Lb. cellobiosus, Lb. Plantarum* *W. confusa* *P. pentosaceus*	*P. orientalis* *S. cerevisiae* *C. pelliculosa* *C. tropicalis*	Mugula et al. 2003
Bushera (Uganda)	Sorghum, millet	*Lb. brevis; E. faecium* *Ln. mesenteroides* subsp. *mesenteroides*	NR	Muyanja et al. 2003
Ogi (Nigeria)	Maize	*Lb. reuteri, Lb. Leichmanii, Lb. plantarum, Lb. casei, Lb. fermentum,*	*S. cereviseae and Candida mycoderma*	Ogunbanwo et al. 2003
		Lb. brevis, Lb. alimentarius, Lb. buchneri, Lb. jensenii.		Odunfa 1985
Pozol (Mexico)	Maize	*Streptococcus* spp., *Lb. plantarum, Lb. fermentum*		Omar and Ampe 2000
		Lb. brevis *Ln. mesenteroides* *Lc. lactis, Lc. raffinolactis*	*C. mycoderma, S. cerevisiae, Rhodotorula* spp.	Nuraida and others 1995
Bordc (Ethiopia)	Various	*Lb. brevis* *W. confusa* *P. pentosaceus*	NR	Abegaz 2002
Idli (India)	Rice and blackgram beans	*Leuconostocs spp.* *Enterococcus faecalis*	*Saccharomyces* spp.	Soni and Sandu 1990

Note: Lb., Lactobacillus; Ln., Leuconostoc; L., Lactococcus; P., Pediococcus, E., Enterococcus; W., Weisella; S., Saccharomyces; I., Isa; NR, Not recorded.

malted flour in a product. The starter should also be able to hydrolyze the cereal protein in order to obtain the amino acids sufficient for rapid growth, and it should produce desirable and product-typical aroma and flavor compounds, but not off-flavors. Some products are characterized by a foaming consistency, and heterofermentative organisms (LAB or yeasts) are required for this property. Bacteriocin-producing strains have also been sought (Holzapfel 2002) in an attempt to increase the microbiological safety of the products. Starter cultures must also be commercially propagable and be able to survive preservation methods without loss of viability, activity, or metabolic traits.

Microbiological and Biochemical Changes in Traditional Fermented Cereals

Few studies have been made on the biochemical changes that take place in traditional fermented cereals. Mugula et al. (2003) analyzed samples of naturally fermented togwa made from sorghum and maize, to which togwa was backslopped and malt was added. The development of groups of microorganisms, organic acids, soluble carbohydrates, and volatile components was studied during the 24-hour fermentation. Maltose and glucose increased during the first part of the fermentation due to the action of cereal amylases, but later were reduced as the growth of LAB and yeasts increased. The pH dropped from around 5.0 to 3.2 in 24 hours, and this was mirrored by a rise in lactic acid to about 0.5%. Ethanol and secondary alcohols and aldehydes increased during the secondary part of the fermentation. Malty flavors are typical for fermented cereal products and may be produced during grain malting. Secondary aldehydes and alcohols are responsible for these flavors and may also originate from microbial metabolism of the branched-chain amino acids leucine, isoleucine, and valine. These compounds are produced by yeasts, some LAB, and probably also by other microorganisms in the product.

Many spontaneously fermented cereals also have a very short shelf life, since fermentation continues in the absence of refrigeration. Off-flavors, in particular vinegary notes, are a common problem. The very low pH in fermented cereal products may be sensorially compensated for by saccharification by β-amylase.

FERMENTED PROBIOTIC CEREAL FOODS

A probiotic food is a live bacterial food supplement, which when ingested, may improve the well-being of the host in a variety of ways by influencing the balance of the host's intestinal flora (Fuller 1989). Most probiotic bacteria have been isolated from the healthy human intestine and are members of the genus *Lactobacillus*, but some products may contain *Bifidobacterium* spp. or the yeast *Saccharomyces boulardii*. While the potential benefits of probiotic bacteria have been generally accepted for many decades, it is only in comparatively recent years that research has been able to scientifically document the beneficial medical effect due to some specific strains (Gorbach 2002). There is now strong scientific evidence that specific strains of probiotic microorganisms are able to:

- show a prophylactic action against and alleviate diarrhea caused by bacterial and viral infections, radiation therapy or the use of antibiotics;
- suppress undesirable bacteria in the gut with beneficial results for patients with conditions such as irritable bowel syndrome and ulcerative colitis;
- influence the immune system, showing positive results for infant atopic eczema and other allergies.

Indeed, in addition to the list above, other effects have been proposed: the lowering of blood cholesterol; the prevention of acute respiratory infections, *Helicobacter pylori* infections, and colonization by potential pathogens in intensive care units in hospitals; relief of constipation; and a protection against the development of various forms of cancer. However, so far, convincing proof for the efficacy of probiotics against these problems has not been obtained. The positive effects that have been documented have led to a great interest from food manufacturers and consumers alike. The main motivation for consuming probiotic products is said to be the developing consumer trend towards healthy living though natural foods and medicines and a trend away from the use of antibiotics and the incorporation of chemical additives in food. As the beneficial effects of probiotic foods become scientifically accepted, there will be increasing pressure from food manufacturers on the authorities to allow health claims to be used in product advertising. Probiotic fermented milks were the first probiotic products to be produced commercially and are available in many countries (Tamime and Marshall 1997).

Some fermented probiotic cereal products are now being prepared and marketed (Table 31.4) and may have an appeal for those who do not consume dairy products.

Oats are a popular basis for probiotic cereal foods. This choice is due to the healthy image of oats with respect to soluble and insoluble fiber content and the potential to reduce blood cholesterol due to β-glucans. A prebiotic is a compound, usually an oligosaccharide, that reaches the colon undigested by the host's enzymes and selectively favors the growth of probiotic bacteria. Such compounds include lactulose, fructooligosaccharides, and inulin. It has been suggested that the best probiotic results may be obtained by using a combination of a prebiotic (such as oats) and a probiotic organism (Charalampopoulos et al. 2002). In this way, the total physiological effect of the food could be increased.

In order for a probiotic product to have a physiological effect, it has been suggested that it should contain at least 10^6 cfu/g product, and that daily intake should be at least 100 g (Sanders and Huis in't Veld 1999). The final acidity in the product has been shown to be of critical importance for the survival of probiotic bacteria during storage (Mårtensson, Öste and Holst 2002). Many probiotic bacteria do not tolerate a pH below 4.0, and fermented cereals frequently reach this pH due to the poor buffering capacity of the substrate. In addition, the physiological state of the probiotic organisms at the time of storage also determines their survival. Organisms that show poor growth during a fermentation period are more likely to die out during cold storage. This necessitates careful formulation of the product as well as selection of the right probiotic culture.

The choice of a substrate for a probiotic food is partially governed by the tolerance of the food towards heat pasteurization or even sterilization before fermentation, and cereal mixtures lend themselves well to this treatment. Probiotic products require fermentation at around 37°C for 8–18 hours, depending on substrate. The suitability of such conditions for the growth of pathogenic organisms necessitates strict adherence to hygiene both before and during fermentation. A fast lactic acid

Table 31.4. Fermented Probiotic Cereal Foods

Type of Product (Commercial Name)	Cereal Constituent	Probiotic Constituent	Reference
Fermented fruit flavored cereal drink (Pro Viva)	Oat + malted barley flour	*Lb. plantarum* 299v	Molin 2001
Fermented cereal drink	Oat "milk"	*LB. reuteri; lb. acidophilus;* *B. bifidus*	Mårtensson, Öste and Holst 2002
Fruit flavored cereal pudding (Yosa)	Oat flour	*Lb. acidophilus; B. bifidus*	Blandino et al. 2003
Cereal-based weaning food	Maize + malted barley flour	*Lb. acidophilus* *Lb. rhamnosus 'GG'* *Lb. reuteri*	Helland et al. 2004

Note: LB = Lactobacillus; B = Bifidobacterium.

development in the product during fermentation is a critical step, and the growth of probiotic organisms in cereal products is greatly stimulated by the addition of malted flour (either of the same grain or of barley malt) or milk, due to the increased availability of fermentable sugars, peptides, and amino acids. For probiotic weaning foods, the use of malt has a further advantage since at a given viscosity, the product has a higher nutritional den-sity.

Probiotic cereal foods are in their infancy, and the future will probably see further development in this type of product. New strains with proven probiotic efficacy and good flavor-forming abilities will increase the range of probiotic products available.

REFERENCES

Abegaz K. 2002. Traditional and Improved Fermentation of *Borde*, a Cereal-Based Ethiopian Beverage. [PhD thesis] Ås, Norway. Agric Univ of Norway. ISBN: 82-575-0508-0.

Abegaz K et al. 2002. Indigenous processing methods and raw materials of *borde*, an Ethiopian traditional fermented beverage. *J Food Technol Africa* 7: 59–64.

Adams MR. 1998. Fermented weaning foods. In: BJB Wood (ed.). *Microbiology of Fermented Foods*. Blackie Academic and Professional, London.

Axelsson LT. 1993. Lactic acid bacteria: classification and physiology. In: S Salminen, A von Wright (eds.). *Lactic Acid Bacteria*. Marcel Dekker, New York, pp. 1–63.

Berg RV et al. 1981. Identification of a growth stimulant for *Lactobacillus sanfrancisco*. *Appl Env Microbiol* 42: 786–788.

Blandino A et al. 2003. Cereal-based fermented foods and beverages. *Food Res Int* 37: 527–543.

Cauvain SP. 1998. Other cereals in breadmaking. In: Cauvain SP, Young LS (eds.), *Technology of Breadmaking*. Blackie Academic and Professional, London, pp. 330–346.

Charalampopoulos D et al. 2002. Application of cereal components in functional foods: a review. *Int J Food Microbiol* 79: 131–141.

Condon S. 1987. Responses of lactic acid bacteria to oxygen. *FEMS Microbiol Rev* 46: 269–280.

Corsetti A et al. 1998. Antimould activity of sourdough lactic acid bacteria: Identification of a mixture of organic acids produced by *Lactobacillus sanfrancisco* CB1. *Appl Microbiol Biotechnol* 50: 253–256.

Corsetti A et al. 1996. Antimicrobial activity of sourdough lactic acid bacteria: Isolation of a bacteriocin-like inhibitory substance from *Lactobacillus sanfrancisco* C57. *Food Microbiol* 13: 447–456.

Di Cagno R et al. 2003. Interactions between sourdough lactic acid bacteria and exogenous enzymes: Effects on the microbial kinetics of acidification and dough textural properties. *Food Microbiol* 20: 67–75.

Di Cagno R et al. 2002. Proteolysis by lactic acid bacteria: Effects on wheat flour protein fractions and gliadin peptides involved in human cereal tolerance. *Appl Env Microbiol* 68: 623–633.

El-Ziney MG et al. 2000. Reuterin. In: AS Naidu (ed.). *Natural Food Antimicrobial Systems*. CRC Press, London, pp. 567–587.

FAO. 1999. Fermented cereals. *A Global Perspective. Food and Agricultural Services Bulletin No. 138*. Food and Agriculture Organization of the United Nations, Rome, Italy.

Fenema OR. 1996. *Food Chemistry*. Marcel Dekker, New York.

Fuller R. 1989. Probiotics in man and animals. *J Appl Bact* 66: 365–378.

Gänzle MG et al. 2000. Characterization of reutericyclin produced by *Lactobacillus reuteri* LTH2584. *Appl Env Microbiol* 66: 4325–4333.

Gänzle MG, Vogel R. 2003. Contribution of reutericyclin production to the stable persistence of *Lactobacillus reuteri* in an industrial sourdough fermentation. *Int J Food Microbiol* 80: 31–45.

Gobbetti M. 1998. The sourdough microflora: Interactions of lactic acid bacteria and yeasts. *Trends Food Sci Technol* 9: 267–274.

Gobbetti M, Corsetti A. 1996. Co-metabolism of citrate and maltose by *Lactobacillus brevis* subsp. *lindneri* CB1 citrate-negative strain: Effect on growth, end-products and sourdough fermentation. *Z Lebensmittel-Untersuchung und -Forschung* 203: 82–87.

Gobbetti M, Corsetti A. 1997. *Lactobacillus sanfrancisco* a key sourdough lactic acid bacterium: A review. *Food Microbiol* 14: 175–187.

Gobbetti M et al. 1995. Volatile compound and organic acid productions by mixed wheat sour dough starters: Influence of fermentation parameters and dynamics during baking. *Food Microbiol* 12: 497–507.

Gobbetti M et al. 1996. The proteolytic system of *Lactobacillus sanfrancisco* CB1: Purification and characterization of a proteinase, dipeptidase and aminopeptidase. *Appl Env Microbiol* 62: 3220–3226.

Gobbetti M et al. 1996. The sourdough microflora. Cellular localization and characterization of proteolytic enzymes in lactic acid bacteria. *Lebensmittel-Wissenschaft und -Technologie* 29: 561–569.

Gorbach SL. 2002. Probiotics in the third millenium. *Digest Liver Dis* 34: S2–7.

Hammes WP, Gänzle MG. 1998. Sourdough breads and related products. In: BJB Wood (ed.). *Microbiology of Fermented Foods*. Blackie Academic and Professional, London, pp. 199–216.

Helland M et al. 2004. Growth and metabolism of selected strains of probiotic bacteria, in maize porridge with added malted barley. *Int J Food Microbiol* 44: 957–965.

Höltzel A et al. 2000. The first low-molecular-weight antibiotic from lactic acid bacteria: Reutericyclin, a new tetramic acid. *Angewandte Chemie Int Ed* 39: 2766–2768.

Holzapfel WH. 2002. Appropriate starter culture technologies for small-scale fermentation in developing countries. *Int J Food Microbiol* 75: 197–212.

Hoseney RC. 1998. Cereal Science and Technology, 2nd edition. St. Paul, Minnesota: American Association of Cereal Chemists.

Jespersen L. 2003. Occurrence and taxonomic characteristics of strains of *Saccharomyces cerevisiae* predominant in African indigenous fermented foods and beverages. *FEMS Yeast Res* 3: 191–200.

Kaiser AL, Montville TJ. 1996. Purification of the bacteriocin bavaricin MN and characterization of its mode of action against *Listeria monocytogenes* Scott A cells and lipid vesicles. *Appl Env Microbiol* 62: 4529–4535.

Kent NL. 1983. *Technology of Cereals*, 2nd edition. Pergamon Press, Oxford.

Kline L, Sugihara TF. 1971. Microorganisms of the San Francisco sourdough bread process. II. Isolation and characterization of undescribed bacterial species responsible for souring activity. *Appl Microbiol* 21: 459–465.

Kunene NF et al. 2000. Characterization and determination of origin of lactic acid bacteria from a sorghum-based fermented weaning food by analysis of soluble proteins and amplified fragment length polymorphism fingerprinting. *Appl Environ Microbiol.* 66: 1084–1092.

Larsen AG et al. 1993. Antimicrobial activity of lactic acid bacteria isolated from sour doughs: Purification and characterization of bavaricin A, a bacteriocin produced by *Lactobacillus bavaricus* MI401. *J Appl Bacteriol* 75: 113–122.

Lavermicocca P et al. 2000. Purification and characterization of novel antifungal compounds from the sourdough *Lactobacillus plantarum* strain 21B. *Appl Env Microbiol* 66: 4084–4090.

Linko Y-Y et al. 1997. Biotechnology of bread baking. *Trends Food Sci Technol* 8: 339–344.

Lotong N. 1998. Koji. In: BJB Wood (ed.). *Microbiology of Fermented Foods*. Blackie Academic and Professional, London.

Maloney DH, Foy JJ. 2003. Yeast Fermentations. In: K Kulp, K Lorenz (eds.). *Handbook of Dough Fermentations*. Marcel Dekker, New York.

Mårtensson O et al. 2002. The effect of yoghurt culture on the survival of probiotic bacteria in oat-based, non-dairy products. *Food Res Int* 35: 775–784.

Martinez-Anaya MA. 2003. Associations and interactions of microorganisms in dough fermentations: Effects on dough and bread characteristics. In: K Kulp, K Lorenz (eds.). *Handbook of Dough Fermentations*. Marcel Dekker, New York: pp. 63–95.

Messens W, De Vuyst L. 2002. Inhibitory substances produced by *Lactobacilli* isolated from sourdoughs – a review. *Int J Food Microbiol* 72: 31–43.

Molin G. 2001. Probiotics in food not containing milk or milk constituents, with special reference to *Lactobacillus plantarum* 299v. *Am J Clinical Nutr* 73: 380S–385S.

Mugula JK. 2001. Microbiology, fermentation and shelf life extension of togwa, a Tanzanian indigenous food. [PhD thesis]. Ås, Norway. Agric Univ of Norway. ISBN 82-575-0457-2.

Mugula JK et al. 2003. Use of starter cultures of lactic acid bacteria and yeasts in the preparation of *togwa*, a Tanzanian fermented food. *Int J Food Microbiol* 83: 307–318.

Muyanja CBK. 2001. Studies on the Fermentation of *Bushera* : A Ugandan Traditional Fermented Cereal-Based Beverage. [PhD thesis]. Ås, Norway, Agric Univ Norway. ISBN 82-575-0486-8.

Muyanja CBK et al. 2003. Production methods and composition of *bushera*, a Ugandan Traditional Fermented Cereal Beverage. *African J Food, Agric, Nutr Development* 3: 10–19.

Muyanja C et al. 2004. Chemical changes during spontaneous and lactic acid starter bacteria starter culture fermentation of bushera. *MURARIK Bulletin* 7: 606–616.

Nuraida L et al. 1995. Microbiology of *pozol*, a Mexican fermented maize dough. *World J Microbiol Biotechnol* 11: 567–571.

Nout MJR, Motarjemi Y. 1997. Assessment of fermentation as a household technology for improving food safety: A joint FAO/WHO workshop. *Food Control* 8: 221–226.

Nout MJR et al. 1989. Effect of accelerated natural lactic fermentation of infant food ingredients on some pathogenic microorganisms. *Int J Food Microbiol* 8: 351–361.

Odunfa SA. 1985. African Fermented Foods. In: BJB Wood (ed.). *Microbiology of Fermented Foods, Vol, 2*. Elsevier Applied Science Publishers, London and New York.

Ogunbanwo ST et al. 2003. Characterization of bacteriocin produced by *Lactobacillus plantarum* F1 and *Lactobacillus brevis* OG1. *African J Biotechnol* 2: 219–227.

Omar benN, Ampe F. 2000. Microbial community dynamics during production of the Mexican fermented maize dough Pozol. *Appl and Env Microbiol* 66: 3644–3673.

Ponte JG, Tsen CC. 1987. Bakery Products. In: LR Beuchat, editor. *Food and Beverage Mycology*, 2nd edition. Van Nostrand Reinhold Company, New York.

Romano AE et al. 1987. Regulation of β-galactoside transport and accumulation in heterofermentative lactic acid bacteria. *J Appl Bacteriol* 169: 5589–5596.

Sanders ME, Huis in't Veld JHJ. 1999. Bringing a probiotic-containing functional food to the market: Microbiological, product, regulatory and labelling issues. In: WN Konings, OP Kuipers, Hui in T'Veld JHJ (eds.). *Lactic Acid Bacteria: Genetics, Metabolism and Applications*. Kluwer Academic Publishing, Dordrecht, The Netherlands, pp. 293–315.

Sanni AI et al. 2002. New efficient amylase-producing strains of *Lactobacillus plantarum* and *L. fermentum* isolated from different Nigerian fermented foods. *Int J Food Microbiol* 72: 53–62.

Soni SK, Sandu DK. 1990. Indian fermented foods. Microbiological and biochemical aspects. *Indian J Microbiol.* 30: 135–157.

Spicher G. 1983. Baked goods. In: HJ Rehm, G Reed (eds.). *Biotechnology, Vol. 5*. Verlag Chemie: Weinheim, pp. 1–80.

Spicher G, Nierle W. 1984. The microflora of sourdough. XVIII. Communication: The protein degrading capabilities of the lactic acid bacteria of sourdough. *Z für Lebensmittel-Untersuchung und-Forschung* 178: 389–392.

Spicher G, Nierle W. 1988. Proteolytic activity of sourdough bacteria. *Appl Microbiol Biotechnol* 28: 487–492.

Stanton WR. 1998. Food fermentation in the tropics. In: BJB Wood (ed.). *Microbiology of Fermented Foods*. Blackie Academic and Professional, London, pp. 696–712.

Stear CA. 1990. *Handbook of Breadmaking Technology*. Barking, Elsevier Applied Science, United Kingdom.

Stolz P. 2003. Biological fundamentals of yeast and lactobacilli fermentation in bread dough. In: K Kulp, K Lorenz (eds.). *Handbook of Dough Fermentations*. Marcel Dekker, New York: pp. 23–42.

Stolz P et al. 1993. Utilization of maltose and glucose by lactobacilli isolated from sourdough. *FEMS Microbiol Lett* 109: 237–242.

Stolz P et al. 1995. Utilization of electron acceptors by lactobacilli isolated from sourdough. II. *Lactobacillus pontis, L. reuteri, L. amylovorus,* and *L fermentum. Z für Lebensmittel-Untersuchung und -Forschung* 201: 402–410.

Todorov S et al. 1999. Detection and characterization of a novel antibacterial substance produced by *Lactobacillus plantarum* ST31 isolated from sourdough. *Int J Food Microbiol* 48: 167–177.

Tamime AY, Marshall VME. 1997. Microbiology and technology of fermented milks. In: BA Law (ed.). *Microbiology and Biochemistry of Cheese and Fermented Milk*. Blackie Academic and Professional, London.

Walker GM. 1998a. *Yeast technology. In: Yeast Physiology and Technology*. John Wiley & Sons, Inc: New York.

Walker GM. 1998b. *Yeast metabolism. In: Yeast Physiology and Technology*. John Wiley & Sons, Inc, New York.

Williams T, Pullen G. 1998. Functional ingredients. In: SP Cauvain, LS Young (eds.). *Technology of Breadmaking*. Blackie Academic and Professional, London.

Wood BJB. 1994. Technology transfer and indigenous fermented foods. *Food Res Int* 27: 269.

Wood BJB. 2000. Sourdough bread. In: RK Robinson, CA Blatt, PD Patel (eds). *Encyclopedia of Food Microbiology*. Academic Press, San Diego, pp. 295–301.

Zehentbauer G, Grosch W. 1998. Crust aroma of baguettes II. Dependence of the concentrations of key odorants on yeast level and dough processing. *J Cereal Sci* 28: 93–96.

32
Starch Synthesis in the Potato Tuber

P. Geigenberger and A.R. Fernie

INTRODUCTION

Starch is the most important carbohydrate used for food and feed purposes and represents the major resource of our diet. The total yield of starch in rice, corn, wheat, and potato exceeds 10^9 tons/year (Kossmann and Lloyd 2000, Slattery et al. 2000). In addition to its use in a nonprocessed form, due to the low cost incurred, extracted starch is processed in many different ways. Processed starch is subsequently used in multiple forms, for example in high-fructose syrup, as a food additive, or for various technical processes based on the fact that as a soluble macromolecule it exhibits high viscosity and adhesive properties (Table 32.1). The considerable importance of starch has made increasing the content and engineering the structural properties of plant starches major goals of both classical breeding and biotechnology over the last few decades (Smith et al. 1997, Sonnewald et al. 1997, Regierer et al. 2002). Indeed, since the advent and widespread adoption of transgenic approaches some 15 years ago gave rise to the discipline of molecular plant physiology, much information has been obtained concerning the potential to manipulate plant metabolism. For this chapter, we intend to review genetic manipulation of starch metabolism in potato (*Solanum tuberosum*). Potato is one of the most important crops worldwide, ranking fourth in annual production behind the cereal species rice (*Oryza sativa*), wheat (*Triticum aestivum*), and maize (*Zea mais*). Although in Europe and North America the consumption of potatoes is mainly in the form of processed foodstuffs such as fried potatoes and chips, in less developed countries it represents an important staple food and is grown by many subsistence farmers. The main reasons for the increasing popularity of the potato in developing countries are the high nutritional value of the tubers combined with the simplicity of its propagation by vegetative amplification (Fernie and Willmitzer 2001). Since all potato varieties are true tetraploids and display a high degree of heterozygosity, genetics have played only a minor role in metabolic studies in this species. However, because the potato is a member of the *Solanaceae* family, it was amongst the first crop plants to be accessible to transgenic approaches. Furthermore, due to its relatively large size and metabolic homogeneity, the potato tuber represents a convenient experimental system for biochemical studies (Geigenberger 2003a).

In this chapter we will describe transgenic attempts to modify starch content and structure in potato tubers that have been carried out in the last two decades. In addition to describing biotechnologically significant results we will also detail fundamental research in this area that should enable future biotechnology strategies. However, we will begin by briefly describing starch, its structure, and its synthesis.

WHAT IS STARCH?

For many years, it has been recognized that the majority of starches consist of two different macromolecules, amylose and amylopectin (Fig. 32.1), which are both polymers of glucose and are organized into grains that range in size from 1 μm to more than 100 μm. Amylose is classically regarded as an essentially linear polymer wherein the glucose units are linked through α-1-4-glucosidic bonds. In contrast, although amylopectin contains α-1-4-glucosidic bonds, it also consists of a high proportion of α-1-6-glucosidic bonds. This feature of amylopectin makes it a more branched, larger molecule than amylose, having a molecular weight of 10^7–10^8 as opposed to 5×10^5–10^6.

Food Biochemistry and Food Processing, Second Edition. Edited by Benjamin K. Simpson, Leo M.L. Nollet, Fidel Toldrá, Soottawat Benjakul, Gopinadhan Paliyath and Y.H. Hui.

Table 32.1. Industrial Uses of Starch

Starch and Its Derivates	Processing Industry	Application
Amylose and amylopectin (polymeric starch)	Food	Thickener, texturants, extenders, low-calorie snacks
	Paper	Beater sizing, surface sizing, coating
	Textile	Wrap sizing, finishing, printing
	Polymer	Absorbents, adhesives, biodegradable plastics
Products of starch hydrolysis, such as glucose, maltose, or dextrins	Food	Sweeteners or stabilizing agents
	Fermentation	Feedstock to produce ethanol, liquors, spirits, beer, etc.
	Pharmaceutical	Feedstock to produce drugs and medicine
	Chemical	Feedstock to produce organic solvents or acids

Source: Adapted from Jansson et al. 1997, with modification.

These molecules can thus be fractionated by utilizing differences in their molecular size as well as in their binding behavior. Moreover, amylose is able to complex lipids, while amylopectin can contain covalently bound phosphate, adding further complexity to their structure. In nature, amylose normally accounts between 20 and 30% of the total starch. However, the percentage of amylose depends on the species and the organ used for starch storage. The proportion of amylose to amylopectin and the size and structure of the starch grain give distinct properties to different extracted starches (properties important in food and industrial purposes; Dennis and Blakeley 2000). Starch grain size is also dependent on species and organ type. It is well established that starch grains grow by adding layers, and growth rings within the grain may represent areas of fast and slower growth (Pilling and Smith 2003); however, very little is known about how these highly ordered structures are formed in vivo. The interested reader is referred to articles by Buleon et al. (1998) and Kossmann and Lloyd (2000).

(A)

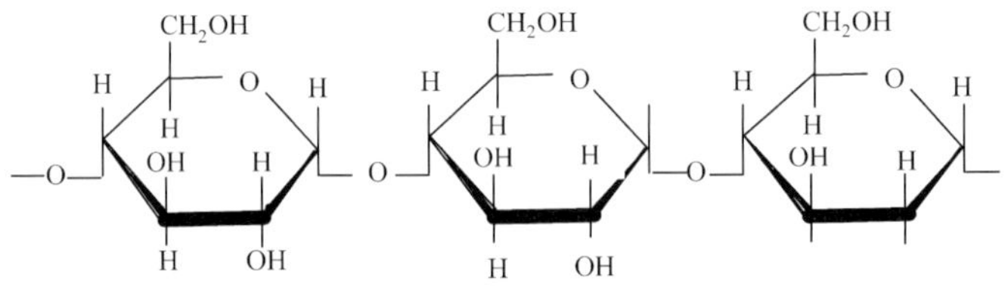

(B)

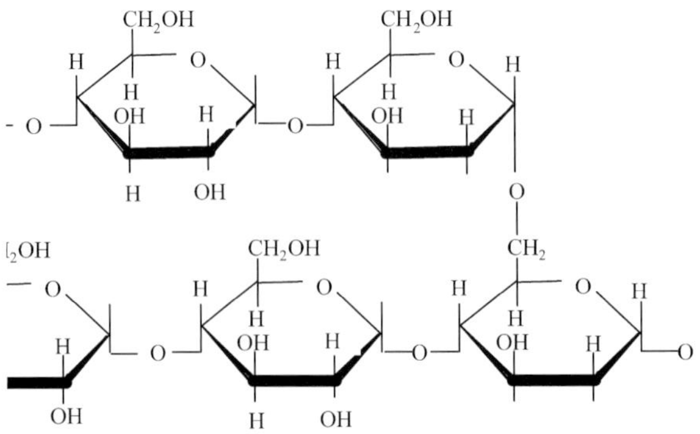

Figure 32.1. Starch structure: Amylose (**A**) is classically regarded as an essentially linear polymer wherein the glucose units are linked through α-1-4-glucosidic bonds. In contrast, although amylopectin contains α-1-4-glucosidic bonds, it also consists of a high proportion of α-1-6-glucosidic bonds (**B**).

ROUTES OF STARCH SYNTHESIS AND DEGRADATION AND THEIR REGULATION

With the possible exception of sucrose, starch is the most important metabolite of plant carbohydrate metabolism. It is by far the most dominant storage polysaccharide and is present in all major organs of most plants, in some instances at very high levels. Due to the high likelihood of starch turnover, its metabolism is best considered as the balance between the antagonistic operation of pathways of synthesis and degradation.

To investigate the regulation of starch synthesis in more detail, growing potato tubers have been used as a model system. Unlike many other tissues, the entry of sucrose into metabolism is relatively simple, in that it is unloaded symplasmically from the phloem, degraded via sucrose synthase to fructose and UDPglucose, which are converted to hexose monophosphates by fructokinase and UDPglucose pyrophosphorylase, respectively (Geigenberger 2003a). In contrast to sucrose degradation, which is localized in the cytosol, starch is synthesized predominantly, if not exclusively, in the plastid. The precise pathway of starch synthesis depends on the form in which carbon crosses the amyloplast membrane (Fig. 32.2). This varies between species and has been the subject of considerable debate (Keeling et al. 1988, Hatzfeld and Stitt 1990, Tauberger et al. 2000). Categorical evidence that carbon enters potato tuber, *Chenopodium rubrum* suspension cell, maize endosperm, and wheat endosperm amyloplasts in the form of hexose monophosphates rather than triose phosphates was provided by determination of the degree of randomization of radiolabel in glucose units isolated from starch following incubation of the various tissues with glucose labeled at the C1 or C6 positions (Keeling et al. 1988, Hatzfeld and Stitt 1990). The cloning of a hexose monophosphate transporter from potato and the finding that the cauliflower homolog is highly specific for glucose-6-phosphate provides strong support for this theory (Kammerer et al. 1998). Further evidence in support of glucose-6-phosphate import was provided by studies of transgenic potato lines in which the activity of the plastidial isoform of phosphoglucomutase was reduced by antisense inhibition, leading to a large reduction in starch content of the tubers (Tauberger et al. 2000). These data are in agreement with the observations that heterotrophic tissues lack plastidial fructose-1,6-bisphosphatase expression and activity (Entwistle and Rees 1990, Kossmann et al. 1992). The results of recent transgenic and immunolocalization experiments have indicated that the substrate for uptake is most probably species specific, with clear evidence for the predominant route of uptake in the developing potato tuber being in the form of glucose-6-phosphate. By contrast, in barley, wheat, oat, and possibly maize, the predominant form of uptake, at least during early stages of seed endosperm development, is as ADP-glucose (Neuhaus and Emes 2000).

Irrespective of the route of carbon import, ADP-glucose pyrophosphorylase (AGPase, EC 2.7.7.27) plays an important role in starch synthesis, catalyzing the conversion of

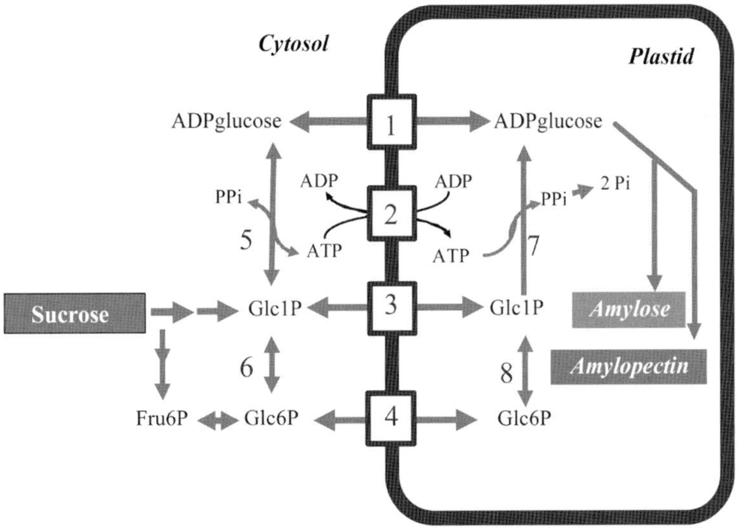

Figure 32.2. Pathways of sucrose to starch conversion in plants: **1**, ADP-glucose transporter; **2**, ATP/ADP translocator; **3**, glucose-1-phoshate (Glc1P) translocator; **4**, glucose-6-phosphate (Glc6P) translocator; **5**, cytosolic ADP-glucose pyrophosphorylase (AGPase); **6**, cytosolic phosphoglucomutase; **7**, plastidial AGPase; **8**, plastidial phosphoglucomutase. In growing potato tubers, incoming sucrose is degraded by sucrose synthase to fructose and UDPglucose and subsequently converted to fructose-6-phosphate (Fru6P) and Glc1P by fructokinase and UDPglucose pyrophosphorylase, respectively (not shown in detail). The conversion of Fru6P to Glc6P in the cytosol is catalyzed by phosphoglucoisomerase, and the cleavage of PP_i to 2 P_i in the plastid is catalyzed by inorganic pyrophosphatase. In potato tubers, there is now convincing evidence that carbon enters the plastid almost exclusively via the Glc6P translocator, whereas in cereal endosperm, the predominant form of uptake is as ADPglucose.

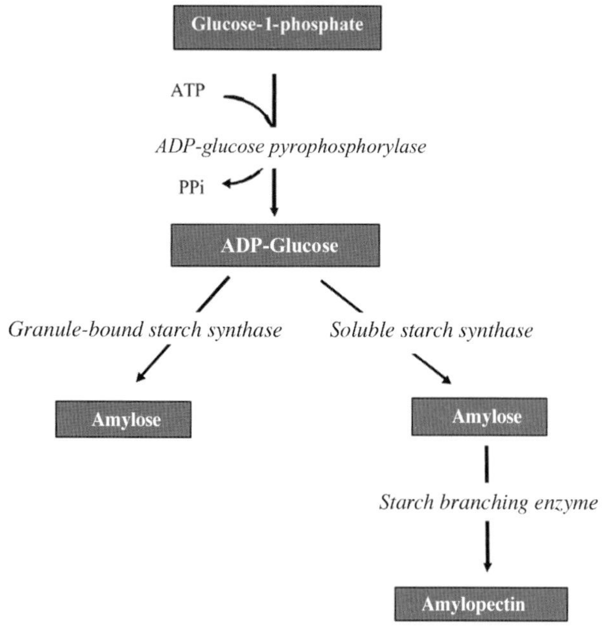

Figure 32.3. Routes of amylose and amylopectin synthesis within potato tuber amyloplasts. For simplicity, the possible roles of starch-degrading enzymes in trimming amylopectin have been neglected in this scheme.

glucose-1-phosphate and ATP to ADP-glucose and inorganic pyrophosphate. Inorganic pyrophosphate is subsequently metabolized to inorganic phosphate by a highly active inorganic pyrophosphatase within the plastid. AGPase is generally considered as the first committed step of starch biosynthesis since it produces ADP-glucose, the direct precursor for the starch polymerizing reactions catalyzed by starch synthase (SS) (EC 2.4.1.21) and branching enzyme (EC 2.4.1.24; Fig. 32.3). These three enzymes appear to be involved in starch synthesis in all species. With the exceptions of the cereal species described above, which also have a cytosolic isoform of AGPase, these reactions are confined to the plastid. The activities of AGPase and SS are sufficient to account for the rates of starch synthesis in a wide variety of photosynthetic and heterotrophic tissues (Smith et al. 1997). Furthermore, changes in the activities of these enzymes correlate with changes in the accumulation of starch during development of storage organs (Smith et al. 1995). AGPase from a range of photosynthetic and heterotrophic tissues is well established to be inhibited by phosphate, which induces sigmoidal kinetics, and to be allosterically activated by 3-phosphoglycerate (3PGA), which relieves phosphate inhibition (Preiss 1988; Fig. 32.4). There is clear evidence that allosteric regulation of AGPase is important in vivo to adjust the rate of starch synthesis to changes in the rate of respiration that go along with changes in the levels of 3PGA, and an impressive body of evidence has been provided that there is a strong correlation between the 3PGA and ADP-glucose levels and the rate of starch synthesis under a wide variety of environmental conditions (Geigenberger et al. 1998, Geigenberger 2003a).

More recently, an important physiological role for posttranslational redox regulation of AGPase has been established (Tiessen et al. 2002). In this case, reduction of an intermolecular cysteine bridge between the two small subunits of the heterotetrameric enzyme leads to a dramatic increase of activity, due to increased substrate affinities and sensitivity to allosteric activation by 3PGA (Fig. 32.4). Redox activation of AGPase *in planta* correlated closely with the potato tuber sucrose content across a range of physiological and genetic manipulations (Tiessen et al. 2002), indicating that redox modulation is part of a novel regulatory loop that directs incoming sucrose towards storage starch synthesis (Tiessen et al. 2003). Crucially, it allows the rate of starch synthesis to be increased in response to sucrose supply and independently of any increase in metabolite levels (Fig. 32.4), and it is therefore an interesting target for approaches to improving starch yield (see later). There are at least two separate sugar signaling pathways leading to posttranslational redox activation of AGPase, one involving an SNF1-like protein kinase (SnRK1), the other involving hexokinase (Tiessen et al. 2003). Both hexokinase and SnRK1 have previously been shown to be involved in the transcriptional regulation of many plant genes. Obviously, the transduction pathway that regulates the reductive activation of AGPase in plastids and the regulatory network that controls the expression of genes in the cytosol share some common components. How the sugar signal is transferred into the plastid and leads to redox changes of AGPase is still unknown, but may involve interaction of specific thioredoxins with AGPase.

Subcellular analyses of metabolite levels in growing potato tubers have shown that the reaction catalyzed by AGPase is far from equilibrium in vivo (Tiessen et al. 2002), and consequently, the flux through this enzyme is particularly sensitive to regulation by the above-mentioned factors. It is interesting to note that plants contain multiple forms of AGPase, which is a tetrameric enzyme comprising two different polypeptides, a small subunit and a large subunit, both of which are required for full activity (Preiss 1988). Both subunits are encoded by multiple genes, which are differentially expressed in different tissues. Although the precise function of this differential expression is currently unknown, it seems likely that these isoforms will differ in their capacity to bind allosteric regulators. If this is indeed the case, then different combinations of small and large subunits should show different sensitivity to allosteric regulation—such as those observed in tissues from cereals.

While there is a wealth of information on the regulation of the plastidial isoforms of AGPase—the only isoform present in the potato tuber—very little is known about cytosolic isoforms in other species. Several important studies provide evidence that the ADP-glucose produced in the cytosol can be taken up by the plastid (Sullivan et al. 1991, Shannon et al. 1998). From these studies and from characterization of mutants unable to transport ADP-glucose, it would appear that this is a predominant route for starch synthesis within maize. Despite these findings, the physiological significance of cytosolic ADP-glucose production remains unclear for a range of species. However, it has been calculated that the AGPase activity of the plastid is insufficient

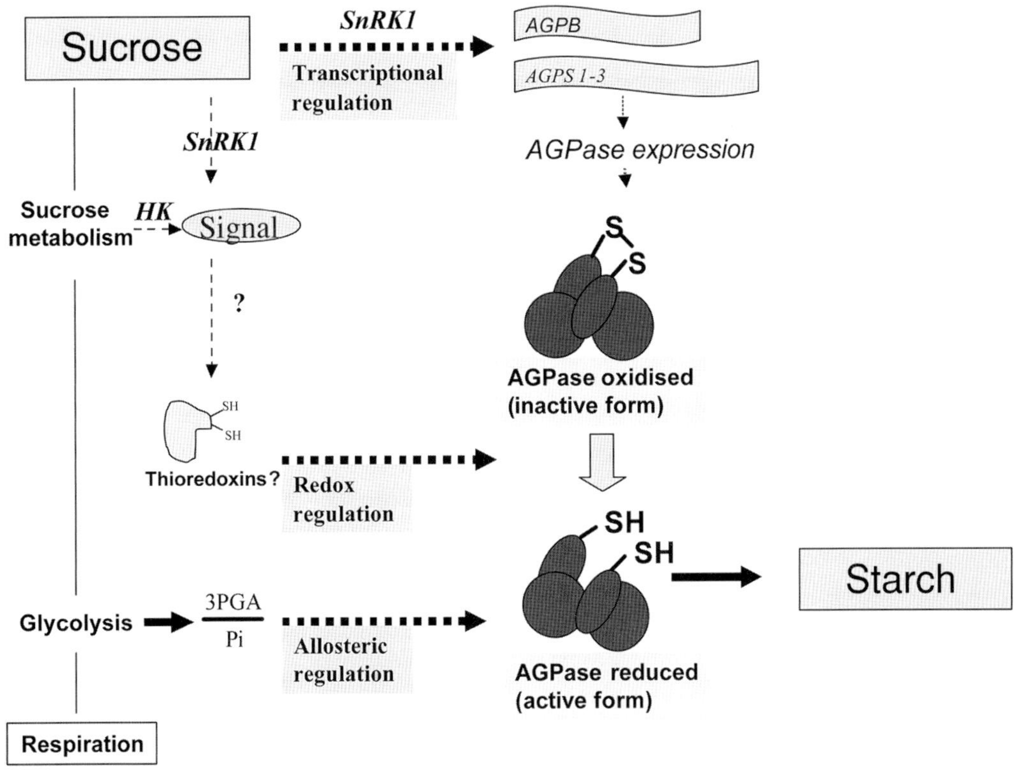

Figure 32.4. Regulation of starch synthesis in potato tubers: ADPGlc-pyrophosphorylase (AGPase) is a key regulatory enzyme of starch biosynthesis. It is regulated at different levels of control, involving allosteric regulation, regulation by posttranslational redox modification and transcriptional regulation. Redox regulation of AGPase represents a novel mechanism regulating starch synthesis in response to changes in sucrose supply (Tiessen et al. 2002). There are at least two separate sugar-sensing pathways leading to posttranslational redox activation of AGPase, one involving an SNF1-like protein kinase (SnRK1), the other involving hexokinase (HK) (Tiessen et al. 2003).

to account for the measured rates of starch synthesis in barley endosperm, suggesting that at least some of the ADP-glucose required for this process is provided by cytosolic production (Thorbjornsen et al. 1996).

It is interesting to note that the involvement of the various isoforms of AGPase in starch biosynthesis is strictly species dependent, whereas the various starch-polymerizing activities are ever present and responsible for the formation of the two different macromolecular forms of starch, amylose and amylopectin (Fig. 32.3). SS catalyze the transfer of the glucosyl moiety from ADP-glucose to the nonreducing end of an α-1,4-glucan and are able to extend α-1,4-glucans in both amylose and amylopectin. There are four different SS isozymes—three soluble and one that is bound to the starch granule.

Starch branching enzymes (SBE), meanwhile, are responsible for the formation of α-1,6 branch points within amylopectin (Fig. 32.3). The precise mechanism by which this is achieved is unknown; however, it is thought to involve cleavage of a linear α-1,4-linked glucose chain and reattachment of the chain to form an α-1,6 linkage. Two isozymes of starch branching enzyme, SBEI and SBEII, are present and differ in specificity. The former preferentially branches unbranched starch (amylose), while the latter preferentially branches amylopectin. Furthermore, in vitro studies indicate that SBEII transfers smaller glucan chains than does SBEI and would therefore be expected to create a more highly branched starch (Schwall et al. 2000). The fact that the developmental expression of these isoforms correlates with the structural properties of starch during pea embryo development is in keeping with this suggestion (Smith et al. 1997). Apart from that, isoforms of isoamylase (E.C. 3.2.1.68) might be involved in debranching starch during its synthesis (Smith et al. 2003).

While the pathways governing starch synthesis are relatively clear, those associated with starch degradation remain somewhat controversial (Smith et al. 2003). The degradation of plastidial starch can proceed via phosphorolytic or hydrolytic cleavage mechanisms involving α-1-4-glucan phosphorylases or amylases, respectively. The relative importance of these different routes of starch degradation has been a matter of debate for many years. The question is whether they are, in fact, independent pathways, since oligosaccharides released by hydrolysis can be further degraded by amylases or, alternatively, by phosphorylases. In addition to this, the mechanisms responsible for the initiation of starch grain degradation in the plastid remain to be resolved, since starch grains have been found to

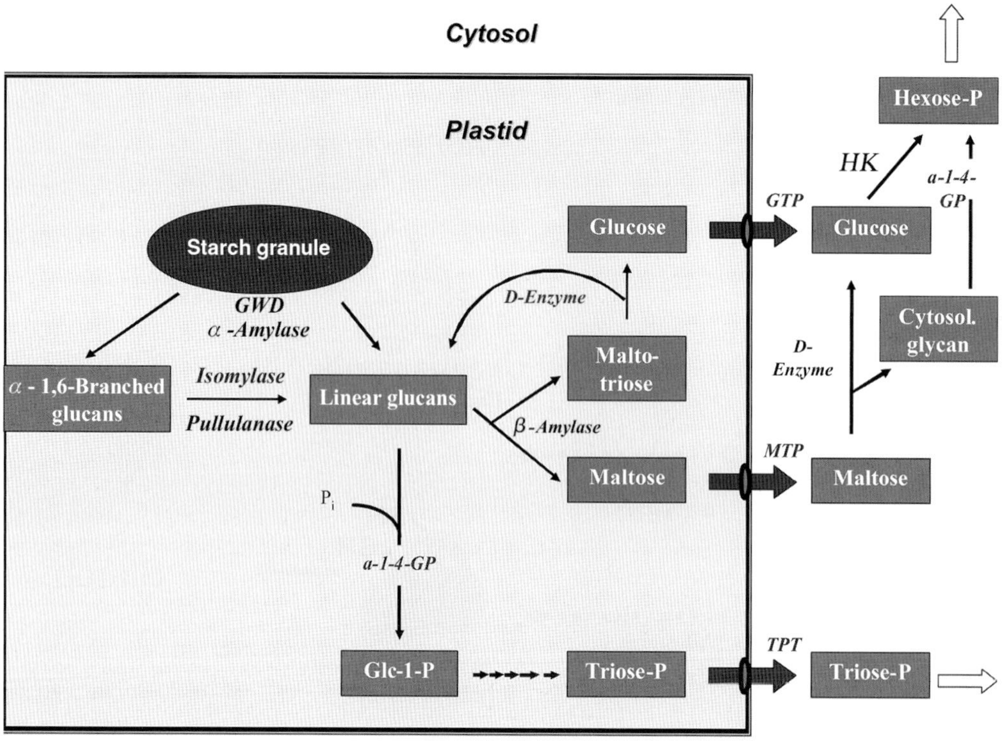

Figure 32.5. Outline of the pathway of plastidial starch degradation. The scheme is mainly based on recent molecular studies in *Arabidopsis* leaves (Ritte et al. 2002, Smith et al. 2003, Chia et al. 2004). After phosphorylation of glucans on the surface of the starch granule via glucan water dikinase (GWD; R1-protein), starch granules are attacked most probably by α-amylase and the resulting branched glucans subsequently converted to unbranched α-1,4-glucans via debranching enzymes (isoamylase and pullulanase). Linear glucans are metabolized by the concerted action of β-amylase and disproportionating enzyme (D-enzyme, glucan transferase) to maltose and glucose. The phosphorolytic degradation of linear glucans to Glc-1-P (glucose-1-phosphate) by α-1-4-glucan phosphorylase (α-1-4-GP) is also possible, but seems of minor importance under normal conditions. Most of the carbon resulting from starch degradation leaves the chloroplast via a maltose transporter in the inner membrane. Subsequent cytosolic metabolism of maltose involves the combined action of glucan transferase (D-enzyme), α-1-4-glucan phosphorylase (α-1-4-GP), and hexokinase (HK). Glucose transporter (GTP) and triose-P/P_i-translocator (TPT) are also shown.

be very stable and relatively resistant against enzymatic action in vitro.

More recently, molecular and genetic approaches have allowed rapid progress in clarifying the route of starch degradation in leaves of the model species *Arabidopsis thaliana* (Fig. 32.5). The phosphorylation of starch granules by glucan water dikinase (GWD, R1) has been found to be essential for the initiation of starch degradation in leaves and tubers (Ritte et al. 2002, Ritte et al. 2004). Transgenic potato plants (Lorberth et al. 1998) and *Arabidopsis* mutants (Yu et al. 2001) with decreased GWD activity showed a decrease in starch-bound phosphate and were severely restricted in their ability to degrade starch, leading to increased starch accumulation in leaves. The underlying mechanisms are still unknown. Phosphorylation of starch may change the structure of the granule surface to make it more susceptible to enzymatic attack or may regulate the extent to which degradative enzymes can attack the granule.

It is generally assumed that the initial attack on the starch granule is catalyzed by α-amylase (endoamylase). This enzyme catalyzes the internal cleavage of glucan chains from amylose or amylopectin, yielding branched and unbranched α-1,4-glucans, which are then subject to further digestion (Fig. 32.5). Debranching enzymes (isoamylase and pullulanase) are needed to convert branched glucans into linear glucans by cleaving the α-1,6 branch points. Further metabolism of linear glucans could involve phosphorolytic or hydrolytic routes. In the first case, α-glucan phosphorylase leads to the phosphorolytic release of Glc1-P, which can be further metabolized to triose-P within the chloroplast and subsequently exported to the cytosol via the triose-P/P_i translocator. Recent results show that the contribution of α-glucan phosphorylase to plastidial starch degradation is relatively small. Removal of the plastidial form of phosphorylase in *Arabidopsis* did not affect starch degradation in leaves of *Arabidopsis* (Zeeman et al. 2004) and potato (Sonnewald et al. 1995). It has been suggested that the phosphorolytic pathway could be more important to degrading starch under certain stress conditions (i.e., water stress; Zeeman et al. 2004). However, no regulatory properties have been described for glucan phosphorylase in plants, other than the effect of changes in the concentrations of inorganic phosphate on the activity of the enzyme (Stitt and Steup 1985).

In the second case, hydrolytic degradation of linear glucans in the plastid can involve the combined four action enzymes: α-amylase, β-amylase, α-glucosidase and disproportionating enzyme (D-enzyme, glucan transferase). There is now direct molecular evidence that β-amylase (exoamylase) plays a significant role in this process (Scheidig et al. 2002). This enzyme catalyzes the hydrolytic cleavage of maltose from the nonreducing end of a linear glucan polymer that is larger than maltotriose. Maltotriose is further metabolized by D-enzyme, producing new substrate for β-amylase and releasing glucose (Fig. 32.5).

Recent studies document that most of the carbon that results from starch degradation leaves the chloroplast in the form of maltose, providing evidence that hydrolytic degradation is the major pathway for mobilization of transitory starch (Weise et al. 2004). Elegant studies with *Arabidopsis* mutants confirmed this interpretation and identified a maltose transporter in the chloroplast envelope that is essential for starch degradation in leaves (Niittylä et al. 2004). The further metabolism of maltose to hexose-phosphates is then performed in the cytosol and is proposed to involve cytosolic forms of glycosyltransferase (D-enzyme; Lu and Sharkey 2004, Chia et al. 2004), α-glucan phosphorylase (Duwenig et al. 1997), and hexokinase, similar to maltose metabolism in the cytoplasm of *E. coli* (Boos and Shuman 1998). It will be interesting to find the potato homolog of the maltose transporter and to investigate its role during starch degradation in tubers.

Despite recent progress in clarifying the route of starch degradation in *Arabidopsis* leaves and potato tubers, the regulation of this pathway still remains an open question. More information is available concerning cereal seeds, where the enzymes involved in starch hydrolysis have been found to be especially active during seed germination, when starch is mobilized within the endosperm, which at this stage of development represents a nonliving tissue. The most studied enzyme in this specialized system is α-amylase, which is synthesized in the surrounding aleurone layer and secreted into the endosperm. This activity and that of α-glucosidase increase in response to the high levels of gibberellins present at germination. A further level of control of the amylolytic pathway is achieved by the action of specific disulphide proteins that inhibit both α-amylase and debranching enzyme. Thioredoxin *h* reduces and thereby inactivates these inhibitor proteins early in germination. Glucose liberated from starch in this manner is phosphorylated by a hexokinase, before conversion to sucrose and subsequent transport to the developing embryo (Beck and Ziegler 1989).

MANIPULATION OF STARCH YIELD

In potato tubers, like all crop species, there has been considerable interest to increase the efficiency of sucrose to starch conversion and thus to increase starch accumulation by both conventional plant breeding and genetic manipulation strategies. Traditional methodology based on the crossing of haploid potato lines and the establishment of a high density genetic map have allowed the identification of quantitative trait loci (QTL) for starch content (Schäfer-Pregl et al. 1998); however, this is outside the scope of this chapter, and the interested reader is referred to Fernie and Willmitzer (2001). Transgenic approaches in potato have focused primarily on the modulation of sucrose import (Leggewie et al. 2003) and sucrose mobilization (Trethewey et al. 1998) or the plastidial starch biosynthetic pathway (see Table 32.2); however, recently more indirect targets have been tested, which are mostly linked to the supply of energy for starch synthesis (see Tjaden et al. 1998, Jenner et al. 2001, Regierer et al. 2002). To date, the most successful transgenic approaches have resulted from the overexpression of a bacterial AGPase (Stark et al. 1991) and the *Arabidopsis* plastidial ATP/ADP translocator (Tjaden et al. 1998), and the antisense inhibition of a plastidial adenylate kinase (Regierer et al. 2002) in potato tubers.

The majority of previous attempts to improve the starch yield of potato tubers concentrated on the expression of a more efficient pathway of sucrose degradation, consisting of a yeast invertase, a bacterial glucokinase, and a sucrose phosphorylase (Trethewey et al. 1998, 2001). However, although the transgenics exhibited decreased levels of sucrose and elevated hexose phosphates and 3-PGA with respect to wild type, these attempts failed. Tubers of these plants even contained less starch than the wild type, but showed higher respiration rates. Recent studies have shown that as a consequence of the high rates of oxygen consumption, oxygen tensions fall to almost zero within growing tubers of these transformants, possibly as a consequence of the high energy demand of the introduced pathway, and this results in a dramatic decrease in the cellular energy state (Bologa et al. 2003, Geigenberger 2003b). This decrease is probably the major reason for the unexpected observation that starch synthesis decreases in these lines. In general, oxygen can fall to very low concentrations in developing sink organs like potato tubers and seeds, even under normal environmental conditions (Geigenberger 2003b, Vigeolas et al. 2003, van Dongen et al. 2004). The consequences of these low internal oxygen concentrations for metabolic events during storage product formation have been ignored in metabolic engineering strategies. Molecular approaches to increase internal oxygen concentrations could provide a novel and exiting route for crop improvement.

Another failed attempt at increasing tuber starch accumulation was the overexpression of a heterologous sucrose transporter from spinach under the control of the CaMV 35S promoter (Leggewie et al. 2003). The rationale behind this attempt was that it would increase carbon partitioning toward the tuber; however, in the absence of improved photosynthetic efficiency this was not the case.

With respect to the plastidial pathway for starch synthesis, much attention has been focused on AGPase. Analysis of potato lines exhibiting different levels of reduction of AGPase due to antisense inhibition have been used to estimate flux control coefficients for starch synthesis of between 0.3 and 0.55 for this enzyme (Geigenberger et al. 1999a, Sweetlove et al. 1999), showing that AGPase is collimating for starch accumulation in potato tubers. In addition to having significantly reduced starch content in the tubers, these lines also exhibit very high tuber sucrose content, produce more but smaller tubers per plant, and

Table 32.2. Summary of Transgenic Approaches to Manipulate Starch Structure and Yield in Potato Tubers

Target	Transgenic Approach	Starch Content/ Yield	Starch Structure	Grain/Tuber Morphology	Application	Reference
Plastidial ATP/ADP translocator	Antisense inhibition	Starch content decreased to 20%	Less amylose	Grains size decreased to 50%, strange tuber shape	—	Tjaden et al. 1998
Plastidial ATP/ADP translocator	Overexpression of AATP	Starch content increased by up to 30%	More amylose	Grains with angular shape	High starch with more amylose	Tjaden et al. 1998, Geigenberger et al. 2001
Plastidial adenylate kinase	Antisense inhibition	Starch content and yield increased by up to 60%	Not determined	Not determined	High starch	Regierer et al. 2003
ADPGlc pyro-phosphorylase	Antisense inhibition of AGPB	Starch content decreased down to 10% of WT-level	Amylose content decreased by 40%	Smaller grains, smaller tubers, increased tuber number	—	Müller-Röber et al. 1992, Lloyd et al. 1999b
ADPGlc pyro-phosphorylase	Overexpression of *glgC*	Starch content increased by up to 30%	Not determined	Not determined	High starch	Stark et al. 1991
Granule bound starch synthase	Antisense inhibition of GBSSI	No effect	Starch is free of amylose	No effect	Amylose free starch	Kuipers et al. 1994

Granule-bound starch synthase	Overexpression of GBSS I	No effect	Amylose increased up to 25.5%	No effect	—	Kuipers et al. 1994
Soluble starch synthase	Antisense inhibition of SS III	No effect	Amylopectin chain length altered, more P in starch	Granules cracked with deep fissures	—	Abel et al. 1996, Marshall et al. 1996; Fulton et al. 2002
Soluble and granule-bound starch synthases	Antisense inhibition of GBSS I, SS II & SS III	No effect	Less amylose, shorter amylopectin chains	Concentric granules	Freeze/thaw- stable starch	Jobling et al. 2002, Fulton et al. 2002
Starch branching enzymes	Antisense inhibition of SBE I and SBE II	Decreased yield	Amylose increased to >70%, low amylopectin, five-fold more P in starch.	Granule morphology altered	High amylose and high P starch	Schwall et al. 2000
Starch water dikinase (R1)	Antisense inhibition of R1	No effect	Less P in starch	Not determined	Low P starch	Lorberth et al. 1998

produce smaller starch grains than the wild type (Müller-Röber et al. 1992; see also later discussion and Table 32.2). To date, one of the most successful approaches for elevating starch accumulation in tubers was that of Stark et al. (1991), who overexpressed an unregulated bacterial AGPase. This manipulation resulted in up to a 30% increase in tuber starch content. However, it should be noted that the expression of exactly the same enzyme within a second potato cultivar did not significantly affect starch levels (Sweetlove et al. 1996), indicating that these results might be highly context dependent. Another more promising route to increase starch yield would be to manipulate the regulatory network leading to posttranslational redox activation of AGPase (see Fig. 32.4). Based on future progress in this field, direct strategies can be taken to modify the regulatory components leading to redox regulation of starch synthesis in potato tubers.

More recent studies have shown that the adenylate supply to the plastid is of fundamental importance to starch biosynthesis in potato tubers (Loef et al. 2001, Tjaden et al. 1998). Overexpression of the plastidial ATP/ADP translocator resulted in increased tuber starch content, whereas antisense inhibition of the same protein resulted in reduced starch yield, modified tuber morphology, and altered starch structure (Tjaden et al. 1998). Furthermore, incubation of tuber discs in adenine resulted in a considerable increase in cellular adenylate pool sizes and a consequent increase in the rate of starch synthesis (Loef et al. 2001). The enzyme adenylate kinase (EC 2.7.4.3) interconverts ATP and AMP into 2 ADP. Because adenylate kinase is involved in maintaining the levels of the various adenylates at equilibrium, it represents an interesting target for modulating the adenylate pools in plants. For this reason, a molecular approach was taken to downregulate the plastidial isoform of this enzyme by the antisense technique (Regierer et al. 2002). This manipulation led to a substantial increase in the levels of all adenylate pools (including ATP) and, most importantly, to a record increase in tuber starch content up to 60% above wild type. These results are particularly striking because this genetic manipulation also resulted in a dramatic increase in tuber yield during several field trials of approximately 40% higher than that of the wild type. When taken in tandem, these results suggest a doubling of starch yield per plant. In addition to the changes described above, more moderate increases in starch yield were previously obtained by targeting enzymes esoteric to the pathway of starch synthesis, for example, plants impaired in their expression of the sucrose synthetic enzyme, sucrose phosphate synthase (Geigenberger et al. 1999b). This enzyme exerts negative control on starch synthesis since it is involved in a futile cycle of sucrose synthesis and degradation and leads to a decrease in the net rate of sucrose degradation in potato tubers (Geigenberger et al. 1997).

Although these results are exciting from a biotechnological perspective and they give clear hints as to how starch synthesis is coordinated in vivo, they do not currently allow us to establish the mechanisms by which they operate. It is also clear that these results, while promising, are unlikely to be the only way to achieve increases in starch yield. Recent advances in transgenic technologies now allow the manipulation of multiple targets in tandem (Fernie et al. 2001), and given that several of the successful manipulations described above were somewhat unexpected, the possibility that further such examples will be uncovered in the future cannot be excluded. One obvious future target would be to reduce the expression levels of the starch degradative pathway since, as described above, starch content is clearly a function of the relative activities of the synthetic and degradative pathways. Despite the fact that a large number of *Arabidopsis* mutants has now been generated that are deficient in the pathway of starch degradation, the consequence of such deficiencies has not been investigated in a crop such as potato tubers. Furthermore, there are no reports to date of increases in starch yield in heterotrophic tissues displaying mutations in the starch degradative pathway.

A further phenomenon in potatoes that relates to starch metabolism is that of cold-induced sweetening, where the rate of degradation of starch to reducing sugars is accelerated. As raw potatoes are sliced and cooked in oil at high temperature, the accumulated reducing sugars react with free amino acids in the potato cell, forming unacceptably brown- to black-pigmented chips or fries via a nonenzymatic, Maillard-type reaction. Potatoes yielding these unacceptably colored products are generally rejected for purchase by the processing plant. If a "cold-processing potato" (i.e., one which has low sugar content even in the cold) were available, energy savings would be realized in potato-growing regions where outside storage temperatures are cool. In regions where outside temperatures are moderately high, increased refrigeration costs may occur. This expense would be offset, however, by removal of the need to purchase dormancy-prolonging chemicals, by a decreased need for disease control, and by improvement of long-term tuber quality. Although such a cold-processing potato is not yet on the market, several manipulations potentially fulfill this criterion, perhaps most impressively the antisense inhibition of GWD, which is involved in the initiation of starch degradation (Lorberth et al. 1998). Further examples on this subject are excellently reviewed in a recent paper by Sowokinos (2001).

While only a limited number of successful manipulations of starch yield have been reported to date, far more successful manipulations have been reported with respect to engineering starch structure. These will be reviewed in Section "Manipulation of Starch Structure."

MANIPULATION OF STARCH STRUCTURE

In addition to attempting to increase starch yield, there have been many, arguably more, successful attempts to manipulate its structural properties. Considerable natural variation exists between the starch structures of crop species, with potato starch having larger granules, less amylose, a higher proportion of covalently bound phosphate, and less protein and lipid content than cereal starches. The level of phosphorylation strongly influences the physical properties of starch, and granule size is another important factor for many applications; for example, determining starch noodle processing and quality (Jobling et al. 2004). In addition, the ratio between the different polymer types can affect the functionality of different starches (Slattery et al. 1998). High amylose starches are used in fried snack products to create crisp and evenly brown snacks, as gelling agents, and in photographic films, whereas high amylopectin starches are useful in the food industry (to improve uniformity, stability, and texture) and in the paper and adhesive industries. Both, conventional breeding and transgenic approaches have been utilized for modification of starch properties (specifically amylose, amylopectin, and phosphate content; Sene et al. 2000, Kossmann and Lloyd 2000). However, in contrast to the situation described previously for yield, the use of natural mutants has been far more prevalent in this instance.

QTL analyses have recently been adopted for identifying the genetic factors underlying starch structure. Recombinant inbred lines were produced from parental maize lines of differing starch structure, and the amylose, amylopectin, and water-soluble fractions were analyzed. The loci linked to these traits were located on a genetic map consisting of RFLP (restriction fragment length polymorphism) markers (Sene et al. 2000). Using this strategy, candidate genes were identified that influenced starch structure. It is clear that application of these techniques may facilitate future breeding approaches to generate modified starch properties.

In addition to plant breeding approaches the use of transgenesis has been of fundamental importance for understanding and influencing starch structure (see Table 32.2). Potato tubers with decreased expression of AGPase (Müller-Röber et al. 1992) had decreased amylose contents, down to about 60% of that found in wild type, and smaller starch granules (Lloyd et al. 1999b). The reduction in amylose content is most likely due to the decreased levels of ADP-glucose found in these plants, since this leads to selective restriction of granule-bound versus soluble SSs, the latter having a higher affinity for ADPGlc (Frydman and Cardini 1967). Similarly, when the plastidial adenylate supply was altered by changes in the expression of the amyloplastidial ATP/ADP translocator, not only the starch content, but also its structural properties were altered (Tjaden et al. 1998, Geigenberger et al. 2001). Tubers of ATP/ADP-translocator overexpressing lines had higher levels of ADP-glucose and starch with higher amylose content, whereas the opposite was true for antisense lines. Furthermore, microscopic examination of starch grains revealed that their size in antisense tubers was considerably decreased (by 50%) in comparison with the wild type.

Antisensing granule-bound SS I (GBSSI) in potato tubers resulted in a starch that was almost free of amylose (Kuipers et al. 1994). This type of starch has improved paste clarity and stability with potential applications in the food and paper industries (Jobling 2004). In contrast to this, the amylose content of potato starch could not be increased above a value of 25.5% on overexpression of GBSS, hinting that this enzyme is not limiting in amylose production (Flipse et al. 1994, 1996). It has additionally been demonstrated that amylose can be completely replaced by a branched material that exhibits properties in between those of amylose and amylopectin following plastid-targeted expression of bacterial glycogen synthases (Kortstee et al. 1998).

The roles of the other isoforms of starch synthase (SS) are, if anything, less clear. A dramatic reduction in the expression of SSI in potato had absolutely no consequences on starch structure (Kossmann et al. 1999). The effects of modifying SSII were observed in pea seeds mutated at the *rug*5 locus, which encodes this protein. These mutants display a wrinkled phenotype that is often caused by decreased embryo starch content. Starch granules in these plants exhibit a striking alteration in morphology in that they have a far more irregular shape and exhibit a reduction in medium chain length glucans coupled to an increase in short- and long-chain glucans. Despite the dramatic changes observed in the pea mutant, the down-regulation of SSII in the potato tuber had only minor effects, including a 50% reduction in the phosphate content of the starch and a slight increase in short chain length glucans (Kossmann et al. 1999, Lloyd et al. 1999a). However, it is worth noting that SSII only contributes 15% of the total soluble SS activity within the tuber. Finally, a third class of SS has also been identified and has been reported to be the major form in potato tubers (Marshall et al. 1996). When this isoform (SSIII) was downregulated by the creation of transgenic potato plants, deep fissures were observed in the starch granule under the electron microscope, most probably due to GBSS making longer glucan chains in this background (Fulton et al. 2002). However, the starch was not altered in its amylose content, nor did it display major differences in chain length distribution of short chains within the amylopectin (Marshall et al. 1996, Lloyd et al. 1999a). However, dramatic changes were observed when the structure of the side chains was studied, with an accumulation of shorter chains observed in the transgenic lines (Lloyd et al. 1999a). In addition to this, the phosphate content of the starch was doubled in this transformant (Abel et al. 1996).

Given that some of the changes observed on alteration of a single isoform of SS were so dramatic, several groups have looked into the effects of simultaneously modifying the activities of more than one isoform by the generation of chimeric antisense constructs. When potato tubers were produced in which the activities of SSII and SSIII were simultaneously repressed (Edwards et al. 1999, Lloyd et al. 1999b), the effects were not additive. The amylopectin from these lines was somewhat different from that resulting from inhibition of the single isoforms, with the double antisense plants exhibiting grossly modified amylopectin consisting of more short and extra long chains but fewer medium length chains, leading to a gross alteration in the structure of the starch granules. In another study, the parallel reduction of GBSSI, SSII, and SSIII resulted in starch with less amylose and shorter amylopectin chains, which conferred additional freeze-thaw stability of starch with respect to the wild type (Jobling et al. 2002). This is beneficial since it replaces the necessity for expensive chemical substitution reactions, and in addition, its production may require less energy, as this starch cooks at much lower temperatures than normal potato starches.

Simultaneously inhibiting two isoforms of starch-branching enzyme to below 1% of the wild-type activities resulted in highly modified starch in potato tubers. In this starch, normal, high molecular weight amylopectin was absent, whereas the amylose content was increased to levels above 70%, comparable to that in the highest commercially available maize starches (Schwall et al. 2000; Table 32.2). There was also a major effect on starch granule morphology. In addition, the phosphorus content of the starch was increased more than five fold. This unique starch, with its high amylose, low amylopectin, and high phosphorus levels, offers novel properties for food and industrial applications. A further example of altered phosphate starch is provided by research into potato plants deficient in GWD (previously known as R1, Lorberth et al. 1998, Ritte et al. 2002)). The importance of starch phosphorylation becomes clear when it is considered that starch phosphate monoesters increase the clarity and viscosity of starch pastes and decrease the gelatinization and retrogradation rate. Interestingly, crops that produce high-amylose starches are characterized by considerably lower yield (Jobling 2004). Additional transgenic or breeding approaches, as discussed in Section "Manipulation of Starch Yield" will be required to minimize the yield penalty that goes along with the manipulation of crops for altered starch structure and functionality.

As discussed previously for starch yield, the importance of starch degradative processes in determination of starch structural properties is currently poorly understood. However, it represents an interesting avenue for further research.

CONCLUSIONS AND FUTURE PERSPECTIVES

In this chapter we have reviewed the pathway organization of starch synthesis within the potato tuber and detailed how it can be modulated through transgenesis to result in higher starch yield or the production of starches of modified structure. While several successful examples exist for both types of manipulation, these are yet to reach the field. It is likely that such crops, once commercially produced, will yield both industrial and nutritional benefits to society. In addition to this, other genetic and biochemical factors may well also influence starch yield and structure, and further research is required in this area in order to fully optimize these parameters in crop species. In this context, transgenic approaches complementary to conventional breeding will allow more "fine-tuning" of starch properties.

REFERENCES

Abel GJW, et al. 1996. Cloning and functional analysis of a cDNA encoding a novel 139 kDa starch synthase from potato (Solanum tuberosum L). *Plant J* 10: 981–991.

Beck E, Ziegler P. 1989. Biosynthesis and degradation of starch in higher plants. *Annu Rev Plant Physiol Plant Mol Biol* 40: 95–117.

Boos W, Shuman H. 1998. Maltose/maltodextrin system of *Escherichia coli*: Transport, metabolism, and regulation. *Microbiol Mol Biol Rev* 62: 204–229.

Bologa KL, et al. 2003. A bypass of sucrose synthase leads to low internal oxygen and impaired metabolic performance in growing potato tubers. *Plant Physiol* 132: 2058–2072.

Buleon A, et al. 1998. Starch granule structure and biosynthesis. *Int J Biol Macromol* 23: 85–112.

Chia T, et al. 2004. A cytosolic glycosyltransferase is required for conversion of starch to sucrose in Arabidopsis leaves at night. *Plant J* (in press).

Dennis DT, Blakeley SD. 2000. Carbohydrate metabolism. In: BB Buchanan, et al. (eds.), *Biochemistry and Molecular Biology of Plants*, American Society of Plant Physiologists, Rockville, pp. 630–675.

Duwenig E, et al. 1997. Antisense inhibition of cytosolic phosphorylase in potato plants (*Solanum tuberosum* L.) affects tuber sprouting and flower formation with only little impact on carbohydrate metabolism. *Plant J* 12: 323–333.

Edwards A, et al. 1999. A combined reduction in the activity of starch synthases II and III of potato has novel effects on the starch of tubers. *Plant J* 17: 251–261.

Entwistle G, Rees T. 1990. Lack of fructose-1,6-bisphosphatase in a range of higher plants that store starch. *Biochem J* 271: 467–472.

Fernie AR, et al. 2001. Simultaneous antagonistic modulation of enzyme activities in transgenic plants through the expression of a chimeric transcript. *Plant Physiol Biochem* 39: 825–830.

Fernie AR, Willmitzer L. 2001. Molecular and biochemical triggers of potato tuber development. *Plant Physiol* 127: 1459–1465.

Flipse E, et al. 1994. Expression of a wild type GBSS introduced into an amylose free potato mutant by *Agrobacterium tumefaciens* and the inheritance of the inserts at the microscopic level. *Theo Appl Gen* 88: 369–375.

Flipse E, et al. 1996. The dosage effect of the wild type GBSS allele is linear for GBSS activity but not for amylose content: Absence of amylose has a distinct influence on the physio-chemical properties of starch. *Theo Appl Gen* 92: 121–127.

Frydman RB, Cardini CE. 1967. Studies on the biosynthesis of starch. *J Biol Chem* 242: 312–317.

Fulton DC, et al. 2002. Role of granule-bound starch synthase in determination of amylopectin structure and starch granule morphology in potato. *J Biol Chem* 277: 10834–10841.

Geigenberger P. 2003a. Regulation of sucrose to starch conversion in growing potato tubers. *J Exp Bot* 54: 457–465.

———. 2003b. Response of plant metabolism to too little oxygen. *Curr Opin Plant Biol* 6: 247–256.

Geigenberger P, et al. 1998. High-temperature perturbation of starch synthesis by decreased levels of glycerate-3-phosphate in growing potato tubers. *Plant Physiol* 117: 1307–1316.

Geigenberger P, et al. 1999a. Contribution of adenosine 5'-diphosphoglucose pyrophosphorylase to the control of starch synthesis is decreased by water stress in growing potato tubers. *Planta* 209: 338–345.

Geigenberger P, et al. 1999b. Decreased expression of sucrose phosphate synthase strongly inhibits the water stress-induced synthesis of sucrose in growing potato tubers. *Plant J* 19: 119–129.

Geigenberger P, et al. 1997. Regulation of sucrose and starch metabolism in potato tubers in response to short-term water deficit. *Planta* 201: 502–518.

Geigenberger P, et al. 2001. Tuber physiology and properties of starch from tubers of transgenic potato plants with altered plastidic adenylate transporter activity. *Plant Physiol* 12: 1667–1678.

Hatzfeld WD, Stitt M. 1990. A study of the rate of recycling of triose phosphates in heterotrophic *Chenopodium rubrum* cells, potato tubers and maize endosperm. *Planta* 180: 198–204.

Jenner HL, et al. 2001. NAD-malic enzyme and the control of carbohydrate metabolism in potato tubers. *Plant Physiol* 126: 1139–1149.

Jansson C, et al. 1997. Cloning, characterisation and modification of genes encoding starch branching enzymes in barley. In: PJ

Frazier, et al. (eds.), *Starch Structure and Functionality*. The Royal Society of Chemistry, Cambridge, pp. 196–203.

Jobling SA. 2004. Improving starch for food and industrial applications. *Curr Opin Plant Biol* 7 (in press).

Jobling SA, et al. 2002. Production of a freeze-thaw-stable potato starch by antisense inhibition of three starch synthase genes. *Nature Biotech* 20: 295–299.

Kammerer B, et al. 1998. Molecular characterization of a carbon transporter in plastids from heterotrophic tissues: The glucose 6-phosphate phosphate antiporter. *Plant Cell* 10: 105–117.

Keeling PL, et al. 1988. Starch biosynthesis in developing wheat grains. Evidence against the direct involvement of triose phosphates in the metabolic pathway. *Plant Physiol* 87: 311–319.

Kortstee AJ, et al. 1998. The influence of an increased degree of branching on the physio-chemical properties of starch from genetically modified potato. *Carb Poly* 37: 173–184.

Kossmann J, et al. 1992. Cloning and expression analysis of the plastidic fructose-1,6-bisphosphatase coding sequence: Circumstantial evidence for the transport of hexoses into chloroplasts. *Planta* 188: 7–12.

Kossmann J, et al. 1999. Cloning and functional analysis of a cDNA encoding a starch synthase from potato (*Solanum tuberosum* L.) that is predominantly expressed in leaf tissue. *Planta* 208: 503–511.

Kossmann J, Lloyd J. 2000. Understanding and influencing starch biochemistry. *Crit Rev Plant Sci* 19: 171–226.

Kuipers AGJ, et al. 1994. Formation and deposition of amylose in the potato tuber starch granule are affected by the reduction of granule bound starch synthase gene expression. *Plant Cell* 6: 43–52.

Leggewie G, et al. 2003. Overexpression of the sucrose transporter SoSUT1 in potato results in alterations in leaf partitioning and in tuber metabolism but has little impact on tuber morphology. *Planta* 217: 158–167.

Lloyd JR, et al. 1999a. Simultaneous antisense inhibition of two starch synthase isoforms in potato tubers leads to accumulation of grossly modified amylopectin. *Biochem J* 338: 515–521.

Lloyd JR, et al. 1999b. The influence of alterations in ADPglucose pyrophosphorylase activities on starch structure and composition in potato tubers. *Planta* 209: 230–238.

Loef I, et al. 2001. Increased levels of adenine nucleotides modify the interaction between starch synthesis and respiration when adenine is supplied to discs from growing potato tubers. *Planta* 212: 782–791.

Lorberth R, et al. 1998. Inhibition of a starch-granule-bound protein leads to modified starch and the repression of cold sweetening. *Nature Biotech* 16: 473–477.

Lu Y, Sharkey TD. 2004. The role of amylomaltase in maltose metabolism in the cytosol of photosynthetic cells. *Planta* 218: 466–473.

Marshall J, et al. 1996. Identification of the major isoform of starch synthase in the soluble fraction of potato tubers. *Plant Cell* 8: 1121–1135.

Müller-Röber B, et al. 1992. Inhibition of ADPglucose pyrophosphorylase in transgenic potatoes leads to sugar storing tubers and influences tuber formation and the expression of tuber storage protein genes. *EMBO J* 11: 1229–1238.

Neuhaus HE, Emes MJ. 2000. Nonphotosynthetic metabolism in plastids. *Annu Rev Plant Physiol Plant Mol Biol* 51: 111–140.

Niittylä T, et al. 2004. A previously unknown maltose transporter essential for starch degradation in leaves. *Science* 303: 87–89.

Pilling E, Smith AM. 2003. Growth ring formation in the starch granules of potato tubers. *Plant Physiol* 132: 365–371.

Preiss J. 1988. Biosynthesis of starch and its regulation. In: J Preiss, (ed.), *The Biochemistry of Plants*, Academic Press, San Diego, pp. 181–254.

Regierer B, et al. 2002. Starch content and yield increase as a result of altering adenylate pools in transgenic plants. *Nature Biotech* 20: 1256–1260.

Ritte G, et al. 2002. The starch-related R1 protein is an alpha-glucan, water dikinase. *Proc Natl Acad Sci USA* 99: 7166–7171.

Ritte G, et al. 2004. Phosphorylation of transitory starch is increased during degradation. *Plant Physiology* (in press).

Schäfer-Pregl R, et al. 1998. Analysis of quantitative trait loci (QTLs) and quantitative trait alleles (QTAs) for potato tuber yield and starch content. *Theo Appl Gen* 97: 834–846.

Scheidig A, et al. 2002. Down regulation of a chloroplast-targeted beta-amylase leads to a starch-excess phenotype in leaves. *Plant J* 30: 581–591.

Schwall GP, et al. 2000. Production of a very-high amylose potato starch by inhibition of SBE A and B. *Nature Biotech* 18: 551–554.

Sene M, et al. 2000. Quantitative trait loci affecting amylose, amylopectin and starch content in maize recombinant inbred lines. *Plant Physiol Biochem* 38: 459–472.

Shannon JC, et al. 1998. Brittle-1, an adenylate translocator, facilitates transfer of extra-plastidial synthesized ADPglucose into amyloplasts of maize endosperms. *Plant Physiol* 117: 1235–1252.

Slattery CJ, et al. 2000. Engineering starch for increased quantity and quality. *Trends Plant Sci* 5: 291–298.

Smith AM, et al. 1995. What controls the amount and structure of starch in storage organs. *Plant Physiol* 107: 673–677.

———. 1997. The synthesis of the starch granule. *Annu Rev Plant Physiol Plant Mol Biol* 48: 65–87.

Smith AM, et al. 2003. Starch mobilisation in leaves. *J Exp Bot* 54: 577–583.

Sonnewald U, et al. 1995. A second L-type isozyme of potato glucan phophorylase: Cloning, antisense inhibition and expression analysis. *Plant Molecular Biology* 21: 567–576.

Sonnewald U, et al. 1997. Expression of a yeast invertase in the apoplast of potato tubers increases tuber size. *Nature Biotech* 15: 794–797.

Stark DM, et al. 1991. Regulation of the amount of starch in plant tissues. *Science* 258: 287–292.

Sowokinos JR. 2001. Biochemical and molecular control of cold-induced sweetening in potatoes. *Am J Pot Res* 78: 221–236.

Stitt M, Steup M. 1985. Starch and sucrose degradation. In: R Douce, DA Day, (eds.), *Encyclopedia of Plant Physiology*, vol. 18, Springer-Verlag, Berlin, pp. 347–390.

Sullivan TD, et al. 1991. Analysis of maize *Brittle*-1 alleles and a defective suppressor mutator inducible allelle. *Plant Cell* 3: 1337–1348.

Sweetlove LJ, et al. 1996. Characterisation of transgenic potato (*Solanum tuberosum*) tubers with increased ADPglucose pyrophosphorylase. *Biochem J* 320: 487–492.

Sweetlove LJ, et al. 1999. The contribution of adenosine 5'-diphosphoglucose pyrophosphorylase to the control of starch synthesis in potato tubers. *Planta* 209: 330–337.

Tauberger E, et al. 2000. Antisense inhibition of plastidial phosphoglucomutase provides compelling evidence that potato tuber amyloplasts import carbon from the cytosol in the form of glucose-6-phosphate. *Plant J* 23: 43–53.

Thorbjornsen T, et al. 1996. Distinct isoforms of ADPglucose pyrophosphorylase occur inside and outside the amyloplast in barley endosperm. *Plant J* 16: 531–540.

Tiessen A, et al. 2002. Starch synthesis in potato tubers is regulated by post-translational redox modification of ADP-glucose pyrophosphorylase: A novel regulatory mechanism linking starch synthesis to the sucrose supply. *Plant Cell* 14: 2191–2213.

Tiessen A, et al. 2003. Evidence that SNF1-related kinase and hexokinase are involved in separate sugar-signalling pathways modulating post-translational redox activation of ADP-glucose pyrophosphorylase in potato tubers. *Plant J* 35: 490–500.

Tjaden J, et al. 1998. Altered plastidic ATP/ADP-transporter activity influences potato (*Solanum tuberosum* L.) tuber morphology, yield and composition of tuber starch. *Plant J* 16: 531–540.

Trethewey RN, et al. 2001. Expression of a bacterial sucrose phosphorylase in potato tubers results in a glucose-independent induction of glycolysis. *Plant Cell Environ* 24: 357–365.

Trethewey RN, et al. 1998. Combined expression of glucokinase and invertase in potato tubers leads to a dramatic reduction in starch accumulation and a stimulation of glycolysis. *Plant J* 15: 109–118.

Van Dongen JT, et al. 2004. Phloem import and storage metabolism are highly coordinated by the low oxygen concentrations within developing wheat seeds. *Plant Physiol* (in press).

Vigeolas, H, et al. 2003. Lipid metabolism is limited by the prevailing low oxygen concentrations within developing seeds of oilseed rape. *Plant Physiol* 133: 2048–2060.

Weise SE, et al. 2004. Maltose is the major form of carbon exported from the chloroplast at night. *Planta* 218: 474–482.

Yu T-S, et al. 2001. The *Arabidopsis sex1* mutant is defective in the R1 protein, a general regulator of starch degradation in plants, and not in the chloroplast hexose transporter. *Plant Cell* 13: 1907–1918.

Zeeman SC, et al. 2004. The role of plastidial α-glucan phosphorylase in starch degradation and tolerance of abiotic stress in *Arabidopsis* leaves. *Plant Physiol* (in press).

33
Biochemistry of Beer Fermentation

Ronnie Willaert

Abstract: The carbohydrate metabolism and flavor formation during yeast primary and secondary fermentation (maturation) is reviewed. Carbohydrate metabolism and ethanol production during the primary fermentation is discussed firstly. Next, the metabolism of the bioflavoring by-product formation, that is, higher alcohols, esters, organic acids, and vicinal diketones, is elaborated. The next step of the fermentation process is the maturation process where the vicinal diketones, acetaldehyde, and hydrogen sulfide concentration needs to be reduced to acceptable levels. The chapter is concluded with a discussion about the use of immobilized cell technology to intensify the fermentation process and its impact on flavor production.

INTRODUCTION

The production of alcoholic beverages is as old as history. Wine may have an archeological record going back more than 7500 years, with the early suspected wine residues dating from early to mid-fifth millennium BC (McGovern et al. 1996). Clear evidence of intentional winemaking first appears in the representations of wine presses that date back to the reign of Udimu in Egypt, some 5000 years ago. The direct fermentation of fruit juices, such as that of grape, had doubtlessly taken place for many thousands of years before early thinking man developed beer brewing and, probably coincidentally, bread baking (Hardwick 1995). The oldest historical evidence of formal brewing dates back to about 6000 BC in ancient Babylonia is a piece of pottery found there, which shows workers either stirring or skimming a brewing vat.

Nowadays, alcoholic beverage production represents a significant contribution to the economies of many countries. The most important beverages today are beer, wine, distilled spirits, cider, sake, and liqueurs (Lea and Piggott 1995). In Belgium ("the beer paradise"), beer is the most important alcoholic beverage, although the beer consumption declined in the last 40 years: from 11,096,717 hL in 1965 to 9,703,000 hL in 2004 (NN 2005). In this time frame, wine consumption doubled from 1,059,964 to 2,215,579 hL. Another trend is the spectacular increase in waters and soft drinks consumption (from 5,215,056 to 26,395,000 hL).

In this chapter, the biochemistry and fermentation of beer is reviewed. First, the carbohydrate metabolism in brewer's yeast is discussed. The maltose metabolism is of major importance in beer brewing, since this sugar is in a high concentration present in wort. For the production of a high-quality beer, a well-controlled

Food Biochemistry and Food Processing, Second Edition. Edited by Benjamin K. Simpson, Leo M.L. Nollet, Fidel Toldrá, Soottawat Benjakul, Gopinadhan Paliyath and Y.H. Hui.

fermentation needs to be performed. During this fermentation, major flavor-active compounds are produced (and some of them are again metabolized) by the yeast cells. The metabolism of the most important fermentation by-products during main and secondary fermentation is discussed in detail. The latest trend in beer fermentation technology is the process intensification using immobilized cell technology (ICT). This new technology is explained and some illustrative applications—on small and large scale—are discussed.

THE BEER BREWING PROCESS

The principal raw materials used to brew beer are water, malted barley, hops and yeast. The brewing process involves extracting and breaking down the carbohydrate from the malted barley to make a sugar solution (called "wort"), which also contains essential nutrients for yeast growth, and using this as a source of nutrients for "anaerobic" yeast growth. During yeast fermentation, simple sugars are consumed, releasing heat and producing ethanol and other flavoring metabolic by-products. The major biological changes, which occur in the brewing process, are catalyzed by naturally produced enzymes from barley (during malting) and yeast. The rest of the brewing process largely involves heat exchange, separation, and clarification, which only produces minor changes in chemical composition when compared to the enzyme catalyzed reactions. Barley is able to produce all the enzymes that are needed to degrade starch, β-glucan, pentosans, lipids, and proteins, which are the major compounds of interest to the brewer. An overview of the brewing process is shown in Figure 33.1, where also the input and output flows are indicated. Table 33.1 gives a more detailed explanation of each step in the process.

CARBOHYDRATE METABOLISM—ETHANOL PRODUCTION

Wort Carbohydrates Uptake and Metabolism

Carbohydrates in wort make up 90–92% of wort solids. Wort from barley malt contains the fermentable sugars sucrose, fructose, glucose, maltose, and maltotriose together with some

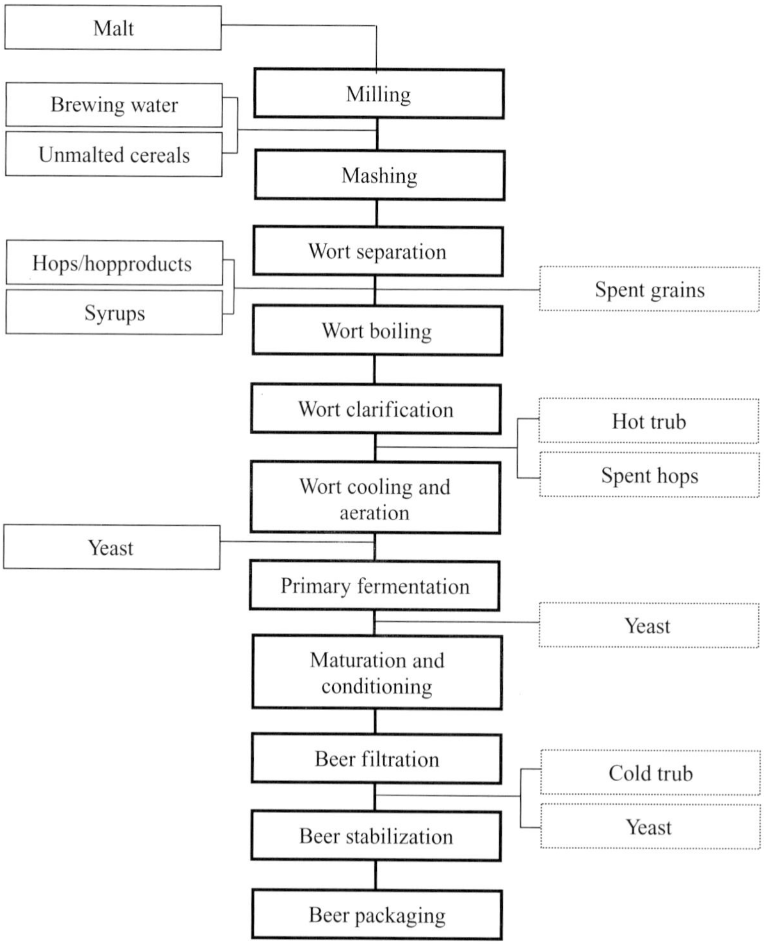

Figure 33.1. Schematic overview of the brewing process (input flows are indicated on the left side and output flows on the right side).

Table 33.1. Overview of the Brewing Processing Steps: From Barley to Beer

Process	Action	Objectives	Time	Temperature (°C)
Malting	Moistening and aeration of barley	Preparation for the germination process	48 h	12–22
Steeping			3–5 d	22
Germination	Barley germination	Enzyme production, chemical structure modification	24–48 h	22–110
Kilning	Kilning of the green malt	Ending of germination and modification, production of flavoring and coloring substances		
Milling	Grain crushing without disintegrating the husks	Enzyme release and increase of surface area	1–2 h	22
Mashing + wort separation	Addition of warm/hot water	Stimulation of enzyme action, extraction and dissolution of compounds, wort filtration, to obtain the desired fermentable extract as quick as possible	1–2 h	30–72
Wort boiling	Boiling of wort and hops	Extraction and isomerization of hop components, hot break formation, wort sterilization, enzyme inactivation, formation of reducing, aromatic and coloring compounds, removal of undesired volatile aroma compounds, wort acidification, evaporation of water	0.5–1.5 h	>98
Wort clarification	Sedimentation or centrifugation	Removal of spent hops, clarification (whirlpool, centrifuge, settling tank)	<1 h	100–80
Wort cooling and aeration	Use of heat exchanger, injection of air bubbles	Preparing the wort for yeast growth	<1 h	12–18
Fermentation	Adding yeast, controlling the specific gravity, removal of yeast	Production of green beer, to obtain yeast for subsequent fermentations, carbon dioxide recovery	2—7 d	12–22 (ale) 4–15 (lager)
Maturation and conditioning	Beer storage in oxygen free tank, beer cooling, adding processing aids	Beer maturation, adjustment of the taste, adjustment of CO_2 content, sedimentation of yeast and cold trub, beer stabilization	7–21 h	−1–0
Beer clarification	Centrifugation, filtration	Removal of yeast and cold trub	1–2 h	−1–0
Biological stabilization	Pasteurization of sterile filtration	Killing or removing of microorganisms	1–2 h	62–72 (past.) −1–0 (filtr.)
Packaging	Filling of bottles, cans, casks, and kegs; pasteurization of small volumes in packings	Production of packaged beer according to specifications	0.5–1.5 h	−1–room temperature

dextrin material (Table 33.2). The fermentable sugars typically make up 70–80% of the total carbohydrate (MacWilliam 1968). The three major fermentable sugars are glucose, α-glucosides maltose, and maltotriose. Maltose is by far the most abundant of these sugars, typically accounting for 50–70% of the total fermentable sugars in an all-malt wort. Sucrose and fructose are present in a low concentration. The unfermentable dextrins play little part in brewing. Wort fermentability may be reduced or increased by using solid (e.g., corn grits, flakes, or rice) or liquid adjuncts (e.g., sugar syrups).

Table 33.2. Carbohydrate Composition of Worts

Origin Type of Wort	Danish Lager 11°P	Canadian Lager 13°P	British Pale Ale 10°P	Canadian Corn Adjunct Wort	All-Malt Wort 18°P
Fructose (g/L) (%)[a]	2.1	1.5	3.3	1.3	3.0
	2.7	1.6	4.8	1.3	
Glucose (g/L) (%)[a]	9.1	10.3	10.0	14.7	13.0
	11.6	10.9	14.5	15.7	
Sucrose (g/L) (%)[a]	2.3	4.2	5.3	1.8	
	2.9	4.5	7.7	1.9	
Maltose (g/L) (%)[a]	52.4	60.4	38.9	62.8	80.0
	66.6	64.2	56.5	67.0	
Maltotriose (g/L) (%)[a]	12.8	17.7	11.4	13.2	24.0
	16.3	18.8	16.5	14.1	
Total ferm. sugars (g/L)	78.7	94.1	68.9	93.8	121.0
Maltotetraose (g/L)	2.6	7.2	2.0		
Higher sugars (g/L)	21.3	26.8	25.2		
Total dextrins (g/L)	23.9	34.0	25.2		
Total sugars (g/L)	102.6	128.1	94.1	117.5	

Source: Patel and Ingledew (1973), Huuskonen et al. (2010), and Hough et al. (1982).
[a]Percent of the total fermentable sugars.

Brewing strains consume the wort sugars in a specific sequence: glucose is consumed first, followed by fructose, maltose, and finally maltotriose. The uptake and consumption of maltose and maltotriose is repressed or inactivated at elevated glucose concentrations. Only when 60% of the wort glucose has been taken up by the yeast, the uptake and consumption of maltose will start. Maltotriose uptake is inhibited by high glucose and maltose concentrations. When high amounts of carbohydrate adjuncts (e.g., glucose) or high-gravity wort are employed, the glucose repression is even more pronounced, resulting in fermentation delays (Stewart and Russell 1993).

The efficiency of brewer's yeast strains to effect alcoholic fermentation is dependent upon their ability to utilize the sugars present in wort. This ability very largely determines the fermentation rate as well as the final quality of the beer produced. In order to optimize the fermentation efficiency of the primary fermentation, a detailed knowledge of the sugar consumption kinetics, which is linked to the yeast growth kinetics, is required (Willaert 2001).

Maltose and Maltotriose Metabolism

The yeast *Saccharomyces cerevisiae* transports the monosaccharides across the cell membrane by the hexose transporters. There are 20 genes encoding hexose transporters (Dickinson 1999, Rintala et al. 2008): 18 genes encoding transporters (*HXT1* to *HXT17*, *GAL2*) and two genes encoding (*SNF3*, *RGT2*). The disaccharide maltose and the trisaccharide maltotriose are transported by specific transporters into the cytoplasm, where these molecules are hydrolyzed by the same α-glucosidase yielding two or three molecules of glucose, respectively (Panchal and Stewart 1979, Zheng et al. 1994a).

Maltose utilization in yeast is conferred by any one of five *MAL* loci: *MAL1* to *MAL4* and *MAL6* (Bisson et al. 1993, Dickinson 1999). Each locus consists of three genes (*MALx1*, where x stands for one of the five loci): gene 1 encodes a maltose transporter (permease), gene 2 encodes a maltase (α-glucosidase), and gene 3 encodes a transcriptional activator of the other two genes. Thus, for example, the maltose transporter gene at the *MAL1* locus is designated *MAL61*. The three genes of a *MAL* locus are all required to allow fermentation. Alternatively, some authors use gene designations such as for the *MAL1* locus: *MAL1T* (transporter = permease), *MAL1R* (regulator), and *MAL1S* (maltase). The genetic and biochemical analysis of maltose fermentation by yeast cells revealed a series of five unlinked telomere-associated multigene *MAL* loci: *MAL1* (chromosome VII), *MAL2* (chromosome III), *MAL4* (chromosome II), and *MAL6* (chromosome VIII). The *MAL* loci exhibit a very high degree of homology and are telomere linked, suggesting that they evolved by translocation from telomeric regions of different chromosomes (Michels et al. 1992). Since a fully functional or partial allele of the *MAL1* locus is found in all strains of *S. cerevisiae*, this locus is proposed as the progenitor of the other *MAL* loci (Chow et al. 1983), as all *S. cerevisiae* strains, and even its closest related yeast species *S. paradoxus*, contain *MAL1* sequences near the right telomere of chromosome VII. The genes in the *MAL* loci show a high degree of sequence and functional similarity, but there can be extensive variability, and several different alleles that determine distinct phenotypes (i.e., *MAL*-inducible and *MAL*-constitutive strains) have been described (Novak et al. 2004). Gene dosage studies performed with laboratory strains of yeast have shown that the transport of maltose in the cell may be the rate-limiting step in the utilization of this sugar (Goldenthal et al. 1987). Constitutive expression of the maltose transporter gene (*MALT*) with high-copy-number plasmids in a lager yeast strain has been found to accelerate the fermentation of maltose during high-gravity (24°P) brewing (Kodama et al. 1995). The constitutive expression of *MALS* and *MALR* had no effect on maltose fermentability.

The control over *MAL* gene expression is exerted at three levels. The presence of maltose induces, whereas glucose represses, the transcription of *MALS* and *MALT* genes (Federoff et al. 1983a, 1983b, Needleman et al. 1984). The constitutively expressed regulatory protein (*MALR*) binds near the *MALS* and *MALT* promotors and mediates the induction of *MALS* and *MALT* transcription (Cohen et al. 1984, Chang et al. 1988, Ni and Needleman 1990). Experiments with *MALR*-disrupted strains led to the conclusion that MalRp is involved in glucose repression (Goldenthal and Vanoni 1990, Yao et al. 1994). Relatively little attention has been paid to posttranscriptional control, that is, the control of translational efficiency, or mRNA turnover, as mechanisms complementing glucose repression (Soler et al. 1987). The addition of glucose to induced cells has been reported to cause a 70% increase in the liability of an mRNA population containing a fragment of *MALS* (Federoff et al. 1983a). The third level of control is posttranslational modification. In the presence of glucose, maltose permease is either reversibly converted to a conformational variant with decreased affinity (Siro and Lövgren 1979, Peinado and Loureiro-Dias 1986) or irreversibly proteolytically degraded depending on the physiological conditions (Lucero et al. 1993, Riballo et al. 1995). The latter phenomenon is called catabolite inactivation. Glucose repression is accomplished by the Mig1p repressor protein, which is encoded by the *MIG1* gene (Nehlin and Ronne 1990). It has been shown that Mig1p represses the transcription of all three *MAL* genes by binding upstream of them (Hu et al. 1995). The *MIG1* gene has been disrupted in a haploid laboratory strain and in an industrial polyploid strain of *S. cerevisiae* (Klein et al. 1996). In the *MIG1*-disrupted haploid strain, glucose repression was partly alleviated; that is, maltose metabolism was initiated at higher glucose concentrations than in the corresponding wild-type strain. In contrast, the polyploid Δ*mig1* strain exhibited an even more stringent glucose control of maltose metabolism than the corresponding wild-type strain, which could be explained by a more rigid catabolite inactivation of maltose permease, affecting the uptake of maltose.

All α-glucoside transport systems so far characterized in yeast are H^+-transporters that use the electrochemical proton gradient to actively transport these sugars into the cell (Crumplen et al. 1996, Stambuk and de Araujo 2001). It has been shown that maltose uptake is the rate-limiting step of fermentation (Kodama et al. 1995, Wang et al. 2002, Rautio and Londesborough 2003). At least three different maltose transporters have been identified in *S. cerevisiae*, and while the *MALx1* transporters (and probably the two *MPH2* and *MPH3* alleles) encode high affinity ($K_m = 2$–4 mmol l^{-1}) maltose permeases, the *AGT1* permease (a gene present in partially functional *mal1g* loci) transports maltose with lower ($K_m \approx 20$ mmol l^{-1}) affinity (Han et al. 1995, Stambuk and de Araujo 2001, Day et al. 2002a, Alves et al. 2007, 2008). *AGT1* is found in many *S. cerevisiae* laboratory strains and maps to a naturally occurring, partially functional allele of the *MAL1* locus (Han et al. 1995). Agt1p is a highly hydrophobic, postulated integral membrane protein. It is 57% identical to Mal61p (the maltose permease encoded at *MAL6*) and is also a member of the 12 transmembrane domain superfamily of sugar transporters (Nelissen et al. 1995). Like Mal61p, Agt1p is a high-affinity, maltose/proton symporter, but Mal61p is capable of transporting only maltose and turanose, while Agt1p transports these two α-glucosides as well as several others including isomaltose, α-methylglucoside, maltotriose, palatinose, trehalose and melezitose. *AGT1* expression is maltose inducible and induction is mediated by the Mal-activator.

Brewing strains of yeast are polyploid, aneuploid, or, in the case of lager strains, alloploid. Recently, Jespersen et al. (1999) examined 30 brewing strains of yeast (5 ale strains and 25 lager strains) with the aim of examining the alleles of maltose and maltotriose transporter genes contained by them. All the strains of brewer's yeast examined, except two, were found to contain *MAL11* and *MAL31* sequences, and only one of these strains lacked *MAL41*. *MAL21* was not present in the 5 ale strains and 12 of the lager strains. *MAL61* was not found in any of the yeast chromosomes other than those known to carry *MAL* loci. Sequences corresponding to the *AGT1* gene (transport of maltose and maltotriose) were detected in all but one of the yeast strains.

Although maltose is easily fermented by the majority of yeast strains after glucose exhaustion, maltotriose is not only the least preferred sugar for uptake by these *Saccharomyces* cells, but many yeasts may not use this α-glucosidase at all (Zheng et al. 1994a, Yoon et al. 2003). Incomplete maltotriose uptake during brewing fermentations results in yeast fermentable extract in beer, material loss, greater potential for microbiological stability and sometimes atypical beer flavor profiles (Stewart and Russell 1993). Maltotriose uptake from wort is always slower with ale strains than with lager strains under similar fermentation conditions. However, the initial transport rates are similar to those of maltose in a number of ale and lager strains. Elevated osmotic pressure inhibits the transport and uptake of glucose, maltose and maltotriose with maltose and maltotriose being more sensitive to osmotic pressure than glucose in both lager and ale strains. Ethanol (5% w/v) stimulated the transport of maltose and maltotriose, due in all probability to an ethanol-induced change in the plasma membrane configuration, but had no effect on glucose transport. Higher ethanol concentrations inhibited the transport of all three sugars.

Maltotriose uptake shows complex kinetics indicating the presence of high- and low-affinity transport activities, and studies on sugar utilization revealed that maltose and maltotriose are apparently transported by different permeases (Zheng et al. 1994b, Zastrow et al. 2001). Two genes are recognized as permease genes for transporting maltotriose in yeasts: *AGT1* transporter from *S. cerevisiae* and *MTY1* (also called *MTT1*) permease from *S. pastorianus*. They are characterized as low-affinity ($K_m \approx 20$ mmol l^{-1}) maltotriose transporters (Stambuk and de Araujo 2001, Dietvorst et al. 2005, Salema-Oom et al. 2005, Alves et al. 2007, 2008).

Recent reports regarding the observed patterns of maltose and maltotriose utilization by yeast cells show contradicting results.

Data indicated that all known α-glucoside transporters present in *S. cerevisiae*, including the maltose permeases *MAL31*, *MAL61*, *MPH2*, and *MHP3*, allowed growth of the yeast cells on both maltose and maltotriose (Day et al. 2002a, 2002b). Their kinetic analysis of maltose and maltotriose uptake indicated that all these transporters, including *AGT1* permease, could transport both sugars with practically the same affinities and capacity. Recently, to better understand maltotriose utilization by yeast strains, an analysis of maltotriose and maltotriose utilization by 52 laboratory and industrial *Saccharomyces* yeast strains was performed (Duval et al. 2009). Microarray comparative genome hybridization (aCGH) was used to correlate the observed phenotypes with copy number variations in genes known to be involved in maltose and maltotriose utilization by yeasts. The results showed that *S. pastorianus* strains utilized maltotriose more efficiently than *S. cerevisiae* strains and highlighted the importance of the *AGT1* gene for efficient maltotriose utilization by *S. cerevisiae* yeasts. The fermentation performance of a lager (*S. pastorianus*) strain was improved when its *AGT1* gene was replaced with the *AGT1* gene of an ale (*S. cerevisiae*) strain, since the *ATG1* gene of the lager strains studied contained a premature stop codon and did not encode functional transporters (Vidgren et al. 2005, 2009). The transformants with repaired *AGT1* had higher maltose transport activity, especially after growth on glucose, which represses endogeneous α-glucoside transporter genes. The sequences of two *AGT1*-encoded α-glucosidase transporters with different efficiencies of maltotriose transport in two *Saccharomyces* strains were compared (Smit et al. 2008). The amino acids Thr^{505} and Ser^{557}, which are respectively located in the transmembrane (TM) segment TM^{11} and on the intracellular segment after TM^{12} of the *AGT1*-encoded α-glucosidase transporters, are critical for efficient transport of maltotriose in *S. cerevisiae*. It was also shown that maltotriose utilization could be improved by attaching a maltase encoded by *MAL32* to the yeast cell surface (Dietvorst et al. 2007).

Glycogen and Trehalose Metabolism

Glycogen and trehalose are the main storage carbohydrates in yeast cells (Panek 1991). The synthesis of both compounds commences with the formation of uridine diphosphate (UDP)-glucose catalyzed by UDP-glucose pyrophosphorylase.

Glycogen is a reserve carbohydrate that is metabolized during periods of starvation. It is a branched polysaccharide composed of linear α-(1,4)-glycosyl chains with α-(1,6)-linkages (similar to starch but with a higher degree of branching). It is synthesized starting from glucose via glucose-6-phosphate and glucose-1-phosphate. Glycogen is synthesized from glucose, via glucose-6-phosphate and glucose-1-phosphate (François and Parrou 2001). UDP serves as a carrier of glucose units and is formed by a two-step reaction, catalyzed by phosphoglucomutase and UDP-glucose phosphorylase. Glycogen synthesis is initiated by glycogenin that produces a short α-(1,4)-glucosyl chain, which is elongated by glycogen synthase. The α-(1,6)-glucosidic bonds are formed by the branching enzyme Glc3. The dissimilation of glycogen occurs through the action of glycogen phosphorylase, which releases glucose-1-phosphate from the nonreducing ends of the glycogen chains, and the debranching enzyme Gdb1, which transfers a maltosyl unit to the end of an adjacent linear α-(1,4) chain and releases glucose by cleaving the remaining α-(1,6)-linkage.

When yeast cells are pitched in aerated wort, an immediate glycogen mobilization is observed (Quain 1988, Boulton 2000). Glycogen accumulates during the exponential growth phase, after oxygen has been consumed. When the yeast growth is ceased toward the end of the primary fermentation, are maximum glycogen levels obtained. In the stationary phase, the glycogen levels decline slowly. Glycogen provides energy for the synthesis of sterols and unsaturated fatty acids during the aerobic phase of the beer fermentation, and energy for the cellular maintanance functions during the stationary phase in the storage phase between cropping and pitching (Quain and Tubb 1982, Boulton et al. 1991). The glycogen content is directly related to subsequent fermentation performance. Therefore, yeast storage occurs best at a low temperature, without agitation and under an atmosphere of nitrogen or carbon dioxide to minimize glycogen breakdown (Murray et al. 1984).

The regulation of the glycogen content is complex and occurs in part by the cAMP/PKA pathway (Smith et al. 1998). Glycogen accumulation is repressed by a high PKA-activity (François and Parrou 2001). The glycogen content increases and the PKA-activity is reduced, when an essential nutrient is progressively consumed from the growth medium.

Trehalose is a disaccharide (α-D-glucopyranosyl-1,1- α-D-glucopyranoside) which contains two molecules of D-glucose (Boulton and Quain 2006). Trehalose biosynthesis is catalyzed by the trehalose synthase complex, which forms trehalose-6-phosphate from UDP-glucose and glucose-6-phosphate, and next dephosphorylates it to trehalose. Trehalose is degradaded by the neutral (Nth1) or the acid (Ath1) trehalase (Panek and Panek 1990, François and Parrou 2001). Like glycogen, trehalose also accumulates in yeast under conditions of nutrient limitation. It has been observed that trehalose rapidly accumulates in response to environmental stress, such as dehydration, heating and osmotic stress (during high gravity brewing) (Majara et al. 1996, Hounsa et al. 1998). Under stress conditions, the higher levels of trehalose protect the cells by binding to membranes and proteins (François and Parrou 2001). Trehalose accumulation in response to stress is, however, a transient phenomenon (Parrou et al. 1997).

In the beginning of the primary fermentation, high glucose levels activate the camp/PKA pathway (Zähringer et al. 2000). This results in posttranslational activation of neutral trehalase and in induction of its *NTH1* gene, and in repression of the trehalose synthase genes, which results in reduced levels of trehalose.

Wort Fermentation

Before the fermentation process starts, wort is aerated. This is a necessary step since oxygen is required for the synthesis of sterols and unsaturated fatty acids, which are incorporated in the yeast cell membrane (Rogers and Stewart 1973). It has been shown that ergosterol and unsaturated fatty acids increase both in concentration as long as oxygen is present in the wort (e.g., Haukeli and Lie 1979). A maximum concentration is obtained in 5–6 hours after pitching, but the formation rate is dependent upon the pitching rate and the temperature. Unsaturated fatty acids can also be taken up from the wort, but all malt wort does not contain sufficient unsaturated lipids to support a normal growth rate of yeast. Adding lipids to the wort, especially unsaturated fatty acids might be an interesting alternative (Moonjai et al. 2000a, 2000b).

The oxygen required for lipid biosynthesis can also be introduced by oxygenation of the separated yeast cells. The use of the preoxygenation technique resulted in more controllable and consistent fermentations, and as a consequence, in a more balanced beer flavor profile (Jakobsen 1982, Ohno and Takahashi 1986a, 1986b, Boulton et al. 1991, Devuyst et al. 1991, Masschelein et al. 1995, Depraetere et al. 2003, Depraetere 2007). Recently, the impact of yeast preoxygenation on yeast metabolism has been assessed (Verbelen et al. 2009). Therefore, expression analysis was performed of genes that are of importance in oxygen-dependent pathways, oxidative stress response and general stress response during 8 hours of preoxygenation. The gene expressions of both the important transcription factors Hap1 and Rox1, involved in oxygen sensing, were mainly increased in the first 3 hours, while *YAP1* expression, which is involved in the oxidative stress response, increased drastically only in the first 45 minutes. The results also show that stress-responsive genes (*HSP12*, *SSA3*, *PAU5*, *SOD1*, *SOD2*, *CTA1*, and *CTT1*) were induced during the process, together with the accumulation of trehalose. The accumulation of ergosterol and unsaturated fatty acids was accompanied by the expression of *ERG1*, *ERG11*, and *OLE1*. Genes involved in respiration (*QCR9*, *COX15*, *CYC1*, and *CYC7*) also increased during preoxygenation. Yeast viability did not decrease during the process, and the fermentation performance of the yeast reached a maximum after 5 hours of preoxygenation. These results suggest that yeast cells acquire a stress response along the preoxygenation period, which makes them more resistant against the stressful conditions of the preoxygenation process and the subsequent fermentation.

Different devices are used to aerate the cold wort: ceramic or sintered metal candles, aeration plants employing venturi pipes, two component jets, static mixers, or centrifugal mixers (Kunze 1999). The principle of these devices is that very small air (oxygen) bubbles are produced and quickly dissolve during turbulent mixing.

As a result of this aeration step, carbohydrates are degraded aerobically during the first few hours of the "fermentation" process. The aerobic carbohydrate catabolism takes typically 12 hours for a lager fermentation.

During the first hours of the fermentation process, oxidative degradation of carbohydrates occurs through the glycolysis and Krebs (TCA) cycle. The energy efficiency of glucose oxidation is derived from the large number of $NADH_2^+$ produced for each mole of glucose oxidized to CO_2. The actual wort fermentation gives alcohol and carbon dioxide via the Embden-Meyerhof-Parnas (glycolytic) pathway. The reductive pathway from pyruvate to ethanol is important since it regenerates NAD^+. Energy is obtained solely from ATP-producing steps of the Embden-Meyerhof-Parnas pathway. During fermentation, the activity of the TCA cycle is greatly reduced, although it still serves as a source of intermediates for biosynthesis (Lievense and Lim 1982).

Lagunas (1979) observed that during aerobic growth of *S. cerevisiae*, respiration accounts for less than 10% of glucose catabolism, the remainder being fermented. Increasing sugar concentrations resulting in a decreased oxidative metabolism is known as the Crabtree effect. This was traditionally explained as an inhibition of the oxidative system by high concentrations of glucose. Nowadays, it is generally accepted that the formation of ethanol at aerobic conditions is a consequence of a bottleneck in the oxidation of pyruvate, for example, in the respiratory system (Petrik et al. 1983, Rieger et al. 1983, Käppeli et al. 1985, Fraleigh et al. 1989, Alexander and Jeffries 1990).

A reduction of ethanol production can be achieved by metabolic engineering of the carbon flux in yeast resulting in an increased formation of other fermentation product. A shift of the carbon flux towards glycerol at the expense of ethanol formation in yeast was achieved by simply increasing the level of glycerol-3-phosphate dehydrogenase (Michnick et al. 1997, Nevoigt and Stahl 1997, Remize et al. 1999, Dequin 2001). The *GDP1* gene, which encodes glycerol-3-phosphate dehydrogenase, has been overexpressed in an industrial lager brewing yeast to reduce the ethanol content in beer (Nevoigt et al. 2002). The amount of glycerol produced by the *GDP1*-overexpressing yeast in fermentation experiments—simulating brewing conditions—was increased 5.6 times and ethanol was decreased by 18% compared to the wild-type strain. Overexpression did not affect the consumption of wort sugars and only minor changes in the concentration of higher alcohols, esters and fatty acids could be observed. However, the concentrations of several other by-products, particularly acetoin, diacetyl, and acetaldehyde, were considerably increased.

Mutants of *S. cerevisiae* strains deficient in tricarboxylic acid cycle genes have been reported as suitable strains for the production of nonalcoholic beer (Navratil et al. 2002). Strains deficient in fumarase and α-ketoglutarate dehydrogenase made nonalcoholic beers with an alcohol content lower than 0.5% (v/v) (Selecky et al. 2008). The low ethanol content was compensated by the considerable increase of organic acids (citrate, succinate, fumarate, and malate). Some of the mutants released high levels of lactic acid, which protects beers against contamination and masks an unacceptable worty off-flavor.

METABOLISM OF BIOFLAVORING BY-PRODUCTS

Yeast is an important contributor to flavor development in fermented beverages. The compounds, which are produced during fermentation, are many and varied, depending on both the raw materials and the microorganisms used. The interrelation between yeast metabolism and the production of bioflavoring by-products is illustrated in Figure 33.2. The important flavor compounds produced by yeast can be classified into five categories: alcohols, esters, organic acids, carbonyl compounds

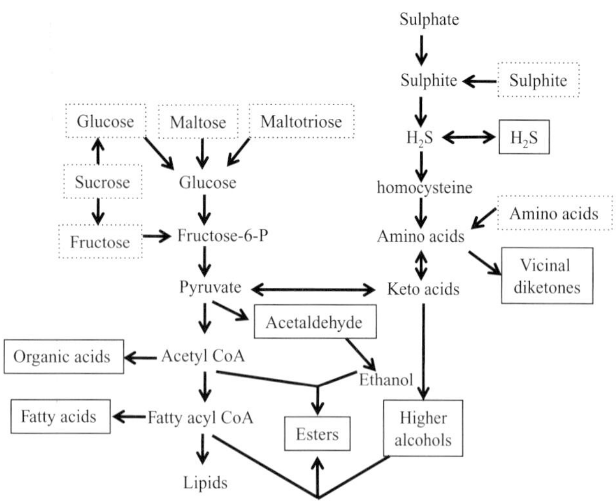

Figure 33.2. Interrelation between yeast metabolism and the production of bioflavoring by-products.

(aldehydes and vicinal diketones), and sulfur-containing compounds (Hammond 1993). In addition, some speciality beers can contain important concentrations of volatile phenol compounds (Vanbeneden et al. 2007).

Biosynthesis of Higher Alcohols

During beer fermentation, higher alcohols (also called "fusel alcohols") are produced by yeast cells as by-products and represent the major fraction of the volatile compounds. More than 35 higher alcohols in beer have been described. Table 33.3 gives the most important compounds, which can be classified into aliphatic (*n*-propanol, isobutanol, 2-methylbutanol (or active amyl alcohol), and 3-methylbutanol (or isoamyl alcohol)), and aromatic (2-phenylethanol, tyrosol, tryptophol) higher alcohols. Aliphatic higher alcohols contribute to the "alcoholic" or "solvent" aroma of beer, and produce a warm mouthfeel. The aromatic alcohol 2-phenylethanol has a sweet rose-like aroma and has a positive contribution to the beer aroma. It is believed that this compound masks the dimethyl sulfide (DMS) perception (Hegarty et al. 1995). The aroma of tyrosol and tryptophol are undesirable, but they are only present above their thresholds in some top-fermented beers.

Higher alcohols are synthesized by yeast during fermentation via the catabolic (Ehrlich) and anabolic pathway (amino acid metabolism) (Ehrlich 1904, Chen 1978, Oshita et al. 1995, Hazelwood et al. 2008). In the catabolic pathway, the yeast uses the amino acids of the wort to produce the corresponding α-keto acid via a transamination reaction. Isoamyl alcohol, isobutanol, and phenylethanol are produced via this route from leucine, valine, and phenylalanine, respectively. An outsider in this pathway is propanol, which is derived from threonine via an oxidative deamination. The excess oxoacids are subsequently decarboxylated into aldehydes and further reduced (alcohol dehydrogenase) to higher alcohols. This last reduction step also regenerates NAD^+.

Dickinson and coworkers looked at the genes and enzymes, which are used by *S. cerevisiae* in the catabolism of leucine to isoamyl alcohol (Dickinson et al. 1997), valine to isobutanol (Dickinson et al. 1998), and isoleucine to active amyl alcohol (Dickinson et al. 2000). In all cases, the general sequence of biochemical reactions is similar, but the details for the formation of the individual alcohols are surprisingly different. The branched-chain amino acids are first deaminated to the corresponding α-ketoacids (α-ketoisocapric acid from leucine, α-ketoisovaleric acid from valine, and α-keto-β-methylvaleric

Table 33.3. Major Higher Alcohols in Beer (Partly Adapted From NN 2000.)

Compound	Flavor Threshold (mg/L)	Aroma or Taste[b]	Concentration Range (mg/L) Bottom Fermentation	Concentration Range (mg/L) Top Fermentation
n-Propanol	600[c], 800[b]	Alcohol	7–19 (12)[a],[f], 6–30[j]	20–45[i]
Isobutanol	100[c], 80–100[g], 200[b]	Alcohol	4–20 (12) [f], 6–32[j]	10–24[i]
2-Methylbutanol	50[c], 50–60[g], 70[b]	Alcohol	9–25 (15) [a], 12–16[j]	80–140[i]
3-Methylbutanol	50[c], 50–60[g], 65[b]	Fusely, pungent	25–75 (46) [a], 30–60[j]	80–140[i]
2-Phenylethanol	5[a], 40[c], 45–50[g], 75[d], 125[b]	Roses, sweetish	11–51 (28) [f], 4–22[g], 16–42[h], 4–10[j]	35–50[g], 8–25[a], 18–45[i]
Tyrosol	10[a], 10–20[e], 20[c], 100[d], [g], 200[b]	Bitter chemical	6–9[a], 6–15[a]	8–12([g], 7–22[g]
Tryptophol	10[a], 10–20[e], 200[d]	Almonds, solvent	0.5–14[a]	2–12[g]

Sources: (a) Szlavko (1973), (b) Meilgaard (1975a), (c) Engan (1972), (d) Rosculet (1971), (e) Charalambous et al. (1972), (f) Values in 48 European lagers (Dufour unpublished date), (g) Reed and Nogodawithana (1991), (h) Iverson (1994), (i) Derdelinckx (unpublished data), (j) immobilised cells (Willaert and Nedovic 2006).
[a]Mean value.

acid from leucine). There are significant differences in the way each α-ketoacid is subsequently decarboxylated. The catabolism of phenylalanine to 2-phenylethanol and of tryptophan were also studied (Dickinson et al. 2003). Phenylalanine and tryptophan are first deaminated to 3-phenylpyruvate and 3-indolepyruvate, respectively, and then decarboxylated. These studies revealed that all amino acid catabolic pathways studied to date use a subtle different spectrum of decarboxylases from the five-membered family that comprises Pdc1p, Pdc5p, Pdc6p, Ydl080cp, and Ydr380wp. Using strains containing all possible combinations of mutations affecting the seven *AAD* genes (putative aryl alcohol dehydrogenases), five *ADH* and *SFA1* (other alcohol dehydrogenase genes), showed that the final step of amino acid catabolism can be accomplished by any one of the ethanol dehydrogenases (Ahd1p, Ahd2p, Ahd3p, Ahd4p, Ahd5p) or Sfa1p (formaldehyde dehydrogenase).

In the anabolic pathway, the higher alcohols are synthesized from α-keto acids during the synthesis of amino acids from the carbohydrate source. The pathway choice depends on the individual higher alcohol and on the level of available amino acids available. The importance of the anabolic pathway decreases as the number of carbon atoms in the alcohol increases (Chen 1978) and increases in the later stage of fermentation as wort amino acids are depleted (MacDonald et al. 1984). Yeast strain, fermentation conditions, and wort composition all have significant effects on the combination and levels of higher alcohols that are formed (MacDonald et al. 1984).

Conditions that promote yeast cell growth such as high levels of nutrients (amino acids, oxygen, lipids, zinc, ...) and increased temperature and agitation stimulate the production of higher alcohols (Engan 1969, Engan and Aubert 1977, Landaud et al. 2001, Boswell et al. 2002). The synthesis of aromatic alcohols is especially sensitive to temperature changes. On the other hand, conditions that restrict yeast growth, such as lower temperature and higher pressure, reduce the extent of higher alcohol production. Higher pressures can reduce the extent of cell growth and, therefore, the production of higher alcohols (Landaud et al. 2001). The yeast strain, fermentation conditions, and wort composition have all significant effects on the pattern and concentrations of synthesized higher alcohols. Supplementation of wort with valine, isoleucine, and leucine induces the formation of isobutanol, amyl alcohol, and isoamyl alcohol, respectively (Äyräpää 1971, Kodama et al. 2001). The overexpression of the branched chain amino acid transferases genes *BAT1* and *BAT2* result in an increased production of isoamyl alcohol and isobutanol (Lilly et al. 2006).

Biosynthesis of Esters

Esters are very important flavor compounds in beer. They have an effect on the fruity/flowery aromas. Table 33.4 shows the most important esters with their threshold values, which are considerably lower than those for higher alcohols. The major esters can be subdivided into acetate esters and C_6–C_{10} medium-chain fatty acid ethyl esters. They are desirable components of beer when present in appropriate quantities and proportions but can become unpleasant when in excess.

Esters are produced by yeast both during the growth phase (60%) and also during the stationary phase (40%) (NN 2000). They are formed by the intracellular reaction between a fatty acyl-coenzyme A and an alcohol:

$$R'OH + RCO - ScoA \rightarrow RCOOR' + CoASH \quad (1)$$

This reaction is catalyzed by alcohol acyltransferases (AATases; or ester synthethases) of the yeast. Since acetyl CoA is also a central molecule in the synthesis of lipids and sterols, ester synthesis is linked to the fatty acid metabolism (see also Fig. 33.2). The majority of acetyl-CoA is formed by oxidative decarboxylation of pyruvate, while most of the other acyl-CoAs are derived from the acylation of free CoA, catalyzed by acyl-CoA synthase.

Several enzymes are involved in the synthesis, the best characterized are AATase I and II that are encoded by *ATF1* and *ATF2*. Alcohol acetyltransferase (AAT) has been localized in the plasma membrane (Malcorps and Dufour 1987) and found to be strongly inhibited by unsaturated fatty acids, ergosterol, heavy metal ions, and sulfydryl reagents (Minetoki et al. 1993). Subcellular fractionation studies conducted during the batch fermentation cycle demonstrated the existence of both cytosolic and membrane-bound AAT (Ramos-Jeunehomme et al. 1989, Ramos-Jeunehomme et al. 1991). In terms of controlling

Table 33.4. Major Esters in Beer (Adapted From Dufour and Malcorps 1994.)

Compound	Flavor Threshold (mg/L)	Aroma	Concentration Range (mg/L) in 48 Lagers
Ethyl actetate	20–30, 30[(a)]	Fruity, solvent-like	8–32 (18.4)[(a)], 5–40[(b)]
Isoamyl acetate	0.6–1.2, 1.2[(a)]	Banana, peardrop	0.3–3.8 (1.72), <0.01–2.8[(b)]
Ethyl caproate (ethyl hexanoate)	0.17–0.21, O.21[(a)]	Apple-like with note of aniseed	0.05–0.3 (0.14), 0.01–0.54[(b)]
Ethyl caprylate (ethyl octanoate)	0.3–0.9, 0.9[(a)]	Apple-like	0.04–0.53 (0.17), 0.01–1.2[(b)]
2-Phenylethyl acetate	3.8[(a)]	Roses, honey, apple, sweetish	0.10–0.73 (0.54)

Source: (a) Meilgaard (1975b) and (b) immobilised cells (Willaert and Nedovic 2006).
[a]Mean value.

ester formation on a metabolic basis, it has further been shown that ester synthesizing activity of AAT is dependent on its positioning within the yeast cell. An interesting feature of this distribution pattern is that specific rates of acetate ester formation varied directly with the level of cytosolic AAT activity (Masschelein 1997).

The *ATF1* gene, which encodes AAT, has been cloned from *S. cerevisiae* and brewery lager yeast (*S. cerevisiae uvarum*) (Fujii et al. 1994). An hydrophobicity analysis suggested that AAT does not have a membrane-spanning region that is significantly hydrophobic, which contradicts the membrane-bound assumption. A Southern analysis of the yeast genomes in which the *ATF1* gene was used as a probe revealed that *S. cerevisiae* has one *ATF1* gene, while brewery lager yeast has one *ATF1* gene and another homologous gene (*Lg-ATF1*). The AAT activities have been compared in vivo and in vitro under different fermentation conditions (Malcorps et al. 1991). This study suggested that ester synthesis is modulated by a repression–induction of enzyme synthesis or processing the regulation of which is presumably linked to lipid metabolism. Other enzymes are Eht1 (ethanolhexanoyl transferase) and Eeb1, which are responsible for the formation of ethyl esters (Verstrepen et al. 2003, Saerens et al. 2006).

The ester production can be altered by changing the synthesis rate of certain fusel alcohols. Hirata et al. (1992) increased the isoamyl acetate levels by introducing extra copies of the *LEU4* gene in the *S. cerevisiae* genome. A comparable *S. cerevisiae uvarum* mutant has been isolated (Lee et al. 1995). The mutants have an altered regulation pattern of amino acid metabolism and produce more isoamyl acetate and phenylethyl acetate.

Isoamyl acetate is synthesized from isoamyl alcohol and acetyl coenzyme A by AAT and is hydrolyzed by esterases at the same time in *S. cerevisiae*. To study the effect of balancing both enzyme activities, yeast strains with different numbers of copies of *ATF1* gene and isoamyl acetate-hydrolyzing esterase gene (*IAH1*) have been constructed and used in small-scale sake brewing (Fukuda et al. 1998). Fermentation profiles as well as components of the resulting sake were largely alike. However, the amount of isoamyl acetate in the sake increased with increasing ratio of AAT/Iah1p esterase activity. Therefore, it was concluded that the balance of these two enzyme activities is important for isoamyl acetate accumulation in sake mash.

The synthesis of acetate esters by *S. cerevisiae* during fermentation is ascribed to at least three acetyltransferase activities, namely, AAT, ethanol acetyltransferase, and isoamyl AAT (Lilly et al. 2000). To investigate the effect of increased AAT activity on the sensory quality of Chenin blanc wines and distillates from Colombar base wines, the *ATF1* gene of *S. cerevisiae* was overexpressed. Northern blot analysis indicated constitutive expression of *ATF1* at high levels in these transformants. The levels of ethyl acetate, isoamyl acetate and 2-phenylethyl acetate increased 3- to 10-fold, 3.8- to 12-fold, and 2- to 10-fold, respectively, depending on the fermentation temperature, cultivar, and yeast used. The concentrations of ethyl caprate, ethyl caprylate, and hexyl acetate only showed minor changes, whereas the acetic acid concentration decreased by more than half. This study established the concept that the overexpression of acetyltransferase genes such as *ATF1* could profoundly affect the flavor profiles of wines and distillates deficient in aroma.

In order to investigate and compare the roles of the known *S. cerevisiae* AATs, Atf1p, Atf2p, and Lg-Atf1p, in volatile ester production, the respective genes were either deleted or overexpressed in a laboratory strain and a commercial brewing strain (Verstrepen et al. 2003). Analysis of the fermentation products confirmed that the expression levels of *ATF1* and *ATF2* greatly affect the production of ethyl acetate and isoamyl acetate. GC-MS analysis revealed that Atf1p and Atf2p are also responsible for the formation of a broad range of less volatile esters, such as propyl acetate, isobutyl acetate, pentyl acetate, hexyl acetate, heptyl acetate, octyl acetate, and phenyl ethyl acetate. With respect to the esters analyzed in this study, Atf2p seemed to play only a minor role compared to Atf1p. The *atf1*Δ*atf2*Δ double deletion strain did not form any isoamyl acetate, showing that together, Atf1p and Atf2p are responsible for the total cellular isoamyl AAT activity. However, the double deletion strain still produced considerable amounts of certain other esters, such as ethyl acetate (50% of the wild-type strain), propyl acetate (50%), and isobutyl acetate (40%), which provides evidence for the existence of additional, as-yet-unknown ester synthases in the yeast proteome. Interestingly, overexpression of different alleles of *ATF1* and *ATF2* led to different ester production rates, indicating that differences in the aroma profiles of yeast strains may be partially due to mutations in their *ATF* genes.

Recently, it has been discovered that the Atf1 enzyme is localized inside lipid vesicles in the cytoplasm of the yeast cell (Verstrepen 2003). Lipid vesicles are small organelles in which certain neutral lipids are metabolized or stored. This indicates that fruity esters are possibly by-products of these processes.

Ester formation is highly dependent on the yeast strain used (Nykänen and Nykänen 1977, Peddie 1990, Verstrepen et al. 2003b) and on certain fermentation parameters such as temperature (Engan and Aubert 1977, Gee and Ramirez 1994, Sablayrolles and Ball 1995), specific growth rate (Gee and Ramirez 1994), pitching rate (Maule 1967, D'Amore et al. 1991, Gee and Ramirez 1994), and top pressure (NN 2000, Verstrepen et al. 2003b). Additionally, the concentrations of assimilable nitrogen compounds (Hammond 1993, Calderbank and Hammond 1994, Sablayrolles and Ball 1995), carbon sources (Pfisterer and Stewart 1975, White and Portno 1979, Younis and Stewart 1998, 2000), dissolved oxygen (Anderson and Kirsop 1975a, 1975b, Avhenainen and Mäkinen 1989, Sablayrolles and Ball 1995), and fatty acids (Thurston et al. 1981, 1982) can influence the ester production rate.

Acetate ester formation in brewer's yeast is controlled mainly by the expression level of the AATase-encoding genes (Verstrepen et al. 2003b). Additionally, changes in the availability of the two substrates for ester production, higher alcohols and acyl-CoA, also influences ester synthesis rates. Any factor that influences the expression of the ester synthase genes and/or the concentrations of substrates will affect ester production accordingly. Perhaps the most convenient and selective way to reduce ester production is applying tank overpressure, if necessary in combination with (slightly) lower fermentation temperatures, low wort free amino nitrogen (FAN) and glucose levels and elevated wort aeration or wort lipid concentration (Verstrepen et al. 2003b). Enhancing ester production is slightly more complicated.

If it is possible, overpressure or wort aeration can be reduced. Otherwise, worts rich in glucose and nitrogen combined with higher fermentation temperatures and lower pitching rates or application of the drauflassen technique may prove helpful.

Biosynthesis of Organic Acids

Over hundred different organic acids have been reported in beer (Meilgaard 1975b). Important organic acids detected in beer include pyruvate, acetate, lactate, succinate, pyroglutamate, malate, citrate, α-ketoglutarate, and α-hydroxyglutarate; and the medium-chain length fatty acids, caproic (C6), caprylic (C8), and capric (C10) acid (Coote and Kirsop 1974, Meilgaard 1975b, Klopper et al. 1986). They influence flavor directly when present above their taste threshold and by their influence on beer pH. These components have their origin in raw materials (malt, hops) and are produced during the beer fermentation. Organic acids, which are excreted by yeast cells, are synthesized via amino acid biosynthesis pathways and carbohydrate metabolism. Especially, they are overflow products of the incomplete Krebs cycle during beer fermentation. Excretion of organic acids is influenced by yeast strain and fermentation vigor. Sluggish fermentations lead to lower levels of excretion. Pyruvate excretion follows the yeast growth: maximal concentration is reached just before the maximal yeast growth and is next taken up by the yeast and converted to acetate. Acetate is synthesized quickly during early fermentation and is later partially reused by the yeast during yeast growth. At the end of the fermentation, acetate is accumulated. The reduction of pyruvate results in the production of D-lactate of L-lactate (most yeast strains produce preferentially D-lactate). The highest amount of lactate is produced during the most active fermentation period.

The change in organic acid productivity by disruption of the gene encoding fumarase (*FUM1*) has been investigated and it has been suggested that malate and succinate are produced via the oxidative pathway of the TCA cycle under static and sake brewing conditions (Magarifuchi et al. 1995). Using a NAD^+-dependent isocitrate dehydrogenase gene (*IDH1*, *IDH2*) disruptant, approximately half of the succinate in sake mash was found to be synthesized via the oxidative pathway of the TCA cycle in sake yeast (Asano et al. 1999).

Sake yeast strains possessing various organic acid productivities were isolated by gene disruption (Arikawa et al. 1999). Sake fermented using the aconitase gene (*ACO1*) disruptant contained a twofold higher concentration of malate and a twofold lower concentration of succinate than that made using the wild-type strain. The fumarate reductase gene (*OSM1*) disruptant produced sake containing a 1.5-fold higher concentration of succinate, whereas the α-ketoglutarate dehydrogenase gene (*KGD1*) and fumarase gene (*FUM1*) disruptants gave lower succinate concentrations. In *S. cerevisiae*, there are two isoenzymes of fumarate reductase (FRDS1 and FDRS2), encoded by the *FRDS* and *OSM1* genes, respectively (Arikawa et al. 1998). Recent results suggest that these isoenzymes are required for the reoxidation of intracellular NADH under anaerobic conditions, but not under aerobic conditions (Enomoto et al. 2002).

Succinate dehydrogenase is an enzyme of the TCA cycle and thus essential for respiration. In *S. cerevisiae*, this enzyme is composed of four nonidentical subunits, that is, the flavoprotein, the iron–sulfur protein, the cytochrome b_{560}, and the ubiquinone reduction protein encoded by the *SDH1*, *SDH2*, *SDH3*, and *SDH4* genes, respectively (Lombardo et al. 1990, Chapman et al. 1992, Bullis and Lamire 1994, Daignan-Fournier et al. 1994). Sdh1p and Sdh2p comprise the catalytic domain involved in succinate oxidation. These proteins are anchored to the inner mitochondrial membrane by Sdh3p and Sdh4p, which are necessary for electron transfer and ubiquinone reduction, and constitute the succinate:ubiquinone oxidoreductase (complex II) of the electron transport chain. Single or double disruptants of the *SDH1*, *SDH1b* (which is a homologue of the *SDH1* gene), *SDH2*, *SDH3*, and *SDH4* genes have been constructed and shown that the succinate dehydrogenase activity was retained in the *SDH2* disruptant and that double disruption of *SDH1* and *SDH2* or *SDH1b* genes is necessary to cause deficiency of succinate dehydrogenase activity in sake yeast (Kubo et al. 2000). The role of each subunit in succinate dehydrogenase activity and the effect of succinate dehydrogenase on succinate production using strains that were deficient in succinate dehydrogenase, have also been determined. The results suggested that succinate dehydrogenase activity contributes to succinate production under shaking conditions, but not under static and sake brewing conditions.

The medium-chain fatty acids account for 85–90% of the fatty acids in beer and impart an undesirable goaty, sweaty, and yeasty flavor (Chen 1980). These fatty acids are produced de novo by yeast during anaerobic fermentation and are not the result of β-oxydation of wort or yeast long-chain fatty acids. As a result, any change in fermentation conditions that promote the extent of yeast growth also favor increased levels of medium-chain fatty acids in beer. Higher temperature, increased wort oxygenation, and possibly elevated pitching rates are all effective in this respect (Boulton and Quain 2006). The presence of these medium-chain fatty acids in beer is also related to yeast autolysis (Masschelein 1981). Yeast autolytic off-flavors are stimulated at high temperatures, high yeast concentrations, and prolonged contact times at the end of primary fermentation and during secondary fermentation.

Biosynthesis of Vicinal Diketones

Vicinal diketones are ketones with two adjacent carbonyl groups. During fermentations, these flavor-active compounds are produced as by-products of the synthesis pathway of isoleucine, leucine, and valine (ILV pathway) (see Fig. 33.3) and thus also linked to amino acid metabolism (Nakatani et al. 1984) and the synthesis of higher alcohols. They impart a "buttery", "butterscotch" aroma to alcoholic drinks. Two of these compounds are important in beer, that is, diacetyl (2,3-butanedione) and 2,3-pentanedione. Diacetyl is quantitatively more important than 2,3-pentanedione. It has a taste threshold in lager beer from 17 μg/L (Saison et al. 2009) to 150 μg/L (Meilgaard

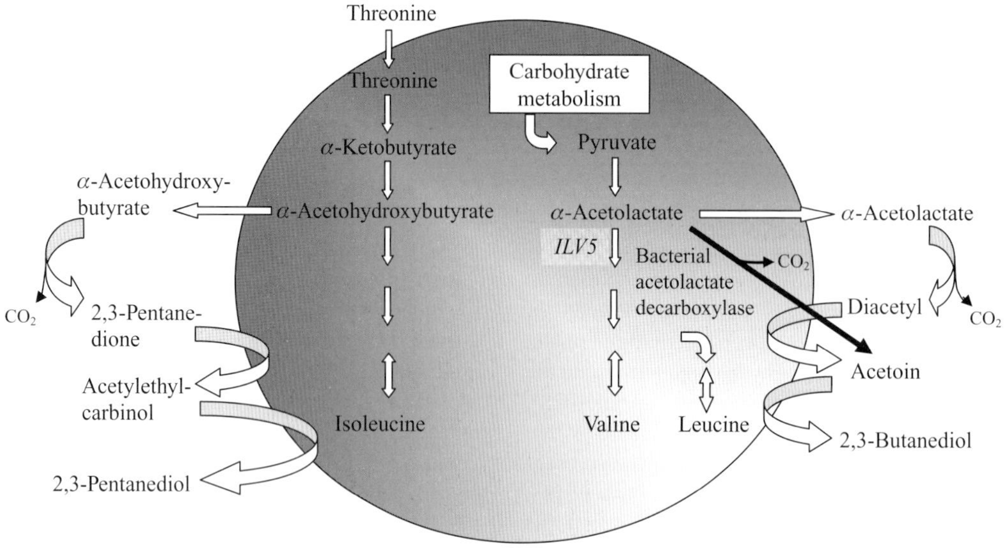

Figure 33.3. The synthesis and reduction of vicinal diketones in *Saccharomyces cerevisiae*.

1975b), approximately 10 times lower than that of pentanedione (Wainwright 1973).

The excreted α-acetohydroxy acids are overflow products of the ILV pathway that are nonenzymatically degraded to the corresponding vicinal diketones (Inoue et al. 1968). Tetraploid gene dosage series for various *ILV* genes have been constructed and the obtained yeast strains were used to study the influence of the copy number of *ILV* genes on the production of vicinal diketones (Debourg et al. 1990, Debourg 2002). It was shown that the *ILV5* activity is the rate-limiting step in the ILV pathway and responsible for the overflow (Fig. 33.3). The nonenzymatic oxidative decarboxylation step is the rate-limiting step and proceeds faster at a higher temperature and a lower pH (Inoue and Yamamoto 1970, Haukeli and Lie 1978). The produced amount of α-acetolactate is very dependent on the used yeast strain. The production increases with increasing yeast growth. For a classical fermentation, 0.6 ppm α-acetolactate is formed (Delvaux 1998). At high aeration, this value can be increased to 0.9 ppm and in cylindro-conical fermentations tanks even to 1.2–1.5 ppm.

It has been shown that valine inhibits the synthesis of α-acetolactate through feedback inhibition (Magee and de Robinson-Szulmajster 1968). This inhibition is directed to the protein Ilv6p, which is the regulatory subunit of acetohydroxy acid synthase (the catalytic subunit encoded by *ILV2*)(Pang and Duggleby 1999, 2001). Because the uptake of valine is delayed in a normal beer fermentation, the suppressive effect of valine accounts for the postponed onset of total diacetyl (sum of actual diacetyl and α-acetolactate), and this effect persists longer in worts with high levels of FAN content. In contrast, low FAN levels give two diacetyl peaks as a result of the requirement for valine biosynthesis. Therefore, a minimum FAN level above the critical value of 50 ppm (Nakatani et al. 1984) or 140 ppm (Pugh et al. 1997) should be maintained during the fermentation to ensure the presence of valine in the fermenting wort.

Yeast cells posses the necessary enzymes (reductases) to reduce diacetyl to acetoin and further to 2,3-butanediol, and 2,3-pentanedione to 2,3-pentanediol (Bamforth and Kanauchi 2004). These reduced compounds have much higher taste thresholds and have no impact on the beer flavor (Van Den Berg et al. 1983). The reduction reactions are yeast strain dependent. The reduction occurs at the end of the main fermentation and during the maturation. Sufficient yeast cells in suspension are necessary to obtain an efficient reduction. Yeast strains that flocculate early during the main fermentation needs a long maturation time to reduce the vicinal diketones. Diacetyl can be complexed using SO_2. These complexes cannot be reduced, but diacetyl can again be liberated at a later stage by aldehydes. This situation is especially applicable to yeast strains, which produce a lot of SO_2. Worts, which are produced using a high content of adjuncts, can be low in free amino acid content. These worts can give rise to a high diacetyl peak at the end of the fermentation.

There are several strategies, which can be chosen to reduce the vicinal diketones amount during fermentation:

1. Since the temperature has a positive effect on the reduction efficiency of the α-acetohydroxy acids, a warm rest period at the end of the main fermentation and a warm maturation are applied in many breweries. In this case, temperature should be well controlled to avoid yeast autolysis.
2. Since the rapid removal of vicinal diketones requires yeast cells in an active metabolic condition, the addition of 5–10% Krausen (containing active, growing yeast) is a procedure, which gives enhanced transformation of vicinal diketones (NN 2000). This procedure can lead to

overproduction of hydrogen sulfide, depending upon the proportions of threonine and methionine carried forward from primary fermentation.

3. Heating up the green beer to a high temperature (90°C) and hold it there for a short period (ca. 7–10 min) to decarboxylate all excreted α-acetohydroxy acids. To avoid cell autolysis, yeast cells are removed by centrifugation prior to heating up. The vicinal diketones can be further reduced by immobilized yeast cells in a few hours (typically at 4°C) (see further).
4. *Adding the enzyme α-acetolactate decarboxylase* (Godtfredsen et al. 1984, Rostgaard-Jensen et al. 1987): This enzyme decarboxylates α-acetolactate directly into acetoin (see Fig. 33.3). It is not present in *S. cerevisiae*, but has been isolated from various bacteria such as *Enterobacter aerogenes*, *Aerobacter aerogenes*, *Streptococcos lactis*, *Lactobacillus casei*, *Acetobacter aceti*, and *Acetobacter pasteurianus*. It has been shown that the addition of α-acetolactate decarboxylase from *L. casei* can reduce the maturation time to 22 hours (Godtfredsen et al. 1983, 1984). An example of a commercial product is Maturex L from Novo Nordisk (Denmark) (Jensen 1993). Maturex L is a purified α-acetolactate decarboxylase produced by a genetically modified strain of *Bacillus subtilis*, which has received the gene from *Bacillus brevis*. The recommended dosage is 1 à 2 kg per 1000 hL wort, to be added to the cold wort at the beginning of fermentation.
5. Using genetic modified yeast strains:
   a. Introducing the bacterial α-acetolactate decarboxylase gene into yeast chromosomes (Fujii et al. 1990, Suihko et al. 1990, Blomqvist et al. 1991, Enari et al. 1992, Linko et al. 1993, Yamano et al. 1994, Tada et al. 1995, Onnela et al. 1996). Transformants possessed a very high α-acetolactate decarboxylase activity that reduced the diacetyl concentration considerably during beer fermentations.
   b. Modifying the biosynthetic flux through the ILV pathway by partially deactivation of *ILV2*. Spontaneous mutants resistant to the herbicide sulfometuron methyl have been selected. These strains showed a partial inactivation of the α-acetolactate synthase activity and some mutants produced 50% less diacetyl compared to the parental strain (Gjermansen et al. 1988).
   c. Increasing the flux of α-acetolactate acid isomerorecuctase activity encoded by the *ILV5* gene (Dillemans et al. 1987). Since α-acetolactate acid isomerorecuctase activity is responsible for the rate-limiting step, increasing its activity reduces the overflow of α-acetolactate. A multicopy transformant resulted in a 70% decreased production of vicinal diketones (Villaneuba et al. 1990), whereas an integrative transformant gave a 50% reduction (Goossens et al. 1993). A tandem integration of multiple *ILV5* copies resulted also in elevated transciption in a polyploidy industrial yeast strain (Mithieux and Weiss 1995). Vicinal diketones production could be reduced by targeting the mitochondrial Ilv5p to the cytosol (Omura 2008).

SECONDARY FERMENTATION

During the secondary fermentation or maturation of beer, several objectives should be realized:

- Sedimentation of yeast cells
- Improvement of the colloidal stability by sedimentation of the tannin–protein complexes
- Beer saturation with carbon dioxide
- Removal of unwanted aroma compounds
- Excretion of flavor-active compounds from yeast to give body and depth to the beer
- Fermentation of the remaining extract
- Improvement of the foam stability of the beer
- Adjustment of the beer color (if necessary) by adding coloring substances (e.g., caramel)
- Adjustment of the bitterness of beer (if necessary) by adding hop products

In the presence of yeast, the principal changes that occur are the elimination of undesirable flavor compounds, such as vicinal diketones, hydrogen sulfide, and acetaldehyde, and the excretion of compounds enhancing the flavor fullness (body) of beer.

Vicinal Diketones

In traditional fermentation lagering processes, the elimination of vicinal diketones required several weeks and determined the length of the maturation process. Nowadays, the maturation phase is much shorter since strategies are used to accelerate the vicinal diketones removal (see the preceding text). Diacetyl is used as a marker molecule. The objective during lagering is to reduce the diacetyl concentration below its taste threshold (<0.10 mg/mL).

Hydrogen Sulfide

Sulfite and hydrogen sulfide are intermediates in the biosynthesis of the sulfur-containing amino acids methionine and cysteine (Van Haecht and Dufour 1995, Duan et al. 2004). Hydrogen sulfide plays an important role during maturation. Inorganic sulfate is taken up by the yeast cells via a permease. Subsequently, it is reduced to sulfide via the intermediates adenylyl sulfate, phosphoadenylyl sulfate and sulfite (see Fig. 33.4). H_2S and SO_2, which are not incorporated in S-containing amino acids, are excreted by the yeast cell during the growth phase (Ryder and Masschelin 1983, Thomas and Surdin-Kerjan 1997). The excreted amount depends on the used yeast strain, the sulfate content of the wort and the growth conditions (Romano and Suzzi 1992). Methionine causes decreased production of sulfur compounds by feedback inhibition, while threonine increases the production (Thomas and Surdin-Kerjan 1997). H_2S and SO_2 can also be formed from the catabolism of S-containing amino acids (Dual et al. 2004). The production of H_2S and SO_2 depends on the yeast strain, sulfate content in the wort, and growth conditions. The production of H_2S could be reduced by the expression of cystathione synthase genes from *S. cerevisiae* in a brewing yeast strain (Tezuka et al. 1992).

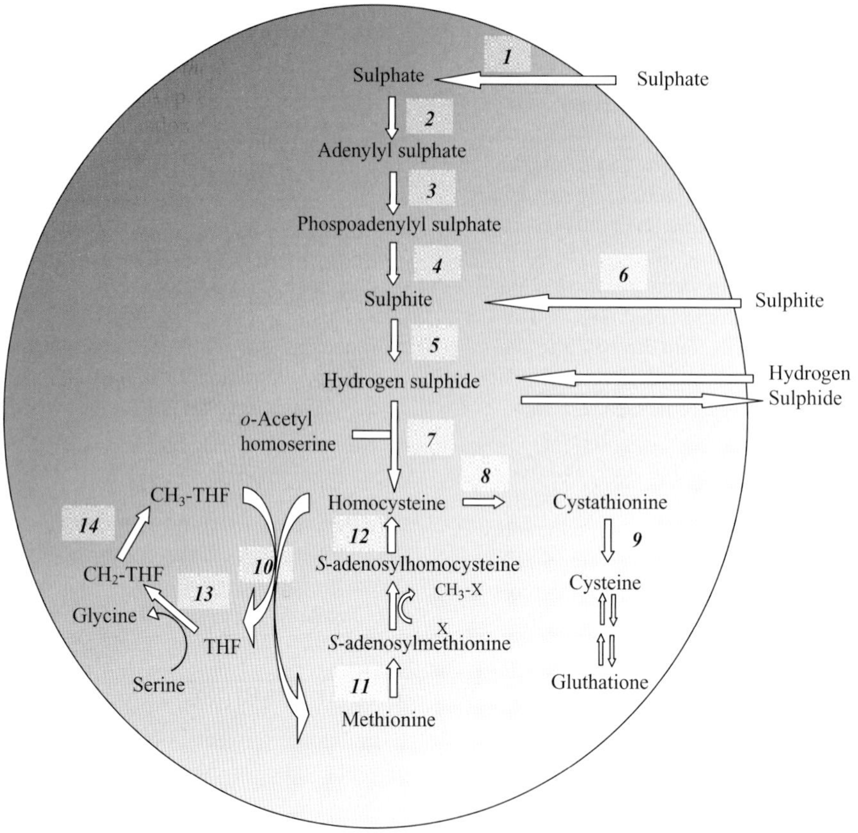

Figure 33.4. The remethylation, transulfuration, and sulfur assimilation pathways. Genes and enzymes catalyzing individual reactions are as follows: 1, sulfate permease; 2, ATP sulfurylase; 3, *MET14*: adenylylsulfate kinase (EC 2.7.1.25); 5, *MET10*: sulfite reductase (EC 1.8.1.2); 6, sulfite permease; 7, *MET17*: *O*-acetylhomoserine (thiol)-lyase (EC 2.5.1.49); 8, *CYS4*: cystathionine β-synthase (CBS; EC 4.2.1.22); 9, *CYS3*: cystathionine γ-lyase (EC 4.4.1.1), 10, *MET6*: methionine synthase (EC 2.1.1.14); 11, *SAM1* and *SAM2*: *S*-adenosylmethionine synthetase (EC 2.5.1.6); 12, *SAH1*: *S*-adenosylhomocysteine hydrolase (EC 3.3.1.1); 13, *SHM1* and *SHM2*: serine hydroxymethyltransferase (SHMT; EC 2.1.2.1); 14, *MET12* and *MET13*: methylenetetrahydrofolate reductase (MTHFR; EC 1.5.1.20). "X" represents any methyl group acceptor; THF, tetrahydrofolate; CH_2-THF, 5,10-methylenetetrahydrofolate; CH_3-THF, 5-methyltetrahydrofolate. (Partly adapted from Chan and Appling 2003.)

At the end of the primary fermentation and during the maturation, the excess H_2S is reutilized by the yeast. A warm conditioning period at 10 à 12°C may be used to remove excessive levels of H_2S.

Brewing yeasts produce H_2S when they are deficient in the vitamin pantothenate (Walker 1998). This vitamin is a precursor of coenzyme A, which is required for metabolism of sulfate into methionine. Therefore, panthothenate deficiency may result in an imbalance in sulfur amino acid biosynthesis, leading to excess sulfate uptake and excretion of H_2S (Slaughter and Jordan 1986).

Sulfite is a versatile food additive used to preserve a large range of beverages and foodstuffs. In beer, sulfite has a dual purpose, acting both as an antioxidant and an agent for masking of certain off-flavors. Some of the flavor stabilizing properties of sulfite is suggested to be due to complex formation of bisulfate with varying carbonyl compounds, of which some would give rise to off-flavors in bottled beer (Dufour 1991). Especially, the unwanted carbonyl *trans*-2-nonenal has received particular attention, since it is responsible for the "cardboard" flavor of some types of stale beer. It has been suggested that it would be better to use a yeast strain with reduced sulfite excretion during fermentation and to add sulfite at the point of bottling to ensure good flavor stability (Francke Johannesen et al. 1999). Therefore, a brewer's yeast disabled in the production of sulfite has been constructed by inactivating both copies of the two alleles of the *MET14* gene (which encodes for adenylylsulfate kinase). Fermentation experiments showed that there was no qualitative difference between yeast-derived and artificially added sulfite, with respect to *trans*-2-nonenal content and flavor stability of the final beer.

The elimination of the gene encoding sulfite reductase (*MET10*) in brewing strains of *Saccharomyces* results in increased accumulation of SO_2 in beer (Hansen and Kielbrandt 1996a). The inactivation of *MET2* resulted into elevated sulfite concentrations in beer (Hansen and Kielbrandt 1996b). Beers produced with increased levels of sulfite showed an improved flavor stability.

Acetaldehyde

Aldehydes—in particular, acetaldehyde (green apple-like flavor)—have an impact on the flavor of green beer. The accumulation of acetaldehyde is dependent on the kinetic properties of enzymes responsible for its formation (pyruvate decarboxylase and acetaldehyde dehydrogenase) and dissimilation (alcohol dehydrogenase) (Boulton and Quain 2006). Acetaldehyde synthesis is linked to yeast growth (Geiger and Piendl 1976). Its concentration is maximal at the end of the growth phase and is reduced at the end of the primary fermentation and during maturation by the yeast cells. As with diacetyl, levels may be enhanced if yeast metabolism is stimulated during transfer, especially by oxygen ingress. Removal also requires the presence of enough active yeast. Fermentations with early flocculating yeast cells can result in too high acetaldehyde concentrations at the end.

Development of Flavor Fullness

During maturation, the residual yeast will excrete compounds (i.e., amino acids, phosphates, peptides, nucleic acids, ...) into the beer. The amount and "quality" of these excreted materials depend on the yeast concentration, yeast strain, its metabolic state and the temperature (NN 2000). Rapid excretion of material is best achieved at a temperature of 5–7°C during 10 days (Van de Meersche et al. 1977).

When the conditioning period is too long or when the temperature is too high, yeast cell autolysis will occur. Some enzymes are liberated (e.g., α-glucosidase), which will produce glucose from traces of residual maltose (NN 2000). At the bottom of a fermentation tank, the amount of α-amino-nitrogen can rise to 40–10,000 mg/L, which account for an increase of 30 mg/L for the total beer volume. The increase in amino acid concentration in the beer has a positive effect on the flavorfullness of the beer. Undesirable medium-chain fatty acids can also be produced in significant amounts if the maturation temperature is too high (Masschelein 1981). Measurement of these compounds indicates the level of autolysis and permits the determination of the most appropriate conditioning period and temperature.

BEER FERMENTATION USING IMMOBILIZED CELL TECHNOLOGY

The advantages of continuous fermentation—such as greater efficiency in utilization of carbohydrates and better use of equipment—led also to the development of continuous beer fermentation processes. Since the beginning of the twentieth century, many different systems using suspended yeast cells have been developed. The excitement for continuous beer fermentation led—especially during the 1950 and 1960s—to the development of various interesting systems. These systems can be classified as (i) stirred *versus* unstirred tank reactors, (ii) single-vessel systems versus a number of vessels connected in series, (iii) vessels that allow yeast to overflow freely with the beer ("open system") versus vessels that have abnormally high yeast concentrations ("closed" or "semiclosed system") (Wellhoener 1954, Coutts 1957, Bishop 1970, Hough et al. 1982,). However, these continuous beer fermentation processes were not commercially successful due to many practical problems, such as the increased danger of contamination (not only during fermentation but also during storage of wort in supplementary holdings tanks that are required since the upstream and downstream brewing processes are usually not continuous), changes in beer flavor (Thorne 1968) and a poor understanding of the beer fermentation kinetics under continuous conditions. One of the well-known exceptions is the successful implementation of a continuous beer production process in New Zealand by Morton Coutts (Dominion Breweries), which is still in use today (Coutts 1957, Hough et al. 1982).

In the 1970, there was a revival in developing continuous beer fermentation systems due to the progress in research on immobilization bioprocesses using living cells. Immobilization gives fermentation processes with high cell densities, resulting in a drastic increase in fermentation productivities compared to the traditional time-consuming batch fermentation processes.

The last 30 years, ICT has been extensively examined and some designs have reached already commercial exploitation. Immobilized cell systems are heterogeneous systems in which considerable mass transfer limitations can occur, resulting in a changed cell yeast metabolism. Therefore, successful exploitation of ICT needs a thorough understanding of mass transfer and intrinsic yeast kinetic behavior of these systems.

Carrier Materials

Cell immobilization can be classified into four categories based on the mechanism of cell localization and the nature of support material: (i) attachment to the support surface, which can be spontaneous or induced by linking agents; (ii) entrapment within a porous matrix; (iii) containment behind or within a barrier; and (iv) self-aggregation, naturally or artificially induced (Karel et al. 1985, Willaert and Baron 1996). Various cell immobilization carrier materials have been tested and used for beer production/bioflavoring. Selection criteria are summarized in Table 33.5. Depending on the particular application, reactor type and operational conditions, some selection criteria will be more appropriate. Examples of selected carrier materials for particular applications are tabulated in Table 33.6.

Cell immobilization by self-aggregation is based on the formation of cell clumps or flocs. Actually, flocs are formed at the end of the primary fermentation when flocculent strains are used. Flocculent strains can also be used in continuous fermentation systems (see "Coutts continuous beer fermentation system" described in the preceding text). Very high yeast concentrations may be achieved by the use of inclined tubes, still zones around outlet pipes and by holding the yeast in a filter (Hough et al. 1982). Growth of the sedimentary yeast may be controlled by the amount of air injected, while carbon dioxide is used to cause some mixing. A flocculent strain has also been used in the tower

Table 33.5. Selection Criteria for Yeast Cell Immobilization Carrier Materials

- High cell mass loading capacity
- Easy access to nutrient media
- Simple and gentle immobilization procedure
- Immobilization compounds approved for food applications
- High surface area-to-volume ratio
- Optimum mass transfer distance from flowing media to centre of support
- Mechanical stability (compression, abrasion)
- Chemical stability
- Highly flexible: rapid start-up after shut-down
- Sterilizable and reusable
- Suitable for conventional reactor systems
- Low shear experienced by cells
- Easy separation of cells and carrier from media
- Readily up-scalable
- Economically feasible (low capital and operating costs)
- Desired flavor profile and consistent product
- Complete attenuation
- Controlled oxygenation
- Control of contamination
- Controlled yeast growth
- Wide choice of yeast

Source: Nedovic et al. (2005) and Verbelen et al. (2010).

fermenter system, which comprises a vertical tube into the base of which wort is pumped (Royston 1966). The sedimentary yeast forms a solid plug at the base of the vessel and through it the wort permeates. Fermentation proceeds as the wort rises, with the rate of wort injection being so adjusted that at the top of the tower the wort is completely fermented.

Yeast flocculation is a reversible, asexual, and calcium-dependent process in which cells adhere to form flocs consisting of thousands of cells (Stratford 1989, Bony et al. 1997, Jin and Speers 1999). Many fungi contain a family of cell wall "adhesines" (which are glycoproteins) that confer unique adhesion properties (Teunissen and Steensma 1995, Guo et al. 2000, Hoyer 2001, Sheppard et al. 2004). These molecules are required for the interactions of fungal cells with each other (flocculation and filamentation) (Teunissen and Steensma 1995, Lo and Dranginis 1998, Guo et al. 2000, Viyas et al. 2003), with inert surfaces such as agar and plastic (Gaur and Klotz 1997, Lo and Dranginis 1998, Reynolds and Fink 2001, Li and Palecek 2003) and with mammalian tissues/cells (Cormack et al. 1999, Staab et al. 1999, Fu et al. 2002, Li and Palecek 2003). They are also crucial for the formation of fungal biofilms (Baillie and Douglas 1999, Reynolds and Fink 2001, Green et al. 2004). The adhesin proteins in *S. cerevisiae* are encoded by *FLO* genes, including *FLO1*, *FLO5*, *FLO9*, *FLO10*, and *FLO11* (Verstrepen et al. 2004). These proteins are called flocculins (Caro et al. 1997) because these proteins promote cell–cell adhesion to form multicellular clumps that sediment out of solution.

The flocculation phenomenon is genetically controlled by 33 genes (Teunissen and Steensma 1995). Recently, the five members of the *FLO*-adhesine family were studied by selectively overexpressing each *FLO* gene in the laboratory strain S288C (Van Mulders et al. 2009). As all *FLO* genes are transcriptionally silent in the S288C-strain background, each *FLO* gene can be activated one by one and each resultant phenotype can be investigated. The *FLO1*, *FLO5*, *FLO9*, and *FLO10* genes share considerable sequence homology. The member proteins of the adhesin family have a modular configuration that consists of three domains (A, B, and C) and an *N*-terminal secretory sequence that must be removed as the protein moves through the secretory pathway to the plasma membrane (Hoyer et al. 1998). The *N*-terminal domain (A) is involved in sugar recognition (Kobayashi et al. 1998). The adhesins undergo several posttranslational modifications, that is, *N*- and *O*-glycosylations. They move from the endoplasmic reticulum (ER) through the Golgi and pass through the plasma membrane and find their final destination in the cell wall, where they are anchored by a glycosyl phosphatidylinositol (GPI) (Teunissen et al. 1993a, 1993b, Bidard et al. 1994, Bony et al. 1997, Hoyer et al. 1998). The GPI anchor is added to the *C*-terminus in the ER, and mannose residues are added to the many serine and threonine residues in domain B in the Golgi (Udenfriend and Kodukula 1995, Bony et al. 1997, Frieman et al. 2002, De Groot et al. 2003). The *FLO1* gene product (Flo1p) has been localized at the cell surface by immunofluorescent microscopy (Bidard et al. 1995). The amount of Flo proteins in flocculent strains increased during batch yeast growth and the Flo1p availability at the cell surface determined the flocculation degree of the yeast. Flo proteins are polarly incorporated into the cell wall at the bud tip and the mother–daughter neck junction (Bony et al. 1997). The transcriptional activity of the flocculation genes is influenced by the nutritional status of the yeast cells as well as other stress factors (Verstrepen et al. 2003a). This implies that during beer fermentation, flocculation is affected by numerous parameters such as nutrient conditions, dissolved oxygen, pH, fermentation temperature, and yeast handling and storage conditions.

Applications of ICT in the Brewing Industry

Beer production with immobilized yeast has been the subject of research for approximately 40 years, but has so far found limited application in the brewing industry, because of engineering problems, unrealised cost advantages, microbial contaminations, and an unbalanced beer flavor (Linko et al. 1998, Brányik et al. 2005, Nedovic et al. 2005, Willaert and Nedovic 2006, Verbelen et al. 2010). The ultimate aim of this research is the production of beer, of desired quality, within 1–3 days. Traditional beer fermentation systems use freely suspended yeast cells to ferment wort in an unstirred batch reactor. The primary fermentation takes approximately 7 days with a subsequent secondary fermentation (maturation) of several weeks. A batch culture system employing immobilization could benefit from an increased rate of fermentation. However, it appears that in terms of increasing productivity, a continuous fermentation system with immobilization would be the best method (Verbelen et al. 2006). An important issue of the research area is to whether beer can be produced by immobilized yeast in continuous culture with the same characteristic as the traditional method.

Table 33.6. Some Selected Applications of Cell Immobilization Systems Used for Beer Production

Carrier Material	Reactor Type	Reference
	Flavor maturation	
Calcium alginate beads	Fixed bed	Shindo et al. (1994)
DEAE-cellulose	Fixed bed	Pajunen and Grönqvist (1994)
Polyvinyl alcohol beads	Fixed bed	Smogrovicová et al. (2001)
Porous glass beads	Fixed bed	Linko et al. (1993), Aivasidis (1996)
Alcohol-free beer		
DEAE-cellulose beads	Fixed bed	Collin et al. 1991, Lomni et al. 1990
Porous glass beads	Fixed bed	Aivasidis et al. (1991)
Silicon carbide rods	Monolith reactor	Van De Winkel et al. (1991)
	Acidified wort	
DEAE-cellulose beads	Fixed bed	Pittner et al. (1993)
	Main fermentation	
Calcium alginate beads	Gas lift	White and Portno (1979), Nedovic et al. (1997)
Calcium pectate beads	Gas lift	Smogrovicová et al. (1997)
κ-Carrageenan beads	Gas lift	Mensour et al. (1996), Decamps et al. (2004)
Ceramic beads	Fixed bed	Inoue (1995)
Corncobs	Gas lift	Brányik et al. (2006a)
Gluten pellets	Fixed bed	Bardi et al. 1997
Polyvinyl alcohol beads	Gas lift	Smogrovicová et al. (2001)
Polyvinyl alcohol Lentikats®	Gas lift	Smogrovicová et al. (2001), Bezbradica et al. (2007)
Polyvinyl chloride granules	Gas lift	Moll et al. (1973)
Porous glass beads	Fixed bed	Virkajärvi and Krönlof (1998)
Porous chitosan beads	Fluidized bed	Unemoto et al. (1998), Maeba et al. (2000)
Silicon carbide rods	Monolith reactor	Andries et al. (1996)
Spent grains	Gas lift	Brányik et al. (2002, 2004)
Stainless steel fibre cloths	Gas lift	Verbelen et al. (2006)
Wood (aspen beech) chips	Fixed bed	Linko et al. (1997), Kronlöf and Virkajärvi (1999), Pajunen et al. (2001)

Started in 1971, one of the first ICT systems for the fermentation and maturation of beer was developed by TREPAL (the R&D Centre of the Konenbourg Brewery and the European Brewery) with INSA (Institut National des Sciences Appliquées, Toulouse, France) and INRA (Institus National des Recherche Agronomique, Dijon, France) (Moll et al. 1973, Moll and Dueurtre 1996, Moll 2006). A continuous pilot installation for fermentation and maturation of beer with a flow rate of 5 L/h was developed and worked for 9 months without any microbial contamination. Around the same time, Narziss and Hellich (1971) developed an ICT process where yeast cells were immobilized in kieselguhr (which is widely used in the brewing industry as a filter aid) and a kieselguhr filter was employed as bioreactor (called the "bio-brew bioreactor"). Their yeast cell immobilization method was based on a method for the immobilization of enzymes (Berdelle-Hilge 1966). From the beginning of the nineteen seventies, various systems have been developed and some have been implemented on an industrial scale.

ICT processes have been developed for (i) the flavor maturation of green beer, (ii) the production of alcohol-free or low-alcohol beer, (iii) the production of acidified wort using immobilized lactic acid bacteria, and (iv) the continuous beer fermentation (Brányik et al. 2005, Verbelen et al. 2006, Willaert and Nedovic 2006, Willaert 2007, Verbelen et al. 2010).

Flavor Maturation of Green Beer

The objective of flavor maturation is the removal of diacetyl and 2,3-pentanedione, and their precursors α-acetolactate and α-acetohydroxybutyrate, which are produced during the main fermentation (see the preceding text). The conversion of α-acetohydroxy acids to the vicinal diketones is the rate-limiting step. This reaction step can be accelerated by heating the beer—after yeast removal—to 80–90°C during a couple of minutes. The resulting vicinal diketones are subsequently reduced by immobilized cells into their less-flavor-active compounds.

The traditional maturation process is characterized by a near-zero temperature, low pH and low yeast concentration, resulting in a very long maturation period of 3–4 weeks. Different strategies have been developed to accelerate diacetyl removal (see Section "Vicinal diketones"). One of the techniques is the use of an ICT process to reduce the maturation period to about 2 hours.

Two continuous maturation systems have been implemented industrially so far: one at Sinebrychoff Brewery (Finland, capacity: 1 million hL per year) (Pajunen 1995) and another system, developed by Alfa Laval and Schott Engineering (Mensour et al. 1997). They are both composed of a separator (to prevent growing yeast cells in the next stages), an anaerobic heat treatment unit (to accelerate the chemical conversion of α-acetolactate to diacetyl, but also the partial directly conversion to acetoin), and a packed bed reactor with yeast immobilized on DEAE-cellulose granules or porous glass beads (to reduce the remaining diacetyl), respectively (Yamauchi et al. 1994). Later on, the DEAE-cellulose carriers were replaced by cheaper wood chips (Virkajärvi 2002). The heat treatment has been replaced by an enzymatic transformation in a fixed bed reactor in which the α-acetolactate decarboxylase is immobilized in special multilayer capsules, followed by the reduction of diacetyl by yeast in a second packed bed reactor (Nitzsche et al. 2001).

Production of Alcohol-Free or Low-Alcohol Beer

The classical technology to produce alcohol-free or low-alcohol beer is based on the suppression of alcohol formation by arrested batch fermentation (Narziss et al. 1992). However, the resulting beers are characterized by an undesirable wort aroma, since the wort aldehydes have only been reduced to a limited degree (Collin et al. 1991, Debourg et al. 1994, van Iersel et al. 1998). The reduction of these wort aldehydes can be quickly achieved by a short contact time with immobilized yeast cells at a low temperature without undesirable cell growth and ethanol production. A disadvantage of this short contact process is the production of only a small amount of desirable esters.

Controlled ethanol production for low-alcohol and alcohol-free beers have been successfully achieved by partial fermentation using DEAE-cellulose as carrier material, which was packed in a column reactor (Collin et al. 1991, Van Dieren 1995). This technology has been successfully implemented by Bavaria Brewery (The Netherlands) to produce malt beer on an industrial scale (150,000 hL/year) (Pittner et al. 1993). Several other companies—that is, Faxe (Denmark), Ottakringer (Austria), and a Spanish brewery—have also implemented this technology (Mensour et al. 1997). In Brewery Beck (Germany), a fluidized-bed pilot scale reactor (8 hL/day) filled with porous glass beads was used for the continuous production of nonalcoholic beer (Aivasidis et al. 1991, Breitenbücher and Mistler 1995, Aivasidis 1996). Yeast cells immobilized in silicon carbide rods and arranged in a multichannel loop reactor (Meura, Belgium) have been used to produce alcohol-free beer at pilot scale by Grolsch Brewery (The Netherlands) and Guinness Brewery (Ireland) (Van De Winkel 1995).

Nuclear mutants of *S. cerevisiae* that are defective in the synthesis of tricarboxylic acid cycle enzymes; that is, fumarase (Kaclíková et al. 1992) or 2-oxoglutarate dehydrogenase (Mockovciaková et al. 1993) have been immobilized in calcium pectate gel beads and used in a continuous process for the production of nonalcoholic beer (Navrátil et al. 2000). These strains produced minimal amounts of ethanol and they were also able to produce much lactic acid (up to 0.64 g/dm^3).

Production of Acidified Wort Using Immobilized Lactic Acid Bacteria

The objective of this technology is the acidification of the wort according to the "Reinheidsgebot", before the start of the boiling process in the brewhouse. An increased productivity of acidified wort has been obtained using immobilized *Lactobacillus amylovorus* on DEAE-cellulose beads (Pittner et al. 1993, Meersman 1994). The pH of wort was reduced below a value of 4.0 after contact times of 7–12 minutes using a packed-bed reactor in downflow mode. The produced acidified wort was stored in a holding tank and used during wort production to adjust the pH.

Continuous Main Fermentation

During the main fermentation of beer, not only ethanol is being produced but also a complex mixture of flavor-active secondary metabolites, of which the higher (or fusel) alcohols and esters are the most important (Verbelen et al. 2010). In addition, diacetyl and some sulfury compounds can cause off-flavors. Since this complex flavor profile is closely related to the amino acid metabolism and consequently to the growth of the yeast cells, differences in the growth metabolic state between freely suspended and immobilized yeast cell systems are most probably responsible for the majority of alterations in the beer flavor. For that reason, it is important that the physiological and metabolic state of the yeast in conventional batch systems is mimicked as much as possible during the continuous fermentation with immobilized yeast. In the continuous mode of operation, cells are not exposed to significant alterations of the environment, influencing the metabolism of the cells and consequently the flavor. Hence, the microbial population of continuous systems lacks the different growth phases of a batch culture. To imitate the batch process as much as possible, plug-flow reactors or a series of reactors can be used. As can be assumed, both the continuous mode of operation and the immobilization of yeast cells can influence the beer flavor.

The Japanese brewery Kirin developed a multistage continuous fermentation process (Inoue 1995, Yamauchi et al. 1994, Yamauchi and Kasahira 1995). The first stage is a stirred tank reactor for yeast growth, followed by packed-bed fermenters, and the final step is a packed-bed maturation column. The first stage ensures adequate yeast cell growth with the desirable FAN consumption. Ca-alginate was initially selected as carrier material to immobilize the yeast cells. These alginate beads were later replaced by ceramic beads ("Bioceramic®"). This system allowed to produce beer within 3–5 days.

The engineering company Meura (Belgium) developed a reactor configuration with a first stage with immobilized yeast cells where partially attenuation and yeast growth occurs, followed by a stirred tank reactor (with free yeast cells) for complete attenuation, ester formation, and flavor maturation (Andries et al. 1996, Masschelein and Andries 1995). Silicon carbide rods are used in the first reactor as immobilization carrier material. The stirred tank (second reactor) is continuously inoculated by free cells that escape from the first immobilized yeast cell reactor.

Labatt Breweries (Interbrew, Canada) in collaboration with the Department of Chemical and Biochemical Engineering at the University of Western Ontario (Canada) developed a continuous system using κ-carrageenan immobilized yeast cells in an airlift reactor (Mensour et al. 1995, 1996, 1997). Pilot scale research showed that full attenuation was reached in 20–24 hours with this system compared to 5–7 days for the traditional batch fermentation. The flavor profile of the beer produced using ICT was similar to the batch fermented beer.

Hartwell Lahti and VTT Research Institute (Finland) developed a primary fermentation system using ICT on a pilot scale of 600 L/day (Kronlöf and Virkajärvi 1999). Woodchips were used as the carrier material that reduced the total investment cost by one-third compared to more expensive carriers. The results showed that fermentation and flavor formation were very similar compared to a traditional batch process, although the process time was reduced to 40 hours.

Andersen et al. (1999) developed a new ICT process in which the concentration of carbon dioxide is controlled in a fixed-bed reactor in such a way that the CO_2 formed is kept dissolved and is removed from the beer without foaming problems. DEAE-cellulose was used as carrier material. High-gravity beer of acceptable quality has been fermented in 20 hours at a capacity of 50 L/h.

A well-explored concept for main beer fermentation is the use of a gas-lift bioreactor system (see Table 33.6). In this system, mixing established by the circulation of liquid and solid phases, provided high liquid circulation rates, low shear environment, and good mass transfer properties (Siegel and Robinson 1992, Vunjak-Novakovic et al. 1992, Chisti and Moo-Young 1993, Baron et al. 1996). Additionally, they possess the following positive characteristics: high loadings of solids, simple construction, low risk of contamination, easy adjustment and control of operational parameters and simple capacity enlargement (Vunjak-Novakovic et al. 1998). Various carrier materials for gas-lift bioreactors have been studied to perform the main fermentation (see Table 33.6).

The optimization of temperature, wort gravity, feed volume, and wort composition seems to be an important tool for the control of the flavor-active compounds formation in immobilized beer fermentation systems (Verbelen et al. 2006, 2010, Willaert and Nedovic 2006, Brányik et al. 2008). Many researchers have concluded that the optimization of aeration during continuous fermentation is essential for the quality of the final beer (Virkajärvi et al. 1999). Oxygen is needed for the formation of unsaturated fatty acids and sterols that needed for growth (Depraetere et al. 2003). However, excess oxygen will lead to low ester production but to excessive diacetyl, acetaldehyde, and fusel alcohol formation (Okabe et al. 1992, Wackerbauer et al. 2003, Brányik et al. 2004). It is possible to adjust the flavor of the produced beer by ensuring the adequate amount of dissolved oxygen by sparging with a mixture of air, nitrogen, or carbon dioxide (Kronlöf and Linko 1992, Brányik et al. 2004). However, it remains difficult to predict the right amount of oxygen because the oxygen availability to the immobilized yeast cells is dependent of external and internal mass transfer limitations (Willaert and Baron 1996, Willaert et al. 2004).

Also, the reactor design, the carrier, and yeast strain can have a dominant effect on flavor formation (Cop et al. 1989, Linko, et al. 1997, Smogrovicová and Dömény 1999, Tata et al. 1999, Virkajärvi and Pohjala 2000).

ACKNOWLEDGMENTS

This work was supported by the Belgian Federal Science Policy Office and European Space Agency PRODEX program, the Institute for the Promotion of Innovation by Science and Technology in Flanders (IWT), and the Research Council of the VUB.

REFERENCES

Aivasidis A. 1996. Another look at immobilized yeast systems. *Cerevisia* 21(1): 27–32.

Aivasidis A et al. 1991. Continuous fermentation of alcohol-free beer with immobilized yeast cells in fluidized bed reactors. Proceedings of the European Brewery Convention Congress, Lisbon, pp. 569–576.

Alexander MA, Jeffries TW. 1990. Respiratory efficiency and metabolite partitioning as regulatory phenomena in yeast. *Enzyme Microb Technol* 12: 2–19.

Alves SL Jr et al. 2007. Maltose and maltotriose active transport and fermentation by *Saccharomyces cerevisiae*. *J Am Soc Brew Chem* 65: 99–104.

Alves SL Jr et al. 2008. Molecular analysis of maltotriose active transport and fermentation by *Saccharomyces cerevisiae* reveals a determinant role for the *AGT1* permease. *Appl Environ Microbiol* 74: 1494–1501.

Andersen K et al. 1999. New process for the continuous fermentation of beer. Proceedings of the European Brewery Convention Congress, Cannes, pp. 771–778.

Anderson RG, Kirsop BH. 1975a. Oxygen as regulator of ester accumulation during the fermentation of worts of high gravity. *J Inst Brew* 80: 111–115.

Anderson RG, Kirsop BH. 1975b. Quantitative aspects of the control by oxygenation of acetate ester formation of worts of high specific gravity. *J Inst Brews* 81: 296–301.

Andries M et al. 1996. Design and application of an immobilized loop bioreactor for continuous beer fermentation. In: RH Wijffels et al. (eds.) *Immobilized Cells: Basics and Applications*. Elsevier, Amsterdam, pp. 672–678.

Arikawa Y et al. 1998. Soluble fumarate reductase isoenzymes from *Saccharomyces cerevisiae* are required for anaerobic growth. *FEMS Microbiol Lett* 165: 111–116.

Arikawa Y et al. 1999. Isolation of sake yeast strains possessing various levels of succinate- and/or malate-producing abilities by gene disruption or mutation. *J Biosci Bioeng* 87: 333–339.

Asano T et al. 1999. Effect of NAD^+-dependent isocitrate dehydrogenase gene (*IDH1*, *IDH2*) disruption of sake yeast on organic acid composition in sake mash. *J Biosci Bioeng* 88: 258–263.

Avhenainen J, Mäkinen V. 1989. The effect of pitching yeast aeration on fermentation and beer flavor. *Proceedings of the European Brewery Convention Congress*, pp. 517–519.

Äyräpää T. 1971. Biosynthetic formation of higher alcohols by yeast. Dependence on the nitrogenous nutrient level of the medium. *J Inst Brew* 77: 266–275.

Baillie GS and Douglas LJ. 1999. Role of dimorphism in the development of *Candida albicans* biofilms. *J Med Microbiol* 48: 671–679.

Bamforth CW and Kanauchi M. 2004. Enzymology of vicinal diketone reduction in brewer's yeast. *J Inst Brew* 110: 83–93.

Bardi E et al. 1997. Room and low temperature brewing with yeast immobilized on gluten pellets. *Process Biochem* 32: 691–696.

Baron GV et al. 1996. Immobilised cell reactors. In: RG Willaert et al. (eds.) *Immobilised Living Cell Systems: Modelling and Experimental Methods*. John Wiley & Sons, Chichester, pp. 67–95

Berdelle-Hilge P. 1966. Verfahren und Vorrichtung zur kontinuerlichen Behandlung von Flüssigkeiten mit Einzymträgern. German patent 1517814.

Bezbradica D et al. 2007. Immobilization of yeast cells in PVA particles for beer fermentation. *Process Biochem* 42: 1348–1351.

Bidard F et al. 1994. Cloning and analysis of a *FLO5* flocculation gene from *Saccharomyces cerevisiae* yeast. *Curr Genet* 25: 196–201.

Bidard F et al. 1995. The *Saccharomyces cerevisiae FLO1* flocculation gene encodes for a cell surface protein. *Yeast* 11: 809–822.

Bishop LR. 1970. A system of continuous fermentations. *J Inst Brew* 76: 172–181.

Bisson LF et al. 1993. Yeast sugar transporters. *Crit Rev Biochem Mol Biol* 28: 259–308.

Blomqvist K et al. 1991. Chromosomal integration and expression of two bacterial α-acetolactate decarboxylase genes in brewer's yeast. *Appl Environ Microbiol* 57: 2796–2803.

Bony M et al. 1997. Localisation and cell surface anchoring of the *Saccharomyces cerevisiae* flocculation protein Flo1p. *J Bacteriol* 179: 4929–4936.

Boswell CD et al. 2002. Studies on the effect of mechanical agitation on the performance of brewing fermentations: fermentation rate, yeast physiology, and development of flavor compounds. *J Am Soc Brew Chem* 60: 101–106.

Boulton CA. 2000. Trehalose, glycogen and sterol. In: KA Smart (ed.) *Brewing Yeast Fermentation Performance*. Blackwell Science, Oxford, pp. 10–19.

Boulton CA et al. 1991. Yeast physiological condition and fermentation performance. Proceedings of the 21st European Brewery Convention, pp. 385–392.

Boulton C, Quain D. 2006. *Brewing Yeast and Fermentation*. Blackwell Science, Oxford.

Brányik T et al. 2002. Continuous primary beer fermentation with brewing yeast immobilized on spent grains. *J Inst Brew* 108: 410–415.

Brányik T et al. 2004. Continuous primary fermentation of beer with yeast immobilized on spent grains—the effect of operational conditions. *J Am Soc Brew Chem* 62: 29–34.

Brányik T et al. 2005. Continuous beer fermentation using immobilized yeast cell bioreactor systems. *Biotechnol Prog* 21: 653–663.

Brányik T et al. 2006a. Continuous immobilized yeast reactor system for complete beer fermentation using spent grains and corncobs as carrier materials. *J Ind Microbiol Biotechnol* 33: 1010–1018.

Brányik T et al. 2008. A review of flavour formation in continuous beer fermentations. *J InstBrew* 114: 3–13.

Breitenbücher K, Mistler M. 1995. Fluidized-bed fermenters for the continuous production of non-alcoholic beer with open-pore sintered glass carriers. EBC Monograph XXIV, EBC Symposium on Immobilized Yeast Applications in the Brewing Industry, pp. 77–89.

Bullis BL, Lamire BD. 1994. Isolation and characterization of the *Saccharomyces cerevisiae SDH4* gene encoding a membrane anchor subunit of succinate dehydrogenase. *J Biol Chem* 269: 6543–6549.

Calderbank J, Hammond JRM. 1994. Influence of higher alcohol availability on ester formation by yeast. *J Am Soc Brew Chem* 52: 84–90.

Caro LH et al. 1997. In silicio identification of glycosylphosphatidylinositol-anchored plasma-membrane and cell wall proteins of *Saccharomyces cerevisiae*. *Yeast* 13: 1477–1489.

Chan SY, Appling DR. 2003. Regulation of S-adenosylmethionine levels in *Saccharomyces cerevisiae*. *J Biol Chem* 278: 43051–43059.

Chang YS et al. 1988. *MAL63* codes for a positive regulator of maltose fermentation in *Saccharomyces cerevisiae*. *Curr Genet* 14: 201–209.

Chapman KB et al. 1992. *SDH1*, the gene encoding the succinate dehydrogenase flavoprotein subunit from *Saccharomyces cerevisiae*. *Gene* 118: 131–136.

Charalambous G et al. 1972. Effect of structurally modified polyphenols on beer flavour and flavour stability, part II. *Tech Quart Master Brew Assoc Am* 9: 131–135.

Chen EC-H. 1978. Relative contribution of Ehrlich and biosynthetic pathways to the formation of fusel alcohols. *J Am Soc Brew Chem* 36: 39–43.

Chen EC-H. 1980. Utilisation of fatty acids by yeast during fermentation. *J Am Soc Brew Chem* 38: 148–153.

Chisti Y, Moo-Young M. 1993. Improve the performance of airlift reactors. *Chem Eng Prog* 89: 33–45.

Chow T et al. 1983. Identification and physical characterization of the maltose of yeast maltase structural genes. *Mol Gen Genet* 191: 366–371.

Cohen JD et al. 1984. Mutational analysis of *MAL1* locus of *Saccharomyces*: identification and functional characterization of three genes. *Mol Gen Genet* 196: 208–216.

Collin S et al. 1991. Yeast dehydrogenase activities in relation to carbonyl compounds removal from wort and beer. Proceedings of the European Brewery Convention Congress, Lisbon, pp. 409–416.

Coote N, Kirsop BH. 1974. The content of some organic acids in beer and in other fermented media. *J Inst Brew* 80: 474–483.

Cop J et al. 1989. Reactor design optimization with a view on the improvement of amino acid utilization and flavour development of calcium-alginate entrapped brewing yeast fermentations. Proceedings of the European Brewery Convention, Zürich, pp. 315–322.

Cormack BP et al. 1999. An adhesin of the yeast pathogen *Candida glabrata* mediating adherence to human epithelial cells. *Science* 285: 578–582.

Coutts MW. 1957. A continuous process for the production of beer. UK Patents 872, 391–400.

Crumplen RM et al. 1996. Characteristics of maltose transporter activity in an ale and large strain of the yeast *Saccharomyces cerevisiae*. *Lett Appl Microbiol* 23: 448–452.

Daignan-Fournier B et al. 1994. Structure and regulation of *SDH3*, the yeast gene encoding the cytochrome b_{560} subunit by respiratory complex II. *J Biol Chem* 269: 15469–15472.

D'Amore T et al. 1991. Advances in the fermentation of high gravity wort. *Proceedings of the European Brewery Convention Congress*, pp. 337–344.

Day RE et al. 2002a. Characterization of the putative maltose transporters encoded by YDL247w and YJR160c. *Yeast* 19: 1015–1027.

Day RE et al. 2002b. Molecular analysis of maltotriose transport and utilization by *Saccharomyces cerevisiae*. *Appl Environ Microbiol* 68: 5326–5335.

Debourg A. 2002. Yeast in action: from wort to beer. *Cerevisia* 27(3): 144–154.

Debourg A et al. 1990. Effets of *ILV* genes doses on the flux through the isoleucine-valine pathway in *Saccharomyces cerevisiae*. Abstract C31, 6th International Symposium on Genetics of Industrial Microorganisms.

Debourg A et al. 1994. Wort aldehyde reduction potential in free and immobilized yeast systems. *J Amer Soc Brew Chem* 52: 100–106.

Decamps C et al. 2004. Continuous pilot plant-scale immobilization of yeast in κ-carrageenan gel beads. *AIChE J* 50: 1599–1605.

De Groot PWJ et al. 2003. Genome-wide identification of fungal GPI proteins. *Yeast* 20: 781–796.

Delvaux F. 1998. Beheersing van gistingsproducten met belangrijke sensorische eigenschappen. *Cerevisia* 23(4): 36–45.

Depraetere SA. 2007. The potential of yeast preoxygenation for application in the brewing industry. Doctoral dissertation, no. 751, Faculty of Bioscience Engineering, Katholieke Universiteit Leuven, Belgium.

Depraetere SA et al. 2003. Evaluation of the oxygen requirement of lager and ale yeast strains by preoxygenation. *MBAA Tech Quart* 40: 283–289.

Dequin S. 2001. The potential of genetic engineering for improved brewing wine-making and baking yeasts. *Appl Microbiol Biotechnol* 56: 577–588.

Devuyst R et al. 1991. Oxygen transfer efficiency with a view to improvement of lager yeast performance and eer quality. Proceedings of the European Brewery Convention Congress, Lisbon, pp. 377–384.

Dickinson JR. 1999. Carbon metabolism. In: JR Dickinson, M Schweizer (eds.) The Metabolism and Molecular Physiology of *Saccharomyces cerevisiae*. Taylor & Francis, London, pp. 23–55.

Dickinson JR et al. 1997. A ^{13}C nuclear magnetic resonance investigation of the metabolism of leucine to isoamyl alcohol in *Saccharomyces cerevisiae*. *J Biol Chem* 272: 26871–26878.

Dickinson JR et al. 1998. An investigation of the metabolism of valine to isobutyl alcohol in *Saccharomyces cerevisiae*. *J Biol Chem* 273: 25751–25756.

Dickinson JR et al. 2000. An investigation of the metabolism of isoleucine to active amyl alcohol in *Saccharomyces cerevisiae*. *J Biol Chem* 275: 10937–10942.

Dickinson JR et al. 2003. The catabolism of amino acids to long chain and complex alcohols in *Saccharomyces cerevisiae*. *J Biol Chem* 278: 8028–8034.

Dietvorst J et al. 2005. Maltotriose utilization in lager yeast strains: *MTT1* encodes a maltotriose transporter. *Yeast* 22: 775–788.

Dietvorst J et al. 2007. Attachment of *MAL32*-encoded maltase on the outside of yeast cells improves maltotriose utilization. *Yeast* 24: 27–38.

Dillemans M et al. 1987. The amplification effect of the *ILV5* gene on the production of vicinal diketones in *Saccharomyces cerevisiae*. *J Am Soc Brew Chem* 45: 81–85.

Duan W et al. 2004. A parallel analysis of H_2S and SO_2 formation by brewing yeast in response to sulphur-containing amino acids and ammonium ions. *J Am Soc Brew Chem* 62: 35–41.

Dufour J-P. 1991. Influence of industrial brewing and fermentation working conditions on beer SO_2 levels and flavour stability. Proceedings of the European Brewery Convention Congress, Lisbon, pp. 209–216.

Dufour J-P, Malcorps P. 1994. Ester synthesis during fermentation: enzyme characterization and modulation mechanism. *Proceedings of the 4th Institute of Brewing Aviemore Conference, Aviemore*, pp. 137–151.

Duval EH et al. 2009. Microarray karyotyping of maltose-fermenting *Saccharomyces* yeasts with differing maltotriose utilization profiles reveals copy number variation in genes involved in maltose and maltotriose utilization. *J Appl Microbiol* 109: 248–259.

Ehrlich F. 1904. Uber das natürliche isomere des leucins. *Berichte der Deutschen Chemisten Gesellschaft* 37: 1809–1840.

Enari T-M et al. 1992. Process for accelerated beer production by integrative expression in the *PGK1* or *ADC1* genes. US Patent 5,108,925.

Engan S. 1969. Wort composition and beer flavour. I. The influence of some amino acids on the formation of higher aliphatic alcohols and esters. *J Inst Brew* 76: 254–261.

Engan S. 1972. Organoleptic threshold values of some alcohols and esters in beer. *J Inst Brew* 78: 33–36.

Engan S, Aubert O. 1977. Relations between fermentation temperature and the formation of some flavour components. Proceedings of the European Brewery Convention Congress, Amsterdam, pp. 591–607.

Enomoto K et al. 2002. Physiological role of soluble fumarate reductase in redox balancing during anaerobiosis in *Saccharomyces cerevisiae*. *FEMS Microbiol Lett* 24: 103–108.

Federoff HJ et al. 1983a. Carbon catabolite repression of maltase synthesis in *Saccharomyces cerevisiae*. *J Bacteriol* 156: 301–307.

Federoff HJ et al. 1983b. Regulation of maltase synthesis in *Saccharomyces carlsbergensis*. *J Bacteriol* 154: 1301–1308.

Fraleigh S et al. 1989. Regulation of oxidoreductive yeast metabolism by extracellular factors. *J Biotechnol* 12: 185–198.

Francke Johannesen P et al. 1999. Construction of *S. carlsbergensis* brewer's yeast without production of sulfite. Proceedings of the European Brewery Convention Congress, Cannes, pp. 655–662.

François J, Parrou JL. 2001. Reserve carbohydrates metabolism in the yeast *Saccharomyces cerevisiae*. *FEMS Microbiol Rev* 25: 125–145.

Frieman MB et al. 2002. Modular domain structure in the *Candida glabrata* adhesin Epa1p, a β-1,6 glucan-cross-linked cell wall protein. *Mol Microbiol* 46: 479–492.

Fu Y et al. 2002. *Candida albicans* Als1p: an adhesin that is a downstream effector of the *EFG1* filamentation pathway. *Mol Microbiol* 44: 61–72.

Fujii T et al. 1990. Application of a ribosomial DNA integration vector in the construction of a brewers' yeast having α-acetolactate decarboxylase activity. *Appl Environ Microbiol* 56: 997–1003.

Fujii T et al. 1994. Molecular cloning, sequence analysis, and expression of the yeast alcohol acetyltransferase gene. *Appl Environ Microbiol* 60: 2786–2792.

Fukuda K et al. 1998. Balance of activities of alcohol acetyltransferase and esterase in *Saccharomyces cerevisiae* is important for production of isoamyl acetate. *Appl Environ Microbiol* 64: 4076–4078.

Gaur NK, Klotz SA. 1997. Expression, cloning, and characterization of a *Candida albicans* gene, ALA1, that confers adherence properties upon *Saccharomyces cerevisiae* for extracellular matrix proteins. *Infect Immun* 65: 5289–5294.

Gee DA, Ramirez WF. 1994. A flavour model for beer fermentation. *J Inst Brew* 100: 321–329.

Geiger E, Piendl A. 1976. Technological factors in the formation of acetaldehyde during fermentation. *MBAA Tech Quart* 13: 51–61.

Gjermansen C et al. 1988. Towards diacetyl-less brewers' yeast. Influence of *ilv2* and *ilv5* mutations. *J Basic Microbiol* 28: 175–183.

Godtfredsen SE et al. 1983. Application of the acetolactate decarboxylase from *Lactobacillus casei* for accelerated maturation of beer. Proceedings of the European Brewery Convention Congress, London, pp. 161–168.

Godtfredsen SE et al. 1984. Occurrence of α-acetolactate decarboxylases among lactic acid bacteria and their utilization for maturation of beer. *Appl Microbiol Biotechnol* 20: 23–28.

Goldenthal MJ, Vanoni M. 1990. Genetic mapping and biochemical analysis of mutants in the maltose regulatory gene of the *MAL1* locus of *Saccharomyces cerevisiae*. *Arch Microbiol* 154: 544–549.

Goldenthal MJ et al. 1987. Regulation of *MAL* gene expression in yeast: gene dosage effects. *Mol Gen Gen* 209: 508–517.

Goossens E et al. 1993. Decreased diacetyl production in lager brewing yeast by integration of the *ILV5* gene. Proceedings of the European Brewery Convention Congress, Oslo, pp. 251–258.

Green CB et al. 2004. T-PCR detection of *Candida albicans ALS* gene expression in the reconstituted human epithelium (RHE) model of oral candidiasis and in model biofilms. *Microbiology* 150: 267–275.

Guo B et al. 2000. A *Saccharomyces* gene family involved in invasive growth, cell-cell adhesion, and mating. *Proc Natl Acad Sci USA* 97: 12158–12163.

Hammond JRM. 1993. Brewer's yeast. In: AH Rose, JS Harrison (eds.) *The Yests*. Academic Press, London, pp. 7–67.

Han E-K et al. 1995. Characterization of *AGT1* encoding a general α-glucoside transporter from *Saccharomyces*. *Mol Microbiol* 17: 1093–1107.

Hansen J, Kielbrandt MC. 1996a. Inactivation of *MET10* in brewers' yeast specifically increases SO_2 formation during beer production. *Nat Biotechnol* 14: 1587–1591.

Hansen J, Kielbrandt MC. 1996b. Inactivation of *MET2* in brewers' yeast increases the level of sulphite in beer. *J Biotechnol* 50: 75–87.

Hardwick WA. 1995. History and antecedents of brewing. In: WA Hardwick (ed.) *Handbook of Brewing*. Marcel Dekker, New York, pp. 37–51.

Haukeli AD, Lie S. 1978. Conversion of α-acetolactate and removal of diacetyl: a kinetic study. *J Inst Brew* 84: 85–89.

Haukeli AD, Lie S. 1979. Yeast growth and metabolic changes during brewery fermentation. *Proceedings of the European Brewery Convention Congress, Berlin*, pp. 461–473.

Hazelwood LA et al. 2008. The Ehrlich pathway for fusel alcohol production: a century of research on *Saccharomyces cerevisiae* metabolism. *Appl Environ Microbiol* 74: 2259–2266.

Hegarty PK et al. 1995. Phenyl ethanol—a factor determining lager character. Proceedings of the European Brewery Convention Congress, pp. 515–522.

Hirata D et al. 1992. Stable overproduction of isoamylalcohol by *S. cerevisiae* with chromosome-integrated copies of the *LEU4* genes. *Biosci Biotechnol Biochem* 56: 1682–1683.

Hough JS et al. 1982. *Malting and Brewing Science—Volume 2: Hopped Wort and Beer*. Chapman and Hall, London.

Hounsa CG et al. 1998. Role of trehalose in survival of *Saccharomyces cerevisiae* under osmotic stress. *Microbiology* 144: 671–680.

Hoyer LL. 2001. The *ALS* gene family of *Candida albicans*. *Trends Microbiol* 9: 176–180.

Hoyer LL et al. 1998. Identification of *Candida albicans ALS2* and *ALS4* and localization of *ALS* proteins to the fungal cell surface. *J Bacteriol* 180: 5334–5343.

Huuskonen A et al. 2010. Selection from industrial lager yeast strains of variants with improved fermentation performance in very-high-gravity worts. *Appl Environ Microbiol* 76: 1563–1573.

Hu Z et al. 1995. *MIG1*-dependent and *MIG1*-independent glucose regulation of *MAL* gene expression in *Saccharomyces cerevisiae*. *Curr Genet* 28: 258–266.

Inoue T. 1995. Development of a two-stage immobilized yeast fermentation system for continuous beer brewing. Proceedings of the European Brewery Convention Congress, Brussels, pp. 25–36.

Inoue T, Yamamoto Y. 1970. Diacetyl and beer fermentation. Proceedings of the Annual Meeting of the American Society of Brewing Chemists, pp. 198–208.

Inoue T et al. 1968. Mechanism of diacetyl formation in beer. Proceedings of the Annual Meeting of the American Society of Brewing Chemists, MN, pp. 158–165.

Iverson WG. 1994. Ethyl acetate extraction of beer: quantitative determination of additional fermentation by-products. *J Am Soc Brew Chem* 52: 91–95.

Jakobsen M. 1982. The effect of yeast handling procedures on yeast oxygen requirement and fermentation. Proc. Inst. Brew. (Aust. N.Z. Section) Conv, Adelaide, pp. 132–137.

Jensen S. 1993. Using ALDC to speed up fermentation. *Brewers' Gardian*, September: 55–56.

Jespersen L et al. 1999. Multiple α-glucoside transporter genes in brewer's yeast. *Appl Environ Microbiol* 65: 450–456.

Jin YL, Speers RA. 1999. Flocculation of *Saccharomyces cerevisiae*. *Food Res Int* 31: 421–440.

Kaclíková E et al. 1992. Fumaric acid overproduction in yeast mutants deficient in fumarase. *FEMS Microbiol Lett* 91: 101–106.

Käppeli O et al. 1985. Transient responses of *Saccharomyces uvarum* to a change of growth-limiting nutrient in continuous culture. *J Gen Microbiol* 131: 47–52.

Karel SF et al. 1985. The immobilization of whole cells: engineering principles. *Chem Eng Sci* 40: 1321–1353.

Klein CJ.L et al. 1996. Alleviation of glucose repression of maltose metabolism by *MIG1* disruption in *Saccharomyces cerevisiae*. *Appl Environ Microbiol* 62: 4441–4449.

Klopper WJ et al. 1986. Organic acids and glycerol in beer. *J Inst Brew* 92: 311–318.

Kobayashi O et al. 1998. Region of Flo1 proteins responsible for sugar recognition. *J Bacteriol* 180: 6503–6510.

Kodama Y et al. 1995. Improvement of maltose fermentation efficiency: constitutive expression of *MAL* genes in brewing yeasts. *J Am Soc Brew Chem* 53: 24–29.

Kodama Y et al. 2001. Control of higher alcohol production by manipulation of the *BAP2* gene in brewing yeast. *J Am Soc Brew Chem* 59: 57–62.

Kronlöf J, Linko M. 1992. Production of beer using immobilized yeast encoding α-acetolactate decarboxylase. *J Inst Brew* 98: 479–491.

Kronlöf J, Virkajärvi I. 1999. Primary fermentation with immobilized yeast. Proceedings of the European Brewery Convention Congress, Cannes, pp. 761–770.

Kubo Y et al. 2000. Effect of gene disruption of succinate dehydrogenase on succinate production in sake yeast strain. *J Biosci Bioeng* 90: 619–624.

Kunze W. 1999. *Technology of Brewing and Malting*. VLB, Berlin.

Landaud S et al. 2001. Top pressure and temperature control of the fusel alcohol/ester ratio through yeast growth in beer fermentation. *J Inst Brew* 10: 107–117.

Lagunas R. 1979. Energetic irrelevance of aerobiosis for *S. cerevisiae* growing on sugars. *Mol Cell Biochem* 27: 139–146.

Lea AH, Piggott JRP. 1995. *Fermented Beverage Production*. Blackie Academic and Professional, Glasgow.

Lee S et al. 1995. Yeast strain development for enhanced production of desirable alcohols/esters in beer. *J Am Soc Brew Chem* 39: 153–156.

Li F, Palecek SP. 2003. *EAP1*, a *Candida albicans* gene involved in binding human epithelial cells. *Eukaryot Cell* 2: 1266–1273.

Lievense JC, Lim HC. 1982. The growth and dynamics of *Saccharomyces cerevisiae*. In: GT Tsao et al. (eds.) *Annual Reports on Fermentation Processes*. Academic Press, New York.

Lilly M et al. 2000. Effect of increased yeast alcohol acetyltransferase activity on flavor profiles of wine and distillates. *Appl Environ Microbiol* 66: 744–753.

Lilly M et al. 2006. The effect of increased branched-chain amino acid transaminase activity in yeast on the production of higher alcohols and on the flavour profiles of wine and distillates. *FEMS Yeast Res* 6: 726–743.

Linko M et al. 1993. Use of brewer's yeast expressing α-acetolactate decarboxylase in conventional and immobilized fermentations. *MBAA Tech Quart* 30: 93–97.

Linko M et al. 1997. Main fermentation with immobilized yeast—a breaktrough? Proceedings of the European Brewery Convention Congress, Maastricht, pp. 385–394.

Linko M et al. 1998. Recent advances in the malting and brewing industry. *J Biotechnol* 65: 85–98.

Lo WS, Dranginis AM. 1998. The cell surface flocculin Flo11 is required for pseudohyphae formation and invasion by *Saccharomyces cerevisiae*. *Mol Biol Cell* 9: 161–171.

Lombardo A et al. 1990. Cloning and characterization of the iron-sulfur subunit gene succinate dehydrogenase from *Saccharomyces cerevisiae*. *J Biol Chem* 265: 10419–10423.

Lomni H et al. 1990. Immobilized yeast reactor speeds beer production. *Food Technol* 5: 128–133.

Lucero P et al. 1993. Catabolite inactivation of the yeast maltose transporter is due to proteolysis. *FEBS Lett* 333: 165–168.

MacDonald J et al. 1984. Current approaches to brewery fermentations. *Prog Ind Microbiol* 19: 47–198.

MacWilliam I.C. 1968. Wort composition. *J Inst Brew* 74: 38–54.

Maeba H et al. 2000. Primary fermentation with immobilized yeast in porous chitosan beads. Pilot scale trial. Proceedings of the 26th Convention of the Institute Brewing, Australia and New Zealand Section, Singapore, pp. 82–86.

Magarifuchi T et al. 1995. Effect of yeast fumarase gene (*FUM1*) disruption on the production of malic, fumaric and succinic acids in sake mash. *J Ferment Bioeng* 80: 355–361.

Magee PT, de Robinson-Szulmajster H. 1968. The regulation of isoleucine-valine biosynthesis in *Saccharomyces cerevisiae*. Properties and regulation of the activity of acetohydroxyacid synthase. *Eur J Biochem* 3: 507–511.

Majara N et al. 1996. Trehalose: a stress protectant and stress indicator compound for yeast exposed to adverse conditions. *J Am Soc Brew Chem* 54: 221–227.

Malcorps P, Dufour J-P. 1987. Proceedings of the European Brewery Convention Congress, Madrid, p. 377.

Malcorps P et al. 1991. A new model for the regulation of ester synthesis by alcohol acetyltransferase in *Saccharomyces cerevisiae*. *J Am Soc Brew Chem* 49: 47–53.

Masschelein CA. 1981. Role of fatty acids in beer flavour. EBC Flavour Symposium, Copenhagen, Denmark. EBC Monograph VII, pp. 211–221, 237–238.

Masschelein CA. 1997. A realistic view on the role of research in the brewing industry today. *J Inst Brew* 103: 103–113.

Masschelein CA, Andries M. 1995. Future scenario of immobilized systems: promises and limitations. EBC Monograph XXIV, EBC Symposium on immobilized yeast applications in the brewing industry, Espoo, pp. 223–241.

Masschelein CA et al. 1995. The membrane loop concept: a new approach for optimal oxygen transfer into high cell density pitching yeast suspensions. Proceedings of the European Brewery Convention Congress, Brussels, pp. 377–386.

Maule DRJ. 1967. Rapid gas chromatographic examination of beer flavour. *J Inst Brew* 73: 351–361.

McGovern PE et al. 1996. Neolithic resinated wine. *Nature* 381: 480–481.

Meersman E. 1994. Biologische aanzuring met geïmmobiliseerde melkzuurbacteriën. *Cerevisia Biotechnol* 19(4): 42–46.

Meilgaard MC. 1975a. Flavour chemistry of beer. Part 1: flavour interactions between principle volatiles. *Tech Quart Master Brew Assoc Am* 12: 107–117.

Meilgaard MC. 1975b. Flavour chemistry of beer. Part 2: flavour and threshold of 239 aroma volatiles. *Tech Quart Master Brew Assoc Am* 12: 151–173.

Mensour N et al. 1995. Gas lift systems for immobilized cell systems. EBC Monograph XXIV, EBC Symposium on Immobilized Yeast Applications in the Brewing Industry, Espoo, pp. 125–133.

Mensour N et al. 1996. Applications of immobilized yeast cells in the brewing industry. In: RH Wijffels et al. (eds.) *Immobilized Cells: Basics and Applications*. Elsevier, Amsterdam, pp. 661–671.

Mensour N et al. 1997. New developments in the brewing industry using immobilized yeast cell bioreactor systems. *J Inst Brew* 103: 363–370.

Michels CA et al. 1992. The telomere-associated *MAL3* locus of *Saccharomyces* is a tandem array of repeated genes. *Yeast* 8: 655–665.

Michnick S et al. 1997. Modulation of glycerol and ethanol yields during alcoholic fermentation in *Saccharomyces cerevisiae* strains overexpressed or disrupted for *GPD1* encoding glycerol-3-phosphate dehydrogenase. *Yeast* 13: 783–793.

Minetoki T et al. 1993. The purification, properties and internal peptide sequences of alcohol acetyltransferase isolated from *Saccharomyces cerevisiae*. *Biosci Biotechnol Biochem* 57: 2094–2098.

Mithieux SM, Weiss AS. 1995. Tandem integration of multiple *ILV5* copies and elevated transcription in polyploid yeast. *Yeast* 11: 311–316.

Mockovciaková D et al. 1993. The *ogd1* and *kgd1* mutants lacking 2-oxoglutarate dehydrogenase activity in yeast are allelic and can be differentiated by the cloned amber suppressor. *Curr Genet* 24: 377–381.

Moll M. 2006. Fermentation and maturation of beer with immobilised yeasts. *J Inst Brew* 112: 346.

Moll M, Duteurtre B. 1996. Fermentation and maturation of beer with immobilised microorganisms. *Brauwelt Int* 14: 248–252.

Moll M et al. 1973. Continuous production of fermented liquids. French Patent 73/23397; US Patent 4009286.

Moonjai N et al. 2000a. Unsaturated fatty acid supplementation of stationary phase brewing yeast. *Cerevisia* 25(3): 37–50.

Moonjai N et al. 2000b. The effects of linoleic acid supplementation of cropped yeast on its performance and acetate ester synthesis. *J Inst Brew* 108: 227–235.

Murray CR et al. 1984. The effect of yeast storage conditions on subsequent fermentations. *MBBA Tech Quart* 21: 189–194.

Nakatani K et al. 1984. Kinetic studies of vicinal diketones in brewing. 2. Theoretical aspects for the formation of total vicinal diketones. *Tech Quart Master Brew Assoc Am* 21: 175–183.

Narziss L, Hellich P. 1971. Ein Beitrag zur wesentlichen Beschleunigung der Gärung und Reifung des Bieres. *Brauwelt* 111: 1491–1500.

Narziss L et al. 1992. Technology and composition of non-alcoholic beers. *Brauwelt Int* 4: 396.

Navrátil M et al. 2000. Fermented beverages produced by yeast cells entrapped in ionotropic hydrogels of polysaccharide nature. *Minerva Biotechnol* 12: 337–344.

Navrátil M et al. 2002. Production of non-alcoholic beer using free and immobilized cells of *Sccharomyces cerevisiae* deficient in tricarboxylic acid cycle. *Biotechnol Appl Biochem* 35: 133–141.

Nedovic VA et al. 1997. Analysis of liquid axial dispersion in an internal loop gas-lift bioreactor for beer fermentation with immobilized yeast cells. Proceedings of the 2nd European Conference on Fluidization, Bilbao, pp. 627–635.

Nedovic V et al. 2005. Beer production using immobilised cells. In: V Nedovic, R Willaert (eds.) *Applications of Cell Immobilisation Biotechnology*. Kluwer Academic Publishers, Dordrecht, pp. 259–273.

Needleman RB et al. 1984. *MAL6* of *Saccharomyces*: a complex genetic locus containing three genes required for maltose fermentation. *Proc Natl Acad Sci USA*. 81: 2811–2815.

Nehlin JO, Ronne H. 1990. Yeast *MIG1* repressor is related to the mammalian early growth response and Wilms' tumour finger proteins. *EMBO J* 9: 2891–2898.

Nelissen B et al. 1995. Phylogenetic classification of the major superfamily of membrane transport facilitators, as deduced from yeast genome sequencing. *FEBS Lett* 377: 232–236.

Nevoigt E, Stahl U. 1997. Reduced pyruvate decarboxylase and increased glycerol-3-phosphate degydrogenase [NAD^+] levels enhance glycerol production in *Saccharomyces cerevisiae*. *Yeast* 12: 1331–1337.

Nevoigt E et al. 2002. Genetic engineering of brewing yeast to reduce the content of ethanol in beer. *FEMS Yeast Res* 2: 225–232.

Ni B, Needleman RB. 1990. Identification of the upstream activating sequence of *MAL* and the binding sites for MAL63 activator of *Saccharomyces cerevisiae*. *Mol Cell Biol* 10: 3797–3800.

Nitzsche F et al. 2001. A new way for immobilized yeast systems: secondary fermentation without heat treatment. Proceedingsof the European Brewery Convention, Budapest, pp. 486–494.

NN. 2000. Fermentation and Maturation. In: *European Brewery Convention Manual of Good Practice*. Getränke-Fachverlag Hans Carl, Nürnberg.

NN. 2005. Dossier sectoroverzicht van de Belgische brouwerijnijverheid. *Het Brouwersblad*. juni: 6–20.

Novak S et al. 2004. Regulation of maltose transport and metabolism in *Saccharomyces cerevisiae*. *Food Technol Biotechnol* 42: 213–218.

Nykänen L, Nykänen I. 1977. Production of esters by different yeast strains in sugar fermentations. *J Inst Brew* 83: 30–31.

Ohno T, Takahashi R. 1986a. Role of wort aeration in the brewing process. Part 1: oxygen uptake and biosynthesis of lipids by the final yeast. *J Inst Brew* 92: 84–87.

Ohno T, Takahashi R. 1986b. Role of wort aeration in the brewing process. Part 2: the optimal aeration conditions for the brewing process. *J Inst Brew* 92: 88–92.

Okabe M et al. 1992. Growth and fermentation characteristics of bottom brewer's yeast under mechanical stirring. *J Ferment Bioeng* 73: 148–152.

Omura F. 2008. Targeting of mitochondrial *Saccharomyces cerevisiae* Ilv5p to the cytosol and its effect on vicinal diketone formation in brewing. *Appl Microbiol Biotechnol* 78: 503–513.

Onnela M-L et al. 1996. Use of a modified alcohol dehydrogenase, *ADH1*, promotor in construction of diacetyl non-producing brewer's yeast. *J Biotechnol* 49: 101–109.

Oshita K et al. 1995. Clarification of the relationship between fusel alcohol formation and amino acid assimilation by brewing yeast using ^{13}C-labeled amino acid. Proceedings of the European Brewery Convention Congress, Brussels, pp. 387–402.

Pajunen E. 1995. Immobilized yeast lager beer maturation: DEAE-cellulose at Synebrychoff. EBC Monograph XXIV, EBC Symposium on Immobilized Yeast Applications in the Brewing Industry, Espoo, pp. 24–40.

Pajunen E, Grönqvist A. 1994. Immobilized yeast fermenters for continuous lager beer maturation. Proceedings of the 23rd Convention Institute of Brewing Australia and New Zealand Section, Sydney, pp. 101–103.

Panchal CJ, Stewart GG. 1979. Utilization of wort carbohydrates. *Brew Digest* June: 36–46.

Pajunen E et al. 2001. Controlled beer fermentation with continuous one-stage immobilized yeast reactor. Proceedings of the European Brewery Convention, Budapest, pp. 465–476.

Panek AD. 1991. Storage carbohydrates. In: AH Rose, JS Harrison (eds.) *The Yeasts, Volume 4, Yeast Organelles*, 2nd edn. Academic Press, London, pp. 655–678.

Panek AD, Panek AC. 1990. Metabolism and thermotolerance function of trehalose in *Saccharomyces cerevisiae*: a current perspective. *J Biotechnol* 14: 229–238.

Pang SS, Duggleby RG. 1999. Expression, purification, characterization, and reconstitution of the large and small subunits of yeast acetohydroxyacid synthase. *Biochemistry* 38: 5222–5231.

Pang SS, Duggleby RG. 2001. Regulation of yeast acetohydroxyacid synthase by valine and ATP. *Biochem J* 357: 749–757.

Patel GB, Ingledew WM. 1973. Trends in wort carbohydrate utilization. *Appl Microbiol* 26: 349–353.

Parrou JL et al. 1997. Effects of various types of stress on the metabolism of reserve carbohydrates in *Saccharomyces cerevisiae*: genetic evidence for a stress-induced recycling of glycogen and trehalose. *Microbiology* 143: 1891–900.

Peddie HAB. 1990. Ester formation in brewery fermentations, *J Inst Brew* 96: 327–331.

Peinado JM, Loureiro-Dias MC. 1986. Reversible loss of affinity induced by glucose in the maltose-H^+ symport of *Saccharomyces cerevisiae*. *Biochim Biophys Acta* 856: 189–192.

Petrik M et al. 1983. An expanded concept for the glucose effect in the yeast *Saccharomyces uvarum*: involvement of short- and long-term regulation. *J Gen Microbiol* 129: 43–49.

Pfisterer E, Stewart GG. 1975. Some aspects on the fermentation of high gravity worts. Proceedings of the European Brewery Convention Congress, Nice, pp. 255–267.

Pittner H et al. 1993. Continuous production of acidified wort for alcohol-free-beer with immobilized lactic acid bacteria. Proceedings of the European Brewery Convention Congress, Oslo, pp. 323–329.

Pugh TA et al. 1997. The impact of wort nitrogen limitation on yeast fermentation performance and diacetyl. *MBAA Tech Quart* 34: 185–189.

Quain DE. 1988. Studies on yeast physiology: impact of fermentation performance and product quality. *J Inst Brew* 95: 315–323.

Quain DE, Tubb RS. 1982. The importance of glycogen in brewing yeast. *MBAA Tech Quart* 19: 19–23.

Ramos-Jeunehomme C et al. 1989. Proceedings of the European Brewery Convention Congress, pp. 513–519.

Ramos-Jeunehomme C et al. 1991. Why is ester formation in brewery fermentations yeast strain dependent? Proceedings of the European Brewery Convention Congress, Lisbon, pp. 257–264.

Rautio J, Londesborough J. 2003. Maltose transport by brewer's yeast in brewer's wort. *J Inst Brew* 109: 251–261.

Reed G, Nogodawithana TW. 1991. *Yeast Technology*. Chapter 3. Van Nostrand Reinhold, New York.

Remize F et al. 1999. Glycerol overproduction by engineered *Saccharomyces cerevisiae* wine yeast strains leads to substantial changes in by-product formation and to stimulation of fermentation rate in stationary phase. *Appl Environ Microbiol* 65: 143–149.

Reynolds TB, Fink GR. 2001. Baker's yeast, a model for fungal biofilm formation. *Science* 291: 878–881.

Riballo E et al. 1995. Catabolite inactivation of the yeast maltose transporter occurs in the vacuole after internalization by endocytosis. *J Bacteriol* 177: 5622–5627.

Rieger M et al. 1983. The role of limited respiration in the incomplete oxidation of glucose by *Saccharomyces cerevisiae*. *J Gen Microbiol* 129: 653–661.

Rintala E et al. 2008. Transcription of hexose transporters of *Saccharomyces cerevisiae* is affected by change in oxygen provision. *BMC Microbiol* 8: 53.

Rogers PJ, Stewart PR. 1973. Mitochondrial and peroxisomal contributions to the energy metabolism of *Saccharomyces cerevisiae* in continuous culture. *J Gen Microbiol* 79: 205–217.

Romano P, Suzzi G. 1992. Production of H_2S by different yeast strains during fermentation. Proceedings of the 22nd Convention of the Institute of Brewing, Melbourne, Institute of Brewing, Adelaide, Australia, pp. 96–98.

Rosculet G. 1971. Aroma and flavour of beer. Part II: the origin and nature of less-volatile and non-volatile components of beer. *Brew Digest* 46(6): 96–98.

Rostgaard-Jensen B et al. 1987. Isolation and characterization of an α-acetolactate decarboxylase useful for accelerated beer maturation. Proceedings of the European Brewery Convention Congress, pp. 393–400.

Royston MG. 1966. Tower fermentation of beer. *Process Biochem* 1: 215–221.

Ryder D, Masschelein CA. 1983. Aspects of metabolic regulatory systems and physiological limitations with a view to the improvement of brewing performance. EBC Symposium on Biotechnology, Monograph IX, Nutfiels, pp. 2–29.

Sablayrolles JM, Ball CB. 1995. Fermentation kinetics and the production of volatiles during alcoholic fermentation. *J Am Soc Brew Chem* 53: 71–78.

Saerens SM et al. 2006. The *Saccharomyces cerevisiae EHT1* and *EEB1* genes encode novel enzymes with medium-chain fatty acid ethyl ester synthesis and hydrolysis capacity. *J Biol Chem* 281: 4446–4456.

Saison D et al. 2009. Contribution of staling compounds to the aged flavour of lager beer by studying their flavour thresholds. *Food Chem* 114: 1206–1215.

Salema-Oom M et al. 2005. Maltotriose utilization by industrial *Saccharomyces* strains: characterization of a new member of the alpha-glucoside transporter family. *Appl Environ Microbiol* 71: 5044–5049.

Selecky R et al. 2008. Beer with reduced ethanol content produced using *Saccharomyces cerevisiae* yeasts deficient in various tricarboxylic acid cycle enzymes. *J Inst Brew* 114: 97–101.

Sheppard DC et al. 2004. Functional and structural diversity in the Als protein family of *Candida albicans*. *J Biol Chem* 279: 30480–30489.

Shindo S et al. 1994. Suppression of α-acetolactate formation in brewing with immobilized yeast. *J Inst Brew* 100: 69–72.

Siegel MH, Robinson CW. 1992. Applications of airlift gas-liquid-solid reactors in biotechnology. *Chem Eng Sci* 47: 3215–3229.

Siro M-R, Lövgren T. 1979. Influence of glucose on the α-glucoside permease activity of yeast. *Eur J Appl Microbiol* 7: 59–66.

Slaughter JC, Jordan B. 1986. The production of hydrogen sulphide by yeast. In: I Campbell, FG Priest (eds.) *Proceedings of the Second Aviemore Conference on Malting, Brewing and Distilling*. Institute of Brewing, London, pp. 308–310.

Smit A et al. 2008. The Thr^{505} and Ser^{557} residues of the *AGT1*-encoded alpha-glucoside transporter are critical for maltotriose transport in *Saccharomyces cerevisiae*. *J Appl Microbiol* 104: 1103–1111.

Smith AU et al. 1998. Yeast PKA represses Msn2p/Msn4p-dependent gene expression to regulate growth, stress response and glycogen accumulation. *EMBO J* 17: 3556–3564.

Smogrovicová D, Dömény Z. 1999. Beer volatile by-product formation at different fermentation temperature using immobilized yeasts. *Process Biochem* 34: 785–794.

Smogrovicová D et al. 1997. Reactors for the continuous primary beer fermentation using immobilised yeast. *Biotechnol Tech* 11: 261–264.

Smogrovicová D et al. 2001. Continuous beer fermentation using polyvinyl alcohol entrapped yeast. Proceedings of the European Brewery Convention Congress, Budapest, pp. 1–9.

Soler AP et al. 1987. Differential translational efficiency of the mRNA isolated from derepressed and glucose repressed *Saccharomyces cerevisiae*. *J Gen Microbiol* 133: 1471–1480.

Staab JF et al. 1999. Adhesive and mammalian transglutaminase substrate properties of *Candida albicans* Hwp1. *Science* 283: 1535–1583.

Stambuk BU, de Araujo PS. 2001. Kinetics of active alpha-glucoside transport in *Saccharomyces cerevisiae*. *FEMS Yeast Res* 1: 73–78.

Stewart GG, Russell I. 1993. Fermentation—the "black box" of the brewing process. *MBAA Tech Quart* 30: 159–168.

Stratford M. 1989. Yeast flocculation: calcium specificity. *Yeast* 5: 487–496.

Suihko M et al. 1990. Recombinant brewer's yeast strains suitable for accelerated brewing. *J Biotechnol* 14: 285–300.

Szlavko CM. 1973. Tryptophol, tyrosol and phenylethanol—the aromatic higher alcohols in beer. *J Inst Brew* 83: 283–243.

Tada S et al. 1995. Pilot scale brewing with industrial yeasts which produce the α-acetolactate decarboxylase of *Acetobacter aceti* ssp. *xylium*. Proceedings of the European Brewery Convention Congress, Brussels, pp. 369–376.

Tata M et al. 1999. Immobilized yeast bioreactor systems for continuous beer fermentation. *Biotechnol Prog* 15: 105–113.

Teunissen AW et al. 1993a. Sequence of the open reading frame of the *FLO1* gene from *Saccharomyces cerevisiae*. *Yeast* 9: 423–427.

Teunissen AW et al. 1993b. Physical localization of the flocculation gene *FLO1* on chromosome I of *Saccharomyces cerevisiae*. *Yeast* 9: 1–10.

Teunissen AW, Steensma HY. 1995. Review: the dominant flocculation genes of *Saccharomyces cerevisiae* constitute a new subtelomeric gene family. *Yeast* 11: 1001–1013.

Tezuka H et al. 1992. Cloning of a gene suppressing hydrogen sulphide production by *Saccharomyces cerevisiae* and its expression in a brewing yeast. *J Am Soc Brew Chem* 50: 130–133.

Thomas D, Surdin-Kerjan Y. 1997. Metabolism of sulfur amino acids in *Saccharomyces cerevisiae*. *Microbiol Mol Biol Rev* 61: 503–532.

Thorne RSW. 1968. Continuous fermentation in retrospect. *Brew Digest* 43(2): 50–55.

Thurston PA et al. 1981. Effects of linoleic acid supplements on the synthesis by yeast of lipid and acetate esters. *J Inst Brew* 87: 92–95.

Thurston PA et al. 1982. Lipid metabolism and the regulation of volatile synthesis in *Saccharomyces cerevisiae*. *J Inst Brew* 88: 90–94.

Udenfriend S, Kodukula K. 1995. How glycosylphosphatidylinositol-anchored membrane proteins are made. *Annu Rev Biochem* 64: 563–591.

Unemoto S et al. 1998. Primary fermentation with immobilized yeast in a fluidized bed reactor. *MBAA Tech Quart* 35: 58–61.

Vanbeneden N et al. 2007. Formation of 4-vinyl and 4-ethyl derivatives from hydroxycinnamic acids: occurrence of volatile phenolic flavour compounds in beer and distribution of Pad1-activity among brewing yeasts. *Food Chem* 107: 221–230.

Van de Meersche J et al. 1977. The role of yeasts in the maturation of beer flavour. Proceedings of the European Brewery Convention Congress, Amsterdam, pp. 561–575.

Van Den Berg R et al. 1983. Diacetyl reducing activity in brewer's yeast. Proceedings of the European Brewery Convention Congress, London, pp. 497–504.

Van De Winkel L. 1995. Design and optimization of a multipurpose immobilized yeast bioreactor system for brewery fermentations. *Cerevisia* 20(1): 77–80.

Van De Winkel L et al. 1991. The application of an immobilized yeast loop reactor to the continuous production of alcohol-free beer. Proceedings of the European Brewery Convention Congress, Lisbon, pp. 307–314.

Van Dieren D. 1995. Yeast metabolism and the production of alcohol-free beer. EBC Monograph XXIV, EBC Symposium on Immobilized Yeast Applications in the Brewing Industry, Espoo, pp. 66–76.

Van Haecht JL, Dufour JP. 1995. The production of sulfur compounds by brewing yeast: a review. *Cerevisiae* 20: 51–64.

van Iersel MFM et al. 1998. Effect of environmental conditions on flocculation and immobilization of brewer's yeast during production of alcohol-free beer. *J Inst Brew* 104: 131–136.

Van Mulders SE et al. 2009. Phenotypic diversity of Flo protein family-mediated adhesion in *Saccharomyces cerevisiae*. *FEMS Yeast Res* 9: 178–190.

Verbelen PJ et al. 2006. Immobilized yeast cell systems for continuous fermentation applications. *Biotechnol Lett* 28: 1515–1525.

Verbelen PJ et al. 2009. The influence of yeast oxygenation prior to brewery fermentation on yeast metabolism and the oxidative stress response. *FEMS Yeast Res* 9: 226–239.

Verbelen PJ et al. 2010. Bioprocess intensification of beer fermentation using immobilised cells. In: NJ Zuidam, VA Nedovic (eds.) *Encapsulation Technologies for Active Food Ingredients and Food Processing*. Springer, New York, pp. 303–325.

Verstrepen KJ. 2003. Flavour-active ester synthesis in *Saccharomyces cerevisiae*. PhD thesis, Katholieke Universiteit Leuven, Belgium.

Verstrepen KJ et al. 2003a. Yeast flocculation: what brewers should know. *Appl Microbiol Biotechnol* 61: 197–205.

Verstrepen KJ et al. 2003b. Flavor-active esters: adding fruitiness to beer. *J Biosci Bioeng* 96: 110–118.

Verstrepen KJ et al. 2003. Expression levels of the yeast alcohol acetyltransferase genes *ATF1*, *Lg-ATF1*, and *ATF2* control the formation of a broad range of volatile esters. *Appl Environ Microbiol* 69: 5228–5237.

Verstrepen KJ et al. 2004. Origins of variation in the fungal cell surface. *Nat Rev Microbiol* 2: 533–540.

Vidgren V et al. 2005. Characterization and functional analysis of the *MAL* and *MPH* Loci for maltose utilization in some ale and lager yeast strains. *Appl Environ Microbiol* 71: 7846–7857.

Vidgren V et al. 2009. Improved fermentation performance of a lager yeast after repair of its *AGT1* maltose and maltotriose transporter genes. *Appl Environ Microbiol* 75: 2333–2345.

Villaneuba KD et al. 1990. Subthreshold vicinal diketone levels in lager brewing yeast fermentations by means of *ILV5* gene amplification. *J Am Soc Brew Chem* 48: 111–114.

Virkajärvi I, Krönlof J. 1998. Long-term stability of immobilized yeast columns in primary fermentation. *J Am Soc Brew Chem* 56: 70–75.

Virkajärvi I, Pohjala N. 2000. Primary fermentation with immobilized yeast: effects of carrier materials on the flavour of the beer. *J Inst Brew* 106: 311–318.

Virkajärvi I et al. 1999. Effects of aeration on flavor compounds in immobilized primary fermentation. *Monatschrift für Brauwissenschaft* 52: 9–12, 25–28.

Virkajärvi I et al. 2002. Productivity of immobilized yeast reactors with very-high-gravity worts. *J Am Soc Brew Chem* 60: 188–197.

Viyas VK et al. 2003. Snf1 kinases with different β-subunit isoforms play distinct roles in regulating haploid invasive growth. *Mol Biol Cell* 23: 1341–1348.

Vunjak-Novakovic G et al. 1992. Flow regimes and liquid mixing in a draft tube gas-liquid-solid fluidized bed. *Chem Eng Sci* 47: 3451–3458.

Vunjak-Novakovic G et al. 1998. Air-lift reactors: research and potential applications. *Trends Chem Eng* 5: 159–172.

Wackerbauer K et al. 2003. Measures to improve long term stability of main fermentation with immobilized yeast. Proceedings of the European Brewery Convention, Dublin, pp. 445–457.

Wainwright T. 1973. Diacetyl—a review. *J Inst Brew* 79: 451–470.

Walker GM. 1998. *Yeast—Physiology and Biotechnology*. John Wiley & Sons, Chichester.

Wang X et al. 2002. Intracellular maltose is sufficient to induce *MAL* gene expression in *Saccharomyces cerevisiae*. *Eukaryot Cell* 1: 696–703.

Wellhoener HJ. 1954. Ein Kontinuierliches Gär- und Reifungsverfahren für Bier. *Brauwelt* 94(1): 624–626.

White FH, Portno AD. 1979. The influence of wort composition on beer ester levels. Proceedings of the European Brewery Convention Congress, Berlin, pp. 447–460.

Willaert R. 2001. Sugar consumption kinetics by brewer's yeast during the primary beer fermentation. *Cerevisia* 26(1): 43–49.

Willaert R. 2007. Cell immobilization and its applications in biotechnology: current trends and future prospects. In: EMT El-Mansi et al. (eds.) *Fermentation Microbiology and Biotechnology*. CRC Press, Boca Raton, FL, pp. 287–361.

Willaert R, Baron GV. 1996. Gel entrapment and microencapsulation: methods, applications and engineering principles. *Rev Chem Eng* 12: 1–205.

Willaert R et al. 2004. Diffusive mass transfer in immobilised cell systems. In: VA Nedovic, R Willaert (eds.) *Fundamentals of Cell Immobilisation Biotechnology*. Kluwer Academic Publishers, Dordrecht, pp. 359–387.

Willaert R, Nedovic VA. 2006. Primary beer fermentation by immobilised yeast—a review on flavour formation and control strategies. *J Chem Technol Biotechnol* 81: 1353–1367.

Yamano S et al. 1994. Construction of a brewer's yeast having α-acetolactate decarboxylase gene from *Acetobacter acetii* ssp. *xylinum* integrated in the genome. *J Biotechnol* 32: 173–178.

Yamauchi Y, Kashihara T. 1995. Kirin immobilized system. EBC Monograph XXIV, EBC Symposium on Immobilized Yeast Applications in the Brewing Industry, pp. 99–117.

Yamauchi Y et al. 1994. Beer brewing using an immobilized yeast bioreactor design of an immobilized yeast bioreactor for rapid beer brewing system. *J Ferment Bioeng* 78: 443–449.

Yao B et al. 1994. Shared control of maltose induction and catabolite repression of the *MAL* structural genes in *Saccharomyces*. *Mol Gen Genet* 243: 622–630.

Yoon SH et al. 2003. Specificity of yeast (*Saccharomyces cerevisiae*) in removing carbohydrates by fermentation. *Carbohydr Res* 338: 1127–1132.

Younis OS, Stewart GG. 1998. Sugar uptake and subsequent ester and higher alcohol production by *Saccharomyces cerevisiae*. *J Inst Brew* 104: 255–264.

Younis OS, Stewart GG. 2000. The effect of wort maltose content on volatile production and fermentation performance in brewing yeast. In: K Smart (ed.) *Brewing Yeast Fermentation and Performance*, vol 1. Blackwell Science, Oxford, pp. 170–176.

Zähringer H et al. 2000. Induction of neutral trehalase Nth1 by heat and osmotic stress is controlled by STRE elements and Msn2/Msn4 transcription factors: variations of PKA effect during stress and growth. *Mol Microbiol* 35: 397–406.

Zastrow CR et al. 2001. Maltotriose fermentations by *Saccharomyces cerevisiae*. *J Ind Microbiol Biotechnol* 27: 34–38.

Zheng X et al. 1994a. Transport kinetics of maltotriose in strains of *Saccharomyces*. *J Ind Microbiol* 13: 159–166.

Zheng X et al. 1994b. Factors influencing maltotriose utilization during brewery wort fermentations. *J Am Soc Brew Chem* 52: 41–47.

34
Rye Constituents and Their Impact on Rye Processing

T. Verwimp, C. M. Courtin, and J. A. Delcour

CLASSIFICATION

Rye (*Secale cereale* L.), member of the grass family (Gramineae) and typically classified into winter and spring varieties, is a diploid, seven-pair chromosome cereal, with tetraploid varieties sometimes being produced artificially. While cultivated rye varieties are less numerous than those of other cereal crops, most varieties are mixed populations because of cross-pollination.

PRODUCTION

Although rye production (15 million tons) is low compared with the 2003 total world production of the three major cereals, wheat, corn, and rice (556, 638, and 589 million tons, respectively) (FAO 2004), rye's low soil and fertilization requirements as well as its relatively good overwintering ability guarantee continuing interest in the cereal, as it can be cultivated in areas generally not suited for other cereal crops. In 2003, high amounts of rye were produced by Russia (4.1 million tons), Poland (3.2 million tons), Germany (2.3 million tons), and Belarus (1.4 million tons) (FAO 2004), making this cereal crop of major importance in parts of

Food Biochemistry and Food Processing, Second Edition. Edited by Benjamin K. Simpson, Leo M.L. Nollet, Fidel Toldrá, Soottawat Benjakul, Gopinadhan Paliyath and Y.H. Hui.

Europe and Asia, but of less importance in America. Production has remained fairly constant, but with a downward trend, during the past 10 years.

THE RYE KERNEL

The main components of the rye kernel are, from the outside to the inside, the hull, pericarp, testa, nucellar epidermis, aleurone and starchy endosperm, and the germ. As the hull threshes free, the pericarp is the actual outer part of the grain, surrounding the seed and consisting of different layers, ranging from the epidermis, hypodermis, and cross cells to tube cells. The latter adhere to the testa or seed coat, which is in turn lined with the nucellar epidermis. The aleurone cells are located beneath the nucellar epidermis, forming the outermost layer of the endosperm tissue, and merge with the scutellum of the germ. The starchy endosperm is composed of three main cell types, that is, the subaleurone or peripheral cells, the prismatic cells, and the central endosperm cells, and contains two major storage reserves. The subaleurone cells contain high levels of storage protein, while the prismatic cells contain high levels of starch granules (Shewry and Bechtel 2001). The pericarp, testa, and nucellar epidermis form the botanical bran and are separated from the starchy endosperm together with the aleurone layer and the germ in the milling process, forming the technical or miller's bran.

RYE CONSTITUENTS

The major rye constituents are starch and nonstarch polysaccharides and protein. Its lipids, vitamins, and minerals are minor constituents (Table 34.1).

STARCH

Starch Composition

Starch granules are mainly composed of two types of polymers, amylose, and amylopectin. In general, amylose is an essentially linear polymer consisting of α-D-glucopyranose residues linked by α-1,4 bonds with few (<1%; Ball et al. 1996) α-1,6 bonds. Amylose occurs in both free and lipid-complexed forms. Lipid-complexed amylose may be present in native starch, but is possibly also formed during gelatinization of starch (Andreev et al. 1999, Morrison 1995). Amylose contents in rye starch vary from 12 to 30% (Andreev et al. 1999, Fredriksson et al. 1998, Mohammadkhani et al. 1998, Wasserman et al. 2001). Berry et al. (1971) reported the rye amylose fraction to have a molecular weight of about 220 k.

Amylopectin is a highly branched polymer consisting of α-D-glucopyranose residues linked by α-1,4 linkages with 4–5% (according to Keetels et al. 1996) α-1,6 linkages. Berry et al. (1971) showed that rye amylopectin has 4.8% branching, with branch points about every 21 glucose units on average. The chains of amylopectin can be classified into three types: A-, B-, and C-chains. The A-chains are unbranched outer chains and are attached to the inner B-chains through their potential reducing end. In their turn, the B-chains are attached to other B-chains or to the single C-chain through their potential reducing end. The C-chain is the only chain carrying a reducing end group. Lii and Lineback (1977) reported an A-chain to B-chain ratio of 1.71–1.81 and an average unit chain length of 26 for rye amylopectin.

Minor components of starch granules are lipids, proteins, phosphorus, and ash.

Starch Structure

Rye starch has a bimodal particle size distribution comprising a major population of large, lenticular A-type granules (>10 μm, 85%), and a minor population of small, spherical B-type granules (>10 μm, 15%) (Schierbaum et al. 1991, Verwimp et al. 2004). Rye A-type starch granules have larger average particle sizes and broader particle size distribution profiles than their wheat counterparts (Fredriksson et al. 1998, Verwimp et al. 2004).

In starch granules, the amylose and amylopectin molecules are radially ordered, with their single reducing end groups towards the center, or hilum. Ordering of crystallites within the starch granule causes optical birefringence, and a Maltese cross can be observed under polarized light. Starch granules are made up of alternating semicrystalline and amorphous growth rings, arranged concentrically around the hilum. The amorphous growth

Table 34.1. Composition of the Rye Kernel

Constituent	Content (%)	References
Starch	57.0–65.6	Aman et al. 1997, Hansen et al. 2004
Nonstarch polysaccharide	13.0–15.0	Härkönen et al. 1997
Arabinoxylan	6.5–12.2	Aman et al. 1997, Hansen et al. 2003, 2004, Henry 1987, Vinkx and Delcour 1996
Mixed-linkage β-glucan	1.5–2.6	Aman et al. 1997, Hansen et al. 2003, Henry 1987
Cellulose	2.1–2.6	Aman et al. 1997, Vinkx and Delcour 1996
Arabinogalactan peptide	0.2	Van den Bulck et al. 2004
Fructan	4.6–6.6	Hansen et al. 2003, Karpinnen et al. 2003
Protein	8.0–17.7	Fowler et al. 1990, Hansen et al. 2004
Lipid	2.0–2.5	Härkönen et al. 1997, Vinkx and Delcour 1996
Mineral	1.7–2.2	Aman et al. 1997, Hansen et al. 2004

ring contains amylose and probably less-ordered amylopectin (Morrison 1995). The semicrystalline growth ring consists of alternating crystalline and amorphous lamellae. The crystalline lamellae are built from amylopectin double helices, whereas the amorphous lamellae contain the branch points of the amylopectin side chains (Andreev et al. 1999, Gallant et al. 1997). Evidence has been provided that the crystalline and amorphous lamellae of the amylopectin are organized into larger, more or less spherical blocklets (Gallant et al. 1997).

The structure of starch can be studied at different levels by several techniques such as optical microscopy, scanning electron microscopy, transmission electron microscopy, and X-ray diffraction. X-ray diffraction provides information about the crystal structure of the starch polymers, but also about the relative amounts of the crystalline and amorphous phases. Cereal starches have an A-type X-ray diffraction pattern, while B-type and C-type X-ray diffraction patterns are characteristic for tuber and root starches and for legume starches, respectively. C-type starches are considered to be mixtures of A-type and B-type starches. In general, rye starches show A-type X-ray diffraction patterns, but some researchers (Gernat et al. 1993, Schierbaum et al. 1991) also reported a significant B-type portion.

Starch Physicochemical Properties

The physicochemical gelatinization, pasting, gelation, and retrogradation properties of starches are important parameters in determining the behavior of cereals in food systems.

When a starch suspension is heated above a characteristic temperature (the gelatinization temperature) in the presence of a sufficient amount of water, the starch granules undergo an order-disorder transition known as gelatinization. The phenomenon is accompanied by melting of the crystallites (loss of X-ray pattern), irreversible swelling of the granules, and solubilization of the amylose. Gelatinization can be studied by measuring the loss of birefringence of the starch granules when observed under polarized light. The end point and the range of temperatures at which gelatinization occurs can be determined. For rye starch, birefringence end point temperatures of 59.6°C (Klassen and Hill 1971) and 58.0°C (Lii and Lineback 1977) were reported. These birefringence end point temperatures were somewhat lower than those observed for wheat starches by the same authors. Another technique often used to study gelatinization-associated phenomena is differential scanning calorimetry (DSC). DSC enables determination of onset, peak, and conclusion gelatinization temperatures, gelatinization temperature ranges, and gelatinization enthalpy. The gelatinization temperature is a qualitative index of crystal structure, whereas gelatinization enthalpy is a quantitative measure of order. The DSC technique also measures the temperature and enthalpy of the dissociation of the amylose-lipid complexes. Rye starches show an onset of gelatinization at rather low temperatures. Gudmundsson and Eliasson (1991) reported onset gelatinization temperatures of 54.8–60.3°C, Radosta et al. (1992) of 53.5°C, and Verwimp et al. (2004) of 49.2–51.4°C. Rye starch has somewhat lower gelatinization temperatures and enthalpies than wheat starch (Gudmundsson and Eliasson 1991, Radosta et al. 1992, Verwimp et al. 2004). Lower dissociation enthalpies of the amylose-lipid complexes for rye starch than for wheat starch were observed by Gudmundsson and Eliasson (1991) and Verwimp et al. (2004), indicating lower levels of lipids in rye starch.

When starch suspensions are heated beyond their gelatinization temperature, granule swelling and amylose and/or amylopectin leaching continue. Eventually, total disruption of the granules occurs, and a starch paste with viscoelastic properties is formed. This process is called pasting. Upon cooling, gelation occurs. During gelation, at sufficiently high starch concentrations (>6%, w/w), the starch paste is converted into a gel. The gel consists of an amylose matrix enriched with swollen granules. During gelation and storage of starch gels, retrogradation occurs. This is a process whereby the starch molecules begin to reassociate in an ordered structure, without regaining the original molecular order. Initially, starch retrogradation is dominated by gelation and crystallization of the amylose molecules. In a longer time frame, amylopectin molecules recrystallize. Rye starch exhibits a unique, typical swelling behavior with a high pasting temperature (79–87°C according to Schierbaum and Kettlitz 1994; 75°C according to Verwimp et al. 2004), a stable consistency during cooking, and a high viscosity increase during cooling (Schierbaum and Kettlitz 1994, Verwimp et al. 2004). It has lower pasting temperatures than wheat starch (Schierbaum and Kettlitz 1994, Verwimp et al. 2004). Fredriksson et al. (1998) and Gudmundsson and Eliasson (1991) showed that rye starches retrograde to a lesser extent than starches from wheat or corn and ascribed this to differences in the structure of the amylopectin polymers.

Nonstarch Polysaccharides

Rye contains, on average, 13.0–15.0% nonstarch polysaccharides, which serve mainly as cell wall components and the majority of which are arabinoxylans (Table 34.1). Other nonstarch polysaccharide constituents of rye are mixed-linked β-glucan, cellulose, arabinogalactan peptides, and fructans (Table 34.1). The variation in nonstarch polysaccharide content and composition is significantly influenced by both harvest year and rye genotype (Hansen et al. 2003, 2004). Most of the nonstarch polysaccharides are classified as dietary fiber because of their resistance to digestion by human digestive enzymes. In view of the major importance of arabinoxylans in rye, these components will be further discussed in detail.

Arabinoxylan Occurrence and Structure

Arabinoxylans consist of a backbone of 1,4-linked β-D-xylopyranose residues, unsubstituted or mono- or disubstituted with single α-L-arabinofuranose residues at the O-2- and/or O-3-position. To some of these arabinose residues, a ferulic acid moiety is ester bound at the O-5 position. In rye, as in other cereals, two types of arabinoxylans can be distinguished: water extractable and water unextractable. Water-unextractable arabinoxylans, representing approximately 75% of the total arabinoxylan population of the rye kernel, are unextractable because they are retained in the cell walls by covalent and noncovalent interactions among arabinoxylans and between arabinoxylans and other cell wall constituents. Water-extractable arabinoxylans are thought to be loosely bound at the cell wall surface. Their content in the rye kernel ranges from 1.5 to 3.0% (Figueroa-Espinoza et al. 2002; Hansen et al. 2003, 2004) and is influenced by environmental and genetic factors but also by the conditions under which extraction occurs (Cyran et al. 2003, Härkönen et al. 1995). While the rye bran and shorts are high in arabinoxylan (14.9–39.5% and 13.5–18.7%, respectively), the endosperm contain lower levels (2.1–4.9%) (Fengler and Marquardt 1988a, Glitso and Bach Knudsen 1999, Härkönen et al. 1997, Nilsson et al. 1997a).

A considerable variation in substitution degree and pattern of xylose residues, arabinoxylan molecular weight, and ferulic acid content has been reported for both water-extractable and water-unextractable arabinoxylan. This variability in structure can, in part, account for the differences in extractability and solubility of the arabinoxylans.

Substitution Degree and Pattern of Xylose Residues Bengtsson et al. (1992a) reported on what could either be two different types of polymers or two different regions in the same molecule, that is, arabinoxylan I and arabinoxylan II. Arabinoxylan I contains mainly un- and monosubstituted xylose residues with on average 50% of the xylose residues substituted at O-3 with an arabinose residue. Only 2% of the xylose residues are disubstituted at O-2 and O-3 with arabinose residues (Bengtsson and Aman 1990). The xylose residues carrying an arabinose side chain occur predominantly as isolated residues (36%) or small blocks of two residues (62%) (Aman and Bengtsson 1991). Arabinoxylan II contains mainly un- and disubstituted xylose residues with on average 60–70% of the xylose residues substituted at O-2 and O-3 with arabinose residues (Bengtsson et al. 1992a). The levels of arabinoxylan I and II in different rye varieties ranges from 1.4 to 1.7% and from 0.6 to 1.0%, respectively (Bengtsson et al. 1992b). In contrast to the two classes of arabinoxylans described by Bengtsson et al. (1992a), Vinkx et al. (1993) concluded that there were a range of rye water-extractable arabinoxylans. The latter authors fractionated rye water-extractable arabinoxylans by graded ammonium sulfate precipitation into several fractions with arabinose:xylose (A/X) ratios of 0.5–1.4, with among them a major fraction containing almost purely O-3 monosubstituted arabinoxylans and a minor fraction consisting of almost purely disubstituted arabinoxylans. All arabinoxylan fractions contained a small amount of xylose residues substituted at O-2 with arabinose (Vinkx et al. 1995a).

Cyran et al. (2003) obtained rye flour arabinoxylans with different structural features after sequential extraction with water at different temperatures. A gradual increase in the degree of substitution in general and disubstitution in particular and a decrease in O-3 monosubstitution were observed from cold to hot water-extractable fractions. Within a water-extractable fraction, subfractions obtained after ammonium sulfate precipitation showed structures analogous to those reported by Bengtsson et al. (1992a) and Vinkx et al. (1993).

Rye water-unextractable arabinoxylan molecules also consist of a range of structures, which can only be studied after alkaline solubilization of the arabinoxylans. This treatment results in the saponification of the ester bonds linking ferulic acid to arabinose, releasing individual arabinoxylan molecules from the cell wall structure. Based on the studies of Hromadkova et al. (1987), Nilsson et al. (1996, 1999) and Vinkx et al. (1995b) (Table 34.2), three different groups of alkali-extractable arabinoxylans can be distinguished in rye bran. A first group shows an intermediate arabinose:xylose ratio (0.54–0.65) and contains mainly unsubstituted (57–64%) and monosubstituted (24–29%) xylose residues. A second group consists of almost pure unsubstituted (89%) arabinoxylans with a low arabinose:xylose ratio (0.20–0.27). A third group is characterized by a high arabinose:xylose ratio (1.08–1.10) and contains approximately 46% monosubstituted and 33% di-substituted xylose residues. In contrast to rye bran water-unextractable arabinoxylans, the rye flour alkali-solubilized arabinoxylans all show similar xylose substitution levels (Cyran et al. 2004, Vinkx 1994) (Table 34.2). However, further fractionation of the alkali-extracted arabinoxylans from rye flour by ammonium sulfate precipitation yields subfractions that differ in structure (Cyran et al. 2004). The arabinoxylans sequentially extracted with alkali from rye bran show lower degrees of branching than those extracted from rye flour by the same extraction solvent (Table 34.2).

In general, for rye bran as well as for rye flour, low yields of alkali-extractable arabinoxylan fractions with low (0.2) and high (1.1) arabinose:xylose ratios are obtained. Glitso and Bach Knudsen (1999) found lower substitution degrees for water-unextractable arabinoxylans in the aleurone layer (0.35) than for those in the starchy endosperm (0.83) and pericarp/testa fraction (1.02). From these results and because in rye milling the endosperm can be contaminated with bran, one can assume that the arabinoxylan fractions with high substitution degrees isolated from rye flour originate from contamination with bran fractions. Similarly, the arabinoxylan fractions isolated from rye bran might be contaminated with endosperm fractions.

Water-unextractable arabinoxylans in bran have a lower degree of branching than their water-extractable counterparts (Glitso and Bach Knudsen 1999), whereas the opposite is the case in the endosperm (Cyran and Cygankiewicz 2004, Glitso and Bach Knudsen 1999).

Molecular Weight Large differences in molecular weight of rye arabinoxylans exist. For rye, whole meal water-extractable arabinoxylans, average molecular weights of 770 k (Girhammar and Nair 1992a) and more than 1000 k (Härkönen et al. 1995), as determined by gel permeation chromatography, have been reported. Whether the water-extractable arabinoxylans from rye bran have a higher (Meuser et al. 1986) or lower (Härkönen et al. 1997) mean molecular weight than those from rye flour is not clear. It is also not clear whether extraction temperature affects the molecular size distribution of rye flour water-extractable arabinoxylans (Cyran et al. 2003) or not (Härkönen et al. 1995, Meuser et al. 1986). For water-extractable arabinoxylans isolated from rye varieties differing in extract viscosity, increasing molecular weights were correlated with increasing rye extract viscosity (Ragaee et al. 2001). Rye water-extractable arabinoxylans have higher average molecular weights than those of wheat (Dervilly-Pinel et al. 2001, Girhammar and Nair 1992a).

The water-unextractable arabinoxylans from rye flour have a higher molecular weight (1500–3000 k) than their

Table 34.2. Substitution Degrees (A/X Ratios) and Xylose Substitution Levels (%) of Alkali-Solubilized Water-Unextractable Arabinoxylans (AS-AX) from Rye Bran and Flour

AS-AX Source	Extraction Solvent[a]	A/X	Un[b]	Mono[b]	Di[b]	Reference
Rye bran	1.1 M NaOH	0.14	88	12	0	Hromadkova et al. 1987
Rye bran	1% NH_4OH	0.78	41	33	26	Ebringerova et al. 1990
Rye bran	Saturated $Ba(OH)_2$	0.65	61	29	10	Vinkx et al. 1995b
	H_2O	0.55	64	24	12	
	1.0 M KOH	0.21	89	8	3	
	Delignification followed by 1.0 M KOH	1.10	21	46	33	
Rye bran	Saturated $Ba(OH)_2$	0.54	ND[c]	ND	ND	Nilsson et al. 1996
	H_2O	0.60	ND	ND	ND	
	4.0 M KOH	0.27	ND	ND	ND	
	2.0 M KOH	1.08	ND	ND	ND	
Rye bran	Saturated $Ba(OH)_2$	0.56	57	29	14	Nilsson et al. 1999
Rye flour	Saturated $Ba(OH)_2$	0.68	46	40	14	Vinkx 1994
	H_2O	0.78	44	34	22	
	1.0 M KOH	0.72	ND	ND	ND	
Rye flour	Saturated $Ba(OH)_2$	0.67	ND	ND	ND	Nilsson et al. 1996
	H_2O	0.93	ND	ND	ND	
	4.0 M KOH	0.63	ND	ND	ND	
	2.0 M KOH	1.10				
Rye flour	Saturated $Ba(OH)_2$	0.69–0.70	51	28–29	20–21	Cyran et al. 2004
	H_2O	0.79–0.84	42–44	32–34	22–26	
	1.0 M NaOH	0.70–0.77	45–48	33	19–22	

[a]Consecutive extraction solvents in one reference indicate sequential alkaline extraction.
[b]Un, mono, and di: percentages of total xylose occurring as unsubstituted, O-2 and/or O-3 monosubstituted, and O-2, O-3 disubstituted xylose residues.
[c]ND = Not determined.

water-extractable counterparts (620–1200 k) (Meuser et al. 1986). For rye bran water-unextractable arabinoxylans, however, Vinkx et al. (1995b) found that the molecular weight of arabinoxylans solubilized with alkali were in the same range as those reported for water-extractable arabinoxylans (Vinkx et al. 1993).

Rye arabinoxylans were reported to have an extended, rodlike conformation (Anger et al. 1986, Girhammar and Nair 1992a). In contrast, Dervilly-Pinel et al. (2001) stated that the water-extractable arabinoxylans from rye have a random coil conformation.

Ferulic Acid Content Ferulic acid residues can link adjacent arabinoxylan chains through formation of diferulic acid bridges, thereby affecting extractability of arabinoxylans.

Ferulic acid and ferulic acid dehydrodimer contents in rye vary from 0.90 to 1.17% and from 0.24 to 0.41%, respectively, and are significantly influenced by both rye genotype and harvest year (Andreasen et al. 2000b). These phenolic compounds are concentrated in the bran of the rye kernel (Andreasen et al. 2000a, Glitso and Bach Knudsen 1999).

The water-extractable arabinoxylans from rye flour contain low levels of ferulic acid (0.03–0.15%) (Figueroa-Espinoza et al. 2002, Vinkx et al. 1993) and ferulic acid dehydrodimers (0.03×10^{-2}%) (Dervilly-Pinel et al. 2001, Figueroa-Espinoza et al. 2002).

The water-unextractable arabinoxylans from rye flour contain higher amounts of ferulic acid (0.20–0.36%) and ferulic acid dehydrodimers (0.35%) than their water-extractable counterparts (Figueroa-Espinoza et al. 2002). After alkaline extraction with saturated barium hydroxide and 1.0 M sodium hydroxide, rye flour arabinoxylans still contain a substantial level of ferulic acid. Further fractionation of these arabinoxylans by ammonium sulfate precipitation reveals increasing levels of ferulic acid with increasing ammonium sulfate concentration (Cyran et al. 2004). This observation might corroborate the above-mentioned hypothesis that the different arabinoxylan fractions with varying structures obtained from rye flour may reflect contamination of the rye flour with bran fractions.

Arabinoxylan Physicochemical Properties

Solubility of a large part of the individual arabinoxylan molecules in water is mainly associated with the presence of arabinose substituents attached to the xylose residues, which prevent intermolecular aggregation of unsubstituted xylose residues. High molecular weight arabinoxylans with arabinose:xylose ratios below 0.3 tend to be insoluble.

Rye arabinoxylans are said to have high water-holding capacities, with water-extractable and water-unextractable arabinoxylans able to absorb 11 and 10 times their weight of water, respectively (Girhammar and Nair 1992a). These values were typically determined by addition of arabinoxylan to flour and measurement of the consecutive rise in Farinograph absorption. However, it is not clear whether the results obtained by this technique may be simply defined as "water-holding capacity." The mechanism behind water binding of discrete cell wall fragments that contain water-unextractable arabinoxylans is bound to be different from that of water-extractable arabinoxylans in solution. It is also unclear whether the increased Farinograph absorption is solely caused by the arabinoxylans. That water-holding values obtained with the Farinograph method have to be interpreted with care is clearly demonstrated by Girhammer and Nair (1992b), who reported a water-holding capacity of 0.47 g/g dry matter for rye water-extractable arabinoxylan, when determined by measuring the level of unfreezable water associated with the water-extractable arabinoxylan using differential scanning calorimetry. These authors stated that the water-holding capacity was not related to the molecular weight of the arabinoxylan. Different milling fractions of rye showed different water absorption capacities in relation to this total arabinoxylan content (Härkönen et al. 1997).

Even at relatively low concentrations, water-extractable arabinoxylans are able to form highly viscous solutions in water (Girhammar and Nair 1992b). Highly positive correlations were found between the viscosity of a rye extract and its content of water-extractable arabinoxylans (Boros et al. 1993; Fengler and Marquardt 1988a; Härkönen et al. 1995, 1997; Ragaee et al. 2001). For different rye milling fractions, extract viscosity correlates with the content of water-extractable arabinoxylans (Fengler and Marquardt 1988a, Glitso and Bach Knudsen 1999, Härkönen et al. 1997). Water extracts from rye are more viscous than those from other cereals (Boros et al. 1993, Fengler and Marquardt 1988a), which can be ascribed to the higher concentration of water-extractable arabinoxylans in rye than in other cereals. Differences in viscosity are related to differences in the structural features and molecular weight of water-extractable arabinoxylans. Although the impact of un-, mono-, and disubstitution is unclear (Bengtsson et al. 1992b, Dervilly-Pinel et al. 2001, Ragaee et al. 2001), molecular weight seems a logical determining parameter (Girhammar and Nair 1992b, Nilsson et al. 2000, Ragaee et al. 2001, Vinkx et al. 1993). Arabinoxylan conformation was also found to have a bearing on viscosity, with a higher viscosity being associated with a larger radius of gyration (Dervilly-Pinel et al. 2001, Ragaee et al. 2001). Viscosity measurement conditions such as shear rate (Bengtsson et al. 1992b, Härkönen et al. 1997, Nilsson et al. 2000) and temperature, pH, and salt concentration (Girhammar and Nair 1992b) also influence the viscosity of rye water-extractable arabinoxylan solutions.

Water-extractable arabinoxylans undergo oxidative gelation upon addition of hydrogen peroxide and peroxidase, resulting in an increase in viscosity and eventually the formation of a gel (Dervilly-Pinel et al. 2001, Vinkx et al. 1991). The phenomenon is caused by the formation of covalent linkages through oxidative coupling of ferulic acid residues esterified to arabinoxylan. The aromatic ring and not the propenoic moiety of ferulic acid is involved in the oxidative gelation (Vinkx et al. 1991). The oxidative gelation capacity of arabinoxylans depends on several parameters including the ferulic acid content, the molecular weight, and the structure of the arabinoxylans. Arabinoxylan fractions with higher ferulic acid content, higher molecular weights, and fewer disubstituted xylose residues have a higher gelation potential (Dervilly-Pinel et al. 2001, Vinkx et al. 1993). Dervilly-Pinel et al. (2001) stated that the intrinsic viscosity of arabinoxylans is the main parameter governing gel rigidity.

PROTEINS

The total protein content in rye is influenced by both environmental and genotypic factors and varies from 8.0 to 17.7% (see Table 34.1). The protein content is higher in the outer layers than in the inner layers of the rye kernel (Glitso and Bach Knudsen 1999; Nilsson et al. 1996, 1997a).

Protein Classification

Rye proteins can be separated by the classical Osborne fractionation procedure. The albumins are extractable with water, the globulins with dilute salt solutions, the prolamins with alcohol, and the glutelins with dilute acid or alkali. Chen and Bushuk (1970) found that the albumin, globulin, prolamin, and glutelin fractions accounted for 34, 11, 19, and 9%, respectively, of the total nitrogen content in a rye flour, with 21% of the protein remaining in the residue. Similar results were obtained by Preston and Woodbury (1975) and Jahn-Deesbach and Schipper (1980), except that the latter authors extracted a much higher proportion of glutelins. Gellrich et al. (2003), however, extracted fewer albumins and globulins and many more prolamins from rye flour. In all cases, rye flour contained higher levels of albumins and globulins and lower levels of prolamins and glutelins than wheat (Chen and Bushuk 1970, Gellrich et al. 2003, Jahn-Deesbach and Schipper 1980).

From a physiological point of view, rye proteins can be divided into two main groups: the storage proteins and the nonstorage proteins. The storage proteins are deposited in protein bodies in the developing starchy endosperm and function solely as a supply of amino acids for use during germination and seedling growth. They include the prolamins, glutelins, and the proteins in the Osborne residue. The nonstorage proteins include, amongst others, enzymes and (in some cases) their inhibitors and correspond with the Osborne albumin and globulin fractions.

Major differences are found in the amino acid compositions of the proteins in the two groups. The nonstorage proteins contain high proportions of lysine and arginine and low proportions of glutamine/glutamate and proline, whereas the opposite was observed for the storage proteins (Dexter and Dronzek 1975, Gellrich et al. 2003, Preston and Woodbury 1975). No significant differences exist in the amino acid composition of the proteins from different rye varieties (Dembinski and Bany 1991,

Table 34.3. Classification of Rye Proteins

Physiological classification	Osborne classification	Technologically important proteins
Nonstorage proteins	Albumins Globulins	Enzymes
		→ α-amylases Endoxylanases Arabinofuranosidases Xylosidases Ferulic acid esterases Proteases
		→ Serine proteases Metalloproteases Aspartic proteases Cysteine proteases
		Enzyme inhibitors
		→ α-amylase inhibitors Endoxylanase inhibitors Protease inhibitors
Storage proteins	Secalins	γ-secalins
		→ 40k γ-secalins 75k γ-secalins
		ω-secalins
		HMW secalins
	Glutelins	γ-secalins
		→ 40k γ-secalins 75k γ-secalins
		HMW secalins

HMW, high molecular weight.

Morey and Evans 1983). The higher proportions of lysine make rye proteins nutritionally superior to wheat proteins (Chen and Bushuk 1970, Dexter and Dronzek 1975, Jahn-Deesbach and Schipper 1980, Morey and Evans 1983).

Table 34.3 gives an overview of the classification of the rye proteins, which are of technological importance and are further discussed in detail below.

Storage Proteins

Secalins The rye alcohol-soluble prolamins, also called "secalins," form the major storage proteins in rye (Gellrich et al. 2003). They have unusual amino acid compositions with high glutamine/glutamate and proline contents and low levels of basic and acidic amino acids (Dexter and Dronzek 1975, Gellrich et al. 2003, Preston and Woodbury 1975). The secalins are highly polymorphic and can be classified into three groups on the basis of their amino acid compositions: S-rich, S-poor, and high molecular weight (HMW) secalins (Fig. 34.1). Some researchers also purified secalins of low molecular weight (LMW) (10–16 k) (Charbonnier et al. 1981; Preston and Woodbury 1975), which is in the same range as that of wheat low molecular weight gliadin and barley A hordein.

The S-rich or γ-secalins are the most abundant secalins and consist of two major groups of polypeptides: 40k γ-secalins (26% of total secalin fraction) and 75k γ-secalins (45% of total secalin fraction) (Gellrich et al. 2003, Shewry et al. 1982). The molecular weights of 40k γ- and 75k γ-secalins are 40 k and 75k, respectively, as determined by SDS-PAGE (sodium dodecyl sulfate–polyacrylamide gel electrophoresis), hence their names (Field et al. 1983, Gellrich et al. 2003, Shewry et al. 1982), and 33 k and 54 k, respectively, as determined by sedimentation equilibrium ultracentrifugation (Shewry et al. 1982) or mass spectrometry (Gellrich et al. 2001, 2003). The two groups differ in amino acid composition, with 75k γ-secalins having higher contents of glutamine/glutamate and proline and lower contents of cysteine than the 40k γ-secalins (Gellrich et al. 2003, Shewry et al. 1982, Tatham and Shewry 1991).

The γ-secalins contain two structural domains: an *N*-terminal domain rich in glutamine and proline and consisting of repetitive sequences (Gellrich et al. 2001, 2004a; Kreis et al. 1985; Shewry et al. 1982), and a *C*-terminal domain with a nonrepetitive structure, which has lower glutamine and proline content and is rich in cysteine (Kreis et al. 1985). The *N*-terminal sequences of the 40k γ- and 75k γ-secalins are identical at 16 or 17 of the first 20 positions (Gellrich et al. 2003, 2004a; Shewry et al. 1982). The *N*-terminal repetitive domains probably have a rod-like conformation (Shewry and Tatham 1990), whereas the *C*-terminal nonrepetitive domains probably have a compact globular conformation (Shewry and Tatham 1990) rich in α-helical structures (Tatham and Shewry 1991). The 75k γ-secalins have a lower content of α-helical structures than the 40k γ-secalins (Tatham

Prolamins (Mainly monomeric, some smaller polymers)				**Glutelins** (Polymeric)
		Rye proteins		
• Secalins				• Glutelins
- 40k γ-, 75k γ-secalins	→	*S*-rich prolamins	←	- 40k γ-, 75k γ-secalins
- ω-, secalins	→	*S*-poor prolamins		
- HMW secalins	→	HMW prolamins	←	- HMW secalins

		Wheat proteins		
• Gliadins				• Glutenins
- α-, β-, γ-gliadins	→	*S*-rich prolamins	←	- B and C LMW-GS[a]
- ω-gliadins	→	*S*-poor prolamins	←	- D LMW-GS*
	→	HMW prolamins	←	- HMW-GS

[a]LMW-GS show strong similarities with gliadins, differing in one very important characteristic, i.e., the occurrence of interchain disulphide bonds in LMW-GS.

Figure 34.1. Schematic representation of the classification of rye and wheat proteins. LMW, low molecular weight; HMW, high molecular weight.

and Shewry 1991), consistent with a higher proportion of repetitive sequences in the former group (Shewry et al. 1982).

The two γ-secalin types differ in aggregation behavior. The 40k γ-secalins predominantly occur as monomers containing intramolecular disulphide bonds, while the 75k γ-secalins are present predominantly as disulphide-linked aggregates (Field et al. 1983, Gellrich et al. 2003, Shewry et al. 1983). A possible explanation for this different aggregation behavior is the presence or absence of cysteine residues in positions that are favorable or unfavorable for formation of intermolecular disulphide bonds. Alternatively, it might result from the specific action of enzymes responsible for the formation of such bonds (Field et al. 1983, Shewry et al. 1983). Gellrich et al. (2001, 2004b) showed that the eight-cysteine residues of the *C*-terminal domain of 75k γ-secalin are linked by intramolecular disulphide bonds and that the cysteine residue located in position 12 of the *N*-terminal domain of the 75k γ-secalin forms an intermolecular disulphide bond with 75k γ-secalins. The cysteine residue in the *N*-terminal domain is unique to the 75k γ-secalins and is likely to be responsible for their aggregative nature (Gellrich et al. 2003, 2004b).

The 40k γ-secalins are homologous to the wheat γ-gliadins and to the B hordeins of barley (Field et al. 1983; Gellrich et al. 2001, 2003; Shewry et al. 1982), whereas the 75k γ-secalins differ from them due to the presence of additional repetitive sequences rich in glutamine and proline and their occurrence as aggregates. The latter are suggested to be homologous to the LMW glutenin subunits (LMW-GS) of wheat (Shewry et al. 1982) (Fig. 34.1).

The S-poor or ω-secalins are quantitatively minor components, accounting for about 19% of the total secalin fraction (Gellrich et al. 2003). They have SDS-PAGE molecular weights of about 48 k to 53 k (Field et al. 1983, Gellrich et al. 2003, Shewry et al. 1983). The amino acid compositions of the ω-secalins are characterized by high levels of glutamine/glutamate, proline, and phenylalanine and the absence of cysteine and methionine (Tatham and Shewry 1991). The ω-secalins consist almost entirely of repetitive sequences (Tatham and Shewry 1991) forming a stiff, wormlike coil (Tatham and Shewry 1995). They are present predominantly as monomers (Field et al. 1983, Gellrich et al. 2003) and are homologous to the ω-gliadins of wheat and to the C hordeins of barley (Field et al. 1983, Gellrich et al. 2003, Tatham and Shewry 1991) (Fig. 34.1).

HMW secalins are also quantitatively minor components, accounting for about 5% of the total secalin fraction (Gellrich et al. 2003). SDS-PAGE molecular weights above 100 k were observed (Field et al. 1982, Gellrich et al. 2003, Kipp et al. 1996, Shewry et al. 1983), whereas sedimentation equilibrium ultracentrifugation showed a molecular weight of 68 k (Field et al. 1982). The HMW secalins have a high content of glycine and glutamine/glutamate and contain lower levels of proline than the other secalin groups (Field et al. 1982, Tatham and Shewry 1991). The conformation of HMW secalins has not been reported in detail, but based on studies with HMW glutenin subunits (HMW-GS) of wheat, it was suggested that they contain central repetitive sequences with a rodlike shape, flanked by nonrepetitive domains having a globular conformation rich in α-helical structure. The nonrepetitive domains contain most of the cysteine residues involved in intermolecular cross-linking (Shewry et al. 1995). The HMW secalins are present only in an aggregated state stabilized by disulphide bonds (Field et al. 1983, Gellrich et al. 2003, Shewry et al. 1983). They are homologous to the HMW-GS of wheat and the D hordeins of barley (Field et al. 1982, 1983; Gellrich et al. 2003; Shewry et al. 1982, 1988; Fig. 34.1).

Glutelins In contrast to the wheat glutelins, which are called "glutenins," no alternative name has been given to the rye glutelin fraction. The glutelins are not soluble in aqueous alcohols. They form HMW polymers stabilized by interchain disulphide bonds. However, the individual subunits of the polymers obtained in the presence of a reducing agent are soluble in alcohol media and rich in proline and glutamine (Field et al. 1982, 1983; Gellrich et al. 2003; Kipp et al. 1996; Shewry et al. 1983). The building blocks of the glutelins are therefore considered to be secalins. The glutelins soluble in alcohol after reduction contain mainly HMW-secalins (26%) and 75k γ-secalins (52%) and small levels of 40k γ-secalins (12%) (Gellrich et al. 2003; Fig. 34.1). When the distribution of the 75k γ-, 40k γ-, ω- and HMW secalins over the storage proteins (secalins + glutelins) is considered, it can be concluded that the major portion of 75k γ-secalins is present in the prolamin fraction (40%) and only 7% in the glutelin fraction. The 40k γ-secalins (23%) and ω-secalins (16%) appear mainly in the prolamin fraction due to their monomeric state. A larger portion of HMW secalins is present in the prolamin (4%) than in the glutelin fraction (3%) (Gellrich et al. 2003).

Gluten Formation The wheat gliadin and glutenin fractions can form a polymeric gluten network by intermolecular noncovalent interactions and disulphide bonds. In contrast to wheat, rye storage proteins do not form gluten. Despite partial homology, storage proteins of rye differ significantly from those of wheat with respect to quantitative and structural parameters that are important for the formation and properties of gluten. Typical for rye are the low content of storage proteins and the high ratio of alcohol-soluble proteins to insoluble proteins (Chen and Bushuk 1970, Gellrich et al. 2003, Preston and Woodbury 1975). The 75k γ-secalins and HMW secalins differ in aggregation behavior from the LMW-GS and HMW-GS of wheat, respectively, due to the absence (Gellrich et al. 2004b) or presence (Köhler and Wieser 2000) of cysteine residues in the *C*-terminal domain in positions favorable for formation of intramolecular disulphide bonds and unfavorable for formation of intermolecular disulphide bonds. It is likely that these structural differences contribute to the lack of ability of rye to form gluten. Another factor hampering the formation of a protein network from rye secalin and glutelin fractions may be a higher degree of glycosylation in rye subunits than in wheat subunits (Kipp et al. 1996).

Functional Proteins

Among the many enzymes and structural proteins in cereals in general, the most important enzymes present in rye grain that break down its major constituents (starch, arabinoxylan, and protein) are the α-amylases, endoxylanases, and proteases, respectively. Such enzymes can be of major importance in rye processing. Rye also contains specific proteins with inhibition activity against these enzymes. Presumably, these enzyme inhibitors contribute to plant defense mechanisms and/or possibly intervene in the complex regulation of plant metabolic processes.

Starch-Degrading Enzymes and Their Inhibitors α-amylases or, in full, 1,4-α-D-glucan glucanohydrolases (E.C. 3.2.1.1) catalyze the hydrolysis of the internal 1,4-α-D-glucan linkages in starch components.

Rye synthesizes two groups of α-amylases: low pI ($\leq$ pI 5.5) and high pI ($\leq$ pI 5.8). The low-pI α-amylases are synthesized during grain development, particularly in the pericarp (Dedio et al. 1975). Their activity decreases during the later stages of grain development and is low at maturity. The high-pI α-amylases are synthesized during germination. In germinated rye, the high-pI groups represent a high proportion of the total amylase activity (MacGregor et al. 1988). Gabor et al. (1991) isolated a high-pI α-amylase from germinated rye, also called "germination-specific" α-amylase, which showed optimal activity at pH 5.5 and is stable at pH 6.0–10.0 after storage at 4°C. The temperature optimum over a 3 minute incubation period was 65°C, but 40% of the activity was lost when the enzyme was incubated without substrate for 4 hours at 55°C. Täufel et al. (1991) also isolated a germination-specific α-amylase from rye and reported a pH optimum of 5.0, pH stability between pH 5.0 and 7.0 and thermal stability up to 55°C.

Variation in α-amylase levels exists between different rye lines (Masojc and Larssonraznikiewicz 1991a). Masojc and Larssonraznikiewicz (1991b) showed the existence of low and high α-amylase genotypes. In contrast, Hansen et al. (2004) stated that the α-amylase activity is mostly influenced by harvest year, attributing only a small effect to genotype.

α-Amylase inhibitors can reduce the activity of one or more α-amylases and are mainly found in two major families, that is, the "cereal trypsin/α-amylase inhibitor" and the "Kunitz" or "α-amylase/subtilisin inhibitor" families (Garcia-Olmedo et al. 1987). The cereal trypsin/α-amylase inhibitor family represents the major part of the albumin and globulin fraction of the endosperm and consists of monofunctional inhibitors, active against either trypsin or exogenous α-amylase. The Kunitz-type inhibitors are bifunctional inhibitors of about 20 k, containing two intramolecular disulphide bridges. They can simultaneously inhibit both subtilisin and endogenous α-amylases, in particular the germination specific, high-pI α-amylase.

Täufel et al. (1991, 1997) purified two proteinaceous inhibitors of α-amylase from rye: a regulation or R-type inhibitor, which only inhibits the endogenous germination-specific α-amylase, and a defense or D-type inhibitor, which only inhibits exogenous α-amylases of animal and human origin. The inhibitors are present in different tissues of the rye kernel (Täufel et al. 1997). The R-type inhibitor predominantly occurs in the bran, whereas the D-type inhibitor is enriched in the germ and the endosperm. The high D-type inhibitor activity in these kernel sections is consistent with its assumed defense function, as it can protect the starch against exogenous α-amylases from insects and rodents. The levels of both types of inhibitor vary with growing conditions (Täufel et al. 1997). Rye varieties cultivated under drought conditions show high D-type inhibitor activities and low R-type inhibitor activities, whereas the opposite is true when they are cultivated under wet conditions. Small variations in α-amylase inhibitor activities exist between sprout-stable and sprout-sensitive rye genotypes (Täufel et al. 1997). Whereas α-amylase activity increases rapidly after 24 hours of germination, the R- and D-type inhibitor activities are stable during the 72 hours of germination (Täufel et al. 1997). The α-amylase inhibitors obviously have no importance at the early stages of growth.

Arabinoxylan-Degrading Enzymes and Their Inhibitors Arabinoxylan structural and physicochemical properties can be impacted by enzymes such as arabinofuranosidases, xylosidases, esterases, and endoxylanases, with the latter being by far the most relevant in cereal processing.

Endoxylanases or, in full, endo-1,4-β-D-xylan xylanohydrolases(E.C. 3.2.1.8) hydrolyze internal 1,4 linkages between β-D-xylopyranose residues of (arabino-) xylan molecules, generating (arabino-) xylan fragments with lower molecular weight and (un)substituted xylooligosaccharides. Their action can thus lead to (partial) solubilization of water-unextractable arabinoxylans and to a decrease in molecular weight of water-extractable and/or solubilized arabinoxylans. The susceptibility of arabinoxylans to endoxylanase attack probably depends on their substitution degree and their substitution pattern as well as on their linkages to other cell wall components. Enzymic hydrolysis also depends on the substrate selectivity of the endoxylanases. Certain endoxylanases preferentially release high molecular weight arabinoxylans from water-unextractable arabinoxylans and slowly degrade water-extractable arabinoxylans; other endoxylanases solubilize arabinoxylans of low molecular weight and/or extensively degrade water-extractable arabinoxylans.

Based on their primary sequence and structure, endoxylanases are classified into two main groups: glycosyl hydrolase family 10 and family 11 (Henrissat 1991). Plant endoxylanases to date have been exclusively classified in glycosyl hydrolase family 10 in contrast to fungal and bacterial endoxylanases, which are present in both glycosyl hydrolase families 10 and 11 (Coutinho and Henrissat 1999).

Arabinofuranosidases (α-L-arabinofuranosidases, E.C. 3.2.1.55) hydrolyze the linkage between arabinose residues and the xylan backbone, rendering the latter less branched and more accessible to depolymerization by endoxylanases. Ferulic acid esterase (feruloyl esterase, E.C. 3.1.1.73) can hydrolyze the ester linkage between ferulic acid and arabinose. Xylosidases (exo-1,4-β-xylosidase, E.C. 3.2.1.37) release single xylose residues from the nonreducing end of arabinoxylan fragments, thereby increasing the level of reducing end sugars. While synergy between these classes of enzymes and endoxylanases can be expected, it is not always observed (Figueroa-Espinoza et al. 2002, 2004).

Endogenous xylanolytic activity, although low, has been reported to be present in ungerminated rye (Rasmussen et al. 2001). The rye endogenous endoxylanase showed optimal activity at pH 4.5 and 40°C and was pH stable but heat labile. Xylosidase and arabinofuranosidase enzymes are also present in ungerminated rye grain (Rasmussen et al. 2001). During germination of rye, endoxylanase activity increases significantly after the first day (Autio et al. 2001).

Endoxylanase inhibitors can affect the enzymic hydrolysis of arabinoxylans by endoxylanases. Two different types of such inhibitors have been found to be present in rye (Goesaert et al. 2002, 2003).

The first type of inhibitor found in rye is the *Secale cereale* L. xylanase inhibitor (SCXI) (Goesaert et al. 2002), representing a family of isoinhibitors (SCXI I–IV) with similar structures and specificities. These inhibitors are basic proteins with isoelectric points of at least 9.0 and have highly homologous *N*-terminal amino acid sequences. They occur in two molecular forms, that is, a monomeric 40 k protein with at least one intramolecular disulphide bridge and, presumably following proteolytic cleaving of this form, a heterodimer consisting of two disulphide linked subunits (30 and 10 k). They specifically inhibit family 11 endoxylanases, while fungal family 10 endoxylanase is not affected (Goesaert et al. 2002). These inhibitors are homologous with wheat *Triticum aestivum* L. xylanase inhibitor I (TAXI I) (Gebruers et al. 2001).

Another family of isoinhibitors in rye, is the XIP-type (xylanase inhibiting protein) endoxylanase inhibitor family (Elliott et al. 2003, Goesaert et al. 2003). These inhibitors are basic, monomeric proteins with a molecular weight of 30 k with pI values of at least 8.5. They show inhibitory activity against fungal glycosyl hydrolase family 10 and 11 endoxylanases (Goesaert et al. 2003). Structural characteristics and inhibition specificities from the rye XIP-type inhibitors are similar to those of wheat XIP-type inhibitors (Gebruers et al. 2002).

Protein-Degrading Enzymes and Their Inhibitors Literature on proteins from different botanical sources (e.g., soy, wheat, rice) reveals that their enzymic hydrolysis can significantly alter their properties (Caldéron de la Barca et al. 2000, Popineau et al. 2002, Shih and Daigle 1997, Wu et al. 1998). This, however, has not been reported for rye protein.

Proteases hydrolyze the peptide bonds between the amino end of one amino acid residue and the carboxyl end of the adjacent amino acid residue in a protein. Endoproteases hydrolyze peptide bonds somewhere along the protein chain, whereas exoproteases attack the ends of the protein chain and remove one amino acid at a time. The latter are called carboxypeptidases when acting from the carboxy terminus or aminopeptidases when acting from the amino terminus. Based on the chemistry of their catalytic mechanism, proteases have been classified into four groups: serine (E.C. 3.4.21), metallo- (E.C. 3.4.24), aspartic (E.C. 3.4.23), and cysteine (E.C. 3.4.22) proteases.

Nowak and Mierzwinska (1978) reported the presence of endopeptidase, carboxypeptidase, and aminopeptidase activities in the embryo and endosperm of rye seeds. Endoproteolytic, exoproteolytic, carboxypeptidase, aminopeptidase, and *N*-α-benzoylarginine-*p*-nitroanilide (BAPA) hydrolyzing activities have also been found in whole meal extracts from ungerminated rye seeds by Brijs et al. (1999). During germination, proteolytic activity increases during the first 3 days to then remain constant (Brijs et al. 2002). Milling fractions of germinated rye (Brijs et al. 2002) have much higher proteolytic activities than those of ungerminated rye (Brijs et al. 1999). The enzymes are mainly present in the bran and shorts fractions. Ungerminated rye mainly contains two classes of proteases (serine and aspartic proteases) (Brijs et al. 1999), whereas all four protease classes are present during germination (Brijs et al. 2002). Germinated rye contains mainly aspartic and cysteine protease activities; serine and metalloproteases are less abundant.

Aspartic proteases have been purified from bran of ungerminated rye (Brijs 2001). Two heterodimeric aspartic proteases were found: a larger 48 k enzyme, consisting of 32 k and 16 k subunits which are probably linked by disulphide bridges and a smaller one of 40 k, consisting of two noncovalently bound 29 k and 11 k subunits. Both proteins have isoelectric points close to pI 4.6. Rye aspartic proteases show optimal activity around pH 3.0 and 45°C and can be completely inhibited by pepstatin A (Brijs 2001). Amino acid sequence alignment showed that the rye aspartic proteases resemble those from wheat (Bleukx et al. 1998) and barley (Sarkkinen et al. 1992). Cysteine proteases were isolated from the starchy endosperm of 3-day germinated rye (Brijs 2001). SDS-PAGE revealed three protein bands with apparent molecular weights of 67 k, 43 k, and 30 k, although it was not established that all three protein bands are cysteine proteases. The rye cysteine proteases are optimally active around pH 4.0 and at 45°C and can be totally inhibited by L-transepoxysuccinyl-L leucylamido-(4-guanidino)butane (or E-64).

Protease inhibitors can reduce the activity of proteases and have also been found in rye. Boisen (1983) and Lyons et al. (1987) isolated a trypsin inhibitor from rye endosperm and rye seed, respectively. The inhibitor had a molecular weight of about 12 to 14 k, contained four disulphide bridges, and was resistant to trypsin and elastase but was completely inactivated by chymotrypsin. The amino acid composition was very similar to that of the barley and wheat trypsin inhibitors.

Mosolov and Shul'gin (1986) purified a subtilisin inhibitor from rye that belongs to the Kunitz-type inhibitor family. The inhibitor consisted of two forms with pI values of 6.8 and 7.1 and had a molecular weight of about 20 k. It inhibited subtilisin (at a molar ration of 1:1) and fungal proteinases from the genus *Aspergillus*, but was inactive against trypsin, chymotrypsin, and pancreatic elastase. The amino acid composition of the rye subtilisin inhibitor was similar to that from its wheat and barley counterparts.

Hejgaard (2001) purified and characterized six serine protease inhibitors, also called "serpins," from rye grain. The serpins each have different regulatory functions based on different reactive center loop sequences, allowing for specific associations with the substrate-binding subsites of the proteases. The reactive center loops of the rye serpins were similar to those of wheat serpins. Five of the six serpins contained one or two glutamine residues at the reactive center bond, which were differently positioned than in the wheat reactive center loops. The rye serpins have a molecular weight of 43 k and inhibit proteases with chymotrypsin-like specificity, but not pancreatic elastase or proteases with trypsin-like specificity.

Protein Physicochemical Properties

The properties of rye proteins, for example, their foam-forming and emulsifying capacities, have hitherto not been studied well. Ludwig and Ludwig (1989) isolated a protein mixture from rye flour that exhibited low foam stability but excellent emulsifying properties at pH 3.5. Meuser et al. (2001) reported on a rye water-soluble protein with foam-forming capacity. It had a molecular weight of 12.3 k and an isoelectric point of 5.97. The optimum pH value for the water-soluble rye protein to form stable foam was pH 6.0. The protein did not appear to be located in specific morphological parts of the rye kernel. Wannerberger et al. (1997) found that secalins are more surface-active than gliadins. The lower molecular weight secalins spread more rapidly at the air-water interface, resulting in a higher surface pressure, than did the higher molecular weight gliadins. The surface pressure developed by both secalin and gliadin decreased with decreasing pH. Different rye milling fractions showed differences in surface behavior in relation to their protein content. The milling fraction with the highest protein content spread fastest and reached the highest surface pressure value.

RYE PROCESSING

While rye processing has gained interest because of its perceived positive health effects (Poutanen 1997), the functional properties of the rye major constituents limit the processing performance of rye in industrial processes (Weipert 1997). Milling and bread making are the most common processes in rye grain processing for human consumption.

Rye Milling

The objective of milling is to separate the bran and the germ from the starchy endosperm and to reduce the starchy endosperm in size, in order to obtain a refined product, that is, flour. Rye milling differs somewhat from wheat milling. Rye kernels are smaller than wheat kernels, and the separation of the endosperm from the bran is poorer in rye than in wheat (Nyman et al. 1984). The extraction rate of rye flour is lower than that of wheat flour, and at any given extraction rate, the ash content of rye flour is always higher than that of wheat flour (Weipert 1997).

Cleaning and Tempering of Rye

Prior to milling, rye grain is cleaned to remove stones, magnetic objects, broken rye kernels, cockle, oats, and ergot. Cleaning is based on a separation between rye and the impurities on the basis of size, density, and shape.

The second step involves tempering of the rye grain by adding a required quantity of moisture to bring it to an optimum condition for milling. Rye is, in most cases, milled at a moisture level of 15%, (which is 1% less than the preferred moisture level for wheat) and, in general, is tempered for a shorter time (2–4 hours with an absolute maximum of 6 hours) than wheat. This is due to the fact that water penetrates the rye kernel faster than the wheat kernel due to rye's weaker cell structure. The tempering process can be eliminated to reduce bacterial counts. However, if rye is milled too dry, the risk of shredding the bran into smaller particle sizes and thus contaminating the rye flour exists.

After tempering, a second cleaning step removes surface dirt, loose hairs, and outer bran layers. Just before milling, an additional 0.5% moisture is added (Zwingelberg and Sarkar 2001).

Milling of the Rye

In Europe, "roller" milling and "pin" milling are the most important rye milling processes. In roller milling, each passage through rolls involves particle size reduction by pressure and shear forces, followed by flour separation by particle size using sieves. Rolls are matched to the product needed. Their size, surface flutes, rotation velocity, and the gap between pairs of rolls rotating in opposite directions at dissimilar speeds, all can be selected or adjusted. Typically, milling starts with prebreak or flatting rolls, followed by six to seven break rolls, six bran finishers, and six reduction rolls. All rolls run at a differential speed of 3:1, except the flatting rolls (1:1 ratio). The first four break and reduction rolls are run dull to dull and the following breaks, sharp to sharp. A complex sieving system with plan sifters and centrifugal sifters is used. In this process, rye is fractionated into different milling streams, a number of which are combined to give the final flour. When practically all streams are used, an extraction rate of approximately 100% is reached and whole meal rye flour is obtained. When an increasing number of the fractions containing outer layer materials are left out of the reconstitution process, extraction rates decrease. For straight-run flour, a typical extraction rate is 58–68%.

In pin milling, rye is milled in impact mills, which work by the principle of impact disintegration. As this process requires a moisture content of 17–18%, rye is tempered a second time. In impact milling, five to seven milling steps are used. The extraction rate, ash content, and flour quality are similar for impact mills and roller mills. The starch damage, however, is lower in rye flour from impact mills than in that from roller mills because of the number of milling steps required.

In North America, the main objective in rye milling is to produce as much flour of the desired granulation as possible. The ash content and color of rye flours are not as critical as for common wheat flours. Therefore, the flow sheet of a rye mill is simple relative to that of a common wheat mill. A typical rye mill contains three to five break rolls, up to two sizings, and four to six reduction rolls. All grinding rolls are corrugated and run at a differential rate of 3:1. To properly remove the bran, bran dusters precede the last two or three break passages rather than coming just at the end. Because rye flour is sticky and difficult to sift, generous sifting surfaces are required (Zwingelberg and Sarkar 2001).

Impact of Rye Constituents on Rye Milling

Rye contains higher levels of arabinoxylans than wheat (Aman et al. 1997) and is accordingly more difficult to mill. Because the arabinoxylans are hygroscopic, the flour absorbs ambient humidity and tends to clump. This interferes with sifting. For this reason, rye is milled at a lower moisture content and requires more sifting surface than wheat (Zwingelberg and Sarkar 2001).

Protein content also influences the performance of rye in the mill. When it exceeds 12%, rye becomes more difficult to grind and has a lower extraction rate (Weipert 1993a).

Rye Bread Making

The bread-making behavior and performance of rye differ considerably from those of wheat (Weipert 1997). The principal reason for this is that rye proteins cannot form a gluten network, which lies at the basis of wheat bread quality. In addition, rye starch gelatinizes at a lower temperature than wheat starch and is therefore more prone to enzymic degradation during the oven phase than wheat starch. The main differences between rye and wheat breads are observed in terms of volume yield, crumb texture, and shelf life. Baking volume of rye breads is normally only about half that of wheat. However, rye bread has a longer shelf life and is richer in taste and aroma (Weipert 1997).

A wide assortment of rye-containing breads exists. They vary in size, shape, formulation, bread-making process, and sensory properties. Rye flour and whole meal rye can be incorporated into rye bread (minimum 90% rye), mixed rye/wheat bread (minimum 50% rye), or mixed wheat/rye bread (minimum 50% wheat). Other ingredients can be crushed rye grains, malted and crushed rye grains, malted rye kernels, or precooked rye kernels (Kujala 2004). Examples of rye bread types are pumpernickel (containing whole kernels), soft sourdough bread, crisp bread, rolls, and buns.

Rye Bread-Making Process

For rye bread production, acidification is a prerequisite to obtain a good quality product. The traditional method of preparing rye bread uses sourdough, although sometimes, direct acidification by lactic acid or other organic acids is used.

Preparation of Sourdough Sourdough is classically made by mixing rye flour or whole meal rye with water, yeast, and a starter and allowing the mixture to ferment (Kariluoto et al.

2004, Meuser et al. 1994, Seibel and Brümmer 1991). As starter, a commercial culture containing lactic acid bacteria or a portion of previous sourdough, equally containing lactic acid bacteria, is used (Seibel and Brümmer 1991). In homofermentative lactic acid fermentation, the lactic acid bacteria ferment glucose mainly into lactic acid; in heterofermentative lactic acid fermentation, considerable levels of acetic acid are formed in addition to lactic acid. The formation of lactic acid and acetic acid during lactic acid fermentation can and must be controlled, as the ratio of lactic acid to acetic acid is quite important for optimal bread quality. Sourdoughs with a high lactic acid content are used mainly for rye and mixed rye breads, while sourdoughs with a high acetic acid content are ideally suited for mixed wheat bread (Seibel and Brümmer 1991). One-, two-, and three-stage sourdough processes are used (Meuser et al. 1994, Seibel and Brümmer 1991). These processes require different fermentation times, dough temperatures, and water to flour ratios, and the resulting doughs have different lactic acid to acetic acid ratios (Seibel and Brümmer 1991). Depending on the proportion of rye flour in the sourdough, various degrees of acidity are required (Seibel and Brümmer 1991).

There are different reasons for using sourdough. Acid conditions have a positive influence on the swelling of rye flour constituents and control the enzyme activity in the dough, thereby preventing early staling (Meuser et al. 1994, Seibel and Brümmer 1991). Rye flours with high enzyme activity require a higher degree of acidification than flours with less enzyme activity (Meuser et al. 1994). Bread sensory properties, such as taste and aroma, can be improved by optimal use of sourdough (Heiniö et al. 2003a, Meuser et al. 1994, Seibel and Brümmer 1991). It contributes to aroma in the bread due to the production of acids, alcohols, and other volatile compounds (Heiniö et al. 2003a, Poutanen 1997). Sourdough bread has good crumb characteristics (Seibel and Brümmer 1991) and nutritional properties (Kariluoto et al. 2004, Meuser et al. 1994, Poutanen 1997). Acidification of rye dough also protects against spoilage by mold growth, thereby improving product shelf life (Meuser et al. 1994, Poutanen 1997, Seibel and Brümmer 1991). When sourdough is used in wheat bread making, its main function is to improve sensory properties and to prolong shelf life (Brümmer and Lorenz 1991).

Mixing and Kneading Rye flour, yeast, salt, and water are optimally mixed and kneaded with sourdough, lactic acid, or lactic and acetic acid to form a viscoelastic dough. The level of water is that yielding the required dough consistency. Rye doughs have to be mixed slowly because of their high arabinoxylan content, which leads to tough doughs when the energy input is excessive. The optimal mixing time depends on the composition and enzymic activities of the rye flour (Seibel and Weipert 2001a).

Important dough quality characteristics are dough yield, that is, the amount of optimally developed dough obtained from 100 g flour, and dough structure.

Fermentation Mixing and kneading of the dough is followed by a floor time. Thereafter, the dough is divided into pieces, which are molded by hand into the desired shape. The doughs are then proofed to their best potential. Fermentation takes place at 30–36°C and 70–85% relative humidity. Fermentation time depends on the composition of the flour, the activity of the flour enzymes, the dough formulation, and the fermentation temperatures.

During fermentation, carbon dioxide is produced, and the gas is retained by the dough, resulting in a volume increase of the dough. Although the rate of gas production in rye flour dough is high, the gas-retaining capacity is low, resulting in dense, compact loaves (He and Hoseney 1991).

Baking After fermentation, the dough is placed in an oven and transformed into bread. Bread quality depends on baking temperature, uniformity of temperature in the oven, and baking time. Optimum baking temperature and time depend on the weight of the dough pieces and the shape of the loaves, which determine the rate of heat transfer. In rye bread making, at the beginning of the baking stage, steam is often used in the oven to develop good crust characteristics (Seibel and Weipert 2001a).

Important bread quality characteristics are specific volume, crumb structure, crumb elasticity, crumb firmness during storage; and sensory properties such as taste, flavor, and odor.

Impact of Rye Constituents on Dough and Bread Quality Characteristics

Different rye varieties show highly different bread-making quality as influenced by harvest year and genotype (Hansen et al. 2004, Nilsson et al. 1997b, Rattunde et al. 1994). This can be attributed to significant differences in the content and structure of the major rye constituents, which are influenced by both factors (Hansen et al. 2003).

Starch Starch gelatinization and α-amylase activity are very important parameters for the bread-making quality of rye flour. Both properties can, for example, be measured by the falling number method or the amylograph test. The falling number is inversely related to α-amylase activity. An important parameter in the amylograph test is amylograph peak viscosity, which is correlated positively with starch content and negatively with α-amylase activity. It should be noted, however, that arabinoxylans can influence falling number and amylograph characteristics of rye flours. At similar α-amylase activity levels, higher falling numbers (Weipert 1993a, 1993b, 1994) and higher amylograph peak viscosities (Nilsson et al. 1997b) can be measured due to higher arabinoxylan contents. The arabinoxylans increase the viscosity, but also protect the starch granules against enzymic breakdown by hindering and retarding the activity of α-amylase (Weipert 1993b, 1994).

For rye bread-making suitability, a minimum starch gelatinization temperature of 63°C and, under specified experimental conditions, a minimum amylograph peak viscosity of 200 amylograph units (AU) are required. Rye shows optimal bread-making quality at a starch gelatinization temperature of 65–69°C and an amylograph peak viscosity of 400–600 AU (Weipert 1993a). Maximum bread-making quality is associated with intermediate values of falling number (Rattunde et al. 1994). Rye flours with high α-amylase activity give a soft, sticky dough, which may be too fluid to process into bread or results in a doughlike crumb. Therefore, preharvest sprouting, which goes hand in hand with high α-amylase activity, is disadvantageous for rye flour quality. Rye flours with low α-amylase activity yield a rigid, stable dough but lead to breads of low volume having a dry and firm crumb, resulting in a short shelf life (Autio and Salmenkallio-Marttila 2001, Fabritius et al. 1997). However, when rye flours with low α-amylase activity contain high levels of total and water-extractable arabinoxylans, better dough and bread characteristics can be observed (Weipert 1993a). It is believed that optimum bread-making performance is reached at an optimal arabinoxylan to starch ratio of 1:16 (Seibel and Brümmer 1991). This ratio provides balanced water-binding capacity and water distribution in the dough and the bread crumb. A shift in favor of more arabinoxylan causes stiffening of the dough (Kühn and Grosch 1989), due to the high water-holding capacity of the arabinoxylans. It also results in breads of greater specific volume and a firmer crumb (Kühn and Grosch 1989). A shift in favor of starch results in a low dough yield, soft dough, a flat bread form, a split and tight crumb, and a low shelf life (Vinkx and Delcour 1996).

Arabinoxylans Although many studies have reported conflicting results concerning the functional role of arabinoxylans in the bread-making process and during storage of breads, they are of considerable importance for rye bread-making quality.

Rye cultivars with high total arabinoxylan content, particularly when they also have a high water-extractable arabinoxylan fraction, perform much better in bread making than those with low arabinoxylan content. This is due to the high water-holding capacity of the arabinoxylans, which results in a delay in starch gelatinization and protects starch from α-amylase degradation, resulting in a higher volume yield, a better crumb elasticity, and a longer bread shelf life (Weipert 1993a, 1994, 1997). Small differences in arabinoxylan content cause differences in water absorption and significant differences in bread-making performance (Hansen et al. 2004, Weipert 1997).

Besides the content of arabinoxylans, their structure also significantly influences rye bread-making performance. Differences in arabinoxylan structure, such as degree of branching and ferulic acid cross-linking, as well as molecular size, may have influence on water absorption (Hansen et al. 2004). A high degree of branching and a high degree of ferulic acid cross-linking of arabinoxylans are related to low water absorption in rye flour (Hansen et al. 2004).

The dough and bread-making properties of rye flour also depend on the type of arabinoxylans present. In general, dough consistency is negatively related to dough water content. Because of their high (and similar) water-holding capacities, both water-extractable and water-unextractable arabinoxylans affect dough consistency (Kühn and Grosch 1989). However, the water-extractable and -unextractable arabinoxylans affect the bread-making properties of rye flour differently. The water-extractable arabinoxylans increase bread specific volume and decrease crumb firmness, whereas the water-unextractable arabinoxylans decrease bread specific volume and form ratio and slightly increase crumb firmness (Kühn and Grosch 1989). Increased water-extractable to water-unextractable arabinoxylan ratios result in flatter breads with higher specific volume, softer crumb, and darker crust (Cyran and Cygankiewicz 2004, Kühn and Grosch 1989).

Enzymic breakdown of the water-unextractable arabinoxylans by endoxylanases to solubilized arabinoxylans with medium to high molecular weight and with limited depolymerization of the water-extractable fraction has been reported to improve both rye (Kühn and Grosch 1988) and wheat (Courtin et al. 1999) bread-making performance. The increase in the level of high molecular weight arabinoxylans in the dough aqueous phase results in an improved water distribution and facilitation of molecular interactions between the dough constituent macromolecules. The subsequent increase in viscosity of the dough aqueous phase is likely to play a positive role in dough rheology and gas retention (Courtin et al. 1999, Rouau et al. 1994). Kühn and Grosch (1988) showed an improved crumb structure, decreased crumb firmness, increased width to height ratio, and increased loaf volume of rye bread after enzymic treatment of the water-unextractable arabinoxylans. Shelf life is also increased, resulting from an initial decrease in crumb firmness (Kühn and Grosch 1988). The endogenous arabinoxylan-degrading enzymes of rye might be active under the conditions prevailing in rye dough (Hansen et al. 2002, Rasmussen et al. 2001) and in the first part of the oven phase, thus prior to enzyme inactivation (Rasmussen et al. 2001). Endogenous enzymes in flour may thus affect crumb properties such as firmness, elasticity, and stickiness (Nilsson et al. 1997b).

According to Figueroa-Espinoza et al. (2004), the enzymic release of high molecular weight arabinoxylans may be positive due to the capacity of the latter to gel and form a polysaccharide network in the dough, thereby increasing water absorption, gas retention, bread volume, and shelf life.

Proteins Proteins are only of secondary importance for rye bread-making quality because of their inability to form a gluten network (Gellrich et al. 2003, 2004b; Kipp et al. 1996; Köhler and Wieser 2000). However, they should not be neglected, as they seem to be important at the dough-mixing step, at least for certain cultivars (Parkkonen et al. 1994). Indeed, although rye proteins do not form a regular gluten network, they have some ability to aggregate (Field et al. 1983) and are surface active (Wannerberger et al. 1997).

Sensory Properties of Rye Bread

Rye bread has a typical strong and slightly bitter flavor. The most dominant sensory attribute of rye bread is its rye-like flavor, but perceptions of sourness and saltiness are also substantially associated with rye bread (Heiniö et al. 1997). The aroma results from a mixture of many different volatile compounds formed by enzymic reactions during fermentation and thermal reactions during baking (Heiniö et al. 2003a, 2003b; Kirchhoff and Schieberle 2001). Besides the generation of aromas during these steps, the flour itself is an important source of dough aroma (Kirchhoff and Schieberle 2002). The specific rye-like flavor can be slightly modified to have a milder taste, without decreasing the high dietary fiber content and other positive health effects of rye, for example, by using the processing techniques of sourdough fermentation, germination prior to use, and extrusion cooking (Heiniö et al. 2003a), or by milling fractionation (Heiniö et al. 2003b).

Rye in Other Food Products

In addition to bread, many other food products such as ginger cake, porridge, pastries, breakfeast cereals, and pasta products are based on rye (Kujala 2004, Poutanen 1997). Rye can also be fermented to produce alcoholic beverages (Seibel and Weipert 2001a).

Rye in Feed

Although the majority of rye is fed to animals, rye is considered to be of inferior quality to other feed grains. The soluble, highly viscous rye arabinoxylans cause reductions in both the rate of digestion and the retention of nutrients in the gastrointestinal tract (Boros et al. 2002, Fengler and Marquardt 1988b). Therefore, rye is generally used in small proportions in mixtures with other grains. The use of enzymes (Boros and Bedford 1999, Boros et al. 2002, He et al. 2003, Lazaro et al. 2003), however, has recently substantially increased the proportion of rye included in animal feed.

Rye Fractionation

Wheat is an important raw material in the industrial separation of gluten and starch because of the abundant applications for gluten and starch in both the food and nonfood sectors. Rye would be an interesting alternative raw material because of the specific properties of the rye constituents. However, in spite of the development of some methods on the laboratory scale to separate starch from the arabinoxylans and proteins in rye (Schierbaum et al. 1991), no such industrial process exists for rye, presumably because of its relatively high content of arabinoxylans with high water-holding capacity and the inability of the rye proteins to form gluten.

Rye in Industrial Uses

Rye and rye products are also used in industrial applications such as production of adhesives and glue, film coating, sludge and oil-well drilling materials, and textiles and paper (Seibel and Weipert 2001b).

RYE AND NUTRITION

Rye products, especially when they contain whole meal rye flour, provide considerable levels of dietary fiber, vitamins, minerals, and phytoestrogens, all of which are considered to have positive health effects (Poutanen 1997). High rye consumption leads to positive effects on digestion and decreased risk of heart disease, hypercholesterolemia, obesity, and non-inulin-dependent diabetes (Hallmans et al. 2003, Smith et al. 2003). It probably also protects against some hormone-dependent cancer types (Hallmans et al. 2003, Smith et al. 2003). Liukkonen et al. (2003) showed that many of the bioactive compounds in whole meal rye are stable during processing and that their levels can even be increased through suitable processing (e.g., sourdough bread making). Rye milling, however, is accompanied by losses in dietary fiber and other phytochemicals, since these compounds are concentrated in the outer layers of the rye kernel (Glitso and Bach Knudsen 1999, Liukkonen et al. 2003, Nilsson et al. 1997a). In overall nutritional value, rye has some advantages over wheat because of the higher content of dietary fiber, vitamins, minerals, phytoestrogens, and lysine in rye.

REFERENCES

Aman P, Bengtsson S. 1991. Periodate oxidation and degradation studies on the major water-soluble arabinoxylan in rye grain. *Carbohydrate Polymers* 15: 405–414.

Aman P, et al. 1997. Positive health effects of rye. *Cereal Foods World* 42: 684–688.

Andreasen MF, et al. 2000a. Ferulic acid dehydrodimers in rye (*Secale cereale* L.). *Journal of Cereal Science* 31: 303–307.

Andreasen MF, et al. 2000b. Content of phenolic acids and ferulic acid dehydrodimers in 17 rye (*Secale cereale* L.) varieties. *Journal of Agricultural and Food Chemistry* 48: 2837–2842.

Andreev NR, et al. 1999. The influence of heating rate and annealing on the melting thermodynamic parameters of some cereal starches in excess water. *Starch* 51: 422–429.

Anger H, et al. 1986. Untersuchungen zur Molmasse und Grenzviskositat von Arabinoxylan (Pentosan) aus Roggen (*Secale cereale*) zur Aufstellung der Mark-Houwink-Beziehung. *Die Nahrung* 30: 205–208.

Autio K, Salmenkallio-Martilla M. 2001. Light microscopic investigations of cereal grains, doughs and breads. *Lebensmittel Wissenschaft und Technologie* 34: 18–22.

Autio K, et al. 2001. Structural and enzymic changes in germinated barley and rye. *Journal of the Institute of Brewing* 107: 19–25.

Ball S, et al. 1996. From glycogen to amylopectin: a model for the biogenesis of the plant starch granule. *Cell* 86: 349–352.

Bengtsson S, Aman P. 1990. Isolation and chemical characterization of water-soluble arabinoxylans in rye grain. *Carbohydrate Polymers* 12: 267–277.

Bengtsson S, et al. 1992a. Structural studies on water-soluble arabinoxylans in rye grain using enzymatic hydrolysis. *Carbohydrate Polymers* 17: 277–284.

Bengtsson S, et al. 1992b. Content, structure and viscosity of soluble arabinoxylans in rye grain from several countries. *Journal of Scientific Agriculture* 58: 331–337.

Berry CP, et al. 1971. The characterization of triticale starch and its comparison with starches of rye, durum and HRS wheat. *Cereal Chemistry* 48: 415–427.

Bleukx W, et al. 1998. Purification, properties and N-terminal amino acid sequence of a wheat gluten aspartic proteinase. *Journal of Cereal Science* 28: 223–232.

Boisen S. 1983. Comparative physico-chemical studies on purified trypsin inhibitors from the endosperm of barley, rye and wheat. *Zeitschrift fur Lebensmittel-Untersuchung und -Forschung* 176: 434–439.

Boros D, et al. 1993. Extract viscosity as an indirect assay for water-soluble pentosan content in rye. *Cereal Chemistry* 70: 575–580.

Boros D, Bedford MR. 1999. Influence of water extract viscosity and exogenous enzymes on nutritive value of rye hybrids in broiler diets. *Journal of Animal and Feed Sciences* 8: 579–588.

Boros D, et al. 2002. Chick adaptation to diets based on milling fractions of rye varying in arabinoxylans content. *Animal Feed Science and Technology* 101: 135–149.

Brijs K, et al. 1999. Proteolytic activities in dormant rye (*Secale cereale* L.) grain. *Journal of Agricultural and Food Chemistry* 47: 3572–3578.

Brijs K. 2001. Proteolytic enzymes in rye: isolation, characterization and specificities [Ph. D. dissertation]. Leuven, Belgium: Katholieke Universiteit Leuven, p. 141.

Brijs K, et al. 2002. Proteolytic enzymes in germinating rye grains. *Cereal Chemistry* 79: 423–428.

Brümmer JM, Lorenz K. 1991. European developments in wheat sourdoughs. *Cereal Foods World* 36: 310–314.

Caldéron de la Barca AM, et al. 2000. Enzymatic hydrolysis and synthesis of soy protein to improve its amino acid composition and functional properties. *Journal of Food Science* 65: 246–253.

Charbonnier L, et al. 1981. Rye prolamins: extractability, separation and characterization. *Journal of Agricultural and Food Chemistry* 29: 968–973.

Chen CH, Bushuk W. 1970. Nature of proteins in triticale and its parental species. I. Solubility characteristics and amino acid composition of endosperm proteins. *Canadian Journal of Plant Science* 50: 9–14.

Courtin CM, et al. 1999. Fractionation-reconstitution experiments provide insight into the role of endoxylanases in bread-making. *Journal of Agricultural and Food Chemistry* 47: 1870–1877.

Coutinho PM, Henrissat B. 1999. Carbohydrate-active enzymes server at URL: http://afmb.cnrs-mrs.fr/CAZY/.

Cyran M, et al. 2003. Structural features of arabinoxylans extracted with water at different temperatures from two rye flours of diverse breadmaking quality. *Journal of Agricultural and Food Chemistry* 51: 4404–4416.

Cyran M, et al. 2004. Heterogeneity in the fine structure of alkali-extractable arabinoxylans isolated from two rye flours with high and low breadmaking quality and their coexistence with other cell wall components. *Journal of Agricultural and Food Chemistry* 52: 2671–2680.

Cyran M, Cygankiewicz A. 2004. Variability in the content of water-extractable and water-unextractable non-starch polysaccharides in rye flour and their relationship to baking quality parameters. *Cereal Research Communications* 32: 143–150.

Dedio W, et al. 1975. Distribution of alpha-amylase in triticale kernel during development. *Canadian Journal of Plant Science* 55: 29–36.

Delcour JA, et al. 1991. Physico-chemical and functional properties of rye nonstarch polysaccharides. II. Impact of a fraction containing water-soluble pentosans and proteins on gluten-starch loaf volumes. *Cereal Chemistry* 68: 72–76.

Dembinski E, Bany S. 1991. The amino acid pool of high and low protein rye inbred lines (*Secale cereale* L.). *Journal of Plant Physiology* 138: 494–496.

Denli E, Ercan R. 2001. Effect of added pentosans isolated from wheat and rye grain on some properties of bread. *European Food Research and Technology* 212: 374–376.

Dervilly-Pinel G, et al. 2001. Water-extractable arabinoxylan from pearled flours of wheat, barley, rye and triticale. Evidence for the presence of ferulic acid dimers and their involvement in gel formation. *Journal of Cereal Science* 34: 207–214.

Dexter JE, Dronzek BL. 1975. Note on the amino acid composition of protein fractions from a developing triticale and its rye and durum wheat parents. *Cereal Chemistry* 52: 587–596.

Ebringerova A, et al. 1990. Structural features of a water-soluble L-arabino-D-xylan from rye bran. *Carbohydrate Research* 198: 57–66.

Elliott, GO, et al. 2003. A wheat xylanase inhibitor protein (XIP-I) accumulates in the grain and has homologues in other cereals. *Journal of Cereal Science* 37: 187–194.

Fabritius M, et al. 1997. Structural changes in insoluble cell walls in wholemeal rye doughs. *Lebensmittel Wissenschaft und Technologie* 30: 367–372.

FAOSTAT data. 2004.

Fengler AI, Marquardt RR. 1988a. Water-soluble pentosans from rye: I. Isolation, partial purification and characterization. *Cereal Chemistry* 65: 291–297.

Fengler AI, Marquardt RR. 1988b. Water-soluble pentosans from rye: II. Effects on rate of dialysis and on the retention of nutrients by the chick. *Cereal Chemistry* 65: 298–302.

Field JM, et al. 1982. The purification and characterization of homologous high molecular weight storage proteins from grain of wheat, rye and barley. *Theoretical and Applied Genetics* 62: 329–336.

Field JM, et al. 1983. Aggregation states of alcohol-soluble storage proteins of barley, rye, wheat and maize. *Journal of the Science of Food and Agriculture* 34: 362–369.

Figueroa-Espinoza MC, et al. 2002. Enzymatic solubilization of arabinoxylans from isolated rye pentosans and rye flour by different endo-xylanases and other hydrolyzing enzymes. Effect of a fungal laccase on the flour extracts oxidative gelation. *Journal of Agricultural and Food Chemistry* 50: 6473–6484.

Figueroa-Espinoza MC, et al. 2004. Enzymatic solubilization of arabinoxylans from native, extruded, and high-shear-treated rye bran by different endo-xylanases and other hydrolyzing enzymes. *Journal of Agricultural and Food Chemistry* 52: 4240–4249.

Fowler DB, et al. 1990. Environment and genotype influence on grain protein concentration of wheat and rye. *Agronomy Journal* 82: 655–664.

Fredriksson H, et al. 1998. The influence of amylose and amylopectin characteristics on gelatinization and retrogradation properties of different starches. *Carbohydrate Polymers* 35: 119–134.

Gabor R, et al. 1991. Studies on the germination specific alpha-amylase and its inhibitor of rye (*Secale cereale*). 1. Isolation and characterization of the enzyme. *Zeitschrift fur Lebensmittel-Untersuchung und -Forschung* 192: 230–233.

Gallant DJ, et al. 1997. Microscopy of starch: evidence of a new level of granule organization. *Carbohydrate Polymers* 32: 177–191.

Garcia-Olmedo F, et al. 1987. Plant proteinaceous inhibitors of proteinases and a-amylases. *Oxford Surveys of Plant Molecular and Cell Biology* 4: 275–334.

Gebruers K, et al. 2001. *Triticum aestivum* L. endoxylanase inhibitor (TAXI) consists of two inhibitors, TAXI I and TAXI II, with different specificities. *Biochemical Journal* 353: 239–244.

Gebruers K, et al. 2002. Affinity chromatography with immobilised endoxylanases separates TAXI- and XIP-type endoxylanase inhibitors from wheat (*Triticum aestivum* L.). *Journal of Cereal Science* 36: 367–375.

Gellrich C, et al. 2001. Isolierung und Charakterisierung der γ-Secaline von Roggen. *Getreide Mehl und Brot* 55: 275–277.

Gellrich C, et al. 2003. Biochemical characterization and quantification of the storage protein (secalin) types in rye flour. *Cereal Chemistry* 80: 102–109.

Gellrich C, et al. 2004a. Biochemical characterization of γ-75k secalins of rye. I. Amino acid sequences. *Cereal Chemistry* 81: 290–295.

Gellrich C, et al. 2004b. Biochemical characterization of γ-75k secalins of rye. II. Disulfide bonds. *Cereal Chemistry* 81: 296–299.

Gernat C, et al. 1993. Crystalline parts of three different conformations detected in native and enzymatically degraded starches. *Starch* 45: 309–314.

Girhammar U, Nair BM. 1992a. Isolation, separation and characterization of water soluble non-starch polysaccharides from wheat and rye. *Food Hydrocolloids* 6: 285–299.

Girhammar U, Nair BM. 1992b. Certain physical properties of water soluble non-starch polysaccharides from wheat, rye, triticale, barley and oats. *Food Hydrocolloids* 6: 329–343.

Glitso LV, Bach Knudsen KE. 1999. Milling of whole grain rye to obtain fractions with different dietary fibre characteristics. *Journal of Cereal Science* 29: 89–97.

Goesaert H, et al. 2002. A family of 'TAXI'-like endoxylanase inhibitors in rye. *Journal of Cereal Science* 36: 177–185.

Goesaert H, et al. 2003. XIP-type endoxylanase inhibitors in different cereals. *Journal of Cereal Science* 38: 317–324.

Gudmundsson M, Eliasson AC. 1991. Thermal and viscous properties of rye starch extracted from different varieties. *Cereal Chemistry* 68: 172–177.

Gudmundsson M, et al. 1991. The effects of water soluble arabinoxylan on gelatinization and retrogradation of starch. *Starch* 43: 5–10.

Hallmans G, et al. 2003. Rye, lignans and human health. *Proceedings of the Nutrition Society* 62: 193–199.

Hansen HB, et al. 2002. Changes in dietary fibre, phenolic acids and activity of endogenous enzymes during rye bread-making. *European Food Research and Technology* 214: 33–42.

Hansen HB, et al. 2003. Effects of genotype and harvest year on content and composition of dietary fibre in rye (*Secale cereale* L.) grain. *Journal of the Science of Food and Agriculture* 83: 76–85.

Hansen HB, et al. 2004. Grain characteristics, chemical composition and functional properties of rye (*Secale cereale* L.) as influenced by genotype and harvest year. *Journal of Agricultural and Food Chemistry* 52: 2282–2291.

Härkönen H, et al. 1995. The effects of a xylanase and a b-glucanase from Trichoderma reesei on the non-starch polysaccharides of whole meal rye slurry. *Journal of Cereal Science* 21: 173–183.

Härkönen H, et al. 1997. Distribution and some properties of cell wall polysaccharides in rye milling fractions. *Journal of Cereal Science* 26: 95–104.

He H, Hoseney RC. 1991. Gas retention of different cereal flours. *Cereal Chemistry* 68: 334–336.

He T, et al. 2003. Performance of broiler chicks fed normal and low viscosity rye or barley with or without enzyme supplementation. *Asian Australian Journal of Animal Sciences* 16: 234–238.

Heiniö RL, et al. 1997. Identity and overall acceptance of two types of sour rye bread. *International Journal of Food Science and Technology* 32: 169–178.

Heiniö RL, et al. 2003a. Relationship between sensory perception and flavour-active volatile compounds of germinated, sourdough fermented and native rye following the extrusion process. *Lebensmittel Wissenschaft und Technologie* 36: 533–545.

Heiniö RL, et al. 2003b. Milling fractionation of rye produces different sensory profiles of both flour and bread. *Lebensmittel Wissenschaft und Technologie* 36: 577–583.

Hejgaard J. 2001. Inhibitory serpins from rye grain with glutamine as P_1 and P_2 residues in the reactive center. *FEBS Letters* 488: 149–153.

Henrissat B. 1991. A classification of glycosyl hydrolases based on amino acid sequence similarities. *Biochemical Journal* 280: 309–316.

Henry RJ. 1987. Pentosan and $(1\rightarrow3, 1\rightarrow4)$-$\beta$-glucan concentrations in endosperm and wholegrain of wheat, barley, oats and rye. *Journal of Cereal Science* 6: 253–258.

Hromadkova Z, et al. 1987. Structural features of a rye bran arabinoxylan with a low degree of branching. *Carbohydrate Research* 163: 73–79.

Jahn-Deesbach W, Schipper A. 1980. Protein-Fraktionen und Aminosäuren in ungekeimten und gekeimten Körnern von Weizen, Gerste, Roggen und Hafer. *Getreide Mehl und Brot* 34: 281–287.

Kariluoto S, et al. 2004. Effect of baking method and fermentation on folate content of rye and wheat breads. *Cereal Chemistry* 81: 134–139.

Karppinen S, et al. 2003. Fructan content of rye and rye products. *Cereal Chemistry* 80: 168–171.

Keetels CJAM, et al. 1996. Recrystallization of amylopectin in concentrated starch gels. *Carbohydrate Polymers* 30: 61–64.

Kipp B, et al. 1996. Comparative studies of high M_r subunits of rye and wheat. I. Isolation and biochemical characterization and effects on gluten extensibility. *Journal of Cereal Science* 23: 227–234.

Kirchhoff E, Schieberle P. 2001. Determination of key aroma compounds in the crumb of a three-stage sourdough rye bread by stable isotope dilution assays and sensory studies. *Journal of Agricultural and Food Chemistry* 49: 4304–4311.

Kirchhoff E, Schieberle P. 2002. Quantitation of odor-active compounds in rye flour and rye sourdough using stable isotope dilution assays. *Journal of Agricultural and Food Chemistry* 50: 5378–5385.

Klassen AJ, Hill RD. 1971. Comparison of starch from triticale and its parental species. *Cereal Chemistry* 48, 647–654.

Köhler P, Wieser H. 2000. Nachweis von Cysteinresten in hochmolekularen Glutelin-Untereinheiten aus Roggen. *Getreide Mehl und Brot* 54: 283–289.

Kreis M, et al. 1985. Molecular evolution of the seed storage proteins of barley, rye and wheat. *Journal of Molecular Biology* 183: 499–502.

Kühn MC, Grosch W. 1988. Influence of the enzymic modification of the nonstarchy polysaccharide fractions on the baking properties of reconstituted rye flour. *Journal of Food Science* 53: 889–895.

Kühn MC, Grosch W. 1989. Baking functionality of reconstituted rye flours having different nonstarchy polysaccharide and starch contents. *Cereal Chemistry* 66: 149–154.

Kujala T. 2004. Rye and health at URL: http://rye.vtt.fi/.

Lazaro R, et al. 2003. Influence of enzymes on performance and digestive parameters of broilers fed rye-based diets. *Poultry Science* 82: 132–140.

Lii CY, Lineback DR. 1977. Characterization and comparison of cereal starches. *Cereal Chemistry* 54: 138–149.

Liukkonen KH, et al. 2003. Process-induced changes on bioactive compounds in whole grain rye. *Proceedings of the Nutrition Society* 62: 117–122.

Ludwig I, Ludwig E. 1989. Untersuchungen zur Extraktion von Roggenproteinen mit grenzflächenaktiven Verbindungen. *Die Nahrung* 33: 761–765.

Lyons A, et al. 1987. Characterization of homologous inhibitors of trypsin and α-amylase from seeds of rye (*Secale cereale* L.). *Biochimica et Biophysica Acta* 915: 305–313.

MacGregor AW, et al. 1988. Multiple a-amylase components in germinated cereal grains determined by isoelectric focusing and chromatofocusing. *Cereal Chemistry* 65: 326–333.

Masojc P, Larssonraznikiewicz M. 1991a. Variations of the levels of alpha-amylase and endogenous alpha-amylase inhibitor in rye and triticale grain. *Swedish Journal of Agricultural Research* 21: 3–9.

Masojc P, Larssonraznikiewicz M. 1991b. Genetic variation of alpha-amylase levels among rye (*Secale cereale* L.) kernels tested by gel-diffusion technique. *Swedish Journal of Agricultural Research* 21: 141–145.

Meuser F, et al. 1986. Chemisch-physikalische Charakterisierung von Roggenpentosanen. *Getreide Mehl und Brot* 40: 198–204.

Meuser F, et al. 1994. Bread varieties in Central Europe. *Cereal Foods World* 39: 222–230.

Meuser F, et al. 2001. Foam-forming capacity of substances present in rye. *Cereal Chemistry* 78: 50–54.

Michniewicz J, et al. 1992. Effect of added pentosans on some properties of wheat bread. *Food Chemistry* 43: 251–257.

Mohammadkhani A, et al. 1998. Survey of amylose content in *Secale cereale, Triticum monococcum, T. turgidum and T. tauschii. Journal of Cereal Science* 28: 273–280.

Morey DD, Evans JJ. 1983. Amino acid composition of six grains and winter wheat forage. *Cereal Chemistry* 60: 461–464.

Morrison WR. 1995. Starch lipids and how they relate to starch granule structure and functionality. *Cereal Foods World* 40: 437–445.

Mosolov VV, Shul'gin MN. 1986. Protein inhibitors of microbial proteinases from wheat, rye and triticale. *Planta* 167: 595–600.

Nilsson M, et al. 1996. Water unextractable polysaccharides from three milling fractions of rye grain. *Carbohydrate Polymers* 30: 229–237.

Nilsson M, et al. 1997a. Content of nutrients and lignans in roller milled fractions of rye. *Journal of the Science of Food and Agriculture* 73: 143–148.

Nilsson M, et al. 1997b. Nutrient and lignan content, dough properties and baking performance of rye samples used in Scandinavia. *Acta Agriculturae Scandinavica* 47: 26–34.

Nilsson M, et al. 1999. Arabinoxylan fractionation on DEAE-cellulose chromatography influenced by protease pre-treatment. *Carbohydrate Polymers* 39: 321–326.

Nilsson M, et al. 2000. Heterogeneity in a water-extractable rye arabinoxylan with a low degree of disubstitution. *Carbohydrate Polymers* 41: 397–405.

Nowak J, Mierzwinska T. 1978. Activity of proteolytic enzymes in rye seeds of different ages. *Zeitschrift fur Pflanzenphysiologie* 86: 15–22.

Nyman M, et al. 1984. Dietary fiber content and composition in six cereals at different extraction rates. *Cereal Chemistry* 61: 14–19.

Parkkonen T, et al. 1994. Effect of baking on the microstructure of rye cell walls and protein. *Cereal Chemistry* 71: 58–63.

Popineau Y, et al. 2002. Foaming and emulsifying properties of fractions of gluten peptides obtained by limited enzymatic hydrolysis and ultrafiltration. *Journal of Cereal Science* 35: 327–335.

Poutanen K. 1997. Rye bread: added value in the world's bread basket. *Cereal Foods World* 42: 682–683.

Preston KR, Woodbury W. 1975. Amino acid composition and subunit structure of rye gliadin proteins fractionated by gel filtration. *Cereal Chemistry* 52: 719–726.

Radosta S, et al. 1992. Studies on rye starch properties and modification. Part II: Swelling and solubility behaviour of rye starch granules. *Starch* 44: 8–14.

Ragaee SM, et al. 2001. Studies on rye (*Secale cereale*) lines exhibiting a range of extract viscosities. 1. Composition, molecular weight distribution of water extracts, and biochemical characteristics of purified water-extractable arabinoxylan. *Journal of Agricultural and Food Chemistry* 49: 2437–2445.

Rasmussen CV, et al. 2001. pH-, temperature- and time-dependent activities of endogenous endo-β-D-xylanase, β-D-xylosidase and α-L-arabinofuranosidase in extracts from ungerminated rye (*Secale cereale* L.) grain. *Journal of Cereal Science* 34: 49–60.

Rattunde HF, et al. 1994. Variation and covariation of milling and baking quality characteristics among winter rye single-cross hybrids. *Plant Breeding* 113: 287–293.

Rouau X, et al. 1994. Effect of an enzyme preparation containing pentosanases on the bread-making quality of flours in relation to changes in pentosan properties. *Journal of Cereal Science* 19: 259–272.

Sarkkinen P, et al. 1992. Aspartic proteinase from barley grains is related to mammalian lysosomal cathepsin-D. *Planta* 186: 317–323.

Schierbaum F, et al. 1991. Studies on rye starch properties and modification. Part I: Composition and properties of rye starch granules. *Starch* 43: 331–339.

Schierbaum F, Kettlitz B. 1994. Studies on rye starch properties and modification. Part III: Viscograph pasting characteristics of rye starches. *Starch* 46: 2–8.

Seibel W, Brümmer JM. 1991. The sourdough process for bread in Germany. *Cereal Foods World* 36: 299–304.

Seibel W, Weipert D. 2001a. Bread baking and other food uses around the world. In: Busuk W, (ed.) *Rye: production, chemistry and technology*. 2nd ed. St. Paul, Minnesota, USA: American Association of Cereal Chemists, pp. 147–211.

Seibel W, Weipert D. 2001b. Animal feed and industrial uses. In: Busuk W, (ed.) *Rye: production, chemistry and technology*. 2nd ed. St. Paul, Minnesota, USA: American Association of Cereal Chemists, pp. 213–233.

Shewry PR, et al. 1982. The purification and characterization of two groups of storage proteins (secalins) from rye (*Secale cereale* L.). *Journal of Experimental Botany* 33: 261–268.

Shewry PR, et al. 1983. Extraction, separation and polymorphism of the prolamin storage proteins (secalins) of rye. *Cereal Chemistry* 60: 1–6.

Shewry PR, et al. 1988. N-terminal amino acid sequences show that D hordein of barley and high molecular weight (HMW) secalins of rye are homologous with HMW glutenin subunits of wheat. *Cereal Chemistry* 65: 510–511.

Shewry PR, Tatham AS. 1990. The prolamin storage proteins of cereal seeds: structure and evolution. *Biochemical Journal* 267: 1–12.

Shewry PR, et al. 1995. Seed storage proteins: structures and biosynthesis. *The Plant Cell* 7: 945–956.

Shewry PR, Bechtel DB. 2001. Morphology and chemistry of the rye grain. In: Busuk W, (ed.) *Rye: production, chemistry and technology*. 2nd ed. St. Paul, MN: American Association of Cereal Chemists, pp. 69–127.

Shih FF, Daigle K. 1997. Use of enzymes for the separation of protein from rice flour. *Cereal Chemistry* 74: 437–441.

Smith AT, et al. 2003. Behavioural, attitudinal and dietary responses to the consumption of wholegrain foods. *Proceedings of the Nutrition Society* 62: 455–467.

Smulikowska S, Nguyen VC. 2001. A note on variability of water extract viscosity of rye grain from north-east regions of Poland. *Journal of Animal and Feed Sciences* 10: 687–693.

Tatham AS, Shewry PR. 1991. Conformational analysis of the secalin storage proteins of rye (*Secale cereale* L.). *Journal of Cereal Science* 14: 15–23.

Tatham AS, Shewry PR. 1995. The S-poor prolamins of wheat, barley and rye. *Journal of Cereal Science* 22: 1–16.

Täufel A, et al. 1991. Studies on the germination specific α-amylase and its inhibitor of rye (*Secale cereale*). 2. Isolation and characterization of the inhibitor. *Zeitschrift fur Lebensmittel-Untersuchung und -Forschung* 193: 9–14.

Täufel A, et al. 1997. Protein inhibitors of alpha-amylase in mature and germinating grain of rye (*Secale cereale*). *Journal of Cereal Science* 25: 267–273.

Van den Bulck K, et al. 2005. Isolation of cereal arabinogalactan-peptides and structural comparison of their carbohydrate and peptide moieties. *Journal of Cereal Science* 41: 59–67.

Vanhamel S, et al. 1993. Physicochemical and functional properties of rye nonstarch polysaccharides. IV. The effect of high molecular weight water-soluble pentosans on wheat-bread quality in a straight-dough procedure. *Cereal Chemistry* 70: 306–311.

Verwimp T, et al. 2004. Isolation and characterization of rye starch. *Journal of Cereal Science* 39: 85–90.

Vinkx CJA, et al. 1991. Physicochemical and functional properties of rye nonstarch polysaccharides. III. Oxidative gelation of a fraction containing water-soluble pentosans and proteins. *Cereal Chemistry* 68: 617–622.

Vinkx CJA, et al. 1993. Physicochemical and functional properties of rye nonstarch polysaccharides. V. Variability in the structure of water-soluble arabinoxylans. *Cereal Chemistry* 70: 311–317.

Vinkx CJA. 1994. Structure and properties of arabinoxylans from rye [Ph. D. dissertation]. Leuven, Belgium: Katholieke Universiteit Leuven, p. 127.

Vinkx CJA, et al. 1995a. Rye water-soluble arabinoxylans also vary in their 2-monosubstituted xylose content. *Cereal Chemistry* 72: 227–228.

Vinkx CJA, et al. 1995b. Physicochemical and functional properties of rye nonstarch polysaccharides. VI. Variability in the structure of water-unextractable arabinoxylans. *Cereal Chemistry* 72: 411–418.

Vinkx CJA, Delcour JA. 1996. Rye (*Secale cereale*) arabinoxylans: a critical review. *Journal of Cereal Science* 23: 1–14.

Wannerberger L, et al. 1997. Interfacial behaviour of secalin and rye flour-milling streams in comparison with gliadin. *Journal of Cereal Science* 25: 243–252.

Wasserman LA, et al. 2001. The application of different physical approaches for the description of structural features in wheat and rye starches. A DSC study. *Starch* 53: 629–634.

Weipert D. 1993a. Verarbeitungswert von deutschen Roggensorten. *Getreide Mehl und Brot* 47(2): 6–12.

Weipert D. 1993b. Messung der Auswuchsfestigkeit bei Roggen. *Getreide Mehl und Brot* 47(4): 3–9.

Weipert D. 1994. Beschreibung des Verarbeitungswertes neuer Roggensorten. *Getreide Mehl und Brot* 4: 3–5.

Weipert D. 1997. Processing performance of rye as compared to wheat. *Cereal Foods World* 42: 706–712.

Wu WU, et al. 1998. Hydrophobicity, solubility and emulsifying properties of soy protein peptides prepared by papain modification and ultrafiltration. *Journal of the American Oil Chemists Society* 75: 845–850.

Zwingelberg H, Sarkar A. 2001. Milling of rye. In: Busuk W (ed.) *Rye: production, chemistry and technology*. 2nd ed. St. Paul, Minnesota, USA: American Association of Cereal Chemists, pp. 129–145.

Part 6
Health/Functional Foods

35
Biochemistry and Probiotics

Claude P. Champagne and Fatemeh Zare

Abstract: The level of viable cells is essential to the functionality of probiotic bacteria in foods. Many enzymatic activities have been identified to be critical in the ability of probiotics cultures to grow in food matrices, remain stable during storage, survive passage to the stomach and duodenum as well as to express activity in the gastro-intestinal tract (GIT). The importance of the following enzymatic activities on these functionalities will be addressed: α-galactosidase for growth in soy and pulses as well as to help prevent gas discomforts in the GIT, β-galactosidase for growth in milk as well as to reduce lactose maldigestion problems in the GIT, proteases for growth in milk and soy as well as to generate bioactive peptides, oxidases and peroxidase to prevent oxygen toxicity during fermentation and storage, proton-translocating ATPases for survival to acid environments in foods and the GIT, bile salt hydrolase for survival in the GIT and bioactivity toward cholesterol metabolism, β-glucosidases for the release of bioactives (isoflavones, quercetin) during fermentation.

INTRODUCTION

Probiotics are live microorganisms which, when administered in sufficient quantity, confer a health benefit to the host (Araya et al. 2002). Many health effects have been suggested: improved carbohydrate digestion in the gastrointestinal (GI) tract, reduction of the incidence of diarrhoea, immune system enhancement, blood cholesterol reduction (Shah 2000, Farnworth 2004). Official recognition of these effects by regulatory agencies is beginning (EFSA 2010). In light of these potential health benefits, digestive health is considered as one of the ten key food trends for 2010 (Mellentin 2010). As a result, the market for foods which contain probiotic bacteria is in constant increase and is predicted to reach US$30 billion by 2015 (Starling 2010).

From a technological and biochemical standpoint, this definition of probiotics implies two important features: (1) the cells must be viable, and (2) a certain quantity must be maintained. However, data suggest that, for some applications, nonviable cells can present biological activity as well (Ouwehand and Salminen 1998). The benefits of probiotics are increasingly linked to specific biogenic metabolites (Stanton et al. 2005) that may not necessarily require viable cells. In the current definition of probiotics, fermented foods that carry bioactive compounds but do not have viable cells are not considered to be as probiotics. If the functionality of the product is nevertheless directly linked to some activity of the probiotic culture, then the concept of "probioactive" may better apply to this health function. It has been proposed that a probioactive be "a bioactive compound influencing health that is synthesized by a probiotic culture or that specifically results from the bioconversion of a food matrix by a probiotic microorganism" (Farnworth and Champagne 2010). Exopolysaccharides, peptides, deconjugated metabolites

Food Biochemistry and Food Processing, Second Edition. Edited by Benjamin K. Simpson, Leo M.L. Nollet, Fidel Toldrá, Soottawat Benjakul, Gopinadhan Paliyath and Y.H. Hui.

(e.g., isoflavones), and enzymes are but a few of the many compounds have been suggested as being linked to functionality of probiotics. In this chapter, the potential contribution of enzymes produced by probiotic bacteria to their functionality will be examined. The critical contribution of enzymes will be addressed from both a technological perspective (ability to grow and remain viable in a food product) and a health perspective (survival to GI environments, effects on human functions).

BIOCHEMISTRY AND GROWTH OF PROBIOTIC CULTURES IN FOODS

Several different metabolic pathways explain the biosynthetic functions of microorganisms that could maintain their life. Each metabolic pathway consists of many reactions that are regulated by different enzyme systems, and so the level of enzymes and their activity sustain and control the functions of the microbial cell (Stanier et al. 1987). Breakdown of nutrients (carbohydrates, proteins, lipids, and other minor constituents) present in the growth medium results in the production of smaller molecules that are subsequently consumed for buildup and division of the microbial cells as well as energy requirements for survival.

In lactic acid bacteria (LAB) (i.e., the lactococci, leuconostoc, lactobacilli, streptococci, and bifidobacteria), energy is mainly supplied by the fermentation of carbohydrates (Lawrence et al. 1976). Hundreds of enzymes are required to sustain life in microorganisms. However, some are recognized as being more critical in affecting growth rates and biomass levels on food substrates. In general, the two most important biochemical reactions are carbohydrate assimilation and protein metabolism. This justifies the focus that will be made of these particular enzymatic activities.

MILK

In milk, "lactase" and "protease" activities of yogurt and probiotic cultures are critical to growth and acidification rates (Tamime and Robinson 1999).

Lactose is the main carbohydrate in milk and it can be metabolized either through the homo- or heterofermentative metabolic pathways. *Streptococcus thermophilus, Lactobacillus delbrueckii subsp. bulgaricus*, and *Lactobacillus acidophilus* ferment lactose homofermentatively, lactic acid being the main metabolite. *Bifidobacterium* spp., on the other hand, ferments the same sugar heterofermentatively, producing lactic and acetic acid typically at a 2:3 ratio. Lactose must enter the cytoplasm to be catabolized. A specific system is involved in lactose transport in the lactococci and certain strains of *L. acidophilus* and sugar is phosphorylated by phosphoenolpyruvate (PEP) during translocation by the PEP dependent phosphotransferase system (PTS) (Kanatani and Oshimura 1994, Marshall and Tamime 1997). This mechanism is known as PEP:PTS. β-Phosphogalactosidase hydrolyses lactose-6-phosphate to its monosaccharide components. The galactose and glucose thus released are then catabolized via Tagatose and Emden–Meyerhof–Parnas (EMP) pathways, respectively (Monnet et al. 1996, Marshall and Tamime 1997). Some organisms including *Bifidobacterium* spp. have an alternative system for lactose transport into the cells that involves cytoplasmic proteins (permeases). This is to translocate lactose without chemical modification and the mechanism could be similar to the lactose permease system in *Escherichia coli*. After lactose enters the cell, it is hydrolyzed by β-galactosidase (β-gal) to nonphosphorylated glucose and galactose. In many yogurt cultures, glucose is catabolized to pyruvate and the galactose is secreted from the cell. Less is known on the lactases of probiotic bacteria. Some bifidobacteria seem to have two types of β-gal and their level seem critical to rapid growth in milk. Indeed, Desjardins et al. (1991) showed that the growth performance of bifidobacteria in milk is linked to their β-gal levels. *L. acidophilus* or *Bifidobacterium* sp. generally grows faster on synthetic media than on milk (Misra and Kuila 1991, Gaudreau et al. 2005). Therefore, it is not surprising that the addition of β-gal to milk enhances the growth rate of *L. acidophilus* (Khattab et al. 1986). In an interesting variation of this concept, ruptured cells of a yogurt culture were added to milk in order to promote the development of *L. acidophilus* and *Bifidobacterium longum* in the fermented product (Shah et al. 1997). Presumably, the liberation of β-gal from the ruptured starter cells contributed to the stimulation. This was also noted for *Lactobacillus rhamnosus* (Gaudreau et al. 2005). However, the ability of a strain to synthesize enzymes essential in the assimilation of a carbohydrate provides an incomplete picture of the ability to grow rapidly on the food matrix. The activity of the enzymes will also influence growth rates. Thus, for a given strain, growth rates in identical media will vary as a function of the carbohydrate (Table 35.1). There are numerous observations of variable enzyme activity levels between strains (Desjardins et al. 1991, Donkor et al. 2007).

The protein fraction in milk is composed of casein and whey proteins and the basic constituents of a protein molecule are compounds from 21 amino acids (Tamime and Robinson 1999). Since LAB cannot synthesize many amino acids (Marshall and Law 1984), proteolytic activity is greatly involved in both nutrition and interactions between the two yogurt cultures as well as with probiotic bacteria. Amino acids and peptones are good supplements for bifidobacteria (Klaver et al. 1993). This would suggest that high proteolytic activities would enhance growth rates and levels. This is not always the limiting factor with

Table 35.1. Effect of the Nature of the Sugar on Growth Rate (h^{-1}) of Two Probiotic Cultures in a Laboratory Growth Medium (Peptones, Salts, Tween)

Carbohydrate	*Lactobacillus helveticus* R0052	*Bifidobacterium longum* R0175
Glucose	0.13	0.06
Stachyose	0	0.06
Lactose	0.15	0.07
Fructose	0.08	0
Sucrose	0.08	0
Raffinose	0	0.06

Source: Champagne et al. 2010.

bifidobacteria (Desjardins et al. 1991). Indeed, some *L. acidophilus* and *Bifidobacterium* cultures have proteolytic activities as high as those of yogurt starters (Shihata and Shah 2000), but still do not multiply as fast in milk. In some instances, mixing a nonproteolytic *Bifidobacterium* strain with a highly proteolytic *L. acidophilus* culture will be helpful (Klaver et al. 1993), but if the LAB grows too fast the fermentation time is shortened and counts in probiotic bacteria can even be lowered (Shihata and Shah 2002).

Proteolytic activities are not only crucial for the development of pure cultures, but also they are also crucial in a symbiotic relationship between the two strains that compose the yogurt cultures. As for probiotics, yogurt starters require an exogenous nitrogen source and utilize peptides and proteins in their growth medium (milk) by more or less complete enzyme systems. Besides the free amino acids, the enzymatic hydrolysis of milk proteins results in the liberation of peptides of varying sizes and soluble nitrogenous compounds. In yogurt, peptides, free amino acid and other released components depends on the type of milk (animal species, season), type of proteolytic enzyme and bacterial strains, heat treatment, manufacturing techniques, and storage conditions (Zourari et al. 1992, Tamime and Robinson 1999). In yogurt, *L. bulgaricus* is the main species responsible for increasing free amino acid (Zourari et al. 1992, El-zahar et al. 2004) that partially explains the associative growth relationship that exists between *S. thermophilus* and *L. bulgaricus*. Therefore, the proteinase activity of *L. bulgaricus* hydrolyzes the casein to yield polypeptides, which are broken down by the peptidases of *S. thermophilus* with the liberation of amino acids (Tamime and Robinson 1999). The proteinase of *L. bulgaricus* is more active on β-casein than on whey proteins (Zourari et al. 1992) and has an optimum pH between 5.2 and 5.8 (Argyle et al. 1976, Ezzat et al. 1987). The importance of peptides for their growth stimulation and their acidification is now well established, especially for *S. thermophilus* (Zourari et al. 1992). According to Accolas et al. (1971), the stimulation of *S. thermophilus* by milk culture filtrate of *L. bulgaricus* was due to the presence of valine, leucine, isoleucine, and histidine. Bracquart et al. (1978) as well as Bracquart and Lorient (1979) reported that reducing the growth medium of valine, histidine, glutamic acid, tryptophan, leucine, and isoleucine results in reduction in stimulation of *S. thermophilus* to 50% (Tamime and Robinson 1999).

Soy and Pulses

Soy beverages contain, by order of importance, sucrose, stachyose, raffinose, glucose, and fructose. Therefore, to grow in milk, the ability to use lactose is critical, but cultures have a variety of options in soy substrates. The assimilation of sucrose requires invertase, while the hydrolysis of stachyose and raffinose demands α-galactosidase (α-gal). In soy substrates, the lactobacilli and the streptococci mainly use sucrose, glucose, and fructose, while the bifidobacteria tend to use stachyose and raffinose (Champagne et al. 2009). However, the ability of probiotic bacteria to synthesize these enzymes varies much between strains and *L. acidophilus*, as well as yogurt starter cultures may possess α-gal (Donkor et al. 2007). Therefore, strain selection must be carried out (Champagne et al. 2009). The α-gal of *B. longum* has an optimal activity at pH 5.8 and between 40–45°C (Garro et al. 1994). In soy, numerous enzymatic substrates are required to assimilate all carbohydrates. Thus, α-gal will hydrolyze stachyose and then raffinose to generate galactose and sucrose (Connes et al. 2004). Invertase is then required for sucrose hydrolysis. Therefore, complete assimilation of the soy carbohydrates requires two sets of hydrolytic enzymes. Thereafter, it is the ability to assimilate the monosaccharides that modulates growth rates. Not all probiotic cultures can metabolize the products of α-gal and invertase (galactose, glucose, and fructose) (Table 35.2), and even then, growth rates vary (Table 35.1).

It was observed that *B. longum* R0175 acidified milk at a much slower rate than *Lactobacillus helveticus* R052 (Fig. 35.1), and that the bifidobacteria grew at half the rate *of L. helveticus* on lactose (Table 35.1). These data suggest that acidification rates of milk by two probiotic bacteria (Fig. 35.1) can potentially

Table 35.2. Fermentation of Some Carbohydrates by Various Probiotic Bacteria

	Food Based								Prebiotics			
Species	Glu	Fru	Suc	Lac	Mal	Sta	Raf	Man	FOS	GOS	Inu	Ltl
Lactobacillus acidophilus	+	V	+	+	V		−	−	V	+	V	+
Lactobacillus casei	+	+	+	−	V		−	+	V	+	V	+
Lactobacillus plantarum	+	+	+	+	+	+	+	+			V	
Lactobacillus reuteri	+	+	+	+	+		+	−				
Lactobacillus rhamnosus	+	+	+	+	+	−	−				−	
Bifidobacterium lactis	+								+	+	+	+
Bifidobacterium longum	+	−	V	+	+	+	+	+	+	+	+	+

Source: From Pokusaeva et al. 2011, Tungland 2000, Champagne et al. 2010, and Dellaglio et al. 1994.
Symbols: + = positive, − = negative, V = variable.
Sugars: Glu, glucose; Fru, fructose; Suc, sucrose; Lac, lactose; Mal, maltose; Sta, Stachyose; Raf, raffinose; Man, mannitol; Inu, inulin; Ltl, lactulose; FOS, fructooligosaccharides; GOS, galactooligosaccharides.

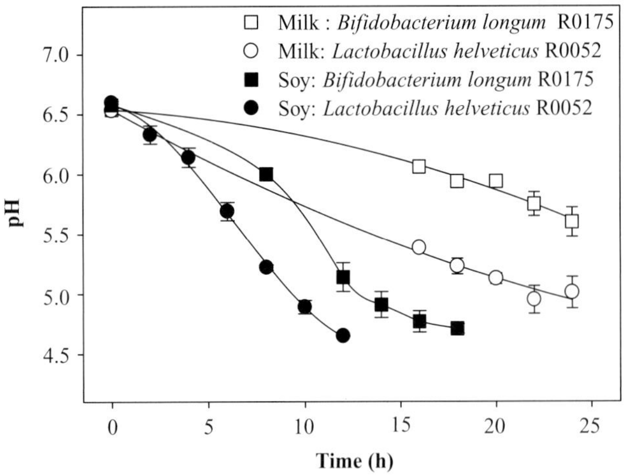

Figure 35.1. Acidification of milk or soy beverages adjusted at identical protein and fat levels by two probiotic cultures. (Redrawn from Champagne et al. 2010.)

be linked to their growth rates on lactose (Table 35.1). It was examined, if this extended to soy fermentation as well. However, it was observed that acidification rates in a soy beverage by two probiotic cultures were much higher than in milk (Fig. 35.1) even though their growth rates on soy carbohydrates were lower that on lactose (Table 35.1). These data show that enzyme activity on carbohydrate is not always the limiting factor on growth rates and fermentative activity. As was the case in milk, amino acids in soy are mainly found in the proteins. Therefore, proteolytic activities are crucial to attain high biomass levels in a soy substrate. A link was indeed established between the ability of bifidobacteria to synthesize galactosidases as well as proteinases and their multiplication in a soy substrate (Donkor et al. 2007).

With respect to soy, another enzymatic system is also of interest: hexanal and pentanal assimilation. These compounds contribute to the "beany" flavor of soy (Desai et al. 2002, Blagden and Gilliland 2005). For some consumers, this is an undesirable trait. Both lactobacilli (Blagden and Gilliland 2005) and bifidobacteria (Desai et al. 2002) have been shown to reduce the level of these metabolites in fermented soy products.

Prebiotics and Other Foods

Prebiotics are nondigestible food ingredients that beneficially affect the host by selectively stimulating the growth and/or activity of one or a limited number of bacteria in the GI system, and thereby confer health benefits to the host (Roberfroid 2007). This definition overlaps with the definition of dietary fiber, with the exception of its selectivity for certain bacterial species and a wider range of health effects. Peptides, proteins, and lipids contain prebiotics characteristics, but some carbohydrates have received the most attention, including lactulose, inulin, and a range of oligosaccharides that supply a source of fermentable carbohydrate for the beneficial bacteria in the colon (Prado et al. 2008).

The nutritional composition of pulses such as complex carbohydrates (Table 35.3), protein, vitamins, and minerals as well as antioxidants, and only very small amounts of unsaturated

Table 35.3. Average of Carbohydrate Content in Different Canadian Pulses and Their Digestibility in Human Body

	Mean of Carbohydrates Content in Pulses (g/100 g dry matter)				
	Sucrose	Raffinose	Stachyose	Verbascose	Oligosaccharides
Digestibility in human body	+	−	−	−	−
Chickpea (desi)	2.03	0.54	1.64	ND[a]	2.18
Chickpea (kabuli)	3.84	0.61	2.20	ND[a]	2.81
Green lentil	2.01	0.43	2.09	0.56	3.07
Red lentil	1.80	0.42	1.94	0.52	2.87
Field pea	2.8	0.7	2.7	1.0	4.4
Navy bean	3.2	0.5	4.0	ND[a]	4.6
Black turtle bean	3.93	0.57	3.50	0.07	4.14
Cranberry bean	4.14	0.23	3.13	0.21	3.58
Dutch brown bean	2.85	0.34	2.97	0.17	3.47
Dark red kidney bean	3.45	0.26	3.80	0.14	4.20
Great Northern bean	5.14	0.54	3.42	0.02	3.98
Light red kidney bean	4.69	0.26	3.44	0.16	3.85
Pink bean	4.54	0.31	3.65	0.02	4.02
Pinto bean	4.40	0.37	3.65	0.04	4.07
Small red bean	4.74	0.45	3.48	0.09	4.02
White kidney bean	3.67	0.22	3.53	0.18	3.93

[a]Not defined.
From Wang and Daun, 2004.

fats made this ingredients as a very good source of prebiotic components for human body as well as yogurt and probiotics bacteria (Boye et al. 2010). Enriching milk with prebiotic supplements (Capela et al. 2006) enhances the stability of probiotics in yogurt, but very few studies have been conducted with soy or pulses. Addition of soy protein isolates in yogurt does not improve stability during storage (Pham and Shah 2009) but some pulses do (Zare et al. 2011).

Food fermentations with probiotics have mainly been conducted on dairy and soy substrates. Pulses contain many of the carbohydrates of soy but there are also various other oligosaccharides (Table 35.3). Therefore, from a carbohydrate perspective, growth of probiotics in pulses should require similar enzymatic profiles to those required in soy; however, a wider range of cultures can theoretically develop in some pulses due to the presence of verbascose or other oligosaccharides.

Cereal-based products also offer many possibilities for the development of probiotic-based (Farnworth 2004). Cereals mainly have starch. Assimilation of starch typically requires amylases and maltases.

Vegetables are also frequently used as substrates for lactic acid fermentations and are thus potential matrices for probiotic cultures (Farnworth 2004). Sucrose is often found in these matrices, but a wide variety of substrates are encountered. In celery, for example, mannitol is the main carbon-based substrate but original polysaccharides are also discovered (Thimm et al. 2002).

It would take too long to examine each plant-based matrix for the required enzymes. Following are the points that must be emphasized:

1. Carbohydrate substrates vary as a function of the food matrix (Table 35.3) and, thus enzymatic requirements vary accordingly.
2. There is variability in the ability of probiotic bacteria to use the carbohydrates (Tables 35.1 and 35.2); a strain selection process is required.

BIOCHEMISTRY AND STABILITY DURING STORAGE IN FOODS

Protection Against Oxygen

Probiotic bacteria are sensitive to many processes (heating, freezing, aeration) associated with food production or storage conditions (Champagne et al. 2005). But they are also sensitive to some conditions that occur during storage. The two most important are the presence of oxygen and high acidity of the food. In both cases, some enzymatic systems may enhance the survival of the cultures to these stressful conditions.

Since the intestines offer an anaerobic environment, many strains are not adapted to growth in aerobic conditions. Technological adaptations can be made: microencapsulation, addition of antioxidants, packaging conditions (Talwalkar and Kailasapathy 2004, Jimenez et al. 2008). But some strains do possess enzymatic systems that enhance their stability in foods containing oxygen. The major problem associated with the presence of oxygen is that hydrogen peroxide is produced in various metabolic pathways. The two major enzyme systems that eliminate this toxic by-product of oxygen are NADH oxidase and NADH peroxidase (Shimamura et al. 1992, Talwalkar and Kailasapathy 2003). These systems are inducible. Thus, when cultures are gradually exposed to peroxide, an increased synthesis of these enzymes occurs. NADH oxidase and NADH peroxidase function optimally at pH 5.0 (Talwalkar and Kailasapathy 2003), which suggests that the protection to oxygen would be higher in acid foods (yogurt, fruit juices) than in neutral foods (unfermented milk, vegetables). Not surprisingly, it was shown that when a probiotic culture was added at the beginning of a yogurt fermentation process, it was more stable during storage than when it was added directly during the finished product (Hull et al. 1984). One hypothesis was that the cultures were gradually exposed to H_2O_2 produced by *L. delbrueckii* ssp. *bulgaricus* during fermentation and that enzymes were synthesized that were subsequently useful during storage. Indeed, when catalase was added in the products having direct inoculation of the probiotic culture in the finished product, an increased stability during storage was noted, while this was not the case in those where the probiotic culture was added earlier in the manufacture (Hull et al. 1984). It can also be argued that the gradual exposure to acid during fermentation would also be involved in the beneficial effect of inoculation at the beginning of the fermentation. Thus, enzymatic adaptation in acid environments needs to be addressed.

Protection Against Acid

Most LAB and probiotic cultures have optimum growth rates between pH 5.5 and 6.6. Consequently, when they are exposed for short or long periods to pH levels below 5.5 (cheese, yogurt, fruit juices, stomach) acid diffuses into the cells and reduces the efficiency of the enzymatic processes. Death may ultimately result. Indeed, numerous studies report high viability loses of probiotics exposed to the acid of the stomach (Mainville et al. 2005) as well as during storage in acid foods (Champagne et al. 2005). In the latter situation, there is often a correlation between the poststorage pH in yogurts and the survival of probiotic bacteria (Kailasapathy et al. 2008).

It is well known that LAB try to maintain an intracellular pH (pH_i) constant in these acid environments, but it is not always possible. As a result, pH_i may drop and generate a variety of responses (O'Sullivan and Condon 1997). In many cases, an acid tolerance response (ATR) occurs that significantly increases the ability of cells to survive a short exposure (a few hours) to high acid environments. In addition, cross-protection occurs and ATR helps protect against short-term stresses caused by heating, H_2O_2, salt, and ethanol (O'Sullivan and Condon 1997). The effect of the matrix medium on pH_i varies between strains and if affected by incubation time (Nannen and Hutkins 1991). Therefore, ATR is not generalized amongst the LAB. Unfortunately, most ATR studies were tested on short exposures to acidity, such as passage through the GI tract, but little is known on the enzymatic factors that will improve stability to weeks of exposure to acid. There seems to be a statistically significant correlation between stability during storage in a fruit drink and the ability of the strains to grow at pH 4.2, but that the relationship was not strong ($R^2 = 0.49$) (Champagne and Gardner 2008). More data are needed for the enzymatic activities associated with resistance to long-term exposure to acid foods.

BIOCHEMISTRY AND HEALTH FUNCTIONALITY

Survival to Gastric Acidity

Many bacteria do not survive passage to the conditions of the GI tract and, as mentioned previously, one of the most important criteria for functionality of probiotic bacteria is the survival to acid conditions of the stomach (Prasad et al. 1998, Mainville et al. 2005).

Proton-translocating ATPases are the enzymes that are principally involved in maintaining a high pH_i, as they contribute to excrete protons outside the cell (De Angelis and Gobbetti 2004). There are other enzymatic systems as well. The arginine deiminase pathway, which involves three enzymes, contributes to maintaining pH_i by producing ammonia as well as ATP for the ATPases (De Angelis and Gobbetti 2004). It must be kept in mind that there must be a source of ATP for the proton-translocating enzymes to function. Therefore, acid resistance of probiotic cells in the stomach can also be linked to the presence of a fermentable sugar in the medium (Corcoran et al. 2005). This shows that numerous enzymatic systems interact to prevent a big drop in pH_i, when the bacteria produce lactic acid during their fermentation or when the cells are exposed to an exocellular environment having a low pH.

As mentioned previously, cells can be adapted to enhance their resistance to acid. This ATR has been shown to be effective on subsequent short-term acid stresses such as those that occur when cells are exposed to gastric solutions. As a result, enzymatic adaptations (H^+-ATPases or others) improve the survival of probiotics to passage in the GI tract (Matto et al. 2006).

In summary, it can be assumed that survival of probiotics to passage in the stomach requires at least two sets of enzymes: (1) those that can assimilate sugar or amino acid substrates and generate ATP and (2) those that excrete protons outside the cell to help maintain an acceptable pH_i.

Lactose Maldigestion

Lactase insufficiency indicates that the concentration of the β-gal in the cells of the small intestine mucosa is very low. As a result, hypolactasia causes insufficient digestion of lactose in the GI tract. This phenomenon is alternatively called lactose malabsorption, lactose maldigestion, or lactose intolerance by various authors. In addition to the intrinsic intestinal lactase activity and its determinants, other parameters that affect lactose digestion include ethnic origin and age. Lactase activity is high at birth, decreases in childhood and adolescence, and remains low in adulthood. The symptoms of lactose intolerance are increased breath hydrogen, flatulence, abdominal pain, and diarrhoea.

An official recognition of the benefit of yogurt in lactose digestion in the GI tract has been granted (EFSA 2010) and evidence points toward a reduction of symptoms of intolerance in lactose maldigesters. The synthesis of β-gal by the yogurt starter culture is considered as the probioactive component involved in the positive effects on intestinal functions and colonic microflora, and reduced sensitivity to symptoms. Unbroken bacterial cell walls act as a mechanical protection of lactase during gastric transit and also affect the discharge of the enzyme into the small intestine; they consequently influence the efficiency of the system. It must also be mentioned that reduced amounts of lactose are often found in cheeses due to lactose utilization by starter microorganisms (Kilara and Shahani 1975, De Vrese et al. 2001).

Lactose catabolism mainly occurs by β-gal derived from the yogurt starters in fermented milk processing. The optimum activity of streptococcal β-gal is at a neutral pH and at 55°C in presence of buffer. Activity of β-gal is stimulated in presence of Mg^{2+} and oxgall (0.15 mL/100mL), while ethylenediaminetetraacetic acid (EDTA) causes inhibition. Thermal denaturation occurs at 60°C, although stability can be enhanced by the addition of bovine serum albumin (Chang and Mahoney 1994). Therefore, it can be expected that yogurt drinks heated at high temperatures to enable storage at room temperature would not contain active β-gal and therefore not present the enzymatic functionality.

In comparison with yogurt cultures, other probiotic bacteria usually promote lactose digestion in the small intestine less efficiently. However, it is suggested that some probiotic bacteria may act by preventing symptoms of intolerance in the large intestine in addition to (or rather than) by improving lactose digestion in the small intestine (De Vrese et al. 2001). It is noteworthy that yogurt cultures are generally not included in lists of probiotic bacteria (Stanton et al. 2003, CFIA 2009), presumably because they generally do highly survive the gastric transit nor grow in the intestines, but some authors do consider them as candidates (Santosa et al. 2008).

Blood Cholesterol Level

High cholesterol in serum is associated with the incidence of human cardiovascular diseases (Othman et al. 2011). One of the health-promoting benefits of probiotics is their ability to reduce blood cholesterol, which was observed in humans (Xiao et al. 2003) and animals (Du Toit et al. 1998, Nguyen et al. 2007). However, some studies have been negative (Simons et al. 2006). This points to a specific biological activity that is variable between cultures. Bile salt hydrolase (BSH) in probiotics renders them more tolerant to bile salts, and it is believed that this activity also helps to reduce the blood cholesterol level of the host (Taranto et al. 1997). Indeed, oral administration of encapsulated BSH reduced serum cholesterol levels in mice by 58% (Sridevi et al. 2009). It is thought that bile salt deconjugation affects its enterohepatic circulation (Kim and Lee 2008), but the BSH hypothesis has not yet been completely demonstrated or elucidated.

Three different types of BSH from various *Bifidobacterium* strains have been identified (Kim et al. 2004, Kim and Lee 2008). Data show that higher BSH activity can be acquired (Noriega et al. 2006). It is unknown if cells that are suddenly exposed to bile salts upon entrance to the duodenum can increase their BSH level and increase their functionality. Thus, BSH activity in the GI tract can potentially be enhanced by culture preparation methodologies prior to consumption.

It must be kept in mind, however, that the production of secondary deconjugated bile salts through BSH activity may have undesirable effects. Contradictory data suggest that colon cancer could be enhanced (Patel et al. 2010), while another study suggests that *Lactobacillus reuteri* could have protective properties through precipitation of the deconjugated bile salts and a physical binding of bile salts by the bacterium, thereby making the harmful bile salts less bioavailable (De Boever et al. 2000). Therefore, more data are needed to better ascertain the desirability of BSH activity.

Gas Discomforts

It is considered that oligosaccharides that are not broken down by the human GI enzymes may be responsible for gas production in the GI tract (Yamaguishi et al. 2009). Since they are not assimilated in the small intestine, these oligosaccharides end up in the colon where they are fermented by the microbiota (LeBlanc et al. 2008). Unfortunately, the consumption of beans has been associated with gas production in the GI tract (Machaiah et al. 1999) if consumed in sufficient amount. Fortunately, symptoms reduce on frequent and continuous consumption (O'Donnell and Fleming 1984), but means to reduce discomforts are nevertheless desirable.

Legumes can be good substrates for the growth of probiotics (Chopra and Prasad 1990), but strain selection is required (Farnworth et al. 2007). Although some probiotic cultures only use the short-chain carbohydrates in legumes (Desai et al. 2002), the prospect of introducing probiotics into a high-fiber bean product is a very promising one. It is unknown at the present time, if the carbohydrates in beans can indeed have a positive impact on survival and growth in the GI tract. In addition to the health benefits, probiotics may also have the potential of reducing problems associated with digestibility.

Various studies in animals (LeBlanc et al. 2008) and in humans (Nobaek et al. 2000, Di Stefano et al. 2004) have shown the potential of probiotics to reduce the symptoms of gas discomforts. There appears to be a link between the synthesis of α-gal, which hydrolyze the oligosaccharides in beans, and the bioactivity of the cultures (LeBlanc et al. 2008). As a result, a product (Proviva Fruit Drink) has received an opinion by the Swedish Nutrition Foundation (SNF 2003) "that an effect to decrease flatulence is reasonably well documented." This was apparently the first such specific health benefit recognition for probiotic bacteria. The culture used was *Lactobacillus plantarum* 299v. However, a similar petition with *L. rhamnosus* GG was not granted (EFSA 2008).

This benefit of probiotics on sugar metabolism in the GI tract through α-gal resembles that of β-gal for lactose maldigestion mentioned previously. It is unknown to what extent live cells are required for the health benefit to occur.

Other Benefits

For probiotic bacteria, the main purpose of proteases is for nutrition. By hydrolysing proteins in the medium, the cells gain access to peptides and amino acids for the synthesis of their own proteins. However, there are instances where peptides show biological activities. Milk caseins have particularly been assessed in this respect (Korhonen and Pihlanto 2006). Thus, milk proteins contain peptidic angiotensin I-converting enzyme (ACE) inhibitors, which can be released by proteolysis during milk fermentation by some strains of *L. helveticus* (Leclerc et al. 2002). Hydrolysis of soy proteins by probiotic lactobacilli has also revealed the presence of ACE inhibitors (Donkor et al. 2005). Through the ACE inhibition, peptides exert antihypertensive activity. It is important to stress that the functionality of probiotics is linked to the substrate on which growth has occurred. That is why the concept of probioactive not only includes bioactive compounds synthesized by the probiotic cells as such, such as exopolysaccharides, but also includes probioactives that are specifically the result of transformation of the food substrate. This link between probiotic functionality and the food matrix may partially explain why there are contradictory data on the benefits of probiotics in the literature. An ACE inhibition activity recorded in fermented milk may unfortunately not extend to fermented fruits or vegetables low in proteins.

The production of peptides is not the only way lactic cultures can affect blood pressure. Certain strains of *Lactococcus lactis* convert glutamic acid to gamma-amino butyric acid (GABA) thanks to glutamate decarboxylase (Inoue et al. 2003, Minervini et al. 2009). High levels of free glutamate are rarely found in foods, and proteolysis is generally required to release glutamate from proteins before GABA production can occur. The activity of glutamate decarboxylase tends to be higher in acid environments (Komatsuzaki et al. 2005). Therefore, fermented milk or soy beverages seem particularly well suited for GABA production.

Foods naturally contain antioxidants and bioactive compounds, but some are linked to sugars or proteins. When combined with sugars or other food ingredients, some of these bioactive compounds have less biological activity. Therefore, deconjugation of the bioactives and their release into the product are desirable. Two examples serve to illustrate this particular concept: (1) isoflavones in soy and (2) quercetin in onions.

Antioxidants are bioactive compounds that are quite varied in chemical. In foods, examples include carotenoids, terpenes, anthocyanins, isoflavones, and flavonoids like quercetin. Quercetin is of interest because of its high antioxidant capacity, that is, 229% higher than that of ascorbic acid (Kim and Lee 2004). In onions, quercetin is mainly found in three forms: (1) quercetin diglucoside (Qdg), (2) quercetin monoglucoside (Qmg), and (3) free quercetin. Fermentation of red onions by lactic cultures substantially increased the proportion of Qmg (Bisakowski et al. 2007), that may have a positive effect as fractions containing higher ratios of Qmg:Qdg have been reported to have higher antioxidant activity (Makris and Rossiter 2001). The enzymes responsible for this useful bioconversion in onions are glucosidases. These enzymes contribute to enhancing the biological

Table 35.4. Some Enzymatic Activities Which May be Linked to the Functionality of Probiotic Bacteria

Enzyme	Function	Functionality	Species
β-Galactosidase	Lactose hydrolysis in food matrix	Improved growth in milk	Wide-ranging
	Lactose hydrolysis in the GI tract	Improved lactose hydrolysis in GI tract for population suffering of lactose maldigestion	Yogurt starters *Streptococcus thermophilus L. delbrueckii ssp. bulgaricus*
NADH oxidase and NADH peroxidase	Eliminate H_2O_2	Improved stability in presence on oxygen	Lactobacilli and bifidobacteria—strain variable
ATPases	Excretion of protons (H^+) from cells to outer medium	Reduced intracellular drop in pH and increased survival to gastric transit	Many LAB (*Streptococcus, Lactococcus and Lactobacillus spp*)
Arginine deiminase pathway	Production of NH_3 and ATP	Reduced intracellular drop in pH and increased survival to gastric transit	Lactobacilli
Bile salt hydrolase	Deconjugation of bile salts	Lowers blood cholesterol level	*Bifidobacterium* ssp. often show high activity, but many *Lactobacillus* ssp. cultures are also active
α-Galactosidase	Hydrolysis of stachyose and raffinose in GI tract	Reduces gas discomforts linked to consumption of beans	*L. plantarum, B. longum*
Proteases	Hydrolysis of caseins, or soy proteins	Produces ACE inhibitory peptides, which are antihypertensive	*L. helveticus*
Glutamate decarboxylase	Conversion of glutamate to GABA	Lowering of blood pressure	*Lactococcus lactis, L. paracasei*
Glucosidases	Deconjugates isoflavones or quercetin from glycosylated forms	Higher levels of the more bioactive forms of antioxidants (cancer, bone metabolism, etc.)	*L. helveticus, L. acidophilus, B. longum, L. plantarum*

GI, gastro-intestinal; ACE, acetyl choline esterase; *L., Lactobacillus; B., Bifidobacterium;* LAB, lactic acid bacteria; GABA, gamma-amino butyric acid.

value of other foods as well. For example, isoflavones are encountered in soy products. The native forms of the isoflavone in soy are mainly daidzein, genistin, and glycitein that are the glycosylated forms. These β-glucosides, are not readily bioavailable in humans as they are unable to be absorbed through the intestinal tract. Hydrolysis of β-glucosides by bacterial β-glucosidases, produces the active and more readily metabolized aglycone forms (Setchell et al. 2002). Many lactobacilli and bifidobacteria possess glucosidases. Thus, the data show that soy fermentation with selected strains contributes to increasing the levels of bioactive aglycons (Champagne et al. 2010).

In summary, many enzymatic activities of lactic cultures contribute to the synthesis, or simply to the release, of probioactives in foods (Table 35.4).

CONCLUSION

Many biochemical processes are critical in the functionality of probiotic bacteria. From a technological perspective, it can be foreseen that cells will be adapted to enhance a specific enzymatic activity that will promote their growth or their stability during storage. Minor acid, osmotic, and thermal stresses have been proposed for this purpose (De Angelis and Gobbetti 2004).

As knowledge of the enzymatic systems involved in health functionality of probiotics increases, the requirement for viable cells might gradually be reduced. Thus, the critical element for the health effect may be limited to the enzymatic components, for example, the quantity of the probioactive. The use of pure enzymes might eventually be considered in substitution to probiotics, but only if the enzymes are stable in the matrix and if they survive passage to the GI conditions. Even if it is enzyme-linked, functionality might still require that the enzymes be delivered through a cell, viable or not, in order to protect them against inactivation in the adverse conditions, they will be exposed to in the food or the GI tract. Therefore, more research is needed not only on the biochemical processes involved in the functionality of probiotic bacteria, but also on factors that will enable the "delivery" of the biochemical activity at the proper site.

REFERENCES

Accolas JP et al. 1971. Study of interactions between various mesophilic and thermophilic lactic bacteria, in connection with the manufacture of cheese. *Lait* 51: 249.

Araya M et al. 2002. Guidelines for the evaluation of probiotics in food. Joint FAO/WHO Working Group Report on Drafting Guidelines for the Evaluation of Probiotics in Food, London (ON, Canada) April 30 and May 1. Available at ftp://ftp.fao.org/docrep/fao/009/a0512e/a0512e00.pdf.

Argyle PJ et al. 1976. Production of cell-bound proteinase by Lactobacillus bulgaricus and its location in the bacterial cell. *J Appl Bacteriol* 41: 175–184.

Bisakowski B et al. 2007. Effect of lactic acid fermentation of onions (Allium cepa) on the composition of flavonol glucosides. *Int J Food Sci Technol* 42: 783–789.

Blagden TD, Gilliland SE. 2005. Reduction of levels of volatile components associated with the 'beany' flavor in soymilk by lactobacilli and streptococci. *J Food Sci* 70: M186–M189.

Boye JI et al. 2010. Pulse proteins: Processing, characterization, functional properties and applications in food and feed. *Food Res Int* 43: 414–431.

Bracquart P, Lorient D. 1979. Effect of amino acids on growth of Streptococcus thermophilus. *Milchwissenschaft* 34: 676.

Bracquart P et al. 1978. Effect of amino acids on growth of Streptococcus thermophilus. II. Study on five strains. *Milchwissenschaft* 33(6): 341–344.

Capela P et al. 2006. Effect of cryoprotectants, prebiotics and microencapsulation on survival of probiotic organisms in yoghurt and freeze-dried yoghurt. *Food Res Int* 39: 203–211.

CFIA (Canadian Food Inspection Agency). 2009. Chapter 8, Section 8.7. Probiotic claims. Available at http://www. inspection.gc.ca/english/fssa/labeti/guide/ch8ae.shtml.

Champagne CP et al. 2005. *CRC Crit Rev Food Sci Nutr* 14(2): 109–134.

Champagne CP, Gardner NJ. 2008. Effect of storage in a fruit drink on subsequent survival of probiotic lactobacilli to gastro-intestinal stresses. *Food Res Int* 41: 539–543.

Champagne CP et al. 2009. Selection of probiotic bacteria for the fermentation of a soy beverage in combination with Streptococcus thermophilus. *Food Res Int* 42: 612–621.

Champagne CP et al. 2010. Effect of fermentation by pure and mixed cultures of *Streptococcus thermophilus* and *Lactobacillus helveticus* on isoflavone and vitamin content of a fermented soy beverage. *Food Microbiol* 27: 968–972.

Chang BS, Mahoney RR. 1994. Thermal denaturation of β-galactosidase (Streptococcus thermophilus) and its stabilization by bovine serum albumin: an electrophoretic study. *Biotechnol Appl Biochem* 19: 169–178.

Chopra R, Prasad DN. 1990. Soymilk and lactic fermentation products, a review. *Microbiol-Aliments-Nutr* 8: 1–13.

Connes C et al. 2004. Towards probiotic lactic acid bacteria strains to remove raffinose-type sugars present in soy-derived products. *Lait* 84: 207–214.

Corcoran BM et al. 2005. Survival of probiotic lactobacilli in acidic environments is enhanced in the presence of metabolizable sugars. *Appl Environ Microbiol* 71: 3060–3067.

De Angelis, M, Gobbetti M. 2004. Environmental stress responses in *Lactobacillus*. A review. *Proteomics* 4: 106–122.

De Boever P et al. 2000. Protective effect of the bile salt hydrolase-active Lactobacillus reuteri against bile salt cytotoxicity. *Appl Microbiol Biotechnol* 53: 709–714.

De Vrese M et al. 2001. Probiotics compensation for lactase insufficiency. *Am J Clin Nutr* 73: 421S–429S.

Dellaglio F et al. 1994. H de Roissard, FM Luquet (eds.) *Bactéries Lactiques*, vol 1. Lorica, Uriage, pp. 25–115.

Desai A et al. 2002. Metabolism of raffinose and stachyose in reconstituted skim milk and of n-hexanal and pentanal in soymilk by bifidobacteria. *Biosci Microflora* 21: 245–250.

Desjardins ML et al. 1991. B-Galactosidase and proteolytic activities of bifidobacteria in milk: A preliminary study. *Milchwissenschaft* 46: 11–13.

Di Stefano M et al. 2004. Probiotics and functional abdominal bloating. *J Clin Gastroenterol* 38(6 Suppl): S102–S103.

Donkor ON et al. 2005. Probiotic strains as starter cultures improve angiotensin-converting enzyme inhibitory activity in soy yogurt. *J Food Sci* 70: M375–M381.

Donkor ON et al. 2007. α-galactosidase and proteolytic activities of selected probiotic and dairy cultures in fermented soymilk. *Food Chem* 104: 10–20.

Du Toit M et al. 1998. Characterisation and selection of probiotic lactobacilli for a preliminary minipig feeding trial and their effect on serum cholesterol levels, faeces pH and faeces moisture content. *Int J Food Microbiol* 40: 93–104.

EFSA (European Food Safety Authority). 2008. Scientific substantiation of a health claim related to LGG® MAX and reduction of gastro-intestinal discomfort pursuant to Article 13(5) of Regulation (EC) No 1924/20061. *EFSA J* 853: 1–15.

EFSA (European Food Safety Authority). 2010. EFSA Panel on Dietetic Products, Nutrition and Allergies (NDA); Scientific Opinion on the substantiation of health claims related to Yoghurt cultures and improving lactose digestion (ID 1143, 2976) pursuant to Article 13(1) of Regulation (EC) No 1924/2006. *EFSA J* 8(10): 1763. Available at www.efsa.europa.eu/efsajournal.htm.

El-zahar K et al. 2004. Proteolysis of ewe'e caseins and whey proteins during fermentation of yogurt and storage. Effect of the starters used. *J Food Biochem* 28: 319–335.

Ezzat N et al. 1987. Partial purification and characterization of a cell wall associated proteinase from *Lactobacillus bulgaricus. Milchwissenchaft* 42: 95–97.

Farnworth E. 2004. The beneficial health effects of fermented foods-potential probiotics around the world. *J Nutraceuticals. Funct Med Foods* 493–117.

Farnworth ER, Champagne CP. 2010. Production of probiotic cultures and their incorporation into foods. In: RR Watson, VR Preedy (eds.) *Bioactive Foods in Promoting Health: Probiotics and Prebiotics*. Elsevier Academic Press, Oxford, pp. 3–17.

Farnworth ER et al. 2007. Growth of probiotic bacteria and bifidobacteria in a soy yogurt formulation. *Int J Food Microbiol* 1: 174–121.

Garro MS et al. 1994. α-D-Galactosidase (EC 3.2.1.22) from *Bifidobacterium longum. Lett Appl Microbiol* 19: 16–19.

Gaudreau H et al. 2005. The use of crude cellular extracts of *Lactobacillus delbrueckii ssp bulgaricus* 11842 to stimulate growth of a probiotic *Lactobacillus rhamnosus* culture in milk. *Enzyme Microb Technol* 36: 83–90.

Hull RR et al. 1984. Survival of Lactobacillus acidophilus in yoghurt. *Australian J Dairy Technol* 39(4): 164–166.

Inoue K et al. 2003. Blood-pressure lowering effect of a novel fermented milk containing g-aminobutyric acid (GABA) in mild hypertensives. *Eur J Clin Nutr* 57: 490–495.

Jimenez AM et al. 2008. On the importance of adequately choosing the ingredients of yoghurt and enriched milk for their antioxidant activity. *Int J Food Sci Technol* 43: 1464–1473.

Kailasapathy K et al. 2008. Survival of Lactobacillus acidophilus and Bifidobacterium animalis ssp. lactis in stirred fruit yogurts. *LWT- Food Sci Technol* 41: 1317–1322.

Kanatani K, Oshimura K. 1994. Isolation and structural analysis of the phospho-β-galactosidase gene from *Lactobacillus acidophilus*. *J Ferment Bioeng* 78: 123.

Khattab AA et al. 1986. Stimulation of Lactobacillus acidophilus and Bifidoacterium adolescentis in milk by β-Galactosidase. *Egyptian J Dairy Sci* 14: 155–163.

Kilara A, Shahani KM. 1975. Lactase activity of cultured and acidified dairy products. *J Dairy Sci* 59: 2031–2035.

Kim DO, Lee CY. 2004. Comprehensive study on vitamin C equivalent antioxidant capacity (VCEAC) of various polyphenolics in scavenging a free radical and its structural relationship. *Crit Rev Food Sci Nutr* 44: 253–273.

Kim GB, Lee BH. 2008. Genetic analysis of a bile salt hydrolase in *Bifidobacterium animalis subsp. lactis* KL612. *J Appl Microbiol* 105: 778–789.

Kim GB et al. 2004. Cloning and Characterization of the Bile Salt Hydrolase Genes (bsh) from *Bifidobacterium bifidum. Strains Appl Environ Microbiol* 70: 5603–5612.

Klaver FAM et al. 1993. Growth and survival of bifidobacteria in milk. *Neth Milk Dairy J* 47: 151–164.

Komatsuzaki N et al. 2005. Production of γ-aminobutyric acid (GABA) by *Lactobacillus paracasei* isolated from traditional fermented foods. *Food Microbiol* 22: 497–504.

Korhonen H, Pihlanto A. 2006. Bioactive peptides: Production and functionality. *Int Dairy J* 16: 945–960.

Lawrence RC et al. 1976. Reviews of the progress of dairy science: cheese starters. *J Dairy Res* 43: 141.

LeBlanc JG et al. 2008. Ability of Lactobacillus fermentum to overcome host alpha-galactosidase deficiency, as evidenced by reduction of hydrogen excretion in rats consuming soya alpha-galacto-oligosaccharides. *BMC Microbiol* 8: 22.

Leclerc PL et al. 2002. Antihypertensive activity of casein-enriched milk fermented by Lactobacillus helveticus. *Int Dairy J* 12: 995–1004.

Machaiah JP et al. 1999. Reduction in flatulence factors in mung beans (Vigna radiata) using low-dose gamma – irradiation. *J Sci Food Agric* 79: 648–652

Mainville I et al. 2005. A dynamic model that simulates the human upper gastrointestinal tract for the study of probiotics. *Int J Food Microbiol* 99: 287–296.

Makris DP, Rossiter JT. 2001. Domestic processing of onion bulbs (Allium cepa) and asparagus spears (Asparagus officinalis): effect on flavonol content and antioxidant status. *J Agric Food Chem* 49: 3216–3222.

Marshall VME, Law BA. 1984. FL Davies, BA Law (eds.) *Advances in the Microbiology and Biochemistry of Cheese and Fermented Milks*. Elsevier, Amsterdam.

Marshall VME, Tamime AY. 1997. In: BA Law (ed.) *Microbiology and Biochemistry of Cheese and Fermented Milk*, 2nd edn. Blackie Academic & Professional, London, pp. 153–192.

Matto J et al. 2006. Influence of processing conditions on Bifidobacterium animalis subsp. lactis functionality with a special focus on acid tolerance and factors affecting it. *Int Dairy J* 16: 1029–1037.

Mellentin. 2010. 10 key trends in food, nutrition and health. New Nutrition Business. Report can be purchased at www.new-nutrition.com/10kt2010.asp. Accessed on October 12, 2010.

Minervini F et al. 2009. Fermented goats' milk produced with selected multiple starters as a potentially functional food. *Food Microbiol* 26: 559–564.

Misra AK, Kuila RK. 1991. Intensified growth of Bifidobacterium and preparation of Bifidobacterium bifidum for a dietary adjunct. *Cultured Dairy Products J* 26(4): 4–6.

Monnet V et al. 1996. In: TM Cogan, J-P Accolas (eds.) *Dairy Starter Cultures*. VCH Publishers, New York, pp. 47–99.

Nannen NL, Hutkins RW. 1991. Intracellular pH effects in lactic acid bacteria. *J Dairy Sci* 74: 741–746.

Nguyen TDT et al. 2007. Characterization of *Lactobacillus plantarum* PH04, a potential probiotic bacterium with cholesterol-lowering effects. *Int J Food Microbiol* 113: 358–361.

Nobaek S et al. 2000. Alteration of intestinal microflora is associated with reduction in abdominal bloating and pain in patients with irritable bowel syndrome. *Am J Gastroenterol* 95: 1231–1238.

Noriega L et al. 2006. Deconjugation and bile salts hydrolase activity by Bifidobacterium strains with acquired resistance to bile. *Int Dairy J* 16: 850–855.

O'Donnell AU, Fleming SE. 1984. Influence of frequent and long-term consumption of legume seeds on excretion of intestinal gases. *Am J Clin Nutr* 40: 48–57.

O'Sullivan E, Condon S. 1997. Intracellular pH is a major factor in the induction of tolerance to acid and other stresses in Lactococcus lactis. *Appl Environ Microbiol* 63: 4210–4215.

Othman RA et al. 2011. Cholesterol-lowering effects of oat beta-glucan. *Nutr Rev* 69(6): 299–309.

Ouwehand AC, Salminen SJ. 1998. The health effects of cultured milk products with viable and non-viable bacteria. *Int Dairy J* 8: 749–758.

Patel AK et al. 2010 Probiotic bile salt hydrolase: current developments and perspectives. *Appl Biochem Biotechnol* 162: 166–180.

Pham TT, Shah NP. 2009. Performance of starter in yogurt supplemented with soy protein isolate and biotransformation of isoflavones during storage period. *J Food Sci* 74: M190–M195.

Prado FC et al. 2008. Trends in non-dairy probiotic beverages. *Food Res Int* 41: 111–123.

Prasad J et al. 1998. Selection and characterisation of Lactobacillus and Bifidobacterium strains for use as probiotics. *Int Dairy J* 8: 993–1002.

Roberfroid MM. 2007. Inulin-type fructans: functional food ingredients. *J Nutr* 137(Suppl 1): 2493S–2502S.

Pokusaeva K. 2011. Carbohydrate metabolism in Bifidobacteria. *Genes Nutr* 6(3): 285–306.

Santosa S et al. 2008. Probiotics and their potential health claims. *Nutr Rev* 64: 265–274.

Setchell KDR et al. 2002. The clinical importance of the metabolite equol-a clue to the effectiveness of soy and its isoflavones. *Am Soc Nutr Sci* 35: 77–84.

Shah NP. 2000. Some beneficial effects of probiotic bacteria. *Biosci Microflora* 19: 99–106.

Shah NP et al. 1997. Improving viability of Lactobacillus acidophilus and Bifidobacterium spp. in yogurt. *Int Dairy J* 7: 349–356.

Shihata A, Shah NP. 2000. Proteolytic profiles of yogurt and probiotic bacteria. *Int Dairy J* 10: 401–408.

Shihata A, Shah NP. 2002. Influence of addition of proteolytic strains of Lactobacillus delbrueckii subsp. bulgaricus to commercial ABT starter cultures on texture of yoghurt, exopolysaccharide production and survival of bacteria. *Int Dairy J* 12: 765–772.

Shimamura S et al. 1992. Relationship between oxygen sensitivity and oxygen metabolism of Bifidobacterium species. *J Dairy Sci* 75: 3296–3306.

Simons LA et al. 2006. Effect of *Lactobacillus fermentum* on serum lipids in subjects with elevated serum cholesterol. *Nutr Metab Cardiovasc Dis* 16: 531–535.

SNF (Swedish Nutrition Foundation). 2003. Statement concerning evaluation of the scientific documentation behind a product specific health claim. Product: Proviva Fruit Drink with Lactobacillus plantarum 299v. Available at http://www.snf.ideon.se/snf/en/rh/pfp/finalreport_proviva.pdf.

Sridevi N et al. 2009. Hypocholesteremic effect of bile salt hydrolase from *Lactobacillus buchneri* ATCC 4005. *Food Res Int* 42: 516–520.

Stanier RY et al. 1987. *The Microbial World*, 5th edn. MacMillan Education, London.

Stanton C et al. 2003. Probiotic health benefits – reality or myth? *Australian J Dairy Technol* 58(2): 107–113.

Stanton C et al. 2005. Fermented functional foods based on probiotics and their biogenic metabolites. *Curr Opin Biotechnol* 16: 198–203.

Starling S. 2010. Global probiotics market approaching $30bn by 2015: Report. NutraIngredients at www.nutraingredients.com/content/view/print/321175. Accessed on January 1, 2012.

Talwalkar A, Kailasapathy K. 2003. Metabolic and biochemical responses of probiotic bacteria to oxygen. *J Dairy Sci* 86: 2537–2546.

Talwalkar A, Kailasapathy K. 2004. A review of oxygen toxicity in probiotic yogurts: influence on the survival of probiotic bacteria and protective techniques. *Compr Rev Food Sci Food Saf* 3: 117–124.

Tamime AY, Robinson RK. 1999. *Yogurt: Science and Technology*, 3rd edn. Woodhead Publishing, Cambridge.

Taranto MP et al. 1997. Bile salts hydrolase plays a key role on cholesterol removal by Lactobacillus reuteri. *Biotechnol Lett* 19: 845–847.

Thimm J-C et al. 2002. Celery (Apium graveolens) parenchyma cell walls: Cell walls with minimal xyloglucan. *Physiol Plant* 116: 164–171.

Tungland BC. 2000. Duncan Crow (ed.) *Inulin: A Comprehensive Scientific Review*. Available at http://members.shaw.ca/duncancrow/inulin_review.html.

Wang N, Daun JK. 2004. *The Chemical Composition and Nutritive Value of Canadian Pulses*. Canadian Grain Commission, Grain Research Laboratory, Canada, pp. 12–13.

Xiao JZ et al. 2003. Effects of milk products fermented by *Bifidobacterium longum* on blood lipids in rats and healthy adult male volunteers. *J Dairy Sci* 86: 2452–2461.

Yamaguishi CT et al. 2009. Biotechnological process for producing black bean slurry without stachyose. *Food Res Int* 42: 4, 425–429.

Zare F et al. 2011. Microbial, physical and sensory properties of yogurt supplemented with lentil flour. *Food Res Int* 44: 2482–2488.

Zourari A et al. 1992. Metabolism and biochemical characteristics of yogurt bacteria. A review. *Lait* 72: 1–34.

36
Biological Activities and Production of Marine-Derived Peptides

Wonnop Vissesangua and Soottawat Benjakul

Abstract: Marine bioactive peptides are known to possess potential health benefits for humans. Beyond their basic nutritional roles, biopeptides are involved into many physiological processes in the living organisms. Many studies have reported that marine-derived bioactive peptides possess angiotensin I converting enzyme (ACE) inhibitory, antioxidant, antimicrobial, calcium binding, anticoagulant, immunomodulatory, and gastrin/CCK-like activity, which can be applied as functional foods and pharmaceuticals. Some of these biological characteristics are attributed to native peptides but also to those generated in vitro by enzymatic hydrolysis and released during processing. The recovery and production of these compounds from various food sources have been investigated for potential commercial applications in the food, health, and allied industries.

INTRODUCTION

Marine animals are emerging as a leading group for identifying and extracting bioactive peptides (Aneiros and Garateix 2004, Barrow and Shahidi 2008, Kim et al. 2008). These peptides have been described as a source of pharmaceutical products with beneficial effects on humans. Beyond their basic nutritional roles, peptides are involved in many processes in the living organisms on the basis of their physiological functions as hormones, neuropeptides, alkaloids, antibiotics, toxins, and regulation peptides. Depending on the sequence of amino acids, these peptides can exhibit diverse activities with physiological significance. Many studies have reported that marine-derived bioactive peptides possess angiotensin I converting enzyme (ACE) inhibitory, antioxidant, antimicrobial, calcium binding, anticoagulant, immunomodulatory, and gastrin/cholecystokinin (CCK)-like activity that can be applied as functional foods and pharmaceuticals. Some of these biological characteristics are attributed to native peptides but also to those generated in vitro by enzymatic hydrolysis and released during processing. A number of marine-derived bioactive peptides have been discovered. Nevertheless, more studies are being performed to explore the sources especially from underutilized marine processing by-products, their bioavailabilities and possible physiological function, and their mechanisms of action. In addition, technological approaches in terms of production and recovery have also been investigated for potential commercial applications in food industry, biotechnologies, cosmetics, or human health.

Food Biochemistry and Food Processing, Second Edition. Edited by Benjamin K. Simpson, Leo M.L. Nollet, Fidel Toldrá, Soottawat Benjakul, Gopinadhan Paliyath and Y.H. Hui.

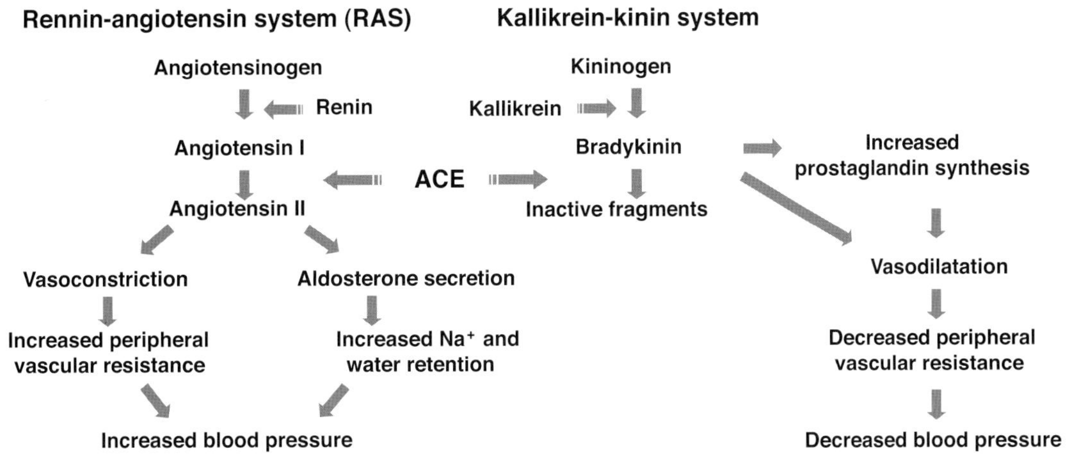

Figure 36.1. Role of angiotensin converting enzyme (ACE) in blood pressure regulation.

BIOLOGICAL ACTIVITIES OF MARINE-DERIVED PEPTIDES

ACE Inhibitory Peptides

ACE inhibition has become an important target in the development of drugs to control high blood pressure, which is a significant health problem worldwide. ACE (EC 3.4.15.1) is a zinc-metallopeptidase that needs zinc and chloride ions for its activity (Erdos and Skidgel 1987). It is widely distributed in mammalian tissues, predominantly as a membrane-bound ectoenzyme in vascular endothelial cells and also in several cell types including absorptive epithelial, neuroepithelial, and male germinal cells (Steve et al. 1988, Sibony et al. 1993). In the renin-angiotensin system (RAS), ACE plays a crucial role in the regulation of blood pressure as well as cardiovascular function (Li et al. 2004). Within the enzyme cascade of the RAS (Fig. 36.1), ACE converts the inactive angiotensin I by cleaving dipeptide from the *C*-terminus into the potent vasoconstricting angiotensin II. This potent vasoconstrictor is also involved in the release of a sodium-retaining steroid, aldosterone, from the adrenal cortex, which has a tendency to increase blood pressure. ACE is a multifunctional enzyme that also catalyses the degradation of bradykinin, a blood pressure lowering nonapeptide in the kallikrein-kinin system (Erdos and Skidgel 1987, Johnston and Franz 1992). Moreover, ACE has been shown to degrade neuropeptides including enkephalins, neurotensin, and substance P, which may interact with the cardiovascular system (Wyvratt and Patchett 1985).

A large number of highly potent and specific ACE inhibitors have been developed as orally active drugs that are used in therapy to reduce morbidity and mortality of patients with hypertension. However, these synthetic drugs, such as captopril, enalapril, alacepril, and lisinopril, which are used extensively, are believed to have certain side effects, such as cough, taste disturbances, and skin rashes angioedema and many other dysfunctions of human organs (Atkinson and Robertson 1979). Therefore, a search for ACE inhibitors from foods and natural sources has become a major area of research.

Marine-derived ACE inhibitory peptides have been discovered in enzymatic hydrolysates of different marine animals (Table 36.1). These peptides are generally short chain, often carrying polar amino acid residues like proline, lysine, or arginine at *C*-terminal residue (Meisel 1997). Cheung and Chushman (1971) suggested that tryptophan, tyrosine, proline, or phenylalanine at the *C*-terminal and branched-chain aliphatic amino acids at the *N*-terminal are suitable for peptides to act as competitive inhibitors by binding with ACE. Furthermore, structure–activity relationships among variety of peptide inhibitors of ACE indicate that binding to ACE is strongly influenced by the *C*-terminal tripeptide sequence of the substrate, and it is suggested that peptides, which contain hydrophobic amino acids at these positions, are potent inhibitors (Qian et al. 2007).

It has been demonstrated that some marine-derived ACE-inhibitory peptides exhibit higher or comparable in vivo antihypertensive activity to captopril (Fujita and Yoshikawa 1999, Lee et al. 2010). By intravenous administration, Fujita and Yoshikawa (1999) reported a reduction in blood pressure by 30 mm Hg in spontaneously hypertensive rats (SHR) by giving Leu–Lys–Pro–Asp–Met, a peptide derived from thermolysis-digest of dried bonito at a dose of 100 μg/kg. A greater reduction by 50 mm Hg was observed when Leu–Lys–Pro, the hydrolyzed product of Leu–Lys–Pro–Asp–Met by ACE, was given at a dose of 30 μg/kg. Furthermore, a study conducted over an 8-week period in humans involving 30 subjects with hypertension produced an interesting result that blood pressure was reduced by 60–66% in individuals fed a thermolysis-digest of dried bonito. In another study (Kawasaki et al. 2002), sardine protein hydrolysate incorporated at a dosage of 0.5 g into a vegetable drink was shown in a randomized double-blind placebo-controlled study on 63 subjects over 13 weeks to lower systolic blood

Table 36.1. Angiotensin Converting Enzyme (ACE) Inhibitory Peptides Derived from Fishery Sources

Source	Enzyme	Peptide Sequence	IC_{50} Value[a]	References
Tuna frame	Peptic enzymes	Gly–Asp–Leu–Gly–Lys–Thr–Thr–Thr–Val–Ser–Asn–Trp–Ser–Pro–Pro–Lys–Try–Lys–Asp–Thr–Pro	11.28 μM	Lee et al. (2010)
Shark meat	Protease	Cys–Phe	1.96 μM	Wu et al. (2008)
		Glu–Tyr	2.68 μM	
		Phe-Glu	1.45 μM	
Bigeye tuna dark muscle	Pepsin	Trp–Pro–Glu–Ala–Ala–Glu–Leu–Met–Met–Glu–Val–Asp–Pro	21.6 μM	Qian et al. (2007)
Anchovy	natural fermentation	Arg–Pro	21 μM	Ichimura et al. (2003)
		Lys–Pro	22 μM	
		Ala–Pro	29 μM	
Salmon	Thermolysin	Val–Trp	2.5 μM	Ono et al. (2003)
		Ile–Trp	4.7 μM	
		Met–Trp	9.9 μM	
		Leu–Trp	17.4 μM	
Cuttlefish	Bacterial protease	Ala–His–Ser–Tyr	11.6 μM	Balti et al. (2010a)
		Gly–Asp–Ala–Pro	22.5 μM	
		Ala–Gly–Ser–Pro	37.2 μM	
		Asp–Phe–Gly	44.7 μM	
Cuttlefish	Proteases	Val–Tyr–Ala–Pro	6.1 μM	Balti et al. (2010b)
		Val–Ile–Ile–Phe	8.7 μM	
		Met–Ala–Trp	16.32 μM	
Shrimp	Peptide enzymes	Leu–His–Pro	1.6 μM	Cao et al. (2010)
Shrimp	*Lactobacillus fermentum* enzymes	Asp–Pro	2.15 μM	Wang et al. (2008b)
		Gly–Thr–Gly	5.54 μM	
		Ser–Thr	4.03 μM	
Hard clam	Protamex	Tyr–Asn	51 μM	Tsai et al. (2008)
Oyster	Pepsin	Val–Val–Tyr–Pro–Trp–Thr–Gln–Arg–Phe	66 μM	Wang et al. (2008a)
Freshwater clam	Protamex and flavourzyme	Val–Lys–Pro	3.7 μM	Tsai et al. (2006)
		Val–Lys–Lys	1045 μM	
Blue mussel	Natural fermentation	Glu–Val–Met–Ala–Gly–Asn–Leu–Tyr–Pro–Gly	19.34 μg/mL	Je et al. (2005d)

[a] IC_{50} value is defined as the concentration of inhibitor required to inhibit 50% of the ACE activity.

pressure by 11 mm Hg, while the control group showed no decrease. Lee et al. (2010) reported that a single oral administration (10 mg/kg of body weight) of peptide from tuna frame hydrolysate displayed a strong suppressive effect on systolic blood pressure of SHR and this antihypertensive activity was similar to captopril. Moreover, no adverse side effects were observed in the rats after administration of antihypertensive peptide. Although, the exact mechanisms underlying this phenomenon have not yet been identified, it was suggested that bioactive peptides have higher tissue affinity and are subjected to a slower elimination than captopril.

Currently, protein hydrolysates from the bonito and sardine are widely available as supplements in Japan (Guérard et al. 2010). Bonito hydrolysate is also being sold in North America as a supplement to lower blood pressure. Similar products have been derived from other proteins, including soy and whey. However, active dosages in these nonmarine products are 1020 g/day, whereas bonito protein hydrolysis, for example, has a recommended daily dosage of only 1.5 g/day. A low daily dose is important for both ingredient cost and ease of formulation. Marine-derived bioactive peptides have potential for use as functional ingredients in nutraceuticals and pharmaceuticals due to their effectiveness in both prevention and treatment of hypertension. However, further studies are needed with clinical trials for these antihypertensive peptides.

Antioxidant Activity

Lipid oxidation is one of the major deteriorations in many types of natural and processed foods, leading to changes in food quality and nutritional value and also production of potentially toxic reaction products. Furthermore, lipid oxidations are major

Table 36.2. Source, Amino Acid Sequence and Molecular Weight of Antioxidative Peptides from Some Aquatic Protein Hydrolysates

Sources	Enzymes	Da	Peptide Sequence	References
Tuna cooking juice	Protease XXIII	751	Pro–His–His–Ala–Asp–Ser	Jao and Ko (2002)
Yellowfin sole frame	Pepsin, Mackerel intestine crude enzyme	1300	Arg–Pro–Asp–Phe–Asp–Leu–Glu–Pro–Pro–Tyr	Jun et al. (2004)
Alaska Pollack frame	Mackerel intestine crude enzyme	672	Leu–Pro–His–Ser–Gly–Tyr	Je et al. (2005c)
Giant squid muscle	Trypsin	747	Asn–Gly–Leu–Glu–Gly–Leu–Lys	Rajapakse et al. (2005b)
		1307	Asn–Ala–Asp–Phe–Gly–Leu–Asn–Gly–Leu–Glu–Gly–Leu–Ala	
Jumbo squid skin gelatin	Trypsin	880	Phe–Asp–Ser–Gly–Pro–Ala–Gly–Val–Leu	Mendis et al. (2005b)
		1242	Asn–Gly–Pro–Leu–Gln–Ala–Gly–Gln–Pro–Gly–Glu–Arg	
Hoki skin gelatin	Trypsin	797	His–Gly–Pro–Leu–Gly–Pro–Leu	Mendis et al. (2005a)
Conger eel	Tryptic enzyme	928	Leu–Gly–Leu–Asn–Gly–Asp–Asp–Val–Asn	Ranathunga et al. (2006)
Tuna backbone	Pepsin	1519	Val–Lys–Ala–Gly–Phe–Ala–Trp–Thr–Ala–Asn–Gln–Gln–Leu–Ser	Je et al. (2007)
Hoki frame	Pepsin	1801	Glu–Ser–Thr–Val–Pro–Glu–Arg–Thr–His–Pro–Ala–Cys–Pro–Asp–Phe–Asn	Kim et al. (2007)
Oyster protein	Pepsin	1600	Leu–Lys–Gln–Glu–Leu–Glu–Asp–Leu–Leu–Glu–Lys–Gln–Glu	Qian et al. (2008b)
Bigeye tuna dark muscle	Pepsin	1222	Leu–Asn–Leu–Pro–Thr–Ala–Val–Tyr–Met–Val–Thr	Je et al. (2008)
Tuna cooking juice	Orientase	1305	Pro–Val–Ser–His–Asp–His–Ala–Pro–Glu–Tyr	Hsu et al. (2009)
		938	Pro–Ser–Asp–His–Asp–His–Glu	
		584	Val–His–Asp–Tyr	
Sardinelle (head, viscera)	Crude extract from sardine viscera	431	Leu–His–Tyr	Bougatef et al. (2010)
Tuna dark muscle by-product	Protease XXIII	756	Pro–Met–Asp–Tyr–Met–Val–Thr	Hsu (2010)
		978	Leu–Pro–Thr–Ser–Glu–Ala–Ala–Lys–Tyr	
Loach muscle	Papain	464	Pro–Ser–Tyr–Val	You et al. (2010)

causes of many serious human diseases, such as cardiovascular disease, cancer, and neurological disorders as well as the aging process (Jittrepotch et al. 2006). To prevent foods from undergoing oxidative deterioration and to provide protection against serious diseases, it is important to inhibit the oxidation of lipids and formation of free radicals occurring in the living body and foodstuffs. In this regard, several antioxidants have been used to maintain food quality. Also, there is keen interest in antioxidants from natural sources for use as food supplements or processing aids. The commonly used antioxidants are chemically synthesized antioxidants such as butylated hydroxyanisole, butylated hydroxytoluene, tert-butylhydroquinone, etc. However, these antioxidants are suspected to pose toxicity problems in the long term and their use in foodstuffs is restricted or prohibited in some countries (Sakanaka et al. 2004, Je et al. 2005a, Pihlanto et al. 2008).

Recently, a number of studies have demonstrated that peptides derived from different marine animal protein hydrolysates act as potent antioxidants (Table 36.2). Some of these antioxidant peptides have exhibited varying capacities to scavenge free radicals. Several studies have indicated that peptides derived from marine fish proteins have greater antioxidant properties than α-tocopherol in different oxidative systems (Jun et al. 2004, Rajapakse et al. 2005b). However, the exact mechanism of action of bioactive peptides as antioxidants is not clearly known. Some peptides are capable of chelating metal ions, which acts as the pro-oxidant (Klompong et al. 2007). The antioxidant activity of protein hydrolysates has been attributed to the presence of certain amino acids in the peptide sequence. High amounts of histidine and some hydrophobic amino acids are related to the antioxidant potency. Dávalos et al. (2005) indicated that tryptophan, tyrosine and methionine showed the highest antioxidant activity among the amino acids, followed by cysteine, histidine, and phenylalanine, while histidine and hydrophobic amino acids were suggested by Peña-Ramos et al. (2004) as the key factors in delaying lipid oxidation. Mendis et al. (2005b) reported that the presence of nonaromatic amino acids such as proline, alanine, valine, and leucine in jumbo squid skin hydrolysate contributed to the higher antioxidative activities, and phenylalanine and leucine residues at *N*- and *C*-terminals of peptide could contribute to the high activity. Wang et al. (2008c) stated that peptides exhibiting good antioxidative activity usually contain certain amino acids such as histidine, proline, tyrosine and lysine. Bougatef et al. (2010) indicated that peptides containing histidine, tryptophan and tyrosine residues possessed antioxidative activity. Guo et al. (2009) concluded that peptides containing tyrosine residues at the *C*-terminus, lysine or phenylalanine residues at the *N*-terminus, and tyrosine residues in their sequences had strong free radical scavenging activity. Moreover, Suetsuna et al. (2000) indicated that some other amino acids such as proline, alanine, and leucine contribute to free radical scavenging activity. Leucine and proline could favor antioxidant activity when they occur at the *C*-terminus end of the sequence (Suetsuna et al. 2000). In general, peptides containing tyrosine tend to exhibit strong free radical scavenging activity due to the phenolic hydroxyl groups, which contribute substantially to scavenging activity toward free radicals via the mechanism of donating a hydrogen atom from their hydroxyl group (Suetsuna et al. 2000, Guo et al. 2009). Other aromatic amino acids, tryptophan and phenylalanine, are generally considered as effective radical scavengers, because they can donate protons easily to electron deficient radicals while at the same time maintaining their stability via resonance structures (Rajapakse et al. 2005b, Zhang et al. 2009). In case of histidine-containing peptides, it is thought to be connected to hydrogen-donating ability, lipid peroxy radical trapping, and/or the metal ion chelating ability of the imidazole group (Mendis et al. 2005a, b). Hydrophobic peptides can help in scavenging of free radicals by keeping close contact with oxidizing substances leading to the rapid scavenging of radicals (Mendis et al. 2005a, b). Some aromatic amino acids and histidine are reported to play a vital role in the observed activity. Gelatin peptides contain mainly hydrophobic amino acids and abundance of these amino acids favors a higher emulsifying ability. Hence, marine-gelatin-derived peptides are expected to exert higher antioxidant effects among other antioxidant peptide sequences (Mendis et al. 2005a). Therefore, marine-derived bioactive peptides with antioxidative properties may have great potential for use as nutraceuticals and pharmaceuticals and a substitute for synthetic antioxidants.

Antimicrobial Activity

Antimicrobial peptides (AMPs) have captured the attention of researchers in recent years because of their efficiency in inhibiting pathogens, bacteria, fungi, and virus. Adding preservative is a common way of preventing or slowing microbial growth, the major reason of food spoilage and poisoning. However, there is a shortage of efficient and safe preservatives as a result of appearance of resistant forms of food pathogens in response to massive use of preservatives. Marine-derived AMPs are found in a wide range of marine animals. These peptides are naturally synthesized as a part of innate host defense mechanisms (Brown and Hancock 2006) and may also be generated by enzymatic hydrolysis of native proteins (Bulet et al. 2004, Liu et al. 2008, Reddy et al. 2004). Among the marine animals, they are well described in fish (Kitts and Weiler 2003, Hwang et al. 2010) and the hemolymph of many marine invertebrates, including spider crab (Stensvag et al. 2008), American lobster (Battison et al. 2008), green sea urchin (Li et al. 2008), oysters (Liu et al. 2008), mussels, scallops, venerid clams and abalone (Cheng-Hua et al. 2009), and shrimp (Destoumieux et al. 1997,Bartlett et al. 2002). For example, the mudfish (*Misgurnus anguillicaudatus*) produces a strongly basic peptide termed misgurin (Park et al. 1997). This 2502 Da peptide contains five arginine and four lysine residues and has antimicrobial activity against a broad spectrum of microorganisms. The tissues of Atlantic salmon (*Salmo salar*) contain an antimicrobial peptide with a molecular mass of 20,734 Da. This purified peptide has been identified as a histone protein and is termed histone H1 (Richards et al. 2001). Three antibacterial basic polypeptides, HLP-1, HLP-2, and HLP-3 have been isolated from acetic extracts of channel catfish (*Ictalurus punctatus*) skin (Robinette et al. 1998). The molecular mass of these peptides were reported to be 15.5 kDa, 30 kDa, and 15.0 kDa, respectively. HLP-1 is the dominant peptide and closely related to histone H2B, with inhibitory effects against both fish bacterial pathogens and *Escherichia coli* D31. Catfish (*Parasilurus asotus*) also contains a strong antimicrobial peptide with the molecular mass of 2000 Da, named parasin I consisting of 19 amino acids that include three arginine and five lysine residues (Park et al. 1998). Because of high homology to *N*-terminal sequence of histone H2A, parasin I may be derived from histone H2A by a specific protease cleavage. These AMPs have strong activity against gram-negative and gram-positive bacteria as well as fungi. Potent antibacterial peptides originating from pardaxin have been isolated from Moses sole fish (*Pardachirus marmoratus*) (Shai 1994). Pardaxin is composed of a helix-hinge-helix structure that is a common motif found in many antibacterial and cytotoxic peptides. The peptide, termed pleurocidin, is found specifically in the general epithelial mucous cells of flounder and has an amphipathic α-helical conformation enabling the binding to anionic phospholipid rich membranes. In addition, two hydrophobic proteins with molecular masses of 31 kDa and 27 kDa, respectively, have also been isolated from skin mucous of carp (*Cyprinus carpio*)(Lemaitre et al. 1996). In addition, marine-derived AMPs, such as urechistachykinins, piscidins, and arenicin-1 exhibited significant antimicrobial activities against human microbial pathogens without remarkable hemolytic effects against human erythrocytes (Hwang et al. 2010).

Of marine invertebrates, antibacterial activity has been reported in the hemolymph of the blue crab, *Callinectus sapidus*, and it was highly inhibitory to gram-negative bacteria (Edward et al. 1996). Although there are several reports on antibacterial activity in seminal plasma, few antibacterial peptides have been reported in mud crab, *Scylla serrata* (Jayasankar and Subramonium 1999). The antimicrobial peptide derived from

American lobster (*Homarus americanus*) exhibited bacteriostatic activity against some gram-negative bacteria and both protozoastatic and protozoacidal activity against two scuticociliate parasites *Mesanophrys chesapeakensis* and *Anophryoides haemophila* (Battison et al. 2008). Furthermore, antimicrobial activity and growth inhibition of bacteria such as *E. coli*, *Pseudomonas aeruginosa*, *Bacillus subtilis*, and fungi such as *Botrytis cinerea*, and *Penicillium expansum* have been reported by antimicrobial peptide, CgPep33, derived from oyster (*Crassostrea gigas*) (Liu et al. 2008). In addition, Lee and Susumu (1998) isolated two peptides, Leu–Leu–Glu–Tyr–Ser–Ile and Leu–Leu–Glu–Tyr–Ser–Leu, from thermolysin hydrolysate of oyster, *C. gigas* that inhibited HIV-1 protease. Moreover, an active peptide, Arg–Arg–Trp–Trp–Cys–Arg–X (where X is an amino acid or an amino acid analog) isolated from the enzymatic hydrolysate of oyster, *C. gigas,* showed high inhibitory activity on herpes virus (Zeng et al. 2008). In general, these AMPs are characterized by relatively short cationic peptides with common features, including (1) α-helices, (2) β-sheet and small proteins, (3) peptides with thio-ether rings, (4) peptides with an overrepresentation of one or two amino acids, (5) lipopeptides, and (6) macrocyclic cystine knot peptides (Epand and Vogel 1999). Their modes of action are based on interaction with the microbial cellular lipid bilayer and eventual disintegration of the cell membrane (Hwang et al. 2010). In addition to antimicrobial action, AMPs have demonstrated diverse biological effects, all of which participate in the control of infectious and inflammatory diseases (Kamysz et al. 2003, Guaní-Guerraa et al. 2010). Therefore, there are several potential applications of AMPs to develop new strategies or substitute the existing ones for more sustainable and promising future of these peptides.

Calcium Binding Peptides

Calcium is an important mineral for the human body. Generally, the principal sources of calcium are milk and other diary products (Anderson and Garner 1996). Casein phosphopeptides (CPP) obtained after intestinal digestion of casein enhance bone calcification (Tsuchita et al. 1993). CPP have the capacity of chelating Ca and preventing the precipitation of Ca phosphate salts, resulting in the increased availability of soluble Ca for absorption (Berrocal et al. 1989, Yuan and Kitts 1994). However, some groups of people do not drink milk due to lactose indigestion and intolerance. Therefore, peptides from other sources, especially from aquatic source can be an alternative for use in food to increase Ca solubility and bioavailability. Bone oligopeptides with high affinity to calcium was prepared from hoki bone with the aid of tuna intestine crude enzyme, which was able to degrade hoki bone matrices comprising collagen, noncollagenous proteins, carbohydrates, and minerals (Jung et al. 2005). Fish bone phosphopeptides (FBP) containing 23.6% phosphorus and a molecular weight (MW) of 3.5 kDa could solubilize calcium (Jung et al. 2005). Fish bone peptide II (FBP II) with a high content of phosphopeptide from hoki bone could inhibit the formation of insoluble Ca salts. The levels of femoral total Ca, bone mineral density, and strength were increased in ovariectomized rats fed with FBP II diet (Jung et al. 2006b). Alaska pollack backbone peptide, Val–Leu–Ser–Gly–Gly–Thr–Thr–Met–Ala–Met–Ala–Met–Tyr –Thr–Leu–Val with the MW of 1442 Da, prepared using pepsin hydrolysis showed affinity for calcium ions on the surface of hydroxyapatite crystals. The peptide could solubilize calcium at levels similar to that by casein phosphopeptide (Jung et al. 2006a). Furthermore, calcium binding peptide derived from pepsinolytic hydrolysate of hoki frame was identified as Val–Leu–Ser–Gly–Gly–Thr–Thr–Met–Tyr–Ala–Ser–Leu–Tyr–Ala–Glu with MW of 1561 Da (Jung and Kim 2007). Therefore, fish bone oligophosphopeptide could be used as the nutraceutical to increase the absorption of calcium.

Anticoagulant Activity

An anticoagulant therapy is a type of medication that may be used to prevent blood from coagulating or clotting. Blood clotting could be life-threatening for patients with atherosclerosis and related cardiovascular diseases, the leading cause of heart attack, stroke, and death (Libby 2002, Levi et al. 2003, Schultz et al. 2003). The mainstays of clinical anticoagulant treatments are heparin, which is a cofactor of plasma-derived and naturally occurring inhibitors of thrombin, and coumarins that antagonize the biosynthesis of vitamin K-dependent coagulation factors. Although effective and widely used, heparins and coumarins have practical limitations because their pharmacokinetics and anticoagulation effects are unpredictable, with the risk of many undesirable side effects, such as hemorrhaging and thrombocytopenia resulting in the need for close monitoring of their use. More seriously, heparins are involved in many aspects of cellular physiology (Kakkar 2003), making their long-term uses as anticoagulants plagued with potential side effects. The anticoagulant marine-derived bioactive peptides have rarely been reported, but have been isolated from marine organisms such as marine echiuroid worm (Jo et al. 2008), starfish Koyama et al. 1998), and blue mussel (Jung and Kim 2009). Moreover, marine anticoagulant proteins have been purified from blood of ark shell (Jung et al. 2001) and yellowfin sole Rajapakse et al. 2005a). The anticoagulant activity of the above peptides has been determined by prolongation of activated partial thromboplastin time (APTT), prothrombin time PT) and thrombin time assays and the activity was compared with heparin, the commercial anticoagulant. The anticoagulant peptide Gly–Glu–Leu–Thr–Pro–Glu–Ser–Gly–Pro–Asp–Leu–Phe–Val–His–Phe–Leu–Asp–Gly–Asn–Pro–Ser–Tyr–Ser–Leu–Tyr–Ala–Asp–Ala–Val–Pro–Arg, isolated from marine echiuroid worm, effectively prolonged the normal clotting time on APTT from 32.3 ± 0.9 to 192.2 ± 2.1 seconds in a dose-dependent manner (Jo et al. 2008). This peptide binds specifically with clotting factor FIXa, a major component of the intrinsic tenase complex. An anticoagulant peptide Glu–Ala–Asp–Ile–Asp–Gly–Asp–Gly–Gln–Val–Asn–Tyr–Glu –Glu–Phe–Val–Ala–Met–Met–Thr–Ser–Lys, derived from blue mussel, showed prolongation of 321 ± 2.1 seconds clotting time on APTT (from 35.3 ± 0.5 seconds of control) and 81.3 ± 0.8 seconds clotting time on TT (from 11.6 ± 0.4 seconds of control), respectively (Jung and Kim 2009). In addition, a protein derived from the blood of arch shell, prolonged the APTT from a 32 seconds control clotting time to 325 seconds. (Jung et al. 2001). These marine-derived anticoagulant peptides have potential for use as functional ingredients in nutraceuticals or pharmaceuticals.

Immunomodulatory Activity

There has been increased interest in the study of the relationship between nutrition and immunity due to the hypothesis that consumption of specific foods may reduce susceptibility for the development and/or progression of immunological diseases. Many immunomodulatory peptides are derived from milk proteins. Some of these peptides have been found to be immunoinhibitory, while others are immunostimulatory in action. Furthermore, it has been found that oral administration of the peptides is a highly effective way to induce the desired immunomodulatory effect, even in the absence of any transport agents such as delivery vehicles. In addition, it has also been found that the amount of peptide required to produce the therapeutic effect by oral delivery can be significantly lower than that required to produce a similar effect when the peptide is delivered by parenteral injection. The immunomodulatory effects of marine-derived bioactive peptides have rarely been reported. Gildberg et al. (1996) reported that four acid peptide fractions from cod stomach hydrolysate were shown to possess an ability to stimulate leucocyte superoxide anion production in Atlantic salmon (*S. salar*). Increasing the production of reactive oxygen metabolites or promoting the phagocytosis and pinocytosis in macrophages can enhance the nonspecific immune defense system. It has been suggested that acid-derived cod protein hydrolysate peptide fractions might be useful as an adjuvant in fish vaccine and as an immune stimulant in fish feed. Immunomodulatory effects of marine oligopeptide preparation (MOP) from Chum Salmon (*Oncorhynchus keta*) were also reported by Yang et al. (2009b). Female ICR mice (6–8 weeks old) were administered the MOP for 4 weeks with the dose of 0, 0.22, 0.45, and 1.35 g/kg/body weight. In comparison with the control group, MOP could significantly enhance the capacity of lymphocyte proliferation induced by the mitogen concanavalin A, the number of plaque-forming cells, natural killer (NK) cell activity, the percentage of $CD4^+$ T helper (Th) cells in spleen and the secretion of Th1 (interleukin (IL)-2, IFN-γ) and Th2 (IL-5, IL-6) type cell cytokines. Nevertheless, no significant differences in weight gain, lymphoid organ indices, and phagocytosis capacity were observed. In irradiation-treated mice (Yang et al. 2010), MOP significantly increased the survival rate and prolonged the survival times for 30 days after irradiation, and lessened the radiation-induced suppression of T- or B-lymphocyte proliferation, resulting in the recovery of cell-mediated and humoral immune functions. This effect may be produced by augmentation of the relative numbers of radioresistant $CD4^+$ T cells, enhancement of the level of immunostimulatory cytokine, IL-12, reduction of the level of total cellular NF-κB through the induction of IκB in spleen, and inhibition of the apoptosis of splenocytes. In addition, it is proposed that MOP can be used as an ideal adjuvant therapy to alleviate radiation-induced injuries in cancer patients.

The oligopeptide-enriched hydrolysates from oyster (*C. gigas*), produced using the protease from *Bacillus* sp. SM98011 possessed antitumor activity and immunostimulating effects of the oyster hydrolysates in BALB/c mice (Wang et al. 2010). The growth of transplantable sarcoma-S180 was obviously inhibited in a dose-dependent manner in BALB/c mice given the oyster hydrolysates. Mice receiving 0.25, 0.5, and 1 mg/g of body weight by oral gavage had 6.8%, 30.6%, and 48% less tumor growth, respectively. Concurrently, the weight coefficients of the thymus and the spleen, the activity of NK cells, the spleen proliferation of lymphocytes and the phagocytic rate of macrophages in S180-bearing mice significantly increased after administration of the oyster hydrolysates. These results demonstrated that oyster hydrolysates produced strong immunostimulating effects in mice, which might result in its antitumor activity. The antitumor and immunostimulating effects of oyster hydrolysates prepared in this study reveal its potential for tumor therapy and as a dietary supplement with immunostimulatory activity.

Gastrin/CCK-Like Activity

Different authors have reported the presence of gastrin/CCK-like molecules in protein hydrolysates from fish by-products (Cancre et al. 1999, Ravallec-Ple and Van Wormhoudt 2003). These molecules are the only known members of the gastrin family in humans, and could have a positive effect on food intake in humans and fish species in aquaculture. Gastrin is a gastric hormone that stimulates postprandial gastric acid secretion and epithelial cell proliferation. In humans, there are two different gastrins, one with 17 and one with 34 amino acids residues. CCK is a group of peptides that controls the emptying of the gallbladder, as well as pancreatic enzyme secretion. It is also a growth factor, and regulates intestinal motility, satiety signaling, and inhibition of gastric acid secretion (Rehfeld et al. 2001). Both gastrin and CCK inhibit food intake and share a common COOH-terminal pentapeptide amide that also includes the sequences essential for biological activity. The incorporation of protein hydrolysates including gastrin/CCK-like molecules in functional foods could be of interest for controlling appetite, food intake, and obesity. The production of protein hydrolysates including CCK-like molecules could also be useful in the treatment of paralytic ileus, for the removal of small concrements from the common bile duct and also for the improvement of pancreatic insufficiency caused by long-term parenteral nutrition or chronic pancreatitis (Rehfeld 2004). Above and beyond the stimulation of satiety through bioactive peptides, lipid-lowering effects as well as increasing metabolic rate are also beneficial in fighting obesity.

PRODUCTION OF BIOACTIVE PEPTIDES

Enzyme Hydrolysis

Marine bioactive peptides have widely been produced by enzymatic hydrolysis of proteins derived from marine animals (Zhao et al. 2007, Je et al. 2008, Sheih et al. 2009, Šližytė et al. 2009). These newly formed peptides can retain the biological properties of the native protein or can show new properties. Using appropriate proteolytic enzymes through the control of process parameters such as pH, time, and enzyme/substrate ratio, it is possible to produce hydrolysates whose components may present some interesting biological properties. Especially for food and pharmaceutical industries, the enzymatic hydrolysis method is preferred because of lack of residual organic solvents or toxic chemicals in the products. Hydrolysis can be generally carried out in a thermostatically stirred-batch reactor in which the hydrolysis conditions (pH, temperature, enzyme concentration, and stirring speed) are adjusted in order to optimize the activity of the enzyme used. An initial mixing is usually done to adjust the pH and temperature to the desired values. To obtain the hydrolysate with high stability with less pro-oxidants, the raw materials can be washed to remove heme protein and lipids (Kristinsson 2007). In addition, antioxidants can be added before hydrolysis. To enhance the hydrolysis process, the raw materials may be subjected to size reduction by homogenization. The bone, scales, or skin, which may interfere with enzymatic hydrolysis, should be removed. Thereafter, the selected protease is added into the mixture containing the protein substrate, which is previously adjusted to pH to the desired value. After hydrolysis, the reaction is terminated by inactivation of protease by heat treatment or pH adjustment. The combination of pH and temperature to denature the protease used is another approach to avoid the harsh condition, which may affect the resulting hydrolysate or peptides. The reaction mixture containing the peptides as well as the unhydrolyzed debris is centrifuged or filtered. The supernatant or filtrate is concentrated or dried. The hydrolysate can be an excellent source of peptides with functionalities and bioactivities, which are determined by the types of protease, pretreatment of raw materials, the condition of hydrolysis, and so on. In order to enhance the bioactivity of peptides produced, several approaches have been implemented such as the use of multistep hydrolysis with different proteases (Phanturat et al. 2010). Moreover, it is possible to obtain serial enzymatic digestions in a system using a multistep recycling membrane reactor combined with ultrafiltration membrane system to separate marine-derived bioactive peptides (Jeon et al. 1999, Byun and Kim 2001, Kim and Mendis 2006). This membrane bioreactor technology equipped with ultrafiltration membranes is recently emerging for the development of bioactive compounds and considered as a potential method to utilize marine proteins as value added nutraceuticals with beneficial health effects.

Choices of Enzyme

Different proteases exhibit varying specificities and reaction rates in the hydrolysis of polypeptide chains. Many authors, in view of the economic interest in the recovering of protein from poorly studied species of fish, have compared some commercial proteases in order to test the most suitable one for the substrate employed. The most common commercial proteases reported are both from plant source such as papain (Quaglia and Orban 1987a, 1987b, Hoyle and Merritt 1994, Shahidi et al. 1995) or from animal origin such as pepsin (Viera et al. 1995) and chymotrypsin or trypsin (Simpson et al. 1998). Enzymes of microbial origin have been also applied to the hydrolysis of fish. In comparison to animal- or plant-derived enzymes, microbial enzymes offer several advantages including a wide variety of available catalytic activities, greater pH, and temperature stabilities (Diniz and Martin 1997). Generally, Alcalase® 2.4 L have been repeatedly favored for fish hydrolysis due to the high degree of hydrolysis (DH) that can be achieved in a relatively short time under moderate conditions compared to neutral or acidic enzymes (Hoyle and Merritt 1994, Shahidi et al. 1994, Diniz and Martin 1996, Martin and Porter 1995, Benjakul and Morrissey 1997).

Alternatively, endogenous proteases in the muscle or internal organs can serve as an important source of proteases, which can be used for production of peptides. Khantaphant and Benjakul (2008) used the ammonium sulphate fraction of fish pyloric caeca extract for preparation of gelatin hydrolysate from brownstripe red snapper skin with 2,2-diphenyl-1-picrylhydrazyl (DPPH) and 2,2-azino-bis (3-ethylbenzothiazoline-6-sulfonic acid) (ABTS) radical scavenging activities. In addition, Phanturat et al. (2010) used the pyloric caeca extract from bigeye snapper (*Priacanthus macracanthus*) for preparation of gelatin hydrolysate with antioxidative activity. The antioxidative peptide of gelatin hydrolysate produced had MW of 1.7 kDa. To maximize autolysis or hydrolysis, optimal pH and temperature are implemented, thereby enhancing the cleavage of peptides. Nevertheless, it could be difficult to control the DH or obtaining the desired peptides by autolysis or the use of endogenous protease.

The impact of the enzyme's specificity is a key factor influencing both the characteristics of hydrolysates and the nature and composition of peptides produced. Proteolysis can operate either sequentially, releasing one peptide at a time, or through the formation of intermediates that are further hydrolyzed to smaller peptides as proteolysis progresses, which is often termed "the zipper mechanisms" (Panyam and Kilara 1996). Depending on the specificity of the enzyme, environmental conditions and the extent of hydrolysis, a wide variety of peptides can be preferentially generated.

Hydrolytic curves reported for enzymatic hydrolysis of different protein substrates by protease generally exhibited an initial fast reaction followed by a slowdown. With regards to the effect of the enzyme concentration, it was found that the DH increased with higher enzyme concentrations. Less important increases were found with enzyme treatment at higher concentration. The exact concentration of enzymes required to hydrolyze the substrate at a required DH could be calculated when log10 (enzyme concentration) versus DH were plotted (Benjakul and Morrissey 1997). Diniz and Martin (1996) used response surface analysis to study the effects of pH, temperature, and enzyme/substrate ratio on the DH of Dogfish proteins. The polynomial model they proposed was well adjusted to the experimental data and was sufficiently accurate for predicting the DH for any combination of independent variables within the ranges studied. However, it should be realized that the higher DH could also negatively affect the biological activity of peptide.

The hydrolytic reaction also depends on the availability of susceptible bonds, on which the primary enzymic attack is concentrated, and on the physical structure of the protein molecule (Raghunath 1993). Very little is known about the factors limiting enzyme activity. The hypothesis explaining the downward curvature could be related to one of the following phenomena: a decrease in the concentration of peptide bonds available for hydrolysis, enzyme inhibition, or enzyme deactivation. To obtain more information about enzyme deactivation or inhibition, the activity enzyme could be assayed during the course of hydrolysis, compared at the beginning and during the course of the hydrolysis. A regular decrease in the areas observed during the course of hydrolysis suggests an enzyme deactivation or an enzyme inhibition by the inhibitory peptides, which are continuously solubilized during hydrolysis. Such behavior also suggests possible deactivation of the enzyme over the time due to a low stability, including the possibility that the enzyme hydrolyzes itself. This observation together with the results of the extra enzyme and substrate additions makes it possible to conclude that the shape of the hydrolysis curves can be explained as a result of the lack of peptide bonds available for hydrolysis combined to a partial enzyme deactivation during the course of hydrolysis.

Choices of Substrates

Muscle of different fish species have been used for the production of protein hydrolysate such as capelin (*Mallotus villosus*) (Amarowicz and Shahidi 1997), Atlantic salmon (*S. salar*) (Kristinsson and Rasco 2000), mackerel (*Scomber austriasicus*) (Wu et al. 2003), herring (*Clupea harengus*) (Sathivel et al. 2003), smooth hound (*Mustelus mustelus*) (Bougatef et al. 2009), yellowstripe trevally (*Selaroides leptolepis*) (Klompong et al. 2007), red salmon (*Oncorhynchus nerka*) (Sathivel et al. 2005), round scad (*Decapterus maruadsi*) (Thiansilakul et al. 2007), tilapia (*Oreochromis niloticus*) (Raghavan et al. 2008), silver carp (*Hypophthalmichthys molitrix*) (Dong et al. 2008), and loach (*M. anguillicaudatus*) (You et al. 2009). Recently, tuna dark muscle has been used for production of hydrolysate containing antioxidative peptides (Hsu 2010). For some fish species, their myofibrillar proteins were isolated as protein isolate prior to enzymatic hydrolysis such as channel catfish (*I. punctatus*) (Theodore et al. 2008). The protein isolate might be more preferable for the proteinase used. As a consequence, hydrolysis could take place at the higher degree, compared with that found in the intact muscle. Moreover, oyster (*C. gigas*) protein (Qian et al. 2008b) and giant squid muscle (Rajapakse et al. 2005b) were also used as source of protein hydrolysate with the antioxidative activity.

By-products obtained from fish-processing plants have been used for the preparation of protein hydrolysate with functional properties and bioactivities. In general, a variety of proteases have been used to cleave the protein, resulting in the peptides with varying functional properties and bioactivity. By-products (heads, viscera, frames, skin, trimmings) of black scabbardfish (*Aphanopus carbo*) (Bougatef et al. 2010) and those (head and viscera) from sardinella (*Sardinella aurita*) were used to produce the protein hydrolysate. The frames from yellowfin sole (Jun et al. 2004), Alaska ollack (Je et al. 2005c), and hoki (*Joohnius belengerii*) (Kim et al. 2007) as well as backbones from tuna (Je et al. 2007) and Atlantic cod (*Gadus morhua*) (Šližytė et al. 2009) have been used for the preparation of protein hydrolysates. Apart from solid by-products, liquid effluent such as cooking juice from tuna (Jao and Ko 2002, Hsu et al. 2009) have been used as the raw material for hydrolysate production.

Protein hydrolysates were also prepared from gelatin extracted from aquatic animals. Gelatin hydrolysates with the antioxidative activity have been produced from gelatin from the skin of Alaska pollack (Kim et al. 2001), hoki (Mendis et al. 2005a), cobia (Yang et al. 2008), and sole (Gimenez et al. 2009). Furthermore, gelatin hydrolysates were also prepared from the skin of jumbo squid (Mendis et al. 2005b), bullfrog (Qian et al. 2008a), and squid (Gimenez et al. 2009). These hydrolysates were mainly prepared with the aid of proteolytic enzymes. Since gelatin can be hydrolyzed at high temperature, thermal hydrolysis was applied to produce the gelatin hydrolysate from cobia and tilapia skins (Yang et al. 2008, 2009a). Recently, a protein hydrolysate from gelatin from the Nile tilapia scale with the antioxidative activity was prepared by Ngo et al. (2010).

Microbial Fermentation

Fermented fishery products are a good source of peptides and amino acids, and rich sources of structurally diverse bioactive compounds. Thus, fishery proteins are of particular interest due to their high protein content and diverse physiological activities in the human organism. A great variety of naturally formed bioactive peptides have been found in fermented fishery products. In fermented marine food sources, enzymatic hydrolysis has already been carried out using microorganisms; thus, bioactive peptides can be purified without further hydrolysis (Je et al. 2005b, Je et al. 2005d).

Proteolysis is an important biochemical process occurring during fermentation. It is likely to be constituted by a very large and complex group of enzymes, which differ in properties such as substrate specificity, active site and catalytic mechanism, pH and temperature optima, and stability profile. In general, enzymes important in fermentation may originate from four general sources: (1) viscera and digestive systems, (2) muscle tissue, (3) plant raw materials added to the fermentation, and (4) microorganisms active in the fermentation.

Proteolytic microorganisms play an important role during fishery fermentation. There are several groups of microorganisms that are well known for excretion of proteolytic enzymes capable of degrading proteins. Many types of microbes excrete proteolytic enzymes, including *Aspergillus spp.*, *Bacillus spp*, halophilic bacteria, halotolerant and lactic acid bacteria (Mackie et al. 1971, Lopetcharat et al. 2001). Careful selection by seeding or controlling the growth environment within the fermentation chamber enables the desired microbes to flourish and produce significant quantities of proteolytic enzymes that help to hydrolyze the fish protein (Wheaton and Lawson 1985).

From our study, ACE inhibitory activity and antioxidative activity of peptides extracted from various Thai traditional

Table 36.3. ACE Inhibitory Activity and Antioxidative Activities of Soluble Fractions from Some Fermented Fishery Products

Type of Products	ACE Inhibitory Activity[a] (%)	DPPH[a] (μmolTE/g protein)	ABTS[a] (μmolTE/g protein)	FRAP[a] (μmolTE/g protein)
Fermented shrimp				
Kapi I	3.36 ± 0.79^{n}	53.67 ± 0.53^{bc}	94.40 ± 0.00^{d}	42.01 ± 0.34^{m}
Kapi II	42.25 ± 1.92^{kl}	49.49 ± 1.19^{c}	75.30 ± 2.17^{fg}	42.23 ± 0.15^{m}
Kapi III	0.00 ± 0.00^{n}	43.25 ± 3.64^{d}	55.70 ± 3.70^{l}	33.39 ± 0.07^{no}
Kapi IV	0.00 ± 0.00^{n}	26.40 ± 1.25^{jkl}	150.55 ± 5.60^{c}	21.98 ± 0.26^{r}
Kung-chom I	53.82 ± 4.70^{i}	37.61 ± 0.69^{ef}	264.17 ± 3.72^{a}	31.02 ± 0.19^{op}
Kung-chom II	52.40 ± 0.08^{i}	36.55 ± 0.66^{ef}	186.40 ± 8.13^{b}	27.53 ± 0.09^{pq}
Kheeuy-naam I	66.38 ± 0.26^{gh}	28.89 ± 0.42^{ijk}	63.89 ± 2.14^{ijk}	14.15 ± 0.08^{s}
Kheeuy-naam II	0.00 ± 0.00^{n}	27.07 ± 0.90^{ijkl}	59.21 ± 1.36^{jkl}	188.10 ± 2.67^{e}
Fermented fish				
Bu-du I	65.83 ± 2.69^{h}	74.14 ± 7.54^{a}	183.49 ± 1.70^{b}	60.86 ± 0.55^{k}
Bu-du II	0.00 ± 0.00^{n}	55.12 ± 0.49^{b}	81.98 ± 2.24^{ef}	450.22 ± 5.86^{c}
Tai-pla I	68.17 ± 0.43^{gh}	31.57 ± 2.50^{hi}	86.14 ± 0.63^{e}	22.81 ± 0.52^{r}
Tai-pla II	45.51 ± 4.96^{kj}	40.63 ± 0.63^{de}	189.10 ± 5.43^{b}	24.69 ± 0.26^{qr}
Pla-ra I	76.66 ± 1.13^{de}	23.44 ± 3.46^{lm}	51.79 ± 3.54^{l}	8.43 ± 0.02^{tu}
Pla-ra II	44.04 ± 1.00^{kl}	25.40 ± 0.13^{klm}	28.82 ± 0.00^{n}	148.43 ± 2.51^{g}
Pla-ra III	74.74 ± 0.09^{ef}	27.18 ± 0.25^{ijkl}	43.05 ± 2.96^{m}	10.96 ± 0.26^{st}
Pla-ra IV	47.10 ± 1.33^{kj}	40.68 ± 1.03^{de}	40.35 ± 5.27^{m}	363.90 ± 4.18^{d}
Pla-chao I	65.77 ± 3.46^{h}	16.42 ± 2.40^{m}	25.66 ± 3.79^{no}	3.54 ± 0.09^{v}
Pla-chao II	87.54 ± 1.55^{b}	$13.17 \perp 0.16^{m}$	16.92 ± 0.95^{p}	35.84 ± 2.55^{n}
Pla-som I	77.31 ± 1.40^{cde}	20.76 ± 0.56^{l}	57.14 ± 6.70^{kl}	10.32 ± 0.43^{st}
Pla-som II	80.74 ± 0.37^{cd}	30.57 ± 2.55^{hij}	71.03 ± 7.27^{ghi}	12.98 ± 0.23^{st}
Pla-som III	97.96 ± 6.24^{a}	15.56 ± 3.54^{m}	53.06 ± 0.00^{l}	5.16 ± 0.29^{uv}
Pla-som IV	81.97 ± 0.89^{c}	34.62 ± 0.55^{fg}	17.48 ± 1.88^{op}	105.47 ± 1.34^{i}
Som-fug I	76.57 ± 1.11^{de}	13.49 ± 0.22^{m}	22.10 ± 1.11^{nop}	53.24 ± 1.74^{l}
Som-fug II	73.12 ± 2.77^{ef}	15.95 ± 0.31^{m}	17.19 ± 3.72^{p}	71.54 ± 1.11^{j}
Pla-chom I	31.27 ± 3.99^{m}	22.78 ± 0.18^{lm}	29.52 ± 1.66^{n}	112.12 ± 3.56^{h}
Pla-chom II	40.28 ± 3.44^{l}	25.28 ± 0.38^{klm}	22.24 ± 2.29^{nop}	181.74 ± 3.54^{f}
Kem-buk-nud I	49.14 ± 0.22^{ij}	50.28 ± 1.52^{c}	65.52 ± 1.06^{hij}	527.56 ± 2.28^{b}
Kem-buk-nud II	49.69 ± 0.00^{ij}	51.58 ± 1.07^{c}	78.84 ± 2.69^{efg}	582.06 ± 4.61^{a}
Fermented mussels				
Hoi-dong I	28.92 ± 0.78^{m}	29.02 ± 0.58^{ijk}	18.26 ± 2.27^{op}	151.78 ± 1.08^{g}
Hoi-dong II	71.11 ± 1.31^{fg}	28.89 ± 1.37^{ijk}	73.04 ± 4.31^{gh}	14.02 ± 0.05^{s}

TE, Trolox equivalents.
[a]Mean ± SD from duplicate determinations.
[a–v]Different superscripts in the same column indicate significant differences ($p < 0.05$). ACE inhibitory activity of soluble fractions from some traditional fermented fishery products was assayed at a final concentration of 3 mg dried extraction/mL.
DPPH, 2,2-diphenyl-1-picrylhydrazyl; ABTS, 2,2-azino-bis (3-ethylbenzothiazoline-6-sulfonic acid); FRAP, ferric reducing antioxidant power.

fermented fishery products were studied (Table 36.3). *Pla-som* had the highest inhibition of ACE, whereas *Bu-du*, *Kung-chom*, and *Kem-buk-nud* exhibited the highest DPPH radical scavenging activity, ABTS radical scavenging activity, and ferric reducing antioxidant power (FRAP) ($p < 0.05$), respectively. On the basis of principal component analysis (PCA), peptides with high ACE inhibitory activities were likely to be found in products with low pH and low salt content. The presence of these peptides was probably related to the use of garlic and raw material used but not affected by the ratio of free and total free α-amino acids (F/T ratio). For antioxidative activity, peptides with high FRAP were likely to be found in products with high F/T ratio. Products with high DPPH radical scavenging activity generally had high ABTS radical scavenging activity but both activities showed no correlation to pH, salt content, and F/T ratio. Therefore, some Thai traditional fermented fishery products might serve as possible protective agents in human diets to reduce damage from high blood pressure and oxidative stress.

The fish silage processing technology developed in the late 1970s revealed the possibility of recovering pepsins and bioactive peptides from fish silage. Silage generally refers to the process of preserving wet fodder, that is, whole fish or parts of fish

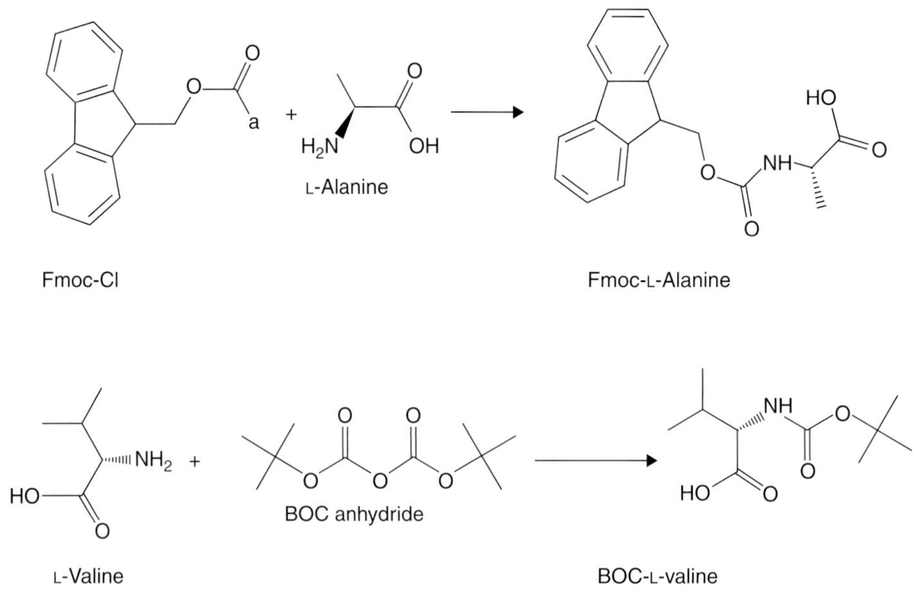

Figure 36.2. Protecting groups commonly used in solid phase peptide synthesis.

and so on, by adding acid or by the anaerobic production of lactic acid by bacteria. Whereas the pepsins are used for gentle bioprocessing of certain fishery products, the peptides occurred by autolysis may be valuable immune stimulants (Gildberg 2004). In vitro and in vivo studies have shown that certain peptide fractions in fish protein hydrolysates may stimulate the nonspecific immune defense system. Both fish sauce and fish silage are protein hydrolysates with immune stimulating properties. Generally, fish sauce is regarded as a typical Asian product made from tropical fish species, but ancient literature reveals that fish sauce was a common food product in Southern Europe more than 2000 years ago. Recent studies have shown that it can be made also from cold water species.

Chemical Peptide Synthesis

On the basis of the peptide sequences available in all published sources and existing peptide databases (Liu et al. 2008), for example, the EROP-Moscow oligopeptide database, SwePep database, peptide of interest can be obtained through chemical synthesis with appreciable purity and reasonable price. Peptide synthesis is a chemical process of coupling of the carboxyl group of one amino acid to the amino group of another amino acid. Usually, chemical techniques are used to synthesize peptides of up to 30–40 amino acids length. Peptide synthesis process can be classified on the basis of used techniques and type of the final product. Liquid-phase peptide synthesis can be divided into two types, step-by-step peptide synthesis with subsequent adding of one amino acid at ones from *C*-terminal to *N*-terminal and block-synthesis with coupling of polypeptide fragments. Liquid-phase peptide synthesis is used in large-scale peptide production for industry. In contrast, solid phase peptide synthesis (SPPS) is a process during which the polypeptide chain is covalently bound via linker to the porous insoluble bed particles (Merrifield 1963). SPPS involves two stages in which the first stage of the technique consists of peptide chain assembly with protected amino acid derivatives on a polymeric support. The second stage of the technique is the cleavage of the peptide from the resin support with the concurrent cleavage of all side chain protecting groups to give the crude free peptides. Currently, two protecting groups are commonly used in solid phase peptide synthesis—Fmoc or 9-fluorenylmethyl carbamate and t-Boc or di-tert-butyl dicarbonate (Fig. 36.2). Fmoc chemistry is known for generating peptides of higher quality and in greater yield than t-Boc chemistry. Impurities in t-Boc-synthesized peptides are mostly attributed to cleavage problems, dehydration, and t-butylation. The advantage of Fmoc is that it is cleaved under very mild basic conditions (e.g., piperidine), but stable under acidic conditions. After base treatment, the nascent peptide is typically washed and then a mixture including an activated amino acid and coupling reagents is placed in contact with the nascent peptide to couple the next amino acid. After coupling, noncoupled reagents can be washed away and then the protecting group on the *N*-terminus of the nascent peptide can be removed, allowing additional amino acids or peptide material to be added to the nascent peptide in a similar fashion. After cleavage from the resin, peptides are usually purified by reverse phase high-performance liquid chromatography using columns such as C-18, C-8, and C-4. The primary advantage of SPPS is its high yield. With modern SPPS instrumentation, coupling and deprotection yields greater than 99.99%, giving an overall yield of greater than 99% for a 50 amino acid peptide.

Production of Peptides by Recombinant DNA Technology

Technologies for heterologous protein expression have been invented and developed exponentially (Terpe 2006). Although there are a variety of expression systems by which a protein or peptide can be recombinantly produced, expression of a peptide in a recombinant host cell has met with limited success. Frequently, peptides are poorly expressed in bacteria because intracellularly the peptide may encounter one or more difficulties including insolubility, instability, and degradation. If a peptide is successfully produced recombinantly and purified, such isolated peptide may have reduced biological activity compared to the same amino acids that are part of the protein from which the peptide sequence is derived. Such reduced activity is typical because the active domain of the peptide is conformational rather than linear; and thus, in solution, the isolated peptide does not occur in proper conformation for full biological activity, or such proper conformation is only one of many alternative structures of the peptide.

To overcome the difficulties encountered in expression of peptides, there have been various attempts to express a peptide in a conformation reflective of the conformation of the peptide when expressed as part of the protein from which it is derived. One of those approaches is to express the peptide as part of a fusion protein (Liu et al. 2007). Typically, a fusion protein consists of a microbial (e.g., bacterial) polypeptide backbone into which is incorporated an amino acid sequence representing one or more heterologous peptide sequences. A step in fusion protein expression comprises inserting into a gene, encoding the microbial polypeptide, a nucleic acid sequence encoding one or more heterologous peptides. Usually, the gene is part of an expression vector, such that when the vector is introduced into a host cell system, the fusion protein is then produced either as remaining host cell-associated, or secreted into the culture medium of the expression system. In the art of recombinant protein expression, there remains a need for new systems, and new compositions for the production and delivery of biologically active and stable peptides for use. For example, necessary control elements for efficient transcription and translation (promoters and enhancer sequences) that are compatible with, and recognized by the particular host system used for expression must be incorporated. In the case that the fusion protein may be lethal or detrimental to the host cells, the host cell strain/line and expression vectors may be chosen such that the action of the promoter is inhibited until specifically induced. A variety of operons such as the trp operon, are under different control mechanisms. The trp operon is induced when tryptophan is absent in the growth media. The P_L promoter can be induced by an increase in temperature of host cells containing a temperature sensitive lambda repressor. In this way, greater than 95% of the promoter-directed transcription may be inhibited in uninduced cells. Thus, expression of the fusion protein may be controlled by culturing transformed or transfected cells under conditions such that the promoter controlling the expression from the fusion sequence encoding the fusion protein is not induced, and when the cells reach a suitable density in the growth medium, the promoter can be induced for expression from the inserted DNA. Accordingly, a recombinant DNA molecule containing a fusion sequence, can be ligated into an expression vector at a specific site in relation to the vector's promoter, control, and regulatory elements so that when the recombinant vector is introduced into the host cell, the fusion sequence can be expressed in the host cell. The recombinant vector is then introduced into the appropriate bacterial host cells, and the host cells are selected, and screened for those cells containing the recombinant vector. Selection and screening may be accomplished by methods known, depending on the vector and expression system used (*E. coli*, *B. subtilis*, yeast, fungi, mammalian, and insect cells). Bacterial expression is the usual starting point for expression of heterologous proteins. Bacterial fermentations are inexpensive and can reach high cell densities resulting in high volumetric yields of the target protein. The use of *B. subtilis* has many advantages, such as its GRAS status and easy and inexpensive culturing methods that can result in very high cell densities. Baculovirus/insect cell systems have found wide application for the expression of highly recalcitrant proteins such as protein tyrosine kinases. As with bacterial expression systems, yeasts (*Saccharomyces cerevisiae* and *Pichia pastoris*) offer relatively inexpensive growth media and high-density fermentation. Furthermore, yeast offers the capability of carrying out limited posttranslational modifications such as disulfide bond formation and glycosylation. Mammalian cell expression has become the system of choice for production of complex, glycosylated biotherapeutic proteins such as antibodies, growth factors, and fertility hormones. Several developed methodologies appropriate for large-scale industrial production of intact recombinant peptides in bacteria have been developed without significant toxicity and a complicated purification method. Using this technology, pilot-scale fermentation can be performed to produce large quantities of biologically active peptides. Together, this new method represents a cost-effective means, compared to chemical peptide synthesis to enable commercial production of peptides in large-scale under good laboratory manufacturing practice (GMP) for application in humans.

Membrane Technology

Membrane processes involve separating components by means of selective permeability through the membrane according to flow under pressure over the surface of membrane. The technology is actually a family of processes that include reverse osmosis, nanofiltration, ultrafiltration, and microfiltration, which can be differentiated by the separation range (Short 1995). Because of the fact that the differences in physicochemical properties of active sequences are often small, the specific separation of one or more peptides from a raw hydrolysate is difficult. Until now, ultrafiltration has provided the best method available for the enrichment of peptides (Korhonen and Pihlanto 2006, Kozlov and Moya 2007), but the selectivity between peptides of close similar MW is poor. Among them, ultrafiltration has been widely used to enrich bioactive peptides from protein hydrolysates by which enzymatic hydrolysis can be performed through either conventional batch hydrolysis or continuous hydrolysis using ultrafiltration membranes.

Ultrafiltration is a pressure-driven membrane process that generally uses membranes with pore sizes in the range of 0.05 μm−1 nm at pressures from 10 to 100 psi to remove both ionic species and low MW solutes from the stream being processed (Short 1995). Ultrafiltration can offer a great advantage to recover and concentrate a small amount of substances from a large volume of liquid phase. Since a positive pressure is employed as a driving force to push the liquid phase through the membrane pores without involving a phase change, ultrafiltration is especially suitable for the separation of sensitive biological substances with activity, such as enzymes. While water and particles smaller than membrane pores pass through, larger molecules, that is, colloids, emulsion droplets, and particulates, are retained and concentrated. Ultrafiltration processes have three distinctive characteristics from the conventional filtration process: (1) they are cross-flow systems in which the solution flows parallel to the membrane surface, clearing away any particles accumulated on the membrane surface, (2) they are critically dependent on membrane materials and their nominal MW cutoff (MWCO), and (3) they depend on membrane geometry in the actual equipment that highly affects the efficiency of the entire process (Belter et al. 1988). The mechanism of separation is complex and is influenced by numerous factors, such as, method of membrane manufacture, composition of the membrane, chemical interactions between the feed stream and the membrane, fluid dynamics of the membrane, pressure, temperature, and velocity of the feed stream (Dziezak 1990). The main problem in many ultrafiltration processes is fouling or the buildup of a layer on the membrane surface until the retained mass offers hydrodynamic resistance and interferes with flux (Paulson and Wilson 1987). Although conventional pressure-driven processes have characteristics that make them capable of performing most of the separations required economically, their fouling problems dramatically reduce their efficiency and selectivity when separating similar sized molecules.

In order to improve the yield and selectivity of peptide separation from hydrolysates, integrative processes based on the use of electrodialysis and filtration membrane (EDFM) have been recently developed (Bazinet et al. 2005, Bazinet and Firdaous 2009). EDFM combines size exclusion capabilities of filtration membranes with the proper MWCO range with the charge selectivity of electrodialysis. With no pressure applied to the ED cell, only the charged molecules migrate under the electrical field and the neutral molecules theoretically stay in the primary solution and do not pass the filtration membrane. In addition, the ED cell can be configured into multicompartment to perform simultaneous sequential separation of different cationic or anionic peptides (Fig. 36.3). The application has been tested for the purification

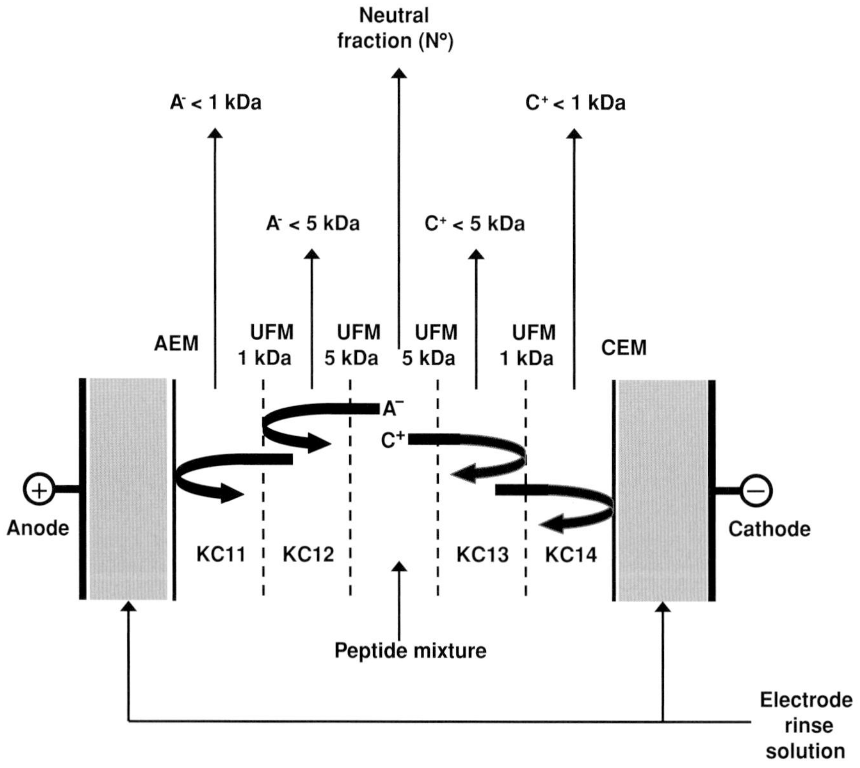

Figure 36.3. Configuration of the electrodialysis (ED) cell and filtration membrane (FM) for simultaneous sequential fractionation of anionic and basic peptides. AEM, anion-exchange membrane; UFM, ultrafiltration membrane; CEM, cation-exchange membrane.

or separation of bioactive peptides (Poulin et al. 2006, 2007, 2008). EDFM represents a reliable and cost-effective separation technique, especially when compared to traditional methods, such as chromatography. Further scale-up developments will be necessary to confirm its feasibility at large scale.

Chromatography

Chromatography offers a variety of methods to separate peptides based on charge, hydrophobicity, size, and molecular recognition. Chromatography is a powerful technique to achieve high degrees of purity. It has been well established in biotechnological industries as a production scale unit operation and as an analytical tool to monitor the quality of raw materials and end products and the purification efficiency of sequential downstream operations. Much effort has been given to develop selective column chromatography methods that can isolate and purify bioactive peptides. However, lesser reported are the advances in preparative chromatography of bioactive peptides. This area will play an increasingly important role in the cost-efficient and environmentally acceptable manufacture of bioactive peptides. Purification of peptides may be performed via different separation techniques used separately or in combination. Some, like crystallization or precipitation, are fairly well established, although they are constantly being improved. Others, like liquid–liquid extraction, chromatography, and electrophoresis, are being developed with a view to being scaled up in the near future.

CONCLUSIONS

Marine-derived bioactive peptides play a vital role in human health and have potential as active ingredients for preparation of various functional foods and pharmaceutical products. The possibilities of designing new functional foods and pharmaceuticals based on marine-derived peptides to support reducing or regulating diet related chronic ailments are promising. These peptides can be naturally formed by hydrolysis by proteolytic enzymes or fermentation. Furthermore, fish processing by-products can be easily utilized for producing bioactive peptides. Until now, most of the biological activities of marine-derived bioactive peptides have been observed in vitro or in mouse model systems. Human clinical studies are limited or nonexistent. Therefore, further research is needed in order to clarify the relevance and potential therapeutic role of bioactive peptides in human subjects. A few commercial food products supplemented with marine-derived bioactive peptides have been launched in limited markets. As part of a normal dietary regimen, they must demonstrate their effects in amounts that are reasonably expected to be consumed. Recently, industrial-scale technologies suitable for the industrial production of bioactive peptides have been developed. The multifunctional properties of marine-derived peptides appear to offer considerable potential for the development of many similar products in the near future.

REFERENCES

Amarowicz R, Shahidi F. 1997. Antioxidant activity of peptide fractions of capelin protein hydrolysates. *Food Chem* 58(4): 355–359.

Anderson JJB, Garner SC. 1996. Calcium and phosphorous nutrition in health and disease. In: JJB Anderson, SC Garner (eds.) *Calcium and Phosphorus in Health and Disease*. CRC Press, New York, pp. 1–5.

Aneiros A, Garateix A. 2004. Bioactive peptides from marine sources: Pharmacological properties and isolation procedures. *J Chromatogr B* 803: 41–53.

Atkinson AB, Robertson JIS. 1979. Captopril in the treatment of clinical hypertension and cardiac failure. *Lancet* 2: 836–839.

Balti R et al. 2010a. Analysis of novel angiotensin I-converting enzyme inhibitory peptides from enzymatic hydrolysates of cuttlefish (*Sepia officinalis*) muscle proteins. *J Agric Food Chem* 58: 3840–3846.

Balti R et al. 2010b. Three novel angiotensin I-converting enzyme (ACE) inhibitory peptides from cuttlefish (*Sepia officinalis*) using digestive proteases. *Food Res Int* 43: 1136–1143.

Barrow C, Shahidi F. 2008. *Marine Nutraceuticals and Functional Foods*. CRC Press, New York.

Bartlett TC et al. 2002. Crustins, homologues of an 11.5-kDa antibacterial peptide, from two species of penaeid shrimp, *Litopenaeus vannamei* and *Litopenaeus setiferus*. *Marine Biotechnol* 4: 278–293.

Battison AL et al. 2008. Isolation and characterization of two antimicrobial peptides from haemocytes of the American lobster *Homarus americanus*. *Fish Shellfish Immunol* 25: 181–187.

Bazinet L, Firdaous L. 2009. Membrane processes and the devices developed for bioactive peptide. *Recent Patents Biotechnol* 3: 61–72.

Bazinet L et al. 2005. Process and system for separation of organic charged compounds. International Patent WO2005082495 A1.

Belter PA et al. 1988. *Bioseparations: Downstream Processing for Biotechnology*. John Wiley & Sons, New York.

Benjakul S. Morrissey MT. 1997. Protein hydrolysates from Pacific whiting solid wastes. *J Agric Food Chem* 45: 3423–3430.

Berrocal R et al. 1989. Tryptic phosphopeptides from whole casein. II. Physiochemical properties related to the solubilization of calcium. *J Dairy Res* 56: 335–341.

Bougatef A et al. 2009. Antioxidant and free radical-scavenging activities of smooth hound (*Mustelus mustelus*) muscle protein hydrolysates obtained by gastrointestinal proteases. *Food Chem* 114: 1198–1205.

Bougatef A et al. 2010. Purification and identification of novel antioxidant peptides from enzymatic hydrolysates of sardinelle (*Sardinella aurita*) by-products proteins. *Food Chem* 118: 559–565.

Brown KL, Hancock REW. 2006. Cationic host defense (antimicrobial) peptides. *Curr Opin Immunol* 18: 24–30.

Bulet P et al. 2004. Anti-microbial peptides: From invertebrates to vertebrates. *Immunol Rev* 198: 169–184.

Byun HG, Kim SK. 2001. Purification and characterization of angiotensin I converting enzyme (ACE) inhibitory peptides from Alaska Pollack (*Theragra chalcogramma*) skin. *Process Biochem* 36: 1155–1162.

Cancre I et al. 1999. Secretagogues and growth factors in fish and crustacean protein hydrolysates. *Marine Biotechnol* 1(5): 489–494.

Cao W et al. 2010. Purification and identification of an ACE inhibitory peptide from the peptic hydrolysate of *Acetes chinensis* and its antihypertensive effects in spontaneously hypertensive rats. *Int J Food Sci Technol* 45: 959–965.

Cheng-Hua L et al. 2009. A review of advances in research on marine molluscan antimicrobial peptides and their potential application in aquaculture. *Molluscan Res* 29: 17–26.

Cheung HS, Chushman DW. 1971. Spectrophotometric assay and properties of the angiotensin-converting enzyme of rabbit lung. *Biochem Pharmacol* 20: 1637–1648.

Dávalos A et al. 2005. Antioxidant properties of commercial grape juices and vinegars. *Food Chem* 93: 325–330.

Destoumieux D et al. 1997. Penaeidins: a new family of antimicrobial peptides in the shrimp, *Penaeus vannamei* Decapoda). *J Biol Chem* 272: 28398–28406.

Diniz FM, Martin AM. 1996. Use of response surface methodology to describe the combined effects of pH, temperature and E/S ratio on the hydrolysis of dogfish (Squalus acanthias) muscle. *Int J Food Sci Technol* 31(5): 419–426.

Diniz FM, Martin AM. 1997. Fish protein hydrolysates by enzymatic processing. *Agro Food Industry Hi-Tech* 8(3): 9–13.

Dong S et al. 2008. Antioxidant and biochemical properties of protein hydrolysates prepared from silver carp (*Hypophthalmichthys molitrix*). *Food Chem* 107: 1485–1493.

Dziezak JD. 1990. Membrane separation technology offers processors ultimated potential. *Food Technol* 9: 108–113.

Edward NJ et al. 1996. Specificity and some physicochemical characteristics of the antibacterial activity from blue crab *Callinectus sapidus*. *Fish Shellfish Immunol* 6: 403–413.

Epand R, Vogel H. 1999. Diversity of antimicrobial peptides and their mechanisms of action. *Biochimca et Biophysica Acta* 1462: 11–28.

Erdos EG, Skidgel RA. 1987. The angiotensin I–converting enzyme. *Lab Invest* 56: 345–348.

Fujita H, Yoshikawa M. 1999. LKPNM: A prodrug-type ACE-inhibitory peptide derived from fish protein. *Immunopharmacology* 44: 123–127.

Gildberg A. 2004. Enzymes and bioactive peptides from fish waste related to fish silage, fish feed and fish sauce production. *J Aquat Food Prod Technol* 13(2): 3–11.

Gildberg A et al. 1996. Isolation of acid peptide fractions from a fish protein hydrolysate with strong stimulatory effect on Atlantic salmon (*Salmo salar*) head kidney leucocytes. *Comp Biochem Physiol B: Biochem Mol Biol* 114(1): 97–101.

Gimenez B et al. 2009. Antioxidant and functional properties of gelatin hydrolysates obtained from skin of sole and squid. *Food Chem* 114: 976–983.

Guaní-Guerraa E et al. 2010. Antimicrobial peptides: General overview and clinical implications in human health and disease. *Clin Immunol* 135: 1–11.

Guérard F et al. 2010. Recent developments of marine ingredients for food and nutraceutical applications: a review. *J Can Sci Halieutiques et Aqatiques* 2: 21–27.

Guo H et al. 2009. Structures and properties of antioxidative peptides derived from royal jelly protein. *Food Chem* 113: 238–245.

Hoyle NT, Merritt JH. 1994. Quality of fish protein hydrolysates from herring (*Clupea harengus*). *J Food Sci* 59(1): 76–79.

Hsu KC. 2010. Purification of antioxidative peptides prepared from enzymatic hydrolysates of tuna dark muscle by-product. *Food Chem* 122: 42–48.

Hsu KC et al. 2009. Antioxidative properties of peptides prepared from tuna cooking juice hydrolysate with orientase (*Bacillus subtilis*). *Food Res Int* 42: 647–652.

Hwang B et al. 2010. Antimicrobial peptides derived from the marine organisms and its mode of action. *Korean J Microbiol Biotechnol* 38: 19–23.

Ichimura T et al. 2003. Angiotensin I-converting enzyme inhibitory activity and insulin secretion stimulative activity of fermented fish sauce. *J Biosci Bioeng* 96(5): 496–499.

Jao CL, Ko WC. 2002. 1,1-Diphenyl-2-picrylhydrazyl (DPPH) radical scavenging by protein hydrolyzates from tuna cooking juice. *Fish Sci* 68: 430–435.

Jayasankar V, Subramonium T. 1999. Antibacterial activity of seminal plasma of the mud crab *Scylla serrata* (forskal). *J Exp Mar Biol Ecol* 236: 253–259.

Je JY et al. 2005a. Preparation and antioxidative activity of hoki frame protein hydrolysate using ultrafiltration membranes. *Eur Food Res Technol* 221: 157–162.

Je JY et al. 2005b. Isolation of angiotensin I converting enzyme (ACE) inhibitor from fermented oyster (*Crassostrea gigas*) *sauce*. *Food Chem* 90: 809–814.

Je JY et al. 2005c. Antioxidant activity of a peptide isolated from Alaska pollack (*Theragra chalcogramma*) frame protein hydrolysate. *Food Res Int* 38: 45–50.

Je JY et al. 2005d. Angiotensin I converting enzyme (ACE) inhibitory peptide derived from the sauce of fermented blue mussel, *Mytilus edulis*. *Bioresour Technol* 96: 1624–1629.

Je JY et al. 2007. Purification and characterization of an antioxidant peptide obtained from tuna backbone protein by enzymatic hydrolysis. *Process Biochem* 42: 840–846.

Je JY et al. 2008. Purification and antioxidant properties of bigeye tuna (*Thunnus obesus*) dark muscle peptide on free radical-mediated oxidation systems. *J Med Food* 11(4): 629–637.

Jeon YJ et al. 1999. Improvement of functional properties of cod frame protein hydrolysates using ultrafiltration membranes. *Process Biochem* 35: 471–478.

Jittrepotch N et al. 2006. Effects of EDTA and a combined use of nitrite and ascorbate on lipid oxidation in cooked Japanese sardine (*Sardinops melanostictus*) during refrigerated storage. *Food Chem* 99: 70–82.

Jo HY et al. 2008. Purification and characterization of a novel anticoagulant peptide from marine echiuroid worm, *Urechis unicinctus*. *Process Biochem* 43: 179–184.

Johnston JI, Franz VL. 1992. Renin-angiotensin system: a dual tissue and hormonal system for cardiovascular control. *J Hypertens* 10: 13–26.

Jun SY et al. 2004. Purification and characterization of an antioxidative peptide from enzymatic hydrolysates of yellowfin sole (*Limanda aspera*) frame protein. *Eur Food Res Technol* 219: 20–26.

Jung WK, Kim SK. 2007. Calcium-binding peptide derived from pepsinolytic hydrolysates of hoki (*Johnius belengerii*) frame. *Eur Food Res Technol* 224: 763–767.

Jung WK, Kim SK. 2009. Isolation and characterization of an anticoagulant oligopeptide from blue mussel, *Mytilus edulis*. *Food Chem* 117: 687–692.

Jung WK et al. 2001. A novel anticoagulant protein from *Scapharca broughtonii*. *J Biochem Mol Biol* 35: 199–205.

Jung WK et al. 2005. Preparation of hoki (*Johnius belengerii*) bone oligopeptide with a high affinity to calcium by carnivorous intestine crude proteinase. *Food Chem* 91: 333–340.

Jung WK et al. 2006a. Recovery of a novel Ca-binding peptide from Alaska pollack (*Theragra chalcogramma*) backbone by pepsinolytic hydrolysis. *Process Biochem* 41: 2097–2100.

Jung WK et al. 2006b. Fish-bone peptide increases calcium solubility and bioavailability in ovariectomised rats. *Br J Nutr* 95: 124–128.

Kakkar AK. 2003. An expanding role for antithrombotic therapy in cancer patients. *Cancer Treat Rev* 29(Suppl 2): 23–26.

Kamysz W et al. 2003. Novel properties of antimicrobial peptides. *Acta Biochim Pol* 50: 461–469.

Kawasaki T et al. 2002. Antihypertensive effect and safety evaluation of vegetable drink with peptides derived from sardine protein hydrolysates on mild hypertensive, high-normal and normal blood pressure subjects. *Fukuoka Igaku Zasshi* 93: 208–18.

Khantaphant S, Benjakul S. 2008. Comparative study on the proteases from fish pyloric caeca and the use for production of gelatin hydrolysate with antioxidative activity. *Comp Biochem Physiol B: Biochem Mol Biol* 151B: 410–419.

Kim SK, Mendis E. 2006. Bioactive compounds from marine processing byproducts: A review. *Food Res Int* 39: 383–393.

Kim SK et al. 2001. Isolation and characterization of antioxidative peptides from gelatin hydrolysate of Alaska pollack skin. *J Agric Food Chem* 49: 1984–1989.

Kim SY et al. 2007. Purification and characterization of antioxidant peptide from hoki (*Johnius belengerii*) frame protein by gastrointestinal digestion. *J Nutr Biochem* 18: 31–38.

Kim SK et al. 2008. Prospective of the cosmeceuticals derived from marine organisms. *Biotechnol Bioprocess Eng* 13: 511–523.

Kitts DD, Weiler K. 2003. Bioactive proteins and peptides from food sources. Applications of bioprocesses used in isolation and recovery. *Curr Pharm Des* 9: 1309–132.

Klompong V et al. 2007. Antioxidative activity and functional properties of protein hydrolysate of yellow stripe trevally (*Selaroides leptolepis*) as influenced by the degree of hydrolysis and enzyme type. *Food Chem* 102: 1317–1327.

Korhonen H, Pihlanto A. 2006. Bioactive peptides: Production and functionality. *Int Dairy J* 16: 945–960.

Koyama T et al. 1998. Analysis for sites of anticoagulant action of plancinin, a new anticoagulant peptide isolated from the starfish *Acanthaster planci*, in the blood coagulation cascade. *Gen Pharmacol* 31: 277–282.

Kozlov M, Moya W. 2007. Membrane surface modification by radiation-induced polymerization. US20070272607 A1.

Kristinsson HG. 2007. Aquatic food protein hydrolysate. In: F Shahidi (ed.) *Maximising the Value of Marine By-products*. CRC Press, New York, pp. 229–248.

Kristinsson HG, Rasco BA. 2000. Biochemical and functional properties of Atlantic salmon (*Salmo salar*) muscle hydrolyzed with various alkaline proteases. *J Agric Food Chem* 48: 657–666.

Lee TG, Susumu M. 1998. Isolation of HIV-1 protease inhibiting peptides from thermolysin hydrolysate of oyster proteins. *Biochem Biophys Res Commun* 253(3): 604–608.

Lee SH et al. 2010. A novel angiotensin I converting enzyme inhibitory peptide from tuna frame protein hydrolysate and its antihypertensive effect in spontaneously hypertensive rats. *Food Chem* 118: 96–102.

Lemaitre C et al. 1996. Characterization and ion channel activities of novel antibacterial proteins from the skin mucosa of carp (*Cyprinus carpio*). *Eur J Biochem* 240(1): 143–149.

Levi M et al. 2003. Infection and inflammation and the coagulation system. *Cardiovasc Res* 60: 26–39.

Li GH et al. 2004. Angiotensin I-converting enzyme inhibitory peptides derived from food proteins and their physiological and pharmacological effects. *Nutr Res* 24: 469–486.

Li C et al. 2008. Strongylocins, novel antimicrobial peptides from the green sea urchin. *Dev Comp Immunol* 32: 1430–1440.

Libby P. 2002. Inflammation in atherosclerosis. *Nature* 420: 868–874.

Liu D et al. 2007. High-level expression of milk-derived antihypertensive peptide in *Escherichia coli* and its bioactivity. *J Agric Food Chem* 55: 5109–5112.

Liu Z et al. 2008. Production of cysteine-rich antimicrobial peptide by digestion of oyster (*Crassostrea gigas*) with alcalase and bromelin. *Food Control* 19: 231–235.

Lopetcharat K et al. 2001. Fish sauce products and manufacturing: review. *Food Rev Int* 17: 65–88.

Mackie IM et al. 1971. Fermented fish products, FAO Fisheries Report No. 100. Rome, Italy.

Martin AM, Porter D. 1995. Food flavors: Generation, analysis and process influence. In: G Charalambous (ed.) *Studies on the Hydrolysis of Fish Protein by Enzymatic Treatment*. Elsevier Science Press, New York, pp. 1395–1404.

Meisel H. 1997. Biochemical properties of regulatory peptides derived from milk proteins. *Biopolymers* 43: 119–128.

Mendis E et al. 2005a. Antioxidant properties of a radical-scavenging peptide purified from enzymatically prepared fish skin gelatin hydrolysate. *J Agric Food Chem* 53: 581–587.

Mendis E et al. 2005b. Investigation of jumbo squid (*Dosidicus gigas*) skin gelatin peptides for their in vitro antioxidant effects. *Life Sci* 77: 2166–2178.

Merrifield RB. 1963. Solid Phase Peptide Synthesis I: the synthesis of a tetrapeptide. *J Am Chem Soc* 85(14): 2149–2154.

Ngo DH et al. 2010. *In vitro* antioxidative activity of a peptide isolated from Nile tilapia (*Oreochromis niloticus*) sacle gelatin in free radical mediated oxidative systems. *J Funct Foods* 2(2): 107–117.

Ono S et al. 2003. Isolation of peptides with angiotensin I–converting enzyme inhibitory effect derived from hydrolysate of upstream chum salmon muscle. *Food Chem Toxicol* 68(5): 1611–1614.

Panyam D, Kilara A. 1996. Enhancing the functionality of food proteins by enzymatic modification. *Trends Food Sci Technol* 7(4): 120–125.

Park CB et al. 1997. A novel antimicrobial peptide from the loach, *Misgurnus anguillicaudatus*. *FEBS Lett* 411: 173–178.

Park IY et al. 1998. Cylexin: A P-selectin inhibitor prolongs heart allograft survival in hypersensitized rat recipients. *Transplant Proc* 30(7): 2927–2928.

Paulson DJ, Wilson RL. 1987. Crossflow membrane technology: Its use in the food industry, recent innovations. In: M Kroger, R Shapiro (eds.) *Changing Food Technology: Innovation and Communication: Partners for Progress in the Food Industry*. Technomic Publishing Co., Lancaster, PA, p. 85.

Peña-Ramos EA et al. 2004. Fractionation and characterization for antioxidant activity of hydrolysed whey protein. *J Sci Food Agric* 84(14): 1908–1918.

Phanturat P et al. 2010. Use of pyloric caeca extract from bigeye snapper (*Priacanthus macracanthus*) for the production of gelatin hydrolysate with antioxidative activity. *LWT-Food Sci Technol* 43: 86–97.

Pihlanto A et al. 2008. ACE-inhibitory and antioxidant properties of potato (*Solanum tuberosum*). *Food Chem* 108: 104–112.

Poulin JF et al. 2006. Simultaneous separation of acid and basic bioactive peptides by electrodialysis with ultrafiltration membrane. *J Biotechnol* 123: 314–328.

Poulin JF et al. 2007. Improved peptide fractionation by electrodialysis with ultrafiltration membrane: Influence of ultrafiltration membrane stacking and electrical field strength. *J Membr Sci* 299: 83–90.

Poulin JF et al. 2008. Impact of feed solution flow rate on peptide fractionation by electrodialysis with ultrafiltration membrane. *J Agric Food Chem* 56: 2007–2011.

Qian ZJ et al. 2007. Antihypertensive effect of angiotensin I converting enzyme-inhibitory peptide from hydrolysates of bigeye tuna dark muscle, *Thunnus obesus*. *J Agric Food Chem* 55: 8398–8403.

Qian ZJ et al. 2008a. Free radical scavenging activity of a novel antioxidative peptide purified from hydrolysate of bullfrog skin, *Rana catesbeiana* Shaw. *Bioresour Technol* 99: 1690–1698.

Qian ZJ et al. 2008b. Protective effect of an antioxidative peptide purified from gastrointestinal digests of oyster, *Crassostrea gigas* against free radical induced DNA damage. *Bioresour Technol* 99: 3365–3371.

Quaglia G, Orban E. 1987a. Enzymic solubilization of proteins of sardine (*Sardina pilchardus*) by commercial proteases. *J Sci Food Agric* 38: 263–269.

Quaglia GB, Orban E. 1987b. Enzymic solubilization of protein of sardine (*Sardina pilchardus*) by commercial enzymes. *J Sci Food Agric* 38: 263–269.

Raghavan S et al. 2008. Radical scavenging and reducing ability of tilapia (*Oreochromis niloticus*) protein hydrolysates. *J Agric Food Chem* 56: 10359–10367.

Raghunath MR. 1993. Enzymatic protein hydrolysate from tuna canning wastes: standardization of hydrolysis parameters. *Fishery Technol* 30: 40–45.

Rajapakse N et al. 2005a. A novel anticoagulant purified from fish protein hydrolysate inhibits factor XIIa and platelet aggregation. *Life Sci* 76: 2607–2619.

Rajapakse N et al. 2005b. Purification and *in vitro* antioxidative effects of giant squid muscle peptides on free radical-mediated oxidative systems. *J Nutr Biochem* 16: 562–569.

Ranaṭhunga S et al. 2006. Purification and characterization of antioxidantative peptide derived from muscle of conger eel (*Conger myriaster*). *Eur Food Res Technol* 222: 310–315.

Ravallec-Ple R, and Van Wormhoudt A. 2003. Secretagogue activities in cod (*Gadus morhua*) and shrimp (*Penaeus aztecus*) extracts and alcalase hydrolysates determined in AR4–2J pancreatic tumour cells. *Comp Biochem Physiol B: Biochem Mol Biol* 134(4): 669–679.

Reddy KV et al. 2004. Antimicrobial peptides: Premises and promises. *Int J Antimicrob Agents* 24: 536–547.

Rehfeld JF. 2004. Clinical endocrinology and metabolism: Cholecystokinin. *Best Practice Res Clin Endocrinol Metab* 18(4): 569–586.

Rehfeld JF et al. 2001. The predominant cholecystokinin in human plasma and intestine is cholecystokinin-33. *J Clin Endocrinol Metab* 86(1): 251–258.

Richards RC et al. 2001. Histone H1: an antimicrobial protein of Atlantic salmon (*Salmo salar*). *Biochem Biophys Res Commun* 284(3): 549–555.

Robinette D et al. 1998. Antimicrobial activity in the skin of the channel catfish Ictalurus punctatus: Characterization of broad-spectrum histone-like antimicrobial proteins. *Cell Mol Life Sci* 54(5): 467–475.

Sakanaka S et al. 2004. Antioxidant activity of egg-yolk protein hydrolysates in a linoleic acid oxidation system. *Food Chem* 86: 99–103.

Sathivel S et al. 2003. Biochemical and functional properties of herring (*Clupea harengus*) byproduct hydrolysates. *J Food Sci* 68: 2196–2200.

Sathivel S et al. 2005. Functional and nutritional properties of red salmon (*oncorhynchus nerka*) enzymatic hydrolysates. *J Food Sci* 70: 401–406.

Schultz MJ et al. 2003. Anticoagulant therapy for acute lung injury or pneumonia. *Curr Drug Targets* 4: 315–321.

Shahidi F et al. 1994. Proteolytic hydrolysis of muscle proteins of harp seal (Phoca groenlandica). *J Agric Food Chem* 42(11): 2634–2638.

Shahidi F et al. 1995. Production and characteristics of protein hydrolysates from capelin (*Mallotus villosus*). *Food Chem* 53: 285–293.

Shai Y. 1994. Pardaxin: Channel formation by a shark repellant peptide from fish. *Toxicology* 87(1–3): 109–129.

Sheih IC et al. 2009. Antioxidant properties of a new antioxidative peptide from algae protein waste hydrolysate in different oxidation systems. *Bioresour Technol* 100: 3419–3425.

Short J. 1995. Membrane separation in food processing, Chapter 9. In: RK Singh, SSH Rizvi (eds.) *Bioseparation Processes in Foods*. Marcel Dekker, New York, pp. 333–350.

Sibony M et al. 1993. Gene expression and tissue localization of the two isomers of ACE. *Hypertension* 21: 827–835.

Simpson BK et al. 1998. Enzymatic hydrolysis of shrimp meat. *Food Chem* 61: 131–138.

Šližytė R et al. 2009. Functional, bioactive and antioxidative properties of hydrolysates obtained from cod (*Gadus morhua*) backbones. *Process Biochem* 44: 668–677.

Stensvag K et al. 2008. Arasin 1, a proline-arginine-rich antimicrobial peptide isolated from the spider crab, *Hyas araneus*. *Dev Comp Immunol* 32: 275–285.

Steve BR et al. 1988. Human intestinal brush border angiotensin-converting enzyme activity and its inhibition by antihypertensive ramipril. *Gastroenterology* 98: 942–944.

Suetsuna K et al. 2000. Isolation and characterization of free radical scavenging activities peptides derived from casein. *J Nutr Biochem* 11: 128–131.

Terpe K. 2006. Overview of bacterial expression systems for heterologous protein production: from molecular and biochemical fundamentals to commercial systems. *Appl Microbiol Biotechnol* 72: 211–222.

Theodore AE et al. 2008. Antioxidative activity of protein hydrolysates prepared from alkaline-aided channel catfish protein isolates. *J Agric Food Chem* 56: 7459–7466.

Thiansilakul Y et al. 2007. Antioxidative activity of protein hydrolysate from round scad muscle using Alcalase and Flavourzyme. *J Food Biochem* 31: 266–287.

Tsai JS et al. 2008. ACE-inhibitory peptides identified from the muscle protein hydrolysate of hard clam (*Meretrix lusoria*). *Process Biochem* 43: 743–747.

Tsai JS et al. 2006. The inhibitory effects of freshwater clam (*Corbicula fluminea*, Muller) muscle protein hydrolysates on angiotensin I converting enzyme. *Process Biochem* 41: 2276–2281.

Tsuchita H et al. 1993. The effect of casein phosphopeptides on calcium utilization in young ovariectomized rats. *Z Ernahrungswiss* 32: 121–130.

Viera GH et al. 1995. Studies on the enzymatic hydrolysis of Brazilian lobster (*Panulirus* spp.) processing wastes. *J Sci Food Agric* 69(1): 61–65.

Wang J et al. 2008a. Purification and identification of an ACE inhibitory peptide from oyster proteins hydrolysate and the antihypertensive effect of hydrolysate in spontaneously hypertensive rats. *Food Chem* 111(2): 302–308.

Wang YK et al. 2008b. Production of novel angiotensin I-converting enzyme inhibitory peptides by fermentation of marine shrimp *Acetes chinensis* with *Lactobacillus fermentum* SM 605. *Appl Microbiol Biotechnol* 79: 785–791.

Wang Y et al. 2008c. Purification and characterization of antioxidative peptides from salmon protamine hydrolysate. *J Food Biochem* 32: 654–671.

Wang YK et al. 2010. Oyster (*Crassostrea gigas*) hydrolysates produced on a plant scale have antitumor activity and immunostimulating effects in BALB/c mice. *Marine Drugs* 8(2): 255–268.

Wheaton FW, Lawson TB. 1985. *Processing Aquatic Food Products*. John Wiley & Sons, New York, pp. 349–390.

Wu HC et al. 2003. Free amino acids and peptides as related to antioxidant properties in protein hydrolysates of mackerel (*Scomber austriasicus*). *Food Res Int* 36: 949–957.

Wu H et al. 2008. Purification and identification of novel angiotensin I-converting enzyme inhibitory peptides from shark meat hydrolysate. *Process Biochem* 43: 457–461.

Wyvratt MJ, Patchett AA. 1985. Recent developments in the design of angiotensin converting enzyme inhibitors. *Med Res Rev* 5: 483–531.

Yang JL et al. 2008. Characteristic and antioxidative activity of retorted gelatin hydrolysate from cobia (*Rachycentron canadum*) skin. *Food Chem* 110: 128–136.

Yang JI et al. 2009a. Process for the production of tilapia retorted skin gelatin hydrolysates with optimized antioxidative properties. *Process Biochem* 44: 1152–1157.

Yang R et al. 2009b. Immunomodulatory effects of marine oligopeptide preparation from Chum Salmon (*Oncorhynchus keta*) in mice. *Food Chem* 113: 464–470.

Yang R et al. 2010. Protective effect of a marine oligopeptide preparation from chum salmon (*Oncorhynchus keta*) on radiation-induced immune suppression in mice. *J Sci Food Agric* 90: 2241–2248.

You L et al. 2009. Effect of degree of hydrolysis on the antioxidant activity of loach (*Misgurnus anguillicaudatus*) protein hydrolysates. *Innovative Food Sci Emerging Technol* 10: 235–240.

You L et al. 2010. Purification and identification of antioxidative peptides from loach (*Misgurnus anguillicaudatus*) protein hydrolysate by consecutive chromatography and electrospray ionization-mass spectrometry. *Food Res Int* 43(4): 1167–1173.

Yuan YV, Kitts DD. 1994. Calcium absorption and bone utilization in spontaneously hypertensive rats fed on native and heat-damaged casein and soybean protein. *Br J Nutr* 71: 583–603.

Zeng M et al. 2008. Antiviral active peptide from oyster. *Chin J Oceanol Limnol* 26(3): 307–312.

Zhang SB et al. 2009. Purification and characterization of radical scavenging peptide from rapeseed protein hydrolysates. *J Am Oil Chem Soc* 86: 959–966.

Zhao Y et al. 2007. Antihypertensive effect and purification of an ACE inhibitory peptide from sea cucumber gelatin hydrolysate. *Process Biochem* 42: 1586–1591.

37
Natural Food Pigments

Benjamin K. Simpson, Soottawat Benjakul, and Sappasith Klomklao

Abstract: Nature has endowed living organisms with different pigments for a plethora of functions. These include enhancing the visual appeal of the source material, a means for camouflage and concealments, as an index of food quality, and even as sex attractants. The interest in natural pigments as food-processing aids derives from the increasing consumer aversion to the use of chemicals and synthetic compounds in foods. The chapter provides information on the major types of natural food pigments and their major sources, their functions and uses in food, their health benefits, and their fate under various processing and storage conditions. The chapter also provides information on their relative advantages over their counterparts perceived as artificial food colorants.

INTRODUCTION

Nature relies on a variety of compounds to impart colors to living organisms and to carry out various functions. The colors may be red, purple, orange, brown, yellow, green, blue, or their various shades. In the case of raw unprocessed foods, these compounds or natural pigments enhance their visual appeal to consumers who associate the colors of food with quality and freshness. These natural compounds are widespread in animals, plants, and microorganisms (including algae, fungi, and yeasts). In all these organisms, these compounds display various shades of black, blue, brown, green, orange, pink, red, or yellow colors. Examples of fresh foods with dark or black colors are blackberries, grapes, black beans, black olives; those with brown colors include seaweeds (brown algae), mushroom, cocoa, coffee, and tea; some others with blue or purple colors include blueberries,

Food Biochemistry and Food Processing, Second Edition. Edited by Benjamin K. Simpson, Leo M.L. Nollet, Fidel Toldrá, Soottawat Benjakul, Gopinadhan Paliyath and Y.H. Hui.

aubergines, raisins, currants, plums, pomegranates, and prunes; the kinds with green colors include asparagus, avocado, broccoli, Brussels sprouts, cabbage, celery, cucumber, green beans, green onion, kiwi fruit, lettuce, okra, spinach, and zucchini; examples of the types with orange or yellow colors are crustacea (crabs, lobster, scampi, and shrimps), salmonids (Arctic char, salmon, and trout), carrots, apricots, yellow peppers, mangoes, lemon, pumpkin, pineapple, squash; and the ones with shades of red colors include several fruits and vegetables (e.g., beets, cranberries, raspberries, red bell peppers, strawberries, tomatoes, and watermelon), meats (e.g., beef, hog, and poultry), and seaweeds (red algae).

Although it is their visual appeal that first attracts the consumers' attention to foods, these compounds also carry out a plethora of key functions in living organisms. In certain species, the compounds function as attractants or mating signals to gain the attention of their mates; in some cases, the colors provide camouflage against predators; some of the compounds participate in metabolic processes (e.g., chlorophylls or carotenoids in biosynthetic reactions and energy generation); some others serve as antioxidants and protect vulnerable biomolecules from oxidative damage; some carry out a transport function (myoglobin (Mb) and hemoglobin (Hb) for oxygen (O_2) transport and energy generation); while others play a role in vision or protect skins and membranes against the damaging effects of sunlight and radiation. Thus, it is not surprising that interest in these compounds extends beyond their visual appeal and consumer preferences in foods to also encompass other aspects such as the health benefits they provide. Each of these interests in itself can form the subject matter of an entire book; thus, the focus in this chapter will be on the major natural food pigments and their sources, structures and functions, properties and uses, and some of their important health benefits.

THE MAJOR NATURAL PIGMENT TYPES

For the purpose of this discourse, the term major natural food pigments is used to encompass the heme pigments (Mb and Hb), carotenoids, chlorophylls, anthocyanins, flavonoids, betalains, melanin, tannins, quinones, and xanthones. The heme pigments are the major pigments found in meats; they are water-soluble red, purplish or brownish compounds and they perform O_2 transport and energy generations functions in animal tissues; the carotenoids are bright red, orange, and yellow fat-soluble and water-insoluble pigments widespread in plants, animals, and microorganisms; the chlorophyll pigments tend to occur with the carotenoids and are found in the plastids of photosynthetic organisms (i.e., plants, algae, and certain bacteria); the anthocyanins are water-soluble red, blue, and violet pigments found in several plants (fruits and vegetables) and microorganisms; flavonoids (also known as anthoxanthins) are water-soluble pigments ranging from colorless to yellowish and are found in plant and microbial sources; the betalains are red and yellow water-soluble pigments found extensively in plants; melanins are shades of brown to dark colored pigments formed by the enzymatic oxidation of polyphenolic compounds and are responsible for the dark colorations in animals, plants, and microorganisms; tannins are colorless to yellow or brown colored pigments found in the barks of certain trees like the oak or sumac; the quinones are a group of pale yellow to dark brown or black pigments also found in plants and microorganisms; and xanthones are yellow-colored compounds found mostly in the plant families of *Bonnetiaceae, Clusiaceae*, and *Podostemaceae*.

THE HEME PIGMENTS (MB AND HB)

STRUCTURES AND FUNCTIONS

The two main heme pigments of interest in foods are Mb and Hb. They are responsible for the red color of meats. The principal function of these two pigments in animal tissues is O_2 transport for the purpose of energy generation. Mb comprises a single polypeptide chain called globin with a molecular weight of approximately 16.4 kDa. This single protein chain is bound to an essential nonprotein four-member ring compound known as porphyrin (or tetrapyrrole) through a central iron (Fe) atom. The Fe can exist in two oxidation states, the ferrous (Fe^{2+}), and the ferric (Fe^{3+}) forms. Both of these forms have a coordination number of six, giving heme the capacity to bind or accept up to six ligands (Fig. 37.1). In both heme pigments (Mb and Hb), the central Fe atom uses four (4) of its coordination positions to bind to the four nitrogen (N) atoms in the porphyrin ring, and its fifth coordination site to bind with a N atom of one of the

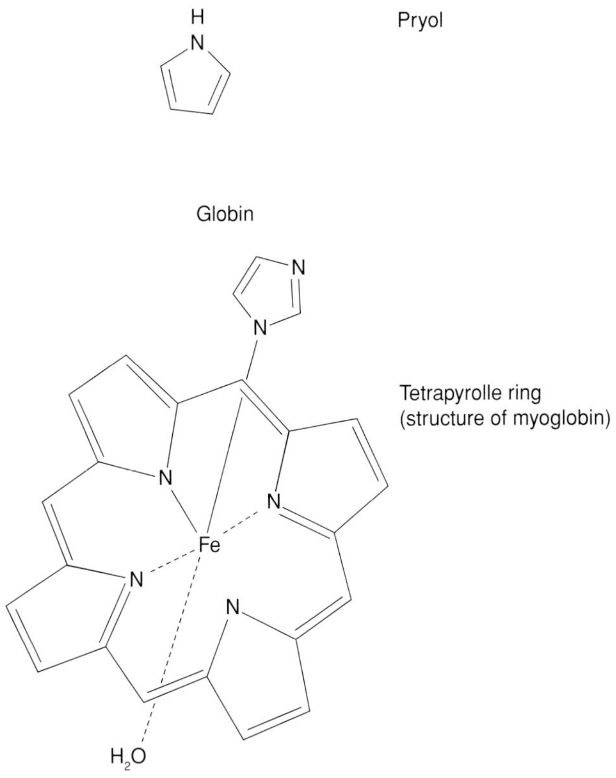

Figure 37.1. Tetrapyrolle ring structure of heme pigments.

histidine residues in globin. The sixth coordination site of the central Fe atom is available for binding with several groups (e.g., CO_2, CO, CN, H_2O, H_2O_2, NO, O_2). Which of these groups (or ligands) can bind to the central Fe atom largely depends on the oxidation state of the Fe. For example, :O_2, :NO, and :CO bind to Fe in the Fe^{2+} state, while -CN, -OH -SH and H_2O_2, bind to Fe in the Fe^{3+} form.

The synthesis of the heme pigments has initial stages that are common to those of the chlorophyll pigments, and lead to the formation of the tetrapyrrole or porphyrin ring structure. The common initial steps (for the heme and chlorophyll pigments) involve the formation of a 5-carbon intermediate compound, 5-aminolevulinic acid (dALA), from smaller molecules like glycine and succinate or glutamate (Beale 1990). Next, two of the dALA molecules combine and cyclicize to form the pyrrol ring compound, porphobilinogen (PBG). Four molecules of the PBGs subsequently form hydroxymethylbilane (HMB) via deamination and complexation reactions, and the HMB molecule formed next undergoes dehydration to form the tetrapyrrole ring compound, uroporphyrinogen (UPP) (Jordan 1989). The heme pigments are formed from UPP via a series of enzyme catalyzed reactions including the insertion of Fe into the tetrapyrrole structure (by ferrochelatase), and its eventual binding to the protein, globin. The chlorophylls, on the other hand, are formed from UPP via a series of enzyme-catalyzed reactions including chelation with magnesium (Mg) by magnesium chelatase, and attachment of the phytol side chain (from phytyl pyrophosphate by chlorophyll synthetase).

In terms of concentrations or levels, Mb is the major pigment of meat muscle. Approximately 80% of muscle pigment is Mb with the balance (approximately 20%) as Hb. Mb traps O_2 in muscle cells as a complex known as oxymyoglobin (MbO_2) for the purpose of producing energy for biological activity in a continuous fashion. Hb, on the other hand, occurs mostly in the blood vessels for O_2 transport. Hb is a tetramer of four Mb units and has a molecular weight of approximately 64 kDa. For this chapter, the focus will be more on Mb than Hb.

Mb and Meat Color

Meats are classified into two types, that is, red (or dark) versus white meats based on the muscle type and the Mb content. The Mb content varies in cells; the higher the Mb content, the more intense the red or dark color of the meat. Thus, red (or dark) meats have relatively higher content of Mb than white meats. This is because red or dark meat comprises muscles that require more uptake of O_2 to be able to generate a constant supply of energy to carry out activities over protracted periods. White meat, on the other hand, has muscle types with lesser Mb content and relies on glycogen as the source of energy for rapid spurts of activity for only brief periods of time. In addition to muscle type, other factors like species, age, sex, and physical activity all influence the Mb levels (and the color) of meat. For example, fresh beef tends to be bright red in color, pork is pinkish, veal is brownish pink, lamb and mutton are pale to brick red, poultry (chicken, ducks, and turkeys) are white to dull red, and most fish tend to range from white to gray. With regards to the physical state of the meat, the muscle protein in its native state imparts a reddish/purplish color while the denatured form is pinkish to brownish. Furthermore, exercised muscles tend to be darker in color than nonexercised muscles. Thus, the color in the muscles of the same animal can vary depending on the state of activity of the animal. In terms of age, the Mb content tends to increase as the animal gets older (Kim et al. 2003), and the color of female foals were reported in a study to be darker than their male counterparts (Sarriés and Beriain 2006), although Carrilho et al. (2009) found male rabbit meat to be more intensely colored than females rabbit meat. The diets of the animal also affect meat color. For example, cattle raised on pasture has a darker meat color than their counterparts raised on concentrates (Priolo et al. 2001), and there appears to be seasonal variations in meat color with the color being darker in the winter season than in the spring and autumn seasons (Kim et al. 2003).

The color of meat also depends to a large extent on the oxidation state of the Fe present in the porphyrin ring, as well as on the nature of the ligand bound to the sixth coordination position of the central Fe atom. When meats are cooked, the color of Mb changes depending on the extent and/or intensity of the cooking. For example, Mb retains its native color when dark meats are cooked "rare." However, medium cooking causes denaturation of the protein group (globin) and oxidation of the central Fe^{2+} to the Fe^{3+} form, and changes the meat to a tan-colored product known as hemichrome. When meats are "well done," the amount of hemichrome formed increases and the meat acquires a darker brown hue. When white meat is cooked, it changes from a translucent pallor to an opaque or whitish color.

The interconversions between Mb, MbO_2, and MetMb are summarized in Figure 37.2. O_2 adds to purple colored Mb (Fe^{2+}) by oxygenation to form the bright red colored MbO_2 (with the central Fe atom still in the Fe^{2+} form). MbO_2 thus formed may be deoxygenated to regenerate Mb, or oxidized via electron loss to form the brown-colored metmyoglobin (MetMb, Fe^{3+}). MetMb may then undergo reduction by electron gain to re-form Mb (Fig. 37.2).

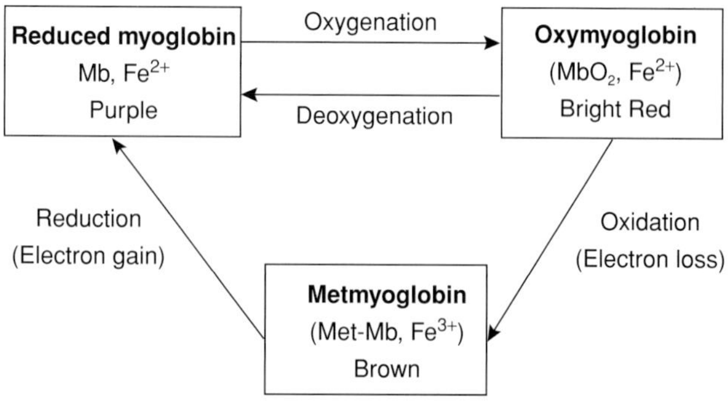

Figure 37.2. Interconversions between myoglobin (Mb), oxymyoglobin (MbO_2), and metmyoglobin (MetMb).

Discoloration of Meat

The color of meats is important because it is used by consumers as an index of quality. In addition to the color changes ascribed previously to the chemical state of the central Fe atom, other factors may also influence the meat color. For example, pale, soft, and exudative (PSE) meats that are characterized by pale colors, soft textures and low water-holding capacities, tend to have considerably more moisture on their surfaces. This is caused by increased glycolysis and a rapid decline in muscle pH that denature meat proteins to adversely impact the texture. Unlike PSE meats, some meats may become dark, firm, and dry due to relatively higher pH levels (>6.0), which make the meat proteins bind water molecules more tenaciously and cause the meat surface to dry out, and reduce light absorption making the meat color to darken. The elevated pH also promotes meat spoilage from microbial proliferation and metabolism. An increase in microbial activity could enhance production of compounds such as amines (via deamination of amino acids) that can react with nitrites (used to "cure" meats) to form nitrosamines in the meats and/or impart a reddish color; or catalase negative bacteria (e.g., lactic acid bacteria) may produce H_2O_2 to give meats a greenish tinge. Exposure of meat to light may cause O_2 to dissociate from MbO_2 in meats to cause fresh meat color to become paler. Antioxidant compounds, for example, vitamin E (α-tocopherol) or vitamin C (ascorbic acid), may be applied to meats to curtail the oxidation of Mb and/or MbO_2 to MetMb, and thereby stabilize meat color. The use of nitrites to "cure" meats is primarily to safeguard against *Clostridium botulinum*, but the process also imparts a pinkish color to meats. Vacuum packaging films also affect the relative proportions of Mb, MbO_2, and MetMb in packaged meats to impact the color from the differences in O_2 permeability or O_2 barrier properties of the packaging films.

Heme Pigments and Health

Red meats are a good source of Fe (by virtue of Fe being part of the heme). When Fe is deficient, it can cause a reduction in cognitive abilities particularly in children, or result in anemia (i.e., Fe deficiency anemia or IDA). The body requires Fe to synthesize red blood cells and to regulate body temperature, among other things. When a person has IDA, the red blood cell levels decline and the victim's health suffers. People with IDA tend to feel cold on a continuing basis, because regulation of the body's temperature is impaired. Because it is part of Hb and Mb, Fe-deficiency causes people to tire easily since their bodies have insufficient O_2 for both biosynthesis and energy generation.

Measurement of Mb

A number of procedures have been described in the literature for the measurements of Mb. These include the optical density method by atomic absorption spectroscopy (Weber et al. 1974), differential scanning calorimetry (Chen and Chow 2001), spectrophotometry (Tang et al. 2004), and NIR spectrophotometry (van Beek and Westerhof 1996).

CAROTENOID PIGMENTS

Structures, Sources, and Functions

Carotenoids are ubiquitous in nature and several different members (>400) have been identified in living organisms. One of the best known naturally occurring carotenoid pigment is the yellow-orange colored compound *β,β*-carotene, which was first crystallized from carrot as far back as in the early 1830s. It is from this source material (carrot) that the name for the entire class of these compounds (carotenoids) is derived. Most carotenoids are water insoluble, but are fat or oil soluble and heat stable. They are found in fruits and vegetables like oranges, tomatoes, and carrots, in the yellow colors of many flowers, in animal species including crustacea (e.g., crab, lobster, scampi, shrimp), fishes (e.g., goldfish and salmonids like Arctic char, salmon, trout) in birds (e.g., canaries, flamingos, and finches), and insects (phasmids, lady bird, and moths); in seaweeds (e.g., *Undaria pinnatifida*, *Laminaria japonica*); in yeasts (e.g., *Rhodotorula rubra*, *Phaffia rhodozyma*); in algae (e.g., *Xanthophyceae*, *Chlorophyceae*); in fungi (e.g., *Blakeslee*, *Xanthophyllomyces*); and in bacteria (e.g., *Corynebacterium autotrophicum*, *Rhodopseudomonas spheroides*). Carotenoids may occur in the free form (e.g., fucoxanthin in seaweeds), or complexed with proteins to form the relatively more stable carotenoproteins as found in several invertebrates (e.g., crustacea). When complexed with proteins, the colors of carotenoids may be altered from orange, red, or yellow to blue, green, or purple. In general, carotenoids are present in low amounts in most organisms, but good sources of this class of pigments include algae and plants (the predominant sources for humans), from fruits (cantaloupe, pumpkin, watermelon, pineapple, citrus fruits, red or yellow peppers, tomatoes, mangoes, papaya, and guava) and vegetables (carrot, rhubarb, sweet potato, kale, parsley, cabbage, collards, and spinach).

Carotenoid compounds are synthesized by plants, bacteria and microalgae from the low molecular weight precursor molecules, pyruvate and acetyl CoA, to initially form a 5-carbon intermediate compound known as isopentenyl pyrophosphate (IPP). IPP has a molecular weight of 246.1 Da and is transformed by a series of enzyme-assisted reactions to form the 40-carbon polyunsaturated hydrocarbon compound known as phytoene ($C_{40}H_{64}$) with a molecular weight of 544.94 Da. Phytoene (Fig. 37.3) subsequently undergoes four desaturation steps to form lycopene ($C_{40}H_{56}$), a hydrocarbon carotenoid with thirteen double bonds (Fig. 37.3). Lycopene thus formed, may isomerize and/or cyclicize to form various other hydrocarbon carotenoids such as α-carotene, β-carotene, and γ-carotene. The hydrocarbon carotenoids may then be hydroxylated by hydrolases to form their oxygenated counterparts such as lutein, astaxanthin, canthaxanthin, cryptoxanthin, and zeaxanthin. Figure 37.3 has the structures of some of the common carotenoid compounds. Various raw materials contain different levels of carotenoids. For example, lutein is abundant in many green plants; carrots and most green fruits and vegetables are rich in α-carotene and β-carotenes; astaxanthin, canthaxanthin, and astacene are the major carotenoids in crustacea and salmonids; astaxanthin is also abundant in mushroom, algae, yeasts, and bird feathers

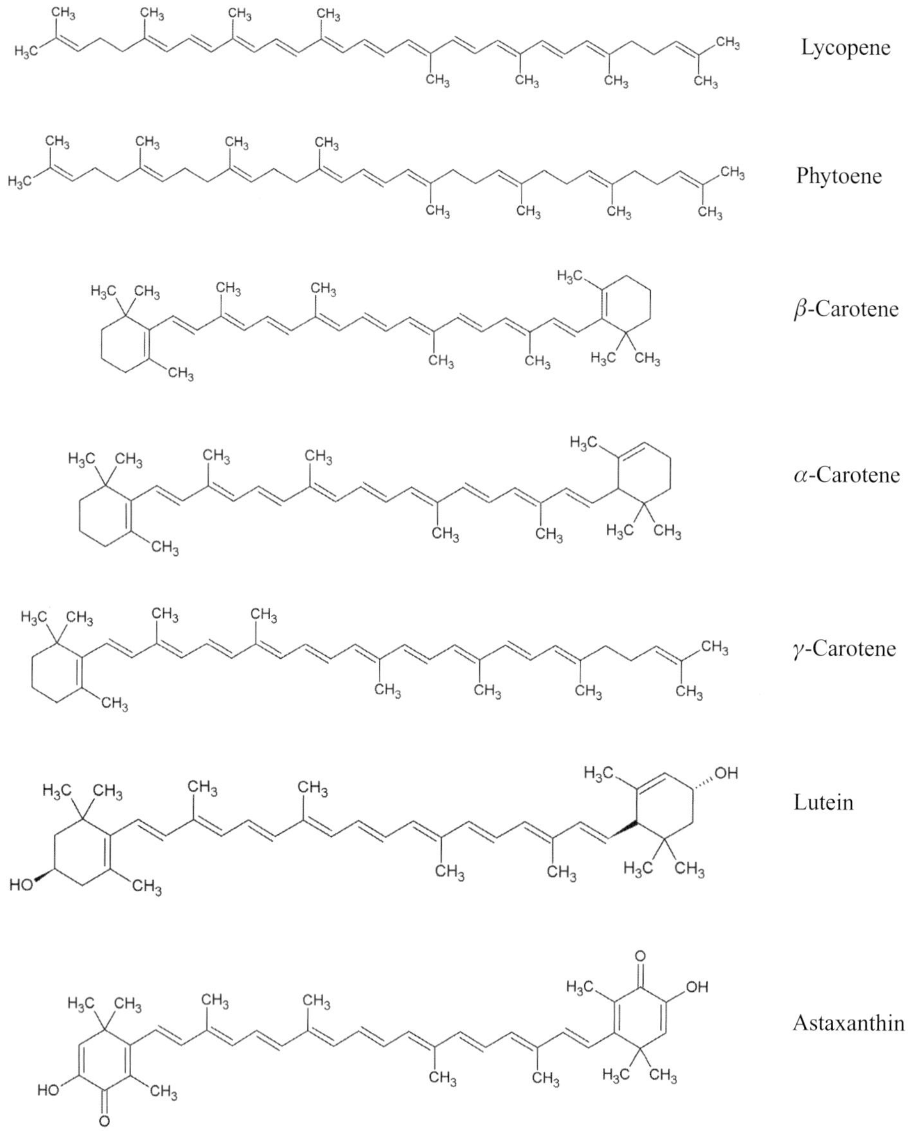

Figure 37.3. Structures of some common carotenoids (*continued*).

(e.g., flamingos and finches); cryptoxanthin, antheraxanthin, and zeaxanthin are the principal ones in maize; cryptoxanthin and zeaxanthin are also predominant in papaya and cantaloupe; red bell peppers have a high content of capsanthin; tomatoes have high levels of phytoene, lycopene, $\beta\beta$–carotene, neurosporene, and phytofluene; and seaweed and many algae have fucoxanthin, fucoxanthinol, and lactucaxanthin as the major ones.

Unlike plants and microorganisms, animals do not synthesize carotenoid compounds de novo, and must derive these compounds from their diets for various processes, although animals may transform dietary carotenoids to other forms for various functions (Tanaka et al. 1976).

Carotenoids like α-carotene, β-carotene, zeaxanthin, and β-cryptoxanthin are important sources of dietary vitamin A. These molecules are cleaved by dioxygenase in the gastrointestinal tract to release, at least a molecule of retinal, which is subsequently reduced to retinol (or vitamin A); thus, such carotenoid compounds are classed as having provitamin A activity. Vitamin A or retinol is needed in the retina of the eye for vision. Other biological functions of carotenoids include their role in light-harvesting and energy transfer reactions in photosynthesis, and in photo-protection of tissues against the ravaging effects of oxidation by exposure to light. The antioxidant role of carotenoids is well known. As antioxidants, carotenoids protect membranes and tissues from oxidative damage and prevent or minimize serious illnesses such as cancer, heart disease, and nutritional blindness. The antioxidant capacity of carotenoids enables them to ward off degenerative diseases by "quenching" harmful free

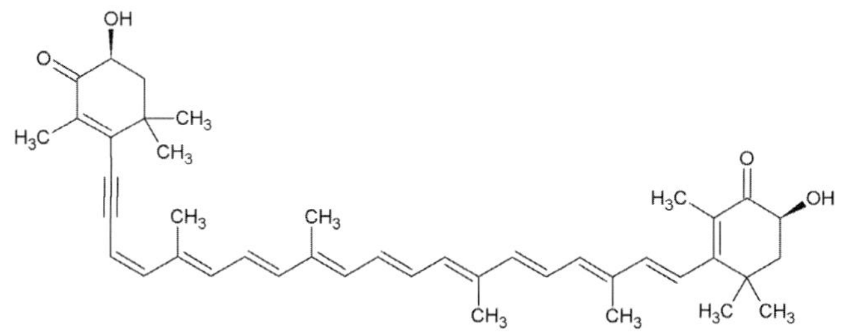

Astacene

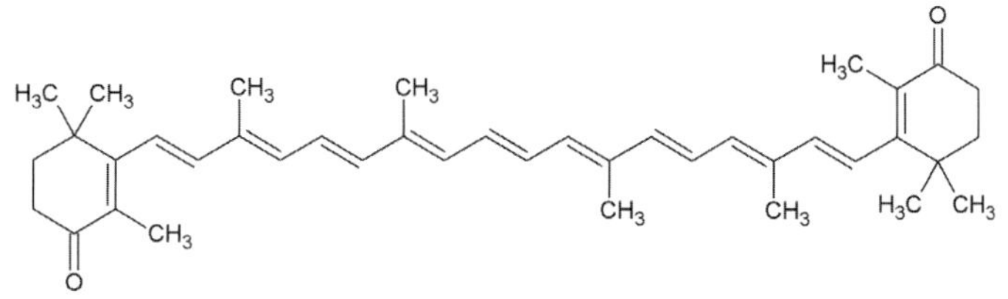

Canthaxanthin

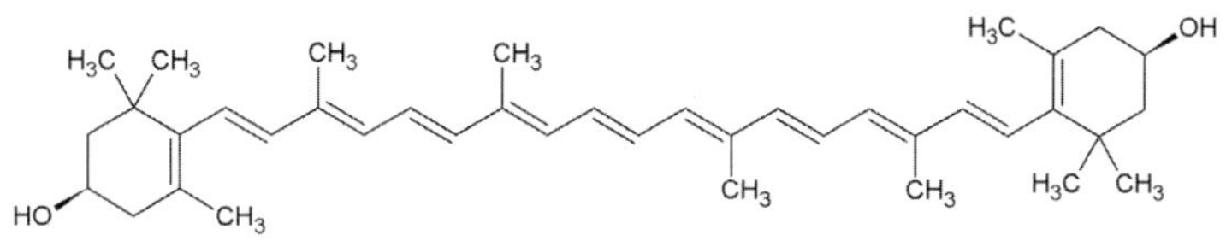

Tunaxanthin

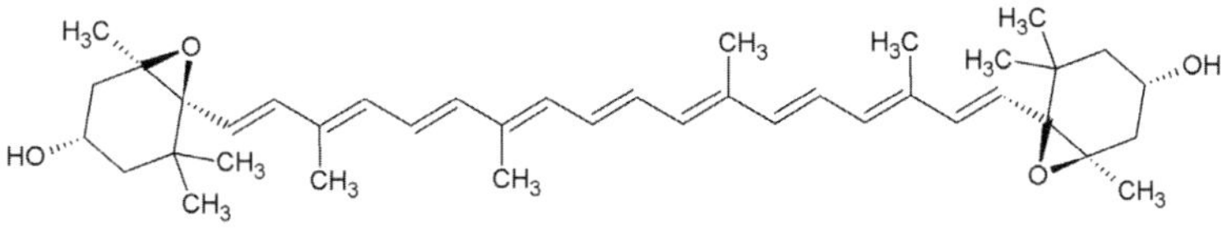

Violaxanthin

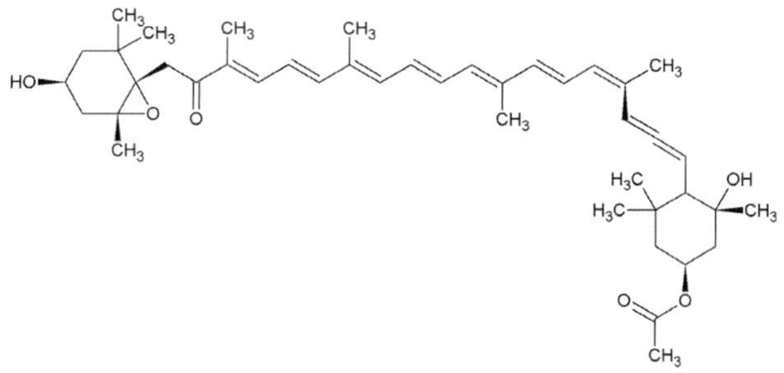

Fucoxanthin

Figure 37.3. (*Continued*)

radicals and reactive O_2 species such as singlet O_2 from causing damage to cells or tissues. Evidence for this behavior by carotenoids includes studies in men that showed that carotenoids could reduce the risks of heart attacks significantly. The same free radical "quenching" behavior of carotenoids enables them to act as anticarcinogens, whereby they act to lower the incidence of certain cancers (e.g., lycopene in tomatoes and prostate cancer). There is also evidence that carotenoids like lycopene suppress the oxidation of low-density lipoprotein (LDL) and helps to lower blood cholesterol levels. Thus, carotenoids play crucial roles in vision and are effective in slowing down age-related eye disorders, and inhibiting oxidation of LDL that promotes platelet formation and aggregation and cardiovascular diseases, as well as free radical scavenging to prevent certain cancers and membrane or tissues damage from light (Hadley et al. 2003). Carotenoids also boost the immune system by increasing the activities of lymphocytes, or protecting macrophages and the immune system from damage by ultraviolet (UV) light and X-rays. However, the efficacy of antioxidants like β-carotene in protecting against cancers has been questioned (Albanes et al. 1996), and it has even been suggested that it may actually increase the risk of lung cancer in heavy smokers, and cause a higher incidence of cardiovascular diseases as well as high mortalities (Albanes et al. 1996).

An oxygenated carotenoid pigment touted for its health benefits is fucoxanthin (Fig. 37.3) found in the brown algae (kelp). Fucoxanthin has a molecular formula of $C_{40}H_{60}O_6$, a molecular weight of 658.9 Da, and studies have shown it to induce apoptosis and/or antiproliferative effects on certain cancer cells (Hosokawa et al. 2004, Kotake-Nara et al. 2005). For these and other reasons, there is intensive research afoot on the antioxidant effects of carotenoids, and their role as regulators of the immune system, and there is considerable interest in the production of carotenoids by biotechnological approaches to increase supplies for the demand for these compounds.

Properties and Uses of Carotenoids

The market for carotenoids is estimated at approximately a billion US$ per annum (Fraser and Bramley 2004). They are used as colorants (e.g., β-carotene, annatto) to impart colors to many processed foods (e.g., margarine, butter, dairy products, confectionery, and baked goods) beverages (e.g., soft drinks), drugs and cosmetics (Britton 1996). They are also used as nutritional supplements and in animal feeds, for example, as colorants for farmed animals (e.g., poultry eggs, salmon, trout, shrimp, and lobster) (Lorenz and Cysewksi 2000). Carotenoids (e.g., astaxanthin, lycopene, and zeaxanthin) can reduce the adverse side effects such as inflammation and/or pain associated with administration of cyclooxygenase inhibitor drugs (Kearney et al. 2006).

Health Benefits of Carotenoids

The health benefits of carotenoids include protection against cardiovascular diseases (by inhibiting the oxidation of LDL cholesterol), certain cancers (e.g., cervical, gastrointestinal tract, lung, skin, uterine), and age-related macular degeneration and cataracts. Carotenoids inhibit the growth of unhealthy cells while promoting the growth of healthy cells to invigorate the body. Most are antioxidants and are able to absorb light/filter the UV rays of the sun to reduce photo-oxidation, and thereby act to protect cells and tissues against oxidative damage potentiated by free radicals and reactive O_2 species (Krinsky and Johnson 2005, Dembinska-Kiec 2005).

Methods for Measuring Carotenoid Pigments

Carotenoid pigments (as astaxanthin) may be measured quantitatively by measuring the absorbance at 485 m in a spectrophotometer (Saito and Regier 1971). Other methods that have been used to study the molecular properties of carotenoids include field desorption mass spectrometry (Takaichi et al. 2003), thin layer chromatography and analysis by gas chromatography (Renstrom and Liaaen-Jensen 1981); high-pressure liquid chromatography (HPLC) (Yuan et al. 1996), or by negative ion liquid chromatography-atmospheric pressure chemical ionization mass spectrometry (negative ion LC-[APC]-MS) (Breithaupt 2004). Other methods described for the measurement of carotenoids include reversed phase HPLC (Burri et al. 2003, Graça Dias et al. 2008), resonance Raman spectroscopy (Bernstein et al. 2004), and by UV/Vis and mass spectra chromatography (Meléndez-Martínez et al. 2007).

CHLOROPHYLLS

Structure and Functions

Chlorophylls are fat and oil soluble green pigments that occur in the plastids of most plants, algae, and certain bacteria. In these organisms, chlorophylls participate in the biosynthesis of complex biomolecules ($C_6H_{12}O_6$) from simpler ones (CO_2 and H_2O) by the process of photosynthesis. They are the most widely distributed natural plant pigments and are present in all green leafy vegetables, and also occur abundantly in the green algae known as chlorophytes. The chemical structure of chlorophylls comprises a tetrapyrrole (or porphyrin) ring system similar to those of Mb and Hb, except that the central atom of chlorophyll is Mg, and the side chain is a 20 carbon hydrocarbon (phytol) group that is esterified to one of the pyrrole rings (while Mb or Hb has the protein, globin, bound to the fifth coordination position of the central Fe atom). Thus, four pyrrole groups are linked together by a central Mg^{+2} ion to form a porphyrin ring, which together with phytol (a 20-carbon hydrocarbon chain) makes the chlorophyll molecule (Fig. 37.4). When foods containing chlorophyll (e.g., vegetables) are cooked, the chlorophyll may undergo changes in color and/or solubility. The colors that form may be dull green, bright green, or brownish. These different colors are due to changes in the chlorophyll molecule such as loss of the phytol side chain, or the removal of the central Mg^{2+}

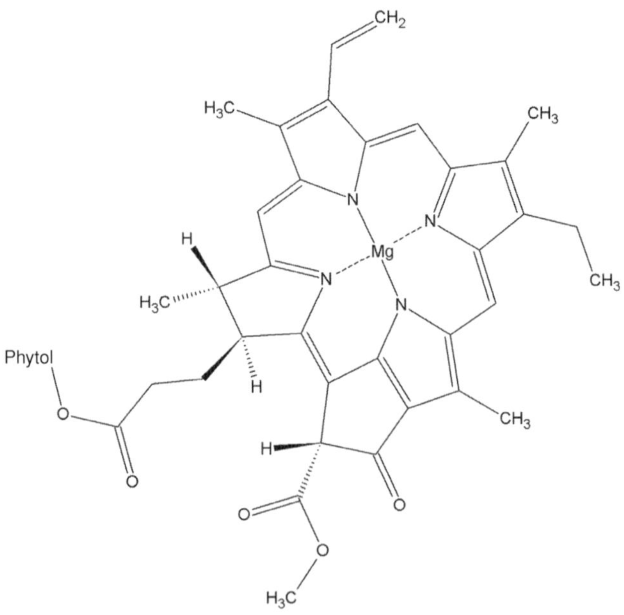

Figure 37.4. Structure of chlorophyll.

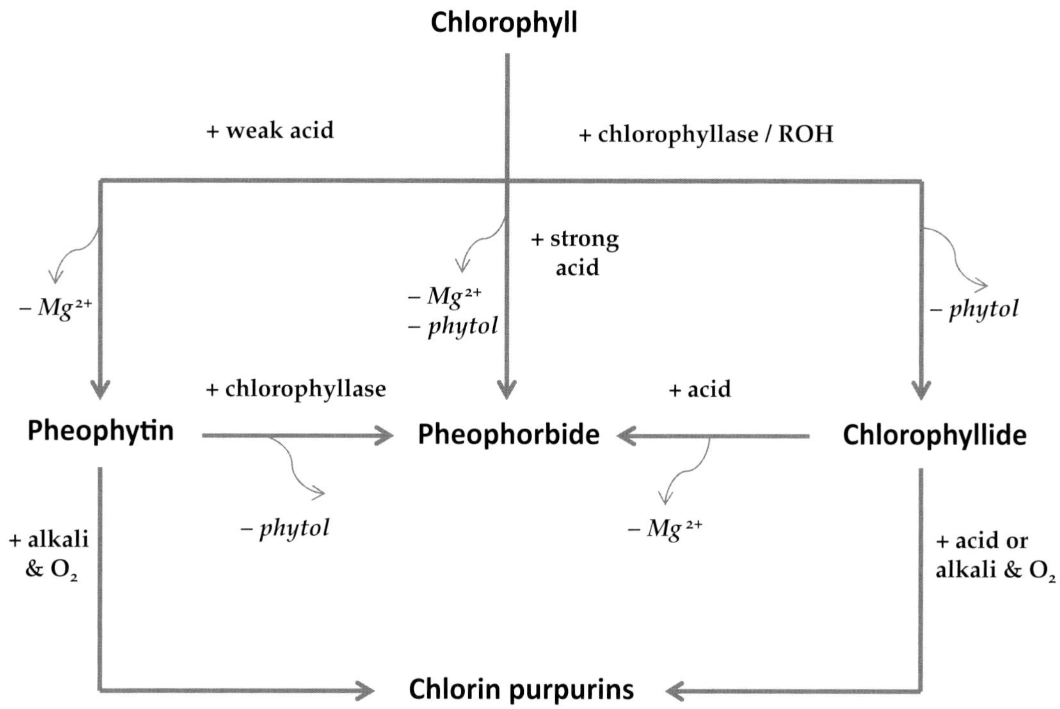

Figure 37.5. Interconversions between chlorophyll pigments.

atom (Fig. 37.5). The phytol side chain may be removed by the action of enzymes (chlorophyllase) and/or acidic conditions, while Mg^{2+} may be removed from exposure to acid and heat treatment. It is the 20 carbon hydrocarbon side chain (phytol) that makes chlorophylls nonpolar and water insoluble. So, its removal from the molecule renders the residual compound water soluble. Alkaline conditions and exposure to light can also induce changes in the chlorophyll molecule (Fig. 37.5).

Because of their light trapping and electron transfer roles in photosynthesis, chlorophyll pigments are considered as the basis of all plant life. Chlorophylls are synthesized via the formation of a porphyrin ring, and the initial steps in the biosynthesis pathway are similar to those of the heme pigments, Mb and Hb, described in Section "Structures and Functions." Chlorophylls exist in nature in different forms (e.g., chlorophylls a, b, c_1, c_2, and d). The differences in these forms arise from subtle changes in either the tetrapyrrole ring itself or in the side chain groups. For example, chlorophylls a and b differ only in the substituent group on C7 of the molecule (i.e., $-CH_3$ for chlorophyll a, and $-CH = O$ for chlorophyll b), while chlorophylls a and b both have $-CH = CH_2$ group at the C3 position unlike chlorophyll d that has a $-CH = O$ group.

Chlorophyll is commonly extracted from plant species like spinach, alfalfa, nettles, or grass using organic solvents (e.g., acetone and hexane), preferably in dim light to minimize degradation of the pigment. Exposure to light, air, heat, and extreme pH all adversely affect its stability. Nevertheless, the stability of chlorophylls against degradation may be enhanced by de-esterification of the chlorophyll and complexation with copper ions.

Properties, Sources, and Uses

Chlorophylls are safe for human consumption and since ancient times humans have ingested chlorophylls in their fruits and vegetables. Commercial sources of chlorophylls include the green algae chlorella, the blue-green algae *Spirulina*, the string lettuce (*Enteromorpha*) and the sea lettuce (*Ulva*); all the sources mentioned previously are used as human food (Ayehunie et al. 1996) in salads and soups. Chlorophylls are used as food colorant and in this regard, can be safely added to several foods either in the pure form or complexed with copper. They are also used in several other applications including cosmetics (soaps, creams, and body lotions), oral hygiene products (mouthwash, toothpaste) as well as confectionary (gums and candies) because of their intense green color. Chlorophyll-colored products are best stored under dry conditions and protection from air, light, and heat to better retain the color. Certain countries have restrictions on the use of copper complexes of chlorophylls in foods and drugs because of the toxicity of copper.

Chlorophyll content in tissues and food materials may be measured spectrophotometrically by the AOCS (2004) official method AK 2–92.

Chlorophylls and Health

Chlorophylls also have several health benefits for humans. They have been shown to be capable of rebuilding the bloodstream (Patek 1936), and to be nontoxic and without harmful side effects even when administered in large doses in various routes (intravenous, intramuscular, or oral). The high Mg content in chlorophyll promotes fertility by increasing the levels and activities of the enzymes that regulate sex hormones. Chlorophylls have antibacterial properties and can be used both inside and outside the body for this purpose (Bowers 1947). They are able to clean out drug deposits and deactivate toxins in the body, cleanse the liver, and reduce problems associated with blood sugar (Colio and Babb 1948). They are also used in deodorizers, to inhibit oral bacterial infections, promote healing of rectal sores, and reduce typhoid fevers (Offenkrantz 1950).

ANTHOCYANINS AND ANTHOCYANIDINS

Properties, Structures, and Functions

Anthocyanins are water-soluble, pH-sensitive, red, purple, and blue pigments found in higher plants. They impart the intricate colors to many flowers, fruits, and vegetables. They also occur in leaves, stems, and roots of plants particularly in the outer cells of these parts of the plant. They are found in almost all the plant families, and in fruits and vegetables the main sources include avocados, blackberries, black currants, blueberries, cherries, chokeberries, cranberries, grapes, prunes, oranges, mangos, aubergines, olives, onions, and sweet potatoes. Anthocyanins (and anthocyanidins) may be considered as members of the flavonoid family of compounds (Section "Flavonoids"), but because of their importance and the extensive attention they have received, a separate section is devoted in this chapter to the discussion of these compounds.

The color of anthocyanins is influenced by acidity; they are red under acidic conditions and turn blue at neutral to alkaline pHs. On the basis of the chemical structures, two types are distinguished; the anthocyanidin aglycones (or anthocyanidins) that are devoid of sugar moieties, and the true anthocyanins are glycosides or sugar esters of the anthocyanidins. They all have a single basic core structure, the flavylium ion (Fig. 37.6) that has seven different side groups (denoted as R in the figure) that may be a hydrogen (H) atom, a hydroxide (—OH) group, or a methoxy (-OCH_3) group. The sugars of anthocyanins are esterified to the different side groups, and because there are seven of these positions and several sugars (monosaccharides, disaccharides, and trisaccharides) in higher plants, it is to be expected that several different anthocyanins would occur in nature. Furthermore, the sugars may be acylated with the organic acids that are naturally present in plants (e.g., acetate, citrate, ascorbate, succinate, propionate, malate, and many others) to form acylated anthocyanins. The sugar residues may be monosaccharides, disaccharides, or trisaccharides. When there is only one monosaccharide residue esterified to the aglycone, the anthocyanin may be referred to as a monoside; where esterification is to either two monosaccharide units or one disaccharide unit, the product is known as a bioside; where there are three monosaccharides, or two disaccharides and one monosaccharide, or a single trisaccharide unit esterified, the compound is referred to as a trioside. Those without sugar molecules are known as aglycones. Some of the common anthocyanidins include: cyanidin, delphinidin, pelargonidin, malvidin, peonidin, and petunidin; and some of the known anthocyanin glycosides are: cyanindin-3-glucoside and pelargonidin-3-glucoside, as well as the glycosides and diglucosides of cyanidin, pelargonidin, delphinidin, and malvidin (Fig. 37.6). Their biological functions include their action as antioxidants whereby they protect plants against the damaging effects of reactive O_2 species and UV radiation, and also serve as attractant for birds and insects to flowers for pollination.

Uses of Anthocyanins

Anthocyanins are used as food colorant in drinks, food spreads (jams and jellies), jellos, pastries, and confectioneries because of their intense blue or red colorations. There are, however, some drawbacks in the use of anthocyanins in foods. This is due to their high solubility in water and vulnerability to pH changes. Thus, red cabbage may exhibit a deep reddish color or a purplish or bluish color depending on the pH of the milieu. Other factors such as exposure to high temperature, oxidizing agents (tocopherols, peroxides, and quinones), air, or light are detrimental to the stability of and colors imparted by anthocyanins. Although the sugars in anthocyanins may provide some protection for the molecule against degradation under ambient conditions, the same sugar moieties could elicit browning reactions in food products when the latter are exposed to high temperatures and/or light. The detrimental effects of temperature and air may be slowed down by low temperature storage, with protection from light and air/O_2.

Anthocyanins and Health

The toxicity of the anthocyanins is not well documented. But they are believed to be nontoxic even at high concentrations. The compounds are potent antioxidants and have been associated with health benefits such as reduced risk of coronary heart disease, improved visual acuity, and antiviral activity. Studies with pelargonidin and delphinidin (both anthocyanins) have shown both of them to be capable of inhibiting aldoreductase, the enzyme that converts glucose into fructose and sorbitol (Varma and Kinoshita 1976). Inhibition of this enzyme is desirable in preventing complications from diabetes because without the inhibition, the sorbitol and fructose formed could diffuse into the myelin sheath along with fluid (by osmotic action) to cause edema of the myelin sheath in nerves (Levin et al. 1980). Thus, substances like pelargonidin and delphinidin that can inhibit the aldoreductase enzymes would be useful in reducing this type of complications for diabetics. The anthocyanins petunidin, delphinidin, and malvidin have also been shown to enhance the activity of glutamate decarboxylase (GAD), the enzyme that catalyzes the decarboxylation of glutamic acid ($HOOC—CH_2—CH_2—CH(NH_2)—COOH$) to γ-amino benzoic acid ($HOOC—CH_2—CH_2—CH_2NH_2$) or GABA (for short). GABA is an important neurotransmitter in the brain, and an

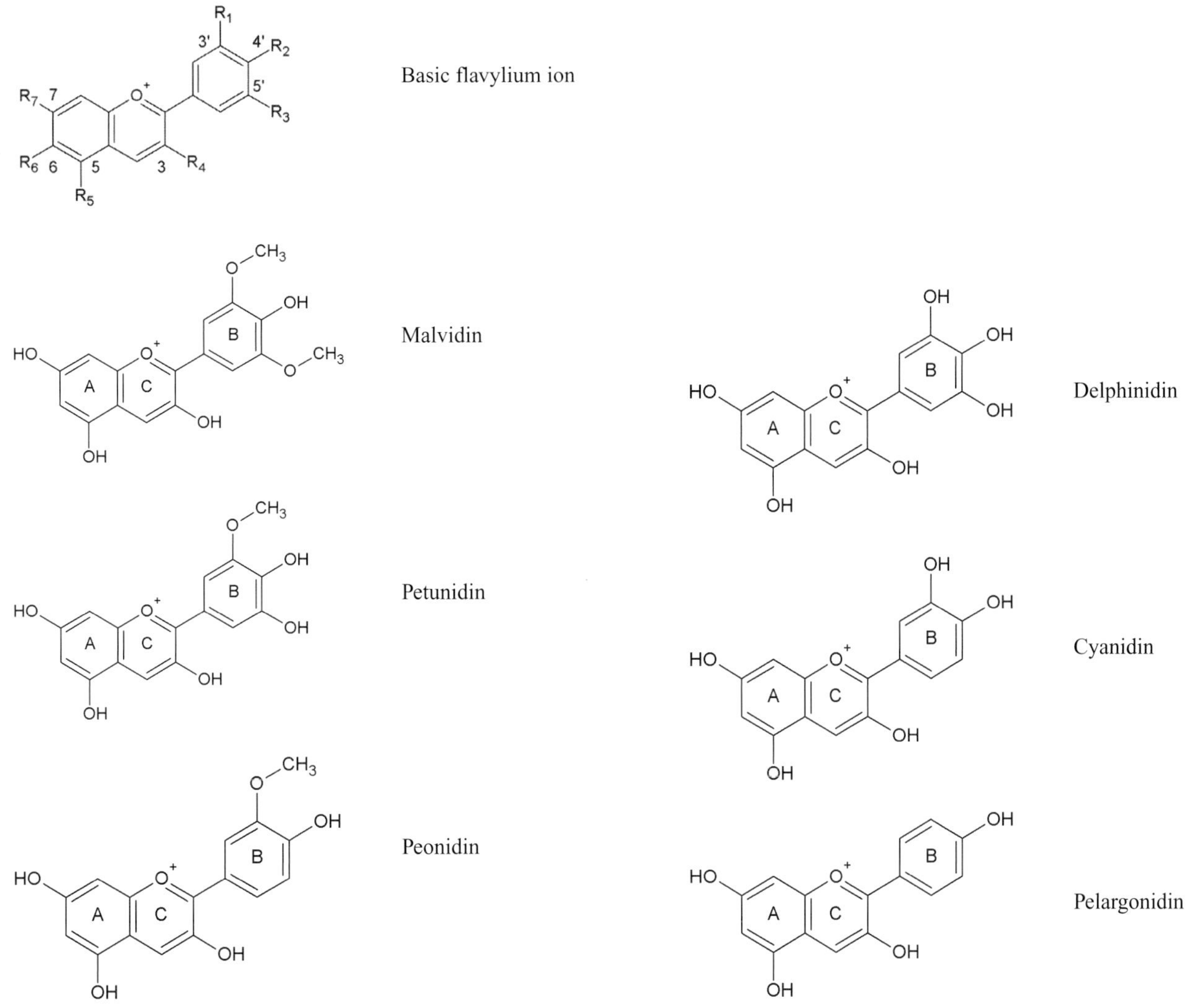

Figure 37.6. Structures of common anthocyanins.

increase in the level of GAD is useful not only because of the formation of GABA, but also because it can maintain insulin production in people with type 1 diabetes (Ludvigsson et al. 2008). Petunidin, delphinidin, and malvidin can also inhibit glycerol dehydrogenase, a key enzyme in glycerolipid metabolism (Carpenter et al. 1967), and malate dehydrogenase that participate in both the tricarboxylic acid (TCA) cycle (i.e., the dehydrogenation of malate to oxaloacetate) and in gluconeogenesis (i.e., the synthesis of glucose from noncarbohydrate carbon sources) (Carpenter et al. 1967). Thus, by inhibiting malate dehydrogenase, antioxidants also act indirectly to regulate gluconeogenesis.

Measurement of Anthocyanins

Several methods have been reported for the measurement of anthocyanins in samples. These include spectrophotometric procedures for measuring anthocyanins in wines (Cliff et al. 2007), and the use of HPLC and ultra performance liquid chromatography to measure the pigment levels in food samples (Hosseinian and Li 2008). Other procedures for measuring anthocyanins include pH-differential measurements (Giusti and Wrolstad 2001) and FTIR (Soriano et al. 2007).

FLAVONOIDS

Properties, Structures, and Functions

The term flavanoids is used here to refer to a family of compounds produced by plants that are sometimes referred to as anthoxanthins. They are water soluble but have lower coloring power compared to the anthocyanins or anthocyanidins; they range in color from colorless to shades of yellow, and they

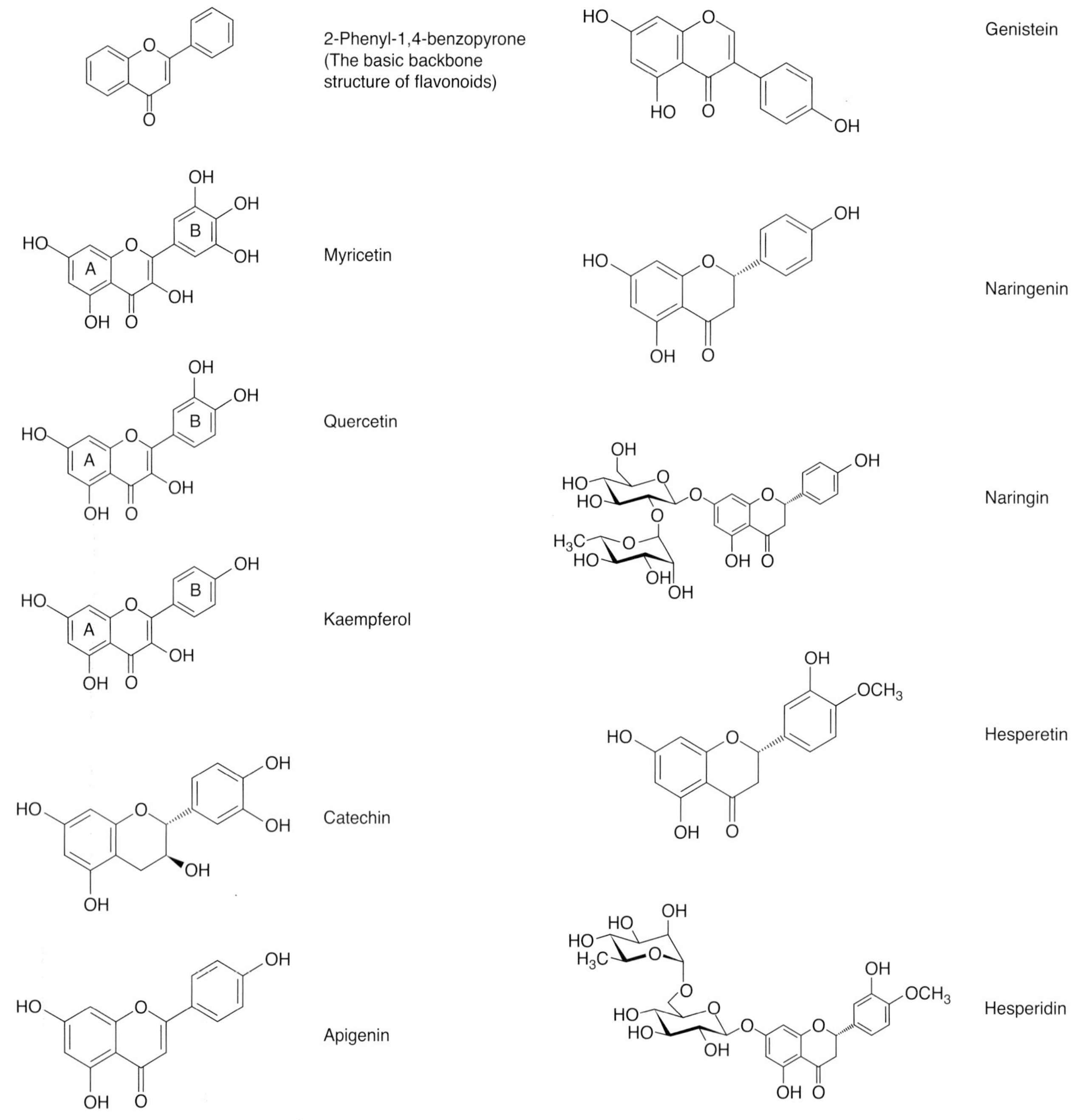

Figure 37.7. Structures of common flavonoids.

have a common core structure, the benzopyrone ring (Fig. 37.7), which bears a structural resemblance to the flavylium ion of anthocyanins or anthocyanidins (Fig. 37.6). Flavonoids are subdivided into five sub-classes: the flavanols, the flavanones, the flavonols, the flavones, and the isoflavones. These five groups of compounds exhibit subtle differences in structure, such as degree of saturation or unsaturation, and/or position of substituent groups (Fig. 37.7). Examples of the flavanols are catechin, epicatechin, epigallocatechin, epicatechin gallate, epigallocatechin gallate, and theaflavins; examples of the flavanones are butin, hesperidin, hesperetin, naringenin and naringin; the flavonols include azaleatin, quercitin, kaempferol, myricetin, morin, and rhamnetin; members of the flavones family include apigenin, baicalein, chrysin, luteolin, tangeritin, and wogonin; and the isoflavones are typified by daidzein, genistein, and glycitein. The flavonoids have lower coloring power compared with the anthocyanins or anthocyanidins and range from colorless to shades of yellow. Some common food sources at flavonoids are apples, broccoli, citrus fruits, chocolate, cocoa, grapes, green tea, oolong tea, parsley, red peppers, white tea, yellow onions, celery, soybeans, and thyme.

Like anthocyanins, some flavonoids are esterified with one or more sugar residues. Apart from flavanols (catechins and proanthocyanidins), the other known flavonoids appear to occur in plants and foodstuffs in the esterified forms. A lot of these flavonoid glycosides appear to survive the cooking process, as well as the preliminary digestion steps in the mouth and stomach. However, in the small intestines, it appears only the aglycone forms and the flavonoid glucosides are absorbed and metabolized to form various secondary metabolites; and colon bacteria metabolize flavonoids to facilitate absorption (Manach et al. 2004). Nevertheless, the flavonoids have low bioavailability due to inadequate absorption and their fast removal from the body. Of the flavonoids, the isoflavones are said to be the most bioavailable while the flavanols are the least bioavailable (Manach et al. 2005).

Biological Functions and Uses

Flavonoids also have antioxidant properties and are effective free radical scavengers (Chun et al. 2003). However, most circulating flavonoids are believed to be metabolized intermediates or by-products that have relatively lower antioxidant capacity than the parent flavonoid compounds. Thus, the relative contribution of dietary flavonoids to antioxidant behavior in the body is suggested to be miniscule (Lotito and Frei 2006). By virtue of their polyphenolic nature, flavonoids are also able to act as metal ion chelators to stop ions such as Cu^{+}/Cu^{2+} and Fe^{2+}/Fe^{3+} from promoting the formation and release of free radicals (Mira et al. 2002). The chelating behavior of flavonoids could also adversely affect intestinal absorption of Fe, thus it is recommended for Fe supplements not to be taken together with beverages high in flavonoids or flavonoids supplements (Zijp et al. 2000).

Flavonoids and Health

Although various suggestions have been made to the effect that high intakes of flavonoids from plant sources do not elicit adverse effects in humans due to their low bioavailability and rapid metabolism and removal from the body, there are some reports indicating that high doses of some flavonoids can cause adverse health effects in some people. For example, some flavonoids like quercitin have been reported to induce nausea, headaches and/or tingling sensations in some people, or renal toxicity or vomiting, sweating, flushing, and dyspnea in certain cancer patients (Ferry et al. 1996, Shoskes et al. 1999). Other reports attribute liver toxicity, abdominal pain, diarrhea insomnia, tremors, dizziness, and confusion sometimes observed from intake of caffeinated green tea extracts to the flavonoids present in tea (Jatoi et al. 2003, Bonkovsky 2006). The flavonoids quercitin and naringenin have also been reported to inhibit cytochrome P450 that is involved in the oxidation of various biomolecules such as lipids, hormones, and xenobiotics (drugs) (Bailey and Dresser 2004). They are also known to inhibit the enzyme 3-hydroxy-3-methylglutaryl-coenzyme A reductase (HMG-CoA reductase), the regulatory enzyme in the mevalonic acid pathway that leads to the formation of isoprenoid compounds, for example, carotenes, cholesterol, and steroids. The effects of these compounds on drug metabolism have been shown to affect the bioavailability and toxicity of drugs such as amiodarone, atorvastatin, cyclosporine, saquinavir, and sildenafil (Bailey and Dresser 2004, Dahan and Altman 2004). It is partly for this reason that patients are cautioned to avoid taking grapefruit juice when they are on some of these drugs, as grapefruit juice has high levels of quercetin and naringenin, as well as furanocoumarins. Copious consumption of purple grape juice has also been found to inhibit platelet aggregation, and the inhibition has been attributed to the high levels of flavonoids in the juice (Polagruto et al. 2003). On the basis of this, it has been speculated that high intake of some of these flavonoids in conjunction with anticoagulants could make some patients more prone to bleeding (Polagruto et al. 2003).

BETALAINS

Properties, Structures, and Functions

The name betalain is derived from beet vegetable (*Beta vulgaris*) from which it was first extracted. They are red or violet and yellow or orange water-soluble aromatic indole pigments found in plants and fungi (*Basidiomycetes*). Betalains are subdivided into two categories based on their colors as betacyanins (reddish to violet) and betaxanthins (yellowish to orange). Examples of the betacyanins are amaranthine, isoamaranthine, betanin, isobetanin, phyllocactin, and iso-phyllocactin; and representatives of betaxanthins include dopaxanthin, miraxanthin, indicaxanthin, portulacaxanthin, portulaxanthin, and vulgaxanthin. The betalains are synthesized from tyrosine via several routes. The amino acid tyrosine may be converted to betalains via either dihydroxy phenylalanine (L-DOPA) or dopaquinone; or through spontaneous combination with betalamic acid to an intermediate portulacaxanthin II (Tanaka et al. 2008). The structures of betanin and vulgaxanthin are shown in Figure 37.8. The betalains are most stable within the pH range of 3.5 and 7.0, but they are susceptible to light, heat and air.

Uses of Betalains

Extracts from betalains have been used to color foods since time immemorial, and beetroot extract is currently approved for use as food color in many countries to impart color to juices and wines. In addition, different varieties of beet have been used as a source of sugar (sugar beet), as vegetable (beetroot), and fodder (fodder beet). Beetroot tuber is cooked and eaten like potatoes, and it is pickled and used in salads.

Because of their high solubility in water and susceptibility to light and heat, the use of betalains to color food may be limited, although the dried form of betanin is quite stable. High sugar content has also been shown to impart protection to betalains, thus they are used in soft drink powders, frozen dairy products, and in confectioneries. Gentile et al. (2004) reported that

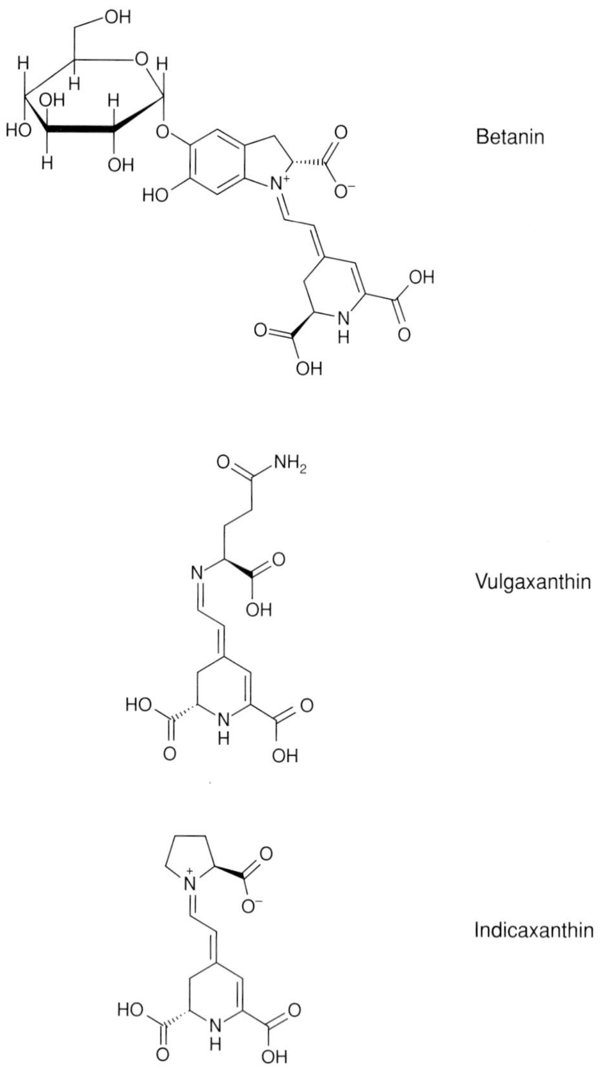

Figure 37.8. Structures of some common betalains.

betalains have anticancer properties by virtue of their antioxidant and free radical scavenging behavior. They have been shown to inhibit redox state alteration induced by cytokines to protect the endothelium layer that line the internal surfaces of blood vessels. The content of betalains in samples may be measured by the HPLC and LC-MS/MS analyses (Stintzing et al. 2006).

MELANINS

Properties and Functions

Melanins are dark brown and black pigments found in plants, animals, and microorganisms. They are formed via the enzymatic oxidation of phenolic compounds, for example, tyrosine, through several intermediates like DOPA and quinones, followed by polymerization to form the large molecular weight melanins. The enzymes catalyzing the oxidation of the phenolic compounds are polyphenoloxidases, also known as phenolases or tyrosinases. In animals and humans, melanins occur in the eyes, hair, skins, and peritoneal lining where they protect tissues from the detrimental effects of factors like light or UV radiation from the sun's rays. Damage from UV radiation is potentiated in various ways such as dimerization of thymine in DNA molecules, or by the effects of reactive O_2 species (e.g., singlet O_2 and superoxide radicals (Jagger 1985, Tyrell 1991). Examples of melanins are eumelanin and pheomelanin. Pheomelanin is dark red-brown in color and is thought to be responsible for red hair and freckles in humans, while eumelanin imparts black and brown coloring to skin and hair (Meredith and Riesz 2004).

Microorganisms synthesize melanins as part of their normal metabolism, from where they are passed along the food chain to higher forms of life. In crustacea, the formation of melanins is sometimes referred to as melanosis or "blackspot" formation, and connotes spoilage to consumers—mainly due to their lack of visual appeal. The "blackspot" phenomenon has been extensively studied to develop strategies to control this undesirable effect and products such as glucose oxidase, glucose, catalase mixes, or 4-hexyl resorcinol were found to be effective in controlling blackspot formation in raw shrimp (Ogawa et al. 1984, Ferrer et al. 1989, Yan et al. 1989, Chen et al. 1993, Benjakul et al. 2005). Nevertheless, these same compounds are desirable in other situations, such as protecting against UV radiation from sunlight to prevent damage to the skin and also to minimize glare in the eyes. This feature of melanin has been exploited by PhotoProtective Technologies (United States) in making a variety of eye wear that filter out colors to reduce the risks of macular degeneration and cataracts (Anon 2005). Melanin has also been incorporated in a hair dye, Melancor-NH, for darkening gray hair in humans.

Melanins and Health

Melanin deficiency in humans has been linked with certain abnormalities and diseases. Examples include oculocutaneous albinism that is characterized by reduced levels of the pigment in the eyes, hair, and skin, and ocular albinism that affects both eye pigmentation and visual acuity (Peracha et al. 2008). Melanin deficiency has also been linked to deafness (Cable et al. 1994) and Parkinson's disease (Nicolaus 2005).

Measurement of Melanins

Melanin measurements may be carried out by spectrofluorometry (Rosenthal et al. 1973) by measuring the fluorescence induced by excitation at 410 nm and emission at 500 nm, or by spectrophotometric measurement of the rate of melanochrome production from 5,6-dihydroxyindole-2-carboxylic acid and 5,6-dihydroxyindole systems at 540 and 500 nm (Blarzino et al. 1999). Melanin in tissues may also be measured based on spectrophotometry, HPLC, and electron spin resonance spectroscopy (Ito 2006, Hu et al. 1995).

TANNINS

Properties and Functions

Tannins are a heterogeneous group of water-soluble polyphenolic compounds present in many plants. The term "tannin" originated from the ancient practice of using plant extracts to "tan" leather, thus the name "tanning" is used to describe the process. This process makes use of the capacity of tannins to form cross-links with other biomolecules such as proteins or polysaccharides to "toughen" the textures of otherwise soft materials. Tannins occur mainly in tea, coffee, immature fruits, leaves, and the bark of trees where their function is believed to be one of defense. They have an astringent and bitter taste that is unpleasant to prospective consumers. The best known members of this family of compounds include tannic acid and its breakdown products (ellagic acid, gallic acid, and propyl gallate; Fig. 37.9). Tannic acid is used as a food additive. Foods known to be high in tannins include pomegranates, persimmons, cranberries, strawberries and blueberries, hazelnuts, walnuts, smoked meats and smoked fish, spices (e.g., cloves, tarragon, thyme, and cinnamon), most legumes, and beverages such as wine, cocoa, chocolate, and tea.

Tannins are classified into two main classes, (1) the hydrolyzable and (2) the condensed (nonhydrolyzable) tannins. The hydrolyzable tannins are those that on heating with dilute mineral acids or bases form carbohydrates and phenolic acids (i.e., gallic or egallic acid). The condensed tannins form phlobaphenes (e.g., phloroglucinol) when heated with dilute acids or bases. It is the condensed tannins that are more important in alcoholic beverages such as wines.

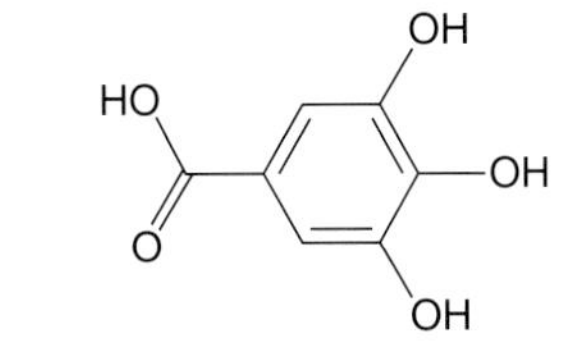

Gallic acid

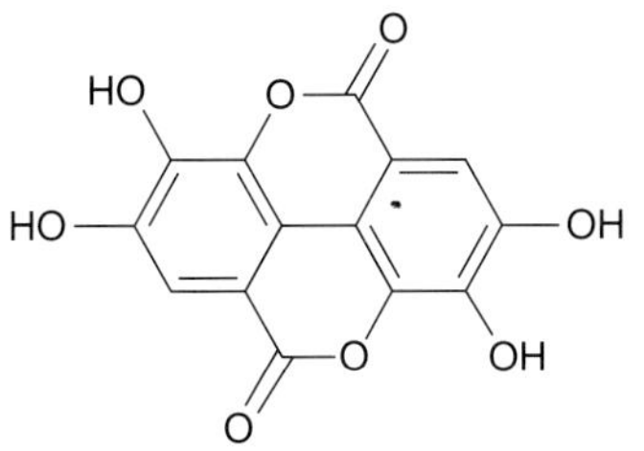

Ellagic acid

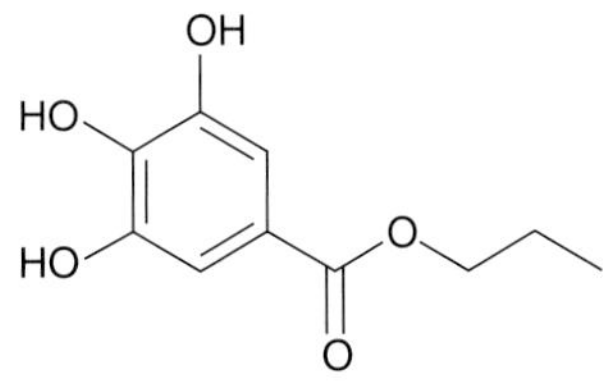

Propyl gallate

Figure 37.9. Structures of some common tannins.

Some tannins such as tannic acid and propyl gallate display antimicrobial activity toward foodborne bacteria, aquatic bacteria and off-flavor-producing microorganisms, e.g., tannic acid reduces staphylococcal activity (Akiyama et al. 2001). Thus, they can also be used in food processing to extend the shelf-life of some food products.

Tannins and Health

Some of the health benefits of *tannins* include their anti-inflammatory, antiseptic, and astringent properties. They also are effective against diarrhea, irritable bowel syndrome, and skin disorders. However, tannins are metal ion chelators and their binding to metal irons make these mineral elements not readily bioavailable. High intake of tannins could hamper the absorption of minerals such as Fe to result in anemia. They can also cause protein precipitation to interfere with the uptake of nutrients in some ruminants. Foods with high levels of tannins are judged to be of poor nutritional value, because they reduce feed intake, protein digestibility, feed efficiency and growth rates, among other things (Chung et al. 1998). These observations have been linked to tannins decreasing the efficiency in converting absorbed nutrients into new body components (Chung et al. 1998). The intake of foods high in tannins such as herbal teas is also thought to increase the incidence of cancers (e.g., esophageal cancer) and this has led to the suggestion that these compounds may be carcinogenic, but other reports attribute the apparent carcinogenic activity observed to constituents associated with tannins and not the tannins per se (Chung et al. 1998). Some of these observations contradict findings alluding to the anticarcinogenic behavior of tannins (Nepka 2009), or their capacity to reduce the mutagenicity of certain compounds (Chen and Chung 2000). The anticarcinogenic and antimutagenic activities of tannins and related compounds have been attributed to their polyphenolic nature that enables them to behave as antioxidants to check lipid peroxidation (Hong et al. 1995) and the damaging effects of reactive O_2 species (Nakagawa and Yokozawa 2002) in biological tissues. Tannins are also reported to reduce blood clotting, blood pressure, and serum lipid levels (Akiyama et al. 2001). They also induce liver necrosis (Goel and Agrawal 1981), and modulate immunoresponses (Marzo et al. 1990).

QUINONES

Properties and Functions

Quinones are a class of heterocyclic organic compounds with structures like cyclic conjugated diones. They have been described as secondary metabolites and are found in several organisms including flowering plants, bacteria, fungi, and to a small extent in animals like insects, sea urchins, corals, and other marine animals (Thompson 1991). The compounds display a plethora of colors from yellow to orange to intense purple to black. Examples of quinones include anthraquinone,

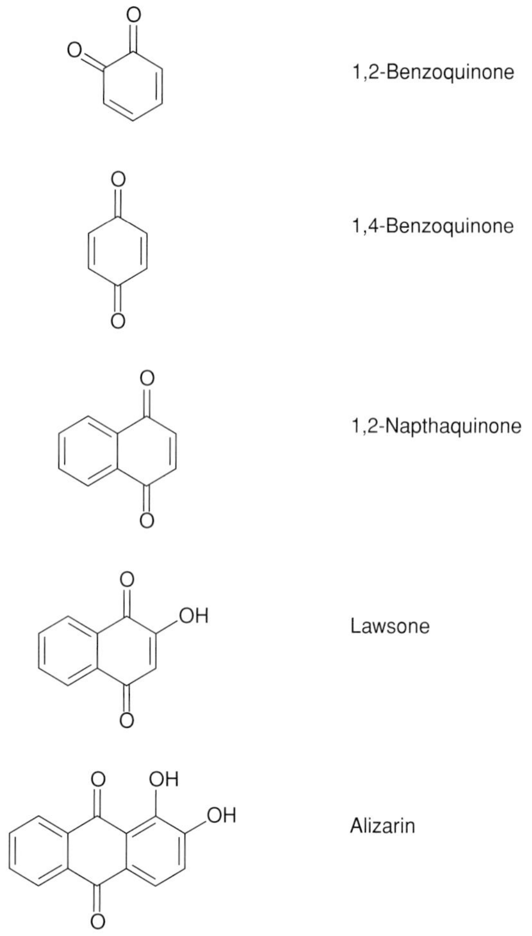

Figure 37.10. Structures of some common quinones.

1,2-benzo- and 1,4-benzo- quinones and naphthaloquinones (Fig. 37.10). Because of their high coloring power, quinones are used as dyes.

Quinones are used for the commercial scale production of hydrogen peroxide (H_2O_2) by the hydrogenation of anthraquinones to hydroquinones, and subsequent transfer of hydrogen (H_2) to O_2 (Schreyer et al. 1972, Drelinkiewicz and Waksmundzka-Góra 2006).

For example:

2-ethyl-9,10-anthraquinone + H_2
$\rightarrow$ 2-ethyl-9,10-anthrahydroquinone

dihydroanthraquinone + $O_2 \rightarrow$ anthraquinone + H_2O_2

Quinones can also react and form conjugates with amino acids, and the products formed can impact the flavor and color of foods (Bittner 2006).

In nature, some quinones serve as important constituents of biomolecules (e.g., vitamin K_1); some like ubiquinone and plastoquinone in mitochondria participate in electron transfer reactions in aerobic respiration, while plastoquinone in chloroplasts are involved in electron transfer reactions between photosystems 1 and 2 in photosynthesis.

Quinones and Health

Quinones and related compounds (e.g., tertiary butyl hydroquinone or TBHQ) have antioxidant properties and the terpenes of hydroquinone are thought to be more potent than the more common antioxidants such as α-tocopherol (Delgado-Vargas and Paredes-López 2003). Some other quinones like the napthoquinones have antibacterial activity and can inhibit various strains of aerobic and anaerobic microorganisms (Didry et al. 1986, Riffel et al. 2002). Some have also been shown to have anticancer or antitumor activities (Smith 1985). For example, dietary intake of certain quinone derivatives was reported to play a role in the prevention of breast cancer (Zhang et al. 2009). Ironically, certain quinones and their derivatives have also been shown to be mutagenic or carcinogenic (Zhang et al. 2009). In this regard, it is reported that extended exposure to some of these compounds increases the incidence of breast cancer. Some metabolic products of quinones are also known to be toxic, and the toxicity appears to arise from either the arylation of sulfhydryls compounds or the production of free radicals and/or reactive O_2 species (Smith 1985).

XANTHONES

Properties and Functions

Xanthones are a group of heat stable organic compounds with 9H-xanthen-9-one (xanthone) as the core structure (Fig. 37.11). Xanthones occur naturally in tropical plants and about two

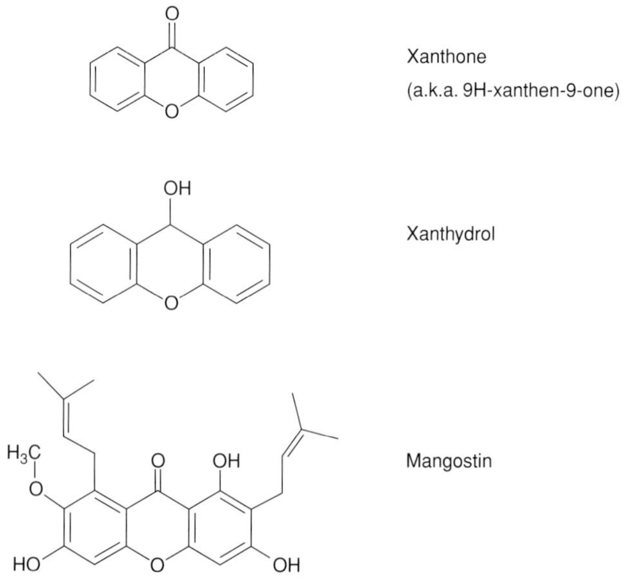

Figure 37.11. Structures of some common xanthones.

hundred known compounds in this group have been identified. Well-known examples are α-mangostin, β-mangostin, garcinone B, and garcinone E from the pericarp of mangosteen fruit. Xanthones are said to enhance the body's immune system and also impart other human health benefits such as promoting healthy cardiovascular or respiratory systems, supporting cartilage and joint function (Weecharangsan et al. 2006). Other health benefits attributed to xanthones include their capacity to reduce: allergies, cholesterol levels, inflammation, skin disorders and fatigue, and they also curb susceptibility to infections and help maintain gastrointestinal health (Weecharangsan et al. 2006). They have antioxidant properties and they do this by deactivating free radicals in the body. In this regard, they are said to be even more potent antioxidants than vitamins E and C.

Some applications of xanthones include their use as insecticide and it has been used in this regard against moth eggs and larvae. Xanthone in the form of xanthydrol is also used to verify urea levels in the blood.

OTHER PIGMENTS

Apart from the pigments described above, there are other lesser known natural pigments that are also found in various living organisms. Examples of these include the blue and red chromoproteins (phycocyanin and phycoerythrin) found in the blue-green algae and cyanobacteria. There are also other pigments such as flavins and pterins.

Phycocyanin and Phycoerythrin

Phycocyanin is blue pigment found in the blue-green algae and cyanobacteria, while phycoerythrin refers to red pigment found in the red algae commonly referred to as rhodophytes. These pigments are conjugated chromoproteins and they comprise of subunits each with a protein backbone that is covalently linked to open chain tetrapyrrole groups. Their molecular weights range from 44,000 daltons (monomers) to 260,000 daltons (hexamers) (Boussiba and Richmond 1979), and they function as light-absorbing substances together with chlorophyll in photosynthesis.

An example of useful red algae is the *nori* or *Porphyra* that is consumed as food and is used as wraps for sushi and in several other Japanese dishes. These pigments are produced commercially from *Spirulina platensis*. *Spirulina* are blue-green algae and they grow well in warm climates and alkaline waters. The two best known members of the *Spirulina* species are *S. platensis* and *Spirulina maxima*. *S. platensis* is cultivated in California while *S. maxima* are cultivated in Mexico. Phycocyanin and related compounds are thought to have antiviral and anticancer activities as well as immune system stimulatory ability. Phycocyanins also enhance the formation and development of red blood cells to reduce the incidence of anemia (Mathew et al. 1995, Jensen and Ginsberg 2000, Mani et al. 2000, Jensen et al. 2001, Samuels et al. 2002, Shih et al. 2003), and also thought to promote the development of healthy skins. Thus, it is used to treat eczema and psoriasis and some other skin disorders. These pigments occur together with carotenoids and chlorophylls in the blue-green and red algae. Phycoerythrin, in particular, is believed to facilitate red seaweed subsistence at greater depths in the ocean where hydrostatic pressures are high, unlike the other seaweed species (e.g., brown and green algae) that subsist in shallow waters.

The antioxidant, free radical scavenging, and anti-inflammatory properties of phycocyanin, as well as antiviral activity of the pigment have been demonstrated in studies with laboratory animals (Bhat and Madyastha 2000). Other studies have also revealed protective effects by phycocyanin against neuronal damage and oxidative damage to DNA (Bhat and Madyastha 2001, Piñero Estrada et al. 2001). Phycocyanin is used as a natural food-coloring agent in food products such as yoghurts, milk shakes and ice creams, beverages, desserts, and also in cosmetic products. Phycocyanin is used in medicine to facilitate selective destruction of atherosclerotic plaques or cancer cells by radiation with little or no damage to the surrounding cells or tissue (Morcos and Henry 1989), while phycoerythrin is used as fluorescent dyes for fluorescence activated cell scanning analysis (Hardy 1986).

Other pigments include curcumin found in turmeric. Curcumin is an oil soluble pigment and tends to fade when exposed to light, but it is heat stable. It is used to impart lemon yellow colors in food products such as curry, soups, and confectionery. Riboflavin (or vitamin B2), is a water-soluble and heat stable compound with an intense yellow color. It is used to color and fortify foods like dairy products, cereals, and dessert mixes. Carmine (or carminic acid, $C_{22}H_{20}O_{13}$) is a water-soluble pigment obtained from the female cochineal insect (*Dactylopius coccus*). It is stable to heat and light, and resistant to oxidation; and is used in alcoholic beverages and processed meat products. It is also used as an antirepellant.

SUMMARY/CONCLUDING REMARKS

For some time, interest in synthetic food colors was high because of the perceived disadvantages with natural pigments. These disadvantages include inconsistency in the product quality and yields from batch to batch, susceptibility to factors such as heat, pH, light, and air, as well as the high cost. However, pressure now seems to be increasing for less use of synthetic pigments as food colorants. Consumers are showing higher preference for natural pigments in foods for health safety reasons, and recent studies published in various scientific journals, including the *Lancet*, linked synthetic colors with hyperactivity in children (Harley et al. 1978, McCann et al. 2007). In addition, food regulations in Europe are promoting the use of natural pigments in foods over their synthetic counterparts. And the United States Food and Drug Administration has been petitioned to ban the use of synthetic colors like Yellow 5 and 6, Red 3 and 40, Blue 1 and 2, Green 3, and Orange B in foods, and to require labels on foods that contain these dyes. In addition to their coloring power, some of these natural pigments have also been shown to have more potent antioxidant and free radical scavenging capabilities than the synthetic ones commonly used in foods (Frankel et al. 1996). It is estimated that the world market for food colors is in excess of a billion dollars (United States) annually, and natural pigments account for about 30% of the market. It is predicted that the market share of natural pigments will continue to grow and surpass that of synthetic colors due to the increasing consumer concerns with the latter.

Thus, what needs to be done is more research to address the disadvantages with the commercial production and use of natural food pigments. Sources of the major food colors (red, blue, and yellow) must be identified; procedures to recover them efficiently and in purer and more stable forms, such as enzyme-assisted treatments (Kim et al. 2005) need to be developed; and treatments and/or conditions to stabilize them such as use in combination with antioxidants need to be exploited.

REFERENCES

Akiyama H et al. 2001. Antibacterial action of several tannins against *Staphylococcus aureus*. *J Antimicrob Chemother* 48: 487–491.

Albanes D et al. 1996. Alpha-tocopherol and beta-carotene supplements and lung cancer incidence in the alpha-tocopherol, beta-carotene Cancer Prevention Study: effects of base-line characteristics and study compliance. *J Nat Cancer Inst* 88: 1560–1570.

Anon. 2005. 1996–2005, MelaninProducts.com – PhotoProtective Technologies.

AOCS. 2004. VC Mehlenbacher et al. (eds.) *Official Methods and Recommended Practices of the American Oil Chemists' Society*, 5th edn. American Oil Chemists Society, Champaign, IL.

Ayehunie S et al. 1996. Inhibition of HIV-1 replication by an aqueous extract of *Spirulina platensis (Arthrospira platensis)*. 7th IAAA Conference, Knysa, South Africa, April 17.

Bailey DG, Dresser GK. 2004. Interactions between grapefruit juice and cardiovascular drugs. *Am J Cardiovasc Drugs* 4(5): 281–297.

Beale SI. 1990. Biosynthesis of the tetrapyrrole pigment precursor δ–aminolevulinic acid from glutamate. *Plant Physiol* 93: 1273–1279.

Benjakul S et al. 2005. Properties of phenoloxidase isolated from the cephalothorax of kuruma prawn (*Penaeus japonicus*). *J Food Biochem* 29: 470–485.

Bernstein PS et al. 2004. Resonance Raman measurement of macular carotenoids in the living human eye. *Arch Biochem Biophys* 430(2): 163–169.

Bhat VB, Madyastha KM. 2000. C-Phycocyanin: A potent peroxyl radical scavenger *in vivo* and *in vitro*. *Biochem Biophys Res Comm* 275: 20–25.

Bhat VB, Madyastha KM. 2001. Scavenging of peroxynitrite by phycocyanin and phycocyanobilin from *Spirulina platensis:* protection against oxidative damage to DNA. *Biochem Biophys Res Comm* 286: 262–266.

Bittner S. 2006. When quinones meet amino acids: chemical, physical and biological consequences. *Amino Acids* 30(3): 205–224.

Blarzino C et al. 1999. Lipoxygenase/H_2O_2-catalyzed oxidation of dihydroxyindoles: synthesis of melanin pigments and study of their antioxidant properties. *Free Radic Biol Med* 26(3–4): 446–453.

Bonkovsky HL. 2006. Hepatotoxicity associated with supplements containing Chinese green tea (*Camellia sinensis*). *Ann Intern Med* 144(1): 68–71.

Boussiba S, Richmond AE. 1979. Isolation and characterization of phycocyanins from the blue-green alga *Spirulina platensis*. *Arch Microbial* 120: 155–159.

Bowers WF. 1947. Chlorophyll in wound healing and suppurative disease. *Am J Surg* 71: 37–50.

Breithaupt DE. 2004. Identification and quantification of astaxanthin esters in shrimp (*Pandalus borealis*) and in a Microalga (*Haematococcos pluvialis*) by liquid chromatography-mass spectrometry using negative ion atmospheric pressure chemical ionization. *J Agric Food Chem* 52: 3870–3875.

Britton G. 1996. Carotenoids. In: GAF Hendry, JD Houghton (eds.) *Natural Food Colorants*, 2nd edn. Blackie Academic & Professional, London, pp. 197–243.

Burri BJ et al. 2003. Measurements of the major isoforms of vitamins A and E and carotenoids in the blood of people with spinal-cord injuries. *J Chromatogr A* 987(1–2): 359–366.

Cable J et al. 1994. Effects of mutations at the W locus (c-kit) on inner ear pigmentation and function in the mouse. *Pigment Cell Res* 7(1): 17–32.

Carpenter JA et al. 1967. Effects of anthocyanin pigments on certain enzymes. *Proc Soc Exp Biol Med* 124: 702–706.

Carrilho MC et al. 2009. Effect of diet, slaughter weight and sex on instrumental and sensory meat characteristics in rabbits. *Meat Sci* 82(1): 37–43.

Chen JS et al. 1993. Effect of carbon dioxide on the inactivation of Florida spiny lobster polyphenol oxidase. *J Sci Food Agric* 61: 253–259.

Chen S-C, Chung K-T. 2000. Mutagenicity and antimutagenicity studies of tannic acid and its related compounds. *Food Chem Toxicol* 38(1): 1–5.

Chen W-L, Chow C-J. 2001. Studies on the physicochemical properties of milkfish myoglobin. *J Food Biochem* 25: 157–174.

Chun OK et al. 2003. Superoxide radical scavenging activity of the major polyphenols in fresh plums. *J Agric Food Chem* 51(27): 8067–8072.

Chung K-T et al. 1998. Tannins and human health: a review. *Crit Rev Food Sci Nutr* 38(6): 421–64.

Cliff MA et al. 2007. Anthocyanin, phenolic composition, colour measurement and sensory analysis of BC commercial red wines. *Food Res Int* 40(1): 92–100.

Colio LG, Babb V. 1948. Study of a new stimulatory growth factor. *J Biol Chem* 174: 405–409.

Dahan A, Altman H. 2004. Food-drug interaction: grapefruit juice augments drug bioavailability-mechanism, extent and relevance. *Eur J Clin Nutr* 58(1): 1–9.

Delgado-Vargas F, Paredes-López O. 2003. *Natural Colorants for Food and Nutraceutical Uses*. CRC Press, Boca Raton, FL.

Dembinska-Kiec A. 2005. Carotenoids: risk or benefit for health. *Biochim Biophys Acta (BBA) - Mol Basis Dis* 1740(2): 93–94.

Didry N et al. 1986. Activité antibacterienne de naphthoquinones d'origine végetale. *Ann Pharm Fr* 44: 73–78.

Drelinkiewicz A, Waksmundzka-Góra A. 2006. Hydrogenation of 2-ethyl-9,10-anthraquinone on Pd/SiO_2 catalysts: The role of humidity in the transformation of hydroquinone form. *J Mol Catal A: Chem* 258(1.2): 1–9.

Ferrer OJ et al. 1989. Phenoloxidase levels in Florida spiny lobster (*Panulirus argus*): relationship to season and molting stage. *Comp Biochem Physiol* 93B: 595–599.

Ferry DR et al. 1996. Phase I clinical trial of the flavonoid quercetin: pharmacokinetics and evidence for in vivo tyrosine kinase inhibition. *Clin Cancer Res* 2(4): 659–668.

Frankel EN et al. 1996. Antioxidant activity of a rosemary extract and its constituents, carnosic acid, carnosol, and rosmarinic acid, in bulk oil and oil-in-water emulsion. *J Agric Food Chem* 44: 131–135.

Fraser PD, Bramley PM. 2004. The biosynthesis and nutritional uses of carotenoids. *Prog Lipid Res* 43: 228–265.

Gentile C et al. 2004. Antioxidant betalains from cactus pear (*Opuntia ficus-indica*) inhibit endothelial ICAM-1 expression. *Ann NY Acad Sci* 1028: 481–486.

Giusti MM, Wrolstad RE. 2001. Anthocyanins. Characterization and measurement with Uv-visible spectroscopy. In: RE Wrolstad, SJ Schwartz (eds.) *Current Protocols in Food Analytical Chemistry*. Wiley, New York, pp. 1–13. A.

Goel BB, Agrawal VP. 1981. Tannic acid-induced biochemical changes in the liver of two teleost fishes, *Clarias batrachus* and *Ophiocephalus punctatus*. *Ecotoxicol Environ Saf* 5(4): 418–423.

Graça Dias, M et al. 2008. Uncertainty estimation and in-house method validation of HPLC analysis of carotenoids for food composition data production. *Food Chem* 109(4): 815–824.

Hadley CW et al. 2003. The consumption of processed tomato products enhances plasma lycopene concentrations in association with a reduced lipoprotein sensitivity to oxidative damage. *J Nutr* 133: 727–732.

Hardy RR. 1986. Purification and coupling of fluorescent proteins for use in flow cytometry. In: DM Weir et al. (eds.), *Handbook of Experimental Immunology*, 4th edn. Blackwell Scientific Publications, Boston, pp. 31.1–31.12.

Harley JP et al. 1978. Synthetic food colors and hyperactivity in children: A double-blind challenge experiment. *Pediatrics* 62(6): 975–983.

Hong CY et al. 1995. The inhibitory effect of tannins on lipid peroxidation of rat heart mitochondria. *J Pharm Pharmacol* 47(2): 138–142.

Hosokawa M et al. 2004. Fucoxathin induces apoptosis and enhances the antiproliferative effect of the PPARγ ligand, troglitazone, on colon cancer cells. *Biochem Biophys Acta* 1675: 113–119.

Hosseinian FS, Li W. 2008. Trust beta measurement of anthocyanins and other phytochemicals in purple wheat. *Food Chem* 109(4): 916–924.

Hu DN et al. 1995. Melanogenesis in cultured human uveal melanocytes. *Invest Ophthalmol Vis Sci* 36: 931–938.

Ito S. 2006. Advances in chemical analysis of melanins. In: JJ Nordlund et al. (eds.), *The Pigmentary System*. Blackwell Publishing, New York, pp. 439–450.

Jagger J. 1985. *Solar UV Actions on Living Cells*. Praeger, New York.

Jatoi A et al. 2003. A phase II trial of green tea in the treatment of patients with androgen independent metastatic prostate carcinoma. *Cancer* 97(6): 1442–1446.

Jensen GS, Ginsberg DI. 2000. Consumption of *Aphanizomenon flos aquae* has rapid effects on the circulation and function of immune cells in humans. *J Amer Nutraceut Assoc* 2: 50–58.

Jensen GS et al. 2001. Blue-green algae as an immunoenhancer and biomodulator. *J Amer Nutraceut Assoc* 3: 24–30.

Jordan PM. 1989. Biosynthesis of 5-aminolevulinic acid and its transformation into coproporphyrinogen in animals and bacteria. In: HA Dailey (ed.) *Biosynthesis of Heme and Chlorophylls*. McGraw-Hill, New York, pp. 55–121.

Kearney PM et al. 2006. Do selective cyclo-oxygenase-2 inhibitors and traditional non-steroidal anti-inflammatory drugs increase the risk of atherothrombosis? Meta-analysis of randomized trials. *BMJ* 332(7553): 1302–1308.

Kim DH et al. 2005. Enhancement of natural pigment extraction using *Bacillus* species Xylanase. *J Agric Food Chem* 53(7): 2541–2545.

Kim YS et al. 2003. Effect of season on color of Hanwoo (Korean native cattle) beef. *Meat Sci* 63(4): 509–513.

Kotake-Nara E et al. 2005. Characterization of apoptosis induced by fucoxanthin in human promyelocytic leukemia cells. *Biosci Biotechnol Biochem* 69: 224–227.

Krinsky NI, Johnson EJ. 2005. Carotenoid actions and their relation to health and disease. *Mol Aspects Med* 26(6): 459–516.

Levin ME et al. 1980. Prevention and Treatment of Diabetic Complications. *Arch Intern Med* 140: 691–696.

Lorenz RT, Cysewksi GR. 2000. Commercial potential for *Haematococcus* microalgae as a natural source of astaxanthin. *TIBTECH* 18: 160–167.

Lotito SB, Frei B. 2006. Consumption of flavonoid-rich foods and increased plasma antioxidant capacity in humans: cause, consequence, or epiphenomenon? *Free Radic Biol Med* 41(12): 1727–1746.

Ludvigsson J et al. 2008. GAD treatment and insulin secretion in recent-onset type 1 diabetes. *N Engl J Med* 359(18): 1909–1920.

Manach C et al. 2004. Polyphenols: food sources and bioavailability. *Am J Clin Nutr* 79(5): 727–747.

Manach C et al. 2005. Bioavailability and bioefficacy of polyphenols in humans. I. Review of 97 bioavailability studies. *Am J Clin Nutr* 81(1): 230S–242S.

Mani UV et al. 2000. Studies on the long-term effect of Spirulina supplementation on serum lipid profile and glycated proteins in NIDDM patients. *J Nutraceut* 2: 25–32.

Marzo F et al. 1990. Effect of tannic acid on the immune response of growing chickens. *J Anim Sci* 68(10): 3306–3312.

Mathew B et al. 1995. Evaluation of chemoprevention of oral cancer with *Spirulina fusiformis*. *Nutr Cancer* 24: 197–202.

McCann D et al. 2007. Food additives and hyperactive behaviour in 3-year-old and 8/9-year-old children in the community: a randomised, double-blinded, placebo-controlled trial. *Lancet* 370(9598): 1560–1567.

Meléndez-Martínez AJ et al. (2007). Geometrical isomers of violaxanthin in orange juice. *Food Chem* 104: 169–175.

Meredith P, Riesz J. 2004. Radiative relaxation quantum yields for synthetic eumelanin. *Photochem Photobiol* 79(2): 211–216.

Mira L et al. 2002. Interactions of flavonoids with iron and copper ions: a mechanism for their antioxidant activity. *Free Radic Res* 36(11): 1199–1208.

Morcos NC, Henry WL. 1989. Medical uses of phycocyanin. US Patent 4886831.

Nakagawa T, Yokozawa T. 2002. Direct scavenging of nitric oxide and superoxide by green tea. *Food Chem Toxicol* 40(12): 1745–1750.

Nepka C. 2009. Chemopreventive activity of very low dose dietary tannic acid administration in hepatoma bearing C3H male mice. *Cancer Lett* 141(1): 57–62.

Nicolaus BJ. 2005. A critical review of the function of neuromelanin and an attempt to provide a unified theory. *Med Hypotheses* 65(4): 791–796.

Offenkrantz W. 1950. Water-soluble chlorophyll in ulcers of long duration. *Rev Gastroent* 17: 359–367.

Ogawa M et al. 1984. On physiological aspects of blackspot appearance in shrimp. *Nip Sui Gakkai* 50: 1763–1769.

Patek A. 1936. Chlorophyll and regeneration of blood. *Arch Int Med* 57: 73–84.

Peracha MO et al. 2008. Ocular manifestations of albinism. Available at http://emedicine.medscape.com/article/1216066-overview.

Piñero Estrada JE et al. 2001. Antioxidant activity of different fractions of *Spirulina platensis* protean extract II. *Farmaco* 56, 5–7 & 497–500.

Polagruto JA et al. 2003. Effects of flavonoid-rich beverages on prostacyclin synthesis in humans and human aortic endothelial cells: association with ex vivo platelet function. *J Med Food* 6(4): 301–308.

Priolo A et al. 2001. Effects of grass feeding systems on ruminant meat color and flavor. A review. *Anim Res* 50: 185–200.

Renstrom B, Liaaen-Jensen S. 1981. Fatty acid composition of some esterified carotenoids. *Comp Biochem Physiol* 69B: 625–627.

Riffel A et al. 2002. *In vitro* antimicrobial activity of a new series of 1,4-naphthoquinones. *Braz J Med Biol Res* 35(7): 811–818.

Rosenthal MH et al. 1973. Quantitative assay of melanin in melanoma cells in culture and in tumor. *Aural Biochem* 56: 91–99.

Saito A, Regier LW. 1971. Pigmentation of brook trout (*Salvelinus fontinalis*) by feeding dried crustacean waste. *J Fish Res Bd Canada* 28: 509–512.

Samuels R et al. 2002. Hypocholesterolemic effect of spirulina in patients with hyperlipidemic nephritic syndrome. *J Med Food* 5: 91–96.

Sarriés MV, Beriain MJ. 2006. Colour and texture characteristics in meat of male and female foals. *Meat Sci* 74(4): 738–745.

Schreyer GG et al. 1972. Process for the production of hydrogen peroxide. US Patent 3,699,217.

Shih SR et al. 2003. Inhibition of enterovirus 71-induced apoptosis by allophycocyanin isolated from a blue-green alga *Spirulina platensis*. *J Med Virol* 70: 119–125.

Shoskes DA et al. 1999. Quercetin in men with category III chronic prostatitis: a preliminary prospective, double-blind, placebo-controlled trial. *Urology* 54(6): 960–963.

Smith MT. 1985. Quinones as mutagens, carcinogens, and anticancer agents: Introduction and overview. *J Toxicol Environ Health A* 16(5): 665–672.

Soriano PM et al. 2007. Determination of anthocyanins in red wine using a newly developed method based on Fourier transform infrared spectroscopy. *Food Chem* 104(3): 1295–1303.

Stintzing FC et al. 2006. Characterization of anthocyanin–betalain mixtures for food colouring by chromatic and HPLC-DAD-MS analyses. *Food Chem* 94(2): 296–309.

Takaichi S et al. 2003. Fatty acids of astaxanthin esters in krill determined by mild mass spectrometry. *Comp Biochem Physiol* 136B: 317–322.

Tanaka Y et al. 1976. The biosynthesis of astaxanthin – XVI. The carotenoids in crustacea. *Comp Biochem Physiol* 54B: 391–393.

Tanaka Y et al. 2008. Biosynthesis of plant pigments: anthocyanins, betalains and carotenoids. *Plant J* 54(4): 733–749.

Tang J et al. 2004. Equations for spectrophotometric determination of myoglobin redox forms in aqueous meat extracts. *J Food Sci* 69: C717–C720.

Thompson RH. 1991. Distribution of naturally occurring quinones. *Pharm World Sci* 13(2): 70–73.

Tyrell RM. 1991. UVA (320–380nm) as an oxidative stress. In: H Sies (ed.) *Oxidative Stress: Oxidants and Antioxidants*. Academic, San Diego, pp. 57–83.

Varma SD, Kinoshita JH. 1976. Inhibition of lens aldose reductase by flavanoids - their possible role in the prevention of diabetic cataracts. *Biochem Pharmacol* 25: 2505–2513.

van Beek JHGM, Westerhof N. 1996. Measurement of the oxygenation status of the isolated perfused rat heart using near-infrared detection. *Adv Exp Med Biol* 388: 147–154.

Weber RE et al. 1974. Functional and biochemical studies of penguin myoglobin. *Comp Biochem Physiol B: Biochem Mol Biol* 49(2): 197–204.

Weecharangsan W et al. 2006. Antioxidative and neuroprotective activities of extracts from the fruit hull of mangosteen (Garcinia mangostana L.). *Med Princ Pract* 15(4): 281–287.

Health shop. Melancor NH Hair Color Restorer & Rejuvenator. Available at www.911healthshop.com/melancornh.html.

Yan X et al. 1989. Studies on the mechanism of blackspot development in Norway lobster (*Nephrops norvegicus*). *Food Chem* 34: 273–283.

Yuan J-P et al. 1996. Separation and identification of astaxanthin esters and chlorophylls in *Haematococcus lacustris* by HPLC. *Biotech Tech* 10: 655–660.

Zhang Q et al. 2009. Balance of beneficial and deleterious health effects of quinones: a case study of the chemical properties of genistein and estrone quinones. *J Am Chem Soc* 131(3): 1067–1076.

Zijp IM et al. 2000. Effect of tea and other dietary factors on iron absorption. *Crit Rev Food Sci Nutr* 40(5): 371–398.

Part 7
Food Processing

38
Thermal Processing Principles

Yetenayet Bekele Tola and Hosahalli S. Ramaswamy

Abstract: Food intended for human consumption must be produced in safe and stable forms to ensure availability, distribution, as well as normal growth and development. Thus, foods are processed in various forms to achieve these desired effects. Thermal processing entails the application of heat energy for food transformation into the desired safe and stable forms. This chapter describes the basic principles of thermal processing and surveys conventional versus novel thermal processing methods currently available for food transformation.

INTRODUCTION

Food processing is used to transform and/or preserve raw ingredients from the farm into various food forms for consumption by human being. Food processing often takes clean, harvested produce or edible portions and uses them to produce attractive and marketable food products. Fresh agricultural products (plant and animal origin), due to their high moisture content, are highly perishable and can be kept only for a short period of time. Especially during peak harvesting seasons, due to large volume of these products, unless and otherwise pertinent measures are taken, the problem leads to spoilage and high economic losses. Furthermore, food products should be available throughout the year in market to fulfill the demand of consumers in all seasons. To do this, we need to minimize postharvest losses and preserve agricultural products in a safe way with no or minimum quality loss. The major emphasis of food processing is shelf life extension by preventing undesirable changes in the wholesomeness, nutritive value, and sensory qualities (Ramaswamy and Marcotte 2006).

Food Biochemistry and Food Processing, Second Edition. Edited by Benjamin K. Simpson, Leo M.L. Nollet, Fidel Toldrá, Soottawat Benjakul, Gopinadhan Paliyath and Y.H. Hui.

Various food processing operations make the food products more attractive, satisfying, safer, and easier to eat. Common food-processing techniques include pasteurization, sterilization, cooking, drying, cooling and freezing, fermentation, addition of preservatives and reduction of water activity, and use of two or more of the above techniques to inhibit or stop chemical, biochemical, and microbiological activities. Most foods available in the market are subjected to some form of thermal processing. Although the canning process started from Nicholas Appert's time in 1809, food processing via the canning process still provides a universal and economic method for preserving and processing foods. Thermal processing of canned foods has been one of the most widely used methods for food preservation during the twentieth century and has contributed significantly to the nutritional well-being of much of the world's population (Teixeira and Tucker 1997).

The two very common industrial conventional thermal food processing technologies in production of canned products are pasteurization and sterilization. Pasteurization is the process of heating liquids and/or solid foods for the purpose of destroying principal pathogens capable of growing under aerobic conditions and reducing the level of spoilage vegetative bacteria, protozoa, and fungi from high acidic foods (pH $<$ 4.5). When pH $>$ 4.5, foods produced by this process should be stored at low temperature to avoid further growth spore forming microorganisms. However, commercial sterilization refers to an intensive heat treatment process that effectively kills or eliminates all pathogens and vegetative microorganisms as well as bulk of spore-forming bacteria from the low acid foods (pH $\geq$ 4.5). Products produced through this method can be shelf stable for up to two years. In order to reduce the process severity, the thermophilic spore formers are not targeted to be completely eliminated, instead the canned product is advised to be stored at temperatures below 30°C to prevent the growth of thermophiles. Because of the high safety implications and severe processing conditions, the conventional thermal sterilization has been accepted to result in considerable product quality degradation. However, many improvements have been implemented to improve thermal processing operations and techniques, and novel processing approaches have been introduced in recent years to improve the quality of thermally processed foods without compromising safety of these products. Therefore, the main focus of this chapter is to describe the basic principles of thermal processing operation in terms of production of safe and better quality products.

THERMAL PROCESSING BASICS

Introduction

The major objective of thermal processing is production of safe and stable products that consumers are willing and able to buy. To achieve this goal, it is necessary to understand the scientific basis on which the process is established. Silva et al. (1992) indicated that commercial thermal processing is a function of several factors, such as product thermo-physical properties (product heating rate), surface heat transfer coefficient, initial food temperature, retort temperature, heating medium (hot water or steam), heating medium come-up time (CUT), temperature resistance of food microorganisms and quality factors, and target degree of lethality or safety level we need to achieve.

The success of thermal processing does not depend on the elimination of the entire microbial population, because this would result in low product quality due to the long heating required. Instead, all pathogenic and most spoilage-causing microorganisms in a hermetically sealed container are destroyed, bulk of the spore formers is killed, and an environment is created inside the container that does not support the growth of remaining spore formers. There are different mechanisms that enable one to control the germination and growth of such type of spores in canned foods. These are based on the microbial growth and inactivation with respect to oxygen requirement, pH preference, and temperature sensitivity.

In general, canned foods have a 200-year history and are likely to remain popular in the foreseeable future owing to their convenience, long shelf life, and low cost of production. The technology is receiving increasing attention from thermal-processing specialists to improve both the economy and quality of some canned foods (Durance 1997). However, the sterilization process not only extends the shelf life of the food, but also affects its nutritional and sensorial qualities. Process optimization is therefore necessary in order to promote better quality retention without sacrificing safety.

Reaction Rate

During thermal processing of foods, several types of chemical reactions occur. Some reactions result in a quality loss and such type of reactions must be minimized, whereas others result in the development and formation of a desirable flavor, taste, or color, and these ones must be optimized to obtain the best product quality (Toledo 2007). In order to maintain quality of food products through optimization of processing conditions, predictive mathematical models are very important. To realize this goal, information is needed on the rates of destruction of microbes as well as quality parameters and their dependence on variables such as temperature, pH, light, oxygen, and moisture content, which can be expressed by mathematical models. A better understanding of kinetics of food products can provide better opportunities for developing food processes to maximize quality parameters and ensuring safety.

Each reaction undergoes on its own rate and the rate of reaction is described by the reaction kinetics. Kinetics is the study of the rate at which compounds react. Reaction kinetics (rate theory) deals to a large extent with the factors that influence the reaction velocity. The rate depends on several factors, including the contact between the reacting components, their concentration, temperature, and pressure at which the reaction takes place. The "collision theory" implies that the molecules need to collide with each other in order for the reaction to take place. If there are a higher number of collisions in a system, there is a greater chance for the reaction to occur. The reaction will go faster, and the rate of the reaction will be higher. In collision theory, two main things are to be considered: the activation energy, which is the minimum energy required to initiate the collision, and the statistical probability for collisions between certain molecules that possess an adequate energy level for the reaction to occur at a given temperature. To measure a reaction rate, it is necessary to monitor the concentration of one of the reactants or products as a function of time. Therefore, the rate of reaction can be expressed as a rate of change in concentration to change in time.

On the basis of this concept, the rate law is an expression relating the rate of a reaction to the concentrations of the chemical species present, which may include reactants, products, and catalysts. Many food reactions follow a simple rate law, which takes the form

$$r = k[A]^a[B]^b[C]^c \quad (1)$$

that is, the rate (r) is proportional to the concentrations of the reactants (A, B, C) each raised to some power. The constant of proportionality, k, is called the rate constant. The power a particular concentration is raised is the order of the reaction with respect to that reactant. Note that the orders do not have to be integers. The sum of the powers in Equation 1 is called the overall reaction order.

Zero-Order Reaction

A zero-order reaction is independent of the concentration of the reactants. A higher concentration of reactants will not speed up the zero-order reaction. This means that the rate of the reaction is equal to the rate constant, k, of that reaction. Zero-order reaction is described as

$$\text{Rate} = r = -\frac{dA}{dt} = k[A]^0 \quad (2)$$

After separating variables and integrating both sides of Equation 2

$$\int_{A_0}^{A} dA = -\int_{t_0}^{t} kdt \quad (3)$$

This provides the integrated form of the rate law

$$[A] = [A_O] - kt \quad (4)$$

where k is the rate constant of the reaction, A is concentration at time t. In a zero-order reaction, when concentration data $[A]$ is plotted versus time (t), the result is a straight line.

First-Order Reaction

A first-order reaction is one where the rate depends on the concentration of the species to the first power. Most of the reactions involved in the processing of foods are of first-order reactions. For a general unimolecular reaction, the decrease in the concentration A over time t can be written as

$$\text{Rate} = r = -\frac{dA}{dt} = k[A]^1 \quad (5)$$

Rearranging the equation

$$-\frac{dA}{A} = kdt \quad (6)$$

Integrate both sides of the equation

$$\int_{A_0}^{A} \frac{dA}{A} = -\int_{t_0}^{t_0} kdt \quad (7)$$

and the linear for of Equation 7 is:

$$\ln[A] = \ln[A_0] - kt \quad (8)$$

where $[A]$ is the concentration at time t, $[A_0]$ is the concentration at time $t = 0$, and k is reaction rate constant (s^{-1}).

Plotting ln $[A]$ with respect to time (t) for a first-order reaction gives a straight line with the slope of the line equal to $-k$, where the rate constant is calculated.

Second-Order Reaction

A second-order reaction is one where the rate depends on the concentration of the species to the second power. The reaction rate expression for a unimolecular second-order reaction is

$$\frac{dA}{dt} = -k[A]^2 \quad (9)$$

Separation of variables and integration of both sides of equations will give us

$$\frac{dA}{[A]^2} = -kdt \quad (10)$$

$$\int_{A_0}^{A} \frac{A}{[A]^2} = -\int_{t_0}^{t} kt \quad (11)$$

$$\frac{1}{[A]} = \frac{1}{[A_0]} + kt \quad (12)$$

Second-order unimolecular reaction is characterized by a hyperbolic relationship between concentration of the reactant or product and time. A linear plot will be obtained if *1/A* is plotted against time.

Second-order bimolecular reactions may also follow the following rate equation:

$$A + B \rightarrow P$$

$$\text{Rate} = \frac{dA}{dt} = -k[A][B] \quad (13)$$

where A and B are the reactants. After separating variables, the differential equation may be integrated by holding B constant to give

$$\frac{dA}{[A]} = -k[B]dt \quad (14)$$

$$\int_{A_0}^{A} \frac{dA}{A} = -\int_{t_0}^{t_0} k'dt \quad (15)$$

$$\frac{1}{[A]} = \frac{1}{[A_0]} - k't \quad (16)$$

k' is a pseudo-first-order rate constant: $k' = kB$.

A second-order bimolecular reaction will yield a similar plot of the concentration of the reactant against time as a first-order unimolecular reaction, but the reaction rate constant will vary with different concentrations of the second reactant (Table 38.1).

Table 38.1. Summary of Zero-, First-, and Second-Order Reaction Rates (*M*-molar Concentration)

Rate Law/Order	Differential Form	Integral Form	Linear Plot to Determine k	Units of Reaction Constant (k)
Zero	$\frac{dA}{dt} = -k$	$[A] = [A_o] - kt$	$[A]$ vs. t	$\frac{M}{s}$
First	$\frac{dA}{dt} = -k[A]^1$	$\ln[A] = \ln[A_o] - kt$	$\ln[A]$ vs. t	$\frac{1}{s}$
Second	$\frac{dA}{dt} = -k[A]^2$	$\frac{1}{[A]} = \frac{1}{[A_o]} + kt$	$\frac{1}{[A]}$ vs. t	$\frac{1}{M.s}$

Common Food Microorganisms

Foods by their very nature involve complex biological molecules that strongly support the survival and growth of food microorganisms. One of the major causes for food deterioration is the growth and activity of microorganisms. Microorganisms can contaminate the food before and after processing from various sources. The major categories of microorganisms involved in food spoilage and deterioration are fungi and bacteria. Each microorganism has its own optimum temperature, pH, and oxygen level to grow. Broadly, food microorganisms can be categorized as pathogenic and spoilage ones. Food pathogenic microorganisms are those groups that can cause illness in human being or animals due to food poisoning or infection. Food poisoning and food infection are different, although the symptoms are similar. True food poisoning or food intoxication is caused by eating food that contains a toxin or poison due to bacterial or fungal growth in food. The bacteria or fungi that produced and excreted the toxic waste products into the food may be destroyed during processing, but the toxin they produced can cause the illness or digestive upset to occur. *Staphylococcus aureus* and *Clostridium botulinum* are two common species of bacteria that cause food poisoning. Food infection is the second type of foodborne illness. It is caused by eating food that contains certain types of live bacteria, which are present in the food. Once the food is consumed, the bacterial cells themselves continue to grow and illness can result. *Salmonellae, C. perifringens, Vibiro parahaemolyticus, Yersinia enterocolitica, and Listeria monocytogenesis* are common infectious bacteria of food.

The other category includes food spoilage microorganisms. Food spoilage can be defined as the process or change leading to a product becoming undesirable or unacceptable for consumption. This can be chemical, physical, or microbial spoilage or the combination of these factors. Food spoilage is more of an economic concern than safety issue.

Food microorganisms can be grouped based up on their oxygen and temperature requirement for survival and growth (Table 38.2). In foods that are packaged under vacuum, low oxygen levels are intentionally achieved. Therefore, the prevailing conditions do not support the growth of microorganisms that require oxygen (obligate aerobes). Likewise, the thermophilic microorganisms require temperatures much higher than 30°C for their growth.

Acidity level of a food is another intrinsic food factor that determines the survival and growth of food microorganisms. From thermal processing standpoint, foods are divided into three pH groups: (i) high-acid foods (pH $<$ 3.7), (ii) acid or medium-acid foods (3.7 $<$ pH $<$ 4.5), and (iii) low-acid foods (pH $>$ 4.5). The most important distinction in the pH classification, with reference to thermal processing, is the dividing line between acid and low-acid foods. Most laboratories concerned with thermal processing have devoted attention to *C. botulinum,* which is a highly heat-resistant, rod-shaped, spore-forming, anaerobic

Table 38.2. Classification of Microorganisms and Bacteria Based on Their Oxygen Demand and Optimum Temperature for Growth

Classification Groups of Microorganisms	Examples of Microorganisms
Oxygen-based classification	
Obligatory aerobes (require oxygen)	Most molds, *Micrococcus, Serratia marcescens, Mycobacterium tuberculosis*
Obligatory anaerobes (require absence of oxygen)	*C. botulinum, C. sporogenes, C. thermosaccharolyticum*
Facultative anaerobes (tolerate some oxygen)	*Bacillus coagulans, Staphylococcus aureus*
Optimum temperature for growth	
Thermophilic (55–35°C)	*B. coagulans; S. thermophilus; C. thermosaccharolyticum, C. nigrificans*
Mesophilic (40–10°C)	*C. pasteurianum, B. mascerans, C. sporogenes, B. licheniformis*
Psychrophilic (35–<5°C)	*Pseudomonas, Micrococcus, C. botulinum* Type E, *S. aureus*

pathogen that produces the deadly botulism toxin. It has been generally recognized that *C. botulinum* does not grow and produce toxin below a pH of 4.5. Hence, the dividing pH between the low-acid and acid groups is set at 4.5. There may be other microorganisms that are more heat-resistant than *C. botulinum*.

Thermal Resistance of Food Microorganisms

As indicated in the above sections, the success of thermal processing depends on the destruction of all pathogenic and most spoilage-causing microorganisms in a hermetically sealed container and creating an environment inside the package that is not conducive to the growth of spoilage-type microorganisms and their spores that resist the thermal process. The microorganisms vary in terms of heat resistance and require different degree of thermal processing. In order to establish a thermal process, first, the target heat-resistant microorganisms of concern need to be identified. *C. botulinum* is the principle target microorganism of concern in low-acid canned foods. The second step is evaluating the thermal resistance of target organism. The thermal resistance data should be well described by an appropriate and reliable mathematical model. Data on the temperature dependence of the microbial destruction rate are also needed to integrate the destruction effect through the temperature profile under processing conditions.

Kinetics of Microbial Death

An understanding of the mechanism and kinetics of thermal death of food microorganisms would be helpful in the practical use of heat in processing of foods. Thermal destruction of bacteria is generally accepted to be exponential with time, and process calculations used in thermal processing are generally based on this assumption (Stumbo 1973). However, there are also nonexponential thermal death patterns indicated in different literatures (Stumbo 1973, Toledo 2007). The theory that thermal death of bacteria follows the first-order kinetics has been well recognized (Stumbo 1973). A first-order reaction is one in which the rate is proportional to the number of microorganisms present.

Decimal Reduction Time and Thermal Death Time

When a suspension of microorganisms is heated at constant temperature, the decrease in number of viable organisms follows logarithmic order of death. In other words, the logarithm of the surviving number of microorganisms following a heat treatment at a particular temperature plotted against heating time will give a straight-line curve (Fig. 38.1). These curves are commonly called survivor curves. The microbial destruction rate is defined as a decimal reduction time or D-value (minutes), which is the heating time in minutes at a given temperature required to result in one decimal or log unit reduction in the surviving microbial population. In other words, D-value represents a heating time that results in 90% destruction of the microbial population from the initial number. The rate of microbial destruction with time at constant temperature expressed as

$$\frac{dN}{dt} = -k[N]^1 \tag{17}$$

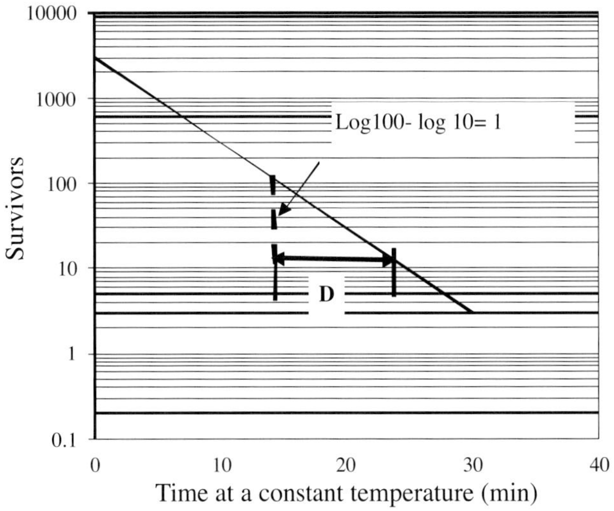

Figure 38.1. Survivor curve that depicts the number of survivor microorganisms versus time at a constant temperature.

where dN/dt is the change in microbial population with time t, k is the reaction rate constant, and 1 is to indicate the order of reaction (first order). After separation of variables and integration, the above equation can be written as

$$\ln N = \ln N_0 - kt \tag{18}$$

where N_0 is initial population of microorganisms.

D is based on common logarithms, in contrast with k, which is based on natural logarithms. But D and k are related as

$$\ln\left(\frac{N}{N_0}\right) = \ln(10)\log\left(\frac{N}{N_0}\right) = -kt \tag{19}$$

From this

$$\log\left(\frac{N}{N_0}\right) = -\frac{k}{\ln(10)}t \tag{20}$$

Thus, the D value is the negative reciprocal of the slope of a plot of log (N) against t, and D and k are related as

$$-\frac{1}{D} = -\frac{k}{\ln(10)}; \quad D = \frac{\ln(10)}{k} \quad \text{or} \quad D = \frac{2.303}{k} \tag{21}$$

Therefore, the D value from survivor curve can be expressed as

$$D = \frac{(t_2 - t_1)}{(\log N_0 - \log N)} \tag{22}$$

where N_0 and N represent the survivors following heating for t_1 and t_2 (minutes), respectively.

Commonly in thermal processing applications, survivor curves are plotted on specially designed semi-log papers for easy handling and interpretation data. The survivor number of microorganisms is plotted directly on the logarithmic y-axis against time on the linear x-axis. From this graph, the D value can be

Table 38.3. Probability of Residual Fraction Survivors after Multiple *D* Values

D Value ($N_o = 10^3$)	Probability of Residual Fraction Survivors After 90% of Destruction
1*D*	10^2
2*D*	10^1
3*D*	10^0
4*D*	10^{-1}
5*D*	10^{-2}
nD	10^{n-3}

calculated from the time interval on *x*-axis between which the straight-line portion of the curve on the *y*-axis shows a reduction by logarithmic unit. For instance, as indicated in Figure 38.1 the number of survivor microorganisms is reduced from 100 (two logarithmic unit) to 10 (one logarithmic unit) in approximate time interval of 10 minutes (24–14 minutes), which is the *D* value at a given constant temperature. However, in today's computer era, the *D* value can be obtained from a log (*N*) versus *t* computer graph on a spreadsheet as the negative reciprocal slope. The logarithmic nature of the survivor curve indicates that complete destruction of microbial population is not theoretically possible; a decimal fraction of the population should remain even after an infinite number of *D* values (Table 38.3).

Thermal death time (TDT) is another concept that is commonly used in thermal processing of food. The concept somewhat contradicts the logarithmic destruction approach. As compared to decimal reduction concept, TDT is the minimum heating time required to cause complete destruction of a microbial population. In sequential study with microbial destruction as a function of time (at a given temperature), TDT then represents a time between the shortest destruction and the longest survival times. The difference between the two is sequentially reduced or geometrically averaged to get an estimate of TDT. The "death" in this instance generally indicates the failure of a given microbial population, after the heat treatment, to show a positive growth in the subculture media. Comparing the TDT approach with the decimal reduction approach, one can easily recognize that TDT value depends on the initial microbial load (whereas *D* value does not). Further, if TDT is always measured with reference to a standard initial load or load reduction, it would simply represent a certain multiple of *D* value. For example, if *TDT* represented the time to reduce the population from $N_0 = 10^2$ to $N = 10^{-1}$, then TDT is a measure of 3*D* values (Figure 38.2). On the other hand, if it is based on 10^4–10^{-6}, it would represent a 10*D* value. Therefore, TDT is a multiple of *D* (TDT = *nD*, were *n* is the number of decimal reductions).

Temperature Dependency of Kinetic Parameters

The *D* value (and TDT) is strongly dependent on the temperature applied during heating of bacterial suspension. Higher temperatures will give shorter *D* values and vice versa. Temperature dependency of kinematic parameters is expressed in terms of *D*-*z* or TDT-*z* models. By plotting the various *D*-values (or TDT) against temperature, again on a logarithmic scale, the *z*-value can be obtained as temperature range for the *D*-value (or TDT) curve to pass through one log cycle (Fig. 38.3). In this concept, the temperature sensitivity indicator is defined as the *z* value. Physically, *z* represents the temperature change need to increase or decrease the decimal reduction time (or TDT) by a factor of 10.

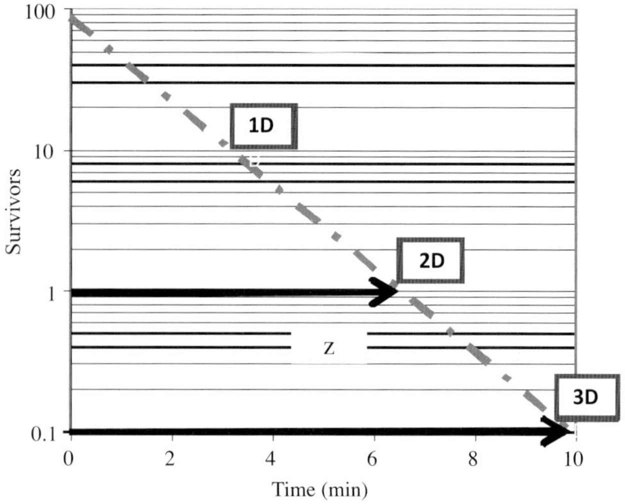

Figure 38.2. Multiple *D* and TDT.

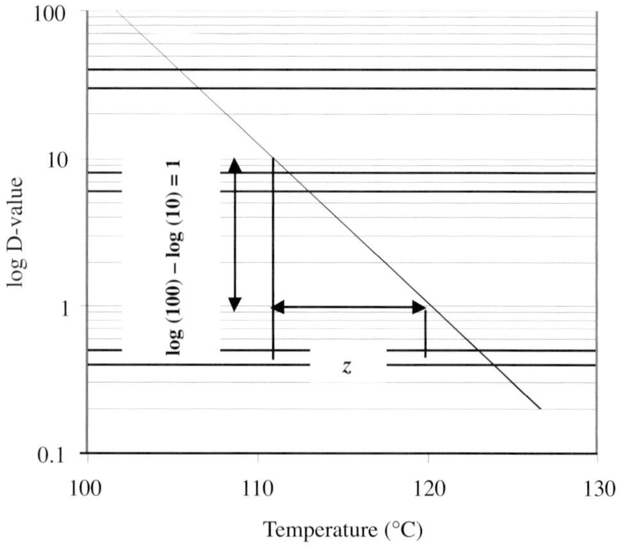

Figure 38.3. *D* value versus temperature curve and *z* value.

Mathematically, using reference temperature (T_r) and reference D value (D_0), z value (°C) is expressed as

$$z = \frac{T_r - T}{\log D - \log D_0} \quad (23)$$

or in between two temperature ranges of T_1 and T_2,

$$z = \frac{T_2 - T_1}{\log D_1 - \log D_2} \quad (24)$$

where D_1 and D_2 are D values at T_1 and T_2, respectively.

A larger z value indicates that the rate of destruction of microorganism is less temperature sensitive. A small z value indicates higher temperature sensitivity and a small change in temperature will result in a significant change on microbial population. Most commonly, quality parameters like nutrients and color have large z values as compared to microorganisms that have small z values. Again, microbial spores will have a higher z as compared to vegetative bacteria.

The D value at any given temperature can be obtained from a modified form of the above equation using the reference D value (D_0 at a reference temperature, $T_{r,}$ usually 121.1°C (250°F) for thermal sterilization and 82.2°C (180°F) for pasteurization).

$$D = D_0 10^{\left(\frac{T_r - T}{z}\right)} \quad (25)$$

Concepts of Process Lethality

Lethality (F_0 value) is a measure of the heat treatment or sterilization processes. To compare the relative sterilizing capacities of heat processes, a unit of lethality needs to be established. For convenience, this is defined as an equivalent heating time of 1 minute at a reference temperature, which is usually taken to be 121.1°C (250°F) for the sterilization processes. The criteria for the adequacy of a process must be based on two microbiological considerations: (i) destruction of the microbial population of public health significance and (ii) reduction in the number of spoilage-causing bacteria.

In the canning industry, the F_0 value is often used for the low-acid canned foods and refers to the lethality value ($F^z_{T_{ref}}$ value) with a z value of 10°C (18°F) and a reference temperature (T_{ref}) of 121.1°C (250°F). The z value of 10°C is used as the thermal characteristic for the pathogenic microorganism, *C. botulinum* spores.

Experimental work and experience have shown that a process that achieves 12 log reductions of *C. botulinum* can be considered commercially sterile (Tucker 2008). If such a process is correctly applied to a product, then the health risk to the consumer will be insignificant. Applying a 12D process reduces the probability of spore survival by a factor of 10^{12}. If cans contain one initial spore, then for trillion cans produced and given a 12D process, only one can would contain a surviving spore.

The $D_{121.1°C}$ ($D_{250°F}$ or D_0) value can be used for establishing such a process. The D_0 for *C. botulinum* is taken as 0.23 minute. Thus, 12D value would represent 12 × 0.23 or 2.76 minutes. The minimum F_0 value should be 2.76 minutes to accomplish the 12D process. In general, if N_0 is the initial spore concentration and N the target spore concentration need to achieve, then for exposure to a theoretical constant process temperature of 121.1°C at the coldest or slowest heating point, the target process would be:

$$F_0 = D_0 \log\left(\frac{N_0}{N}\right) \quad (26)$$

However, several low-acid foods are processed beyond the minimum F_0 value of 2.76 minutes in order to deal with spoilage-causing bacteria of much greater heat resistance. For these organisms, acceptable levels of spoilage probability are usually dictated by economic considerations. Most food companies accept a spoilage probability of 10^{-5} (5D = TDT) from mesophilic spore formers (organisms that can grow and spoil food at room temperature). The organism most frequently used to characterize this classification of food spoilage is a strain of *C. sporogenes*, known as PA 3679, with a maximum $D_{121.1°C}$ value of 1.00 minute. Therefore, the 5D minimum value would be 5 minutes and hence a F_0 of 5 minutes is more commonly used for these foods. Where thermophilic spoilage is a problem, more severe processes may be necessary because of the high heat resistance of thermophilic spores. Fortunately, most thermophiles do not grow readily at room temperature and require incubation at unusually high storage temperatures (>38°C) to cause food spoilage.

Sterilization time required at temperature (T) for the same value target microorganism to deliver equal degree of sterility is mathematically expressed as

$$F^z_T = F_0 * 10^{\left(\frac{T_{ref} - T}{z}\right)} \quad (27)$$

The same general relationships as were discussed under sterilization apply to pasteurization. A combination of temperature and time that is sufficient to inactivate the particular species of bacteria must be used. Fortunately, most of the pathogenic organisms, which can be transmitted from food to the person who eats it, are not very resistant to heat. For thermal process calculation of high-acid canned foods with an extended refrigerated shelf life, for shelf-stable low pH < 3.9 products (such as fruit juices) and acidified foods under normal storage conditions, food industries commonly use the pasteurization value (P_o with z value of 10°C and T_{ref} of 82.2°C) instead of the sterilization value (F_o *value*), since foods generally are processed below or around 100°C.

THERMAL PROCESS DETERMINATION METHODS

Thermal process evaluation is a process of determining the desired processing time at a given heating condition to achieve desired process lethality or vice versa. Determination of process lethality (F_0) can be done in two ways: using calculation method (Equation 27) and microbiological data (Equation 26). Different calculation methods have been used to determine the required processing time to achieve the desired lethality. The reliability of the method depends on how accurately it integrates the lethality effect of transit temperature response of the food undergoing thermal process, with respect to test microorganism of public health concern. This implies that accurate determination of microbial kinetic data and reliable measurement of temperature profile of the product at the worst case or slowest heating point within the container are important inputs to determine or evaluate the reliability of a given thermal process.

The desired degree of lethality in terms of minimum equivalent time at a given reference temperature is generally preestablished for a given product considering kinetic data of target microorganism, and processes are designed to deliver a minimum of this preset value at the thermal center. There are essentially two widely accepted categories of methods for using these data to perform thermal process calculations. The first of category is the General Method of process calculation based on Bigelow et al. (1920) method of calculation, and the second is the Formula Method of process calculation based on concepts of Ball (1923) or Stumbo (1973).

General Method

The General Method developed by Bigelow et al. (1920) is a graphical procedure to calculate thermal processes. It is a simple method and can be applied to any product and process situation. The method is used to determine sterility equivalents at known product temperatures measured directly by the use of thermocouples positioned within the slowest heating test container at the slowest heating spot in the container. It is the most accurate method for evaluating under a given set of heat penetration data gathering. It is universally applicable to essentially any type of thermal processing situation and used to get the baseline target value upon which all other experimental and formula calculation method performances are compared.

However, the time–temperature data is at the coldest spot and the method is only used to calculate process times for the given heat profile data. Because of this main limitation of the method, it lacks predictive power for different processes. There are two General Method approaches: Original General Method and Improved General Method.

Original General Method

The Original General Method uses the time–temperature data at slowest heating point of the product and converts this data to destruction rate graph to determine sterilization value of the process (area under the curve in the graphical procedure). The procedure is to plot the inverse of TDT against time to produce a destruction rate curve. The area beneath the curve corresponds to the accumulated sterilization value delivered by the process during heating and cooling (Fig. 38.4). Conversion of available time–temperature data to destruction rate is done using Equation 27. From gathered time–temperature data, TDT is calculated and then the reciprocal of TDT (TDT^{-1}) is computed for corresponding temperature (Table 38.4 columns 3 and 4). The reciprocal of TDT is known as the destruction rate achieved at that particular time–temperature combination. Using destruction rate versus time (1/TDT vs. Time), the graph is drawn and the area under the curve is determined by counting the squares, or using planimeter. A unit of sterilization area on the curve is equal to the product of destruction rate and time of unity, which is the minimum processing requirement to achieve desired degree of commercial sterilization value (SV) with respect to target microorganism. The sterilization value of the overall process can be calculated by considering the area under the curve divided by the unit sterilization area.

Graphical method of process calculation is summarized in Table 38.4 for *C. botulinum* with D_0 value of 0.21 minute to achieve TDT value of 12 D_0 (F_0) with z value of 18°F.

In order to get a target sterilization value, the process (steam off) time is appropriately increased or decreased, as the case may be by shifting cooling curve to right or left direction on

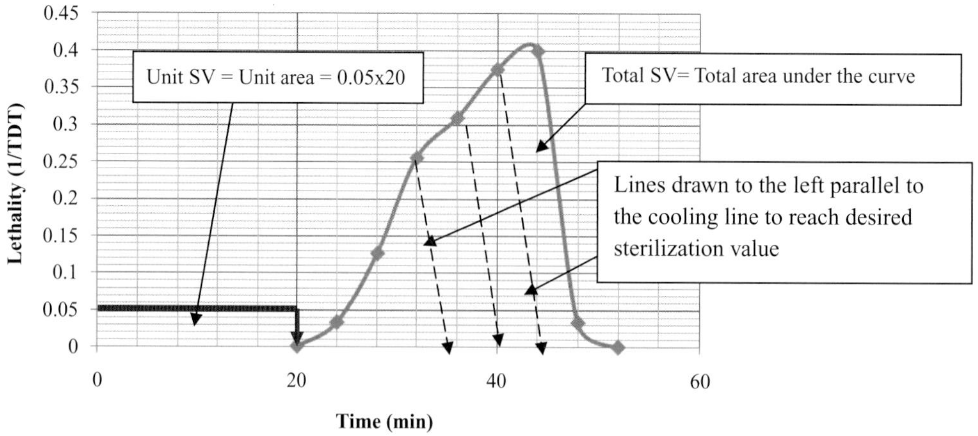

Figure 38.4. Graphical approach of process calculation. Broken lines are parallel lines drawn to the left after carrying out over-processing in order to get desired SV.

Table 38.4. Process Calculation by General Graphical and Numerical Integration Method

Time (min)	Temperature (°F)	$F = TDT = F_o{}^*10^{(Tr-T/z)}$	$1/TDT$	$1/TDT^*\Delta t$
0	62	7.01E+10	1.4E−11	0
4	71	2.22E+10	4.5E−11	1.8E−10
8	99	6.17E+08	1.6E−09	6.5E−09
12	134	7.01E+06	1.4E−07	5.7E−07
16	171	6.17E+04	1.6E−05	6.5E−05
20	202	1.17E+03	8.5E−04	3.4E−03
24	231	2.86E+01	3.5E−02	1.4E−01
28	241	7.97E+00	1.3E−01	5.0E−01
32	247	3.70E+00	2.7E−01	1.1E+00
36	248	3.25E+00	3.1E−01	1.2E+00
40	250	2.52	4.0E−01	1.6E+00
44	250	2.52	4.0E−01	1.6E+00
48	231	2.86E+01	3.5E−02	1.4E−01
52	181	1.72E+04	5.8E−05	2.3E−04
56	137	4.78E+06	2.1E−07	8.4E−07
60	97	7.97E+08	1.3E−09	5.0E−09
			SV = Σ(1/TDT)Δt =	**6.27**

a trial-and-error basis until the desired lethality achieved. In other words, this whole process is repeated until the desired SV is obtained, and the corresponding processing time is known. From food safety point of view and commercial processing conditions, the SV should be more than unity to ensure complete destruction of pathogens and reduction of spoilage microorganisms to the level that do not cause spoilage in postprocessing duration.

However, this graphical method is a tedious and cumbersome approach and requires a graphical trial-and-error works. Because of this reason, several approaches have been attempted to develop a simple approach in order to get similar information from gathered time–temperature data. One of the methods is the numerical integration method (Tabular Method) proposed by Patashnik (1953) (Table 38.4 column 5). In this method, the corresponding SV of each temperature is numerically calculated for each time interval, and the total SV is obtained for equal (Equation 28) and unequal (Equation 29) interval of time.

$$\text{Sterilization value} = \sum \left({}^{1}\!/_{\text{TDT}}\right) \Delta t \tag{28}$$

$$\text{Sterilization value} = \sum \left({}^{1}\!/_{\text{TDT}\Delta t}\right) \tag{29}$$

In Table 38.4, we obtained a SV of 6.14, which can be interpreted that the given heat treatment is 6.14 times more than what is required to destroy a given target microorganism. This is definitely an overprocessing and will have an impact on the quality of the product. Therefore, SV of 6.14 should be reduced to the value close to unity in order to achieve the desired degree of lethality of target microorganism.

Improved General Method

Like the original graphical approach, this method also requires heat penetration data and the conversion of product temperature to lethal rate. For the development of Improved General method, the two main contributions of Ball (1928) played a significant role. The first important point was the construction of a hypothetical reference TDT curve passing through 1 minute at reference temperature of 121.1°C (250°F) having a given z value. In addition to this, Ball (1928) introduced the term lethal rate (L) for a given temperature according to the following Equation 30 on the basis of Equation 27.

$$\frac{F_T^z}{F_0} = 10^{\left(\frac{T_{\text{ref}}-T}{z}\right)} \tag{30}$$

For commercial sterilization process, the numerator on left side of Equation 30 indicates TDT at temperature T, which is equivalent to thermal treatment of 1 minute at reference temperature of 121.1°C (250°F) ($F^{z=10}_{T_{\text{ref}}=121.1^\circ\text{C}}$). Therefore, based upon this concept, Equation 30 can be written as

$$\frac{F_T^z}{1} = 10^{\left(\frac{T_{\text{ref}}-T}{z}\right)},$$

where $F_T^z = \text{TDT}_\text{T}$ and after rearranging, we get

$$\frac{1}{\text{TDT}_\text{T}} = L = 10^{\left(\frac{T_{T}-\text{ref}}{z}\right)} \tag{31}$$

The left-hand side term refers lethal rate (L) and L is calculated as a function of temperature versus time from heat penetration data at a specific temperature. The lethal effects at the different time–temperature combinations in a thermal process are integrated so as to account for the total accumulated lethality, since each temperature is considered to have a sterilizing

Table 38.5. Example of Improved General Method for Process Evaluation at Reference Temperature of 250°F and z Value of 18°F

Time (min)	Temperature (°F)	Lethality Rate (min) $L = 10^{(T-Tr/z)}$	Accumulated lethality (min) $L * \Delta t$
0	62	3.59E−11	0.0E+00
4	71	1.14E−10	4.5E−10
8	99	4.08E−09	1.6E−08
12	134	3.59E−07	1.4E−06
16	171	4.08E−05	1.6E−04
20	202	2.15E−03	8.6E−03
24	231	8.80E−02	3.5E−01
28	241	3.16E−01	1.3E+00
32	247	6.81E−01	2.7E+00
36	248	7.74E−01	3.1E+00
40	250	1.00E+00	4.0E+00
44	250	1.00E+00	4.0E+00
48	231	8.80E−02	3.5E−01
52	181	1.47E−04	5.9E−04
56	137	5.27E−07	2.1E−06
60	97	3.16E−09	1.3E−08
		Accumulated lethality = 15.80	

value during the entire heating and cooling phases (Equation 32; Table 38.5).

$$\text{Accumulated lethality} = F_0 = \int_0^t 10^{\left(\frac{T-T_r}{z}\right)} dt = \int_0^t L = \sum_0^t L\Delta t \qquad (32)$$

Note that general methods do not have any specific requirement about the shape of the time–temperature curve. They are therefore accurate and they should be the methods of choice for handling "complex" temperature data (Ball 1928). In 1980s, computerized mathematical models for predicting time–temperature data using improved general method were developed by different researchers (Teixeira and Manson 1982, Datta et al. 1986, Sastry 1986, Chandarana and Gavin 1989, Chang and Toledo 1989, Lee et al. 1990) for F_0 value calculations.

Even though the General Method is the base for comparing the accuracy of other formula methods, the process is critically dependent on product, container, and process delivery system conditions as studied at the time of the test. Extrapolation outside the given process condition yields undefined, invalid lethality results. Additional testing must be performed to verify the sterilizing value delivered under the altered conditions.

Formula Methods

The formula method of process calculation simplifies the tedious steps associated in calculating process time using general methods. However, it requires characterization of the nonlinear heat penetration data to compute the key heat penetration parameters.

Characterization of Heat Penetration Data

Heat penetration data are time–temperature combination data that are collected after exposing a given food for a given period of time under a given heating condition. For products in metal cans, which are mainly heated by conduction, the slowest heating point is the center of the container, and for products whose mode of heating predominated by convection, the slowest heating point is 1/10th above the bottom of the can along the central vertical axis. Accurate determination of heating profile of a given system is very important for accurate establishment of process schedule. The profile should be collected under conditions that simulate the actual process situations. Typical heat penetration profiles for conduction and convention heating products is shown in Figure 38.5.

Because of the limitations of the General Method, a more robust Formula Method was developed and proposed by Ball (1923). It makes use of the fact that the difference between retort and cold spot temperature decays exponentially over process time after an initial lag period. Therefore, a semi-logarithmic plot of this temperature difference over time appears as a straight line that can be described mathematically by a simple formula and related to lethality requirements by a set of tables that must be used in conjunction with the formula. The originally developed method was later modified by Ball and Olson (1957), which has been still widely used in the canning industry.

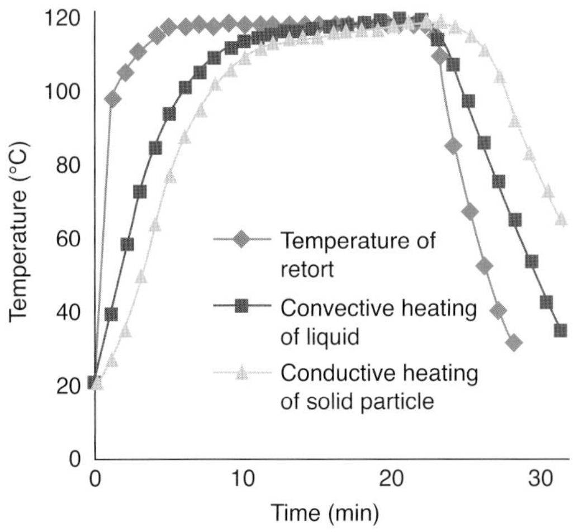

Figure 38.5. Heat penetration curves foods in container in relation to retort temperature (◇), convective heating (■), and conductive heating (▲).

Equation 33 derived from the heat penetration curve is used to estimate the process time, B (minutes).

$$B = f_h \log \frac{j_h I_h}{g} \quad (33)$$

In order to understand the contents of this equation and determine the process time (B), it is necessary to recognize heat penetration parameters, retort CUT, other several inputs, and assumptions of Ball (1923) as indicated in the following section.

Heat Penetration Parameters

The following data are normally obtained from the heat penetration data and heating conditions for calculation purposes as well:

f_h—Heating rate index. It is the time required for the straight-line portion of the heating curve to pass through one log cycle. It is also the negative reciprocal slope of the heating rate curve.

j_h—Heating rate lag factor. This is a factor that, when multiplied by I_h, locates the intersection of the extension of the straight-line portion of the semi-log heating curve and the vertical line representing the effective beginning of the process (0.58 CUT).

$$j_{ch} = \frac{T_r - T_{pih}}{T_r - T_{ih}} \quad (34)$$

I_{ih}—Difference between the retort temperature and food temperature at the start of the heating process ($T_r - T_{ih}$).

T_r—Retort temperature.

T_{pih}—Pseudo-initial temperature during heating. It is the temperature indicated by the intersection of the extension of the heating curve (original T_{pih}) and the vertical line representing the effective beginning of the process (0.58 CUT) (New T_{pih}).

T_{ih}—Initial food temperature when heating is started.

CUT—Come-up time. In batch processing operations, the retort requires some time for reaching the operating condition. The time from steam on to when the retort reached T_r is called the CUT.

0.58 CUT—Effective beginning of the process. Implies 58% of CUT from the start, which do not significantly contributes for process lethality. Ball recognized that the last 42% of the CUT only has a significant contribution on sterilization process.

g—Difference between the retort temperature (T_r) and the maximum temperature (T) reached by the food at the slowest heating point ($T_r - T$).

T_{ic}—Food temperature when cooling is started.

T_{cw}—Cooling water temperature.

I_c—Difference between the cooling water temperature and food temperature at the start of the cooling process ($T_{ic} - T_w$).

g_c—The value of g at the end of heating or beginning of cooling ($T_r - T_{ic}$).

f_c—Cooling rate index. The time required for the straight-line portion of the cooling curve to pass through one log cycle. It is also the negative reciprocal slope of the cooling rate curve.

T_{pic}—Pseudo-initial temperature during cooling. It is the temperature indicated by the intersection of the extension of the cooling curve and the vertical line representing the start of cooling.

j_{cc}—Cooling rate lag factor. This is a factor that, when multiplied by I_c, locates the intersection of the extension of the straight-line portion of the semi-log cooling curve and the vertical line representing the start of the cooling process.

$$j_{cc} = \frac{T_{cw} - T_{pic}}{T_{cw} - T_{ic}} \quad (35)$$

Formula methods make use of heat penetration data obtained from a semi-logarithmic plot of the temperature difference ($T_r - T$) on log scale against time (on linear abscissa).When the temperature difference is greater than one, three-cycle log paper is recommended (Fig. 38.6), and when the difference is below one, four-cycle log paper is required.

Gathered time–temperature data may be plotted in two different ways. In the first case, log ($T_r - T$) versus time is plotted on the paper (spread sheet), and the slope is obtained from the linear portion of the negative reciprocal slope of the graph. The lag factor is calculated from the intercept of the equation of the line. In the second case, the semi-log paper is rotated 180° (Jackson plot) and the top line of the paper is marked 1° less than retort temperature ($T_r - 1$°C), the next log cycle is marked 10° less ($T_r - 10$°C), and then the third cycle 100°C less ($T_r - 100$°C). Using this interval, the product temperatures are directly numbered without conversion on semi-log graphing paper to produce the linear plot needed to determine heating rate index (f_h) and lag factor (j_h) (Fig. 38.7).

The Retort CUT, f_h and j_h Values

Once heating medium (hot water or steam) is injected into the retort, the retort is not reached immediately to the specified operating temperature of 121.1°C. The time from introduction of

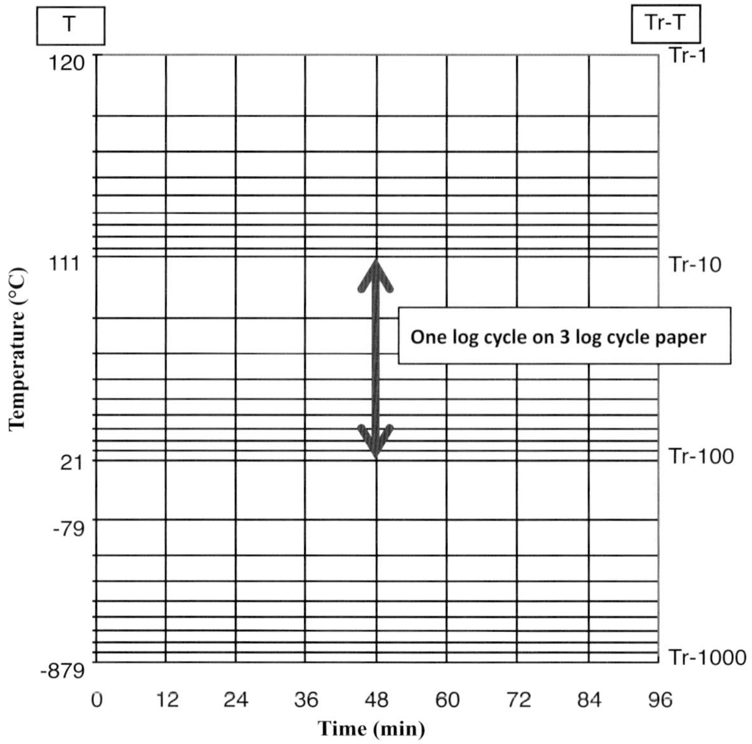

Figure 38.6. Diagram showing how the axis of semi-logarithmic graphing paper (rotated 180°C) is marked for plotting heat penetration data for retort temperature of 121°C.

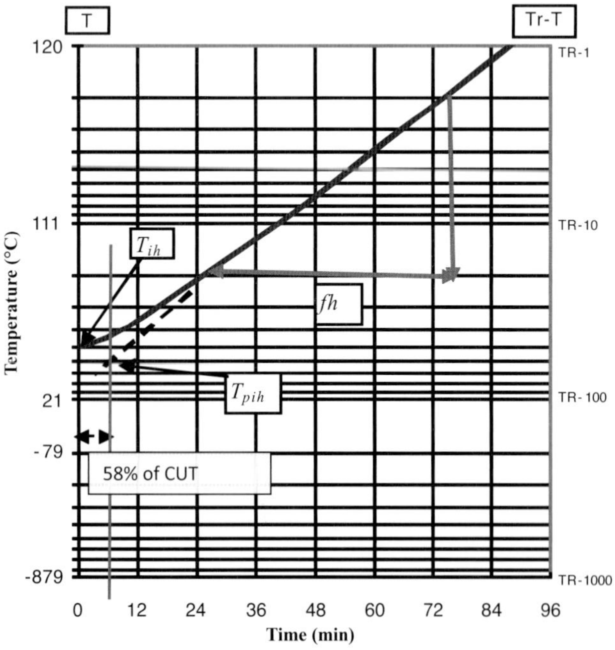

Figure 38.7. Heat penetration curve on semi-log paper during heating phase.

heating medium to when processing temperature (T_r) is reached is the retort CUT. Among his several assumptions, Ball (1923) assumed that 58% of the retort CUT has no significant heating or lethality value; therefore, heating starts from a pseudo-initial time (t_{pih}), which is 058% of CUT (0.58 CUT). According to Figure 38.7, the total CUT is assumed as 12 minutes. Therefore, the corrected zero time or pseudo-initial time (t_{pih}) is 7 minutes (0.58∗12). From hypothetical data of Figure 38.7, the heating parameters of j_h and f_h are

$$j_h = \frac{T_r - T_{pih}}{T_r - T_{ih}} = \frac{121 - 61}{121 - 71} = 1.2 \quad \text{and,}$$

$$f_h = 74.4 - 14.4 = 60 \text{ minutes}$$

The heating rate index (f_h—value) is obtained as the time for the curve to traverse one logarithmic cycle (Fig. 38.7).

For cooling part, to plot the cooling curve on semi-log paper (in this case you do not need to rotate 180°), the bottom line is marked 1°C, the second log cycle 10°C, and the third log cycle 100°C above the cooling water temperature (Figure 38.8). Then, temperatures are plotted directly on the y-axis of semi-log paper and time of cooling on x-axis. Cooling parameters, f_c and j_{cc}, (Equation 35) are obtained likewise during heating, except there are no come-down time considerations during cooling.

In addition to this, the cooling curve can be obtained by plotting a cooling curve graph from log difference of temperature of a product to that of cooling water temperature ($\log(T - T_{cw})$)

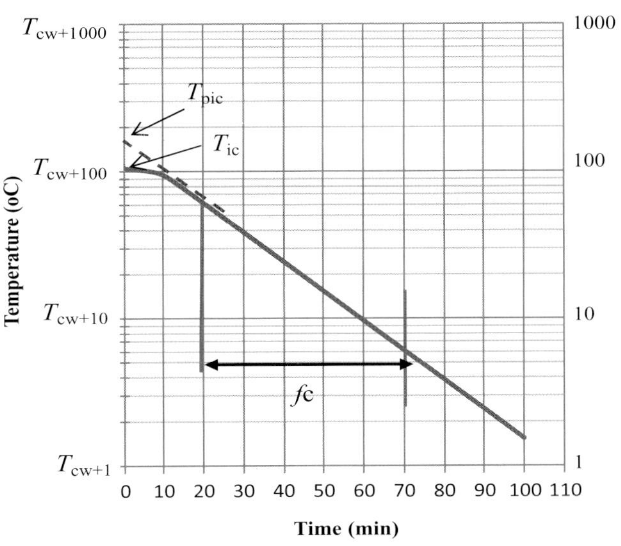

Figure 38.8. Heat penetration curve on semi-log paper during cooling phase.

versus time. The beginning of cooling time (time zero) is considered the end of heating time. The cooling rate index (f_c) is calculated from the negative reciprocal slope of cooling curve and the cooling lag factor (j_{cc}) from the intercept of the equation.

Ball calculated the process time based on straight-line equation of heating curve and integrating the effect of process time over the kinetic data of microbial destruction. He developed tables (Table 38.6) and graphs to relate process time and process lethality. These were based on relating two parameters: g (temperature difference between heating medium and product at the end of heating) as a measure of process time and f_h/U (ratio of heating rate to sterilization value) as the measure of process lethality.

Table 38.6. f_h/U Versus Log g Values for the Ball Method

f_h/U	Log g	f_h/U	Log g
0.4	−1.79	5	0.742
0.5	−1.29	6	0.805
0.6	−0.949	8	0.894
0.7	−0.736	10	0.955
0.8	−0.544	20	1.112
0.9	−0.392	30	1.187
1	−0.273	40	1.235
1.2	−0.09	50	1.27
1.3	−0.019	60	1.296
1.4	0.042	70	1.318
1.5	0.097	80	1.336
1.6	0.146	90	1.352
1.7	0.183	100	1.365
1.8	0.229	120	1.388
1.9	0.265	140	1.406
2	0.298	160	1.422
3	0.525	180	1.435
4	0.655	200	1.447

The lag factor for cooling cycle is important, because a significant contribution of sterilization takes place during the early part of the cooling period. Ball (1923) assigned a value of 1.41 for j_{cc} and also assumed $f_h = f_c$ so that the needed information can be obtained just from the heating curve. The values for f_c tend to be larger f_h because of the slower heat transfer with water compared to steam. Considering these factors, Ball related dimensionless ratio f_h/U versus g and expressed their relationship in the form of tables and figures. With the above assumptions, and experimentally obtained heating parameters (f_h, j_{ch}, I_{ch}), process calculations can easily be accomplished. Table 38.7 is an illustration of how process time (B) can be calculated. It can be seen it is necessary to get the input from Table 38.7 for finding the value of log g from f_h/U. As discussed before, the retort does not reach the desired retort temperature immediately after the steam is turned on, rather needs a finite heating time to come to operating temperature (CUT). On the basis of the assumed 42% effectiveness for the CUT, the operator's process time (P_t) is obtained from the Ball process time (B) (Equation 36).

$$B = P_t + 0.42\text{CUT} \tag{36}$$

The operator's process time P_t is the time interval from the time the retort reaches the desired process temperature to the time the steam is turned off. The CUT correction concept is only applied in Formula Methods, because in General Methods, the effect of the length of the CUT will be automatically included in the calculated lethality value because all the temperatures used in the calculation will reflect the effect of heat flowing into the product during the CUT. The stepwise procedure for calculating process lethality using Ball formula method is shown in Table 38.8.

Stumbo's Method

Stumbo's f_h/U versus log g tables (Stumbo 1973) includes the contribution from both the heating and cooling lags. These tables were developed after considering some restrictive assumptions made by Ball (1923).

Generally, the Formula Method assumptions were

1. Ball assumed a constant hyperbolic lag factor of 1.41 ($j_{cc} = 1.41$) for the cooling conditions, which is not the same in many cases. Early part of the cooling curve contributes significantly to the process lethality and hence j_{cc} is very important. Stumbo and Longley (1966) published tables that considered variations in cooling lag factor instead of assuming it to be constant (Table 38.9).
2. This is one of the assumptions made by Ball (1923) assuming the heating rate is equal to cooling rate during thermal processing time at the slowest heating part. Commonly under practical condition $f_h < f_{cc}$, which is more

Table 38.7. Procedure in Calculation of Process Time Using Ball Method for Conductive Heating Food

		Process Time Calculation
1	j_h	1.41
2	f_h	60 min
3	Retort temperature (T_r)	245°F
4	z (°F) value	18°F
5	Lethality(F_o)	5 min
6	*CUT*	8 min
7	Initial product temperature (T_i)	170°F
8	$I_h = T_r - T_i$	75°F =245−170
9	$j_h{}^*I_h$	105.75 = 1.41*75
10	Log ($j_h{}^*I_h$)	2.024
11	$F_i = 10^{\left(\frac{250-T_r}{z}\right)}$	$1.9 = 10^{\left(\frac{250-245}{18}\right)}$
12	$f_h/U = f_h/(F_o{}^*F_i)$	f_h/U = 60/(1.9*5)=6.32
13	Log g	From Ball Table 38. 6 by interpolation = 0.821
14	Log($j_h.I_h$) − log g	1.203 = 2.024−0.821
15	$B = f_h\ [\log(j_h{}^*I_h) - \log g)]$	72.18 min =60*1.203
16	$Pt = B - 42\%\ CUT$	68.82 min = 72.18 − 0.42*8

conservative assumption from safety point of view but from quality point of view it has an over processing effect. In practice, most of the heating is done with steam or agitated water, while cooling is normally done in slowly circulating water. Therefore, generally $f_h < f_c$ and $f_h = f_c$ gives a more conservative processing conditions.

3. The temperature difference between the retort temperature and cooling water temperature ($T_r - T_{cw}$) is the driving force. This is expressed in terms of I_c, which is ($I_c = T_{ic} - T_{cw}$). Ball assumed that the sum of $I_c + g = T_r - T_{cw}$ = 180°F (for steam) or 130°F (for full immersion water overpressure). Stumbo (1973) discovered that, a 10°F change in $T_r - T_{cw}$ can result in 1% difference in calculated lethality (F_0) and can be corrected if required.
4. CUT correction for j_h value. Ball noticed that only about 42% of the CUT can be considered as effective contribution to the process time.
5. No further product heating after cooling starts. However, this is not always valid for the critical point of conduction in heated canned foods. In such cases, considerable overprocessing can occur.

Table 38.8. Calculation of Process Lethality Using Ball Method From Hypothetical Given Data

		Process Time Calculation
1	j_h	2.05
2	f_h	34.9 min
3	*CUT*	8 min
4	*Pt*	45 min
	Process time	$B = Pt + 42\%\ CUT$
5	($B = Pt + 42\%CUT$)	B =45+0.42*8 = 48.36 *min*
6	Retort Temperature (T_r)	260°F
7	Initial temperature(T_i)	100°F
8	$I_{ch} = Tr - Ti$	160 °F =260−100
9	$j_h{}^*I_{ch}$	328 °F = 2.05 *160
10	log ($j_h{}^*I_h$)	2.52
11	z value	18°F
12	$F_i = 10^{\left(\frac{250-T_r}{z}\right)}$	0.278
13	B/f_h	1.39 = 48.36/34.9
14	log g = log($j_h{}^*I_h$)−B/f_h	log g= 1.13 = 2.52−1.39
15	f_h/U	From Ball Table 38.6 by interpolation = 22.4
16	$F_o = \frac{fh}{(fh/U*F_i)}$	F_o = 34.9/(22.4 *0.278) = 5.60 min

Table 38.9. f_h/U Versus g Relationships When $z = 18°F$ for Stumbo Method (1973)

	Values of g When *j* of Cooling Curve is								
f_h/U	0.40	0.60	0.80	1.00	1.20	1.40	1.60	1.80	2.00
0.20	4.09–05	4.42–05	4.76–05	5.09–05	5.43–05	5.76–05	6.10–05	6.44–05	6.77–05
0.30	2.01–03	2.14–03	2.27–03	2.40–03	2.53–03	2.66–03	2.79–03	2.93–03	3.06–03
0.40	1.33–02	1.43–02	1.52–02	1.62–02	1.71–02	1.80–02	1.90–02	1.99–02	2.09–02
0.50	4.11–02	4.42–02	4.74–02	5.06–02	5.38–02	5.70–02	6.02–02	6.34–02	6.65–02
0.60	8.70–02	9.43–02	1.02–01	1.09–01	1.16–01	1.23–01	1.31–01	1.38–01	1.45–01
0.70	0.150	0.163	0.176	0.189	0.202	0.215	0.228	0.241	0.255
0.80	0.226	0.246	0.267	0.287	0.308	0.328	0.349	0.369	0.390
0.90	0.313	0.342	0.371	0.400	0.429	0.458	0.487	0.516	0.545
1.00	0.408	0.447	0.485	0.523	0.561	0.600	0.638	0.676	0.715
2.00	1.53	1.66	1.80	1.93	2.07	2.21	2.34	2.48	2.61
3.00	2.63	2.84	3.05	3.26	3.47	3.68	3.89	4.10	4.31
4.00	3.61	3.87	4.14	4.41	4.68	4.94	5.21	5.48	5.75
5.00	4.44	4.76	5.08	5.40	5.71	6.03	6.35	6.67	6.99
6.00	5.15	5.52	5.88	6.25	6.61	6.98	7.34	7.71	8.07
7.00	5.77	6.18	6.59	7.00	7.41	7.82	8.23	8.64	9.05
8.00	6.29	6.75	7.20	7.66	8.11	8.56	9.02	9.47	9.93
9.00	6.76	7.26	7.75	8.25	8.74	9.24	9.74	10.23	10.73
10.00	7.17	7.71	8.24	8.78	9.32	9.86	10.39	10.93	11.47
15.00	8.73	9.44	10.16	10.88	11.59	12.31	13.02	13.74	14.45
20.00	9.83	10.69	11.55	12.40	13.26	14.11	14.97	15.82	16.68
25.00	10.7	11.7	12.7	13.6	14.6	15.6	16.5	17.5	18.4
30.00	11.5	12.5	13.6	14.6	15.7	16.8	17.8	18.9	19.9
35.00	12.1	13.3	14.4	15.5	16.7	17.8	18.9	20.0	21.2
40.00	12.8	13.9	15.1	16.3	17.5	18.7	19.9	21.1	22.3
45.00	13.3	14.6	15.8	17.0	18.3	19.5	20.8	22.0	23.2
50.00	13.8	15.1	16.4	17.7	19.0	20.3	21.6	22.8	24.1
60.00	14.8	16.1	17.5	18.9	20.2	21.6	22.9	24.3	25.7
70.00	15.6	17.0	18.4	19.9	21.3	22.7	24.1	25.6	27.0
80.00	16.3	17.8	19.3	20.8	22.2	23.7	25.2	26.7	28.1
90.00	17.0	18.5	20.1	21.6	23.1	24.6	26.1	27.6	29.2
100.00	17.6	19.2	20.8	22.3	23.9	25.4	27.0	28.5	30.1
150.00	20.1	21.8	23.5	25.2	26.8	28.5	30.2	31.9	33.6
200.00	21.7	23.5	25.3	27.1	28.9	30.7	32.5	34.3	36.2
250.00	22.9	24.8	26.7	28.6	30.5	32.4	34.3	36.2	38.1
300.00	23.8	25.8	27.8	29.8	31.8	33.7	35.7	37.7	39.7
350.00	24.5	26.6	28.6	30.7	32.8	34.9	37.0	39.0	41.1
400.00	25.1	27.2	29.4	31.5	33.7	35.9	38.0	40.2	42.3
450.00	25.6	27.8	30.0	32.3	34.5	36.7	38.9	41.2	43.4
500.00	26.0	28.3	30.6	32.9	35.2	37.5	39.8	42.1	44.4
600.00	26.8	29.2	31.6	34.0	36.4	38.8	41.2	43.6	46.0
700.00	27.5	30.0	32.5	35.0	37.5	39.9	42.4	44.9	47.4
800.00	28.1	30.7	33.3	35.8	38.4	40.9	43.5	46.0	48.6
900.00	28.7	31.3	34.0	36.6	39.2	41.8	44.4	47.0	49.7
999.99	29.3	31.9	34.6	37.3	39.9	42.6	45.3	47.9	50.6

6. Also, the Stumbo method accommodates different values of z and can be used for not only cold-point values but also for integrated lethality computations. The use of this yield more accurate process times (Smith and Tung 1982).

Typical calculations of process time and process lethality using Stumbo method are illustrated in Tables 38.10 and 38.11.

QUALITY OPTIMIZATION

The effect of the thermal sterilization process on quality and nutrient retention of food has been of major concern for food processors since the invention of canning. During sterilization of foods, heat has two simultaneous effects: sterilization and cooking. Therefore, it is necessary to optimize between these

Table 38.10. Calculation of Process Time (B) Using Stumbo Method

1	j_h	0.96
2	f_h	12.6 min
3	Process lethality (F_o)	5 min
4	Retort temperature (T_r)	245°F
5	Initial temperature (T_i)	175°F
6	$I_h = T_r - T_i$	70°F = 245−175
7	$j_h{}^*I_h$	67.2
8	$\text{Log}(j_{ch}{}^*I_h)$	1.75
9	Z value	18°F
10	$F_i = 10^{\left(\frac{250-T_r}{z}\right)}$	1.9
11	$f_h/U = f_h/(F_o{}^*F_i)$	1.33
12	j_{cc}	1.6
	From Stumbo's table for $z = 18$°F (18°F) ($j_{cc} = 1.6$), obtain g value by interpolation f_h/U — g value 1.00 — 0.638 1.33 — ? 2.00 — 2.34 Interpolated g value is 1.2 for f_h/U value of 1.33	
13	$\log g$	0.0792
14	$B = f_h\,[\log(j_{ch}\,I_h) - \log g)]$	21.68 min

Table 38.11. Calculation of Process Lethality (F_0) Using Stumbo Method

1	j_h	1.9
2	f_h	36.2 min
3	Operator process time (P_t)	42 min
4	CUT	10 min
5	Retort temperature (T_r)	255°F
6	Initial temperature (T_i)	160°F
7	$I_h = T_r - T_i$	95°F
8	$j_{ch}{}^*I_h$	180.5
9	$\text{Log}(j_{ch}{}^*I_h)$	2.26
10	z value	18°F
11	$F_i = 10^{\left(\frac{250-T_r}{z}\right)}$	0.53
12	$B = P_t + 42\%\ CUT$	46.2 min
12	B/f_h	1.28
13	$\log g = \log(j_{ch}{}^*I_h) - B/f_h$	0.98
14	g	9.55
15	j_{cc}	0.8
	From Stumbo's table for $z = 18$°F ($j_{cc} = 0.8$). Obtain f_h/U value by interpolation f_h/U — g value 10.00 — 8.24 ? — 9.55 15.00 — 10.16 Interpolated f_h/U value is 13.41 for g value of 9.55	
16	$F_0 = f_h/[(f_h/U)^*F_i]$	5.1 min

two effects to get a better quality safe product. This goal can be achieved by optimizing thermal process conditions. In order to achieve proper process optimization, accurate determination of kinetic parameters of microorganisms, spoilage enzymes, and quality factors are crucial. The need to optimize processing conditions arises when the kinetic behavior of the different components is considered because the rate of a chemical reaction generally doubles for a 10°C rise in temperature, whereas rates of bacterial destruction increase tenfold under similar conditions (Holdsworth 1985). Gathered kinetic parameters are used to develop reliable predictive models that will ensure to produce better quality products without compromising food safety.

There are two key issues that should be explored to produce better-quality safe products. The first choice may focus on reducing thermal intensity levels, through reducing of the heating, and cooling times associated with the process through improving heat transfer property of processing condition and modification of geometry of packaging material. The other option is to look for novel thermal or nonthermal food processing technologies that will ensure better quality products with relatively same degree of safety. New preservation technologies are utilized because of their expected potential to inactivate microorganisms and spoilage enzymes with a very little damage on product quality.

High-Temperature Short-Time and Ultra-High Temperature Processing

As the names implies, high-temperature short-time (HTST) and ultra high temperature (UHT) processes use higher temperatures and shorter processing times than conventional thermal processes to improve quality of foods and beverages. Because the products are exposed to high temperatures for short times, there is reduced degradation of quality factors of the products while causing a greater effect on destruction of food microorganisms.

The principle of HTST or UHT process is easily understood by comparing the heat resistance of nutrient components, microbial spores, and vegetative bacteria (Figure 38.9). The D value at reference temperature for vegetative bacteria are in fraction of seconds ($z = 5$°C); microbial spores, 0.2 minute ($z = 10$°C); and nutrient components, 100–150 minutes ($z = 25$–30°C). This variation gives an opportunity to optimize processing condition for better quality production of foods and beverages.

From destruction of vegetative bacteria point of view, area (A, B, and C) beneath the dotted line of the graph are not acceptable (Figure 38.9). In these sections, the vegetative bacteria are not yet destroyed. The same is true for section D, E, and F, where the product is cooked but spore formers still survive the heat treatment. All sections of G, H, and I above the bold line gives sterile product. However, the question that comes here is that, to what extent the product is cooked with respect to the quality factors. The top and bottom broken D lines represent 90% and 10% destruction of a nutrient component, respectively. Particularly, section I of the graph gives a safe product with less than 10% destruction of nutrient. This is the region corresponding to HTST or UHT treatment as indicated by the arrows.

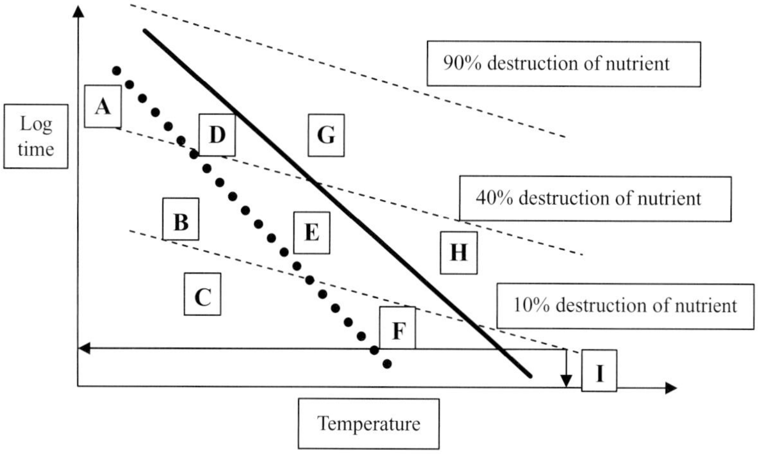

Figure 38.9. A diagram of resistance of nutrient (- - -), microbial spores (–), vegetative bacteria (•••), and HTST heating principle.

For instance, in UHT sterilization process, milk is exposed for 2–5 seconds at high temperatures of 140–145°C to kill bacteria spores. UHT treatment limits browning of the milk, development of cooked flavor, and denaturation of proteins. In HTST pasteurization, milk is subjected to a temperature of 71.1°C for 15–20 seconds. HTST or UHT approaches are not always beneficial for conduction heating food products that heat relatively slowly, exhibiting large temperature gradients between surface and center of the container. However, they are best for liquid foods and liquids containing small particles that can heat rapidly while subjected to in-container thermal processing or in-heat exchangers, as an aseptic processing.

Agitation Processing

Some retorts agitate the cans during processing in order to increase the rate of heat penetration in cans. As compared to static retorts, the process time may be reduced by 80% because the contents are heated up faster and more evenly. Agitation processing is mainly groped into axial and end-over-end (EOE) type (Fig. 38.10). EOE involves the containers being loaded vertically and a crate rotating around a central horizontal axis. During axial rotation, cans are rotated individually in the horizontal plane.

As the containers are agitated inside a retort, the contents of the containers are mixed uniformly; this eliminates cold spots and increase heat penetration rate. Mixing largely is due to the movement of the headspace bubble during agitation, and to be effective, there must be sufficient headspace in cans. Both small and large headspace may result in lower heat transfer, leading to underprocessing because of limitation of mixing of food components in cans. In addition to headspace, fill of the container, solid to liquid ratio, consistency of the product, and the speed of agitation are the crucial factors to be standardized in agitating processing. Unless and otherwise these conditions are properly optimized for a given product and processing condition, the required fast heat penetration may not be achieved.

Aseptic Processing

The word "asepsis" implies the process of removing pathogenic microorganisms or protecting against infection by such organisms. It can be defined as a state of control attained by using an aseptic work area and performing activities in a manner that precludes microbiological contamination of the exposed sterile product. In food processing, aseptic processing involves a section that precludes microbiological contamination of the

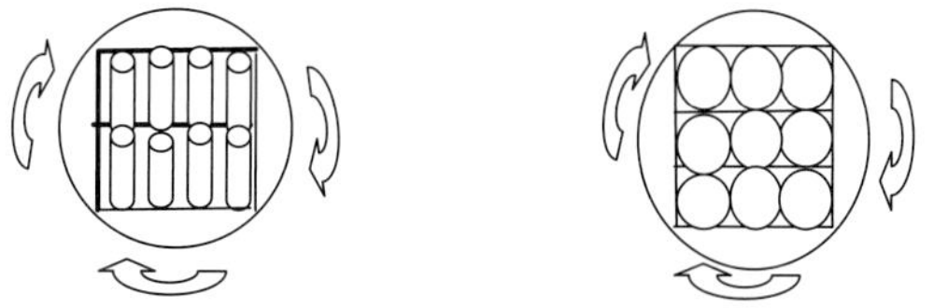

Figure 38.10. End-over-end and axial agitation orientation of cans in retort cage.

final sealed product. During aseptic processing, the product is exposed to desired treatment temperature to eliminate food microorganisms and finally packed in sterile container in an aseptic environment. This processing method is more efficient for liquid products and liquid products containing small particles. Aseptic processing sterilizes food and beverages in a way that puts the least amount of thermal stress on a product, so nutrients and natural flavors, colors, and textures are maintained while sterility is ensured.

The aseptic process begins with sterilization of both the processing system and the filler. Generally, this is done with hot water or saturated steam. Food is then pumped into the aseptic processing system for sterilization. The flow of food is controlled via a timing pump, so that no food flows too fast or too slow.

This process consists of heating, holding, cooling, and packing steps. First, the product is heated and held for some time to attain desired degree of sterilization. Once the food leaves the hold tube, it is sterile and subject to contamination if microorganisms are permitted to enter the system. The best way to keep the product sterile is to keep it flowing and pressurized. Heating step is followed by cooling of the product. The next step is moving the product into an aseptic surge tank to hold the product just before packaging. Meanwhile, the packaging material is sterilized from the other side of the aseptic environment. Packaging material is sterilized by liquid hydrogen peroxide at a high temperature. After the food and package are sterilized, the sterile package is filled with the food, closed, and sealed in a sterile chamber.

The direct and indirect heating processes used in aseptic packaging can affect the final taste of the food. Indirect heating, the food comes into contact with either a metal plate or tube, which can give food, such as milk, a burnt flavor. There are three types of indirect heating processes: plate heat exchangers, tubular heat exchangers, and scraped-surface heat exchangers. The former two are mainly used for liquid foods and the latter one is for more viscous and particulate foods to prevent fouling due to high temperature effect. All use a physical separation between the product and the heating medium and transfer heat through either a plate or tube to the product.

Direct heating uses steam injection or steam infusion and minimizes the burnt flavor of the product by letting the food come into direct contact with the heat source. With steam injection, product and steam are pumped through the same chamber; while with steam infusion, the product is pumped through a steam-filled infusion chamber. In all these heat exchangers, since the product is directly exposed to heating medium or subjected to thin profile mode of heating, it allows faster heat penetration within short period of time, which destroys microorganisms with limited effect on quality of the product. That is why aseptic processing is considered as HTST processing in terms of optimization of quality with desired degree of safety. However, for particulate foods, the rate of heat penetration in the center of the slowest heating point of the product should be determined to ensure desired degree of pasteurization/sterilization. In particulate foods, first the heat is exchanged between the heating medium and liquid food, and then transferred to the particle. In this case, it is difficult to measure the temperature of the moving particle. The time–temperature profile in the particle can be estimated using mathematical models. Based upon the data of the model used, the particle should be held at appropriate temperature for sufficient time to achieve the required sterilization value at the center.

Thin Profile and Retort Pouch Processing

Quality optimization is mainly achieved through enhancing the heat transfer rate from the heating medium to the product. Any resistance to rapid heat penetration slows down the heat penetration rate and exposed the product for longer heating time. The nature of packaging material, thickness, and its overall thermal conductivity determine the heat transfer rate. In conventional thermal processing, metal cans and glass containers are the main type of containers used. However, the thickness of these containers limits fast transfer of heat between heating medium and product. Modification of geometric configuration of the packaging materials from material thickness and geometry point of view is additional opportunity to optimize quality through improving heat transfer rate. Foods in rigid polymer trays or flexible pouches heat more rapidly, owing to the thinner material and smaller cross-section of the container.

Much of the retort pouch development was conducted by the U.S. Army Natick Research and Development Center for use in the Meal Ready-to-Eat (MRE), having relatively light weight as compared to conventional metal containers. However, the thin thickness of the pouch and its flat shape allows fast heat penetration and contributes much on improvement of quality of foods as compared to conventional packaging materials. The retort pouch is a flexible, heat sealable container that is thermally processed like a can and used to produce shelf stable, commercially sterile food products. It is constructed of a 3-ply laminate composed of an outer layer of polyester film, a middle layer of aluminum foil, and an inner layer of polypropylene. The layers are bonded together with a special adhesive. The tri-laminate material provides seal integrity, toughness, puncture resistance, printability, and superior barrier properties for long shelf life. It also withstands the rigors of thermal processing up to 135°C.

Retort pouches are filled with wet foods sealed and then heat-treated in steam/hot water retort kettles to achieve commercial sterilization (for shelf-stable foods) or pasteurization (for refrigerated foods). Because processing time typically is faster in the pouch than in metal, glass, or rigid plastic containers, the product tends to end up with better quality and safety. Pouches also offer lower shipping and storage costs (pouch material is lighter than cans, and pouches take up less storage space than cans); easier, safer handling (no can openers or thawing time required); and reduced product waste and reduced volumes of disposed packaging waste material as compared to cans.

Some of the disadvantages include a lack of physical durability and slow production rates due to slow filling and sealing rate as compared to cans or glass jars. Furthermore, it needs an overpressure processing to protect the integrity of the packages during processing. Pouches are more easily punctured; they can overwrap for safe distribution.

Novel Thermal Food Processing Technologies

Microwave and Radio Frequency Heating

Microwave (MW; 300–300,000 MHz) and Radio Frequency (RF) waves (0.003–300 MHz) are a part of the electromagnetic spectrum. MW and RF energy generates heat in dielectric materials such as foods through dipole rotation and/or ionic polarization. MW ovens are now common household appliances. Popular industrial applications of MW heating in food processing operations include tempering meat or fish blocks and precooking bacon or meat patties, while RF heating is commonly used in finishing drying of freshly baked products. Such applications shorten processing times, reduce floor space, and improve product qualities compared to conventional methods.

Extensive research has been carried out over the past 50 years on MW and RF energy in pasteurization, sterilization, drying, rapid extraction, enhanced reaction kinetics, selective heating, disinfestations, etc., but with limited applications. Technological challenges remain and further research is needed for those applications. MW/RF sterilization applications demand more thorough and systematic studies as compared to other applications. These studies will have far reaching impacts to the food industry and research communities.

Several commercial 2450 MHz MW sterilization systems produce shelf-stable packaged foods in Europe (e.g., Tops Foods, Olen, Belgium) and Japan (Otsuka Chemical Co., Osaka), but these systems are designed with multi-mode MW cavities. Generally, MW/RF heating is a promising alternative to conventional methods of heat processing as it is regarded as a volumetric form of heating in which heat is generated within the product, which reduces cooking times and could potentially lead to a more uniform heating. Reduction in processing time and uniform heating results in getting high-quality product in terms of its nutrient content, desired flavor, texture, color, and taste.

Ohmic Heating

Ohmic heating (OH) is based on the passage of alternating electrical current through a food product that serves as an electrical resistance. Because of the current passing through the food sample and its resistance to the flowing current, relatively rapid heating occurs. OH has good energy efficiency since almost all of the electrical power supplied is transformed into heat. Many factors affect the heating rate of foods undergoing OH: electrical conductivities of fluid and particles, the product formulation, specific heat, particle size, shape, and concentration as well as particle orientation in the electric field.

As compared to conventional heating technologies, internal heat generation by OH eliminates the problems associated with the heat conduction in food materials and then prevents the problems associated with overcooking. Aseptic processing has been used commercially for long time for liquid foods, but for products containing particulates, the use of conventional heat-transfer techniques leads to overprocessing of the liquid phase to ensure that the center of a particulate is sterilized. This can result in destruction of flavors and nutrients, and mechanical damage to the particulate. However, OH-treated product is clearly superior in quality than those processed by conventional technologies. This is mainly due to its ability to heat materials rapidly and uniformly, leading to a less-aggressive thermal treatment. Hence, OH can be considered as a HTST aseptic process (Castro et al. 2004).

Novel Nonthermal Processing Technologies

High-Pressure Processing

High-pressure Processing (HPP) is an innovative technological concept that has a great potential for extending the shelf life of foods with minimal or no heat treatment. It is a process aimed at controlling deteriorative changes such as microbial and enzymatic activity without subjecting the product to drastic thermal processing and mass (drying) transfer techniques such that the original quality is retained.

The application of hydrostatic pressure to food results in the instantaneous and uniform transmission of the pressure throughout the product independent of the product volume. The treatment is unique in that the effects neither follow a concentration gradient nor change as a function of time. A significant advantage is the possibility of operation at low or ambient temperatures so that the food is essentially raw. Gelation, gelatinization, and texture modification can be achieved without the application of heat. Apparently, HPP is a physical treatment and is not expected to cause extensive chemical changes in food system. Once the desired pressure is reached, the pressure can be maintained without the need for further energy input. Liquid foods can be pumped to treatment pressures, held, and then decompressed aseptically for filling as with other aseptic processes.

HPP is a novel technique for processing of foods and has attracted considerable attention in recent years. It has been commercialized for a variety of acid and acidified food products. For low-acid foods, it has been used only as a temporary measure of extending the shelf life under refrigerated storage conditions. It has also been used for several other purposes, including control of some pathogens and viruses, for inducting functional changes, as well as improving nutritional and sensory quality of foods. HPP has been used mainly for refrigerated and high-acid foods. Pasteurization by HPP can be carried out at pressures in the range of 400–600 MPa at relatively moderate temperatures (20–50°C). Under these conditions, HPP can be effective in inactivating most vegetative pathogens and spoilage microorganisms, but the quality factors remain unaffected.

HPP for producing shelf-stable low-acid foods is still a topic of considerable controversy. It has the potential to produce better quality foods than possible from the use of processing novelties such as MW, RF, or OH techniques in combination with aseptic processing. This is because HPP allows the product temperature to increase very rapidly (due to adiabatic heating) from around 90–100°C to the sterilization zone (120–130°C; in about 2 minutes) and bring it back almost instantaneously by depressurization. The process can be formulated either as a pressure-assisted thermal processing. A better understanding of the inactivation kinetics of pathogens or their surrogates under pressure-assisted thermal process conditions is the key for the success of the HPP and for regulatory approval.

The critical factors in the HPP include pressure, time to achieve treatment pressure, time at pressure, depressurization time, treatment temperature (including adiabatic heating), product initial temperature, vessel temperature distribution at pressure, product pH, product composition, product water activity, packaging material integrity, and concurrent processing aids. Although HP processing related research work has increased tremendously in the last decade, there is serious lack of information in this area to permit establishing a reliable process.

Pulsed Electric Field

Pulsed electric field (PEF) processing is a nonthermal method of food preservation that uses short bursts of electricity for microbial inactivation and causes minimal or no detrimental effect on food-quality attributes. PEF can be used for processing liquid and semi-liquid food products. PEF processing involves treating foods placed between electrodes by high-voltage pulses in the order of 20–80 kV (usually for a couple of microseconds). The applied high-voltage results in an electric field that causes microbial inactivation. A series of short, high-voltage pulses breaks the cell membranes of vegetative microorganisms in liquid media by expanding existing pores (electroporation) or creating new ones. The membranes of PEF-treated cells become permeable to small molecules; permeation causes swelling and eventual rupture of the cell membrane. The treatment is applied for less than one second, so there is little heating of the food, and it maintains its "fresh" appearance, shows little change in nutritional composition, and has a satisfactory shelf life (Castro et al. 1993, Kozempel et al. 1998). Since it preserves foods without using heat, foods treated this way retain their fresh aroma, taste, and appearance.

RETORT TYPES FOR COMMERCIAL APPLICATION

Retorts generally are either batch or continuous types. Their configuration may be either vertical or horizontal. Horizontal retorts are commonly preferred due to their ease of loading and unloading facilities as compared to vertical retorts.

Batch Retorts

Steam Heating Retort

In this type of retort (static or rotary), saturated steam is used as a heating medium. Latent heat is transferred from steam to the cans when saturated steam condensed at the surface of the cans. The saturated steam process is the oldest method of in-container sterilization. Since air is considered an insulating medium, saturating the retort vessel with steam is a requirement of the process. If air enters in the retort, it forms insulation layer of the surface of the can and prevents the condensation of steam and causes underprocessing of the product. It is inherent in the process that all air be evacuated from the retort by flooding the vessel with steam and allowing the air to escape through vent valves. There is no overpressure during the sterilization phases of this process, since air is not permitted to enter the vessel at any time during any sterilization step. However, there may be air-overpressure applied during the cooling steps to prevent container deformation. This is because during cooling, the steam rapidly condenses in the retort and the food cools more slowly and pressure in the container remains high. When the food temperature becomes below 100°C, overpressuring is stopped and the food is allowed to continuously cool up to 40°C.

Water Heating Retort (Immersion and Spray Modes)

The water immersion process is the most widely accepted method of sterilizing product using an overpressure process. The water immersion process is similar to a saturated steam process in that the product is totally isolated from any influence of cooling air. The product is totally submerged in preheated water at preset temperature. But it is different from saturated steam in that air can be introduced into the vessel during sterilization. The water is preheated in different chamber with steam to desired temperature and pumped to retort chamber. Overpressure is provided by introducing air (or steam) on top of the water. In some instances, air is added to the steam (which then heats the air). The heated air agitates the water as it flows to the surface and serves to pressurize the process load. Because it is an overpressure process, the machine can handle most, if not all, of the fragile containers. For cooling, the water present in the retort passes through a heat exchanger, where it is gradually cooled by fresh water circulating in the service side.

The water spray process is also an overpressure process, like water immersion, except that the product is exposed to the influence of the overpressure air. In this retort, low amount of water is stored in the bottom of the retort and circulated by a pump with high flow rate and sprayed on containers. It is similar to the saturated steam process in that steam is the driving force for reaching the center of the load. But the water spray process is different from saturated steam in that air can be introduced into the vessel during sterilization. Overpressure is provided by introducing air (or steam) into the retort. To overcome the insulating effects of the air, the spray nozzles vaporize the steam and mix the steam with the air. The condensates are automatically evacuated by a drainer and can be returned to the boiler.

Steam-Air Heating Retort

The steam-air process is an overpressure process, like water immersion, except that the product is exposed to the influence of the overpressure air. Steam is directly injected inside the retort chamber and distributed and mixed uniformly with air with the help of a fan to prevent cold spots in the retort. It is a "ventless" process, resulting in significant energy savings. Furthermore, a precise control of the steam/air process results in a shortened cycle time for maximum production and consistent quality, and reduces the process time and optimal heat transfer characteristics, which allow foods to retain more of their natural qualities. The opening of the steam inlet valve is automatically monitored, depending on the programmed temperature parameters. The pressure is also independently controlled from the temperature by injecting or venting compressed air.

Continuous Retorts

In continuous retort, filled and sealed containers are continuously moved from the atmospheric condition into steam-pressurized environment for desired degree of processing. In this process, cans enter a continuous retort and are instantaneously at the processing temperature; therefore, the process time is exactly the residence time of the cans within the retort.

REFERENCES

Ball CO. 1923. Thermal process time for canned food. *Bull Nat Res Council* 7(37): 9–76.

Ball CO. 1928. *Mathematical Solution of Problems on Thermal Processing of Canned Food*, vol 1, No 2. University of California Press, Berkeley, CA.

Ball CO, Olson FCW. 1957. General method of process calculation. In: CO Ball, FCW Olson (eds.) *Sterilization in Food Technology*. McGraw-Hill, New York, pp. 291–312.

Bigelow WD et al. 1920. Heat Penetration in Processing Canned Foods. National Canners Association Bulletin No. 16L, Washington, DC.

Castro AJ et al. 1993. Microbial inactivation of foods by pulsed electric fields. *J Food Process Pres* 17: 47–73.

Castro I et al. 2004. Ohmic heating of strawberry products: electrical conductivity measurements and ascorbic acid degradation kinetics. *Innov Food Sci Emerg Technol* 5: 27–36.

Chandarana DI, Gavin A III. 1989. Establishing thermal processes for heterogeneous foods to be processed aseptically: a theoretical comparison of process development methods. *J Food Sci* 54: 198–204.

Chang SY, Toledo RT. 1989. Heat transfer and simulated sterilization of particulate solids in a continuously flowing system. *J Food Sci* 54: 1017–1030.

Datta AK et al. 1986. Computer based retort control logic for on-line correction of process deviations. *J Food Sci* 51: 480–507.

Durance TD. 1997. Improving canned food quality with variable retort temperature process. *Trends Food Sci Technol* 8: 113–118.

Holdsworth SD. 1985. Optimization of thermal processing—a review. *J Food Eng* 4: 89.

Kozempel MF et al. 1998. Inactivation of microorganisms with microwaves at reduced temperatures. *J Food Prot* 61: 582–590.

Lee JH et al. 1990. Determination of lethality and processing time in a continuous sterilization system containing particulates. *J Food Eng* 11: 67–92.

Patashnik M. 1953. A simplified procedure for thermal process evaluation. *Food Technol* 7: 1–6.

Ramaswamy HS, Marcotte M. 2006. *Food Processing: Principles and Practice*. CRC/Taylor & Francis, Boca Raton, FL, pp. 67–168.

Sastry SK. 1986. Mathematical evaluation of process schedules for aseptic processing of low-acid foods containing discrete particulates. *J Food Sci* 51: 1323–1332.

Smith T, Tung MA. 1982. Comparison of formula methods for calculating thermal process lethality. *J Food Sci* 47: 626.

Silva C et al. 1992. Optimal sterilization temperatures for conduction heating foods considering finite surface heat transfer coefficients. *J Food Sci* 57(3): 743–748.

Stumbo CR, Longley RE. 1966. New parameters for process calculation. *Food Technol* 20(3): 341.

Stumbo CR 1973. *Thermobacteriology in Food Processing*, 2nd edn. Academic Press, New York.

Teixeira AA, Tucker GS. 1997. On-line retort control in thermal sterilization of canned foods. *Food Control* 8(1): 13–20.

Teixeira AA, Manson JE. 1982. Computer control of batch retort operations with on-line correction of process deviations. *Food Technol* 36: 85–89.

Toledo RT. 2007. *Fundamentals of Food Process Engineering*, 3rd edn. Springer Science, New York, pp. 285–299.

Tucker GS. 2008. History of the minimum *C. botulinum* cook for low-acid canned foods. Campden BRI R and D Report No. 260.

39
Minimally Processed Foods

Michael O. Ngadi, Sammy S.S. Bajwa, and Joseph Alakali

Abstract: The widespread, strong and consistent demands for fresh-like and safe food products have been among the major factors driving innovative food processing technologies. The sometimes divergent requirements for food quality and safety have continued to challenge food scientists and engineers. Minimally processed foods are produced without or with mild heat treatment that result in safe products without quality degradation. To produce these products, nonthermal technologies are being introduced to replace or complement the well-established thermal processes. Techniques that combine different mild hurdles to restrict the growth of food spoiling microorganisms are also used. The hurdles are typically applied in a logical sequence to take advantage of possible synergistic interactions between them to enhance safety and quality of food products. The various techniques of minimal processing of foods are reviewed in this chapter along with the required packaging considerations that are needed for the processed products.

INTRODUCTION

The normal objective of food processing is to transform a product to ensure that it has adequate nutrient content, acceptable to consumers with respect to sensory and functional properties, and that it is microbiologically and chemically safe to consume. Thus, humans have traditionally embarked on series of activities and have taken advantage of various techniques, including various types of cooking, smoking, fermenting, sun drying, and salt preservation to transform food products. Nicolas Appert's introduction of canning and thermal processing revolutionized and set the stage for the modern food processing industry. Despite the huge market for canned and processed foods, there is strong and increasing demand for fresh and minimally processed health-centered foods. This demand for products with fresh-like characteristics has changed the traditional food processing mantra. New and lucrative markets are being developed for "fresh or fresher foods" that are "high" or "low" in certain target constituents (e.g., less salt, fat and sugar; and more in fiber and health-promoting chemicals). This rise in consumer health consciousness can be attributed to wide societal and scientific reawakening to the benefits of fresh products, including fruits and vegetables (Allende et al. 2006). Fresh fruits and vegetables are known to be of higher quality and rich in antioxidants, which are crucial in the fight against cancer (Flood et al. 2002). However, the recent and increasing incidences of food-borne pathogens in food products have heightened the alert for increased food safety. Although Canada has one of the best

Food Biochemistry and Food Processing, Second Edition. Edited by Benjamin K. Simpson, Leo M.L. Nollet, Fidel Toldrá, Soottawat Benjakul, Gopinadhan Paliyath and Y.H. Hui.

food safety systems in the world, outbreaks of illness that are linked to consumption of fresh fruits and vegetables have been reported. The Public Health Agency of Canada, Health Canada, and the Canadian Food Inspection Agency have been coordinating to identify gaps, develop and enforce solutions to improve the Government's response to outbreaks, and to ensure Canadians are provided with the highest levels of food safety protection.

Food quality and safety are not necessarily complementary. In fact, they are opposing in some cases. The balance and optimization of the two parameters has been a major challenge in food processing. The use of intense and broad-spectrum technologies such as thermal treatment for microbial inactivation and shelf life extension are known to be effective but unfortunately result in accelerated degradation of sensory, nutritive, and functional quality of food products. There is also the aspect of excessive energy use with some of the food processing technologies. More recent technological advancements have allowed the application of mild processing of food with the promise of safe products without the downside of quality degradation. These technologies are termed as minimal processing technologies, and food products obtained by their applications are referred to as minimally processed foods. Thus, minimally processed products have not undergone very extensive transformation, particularly thermal processing as compared to conventional foods. The aim of each minimal processing technique is to minimize the rate of deterioration within the food matrix.

OVERVIEW OF THERMAL PROCESSING TECHNIQUES

Thermal processing involves the application of adequate heat to inactivate undesired spoilage microorganisms, degrading enzymes, and antinutrients, and prepare the food product to be consumed. It is robust and widely used in processing and preserving foods. Sufficient heat destroys microorganisms and food enzymes but also impacts both desirable (textural changes, protein coagulation, and release of aroma) and undesirable changes (nutrient loss, loss of freshness) in the food. Thus, in practice, designing appropriate thermal process typically involves consideration of time–temperature combinations that are required to inactivate the most heat-resistant pathogens and spoilage organisms in a given food matrix. There is often the need to evaluate heat penetration characteristics in a given food product, including can or container in order to evaluate temperature distributions. The choice of process depends on the minimum possible heat treatment that should guarantee freedom from pathogens and toxins and give the desired quality and storage life.

High Temperature Short Time Process

The mechanism of heat destruction of microorganisms has been attributed to thermally induced changes in the chemical structure of proteins in the cells. The phenomenon can be described by a first-order reaction equation.

$$\frac{dN}{dt} = -k_o N e^{\left(-\frac{E}{RT}\right)} \quad (1)$$

where N is concentration, t is process time, k_o is the activation factor, E is activation energy, and T is absolute temperature. Along with the microbial inactivation, other chemical reactions that occur during thermal processing lead to degradation of various quality attributes (sensory, vitamin C, thiamine, flavor, etc.). These degradation reactions can also be described using kinetic equations similar to Equation 1 with their relevant activation factors and activation energies. Several studies have confirmed that at high temperatures, the destruction rate of pathogenic bacteria accelerated more rapidly than the degradation rate of nutrients (Holdsworth 1985). The z-value (defined as change in the death or degradation rate based on temperature, i.e., the change in temperature required for tenfold change in microbial or chemical resistance) is typically around 10°C for microbial death, whereas the value for chemical degradation is around 30°C. Thus, the interaction between safety and quality can be examined by considering microbial death kinetics and chemical degradation kinetics.

High temperature short time (HTST) process takes advantage of the differences in the microbial inactivation and quality degradation kinetics to improve retention of nutrients (higher quality) while ensuring adequate microbial inactivation. Temperature–time combinations used for HTST may range from about 72°C for 15 seconds for pasteurization of milk to much higher temperatures for shorter times. It is typically applied at about 132–143°C for different times depending on the required F_0 values (the amount of time a food is maintained at the reference temperature of 121.11°C; used as a benchmark and indicating the severity of a process) to sterilize low-acid foods and beverages. The use of temperatures beyond 138°C is usually referred to as ultra high temperature (UHT) processing. A major limitation associated with HTST lies in its application to products that are highly viscous, solid, or that contain solid particles. Safety considerations require that sufficient heat is applied to sterilize the fastest moving (which is really the slowest heating) particle in the system. This implies that other slower moving particles are overprocessed, since they receive the high temperature but not the short time, resulting in degraded quality of the product. There are also limitations of heat transfer rate in solids, which may lead to surface heating and reduced quality. These limitations create a need for other technologies for effective delivery of energy to the product to enhance both safety and quality.

Aseptic Processing

Aseptic processing is an answer to the limitations of HTST. In this technique, processed food products and package are brought together aseptically (i.e., in a sterile environment). Liquid food product is initially sterilized by HTST or UHT and then heat is transferred to the solid particulate components as they are transported by the liquid and filled into presterilized containers. This revolutionary packaging system first appeared in US supermarkets in the 1970s (Fellows 1988, Holdsworth 1992, Welch and Mitchell 2000). In 1989, the Institute of Food Technologist selected aseptic packaging as "the most significant food science innovation in the past fifty years." An aseptic process for a low-acid product containing particulates (potato soup) was filed with the Food and Drug Administration in 1997 by Tetra Pak, following two workshops conducted in 1995 and 1996 by Center for Advanced Processing and Packaging Studies (multi-University partnership of The Ohio State University, North Carolina State University, and University of California-Davis) and the National Center for Food Safety and Technology.

Aseptically processed products have shelf life of 1–2 years and they have superior sensory and nutritional characteristics compared to conventional thermally processed products. Some of the difficulties associated with the techniques include variation of residence time of particulates in different parts of the processing equipment, control of heat transfer rate between heated liquid and solid particulates, fouling in heat exchangers, potential survival of enzymes, and potential damage to solids by pumping.

Sous-Vide Processing

Sous vide is a French word, which essentially means under vacuum. Chef Georges Pralus is credited with using sous-vide cooking for the first time to reduce shrinkage on a foie gras terrine by cooking in a laminated plastic film sachets in France in 1967. In this technique, ingredients or the raw food materials are put in a plastic pouch, vacuum sealed and pasteurized at varying time–temperature combinations (usually below 100°C and time longer than traditional cooking). The products can be regarded as minimally processed since they are typically cooked at lower temperatures. After being cooked, the food is cooled to temperatures in the range of 1–8°C. Sous vide is increasingly being used to process convenience foods (ready-to-eat meals) as it is reputed to produce superior quality product because of the mild heat processing and the absence of oxygen in the pack (Creed 2001). Pasteurized sous-vide pouches when held below 3.3°C remain safe and palatable up to 3–4 weeks (Armstrong and McIlveen 2000, Nyati 2000, Rybka-Rodgers 2001, González-Fandos et al. 2005, Peck and Stringer 2005). Sous-vide processed products retain fresh-like textures and superior flavors. They also tend to show higher mineral, vitamin, and color retention. However, inappropriate heating, refrigeration, and storage temperatures can be the breeding ground for *Clostridium botulinum* and more specifically the type A, B, and E spores, which can produce toxins dangerous for human health. After *Clostridium botulinum*, *Listeria* is the most heat-resistant non-spore forming pathogen that has the ability to grow at refrigerator temperatures. Therefore, there is greater risk for food poisoning, and it is important to achieve 6D reduction (Rybka-Rodgers 2001). It is vital to use the sous-vide techniques within systems that minimizes contamination. At industrial level, there is need to implement HACCP practices to get high-quality and microbiologically safe minimally processed food.

Electrical Heating of Foods

In electrical heating, electromagnetic energy is applied to increase the temperature of a product in such a way that the entire volume and not just a part of the product is heated rapidly in order to minimize quality losses. The technique was developed to address the limitation of conventional surface conduction-based methods. As a result of the rapid heating, quality changes in electrical processed foods are generally better and are more controlled compared to HTST processing. Electric heating of different foods can be achieved by using different methods, including ohmic heating, microwave, radio frequency (RF), and infrared heating.

Ohmic Heating

Ohmic heating involves passing electric currents (primarily alternating currents) through a food product for the purpose of heating it. The process is also sometimes called Joule heating, electrical resistance heating, direct electroheating, and electroconductive heating. Heat is generated internally in the product as a result of its electrical resistivity. The electrical resistance of the food determines the extent of current or voltage and the relative amount of heat generated. It is a promising method of food processing as the food heats volumetrically and provides the potential to reduce overprocessing problems associated with typical inside–outside heating pattern of conventional heat transfer (Lima 2007, Wang et al. 2007b). The process is efficient for foods with large particulates, which are not easy to process by conventional thermal processing techniques. Rapid heat rates greater than 1°C/s can be obtained along with considerable energy saving with better control of process parameters (temperature and time).

Ohmic heating systems typically use electrodes contacting the food. Applications of the technology include evaporation, blanching, dehydration, and extraction. Zhong and Lima (2003) studied the effect of ohmic heating on drying rate of sweet potatoes. The authors reported that ohmic heating of potato increased the vacuum drying rate and 24% drying time was saved, which is crucial in controlling economics of the process. The technique is being used in the United States and Japan for processing strawberries and different fruits to be used in yogurt preparation (Sastry and Barach 2000, Castro et al. 2003). Red apple, golden apple, peach, pear, pineapple, and strawberry have been heated ohmically in the temperature range of up to 25–140°C (Sarang et al. 2008). It has been reported that the application of electric current has no impact on the rheological, color, and chemical content of ohmically treated liquid juice (Yildiz et al. 2009). Quality of carrot pieces that had received varied thermal treatments, such as blanching at low and high temperatures, were tested for qualitative changes when three different heating methods, namely conventional, microwave, and ohmic heating, were applied. It was observed that time/temperature of thermal pretreatment rather than method of heating influenced quality of products (Lemmens et al. 2009).

Dielectric Heating

RF heating and microwave (MW) heating are based on the use of electromagnetic energy to quickly heat foods. When poor conducting dielectric food products are brought into a rapidly altering electrical field, the H^+ and OH^- components of water molecule separate to form an electric dipole and orient themselves in the direction of the applied varying electric field. Thus, there is dipole rotation and ionic polarization resulting in volumetric heat generation in the food (Metaxas and Meredith 1993). The extent of energy absorption by the food sample depends on its dielectric loss factor. The higher the loss factor, the more energy can be absorbed in the material. Chemical composition, including water and salt, and process temperature are among the factors that determine the materials loss factor. Although RF and MW heating are forms of dielectric heating, they are distinct from technological point of view. Both systems are operated at specific bands, namely frequencies between 0.003 and 300 MHz for RF and between 300 and 300,000 MHz for microwaves (Ramaswamy and Tung 2008). RF systems have longer wavelengths compared to MW systems. Thus, the penetration depth is typically smaller for MW heating. In general, the efficiency of power utilization is far lower in a RF generator than a microwave unit, although the initial capital cost per KW of power output may be higher. Selection of RF or MW heating usually depends on product physical properties and required process conditions for a particular application. For large products where greater penetration depth is required and control of uniformity of heating is not a major issue, RF offers a good solution. However, microwave systems are preferred where uniformity of drying and moisture control is essential.

The earlier applications of RF heating in fruit and vegetable processing was carried out for vegetable dehydration and quality assessment of juice from orange, peach, and quince (Kinn 1947, Moyer and Stotz 1947, Demeczky 1974). Nelson and Trabesli (2009) explained how dielectric properties of fruits affect their heating characteristics. RF-based treatments have been widely used in postharvest operations (Ikediala et al. 2000, Wang et al. 2006, 2009a). Wang et al. (2006) proposed the use of radio frequencies at various levels to control insect infestation without compromising quality aspects. Sosa-Morales et al. (2009) noted that RF and MW treatments can be highly efficient in reducing the heating time required for pest control, as compared to conventional thermal treatments. Water-assisted RF heating treatment was used to control the infestation of apples by codling moth, and it was reported to be an efficient technique, which also sustained the fruit's quality (Wang et al. 2005, Wang et al. 2006). Further, RF electric field has been used in deactivating *E. coli* bacteria in apple and orange juice (Geveke et al. 2007). Despite the clear advantages of RF heating, the problem of uneven heating imposes a big hurdle in its wide acceptance. There is ongoing effort to mitigate some of these problems using computer modeling (Birla et al. 2008). New and advanced RF systems are being developed to improve heating applications.

MW-based processing systems are more widely used for either domestic or commercial food preparations. There has been considerable advance in the design, power, and cost features of these systems since their introduction for small scale heating applications. The reliability of MW-assisted treatment of fruits and vegetables has been rated high over the conventional heating methods (Vadivambal and Jayas 2007). MW-assisted drying has proved to be an excellent technique for processing the raw fruits and vegetables. The technique was used to study drying behavior of various fruits like apples, mangoes, and carrots (Wang and Xi 2005, Orsat et al. 2006, Zhang et al. 2006, Vadivambal and Jayas 2007). Various fruits and vegetables have been reviewed to assess the effect of varying microwave conditions on banana slice, cranberries, and carrots (Drouzas and Schubert 1996, Yongsawatddigul and Gunasekaran 1996a, 1996b, Tein et al. 1998). Also, there has been use of vacuum for assisting microwave drying to remove the moisture. Chou and Chua (2001) reported that it was necessary to combine MW drying with vacuum in order to improve drying performance. Many other authors (Table 39.1) have combined MW with other heating techniques in order to reduce process energy and produce high-quality products.

There are some indication of nonthermal and enhanced thermal effects of microwave processing (Ramaswamy et al. 2002, Datta and Anantheswaran 2004). However, it is generally

Table 39.1. The Use of Microwave Drying on Different Fruits and Vegetables

Fruit or Vegetable	Type of Treatment	References
Carrots	MW & vacuum	Regier 2005, Wang and Xi 2005, Orsat et al. 2007
Potato	MW & convective, MW & vacuum	Bondrauk et al. 2007, Reyes et al. 2007, Markowski et al. 2009, Song et al. 2009
Strawberry	MW & vacuum, MW & convective	Bohm et al. 2006, Charangue et al. 2008, Contreras et al. 2008, Wojdylo et al. 2009
Mushroom	MW & vacuum, MW & convective	Orsat et al. 2007, Yang et al. 2008, Askari et al. 2009, Giri and Prasad 2009
Spinach	MW, MW & convective	Ozkan et al. 2007, Dadali et al. 2008, Karaaslan et al. 2008, Dadali and Ozbek 2009
Peach	MW	Wang and Sheng 2006
Orange	MW, MW & vacuum, MW & convection	Ruiz et al. 2003, De Pilli et al. 2008, Guo et al. 2009
Apple	MW, MW & vacuum, MW & convection	Han et al. 2009, Huang et al. 2009, Tarko et al. 2009, Witrowa and Rzaca 2009, Li et al. 2010
Pumpkin	MW, MW & vacuum, MW & convection	Alibas 2007, Wang et al. 2007a, Nawirska et al. 2009
Banana	MW, MW & vacuum, MW & convection	Mousa and Farid 2002, Kar et al. 2003, Pereira et al. 2007

Note: The objectives of researches shown may differ.

believed that the kinetics of microbial inactivation in microwave processing is essentially same as the inactivation kinetics of conventional thermal processing for the same heating rate. Recent technological advancement, including development of new chemical marker system for process monitoring and identification of cold spots, techniques for achieving temperature uniformity, computer modeling, and microbial inactivation validation, has led to US FDA approval of microwave sterilization in 2009. This clears the path for commercialization of the process. There is no report of industrial application of RF sterilization (Ramaswamy and Tang 2008). There is a need to speed up research and development efforts to enhance MF- and RF-based industrial sterilization processes.

NONTHERMAL PROCESSING TECHNOLOGIES

Traditional thermal processing technologies are well established and proven to inactivate microorganism and prolong shelf life in foods. However, some of the challenges associated with maintaining product quality and freshness have been discussed in the preceding sections. Nonthermal technologies that inactivate microorganisms by means other than heat are being introduced to mitigate some of the shortcomings of traditional processes. These technologies include high pressure, pulsed electric fields, irradiation and pulsed ultraviolet (UV) light, etc.

High-Pressure Processing

High-pressure processing (HPP) of foods is the application of ultra high pressure (100–1000 MPa) on foods for short duration, which may range from few seconds to minutes. The most important aspect of HPP lies in its effectiveness in microbial control in food products. HPP has shown its potential in keeping food fresh for long time without the use of preservatives, in inactivating food-spoiling microbes, and most importantly in maintaining minimally processed fresh-like texture. The technique was initially commercialized in Japan in 1990s, but, historically, the United States was the first country to use the HPP technology for preserving products like juice, some fruits, milk, etc. The monetary constraints and lack of expertise in high-pressure equipment and package manufacturing emerged as a roadblock until Japan reinitiated use of this technology in 1980s. There have been an increasing number of HPP systems installed throughout the world. In 2008, there were four systems (71 overall) systems in North America (Campus 2010). A wide range of products including juices, fruits and fruit related products, seafood, and meat products have been successfully processed using the technology.

HPP operations can be conducted in three modes, namely batch, semi-continuous, and continuous modes. Semi-continuous or continuous modes typically apply to pumpable products, in which case aseptic packaging is required. Products can be processed in-package in batch mode. A flexible package like a retort pouch or plastic bottle is introduced in a high-pressure chamber, and pressure is transmitted to the food via its package by the use of a hydraulic liquid. Water is preferred as the hydraulic liquid because of operational ease and high compatibility with the food components. HPP packages must be airtight and capable of allowing about 19% decrease in volume without losing seal integrity. Juliano et al. (2010) have provided a good review on packaging concepts for HPP. Since pressure is transmitted uniformly in all directions, the texture of processed product, as well as nutritional and sensory attributes of the food product, is preserved in HPP processing (Fernandez et al. 2001).

Typical HPP system comprises of four components, namely pressure vessel, pressure generating device, pressure and temperature control, and material handling units (Mertens 1995). Pressure vessels are designed carefully to meet very strict safety and operational standards of the American Society of Mechanical Engineers (ASME) boiler and pressure vessel codes. To withstand the high pressures used in HPP, prestressed multilayer or wire-wound vessels are used. High pressure can be generated either by direct or indirect compression or simply by heating the pressure fluid. In the more widely used indirect compression, water (or other pressure transmitting fluid) is pumped into a closed high-pressure vessel until the desired pressure is achieved (Mertens 1995). Technological developments on design of pressure vessels and intensifiers are the major driving advances in the HPP technology. Early HPP systems were operated with vertical vessels. Recently in 2000, NC Hyperbaric, one of the leading HP manufactures introduced the first horizontal HPP vessel. Information from the company suggests that installation and product handling in a horizontal system is reported to be much easier and less expensive than the vertical system (www.nchyperbaric.com). In particular, loading and unloading motions in the horizontal machines are made at the waist level and can be more easily done either manually or automatically than in the vertical systems, which are typically several meters high.

High pressure inactivates microorganisms by increasing the permeability of the cell membrane, causing loss of intracellular fluids, thereby inhibiting biochemical reactions within the cell and hampering cell growth (Cheftel 1995, Ananta et al. 2001). On application of high pressure, the acyl chains get tightly packed with the phospholipid bilayer of membranes, and it undergoes transition from liquid crystalline to gel phase, hence modifying the internal microbial resistance. Variation in time–pressure conditions affects the cytokinetic and mitotic activity of the cell (Lopes et al. 2010). Presence of at least 40% free moisture content is important for effective microbe destruction (Earnshaw 1996). Gill and Ramaswamy (2008) used a pressure of about 600 MPa with holding time of about 3 minutes on two RTE meats (Hungarian salami and All beef salami) inoculated with food-borne pathogen *E. coli* O157. It was observed that there was reduction in *E. coli* count in both samples. However, when samples were enriched with nutrient medium to recover *E. coli*, there was increase in the *E. coli* count of Hungarian salami, whereas stagnant *E. coli* count was found in all beef salami that was conditioned in regulated pH and water activity. Bacterial spores are of small size, low in moisture content, and impermeable to water, which makes them highly resistive to intense pressure conditions. They are considered as potential danger in postprocessing storage, as they may germinate and cause food toxification. Spores can tolerate pressure up to

1000 MPa (Cheftel 1995). There is a long list of spore-forming bacterium that spoils food such as milk, cheese, meat, juices, etc. The spores of *Bacillus* and *Clostridium* species have been cited as the common reason for food-borne diseases (Shapiro et al. 1998, Salkinoja-Salonen et al. 1999, Borge et al. 2001, Brown 2000). An elaborate review of studies on spore inactivation has been given by Black et al. (2007). Although HPP has shown great potential as a reliable processing technique, very high pressures may result in poor quality. Thus, in some cases, combination of HPP with other factors (such as temperature, pH, water activity, etc.) may be beneficial. There is increasing interest in the high pressure, high temperature or Pressure-Assisted Thermal Sterilization or Pressure-Assisted Thermal Processing (PATP) for development of shelf-stable low-acid food products (Barbosa-Canovas and Juliano 2007, Zhu et al. 2008).

Pulse Electric Field Processing

Pulse electric field (PEF) processing of foods is growing in importance as a technique for producing microbiologically safe, nutritious, fresh-like, and high-quality products (Mittal 2009). There are also economic and energy-saving advantages associated with the technology. It has been used for processing liquid food products like fruit juices, alcoholic beverages, soups, liquid eggs, and milk. It can be used in continuous for pumpable fluids or in batch modes for both liquids and solids.

PEF processing is based on the principle of application of electric field (20–80 kV/cm) on a food product contained between two electrodes in short pulses (1–100 μs). Microbial cells present in the food are inactivated and drastically reduced in count. Electric breakdown of cells and electroporation are the two mechanisms of microbial destruction during PEF (Zimmermann 1986, Harrison et al. 1997, Barbosa-Canovas and Sepulveda 2005). In electric breakdown mechanism, application of external electric field causes development of electrical potential difference across cell membrane, also called as transmembrane potential. When the transmembrane potential is higher than the natural potential of the cell (nearly 1 V), pores are formed, reducing the thickness of cell membrane. With higher potential, the cell membrane is eventually disrupted permanently, leading to cell destruction. With respect to the electroporation mechanism, lipid bilayers and proteins are destabilized, increasing membrane permeability and number of pores. The pores cause the cell membrane to rupture, cytoplasmic material oozes out, and finally cell dies (Vega-Mercado et al. 1997). Effective application of PEF depends on the type of pulse (monopolar and bipolar) and waveform (sinusoidal, square, or exponentially decaying) used (Ho and Mittal 2000, Ngadi and Bazhal 2004).

Earlier application of PEF in food processing in North America focused on pasteurization of liquid or semi-liquid products. Some of the foods that have been processed using the technology include milk (Evrendilek and Zhang 2005), apple juice (Vega-Mercado et al. 1997, Sanchez-Vega et al. 2009), orange juice (Rivas et al. 2006), liquid egg (Martin-Belloso et al. 1997, Amiali et al. 2006, Amiali et al. 2007), and egg yolk (Bazhal et al. 2006). These studies and others showed that PEF can be used to successfully inactivate pathogenic and food spoilage microorganisms as well as selected enzymes, resulting in better retention of flavors, nutrients, and fresher taste compared to heat-pasteurized products. Mosqueda-Melgar et al. (2007) reported efficacy of PEF in decreasing *Salmonella* spp., *E. coli* and *L. monocytogenes* in melon and watermelon fruit juices. A reduction in the bacterial yeast and mould count in naturally contaminated orange juice was observed. The color, aroma, and flavor features were better than as in conventional HTST treatments. PEF was also effective in inactivating yeast (*Saccharomyces cerevisae*) and bacteria (*Kloeckera apiculata, Lactobacillus plantarum, Lactobacillus hilgardii, and Gluconobacter oxydans*) present in grape juice (Marsellés-Fontanet et al. 2009). Many factors influence effectiveness of PEF processing. These include process factors (electric field strength, number of pulses, treatment time, treatment temperature, pulse shape, pulse width, pulse polarization, frequency, specific energy, treatment chamber design, etc.), microbial factors (cell concentration, microorganism resistance to environmental factors, type and species of the microorganism, as well as growth conditions such as medium composition, temperature and oxygen concentration and other stress conditions), and product parameters (product composition, presence or absence of particles, sugars, salt and thickeners, conductivity, ionic strength, pH, water activity, etc.). Optimal application of PEF often involves appropriate control of these factors. There has been effort to develop mathematical models that will describe the complex mechanism of microbial inactivation using PEF. Research is still ongoing to discover suitable models. However, typical inactivation characteristics tend to follow the sigmoidal shape and may be described using kinetic-based or probability-based models. Table 39.2 indicates how different parameters might influence microbial inactivation using PEF.

Although the influence of PEF on microbial inactivation is well established, there are varying reports on its effect on quality of processed liquids. There was high retention of lycopene and vitamin C in water melon juice (Oms-Oliu et al. 2009), higher vitamin A in orange–carrot juices (Torregrosa et al. 2006), higher phenolic content and antioxidant capacity in strawberry juice (Odriozola-Serrano et al. 2008), and no change in phenolic content of tomato juice (Odriozola-Serrano et al. 2009) with PEF processing. A new orange juice-milk beverage was developed by insertion of bioactive components (n-3 fatty acids and oleic acid) as an alternative to soft drinks. The effect of PEF at two levels, namely 35 and 40 kV/cm, was used to study effects on physicochemical properties, pH, degree Brix, and peroxide index. The study concluded that there was no significance effect of PEF treatment on saturated and unsaturated fatty acid contents, with no detection of peroxide and tolerable levels of furfurals (Zulueta et al. 2007).

PEF treatment may influence physical and chemical properties of products. The nature and extent of PEF influence on quality changes are still being actively discussed. Ngadi et al. (2010) provided an excellent review of quality aspects of PEF processing. Barsotti et al. (2002) indicated that PEF treatment of model emulsions and liquid dairy cream may result in dispersal of oil droplets and dissociation of fat globule aggregates. Qin et al. (1995) reported no apparent change in the physical and chemical attributes of PEF-processed milk. PEF

Table 39.2. Parameters of Weibull Model for Microbial Inactivation Using Pulsed Electric Field (Ngadi and Dehghannya 2010)

Control Parameter	Product or Medium	Microorganism	pH	E (kV/cm)	α	β	Reference
Treatment time	Citrate–phosphate McIlvaine buffer	*Yersinia enterocolitica*	7.0	15	1.67	0.33	Alvarez et al. (2003a)
				25	0.02	0.22	
		Lactobacillus plantarum	7.0	25	234.90	0.99	Gomez et al. (2005a)
			6.5	25	41.98	0.98	
			5.0	25	14.54	0.99	
			3.5	25	12.81	0.98	
		Listeria monocytogenes	7.0	15	2101.0	0.85	Gomez et al. (2005b)
				25	111.40	0.62	
			6.5	15	2514.0	0.57	
				25	57.24	0.61	
			5.0	15	108.20	0.36	
				25	4.31	0.39	
			3.5	15	17.64	0.34	
				25	0.95	0.43	
	Orange juice-milk beverage	*Escherichia coli* CECT 516 (ATCC 8739)	4.05	15	0.68	0.31	Rivas et al. (2006)
				25	0.17	0.27	
		Lactobacillus plantarum	4.05	15	29.07	0.49	Sampedro et al. (2006)
				25	0.03	0.16	
Specific energy	Citrate–phosphate McIlvaine buffer	*Yersinia enterocolitica*	7.0	15	0.71	0.32	Alvarez et al. (2003a)
				25	0.03	0.22	
		Listeria monocytogenes		15	1135.0	0.85	Alvarez et al. (2003b)
				25	158.2	0.62	

treatment of various liquid foods, including apple juice, orange juice, and milk, has not shown any significant physicochemical changes. PEF-processed yogurt-based drink retained its physical and chemical characteristics (Evrendilek et al. 2004). There was a slight decrease in vitamin C content in PEF-treated orange juice compared to heat-treated orange juice (Zhang et al. 1997). In a comparative study of orange juice pasteurized by ultra-high temperature (processing at 110°C, 120°C, and 130°C for 2 and 4 s) and PEF (20 and 25 kV/cm for 2 ms), Gallardo-Reyes et al. (2008) concluded that although there was no difference in pH and soluble solids obtained with both treatments and freshly squeezed control samples, the color of PEF-treated sample was closer to the control. PEF inactivates various types of enzymes. A reduction of about 95% in the residual activity of polyphenol oxidase (PPO) in apple juice was observed when treated with PEF. Decreased pectin methylesterase (PME) and PPO activity and nonenzymatic browning were observed in the carrot juice (Luo et al. 2008). When treated with PEF, strawberry, tomato, and water melon juice noticed a decreased level of nonenzymatic browning (Aguilo-Aguayo et al. 2009a, 2009b). The activity of lipoxygenase and β-glucosidase in strawberry juice was reduced effectively (Aguilo-Aguayo et al. 2008). There are indications that as a result of its influence on enzymes and bioactive compounds, PEF may affect quality of cheese produced using PEF-processed milk. Electric field intensity (E) and treatment temperature significantly affected rennet coagulation properties of milk in terms of curd firmness (Yu et al. 2010). The result implied that treating milk with PEF and mild temperature impact less changes in terms of milk coagulation properties compared to conventional heat pasteurization. This result is consistent with the finding of Dunn (1996), who studied the PEF-treated raw milk and concluded that no significant physicochemical changes were observed.

Although the interest is increasing rapidly, there is much less information on application of PEF on solid food materials. Bazhal et al. (2003) applied 300 μs pulses of 1 kV/cm at the rate of 1 Hz on apple tissues and studied the resulting structural and morphological changes. The authors reported an increase of porosity from 63% to 69.4% after PEF electroplasmolysis. The sizes of the induced pores were smaller compared to the pores of untreated samples and were comparable with the tissue cell-wall thickness. When PEF-treated apple slices were compressed, it was observed that there was a relationship between mechanical failure stress and degree of electroplasmolysis as shown in Figure 39.1. Influences of PEF on physical and textural properties of food materials have been reported (Rastogi et al. 1999, Angersbach et al. 2000, Gadmundsson and Hafsteinsson 2001, Taiwo et al. 2001, Gachovska et al. 2009, Grimi et al. 2009). Several other authors have exploited this phenomenon in using PEF to enhance extraction of juice from different food materials (Bazhal and Vorobiev 2000, Eshtiaghi and Knorr 2002, Gachovska et al. 2006, Gachovska et al. 2008, Gachovska et al. 2010, Hou et al. 2010, Loginova et al. 2010). The various studies indicate improved pressing efficiency, faster extraction processes, and improvement in drying processes as advantages of PEF-enhanced processes. Despite the enormous progress that has been made on PEF processing of food, some of the challenges of the technology include difficulty in comparing different systems and process parameters. Different equipment and treatment chambers may present different treatment conditions. There is need to standardize various parameters and systems in order to facilitate effective evaluation of various processes reported by various authors.

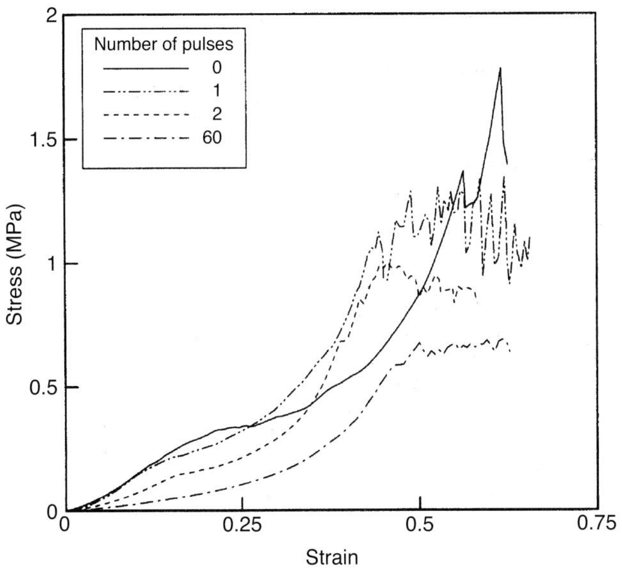

Figure 39.1. Typical stress–strain curves from the compression test on apple samples treated by electric field pulses (field strength, E = 1 kV/cm; treatment time = 300 μs; frequency = 1 Hz). Control sample has no treatment ($n = 0$) (Bazhal et al. 2004).

Low-Dose Irradiation

Ionizing radiations have been used to control spoilage microorganisms and extend food shelf life. As it involves no heating, the food retains most of its organoleptic features. The technique involves application of precisely controlled ionizing radiation to either bulk or packaged product for a preestablished duration in order to destroy microorganisms, insects, or other pests; minimize postharvest losses; decrease or inhibit sprouting; or other specific purposes. The form of ionizing radiations used in food processing includes gamma rays from radioisotopic sources such as Cobalt-60 or Cesium-137, high-energy electrons from electron beam devices, and X-rays from electron beam accelerators (Patterson and Loaharanu 2000). The use of gamma rays or X-rays of energies no greater than 5 MeV is generally recommended for food processing, whereas for accelerated electrons, energy levels below 10 MeV are suitable (Wilkinson and Gould 1998). The mechanism of microbial inactivation by ionizing radiations has been ascribed to direct interaction of the radiation with cell components and food molecules or indirect action from radiolytic products such as water radicals $H+$, $OH-$, and e_{aq} (Wilkinson and Gould 1998, Morris et al. 2007). It primarily targets the cell's chromosomal DNA and exerts a secondary effect on the cytoplasmic membrane, either of which can result in microbial inhibition or inactivation. Energy from the radiation sources may be sufficient to dislodge electrons from food molecules, converting them to electrically charged particles or ions. However, the ionizations are too low to induce radioactivity in food products. To irradiate the food, it is important to use the radiations that can reach the core of the irradiated food in order to achieve uniform treatment. Gamma rays have high penetrating power and thus can be used to treat foods in large packages. The energy decreases exponentially with depth of the absorbing product. However, accelerated electron beams continuously lose energy in a series of interactions with orbital electrons in the absorbing medium. Their penetration depth is low such that 10 MeV electrons may only penetrate about 4 cm (Wilkinson and Gould 1998).

Microorganisms, enzymes, insects, and vegetable sprouting have different degrees of sensitivity to radiation. Required irradiation doses for a specific goal depend on the rays (type, quantity, and radiation time) and the irradiated environment (absorption capacity, physical, chemical and biological modifications, and secondary reactions). Irradiation dose is measured in grays (Gy), defined as the absorption of 1 J of ionizing radiation by 1 kg of matter (i.e., 1 Gy = 1 J/kg). High-radiation dose in the range of 10–74 kGy is usually applied for microbial sterilization (Morris et al. 2007). Milder doses less than 10 kGy have been applied for pasteurization applications (Patterson and Loaharanu 2000). Although high irradiation doses have intense and effective microbial inactivation action, they are not preferred since they also destroy organoleptic features of the food material. Thus, low doses of radiation that are severe enough to reduce microbial populations and maintain sensorial and nutritional quality of foods are generally applied. FDA regulations specify the maximum radiation dose for control of food-borne pathogens in fresh or frozen, uncooked poultry products (3 kGy); refrigerated, uncooked meat products (4.5 kGy); and frozen, uncooked meat products (7 kGy) (FDA 2005, Komolprasert 2007). Low dose (1 kGy maximum) is imposed for control of growth and maturation inhibition in fresh foods. However, much higher dose of 30 kGy is imposed for microbial disinfection of dry or dehydrated spices.

Many studies have demonstrated that low-dose radiation is useful in improving microbiological safety of fresh-cut fruits and vegetables without significant change in appearance, texture, flavor, or nutrition quality (Kim et al. 2006, Niemira and Fan 2009). Gamma radiation of broccoli, cabbage, tomatoes, and mung bean sprouts at the dose of 1 kGy resulted in more than 4 log reduction in *Listeria monocytogenes* without significant changes in appearance, color, taste, and overall acceptability (Bari et al. 2005). Usage of low energy X-ray irradiation (1 kGy surface dose) on lettuce caused 5 D reduction in *E. coli* O157:H7 (Jeong et al. 2010). Reduction in the *E. coli* O157:H7 count without any effect on quality of spinach leaves was noticed when gamma irradiation (doses of 0.3–1 kGy) was used (Gomes et al. 2008). Along with the effect on microorganism, radiation also denatures enzymes in foods. Latorre et al. (2010) reported that when red beets were treated with 1–2 kGy gamma radiation, peroxidase (POX) activity increased significantly with the increasing radiation dose, whereas PPO activity increased only for a radiation dose of 2 kGy. The trend was attributed to the higher stiffness and to the rise in tissue elasticity after irradiation. As a result, color pigments such as betacyanin and betaxanthin did not change after 1 kGy radiation but decreased sharply at 2 kGy. A detailed review on the use of irradiation in the fruits and vegetables has been given by Arvanitoyannis et al. (2009).

There are challenges associated with producing and marketing irradiated food products. Although it is well established that food products processed using appropriate low dose of radiation is safe to consume, there continues to be some widespread hesitation by consumers. There is a strong opposition in the European Union to irradiated foods, but as food safety takes center stage internationally and people prioritize food safety, these attitudes may shift. In the United States, a large number of irradiated food products are available and without any significant opposition from consumers. There are strict regulations limiting the marketing of irradiated products in the United States. The Federal Food, Drug and Cosmetic Act deems a food product to be adulterated if it has been intentionally irradiated (unless the irradiation is carried out in compliance with an applicable regulation under the prescribed conditions of use specified in the regulation). Irradiated products are required to be adequately labeled. Further, since most foods are generally prepackaged in their final form before irradiation, there are serious constrains on the type of packaging materials that can be used. Komolprasert (2007) discussed some of the constraints of irradiation on packaging materials. To ensure that components of the packaging materials that have been irradiated at a given level do not migrate into the food, the regulation stipulates that the use of packaging materials for irradiated food is considered a new use and is subject to premarket safety evaluation and approval. In general, development of new methods of irradiation detection and chemical analysis of trace elements in foods could expedite introduction of safe and quality irradiated products and enhance consumer confidence in accepting the products.

Ultrasound

Ultrasound processing is the application of high-intensity sound waves (20–100 kHz) on foods. The technology can be used mainly in two ways: as a diagnostic tool for nondestructive quality evaluation and as energy for processing (Mason et al. 2005). In the past, ultrasound has been used in quality assessment of agricultural produce (preharvest and postharvest) but is now increasingly being applied in processing. Ultrasound can be applied for various operations such as cell disintegration, extraction (phenolic compounds, pigments, lipids, proteins), mixing, acceleration of enzyme activity and microbial fermentation, emulsification, fruit juice processing, and many other uses. Further, ultrasound has also been used for enzyme inactivation and microbial reduction in food. The effectiveness of ultrasound in its antimicrobial action is based on controlling critical factors such as frequency and intensity of wave and time of exposure. Actual application depends on the type of microorganism, temperature, and the nature of the food. In fluid systems, ultrasound induces rapid appearance, growth, and collapse of bubbles or cavitations, leading to considerable agitation, localized "hot spots," and increased bulk transport within the mass (Mulet et al. 2003). In solids, ultrasound produces series of rapid compression and expansions of the material comparable to repeated squeezing and releasing of a sponge, which enhances mass transfer and creates microchannels for fluid movement (Floros and Liang 1994). Thus, when applied to solid–fluid systems, both internal and external resistances to mass transfer between the solid and liquid phases are affected. The interaction of ultrasound in solid–fluid interfaces can produce a microagitation in the immediate vicinity of the solid surface, resulting in reduction of diffusion boundary layer thickness (Mulet et al. 2010). It was reported that the extractive value of herbs such as for fennel, hops, marigold, and mint increased by 34%, 18%, 2%, and 3%, respectively, in water, whereas in ethanol, there was an increase of 34%, 12%, 3%, and 7%, respectively, when ultrasound extraction was applied when compared to conventional methods (Vinatoru 2001). Different solvents may present varying effect on extraction effectiveness of ultrasound. For extraction of carnosic acid from rosemary, ultrasound improved the relative performance of ethanol such that it was comparable to butanone and ethyl acetate alone. Thus, ultrasonication may reduce the dependence on harsh solvents and enable use of environmentally benign solvents (Albu et al. 2004). Knorr et al. (2004) and Vilkhu et al. (2008) provided good review of applications of ultrasound in the food industry.

More recent reviews on the technology are Soria and Villamiel (2010) and Chemat et al. (2010).

Different bacteria exhibit different sensitivities to ultrasonic treatment in different media (Wang et al. 2010). Early applications of ultrasound showed relatively low microbial inactivations. However, there has been considerable progress in equipment development that has resulted in increased inactivations but typically below 5 logs. Application of ultrasound alone is not very effective for microbial inactivation in commercial processing, but the technique can be effective when used in combination with other treatments (Raso et al. 1998). Thus, three different techniques, namely thermosonication, manosonication, and manothermosonication have been promoted as a result of the synergistic actions of the different treatments on microbes (Lee et al. 2009). Manothermosonication is an emerging technique that combines heat and ultrasound at elevated pressure. Based on intensity, amplitude, and time of treatment, manothermosonication can be 6–30 times more effective in killing microorganisms (*Bacillus* species, *Sacchromyces cerevisae*) compared to thermal treatment given at same temperature. Different enzymes such as POX, lipase, lipoxygenase, protease, and pectin methylesterase have been tested for inactivation by manothermosonication (Demirdoven and Baysal 2009). Manothermosonication has been used to enhance the textural and functional properties of tomato juice and milk proteins (Lopez and Burgos 1995, Vercet et al. 2002). Ultrasound has also been used in combination with chlorine, and a strong bactericidal effect was observed (Blume and Neis 2005). Chlorine dioxide, when used with heat and ultrasound, destroyed *Salmonella* and *E. coli* cells in alfalfa seeds (Scouten and Beuchat 2002). The full potentials of ultrasound in food processing have yet to be tapped. There will continue to be progress made on design of improved and efficient equipment. Better understanding of the effect of the process on technological and functional properties will be crucial in identifying niche applications of the technology.

UV Irradiation

The nonionizing UV radiation in the wavelength range of 100–400 nm is widely used in food processing. The electromagnetic spectrum is classified into three groups, namely UV-A (315–400 nm), UV-B (280–315 nm), and UV-C (less than 280 nm). UV-C is particularly used because of its effective germicidal capacity. The wavelength of 253.7 nm is known to have the most lethal effect on microorganisms since photons are absorbed most by the DNA of microorganisms at this wavelength (Labas et al. 2005). UV light can be generated from various sources. The low pressure mercury vapor UV lamps are widely used as reliable and low cost sources of UV light (Ngadi et al. 2003). These lamps operate at the nominal total gas pressures of 102–103 Pa and their UV output is in the range of 0.2–0.3 W/cm (Koutchma 2009). More recently, high intensity lamps with enhanced potential for UV microbial inactivation are being developed. Pulsed UV systems (PUV) have been developed and have been shown to be more effective in inactivating bacteria. In these systems, alternating current is stored in a capacitor and the energy is discharged through a high-speed switch to form a pulse of intense emission of light of about 100 μs durations. Some recent studies have reported application of PUV light for surface treatment of food products such as fresh-cut fruit, meats, and fish (Woodling and Moraru 2005, Ozer and Demirci 2006, Alothman et al. 2009, Oms-Oliu et al. 2009, Pombo et al. 2009). A prime goal of UV application in food processing is to reduce microbial load and achieve high-quality product with improved shelf life while maintaining sensory attributes. Some fresh-cut fruits and vegetables such as cantaloupe and fresh-cut melon treated with UV light yield better quality retention, since the treatment was effective in reducing microbial populations (Lamikanara et al. 2005, Artés-Hernández et al. 2010). The efficiency of the treatment depends on the structure of the fruit surface (Koutchma 2008). A reduction in microbial load was observed in juices from apple, guava, pineapple, and orange when treated with UV radiation (Keyser et al. 2008). Application of UV at 24 mW/cm^2 on apples inoculated with *Salmonella* and *E. coli* O157:H7 resulted in 3.3 log reduction (Yaun et al. 2004). In order to achieve high microbial inactivation, UV light should be applied for a sufficient time. Combination of UV and other minimal processing technologies can be used to improve microbial inactivation efficiency. Using modified atmosphere packaging (MAP; high concentrations of carbon dioxide) and UV together significantly reduced microbial population (Artés et al. 2009). Apart from antimicrobial action, UV may also influence the antioxidant activity of fresh-cut fruits and vegetables. An increase in the total phenol contents of fresh-cut tropical fruits such as banana and guava (Alothman et al. 2009), and enhanced flavonoid and antioxidant levels were observed in blueberries (Perkins-Veazie et al. 2008, Wang et al. 2009c) when the products were treated with UV irradiation. Gomez et al. (2010) reported changes in the color and mechanical compression behavior of apple cuts exposed to UV irradiation. Treated cuts showed accelerated browning, which the authors attributed to breakage of cellular membranes (plasmalemma and tonoplast). The phenomenon was also used to explain decrease in rupture stress and deformability indices for UV-treated samples during storage. This observation is curious as the action of UV on biological cells is traditionally attributed to changes in DNA modification. Pretreatment of apple cuts blanching or dipping in anti-browning solution were reported as effective in reducing quality changes and maintaining the original color of apple slices after UV treatment. Thus, the influence of UV on biological cells may be more complicated as previously thought. Improved understanding of the effect may open doors to innovative application of the technology.

Other Techniques

Various other novel techniques, photochemical (intense light pulses, etc.) and nonphotochemical processes are being used for minimal processing of fruits and vegetables. Electrolyzed water has been used to disinfect food surfaces. The pH of electrolyzed water can be raised or lowered by adding hydroxyl or hydrogen ion concentration. Decreased bacterial growth in fresh-cut cabbage was observed when sanitized with slightly acid electrolyzed water (Koide et al. 2009). Neutral Electrolyzed Water (NEW) was used to demonstrate the efficacy of NEW over Sodium hypochlorite in inhibiting *Listeria monocytogenes*, *Salmonella*, *E. coli* O157:H7, and the bacteria *Erwinia carotovora* in lettuce and fresh-cut products (Abadias et al. 2008). Comparison of bacterial reduction potency of strongly acidic electrolyzed water (SAEW), sodium hypochlorite solution (NaOCl), and slightly acidic electrolyzed water (StAEW) yielded the following order: StAEW > NaOCl > SAEW (Issa-Zacharia et al. 2010). The level of implementation of these techniques at commercial scale has been limited.

QUALITY AND SAFETY CONSIDERATIONS: THE HURDLE CONCEPT

Although food safety is critical, it is known that quality is a top-of-mind consideration when consumers purchase food products. A successful processing technique must therefore not only ensure that the product is safe, it must also maintain appropriate quality attributes that can be acceptable by consumers. Thermal treatment is a broad-spectrum antimicrobial process. However, novel nonthermal or minimal processing techniques have been developed due to the potential of thermal processing to degrade nutritive quality and functional properties of foods. The individual use of most of the new techniques may not provide the inactivation level required for commercial processing. Recently, combined inactivation techniques of microorganisms have been widely investigated since a combination of different "hurdles" is a more effective means of inhibiting microorganisms than using each "hurdle" alone (Alakomi et al. 2002, Leistner 2000). The hurdle concept is based on the imposition of certain hurdles to restrict the growth of food spoiling microorganisms as illustrated in Figure 39.2. Any microbial inactivation factor can potentially be adapted as a hurdle. Nearly 60 potential hurdles have been identified for food preservation. The hurdles must be applied in a logical sequence to ensure safety of the food product. The resistance of different pathogens should be considered when a hurdle is the lone preservation factor. For instance, PEF as well as high-pressure treatment effectively inactivate bacteria and yeast but are ineffective against bacterial and mold spores and some enzymes. Ultrasound is ineffective against wide-ranging pathogens. Although heating is the most broad-spectrum inactivation method, cells of *Clostridium botulinum* are highly resistant even to thermal treatment. Moreover, there is an inactivation threshold after which further hurdle application and energy input do not inactivate microorganisms or degrade nutritive quality of the treated foods.

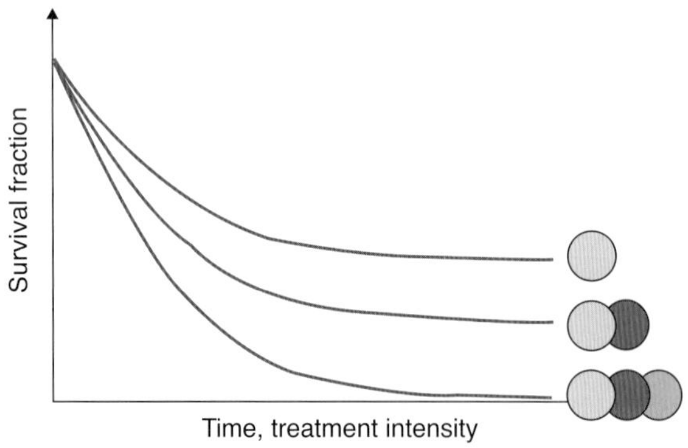

Figure 39.2. Schematic of enhanced microbial inactivation using different hurdles in food processing.

A synergistic relationship between different nonthermal physical and chemical hurdles has been observed for both foods (Fielding et al. 1997, Raso et al. 1998, Dutreux et al. 2000, Heinz and Knorr 2000, Aronsson and Rönner 2001, Fernanda et al. 2001) as well as nonfoods liquids (wastes, poultry chiller water, and others) (Liltved et al. 1995, Unal et al. 2001, Larson and Mariñas 2003). Although the simultaneous application of different nonthermal technologies has been shown to have a significant bactericidal effect, thermal treatment may still be required to achieve the level of inactivation of different microorganisms necessary for practical use. Thus, combination of thermal treatment with PEF (thermoelectrical treatment), pressure (manothermal treatment), ultrasound (thermoultrasonication), irradiation (thermoradiation), and other hurdles have been reported (Raso et al. 1998, Hoover 2000, Aronsson and Rönner 2001, Kim et al. 2001, Ohshima et al. 2002). Synergistic bactericidal effects have also been observed between heat and ultrasound (Wrigley and Llorca 1992), heat and radiation (Schaffner et al. 1989), heat and nisin (Knight et al. 1999), high hydrostatic pressure and nisin (Ponce et al. 1998), and PEF and nisin (Calderón-Miranda et al. 1999). Some general approaches for understanding hurdle concept can be found in Barbosa-Canovas et al. (1998) and Leistner and Gorris (1995).

Microbial Stress

When microorganisms present in food encounter stress in the form of various hurdles, they undergo homeostasis. The organisms try to prevail over inclement conditions caused by hurdles, but metabolic exhaustion due to repair action leads to death of the microorganisms. However, when bacteria are under stress, they can become more resistant and synthesize stress shock proteins. These proteins play the role of molecular chaperons by folding distorted proteins into a shape that retains the cell functionality under stress (Hightower 1991). But if the food is exposed to different stresses simultaneously, the microorganism generates more shock protein using all the cell energy and dies due to metabolic exhaustion (Leistner 2000).

Multitarget Preservation of Foods

Hurdle technology emphasizes intelligent combination of various preservation techniques. Leistner (1995) proposed concept of multitarget preservation of foods and suggested that different hurdles rather than having an accruing effect may have synergistic action. It has been suggested that different preservative factors of variable intensity be used for the synergistic action, instead of using a single high-impact preservative. These targets include cell components, enzymes, pH, water activity, redox potential, etc. Barbosa-Canovas et al. (1998) suggested that the applicability of multitarget preservation approach is not only limited to traditional methods of food preservation, but also valid for emerging nonthermal technologies like HPP, PEF, etc. A lot of new research is being done using newer technologies, which are based upon the central idea of multitarget preservation.

Limitations of Hurdle Concept

Combination of two or more hurdles results in either additive, synergistic, or antagonistic effect. Addition and synergism justify the use of hurdle technology but some studies have shown the negative effect of using some hurdle combinations (Jordan et al. 1999, Casey and Condon 2002). The antagonistic phenomenon is related to the type of conditions, intensity of preservative action of each hurdle, and the type and nature of food to be minimally processed.

PACKAGING TECHNIQUES FOR MINMALLY PROCESSED FOODS

Map

MAP is a method of preserving the fruits and vegetables by changing the composition of the air surrounding the food in a package. Different gas mixtures in varying concentrations are used in modifying the atmosphere inside the package. Generally, oxygen and carbon dioxide are used for packaging of minimally processed fruits and vegetables, but the potentials of other gases like nitrogen, carbon monoxide, and noble gases (Helium, Argon, and Neon) have also been realized (Sandhya 2010). Higher carbon dioxide and reduced oxygen levels have been found efficient in enhancing the shelf life and preventing the problem of enzymatic browning in fruits and vegetables. Generally, 15–20% CO_2 has been considered effective in preventing decay in fresh fruits and vegetables. In case of meat products, oxygen is used for retaining the red color of oxymyoglobin, but oxygen levels are reduced in other products to prevent oxidative rancidity and spoilage due to microbes. Very low oxygen or high concentration of carbon dioxide can initiate anaerobic respiration in the package, which may lead to formation of certain undesirable metabolites, harming the product's physiology (Soliva-Fortuny and Martin-Belloso 2003).

The essential characteristics for MAP packaging material are the gas permeability and water vapor transmission rate. Most common packaging materials include polyvinyl chloride, polypropylene, polyethylene, and polyethylene terephthalate (Mangaraj and Goswami 2009). These days, laminates or coextruded films are used for packaging. The traditional gas mixture configuration is not enough to prevent the deteriorative reaction if fruits and vegetables get wounded. Minimal processing up to some extent is responsible for initiating the tissue damaging. Also, the packaging materials used for MAP are prone to some limitations pertaining to textural, color, and permeability changes. Hence, as a solution, edible coatings are being seen as potential alternative to the MAP technique (Rojas-Grau et al. 2009). The advantages of edible coating are numerous, and constantly improvements are being made by incorporating active ingredients like antioxidant, antimicrobials, antibrowning agents, etc. Edible coatings have been explained later in the chapter.

Active and Edible Packaging

Active packaging implies incorporation of certain additives that can enhance the shelf life, flavor, texture, etc. by interacting with the food product inside the package. These additives can be oxygen scavengers, carbon dioxide absorbers or generators, ethanol emitters, ethylene absorbers, and moisture absorbers (Ohlsson and Bengtsson 2002). Antioxidants and antimicrobial compounds are also used in the active packaging.

Inclusion of chemical or physical additives in packages may become a hindrance for consumer acceptance of packaged foods. The development of packaging techniques that use natural materials like edible coating is a potential alternative to the chemical-based packaging methods. Edible coating is applied on the surface of food by spraying, dipping, or brushing. Sources of edible coatings are polysaccharides, proteins, and lipids (Lin and Zhao 2007). Edible coatings help reduce water loss and delay ageing by allowing controlled and selective gas permeability through product. They are also environmentally friendly as they reduce synthetic packaging waste.

CONCLUSION

Minimally processed foods have become very popular with consumers. All the aspects related to minimally processed food, from processing to packaging, are witnessing an unprecedented continuous improvement. The combined treatment methods using thermal and nonthermal technologies have shown promising results. Updated knowledge of the different emerging techniques and how they might interact when combined may be critical in developing new and innovative processing strategies.

REFERENCES

Abadias M et al. 2008. Efficacy of neutral electrolyzed water (NEW) for reducing microbial contamination on minimally processed vegetables. *Int J Food Microbiol* 123(1–2): 151–158.

Aguilo-Aguayo I et al. 2008. Influence of high-intensity pulsed electric field processing on lipoxygenase and β-glucosidase activities in strawberry juice. *Innov Food Sci Emerg Technol* 9(4): 455–462.

Aguilo-Aguayo I et al. 2009a. Changes in quality attributes throughout storage of strawberry juice processed by high-intensity pulsed electric fields or heat treatments. *LWT Food Sci Technol* 42(4): 813–818.

Aguilo-Aguayo I et al. 2009b. Avoiding non-enzymatic browning by high-intensity pulsed electric fields in strawberry, tomato and watermelon juices. *J Food Eng* 92(1): 37–43.

Alakomi H et al. 2002. The hurdle concept. In: T Ohlsson, N Bengtsson (eds.) *Minimal Processing Technologies in the Food Industry*. Cambridge Woodhead Publishing, Cambridge, UK, pp. 175–195.

Albu S et al. 2004. Potential for the use of ultrasound in the extraction of antioxidants from *Rosmarinus officinalis* for the food and pharmaceutical industry. *Ultrason Sonochem* 11: 261–265.

Alibas I. 2007. Microwave, air and combined microwave-air-drying parameters of pumpkin slices. *LWT Food Sci Technol* 40(8): 1445–1451.

Allende A et al. 2006. Minimal processing for healthy traditional foods. *Trends Food Sci Technol* 17(9): 513–519.

Alothman M et al. 2009. UV radiation-induced changes of antioxidant capacity of fresh-cut tropical fruits. *Innovative Food Sci Emerg Technol* 10(4): 512–516.

Alvarez I et al. 2003a. Inactivation of *Yersinia enterocolitica* by pulsed electric fields. *Food Microbiol* 20(6): 691–700.

Alvarez I et al. 2003b. The influence of process parameters for the inactivation of *Listeria monocytogenes* by pulsed electric fields. *Int J Food Microbiol* 87(1–2): 87–95.

Amiali M et al. 2006. Inactivation of Escherichia coli O157:H7 and *Salmonella enteritidis* in liquid egg white using pulsed electric field. *J Food Sci* 71: 88–94.

Amiali M et al. 2007. Synergistic effect of temperature and pulsed electric field on inactivation of *Escherichia coli* O157:H7 and *Salmonella Enteritidis* in liquid egg yolk. *J Food Eng* 79: 689–694.

Ananta E et al. 2001. Kinetic studies on high-pressure inactivation of *Bacillus stearothermophilus* spores suspended in food matrices. *Innov Food Sci Emerg Technol* 2(4): 261–272.

Angersbach A et al. 2000. Effects of pulsed electric fields on cell membranes in real food systems. *Innov Food Sci Emerg Technol* 1: 135–149.

Armstrong GA, McIlveen H. 2000. Effects of prolonged storage on the sensory quality and consumer acceptance of sous vide meat-based recipe dishes. *Food Qual Prefer* 11: 377–385.

Aronsson K, Ronner U. 2001. Influence of pH, water activity and temperature on the inactivation of *Escherichia coli* and *Saccharomyces cerevisiae* by pulsed electric fields. *Innov Food Sci Emerg Technol* 2(2): 105–112.

Artes F et al. 2009. Sustainable sanitation techniques for keeping quality and safety of fresh-cut plant commodities. *Postharv Biol Technol* 51(3): 287–296.

Artés-Hernández F et al. 2010. Low UV-C illumination for keeping overall quality of fresh-cut watermelon. *Postharv Biol Technol* 55(2): 114–120.

Arvanitoyannis IS et al. 2009. Irradiation applications in vegetables and fruits: a review. *Crit Rev Food Sci Nutr* 49(5): 427–462.

Askari GR et al. 2009. An investigation of the effects of drying methods and conditions on drying characteristics and quality attributes of agricultural products during hot air and hot air/microwave-assisted dehydration. *Drying Technol* 27(7): 831–841.

Barbosa-Canovas GV, Juliano P. 2007. Food sterilization by combining high pressure and heat. In: GF Gutierrez-Lopez, G Barbosa-Canovas, J Welti-Chanes, E Paradas-Arias (eds.) *Food Engineering Integrated Approaches*. Springer, New York, pp. 9–46.

Barbosa-Cánovas GV, Sepulveda DR. 2005. Present status and the future of PEF technology. In: GV Barbosa-Cánovas, MS Tapia, MP Cano (eds.) *Novel Food Processing Technologies*. CRC Press, Boca Raton, FL.

Barbosa-Cánovas GV et al. 1998. In: *Nonthermal Preservation of Foods*. Marcel Dekker, New York, p. 276.

Bari ML et al. 2005. Effectiveness of irradiation treatments in inactivating Listeria monocytogenes on fresh vegetables at refrigeration temperature. *J Food Prot* 68(2): 318–323.

Barsotti L et al. 2002. Effects of high voltage electric pulses on protein-based food constituents and structures. *Trends Food Sci Technol* 12: 136–144.

Bazhal MI, Vorobiev E. 2000. Electrical treatment of apple cossettes for intensifying juice pressing. *J Sci Food Agric* 80: 1668–1674.

Bazhal MI et al. 2003. Minimal processing of foods using hurdle technologies. Paper read at the CSAE/SCGR meeting, Montreal, Quebec.

Bazhal MI et al. 2004. Modeling compression of cellular systems exposed to combined pressure and pulsed electric fields. *Trans ASAE* 47(1): 165–171.

Bazhal MI et al. 2006. Inactivation of *Escherichia coli* O157:H7 in liquid whole egg using combined pulsed electric field and thermal treatments. *LWT Food Sci Technol* 39: 420–426.

Birla SL et al. 2008. Characterization of radio frequency heating of fresh fruits influenced by dielectric properties. *J Food Eng* 89(4): 390–398.

Black EP et al. 2007. Response of spores to high-pressure processing. *Compr Rev Food Sci Food Safety* 6(4): 103–119.

Blume T, Neis U. 2005. Improving chlorine disinfection of wastewater by ultrasound application. *Water Sci Technol* 52(10–11): 139–144.

Böhm V et al. 2006. Improving the nutritional quality of microwave-vacuum dried strawberries: a preliminary study. *Food Sci Technol Int* 12: 67–75.

Bondaruk J et al. 2007. Effect of drying conditions on the quality of vacuum-microwave dried potato cubes. *J Food Eng* 81: 164 –175.

Borge GIA et al. 2001. Growth and toxin profiles of *Bacillus cereus* isolated from different food sources. Int *J Food Microbiol* 6: 237–246.

Brown KL. 2000. Control of bacterial spores. *Br Med Bull* 56: 158–171.

Calderon-Miranda ML et al. 1999. Inactivation of *Listeria innocua* in skim milk by pulsed electric fields and nisin. *Int J Food Microbiol* 51: 19–30.

Campus M. 2010. High pressure processing of meat, meat products and seafood. *Food Eng Rev* 2: 256–273.

Casey P, Condon S. 2002. Sodium chloride decreases the bacteriocidal effect of acid pH on *Eshcerichia coli* O157:H45. *Int J Food Microbiol* 76: 199–206.

Castro I et al. 2003. The influence of field strength, sugar and solid content on electrical conductivity of strawberry products. *J Food Process Eng* 26(1): 17–30.

Changrue V et al. 2008. Osmotically dehydrated microwave-vacuum drying of strawberries. *J Food Process Presv* 32(5): 798–816.

Cheftel JC. 1995. Review: high pressure, microbial inactivation, and food preservation. *Food Sci Technol Int* 1: 75–90.

Chemat F et al. 2010. Applications of ultrasound in food technology: processing, preservation and extraction. *Ultrason Sonochem* 18(4): 813–835.

Chou SK, Chua KJ. 2001. New hybrid drying technologies for heat sensitive foodstuffs. *Trends Food Sci Technol* 12: 359–369.

Contreras C et al. 2008. Influence of microwave application on convective drying: effects on drying kinetics, and optical and mechanical properties of apple and strawberry. *J Food Eng* 88(1): 55–64.

Creed PG. 2001. The potential of foodservice systems for satisfying consumer needs. *Innov Food Sci Emerg Technol* 2(3): 219–227.

Dadali G, Ozbek B. 2009. Kinetic thermal degradation of vitamin C during microwave drying of okra and spinach. *Int J Food Sci Nutr* 60(1): 21–31.

Dadali G et al. 2008. Effect of drying conditions on rehydration kinetics of microwave dried spinach. *Food Bioprod Process* 86(4): 235–241.

Datta AK, Ananantheswaran RC. 2004. *Handbook of Microwave Technology for Food Applications*. Book News, Portland, OR.

De Pilli T et al. 2008. Study on operating conditions of orange drying processing: comparison between conventional and combined treatment. *J Food Process Pres* 32(5): 751–769.

Demeczky M. 1974. Continuous pasteurization of bottled fruit juices by high frequency energy. *Proceedings of IV International Congress on Food Science and Technology IV*, pp. 11–20.

Demirdove A, Baysal T. 2009. The use of ultrasound and combined technologies in food preservation. *Food Rev Int* 25(1): 1–11.

Drouzas AE, Schubert H. 1996. Microwave application in vacuum drying of fruit. *J Food Eng* 28(2): 203–209.

Dunn J. 1996. Pulsed light and pulsed electric field for foods and eggs. *Poultry Sci* 75(9): 1133–1136.

Dutreux N et al. 2000. Pulsed electric fields inactivation of attached and free-living *Escherichia coli* and *Listeria innocua* under several conditions. *Int J Food Microbiol* 54: 91–98.

Earnshaw R. 1996. High pressure processing. *Nutr Food Sci* 2: 8–11.

Eshtiaghi MN, Knorr D. 2002. High electric field pulse pretreatment: potential for sugar beet processing. *J Food Eng* 52(3): 265–272.

Evrendilek GA, Zhang QH. 2005. Effects of pulse polarity and pulse delaying time on pulsed electric fields-induced pasteurization of *E. coli* O157:H7. *J Food Eng* 68(2): 271–276.

Evrendilek GA et al. 2001. Shelf-life evaluations of liquid foods treated by pilot plant pulsed electric field system. *J Food Process Pres* 25: 283–297.

FDA. 2005. Irradiation in the production, processing, and handling of food. *Fed Regist Final Rule* 70(157): 48057–48073.

Fellows PJ. 1988. *Food Processing Technology, Principles and Practice*. Elks Horwood, Chichester.

Fernanda SMM et al. 2001. Inactivation effect of an 18-T pulsed magnetic field combined with other technologies on *Escherichia coli*. *Innov Food Sci Emerg Technol* 2(4): 273–277.

Fernandez GA et al. 2001. Antioxidative capacity, nutrient content and sensory quality of orange juice and an orange lemon carrot juice product after high-pressure treatment and storage in different packaging. *Eur Food Res Technol* 213(4–5): 290–296.

Fielding LM et al. 1997. The effect of electron beam irradiation, combined with acetic acid, on the survival and recovery of *Escherichia coli* and *Lactobacillus curvatus*. *Int J Food Microbiol* 35(3): 259–265.

Flood A et al. 2002. Fruit and vegetable intakes and the risk of colorectal cancer in the Breast Cancer Detection Demonstration Project follow-up cohort. *Am J Clin Nutr* 75(5): 936–943.

Floros JD, Liang H. 1994. Accoustically assisted diffusion through membranes and biomaterials. *Food Technol* 48(12): 79–84.

Gachovska TK et al. 2006. Pulsed electric field assisted juice extraction from alfalfa. *Can Biosyst Eng* 48: 3.33–3.37.

Gachovska TK et al. 2008. Drying characteristics of pulsed electric field-treated carrot. *Drying Technol* 26(10): 1244–1250.

Gachovska TK et al. 2009. Pulsed electric field treatment of carrots before drying and rehydration. *J Sci Food Agric* 89(14): 2372–2376.

Gachovska T et al. 2010. Enhanced anthocyanin extraction from red cabbage using pulsed electric Field processing. *J Food Sci* 75(6): E323–E329.

Gallardo-Reyes ED et al. 2008. Comparative quality of orange juice as treated by pulsed electric fields and ultra high temperature. *Agro Food Ind Hi-tech* 19(1): 35–36.

Geveke DJ et al. 2007. Radio frequency electric fields processing of orange juice. *Innov Sci Emerg Technol* 8(4): 549–554.

Gill A, Ramaswamy HOS. 2008. Application of high pressure processing to kill *Escherichia coli* O157 in ready-to-eat meats. *J Food Prot* 71(11): 2182–2189.

Giri SK, Prasad S. 2009. Quality and moisture sorption characteristics of microwave-vacuum, air and freeze-dried button mushroom (*Agaricus bisporus*). *J Food Process Pres* 33(1): 237–251.

Gomes C et al. 2008. E-beam irradiation of bagged, ready-to-eat spinach leaves (*Spinacea oleracea*): an engineering approach. *J Food Sci* 73(2): E95–E102.

Gomez N et al. 2005a. A model describing the kinetics of inactivation of *Lactobacillus plantarum* in a buffer system of different pH and in orange and apple juice. *J Food Eng* 70(1): 7–14.

Gomez N et al. 2005b. Modelling inactivation of *Listeria monocytogenes* by pulsed electric fields in media of different pH. *Int J Food Microbiol* 103(2): 199–206.

Gómez PL et al. 2010. Effect of ultraviolet-C light dose on quality of cut-apple: microorganism, color and compression behaviour. *J Food Eng* 98(1): 60–70.

González-Fandos E et al. 2005. Microbiological safety and sensory characteristics of salmon slices processed by the sous vide method. *Food Control* 16: 77–85.

Grimi N et al. 2009. Compressing behavior and texture evaluation for potatoes by pulsed electric field. *J Texture Stud* 40(2): 208–224.

Gudmundsson M, Hafsteinsson H. 2001. Effect of electric field pulses on microstructure of muscle foods and roes. *Trends Food Sci Technol* 12: 122–128.

Guo Y et al. 2009. Fast determination of essential oil extracted from fresh orange peel by improved solvent-free microwave extraction. *Fenxi Huaxue/Chin J Anal Chem* 37(3): 407–411.

Han Q et al. 2009. Effects of microwave vacuum drying technology on sensory quality of apple slices. *Nongye Jixie Xuebao/Trans Chin Soc Agric Machinery* 40(3): 130–134.

Harrison SL et al. 1997. *Saccharomyces cerevisiae* structural changes induced by pulsed electric field treatment. *LWT Food Sci Technol* 30(3): 236–240.

Heinz V, Knorr D. 2000. Effect of pH, ethanol addition and high hydrostatic pressure on the inactivation of *Bacillus subtilis* by pulsed electric fields. *Innov Food Sci Emerg Technol* 1(2): 151–159.

Hightower LE. 1991. Heat shock, stress proteins, chaperones, and proteotoxicity. *Cell* 56: 191–197.

Ho S, Mittal GS. 2000. High voltage pulsed electrical field for liquid food pasteurization. *Food Rev Int* 16(4): 395–434.

Holdsworth SD. 1985. Optimisation of thermal processing—a review. *J Food Eng* 4: 89–116.

Holdsworth SD. 1992. *Aseptic Processing and Packaging of Food Products*. Elsevier, London.

Hoover DG. 2000. Ultrasound. *J Food Sci* 65(l): 93–95.

Hou J et al. 2010. A method of extracting ginsenosides from panax ginseng by pulsed electric field. *J Sep Sci* 33(17–18): 2707–2713.

Huang L-L et al. 2009. Studies on decreasing energy consumption for a freeze-drying process of apple slices. *Drying Technol* 27(9): 938–946.

Ikediala JN et al. 2000. Dielectric properties of apple cultivars and codling moth larvae. *Trans Amer Soc Agric Eng* 43(5): 1175–1184.

Issa-Zacharia A et al. 2010. Sanitization potency of slightly acidic electrolyzed water against pure cultures of *Escherichia coli* and *Staphylococcus aureus*, in comparison with that of other food sanitizers. *Food Control* 21(5): 740–745.

Jeong S et al. 2010. Inactivation of *Escherichia coli* O157:H7 on lettuce, using low-energy X-ray irradiation. *J Food Prot* 73(3): 547–551.

Jordan KN et al. 1999. Survival of low pH stress by *Escherichia coli* O157:H7: a correlation between alterations in the cell envelope and increased acid tolerance. *Appl Environ Microbiol* 65: 3048–3055.

Juliano P et al. 2010. Polymeric-based food packaging for high-pressure processing. *Food Eng Rev* 2: 274–297.

Kar A et al. 2003. Comparison of different methods of drying for banana (Dwarf Cavendish) slices. *J Food Sci Technol* 40(4): 378–381.

Karaaslan SN, Tunçer IK. 2008. Development of a drying model for combined microwave-fan-assisted convection drying of spinach. *Biosyst Eng* 100(1): 44–52.

Keyser M et al. 2008. Ultraviolet radiation as a non-thermal treatment for the inactivation of microorganisms in fruit juice. *Innov Food Sci Emerg Technol* 9(3): 348–354.

Kim JH et al. 2006. Effect of gamma irradiation on *Listeria ivanovii* inoculated to iceberg lettuce stored at cold temperature. *Food Control* 17: 397–401.

Kim YS et al. 2001. Effects of combined treatment of high hydrostatic pressure and mild heat on the quality of carrot juice. *J Food Sci* 66(9): 1355–1360.

Kinn. 1947. Basic theory and limitations of high frequency heating equipment. *Biosyst Eng* 1: 161–173.

Knight KP et al. 1999. Nisin reduced thermal resistance of *Listeria monocytogenes* Scott A in liquid whole egg. *J Food Prot* 62: 999–1003.

Knorr D et al. 2004. Applications and potential of ultrasonics in food processing. *Trends Food Sci Technol* 15: 261–266.

Koide S et al. 2009. Disinfection efficacy of slightly acidic electrolyzed water on fresh cut cabbage. *Food Control* 20(3): 294–297.

Komolprasert V. 2007. Packaging for foods treated with ionizing radiation. In: JH Han (ed.) *Packaging for Non Thermal Processing of Food*. Blackwell Publishing, Ames, IA.

Koutchma T. 2008. UV light for processing foods. *Ozone: Sci Eng* 30(1): 93–98.

Koutchma T. 2009. Advances in ultraviolet light technology for non-thermal processing of liquid foods. *Food Bioprocess Technol* 2(2): 138–155.

Labas MD et al. 2005. Kinetics of bacteria disinfection with UV radiation in an absorbing and nutritious medium. *Chem Eng J* 114: 87–97.

Lamikanra O et al. 2005. Effect of processing under ultraviolet light on the shelf life of fresh-cut cantaloupe melon. *J Food Sci* 70(9): C534–C539.

Larson MA, Mariñas BJ. 2003. Inactivation of *Bacillus subtilis* spores with ozone and monochloramine. *Water Res* 37(4): 833–844.

Latorre ME et al. 2010. Effects of gamma irradiation on biochemical and physico-chemical parameters of fresh-cut red beet (*Beta vulgaris* L. var. *conditiva*) root. *J Food Eng* 98(2): 178–191.

Lee H et al. 2009. Inactivation of *Escherichia coli* cells with sonication, manosonication, thermosonication, and manothermosonication: microbial responses and kinetics modeling. *J Food Eng* 93(3): 354–364.

Leistner L. 2000. Basic aspects of food preservation by hurdle technology. *Int J Food Microbiol* 55(1–3): 181–186.

Leistner L, Gorris LGM. 1995. Food preservation by hurdle technology. *Trends Food Sci Technol* 6: 43–46.

Lemmens L et al. 2009. Thermal pretreatments of carrot pieces using different heating techniques: effect on quality related aspects. *Innov Food Sci Emerg Technol* 10(4): 522–529.

Li Z et al. 2010. Apple volatiles monitoring and control in microwave drying. *LWT Food Sci Technol* 43(4): 684–689.

Liltved H et al. 1995. Inactivation of bacterial and viral fish pathogens by ozonation or UV irradiation in water of different salinity. *Aquacult Eng* 14(2): 107–122.

Lima M. 2007. Ohmic heating: quality improvements. *Encyclopedia Agric Food Biol Eng* 1: 1–3.

Lin D, Zhao Y. 2007. Innovations in the development and application of edible coatings for fresh and minimally processed fruits and vegetables. *Compr Food Sci Food Saf* 6(3): 60–75.

Loginova KV et al. 2010. Pilot study of countercurrent cold and mild heat extraction of sugar from sugar beets, assisted by pulsed electric fields. *J Food Eng* 102(4): 340–347.

Lopes MLM et al. 2010. High hydrostatic pressure processing of tropical fruits: importance for maintenance of the natural food properties. *Ann N Y Acad Sci* 1189: 6–15.

Lopez P, Burgos J. 1995. Lipoxygenase inactivation by manothermosonication: effects of sonication physical parameters, pH, KCl, sugars, glycerol, and enzyme concentration. *J Agric Food Chem* 43: 620–625.

Luo W et al. 2008. Investigation on shelf life of carrot juice processed by pulse electric field. Paper read at International Conference on High Voltage Engineering and Application, art no.4774040, pp. 735–741.

Mangaraj S, Goswami TK. 2009. Modified atmosphere packaging—an ideal food preservation technique. *J Food Sci Technol* 46(5): 399–410.

Markowski M et al. 2009. Rehydration behavior of vacuum-microwave-dried potato cubes. *Drying Technol* 27(2): 296–305.

Marsellés-Fontanet ÀR et al. 2009. Optimising the inactivation of grape juice spoilage organisms by pulse electric fields. *Int J Food Microbiol* 130(3): 159–165.

Martín-Belloso O et al. 1997. Inactivation of *Escherichia coli* suspended in liquid egg using pulsed electric fields. *J Food Process Eng* 20: 317–336.

Mason T et al. 2005. Aplications of ultrasound. In: DW Sun (ed.) *Emerging Technologies for Food Processing*. Elsevier Academic Press, London, Chap. 13, pp. 323–351.

Mertens B. 1995. Hydrostatic pressure treatment of foods: equipment and processing. In: GW Gould (ed.) *New Methods of Food Preservation*. Blackie Academic and Professional, New York, p. 135.

Metaxas AC, Meredith RJ. 1993. *Industrial Microwave Heating*. Peter Peregrinus, London.

Mittal GS. 2009. Non-thermal food processing with pulsed electric field technology. *Resou: Eng Technol Sustain World* 16(2): 6–8.

Morris C et al. 2007. Non-thermal food processing/preservation technologies: a review with packaging implications. *Packag Technol Sci* 20: 275–286.

Mosqueda-Melgar J et al. 2007. Influence of treatment time and pulse frequency on *Salmonella enteritidis*, *Escherichia coli* and *Listeria monocytogenes* populations inoculated in melon and watermelon juices treated by pulsed electric fields. *Int J Food Microbiol* 117(2): 192–200.

Mousa N, Farid M. 2002. Microwave vacuum drying of banana slices. *Drying Technol* 20(10): 2055–2066.

Moyer JC, Stotz E. 1947. The blanching of vegetables by electronics. *Biosyst Eng* 1: 252–257.

Mulet A et al. 2010. Ultrasound-assisted hot air drying of foods. In: H Feng, G Barbosa-Canovas, J Weiss (eds.) *Ultrasound Technologies for Food and Bioprocessing Series: Food Engineering Series*. Springer, New York.

Mulet A et al. 2003. New food drying technologies—use of ultrasound. *Food Sci Technol Int* 9(3): 215–218.

Nawirska A et al. 2009. Drying kinetics and quality parameters of pumpkin slices dehydrated using different methods. *J Food Eng* 94(1): 14–20.

Nelson SO, Trabelsi S. 2009. Dielectric properties of agricultural products and applications. American Society of Agricultural and Biological Engineers Annual International Meeting 5, pp. 2901–2919.

Ngadi M, Bazhal M. 2004. Emerging food processing technologies: food quality and safety. In: F Shahidi, BK Simpson (eds.) *Seafood Quality and Safety: Advances in the New Millennium*. ScienceTech Publishing, St. John's, NL.

Ngadi M, Dehghannya J. 2010. Modeling microbial inactivation by a pulsed electric field. In: MF Mohammed (ed.) *Mathematical Modeling of Food Processing*. CRC Press, Boca Raton, FL, pp. 559–573. DOI: 10.1201/9781420053548-c20.

Ngadi M et al. 2003. Kinetics of ultraviolet light inactivation of *Escherichia coli* O157:H7 in liquid foods. *J Sci Food Agric* 83(15): 1551–1555.

Ngadi MO et al. 2010. Food quality and safety issues during pulsed electric field processing. In: E Ortega-Rivas (ed.) *Processing Effects on Safety and Quality of Foods*. CRC Press, Boca Raton, FL, pp. 441–479. DOI: 10.1201/9781420061154-c16.

Niemira BA, Fan X. 2009. Irradiation enhances quality and microbial safety of fresh and fresh-cut fruits and vegetables. In: E Feeherry, RB Gravani (eds.) *Microbial Safety of Fresh Produce*. Wiley-Blackwell, Oxford, UK. Doi: 10.1002/9781444319347.ch10.

Nyati H. 2000. An evaluation of the effect of storage and processing temperatures on the microbiological status of sous vide extended shelf-life products. *Food Control* 11: 471–476.

Odriozola-Serrano I et al. 2008. Phenolic acids, flavonoids, vitamin C and antioxidant capacity of strawberry juices processed by high-intensity pulsed electric fields or heat treatments. *Eur Food Res Technol* 228(2): 239–248.

Odriozola-Serrano I et al. 2009. Carotenoid and phenolic profile of tomato juices processed by high intensity pulsed electric fields compared with conventional thermal treatments. *Food Chem* 112(1): 258–266.

Ohlsson T, Bengtsson N. 2002. The hurdle concept. In *Minimal Processing Technologies in the Food Industry*. CRC Press, New York, Chapter 7 pp. 175–195.

Ohshima T et al. 2002. Effect of culture temperature on high-voltage pulse sterilization of *Escherichia coli*. *J Electrostatics* 55(3–4): 227–235.

Oms-Oliu G et al. 2009. Effects of high-intensity pulsed electric field processing conditions on lycopene, vitamin C and antioxidant capacity of watermelon juice. *Food Chem* 115(4): 1312–1319.

Orsat V et al. 2006. Microwave drying of fruits and vegetables. *Stewart Postharv Rev* 2(6): 1–7.

Orsat V et al. 2007. Microwave-assisted drying of biomaterials. *Food Bioprod Process* 85(3C): 255–263.

Ozer NP, Demirci A. 2006. Inactivation of *Escherichia coli* O157:H7 and *Listeria monocytogenes* inoculated on raw salmon fillets by pulsed-UV light treatment. *Int J Food Sci Technol* 41(4): 354–360.

Ozkan I.A et al. 2007. Microwave drying characteristics of spinach. *J Food Eng* 78(2): 577–583.

Patterson MF, Loaharanu P. 2000. Irradiation. In: BM Lund, TC Baird-Parker, GW Gould (eds.) *The Microbiological Safety and Quality of Food,* Vols. 1–2. Aspen Publishers, Springer–Verlag, Gaithersburg, Chapter 4, pp. 65–100.

Peck MW, Stringer SC. 2005. The safety of pasteurised in-pack chilled meat products with respect to the foodborne botulism hazard. *Meat Sci* 70: 461–475.

Pereira NR et al. 2007. Effect of microwave power, air velocity and temperature on the final drying of osmotically dehydrated bananas. *J Food Eng* 81(1): 79–87.

Perkins-Veazie P et al. 2008. Blueberry fruit response to postharvest application of ultraviolet radiation. *Postharv Biol Technol* 47(3): 280–285.

Pombo MA et al. 2009. UV-C irradiation delays strawberry fruit softening and modifies the expression of genes involved in cell wall degradation. *Postharvest Biol Technol* 51(2): 141–148.

Ponce E et al. 1998. Combined effect of nisin and high hydrostatic pressure on destruction of *Listeria innocua* and *Escherichia coli* in liquid whole egg. *Int J Food Microbiol* 43(1–2): 15–19.

Qin B et al. 1995. Nonthermal inactivation of *S. cerevisiae* in apple juice using pulsed electric fields. *LWT Food Sci Technol* 28: 564–568.

Ramaswamy H, Tang J. 2008. Microwave and radio frequency heating. *Food Sci Technol Int* 14: 423–427.

Ramaswamy HS et al. 2002. Enhanced thermal effects under microwave heating conditions. In: J Welti-Chanes, G Barbosa-Canovas, JM Aguilera (eds.) *Engineering and Food for the 21st Century*. CRC Press, Boca Raton, FL, pp. 739–762.

Raso J et al. 1998. Influence of temperature and pressure on the lethality of ultrasound. *Appl Environ Microbiol* 64: 465–471.

Rastogi NK et al. 1999. Accelerated mass transfer during osmotic dehydration of high intensity electrical field pulse pretreated carrots. *J Food Sci* 64(6): 1020–1023.

Regier M et al. 2005. Influences of drying and storage of lycopene-rich carrots on the carotenoid content. *Drying Technol* 23: 989–998.

Reyes A et al. 2007. A comparative study of microwave-assisted air drying of potato slices. *Biosyst Eng* 98(3): 310–318.

Rivas A et al. 2006a. Effect of PEF and heat pasteurization on the physical–chemical characteristics of blended orange and carrot juice. *LWT Food Sci Technol* 39: 1163–1170.

Rivas A et al. 2006b. Nature of the inactivation of *Escherichia coli* suspended in an orange juice and milk beverage. *Eur Food Res Technol* 223(4): 541–545.

Rojas-Grau MA et al. 2009. Edible coatings to incorporate active ingredients to fresh-cut fruits: a review. *Trends Food Sci Technol* 20(10): 438–447.

Ruiz-Diaz G et al. 2003. Modelling of dehydration-rehydration of orange slices in combined microwave/air drying. *Innovative Food Sci Emerging Technol* 4(2): 203–209.

Rybka-Rodgers S. 2001. Improvement of food safety design of cook-chill foods. *Food Res Int* 34: 449–455.

Salkinoja-Salonen MS et al. 1999. Toxigenic strains of *Bacillus licheniformis* related to food poisoning. *Appl Environ Microbiol* 65: 4637–4645.

Sampedro F et al. 2006. Effect of temperature and substrate on PEF inactivation of *Lactobacillus plantarum* in an orange juice-milk beverage. *Eur Food Res Technol* 223(1): 30–34.

Sanchez-Vega R et al. 2009. Enzyme inactivation on apple juice treated by ultrapasteurization and pulsed electric fields technology. *J Food Process Pres* 33(4): 486–499.

Sandhya. 2010. Modified atmosphere packaging of fresh produce: current status and future needs. *LWT - Food Sci Technol* 43(3): 381–392.

Sarang S et al. 2008. Electrical conductivity of fruits and meats during ohmic heating. *J Food Eng* 87(3): 351–356.

Sastry SK, Barach JT. 2000. Ohmic and inductive heating. *J Food Sci* 65(4): 42–46.

Schaffner DF et al. 1989. *Salmonella* inactivation in liquid whole egg by thermoradiation. *J Food Sci* 54(4): 902–905.

Scouten AJ, Beuchat LR. 2002. Combined effects of chemical, heat and ultrasound treatments to kill *Salmonella* and *Escherichia coli* O157:H7 on alfalfa seeds. *J Appl Microbiol* 92: 668–674.

Shapiro RL et al. 1998. Botulism in the United States: a clinical and epidemiologic review. *Ann Intern Med* 129: 221–228.

Soliva-Fortuny RC, Martin-Belloso O. 2003. New advances in extending the shelf-life of fresh-cut fruits: a review. *Trends Food Sci Technol* 14: 341–353.

Song X-J et al. 2009. Drying characteristics and kinetics of vacuum microwave-dried potato slices. *Drying Technol* 27(9): 969–974.

Soria AC, Villamiel M. 2010. Effect of ultrasound on the technological properties and bioactivity of food: a review. *Trends Food Sci Technol* 21(7): 323–331.

Sosa-Morales ME et al. 2009. Dielectric heating as a potential post-harvest treatment of disinfesting mangoes, Part I: relation between dielectric properties and ripening. *Biosyst Eng* 103(3): 297–303.

Taiwo KA et al. 2001. Effects of pretreatments on the diffusion kinetics and some quality parameters of osmotically dehydrated apple slices. *J Agric Food Chem* 49: 2804–2811.

Tarko T et al. 2009. The influence of microwaves and selected manufacturing parameters on apple chip quality and antioxidant activity. *J Food Process Pres* 33(5): 676–690.

Tein ML et al. 1998. Characterization of vacuum microwave, air and freeze dried carrot slices. *Food Res Int* 31(2): 111–117.

Torregrosa F et al. 2006. Ascorbic acid stability during refrigerated storage of orange-carrot juice treated by high pulsed electric fields and comparison with pasteurized juice. *J Food Eng* 73: 339–345.

Unal U et al. 2001. Inactivation of *Escherichia coli* O157:H7, *Listeria monocytogenes*, and *Lactobacillus leichmannii* by combinations of ozone and pulsed electric field. *J Food Prot* 64: 777–782.

Vadivambal R, Jayas DS. 2007. Changes in quality of microwave-treated agricultural products—a review. *Biosyst Eng* 98: 1–16.

Vega-Mercado H et al. 1997. Non-thermal food preservation: pulsed electric fields. *Trends Food Sci Technol* 8(5): 151–157.

Vercet A et al. 2002. The effects of manothermosonication on tomato pectic enzymes and tomato paste rheological properties. *J Food Eng* 53: 273–278.

Vilkhu K et al. 2008. Applications and opportunities for ultrasound assisted extraction in the food industry—a review. *Innov Food Sci Emerg Technol* 9(2): 161–169.

Vinatoru M. 2001. An overview of the ultrasonically assisted extraction of bioactive principles from herbs. *Ultrason Sonochem* 8: 303–313.

Wang CY et al. 2009c. Changes of flavonoid content and antioxidant capacity in blueberries after illumination with UV-C. *Food Chem* 117(3): 426–431.

Wang J, Sheng K. 2006. Far-infrared and microwave drying of peach. *LWT Food Sci Technol* 39(3): 247–255.

Wang J, Xi YS. 2005. Drying characteristics and drying quality of carrot using a two-stage microwave process. *J Food Eng* 68(4): 505–511.

Wang J et al. 2007a. Microwave drying characteristics and dried quality of pumpkin. *Int J Food Sci Technol* 42(2): 148–156.

Wang J et al. 2010. Kinetics models for the inactivation of *Alicyclobacillus acidiphilus* DSM14558T and *Alicyclobacillus acidoterrestris* DSM 3922T in apple juice by ultrasound. *Int J Food Microbiol* 139: 177–181.

Wang LJ et al. 2007b. Application of two-stage ohmic heating to tofu processing. *Chem Eng Process* 46(5): 486–490.

Wang R et al. 2009b. Non-thermal processing of orange juice by PEFP in continuous chamber. *Jiangsu Daxue Xuebao (Ziran Kexue Ban)/J Jiangsu Univ (Nat Sci Ed)* 30(2): 169–173.

Wang S et al. 2005. Temperature-dependent dielectric properties of selected subtropical and tropical fruits and associated insect pests. *Trans Amer Soc Agric Eng* 48(5): 1873–1881.

Wang S et al. 2006. Postharvest treatment to control codling moth in fresh apples using water assisted radio frequency heating. *Postharv Biol Technol* 40(1): 89–96.

Welch RW, Mitchell PC. 2000. Food processing: a century of change. *Br Med Bull* 56(1): 1–17.

Wilkinson VM, Gould GW. 1998. *Food Irradiation: A Reference Guide*. Woodhead Publishing, Abington, Cambridge, UK.

Witrowa-Rajchert D, Rzaca M. 2009. Effect of drying method on the microstructure and physical properties of dried apples. *Drying Technol* 27: 903–909.

Wojdylo A et al. 2009. Effect of drying methods with the application of vacuum-microwave on the phenolic compounds, colour and antioxidant activity in strawberry fruits. *J Agri Food Chem* 57(4):1337–1343.

Woodling SE, Moraru CI. 2005. Influence of surface topography on the effectiveness of pulsed light treatment for the inactivation of *Listeria innocua* on stainless steel surfaces. *J Food Sci* 70(7): 345–351.

Wrigley DM, Llorca NG. 1992. Decrease of *Salmonella typhimurium* in skim milk and egg by heat and ultrasonic wave treatment. *J Food Prot* 55: 678–680.

Yang W et al. 2008. Comparative experiment on hot-air, microwave-convective and microwave-vacuum drying of mushroom. *Nongye Jixie Xuebao/Trans Chin Soc Agric Mach* 39(6): 102–104, 112.

Yaun BR et al. 2004. Inhibition of pathogens on fresh produce by ultraviolet energy. *Int J Food Microbiol* 90(1): 1–8.

Yildiz H et al. 2009. Ohmic and conventional heating of pomegranate juice: effects on rheology, color, and total phenolics. *Food Sci Technol Int* 15(5): 503–512.

Yongsawatddigul J, Gunasekaran S. 1996a. Microwavevacuum-drying of cranberries, part I: energy use and efficiency. *J Food Process Pres* 20(2): 121–143.

Yongsawatdigul J, Gunasekaran S. 1996b. Microwave-vacuum drying of cranberries, part II: quality evaluation. *J Food Process Pres* 20(2): 145–156.

Yu LJ et al. 2010. Proteolysis of cheese slurry made from pulsed electric field-treated milk. *Food Bioprocess Technol,* Published Online March 31, 2010.

Zhang H et al. 2006. Postharvest control of blue mold rot of pear by microwave treatment and *Cryptococcus laurentii*. *J Food Eng* 77: 539–544.

Zhang QH et al. 1997. Recent development in pulsed processing. In: *New Technologies Yearbook*. National Food Processors Association, Washington, DC, pp. 31–42.

Zhong T, Lima M. 2003. The effect of ohmic heating on vacuum drying rate of sweet potato tissue. *Bioresource Technol* 87(3): 215–220.

Zhu S et al. 2008. High-pressure destruction kinetics of *Clostridium sporogenes* spores in ground beef at elevated temperatures. *Int J Food Microbiol* 126(1–2): 86–92.

Zimmermann U. 1986. Electrical breakdown, electropermeabilization and electrofusion. *Rev Physiol Biochem Pharmacol* 105: 175–256.

Zulueta A et al. 2007. Fatty acid profile changes during orange juice-milk beverage processing by high-pulsed electric field. *Eur J Lipid Sci Technol* 109(1): 25–31.

40
Separation Technology in Food Processing

John Shi, Sophia Jun Xue, Xingqian Ye, Yueming Jiang, Ying Ma, Yanjun Li, and Xianzhe Zheng

INTRODUCTION

Separation of one or more components from a complex mixture is a requirement for many operations and is extensively used in various applications in the food and biotechnology industries. Application of separation technologies is used either to recover high-value components from agricultural commodities, being an important operation for the production of food products such as oil and proteins, or especially for the development of health-promoting food ingredients and high value-added food products such as antioxidants and flavors. Separation processes are also used to remove contaminants and impurities from food materials. Separation, often known as "downstream processing," is totally devoted to the science and engineering principles of separation and purification. Over the last 20 years, separation

Food Biochemistry and Food Processing, Second Edition. Edited by Benjamin K. Simpson, Leo M.L. Nollet, Fidel Toldrá, Soottawat Benjakul, Gopinadhan Paliyath and Y.H. Hui.

technologies in the food processing area have undergone an explosive growth. The competitive nature of the biotechnologies applied to the pharmaceutical and food industries for cost-effective manufacturing have provided much impetus for the development and use of new separation techniques on a large scale, but at a lower cost. Aside from the conventional separation techniques such as solvent and water extraction (solid–liquid contacting extraction or leaching), crystallization, precipitation, distillation, and liquid–liquid extraction, etc., have already been incorporated in basic food processing and well established and commercialized. A number of newer separation techniques as promising alternative methods for improved application in food engineering have been implemented on the commercial scale, including supercritical CO_2 fluid extraction, membrane-based separation, molecular distillation, and pressured low-polarity water extraction procedures. These newer techniques have made impressive advances in obtaining adequate segregations of components of interest with maximum speed, minimum effort, and minimum cost at as large a capacity as possible in production-scale processes. These techniques have been implemented for the purification of proteins, characterization of aromas, whey protein removal from dairy products, extraction of health-benefiting fish oil, and clarification of beverages including beer, fruit juices, and wine. The separation processes and technologies for high value-added products are based on their polarity and molecular size. Many potential high-value products can be developed from natural resources by different separation technologies and processes. Carotenoids, including lycopene, β-carotene, astaxanthins, and lutein, make up a world market nearing $1 billion with a growth rate of about 3%. Therefore, efforts to utilize natural agricultural materials for the production of high value-added products, especially health-promoting foods and ingredients, are of great interest to the food and biotechnology industries.

MAIN SEPARATION PROCESSES IN FOOD INDUSTRIAL APPLICATIONS

Separation processes such as extraction, concentration, purification, and fractionation of nutrients or bioactive components from agricultural materials are the main processes used to obtain high-value end products. All separations rely on exploiting differences in physical or chemical properties of mixture of components. Some of the more common properties involved in separation processes are particles or molecular size and shape, density, solubility, and electrostatic charge. In some operations, more than one of these physical and chemical properties is involved. As a unit operation, the separation process plays a key function in the whole procedure for value-added food processing. The science of separation consists of a wide variety of processes, including mechanical, equilibrium, and chromatographic methods.

Mechanical Separation Processes

Centrifugation

The centrifugation process works to separate immiscible liquids or solids from liquids by the application of centrifugal force. A centrifuge is a spinning settling tank. The rapid rotation of the entire unit replaces gravity by a controllable centrifugal or radial force. Centrifugal separation is used primarily in solid–liquid or liquid–liquid separation processes, where the process is based on density difference between the solid or liquids. Centrifugation is typically the first step in which the suspended solids or liquids are separated from the fluid phase. Various designs of centrifuge are used in the food industry such as removal of solids from juice, beverage, fermentation broths, or dewatering of food materials. Most centrifugal processes are carried out on a batch basis. However, some automatic and continuous centrifuge equipment and process are applied in the food and biotechnology industry.

Filtration

Filtration is the separation of two phases, solid particles or liquid droplets, and a continuous phase such as liquid or gas, from a fluid stream by passing the mixture through a porous medium. Filtration finds applications through the food and biotechnology industries. Filtration is employed at various stages in food manufacture, such as the refining of edible oils, clarification of sugar syrups, fruit juice, vinegar, wine, and beer, as well as yeast recovery after fermentation. Filtration is also carried out to clarify and recover cells from fermentation broths in the biotechnology area.

Filtration can be classified into conventional and nonconventional filtrations. The conventional filtration process uses filtration media with coarse selectivity and cannot separate similar size particles. It usually refers to the separation of solid, immiscible particles from liquid streams. Conventional filtration is typically the first step in which the suspended solids are separated from the fluid phase.

Membrane Separation

Nonconventional filtration processes use membrane separation technology. This nonconventional filtration has evolved into a quite sophisticated technique with the advent of membranes whose pore size and configuration can be controlled to such a degree that the filtration area is maximized while keeping the total volume of the unit small. A membrane separation process is based on differences in the ability to flow through a selective barrier (membrane) that separate two fluids. It should permit passage of certain components and retain certain other components of the mixture. Membrane separation processes are classified as microfiltration, ultrafiltration, and reverse osmosis according to the pore size of the membranes.

Ion-Exchange and Electrodialysis

Ion-exchange and electrodialysis are distinct separation processes, but both processes are based on the molecular electrostatic charge.

Electrodialysis

The electrodialysis process is based on the use of ion-selective membranes, which are sheets of ion-exchange resin. The process permits the separation of electrolytes from nonelectrolytes in solution and the exchange of ions between solutions. The electrodialysis process is used to separate ionic species in the food and biotechnology industries. In the electrochemical separation process, a gradient in electrical potential is used to separate ions with charged, ionically selective membranes. The electrodialysis process is used widely to desalinate brackish water to produce potable water. The process is also used in the food industry to deionize cheese whey and in a number of pollution-control applications.

Ion-Exchange

The ion-exchange process is defined as the selective removal of charged molecules from one liquid phase by transfer to a second liquid phase by a solid ion-exchange material. The mechanism of absorption is electrostatic opposite charges on the solutes and ion-exchanger. The feed solution is washed off, followed by desorption, in which the separated species are recovered back into solution in a purified form. The main areas using it are sugar, dairy (separation of protein, amino acids), and water purification. Ion-exchange processes are also widely used in the recovery, separation, and purification of biochemicals, monoclonal antibodies, and enzymes.

Solid/Solid Separation

Solid/solid separation is a mechanical separation such as in a milling separation facility. Solid/solid separation can be achieved on the basis of particle size from the sorting of large food units down to the molecular level. Shape, moisture content, electrostatic charge, and specific gravity may affect the separation process. Screening of materials through perforated beds (wire mesh, cloth screen) produces materials of more uniform particle size. For example, screening is usually used to sort and grade many food materials such as fruits, vegetables, and grains. A wide range of geometric designs of flat bed and rotary fixed aperture screens are used in the food industry.

Sedimentation (Precipitation or Settling)

The sedimentation process is based on gravitational settling of solids in liquids. Sedimentation processes are slow and are widely used in water and effluent treatment processes.

Magnetic Separation

Magnetic separation relies on the behavior of individual particles under the influence of magnetic forces. When exposed to a magnetic field, ferromagnetic materials are attracted along lines of magnetic force from points of lower magnetic field intensity to points of higher magnetic field intensity.

Equilibration Separation Processes

The following processes are the unit operations commonly observed in the food industry that involve equilibrium between solid and liquid phases.

Adsorption

Absorption is a process whereby a substance (absorbate or sorbate) is accumulated on the surface of a solid. Absorption process is used to recover and purify the desired products from a mixture of liquids. Adsorption is used for decolorization and enzyme and antibiotic recovery. Liquid-phase adsorption is usually used for the removal of contaminants present at low concentrations in process streams.

Evaporation

Evaporation is the concentration of a solution by evaporating water or other solvents. The process is largely dependent on the heat sensitivity of the material. Boiling temperatures can be lowered under vacuum. Most commercial evaporators work in the range 40–90°C and minimize the residence time in the heating zone, for the concentration of heat-sensitive liquid foods and volatile flavor and aroma solutions. The final product after an evaporation process is usually in the liquid form.

Crystallization

Crystallization processes can be used to separate a liquid material from a solid. Crystallization separates materials and forms solid particles of defined shape and size from a supersaturated solution by creating crystal nuclei and growing these nuclei to the desired size. Crystallization is often used in a high-resolution, polishing, or confectioning step during the separation of biological macromolecules. Crystallization can be affected by either cooling or evaporation to form a supersaturated solution in which crystal nuclei formation may occur. Sometimes, it is necessary to seed the solution by addition of solute crystals. Batch and continuous operations of crystallization processes are used in commercial food production.

Dewatering (Dehydration or Drying)

Dewatering is the process applied to separate water (or volatile liquids) from solids, slurries, and solutions to yield solid products.

Extraction

Solvent extraction is a method for a solvent–solvent or solvent–solid contacting operation.

Solid–Liquid Extraction, e.g., Organic–Aqueous Extraction

The solid is contacted with a liquid phase in the process called solid–liquid extraction or leaching in order to separate the desired solute constituent or to remove an unwanted component from the solid phase. Solid–liquid extraction or leaching is a separation process affected by a fluid involving the transfer of solutes from a solid matrix to a solvent. It is an operation extensively used to recover many important high-value food components such as oil from oilseeds, protein from soybean meal, phytochemicals from plants, etc. Solid–liquid extraction is also used to remove undesirable contaminants or toxins from food materials.

Liquid–Liquid Extraction, e.g., Two-Phase Aqueous Extraction

Liquid–liquid extraction separates a dissolved component from its solvent by transfer to a second solvent, mutually nonmiscible with the carrier solvent. The liquid–liquid extraction as a technology has been used in the antibiotics industry for several decades and now is recognized as a potentially usefully separation step in protein recovery on a commercial production.

Supercritical Fluid Extraction

Supercritical CO_2 fluid extraction is mostly for high value-added products that are sensitive to heat, light, and oxygen. The extraction process is implemented in a supercritical region where the extraction fluid (e.g., CO_2) has liquid-like densities and solvating strengths but retains the penetrating properties of a gas.

Distillation

Distillation is a physical separation process used to separate the components of a solution (mixture) that contains more components in the liquid mixture, by the distribution of gas and liquid in each phase. The principle of separation is based on the differences in composition between a liquid mixture and the vapor formed from it, because each substance in the mixture has its own unique boiling point.

Chromatographic Methods

Chromatography is a separation technique based on the uneven distribution of analytes between a stationary (usually a solid) and a mobile phase (either a liquid or a gas). Chromatography can be conducted in two or three dimensions. Two-dimensional chromatography, for example paper chromatography, is almost solely used for analytical purposes. Column chromatography separation is the most common form of chromatography used in the food processing field for further purification of high value-added components. Many variations exist on the basic chromatography methods. Some well-established methods are listed below.

Liquid–Solid Absorption Chromatography (e.g., Hydroxyapatite)

In absorption, molecules in a fluid phase concentrate on a solid surface without any chemical change. Physical surface forces from the solid phase attract and hold certain molecules (substrate) from the fluid surrounding the surface.

Affinity Chromatography

Affinity chromatography exploits the natural, biospecific interactions that occur between biological molecules. These interactions are very specific and because of this affinity separation process are very high resolution methods for the purification of food products such as proteins. Affinity chromatography has a gel surface that is covered with molecules (ligand) binding a particular molecule or a family of molecules in the sample.

Size-Exclusion Chromatography

Size-exclusion chromatography, also known as gel permeation chromatography when organic solvents are used, is used to control the pore size of the stationary phase so as to separate molecules according to their hydrodynamic volume and molecular size.

Ion-Exchange Chromatography

Ion-exchange chromatography and a related technique called ion-exclusion chromatography are based on differences in the electric charge density of their molecules during charge–charge interaction of molecules. It is a variation of absorption chromatography in which the solid adsorbent has charged groups chemically linked to an inert support. Ion-exchange chromatography is usually used for recovery and purification of amino acids, antibiotics, proteins, and living cells.

Hydrophobic Interaction Chromatography

Hydrophobic interaction chromatography is based on a mild adsorption process, and separates solutes by exploiting differences in their hydrophobicity.

Reverse Phase Chromatography

Reverse phase chromatography is based on the similar principles of hydrophobic interaction chromatography. The difference is that the solid support is highly hydrophobic in reverse phase chromatography, which allows the mobile phase to be aqueous.

ENGINEERING ASPECTS OF SEPARATION PROCESSES

Much of the recent process development for the separation of valuable components from natural materials has been directed toward high value-added products. For the recovery of valuable components from raw materials, applications of separation technologies to recover major or minor components from agricultural commodities is usually a solid–liquid contacting operation. As research moves into commercial production, there is a great need to develop scalable and cost-effective methods of separation technologies. The successful and effective development of separation technologies is a critical issue in the chain of value-added processing of agricultural materials such as developments in functional food ingredients, nutrients, and nutraceuticals.

Separation operations are interphase mass transfer processes because they involve certain heat, mass, and phase transfers, as well as chemical reactions among food components. The engineering properties of the targeted food components via separation systems include separation modeling, simulations, optimization control studies, and thermodynamic analyses. The principles of mass conservation and component transfer amounts are used to analyze and design industrial processes. Molecular intuition and a thermodynamic approach constitute powerful tools for the design of a successful separation process. The following issues of engineering properties are important for process optimization and simulation.

1. Chemical equilibration—binary, ternary, and multicomponent systems in solid–liquid contacting operation,
2. Diffusivity (pressure diffusion, thermal diffusion, gaseous diffusion) and convection,
3. Solubility of targeted components under different separation operating conditions,
4. Iso-electric points and charge dependence on pH,
5. Chemical interaction kinetics (colloid formation and affinity),
6. Physical properties of particles of the food material,
7. Flux and fouling properties in membrane separation processes,
8. Solvent selection, recycling, and management,
9. Nature of solvents, optimum composition of mixed solvents for certain nutrient separations, and solvent residues in food products are factors to be optimized.
10. Some solvent treatments such as evaporation, concentration, de-watering, de-coloring, toxicological analyses, waste minimization, recycling, and disposal are necessary.

SEPARATION SYSTEM DESIGN

Technical Request for a Separation System

An important consideration in determining the appropriateness of a separation technique and system is the actual purity requirement for the end products. In design of organic solvent (toxic chemical)-free separation technologies and systems, there are several essential technical approach requirements for technology advances, such as combination of new techniques available for system optimization, product design and reasonable separation and purification steps, environmentally friendly process, less air pollution and industrial waste (e.g., energy, greenhouse gases emission, reduction of waste water production), and economic feasibility and raw material selection.

Food Quality and Separation System

The major issues related to product quality after separations are the impacts of processing on bioactive compounds and the nutritional aspects of foods, as well as the quality characteristics. To meet the food safety regulations, no toxic chemical solvent residues are permitted in the end food products, e.g., "green" food products. Nutrition and health regulations must be met. Also important are a high stability of nutrients and bioactive components, processes operating at low temperatures to reduce thermal effects, processes that exclude light to reduce light induced (UV) irradiation effects, and processes that exclude oxygen to reduce oxygen effects. And the final product must maintain uniformity and quality consistency, and purity can meet food grade or pharmaceutical grade requirements.

Scaling Up Technology for Industrial Production

Scaling up of a natural product separation process is by no means a simple affair. The enormous variations from process to process necessitate careful attention to details at all stages of product development. When the technology in a food process is designed for industrial-scale production, an important area for consideration is the balance of capital and operating costs as the scale of the separation operation increases. Scale up of separation technology also involves optimization with respect to increasing the efficiency of each stage, giving rise to increasing demands on the accuracy of the assay system.

NEW TECHNOLOGY DEVELOPMENT

Extraction of health-promoting components from plant materials has usually been accomplished by conventional extraction processes such as solid–liquid extractions employing methanol, ethanol, acetone, or hexane and also through steam distillation or evaporation processes to remove solvents from extracts. Currently, the demand for natural bioactive compounds is increasing due to their use by the functional food and pharmaceutical industry. Thus, there has been increasing interest in the use of environmentally friendly "green" separation technologies able to provide high quality–high bioactivity extracts while precluding any toxicity associated with the solvent. Some of the motivations to employ "green" separation technologies as a viable separation technique are (a) tightening government regulations on toxic-chemical solvent residues and pollution control, (b) consumers' concern over the use of toxic chemical solvents in the processing, and (c) increased demand for higher quality products that traditional processing techniques cannot meet.

One of the most important considerations in developing new extraction processes is the safety aspect. In this sense, a variety of processes involving extractions with supercritical CO_2 fluid extraction, membrane-based separation, molecular distillation, and pressured low-polarity water extraction, etc., are generally recognized as "green" separation technology and are considered clean and safe processes to meet the requirements (Ibáñez et al. 1999, Fernandez Perez et al. 2000, Herrero et al. 2006, Chang et al. 2008). They have been developed and are regarded as innovative emerging separation technologies that meet food quality and safety requirements. These processes can be used to solve some of the problems associated with conventional organic solvent-oriented separation processes. Operation parameters and other factors related to the quality of the original plant, its geographic origin, the harvesting date, its storage, and its pretreatment process prior to extraction also influence the separation operations and the final composition of the extracts obtained.

Supercritical CO_2 Fluid Technology

Changes in food processing practices and new opportunities for innovative food products have spurred interest in supercritical CO_2 fluid extraction. Supercritical CO_2 fluid technology has been widely used to extract essential oils, functional fatty acids, and bioactive compounds, and also been applied in recently developed extraction and fractionation for carbohydrates (Glisic et al. 2007, Shi et al. 2007a, Montañés et al. 2008, Mitra et al. 2009, Montañés et al. 2009, Sanchez-Vicente et al. 2009, Shi et al. 2010a, 2010b). Supercritical CO_2 fluid extraction is a novel separation technique that utilizes the solvent properties of fluids near their thermodynamic critical point. Supercritical CO_2 is being given a great deal more attention as an alternative to organic chemical industrial solvents, and increased governmental scrutiny and new regulations restricting the use of common industrial solvents such as chlorinated hydrocarbons. It is one of the "green" separation processes that provides nontoxic and environmentally friendly attributes and leaves no traces of any toxic chemical solvent residue in foods.

When CO_2 fluid is forced into a pressure and temperature above its critical point (Fig. 40.1), CO_2 becomes a supercritical fluid. Under these conditions, various properties of the fluid are placed between those of a gas and those of a liquid. Although the density of a supercritical CO_2 fluid is similar to a liquid and its viscosity is similar to a gas, its diffusivity is intermediate between the two states. Thus, the supercritical state of a fluid has been defined as a state in which liquid and gas are indistinguishable from each other or as a state in which the fluid is compressible (i.e., similar behavior to a gas), even though possessing a density similar to that of a liquid and with similar solvating power. Because of its different physicochemical properties, supercritical CO_2 provides several operational advantages over traditional extraction methods. Because of their low viscosity and relatively high diffusivity, supercritical CO_2 fluids have better transport properties than liquids. They can diffuse easily through solid materials and therefore give faster extraction yields. One of the main characteristics of a supercritical fluid is the possibility of modifying the density of the fluid by changing its pressure and/or its temperature. Since density is directly related to solubility (Raventós et al. 2002, Shi and Zhou 2007, Shi et al. 2009a), by altering the extraction pressure, the solvent strength of the fluid can be modified.

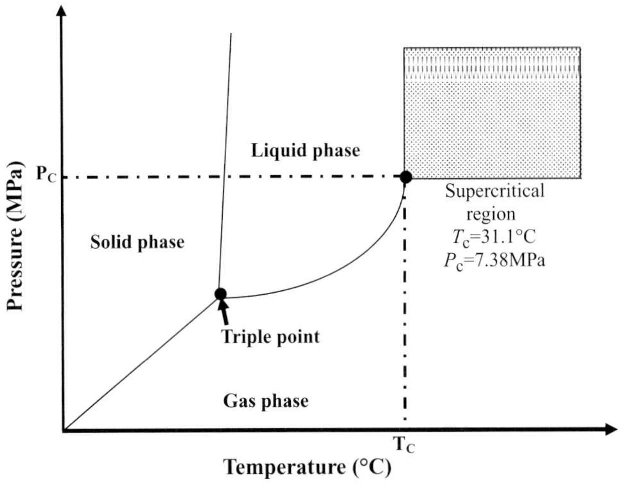

Figure 40.1. Supercritical pressure–temperature diagram for carbon dioxide.

The characteristic traits of CO_2 are inertness, nonflammability, noncorrosiveness, inexpensive, easily available, odorless, tasteless, and environmentally friendly. Its near-ambient critical temperature makes it ideal for thermolabile natural products (Mendiola et al. 2007). CO_2 has its favorable properties and the ease of changing selectivity by the addition of a relatively low amount of modifier (co-solvent) such as ethanol and other polar solvents (e.g., water). CO_2 may be considered the most desirable supercritical fluid for extracting natural products for food and medicinal uses (Shi et al. 2007b, Kasamma et al. 2008, Shi et al. 2009b, Yi et al. 2009). Other supercritical fluids, such as ethane, propane, butane, pentane, ethylene, ammonia, sulphur dioxide, water, chlorodifluoromethane, etc., are also used in supercritical fluid extraction processes.

In a supercritical CO_2 fluid extraction process, supercritical CO_2 fluid has a solvating power similar to organic liquid solvents and a higher diffusivity, with lower surface tension and viscosity. The physicochemical properties of supercritical fluids, such as the density, diffusivity, viscosity, and dielectric constant, can be controlled by varying the operating conditions of pressure and temperature or both in combination (Tena et al. 1997, Shi et al. 2007a, 2007b, 2007e, Kasamma et al. 2008, Shi et al. 2009b). The separation process can be affected by simply changing the operating pressure and temperature to alter the solvating power of the solvent. After modifying CO_2 with a co-solvent, the extraction process can significantly augment the selective and separation power and in some cases extend its solvating powers to polar components (Shi et al. 2009a).

Supercritical CO_2 fluid extraction is particularly relevant to food and pharmaceutical applications involving the processing and handling of complex, thermo-sensitive bioactive components, and an increased application in the areas of nutraceuticals, flavors, and other high-value items such as in the extraction and fractionation of fats and oils (Reverchon et al. 1992, Rizvi and Bhaskar 1995), the purification of a solid matrix, separation of tocopherols and antioxidants, removal of pesticide residues from herbs, medicines, and food products, the detoxification of shellfish, the concentration of fermented broth, fruit juices, essential oils, spices, coffee, and the separation of caffeine, etc. (Perrut 2000, González et al. 2002, Miyawaki et al. 2008, Martinez et al. 2008, Liu et al. 2009a, 2009b).

This technology has been successfully applied in the extraction of bioactive components (antioxidants, flavonoids, lycopene, essential oils, lectins, carotenoids, etc.) from a variety of biological materials such as hops, spices, tomato skins, and

Table 40.1. Comparison of Physical Properties of Supercritical CO_2 at 20 MPa and 55°C with Some Selected Liquid Solvent at 25°C

Properties	CO_2	*n*-Hexane	Methylene Chloride	Methanol
Density (g/mL)	0.75	0.66	1.33	0.79
Kinematic viscosity (m^2/s)	1.0	4.45	3.09	6.91
Diffusivity (m^2/s)	6.0×10^9	4.0×10^9	2.9×10^9	1.8×10^9
Cohesive energy density (δ (cal/cm^3))	10.8	7.24	9.93	14.28

Source: Modified from King et al. 1993.

other raw or waste agricultural materials (Kasamma et al. 2008, Shi et al. 2009a, Yi et al. 2009, Huang et al. 2010, Shi et al. 2010a, 2010b, Xiao et al. 2010). The scale up of some supercritical extraction processes have already proven to be possible and are in industrial use.

The process requires intimate contact between the packed beds formed by a ground solid substratum (fixed-bed of extractable material) with supercritical CO_2 fluid. During the supercritical extraction process, the solid phase comprising of the solute and the insoluble residuum (matrix) is brought into contact with the fluid phase, which is the solution of the solute in the supercritical CO_2 fluid (solvent). The extracted material is then conveyed to a separation unit.

Process Concept Schemes

The physicochemical properties of the supercritical fluids are crucial to the understanding of the process design calculation and modeling of the extraction process. Therefore, selectivity of solvents to discriminate solutes is a key property for the process engineer. Physical characteristics such as density and interfacial tension are important for separation to proceed; the density of the extract phase must be different from that of the raffinate phase and the interfacial properties influence coalescence, a step that must occur if the extract and raffinate phase are to separate. The supercritical state of the fluid is influenced by temperature and pressure above the critical point. The critical point is the end of the vapor–liquid coexistence curve as shown on the pressure–temperature curve in Figure 40.1, where a single gaseous phase is generated. When pressure and temperature are further increased beyond this critical point, it enters a supercritical state. At this state, no phase transition will occur, regardless of any increase in pressure or temperature, nor will it transit to a liquid phase. Hence, diffusion and mass transfer rates during supercritical extraction are about two orders of magnitude greater than those with solvents in the liquid state.

Substances that have similar polarities will be soluble in each other, but increasing deviations in polarity will make solubility increasingly difficult. Intermolecular polarities exist as a result of van der Waals forces, and although solubility behaviors depend on the degree of intermolecular attraction between molecules, the discriminations between different types of polarities are also important. Substances dissolve in each other if their intermolecular forces are similar or if the composite forces are made up in the same way. Properties such as the density, diffusivity, dielectrical constant, viscosity, and solubility are paramount to supercritical extraction process design. The dissolving power of SCF depends on its density and the mass transfer characteristic, and is superior due to its higher diffusivity and lower viscosity and interfacial tension than liquid solvents.

Although many different types of supercritical fluids exist and have many industrial applications, CO_2 is the most desirable for SCE of bioactive components. Table 40.1 shows some physical properties of compressed (20 MPa) supercritical CO_2 at 55°C as compared to some condensed liquids that are commonly used as extraction solvents at 25°C. It should be noticed that supercritical CO_2 exhibits similar density as those of the liquid solvents, while being less viscous and more highly diffusive. This fluid-like attribute of CO_2 coupled with its ideal transport properties and other quality attributes outlined above make it a better choice over other solvents.

The specific heat capacity (C_p) of CO_2 rapidly increases as the critical point (31.1°C temperature, 7.37 MPa pressure, and 467.7 g/L flow rate) is approached. Like enthalpy and entropy, the heat capacity is a function of temperature, pressure, and density (Mukhopadhyay 2000). Under constant temperature, both the enthalpy and entropy of supercritical CO_2 decreases with increased pressure and increases with temperature at constant pressure. The change in specific heat as a result of varying the pressure and temperature is also dependent on density. For example, under constant temperature, specific heat increases with increasing density up to a certain critical level. Above this critical level, any further increase in density reduces the specific heat.

Process System

The supercritical CO_2 fluid extraction process is governed by four key steps: extraction, expansion, separation, and solvent conditioning. The steps are accompanied by four generic primary components: extractor (high pressure vessel), pressure and temperature control system, separator, and pressure intensifier (pump; Fig. 40.2). Raw materials are usually ground and charged into a temperature-controlled extractor forming a fixed bed, which is usually the case for a batch and single-stage mode (Shi et al. 2007a, 2007b, Kasamma et al. 2008).

The supercritical CO_2 fluid is fed at high pressure by means of a pump, which pressurizes the extraction tank and also

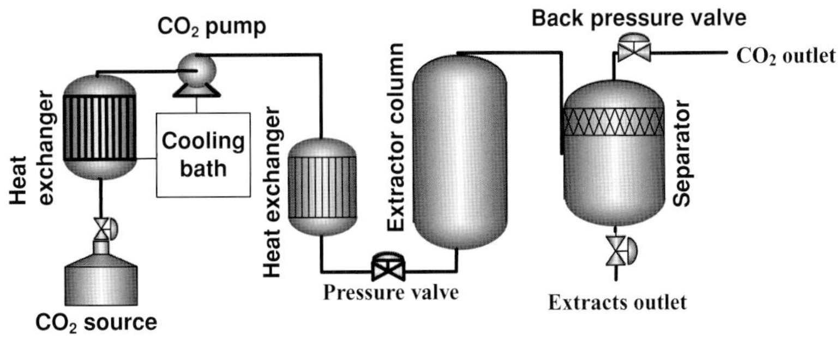

Figure 40.2. Schematic diagram of a typical single-stage supercritical fluid extraction system with CO_2.

circulates the supercritical medium throughout the system. Figure 40.2 shows an example of a typical single-stage supercritical CO_2 fluid extraction system. Once the supercritical CO_2 and the feed reach equilibrium in the extraction vessel, through the manipulation of pressure and temperature to achieve the ideal operating conditions, the extraction process proceeds. The mobile phase consisting of the supercritical CO_2 fluid and the solubilized components is transferred to the separator where the solvating power of the fluid is reduced by raising the temperature and/or decreasing the pressure of the system. The extract precipitates in the separator while the supercritical CO_2 fluid is either released to the atmosphere or recycled back to the extractor. In the case where highly volatile components are being extracted, a multistage configuration may have to be employed as shown in Figure 40.3 (Shi et al. 2007a, 2007b, Kasamma et al. 2008).

The processes described above are semi-batch continuous processes where the supercritical CO_2 flows in a continuous mode while the extractable solid feed is charged into the extraction vessel in batches. In commercial-scale processing plants, multiple extraction vessels are sequentially used to enhance process performance and output. Although the system is interrupted at the end of the extraction period, when the process is switched to another vessel prepared for extraction, the unloading and/or loading of the spent vessels can be carried out while extraction is in progress, reducing the downtime and improving the production efficiency.

A semi-continuous approach on a commercial scale uses a multiple stage extraction processes that involve running the system concurrently by harnessing a series of extraction vessels in tandem, as shown in Figure 40.4. In this system, the process is not interrupted at the end of extraction period for each

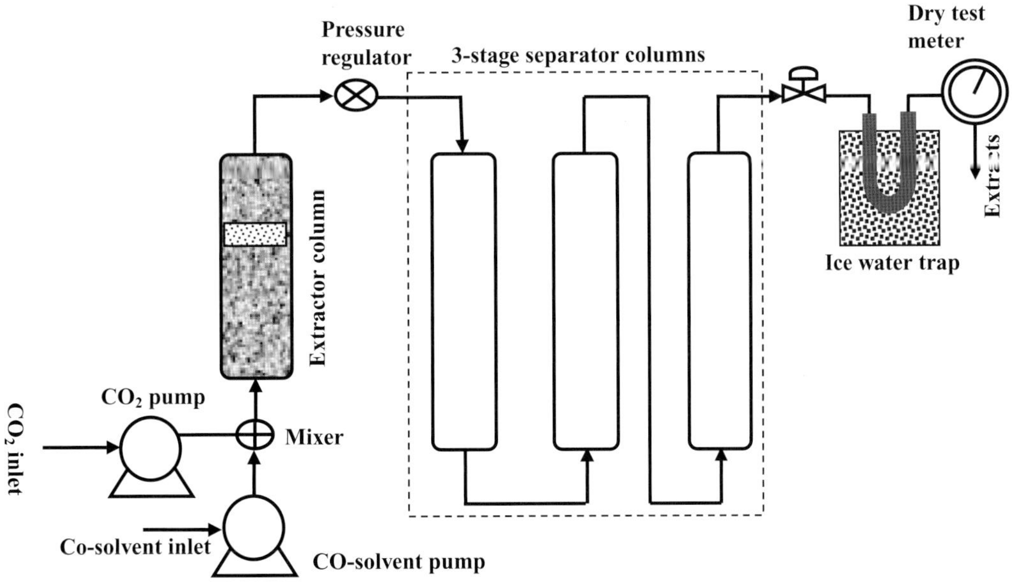

Figure 40.3. Schematic diagram of supercritical fluid extraction system used to fractionate bioactive components from plant matrix using supercritical carbon dioxide.

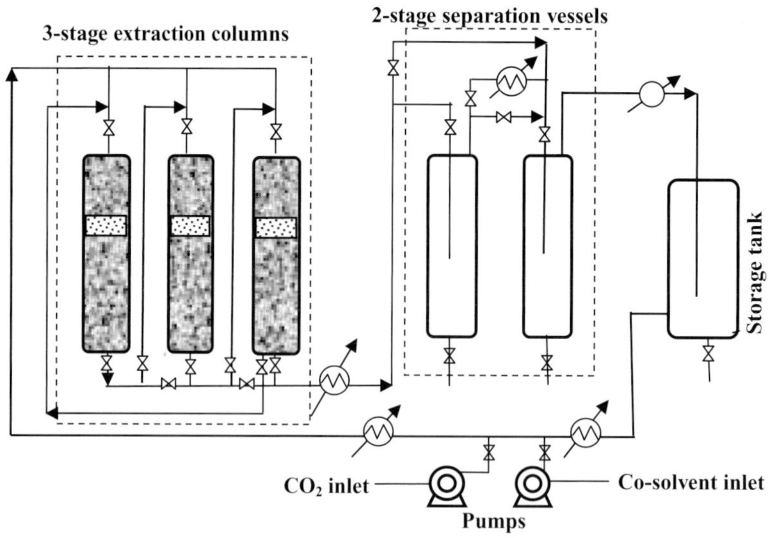

Figure 40.4. Schematic diagram of commercial scale multistage supercritical fluid extraction system used to fractionate bioactive components. The symbol "⧖" is pressure valves and "⊛" is heat exchangers.

vessel, because the process is switched to the next prepared vessel by control valves for extraction while unloading and/or loading the spent vessels. Thus, supercritical CO_2 technology is available in the form of single-stage batch that could be upgraded to multistage semi-continuous batch operations coupled with a multi-separation process. The need to improve the design into truly continuous modes is growing. Supercritical CO_2 fluid extraction could be cost-effective under large-scale production.

Industrial Applications

Large-scale supercritical CO_2 fluid extraction has become a practical process for the extraction of high-value products from natural materials. The solvating power of supercritical CO_2 fluids is sensitive to temperature and pressure changes; thus, the extraction parameters may be optimized to provide the highest possible extraction yields with maximum antioxidant activity for health-promoting components in bioactive extraction production (Kasamma et al. 2008, Chen et al. 2009, Yi et al. 2009b).

A supercritical CO_2 fluid extraction process offers the unique advantage of adding value to agricultural waste by extracting bioactives from agricultural by-products, which are then used for the fortification of foods and other applications. Its drawbacks are the difficulties in extracting polar compounds and its difficulty of extracting compounds from a complex matrix where the phase interaction with the intrinsic properties of the product inhibits its effectiveness. Some drawbacks can be ameliorated by using small amounts of food-grade co-solvents (less than 10%) to approach the high extraction efficiency (Shi et al. 2009a). However, much investigation is required to understand the solvation effects on the targeted bioactive components being extracted. The CO_2 density, pressure, and temperature have been noted to have great impacts on the results of the extraction process. By understanding the effect of the parameters that influence the extraction process, the conditions may be set to optimize yield and cost efficiency. When determining the parameters that should be used to maximize yields and solubility of the targeted components, many researchers attempted to use conditions that may be applicable in large-scale applications (Shi et al. 2007f, Kasamma et al. 2008). For example, nontoxic co-solvents and modifiers could be acceptable for food processing; therefore, a number of researchers have opted to use food-grade co-solvents and modifiers in extraction processes (Shi et al. 2009a). The nature of the material used as a source of high-value components, such as health-promoting components, governs the availability of the compounds for the extraction process. The presence of other components such as lipids may impede the process or elevate costs due to an elongated extraction time.

Although a high temperature in the extraction process generally increases the solubility of components in supercritical CO_2 fluids, the conditions under which thermally labile targeted compounds are negatively affected should be considered (Shi et al. 2007a, 2007b, 2007e). The intensity and the length of heat processing affect the health-promoting properties of bioactives. Therefore, ideally, the extraction time and temperature should be minimized. Minimizing such conditions also leads to a more economically viable process (Shi et al. 2007f, Kasamma et al. 2008). Excessively high flow rates may reduce the contact time between the solute and the solvent and restrict the fluid flow in the sample if it becomes compacted. The optimal flow rate appears to vary with the targeted molecule, relatively high flow rates having a negative effect on some components. Raising the pressure increases extraction yields.

Sample matrix is an important parameter that influences the solubility and mass transfer process during SCE. Properties such as particle shape and size distribution, porosity and pore size distributions, surface area, and moisture content influence solubility and mass transfer. The presence of water (moisture content) in the sample matrix during supercritical extraction also has an effect on the extraction outcome. In order to improve the yield and quality of the extracted high-value food components from raw material, a pretreatment of the raw material is an essential process (Yang et al. 2008, Zheng et al. 2009, Nagendra et al. 2010). Cell disruption is the most important pretreatment, and this procedure can be conducted by several processes such as mechanical, ultrasonic, high electronic field pulse, and nonmechanical treatments. With improved processing conditions and reduced cost, high-value components extracted from natural materials by supercritical CO_2 extraction process will become even more economical at high throughput.

Membrane-Based Separation Technology

Membrane-based technology is a rising separation technology. At present, more and more fields use membrane technology to separate the fluids and get good results. With the principal development of membrane technology, the applications are expanding much wider and the technology brings remarkable economic benefits. Now, many countries in the world have noticed the importance of membrane technology in the food processing field, especially since we are short of energy and resources and the deteriorating environment all exist in our lifetime. So, industries regard membrane-based separation technologies as important technologies in food processing areas.

The common membrane-based separation process is usually run under pressure. A membrane can be described as a thin barrier between the two bulk phases, and it is either a homogeneous phase or a heterogeneous collection of phases. In membrane separations, each membrane has the ability to transport one component more readily than the others because of differences in physical and chemical properties between the membrane and the permeating components. Furthermore, some components can freely permeate through the membrane, while others will be retained. The stream containing the components that pass through the membrane is called the permeate, and the stream containing the retained components is called the retentate. The flow of material across a membrane is kinetically driven by the application of pressure concentration, vapor pressure, hydrostatic pressure, or electrical potential (Mulder 1996, Cheryan 1998).

Membrane separation technology adopts a selective multihole membrane as the separation medium. The separated fluid, driven by the outside pressure, goes through the surface of the membrane with different pore sizes. The molecules will pass the membrane but larger molecules will be rejected. Therefore, the fluid can be separated into different molecular weights with high efficiency (Cheryan 1998).

Compared with the traditional separation process technologies, membrane-based separation processes can run under normal temperatures, so are especially good for separating and concentrating heat-sensitive materials such as juices, enzymes, or phytochemicals (Vaillant et al. 2001, Enevoldsen et al. 2007, Pouliot 2008). Moreover, membrane-based processes are energy-efficient processes that do not involve phase change or heat input; thus, the processes do not require ancillary equipment such as heat generator, evaporator, and condenser. They offer ease of operation and great flexibility, and do not require the addition of any chemical agents. The process also provides for minimal thermal degradation, occurs at ambient temperatures, and is used both to filter molecular-sized particulates and also to concentrate an isolate of interest, such as lycopene. Determination of the optimum operating conditions is vital importance.

Membrane-based processes are usually more energy efficient than distillation, adsorption, and chromatography. Furthermore, membrane-based separation has the advantage of compatibility with a wide range of solvents and chemical products, an ability to process thermally sensitive compounds and easy amenability to automation. These advantages open up several possibilities of membrane application in the production of bioactive compounds in the areas of energy-efficient pre-concentration of dilute solutions, fractionation of diverse classes of compounds from complex mixtures, recovery of intermediates, and recycling of solvents. The challenges posed by a membrane-based separation process are limited selectivity, fouling leading to performance decline, and a necessary and occasionally difficult periodic cleaning process. The performance of a membrane can be distinguished by two simple factors: flux (or product rate) and selectivity through the membrane. Flux is defined as the permeation capacity that refers to the quantity of fluid permeating per unit area of membrane per unit time. Flux depends linearly on both the permeability and the driving force. The flux also depends inversely upon the thickness of the membrane (Yang et al. 2001, Bhanusali and Bhattacharyya 2003, Kertész et al. 2005).

Figure 40.5 shows a classification of various separation-based processes that are based upon particle or molecular size. The five major membrane separation processes, including microfiltration, ultrafiltration, nanofiltration, reverse osmosis, and electrodialysis, cover a wide range of particle sizes according to the

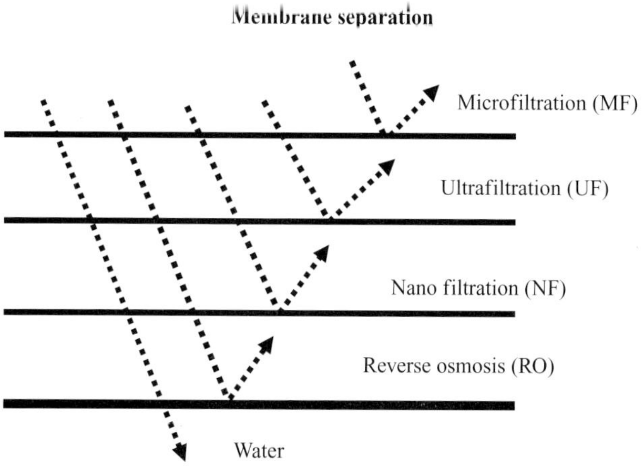

Figure 40.5. Schematic diagram of membrane molecular cut property.

Table 40.2. Size of Materials Retained, Driving Force, and Type of Membrane

Process	Size of Materials Retained	Driving Force	Type of Membrane
Microfiltration	0.2–10 μm (microparticles)	Pressure difference (<2 bar)	Porous
Ultrafiltration	1–100 nm MWCO 10^3–10^6 Da (macromolecules)	Pressure difference (1–10 bar)	Microporous
Nanofiltration	0.5–5 nm (molecules)	Pressure difference (10–100 bar)	Microporous
Reverse osmosis	<2 nm MWCO 10^3 Da (molecules)	Pressure difference (10–100 bar)	Nonporous
Dialysis	1–3 μm (molecules)	Concentration difference	Nonporous or microporous
Electrodialysis	<1 nm (molecules)	Electrical potential difference	Nonporous or microporous

separation medium. The technology involves separating components from fluid streams by means of forcing the stream to flow under pressure over the surface of the membrane. Membrane properties (pore size, membrane material, and membrane configuration) and operating conditions (pressure, temperature, and feed velocity) play important roles for successful separation (Table 40.2). The introduction of cross-flow membrane system made large-scale continuous separation possible. The technology enables processors to concentrate, fractionate, and purify products simultaneously. The operation is usually conducted at ambient temperature, although they also have the capability to be run at higher or lower temperatures.

Microfiltration

The microfiltration membrane separation process is the most common pressure-driven membrane separation process and is defined as the separation of a retained particle size between 0.2 and 10 μm by the membrane. Two principal types of membrane filters are used: depth filters and screen filters, running at very low pressures owing to the open structure of the membranes. On the basis of the multi-hole membrane, microfiltration can effectively remove suspended particles, bacteria, colloids, and solid proteins. The common membrane modules include spiral-wound membranes, plate and frame membranes, tubular membranes, and hollow fiber membranes.

Ultrafiltration

Ultrafiltration is a pressure-driven filtration separation occurring on a molecular scale and can separate high-molecular weight solutes from a solvent, typically retaining macromolecules with a molecular weight cutoff of more than 1000 Da. The pore size of an ultrafiltration membrane normally is from 0.001 to 0.02 μm in order to provide purification, separation, and concentration. The membranes used in this process have a much smaller pore size for solvent passage than do microfiltration membranes, with a relatively small pore area per membrane surface area.

By using an ultrafiltration membrane with a different pore size, we can separate the contents in the solution with different molecular weight and shape, such as the purification and concentration of enzymes, proteins, cells, pathogenic organisms, and polysaccharides, and the clarification and decolorizations for antibiotic fermentation (Li et al. 2004, Feins and Sirkar 2005, Krstić et al. 2007, Susanto et al. 2008, Cuellar et al. 2009). Ultrafiltration mainly has the following advantages: steady high permeated flux, resists free chlorine, wide range for pH and temperature, easy operation, low energy and operation cost, less pollution discharge, and compact equipment.

Ultrafiltration is used to separate solvents from solutions of very large molecules, as well as the liquid from a suspension of colloidal solids. Ultrafiltration membranes are commercially fabricated in sheet, capillary, and tubular forms. The liquid to be filtered is forced into the assemblage and dilute permeate passes perpendicularly through the membrane while concentrate passes out the end of the media. This may prove useful for the recovery and recycle of suspended solids and macromolecules. A key factor determining the performance of an ultrafiltration membrane is concentration polarization due to macromolecules being retained on the membrane surface.

Nanofiltration

A nanofiltration membrane is between a reverse osmosis membrane and an ultrafiltration membrane in pore size, which can remove NaCl under 90% concentration. A reverse osmosis membrane has a high removal rate for nearly all solutes. But a nanofiltration membrane only has a high removal rate for the special contents. Nanofiltration membranes mainly remove the particle whose diameter is near 1 nm, with a molecular weight cutoff of 100~1000 Da. In the drinking water area, nanofiltration is used to deal with quite large inorganic ions such as Ca^{2+}, Mg^{2+}, peculiar smells, pigments, pesticides, synthesized surfactants, dissoluble organics, and vaporized rudimental materials (Bates 2000, Braeken et al. 2004, Baisali et al. 2007). The character of nanofiltration is that it holds the charge itself, so under low pressure, it can also have a high desalination rate. The greatest field for the nanofiltration is to soften and desalt brine water. Nanofiltration has some advantages, including good chemical stability, long life, and high rejection selectivity.

Reverse Osmosis

Reverse osmosis is a liquid/liquid separation process that uses a dense semipermeable membrane, highly permeable to water. A pressurized feed solution is passed over one surface of the membrane, as long as the applied pressure is greater than the osmotic pressure of the feed solution. Reverse osmosis membranes have been widely applied in water treatment, such as desalting, pollution control, pure water treatment, and wastewater treatment. Reverse osmosis operates by diffusion from a solution. Under the high differential pressure across the membrane, the solvent from the solution actually dissolves in the material of the membrane, diffuses across it, and transfers out into the clean solvent on the other side. It is not a perfect separation, because the dissolved species from the feed solution have definite abilities to diffuse through it as well, but the diffusion coefficient for the solvent is so much higher than that for the solute that the separation is virtually complete. Reverse osmosis is often used to remove dissolved organics and metals. Thus, reverse osmosis, as a device solely required to separate sodium chloride from water in the early days, is now being used to reject a wider range of solutes.

Electrodialysis

Electrodialysis, an electrically driven membrane process, is commonly used to desalt solutions. Ions in aqueous solution can be separated using a direct current electrical driving force on an ion-selective membrane. Electrodialysis usually uses many thin compartments of solution separated by membranes that permit passage of either positive ions (cations) or negative ions (anions) and block passage of the oppositely charged ion. Cation-exchange membranes are alternatively stacked with anion-exchange membranes placed between two electrodes. The solution to be treated is circulated through the compartments and a direct current power source is applied. All cations gravitate toward the cathode (negatively charged terminal) and transfer through one membrane, while anions move in the opposite direction, thereby concentrating in alternating compartments. Electrodialysis is commonly used to recover spent acid and metal salts from plating rinse. It obviously is not effective for nonpolar solutions.

Membrane-Based Separation in Industrial Applications

Membrane-based processes offer many advantages in terms of lower energy consumption, higher selectivity, faster separation rate, and lower recovery costs (Cassano et al. 2008). There has been increasing interest in the use of membrane-based technologies in the food industry. Various designs of membrane-based separation equipment are used in the food industry (Figure 40.6). In the dairy industry, ultrafiltration is widely used to fractionate cheese whey and pre-concentrate milk for cheese making (Pouliot 2008). In sequential membrane processes, ultrafiltration may also be used as a membrane reactor to produce protein hydrolysate from protein isolate (Feins and Sirkar 2005, Charoenvuttitham et al. 2006, Sarkar et al. 2007, Annathur et al. 2010) and as a membrane fermenter to produce fuels and chemicals from reverse osmosis concentrates (Greenlee et al. 2009, Qu et al. 2009, Saxena et al. 2009). In sugar and corn syrup processing, microfiltration and ultrafiltration are used for clarification to remove impurities (Singh and Cheryan 1994, Jaeger de Carvalho et al. 2008). In juice, wine, and beer processes, microfiltration and ultrafiltration not only remove insoluble solids to produce sparkling clear product, but also remove spoilage microorganisms, eliminating the need for pasteurization (Massot et al. 2008, Vaillant et al. 2008, Cassano et al. 2010). Ultrafiltration is also used to concentrate natural colorants and flavors (Nawaz et al. 2005, Nawaz et al. 2006, Shi et al. 2007d, Galaverna et al. 2008, Ren et al. 2008a, 2008b, 2008c).

Figure 40.6. The pilot scale membrane separation system.

The main operating problem of all membrane separation processes is that the membrane material eventually plugs, called fouling, causing the resistance to flow to increase. The plugging process is accentuated by the concentration polarization that occurs in the relatively quiescent fluid zone close to the membrane surface, as the species separated from previously processed fluids build up in this zone and interfere with fresh material trying to get to the surface.

Membrane-based separation is still evolving and finding more and more applications in a broad range of fields, and the development of new membrane materials will strongly influence separation processes in the future. Because of the great advantages of membrane separation over conventional separation practices, the availability of the choice of membrane size (microfiltration, ultrafiltration, nanofiltration, reverse osmosis, pre-evaporation membrane, and distillation membrane), materials (polymeric and ceramic, hydrophilic and hydrophobic, symmetric and asymmetric), configurations (spiral wound, hollow fiber, and plate and frame), operation modes (dead-end and cross-flow, batch, semi-batch, and continuous), and membrane technology offers more selective, flexible, and efficient separations over a wide range of compounds. This technology will continue to gain recognition and acceptance in the food industry.

Pressurized Low-Polarity Water Extraction

Pressurized low-polarity water extraction, also known as subcritical water extraction (or hot water extraction, pressurized hot water extraction, superheated water extraction, or high-temperature water extraction), that is, extraction using hot water under pressure, has recently become a popular green processing technology and emerges as a promising extraction and fractionation technique for replacing the traditional extraction methods. The pressurized low-polarity water extraction is also used in sample preparation to extract organic contaminants from foodstuff for food safety analysis and solids/sediments for environmental monitoring purpose.

The pressurized low-polarity water extraction process is an environmentally friendly technique that can provide higher extraction yields from solid plant materials (Luque de Castro and Jiménez-Carmona 1998). Pressurized low-polarity water extraction is based on the use of water as an extractant in a dynamic mode, at temperatures between 100°C and 374°C (critical point of water, 221 bar and 374°C) and under pressure high enough to maintain the liquid state. The critical temperature and pressure of water are shown as a phase diagram in Figure 40.5 (T_c = 374°C, P_c = 221 bar or 22 MPa). The pressurized low-polarity water extraction process can maintain the water in the liquid form up to a temperature of 374°C and a pressure of 22.1 MPa (221 bar) (Haar et al. 1984, Hawthorne et al. 2000). A pressure of 5 MPa would be high enough to prevent the water from vaporizing at temperatures from 100°C to 250°C. Once pressure is high enough to keep water in a liquid state, additional pressure is not necessary as it has limited influence on the solvent characteristics of water. Increasing the water temperature from 25°C to 250°C causes similar changes in dielectric constant, surface tension, and viscosity (Kronholm et al. 2007, Brunner 2009). Pressurized low-polarity water extraction can easily solublize organic compounds such as phytochemicals, which are normally insoluble in ambient water.

Pressurized low-polarity water extraction has the ability to selectively extract different classes of compounds, depending on the temperature used. The selectivity of subcritical water extraction allows for manipulation of the composition of the extracts by changing the operating parameters, with the more polar ones extracted at lower temperatures and the less polar compounds extracted at higher temperatures (Basile et al. 1998, Ammann et al. 1999, Clifford et al. 1999, Miki et al. 1999, Kubatova et al. 2001, Soto Ayala and Luque de Castro 2001).

Process System

As shown in Figure 40.7, the instrumentation consists of a water reservoir coupled to a high-pressure pump to introduce the pressurized low-polarity water into the system, an oven, where the extraction cell is placed and extraction takes place, and a restrictor or valve to maintain the pressure. Extracts are collected in a vial placed at the end of the extraction system. In addition, the system can be equipped with a coolant device for rapid cooling of the resultant extract. As the unique properties of pressurized low-polarity water, the pressurized low-polarity water extraction has a disproportionately high boiling point for its mass, a high dielectric constant, and high polarity. As the temperature rises, there is a marked and systematic decrease in permittivity, an increase in the diffusion rate, and a decrease in the viscosity and surface tension. In consequence, more polar target materials with high solubilities in water at ambient conditions are extracted most efficiently at lower temperatures, whereas moderately polar and nonpolar targets require a less-polar medium, induced by elevated temperature.

Water changes dramatically when its temperature rises, because of the breakdown in its hydrogen-bonded structure with temperature. The high degree of association in the liquid causes its relative permittivity (more commonly called its dielectric constant) to be very high at ca. 80 under ambient conditions. But as the temperature rises, the hydrogen bonding breaks down and the dielectric constant falls, as shown in Figure 40.5. The most outstanding feature of this leaching agent is the easy manipulation of its dielectric constant (ε). In fact, this parameter can be changed within a wide range just by changing the temperature under moderate pressure. Thus, at ambient temperature and pressure, water has a dielectric constant of ca. 80, making it an extremely polar solvent. This parameter is drastically lowered by raising the temperature under moderate pressure. For example, subcritical water at 250°C and a pressure over 40 bar has $\varepsilon = 37$, which is similar to that of ethanol and allows for the leaching of low-polarity compounds. By 250°C, its dielectric constant has fallen so that it is equal to that for methanol (i.e., 33) at ambient temperature. Thus, between 100°C and 200°C, superheated water is behaving like a water–methanol mixture. Partly because of its fall in polarity with temperature, superheated water can dissolve organic compounds to some extent, especially if they are slightly polar or polarizable like aromatic compounds. Therefore, water can be used as extraction solvent to extract the polar, the moderately polar, and the nonpolar compounds by adjusting the extraction temperature from range of 50°C to 275°C.

The solubility of an organic compound is often many orders of magnitude higher than its solubility in water at ambient temperature for two reasons. One is the polarity change and the other is that of a compound with low solubility at ambient temperature. Pressurized low-polarity water will have a high positive enthalpy of solution and thus a large increase in solubility with temperature. Because of the greater solubility of some

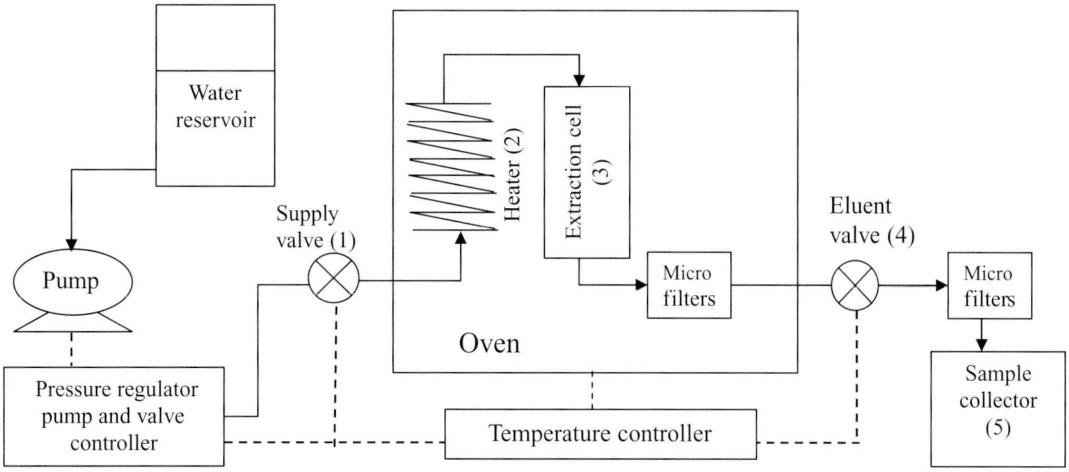

Figure 40.7. Diagram of pressurized low-polarity water extractor. The electrical connections are marked by dashed lines, while the path of subcritical water is shown by solid line with arrow. The high-pressured water passes through a supply vale (1) into a heating coil (2) and into an extraction cell (3). The microfilters are placed before and after an eluent valve (4). The extract is collected in a sample collector (5).

organic compounds in superheated water, this medium can be considered for the extraction and other processes, to replace conventional organic solvents. But some additional reactions of the compounds being processed may also occur, by hydrolysis, oxidation, etc.

Industrial Applications

Using pressurized low-polarity water provides a number of advantages over traditional extraction techniques (i.e., hydrodistillation, organic solvents, solid–liquid extraction). These are mainly shorter extraction times, higher quality of the extracts (mostly for essential oils), lower costs of the extracting agent, and an environmentally compatible technique. Since water is perhaps the most environmentally friendly solvent available in high purity and at low cost, it has been exploited for the extraction of avoparcin in animal tissue (Curren and King 2001), fungicides in agricultural commodities (Pawlowski and Poole 1998), fragrances from cloves (Rovio et al. 1999), antioxidative components from sage (Ollanketo et al. 2002), anthocyanins and total phenolics from dried red grape skin (Ju and Howard 2003), saponins from cow cockle seed (Güçlü-Üstündag et al. 2007), and other bioactive components from plant materials (Ong and Len 2003). Some additional successful applications of this technique are for the extraction of essential oils from various plant materials (Khajenoori et al. 2009, Mortazavi et al. 2010), extraction of sweet components from *Siraitia grosvernorii*, extraction of lactones from kava roots, extraction of antioxidant compounds from microalgae *S. platensis* (Ibáñez et al. 1999, Ibáñez et al. 2003, Herrero et al. 2004), extraction of *Ginkgo biloba*, and of biophenols from olive leaves (Japón-Lujána and Luque de Castro 2006).

The quality of the oil obtained is therefore better than that from steam distillation, as it contains more of the oxygenated compounds and lower terpene content. The yield is also slightly higher than from steam distillation, in spite of the fact that all the terpenes are not extracted. This may be because, at the higher temperatures and under pressure, the plant material is more effectively penetrated. However, about twice the amount of water is required than for steam distillation. Energy costs are much less than for steam distillation. The energy required to heat a given mass of water from 30°C to 150°C under pressure is one-fifth of that needed to boil water at atmospheric pressure from 30°C. Furthermore, it is possible to recycle most (three-quarters) of the heat, whereas it is difficult to recycle heat in steam distillation. Thus, in spite of the fact that twice as much water is needed, only one-tenth of the energy of a steam distillation is required. Pressurized low-polarity water extraction has been suggested as a method to extract valuable health-promoting compounds from plant materials.

Molecular Distillation

Molecular distillation is a unit operation that is a peculiar case of evaporation, which happens under extremely low pressures and low temperatures and is used for the separation of constituents from mixtures by partial evaporation. It is based on the fact that the vapor is relatively richer in the component with the highest vapor pressure, that is, the more volatile component. Distillation is a process of heating a substance until the most volatile constituents change into the vapor phase, and then cooling the vapors to recover the constituents in liquid form by condensation. The main purpose of distillation is to separate a mixture into individual components by taking advantage of their different level of volatilities. Distillation is one of the main methods of extracting essential oils from plants. It can be carried out either as simple distillation or fractional distillation (rectification). In simple distillation, the vapors are recovered by condensation. In rectification, successive vaporization and condensation are carried out simultaneously and a part of the condensed liquid, called the reflux, flows down the column countercurrent to the flow of vapors for isolating components from a mixture based on differences in boiling points. The percentage of each constituent in the vapor phase usually depends on its vapor pressure at a certain temperature. The principle of vacuum distillation may be applied to substances, such as oils, that would be damaged by overheating by the conventional method (Liu et al. 2008). Several new methods have been developed for the separation and recovery of minor components from vegetable oils such as palm oil (Rodríguez et al. 2007, Shi et al. 2007c).

Distillation and its companion processes, azeotropic and extractive distillations, are by far the most widely used separation processes for mixtures that can be vaporized. Vapors are generated from liquids or solids by heating and are then condensed into liquid products. However, many mixtures exhibit special states, known as azeotropes, at which the composition, temperature, and pressure of the liquid phase become equal to those of the vapor phase. Thus, further separation by conventional distillation is no longer possible. By adding a carefully selected other component as an entrainer to the mixture, it is often possible to "break" the azeotrope and thereby achieve the desired separation. In azeotropic distillation, a compound is added to form an azeotrope with at least one of the components of the mixture. That component can then be more readily separated from the mixture because of the increased difference between the volatilities of the components. Extractive distillation combines continuous fractional distillation with absorption. A relatively high-boiling solvent is used to selectively scrub one or more of the components from a mixture of components with similar vapor pressures. Distillation processes are also widely used for the separation of organic chemicals, usually at cryogenic temperatures.

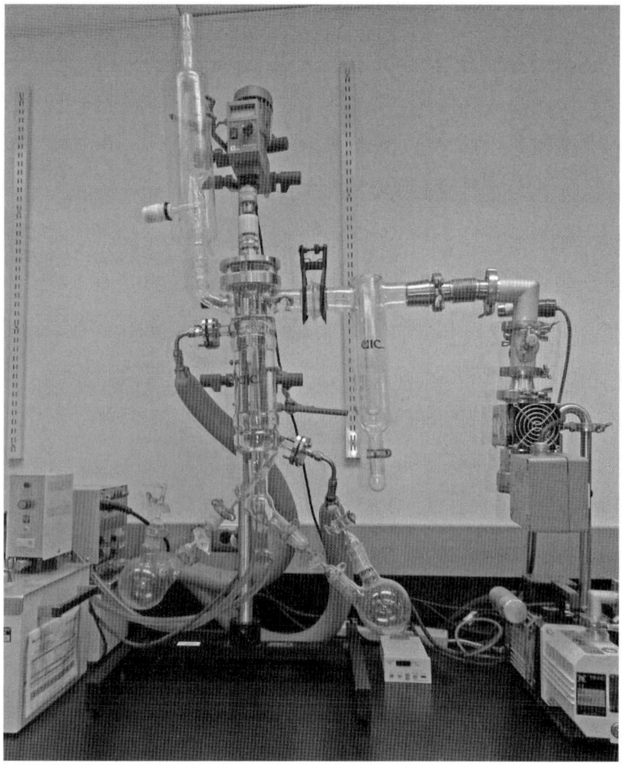

Figure 40.8. Molecular distillation system.

Separation Processes in Distillation

When vacuum is applied, there are three major reduced pressure ranges that can be utilized for distillation: (a) distillation at moderate vacuum or equilibrium distillation, (b) unobstructed path distillation, and (c) molecular distillation. Distillation at moderate vacuum is characterized by the use of conventional distillation equipment as shown in Figure 40.8. Its lowest pressure limit is of the order of 1 Torr, that is, 1 mm Hg. Unobstructed path distillation is defined as distillation in which the path between the evaporator and the condenser is not blocked, in other words there is a free transfer of molecules (Eckles and Benz 1992). When the transfer distance is comparable with the mean free path of the vapor molecules, the distillation is known as molecular distillation.

Mean free path is defined as the average distance a molecule will travel in the vapor phase without colliding with another vapor molecule (Eckles et al. 1991). This implies that, in molecular distillation, the vapor molecules can reach the condenser without intermolecular collisions. A dynamic equilibrium can not be established between the vapor and the liquid. An individual molecule that has evaporated will be able to travel any distance without a collision.

Molecular distillation occurs at low temperatures and, therefore, reduces the problem of thermal decomposition. High vacuum also eliminates oxidation that might otherwise occur in the presence of air. Unobstructed path and molecular distillations are often classified together as short-path or high-vacuum distillation. The difference is in the dimensions and operating conditions. Unobstructed path distillation is carried out at pressures as low as 10^{-2} Torr, while in molecular distillation pressures of 10^{-3} Torr, that is, a mTorr, are used. Another useful distinction between the methods of vacuum distillation can be made with reference to the way in which the vapor phase is formed: (a) Ebullition, (b) evaporative distillation, and (c) molecular distillation.

Ebullition is accompanied by the formation of bubbles when the saturated vapor pressure exceeds the pressure of the surrounding gas. The only limit to the rate of evaporation in this case is the rate at which heat can be transferred to the liquid. Evaporative distillation occurs when the pressure of the surrounding gas is higher than the vapor pressure of the liquid at a given temperature, so that bubbles are not formed. The rate of evaporation is controlled by the temperature of the liquid and the conditions above the liquid surface. In molecular distillation, the rate of evaporation is controlled by the rate at which the molecules escape from the free surface of the liquid and condense on the condenser. The condenser should be in the immediate vicinity of the evaporating surface. When an appreciable number of collisions can occur in the vapor space, some of the molecules will return to the liquid. This leads to a decrease in the number of molecules that reach the condensation surface.

High-vacuum distillation may be used for certain classes of chemical compounds that decompose, polymerize, react, or are destroyed by conventional distillation methods. Low cost per pound and high throughput may be obtained on certain groups of compounds such as vitamins, epoxy resins, highly concentrated pure fatty acids, plasticizers, fatty acid nitrogen compounds, and a host of other heat-sensitive materials, which may require only deodorizing and decolorizing (Spychaj 1986, Batistella et al. 2002a,b, Cermak et al. 2007, Shao et al. 2007, Compton et al. 2008). High-vacuum distillation is a safe process to separate mixtures of organic or silicon compounds, most of which can not withstand prolonged heating without excessive structural change or decomposition. With short residence times and lower distilling temperatures, thermal hazards to the organic materials are greatly reduced. Purity of the distillate also depends on the film thickness. Controlling positive pressure and supply to the heated evaporator surface will usually provide a uniform film throughout the distillation. The absence of air molecules in the high-vacuum distillation column permits most of the distilling molecules to reach the condenser with relatively few molecules returning to the liquid film surface in the evaporator. Experimental results show a relationship between the molecular weight and distillation temperature for a broad range of different materials.

There are several basic design variations with short-path evaporators. These are the short-path falling film evaporator, the centrifugal molecular still, and the short-path, wiped-film evaporator. They operate at the lowest pressure of any system and are capable of high throughput per unit size, due to their continuous nature. They also offer the shortest thermal exposure to any process (Eckles and Benz 1992). Short-path, falling film evaporators can handle materials with viscosities up to 5000 cP, and the residence time, temperature, and pressure can be controlled. This is the simplest of the short-path stills, but more current designs use either centrifugal force or roller-wipers to spread the feed material on the evaporator surface. In the centrifugal molecular still, feed material is fed into the center of a heated spinning rotor. The material is evenly spread towards the edge and condensed in front of the rotor.

Molecular Distillation Applications

Molecular distillation, with its important characteristics of low pressure and low temperature, gives a high potential for this process in the separation, purification, and concentration of natural products, which usually consist of complex and thermally sensitive molecules. Especially under high vacuum for short operating times, while using no solvents, it avoids any toxicity problems. The effects of feed flow rate and distillation temperature on the extraction of minor components are related to the yield, purity, and rate of evaporation in terms of concentrations, distribution coefficients, and relative volatilities. Molecular distillation is a valuable processing method that is often used to separate or purify high-boiling materials that decompose, oxidize, or polymerize at elevated temperatures. This process can be considered for industrial uses if both distillate and residue become high value-added products. Some molecular distillation processes are used in the production of biologically active substances such as vitamins, sterols, and antioxidants from natural oils and fats such as the extraction of vitamin A from fish liver and whale oil and carotenoid recovery from esterified palm oil (Batistella and Wolf Maciel 1998, Mutalib et al. 2003, Martins et al. 2006, Bettini 2007, Fregolente et al. 2007, Shi et al. 2007c), separation of mono and diglycerides in partially saponified fats (Holló and Kurucz 1968), and the refining of crude animal and vegetable oils (Martins et al. 2006, Behrenbruch and Dedigama 2007, Posada et al. 2007). For instance, a Malaysian company produces *Carotino*®, an edible red palm oil, using a process that involves a pretreatment of crude palm oil (i.e., degumming with phosphoric acid and treatment with bleaching earth) followed by deacidification and deodorization using molecular distillation (Ooi et al. 1996; Fig. 40.9). Recovery of special components used in the nutritional, pharmaceutical, and cosmetic areas from natural products by molecular distillation

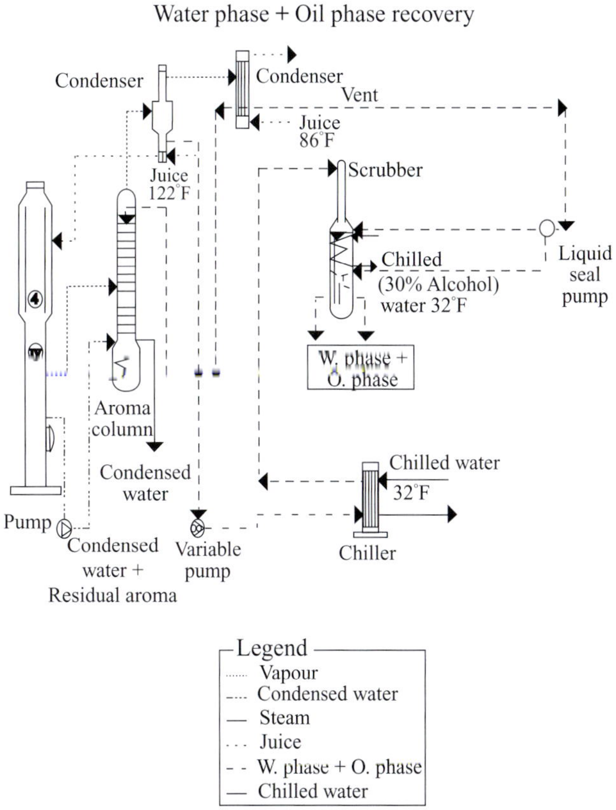

Figure 40.9. Recovery system of water phase and oil phase in the orange juice industry (Bettini 2007).

processes have gained wide applications for products such as those derived from refined vegetable oils, e.g., deodorizer distillate of vegetable oils, palm oil for obtaining tocotrienols and tocopherols, rice oil for oryzanol recovery, monoglyceride concentration, carotenoid recovery from palm oil, heavy petroleum characterization, and herbicides (Barnicki et al. 1996, Sakiyama et al. 2001, Batistella et al. 2002a,b, Fregolente et al. 2005, Ito et al. 2006, Shi et al. 2007c). Recently, a new process of molecular distillation was developed for recovery of tocotrienols and tocopherols from rapeseed by a combination of acid-catalyzed methyl esterification and crystallization followed by fractional distillation of derived products and for the recovery of orange peel oil and essence products rich in aldehydes, esters, and other special volatile compounds (Lutisan et al. 2002, Jiang et al. 2006). An industrial production of a typical industrial fractional vacuum distillation plant is shown in Figure 40.10. In order to increase the yield and quality of the final products, the distiller is operated continuously.

Figure 40.10. The vacuum fractional molecular distillation column.

SUMMARY

One of the most important trends in the food industry today is the demand for "natural" foods and ingredients that are free from toxic chemical additives. The growing interest in natural foods has raised the demand for natural health-promoting products of nonsynthetic origin. High-value functional substances can be obtained from biological materials by various purification and separation methods from plant materials or by-products. Among the separation processes being used or developed are physical and chemical processes such as centrifugation, filtration, membrane separation, precipitation, chromatography, solvent extraction, supercritical fluid extraction, crystallization, evaporation, and distillation.

The separation problems presented by the production of soluble material have a number of aspects that influence the nature of the extraction technique chosen. An important factor is the stability of the extracted product. In the case of unstable products, a rapid separation process is necessary in order to avoid significant loss of products. Extraction with organic solvents by traditional methods is a well-established method for the selective separation of specific constituents from raw materials. The conventional procedures for preparing concentrated extracts are steam distillation or solvent extraction with an organic solvent. Extractants with a low boiling point, such as ethyl acetate, methanol, dichloromethane, etc., are traditionally used for isolating valuable constituents from such products as hops, spices, and oil seeds and also for removing or reducing the level of less-desirable accompanying substances (such as nicotine, caffeine).

But a very important step in the process is the complete removal of the solvents from extracts and its residue as they are in part toxicologically objectionable. In addition, organic solvents have a low selectivity. Extraction must be followed by removal of the solvent from the extract, for example by distillation. The challenges in the separation processes are to both meet food regulations and to conduct the separation effectively and economically. Many compounds in the concentrated extract will be thermally labile and prone to degradation if heated too vigorously. The removal of the solvent can pose serious problems since the residual level of the solvent must be minimized, but not at the expense of the degradation of the final extract. Public health, environment, and safety issues are the major concerns in the use of organic solvents in food processing. The possibility of solvent residues remaining in the final product has been a growing concern to consumers, thus warranting stringent environmental regulations. The extent of residual solvent levels and also the composition of the extraction solvents used in the food industry are well regulated within the existing laws and regulations. The solvents used for this purpose must meet the requirements of foodstuffs legislation, which can vary from country to country. In recent years, national and international bodies (such as the Food and Drug Administration, USA; the EC Codex Committee; and the FAO/WHO, Geneva) have increasingly emphasized the importance of complying with such criteria.

The demand for ultra-pure and high value-added products is redirecting the focus of the food and pharmaceutical industries into seeking the development of new and clean technologies for their products. To solve these problems, some innovative emerging "green" processes such as supercritical CO_2 fluid extraction, membrane-based separations, molecular distillation, and pressurized low-polarity water extraction have provided excellent alternatives to the conventional organic solvent extraction methods. Extracts from natural sources are key elements in the manufacturing of health-promoting functional foods and ingredients. Especially such "green" separation processes and technologies are widely employed in the development of bioactive components for use as supplements for health-promoting foods.

Development of new and efficient separation processes will be based on more effectively exploiting differences in the actual physicochemical properties of food products. Liquid membrane extraction is one of the relatively new separation technologies that has significant potential for the selective separation and concentration of low-molecular weight chemicals used in food-processing industry.

REFERENCES

Ammann A et al. 1999. Superheated water extraction, steam distillation and SFE of peppemint oil. *Fresenius' J Anal Chem* 364: 650–653.

Annathur GV et al. 2010. Ultrafiltration of a highly self-associating protein. *J Membr Sci* 353: 41–50.

Baisali SN et al. 2007. Treatment of pesticide contaminated surface water for production of potable water by a coagulation-adsorption-nanofiltration approach. *Desalination* 212: 129–140.

Barnicki SD et al. 1996. Process for the production of tocopherol concentrates. US Patent No.5512691: 1–38.

Basile A et al. 1998. Extraction of rosemary by superheated water. *J Agric Food Chem* 46: 5204–5209.

Bates AJ. 2000. Water as consumed and its impact on the consumer—do we understand the variables? *Food Chem Toxicol* 38: S29–S36.

Batistella CB, Wolf Maciel MR. 1998. Recovery of carotenoids from palm oil by molecular distillation. *Comput Chem Eng* 22: S53–S60.

Batistella CB et al. 2002a. Molecular distillation process for recovering biodiesel and carotenoids from palm oil. *Appl Biochem Biotechnol* 98–100: 1149–1159.

Batistella CB et al. 2002b. Molecular distillation: rigorous modeling and simulation for recovering vitamin E from vegetal oils. *Appl Biochem Biotechnol* 98–100: 1187–1206.

Behrenbruch P, Dedigama T. 2007. Classification and characterization of crude oils based on distillation properties. *J Pet Sci Eng* 57: 166–180.

Bettini MFM. 2007. Purification of orange peel oil and oil phase by vacuum distillation. In: J Shi (ed.) *Functional Food Ingredients and Nutraceuticals: Processing Technology*. CRC Press, Boco Raton, FL, pp. 157–172.

Bhanusali D, Bhattacharyya D. 2003. Advances in solvent resistant nanofiltration membranes: experimental observations and applications. *Ann NY Acad Sci* 984: 159.

Braeken L et al. 2004. Regeneration of brewery waste water using nanofiltration. *Water Res* 38: 3075–3082.

Brunner G. 2009. Near critical and supercritical water. Part I. Hydrolytic and hydrothermal processes. *J Supercrit Fluids* 47(3): 373–381.

Cassano A et al. 2008. Recovery of bioactive compounds in kiwi fruit juice by ultrafiltration. *Innov Food Sci Emerg Technol* 9: 556–562.

Cassano A et al. 2010. Physico-chemical parameters of cactus pear (*Opuntia ficus-indica*) juice clarified by microfiltration and ultrafiltration processes. *Desalination* 250: 1101–1104.

Cermak SC et al. 2007. Enrichment of decanoic acid in cuphea fatty acids by molecular distillation. *Ind Crops Products* 26: 93–99.

Chang CH et al. 2008. Relevance of phenolic diterpene constituents to antioxidant activity of supercritical CO_2 extract from the leaves of rosemary. *Nat Product Res* 22: 76–90.

Charoenvuttitham P et al. 2006. Chitin extraction from black tiger shrimp (*Penaeus Monodon*) waste using organic acids. *Sep Sci Technol* 41: 1135–1153.

Chen J et al. 2009. Effects of supercritical CO_2 fluid parameters on chemical composition and yield of carotenoids extracted from pumpkin. *LWT Food Sci Technol* 43: 39–44.

Cheryan M. 1998. *Ultrafiltration and Microfiltration Handbook*. Technomic Publishing Co., Lancaster, PA.

Clifford AA et al. 1999. A comparison of the extraction of clove buds with supercritical carbon dioxide and superheated water. *Fresenius' J Anal Chem* 364: 635–637.

Compton DL et al. 2008. Purification of 1,2-diacylglycerols from vegetable oils: comparison of molecular distillation and liquid CO_2 extraction. *Ind Crops Products* 28(2): 113–121.

Cuellar MC et al. 2009. Model-based evaluation of cell retention by crossflow ultrafiltration during fed-batch fermentations with *Escherichia coli*. *Biochem Eng J* 44: 280–288.

Curren MSS, King JW. 2001. Ethanol-modified subcritical water extraction combined with solid-phase microextraction for determining atrazine in beef kidney. *J Agric Food Chem* 49: 2175–2180.

Eckles A, Benz PH. 1992. The basics of vacuum processing. *Chem Eng* 99: 78–86.

Eckles A et al. 1991. When to use high-vacuum distillation. *Chem Eng* 98: 201–203.

Enevoldsen AD et al. 2007. Electro-ultrafiltration of amylase enzymes: process design and economy. *Chem Eng Sci* 62: 6716–6725.

Feins M, Sirkar KK. 2005. Novel internally staged ultrafiltration for protein purification. *J Membr Sci* 248(1–2): 137–148.

Fernandez Perez V et al. 2000. An approach to the static-dynamic subcritical water extraction of laurel essential oil: comparison with conventional techniques. *Analyst* 125: 481–485.

Fregolente LV et al. 2005. Response surface methodology applied to optimization of distilled monoglycerides production. *J Am Oil Chem Soc* 82(9): 673–678.

Fregolente LV et al. 2007. Effect of operating conditions on the concentration of monoglycerides using molecular distillation. *Chem Eng Res Des* 85: 1524–1528.

Galaverna G et al. 2008. A new integrated membrane process for the production of concentrated blood orange juice: effect of bioactive compounds and antioxidant activity. *Food Chem* 106: 1021–1030.

Glisic SB et al. 2007. Supercritical carbon dioxide extraction of carrot fruit essential oil: chemical composition and antimicrobial activity. *Food Chem* 105(1): 346–352.

González JC et al. 2002. Basis for a new procedure to eliminate diarrheic shellfish toxins from a contaminated matrix. *J Agric Food Chem* 50(2): 400–405.

Greenlee LF et al. 2009. Reverse osmosis desalination: water sources, technology, and today's challenges. *Water Res* 43(9): 2317–2348.

Güçlü-Üstündag O et al. 2007. Pressurized low polarity water extraction of saponins from cow cockle seed. *J Food Eng* 80: 619–630.

Haar L et al. 1984. *National Bureau of Standards/National Research Council Steam Tables*. Hemisphere Publishing, Bristol, PA.

Hawthorne SB et al. 2000. Comparisons of Soxhlet extraction, pressurized liquid extraction, supercritical fluid extraction and subcritical water extraction for environmental solids: recovery, selectivity and effects on sample matrix. *J Chromatogr A* 892: 421–433.

Herrero M et al. 2004. Pressurized liquid extracts from Spirulina platensis microalga: Determination of their antioxidant activity and preliminary analysis by micellar electrokinetic chromatography. *J Chromatogr A* 1047(2): 195–203.

Herrero M et al. 2006. Sub- and supercrtical fluid extraction of functional ingredients from different natural sources: plants, food by-products, algae and microalgae. A review. *Food Chem* 98: 136–148.

Holló J, Kurucz E. 1968. Molecular distillation. *Process Biochem* 3: 23–24.

Huang W et al. 2010. Purification and characterization of an antioxidant protein from ginkgo biloba seeds. *Food Res Int* 43(1): 86–94.

Ibáñez E et al. 1999. Supercrtical fluid extraction and fractionation of different pre-processed rosemary plants. *J Agric Food Chem* 47: 1400–1404.

Ibáñez E et al. 2003. Subcritical water extraction of antioxidant compounds from rosemary plants. *J Agric Food Chem* 51: 375–382.

Ito VM et al. 2006. Natural compounds obtained through centrifugal molecular distillation. *Appl Biochem Biotechnol* 129–132: 716–726.

Jaeger de Carvalho LM et al. 2008. A study of retention of sugars in the process of clarification of pineapple juice (*Ananas comosus* L. *Merril*) by micro- and ultrafiltration. *J Food Eng* 87: 447–454.

Japón-Lujána R, Luque de Castro MD. 2006. Superheated liquid extraction of oleuropein and related biophenols from olive leaves. *J Chromatogr A* 1136: 185–191.

Jiang ST et al. 2006. Molecular distillation for recovering tocopherol and fatty acid methyl esters from rapeseed oil deodorizer distillate. *Biosyst Eng* 93(4): 383–391.

Ju ZY, Howard LR. 2003. Effects of solvent and temperature on pressurized liquid extraction of anthocyanins and total phenolics from dried red grape skin. *J Agric Food Chem* 51: 5207–5213.

Kasamma L et al. 2008. Response surface analysis of lycopene yield from tomato using supercritical CO_2 fluid extraction. *Sep Purif Technol* 60: 278–284.

Kertész R et al. 2005. Membrane-based solvent extraction and stripping of phenylalanine in HF contactors. *J Membr Sci* 257: 37–41.

Khajenoori M et al. 2009. Proposed models for subcritical water extraction of essential oils. *Chinese J Chem Eng* 17: 359–365.

King JW et al. 1993. Analytical supercritical fluid chromatography and extraction. In: *Supplement and Cumulative Index*. John Wiley & Sons, New York, pp. 1–83.

Kronholm J et al. 2007. Analytical extractions with water at elevated temperatures and pressures. *Trends Anal Chem* 26(5): 396–412.

Krstić DM et al. 2007. The possibility for improvement of ceramic membrane ultrafiltration of an enzyme solution. *Biochem Eng J* 33: 10–15.

Kubatova A et al. 2001. Selective extraction of oxygenates from savoury and peppermint using subcritical water. *Flavour Frag J* 16: 64–73.

Kumar A. 2007. Membrane separation technology in processing bioactive components. In: John Shi (ed.) *Functional Food Ingredients and Nutraceuticals: Processing Technology*. CRC Press, Boca Raton, FL, pp. 193–210.

Li SZ et al. 2004. Application of ultrafiltration to improve the extraction of antibiotics. *Sep Purif Technol* 34: 115–123.

Liu D et al. 2008. Separation of tocotrienols from palm oil by molecular distillation process. *Food Rev Int* 24: 376–391.

Liu G et al. 2009a. Supercritical CO_2 extraction optimization of pomegranate (*Punica granatum* L.) seed oil using response surface methodology. *LWT Food Sci Technol* 42: 1491–1495.

Liu X et al. 2009b. Antimicrobial and antioxidant activity of emblica extracts obtained by supercritical carbon dioxide extraction and methanol extraction. *J Food Biochem* 33(3): 307–330.

Luque de Castro MD, Jiménez-Carmona MM. 1998. Potential of water for continuous automated sample-leaching. *Trends Anal Chem* 17: 441–447.

Lutisan J et al. 2002. Heat and mass transfer in the evaporating film of a molecular evaporator. *Chem Eng J* 85: 225–234.

Martinez ML et al. 2008. Pressing and supercritical carbon dioxide extraction of walnut oil. *J Food Eng* 88: 399–404.

Martins PF et al. 2006. Free fatty acid separation from vegetable oil deodorizer distillate using molecular distillation process. *Sep Purif Technol* 48(1): 78–84.

Massot A et al. 2008. Nanofiltration and reverse osmosis in winemaking. *Desalination* 231: 283–289.

Mendiola JA et al. 2007. Use of compressed fluids for sample preparation: food applications. *J Chromatogr A* 1152(1–2): 234–246.

Miki W et al. 1999. Process for producing essential oil via treatment with supercritical water and essential oil obtained by treatment with supercritical water. Int. Patent App. WO99/53002A1.

Mitra P et al. 2009. Pumpkin (*Cucurbita maxima*) seed oil extraction using supercritical carbon dioxide and physicochemical properties of the oil. *J Food Eng* 95: 208–213.

Miyawaki T et al. 2008. Development of supercritical carbon dioxide extraction with a solid phase trap for dioxins in soils and sediments. *Chemosphere* 70(4): 648–655.

Montañés F et al. 2008. Selective fractionation of carbohydrate complex mixtures by supercritical extraction with CO_2 and different co-solvents. Selective fractionation of carbohydrate complex mixtures by supercritical extraction with CO_2 and different co-solvents. *J Supercrit Fluids* 45(2): 189–194.

Montañés F et al. 2009. Supercritical technology as an alternative to fractionate prebiotic galactooligosaccharides. *Sep Purif Technol* 66(2): 383–389.

Mortazavi SV et al. 2010. Extraction of essential oils from *Bunium persicum Boiss*. using superheated water. *Food Bioprod Process* 88(2–3): 222–226.

Mukhopadhyay M. 2000. *Natural Extracts Using Supercritical Carbon Dioxide*. CRC Press, Boca Raton, FL.

Mulder M. 1996. *Basic Principles of Membrane Technology*, 1st edn. Kluwer Academic Publishers, Dordrecht, The Netherlands.

Mutalib MSA et al. 2003. Palm-tocotrienol rich fraction (TRF) is a more effective inhibitor of LDL oxidation and endothelial cell lipid peroxidation than α-tocopherol in vitro. *Food Res Int* 36: 405–413.

Nagendra PK et al. 2010. Enhanced antioxidant and antityrosinase activities of longan fruit pericarp by ultra-high-pressure-assisted extraction processing. *J Pharm Biomed Anal* 51(2): 471–477.

Nawaz H et al. 2005. Extraction of polyphenolics from plant material for functional foods—engineering and technology. *Food Rev Int* 21: 139–166.

Nawaz H et al. 2006. Extraction of polyphenolics from grape seeds and concentration by ultrafiltration. *Sep Sci Technol* 48: 176–181.

Ollanketo M et al. 2002. Extraction of sage (*Salvia officinalis* L.) by pressurized hot water and conventional methods: antioxidant activity of the extracts. *Eur Food Res Technol* 215: 158–163.

Ong ES, Len SM. 2003. Pressurized hot water extraction of berberine, baicalein and glycyrrhizin in medicinal plants. *Anal Chim Acta* 482: 81–89.

Ooi CK et al. 1996. Refining of red palm oil. *Elaeis* 8: 20–28.

Pawlowski TM, Poole CF. 1998. Extraction of chiabendazole and carbendazim from foods using pressurized hot (subcritical) water for extraction: a feasibility study. *J Agric Food Chem* 46: 3124–3132.

Perrut M. 2000. Supercritical fluid applications: industrial developments and economic issues. *Ind Eng Chem Res* 39(12): 4531–4535.

Posada LR et al. 2007. Extraction of tocotrienols from palm fatty acid distillates using molecular distillation. *Sep Purif Technol* 57: 220–229.

Pouliot Y. 2008. Membrane processes in dairy technology—from a simple idea to worldwide panacea. *Int Dairy J* 18(7): 735–740.

Qu D et al. 2009. Integration of accelerated precipitation softening with membrane distillation for high-recovery desalination of primary reverse osmosis concentrate. *Sep Purif Technol* 67: 21–25.

Raventós M et al. 2002. Application and possibilities of supercritical CO_2 extraction in food processing industry: an overview. *Food Sci Technol Int* 8: 269–284.

Ren J et al. 2008a. Comparison of the phytohaemagglutinin from red kidney bean (*Phaseolus vulgaris*) purified by different affinity chromatography. *Food Chem* 108: 394–401.

Ren J et al. 2008b. phytohemagglutinin isolectins extracted and purified from red kidney bean and its cytotoxicity on human H9 lymphoma cell line. *Sep Purif Technol* 63: 122–128.

Ren J et al. 2008c. Purification and identification of antioxidant peptides from grass carp muscle hydrolysates by consecutive chromatography and electrospray ionization-mass spectrometry. *Food Chem* 108: 727–736.

Reverchon E et al. 1992. Extraction of essential oils using supercritical CO_2: effects of some process and preprocess parameters. *Italian J Food Sci* 3: 187–194.

Rizvi SSH, Bhaskar AR. 1995. Supercritical fluid processing of milk fat: fractionation, scale-up, and economics. *Food Technol* 49(2): 90–97.

Rodríguez PL et al. 2007. Extraction of tocotrienols from palm fatty acid distillates using molecular distillation. *Sep Purif Technol* 57: 220–229.

Rovio S et al. 1999. Extraction of clove using pressurized hot water. *Flavour Frag J* 14: 399–404.

Sakiyama T et al. 2001. Analysis of monoglyceride synthetic reaction in a solvent-free two-phase system catalyzed by a monoacylglycerol lipase from *Pseudomonas* sp. *J Biosci Bioeng* 91: 88–90.

Sanchez-Vicente Y et al. 2009. Supercritical fluid extraction of peach (*Prunus persica*) seed oil using carbon dioxide and ethanol. *J Supercrit Fluids* 49: 167–173.

Sarkar P et al. 2007. Effect of different operating parameters on the recovery of proteins from casein whey using a rotating disc membrane ultrafiltration cell. *Desalination* 249(1): 5–11.

Saxena A et al. 2009. Membrane-based techniques for the separation and purification of proteins: an overview. *Adv Colloid Interface Sci* 145: 1–22.

Shao P et al. 2007. Optimization of molecular distillation for recovery of tocopherol from rapeseed oil deodorizer distillate using response surface and artificial neural network models. *Food Bioprod Process* 85: 85–92.

Shi J, Zhou X. 2007. Solubility property of bioactive components on recovery yield in separation process by supercritical fluid. In: John Shi (ed.) *Functional Food Ingredients and Nutraceuticals: Processing Technology*. CRC Press, Boca Raton, FL, pp. 45–74.

Shi J et al. 2007a. Correlation of mass transfer coefficient in the extraction of plant oil in a fixed bed for supercritical CO_2. *J Food Eng* 78: 33–40.

Shi J et al. 2007b. Solubility of carotenoids in supercritical CO_2. *Food Rev Int* 23: 341–371.

Shi J et al. 2007c. Molecular distillation of palm oil distillates: evaporation rates, relative volatility and distribution coefficients of tocotrienols and other minor components. *Sep Sci Technol* 42(14): 3029–3048.

Shi J et al. 2007d. Isolation and partial characterization of lectins from dry raw and canned kidney beans (*Phaseolus vulgaris*). *Process Biochem* 42: 1436–1442.

Shi J et al. 2007e. Correlation of mass transfer coefficient in separation process with supercritical CO_2. *Drying Technol Int* 25: 335–339.

Shi J et al. 2007f. Supercritical fluid technology for extraction of bioactive components. In: John Shi (ed.) *Functional Food Ingredients and Nutraceuticals: Processing Technology*. CRC Press, Boca Raton, FL, pp. 3–44.

Shi J et al. 2009a. Solubility of lycopene in supercritical CO_2 fluid affected by temperature and pressure. *Sep Purif Technol* 66: 322–328.

Shi J et al. 2009b. Effects of modifier on lycopene extract profile from tomato skin using supercritical-CO_2 fluid. *J Food Eng* 93: 431–436.

Shi J et al. 2010a. Supercritical-fluid extraction of lycopene from tomatoes. In: S Rizvi (ed.) *Separation, Extraction and Concentration Processes in the Food, Beverage and Nutraceutical Industries*. Woodhead Publishing Limited, UK, pp. 619–639.

Shi J et al. 2010b. Effects of supercritical CO_2 fluid parameters on chemical composition and yield of carotenoids extracted from pumpkin. *LWT Food Sci Technol* 43(1): 39–44.

Singh N, Cheryan M. 1994. Fouling of a ceramic microfiltration membrane by corn starch hydrolysate. *J Membr Sci* 135(2): 195–202.

Soto Ayala R, Luque de Castro MD. 2001. Continuous subcritical water extraction as a useful tool for isolation of edible essential oil. *Food Chem* 75: 109–113.

Spychaj T. 1986. Fractionation of low molecular weight epoxy resins. *Prog Org Coat* 13: 421–430.

Susanto H et al. 2008. Ultrafiltration of polysaccharide–protein mixtures: elucidation of fouling mechanisms and fouling control by membrane surface modification. *Sep Purif Technol* 63: 558–565.

Tena MT et al. 1997. Supercritical fluid extraction of natural antioxidants from rosemary: comparison with liquid solvent sonication. *Anal Chem* 69: 521–526.

Vaillant F et al. 2001. Strategy for economical optimization of the clarification of pulpy fruit juices using crossflow microfiltration. *J Food Eng* 48(1): 83–90.

Vaillant F et al. 2008. Turbidity of pulpy fruit juice: a key factor for predicting cross-flow microfiltration performance. *J Membr Sci* 325(1): 404–412.

Xiao J et al. 2010. Supercritical fluid extraction and identification of isoquinoline alkaloids from leaves of Nelumbo nucifera Gaertn. *Eur Food Res Technol* 231(3): 407–414.

Yang B et al. 2008. Effects of ultrasonic extraction on the physical and chemical properties of polysaccharides from longan fruit pericarp. *Polym Degrad Stab* 93(1): 268–272.

Yang XJ et al. 2001. Experimental observations of nanofiltration with organic solvents. *J Membr Sci* 190: 45–51.

Yi C et al. 2009. Effects of supercritical fluid extraction parameters on lycopene yield and antioxidant activity. *Food Chem* 113(4): 1088–1094.

Zheng X et al. 2009. Application of response surface methodology to optimize microwave-assisted extraction of silymarin from milk thistle seeds. *Sep Purif Technol* 70: 34–40.

Part 8
Food Safety and Food Allergens

41
Microbial Safety of Food and Food Products

J. A. Odumeru

PREFACE

Globally, food safety issues are of top priorities to the food industry, government food safety regulators, and consumers as a result of a significant increase in the number of foodborne disease cases and outbreaks reported worldwide in the twentieth century. These issues led to the proliferation of several food safety programs designed to reduce the incidence of foodborne illness. Although a number of producers and processors have implemented a variety of food safety programs, the occurrence of foodborne illness from emerging and existing pathogens remains a challenge to the food industry and food safety regulators. Food safety begins on the farm and continues through processing, transportation, and storage until the food is consumed. Food safety programs such as Good Manufacturing Practices (GMP), Sanitation, Food Quality, and Safety Tests, and Hazard Analysis Critical Control Points (HACCP) are examples of food safety programs that are commonly used to control and monitor microbial contamination of food.

The three main categories of food safety concerns in the food industry include microbiological, chemical, and physical hazards. The microbiological hazards are those involving foodborne pathogens; chemical hazards include concerns related to antibiotics, pesticides, and herbicides; and physical hazards are those related to foreign objects in foods that can result in injury or illness when consumed with foods. Although this chapter addresses issues related to microbial hazards, food safety programs, which provide protection against these three types of hazards, especially during food processing, will be discussed. Foodborne organisms, sources of microbial contamination of foods and emerging pathogens, will be reviewed in relation to food safety issues.

INTRODUCTION

Food safety concerns are currently at an all-time high due to worldwide publicity about cases and outbreaks of foodborne illness. These concerns are now of top priority in the political and economic agendas of governments at various levels. One of the worst nightmares for food producers or processors is to have the name of their company show up in a news report as the source of a foodborne illness. Apart from the loss of consumer confidence and loss of sales, there are also legal aspects about which food companies must be concerned (Odumeru 2002). An estimated 76 million cases of foodborne illness per year occur in the United States, resulting in 325,000 hospitalizations and 5000 deaths (Mead et al. 1999). The economic impact of these illnesses is estimated at $5 billion or more. A number of food safety programs are currently in place in the food industry in an attempt to reduce the incidence of foodborne illness, which has been on

Food Biochemistry and Food Processing, Second Edition. Edited by Benjamin K. Simpson, Leo M.L. Nollet, Fidel Toldrá, Soottawat Benjakul, Gopinadhan Paliyath and Y.H. Hui.

the rise in the last two decades (Maurice 1994). The increase in the incidence of foodborne disease has been attributed to a combination of factors. These include changes in food production and processing practices, changes in retail distribution, social changes including consumer preferences and eating habits, lack of experience of mass kitchen personnel, changes in population demographics, and increases in population mobility worldwide as a result of increases in international trade and travel (McMeekin and Olley 1995, Baird Parker 1994). Furthermore, advances in sciences in the area of analytical methods development has led to the availability of better detection methods for the diagnosis of foodborne illness and a subsequent increase in the number of cases reported. Other factors such as better reporting systems and increases in the occurrence of emerging pathogens have contributed to the increase in cases of foodborne diseases reported. Issues related to emerging pathogens will be discussed in a subsection of this chapter. A number of scientific tools are now available to the food industry and food safety regulators for implementing food safety programs designed to reduce the incidence of foodborne illness. These include the use of risk assessment of foods and food ingredients, to determine the risks associated with various types of foods under certain processing conditions; predictive modeling, which estimates the growth and survival of pathogens and spoilage organisms under specified conditions; and rapid methods for screening foods for quality and safety during and after production. This chapter will provide an overview of issues related to microbial safety of food and food products, food preservation technologies, emerging pathogens, food safety programs to control microbial contamination, and future perspectives on food safety.

SHELF LIFE OF FOODS AND FOOD INGREDIENTS AND FOOD SAFETY

The shelf life of a food product generally refers to the keeping quality of the food. An estimated 25% of the food supplies worldwide are lost as a result of spoilage; hence, it is economically beneficial to maintain the quality of food products at various stages of food production and storage. There are two categories of foods in relation to shelf life: shelf stable and perishable. Whether a particular food product is shelf stable or perishable depends on the intrinsic properties of the food (e.g., pH, water activity, and structure). Shelf-stable foods usually have low water activity, low pH, or a combination of both, while perishable foods tend to have high water activity and high pH. The structure or texture of the food is also an important factor in shelf stability. Extrinsic factors such as storage temperature, gaseous atmosphere, and relative humidity also determine the shelf stability of food products (McMeekin and Ross 1996). These intrinsic and extrinsic factors influence the survival and growth not only of spoilage organisms but also of pathogenic organisms in foods. Food spoilage occurs as a result of physical or chemical changes in the food or of the by-products of spoilage microorganisms growing in the food product. Pathogens present in low levels may not produce identifiable changes in the food; hence, the presence of pathogens cannot be determined using noticeable changes in the food as an indicator.

Although shelf-stable foods are less likely to be implicated in foodborne illness than perishable foods, cross-contamination of shelf-stable or perishable foods by pathogens can be a source of foodborne illness. A number of preservation applications used in the food industry are designed to extend the shelf life of the food product by reducing microbial growth; however, pathogens that are able to survive or even grow under preservation techniques such as refrigeration can cause foodborne illness. Effective strategies for controlling the presence of spoilage and foodborne pathogens in foods should include elimination of sources of contamination combined with food preservation technologies such as drying, freezing, smoking, curing, fermenting, refrigeration (Baird-Parker 2000) and modified-atmosphere packaging (Farber 1991).

CATEGORIES OF FOODBORNE ORGANISMS

Microorganisms that can be transmitted to humans or animals through food are referred to as foodborne organisms. There are three main categories of foodborne organisms: spoilage, pathogenic, and beneficial. Spoilage organisms can grow and produce physical and chemical changes in foods, resulting in unacceptable flavor, odor, formation of slime, gas accumulation, release of liquid exudates or purge, and changes in consistency, color, and appearance. Also, extracellular or intracellular enzymes released by spoilage organisms can result in deterioration of food quality. Growth of microorganisms to high numbers is usually required before spoilage becomes noticeable. Hence, control of growth of spoilage organisms is required to impede microbial spoilage. The presence of foodborne pathogens in foods in low concentrations can render foods harmful to humans if consumed. Because pathogenic organisms at low levels may not produce noticeable changes in foods, consumers may not have advance warning signals of the danger associated with consumption of contaminated foods. "Beneficial" or "useful" organisms include microorganisms used in various food fermentation processes. These organisms are either naturally present in such foods or added to produce the desired by-product of fermentation. Various types of foods such as fruits and vegetables, pickles, dairy products, meats, sausages, cheeses, and yogurt are common types of fermented products involving the use of beneficial organisms. Beneficial organisms include organisms in the group of lactic acid bacteria, yeasts, and molds. Bacteria species from 10 genera are included in the group of lactic acid bacteria. These include *Lactococcus, Leuconostoc, Streptococcus, Lactobacillus, Pediococcus, Carnobacterium, Tetragenococcus, Aerococcus, Vagococcus*, and *Enterococcus*. The most important type of yeast used for fermentation of food and alcohol is *Saccharomyces cerevisiae*. This yeast is used for leavening bread and production of beer, wine, and liquors. It is also used for food flavor. Non-mycotoxin-producing molds from the genera *Penicillium* and *Aspergillus*, and some in the *Rhizopus* and *Mucor* genera, have been used for beneficial purposes in food preparation (Bibek 1996).

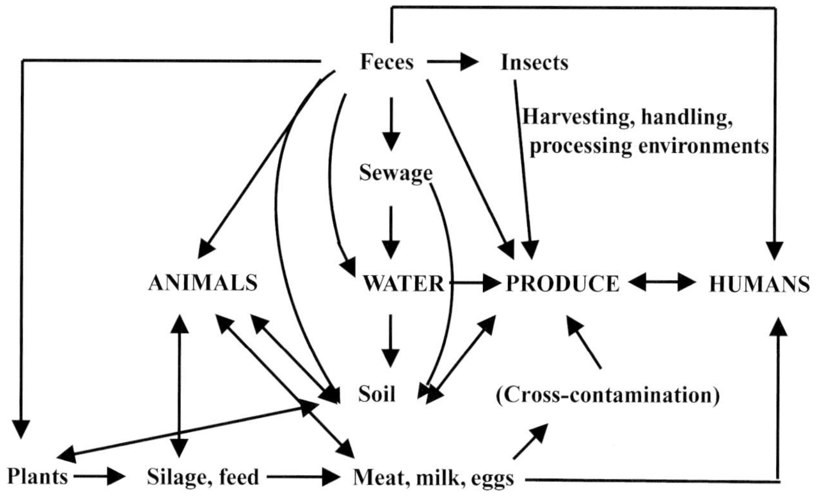

Figure 41.1. On-farm sources of microbial contamination of water and food products of plant and animal origin. (Adapted from Beuchat 1995.)

SOURCES OF FOODBORNE PATHOGENS

In order to determine and implement effective control measures for pathogens in foods, it is important to identify potential sources of contamination. Plants and animals are the main source of human food supply. The exterior and, in some cases, the interior of plants and animals harbor microorganisms from external sources such as soil, water, and air. These environmental sources contain a wide variety of microorganisms, some of which are pathogenic to man. Contamination of the food supply can occur at various stages of production, processing, transportation, and storage. For example, fruits and vegetables can be contaminated at the farm level as well as during harvesting, transportation, and processing. Potential sources of on-farm contamination of fruits and vegetables are summarized in Figure 41.1. Meats can be contaminated at the time of slaughter, processing, and storage. Microbial contamination can come from slaughtered animals, water, equipment, utensils, the slaughterhouse environment, and workers. Thus, intervention strategies to control microbial contamination of the food supply must be implemented at various stages of food production, processing, transportation, and storage, and also at the consumer end of food preparation.

FOODBORNE DISEASE CASES AND OUTBREAKS

Foodborne disease in a susceptible host can result from consumption of food or water contaminated with pathogenic organisms. A single or sporadic case of foodborne illness refers to an instance when an illness that is unrelated to other cases occurs as a result of consumption of contaminated food or water. An outbreak, on the other hand, refers to an incident in which two or more persons become ill after consuming the same food or water from the same source. The occurrence of a foodborne illness depends on a number of risk factors such as (1) type and number of pathogenic microorganisms in the food, (2) effect of food product formulation or processing on the viability of the pathogen, (3) storage conditions of the food that may promote contamination, growth, and survival of the pathogen, and (4) the susceptibility of the individual to foodborne illness. It is believed that the number of reported outbreaks represents only 10% of the real incidence of foodborne disease, even in countries with well-established surveillance systems (Baird-Parker 2000). Foodborne illness resulting from severe infections such as hemolytic uremic syndrome, botulism, and listeriosis often require hospitalization and are more likely to be reported, while self-limiting foodborne illness such as salmonellosis, campylobacteriosis, and *S. aureus* enterotoxin–related infections are less likely to be reported. It is believed that viral foodborne infections account for a large portion of cases of foodborne illness (Caul 2000). However, the extent of the problem on a global scale is difficult to assess as a result of lack of surveillance data in most parts of the world, coupled with the fact that most viral infections are self-limiting. Although viruses and bacteria-related infections account for the majority of foodborne diseases, certain groups of parasites and fungi are also etiologic agents of foodborne diseases.

Foodborne Bacterial Infections

Several types of foodborne bacterial pathogens are implicated in foodborne diseases. Examples of bacterial genera most commonly implicated in foodborne infections, onset and duration of the symptoms of the disease, types of foods that are likely to be contaminated by these groups of bacteria, and potential sources of contamination are summarized in Table 41.1. Bacterial pathogens including *Campylobacter, Salmonella, Listeria monocytogenes*, and verocytotoxigenic *E. coli* are the top causes of bacterial foodborne infections in westernized countries, based on the number of reported cases (Sharp and Reilly 2000). The

Table 41.1. Foodborne Bacterial Pathogens, Symptoms of Disease, Target Populations, and Potential Sources of Food Contamination

Organism	Symptoms	Onset	Duration	Target Population	Source	Suspect Foods
Bacillus cereus (*B. cereus*)	Two forms: 1. Diarrheal illness including abdominal cramps, nausea (vomiting rare) 2. Emetic illness including nausea and vomiting	Diarrheal: 6–15 hours Emetic: 30 minutes to 5 hours	Diarrheal: 12–24 hours Emetic: 6–24 hours	Everyone	Soil, Dust	Diarrheal: meat, milk, vegetables, fish, soups Emetic: rice products, potatoes, pasta, cheese products
Campylobacter jejuni	Diarrhea (may contain blood), fever, nausea, vomiting, abdominal pain, headache, muscle pain	2–5 days	2–10 days	Children under five, young adults 15–29	Cattle, chicken, birds, flies, stream, pond water	Raw chicken, turkey, raw milk, beef, shellfish, water
Clostridium botulinum	Fatigue, weakness, double vision, respiratory failure	18–36 hours, varies 4 hours to 8 days	Months	Everyone	Soil, sediment, intestinal tracts of fish and mammals	Canned foods, smoked and salted fish, chopped bottled garlic, sautéed onions, honey
Clostridium perfringens	Diarrhea, cramps, nausea (vomiting rare)	6–22 hours	Under 24 hours, may persist 1–2 weeks	Everyone, but young and elderly are most affected	Soil, feces	Meats, meat products, poultry, gravy
Escherichia coli (Traveller's diarrhea)	Watery diarrhea, abdominal cramps, low-grade fever, nausea, malaise	1–3 days	Days	Everyone	Water, Human sewage	Dairy products
E. coli O157:H7	Severe cramping, watery diarrhea becoming bloody, low-grade fever	12–60 hours	2–9 days to weeks	Children	Cattle, deer	Undercooked or raw hamburger, raw milk, unpasteurized apple cider
Listeria monocytogenes	Septicemia, maningoen cephalitis, spontaneous abortions, stillbirths, influenza-like symptoms	A few days to 6 weeks	Days to weeks	Pregnant woman, immunocompromised persons, cancer and AIDS patients and those with chronic diseases, elderly, (fatality rate is as high as 70%)	Soil, improperly made silage	Raw milk, soft ripened cheeses, ice cream, raw vegetables, fermented raw meat, sausages, hot dogs, luncheon meats, raw and cooked poultry, raw and smoked fish

Salmonella	Nausea, chills, vomiting, cramps, fever, headache, diarrhea, dehydration	6–48 hours	1–4 days	Everyone, but more severe symptoms in infants, elderly, infirm, and AIDS patients	Water, soil, insects, animal feces, raw meat, raw poultry, raw seafood	Raw meats, poultry, eggs, milk, dairy products, fish, shrimp, frogs legs, yeast, coconut, sauces and salad dressings, cake mixes, cream-filled desserts and toppings, dried gelatin, peanut butter, cocoa, chocolate
Shigella	Abdominal pain, diarrhea, fever, vomiting, sometimes cramps, nausea	1–7 days	4–7 days	Infants, elderly, infirm, AIDS patients	N/A	Salads (Potato, tuna, shrimp, macaroni and chicken), raw vegetables, milk and dairy products, poultry
Staphylococcus aureus	Nausea, vomiting, cramps, diarrhea, prostrations	1–6 hours	1–2 days	Everyone	Air, dust, sewage, water, milk, food equipment, humans, animals, nasal passages and throat, hair, skin, boils	Meat and meat products, poultry and poultry products, salads, (egg, tuna, chicken, potato, macaroni) cream-filled pastries, cream pies, milk and dairy products
Vibrio cholera (Cholera)	Mild, watery diarrhea, abdominal cramps, nausea, vomiting, dehydration, shock	6 hours to 5 days	Days	Everyone, but especially those with immunocom-promised systems, reduced gastric acidity, malnutrition	Raw shellfish, water, poor sanitation	Seafood, water

Source: Adapted from American Academy of Microbiology: Food Safety Current Status and Future Needs 1999, research paper by Stephanie Doores.

Table 41.2. The Incidence of the Four Most Frequent Foodborne Disease-Causing Pathogens in Canada and the United States

	Estimated Cases/Year (Based on Population Ratio)	
Bacteria	Canada	United States[a]
Campylobacter	110,000–700,000	1,100,000–7,000,000
Listeria	93–177	928–1767
Salmonella	69,600–384,000	696,000–3,840,000
Verotoxigenic *E. coli* (Including *E. coli* O157:H7)	1600–3200	16,000–32,000
Total	181,3000–1,087,400	1,812,900–10,873,800

[a]Adapted from Buzby and Roberts 1997.

number of reported cases of *Salmonella, Campylobacter*, and *E. coli* O157 increased from 2- to 40-fold between 1982 and 1994 (Sharp and Reilly 2000). The increase in the number of cases was attributed to heightened awareness of foodborne related illness, improved diagnostic methods, and improved reporting and surveillance procedures. *Campylobacter* is currently the leading cause of foodborne disease in the United States and Canada, followed by *Salmonella*-related infections. The numbers of estimated cases of foodborne illness per year in the United States and Canada countries are listed in Table 41.2. Although the number of *Listeria monocytogenes*–related infections is considerably lower than those of other pathogens, foodborne infections caused by this organism are characterized by a high case mortality ratio, especially in certain high risk groups such as the elderly, the young, immunocompromised individuals, and pregnant women. Infections caused by *E. coli* O157:H7 often result in a high morbidity and mortality rate. For example, complications such as hemolytic uremic syndrome can result in kidney failure and death (Riley et al. 1983, Karmali 1989). Sources of bacterial infections are not limited to contaminated foods; contaminated water and person-to-person transmission have also been implicated. For example, in a waterborne outbreak of *E. coli* 0157:H7 in Walkerton, Ontario, 1700 cases and seven deaths were reported. This led to changes in the Ontario Drinking Water Regulations to include strict water treatment procedures and zero tolerance for the presence of coliforms and *E. coli* in drinking water. Current food safety programs in the food industry must now document the use of pathogen-free water in food production and processing facilities.

Campylobacter Enteritis

Campylobacter jejuni and *C. coli* are the species of *Campylobacter* implicated in most *Campylobacter* enteritis infections. Other species of *Campylobacter* such as *C. lari* and *C. upsaleusis* may also cause *Campylobacter* enteritis. Symptoms of *Campylobacter* infection in humans are watery and/or bloody diarrhea, abdominal pain, fever, and general malaise. The disease is usually self-limiting, and antibiotics are required only when complications occur. It is estimated that *Campylobacter* enteritis accounts for 10% of cases of foodborne illness, and death is rarely reported. Most cases of this disease are sporadic, and foods of animal origin, particularly poultry, are largely responsible for most infections. Contaminated fruits and vegetables have also been implicated in *Campylobacter* enteritis cases.

Salmonellosis

Salmonellosis infection ranks second in incidence to *Campylobacter* enteritis. This organism is ubiquitous (present everywhere) in the environment, especially in the feces of most food-producing animals; hence, a variety of foods are readily contaminated by this organism. *Salmonella* are present in animals without causing apparent illness. However, certain serotypes of *Salmonella* such as *S. enteritidis* can penetrate poultry reproductive organs, resulting in contamination of egg content and posing a health risk to consumers. Salmonellosis symptoms include watery diarrhea, abdominal pain, nausea, fever, headache, and occasional constipation. Hospitalization may be required in cases of severe infections. Foods that can become contaminated with *Salmonella* include meat, raw milk, poultry, eggs, dairy products, and other types of foods that can become contaminated with fecal material. An increase in the number of cases of Salmonellosis linked to consumption of contaminated fruits and vegetables such as bean sprouts, raw tomatoes, melons, and cantaloupes has been reported. In addition to fecal contamination, cross-contamination of foods by *Salmonella* during food preparation can be a source of foodborne illness.

Listeriosis

Listeriosis is caused by *Listeria monocytogenes*. The ubiquitous nature of this organism contributes to the widespread incidence of the organism in foods. It is psychrotrophic in nature and is able to grow at refrigeration temperatures. Hence, this organism is of concern in refrigerated foods with extended shelf life. Listeriosis infections often result in septicemia and/or meningitis. The case mortality ratio for Listeriosis infections is much higher than those of *Salmonella*, *Campylobacter,* and *E. coli* 0157:H7 infections. An estimated 500 deaths associated with Listeriosis are reported annually in the United States. Individuals most susceptible to Listeriosis include the elderly, the young, immunocompromised patients, and pregnant woman. A wide variety of foods including processed meats, milk and dairy products, fruits, and vegetables have been implicated in a number of Listerioses outbreaks. In 1999, a multistate outbreak involving 1000 cases of Listeriosis in the United States was linked to consumption of contaminated hot dogs and delicatessen meats, resulting in a recall of 35 million pounds of these food products (MMWR 1999).

Verotoxigenic *Escherichia coli* (VTEC) Infections

Verotoxigenic *Escherichia coli* (VTEC) belonging to the O157 serogroup is the most common VTEC serogroup implicated in foodborne disease outbreaks. This organism was first recognized as a cause of hemorrhagic colitis in 1982. VTEC O157 and serogroups O5, O26, and O111 are of high prevalence in feces of healthy cattle. Thus, foods of animal origin such as meat, milk, and dairy products can be contaminated with these organisms. Other types of foods such as salads, vegetables, fruits, sandwiches, and cooked meat can be contaminated during preparation. Contaminated water can also be a source of infection by this organism. The majority of food- and waterborne outbreaks had links with contamination of fecal origin. Infection by VTEC can result in hemorrhagic colitis (HC) and hemolytic uremic syndrome (HUS), abdominal pain, and watery diarrhea (bloody or nonbloody). The HC symptoms may lead to complications of the HUS and subsequent renal failure. Although HUS can occur in any age group, children are more susceptible (Karmali 1989). In adults, 50% of thrombocytopenic purpura (TP) cases are caused by HUS.

Other bacterial foodborne infections can occur as a result of ingestion of food contaminated with *Staphylococcus aureus* toxin, *Clostridium perfringens, Clostridium botulinum* toxin, *Shigella* spp., *Yersinia enterocolitica, Brucella* spp., *Vibrio cholera, Vibrio paraheamolylicus*, and *Bacillus cereus* enterotoxin.

Foodborne Parasite Infections

Foodborne parasite infections are caused by certain groups of protozoa. It is believed that foodborne infections related to parasites are underreported by an estimated factor of 10 or more (Casemore 1991), and the etiological agent of foodborne outbreaks is identified in less than 50% of the cases (Bean et al. 1990, Cliver 1987). Parasites commonly associated with foodborne disease and possible food and water sources are listed in Table 41.3.

Cryptosporidium parvum is the species of *Cryptosporidium* associated with infection in humans (Current and Blagburn 1990). It is a common cause of gastrointestinal infections in immunocompromised individuals and is an opportunistic pathogen associated with infections in patients with acquired immune deficiency syndrome (AIDS), but it affects healthy people as well. Although drinking water contaminated with human or animal feces is the usual source of transmission of this organism to humans, epidemiological links between the consumption of contaminated foods such as raw sausage, offal, and raw milk and cryptosporidiosis have been reported (Casemore 1990, Casemore et al. 1986). The largest outbreak of waterborne cryptosporidiosis was reported in Milwaukee, Wisconsin, with 403,000 persons ill. Reported outbreaks in most countries are associated with contaminated drinking water or contaminated swimming pools.

Table 41.3. Foodborne Parasites, Fungi, and Viruses Associated with Food- or Waterborne Illness

Pathogen	Sources of Pathogen or Toxin
Parasites	
Cryptosporidium	Contaminated surface water or foods in contact with contaminated water
Cyclospora	Raspberry, mesculun lettuce, and basil
Sarcocystis	Raw or undercooked meat
Toxoplasma	Raw or undercooked meat
Giardia	Contaminated surface water
Fungi	
Penicillium	Mycotoxins in apple juice, walnuts, corn and cereals
Aspergillus	Mycotoxins in groundnuts, corn, figs and tree nuts
Fusarium	Mycotoxins in cereals, corn, wheat and barley
Viruses	
Hepatitis A and E	Contaminated water, fruits and vegetables, raw shellfish, raw oysters
Norwalk/Norwalk–like viruses	Variety of foods and water
Calcivirus	Water and contaminated foods
Astrovirus	Water and contaminated foods

Cyclospora is an emerging pathogen, which causes diarrheal infections in humans. The first reported outbreak of *Cyclospora* infection occurred in 20 US states and two provinces in Canada, Ontario and Quebec, during the months of May and June, 1996, and April and May, 1997. Consumption of contaminated imported raspberries was implicated in many of these outbreaks (Herwaldt and Ackers 1996, Herwaldt and Beach 1997). Contaminated mesclun lettuce and basil have also been linked to *Cyclospora* outbreaks (MMWR 1997).

Sarcocystis is another coccidian parasite, which is prevalent in livestock. Human infection occurs as a result of ingestion of raw or undercooked meat containing mature sarcocysts (Tenter 1995).

Toxoplasma gondii is an intracellular parasite, which is prevalent in man and animals (Fayer 1981). Ingestion of raw or undercooked meat from livestock and game animals is often implicated in human toxoplasmosis infections. Another parasite commonly implicated in food and waterborne illness is *Giardia*.

Outbreaks of food and waterborne illness have been reported in the United States and the United Kingdom (Craun 1990, Porter et al. 1990). Infection with this parasite is associated with unsanitary conditions, and contaminated water is the most common source of infection.

Foodborne Fungi

Fungi most commonly associated with foodborne intoxications are *Penicillium*, *Aspergillus*, and *Fusarium*. The types of foods that may contain mycotoxin contamination are listed in Table 41.3. A foodborne fungi outbreak associated with consumption of corn contaminated with *Aspergillus flavus* was reported in India. The outbreak involved 1000 cases and 100 deaths (Krishnamachari et al. 1975). Mold growth can occur in foods stored under high temperature or humidity, resulting in the production of mycotoxin, which can be harmful to humans.

Foodborne Virus Infection

Viruses from human fecal origin can result in illness if ingested with food or water. Food or waterborne viruses commonly associated with human illness are listed in Table 41.3. Viral gastroenteritis is caused by Norwalk-like viruses and in some cases by the calicivirus and astrovirus groups. The symptoms often include acute but short self-limiting episode of diarrhea and vomiting. The viruses can be transmitted from person to person via contaminated utensils and foods. Foodborne viral hepatitis in humans is caused by the hepatitis A or hepatitis E virus. The onset of hepatitis may be preceded by anorexia, fever, fatigue, nausea, and vomiting. Infected individuals shed the organism in feces, which if allowed to contaminate food or water, can result in person-to-person transmission (Caul 2000).

EMERGING PATHOGENS AND FOOD SAFETY

Emerging infectious diseases have increased in the last two decades and are likely to increase in the future. Hence, the characteristics of the etiologic agents of these diseases must be considered when designing control measures to ensure food safety. There are two types of emergence: true emergence and reemergence. A true emergence involves the occurrence of microbial agent not previously identified as a public health threat. A reemergence involves the occurrence of a microbial agent causing disease in a new way not previously reported or the reemergence of a human disease after a decline in incidence. Emergence may be due to the introduction of a new agent, to recognition of an existing disease that was not previously detected, or to environmental pressures resulting in occurrence of a pathogen that can cause human disease. For example, the occurrence of enterohemorrhagic *E. coli* O157:H7 as a pathogen in 1982 is believed to be due to introduction of a new agent. The organism was involved in two outbreaks that year in the United States, and these outbreaks were associated with consumption of undercooked hamburgers from a fast-food restaurant chain. Several countries worldwide have reported outbreaks of infection caused by this organism. The appearance of *Salmonella typhimurium* phage type 104 is another example of a new agent. Increases in the use of antibiotics in humans and animals may have provided the environmental pressures resulting in the occurrence of *Salmonella typhimurium* phage type 104 with multiresistance to five antibiotics: ampicillin, chloramphenicol, streptomycin, sulphonamides, and tetracycline. This multiresistant characteristic leaves little choice of antibiotics for treatment of disease caused by this organism. The occurrence of *Listeria monocytogenes* as a foodborne agent is an example of reemergence. The organism is a well-known infectious agent; however, its role as a foodborne organism was not detected until the early 1980s. Examples of emerging foodborne pathogens and possible causes of emergence are provided in Table 41.4. The challenge posed by the appearance of emerging pathogens is that these organisms may not always behave as traditional pathogens. Therefore, new control measures to ensure that foods are free of these pathogens may be required.

Table 41.4. Emerging Pathogens and Suspected Causes of Emergence

Pathogens	Cause of Emergence
Salmonella DT104	Resistance to antibiotics
E. coli O157:H7	Development of a new pathogen
Cyclospora cayetanensis	Development of a new pathogen
Cryptosporidium parvum	Development of a new watershed areas
Hepatitis E virus	Newly recognized
Norwalk Virus	Increased recognition
Aeromonas Spp.	Immunosuppression and improved detection
C. jejuni	Increased recognition, consumption of uncooked poultry
L. monocytogenes	Increased awareness
Helicobacter pylori	Increased recognition
Vibrio vulnificus	Increased recognition

CONTROL MEASURES FOR MICROBIAL CONTAMINANTS

A combination of factors is normally responsible for occurrence of an incident of foodborne illness. The pathogen must first reach the food involved; the organism must survive until food is ingested; in many cases, the organism must multiply to an infectious level or produce toxins; and lastly, the host must be susceptible to the level of organisms ingested with the food. Control measures to ensure food safety include (1) prevention of contamination of foods by pathogenic organisms (2) inhibition of growth or elimination of pathogens in foods and food products. The first stage of control measures is to prevent contamination of food animals and plants during the production stage. Production practices such as the use of manure and other organic fertilizer materials can provide the vehicle for contamination of food crops; hence, it is important to ensure that this type of fertilizer is pathogen-free prior to use. The second stage of control measure is to prevent contamination and growth of pathogenic organisms during harvesting and transportation of food products. It is generally believed that control of microbial contamination early in production is more effective than control measures applied at a later stage of production. However, in cases when the presence of pathogens cannot be eliminated during the production stage, processing procedures that are designed to control the presence of pathogens must be implemented. A combination of intrinsic factors, (i.e., factors associated with the properties of the food) and extrinsic factors (i.e., factors associated with external conditions) is often used during processing and storage to control microbial growth or survival in foods (Bibek 1996). Intrinsic factors such as the pH, water activity, and food components can promote or inhibit microbial growth or survival. These properties can be used in combination with extrinsic factors such as heat, preservatives, irradiation, and storage conditions (e.g., refrigeration, relative humidity, and gaseous atmosphere).

Currently, there are several food preservation technologies available for controlling microbial growth and survival. The most commonly used technologies for controlling microbial growth include application of low temperatures such as refrigeration or freezing, reduction of water activity of the food by drying, curing with salt or increased sugar level, pH reduction by acidification or fermentation, use of food preservatives, and modified atmosphere techniques. Other preservation technologies designed to inactivate foodborne organism include heat treatment such as sterilization, cooking, retorting, pasteurization, irradiation, and the use of hydrostatic pressure and pulse light. Also, use of packaging materials and adequate sanitation of the food production or processing environment can be used to limit entry of microorganisms into the food product and prevent recontamination of processed foods (Gould 1989, Hall 1997). The use of a combination of preservation techniques, often referred to as the hurdle concept, can be very effective in controlling microbial growth. A combination of suboptimal levels of the growth-limiting factors can be very useful where higher levels of one of the factors can be detrimental to the quality of the product. The most important hurdles used in food preservation include high or low temperature, water activity (*a*w), redox potential (*E*h), acidity (pH), preservatives such as sulfite, nitrite, and sorbate, and the use of competing microorganisms such as lactic acid bacteria. The application of an appropriate combination of these hurdles can improve the microbial stability, sensory and nutritional qualities, and safety of the foods.

FOOD SAFETY PROGRAMS

Traditional approaches to controlling the safety and quality of foods involve inspection of foods after production or processing for compliance with general hygienic practice, and where appropriate, foods are sampled for laboratory testing. This approach does not ensure food safety since reliance on visual inspection and testing of finished products cannot guarantee the absence of harmful pathogens in food. A more effective food safety control program, called the hazard analysis critical control point (HACCP) program, was introduced to the food industry in the early 1970s, and various food safety regulations and trade bodies worldwide endorsed its use as an effective and rational approach to assurance of food safety. HACCP is a systematic approach to hazard identification, assessment, and control. The benefits of the HACCP approach to ensuring food safety include the following:

- The food industry has a better proactive tool for ensuring the safety of foods produced.
- Potential food safety problems can be detected early.
- Food inspectors can focus more on verifying plant controls.
- More effective use is made of resources by directing attention to where the need is greatest.
- A significant reduction in the cost of end product testing is achieved.

There are seven key principles of HACCP. Principle No. 1 includes hazard analysis to identify the microbiological, chemical, and physical hazards of public health concern. Raw materials processing procedures, including packaging and storage, are assessed for microbiological hazards. The second principle is to determine the procedures or points in the food operation where hazards can be controlled effectively. These are called the critical control points (CCPs). The third principle includes the establishment of critical limits that separate acceptable from nonacceptable limits. The fourth principle involves the development of a system to monitor the CCPs so that the limits are not exceeded. The fifth principle is to determine what corrective actions to take when the CCPs are exceeded. The sixth principle includes the establishment of procedures for verifying that the HACCP program is working as expected. The seventh principle involves the documentation procedures and records for all aspects of these six principles. These HACCP principles have received worldwide recognition, by governments and the food industry. Many industrial organizations have adopted this food safety program as a means of controlling food safety hazards.

One of the most successful food safety programs at the consumer level is the FightBac program developed by the Partnership for Food Safety Education (PFSE), formed in 1997. This nonprofit organization provides education on safe handling of food to the public. The key features of the information on safe handling of food in the FightBac program are summarized in Table 41.5. Further information on this program can be found in the FightBac website (www.fightbac.org). Implementation of these safe food-handling procedures by consumers can prevent the occurrence of foodborne illness.

FUTURE PERSPECTIVES ON FOOD SAFETY

Foodborne disease is preventable, provided control measures to prevent contamination of the food supply are implemented at various stages of the food chain, from the farm to the point of consumption. On-farm food safety programs include control measures for preventing contamination of plants, animals, and plant and animal products that are produced for human consumption. Implementation of effective environmental sanitation, good

Table 41.5. Four Simple Steps for Consumers to Ensure Food Safety (FightBac!^{TM})

Clean: Wash hands and surfaces often
- Wash your cutting boards, dishes, utensils, and counter tops with hot soapy water after preparing each food item and before you go onto the next food
- Use plastic or other nonporous cutting boards
- Consider using paper towels to clean up kitchen surfaces

Separate: Don't cross-contaminate
- This is especially true when handling raw meat, poultry, and seafood
- Never place cooked food on a plate that previously held raw meat, poultry, or seafood

Cook: Cook to proper temperatures
- Use a clean thermometer that measures the internal temperatures of cooked foods to make sure meat, poultry, casseroles, and other foods are cooked all the way through
- Cook roasts and steaks to at least 145°F. Whole poultry should be cooked to 180°F for doneness
- Cook ground beef, where bacteria can spread during processing, to at least 160°F
- Don't use recipes in which eggs remain raw or only partially cooked

Chill: Refrigerate promptly
- Refrigerate or freeze perishables, prepared foods, and leftovers within 2 hours or sooner
- Never defrost at room temperatures. Thaw food in the refrigerator, under cold running water, or in the microwave. Marine foods in the refrigerator
- Divide large amounts of leftovers into small, shallow containers for quick cooling in the refrigerator

Source: FightBac!, Partnership for Food Safety Education. Available at www.fightbac.org.

manufacturing practices (GMPs) and hazard analysis critical control points (HACCP) programs are key control measures for microbial, chemical, and physical hazards, not only at the farm level, but also during food processing, transportation, storage, and final preparation of foods prior to consumption. Furthermore, education and training of producers, processors, retailers, restaurant personnel, and consumers are important in the implementation of control measures to ensure the safety of foods consumed. Also, adequate inspection of food processing facilities and operations is very important to ensure that sanitation, GMP and HACCP protocols are being followed on a consistent basis. In order to implement a HACCP program successfully, the management of food operations must be committed to the program, by being aware of the benefits the program offers, and be prepared to invest time and money into the program. The use of microbiological testing as a verification tool for food safety programs including HACCP is important in providing evidence that indeed the control measures in place are effective. Finished product testing alone must not be relied on as means of ensuring food safety. The application of rapid methods for microbiological tests provides a useful tool for ensuring production of a safe food product at various points in the HACCP implementation. Microbiological tests can be performed for (1) the raw materials required for processing, (2) monitoring critical control points, and (3) HACCP verification. The reliance on testing of finished food products alone is a thing of the past. The challenge to food microbiologists now is to develop on-line food safety testing tools that can be used during food production. The limitation of current rapid tests for foodborne pathogens is that they require 24 hours or more to complete because of the requirement for amplification of pathogen numbers to detectable levels. The ability to detect pathogens at very low levels within seconds, minutes, or in less than 24 hours is an important future development for food microbiologists.

The likelihood of future increases in the incidence of foodborne illness is high as a result of increases in the number of susceptible aging populations, the treatment of disease with immunosuppressing drugs, and increases in the availability of ready-to-eat foods requiring no further processing prior to consumption. Unless steps are taken to implement food safety programs, which can reduce microbial contamination of foods at all levels of food production, processing, and retailing, and to educate restaurant personnel and consumers about the safe handling of foods, outbreaks of foodborne illness will continue to occur.

REFERENCES

Baird-Parker AC. 1994. Food and microbiological risks. *Microbiology* 140: 687–695.

Baird-Parker TC. 2000. Chapter 1. The production of microbiologically safe and stable foods. In: BM Lund, TC Baird Parker, GW Gould, (eds.), *The Microbiological Safety and Quality of Food*, vol. 1, Aspen Publishers, Gaithersburg, MD, pp. 3–18.

Bean NH. et al. 1990. Foodborne disease outbreaks, five-year summary, 1983–1987. *Morbility and Mortality Weekly Report* 39: 15–57.

Bibek R. 1996. Chapter 6. *Fundamental Food Microbiology*, CRC Press, Boca Raton, FL, pp. 61–72.

Beuchat LR. 1995. Pathogenic microorganisms associated with fresh produce. *Journal of Food Protection* 59: 204–216.

Buzby FL, Roberts T. 1997. Guillain-Barre Syndrome increases foodborne disease costs. *Food Review* 20(3): 36–42.

Casemore DP. 1990. Epidemiological aspects of human cryptosporidiosis. *Epidemiology and Infection* 104: 1–28.

Casemore DP. 1991. Foodborne protozoal infection. In: WM Waites, JP Arbuthnott, (eds.) *Foodborne Illness. A Lancet Review*, Edward Arnold, London, pp. 108–119.

Casemore DP et al. 1986. *Cryptosporidium* plus *Campylobacter*: An outbreak in a semi-rural population. *Journal of Hygiene* 96: 95–105.

Caul EO. 2000. Chapter 52. Foodborne viruses. In: *The Microbiological Safety and Quality of Food*, vol. 1, Aspen Publishers, Gaithersburg, MD, pp. 1457–1489.

Cliver DO. 1987. Foodborne disease in the United States, 1946–1986. *International Journal of Food Microbiology* 4: 260–277.

Craun GF. 1990. Waterborne giardiasis. In: EA Meyer (ed.) *Giardiasis*, Elsevier, New York, pp. 305–313.

Current WL, Blagburn BL. 1990. Cryptosporidium: infections in man and domestic animals. In: PL Long, editor. *Coccidiosis of Man and Domestic Animals* CRC Press, Boca Raton, FL, pp. 155–185.

Farber JM. 1991. Microbiological aspects of modified atmosphere packaging technology: A review. *Journal of Food Protection* 54: 58–70.

Fayer R. 1981. Toxoplasmosis and public health implications. *Canadian Veterinary Journal* 22: 344–352.

Gould GW. 1989. Introduction. In: *Mechanisms of Action of Food Preservation Procedures*. Elsevier Applied Science, London, pp. 1–10.

Hall RL. 1997. Foodborne illness: Implications for the future. *Emerging Infectious Diseases* 3: 555–559.

Herwaldt BL, Ackers M-L. 1996. Cyclospora Working Group. An outbreak in 1996 of cyclosporiasis associated with imported raspberries. *N England Journal of Medicine* 336: 1548–1556.

Herwaldt BL, Beach MJ. 1997. Cyclospora Working Group. The return of *Cyclospora* in 1997: Another outbreak of cyclosporiasis in North America associated with imported raspberries. *Annals of Internal Medicine* 130: 210–220.

Kaferstein FK et al. 1997. Foodborne disease control: A transnational challenge. *Emerging Infectious Diseases* 3: 503–510.

Karmali MA. 1989. Infection by verocytotoxinproducing *Echerichia coli*. *Clinical Microbiological Review* 2: 15–38.

Krishnamachari KAVR et al. 1975. Hepatitis due to aflatoxicosis. *Lancet* 1: 1061–1063.

Maurice J. 1994. The rise and rise of food poisoning. *New Scientist* 144: 28–33.

Mead PS et al. 1999. Food-related illness and death in the United States. *Emerging Infectious Disease* 5: 607–624.

McMeekin TA, Olley J. 1995. Predictive microbiology and the rise and fall of food poisoning. *ATS Focus*, pp. 14–20.

McMeekin TA, Ross T. 1996. Shelf life prediction: Status and future possibilities. *International Journal of Food Microbiology* 28: 65–83.

Morbidity and Mortality Weekly Report. 1997. Update: Outbreaks of Cyclosporiasis—United States and Canada. *MMWR* 46: 521–523.

Morbidity and Mortality Weekly Report. 1999. Centre of Disease Control and Prevention. Update: Multistate outbreak of listeriosis—United States, 1998–1999. *MMWR* 47: 1117–1118.

Odumeru J. 2002. Inside Microbiology: Current microbial concerns in the dairy industry. *Food Safety Magazine* 8(1): 20–23.

Porter JDH et al. 1990. Foodborne outbreak of *Giardia lamblia*. *American Journal of Public Health* 80: 1259–1260.

Riley LW et al. 1983. Hemorrhagic colitis associated with a rare *Echerichia coli* serotype. *New England Journal of Medicine* 308: 681–685.

Sharp JCM, Reilly W. 2000. Chapter 37. Surveillance of foodborne disease. In: BM Lund, TC Baird-Parker, GW Gould (eds.) *The Microbiological Safety and Quality of Food*, vol. 2., Aspen Publishers, Gaithersburg, MD, pp. 975–1010.

Tenter AM. 1995. Current research on *Sarcocystis* species of domestic animals. *International Journal of Parasitology* 25: 1311–1330.

42
Food Allergens

J. I. Boye, A. O. Danquah, Cin Lam Thang, and X. Zhao

Abstract: The management of allergens along the food value chain and the diagnosis of food allergic diseases continue to pose serious challenges to the food industry as well as health care professionals. Of the over 170 foods known to provoke allergic reactions, nine foods (and their derived products) are today considered to be major allergens accounting for over 90% of all food allergic reactions. These priority allergens include milk, eggs, soya beans, peanuts, tree nuts (e.g. almonds, walnuts, pecans, cashews, Brazil nuts, hazel nuts, pistachios, pine nuts, macadamia nuts, chestnuts and hickory nuts), seafood such as fish (i.e. both saltwater and freshwater

Food Biochemistry and Food Processing, Second Edition. Edited by Benjamin K. Simpson, Leo M.L. Nollet, Fidel Toldrá, Soottawat Benjakul, Gopinadhan Paliyath and Y.H. Hui.

finfish), crustaceans (e.g. shrimp, prawns, crab, lobster and crayfish) and molluscs (e.g. snails, oysters, clams, squid, octopus and cuttlefish), gluten-containing cereals (i.e. wheat, rye, barley and their hybridised strains and products), sesame and mustard. Symptoms of food allergic reactions and the threshold dose required to provoke allergic reaction markedly vary among sensitised individuals. Food production practices, processing conditions and matrix effects can also modify the molecular structure of food allergens and their potential immunogenic properties which can make allergens difficult to detect even when present in foods. This chapter provides an overview of the different types of food hypersensitivities including a distinction between food allergies and food intolerance, the properties of the nine priority food allergens, current approaches for the management of food allergens and a summary of some of the current methods used in food allergen detection.

INTRODUCTION

Humans require food to survive. In addition to basic nutrition, food consumption provides a sense of satisfaction and culinary pleasure and also serves as a source of social entertainment. However, for a small percentage of the population, consumption of certain foods even in small quantities can result in life-threatening allergic reactions. Approximately 4–8% of young children and 2–4% of adults in developed countries suffer from food allergies (Kanny et al. 2001, Munoz-Furlong et al. 2004, Sampson 2004, Sicherer et al. 2004, Breiteneder and Mills 2005). Although over 170 foods are known to cause food allergies, nine foods (and their derived products) are today considered to be major allergens accounting for over 90% of all food allergic reactions. These priority allergens include milk, eggs, soya beans, peanuts, tree nuts (e.g. almonds, walnuts, pecans, cashews, Brazil nuts, hazel nuts, pistachios, pine nuts, macadamia nuts, chestnuts and hickory nuts), seafood such as fish (i.e. both saltwater and freshwater finfish), crustacea (e.g. shrimp, prawns, crab, lobster and crayfish) and molluscs (e.g. snails, oysters, clams, squid, octopus and cuttlefish), gluten-containing cereals (i.e. wheat, rye, barley and their hybridised strains and products), sesame and mustard.

Food allergy is thus emerging as a growing public health problem. Unfortunately, the management of food allergens along the food value chain and the diagnosis of food allergic diseases continue to pose a challenge to the food industry as well as health care professionals. The factors responsible for this are varied. Multiple foods can induce food allergic reactions, and the threshold dose required to provoke allergic reaction markedly varies for different patients; thus, small quantities of foods can cause severe reactions in sensitised individuals, whereas other sensitised individuals can tolerate quantities of the allergen that are orders of magnitude higher. Additionally, symptoms of food allergies vary tremendously among sensitised individuals, and food production practices, processing conditions and matrix effects can alter the molecular structure of food allergens and their potential immunogenic properties and detection. In this chapter, we will provide an overview of the different types of food hypersensitivities, including a distinction between food allergies and food intolerance, the properties of the nine priority food allergens, current approaches for the management of food allergens and a summary of some of the current methods used for food allergen detection.

FOOD HYPERSENSITIVITY

Food Allergy

Food allergy is an immunological reaction resulting from the ingestion, inhalation or atopic contact of food. Immunological reactions can be mediated by IgE antibodies or other immune cells such as T cells. Some workers define food allergy specifically as being those immunological responses mediated by immunoglobulin E (IgE) antibodies, whereas others use the broader definition of immunological response, which include T-cell mediated responses as well. In this chapter, the latter definition will be used.

In general, any protein-containing food can elicit an allergic response in sensitised individuals. Allergenic proteins in foods may be enzymes, enzyme inhibitors, structural proteins or binding proteins with varied biological functions (Stewart and Thompson 1996, Valenta et al. 1999, Chapman et al. 2000, Martin and Chapman 2001). The pathogenesis of food allergy begins with a sensitisation phase during which time the body recognises one or more proteins in a particular food source as a foreign invader and begins to mount an immune-defensive response. Subsequent consumption of the offending food can result in an allergic response that may manifest in one of two forms, that is, immediate or delayed response as described below.

Immediate Hypersensitivity Reactions

Immediate hypersensitivity reactions are mediated by a specific class of antibodies known as IgE. Symptoms of an IgE-mediated food allergy develop within a few minutes to a few hours after an individual has ingested the offending food (Taylor 2000). During the sensitisation phase, the body produces IgE antibodies, which recognise allergenic fragments from the offending food. After production, IgE antibodies circulate in the blood and bind to basophiles and to the surface of mast cells (Taylor 2000). Mast cells and basophiles contain granules packed with potent inflammatory mediators such as histamine, cytokines and other chemotactic factors. When an allergic person encounters an allergen to which the person has previously produced IgE antibodies, the allergen combines with the IgE antibody on the surface of the mast cells and basophiles. This triggers a complex series of reactions that result in the release of histamine and other chemical mediators (prostaglandin D2, tryptase, heparin, etc.) from the granules inside the mast cells and basophiles (Taylor and Lehrer 1996, Lu et al. 2007). Histamine as well as the other chemical mediators has been shown to be the agents responsible for producing symptoms of allergy. The release of some of these mediators occurs rapidly, within 5 minutes after the interaction between the allergen and the IgE antibody (Taylor 2000). Once released, these compounds enter the blood stream and bind to 'receptors' on other cells in the body, causing typical allergic symptoms.

The severity of an allergic reaction depends on how sensitised the person is and the amount of the allergenic component ingested. Histamine released from mast cells is usually complete within 30 minutes after the allergen–IgE antibody interaction. The release of mediators (other than histamine) is slower and their effects are more prolonged. Allergic reactions may, thus, sometimes occur in two phases. The first stage or first symptoms disappear on their own or with medication only to recur in 4–6 hours. Reactions may range from mild to severe life-threatening conditions, and may manifest as gastrointestinal disorders (nausea, vomiting, diarrhoea and abdominal cramping), or involve the skin, leading to urticaria or hives, dermatitis, eczema, angioderma, pruritis or itching. It may also involve the respiratory tract, in which case the individual may suffer from rhinitis, asthma or laryngeal oedema. Systemic anaphylaxis represents the most dramatic and potentially catastrophic manifestation of immediate hypersensitivity. Virtually, every organ in the body can be affected, although reactions involving the pulmonary, circulatory, cutaneous, neurological and gastrointestinal tract are the most common (Anderson 1986).

About 1–2% of the total population are affected by IgE-mediated food allergies (Chafen et al. 2010). The most vulnerable group are infants and young children, with 5–8% infants under 3 years being affected (Sampson and McCaskill 1985, Motala and Lockey 2004, Sicherer 2010). More than 170 different foods have been implicated in immediate hypersensitivity reactions (Hefle et al. 1996, Taylor 2000).

Delayed Hypersensitivity Reactions

Unlike the immediate hypersensitivity reactions described above, symptoms associated with delayed hypersensitivity induced by T-cells may appear after several hours to even days following consumption of the offending food. Delayed allergic reactions to food are relatively more difficult to identify unless the suspected food is eliminated from the diet for at least several weeks and then slowly reintroduced followed by monitoring for the onset of any physical, emotional or mental changes (Anderson 1986). These reactions are mediated by tissue-bound immune cells, and the acuteness of the reaction may be less than that of the immediate reactions (Taylor 2000). An example of delayed hypersensitivity reaction to food is celiac disease (CD), which is described in greater detail below.

Food Intolerance

Food intolerances are abnormal reactions to food or food components that do not involve the immune system (Breneman 1987). They are generally less severe, shorter in duration and more localised than immunological reactions. Food intolerances are the most common types of food sensitivities and specific examples include the metabolic food disorders and anaphylactoid responses described below.

Metabolic Food Disorders

Metabolic food disorders arise from inherited genetic deficiencies that reduce the capacity of afflicted individuals to efficiently metabolise food components. One such example is lactose intolerance, which is caused by an inherited deficiency of the digestive enzyme lactase, which hydrolyses lactose in milk and milk products into galactose and glucose for further processing within the body. As a result, the undigested lactose cannot be absorbed by the small intestine and passes into the colon where bacteria metabolise it into carbon dioxide (CO_2) and water (H_2O), leading to bloating, abdominal cramping and frothy diarrhoea (Taylor 2000).

Another example is phenylketonuria, a metabolic disorder in which afflicted individuals are deficient in the hepatic enzyme phenylalanine hydroxylase, which metabolises phenylalanine into tyrosine. Thus, phenylalanine and phenylpyruvate accumulate in the body, and if left untreated, it could affect the central nervous system and cause mental retardation and/or brain damage in infants and children.

A third example of food intolerance due to a genetic deficiency is favism, where individuals are intolerant to fava beans or the pollen from the *Vicia faba* plant. Persons with this intolerance have an inherited deficiency of the erythrocyte glucose-6-phosphate dehydrogenase (G6PDH) enzyme, which is essential to protect erythrocyte membranes against oxidative damage (Taylor 2000). This deficiency is crucial as several endogenous oxidants present in broad beans, for example, vicine and convicine, are capable of damaging erythrocyte membranes in G6PDH-deficient individuals, resulting in acute haemolytic anaemia with pallor, fatigue, dyspnoea, nausea, abdominal and/or back pain, fever and chills. It may even lead to more serious symptoms such as haemoglobinuria, jaundice and renal failure, but these situations are rare. Symptoms occur quite rapidly, usually within 5–24 hours following consumption of fava beans.

Anaphylactoid Responses

Anaphylactoid responses are due to the non-immunologic release of chemical mediators, such as histamine, from mast cells (Sampson et al. 1992). The specific substances that cause this reaction are not well known, and the reactions are often confused with true food allergies because they display similar symptoms. Anaphylactic shock is one of the most startling symptoms associated with food allergies, and such food intolerances, and may affect the gastrointestinal tract, skin, respiratory tract and the cardiovascular system. It can cause severe hypotension, and if not treated properly, it may lead to death within minutes after ingesting the offending food (Taylor 2000).

The rest of this chapter will focus on describing some of the properties of the nine most common food allergens and methods for detecting allergens.

MILK ALLERGENS

Milk is a nutritional biological fluid secreted from the mammary gland of female mammalians, and it is primarily intended to provide nutritional requirements of neonates. Human milk and the milks of dairy animals, such as cows, goats, sheep and buffaloes, have been extensively studied. Humans are the only known mammals that consume the milk of other mammals, particularly from cows. In fact, cow's milk and cow's milk-associated food have been deeply rooted as an important part of the human diet, and the dairy industry plays a huge role in supporting nutrition of humans and continues to be a backbone of the agri-food sector in many countries.

In developed countries, bovine milk and milk-derived products contribute about 19% of total dietary protein intake and 73% of calcium intake (Tome et al. 2004). However, for some people, consumption of cow's milk and cow's milk-derived food has to be avoided due to milk allergy. Recent studies have indicated that prevalence and persistence of cow's milk allergy (CMA) in industrialised countries may be increasing. In North America, incidence of CMA is estimated at 2.5% in children and about 1% in the adult population with a 75% outgrowing rate at the age of 16 (Sicherer and Sampson 2010).

Milk Protein

Milk protein is a very complex mixture of molecules. Thanks to advanced analytical techniques, over 200 types of proteins have been identified in bovine milk (Ng-Kwai-Hang 2002). These proteins can be arranged into five groups: caseins, whey proteins, milk fat globule proteins, enzymes and minor miscellaneous proteins. Caseins and whey proteins make up almost all of the total milk protein composition, whereas the other 3 groups are found in trace amounts (Table 42.1). Caseins can be separated from the other proteins by acid precipitation at pH 4.6 to form a coagulum. However, the ratio of casein and whey in the milk varies among different species. In human milk, the ratio is 40:60, in equine the ratio is 50:50 and in major dairy animals (cows, goats, sheep and buffaloes) the ratio is 80:20.

Major Milk Allergens

β-Lactoglobulin (BLG, not found in human milk) and caseins, α-s1-casein in particular, are regarded as the principal allergens in cow's milk (Kaminogawa and Totsuka 2003). A recent study of 115 CMA Japanese children identified casein as a major milk allergen with 107 (97.3%) children reacting positively to casein-specific IgE, whereas 51 patients (46.6%) were positive to BLG-specific IgE (Nakano et al. 2010). However, studies on large populations show that a high proportion of CMA patients are sensitised to multiple milk proteins, including proteins present in very low quantities, such as bovine serum albumin (BSA), immunoglobulins and lactoferrin, suggesting that all milk proteins could be potentially allergenic. Sensitisation to caseins, BLG, and α-lactalbumin (ALA) appear to be closely linked. However, sensitivity to BSA seems to be completely independent. Epitope mapping of major milk allergens has revealed multiple allergenic epitopic regions within each protein (Busse et al. 2002, Cocco et al. 2003).

Caseins

Casein (CN) forms the main fraction of milk proteins and is subdivided into a number of families, including α-S1-, α-S2-, β-, κ- and γ-caseins.

The α-S1-casein (α-S1-CN) is a single-chain phosphoprotein of 199 amino acid residues and represents about 40% of total casein. α-S1-CN is characterised by a high content of proline residues without disulfide bonds and the presence of a small amount of secondary structure, such as α-helix or β-sheets. Several epitope regions of α-S1-CN recognised by human IgE antibodies have been identified. These regions include amino acid (aa) 19–30, aa86–103 and aa141–150 (Spuergin et al. 1996), aa181–199 (Nakajima-Adachi et al. 1998) and nine regions from amino acid residues 17 through 194 (Chatchatee et al. 2001). Reasons for the differences in the dominant

Table 42.1. Characteristics of the Major Proteins in Human and Cow's Milk

Proteins	Human (mg/mL)	Cow (mg/mL)	Molecular Weight (kDa) of Cow's Milk Proteins	Number of Amino Acids in Cow's Milk
Whole caseins				
α-s1-Casein	0	11.6	23.6	199
α-s2-Casein	0	3	25.2	207
β-Casein	2.2	9.6	24	209
κ-Casein	0.4	3.6	19	169
γ-Casein	0	1.6	11.6–20.5	
Whey proteins				
β-Lactoglobulin	0	3	5.3	162
α-Lactalbumin	2.2	1.2	4.8	123
Immunoglobulins	0.8	0.6	<150	
Serum albumin	0.4	0.4	66.4	582
Lactoferrin	1.4	0.3	76.2	703
Other	1.3	0.6		

epitopic regions reported by different researchers may be due to racial (Kaminogawa and Totsuka 2003), dietary, breed and environmental differences.

The α-s2-casein is comprised of 207 amino acids and has one disulfide bond (Wal 1998). The α-S2-CN family accounts for 12.5% of the casein fraction and is the most hydrophilic among all caseins due to the presence of clusters of anionic groups. Ten IgE-binding regions have been identified between amino acid positions 31–200 of α-s2-casein (Busse et al. 2002).

β-Casein (β-CN) represents 35% of the total caseins and is quite complex because of the action of native milk protease plasmin. Plasmin cleaves β-CN and thereby generates γ1-, γ2-, and γ3-CN fragments. β-CN is the most hydrophobic component among casein fractions. Six major and three minor IgE-binding epitopes, as well as eight major and one minor IgG-binding regions, have been identified on β-CN (Chatchatee et al. 2001).

κ-CN accounts for 12.5% of the total casein fraction. Eight major IgE-binding epitopes, as well as two major and two minor IgG-binding epitopes, have been detected in κ-casein (Chatchatee et al. 2001).

β-Lactoglobulin

Bovine BLG is the most abundant whey protein and represent 50% of total whey proteins. It has no homologous counterpart in human milk. It possesses three disulfide bridges. BLG is relatively resistant to proteases and acid hydrolysis. BLG belongs to the lipocalin superfamily and is capable of binding a wide range of molecules, including retinol, β-carotene, saturated and unsaturated fatty acids and aliphatic hydrocarbons (Breiteneder and Mills 2005). BLG epitopes reported as markers for persistent CMA include aa1–16, aa31–48, aa47–60, aa67–78 and aa75–86 (Jarvinen et al. 2001, Inoue et al. 2001).

α-Lactalbumin

Bovine ALA is characterised by four disulfide bridges and possesses a high-affinity binding site for calcium, which stabilises its secondary structure.

Bovine ALA shows a 72% amino acid sequence homology to human ALA and is, thus, an ideal protein for the nutrition of human infants. Four different linear IgE-binding peptides, aa1–16, aa3–26, aa47–58 and aa93–102, have been identified by epitope mapping in children who outgrow CMA later in life (Jarvinen et al. 2001).

Bovine Serum Albumin

BSA accounts for about 5% of total whey proteins and is physically and immunologically very similar to human blood serum albumin. BSA has 17 disulfide bonds and most of the disulfide bonds are protected in the core of the protein and are therefore not easily accessible (Restani et al. 2004). This may be the reason for its relatively stable tertiary structure. IgE-binding epitopes identified for BSA have been inconsistent among studies (Karjalainen et al. 1992).

Lactoferrin

Lactoferrin (LF) is a milk-specific iron-binding protein. Although LF in cow's milk is homologous to human LF, the content is lower than that of human's milk. Its main function is to defend the host against infections and inflammations due to its ability to sequester iron from the environment, thereby removing this essential nutrient for bacterial growth (Ward et al. 2002). Despite its low concentration in bovine milk, LF-specific IgE have been detected in 45% of CMA patients (Wal 1998).

Milk Allergen Cross-Reactivities

Different species of milk-producing ruminants have the same or closely related milk proteins with relatively similar ratios of casein and whey (Monaci et al. 2006). Furthermore, considerably high amino acid sequence homology (varying from 87–96%) exists among major milk proteins of cow, goat and ewe (Table 42.2). As a result, IgE cross-reactivity among the milk of goat, ewe and cow has been reported for most CMA patients. IgE sensitisation to sheep and goat casein has been found to be as high as 93–98% in children with IgE-mediated CMA (Besler et al. 2002a). In general, the milk of other ruminants is not a good alternative for CMA patients.

Cross-reactivity between cow's milk allergens and soya bean proteins has also been reported. Allergic reactions to soya bean was observed in about 17–47% of children with CMA (Hill et al. 1999). Cross-reactivity between soya bean and casein was confirmed by Rozenfeld et al. (2002), who showed that glycinin-like protein (two polypeptides, A5-B3) from soya bean were able to bind casein-specific monoclonal antibodies.

Threshold Dose

There is limited data on the threshold dose required to provoke a milk allergy reaction. Several research groups using double blind placebo controlled food challenges with milk protein have used doses varying from 0.6 to 180 mg to induce allergic reactions in CMA patients (Monaci et al. 2006). However, these studies were primarily intended for diagnostic purpose and not necessarily for determining the lowest provoking dose. Severe adverse reactions

Table 42.2. Comparisons of the Amino Acid Sequence Homology of the Major Proteins Found in Cow's Milk and That of Goat and Ewe

	Sequence Homology (%)	
Milk Proteins	Cow vs Goat	Cow vs Ewe
β-Lactoglobulin	96	96
α-Lactalbumin	95	94
α-s1-Casein	87	89
α-s2-Casein	88	89
β-Casein	90	90
κ-Casein	85	84

Source: Adapted from Wal 2004.

have been reported after eating food contaminated with trace amounts of milk protein, including breast milk from mothers who consume cow's milk (Gerrard and Shenassa 1983), meat treated with casein to enhance texture (Yman et al. 1994), frozen desserts that contain trace amounts of whey protein (Laoprasert et al. 1998) and lactose-containing medications that contained residual milk protein (Nowak-Wegrzyn et al. 2004).

Bindslev-Jensen et al. (2002) developed a statistical approach to estimate threshold levels of four major food allergens. They reported that the threshold dose required for a person to develop an adverse reaction using a million CMP susceptible population is 0.005 mg of cow's milk or 7×10^{-5} mg of milk protein (Bindslev-Jensen et al. 2002). Other workers have reported threshold levels ranging between 3 and 180 mg (Morisset et al. 2003a, Taylor et al. 2002).

Effect of Processing on the Allergenicity of Cow's Milk Proteins

Processing techniques are routinely applied to raw milk in order to reduce or eliminate microorganisms and enhance shelf life. Some of the techniques such as homogenisation, pasteurisation and sterilisation do not modify protein structure significantly, whereas others such as hydrolysis and irradiation do. In general, results from several studies have shown that cow's milk allergenicity could be decreased, increased or unchanged by processing treatments such as pasteurisation or sterilisation and homogenisation (Host and Samuelsson 1988).

Heat treatment

The effect of heating on milk protein allergenicity remains controversial. Pasteurised milk has been reported to have higher allergenicity than raw or homogenised milk (Host and Samuelsson 1988). Caseins are more thermostable, whereas BLG manifests a thermolabile behaviour (Wal 2004). However, BLG may be protected from denaturation when heated due to possible interactions with caseins. Currently, heat denaturation is not accepted as a satisfactory process to reduce the allergenicity of milk protein. On the contrary, application of heat treatment could lead to the formation of neo-allergens.

Hydrolysis

Hydrolysis of milk proteins reduces their allergenicity to varying degrees. Several cow's milk proteins, for example BLG, are relatively resistant to degradation by proteolytic enzymes, while others are considered very labile (e.g. caseins). Boza et al. (1994) obtained an extensively hydrolysed hypoallergenic infant formulae by using a combination of ultrafiltration process and hydrolysis with enzymes from bacterial/fungal origin with broad specificity. Extensively hydrolyzed formulas (eHF) successfully protect the development of allergy symptoms in the majority of CMA infants (Walker-Smith 2003). However, even eHFs (molecular weight (MW) less than 1.5 kDa) can still induce allergic response in infants with atopic family background (Nentwich et al. 2001) and CMA infants (De Boissieu et al. 1997). Moreover, eHF have also been criticised for their poor functionality (Crittenden and Bennett 2005).

Radiation

Lee et al. (2001) reported that the application of gamma irradiation to α-casein and BLG altered the epitope structure of both milk proteins, probably due to agglomeration of the milk proteins and consequent decrease in solubility. Further research is needed to clearly elucidate the effect of irradiated milk on allergenicity.

High-Pressure Treatment

High pressure may reveal potentially immunogenic hydrophobic regions to enzymes, resulting in better hydrolysis. Bonomi et al. (2003) reported that enzymatic hydrolysis under high pressure (600 MPa) produced non-immunogenic peptides. Other workers have, however, reported an increase in antigenicity of milk proteins on high-pressure treatment in the absence of hydrolysis (Kleber et al. 2004, 2007). Thus, as with irradiation, further research will be useful to ascertain the effect of high pressure on milk protein allergenicity.

EGG ALLERGENS

Egg protein is used as a protein nutritional standard because it is highly nutritional and has all the essential amino acids in the right amounts required by the human body. Moreover, with the exception of vitamin C, eggs serve as a good source of vitamins A, D, E, K, as well as the B vitamins. Whole egg and egg-derived ingredients also possess excellent functional properties (e.g. gelation, emulsification and foaming), which has resulted in their extensive use in the formulation of various food products.

Egg consists of the white and yolk, both of which contain allergenic proteins. Whole egg contains 12.8–13.4% protein, 10.5–11.8% fat, 0.3–1% carbohydrate and 0.8–1.0% ash on a wet basis (Breeding and Beyer 2000). Proteins found in egg white are ovalbumin, conalbumin, ovomucoid, lysozyme, ovomucin and other minor albumen proteins such as avidin, ovoglobulins, flavoprotein and ovoinhibitors. The egg yolk proteins are rich in lipoproteins and phosphoproteins and primarily comprise of lipovitellin, lipovitellenin, vitellin, vitellenin, phosvitin, transferrin, γ-globulin, serum albumin and α_2-glycoprotein.

Prevalence, Symptoms and Thresholds

Although the majority of people can tolerate eggs in their diet, a small percentage of the population, mostly children, suffer severe allergic reactions after consuming egg. Symptoms of egg allergic reactions include vomiting, diarrhoea, gastrointestinal pain, urticaria, angiodema, atopic dermatitis, asthma and rhicoconjunctivitis (Martorell Aragonés et al. 2001).

Egg allergy prevalence in the general population is estimated between 1.6 and 3.2%, making egg the second most important cause of food allergic reactions next to peanut (Arede et al. 2000, Mine and Yang 2008 and references within). In children, egg allergy is the most commonly occurring food hypersensitivity with symptoms usually becoming evident in the child's first year.

Major Egg Allergens

Egg white is more allergenic than the egg yolk. Ovalbumin and ovocumoid, the two most potent egg allergens together make up about 65% of the composition of egg white proteins. Ovalbumin has a MW of 44.5 kDa and contains 385 amino acids. Using pooled sera from 18 egg-allergic patients, Mine and Rupa (2003) identified five IgE binding linear epitopes in the primary structure of the protein. Four of the five epitopes were exposed at the surface of the protein.

Ovomucoid, the other major allergen, is a 28 kDa MW protein comprising 186 amino acids. The protein has three structurally distinct domains (Domain I, II and III), which are cross-linked by disulfide bonds. Several linear sequences representing allergenic epitopes recognised by both IgE and IgG antibodies have been identified (Zhang and Mine 1998, Mine and Zhang 2002). Ovomucoid also contains 20–25% carbohydrate moieties. Most reports indicate that the carbohydrate moieties are not immunogenic; however, some studies suggest that the carbohydrate region in the third domain is immunologically active (Mine and Rupa 2004). Ovomucoid is relatively stable to digestion, and heat and allergenic fragments have been detected even after digestion by pepsin.

Other reported egg allergens are ovotransferrin (MW 76 kDa) and lysozyme (MW 14.3 kDa). Aabin et al. (1996) reported a higher frequency of reactivity for ovotransferrin (53%) than for ovomucoid (38%) and ovalbumin (32%), suggesting that it may be a more important allergen than the latter two. The same authors reported frequency of reactivity of 15% for lysozyme. Very few studies have reported on the allergenicity of egg yolk proteins; however, some of the allergenic proteins identified in egg yolk include apovitellenin I, apovitellenin VI and phosvitin (Walsh et al. 1988, 2005).

Thresholds of Clinical Reactivity to Eggs

Reported thresholds of clinical reactivity to eggs range between 0.13 and 6.5 mg of hen's egg protein (Morisset et al. 2003a, Taylor et al. 2004). As with other allergens, there are currently no cures for egg allergy, and the best management tool is the reading of food labels and avoidance of foods containing egg or egg-derived ingredients. The use of multiple definitions for eggs and egg ingredients (e.g. albumin, albumen, conalbumin, egg nog, egg yolk, egg white, lecithin, livetin, lysozyme, ovalbumin, ovoglobulin, ovomacroglobulin, ovomucin, ovomucoid, ovomucin, ovovitellin, ovotransferrin and vitellin) in the past made egg avoidance challenging. Today, most countries in North America as well as in Europe require the use of the common names of priority allergens when they are used as ingredients in foods or are likely to be present in foods in spite of best efforts to control their presence. Fortunately, most children outgrow their egg allergy by school age (Heine et al. 2006). Egg allergy, however, tends to persist in children with positive skin tests, multiple allergies and/or who have more severe reactions (e.g. respiratory symptoms, angiodema and multisystemic reactions) or other atopic diseases such as asthma (Ford and Taylor 1982, Mine and Yang 2008).

SOYA BEAN ALLERGENS

Soya bean (*Glycine max*) is a legume belonging to the *fabaceae* family and the *glycine* genus. On a dry basis, soya beans contain 35–40% protein, 17–23% lipid, 31% carbohydrate and 4–5% minerals. Soya bean is considered as one of the most nutritional plant sources of food providing a well-balanced amino acid profile and good supplies of omega 3 and omega 6 fatty acids. The history of the seed dates as far back as 3000–5000 years, with its origins somewhere in Asia.

Consumption of soya bean and soya bean foods has been linked to many health benefits. Various studies have reported associations between soya bean consumption and reduced risk of cardiovascular disease, cancer, diabetes, bone loss and menopausal symptoms amongst others (Friedman and Brandon 2001, Hori et al. 2001, Chen et al. 2003, Zhang et al. 2003, Stephenson et al. 2005, Anderson 2008). These reports have fuelled the growth of soya foods in Western countries. In October 1999, the Food and Drug Administration of the United States approved a health claim linking the consumption of 25 grams of soya protein a day (as part of a diet low in saturated fatty acids) to reduced cardiovascular disease risk (FDA 1999). For manufacturers to make this claim, products must provide at least 6.25 grams of soya protein per serving. Soya flour and soya protein ingredients are therefore increasingly being used in many food products.

Unfortunately, soya bean is listed in Canada, the United States, Australia and the European Union as a priority allergen, requiring labelling when it is used as an ingredient in foods. Symptoms of soya bean allergy are similar to the other major allergens and include cutaneous, respiratory as well as gastrointestinal responses. Although the majority of allergic responses occur on ingestion, allergic reactions on inhalation of soya bean and soya bean byproducts has also been reported (Gonzalez et al. 1992, 1995, Codina et al. 1997).

Major Soya Allergens

Over 17 different allergens have been identified in soya bean. These include soya bean glycinin (11S), β-conglycinin (7S), soya bean vacuolar protein (Gly m Bd 30K or P34), the Kunitz trypsin inhibitor (KTI), Gly m Bd 28K, soya bean profilin (Gly m 3), soya bean hull proteins (Gly m 1.0101, Gly m 1.0102, Gly m 2) and the pathogensis-related (PR) soya bean protein SAM22 (Gly m 4) (Wilson et al. 2005, L'Hocine and Boye 2007, Boye et al. 2010).

Glycinin and β-conglycinin have been studied extensively as they are the major soya bean storage proteins and represent over 70% of the proteins found in soya (Liu 1997). They are both globulins belonging to the cupin superfamily (Breiteneder and Radauer 2004) and have complex quaternary structures with conserved amino acid sequence homology. Glycinin

(MW ~ 300 kDa) has a hexameric structure with each monomer comprising an acidic and basic subunit linked by disulfide bonds. β-conglycinin (MW ~ 150 kDa) has a trimeric structure with the three major subunits α, α' and β having MWs of 76, 72 and 53 kDa, respectively (Liu 1997, Krishnan 2000). Various studies have shown that all glycinin and β-conglycinin subunits bind IgE from soya bean-allergic patients although a few studies have found an absence of binding to some subunits (Pedersen and Djurtoft 1989, Krishnan et al. 2009, Boye et al. 2010).

The soya bean vacuolar protein (P34; MW of 34 kDa), which is an oil body-associated protein, is also a major soya allergen. Over 65% of soya bean-allergic patients reportedly react to it (Ogawa et al. 1991, 1993). The KTI, which is a 21 kDa belonging to the plant defence system, is another important allergen. KTI exerts its effect by inhibiting the activity of proteases such as trypsin. Although the major allergenic epitopes have not been identified, reports suggest that it may be a culprit in soya bean-induced occupational respiratory disorders (Baur et al. 1996) along with the soya bean hydrophobic lipid transfer proteins identified in soya bean hulls (Gonzalez et al. 1992, 1995).

Prevalence, Symptoms and Threshold

Prevalence rates of soya allergy range between 0.3% and 0.4% (Becker et al. 2004, Sicherer and Sampson 2006). The exact amount of soya proteins required to induce an allergic response is, however, not known. Reported threshold levels vary significantly and range between 0.0013 and 500 mg of soya protein (Bindslev-Jensen et al. 2002, Ballmer-Weber et al. 2007).

Effect of Processing on Soya Allergens

In general, processing treatments such as thermal treatment, microwaving, irradiation, hydrolysis and fermentation have not been shown to remove soya bean allergenicity. Some treatments may cause modifications in proteins structure and hide allergenic epitopes reducing but not eliminating their immunogenicity, whereas other treatments cause unfolding of the proteins, allowing greater exposure of hidden allergenic epitopes and increasing the immunogenic properties of the proteins (Davis and Williams 1998, Soler-Rivas and Wichers 2001).

Soya bean is processed into products such as soya beverages, tofu, edamame, miso, natto and tempeh. The beans can also be processed to obtain soya oil, soya flour, soya lecithin, soya fibre, soya protein concentrate (>65% protein on dry basis), soya protein isolate (>90% protein on dry basis), textured soya products and soya hydrolysates or hydrolysed vegetable protein, which are used in a wide variety of products. Other novel applications include soya ice cream and soya yogurt. Allergenicity of soya foods is linked to the presence of residual soya proteins in these foods, so unless the soya ingredient used is highly refined (e.g. refined soya bean oil), it is likely to contain allergenic protein and must be labelled when used in foods. Studies conducted on soya lecithin have generally reported little or no allergy risk to sensitised individuals (Awazuhara et al. 1998); however, hydrolysed and fermented soya products tend to retain some of their allergenic properties (Hefle et al. 2005, L'Hocine and Boye 2007). Appropriate caution therefore needs to be exercised in the consumption of processed soya foods.

Soya Allergen Cross-Reactivities

Although not confirmed clinically, a high cross-reactivity has been reported for different legumes (soya bean, peanut, lentils and beans; Yunginger 1990, Eigenmann et al. 1996, Kalogeromitros et al. 1996). This may be explained by the high amino acid sequence homology, which frequently occurs in plant proteins belonging to the same family. Clinical co-reactivity rates in peanut-allergic patients for soya using placebo-controlled challenges range between 1% and 6.5% (Burks et al. 1988). Additionally, coexisting clinical reactivity has been reported between soya bean and cow's milk for some allergic patients ranging between 5% and 50% depending on the specific group of patients studied (Host and Halken 1990, Burks et al. 1994).

PEANUT AND TREE NUT ALLERGENS

Peanuts and tree nuts contain some of the most potent food allergens. Prevalence rates of 0.8–1.5% for peanut allergy in the UK and US population and about 0.6% for tree-nut allergy in the US population have been reported (Grundy et al. 2002, Sicherer et al. 2003). Nuts and tree nuts belong to different plant species; however, they are frequently discussed together as their presence, handling and use in the food chain as well as the allergic responses they induce are often similar.

Peanut (*Arachis hypogea*) is a legume belonging to the family *Leguminosae*. It grows under the ground in peanut pods containing the peanut seed. Tree nuts, on the other hand, are edible seeds that grow on trees. Examples of tree nuts that are of most concern as allergens are hazelnut (*Corylus avellana*), almond (*Prunus dulcis*), pistachio (*Pistachia vera*), macadamia nuts (*Macadamia integrifolia*), cashew (*Anacardium occidentale*), walnut (*Juglans regia*), pine nut (*Pinus pinea*), pecans (*Carya illinoinensis*) and Brazil nut (*Bertholetia excelssa*).

Prevalence and Threshold

Prevalence rates of peanut and tree nut allergies vary for different populations but appear to be higher in Western societies such as in Europe and North America. Reported rates vary from 0.2 to 1.7 for peanut and 0.1 to 1.4 for tree nuts (Sicherer et al. 2003). Allergic responses following ingestion of peanut and tree nuts by allergic patients range from oral pruritus, nausea, vomiting, urticaria, angiodema, bronchospasm, bronchitis, hypotension, anaphylaxis and death in some instances. The reaction often occurs within minutes to a few hours after food consumption, and the severity of the response may be exacerbated by preexisting asthma (Sampson 2002, Sicherer et al. 2003).

Threshold doses for clinical reactivity reported in the literature range from 100 μg to 10 mg for peanut and 20 μg to 7.5 mg for tree nuts (Hourihane et al. 1997, FDA 2006). Cross-reactivities between peanut, soya bean as well as other tree nuts have been reported (de Leon et al. 2003).

Major Peanut and Tree Nut Allergens

The major allergens identified in peanut are Ara h 1 (glycoprotein, vicilin, MW 63.5 kDa), Ara h 2 (glycoprotein, conglutin, MW 17.5 kDa), Ara h 3 (legumin, MW ~ 60 kDa), Ara h 4 (legumin, MW 37 kDa), Ara h 5 (profilin, MW 14–15 kDa), Ara h 6 (conglutin, MW 14.5 kDa), Ara h 7 (conglutin, MW 15.8 kDa) and Ara h 8 (pathogenesis-related protein, MW 16.9 kDa; Wen et al. 2007, Rajamohamed and Boye 2010). Ara h 1 and Ara h 2 are classified as major allergens and are recognised by the sera of >90% of peanut-allergic patients. Ara h 3, Ara h 4, Ara h 5, Ara h 6, Ara h 7 and Ara h 8 are less frequently recognised by the sera of peanut allergic individuals and are classified as minor allergens (Wen et al. 2007, Rajamohamed and Boye 2010).

Allergenic proteins in tree nuts vary depending on the type of nuts. In a voluntary survey report on tree nut allergy conducted by Sicherer et al. (2001), 46% of tree-nut-allergic individuals reacted to multiple tree nuts, and 54% reacted to single tree nuts with the highest reactions being reported toward walnut (34%), cashew (20%) and almond (15%) and lower allergic responses to pecan (9%), pistachio (7%), hazelnut, Brazil nut, macadamia nut, pine nut and hickory (less than 5% each). Table 42.3 provides a list of some of the major proteins identified in tree nuts and their properties (Rajamohamed and Boye 2010).

Processing-Induced Changes in Peanut and Tree Nut Allergenic Proteins

Processing induces changes in peanut and tree nut proteins, which can modify their allergenic properties. Many research studies have found, for example, that roasting increases the immunogenic properties of peanut compared to frying and boiling. Using sera of peanut-allergic patients, Beyer et al. (2001) found lower IgE-binding intensities of Ara h 1, Ara h 2 and Ara h 3 in fried and boiled peanuts compared to roasted peanuts. Similarly, Maleki et al. (2000) found significant increases in the allergenic properties of roasted peanut compared to raw peanut. On the contrary, Koppelman et al. (1999) reported no change in the allergenicity of Ara h 1on heat treatment. Hansen et al. (2003) also reported that dry roasting of hazelnut reduced its allergenicity compared to raw hazel nut. Differences in the effect of the thermal treatment on the molecular structure of the proteins and their solubility may explain the variations in the responses reported.

FISH AND SHELLFISH ALLERGENS

Fish and shellfish represent one of the most common sources of food allergens in the adult population. Fish and shellfish species known to cause allergic reactions include but are not limited to cod, flounder, grouper, haddock, halibut, hake, herring, mackerel, pike, sole, snapper, trout, crabs, lobsters, prawns, shrimps, crayfish, octopus, squid, clams, mussels, oysters, scallops and snails.

The major allergen in fish is parvalbumin (Gad c 1), a 12 kDa protein (O'Neil et al. 1993). Tropomyosin, with a MW of ~36 kDa, is the major allergen found in shrimp, lobster, crab and molluscs such as squid, oyster, snail, mussels, clam and scallops (Daul et al. 1993a, 1993b). Both parvalbumin

Table 42.3. Major Tree-Nut Allergens and Their Characteristics

Tree Nuts	Allergen	Molecular Weight (kDa)	Protein Family	Allergen Type	Identified Epitopes	Allergen Stability	References
Cashew	Ana o 1	50	Vicilin (7S)	Major	11	Thermostable	51, 54
	Ana o 2	33 and 53	Legumin (11S)	Major	22	Thermostable	53, 54
	Ana o 3	12	Albumin (2S)	Major	16	Thermostable	52, 54
Walnut	Jug r 1	14	Albumin (2S)	Major	3	NR	55, 56
	Jug r 2	44–47	Vicilin (7S)	Major	NR	NR	57
	Jug r 3	9	LTP	NR	NR	NR	58
	Jug r 4	NR	Legumin (11S)	NR	NR	NR	58, 76
Hazelnut	Cor a 1	18	PR-10	Major	NR	Thermolabile	59, 61
	Cor a 2	14	Profilin	Major	NR	NR	59
	Cor a 8	9	LTP	Major	NR	NR	63
	Cor a 9	35–40	Legumin (11S)	Major	NR	NR	63
	Cor a 11	47	Vicilin (7S)	NR	NR	NR	63
Brazil nut	Ber e 1	9	Albumin (2S)	Major	NR	Thermostable and resistant to proteolysis	64, 65
	Ber e 2	22 and 35	Legumin (11S)	Minor	NR	NR	67
Almond	NR	45	Vicilin (7S)	Major	NR	NR	70
	NR	20–22 and 38–42	Legumin (11S)	Major	NR	NR	68, 69
	NR	12	Albumin (2S)	Major	NR	NR	70

Source: Adapted from Rajamohamed and Boye 2010.
MW, molecular weight; NR, - not reported; PR, pathogenesis-related protein family; LTP, lipid transfer protein.

and tropomyosin are muscle cell proteins. To date, eight major IgE-binding epiopes have been identified in shrimp tropomyosin (Lehrer et al. 2003).

Exposure, including handling, consumption and inhalation of air-borne particles from fish and fish ingredients, can induce allergic reaction in sensitised individuals. Seafood-induced allergic reactions are generally similar to responses induced by many of the other allergenic foods. A study conducted using 30 shrimp-sensitive and 37 fish-allergic individuals reported allergic symptoms ranging from generalised itching, urticaria to swelling of the lips and tongue (Lehrer et al. 2003). Other reported symptoms include difficulty breathing, gastrointestinal distress and anaphylactic shock (O'Neil et al. 1993, Daul et al. 1993a, 1993b).

Additionally, occupational reactions among a variety of seafood workers (e.g. fish and prawn workers, seafood processing workers, fishermen, canners, restaurant cooks and other workers in the seafood industry) have been reported. Cartier et al. (1984) showed that workers in the seafood industry were exposed to occupational allergens through direct contact with seafood products as well as inhalation of bits of seafood or water droplets generated during processing. Out of the 303 crab workers investigated, 18% reported rhinitis or conjunctivitis, about 24% some sort of skin rash and over a third reported of asthma.

Very little work has been done on the effect of processing on the allergenicity of seafood. As with other allergens, highly refined products from seafoods that do not contain residual proteins (e.g. refined fish oils, gelatine and isinglass) do not pose a risk to allergic consumers. However, processing techniques that leave seafood protein fragments in the finished product may pose serious allergenic risk to sensitised individuals. As consumers and the food industry become increasingly aware of the health benefits of fish, consumption and utilisation of fish and fish products is likely to increase, which could increase the allergen risk for fish- and shellfish-sensitised individuals.

ALLERGENS IN CEREALS

Certain cereal grains contain proteins that induce immune-mediated responses in individuals who are predisposed to CD or who have specific cereal allergy. As the mechanisms involved in celiac disease are distinctly different from those involved in IgE-mediated cereal allergy, they will be discussed separately.

Celiac Disease

CD also sometimes known as gluten-sensitive enteropathy or gluten intolerance is an abnormal immunological response to gluten/gliadin, which frequently results in a diseased state characterised by damage of the lining of the gut (villous atrophy). In these individuals, the T lymphocytes in the small intestines respond abnormally to gluten, causing inflammation and damage to the absorptive epithelium of the small intestine, resulting in malabsorption and disorders such as diarrhoea, bloating, weight loss, anaemia, weakness and muscle cramps. In children, CD leads to growth retardation and underweight. Symptoms linger for some days even after the offending food is avoided due to the fact that the damaged intestine requires time to heal. Mortality rate has not been reported but patients are likely to develop malignant lymphomas (Ferguson 1997).

CD occurs more commonly in Caucasians than in Blacks, Asians and Hispanics according to present knowledge. Whether this is due to under-diagnosis or true biosocial/genetic difference is not clear. Reported prevalence rates are 1:200–400 in Europe, 1:133 in America, 1:100–300 in a UK study and 1:120 in a Belfast study (Rostami et al. 1999, Gomez et al. 2001, Fasano and Catassi 2001, Fasano et al. 2003, García Novo et al. 2007). CD appears to be genetic with 10% prevalence rates reported among first-degree relatives of CD patients and 70–100% concordance rates amongst twins. Higher prevalence rates are reported in women likely due to higher rates of diagnoses.

The primary offending foods for celiacs are wheat, barley and rye. The major proteins present in these cereals are albumins, globulins, gliadin (prolamin) and glutenin (glutelin) and the offending protein for celiacs is the gluten fraction in these cereals, which are the prolamins and glutelins, particularly the prolamins (i.e. hordein (barley), secalin (rye) and gliadin (wheat)). Several repeating peptide sequences (e.g. QQPFP, QQQP, QQPY, QPYP, PSQQ) in the primary structure of these proteins have been blamed (Osman et al. 2001, Kahlenberg et al. 2006, Darewicz et al. 2008). Although the mechanism involved in the pathogenesis of the disease is unclear, tissue transglutaminase is believed to play a key role in the deamidation of glutamine converting it to glutamic acid, which allows the immune cells to bind, provoking continued immune response (Anderson et al. 2000, Mazzeo et al. 2003). The principal organ targeted is the gut (i.e. small intestine); however, damage to other parts of the body such as the skin (dermatitis herpetiformis), the teeth and the liver has been reported (Lohi 2010). Severity of the disease increases significantly with delays in diagnosis (i.e. age of diagnosis) and the degree of susceptibility increase with the rate of gluten consumption (quantity). Tolerance thresholds for the general population of CD patients are not known but some workers have reported values ranging between 10 and 100 mg gluten (Collin et al. 2004, Hischenhuber et al. 2006, Catassi et al. 2007).

IgE-Mediated Cereal Allergy

IgE-mediated cereal allergy is distinctly different from CD. This type of allergy is an immediate-type hypersensitivity occurring minutes to hours after consumption of the offending food. Symptoms are similar to those described for the other allergens and include oral allergy syndrome (e.g. swelling of lips), respiratory difficulties (e.g. asthma), skin reaction (e.g. eczema, atopic dermatitis) and gastrointestinal distress (e.g. nausea, diarrhoea, vomiting and cramps). There is no accurate data available on the prevalence, but as is the case for the other allergens, genetic susceptibility has been suggested (Becker et al. 2004). Severity of IgE-mediated cereal allergy depends on the immune state of the patient (degree of sensitisation/tolerance) and the concentration of antigen in the food consumed. The proteins responsible for this type of allergy can be the albumins, globulin, gliadins or glutenins and will vary depending on the individual and the specific allergenic food. Offending foods include buckwheat, rice, corn, millet, wheat, oats, rye and barley (Cantani 2008). Buckwheat and rice allergies are more frequently observed in Asia than in Europe or North America (Taylor and Hefle 2001, Kumar et al. 2007). The pathway for sensitisation and cereal allergy elicitation is by ingestion (mouth) and/or inhalation (nose) (e.g. baker's asthma). Wheat allergy is responsible for up to 30% of occupational asthma in the bakery industry.

Another type of cereal allergy is exercise-induced cereal allergy. Symptoms in this case appear only after food allergen consumption is followed by exercise. Wheat, as well as shellfish and nuts, is mostly associated with this type of allergy (Romano et al. 2001, Beaudouin et al. 2006, Porcel et al. 2006).

Foods to Avoid for Gluten-Sensitive Enteropathy Patients and IgE-Mediated Cereal-Allergic Patients

As with the other allergens, there are presently no cures for CD. Major foods to avoid include barley, wheat, durum, farina, kamut, rye, semolina, spelt and triticale. Oats was previously included in the list of gluten-containing foods; however, several recent studies suggest that quantities of oats of up to 50 g/day are harmless to the majority of gluten-sensitive individuals (Janatuinen et al. 1995, 2000, 2002, Lundin et al. 2003). A major challenge for the industry is that commercial oats is very frequently contaminated with high amounts of wheat, barley or rye (Thompson 2004, 2005) as these crops are often grown in the same regions and in close proximity. Furthermore, a small percentage of CD patients may react to oat proteins. The mechanisms at play in this instance and the specific proteins in oats to which these patients react still remain to be clarified. As gluten-containing cereals are often used as ingredients in food formulation, other foods to avoid include hydrolyzed vegetable protein, flavouring, malt, maltodextrin, malted barley, malt vinegar and starch from gluten sources, especially if these contain residual amounts of gluten proteins.

Many countries have adopted the gluten-free Codex Alimentarius Standards (Joint FAO/WHO Food Standards Programme and Codex Alimentarius Commission 2008), which sets a maximum limit of 20 ppm for gluten-free foods that are naturally free from gluten and 100 ppm for gluten-free foods that have been rendered gluten-free through processing.

For IgE-mediated cereal allergy, foods to avoid will depend on the particular allergy. Major challenges for cereal-allergic individuals are cross-reactivity and cross-contamination of foods. Food ingredients that are particularly problematic if not properly labelled are starches, spices, seasonings, sauces, flavourings, colourings, some vinegars, hydrolysed plant protein, syrups (e.g. brown rice syrup), beverages (e.g. beer, ale, etc.). Other potential hidden sources of cereals or cereal ingredients are cosmetics, pillows, toy stuffings and certain medications when products derived from these are used as ingredients.

SESAME AND MUSTARD ALLERGENS

Sesame Allergy

Sesame (*Sesamum indicum*) is a herbaceous plant of the *Pedaliaceae* family originating from India, which is now grown in many countries. It is also known as Benne, Gingelly, Til or Teel, Simsim and Ajonjoli and is now a common ingredient used extensively in everyday foods because of its high nutritional value (Perkins 2000). Sesame proteins are rich in methionine (Dalal et al. 2002, Wolff et al. 2003). Common sesame products include biscuits, crackers, breadsticks, rice cakes, etc., as well as prepacked delicatessen and processed foods such as noodles, dips, soups, sausages, samosas, processed meats, vegeburgers, chutneys, etc., (Perkins 2000, Allergyexpert 2010).

Sesame seeds, which may be used whole or crushed, are extremely potent allergens, causing severe allergic reactions in susceptible individuals. The first case of sesame allergy was reported in 1950 (Gangur et al. 2005). More recently, a study of Australian children showed that allergic reactions to sesame ranked fourth behind reactions to egg, milk, and peanuts, and sesame was also found to be the third most common allergy-inducing food in Israeli children (Gangur et al. 2005). Another recent study showed that sesame allergy in Israeli children was more common than peanut allergy (More 2009). Sesame allergy seems to affect people of all ages, which imply that this food allergy is not commonly outgrown. The symptoms of sesame allergy can include urticaria, angioedema, asthma, atopic dermatitis, oral allergy syndrome, allergic rhinitis and anaphylaxis.

One of the major sesame seed allergens is the 9 kDa, 2S albumin (Pastorello et al. 2001). Beyer et al. (2002) also identified 10 IgE-binding proteins in sesame, four of which had MWs of 7, 34, 45 and 78 kDa. Wolff et al. (2003) have also reported a 14 kDa sesame allergenic protein belonging to the 2S albumin family.

Sesame products are used in a wide variety of food products and may represent hidden allergens in foods. Fatal anaphylactic reactions have occurred as a result of consuming sesame (Gangur et al. 2005). Unrefined sesame oil may be used in food products, which may trigger allergic reactions in susceptible individuals. The oil resists rancidity and is popular with Oriental chefs. In the bakery industry, workers have reported allergic reactions, which include asthma, to sesame products. In addition to its use in the food industry, sesame and sesame products are used in the pharmaceutical and cosmetic industries. Sesame products used in cosmetics and ointments may cause allergic dermatitis, an inflammatory condition of the skin, in sensitised individuals. Contact dermatitis as a result of direct exposure to cosmetics or pharmaceutical products containing sesame allergens has been reported. Specific instances of sesame allergy resulting in skin rashes and inflammation after baking with sesame seeds and skin rashes from cosmetics that contain sesame oil have also been reported (Stoppler and Marks 2005).

As sesame allergens are similar in biochemical structure to peanut allergens, people with sesame allergy may be at risk of having allergic reactions as a result of eating peanuts and vice-versa (Gangur et al. 2005). Cross-reactivity may also exist with rye, kiwi, poppy seed and various tree nuts (such as hazelnut, black walnut, cashew, macadamia and pistachio), but clinical studies are lacking.

While the number of reports of sesame allergy has steadily increased, it is still not clear whether this increase is in the number of reactions or an increased rate of detection and reporting of these allergic reactions (Stoppler and Marks 2005).

Mustard Seed Allergy

The mustard plant belongs to the family *Brassicaceae* with cabbage, cauliflower, broccoli, Brussel sprout, turnip and radish (Monreal et al. 1992). Mustard is often consumed as a condiment prepared from the mustard seed. The powder is made from a mixture of two species, *Sinapis alba* L. (yellow mustard) and *Brassica juncea* L. (oriental mustard). The varieties *Brassica nigra* and *Brassica juncea* are used for food products. In addition to mustard powder, mustard is also usually found in salad dressing, mayonnaise, soups and sauces (Ensminger et al. 1983).

Mustard allergy is now considered to be a very common food allergy, accounting for about 1.1% of food allergies in children (Morisset et al. 2003b) and ranks fourth in children's food allergies after eggs, peanuts and cow's milk (Rance et al. 2000). The major allergen of yellow mustard is Sin a 1, which has been found to be resistant to heating and proteolysis (Dominguez et al. 1990, Gonzalez de la Pena et al. 1996, Rance et al. 2000). Bra j 1 is the major allergen in oriental mustard with a structure similar to that of Sin a 1 (Gonzalez de la Pena et al. 1991, Caballero et al. 1994).

Mustard allergy can cause a wide array of symptoms. Rance et al. (2000) reported that allergic reactions to mustard start early in life and are probably linked to early consumption of baby foods. Mustard-allergic patients will react to any food that comes from the mustard plant, including jars of mustard, mustard powder, mustard leaves, seeds and flowers, sprouted mustard seeds, mustard oil and foods that contain these (Bock 2008).

Common symptoms of mustard allergy include difficulty in breathing, shortness of breath and other breathing complications, a rash or hives, itchy skin or general skin irritation. In some severe cases, it can lead to anaphylaxis, and if left untreated, anaphylaxis can lead to anaphylactic shock and even death. Some of the common symptoms of anaphylaxis include constriction of airways in the throat and lungs, anaphylactic shock, severe drop in blood pressure, heightened pulse and heavy heartbeat, dizziness, nausea and abdominal pain, confusion and disorientation, and loss of consciousness. Panconesi et al. (1980) reported the first case of mustard-induced anaphylaxis after the subject had consumed pizza.

Incidents of cross-reactions have been rarely reported. However, single cases have been described of cross-allergy to cauliflower, broccoli, cabbage and Brazil nuts, which may be linked to sequence homology of some proteins (Moneret-Vautrin 2006). One of the major challenges of mustard allergy is that many foods contain mustard even when it would seem unlikely (e.g. lunchmeat and hot dogs). Thus, careful reading of labels on processed foods is important for mustard-allergic patients.

MINOR FOOD ALLERGENS

As previously indicated, over 170 foods are known to provoke allergic reactions in humans (Taylor 2000). In addition to the nine major priority allergens, other minor food allergens and/or emerging allergens include lupin, pea, chickpea, lentil, fruits (e.g. apple, apricot, avocado, banana, cherry, grape, kiwi, mango, melon, peach, pear, pineapple and strawberry) and vegetables (celery, carrot, eggplant, lettuce, potato, pumpkin and tomato). The reader is referred to the following references for further reading (Pereira et al. 2002, Fernández-Rivas 2003, Fernández-Rivas et al. 2008, Harish Babu et al. 2008, Towell 2009, Skypala 2009, Jappe and Vieths 2010).

MANAGEMENT OF FOOD ALLERGY

Many of the priority food allergens such as milk, eggs, nuts and soya bean are commonly used in food processing (e.g. processed beef, sausages, salad dressings, breads, cakes, soups and sauces) and pharmaceutical products (e.g. casein hydrolysates as "drug" carriers) due to their desirable properties (Monaci et al. 2006). Extensive use of these foods as ingredients in various products increases the chances of their presence as hidden allergens, particularly when they are undeclared or present as a result of cross-contact.

As there are currently no cures for food allergy, the best management tool is the reading of food labels and avoidance of foods containing allergens or allergen-derived ingredients. As a result, many countries presently require the use of the common names of priority allergens when they are used as ingredients in foods.

A major development in the last decade has been the use of precautionary allergen labelling to warn allergic consumers of the likely presence of allergens. Unfortunately, consumers with food allergy have become less avoidant to products with advisory labels, such as "may contain" or "shared equipment" due to their misuse (Hefle et al. 2007). Pieretti et al. (2009) studied the use of advisory labels in the United States and found that 17% of 20,241 manufactured foods contained advisory labels. Thus, the unregulated use of advisory labels has become a source of confusion and frustration, which could eventually pose challenges for allergic consumers (Pieretti et al. 2009).

Recent emergence of the hygiene hypothesis (Yazdanbakhsh et al. 2002) brings a new approach for the management of allergic diseases. Several epidemiological studies have suggested that allergic diseases are more common in industrialised nations and urban areas compared to developing countries and rural areas. Less frequent microbial exposures in the developed world has been regarded as an important predisposing factor for having higher allergic population in these regions. To compensate the inadequate exposure of microbial load, probiotics (live bacteria, usually *Lactobacillus* and *Bifidobacterium*) are now being added in some infant formula along with oligosaccharides (prebiotics; Matricardi et al. 2003). The goal is to achieve the development of regulatory T cell or balanced Th1/Th2 activity, which could eventually prevent allergy development (Cross et al. 2001, Prioult et al. 2004). Further research will be required to confirm the purported effects.

Another area of interest is specific oral tolerance induction (SOTI). Taking milk as an example, about 15% of children with milk allergy maintain the susceptibility permanently in their life. Strict avoidance of cow's milk and cow's milk derivatives remains the gold standard for allergen management for such CMA patients. For these patients, however, the likelihood of exposing the offending allergen unintentionally always remains and thus total avoidance cannot be guaranteed. Similar concerns exist for patients with persistent peanut and tree nut allergies. SOTI is a promising approach particularly for patients with persistent food allergy. Staden et al. (2007) reported that SOTI treatment (using a daily dose of CMP, starting from 0.002 mg CMP) remarkably increased the threshold dose for allergic reaction in CMA patients. As a result, the patient could be protected against potentially fatal allergic reactions when small amounts of the hidden allergen are accidentally ingested. Since this approach uses the specific offending allergen to induce tolerance, development of mild to moderate side effects (such as itching in the mouth, nausea and wheals, etc.) is very common. Close supervision by an experienced professional is, therefore, necessary for SOTI treatment. Similar promising studies have been reported for peanut (Blumchen et al. 2008, Clark et al. 2009, Jones et al. 2009).

For the food industry, food allergen testing has become an increasingly important tool for the management and control of allergens during food production and processing. The rest of the chapter provides a brief summary of some of the methods currently available for allergen detection.

METHODS FOR DETECTING ALLERGENS

Table 42.4 provides examples of some of the commonly used methods for food allergen detection. Food allergens may be detected either by the presence of the target allergen itself or by using a marker in the target food. The method of choice will depend on factors such as food matrix interference, nature and quantity of the target allergen, the desired level of detection, specificity and time and resources required for running the assay. The threshold dose for sensitised individuals to manifest allergic reactions is quite low and variable among individuals (Bindslev-Jensen et al. 2002). However, it is commonly agreed that food allergen detection methods should be sensitive enough to detect 1–100 mg of analyte (allergen) per kg of processed food (Poms et al. 2004). Increasing numbers of rapid and user-friendly test kits to detect different food allergens are now commercially available with sensitivities ranging from 0.05 to 10 mg/kg (Table 42.5).

Protein-based food allergen detection methods include immunoblotting, rocket immunoelectrophoresis, enzyme allergosorbent test, enzyme-linked immunosorbent assay (ELISA), protein microarray and biosensors. More recently developed DNA-based methods are available as supplementary and complementary methods to protein-based methods and are particularly useful in species differentiation and detection of genetically modified food (Mustorp et al. 2008). This chapter will focus on ELISA-based detection, lateral-flow assays (LFA) and dipstick test, proteomic approaches and DNA-based allergen detection methods.

ELISA-Based Detection Methods

Among the growing number of food allergen detection techniques, ELISA has gained prominence and has become an extensively used technique due to its high sensitivity and specificity, availability of automation and user friendliness. Some ELISAs are designed to detect specific allergens (e.g. BLG, Ara h1 (a major peanut allergen) and shrimp tropomyosin), whereas other ELISAs detect mixtures of proteins from the allergenic source (e.g. total milk, egg, peanut and almond-soluble proteins; Taylor et al. 2009). Detection is based on binding of an allergen or a specific marker protein with an antibody specifically generated to recognise and bind to these proteins. This binding complex is visualised by a colourimetric assay when the enzymes, which are labelled to the allergen specific antibody, interact with the substrate solution. The concentration of the allergen can be quantified by obtaining optical density (OD) values with a microplate reader and plotting these values using a standard curve.

There are certain limitations of ELISA-based detection methods, however (Yeung 2006). ELISAs are based on aqueous systems and they do not work well in detecting insoluble proteins or allergens derived from edible oil-producing foods. In addition, ELISA may not detect oleosins, a family of protein involved in the formation of oil bodies in peanut and sesame (Pons et al. 2002 and Leduc et al. 2006) and soya lecithin containing more than 50 ppm residual protein (Taylor et al. 2009). ELISAs also fail to distinguish certain closely related foods. For example, walnut antisera react with pecan (Niemann et al. 2009) and mustard antisera react to rapeseed (Lee et al. 2008). These may be caused by the presence of cross-reactive epitopes or proteins between two foods (Taylor et al. 2009). Regardless of the above-mentioned limitations, ELISA is still a very powerful tool and remains the method of choice for food allergen detection. ELISA-based food allergen detection methods are developed using either the sandwich ELISA or competitive ELISA format.

Sandwich ELISA

In this assay, as indicated by its self-explanatory name, an allergen of interest is captured between two allergen-specific antibodies. The first allergen-specific antibody is immobilised on the solid phase to capture the allergen, and the second allergen-specific antibody is labelled with enzyme to detect the captured allergen. Allergens need to have enough binding sites, for example, large molecules such as proteins, to allow binding of the two allergen-specific antibodies (Schubert-Ullrich et al. 2009).

Competitive ELISA

This assay is based on the competitive binding of unknown analyte (allergen) in a sample of interest and a known analyte (allergen) to the allergen-specific antibody. The antigen of interest in this case is incubated in the presence of the unlabelled antibody. Antibody/antigen complexes are formed and the solution is then

Table 42.4. Commonly Used Food Allergen Detection Methods

	Principle	Limit of Detection	Advantages/Disadvantages	References
Protein-based methods				
Rocket immunoelectrophoresis (RIE)	Binding of allergen to human IgE or antibodies raised in animal followed by observation in gel	2.5 mg/kg	Not widely used due to the laborious procedure. Semi-quantitative.	Besler et al. 2002b
EAST inhibition	Binding of specific IgE antibodies to food allergens bound to a solid phase followed by measurement	1 mg/kg	Mainly used for clinical diagnosis. Limited commercial products available due to reliance on human sera. Quantitative.	Nordlee and Taylor 1995
SDS-PAGE and immunoblotting	Proteins are denatured and separated according to their molecular mass. The proteins are then transferred onto a membrane and detected with labelled human IgE or antibody raised in animals.	5 mg/kg	Complex and time-consuming procedure. Reliance on human sera.	Scheibe et al. 2001
Dot immunoblotting	Reaction between sample protein extracts and enzyme-labelled specific antibodies on a nitrocellulose membrane.	Intensity of dot is proportional to the amount of allergen.	Semi-quantitative but simple and inexpensive.	Blais and Phillippe 2000
ELISA	Detects allergens by colourimetric reaction formed by the binding of allergen and specific enzyme-labelled antibody.	0.05–10 mg/kg	Most widely used method. High specificity and sensitivity. Rapid ELISA kits for major allergens are commercially available.	Schubert-Ullrich et al. 2009
Flow-through Microarray	Immobilised allergen specific antibodies on a microarray chip react with allergens.	1.3 ng/mL for ovalbumin	Powerful tool but still in infancy stage for food allergen detection.	Shriver-Lake et al. 2004
Optical biosensors, e.g., surface plasmon resonance (SPR)	Interaction of immobilised molecules (e.g. allergen-specific antibodies) on a sensor surface and a target molecule (e.g. allergens) in solution.	1–12.5 mg/kg	Interaction is measured by either a resonance angle or refractive index value. No need to use labelled molecules.	Yman et al. 2006
DNA-based methods				
PCR with gel electrophoresis	Amplification of a specific DNA fragment followed by agarose gel electrophoresis and visualisation by staining or southern blotting.	<10 mg/kg	Qualitative. Target DNA is less affected than protein during processing and extraction from food matrices. Results do not truly represent actual allergen exposure.	Poms et al. 2001
Real-time PCR	Uses a target-specific oligonucleotide probe with a reporter dye and a quencher dye attached.	<10mg/kg	Requires costly laboratory equipments. Allows for gel-free detection. RT-PCR kits are available for various food allergens.	Oliver and Vieths 2004
DNA-ELISA	Amplified product of specific DNA fragment is linked with a specific protein labelled DNA probe coupled with a specific enzyme-labelled antibody.	<10mg/kg	Semi-quantitative.	Poms et al. 2004

Source: Adapted from Poms et al. 2004.

Table 42.5. Examples of Some Commercially Available Test Kits for Food Allergen Detection

Allergenic Food	Target Allergens	Test Format and LOD (mg/kg food)	Testing Time (min)	Manufacturer
Milk	Milk protein	ELISA (<5)	30	Neogen
	Casein	ELISA (0.5)	30	R-Biopharm
	β-Lactoglobulin	ELISA (0.1)	45	ELISA Systems
Eggs	Egg-white protein	ELISA (0.6)	35	R-Biopharm
	Egg-white protein	LFA (LOD: not specified)	<10	Tepnel BioSystems
	Egg protein	ELISA (<5)	30	Neogen
Peanuts	Peanut protein	LFA (<5)	10	Neogen
	Peanut protein	LFA (1)	<10	Tepnel BioSystems
	Ara h1	LFA (5)	10	R-Biopharm
Nuts	Almond protein	LFA (1)	<10	Tepnel BioSystems
	Hazelnut protein	LFA (5)	10	R-Biopharm
Soya beans	Soya protein	ELISA (<5)	30	Neogen
Crustacean	Tropomyosin	ELISA (1)	60	ELISA Systems
Cereals	Gliadin, secalins, hordeins	ELISA (2)	30	R-Biopharm
	Gluten	Dipstick (10–20)	<15	Hallmark
	Gliadin	LFA (2.5)	5	R-Biopharm
Molluscs	Not specified	LFA (5)	<10	Tepnel BioSystems

Source: Adapted from Schubert-Ullrich et al. 2009.
LOD, limit of detection; ppm, milligram of allergenic food/kg of target food sample.

added to an antigen-coated well followed by washing to remove any unbound material. The higher the concentration of antigen in the sample, the lower will be the amount of free antibody able to bind to the antigen in the well. The allergen concentration in the sample in this assay is therefore inversely proportional to OD values obtained after adding the substrate. Competitive ELISA assay is known for its ability to detect relatively small proteins (Poms et al. 2004). Today, continuous developments in technology allows the possibility of enzyme labelling of either the analyte or the analyte-specific antibody.

LFA and Dipstick Tests

Both of these tests are modified versions of ELISA and are designed to meet the growing need of portable, reliable, user friendly and cost-effective methods that can detect trace amount of allergens in suspected foods within a short period of time (Schubert-Ullrich et al. 2009). Although quantification is possible in principle by using special strip test readers, commercially available LFA and dipstick test are primarily intended for qualitative or semi-quantitative purposes (yes or no). Table 42.5 shows some of the commercially available tests based on ELISA, LFA and dipstick methods.

LFAs or strip tests are immunochromatographic tests with a mobile phase that allows movement of the allergen–antibody complex and/or the sample along a test strip. This technique is based on an immunochromatographic procedure that utilises antigen–antibody properties and enables rapid detection of the analyte (Rong-Hwa et al. 2010). For LFAs using a sandwich ELISA, an allergen-specific detection antibody, deposited on the membrane, is solubilised upon introduction of a liquid sample and moves along with the sample until it reaches the zone where a capture antibody has been immobilised. At that point, the allergen–antibody complex is trapped. The presence or absence of allergen is determined with the colour development pattern. Competitive ELISA format can also be applied in LFA tests. LFAs results can be obtained within 3–15 minutes after running the assay (Schubert-Ullrich et al. 2009).

Dipstick tests are also based on a similar principal as LFA but with no mobile phase. Instead, an incubation phase is required in the dipstick test. Oliver et al. (2002) used dipstick assay to detect peanut and hazelnut allergens and reported that results of the dipstick assays and the corresponding ELISA were in concordance. Although most dipstick tests are based on sandwich ELISA format, application of competitive ELISA has been reported. Depending on the time needed for incubation, food allergen detection time, using dipstick test, could vary from 10 minutes to 3 hours (Schubert-Ullrich et al. 2009).

Proteomic Approach

Recently, proteomics has gained much attention especially for the detection and identification of IgE-binding allergens in complex mixtures. Proteomics is useful for identification and characterisation of proteins, including their post-translational modifications, protein conformations (native, denatured, folding intermediates) and protein–protein interactions (Leadbeater and Ward 1987, Alomirah et al. 2000).

A proteomic analysis usually consists of sample collection and preparation, protein separation using electrophoresis or chromatography and protein identification and characterisation using mass spectrometry. Successful identification of proteins depends on many factors such as the sensitivity of the mass spectrometer, the completeness of the database, post-translational modifications and more (Monaci et al. 2006). Many technologies, including protein separation via two-dimensional electrophoresis or liquid chromatography (LC), mass spectrometry and arrays (to map protein–protein interactions), are applied in proteomic-based studies. Shefcheck and Musser (2004) showed useful application of LC-based proteins separation with MS analysis (LC-MS/MS) to detect peanut allergen Ara h1 in ice cream with a detection level of 10 mg/kg. Proteomic studies can offer more detailed information about allergens (Shefcheck and Musser 2004), help in the identification of new allergens (Yu et al. 2003) and provide objective evaluation of changes induced by genetic modification (Herman et al. 2003). However, the requirement of a highly skilled operator, the high cost of mass spectrometers, the length of time required for sample preparation and analysis, and the need for separate protocols and optimisation for each sample prevent current proteomic approaches to be used for routine detection of food allergens. Nevertheless, proteomics remains a promising tool in allergen studies due to its great potential for detection and identification of food allergens in complex mixtures. Recent developments in mass spectrometer instrumentation and software offer ultra-high throughput, high sensitivity, and high-resolution capabilities (Chassaigne et al. 2009). Currently, attempts to couple more efficient protein separation (microfluidic) and detection technologies with MS analysis are progressing. Proteomic approaches could therefore soon be widely applicable for detection and identification of food allergens.

DNA-Based Allergen Detection Methods

Detection of allergens in food, by traditional protein-based detection methods, can sometimes be very difficult, as they are often present in minute amounts and are masked by the food matrix (Keck-Gassenmeier et al. 1999). DNA-based allergen detection methods have emerged as an alternative and complementary method to detect allergens where effective protein-based allergen detection methods are unavailable. DNA-based tests also possess high accuracy and sensitivity and, in general, are able to withstand harsh processing conditions better than corresponding allergenic proteins (Mustorp et al. 2008). On the other hand, in some food, DNA are partially or totally degraded and consequently false-negative results could occur, while the food may still contain the allergen. In addition, some food matrix components may reduce DNA amplification efficiency. DNA-based methods, in comparison to ELISA-based methods, are relatively more complicated, require more equipment and have lengthier protocols. For this reason, DNA-based methods have been considered impractical for routine detection. They may, however, be desirable for use by regulatory agencies and in institutions where there is easy access to appropriate equipment.

The sensitivity of the DNA-based methods strongly depends on the amount and on the quality of the template DNA since the DNA encoding the allergen in the food samples may be present in low amounts and may also be degraded as a result of processing (Holzhauser et al. 2006). Three DNA-based approaches are currently available to detect food allergens.

Polymerase Chain Reaction (PCR) with Gel Electrophoresis

The assay involves DNA extraction from food, amplification of specific DNA fragment using primers that bind selectively to complementary parts of the DNA to be amplified, followed by agarose gel electrophoresis (Besler 2001). PCR consists of three steps, determined by different temperatures in every cycle of amplification: denaturation of double strands DNA, annealing of the primers and extension of the primers by the enzyme Taq polymerase (Poms et al. 2004). The amplified product is loaded to run agarose gel electrophoresis and visualised by staining the gel. PCR with gel electrophoresis approach is good for qualitative detection of allergen in the food.

PCR with ELISA

This assay is a combination of a highly specific DNA-based method with the relatively simple and highly sensitive ELISA assay for semi-quantitative analysis (Poms et al. 2004). As in other PCR-based methods, extraction and purification of DNA from a food sample and amplification of specific DNA fragment encoding to suspected allergen using primers in a thermocycler are prerequisite for PCR-ELISA assay. The ELISA component of the test involves immobilisation of the amplified PCR product on a solid phase (a microplate), denaturation of double strand, hybridisation with specific probe, detection of the probe with enzyme-conjugated antibody and observation of the enzymatic colour reaction after addition of substrate (Holzhauser et al. 2002). The DNA concentration can be determined via absorbance obtained from an enzyme–substrate reaction.

Real-Time PCR

Real-time PCR allows gel-free detection 'in real time' with high accuracy by using a target-specific oligonucleotide probe with a reporter dye and a quencher dye attached (Desjardin et al. 1998). Oliver and Vieths (2004) reported real-time PCR method to be as sensitive as sandwich ELISA in detecting peanut allergens in 33 food samples. Many different types of real-time PCR assay kits are commercially available to detect many of the priority food allergens (Poms et al. 2004), including kits for detecting allergen encoding genes from mustard, sesame and celery (Mustorp et al. 2008).

CONCLUSION

Food allergy will likely continue to be an important food safety issue in the coming decades. Although substantial progress has been made in the field in the last 20 years, much still remains unknown. One area still requiring significant work is the question of the minimum eliciting dose required to provoke reactions in sensitised individuals. Increasing prevalence rates along with the possibility of severe allergic reactions after exposure to very low level of allergens in sensitised consumers continue to be a concern. Although exact threshold doses for food allergens are not known at present, levels of detection offered by current commercial allergens detection kits appear to be adequate for allergic consumer protection. There continues to be a need, however, for more sensitive and easily applicable analytical methods and improved understanding of the sources of cross-contact along the food value chain.

Currently, ELISA-based rapid allergens detection kits including LFA and dipstick are extensively used to detect suspected major allergens in processed foods. They will likely remain the method of choice in the food industry for a while, because of their fast and easy usage. Other methods such as surface plasmon resonance on automated platforms or protein microarrays in combination with immunochemical analytical approaches can provide rapid, efficient and simultaneous detection of multiple allergens in multiple samples (Moreno-Bondi et al. 2003 and Schubert-Ullrich et al. 2009). However, application of these novel techniques to food allergen analyses is still limited and will hopefully grow. Similarly, use of proteomics approach to identify and detect food allergens is expected to increase as microfluidic system-based protein separation and analysis systems become available as handheld devices in the near future. It is also anticipated that DNA-based food allergen detection methods will continue to be used as a valuable reinforcement tool in the ongoing effort to manage food allergens in the food supply.

ACKNOWLEDGEMENT

The help of Dr. Sahul H. Rajamohamed in the preparation and revision of this manuscript is gratefully acknowledged.

REFERENCES

Aabin B et al. 1996. Identification of IgE-binding egg white proteins: comparison of results obtained by different methods. *Int Arch Allergy Immunol* 109(1): 50–57.

Allergyexpert. Available at www.allergyexpert.com. Accessed on March 3, 2010.

Alomirah HF et al. 2000. Applications of mass spectrometry to food proteins and peptides. *J Chromatogr A* 893: 1–21.

Anderson JA. 1986. The establishment of common language concerning adverse reactions to food and food additives. *J Allergy Clin Immunol* 78(1): 140–143.

Anderson JW. 2008. Beneficial effects of soy protein consumption for renal function. *Asia Pacific J Clin Nutr* 17(Suppl 1): 324–328.

Anderson RP et al. 2000. In vivo antigen challenge in celiac disease identifies a single transglutaminase-modified peptide as the dominant A-gliadin T-cell epitope. *Nature Med* 6(3): 337–342.

Arede A et al. 2000. Food allergy pattern in children from Southern Europe. *Allergy* 55(63): 52–57.

Awazuhara H et al. 1998. Antigenicity of the proteins in soy lecithin and soy oil in soybean allergy. *Clin Exp Allergy* 28(12): 1559–1569.

Ballmer-Weber BK et al. 2007. Clinical characteristics of soybean allergy in Europe: a double-blind, placebo-controlled food challenge study. *J Allergy Clin Immunol* 119(6): 1489–1496.

Baur X et al. 1996. Characterization of soybean allergens causing sensitization of occupationally exposed bakers. *Allergy* 51(5): 326–330.

Beaudouin E et al. 2006. Food-dependent exercise-induced anaphylaxis – update and current data. *Eur Ann Allergy Clin Immunol* 38(2): 45–51.

Becker W et al. 2004. Opinion of the scientific panel on dietetic products, nutrition and allergies on a request from the commission relating to the evaluation of allergenic foods for labelling purposes (Request nr EFSA-Q-2003-016). *Eur Food Saf Authority J* 32: 1–197.

Besler M. 2001. Determination of allergens in foods. *Trends Anal Chem* 20: 662–672.

Besler M et al. 2002a. Allergen data collection: goat's milk (*Capra* spp.). Available at www.food-allergens.de/. *Internet Symp Food Allergens* 4(2): 119–124.

Besler M et al. 2002b. Determination of hidden allergens in foods by immunoassays. Available at www.food-allergens.de/. *Internet Symp Food Allergens* 4(1): 1–18.

Beyer K et al. 2001. Effects of cooking methods on peanut allergenicity. *J Allergy Clin Immunol* 107: 1077–1081.

Beyer K et al. 2002. Identification of sesame seed allergens by 2-dimensional proteomics and Edman sequencing: seed storage proteins as common food allergens. *J Allergy Clin Immunol* 110: 154–159.

Bindslev-Jensen C et al. 2002. Can we determine a threshold level for allergenic foods by statistical analysis of published data in the literature? *Allergy* 57: 741–746.

Blais BW, Phillippe LM. 2000. A cloth-based enzyme immunoassay for detection of peanut proteins in foods. *Food Agric Immunol* 12: 243–248.

Blumchen K et al. 2008. Rush specific oral tolerance induction in peanut allergic patients with high risk of anaphylactic reactions. *J Allergy Clin Immunol* 121(2): S136.

Bock T. 2008. What are the symptoms of mustard allergy. Available at http://ezinearticles. com/?What-Are-the-Symptoms-of-Mustard-Allergy?andid=1682496. Accessed on March 2, 2010.

Bonomi F et al. 2003. Reduction of immunoreactivity of bovine β-lactoglobulin upon combined physical and proteolytic treatment. *J Dairy Res* 70: 51–59.

Boye JI et al. 2010. Processing foods without soybean ingredients. In: JI Boye, S Godefroy (eds.) *Allergen Management in the Food Industry*, Chapter 13. John Wiley & Sons, New York.

Boza JJ et al. 1994. Nutritional value and antigenicity of two milk protein hydrolysates in rats and guinea pigs. *J Nutr* 124: 1978–1986.

Breeding CJ, Beyer RS. 2000. Eggs. In: L Genevieve et al. (eds) *Food Chemistry: Principles and Applications*. Science Technology Systems, West Sacramento, CA, pp. 421–431.

Breiteneder H, Mills EN. 2005. Molecular properties of food allergens. *J Allergy Clin Immunol* 115: 14–23.

Breiteneder H, Radauer CA. 2004. Classification of plant food allergens. *J Allergy Clin Immunol* 113(5): 821–830.

Breneman JC. 1987. *Handbook of Food Allergies*. Marcel Dekker, New York.

Burks AW Jr et al. 1988. Allergenicity of major component proteins of soybean determined by enzyme-linked immunosorbent assay (ELISA) and immunoblotting in children with atopic dermatitis and positive soy challenges. *J Allergy Clin Immunol* 11(6): 1135–1142.

Burks AW et al. 1994. Prospective oral food challenge study of two soybean protein isolates in patients with possible milk or soy protein enterocolitis. *Pediatr Allergy Immunol* 5(1): 40–45.

Busse PJ et al. 2002. Identification of sequential IgE-binding epitopes on bovine alpha (s2)- casein in cow's milk allergic patients. *Int Arch Allergy Immunol* 129: 93–96.

Caballero T et al. 1994. Relationship between pollinosis and fruit or vegetable sensitization. *Pediatr Allergy Immunol* 5: 218–222.

Cantani A. 2008. *Pediatric Allergy, Asthma and Immunology*. Springer-Verlag, Berlin, Heidelberg and New York, p. 628.

Cartier A et al. 1984. Occupational asthma in snow crab-processing workers. *J Allergy Clin Immunol* 74(3 Pt 1): 261–269.

Catassi C et al. 2007. A prospective, double-blind, placebo-controlled trial to establish a safe gluten threshold for patients with celiac disease. *Am J Clin Nutr* 85(1): 160–166.

Chafen JJ et al. 2010. Diagnosing and managing common food allergies: a systematic review. *JAMA* 12(18): 1848–1856.

Chapman MD et al. 2000. Recombinant allergens for diagnosis and therapy of allergic disease. *J Allergy Clin Immunol* 106: 409–418.

Chassaigne H et al. 2009. Resolution and identification of major peanut allergens using a combination of fluorescence two-dimensional differential gel electrophoresis, Western blotting and Q-TOF mass spectrometry. *J Proteomics* 72: 511–526.

Chatchatee P et al. 2001. Identification of IgE and IgG binding epitopes on β- and κ-casein in cow's milk allergic patients. *Clin Exp Allergy* 31: 1256–1262.

Chen ST et al. 2003. Effect of dietary protein on renal function and lipid metabolism in five-sixths nephrectomized rats. *Br J Nutr* 89(4): 491–497.

Clark AT et al. 2009. Successful oral tolerance induction in severe peanut allergy. *Allergy* 64(8): 1218–1220.

Cocco RR et al. 2003. Mutational analysis of major, sequential IgE-binding epitopes in alpha s1-casein, a major cow's milk allergen. *J Allergy Clin Immunol* 112: 433–437.

Codina R et al. 1997. Purification and characterization of a soybean hull allergen responsible for the Barcelona asthma outbreaks. II: purification and sequencing of the Gly m 2 allergen. *Clin Exp Allergy* 27(4): 424–430.

Collin P et al. 2004. The safe threshold for gluten contamination in gluten-free products. Can trace amounts be accepted in the treatment of coeliac disease? *Aliment Pharmacol Ther* 19(12): 1277–1283.

Crittenden RG, Bennett LE. 2005. Cow's milk allergy: a complex disorder. *J Am Coll Nutr* 24: S582–S591.

Cross M et al. 2001. Anti-allergy properties of fermented foods: an important immunoregulatory mechanism of lactic acid bacteria? *Int Immunopharmacol* 1: 891–901.

Dalal I et al. 2002. Food allergy is a matter of geography after all: sesame as a major cause of severe IgE-mediated food allergic reactions among infants and young children in Israel. *Allergy* 57: 362–365.

Darewicz M et al. 2008. Celiac disease – background, molecular, bioinformatics and analytical aspects. *Food Rev Int* 24: 311–329.

Daul CB et al. 1993a. Hypersensitivity reactions to crustacean and molluscs. *Clin Rev Allergy* 11: 201–222.

Daul CB et al. 1993b. Common crustacean allergens: identification with shrimp-specific allergens. In: D Kraft, A Sehon (eds.) *Molecular Biology and Immunology of Allergens*. CRC Press, Baton Rouge, pp. 291–293.

Davis PJ, Williams SC. 1998. Protein modification by thermal processing. *Allergy* 53(46, Suppl): 102–105.

De Boissieu D et al. 1997. Allergy to extensively hydrolyzed cow's milk proteins in infants: identification and treatment with an amino acid-based formula. *J Pediatr* 131: 744–747.

de Leon MP et al. 2003. Immunological analysis of allergenic cross reactivity between peanut and tree nuts. *Clin Exp Allergy* 33: 1273–1280.

Desjardin LE et al. 1998. Comparison of the ABI 7700 system (TaqMan) and competitive PCR for quantification of IS6110 DNA in sputum during treatment of tuberculosis. *J Clin Microbiol* 36: 1964–1968.

Dominguez J et al. 1990. Purification and characterization of an allergen of mustard seed. *Ann Allergy* 64: 352–357.

Eigenmann PA et al. 1996. Identification of unique peanut and soy allergens in sera adsorbed with cross-reacting antibodies. *J Allergy Clin Immunol* 98(5 Pt 1): 969–978.

Ensminger AH et al. 1983. *Foods and Nutrition Encyclopedia*. Pegus Press, Clovis, CA.

Fasano A, Catassi C. 2001. Current approaches to diagnosis and treatment of celiac disease: an evolving spectrum. *Gastroenterology* 120(3): 636–651.

Fasano A et al. 2003. Prevalence of celiac disease in at-risk and not-at-risk groups in the United States: a large multicenter study. *Arch Intern Med* 163(3): 286–292.

FDA. 1999. Food labeling: health claims; soy protein and coronary heart disease. The United States Food and Drug Adminstration, Department of Health and Human Services. *Federal Register* 64: 57699–57733.

FDA. 2006. The United States Food and Drug Administration, Department of Health and Human Services, Threshold Working Group. Approaches to establish thresholds for major food allergens and for gluten in food. Available at http://www.fda.gov/Food/LabelingNutrition/FoodAllergensLabeling/GuidanceComplianceRegulatoryInformation/ucm106108.htm. Accessed on November 28, 2011

Ferguson A. 1997. Gluten-sensitive enteropathy (celiac disease). In: DD Metcalfe et al. (eds.) *Food Allergy – Adverse Reactions to Foods and Food Additives*, 2nd edn. Wiley-Blackwell, Malden, MA, pp. 287–301.

Fernández-Rivas M. 2003. Cross-reactivity between fruit and vegetables. *Allergol Immunopathol* 31(3): 141–146.

Fernández-Rivas M et al. 2008. Allergies to fruits and vegetables. *Pediatr Allergy Immunol* 19(8): 675–681.

Ford RP, Taylor B. 1982. Natural history of egg hypersensitivity. *Arch Dis Childhood* 57: 649–652.

Friedman M, Brandon DL. 2001. Nutritional and health benefits of soy proteins. *J Agric Food Chem* 49(3): 1069–1086.

Gangur V et al. 2005. Sesame allergy: a growing food allergy of global proportions. *Ann Allergy Asthma Immunol* 95: 4–11.

García Novo MD et al. 2007. Prevalence of celiac disease in apparently healthy blood donors in the autonomous community of Madrid. *Revista Española de Enfermedades Digestivas* 99(6): 337–342.

Gerrard JW, Shenassa M. 1983. Sensitization to substances in breast milk: recognition, management and significance. *Ann Allergy* 51: 300–302.

Gomez JC et al. 2001. Prevalence of celiac disease in Argentina: screening of an adult population in the La Plata area. *Am J Gastroenterol* 96(9): 2700–2704.

Gonzalez de la Pena MA et al. 1991. Isolation and characterization of a major allergen from oriental mustard seeds, Brajl. *Int Arch Allergy Appl Immunol* 96: 263–270.

Gonzalez de la Pena MA et al. 1996. Expression in *Escherichia coli* of Sin a 1, the major allergen from mustard. *Eur J Biochem* 237: 827–832.

Gonzalez R et al. 1992. Purification and characterization of major inhalant allergens from soybean hulls. *Clin Exp Allergy* 22(8): 748–755.

Gonzalez R et al. 1995. Soybean hydrophobic protein and soybean hull allergy. *Lancet* 346(8966): 48–49.

Grundy JR et al. 2002. Rising prevalence of allergy to peanut in children: data from 2 sequential cohorts. *J Allergy Clin Immunol* 110(5): 784–778.

Hansen KS et al. 2003. Roasted hazelnuts – allergenic activity evaluated by double-blind, placebo-controlled food challenge. *Allergy* 58: 132–138.

Harish Babu BN et al. 2008. A cross-sectional study on the prevalence of food allergy to eggplant (*Solanum melongena* L.) reveals female predominance. *Clin Exp Allergy* 38(11): 1795–1802.

Hefle SL et al. 1996. Allergenic foods. *Crit Rev Food Sci Nutr* 36: S69–S89.

Hefle SL et al. 2005. Soy sauce retains allergenicity through the fermentation production process. *J Allergy Clin Immunol* 115(2, Suppl): S32.

Hefle SL et al. 2007. Consumer attitudes and risks associated with packaged foods having advisory labelling regarding the presence of peanuts. *J Allergy Clin Immunol* 120: 171–176.

Heine RG et al. 2006. The diagnosis and management of egg allergy. *Curr Allergy Asthma Reports* 6: 145–152.

Herman EM et al. 2003. Genetic modification removes an immunodominant allergen from soybean. *Plant Physiol* 132: 36–43.

Hill DJ et al. 1999. The natural history of intolerance to soy and extensively hydrolyzed formula in infants with multiple food protein intolerance (MFPI). *J Pediatr* 135: 118–121.

Hischenhuber C et al. 2006. Review article: safe amounts of gluten for patients with wheat allergy or coeliac disease. *Aliment Phar macol Ther* 23(5): 559–575.

Holzhauser T et al. 2002. Detection of potentially allergenic hazelnut (*Corylus avellana*) residues in food: a comparative study with DNA PCR-ELISA and protein sandwich-ELISA. *J Agric Food Chem* 50: 5808–5215.

Holzhauser T et al. 2006. Polymerase chain reaction (PCR) methods for the detection of allergenic foods. In: SJ Koppelman, SL Hefle (eds.) *Detecting Allergens in Food*. Woodhead Publishing, Cambridge, pp. 125–143.

Hori G et al. 2001. Soy protein hydrolyzate with bound phospholipids reduces serum cholesterol levels in hypercholesterolemic adult male volunteers. *Biosci Biotechnol Biochem* 65(1): 72–78.

Host A, Halken S. 1990. A prospective study of cow milk allergy in Danish infants during the first 3 years of life: clinical course in relation to clinical and immunological type of hypersensitivity reaction. *Allergy* 45(8): 587–596.

Host A, Samuelsson EG. 1988. Allergic reactions to raw, pasteurized, and homogenized/pasteurized cow milk: a comparison. A double-blind placebo-controlled study in milk allergic children. *Allergy* 43: 113–118.

Hourihane JO et al. 1997. An evaluation of the sensitivity of subjects with peanut allergy to very low doses of peanut protein: a randomized, double-blind, placebo-controlled food challenge study. *J Allergy Clin Immunol* 100: 596–600.

Inoue R et al. 2001. Identification of beta-lactoglobulin-derived peptides and class II HLA molecules recognized by T cells from patients with milk allergy. *Clin Exp Allergy* 31: 1126–1134.

Janatuinen EK et al. 1995. A comparison of diets with and without oats in adults with celiac disease. *N Engl J Med* 333(16): 1033–1037.

Janatuinen EK et al. 2000. Lack of cellular and humoral immunological responses to oats in adults with coeliac disease. *Gut* 46(3): 327–331.

Janatuinen EK et al. 2002. No harm from five year ingestion of oats in coeliac disease. *Gut* 50(3): 332–335.

Jappe U, Vieths S. 2010. Lupine, a source of new as well as hidden food allergens. *Mol Nutr Food Res* 54(1): 113–126.

Jarvinen KM et al. 2001. IgE and IgG binding epitopes on alphalactalbumin and beta-lactoglobulin in cow's milk allergy. *Int Arch Allergy Immunol* 126: 111–118.

Joint FAO/WHO Food Standards Programme and Codex Alimentarius Commission. 2008. Report of the 30th Session of the Codex Committee on Nutrition and Foods for Special Dietary Uses, 30 June – 5 July 2008, Geneva, Switzerland. ALINORM 08/31/26. Available at http://www. codexalimentarius.net/download/report/687/al08_26e.pdf. Accessed on 28 November 2011

Jones SM et al. 2009. Clinical efficacy and immune regulation with peanut oral immunotherapy. *J Allergy Clin Immunol* 124(2): 292–300.

Kahlenberg F et al. 2006. Monoclonal antibody R5 for detection of putatively coeliac-toxic gliadin peptides. *Eur Food Res Technol* 222(1–2): 78–82.

Kalogeromitros D et al. 1996. Anaphylaxis induced by lentils. *Ann Allergy Asthma Immunol* 77(6): 480–482.

Kaminogawa S, Totsuka M. 2003. Allergenicity of milk proteins. In: PF Fox (ed.) *Advanced Dairy Chemistry, Volume 1, Proteins*. Elsevier Science Publications, London, pp. 648–674.

Kanny G et al. 2001. Population study of food allergy in France. *J Allergy Clin Immunol* 108: 133–140.

Karjalainen J et al. 1992. A bovine albumin peptide as a possible trigger of insulin-dependent diabetes mellitus. *N Engl J Med* 327: 302–307.

Keck-Gassenmeier B et al. 1999. Determination of peanut traces in food by a commercially available ELISA test. *Food Agric Immunol* 11: 243–250.

Kleber N et al. 2004. The antigenic response of β-lg is modulated by thermally induced aggregation. *Eur Food Res Technol* 219(2): 105–110.

Kleber N et al. 2007. Antigenic response of bovine beta lactoglobulin influenced by ultra high pressure treatment and temperature. *Innovative Food Sci Emerg Technol* 8(1): 39–45.

Koppelman SJ et al. 1999. Heat induced conformational changes of Ara h 1, a major peanut allergen, do not affect its allergenic properties. *J Biol Chem* 274(8): 4770–4777.

Krishnan HB. 2000. Biochemistry and molecular biology of soybean seed storage protein. *J New Seeds* 2(3): 1–25.

Krishnan HB et al. 2009. All three subunits of soybean β-conglycinin are potential food allergens. *J Agric Food Chem* 57(3): 938–943.

Kumar R et al. 2007. Rice (*Oryza sativa*) allergy in rhinitis and asthma patients: a clinico-immunological study. *Immunobiology* 212(2): 141–147.

L'Hocine L, Boye JI. 2007. Allergenicity of soybean: new developments in identification of allergenic proteins, cross-reactivities and hypoallergenization technologies. *Crit Rev Food Sci Nutr* 47(2): 127–143.

Laoprasert N et al. 1998. Anaphylaxis in a milk-allergic child following ingestion of lemon sorbet containing trace quantities of milk. *J Food Prot* 61: 1522–1524.

Leadbeater L, Ward FB. 1987. Analysis of tryptic digests of bovine β-casein by reversed-phase high-performance liquid chromatography. *J Chromatogr* 397: 435–443.

Leduc V et al. 2006. Identification of oleosins as major allergens in sesame seed allergic patients. *Allergy* 61: 349–356.

Lee JW et al. 2001. Effects of gamma-radiation on the allergenic and antigenic properties of milk proteins. *J Food Prot* 64: 272–276.

Lee PW et al. 2008. Sandwich enzyme-linked immunosorbent assay (ELISA) for detection of mustard in foods. *J Food Sci* 73: T62–T68.

Lehrer SB et al. 2003. Seafood allergy and allergen: a review. *Mar Biotechnol* 5: 339–348.

Liu K. 1997. *Soybeans: Chemistry, Technology and Utilization*. Chapman and Hall, New York, pp. 28–35.

Lohi S. 2010. Prevalence and prognosis of coeliac disease: a special focus on undetected condition. Ph.D. Dissertation, Submitted to Faculty of Medicine of the University of Tampere, Finland.

Lu Y et al. 2007. Immunological characteristics of monoclonal antibodies against shellfish major allergen tropomyosin. *Food Chem* 100: 1093–1099.

Lundin KE et al. 2003. Oats induced villous atrophy in coeliac disease. *Gut* 52(11): 1649–1652.

Maleki SJ et al. 2000. The effects of roasting on the allergenic properties of peanut proteins. *J Allergy Clin Immunol* 106: 763–768.

Martin D, Chapman M. 2001. Prediction of the allergenicity of a molecule. *J Allergy Clin Immunol* 41: 13–16.

Martorell Aragonés A et al. 2001. Allergy to egg proteins. *Allergol Immunopathol* 29: 84–95.

Matricardi PM et al. 2003. Microbial products in allergy prevention and therapy. *Allergy* 58: 461–471.

Mazzeo MF et al. 2003. Identification of transglutaminase-mediated deamidation sites in a recombinant alpha-gliadin by advanced mass-spectrometric methodologies. *Protein Sci* 12(11): 2434–2442.

Mine Y, Rupa P. 2003. Fine mapping and structural analysis of immunodominant IgE allergenic epitopes in chicken egg ovalbumin. *Prot Eng* 16: 747–752.

Mine Y, Rupa P. 2004. Immunological and biochemical properties of egg allergens. *World Poult Sci J* 60: 321–330.

Mine Y, Yang M. 2008. Recent advances in the understanding of egg allergens: basic, industrial, and clinical perspectives. *J Agric Food Chem* 56: 4874–4900.

Mine Y, Zhang JW. 2002. Identification and fine mapping of IgG and IgE epitopes in ovomucoid. *Biochem Biophys Res Commun* 292: 1070–1074.

Monaci L et al. 2006. Milk allergens, their characteristics and their detection in food. *Eur Food Res Technol* 223: 149–179.

Moneret-Vautrin A. 2006. Allergenic foods: mustard. Available at http://www.anaphylaxis.eu/uk_mustard.html.

Monreal P et al. 1992. Two anaphylactic reactions to ingestion of mustard sauce. *Ann Allergy* 69: 317–320.

More D. 2009. Sesame seed allergy. Available at http://allergies.about.com/od/otherfoodallergies/a/sesameallergy.htm. Accessed on November 28, 2011.

Moreno-Bondi MC et al. 2003. Multi-analyte analysis system using an antibody-based biochip. *Anal Bioanal Chem* 375: 120–124.

Morisset M et al. 2003a. Thresholds of clinical reactivity to milk, egg, peanut and sesame in immunoglobulin E-dependent allergies: evaluation by double-blind or single-blind placebo controlled oral challenges. *Clin Exp Allergy* 33: 1046–1051.

Morisset M et al. 2003b. Prospective study of mustard allergy: first study with double-blind placebo-controlled food challenge trials (24 cases). *Allergy* 58(4): 295–299.

Motala C, Lockey R. 2004. Food Allergy. World Allergy Organization Allergic Diseases Resource Centre. Available at http://www.worldallergy.org/professional/allergic_diseases_center/foodallergy/

Munoz-Furlong A et al. 2004. Prevalence of self-reported seafood allergy in the U.S. *J Allergy Clin Immunol* 113: S100.

Mustorp D et al. 2008. Detection of celery (*Apium graveolens*), mustard (*Sinapis alba, Brassica juncea, Brassica nigra*) and sesame (*Sesanum indicum*) in food by real time PCR. *Eur Food Res Technol* 226: 771–778.

Nakajima-Adachi H et al. 1998. Determinant analysis of IgE and IgG antibodies and T cells specific for bovine alpha s1-casein from the same patients allergic to cow's milk: existence of alpha s1-casein-specific B cells and T cells characteristic in cow's milk allergy. *J Allergy Clin Immunol* 101: 660–671.

Nakano T et al. 2010. Sensitization to casein and β-lactoglobulin (blg) in children with cow's milk allergy (CMA). *Arerugi* 59: 117–122.

Nentwich I et al. 2001. Cow's milk-specific cellular and humoral immune responses and atopy skin symptoms in infants from atopic families fed a partially (pHF) or extensively (eHF) hydrolyzed infant formula. *Allergy* 56: 1144–1156.

Ng-Kwai-Hang KF. 2002. Heterogeneity, fractionation and isolation. In: H Roginski et al. (eds.) *Encyclopaedia of Dairy Sciences*. Academic Press, London, pp. 1881–1894.

Niemann L et al. 2009. Detection of walnut residues in foods using an enzyme-linked immunosorbent assay. *J Food Sci* 74: T51–T57.

Nordlee JA, Taylor SL. 1995. Immunological analysis of food allergens and other food proteins. *Food Technol* 2: 129–132.

Nowak-Wegrzyn A et al. 2004. Contamination of dry powder inhalers for asthma with milk proteins containing lactose. *J Allergy Clin Immunol* 113: 558–560.

O'Neil C et al. 1993. Allergic reactions to fish. *Clin Rev Allergy* 11: 183–200.

Ogawa T et al. 1991. Investigation of the IgE binding proteins in soybeans by immunoblotting with the sera of the soybean-sensitive patients with atopic dermatitis. *J Nutr Sci Vitaminol* 37(6): 555–565.

Ogawa T et al. 1993. Identification of the soybean allergenic protein, *Gly m* Bd 30K, with the soybean seed 34-kDa oil-body-associated protein. *Biosci Biotechnol Biochem* 57(6): 1030–1033.

Oliver S, Vieths S. 2004. Development of a real-time PCR and a sandwich ELISA for detection of potentially allergenic trace amounts of peanut (*Arachis hypogaea*) in processed foods. *J Agric Food Chem* 52: 3754–3760.

Oliver S et al. 2002. Development and validation of two dipstick type immunoassays for determination of trace amounts of peanut and hazelnut in processed foods. *Eur Food Res Technol* 215: 431–436.

Osman AA et al. 2001. A monoclonal antibody that recognizes a potential coeliac-toxic repetitive pentapeptide epitope in gliadins. *Eur J Gastroenterol Hepatol* 13(10): 1189–1193.

Panconesi E et al. 1980. Anaphylactic shock from mustard after ingestion of pizza. *Contact Dermatitis* 6: 294–295.

Pastorello EA et al. 2001. The major allergen of sesame seeds (*Sesamum indicum*) is a 2S albumin. *J Chromatogr B Biomed Sci Appl* 756: 85–93.

Pedersen HS, Djurtoft R. 1989. Antigenic and allergenic properties of acidic and basic peptide chains from glycinin. *Food Agric Immunol* 1(2): 101–109.

Pereira MJ et al. 2002. The allergenic significance of legumes. *Allergol Immunopathol* 30(6): 346–353.

Perkins MS. 2000. Sesame allergy. Available at www.allergyuk.org/fs_sesame.aspx. Accessed on February 24, 2010.

Pieretti MM et al. 2009. Audit of manufactured products: use of allergen advisory labels and identification of labeling ambiguities. *J Allergy Clin Immunol* 124: 337–341.

Poms RE et al. 2001. Increased sensitivity for detection of specific target DNA in milk by concentration in milk fat. *Eur Food Res Technol* 213: 361–365.

Poms RE et al. 2004. Methods for allergen analysis in food: a review. *Food Addit Contam* 21: 1–31.

Pons L et al. 2002. The 18 kDa peanut oleosin is a candidate allergen for IgE-mediated reactions to peanuts. *Allergy* 57(72): 88–93.

Porcel S et al. 2006. Food-dependent exercise-induced anaphylaxis to pistachio. *J Investig Allergol Clin Immunol* 16(1): 71–73.

Prioult G et al. 2004. Stimulation of interleukin-10 production by acidic beta-lactoglobulin-derived peptides hydrolyzed with *Lactobacillus paracasei* NCC2461 peptidases. *Clin Diagn Lab Immunol* 11: 266–271.

Rajamohamed SH, Boye JI. 2010. Processing foods without peanuts and tree nuts. In: JI Boye, S Godefroy (eds.) *Allergen Management in the Food Industry*, Chapter 11. John Wiley & Sons, New York.

Rance F et al. 2000. Mustard allergy in children. *Allergy* 55(5): 496–500.

Restani P et al. 2004. Characterization of bovine serum albumin epitopes and their role in allergic reactions. *Allergy* 59: 21–24.

Romano A et al. 2001. Food-dependent exercise-induced anaphylaxis: clinical and laboratory findings in 54 subjects. *Int Arch Allergy Immunol* 125(3): 264–272.

Rong-Hwa S et al. 2010. Gold nanoparticle-based lateral flow assay for detection of staphylococcal enterotoxin B. *Food Chem* 118: 462–466.

Rostami K et al. 1999. High prevalence of celiac disease in apparently healthy blood donors suggests a high prevalence of undiagnosed celiac disease in the Dutch population. *Scand J Gastroenterol* 34(3): 276–279.

Rozenfeld P et al. 2002. Detection and identification of a soy bean component that cross reacts with caseins from cow's milk. *Clin Exp Immunol* 130: 49–58.

Sampson HA. 2002. Peanut allergy. *N Engl J Med* 346(17): 1294–1299.

Sampson HA. 2004. Update on food allergy. *J Allergy Clin Immunol* 113: 805–819.

Sampson HA, McCaskill CC. 1985. Food hypersensitivity and atopic dermatitis: evaluation of 113 patients. *J Pediatr* 107: 669–675.

Sampson HA et al. 1992. Fatal and near-fatal anaphylactic reactions to food in children and adolescents. *N Engl J Med* 327(6): 380–384.

Scheibe B et al. 2001. Detection of trace amounts of hidden allergens: hazelnut and almond proteins in chocolate. *J Chromatogr B Biomed Sci Appl* 756: 229–237.

Schubert-Ullrich P et al. 2009. Commercialized rapid immunoanalytical tests for determination of allergenic food proteins: an overview. *Anal Bioanal Chem* 395: 69–81.

Shefcheck KJ, Musser SJ. 2004. Conformation of the allergenic protein Ara h 1 in a model matrix using liquid chromatography/tandem mass spectrometry. *J Agric Food Chem* 52: 2785–2790.

Shriver-Lake LC et al. 2004. Applications of array biosensor for detection of food allergens. *J AOAC Int* 87: 1498–1502.

Sicherer SH. 2010. Food allergies. eMedicine. Available at http://emedicine.medscape.com/article/135959-overview.

Sicherer SH, Sampson HA. 2006. 9. Food allergy. *J Allergy Clin Immunol* 117(2 Suppl Mini-Primer): S470–S475.

Sicherer SH, Sampson HA. 2010. Food allergy. *J Allergy Clin Immunol* 125: S116–S125.

Sicherer SH et al. 2001. A voluntary registry for peanut and tree-nut allergy: characteristics of the first 5149 registrants. *J Allergy Clin Immunol* 108: 128–132.

Sicherer SH et al. 2003. Prevalence of peanut and tree nut allergy in the United States determined by means of a random digit dial telephone survey: a 5-year follow-up study. *J Allergy Clin Immunol* 112(6): 1203–1207.

Sicherer SH et al. 2004. Prevalence of seafood allergy in the United States determined by a random telephone survey. *J Allergy Clin Immunol* 114: 159–165.

Skypala I. 2009. Fruits and vegetables. In: I Skypala, C Venter (eds.) *Food Hypersensitivity: Diagnosing and Managing Food Allergies and Intolerance*. Blackwell Publishing, Oxford, pp. 147–165.

Soler-Rivas C, Wichers HJ. 2001. Impact of (bio)chemical and physical procedures on food allergen stability. *Allergy* 56(67): 52–55.

Spuergin P et al. 1996. Allergenic epitopes of bovine alpha s1-casein recognized by human IgE and IgG. *Allergy* 51: 306–312.

Staden U et al. 2007. Specific oral tolerance induction in food allergy in children: efficacy and clinical patterns of reaction. *Allergy* 62: 1261–1269.

Stephenson TJ et al. 2005. Effect of soy protein-rich diet on renal function in young adults with insulin-dependant diabetes mellitus. *Clin Nephrol* 64(1): 1–11.

Stewart GA, Thompson PJ. 1996. The biochemistry of common aeroallergens. *Clin Exp Allergy* 26: 1020–1044.

Stoppler MC, Marks JW. 2005. Sesame seed allergy: a growing problem? Available at www.medicinenet.com/script/main/art.asp. Accessed on February 24, 2010.

Taylor SL. 2000. Emerging problems with food allergens. *Food Nutr Agric* 26: 14–29.

Taylor SL, Hefle SL. 2001. Food allergies and other food sensitivities – an application of the Institute of Food Techologists' expert panel on food safety and nutrition. *Food Technol* 55(9): 68–83.

Taylor SL, Lehrer SB. 1996. Principles and characteristics of food allergens. *Crit Rev Food Sci Nutr* 36: S91–S118.

Taylor SL et al. 2002. Factors affecting the determination of threshold doses for allergenic foods: how much is too much? *J Allergy Clin Immunol* 109(1): 24–30.

Taylor SL et al. 2004. Consensus protocol for the determination of the threshold doses for allergenic foods: how much is too much? *Clin Exp Allergy* 34: 689–695.

Taylor SL et al. 2009. Allergen immunoassays – considerations for use of naturally incurred standards. *Anal Bioanal Chem* 395: 83–92.

Thompson T. 2004. Gluten contamination of commercial oat products in the United States. *N Engl J Med* 351(19): 2021–2022.

Thompson T. 2005. Contaminated oats and other gluten-free foods in the United States. *J Am Diet Assoc* 105(3): 348–349.

Tome D et al. 2004. Nutritional value of milk and meat products derived from cloning. *Cloning Stem Cells* 6: 172–177.

Towell R. 2009. Peanuts, legumes, seeds and tree nuts. In: I Skypala, C Venter (eds.) *Food Hypersensitivity: Diagnosing and Managing Food Allergies and Intolerance*. Blackwell Publishing, Oxford, pp. 166–182.

Valenta R et al. 1999. The recombinant allergen-based concept of component resolved diagnostics and immunotherapy (CRD and CRIT). *Clin Exp Allergy* 29: 896–904.

Wal JM. 1998. Cow's milk allergens. *Allergy* 53: 1013–1022.

Wal JM. 2004. Bovine milk allergenicity. *Ann Allergy Asthma Immunol* 93: 2–11.

Walker-Smith J. 2003. Hypoallergenic formulas: are they really hypoallergenic? *Ann Allergy Asthma Immunol* 90: 112–114.

Walsh BJ et al. 1988. New allergens from hen's egg white and egg yolk: *in vitro* study of ovomucin, apovitellenin I and VI, and phosvitin. *Int Arch Allergy Appl Immunol* 87: 81–86.

Walsh BJ et al. 2005. Detection of four distinct groups of hen egg allergens binding IgE in the sera of children with egg allergy. *Allergol Immunopathol* 33: 183–191.

Ward PP et al. 2002. Lactoferrin and host defense. *Biochem Cell Biol* 80: 95–102.

Wen HW et al. 2007. Peanut allergy, peanut allergens, and methods for the detection of peanut contamination in food products. *Compr Rev Food Sci Food Saf* 6(2): 47–58.

Wilson S et al. 2005. Allergenic proteins in soybean: processing and reduction of P34 allergenicity. *Nutr Rev* 63(2): 47–58.

Wolff N et al. 2003. Allergy to sesame in humans is associated primarily with IgE antibody to a 14 kDa 2S albumin precursor. *Food Chem Toxicol* 41(8): 1165–1174.

Yazdanbakhsh M et al. 2002. Allergy, parasites, and the hygiene hypothesis. *Science* 296: 490–494.

Yeung J. 2006. Enzyme-linked immunosorbent assays (ELISAs) for detecting allergens in foods. In: K Stef, H Sue (eds.) *Detecting Allergens in Food*. Woodhead Publishing, Cambridge, pp. 109–124.

Yman IM et al. 1994. Analysis of food proteins for verification of contamination or mislabelling. *Food Agric Immunol* 6: 167–172.

Yman IM et al. 2006. Food allergen detection with biosensor immunoassays. *J AOAC Int* 89: 856–861.

Yu CJ et al. 2003. Proteomics and immunological analysis of a novel shrimp allergen (Pen m 2). *J Immunol* 170: 445–453.

Yunginger JW. 1990. Classical food allergens. *Allergy Asthma Proc* 11(1): 7–9.

Zhang JW, Mine Y. 1998. Characterization of IgE and IgG epitopes on ovomucoid using egg white allergic patient's sera. *Biochem Biophys Res Commun* 253: 124–127.

Zhang X et al. 2003. Soy food consumption is associated with lower risk of coronary heat disease in Chinese women. *J Nutr* 133(9): 2874–2878.

43
Biogenic Amines in Foods

Angelina O. Danquah, Soottawat Benjakul, and Benjamin K. Simpson

Abstract: Biogenic amines (also known as biologically active amines) are low-molecular weight organic compounds produced in biological systems by enzymatic decarboxylation of certain amino acids such as histamine and tyrosine. Examples include dopamine, histamine, norepinephrine, serotonin, and tyramine. They function in the body as neurotransmitters and relay signals between neurons across synapses to impact on mental functions, blood pressure, body temperature, appetite, and several other physiological processes. The levels of biogenic amines in the body are regulated for proper functioning of the various physiological processes that they are associated with. Certain foods contain biogenic amines that can add to the amounts naturally present in the body. High intake of these foods could upset the balance of biogenic amines in the body to cause health problems such as hypotension, hypertension, gastrointestinal distress, headaches and migraine, and others. Thus, it is crucial to curtail the formation and levels of these compounds in certain food products. This chapter discusses the formation and occurrence of biogenic amines in selected food products, their effects of human health and significance to the food industry, their detection, and their fate during processing and storage. The chapter also covers regulation of biogenic amines in foods.

INTRODUCTION

Biogenic amines, also known as biologically active amines, are nitrogen-containing low-molecular weight organic compounds that are naturally present in food products such as cheese, wine, beer, plant food, fish, and meat (Tarjan and Janossy 1978). These substances are described as biogenic because they are formed by the action of living organisms. In foodstuffs, they are perceived as undesirable anti-nutritional factors of public health concern since they have been associated with food poisoning (Shalaby 1996), particularly when ingested in large quantities, or where there is inhibition of their breakdown in humans. Examples of biogenic amines are the catecholamines and indoleamines. The prominent ones include acetylcholine (ACh), histamine, tyramine, dopamine (Fig. 43.1), serotonin (Fig. 43.2), norepinephrine (NE; also known as noradrenalin (NA)), and epinephrine (or adrenalin; Fig. 43.3).

Acetylcholine, also known as 2-acetoxy-*N,N,N*-trimethylethanaminium, is a neuromodulator found in both the peripheral and the central nervous systems (PNS and CNS). The compound has a chemical formula of $CH_3COOCH_2CH_2N^+(CH_3)_3$ and is formed from the esterification of choline with acetic acid in a reaction catalyzed by

Food Biochemistry and Food Processing, Second Edition. Edited by Benjamin K. Simpson, Leo M.L. Nollet, Fidel Toldrá, Soottawat Benjakul, Gopinadhan Paliyath and Y.H. Hui.

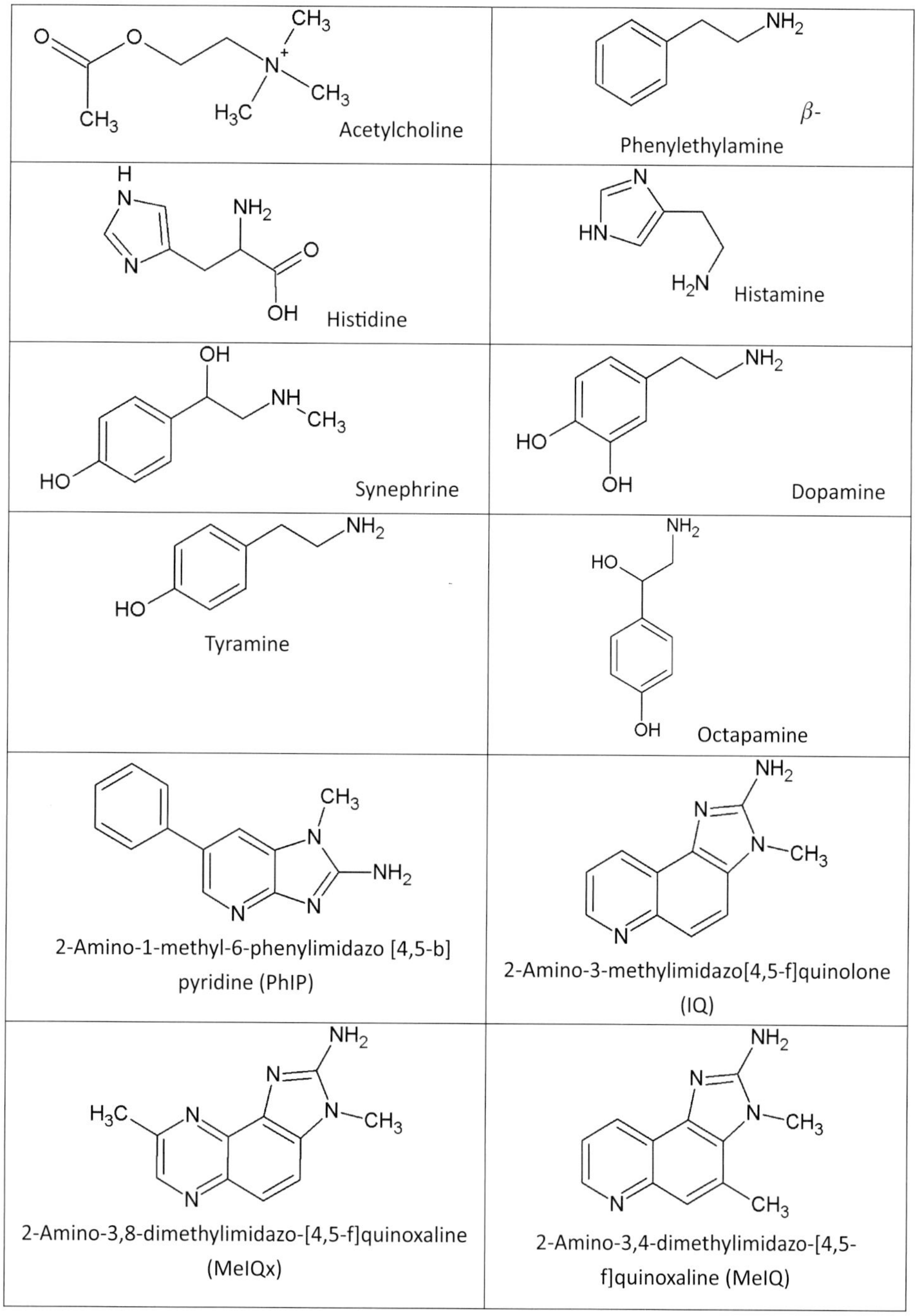

Figure 43.1. Structures of some common biogenic amines.

the enzyme choline acetyl transferase. In the PNS, ACh acts to activate muscles, while in the CNS, it elicits excitatory actions.

Histamine is derived from the amino acid histidine via decarboxylation by histidine decarboxylase. It is a neurotransmitter and is produced by basophiles and mast cells, and functions in immune response and the regulation of physiological functions in the gut (Marieb 2001). Histamine enhances the permeability of capillaries to white blood cells and protein antibodies, which are needed to destroy foreign substances and nullify their harmful effects. It also acts in pro-inflammatory response to tissue damage or allergic reactions and also enhances the secretion of gastric HCl through histamine receptors.

Tyramine is formed from the amino acid tyrosine by decarboxylation, and it occurs in many common foods that humans

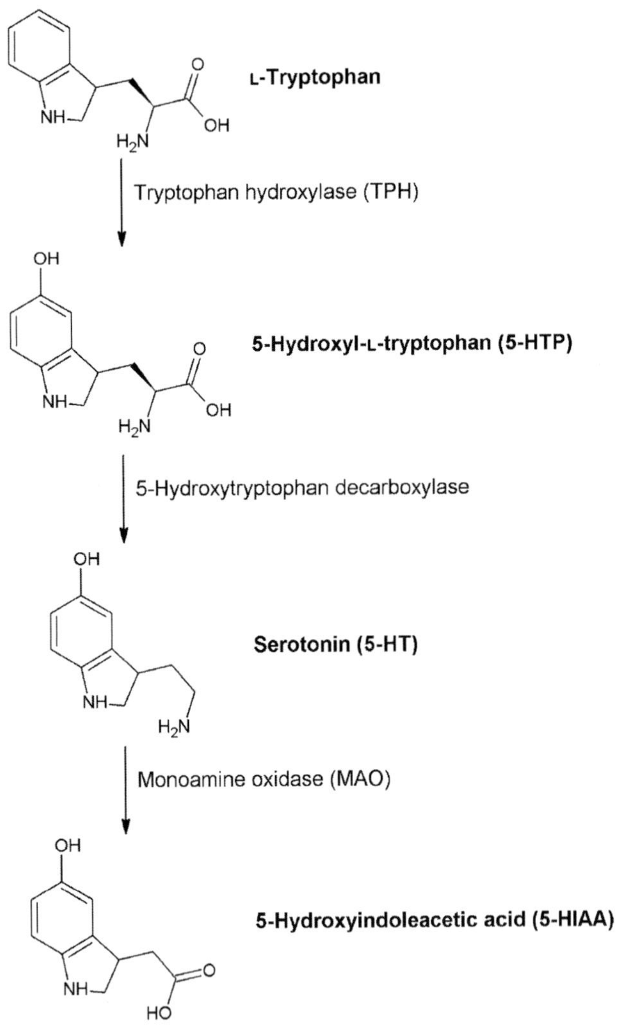

Figure 43.2. Scheme for the biosynthesis of serotonin from tryptophan.

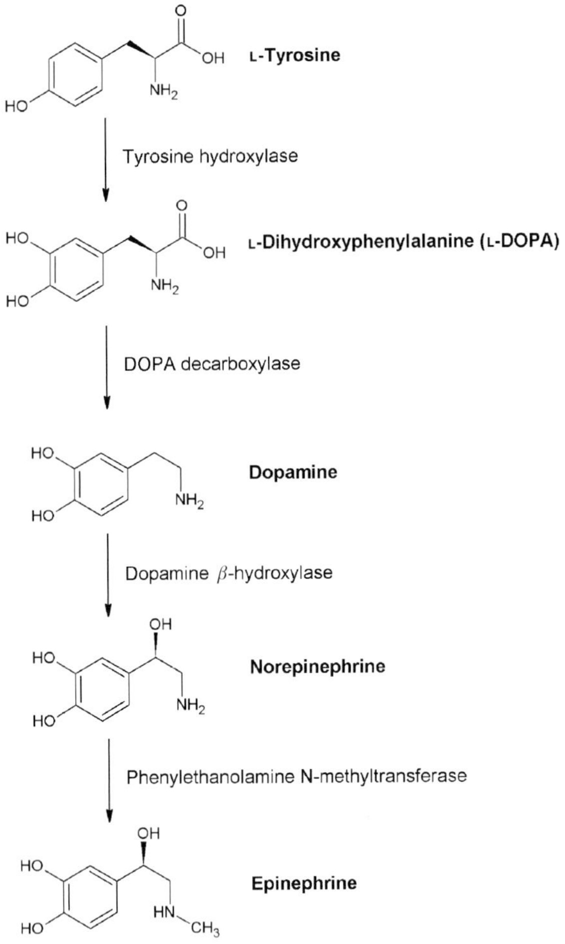

Figure 43.3. Scheme for the synthesis of adrenaline (epinephrine) from tyrosine.

ingest or consume. Well-known food sources of tyramine include fermented, pickled and aged foods, marinated foods, smoked foods, chocolate, alcoholic beverages, and spoiled foods. Its production and accumulation in the body have been linked to elevated blood pressure and headaches.

Serotonin, known systematically as 5-hydroxytryptamine, is synthesized from L-tryptophan by a series of reactions (Fig. 43.2), including hydroxylation, decarboxylation, and deamination steps. These reactions involve the participation of several enzymes, including tryptophan hydroxylase (TPH), 5-hydroxytryptophan decarboxylase, and monoamine oxidase (MAO). Serotonin is a neurotransmitter found mostly in the gastrointestinal tract (GIT), the CNS, and platelets of animals including humans. In the GIT, serotonin acts to regulate intestinal movements; in the CNS, serotonin performs various functions, including the regulation of appetite, mood, and sleep, as well as muscle contraction and some cognitive functions (Berger et al. 2009). Platelet-related serotonin is involved in haemostasis and liver regeneration (Lesurtel et al. 2006). In plants, serotonin is thought to be produced as a mechanism for obviating the build-up of toxic ammonia. This involves the coupling of NH_3 to the indole of L-tryptophan followed by decarboxylation into tryptamine and subsequently by cytochrome P450 (Cyt P_{450}) monooxygenase-mediated hydroxylation to serotonin. Plant sources of serotonin include mushroom, tomato, plantain, banana, plum, kiwifruit, walnut, and hickory. Excessive levels of serotonin can be toxic to humans and is manifested by specific cognitive, autonomic, and somatic effects that may range from nondetectable to fatal (Boyer and Shannon 2005).

Tryptamines are a group of alkaloid compounds found in plants, animals, and fungi. Well-known tryptamines include serotonin and melatonin. These compounds are known for their psychotropic effects and this forms the basis of their use by some people; examples include psilocybin (from "magic mushrooms") and dimethyltryptamine (Jones 1983). There are also synthetic counterparts of tryptamines, and these include the drug sumatriptan and its derivatives used to treat migraines (Yen and Hsieh 1991).

NE or NA is synthesized in the adrenal medulla and the neurons of the sympathetic nervous system from tyrosine via several enzyme-catalyzed reactions including hydroxylation followed by decarboxylation to dopamine and a final oxidation step. The enzymes involved in the reactions are tyrosine hydroxylase, DOPA decarboxylase, and dopamine β-hydroxylase, respectively (Aston-Jones et al. 2002). The NE (or NA) compound serves multifunctions, as a stress hormone, a neurotransmitter, and as a drug. As stress hormone, it is involved in triggering responses such as increasing heart rate, facilitating release of glucose from storage depots, and increasing blood flow to the skeletal muscle; as a neurotransmitter, it is released in the brain to mitigate inflammations; and as a drug, it can increase vascular tone and blood pressure.

Epinephrine or adrenaline is synthesized in the adrenal gland from the amino acids phenylalanine and tyrosine via a series of intermediates (Fig. 43.3). In the first step, tyrosine is oxidized to L-DOPA, followed by decarboxylation to dopamine, which is subsequently oxidized to NE and then methylated to form adrenaline or epinephrine. It functions both as a hormone and as a neurotransmitter and carries out manifold functions in the body. For instance, it increases heart rate, modulates dilation of blood vessels and air passages, and participates in the fight/flight response of the sympathetic nervous system. It also participates in the contraction of the smooth muscles; inhibits insulin secretion while enhancing glucagon secretion in the pancreas; and also stimulates glycolysis in the muscle, lipolysis in adipose tissues, and glycogenolysis in the liver and the muscle (Cannon 1929).

Dopamine, also known as 4,2-aminoethyl benzene-1,2-diol, has a molecular formula of $C_6H_3(OH)_2(CH_2)_2NH_3$. It is synthesized by nervous tissues and the adrenal medulla from tyrosine (L-Tyr) by hydroxylation to form dihydroxyphenyl alanine (L-DOPA) in a reaction catalyzed by tyrosine hydroxylase, followed by a decarboxylation step catalyzed by DOPA decarboxylase (Fig. 43.3). Dopamine thus formed may be processed further into NE by the enzyme dopamine β-hydroxylase and then methylated into epinephrine (Fig. 43.3). Dopamine is involved in motivation, addictions, behavioral manifestations, and coordination of bodily movements (Yen and Hsieh 1991).

Other commonly known biogenic amines and their precursors include cadaverine, agmatine, β-phenylethylamine, and putrescine derived, respectively, from lysine, arginine, phenylalanine, and ornithine. Polymerization of putrescine forms the polyamines spermine and spermidine (Pegg and McCann 1982). These polymeric products are highly resistant to degradation during processing and are capable of withstanding procedures like canning, freezing, smoking, and cooking (Etkind and Wilson 1987). Some biogenic amines are present at low levels in fresh foods; however, these levels may increase via microbial decarboxylation of free amino acids during processes such as fermentation, storage, and in food spoilage. Various factors such as pH, temperature, salt concentration, and glucose availability all influence the formation of biogenic amines in foodstuffs.

The importance of biogenic amines to the food industry is due to two main reasons. First, if ingested in high quantities, they may induce toxic effects. For example, consumption of tyramine and histamine has been associated with disorders such as a burning sensation in the throat, flushing, headaches, nausea, hypertension, numbing and tingling of the lips, rapid pulse, and vomiting (De Vries 1996). Certain biogenic amines have also been implicated in the formation of carcinogenic nitrosamines, and the variation in biogenic amine levels has been suggested as a useful chemical indicator of food quality or spoilage. Thus, the food industry has developed several techniques for determining biogenic amine concentrations in food products for various purposes, including research, health and safety, or quality control. The effects of food processing on biogenic amine formation or levels have also been extensively studied in order to develop methods of limiting biogenic amine formation in food products (Stratton et al. 1991, Shalaby 1996). Because of its potential health concerns, governments and related regulatory agencies have established legal limits and recommendations for these compounds in certain products (FDA 2001).

Basic Structure and Formation

Biogenic amines may be classified into three main groups based on their molecular structures as aliphatic, aromatic, and heterocyclic. Examples of aliphatic biogenic amines include methylamine, dimethyl amine, ethylamine, diethyl amine, butyl amine, cadaverine, putrescine, spermidine, and spermine; examples of the aromatic amines are tyramine, octapamine, dopamine, and synephrine; and examples of the heterocyclic biogenic amines are 2-amino-2-methylimidazolo[4,5-*f*]quinolone (IQ), 2-amino-3,8-dimethylimidazo[4,5-*f*]quinoxaline and 2-amino-1-methyl-6-phenylimidazo[4,5-b]pyridine (Fig. 43.1).

Biogenic amines may be formed endogenously during normal metabolic processes in living cells from the degradation of biological molecules like proteins and/or decarboxylation of certain amino acids (e.g., tyrosine, phenylalanine, arginine, and histidine) by microbial enzymes. Endogenous amines formed from cellular metabolism play a plethora of important roles within the cardiovascular, nervous, and digestive systems. They are produced in many different tissues, for example, adrenaline is produced in the adrenal medulla, while histamine is formed in mast cells and liver and is subsequently transmitted locally or via the blood stream to perform various functions as in the modulation of growth and proliferation of eukaryotic cells by spermine and spermidine (Pegg and McCann 1982). Biogenic amines may also be formed exogenously as intermediates in the chemical and pharmaceutical industries, or released into the atmosphere from livestock breeding operations, animal feeds, waste treatment/incineration, and in automobile exhaust fumes. The exogenous biogenic amines are directly absorbed in the intestine, a process that is enhanced by intake of alcohol.

The amino acid decarboxylase enzymes of microorganisms carry out the dual functions of producing the biogenic amines for normal functions in the organism or removing the substrate (parent amino acid) to prevent the excessive accumulation of these substrate molecules. Several of the decarboxylase enzymes act on the α-carboxyl group of the appropriate L-amino acid to produce the corresponding biogenic amine and carbon dioxide. The enzymatic decarboxylation process depends on various factors such as the availability of the substrate in free form, the presence of decarboxylase-producing microorganisms, and the reaction medium conditions. The amino acid substrates for the decarboxylations are either naturally present in the food material or produced via proteolysis by endogenous proteases in the raw material. Also, decarboxylase enzyme-producing microorganisms may be present as part of the natural micro flora of a food product, and they may be introduced by contamination or added as part of a starter culture. Medium conditions that favor the growth of the bacteria (i.e., carbon source, pH, temperature, availability of O_2, and light) and production of the decarboxylase enzymes in active forms all promote the formation of biogenic amines.

Biogenic amines may also be broken down by the MAO enzyme system to control the levels of these compounds. Proteolysis to form free amino acids may also be potentiated by the action of exogenous proteases that are intentionally added for the modification of foodstuffs.

In foods and the food industry, it is the exogenously formed biogenic amines that are of particular interest. Heterocyclic amines (HCAs) are formed in muscle foods (beef, pork, poultry, and fish) during high temperature cooking procedures such as frying, broiling, and barbecuing. The amino acids from food proteins (meats) condense with creatinine to result in the HCAs. HCAs can be potent mutagens and carcinogens and may increase the risk of cancer for humans (Adamson and Thorgeirsson 1995). Nonetheless, biogenic amines are also used to treat depression and to promote positive mood.

OCCURRENCE IN FOOD PRODUCTS

Fermented Food Sources

Fermentation is a biological method of processing foodstuff to preserve its quality or to transform foods into stable and useful forms. The fermentation process invariably makes finished products with characteristic flavors and textural properties. The process is potentiated mostly by microbial activity and metabolism and by enzymes. Several different microorganisms including *Lactococcus*, *Lactobacillus*, *Leuconostoc*, *Streptococcus*, and *Pediococcus* species have been shown to participate in various food fermentations. These microorganisms may be naturally present in the food material or added to foods as starter culture, and they secrete their enzymes (including various decarboxylases and hydrolases) in foods for their transformation. As a result, certain fermented foods may accumulate large amounts of biogenic amines through the decarboxylation of amino acids by microbial decarboxylases or via proteolysis by proteases to generate intermediates (free amino acids) that may condense with creatinine to form heterocyclic biogenic amines. For example, products such as the Chinese mitten crab, beer, cheese, and other fermented foods have all been shown to accumulate high levels of histamine during storage (Stratton et al. 1991, Izquierdo-Pulido et al. 1999, Xu et al. 2009). Although the presence of biogenic amines may not always indicate spoilage, their presence in food products is not desired due to their potential health effects.

Cheese

Cheese is a commonly consumed fermented food product. During fermentation and ripening of cheeses, milk casein is hydrolyzed by proteases and peptidases into a mixture of smaller peptides and free amino acids. Some of the free amino acids produced become substrates for the decarboxylases of microorganisms associated with cheeses to form biogenic amines (mainly, putrescine, cadaverine, histamine, and tyramine) in the products (ten Brink et al. 1990). As well, free amino acids from proteolysis may form heterocyclic biogenic amines with creatinine.

Some microorganisms have been implicated in the formation of histamine, tyramine, putrescine, and cadaverine in cheeses. For example, *Lactobacillus buchneri* is known to produce vast amounts of histamine in cheeses, which can potentially elicit histamine poisoning in consumers, while *L. brevis* and *Enterococcus faecalis* have also been implicated in the formation of tyramine in certain cheese products (Stratton et al. 1991, Shalaby 1996).

Certain conditions can promote or inhibit the formation of biogenic amines in foods by microbial enzymes. The increase of free amino acids during fermentation is one of the main promoters of biogenic amines formation since the amino acids are the substrates for the enzymatic decarboxylase action, leading to the formation of the amines and carbon dioxide. As a result, cheeses with longer ripening times and higher degrees of proteolysis may have higher amounts of biogenic amines. The use of high temperatures during cheese making and/or storage may also increase biogenic amine levels in the products. When cheeses are stored or transported under different temperature conditions, the products handled under proper refrigeration conditions (4°C) tend to have lower levels of biogenic amines than their counterparts stored or transported at elevated temperatures (e.g., 25°C; Shalaby 1996, Pinho et al. 2001). On the other hand, high-temperature pasteurization of milk prior to its use for cheese-making can curtail biogenic amine formation in the product, and when high-temperature pasteurized milk is used in combination with particular starter cultures, it could result in the formation of considerably lower levels of biogenic amines in the cheese product; the use of appropriate levels of acid and salt may also slow down biogenic amine production in cheese products (Ordóñez et al. 1997, Valsamaki et al. 2000, Martuscelli et al. 2009).

Alcoholic Beverages

Alcoholic beverages, including beer and wines, are produced from plant materials by microbial fermentation process (e.g., red wine is produced by fermentation of grape must) and are common sources of biogenic amines. Beer and red wines have been implicated in outbreaks of histamine and tyramine poisoning. Biogenic amines in alcohol act synergistically to elicit various adverse health symptoms such as nausea, respiratory problems, palpitations, headaches, rashes, and blood pressure problems (Bodemer et al. 1999, Du Toit 2005). In the process, fermentation microorganisms such as *Saccharomyces cerevisiae* and *S. ellipsoideus* hydrolyze carbohydrates into simpler sugars that are further degraded into ethanol and a variety of flavor compounds. The beverages are then stored and aged for various time periods. In wines, the most common biogenic amines are histamine, tyramine, and putrescine, which are formed by bacterial strains such as *Pediococcus spp., L. brevis, L. buchneri, Leuconostoc mesenteroides*, and *Oenococcus oeni* (Guerrini et al. 2002, Moreno-Arribas et al. 2003).

The production of beer entails an initial mashing step to produce the wort, which is subsequently inoculated with yeast (e.g., *S. carlsbergensis* or *S. cerevisiae*) to potentiate the fermentation. Histamine, tyramine, putrescine, and cadaverine have been found in certain beers and because alcohol enhances the absorption of biogenic amines in humans, a legal upper limit for biogenic amine content of 2 mg per alcoholic beverage has been established as guide for manufacturers of the products (Moreno-Arribas et al. 2003).

The formation and accumulation of biogenic amines in beer is due to several factors, such as the quality of the raw material and microbial contaminations. Thus, lactic acid bacteria form putrescine, cadaverine, histamine, and tyramine during fermentations; storage and aging may lead to further increases in the levels of these compounds in the food products (Tabor et al. 1958, Bodemer et al. 1999).

The formation and/or accumulation of biogenic amines in foods may be controlled by adopting proper good sanitary and hygienic practices during food handling and processing, to control microbial contaminations and prevent temperature abuse during storage and transportation.

Meat Products

Fermentation is also used as a traditional preservation method to extend the shelf life of meat. Examples of fermented meats include ham, salami, pepperoni, chorizo, thuringer, bologna, and cervelat (Potter and Hotchkiss 1998). Sausages are classified as semi-dry and dry sausages, based on fermentation techniques. The semi-dry sausages are characterized by a relatively higher water activity ($A_w = 0.90$–0.94) and a relatively higher pH (≤ 5.0), as opposed to A_w of 0.85–0.91 and a pH of ≤ 4.5 for dry sausage. The semi-dry sausages are usually produced using a lactic acid bacteria starter culture over a relatively shorter fermentation period, while dry sausages are fermented over much longer time periods using the endogenous microflora associated with the raw material. The biogenic amines histamine, tyramine, tryptamine, cadaverine, putrescine, 2-phenylethyamine, spermidine, and spermine have all been found in dry sausages. However, their levels and distribution vary depending on factors such as quality of the raw material and the availability of precursor molecules to serve as substrates (Santos 1996, Suzzi and Gardini 2003).

Raw meats tend to have lower amounts of biogenic amines such as spermine and spermidine unlike fermented products, and some industry-derived commercial products tend to have relatively higher amounts of biogenic amines than their nonindustry-produced traditional products (Lakritz et al. 1975). The microorganisms responsible for biogenic amines production in fermented meats include *Enterobacteriaceae, Enterococci, Lactobacilli, Pediococcus,* and *Pseudomonads*. The lactic acid bacteria strain, *Lactobacilli* and *Enterococci,* are more associated with tyramine production, while *Enterobacteriaceae* is more connected with high putrescine and cadaverine levels in fermented meat products (Suzzi and Gardini 2003).

It is possible to have the microbial load associated with fermented meat products to be low; however, the enzymes (proteases and decarboxylases) released by microorganisms into the products may survive the fermentation process and cause continued production of biogenic amines in the finished products. Storage of the raw materials intended for fermentation at low temperatures is a good practice to control both microbial proliferation and enzymatic activity that lead to biogenic amines formation in the products. Fermented meat products made from raw material that have been exposed to high temperatures tend to have significantly high contents of biogenic amines such as tyramine, cadaverine, and putrescine.

In general, high temperature, high pH, and low salt content can enhance biogenic amine formation. Higher temperatures favor proteolytic, lipolytic, and decarboxylation reactions. For instance, the amount of biogenic amine content is higher in products prepared at 15°C compared with similar products made at 4°C, and more histamine is formed during ripening at 18°C as compared with processing at 7°C; biogenic amine formation is also less at lower pHs, because the decarboxylation of histidine tends to be relatively low at acidic pH (Maijala et al. 1995). Thus, amine-negative lactic acid bacteria are preferred as starter culture to rapidly produce acids to decrease the pH in the product(s) and suppress biogenic amine formation (Battcock and Azam-Ali 1998). A high salt content also generally inhibits microflora growth because the reduced water activity makes the product less conducive for the growth and proliferation of microbes; consequently, low levels of biogenic amine are formed.

Other factors may also contribute to the biogenic amine accumulation. For example, bacteria growth and metabolism are enhanced by the presence of glucose, while the addition of sulphites or nitrites suppresses the formation of cadaverine, putrescine, and cadaverine. In the case of sausages, the diameter of the product may affect biogenic amine formation, because it impacts the salt content and water activity of the product. Thus, sausages with bigger diameters tend to have greater tendency for biogenic amine production and accumulation than smaller-sized sausages (Battcock and Azam-Ali 1998).

In addition to meats, fish are also fermented into products such as faseekh, fish paste (e.g., bagoong, patis, and shrimp paste), and fish sauce (e.g., joetgal, shottsuru, garum, nuoc mam, and nam pla). Like their meat counterparts, these products also may contain biogenic amines. Histamine, putrescine, cadaverine, tyramine, spermidine, and spermine have all been detected in fermented fish products and their levels increased with storage (Mah et al. 2002). However, the amounts of biogenic amines found in these products (0–70 mg/kg sample) are considered as safe for human consumption, if they are handled and stored properly.

Vegetable Products

Vegetables may also be fermented to preserve them or form modified products with characteristic flavors and textures. Fermented vegetable products include pickles, ripe olives, sauerkraut, doenjang, shoyu, stinky tofu, miso, tempeh, injera, kimchi, koji, natto, soy sauce, brandy, cider, sake, and vinegar; they are produced from sources such as beans, grains, cucumbers, lettuce, olives, cabbage, turnips, fruits, and rice. In general, biogenic amine content is very low in fresh vegetables; however, the concentration increases during the fermentations process and throughout storage. Sauerkraut is made from shredded cabbage by fermentation with lactic acid bacteria. In the case of fermented vegetable products, there does not appear to be a significant correlation between biogenic amine levels and spoilage unlike the situation with fermented meat products, and compounds like tyramine, putrescine, and cadaverine are commonly found in fermented vegetable products (Potter and Hotchkiss 1998, Kalac et al. 1999, 2000a, 2000b).

The amounts of biogenic amines present in fermented vegetable products also vary and depend on factors such as temperature, salt content, and pH/acidity, as well as starter culture type. These factors all affect the growth and metabolism of the organism, which influence the formation of the biogenic amines. Low temperatures generally minimize microbial growth and proliferation, and enzyme activities to reduce biogenic amine formation, while elevated temperatures tend to do the reverse.

Salt plays an important role in the fermentation of vegetables. By modulating the A_w levels, salting can influence the texture of fermented vegetable products. For example, under dry salted fermentation conditions, product firmness is enhanced with increasing salt content. Although a high salt may suppress the activity of certain (desirable) bacteria, there are halophilic bacteria that are salt tolerant, and advantage may be taken of this fact to use halophilic bacteria such as *Enterococci* to overcome the growth and proliferation of a moderately salt tolerant *Lactobacilli* species in fermented vegetable products (Kalac et al. 1999).

Acidity and pH are also important in the fermentation of vegetables. Hitherto, the use of starter cultures helps to exclude undesirable microorganisms by generating an acidic milieu that is not conducive for the growth of certain microorganisms. Thus, the pH level in a product can determine the types of microorganisms that would be dominant in the course of the fermentation process. For instance, *Lactobacillus* and *Streptococcus* species are acid tolerant, and when they are used in mixed cultures, species like *Leuconostoc mesenteroides* would initiate the process and produce an acidic environment, and then be replaced by *L. plantarum* till the lactic acid level rises sufficiently to kill the *L. plantarum* and be replaced by *L. pentoaceticus*. *L. plantarum* is a typical tyramine producer; thus, products formed with such cultures would be expected to accumulate biogenic amines (Battcock and Azam-Ali 1998, Suzzi and Gardini 2003).

Fermented vegetable products also accumulate biogenic amines during storage. For example, the levels of tyramine and putrescine in sauerkraut increase during storage, and the longer the storage period, the more of these biogenic amines accumulate (Kalac et al. 2000a, 2000b).

Non-fermented Foods

Biogenic amines occur in non-fermented foods also. Fresh fish and meats in particular are known to contain biogenic amines due to their high protein and free amino acid contents and their high perishability. These food products have also been associated with histamine poisoning, which also attests to the presence of these compounds. Biogenic amines in non-fermented foods are also formed by the action of decarboxylating and proteolytic enzymes exuded by microorganisms naturally present as part of the native microbial flora of fish and meats, or microorganisms purposefully added as starter culture, or added through contamination. High levels of biogenic amines in foodstuffs may be used as indices of undesirable microbial growth and spoilage.

Fish

Biogenic amines in fish are of particular health concern because they cause histamine poisoning in humans (Sanchez-Guerrero et al. 1997). Histamine poisoning or scombroid fish poisoning (scombrotoxicosis) is the most widespread form of seafood poisoning in fish-producing and fish-consuming communities. Fish in the *Scombridae* and *Scomberesocidae* families as well as non-scombroid fish have both been implicated in biogenic amine poisoning of humans. These include fish species like bluefish (*Pomatomus* spp.), bonito (*Sarda* spp.), mackerel (*Scomber* spp.), mahi mahi (*Coryphaena* spp.), marlin (*Makaira* spp.), pilchards (*Sardina pilchardus*), sardines (*Sardinella* spp.), saury (*Cololabis saida*), sockeye salmon (*Oncorhynchus nerka*), tuna (*Thunnus* spp.), and yellowtail (*Seriola lalandii*) (Lehane and Olley 2000, Sarkadi 2009). These fish species tend to have high levels of free histidine in their tissues, which may undergo decarboxylation to form the toxic histamine. The levels of biogenic amines in fish tissues vary greatly and are influenced by various factors including muscle type, the native microflora, harvesting, and postharvest management practices (including handling, processing, transportation and storage; Veciana-Nogues et al. 1997a, 1997b, Gloria et al. 1999).

Of the biogenic amines found in fish tissues, histamine usually tends to be the most abundant, followed by putrescine and cadaverine in that order. For example, tuna fish samples stored at 20°C produced histamine, cadaverine, and putrescine levels of 3103 μg/g, 44.2 μg/g, and 3.0 μg/g, respectively, after 48 hours of storage (Veciana-Nogues et al. 1997a, 1997b). Nonetheless, these biogenic amines may not occur in all fish tissues, as is the case for tyramine, which is absent in sea bream (*Sparus aurata*) but present in anchovies (*Engraulis mordax*; Koutsoumanis et al. 1999).

The relative amounts of biogenic amines and their distribution in fish tissues depend to a large extent on the types of microorganisms associated with the raw material and their capacity to produce decarboxylases and hydrolytic enzymes like proteases and lipases. Generally, when the microbial load is high, the biogenic amine levels also tend to be high (Du et al. 2002).

Approximately, one-third of the microorganisms isolated from spoiled tuna and mackerel are decarboxylase enzyme producers. These include *Acinetobacter iwoffi*, *Aeromonas hydrophila*, *Clostridium perfringens*, *Enterobacter aerogenes*, *Hafnia alvei*, *Morganella morganii*, *Proteus mirabilis*, *Proteus vulgaris*, *Pseudomonas fluorescens*, *Pseudomonas putrefaciens*, and *Vibrio alginolyticus*, and they are members from both mesophilic and psychrophilic organisms. In this regard, the mesophilic types tend to play a relatively more significant role in histamine production and fish flesh deterioration than their psychrophilic counterparts, and species like *Morganella morganii*, as well as *Enterobacteriaceae*, *Clostridium*, and *Lactobacillus* are of particular importance in terms of histamine production in fish (Omura et al. 1978, Taylor 1986, Middlebrooks et al. 1988, Lakshmanan et al. 2002, Kim et al. 2006).

Environmental factors (e.g., temperature, pH, and salt content) also play a role in biogenic amines formation and accumulation in fish tissues, and these factors may be manipulated to slow down biogenic amine formation.

Fish muscle type also influences the amounts of biogenic amines that form and/or accumulate in fish tissues. In this connection, red-fleshed fish tend to have higher levels of histamine in their tissues than their white-fleshed counterparts. This is due to the relatively higher content of the precursor molecule, histidine, in red-fleshed fish tissues, as well as higher amounts of carbohydrates and lipids, and a slightly more acidic environment, all conditions that favor the growth and proliferation of decarboxylase positive microorganisms.

The importance of temperature as a foremost manipulative factor to limit the formation of biogenic amines in fish tissues is due to the fact that several of the major histidine-decarboxylase-positive bacteria associated with fish are mesophilic. Thus, a decrease in temperature slows down their growth and metabolism (Ababouch et al. 1991). For example, icing decreases the initial mesophilic microorganism population while increasing the lag phase of growth. Nevertheless, decreasing product temperature is not an absolute means of control. This is because certain psychrophilic bacteria can also produce decarboxylases that can lead to the formation of biogenic amines, even at proper refrigeration temperatures. Furthermore, the decarboxylase enzymes released in fish tissues are still functional at low temperatures although at a reduced rate. Thus, proper postharvest fish management is crucial to control formation and accumulation of biogenic amines in fish flesh.

Meat

Fresh meats (beef, pork, and poultry) are also non-fermented foods known to have biogenic amines. Their high proteinaceous nature makes them prone to proteolysis to form free amino acids that may subsequently be decarboxylated by microbial enzymes into the biogenic amines. Spermine and spermidine are believed to be present in all fresh meats (beef, pork, and poultry) at fairly constant levels, unlike the other biogenic amines that occur in varying amounts. Other predominant biogenic amines in fresh meats are putrescine and cadaverine, together with relatively lower levels of histamine, tryptamine, and tyramine (Paulsen and Bauer 2007, Galgano et al. 2009). For beef, tyramine or a combination of the three most prevalent biogenic amines, tyramine, putrescine, and cadaverine, have been recommended for use as index of quality or spoilage (Vinci and Antonelli 2002).

Fresh poultry meat can also have significant amounts of biogenic amines and the accumulation of compounds such as spermidine, putrescine, cadaverine, histamine, and tyramine tend to increase during storage. In poultry, putrescine and cadaverine tend to be the predominant biogenic amines and may attain total levels of 50–100 mg/kg when stored over extended periods. Biogenic amines also accumulate more rapidly in raw poultry meat because of the higher susceptibility of white muscle fibers to proteolysis versus the darker muscle fibers of red meats, beef, or pork (Perez et al. 1998, Balamatsia et al. 2006, Patsias et al. 2006).

Enterobacteriaceae species are believed to be primarily responsible for putrescine, cadaverine, and histamine formation in poultry meat. Other microorganisms that have been identified as major biogenic amine producers include *Brochothrix thermosphacta*, *Carnobacterium divergens*, *Carnobacterium piscicola*, *Citrobacter freundii*, *Enterobacter agglomerans*, *Escherichia coli*, *Escherichia vulnaris*, *Hafnia alvei*, *L. curvatus*, *Morganella morganii*, *Proteus alcalifaciens*, and *Serratia liquifaciens* (Masson et al. 1996, Pircher et al. 2007).

Biogenic amine formation and/or accumulation in non-fermented meats are influenced by essentially the same factors as for fermented meats. For example, the type of meat is important, as it determines the availability of the substrate or precursor molecules. Relatively larger quantities of cadaverine are produced in both red and white meat due to the high levels of free lysine in both muscle types; however, red meats tend to have higher tyramine contents from their higher free tyrosine contents.

In fresh meats, temperature, salt content, and pH all play an important role in biogenic amine production. However, this attribute varies from species to species. For example, low pHs enhance the production of histamine, tyramine, and tryptamine in poultry meats by lactic acid than at a higher pH, while *Enterobacteriaceae* produced less cadaverine at low pHs. High salt concentrations also has a greater inhibitory effect on the formation of biogenic amines, and this is due to inhibition of microbial growth by the reduced A_w from the high salt content; higher temperatures promote more biogenic amines formation in beef, pork, and poultry samples than frozen temperatures. Thus, a higher salt content or lower storage temperature (or both) is recommended for curtailing biogenic amine formation and extending the shelf life of meat products (Chander et al. 1989, Karpas et al. 2002).

SIGNIFICANCE TO THE FOOD INDUSTRY

Biogenic amines are of particular interest to the food industry due to their toxicity, and their use as indicators of food quality or spoilage (Ruiz-Capillas and Moral 2004). For example, histamine poisoning occurs following ingestion of foods that have large quantities of histamine (Stratton et al. 1991). Scombroid fish if not handled properly can accumulate high quantities of histamine, and such products can pose a health hazard when consumed. Other fermented foods, such as cheese, wine, dry sausage, sauerkraut, miso, and soy sauce have also been linked with biogenic amines poisoning. Histamine and tyramine toxicities have been extensively investigated, and biogenic amines such as β-phenylethylamine and tyramine have been linked to hypertensive crises (Shalaby 1996) and migraines. Other biogenic amines such as the HCAs have also been implicated in the formation of mutagenic or carcinogenic nitrosamines (Stratton et al. 1991), and the potential health risks of these compounds are areas of active research to better understand the extent of the health hazards posed by these compounds in food products.

Health Effects

Biogenic amines participate in various functions connected with cell growth and metabolism in humans. For example, spermine and spermidine are involved in free radicals scavenging, cell membrane channels modulation, and in DNA synthesis. However, the compounds must be present in minute quantities to carry out useful functions, because they can be toxic in high quantities. Other biogenic amines, such as histamine and tyramine, occur naturally in small quantities and participate in allergic and inflammatory responses (Khan et al. 1992, Ha et al. 1998, Schneider et al. 2002). The metabolism of histamine entails absorption by active or passive transport through the lumen of the GIT. During this passage, most of the histamine is broken down by the enzymes diamine oxidase (DAO) and histamine-*N*-methyltransferase (HNMT) into useful intermediates like imidazoleacetic acid and *N*-methyl histamine. These molecules together with the residual histamine are transported in the blood to the liver where they are degraded further. In the body, histamine binds to receptors to initiate various responses, including smooth muscle contraction, vasodilation, alteration of blood pressure, and enhancing vascular permeability. Nevertheless, high levels of histamine are potentially toxic and this intoxication manifests itself within 2 hours after ingestion and may last up to 16 hours. The symptoms include allergic-type reactions and gastrointestinal problems in the form of nausea, cramps, vomiting, and diarrhea (Izquierdo-Pulido et al. 1999, Attaran and Probst 2002, Maintz and Novak 2007). Histamine intoxication can also manifest as neurological and cutaneous disorders in the form of dizziness, flushing, headache, rash, and itchy eyes. These symptoms are attributed to the release of nitric oxide from vascular endothelial cells to bring about vasodilation (Becker et al. 2001, Predy et al. 2003, Borade et al. 2007). In rare cases, histamine poisoning could cause severe hypotension, atrial arrhythmias, and loss of consciousness (McInerney et al. 1996).

Various factors may influence histamine toxicity. These include the use of drugs that inhibit mono- and DAO enzymes, thereby preventing detoxification of histamine by deamination, gastrointestinal diseases, alcohol consumption, and bacteria endotoxins. Other biogenic amines like putrescine and cadaverine can also compete for the same catabolic enzymes as histamine, thus inhibiting the effectiveness of histamine detoxification and making the toxic compounds accumulate highly in the plasma.

Although histamine intoxication is the type most commonly associated with biogenic amines, tyramine also has high toxicity. Ingestion of large quantities of tyramine-containing cheeses, alcohol, chocolate, and other dairy products can elicit tyramine intoxication, which manifests as headaches and elevated blood pressure (Millichap and Yee 2003, Panconesi 2008). Tyramine is catabolized rapidly in the small intestine by MAOs, and monitoring tyramine ingestion is crucial for individuals on MAO-inhibiting drugs as their ingestion could bring on hypertensive crisis (Schmidt and Ferger 2004, Vlieg-Boerstra et al. 2005). Tyramine poisoning can also cause nausea, vomiting, and even death from intracranial hemorrhage. There are also emerging concerns with nitrosamines because they can form carcinogenic nitrosamines with cadaverine and putrescine by reacting with nitrites (used to cure meats) or by serving as precursor molecules (Bover-Cid et al. 1999).

Certain drugs, for example, ephedrine and related compounds (amphetamines, phentermine, mazindol, and fenfluramine) can block the uptake of the biogenic amines NE and serotonin. Thus, they are able to suppress appetite and food intake, and there is interest in the development of thermogenic drugs that could act on appropriate biogenic amines to control obesity safely without increasing heart rate and blood pressure.

Quality Indicators

Biogenic amines are of interest and use to food scientists/food technologists due to their use as food quality indicators. Aqueous solutions of putrescine and cadaverine impart discernible and objectionable odors at levels of 22 ppm and 190 ppm, respectively. The relative contribution of flavor from biogenic amines to the overall food quality is not well established; rather, the main focus has been on their possible use as chemical indicators of food quality. The relationship between the amounts of particular biogenic amines in a food product and the extent of food spoilage has been used to estimate a parameter known as the chemical quality index (CQI) or the biogenic amines index (BAI) for fish, and this takes into account the concentrations of putrescine, cadaverine, histamine, spermine, and spermidine in the fish sample (Mietz and Karmas 1977). On the basis of this, the CQI is calculated as

$$\text{Chemical Quality Index [CQI]} = \frac{\text{Histamine} + \text{Putrescine} + \text{Cadaverine}}{1 + \text{Spermine} + \text{Spermidine}}$$

and a CQI value below 1 denotes a good quality product, a value between 1 and 10 is considered mediocre, while a value ≥ 10 connotes spoilage. The CQI or BAI concept was extended further as the beer BAI by taking into consideration the concentrations of other biogenic amines, tyramine, β-phenylethylamine, and agmatine that occur in products like beer (Loret et al. 2005). The beer BAI is calculated as

$$\text{Beer BAI} = \frac{\text{Cadaverine} + \text{Histamine} + \text{Tyramine} + \text{Putrescine} + \text{Phenylethylamine} + \text{Trytamine}}{1 + \text{Agmatine}}$$

Hitherto, a beer BAI value below 1 indicates a good quality product, a value between 1 and 10 is considered as mediocre quality, while a value $\geq$10 connotes spoilage. For meats, the total content of putrescine, cadaverine, histamine, and tyramine may be used to determine freshness, with levels $\leq$5.0 μg/g meat indicating high quality products (Hernandez-Jover et al. 1997).

ANALYSIS METHODS

There are several analytical methods available for the detection, isolation, and quantification of biogenic amine levels in food products. These include chromatographic, fluorometric, and enzyme-linked immunosorbent assay (ELISA) methods.

Extraction and Chromatography

Various chromatographic methods have been used to analyze biogenic amines. These methods all entail an initial extraction step to recover the amines from the food product using organic solvents, together with perchloric and trichloroacetic acids. The extract thus obtained may then be subjected to chromatographic separation methods such as ion-exchange column chromatography, paper chromatography, thin layer chromatography (TLC), gas liquid chromatography, and high-pressure liquid chromatography (HPLC; Moret and Conte 1996, Cinquina et al. 2004, Paleologos and Kontominas 2004). TLC is considered as the simplest of the chromatographic methods for the separation and quantification of biogenic amines (Shalaby 1996). It is effective, accurate, fairly rapid, relatively inexpensive, and can be applied to large sample sizes. HPLC methods are commonly used method for biogenic amine analysis because they are highly sensitive and reproducible, although they require relatively more sophisticated and expensive technology, and are more time-consuming and tedious to carry out compared with TLC (Shakila et al. 2001, Lange et al. 2002).

Fluorometric Methods

Biogenic amines may also be determined by fluorometric methods. These methods also involve extraction, purification, and separation of the compounds as in the chromatographic procedures. However, in order to quantify particular biogenic amines, the eluant is mixed with a molecule to form a fluorescent product, and the degree to which it fluoresces at particular wavelengths is determined and used as a measure of the amount of the biogenic amine in a food product by comparing with known standards (Du et al. 2002).

Biosensors

This technology is based on the recognition of biogenic amines by specific ligands such as enzymes, antibodies, receptors, or microorganisms. The binding of the amines to these ligands is "sensed" by electrochemical, mass, optical, or thermal sensors, and the degree of response is measured and correlated with the amount of the substrate (amines) present in the raw material or sample (Tombelli and Mascini 1998, Sarkadi 2009). Commonly used biosensors for biogenic amines include putrescine oxidase and DAOs (Carelli et al. 2007). These enzymes utilize O_2 to react with the amines and produce H_2O_2. The change in concentration (of O_2 or H_2O_2) is typically measured as changes in electric signal by electrodes (Prodromidis and Karayannis 2002).

Enzyme-Linked Immunosorbent Assay

Biogenic amines in foods may also be measured by the ELISA. This method is based on the detection of *N*-amino derivatives of histamine using antibodies. The *N*-amino derivatives are synthesized from histamine before analysis using compounds such as *p*-benzoquinone or propionic acid esters. There are presently commercial ELISA test methods that are available for the analysis of histamine levels in wine, cheese, and fish (Serrar et al. 1995, Aygun et al. 1999, Rupasinghe and Clegg 2007). The ELISA method used to analyze wine sometimes yields slightly higher results than the HPLC method; however, the ELISA method has the advantage of being more rapid and less tedious to perform than the HPLC methods.

EFFECTS OF FOOD PROCESSING AND STORAGE

Food processing is an essential mechanism for controlling the biogenic amine levels in foods. Because of their high thermal stability, once amines are formed, their concentrations will not be significantly decreased during cooking and heating processes. Processing techniques such as evisceration, postharvest handling, freezing, salting, and smoking can greatly affect the biogenic amine content and quality of the final product.

Evisceration

Evisceration can help control fish quality. For example, Atlantic croacker and trout eviscerated directly after capture had a longer shelf life than their iced un-eviscerated counterparts, and the cadaverine and histamine concentrations were significantly less in the eviscerated samples versus those of un-gutted fish samples (Paleologos et al. 2004). This observation is due to a decrease in the load of microorganisms associated with the eviscerated fish compared with the un-eviscerated ones.

Temperature Control

Biogenic amine levels in food products are also greatly influenced by storage temperature. Postharvest handling techniques like immediate icing are important in controlling biogenic amine levels in fish because at the reduced temperatures, both enzymatic and microbial activities are considerably reduced, so the rates of formation of the biogenic amines and their accumulation in the products are correspondingly reduced.

Salting

Salting can effectively inhibit biogenic amine formation particularly at high levels. In general, a higher salt content results in a reduced biogenic amine formation. This is because the high salt lowers the A_w of the milieu, and this can inhibit both the microbial and enzymatic activities that are required for the biogenic amine formation.

Smoking

Smoking is a traditional method used to preserve fish. The fish is smoked, then frozen, and transported. Histamine content is found to increase during the smoking process and continues to accumulate during subsequent freezing (Zotos et al. 1995). Even though smoking may induce histamine production and nitrosamines formation, the high temperature applied in smoking can also curtail the growth and proliferation of harmful and/or spoilage microorganisms (Martuscelli et al. 2009).

REGULATIONS

Since high level of biogenic amines, for example, histamine and tyramine, are associated with food poisoning, their intake has to be limited. Even though not all amines are equally toxic, and toxicological levels are hard to ascertain due to synergetic effects between amines, permissible limits have still been proposed. Thus, fermented products that are prepared using good manufacturing practices may be expected to contain histamine, tyramine, and β-phenylethylamine concentrations of 50–100 , 100–800, and 30 mg/kg, respectively, and still considered safe and acceptable for human consumption (Nout 1994). Levels of histamine, cadaverine, putrescine, tyramine, and β-phenylethylamine in sauerkraut must not exceed 10, 25, 50, 20, and 5 mg/kg, respectively, and the total amount of biogenic amines in fish, cheese, and sauerkraut must be less than 300 mg/kg (Shalaby 1996). Furthermore, biogenic amine intake greater than 40 mg in a meal is potentially toxic.

As biogenic amines are not extremely toxic or associated with many fatal incidences, there are few general regulations controlling their concentration. Histamine, however, is a common concern and different countries have set up their own standards for these compounds in foods. For example, a level of 10 mg histamine for every liter of wine is considered as acceptable in Switzerland. In the United States, 20 mg histamine per 100 g canned fish is considered unsafe for human consumption, and 50 mg histamine per 100 g canned fish is considered as a health hazard by the Food and Drugs Administration (FDA 2001); the European Economic Community has set the acceptable level of histamine content in fish as 10–20 mg/100 g (Shalaby 1996); the Canadian Food Inspection Agency, has set the action level of histamine as 20 mg/100 g in fermented products and 10 mg/100 g in scombroid fish products (CFIA 2005, Moret et al. 2005).

CONCLUSION

Biogenic amines are important food components found in many fermented and non-fermented food production. They are present as by-products of microbial activity in foods. The microorganisms are either present naturally, added for fermentation purposes, or are introduced through contamination. It is desirable to minimize the formation of biogenic amines due to their adverse effects on human health and their contribution to food spoilage and food losses. In particular, the levels of tyramine and histamine need to be below a certain threshold level in order to prevent toxic responses.

The study of biogenic amines is an ongoing field of research, with several foci. Analytical methods are constantly being developed and/or improved in order to quantify their presence more rapidly and accurately. Further studies are needed to generate the basic knowledge on the effects of ingestion, toxic limits, and interactions with other biological molecules, to facilitate the rationalization and formulation of more useful recommendations and regulations regarding their safe levels in food products.

REFERENCES

Ababouch L et al. 1991. Quantitative changes in bacteria, amino acids and biogenic amines in sardine (*Sardina pilchardus*) stored at ambient temperature (25–28°C) and in ice. *Int J Food Sci Technol* 26: 297–306.

Adamson RH, Thorgeirsson UP. 1995. Carcinogens in foods: heterocyclic amines and cancer and heart disease. *Adv Exp Med Biol* 369: 211–220.

Aston-Jones G et al. 2002. Prominent projections from the orbital prefrontal cortex to the locus coeruleus in monkey. *Soc Neurosci Abstr* 28: 86–89.

Attaran RR, Probst F. 2002. Histamine fish poisoning: a common but frequently misdiagnosed condition. *Emerg Med J* 19: 474–475.

Aygun O et al. 1999. Comparison of ELISA and HPLC for the determination of histamine in cheese. *J Agric Food Chem* 47: 1961–1964.

Balamatsia CC et al. 2006. Correlation between microbial flora, sensory changes and biogenic amines formation in fresh chicken meat stored aerobically or under modified atmosphere packaging at 4°C: possible role of biogenic amines as spoilage indicators. *Antonie van Leeuwenhoek* 89: 9–17.

Battcock M, Azam-Ali S. 1998. *Fermented Fruits and Vegetables: A Global Perspective*. Food and Agruiculture Organization of the United Nations, Rome.

Becker K et al. 2001. Histamine poisoning associated with eating tuna burgers. *J Am Med Assoc* 285: 1327–1330.

Berger M et al. 2009. The expanded biology of serotonin. *Annu Rev Med* 60: 355–366.

Bodemer S et al. 1999. Biogenic amines in foods: histamine and food processing. *Inflamm Res* 48: 296–300.

Borade PS et al. 2007. A fishy case of sudden near fatal hypotension. *Resuscitation* 72: 158–160.

Bover-Cid S et al. 1999. Relationship between biogenic amine contents and the size of dry fermented sausages. *Meat Sci* 51: 305–311.

Boyer EW, Shannon M. 2005. The serotonin syndrome. *N Engl J Med* 352(11): 1112–1120.

Canadian Food Inspection Agency. 2005. Fish products standards and methods manual. Canadian guidelines for chemical contaminants and toxins in fish and fish products. Available at: http://www.inspection.gc.ca/english/fssa/fispoi/man/samnem/smnmalle.pdf. Accessed on June 20, 2011.

Cannon WB. 1929. The interrelations of emotion as suggested by recent physiological researches. *Am J Physiol* 25: 256–282.

Carelli D et al. 2007. An interference free amperometric biosensor for the detection of biogenic amines in food products. *Biosens Bioelectron* 23: 640–647.

Chander H et al. 1989. Factors affecting amine production by a selected strain of *Lactobacillus bulgaricus*. *J Food Sci* 54: 940–942.

Cinquina AL et al. 2004. Determination of biogenic amines in fish tissues by ion-exchange chromatography with conductivity detection. *J Chromatogr A* 1032: 73–77.

De Vries J (ed.). 1996. *Food Safety and Toxicity*. CRC Press, Boca Raton, FL.

Du WX et al. 2002. Development of biogenic amines in yellowfin tuna (*Thunnis albacares*): effect of storage and correlation with decarboxylase-positive bacterial flora. *J Food Sci* 67: 292–301.

Du Toit M. 2005. Biogenic amine production in wine. Wynboer. Available at http://www.wynboer.co.za/recentarticles/200506bio.php3 (Accessed in October 2009).

Etkind P, Wilson ME. 1987. Bluefish-associated scombroid poisoning: an example of the expanding spectrum of food poisoning from seafood. *J Am Med Assoc* 258: 3406–3410.

FDA. 2001. *Fish and Fishery Products Hazards and Controls Guide*, 3rd edn. Office of Seafood, Washington, DC, p. 326.

Galgano F et al. 2009. Role of biogenic amines as index of freshness in beef meat packed with different biopolymeric materials. *Food Res Int* 42: 1147–1152.

Gloria M et al. 1999. Histamine and other biogenic amines in albacore tuna. *J Aquat Food Product Technol* 8: 55–69.

Guerrini S et al. 2002. Biogenic amine production by *Oenococcus oeni*. *Curr Microbiol* 44: 374–378.

Ha HC et al. 1998. The natural polyamine spermine functions directly as a free radical scavenger. *Proc Nat Acad Sci USA* 95: 11140–11145.

Hernandez-Jover M et al. 1997. Biogenic amine and polyamine contents in meat and meat products. *J Agric Food Chem* 45: 2098–2102.

Izquierdo-Pulido M et al. 1999. Polyamine and biogenic amine evolution during food processing. In: S Bardocz, A White (eds.) *Polyamines in Health and Nutrition*. Kluwer Academic Publishers, Norwell, MA, pp. 3–26.

Jones RSG. 1983. Trace biogenic amines: a possible functional role in the CNS. *Trends Pharmacol Sci* 4: 426–429.

Kalac P et al. 1999. Concentrations of seven biogenic amines in sauerkraut. *Food Chem* 67(3): 275–280.

Kalac P et al. 2000a. Changes in biogenic amine concentrations during sauerkraut storage. *Food Chem* 69(3): 309–314.

Kalac P et al. 2000b. The effects of lactic acid bacteria inoculants on biogenic amines formation in sauerkraut. *Food Chem* 70(3): 355–359.

Karpas Z et al. 2002. Determination of volatile biogenic amines in muscle food products by ion mobility spectrometry. *Anal Chim Acta* 463: 155–163.

Khan AU et al. 1992. A proposed function for spermine and spermidine: Protection of replicating DNA against damage by singlet oxygen. *Proc Natl Acad Sci USA* 89: 11426–11427.

Kim SH et al. 2006. Histamine production by *Morganella morganii* in mackerel, albacore, mahi-mahi, and salmon at various storage temperatures. *J Food Sci* 67: 1522–1528.

Koutsoumanis K et al. 1999. Biogenic amines and sensory changes associated with the microbial flora of Mediterranean gilt-head sea bream (*Sparus aurata*) stored aerobically at 0, 8, 15°C. *J Food Prot* 62: 398–402.

Lange J et al. 2002. Comparison of a capillary electrophoresis method with high-performance liquid chromatography for the determination of biogenic amines in various food samples. *J Chromatogr B* 889: 229–239.

Lakritz L et al. 1975. Determination of amines in fresh and processed pork. *J Agric Food Chem* 23: 344–346.

Lakshmanan R et al. 2002. Survival of amine-forming bacteria during the ice storage of fish and shrimp. *Food Microbiol* 19: 617–625.

Lehane L, Olley J. 2000. Histamine fish poisoning revisited. *Int J Food Microbiol* 58: 1–37.

Lesurtel M et al. 2006. Platelet-derived serotonin mediates liver regeneration. *Science* 312: 104–107.

Loret S et al. 2005. Levels of biogenic amines as a measure of the quality of the beer fermentation process: data from Belgian samples. *Food Chem* 89: 519–525.

Mah J-H et al. 2002. Biogenic amines in Jeotkals, Korean salted and fermented fish products. *Food Chem* 79(2): 239–243.

Maintz L, Novak N. 2007. Histamine and histamine intolerance. *Am J Clin Nutr* 85: 1185–1196.

Maijala R et al. 1995. Influence of processing temperature on the formation of biogenic amines in dry sausages. *Meat Sci* 39(1): 9–22.

Marieb EN. 2001. *Human Anatomy and Physiology*, 5th edn. Addison Wesley Longman, New York.

Martuscelli M et al. 2009. Effect of intensity of smoking treatment on the free amino acids and biogenic amines occurrence in dry cured ham. *Food Chem* 116: 955–962.

Masson F et al. 1996. Histamine and tyramine production by bacteria from meat products. *Food Microbiol* 32: 199–207.

McInerney J et al. 1996. Scombroid poisoning. *Ann Emerg Med* 28: 235–238.

Middlebrooks BL et al. 1988. Effects of storage time and temperature on the microflora and amine development in Spanish Mackerel (*Scomberomorus maculates*). *J Food Sci* 53: 1024–1029.

Mietz JL, Karmas E. 1977. Chemical quality index of canned tuna as determined by high-pressure liquid chromatography. *J Food Sci* 42: 155–158.

Millichap JG, Yee MM. 2003. The diet factor in pediatric and adolescent migraine. *Pediatr Neurol* 28: 9–15.

Moreno-Arribas MV et al. 2003. Screening of biogenic amine production by lactic acid bacteria isolated from grape must and wine. *Int J Food Microbiol* 84: 117–123.

Moret S, Conte LS. 1996. High-performance liquid chromatographic evaluation of biogenic amines in foods: an analysis of different methods of sample preparation in relation to food characteristics. *J Chromatogr A* 729: 353–369.

Moret S et al. 2005. A survey on free biogenic amine content of fresh and preserved vegetables. *Food Chem* 89: 355–361.

Nout MJR. 1994. Fermented food and food safety. *Food Res Int* 27: 291–298.

Omura Y et al. 1978. Histamine-forming bacteria isolated from spoiled skipjack tuna and jack mackerel. *J Food Sci* 43: 1779–1781.

Ordóñez AI et al. 1997. Formation of biogenic amines in Idiazábal ewe's-milk cheese: effect of ripening, pasteurization, and starter. *J Food Prot* 60: 1375–1380.

Paleologos EK, Kontominas MG. 2004. On-line solid-phase extraction with surfactant accelerated on-column derivatization and micellar liquid chromatographic separation as a tool for the determination of biogenic amines in various food substrates. *Anal Chem* 76: 1289–1294.

Paleologos EK et al. 2004. Biogenic amines formation and its relation to microbiological and sensory attributes in ice-stored whole, gutted and filletes Mediterranean sea bass (*Dicentrarchus labrax*). *Food Microbiol* 21: 549–557.

Panconesi A. 2008. Alcohol and migraine: trigger factor, consumption, mechanisms—a review. *J Headache Pain*. 9: 19–27.

Patsias A et al. 2006. Relation of biogenic amines to microbial and sensory changes of precooked chicken meat stored aerobically and under modified atmosphere packaging at 4°C. *Eur Food Res Technol* 223: 683–689.

Paulsen P, Bauer F. 2007. Spermine and spermidine concentrations in pork loin as affected by storage, curing and thermal processing. *Eur Food Res Technol* 225: 921–924.

Pegg AE, McCann PP. 1982. Polyamine metabolism and function. *Am J Physiol Cell Physiol* 243: C212–C221.

Perez ML et al. 1998. Effect of calcium chloride marination on calpain and quality characteristics of meat from chicken, horse, cattle and rabbit. *Meat Sci* 48: 125–134.

Pinho O et al. 2001. Effect of temperature on evolution of free amino acid and biogenic amine contents during storage of Azeitao cheese. *Food Chem* 75: 287–291.

Pircher A et al. 2007. Formation of cadaverine, histamine, putrescine and tyramine by bacteria isolated from meat, fermented sausages and cheeses. *Eur Food Res Technol* 226: 225–231.

Potter NN, Hotchkiss JH. 1998. *Food Science*. Chapman and Hall, New York.

Predy G et al. 2003. Was it something she ate? Case report and discussion of scombroid poisoning. *Can Med Assoc J* 168: 587–588.

Prodromidis MI, Karayannis MI. 2002. Enzyme-based amperometric biosensors for food analysis. *Electroanalysis* 14: 241–261.

Ruiz-Capillas C, Moral Q. 2004. Free amino acids and biogenic amines in red and white muscle of tuna stored in controlled atmospheres. *Amino Acids* 26: 125–132.

Rupasinghe HPV, Clegg S. 2007. Total antioxidant capacity, total phenolic content, mineral elements, and histamine concentrations in wines of different fruit sources. *J Food Comp Anal* 20: 133–137.

Sanchez-Guerrero IM et al. 1997. Scombroid fish poisoning: a potentially life-threatening allergic-like reaction. *J Allergy Clin Immunol* 100: 433–434.

Santos MHS. 1996. Biogenic amines: their importance in foods. *Int J Food Microbiol* 29: 213–231.

Sarkadi LS. 2009. Biogenic amines. In: RH Stadler, DR Lineback (eds.) *Process-Induced Food Toxicants: Occurence, Formation, Mitgation, and Health Risks*. John Wiley & Sons, Hoboken, pp. 321–361.

Schmidt N, Ferger B. 2004. The biogenic trace amine tyramine induces a pronounced hydroxyl radical production via a monoamine oxidase dependent mechanism: an in vivo microdialysis study in mouse striatum. *Brain Res* 1012: 101–107.

Schneider E et al. 2002. Trends in histamine research: new functions during immune response and hematopoiesis. *Trends Immunol* 23: 255–263.

Serrar D et al. 1995. The development of a monoclonal antibody-based ELISA for the determination of histamine in food: application to fishery products and comparison with the HPLC assay. *Food Chem* 54: 85–91.

Shakila RJ et al. 2001. A comparison of the TLC-densitometry and HPLC method for the determination of biogenic amines in fish and fishery products. *Food Chem* 75: 255–259.

Shalaby AR. 1996. Significance of biogenic amines to food safety and human health. *Food Res Int* 29: 675–690.

Stratton JE et al. 1991. Biogenic amines in cheese and other fermented foods. *J Food Prot* 54(6): 460–470.

Suzzi G, Gardini F. 2003. Biogenic amines in dry fermented sausages: a review. *Int J Food Microbiol* 88: 41–54.

Tabor H et al. 1958. The biosynthesis of spermidine and spermine from putrescine and methionine. *J Biol Chem* 233: 907–914.

Tarjan V, Janossy G. 1978. The role of biogenic amines in foods. *Die Nahrung* 22: 285–289.

Taylor SL. 1986. Histamine food poisoning: toxicology and clinical aspects. *CRH Crit Rev Toxicol* 17: 91–128.

ten Brink B et al. 1990. Occurrence and formation of biologically active amines in foods. *Int J Food Microbiol* 11: 73–84.

Tombelli S, Mascini M. 1998. Electrochemical biosensors for biogenic amines: a comparison between different approaches. *Anal Chim Acta* 358: 277–284.

Valsamaki K et al. 2000. Biogenic amine production in feta cheese. *Food Chem* 71: 259–266.

Veciana-Nogues MT et al. 1997a. Biogenic amines in fresh and canned tuna: Effects of canning on biogenic amine contents. *J Agric Food Chem* 45: 4324–4328.

Veciana-Nogues MT et al. 1997b. Biogenic amines as hygienic quality indicators of tuna: relationships with microbial counts, ATP-related compounds, volatile amines, and organoleptic changes. *J Agric Food Chem* 45: 2036–2041.

Vinci G, Antonelli ML. 2002. Biogenic amines: quality index of freshness in red and white meat. *Food Control* 13: 519–524.

Vlieg-Boerstra BJ et al. 2005. Mastocytosis and adverse reactions to biogenic amines and histamine-releasing food: what is the evidence? *Neth J Med* 63: 244–249.

Yen GC, Hsieh CL. 1991. Simultaneous analysis of biogenic amines in canned fish by HPLC. *J Food Sci* 56: 158–160.

Xu Y et al. 2009. Biogenic and volatile amines in Chinese mitten crab (*Eriocheir sinensis*) stored at different temperatures. *Int J Food Sci Technol* 44(8): 1547–1552.

Zotos A et al. 1995. The effect of frozen storage of mackerel (*Scomber scombrus*) on its quality when hot-smoked. *J Sci Food Agric* 67: 43–48.

44
Emerging Bacterial Food-Borne Pathogens and Methods of Detection

Catherine M. Logue and Lisa K. Nolan

Abstract: Bacterial food-borne pathogens cause significant human illness and misery worldwide. In this review, we focus on an examination of the bacterial pathogens associated with food-borne illness that are common, those that are emerging and those that are reemerging. This overview will provide the reader with an understanding of the importance of selected pathogens, the type of disease they cause and why they should still be considered as emerging pathogens. In this review, we also focus on some of the newer generation pathogens such as shiga toxin-producing *Escherichia coli*, *Arcobacter*, *Mycobacterium*, *Cronobacter*, and antimicrobial-resistant pathogens. The second portion of this review provides the reader with an overview of methods for the detection of pathogens from culture based to automated methods as well as immunologic and nucleic acid-based techniques. Finally, this chapter includes some overview of the next-generation technology for the detection and characterization of pathogens including microarray, biosensor, mass sensitive biosensors, and nanotechnology. We recognize that the world of food-borne pathogens is ever changing and as the pathogen poses greater challenges to the host so too does our ability to detect and identify it in shorter timeframes.

INTRODUCTION

The costs of food-borne illnesses in terms of human health and economic loss remain relatively unknown, but are likely to be substantial. However, data, as it pertains to the industrialized countries, do exist but are typically limited to a select few microorganisms; globally, the burden of food-borne illness remains poorly understood and presents a significant challenge to society in how to monitor and assess disease trends.

The World Health Organization seeks to respond to this data gap by estimating the global burden of food-borne illness. In the United States, estimates of food-borne illness have been reported as 48 million cases, 128,000 hospitalizations, and 3000 deaths annually (CDC 2011) placing an economic burden of US$152 billion on the nation, with an average cost of US$1850 per

Food Biochemistry and Food Processing, Second Edition. Edited by Benjamin K. Simpson, Leo M.L. Nollet, Fidel Toldrá, Soottawat Benjakul, Gopinadhan Paliyath and Y.H. Hui.

illness (Schraff 2010). The Centers for Disease Control (CDC) annual food-borne disease report posted through the FoodNet data site reports incidences of illness based on data from ten sentinel sites across the United States. The data cover approximately 46 million people representing approximately 15% of the population. Current data suggest that the incidence of infections caused by *Campylobacter*, *Salmonella*, *Listeria*, and shiga toxin-producing *Escherichia coli* (STEC), *Yersinia*, and *Shigella* (CDC 2010) is declining.

While focusing on food-borne pathogens in this review it is important to define what this review means by an "emerging food-borne pathogen." Mor-Mur and Yuste (2010) suggest that there are pathogens not previously known (new pathogens), others that have arisen as food-borne (emerging pathogens), and others that have become more potent with other products (evolving pathogens). Emerging pathogens are therefore present strains that are adapting to stresses in new environments. Emerging pathogens that have newly arisen have been recognized as a pathogen for many years and are now associated with food-borne transmission (Meng and Doyle 1997, 1998, Sofos 2008). While a considerable number of pathogens are present and capable of causing disease in foods, there are over 200 recognized natural chemical and physical agents (Acheson 1999). Perhaps the most important and emergent are *Campylobacter jejuni*, *Salmonella typhimurium* DT104, *E coli* O157:H7, and other enterohemorrhagic *E. coli* (EHEC) and STEC, *Listeria monocytogenes*, *Arcobacter butzleri*, *Mycobacterium avium* sbsp *paratuberculosis*, *Aeromonas hydrophila*, and *Cronobacter sakazakii*. Other agents of emergence include prions, viruses, (noroviruses, hepatitis A, rotavirus), severe acute respiratory syndrome (SARS), and parasites such as *Ascaris*, *Cryptosporidium*, and *Trichinella*. Also of concern is the emergence of antibiotic-resistant strains in pathogens such as *Salmonella*, *Campylobacter*, *Shigella*, *Vibrio*, *Staphylococcus aureus*, *E. coli*, and *Enterococci*.

There are a number of factors that contribute or have had an influence on the changing trends in food-borne disease; Newell et al. (2010) provides an excellent overview of changes that have occurred and contribute to the changing trends in food-borne disease. Some of the highlights include: a rapid increase in the human population and a shift towards an aging population; an increasing global market for foods, transportation, travel, changes in eating habits, a greater population of immunocompromised, and changes in farming practices and so on.

Ray (2004a) cites other factors including some which overlap with Newell et al. (2010) that can be attributed to our recognition of how we can define an emerging food-borne pathogen. *There is an overall better knowledge of food-borne pathogens*: previously, outbreaks linked to foods were screened for a range of standard pathogens but over the years we have expanded the repertoire to include new pathogens. Perhaps our abilities to test for new pathogens are based on our understanding of them. Technologies incorporating molecular analysis and the use of genomics have expanded our capabilities of what to search for. So too does the technology available to rapidly screen and identify greatly expand our knowledge.

Improvements in regulations and the actions of regulatory authorities have enhanced our capabilities to rapidly identify and monitor for emerging threats. Regulatory authorities at the state and national levels are able to collate data regarding specific outbreaks and the modes of transmission or vehicles implicated in food-borne outbreaks. On the basis of this type of criteria it is possible to make recommendations regarding criteria for what is acceptable and what is not, as well as methods to detect pathogens of concern. These agencies have also developed or guide the development of appropriate sanitation procedures to reduce the risk of human disease or educate consumers in safe handling practices. In the United States, agencies such as United States Department of Agriculture Food Safety and Inspection service (USDA FSIS), and the Food and Drugs Administration (FDA) direct some of these activities, while in Europe, the European Food Safety Authority (EFSA) has a similar role.

Changes in the lifestyle habits of consumers have changed significantly in the last 50 years. These have included traveling, (domestic and international), changes in consumer eating habits, (close to half of every dollar earned is spent eating out), the types of food consumed has changed, and there are increases in seafood consumption, exotic foods, cheeses, and minimally processed or more fresh or natural foods as well as greater access to a larger variety of fresh and raw foods that are imported.

New processing technologies have led to faster and more efficient ways to produce foodstuffs for larger retail markets and consequently there is considerably more product distributed over a wide area—examples of the effects of this have been seen in the outbreak of listeriosis associated with the consumption of contaminated hotdogs distributed nationally (CDC 1998, Donnelly 2001).

Some additional factors that may contribute to the emergence of food-borne pathogens include the recognition of illness associated with multiple strains of *E. coli* termed STEC (Eblen 2007), or strains that appear to be emergent as a result of transfer of resistance and other genetic virulence properties (Fricke et al. 2009), additional factors that warrant inclusion here are also related to the transfer of antimicrobial resistance and the selection of drug-resistant strains, which can become dominant in a food system (Endtz et al. 1991, Aarestrup and Engberg 2001, Engberg et al. 2001, Nielsen et al. 2006, Logue et al. 2010) and have potential links to disease that is difficult to treat. Also of interest are the capabilities of certain pathogens to grow at refrigeration temperatures (*Listeria*, *Yersinia*, *Aeromonas*) and the ability to survive in low-pH (acid) foods (*E. coli* O157:H7, *Salmonella*, *Listeria*).

In this chapter, we address emerging food-borne pathogens of human disease and their potential impacts on human health and methods for their detection. In general, trends in addressing disease have shown a reduction (downward trends). The latest data from the CDC shows prevalence in sentinel states has been decreasing (CDC 2010).

Emerging Food-Borne Pathogens

Salmonella

Nontyphoidal *Salmonella* is a major cause of food-borne disease in industrialized countries. In the United States alone, it is estimated that nontyphoidal *Salmonella* was responsible for 7039 illnesses per 100,000 people in 2008 (CDC 2010). The organism colonizes a range of hosts of food animals including poultry, cattle, and pigs and rarer animals such as bison (Li et al. 2004, Li et al. 2006, Mor-Muir and Yuste 2010). *Salmonella* has also been found in a range of other habitats including soil, water, sewage, and so on.

Salmonella outbreaks are relatively common and have been associated with a range of foods and meats including poultry, beef, pork, and so on. Of interest are outbreaks of illness associated with more unusual foodstuffs such as chocolate, (Werber et al. 2005), shell eggs (Gantois et al. 2009), produce (Franz and van Bruggen 2008), tomatoes (CDC 2005a), and peanut butter (CDC 2009). Currently, there are over 2000 serotypes of *Salmonella* recognized (Ray 2004b) and annual trends from the CDC report common serotypes implicated in food-borne illness. The top two serotypes implicated include *S. typhimurium* and *Salmonella enteritidis* (CDC 2010) http://www.cdc.gov/ncidod/dbmd/phlisdata/*Salmonella*.htm; some novel strains also emerge on a regular basis.

Symptoms of human salmonellosis are usually observed as uncomplicated enteritis and enteric (typhoid) fever, which can result in a more complicated disease involving diarrhea, fever, abdominal pain, and headache. Systemic infection may also occur resulting in a chronic reactive arthritis (Smith 1994). *Salmonella* invades the mucosa of the small intestine and can proliferate in the epithelium, and produce a toxin, which causes inflammation and fluid accumulation in the intestine. The infectious dose required to cause illness is relatively high, about 10^5–10^6 organisms. Most individuals develop symptoms within 24–36 hours but the incubation time may be longer if the dose ingested is lower. Symptoms usually last 2–3 days but patients may remain carriers for periods of months after recovery.

Of greatest concern are *S. typhimurium* DT104 strains exhibiting multiple antimicrobial resistance to ampicillin, chloramphenicol, streptomycin, sulfonamides, and tetracycline (ACSSuT), most of the resistance appears to be chromosomally encoded (Threlfall 2000). These strain types have been recognized worldwide and have also caused infection in humans through consumption of contaminated foods including chicken, beef, pork sausages, and meat paste and also through occupational contact with cattle (Wall et al. 1995, Mor-Mur and Yuste 2010). In recent years, the incidence of *S. typhimurium* DT104 appears to be declining, while resistance to other antimicrobials including extended spectrum β-lactamases appears to be increasing (Newell et al. 2010) and have been detected in both humans and animals (Hasman et al. 2005).

Campylobacter

Next to *Salmonella*, *Campylobacter* is the most common cause of human food-borne disease annually in developed countries. In the European Union, *Campylobacter* is the most common cause of acute food poisoning; over 20,000 cases were reported in 2007 compared to 150,000 cases of salmonelosis (Westrell et al. 2009). In the United States, *Campylobacter* is the second most common cause of disease (gastroenteritis) responsible for approximately 845,000 illnesses annually and almost 100 deaths (CDC 2011). *Campylobacter* gastroenteritis is typically associated with *C. jejuni* and *Campylobacter coli*.

Campylobacter is found in a range of animal hosts including cattle, lamb, pigs with poultry being implicated as a primary source (Niesen et al. 2006, Nachamkin 2007, Mor-Mur and Yuste 2010). Additional sources of *Campylobacter* include water, sewage, raw milk, raw meats, and vegetables (Meng and Doyle 1998, Chan et al. 2001, Inglis et al. 2004, Nachamkin 2007). Prevalence levels of *Campylobacter* in food animals vary significantly, while *Campylobacter* levels on pork products are relatively low and may be related to postslaughter chill and drying effects, which significantly reduce the populations (Snijers and Collins 2004). Raw and undercooked poultry appear to be primary sources of *Campylobacter* associated with food-borne illness (Olson et al. 2008, Mor-Mur and Yuste 2010, Newell et al. 2010).

Campylobacter is a gram-negative organism and is a microaerobe with a requirement for low oxygen (microaerophilic) and a temperature range of 35–42°C for growth. The organism is generally a poor competitor, is sensitive to low pH, temperatures less than 30°C, pasteurization temperatures, and drying (Doyle and Jones 1992). It does, however, appear to survive under refrigeration temperatures and freezing may cause cell reduction but low levels of pathogen have been recovered (Blaser et al. 1980, Jacobs-Reitsma 2000). The organism does not survive well in foods and is easily destroyed by adequate cooking temperatures (Meng and Doyle 1998, Nachamkin 2007). *Campylobacter* is not capable of growth outside the host, but studies have demonstrated the survival of this pathogen on kitchen surfaces (Humphrey 2001).

The two major strains of *Campylobacter* implicated in the majority of human *Campylobacter* enteritis are *C. jejuni* and *C. coli*. Typical cases of human campylobacteriosis include diarrhea, abdominal cramps, profuse diarrhea, nausea, and vomiting; other symptoms include fever, headache, and chills, in some cases bloody diarrhea can occur, recurrence of symptoms (relapse) is often reported. Symptoms of illness are usually observed within 2-5 days of ingestion of contaminated food and can persist for up to 2 weeks. The infectious dose required to cause illness is relatively low and is estimated at approximately 500 organisms (Robinson 1981, Black et al. 1988). Additional complications recognized include reactive arthritis, pancreatitis, meningitis, endocarditis, and Guillain–Barré syndrome a postinfection complication of campylobacteriosis that causes peripheral nervous system paralysis resulting in a flaccid paralysis of the limbs, which can be life threatening (Blaser and Engberg 2008, Jacobs et al. 2008).

Outbreaks of campylobacteriosis typically occur in the summer months with most being sporadic in nature (Olson et al. 2008). Most outbreaks have been linked with improperly prepared or consumption of cross-contaminated, poorly handled poultry (Alketruse et al. 1997, Friedman et al. 2000).

Of significant importance is the prevalence of antibiotic resistance to agents such as fluoroquinolones and macrolides, two of the primary agents used to treat human disease (Engberg et al. 2001, Lutgen et al. 2009). Studies have reported fluoroquinolone resistance in human isolates and those of production animals (Endtz et al. 1991, Engberg et al. 2001), while resistance to other antimicrobials has also been recognized including macrolides, tetracycline, aminoglycosides, β-lactams, and so on (Olah et al. 2006, Zhang and Plummer 2008, Lunagtongkum et al. 2009, Logue et al. 2010).

Resistances observed have been linked to the use of antimicrobial agents in food production animals, which can result in the selection or creation of drug-resistant strains (Endtz et al. 1991, Engberg et al. 2001, Logue et al. 2010). In *Campylobacter*, antimicrobial resistance is associated with a range of mechanisms including efflux pumps and point mutations (Lin et al. 2002, Taylor and Tracz 2005, Olah et al. 2006, Zhang and Plummer 2008, Logue et al. 2010, Guo et al. 2010). For detailed reviews of antimicrobial resistance mechanisms in *Campylobacter*, see Taylor and Tracz (2005), Zhang and Plummer (2008), and Lunagtongkum et al. (2009).

Pathogenesis in *Campylobacter* is associated with the ability to invade host cells, motility, and the production of cytolethal distending toxins which causes diarrhea (Nachamkin 2007); the presence of the pVir plasmid also appears to have a role in pathogenesis (Larsen and Guerry 2005). While *Campylobacter* appears to have emerged as a pathogen of human concern, its recognition as an agent of significant importance in human disease and its standing as the primary or secondary most common food-borne pathogen warrants its continued study.

Escherichia coli Including STEC

While *E. coli* is a common contaminant of meats, of significance are strains implicated in human disease including EHEC and other STEC. EHEC have been recognized as a significant cause of hemorrhagic colitis and complications associated with infection including hemolytic uremic syndrome (HUS) and thrombocytic purpura TTP (Ray 2004d). The organism was first recognized in 1982 as the cause of two major outbreaks of hemorrhagic colitis (Meng et al. 2007). Since that time, other serotypes associated with hemorrhagic colitis have been recognized and the list continues to grow. More than 600 serotypes of STEC are recognized and about 130 different types of EHEC have been recovered from patients presenting with hemorrhagic colitis (Meng et al. 2007).

One of the most common agents of EHEC is *E. coli* O157:H7, which has caused devastating disease worldwide. O157 strains are typically associated with beef cattle, beef, and beef products; however, other sources of the organism are now recognized including domestic and wild animals as well as meats, milk, vegetables, and so on. (Ray 2004d, Meng et al. 2007, Mor-Mur and Yuste 2010). Sources typically implicated include beef, beef products, luncheon meats, salami, and more recently outbreaks have been linked to unusual foods such as spinach and apple juice (CDC 1997b, CDC 2002, CDC 2006, Mor-Mur and Yuste 2010).

Of importance with O157 is the low dose ($<$100 organisms) required to cause human illness, which has been reported worldwide (Meng and Doyle 1998, Ray 2004d, Meng et al. 2007). The pathogenesis of *E. coli* O157 and other STEC is related to its ability to adhere to the host cell membrane, colonize the large intestine, and produce cytotoxic factors and cytolethal toxins including Shiga toxins. Toxins produced can become absorbed into the bloodstream where they cause further damage of blood vessels of the intestine, kidneys, and brain.

Symptoms of infection usually occur within 3–9 days after consumption of contaminated foods and generally last 4 days. Colitis symptoms usually include sudden onset of abdominal cramps, watery diarrhea, which in some cases becomes bloody (about 35–75% of cases can become bloody), followed by vomiting. The presence of toxins in the bloodstream cause breakdown of red blood cells and can cause clotting in small blood vessels of the kidney, which can cause kidney damage and failure, thus resulting in HUS. *E. coli* O157 infections can be fatal in young children. In addition, *E. coli* O157:H7 can cause brain damage as a result of TTP where blood clots in the brain lead to seizures and coma or death.

The largest outbreaks of *E. coli* O157 have been associated with ground meats (e.g., CDC1997a, CDC 2002); other sources of the organism have included raw milk, apple cider, fermented meats such as salami and pepperoni, sprouts, spinach, salads (CDC 1995a, 1995b, CDC 1997b, 1997c, CDC 2005b, CDC 2006, CDC 2008).

E. coli O157:H7 does not grow well at temperatures more than 44.5°C but does appear to have increased acid tolerance. The pH range for growth is 4–4.5 (Meng et al. 2007). *E. coli* O157 strains also display resistance to antimicrobials including streptomycin, tetracycline, and sulfisoxazole, which appear to be the most common resistance profiles observed (Meng et al. 2007) and care should be take when prescribing antimicrobials for treatment of disease as it may exacerbate illness. The importance of *E. coli* O157:H7 as an emerging (and continuing to emerge) pathogen that warrants its continued status as a significant pathogen of human concern; the observation of novel foods that have been implicated in human illness is ever changing, warranting careful monitoring.

Listeria monocytogenes

L. monocytogenes is a gram-positive organism that is ubiquitous in nature and found in a range of environments. Of all the *Listeria* species, *L. monocytogenes* is of primary importance in human disease. The organism can be a significant health risk for the unborn, newborns, infants, the elderly, pregnant women, and immunocompromised individuals.

Listeria is a psychrotroph, capable of growth at refrigeration temperatures, which poses a significant health risk and has been linked to food-borne outbreaks (Lianou and Sofos 2007, Swaminathan and Gerner-Smidt 2007). Therefore, contamination of postprocessed foods may pose a significant health risk to susceptible individuals.

While the incidence of listeriosis in the United States is relatively low, the CDC estimates about 2500 cases occur per year with 500 deaths. *Listeria* is a significant public health concern because of the severity of disease in susceptible populations. Individuals exposed to *Listeria* usually through the consumption of contaminated foods may or may not present with symptoms of infection. Incubation times of 2–70 days after eating contaminated foods have been reported (Bortolussi 2008). Often, the symptoms appear as a mild flu-like illness with slight fever, abdominal cramps, and diarrhea. In healthy individuals, the symptoms usually subside with no significant side effects; patients do, however, continue to shed the pathogen for considerable periods postinfection (Marth 1988). In susceptible individuals, the initial listeriosis infection results in enteric symptoms with nausea, vomiting, abdominal cramps, diarrhea, fever, and headache being common. However, as the pathogen leaves the gut resulting in a systemic infection the symptoms change. In pregnant women, infection of the fetus can occur *in utero* resulting in abortion or preterm delivery; alternatively, the infant may present with symptoms of infection early (<7 days) or late (>7 days) postdelivery. Infants diagnosed early (<7 days) are most likely infected transplacentally, while those presenting with listeriosis late (>7 days) do not appear linked to maternal infection but the sources of infection are poorly understood (Bortolussi 2008). Symptoms of infection in infants and the immunocompromised include bacteremia, meningitis, encephalitis, and endocarditis. Fatality rates in unborn and newborn infants may be as high as 50% and has an average case fatality rate of 20–30% in others (Swaminathan and Gerner-Smidt 2007).

The infective dose to cause illness is considered to be 100–1000 cells; however, individuals that are compromised may be susceptible to lower doses (Ray 2004c). Cooked and ready-to-eat meat and poultry have been implicated as sources of the pathogen, and the growth of the organism has been demonstrated in refrigerated ready-to-eat meats and raw foods that may pose as significant health risk for consumers (Lianou and Sofos 2007, Swaminathan and Gerner-Smidt 2007).

The pathogenesis of *Listeria* is associated with the invasive ability of the organism, its ability to evade the host defenses, the ability to produce a hemolysin called listeriolysin O, and cross the blood–brain barrier (Drevets and Bronze 2008). Despite our knowledge of *Listeria* and methods of its control, the pathogen is still a significant risk to populations within the community and education of consumers is essential to ensure the safety of certain foods such as ready-to-eat meats and vegetables.

Arcobacter

The genus *Arcobacter* was first introduced in 1991, when it was identified as a distinct member of the family Campylobacteraceae (Vandamme et al. 1991, Vandamme and De Ley 1991). There are currently six recognized members of the genus including *A. butzleri*, *Arcobacter cryaerophilus*, *Arcobacter skirrowii*, *Arcobacter cibarius*, *Arcobacter halophilus*, and *Arcobacter nitrofigilis* (McClung et al. 1983, Houf et al. 2005). Although the organism still remains relatively unknown, *Arcobacter* is recognized as an opportunistic pathogen and a commensal and an emerging pathogen of human concern (Ho et al. 2006, Snelling et al. 2006).

To date, studies have found *Arcobacter* is a contaminant of water and foods of animal origin. *Arcobacter* have been associated with abortion in bovine, porcine, and ovine (Vandamme et al. 1992a,b, On et al. 2002). On et al. (2002) suggests that some strains of *Arcobacter* play a primary role in abortions and reproductive disorders, while other stains may be opportunistic pathogens. Regardless, *Arcobacter* has frequently been associated with production animals and it appears that animals can be healthy carriers (Hume et al. 2001). The organism has been recovered from the feces of healthy animals including poultry (Houf et al. 2005), cattle (Wesley et al. 2000, Kabeya et al. 2003), pigs (Hume et al. 2001, Kabeya et al. 2003, Van Driessche et al. 2004), lamb, and equines (Van Driessche et al. 2003). Poultry appear to be a significant source of the pathogen with contamination occurring on carcass surfaces. Gude et al. (2005) reported that *A. butzleri* was commonly found in poultry abattoirs with carcasses being contaminated during processing. Atabay and Corry (1997) suggest that although there is a high prevalence of *Arcobacter* on poultry, the organism does not appear to be common in the gut of poultry suggesting that contamination of carcasses may arise from other sources during processing.

Arcobacter is a gram-negative organism capable of growth under aerobic and anaerobic conditions; however, optimal growth appears to occur under microaerobic conditions (3–10% O_2). The organism is considered a potential zoonotic agent due to its association with animals and the detection of *Arcobacter* in foods of animal origin including chicken, pork, lamb, and beef (Kabeya et al. 2004, Rivas et al. 2004, Morita et al. 2004, Lehner et al. 2005). *Arcobacter* have also been associated with water including a water reservoir, river or surface water, ground water, and sewage (Diergaardt Jacob et al. 1993, Rice et al. 1999, Stampi et al. 1999, Moreno et al. 2003, Morita et al. 2004, Diergaardt et al. 2004).

Although the organism is relatively unknown, it has been recognized as a potential agent of enteric disease and extraintestinal invasive diseases such as septicemia and bacteremia (Ho et al. 2006). In animals, *Arcobacter* has been recognized as a cause of diarrhea and hemorrhagic colitis (Vandamme et al. 1992a, Vandamme 2000). While in humans, *Arcobacter* was reported as the fourth most common *Campylobacter*-like organism isolated from the stool of human patients in Belgium and France (Vandenberg et al. 2004, Prouzet-Mauleon et al. 2006).

As mentioned above, *Arcobacter* is capable of growth under aerobic and anaerobic atmospheres with optimum growth under microaerobic conditions. The organism has a growth range of 15–37°C with an optimum of 30°C. This strict growth range may explain the lack of detection of the organism in the gut of poultry where the temperature is about 42°C. *Arcobacter* does also possess distinct abilities for growth at 15°C, which differentiates it from *Campylobacter* spp. *Arcobacter* can grow at a pH range of 5 and 8.5 with an optimum of 6 and 8 at 30°C (Hilton et al. 2001). Storage at 4°C resulted in a 4 log reduction over 21 days and freezing at −21°C resulted in a 2 log loss after 24 hours but remained constant thereafter.

Evidence suggests that there is an association between *Arcobacter* and human disease with food playing a role; however, the factors associated with *Arcobacter* and human disease are not well understood and warrant further investigation of the epidemiology, pathogenesis of and nature of this organism.

Mycobacterium avium Subsp *paratuberculosis*

M. avium sub species *paratuberculosis* (MAP) is a relatively new emerging disease of human illness. The organism MAP is part of a complex group of related organisms (Motiwala et al. 2006). The organisms within the group have high-level identity at the genetic level but differ in host source, disease phenotypes, and pathogenicity. Members of the group identified as *M. avium* are associated with disease in both animals and humans. Of interest is the association between *M. avium* and immunocompromised patients (AIDS) and its links to Chron's disease (*Mycobacterium paratuberculosis*) (Harris and Lammerding 2001, Greenstein 2003, Ghadiali et al. 2004, Karakousis et al. 2004). *M. paratuberculosis* is recognized as the causative agent of Johne's disease in ruminants and is recognized by chronic gastroenteritis, which can be debilitating.

In humans, Chron's causes a similar type of disease to that observed in animals and may be linked with MAP (Harris and Lammerding 2001); the association between Chron's disease and MAP does, however, remain controversial. Chron's disease appears to occur in a range of age groups of the population with cases observed in adolescents and in patients older than 50 years.

Recognized sources of MAP include the feces of infected animals, the environment, milk, foods, and contaminated drinking water (Glover et al. 1994, Grant et al. 1998, Yoder et al. 1999, Argueta et al. 2000, Grant et al. 2000, Raizman et al. 2004, O'Reilly et al. 2004). Methods using polymerase chain reaction (PCR) are valuable for the detection of MAP because of the extended periods of incubation required to detect the pathogen and the state of the cells that may make the organism difficult to detect by standard culture methods.

MAP may also be linked to milk. Studies have demonstrated that the organism survives pasteurization and may be present in raw and pasteurized milk. Grant et al. (1998, 2000) reported on the detection of MAP in milk using immunomagnetic separation (IMS). More recent studies have found the prevalence of MAP to be 12% in raw milk and 9.8% of pasteurized using IMS PCR. However, no viable cells were found, only DNA was detected suggesting that the pasteurization process destroyed cell viability. In the same study, only one raw milk sample was positive for MAP (O'Reilly et al. 2004). Others have assessed the prevalence of MAP in a range of produce from retail stores and supermarkets finding that of 129 foods sampled, 25 were positive for *Mycobacteria* with half of all isolates recovered identified as *M. avium* (Argueta et al. 2000). Yoder et al. (1999) demonstrated the relatedness of strains recovered from human disease with those found in foods, supporting the role of foods as a source of *Mycobacteria* for human disease.

In human cases of Crohn's disease, a chronic condition resulting in significant intestinal disease; it is recognized in three forms including inflammatory, structuring, and stenosing and fistulizing (Andres and Friedman 1999, Kirsner 1999). The primary location of disease is the distal ileum and colon. Currently, there is no known cure for Crohn's, with up to 74% of patients requiring surgery (Andres and Friedman 1999). Links between Chron's and MAP varies significantly and the epidemiology of the disease is not well understood although the organism has been recovered from the tissues of Crohn's patients (Sechi et al. 2004). Treatment of Crohn's uses combinations of anti-inflammatories, steroids, and antibiotics until the patient experiences relief. Some antimycobacterial drugs used include rifabutin-macrolide-clofazimine (Shafran et al. 2000). Although the data on MAP are limited, its potential as an emerging food-borne pathogen at this time cannot be discounted and warrants further epidemiological studies.

Aeromonas hydrophila

The genus *Aeromonas* is a member of the family Aeromonadaceae and consists of 24 different species (Janda and Abbott 2010); this number is, however, continuously expanding with over 3500 new species proposed (Janda and Abbott 2010). *Aeromonas* is a gram-negative facultative anaerobe that appears to be ubiquitous in water and foods. The role of *Aeromonas* in food-borne disease is not well understood but is suspected and there is evidence of the pathogen in reported outbreaks. *Aeromonas* is capable of growth at refrigeration temperatures with an optimum growth range of 28°C and in a pH range of 5.5–9.

Most aeromonads are considered opportunistic in aquatic and terrestrial animals but can also be a pathogen of man and has been recognized as so since 1984 (Tsai and Chen 1996). Epidemiological evidence suggests that aeromonads can cause gastroenteritis in humans (Deodhar et al. 1991, Kuhn et al. 1997, von Grawvenitz 2007) and other complications such as wound infections, septicemia, and endocarditis (Janda et al. 1994, Broqui and Raoult 2001). *Aeromonas* appears to act as both an infectious agent and an enterotoxigenic pathogen (Kingombe et al. 1999). Members of the species of interest in food-borne disease appear to be associated with strains identified as *A. hydrophila*, *Aeromonas sobria*, and *Aeromonas caviae* (Gavriel et al. 1998). The relationship between motile aeromonads and food-borne illnesses is tentative; a few reports have linked illness to foods consumed such as shrimp cocktail (Janda and Abbott 1999) and oysters (Abeyta et al. 1990).

In humans, food-related *Aeromonas* illness is associated with gastroenteritis, with most cases affecting children, the elderly, and the immunocompromised (Isonhood and Drake 2002); annually it is estimated that *Aeromonas* may be responsible for up to 13% of all cases of gastroenteritis in the United States (Kingombe et al. 1999). The infectious dose of *Aeromonas* to cause human illness is unknown as is the mechanism of virulence; however, gastroenteritis, wound infections, and septicemia appear to be recognized as consequences of infection (Janda and Abbott 2010).

Aeromonas has been isolated from a range of foods and environments including retail vegetables, seafood, meats, cheese, and milk (Okrend et al. 1987, Abeyata et al. 1990, Kirov et al. 1993a, 1993b, Pon et al. 1995, Saad et al. 1995, Tsai and Chen 1996, Kuhn et al. 1997, Gavriel et al. 1998, Yadav and Verma 1998).

As with other emerging pathogens, our recognition of *Aeromonas* as an emerging pathogen of significant human concern warrants further attention into the epidemiology of this pathogen and its relationship with human food-borne illness.

Cronobacter sakazakii (*Enterobacter sakazakii*)

Originally discovered in the early 1960s, this organism has primarily been associated with neonatal meningitis of newborns (Chenu and Cox 2009). Since then, the organism has gained increased attention as a pathogen of infants and newborns and was named *E. sakazakii* in the 1980s; as a consequence of genetic diversity, a new genus for the species was proposed in 2008 and named *Cronobacter*. This emerging pathogen has gained considerable attention in the past 10 years or so with much still to be learned about its epidemiology and pathogenesis.

Studies to date have implicated powdered infant formula (dried milk powder or a modified version) as the vehicle of transmission of the organism in some cases of neonatal meningitis (Simmons et al. 1989, Van Acker et al. 2001, Himelright et al. 2002, Caubilla-Barron et al. 2007, Hunter et al. 2008).

Cronobacter is a gram-negative rod, motile by peritrichous flagella and a member of the family Enterobacteriaceae. The genus consists of six species (Iversen et al. 2008). The organism has a growth range of 6–45°C with an optimum temperature of 37–43°C; *Cronobacter* is capable of growth at refrigeration temperatures, which may be a significant health risk (Iversen et al. 2004).

Clinical manifestations of illness have previously been associated with neonates and present as life-threatening meningitis, necrotizing enterocolitis, bacteremia, and septicemia (Muytjens et al. 1983, Van Acker et al. 2001, Lai 2001, Himelright et al. 2002). Case fatality rates as high as 80% have been reported; the survivors often suffer severe neurological disorders (Lehner and Stephan 2004) and long-term complications. *Cronobacter* illness has also been recognized in adults primarily among adults with underlying disease, the immunocompromised, stroke patients, and the elderly (See et al. 2007, Healey et al. 2010).

The pathogenesis of *Cronobacter* is not well understood but studies suggest that enterotoxins and the outer membrane protein OmpA may play a role in virulence and the ability of the organism to cross the blood–brain barrier (Iversen and Forsythe 2004, Townsend et al. 2007, Singamsetty et al. 2008, Mittal et al. 2009, Mohan Nair et al. 2009).

Cronobacter appears to have an environmental association; it has been postulated that plant material may be a source of contamination (Healey et al. 2010); new reservoirs of *Cronobacter* are being reported on a regular basis. Although the association between the gut of animals and *Cronobacter* is questionable, the organism has been found in water, soils, and plant materials (Iversen and Forsythe 2003). *Cronobacter* has also been recovered from a range of foods including meats, poultry, rice and other grains, vegetables, herbs and spices, dry food ingredients, breads, ultra-high temperature processed milk, cheese, sausage, tofu, keifir, and sour tea (Iversen and Forsythe 2004, Kandhai et al. 2004, Gurtler et al. 2005, Mullane et al. 2007, Friedemann 2007). Baumgartner et al. (2009) identified *Cronobacter* in ready-to-eat foods with 61% of sprout or fresh herbs and 27% of dried herbs or spices being positive for the organism. O'Brien et al. (2009) detected *Cronobacter* in follow-on infant formula but not in powdered milk products suggesting that the risk of illness is low but monitoring of raw ingredients is necessary to establish sources of contamination.

One of the primary sources of illness in neonates appears to be associated with powdered infant formula. As the powder is nonsterile, it may become contaminated during production or reconstitution (Simmons et al. 1989, Van Acker et al. 2001, Caubilla-Barron et al. 2007). Also of concern is the ability of *Cronobacter* to produce biofilms, which may enhance its abilities to resist the killing action of disinfectants (Kim et al. 2007a). Regardless of the narrow scope and sources of *Cronobacter*, its recognition as an emerging pathogen warrants attention in light of its epidemiology and pathogenesis being poorly understood.

Clostridium difficile

Clostridium difficile is a gram-positive anaerobic spore former that has gained increased attention as a pathogen of human concern (Weese 2010). Originally, this organism was recognized as a pathogen associated with diarrhea as a result of antimicrobial treatment of patients. Now, however, there is concern that this organism may also be an agent of zoonoses (Songer 2010). Indra et al. (2009) reported that *C. difficile* is primarily considered a nosocomial pathogen causing diarrhea and pseudomembranous colitis, yet others have found that *C. difficile* is found in food production animals suggesting a potential relationship between humans and animals (Songer 2004, Weese 2009a).

The pathogenesis of *C. difficile* is usually associated with overgrowth of the organism in the gut resulting in mild to life-threatening pseudomembranous colitis, toxic megacolon, and/or intestinal perforation (Borriello 1998); changes in the gut microflora also appear to be linked to the overgrowth of *C. difficile* (Bartlett and Perl 2005). Two toxins are also recognized as contributing to pathogenesis—tox A and tox B and a third toxin CDT is also a contributing factor (Rupnik et al. 2005). In human disease, there has been an increase in the incidence of *C. difficile* associated with ribotype 027, which has been recognized as an epidemic strain worldwide (Goorhuis et al. 2008).

Recently, concern for *C. difficile* has escalated due to studies demonstrating the prevalence of the organism in animals and foods of animal sources, suggesting that community acquired *C. difficile* may contribute to human outbreaks. Studies to date are, however, limited and have used varying approaches to detect the pathogen. Indra et al. (2009) assessed the prevalence of *C. difficile* in cattle, pigs, and broilers demonstrating that 4.5% of cattle, 3.3% of pigs, and 5% of broilers were positive for the organism. A similar assessment of retail meat samples (ground beef, pork, and chicken) failed to detect the pathogen. In contrast, Songer et al. (2009) reported the prevalence of *C. difficile* in ready-to-eat meats and raw meats ranging from 14% to 50%. Of significance in the study was the detection of ribotype 027, a strain type that is almost exclusively associated with human disease worldwide and has been identified as an epidemic strain (Goorhuis et al. 2008). This ribotype has also been identified as possessing resistance to the fluoroquinolone class of antimicrobials (Razavi et al. 2007). *C. difficile* has also been found in other food types including vegetables (Al Saif and Brazier 1996), ready-to-eat salads (Bakri et al. 2009), vacuum-packaged beef (Bouttier et al. 2010), retail ground meats (Jobstl et al. 2009), red meats, ground products, pork and chicken products (Rodriguez-Placios et al. 2007, 2009, Norman et al. 2009, Von Abercron et al. 2009, Weese et al. 2009b, Weese et al. 2010), and in swine operations with the highest prevalence levels observed in young animals (Norman et al. 2009). Speculation as to the sources of *C. difficile* on meat have included animal hides, processing equipment, the environment and employees, and noting that the spores are probably able to survive processing, cleaning, and disinfection (Weese 2009a).

Rupnik (2007) strongly suggests that the current data provide preliminary evidence or support speculation that animal reservoirs and transmission via foods may be a source of community-associated infections. Regardless, there are currently considerable knowledge gaps in our understanding of *C. difficile* and its links to food-borne disease; to date, there is no direct evidence of illness and an identified source, which suggests that a greater understanding of the epidemiology of *C. difficile* is warranted.

Antimicrobial-Resistant Pathogens

Although mentioned in regards to some of the pathogens discussed above, the emergence of antimicrobial resistance in food-borne pathogens warrants further review in this chapter. Doyle and Erickson (2006) consider that antimicrobial resistance in pathogens is a concern and will continue to be an issue for the future. Over the last 50 years or more, the use of antimicrobials has been a mainstay for the treatment of infectious disease in humans and animals. However, over the same period, the use of antimicrobials has led to the emergence of antimicrobial resistance in pathogens. Newell et al. (2010) considers that any kind of antimicrobial can select for emergence of resistance and promote the dissemination of resistant bacteria and resistance genes. As a consequence, resistance can be selected in the agricultural sector and can contribute to the public health burden. Woteki and Kineman (2003) and Teale (2002) suggest that the overuse of antimicrobials in agriculture and healthcare have been identified as contributing factors to drug-resistant strains. Food, therefore, can be both a source of antimicrobial-resistant bacteria and resistance genes; therefore, foods harboring antimicrobial-resistant bacteria can be a significant threat to human illness. Studies have highlighted the presence of resistant *Salmonella*, *Campylobacter*, *E. coli*, and other such pathogens in the food chain potentially contributing to disease.

Indirectly, these antimicrobial-resistant organisms can also be an indirect source of bacterial resistance by transferring resistance genes to other organisms or commensals. Studies have highlighted the role of mobile genetic elements in the transfer of resistance (Barlow 2009). Kruse and Sorum (1994) highlighted the transfer of resistance using plasmids. Regardless of how resistance occurs, and its complexities, antimicrobial resistance appears to have emerged in food-borne pathogens, thus posing a new challenge in understanding its epidemiology and the means to control it.

Some of the important antimicrobial-resistant pathogens include: *Salmonella*, there are well-documented studies of food-borne illness associated with drug-resistant *Salmonella*. The pathogen has been associated with a variety of foods including meats, eggs, produce, and so on. Studies have highlighted the movement of drug-resistant strains from food to humans (Endtz et al. 1991, Engberg et al. 2001). Other studies have linked drug use at the farm with the emergence of drug-resistant strains (Logue et al. 2003, Lin et al. 2007, Logue et al. 2010, Newell et al. 2010), thus leading to potential transfer to humans via foods or direct contact with animals.

In *Salmonella*, some species appear to have the ability to resist multiple antimicrobial agents. Multidrug-resistant *S. typhimurium* DT104 is an example of a strain possessing resistance to five different antimicrobials (Threlfall 2000). Newly emerging resistance to extended spectrum β-lactam antimicrobials has also become an issue (Patterson 2006, Carattoli 2008).

In *Campylobacter*, a similar situation has also been observed with poultry being a primary source of the pathogen and the intensive nature of poultry rearing has led to the emergence of drug-resistant *Campylobacter*. In recent years, the emergence of fluoroquinolone-resistant *Campylobacter* has been observed and in some countries, the emergence of resistance has been linked to the use of fluoroquinolones in poultry production (Aarestrup and Engberg 2001, Humphrey et al. 2005) and a consequent increase in fluoroquinolone-resistant strains occurring in humans (Endtz et al. 1991, Engberg et al. 2001). In light of this evidence and the potential risks for food-borne disease, the use of fluoroquinolones for use in poultry production was banned in 2005. Regardless of the ban, however, it would appear that fluoroquinolone-resistant *Campylobacter* are still present in production birds and meat (Price et al. 2007, Han et al. 2009a). In addition, the emergence of macrolide-resistant *Campylobacter* is also of concern. Lin et al. (2007), Lunagtongkum et al. (2009), and Logue et al. (2010) highlighted the effect of dosing to select resistant strains, however, in the absence of selective pressure, levels of resistant strains appear to drop in favor of nonresistant strains, suggesting that the fitness cost associated with resistance maintenance may be too high to promote persistence (Han et al. 2009b, Hao et al. 2009, Logue et al. 2010).

In *E. coli*, the emergence of multidrug-resistant strains has also been evident. Commensal *E. coli* may also be a source of resistance to other *E. coli* and pathogens. Studies by our colleagues have highlighted resistance carried on plasmids that form islands of resistance (Johnson et al. 2004, Johnson et al. 2010) and appear to be co-selected with virulent strain types, suggesting that virulence and resistance are complementary (Johnson and Nolan 2009). Travers and Barza (2002) noted that drug-resistant strains appear to be more virulent than drug susceptible strains. In addition, transfer of resistance genes from *E. coli* to other pathogens has been demonstrated and has been shown in both the animal and human gut (Gast and Stephens 1986, Su et al. 2003, Poppe et al. 2005, Yan et al. 2005) and in foods (Kruse and Sorum 1994, Walsh et al. 2008).

In gram-positive organisms, methicillin-resistant *S. aureus* (MRSA) has emerged as a significant problem; once thought of as a nosocomial infection, community acquired MRSA have been identified as a potential zoonoses following detection of the pathogen in a range of animal hosts. MRSA strain types have been isolated from pigs (De Boer et al. 2009) and a prevalence of MRSA in retail meats of 11.9% was also observed with highest prevalence in poultry meats.

Resistance of food-borne pathogens is currently monitored in the United States through the National Antimicrobial Resistance Monitoring System. There are two branches that follow and assess drug resistance in animal production and foods of animals and also the clinical side of disease. Trends in resistance are released annually and indicate emerging resistance types and changing profiles for *Salmonella*, *E. coli*, *Campylobacter*, and some gram-positive organisms (see http://www.cdc.gov/narms/ and http://www.fda.gov/AnimalVeterinary/SafetyHealth/AntimicrobialResistance/NationalAntimicrobialResistanceMonitoringSystem/default.htm). For further discussion on antimicrobial resistance in relation to food-borne pathogens the authors recommend the review by Cleveland McEntire and Montville (2007).

METHODS FOR THE DETECTION OF FOOD-BORNE PATHOGENS

In this section, we discuss some of the approaches used in detection of food-borne pathogens. The review discusses methods from the classical to the rapid to the next generation of technologies. While this section provides an overview of the most common methods used it is not comprehensive, as the field of rapid technologies is changing fast and newer technologies have taken advantage of nano sciences for their applications. This section covers detection methods; for detail on typing (see Logue and Nolan (2009)).

Culture-Dependent Methods

Detection of food-borne pathogens in a food sample usually involves increasing the population of the target organism present to levels that can be detected by conventional or other means. Typically, culture-dependent methods involve a series of enrichment phases designed to increase the population of the pathogen present, which can be followed by a second selective enrichment designed to selectively enrich the target pathogen, while inhibiting or suppressing the growth of contaminant flora that may also be present. The method for culture enrichment is relatively straightforward and for the primary enrichment usually involves volumes about 25 g of the suspect food in 225 mL of a nonselective enrichment broth (a 1:10 ratio of sample to enrichment broth is typically recommended). The sample is homogenized and then incubated at a temperature optimum for the target pathogen with an incubation time in the region of 18–24 hours. Following the primary enrichment, a secondary enrichment designed to select the pathogen and suppress contaminants is usually carried out. In this case, the selective enrichment phase, media, and ratios of sample to selective broth are significantly reduced and volumes of 0.1–0.5 mL (or at slightly higher ratios if needed) of the primary enrichment broth are transferred to 9.9–9.5 mL volumes of the secondary broth. Secondary enrichment broths are usually incubated for a further 18–24 hours at the optimum temperature of the pathogen. Incubation temperature for the selective broth can be the same or different from the temperature used for the primary enrichment broth. Following secondary enrichment, the sample is typically struck to a selective or differential agar designed for specific isolation of the target pathogen. Incubation of selective agar plates to recover the target can take an additional 24–48 hours (or longer for some pathogens) to determine presence of a suspect pathogen. Further confirmation and identification of a suspect pathogen from selective plates to the genus or species level may take an additional 4–5 days to complete using conventional biochemical methods, morphological analysis, serological analysis, or other such tests. Some of the post isolation analysis can be considerably shortened with the application of automated or molecular testing. While the methods such as those described here are considered the standard for detection of a target pathogen, they are recognized by federal agencies as the gold standard for the detection of bacterial contamination. Some examples of agencies who provide standardized methods online for use in the detection of pathogens in foodstuffs include the FDA Bacteriological Analytical Manual http://www.fda.gov/Food/ScienceResearch/LaboratoryMethods/BacteriologicalAnalyticalManualBAM/default.htm; the USDA FSIS Microbiology Laboratory Guidebook http://www.fsis.usda.gov/science/microbiological_Lab_Guidebook/; other guides of importance include the American Association of Analytical Chemists (AOAC) and the International Standards Organization (ISO); additional standards and protocols are recommended on a national level for individual countries and the reader is advised to refer to the protocols recommended on a national level. While the use of standard methods is of value in pathogen detection, they have considerable drawbacks in terms of time and labor required to obtain a result and cost issues associated with isolation and detection; some of these methods also suffer in their detection capabilities (sensitivity and specificity) and may fail to provide adequate detection of the target pathogen if the medium is too selective for a particular strain type or the state of the organism may not allow its detection prior to application of the protocol. Other problems also complicating pathogen detection include nonuniform distribution of the pathogen in a sample, which can mean that dependent on where the sample is collected it may not be representative of the whole sample. Also, a low level of pathogen may be present in the sample, which may complicate its detection in the presence of high levels of natural microflora. The heterogeneous nature of the food matrix can also interfere with the growth or detection of the target pathogen. Finally, injured cells as a result of processing may result in inability to detect target pathogens using selective media.

Culture methods for the detection of a pathogen are therefore subject to limitations such as the initial number of organisms in a sample: (1) if the number of organisms in a sample is very low and the sample is enriched, competition from other naturally occurring microflora present in the sample may outcompete the target organism, other issues associated with competition may not only be competition for space and nutrients but also production of by-products of metabolism that have a negative effect on the target organism such as acids and other inhibitory substances; (2) the state of the organism in the sample (injury status); studies have demonstrated that injured cells may need additional time to recover prior to any standard enrichment as they may not be able to grow in the presence of a selective agent, and so on, therefore, there is a risk in missing detection of the target; (3) selectivity of the enrichment media used--authors have noted that not all standard protocols may be adequate for detection of the target organism, there is evidence to suggest that some selective enrichment media may not be appropriate for detection of all target populations and that some strain types may be inhibited or missed as a result of the enrichment process, thus there is a risk that the protocol may not truly be reflective of the population in a sample. Wu et al. (2001) demonstrated some of the challenges posed in the detection and enumeration of *E. coli* O157:H7 in alfalfa seeds, an unusual food product that has been linked to outbreaks of human illness (Mohle-Boetani et al. 2001). Regardless of the approach taken, readers of this article should consider carefully the type of food matrix that requires testing and how it in itself may pose unique challenges to detection of the target pathogen.

In recent years, methods to automate and significantly shorten the detection time required have been implemented for some protocols used in the detection of pathogens and these can significantly reduce labor and time of detection. They do, however, come at a cost in terms of technology or specialized systems for detection, which may not be appropriate for all laboratory or production facilities.

Automated Methods to Detect Pathogens

Commercial companies have considerably expanded their offerings in automated and rapid methods for the detection and identification of food-borne pathogens, while in some instances the methods are proprietary, the basic principle of the methods are relatively similar—examples of some of the more common commercial detection systems currently available include the VIDAS® system (BioMérieux), which has application for the detection of pathogens such as *E. coli* O157:H7, *Campylobacter*, *Listeria*, *Salmonella*, and so on in food systems. The VIDAS system is based on an immunoassay type reaction that can detect the pathogen in an enrichment culture in as little time as 45 minutes. The protocol does, however, require enrichment prior to detection and this can entail time frames of 6–24 hours.

The BAX® system (DuPont and Qualcon) is another automated system for the detection of pathogens such as *Campylobacter*, *E. coli* O157:H7, *Salmonella*, *Listeria*, *S. aureus*, and so on in foodstuffs using a real-time PCR-based assay. The BAX system can detect the pathogen directly in a sample (if heavy contamination is suspected) or following an enrichment phase. The BAX system uses a PCR-based protocol to amplify genes specific to the species of interest and detect presence of the organism through fluorescence of the amplified product. The system is also capable of assessing the number of cells in a sample—the minimum detection limit for the technology is about 10^4 colony forming units (CFU)/mL, and can detect target pathogens in as little as 90 minutes. This technology has recently received AOAC approval for application in the detection of a range of pathogens.

Another rapid technology for the detection of pathogens includes the simple and relatively low-cost alternative called lateral flow kits. Lateral flow kits (Dipstick, DuPont) work on the principle of an antigen antibody reaction and are designed to detect the presence of antigens in a sample. The system uses specific enrichment media to increase the population of the target pathogen and then a test strip is added to an aliquot of the enriched sample, the strip employs a combination of antibodies specific to the target pathogen and gold colloidal particles on the surface of the membrane. The liquid sample wicks up the test strip by capillary action and if the target pathogen is present it reacts with the antibodies present to result in a color change and a visible line in the strip. This method has been available from a number of vendors and has proven successful for rapid screening of samples for target pathogens. Of concern, however, is the specificity of the antibodies used in the strip that may not be able to detect all strain types present in a sample. Fakhr et al. (2006) found that the a lateral flow kit can fail to detect *Salmonella* in some poultry samples that were positive for *Salmonella* by other methods including real-time PCR and selective culture, suggesting that specificity of the antibodies used in the assay may not always detect the target.

Automated Methods of Identification

Although not strictly part of this review, it is important to discuss the differences between detection of a pathogen and its identification. While there are a range of technologies available for the detection of pathogens, there are also a significant number of technologies available for the rapid identification of pathogens once detected. These include a range of automated systems that are based on biochemical tests or metabolic profiles Vitek – (BioMérieux) and Microstation (Biolog), or cellular fatty acid analysis systems (gas chromatograph based, Sherlock Microbial Identification system). In addition, there are multiple smaller rapid identification kits that can successfully identify a pathogen in a period of a few minutes to a few hours, these include systems such as microID, Trek's GNID, API, and so on; the majority of these kits are visual and based on biochemical reactions leading to changes in color (or fluorescence if read automatically, TREK GNID). In addition, DNA-based methods can successfully identify a potential pathogen by amplification of single or multiple genes associated with a particular species.

Immunologic-Based Methods of Pathogen Detection

Immunological-based methods for pathogen detection rely on the binding of a bacterial antigen to a monoclonal or polyclonal antibody. A range of different immunoassays are available for detection of pathogens these include the enzyme linked immunosorbent assay (ELISA), immunochromatic assay, and immuofluorescent assay. The sandwich ELISA is a commonly used technique for the detection of pathogens and is based on the application of an immobilized antibody on a solid surface (usually, a microtiter plate) followed by exposure of the antibody to a sample containing the target cells. The antibodies bind or capture the target cells. Once captured, the sample is washed to remove unbound cells and then a second antibody conjugated to an enzyme is added followed by addition of an enzyme substrate. A positive result is usually observed through the colorimetric detection of the product of the reaction between the enzyme and the substrate (see Fig. 44.1).

Immunoprecipitation (*Dipstick/Lateral* flow)

The lateral flow kit is an immunochromatographic method for the detection of pathogens. The sample is usually applied to a well and the sample is wicked from the sample port to an area containing antibodies conjugated to a precipitable material such as colloidal gold. If the target pathogen is present in the sample the antigen antibody complex moves to a second region where binding to a second antibody occurs resulting in an antibody – antigen – antibody complex which usually results in a colored immunoprecipitate (see Fig. 44.2).

Immunomagnetic Separation

Another application of immunologic-based detection is the use of antibody coated beads for the detection of pathogens in enrichment samples using immunomagnetic separation (IMS). Specially designed beads are coated with antibody specific to the pathogen of interest. The magnetic beads themselves are paramagnetic meaning that they are only magnetic in the presence of a magnetic field. Usually, the beads are mixed with a sample of the enrichment broth containing the target pathogen. Following a period of incubation to allow the antibody to react with the antigen of the target organism the beads are exposed to a magnetic field which captures the beads at the side of the tube. The remaining liquid is removed containing unbound organism, food components and remaining enrichment broth. The beads containing the bound cells is usually washed and then suspended in a small volume of diluent, which can then be plated directly on selective media to detect the target organism or further processed into a PCR reaction to detect the target pathogen or examined by other means such as immunofluorescence, and so on.

While the use of immunologic-based assays significantly enhances detection of pathogens in foods, there are however, some drawbacks to the use of immunologic-based assays; the first is the risk of cross-reactivity, due to the ability of the antibodies used to cross-react with nontarget antigens, thus leading to false positive results. In addition, most antibody-based assays have a minimum detection limit of about 10^4 CFU/ml; therefore, levels of the target antigen below this threshold may not be easily detected leading to false negative results.

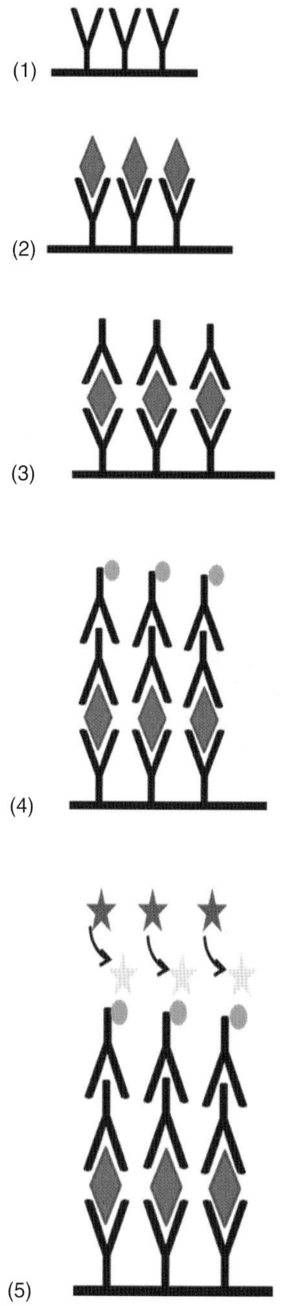

Figure 44.1. Direct enzyme-linked immunosorbent assay for the detection of pathogens. **1.** The plate is coated with capture antibody. **2.** The test sample is added and the antigen (or target organism) binds to the capture antibody. **3.** An enzyme linked capture antibody is added as a detecting antibody and binds to the antigen. **4.** A substrate is added. **5.** The substrate is converted by enzyme to a detectable form (usually, a color reaction).

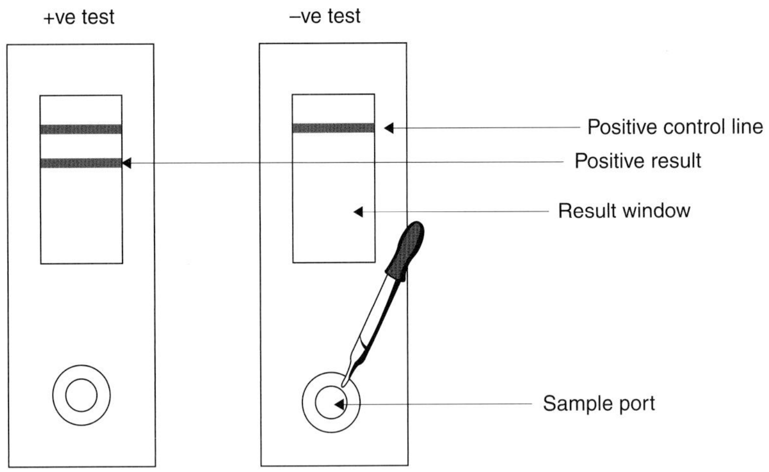

Figure 44.2. Rapid detection of pathogens using a lateral flow kit.

Latex Agglutination Assays

Latex agglutination assays use antibody coated colored latex beads or colloidal gold particles that will clump in the presence of a specific antigen. The test is based on the insoluble bacterial cells clumping in the presence of the antibody. In general, for latex agglutination assays to be accurate they require a pure culture for the specific assay as the sensitivity of the antibodies can be relatively low, thus resulting in cross-reactions. The test is however relatively fast, easy to perform and visual interpretation relatively straightforward. Latex assays have been developed for the detection of a range of organisms including *E. coli* O157:H7 and other STEC members, *Salmonella* spp, *Campylobacter* spp., *Listeria*, *Shigella*, *S. aureus*, *Vibrio* and *Yersinia*. One of the drawbacks of the latex agglutination assays is the potential for the particles to cross-react with other members of the same genus because of structural similarities in the lipopolysaccharide layer (Feng 2007).

A modification of the latex agglutination assay is the reverse passive latex agglutination (RPLA), in this assay the antibody is used to detect a soluble component of the bacterium – the toxin. Usually, toxins are soluble and will form a diffuse lattice network or layer in the presence of the antibody indicating a positive test. RPLA is typically carried out in a microplate to detect toxins in a food extract or pure culture suspension (Feng 2007).

Gel Immunodiffusion

The gel immunodiffusion assay was developed for the rapid detection of *Salmonella* in a food system. The test is based on the use an L-shaped chamber containing motility agar and antibodies against *Salmonella* flagella on one side of the chamber and on the other side, selective enrichment media for enhancing the growth of *Salmonella*. A sample is added to the enrichment medium side and growth of the pathogen will occur, over time, the pathogen will migrate through the motility agar where it cross-reacts with the anti flagellar antibodies resulting in a visible layer of precipitation in the motility agar. The test is relatively specific and has AOAC approval for use in the detection of *Salmonella* species. The test, however, is limited to motile *Salmonella* species and will not detect nonmotile strains. The test is marketed as a *Salmonella* 1-2 kit by BioControl.

NUCLEIC ACID-BASED DETECTION

Pathogen detection using nucleic acid-based approaches are usually designed to target DNA, RNA, or both. In recent years, there has been a rapid increase in the use of such approaches for not only detection of a pathogen but also for identification and characterization.

Hybridization Techniques

DNA/Colony Hybridization

Colony hybridization is a relatively low-cost method for pathogen detection where colonies of a suspect target are selected from a plate and transferred to a solid support (e.g., membrane). The cells of the sample are then lysed using an enzyme to release the DNA. The DNA is then denatured to allow linkage to the support membrane and then hybridized to a gene probe with complementary sequence to the target DNA. Usually, the gene probe is labeled with a fluorescent tag or other color marker which allows easier detection of the pathogen. In recent years, the use of the solid support for detection of target DNA has changed slightly and the application of hybridization can occur through the use of a dip stick approach where the DNA is hybridized to probes embedded on a solid surface. Hybridization can also be achieved in solution. One such system that is commercially available with success in the detection of pathogens such as *L. monocytogenes* and *Salmonella* is the Gene-Trak kit marketed by Neogen.

Polymerase Chain Reaction (PCR)

PCR has revolutionized molecular studies and our approaches in detection and characterization of food-borne pathogens. The basic principle of the technique is that a thermostable DNA polymerase is used to amplify a specific region of DNA to detectable levels from template DNA found in the target pathogen. In conventional PCR methodology, the amplicons generated are then separated out by gel electrophoresis, the DNA stained, and the size of the DNA bands determined by comparison to standard DNA and a control strain. The use of such amplification techniques has expanded considerably over recent years and is used not only in the detection of a target pathogen but also for identification of pathogens in foods and other substrates, detection of genes associated with pathogenesis, virulence or toxins, and detection of viruses or fungi (Auvray et al. 2007, Kim et al. 2007b, Baert et al. 2008, Niessen 2008).

Newer PCR techniques are emerging, making this technique ever more useful for rapid assessment of a food. For example, PCR reactions are commonly configured to amplify multiple gene targets simultaneously; therefore, the application of PCR-based approaches has the capability to detect multiple pathogens concurrently. To work well, multiplex PCR protocols should use primer sets which ensure that amplification products are at least 30-50 bp apart, thus allowing adequate resolution if the amplification is followed by gel electrophoresis. Multiplex PCR reactions have been used in a number of ways including the simultaneous detection of multiple target pathogens in a mixed sample or matrix, for example, the detection of pathogens such as *Salmonella*, *Shigella*, *Listeria*, *E. coli O157:H7*, and so on in artificially inoculated foods (Alarcon et al. 2004, Kim et al. 2005, Kim et al. 2007b, Yuan et al. 2009, Kawasaki et al. 2010) and detection of multiple virulence or resistance genes in *E. coli* and *Salmonella* (Skyberg et al. 2003, Skyberg et al. 2006, Li et al. 2006, Johnson et al. 2007, Nde and Logue 2008). Martinez et al. (2006) used multiplex PCR for the simultaneous detection of three genes associated with the production of cytolethal distending toxin (CDT) in *C. jejuni* and found the prevalence of the genes to be 98% in human and animal isolates recovered from a wide European geographic origin. In a similar study, Asakura et al. (2008) developed a multiplex for *cdt* genes in *C. jejuni* and *C. coli*, with a minimum detection limit of 10–100 CFU.

When PCR is used in the detection of target genes from food matrices or mixed culture, the quality of the DNA that serves as template for amplification is of concern. Cultures and food matrices can contain substances which inhibit PCR (Fakhr et al. 2006); these include components such as carbohydrates, fats, polysaccharides and proteins; commercial PCR clean-up kits are available for reducing these effects. The efficiency of, or utility of PCR may be enhanced using different enzymes and labeling agents; specialty *Taq* polymerases can increase the efficiency of the PCR reaction (Fratamico and Bales 2005), and fluorescent-labeled primers has been shown to lower detection limits. Chen et al. (1998) used fluorescent-labeled primers to detect STEC in foods. Primers were designed to detect the presence of the shiga toxin genes *stx1*, *stx2*, and *stxe*. Detection limits of the method were 1-5 CFU per PCR mixture (pure culture), and 3 CFU per 25g of food. Fanning et al. (1995) used a similar approach to develop a color amplified PCR system to detect the heat stable toxin gene (ST) in enterotoxigenic *E. coli* (ETEC). The method was found to have a lower detection limit of 10 fg of purified DNA and was capable of detecting 270 CFU of an ETEC strain possessing the ST gene.

Niessen (2008) described the use of PCR for the diagnosis and quantification of mycotoxin-producing fungi in foods and other commodities, while Zhang et al. (2008) reported a modification of PCR using an immuno-PCR assay for the detection of shiga toxin 2 (*stx2*) in culture which reduced the sensitivity to a level of 10pg/ml, when compared with commercial enzyme immuno assays which had a limit of detection of 1 ng/mL. Ge et al. (2002) used a combination of PCR with ELISA to detect STEC in food and found detection limits of the assay were 0.1–10 CFU, depending on the strain type. Another modification of PCR combined restriction fragment length polymorphism (RFLP) (Atanassova et al. 2001) for the detection of *S. aureus* and staphylococcal enterotoxins in pork and pork products. This method was found to be more sensitive than standard culture techniques with a detection rate of 28.6–34.8% compared to 11.1% by standard culture. Other modifications of PCR include restriction site specific PCR (RSS-PCR) (Kimura et al. 2000) and the ramification assay for detection of *E. coli* O157:H7 and STEC (Li et al. 2005).

Nested PCR

Nested PCR is a refined method for enhancing the specifics of PCR for pathogen detection. It uses a primary set of primers to amplify a specific region of DNA, then a secondary set of primers is used to amplify a product that lies (i.e., nested) within the initial product. Nested PCR is designed to increase the accuracy of the PCR reaction; that is, the dual amplification strategy ensures that if the wrong target is amplified in the primary reaction, the odds of a second successful amplification is low, as the second pair of primers are smaller and designed to amplify only within a region of the first PCR product. Therefore, the odds of amplifying such a product in an error sequence become low and the accuracy of the results are likely increased over standard methods. Nested PCR has been used in the detection of *Shigella* in food (Lindqvist 1999, Warren et al. 2006) and for the detection of toxin-associated genes in some pathogens. Miwa et al. (1996) described the use of nested PCR for the detection of enterotoxigenic *Clostridium perfringens* in animal feces and meat using a pair of nested primers homologous to the *C. perfringens* enterotoxin gene. The sensitivity of the test was about 10^3 fold greater than standard PCR.

Real-time PCR

Real-time PCR is a modification of conventional PCR that eliminates the need for gel electrophoresis, thus providing the user results more quickly. A recent review by Mackay (2004) provides good information into the detail of the technology. Whereas, its underlying principle is based on standard PCR, a fluorescent tag is added to the primers so that the amplicons can be detected on a real-time basis by monitoring fluorescence. Thus, a detector in the thermocycler detects amplified product as it is produced. Real-time PCR protocols can be designed to detect multiple products simultaneously using a similar approach to multiplex PCR—typically, four or five fluorescent tags can be detected simultaneously as with conventional multiplex PCR. It also means that products of the similar size can be amplified easily without the challenges posed by separation on a standard electrophoresis gel, as each amplicon will be amplified with a different fluorescent marker. The advantages of real-time PCR over conventional PCR include: easy resolution of product, a faster run time as there is no post-PCR gel analysis, and increased sensitivity. The latter is greater than conventional PCR, as real-time PCR technology relies on the detection of a signal and its quantification with the release of light being in proportion to amplified product formed. Fluorescent dyes such as HEX, FAM, ROX, and SYBR green are some of the commonly available fluorescent dyes, which can be used simultaneously in a multiplex real-time-PCR protocol (Huang et al. 2007, Wang et al. 2007, Nde et al. 2008b). Multiple commercial thermocyclers are available for use in performing real-time PCR offering capabilities from single tubes to 96 and 384 well formats and have the capability to detect up to five different fluorescent targets simultaneously. Despite the practical advantages of this method, the expense of the requisite thermocyclers is high that may curtail the use of real-time PCR in some instances. Potential applications of real-time PCR include detection of genes or strains in a range of media (Rodriguez-Lazaro et al. 2004, Fakhr et al. 2006, Bohaychuk et al. 2007, Wang et al. 2007, O'Grady et al. 2008), and diarrheagenic *E. coli* (Vidal et al. 2005). Horsmon et al. (2006) used real-time fluorogenic PCR to detect *entA*, encoding the staphylococcal enterotoxin A (SEA). The method detected SEA at levels as low as 1–13 gene copies. Fykse et al. (2007) used molecular beacon real-time nucleic acid sequence-based amplification for *Vibrio cholerae* by detecting the cholera toxin gene (*ctxA*) and the genes *tcpA*, *toxR*, *hlyA*, and *groEL* and was able to detect the organism at a level of 50 CFU/mL, the method could also differentiate toxigenic from nontoxigenic *Vibrio* strains by amplification of the toxin genes *tcpA* and *ctxA*. Grant et al. (2006) modified the real-time PCR technique to simultaneously detect heat stable and labile toxin genes of enterotoxigenic *E. coli* with threshold cycles of 25.2–41.1.

A primary limitation of PCR has been that it could not distinguish between live and dead cells. However, newly designed PCR protocols can target viable cells by detection of mRNA, which is a marker of viability (Klein and Juneja 1997, McIngvale et al. 2002, Morin et al. 2004).

Overall, PCR has tremendous power as a molecular detection and profiling tool especially as a means to detect a pathogen as well as determine a pathogen's traits or to define its pathotype. It can also be useful for clustering gene traits and can provide significant information when bundled with analysis software to sort organisms into cluster groups or by trait possession (Rodriguez-Siek et al. 2005a, 2005b).

NEXT-GENERATION TECHNOLOGIES

The application of sensitive and rapid detection technologies has become of paramount interest in advancing methods for rapid detection and identification of pathogens of concern. In this section, we review some of the newer technologies emerging for the future and their potential for enhancing methods to detect foodborne pathogens. As we write this section, however, the reader should be aware that the technology is rapidly changing, some recent reviews in relation to the use of biosensors include articles by Kaittanis et al. (2010), Tallury et al. (2010), and Velusamy et al. (2010). Two of the overriding factors in sensor detection methodologies are the ability to obtain a result in a faster time frame and a preference to automate the system (Anderson and Taitt 2005). Biosensors are detection devices that use biological molecules to recognize and quantify analytes of interest—these analytes include antibodies, receptors, enzymes, nucleic acids, oligosaccharides, peptides, and so on. A recognition event between a biomolecule and an agent (e.g., nucleic acid, receptor, and so on) results in an output that can be measured either optically, chemically, or electronically. Biosensors should be able to monitor a sample continuously for analyte and provide a quantitative readout (such as measure of concentration).

Adenosine Triphosphate Detection

Although not a strict biosensor per se, detection of adenosine triphosphate (ATP) from living cells is used as a means to assess the quality of cleaning or disinfection of surfaces and utensils. ATP detection usually involves analysis of a sample swab of an area; the swab is then combined with a mixture of the enzyme luciferase and its substrate luciferin. The reaction of luciferin and luciferase is dependent on the presence of ATP to catalyze the reaction, which is only found in living cells. The ATP combines with luciferin to form luciferyl adenylate and pyrophosphate.

$$\text{Luciferin} + \text{ATP} \rightarrow \text{Luciferyl} + \text{PP}_i$$

The luciferyl adenylate reacts with oxygen to form oxyluciferin and adenosine monophosphate with the release of light.

$$\text{Luciferyl adenylate} + O_2 \rightarrow \text{Oxyluciferin} + \text{AMP} + \text{Light}$$

The enzymatic process results in the release of light, which can be measured and is proportional to the amount of ATP present. There are a number of handheld devices that can be used for this type of testing and are available commercially. Typical levels of detection are 1–200 CFU. ATP assays are nonspecific and in general provide information of a quality nature; it cannot detect for specific pathogens or identify the source of ATP and typically are used as an indicator of hygiene quality or successful sanitization. In addition, the system may have errors associated with the sampling technique resulting in false-positives and false-negatives (Carrick et al. 2001). Modifications of ATP detection have incorporated the use of fluorescence for detection of fluoroimmuno-stained cells; bioluminescence with IMS (Takahashi et al. 2000, Tu et al. 2000, Squirrel et al. 2002).

DNA Microarrays

DNA microarrays consist of gene probes arrayed on a substrate. Such arrays may be limited to select genes of interest or be multigenome-wide in scope. Among other things, they can be used for comparative genomic hybridization (also called genomotyping), to study genome-wide gene expression or for the detection of pathogens in a sample. They have quickly become a powerful tool in genomic analysis of pathogens. The microarray itself (often called a gene chip or biochip) consists of a series of DNA molecules of known sequence called probes that are fixed to a substrate (usually, a special type slide). These probes consist of partial gene sequences, generated from PCR, full-length cDNA, or oligonucleotides (Pagotto et al. 2005) of the pathogens of interest. Such microarrays can be used to detect a large number of pathogens in a sample or alternatively, they can be used to perform expression studies of the whole genome or select genes; for the purposes of this chapter we will focus on the use of microarrays as a detection technology only; for more detail regarding other applications of microarrays the reader is directed to reviews on the subject. A recent review by Ojha and Kostrzynska (2008) highlights the application of microarray technology in the field of veterinary research for pathogen infection investigations, diagnostics, and studies of host pathogen interactions. Similar research by Jin et al. (2005) used microarray technology to investigate *E. coli* O157:H7.

Microarrays can also be used to assess similarities between strains, characterize strains or subtype strains. Boyd et al. 2003, Gaynor et al. (2004), Hain et al. (2006), Malik-Kale et al. (2007), Parker et al. (2007), and Raengpradub et al. (2008) have described the use of microarray technology to analyze strains at the genetic level for comparative purposes. Volokhov et al. (2002), (2003), Chen et al. (2005), Reen et al. (2005), Garaizar et al. (2006), Anjum et al. (2007), Yoshida et al. (2007), Zhang et al. (2007), Batchelor et al. (2008) have used microarrays to identify various bacteria to species or subspecies level, detection of virulence or antimicrobial resistance genes and Call et al. (2001), Chandler et al. (2001), Keramas et al. (2004), Kostrzynska and Bachand (2006), Kostic et al. (2007), Quinones et al. (2007) have used them for detection, identification, and characterization of pathogens in a range of samples. Given the unlimited amount of information available, microarrays can be designed and built for a range of purposes such as, determining expression of specific virulence or antimicrobial resistance genes or for more specific processes, such as study of invasion, flagellar production, growth processes, or biofilm production. DNA microarrays offer much promise for future studies in understanding pathogens, hosts, and production systems—they can be used to model host–pathogen interactions and the effect of various drugs or vaccines on a host or pathogen. Therefore, it is likely that microarray technology will provide needed insight into faster methods for detection and understanding the mechanisms of pathogenesis used by food-borne pathogens that can be exploited to make food safer.

Immunosensors or Biosensors

Biosensors are analytical devices that convert a biological response into an electrical signal. Biosensors work on the principle of a biological component that is coupled to a physiochemical type transducer that collects signal and converts it to an electronic readout. A range of bioreceptors currently recognized for use in pathogen detection include antibodies, enzymes, nucleic acids, cellular, biomimetic, and bacteriophages (Velusamy et al. 2010). Biosensor technology has rapidly expanded in the last 10 years and as this chapter is completed, there will no doubt be many more new applications emerging. Some of the more promising technologies at this point are based on fiber optics and electrochemical reactions. Velusamy et al. (2010) provide a comprehensive overview of sensors for application in pathogen detection.

Fiber Optic Biosensor

Optical fibers for the detection of pathogens work on the principle of a fiber optic taper that sends excitation laser light to a detection surface and receives emitted light. As such, fiber optic technology has been reported to be very useful in the detection of food-borne pathogens. Optical biosensors measure changes in refractive index, fluorescence emission or quenching, chemiluminescence, and fluorescence energy transfer. One such fiber optic biosensor operates by using an antibody sandwich format on optic fiber to detect the capture of antigens using fluorescently tagged conjugates. Geng et al. (2004) used a fiber optic-based detection system to detect *Listeria* in mixed cultures, and meat sample enrichments with a detection level as low as $10–10^3$ cells/mL. In a similar type approach, Kramer and Lim (2004) developed a rapid automated fiber-based biosensor for the detection of *Salmonella* in sprout rinse water. The sensor was capable of detecting *Salmonella* when counts as low as 50 CFU/g were inoculated into the seeds. Fiber optics have been used in *Listeria*, *Salmonella*, *E. coli* O157:H7, and *Clostridium botulinum* toxin detection (Ogert et al. 1992, Strachan and Gray 1995, Simpson and Lim 2005, Ko and Grant 2006) and for the detection of *S. aureus* (Chang et al. 1996).

Raman and Fourier Transform Spectroscopy

This technology is frequently used in whole organism fingerprinting but also has application in analysis of a sample of interest. The technique does, however, depend on increasing the population so that there is sufficient amount for biomass analysis. Raman spectroscopy is an optical technique that uses light scattering to detect pathogens (Schmilovitch et al. 2005). Modifications of Raman spectroscopy using surface enhanced technologies (Grow et al. 2003, Kalasinsky et al. 2007) have also been developed (see below).

Fourier Transform Infrared Spectroscopy

Fourier transform infrared spectroscopy (FTIR) is a nondestructive technique for pathogen detection that measures infrared loss after passing through a sample, the data collected is analyzed using Fourier transformation and the resulting output results in a spectrum identical to the conventional infrared spectroscopy. FTIR has been used to differentiate and quantify *Salmonella*, *E. coli* O157:H7, and *Listeria* in various media or substrates (Lin et al. 2004, Yu et al. 2004, Al-Holy et al. 2006). Ellis et al. (2002) reported the use of FTIR directly on food to detect unique biochemical fingerprints of pathogens and provided a measure of bacterial loads.

Surface Plasmon Resonance

Surface plasmon resonance (SPR) uses reflectance spectroscopy to detect pathogens. The principle of the technique is based on the technology being able to detect minor changes in refractive index, which occur as cells of the target bind to receptors immobilized on a transducer surface. Changes in the angle of reflected light are measured as a function of changes in density of the medium versus time. SPR sensors have been used by a number of researchers for the detection of a range of pathogens including *L. monocytogenes*, *Salmonella*, *E. coli* O157:H7, and *C. jejuni* (Koubova et al. 2001, Bokken et al. 2003, Bhunia et al. 2004, Leonard et al. 2004, Meeusen et al. 2005, Subramanian et al. 2006, Taylor et al. 2006, Waswa et al. 2007).

SPR technologies use antibodies that are immobilized on a gold electrode surface that can measure miniscule changes in resonance frequency as a result of antibody antigen binding occurring (Feng 2007). Su and Li (2004) incorporated SPR with a piezoelectric system for the detection of *E. coli* O157:H7 at levels ranging from 10^3 to 10^8 cells; a similar approach by Oh et al. (2004) could detect *S. typhimurium* at levels ranging from 10^2 to 10^9 CFU/mL and Leonard et al. (2004, 2005) report detection of 10^5 cells/mL of *L. monocytogenes* in less than 30 minutes. Oh et al. (2004) report the development of an SPR chip capable of detecting multiple organisms simultaneously including *E. coli* O157:H7, *S. typhimurium*, *Legionella pneumophila*, and *Yersinia enterocolitica*.

Mass Sensitive Biosensors

Mass sensitive biosensors work on the principle of detection of minute changes in mass. The analysis depends on the use of piezoelectric crystals that vibrate at a specific frequency, as a result of electrical input at the same frequency. The frequency depends on electrical input and the mass of the crystal. If the mass of the crystal increases due to binding of agents, the weight of the crystal will change and therefore the oscillation frequency. The difference in oscillation results in a change, which can be measured electronically. Quartz is one of the most common piezoelectric material and the main types of sensors are quartz crystal microbalance (QCM) or surface acoustic wave (SAW). When the surface of the sensor is coated with an antibody and exposed to liquid containing the target pathogen—the pathogens will bind to the antibody resulting in a change in the weight of the crystal, and a shift in its frequency oscillation. QCM have been used in the detection of *Salmonella* (Su and Li 2005), *L. monocytogenes* (Vaughn et al. 2001), and *E. coli* O157:H7 (Berkenpas et al. 2006). Modifications of the microbalance principle include piezoelectric excited millimeter sized cantilever and magnetoelastic sensors (Ruan et al. 2004, Mutharasan and Campbell 2008).

Electronic Nose Sensors

Electronic nose technology for pathogen detection works on the principle of detection of volatiles produced by active organisms in a food sample. The system is based on adsorption of volatiles to a series of conducting organic polymers; once the volatiles are adsorbed there is a change in the resistance of the sensors that can be equated to the presence of an organism or a population of organisms. Balasubramanian et al. (2005) used a commercial electronic nose system to detect *Salmonella* in inoculated beef samples. The system was able to differentiate *Salmonella* contaminated from noncontaminated meat. Similar systems have also been used in the detection of *E. coli* O157:H7 Younts et al. (2002, 2003) assessed the use of gas sensors for detection of *E. coli* O157 and non-O157 strains. Sensitivity and specificity ranged from 41.7 to 50% for nonnormalized data, which contrasted to data normalized using artificial neural network (ANN) classification that increased sensitivity from 91.7% to 100% but specificity ranged from 37.5% to 50%. Magan et al. (2001) used a commercial system for detection of *Pseudomonas* and *Bacillus* species in milk; Muhamed-Tahir and Alocilja (2003) developed a portable sensor system for detection of *E. coli* O157:H7 with a minimum detection limit of 7.8×10^1 CFU/mL. Arshak et al. (2009) used a series of conducting polymers to detect a range of food-borne pathogens including *Salmonella*, *Bacillus cereus* and *Vibrio parahaemolyticus*, and was able to differentiate each based on its signal.

Nanotechnology for Pathogen Detection

As one of the latest technologies, nanotechnology presents a great opportunity for rapid detection, diagnosis, and identification of pathogens. Nanoparticles, in particular, gold and silver have electronic and optical properties that make them useful in next generation detection. When the particles are coupled to affinity ligands they can be useful in pathogen detection; for example, gold nanoparticles coupled with specific oligonucleotides can detect complementary DNA. Other types of nanoparticles including quantum dots and carbon nanotubes have been used in assays for detection of pathogens, toxins, DNA, and in immunoassay development (Kaittanis et al. 2010). Recent articles by Kaittanis et al. (2010) and Tallury et al. (2010) provide some interesting overviews of nanotechnology applications in pathogen detection.

SUMMARY

Pathogens associated with human disease are constantly changing, what was old is now new and what was new has emerged in new ways not previously recognized as a means to cause disease. As fast as the pathogens are emerging, technologies and applications for their detection have become the next race to find the ultimate technology that can detect all in a user-friendly manner, in as short a time as possible and with the ultimate sensitivity and specificity.

REFERENCES

Aarestrup FM, Engberg J. 2001. Antimicrobial resistance of thermophilic *Campylobacter*. *Vet Res* 32: 311–321.

Abeyta C et al. 1990. Incidence of motile aeromonads from United States west coast shellfish growing estuaries. *J Food Prot* 53: 849–855.

Acheson DW 1999. Foodborne Infections. *Curr Opin Gastroenterol* 15: 538–545.

Alarcon B et al. 2004. Simultaneous and sensitive detection of three foodborne pathogens by multiplex PCR, capillary gel electrophoresis, and laser-induced fluorescent. *J Agric Food Chem* 52: 7180–7186.

Al Saif N, Brazier JS. 1996. The distribution of *Clostridium difficile* in the environment of South Wales. *J Med Microbiol* 45: 133–137.

Al-Holy MA et al. 2006. The use of fourier transform infrared spectroscopy to differentiate *Escherichia coli* O157: H7 from other bacteria inoculated into apple juice. *Food Microbiol* 23: 162–168.

Alketruse SF et al. 1997. Emerging foodborne diseases. *Emerg Infect Dis* 3: 285–293.

Anderson GP, Taitt CR. 2005. Biosensor-based detection of foodborne pathogens. In: PM Fratamico, et al. (eds.) *Foodborne Pathogens Microbiology and Molecular Biology*. Caister Academic Press, Norfolk, pp. 33–50.

Andres PG, Friedman LS. 1999. Epidemiology and the natural course of inflammatory bowel disease. *Gastroenterol Clin North Am* 28: 255–281.

Anjum MF et al. 2007. Pathotyping *Escherichia coli* by using miniaturized DNA microarrays. *Appl Environ Microbiol* 73: 5692–5697.

Argueta C et al. 2000. Isolation and Identification of nontuberculosis mycobacteria from foods as possible exposure sources. *J Food Prot* 63: 930–933.

Arshak K et al. 2009. Conducting polymers and their applications to biosensors: emphasizing on foodborne pathogen detection. *IEEE Sensors Journal* 9: 1942–1951.

Asakura M et al. 2008. Development of a cytolethal distending (cdt) gene –based species-specific multiplex PCR assay for the detection and identification of *Campylobacter jejuni*, *Campylobacter coli* and *Campylobacter fetus*. *FEMS Immunol Med Microbiol* 52: 260–266.

Atabay HI, Corry JE. 1997. The prevalence of *Campylobacter*s and arcobacters in broiler chickens. *J Appl Microbiol* 86: 619–626.

Atanassova V et al. 2001. Prevalence of *Staphylococcus aureus* and staphylococcal enterotoxins in raw pork and uncooked smoked ham – a comparison of classical culturing detection and RFLP-PCR. *Int J Food Microbiol* 15: 105–113.

Auvray F et al. 2007. Detection, isolation and characterization of shiga toxin-producing *Escherichia coli* in retail-minced beef using PCR-based techniques, immunoassays, and colony hybridization. *Lett Appl Microbiol* 45: 646–651.

Baert L et al. 2008. Evaluation of viral extraction methods on a broad range of ready-to-eat foods with conventional and real-time RT-PCR for norovirus GII detection. *Int J Food Microbiol* 31: 101–108.

Bakri M et al. 2009. *Clostridium difficile* in ready to eat salads, Scotland. *Emerg Infect Dis* 15: 817–818.

Balasubramanian S et al. 2005. Identification of *Salmonella* inoculated beef using a portable electronic nose system. *Journal of Food Safety* 13: 71–95.

Barlow M. 2009. What antimicrobial resistance has taught us about horizontal gene transfer. *Methods Mol Biol* 532: 397–411.

Bartlett JG, Perl TM. 2005. The new *Clostridium* – what does it mean? *N Engl J Med* 353: 2503–2505.

Batchelor M et al. 2008. Development of a miniaturized microarray-based assay for the rapid identification of antimicrobial resistance genes in gram-negative bacteria. *Int J Antimicrob Agents* 31: 440–451.

Baumgartner A et al. 2009. Detection and frequency of *Cronobacter* spp. (*Enterobacter sakazakii*) in different categories of ready-to-eat foods other than infant formula. *Int J Food Microbiol* 31: 189–192.

Berkenpas E et al. 2006. Detection of *Escherichia coli* O157: H7 with langasite pure shear horizontal surface acoustic wave sensors. *Biosens Bioelectron* 21: 2255–2262.

Bhunia AK et al. 2004. Optical immunosensors for detection of *Listeria monocytogenes* and *Salmonella enteritidis* from food. In: BS Bennedsen, et al. (eds.) *SPIE-International Society for Optical Engineering Providence, RI*, pp. 1–6.

Black RE et al. 1988. Experimental *Campylobacter jejuni* infection in humans. *J Infect Dis* 157: 472–479.

Blaser MJ, Engberg J. 2008. Clinical aspects of *Campylobacter jejuni* and *Campylobacter coli* infections. In: I Nachamkin, et al. (eds.) Campylobacter, 3rd edn. ASM Press, Washington, DC, pp. 99–121.

Blaser MJ et al. 1980. Survival of *Campylobacter fetus* subsp. *jejuni* in biological milieus. *J Clin Microbiol* 11: 309–313.

Bohaychuk VM et al. 2007. A real-time PCR assay for the detection of *Salmonella* in a wide variety of food and food-animal matrices. *J Food Prot* 70: 1080–1087.

Bokken G et al. 2003 Immunochemical detection of *Salmonella* group B, D and E using an optical surface Plasmon resonance biosensor. *FEMS Microbiology Letters* 222: 75–82.

Borriello S. 1998. Pathogenesis of *Clostridium difficile* infection. *J Antimicrob Chemother* 41(suppl C): 13–19.

Bortolussi R. 2008. Listeriosis: a primer. *Can Med Assoc J* 179: 795–797.

Bouttier S et al. 2010. *Clostridium difficle* in ground meat, France. *Emerg Infect Dis* 16: 733–735.

Boyd EF et al. 2003. Differences in gene content among *Salmonella enterica* serovar typhi isolates. *J Clin Microbiol* 41: 3823–3828.

Broqui P, Raoult D. 2001. Endocarditis due to rare and fastidious bacteria. *Clin Microbiol Rev* 14: 177–207.

Call DR et al. 2001. Detecting and genotyping of *Escherichia coli* O157:H7 using multiplexed PCR and nucleic acid microarrays. *Int J Food Microbiol* 20: 71–80.

Carattoli A. 2008. Animal reservoirs for extended spectrum beta-lactamase producers. *Clin Microbiol Infect* 1: 117–123.

Carrick K et al. 2001. The comparison of four bioluminators and their swab kits for instant hygiene monitoring and detection of microorganisms in the brewery. *J Inst Brew* 107: 31–37.

Caubilla-Barron J et al. 2007. Genotypic and phenotypic analysis of *Enterobacter sakazakii* strains from an outbreak resulting in fatalities in a neonatal intensive care unit in France. *J Clin Microbiol* 43: 3978–3985.

CDC. 1995a. Community outbreak of hemolytic uremic syndrome attributable to *Escherichia coli* O111:NM – South Australia 1995. *Morb Mortal Wkly Rep* 44: 550–551.

CDC. 1995b. *Escherichia coli* O157:H7 outbreak linked to commercially distributed dry-cured salami-Washington and California 1994. *Morb Mortal Wkly Rep* 44: 157–160.

CDC. 1997a. *Escherichia coli* O157:H7 infections associated with eating a nationally distributed commercial brand of frozen beef patties and burgers – Colorado 1997. *Morb Mortal Wkly Rep* 46: 777–778.

CDC. 1997b. Outbreaks of *Escherichia coli* O157:H7 infection and cryptosporidiosis associated with drinking unpasteurized apple cider – Connecticut and New York, October 1996. *Morb Mortal Wkly Rep* 46: 4–8.

CDC. 1997c. Outbreaks of *Escherichia coli* O157:H7 infection associated with eating alfalfa sprouts – Michigan and Virginia, June-July 1997. *Morb Mortal Wkly Rep* 46: 741–744.

CDC. 1998. Multistate outbreak of listeriosis – Unites States, 1998. *Morb Mortal Wkly Rep* 47: 1085–1086.

CDC. 2002. Multistate outbreak of *Escherichia coli* O157:H7 infections associated with eating ground beef – United States, June-July 2002. *Morb Mortal Wkly Rep* 51: 637–639.

CDC. 2005a. Outbreaks of *Salmonella* infections associated with eating Roma tomatoes – United States and Canada 2004. *Morb Mortal Wkly Rep* 54: 325–328.

CDC. 2005b *Escherichia coli* O157:H7 infections associated with drinking raw milk – Washington and Oregon, November – December 2005. *Morb Mortal Wkly Rep* 56: 165–167.

CDC. 2006. Ongoing multistate outbreak of *Escherichia coli* serotype O157:H7 infections associated with consumption of fresh spinach – United States, September 2006. *Morb Mortal Wkly Rep* 55: 1–2.

CDC. 2008. *Escherichia coli* O157:H7 infections in children associated with raw milk and raw colostrums from cows – California 2006. *Morb Mortal Wkly Rep* 57: 625–628.

CDC. 2009. Multistate outbreak of *Salmonella* infections associated with peanut butter and peanut butter-containing products – United States, 2008–2009. *Morb Mortal Wkly Rep* 58: 85–90.

CDC. 2010. Preliminary FoodNet data on the incidence of Infection with pathogens transmitted commonly through food – 10 States 2009. *Morb Mortal Wkly Rep* 59: 418–422.

CDC. 2011. Estimates of Foodborne Illness in the United States. Available at http://www.cdc.gov/foodborneburden/2011-foodborne-estimates.html.

Chan KF et al. 2001. Survival of clinical and poultry derived isolates of *Campylobacter jejuni* at a low temperature (4C). *Appl Environ Microbiol* 67: 4186–4191.

Chandler DP et al. 2001. Automated immunomagnetic separation and microarray detection of *E. coli* O157:H7 from poultry carcass rinse. *Int J Food Microbiol* 70: 143–154.

Chang YH et al. 1996. Detection of protein a produced by *Staphylococcus aureus* with a fiber-optic-based biosensor. *Biosci Biotechnol Biochem* 60: 1571–1574.

Chen S et al. 1998. An automated fluorescent PCR method for detection of shiga toxin-producing *Escherichia coli* in foods. *Appl Environ Microbiol* 64: 4210–4216.

Chen S et al. 2005. A DNA microarray for identification of virulence and antimicrobial resistance genes in *Salmonella* serovars and *Escherichia coli*. *Mol Cell Probes* 19: 195–201.

Chenu JW, Cox JM. 2009. *Cronobacter* ('*Enterobacter sakazakii*'); current status and future prospects. *Lett Appl Microbiol* 49: 153–159.

Cleveland McEntire J, Montville TJ. 2007. Antimicrobial resistance. In: M.P. Doyle, L.R. Beuchat (eds.) *Food Microbiology: Fundamentals and Frontiers*. ASM Press, Washington, DC, pp. 23–34.

De Boer E et al. 2009. Prevalence of methicillin-resistant *Staphylococcus aureus* in meat. *Int J Food Microbiol* 134: 52–56.

Deodhar LP et al. 1991. *Aeromonas spp*. and their association with human diarrheal disease. *J Clin Microbiol* 29: 853–856.

Diergaardt SM et al. 2004. The occurrence of Campylobacters in water sources in South Africa. *Water Res* 38: 2589–2595.

Donnelly, C.W. 2001. *Listeria monocytogenes*: a continuing challenge. *Nutr Rev* 59: 183–194.

Doyle MP, Erickson MC. 2006. Emerging microbiological food safety issues related to meat. *Meat Sci* 74: 98–112.

Doyle MP, Jones DM. 1992. Foodborne transmission and antibiotic resistance of *Campylobacter jejuni*. In: I Nachamkin, et al. (eds.) *Campylobacter jejuni: Current Status and Future Trends*. ASM Press, Washington, DC, pp. 45–48.

Drevets DA, Bronze MS. 2008. *Listeria monocytogenes*: epidemiology, human disease and mechanisms of brain invasion. *FEMS Immunol Med Microbiol* 53: 151–165.

Eblen DR. 2007. Public health importance of non-O157 Shiga toxin-producing *Escherichia coli* (non-O157 STEC) in the US food supply. USDA FSIS OPHS White Paper. Available at http://origin-www.fsis.usda.gov/PDF/STEC_101207.pdf.

Ellis DI et al. 2002. Rapid and quantitative detection of the microbial spoilage of meat by Fourier transform infrared spectroscopy and machine learning. *Appl Environ Microbiol* 68: 2822–2828.

Endtz HP et al. 1991. Quinolone resistance in *Campylobacter* isolated from man and poultry following the introduction of fluoroquinolones in veterinary medicine. *J Antimicrob Chemother* 27: 199–208.

Engberg J et al. 2001. Quinolone and macrolide resistance in *Campylobacter jejuni* and *C. coli*: resistance mechanisms and trends in human isolates. *Emerg Infect Dis* 7: 24–34.

Fakhr MK et al. 2006. Adding a selective enrichment step to the iQ-Check real-time PCR improves the detection of *Salmonella* in naturally contaminated retail turkey meat products. *Lett Appl Microbiol* 43: 78–83.

Fanning S et al. 1995. Detection of the heat-stable toxin encoding gene (ST-gene) in enterotoxigenic *Escherichia coli*: development of a color coding amplified PCR detection system. *Br J Biomed Sci* 52: 317–320.

Feng, P. 2007. Rapid methods for the detection of foodborne pathogens: current and next generation technologies. In: MP Doyle, LR Beuchat (eds.) *Food Microbiology, Fundamentals, and Frontiers*. ASM Press, Washington, DC, pp. 911–934.

Franz E, van Bruggen AH. 2008. Ecology of *E. coli* O157:H7 and *Salmonella enterica* in the primary vegetable production chain. *Crit Rev Microbiol* 34: 143–161.

Fratamico PM, Bales DO. 2005. Molecular approaches for detection, identification, and analysis of foodborne pathogens. In: PM Fratamico, et al. (eds.), *Foodborne Pathogens Microbiology and Molecular Biology*. Caister Academic Press, Norfolk, pp. 1–13.

Fricke WF et al. 2009. Antimicrobial resistance-conferring plasmids with similarity to virulence plasmids from avian pathogenic *Escherichia coli* strains in *Salmonella enterica* serovar *Kentucky* isolates from poultry. *Appl Environ Microbiol* 75: 5963–5971.

Friedemann M. 2007. *Enterobacter sakazakii* in food and beverages (other than infant formula and milk powder). *Int J Food Microbiol* 116: 1–10.

Friedman CR et al. 2000. Epidemiology of *Campylobacter* infections in the United States and other industrialized nations. In: I. Nachamkin, MJ Blaser (eds.) *Campylobacter*, 2nd edn. ASM Press, Washington, DC, pp. 121–138.

Fykse EM et al. 2007. Detection of *Vibrio cholerae* by real time nucleic acid sequence based amplification. *Appl Environ Microbiol* 73: 1457–1466.

Gantois I et al. 2009. Mechanisms of egg contamination by *Salmonella enteritidis*. *FEMS Microbiol Rev* 33: 718–738.

Garaizar J et al. 2006. DNA microarray technology: a new tool for the epidemiological typing of bacterial pathogens? *FEMS Immunol Med Microbiol* 47: 178–189.

Gast RK, Stephens JF. 1986. In vivo transfer of antibiotic resistance to a strain of *Salmonella* arizonae. *Poultry Science* 65: 270–279.

Gavriel AA et al. 1998. Incidence of mesophilic *Aeromonas* within a public drinking water supply in north-east Scotland. *J Appl Microbiol* 84: 383–392.

Gaynor EC et al. 2004. The genome-sequenced variant of *Campylobacter jejuni* NCTC 11168 and the original clonal clinical isolate differ markedly in colonization, gene expression, and virulence associated phenotypes. *J Bacteriol* 186: 503–517.

Ge B et al. 2002. A PCR-ELISA for detecting Shiga toxin-producing *Escherichia coli*. *Microbes Infect* 4: 285–290.

Geng T et al. 2004. Detection of Low levels of *Listeria monocytogenes* cells by using a fiber-optic immunosensor. *Appl Environ Microbiol* 70: 6138–6146.

Ghadiali AH et al. 2004. *Mycobacterium avium subsp. Paratuberculosis* strains isolated from Crohn's disease patients and animal species exhibit similar polymorphic locus patterns. *J Clin Microbiol* 42: 5345–5348.

Glover N et al. 1994. The isolation and identification of *Mycobacterium avium* complex (MAC) recovered from Los Angeles potable water, a possible source of infection in AIDS patients. *Int J Environ Health Res* 4: 63–72.

Goorhuis A et al. 2008. Emergence of *Clostridium difficile* infection due to a new hypervirulent strain, polymerase chain reaction ribotype 078. *Clin Infect Dis* 47: 1162–1170.

Grant IR et al. 1998. Isolation of *Mycobacterium paratuberculosis* from milk by immunomagnetic separation. *Appl Environ Microbiol* 64: 3153–3158.

Grant IR et al. 2000. Improved detection of *Mycobacterium avium subsp. Paratuberculosis* in milk by immunomagentic PCR. *Vet Microbiol* 77: 369–378.

Grant MA et al. 2006. Multiplex real-time PCR detection of heat-labile and heat-stable toxin genes in enterotoxigenic *Escherichia coli*. *J Food Prot* 69: 412–416.

Greenstein, R.J. 2003. Is Crohn's disease caused by a Mycobacterium? Comparisons with leprosy, tuberculosis, and Johne's disease. *Lancet Infect Dis* 3: 507–514.

Grow AE et al. 2003. New biochip technology for label-free detection of pathogens and their toxins. *J Microbiol Methods* 53: 221–233.

Gude A et al. 2005. Ecology of *Arcobacter* species in chicken rearing and processing. *Lett Appl Microbiol* 41: 82–87.

Guo B et al. 2010. Contribution of the multidrug efflux transporter CmeABC to antibiotic resistance in different *Campylobacter* species. *Foodborne Pathogens Dis* 7: 77–83.

Gurtler JB et al. 2005. *Enterobacter sakazakii*: a coliform of increased concern to infant health. *Int J Food Microbiol* 104: 1–34.

Hain, T., Steinweg, C., Chakraborty, T. 2006. Comparative and functional genomics of *Listeria spp*. *J Biotechnol* 20: 37–51.

Han F et al. 2009a. Prevalence and antimicrobial resistance among *Campylobacter* spp. in Louisiana retail chickens after the enrofloxacin ban. *Foodborne Pathog Dis* 6: 163–171.

Han F et al. 2009b. Fitness cost of macrolide resistance in *Campylobacter jejuni*. *Int J Antimicrob Agents Chemother* 34: 462–466.

Hao H et al. 2009. 23S rRNA mutation A2074C conferring high-level macrolide resistance and fitness in *Campylobacter jejuni*. *Microb Drug Resist* 15: 239–244.

Harris JE, Lammerding AM. 2001. Crohn's disease and *Mycobacterium avium subsp. Paratuberculosis*: Current Issues. *J Food Prot* 64: 2103–2110.

Hasman H et al. 2005. Beta-lactamases among extended-spectrum beta-lactamase (ESBL)-resistant *Salmonella* from poultry, poultry products and human patients in The Netherlands. *J Antimicrob Chemother* 56: 115–121.

Healey B et al. 2010. *Cronobacter* (*Entrobacter sakazakii*): An opportunistic foodborne pathogen. *Foodborne Pathog Dis* 7: 339–350.

Hilton CL et al. 2001. The recovery of *Arcobacter butzleri* NCTC 12481 from various temperature treatments. *J Appl Microbiol* 91: 929–932.

Himelright I et al. 2002. From the centers for disease control and prevention *Enterobacter sakazakii* infections associated with the use of powdered infant formula – Tennesee 2001. *JAMA* 287: 2204–2205.

Ho HTK et al. 2006. *Arcobacter*, what is known and unknown about a potential foodborne zoonotic agent! *Vet Microbiol* 115: 1–13.

Horsmon JR et al. 2006. Real-time fluorogenic PCR assays for the detection of *entA*, the gene encoding staphylococcal enterotoxin A. *Biotechnol Lett* 28: 823–829.

Houf K et al. 2005. *Arcobacter cibarius* sp. nov., isolated from broiler carcasses. *Int J Syst Evol Microbiol* 55: 713–717.

Huang Q et al. 2007. Identification of 8 foodborne pathogens by multicolor combinational probe coding technology in a single real-time PCR. *Clin Chem* 53: 1741–1748.

Hume ME et al. 2001. Genotypic variation from a farrow-to-finish swine facility. *J Food Prot* 64: 645–651.

Humphrey T. 2001. The spread and persistence of *Campylobacter* and *Salmonella* in the Domestic Kitchen. *J Infect* 43: 50–53.

Humphrey TJ et al. 2005. Prevalence and subtypes of ciprofloxacin-resistant *Campylobacter* spp. in commercial poultry flocks before, during, and after treatment with fluoroquinolones. *Antimicrob Agents Chemother* 49: 690–698.

Hunter CJ et al. 2008. *Enterobacter sakazakii*: an emerging pathogen in infants and neonates. *Surg Infect (Larchmt)* 9: 533–539.

Indra A et al. 2009. *Clostridium difficile*: a new zoonotic agent? *Weiner Klinische Wochenschrift* 121: 91–95.

Inglis GD et al. 2004. Chronic shedding of *Campylobacter* species in beef cattle. *J Appl Microbiol* 97: 410–420.

Isonhood JH, Drake M. 2002. *Aeromonas* species in foods. *J Food Prot* 65: 575–582.

Iversen C, Forsythe S. 2003. Risk profile for *Enterobacter sakazakii*, an emergent pathogen associated with infant milk formula. *Trends Food Sci Technol* 14: 443–454.

Iversen C, Forsythe S. 2004. Isolation of *Enterobacter sakazakii* and other enterobacteriaceae from powdered infant formula milk and related products. *Food Microbiol* 21: 771–777.

Iversen C et al. 2004 The growth profile, thermotolerance and biofilm formation of *Enterobacter sakazakii* grown in infant formula. *Lett Appl Microbiol* 38: 378–382.

Iversen C et al. 2008. *Cronobacter gen. nov.*, a new genus to accommodate the biogroups of *Enterobacter sakazakii* and proposal of *Cronobacter sakazakii gen.nov.*, comb., nov., *Cronobacter malonaticus sp. nov.*, *Cronobacter genomospecies* 1, and three subspecies, *Cronobacter dublinensis, subsp dublinensis subsp. nov.*, and *Cronobacter dublinensis subsp. lausannensis subsp. nov.* and *Cronobacter dublinensis* subsp. lactardi subsp.nov. *Int J Syst Evol Microbiol* 58: 1442–1447.

Jacob J et al. 1993. Isolation of *Arcobacter butzleri* from a drinking water reservoir in Eastern Germany. *Zentralbl Hyg Umweltmed* 193: 557–562.

Jacobs BC et al. 2008. Guillain-Barre syndrome and *Campylobacter* infection. In: I. Nachamkin, et al. (eds.) *Campylobacter*, 3rd edn. ASM Press, Washington, DC, pp. 245–261.

Jacobs-Reitsma W. 2000. *Campylobacter* in the food supply. In: I Nachamkin, MJ Blaser (eds.) *Campylobacter* 2nd edn. ASM Press, Washington, DC, pp. 467–481.

Janda JM, Abbott SL. 1999. Unusual food-borne pathogens. *Listeria monocytogenes*, *Aeromonas*, *plesiomonas*, and *Edwardsiella* species. *Clin Lab Med* 19: 553–582.

Janda JM, Abbott SL. 2010. The Genus *Aeromonas*: taxonomy, pathogenicity, and infection. *Clin Microbiol Rev* 23: 35–73.

Janda JM et al. 1994. *Aeromonas* species in septicemia: laboratory characteristics and clinical observations. *Clin Infect Dis* 19: 77–83.

Jin HY et al. 2005. Microarray analysis of *Escherichia coli* O157:H7. *World J Gastroenterol* 7: 5811–5815.

Jobstl M et al. 2009. *Clostridium difficile* in raw products of animal origin. *Int J Food Microbiol* 138: 172–175.

Johnson TJ et al. 2010. Sequence analysis and characterization of a transferable hybrid plasmid encoding multidrug resistance and enabling zoonotic potential for extraintestinal *Escherichia coli*. *Infect Immun* 78: 1931–942.

Johnson TJ, Nolan LK. 2009. Pathogenomics of the virulence plasmids of *Escherichia coli*. *Microbiol Mol Biol Rev* 73: 750–774.

Johnson TJ et al. 2004. Multiple antimicrobial resistance region of a putative virulence plasmid from an *Escherichia coli* isolate incriminated in avian colibacillosis. *Avian Dis* 48: 351–360.

Johnson TJ et al. 2007. Plasmid replicon typing of commensal and pathogenic *Escherichia coli* isolates. *Appl Environ Microbiol* 73: 1976–1983.

Kabeya H et al. 2003. Distribution of *Arcobacter* species among livestock in Japan. *Vet Microbiol* 93: 153–158.

Kabeya H et al. 2004. Prevalence of *Arcobacter* species in retail meats and antimicrobial susceptibility of the isolates in Japan. *Int J Food Microbiol* 90: 303–308.

Kaittanis C et al. 2010. Emerging nanotechnology-based strategies for the identification of microbial pathogenesis. *Adv Drug Deliv Rev* 18: 408–423.

Kalasinsky KS et al. 2007. Raman chemical imaging spectroscopy reagentless detection and identification of pathogens: signature development and evaluation. *Anal Chem* 79: 2658–2673.

Kandhai MC et al. 2004. Occurrence of *Enterobacter sakazakii* in food production environments and households. *Lancet* 363: 39–40.

Karakousis PC et al. 2004. *Mycobacterium avium* complex in patients with HIV infection in the era of highly active anti-retroviral therapy. *Lancet Infect Dis* 4: 557–565.

Kawasaki S et al. 2010. Multiplex real-time polymerase chain reaction assay for simultaneous detection and quantification of *Salmonella* species, *Listeria monocytogenes* and *Escherichia coli* O157:H7 in ground pork samples. *Foodborne Pathog Dis* 7: 549–554.

Keramas G et al. 2004. Use of culture, PCR analysis, and DNA microarrays for detection of *Campylobacter jejuni* and *Campylobacter coli* from chicken feces. *J Clin Microbiol* 42: 3985–3991.

Kim J-Y et al. 2005. Isolation and identification of *Escherichia coli* O157:H7 using different detection methods and molecular determination by multiplex PCR and RAPD. *J Vet Sc* 6: 7–19.

Kim JS et al. 2007a. A novel multiplex PCR assay for rapid and simultaneous detection of five pathogenic bacteria: *Escherichia coli* O157:H7, *Salmonella*, *Stapylococcus aureus*, *Listeria monocytogenes* and *Vibrio parahaemolyticus*. *J Food Prot* 70: 1656–1662.

Kim H et al. 2007b. Effectiveness of disinfectants in killing *Enterobacter sakazakii* I suspension dried on the surface of stainless steel, and in a biofilm. *Appl Environ Microbiol* 73: 1256–1265.

Kimura R et al. 2000. Restriction site specific PCR as a rapid test to detect enterohemorrhagic *Escherichia coli* O157:H7 strains in Environmental samples. *Appl Environ Microbiol* 66: 5213–2519.

Kingombe CIB et al. 1999. PCR detection, characterization, and distribution of virulence genes in *Aeromonas* spp. *Appl Environ Microbiol* 65: 5293–5302.

Kirov SM. 1993a. The public health significance of *Aeromonas* spp. In foods. *Int J Food Microbiol* 20: 179–198.

Kirov SM et al. 1993b. Milk as a potential source of *Aeromonas* gastrointestinal infection. *J Food Prot* 56: 306–312.

Kirsner JB. 1999. "Nonspecific" inflammatory bowel disease (ulcerative colitis and Crohn's disease) after 100 years – what next?. *Ital J Gastroenterol Hepatol* 31: 651–658.

Klein PG, Juneja VK. 1997. Sensitive detection of viable *Listeria monocytogenes* by reverse transcription-PCR. *Appl Environ Microbiol* 63: 4441–8.

Ko SH, Grant SA. 2006. A novel FRET-based optical biosensor for rapid detection of *Salmonella typhimurium*. *Biosens Bioelectron* 21: 1283–1290.

Kostic T et al. 2007. A microbial diagnostic microarray technique for the sensitive detection and identification of pathogenic bacteria in a background of nonpathogens. *Anal Biochem* 15: 244–254.

Kostrzynska M, Bachand A. 2006. Application of DNA microarray technology for detection, identification, and characterization of food-borne pathogens. *Can J Microbiol* 52: 1–8.

Koubova V et al. 2001. Detection of foodborne pathogens using surface Plasmon resonance biosensors. *Sens Actuators B Chem* 74: 100–105.

Kramer MF, Lim DV. 2004. A rapid and automated fiberoptic-based biosensor assay for the detection of *Salmonella* in spent irrigation water used in the sprouting of sprout seeds. *J Food Prot* 67: 46–52.

Kruse H, Sorum H. 1994. Transfer of multiple drug resistance plasmids between bacteria of diverse origins in natural microenvironments. *Appl Environ Microbiol* 60: 4015–4021.

Kuhn I et al. 1997. Characterization of *Aeromonas* spp. Isolated from humans with diarrhea, from healthy controls and from surface water in Bangladesh. *J Clin Microbiol* 35: 369–373.

Lai KK. 2001. *Enterobacter sakazakii* infections among neonates, infants, children and adults: case reports and review of the literature. *Medicine (Baltimore)* 80: 113–122.

Larsen JC, Guerry P. 2005. Plasmids of *Campylobacter jejuni*. In: JM Ketley, ME Konkel (eds.) *Campylobacter: Molecular and Cellular Biology*. Horizon Bioscience, Norfolk, pp. 181–192.

Lehner A, Stephan R. 2004. Microbiological, epidemiological, and food safety aspects of *Enterobacter sakazakii*. *J Food Prot* 67: 2850–2857.

Lehner A et al. 2005. Relevant aspects of *Arcobacter* spp. as potential foodborne pathogen. *Int J Food Microbiol* 102: 127–135.

Leonard P et al. 2004. A generic approach for the detection of whole *Listeria monocytogenes* cells in contaminated samples using surface plasmon resonance. *Biosens Bioelectron* 19: 1331–1335.

Leonard P et al. 2005. Development of a surface plasmon resonance-based immunoassay for *Listeria monocytogenes*. *J Food Prot* 68: 728–735.

Li Q et al. 2004. The prevalence of *Listeria*, *Salmonella*, *Escherichia coli* and *E. coli* O157:H7 on bison carcasses during processing. *Food Microbiol* 24: 791–799.

Li F et al. 2005. Use of a ramification amplification assay for detection of *Escherichia coli* O157:H7 and other *E. coli* shiga toxin-producing strains. *J Clin Microbiol* 43: 6086–6090.

Li Q et al. 2006. Antimicrobial susceptibility and characterization of *Salmonella* isolates from processed bison carcasses. *Appl Environ Microbiol* 72: 3046–9.

Lianou A, Sofos JN. 2007. A review of the incidence and transmission of *Listeria monocytogenes* in ready-to-eat products in retail and food service establishments. *J Food Prot* 70: 2172–2198.

Lin J et al. 2002. CmeABC functions as a multidrug efflux system in *Campylobacter jejuni*. *Antimicrob Agents Chemother* 46: 2124–2131.

Lin MS et al. 2004. Discrimination of intact and injured *Listeria monocytogenes* by fourier transform infrared spectroscopy and principal component analysis. *J Agric Food Chem* 52: 5679–5772.

Lin J et al. 2007. Effect of macrolide usage on the emergence of erythromycin-resistant *Campylobacter* isolates in chickens. *Antimicrob Agents Chemother* 51: 1678–1686.

Lindqvist R. 1999. Detection of *Shigella* spp. in food with a nested PCR method-sensitivity and performance compared with a conventional culture method. *J Appl Microbiol* 86: 971–8.

Logue CM et al. 2003. The incidence of antimicrobial-resistant *Salmonella* spp on freshly processed poultry from US Midwestern processing plants. *J Appl Microbiol* 94: 16–24.

Logue CM, Nolan LK. 2009. Molecular analysis of pathogenic bacteria and their toxins. In: F Toldra (ed.) *Safety of Meat and Processed Meats*. Springer, New York.

Logue CM et al. 2010. Repeated therapeutic dosing selects macrolide resistant *Campylobacter* spp. in a turkey facility. *J Appl Microbiol* 109: 1379–1388.

Lunagtongkum T et al. 2009. Antibiotic resistance in *Campylobacter* : emergence, transmission and persistence. *Future Microbiol* 4: 189–200.

Lutgen EM et al. 2009. Antimicrobial resistance profiling and molecular subtyping of *Campylobacter* spp. from processed turkey. *BMC Microbiol* 21: 203.

Mackay IM. 2004. Real-time PCR in the microbiology laboratory. *Clin Microbiol Infect* 10: 190–212.

Magan N, et al. 2001. Milk-sense: a volatile sensing system recognizes spoilage bacteria an yeasts in milk. *Sens Actuators B Chem* 72: 28–34.

Malik-Kale P et al. 2007. Characterization of genetically matched isolates of *Campylobacter jejuni* reveals that mutations in genes involved in flagellar biosynthesis alter the organism's virulence potential. *Appl Environ Microbiol* 73: 3123–3136.

Marth EH. 1988. Disease characteristics of *Listeria monocytogenes*. *Food Technol* 42: 165–168.

Martinez I et al. 2006. Detection of *cdtA*, *cdtB*, and *cdtC* genes in *Campylobacter* by multiplex PCR. *Int J Med Microbiol* 296: 45–48.

McClung CR et al. 1983 *Campylobacter* nitrofigilis sp. Nov., a nitrogen-fixing bacterium associated with roots of Spartina alterniflora Loisel. *Int J Syst Bacteriol* 33: 605–612.

McIngvale SC et al. 2002. Optimization of reverse transcriptase PCR to detect viable Shiga-toxin-producing *Escherichia coli*. *Appl Environ Microbiol* 68: 799–806.

Meeusen CA et al. 2005. Detection of *E. coli* O157:H7 using a miniaturized surface Plasmon resonance biosensor. *Trans ASAE* 48: 2409–2416.

Meng J, Doyle MP. 1997. Emerging issues in microbiological food safety. *Annu Rev Nutr* 17: 255–275.

Meng J, Doyle, MP. 1998. Emerging and Evolving microbial food-borne pathogens. *Bull Inst Pasteur* 96: 151–164.

Meng J. et al. 2007. Enterohemorrhagic *Escherichia coli*. In: MP Doyle, LR Beuchat (eds.) *Food Microbiology: Fundamentals and Frontiers* 3rd edn. ASM Press, Washington, DC, pp. 249–269.

Mittal R. et al. 2009. Brain damage in a newborn rat model of meningitis by *Enterobacter sakazakii*: role for outer membrane protein A. *Lab Invest* 89: 263–377.

Miwa N et al. 1996. Nested polymerase chain reaction for detection of low levels of enterotoxigenic *Clostridium perfringens* in animal feces and meat. *J Vet Med Sci* 58: 197–203.

Mohan Nair MK et al. 2009. Outer membrane protein A (OmpA) of *Enterobacter sakazakii* binds fibronectin and contributes to invasion of human brain microvascular endothelial cells. *Foodborne Pathog Dis* 6: 495–501.

Mohle-Boetani JC et al. 2001. *Escherichia coli* O157 and *Salmonella* infections associated with sprouts in California 1996–1998. *Ann Intern Med* 135: 239–247.

Moreno Y et al. 2003. Specific detection of Arcobacter and *Campylobacter* strains in water and sewage by PCR and fluorescent in situ hybridization. *Appl Environ Microbiol* 69: 1181–1186.

Morin NJ et al. 2004. Reverse transcription-multiplex PCR assay for simultaneous detection of *Escherichia coli* O157:H7, *Vibrio cholerae* O1, and *Salmonella Typhi*. *Clin Chem* 50: 2037–2044.

Morita Y et al. 2004. Isolation and phylogenetic analysis of *Arcobacter* spp. In ground chicken and environmental water in Japan and Thailand. *Microbiol Immunol* 48: 527–533.

Mor-Mur M, Yuste J. 2010. Emerging bacterial pathogens in meat and poultry: an overview. *Food Bioprocess Technol* 3: 24–35.

Motiwala AS et al. 2006. Current understanding of the genetic diversity of *Mycobacterium avium subsp. Paratubersulosis*. *Microbes Infect* 8: 1406–1418.

Muhamed-Tahir Z, Alocilja EC. 2003. Fabrication of a disposable biosensor for *Escherichia coli* O157:H7 detection. *IEEE Sens J* 3: 345–351.

Mullane N et al. 2007. *Enterobacter sakazakii*: an emerging bacterial pathogen with implications for infant health. *Minerva Pediatr* 59: 137–148.

Mutharasan R, Campbell GA. 2008. Flow cell for receiving piezoelectric-excited millimeter sized cantilever sensor, to detect e.g. pathogens, in media, comprises reservoir portion having flow inlet and outlet apertures to allow flow of media across sensing surface of sensor. Patent # US 2008034840-A1; WO 2008021187-A2.bib>

Muytjens HL et al. 1983. Analysis of eight cases of neonatal meningitis and sepsis due to *Enterobacter sakazakii*. *J Clin Microbiol* 18: 115–120.

Nachamkin I. 2007. *Campylobacter jejuni*. In: MP Doyle, LR Beuchat (eds.) *Food Microbiology: Fundamentals and Frontiers*. ASM Press, Washington, DC, pp. 237–248.

Nde CW et al. 2008. An evaluation of conventional culture, *invA* PCR, and the real-time PCR IQ-Check kit as detection tools for *Salmonella* in naturally contaminated premarket and retail turkey. *J Food Prot* 71: 386–391.

Nde CW, Logue CM. 2008. Characterization of antimicrobial susceptibility and virulence genes of *Salmonella* serovars collected at a commercial turkey processing plant. *J Appl Microbiol* 104: 215–223.

Newell DG et al. 2010. Food-borne diseases – the challenges of 20 years ago still persist while new ones continue to emerge. *Int J Food Microbiol* 139(suppl1): S3–S15.

Niesen EM et al. 2006. Most *Campylobacter* subtypes from sporadic outbreaks can be found in retail poultry products and food animals. *Epidemiol Infect* 134: 758–767.

Niessen L. 2008. PCR-based diagnosis and quantification of mycotoxin-producing fungi. *Adv Food Nutr Res* 54: 81–138.

Norman KN et al. 2009. Varied prevalence of *Clostridium difficile* in an integrated swine operation. *Anaerobe* 15: 256–260.

O'Brien S et al. 2009. Prevalence of *Cronobacter species* (*Enterobacter sakazakii*) in follow-on infant formulae and infant drinks. *Lett Appl Microbiol* 48: 536–541.

O'Grady J et al. 2008. Rapid real-time PCR detection of *Listeria monocytogenes* in enriched food sample based on the *ssrA* gene, a novel diagnostic target. *Food Microbiol* 25: 75–84.

O'Reilly CE et al. 2004. Surveillance of bulk raw and commercially pasteurized cows' milk from approved Irish liquid-milk pasteurization plants to determine the incidence of *Mycobacterium paratuberculosis*. *Appl Environ Microbiol* 70: 5138–5144.

Ogert RA et al. 1992. Detection of *Clostryidium botulinum* toxin-a using a fiber optic based biosensor. *Anal Biochem* 205: 306–312.

Oh B-K et al. 2004. Surface plasmon resonance immunosensor for the detection of *Salmonella* typhimurium. *Biosens Bioelectron* 19: 1497–1504.

Ojha S, Kostrzynska M. 2008. Examination of animal and zoonotic pathogens using microarrays. *Vet Res* 39: 4.

Okrend AJG, et al. 1987. Incidence and toxigenicity of *Aeromonas* species in retail poultry, beef and pork. *J Food Prot* 50: 509–513.

Olah PA et al. 2006. Prevalence of the *Campylobacter* multi-drug efflux pump (CmeABC) in *Campylobacter* spp. isolated from freshly processed turkeys. *Food Microbiol* 23: 453–460.

Olson CK et al. 2008. Epidemiology of *Campylobacter jejuni* in industrialized nations. In: I. Nachamkon, et al. (eds.) *Campylobacter* 3rd edn. ASM Press, Washington, DC, pp. 163–189.

On SL et al. 2002. Prevalence and diversity of *Arcobacter* spp. Isolated from the internal organs of spontaneous porcine abortions in Denmark. *Vet Microbiol* 85: 159–167.

Pagotto F et al. 2005. Molecular typing and differentiation of foodborne bacterial pathogens. In: P Fratamico, et al. (eds.) *Foodborne Pathogens Microbiology and Molecular Biology*. Caister Academic Press, Norfolk, Chapter 4, pp. 51–75.

Parker CT et al. 2007. Common genomic features of *Campylobacter jejuni* subsp. *doylei* strains distinguish them from *C. jejuni* subsp. *jejuni*. *BMC Microbiol* 7: 50.

Patterson DL. 2006. Resistance in gram-negative bacteria: Enterobacteriaceae. *Am J Infect Control* 34: S20–S28.

Pon C, et al. 1995. Differences in production of several extracellular virulence factors in clinical and food *Aeromonas* spp. strains. *J Appl Bacteriol* 78: 175–179.

Poppe C et al. 2005. Acquisition of resistance to extended-spectrum cephalosporins by *Salmonella* enteric serovar Newport and *Escherichia coli* in the turkey poult intestinal tract. *Appl Environ Microbiol* 71: 1184–1192.

Price B et al. 2007. The persistence of fluoroquinolone-resistant *Campylobacter* in poultry production. *Environ Health Perspect* 115: 1035–1039.

Prouzet-Mauleon V et al. 2006. *Arcobacter butzleri*: underestimated enteropathogen. *Emerg Infect Dis* 12: 307–309.

Quinones B et al. 2007. Detection and genotyping of *Arcobacter* and *Campylobacter* isolates from retail chicken samples by use of DNA oligonucleotide arrays. *Appl Environ Microbiol* 73: 3645–3655.

Raengpradub S et al. 2008. Comparative analysis of the sigma B-dependent stress responses in *Listeria monocytogenes* and *Listeria innocua* strains exposed to selected stress conditions. *Appl Environ Microbiol* 74: 158–171.

Raizman EA et al. 2004. The distribution of *Mycobacterium avium subsp. paratuberculosis* in the environment surrounding Minnesota dairy farms. *J Dairy Sci* 87: 2959–2966.

Ray B. 2004a. New and emerging foodborne pathogens. In: B Ray (ed.) *Fundamental Food Microbiology*, 3rd edn. Boca Raton, FL, Chapter 28, pp. 417–428.

Ray B. 2004b. Salmonellosis by *Salmonella* In: B Ray (ed.) *Fundamental Food Microbiology*, 3rd edn. Boca Raton, FL, Chapter 25, pp. 362–367.

Ray B. 2004c. Listeriosis by *Listeria monocytogenes*. In: B Ray (ed.) *Fundamental Food Microbiology*, 3rd edn. Boca Raton, FL, Chapter 25, pp. 367–371.

Ray B. 2004d. Pathogenic *Escherichia coli*. In: B Ray (ed.) *Fundamental Food Microbiology*, 3rd edn. Boca Raton, FL, Chapter 25, pp. 371–374.

Razavi B et al. 2007. *Clostridium difficile*: Emergence of hypervirulence and fluoroquinolone resistance. *Infect* 35: 300–307.

Reen FJ et al. 2005. Genomic comparisons of *Salmonella enterica* serovar Dublin, Agona, and Typhimurium strains recently isolated from milk filters and bovine samples from Ireland, using *Salmonella* microarray. *Appl Environ Microbiol* 71: 1616–1625.

Rice EW et al. 1999. Isolation of *Arcobacter butzleri* from ground water. *Lett Appl Microbiol* 28: 31–35.

Rivas L et al. 2004. Isolation and characterization of *Arcobacter butzleri* from meat. *Int J Food Microbiol* 91: 31–34.

Robinson DA. 1981. Infective dose of *Campylobacter jejuni* in milk. *Br Med J* 282: 1584.

Rodriguez-Lazaro D et al. 2004. Rapid quantitative detection of *Listeria monocytogenes* in meat products by real-time PCR. *Appl Environ Microbiol* 70: 6299–6301.

Rodriguez-Placios A et al. 2007. *Clostridium difficile* in retail ground meat, Canada. *Emerg Infect Dis* 13: 485–487.

Rodriguez-Placios A et al. 2009. Possible seasonality of *Clostridium difficile* in retail meat, Canada. *Emerg Infect Dis* 15: 802–805.

Rodriguez-Siek KE et al. 2005a. Comparison of *Escherichia coli* isolates implicated in human urinary tract infection and avian colibacillosis. *Microbiol* 151: 2097–2110.

Rodriguez-Siek Giddings CW et al. 2005b. Characterizing the APEC pathotype. *Vet Res* 36: 241–256.

Ruan CM et al. 2004. A staphylococcal enterotoxin B magnetoelastic immunosensor. *Biosens Bioelectron* 20: 585–591.

Rupnik M. 2007. Is *Clostridium difficile* –associated infection a potentially zoonotic and foodborne disease? *Clin Microbiol Infect* 13: 457–459.

Rupnik M et al. 2005. Revised nomenclature of *Clostridium difficile* toxins and associated genes. *J Med Microbiol* 54: 113–117.

Saad SMI et al. 1995. Motile *Aeromonas* spp. In retail vegetables from Sao Paulo, Bazil. *Revista de Microbiologica* 26: 22–27.

Schmilovitch Z et al. 2005. Detection of bacteria with low-resolution Raman spectroscopy. *Transactions of the ASABE* 48: 1843–1850.

Schraff RL. 2010. Health-related costs from foodborne illness in the United States. Avialable at http://www.producesafetyproject.org/admin/assets/files/Health-Related-Foodborne-Illness-Costs-Report.pdf-1.pdf.

Sechi LA et al. 2004. *Mycobacterium avium sub. Paratuberculosis* in tissue samples of Crohn's disease patients. *New Microbiol* 27: 75–77.

See KC et al. 2007. *Enterobacter sakazakii* bacteremia with multiple splenic abscesses in a 75 year old woman : a case report. *Age Ageing* 36: 595–596.

Shafran I et al. 2000. Rifabutin and macrolide antibiotic treatment in Crohn's patients identified serologically positive for *Mycobacterium avium ss. paratuberculosis*. *Gastroenterology* 18: A762.

Simmons BP et al. 1989. *Enterobacter sakazakii* infections in neonates associated with intrinsic contamination of a powdered infant formula. *Infect Control Hosp Epidemiol* 10: 398–401.

Simpson JM, Lim DV. 2005. Rapid PCR confirmation of *E. coli* O157:H7 after evanescent wave fiber optic biosensor detection. *Biosens Bioelectron* 21: 881–887.

Singamsetty VK et al. 2008. Outer membrane protein A expression in *Enterobacter sakazakii* is required to induce microtubule condensation in human brain microvascular endothelial cells for invasion. *Microb Pathog* 45: 181–191.

Skyberg JA et al. 2003. Characterizing avian *Escherichia coli* isolates with multiplex polymerase chain reaction. *Avian Dis* 47: 1441–1447.

Skyberg JA et al. 2006. Virulence genotyping of *Salmonella* spp. with multiplex PCR. *Avian Dis* 50: 77–81.

Smith JL. 1994. Arthritis and foodborne bacteria. *J Food Prot* 57: 935–941.

Snelling WJ et al. 2006. Under the microscope: *Arcobacter*. *Lett Appl Microbiol* 42: 7–14.

Snijers JMA, Collins JD. 2004. Research update on major pathogens associated with the processing of pork and pork products. In: JM Smulders, JD Collins (eds.) *Safety Assurance During Food Processing*. Wageningen Academic, Wageningen, pp. 99–113.

Sofos JN. 2008. Challenges to meat safety in the 21st century. *Meat Science* 78: 3–13.

Songer JG. 2004. The emergence of *Clostridium difficile* as a pathogen of food animals. *Anim Health Res Rev* 5: 321–326.

Songer JG. 2010. Clostridia as agents of zoonotic disease. *Vet Microbiol* 140: 399–404.

Songer JG et al. 2009. *Clostridium difficile* in retail meat products, USA, 2007. *Emerg Infect Dis* 15: 819–821.

Squirrel DJ et al. 2002. Rapid and specific detection of bacteria using bioluminescence *Anal Chim Acta* 457: 109–114.

Stampi S et al. 1999. Occurrence, removal and seasonal variation of thermophilic Campylobacters and *Arcobacter* in sewage sludge. *Zentralbl Hyg Umweltmed* 202: 19–27.

Strachan NJC, Gray DI. 1995. A rapid general-method for the identification of PCR products using a fibreoptic biosensor and its application to the detection of *Listeria*. *Lett Appl Microbiol* 21: 5–9.

Su L-H et al. 2003. In vivo acquisition of ceftriaxone resistance in *Salmonella* enteric serotype Anatum. *Antimicrob Agents Chemother* 47: 563–567.

Su X-L, Li Y. 2004. A self-assembled monolayer-based piezoelectric immunosensor for rapid detection of *Escherichia coli* O157:H7. *Biosens Bioelectron* 19: 563–574.

Subramanian A et al. 2006. A mixed self-assembled monolayer-based surface plasmon immunosensor for detection of *E. coli* O157:H7. *Biosens Bioelectron* 21: 998–1006.

Swaminathan B, Gerner-Smidt P. 2007. The epidemiology of human listeriosis. *Microbes Infect* 9: 1236–1243.

Takahashi T et al. 2000. A new rapid technique for detection of microorganisms using bioluminescence and fluorescence microscope method. *J Biosci Bioeng* 89: 509–513.

Tallury P et al. 2010. Nanobiomaging and sensing of infectious diseases. *Adv Drug Deliv Rev* 18: 424–437.

Taylor AD et al. 2006. Quantitative and simultaneous detection of four foodborne bacterial pathogens with a multichannel SPR sensor. *Biosens Bioelectron* 22: 752–758.

Taylor DE, Tracz DM. 2005. Mechanisms of antimicrobial resistance in *Campylobacter*. In: JM Ketley, ME Konkel (eds.) *Campylobacter, Molecular and Cellular Biology*. Horizon Bioscience, Norfolk, pp. 193–204.

Teale, C.J. 2002. Antimicrobial resistance and the food chain. *J Appl Microbiol* 92: 85S-89S.

Threlfall, E.J. 2000. Epidemic *Salmonella* typhimurium DT104 – a truly international multiresistant clone. *J Antimicrob Chemother* 46: 7–10.

Townsend S et al. 2007. The presence of endotoxin in powdered infant formula milk and the influence of endotoxin and *Enterobacter sakazakii* on bacterial translocation in the infant rat. *Food Microbiol* 24: 67–74.

Travers K, Barza M. 2002. Morbidity of infections caused by antimicrobial-resistant bacteria. *Clin Infect Dis* 34: S131–S134.

Tsai GJ, Chen TC. 1996. Incidence and toxigenicity of *Aeromonas hydrophila* in seafood. *Int J Food Microbiol* 31: 121–131.

Tu SI et al. 2000. Detection of *Escherichia coli* O157:H7 using imunomagentic capture and luciferin-luciferase ATP measurement. *Food Res Int* 33: 375–380.

Van Acker J et al. 2001. Outbreak of necrotizing enterocolitis associated with *Enterobacter sakazakii* in powdered milk formula. *J Clin Microbiol* 39: 293–297.

Van Driessche E et al. 2003. Isolation of *Arcobacter* species from animal feces. *FEMS Microbiol Lett* 229: 243–248.

Van Driessche, E et al. 2004. Occurrence and strain diversity of *Arcobacter* species isolated from healthy Belgian pigs. *Res Microbiol* 155: 662–666.

Vandamme P. 2000. Taxnomy of the family Campylobacteriaceae. In: I Nachamkin, MJ Blaser (eds.) *Campylobacter*, 2nd edn. ASM Press, Washington, DC, pp. 3–26.

Vandamme P, De Ley, J. 1991. Proposal for a new family Campylobacteraceae. *Int J Syst Bacteriol*. 41: 451–455.

Vandamme P et al. 1991. Revision of *Campylobacter*, *Helicobacter* and *Wolinella* taxonomy:emendation of generic descriptions and proposal of *Arcobacter gen. nov. Int J Syst Bacteriol* 41: 88–103.

Vandamme P et al. 1992a. Outbreak of recurrent abdominal cramps associated with *Arcobacter butzleri* in an Italian school. *J Clin Microbiol* 30: 2355–2337.

Vandamme P et al. 1992b. Polyphasic taxonomic study of the emended genus *Arcobacter* with *Arcobacter butzleri comb. Nov.* and *Arcobacter skirrowii sp. nov.* an aertolerant bacterium isolated from veterinary specimens. *Int J Syst Bacteriol* 42: 344–356.

Vandenberg O et al. 2004. *Arcobacter* species in humans. *Emerg Infect Dis* 10: 1863–1867.

Vaughn RD et al. 2001. Development of a quartz crystal microbalance (QCM) immunosensor for the detection of *Listeria monocytogenes*. *Enzyme Microb Technol* 29: 635–638.

Velusamy V et al. 2010. An overview of foodborne pathogen detection: In the perspective of biosensors. *Biotechnol Adv* 28: 232–254.

Vidal M et al. 2005. Single multiplex PCR assay to identify simultaneously the six categories of diarrheagenic *Escherichia coli* associated with enteric infections. *J Clin Microbiol* 43: 5362–5365.

Volokhov D et al. 2002. Identification of *Listeria* species by microarray-based assay. *J Clin Microbiol* 40: 4720–4728.

Volokhov D et al. 2003. Microarray-based identification of thermophilic *Campyloabcter jejuni*, *C. coli*, *C. lari*, and *C. upsaliensis*. *J Clin Microbiol* 41: 4071–4080.

Von Abercron SM et al. 2009. Low occurrence of *Clostridium difficile* in retail ground meat in Sweden. *J Food Prot* 72: 1732–1734.

Von Grawvenitz A. 2007. The role of *Aeromonas* in diarrhea: a review. *Infection* 2: 59–64.

Wall PG et al. 1995. Transmission of multi-resistant strains of *Salmonella typhimurium* from cattle. *Vet Rec* 136: 591–592.

Walsh C et al. 2008. Transfer of ampicillin resistance from *Salmonella* Typhimurium DT104 to *Escherichia coli* K12 in food. *Lett Appl Microbiol* 46: 210–215.

Wang L et al. 2007. Rapid and simultaneous quantitation of *Escherichia coli* O157:H7, *Salmonella*, and *Shigella* in ground beef by multiplex real-time PCR and immunomagnetic separation. *J Food Prot* 70: 1366–1372.

Warren BR et al. 2006. *Shigella* as a foodborne pathogen and current methods for detection in food. *Crit Rev Food Sci Nutr* 46: 551–567.

Waswa J, et al. 2007. Direct detection of *E. coli* O157:H7 in selected food systems by a surface Plasmon resonance biosensor. *LWT – Food Sci Technol* 40: 187–192.

Weese JS. 2010. *Clostridium difficile* in food – innocent bystander or serious threat? *Clin Microbiol Infect* 16: 3–10.

Weese JS. 2009a. *Clostridium difficile* in food-innocent bystander or serious threat? *Clin Microbiol Infect* 16: 3–10.

Weese JS et al. 2009b. Detection and enumeration of *Clostridium difficile* spores in retail beef and pork. *Appl Environ Microbiol* 75: 5009–5011.

Weese JS et al. 2010. Detection and Characterization of *Clostridium difficile* in retail chicken. *Lett Appl Microbiol* 50: 362–365.

Werber D et al. 2005. International outbreak of *Salmonella Oranienberg* due to German chocolate. *BMC Infect Dis* 3: 7.

Wesley IV et al. 2000. Fecal shedding of *Campylobacter* and *Arcobacter* spp. In dairy cattle. *Appl Environ Microbiol* 66: 1994–2000.

Westrell T et al. 2009. Zoonotic infections in Europe in 2007: a summary of the EFSA-ECDC annual report. *Eurosurveillance* 14: 1–3.

Woteki CE, Kineman BD. 2003. Challenge and approaches to reducing foodborne illness. *Annu Rev Nutr* 23: 315–344.

Wu FM et al. 2001. Factors influencing the detection and enumeration of *Escherichia coli* O157:H7 on alfalfa seeds. *Int J Food Microbiol* 4: 93–99.

Yadav AS, Verma SS. 1998. Occurrence of enterotoxigenic *Aeromonas* in poultry eggs and meat. *J Food Sci Technol India* 35: 169–170.

Yan J-J et al. 2005. Cephalosporin and ciprofloxacin resistance in *Salmonella*, Taiwan. *Emerg Infect Dis* 11: 947–950.

Yoder S et al. 1999. PCR comparison of *Mycobacterium avium* isolates obtained from patients and foods. *Appl Environ Microbiol* 65: 2650–2653.

Yoshida C et al. 2007. Methodologies towards the development of an oligonucleotide microarray for determination of *Salmonella* serotypes. *J Microbiol Methods* 70: 261–271.

Younts S et al. 2002. Differentiation of *Escherichia coli* O157:H7 from non O157:H7 *E. coli* serotypes using a gas sensor-based, computer-controlled detection system. *Transactions of the ASAE* 45: 1681–1685.

Younts S et al. 2003. Experimental use of a gas sensor-based instrument for differentiation of *Escherichia coli* O157:H7 from non O157:H7 *Escherichia coli* field isolates. *J Food Prot* 66: 1455–1458.

Yu CX et al. 2004. Spectroscopic differentiation and quantification of microorganisms in apple juice. *J Food Sci* 69: S268–S272.

Yuan Y et al. 2009. Universal primer-multiplex PCR approach for simultaneous detection of *Escherichia coli*, *Listeria monocytogenes* and *Salmonella* spp. in food samples. *J Food Sci* 74: M446–M452.

Zhang Q, Plummer P. 2008. Mechanisms of antibiotic resistance in *Campylobacter*. In: I Nachamkin, et al. (eds.) *Campylobacter*, 3rd edn. ASM Press, Washington, DC, pp. 263–276.

Zhang W et al. 2008. A new immuno-PCR assay for the detection of low concentration of shiga toxin 2 and its variants. *J Clin Microbiol* 46: 1292–1297.

Zhang Y et al. 2007. Genome evolution in major *Escherichia coli* O157:H7 lineages. *BMC Genomics* 16: 121.

45
Biosensors for Sensitive Detection of Agricultural Contaminants, Pathogens and Food-Borne Toxins

Barry Byrne, Edwina Stack, and Richard O'Kennedy

Abstract: Immunosensors permit the rapid and sensitive analysis of a range of analytes. Here, we provide a critical assessment of how such formats can be implemented, with emphasis on the detection of bacterial and fungal pathogens, agricultural contaminants (e.g. pesticides and herbicides) and toxins.

INTRODUCTION

The monitoring of quality of food destined for human consumption is a key consideration for farmers, the food industry, legislators and, most importantly, for consumers (Karlsson 2004). Hence, it is an absolute necessity to ensure that any contaminants that may have a deleterious effect on human health are monitored qualitatively and quantitatively in a sensitive and reliable manner. For example, herbicides and pesticides are extremely effective at suppressing the growth of plant and insect populations on agricultural produce, such as tomatoes and strawberries. However, the extensive use of such potentially toxic compounds may compromise the quality of the product, and prolonged exposure may manifest itself as chronic toxicity in human hosts (Keay and McNeil 1998). Furthermore, bacterial strains such as *Salmonella typhimurium* and *Listeria monocytogenes*, which are causative agents of salmonellosis and listeriosis, respectively, can act as opportunistic pathogens and cause death through the ingestion of contaminated produce. Consequently, the development of suitable methods for their rapid detection is an absolute necessity. Finally, there is also an urgent need to accurately monitor the distribution of toxins, including fungal (e.g. mycotoxins) and water-borne toxins (e.g. phycotoxins), which cause severe illness through the consumption of contaminated food (e.g. nuts, shellfish meat). In summary, rapid, sensitive and accurate methodologies are essential for the evaluation of product quality and for satisfying legislative requirements.

CONTAMINANT MONITORING

TRADITIONAL MEANS OF ASSESSMENT

There are several standard methods that are currently used to monitor the quality of food. As an example, fruit and vegetable produce may be inspected by monitoring the colour, gloss, firmness, shape and size of the product, as well as noting the presence or absence of visible defects. This visual inspection may be performed alongside more invasive methods, including the analysis of the soluble solid content of the product and determining the acidity, and is particularly useful for produce such as apples, pears and berries. The main advantage of such tests relates to the fact that they may be carried out immediately post-harvest

Food Biochemistry and Food Processing, Second Edition. Edited by Benjamin K. Simpson, Leo M.L. Nollet, Fidel Toldrá, Soottawat Benjakul, Gopinadhan Paliyath and Y.H. Hui.

by the farmer, and at a minimal cost (Mitchum et al. 1996). As further examples, grain and nuts may be inspected for the presence of fungal contamination, while bacterial spoilage may be indicated by the presence of an uncharacteristically strong odour, such as in coleslaw and milk. However, these tests are not sufficient for providing confirmation to the consumer that the product satisfies regulations with respect to acceptable maximum residue limits (MRLs). Furthermore, it is not possible to provide accurate quantitative or qualitative analysis of individual contaminants through these methodologies, including those that cannot be seen by visual inspection (e.g. mycotoxins). Hence, it is common practice for food samples to be removed and sent to an external laboratory where comprehensive in situ analysis may be performed (Giraudi and Baggiani 1994).

Instrumentation-Based Analysis

There is a selection of different methodologies available for quality determination, and some of the more relevant examples are discussed in this section. Product firmness can be determined through the implementation of the Magness-Taylor test, namely a destructive method that assesses the maximum force required to perforate the product in a specific way (Abbott 2004). This has been applied for the analysis of fruit, including pears (Gómez et al. 2005). The non-destructive determination of elasticity may also be permitted through the measurement of acoustic responses, with the signal being interrogated using fast Fourier transform-based analysis. Shmulevich et al. (2003) demonstrated the efficacy of this approach for evaluating the firmness of apples, and monitored product softening over time in a controlled atmosphere environment. While these methods are suitable for monitoring the structural properties of the product in question, they do not permit rigorous quality evaluation.

More suitable analytical methods include near-infrared (IR) spectroscopy, implemented by Berardo et al. (2005) for the detection of mycotoxigenic fungi and associated toxic metabolites, and scanning electron microscopy, applied for the inspection of ultrastructural changes of the epicuticular layer of oranges treated with fludioxonil, a pesticide used to control the growth of two *Penicillium* species (*Penicillium digitatum* and *Penicillium italicum*) (Schirra et al. 2005). Furthermore, gas chromatography (GS) or mass spectrometry (MS) and high-performance liquid chromatography (HPLC) are accurate and highly sensitive methods for the detection of an array of contaminants, including pesticide, herbicide and toxins residues. The latter method may also be used to detect the presence of indicator molecules that are representative of product freshness, including flavonoids (MacLean et al. 2006), while liquid chromatography coupled with mass spectrometry (LC-MS) can accurately monitor product bitterness. This was demonstrated by Dourtoglou et al. (2006) for the analysis of olives (*Olea europaea*). In spite of the efficacy of these analytical platforms, the instrumentation needed to perform this analysis is expensive and bulky and may require extensive operator training. In addition, analysis times may also be extensive, as many contaminants require lengthy sample pre-treatment prior to assessment.

Here, we focus on the application of biosensor-based platforms that are rapid, sensitive and reliable and are frequently used for the detection of herbicide and pesticide residues, bacterial and fungal pathogens and toxins (fungal and water-borne).

BIOSENSORS

A biosensor can be defined as an analytical device that incorporates a biological element for promoting biorecognition of an analyte of interest (e.g. herbicide, bacterial cell or toxin). A schematic representation of a biosensor, illustrating the three main components of the system, namely a bioligand, a transducer and a readout device, is shown in Figure 45.1. Biosensor-based platforms can use a wide selection of different recognition elements, including nucleic acid probes, lectins and antibodies (Table 45.1). The focus in this chapter will be placed on the use of antibodies for detection of contaminants that are of interest to the food industry with examples of enzyme- and nucleic acid-based detection also provided.

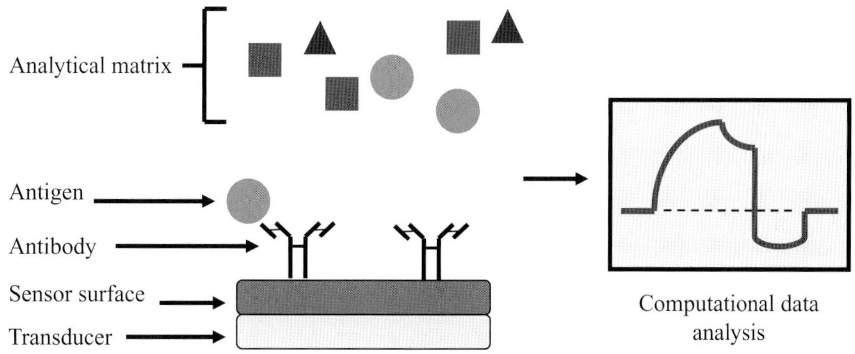

Figure 45.1. General format of a biosensor. The biorecognition element is in contact with the transducer, which converts the signal to an output shown on the computer. For illustrative purposes, an antibody-based platform is shown, with non-specific antigens represented by squares and triangles, respectively.

Table 45.1. The Recognition Elements Commonly Used in Sensor Systems

1. Antibodies and antibody fragments: Derived by enzyme digestion or genetic engineering (Fab fragment, scFv and diabody).
2. Lectins, namely carbohydrate-binding proteins. Only suitable for the detection of glycosylated entities, such as glycoproteins.
3. Enzymes: e.g. those specific for one particular substrate (i.e. horseradish peroxidase for hydrogen peroxide).
4. Cell membrane receptors.
5. A living cell: eukaryotic or prokaryotic.
6. Nucleic acid-based probes: DNA and RNA or peptide nucleic acid.
7. Aptamers.
8. Chemically generated recognition surfaces: Including plastibodies (artificial antibodies or molecularly imprinted polymers).

Fab, fragment antigen binding; scFv, single-chain variable fragment.

A transducer is a device, such as a piezoelectric crystal or photoelectric cell that converts input energy of one form into output energy of another, with the output signal generated proportional to the concentration of the target analyte (Luong et al. 1995). There are many different types of transducers that can be used in biosensor-based platforms, and a selection of these is shown in Table 45.2. Finally, a readout device typically consists of a computer-linked monitor that presents the data in a form that can easily be interpreted by the end-user.

Many different biosensor platforms have dedicated software packages that can facilitate in the interpretation of biomolecular interactions. For example, Biacore, a frequently used commercial optical biosensor based on surface plasmon resonance (SPR), which is discussed in more detail later, uses Biaevaluation software that presents data in the form of a sensorgram (Fig. 45.2). Here, interactions between immobilised and free entities can be easily visualised by monitoring changes in refractive index (RI). This correlates to a change in mass imparted by the interaction between the two binding elements (e.g. antibody and cognate antigen), with units of measurement referred to as response units (RU). In Biacore analysis, a response of 1000 RU is representative of a change in resonance angle of 0.1°, which corresponds to an alteration in the surface coverage of the sensor surface of approximately 1 ng/mm^2.

While Biacore is an excellent example of a biosensor that is commonly used for the evaluation of quality of food (which is discussed in more detail later) other biosensor platforms are also applicable that are based on electrochemical, piezoelectric, magnetic and thermal detection (Byrne et al. 2009). Many of these platforms are developed 'in-house' and have their own dedicated software packages for the interpretation of data, and key examples of these novel sensors are also described later. Finally, for a biosensor to be applicable in any field, including in the monitoring of quality of food produce, it must have certain characteristics. These are listed in Table 45.3.

Biacore

Biacore (GE Healthcare) uses an optical-based transducer system for the measurement of analytes based on the principle of SPR. SPR works on the principle of total internal reflection (TIR), a phenomenon that occurs at the interface between two non-absorbing materials, such as water and a solid. When a source of light is directed at such an interface from a medium with a higher RI to a medium of lower RI (such as light travelling through glass and water), the light is refracted to the interface

Table 45.2. Examples of Transducers Commonly Used in Biosensor Systems

Type	Example	Principle of Use
Electrochemical	Conductimetric	Solutions containing ions conduct electricity. Depending on the reaction, the change in conductance is measured.
	Potentiometric	Measurement of the potential of a cell when there is no current flowing to determine the concentration of an analyte.
	Voltammetric	A changing potential is applied to a system and the resulting change in current is measured.
Field effect transistor-based	Field effect transistor	A current flows along a semi-conductor from a source gate to a drain. A small change in gate voltage can cause a large variation in the current from the source to the drain.
Optical	Surface plasmon resonance	Surface plasmon resonance (a detailed explanation is given in this chapter).
Thermal	Calorimetry	Heat exchange is detected by thermistors and related to the rate of a reaction.
Surface acoustic wave	Rayleigh surface wave	An immobilised sample on the surface of a crystal affects the transmission of a wave to a detector.
Piezoelectric	Electrochemical quartz crystal microbalance	A vibrating crystal generates current that is affected by a material adsorbed onto its surface.

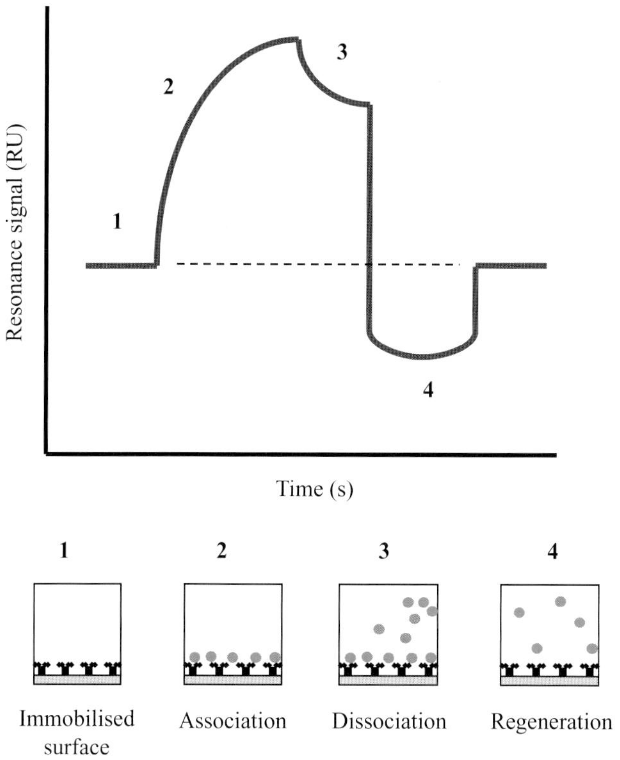

Figure 45.2. Schematic representation of a Biacore sensorgram, which provides information relating to the biomolecular interaction between the immobilised antibody and its cognate antigen (relevant analyte).

(Markey 2000). However, when the light is above a particular angle of incidence, no light is refracted across the interface and TIR occurs. Even though the incident light is reflected back from the interface, an electromagnetic field (called an evanescent wave) penetrates a distance of the order of one wavelength travelling into the less optically dense medium.

A key advantage of SPR over other optical methods of detection is that SPR measures the interaction between immobilised molecules on a surface (e.g. an antigen, or an antibody immobilised onto a sensor chip) and the corresponding ligand in solution passing over this matrix. This means that the reaction can be measured in a coloured solution or in a turbid complex matrix, such as in fruit juice or vegetable preparations. Hence, there is no requirement for food samples to be pre-treated to remove contaminating coloured species prior to analysis.

The Biacore system has a lower limit of detection (LOD) of approximately 10 RU (which is approximately 10 pg/mm^2), which is suitable for the detection of trace amounts of analytes of interest, such as pesticide or herbicide residues. There are several platforms available for use. The Biacore Q instrument has major potential for quality control analysis, whereas the Biacore 3000 is particularly useful for high sensitivity antibody-based analyte detection and kinetic analysis. The T100 and the more recently developed A100, permit high-throughput analysis of protein–protein and protein–ligand interactions that may be useful in the judicious selection of antibodies for specific diagnostic applications.

Table 45.3. The Ideal Characteristics of a Biosensor

1.	The biorecognition element must be highly specific for the substrate or antigen.
2.	Re-usable, with multiple readings permitted on a single device.
3.	High sensitivity.
4.	Cost-effective.
5.	Permit rapid or 'real-time' analysis of biomolecular interactions.
6.	Use of 'on-line' or 'in-situ' measurement.
7.	Good signal to noise ratio.
8.	The device should be robust.
9.	The ability to measure samples in a high throughput fashion, if required.
10.	Fast turnaround time on analysis.
11.	Ease of use.

Sensor Surfaces

When selecting a biosensor-based platform, such as Biacore, for the evaluation of quality of food produce, several key considerations should be made. The first of these relates to the nature of the analytical surface that will be used in the assay. One of the primary reasons why Biacore is commonly selected for food analysis (aside from the excellent sensitivity and flexibility for designing a suitable assay) relates to the fact that many different surface chemistries are available for conjugating molecules onto the sensor chip surface to perform the assay. Biacore uses sensor chips, such as those mentioned in Table 45.4, and the choice of chip chemistry is dependent on a number of different factors, including the desired application of the assay, the physical and chemical properties of the binding partners and the nature of the biomolecular interaction.

The CM5 chip is the most versatile Biacore chip currently available and is the most frequently used for food analysis. Its matrix consists of a completely modified carboxy-methylated dextran covalently attached to a gold surface, and is ideally suited for the analysis of a variety of different binding events, ranging from those involving small organic molecules to multi-domain proteins. Analytes of interest can readily be coupled to the sensor chip surface through the application of a number of different cross-linking chemistries (e.g. amine, thiol or aldehyde coupling), and a stable surface can be produced that permits accurate and repeated analysis on a single surface (Dillon et al. 2005) as non-covalently bound entities can readily be removed from the surface by a process termed regeneration. Typically, this is achieved with low concentrations of acid (HCl) or base (NaOH). As an example, for the detection of a bacterial strain in a complex sample matrix (e.g. coleslaw), a pathogen-specific antibody may be immobilised on a CM5 surface by amine

Table 45.4. The Surface Chemistries of Available Biacore Chips

Chip Type	Modification Type	Applications
CM5	100% carboxylation of dextran surface.	General use. Routinely selected for antibody-antigen and protein-protein interaction analysis.
CM4	30% carboxylation of dextran surface.	Serum, cell extracts.
CM3	100% carboxylation of dextran surface.	Serum, cell extracts.
C1	100% carboxylation of dextran surface.	Permits binding events to occur closer to the sensor surface, which is advantageous in situations where multivalent interactions occur, or where analytes are large.
L1	Lipophilic.	Lipid capturing.
SA	Streptavidin surface.	Detection of biotin-containing molecules. Useful for immobilisation.
NTA	Nickel – nitrilotriacetic acid.	Detection of histidine-tagged molecules.
HPA	Flat hydrophobic surface.	Used for membrane-associated interactions.
Au and SIA	None.	Useful for self-assembled monolayer-based interactions analysis.

coupling. This covalently captured immunoglobulin may then be used as a bioligand to capture the pathogen from the food sample, which is injected over the sensor surface and, therefore, is in solution. The change in mass resulting from the biomolecular interaction introduces a change in RI, as seen by an increase in RU, and regeneration can subsequently be used to liberate the bacterial cell(s), so that the immobilised antibody is available for subsequent analysis. This method of analysis can be used for detection purposes and can be used to establish LODs. Furthermore, where required, kinetic analysis can be used to determine the affinity of the antibody for its cognate antigen.

Assay Configuration

For food-based biosensor analysis, analytes of interest range in size from large (intact bacterial and fungal cells) to small (pesticide and toxin residues), requiring that the assay format selected must be capable of providing accurate and quantitative detection. Biacore-based assays that monitor interactions between large biomolecules, such as antibodies and proteinaceous antigens, can be developed by immobilising either the antibody or antigen on the surface, as the mass change introduced by the binding of the other entity is sufficient to cause a recordable change in RU that can be seen on a sensorgram. However, as contaminants such as herbicides, pesticides and toxins have low molecular weights, it is often necessary to employ an indirect measurement method where the analyte is immobilised directly on the sensor surface, and the larger of the binding elements is subsequently introduced. The resultant change in mass introduced by the binding of the larger element (in solution) to the smaller, immobilised ligand can therefore be easily seen as a change in RU. If the assay was to be performed in reverse (e.g. the larger entity is immobilised and the small hapten or toxin is free in solution), a small change in mass would be introduced during an interaction event, which may be difficult to detect and quantify reliably.

Many of the examples that are discussed in this chapter employ antibody-based competition or inhibition assay formats (Fig. 45.3). An inhibition assay involves the combination of the sample of interest with the specific antibody before injection onto a biosensor chip containing an immobilised target molecule. There is competition between the immobilised and free antigen (from the sample to be analysed) for antibody binding. A change in signal (e.g. RU) is recorded, and this is inversely proportional to the amount of target analyte that remains free in solution. An alternative method to detect analytes of interest requires a competitive assay format. In this format, the antibody, specific to the analyte of interest, is immobilised on the surface. The sample, containing the analyte to be determined, is mixed with a known concentration of standard consisting of the target analyte that has been conjugated to a large carrier protein. This results in the analyte and the conjugated standard competing for the immobilised antibody on the surface of the biochip. An increase in signal is caused by the binding of the large analyte-carrier conjugate and the data generated are similar to the inhibition assay since the signal recorded is inversely proportional to the amount of target analyte present in the sample. However, to perform these assays, it is necessary to have a suitable antibody, and methods for antibody production are discussed in Section 'Antibody Production Strategies'.

ANTIBODIES

Antibodies are the key biorecognition elements of the immune system. A large variety of antibodies and antibody-derived fragments have been produced with the capability of detecting an array of structurally diverse analytes, ranging from proteins to haptens, and these have been implemented in a number of different biosensor formats. Antibodies are globular glycoproteins (sugar-containing proteins), and five main classes (or serotypes) exist in nature, namely IgA, IgM, IgE, IgD and IgG. Single antibody molecules typically have molecular weights of approximately 150–200 kilodaltons (kDa). IgG antibodies (Fig. 45.4) have a Y-shaped backbone with four polypeptide chains located in two identical chains that are covalently attached through disulphide bonds. The innermost chains are referred to as the heavy chains because they are approximately double the molecular weight of the outer arms (termed the light chains). The

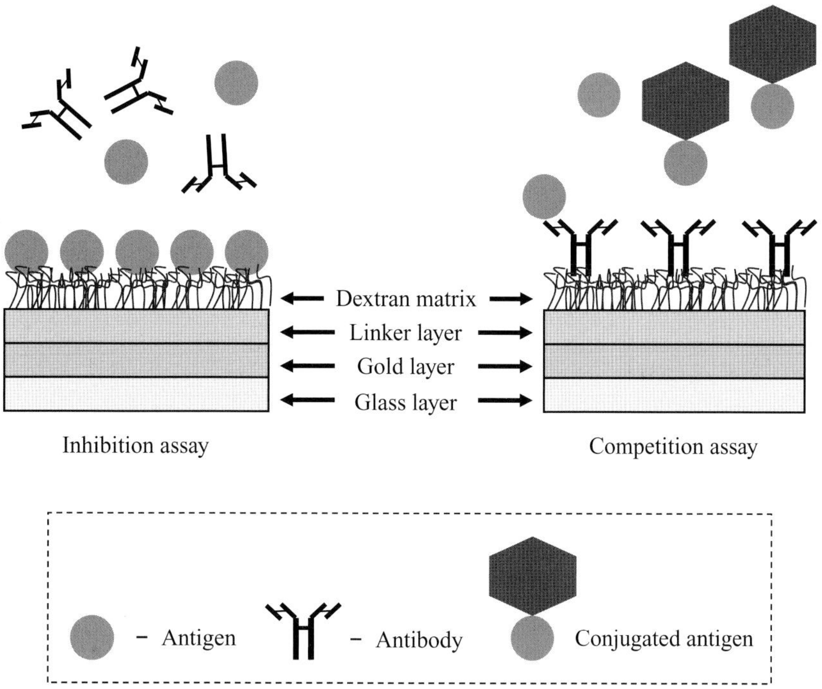

Figure 45.3. Formats for inhibition and competition assays. For illustration, a Biacore surface is represented.
Inhibition: Free analyte inhibits binding of the antibody to the immobilised analyte on the chip. The signal generated when the antibody binds to the immobilised analyte is inversely proportional to the concentration of free analyte in the sample.

Competition: Free (●) and conjugated (⬢●) analyte compete for binding to the immobilised antibody. The signal generated is inversely proportional to the amount of free analyte in the sample.

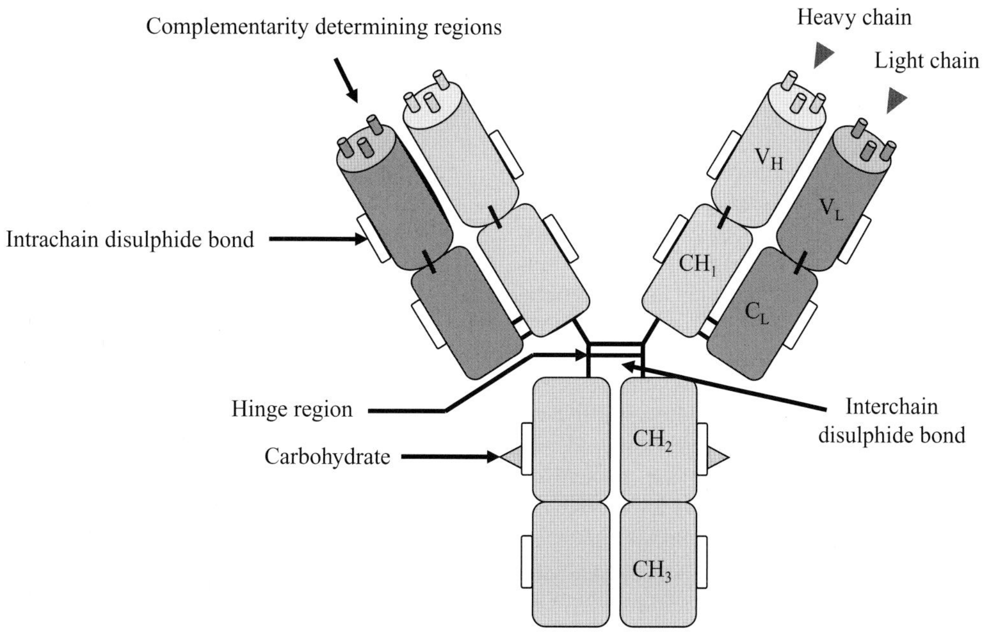

Figure 45.4. Structure of an IgG antibody. CH_{1-3} refers to constant heavy regions 1 to 3, respectively. V_H, variable heavy; V_L, variable light.

recognition sites of the antibody, which interact with an epitope on an antigen of interest, are located at the ends of the variable heavy (V_H) and variable light (V_L) regions of the heavy and light chains, respectively. They are commonly referred to as the complementarity determining regions, or CDRs. Each arm of an antibody can bind to one antigen, so one IgG molecule can theoretically bind to two antigens.

Antibody Production Strategies

There are three main methods for generating antibodies that may be used in biosensor-based platforms for quality evaluation, and these are discussed in this section. Polyclonal antibodies are produced through the immunisation of animal hosts with a particular antigen. The immunogen is typically administered in the presence of a suitable adjuvant, which elicits an immune response in the host. Small molecules, such as toxins or haptens, may have to be conjugated to larger carrier molecules, such as bovine serum albumin (BSA), to enhance immunorecognition. Serum titres are typically analysed by enzyme-linked immunosorbent assay (ELISA) to quantify the host-based response. If this is deemed to be suitable, blood samples are subsequently collected and the antibodies generated are purified from the serum. Polyclonal antibodies typically consist of a variety of different serotypes with varying affinities/specificities towards the analyte in question. Animals often used for polyclonal antibody generation include guinea pigs, rabbits, goats, sheep and donkeys (Leenaars and Hendriksen 2005).

The second method involves the use of hybridoma technology to produce monoclonal antibodies (Köhler and Milstein 1975, Nelson et al. 2000, Hudson and Souriau 2003). These are generated by immunising an animal (typically, a mouse) with the antigen of interest in the presence of a suitable adjuvant. Once a sufficient immune response is detected by ELISA, the spleen, bone marrow from long bones (femur and humerus) or primary lymphoid organs (such as lymph nodes) are removed from the sacrificed animal and the antibody-producing B-cells are harvested. These cells can then be fused to immortal myeloma cells by using an electrical current or polyethylene glycol. The resulting hybrid cells (hybridomas), which secrete antibodies that are directed towards the desired antigen, are then selected and cloned out to ensure monoclonality. The advantage of this approach is that there is a constant supply of the antibody that is required for analysis. However, there is a significant cost involved in the production and the screening of these antibodies.

Recombinant antibodies are the third form of antibodies that are increasingly being used. They are often produced in bacterial strains such as *Escherichia coli* and are expressed in a phage display format. Libraries, with the capacity to express a large number of antibodies, are generated, and they are referred to as naive, synthetic or immune depending on their mode of production (Bradbury and Marks 2004). Immune libraries are constructed though the administration of the immunogen of interest to a suitable host (e.g. mouse, rabbit, chicken), which is monitored for antibody production. Antibody-encoding nucleic acid is then purified from lymphoid organs, such as the spleen, and cloned into a phage or phagemid vector that, in turn, is

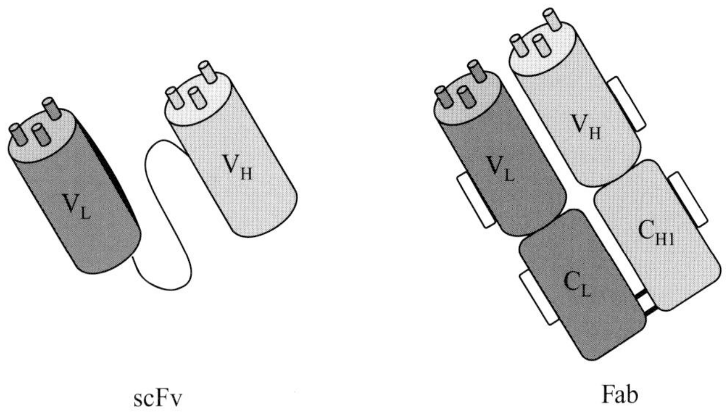

Figure 45.5. Different antibody fragments available for biosensor-based analysis. CH_{1-3} refers to constant heavy region 1; V_H, variable heavy; V_L, variable light; scFv, single-chain variable fragment; Fab, fragment antigen binding.

propagated in *E. coli*. Phage display libraries can subsequently be screened against targets of interest by biopanning. This is feasible since the phage express the active recombinant antibodies on their surface and those that bind to the specific antigen immobilised on a capture surface can be differentiated from non-binding antibodies, and characterised further to determine their affinity for the target antigen. A key advantage with recombinant antibodies is the ability to increase their affinity for an antigen of interest through site-directed mutagenesis or other approaches such as chain shuffling or error-prone PCR, which is not possible for monoclonal or polyclonal antibodies (Conroy et al. 2009, O'Kennedy et al. 2010).

The main types of antibody fragments produced by phage display are the fragment antigen binding (Fab) and single-chain variable fragment (scFv), whose structures are shown in Figure 45.5. The presence of the constant regions in a Fab is thought to aid in the stabilisation of the antibody variable regions, which might not function efficiently when expressed in the monomeric scFv format (Röthlisberger et al. 2005). It was previously shown in our laboratory that the Fab antibody format is the most reliable and sensitive for use in small molecule competition biosensor assays involving haptens. The strict monovalency of this format can lead to a significant enhancement in assay sensitivity in both ELISA and competition SPR assays (Townsend et al. 2006).

The scFv is the most widely used antibody fragment. The variable regions (V_H and V_L) of the antibody are linked by a flexible peptide linker. The most frequently used are based on glycine-serine repeat structures, with the length of the linker related to the intended valency of the molecule. When a short linker is used, the stability and folding of the scFv does not occur properly. This is caused by the insufficient juxtaposing of the V_H and V_L regions in the single chain for the monomer to function. ScFvs selected in this format invariably form bivalent dimers, or diabodies, which often have increased avidity for an antigen over the monomeric forms typically observed when long-linker systems are used (Holliger et al. 1993, Kortt et al. 1997, Atwell et al. 1999). Long linkers (ranging from 18–21 amino acids) favour the production of scFv formats, which are predominantly monomeric (Holliger et al. 1993, Perisic et al. 1994, McGuinness et al. 1996). In summary, recombinant antibodies are an excellent alternative to polyclonal and monoclonal antibodies for the detection of analytes of interest, and have great potential for application in the monitoring of quality in the food industry (Hudson and Souriau 2003, O'Kennedy et al. 2010).

Optical Immunosensors for Quality Determination

There have been several excellent examples of biosensor-based analysis for the detection of pesticides, herbicides, toxins and bacterial cells, and some of the more pertinent observations that employ antibody-based recognition (immunosensors) are discussed in this section. One of the earliest examples demonstrating the use of antibody-based biosensing for detecting herbicide residues was described by Minunni and Mascini (1993). Here, Biacore was selected as a platform to facilitate the detection of traces (50 pg/mL) of the herbicide atrazine in water samples. Another optical biosensor was developed to detect and quantify carbamate residues in vegetables. It was observed that changes in the concentration of carbamate could be monitored using chlorophenol red (Xavier et al. 2000). Moran et al. (2002) used SPR to characterise antibodies that were subsequently used to detect the presence of 2-(4-thiazolyl)benzimidazole, a molecule that is used as a food preservative and an agricultural fungicide. Schlecht et al. (2002) implemented a C1 four-channel sensor chip in a study to quantifiably detect the presence of 2,4-dichlorophenoxyacetic acid (2,4-D), an organochlorine herbicide, and monitor cross-reactivity of polyclonal antibodies with a structurally related analogue, namely 2,4,5-trichlorophenoxyacetic acid (2,4,5-T). This assay was permitted by immobilising 2,4-D analogues onto the surface of a C1 sensor chip surface through a thiol-carboxyl group reaction, and had a sensitivity of 0.1μg/mL. Finally, Caldow et al. (2005) used a Biacore Q instrument to detect the bacteriostatic antibiotic, tylosin, in bees' honey. This polyketide is active against most gram-positive bacteria, mycoplasma, and certain gram-negative bacteria. They were able to detect tylosin at the level of 2.5μg/kg in honey, demonstrating the ability of biosensor platforms, such as Biacore, to detect analytes in complex sample matrices. These five key examples demonstrate early applications of using immunosensors for the detection of low-molecular weight analytes that have a deleterious effect on the quality of agricultural produce.

More recently, miniaturised biosensor platforms have been developed for in situ analysis of pesticide and herbicide residues. A portable immunosensor for the detection of 2,4-D was described by Kim et al. (2007). Here, murine hosts were immunised with a 2,4-D-BSA conjugate, and the resultant monoclonal antibodies were implemented for detection purposes. When tested on spiked river water samples, this assay format had excellent sensitivity (0.1 ppb of 2,4-D) and permitted the parallel analysis of multiple samples. This example also demonstrates how modification of the assay format can greatly improve sensitivity, which is a key consideration for the detection of analytes, such as 2,4-D, which may reside in food or water samples in trace amounts. A sandwich assay format, incorporating a second biotinylated antibody, was subsequently developed whose sensitivity was significantly enhanced (0.1 ppt of 2,4-D).

Competitive and inhibition assay formats can be utilised effectively in a number of different biosensor formats for the detection of pesticide and herbicide residues. In some cases, this can be applied where the direct monitoring of the interaction between an antigen and its cognate antibody is not sufficiently sensitive. To illustrate this, Gouzy et al. (2009) developed a Biacore-based SPR competition assay for the detection of the herbicide isoproturon, as employing a direct detection assay was deemed to be inadequate. Here, a rat-derived anti-isoproturon monoclonal antibody was selected for biorecognition, and the competition assay had a good LOD (0.1μg/L). Salmain et al. (2008) developed an indirect competition immunoassay format for the detection of atrazine. The biosensor format implemented was an IR optical platform that had nanomolar sensitivity.

Salmain et al. (2008) subsequently performed comparative sensitivity analysis in an ELISA-based assay format, and similar observations were made. ELISA assays are routinely selected for immunodetection purposes and, depending on the quality of the antibody, have excellent sensitivity. However, a major drawback relates to the fact that analysis times are often lengthy, with multiple incubation and washing stages required for assay completion. This is in contrast to biosensor-based assays that permit rapid analysis and facilitate the detection of multiple analytes on a single sensor surface, as opposed to using multiple wells. Therefore, ELISA formats may be initially used to validate an assay format before transferring this to a biosensor platform. This has been demonstrated by Herranz et al. (2008) for the evaluation of leporine polyclonal antibodies specific for simazine derivatives prior to their implementation in an immunosensor platform. The resultant assay had a LOD of 1.3 ng/L in contaminated water samples, and results were obtained in 30 minutes.

In the biosensor assay described by Herranz et al. (2008), it was also possible to monitor for cross-reactivity with other molecules, including structurally-related triazines (propazine and atrazine), and demonstrate the absence of non-specific binding to unrelated entities, such as 2,4-D. Biosensors permit rapid cross-reactivity analysis to be performed, where multiple analytes may be tested on a single antibody-immobilised surface. Alternatively, where small analytes are to be detected, numerous structural analogues may be immobilised on different surfaces (e.g. in a CM5 sensor chip, four individual flow cells are available) and tested with a panel of antibodies. This parallel immunosensing approach was recently demonstrated by Gao et al. (2009) for the detection of atrazine and four additional chemicals, namely paraverine, 17-β-estradiol, chloramphenicol and nonylphenol. The biorecognition elements implemented during this analysis were either monoclonal (anti-atrazine, anti-17-β-estradiol and anti-chloramphenicol) or polyclonal (anti-nonylphenol and anti-paraverine) antibodies generated in murine and leporine hosts, respectively. Optical biosensor-based platforms that can sensitively detect multiple contaminants, such as herbicide or pesticide residues, suggest the way forward for

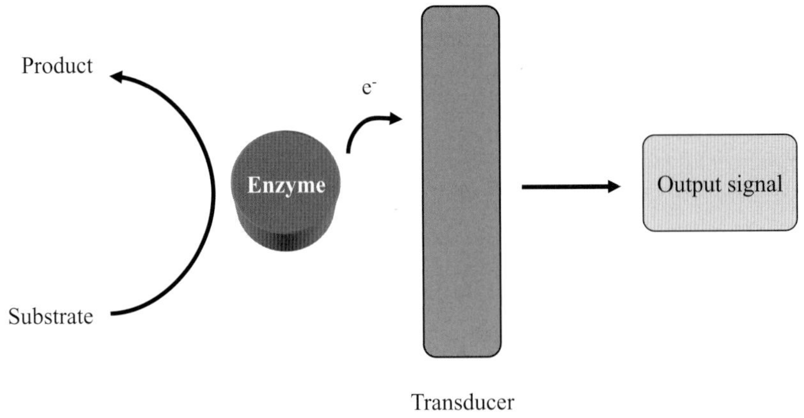

Figure 45.6. Schematic diagram of an enzyme-transducer biosensor. When the redox enzyme goes through its catalytic cycle (going from an oxidised to reduced state and back to its resting state) the redox action of the enzyme is detected by the transducer and the change in electrical state is recorded as a change in the output signal. An electron is represented by e^-.

monitoring the quality of agricultural produce. Furthermore, the possibility of translating these methodologies onto portable microdevices will permit 'on-site' analysis to be performed in a rapid, reliable and sensitive manner.

Electrochemical Sensors

Electrochemical sensors have also been used extensively to detect analytes of interest in agricultural produce. These platforms are based on four different transducer types, namely amperometric, impedimetric, potentiometric and conductimetric (Conroy et al. 2009, Byrne et al. 2009). The biorecognition element in these sensors is in direct contact with a transducer, and the resulting signal that is generated is converted from a biochemical signal to an electrical signal. Similar to optical immunosensors, antibodies and enzymes are commonly used for biorecognition purposes. Enzymes that belong to the oxidoreductase class (enzyme classification (EC) 1) are frequently selected as they alternate between oxidised and reduced states that can be measured electrochemically and, therefore, can be exploited in these analytical devices.

Equation 1 illustrates the generation of an electron through the redox cycling of an enzyme. When the enzyme is located in close proximity to the surface of the transducer, electron transfer can occur directly (as shown in Fig. 45.6). However, as is the case with many naturally occurring enzymes, they are surrounded by a layer of carbohydrate or lipid. This increases the distance that electrons have to traverse, thus, causing a decrease in the signal recorded by the transducer. In situations where this occurs, electron mediators, such as ferrocene, are used (Fig. 45.7).

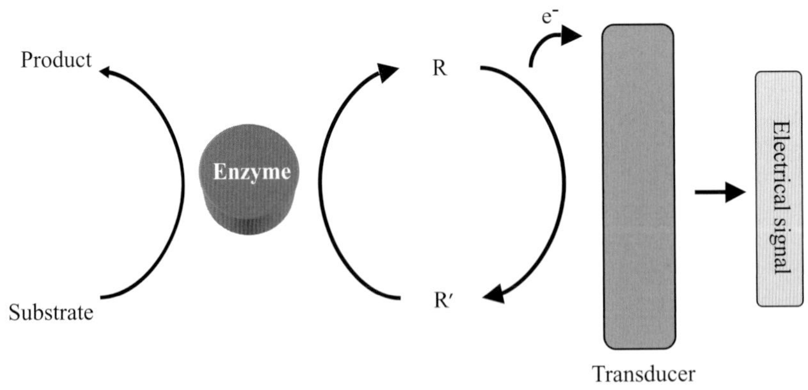

Figure 45.7. An enzyme-based electrochemical biosensor with an electron mediator. The mediator shuttles the electron (e^-) from the enzyme to the surface of the transducer where it is converted from a chemical signal to an electrical signal. R and R′ represent the oxidised and reduced forms, respectively, of an electron mediator.

Table 45.5. The Ideal Characteristics and Properties of an Electron Mediator

1.	Exhibits reversible kinetics.
2.	Reacts readily with the reduced form of an enzyme.
3.	Has a low oxidation potential and is mediator activity is pH independent.
4.	Is stable in both redox forms.
5.	Easily retained at the surface of an electrode.
6.	Unreactive towards oxygen.
7.	Chemically unreactive with the immobilised biological material.

Table 45.5 illustrates the characteristics that are favourable in choosing a specific mediator (Cassidy et al. 1998).

Mediators are usually low molecular redox couples that shuttle the electrons from the enzyme's active site to the surface of the transducer. Equation 2 illustrates the reaction of the mediator and the subsequent generation of an electron resulting electrochemical cycling.

$$H_2O_2 + 2H^+ + 2e^- \rightarrow 2H_2O \quad (1)$$

$$\text{Mediator}_{\text{red}} \rightarrow \text{Mediator}_{\text{ox}} + e^- \quad (2)$$

Equations 3, 4, 5 and 6 illustrate a peroxidase-mediated reaction on a biosensor surface. Equation 3 shows the resting state peroxidase (in its reduced form) reacting with hydrogen peroxide and two hydrogen ions to form an intermediary oxidised-peroxidase compound and two water molecules. Equation 4 shows the reaction of the intermediary oxidised-peroxidase compound with the reduced form of a mediator to produce the resting reduced-state peroxidase and an oxidised mediator. Equation 5 shows the oxidised mediator reacting with two electrons, thus, reverting to the reduced form of the mediator. It is these electrons that the transducer detects and converts into an electrical signal. The final Equation 6 shows the overall reaction, whereby one hydrogen peroxide molecule in the presence of two electrons and two hydrogen ions is converted to two water molecules (Ryan et al. 2006).

$$H_2O_2 + 2H^+ + PO_{\text{red}} \rightarrow PO_{\text{ox}} + 2H_2O \quad (3)$$

$$PO_{\text{ox}} + \text{Mediator}_{\text{red}} \rightarrow PO_{\text{red}} + \text{Mediator}_{\text{ox}} \quad (4)$$

$$\text{Mediator}_{\text{ox}} + 2e^- \rightarrow \text{Mediator}_{\text{red}} \quad (5)$$

$$H_2O_2 + 2e^- + 2H^+ \rightarrow 2H_2O \quad (6)$$

It is often possible to use a multi-enzyme electrochemical biosensor system to detect the presence and determine the concentration of a particular compound in a matrix of interest. This method employs two or more enzymes that are in proximity with each other and in contact with a mediator or directly in contact with the transducer. As illustrated in Figure 45.8, enzyme 1 (glucose oxidase (Gox)) generates hydrogen peroxide (H_2O_2) as a by-product through the catalysis of glucose by Gox. The H_2O_2 generated is catalysed by a peroxidase enzyme and the redox cycling of the mediator. The cycling of the peroxidase enzyme in the presence of H_2O_2 generates electrons that are detected by the transducer. This in turn enables the concentration of the target analyte to be determined, thereby providing quantitative determination.

There are several well-characterised electrochemical biosensor devices that have been applied in the fruit and vegetable industry for the detection of a number of structurally diverse compounds, including pesticides, herbicides, insecticides, organophosphates, organochlorines and carbamates. With reference to the detection of pesticide residues, there are many available options that can be implemented by the end user. More specifically, enzymes that are directly affected by the presence of a pesticide can be incorporated into a biosensor-based platform permitting detection (Amine et al. 2006). As an example, acetylcholinesterase (AChE) presents in muscles, red blood cells

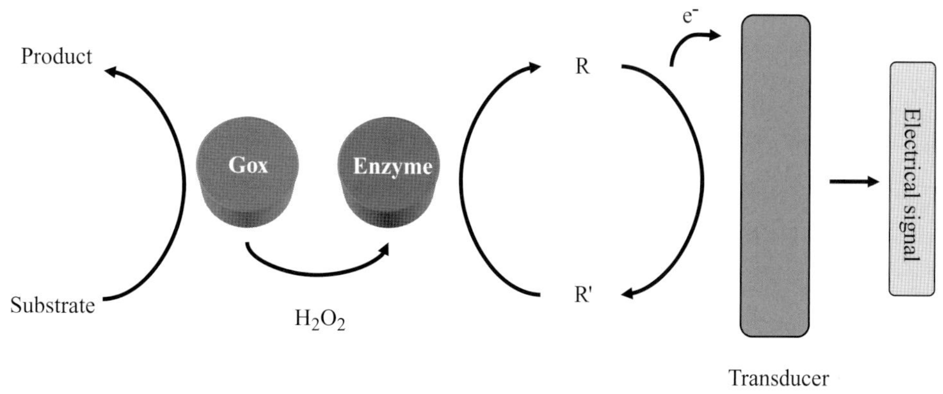

Figure 45.8. A representation of a sensor with two enzymes operating together. The first enzyme catalyses the substrate of interest and, as a by-product, H_2O_2 is generated. This is then catalysed by a peroxidase enzyme. The redox activity of the peroxidase enzyme is detected by the transducer. Gox is glucose oxidase; H_2O_2 is hydrogen peroxide; R and R′ represent the oxidised and reduced forms, respectively, of an electron mediator and e^- is an electron.

and nerve tissue, catalyses the hydrolysis of the acetylcholine, a neurotransmitter, to yield choline and acetic acid. This enzyme is associated with cognition in mammalian hosts. Compounds that effect AChE activity include organophosphates and parathion, which is routinely used as an acaricide and an insecticide. Botulinum toxin, produced by the bacterial strain *Clostridium botulinum*, also suppresses the release mechanism of AChE (Rang et al. 1998). The suppression of activity of this enzyme results in the accumulation of acetylcholine in the host, which can cause an excessive over-stimulation of the cholinergic nerves. Death usually results from the failure of the circulatory and respiratory systems (Timbrell 1991). Schulze et al. (2002) developed an amperometric AChE biosensor that was used for the detection of carbamate and organophosphate residues in a number of fruit-containing baby foods and fruit and vegetable samples taken from four different geographical locations. The protocol involved the printing of thick film electrodes onto sheets of polyvinylchloride and 'curing' for 30 minutes at 90°C prior to the introduction of AChE by glutaraldehyde coupling. The activity of AChE was analysed by monitoring the formation of thiocholine by the enzymatic hydrolysis of acetylcholine chloride. This sensor format permitted the detection of trace levels of these analytes (lower than 5 μg/kg), which included carbofuran, carbaryl and chlorpyrifos.

Wheat is a common component of food produce that should be monitored for the presence of pesticide residues. Del Carlo et al. (2005) quantified the amount of a phosphothionate insecticide (pirimiphos-methyl) present in durum wheat using an electrochemical biosensor. They used an AChE-inhibition assay and obtained a calibration curve between 25–1000 ng/mL with a detection limit of 38 ng/mL. When real samples of durum wheat were analysed, the LOD increased to 65–133 ng/mL, presumably because of the complexity of the sample matrix.

Cell-based biosensors that are also based on the inhibition of AChE are useful alternative formats for detecting chemical contaminants. Here, interactions between mammalian cells and chemicals, such as pesticides, are conducive to the disruption of cell membrane permeability, which can be quantified. To demonstrate how this assay format may be applied, Flampouri et al. (2010) recently prepared neuroblastoma and fibroblast mammalian cells by standard cell culture methodologies and immobilised these on a working electrode through entrapment in sodium alginate beads. Two chemicals of interest were selected for investigation, namely diazinon (an insecticide) and propineb (a fungicide), and were detected at low nanomolar levels on this platform through the monitoring of changes in membrane potential. Of particular interest in this study was the ability of the sensor format to subsequently differentiate between organic ($n = 5$) and pesticide-treated ($n = 8$) tomato samples that had 14 detectable pesticide residues, including diazinon and thiabendazole. The principle behind this particular assay was the monitoring of cytosolic calcium accumulation in the neuroblastoma cells, with pesticide residues inducing cell membrane depolarisation. This interesting observation demonstrates the potential of using biosensors to differentiate between organic and non-organic agricultural produce, which is of particular interest to the consumer. Furthermore, with the recent trends in biosensor miniaturisation, it is likely that future sensors will permit this differentiation to be performed 'on-site'.

Many other examples demonstrate the use of electrochemical biosensors for monitoring the presence of indicator molecules that are representative of contamination. In an early example, Voss and Galensa (2000) used an electrochemical biosensor to monitor the presence and concentration of amino acids in fruit juice. Amino acids were separated on a lithium cation-exchange HPLC column, and an amperometric platform was subsequently used to monitor the production of H_2O_2. This assay could detect amino acid concentrations ranging from 0.1 mg/mL to 5 mg/mL (D-proline and D-methionine to L-alanine, respectively) in a range of different analytical samples, including fruit juice, wine and beer. D-alanine was detected at a concentration of 0.5 mg/mL, and the authors suggested that the presence of this amino acid was a result of bacterial contamination. Kriz et al. (2002) used a SIRE (sensors based on injection of the recognition element)-based biosensor to monitor L-lactate content in baby food and tomato paste, whereby a small amount of enzyme (lactate oxidase) was injected into an internal delivery flow system and was held in direct spatial contact with an amperometric transducer by a semi-permeable membrane. Measurements were determined by the enzymatic conversion of L-lactate to pyruvate and H_2O_2, and the range of detection of L-lactate was between 0.10 and 2.51 mM. Here, all assay measurements were compared to an established spectrophotometric assay, with the contributors stating that the biosensor method employed during this analysis had a distinct advantage, as the measurement could be performed in less than 3 minutes (as opposed to 30–35 minutes for spectrophotometry-based analysis).

Additional examples of electrochemical sensor-based detection of contaminants include the development of an immunosensor for the detection of 2,4-D based on electrochemical impedance spectroscopy (Navrátilová and Skádal 2004) and, more recently, a sensor implementing a molecularly-imprinted conducting polymer, namely poly(3,4-ethylenedioxythiopene-co-thiopene-acetic acid), for the detection of atrazine (Pardieu et al. 2009). These platforms had sensitivities of 45 nmol/L and 10^{-7} mol/L for 2,4-D and atrazine, respectively, and demonstrate the efficacy of using electrochemical detection as an alternative to optical platforms, such as Biacore, for quality determination.

In recent years, there is a steadily increasing trend for the development of biosensors in an array format, where multiple analytes can be measured simultaneously on a single device. Examples of optical platforms that permit this analysis have been discussed earlier by Herranz et al. (2008) and Gao et al. (2009), and many electrochemical platforms are also available. An enzyme-based three-electrode biosensor for detecting the presence of markers of maturity and quality in tropical fruits was developed by Jawaheer et al. (2003). The assay was developed to quantifiably determine the presence of β-D-glucose, total D-glucose, sucrose and ascorbic acid in pineapples, mangos and papaya fruit. These markers are indicative of maturity and quality in selected fruit products and were detected in pectin (a natural polysaccharide present in plant cells) extracted from these samples. A fabrication format was developed that permitted the integration of the individual sensors into a multi-sensor array, and analytes were measured in this matrix to enhance the enzymatic responses over analyte ranges of 0–7 mM. Interferences normally related to electrochemically active compounds present in fruits were minimised by including a membrane made out of cellulose acetate.

Electrochemical assay formats may be further enhanced by electrochemiluminescence (ECL). ECL reactions are of interest because of their versatility for a range of different types of immunoassay. The principle of ECL is the co-oxidation of luminol and a substrate (called an enhancer) by H_2O_2 in the presence of the enzyme horseradish peroxidase (HRP). The resulting amperometric signal is detected by the transducer and a quantifiable electrical signal is generated. ECL as a detection system has advantages over other methods, including high sensitivity and a reduced assay time. In experiments carried out by Rubtsova et al. (1998), specific antibodies against atrazine were covalently immobilised on photo-activated nylon. A chemiluminescence-based assay was subsequently used to measure 2,4-D with a detection limit of 0.2 μg/L.

Portable electrochemical biosensors are also applicable for the detection of herbicide and pesticide residues. A portable and disposable immunomembrane-based electrochemical biosensor was developed for the detection of picloram in spiked lettuce, rice and water samples, and the latter extracted from paddy fields. The competitive assay format developed here by Tang et al. (2008) utilised a polyclonal anti-picloram antibody, purified from serum extracted from immunised rabbits, and had a sensitivity of 5 ng/mL.

Alternative Biosensor Formats

Other types of biosensors have also been used for post-harvest analysis. Pogačnik and Franko (2003) developed a photothermal biosensor to measure organophosphate and carbamate compounds in salads, lettuce and onions. This approach used thermal lens spectrometry, a technique that depends upon the adsorption of optical radiation in the sample generating heat. This introduces a change in the RI, and, through this analysis, concentrations can be determined. Paraoxon was detected in all of the samples tested by this protocol. Recently, a piezoelectric antibody-based sensor for the detection of pesticide residues in fruit juice matrices was described by March et al. (2009). Piezoelectric, or mass-based, sensors operate on the principle that a biorecognition event, such as the interaction between an antibody and an agricultural contaminant, results in a change in mass. This is conducive to a change in, for example, resonance frequency that can readily be detected by the end-user. Piezoelectric sensors typically use quartz crystals as transducers, and are a cost-effective alternative to optical sensors, which are often expensive (Byrne et al. 2009, Conroy et al. 2009). The immunosensor described by March et al. was based on the implementation of a quartz-crystal microbalance for detecting carbaryl and 3,5,6-trichloro-2-pyridinol (TCP) residues. Here, a rapid (20 minute) inhibition assay format was adopted, using antigen-specific monoclonal antibodies for biorecognition, and the piezoelectric sensor had good detection limits for carbaryl (11 μg/L) and TCP (7 μg/L) in spiked fruit juice samples, although the authors suggested that assay sensitivity could be improved by incorporating crystals with enhanced resonance frequencies. However, this example does emphasise the efficiency of using biosensors for detecting analytes of interest in complex and coloured sample matrices without the need for sample pre-preparation, which significantly reduces analysis times.

PATHOGENS

Bacterial Pathogens

Several bacterial strains are of great significance for the food and horticultural industries, including *L. monocytogenes*, *S. typhimurium* and *E. coli* O157:H7, and Table 45.6 show some of the earlier examples of using SPR instruments to detect the presence of bacterial contaminants in food. The bacterial strain *E. coli* 0157:H7 is a major pathogen of interest in fruit and vegetables, as it causes severe illness and can be fatal in the infants, the elderly and the immunocompromised (Byrne et al. 2009). Hence, the detection and enumeration of this and other bacterial pathogens are absolute requirements for the food industry. A disposable conductometric electrochemical immunosensor, based on a lateral flow strip connected to an ohmmeter, was described by Muhammad-Tahir and Alocilja (2004). Anti-*E. coli* 0157:H7 antibodies, labelled with polyaniline (PANI), were immobilised onto the nitrocellulose strip and a sample was allowed to migrate up the strip. PANI is an excellent conducting polymer frequently selected for use in chemical sensor platforms. A drop in resistance, proportional to the concentration of *E. coli* 0157:H7 cells binding to the antibodies, indicated a decrease in the electron transfer from the PANI-conjugated antibody.

L. monocytogenes is a highly infectious pathogen that has very serious implications if present in food or food products due to the

Table 45.6. Examples of Surface Plasmon Resonance-Based Analysis of Bacterial Pathogens

Analyte	Limit of Detection	Reference
Salmonella enteritidis, *Listeria monocytogenes*	106 cells/mL	Koubová et al. 2001.
Salmonella groups B, D and E	1.7 × 103 CFU/mL	Bokken et al. 2003.
Staphylococcal enterotoxin B	0.5 ng/mL	Homola 2003.
Staphylococcal enterotoxin B	1.0 ng/mL	Nedelkov et al. 2000.
Salmonella typhimurium	10^2–10^9 CFU/mL	Oh et al. 2004.

CFU, colony forming unit.

associated risk of fatality (Hearty et al. 2006). While the natural ecosystem of this bacterial strain includes soil, water, plant material and decaying plant detritus (Suihko et al. 2002), numerous foods, such as raw vegetables, fruits, and horticultural samples, are also prone to infection. Furthermore, the psychrophilic nature of this strain, conferring the ability to grow at refrigeration temperatures, and the ability of this bacterium to withstand high salt concentrations and tolerate a wide pH range suggest that contamination of foodstuffs such as coleslaw is a frequent occurrence. Leonard et al. (2005) devised a rapid SPR-based immunosensor assay for the detection of *L. monocytogenes*, which incorporated a polyclonal antibody specific for internalins. Internalin B (InlB), a protein on the surface of *L. monocytogenes*, participates with Internalin A (InlA) in the invasion of mammalian cells. In this assay format, the antibody was immobilised on a CM5 sensor chip and varying concentrations of cells were then injected over the surface. A detection limit level of less than 2×10^5 cells/mL of sample was reported. Hearty et al. (2006) generated a monoclonal antibody that specifically interacted with the InlA surface protein, and demonstrated the efficacy of this antibody by developing a Biacore-based assay capable of detecting 1×10^7 cells/mL. Finally, Tully et al. (2006) described the use of quantum dot-labelled antibodies, specific for InlA, for the immunostaining of *L. monocytogenes* cells, and demonstrated that these have major potential for use with fluorescence-based sensor formats. These three examples demonstrate the efficacy of detecting bacterial pathogens through the identification of cell surface epitopes and subsequent generation of antibodies. Bacterial cells, in particular, present an array of different antigenic determinants, including capsular, flagellar and surface antigens. While carbohydrate elements may also, in theory, be targeted, these typically have lower immunogenic potential than their proteinaceous counterparts, and therefore are rarely used for biorecognition.

Many of the more recent examples of biosensor-based bacterial pathogen detection have focused on the use of electrochemical platforms, including impedimetric assays. Tully et al. (2008) described an impedance-based assay for the analysis of the *L. monocytogenes* InB protein, which used a leporine host-derived polyclonal antibody as a bioligand. The assay format developed, which used planar screen-printed carbon electrodes modified with PANI for antibody capture, had an excellent LOD for InlB (4.1 pg/mL). Additional electrochemical platforms have also been developed for the detection of *S. typhimurium*, a pathogen typically transmitted to human hosts through the ingestion of contaminated animal produce, such as meat, eggs or milk. Nandakumar et al. (2008) devised an antibody-based impedimetric assay capable of detecting 500 colony forming unit (CFU)/mL of *S. typhimurium* in less than 10 minutes. Interestingly, the authors suggested that the developed methodology would be of particular use in portable pathogen detectors, due to the relatively low complexity of the sensor format and the necessity for a small number of data samples (30) to provide accurate pathogen detection.

Biosensors are particularly useful for the rapid detection of analytes of interest in complex sample matrices, and this also applies to the detection of bacterial pathogens where additional methodologies may be used in conjunction with immunosensing for pathogen retrieval. Immunomagnetic separation (IMS) may be applied to initially pre-concentrate bacterial cells from a food sample prior to biosensor-based analysis (Byrne et al. 2009) and, as with all immunodetection-based strategies, the efficacy of this approach is dependent on the quality of the antibody that is selected for biorecognition. Liébana et al. (2009) used a combination of IMS and magnetic electrode-based electrochemical biosensing for detecting *S. typhimurium* cells in milk, with the antibody capable of selectively differentiating between target and *E. coli* cells in this matrix. Detection was verified through the introduction of a second HRP-labelled anti-*Salmonella* antibody, thereby creating a sandwich assay format on the sensor surface that could be completed in less than one hour. Interestingly, the sensitivity of this assay was influenced by the nature of the samples selected for analysis. When *S. typhimurium* cells were propagated in Luria Bertani (LB) medium, the LOD was 5×10^3 CFU/mL. In spiked milk samples diluted tenfold in LB medium, the sensitivity was 7.5×10^3 CFU/mL. However, when the milk was pre-enriched for six and eight hours, the limits of detection could be significantly reduced to 1.4 CFU/mL and 0.1 CFU/mL, respectively, suggesting that sample preparation has a significant impact on assay sensitivity.

Salam and Tothill (2009) also developed an electrochemical immunosensor for *S. typhimurium* detection that was applied for the analysis of chicken breast samples. The assay format consisted of a murine monoclonal antibody captured on a screen-printed gold electrode via amine coupling. Similarly to the example described above by Liébana et al. (2009), a secondary HRP-labelled polyclonal antibody was selected for enhancing specificity. The assay format permitted rapid cross-reactivity analysis to be performed with additional bacterial strains, including *Staphylococcus aureus*, *Pseudomonas* spp. and *Klebsiella pneumonia* (all at a concentration of 1×10^9 CFU/mL). This is an important consideration for pathogen detection, and the ability to develop a species-specific immunoassay and subsequently, screen for non-specific binding to additional bacterial strains/pathogens is essential.

Liao and Ho (2009) describe a novel electrochemical sensor for the detection of *E. coli* O157:H7. Instead of selecting an epitope on the bacterial cell for immunorecognition, a gene encoding an enzyme that participates in O-antigen assembly (*rfb*E) was targeted by a single-stranded DNA probe, and the resultant assay had excellent sensitivity (0.75 amol) for the target gene. In another excellent example, Koo et al. (2009) developed an 'antibody-free' biosensor platform for the detection of *L. monocytogenes*. Here, the biomolecular interaction between the *Listeria* adhesion protein (LAP, also referred to as alcohol acetaldehyde dehydrogenase) and biotin-tagged heat shock protein 60 (HSP60) was used to facilitate the detection of this opportunistic pathogen in a matrix containing additional pathogenic bacterial strains, including *Bacillus cereus* and *Proteus vulgaris*. Of particular interest in this study was the observation that the efficiency of detection observed through the use of the HSP biorecognition element was over 80-fold higher than for an anti-*L. monocytogenes* monoclonal antibody previously developed in the same laboratory.

Fungal Pathogens

The sensitive detection of fungal strains is also of great importance, due to their association with crop spoilage and inherent ability to act as human pathogens. A key consideration when analysing fungal cells and spores by optical biosensing relates to the fact that they are significantly larger than their bacterial counterparts. Fungal spores are often over 40 μM in diameter, which complicates analysis in platforms, such as Biacore, where system blockages may result from their direct injection. Hence, it is possible to utilise modified assay formats, such as a subtraction inhibition assay (SIA), for the detection of these large fungal spores (Byrne et al. 2009). The principle behind the SIA involves the pre-incubation of antigen with a target-specific antibody. Unbound antibody is subsequently retained (e.g. by centrifugation or by filtration) and enumerated by capturing on a biosensor surface immobilised with an antibody-specific capture immunoglobulin. The observed response is inversely proportional to the amount of pathogen present in the original sample or culture, and the major benefit of the SIA assay format is that the system is not exposed to potentially harmful pathogens. This Biacore SIA format is also applicable for the monitoring of bacterial cells, including *L. monocytogenes* (Leonard et al. 2004).

SIA-based detection of fungal pathogens was demonstrated by Skottrup et al. (2007a), who pre-incubated a murine monoclonal antibody with sporangia of the fungal strain *Phytophthora infestans* and used a centrifugation-based protocol for separating free and bound antibodies. Free immunoglobulin was subsequently passed over a Biacore CM5 surface containing immobilised goat anti-mouse polyclonal antibody, which was used to quantitate free antibody over a pre-determined dilution series. The LOD was 2.2×10^6 sporangia/mL, and no cross-reactivity was seen when other fungal strains, such as *Botrytis cinerea* and *Melampsora euphorbiae* were assayed in parallel. A similar assay was also selected for detecting spores of *Puccinia striiformis*, a well-characterised plant pathogen. Here, a polyclonal rabbit anti-mouse IgM antibody was used for the capture of a monoclonal antibody (IgM), whose production was elicited in a murine host immunised with whole urediniospores. The resultant assay took approximately 45 minutes to complete and permitted the detection of 3.1×10^5 urediniospores/mL (Skottrup et al. 2007b). These two important examples demonstrate the applicability of using biosensor-based analysis for the detection of fungal spores, which is another key consideration for the evaluation of quality.

TOXINS

Mycotoxins

Mycotoxins are toxic compounds that are synthesised as secondary metabolites by fungal strains. For example, the fungus *Fusarium moniliforme* produces the carcinogenic mycotoxin, fumonisin B_1. Mullett et al. (1998) developed a rapid (approximately 10 minutes) SPR method to detect this mycotoxin in spiked samples. Here, polyclonal fumonisin B_1-specific antiserum (2 mg/mL) was diluted (1:5,000, 1:10,000, 1:1,500 and 1:2,000) and immobilised onto gold layers. Next, fumonisin B_1, in concentrations of 0.0–100μg/mL, was subsequently passed over each surface, prior to measuring the change in SPR. Their assay had a detection limit of 50 ng/mL. This research group state that the 'lab-made' SPR device used in their research could be implemented as an early stage method in the screening of large numbers of potentially contaminated food samples. Positive samples could then be analysed further by more sensitive methods of analysis, such as liquid chromatography or electrospray ionisation mass spectrometry (Hartl and Humpf 1999). More recently, van der Gaag et al. (2003) developed a biosensor-based assay capable of detecting fumonisin B_1 and three other analytes, namely zearalenone, ochratoxin A and aflatoxin B_1.

The monitoring of the presence of aflatoxins, which are naturally occurring mycotoxins produced by several strains of *Aspergillus* spp., in fruit, vegetable and food produce, is also of critical significance. Produce prone to contamination includes nuts (almonds, walnuts), cereals (rice, wheat, maize) and oilseeds (soybean and peanuts). Aflatoxin monitoring is also important as consumption of infected produce is implicated in carcinoma of the liver (Bhatnagar et al. 2002, Bennett and Klich 2003). While approximately 16 structurally diverse aflatoxins have been reported, aflatoxins B_1, B_2, G_1 and G_2 and M represent the greatest danger to human health (Keller et al. 2005). Many of these compounds may be generated in *Aspergillus parasiticus*, while *Aspergillus flavus* can synthesise aflatoxins B_1 and B_2 (Yu et al. 2004).

Several biosensor-based platforms have been developed to permit the detection of trace levels of different aflatoxins (Lacy et al. 2006). Daly et al. (2000) used a rabbit-derived polyclonal antibody to detect AFB_1, which was conjugated to BSA and immobilised onto a CM5 Biacore chip. A competition assay between free and bound AFB_1 had a linear range of detection (3–98 ng/mL). Daly et al. (2002) subsequently generated murine scFvs against AFB_1 by using a phage display format and incorporated these antibodies into a Biacore-based inhibition assay. Dunne et al. (2005) developed a unique SPR-based inhibition assay that incorporated monomeric and dimeric scFv antibody fragments that could detect AFB_1. Monomeric scFvs could detect between 390 and 12,000 ppb, while the dimeric scFv was more sensitive, detecting between 190 and 24,000 ppb.

Several other research groups have developed similar rapid analytical biosensor-based platforms for aflatoxin detection. Sapsford et al. (2006) developed an indirect competitive immunoassay on a fluorescence-based biosensor that permitted rapid detection of AFB_1 in spiked corn (cornflakes, cornmeal) and nut (peanuts, peanut butter) products. Mouse monoclonal antibodies were labeled with the fluorescent dye CY5, with detectable signals inversely proportional to the concentration of AFB_1 present. Limits of detection for nut and corn products were 0.6–1.4 ng/g and 1.5–5.1 ng/g, respectively. Adanyi et al. (2007) also recently described a unique protocol for permitting aflatoxin detection by using optical wavelength lightmode spectroscopy (abbreviated to OWLS). Integrated optical wavelength sensors were selected for use in this experiment with the detection range for a competitive assay being between 0.5 and 10 ng/mL. An indirect screening protocol was subsequently applied to wheat and barley that permitted the detection of AFB_1 and ochratoxin A. A more recent example of biosensor-based detection of the latter toxin is provided by Yuan et al. (2009), who selected a Biacore Q-based platform for their analysis. In this study, a murine-derived monoclonal antibody was used for the detection of ochratoxin residues in spiked cereal, beverages, juices (apple and grape) and wine samples in a competition assay format. Sensitivity was significantly enhanced by using gold nanoparticles.

Not all biosensor-based technologies for the evaluation of quality use antibodies for biorecognition, and this is also applicable to the detection of aflatoxins. An example is the development of an enzyme-based SPR assay for the detection of AFB_1 and AFG_1. The principle behind this assay was the ability of these aflatoxins to bind to porcine neutrophil elastase. The efficacy of using this approach was demonstrated by the excellent sensitivity that was observed (low ppb for both analytes), which demonstrates the potential of using enzymatic biosensors as alternatives to immunosensors for detection purposes (Cuccioloni et al. 2008).

Water and Marine Toxins

Marine toxins pose a considerable threat to human health because of their potential to cause respiratory, neurological and gastrointestinal problems, including mortalities at very low toxin concentrations. This is primarily linked to the consumption of contaminated shellfish meat. Hence, a number of countries have established regulations and specific concentration limits to control the acceptable levels of these phycotoxins in seafood (FAO 2004).

Phycotoxins include a variety of toxins from different phytoplankton groups. They vary widely in their physical properties, chemical structure and toxic mechanisms. The type of poisoning and, therefore, the phycotoxin groups are classified according to the associated symptoms: paralytic shellfish poisoning (PSP), neurotoxic shellfish poisoning, amnesic shellfish poisoning (ASP), diarrheic shellfish poisoning (DSP) and azaspiracid shellfish poisoning associated with shellfish; and ciguatera poisoning associated with finfish. ASP results from contamination with the amino acid molecule domoic acid and its derivatives. Domoic acid is a potent kainoid, neuroexcitatory toxin produced by the marine diatom *Pseudo-nitzschia pungens* (Pan et al. 2001). Many species of shellfish bioaccumulate domoic acid through the consumption of toxin-containing phytoplankton. Furthermore, shellfish known to have been contaminated with domoic acid include anchovies, crabs, mussels, clams, scallops, gastropods, mackerel and oysters. It was shown that the first gastrointestinal (nausea, vomiting, diarrhoea and abdominal cramps) and neurological symptoms (memory loss) will appear 15 minutes post-consumption of contaminated shellfish. Vision problems may also occur, including diplopia, disconjugate and ophthalmoplegia. Severe intoxication will cause significant neurological issues, including confusion and seizures, sometimes leading to coma and death.

SPR has been applied for the detection of domoic acid in an indirect competitive immunosensor format (Yu et al. 2005). Domoic acid was covalently linked onto a mixed self-assembled monolayer (SAM)-modified SPR chip and competition occurred between free domoic acid and anti-domoic acid monoclonal antibodies in solution. The detection limit was 0.1 ng/mL, which was lower than what was initially obtained when normal colorimetric ELISA analysis was performed. This was due to the lack of non-specific antibody adsorption on the SAM and the high sensitivity of the SPR instrument. Stevens et al. (2007) subsequently developed a portable six-channel SPR system for domoic acid detection in clam extracts with a LOD of 3 ng/mL and a quantitative detection range of 4–60 ng/mL. The portability of this biosensor is a key advantage over other SPR-based sensors, as both researchers and governmental shellfish monitoring programmes have highlighted the need for rapid in situ methods for monitoring biotoxin contamination (Kröger et al. 2002).

DSP results from the consumption of molluscs contaminated with the polycyclic ether toxins okadaic acid (OA), dinophysis toxin-1 (DTX1) and pectenotoxins (PTX). Yessotoxin (YTX) is also classified under this grouping as it was first isolated in 1987 from scallops associated with a DSP poisoning event, although the pharmacological activity of YTXs is different to that of DSP toxins (Bowden 2006). OA, DTX and PTX are all produced by dinoflagellates belonging to the *Dinophysis* and *Prorocentrum* species, whereas YTXs are produced by *Protoceratium reticulatum* (Satake et al. 1999). The known marine organisms susceptible to infection by *Dinophysis* and *Prorocentrum* include clams, mussels, oysters and scallops. The main symptom associated with DSP poisoning is diarrhoea, but other symptoms such as abdominal pain, nausea and vomiting may also occur. Hospitalisation is not normally required, and no human deaths have been documented to date.

Llamas et al. (2007) developed a unique SPR-based immunosensor for the detection of OA in shellfish extracts. They utilised a polyclonal antibody generated through repeated immunisation of an OA-bovine thyroglobulin (BTG) conjugate into a rabbit host and an amine sensor chip surface functionalised with an OA derivative, OA-*N*-hydroxysuccinimide. In this competition assay, 20 mussel samples, negative for OA, were spiked with 160 ng/g of OA. When compared to LC-MS, the immunosensor demonstrated a slight overestimation of OA. Furthermore, the polyclonal antibody was found to have 40% cross-reactivity with DTX-1, but when applied to a matrix environment no cross-reactivity was observed. Subsequently, a monoclonal antibody was developed against an OA-BTG conjugate (Stewart et al. 2009) that showed a cross-reactivity profile for both DTX-1 and DTX-2 that corresponded with the toxin potency of all the OA toxins. The antibody was selected using a novel approach in which hybridoma screening was carried out in an ELISA competition format, a direct ELISA format and a SPR biosensor format, using an OA-immobilised CM5 sensor chip.

PSP causes a significant risk to public health due to its high mortality rate and considerable economic damage. PSPs are water soluble, thermostable, tetrehydropurine molecules, which are subdivided into four structural categories; carbamates, *N*-sulfo-carbamoyls, decarbamoyls and deoxydecarbamoyls. The main phycotoxins associated with PSP include the neurotoxins, gonyatoxin (GTX) and saxitoxin (STX). The carbamate STX is the most toxic and over 20 different analogues have been described (Fusetani and Kern 2009), with a range of associated toxicity levels. PSPs are known to be produced by *Alexandrium, Gymnodinium* and *Pyrodinium* species. Infected organisms can include clams, crabs, cockles, cods, gastropods, herrings, lobsters, mackerel, mussels, oysters, pufferfish and scallops. Symptoms can be neurological and, less frequently, gastrointestinal, including nausea, vomiting and diarrhoea. These symptoms can appear from 15 minutes to 10 hours after ingestion of contaminated shellfish, any may also be associated with the onset of a tingling, a prickly feeling in the toes and fingertips, a burning sensation of the lips and skin, dizziness, numbness in the mouth and extremities, headache, ataxia and fever. Individuals may also experience a feeling of calmness and serenity. Severe poisoning can lead to lack of coordination and respiratory distress that can lead to death within 2–25 hours of intoxication.

Fonfría et al. (2007) developed a SPR inhibition assay using an anti-GTX2/3 antibody and a STX-immobilised CM5 Biacore surface. Here, STX in solution was detected by competition for binding to the anti-GTX2/3 antibody. The biosensor detected a number of different saxitoxin analogues, including dcSTX, GTX2/3, dcGTX2/3, GTX5 and C1/2 (with slightly higher sensitivity than STX). Significantly, the antibody had low cross-reactivity for neoSTX, GTX1/4 or decarbamoyl neosaxitoxin. The biosensor platform was subsequently shown to be effective in a number of different shellfish matrices, including mussels, clams, cockles, scallops and oysters, with a detection limit between 2 and 50 ng/mL observed. More recently, Taylor et al. (2008) devised a novel SPR immunosensor for the low molecular weight toxin, tetrodotoxin (TTX), using a polyclonal anti-TTX antibody as a biorecognition element. TTX was chemically immobilised on a gold film with a mixed SAM, which reduced non-specific binding at the surface. The inhibitory concentration causing 20% displacement (IC_{20}) for the assay was approximately 0.3 ng/mL, and the corresponding concentration for 50% displacement (IC_{50}) was approximately 6 ng/mL. In summary, these examples clearly demonstrate how biosensors can be used to monitor potentially fatal contaminants in water and marine life, which has a direct impact on the food industry.

LEGISLATION

In this chapter, a variety of different biosensor-based platforms used to detect a range of different molecules such as pesticides, herbicides, mycotoxins and phycotoxins are described. It is important to consider the legislation associated with the monitoring of quality. The Food and Agriculture Organisation of the United Nations (FAOSTAT) (http://faostat.fao.org/) is an international body whose overall aim is to protect the health of consumers. FAOSTAT has an online database (The Codex Alimentarius: Pesticide Residues in Food Maximum Residue Limits; see 'Websites of Interest'), which details the MRLs for pesticides in commodities such as fruit and vegetables. The establishment of a MRL (usually expressed as mg/kg) is dependent on good agricultural practice for pesticides. This relates to where the highest detectable residues anticipated (when a pesticide-containing product is applied to a commodity to remove contaminants) are intended to be toxicologically acceptable. Under these arrangements, the important point is that residue levels do not pose unacceptable risks for consumers by being present in values that exceed these levels.

Similar legislation exists within the European Union (EU). Within the EU Directive (91/414/EEC), the Plant Protection Products Directive ('The Authorisations Directive': 1991) was proposed with the aim of harmonising the overall arrangements for authorisation of plant protection products within the EU. However, legislation relating to MRLs of pesticides in cereals, fruit, vegetables, foods of animal origin and feeding stuffs was substantially amended several times since this legislation was developed. A single act has replaced the amended original Directive Regulation (EC) No. 396/2005. This establishes maximum levels of pesticide residues permitted in or on food and feed of plant and animal origin. These MRLs include levels that are specific to particular foodstuffs that are intended for human or animal consumption, and a general limit that applies where no specific MRL has been established.

As infants have developing immune systems, fruit and vegetable-based baby foods need to be rigorously monitored, as the presence of low concentrations of pesticides could be of significance. The EU has set an MRL for any given pesticide in infant foods of a concentration not exceeding 10 μg/mL. This information is very significant for the use and future applications of biosensor-based detection methods.

The EU has established an upper safe limit of 20 mg/kg for total ASP toxin content in the edible part of the mollusc (EEC 1991), with HLPC selected as a reference method. The European Commission have set tolerable levels for OA, DTXs and PTXs present in edible tissues at 160 μg OA equivalent/kg shellfish and a maximal level of YTXs of 1 mg YTX equivalent/kg shellfish (European Commission Decision 2002). The mouse or rat bioassay is the reference method for DSP detection. The EU directive 91/492/EEC allows a maximum level of PSP toxin of 80 μg STX eq/100g shellfish meat. The European Commission also decided in 2002 (2002/226/EC) to allow a conditional harvesting of scallops with a concentration of domoic acid (DA) in the whole body exceeding 20 mg/kg but lower than 250 mg/kg, based on the limit of ASP, imparted by Council Directive 91/492/EEC (1991). More recently (2009), the European Food Safety Authority (EFSA) recommended that it was safe to consume shellfish containing less than 4.5 mg domoic acid/kg, so as not to exceed the acute reference dose of 30 μg DA/kg body weight (equivalent to 1.8 mg DA per portion for a 60 kg adult) (EFSA 2009a, EFSA 2009b).

CONCLUSION

This chapter has provided a concise review of how biosensors may be applied to monitoring the quality of fruit and vegetable produce, and also for the detection of pathogens (bacterial and fungal) and associated toxins. Examples are given where such applications have been successfully employed, while recent developments in the area have focused on the development of low and high-density arrays capable of multiple analyte monitoring and portable or disposable platforms for in situ monitoring. With the advent of such technology in a user-friendly format, the scope for analysis will be greatly expanded and simplified, leading to a very significant improvement in food quality analysis.

ACKNOWLEDGEMENTS

The financial support of the Centre for Bioanalytical Sciences and Biomedical Diagnostics Institute, Dublin City University, the Industrial Developmental Agency, Ireland, Science Foundation Ireland (SFI, grant no: 05/CE3/B754), Enterprise Ireland, *Safe*food Biotoxin Research Network, the EU 7th Framework Programmes (Biolisme, research for benefit of Small Medium Enterprises, grant no. 232037 and the PHARMATLANTIC knowledge transfer Network, project number 2009–1/117), the US-Ireland partnership program, Biosafety for Environmental Contaminants using Novel Sensors (BEACONS; 08/US/1512), supported by SFI and the Beaufort Marine Award carried out under the Sea Change Strategy and the Strategy for Science Technology and Innovation (2006–2013), are gratefully acknowledged.

WEBSITES OF INTEREST

1. Food and Agriculture Organisation of the United Nations. Available at http://faostat.fao.org/
2. Pesticide Residues in Food (MRLs/EMRLs). Available at http://www.codexalimentarius.net/mrls/pestdes/jsp/pest_q-e.jsp?language = EN&version = ext&hasbulk = 0
3. The Pesticides Safety Directorate. Available at http://www.pesticides.gov.uk/
4. Irish National Food Residue Database. Available at http://nfrd.teagasc.ie/db_search_crit.asp
5. Environmental Health and Toxicology Database. Available at http://sis.nlm.nih.gov/enviro.html
6. Toxicology Data Network. Available at http://toxnet.nlm.nih.gov/cgi-bin/sis/htmlgen?TOXLINE
7. Marine Biotoxin Database. Available at http://www.fao.org/docrep/007/y5486e/y5486e00.htm#Contents

REFERENCES

Abbott JA. 2004. Textural quality assessment for fresh fruits and vegetables. *Adv Exp Biol Med* 542: 265–279.

Adanyi N et al. 2007. Development of immunosensor based on OWLS technique for determining Aflatoxin B1 and Ochratoxin A. *Biosens Bioelectron* 22(6): 797–802.

Amine A et al. 2006. Enzyme inhibition-based biosensors for food safety and environmental monitoring. *Biosens Bioelectron* 21(8): 1405–1423.

Atwell JL et al. 1999. ScFv multimers of the anti-neuraminidase antibody NC10: length of the linker between VH and VL domains dictates precisely the transition between diabodies and triabodies. *Protein Eng* 12(7): 597–604.

Bennett JW, Klich M. 2003. Mycotoxins. *Clin Microbiol Rev* 16(3): 497–516.

Berardo N et al. 2005. Rapid detection of kernel rots and mycotoxins in maize by near-infrared reflectance spectroscopy. *J Agric Food Chem* 53(21): 8128–8134.

Bhatnagar D et al. 2002. Toxins of filamentous fungi. *Chem Immunol* 81: 167–206.

Bokken GC et al. 2003. Immunochemical detection of Salmonella group B, D and E using an optical surface plasmon resonance biosensor. *FEMS Microbiol Lett* 222(1): 75–82.

Bowden BF. 2006. Yessotoxins–polycyclic ethers from dinoflagellates: relationships to diarrheic shellfish toxins. *Toxin Rev* 25(2): 137–157.

Bradbury AR, Marks JD. 2004. Antibodies from phage antibody libraries. *J Immunol Methods* 290(1–2): 29–49.

Byrne B et al. 2009. Antibody-based sensors: principles, problems and potential for detection of pathogens and associated toxins. *Sensors* 9: 4407–4445.

Caldow M et al. 2005. Development and validation of an optical SPR biosensor assay for tylosin residues in honey. *J Agric Food Chem* 53(19): 7367–7370.

Cassidy JF et al. 1998. Principles of chemical and biological sensors. In: D Diamond (ed.) *Amperometric Methods of Detection*. Wiley-Interscience, New York, pp. 73–132.

Conroy PJ et al. 2009. Antibody production, design and use for biosensor-based applications. *Semin Cell Dev Biol* 20(1): 10–26.

Cuccioloni M et al. 2008. Biosensor-based screening method for the detection of aflatoxins B1-G1. *Anal Chem* 80(23): 9250–9256.

Daly SJ et al. 2002. Production and characterisation of murine single chain Fv antibodies to aflatoxin B-1 derived from a pre-immunised antibody phage display library system. *Food Agric Immunol* 14(4): 255–274.

Daly SJ et al. 2000. Development of surface plasmon resonance-based immunoassay for aflatoxin B_1. *J Agric Food Chem* 48(11): 5097–5104.

Del Carlo M et al. 2005. Determining pirimiphos-methyl in durum wheat samples using an acetylcholinesterase inhibition assay. *Anal Bioanal Chem* 381(7): 1367–1372.

Dillon PP et al. 2005. Novel assay format allowing the prolonged use of regeneration-based sensor chip technology. *J Immunol Methods* 296(1–2): 77–82.

Dourtoglou VG et al. 2006. Storage of olives (*Olea europaea L.*) under CO_2 atmosphere: liquid chromatography-mass spectrometry characterisation of indices related to changes in polyphenolic metabolism. *J Agric Food Chem* 54(6): 2211–2217.

Dunne L et al. 2005. Surface plasmon resonance-based immunoassay for the detection of alfatoxin B-1 using single-chain antibody fragments. *Spectrosc Lett* 38(3): 229–245.

EEC Commission Decision. 1991. Laying down the health conditions for the production and the placing on the market of fishery products. Official Journal of the European Communities. (91/493/EEC) No. L. 268, 1–15. (PSP, ASP).

EFSA. 2009a. Scientific opinion of the panel on contaminants in the food chain on a request from the European Commission on marine biotoxins in shellfish – summary on regulated marine biotoxins. *EFSA J* 1306: 1–23.

EFSA. 2009b. Scientific opinion of the panel on contaminants in the food chain on a request from the European Commission on marine biotoxins in shellfish – domoic acid. *EFSA J* 1181: 1–61.

European Commission Decision. 2002. Laying down rules for the implementation of Council Directive 91/492/EEC as regards the maximum levels and the methods of analysis of certain marine biotoxins in bivalve molluscs, echinoderms, tunicates and marine gastropods. Official Journal of the European Communities (2002/225/EC) No. L.75, 62–63. (DSP).

FAO 2004. Marine biotoxins: FAO food and nutrition. 2004. Paper 80.

Flampouri K et al. 2010. Development and validation of a cellular biosensor detecting pesticide residues in tomatoes. *Talanta* 80(5): 1799–1804.

Fonfría ES et al. 2007. Paralytic shellfish poisoning detection by surface plasmon resonance-based biosensors in shellfish matrixes. *Anal Chem* 79(16): 6303–6311.

Fusetani N, Kern W. 2009. Marine toxins: an overview. *Prog Mol Subcell Biol* 46: 1–44.

Gao Z et al. 2009. Immunochip for the detection of five kinds of chemicals: atrazine, nonylphenol, 17-beta estradiol, paraverine and chloramphenicol. *Biosens Bioelectron* 24(5): 1445–1450.

Giraudi G, Baggiani C. 1994. Immunochemical methods for environmental monitoring. *Nucl Med Biol* 21(3): 557–572.

Gómez AH et al. 2005. Impulse response of pear fruit and its relation to Magness-Taylor firmness during storage. *Postharvest Biol Technol* 35(2): 209–215.

Gouzy MF et al. 2009. A SPR-based immunosensor for the detection of isoproturon. *Biosens Bioelectron* 24(6): 1563–1568.

Hartl M, Humpf HU. 1999. Simultaneous determination of fumonisin B1 and hydrolyzed fumonisin B1 in corn products by liquid chromatography/electrospray ionization mass spectrometry. *J Agric Food Chem* 47(12): 5078–5083.

Hearty S et al. 2006. Production, characterisation and potential application of a novel monoclonal antibody for rapid identification of virulent *Listeria monocytogenes*. *J Microbiol Methods* 66(2): 294–312.

Herranz S et al. 2008. Preparation of antibodies and development of a sensitive immunoassay with fluorescence detection for triazine herbicides. *Anal Bioanal Chem* 391(5): 1801–1812.

Holliger P et al. 1993. "Diabodies": small bivalent and bispecific antibody fragments. *Proc Natl Acad Sci USA* 90(14): 6444–6448.

Homola J. 2003. Present and future of surface plasmon resonance biosensors. *Anal Bioanal Chem* 377(3): 528–539.

Hudson PJ, Souriau C. 2003. Engineered antibodies. *Nat Med* 9(1): 129–134.

Jawaheer S et al. 2003. Development of a common biosensor format for an enzyme-based biosensor array to monitor fruit quality. *Biosens Bioelectron* 18(12): 1429–1437.

Karlsson R. 2004. SPR for molecular interaction analysis: a review of emerging application areas. *J Mol Recognit* 17(3): 151–161.

Keay RW, McNeil CJ. 1998. Separation-free electrochemical immunosensor for rapid determination of atrazine. *Biosens Bioelectron* 13(9): 963–970.

Keller NP et al. 2005. Fungal secondary metabolism – from biochemistry to genomics. *Nat Rev Microbiol* 3(12): 937–947.

Kim SJ et al. 2007. Novel miniature SPR immunosensor equipped with all-in-one multi-microchannel sensor chip for detecting low-molecular-weight analytes. *Biosens Bioelectron* 23(5): 701–707.

Köhler G, Milstein C. 1975. Continuous cultures of fused cells secreting antibody of predefined specificity. *Nature (Lond)* 256(5517): 495–497.

Koo OK et al. 2009. Targeted capture of pathogenic bacteria using a mammalian cell receptor coupled with dielectrophoresis on a biochip. *Anal Chem* 81(8): 3094–3101.

Kortt AA et al. 1997. Single-chain Fv fragments of anti-neuraminidase antibody NC10 containing five- and ten-residue linkers form dimers and with zero-residue linker a trimer. *Protein Eng* 10(4): 423–433.

Koubová V et al. 2001. Detection of foodbourne pathogens using surface plasmon resonance biosensors. *Sensors Actuators* 74(1–3): 100–105.

Kriz K et al. 2002. Amperometric determination of L-lactate based on entrapment of lactate oxidase on a transducer surface with a semi-permeable membrane using a SIRE technology-based biosensor. Application: tomato paste and baby food. *J Agric Food Chem* 50(12): 3419–3424.

Kröger S et al. 2002. Biosensors for marine pollution research, monitoring and control. *Mar Pollut Bull* 45(1–12): 24–34.

Lacy A et al. 2006. Rapid analysis of coumarins using surface plasmon resonance. *J AOAC Int* 89(3): 884–892.

Leenaars M, Hendriksen CFM. 2005. Critical steps in the production of polyclonal and monoclonal antibodies: evaluation and recommendations. *ILAR J* 46(3): 269–279.

Leonard P et al. 2004. A generic approach for the detection of whole *Listeria monocytogenes* cells in contaminated samples using surface plasmon resonance. *Biosensors Bioelectron* 19(10): 1331–1335.

Leonard P et al. 2005. Development of a surface plasmon resonance-based immunoassay for *Listeria monocytogenes*. *J Food Prot* 68(4): 728–735.

Liao WC, Ho JA. 2009. Attomole DNA electrochemical sensor for the detection of *Escherichia coli* O157:H7. *Anal Chem* 81(7): 2470–2476.

Liébana S et al. 2009. Rapid detection of *Salmonella* in milk by electrochemical magneto-immunosensing. *Biosens Bioelectron* 25(2): 510–513.

Llamas NM et al. 2007. Development of a novel immunobiosensor method for the rapid detection of okadaic acid contamination in shellfish extracts. *Anal Bioanal Chem* 389(2): 581–587.

Luong JHT et al. 1995. Enzyme reactions in the presence of cyclodextrins: biosensors and enzyme assays. *Trends Biotechnol* 13(11): 457–463.

MacLean DD et al. 2006. Postharvest variation in apple (*Malus x domestica Borkh.*) flavonoids following harvest, storage, and 1-MCP treatment. *J Agric Food Chem* 54(3): 870–878.

March C et al. 2009. A piezoelectric immunosensor for the determination of pesticide residues and metabolites in fruit juices. *Talanta* 78(3): 827–833.

Markey F. 2000. Principles of surface plasmon resonance. In: K Nagata, KH Handa (eds.) *Real-Time Analysis of Biomolecular Interactions*. Springer, Tokyo, pp. 13–22.

McGuinness BT et al. 1996. Phage diabody repertoires for selection of large numbers of bispecific antibody fragments. *Nat Biotechnol* 14(9): 1149–1154.

Minunni M, Mascini M. 1993. Detection of pesticides in drinking water using real-time biospecific interaction analysis (BIA). *Anal Lett* 26(7): 1441–1460.

Mitchum B et al. 1996. Methods for determining quality of fresh commodities. *Perishables Handle Newsletters* 85: 1–5.

Moran E et al. 2002. Methods for generation of monoclonal antibodies to the very small drug hapten, 5-benzimidazolecarboxylic acid. *J Immunol Methods* 271(1–2): 65–75.

Muhammad-Tahir Z, Alocilja EC. 2004. A disposable biosensor for pathogen detection in fresh produce samples. *Biosyst Eng* 88(2): 145–151.

Mullett W et al. 1998. Immunoassay of fumonisins by a surface plasmon resonance biosensor. *Anal Biochem* 258(2): 161–157.

Nandakumar V et al. 2008. A methodology for rapid detection of *Salmonella typhimurium* using label-free electrochemical impedance spectroscopy. *Biosens Bioelectron* 24(4): 1045–1048.

Navrátilová I, Skládal P. 2004. The immunosensors for measurement of 2,4-dichlorophenoxyacetic acid based on electrochemical impedance spectroscopy. *Bioelectrochemistry* 62(1): 11–18.

Nedelkov D et al. 2000. Multitoxin biosensor–mass spectrometry analysis: a new approach for rapid, real-time, sensitive analysis of staphylococcal toxins in food. *Int J Food Microbiol* 60(1): 1–13.

Nelson PN et al. 2000. Monoclonal antibodies. *Mol Pathol* 53(3): 111–117.

O'Kennedy R et al. 2010. Speedy, small, sensitive and specific – reality of myth for future analytical methods. *Anal Lett* 43: 1630–1648.

Oh B et al. 2004. Surface plasmon resonance immunosensor for the detection of *Salmonella typhimurium*. *Biosens Bioelectron* 19(11): 1497–1504.

Pan Y et al. 2001. *Pseudo-nitzschia sp.* of *Pseudodelicatissima*—a confirmed producer of domoic acid from the northern Gulf of Mexico. *Mar Ecol Prog Ser* 220: 83–92.

Pardieu E et al. 2009. Molecularly imprinted conducting polymer-based electrochemical sensor for detection of atrazine. *Anal Chim Acta* 649(2): 236–245.

Perisic O et al. 1994. Crystal structure of a diabody, a bivalent antibody fragment. *Structure* 2(12): 1217–1226.

Pogačnik L, Franko M. 2003. Detection of organophosphate and carbamate pesticides in vegetable samples by a photothermal biosensor. *Biosens Bioelectron* 18(1): 1–9.

Rang HP et al. 1998. Cholinergic transmission. In: *Pharmacology*, 3rd edn. Churchill Livingstone, Edinburgh.

Röthlisberger D et al. 2005. Domain interactions in the Fab fragment: a comparative evaluation of the single-chain Fv and Fab format engineered with variable domains of different stability. *J Mol Biol* 347(4): 773–789.

Rubtsova MY et al. 1998. Chemiluminescent biosensors based on porous supports with immobilised peroxidase. *Biosens Bioelectron* 13(1): 75–85.

Ryan BJ et al. 2006. Horseradish and soybean peroxidases: comparable tools for alternative niches? *Trends Biotechnol* 24(8): 355–363.

Salam F, Tothill IE. 2009. Detection of *Salmonella typhimurium* using an electrochemical immunosensor. *Biosens Bioelectron* 24(8): 2630–2636.

Salmain M et al. 2008. Infrared optical immunosensor: application to the measurement of the herbicide atrazine. *Anal Biochem* 373(1): 61–70.

Sapsford KE et al. 2006. Indirect competitive immunoassay for detection of aflatoxin B_1 in corn and nut products using the array biosensor. *Biosens Bioelectron* 21(12): 2298–2305.

Satake M et al. 1999. Confirmation of yessotoxin and 45, 46, 47-trinoryessotoxin production by *Protoceratium reticulatum* collected in Japan. *Nat Toxins* 7(4): 147–150.

Schirra M et al. 2005. Residue level, persistence, and storage of citrus fruit treated with fludioxonil. *J Agric Food Chem* 53(17): 6718–6724.

Schlecht U et al. 2002. Reversible surface thiol immobilization of carboxyl group containing haptens to a BIAcore biosensor chip enabling repeated usage of a single sensor surface. *Bioconjugate Chem* 13(2): 188–193.

Schulze H et al. 2002. Development, validation, and application of an acetylcholinesterase-biosensor test for the direct detection of insecticide residues in infant food. *Biosens Bioelectron* 17(11–12): 1095–1105.

Shmulevich I et al. 2003. Sensing technology for quality assessment in controlled atmospheres. *Postharv Biol Technol* 29(2): 145–154.

Skottrup P et al. 2007a. Rapid determination of *Phytophthora infestans* sporangia using a surface plasmon resonance immunosensor. *J Microbiol Methods* 68(3): 507–515.

Skottrup P et al. 2007b. Detection of fungal spores using generic surface plasmon resonance immunoassay. *Biosens Bioelectron* 22(11): 2724–2729.

Stevens RC et al. 2007. Detection of the toxin domoic acid from clam extracts using a portable surface plasmon resonance biosensor. *Harmful Algae* 6(2): 166–174.

Stewart LD et al. 2009. Development of a monoclonal antibody binding okadaic acid and dinophysistoxins-1, -2 in proportion to their toxicity equivalence factors. *Toxicon* 54(4): 491–498.

Suihko ML et al. 2002. Characterisation of *Listeria monocytogenes* from the meat, poultry and seafood industries by automated ribotyping. *Int J Food Microbiol* 72(1–2): 137–146.

Tang L et al. 2008. Rapid detection of picloram in agricultural field samples using a disposable immunomembrane-based electrochemical sensor. *Environ Sci Technol* 42(4): 1207–1212.

Taylor AD et al. 2008. Quantitative detection of tetrodotoxin (TTX) by a surface plasmon resonance (SPR) sensor. *Sensors and Actuators B: Chemical*. 130(1): 120–128.

Timbrell JA. 1991. Biochemical mechanisms of toxicity: specific examples. In: *Principles of Biochemical Toxicology*, 2nd edn. Taylor and Francis, London.

Townsend S et al. 2006. Optimizing recombinant antibody function in SPR immunosensing. The influence of antibody structural format and chip surface chemistry on assay sensitivity. *Biosens Bioelectron* 22(2): 268–274.

Tully E et al. 2006. The development of rapid fluorescence-based immunoassays, using quantum dot-labelled antibodies for the detection of *Listeria monocytogenes* cell surface proteins. *Int J Biol Macromol* 39(1–3): 127–134.

Tully E et al. 2008. The development of a 'labeless' immunosensor for the detection of *Listeria monocytogenes* cell surface protein, Internalin B. *Biosens Bioelectron* 23(6): 906–912.

Van der Gaag B et al. 2003. Biosensors and multiple mycotoxin analysis. *Food Control* 14: 251–254.

Voss K, Galensa R. 2000. Determination of L- and D-amino acids in foodstuffs by coupling of high-performance liquid chromatography with enzyme reactors. *Amino Acids* 18(4): 339–352.

Xavier MP et al. 2000. Fiber optic monitoring of carbamate pesticides using porous glass with covalently bound chlorophenol red. *Biosens Bioelectron* 14(12): 895–905.

Yu J et al. 2004. Clustered pathway genes in aflatoxin biosynthesis. *Appl Environ Microbiol* 70(3): 1253–1262.

Yu Q et al. 2005. Detection of low-molecular-weight domoic acid using surface plasmon resonance sensor. *Sensors Actuators* 107(1): 193–201.

Yuan J et al. 2009. Surface plasmon resonance biosensor for the detection of ochratoxin A in cereals and beverages. *Anal Chim Acta* 656(1–2): 63–71.

Glossary

This illustrates some of the compounds described in the text that are of relevance to the food industry. These are listed below in order of their appearance in the text.

SMALL MOLECULAR WEIGHT CONTAMINANTS

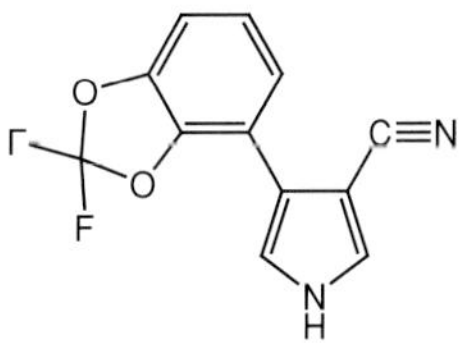

Fludioxonil

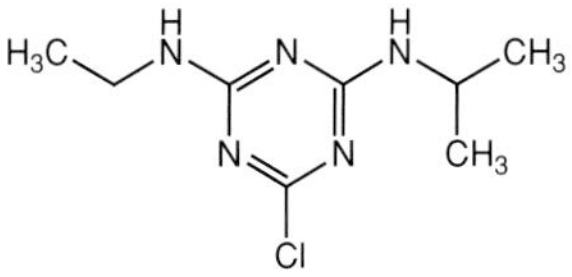

Atrazine

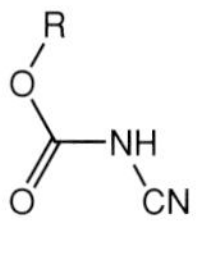

General carbamate structure

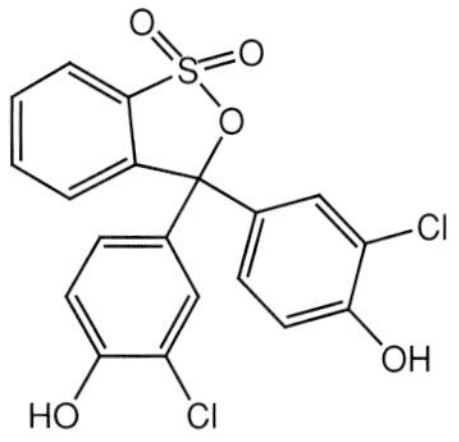

Chlorophenol red

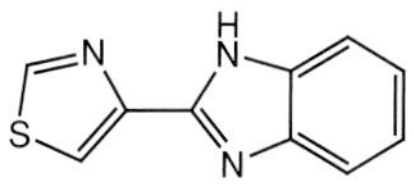

2-(4-Thiazolyl)benzimidazole

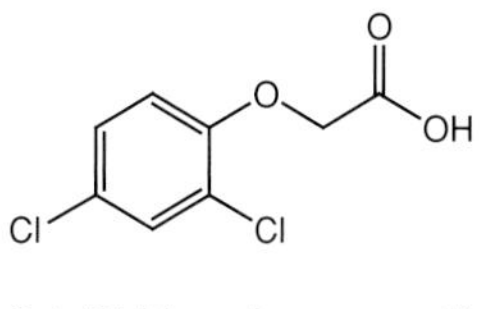

2,4-Dichlorophenoxyacetic acid

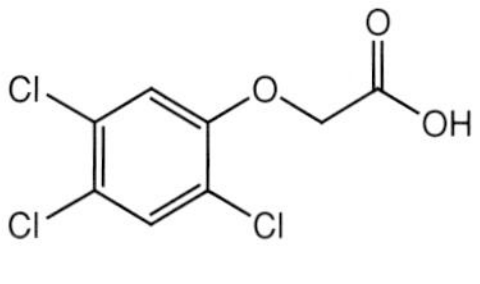

2,4,5-Trichlorophenoxyacetic acid (2,4,5-T)

Food Biochemistry and Food Processing, Second Edition. Edited by Benjamin K. Simpson, Leo M.L. Nollet, Fidel Toldrá, Soottawat Benjakul, Gopinadhan Paliyath and Y.H. Hui.

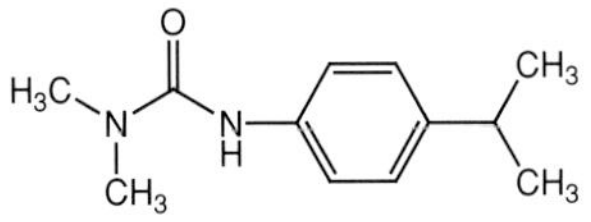

Isoproturon

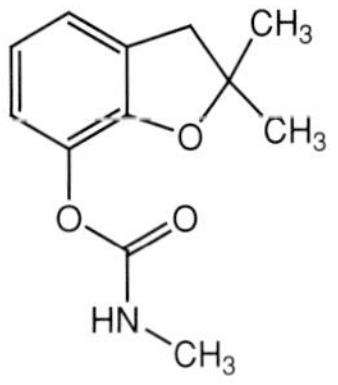

Carbofuran

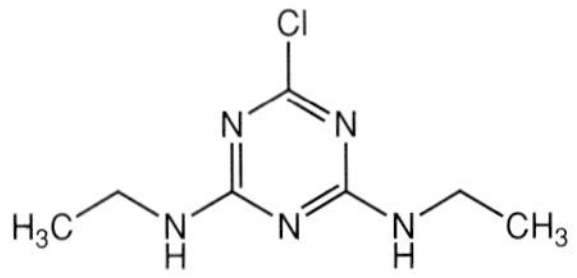

Simazine

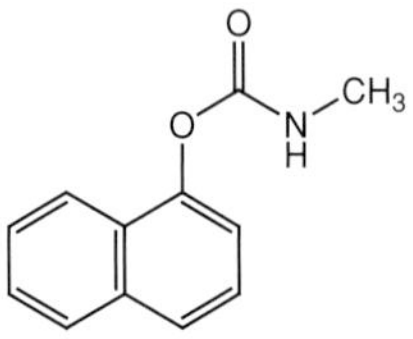

Carbaryl

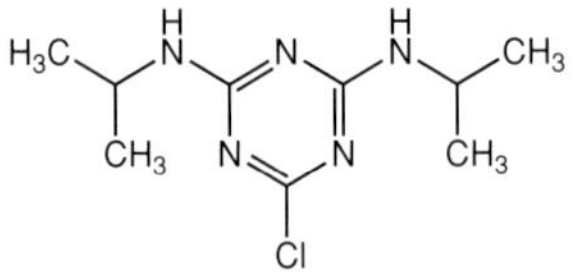

Propazine

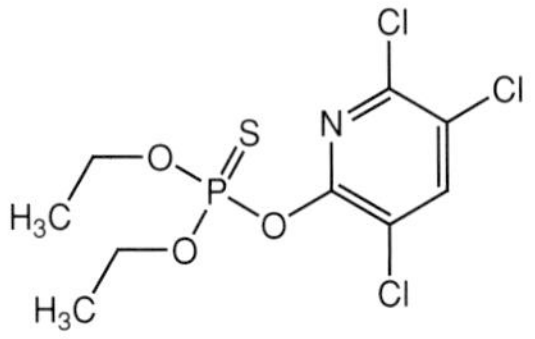

Chlorpyrifos

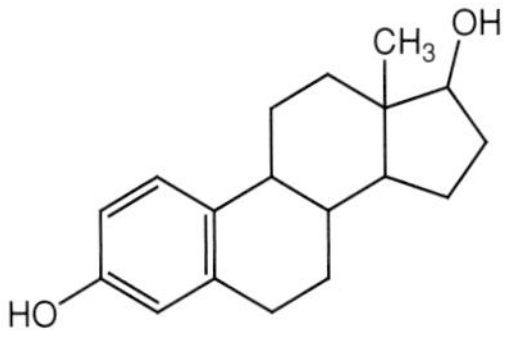

17-β-estradiol

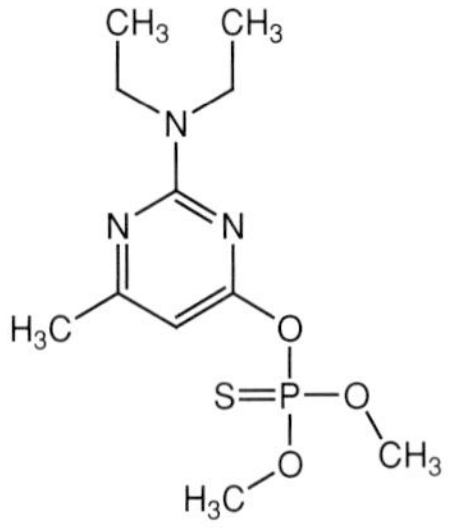

Pirimiphos-methyl

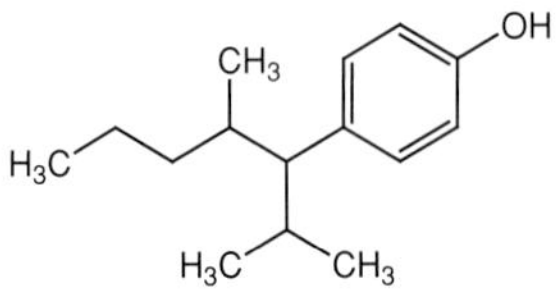

Nonylphenol

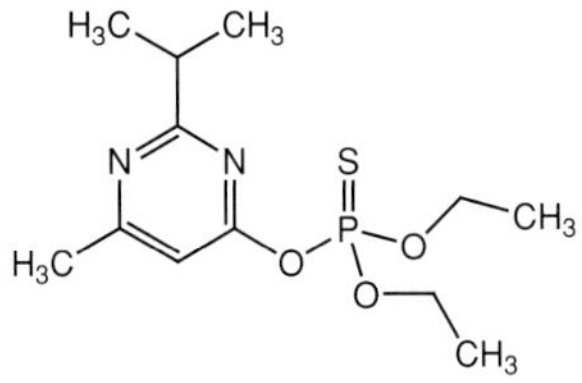

Diazinon

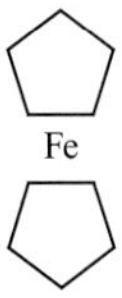

Ferrocene

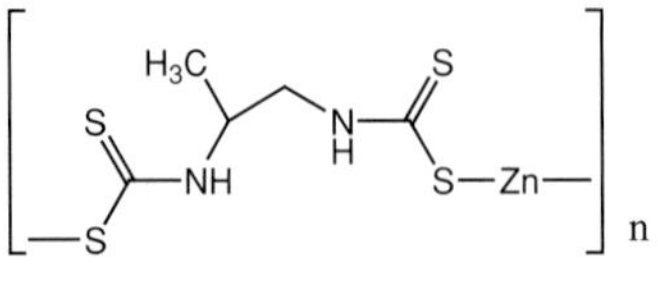

Propineb

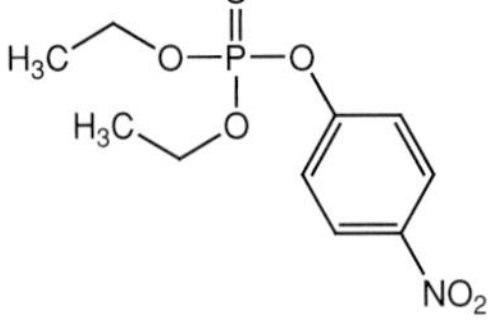

Parathion

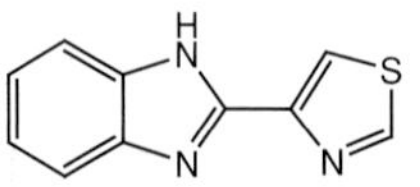

Thiabenzadole

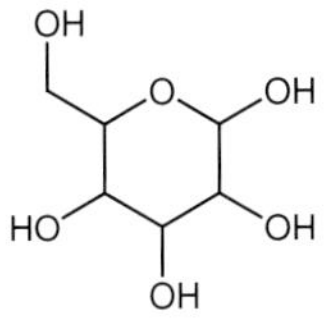

D-glucose

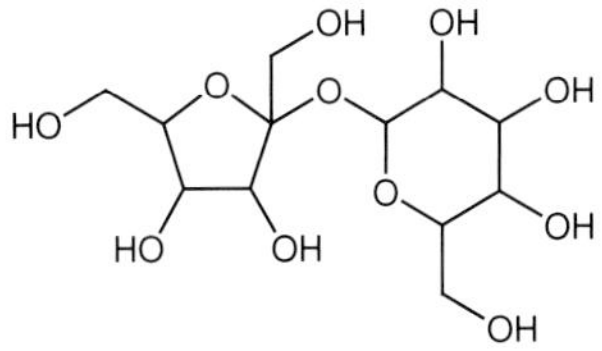

Sucrose

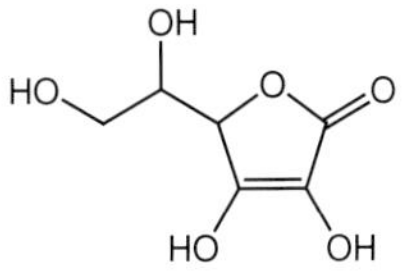

Ascorbic acid

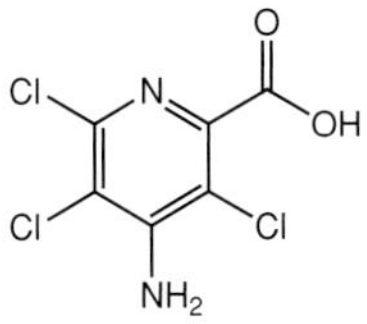

Picloram

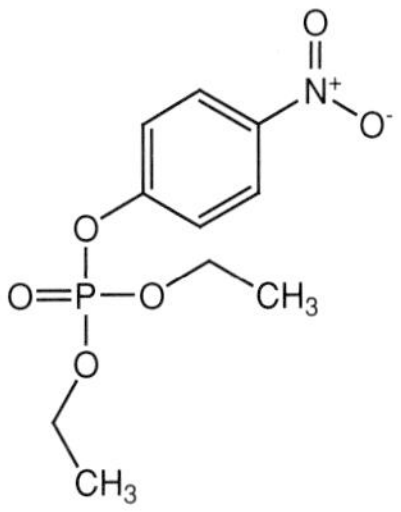

Paraoxon

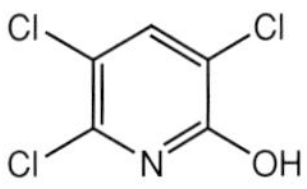

3,5,6-Trichloro-2-pyridinol

POLYKETIDES

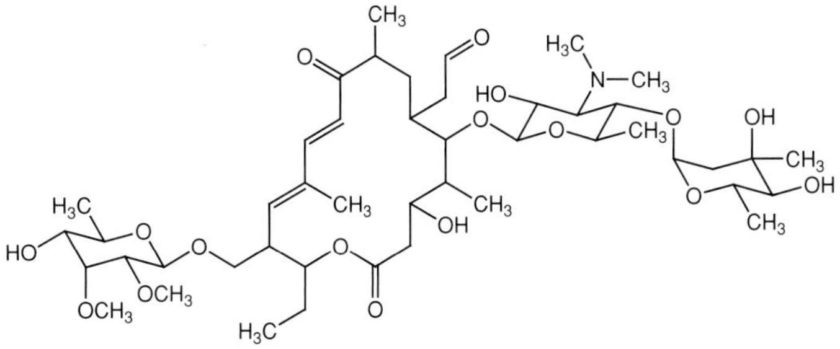

Tylosin

MYCOTOXINS

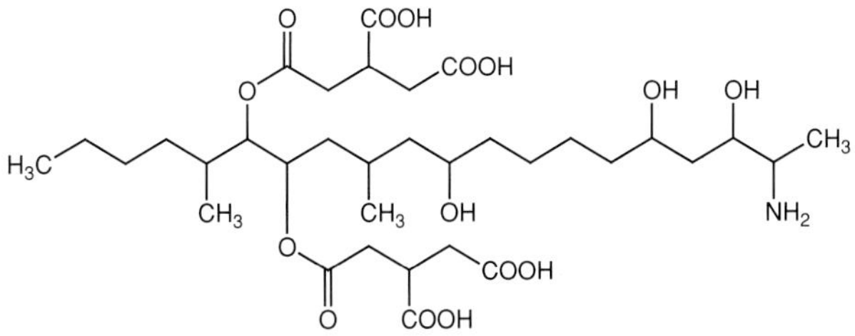

Fumonisin B_1

AFLATOXINS

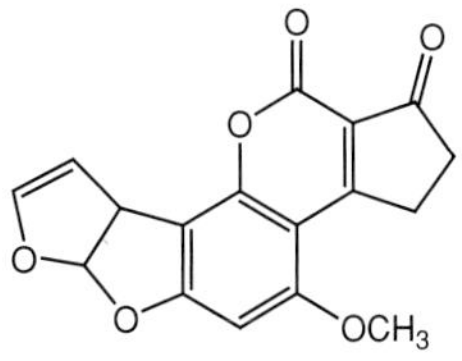

Aflatoxin B_1

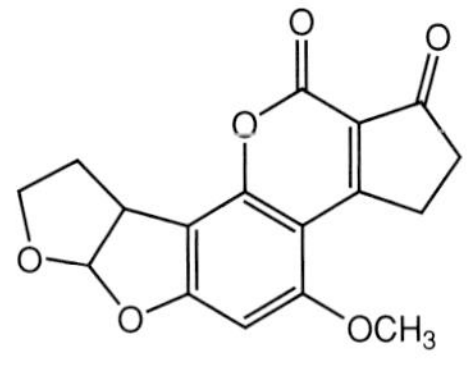

Aflatoxin B_2

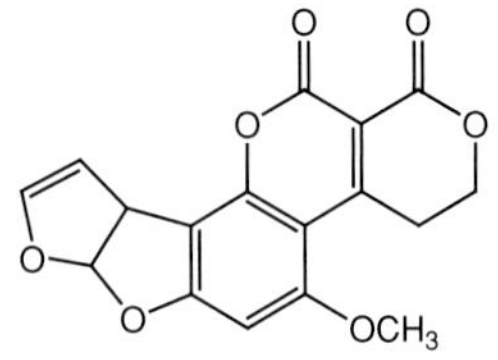

Aflatoxin G_1

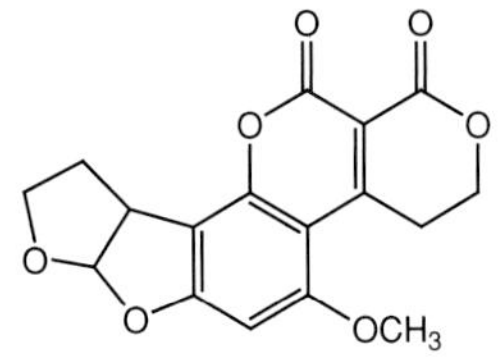

Aflatoxin G_2

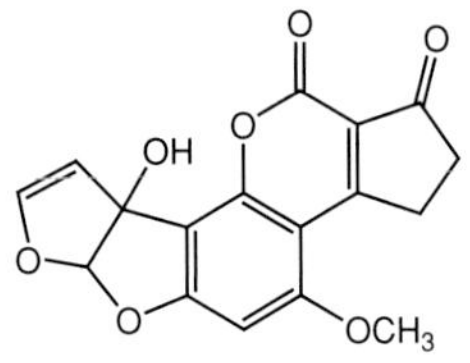

Aflatoxin M_1

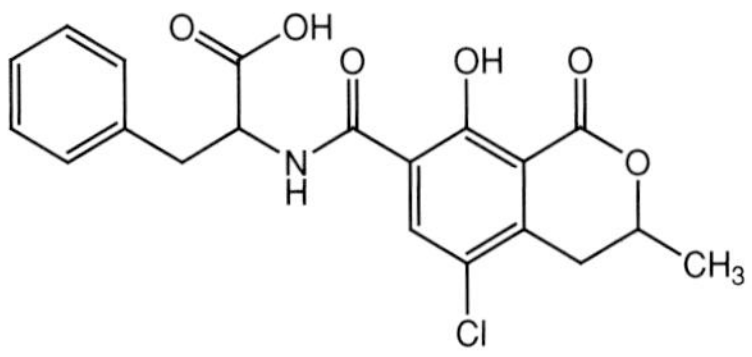

Ochratoxin

PHYCOTOXINS

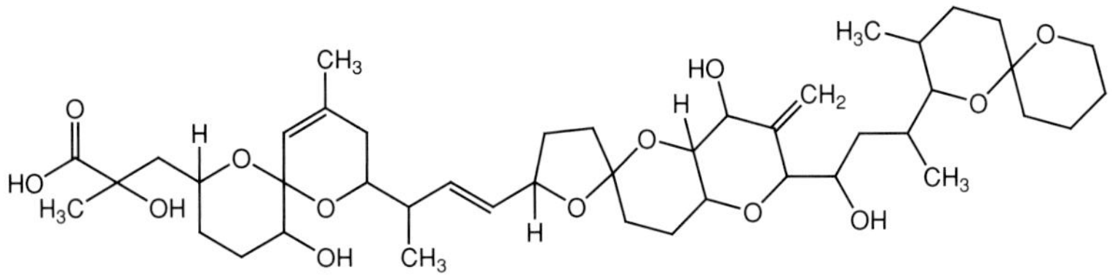

Okadaic acid

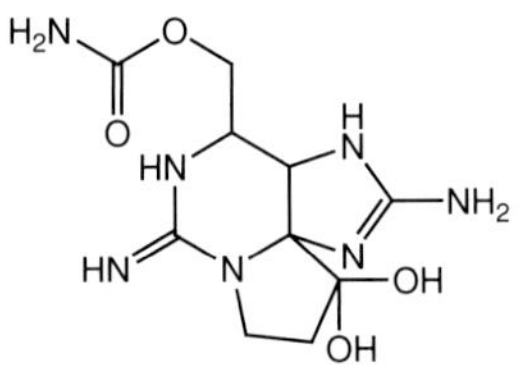

Saxitoxin

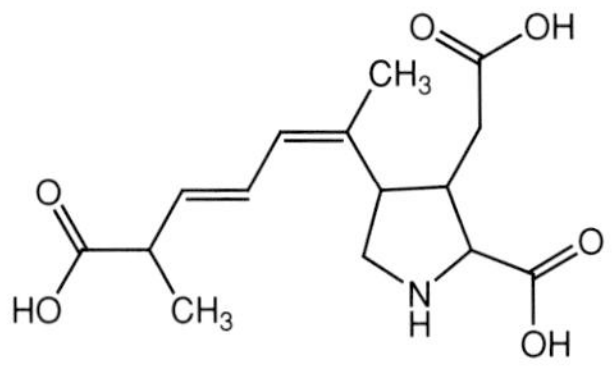

Domoic acid

Index

Food Biochemistry and Food Processing, Second Edition. Edited by Benjamin K. Simpson, Leo M.L. Nollet, Fidel Toldrá, Soottawat Benjakul, Gopinadhan Paliyath and Y.H. Hui.